PROCEEDINGS

OF THE

1991 INTERNATIONAL CONFERENCE

ON

PARALLEL PROCESSING

August 12-16, 1991

Vol. I Architecture
Dr. Chuan-lin Wu, Editor
University of Texas

Sponsored by
THE PENNSYLVANIA STATE UNIVERSITY

CRC Press
Boca Raton Ann Arbor Boston London

The papers appearing in this book comprise the proceedings of the meeting mentioned on the cover and title page. They reflect the authors' opinions and are published as presented and without change in the interest of timely dissemination. Their inclusion in this publication does not necessarily constitute endorsement by the editors, CRC Press, or the the Institute of Electrical and Electronics Engineers, Inc.

Catalog record is available from the Library of Congress
ISSN 0190-3918
ISBN 0-8493-0190-4 (set)
ISBN 0-8493-0191-2 (vol. I)
ISBN 0-8493-0192-0 (vol. II)
ISBN 0-8493-0193-4 (vol. III)
IEEE Computer Society Order Number 2355

Additional copies may be obtained from:

IEEE Computer Society Press	IEEE Computer Society	IEEE Computer Society
Customer Service Center	13, Avenue de l'Aquilon	Ooshima Building
10662 Los Vaqueros Circle	B-1200 Brussels	2-19-1 Minami-Aoyama,
P.O. Box 3014	BELGIUM	Minato-Ku
Los Alamitos, CA 90720-1264		Tokyo 107, JAPAN

PREFACE

The 1991 version of the International Conference on Parallel Processing is the 20th in this long and distinguished series of conferences. Over the past twenty years, the conference has grown enormously, both in terms of numbers of technical papers presented and in terms of topics and areas covered. The banquet on the Monday night before the conference celebrates the continuation of ICPP. This banquet was arranged by Professor Ming T. Liu, of Ohio State University, and features an address by Professor Duncan Lawrie, of the University of Illinois at Urbana-Champaign.

A special event on the program is the panel entitled Toward Teraflops Computing. We want to thank the six panelists for taking the time to present their ideas on this important topic. The six panelists are Chris Hsuing, Cray Research, Dave Patterson, University of California Berkeley, Justin Ratner, Intel, Burton Smith, Tera Computer, Guy Steele, Thinking Machines, and Steve Wallach, Convex Computer. This promises to be an outstanding part of the technical program.

The three Co-Chairs for the Technical Program have worked hard, to try to assure the best possible program. For the third time in the history of the conference, the number of submitted papers has exceeded 500. These papers were submitted to one of three tracks: architecture, software or applications/algorithms. Each paper was handled by one of the three Co-Chairs and was reviewed by at least two external referees. The accepted papers were classified as either a regular paper, a concise paper or a poster session paper. The brief table below summarizes the papers handled in each of the three tracks.

Area	Submitted	Regular	Accepted Concise	Poster
Architecture	253	42 (16%)	66 (26%)	45 (17%)
Software	113	25 (22%)	14 (12%)	24 (21%)
Applications/Algorithms	140	18 (13%)	30 (21%)	27 (19%)
All Areas	506	85 (17%)	110 (22%)	96 (19%)

There were 126 papers submitted from 25 foreign countries. Thus, the word "International" in the title is well justified. This participation by foreign authors and speakers is an important aspect of the broad nature of the conference.

This kind of conference depends on many volunteers. In addition to the three Co-Chairs for the Technical Program, literally hundreds of individuals have served as referees. This is a thankless, but absolutely essential, job in the structuring of a technical conference. We are very grateful to the many referees who agreed to review from one paper, to in some cases five or six papers, within our fairly tight deadlines. The degree of response helped give the conference its high degree of technical quality.

Finally, we have to thank *all* of the authors: those whose papers were accepted and those whose papers were not accepted. The authors of accepted papers had to, in many ases, alter their papers, to fit within the page limitations of concise and poster papers. Again, tight deadlines were an additional problem.

Each of the Co-Chairs needs to thank several individuals. First, Dr. Tse-yun Feng, the founder and sustainer of this conference has contributed much in the areas of local arrangements, publicity, and the overall management of this large conference. Next, Chita R. Das, of Pennsylvania State University, handled the papers submitted by authors with affiliations which created conflicts of interest. And finally, in the Architecture Area, Y.-C. Huange, R. G. Prasadh, D.-Z. Ju, P. Zievers, C.-C. Tong and A. Chandra have been of great assistance in processing submitted papers and preparing the proceedings. The support provided by my home department is also greatly appreciated.

The Program Co-Chairs hope that each attendee finds at least one item of value in the technical program. If each of you are able to do that, it was all worthwhile.

C. L. Wu
1991 Program Co-Chair: Architecture
Department of Electrical and Computer Engineering
The University of Texas
Austin, TX 78712

LIST OF REFEREES
VOLUME I - ARCHITECTURE/HARDWARE

Abraham, J.A.	Univ. of Texas at Austin
Abraham, S.	Purdue Univ.
Adams, G.B.	Purdue Univ.
Agrawal, D.P.	North Carolina State Univ.
Agre, J.	Rockwell Intl.
Aker, S.B.	Univ. of Massachusetts
Alam, M.S.	Univ. of Pittsburgh
Algudady, M.S.	Pennsylvania State Univ.
Anderson, T.	Univ. of Washington
Andrescavage, M.J.	GE Aerospace
Antonio, J.	Purdue University
Arabnia, H.R.	The Univ. of Georgia
Baer, J.-L.	Univ. of Washington
Bagherzadeh, N.	Univ. of California
Barroso, L.A.	Univ. of Southern California
Batcher, K.E.	Kent State Univ.
Beckmann, C.J.	Univ. of Illinois at Urbana-Champaign
Bennett, D.B.	Unisys EISG
Bhattacharya, S.	Univ. of Minnesota
Bhausar, V.C.	Univ. of New Brunswick
Bhuyan, L.N.	Texas A&M Univ.
Bic, L.	University of California
Blough, D.M.	Univ. of California
Boppana, R.V.	Univ. of So. California
Bourbakis, N.G.	IBM San Jose
Brice, R.	MCC
Brooks III, E.D.	Lawerence Livermore National Lab.
Browne, J.C.	The Univ. of Texas at Austin
Butner, S.E.	University of California-Santa Barbara
Byrd, G.T.	Digital Equipment Corporation
Canning, J.T.	Univ. of Lowell
Carlson, W.W.	Supercomputing Research Center
Chalasani, S.	Univ. of So. California
Chaudhary, V.	Univ. of Texas at Austin
Chen, A.C.	The Univ. of Texas at Austin
Chen, C.Y.R.	Syracuse Univ
Chen, S.-K.	University of Illinois
Cheng, J.B.	Penn. State Univ.
Cheng, K.H.	Univ. of Houston
Chi, C.H.	Philips Lab.
Chiang, C.Y.	The Univ. of Texas at Austin
Chien, M.V.	Univ. of Maryland
Chittor, S.	Michigan State University
Chiu, G.-M.	Univ. of Southern California
Choi, H.-A.	The George Washington Univ.
Choi, Y.H.	Univ. of Minnesota
Chowdhury, S.	Duke Univ.
Chuang, H.Y.H.	University of Pittsburgh
Cypher, R.	IBM Almaden Research Center
Dandamudi, S.	Carleton University
Das, C.R.	Pennsylvania State Univ.
DeGroot, D.	Texas Instrument
DeMara, R.	Univ. of Southern California
Delagi, B.	Digital Equipment Corporation

Deshmukh, R.G.	Florida Institute of Technology
Ding, J.	Texas A&M Univ.
Douglass, B.G.	Texas A&M Univ.
Dowd, P.W.	SUNY at Buffalo
Du, D.H.C.	University of Minnesota
Dubois, M.	Univ. of Southern California
Dugan, J.B.	Duke Univ.
Eggers, S.J.	University of Washington
El-Amawy, A.	Louisiana State University
Fang, J.	CONVEX Computer Corp.
Felten, E.	Univ. of Washington
Flynn, M.J.	The Johns Hopkins Univ.
Fuchs, W.K.	Univ. of Illinois
Ganapathy, K.N.	Univ. of Illinois
Garg, V.	The Univ. of Texas at Austin
Gharachorloo, K.	Stanford Univ.
Ghosal, D.	Bellcore
Ghose, K.	State Univ. of New York
Ghosh, J.	The Univ. of Texas at Austin
Gonzalez, M.J.	The Univ. of Texas
Gouda, M.	The Univ. of Texas at Austin
Grady, W.M.	Univ. of Texas at Austin
Granston, E.D.	Univ. of Illinois
Guo, Z.	University of Pittsburgh
Gupta, A.	Stanford University
Gupta, M.	University of Illinois
Han, J.-Y.	Illinois Institute of Tech.
Hayes, J.P.	University of Michigan
Hennessy, J. L.	Stanford Univ.
Herath, J.	Drexel Univ.
Higuchi, T.	Carnegie Mellon Univ.
Ho, C.-T.	IBM Almaden Research Center
Ho, K.M.	San Jose State Univ.
Hoag, J.E.	Lawrence Livermore National Laboratory
Holliday, M.A.	Duke Univ.
Hsiung, C.C.	Cray Research Inc.
Hsu, B.	Univ. of Illinois at Urbana-Champaign
Hsu, W.T.-Y.	Univ. of Illinois
Hsu, W.-J.	Michigan State Univ.
Hsu, Y.	IBM T.J. Watson Research Center
Huang, C.M.	Ohio State Univ.
Huang, K.-S.	IBM
Huang, Y.-C.	Univ. of Texas at Austin
Hughey, R.	Brown Univ.
Hulina, P.T.	Pennsylvannia State Univ.
Hurson, A.R.	The Pennsylvania State University
Hwang, K.	University of Southern California
Jain, R.	The Univ. of Texas at Austin
Jang, J.W.	Univ. of Southern California
Janssens, B.	Univ. of Illinois
Jiang, H.	Texas A&M Univ.
Jin, L.	California State Univ. - Fresno
Joshi, C.S.	Tandem Computer
Ju, D.C.	The Univ. of Texas at Austin

Kanawati, A.N.	The Univ. of Texas at Austin	Padmanabhan, K.	AT&T Bell Lab.
Kang, B.-C.	Univ. of Minnesota	Padua, D.	Univ. of Illinois at Urbana-Champaign
Karthik, S.	The Univ. of Texas at Austin	Pan, Y.	University of Pittsburgh
Kim, J.	Univ. of Michigan	Panda, D.K.	University of Southern California
Kim, J.H.	Univ. of S.W. Louisiana	Park, J.S.	Penn. State Univ.
Kim, K.H.	Univ. of California	Patel, J.H.	Univ. of Illinois
Kim, K.	University of Southern California	Peir, J-K	IBM
Kim, Y.	Univ. of Washington	Peris, V.	Univ. of Maryland
Kogge, P.M.	IBM System Integration Division	Picano, S.	Purdue Univ.
Koren, I.	Univ. of Massachusetts	Pinski, E.	Boston Univ.
Kothari, S.C.	Iowa State Univ.	Pradhan, D.K.	Univ. of Massachusetts
Kreulen, J.T.	Penn. State Univ.	Prasadh, R.G.	The Univ. of Texas at Austin
Kuck, D.J.	Univ. of Illinois at Urbana-Champaign	Qiao, C.	University of Pittsburgh
Kumar, V.	Univ. of Minnesota	Qiu, K.	Queens University
Kumar, V.K.P.	University of Southern California	Radivojevic, I.	Drexel Univ.
Kung, S.Y.	Princeton Univ.	Raghavan, R.	IBM Myers Corners Lab.
Kuo, S.-Y.	The Univ. of Arizona	Raghavendra, C.S.	Univ. of Southern California
Kurian, L.	The Pennsylvania State University	Rahmeh, J.T.	The Univ. of Texas at Austin
Latifi, S.	UNLV	Rajasekaran, S.	Penn. UNIV.
Laudon, J	Stanford Univ.	Ramanathan, J.	Michigan State Univ.
Lee, B.	Penn. State Univ.	Ramanathan, P.	The Univ. of Michigan
Lee, C.-Y.	Univ. of Maryland	Robbins, K.A.	Univ. of Texas at San Antonio
Lee, D.C.	Penn. State Univ.	Rotenstreich, S.	The George Washington Univ.
Lee, S.Y.	Cornell Univ.	Rowley, R.	Oregan State Univ.
Lee, Y.-H.	University of Florida	Saha, A.	The Univ. of Texas
Li, G.-j.	National Research Center, China	Sahni, S.	University of Florida
Li, H.F.	Concordia Univ.	Sarkar, D.	University of Miami
Li, S.-Q.	The Univ. of Texas at Austin	Scherson, I.D.	Univ. of California
Lilja, D.	Univ. of Illinois at Urbana-Champaign	Schneck, P.B.	Supercomputing Research Center
Lin, W.	Univ. of Hawaii at Manoa	Shang, W.	Univ. of SW Louisiana
Lipovski, J.	The Univ. of Texas at Austin	Shaout, A.	The Univ. of Michigan - Dearborn
Liu, J.C.	Texas A&M Univ.	Sharif, H.	Univ. of Nebraska at Omaha Campus
Liu, M.T.	Ohio State Univ.	Sharma, N.K.	University of Akron
Lopresti, D.L.	Brown Univ.	Sheu, T.L.	IBM Corp.
Lu, M.	Texas A&M Univ.	Shih, C.J.	Kent State Univ.
Marinos, P.N.	Duke Univ.	Shin, K.G.	The Univ. of Michigan
McCarthy, A.	MIT	Shing, H.	Michigan State Univ.
Mehra, P.	Univ. of Illinois	Sibai, F.N.	University of Akron
Melhem, R.	Univ. of Pittsburg	Siegel, H.J.	Purdue University
Menezes, B.L.	Univ. of Maryland	Siewiorek	CMU
Michailidis, P.	Univ. of Maryland	Singhal, M.	The Ohio State University
Mohapatra, P.	Penn. State Univ.	Sohi, G.S.	Univ. of Wisconsin
Moldovan, D.	Univ. of Southern California	Somani, A.K.	Univ. of Washington
Mudge, T.N.	University of Michigan	Song, S.P.	Advanced Micro Devices
Mukhamala, R.	Old Dominion Univ.	Sood, A.	George Mason Univ.
Muppala, J.K.	Duke Univ.	Stankovic, J.A.	Univ. of Massachusetts
Murata, T.	Univ. of Illnois at Chicago	Starzyk, J.A.	Ohio Univ.
Nanda, A.K.	Texas A&M Univ.	Stirpe, P.A.	Boston University
Narahari, B.	The George Washington Univ.	Stone, H.S.	IBM T.J. Watson Research Center
Ni, L.	Michigan State Univ.	Stout, Q.F.	University of Michigan
Oner, K.	Univ. of Southern California	Su, H.-M.	Univ. of Illinois at Urbana-Champaign
Opitz, D.W.	Univ. of Wisconsin - Madison	Suri, N.	Univ. of Massachusetts
Oruc, A.Y.	Univ. of Maryland	Swartzlander, E.	The Univ. of Texas
Oyang, Y.J.	National Taiwan Univ.	Szymanski, T.	Columbia Univ.

Taylor, J.M.	New Mexico State Univ.	Xu, Z.	Polytechnic Univ.
Thapar, M.	Stanford Univ.	Yang, G.-C.	Univ. of Illinois at Urbana-Champaign
Torng, H.C.	Conell Univ.	Yang, M.K.	Pennsylvania State Univ.
Troffer, J.	AT&T Bell Lab.	Yang, Q.	The Univ. of Rhode Island
Tzeng, N.F.	Univ. of S.W. Louisiana	Yang, W.	AT&T Bell Lab.
Veidenbaum, A.V.	Univ. of Illinois	Yang, X.D.	University of Regina
Wah, B.W.	Univ. of Illinois	Yao, Y.-W.	Ohio State Univ.
Waldecker, B.	The Univ. of Texas at Austin	Yeh, Y.-M	Penn State Univ.
Wang, J.C.	Tandem Computers Inc.	Youn, H.Y.	Univ. of North Texas
Wang, W.-H.	IBM Austin	Youssef, A.	The George Washington Univ.
Wang, Y.-M.	Univ. of Illinois	Yu, C.S.	Penn. State Univ.
Warhor, C.	CONVEX Computer Corp.	Yu, P.S.	IBM T.J. Watson Research Center
Wei, S.	Rutgers Univ.	Zhang, C.N.	University of Regina
Weiss, S.	Univ. of Maryland	Ziever, P.	The Univ. of Texas at Austin
Werth, J.	The Univ. of Texas at Austin	Zucker, R.	Univ. of Washington
Xu, X.	University of Minnesota		

AUTHOR INDEX -- FULL PROCEEDING

Volume I -- Architecture
Volume II -- Software
Volume III -- Algorithms and Applications

TABLE OF CONTENTS

VOLUME I: ARCHITECTURE

(R): Regular Paper
(C): Concise Paper
(P): Poster Paper

SESSION 1A: System Architectures

SESSION 1B: Resource Allocation

SESSION 1C: Cedar System

xiv

POSTER SESSION

MULTIPLE INTERLEAVED BUS ARCHITECTURES

Jonathan Bertoni[1] and Wen-Hann Wang
Future Systems Technology
IBM Advanced Workstation Division
Austin, TX 78758

Abstract – In this paper, we propose and analyze a new multiple bus scheme, called *multiple interleaved busses*. This scheme is a generalization of previous approaches, and attempts to balance performance and cost tradeoffs in a snoopy-cache multiprocessor environment. A queuing network model is used to derive performance figures. Our results show that multiple interleaved busses perform almost as well as multiple independent busses, but with simpler and less costly implementation. Furthermore, multiple interleaved busses are shown to deliver much better performance than interleaved busses when the skew of accesses across the interleaves is large. Overall, the multiple interleaved bus approach is shown to be a good choice in high-performance shared bus systems.

1. Introduction

Memory bandwidth demands for shared-bus multiprocessor systems are growing faster than available bus bandwidth. Even as increased bandwidth is needed, low latency remains important to system performance. Two-level caches and higher bus transaction rates have been suggested as ways of dealing with these problems. The use of multiple busses for improving system bandwidth and latency also has been widely discussed in the literature. These approaches can complement the use of faster bus mechanisms and two-level caches. In this paper, we will introduce and study a generalization of the multiple bus approaches, and try to provide some insight into the tradeoffs in the design of multiple bus systems.

We categorize the systems described in previous literature into three groups: wide busses, independent busses, and interleaved busses. A *wide bus* [6] is simply a single bus with a large number of data lines. It offers higher memory bandwidth than a narrow bus by fetching more data in parallel from sequential locations in memory. For example, Sequent moved from a 32-bit system bus to a 64-bit bus when they upgraded their multiprocessor product from the Balance series to the Symmetry series[9].

Despite its advantage in simplicity, the wide bus approach has several limitations. A wide bus does not scale well. The width of a bus is limited by the cache line size. The line size cannot be too large because that can increase the degree of false sharing in a parallel processing environment[3]. Furthermore, to utilize a wide bus fully the memory system must be capable of fetching a cache line in one memory cycle. If this cannot be done, the data must be fetched in a series of cycles and queued if any use of the bus width is to be made. Fetching larger amounts of data at a time requires either larger chip counts per memory banks or memory chips that return more bits in parallel. Thus, we cannot increase the bus width arbitrarily.

Another group of proposals advocate the use of *multiple independent busses* [11, 5, 4, 15, 14, 10, 2, 8] to increase system performance. In such a system, more than one bus is provided and each can be used independently by any processor to communicate with any memory module. The SYNAPSE N+1 System[4], for example, used two independent busses to boost performance.

Although multiple independent busses have

[1]This author was supported in part by an ATT Bell Laboratories Fellowship.

been shown to deliver good performance, the cost of such a system can be quite high, depending on the number of busses to be supported. For a machine using a snoopy bus hardware cache coherence protocol [9], the cost is especially high because snooping must be done on all busses. For example, in a system with four processors and four busses, the cache for each processor must snoop all four of the busses, for a total of sixteen snoop interfaces. Using the coherence methods described in [9], [7] and [4], each interface would require a complete directory for the cache and other supporting circuits, resulting in a large degree of complexity that not only dramatically increases the design cost but also can greatly degrade the cache performance[7].

The final alternative to be discussed is *interleaved busses*, which have been used to increase bandwidth without increasing overhead costs as much as the multiple independent busses scheme. Using this method, each bus handles accesses to a certain subset of memory, herein called a *cache line group*. The IBM TOP-1 multiprocessor [12], for example, has two busses, one for even-addressed cache lines and the other for odd-addressed cache lines. The cache is interleaved accordingly, and each cache bank only requires a snoop interface to its associated bus. Therefore, the cost of this scheme can be considerably less than that of multiple independent busses.

An interleaved bus works well if the memory accesses are distributed evenly across the cache line groups. Our simulations show that a skew across memory addresses can occur, which results in a high degree of contention on one of the busses and degrades the overall performance.

All of the above schemes have serious limitations. Wide busses do not scale, multiple independent busses are too complex and too costly, and interleaved busses can suffer from skewed memory reference patterns. This leads us to explore a wider design space and propose a compromise scheme called *multiple interleaved busses*. This scheme attempts to balance performance and cost tradeoffs in a snoopy-cache multiprocessor environment. Our performance studies show that multiple interleaved busses perform almost as well as multiple independent busses, but with simpler and less costly implementation. Furthermore, multiple interleaved busses are shown to deliver much better performance than interleaved busses when the skew of accesses across the interleaves is large. Overall, the multiple interleaved busses approach seems to be a good choice for high-performance shared-bus systems.

Section 2 describes the system we propose, and in Section 3 the modeling methodology is presented. In Section 4, we outline the configurations and parameters for the simulations, and in Section 5 we present and discuss the results of the analysis. Section 6 summarizes our findings and draws some conclusions.

2. The Proposed Architecture

In this section, we describe our proposal of organizing multiple busses, which we call *multiple interleaved busses*. In this scheme we consider using a small number of independent busses (two, for example), each of which is itself a collection of interleaved busses. The small degree of independence helps maintain performance at a reasonable level for programs that demonstrate substantial skew in their memory reference patterns. Furthermore, by using interleaving, a designer can reduce the cost and complexity of the system substantially as compared to a fully independent scheme, but still allow a high degree of parallelism. The optimal choice of the degree of independence and interleaving for a system can be chosen by considering the cost-performance ratios under the implementation technologies available. Such an analysis is beyond the scope of this paper.

We denote each bus configuration with an ordered pair, (m, n), where m is the number of independent busses and n is the degree of interleaving of each independent bus. For example, a $(2, 4)$ system has two independent busses, each of which is implemented using four interleaved busses. Thus, a multiple independent bus scheme is a $(m, 1)$ system in this notation, where m is the number of busses. Likewise, a purely interleaved system is denoted $(1, n)$, where n is the level of interleaving.

In Figure 1, a $(2, 2)$ system with one CPU per second-level cache is shown. Each line represents a collection of physical wires, all forming paths of the same width. A total of four

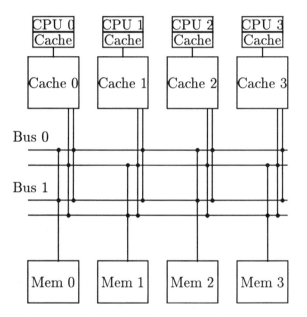

Figure 1: A $(2, 2)$ Multiple Interleaved Bus System

caches are shown, although this number could be increased or decreased without changing the bussing system. There are a total of four busses: two independent busses, labeled "Bus 0" and "Bus 1", each consisting of two interleaves. The memory banks are single-ported and are assigned to alternate bus interleaves.

As stated earlier, this family offers the same possibilities for parallelism as previously proposed systems, but covers a wider range of cost-performance tradeoffs. Another advantage of our scheme is that in a multiprocessor with a shared second-level cache [1], each second-level cache can be interleaved. Such a design

would allow simultaneous access by the first-level caches, provided that each access was to a different cache line group. This capability allows even more interprocessor parallelism, potentially reducing the cost of providing second-level cache bandwidth.

In the following sections, we will attempt to provide some quantitative data to show the performance of our scheme versus other bussing structures. We will also briefly present some results for systems with interleaved shared second-level caches.

3. The Modeling Methodology

In order to analyze the performance of these systems, a queuing network model was developed and analyzed with various parameters. We will discuss first the hardware model, and then the transactions performed on that model.

The Hardware Model

The hardware model follows Figure 1 closely. In Figure 2 a diagram of a sample model shows the queuing resources, delay centers, and transaction routes. Delay centers are shown in double boxes; queuing resources are shown using single boxes. A $(2, 2)$ system is shown. It has three processor modules, with three second-level caches. Each second-level cache is interleaved and so requires two queuing resources. Likewise, each independent bus is modeled using two queuing resources, one for each interleave. The labels $i0$ and $i1$ are used to mark each independent half of these interleaved resources. A total of two memory banks are attached to this sample system.

There is one queue for each queuing resource, except that the two corresponding interleaves of each independent bus share a single queue, i.e., Bus 0 interleave $i0$ and Bus 1 interleave $i0$ share one queue. The CPU delay centers are assigned one client for each CPU attached to the corresponding second-level cache. For example, if two customers are assigned to each CPU delay

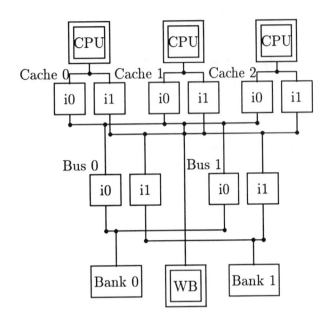

Figure 2: The Queuing Network Model

center, the system contains a total of six processors, two assigned to each second-level cache. Likewise, the writeback delay center is assigned one customer for each processor in the system.

The modeled systems all have interleaved second-level caches. These second-level caches implement a write-back policy. They have a two-cycle latency for reads and require six cycles total for a cache read, five for a write. The line size is 32 bytes, and data can be read from the bus and placed in a cache as fast as the bus can deliver data. A cache miss requires one cycle before the fill can be started.

Each interleave of each bus is modeled as an independent resource. Bus address and data lines are assumed to be shared, so that data transfer and address transfers may not proceed in parallel on the same bus. Invalidations require two cycles on the bus, but do not require any other resources. Data and addresses each require one cycle on the bus to be communicated. The bus width studied is 64 bits and there are four busses in the systems being studied.

A memory bank is modeled as having a single port and requires four cycles to fetch a line.

Subsequent 64-bit portions are delivered one per cycle. Placing the address of the memory on the bus requires one cycle, and the data is then sent during the following four cycles. Thus 11 cycles total are required for a cache line fill if no queuing time is needed: two to deliver the address to memory, and nine to return the data. Each memory bank contains write buffers so that the bank need not be allocated on write operations. Likewise, the memory units always can buffer data just read until a bus can be obtained to return it to a processor, although for reads the memory bus is allocated for the entire fetch cycle.

The Transaction Types

The model supports five types of transactions: a second-level cache read, a second-level cache write, a memory read, a cache coherence invalidation operation, and a write-back operation. The second-level cache operations correspond to misses in the first-level cache that hit in the second-level cache and need not use the bus. The memory read operation fetches a line from memory to a cache on a cache miss. The write-back operation represents a delayed write-back from a write buffer [13]. A cache coherence transaction is used to represent the load from the hardware cache coherence protocol [3]. Our study considered only invalidation-based protocols.

The processor computation and first-level cache phase is represented by a delay center, labeled *CPU* in Figure 2. Upon leaving this delay center, the customer must acquire a cache in order to perform any operation. For operations that correspond to second-level cache hits, the second-level cache interleave is the only resource that need be acquired. After delaying for the time required for a first-level cache fill, the cache is released and the processor computation phase is reentered.

Second-level cache misses must proceed to the bus subsystem. Because the model is for a write-back system, all misses are read misses. Once an appropriate bus has been acquired, the customer places the the memory address on the bus, represented by a delay center. Next the specific transaction type is chosen randomly, using the distribution function given as a parameter. For a cache coherence transaction, the customer simply transfers the address to the bus, holds the bus for two cycles to permit broadcast and acknowledgement. Then, it releases the cache and returns to the computation phase.

A memory read transaction is more complicated. Although buffering can remove the necessity of waiting when a memory write is performed, on a read miss computation cannot proceed until the data has been returned. For this reason, the model simulates each memory bank explicitly, and allows at most one read to occur per memory bank at a time. Split read transactions are assumed. After placing the memory address on the bus, the bus is relinquished, because we assume the address is buffered in the memory controllers.

Once the appropriate memory bank has been chosen, based on the memory reference distribution the customer queues for access. Upon receiving it the customer enters a delay center that represents the memory read time. When the data has been read from the memory array into a buffer in the memory bank, the bank is released and the customer arbitrates for a bus back to the processor, and then transfers the data and finally releases the bus. The customer then reenters the compute delay center.

In addition to the processor model transactions, there is a write buffer transaction type to simulate the bus traffic generated by actual memory write operations. Such transaction begin in a delay center, labeled *WB* in Figure 2, representing the write buffers. Upon leaving that center, the customer obtains the bus and writes the data to the memory bank buffers, and then release the bus and return to the delay center.

4. The Modeling Data

Here we present the configurations simulated and the data used as parameters to the model.

Hardware Configurations

Many configurations are possible, and only a small subset of them could be simulated. We chose to study groups of systems with the same number of busses, and study how performance was affected by the various combinations of interleaving and independent busses. We compared systems that have a total of 256 data lines on the bus. Table 1 shows the systems studied. All were studied with 8, 16, 32, and 64 caches

Table 1: Bussing Systems Simulated

Independent Busses	Interleaving	Bus Widths
1	4	64
2	2	64
4	1	64

attached, with one processor per second-level cache. In addition, systems with 4, 8, 16, and 32 caches were studied with a load of two processors per second-level cache. The number of memory banks per system was fixed at four for all simulations.

Memory Reference Data

As discussed previously, some simulations have shown that memory transactions are not distributed evenly across the address space. Three different distributions of transactions across the line groups were studied. They are taken from three programs of the SPEC benchmark suite, namely, the Lisp interpreter (*li*), a matrix multiplication routine (*matrix300*), and a switching function minimization program (*eqntott*). The distributions are shown in Table 2. Of the three benchmark we studied, *matrix300* has almost no skew, *eqntott* shows moderate skew, and *li* has significant skew.

We fix the transaction load by simulating a 93% hit rate on first-level data references, 98%

Table 2: Memory Transaction Distributions

Program	Group			
	0	1	2	3
li	0.025	0.656	0.242	0.077
matrix300	0.249	0.257	0.247	0.247
eqntott	0.317	0.212	0.252	0.219

on first-level instruction references, and a 50% hit rate for the second-level cache. Loads were set at 66% of all first-level cache operations, stores at 34%. One of every 3.5 instructions was modeled as making a data reference. The write-back rate was set at a tenth of the level two cache miss rate. Invalidations were set at 1% of all bus transactions.

5. Simulation Results

Tables 3, 4, and 5 show the second-level cache fill times required by various system con-

Table 3: Memory Read Performance on *li*

System	CPU's Per System			
	8	16	32	64
(1,4)	12.3	15.8	59.6	205.0
(2,2)	11.4	12.1	17.5	63.0
(4,1)	11.4	12.0	16.6	61.2

Table 4: Memory Read Performance on *matrix300*

System	CPU's Per System			
	8	16	32	64
(1,4)	11.6	12.5	15.3	25.4
(2,2)	11.2	11.6	13.1	18.8
(4,1)	11.2	11.4	12.3	15.4

figurations. This is a measure of the response time of memory reads for the three skew rates from the SPEC benchmarks. All simulations ran to at least $150,000$ transactions.

As can be seen from Tables 3, 4, and 5, a $(2,2)$ system performs almost as well as the $(4,1)$ system. On the other hand, the $(1,4)$ system shows serious degradation on programs

Table 5: Memory Read Performance on *eqntott*

System	CPU's Per System			
	8	16	32	64
(1,4)	11.6	12.6	15.8	33.3
(2,2)	11.2	11.6	13.4	22.3
(4,1)	11.2	11.4	12.3	16.1

with large amounts of skew, and somewhat degraded performance with moderate skew. This degradation occurs primarily with larger systems when the queuing time grows large. The full results for the *li* benchmark show that the memory bank becomes the major bottleneck to system performance, and this is the reason the $(4,1)$ system does not perform much better than the $(2,2)$ system.

Shared-Cache Extension

As stated in Section 2, another advantage of our scheme is that the second-level cache is interleaved and therefore has the bandwidth potential to serve multiple processors. To assess this shared second-level cache potential we extended our performance model to derive the portion of cycles per instruction contributed by cache hierarchy access.

Table 6 shows the cache cycles per instruction for a $(2,2)$ system under various skew rates, across different numbers of CPU per system. From Table 6 we observe that allowing two CPU's per cache does not degrade performance excessively, and can be a good price-performance choice. Note that for larger systems, again, the bus and memory banks are the primary bottlenecks, and so systems with two CPU's per cache perform almost as well as systems with one CPU per cache.

6. Conclusions and Future Work

In this paper, we have proposed and analyzed a new multiple bus scheme, called *multiple interleaved busses*. This scheme is a generalization of previous approaches, and attempts

Table 6: Cache CPI For One and Two CPU's Per Cache

Skew	CPU's / Cache	CPU's Per System			
		8	16	32	64
li	1	0.54	0.56	0.69	1.80
	2	0.60	0.62	0.76	1.81
matrix300	1	0.53	0.54	0.58	0.76
	2	0.59	0.59	0.63	0.79
eqntott	1	0.53	0.54	0.59	0.86
	2	0.58	0.59	0.64	0.86

to balance performance and cost tradeoffs in a snoopy-cache multiprocessor environment. A queuing network model was used to derive performance figures. Our results show that multiple interleaved busses perform almost as well as multiple independent busses, but with simpler and less costly implementation. Furthermore, multiple interleaved busses are shown to deliver much better performance than interleaved busses when the skew of accesses across the interleaves is large. Overall, the multiple interleaved bus approach is shown to be a good choice in high-performance shared bus systems.

Further studies of the number of CPU's per cache, skew rates for a wider range of applications, and larger systems (with more processors and bus lines) are being conducted to illuminate the full design space. Also, the use of wider busses are being examined, as well as the interaction of cache line size and miss and invalidate rates. We also plan to investigate a range of hit ratios and memory bank architectures, including dual-porting, to improve performance on programs with large amount of skew.

Acknowledgements

We thank Jean-Loup Baer for his advice on this study. We are grateful to Hank Chang, Edward MacNair, and Paul Loewner for their technical support for the RESQ queuing package. Jim van Fleet helped with the vagaries of VM/CMS and REXX, as well as contributing greatly to moral support. Khoa Nguyen supplied encouragement and freedom to explore our ideas. Finally, we would like to thank Kimming So and Doug Steves for their helpful comments.

References

[1] J.-L. Baer and W.-H. Wang. Multi-level cache hierarchies: Organizations, protocols and performance. *Journal of Parallel and Distributed Computing*, 6(3):451–476, 1989.

[2] C. Das and L. Bhuyan. Bandwidth availability of multiple-bus multiprocessors. *IEEE Transaction on Computers*, 34:918–926, October 1985.

[3] Eggers, S. J. and R. H. Katz. The effect of sharing on the cache and bus performance of parallel programs. In *ASPLOS-III Proceedings*, pages 257–270, April 1989.

[4] Frank, S. and A. Inselberg. Synapse tightly-coupled multiprocessors: A new approach to solve old problems. In *Proc. National Computer Conference*, pages 164–169, 1984.

[5] Goyal, A. and T. Agerwala. Performance analysis of future shared storage systems. *IBM Journal of Research and Development*, 28:95–108, January 1984.

[6] Hopper, A., A. Jones, and D. Lioupis. Multiple vs wide shared bus multiprocessors. In *Proc. 16th Symposium on Computer Architecture*, pages 300–306, 1989.

[7] Katz, R., S. Eggers, D. Wood, C. Perkins, and R. G. Sheldon. Implementing a cache coherence protocol. In *Proc. 12th Symposium on Computer Architecture*, pages 276–283, 1985.

[8] Lang, T., M. Mateo and I. Alegre. Bandwidth of crossbar and multiple-bus connections for multiprocessors. *IEEE Transaction on Computers*, 31:1227–1234, December 1982.

[9] Lovett, Tom and S. Thakkar. The Symmetry multiprocessor system. In *Proc. 17th International Conference on Parallel Processing*, pages 303–310, 1988.

[10] Marsan, M. A. and M. G. Gerla. Markov models for multiple bus multiprocessor system. *IEEE Transaction on Computers*, 31:239–248, March 1982.

[11] Mudge, T. N., J. P. Hays, G. D. Buzzard, and D. C. Winsor. Anaylsis of multiple-bus interconnection networks. *J. Parallel and Distributed Computing*, 3:328–343, September 1986.

[12] Oba, N., A. Moriwaki, and S. Shimizu. Top-1: A snoop-cache-based multiprocessor. In *IEEE International Phoenix Conference on Computers and Communication*, pages 101–108, March 1990.

[13] Przybylski, S. and J. Hennessy. Performance tradeoffs in cache design. In *Proc. 15th Symposium on Computer Architecture*, pages 290–298, 1988.

[14] Yang, Q. and L. Bhuyan. A queueing network model for a cache coherence protocol on multiple-bus multiprocessors. In *Proc. 17th International Conference on Parallel Processing*, pages 130–137, 1988.

[15] Yang, Q. and L. Bhuyan. Performance of multiple-bus interconnections for multiprocessors. *J. Parallel and Distributed Computing*, 8:267–273, January 1990.

The Performance of Hierarchical Systems with Wiring Constraints

William Tsun–yuk Hsu and Pen–Chung Yew
Center for Supercomputing Research and Development
University of Illinois at Urbana–Champaign
104 S. Wright Street, Urbana IL 61801

Abstract

Packaging constraints, such as package pinouts and system wiring, have major impacts on the construction and performance of large multiprocessor networks. Packaging imposes a physical hierarchy on a multiprocessor system. We examine ways to exploit this hierarchy by grouping processors into clusters. We combine previous analyses on clustering and wiring constraints to compare the effects of packaging on the performance of hierarchical and non–hierarchical system organizations, focusing on clustered and flat hypercubes. We find that clusters have better performance than previously predicted. Under uniform traffic, clustered hypercubes have lower saturation loads than unclustered hypercubes, but substantially better queueing performance prior to saturation. If a significant proportion of traffic is directed to processors within the same cluster, clustered systems have comparable saturation loads and superior queueing performance. Depending on the expected operating environments, clustering is an attractive option when packaging constraints are considered.

1. Introduction

In recent years, there has been increasing experience with the construction of large multiprocessor systems. For example, the NCUBE/ten [HaMS86] has 1024 processors connected as a hypercube, and the Connection Machine [Hill85] has up to 64k processors. According to researchers such as Dally [Dall90], systems based on VLSI components are wire–limited, and wiring (instead of logic) will be a major component of system cost.

With higher degrees of integration, packaging constraints have become very influential in determining system performance. In addition to wiring cost, the limited numbers of pins available on chips and PC boards restrict the number of channels and channel widths (and hence, bandwidth) of network switches. A message has to be broken into many nibbles, and a narrow channel means that a message has to be sent in many clock cycles. For example, the NCUBE/ten has 64 nodes on a single board. If the system is a 1024–node hypercube, and each board has 512 pins for network connections, there are 64*4*2=512 uni-

directional channels that connect to other boards, and each channel is only one–bit wide. Traditional network analysis has mostly assumed a constant channel width and single–nibble messages. Packaging constraints demand a different approach.

Researchers such as Dally [Dall90] and Abraham [AbPa90] have tried to incorporate packaging constraints into network analysis. Dally used *bisection width* [Thom79] as an estimate of wiring area, and compared networks of constant bisection width, but variable channel width and channel degree. *Channel width* is the number of bits in a network channel, and hence the number of bits in a message nibble. *Channel degree* is the number of channels at a node. The *pinout* of a node (or switch or board) is the total number of pins that are available for data paths out of the node (or switch or board); for simplicity, we ignore pins for control signals. Abraham performed detailed queueing analysis, first keeping constant switch pinout, then bisection width.

Both Dally and Abraham studied "flat" systems, i.e., the network is non–hierarchical, and nodes are completely symmetric. However, when studying the impact of packaging constraints, we observe that the physical packaging of most large–scale digital systems immediately imposes a hierarchy on the system. For example, one level of the hierarchy could be the processors that share the same board, and the next level would be all the boards that comprise the system. It seems natural to organize processors on the same board together as a cluster. In this paper, we will look at ways to exploit this hierarchy.

Hierarchical or clustered systems have been studied in the literature. [WuLi81] analyzed clusters of processors using shared buses. [AgMa85] studied clusters of processors with crossbar switches. [Carl85] proposed connecting processors locally as meshes, and using a global mesh to connect the clusters. [DaEa90] studied two–level systems, and compared a variety of network topologies for each of the levels. However, these studies have not incorporated into their analyses the effects of packaging constraints on system performance, and have often restricted the discussion to traffic patterns with high locality. In this paper, we study hierarchical systems with packaging constraints, using wiring costs as a basis for cost comparisons. We found that when packaging constraints are taken into account, hierarchical systems give better performance than previously predicted, even in environments where there is little locality in traffic.

This work was supported in part by the **National Science Foundation** under Grants No. US NSF MIP-8410110 and US NSF (hy89-20891, the **US Department of Energy** under Grant No. US DOE DE-FG02-85ER25001, the **NASA Ames Research Center** under Grant No. NASA NCC 2-559, and Sun Microsystems.

In Section 2, we define the hierarchical structures studied in this paper. In Section 3, we compare the performance of flat and hierarchical hypercubes, under uniform and non–uniform traffic. Conclusions are drawn in Section 4.

2. System configurations

For the purposes of this paper, we will focus on a specific approach to clustering, and limit our discussion to hypercube–like structures. Consider a *flat hypercube (FH)*: each processor on a board has its own channels and switch connections to processors on other boards. For example, in a 1024–processor FH with 64 processors on each board and board pinout of 512, there can be 128 1–bit wide channels between any two directly connected boards, because wiring space and pin availability are limited.

Instead of letting each processor have its own dedicated connections, we make all processors on a board share a single global switch that connects to global switches on different boards. We call these *clustered hypercubes (CH)*. In the example above, an analogous CH would have two 64–bit channels between two boards. Note that the network topology and board pinout do not change. We are using both Abraham's constant pinout constraint [AbPa90] and Dally's constant bisection width constraint [Dall90], and comparing systems with similar wiring costs.

A high bandwidth shared global switch can be constructed by dividing the data path into slices, and having a single crossbar chip handle each slice. One slice of the data path determines the setting of the switch, and broadcasts the state information to all the other slices. This avoids the problems discussed in [FrWT82].

2.1 Definitions

Let N_1 be the number of processors in the system, and $n_1 = \log N_1$. The processors are numbered from 0 to $N_1 - 1$, and $p_{n_1-1} \cdots p_0$ is the binary representation of a processor address. In a hypercube, processors that differ in only one bit of their binary addresses are connected directly. Let N_0 be the number of processors in a cluster (or on a board), and $n_0 = \log N_0$.

In an FH, assume that a subcube of size N_0 is placed on a board. If two processors are on the same board, the top $n_1 - n_0$ bits of their addresses are identical. Each processor has associated with it a crossbar switch of fanout n_1. n_0 of these channels go to processors on the same board, and $n_1 - n_0$ go to processors on different boards. Figure 1a shows a 16–processor FH with four processors on a board.

Let W be the board (or cluster) pinout, L the message length (in bits), t_m the total number of nibbles in a message, and t_h the number of nibbles in the message header. We assume that the message header is

n_0 bits long, since that is the length of a processor address. The number of channels going outside a board is $2N_0(n_1 - n_0)$. The channel width is $C = \left\lfloor W/(2N_0(n_1 - n_0)) \right\rfloor$. The number of crosspoints in the system is $CN_1(n_1 + 1)^2$.

In a CH, each board or cluster of N_0 processors has associated with it a crossbar switch of fanout $n_1 - n_0$. The upper $n_1 - n_0$ bits of the processor number are the address of the cluster, and the lower n_0 bits represent the address of the processor within that cluster. The N_0 processors on a board are connected to the global switch via a system of k buses. The use of buses ensures that processors have equal access to the global switch, and no single connection in a cluster becomes a bottleneck.

There are many possible ways to configure the buses. We arbitrarily connect processor P to bus K if P mod k = K. Each subcluster of N_0/k processors has its own cluster–to–global and global–to–cluster buses, and each bus has its own port to the global switch. Traffic local to the cluster would still go through the global switch, from cluster bus to cluster bus, but does not route through the global network. Figure 1b shows a 16–processor CH with four processors on a board, k=2 buses, subclusters of size 2. Processors x_1x_000, x_1x_001, x_1x_010, and x_1x_011 are grouped together as cluster x_1x_0. Processors x_1x_000 and x_1x_010 are on bus 0, and x_1x_001 and x_1x_011 are on bus 1. Cluster x_1x_0 is connected via hypercube connections to clusters $\bar{x}_1x_0$ and $x_1\bar{x}_0$. Hence, the system might be viewed as a "super–hypercube" with N_1/N_0 "supernodes", with a cluster of N_0 processors at each supernode.

In a CH, each cluster has a crossbar switch of size $n_1 - n_0 + k$, where k is the number of ports from the cluster to the global switch. There are $2(n_1 - n_0)$ channels going outside a board. The channel width is $C = \left\lfloor W/(2(n_1 - n_0)) \right\rfloor$. The number of crosspoints in the global switch is $CN_1(n_1 - n_0 + k)^2/N_0$. The number of crosspoints in the buses is $Ck*2N_0/k = 2CN_0$.

Figure 2 compares the crosspoint costs of FHs and CHs, with constant wiring cost and constant board pinout, for $N_1 = 1024$ (trends are similar for other system sizes). For small cluster sizes, the crosspoint cost of an FH is comparable to that of a CH. For large cluster sizes, the FH actually has more crosspoints than a CH.

Our approach to clustering is also applicable to other network topologies where processors are grouped together in a pin–limited package. For example, a flat mesh has $8\sqrt{N_0}$ channels per board, but a clustered mesh would have only 8 channels, and each channel would be $\sqrt{N_0}$ times as wide.

In the traditional *uniform traffic* model, each processor generates a message per cycle with probability

m_g, and the message is directed at all other processors with equal probability. To model locality of traffic, let α be the probability that a generated message has its destination within the same cluster. Hence, in a 1024 processor system with 16-processor clusters with $m_g=0.1$ and $\alpha=0.4$, a processor generates a local message with probability 0.04, and a global message with probability 0.06. If we set $\alpha=(N_0-1)/(N_1-1)$, we reduce to uniform traffic.

2.2 Preliminary performance estimates

We will refer to the channels between different boards as *global channels*. A bus that routes traffic from the cluster to the global switch is called a *cluster-to-global bus*, and a bus that routes traffic from the global switch to the cluster is called a *global-to-cluster bus*. For FHs, there are *local channels* that connect processors on the same board; these are identical to global channels in the FH. Let m be the traffic rate (in nibbles per cycle) on a global channel. Let P_t be the probability of termination, i.e., the probability that a message received on a channel will terminate at that node. For the flat systems, m is also the traffic rate on a local channel. For an FH, under uniform traffic, from [AbPa90], we have $P_t=(2N_1-2)/(n_1 N_1)$, which is approximately $2/n_1$. $t_m=\lceil L/C \rceil$, and $m=t_m m_g/(n_1 P_t)$. For a CH, under uniform traffic, the total rate of cluster-to-global traffic is $m'_g=m_g N_0 \dfrac{N_1-N_0}{N_1-1}$.

$P_t=\dfrac{2N_1/N_0-2}{(n_1-n_0)N_1/N_0}$, which is approximately $2/(n_1-n_0)$. $t_m=\lceil L/C \rceil$, and $m=t_m m'_g/((n_1-n_0)P_t)$. Note that for both systems, P_t is approximated by the reciprocal of the average number of hops taken by a message in the network.

If we set $m=1$ nibble per cycle for saturation, the traffic rate from the cluster processors to the global switch is approximately $m_s=t_m m_{g,sat} N_0 = 2(N_1-1)/N_1$. $1 < m_s < 2$ nibbles per cycle will saturate a single channel with the same bandwidth as the global channels. This is why we need $k>1$ buses to connect the cluster to the global switch.

We will model each bus as a single queue, and queueing delays approximate closely bus contention delays. We will use the terms "bus" and "queue" interchangeably for the cluster/global connections. We assume that all the channels of the global switch have the same bandwidth.

We will adapt the analyses in [AbPa90] for message switching and partial cut-through, in clustered systems. Partial cut-through is similar to the "wormhole" routing in [Dall90]. In message switching for multiple-nibble messages, an entire message has to arrive at the output port of a switch before it can be forwarded. Partial cut-through allows a message with enough routing information to "cut-through" to the next switch, even if part of the message has not arrived [AbPa90].

In [AbPa89], two types of routing schemes for hypercubes are studied. In *random routing*, a message at a switch takes at random any one of the possible intermediate routes. In *left-to-right (LR) routing*, a message takes the intermediate route that corresponds to the leftmost (highest order) bit of its destination address. [AbPa89] showed that LR routing performs slightly better than random routing, at heavy network loads. For convenience of analysis, we have assumed random routing for our CHs, and LR routing for our FHs. (Hence, our analysis is slightly biased in favor of FHs.)

3. Performance evaluation

3.1 CH: performance analysis

Consider a cluster split into k subclusters, each with its own bus to and from the global switch. Let $\dbinom{i}{j}$ be the binomial coefficient in i,j. First consider queueing delays in the global switch. Let B be the number of global channels routing to other clusters (i.e., $B=n_1-n_0$). The total traffic rate (in nibbles per cycle) on each port is $m_s=t_m m_g N_0/k$, and the traffic rate on a global channel $m=(1-\alpha)k m_s/(BP_t)$. The probability of i arrivals from the cluster at a global channel is

$$c_i=\binom{k}{i}(1-(1-\alpha)m_s/B)^{k-i}((1-\alpha)m_s/B)^i$$

for $i \leq k$, and 0 otherwise. The probability of i arrivals from global channels at a global channel is

$$d_i=\binom{B-1}{i}\left(\frac{m(1-P_t)}{B-1}\right)^i\left(1-\frac{m(1-P_t)}{B-1}\right)^{B-1-i}$$

for $i \leq B-1$, and 0 otherwise. The probability of i total arrivals at a global channel will be $a_i=\sum\limits_{j=0}^{i}c_j d_{i-j}$. The probability of having i items in a global queue is $b_i=b_0 a_i+\sum\limits_{j=0}^{i}a_j b_{i-j+1}$, and $m=1-b_0$.

Next look at the arrivals at each of the cluster-to-global buses, which we model as queues. The probability of i arrivals at each queue is

$$e_i=\binom{N_0/k}{i}(t_m m_g)^i(1-t_m m_g)^{N_0/k-i}.$$

The probability of having i items in the queue is $f_i=e_i f_0+\sum\limits_{j=0}^{i}e_j f_{i-j+1}$, and $m_s=1-f_0$.

Consider arrivals at each of the global-to-cluster queues. At each global-to-cluster queue, the probability of i arrivals from the cluster itself will be

$$x_i = \binom{k}{i}(\alpha m_s/k)^i(1-\alpha m_s/k)^{k-i}.$$

The probability of i arrivals at each queue from a global channel is

$$g_i = \binom{B}{i}(mP_t/k)^i(1-mP_t/k)^{B-i}.$$

The probability of i total arrivals at a queue is then $y_i = \sum_{j=0}^{i} x_j g_{i-j}$. The probability of i items in the system is

$$h_i = y_i h_0 + \sum_{j=0}^{i} y_j h_{i-j+1}.$$ As in [AbPa90], the average message delay is

$$T = t_h(1 + \frac{1-\alpha}{P_t}) + t_m(1 + \frac{(1-\alpha)}{mP_t}\sum_i ib_i + \frac{1}{m_s}\sum_i if_i + \frac{1}{m_s}\sum_i ih_i),$$

for message switching.

Figures 3a–6a compare message delays (in network cycles) obtained from analysis and simulations, for various system parameters, under uniform traffic. For most network loads, the difference between analytical and simulation results is around 5%.

We can obtain network performance under partial cut–through with some simple adjustments to the analysis. Partial cut–through pipelines the message through the network. In each hop, the message has to wait for a period equal to only its queueing delay plus t_h, and not its queueing delay plus its service time (which is t_m+t_h). Since a message takes an average of $1/P_t$ hops, and only $1-\alpha$ of these messages traverse the global network, the average savings from partial cut–through is $(1-\alpha)t_m/P_t$. The average message delay for partial cut–through is just $T-(1-\alpha)t_m/P_t$. This is much simpler than in [Abra90], which involves the computation of queueing probabilities. The new method also gives comparable results to the simulations.

Figures 3b–6b compare message delays (in network cycles) obtained from analysis and simulations, for various system parameters under uniform traffic, using partial cut–through. Figure 7 shows some results with non–uniform traffic. For most network loads, the error between analysis and simulations is less than 10%.

3.2 Modeling hypercubes with locality

Recall that we use left–to–right (LR) routing in the FH. The LR routing analysis in [Abra90] assumes that a processor can send messages to itself, which is inconsistent with the random routing analysis. The effects of the inconsistency is small with flat systems, but could be appreciable if we considered cluster locality. We have to make some simple modifications to Abraham's equations.

For traffic with locality, we divide the dimensions into local ($< n_0$) and global ($\geq n_0$).

$$m_{n_1-1} = \frac{m_g t_m}{2}(1-\alpha)\frac{N_1}{N_1-N_0}$$

$$m_i = (1-\alpha)\frac{m_g t_m}{2^{n_1-i}}\frac{N_1}{N_1-N_0} + \sum_{i<j<n_1}\frac{m_j}{2^{j-i}} \quad \text{for } i \geq n_0$$

$$m_i = \alpha\frac{m_g t_m}{2^{n_0-i}}\frac{N_0}{N_0-1} + \sum_{i<j<n_1}\frac{m_j}{2^{j-i}} \quad \text{for } i < n_0$$

The $N_1/(N_1-N_0)$ factor accounts for the fact that non-local traffic does not go to the same cluster. The $N_0/(N_0-1)$ factor accounts for the fact that processors do not send messages to themselves.

Let the probability that channel i has a message for channel j be $c_{i,j} = m_i/2^{i-j}$ if $i > j$, and zero otherwise. Let the probability of i arrivals at channel j from other channels be $d_{i,j}$. Let the probability of i total arrivals at channel j be

$$a_{i,j} = (1-\alpha)\frac{m_g t_m}{2^{n_1-j}}\frac{N_1}{N_1-N_0}d_{i-1,j} + (1-(1-\alpha)\frac{m_g t_m}{2^{n_1-j}}\frac{N_1}{N_1-N_0})d_{i,j}$$

for $j \geq n_0$, and

$$a_{i,j} = \alpha\frac{m_g t_m}{2^{n_1-j}}\frac{N_0}{N_0-1}d_{i-1,j} + (1-\alpha\frac{m_g t_m}{2^{n_1-j}}\frac{N_0}{N_0-1})d_{i,j}$$

for $j < n_0$. The probability of having i items in queue j is $b_{i,j} = a_{i,j}b_{0,j} + \sum_{k=0}^{i} a_{k,j}b_{i-k+1,j}$. To obtain the average queueing delay, recall that $0.5N_1/(N_1-1)$ of the messages take the lower order n_0 channels, and $0.5(1-\alpha)N_1/(N_1-N_0)$ of the messages take the higher order n_1-n_0 channels. Hence, the average message delay is

$$T = (1 + \frac{0.5N_1}{N_1-1}\sum_{j=0}^{n_0-1}\frac{1}{m_j}\sum_i ib_{i,j} + \frac{0.5(1-\alpha)N_1}{N_1-N_0}\sum_{j=n_0}^{n_1-1}\frac{1}{m_j}\sum_i ib_{i,j})t_m$$
$$+ (1 + (1-\alpha)\frac{N_1(n_1-n_0)}{2(N_1-N_0)} + \frac{n_0 N_1}{2(N_1-1)})t_h.$$

For uniform traffic, $\alpha = N_0-1/(N_1-1)$. For partial cut–through, the average message delay is

$$T - (1-\alpha)t_m\frac{N_1(n_1-n_0)}{2(N_1-N_0)} - t_m\frac{n_0 N_1}{2(N_1-1)}.$$

For cluster–local traffic, local channels have heavier loads than global channels. Since we assume local and global channels have the same bandwidth, local channels will saturate before the global channels. If we set $m_{n_0-1}=1$, $m_{g,sat} = \frac{2}{t_m(1-\alpha+\alpha N_0/(N_0-1))}$, which is very close to the saturation value for uniform traffic, $2/t_m$, for $N_1 \gg N_0 \gg 1$. Hence, for systems with heavy local traffic, it is useful to build a local network that has a higher bandwidth than the global network.

3.3 Uniform traffic performance: results

Figures 3a–6a show average message delays for various systems using message switching. (In Figures 3–5, the curve for k=1 has been clipped before saturation for a cleaner presentation.) As predicted earlier, if only one cluster/global bus is used in the CH, the system saturates at $m_g=1/t_m$, because of the saturation of the cluster/global bus. This is well below the saturation load of the global network. If we split the cluster into two subclusters, each with its own cluster/global bus, the system is more balanced and saturates only when the global network saturates. Splitting the cluster into four or more subclusters gives little performance improvement at low traffic; message delay decreases significantly only close to saturation.

Figures 3b–6b show the same systems but with partial cut–through. The relative performance does not change until the systems are very close to saturation. Partial cut–through saves more time in FHs, because the FHs have more nibbles in a message. It has no effect on saturation load.

For all the system parameters considered, the FH has a higher saturation load than the CH. To get saturation values of m_g, we set the traffic rate on the global link, m, to 1 nibble per cycle. We can estimate the ratio of the two saturation loads. Let $R=m_{g,satFH}/m_{g,satCH}$, where $m_{g,satFH}$ and $m_{g,satCH}$ are the respective message generation rates of the FH and CH at saturation. If we assume the channels in the CH are N_0 times as wide as those in the FH, R is approximately $\lceil L/CN_0 \rceil N_0 / \lceil L/C \rceil$. Assume L divides C; hence $L/C=t_{m,FH}$. Since $\lceil t_{m,FH}/N_0 \rceil N_0 - t_{m,FH} \leq N_0-1$, the maximum value of R is $(t_{m,FH}+N_0-1)/t_{m,FH}$. For a worst case $R_{max} \leq 120\%$, $t_{m,FH} \geq 5N_0$. This is not an unreasonable condition for most system parameters, since large cluster sizes consume more pinouts, and messages have to be broken into many nibbles.

Next consider queueing performance and buffering requirements. When the CH is operating at loads lower than approximately 90% of saturation, it has a better queueing performance than the FH by a substantial margin. The difference in average message delay is mainly because of the wider channel widths in the CHs. A message can be routed in fewer cycles in a CH, and the waiting time at each intermediate hop is greatly reduced. A less important factor is the slightly reduced network diameter in the CHs.

Since we assumed infinite buffering, the extremely long message delay in the FH implies that many buffers have to be reserved at each switch. For network loads below saturation, the delays in the FH are often 2 to 10 times the delays in the CH. This means buffers in the FH must be many times deeper than in the CH. Buffers in the CH are wider by a factor of N_0, but recall that there are N_0 as many switches in the FH, and each FH switch has a greater degree. When operating the FH beyond the saturation point of the comparable CH, a huge number of buffers is required to attain reasonable performance.

We have assumed that board pinouts and system wiring determine the channel widths in the system. Local and global channels are the same width, and there are no pin constraints on local structures. Pin constraints on local structures can usually be circumvented by implementing slices of the data path on different chips, while data path slicing is difficult on the board level. The performance of the FH would improve if we built local channels that are wider than global channels.

It appears that while the FH is able to operate in a larger range of loads than the CH, its performance for much of that range (below saturation) is much less attractive than the CH. The performance of the FH in that extra range is also highly unpredictable, since average message delay increases wildly with a small increase in m_g. Furthermore, the hardware cost for maintaining the extra range of operation may not be realistic.

3.4 Non–uniform traffic performance: results

As in the last section, we compared FHs and CHs, picking pairs of systems that have similar wiring costs and board pinouts. Since relative performance is similar across different system parameters, we only show the results (in Figure 7) for the same system as in Figure 4b. We examine $\alpha=8(N_0-1)/(N_1-1)$ and $16(N_0-1)/(N_1-1)$ (for the system considered, α is approximately 12% and 23%.) For k=2, the relative performance of the systems do not change with locality of traffic. As with uniform traffic, the FHs have higher saturation loads, but the CHs have better queueing performance until the network loads approach saturation. However, at high loads and higher degrees of locality, the CH with k=4 is able to utilize the full bandwidth of the global channels, whereas the *local* channels of the FHs saturate. For example, in Figure 7, the CH with k=4 has both better queueing performance and higher saturation loads than the FH, for higher values of α.

4. Conclusions

Packaging constraints are very important considerations in the construction of large multiprocessors. We examined an approach to clustering, which exploits the physical hierarchy imposed by packaging constraints. By combining techniques from research on cluster–based systems and queueing analysis with wiring constraints, we were able to study in detail the impact of wiring cost on large systems, and compare its effects on flat and clustered hypercubes.

Our analyses and simulations suggest that the benefits of clustering are greater than previously predicted. When we compare clustered hypercubes (CHs) and flat hypercubes (FHs) with similar wiring costs and comparable crosspoint costs, at uniform traffic, CHs saturate at lower network loads, but have much better queueing performance before they saturate. The FH also requires more buffers for reasonable performance.

When message traffic is heavy and highly local, the CH with a small number of subclusters (e.g., k=4) is able to utilize the full bandwidth of both its local and global channels, whereas the local channels of the FH saturates at the same loads, because of the heavy local traffic. In these situations, the CH has both better queueing performance and higher saturation loads than the FH.

When building large multiprocessor systems where pin-count and wiring cost are major limitations, clustering is an attractive option. While we have focused on hypercube-based systems in this paper, our methods are also applicable to systems based on other topologies.

REFERENCES

[AbPa89] S. Abraham and K. Padmanabhan, "Performance of the Direct Binary n-Cube for Multiprocessors", IEEE Trans. on Computers, Vol. C-38, No. 7, July 1989.

[AbPa90] S. Abraham and K. Padmanabhan, "Constraint Based Evaluation of Multicomputer Networks," 1990 Int. Conf. on Parallel Processing, Aug. 1990.

[Abra90] S. Abraham, *Issues in the Architecture of Direct Interconnection Schemes for Multiprocessors*, PhD Thesis, University of Illinois at Urbana-Champaign, 1990.

[AgMa85] D. Agrawal and I. Mahgoub, "Performance analysis of cluster-based supersystems," 1st Int. Conf. on Supercomputer Systems, IEEE Comp. Soc. Press, 1985.

[Carl85] D. Carlson, "The mesh with a global mesh: a flexible, high-speed organization for parallel computation," 1st Int. Conf. on Supercomputer Systems, IEEE Comp. Soc. Press, 1985.

[Dall90] W. Dally, "Performance Analysis of k-ary n-cube Interconnection Networks," IEEE Trans. on Computers, Vol. C-39, No. 6, June 1990.

[DaEa90] S. Dandamudi and D. Eager, "Hierarchical Interconnection Networks for Multicomputer Systems," IEEE Trans. on Computers, Vol. C-39, No. 6, June 1990.

[FrWT82] M. Franklin, D. Wann, and W. Thomas, "Pin Limitations and Partitioning of VLSI Interconnection Networks," IEEE Trans. on Computers, Vol. C-31, No. 11, Nov. 1982.

[HaMS86] J.P. Hayes, T.N. Mudge, and Q.F. Stout, "Architecture of a Hypercube Supercomputer", 1986 Int. Conf. on Parallel Processing, Aug. 1986.

[Hill85] W.D. Hillis, *The Connection Machine,* MIT Press, Cambridge, MA, 1985.

[Seit85] C.L. Seitz, "The Cosmic Cube", CACM, Jan. 1985.

[Thom79] C.D. Thompson, "Area-Time Complexity for VLSI," Ann. Symp. on Theory of Computing, May 1979.

[WuLi81] S. Wu and M. Liu, "A cluster structure as an interconnection network for large multiprocessor systems," IEEE Trans. on Computers, Vol. C-30, No. 4, Apr. 1981.

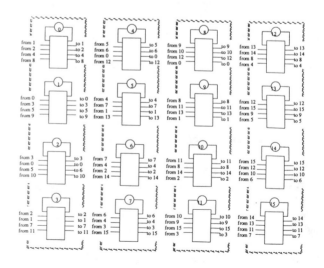

Figure 1a: Flat hypercube
16 processors, 4 per cluster

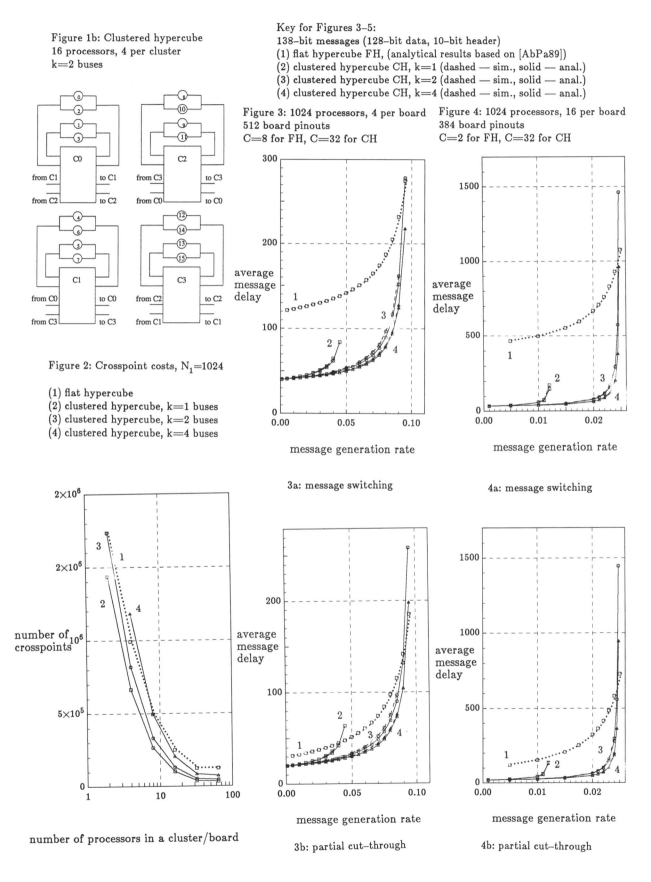

Figure 1b: Clustered hypercube
16 processors, 4 per cluster
k=2 buses

Figure 2: Crosspoint costs, N$_1$=1024

(1) flat hypercube
(2) clustered hypercube, k=1 buses
(3) clustered hypercube, k=2 buses
(4) clustered hypercube, k=4 buses

Key for Figures 3–5:
138–bit messages (128–bit data, 10–bit header)
(1) flat hypercube FH, (analytical results based on [AbPa89])
(2) clustered hypercube CH, k=1 (dashed — sim., solid — anal.)
(3) clustered hypercube CH, k=2 (dashed — sim., solid — anal.)
(4) clustered hypercube CH, k=4 (dashed — sim., solid — anal.)

Figure 3: 1024 processors, 4 per board
512 board pinouts
C=8 for FH, C=32 for CH

Figure 4: 1024 processors, 16 per board
384 board pinouts
C=2 for FH, C=32 for CH

3a: message switching

4a: message switching

number of processors in a cluster/board

3b: partial cut–through

4b: partial cut–through

Figure 5: 4096 processors, 32 per board
896 board pinouts
C=2 for FH, C=64 for CH

Figure 6: 1024 processors, 16 per board
384 board pinouts, $m_g = 0.007$
C=2 for FH, C=32 for CH
(1) flat hypercube (anal.)
(2) clustered hypercube, k=2 (anal.)

Figure 7: 1024 processors, 16 per board
partial cut-through

384 board pinouts, 138-bit messages
(128-bit data, 10-bit header)
C=2 for FH, C=32 for CH
(1) flat hypercube, $\alpha = 11.7\%$
(2) flat hypercube, $\alpha = 23.5\%$
clustered hypercube, k=2
(3) $\alpha = 11.7\%$
(4) $\alpha = 23.5\%$
clustered hypercube, k=4
(5) $\alpha = 11.7\%$
(6) $\alpha = 23.5\%$

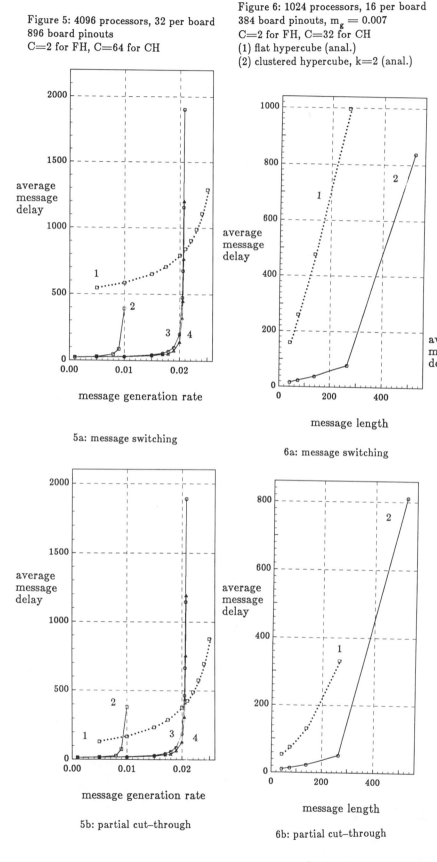

5a: message switching

5b: partial cut-through

6a: message switching

6b: partial cut-through

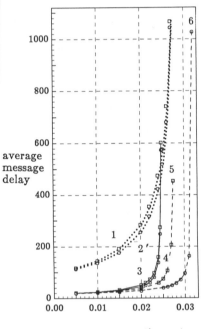

A Hybrid† Architecture and Adaptive Scheduling
for Parallel Execution of Logic Programs ††

Surapong Auwatanamongkol

Division of Computer Science
Asian Institute of Technology
P.O.Box 2754, Bangkok 1051
Thailand

Prasenjit Biswas

Computer Systems Laboratory
Texas Instruments
P.O.Box 655474, M/S 238
Dallas, Texas 75265.

Abstract

In this paper we describe a stream based dynamic data-driven model for OR-Parallel execution of logic programs and an architecture for efficient implementation. The architecture could be considered as representing a hybrid of von Neumann and data driven models of execution. A logic program is compiled into dataflow graphs with variable resolution macro-actors. The macro actors are represented by threads of simple instructions of a von Neumann processing element. Dynamic scheduling of threads plays a major role in the performance of logic programs on such an architecture. We report an adaptive scheduling strategy that caters to the conflicting requirements of data-locality and uniform load sharing in the proposed distributed memory architecture. The performance results in this paper are based on extensive simulation of the multi-ring architecture with hypercube interconnection.

1. Introduction

It has been shown that inherent parallelism in logic programs is naturally exploited by a dataflow execution model. A number of execution models have been proposed in the literature. Most of these have not considered details of implementation and performance issues. LogDf model [BiTs88,TsBi88] is a stream based data-driven model for OR-Parallel and Restricted-AND parallel execution of logic programs. For brevity of presentation, we will only consider OR-parallel execution in this paper. The stream-based data-driven model naturally provides AND-pipelined execution. Detailed simulation and experiments with the LogDf model/architecture over the last few years uncovered several issues related to some of the inefficiencies of the model and its implementation. These observations have changed the execution model significantly. In this paper we consider the modified model and an architecture to support its efficient implementation. The proposed hybrid architecture incorporates extensions to von Neumann model to support data driven execution [BuEk87].

It was realized quite early in the LogDf project that a dataflow model with explicit synchronization at the fine-grain instruction level was not an efficient scheme for parallel execution of Logic programs. This decision was based on the fact that data-driven instruction sequencing for inherently sequential threads of computation did not seem to provide any performance advantage. This led to the definition of the LogDf with macro-actors to represent sequential threads of computation and explicit data-driven synchronization for enabling the execution of these macro-actors.

Experiments with the earlier version of the LogDf model and the architecture revealed that a definition of a fixed set of macro-operators and corresponding function units (at the machine level) denied opportunities for several compile time optimizations that could normally be performed in a Warren Abstract Machine based parallel implementation on von Neumann processors [LWHB88]. The current version of the multi-ring hybrid architecture (LogDf-H) is based on multiple conventional RISC-like processing elements with the additional capability to support split phase memory transactions with I-structure stream memories [ArNi90]. The variable resolution macro-actors representing sequential threads are compiled for these processing elements using standard register optimization techniques. Dynamic scheduling of threads plays a major role in the performance of logic programs on such an architecture. In this paper we introduce an adaptive scheduling strategy that caters to the conflicting requirements of data locality and uniform load sharing.

2. Abstract Execution Model

We assume the readers to have basic familiarity with the fundamentals of logic programming. We will review a few terminologies for the sake of clarity. The subset of clauses in the program with similar head literals (i.e. same predicate name and number of arguments) will be referred to as *candidate clause* for a procedure call. An entry to a procedure would take place on successful unification of the goal literal with the head literal of the clause representing the procedure [Hogg84]. A successful unification causes activation of the body of the clause and procedure calls corresponding to the body literals are allowed to proceed. OR-parallel execution implies parallel activation of all the candidate clauses corresponding to a call. In this paper we assume the calls corresponding to the body literals of a clause are executed in a left to right sequential order. A logic program is compiled into dataflow graphs for execution on the

LogDf architecture. The compilation scheme for the original LogDf model was explained in [BiTs88]. The compilation scheme has been appropriately modified for the hybrid architecture [AuBi91].

We will outline the functions of the major macro-actors and briefly introduce only those aspects of the execution model which are relevant to this paper. The macro actors constituting the dataflow graphs are as follows:

(1)   Unify actor : This actor has 1-1 correspondence with the head of a clause and is responsible for unification of an instance of a goal literal and the head literal. On receipt of a token containing pointers to arguments of a goal literal, and input bindings, the actor starts an invocation of the corresponding clause. If the unification succeeds, local bindings for the candidate clause are created at an I-structure memory and the pointer to the bindings is passed on to an Activate actor for parallel invocations of the first literal in the body of a clause.

(2)   Activate actor : An Activate actor in the graph corresponds to a subgoal literal in the body of a clause. On receipt of a token containing a pointer to input bindings, the actor starts parallel invocations, one for each of the candidate clauses of the corresponding subgoal literal.

(3)   Return actor : This actor terminates an invocation of a candidate clause of a literal after returning solution bindings to the calling level of the invocation. Before returning the solution bindings, the Return actor performs a closing operation on the bindings to make it local to the calling level.

Besides the basic functions of the actors as described above, each actor maintains its solution bindings in the cell of a stream representing a set of solutions from a procedure call for a clause or a goal literals. The solutions in the stream are maintained in a strict order to correspond to execution of Prolog based on its sequential semantics. The strict ordering is necessary to support implementation of cut and side-effects in the model. Details of the implementation is beyond the scope of this paper and were discussed elsewhere [AuCB90].

To see how a logic program can be compiled into a dataflow graph using the above mentioned actors, let us consider a set of clauses as shown in Figure 1 (arguments of each literal are not shown).

Figure 1 shows a tree structured dataflow graph representing a set of nested calls for the execution of a goal p. In this graph arguments to predicates have been intentionally left out for clarity, "success" and "failure" of unifications have been arbitrarily assumed to create the example. The floating input arcs for Unify and Activate nodes are associated with constant templates representing a head literal and body literal respectively [BiTs88]. A success leaf is reached when there is no more goal literal to be solved. For each success leaf, there is a corresponding stream cell, e.g. cell a, b, c, d, e and f, which contains a pointer to solution bindings. These cells are linked together and form a stream of solutions for the goal p. If a unification required to enter a procedure for a candidate clause does not succeed, a failure leaf is reached. Each Return actor in the graph has a corresponding Unify actor. Contours are shown in the figure to clarify the correspondence.

In a dynamic dataflow execution model, different activations of the same dataflow graph are distinguished by coloring of tokens flowing in the graph [ArNi90]. Different colors are assigned to identify distinct invocations. A brief description of the token coloring scheme for the LogDf is essential for understanding how the return address of a call can be passed from a Unify actor to any of its corresponding Return actors. In LogDf, each successful unification creates a descriptor and the address of a descriptor is used as a color of a token in the dynamic dataflow execution [BiTs88].

3. The dataflow/von Neumann hybrid architecture (LogDf-H)

As mentioned earlier, the proposed scheme supports variable resolution macro-dataflow execution and macro-dataflow nodes are represented by threads of simple instructions, which are executed by conventional RISC-like processing units. The dataflow graph of a logic program is represented by a set of instruction sequences, a sequence per macro-actor of the dataflow graph representation. This provides the flexibility of representing specialized node operations and possibilities of several optimization by generating appropriate instruction sequences for the execution of the nodes. Finally,

† Represents a hybridization of dataflow and von Neumann style execution

†† This research was supported by Texas Advanced Technology Program Contract 3128 (1988).

appropriate instructions are provided to initiate split-phase memory transactions at any arbitrary point in the sequence. Each instruction sequence could be viewed as a thread or lightweight process. A sequential chain of macro-actors is also compiled as a single thread without loss of any effective parallelism. The architecture and the execution scheme supports fast switching of lightweight contexts.

Figure 2 shows the organization of a ring of the proposed multi-ring hybrid architecture. The computing section consists of 2 to 4 processors and local memory. The data-driven execution of the original LogDf model is supported by split-phase memory transactions on I-Structure memories (Descriptor, S-Stream, Vframe, and List/Structure memories as shown in the figure). These split phase transactions on I-Structure memories provide the necessary synchronization between producer and consumer nodes and also allow context switching during deferred read operations. This in turn provides the capability of hiding memory latency and facilitates scalability of the architecture. In the following paragraphs we provide a brief overview of the execution scheme, for architectural details reader could refer to [AuBi91].

Interactions between threads and between a thread and I-Structure memory are through tokens of various types. There are essentially four types of tokens: (i) activation, (ii) read (for I-structure memory), (iii) write (for I-structure memory), (iv) reactivation (returning data from I-structure memory; and are identified by appropriate tags. The tokens are issued by the threads using special instructions. Any token generated by a thread goes to the output token queue. Depending on the tag and destination of the token, it is routed by the communication section to either the input queue or I-Structure memory controller at the destination ring. This token based activation of threads and I-Structure transactions is depicted in Figure 3. An arbitrary sequence of events and processor allocations are shown in the figure to demonstrate functions of various tokens (all the arguments with each token type are not shown). Producer-consumer synchronization of threads T_0 and T_1 using the I-Structure store indicates the context switching at processor PE_2. When the read request from T_1 is deferred, context switch allows initiation of T_2 at PE_2. The 'read' token carries a *continuation* to I-Structure store via the output token queue. When data is available from T_0, the I-Structure controller returns read data and the context back to the input token queue and thread T_1 is reactivated at a different processor PE_1.

Activation of new threads T_3 and T_4 by T_1 is also shown in the figure. This is similar to activations of two unification actors by an Activate actor in the dataflow graph shown in Figure 1. The activation of threads T_3 and T_4 is done by generation of 'activation' tokens by T_1. The activation is equivalent to initiating a sequence of instructions and thus the 'activation' tokens carries the program counter values PC_1 and PC_2 to indicate the starting points of the instruction sequences for threads T_3 and T_4 respectively.

4. Adaptive Scheduling

There have been several proposals for dynamic load balancing for loosely coupled distributed memory multiprocessors. In the realm of multi-ring dataflow architectures, there have been some reports on heuristic techniques for dynamic load distribution [BaGu86]. All these papers deal with distributed or centralized algorithms for evenly distributing the total workload on a network of processors to achieve an optimal load balance (pure load balancing). These proposals for load distribution do not address the issue of locality of data and the increased communication overhead of executing data dependent processes at different nodes of the distributed memory multiprocessor. Uniform distribution of load and minimization of communication overhead are the two conflicting criteria that need to be considered simultaneously, to achieve optimal performance.

Some of the recent proposals for static scheduling [BeSn87,Lo88] consider both the above mentioned criteria in design of the partitioning and/or scheduling algorithms for predictable workload. For search problems with non-deterministic behaviors, e.g. execution of logic programs, the execution graph can not be fully instantiated at compile time; therefore, static partitioning and/or scheduling methods are not applicable.

Dynamic scheduling algorithms for loosely coupled systems differ on the basis of transfer and location policies [EaLa86]. In the proposed scheme, the decision heuristic of the transfer policy is adaptive to the state of the system. It is adaptive in the sense that there is no fixed threshold and the decision is based on an inequality. The state of the system is characterized as waiting delay dominated or communication delay dominated. The measures and heuristic proposed in this section are used to estimate the state of the sytem and provide a dynamic decision making capability for the transfer policy. The location policy is based on a simple round robin scheme.

4.1 Rationale

To see how nodes of a LogDf dataflow graph for a logic program can be scheduled in the hybrid architecture, consider the dataflow graph for the execution of goal p shown in Figure 1. To reduce the scheduling overhead as well as the communication cost between nodes, it is natural to distribute only those nodes that contribute to useful parallelism. The best candidate nodes for the scheduling would be the Unify nodes of candidate clauses which are

activated in parallel by an Activate node. Each of the Unify nodes, once scheduled at a particular ring, enables a serial sequence of dataflow nodes that would be executed at the same ring. This serial sequence of nodes will be referred to as a scheduling grain (SG) which represents a unit of work to be scheduled. The introduction of the notion of SG increases both granularity and data locality of execution without loss of any parallelism.

Now, consider any path from a root node to any other node of the dataflow graph. The total execution time of the path would be equal to the total processing time of nodes in the path plus the total delay (T_d) during the execution. The total delay T_d (on a path) is composed of two main components:

1) The total waiting delay (T_w), which is the accumulated time that all nodes in the path have to spend in resource queues before they can gain access to the resources. T_w depends much on how workload is distributed in the system.

2) The total communication delay (T_c), is the accumulated time of communication that occurs between any nodes along the path. T_c is at a minimum when all SG's in the dataflow graph are assigned to the same ring.

Since the total processing time on a path is fixed, the total execution time would depend solely on T_d. In order to achieve optimal performance, T_d or the sum of the two components ($T_w + T_c$) must be minimized. The minimum of T_d can not always be achieved by distributing SG's to simply balance workload in the system. In other words, there is an operating point where some SG's are kept at their source rings in spite of load unbalance, yet achieving minimum T_d. The operating point could be characterized by a load distribution factor Θ as follows:

$$\Theta = N_d / N_t$$

where

- N_d is the total number of SG's that are considered by a load balancing scheme for migration from their source rings to other rings to balance the workload in the system,

- N_t is the total number of SG's that were ready to be scheduled for execution.

It should be noted that the decision of the load balancer could be to retain some SG's in the source rings, but these SG's are still counted as part of N_d. The value of Θ will range from 0, where all SG's are kept at one ring or there is no load distribution, to 1 where all SG's are considered for distribution according to a pure load balancing scheme. The combination of the influences of the two delays on the T_d can lead to two situations:

1. Waiting Delay Dominance is a situation when T_w delay has more influence on T_d than that of T_c. This situation, for instance, can occur when a program has very high parallelism, and the system is short of resources to support such parallelism.

2. Communication Delay Dominance is a situation where T_c has more influence on T_d than that of T_w. This situation, for instance, can occur when a program has low parallelism and the system has an abundance of resources.

4.2 Scheduling Heuristic

To achieve optimal performance, the system must operate around an optimal operating point Θ^* or a specific value of Θ where T_d is minimum. In the case of *waiting delay dominance*, T_d is decreased as Θ approaches 1, so the optimal operating point would be at $\Theta = 1$. In the case of *communication delay dominance*, T_d is at a minimum at an optimal operating point where $\Theta < 1$. Since the optimal operating point depends on several hardware and software factors, it is not likely that the value of the optimal operating point can be estimated in advance and then be used to operate the system around that point. An alternative is to use a a dynamic scheduling strategy that would adaptively adjust to make the system operate around the optimal point. In order to be able to adjust the operating point at runtime, a heuristic is used to decide whether an SG along a path of an execution tree should be sent out to another ring to balance workload or be kept at its source ring to reduce communication delay.

The heuristic for the decision making is based on runtime information about waiting and communication delays accrued during a time interval corresponding to execution of each SG. This would allow the scheduling to be adaptive and adjust the system according to the current state of the system. Two measures are used to represent the two types of delays for the heuristic decision making.

1. δW_m = Total waiting delay during the execution of the current SG_m being considered

$$= \sum_{i=1}^{i=k} w_i$$

where, w_i is waiting delay that occurs at the suspension point i of the execution of SG_m; the suspension point is a point of discontinuity in the execution of the SG caused by either queuing of a node of the SG at a processing resource or split-phase I-structure memory transaction, k is the total number of suspension points during the execution of the SG_m.

2.
$$\delta C_m = \delta TC_m + \delta EC_m$$
where,

δTC_m is total communication delay the execution of the current SG_m being considered, so

$$\delta TC_m = \sum_{i=1}^{i=k} c_i$$

where,

c_i is communication delay that occurs at the suspension point i of the execution of SG_m; the suspension is typically caused by a remote I-structure memory operation; k is the total number of suspension points during the execution of the SG_m

δEC_m is the expected communication cost of initiating the execution of the child SG at a destination ring (the ring being decided according to a pure load balancing scheme, cf. section 4.3), i.e. cost of sending activating binding environment from the source ring to the destination ring.

Let us suppose that at the time the child SG of SG_m has to be scheduled, $(\delta C_m - \delta W_m) < 0$. This indicates that the system might be in a state where the current workload in the system is so high that waiting delays dominate communication delays. In this situation the child SG should be sent out to another ring to improve the load balance in the system and to reduce the influence of the waiting delays on T_d. On the other hand, if $(\delta C_m - \delta W_m) > 0$, the system might be in a state where the workload in the system is so low that the communication delay becomes more prominent as compared to the waiting delay. In this situation the child SG should be kept at its source ring to reduce the influence of communication delays on T_d. In the case of a tie between the two measurements, the decision of the scheduling can be either one.

Notice that the expected communication cost is included in the second measure (δC_m) in order to consider the impact of communication costs between the current SG and its child. If a packet switching interconnection network is assumed, the expected communication cost between the current SG and its child SG could be estimated using the size of the data (estimated from static analysis [AuBi 91]), bandwidth of the interconnection network and the distance from source ring to destination ring.

Two of the measures used in the heuristic, δW_m and δTC_m, have to be acquired at runtime. It is important that the computation of these two quantities should not introduce too much overhead that could undermine the performance of the scheduling algorithm. In order to minimize the overhead, the measurements of the two delays are not accumulated individually. The difference -- $\delta D_m = \delta TC_m - \delta W_m$, is accumulated instead.

If δD_m is obtained, the comparison of δC_m and δW_m required for evaluating the scheduling criterion can be indirectly performed, as $- \delta C_m - \delta W_m = \delta D_m + \delta EC_m$.

The initial value of δD_m is reset to zero and is sent along with a token that activates a child SG. At any suspension point j of the SG, the current value of δD_m is also sent along with an outgoing token from the SG. Special hardware units, (perodically) synchronized clock among the rings and time-stamping provide the value of c_j and w_j and thus δD_m is obtained [AuBi91]. The addition and subtraction of the time units to update δD_m are performed by special hardware units that perform these operations in parallel with work being performed at memories or processing resources. So the overhead associated with the updating can be hidden by overlapping the operations with useful work.

4.3 Simulation of the Scheduling Strategy

The scheduling strategy was simulated for the multi-ring hybrid architecture. Hypercube interconnection of the rings was assumed as a topology for organization of the rings. The simulator was designed to test performance of LogDf-H with this scheduling strategy for several programs with varying hardware and software factors influencing the optimal operating point. In the case studies for determining the efficiency of the heuristic, a load distribution strategy is assumed to find tentative destinations of SG's to be scheduled. The assumed load distribution strategy prescribes the destination ring for an enabled SG; but the SG is not scheduled to execute at that destination ring unless the heuristic scheduling decision dictates so. If there is a disagreement, the enabled SG is scheduled to the same ring as that of the activating SG.

In each test case, results of performance using the same load distribution strategy, with and without the heuristic, are compared. A local round robin strategy is used as a load distribution strategy in the simulation. The assumed load distribution algorithm is fully distributed and requires very little scheduling overhead. In this scheme, each ring is initially assigned a distinct seed number R which is the same as the ring number. When a token corresponding to a Unify node (that starts the execution of an SG) needs to be scheduled, a new seed number designating the destination of the

token is generated from the current seed number using the following function: $R_{new} = (R_{old} + 1) \bmod N$, where N is the total number of rings in the system and the range of R is from 0 to $N-1$.

The following figures show results from simulations of the the proposed hybrid architecture and the dynamic scheduling scheme based on the following system parameters: (1) 10 MHz processor clock, (2) 4 processors/ring, (3) communication delay between two adjacent rings in the hypercube interconnection network is 4 clock cycles/word.

For brevity of presentation, we provide performance results for two well known bench mark programs. The relative speedups shown in the figures are with respect to the performance on a single ring with one processor.

The simulation results indicate that the heuristic can improve the speedup of the program significantly. They show that load balancing with the use of the heuristic can adaptively adjust the system to operate around an optimal operating point. This is confirmed by the improvement of the performance over varying hardware factors. For instance, the scheduling with the heuristic performs much better than the pure load balancing without the heuristic as the communication delay becomes much more dominant over the waiting delay. The scheduler with the heuristic is able to maintain the locality of data references by not operating the system at full 100 % of Θ.

5. Conclusion

In this paper, we presented a new LogDf model for data-driven execution of logic programs on a hybrid architecture. Due to brevity of presentation, it was not possible to elaborate some of interesting features of the model, namely, stream-based eager evaluation for AND-pipelined execution and the scheme for handling cut and side-effect predicates [AuCB90]. Detailed simulation of LogDf-H, at the instruction level, is operational. The test programs had to be hand-coded in the assembly language of the machine. Development of the compiler is in progress. We have also developed a token prioritization scheme associated with the dynamic scheduler. Token prioritization based on textual ordering of the source program shows impressive performance gain and reduction of speculative computation.

References

[ArNi90] Arvind, and Rishiyur S. Nikhil, *Executing a Program on the MIT Tagged-Token Dataflow Architecture*, IEEE Transaction on Computer, Vol. 39, No. 3, March 1990, pp. 300-318.

[AuBi91] Auwatanamongkol, Surapong, and Biswas, Prasenjit, *A Hybrid Architecture and Adaptive Scheduling for Parallel Execution of Logic Programs*, Technical Report, Department of Computer Science and Engineering, Southern Methodist University, January, 1991.

[AuCB90] Auwatanamongkol, Surapong, Ciepielewski, Andrzej, and Biswas, Prasenjit, *Cut and Side-Effects in a Data-Driven Implementation of Prolog*, (under revision), New Generation Computing Journal, 1990.

[BaGu86] Barahona, P., and Gurd, J. R., *Processor Allocation in a Multi-ring Dataflow Machine*, Journal of Parallel and Distributed Computing, Vol. 3, No. 3, September, 1986, pp. 305-327.

[BeSn87] Berman, F., and Snyder, L., *On Mapping Parallel Algorithms into Parallel Architectures*, Journal of Parallel and Distributed Computing, Vol. 4, No. 5, October, 1987, pp. 439-458.

[BiTs88] Biswas, Prasenjit, and Tseng, Chien-Chao, *LogDf : A Data Driven Abstract Machine Model for Parallel Execution of Logic Programs*, Proceeding of the International Conference on Fifth Generation Computer Systems, ICOT, 1988, pp. 1057-1070.

[BuEk87] Buehrer, R., and Ekanadham, K., *Incorporating Dataflow Ideas into von Neumann Processors for Parallel Execution*, IEEE Transaction on Computers, December, 1987, pp. 1515-1522.

[EaLa86] Eager, D. L., Lazoska, and Zahorjan, J., *Adaptive Load Sharing in Homogeneous Distributed Systems*, IEEE Transaction on Software Engineering, May, 1986, pp. 662-675.

[Hogg84] Hogger, C. J., *Introduction to Logic Programming*, Academic Press, 1984.

[Lo88] Lo, V., *Algorithm for Static Task Assignment and Symmetric Contraction in Distributed Systems*, Proceedings of the 1988 International Conference on Parallel Processing, August, 1988, pp. 239-244.

[LWHB88] Lusk, E., Warren, D. H. D., Haridi, S., Butler, R., Calderwood, A., Disz, T., Olson, R., Overbeek, R., Stevens, R., Szeredi, P., Brand, P., Carlsson, M., Ciepielewski, A., and Hausmann, B., *The AURORA OR-Parallel Prolog System*, Proceedings of International Conference on Fifth Generation Computer Systems, ICOT, 1988, pp. 819-830.

[TsBi88] Tseng, C. C., and Biswas, P., *A Data-Driven Parallel Execution Model for Logic Programs*, Proceedings of 5th International Conference Symposium on Logic Programming, August, 1988, pp. 1204-1222.

cl 1: p :- q, r.
cl 2: p :- s.
cl 3: q.
cl 4: q.
cl 5: r.
cl 6: r :- t.
cl 7: s.
cl 8: s.
cl 9: t.
cl10: t.

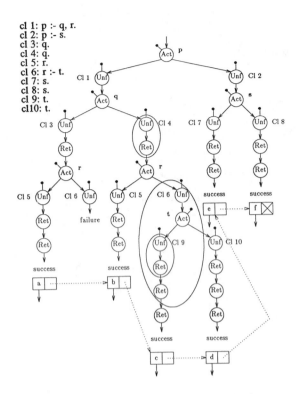

Figure 1 : A dataflow graph corresponding to the execution of goal p
(The contours show levels of nesting)

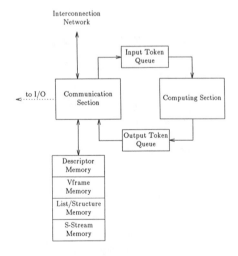

Figure 2 : A ring of LogDf-H.

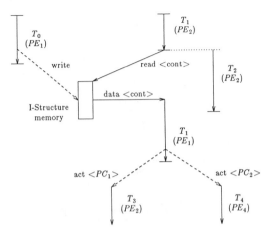

Figure 3 : Token-based Synchronization

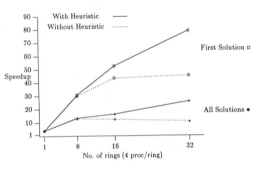

Figure 4 : Speedup for the 7-Queen Program

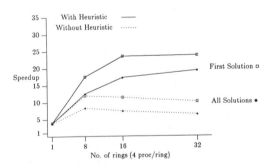

Figure 5 : Speedup for the Naive sort program

BUS CONFLICTS FOR LOGICAL MEMORY BANKS
ON A CRAY Y-MP TYPE PROCESSOR SYSTEM[†]

K. A. Robbins
krobbins@ringer.cs.utsa.edu

S. Robbins
srobbins@ringer.cs.utsa.edu

Division of Mathematics, Computer Science, and Statistics
The University of Texas at San Antonio
San Antonio, TX 78249

Abstract — A memory subsystem design for the Cray Y-MP in which the physical banks are replaced by logical banks is shown to result in significant improvements in memory performance. A complete path simulation of data flow for the logical bank system including subsections and bus return conflicts is given. It is shown that return path conflicts on reads, which result from the introduction of logical banks, do not significantly affect memory performance under moderate loading provided that each physical bank has a single-slot output buffer. The results of this study establish that logical banks are a viable approach to the design of a high-performance shared memory capable of supporting up to 64 Cray Y-MP processors.

1 Introduction

Commercial multiprocessor supercomputers with 2 to 8 processors and shared multiport memories have been shown to give reasonable performance [5], but designs for systems with larger numbers of processors have suffered from reduced memory performance due to interprocessor memory conflicts. A number of proposals to reduce memory conflicts have been made. Skewing and related techniques [6, 7, 8] have been shown to be effective in reducing intraprocessor conflicts. These techniques are not effective when conflicts between processors dominate memory performance. Buffering has been proposed in [4, 6] to reduce memory conflicts in pipelined processors. However, a number of problems arise because of synchronization and consistency requirements.

In this paper a complete data-path simulation for a logical bank memory in a Cray-based multiprocessor system is presented. The memory efficiency, E, is used as the primary measure of memory performance. E is defined by:

$$E = \frac{\text{Number of successful memory references}}{\text{Number of memory reference attempts}}$$

which reflects the fact that a memory reference may be attempted many times before it is successful. It is shown that buffering on the front end of physical memory (input buffering) results in substantial improvements in memory efficiency, while buffering on the back end (output buffering) reduces the difference in performance between reads and writes.

† This work was supported by Cray Research Inc. Computational support was provided by the University of Texas Center for High Performance Computing.

2 Cray Y-MP Memory Design

The Cray Y-MP vector supercomputer contains a complex interconnection network between its eight processors and its 512-megabyte central memory [3]. Each processor has four ports (two read ports, one write port, and an I/O port) which are connected via a crossbar-like switch to four memory lines. The memory is divided into four sections and each section is accessed by a single line from each processor.

Each section is divided into eight subsections and the individual subsections are further subdivided into eight physical banks. The physical bank cycle time, T_c, is five cycles. A processor can access a particular subsection once every T_c cycles. References from different processors to the same subsection can proceed without conflict provided that they are addressed to banks which are not already in use. Since each processor can access each subsection independently, the complexity of a subsection switch is nl where n is the number of processors and l is the number of banks within the subsection. The complexity of the port-to-line switch at each processor is pr where p is the number of ports per processor and r is the number of lines per processor. If s is the number of subsections per section, the interconnection cost is proportional to $npr + rsnl = nr(p + sl)$. For the Cray Y-MP, $r = 4$ and $p = 4$ and $s = l = 8$. The cost is essentially the number of processors times the total number of physical banks in the system.

3 Logical Memory Banks

In the proposed design, the individual physical banks within a subsection are replaced by logical banks in order to accommodate more processors and/or to allow the use of slower memory without affecting performance.

A logical bank consists of a number of physical subbanks each with an input buffer of pending requests as shown in Figure 1. Each physical bank may also have an output buffer which is used to buffer read return values before they enter the fan-in network. When a request is made by a particular processor, the line, subsection, and logical bank must be free. Only one reference to a logical bank can occur during an interval T_l. T_l is called the logical bank cycle time and is assumed to be one. If two lines (from different processors) attempt to access the same logical bank during T_l, a *logical bank conflict* occurs and one reference is delayed.

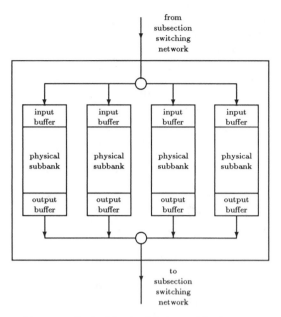

Figure 1: Logical bank with four subbanks.

The reference to a logical bank will be directed to a particular subbank within that logical bank. If the subbank is busy and there is still room in its input buffer, the request is queued for later service. In either case, the reference is considered to be made successfully. An unsuccessful reference causes the logical bank to be unavailable until the next cycle. If a reference is made successfully, the subsection for that processor will only be busy for one cycle rather than T_c cycles as in the original Cray Y-MP design.

Logical banks were introduced by Seznec and Jegou to support the Data Synchronized Pipeline Architecture [11]. Their design includes a reordering unit so that data flows out of the logical bank in chronological order. In the scheme proposed in this paper, a reordering unit is not required at the logical bank, because reordering occurs at the processor. However, a data bus arbitration scheme, which is not needed when no buffers are present, must be introduced to handle port conflicts on the data bus for reads.

4 Analytical Model for Writes

There is a distinction between reads and writes in this system. Read operations must navigate a data fan-in return network while write operations complete when they reach the subbank. A simple analytical model for writes [10] to logical banks gives surprisingly accurate results and provides insight into the mechanism by which buffering improves performance.

The model is based on the observation that as long as the buffer for the logical bank has available buffer slots, the processor sees a memory consisting of l logical banks with a cycle time of T_l, the logical bank cycle time. The efficiency in this case is given by E_l. When the buffer saturates, the memory behaves as though the system had no logical banks and a memory cycle time of T_c. The efficiency in this case is denoted by E_p. The effective efficiency is then a weighted average of the two cases depending on the probability that there are buffer slots available.

Let P be the probability that the subbank input buffer is not full at the time a memory reference is first initiated. A successful reference will take $1/E_l$ cycles with probability P and $1/E_p$ cycles with probability $1 - P$. The average number of cycles for a successful reference is:

$$\frac{1}{E} = \frac{P}{E_l} + \frac{1 - P}{E_p}$$

where E is the combined or effective efficiency. This relationship can be written as:

$$E = \frac{E_l E_p}{P E_p + (1 - P) E_l}$$

The probability, P, that the buffer is not full can be estimated by considering each logical bank as a system of independent queues under the $M/D/1/\infty$ discipline. The distribution depends on the parameter $\rho = \lambda T_c$ where λ is the average arrival rate and T_c is the queue service time. λ can be approximated by npq/b, where q is the probability that a port initiates a reference, n is the number of processors, p is the number of ports per processor, and b is the total number of subbanks. Expressions for E_p and E_l have been derived in [10].

The key observation in the analysis is that as long as the reference arrival rate is less than the service time, the memory appears to have a cycle time T_l (one cycle) independent of the actual bank cycle time of the physical memory. When the arrival rate is greater than the service time, the buffers saturate and the memory appears as though there were no buffering at all. This principle holds for both scalar and vector reference streams.

5 Simulation Model for Data Bus Conflicts

While the above model provides insight into the mechanism which makes logical banks effective, detailed issues of performance must be resolved by simulation. The simulation presented in this paper emulates the operation of a Cray Y-MP type system as closely as possible. The processors are assumed to have three ports connected to four lines. The ports are specifically designated as either read ports or write ports since the implementations of the two types of ports are different. (The I/O port usually does asynchronous block transfers and is not considered in this work.) Each processor also has a certain number of vector registers. When a port is ready to initiate a vector reference, it randomly selects a free vector register to perform a read or write. If no register is available, the operation holds until a register becomes available.

All memory operations are assumed to be vector operations with an associated stride and length. A stride of one is assumed to be the most probable (0.75) with other strides up to a maximum stride being equally probable. The maximum length of the vector operations is determined by the length of the vector registers in the processor. All possible vector lengths are assumed to be equally probable, except the maximum length which will occur more frequently. In this paper the maximum vector length is assumed to be 64, and the probability of a maximum length length vector is assumed to be 0.75.

When a particular memory location is referenced, the line, subsection, logical bank, and subbank numbers are calculated. If the line is free, it is reserved for T_r cycles and the subsection is checked. If the subsection is free, it is reserved for T_s cycles, and the logical bank is checked. If the logical bank is free, it is reserved for T_l cycles, and the subbank is checked. If the subbank is free, the operation is initiated. Otherwise if the subbank is in use but it has space in its input buffer, the operation is queued. In either case the reference is said to be *issued*. The reference generates a hold and fails to issue if the subbank is busy and there is no room in its input buffer. It takes T_c clock cycles for a reference to be processed by a subbank. If the operation is a write, the subbank is free to accept another reference after T_c cycles. If there are any pending requests in the subbank input buffer, these will be processed before additional outside references.

A vector write reference has completed when the last element operation has been issued. A vector read reference has not completed until all of the element values have arrived at the register. Read data values must be routed on the line data bus from the physical memory bank to the appropriate processor vector register. Additional conflicts may occur here because more than one value may become available on a particular cycle. Each logical bank has a single output latch. If the latch is free, the value is moved from the subbank to the latch and the subbank is freed. If the latch is busy, the subbank must wait until the latch is free before accepting another value. If an output buffer is included for each subbank as shown in Figure 1, the value is moved from the subbank to the output buffer and blocking of the subbank on reads is less likely to occur.

All of the data values latched for a particular processor line compete for processing on the return data bus. The return data bus is modeled as a pipeline which can accept one value on each cycle. When the last value for a vector read has emerged from the pipeline, the read is considered to be complete and the vector register is freed.

The system load is determined by the operation initiation rate. When a port is free there is a certain probability, p_f, that on the current cycle a memory operation will be initiated. A vector initiation rate of $p_f = 0.01$ is chosen for this paper corresponding to a base efficiency of 0.67 for the original Cray Y-MP design. This value is similar to the value used by Bailey in his simulations of the Cray X-MP [1] and corresponds to a moderately loaded system.

6 Results of the Simulation

In previous work [9] a simulation which did not account for subsections or return conflicts was performed. Under these assumptions it was shown that with logical banks the number of processors could be tripled from eight to 24 without decreasing the memory efficiency. The current simula-

tion introduces subsections and accounts for flow through the full data path.

Figure 2 shows the effect of buffering on efficiency as the bank cycle time is varied for parameters similar to the existing Cray Y-MP architecture. The efficiency remains relatively flat as long as the buffers don't saturate which confirms the observations of Section 4. When the bank cycle time is tripled from 5 to 15, the efficiency remains above the base efficiency of 0.67.

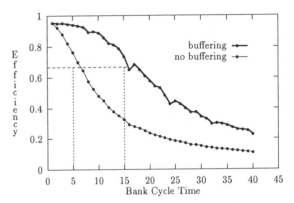

Figure 2: Effect of buffering on memory efficiency for the current 8-processor Cray-YMP architecture.

In Figure 3 the efficiency is shown as a function of the number of processors when the bank cycle time is five. Five cases are considered. The dotted line is the base efficiency for the original Cray Y-MP under the specified loading. Cases a), b), and d) use the buffering methods proposed here. Cases c) and e) do not use buffering.

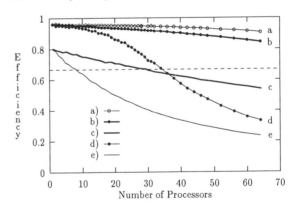

Figure 3: Efficiency as a function of the number of processors for a bank cycle time of five.
a) 1024 logical banks, 1 subbank per logical bank
b) 256 logical banks, 4 subbanks per logical bank
c) 1024 banks, no buffering
d) 256 logical banks, 1 subbank per logical bank
e) 256 banks, no buffering

Cases a) and b) each have 1024 physical banks. a) has 1024 logical banks each with a single subbank, while b) has 256 logical banks each with four subbanks. For up to

64 processors, there is little difference in performance although b) has a much smaller interconnection cost. Case b) achieves an operating efficiency of 0.844 with 64 processors which is above the base efficiency for the 8-processor system under the same per processor loading. The rapid fall-off in efficiency for case d) at 20 processors corresponds to a saturation of the subbank buffers.

Case c) corresponds to adding unbuffered banks to the current Cray Y-MP architecture. Although this design has a per processor interconnection cost which is comparable to case a), it cannot adequately support more than 24 processors. The performance when processors are added to the current Cray Y-MP architecture is shown in case e) for comparison.

As mentioned in Section 5 reads and writes are handled by a different mechanism since read values must traverse a fan-in return network before the operation is complete. The presence of a single slot output buffer at each subbank is sufficient to eliminate the performance difference between reads and writes for the specified loading as shown in Figure 4. The case of writes without output buffering and the case of two read ports and one write port with output buffering have similar memory performance.

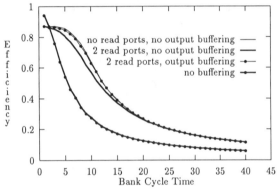

Figure 4: Efficiency as a function of bank cycle time with various types of input and output buffering, 64 processors, 256 logical banks, and 4 physical subbanks per logical bank.

7 Discussion

The addition of logical banks in a Cray Y-MP architecture is a cost-effective method for achieving high performance in a shared memory system with 32 to 64 vector processors. The addition of logical banks results in an increase in hardware memory costs. This cost is due to the additional complexity of the banks, the requirement for return bus arbitration, and increased bus width for tags. For the same interconnection cost per processor as a current Cray Y-MP, a memory efficiency of 0.844 can be achieved for a system with 64 processors for moderate loadings. For maximally loaded systems with a large number of banks, return conflicts can become a problem and additional output buffering may be required.

The Cray memory system is based on *weak ordering* of events as described in [4]. Weak ordering relies on the com-

piler to enforce critical sections in software by appropriate use of synchronization variables which are provided by shared registers in the Cray architecture. References to individual subbanks are performed on a first-in-first-out basis, and the system with logical banks is still weakly ordered. No caches or virtual memory are used, so consistency problems which usually arise with buffering and shared memory do not occur in this context.

Acknowledgments: The authors thank Xiaodong Zhang for his comments on this paper.

References

[1] D. H. Bailey, "Vector computer memory bank contention," *IEEE Trans. on Computers*, vol. C-36, (1987), pp. 293–298.

[2] F. A. Briggs, "Effects of buffered memory requests in multiprocessor systems," *Proc. ACM-SIGMETRICS Conf. on Simulation, Measurements, and Modeling of Computer Systems*, (1979), pp. 434–442.

[3] Cray Research, *Cray Y-MP Computer Systems Functional Description Manual*, Publication HR-4001A, (1988).

[4] M. Dubois, C. Scheurich and F. Briggs, "Memory access buffering in multiprocessors," *13th Intl. Symp. on Computer Architecture*, (1988), pp. 434–422.

[5] G. J. Faanes and J. L. Schwarzmeier, "Comparing the performance of Cray Y-MP and Cray X-MP computer systems," *Cray Channels*, vol. 10, no. 4, (1989), pp. 26–30.

[6] D. T. Harper and J. R. Jump, "Vector access performance in parallel memories using a skewed storage scheme," *IEEE Trans. on Computers*, vol C-36, (1987), pp. 1440–1449.

[7] D. T. Harper, "Address transformations to increase memory performance," *Proc. 1989 Intl. Conf. on Parallel Processing*, (1989), pp. 237–241.

[8] D. Lee, "Scrambled storage for parallel memory systems," *Proc. 15th Intl. Conf. on Computer Architecture*, (1988), pp. 232–239.

[9] K. A. Robbins and S. Robbins, "Logical memory banks for the Cray Y-MP," *Proc. MIDCON/90*, (1990), pp. 262–265.

[10] K. A. Robbins and S. Robbins, "Logical memory banks in vector multiprocessor systems," University of Texas at San Antonio, Technical Report, UTSA-CS-90-101, (1990).

[11] A. Seznec and Y. Jegou, "Optimizing memory throughput in a tightly coupled multiprocessor," *Proc. 1987 Intl. Conf. on Parallel Processing*, (1987), pp. 344–346.

Prime Cube Graph Approach for Processor Allocation in Hypercube Multiprocessors[*]

Hong Wang and Qing Yang

Dept. of Electrical Engineering

The University of Rhode Island

Kingston,RI 02881

ABSTRACT

This paper proposes a new approach for subcube and non-cubic processor allocation for hypercube multiprocessors. The main idea of the new approach is to represent available processors in the system by means of a *prime cube graph* (PC-graph). With the availability of existing linear graph algorithms, cubic and non-cubic processor allocations are efficiently performed by manipulating the PC-graph. In addition to having the full subcube recognition ability, the PC-graph also maintains the inter-relationships between free subcubes. By means of biconnectivity analysis, the algorithm can significantly reduce both internal and external processor fragmentations. Our simulation results show that the new approach can reduce the average task waiting time by 35% to 50% compared to the existing allocation strategies when tasks require random number of processors for execution. In the worst case scenario, the PC-graph strategy performs at least as well as the free list strategy [1] although the new scheme does not possess static optimality.

1 Introduction

In hypercube computers, user programs generally consist of a number of tasks, each of which can be executed by several processors in parallel. The problem of assigning available (free) processors to an arriving task, called processor allocation, is of essential importance for effective parallel processing and high resource utilization. A number of processor allocation schemes have appeared in the literature [1, 2, 3, 4]. In these studies, an incoming task that is assumed to require N ($N = 2^k$) processors is allocated to an available subcube of dimension k for execution. Upon completion of the execution, the subcube occupied by the task is deallocated and made available to other incoming tasks. A good discussion of various allocation strategies can be found in [1, 4].

As pointed out by Chen and Shin in their recent study [5], there exists a *fragmentation* problem as a result of processor allocation and deallocation in the hypercube computers. First of all, due to the dynamic nature of processor allocation and deallocation, available processors are separated by busy processors. Even if a sufficient number of processors are available, they may not form a subcube that is big enough to accommodate an arriving task. When the system is in this situation, the external fragmentation is said to occur. Secondly, many applications such as embeddings in hypercubes for binary tree, incomplete binary tree, rectangular grids and solving large number of nonlinear equations require the number of processors that is not a power of 2, or the size of a complete subcube [6, 7, 8]. Thus, the complete cube allocation strategies may also result in internal fragmentations. Solving these two fragmentation problems is essential to effectively utilizing hypercube computers and achieving high performance.

The external fragmentation problem can be solved by means of task relocation. Chen and Shin are among the first to realize the fragmentation problem and have proposed a full compaction scheme based on Gray Code strategy [5]. Recently, Huang and Juang proposed a partial compaction scheme [9] in which only a small portion of running tasks are moved to generate a new free subcube that is big enough to accommodate the requesting task. However, any relocation scheme has to pay penalty for delaying individual tasks which are subjected to suspension from execution, and for high communication costs incurred during migrating tasks to their new sites.

Internal fragmentation mainly results from the large gap between two consecutive allowable cubic sizes, which have to be a power of 2. This problem can be lessened by means of noncubic allocation. The importance and the potential applications of noncubic structures are well discussed in [7, 10]. To effectively utilize noncubic topology on hypercubes requires a processor allocation scheme which is capable of finding a set of subcubes that are physically adjacent to each other and form a noncubic structure like the incomplete cube for a requesting task. To minimize the cost of inter-process communication, a noncubic allocation scheme is anticipated to be able to recognize subcubes as well as interconnections between them so as to efficiently identify an available noncubic structure in a fragmented configuration in the dynamic environment of allocation and deallocation.

In this paper, we propose a new processor allocation method which can efficiently recognize the physical relationships between available subcubes in addition to having a full subcube recognition ability. The distinctive feature of our allocation scheme is that it treats subcube allocation and non-cubic allocation in a unified way. The scheme is referred to as *prime cube graph* since it maintains an intersection graph of all free prime cubes (or *a complete sum*) of a configuration in the hypercube system. The new method differs from the usual prime implicant technique in that it is a top-down approach [1] in both finding a required subcube and updating the *prime cube graph*. Dutt and Hayes have also proposed a graphic approach for subcube allocation in their recent study [3]. Their approach aims at deriving the *maximal set of subcubes* (MSS) by using consensus graph to reduce external processor fragmentation. Our approach differs from theirs in that the PC-graph scheme emphasizes on recognizing the inter-relationship between available subcubes so that both internal and external fragmentations can be minimized in an efficient way. Our simulation results show that the new approach performs at least as well as the *free list* scheme for complete subcube allocation in terms of processor utilization and task completion time, though *static optimality* is not guaranteed theoretically. The advantages of our scheme become apparent when non-cubic allocation is taken into account to avoid internal fragmentation for incoming tasks requiring a random number of processors rather than just a power of 2. With the availability of connectivity information among free subcubes in *prime cube graph*, the new scheme can identify and utilize

[*]This research is supported by National Science Foundation under grants No. CCR-8909672

smaller subcubes which are physically adjacent to each other for non-cubic allocations in contrast to breaking up larger cubes used by other schemes [1].

The paper is organized as follows. Section 2 introduces necessary definitions, notations and background. The *prime cube graph* approach for cubic and noncubic processor allocation and deallocation is presented in Section 3. In section 4, we present performance analysis and comparison between our approach and existing approaches by means of simulation. Section 5 concludes the paper.

2 Background

2.1 Definitions and Notations

In an n-dimensional complete hypercube, Q^n, we define a *configuration* C_f, as a set of all *free* nodes in the system at a time of interest. Let $\sum$ be a ternary set $\{0, 1, *\}$, where $*$ represents the *don't care* symbol. A subcube α can be uniquely represented by an element in $\sum^n$. We call such n-bit ternary string the *address* of subcube α. The number of $*$'s in subcube α's address is called its dimension, denoted by $dim(\alpha)$. For convenience, we'll use α to interchangeably denote a subcube and its address, while $\alpha[i]$ $(1 \leq i \leq n)$ represents the symbol at the ith bit of α's address with i being counted from left to right. For clarity in description, we call a processor in Q^n a *node*, and a connection between two adjacent processors a *link*. The terms *vertex* and *edge* will be reserved for describing our PC-graph only. Also, cubic allocation is interchangeable with complete cube allocation.

Definition 2.1 *The minimum distance, or Hamming distance H, between $\forall \alpha, \beta \in \sum^n$ is defined as* $\mathrm{H}(\alpha, \beta) = \sum_{i=1}^{n} h(\alpha[i], \beta[i])$, *where,* $h(\alpha[i], \beta[i]) = 1$ *iff* $\alpha[i] = 0$ *and* $\beta[i] = 1$, *or* $\alpha[i] = 1$ *and* $\beta[i] = 0$; *otherwise,* $h(\alpha[i], \beta[i]) = 0$.

The maximum distance, D, between $\forall \alpha, \beta \in \sum^n$ is defined as $\mathrm{D}(\alpha, \beta) = \sum_{i=1}^{n} d(\alpha[i], \beta[i])$, *where* $d(\alpha[i], \beta[i]) = 0$ *if* $\alpha[i] = \beta[i] \neq *$; *and* $d(\alpha[i], \beta[i]) = 1$ *otherwise.*

Two subcubes α and β are said to be *adjacent* or *neighbors* to each other iff $H(\alpha, \beta) = 1$.

To further describe relationship between subcubes, let $\perp$ represent *undefined* symbol, and $\diamond$ represent a quadruplicate symbol set $\sum \cup \{\perp\}$. A partial ordering $\succ$ is defined on $\diamond$ by letting $* \succ 1$, $* \succ 0$, $1 \succ \perp$, $0 \succ \perp$, and $\succeq, \prec, \preceq$ will have the obvious meanings. $\forall x, y \in \diamond$, both *least upper bound* (lub) and *greatest lower bound* (glb) exist within $\diamond$, define $x \vee y = \mathrm{lub}(x, y)$, $x \wedge y = \mathrm{glb}(x, y)$. For example, $* \wedge * = *$, $0 \wedge * = 0$, $* \wedge 1 = 1$, $1 \wedge 0 = \perp$; $* \vee * = *$, $0 \vee * = *$, $1 \vee 0 = *$, $\perp \vee 1 = 1$.

On basis of $\succeq$, we define several other relationships between subcubes in $\sum^n$ as follows,

Definition 2.2 $\forall \alpha, \beta \in \sum^n$, α *is said to* intersect β, *if* $\forall i \in (1 \cdots n)$, $\alpha[i] \wedge \beta[i] \in \sum$. *If* α *intersects* β, *the intersection of two subcubes can be represented by* $\gamma = \alpha \wedge \beta$, *where* $\gamma[i] = \alpha[i] \wedge \beta[i]$, $\forall i \in (1 \cdots n)$.

Definition 2.3 $\forall \alpha, \beta \in \sum^n$, α *is said to* cover β, *denoted by* $\alpha \sqsupseteq \beta$, *if* $\forall i \in (1 \cdots n)$, $\alpha[i] \succeq \beta[i]$. *Let* $\sqsubseteq, \sqsupset$, *and* $\sqsubset$ *have their obvious meanings. Clearly,* $\forall i \in (1 \cdots n)$, $\alpha[i] \succeq \beta[i]$ *iff* $\alpha \wedge \beta = \beta$.

Definition 2.4 *In a given set of nodes S in a hypercube, a subcube α is a* prime cube *if there is no other subcube β in S such that $\beta \sqsupset \alpha$.*

For a given set of nodes S in the hypercube, let $\mathcal{P}(S)$ denote the set of all prime cubes for these nodes.

Adopting the terminology from switching theory, a node of Q^n is analogous to a *miniterm* of an n-variable boolean function, C_f corresponds to a *canonical sum* while the set $\mathcal{P}(C_f)$ corresponds to a *complete sum*. Note that prime cubes in a complete sum are not necessarily mutually disjoint.

Example 2.1 *For $C_f = (0111, 1111, 1101, 1001, 1011)$ of a Q^4, subcube $(1 * *1)$ intersects $(*111)$, and they are prime cubes since no other cubes in C_f covers any of them. Subcube $(1 * 11)$ is not a prime cube because $(1 * *1) \sqsupset (1 * 11)$.*

Definition 2.5 *The consensus (also called star product) of two subcubes α and β, denoted by $\gamma = \alpha \star \beta$, is defined as: γ is nonempty iff $\mathrm{H}(\alpha, \beta) \leq 1$, and $\gamma = \alpha \wedge \beta$, when $\mathrm{H}(\alpha, \beta) = 0$; $\gamma[i] = \alpha[i] \wedge \beta[i]$, with the "$\perp$" symbol being substituted by "$*$", when $\mathrm{H}(\alpha, \beta) = 1$.*

Definition 2.6 *The sharp product of two subcubes α and β, denoted by $\alpha\#\beta$, is the set of nodes belonging to α but not to β.*

Notice that $\alpha\#\beta$ may not equal to $\beta\#\alpha$.

Example 2.2 *The consensus of $(1 * *1)$ and $(*100)$: $(1 * *1) \star (*100) = (110*)$ and the consensus of $(00 * *)$ and $(010*)$: $(00 * *) \star (010*) = (0 * 0*)$. The sharp product $(1 * *1)\#(*111)$ of $(1 * *1)$ with respect to $(*111)$ is $\{(1 * 01), (10 * 1)\}$, while $(*111) \# (1 * *1) = \{(0111)\}$.*

2.2 Incomplete Cubes and Non-cubic Structure

An *incomplete hypercube*, which was first studied by Katseff[7], is a special non-cubic structure. In an incomplete hypercube with N nodes, all nodes can be numbered from 0 to $N-1$ with the property that there is a link between a pair of nodes whose binary representations differ by exactly one bit. This property leads to a simple and deadlock free routing algorithm[7]. In [10], Tzeng further analyzed the structural properties of a special incomplete hypercube structure I_k^p, which consists of two adjacent subcubes $(0*^p)$ and $(10^{p-k}*^k)$ with nodes numbered from 0 to $(2^p + 2^k - 1)$.

The non-cubic structure to be considered in this paper, denoted by $NC(p, q, k)$, consists of two intersecting (prime) cubes: a p-cube $(0^{q-k}*^p)$, a q-cube $(*^{q-k}0^{p-k}*^k)$, and their intersection k-cube $(0^{p+q-2k}*^k)$. Without loss of generality, we assume $p \geq q > k$ throughout the paper. Obviously, I_k^p is a special case of $NC(p, q, k)$, when $q = k + 1$. Unlike in incomplete cube, it's not always possible to number all nodes in an $NC(p, q, k)$ from 0 to $2^p + 2^q - 2^k - 1$ while guaranteeing that there is a link between a pair of nodes if their binary representations differ by exactly one bit. Take $NC(2, 2, 0)$ for example, it's impossible to find such numbering. Figure 1 shows some examples of NC structures.

3 Prime Cube Graph Strategy

3.1 Prime Cube Graph

As mentioned in section 2, $\mathcal{P}(C_f)$, the set of all prime cubes in a given configuration C_f is analogous to a *complete sum* of an n-variable boolean function. Accordingly, $\mathcal{P}(C_f)$ is *unique* for each C_f. A graph representing $\mathcal{P}(C_f)$ is introduced below. Notice that $\mathcal{P}$ and $\mathcal{P}(C_f)$ are interchangeably used for simplicity.

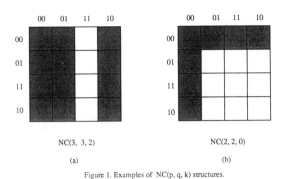

Figure 1. Examples of NC(p, q, k) structures.

Definition 3.1 *For a given configuration C_f, a prime cube graph, $G(\mathcal{P}) = (V(\mathcal{P}), E(\mathcal{P}))$, is a labeled undirected graph as defined below,*

- *$V(\mathcal{P}) = \mathcal{P}$, or each vertex v in $G(\mathcal{P})$ corresponds to a prime cube in $\mathcal{P}$.*

- *$\forall v, w \in V(\mathcal{P})$, $v \neq w$, if v, w intersect, then there exists an edge $e = (v, w)$, $e \in E(\mathcal{P})$.*

- *Every vertex or edge has a label, or a tuple of weights, which in general contains information about individual prime cube and topological relation between different prime cubes.*

A typical vertex label for $\forall v \in \mathcal{P}$ includes subcube address of v, $degree(v)$ and biconnectivity information. And an edge label includes size of intersecting cubes and biconnectivity information on this edge.

The prime cube graph defined above can *uniquely* represent a configuration of the hypercube system in a dynamic environment. Unlike other top-down approaches, the prime cube graph is able to keep more information about both individual subcubes and their connectivity relationships. The problem of cubic and non-cubic processor allocation boils down to the manipulation of the prime cube graph.

Our main objective in developing the new algorithm is to reduce processor fragmentation. In a hypercube structure, external processor fragmentation can be partially described by the prime cube graph. Consider two configurations: C_{f_1} and C_{f_2}, both of which contain the same number of free nodes. C_{f_1} is *more fragmented* than C_{f_2} if C_{f_1} has more disconnected subsets of nodes or more connected components than C_{f_2} does. Unless there is no other choice, an allocation scheme should avoid allocating an articulation point of the PC-graph in order to maintain higher connectivity of the PC-graph. In this way, the external fragmentation can be reduced to certain extent, especially for noncubic allocation. We will use such analysis in our allocation algorithm to minimize processor fragmentation.

Besides biconnectivity information which gives a coarse measure on fragmentation, analysis on vertex adjacency is another important feature to gauge fragmentation. Consider the following example.

Example 3.1 *In a configuration C_f, $\mathcal{P}(C_f) = \{\alpha, \beta, \gamma\}$, where $\alpha = (*1**)$, $\beta = (1*00)$, $\gamma = (0*10)$. An incoming task requests a 2-cube. Obviously, only $\alpha = (*1**)$ is large enough to accommodate this request though it's an articulation point in $G(\mathcal{P})$. However, different ways of decomposing α may result in different remaining configurations which in turn give different performance. Care has to be taken to do the decomposition, if connectivity is considered as a measure of fragmentation. Subcube $(*1*1)$ should be allocated so that all the remaining nodes are in one connected component. If the allocation environment allows noncubic request and the*

next request needs 6 nodes, no delay will be incurred if the connectivity is retained.

In the PC-graph $G(\mathcal{P})$ for a given configuration C_f, a *sharing density vector* with respect to vertex v, denoted SD_v, is defined as an n-bit vector that is evaluated through bit-wise comparison with all $degree(v)$ cubes intersecting v, w_j ($j = 1, ..., degree(v)$). $SD_v[i] = SD_v[i] - 1$, if $v[i] = *$ and $w_j[i] = 0$; $SD_v[i] = SD_v[i] + 1$, if $v[i] = *$ and $w_j[i] = 1$; otherwise, $SD_v[i]$ remains unchanged. The *sharing density vector* for all vertices in a PC-graph can be computed in the course of linear biconnectivity analysis. With sharing density vector, connectivity relationship between intersecting prime cubes can be further represented. Consequently, subcube decomposition can be guided wisely. For instance, the sharing density vector for subcube α in the last example is $SD_\alpha = \overline{0}00(-2)$, which means that the intersecting cubes (here are β and γ) share more 0's or less 1's along the $4th$ dimension. So, by fixing the $4th$ dimension of α by 1, the split subcube $(*1*1)$ will have no common part with β and γ. In other words, the previous connectivity is retained.

3.2 Cubic Allocation Algorithm

The formal algorithm for cubic processor allocation is given below.

Allocation Algorithm

1. Set $k := |I_j|$, where $|I_j|$ is the dimension of a subcube required to accommodate the request I_j.

2. Apply biconnectivity analysis on prime cube graph and select a vertex $\alpha \in V(\mathcal{P})$ (or $\mathcal{P}$) based on size, connectivity information (explained later).
   If none of prime cubes in $\mathcal{P}$ is bigger than or equal to k, then request I_j is kept in the waiting queue and skip over remaining steps.

3. (a) If $dim(\alpha) = k$, allocate α to I_j.

   (b) If $dim(\alpha) > k$, split a k-cube γ out of α based on SD_α, allocate γ to I_j, and put remaining $(dim(\alpha) - k)$ cubes into a working queue.

4. Update prime cube graph.
   For each cube β in $\mathcal{P}$ which intersects α (including α itself)

   (a) Remove β from $\mathcal{P}$.

   (b) Append $(\beta \# \alpha)$ to the working queue.

5. If working queue is empty, then finished;
   else

   (a) $\delta \leftarrow$ dequeue first element in the working queue.

   (b) Deallocate δ.

   (c) Go to step 5.

Step two of the allocation algorithm uses the linear depth-first search algorithm to do biconnectivity analysis. As a result of applying this algorithm, one candidate tagged as articulation vertex or nonarticulation vertex is selected from each biconnected component, provided that there is at least one vertex in the component whose dimension is higher than or equal to k. During the candidate selection, the following reasonable weight evaluation order can be used: 1) subcube size, 2) articulation information, 3) vertex degree. A global selection takes place right after biconnectivity analysis is over. Vertex representatives from all components are compared respectively within articulation and non-articulation categories. After global candidates in both categories have been obtained, one of them is selected as the α. Step 3 does processor allocation by assigning the selected subcube or a part of the selected subcube to the requesting task. If $dim(\alpha) > k$, α is split based on its sharing density vector as discussed in the last subsection. At step 4, all prime cubes (including α itself) in $\mathcal{P}$ which intersect α are removed and the corresponding sharp product cubes with respect to α are appended to the waiting queue. With deallocation algorithm which will be given shortly, all cubes in the waiting queue are deallocated back to the prime cube graph at step 5. When the algorithm is finished, the updated prime cube graph represents the complete sum for the new configuration.

The time and space complexity of the allocation algorithm will be analyzed in details in the next section. One can easily observe that the time complexity of step 2 is $O(l, E(G(\mathcal{P})))$, or linear to the size of the prime cube graph, where l is the number of vertices in the graph. Therefore, the time needed to allocate processors to a requesting task is linear to l. From our simulation experiment, we observed that the average number of vertices in the PC-graph is on the same order of the hypercube size n ($n \leq 10$).

Example 3.2 *In a 6-cube, suppose the current configuration is represented by the PC-graph shown in Figure 2(a). The corresponding prime cubes are, $1 : (01**01), 2 : (**00**), 3 : (10****), 4 : (****10)$. An incoming request needs a 3-cube. At step 2 of the allocation algorithm, vertices 2 and 3 are selected as candidates from two biconnected components $\{1,2\}$ and $\{2,3,4\}$, respectively, as shown in Figure 2(b). When the global selection takes place, although both candidates 2 and 3 have the same size, vertex 3 is selected for allocation because it is an non-articulation point. Vertex 3 is decomposed into two 3-dimensional cubes $(101***)$ and $(100***)$. The cube $(101***)$ is assigned to the request. In making this decomposition, the sharing density vector is used (as discussed previously) to select the dimension along which the splitting is done. The PC-graph is updated at step 4 and 5, and resulting PC-graph is shown in Figure 2(c).*

Deallocation Algorithm
Let α denote the released k-cube.

1. Find α's neighborhood from PC-graph.
   For each neighbor β

   (a) Compute a consensus $\gamma = \alpha \star \beta$.

   (b) If α is covered by γ, then *mark* α.

   (c) If β is covered by γ, then *mark* β.

   (d) Insert γ into a queue L.

2. While $L \neq \emptyset$,

   (a) $\gamma \leftarrow$ dequeue the first element in L.

   (b) Compute $\delta \leftarrow prime(\gamma)$ (explain later).

   (c) Put δ into PC-graph, and delete all the covered cubes.

   (d) Go to step 2.

3. If α is not marked, then put α into the PC-graph.

4. For all remaining vertices in the PC-graph, make edge connections between all pairs of intersecting vertices.

Algorithm to form prime cube $prime(\sigma)$

1. If there is no *unmarked neighbor* for σ in PC-graph and L

then skip to step 2.
else

   (a) find next unmarked neighbor β.

   (b) $\gamma \leftarrow \sigma \star \beta$.

   (c) If σ is covered by γ
       then mark σ and skip to step 2.
       else go to step 1.

2. If σ is marked
   then return $prime(\gamma)$ where γ is the consensus computed at step 1(b).
   else return σ.

Example 3.3 *Consider that a newly finished task relinquishes a cube $\alpha = (01**)$ to a configuration C_f which is represented by $\mathcal{P} = \{\beta_1, \beta_2, \beta_3, \beta_4, \beta_5, \beta_6\}$, where $\beta_1 = (1*00), \beta_2 = (110*), \beta_3 = (11*1), \beta_4 = (1*11), \beta_5 = (*01*)$ and $\beta_6 = (10*0)$. See Figure 3(a). Follow the deallocation algorithm, at step 1, $\{\beta_1, \beta_2, \beta_3, \beta_4, \beta_5\}$ are first found to be α's neighborhood, and corresponding consensuses computed at 1(a) are then put into list L, which therefore becomes $\{(*100), (*10*), (*1*1), (*111), (0*1*)\}$. α is not marked since none of the consensuses in L cover α. On the other hand, β_2 and β_3 are marked, for they are covered by $(*10*), (*1*1)$ respectively. At step 2, each element in L will be used to generate a prime cube for new configuration. After step 2 is finished, the PC-graph will be $\{(1*00), (*01*), (10*0), (*10*), (*1*1), (**11), (0*1*)\}$. Notice the first 3 cubes which are originally in C_f's PC-graph remain in the new PCG. α is then put into the new PCG at step 3, since it's not marked. After edge connections between all pairs of cubes in the new PCG are made at step 4, the new PC-graph is obtained. See Figure 3(b).*

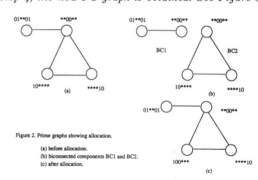

Figure 2. Prime graphs showing allocation.

(a) before allocation.
(b) biconnected components BC1 and BC2.
(c) after allocation.

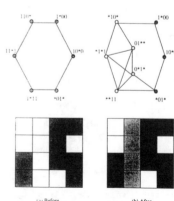

(a) Before   (b) After

Figure 3. Deallocation Illustrations.

3.3 Non-cubic Allocation

The non-cubic allocation can be done in the similar way as for the cubic allocation described in the last subsection. That is, by manipulating the PC-graph. The only difference of the non-cubic allocation from the cubic allocation is that it extracts edge information in addition to vertex information. The formal algorithm is given below:

Noncubic Allocation Algorithm

1. Let k be the number of processors required by the request I_j.

2. Apply biconnectivity analysis on PC-graph and select a vertex $\alpha \in \mathcal{P}$ or an edge e: $(\beta, \gamma) \in E(\mathcal{P})$ such that the number of processors in α or in e is greater than or equal to k.
   If neither vertex nor edge has been found larger than or equal to k, then I_j is kept in waiting queue and skip remaining steps.

3. If a vertex $\alpha \in \mathcal{P}$ is selected, skip remaining steps, and do the same as steps 3 and 4 of cubic allocation.

4. If an edge $(\beta, \gamma) \in E(\mathcal{P})$ is selected (suppose $|\beta| \geq |\gamma|$) then

   (a) For each cube δ in $\mathcal{P}$ intersecting β, or γ
       Remove δ from PC-graph,
       Append $(\delta \# \beta)$ or $(\delta \# \gamma)$ to a working queue.

   (b) Completely allocate β, and allocate cubes in $\gamma \# \beta$ until I_j is totally satisfied. Unused cube(s) are put into the working queue.

   (c) Deallocate all cubes in the working queue to PC-graph.

The candidate selection process of step 2 is performed in the same way as for cubic allocation. The selection rule used there can be easily extended for the edge selection. In addition to the 2 categories of vertex representatives (articulation and nonarticulation), a selected edge representatives can be of (1) articulation edge; (2) semi-articulation edge; (3) non-articulation edge. Inter-cube maximal distance should be taken into account during edge selection process. At the end of the selection, only one vertex representative and one edge representative are left for final selection. The vertex will be selected for allocation unless the edge has smaller size. In this context, size of a vertex α is measured by $2^{dim(\alpha)}$, while size of an edge (β, γ) is measured by $2^{dim(\beta)} + 2^{dim(\gamma)} - 2^{dim(\beta \wedge \gamma)}$.

Example 3.4 *A configuration C_f of a 4-cube system is given by $\mathcal{P} = \{\alpha, \beta, \gamma\}$, where $\alpha = (00**)$, $\beta = (**00)$, and $\gamma = (110*)$. An incoming task is requesting 7 nodes. Though in the current configuration no complete cube is big enough to accommodate the request alone, an edge (α, β) can satisfy the request perfectly.*

This example illustrates that our new approach can integrate a pair of small cubes to accommodate a request, thus result in processor utilization improvement.

The deallocation algorithm is the same as the cubic deallocation algorithm except that the released nodes may not form a complete cube.

3.4 Validity and Efficiency of the PC-graph Scheme

Since in the allocation algorithm, the deallocation algorithm is employed to update the prime cube graph after allocation of one task, the correctness and efficiency of deallocation are essential to the new approach.

In this subsection, we first prove that the deallocation algorithm can correctly form a PC-graph, PCG2, corresponding to a new configuration C_f^2 after a cube α is released to a configuration C_f^1 which is represented by PCG1.

After cube α is released, all prime cubes in configuration C_f^2, can be partitioned into 2 categories

- G_1, containing all prime cubes intersecting α, see the shaded vertices in the graph of Figure 3(b). (Notice that covering is a special intersecting).

- G_2, containing all prime cubes not intersecting α, see the unshaded vertices in the graph of Figure 3(b).

Let $N(\alpha)$ denote the set of all cubes that are neighbor cubes of α in C_f^1 (or PCG1). Let $C(\alpha)$ denote the set of all consensuses computed from α and cubes in $N(\alpha)$. For any prime cube δ in C_f^2 (or PCG2), let $CV(\delta)$ denote the set of all cubes in $C(\alpha)$ which are covered by δ.

Lemma 3.1 *Any prime cube in G_2 must be a prime cube in C_f^1 too.*

Proof: Let γ be any cube in G_2. According to G_2's definition, all nodes covered by γ must be in C_f^1 too. If γ is not a prime cube in C_f^1, then there must be a prime cube δ in C_f^1 (or PCG1), which covers γ. In C_f^2, δ either remains prime or is covered by a prime cube other than γ in PCG2. Both cases are in contradiction with the assumption that γ is a prime cube in C_f^2. $\square$

Following this lemma, the deallocation algorithm is correct provided that it can correctly generate all cubes in G_1.

Lemma 3.2 *1. $\forall \delta \in G_1$, if $\delta \neq \alpha$, then $CV(\delta) \neq \emptyset$, or δ must cover one or more cubes in $C(\alpha)$, where α is the released cube.*

2. For any two prime cubes δ_1, $\delta_2 \in G_1$, $CV(\delta_1)$ and $CV(\delta_2)$ are incompatible, i.e., neither of the following conditions holds: $CV(\delta_1) \supseteq CV(\delta_2)$, $CV(\delta_2) \supseteq CV(\delta_1)$.

Proof:

1. Suppose $CV(\delta) = \emptyset$, then either δ does not have intersection with α, or there is another cube covering δ, in either case, δ is not in G_1, which contradicts the assumption on δ.

2. Suppose $CV(\delta_1) \supseteq CV(\delta_2)$, then with any expansion to generate δ_1, δ_2 can also be generated, i.e., $\delta_1 \sqsupseteq \delta_2$. This contradicts the assumption that δ_2 is a prime cube.

$\square$

This lemma indicates that (1) G_1 has at most $|C(\alpha)|$ prime cubes, (2) CV set of any prime cube in G_1 has at least one element that is distinct from all elements in the CV set of another prime cube in G_1. The importance of (2) is that no matter in what order prime cubes in C_f^1 are organized or previous prime cubes of G_1 are generated, a prime cube in G_1 yet to be generated can always be considered by the deallocation algorithm, because its CV set cannot be empty unless it has been generated.

Lemma 3.3 *$\forall \alpha, \beta, \gamma$, if $\alpha \sqsubseteq \beta$, and $H(\alpha, \gamma) = 1$, then either $\alpha \star \gamma \sqsubseteq \beta \star \gamma$ or $\alpha \star \gamma \sqsubseteq \beta$.*

Proof: Suppose α and γ differ at bit i, strictly following the definition of consensus, we have (1) $\alpha \star \gamma \sqsubseteq \beta \star \gamma$, when $\alpha[i] = \beta[i]$; (2) $\alpha \star \gamma \sqsubseteq \beta$, when $\beta[i] = *$.□

This lemma assures the validity of deletion operation incurred at step 2(c) of the deallocation algorithm.

Lemma 3.4 *For each* $\sigma \in C(\alpha)$, $prime(\sigma)$ *will generate one prime cube* δ *in* G_1 *in* $dim(\delta) - dim(\sigma)$ *steps.*

Proof: According to the algorithm, the resulting cube is generated by iteratively expanding a cube in $C(\alpha)$ until it cannot expand any more. If the resulting cube δ can be covered by a prime cube in C_f^2, then a neighbor in either PC-graph or L must be found at step 1 and a consensus of higher dimension can be therefore formed at step 2.

Since a consensus can expand at most one more dimension in each recursion, the total number of expansions is therefore equal to the dimension difference of two cubes, i.e., $dim(\delta) - dim(\sigma)$. □

Theorem 3.1 *The deallocation algorithm can correctly generate PCG2 for* C_f^2 *with complexity of* $O(\max(|PCG1| * n, |PCG1|^2))$, *where* n *is the dimension of hypercube system.*

Proof: **Correctness** The correctness follows from the above lemmas.

At step 1, $C(\alpha)$ is found out, and

At step 2, all prime cubes are generated.

Following Lemma 3.2, for any pair of prime cubes in G_1, their cover sets are incompatible. Thus, the order of cubes appearing in prime cube graph doesn't affect the correctness of the deallocation algorithm. Lemma 3.3 guarantees the validity of deleting covered cubes from the PC-graph. Following lemma 3.4, when the algorithm is accomplished, all cubes in G_1 will be generated. According to lemma 3.1., those cubes from C_f^1 remaining intact in the resulting PC-graph are G_2. Therefore, the resultant prime cube graph is PCG2 for the new configuration C_f^2.

Complexity Analysis Let l be the size of resulting PCG2, $l = |G_1| + |G_2|$. Since $|G_1| \leq |C(\alpha)| = |N(\alpha)| \leq |PCG1|$, $|G_2| \leq |PCG1|$, we have $l = O(|PCG1|)$.

According to deallocation algorithm,

at step 1, the $C(\alpha)$ is generated using $O(|PCG_1|)$ time.

at step 2, all prime cubes are generated by expanding $C(\alpha)$. In each iteration,

step 2(b), according to lemma 3.4, takes $\max(dim(\delta) - dim(\sigma))$ time, where $\delta \in G_1$, and $\sigma \in CV(\delta) \subset C(\alpha)$,

step 2(c) takes at most l.

Notice $\forall \delta \in G_1, \max(dim(\delta) - dim(\sigma)) < n$, where n is the dimension of hypercube system, or the size of the largest cube. So, totally, step 2 takes

$$O(\Sigma_{\delta \in G_1, \sigma \in CV(\delta)} \max(l, \max(dim(\delta) - dim(\sigma))))$$

or, $O(\max(|PCG1|^2, |PCG1| \cdot n)$.

step 3 takes $O(1)$.

step 4 takes $O(l^2)$ time.

Therefore, the time complexity of the deallocation algorithm is $O(\max(|PCG1| \cdot n, |PCG1|^2))$.

Space Complexity Since only new prime cubes are inserted into the graph, the space complexity is $O(|PCG1| + |G_1|)$ or $O(|PCG1|)$. □

In the cubic allocation algorithm, the deallocation algorithm is used to modify the PC-graph, it is therefore obvious that the allocation algorithm can also correctly form the PC-graph (a complete sum) for the new configuration after an allocation. Similar to deallocation complexity analysis, we see the last step of allocation algorithm accounts for the algorithm's time complexity. The length of working queue can be maximally equal to $O(|PCG1| \cdot n)$, so in worst case, the allocation complexity would be $O(\max(|PCG1|^2 \cdot n^2, |PCG1|^3 \cdot n))$.

Here, the time complexity of the deallocation and allocation algorithms are expressed in terms of number of prime cubes in a PC-Graph. From our simulation experiment, we observed that the average number of vertices in the PC-graph is on the same order of the total dimension of the hypercube system, n ($n \leq 10$). Notice that before the first allocation in the system, the PC-graph has only one cube (all *'s) corresponding to the whole hypercube system.

Theorem 3.2 *The allocation algorithm has full subcube recognition ability.*

Proof: Because the PC-graph represents all prime cubes in each configuration, and any subcube, if exists, must be covered in one of the prime cubes, therefore, this theorem holds naturally. □

Notice that the allocation algorithm described above is not statically optimal, although it can minimize processor fragmentations in a dynamic environment. Consider the following example.

Example 3.5 *Initially, a* Q^4 *system is idle, the sequence of requests require 1, 0, 2, 0, 2, and 2 dimensions respectively.*

After the first 5 requests are allocated by the above allocation algorithm, the resulting free prime cubes are $\{0101, *110, 1 * 10\}$. *The last task can not be allocated since there is no available 2-cube in the configuration. Therefore, although the total number of processors required by six incoming tasks is equal to the system size, the last task cannot be allocated.*

Even though the PC-graph allocation scheme is not statically optimal, our simulation results show that our scheme performs as well as other schemes under dynamic environment. Such result corroborates again the correctness of Tzeng's speculation that all allocation strategies with full subcube recognition ability would lead to almost identical performance for complete cube allocation[4]. One interesting topic as a future research is to develop an adaptive scheme using free list (or multiple gray code, tree collapsing, or maximum subcube sets) and the PC-graph strategies together to obtain optimal performance.

4 Performance Analysis

In order to study the performance of our PC-graph processor allocation scheme, we have developed a simulator written in C language and run on Sparc station 1+. In the simulation, processor allocation is carried out by a host computer exclusive of the simulated hypercube computer so that the hypercube computer is used solely for user tasks. Tasks (or requests) are either generated in the beginning of the simulation program or generated at each time step (time unit) during a simulation run. The generated tasks (requests) are put in a central task queue waiting to be allocated. We consider both cubic and noncubic requests. The required dimension by a cubic request is assumed to follow a uniform distribution between 0 and n, while the required number of processors by a noncubic request is assumed to follow a uniform distribution between 1 and 2^n, where n is the dimension of the hypercube. At each time unit, the host computer tries to schedule the task at the head of the queue. If it fails to find an available subcube of desired size, the host waits until a deallocation event occurs. It will not allocate the next task until the first one in the queue has been allocated. Once a task is assigned to a subcube, it will execute for a random amount of time of given distribution.

Our first experiment is to examine the performance of our PC-Graph processor allocation scheme under the situation that all tasks require complete subcubes. For this purpose, we use our simulator in which a new task arrives at each time unit and observe performance results after the simulation has run for 100, 200 and 300 time units, respectively. The results are averaged over 100 independent runs. Table 1 lists our simulation results and the results previously reported [1]. In this table, we compare the performance of the PC-graph strategy only with the free list strategy, since typical allocation strategies such as multiple-gray-code, free list and tree collapsing show similar performance as observed by Chuang and Tzeng [4]. As shown in the table, the PC-Graph strategy gives almost the same performance as the free list strategy although the PC-Graph scheme is not theoretically statically optimal. We have also used the simulator to do experiment based on exactly the same assumption used in [4] and collected data as shown in Table 2. In this case, 100 tasks are generated in the beginning of the simulation and no tasks are generated during the simulation run. Again, the results match with that of free list. It is interesting to note that all the 4 allocation strategies (free list, multiple gray code, tree collapsing and PC-graph) have virtually identical performance in terms of the completion time (the time needed to finish all 100 tasks) and processor utilization. More importantly, from our experiment and previous published work we speculate that the statically optimality does not play an important role in the dynamic environment. Rather, the fragmentation problems, both internal and external fragmentations, play a key role in performance in the dynamic environment as will be evidenced later in this section.

Now let us consider the effects of the internal fragmentation on the performance of the hypercube computers. We still use the simulator but the number of processors requested by tasks is assumed to be random number uniformly distributed between 1 and 2^n. A new task is generated at each time unit in the simulation and performance results are collected at different observation points (100, 150, ...,300). The results are averaged over 100 independent runs. We first compare the performance of cubic PC-graph allocation with that of non-cubic PC-graph allocation to see the effect of internal fragmentation. Figure 4 shows the total task waiting time (delay) until the last request was assigned. It can be seen from this figure that the performance difference between the two schemes is significant, indicating that the internal fragmentation affect the system performance to a large extent. We expect that the effect of the internal fragmentation increases as the system size increases. To show this fact, we have run our simulation for different system sizes ranging from 2-dimension to 10-dimension hypercubes. Figure 5 plots the total delay as a function of system size. The results in the figure are observed at 200th time unit after the simulation starts. As shown in the figure, the fragmentation is getting severer as system size increases from $n = 2$ to $n = 6$ indicating that non-cubic allocation schemes must be used in order to effectively utilize the system resources. The difference decreases after dimension 6

indicating that the system size is sufficiently large to accommodate most of tasks. Notice that the absolute value of the total waiting time is getting smaller as the system size increases.

Next we compare results of our non-cubic allocation scheme with the free list non-cubic allocation strategy. Figure 6 plots the processor utilization observed at different simulation points. The non-cubic allocation scheme based on the free list strategy shows better processor utilization than cubic free list strategy. Our PC-graph strategy performs significantly better than both cubic and non-cubic free list allocation strategies. This performance improvement can be

attributed mainly to the effective utilization of small available subcubes in a fragmented situation. Our algorithm tries to use small subcubes that are physically close to each other for non-cubic allocation in contrast to breaking up large cubes for non-cubic allocation which will worsen the already fragmented system. Figure 7 shows the total task delay versus different observation time in the simulation. As shown in the figure, the PC-graph scheme has the lowest delay compared to both cubic and non-cubic free list schemes. The performance improvement ranges from 35% to 50%. Similar results are obtained in terms of completion time as shown in Figure 8. The main advantage of the PC-graph allocation scheme is that it minimizes both the internal and external processor fragmentations in an efficient way.

5 Conclusion

In this paper, we have presented a new graphic approach for processor allocation on hypercube multiprocessors, called *prime cube graph* scheme (PC-graph). The distinctive feature of our new allocation scheme is that the problems of cubic and non-cubic processor allocation and deallocation are treated in a unified way. It has been shown that the PC-graph strategy has full subcube recognition ability. Moreover, the new resource representation using the PC-graph can facilitate the recognition of inter-relationships between available subcubes in the system. Thus, both internal and external processor fragmentations can be reduced by using the new allocation scheme. We have proposed techniques to treat the two fragmentation problems. By using some existing efficient, linear graph algorithms, cubic allocation and non-cubic allocation can be done by manipulating the PC-graph with low complexity. A simulation experiment has been carried out to study the performance of our new allocation scheme as well as some existing strategies. Simulation results have shown that our new allocation strategy improve system performance in terms of average task waiting time by 35% to 50% under certain situations. For complete subcube allocations, the PC-graph performs at least as well as free list allocation scheme although static optimality is not guaranteed. From our experiment, we observe that the static optimality of an allocation strategy does not play an important role in a dynamic environment. Rather, processor fragmentations are the key factor that affects the performance of the hypercube systems.

References

[1] J. Kim, C. R. Das, and W. Lin, "A top-down processor allocation scheme for hypercube computers," *IEEE Tran. on Prarallel and Distr. Sys.*, vol. 2, pp. 21–30, Jan. 1991.

[2] M. Chen and K. G. Shin, "Processor allocation in an n-cube multiprocessor using gray codes," *IEEE Trans. on Comput.*, vol. C-36, pp. 1396–1407, Dec. 1987.

[3] S. Dutt and J. P. Hayes, "On allocating subcubes in a hypercube multiprocessor," in *Proc. Third Conf. on Hypercube Comp. and Appl.*, pp. 801–810, Jan. 1988.

[4] P.-J. Chuang and N.-F. Tzeng, "Dynamic processor allocation in hypercube computers," in *Proc. 17th Annu. Int'l Symp. on Comp. Arch.*, May 1990.

[5] M. S. Chen and K. G. Shin, "Subcube allocation and task migration in hypercube multiprocessors," *IEEE Tran. on Comput.*, vol. 39, pp. 1146–1155, Sep. 1990.

[6] N.-F. Tzeng, H. L. Chen, and P. J. Chuang, "Embeddings in incomplete hypercube," in *Proc. 1990 Int'l Conf. on Parallel Processing.*, August 1990.

[7] H. P. Katseff, "Incomplete hypercubes," *IEEE Tran. on Computer*, vol. 37, pp. 604–608, May 1988.

[8] C. Hu, M. Bayoumi, B. Kearfott, and Q. Yang, "A parallelized algorithm for the preconditioned interval newton method," in *Proc. of 5th SIAM Conf. on Parallel Processing for Sci. Comp.*, March 1991.

[9] C.-H. Huang and J.-Y. Juang, "A partial compaction scheme for processor allocation in hypercube multiprocessors," in *Proc. 1990 International Conf. Parallel Proc.*, pp. I221–217, August 1989.

[10] N.-F. Tzeng, "Structural properties of incomplete hypercube computers," in *Proc. 1990 Int'l Conf. on Dist. Comp. Sys.*, May 1990.

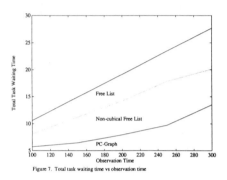

Figure 6. Processor utilization vs. observation time

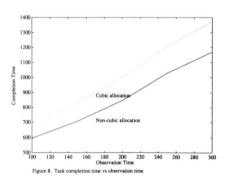

Figure 7. Total task waiting time vs observation time

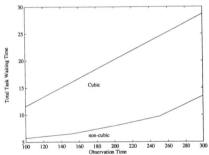

Figure 4. Total task waiting time until the last request assigned vs observation time

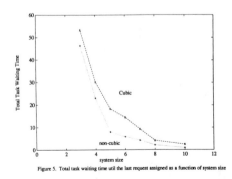

Figure 5. Total task waiting time util the last request assigned as a function of system size

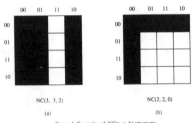

Figure 8. Task completion time vs observation time

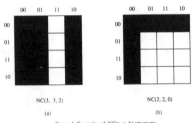

Figure 1. Examples of NC(p, q, k) structures.

COMMUNICATION-EFFICIENT VECTOR MANIPULATIONS ON BINARY N-CUBES

Woei Lin
Department of Electrical Engineering
University of Hawaii at Manoa
Honolulu, HI 96822

ABSTRACT

This paper describes efficient vector manipulations on the binary n-cube structure. A vector is defined to be a set of ordered elements stored in consecutive processors, one element per processor. Vectors may have arbitrary starting points and lengths. It is shown that the dimension order is a key to efficient vector manipulations on the binary n-cube. The use of proper dimension orders can completely eliminate link collisions for the vector manipulating functions. We focus on six most common vector manipulating functions used in image processing and parallel solutions of linear systems: they are merge, split, rotation, reverse, compression and expansion. These six functions form a new set of data communication problems of the binary n-cube. They are different from such data permutations as bit reversal in that their communication patterns can not be specifically described by interchanging bits of binary addresses. Some manipulating functions, e.g. merge, are not even permutations per se, because they may involve overlapped vector operands. A formal network model is developed for determining whether a dimension order may cause link collisions. With the aid of this network model, we show that among various dimension orders, the high-to-low and low-to-high orders and their variants suffice to achieve collision-free communications for the six manipulating functions. As a consequence, these functions can be completed in single network cycles. That the two orders require considerably straightforward encoding and extraction of routing information lends themselves to the feasibility of physical implementation.

Index terms: vector manipulation, parallel processing, network routing, binary n-cube, data communication.

I. INTRODUCTION

We are concerned with manipulating vectors on a class of array processors connected by the binary n-cube structure. In the context of binary n-cubes, we define a vector to be a set of ordered elements which are stored in sequential locations in the array; it may start at any location and contains an arbitrary number of elements. Vector processing is an essential means of exploiting massive parallelism in scientific and engineering computing. We examine most common vector manipulating functions: merge, split, rotation, reverse, compression and expansion. They are frequently used for manipulating bit vectors and numerical vectors in linear system problems and image processing. Our objective is to find efficient ways of controlling, minimizing network communications required by these manipulating functions on binary n-cubes.

The efficiency of a vector processing computer is primarily determined by its capability of handling vectors, arrays and matrices. On array processors vector manipulations are done mainly through network communication. Data elements need be moved across network links in order to engage in a computational rendezvous. Data movement plays a crucial role in vector processing on array processors. It often accounts for a significant proportion of the execution time.

One possible way of manipulating vectors on binary n-cubes is to remap some permutations of a class of multistage interconnection networks [1] into the binary n-cube structure and to use them as primitive operations for synthesizing vector manipulating functions of interest. These networks are often referred to as *indirect* binary n-cube networks. There exists a transformation by which one can unroll a *direct* binary n-cube network in space into an indirect one [3-5]. It is widely known that these multistage interconnection networks are capable of performing such commonly-used data permutations as barrel shift and flip. The essential similarity in structure allows both direct and indirect networks to use a conceptually simple cube routing scheme to establish paths toward destinations. This enables the direct binary n-cube network to accomplish permutations in just a single pass as its indirect counterpart. So, one can make use of these permutations as primitives to manipulate vector operands on the binary n-cube. For instance, consider a unit right rotation of a vector $(7, 6, 5, 4, 3)$ on a 3-cube. The rotation can be realized with two shifts; we first shift 3 left by four positions and then shift the remainder right by one position, as is done with two passes on an 8x8 omega network [5]. Note that although the omega network can perform barrel shift in one pass, it can not accomplish such a rotation in one pass. An attempt to do so will result in link contentions between shifts of 3 to 7 and 7 to 6. In fact, this is true of other vector manipulating functions. It has been pointed out that the multistage interconnection networks can not perform all vector manipulating functions with arbitrary vector configurations in one pass [6]. If taking this remapping approach, we need to use multiple passes, each with a particular permutation realizable by the multistage interconnection networks, to achieve a desired vector manipulating function.

The binary n-cube network and the multistage interconnection network have similarities in structure. However, this does not mean that they are completely identical in functionality. There are numerous cases where permutations can be accomplished on the binary n-cube network, but not on the multistage interconnection network. For instance, later on we will show that the right unit-rotation mentioned previously can be easily realized on the binary n-cube network in just one pass. Their difference in network functionality mainly lies in that on the $2^n \times 2^n$ multistage interconnection network, paths always have to go through the n stages of links toward output terminals, even though their routing tags may have been turned into the destination addresses in some middle stage. But on the binary n-cube, once the routing tag of a path is changed into the destination address, routing for the path is completed and no further link traversals are required. This greatly reduces the possibility of link collisions on the binary n-cube. (A formal network model for describing routing and link collisions on the binary n-cube will be presented in the next section.) We intend to employ this valuable network property to design a set of communication-efficient vector manipulating functions.

Data communication is crucial to the binary n-cube performance. A variety of techniques have been devised for improving data communication efficiency of the n-cube. Considerable efforts are devoted to multicast communication [7-9]. It is a fundamental communication operation needed by many parallel algorithms using iterative relaxation methods. Rearrangeability of the binary n-cube is studied in [10]. A heuristic backtracking depth first search algorithm for routing arbitrary permutations is presented. In addition, a proof shows that the circuit-switched 3-cube is always rearrangeable. However, whether 4-cubes and higher are rearrangeable remains an open question. Several general techniques for improving communication efficiency are presented in [11]. These techniques include communication overlap, quadruple buffering, and global exchange. A deadlock-free routing algorithm for n-cube structures is described in [12]. The routing algorithm uses virtual channels to break potential communication deadlocks.

II. PRELIMINARIES

We are interested in a class of synchronous binary n-cubes. A notable example is the Connection Machine by the Thinking Machines Corporation [13]. In a synchronous n-cube, all processor nodes derive the timing information for message transfers from a common clock source. The common clock source broadcasts equally spaced cycles, called *network cycles*, for processor nodes to transfer messages. As shown in Figure 1, each network cycle is further divided into n *dimension cycles*, one for each dimension of a binary n-cube. We assume that in a network cycle each dimension appears only once but ordering of the n dimensions is a changeable state. In theory there are n! ways of ordering the dimensions of an n-cube. But most of them are not feasible for implementation because of their difficulties in specifying the orderings. We focus on two types of feasible dimension orders: the high-to-low order (HL) and the low-to-high order (LH). Routing information based on these two orders can be easily encoded and extracted by shift register operations. We also consider variations of the two dimension orders. They provide alternatives for achieving efficient vector manipulations as well .

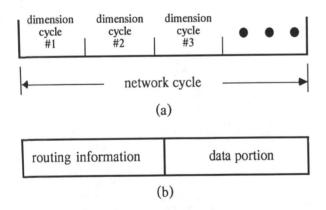

(a)

routing information	data portion

(b)

Figure 1. Network cycle and data message format.

During a network cycle messages are moved across each of the n dimensions in the specified order. A processor node

initiates a message by sending a valid message packet through its I/O ports. A message packet has a routing tag. The routing tag contains the routing information for delivering information to its destination. At the initiating node it is generated by exclusive-oring the binary addresses of the source node and the destination node. The resulting pattern indicates the dimension links over which the packet has to traverse. A message may be relayed for several times, before it reaches the destination node. This is determined by the number of 1 bits in the routing tag. For instance, a message with a routing tag 010011 must move across three links along dimensions 0, 1, and 4 in case that the LH order is used. Each time it is moved from one node to another, the routing tag is updated to maintain its relative address with respect to its destination. Because each bit of the resulting tag corresponds to one dimension of the n-cube, each time a message moves along a dimension the corresponding bit must be changed. When it becomes 0, the message arrives at the destination.

We denote a path by an ordered pair (X,Y); X represents the source node and Y represents the destination node. The following definitions develop a notion for describing some relationships among paths. Let (A,B) and (C,D) be two paths on an n-cube. Each node is represented by an n-bit sequence, for example, $A=a_{n-1}a_{n-2}...a_0$.

Definition $\lambda(A,C)$ indicates the first low-order bit position where A and C differ. $\eta(B,D)$ indicates the first high-order bit position where B and D differ.

Definition (A,B) and (C,D) are said to have a *collision* if they compete a communication link on some dimension cycle. On the other hand, (A,B) and (C,D) do not have a collision if they traverse different links on every dimension cycle.

When there are several messages present in an n-cube simultaneously, link collisions may cause extra communication delay. On a processor node, messages with the kth bit set to 1 have to use the cycle corresponding to the kth dimension to move into the next node; they all need to travel over the same link. But only one of them is chosen and the remaining messages must wait for the next opportunity. To use this jammed dimension, a blocked message has to wait for at least a complete network cycle. As a result, the entire routing takes more than one network cycle to complete.

A wise choice of dimension orders may eliminate potential link collisions among paths. It is possible that two paths have a collision when the HL order is used, but no collision occurs when the LH is used, and vice versa. For example, in a 4-cube, (4,9) and (6,1) have no collisions throughout the routing using the HL order, but they have a collision on the third dimension cycle if the LH order is used. The conditions in which two paths have collisions in the two orders are formally stated as follows. Consider A, B, C, and D as n-bit numbers and let $A = a_{n-1}a_{n-2}...a_0$, etc.

Theorem 1 (*collision conditions*) (A,B) and (C,D) have a collision in the HL order if and only if there exists a bit position x, $\eta(B,D) < x < \lambda(A,C)$ and $a_x \neq b_x$. (A,B) and (C,D) have a collision in the LH order if and only if there exists a bit position x, $\eta(A,C) < x < \lambda(B,D)$ and $c_x \neq d_x$.

Proof First let us consider the HL order. When the HL order is used, there are two situations where (A,B) and (C,D) have no collision. They are: (1) during each dimension cycle the two paths traverse two different nodes, and (2) if they have to pass through a common intermediary node during the course of traversal, they do not leave the node simultaneously on the subsequent dimension cycles. The first situation can be described by that $\eta(B,D) > \lambda(A,C)-1$. The second one corresponds to that if $\eta(B,D) \leq \lambda(A,C)-1$, then for every position x, $\eta(B,D) < x < \lambda(A,C)$, we always have $a_x = b_x$. This implies that if coming to the same intermediary node, the two paths stay until the cycle for dimension $\eta(B,D)$ and then only one of them leaves.

In summary, if none of these two situations happens, then (A,B) and (C,D) must come across a collision. On the other hand, if they incur a collision in the HL order, then there must exist a bit position x, $\eta(B,D) < x < \lambda(A,C)$ and $a_x \neq b_x$. The collision condition for the LH order can be derived in a similar way. □

Many vector manipulating functions to be examined are inverse functions of each other; for instance, split is the inverse of merge. There exists an interesting duality property of dimension orders for a function and its inverse. If one function using the HL order does not cause any collision, then its inverse using the LH order is also collision-free. On the other hand, if one function using the HL order experiences collisions during routing, then the inverse using the LH order must cause collisions. Central to this is the duality theorem described below. We would like to use this theorem to boil down our later discussion on the vector manipulating functions.

Theorem 2 (duality theorem) If (A,B) and (C,D) can be routed without collisions by using the HL order, then (B,A) and (D,C) can be routed without collisions by using the LH order.

Proof From the proof of Theorem 1, we know that if (A,B) and (C,D) incur no collision using the HL order, then either $\eta(B,D) > \lambda(A,C)-1$ or for every position x, $\eta(B,D) < x < \lambda(A,C)$, $a_x = b_x$. Now we reverse directions of the two paths; that is, A and B become destinations and C and D become sources. The two situations also hold for the two reversed paths, (B,A) and (D,C), if the LH order is used. Therefore, using the LH order the two reversed paths do not cause any collision. □

In the subsequent discussion, we formally denote a vector as a 2-tuple $<X, k>$, where X specifies the starting address of the vector and k indicates the vector length. We are primarily interested in unit-cycle vector manipulations. A vector manipulating function can be completed in one network cycle if paths due to the manipulation are free from collisions during all the n dimension cycles arranged in the specified order. So, we use "unit-cycle" and "collision-free" as two interchangeable terms in our discussion.

III. SPLIT AND MERGE

Split and merge are two manipulating functions useful in combining and dividing vectors. Odd-even merge sorting heavily relies on these two functions for manipulating operands. Split requires specifications of one source vector and two destination vectors. Merge, on the other hand, needs specifications of two source vectors and one destination vector. We make no assumptions about starting addresses and lengths of the three vectors, except that sometimes they have to be non-overlapped vectors. Split and merge are two functions with the duality property described in Theorem 2. In the following discussion, we only provide proofs for vector split. Main results of vector merge are summarized in corollaries without proofs.

THEOREM 3 Using the HL dimension order, in a network cycle we can split a vector $<X, 2k>$ into two vectors, $<Y, k>$ and $<Z, k>$, which contain the odd-number and even-number elements of $<X, 2k>$, respectively.

Proof We will show that any two paths due to the split, (A,B) and (C,D), incur no collisions if the HL order is used for arranging the dimensions in a network cycle. Two cases are considered:

(1) Both B and D are in the same resulting vectors. Assume that B and D are in $<Y, k>$. A and C must be even numbers or odd numbers at the same time. Therefore, we have $\lambda(A,C) > 0$. Given X and Y, we can compute B and D in terms of A and C:

$$B = (A - X)/2 + Y$$
$$D = (C - X)/2 + Y.$$

Note that "/" denotes integer divide without rounding. Dividing by two is equivalent to right shift by one. Thus, we obtain

$$\lambda(A/2,C/2) = \lambda(A,C)-1$$
$$= \lambda((A-X)/2+Y,(C-X)/2+Y)$$
$$= \lambda(B,D)$$
$$\leq \eta(B,D).$$

The above equation is based on the fact that adding (or subtracting) constants does not change λ. As can be seen, it is impossible for us to find a bit position x, $\eta(B,D) < x < \lambda(A,C)$. Consequently, the collision condition stated in Theorem 1 never holds.

(2) B and D are in different resulting vectors. We always have $\lambda(A,C)=0$. It is clear that such a bit position x that $\eta(B,D) < x < \lambda(A,C)$ does not exist. Therefore, if the HL order is used, (A,B) and (C,D) do not cause collisions during routing. □

COROLLARY 1 Using the LH dimension order, in a network cycle we can merge two vectors, $<X, k>$ and $<Y, k>$, into a vector $<Z, 2k>$ where elements of the two source vectors are stored in alternate locations.

From the preceding proof, we can observe that vectors to be merged or split can be arranged in either overlapped or non-overlapped manner. This provides us with a great flexibility in merging and splitting vectors. Figure 2 shows a merge of $<1, 3>$ and $<3, 3>$ into $<2, 6>$ on a 3-cube and paths created for the merge. Next, we look into unit-cycle group split and merge. By group we mean that each vector is made up of a number of units each of which contains power-of-two elements and that these units form the basic objects of manipulation. However, to ensure collision-free routing, we have to arrange vectors in a non-overlapped manner. In some cases, they need be separate from each other at certain safeguard distances.

THEOREM 4 Let $<X, k \cdot 2^{i+1}>$ be a vector of 2k units, each with 2^i elements, $i > 0$ and let $<Y, k \cdot 2^i>$ and $<Z, k \cdot 2^i>$ be the two vectors which contain the alternate units split from $<X, k \cdot 2^{i+1}>$. If $Z = Y+k \cdot 2^i$ or $Z \geq Y+(k+1) \cdot 2^i-(Y \bmod 2^i)$, then we can complete the split in a network cycle, using the HL dimension order.

Proof Suppose that (A,B) and (C,D) are two paths created for the split. Consider the following two cases:

(1) Both B and D are located in the same destination vector. If $\lambda(A,C) < i$, we must have $\eta(B,D) \geq \lambda(B,D) =$

$\lambda(A,C)$. If $\lambda(A,C) \geq i$, then there exists a number j such that $|A-C| = j \cdot 2^i$ and $|B-D| = j \cdot 2^{i-1}$. In other words, we have $\eta(B,D) \geq \lambda(B,D) = \lambda(A,C)-1$.

(2) B and D are located in different destination vectors. It must be true that $\lambda(A,C) \leq i$. (Otherwise, both A and C belong to either odd-number units or even-number units simultaneously. And B and D would be located in the same destination vector.) Now consider the following two subcases: (i) $Z = Y+k \cdot 2^i$. That is, $<Z, k \cdot 2^i>$ immediately follows $<Y, k \cdot 2^i>$. If $\lambda(A,C) < i$, then we have $\eta(B,D) \geq \lambda(B,D) = \lambda(A,C)$. If $\lambda(A,C) = i$, then $|B-D| \geq 2^i$. Therefore, $\eta(B,D) \geq i$; (ii) $Z \geq Y+(k+1) \cdot 2^i-(Y \bmod 2^i)$. That is, Z and the last element of $<Y, k \cdot 2^i>$ are separate at a minimum distance of $2^i-(Y \bmod 2^i)$. This minimum distance guarantees that $\eta(B,D) \geq i$.

In either case, there does not exist such a bit position x that $\eta(B,D) < x < \lambda(A,C)$. By Theorem 1, we conclude that (A,B) and (C,D) incur no collisions if the HL order is used. □

COROLLARY 2 Let $<X, k \cdot 2^i>$ and $<Y, k \cdot 2^i>$ be two vectors each with k units of 2^i elements. If $Y = X+k \cdot 2^i$ or $Y \geq X+(k+1) \cdot 2^i-(X \bmod 2^i)$, then using the LH dimension order we can merge the two vectors, in a network cycle, into a 2k-unit vector where units of the two source vectors are arranged in alternate locations.

IV. SHIFT, ROTATION AND REVERSE

In this section, we examine three monadic manipulation functions: shift, rotation and reverse. Each of them requires a

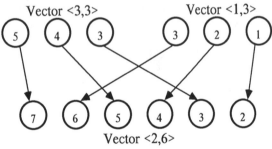

Note: numerals within circles indicate locations where elements are stored.

(a)

	Cycle #1	Cycle #2	Cycle#3		
(1)	001 ⟶	000 ⟶	010	010	(2)
(2)	010	010 ⟶	000 ⟶	100	(4)
(3)	011 ⟶	010	010 ⟶	110	(6)
(3)	011	011	011	011	(3)
(4)	100 ⟶	101	101	101	(5)
(5)	101	101 ⟶	111	111	(7)

(b)

Figure 2. A merge of $<1,3>$ and $<3,3>$ into $<2,6>$ using the LH order.

vector as the operand. Shift uniformly moves elements of a vector by a certain displacement. Rotation cyclically shifts elements of a vector by a designated number of positions, either toward the right end or toward the left end. Reverse refers to interchanging vector elements with respect to the center of a vector.

Shift is a fundamental vector manipulating function. Typical applications of shift include inserting elements into a vector and deleting elements from a vector. For instance, deleting a number of successive elements from a vector requires moving all the elements in the rest of the vector forward by the number of deleted elements. Shift exhibits a highly interesting property that it is always collision-free no matter what dimension order is used. For a proof, see [14].

Rotation is frequently needed in parallel solutions to linear system problems. For example, Jocobi's method and other plane-rotation algorithms use it to move rows and columns of matrices to proceed computations in parallel. Let us first take a look at some cases in which a single dimension order is sufficient for collision-free rotation. Then we will examine a general solution that may require the combined use of two different orders.

THEOREM 5 Using the LH order, we can rotate $<X, i>$ right by one position in a network cycle.

Proof The right unit-rotate requires two types of paths. The first contains a single path (X, X+i-1) and the other consists of paths (Y, Y-1), where $X+i-1 \geq Y \geq X+1$. It is obvious that any two paths of the second type do not incur collision because they can be considered two paths of unit right shift of $<X+1, i-1>$ into $<X, i-1>$. Now we need to show that (X, X+i-1) and any path (Y, Y-1), where $X+i-1 \geq Y \geq X+1$, are collision-free by using the LH order. Let $X = x_{n-1}...x_1x_0$ and $Y = y_{n-1}...y_1y_0$. Suppose $\eta(X,Y) = k$. Then we must have $x_k = 0$ and $y_k = 1$. (Otherwise, we would have $X > Y$.) Now let $Z = Y-1$ and $Z = z_{n-1}...z_1z_0$. Due to the decrement of Y by 1, z_k may become 0. However, other bits z_i, $n-1 \geq i \geq k+1$, should remain unchanged; that is, we have $y_i = z_i$, for all i, $n-1 \geq i \geq k+1$. This leads to that there never exists a bit position i, $\eta(X,Y) < i < \lambda(X+i-1,Z)$, such that $y_i \neq z_i$. By Theorem 1, we conclude that these two paths are collision-free if the LH order is used for routing. □

COROLLARY 3 Using the HL order, we can rotate $<X, i>$ left by one position in a network cycle.

Figure 3 shows an example of left unit-rotation of $<3, 5>$ by the HL order and paths required by the rotation. When vector length is a power-of-two number, we can perform collision-free rotations with more flexibilities in choosing dimension orders and rotation positions.

THEOREM 6 Using either the LH or the HL order, we can rotate $<X, 2^i>$ right by an arbitrary number of positions in a network cycle.

Proof Let k be the number of positions to be rotated. Paths created for the rotation fall into two groups, one with sources in $<X, k>$ and the other with sources in $<X+k, 2^i-k>$. It has been shown previously that two paths of the same group do not cause collision by either the LH or the LH orders. Let us consider the other situation where they belong to different groups. Suppose that (A, B) and (C, D) are two paths, where $X \leq A \leq X+k-1$ and $X+k \leq C \leq X+2^i-1$. Thus, we have $B = A-k$ and $D = C+2^i-k$. Since $|A-C| < 2^i$, we have $\lambda(A,C) < i$. Because of this,

$$\lambda(A,C) = \lambda(A-k,C-k)$$

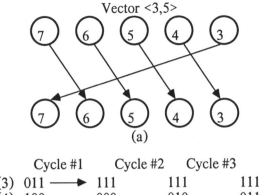

Vector <3,5>

(a)

Cycle #1   Cycle #2   Cycle #3

(3) 011 ⟶ 111   111   111 (7)
(4) 100 ⟶ 000 ⟶ 010 ⟶ 011 (3)
(5) 101   101   101 ⟶ 100 (4)
(6) 110   110 ⟶ 100 ⟶ 101 (5)
(7) 111   111   111 ⟶ 110 (6)

(b)

Figure 3. A unit right rotation on <3,5>
using the HL order.

$$= \lambda(A-k, C+2^i-k)$$
$$= \lambda(B,D).$$

Consequently, both $\eta(B,D) \geq \lambda(A,C)$ and $\eta(A,C) \geq \lambda(B,D)$ become true. According to Theorem 1, we know that (A,B) and (C,D) do not cause collisions if the HL order or the LH order is used. □

An interesting question arises: Is it possible to use either the HL order or the LH order to rotate a vector of arbitrary length by any number of positions? Unfortunately, the answer is no. The HL order and the LH order reach their limit in this respect. We have found that there exist many situations in which both the HL order and the LH fail to accomplish collision-free rotations. In the subsequent discussion, we present an alternative solution to this problem. We use instead two different dimension orders to guide paths for arbitrary rotations. One of them can just be the HL order and the other one is a variation of the HL order with a phase lag. The use of these two dimension orders, however, necessitates a hardware enhancement in switching capability of processor nodes: Each node is capable of processing two simultaneous messages along two different dimensions in a dimension cycle.

DEFINITION The HLC order is a high-to-low arrangement of the n-cube dimensions with a cyclic shift to the right.

As an example, the HLC order for 6-cube is of the form (0, 5, 4, 3, 2, 1). That is, a network cycle in the HLC order starts with dimension 0, followed by dimensions 5, 4, 3, etc.

THEOREM 7 Let <X, i> be a vector to be rotated right k positions, $0 \leq k < i$. We can complete the rotation in a network cycle by applying the HL order to <X, k> and the HLC order to <X+k, i-k>.

Proof Dividing <X, i> into two subvectors, <X, k> and <X+k, i-k>, results in two groups of shifts. For <X, k> it is a left shift by i-k positions, and for <X+k, i-k> it is a right shift by k positions. It is known that any two paths of the same

group are collision-free. Since there exists a phase lag between the HL order and the HLC order, their dimension numbers are different in each of the n dimension cycles. Consequently, any two paths of <X, k> and <X+k, i-k> in the two orders never compete the same link throughout the n dimension cycles. Therefore, they can be completed in a network cycle. □

As can be seen, the key technique we use above is to treat a rotation as two shifts with the HL and the HLC orders. A combination of these two orders gives rise to that every dimension cycle has two distinct dimension numbers. Based on this, we can derive many other combinations of dimension orders than the HL and HLC orders for collision-free arbitrary rotations. Next, we turn our attention to vector reverse.

THEOREM 8 Using either the HL order or the LH order, we can reverse <X, i> in a network cycle.

Proof Let A and C be two elements of <X, i> and let B and D be the reversed elements of A and C. Then, we have B = Y+X-A-1-i and D = Y+X-C-1-i. Let A' and C' denote the one's-complemented numbers of A and C. Then, we have

$$\begin{aligned}\lambda(A,C) &= \lambda(A',C') \\ &= \lambda(A'+1,C'+1) \\ &= \lambda(-A,-C) \\ &= \lambda(Y+X-A-1-i, Y+X-C-1-i) \\ &= \lambda(B,D).\end{aligned}$$

This suggests that $\eta(B,D) \geq \lambda(A,C)$ and $\eta(A,C) \geq \lambda(B,D)$. Therefore, the reverse of <X, i> can be done without collision by either the HL or the LH orders. □

V. COMPRESSION AND EXPANSION

Compression selects elements from a vector and packs them into another vector. Expansion is the inverse function of compression by taking the reverse steps. They are two vector manipulating functions useful in image warping, magnification, reduction and sampling. In this section, we consider both regular and irregular vector manipulations of compression and expansion. In cases of regular compression, elements at uniform intervals are gathered and moved into contiguous locations of another vector. On the other hand, irregular compression deals with elements at nonuniform intervals. In the following discussion, we first consider arbitrary compression and expansion. Actually, they include both regular and irregular manipulations and results can be equally applied to both. Then, we give additional treatments on regular compression and expansion. There are alternatives for collision-free compression and expansion, when intervals of elements become uniform.

THEOREM 9 Using the LH order, we can arbitrarily compress <X, i> into <Y, j>, $i \geq j$, in a network cycle.

Proof Suppose that A and C are two elements selected from <X, i> for compression and that B and D are their destinations in <Y, j>. Let d = |A-C|. Then we have $\eta(A,C) \geq \lfloor \log_2 d \rfloor$. (If $\eta(A,C) < \lfloor \log_2 d \rfloor$, then $|A-C| < 2^{**\lfloor \log_2 d \rfloor} < d$. This contradicts d = |A-C|.) Now let k = |B-D|. We must have $\lambda(B,D) \leq \lfloor \log_2 k \rfloor$. (If $\lambda(B,D) > \lfloor \log_2 k \rfloor$, then $|B-D| \geq 2^{**(\lfloor \log_2 k \rfloor+1)} > k$. This contradicts |B-D| = k.) Since this is vector compression, $|A-C| \geq |B-D|$. Putting these all together, we have $\eta(A,C) \geq \lfloor \log_2 d \rfloor \geq \lfloor \log_2 k \rfloor \geq \lambda(B,D)$. According to Theorem 1, (A,B) and (C,D) do not cause link collisions if the LH order is used for routing. □

COROLLARY 4 Using the HL order, we can arbitrarily expand $<X, i>$ into $<Y, j>$, $i \leq j$, in a network cycle.

For instance, Figure 4 shows an irregular compression from $<1,7>$ to $<2, 4>$. Next, we consider compression and expansion with regular intervals. We find that in addition to the LH order, the HL order is good for compression when elements are sampled at regular odd intervals.

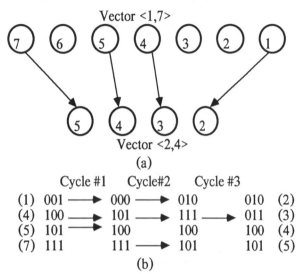

Vector $<1,7>$

Vector $<2,4>$

(a)

Cycle #1		Cycle#2		Cycle #3		
(1)	001 →	000 →	010	010	(2)
(4)	100 →	101 →	111 →	011	(3)
(5)	101 →	100	100	100	(4)
(7)	111	111 →	101	101	(5)

(b)

Figure 4. A compression of $<1,7>$ into $<2,4>$ using the LH order.

THEOREM 10 Suppose that elements of $<X, i>$ at uniform odd intervals are selected for vector compression into $<Y, j>$, $i \geq j$. Then, we can use the HL order to complete the compression in a network cycle.

Proof Without loss of generality, assume that X is the first element selected for the compression and that every other d elements, d is an odd number, is selected subsequently. Let A and C be such two elements and let B and D be their destinations in $<Y, j>$. Since we select every other d elements, the distance between A and C can be expressed in terms of a multiple of d; that is, we have $|A-C| = m \cdot d$, $m > 0$. In addition, we should have $|B-D| = m$. Now let us factor m into two parts, $m = k \cdot 2^p$, where k is an odd number. (We can always do this.) p determines the value of $\lambda(B,D)$. That is, $p = \lambda(B,D)$. Likewise, if we factor $|A-C|$ into an odd number and a power-of-two number, the exponent of the latter must equal p. Thus, we have $\lambda(A,C) = p$. As a result, $\eta(B,D) \geq \lambda(B,D) = \lambda(A,C)$. This means that if the HL order is used, (A,B) and (C,D) never have link collisions. □

COROLLARY 5 Suppose that $<X, i>$ is a vector to be expanded into $<Y, j>$, $i \leq j$, at uniform odd intervals. Then using the LH order we can complete the vector expansion in a network cycle.

We are also interested in compression and expansion with power-of-two intervals and by different dimension orders than those described in Theorem 9 and its corollary. From Theorem 3, we observe that compressing elements at an interval of two is a subfunction of vector split, which moves alternate elements to form two separate vectors. Therefore, we may also use the HL order to compress a vector at intervals of

two. We intend to generalize this result for other power-of-two cases. Two dimension orders are introduced for this purpose.

DEFINITION The HLS order refers to a class of high-to-low orderings of n-cube dimensions with stride s, $s \geq 2$. An HLS order with stride s starts with dimension n-1 and a dimension i in the order is succeeded by dimemension (i-s) mod n; but if n is divisible by s, the successor of a dimension i, $s-1 \leq i \leq 0$, is dimension (i-s-1) mod n.

For example, an HLS order with stride 3 of 7-cube is (6, 3, 0, 4, 1, 5, 2). The HLS order in fact defines a permutation on n-cube dimensions in the HL order. Let σ_s denote an HLS order with stride s; for any i, $n-1 \geq i \geq 0$, $\sigma_s(i)$ indicates the dimension position by the HLS with s. In the previous example, we have $\sigma_3(5) = 1$. Similarly, we define the LHS order as follows.

DEFINITION The LHS order refers to a class of low-to-high orderings of n-cube dimensions with stride s, $s \geq 2$. An LHS order with stride s starts with dimension 0 and a dimension i in the order is succeeded by dimension (i+s) mod n; but if n is divisible by s, the successor of a dimension i, $n-1 \leq i \leq n-s$, is dimension (i-s+1) mod n.

THEOREM 11 Suppose that elements of $<X, i>$ at uniform 2^k intervals, $k \geq 0$, are selected for vector compression into $<Y, j>$, $i \geq j$. Then using an HLS order with stride k we can complete the compression in a network cycle.

Proof Suppose that A and C are two elements selected for the compression and that B and D are their destinations in $<Y, j>$. And suppose that A', B', C' and D' are the bit-permuted numbers of A, B, C, and D by the HLS order with stride k. Due to the uniform compression, we have $\lambda(A,C) \geq k$ and $\lambda(B,D) \leq \lambda(A,C)-k$. This leads to $\lambda(A',C') \leq \sigma_k(\lambda(A,C))$. In addition, we have $\lambda(B',D') \leq \sigma_k(\lambda(A,C))-1$. Therefore, $\eta(B',D') \geq \sigma_k(\lambda(A,C))-1 \geq \lambda(A',C')-1$. By Theorem 1, this indicates that if the HLS order is used, (A,B) and (C,D) do not cause link collision. □

COROLLARY 6 Suppose that $<X, i>$ is a vector to be expanded into $<Y, j>$, $i \leq j$, at uniform 2^k intervals, $k > 0$. Then using an LHS order with stride k we can complete the vector expansion in a network cycle.

VI. RELATIONS WITH SOME PERMUTATIONS

There are a class of familiar data permutations, such as flip and perfect shuffle, whose functions can be described in terms of bit permutations on binary addresses. This section provides a different perspective of these data permutations through the aforementioned vector manipulating functions. When certain restrictions are placed on vectors, a vector manipulating function may turn into a data permutation of this class. We intend to identify such situations and the correspondences between them. Our findings coincide with those optimal routing algorithms designed previously by Nassimi and Sahni [15].

In the following discussion, we deal with vectors which form cubes or subcubes. That is, a k-cube vector starts with 0 and has 2^k consecutive elements. In case of subcubes, we use local addresses for representing vector elements. Let us start with perfect shuffle. Let $X = x_{k-1}x_{n-2}...x_0$ be an element of a k-cube vector. This permutation corresponds to a unit circular left shift of the binary representation of x: shuffle(X) = $x_{n-2}...x_0x_{n-1}$. The perfect shuffle can be performed by dividing the vector into two subvectors and interleaving them, as in perfect card shuffle. We may consider this permutation as a merge of two adjacent (k/2)-subcubes. According to the corollary of Theorem 3, we know that it can be done by the LH order. In a similar way, we can say that the perfect unshuffle is a split of a k-cube vector into two adjacent (k/2)-subcube vectors.

The flip permutation is defined over the binary representation of X by complementing every bit of X: flip(X) = $x'_{k-1}x'_{n-2}...x'_0$. The apostrophe denotes the complement of a given bit. The flip permutation can be considered as the reverse of a k-cube vector. According to Theorem 8, we know that it can be accomplished by either the HL order or the LH order.

Now, let us take a look at the transposition permutation. This permutation is concerned with transposing a $2^p \times 2^q$ matrix whose elements are arranged in row-major order. Let k = p+q. It can be formally defined as: transposition(X) = $x_{q-1}...x_0x_{k-1}...x_{k-p}$. (Note that k-p = q.) The transposition permutation can be treated as multiple compressions of a k-cube vector into 2^q adjacent 2^p-subvectors. All the compressions have the same sampling intervals of 2^q, but each starts with a different elements of $0, 1, 2,...,2^{q-1}$. For instance, the second compression works on elements, $1, 2^q+1, 2^{q+1}+1$, etc. And they are packed into a 2^p-cube vector starting at 2^p. We can use an HLS order with stride q, described in Section 5, to perform such compressions. It can be observed that an HLS order with stride q consists of q subsequences each of which starts with a number i, $k-1 \le i \le k-q$, followed by nonnegative decrements of q in decreasing order. From the proof of Theorem 11, we find that for a single vector compression, cascading these q subsequences with the "modulo q" operation in fact is not a must. But when handling multiple compressions, we must follow the HLS order to move elements across links. The "modulo q" operation prevents paths due to the permutation from meeting the collision condition specified in Theorem 1. Each time we flip routing bits along a dimension i, $q-1 \le i \le 0$, it is replaced by the preceding dimension i+q in the HLS order. By the "modulo q" operation, dimension i reappears immediately on the next cycle. Because of this, all the paths traverse disjoint links on every dimension cycle and never compete for using the same cycles.

VII. CONCLUSION

We have been concerned with performing efficient vector manipulating functions on the binary n-cube array processor. The key idea is to use different dimension orders of the binary n-cube to achieve collision-free communications for vector manipulations. To demonstrate this idea, we have examined six most common vector manipulation functions: merge, split, rotation, reverse, compression and expansion. We assume that a vector may start at any location and contains an arbitrary number of elements. We focus on two major dimension orders - the HL order and the LH order - and their variations. These dimension orders are featured by their simplicity in encoding and extraction of routing information. A formal network model is developed for characterizing link collisions. Using this model, we have shown that the two orders as well as variants are sufficient for the six manipulating functions to accomplish unit-cycle routing. Alternative solutions have been sought for manipulating vectors of different forms in various situations. Our findings have a significant impact on reducing communication time for vector processing on the binary n-cube. The proposed techniques are cost-effective and easy to implement in hardware.

REFERENCES

[1] C. L. Wu and T. Y. Feng, "On a class of multistage interconnection networks," *IEEE Trans. on Computers*, Vol. C-29, No. 8, Aug. 1980, pp. 694-702.

[2] L. N. Bhuyan and D. P. Agrawal, "Generalized hypercube and hyperbus structures for a computer network," *IEEE Trans. on Computers*, Vol. C-33, No. 4, April 1984, pp. 323-332.

[3] C. L. Seitz, "Concurrent VLSI architectures," *IEEE Trans. on Computers*, Vol. C-33, No. 12, Dec. 1984, pp. 1247-1265.

[4] K. Padmanabhan, "Cube structures for multiprocessors," *Comm. of ACM*, Vol. 33, No. 1, Jan. 1990, pp.43-52.

[5] D. H. Lawrie, "Access and alignment of data in array processors," *IEEE Trans. on Computers*, Vol. C-24, No. 12, Dec. 1975, pp. 1145-1155.

[6] D. H. Lawrie and C. R. Vora, "The primary memory system for array access," *IEEE Trans. on Computers*, Vol. C-31, No. 5, May 1982, pp. 435-442.

[7] Y. Lan and L. Ni, "Relay approach message routing in hypercube multiprocessors," *3rd Int'l Conf. on Supercomputing*, May 1988, Vol. III, pp. 174-182.

[8] S. Saad and M. H. Schultz, " Data communication in hypercubes," *J. of Parallel and Distributed Computing*, Vol. 6, No.1, Feb. 1989, pp. 115-136.

[9] C. T. Ho and S. L. Johnsson, "Distributed routing algorithms for broadcasting and personalized communication in hypercubes," *Proc. of 1986 Int'l Conf. on Parallel Proc.*, Aug. 1986, pp. 640-648.

[10] T. Szymansky, "On the permutation capability of a circuit-switched hypercube," *Proc. of 1989 Int'l Conf. on Parallel Proc.*, Vol. 1, Aug. 1989, pp. 103-110.

[11] J. L. Gustafson, G. R. Montry and R. E. Benner, "development of parallel methods for a 1024-processor hypercube," *SIAM J. Sci. Stat. Comput.*, Vol.9, No.4, July 1988, pp. 609-638.

[12] W. J. Dally and C. L. Sietz, "Deadlock-free routing in multiprocessor interconnection networks," *IEEE Trans. on Computers*, Vol. C-36, No. 35, May 1987. pp. 547-553.

[13] L. W. Tucker and G. G. Robertson, "Architecture and applications of the connection machine," *Computer*, Aug. 1988, pp.26-38.

[14] W. Lin, "A dimension-scrambling approach to fast hypercube data permutations," *Proc. of 1990 Int'l Conf. on Parallel Proc.*, Aug. 1990, Vol. III, pp. 119-122.

[15] D. Nassimi and S, Sahni, "Optimum BPC permutation on a cube-connected SIMD computer," IEEE Trans. on Computers, Vol. 31, No. 4, April 1982, pp. 338-341.

A General Model for Scheduling of Parallel Computations And Its Application to Parallel I/O Operations*

Ravi Jain, John Werth and J. C. Browne
Department of Computer Sciences
University of Texas at Austin
Austin, TX 78712

Abstract

The motivation for the research presented here is to develop algorithms for scheduling I/O operations in parallel computer systems. In order to identify existing applicable algorithms, a general model for specifying scheduling problems is first developed. The model is of interest in its own right; for instance, we have demonstrated its coverage of algorithms for scheduling parallel tasks with precedence constraints on multiprocessors.

The model is used to establish an algorithm for scheduling of I/O operations across a multiple bus I/O architecture for a multiprocessor system such as the IBM RP3, by adapting an algorithm used for scheduling of satellite-switched time-division multiplexed communication systems with tree structured architectures. The model then helps recognize that this problem is equivalent to a more general and previously unstudied problem, namely scheduling parallel I/O operations in the presence of a limited class of mutual exclusion constraints, thus providing an optimal algorithm for the latter problem.

1   Introduction

The motivation for the research presented herein was to develop effective and generally applicable methods of scheduling I/O operations for parallel execution. The problem of scheduling I/O in parallel computers is of significant practical interest as the speed of many parallel applications is limited by the speed of I/O rather than computation [4]. In general the discrepancy between processing speeds at different levels of the memory hierarchy is increasing, and appears likely to continue doing so. Thus the speed of data transfers between different levels of the memory hierarchy is an increasingly important bottleneck [2], and it is attractive to schedule these transfers so as to minimize program completion time. Nonetheless, this problem has received relatively little attention [9].

Since there has been a great deal of previous work on algorithms for development of parallel schedules for other application areas, the initial approach we took was to see if algorithms used in other applications could be adapted to the problem of parallel schedules for I/O operations. In order to perform a systematic analysis we created a model for specifying scheduling problems in a common framework at a higher level of abstraction. This general scheduling model is of interest in its own right since almost all research in scheduling algorithms for parallel computations has been done in the context of some particular application. This has led to a profusion of application specific terminology and notation which has made it difficult to apply results across application boundaries. A unification may produce insights leading to more generally applicable formulations of important algorithms. The next section of this paper presents our first attempt at such a unified model.

Application of this approach led immediately to the recognition that a scheduling algorithm used in the domain of satellite-switched time-division multiplexed (SS/TDM) communication systems with tree structured architectures [10] can be adapted to generate parallel schedules for I/O operations on multiple bus architectures similar to the IBM RP3 [15]. Section 3 of this paper gives the transformation leading to this transfer of algorithms across application areas.

Casting the I/O application in more abstract terms has also led to the recognition that both it and the SS/TDM application are special cases of a more general problem, namely scheduling parallel I/O in the presence of mutual exclusion constraints. Mutual exclusion constraints are logically weaker than precedence constraints, and represent synchronization requirements among the operations to be scheduled. It appears that virtually all previous research on algorithms for generating parallel schedules has used only precedence constraints which induce a total order among the operations to be scheduled. We thus obtain what is, to our knowledge, the first result on parallel scheduling in the presence of limited mutual exclusion constraints, which is discussed in Section 4.

2   The Scheduling Model

2.1   Basic Definitions

We take certain notions as primitive. We shall assume the existence of primitive objects called *resources*; intuitively these correspond to machines, parts, communications links, disks etc. We also assume the existence of primitive objects called *units of computation*; intuitively these correspond to computer instructions, industrial processes, etc.

*This research was supported by the IBM Corporation through grant 61653 and by the State of Texas through TATP Project 003658-237.

Def. A *task* is a unit of computation that requires a fixed set of resources.

In practice a complex process requiring different sets of resources at different times may be represented as a sequence of tasks. Discrete time is represented as the set of natural numbers.

Def. The *length* of a task for a given resource is the amount of time the task requires that resource.

In practice we may specify the task length in units such as machine instructions or machine cycles, from which the task length in units of time can be readily calculated knowing the speed of the resource.

We assume that resources are partitioned into a collection of disjoint sets; the set to which a resource belongs is called its *type*.

The resources a task requires are specified in terms of *virtual resources*. Part of the process of computing a schedule is to map the virtual resource requirements of each task to the resources actually available; for some problems this step may be trivial, while for others it is a significant difficulty. Virtual resources have types drawn from the same set of types as (real) resources.

Notation. Let T denote the set of tasks, R the set of resources, VR the set of virtual resources, and RT the set of resource types. Let τ denote the set of natural numbers representing time and N the set of natural numbers.

Def. An *allocation* is a function $al : T \to N^{|RT|}$ specifying the number of resources of each type required by a task.

Def. An *assignment* is a function $as : T \to 2^R$ specifying the resources required by each task.

The notation and definitions above pertain to the specification of a scheduling problem, i.e., the "input" to the scheduling process. In the following definitions we define the schedule and its characteristics, which is the solution to the scheduling problem, i.e., the "output" of the scheduling process.

Def. A *schedule* is a function $s : T \to 2^R \times 2^{\tau \times N}$ specifying for each task the resources and the times τ_i and durations n_i for which they are held, i.e., for each task $t \in T$, there exists $k \in N$ such that
$$s(t) = R(t) \cup \{(\tau_1, n_1), (\tau_2, n_2), ..., (\tau_k, n_k)\}$$
where, for $1 \le i \le k, R(t) \subseteq R, \tau_i \in \tau, \tau_{i+1} \ge \tau_i + n_i$, and $n_i \in N$. Note that if schedule s is non-preemptive then for all $t \in T, k = 1$.

Def. Given a schedule $s(t)$ as defined above we can define the following auxiliary functions.
$$start(s, t) = \tau_1$$
$$stop(s, t) = \tau_k + n_k$$
$$makespan(s) = max\{stop(s, t) : t \in T\}$$
$$- min\{start(s, t) : t \in T\}$$
$$active(s, t) = \{(\tau_i, \tau_i + 1) : \text{for some } j, 1 \le j \le k,$$
$$\tau_j \le \tau_i \wedge \tau_i + 1 \le \tau_j + n_j\}$$

2.2 Definition of the Model

Def. A *scheduling problem* is a tuple $SP = (PG, AG, VRG,$ $f, Preempt)$ specifying constraints on tasks and resources, where PG, AG and VRG are the precedence, architecture, and virtual resource graphs defined below, f is the objective function, and $Preempt$ is true iff the schedule may be preemptive.

The precedence graph PG and architecture graph AG are fairly familiar from previous work in compilers, parallel architectures and parallel programming. The precedence graph captures the relationships between tasks, and the architecture graph captures the interconnections of resources. The virtual resource graph VRG captures the resource requirements of each task.

Terminology. A hyperedge is an undirected connection between one or more vertices. A hyperedge on one vertex is called a self-loop. A hyperedge on two vertices is called an edge. An arc is a directed edge. A path is a sequence of arcs or hyperedges in which consecutive arcs or hyperedges share a vertex and no vertex is included twice.

Def. A *precedence graph* $PG = (T, Ep, Lp)$ consists of a set T of vertices representing tasks, a set Ep of arcs and hyperedges, and a labeling function Lp where
$$Lp(x) = \begin{cases} \text{length of task } x, \text{ if } x \in T \\ \text{communication cost from task } u \text{ to v,} \\ \quad \text{if } x = (u, v) \text{ is an arc} \\ 0, \text{ if } x \text{ is a hyperedge} \end{cases}$$
Informally, a hyperedge specifies that for all schedules the tasks connected by the edge are to be mutually exclusive, i.e., no two may be *active* simultaneously. An arc (u, v) specifies that for all schedules task u must complete before task v may begin.

Def. An *architecture graph* $AG = (R, Ea, La)$ consists of a set R of vertices representing resources, a set Ea of arcs and hyperedges, and a labeling function La:
$$La(x) = \begin{cases} \text{processing speed, if } x \in R \\ \text{bandwidth, if } x \in Ea \end{cases}$$
An arc or hyperedge in AG represents interconnection of resources. Typically hyperedges represent buses and arcs represent unidirectional communication links.

The virtual resource graph captures, for each task, the virtual resources it requires. Arcs and hyperedges correspond to tasks, and the vertices they connect correspond to virtual resources.

Def. A *virtual resource graph* for a given set of tasks T, $VRG = (VR, Er, Lr)$ consists of a set VR of vertices representing virtual resources, a set Er of arcs and hyperedges, and a labeling function Lr. Arcs and hyperedges correspond to tasks in T, and $|Er| = |T|$. For all arcs and hyperedges $e \in Er, Lr(e) = t$ specifies that task $t \in T$ must simultaneously possess all virtual resources connected by e. In addition, if $e = (r, s)$ is an arc then t involves transfer of information from r to s.

Def. A resource function $g : VRG \to AG$
1. provides a mapping from VR to R
2. if $e = (u, v) \in Er$ is an arc in VRG then there is a path from $g(u)$ to $g(v)$ in AG
3. ignores all hyperedges in Er
4. ignores the labeling function Lr

Def. A schedule s satisfies a scheduling problem $SP = (PG, AG, VRG, f, Preempt)$ iff

1. If $Preempt = false$ then for all $t \in T$, for some $\tau' \in \tau, n \in N$, $s(t) = R(t) \cup (\tau', n)$
2. If $(u, v) \in Ep$ is an arc in PG then task u stops before task v may begin, i.e., $stop(s, u) < start(s, v)$
3. If $e \in Ep$ is a hyperedge in PG then no two tasks connected by the hyperedge are active simultaneously, i.e.,
   $$\forall u, v \in e, active(s, u) \cap active(s, v) = \{\}$$
4. there exists a resource function $g : VRG \to AG$.

Remark. For some scheduling problems the arcs in the VRG will correspond directly to arcs in the AG, i.e., the resource function g will preserve arcs. In addition, for some problems g may provide a bijection from VR to R. Thus for some scheduling problems g will be very simple.

Def. A schedule s is *optimal* if the objective function f is minimized.

We have shown that the model covers traditional problem classes such as flow-shop, job-shop, and multiprocessor scheduling [9]. The algorithms develped for these problems, as well as previous models and classification schemes, e.g. [7] and [12], are too restricted for our purpose as in general, unlike the I/O problem, they assume that jobs require only a single instance of a single resource type at any given time. In [11] we discuss strengths and weaknesses of the model.

3   Parallel I/O Scheduling

In this section we describe the scheduling problem in which we are primarily interested: the scheduling of I/O operations in a parallel computer system. Defining the problem in the scheduling model leads to the recognition that it is closely related to problems in two other application areas: the scheduling of file transfers in a network, and scheduling packet transfers in a communications switch. We show how an algorithm used for time slot assignment in satellite-switched time-division multiplexed (SS/TDM) communications systems with tree-structured architectures can be adapted to solve a version of the parallel I/O scheduling problem.

3.1   I/O in Parallel Computers

We describe simple I/O scheduling *(IOS)*, stated in terms of processors and memory units for convenience. Informally, the problem is: Given a set of m I/O transfers, where

1. each transfer requires a fixed but possibly distinct time,
2. each transfer requires a specified processor-memory pair from two given sets of processors and memories,
3. each processor can communicate via a direct dedicated link with each memory, and
4. there exists no partial order in which the transfers are to occur,

is there a preemptive schedule for performing the transfers whose total length is at most some given bound?

The formal specification of *IOS* consists of a precedence graph with no edges, a complete bipartite architecture graph, and a bipartite virtual resource graph representing the transfers [11].

Application 1: Simple I/O scheduling *(IOS)* with the following extension (**E1**) to the architecture graph occurs in multiple-bus I/O systems such as the IBM RP3 parallel computer [15], and has been studied in [17].

E1. A fixed maximum number of data transfers between processors and memories may take place at any given time.

3.2   I/O Scheduling for Tree-Structured Architectures

More realistic cases of the I/O scheduling problem do not assume a direct dedicated link from every processor to every memory. In particular, we define the following generalized I/O scheduling problem which differs from *IOS* in allowing tree-structured interconnection networks of the type shown in Fig. 1 and defined formally as AG below.

$TreeIOS = (PG, AG, VRG, f, Preempt)$ where $Preempt = true$, f is makespan, and,

$PG = (T, Ep, Lp)$ with $|T| = m$, $Ep = \{\}$, and $Lp(t) \geq 0$ for all $t \in T$ are the task lengths.

$VRG = (VR, Er, Lr)$ is a bipartite graph, where $VR \in R$, Er is a set of arcs from vertices of type $SUSER$ (processors) to those of type $RUSER$ (memories) representing the I/O operations to be scheduled, and Lr is a bijection from Er to T.

$AG = (R, Ea, La)$ and (see Fig. 1) $RT = \{SUSER, RUSER, MUX, DMUX, NULL\}$; $|R| = n + m + 1$, n is the number of resources of type $SUSER$ or $RUSER$, m is the number of type MUX and $DMUX$, and there is one resource of type $NULL$. Ea is a directed tree whose root is the resource of type $NULL$. The root has a left subtree MT called the multiplexer subtree with interior nodes of type MUX, leaves of type $SUSER$, and arcs directed towards the root. (The definition of the right subtree DT follows by analogy). Now $La(r) = 0$ if $r \in R$, and $La(e)$ is the capacity of arcs e (in packets per second) for $e \in Ea$. We assume all interior vertices have degree at least 3. We also assume that for vertices in MT the sum of $La(e)$ for incoming arcs e is at least the value of $La(e')$ for the single outgoing arc e'. There is an analogous assumption for DT, while for the root the capacity of the incoming arc equals that of the outgoing arc.

Having cast this extended I/O scheduling problem in our model, we see that it is a generalization of the following problems.

Application 1: Simple I/O scheduling. Clearly *IOS*, extended with **E1**, is a special case of *TreeIOS*. The algorithm given in [17] applies when the following additional restrictions are placed on AG:

1. AG is a tree with exactly three levels
2. arcs connected to the root have capacity k

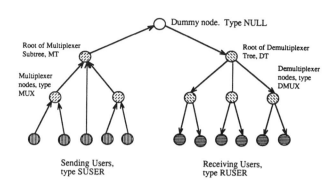

Figure 1: Model of a tree-structured architecture

3. arcs connected to leaves (users) have a capacity of exactly 1 packet per second.

Application 2: Hierarchical Satellite Switching Systems. Hierarchical time division multiplexed (TDM) switching systems have recently been proposed [8] that connect sending users via a bank of multiplexers, followed by the satellite acting as a TDM switch, followed by a bank of demultiplexers, to receiving users, as in Fig. 2. Hence, the switching system has three stages. The scheduling problem in such systems have been studied in [1], [13] and [5]; the algorithms of [1, 13] run in time $O(n^5)$. The problem is a version of *TreeIOS* with the following additional restrictions on AG:

1. AG is a tree with exactly four levels
2. arcs connected to leaves have a capacity of exactly 1 packet per second.

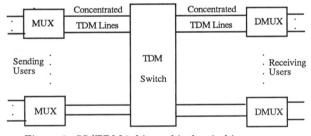

Figure 2: SS/TDMA hierarchical switching system

Application 3: Data Transfers in Networks. In [6] the authors consider a set of data transfers that are to be scheduled between nodes of a communication network where each node has a number of transmitters and receivers. Their algorithm runs in time $O(n^5C + n^4C^2)$, where C is the average capacity of the arcs connecting the leaves in AG. As pointed out in [10], this problem is similar to **Application 2**, and is also a version of *TreeIOS* where:

1. AG is a tree with exactly four levels
2. arcs connected to the root may have capacity k
3. preemption is allowed at non-integer boundaries
4. arcs connected to leaves have capacity **either** exactly 1 packet per second (the case called *speedup*), **or** equal to the number of ports per user (*no speedup*).

3.3 An Optimal Algorithm for I/O Scheduling in Tree Architectures

Definition of the generalized I/O scheduling problem, *Tree-IOS*, in the model leads to the recognition that it can be solved by a algorithm recently developed for time-slot assignment in SS/TDM hierarchical switching systems [10]. The problem in [10] specified in our model is precisely equivalent to *TreeIOS*. Hence the algorithm in [10] can be used to solve *TreeIOS*; the algorithm runs in time $O(n^4C)$, where as before C is the average capacity of user arcs, and represents a generalization and improvement in time complexity over previous results ([1, 13]).

In addition, the definition of *TreeIOS* opens the possibility that the solution may be applicable to I/O scheduling in a variety of tree-structured parallel computer architectures. Examples include the VLSI tree-structured database machine of [18], the relational database machine REPT [16], and interconnection networks such as Kyklos [14].

4 I/O Scheduling With Mutual Exclusion Constraints

A natural extension to the I/O problem *IOS* is to allow the specification of mutual exclusion constraints between tasks, since mutual exclusion constraints are a natural and common means of expressing synchronization requirements in parallel programs (e.g. in the CODE parallel programming environment [3, 19]). To our knowledge mutual exclusion constraints have not been studied in the context of scheduling parallel programs. Previous work has concentrated on scheduling in the presence of precedence constraints, which are logically stronger than mutual exclusion constraints, or "exclusion constraints" [20] where a task executing on a single processor may not be preempted by another.

An advantage of the model over simpler classification schemes is the ability to perform graph transformations to problem specifications, so as to prove that a solution to one problem is applicable to a seemingly different problem. We have demonstrated this by showing the I/O scheduling problem for systems with tree-structured architectures *(TreeIOS)* is related to scheduling I/O in a parallel computer system in the presence of limited mutual exclusion constraints. The limited mutual exclusion constraints we allow are a superset of those permitted in the CODE 1.2 parallel programming environment [19]. (For proofs and details see [11]). We thus obtain an algorithm producing optimal-length schedules for I/O scheduling in the presence of limited mutual exclusion constraints, to our knowledge the first such result in this area.

5 Conclusions

We have defined a general model for the problem of scheduling parallel computations that serves as a consistent framework for specifying scheduling problems and for applying results across application boundaries.

The model was used to show that a simple class of I/O scheduling problems, a class of communications switch scheduling problems, and a class of network file transfer problems are special cases of the problem of scheduling I/O in tree-structured architectures. This recognition allowed us to adapt a recent algorithm for the satellite-switched TDM communications scheduling problem to the problem of I/O in tree-structured architectures. Finally, we obtained a fast optimal algorithm for the problem of parallel I/O scheduling under limited mutual exclusion constraints by transforming it to I/O scheduling in tree-structured architectures.

We are continuing to investigate the parallel I/O scheduling problem both theoretically and experimentally, particularly for cases in which the architecture and precedence graphs are more general, and correspond to situations of practical interest.

Acknowledgement. The first author would like to thank Prof. Galen Sasaki for many instructive discussions.

References

[1] M. A. Bonuccelli. A fast time slot assignment algorithm for TDM hierarchical switching systems. *IEEE Trans. Comm.*, 37:870–874, Aug. 1989.

[2] H. Boral and D. J. DeWitt. Database machines: An idea whose time has passed? A critique of the future of database machines. In *Third Intl. Workshop on Database Machines*, pages 166–187, 1983.

[3] J. C. Browne, Muhammad Azam, and Stephen Sobek. CODE: A unified approach to parallel programming. *IEEE Software*, page 11, July 1989.

[4] J. C. Browne et al. Design and evaluation of external memory architectures for multiprocessor computer systems. Technical report, Univ. Texas at Austin, Dept. of Comp. Sci., 1987.

[5] S. Chalasani and A. Varma. Fast parallel time-slot assignment algorithms for TDM switching. In *Proc. Int'l Conf. Par. Proc.*, volume III, page 154, 1990.

[6] H.-A. Choi and S. L. Hakimi. Data transfers in networks. *Algorithmica*, 3:223–245, 1988.

[7] R. W. Conway, W. L. Maxwell, and L. W. Miller. *Theory of Scheduling*. Addison-Wesley, 1967.

[8] K. Y. Eng and A. S. Acampora. Fundamental conditions governing TDM switching assignments in terrestial and satellite networks. *IEEE Trans. Comm.*, COM-35:755–761, 1987.

[9] Ravi Jain. Scheduling I/O in parallel computing environments. Unpublished manuscript, Dec. 1990.

[10] Ravi Jain and Galen Sasaki. Scheduling packet transfers in a class of TDM hierarchical switching systems. In *Proc. Intl. Conf. Comm.*, 1991.

[11] Ravi Jain, John Werth, and J. C. Browne. A general model for the scheduling problem. Submitted for publication, Apr. 1991.

[12] E. L. Lawler, J. K. Lenstra, and A. H. G. Rinnooy Kan. Recent developments in deterministic sequencing and scheduling: A survey. In *Deterministic and Stochastic Scheduling*, pages 35–73. D. Reidel Publishing, 1982.

[13] S. C. Liew. Comments on "Fundamental conditions governing TDM switching assignments in terrestrial and satellite networks". *IEEE Trans. Comm.*, 37:187–189, Feb. 1989.

[14] B. Menezes and R. Jenevein. KYKLOS: A linear growth fault-tolerant interconnection network. In *Proc. Int'l Conf. Par. Proc.*, pages 498–502, 1985.

[15] G. Pfister et al. The IBM research parallel processor (RP3): Introduction and architecture. In *Proc. Int'l Conf. Par. Proc.*, pages 764–771, 1985.

[16] R. K. Schultz and R. J. Zingg. Response time analysis of multiprocessor computers for database support. *ACM Trans. Database Sys.*, pages 14–17, 1984.

[17] Kiran Somalwar. Data transfer scheduling. Technical Report TR-88-31, Univ. Texas at Austin, Dept. of Comp. Sci., 1988.

[18] S. W. Song. A highly concurrent tree machine for database applications. In *Proc. Int'l Conf. Par. Proc.*, pages 259–268, 1980.

[19] John Werth, J. C. Browne, Steve Sobek, T. J. Lee, Peter Newton, and Ravi Jain. The interaction of the formal and practical in parallel programming environment development: CODE. Technical Report TR-91-09, Univ. Texas at Austin, Dept. of Comp. Sci., 1991.

[20] J. Xu and D. L. Parnas. Scheduling processes with release times, deadlines, precedence, and exclusion relations. *IEEE Trans. Soft. Eng.*, 16:360–369, Mar. 1990.

Efficient Interprocessor Communication on Distributed Shared-Memory Multiprocessors *

Hong-Men Su   and   Pen-Chung Yew

Center for Supercomputing Research and Development

University of Illinois at Urbana-Champaign

Urbana, Illinois 61801

hong@csrd.uiuc.edu, yew@csrd.uiuc.edu

Abstract

Interprocessor communication is one of the major sources of overheads in large-scale shared memory multiprocessors. Overheads resulting from interprocessor communication include communication latency and busy waiting. In this paper, we show that the distributed shared memory (DSM) multiprocessors utilizing their unique nearest shared memory (NeSM) as communication buffers offer a better solution for interprocessor communication. Our approach is to use **direct communication** with **static message passing**. Preliminary results show that this approach is promising.

1   Introduction

Interprocessor communication is one of the major sources of overheads in large-scale shared memory multiprocessors. This is especially true when it happens within loops, such as Doacross loops [4]. In this paper, we examine the problems in interprocessor communication and show how the distributed shared memory (DSM) multiprocessors offer a better solution. We present the programming paradigm for achieving the goal. Readers interested in the compiler algorithms that transform sequential code into such paradigm should consult [14].

The paper is organized as follows. Section 2 discusses the motivation of this work. Section 3 presents a model of DSM multiprocessors and their unique feature, the nearest shared memory (NeSM). Section 4 gives the mechanisms, advantages, and primitives for utilizing the NeSM as the buffers for interprocessor communication. Section 5 describes the simulation and preliminary results. Section 6 concludes this paper.

2   Motivation

Consider the generic interprocessor communication scenario in shared memory multiprocessors shown in Figure 1. The *producer* updates the value of variable **data** which the *consumer* is to read. To order these two events, the producer can send a completion signal by turning on a flag after it updates the data (assume the flag is initially off), and the consumer

producer	consumer
S_1: data= in;	S_3: while(flag!=on);
S_2: flag= on;	S_4: out= data;

Figure 1: Interprocessor communication scenario 1

can busy-wait on the flag until it becomes on, then reads the data. We refer to the part of interprocessor communication involving the flag as **synchronization** (the flag as the synchronization variable), and to that involving the data as **data transfer** when distinction is necessary. To ensure the correctness of data transfer (S_1 and S_4) and synchronization (S_2 and S_3), it requires that the completion of data updating precede the completion of signal sending and the completion of busy-waiting occur before the completion of data reading, or $S_1 \rightarrow S_2$ and $S_3 \rightarrow S_4$, respectively [5].

The scenario is quite general because it shows the major overheads in the frequently used interprocessor communications, namely, **producer stall**, **busy-waiting**, and **remote read**. It implements exactly the producer-consumer communication. It can be viewed as parts of **critical region** code in that signal sending (S_2) is **unlock**, and busy-waiting (S_3) is a part of **lock**. It can also be viewed as parts of codes using **barrier synchronization**, in that signal sending is to notify the barrier completion by the last processor, and busy-waiting is to wait for the barrier completion by other processors. Although the code in Fig.1 does not fully implement critical region and barrier synchronization codes, it does show the major overheads in these two codes.

Let us consider the three major overheads in interprocessor communication in order. In a weakly-ordered system without write-back coherent caches,[1] the producer may need to stall at S_2 to wait for the completion of S_1 in the following cases:

- the system has multiple-pathed networks and S_1 and S_2 cannot be executed atomically; or

- the system has uni-pathed networks, and **data** and **flag** are not allocated in the same memory module.

In a weakly-ordered system with write-back coherent caches, the producer may need to stall at S_2 to wait for obtaining the ownership of the cache line containing **data**. One exception is that a system supporting **release consistency**

*This work is supported in part by the National Science Foundation under Grant No. US NSF MIP 8410110, the U.S. Department of Energy under Grant No. US DOE DE-FG02-85ER25001, NASA NCC 2-559, and Cray Research Inc.

[1]We refer to a system without coherent caches or with write-through coherent caches. We use the term *coherent caches* for **hardware** coherent caches only.

can delay the signal sending in a write buffer, and the producer can immediately execute instructions following S_2. In all cases, however, the signal sending of S_2 is delayed until some global operations are finished, and this means a longer busy-waiting by the consumer. In a system without coherence caches, **busy waiting** is performed on shared memory and repeatedly generates **global traffic**. The overheads of **remote reads** for data and (the last check for) flag can be measured by the number of trips required through the networks. In a system without write-back coherent caches, each remote read requires *two network trips*. In a system with **directory-based** write-back coherent caches, it requires *four network trips* resulting from a sequence of five events: cache miss, home directory lookup, obtaining the exclusive copy, write-back to home, and returning the cache line.

Hiding the latency by inserting unrelated instructions between S_1 and S_2, or S_3 and S_4 is one way to reduce the overheads. However, there may not always be unrelated instructions. Even if unrelated instructions are available, inserting appropriate instructions is not trivial owing to unpredictable memory access time. Using prefetching for remote read of data is bounded by S_3.

3 DSM Multiprocessors and Nearest Shared Memory

We define a **distributed shared memory (DSM)** multiprocessor as a shared memory multiprocessor whose shared memory modules are **physically distributed** across the system so that some shared memory module is closer to a processor than others. DSM multiprocessors belong to the non-uniform memory access (NUMA) architecture, which also includes distributed (non-shared) memory multiprocessors.[2] Examples of DSM systems include CMU's Cm* [6], the BBN Butterfly [2], the IBM RP3 [12], Stanford's DASH [9], CMU's PLUS [3], MIT's ALEWIFE [1] and the Horizon/Tera [8]. In a DSM system, we call the shared memory module that is closest to a processor its **nearest shared memory (NeSM)**, and other shared memory modules its **remote memory**. A processor is the **owner** of its NeSM. We use the term NeSM instead of "local memory" because (1) we want to emphasize the concept of "shared" and to avoid the possible confusion that local memory is private; (2) we want to include the machines without "local memory" such as the Horizon/Tera and the DASH. In the Horizon/Tera, a NeSM is just another node in the 3-D torus as its owner processor, and there may be extra shared memory modules that are not NeSM of any processor. In the DASH, processors are clustered, and each cluster has a shared memory module as the NeSM of all its processors. In other DSM systems mentioned above, a NeSM is attached to its owner processor, and the union of all NeSM's constitutes the whole shared memory. This is the model that we use throughout this paper. We also assume that accesses to a processor's own NeSM generate no global traffic, and accesses to all remote memory take equal time.

The DSM architecture has been claimed for its scalability to a large system. It has been shown in [11] that the

synchronization algorithms, which make processors busy-wait on synchronization variables allocated in their corresponding NeSM's, can outperform the algorithms that are not designed to do so. The proposed algorithms include a queue-based spin-lock and a tree-based barrier. However, they did not try to integrate the synchronization and data transfer. Thus, only global busy-waiting is removed from the overheads of interprocessor communication.

We have proposed a scheme for interprocessor communication which uses the NeSM not only to *eliminate the global busy-waiting*, but also to *remove the processor stall and the remote read*. In this scheme, the interprocessor data transfer latency is effectively reduced. We have also proposed compiler algorithms to generate communication primitives for Doacross loops, loops with cross-iteration dependences that require interprocessor communication [14]. In this paper, we will only show the basic techniques of how this can be done systematically.

4 Nearest Shared Memory as Communication Buffers

In a DSM multiprocessor, interprocessor communication can be implemented efficiently by using the NeSM as communication buffers. The producer first puts the new value of data (i.e. in) *directly in its consumer's NeSM*, and then sets a flag there. The consumer busy-waits on the flag in its NeSM, and gets the value of data. This is called **direct communication**, whereas in Fig.1, the information flow from the producer to the remote memory then to the consumer is **indirect communication**. Two immediate advantages of direct communication can be identified. First, the consumer can perform **local busy-waiting** without the need for coherent caches. Second, the remote read by the consumer is replaced by a much **faster NeSM read**. Further, because the values of data and flag go to the same memory module, **producer stall** at S_2 is not necessary if the system has uni-pathed networks or the data transfer and synchronization can be performed atomically. This also means a shorter busy-waiting. We will assume the availability of **atomic operations**. Because the flag has only two states, on and off, binary semaphore operations such as HEP's FULL/EMPTY bit synchronization will suffice [13]. The flag will be referred to as the **sync bit**, which is associated with each location.

The data and its sync bit passed from a producer to its consumers are called a **message**, and the storage (or variable) allocated in a consumer's NeSM to hold messages temporarily is a **message buffer**. The message passing used is static in that the message buffer location for a particular message is determined statically. In the **static message passing**, a producer can directly store a message in a buffer without performing message enqueuing, and a consumer knows where a particular message is stored, so it can extract the message without the need for message dequeuing. Note that the message buffers are auxiliary storage used temporarily, and each data may need to maintain an up-to-date value, if required, at a fixed location called **home**. If two messages are not *live* simultaneously,[3] they can share the same mes-

---

[2]We distinguish DSM multiprocessors from the Shared Virtual Memory (SVM) systems built on top of distributed memory multiprocessors such as [10].

[3]One example is that the messages received before a barrier and those sent after the barrier are not live simultaneously.

```
producer (p1)                    consumer (p2)
S1:  put(in,nsm(p2,mb));         ........
S2:  data= in;                   S3:  out= get*(mb);
```

Figure 2: Interprocessor communication scenario 2

sage buffer location without sacrificing parallelism [14]. The message buffer sharing problem is analogous to the register allocation problem.

We use "NeSM mb" to declare **mb** as a variable allocated in the NeSM of the executing processor. In this case, if the executing processor is **p**, processor **p** accesses the variable by simply specifying **mb**, and other processors access it by specifying **nsm(p,mb)**. Other variables are referenced in the ordinary way. The primitives performing the required operations on a message are defined as follows:

- **put(D, X)**: write the value of **D** to variable **X**, and set **X**'s sync bit to FULL.

- **get(X)**: return the sync bit and the value of variable **X**, and set the sync bit to EMPTY.

- **get*(X)**: execute **get(X)** repeatedly until the sync bit is FULL, then return the value of **X**.

The primitives can be extended to operate on a block of data.

Using the primitives, the example in Fig.1 is rewritten in Fig.2. The total latency for the interprocessor communication is now reduced to one network trip, and the producer does not need to stall after S_1. The write to data in S_2 is needed to keep an up-to-date value.

The above technique works effectively for DSM systems without coherent caches and for those with coherent caches that can be bypassed by **put** primitive. Note that the technique requires that the data producer know who its consumer is. This is not a strict constraint, as in many applications the producer/consumer relation is easily determined.

5 Preliminary Results

We have done the detailed simulation on the code shown in Fig.3, which contains the dependence patterns in 5-point relaxation code, image smoothing code, and dynamic programming code. It can be transformed by the algorithm in [14] into the single-program-multiple-data (SPMD) execution [7] as shown in Fig.4.[4] The outer loop is cyclically scheduled among processors, where P is the number of processors and mypid∈[1,P] is the executing processor's id. The inner loop is executed sequentially. Let (I, J) be the iteration where $i = I$ and $j = J$. The interprocessor communication is from the write of **a[i,j]** in (I, J) to the read of **a[i-1,j]** in $(I+1, J)$. The declaration statement gives each processor a copy of the message buffer, array **mb[N]**, in its NeSM. The message buffer for a processor is re-used in each iteration executed by the processor, and there is no race involved [14].

The simulated DSM system contains forward and backward omega networks which connect processors to and from

----

[4]The initialization code is ignored.

```
doacr i= 1, N
    do j= 1, N
         a[i,j]=(a[i,j]+a[i,j-1]+a[i-1,j])/3;
    end do
end doacr
```

Figure 3: A Doacross example

```
NeSM mb[N];
consumer= (mypid+1) modulus P;
do i= mypid, N, P
    do j= 1, N
         tmp=(a[i,j]+a[i,j-1]+get*(mb[j]))/3;
         if(i < N) put(tmp,nsm(consumer,mb[j]));
         a[i,j]= tmp;
    end do
end do
```

Figure 4: SPMD code for the Doacross example

remote shared memory. There are no coherent caches. However, prefetching is available to hide the latency for accessing **a[i,j]** and **a[i,j-1]** as they are available before the start of the loops. Both direct and indirect communication are simulated. For direct communication, accesses to a processor's own NeSM variables have the effect described above. For indirect communication, all accesses are forced to go through the networks. We assume atomic operations are available for both schemes. The speedups of direct communication versus indirect communication are shown in Fig.5, where T_c is the time for one network trip. For the systems with the same P, the larger T_c is, the better the direct communication is performed relative to the indirect one. The ups and downs of each curve are mainly the result of the ceiling function in T_c. Fig.5 shows the performance gains for using direct communication.

6 Conclusions

In this paper, we point out the overheads in interprocessor communication. DSM multiprocessors with their unique feature, the NeSM, can offer a solution that greatly reduces the overheads and requires little hardware support. The solution that we propose is to use direct communication with static message passing. Direct communication eliminates the global busy waiting and shortens the latency in synchronization and data transfer. Static message passing makes direct communication more efficient without resort to message enqueuing,

dequeuing, and copying in dynamic message passing. However, static message passing needs extra storage. We have proposed compiler algorithms to transform Doacross loops into the execution scheme and derived conditions that allow some messages to share the same buffer without sacrificing parallelism [14]. As supported by the preliminary results, we believe that such an approach can substantially improve the efficiency of a parallel system.

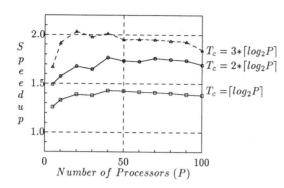

Figure 5: Speedups of direct vs. indirect communication (N=200)

References

[1] A. Agarwal, B.-H. Lim, D. Kranz, and J. Kubiatowicz. APRIL: a processor architecture for multiprocessing. In *Int. Sym. Computer Architecture*, pages 104–114, May 1990.

[2] BBN Advanced Computers. *Butterfly products overview*, 1987.

[3] R. Bisiani and M. Ravishankar. PLUS: a distributed shared-memory system. In *Int. Sym. Computer Architecture*, pages 115–124, May 1990.

[4] R. Cytron. Doacross: Beyond vectorization for multiprocessors. In *Int. Conf. on Parallel Processing*, pages 836–845, Aug. 1986.

[5] M. Dubois, C. Scheurich, and F. Briggs. Memory access buffering in multiprocessors. In *Int. Sym. on Computer Architecture*, pages 434–442, May 1986.

[6] A. Jones and P. Schwarz. Experience using multiprocessor systems: A status report. *ACM Comput. Surveys*, pages 121–166, June 1980.

[7] A. H. Karp. Programming for parallelism. *Computers*, pages 43–57, May 1987.

[8] J. Kuehn and B. Smith. The Horizon supercomputing system: Architecture and software. *Supercomputing*, pages 28–34, Nov. 1988.

[9] D. Lenoski, K. Laudon, K. Gharachorloo, A. Gupta, and J. Hennessy. The directory-based cache coherence protocol for the DASH multiprocessor. In *Int. Sym. Computer Architecture*, pages 148–159, May 1990.

[10] Kai Li. IVY: A shared virtual memory system for parallel computing. *Int. Conf. on Parallel Processing*, 2:94–101, August 1988.

[11] J. Mellor-Crummey and M. Scott. Synchronization without contention. In *ACM Int. Conf. on Architectural Support for Programming Languages and Operating Systems*, pages 269–278, April 1991.

[12] G. Pfister, W. Brantley, D. George, S. Harvey, W. Kleinfelder, K. McAuliffe, E. Melton, V. Norton, and J. Weiss. The IBM research parallel processor prototype (RP3): Introduction and architecture. *Int. Conf. on Parallel Processing*, pages 764–771, Aug. 1985.

[13] B. J. Smith. A pipelined, shared resource MIMD computer. In *Int. Conf. on Parallel Processing*, pages 6–8, Aug. 1978.

[14] H. M. Su and P. C. Yew. Efficient Doacross execution for distributed shared memory multiprocessors. Technical Report No. 1072, Center for Supercomputing Research and Development, University of Illinois at Urbana-Champaign, Jan. 1991.

THE ORGANIZATION OF THE CEDAR SYSTEM

J. Konicek   T. Tilton   A. Veidenbaum   C. Q. Zhu   E. S. Davidson
R. Downing   M. Haney   M. Sharma   P. C. Yew   P. M. Farmwald
D. Kuck   D. Lavery   R. Lindsey   D. Pointer   J. Andrews   T. Beck
T. Murphy   S. Turner   N. Warter

Center for Supercomputing Research and Development *
University of Illinois at Urbana-Champaign
Urbana, Illinois, 61801

Abstract

The Cedar multiprocessor project at the Center for Supercomputing Research and Development at the University of Illinois is a research project to study issues in parallel processing. The goal of the project has been to design and study a scalable high-performance multiprocessor system for execution of parallel programs. This paper describes the organization of the major components of the Cedar system and their implementation, and demonstrates how the design objectives were met. The scalability of the shared memory system is also discussed.

Introduction

The Cedar multiprocessor project at the Center for Supercomputing Research and Development (CSRD) at the University of Illinois is a research project to study issues in parallel processing [1]. The goal of the project has been to design and study a scalable high-performance multiprocessor system for the execution of parallel programs. Currently, a 32-processor system is operational and is running Xylem, the Cedar operating system, and a Fortran restructuring compiler described in the companion papers.

The main objectives of the project can be summarized as follows:

1. Efficient execution of parallel programs

2. An architecturally scalable design

3. Use of memory and control hierarchy

4. Utilization of Omega networks [2] for high-bandwidth interconnect

5. Efficient synchronization

6. Integration of performance evaluation facilities into the system

Implementation of the Cedar architecture began in 1985 with the selection of the Alliant FX/8 system [3] as a building block for our machine. This system is used as a *cluster* of processors and a shared memory system is added linking 4 clusters together for a total of 32 vector processors. The high level organization of the Cedar system is shown in Figure 1.

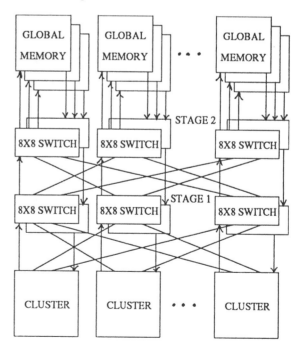

Figure 1: The Cedar System Organization.

The Cedar system has a memory hierarchy to

---

*At the time of this writing, T. Beck, E. S. Davidson, R. Downing, M. Haney, M. Sharma, and C. Q. Zhu are no longer with CSRD but have been included here because of their significant contributions to the implementation of the Cedar system while at CSRD.

exploit locality and the varying degrees of sharing exhibited by instructions and data. The memory in each Alliant system forms one such level shared by 8 processors. The shared global memory of Cedar is the second level in the hierarchy and is shared by all processors. The control hierarchy is similarly divided into cluster and global components.

The shared memory bandwidth requirements of 32 vector processors are supported by a 32-way interleaved memory and the use of two unidirectional packet-switched Omega networks, called the forward and the reverse networks, for connecting the processors to the shared memory. The forward network routes requests from the processors to the memory modules, and the reverse network returns the responses.

The fast synchronization required by parallel programs is supported by implementing synchronization instructions in the memory modules. This provides indivisibility and saves multiple memory accesses on synchronization requests.

An integrated hardware and software performance analysis environment allows measurements to be made in both single-user and multi-user modes. Collected data is filtered and sent to workstations where visualization tools are used to display the data.

The remainder of this paper presents the organization of the major components of the Cedar system and their implementation. The following sections describe the processor cluster, the shared memory system, the implementation technology, and finally, the scalability of the shared memory system.

Processor Cluster

Each Cedar processor cluster consists of an Alliant FX/8 system augmented with the Global Interface hardware (described below). The cluster organization can be seen in Figure 2. In the remainder of this section, we will describe the pertinent features of the Alliant FX/8 architecture.

The FX/8 system is an 8-processor shared memory multiprocessor. Each processor has a vector unit with eight 32-word vector registers and multiple vector functional units. The processor is fully pipelined with a cycle of 170ns. The processor is rated at 11.8 MFLOPs peak (vector) performance.

The processors are connected through a crossbar switch to a 4-way interleaved, direct-mapped, shared, 512KB cache. The shared cache is connected to the interleaved cluster memory through a split transaction memory bus. The data busses

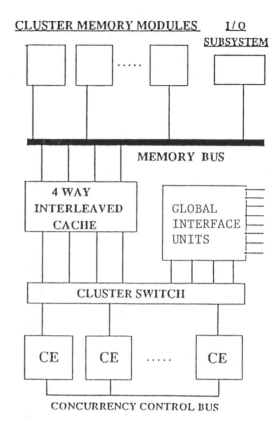

Figure 2: Cluster Organization.

are 64 bits wide. The I/O caches are also connected to the memory bus and are kept coherent with the shared cache by hardware.

Each processor contains a Concurrency Control Unit which is connected to the concurrency control bus. Instructions for starting and synchronizing the computation on multiple processors are implemented by this hardware. These instructions support execution of parallel Doall and Doacross style loops with self-scheduling of the processors.

The FX/8 supports data types ranging from 1 to 8 bytes. For reads, a processor can issue 2 requests before stalling the pipeline. Any number of writes can be outstanding. The memory is byte-addressable. Read requests always bring an 8-byte data word into the processor. Furthermore, unaligned memory accesses are allowed and the processor issues two requests if a word boundary is crossed.

The Shared Memory System

The four Cedar clusters are connected to the shared memory. The shared memory system includes the networks and the interleaved memory. Several major features of the shared memory system are described first, followed by a description of its major sub-systems.

Memory interleaving   The Cedar shared memory consists of 32 independent modules. The word size in the shared memory is 8 bytes. The address interleaving is such that consecutive word addresses access successive memory modules. The memory is byte-addressable.

Address space partitioning   The physical address space of each Cedar cluster is partitioned in two halves. The lower half describes the cluster memory address space and the upper half describes the shared memory space. Four separate cluster addresses spaces exist in the system but there is only one shared global address space. Bit 31 of the physical address distinguishes the global and cluster addresses. The virtual memory system maps the different parts of the physical address space accordingly.

Memory request ordering   To utilize pipelining in the shared memory system weak ordering of memory accesses [4] is used on Cedar. Cedar can have multiple outstanding memory requests from each processor. The only thing guaranteed by the forward Omega network and memory is that requests from a processor to a given memory address are performed in the same order as issued by the processor. Requests to different addresses by the processor are satisfied by different memory modules and can occur in any order. Requests from different processors to a given address can reach the memory module in any order.

When a synchronization instruction to shared memory is executed, the processor is stalled and the synchronization request is not issued until all read and write requests by the processor have been performed by the requested memory modules.

Synchronization   Cedar implements a set of fast, indivisible synchronization operations in shared memory based on the Zhu-Yew synchronization primitives [5]. The multiple disjoint paths in the interconnection network make it impossible to perform synchronization operations using a read-modify-write involving the processor. If the memory implemented only a TestAndSet instruction, a sequence of memory requests would be required to do the work of a Zhu-Yew primitive, increasing the time needed for synchronization and creating extra

memory traffic. For these reasons, the Cedar synchronization instructions are performed by a special processor in each memory module.

Data prefetch   The restriction on the number of outstanding read requests by each Alliant processor makes it impossible to fully utilize the shared memory system pipeline. Data prefetching is implemented to overcome this limitation. It allows the processor to initiate a block move from the shared memory and continue the execution regardless of how many read requests are outstanding.

Interconnection Networks

Each Cedar network is a two-stage packet-switched Omega network composed of 8 crossbar switches and is shown in Figure 3. The network data path

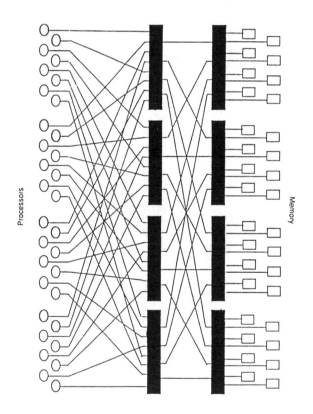

Figure 3: The Cedar Interconnection Network.

is one 64-bit word wide. A processor read request consists of one word in the forward network (address/control) and two words in the reverse network (address/control, and data). A write request is 2 words in the forward and 1 word in the reverse network. The synchronization request can be up to

3 words long in the forward and up to 2 words in the reverse network. A multiword request is sent as a single *packet* through the network.

The 8x8 crossbar switch has a 64-bit wide data path and provides a 2-deep queue per input port and a 2-deep queue per output port. Any crossbar input can be switched to any output in one system clock (85ns) when there are no conflicts. Each output port contains priority resolution logic for selecting which of the input ports simultaneously requesting the output port will be serviced.

The first word of a packet is an address/control (A/C) word. This word contains the destination address of the output port. A switch receives a packet and determines if any other input ports are requesting the same output port. If not, the packet is routed to the output port at the rate of one word per clock. The packet waits in the input queue if the requested output port is already used by another request or if this input port is not selected by priority resolution logic for routing. Once an input port queue is filled, the input port sends a *Busy* signal back to the sender until such time that the input queue has an empty slot.

For multiword packets the A/C word asserts a *Hold* signal which informs the switch to hold the route established by the A/C word for one more clock. In this way, multiple data words are transferred to the same destination. In general, for a packet of N words, *Hold* must be active for the first N-1 words. It should also be noted that a packet is indivisible. Should another request be received that requires the same output port, the current request must finish in its entirety before the new request can be honored.

A fixed-priority, non-blocking conflict resolution scheme is used in the crossbar. The priority of an input port is given by its number. When simultaneous requests conflict for a given output port, the highest priority request will be serviced first, then the next priority, and so on. However, all currently conflicting requests must be resolved before any new request can enter the arbitration. In this way, the highest priority input port cannot "starve" other input ports and prevent them from accessing a given output.

Global Interface Units

The Global Interface Unit (GIU) connects a processor to an input port of the forward network and to an output port of the reverse network. At each port, there is a queue on the GIU which can hold up to four requests. The GIU performs data prefetching and works together with the shared memory modules to execute the Cedar synchronization operations. Each GIU services one processor, so 8 GIUs are added to each cluster.

In the Alliant FX/8, memory requests are transferred between the eight processors and four cache banks by an 8 x 4 crossbar switch. For Cedar, this switch is expanded to an 8 x 8 crossbar. Each of the four new switch ports is shared by two GIUs.

Since each processor can have up to two pending requests, each GIU provides logic for managing two outstanding shared memory operations and works together with the Alliant cache to make sure that data transfers take place in the order requested by the processor. Read operations are completed when the GIU returns data to the processor. However, for write operations, the processor can continue as soon as the data is transferred to the GIU. Hence, there can be many outstanding writes in the shared memory system.

Vector Prefetch Unit   Each GIU contains a Vector Prefetch Unit (VPU). The VPU is a programmable address generator that fetches data from global memory into a direct-mapped 512-word memory called the prefetch data buffer. The VPU can issue a shared memory address every 170ns or two clocks and can have up to 512 outstanding requests.

The VPU receives the stride, mask, and length of the vector from the processor. Then the processor has to supply the starting address of the vector and the VPU fetches the specified vector into the data buffer. The processor can then read the vector using a vector instruction and the original shared memory address. The GIU detects that the shared memory operand is in the buffer and retrieves it from there.

The processor does the virtual address translation and sends physical addresses to the GIU. Therefore, when the VPU reaches a page boundary, it waits for the processor to emit the physical address of the first element to be fetched in the next page. Assuming that the processor will finish all accesses to the current page before issuing an address to the next page, the prefetch data buffer need only be one page long. For Cedar that size is 512 words or 4KB.

Each location in the prefetch data buffer has a valid bit associated with it. When data arrives from shared memory and is written to the buffer, the corresponding valid bit is set. Processor requests for elements which are not yet valid are held pending until the data arrives. All the data in the buffer for the current page is invalidated when the VPU starts to fetch the next page or when prefetching is turned off.

There are two ways for the processor to supply the starting address and begin the prefetch. It can be supplied implicitly by using a vector instruction to access the memory address of the first element in a vector. This access is held pending until the element arrives in the buffer and all successive shared memory accesses from the processor are directed to the data buffer. The second method is to supply the address explicitly using a special processor instruction. A vector instruction issued later will have its accesses directed to the data buffer. The advantage of explicit prefetching is that the processor is not held waiting until the data begins arriving from the shared memory. After starting the prefetch, it is free to do other work while waiting for the data buffer to fill. Thus, explicit prefetches can be moved up in the code to hide the shared memory latency.

Synchronization Control Unit   The Cedar synchronization instructions are performed in the shared memory. A memory module receives the instruction and operands from the processor. The result of the operation is returned to the processor. Controlling these data transfers is the job of the GIU's Synchronization Control Unit (SCU).

To initiate a synchronization operation, the opcode and operands are written to registers on the GIU by the processor. The writing of the opcode arms the SCU. Then, when the processor issues a TestAndSet instruction to a shared memory address, the SCU sends the opcode and the necessary operands to the shared memory, initiating the synchronization operation. It then waits for the result to come back. When the result returns, the processor is informed of the outcome (true or false) of the comparison operation performed by the shared memory module as described in the next section. Some synchronization operations also return additional data. This data is written to a register on the GIU for the processor to retrieve later. If the SCU is not armed when the TestAndSet is issued, then a simple TestAndSet request is sent to the shared memory.

To maintain weak ordering of memory requests, a synchronization operation cannot be initiated until all of the previously initiated writes have been completed. The GIU uses a counter to keep track of pending writes. Synchronization operations are held at the GIU until all writes have been acknowledged.

Shared Memory Modules

A memory module consists of a forward network (input) interface, memory control, a memory array with error correction circuitry, a synchronization processor, and a reverse network (output) interface. The memory has an 85ns cycle and is fully pipelined. The longest pipeline segment is the memory array which takes four 85ns cycles. Each memory module contains 2MB of memory.

The memory array and its control perform 64-bit reads, and 64-, 32-, 16-, or 8-bit writes. For unaligned or short writes it performs a read-modify-write access. Memory control also coordinates the execution of synchronization requests with the synchronization processor. Additional queuing is provided on the input and output of the memory array to buffer incoming or outgoing requests.

The input interface is responsible for receiving requests from the network, checking for errors, and queuing up to 2 words if the memory array is busy. The flow control between the forward network output port and the input interface is identical to the one used in the network.

The output interface receives a request from the memory array and sends it to the reverse network. It can queue up to 2 words if the network input port is busy.

Synchronization Processor   The processor consists of a 32-bit ALU, registers, and control. It has a cycle time of 85ns and can perform an addition, subtraction, or logical operation in one cycle. It performs the TestAndSet instruction and the Cedar synchronization instructions described below.

A Cedar synchronization instruction is an indivisible memory operation on either a 32-bit aligned word of Cedar shared memory called a *Key* or on a 64-bit word containing a *Key* and a 32-bit datum (*Key/Data* pair). Its indivisibility is due to the fact that the memory module in which the word resides is not available for other accesses until the synchronization operation is completed.

A synchronization operation is executed as follows:

1. Read the memory word at the specified address

2. Perform the specified logical test

3. If test fails terminate and notify the processor

4. Else perform the specified action

The actions can be a read or write for a *Key* or a *Key/Data* pair. An arithmetic or logical operation can be performed on a *Key* only.

The test is a comparison between a 32-bit operand (*OPERAND1*) sent by the GIU and the addressed memory word (*MWORD*). The test is *MWORD rel_op OPERAND1*, where *rel_op* can be one of the following: $<$, $<=$, $=$, $<>$, $>$, $>=$, TRUE (test always successful).

The instructions which specify an arithmetic or logical operation on the *Key* either return the *Key* value before the operation, after the operation, or just return an acknowledgment to the GIU, as specified by the instruction (e.g. pre-increment or post-increment is possible). The available synchronization processor operations are:

Arithmetic: Increment or decrement; add or subtract a 20-bit immediate value; add or subtract a 32-bit value.

Logical: OR, AND, XOR, XNOR (possibly after complementing one of the inputs) and NOT.

Other: IDENTITY (pass either operand through the ALU), SET to 0, SET to -1, NO-OP.

A synchronization instruction can have two formats: a 32-bit instruction word followed by one or two 32-bit operand words. The first operand is used in the test, the second operand is used in a 32-bit arithmetic or logical operation. The arithmetic operation can be performed without the second operand using a 20-bit immediate value. The advantage of this is that the synchronization packet from is GIU is only two words long, consisting of the address/control word plus a data word containing the instruction and first operand. 32-bit operations require a third word in the network.

Performance Monitoring

The Cedar performance monitoring system provides tracing and histogramming measurement tools for program and hardware performance analysis. It is composed of sub-systems each containing a Sun CPU board running UNIX and up to 18 CSRD-designed Tracers or Histogrammers. The GIUs, crossbar switches and memory modules have built-in performance monitoring points and can be connected to the Tracers and Histogrammers. Active probes allow connections to the cluster backplanes so that cluster performance measurements can be made. The performance monitoring hardware plugs into a VME bus [6] and is linked to the CSRD Ethernet.

The Tracer allows 20 bits of data to be collected and timestamped [7]. Tracing can be enabled or disabled through special control signals. Data can be acquired at a rate of 20 bits every 170ns. The timestamp is 40 bits and is incremented every 50ns. The timestamps of multiple Tracers can be synchronized. Tracer modules can be used in parallel to increase the trace width. Up to 4 modules can be interleaved to increase the acquisition rate and/or trace buffer depth by a factor of 4.

The Histogrammer can count the number of occurrences of 64K events. A 64K by 32-bit static RAM array stores the number of occurrences of the events. Event detection logic accepts the module's 32 data inputs and 8 data enable inputs and performs a function which generates the RAM address associated with a specific event type. The data at that location is read from RAM, loaded into a 32-bit counter, incremented, and written back to RAM. Built-in functions for generating an address are selected by programming control registers through the VME interface. Histogramming events can be acquired every 170ns. Up to 4 histogrammer modules can be interleaved to increase the acquisition rate to 42.5ns.

Hardware Implementation

Cedar is designed using a synchronous approach. A system clock of 85ns is distributed to each cluster, memory module, GIU, and crossbar switch. The clock distribution logic is implemented using 100K ECL logic and guarantees minimum skew between different boards.

In order to build very fast networks that can be scaled up in size we use twisted pair cables and differential ECL to interconnect boards. ECL is also used in the crossbar switch construction and for the control logic in the memory module where speed is of essence. The rest of the logic is implemented in TTL.

Cedar is implemented using a mixture of surface-mounted and through-hole components. Printed-circuit boards of 11, 15, and 17 layers are used and are connected to passive backplanes using 300 and 400 pin connectors.

Diagnostic hardware on each board is connected to a set diagnostic processors for error monitoring. The errors are reported to a master console. A set of diagnostic routines can exercise and isolate errors in the shared memory hardware.

GIU The GIU is implemented using mostly TTL technology with ECL used for cable drivers and receivers. The data buffer is implemented using fast SRAMs, with the valid bits stored in resetable

SRAMs. Each GIU is a single printed circuit board and resides in the modified Alliant chassis.

Crossbar Switch The crossbar is designed in 10KH ECL. Two boards implement the switch due to a large number of wires. The first switch board is designed to accept differential cable input (1280 wires) and present single-ended output (640). A second switch board provides one added level of buffering for total queue depth of 4 per stage and converts the single-ended outputs of the first board to differential cable drive.

Another component of the switch is a custom silicon device implementing the crossbar data path. This device, FXBAR, is roughly 2000 ECL gates packaged in a 148 pin PGA. It is a 4-bit wide slice of an 8x8 crossbar connection matrix with a 2-deep input queue and a one word output buffer (queue) on each port.

Memory Module Each memory module is a single printed circuit board containing 2MB of memory and the synchronization processor. The memory is implemented using 256K DRAMS and a standard single-bit-correct/double double-bit-detect error correction scheme. The synchronization processor is implemented using 4-bit wide ALUs. The data path is 64 bits wide except for the synchronization processor where it is 32 bits. The control logic and parts of input and output interface are implemented using the 10KH ECL logic. The remainder of the board is implemented using TTL logic.

Tracer and Histogrammer The Tracer is implemented on two boards; the Tracer board and the Tracer I/O board. The Tracer I/O board connects to the Tracer board through VME connectors. The Tracer board contains a VME interface and 3 tracer modules. The trace buffer of a module has 32-bit words and is populated with 1- or 4-megabit DRAMS, providing a 1- or 4-megaword trace buffer.

The Histogrammer is also implemented on two boards. Histogrammer I/O board receives and conditions all performance signals for the Histogrammer. The Histogrammer board contains a VME interface and 2 histogrammer modules.

Scalability Prediction

The performance scalability of the Cedar shared memory system can be gleaned from the simulation results presented below. The simulator uses a register-transfer level description of the actual design and varies the number of processors [8].

The experiments were run using a traffic model which consists of vectors of 32 requests issued with stride one starting from a random base address. Two-thirds of the vectors are entirely read requests, with the other one-third write requests. Each processor issues 128 such vectors. While a vector request is outstanding in the network, processors do not issue any further requests. Once all outstanding words in a vector have been satisfied then a processor issues another vector request after waiting a length of time determined by a random number selected from an exponential distribution with mean given by $1/BIR$, where BIR is the burst issue rate and ranges from 0.001 to 1.0. The BIR approximates computational delays between vector requests. For BIR=1.0 there is no computation, just vector accesses.

The systems simulated are a close approximation of the Cedar shared memory system. The networks use 8x8 switches but in some cases a stage may only use 4 of the 8 crossbar ports. The sizes of the switches used in each stage of the forward network for different configurations are listed in Table 1. The configurations for the reverse network are the same except that the memories are connected to the first stage, and the processors to the last stage.

Table 1: Network Configurations for Simulation.

| Number of Processors | Switch Size | | | Number of Memories |
	1st Stage	2nd Stage	3rd Stage	
32	8x8	4x4		32
64	8x8	8x8		64
128	8x8	4x4	4x4	128
256	8x8	8x8	4x4	256
512	8x8	8x8	8x8	512

As can be seen in Figure 4, the observable memory bandwidth scales up almost linearly with the increase in the number of processors and memory modules. The slight loss of bandwidth that is observed is due to the limited queueing in the memory and network and can be further reduced by expanding the queues. Bandwidth increases with increasing BIR until the shared memory system is "saturated" at BIR values of .032 and higher.

The latency (Figure 5) and the interarrival time (Figure 6) represent the shared memory system

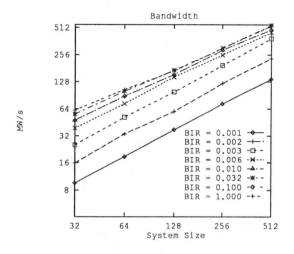

Figure 4: Bandwidth vs. System Size

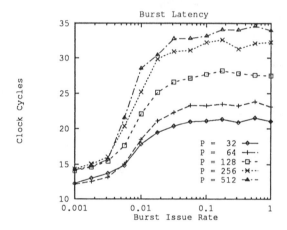

Figure 5: Burst Latency vs. Burst Issue Rate

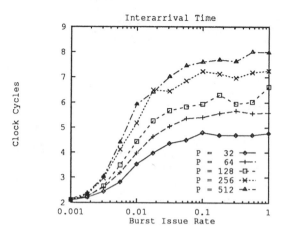

Figure 6: Interarrival Time vs. Burst Issue Rate

performance for a vector request. The latency is paid on the first element of the vector. Additional words requested arrive at intervals given by the interarrival time.

Acknowledgments

This work is supported in part by the National Science Foundation under Grant No. US NSF MIP84-10110, the U.S. Department of Energy under Grant No. US DOE DE-FG02-85ER25001.

The authors would like to thank Frank Dale, Greg Deffley, Steve Look, and Sally Hale for their excellent technical support during the construction of Cedar, and many others at CSRD who contributed directly or indirectly to the completion of this project.

References

[1] D. J. Kuck, E. S. Davidson, D. H. Lawrie, and A. H. Sameh, "Parallel supercomputing today and the Cedar approach," *Science*, vol. 231, pp. 967–974, Feb. 1986.

[2] D. H. Lawrie, "Access and alignment of data in an array processor," *IEEE Trans. Comput.*, vol. c-24, pp. 1145–1155, Dec. 1975.

[3] Alliant Computer Systems Corporation, Littleton, Massachusetts, *FX/Series Product Summary*, Oct. 1986.

[4] C. Scheurich and M. Dubois, "Correct memory operation of cached-based multiprocessors," in *Proc. 14th Annual Int'l Symp. on Computer Architecture*, June 1987.

[5] C.-Q. Zhu and P.-C. Yew, "A scheme to enforce data dependence on large multiprocessor systems," *IEEE Trans. Softw. Eng.*, vol. SE-13, no. 6, pp. 726–739, June 1987.

[6] IEEE Standard P1014, *The VMEbus Specification, Rev. C.1*, 1985.

[7] J. B. Andrews, "A hardware tracing facility for a multiprocessing supercomputer," CSRD Rep. 1009, University of Illinois, Urbana-Champaign, Illinois, May 1990.

[8] E. D. Granston, S. W. Turner, and A. V. Veidenbaum, "Design of a scalable, shared-memory system with support for burst traffic," in *Proc. 1st Annual Workshop on Scalable Shared-Memory Architectures*, Kluwer & Associates, Feb. 1991.

Restructuring Fortran Programs for Cedar *

Rudolf Eigenmann, Jay Hoeflinger, Greg Jaxon, Zhiyuan Li, David Padua
Center for Supercomputing Research & Development
University of Illinois at Urbana-Champaign
Urbana, Illinois 61801

Abstract

This paper reports on the status of the Fortran translator for the Cedar computer at the end of March, 1991. A brief description of the CEDAR FORTRAN language is followed by a discussion of the FORTRAN77 to CEDAR FORTRAN parallelizer that describes the techniques currently being implemented. A collection of experiments illustrate the effectiveness of the current implementation, and point toward new approaches to be incorporated into the system in the near future.

1 Introduction

The University of Illinois has been a pioneer in the development of program translation techniques for vector and parallel computers since the late 1960s, when Illiac IV was developed. It is therefore natural for automatic parallelization, together with language design and machine code generation, to become one of the major concerns of the Cedar project.

The most frequent use of supercomputers today is in the execution of scientific programs that are dominated by numerical algorithms. Furthermore, parallel numerical algorithms have been studied extensively and are relatively well understood. For these reasons we have used, and plan to use in the near future, mostly numerical applications to study Cedar's behavior and performance. Because Fortran is the dominant language in numerical computing today, most of the compiler effort has been devoted to the design and implementation of Fortran translators.

Even though we emphasize numerical computing, it is not our only interest. We believe that Cedar is a general-purpose computer that should also perform effectively on non-numerical problems. Therefore, some effort has been devoted to the implementation and study of the behavior of parallel symbolic programs and more work in this area is planned. To support this effort we are developing parallelizing compilers for symbolic computing languages such as LISP [10] and PROLOG, as well as for C.

Our Fortran translation system, which is shown in Figure 1, consists of two components. The back-end compiler, a modified version of the Alliant Fortran compiler,

* This work was supported by the U.S. Department of Energy under Grant No. DOE DE-FG02-85ER25001.

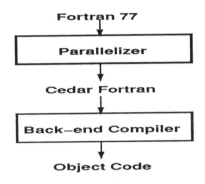

Figure 1: Fortran translation in Cedar

generates machine code for Cedar from programs written in CEDAR FORTRAN, a parallel programming dialect described in Section 2. CEDAR FORTRAN gives the programmer access to the main architectural features of the machine, including its different levels of parallelism and memory hierarchy.

A programmer who is developing a system from scratch, is also concerned with performance and is knowledgeable about parallelism and the target machine could do all the programming in CEDAR FORTRAN. In fact, many of the programs developed by the applications and algorithms researchers on the Cedar project have been written in CEDAR FORTRAN.

Some programmers, however, find it more desirable or even necessary to write in a conventional programming language such as FORTRAN77, because they are not interested in learning the machine details or investing the extra time required to develop a parallel program. Another possible reason is that the programmer may want to use existing FORTRAN77 code (sometimes called "dusty decks") instead of writing a new program, or he or she may want to build a new program using a large fraction of existing sequential code. Still another reason is that parallel source programs are cumbersome to port, in part because of the lack of widely accepted standards. This is in contrast with FORTRAN77 programs, which can be ported relatively easily to most machines (especially if portability was taken into consideration when the program was written).

The parallelizer (also referred to in this paper as the *restructurer*) in Figure 1 was developed for programmers who prefer or need to program Cedar using FORTRAN77. It is based on a 1988 version of KAP that we modified, as discussed in Section 3, to take into account those architectural characteristics of Cedar that distinguish it from other shared-memory multiprocessors. As discussed in Section 4, we are in the process of evaluating the present version of the parallelizer and studying approaches to make it more effective. Some positive results have already been obtained, but we are still far from the goal of succeeding most of the time in automatically producing effective Cedar code from sequential FORTRAN77 programs.

2 Cedar Fortran

2.1 Language Description

CEDAR FORTRAN was designed with two purposes in mind: to be the output language for the Cedar restructurer and to be a programming language for expressing parallel programs. The result is a language with only minor syntactic extensions to FORTRAN77, yet with the expressive power to make full use of the architectural features of the Cedar machine [9].

Vector operations are provided in CEDAR FORTRAN. Assignment and all arithmetic operators work for vectors as well as for scalars. Some vector reduction intrinsics are also provided, such as sum and dotproduct. Also part of the language is the FORTRAN90 WHERE statement used for doing one or more masked vector assignments.

Two types of parallel loops exist in CEDAR FORTRAN: DOACROSS and DOALL. A DOACROSS loop is called an *ordered* parallel loop, because its iterations start in the same order as they would if the loop were sequential. This guarantees correct execution if *cascade synchronization* is used in the DOACROSS loop body. In cascade synchronization, signals are passed from earlier iterations to later ones, for the purpose of maintaining a sequential ordering of the execution of a particular portion of the loop body. An example of using cascade synchronization in a DOACROSS loop may be seen in Figure 3.

The DOALL loops are called *unordered* parallel loops, because no assumptions can be made about the order in which the iterations will be executed.

The syntactic form of all these loops is similar to that of a FORTRAN77 DO loop, but they have a few extra parts as shown in Figure 2.

There are three classes of loops in CEDAR FORTRAN: cluster loops using the prefix C, spread loops using the prefix S, and cross-cluster loops using the prefix X.

CDOALL and CDOACROSS loops cause all processors on a single cluster to join in the execution of the loop body. SDOALL loops cause a single processor on each active Cedar cluster to begin execution of the loop body. CDOALL loops are often nested inside SDOALL loops to engage all processors on each active cluster. An XDOALL loop causes all processors on all active clusters to begin executing the loop body.

$$\left\{ \begin{array}{c} C \\ S \\ X \end{array} \right\} \left\{ \begin{array}{c} \texttt{DOALL} \\ \texttt{DOACROSS} \end{array} \right\} \quad index \; = \; start, \; end\,[, \; incr]$$

$$[local \; declarations]$$

$$\left[\begin{array}{c} preamble \\ \texttt{LOOP} \end{array} \right]$$

$$body$$

$$\left[\begin{array}{c} \texttt{ENDLOOP} \\ postamble \end{array} \right] \qquad \text{(only SDO or XDO)}$$

$$\texttt{END} \left\{ \begin{array}{c} C \\ S \\ X \end{array} \right\} \left\{ \begin{array}{c} \texttt{DOALL} \\ \texttt{DOACROSS} \end{array} \right\}$$

Figure 2: Concurrent loop syntax

Data declared local to CDO and XDO loops (see Figure 2) is visible to only a single processor. Each processor has its own copy of this data. Data declared local to an SDO loop is visible to all processors of a single cluster. The *preamble* of the loop is executed once, prior to execution of the loop body, by each processor that joins the loop. The loop *postamble* is not available for CDO loops, but for SDO and XDO loops. It is executed once by each processor after it finishes all its work on the loop.

```
CDOACROSS i=1,n

  c(i) = d(i) + e(i)
  g(i) = f(i) * h(i)

  call await(1,1)
  b(i) = a(i) + b(i-1)
  call advance(1)

END CDOACROSS
```

Figure 3: Cascade synchronization in DOACROSS

By default, data used in CEDAR FORTRAN programs is visible to all processors on a single cluster. CEDAR FORTRAN provides statements for explicitly declaring data to be visible to all processors on all clusters, or all processors on a single cluster.

The GLOBAL and PROCESS COMMON statements (see Figure 4) declare data that is visible to all processors on all clusters. A single copy of this data exists in global memory, and any processor with the address of one of these data items may access it.

The CLUSTER and COMMON statements declare data that is visible to all processors on a single cluster. A separate copy of this data exists in each cluster task participating in the execution of the program.

```
GLOBAL  var  [ ,  var ]  ...
CLUSTER  var  [ ,  var ]  ...

PROCESS COMMON  /  name  /  var  [ ,  var ]  ...

COMMON  /  name  /  var  [ ,  var ]  ...
```

Figure 4: CEDAR FORTRAN data declaration statements

2.2 Implementing Cedar Fortran on Cedar

2.2.1 Parallel loops

The parallel loops of CEDAR FORTRAN are all self-scheduled through a technique called microtasking [2]. CDO loops use microtasking supported by special concurrency hardware within the Alliant FX/8. This is used for dispatching iterations of CDO loops and for synchronizing between iterations of CDOACROSS loops.

SDOALL and XDOALL loops use microtasking supported by the CEDAR FORTRAN runtime library. The library starts a requested number of *helper tasks* ("implicit tasks" in IBM terminology [12]) which remain idle until an SDOALL or XDOALL loop starts. At that time, the helper tasks begin competing with the main task for iterations of the loop.

2.2.2 Tasking

Subroutine-level tasking is also supported by the CEDAR FORTRAN runtime library. In subroutine-level tasking, a new execution thread is formed for running a subroutine. When the subroutine returns, the thread ends. Two ways of doing subroutine-level tasking are available: via a new cluster task built by the operating system at the time the thread is started (via a ctskstart call), or via an already-existing helper task (the thread is started via an mtskstart call).

The ctskstart mechanism involves much higher overhead, but it allows unrestricted forms of synchronization. On the other hand, synchronization instructions are not allowed in threads started with mtskstart because of the possibility of deadlock in our implementation of the microtasking approach. This deadlock potential arises from the fact that a helper task remains associated with a thread until it completes its execution. Because no context switching is allowed, when the number of helper tasks is smaller than the number of threads, waiting for threads that have not been scheduled on any helper task may produce deadlock. On the other hand, in the right situations, the mtskstart mechanism provides a low-overhead mechanism for subroutine-level tasking, making possible the use of a finer grain of parallelism.

2.2.3 Vector prefetch from global memory

The Cedar machine provides hardware to prefetch data from global memory. The back-end compiler generates instructions to trigger the prefetch mechanism prior to the vector fetching of global data. The data is prefetched into a special buffer attached to the processor that issued the prefetch request. Once data is in the prefetch buffer, it is available to the processor at cache speed. Ideally, the prefetch trigger instruction should be placed as early in the instruction stream as possible, so that the data is in the buffer when it is needed. The back-end compiler generates a prefetch instruction for 32 elements before each vector register load instruction whose source is in global memory.

3 Automatic Parallelization

More than two decades of research in parallelization have produced many commercially available parallelizers. Some are imbedded within machine-specific compilers whereas others, notably VAST and KAP, are machine independent and have been targeted for many different machines.

These parallelizers convert sequential programs into vector/concurrent code that in some cases runs significantly faster than the original version. However, there is still much room for improvement, and as a consequence there is a need for new analysis and restructuring techniques and for a more comprehensive evaluation of the capabilities and limitations of today's techniques [14, 18].

New developments in computer architecture have spurred corresponding adaptations in parallelization. Recent examples address the Very Long Instruction Word (VLIW) and distributed memory multiprocessor architectures. Cedar extends this effort to hierarchical-memory multiprocessors. The organization of the processors into clusters and the presence of three levels of parallelism (across clusters, inside clusters, and the vector pipelined parallelism of the processors) have influenced our parallelization strategies.

Particularly important is the existence of cluster memory that, to be effectively exploited, requires the application of domain decomposition techniques automatically or by means of user assertions. Domain decomposition involves partitioning data structures and storing the parts in memory modules accessible by only a subset of the processors. Domain decomposition techniques can also be used when compiling for distributed-memory computers.

Thanks to the global memory, domain decomposition is not as critical in Cedar as in the distributed memory machines and, at least in some cases, simple data allocation strategies (e.g. the privatization algorithms discussed below) are sufficient to produce effective parallelization for Cedar.

3.1 The Cedar Fortran restructurer

CEDAR FORTRAN's restructurer is built on KAP, a proprietary Fortran restructurer from Kuck & Associates, Inc. [13]. It accepts FORTRAN77 extended with a subset of the vector operations in the FORTRAN90 standard. Our modified version also accepts assertions indicating whether a variable or array is to be stored in global or cluster memory, and whether a COMMON block is to be visible to all cluster tasks or only to one of them. The restructurer produces CEDAR FORTRAN source code as output.

The following subsections present the restructuring techniques we incorporated into KAP. Besides implementing these techniques, we further modified KAP by rewriting or extending several transformations including scalar expansion, stripmining, DOACROSS synchronization, IF to WHERE conversion, recurrence and reduction recognition, inline subroutine expansion, floating of loop bound-related calculations, and last-value assignments.

3.2 Stripmining, Globalizing, and Privatizing

The general restructuring scheme is illustrated by the following simple loop:

```
DO i=1,n
   a(i) = b(i)
END DO
```

which gets restructured into

```
GLOBAL a,b,strip,n
XDOALL i=1,n,strip
   a(i:MIN(i+strip-1,n))= b(i:MIN(i+strip-1,n))
END XDOALL
```

After detecting that the loop iterations are independent, the restructurer generates a parallel version that can be executed on all processors in Cedar. Furthermore, the iteration space has been *stripmined* [15, 18] so that in each iteration a separate *strip* of data is processed in vector form. For a given loop the optimal strip length depends on the total number of iterations and the number of processors that participate. When these quantities are not known at compile time, we use default values. Future versions of the parallelizer will be able to use strip lengths computed at runtime. In general, XDOALL and stripmining are used when only one loop in a nest is parallelized. When several nested loops are parallelized, the outermost loop is transformed into a SDOALL, and the second to the outermost is transformed into a CDOALL loop. If there are only two nested parallel loops, the innermost is also stripmined to generate vector statements.

In the translated version above, the GLOBAL declaration makes a,b,strip and n visible to all processors. This statement is generated by the *globalization* pass, which identifies the variables used in parallel loops involving processors from different clusters and then marks them as GLOBAL. Any variable used by the processors in a single cluster is marked as CLUSTER by the globalization pass.

The *privatization* pass looks for scalar variables that are not shared across iterations of a loop and marks them as local to the loop. It is worth noticing that some of the storage introduced by the compiler such as the bounds computed for the inner loop after stripmining can be kept private. An example of privatization is shown next.

```
DO i=1,n
   t = b(i)
   a(i) = sqrt(t)
END DO
```

⇓

```
GLOBAL a, b, strip, n
XDOALL i=1,n,strip
   INTEGER upper, i3
   REAL t(strip)
   i3 = MIN(strip,n-i+1)
   upper = MIN(i+strip-1,n)
   t(1:i3) = b(i:upper)
   a(i:upper) = sqrt(t(1:i3))
END XDOALL
```

Privatization is related to *scalar expansion* [20] which expands a scalar into an array if all references to the scalar in iteration i of the loop can be replaced by references to the ith element of the array. Privatizing expands the storage for the scalar to one cell per processor. The restructurer searches for the privatizable usage pattern at every level in a loop nest. It creates temporary storage using a combination of privatization and scalar expansion, which can result in local arrays whose dimensions are computed at runtime, as in the above example.

Two improvements on this technique over the present implementation will be incorporated into the restructurer in the next version. One is *array expansion* where the parallelizer recognizes that the original program's use of an array is privatizable. The other is *multilevel expansion* which privatizes or expands a variable to break storage conflicts among several levels of one loop nest. The current version of the Cedar restructurer recognizes multilevel privatizations, but can only parallelize the innermost level.

Both the globalization and privatization passes cooperate with the code that judges the best execution mode for each loop. However, an inherent difficulty of statically deciding the placement of a data item is that the decision affects the execution time of all parts of the program where the data item is used. Placing an array in global memory may benefit some parallel loops, but slow down some serial loops that cannot take advantage of the vector prefetch facilities. In most cases, these costs and benefits cannot even be calculated at compile time, yet placement must be done for every data item. Data placement choices are also complicated by EQUIVALENCE and COMMON block relations between variables.

Often the placement analysis must span procedure boundaries. The Cedar restructurer provides *inline expansion* of subroutine calls as an option to reduce the number of routine boundaries and meet some interprocedural analysis needs.

To simplify the static placement problem, there is a user-settable (global or cluster) default allocation for all data whose usage may cross a routine boundary, which we call *interface data*; this includes COMMON blocks and all formal and actual parameters in subroutine and function calls. Where no single choice is satisfactory, the programmer can force the placement of particular variables using the assertions mentioned above.

3.3 Reductions, Recurrences, and Synchronization

The Cedar restructurer recognizes loops that can be parallel if the order of their arithmetic or logical operations is allowed to change. Loops such as dot products, linear recurrences (e.g., `X(i) = X(i-1)*B(i) + C(i)`) and minimum/maximum searches are replaced by calls into a library of Cedar-optimized functions. For example, a dot product can be distributed to all Cedar processors, its partial results being summed up in two steps: within each cluster, then across the clusters. When a parallel `dotproduct` routine was used in the Conjugate Gradient algorithm [16], it improved performance by 50% over the version of the program that used `dotproduct` vectorized on one processor only.

To make use of a library routine, the restructurer must often distribute an original loop to isolate those computations done by library code, which adds loop control overhead, reduces the average grain size of parallel activity, and reduces the effectiveness of the machine's registers. The payoff comes from the wealth of algebraic and programming insight that library authors use to reduce operation counts and memory references [4].

Loops where different iterations may use the same storage cell can usually be concurrentized as `DOACROSS` loops. Uses of the shared store are serialized by the `await` and `advance` functions in the concurrency control hardware, while the rest of the loop executes in parallel. The Cedar restructurer inserts the smallest set of synchronization instructions that will suffice [17].

When considering a `DOACROSS` loop version, the restructurer lowers its estimate of the benefit owing to parallel execution by a *synchronization delay* factor. Intuitively this is the size of the synchronized region (as a fraction of one iteration) divided by the number of processors that may be executing it concurrently. The delay is reduced by reordering statements to compact the synchronized code.

3.4 Optimization Alternatives

Once parallelism has been recognized, there are still many ways that concurrent activity can be scheduled. Cedar's cluster architecture makes interprocessor communication cost less *within* a cluster than between clusters. For some loops it is not certain that a `DOALL` form could activate other clusters quickly enough to be of benefit. In others, the compiler must guess whether a `DOACROSS` could pass a synchronization signal through 8 or 32 processors fast enough to outperform the same loop distributed into serial `DO`s and parallel `DOALL`s. An understanding of how Cedar's many components really interact on significant programs is still taking shape.

To find the right match between loop levels and hardware levels, the restructurer considers a whole loop nest at one time. A central coordinator tries out many potential transformations such as how loops in a nest might be interchanged, parallelized, or stripmined, and which data must then be placed in global memory. The many sources of parallelism and synchronization in Cedar can make the number of alternatives to consider become quite large.

Currently, the restructurer uses simple heuristics to identify transformed program versions worth further consideration. A user-settable hard limit (50 by default) keeps the number of candidate versions manageable. We believe that as the number of alternatives increases, so does the number of near-optimal ones; this should allow us to keep the heuristics simple and still be confident of finding a good translation of a loop.

4 Experiments

In this section we discuss some of the experience we have accumulated in the automatic parallelization of the Perfect Benchmarks^{TM} programs as well as some linear algebra routines from *Numerical Recipes* [19]. The work reported below is part of an ongoing study whose goal is to learn about the limitations of the current version of the parallelizer as well as to develop new automatic techniques that, once incorporated in the parallelizer, would overcome these limitations.

This section is divided into two parts. In the first part we study the general ability of the restructurer to detect parallelism. This is a summary of work we have reported in [5, 6]. The results in this part are not restricted to the Cedar machine. In the second part we address some issues specific to compiling for the Cedar architecture.

4.1 Parallelism Detection

After the preliminary version of the parallelizer as described in Section 3 was completed, we started to study its effectiveness on small routines and synthetic loops. The initial results have been encouraging. Table 1 shows the speedup results for a set of linear algebra routines. The first routine is a *conjugate gradient* algorithm [16]; the other routines are from *Numerical Recipes*. The columns show an increasing input data set size which, in most routines, represents the number of rows and columns of the input matrices. The numbers report speedups of the parallelized version run on Cedar over the routine in its original serial (scalar) form. In many cases satisfactory speedups are achieved. In fact, in all but two of the routines the compiler was able to parallelize all major loops. The size of the input data set has a great influence on performance and speedup, because, as the amount of computation grows, it overcomes the negative effect due to the parallel loop overhead and to the fetching of data from global memory. Some of the routines exhibit very good speedups with relatively small data sets. Other routines start low, and their speedup is still improving when N reaches 1000. One particularly interesting case is that of routine *mprove*, which has a sharp increase in speedup when N reaches 1000. The reason for this increase is that for values of N greater than 800, the serial version, which has its data in cluster memory, exceeds the physical memory and thrashes, whereas the data of the parallel version fits in the larger global memory. Similar effects contribute to the high speedups of the CG algorithm.

Routine	Data size					
	100	200	400	600	800	1000
CG	52	87	163			
ludcmp	0.51	0.69	2.2	3.5	5.0	9.2
lubksb	0.5	1.0	2.7	4.1	5.6	6.8
sparse	7.7	9.0	8.6	23	29	
gaussj	1.3	2.8	7.0	10		
svbksb	9.0	32				
svdcmp	3.1	7.2				
mprove	0.2	1	1.2	2	8	1079
toeplz	0.6	2.8	1.1	1.0	1.3	
tridag	0.7	0.8	2.3	3.6	2.1	

Table 1: Speedups of linear algebra routines on Cedar

As part of this study, we also ran several of the programs in the Perfect Benchmarks suite, which are complete applications ranging in size from a few hundred to a few thousand lines of code. The results obtained were less satisfactory, as shown in Table 2, where the speedups obtained on the Alliant FX/80 are presented.

Program	Automatically compiled	Manually improved
ARC2D	8.0	
FLO52	10.5	16
BDNA	3.7	
DYFESM	3.6	
ADM	1.2	6.5
MDG	1.2	5.5
MG3D	1.6	
OCEAN	1.5	8.3
TRACK	1.0	5.1
TRFD	2.2	14.3

Table 2: Speedups of Perfect Benchmarks programs on Alliant FX/80

As can be seen from Table 2, in several cases practically no speedup was obtained. Our experience with these codes corresponds to that reported by many computer vendors, who have obtained for these programs a performance far below their machine's theoretical peak. By hand analyzing the restructured codes, we have found that many of the difficulties result from the general weakness of the existing restructuring technology and not from the target architecture or the algorithms used.

In our hand analysis, we examined the loops of the restructured program. If a loop was not parallelized by the restructurer, we studied the reason. If the problem resulted from limitations of the parallelizer, we tried to use more aggressive strategies and hand-parallelize the loop when possible. Throughout this process, we limited ourselves to automatable analyses and transformations rather than

pursuing a complete analysis of the application problems and their numerical solutions. We restricted our work to automatable techniques because our goal is to improve the parallelizer. Those automatable techniques that we found successful will be incorporated into later versions of the parallelizer.

Because hand-simulated mechanical analyses and transformation are cumbersome, we have run our experiments mainly on an Alliant FX/80 multiprocessor. This machine is the main building block of Cedar and has a simpler organization than the complete system. For example, codes parallelized for the FX/80 have CDOALL and CDOACROSS loops but no SDOALL and SDOACROSS loops.

Preliminary results from our experiment are encouraging. For five of the programs we have analyzed so far, the hand-transformed codes perform about five times better than the automatically restructured codes. Resulting speedup numbers for all programs range from 5 to 16 on the eight-processor machine. (Each processor of the FX/80 is equipped with pipelined vector processing units. Hence, a speedup over eight is possible.) We have ported some of these hand-transformed programs to the Cedar system. The performance was also quite encouraging: resulting speedups were up to 43 (TRFD) on the 32-processor system. The rest of this section discusses some of the automatable techniques we applied by hand to obtain the improved performance shown in Table 2.

4.1.1 Interprocedural analysis

In most of the Perfect Benchmarks programs, we found interprocedural analysis useful or even necessary. We need this type of analysis for two reasons. First, if a loop contains subroutine calls, the compiler has to assume the worst-case situation in flow analysis and data dependence analysis, which usually leads to serializing the loop. Second, the values determining the existence of dependences in one subroutine may be passed as arguments. Our compiler currently relies on inlining [11], which replaces call statements with the text of the called subroutines. However, in many cases, subroutine calls are so deeply nested in loops that the results of inlining are extremely long subroutines. Such long subroutines not only create a memory requirement problem for the compiler but also increase the complexity dramatically. However, those problems can be alleviated by using interprocedural analysis to summarize the effects of subroutine calls.

We performed interprocedural analysis manually on five programs (ADM, MDG, OCEAN, TRFD, and TRACK). One important finding is that none of the programs can be transformed into more parallelized codes by interprocedural data dependence analysis alone. Rather, an interprocedural use/def analysis [1] on both scalars and arrays is required for variable privatization or expansion, without which most of the loops containing subroutine calls cannot be parallelized. After this enhanced interprocedural analysis, we are able to parallelize some loops which consume significant portions of serial execution time of ADM,

MDG, OCEAN, and TRFD. For other important loops, including three major loops in TRACK, we used the techniques described in the following subsections.

4.1.2 DOACROSS loops and critical sections

Techniques for executing parallel loops with cross-iteration dependences have been around for many years. The Cedar restructurer can generate `await` and `advance` synchronization instructions to preserve cross-iteration dependences of simple types, thus allowing `DOACROSS` parallel loop execution. In the TRACK program, we meet a variety of problems of how to execute `DOACROSS` loops efficiently. For instance, in one loop, we find it more efficient to perform `await` synchronization conditional upon a runtime dependence test.

CEDAR FORTRAN also provides locking and unlocking to protect *unordered* critical sections. Little has been published in the literature about compiler recognition and protection of unordered critical sections. However, in at least three programs (TRACK, MDG, and TRFD) we could parallelize the most time-consuming loops using unordered critical sections.

4.1.3 Generalized induction variables

In Fortran DO loops, array subscripts often use the values of induction variables [1] which are updated in each iteration in the form of $V = V$ *op* K. Such a recursive assignment causes cross-iteration flow dependences. If a compiler can solve such recursion and rewrite each induction variable in terms of the loop indices, for example, $V = A$ op_1 I op_2 B op_3 J, where I and J are loop indices and A and B are loop invariants, then the appearance of V in array subscripts can be replaced by the expression A op_1 I op_2 B op_3 J. The recursive assignment can be eliminated as a result. There are well-known compiler techniques for recognizing and replacing an induction variable whose values form an arithmetic progression. These techniques typically deal with induction variables assigned in the form of $V = V + K$.

In our experiment with the Perfect Benchmarks code, we found induction variables whose values do not constitute arithmetic progressions. Here we call them *generalized induction variables* or GIVs. We found two types of GIVs. The first type is updated using multiplication instead of addition, thus forming a geometric progression. The second type is updated using addition, but forms no arithmetic progression nonetheless because the loops are *triangular*, that is, an inner loop limit depends on the value of an outer loop index. For both types of induction variables that we found in the Perfect code, we were able to determine the closed form expression for the value of the GIV. In the program OCEAN, one loop could be parallelized and sped up by 8.1, thanks to the recognition of the first type of induction variables (that loop performs 40% of the operations in the whole program). In the program TRFD, we found some generalized induction variables of the second type.

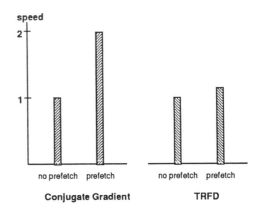

Figure 5: The effect of compiler-inserted prefetch instructions

4.1.4 Run-time dependence test

In TRFD and OCEAN we used aggressive data-dependence tests to achieve significant speedups. In several loops we used a test more accurate than the GCD test and Banerjee's test. This involved a run-time test that chooses between the serial and parallel versions of a loop.

4.2 Important optimization techniques for Cedar programs

In this section we discuss some compiler issues that are specific to the Cedar machine. Several optimization techniques are discussed, and their effects on a few programs are presented. Some of these techniques have already been implemented; while others are being studied for future inclusion in the parallelizer.

4.2.1 Prefetching data from global memory

The memory hierarchy is one of the most significant characteristics of the Cedar architecture. In the presence of such a hierarchy it is important to store and fetch data in such a way that memory access cost is kept low. This holds in particular for referencing data from the Cedar global memory.

A straightforward approach to reducing global memory access costs is to combine data requests and issue them as a block transfer to take advantage of the prefetch facility of Cedar. In CEDAR FORTRAN a natural program entity that refers to a block of data is the vector operation. Section 2.2.3 has described how the compiler inserts prefetch instructions for vector operations. Figure 5 shows the effects of this optimization in two programs, the *Conjugate Gradient(CG)* Algorithm [16] and the Perfect Benchmarks code *TRFD*.

Although there is an improvement of up to 100% in CG, TRFD exhibits only a 15% gain, primarily because vector lengths are large in CG and small in TRFD. There are many additional issues related to prefetching that we plan to study in the near future. For example, what is the effect of an aggressive floating of prefetching instructions [8]?

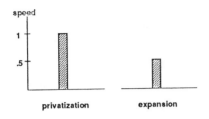

Figure 6: Data privatization vs expansion

The strategy used today in the CEDAR FORTRAN compiler is to generate prefetch code to precede each vector register load from global memory without any code motion optimizations.

4.2.2 Data privatization

Prefetching data reduces the latency for reading "remote" data. Another source of performance loss stems from contention in the shared global memory. An important architectural idea in Cedar is to overcome this problem by providing a local cluster memory, which grants faster and less contended access to data that need be seen by the given cluster only. In addition, cluster data references can benefit from the cache. A major challenge for the compiler is to find data that can be placed at this level of the memory hierarchy.

In Section 3.2 we described the compilation scheme to find data that can be *privatized* to a given loop. All privatized data gets placed in cluster memory. We have found important code sections in the Perfect Club programs where this transformation is sufficient to balance the load of the memory hierarchy. For example, in Figure 6 two variants of the major loop in program MDG are measured. The first variant has privatized array data. In the second variant the same data elements were expanded and put in global memory. The figure shows a 50% slow down of the non-privatized version. The performance loss is not only attributable to the memory placement of the data, but also to the more costly addressing mode of the data which are now expanded by one array dimension. Although the measurement does not discriminate these two sources of performance loss, it clearly demonstrates the execution speed advantage of the privatization transformation.

4.2.3 Data partitioning and distribution

Data can be privatized when its life is confined to a loop iteration. When the lifetime spans several loops, one can attempt to place data partitions onto each cluster memory and assign corresponding subsets of the loop iteration spaces to the cluster processors. This works without further communication for data that is read-only or that is read by the same cluster on which it was written. Figure 7 shows the performance of the Conjugate Gradient algorithm before and after we have applied such a simple data partitioning and privatization strategy.

The figure shows the speed of the CG relative to a

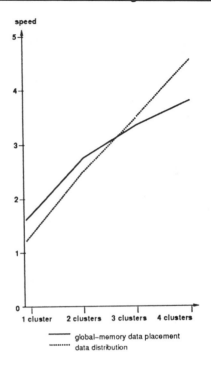

Figure 7: Data partitioning in the Conjugate Gradient Algorithm

program variant that was optimized for a 1-cluster execution and which has its data in cluster memory. The solid curve corresponds to the automatically compiled algorithm. Most data is placed in global memory. On one cluster this causes a 1.6 performance gain because of the high access bandwidth and prefetch. On two clusters the performance is nearly twice the one-cluster performance; however, on 3 and 4 clusters the speed improves less and less. We attribute this effect to the program accessing global data near the maximum transfer rate of the global memory system. The dashed curve represents the data-partitioned implementation variant. This variant has 50% of its data references localized to the cluster memory. On one cluster the speed is less than the global-data version, but then it achieves a near-linear speedup through four clusters.

We are about to implement the sketched scheme in the compiler. Yet our experiment covers a small part of what is needed to do successful data partitioning and distribution for a large range of programs. This area of research is not mature and the practical value of proposed techniques is yet unclear [3, 7, 21]. More experiments are needed. We have found that the Cedar architecture is a useful testbed for this purpose. It allows us to combine shared-memory and distributed-memory programming schemes. It lets us take advantage of newly explored data placement strategies while retaining in shared memory data whose distribution would cause intolerable communication overhead.

4.2.4 Making large concurrent loops

The cluster structure of Cedar may be considered a side effect of the memory hierarchy. However, it is important to note that within a cluster fast communication hardware is available to start, end, and synchronize parallel activities, whereas the global memory is the only mediator for inter-cluster communication. This raises the issue of providing the appropriate large grain parallelism at the program level. In loop-oriented programs, large granularity means loops with a high number of iterations and a large loop body.

Large loop iteration counts can often be obtained with large input data sets. Although current "real program" benchmark suites, such as the Perfect Benchmarks, exhibit relatively small iteration counts for many of their crucial loops, it is commonly accepted for novel highly parallel systems to refer to much larger data sets. We have shown above that linear algebra routines working on matrices of size 1000 by 1000 run quite efficiently on Cedar.

The issue of finding large bodies for concurrent loops is a challenge to the compiler. Figure 8 shows the effect of restructuring techniques that increase the size of parallel loops in the Perfect code Flo52. The major subroutine of this program consists of two loops each having a sequence of small inner loops. The first version of our compiler parallelized the inner loops only, which is represented by variant A. Variant B shows a program where the two outer loops were parallelized. In variant C these two loops were fused, thus the whole subroutine becomes one parallel loop.

The parallel loops were stripmined into CDOALL / vector loops for the Alliant FX/80 and into SDOALL / CDOALL / vector loops for Cedar.

On the Alliant FX/80 architecture the resulting performance gain amounts to 50%, whereas on Cedar, a 100% speedup results, which illustrates the difference in startup latencies between the CDO and SDO loops and shows that compiling a structure of multiple small SDOALL loops into a single SDOALL is a significant improvement on Cedar.

Our current compiler is often able to find large concurrent loops or to interchange loops to an outer position (see Section 3.4). In other cases it fails because too many potential data dependences are detected or the outer loop is in a calling subroutine. These problems constitute important issues for the ongoing project.

5 Conclusions

We have designed and implemented the CEDAR FORTRAN language. The compiler and language support software have operated reliably since the first Cedar configuration came up in mid-1988.

We have retargeted a parallelizing restructurer to automatically translate FORTRAN77 programs into CEDAR FORTRAN programs. We have extended the restructurer to cope with the challenges presented by the Cedar machine. The modified restructurer performs quite well on some linear algebra routines and synthetic loops. However, it does

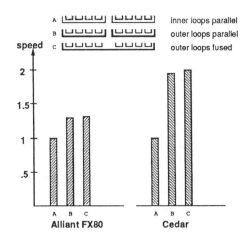

Figure 8: combining multiple parallel loops into a single one

not perform as well on some large application programs. We have engaged in an effort to study how to improve the current techniques for automatic parallelization, and in particular, how to improve our restructurer. In many cases, we find extensions to current techniques that can substantially improve the performance of the restructured code. We hope that, when those extensions are incorporated in later versions of our restructurer, we will be able to generate automatically efficient parallel code for many of the existing sequential application programs that are written in FORTRAN77.

References

[1] Alfred V. Aho, Ravi Sethi, and Jeffrey D. Ullman. *Compilers: Principles, Techniques, and Tools.* Addison-Wesley, Reading, Mass., 1986.

[2] M. Booth and K. Misegades. Microtasking: A New Way to Harness Multiprocessors. *Cray Channels,* pages 24–27, 1986.

[3] David Callahan and Ken Kennedy. Compiling programs for distributed-memory multiprocessors. *Journal of Supercomputing,* 2(2):151–169, October 1988.

[4] Steven Chen and David Kuck. Time and Parallel Processor Bounds for Linear Recurrence Systems. *IEEE Trans. on Computers,* pages 701–717, July 1975.

[5] R. Eigenmann, J. Hoeflinger, Zhiyuan Li, and D. Padua. Experiments in parallelization. Technical Report 1114, Univ. of Illinois at Urbana-Champaign, Center for Supercomp. R&D, 1991.

[6] Rudolf Eigenmann and William Blume. An effectiveness study of parallelizing compiler techniques. In *International Conference on Parallel Processing,* 1991.

[7] Kyle Gallivan, William Jalby, and Dennis Gannon. On the problem of optimizing data transfers for complex memory systems. *Proc. of 1988 Int'l. Conf. on Su-*

percomputing, *St. Malo, France*, pages 238–253, July 1988.

[8] Edward H. Gornish, Elana D. Granston, and Alexander V. Veidenbaum. Compiler-directed data prefetching in multiprocessors with memory hierarchies. *Proc. of ICS'90, Amsterdam, The Netherlands*, 1:354–368, May 1990.

[9] Mark D. Guzzi, David A. Padua, Jay P. Hoeflinger, and Duncan H. Lawrie. Cedar Fortran and other vector and parallel Fortran dialects. *Journal of Supercomputing*, pages 37–62, March 1990.

[10] W. Ludwell Harrison III and David Padua. Parcel: Project for the automatic restructuring and concurrent evaluation of lisp. *Proc. of 1988 Int'l. Conf. on Supercomputing, St. Malo, France*, pages 527–538, July 1988.

[11] Christopher Alan Huson. An in-line subroutine expander for parafrase. Master's thesis, Univ. of Illinois at Urbana-Champaign, Dept. of Computer Sci., Dec., 1982.

[12] International Business Machines Corporation. *Parallel FORTRAN: Language and Library Reference*, 1988. SC23-0431-0.

[13] Kuck & Associates, Inc., Champaign, Illinois. *KAP User's Guide*, 1988.

[14] D. J. Kuck, R. H. Kuhn, D. A. Padua, B. Leasure, and M. Wolfe. Dependence graphs and compiler optimizations.

[15] David B. Loveman. Program Improvement by Source-to-Source Transformation. *Journal of the ACM*, 24(1):121–145, January 1977.

[16] Ulrike Meier and Rudolf Eigenmann. Parallelization and Performance of Conjugate Gradient Algorithms on the Cedar hierarchical-memory Multiprocessor. In *Symposium on Principles and Practice of Parallel Programming, PPoPP*, 1991.

[17] Samuel Midkiff and David Padua. Compiler Algorithms for Synchronization. *IEEE Trans. on Computers*, C-36(12), December 1987.

[18] David A. Padua and Michael J. Wolfe. Advanced Compiler Optimizations for Supercomputers. *Communications of the ACM*, 29(12):1184–1201, December 1986.

[19] William H. Press, Brian P. Flannery, Saul A. Teukolsky, and William T. Vetterling. *Numerical Recipes: The Art of Scientific Computing (FORTRAN Version)*. Cambridge University Press, 1989.

[20] Michael J. Wolfe. *Optimizing Compilers for Supercomputers*. PhD thesis, University of Illinois, October 1982.

[21] Hans P. Zima and Michael Gerndt. SUPERB: A tool for semi-automatic MIMD/SIMD parallelization. *Parallel Computing*, 6:1–18, 1988.

Before mentioned:
Proc. of the 8th ACM Symp. on Principles of Programming Languages (POPL), pages 207–218, Jan., 1981.

The Xylem Operating System

Perry A. Emrath, Mark S. Anderson, Richard R. Barton, Robert E. McGrath

Center for Supercomputing Research and Development
University of Illinois at Urbana-Champaign
104 S. Wright, Urbana, Illinois 61801

Abstract

The Cedar machine, described in another paper in this session, provides three levels of parallelism [1]. Vectors instructions and intracluster parallelism are supported by the hardware and are directly accessible by the programmer using assembly language or explicit constructs in Cedar Fortran [2]. Intercluster parallelism is managed by the operating system, called *Xylem*, along with the C and Cedar Fortran run-time libraries. This paper will focus on the design and development of Xylem and overview its features supporting intercluster parallelism and the management of the Cedar memory hierarchy. Also, some library level tools for performance measurement and debugging will be discussed.

Introduction and Background

The Cedar system is intended to be general purpose in nature. A multiprogramming operating system is desired to effectively utilize the system in an environment for software and applications development. An evolutionary development path was chosen for the operating system. Xylem is based on Alliant Computer System's operating system, Concentrix. Concentrix is based on Berkeley's implementation of Unix. In addition to providing the desired environment, a system that users are already familiar with, Concentrix (henceforth called Unix) provides software support for multiprocessing, devices, and memory subsystems of the Alliant machine used as a Cedar building block (one cluster).

The primary goal in the design of Xylem was to provide the ability for a single application, written in the form of a single program, to effectively utilize the entire Cedar system at one time. This involves allocating processors across all clusters and providing the program with access to cluster memory as well as shared access to global memory. Meanwhile, an environment that would support much of the software base available for Unix had to be provided.

Initially, two Alliant systems were acquired. These eventually became two clusters in the Cedar machine. The system with the larger configuration was used for performance experiments and applications

software development. It quickly became evident that compatibility with the Alliant system needed to be maintained. This forced us in the direction of using the bulk of the Alliant kernel along with its scheduler. As a result, Xylem primarily consists of: 1) a number of new system calls and virtual memory functions added to Alliant's kernel, 2) conditional tests to take Xylem specific actions at strategic points, and 3) a system process that handles intercluster communications and global scheduling decisions. While each cluster services Xylem programs, they also appear to be independent Alliant systems, supporting an entire mix of Unix processes, such as shells, mailers, and networking utilities.

This approach proved successful in a number of ways. The first version of Xylem was brought up on an unmodified Alliant machine in which part of the Alliant memory was used to simulate Cedar global memory. The system behaved like a one cluster Cedar. This made it possible to debug most of the Xylem code before any of the Cedar hardware was actually available. This in turn made it possible to develop and debug the compiler, run-time support, and multi-cluster applications in a native Cedar multi-user environment.

Processes and Tasks

In Xylem, the abstraction of a Unix *process* is modified to allow a process to consist of a number of independently schedulable *cluster-tasks*. A Xylem process, akin to a Unix process, essentially represents the entire state of execution of a program. However, the process itself doesn't execute; individual cluster-tasks execute, ideally on different physical clusters. So for scheduling purposes, the cluster-task (*task*) abstraction also closely resembles a Unix process; both are schedulable and include the state of each processor in a cluster. The per-processor information includes the usual register complement, namely: address, data, floating point, vector, status, and memory management registers.

Each cluster runs a copy of the same Xylem kernel, with all the processors in a cluster sharing that copy. The original Unix variables and structures remain in cluster memory, while most Xylem variables and structures are placed in global memory, sharable by all Cedar processors. A Unix process *(proc)* table remains in each cluster. A Xylem process *(xproc)* table and xylem task *(xtask)* table reside in global memory. A Xylem process is created when the *exec()* system call is invoked on a

This work was supported in part by the U. S. Department of Energy under grant number DOE-DE-FG02-85ER25001 and the U. S. Air Force under grant number AFOSR 90-0044.

Xylem format binary file (described in the next section). The Unix process which does the exec() is effectively promoted to a Xylem process by allocating xproc and xtask table entries and moving information to global memory.

The Xylem process begins as a single task, which can then create more tasks. These become peers of the creator, all belonging to the same Xylem process; there is no hierarchy between them. They are identified by small integers, unique only within the process. Xylem system calls are available to allow any task to stop, start, wait for, or delete any other task in the process. Starting a task involves supplying it with a subroutine name to be called (program counter), an argument (typically a pointer to shared memory), and a fresh stack area. Stopping a task does not remove it from the system, but leaves it reusable via another start call. Waiting for a task causes the caller to block until the target task stops itself or is forcibly stopped.

Xylem processes and tasks, first described in [3], are similar in nature to tasks and threads in Mach [4]. However, a Xylem task includes the states of 8 independent processors, and each Xylem task may have its own private memory, inaccessible by other tasks. This design is to a large extent dictated by the nature of the Cedar architecture.

Memory Management

Like Alliant's Unix, Xylem provides a paged virtual memory environment for each process. Unlike Unix, each page (4 Kbytes) in the Xylem virtual memory system has *location* and *sharability* attributes. The location of a page indicates in which level of the physical memory hierarchy the page should reside. This can be *global* or *cluster*. The sharability of a page specifies whether the page is *shared* between all the tasks of a process, or is *private* to just one task.

The global memory is used as backing store for cluster memory, while disk is used as backing store for global memory. When a page fault occurs on a *cluster* page that has been backed out to global memory, the kernel loads it back into cluster memory before restarting the faulting instruction, rather than simply pointing the page table at the global copy. As a result, the programmer (or compiler) has explicit control over the placement of any data item.

The Xylem executable file provides the information necessary to build virtual memory structures when a Xylem process is created. The assembler and linker were modified to handle new directives and construct Xylem object and executable files. Essentially, code and variables are grouped in sections according to their attributes. Each section is a contiguous area of virtual address space that is paged into memory by the kernel according to its attributes.

If a page is shared, then every task in a process will see the same physical page at the same virtual address. A page that is private (and writable) is never shared between tasks. When a new task is created, it gets all the *shared* pages belonging to the process. When a task is started, fresh copies of private pages are loaded from the executable file, and processor stacks are re-initialized.

Pages of virtual memory can be (de)allocated dynamically via the *memctl()* system call, which takes a request code and address range as arguments. A program can create a shared heap anywhere in the virtual address space. Any task can extend the area as necessary. Other tasks automatically get access to newly allocated shared pages, and addresses can be passed between tasks since the same virtual address will always map to the same physical memory for shared pages. Task private heaps can also be allocated as long as they don't conflict with any existing shared space. More details of the Xylem virtual memory system can be found in [5].

It is also possible to use cluster memory to cache shared pages, and maintain coherency by software means. Such pages would have the shared and cluster attributes. The kernel will allow one task to access such a page, but the second task that tries to access the page while it is in the cluster memory of the first would get a memory management fault. This fault can be caught and recovery might be to change the page from cluster to global, thereby making it a single shared copy.

The notion of partial sharing, that is, allowing a page to be shared between a proper subset of the tasks in a process, while protected from the other tasks, was considered. Having the same physical memory (pages) accessible by different tasks using different virtual addresses (or access permissions) in each task was also considered. Both of these ideas have been seen elsewhere and are feasible on the Cedar hardware. However, we felt that little, if any, performance enhancement could be achieved with these features, and that they came at the expense of additional complexity, memory requirements, and implementation time. It is felt that this simpler memory model is sufficiently powerful and easier for programmers to work with, particularly when it comes to debugging.

Scheduling and the Xylem Server

The Alliant kernel has a complicated scheduler in which each processor works on processes belonging to specific classes. The use of classes ensures that a process executes only on processors meeting the requirements of the process. The process classes that a particular processor works on is table driven, and can be prioritized. Normally, information in the executable file will determine what class a process will be in. This flexibility was taken advantage of by adding a number of new classes for Xylem tasks.

There is a Xylem server process that has one task bound to each cluster. These tasks are alone in the highest priority class, and are scheduled periodically (currently 200 milliseconds). Their purpose is to update cluster status information in global memory, and to act on state changes observed in global memory. For example, a Xylem task cannot execute until a server task sees it *ready* in global memory and assigns it to its cluster. This is done with global knowledge in order to control load balance across clusters.

To schedule a task, the server allocates a Unix proc structure in the cluster, fills it with information from the Xylem process and task data stored in global memory, and puts it on the cluster ready queue. At this point, the task is scheduled by the Alliant scheduler along with existing Unix processes and previously assigned Xylem tasks.

Synchronization

The Alliant and Cedar hardware provides a test-and-set instruction which can be used on any byte in memory (cluster or global). The Cedar global memory hardware also provides special synchronization operations, described in [1]. The Xylem kernel does not presently use any of the Cedar operations, but does save and restore the necessary registers so that user programs can use them.

The Alliant kernel has internal *lock* and *unlock* functions that use the test-and-set instruction to implement critical sections. If a lock is already held, the processor spin waits. Lock structures in the kernel are prioritized to avoid deadlock. Held locks are saved on a list and are automatically unlocked and reacquired by the process context switcher, so system functions may call *sleep()* to block the current process (or task) without having to know what locks are held. These functions were extended to handle kernel lock structures in Cedar global memory, and Xylem uses these same mechanisms for implementing Cedar-wide kernel critical regions.

User programs can perform interprocessor synchronization in a number of ways. There are system calls in Xylem that provide locking and event style synchronization. These calls will block the task and allow the cluster to context switch if the lock is not available or the event is not posted. A process is aborted if all tasks block on synchronization.

Alternatively, there are similar functions in the C library, where the lock function uses the test-and-set instruction. These library routines use a combination of spinning and a system call to allow rescheduling when waiting is required. Usually, they are much more efficient than using the kernel synchronization primitives, but they can lead to wasting CPU cycles by spin waiting. Of course, a user can also explicitly use test-and-set or Cedar synchronization operations in assembly language.

Intercluster Interrupts

Xylem was initially designed to run on a Cedar system that did not have the capability to send hardware interrupts between clusters. Kernel functions were implemented to send and receive intercluster interrupts, by having software place a structure in a global message queue and set a flag in global memory for the target cluster. The flag is checked whenever the kernel is returning from supervisor mode to user mode (i.e. as part of asynchronous system trap (AST) processing).

Since the target cluster may be executing user code when the AST flag is set, the request isn't seen until the next system call or a timer driven reschedule event occurs on the target cluster. When the trap handler sees the flag set, it schedules the Xylem server. The Xylem server finds and dequeues the message and performs the requested action.

This mechanism was useful as soon as the global memory hardware became available. Now that the hardware actually supports intercluster interrupts, the functions work the same way except that they can send an interrupt to the target cluster. The target cluster processes the request after a small predictable latency.

From a user program, there is a family of system calls available for sending interrupts to other tasks. This is a software mechanism built on top of the kernel mechanisms described above. The primary function in this family is *intr_ctask()*, a request to send an interrupt. Arguments specify the target task number (-1 to broadcast), the function to be called in the target task, and an arbitrary argument for that function. The other functions are for enabling and disabling these interrupts, and for blocking until an interrupt is received (much like *sigpause()*).

Within the C run-time library, there are two areas where the intertask interrupt facilities are used. One is in *stdio* and the other is in a debugging package discussed in the next section.

In the current version of Xylem, I/O requests are restricted to the cluster where the Xylem process was created. This is in part because I/O devices are physically connected to cluster memories and controlled by processors that cannot access global memory. Major and minor device numbers, and file and inode pointers are cluster specific and must be used on the correct cluster.

A simple user level approach was taken to address this problem, involving the modification of stdio routines. Stdio buffers are placed in shared memory. The *flsbuf()* and *filbuf()* routines check the current cluster number, and if not on the original cluster, an intertask interrupt is sent to the correct cluster where the *read()* or *write()* call can be made. An acknowledgement interrupt is sent back after the kernel request completes. Locking is done on all requests except *getc* and *putc*. The locking routines are also available to the user, so that concurrent stdio requests can be sequenced correctly.

Performance and Debugging Tools

There are three Xylem tools to be described in this section. Two of them are performance measurement tools, the third is a debugging tool. The performance tools use a high-resolution clock (10 microsecond, 64 bit counter) accessible on the Alliant machines.

Xylem keeps track of real time spent by a user program and breaks it down into three categories: running, ready, and not ready. Running time is further broken down into user, system, and overhead time. System time is accumulated while in supervisor mode on behalf of the user's program, for example in the filing system. Overhead time counts that which is not directly attributable to the user's program, such as in page fault handlers. All of these timers are stored in a page that is mapped into the user's address space so that they don't need to be copied. A system call is available to ensure they are up to date.

The second performance measurement tool is a facility for timestamping each context switch a task makes. This is enabled by having the program supply a buffer to the kernel. The kernel will then record the high-resolution real-time clock each time the task is switched on or off the cluster. A separate context-switch trace can be made for each task in the program. A post-processor is used to correlate these traces so that the user can study the concurrency behavior of the program.

The debugging tool is a group of functions available in the C run-time library and an auxiliary library. During program startup, the program's signal vectors are initialized to catch exceptions such as addressing errors, illegal instructions, and floating point exceptions. When a signal is caught, a debugging monitor is entered. The first task to enter the debugger acquires a lock, and then broadcasts an interrupt to all other tasks in order to freeze them as soon as possible.

The task that acquired the debugging lock prints an error message and becomes interactive, allowing the user to examine registers and memory. When linked with the auxiliary library, the program's symbol table can be loaded, allowing symbolic (assembler level) interaction. Control can be passed to any of the other frozen tasks via an intertask interrupt. The Unix QUIT signal is also caught and sends the process into the debugging routines. This has proven useful in finding synchronization bugs that lead to infinite loops or livelock.

Initially, core dumping was modified to generate a directory of core files when a program crashed, a file for each task and a file containing shared memory. Unfortunately, no program for examining these dumps was produced, nobody seemed interested in these core files, and they tended to be very wasteful of disk space. Rather than rewrite *adb* to make it aware of multi-tasking and this new core file format, the debugging tool described above was developd to work in the context of the program itself. While this debugger is not completely isolated from user code, it requires no special support from the kernel, such as core dumps or *ptrace()*. This approach to multi-tasking debugging is proving to be quite effective.

Conclusion

The Cedar machine is a unique system today, and it requires an operating system geared toward its architecture in order to make it usable. The evolution of Unix into Xylem, described in this paper, has to some extent fulfilled this need. Since most users are already familiar with Unix, Xylem provides a comfortable, conventional environment rather than a rudimentary kernel requiring new understandings and ways of thinking.

However, because of this approach, the full performance of the Cedar machine is somewhat inhibited. In particular, effects of the virtual memory system are the least understood by developers when attempting to maximize performance. Also, the overhead involved in managing and synchronizing tasks at the operating system level limits the granularity of work that can be done in parallel effectively. Having made this tradeoff, effort continues to better characterize overall system performance, reduce overheads, and improve the memory system.

References

[1] J. Konicek, et.al., "The Organization of the Cedar System," *Proc. 1991 International Conference on Parallel Processing*.

[2] R. Eigenmann, et.al., "Restructuring Fortran Programs for Cedar," *Proc. 1991 International Conference on Parallel Processing*.

[3] P. Emrath, "Xylem: An Operating System for the Cedar Multiprocessor," *IEEE Software* (July 1985) pp 30-37.

[4] R. Rashid, et.al., "Mach: A New Kernel Foundation for Unix Development," *Proceedings of Summer Usenix 1986*.

[5] R. McGrath and P. Emrath, "Using Memory in the Cedar System," *Proc. 1987 International Conference on Supercomputing*, Springer-Verlag.

Preliminary Performance Analysis of the Cedar Multiprocessor Memory System*

K. Gallivan   W. Jalby   S. Turner   A. Veidenbaum

H. Wijshoff

Center for Supercomputing Research and Development
University of Illinois, Urbana, IL 61801

Abstract

In this paper we present some preliminary basic results on the performance of the Cedar multiprocessor memory system. Empirical results are presented and used to calibrate a memory system simulator which is then used to discuss the scalability of the system.

Introduction

Given the description of Cedar in an accompanying paper in this volume, [1], it is clear that exploiting the memory system efficiently is crucial if a reasonable fraction of the peak performance of Cedar is to be achieved in practice. This paper presents some of the results of a global memory performance characterization effort presently underway. It includes the use of a simulator, which has been calibrated based on empirical observations of Cedar memory performance, to investigate the scalability of the approach. Due to space limitations only a brief summary is given here. For more details see [2].

Memory system experiments

Description of experiments

The memory system has been explored by a generalization to the Cedar architecture of earlier work done within a cluster for the purposes of characterization and performance prediction, [3, 4]. This approach makes use of a set of parameterized memory access kernels informally referred to as the LOAD/STORE kernels. In order to predict performance on Cedar these memory system kernels must be augmented with a similar set of kernels which isolate the effect of the control constructs available on Cedar. The main parameters that were varied during the experiments are: number of computational elements (CE's) and clusters; type of request (LOAD, STORE, LOAD-STORE, etc.); mode of request (scalar, vector, etc.); temporal distribution of requests; spatial distribution of requests; and scheduling. Our study is restricted to a steady state analysis, i.e., each CE loops around a large number of times around the same piece of code (which by construction will have exactly the same pattern of memory requests). There are ways to approximate the effect of transient behavior on Cedar for performance prediction but they are beyond the scope of this paper.

*This work was supported by the Department of Energy under Grant No. DE-FG02-85ER25001, the National Science Foundation under Grant No. NSF 89-20891, the NASA Ames Research Center under Grant No. NASA NCC 2-559, Cray Research Inc. and Alliant Computer Systems.

Table 1: Parameters for experiments.

N_i	512
S_i	512
S_{bf}	32, 64, 128, 256, 512
I_{nop}	0, 128, 256, 384, 512, 640, 768, 896, 1024
	1536, 2048, 2560, 3072, 3584, 4096

The code which implements the vector-concurrent block prefetch version of the kernel is typical and comprises several nested loops. The other forms are simple modifications. The outermost loop distributes the work among the clusters, each cluster will run exactly the same code. (This loop has been implemented to minimize as much as possible the associated overhead.) Inside each cluster, a loop over N_i iterations is executed; an iteration is the basic unit of work each CE will execute and consists of the processor performing S_i memory accesses. The accesses are performed using a prefetch block size of S_{bf} and temporal distribution is controlled by the insertion of null instructions NOPs at key points in the iteration. The values of the parameters used in the experiments is given in Table 1.

Key considerations in interpretting the results below are the underlying performance limits implied by the components of the memory system and the experimental set up. Since each CE has one port to the cluster memory system, a cluster cache is designed to have a hardware limiting access rate of 8 64-bit words per 170 ns. cycle or 47 MW/s. The cluster memory bandwidth has a hardware limiting access rate of 23.5 MW/s. As a result, computations with limited locality that cannot exploit the cluster memory hierarchy tend to level off in performance when using between 4 and 5 CE's.

The components of the global memory system are such that, at least in terms of hardware limits, they balance fairly well with the cluster memory hierarchy. For a 32 CE/memory bank Cedar configuration we have the following hardware limits: the aggregate bandwidth from cluster cache is 188 MW/s, the aggregate bandwidth from cluster memory is 94 MW/s, and the bandwidth from global memory is 94 MW/s and 32 MW/s with and without prefetch respectively. (See [2] for an explanation of the derivation of the values.) If the cluster cache is taken as the reference value of 1 then the slowdown of bandwidth for the more remote portions of the hierarchy is 2 for cluster memory and 6 for global memory accessed without prefetch. If prefetch is used this is improved to 2 thereby restoring a similarity between cluster and global memory.

Another source of performance limitation is the control overhead within the CE due to experimental setup. In our experiments the multicluster control overhead has been reduced to negligible levels due to our steady state assumptions. However, the control overhead that controls the dynamic scheduling of the iterations within cluster and the looping within a CE represent overhead that is *real* in the sense that it is unavoidable any time the memory system is used in practice. We have kept this as small as possible so that any further overhead incurred in a particular practical situation may be modeled by the NOP parameter. This overhead has been estimated in two ways. The first is a static cycle count of the code generated for the experiments and the second is an empirical test using the kernels with a simple modification which eliminates multiprocessor contention and keeps only control overhead and the effect of latency on the first request of a block. Table 2 shows the empirically determined CE overhead limits to achievable bandwidths. It is these bandwidths that should be used to determine how well the global memory is doing when it is not saturated. The cycle count predictions are more optimistic but very similar.

Table 2: Basic global memory vector read overhead bandwidths in MW/s.

Total CE's	Clusters	pf=32	pf=512
1	1	3.3	4.6
2	2	6.5	9.2
3	3	9.8	13.7
4	4	13.0	18.2
8	1	26.0	35.0
16	2	51.0	70.0
24	3	75.4	105.0
32	4	90.1	140.0

Results

In this section, we present some of the results of the basic performance of the memory system. We concentrate on the LOAD kernel from the various locations in the memory system.

Basic memory performance The performance of the cluster memory hierarchy has been characterized in detail and attendant performance prediction strategies developed elsewhere, [3, 4], and will not be repeated in detail here. The observed performance of a single cluster cache is 4.5, 18 and 34 MW/s for 1, 4, and 8 CE's respectively; a single cluster memory produces 2.3, 9.3, and 12 MW/s for 1, 4, and 8 CE's respectively. This is for a stride 1 access using an iteration block of 512 (as is used for the global memory experiments) and 0 NOPS. The cache bandwidths are much more sensitive to the variation of the experimental parameters than the cluster memory bandwidths and the value of 34 MW/s is somewhat optimistic in practice. Rates in the high 20's are more typical. Note that, not surprisingly, the achieved performance is significantly

lower than the hardware limits; a factor of 2 for the memory and between 1.4 and 2 for the cache.

The performance for various Cedar configurations of the basic experimental setup, i.e., iteration block size of 512 and 0 NOPS, using vector accesses without prefetch is almost exactly as expected. For a single CE .98 MW/s is achieved and virtually ideal speedup is achieved as the number of CE's increase to 32 – 15.7 for 16 CE's and 30.0 for 32 CE's. The performance degradation relative to the observed performance of the same number of processors in the same number of clusters accessing their local caches varies from 4.4 to 4.6. Note that this is less than the degradation factor of approximately 6 implied by the hardware limits. However, due to the sensitivity of the cache access to the experimental parameters a figure of 3.5 to 4 is probably more reasonable in practice.

As the hardware limits indicate, the best way to access global memory is to make use of the implicit prefetch capabilities of the GIB/CE. The simplest mode of this type of access is when all of the CE's access locations starting in memory bank 0 and proceed using a stride of 1. Table 3 contains the bandwidths and speedups for the basic experimental configuration for the two extreme prefetch block sizes. Note that for a small number of CE's the speedup is essentially linear, as one would expect, and increasing the size of the prefetch block used can improve performance significantly. As the number of processors involved increases contention increases degradation and the influence of the size of the prefetch block diminishes rapidly. Notice also that the distribution of the CE's is important when considering the prefetch block size. When 1 CE per cluster is used the influence of the prefetch block size reduces very quickly – at 4 CE's the difference between 32 and 512 is negligible. But, when 8 CE's are used within a single cluster the difference in performance between 32 and 512 is still significant. The is due to the initial shuffle which is present in the network between the CE's and the first stage crossbars. The different distributions of a fixed number of CE's vary considerably in their amount of first stage contention.

Table 3: Basic global memory vector read bandwidths in MW/s (speedups).

Total CE's	Clusters	pf=32	pf=512
1	1	3.2 (1.0)	4.4 (1.4)
2	2	6.3 (1.9)	8.8 (2.8)
3	3	9.4 (2.9)	11.0 (3.4)
4	4	12.0 (3.8)	13.0 (4.1)
8	1	21.5 (6.7)	30.9 (9.7)
16	2	37.2 (11.6)	46.2 (14.4)
24	3	48.3 (15.0)	52.1 (16.3)
32	4	54.7 (17.1)	54.4 (17.0)

It is clear from this data that for a small number of CE's the network is not a problem and the global memory can approach the rate of the cluster cache – note 1 CE for a prefetch block of 512. (Data in the prefetch buffer on each GIB can be accessed in essentially the same amount of time as in a cluster cache.) As the CE's increase contention begins to play a role and the performance falls between that of cluster cache and cluster memory. For some points it is still significantly better than cluster memory performance, e.g., 8, 16 and 24 CE's have slowdowns relative to cache of 1.1, 1.5, and 1.9 respectively. As the number of CE's approach 32 the performance is only slightly better than cluster memory. This clearly demonstrates that in practice the prefetch capabilities of the architecture can be used effectively to offset global memory latency. One must keep in mind however, that the speed of access to global memory is not without its constraints. The prefetch unit must have a simple linear address function to work with in vector mode, i.e., vector gathers and scalar accesses can not use the prefetch buffer address generation capabilities. In this respect the cluster memory is more flexible.

Degradation mitigation   One of the issues that the experimental configuration attempts to address is the effect of the density of requests on the performance of the global memory system. We might, for example, want to answer the question: *What ratio of global and cluster accesses is needed in order to see near linear speedup in global memory bandwidth as the number of CE's increase?* or *For a fixed density of requests how many processor can the global memory system support before contention causes significant performance degradation?* This is modeled by the NOPS parameters in the templates. As the number of NOPS per iteration block increases the density of requests on the network decreases. These NOPS can be used to model a CE performing work involving data from cluster memory or from registers.

There are several ways to look at bandwidth and speedup as the sparsity of requests and the configuration of Cedar change, each of which yield a particular piece of the puzzle (see [2]). The simplest is shown in Figure 1 which plots bandwidth against CE's for various sparsity values ρ. The sparsity, ρ, is NOPS/512 and measures the number of extra cycles added per data element fetch in the block of 512 elements that a CE fetches on a single iteration. If $\rho = 0$ then there are no added dead cycles, i.e., the access is as dense as possible. As ρ increases the sparsity of the access increases and the amount of contention is decreased. To answer the first question above one would look for the value of ρ on beyond which the bandwidth increased nearly linearly. The more linear the curve the less degradation due to contention. Clearly, as ρ approaches 2 the curves have an essentially linear profile against the number of CE's. The second question is answered by noting the number of processors on each curve, P_{crit}, where the bandwidth begins to level out. The expected trend of P_{crit} being a decreasing function of ρ is evident in the figure.

The results above address the issue of balance between global and cluster memory activity. However, for a given code running a particular problem this is a fixed ratio. We next address the issue of manipulating contention in this case. We will take the worst case sparsity from above, i.e., $\rho = 0$ and attempt to improve performance of the steady state global memory bandwidth. Of course, we must still access the same locations, but we do have degrees

of freedom concerning what elements each CE will fetch. This can be controlled via the offset of the initial element accessed by a CE and the stride used for the remaining data elements. The experiments above all assumed stride 1 and an offset of 0 for all CE's. For the given network, it is easy to determine an offset for processor p of cluster c, where $0 \le p \le 7$ and $0 \le c \le 3$, such that the initial conditions of the access have minimum contention. This offset is $\sigma = 8c + p$. The stride parameter then determines how this initial condition contention evolves over time. (See [2] for a more detailed explanation of the choice of stride and offset.) Given that we have 32 CE's and memory banks, a stride of 32 would maintain this initial condition throughout the accesses and minimize contention. Unfortunately, there is a compensating effect that occurs to degrade performance as stride increases. The prefetch buffer works only with physical addresses; therefore when a prefetch buffer crosses a page boundary it must halt and wait for the CE to provide the next physical address before proceeding. The larger the stride the more page boundary crossings and the more prefetch unit stalls. Figure 2 plots the bandwidth against prefetch block size for different strides using σ as the offset in each CE. The effect of these competing trends are clear. Note the superiority of strides of 8 and 16 which strike a reasonable compromise between contention and page boundary crossing.

The performance of the global memory can also be improved by exploiting the fact that the prefetch buffer is a direct map cache with some additional restrictions. For the original 0 offset, stride 1, $p = 32$ experiments and those modified to access every prefetch block a second time after it has been loaded into the buffer we have 87 MW/s and 97.4 MW/s for prefetch block sizes of 32 and 512 respectively when the cache effect is exploited compared to approximately 54 MW/s when it is not. The results show the expected trend of a significant reduction in the cost of the second access which increases with the prefetch block size.

A global memory simulator

In this section, performance estimates obtained through simulation are used to determine the scalability of systems based on the Cedar shared-memory network. The term *scalable* is used here to describe systems whose per-processor performance is roughly constant across a range of system sizes. In our experiments, systems consisting of 32 to 512 processors connected to an equal number of memories are examined.

A register transfer level simulator was used. It was driven with traffic patterns which reflect the behavior of the LOAD/STORE kernels. This simulator has been described before, and results for a more general traffic model presented, in [5]. The simulator models the data path of Cedar's global memory system in detail. This includes the GIBs, switches in both network, and memory modules. The CEs and cluster memory systems are not modeled in detail, instead GIB prefetch block requests are used to generate the traffic patterns which load the network.

The simulation results presented are based on the vector block prefetch: all CE's participate, LOAD

Table 4: Network configuration versus system size

| Number of: | | Switch Sizes for: | |
Procs.	Mems.	Forward Network	Reverse Network
32 Cedar	32	8x4, 4x8	8x4, 4x8
32	32	8x8,4x4	8x8,4x4
64	64	8x8,8x8	8x8,8x8
128	128	8x8,8x8,2x2	8x8,8x8,2x2
256	256	8x8,8x8,4x4	8x8,8x8,4x4
512	512	8x8,8x8,8x8	8x8,8x8,8x8

Table 5: Performance results for sparsity= 0

System Size	Bandwidth (MW/s)	Block Latency	Interarrival Time
Cedar	47	25.2	5.4
32	78	15.5	2.8
64	107	18.3	4.0
128	178	24.1	4.8
256	305	32.4	5.4
512	527	31.7	6.7

requests are considered, and a block prefetch length of 32 words is used. The timing of the GIB prefetch block requests is determined by analysis of the LOAD STORE kernels. The delay between the completion of one prefetch request and the initiation of the next is determined by the loop overhead in the kernels, and the number of sparsity NOPS. The spatial distributions have no offset and a stride of one. The simulator was calibrated by comparing the bandwidth measured in several of the experiments presented previously to that obtained by simulation of the same traffic patterns on an identically configured system. For the cases examined the simulated results match the measured values with less than 10% deviation.

The configurations of the networks used, in terms of the size of the switches at each stage, are given in Table 4. Note that Cedar does not use the same switch configuration as the 32 processor system studied here. Cedar has so far implemented only part of the full network connectivity, and in the interests of generality, the full network is explored in detail here.

Three metrics are shown here. As before, aggregate bandwidth and speedups are shown to evaluate overall performance. In addition, block latency and message interarrival time are used to describe the behavior of individual block fetches. These latter metrics are determined for each individual message and the arithmetic mean over all elements is then calculated. Block latency is defined as the number of clock cycles that has elapsed between the time a processor begins issuing a block and the time that the first message in the block is received by the processor. Interarrival time is the delay between the return of successive packets in a block to the processor that issued them.

The performance results for various systems performing a vector block prefetch with a sparsity of zero are shown in Table 5. In theory, the 32 to 16 fan-in, and 16 to 32 fan-out should not limit the performance of the network, because the memories can only service requests at half the rate at which GIBs generate them. In practice, this can be seen to be untrue. By doubling the number of connections in the middle of the forward and backward networks, achieved bandwidth increases by over 50%. Only the "full" network case will be considered further.

We now consider traffic with more sparsity. In Figure 3, curves are shown which correspond to those shown in [2] for the Cedar system. All the systems achieve a good fraction of the ideal speedup for some value of sparsity, but the larger systems require larger amounts of sparsity (equivalent to locality) in order to offset their increased contention. In Figures 4 and 5 a similar trend can be seen in the interarrival time and latency values. Here the x-axis shows the size of the system on a logarithmic scale to emphasize the fact that for low sparsity performance degrades as the log of system size. As sparsity increases, the system scales better: for very high sparsity, the per-processor performance is almost constant across system sizes. The jump in latency seen as the size increases from 64 to 128 processors is due to the extra network stage needed to construct the larger systems (see Table 4).

References

[1] J. KONICEK, T. TILTON, ET AL., *The organization of the cedar system*, in Proc. 1991 International Conference on Parallel Processing, Penn State UNiversity Press, 1991.

[2] K. GALLIVAN, W. JALBY, S. TURNER, A. VEIDENBAUM, AND H. WIJSHOFF, *Preliminary performance analysis of the cedar multiprocessor memory system*, Tech. Rep. 1116, CSRD, University of Illinois at Urbana-Champaign, Urbana, Illinois, 1991.

[3] K. GALLIVAN, D. GANNON, W. JALBY, A. MALONY, AND H. WIJSHOFF, *Behavioral characterization of multiprocessor memory systems*, in Proc. 1989 ACM SIGMETRICS Conf. on Measuring and Modeling Computer Systems, New York, 1989, ACM Press, pp. 79–89.

[4] K. GALLIVAN, W. JALBY, A. MALONY, AND H. WIJSHOFF, *Performance prediction of loop constructs on multiprocessor hierarchical memory systems*, in Proc. 1989 Intl. Conf. Supercomputing, New York, 1989, ACM Press, pp. 433–442.

[5] E. D. GRANSTON, S. W. TURNER, AND A. V. VEIDENBAUM, *Design and analysis of a scalable shared-memory system with support for burst traffic*, in Proceedings of the 2nd Annual Workshop on Shared-memory Multiprocessors, ISCA-90, Kluwer and Assocs., 1991.

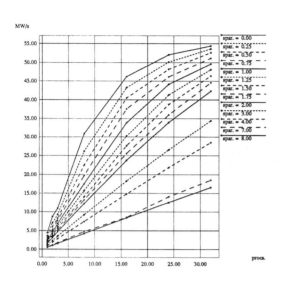

Figure 1: Implicit prefetch vector read bandwidth vs. CE's, $pf = 512$.

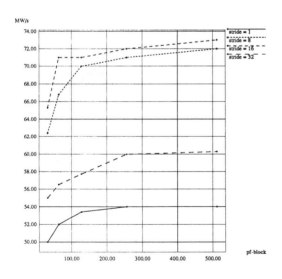

Figure 2: Bandwidth vs. prefetch size, 0 NOPS $p = 32$.

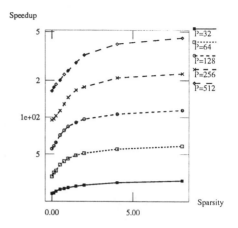

Figure 3: Speedup vs. Sparsity

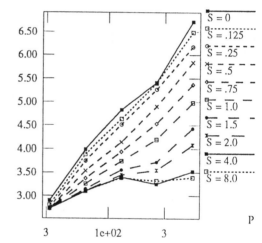

Figure 4: Interarrival Time vs. System Size

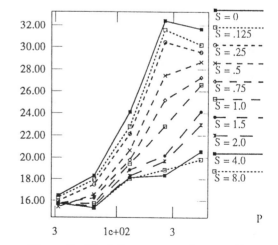

Figure 5: Block Latency vs. System Size

The Concurrent Execution of Multiple Instruction Streams
On
Superscalar Processors[1]

George E. Daddis Jr. and H.C. Torng

School of Electrical Engineering
Cornell University
Ithaca, New York 14853

ABSTRACT

We propose and evaluate an issuing mechanism to boost superscalar performance with multiple and out-of-order instruction dispatching from two or more independent instruction streams. This mechanism, an instruction window, expands and enhances the multiple instruction stream dispatching implemented in HEP.

Simulation of the MIMD and single stream processors were performed on both numeric and general computation benchmarks. The effects of instruction window size, maximum fetch count, and instruction execution times are examined.

We demonstrate through simulation a 90% - 100% increase in the processor throughput when interleaving the execution of just two instruction streams relative to the performance of an analogous single stream processor. A second benefit discovered in this MIMD processor is the increased utilization of processor resources - turbulences in an instruction stream caused by conditional branches and instruction cache faults are masked by the concurrent execution of other independent instruction streams.

The hardware required to implement the instruction window and support the execution of multiple instruction streams in a superscalar processor is outlined.

---

[1] The research reported herein has been supported in part by an IBM Graduate Fellowship and under the Joint Services Electronics Program, Contract Number F49620-90-C-0039.

I. Introduction

In this paper, we propose and evaluate an instruction issuing mechanism to boost superscalar performance: the processing of two or more independent instruction streams on a superscalar processor, creating an MIMD processor with a higher throughput than the analogous single stream processor.

The presence of multiple functional units in superscalar processors facilitates the implementation of fine-grained parallelism [1] found in an instruction stream. The exploration for concurrencies in instruction execution can be carried out statically and/or dynamically. Much work has been done on both approaches [2-9].

The focus of these investigations has been on developing efficient software and hardware mechanisms for detecting and resolving dependencies among instructions in a single instruction stream. One of the keys to enhancing the performance of such a processor is that the workload is endowed with instructions that can be executed concurrently.

A fine-grained MIMD multiprogramming approach has been implemented in HEP [10].

By interleaving the fetching and processing of two or more independent instruction streams, one would increase the average density of "independent" instructions in the instruction window. Furthermore, turbulences in an instruction stream due to conditional branches and other interrupting sources can be "masked" by other concurrently executing streams.

The presence of instructions from several instruction streams in an instruction window makes it possible to

1. attempt multiple and out-of-order instruction issuance;
2. allocate dynamically execution resources to instruction streams.

These are features which are not found in HEP implementations.

The resulting MIMD processor would consequently enjoy an increase in processor throughput. Two instruction streams are mutually independent if their data sets are non-overlapping: the streams' instructions access different portions of memory and their working register sets are separate.

A program counter for each stream processed is necessary. Also, a full register set is required for each instruction stream to be processed concurrently. Alternatively, one can pursue a dynamically allocated register scheme; in this approach, a single pool of dynamically assigned registers services all concurrently processed instruction streams. In this paper, we consider only the case of a seperate register set for each instruction stream.

In section II, we present the multiple stream dispatching in greater detail, and describe the additional hardware necessary in a typical superscalar processor. Section III provides a brief description of the instruction window [2]. Section IV contains the results of our analysis and some discussion. In section V, we conclude the paper with a summary.

II. The Multiple Independent Instruction Stream Technique

In Figure 1, we present an illustrative block diagram of a typical superscalar processor with the hardware necessary to process two independent instruction streams A and B. We provide a full set of registers for each stream. Generalization of the structure in Figure 1 to one that processes N instruction streams is clear.

The fetch unit fetches multiple instructions from the memory through the instruction cache each cycle and delivers them to the instruction window. Instructions free of dependencies are issued to free functional units where they are executed. For a full discussion of dependencies between instructions, see [11]. The functional units interact with the register sets through the interconnection network to obtain operands and store results. A possibly specialized functional unit handles accesses to memory. Conditional branch results are communicated by the functional units to the fetch unit.

One mode of operation is that instructions are fetched from both streams in an interleaved fashion: fetching alternates between stream A and B each cycle; other fetching modes are possible to accommodate different requirements for the streams. A program counter is provided for each stream, along with some control hardware. If a control dependency such as an outstanding conditional branch or an instruction cache fault suspends fetching instructions from one stream, the other stream is processed, even if it was processed in the previous cycle. If both streams are suspended, fetching is inhibited. Figure 2 provides a functional diagram of the control hardware necessary to generate the signals directing

this operation.

If one instruction stream causes an instruction cache miss, the other stream will usually provide useful work to the processor during the memory access, a situation not possible in a single stream processor. Also, when the program counter of stream A is updated by a conditional branch execution or a subroutine call, stream B may continue providing instructions. Further, in a processor with branch prediction, encountering a branch instruction leads to the access of a branch cache or Branch Target Buffer [12]. The predicted path must be determined and the program counter must be updated. All this may occur while fetching proceeds along the other stream.

As illustrated in Figure 1, a full register set is provided for each instruction stream concurrently processed. As each instruction stream is compiled for the same logical register space, a mechanism must be provided to map one stream to the second set of registers. To distinguish and easily address the registers of each set, we number the registers of the first set from 0 to N-1. The registers of the second set are numbered from N to 2N-1, where N is the number of registers in the original, single stream processor. The instruction preprocessor in the fetch unit appends a bit to the register specifiers in each instruction fetched. If the instruction was fetched from stream B, the bit is set to 1. Otherwise, the bit is set to 0. This new bit is to be interpreted as the highest order bit of the register specifier and effectively directs the register accesses of the instruction to the appropriate register set. The alteration of the register specifiers is a simple one, and may possibly be performed without the introduction of another delay or stage in the fetching pipeline.

Disadvantages with this scheme include the doubling of hardware dedicated to the register file and the increased complexity in the interconnection network between the functional units and the registers.

III. The Instruction Window

An instruction window [2] is installed in the MIMD processor. It was designed to achieve two goals:

    1. to allow multiple instruction issue at each clock cycle;
    2. to allow out-of-order instruction issue.

The Instruction Window, represented as the Instruction Stack in Figure 1, dispatches executable instructions to the functional units where they are executed. A functional unit accesses the register file and main memory for operand reads and stores, and notifies the instruction stack of completed executions. The instruction stack is fed by main memory, possibly through an instruction cache. The design of the memory-cache complex is not part of the investigation reported here.

The instruction stack is a stack of uncompleted or blocked instructions. In the stack, instructions maintain their order within the dynamic instruction stream; oldest instructions are placed at the top. Each entry in the stack is of the form:

    tag: op s1 s2 d

where "tag" is an instruction label, "op" is the operation code, "s1" and "s2" are source register specifiers or literal values, and "d" is the destination register specifier. Also present in the operand fields must be some indication as to whether the field contains a register specifier or a literal value. A field may also be unused in an instruction. The Instruction Window, or equivalently the Instruction Stack, is depicted in Figure 3.

The operation of the Instruction Stack is split into two phases: issue and compression/fetch. During the issue phase, the source and destination register specifiers of each instruction are checked against those of preceding instructions in the stack for Write-after-Read (WAR), Write-after-Write (WAW), and Read-after-Write (RAW) dependencies.

Note that the top (oldest) instruction in the stack is always executable. This dependency checking is done in hardware [13]. It is simplified by linearly addressing the registers: for each register, one address bit is provided in each operand field. Any instructions found to be free of dependencies are assigned to available functional units for execution. This completes the issue phase.

In the compression/fetch phase, completed instructions are deleted from the stack, the stack is shifted upwards to remove empty spaces, and new instructions are brought in through the instruction cache. Because instructions may not retain their original positions in the stack during their execution, their tags are provided to the functional units so that a unit may identify the instruction after execution to delete it from the stack. The two phases of issue and compression/fetch are to be executed each cycle.

When a branch instruction is encountered, the Instruction Stack stops fetching instructions from that stream. After the branch is executed and the location of the target instruction following the branch is installed into the program counter, fetching may resume.

We intend to study the maximum performance of the instruction issue mechanism in a superscalar processor fed by multiple instruction streams. Therefore, we have made the following idealistic assumptions in the emulation of the architecture described above.

First, we assume complete availability of functional units: an instruction found free of dependencies by the instruction stack always finds a free functional unit on which to execute.

Second, no blocking in the interconnection network between the functional units and the register file exists. Further, there is no possibility of blocking between the instruction stack and the functional units.

A parameter called "maximum fetch count" is set to determine the maximum number of instructions that may be fetched each cycle from main memory through the instruction cache. The instruction stack will always request as many instructions as it has empty slots, up to the maximum fetch count. A second parameter of the Instruction Stack is its size. These two parameters must be set for every performance test.

The instruction set is a simple RISC type; it assumes a register-to-register architecture. Therefore, memory accesses are made only through loads and stores. We consider two sets of execution latencies for these instructions. The "ideal" set of execution times specifies

We compare the performance of a superscalar processor executing two independent instructions streams against that of the same processor executing a single stream. To form two independent instruction streams, two benchmark programs of the same type (numeric or general) were selected and run in the interleaved fashion described in section II. Each stream accessed a different portion of memory and used logically independent register sets.

The graphs of Figures 4 and 5 represent the processor throughput plotted against the instruction window size for the maximum fetch counts of 2, 4, 6, and 8. In Figure 4, we averaged the results of the 14 Livermore Loops. There is clearly a significant increase in processor throughput over the single stream numeric workload. The graphs of Figure 5 show the analogous results for the Dhrystone Program. Again, the multiple instruction stream approach results in a tangible increase in performance.

We now define the processor speedup due to the increase in performance of the multiple instruction stream technique over the single stream case to be:

$$\text{Speedup} = \frac{\text{Throughput(Multiple Stream)} - \text{Throughput(Single Stream)}}{\text{Throughput(Single Stream)}}$$

one processor cycle for the execution of all instructions. The "real", but not real enough, set of execution times attempts to incorporate some sense of the complexity of the operations to be performed. The instruction set and the two sets of execution times are presented in the Appendix.

IV. Results and Discussion

In this section, we present the results of our computer simulation performance study. We use the 14 Livermore Loops [14] to represent a numerical computation workload and the Dhrystone Program [15] to represent a general computation workload. Unless specifically stated, all results assume an "ideal" execution latency of one cycle for all instructions. We examine the processor throughput as two parameters are varied: the instruction window size and the maximum fetch count. The processor throughput is defined as the average number of instructions executed per cycle.

In Figure 6, we plot processor speedup as a function of instruction window size for maximum fetches of 2, 4, 6, and 8, with both numeric and general computation workloads. Notice that for sufficiently large window sizes and maximum fetch count, the speedup is almost 1.0 - when processing two independent streams in an interleaved fashion, the processor executes instruction at almost twice the rate of a single stream processor.

Put in another way, if we call the time to execute instruction stream A (B) on single stream processor t_A (t_B), and the time to execute streams A and B in an interleaved fashion on a multiple instruction stream processor t_{AB}, then we have $t_{AB} \sim (t_A + t_B)/2$. In other words, the time to execute programs A and B one after another is almost twice the time to execute them both in an interleaved fashion.

Three characteristics of the instruction flow in the MIMD processor combine to produce the increase in performance over the single stream processor.

First, for a superscalar processor that executes two independent instruction streams, any single given instruction in the instruction window will be independent of roughly half the instructions in the window - those originating in the other stream, in addition to several instructions within it's own stream. This increase in the density of independent instructions in the window is clearly an important factor in the processor speedups observed. Multiple and out-of-order instruction issuances, which an HEP-like machine does not do, help to take advantage of these opportunities.

Second, we turn our attention to dependencies that exist within an instruction stream. The fetching of instructions from two or more streams inserts delays between dependent instructions, allowing the preceding instruction time to execute and therefore eliminate the dependency before the later instruction is fetched and examined. The code segment below will illustrate an example:

```
I0:     ADD     R0    #1    R0
I1:     ADD     R0    R2    R2
```

There exists a Read-After-Write dependency between instructions I0 and I1. If I0 was fetched on a given cycle, an instruction from the other stream fetched during the next cycle, and finally I1 fetched in the third cycle, then one cycle of delay has been inserted in the original instruction stream between I0 and I1. This allows time for I0 to begin and possibly finish execution before I1 enters the instruction window (assuming I0 itself is not blocked due to dependencies with previous instructions). When I1 is fetched, it may now be possible to begin it's execution

immediately. Since the processor was working on the other stream during this delay cycle, this time was spent doing useful work. The overall result is an increase in the throughput of the processor.

Lastly, as discussed in section II, turbulences in the flow of instructions of a stream caused by interruptions such as conditional branches

or instruction cache misses are masked by the concurrent execution of the other instruction stream. Cycles which would have been lost waiting for a resolution of the interruption in a single stream processor may be filled with useful work in the MIMD superscalar processor.

A close examination of the graph of the MIMD processor throughput in Figure 4 reveals occasional, slight drops in performance with increasing instruction window size. This unsteady rise in throughput with increasing window size is also reflected in the irregular shape of the speedup graph in Figure 6. These drops in performance are counter-intuitive : in the single stream processor we are guaranteed equal or better throughput with increasing instruction window size. The basis of the proof of this statement lies in the fact that a larger window size can only allow more instructions per cycle to be examined and issued each cycle. This increases the flow of instructions into the processor. What characteristics of the MIMD processor causes these occasional decreases in performance with increasing window size?

With certain window sizes and maximum fetch counts, blocks of instructions large in size relative to the window size from a single instruction stream can be fetched in one cycle. Many of the instructions fetched in such a block will not be executable for many cycles

due to dependencies on earlier instructions. These instructions sit for many cycles in the instruction window, preventing the fetching of executable instructions from the other instruction stream.

Further, when fetching large blocks of instructions, the processor forfeits its ability to split up the fetching of long dependent chains of instructions between two or more instruction streams. The beneficial delay cycles discussed above become infrequent.

These two effects combine to decrease the performance of the MIMD processor at certain combinations of instruction window size and maximum fetch count. However, in every case, the performance drops less than 5%. These decreases in performance are not as prevalent in the general computation benchmark of Figure 5 because the lengths of dependent chains of instructions in this benchmark are, on the average, smaller than those of the numeric benchmarks. Smaller dependent chains lessen the negative effects of fetching large blocks of instructions in one cycle and infrequent delay cycles inserted between consecutive instructions in a stream.

In Figure 7, the processor speedup is plotted against instruction window size in the case where increased or "real" instruction execution times are assumed. Again, note that speedups approach 1.0 for sufficiently large window sizes on both benchmark types. We conclude that the increased throughput obtained in the MIMD processor over the single stream processor is reasonably independent of the instruction execution times.

V. Conclusion

We have formulated and evaluated an instruction issuing meachanism to enhance the performance of superscalar processors; it entails processing concurrently multiple independent instruction streams. This mechanism expands and enhances the multiple instruction stream issuing capabilities that were implemented in HEP. We have demonstrated increases in processor throughput of up to 100% in the resulting MIMD processor executing two instruction streams over an analogous single instruction stream processor. In order to isolate the operation of the instruction issue mechanism, we assumed complete availability of functional units and intra-processor communication resources.

It was found that the additional fetching unit hardware necessary to process multiple instruction streams was minimal. A full register set may be provided to each instruction stream processed in the MIMD architecture. [Alternatively, a dynamic register allocation scheme may be implemented to reduce the cost incurred duplicating register sets.] No other changes in the architecture are necessary. Superscalar processors can be naturally applied to MIMD applications with significant gains in processor throughput over the original single stream processor.

Appendix

The Simple Instruction Set

Instruction	Description	Ex. Times: Real, Ideal
SUB S1 S2 D	D <- S1 - S2	1,2
ADD S1 S2 D	D <- S1 + S2	1,2
MULT S1 S2 D	D <- S1 * S2	1,4
AND S1 S2 D	D <- S1 & S2	1,2
OR S1 S2 D	D <- S1 | S2	1,2
LOAD S1 S2 D	D <- M(S1+S2)	1,2
LOADIM value D	D <- value	1,2
STORE S1 S2	M(S2)<-S1	1,2
PRINT S1	Output(S1)	1,2
JUMP S1	PC <- S1	1,2
BRNCH S1 S2 D	PC <- D if Test(S1,S2) is true. *	1,1
JSBR D	PC <-D, Return address put on stack.	1,1
RETN	PC <- Return address popped off stack	1,1
HALT	Stop fetching	1,1

Notes:

S1, S2, and D are register specifiers.

Execution times are given in processor cycles.

* The test may be specified to be Less Than, Less Than or Equal, Equal, Not Equal, Greater Than, or Greater Than or Equal.

References

[1] G.S. Tjaden and M.J. Flynn, "Detection and Parallel Execution of Independent Instructions," IEEE Transactions on Computers, Vol. C-19, No. 10, pp. 889-895, October 1970.

[2] R.D. Acosta, J. Kjelstrup, and H.C. Torng, "An Instruction Issuing Approach to Enhancing Performance in Multiple Function Unit Processors," IEEE Transactions on Computers, Vol. C-35, No. 9, pp. 815-828, September 1986.

[3] L. Wang and C. L. Wu, "Distributed Instruction Set Architecture", private communication, to appear in IEEE Transactions on Computers.

[4] W. Hwu and Y.N. Patt, "HPSm, a High Performance Restricted Data Flow Architecture Having Minimal Functionality," 13th International symposium on Computer Architectures Conference Proceedings, pp. 297-306, June 1986.

[5] - , "80960 Hardware Designer's Reference Manual," Intel Corporation, 1988.

[6] J.A. Fisher, "Very Long Instruction Word Architectures and the ELI-512," Proceedings of the 10th Annual International Symposium on Computer Architectures," pp. 140-150, June 1983.

[7] K. Ebcioglu, "Some Design Ideas for a VLIW Architecture for Sequential-Natured Software," Proceedings of the IFIP WG 10.3 Working Conference on Parallel Processing, Pisa Italy, pp. 3-21, April 1988.

[8] H. B. Bakoglu, G. F. Grohoski, and R. K. Montoye, "The IBM RISC System/6000 processor: Hardware overview", IBM Journal of Research and Development, 34, 1, pp. 12-22, January 1990.

[9] K. Murakami, N. Irie, M. Kuga, and S. Tomita, "SIMP (Single Instruction Stream/Multiple Instruction Pipelining): A Novel High-Speed Single-Processor Architecture," 16th International Symposium on Computer Architectures Conference Proceedings, pp. 78-85, 1989.

[10] H. F. Jordan, "Performance Measurements on HEP - A Pipelined MIMD Computer", The 10th International Symposium on Computer Architectures Conference Proceedings, pp. 207-212, June, 1983.

[11] R.M. Keller, "Parallel Program Schemata and Maximal Parallelism I. Fundamental Results," Journal of the Association for Computing Machinery," Vol. 20, No. 3, pp. 514-537, July 1973; "Parallel Program Schemata and Maximal Parallelism II. Construction of Closures," Journal of the Association for Computing Machinery," Vol. 20, No. 4, pp. 696-710, October 1973.

[12] J.K.F. Lee and A. J. Smith, "Branch Prediction Strategies and Branch Target Buffer Design," IEEE Computer, Vol. 17, No. 1, pp. 6-22, January 1984.

[13] H. Dwyer and H. C. Torng, "A Fast Instruction Despatch Unit for Multiple and Out-of-Sequence Issuances," Technical Report EE-CEG-87-15, School of Electrical Engineering, Cornell University, Ithaca, NY, November 1987.

[14] E.M. Riganati and P.B. Schnek, "Supercomputing," IEEE Computer, Vol. 17, No. 10, pp. 97-113, October 1984.

[15] R.D. Weicker, "Dhrystone, A Synthetic systems Programming Benchmark," Communications of the ACM, Vol. 27, No. 10, pp. 1013-1030, October 1984.

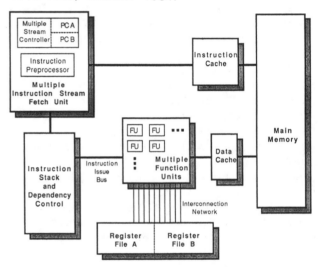

Figure 1. Out-of-Order Execution, Multiple Function Unit Processor Block Diagram

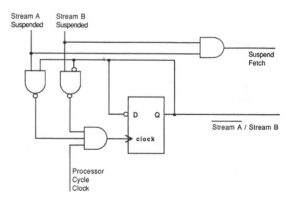

Figure 2. Multiple Stream Controller Hardware

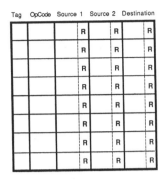

Figure 3. The Dispatch Stack Instruction Stack

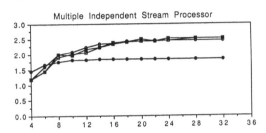

Figure 4. Multiple and Single Stream Throughputs on Numeric Computation

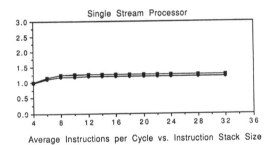

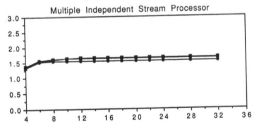

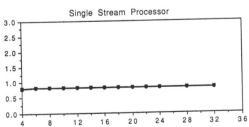

Figure 5. Multiple and Single Stream Throughputs on General Computation

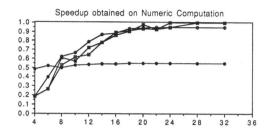

Figure 6. Speedups on General and Numeric Computations

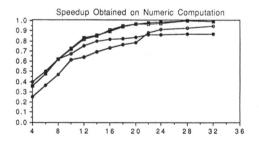

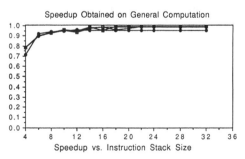

Figure 7. Speedups on General and Numeric Computations with Increased Instruction Execution Times

A Benchmark Evaluation of a Multi-Threaded RISC Processor Architecture*

R.Guru Prasadh and Chuan-lin Wu
Dept. of Electrical and Computer Engineering
The University of Texas at Austin
Austin, TX 78712

Abstract: *This paper presents results from a benchmark evaluation of a multi-threaded RISC architecture model. The RISC processor supports multiple instruction threads of a single program or multiple programs in a uniprocessor environment. The architecture model consists of pipelined, dedicated functional units supporting multiple instruction issuing in every clock cycle. The novelty of the architecture is its dynamic interleaving feature whereby instructions from different threads are interleaved at run time and issued to the functional units, allowing for concurrent thread execution. A performance evaluation indicates that a four-threaded, eight functional unit processor is capable of achieving up to a factor of eight speed-up over a sequential machine.*

Keywords and phrases: dynamic interleaving, multi-threaded execution, MIMD, RISC architecture, parallel instructions.

1. Introduction

The performance of microprocessors has been improved significantly by reducing the instruction execution time and by concurrently executing more than one instruction per cycle, as many RISC superscalar microprocessors are currently designed to do. Today's commercial microprocessors have already achieved a good fraction of the performance level available through supercomputer systems. Furthermore, a significant increase in the clock rate and gate counts in a single chip expected in the near future make microprocessor technology a very unique tool for obtaining additional supercomputation power in a cost effective way. However, the single-threaded pipelined instruction issuing architectures used in the current superscalar microprocessors, like the i860 [1] and the MC88100 [2], will no longer significantly increase the computational power. First, utmost one instruction is issued per cycle in current microprocessors. It is also an established fact that the amount of parallelism which exists in a single instruction thread is limited by the data and control dependencies between the instructions. The dependencies slow down the instruction issuing rate and lead to a poor functional unit utilization. While one functional unit is busy, others may be idle, waiting for the results from the busy unit. One way of improving the utilization is to interleave instruction threads and to dispatch them to the functional units. To this end, we came up with the current approach of multiple-instruction-thread execution in which instruction threads generated from a single program or from different programs are executed concurrently on a single processor.

Studies on multiple-context execution and multi-threaded architectures proposed in the past relate to multiprocessors with homogeneous CPUs [3,4,5] and are language specific [6]. Our architecture, on the other hand, is a general purpose RISC architecture adapted for multiple instruction threads and multiple instruction issue. It uses a software dataflow control method to parallelize and schedule code in a manner similar to the VLIW approach [7,8]. The novelty of the architecture, however, lies in its *dynamic interleaving* feature in which instructions from different threads are combined at run time and issued concurrently. The goal of the current study is to evaluate this multi-threaded superscalar processor architecture using benchmark programs. The simulation results will help to tune the parameters of the processor design.

The paper is organized as follows. The next section describes the processor architecture. Section 3 explains the dynamic interleaving scheme employed by the architecture. Then section 4 details the simulation model employed in evaluating its performance and reports the results of the simulation experiments. Section 5 discusses related works. Finally, section 6 concludes the paper.

2. Processor Architecture Model

The upper bound on the instruction issue rate for a program is related to its dataflow limit and is generally unachieveable for most machine organizations [9]. Satisfying the dataflow instruction issue rate would, no doubt, lead to higher throughput and performance but would require an extensive set of resources resulting in poor resource utilization for most applications. On the other hand, higher resource utilization can be achieved using fewer resources but at the cost of reduced performance. Our architecture works out a compromise between system throughput (performance) and resource utilization. By providing multiple functional units, the instruction issue rate and, hence, the system throughput is enhanced. To improve the resource utilization, multiple instruction threads are supported. Using a *dynamic interleaving* technique, instructions from different threads are combined at run-time and issued to the functional units. This keeps the functional units busy and improves the resource utilization.

The processor under study is a multi-functional unit RISC system that supports multiple instruction issues per cycle. In our architecture, the functions of the CPU are decoupled and distributed among a set of heterogeneous functional units (see figure 1). The functional units operate independent of one another and are dedicated to performing exclusive tasks. Instruction scheduling at the various functional units is done in

---

* This work is supported in part by the National Science Foundation under grant MIP-8819652 and by Cray Research Incorporated.

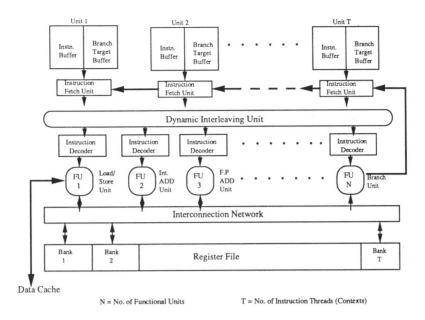

Figure 1. The Processor Architecture Model

software by a parallelizing compiler in such a way that instructions are issued to the functional units as soon as the operands become available. Thus our architecture presents a superscalar dataflow model of computation with decentralized control.

Instruction Sequencing and Execution

The Dynamic Interleaving Unit, DIU, controls the sequencing and issue of instructions to the functional units (see figure 1). Its operation is discussed in a later section. Dynamic interleaving requires the simultaneous fetching of instructions from different threads. To facilitate this, a separate instruction buffer is provided for each of the threads. The Branch Target Buffer, BTB, holds the first few lines of a branch target to reduce branching delays. The compiler schedules the instructions statically at the various functional units. The instruction buffer thus contains pre-sorted and temporally ordered instructions. In every clock cycle, the instruction fetch unit, under the control of the DIU and the branch unit, fetches the next instruction word from the instruction buffers.

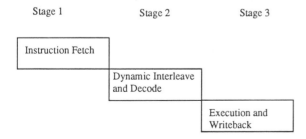

Figure 2. The Instruction Execution Pipeline

Each instruction passes through three main pipelined stages: the instruction pre-fetch stage, the dynamic interleave/decode stage, and the execution/writeback stage (see figure 2). In the pre-fetch stage, the instruction is fetched from the instruction buffer. In the dynamic interleave/decode stage, it is interleaved with instructions from other streams, decoded, and then issued to the functional units. In the third stage, the instruction is executed by the functional unit and the results are written back into the register file. For instructions requiring multiple clock cycles to execute the corresponding functional units are pipelined to accept a new instruction in every cycle. The compiler, by judicious code scheduling, prevents register *write* contentions between instructions in a thread.

Pipeline synchronization is maintained using exceptions. Since instructions from a thread are distributed among the various functional units, if any of the instructions cause an exception, all the functional unit states need to be frozen in order to maintain synchronization. The exception handler identifies the thread that caused the exception, saves the states of all the partially executed instructions of that thread, and removes them from the functional unit pipelines. After making the necessary corrections, the exception handler restarts the functional unit pipelines. Depending upon the type of exception taken, the exception handler may suspend the thread causing the exception and introduce other threads, or allow it to continue execution. For example, on a cache miss, the load/store unit signals a cache miss exception and indicates the thread causing it. The corresponding handler then saves the state of the partially executed instructions of that thread and suspends its execution. Execution of the other threads is restarted. When the requested data becomes available in the cache, the suspended thread is re-introduced in the functional unit pipelines.

Synchronization between threads generated from the same program is achieved through the use of fork/join primitives. Whenever a *fork* instruction is encountered, a new thread of that program is introduced in the execution pipeline. A *join* instruction, on the other hand, terminates that thread.

3. Dynamic Interleaving Model

Dynamic Interleaving is a technique to improve the rate of instruction issues to the functional units. In every clock cycle the Dynamic Interleaving Unit, DIU (see figure 1), attempts to issue the maximal number of instructions to the functional units by interleaving instructions from different threads. An instruction thread is defined as a set of instructions belonging to a particular context that can be independently executed of other instruction threads. Instruction threads can be generated from a single program that exhibits sufficient parallelism or from different programs. Data and control dependencies between instructions in a single thread prevent the simultaneous issue of instructions to all the functional units in every clock cycle. However, instructions from different threads are independent of each other and can be issued concurrently. *Dynamic Interleaving* is based on this main idea.

The dynamic interleaving process is described below. The DIU selects a thread according to a priority scheme (usually round-robin). It examines the instruction word of that thread and replaces, if possible, every one of its "NOOP" instructions (introduced as a result of static scheduling by the compiler) with a corresponding "non-NOOP" instruction from another thread. The newly assembled instruction word is now sent to the decode stage. The instruction decoder at each of the functional units identifies the thread to which the instruction belongs and generates the appropriate control signals. At the end of a dynamic interleaving operation, one or more instruction words, pre-fetched earlier into the DIU, is issued to the functional units. The DIU signals to the instruction fetch units of the corresponding threads to fetch the next instruction words in the following clock cycle. For all other threads, instruction words fetched earlier are still present in the DIU and are used in the dynamic interleaving process in subsequent clock cycles.

Figure 3 shows an example illustrating *dynamic interleaving*. In the example, four functional units and three instruction threads are assumed for simplicity. The instructions shown in the instruction buffers are scheduled by the compiler. We assume that a round-robin strategy is employed in selecting the threads for interleaving, and that the threads have their own register set. At CK=1, the first instruction word from the three threads is fetched from its buffer into the DIU. Following the round-robin strategy, thread 1 is selected first (shown in bold letters in the figure). The ADD2 instruction of this thread is sent to the integer unit decoder. There are no more instructions in thread 1. Now thread 2 is selected. Since the ADD2 instruction of thread 1 has already been issued to the integer unit, the issue of the ADD2 instruction of thread 2 is deferred until the next clock cycle. However, the logic unit is free. Hence, the SHLL2 instruction of thread 2 is sent to the logic unit decoder. Since there are no more instructions in this thread, the third thread is selected. The FMOVEF instruction of this thread is issued to the decoder of the FP<->Int convert unit since it is free. No more instructions can be issued now. Thus, at CK=1, an ADD2 instruction from thread 1, a SHLL2 instruction from thread 2, and a FMOVEF instruction from thread 3 are issued simultaneously to the functional unit decoders.

At the end of the first clock cycle, threads 1 and 3 have no more instructions in the DIU buffer. Hence, the next instruction words from these threads are fetched from their respective instruction buffers. Thread 2, on the other hand, still has an ADD2 instruction to be issued and, therefore, the

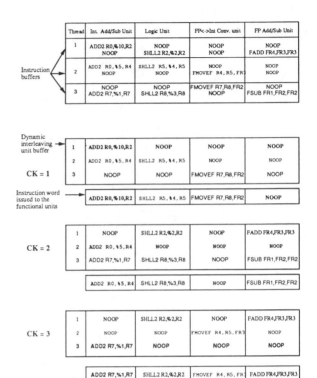

Figure 3. Dynamic Interleaving Example

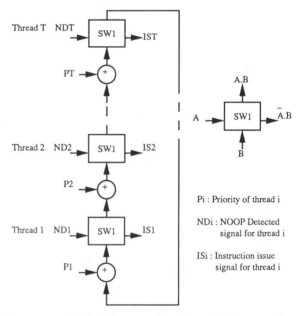

Figure 4. Logic Diagram for Dynamic Interleaving

next instruction word is not fetched. At CK=2, thread 2 is selected first, following the round-robin strategy. Its ADD2 instruction is sent to the integer unit decoder. Next, thread 3 is selected and the SHLL2 and FSUB instructions from this thread are issued to the logic and FP Add/Sub units, respectively. Finally, thread 1 is selected; but no instruction from thread 1 can be issued since the required functional unit decoders are full. Proceeding in a similar fashion, we see that by the end of the third clock, all the instructions from the three threads are issued. In the absence of *dynamic interleaving*, it would take six clock cycles to issue the instructions in the example. *Dynamic interleaving*, thus, improves the instruction issue rate by a factor of two in the example.

We now explain how to implement dynamic interleaving. The basic operation, performed in the DIU, is the partial decoding of an instruction to see if it is a NOOP instruction. A non-NOOP instruction is issued to its functional unit. But if the instruction is a NOOP, then the instruction from the next thread has to be examined. The checking is continued (in a domino fashion) until a non-NOOP instruction is encountered or until all the threads have been exhausted. Figure 4 describes a logic circuit to achieve this. In the figure, SW1 is a logic switch whose function is as shown. ND is the "NOOP Detected" signal, the result of partial instruction decoding. P is the priority signal. In any clock cycle, only one thread has its priority signal high. IS is the instruction issue signal. When high, it indicates that the instruction from the corresponding thread will be issued to its functional unit. The DIU has a logic circuit, like the one in figure 4, for every functional unit.

Table I. Functional Units and their Latencies

SN	Functional Unit	Latency (cycles)
1	Integer Add/Sub Unit	1
2	Logic Unit	1
3	Load/Store Unit	3+WC*
4	FP <-> Int Convert Unit	5
5	FP Add/Sub Unit	5
6	FP Multiply Unit **	6
7	FP Divide Unit **	59
8	Branch Unit	1

* WC - Wait Cycles in case of a cache miss

** Integer Multiply and divide also included

4. Benchmark Evaluation

Simulation Model

The Processor
The processor is modelled as a RISC machine composed of eight functional units (see table I). These include an integer unit that performs data move and integer add and subtract operations; a logic unit responsible for bit-field operations; a load/store unit that performs memory reads and writes; an integer/floating-point convert unit to do data-type conversions; a floating-point adder; a floating-point multiplier; a floating-point divide unit; and a branch unit. All the units are effectively pipelined and are capable of accepting a new instruction in every cycle. The simulator provides for the existence of multiple instances of a particular functional unit in a processor configuration (see table II).

The Register Set
Every instruction thread has a private register set consisting of thirty-two 32-bit integer and sixteen 64-bit floating-point registers. The integer, logic, load/store, and branch units can access only the integer registers while the floating-point units are restricted to using only the floating-point registers. Only the integer/floating-point convert unit can access registers of either type and performs all data transfers between the integer and floating-point registers.

The Instruction Set
The instruction set is a subset of the RISC instruction set developed for the Distributed Instruction Set Computer [10]. It consists of forty-nine instructions defined orthogonally in three formats, e.g., 3-operand, 2-operand, and 1-operand. All the instructions are 32-bits long.

The Compiler
The compiler is composed of two parts. Its front end is a gnu-generated 'C' compiler that transforms a 'C' program into sequential assembly code. The back end is a parallelizing compiler that converts the sequential code into parallel code vectors. The parallel code vector is composed of different sections where each section corresponds to an instruction specific to a functional unit. The compiler generates the parallel code vector by combining instructions that do not have data dependencies between them. Thus, the parallel code vector is

Table II. Processor Configurations used in the Simulation Experiments

Functional Unit	Processor Configuration												
	8	9.1	9.2	. 9.3	9.4	9.6	9.8	10.1	10.11	10.12	10.15	10.17	10.35
Integer Add/Sub Unit	1	2	1	1	1	1	1	3	2	2	2	2	1
Logic Unit	1	1	2	1	1	1	1	1	2	1	1	1	1
Load/Store Unit	1	1	1	2	1	1	1	1	1	2	1	1	2
FP <-> Int Convert Unit	1	1	1	1	2	1	1	1	1	1	1	1	1
FP Add/Sub Unit	1	1	1	1	1	1	1	1	1	1	1	1	1
FP Multiply Unit	1	1	1	1	1	2	1	1	1	1	2	1	1
FP Divide Unit	1	1	1	1	1	1	1	1	1	1	1	1	1
Branch Unit	1	1	1	1	1	1	2	1	1	1	1	2	2

formed by concatenating data-independent instructions that can be issued in the same clock cycle. In the process of parallelization, the compiler is forced to insert NOOP instructions in the parallel code vector.

Dynamic Interleaving

In every clock cycle, the dynamic interleaving unit examines parallel instruction vectors from every thread and tries to combine them into a single instruction vector to be issued to the functional units in the next clock cycle. It does this by replacing NOOP instructions in an instruction vector belonging to a thread with corresponding sections from instruction vectors of other threads. A round-robin strategy is adopted in selecting the threads for instruction issue to prevent the starvation of any particular thread.

Caches and Memory

Separate caches exist for instructions and data. For fast instruction fetching, a separate instruction cache is provided for each thread. The data cache, however, is shared between the threads. An infinite cache is assumed in our simulation experiments. The load/store unit performs memory accesses with a total latency of three clock cycles. However, being pipelined, the load/store unit is capable of initiating a new operation (data fetch or store) in every clock cycle.

Table III. The Benchmark Test Suite

Numerical Benchmarks (Livermoore loops)	
loop1	Hydro excerpt
loop2	Inner product (loop unrolling)
loop3	Inner product (no loop unrolling)
loop4	Banded linear equations
loop5	Tri-diagonal elimination, lower
loop6	Tri-diagonal elimination, upper
loop7	Equation of state excerpt
loop8	P.D.E. integration
loop9	Integrate predictors
loop10	DIfference predictors
loop11	First sum
loop12	First difference
loop13	2-D particle pusher
loop14	1-D particle pusher
loop15	Modified livermoore loop 22
Non-numerical Benchmarks	
bubble	Bubble sort of 100 elements
queen	Eight queens chess problem
qsort	Quick sort of 1000 elements
tree	A binary tree search program

Benchmark Programs

As Table III shows, the original fourteen Lawrence Livermore loops are used as benchmark programs to evaluate the numerical performance [11]. They represent computation intensive applications. A modified version of Livermore loop 22 is used to evaluate the performance of the floating-point divide functional unit. In addition, four non-computation intensive programs form the benchmark test suite for the non-numerical performance. Of these, two are bubblesort and quicksort programs that sort 100 and 1000 random numbers, respectively. The other two non-numerical benchmarks include a recursive binary tree search and a program to solve the eight queens chess problem. The non-numerical test suite represents control-intensive applications, which are highly recursive programs. Multi-threaded execution is simulated by executing the benchmark programs in various combinations.

Results

Simulations were performed on thirteen processor configurations, as table II shows. For each processor configuration, the benchmarks were compiled separately. The simulations were run on a Sun SPARC station 1.

Functional Unit Utilization

Functional unit utilization is reflected in the instruction issue rate. Higher instruction issue rates imply better functional unit utilization. Figure 5 plots the variation in the instruction issue rate for numerical benchmarks as the number of instruction threads increases. The instruction issue rate and hence the functional unit utilization is an almost linear function of the number of threads. However, as the number of threads increases, the instruction issue rate begins to saturate. This is because of insufficient machine parallelism in the processor and instruction parallelism in the benchmark programs. Instruction issue rates using non-numerical benchmarks show a similar variation (see figure 6).

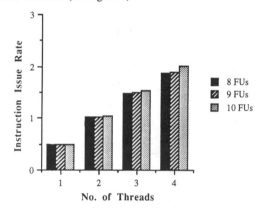

Figure 5. Instruction Issue Rate versus Number of Threads for Numerical Benchmarks

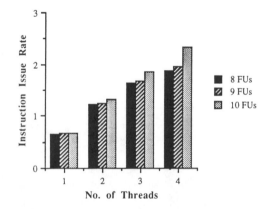

Figure 6. Instruction Issue Rate versus Number of Threads for Non-Numerical Benchmarks

The addition of extra functional units, the 9 and 10 configurations of table II, does not improve the functional unit utilization significantly. This indicates the absence of sufficient parallelism in the benchmark programs. It is interesting to note that the inclusion of additional integer add/sub units produces the maximum increase in the instruction issue rate. This is because the integer add/sub and move instructions are the most frequently occurring instructions in the program code and, hence, demand more resources of its kind [12].

Speed-up

We define speed-up as the ratio of the number of clock cycles required by a program to run on a sequential machine to the number of cycles on our processor. The Sun SPARC station 1, running at a clock rate of 20Mhz, was modelled as our sequential machine. For a fair comparison, a clock rate of 20Mhz was also assumed for all the processor configurations. Figures 7 and 8 show the variations in the speed-up for the numerical and non-numerical benchmarks, respectively. The speed-ups, like the instruction issue rates, have an almost linear relationship with the number of threads supported by the processor. The slight saturation in the non-numerical speed-up is related to the instruction issue rate (figure 8) and is caused for the same reasons.

Speed-ups for non-numerical benchmarks are generally lower than those for numerical benchmarks due to the recursive nature of the former. Sub-routine calls are handled by the SPARC architecture in an efficient manner using register windows that result in a significant improvement in performance. Our architecture, on the other hand, does not support register windows within a thread and, hence, suffers from a non-optimal performance in this regard. This architectural difference results in a lower speed-up.

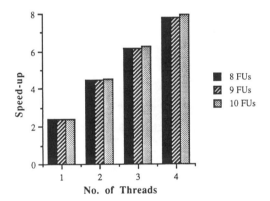

Figure 7. Performance Speed-up versus Number of Threads for Numerical Benchmarks

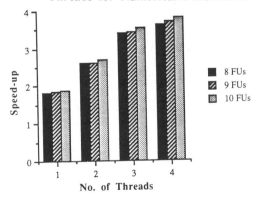

Figure 8. Performance Speed-up versus Number of Threads for Non-Numerical Benchmarks

Register File

One of the important design parameters is the number of read and write ports of the register file. This parameter limits the number of instructions that can be issued concurrently. Ideally, one would like to have as many sets of read/write ports as there are functional units, in order to utilize the full potential of the functional units. But, as figures 9-11 show, this is not required; the amount of program parallelism available is far too little to demand a large number of ports. The data presented in the tables are averaged over the different benchmark runs and correspond to processor configuration 8.

Figures 9 and 10 show the dynamic percentage distribution of the instructions from different threads accessing their respective integer registers for operands simultaneously. As an example, consider numerical benchmarks run on the processor supporting three threads. This processor configuration has a total of three integer register sets and three floating-point register sets corresponding to the three threads. From figure 9 we see that, on average, 72% of the instructions issued from each thread access their respective integer register sets singly, thus requiring only two read ports; whereas 17% of the instructions are issued two at a time and require four read ports in their respective register sets for a simultaneous operand fetch. Finally, less than 1% of the instructions are issued three at a time, requiring up to six read ports. The absence of sufficient parallelism in the benchmarks limits the number of simultaneous register set accesses to three.

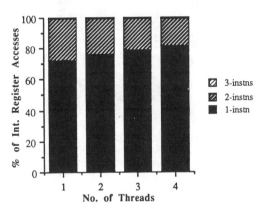

Figure 9. Distribution of Simultaneous Integer Register File Accesses by Individual Threads for Numerical Benchmarks

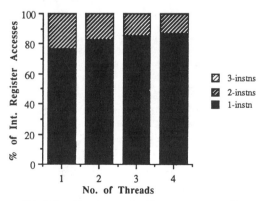

Figure 10. Distribution of Simultaneous Integer Register File Accesses by Individual Threads for Numerical Benchmarks

It is interesting to note that as the number of threads increases, the percentage of simultaneous multiple register accesses (within a thread) decreases. This is explained as follows. During the static compiling process, the compiler is forced to introduce NOOP instructions in the horizontal instruction word as well as in successive instruction words because of data dependencies between instructions. The latter NOOP instructions effectively introduce wait cycles when an instruction is waiting for results from other instructions. Dynamic interleaving makes use of these wait cycles and other NOOP instructions to introduce instructions from other threads. In doing so, interleaving essentially breaks down the statically scheduled horizontal instruction word (see figure 3) and, thus, reduces the number of simultaneous instruction issues from that thread. This results in a lower percentage of instructions accessing the register sets concurrently. For example, in figure 3 at CK=1, thread 2 has two instructions, ADD2 and SHLL2, accessing the integer register set of thread 2 simultaneously. However, because of dynamic interleaving, only one instruction, SHLL2, is issued at CK=1. The ADD2 instruction is issued only at CK=2, thereby reducing the multiplicity of the register set access of thread 2 from two to one.

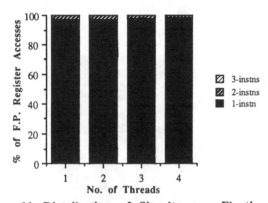

Figure 11. Distribution of Simultaneous Floating-Point Register File Accesses by Individual Threads for Numerical Benchmarks

Figure 11 shows the distribution of floating-point instructions accessing their respective floating-point register sets. There is no table for non-numerical benchmarks since they do not have any floating-point instructions. For numerical benchmarks, most of the floating-point register accesses are single and the percentages do not show much variation with the number of threads. This is because floating-point instructions from the benchmark programs are sequential and have large latencies.

We infer from figures 9-11 that for a processor supporting up to four threads, an integer register file having four read ports and a floating-point register file having two read ports are sufficient. The compiler has to take this into account while scheduling code; no more than two integer instructions and one floating-point instruction from the same thread can be scheduled in a horizontal instruction word. Simulations to determine the number of write ports need to take memory latency into account by suitably modelling caches and main memory.

5. Discussion

This paper presents simulation results of a multi-threaded RISC processor architecture model. The RISC processor architecture model is designed to support multiple threads or contexts, generated from a single program or multiple programs, in a uniprocessor environment. We discuss only the architectural aspects, since we are still researching implementation and realization issues of the model. In our simulation experiments using the Livermore loops and other non-numerical programs as benchmarks, a speed-up factor of approximately eight for a four-threaded, eight functional unit processor and better resource utilization over sequential machines were achieved. A brief discussion of contemporary superscalar and multi-threaded architectures, in contrast to ours, is presented below.

The i860 [1] is a 64-bit RISC microprocessor supporting up to three operations per clock: one integer or control instruction and up to two floating-point instructions. Parallel execution is achieved by incorporating multiple functional units that include a RISC core, a floating-point adder, and a floating-point multiplier. The i860 specifies a dual instruction mode in which two instructions can be issued simultaneously: one to the RISC core unit and one to either of the floating-point units. Even though the peak instruction issue rate is two, this is rarely achieved in view of the limited parallelism in a program. Our architecture, on the other hand, is able to keep the functional units busy most of the time, since it supports multiple threads.

The MC88100 [2] architecture consists of four independent functional units but does not support multiple instruction issue per cycle. As a result, when there is a data dependence causing the issuing unit to hold back an instruction, the rest of the instructions must be held back as well. In comparison, our architecture interplays multiple instruction threads with multiple instruction issuing capability so that the functional units can be utilized more efficiently.

VLIW [7] machines are single threaded, superscalar machines characterized by very long instruction words. Instruction scheduling is done by a trace scheduling compiler [8] that exploits statistical information about program branching by analyzing program traces to allow searching beyond basic blocks for program parallelism. Our architecture resembles the VLIW architecture in that both employ VLIW instructions and static instruction scheduling. The difference is that our architecture derives its parallelism by dynamically interleaving instruction threads at run time and, hence, requires a relatively simple compiler.

The HEP [15], intended as a supercomputer for solving large scientific problems, contains one or more Process Execution Modules (PEMs). Each PEM has an eight-stage pipeline, 2048 registers and sufficient state information to operate upto 128 instruction streams (processes) organized in up to 16 'tasks'. Data memory is external to the PEM and is shared among PEMs, with a pipelined switching network connecting the various data memory modules and PEMs. Independence of the instructions in progress is ensured by interleaving instructions from different instruction streams. No further instruction from the same process is allowed to begin until an instruction in progress from the same stream has completed execution. This is quite different from our approach where any non-conflicting instruction from the same instruction stream is allowed to be issued.

P-RISC [3] is a RISC multiprocessor that incorporates dataflow concepts to support multiple thread execution. There are certain basic differences between P-RISC and our architecture. The former is a multiprocessor architecture modified to support multiple threads. It uses separate memory frames as register sets for the different threads. In addition, instruction scheduling is done in hardware. Our architecture, on the other hand, is a superscalar architecture designed to support multiple threads. Every thread is allocated a separate register set and instruction scheduling is done both in software (by the parallelizing compiler) as well as in hardware (by the dynamic interleaving unit).

6. Conclusion

It is probable that only advances in VLSI technology can produce significant speed-ups in single-threaded processors by increasing the clock rate. Alternatively, the concurrent execution of multiple instruction threads can improve a system's performance using current technology. This study explores architectural features for issuing multiple instructions in a cycle and multi-threaded execution in a uniprocessor environment to further enhance the speed-up. Our approach is different from others in that they are based on multiprocessing at the CPU level. On the other hand, we explore the performance improvements obtained by moving multiprocessing from the CPU level to the ALU level.

The performance of a parallel processor depends on the amount of machine parallelism available and its ability to efficiently use the parallelism within a program. Our architecture provides machine parallelism by incorporating heterogeneous and independently functioning execution units as opposed to a homogeneous CPU. Doing so, we achieve a speed-up factor of nearly 2.5 in program execution time over a sequential processor for the single-threaded case. Program parallelism is enhanced by executing multiple instruction threads. Concurrent thread execution is made possible by dynamically interleaving instructions from various threads. *Dynamic Interleaving* produces an order of magnitude speed-up in program execution time when four instruction threads are concurrently issued. Moreover, *Dynamic Interleaving* increases the instruction issue rate to the functional units, thereby improving the resource utilization. Incorporating four instruction threads increases the functional unit utilization almost 3.5 times. Thus, both the execution speed-up and the instruction issue rate increase almost linearly with the number of instruction threads. Our study also shows that *Dynamic Interleaving* can reduce contention in register file access. This means that the register files can have fewer access ports.

Overall the evaluation study shows that the proposed architecture is feasible. A microprocessor based on our architecture will have great potential in building workstations supporting multiple window environment and in distributed computing requiring remote processing. In addition, multi-user systems and advanced scientific processors performing parallel simulations can greatly benefit from this architecture. We are currently working on the details of the dynamic interleaving scheme, the instruction issuing mechanism, and the register file interconnection network.

References

[1] "i860: 64-bit microprocessor user's manual," Intel Corp., 1989.

[2] "MC88100: RISC microprocessor user's manual," Motorola Inc., 1988.

[3] R. S. Nikhil and Arvind, "Can dataflow subsume von neumann computing?" in *Proc. 16th Annual Intl. Symp. on Comp. Arch.,* 1989, p. 262.

[4] R. A. Iannucci, "Towards a dataflow/von neumann hybrid architecture," in *Proc. 15th. Annual Intl. Symp. on Comp. Arch.,* 1988.

[5] W. D. Weber and A. Gupta, "Exploring the benefits of multiple hardware contexts in a multiprocessor architecture: preliminary results," in *Proc. 16th Annual Intl. Symp. on Comp. Arch.,* 1989, p. 273.

[6] R. H. Halstead and T. Fujita, "MASA: A multithreaded processor architecture for parallel symbolic computing," in *Proc. 15th. Annual Intl. Symp. on Comp. Arch.,* 1988, p. 443.

[7] R. P. Colwell, R. P. Nix, J. J. O'Donnell, D. B. Papworth and P. K. Rodman, "A VLIW architecture for a trace scheduling compiler," *IEEE Trans. on Computers,* vol. 37, Aug. 1988.

[8] J. A. Fisher, "Trace scheduling: a technique for global microcode compaction," *IEEE Trans. on Computers,* vol. 30, Jul. 1981.

[9] A.R.Pleszkun and G.S.Sohi, "The performance potential of multiple functional unit processors," in *Proc. IEEE Symposium on Computer Architecture,* 1988, p. 37-44.

[10] L.Wang and C.L.Wu, "Distributed instruction set computer architecture," *IEEE Trans. on Computers,* in press, 1991.

[11] F. H. McMahon, "The livermore fortran kernels: a computer test of the numerical performance range," UCRL-53745, Lawrence Livermore National Labs., Dec. 1986.

[12] D. W. Clark and H. M. Levy, "Measurement and analysis of instruction use in the VAX-11/780," in *Proc. 9th Annual Intl. Symp. on Comp. Arch.,* 1982, p. 9-17.

[13] C. L. Wu and R. G. Prasadh," Exploitation of multiple-instruction-stream superscalar microprocessor architecture," in *Proc. 1st Workshop on Parallel Processing,* Dec. 1990.

[14] G. S. Sohi, "Instruction issue logic for high-performance, interruptible, multiple functional unit, pipelined computers," *IEEE Trans. on Computers,* vol. 39, pp. 349-359, Mar. 1990.

[15] B. J. Smith, "A pipelined shared resource MIMD computer," in *Proc. Intl. Conference on Parallel Processing,* 1978, pp. 6-8.

Multiple Stream Execution on the DART Processor

Joy Shetler
Steven E. Butner

Electrical and Computer Engineering Department
University of California – Santa Barbara
Santa Barbara, CA 93106

Abstract

This paper presents early simulation results of single-processor multiple stream execution on the experimental DART multiprocessor. This research prototype design seeks to investigate the effects of a fully decoupled organization: specifically, to overlap useful processing while awaiting memory accesses, to develop architectural approaches to achieve built-in fault tolerance, and to arrive at a machine organization that can utilize high performance technologies even though a substantial speed mismatch exists between processor, network, and memory access times.

The design space of possible decoupled processor organizations is large, and we have just begun to explore it. This paper presents early simulation results concerning the multiple stream performance within a single DART processor. The processor's internal design is discussed and models of its execution on several programs are given. The performance ranges from 28 cycles per instruction (CPI) to approximately 3.9 CPI and the processor sustains a utilization as high as 95% on its internal units. The design allows internal subsystems to be implemented in high performance technologies, e.g. GaAs. With high clock rates and 95% pipeline utilization, processor performance would be substantial. The highest throughput observed for a single DART processor in the results presented here occurs with 24 concurrent processes. In a multiprocessor DART system, each processor is expected to offer similar performance.

I. Introduction

Memory latency is becoming increasingly difficult to deal with as the clock rates of modern high-performance processors approach and exceed 100 MHz. It has been predicted [1] that as system clock rates and main memory capacities increase we will see a growing gap — perhaps as large as a few hundred clocks — between processor speed and memory subsystem latency. Pipelining would seem to hold the answer to this dilemma though the data and control-flow dependencies in programs often severely limit a pipeline's effectiveness.

This paper presents early results from the DART project (DART = Decoupled Architecture for Reliable Throughput), which is investigating the feasibility of using highly decoupled hardware structures (with independent locally synchronous but globally asynchronous clocks and multiple interleaved instruction streams) to overcome the memory latency problem. This project is a follow-on to the circulated context multiprocessor (CCMP) research of Staley and Butner [2-4]. The goals of the DART project are to:

- Develop a new multiprocessor structure that uses a fully decoupled strategy to gain high throughput, reliability, and scalability.

- Learn to design and interconnect locally synchronous, globally asynchronous subsystems with very high clock rates. Such structures will be useful for building computers from the emerging high-performance technologies, (e.g. GaAs, HBTs, HEMTs) where processor clock rates in the GHz range are achievable [5].

- Include error detection, reconfiguration, and fault tolerance within the architecture. These critical capabilities, necessary for achieving reliability and dependability in a scalable multi-processor environment, are built in at the most fundamental level.

- Investigate software structuring methods that can take advantage of the multiple interleaved instruction streams available under our fully decoupled philosophy.

In a fully decoupled organization, we are making a deliberate trade of increased latency in exchange for breakless pipelines (i.e. in exchange for throughput). Much of the cost (in terms of increased single-stream laten-

cy) can be hidden since the DART model is intended to be used on extremely high clock-rate machines. Though any given processor may only be executing instructions at a rate of 4-28 clocks per instruction, there are potentially a large number of processors simultaneously delivering results at the same rate. Because all memory transactions (in fact, *all* activities) are split-phase, this organization can tolerate hundreds of clocks of memory latency. The cycles between the request and response of memory and other units are fully utilized for the processing of other instruction streams. Breakless pipelining is guaranteed since although many independent instruction streams are simultaneously active in any DART processor (and there may be up to 1024 such processors in a fully-scaled DART system) there is only one instruction from any given stream active in any part of the machine at any instant. Thus, the machine's pipelines are kept filled with independent instructions from separate instruction streams. There are never dependencies among these instruction streams so the pipelines can remain full and deliver their full potential.

We have found the realization of our 'pipelines-full' goal to be quite a challenge, particularly in light of the scalability and variable latency issues of the DART system. Maintaining efficient flows through our interconnection network and balancing the load across the resources of the multiprocessor are key elements of the approach. DART processors each contain a 4 Mbyte segment of a global, shared memory space. Reference to non-local memory involves memory request and response packets flowing in the ICN. The use of dynamically changeable mid-order interleaving provides a degree of load-spreading by giving a part of the address space of a process to each processor. Since both instructions and data are interleaved across the system, execution moves from processor to processor according to the current program counter and virtual memory segment register settings. This action is expensive in ICN bandwidth in that a portion of the processor's context must be moved each time execution migrates. But, with suitable error detection and correction in the ICN, the payoff is a load-balanced and highly reliable, high throughput system. By moving from processor to processor during execution, the process is automatically check-pointed and recovery from errors is facilitated. Additionally, since processors are identical and process migration is a primitive operation, it is possible for a processor to stand-in for another failed processor. This action can be made transparent to the user process.

II. The DART System

The basic philosophy of a decoupled architecture is to allow several functional units to operate independently under decentralized control at independent clock rates in order to process multiple instruction streams. Multiple independent instruction streams provide the interleaved, pipelined hardware with the type of workload needed to keep its pipelines full, thus achieving extremely high efficiency and throughput. Such systems are decoupled by inserting FIFO queues between units to provide a latency-absorbing spring-like effect, allowing major subsystems to operate on independent clocks. Each processor contains a portion of an interleaved, global memory which all other processors can access via an interconnection network. Figure 1 shows a block diagram of a DART processor.

In a DART system, up to 1024 processors can be connected through the tree-structured SHUNT interconnection network [6]. Packets containing process state migrate between processors under direction of a common operating system to solve computational problems. Though the interconnection network has a significant effect on the performance of the overall system, it is not the subject of this paper. The design and implementation of SHUNT is described in [6]. To utilize high speed technologies such as

GaAs, it is necessary to address the speed mismatch between processors, network, and memories at the architectural level. Appropriate partitioning of a decoupled system can leverage the strengths of one technology without making compromises which result in performance loss in other areas of the system.

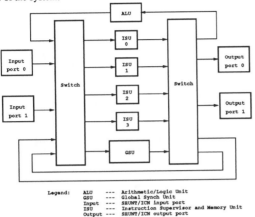

Legend:
ALU --- Arithmetic/Logic Unit
GSU --- Global Synch Unit
Input --- SHUNT/ICN input port
ISU --- Instruction Supervisor and Memory Unit
Output --- SHUNT/ICN output port

⟶ = 32-bit data path with FIFO queue

Figure 1: DART Processor Block Diagram

III. Instruction Execution Models for DART

Using the circulating context schema, instruction flow diagrams (Fig. 2) were used to specify, manipulate, and study the functions to be performed by the processor. Since the processor is part of a shared memory, multiprocessing system, additional steps for virtual memory translation and global synchronization are included in the diagrams. Each block represents a separate operation in the instruction execution. The parallel paths in these flow diagrams represent operations which are performed concurrently. Separate functional units operate concurrently on the same instruction stream, the only constraints being that the operations performed in parallel are data independent and that the hardware is able to resynchronize the instruction stream. Resynchronization may require extra copies of functional units or special resynchronization queues to be inserted that increase the cost of implementing the hardware structures. In some instances this resynchronization would have minimal impact on the design. For the path through the Global Synchronization Unit, resynchronization only occurs in rare cases, when a message is pending from the operating system for that process.

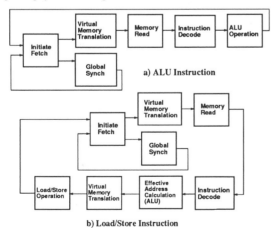

a) ALU Instruction

b) Load/Store Instruction

Figure 2: Sample Instruction Flow Diagrams

For some operations (such as accessing data in other processors, I/O operations, or the migration of processes) the instruction stream must pass a packet through the Interconnection Network (ICN). Two options exist for data accesses and I/O operations, either allow the instruction stream to

execute in the other processor or place the stream on a queue while it waits for the results. The advantage of the first method is not having to use special queues, while the advantage of the second method is that the packets transferred between the processors may be smaller. In the DART system, no extra words are needed for the stream to execute remotely on another processor and the context for the stream is already stored in the originating processor. So, the stream executes remotely in the other processor just as though the stream had been assigned to that processor. Upon completing the operation, the packet is sent back to the originating processor and the stream continues execution on that processor. In the short term, these accesses cause increased usage of the functional units in other processors but lessen the usage in the local processor. Over a long interval, this usage is averaged across all the processors by interleaving data segments for several streams across the system.

IV. DART Processor Design

The traditional approach to designing a processor for this system would be to use a long pipeline, each stage of the pipeline performing the operations detailed in the instruction flow diagrams. Several different types of instructions would map to the same pipeline. Since the same pipeline must be used for all instructions, either some stages would be unused for some instructions or an instruction stream might need to enter the pipeline more than once to execute an instruction. If an instruction stream stalls because of accessing data in other processors or I/O devices, the whole pipeline could backup. To avoid this situation, the instruction stream could be placed on a wait queue until the data or device becomes available. This has the effect of introducing a bubble into the pipeline. In a decoupled system, this effect is minimized since queues are placed between stages to absorb bubbles. Frequent stalling or stalling that occurs for periods of time longer than the number of instruction streams that can be enqueued may still cause some bubbles to be created. One difficulty with having processes re-enter the pipeline is that other instruction streams may be using the pipeline, so either the re-entering process must wait for a gap or else the instruction stream in the previous pipeline stage must wait. If several processes are re-entering at a particular point, the queue behind this point can fill rapidly with streams waiting to execute.

Reconfigurable pipelines can be used to allow an instruction stream to re-enter a pipeline at specific points. The main benefit of this approach is that fewer copies of specific functional units are required to implement the pipeline. This approach assumes that each pipeline stage is unique and is assigned as a resource. The use of reconfigurable pipelines can introduce some contention for resources and can require additional complex hardware mechanisms such as reservation tables [7,8] to allow several instruction streams to use the pipeline without colliding. Even with these complex mechanisms, some stages remain unused during execution to prevent collisions, thus introducing bubbles into the pipeline. This method may still not adequately compensate for long delays associated with accessing data in other processors or I/O devices.

In another approach used in the ZS-1 [9], the pipeline is split, forming two dual concurrent pipelines. The dual concurrent pipelines are both decoupled at the stage following the split. No other stages are decoupled. One path performs fixed point and memory operations while the other path performs floating point operations. A 64 bit machine word is interpreted as either one 64 bit instruction or two 32 bit instructions. For some instructions, only one of the dual pipelines is used. The ZS-1 executes one instruction stream at a time, allowing several instructions to be concurrently executed as a basic block in the pipeline. Dynamic scheduling is used to cross over the basic block boundary, decreasing the number of bubbles normally introduced in the pipeline at the boundary. Complex mechanisms are used to schedule instructions to avoid data and/or control dependencies within the basic blocks. Otherwise this approach is fairly similar to the long pipeline approach. The pipeline must be entered at the first stage and only one instruction enters the pipeline at a time. In some cases, queues are used to hold instructions waiting for data to become available at intermediate stages in the pipeline. The main benefit of this design seems to be the ability to perform two concurrent operations, one in the fixed point/memory pipe and the other in the floating point pipe. Bubbles can still appear in the pipes if the stream doesn't use the arithmetic functions or if the mix accesses one pipeline and not the other.

Our approach is based upon using several identical copies of multipurpose functional units in each processor. Some of the functional units can perform several of the operations detailed in the instruction flow graph and are the main controllers. One multipurpose functional unit contains the context for the instruction stream. The functional units are interconnected to allow instruction streams to make fast and simple transfers between functional units. Instruction streams do not stall waiting for data or results. Some operations take more clocks to be performed than others and this effect is cushioned by the queues decoupling the units.

An instruction stream accesses the interconnected modules as needed by the particular instructions that the stream is executing. One functional unit handles both the memory and context storage functions. No complex mechanisms are needed to schedule the use of the functional units and the number of diverse designs is reduced. The number of unused stages is reduced and decoupling is used to provide a spring-like effect between functional units. In the event of accessing data in other processors or I/O devices, this approach allows the instruction stream to enter or re-enter the pipeline at several points. The instruction stream can actually move to perform a data access on another processor and then return. At the completion of an operation, the next destination for the stream is determined and its address placed in the destination field in the packet control word. This field is used to route the packet to the next functional unit as specified by the instruction flow graph. Changes in the instruction flow graph merely require changing the distributed control mechanism in the affected units. The main disadvantage of this approach is that the functional units designed to perform many operations may be more complex than functional units designed to perform a single operation. Since distributed control mechanisms exist in each functional unit in a decoupled system, the impact of increased complexity is less of a problem. By judiciously selecting the operations to be assigned to each functional unit, this increase in complexity can be minimized.

V. DART Functional Units

The instruction flow diagrams were used to map each function into hardware. The functional units which evolved from this mapping were: an Instruction Supervisor Unit (ISU), an ALU, a Global Synchronization Unit (GSU), and Input and Output ports for the Interconnection Network. Each of these units performs a different set of steps in the instruction flow graph. Each unit acts independently based upon the information received in the control word of the packet. If a unit receives a control word with a step that is not implemented for that unit, an error condition is set. Additional fields in the control word specify the op code, length of the packet, address of the context slot in the ISU bank, the ISU bank, and functional unit or destination for the the next step. Currently, sixteen separate values have been encoded to represent different steps in the instruction flow graphs, I/O requests and global synchronization.

Each of the functional units in the DART processor is pipelined. This increases the latency for each instruction but also allows higher clock rates in the pipelined units. The number of pipeline stages and average packet length determine the latency, utilization of the stages, and average CPI for the system. For maximum throughput within a single processor, the average packet length would ideally be 1. Paths several words wide are required between the functional units to achieve an average packet length of 1. To minimize hardware costs for the prototype, the path width between functional units has been limited to one word (32 bits). This adversely affects performance but allows a relatively inexpensive prototype to be developed.

Instruction Supervisor Unit (ISU)

The Instruction Supervisor Unit (ISU) contains the memory and context of an instruction stream. The ISU performs the steps of instruction fetch, virtual memory translation, data fetch, and receiving results. There are four copies of the ISU hardware in the processor, each capable of storing context for independent streams and acting as a memory bank. The ISU context slot contains the general purpose registers, virtual memory translation registers, process i.d. and instruction counter. Streams are assigned a slot in one of the four ISUs upon migrating into the processor. Each of the four ISUs can be used as an entry or re-entry point in the instruction

flow graph. Four separate streams can simultaneously be entering the same step or different steps. There will, of course, be contention if two streams simultaneously try to enter the same ISU. This contention is resolved in the internal interconnection switches in the processor.

By combining the memory and control operations in the ISU, the memory is incorporated into the processor. Most traditional machines have separated the memory and processor through buses. As larger memory chips become available and systems increase in speed, this separation adversely affects performance. Advances in processing technology have made the addition of random logic to large memory chips feasible. This can result in a memory functional unit that can be classified either as a smart memory or as a processor which has a large on-chip memory. Designs with larger memories combined with processor functions are typical of the direction in which future multiprocessing system development is proceeding. Since one goal of the DART system is to reduce the part count, this integration of functions requires fewer chips and make more efficient use of the area on the chips.

Arithmetic and Logical Unit (ALU)

The ALU performs effective address calculation as well as standard arithmetic and logical operations for the execution phase of an instruction. Since all instructions except Load/Store are register-to-register operations, effective address calculations are only performed for Load/Store instructions and the ALU operations are only performed for other instructions. This mutual exclusivity means that only one copy of the ALU is needed in the processor to ensure that no bottlenecks develop in accessing the ALU.

Currently, the ALU performs binary, fixed-point and floating-point operations. Packets are received in a word-serial manner (32-bits at a time) and when all the words of a packet have been loaded into the ALU, the operation in the op code field of the packet is performed. Some operations may take several clocks to complete such as multiply and divide operations. The current design assumes that these operations are performed during the time required for the next packet to be loaded into the ALU.

Global Synchronization Unit (GSU)

The Global Synchronization Unit (GSU) is used by the operating system as a mailbox to locate and notify executing processes to respond. There is no definite time associated with message delivery, only that the message will be delivered. A mailbox address exists in each GSU for every process. To communicate with a process, the operating system sends a message to every GSU in the system. Additionally, the GSU can monitor the execution of a stream by incrementing a counter associated with each stream as each instruction is executed. Currently, for the simulation model, this function is not implemented.

Interconnection Network Ports

The Interconnection Network Ports (ICN ports) are designed to interface to the SHUNT network [6]. SHUNT is a high-speed, circuit-switched network designed to optimize accesses to processors in the same cluster. To introduce some fault tolerance and allow the simultaneous transfer of packets to different destinations, the SHUNT network has two 16-bit input ports to the processor and two 16-bit output ports coming from the processor. Two packets can be received and two packets sent simultaneously. The processor internally uses a 32-bit word width. So the individual packet will see a 2-to-1 reduction in speed upon entering the ICN. With two ports operating concurrently, the ICN can transfer up to one word per clock. The SHUNT network has extensive error detection and correction capabilities. In the processor, parity checking is performed on each incoming 16-bit halfword. If the incoming packet is a migrating process, bank assignment logic assigns the incoming process to a context slot in the ISU. Comparators determine which bank has the most available slots (if any) and the process is assigned to that bank. This provides a degree of load balancing across the ISU banks.

Internal Switches

The functional units are connected within the processor by switches. The size and design of each switch is based upon commercially available parts. A crossbar switch provides the most bandwidth for parallel transfers of data from several sources to several destinations. The current processor configuration uses two crossbar switches to offset the effect that packets encounter trying to pass through the controller of a crossbar. With several packets trying to access the controller simultaneously, a two-level system reduces the amount of contention through the controller over that of a one-level system. Contention is also reduced since the number of sources and destinations for each crossbar switch is low.

VI. DART Pipeline Performance

Several measures are typically used to judge the performance of a processor. For the DART system, these measures are Cycles Per Instruction (CPI), stage utilization, run time for a program, number of instruction streams that can execute simultaneously, and queue lengths. The CPI inversely measures the rate of work being done by the processor. Stage utilization indicates the percentage of the time a particular stage is in use. The run time for a program provides insight into the amount of latency encountered during execution. For a single instruction stream executing in the processor, the CPI is at a maximum, the stage utilization at a minimum and the run time at a minimum. Single stream execution is the worst case in DART because the full latency is visible with no overlap. As more instruction streams begin executing, the CPI begins to decrease as more instructions complete per clock. The stage utilization begins to increase as more streams enter the pipeline. The run time begins to increase, though minimally, as contention occurs for some resources. The number of instruction streams that can execute simultaneously provides an indication of how many processes can be initiated by the operating system to maximize system performance. If fewer streams are executing, the CPI and stage utilization will be adversely affected. The queue lengths between each stage are a useful indicator of where contention problems exist and of which resources are overburdened.

Test Environment

Register Transfer Level simulation models have been written in n.2 [10] to determine the above parameters. The simulation model is modular, with a separate ISP' procedure running for each stage in the processor. Due to the modular design, the stages communicate through interface queues. The design of a stage can be altered without affecting the design of other stages. A new design can be plugged into the system by merely replacing the appropriate procedures. Clock rates within each stage can be adjusted independently. This allows us to simulate a system where the individual stages are realized using different technologies.

In the simulation model, programs are downloaded into memory from a "host" processor such as will actually happen in the prototype. Packets enter the processor through the ICN Input Ports and are loaded into the memory using the same control words that the host processor will use. After the loading is complete for a particular process, a context packet is sent from the host processor just as will occur during a process migration.

Test Programs

Several programs were run on the simulator including an iterative loop program, a bubble sort, and a Fibonacci series calculation program. Further testing was done on this subset to establish the basic performance characteristics of the processor and fine tune the design of the pipeline stages. This choice was based on the size and simplicity of the program and that these programs are commonly used and understood. The programs were run in several ways to gauge the effectiveness of the system. First, each program was run singly with just one stream executing on the system. This allowed us to calibrate the CPI, run time and stage utilization for one copy of the program running without any contention or effects due to interactions between the streams. Then for each program, more streams running other copies of that program were added to the simulation until the system stabilized. This gave us results for instruction mixes that were mostly composed of certain types of memory access patterns. Finally, simulations with multiple streams running different programs were performed to see what effect having a diverse instruction mix with diverse memory access patterns would have on the system performance.

CPI Results

The average CPI is plotted below as a function of the number of active streams. Several things are clear from this plot. First, the CPI decreases rapidly from the worst case single-stream value (29) to the minimum of 3.9 as the number of active processes increases.

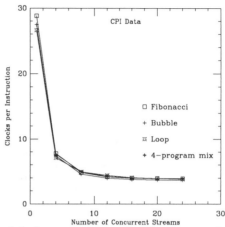

Though the three programs have quite different patterns of memory access, branch, and compute cycles, on average they behave in a very similar way. The machine runs within a factor of 2 of its best CPI with as few as 4 active streams. With 8 streams, the DART processor is running at a CPI less than 5.

By using multiple streams, each running the same program, we tested the response of the system to each of the various patterns. A more realistic scenario would be that the processor is executing many diverse processes, with unrelated mixtures of memory and processor activity. To simulate such a case, we created a four program mix consisting of the three test programs used previously plus an additional small loop-type program. By running multiple copies of this 4-stream set, we obtained data for 4, 8, 12, 16, 20, and 24 streams. The data points are labeled "4-program mix". The plot above shows that CPI is not strongly dependent upon the type of programs running.

Packet lengths and maximum queue lengths were monitored during all simulations. The average packet length across all simulations was 2.09 words with the smallest and largest execution-related packets being 1 and 5 words respectively. During initial program load, packet lengths of 21 words were recorded. With 4 processes running, maximum queue depths of 3 were observed, increasing to 4 as the CPI began to level off. All simulations were run with 4-word deep queues. When deeper queues were tried, a speedup of less than 1% was observed.

Unit Utilization Results

The "ISU Utilization" plot (next page) depicts the average ISU utilization as a function of the number of active streams. This plot reveals the relative differences in memory versus compute cycles for the three simulated programs. The simple loop program is primarily composed of register-to-register instructions while the bubble sort is shuffling integers within the memory. In all cases, the processor is able to operate with unit utilizations near or above 90%.

Simulation studies reveal that the ISU is the hotspot of activity in the DART processor. Previous designs separated the control aspects of the instruction supervisor from the memory-access aspects, i.e. there were distinct ISU units and memory bank units. The separation caused increased intra-processor packet traffic. The ISU and memory functions were subsequently combined into the current form to minimize the amount of traffic during instruction execution. Because the queues buffer memory requests, ISU utilization can be very high. For the simple loop program, the ISU is 100% utilized with 24 active streams.

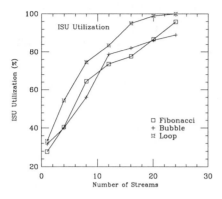

ALU utilization is plotted below for each of the three test programs as well as for the 4-program mix. It is apparent that another portion of the machine, rather than the ALU, is the limiting factor for the programs modeled, since ALU utilization levels off for most cases well below 70%. Bubble sort levels off at a slightly higher level (≈83%).

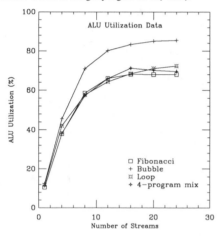

The plot below depicts unit utilizations for the four program mix. Note that even with 24 concurrent streams, the system has not yet reached saturation. This mixed load appears to be more varied than the individual test program streams, utilizing the processor's resources more smoothly.

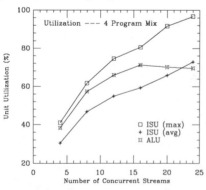

One conclusion reached from the simulation studies was that the internal connections between functional units were the main bottleneck in the processor. Based on this information, further research hass been undertaken to design higher bandwidth internal interconnections. The internal connection scheme must maintain a transfer rate of one packet per clock for each output, be able to queue packets headed for the same destination, and be able to receive one packet per clock at the inputs.

Higher-level simulation models have demonstrated the feasibility of the DART system, but to maximize performance, it is especially important to incorporate the processor as part of the DART multiprocessing system using RTL models. We intend to run considerably more diverse programs and establish more calibrations as our software support development, processor development, and ICN development continue.

VII. Summation

This paper has discussed the internal organization and performance of a single DART processor. Since the decoupling philosophy allows subsystems to use independent clocks, DART subsystems can be implemented in different technologies to obtain the best arithmetic speed as well as the densest memories. All transactions are split-phase, allowing large memory latencies to be absorbed and overlapped. Multiple concurrent instruction streams provide pipelines with independent transactions, thus allowing the pipelines to be breakless. The processor can sustain utilizations of 95% or higher on its internal pipelines. Though the fault tolerance of the system is only mentioned briefly, its incorporation at the most basic level is critical. This provides the correct primitives to allow detection and, in some cases, correction of errors, as well as reconfiguration around failures.

There are many parts of the decoupled system design space still unexplored. Other researchers [9,11] have investigated the benefits of decoupling — breaking the hard synchronization — in selected places, such as between major subsystems. The DART processor in its present form, on the other hand, is *completely* decoupled. This provides the best opportunity for observing, understanding, and improving the flow of transactions within the system.

VIII. Acknowledgements

This research was supported by DARPA under contract N00014-88-K-0497. We gratefully acknowledge our sponsor as well as many fruitful interactions with the high-performance computer architecture research group at UCSB.

IX. References

[1] J.L. Hennessy, D.A. Patterson, *Computer Architecture: A Quantitative Approach*, Morgan Kaufmann Publ., 1990

[2] S. Butner, "The Circulating Context Multiprocessor, An Architecture for Reliable Systems", *Int'l Symposium on Mini and Microcomputers (ISMM)*, pp.50-52, June 1984.

[3] S. Butner, C. Staley, "A RISC Multiprocessor Based on Circulating Context", *IEEE Phoenix Conference on Computers and Communications*, March 1986.

[4] C.A. Staley, "Design and Analysis of the CCMP: A Highly Expandable Shared Memory Parallel Computer", PhD dissertation, ECE Dept. UCSB, Santa Barbara, CA, August 1986.

[5] S.I. Long, S. E. Butner, *Gallium Arsenide Digital Integrated Circuit Design*, McGraw-Hill, 1990.

[6] S. Butner, S. Bordelon, L. Endres, J. Dodd, J. Shetler, "A Fault Tolerant GaAs/CMOS Interconnection Network for Scalable Multiprocessors", *IEEE Journal of Solid-State Circuits*, May 1991.

[7] J.P. Hayes, *Computer Architecture and Organization*, 2nd Edition, McGraw-Hill, 1988, pp. 602-608

[8] E.S. Davidson, "The Design and Control of Pipelined Function Generators," Proc. IEEE Conf. Systems, Networks and Computers, Oaxtepec, Mexico, January 1971, pp. 19-21.

[9] J.E. Smith, "Dynamic Instruction Scheduling and the Astronautics ZS-1," *IEEE Computer*, Vol. 22, No.7, July 1989, pp. 21-35

[10] "N.2 ISP' User's Manual", Document #101, version 1.14, Zycad Corp., 1988.

[11] J.E. Smith, S. Weiss, and N. Pang, "A Simulation Study of Decoupled Architecture Computers", *IEEE Transactions on Computers*, Vol. C-35, Aug. 1986, pp. 692-702

Performance Advantages of Multithreaded Processors

Won Woo Park*, Donald S. Fussell and Roy M. Jenevein

Department of Computer Sciences

*Department of Electrical and Computer Engineering

The University of Texas at Austin

Austin, TX 78712

Abstract

The ever-increasing silicon budgets available for single-chip integrated systems provides an ongoing challenge to designers to maximize performance for a given chip area. This has motivated moving processor subsystems such as floating point units, caches, memory management units, etc. onto the same chip with the integer core and led to various schemes for exploiting instruction-level parallelism such as superscalar processing. Further increases in single-chip integration may require a different approach, however. We argue that in a modern parallel environment, multithreaded processors provide a cost-effective way to exploit increasing silicon resources to achieve higher performance. We present a parametrized multithreaded architecture model in which multiple resident execution threads share expensive functional blocks. We compare the performance of a single-chip multithreaded processor with single-threaded designs using comparable silicon resources. Our results indicate that multi-threading can provide significantly higher performance in such environments than alternative strategies which devote the same silicon area to other purposes.

1. Introduction

Feature sizes on VLSI chips have been decreasing steadily over the last decade and promise to continue doing so for some time to come. Processor designers have sought to exploit this trend to improve the performance of single-chip processors, with impressive results. Modern systems based on single-chip processors like the Intel i860 boast performance exceeding 50 MIPS and 20 MFLOPS at a reasonable cost. This performance level has been achieved through two approaches to exploiting silicon resources. First, increasing integration of processor subsystems such as floating point units, cache memories, memory management units, I/O controllers, etc. combines with decreasing feature sizes to allow higher clock rates. Second, the ability of the machine to execute multiple instructions per cycle from a single sequential instruction stream allows instruction rates to exceed the clock rate.

It is not clear that these approaches will continue to be effective as silicon budgets continue to increase. Today's single chip processors contain all the subsystems that once comprised board level processors, with the exception of large instruction and data caches.

It is generally conceded that the performance gains possible through speeding up serial execution rates are very limited. Ultimately, further exploitation of parallelism in programs will be needed for continued performance improvements. For most nonscientific applications, the available instruction-level parallelism for general-purpose applications is around 2.5 instructions/cycle [3, 4, 10]. Approaches such as superscalar processors have already reached this limit in today's machines. Performance beyond this cannot be expected with a single instruction stream. This implies that parallelism must be built explicitly into programs to obtain further performance gains. As programming languages and compiler techniques advance, we can expect more task-level parallelism, particularly if these tasks are sufficiently low in overhead to achieve relatively fine-grained parallelism.

For environments with a high degree of task-level parallelism, multithreaded processors [9, 2, 8, 11, 1] used in place of single-threaded processors as the basic building blocks of multiprocessor systems, can increase the utilization of the processing hardware by interleaving the execution of different serial instruction streams, or *threads*. A single-chip multithreaded processor would provide hardware support for resident threads for each of the interleaved threads, and could potentially share computational resources which are expensive in terms of silicon area, thus achieving the performance of several independent single-threaded processors at a fraction of the silicon cost.

Shared resources must have a low enough utilization by individual threads that sharing will not degrade the performance of an individual thread. On the other hand, effective masking of latencies such as that due to memory access or interprocess communication, requires enough concurrent threads to assure that useful work is available to cover latent cycles. An efficient multithreaded design will depend on a careful analysis of the requirements of individual threads, and a careful determination of the best degree of sharing for each resource. The silicon costs of the resulting design decisions must be compared with performance increases possible by devoting the same silicon resources to the enhancement of single-threaded performance to determine whether multithreading is preferable.

In this paper, we present the results of a cost-performance study of a multithreaded processing unit in an environment where the interprocess communication is negligible and memory latency is relatively small. This type of concurrent environment can be viewed as a worst case insofar as the latency masking advantages of multithreading are minimized with these assumptions. Thus we did not include interprocess communication effects in the simulation which was used for performance measurement. In spite of this worst case assumption, our results indicate that significant performance advantages can be obtained through multithreading when compared against other architectural strategies with the same silicon cost. This is particularly true when the total silicon budget begins to exceed that of the i860, which indicates that multithreading may well be a significant component of post-superscalar processor architectures.

2. Multithreaded Processor Architecture Model

As with any processor architecture, there are many possible ways to allocate silicon resources to various aspects of the machine. Many design decisions such as the number of registers per thread, the size of on-chip cache, the type of floating-point unit, the sizes of various internal buffers, the number of simultaneous instructions per cycle to be supported, etc. must be made. In our model, the parameters used for the single-threaded basic processor, which is replicated to obtain the multithreaded machine, were fixed at optimal values through extensive simulation using a wide variety of benchmark programs. The details of this process will not be discussed here due to lack of space. Basically, a multithreaded processing unit can be divided into shared and non-shared blocks. The non-shared block consists of multiple copies of an integer core with its private caches, and the shared block includes the floating point functional unit (FPU) and memory bus interface.

2.1. The Non-Shared Block

The non-shared block contains separate hardware to maintain the context of each thread as well as an integer execution unit for each thread. This includes the integer core, instruction cache, data cache, and the branch target buffer (BTB), assuming that we use a branch prediction scheme. The integer core includes the instruction decoder, instruction issue unit, register file, and integer functional units. In the instruction unit, instructions are prefetched and decoded. The data dependency resolution and instruction issue scheduling are done in the instruction unit. There are two pipe stages in the instruction unit. In the first pipe stage, a block of instructions is delivered from the instruction cache to the instruction prefetch buffer. In the next pipe stage, the prefetched instructions are decoded by one or more instruction decoders, and the initial data conflict resolution is made for the decoded instructions. Each thread has a private conflict register set (CRS), which performs the function of a reorder buffer [7, 12], and instruction issue window (IIW). Selection of the instructions to be issued in the next cycle among conflict-free instructions is done dynamically in the instruction decode pipe stage for each thread.

In the integer execution unit, there are one or more arithmetic and logic units (ALUs). The results of integer ALU instructions are sent to the CRS, and the calculated effective addresses of memory instructions are forwarded to the memory data dependency resolution logic. The new branch target address calculated in the integer ALU is sent to the CRS, and when a branch instruction turns out to be incorrectly predicted, instruction prefetching and decoding is stopped. When the incorrectly predicted branch instruction is at the top of the CRS, the CRS, the IIW, and the instruction prefetch buffer are flushed and subsequent instructions are fetched from the correct address. The number of parallel ALUs for a single thread can be varied, depending on application requirements and chip area limitations.

2.2. The Shared Block

The floating point functional unit (FPU) consists of a floating point adder and a floating point multiplier and is shared by all threads. Threads that have ready floating point instructions assert their *floating point instruction issue request* signal lines. The thread selector associated with the FPU selects a thread among requesting threads. If the selected thread has more than one ready floating point instruction, the most first instruction of the thread in program order is selected.

The memory bus interface is also shared. The threads share a single set of I/O pads and drivers for instruction and data transfers between processor and cache. In principle, the cache should also be a shared resource. In the current implementation, we have logically divided the cache into separate regions for each thread for simplicity. There are reasons to expect superior performance from a dynamically shared single cache under some circumstances, but further discussion of this is beyond the scope of this paper.

3. Area Estimation Model

Processor chips are designed to provide the highest performance possible for a given cost. Cost can be measured many ways: manufacturing cost, silicon area, pinout requirements, power consumption, or a combination of these. In general, it is difficult to define precisely what constitutes cost in a real-world sense for a study of this kind. Without specific justification for a more complex strategy, we choose silicon area as a simple measure of cost. We will maintain the pinout

Functional block	Area
Pads & bus driver and receiver	0.222
FPU	0.261
Graphic unit	0.030
Caches	0.260
Integer core	0.086
TLB	0.025
Bus controller	0.025
Global wires	0.091
Total	1.000

Table 1: Area of functional blocks of the i860 processor chip.

STP	1	2	3	4
Register File	0.057	0.081	0.104	0.158
CRS(16 entries)	0.019	0.033	0.052	0.081
IIW(8 entries)	0.006	0.009	0.013	0.019
Instr. Decoder & Execution Unit	0.075	0.151	0.237	0.302
Total Cost	0.157	0.274	0.406	0.560

Table 2: Area of the integer core as a function of STP Area is represented with respect to the area of the i860 processor chip.

requirements approximately equal for our various designs, and we will compare designs which have equal silicon area requirements.

Since the Intel i860 processor chip includes most functional blocks on a single chip and the area taken for each functional block can be measured from the layout, we have based our area estimation model on the i860 processor layout [5].

Table 1 shows the area in units of the i860 devoted to various functional blocks. In practice, the area taken for any functional block can vary widely, depending on the design style. Using i860 blocks at least lets us assume a common design style for all of them, so we have a reasonable basis for uniform area comparisons among blocks of different types.

3.1. Area Estimation of the Integer Processor Core

The integer processor core includes the instruction decoder, instruction issue unit, register file, and integer functional units. The size of the register file is assumed to be 32, 32-bit registers each for integer operations and floating point operations. We used the SPARC instruction set for simulation due to the convenient availability of SPARC software tools. Single-threaded parallelism (STP), the number of integer instructions simultaneously executable in a single machine cycle, is a problem orthogonal to the instruction set, so the configuration of the execution unit can be different from the SPARC processor. The number of read/write ports of the register file and the CRS must be increased with increasing STP. The number of instruction decoders and the number of parallel functional units also increase with increasing STP.

Table 2 shows the costs of the integer core for several different machine configurations [7]. The area of an integer core capable of 2 inst/cycle is about 75% larger than that of a single instruction machine. In general, the area of the integer core increases substantially with increasing STP.

3.2. Branch Target Buffer Area Estimation

Since the cache unit of the i860 processor chip consists of 4K byte instruction and 8K byte data caches which are 2-way set-associative with a 32-byte line size, the total number of storage cells in the cache unit is 110,720. If we assume that the area taken by the cache control, address decoder, and global wiring is proportional to the number of storage cells, the function for the area of a cache of m storage cells is given by $A_{cache}(m) \simeq 1.07 \times 10^{-7} \times m$.

The branch target buffer (BTB) is a cache memory associated with branch instructions [6]. When the BTB has m cells, the area of the BTB is given by the same expression.

4. Simulation

We have developed a trace driven register transfer level (RTL) simulator, which simulates dynamic instruction scheduling, pipelining, cache, branch target buffer, memory bus, and memory. Instruction traces are obtained using a modified GDB debugger on a SPARC processor.

Performance of a processor is evaluated in terms of the execution rate, which is the average number of useful instructions executed per cycle. The instructions that are aborted or discarded due to branch prediction misses or data hazards are not counted as useful instructions.

The operation latency for each instruction class is assumed to be one cycle for ALU class instructions, two cycles for load, store, and branch class instructions, and three cycles for floating point (FPT) class instructions.

As a representative sample of general-purpose C applications, we used ten different programs as benchmarks. All are written in C and compiled using the GNU SPARC compiler.

Table 3 shows the distribution of instruction classes for each benchmark program. Almost half of the instruction instances are ALU class instructions, which include move, logic, shift, and integer arithmetic instructions. The FPT class includes floating point addition, multiply, divide, and

Benchmark	Instruction Classes (%)				
Programs	ALU	FPT	ST	LD	BR
Assembler	53.13	0.00	5.99	21.09	19.78
Dhrystone	53.87	0.00	10.11	19.68	16.33
Grep	57.30	0.00	4.30	18.12	20.28
Newt-Raph	32.78	11.44	10.32	30.56	14.92
Sim-ann	49.65	2.35	9.47	27.62	10.93
Linpack	27.21	16.34	17.61	33.44	5.41
Quicksort	47.92	0.00	5.14	28.15	18.79
Proc-sim	41.98	0.00	17.73	21.83	18.46
Towers	54.46	0.00	5.40	22.47	17.67
Whetstone	34.60	20.66	6.93	35.26	2.56
Average	45.29	5.08	9.30	25.82	14.51

Table 3: Instruction distributions of benchmarks. Sim-ann is layout using simulated annealing and proc-sim is a processor simulator.

Machine parallelism	1	2	3
Number of decoders	1	2	4
Number of integer ALUs	1	2	3

Table 4: Varying the single-thread parallelism (STP)

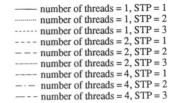

- —— number of threads = 1, STP = 1
- —— number of threads = 1, STP = 2
- ---- number of threads = 1, STP = 3
- - - - number of threads = 2, STP = 1
- — — number of threads = 2, STP = 2
- —— number of threads = 2, STP = 3
- ·—·— number of threads = 4, STP = 1
- ·—·— number of threads = 4, STP = 2
- – – – number of threads = 4, STP = 3

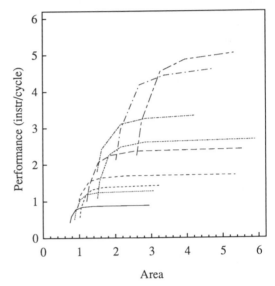

Figure 1: Performance as a function of area for multi-threaded systems.

conversion instructions. The load and store instructions (LD and ST) together account for one third of all instruction instances on average. Thus the memory traffic for data only utilizes one third of the I/O bandwidth if we allow one data transaction per processor cycle. The proportion of FPT class instructions varies greatly depending upon the application, peaking at 20.66% for whetstone. Both I/O interface and floating point unit have spare capacity that could be shared by multiple threads.

For the simulation of a system with multiple threads, traces of the 10 benchmark programs are supplied for each thread one by one, but the sequences of traces supplied to each thread are different. Thus different traces are processed in each thread at any time.

Configuration	Execution rate (instr/cycle)	Relative perf. rate
Singlethreaded	1.4	1.0
Dual-processor	2.4	1.7
Multithreaded	3.6	2.6

Table 5: Performance of three different configurations that have the cost of 2.4.

5. Performance of a Multithreaded Processor

Figure 1 shows performance in useful instructions executed per cycle as a function of area for systems with a single thread, two threads, and four threads respectively, varying two design parameters: the cache size and the STP. In the variation of cache sizes, the data cache size and the BTB size were assumed to be one fourth of the instruction cache size because this shows the highest marginal performance/area rate. The variations in STP are shown in Table 4. The sizes of the CRS and the IIW for each thread were assumed to be 16 and 8 respectively, again based on maximal performance/area results.

In the figure, the x-axis represents silicon area required for the given configuration in units of fractions of the i860.

Increasing any design parameter increases area and performance. However, the marginal performance/area rates are quite different depending on the current configuration and varied design parameter.

When the number of threads is included in the design space, the optimal configuration at any given area is different from the optimal configuration in a system with a single thread. For example, the optimal configuration of a single-threaded system at a cost of 1.2 is the configuration in which the STP is two with a 4K byte instruction cache, but the optimal configuration of a multithreaded system is the configuration where the number of threads is two, the STP is one, and the total instruction cache size is 2K bytes. In this case the multithreaded optimal configuration has 18% higher performance than the single-threaded optimal configuration.

In Table 5, the performance for three configurations that have a cost of 2.4 is given. Each configuration represents a locally-optimal configuration in the given design space. The single-threaded machine has STP of three and a 32K byte instruction cache. The dual-processor system includes two single-threaded processors, each of which has STP of two and a 2K byte instruction cache. The multithreaded system has four threads, STP of two per thread and a total 16K shared instruction cache. The dual-processor system has 1.7 times the performance of the single-threaded machine. The multithreaded system has 2.6 times the performance of the single-threaded machine and 1.5 times the performance of the dual-processor system. This clearly indicates a significant performance advantage for the multithreaded system over the single-threaded and dual-processor alternatives when the silicon area used is about twice that of the i860. We conclude that the sharing of the expensive resources, the I/O pads, the FPU and the cache, among this small number of threads is an efficient strategy for performance optimization in machines using post i860 silicon technology which are intended for use in parallel, general purpose environments.

6. Conclusions

In this paper we have presented a model configuration for a single-chip multithreaded processor in which multiple threads of execution share expensive functional blocks by interleaving instructions. In this model, the non-shared integer cores operate concurrently and independently. Our analysis shows that significant performance advantages can be obtained using multithreading as compared with superscalar machines or even multiple independent processors using the same silicon area. This is particularly true for large silicon budgets, i.e. those exceeding the area of a processor comparable to the i860. The performance advantages of a multithreaded system over a singlethreaded system of the same size come from the increased available parallelism and fine-grained hardware sharing which increases the utilization of otherwise little-used resources.

The SPARC-based model is probably not the best choice for the base architecture for a multithreaded machine. The 32-bit instruction sets of SPARC and other RISC processors, while providing simple, fast decoding and a reasonable addressing model, are not the most compact encodings of instructions available. Since I/O pins on the chip are one of the shared expensive resources in a multithreaded model, performance potential is maximized by making the best possible use of the bandwidth through the pins shared by the executing threads. This argues for a more compact instruction set, even at the expense of some limited additional decoding complexity or a limited code expansion. A thorough study including these architectural considerations is beyond the scope of this paper. We can note, however, that after several years of experimenting with different detailed designs of the METRIC multithreaded processor [8], we have seen evidence that compact instruction sets and smaller register banks provide advantages in the design of high-performance multithreaded machines using limited silicon area and pins.

References

[1] A. Agarwal, B. Lim, D. Kranz, and J. Kubiatowicz, "APRIL: A Processor Architecture for Multiprocessing," *17th Annual Inter. Symp. on Comput. Arch.*, pp. 104-114, 1990.

[2] R. H. Halstead, Jr. and Tetsuya Fujita, "MASA: A Multithreaded Processor Architecture for Parallel Symbolic Computing," *15th Annual Inter. Symp. on Comput. Arch.*, pp. 443-451, 1988.

[3] N. P. Jouppi, D. W. Wall, "Available Instruction-Level Parallelism for Superscalar and Superpipelined Machines," *Third International Conference on Architectural Support for Programming Languages and Operating Systems*, pp. 272-282, 1989.

[4] N. P. Jouppi, "The Nonuniform Distribution of Instruction-level and Machine Parallelism and its Effect on Performance," *IEEE Trans. Comput.*, Vol. 38, No. 12, pp. 1645-1658, Dec., 1989.

[5] Leslie Kohn and Sai-Wai Fu, "A 1,000,000 Transistor Microprocessor," *1989 IEEE International Solid-State Circuits Conference*, pp.54-55, 1989.

[6] J. K. F. Lee and A. J. Smith, "Branch Prediction Strategies and Branch Target Buffer Design," *IEEE Computer*, Vol. 17, No. 1, pp. 6-22, Jan., 1984.

[7] W. W. Park, "Performance-Area Tradeoffs in a Multithreaded Processing Unit," PhD. Dissertation, The University of Texas at Austin, 1991.

[8] R. O. Simpson, "METRIC Context Unit Architecture," PhD. Dissertation, The University of Texas at Austin, 1988.

[9] B. J. Smith, "A Pipelined, Shared Resource MIMD Computer," *Proc. International Conference on Parallel Processing*, 1978.

[10] M. D. Smith, M. Johnson, and M. A. Horowitz, "Limits on Multiple Instruction Issue," *Third International Conference on Architectural Support for Programming Languages and Operating Systems*, pp. 290-302, 1989.

[11] W. Weber and A. Gupta, "Exploring the Benefits of Multiple Hardware Contexts in a Multiprocessor Architecture: Preliminary Results," *16th Annual Inter. Symp. on Comput. Arch.*, pp. 273-280, 1989.

[12] S. Weiss and J. E. Smith, "Instruction Issue Logic in a Pipelined Supercomputer," *IEEE Trans. Comput.*, Vol. C-33, No. 11, pp. 1013-1022, Nov. 1984.

A Mapping Strategy For MIMD Computers*

Jiyuan Yang, Lubomir Bic, Alexandru Nicolau
Department of Information and Computer Science
University of california, Irvine,
Irvine, CA 92717

Abstract — In this paper, a heuristic mapping approach which maps parallel programs, described by precedence graphs, to MIMD architectures, described by system graphs, is presented. The complete execution time of a parallel program is used as a measure, and the concept of critical edges is utilized as the heuristic to guide the search for a better initial assignment and subsequent refinement. An important feature is the use of a termination condition of the refinement process. This is based on deriving a lower bound on the total execution time of the mapped program. When this has been reached, no further refinement steps are necessary. The algorithms have been implemented and applied to the mapping of random problem graphs to various system topologies, including hypercubes, meshes, and random graphs. The results show reductions in execution times of the mapped programs of up to 77 percent over random mapping.

1 Introduction

In order to effectively utilize large-scale parallel computers, the scheduling problem is one of crucial importance. Many researchers have addressed task scheduling in various approaches [1], [2], [6], [7], [8], [13], [14], [16]. The scheduling problems are usually classified into static and dynamic methods. This paper addresses static task scheduling.

A parallel program is represented by a *problem graph*. The parallel computer system on which the parallel program is to be executed is referred to as a *system graph*. The purpose of the static task scheduling presented in this paper is to minimize the complete execution time of the parallel program.

In the problem graph, each task is dependent on its predecessors. A task can only be executed after all its predecessors have completed. Weights are associated with nodes and edges in the graph to represent execution and communication times, respectively.

Usually, the number of nodes in the problem graph, np, is much larger than the number of nodes in the system graph ns, $(np \gg ns)$. In order to simplify the scheduling problem, it can be divided into two steps. The first step, called *clustering*, combines np problem nodes into na groups, where $na = ns$. The edges connecting problem nodes within the same group are removed. The resulting graph is called a *clustered problem graph*. The second step, refered to as *mapping*, then maps the na clusters to the ns system nodes. Here, each cluster is treated as a single

abstract node and edges connecting two abstract nodes are combined into one abstract edge. Under this abstraction, the second step only deals with graphs having the same number of nodes.

In this paper, we present an approach for performing the mapping of a clustered problem graph onto a system graph. In other words, we assume that an existing technique is first applied to produce a clustering from a given problem graph. The resulting clustered problem graph is then used as input to our algorithms. Note that the problem graph has the same number of nodes as the system graph, as has been done with other approaches. However, in our case, we still use the information about individual tasks within each cluster and their communication.

Since the mapping problem is NP-Complete, various heuristic algorithms have been developed in the past [1], [2]. They focus primarily on minimizing the communication overhead. Bokhari [1] describes a mapping strategy, where the cardinality, defined as the number of the problem edges that fall on system edges, is used as the measure for evaluating a mapping. Unfortunately, the edges that don't fall on system edges can have a significant effect on the system's performance. Another limitation of this strategy is that all problem edges are assumed to have the same weight. However, in a general problem graph, the communication load carried by the different problem edges may vary significantly. Furthermore, the algorithm assumes $np \leq ns$, which imposes a serious limitation on the number of problems this method can be applied to.

Lee describes another mapping strategy which takes the phase for each problem edge into account, and uses actual distances between the system nodes rather than their nominal distances [2]. However, the assumption that all problem edges have to be activated simultaneously, i.e., all communications in one phase must start at the same time, is too restrictive for most applications. Similar to Bokhari, he also assumes that the number of nodes in the system graph must be equal to or greater than the number of nodes in the problem graph, i.e., $np \leq ns$. In most actual cases, the number of nodes in the problem graph is much larger than the number of nodes in the system graph.

The main drawback of both of the above mapping strategies is that they only consider communication cost but ignore execution time. We have shown in [4] that an optimal communication cost may still result with a non-optimal complete execution time.

Another limitation is inherent to the process of deriving a solution using these approaches, which is based on iterative improvement, i.e., repeatedly modifying assignments and

*This material is based upon work supported by NSF under grant No. CCR-8709817

comparing their results. Unfortunately, this process can't be terminated until a predetermined number of moves have been performed. Hence the search may continue long after the optimal solution has already been found.

Finally, neither approach considers any data dependencies among nodes. The assumed problem graphs are not directed, which means that they only consider the communications among the tasks but not their precedences.

The above limitations indicate the need for a better mapping strategy, which would consider data dependency, had a more realistic measure of how good a mapping is, and a better termination method. In this paper, we present such a mapping strategy. The complete execution time is used as the measure and the data dependencies are taken into account in the mapping process. The most important merit of this strategy is that, in some cases, it can detect when the optimal solution has been reached and thus no further refinement attempts are necessary. This reduces the total search space and time.

The paper is organized as follow. The outline of the approach is introduced in section 2. The internal problem representation and mapping algorithms are described in sections 3 and 4, respectively. Experimental results are discussed in section 5. Finally, conclusions are given in section 6.

2 Outline of Approach

The algorithms will use the following five graphs:
A *problem graph* $G_p = \{V_p, E_p\}$,
A *clustered problem graph* $G_c = \{V_c, E_c\}$,
An *abstract graph* $G_a = \{V_a, E_a\}$,
An *ideal graph* $G_i = \{V_i, E_i\}$,
A *system graph* $G_s = \{V_s, E_s\}$,
where V_p, V_c, V_a, V_i, V_s are sets of nodes and E_p, E_c, E_a, E_i, E_s are sets of edges in the respective graphs. The numbers of nodes in each graph are given by $np = |V_p|$, $nc = |V_c|$, $na = |V_a|$, $ni = |V_i|$ and $ns = |V_s|$, where $np = nc = ni$ and $na = ns$.

The relationships among the graphs are shown in Fig. 1. The *problem graph* describes the tasks and their interactions. Fig. 2 is an example of a problem graph, where each node has an ID and a weight to indicate the number of time units for executing the task. Each edge also has a weight which represents the communication time. The *clustered problem graph* is derived from the problem graph by combining the problem nodes into groups. An example of a clustered problem graph derived from the problem graph in Fig. 2 is shown in Fig. 3. As mentioned earlier, we assume that an existing technique for clustering a given problem graph is used. Examples of such techniques may be found in [9], [10], [11], [12].

The *abstract graph* is the result of treating each cluster as one abstract node and collapsing edges between the same abstract nodes into one. Fig. 4 is the abstraction of the clustered problem graph in Fig. 3. In particular, we need to know if a given abstract edge is critical. This information is used to guide the mapping.

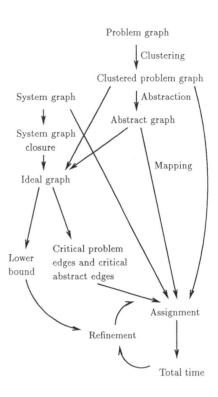

Fig. 1 Relationships among the graphs

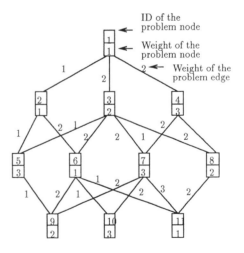

Fig.2 Problem Graph

The *system graph* describes the topology interconnecting homogeneous processing elements of a parallel computer system. A *system graph closure* is the fully connected superset of the system graph. Fig. 5-b shows the closure for the system graph of Fig. 5-a. The concept of the system graph closure is used to derive the *ideal graph*, which is a mapping of the clustered problem graph onto a fully connected system graph closure. This mapping is unique and is easily derivable, since the graphs have the same numbers of nodes and the communication cost between any two system nodes is identical (due to its full connectivity).

The purpose of deriving the ideal graph is to obtain a *lower bound* on the complete execution time of the parallel program. It is also used to derive the *critical problem edges* and *critical abstract edges* (see below) which are used to guide the mapping of the clustered problem graph to the actual system graph.

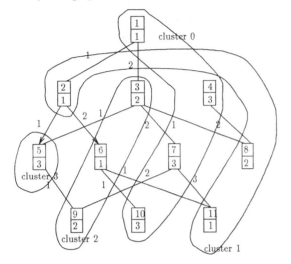

Fig. 3 Clustered problem graph

Fig. 4 Abstract graph

Fig. 5-a System graph

Fig. 5-b System graph closure

An example of the ideal graph resulting from mapping the clustered problem graph of Fig. 3 to the closure of Fig. 5-b is shown in Fig. 6. Note that the ideal graph carries the same information as the clustered problem graph, but is depicted in a different format to visually capture the time line of execution. In particular, the node weights are unchanged, each shown inside the corresponding node. The edge weights, on the other hand, are not shown explicitly as numbers attached to edges but as the time units that separate the nodes on the vertical axis. For example, the weight on the edge (1,3) is 2 (Fig. 3), which corresponds

to a 2-time-unit delay between the end of node 1 and the beginning of node 3 (Fig. 6).

The main difference between the clustered problem graph and the ideal graph, however, is that some edge weights become longer due to data dependencies. For example, the edge (6,11) carries the weight 1 in Fig. 3, but it results in 7 time units (weight 7) in Fig. 6. This is due to its dependency on node 7, which, in turn, depends on node 3, and so on. In other words, the ideal graph may be viewed as the topologically sorted form of the clustered problem graph.

When the abstract graph is mapped to the system graph, rather than its closure, an *assignment* is produced. Based on the assignment and the information in the clustered problem graph, the complete execution time of the parallel program can be derived. If the result is not equal to the lower bound, a refinement of the assignment is attempted, as illustrated in Fig. 1.

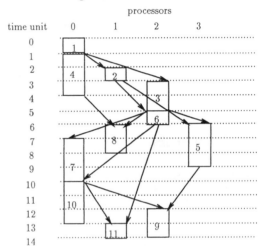

Fig. 6 Ideal graph

We assume that the execution time of each task (problem node) and the communication times are measured in time units. For any two problem nodes connected by an edge, we assume the worst case where each communication takes place between the end of the sending task and the beginning of the receiving task.

We will define several terms based on the graphs mentioned before, which will be used by the algorithms of section 4.

1. *Latest task* — This is the task (problem node) which terminates last. For example, in Fig. 6, tasks 9 and 11 are the latest tasks.

2. *Critical problem edge* — An edge in the ideal graph is critical if increasing the weight of the corresponding edge in the clustered problem graph by any amount will lengthen the complete execution time of the ideal graph. For example, the problem edge e_{i79} in Fig. 6 is critical, since any increase in the weight of the clustered problem edge e_{c79} (Fig. 3) must increase the weight of the ideal edge e_{i79} and thus delay the start time of the latest task 9. On the other hand, edge e_{i59} is not criti-

cal, since increasing the weight of the clustered problem edge e_{c59} will not necessarily increase the weight of the ideal edge e_{i79} (Only when the increase is by more than 2, will the ideal graph edge be affected).

3. *Critical abstract edge* — An abstract edge (v_{a1}, v_{a2}) is critical if there is at least one critical problem edge (v_{i1}, v_{i2}), where (v_{i1}) is mapped onto (v_{a1}) and (v_{i2}) is mapped onto (v_{a2}). The abstract edges $(v_{a0}, v_{a1}), (v_{a0}, v_{a2})$ in Fig. 4, for example, are critical.

4. *Critical degree of an abstract node* — This is the sum of the weights of all critical abstract edges directly connected to that abstract node. For example, there are two critical abstract edges $(v_{a0}, v_{a1}), (v_{a0}, v_{a2})$ that connect to abstract node v_{a0}. Hence, the critical degree of v_{a0} is the sum of the weights of the two critical abstract edges.

5. *Critical abstract node* — An abstract node is critical if it is connected to a critical abstract edge which has been mapped to a single system edge.

6. *Total time* — This is the complete execution time of the parallel program.

3 Internal Representation

So far we have talked about all graphs of Fig. 1 at only a conceptual level. In this section we introduce the internal representation of each graph, and other auxiliary data structures, needed by the mapping algorithms presented in subsequent sections.

1. For problem graph, matrices $prob\_edge[np][np]$, and $task\_size[np]$ are used to describe the weights of the edges and tasks in the problem graph.

2. For clustered problem graph, matrices $clus\_edge[np][np]$, $clus\_pnode[na][np]$ represent clustered problem edges and which problem nodes are in which cluster respectively.

3. For abstract graph, $abs\_edge[na][na]$, $c\_abs\_edge[na][na+1]$, $mca[na]$ describe abstract edges, critical abstract edges in which the last element of each row is the *critical degree*, defined as the sum of all the numbers in that row (see section 2), and communication intensity of the abstract nodes which is the sum of the weights of all clustered problem edges directly connected to an abstract node.

4. For system graph, $sys\_edge[ns][ns]$, $shortest[ns][ns]$, and $deg[ns]$, are used to depict system edges, shortest path between two system nodes and degree of each system node.

5. Since the system graph closure is fully connected, its matrix contains all ones except the diagonal elements. Hence no explicit representation is required.

6. For ideal graph, $i\_edge[np][np]$, $i\_start[np]$ and $i\_end[np]$, and $crit\_edge[np][np]$ represent the ideal

graph edges, start and end times of each task, and the critical problem edges.

7. For assignment, matrix $assi[ns]$ expresses the assignment from abstract nodes to system nodes. A matrix $comm[np][np]$ describes the communication between any pair of problem nodes under a given assignment, which is the product of the weight of a problem edge and the length of the shortest path between two system nodes. Two matrices $start[np]$ and $end[np]$ represent the start time and end time of each task, respectively.

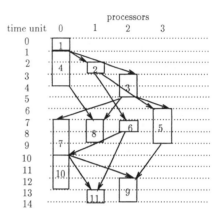

Fig.7

4 Mapping Algorithms

In this section, we will introduce algorithms which perform the transitions of Fig. 1.

4.1 Deriving the Ideal Graph

I Algorithm for deriving the start and end time of each task in ideal graph

Do the following until all tasks have been visited:

1. For an unvisited task i, find its predecessors from matrix $prob\_edge[np][np]$.

2. If it has a predecessor, go to 3. Otherwise, do the following:

   (a) $i\_start[i] = 0$

   (b) $i\_end[i] = i\_start[i] + task\_size[i]$

   (c) Mark task i as visited.

3. If any one of the end times of all the predecessors of task i is unknown, go to 1. Otherwise, do the following:

   (a) For each predecessor j, derive $i\_start[i] = max_j(i\_end[j] + clus\_edge[j][i])$

   (b) $i\_end[i] = i\_start[i] + task\_size[i]$

   (c) Mark task i as visited.

II Algorithm for deriving the lower bound

$$lower\_bound = i\_end[l]$$

where l is the latest task.

After we obtain the start time and end time of each task, it is easy to derive the ideal edge matrix $i\_edge[np][np]$ which will be used to derive critical edges.

III Algorithm for finding ideal edge matrix $i\_edge[np][np]$

Do the following for each pair of tasks i and j:

1. If there is an edge between i and j in the clustered problem graph, and j is the predecessor of i, i.e., $clus\_edge[j][i] > 0$, then $i\_edge[j][i] = i\_start[i] - i\_end[j]$.

2. All other elements remain 0.

4.2 Finding Critical Abstract Edges

To find the critical abstract edges we first need to find all critical problem edges. Following is the basic idea needed for finding the latter.

Theorem 1 An edge e_{iij} in the ideal graph is critical if its weight, $i\_edge[i][j]$, is equal to the weight of the corresponding edge, $clus\_edge[i][j]$, in the clustered problem graph and it directly connects to the latest task.

Theorem 2 An edge in the ideal graph, $i\_edge[i][j]$, is critical if its weight is equal to the weight of the corresponding edge in the clustered problem graph, $clus\_edge[i][j]$, and it is the predecessor of a critical problem edge $i\_edge[j][k]$.

Based on above theorems, we can find all critical problem edges in an ideal graph recursively. This is the main idea of the following algorithm. (The proof of the Theorems may be found in [4].)

I Algorithm for finding critical problem edges

1. Find the set, LS, of latest tasks v_i in the ideal graph from the end time matrix $i\_end[np]$. LS can have one or more elements.

2. Repeat until LS is empty.

   For each v_i in LS do the following:

   (a) Find the predecessors of v_i in the matrix $clus\_edge[np][np]$.

   (b) For each predecessor v_j, compare the weight of the corresponding edge in the clustered problem graph, $clus\_edge[j][i]$, with the weight of the edge in the ideal graph, $i\_edge[j][i]$. If they are equal, then the edge $i\_edge[j][i]$ is critical, and $crit\_edge[j][i] = clus\_edge[j][i]$.

   (c) Include all predecessors of v_i that are connected to v_i by a critical problem edge(s) in LS.

3. All other elements of $crit\_edge[np][np]$ remain 0.

From the definition of the abstract edge, the algorithm for finding the critical abstract edge is to detect whether there is a critical problem edge in an abstract edge.

II Algorithm for finding the critical abstract edges

1. Find all critical problem edges $crit\_edge[i][j]$ (above algorithm).

2. For abstract nodes v_{al}, v_{am}, where $(0 \leq l, m \leq na - 1)$, find all critical problem edges $crit\_edge[i][j]$ where, problem nodes i, j are included in abstract nodes v_{al}, v_{am}, respectively. Assign the

sum of the weights of the critical problem edges as the weight of this critical abstract edge.

$$c\_abs\_edge[v_{al}][v_{am}] = \sum (crit\_edge[i][j] + crit\_edge[j][i])$$

From the definition of the critical degree, we have the following:

III Algorithm for finding the critical degree of each abstract node

For all i $(0 \leq i \leq na - 1)$, do

$$c\_abs\_edge[i][na] = \sum_{j=0}^{na-1} c\_abs\_edge[i][j]$$

4.3 The Mapping Algorithm

The mapping is performed in two stages: initial assignment and refinement. How to refine the initial solution and when to stop refinement are the two most important aspects.

4.3.1 The Termination Condition

The system graph closure provides a means to derive a lower bound on the total time of the problem program. This is based on the following theorem.

Theorem 3 If the total time of any assignment is equal to the total time of the ideal graph, this assignment is an optimal mapping. (The proof may be found in [4].)

From theorem 3 we derive the following termination condition:

Termination Condition If the total time of an assignment is equal to the total time of the ideal graph, terminate the refinement process.

4.3.2 Initial Assignment

Initial assignment tries to achieve the smallest possible total time of a given clustered problem graph and a system graph. As mentioned before, the critical edges influence the goodness of the mapping most significantly. Hence the basic idea of the initial assignment algorithm is to map the critical edges to neighboring system nodes or at least as close as possible. This usually yields a very good initial assignment, as has been verified by our experiments.

The initial assignment algorithm consists of the following three steps:

1. (a) Using the matrix $deg[i]$, select node v_s from V_s such that $deg[v_s]$ has the maximum degree. In the event of a tie, select any qualifying node arbitrarily. Mark v_s as visited.

   (b) Select node v_a from V_a such that $c\_abs\_edge[v_a][na]$ has the maximum critical degree (see section 3.3.(b)). In the event of a tie, select any qualifying node arbitrarily. Mark v_a as visited.

   (c) $assi[v_s] = v_a$. Mark v_a as a critical abstract node.

2. Repeat until all the abstract nodes that have critical abstract edges have been visited.

   (a) Using the matrix $c\_abs\_edge[i][j]$, select node v_a from V_a such that v_a is unvisited, $c\_abs\_edge[v_a][na]$ has the maximum critical degree, v_a is a neighbor of some marked node

v'_a in V_a and the abstract edge $abs\_edge[v_a][v'_a]$ is critical, i.e., $c\_abs\_edge[v_a][v'_a] > 0$. Mark v_a as visited.

(b) If a node v_s from V_s can be selected such that v_s is unvisited, $deg[v_s]$ has the maximum degree and v_s is a neighbor of some marked node v'_s in V_s ($sys\_edge[v_s][v'_s] > 0$), such that $assi[v'_s] = v'_a$, then $assi[v_s] = v_a$, mark v_s as visited, mark v_a as critical abstract node and go to (a). Otherwise, go to (c).

(c) Select node v_s from V_s such that v_s is unvisited, v_s is the closest node from some marked node v'_s in V_s, i.e., $shortest[v_s][v'_s]$ is smallest, and $assi[v'_s] = v'_a$, then $assi[v_s] = v_a$ and mark v_s as visited.

3. Repeat until all nodes in V_a have been visited.

(a) Select node v_a from V_a such that v_a is unvisited, $mca[v_a]$ has the largest communication intensity and v_a is a neighbor of some marked node v'_a in V_a ($abs\_edge[v_a][v'_a] > 0$). Mark v_a as visited.

(b) If a node v_s from V_s can be selected such that v_s is unvisited, $deg[v_s]$ has the largest degree and v_s is a neighbor of some marked node v'_s in V_s, ($sys\_edge[v_s][v'_s] > 0$), such that $assi[v'_s] = v'_a$, then $assi[v_s] = v_a$, mark v_s as visited and go to (a). Otherwise go to (c).

(c) Select node v_s from V_s such that v_s is unvisited, v_s is the closest node of some marked node v'_s in V_s, i.e., $shortest[v_s][v'_s]$ is smallest, and $assi[v'_s] = v'_a$, then $assi[v_s] = v_a$, and mark v_s as visited.

4.3.3 Refinement

The initial assignment which uses the critical abstract edges to guide the mapping process is usually quite good, but subsequent refinement is likely to improve the mapping further.

An iterative improvement technique is chosen to refine the mapping.

1. Derive an initial assignment A1 (algorithm in section 4.3.2).

2. Evaluate the total time (see section 4.3.4).

3. If the total time of A1 is equal to the total time of the ideal graph, stop. Otherwise, go to 4.

4. Repeat the following ns times

(a) Randomly assign the non-critical abstract nodes to the system nodes which are not occupied by critical abstract nodes.

(b) Evaluate the total time of the changed assignment, A2.

(c) If the total time of the changed assignment A2 is equal to the total time of the ideal graph, stop; the optimal solution has been reached.

(d) If the total time of A2 is less than that of A1, assign A2 to be the current assignment; else keep A1.

4.3.4 Evaluating Total Time

Mapping of the clustered problem graph to the system graph closure is similar to mapping it to the system graph. Hence, evaluating the total time is similar to deriving the lower bound, as described in section 4.1. The main difference is that the system graph usually is not fully connected. To derive the start and end time matrices, we first generate the communication matrix $comm[np][np]$, which describes the communication between any pair of problem nodes under a given assignment.

I Algorithm for finding the communication matrix

1. Find the shortest path between any pair of system nodes $shortest[ns][ns]$ (use some existing algorithm [17]).

2. Do the following for each pair of the clustered problem nodes i and j:

If i and j are in different abstract nodes v_{al}, v_{am}, and $assi[v_{sl}] = v_{al}, assi[v_{sm}] = v_{am}$, then

$$comm[i][j] = clus\_edge[i][j] \times shortest[v_{sl}][v_{sm}]$$

Using the matrix $comm[i][j]$, we can derive the start time and end time of each task by using the following algorithm, which is similar to that used for the ideal graph.

II Algorithm for deriving the start time and end time of each task under a given assignment

Do the following until all tasks have been visited:

1. For an unvisited task i, find its predecessors from matrix $prob\_edge[np][np]$.

2. If it has predecessors, go 3. Otherwise, do the following:

(a) $start[i] = 0$

(b) $end[i] = start[i] + task\_size[i]$

(c) Mark task i as visited.

3. If the end time of any of the predecessors of task i is unknown, go to 1. Otherwise, do the following:

(a) For each predecessor j, derive $start[i] = max_j(end[j] + comm[j][i])$

(b) $end[i] = start[i] + task\_size[i]$

(c) Mark task i as visited.

III Algorithm for deriving the total time of the program under an assignment

Assign the maximum end time to the total time of the program, the node with maximum end time is the latest task.

$$total\_time = max_j(end[j])$$

Fig. 7 shows the result of mapping the clustered problem graph from Fig. 3 onto the system graph in Fig. 5-a with the assignment (0, 1, 3, 2) by using the preceding algorithms. Since the total time of this initial assignment is equal to that of the ideal graph, it is an optimal mapping and no further refinement is needed.

5 Experiments

It is hard to compare one heuristic approach with other heuristic approaches, because different approaches make different assumptions. To avoid criticism for having used only several special examples particularly suited to our approach, random mapping was chosen to be compared with our mapping strategy. For this purpose, a random problem graph generator was created and a random clustering program was developed. The weights of the problem nodes and the weights of the problem edges are also produced randomly. The numbers of nodes in a problem graph range from 30 to 300, while the numbers of nodes in a system graph range from 4 to 40. The system topologies are hypercubes, meshes, and random graphs. All algorithms were implemented in C++ and run on a SUN-4 workstation.

5.1 Mapping to Hypercube

The comparison between mapping randomly produced abstract graphs to a hypercube topology using our mapping strategy versus random mapping is shown in Fig. 8.

Each point on the horizontal axis corresponds to one problem graph, i.e., the figure is a histogram. The lower end of each vertical solid line shows the result of a mapping using our mapping strategy; the higher end shows the random mapping result. The difference between higher end and lower end is the improvement by using our mapping strategy over the random mapping. For example, a lower end value of 110 and an upper end value of 160 mean that a program mapped by using our approach requires only 10% more time than the lower bound, while a random mapping would result in a 60% increase in total time. The results demonstrate that the improvement between the results of our approach and those of the random mapping ranges from 29 percent to 63 percent. In 2 out of 10 cases, our results reached the lower bound. It should be noted that optimal mapping can never be better than the lower bound. It is worse than, at most equal to, the lower bound.

5.2 Mapping to Mesh and Random Topologies

Fig. 9 and 10 are analogous to Fig. 8. They show the results of mapping random problem graphs to mesh architectures and to randomly produced system architectures, respectively. They show improvement between the results of our approach and those of the random mapping ranging from 33 percent to 77 percent for the total time. The experiments also demonstrate that the termination condition works well. In Fig. 10 there are 5 out of 17 cases where our mapping stops the refinement by the termination condition. In Fig. 9, there are 7 out of 11 such cases.

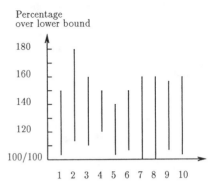

Fig.8 Mapping to Hypercube

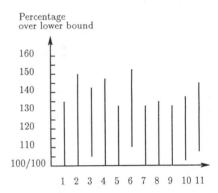

Fig.9 Mapping to Meshes

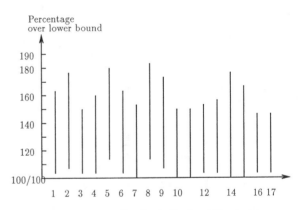

Fig.10 Mapping to Randomly Produced Topologies

6 Conclusion

In this paper, we presented a mapping strategy which maps a clustered problem graph to a system graph. This strategy uses the complete execution time of a parallel program, represented by a clustered problem graph, as the measure to evaluate the goodness of the mapping. Through the analysis of the critical edges based on the mapping of the abstract graph to the system graph closure, we obtain two important concepts: critical abstract edges and a lower bound. The former is used to guide the mapping by attempting to assign critical edges to a single system edge each. The latter allows us to derive a termination condition which stops unnecessary refinement and reduce both searching space and mapping time. The algorithms presented in this paper make it possible to map np problem nodes to ns system nodes where $np > ns$. The effectiveness of this approach has been verified empirically, by deriving the mappings of different, randomly generated problem graphs onto hypercube, mesh-connected, and random system graphs. The results have shown improvements ranging from 29 to 77% in total execution time over random mappings.

Acknowledgement

The authors would like to thank Meng-lai Yin for useful discussions.

References

[1] S. H. Bokhari, "On the Mapping Problem", IEEE Trans. on Computers, vol. V-30, (March 1981), pp. 207-214.

[2] S.-Y. Lee, J.K. Aggarwal, "A Mapping Strategy for Parallel Processing", IEEE Trans. on Computers, vol. V-36,(April, 1987), pp. 433-442.

[3] S.Kirkpatrck, C.D.Gelatt, M.P. Vecchi, "Optimization by Simulated Annealing", Science , vol. V220, (May 13, 1983) pp.671-680.

[4] J. Yang, L. Bic, A. Nicolau, "A Mapping Strategy for MIMD Computers", Technical Report, Department of Information and Computer Science, UC Irvine, (1990).

[5] L.M. Ni and K. Hwang, "Optimal Load Balancing in a Multiple Processor System with Many Job Classes", IEEE Trans. on Software Eng., vol. SE-11, (May 1985), pp.491-496.

[6] K. Fukunaga, S. Yamada, T. Kasai. "Assignment of Job Modules onto Array Processors", IEEE Trans. on Computers, vol. V-36, (July 1987), pp. 888-891.

[7] F. Berman, M. Goodrich, C. Koelbel, W.J. Robison, K. Showell, "Prep-P: A Mapping Processor for CHiP Computers", Int'l Conf. on Parallel Processing, (1985), pp. 731-733.

[8] P. Sadayappan, F. Ercal, "Nearest-Neighbor Mapping of Finite Element Graphs onto Processor Meshes," IEEE Trans. on Computers, vol. V-36, (Dec., 1987), pp. 1408-1424.

[9] A. Gerasoulis, S. Venugopal, T. Yang, "Clustering Task Graphs for Message Passing Architectures", ACM International Conference on Supercomputing, (June 11-15, 1990), Amsterdam, Holand.

[10] K. Efe, "Heuristic Models of Task Assignment Scheduling in Distributed systems", IEEE Computer, 15(6), (1982), pp.50-56.

[11] A. Gerasoulis, I. Nelken. "Static Scheduling for Linear Algebra DAGs." HCCA4(1989).

[12] M. Cosnard, M. Marrakchi, Y. Robert and D. Trystram. "Parallel Gaussian Elimination on an MIMD Computer." Parallel Computing, (6, 1988), pp. 275-296.

[13] J. Baxter, L. H. Patel, "The LAST Algorithm: A Heuristic-Based Static Task Allocation Algorithm", IEEE Int'l Conf. on Parallel Processing, (Aug., 1989), pp. II217-222.

[14] L. Kim, C. R. Das and W. Lin, "A Processor Allocation Scheme for Hypercube Computers", IEEE Int'l Conf. on Parallel Processing, (Aug., 1989), pp. II-231 - II-238.

[15] C. Lee, L. Bic, "Comparing Quenching and Slow Simulated Annealing in the Mapping Problem", IEEE Third Annual Parallel Processing Symposium, (April, 1989).

[16] D.T. Peng, K.G. Shin, "Static Allocation of Periodic Tasks With Precedence Constraints in Distributed Real-Time Systems", 9th Int'l Conf. on Distri. Compt. Syst. , (June 5-9, 1989), pp. 190-198, CA.

[17] S. Baase, Computer Algorithms: Introduction to Design and Analysis, Second Edition, Addison-Wesley Publishing company, (1988).

Impact of Temporal Juxtaposition on the Isolated Phase Optimization
Approach to Mapping an Algorithm to Mixed-Mode Architectures

Thomas B. Berg[*]   Shin-Dug Kim   Howard Jay Siegel
tbb@sequent.com   sdkim@ecn.purdue.edu   hj@ecn.purdue.edu

Parallel Processing Laboratory
School of Electrical Engineering
Purdue University
West Lafayette, IN 47907-0501 USA

Abstract -- Mixed-mode parallel processing systems are capable of executing in either SIMD (synchronous) or MIMD (asynchronous) modes of parallelism. The ability to switch between the two modes at instruction level granularity with very little overhead allows the parallelism mode to vary for each portion of an algorithm. To fully exploit the capability of intermixing both SIMD and MIMD operations within a single program, one must determine the optimum mapping of an algorithm to the mixed-mode architecture. The phase optimization technique, where the programmer makes an implicit assumption that by combining the best version of each phase the optimal implementation of the entire program will be achieved, generally works in a serial computer environment. The application of this approach to the selection of a mode of parallelism for each phase is investigated by presenting a detailed study of a practical image processing application, the Edge Guided Thresholding algorithm, and its mapping to a mixed-mode parallel architecture. The six functional phases of the algorithm, as well as their temporal juxtaposition, are analyzed along with experimental performance measurements obtained from the PASM parallel processing prototype, a mixed-mode system. The results discussed here demonstrate a situation where the advantages of a mixed-mode approach are limited.

1. Introduction

One significant issue in the field of massively parallel computing concerns the mapping of algorithms to parallel computer architectures. To fully exploit the benefit of a parallel processing system, the programmer must understand how to structure an algorithm to take advantage of the target architecture.

Adaptive parallel systems add a new dimension by allowing the user to configure the target machine to best match the algorithm. Although such systems offer potentially higher performance for a wider range of applications, the flexibility places a burden on the programmer to chose for each algorithm those system parameters that will result in the highest performance. Ideally a compiler would remove the burden from the programmer, but usually to achieve peak performance the parameters of the target architecture must be considered throughout the algorithm design process.

In a **mixed-mode** parallel system, processors are capable of executing instructions in both the **SIMD** (Single Instruction stream - Multiple Data stream) and **MIMD** (Multiple Instruction stream - Multiple Data stream) modes of parallelism [14], and switching between these two modes at instruction level granularity. With such flexibility, it is possible for SIMD and MIMD mode operations to be intermixed within a single algorithm. Thus, when implementing a given task, the programmer or compiler has an additional parameter to consider. The parallelism mode can be varied on the task, sub-task, or even instruction level.

A mixed-mode machine must have a control unit to broadcast instructions, as in an SIMD machine, and each processor must be able to decode instructions into control signals as in an MIMD machine. This research is part of an ongoing series of studies to model and evaluate the worth of the mixed-mode architectural approach to parallelism. Earlier works have demonstrated strengths of this approach, e.g., [7, 12, 13]. This work illustrates a potential weakness.

The optimum mapping of multi-phase algorithms to mixed-mode parallel processing systems can be determined through a process of experimentation. One straightforward approach starts by applying **phase optimization**, a common technique in serial computing. The program is divided into functional phases and, through experimentation, the implementation of each phase is optimized. Several versions of each phase are created, possibly utilizing different parallelism modes, and their performance empirically measured. The final program consists of the implementation of each phase that had the best performance when executed in isolation.

When applying the phase optimization technique, the programmer makes an implicit assumption that by combining the best version of each phase the optimal implementation of the entire program will have been achieved. In uniprocessor systems, this assumption may not be valid if input/output formats are mismatched (i.e., each phase requires a different data format). It will be shown that when programming some parallel processing systems, even if the input/output formats match, the phase optimized approach may not yield the optimal solution due to aspects, not found in typical uniprocessor machines, that complicate the interaction among phases.

To demonstrate the process of mapping algorithms to mixed-mode processors, a practical, multiple phase image processing application, the **edge-guided thresholding (EGT)** algorithm, is discussed in detail. After background information is given in Section 2, the EGT algorithm is introduced in Section 3. In Section 4, several implementations for each phase of the EGT algorithm are presented and thoroughly analyzed. The trade-offs among the implementations are studied along with experimental data obtained from a mixed-mode computer, the PASM parallel processing prototype. Section 5 contains an analysis of performance measurements for several versions of the entire algorithm, where each version consists of a different combination of phase implementations. These analyses attempt to concentrate on inherent relative differences among SIMD, MIMD, and mixed-mode processing in an implementation independent way.

The goal of this paper is to contribute to the development of an overall theory of mixed-mode parallelism, which involves exploring both the strengths and the weaknesses of the approach. The EGT algorithm used in this study was chosen because it is a practical algorithm and has computational characteristics that lead to interesting results in terms of mixed-mode mapping. How well the EGT algorithm compares to other similar image processing techniques is not the issue here; it is its computational characteristics that are important for this study, not its image processing performance. By recognizing how aspects of the com-

This research was supported by the Naval Ocean System Center under the High Performance Computing Block, ONT, by the Office of Naval Research under grant number N00014-90-J-1937, and by the National Science Foundation under grant number CDA-9015696.

[*] Author's current address: Sequent Computer Systems, Inc. 15450 S. W. Koll Parkway Beaverton, OR 97006-6063.

putational structure of this algorithm interact with mixed-mode parallel architectures, insights can be gained that will help further the general study of mapping algorithms onto massively parallel architectures and the development of techniques for automatic compilation of parallel code. The results discussed here demonstrate a situation where the advantages of a mixed-mode approach are limited.

2. Background and Mixed-Mode Architectural Model

There have been at least three mixed-mode parallel prototypes built. TRAC [19], designed and built at The University of Texas at Austin, implemented a mixed-mode architecture through a special SW-baynan interconnection network. OPSILA [1, 2, 10], from Laboratoire de Signaux et Systemes in Nice, France, has the capability to execute instructions in either SIMD or **SPMD** (Single Program - Multiple Data stream) mode, a sub-class of MIMD.

The mixed-mode architectural model that has been assumed for this research is based on the PASM (**p**artitionable SIMD/MIMD) design for a large-scale parallel processing system [23]. A small-scale 30-processor prototype has been built at Purdue with a computation engine of 16 processing elements (**PEs**), each consisting of a processor-memory pair [24]. The PEs can communicate with each other through a circuit-switched, extra-stage cube network, a variation of the multistage cube network. The PE processors are MC68000s and perform the actual SIMD and MIMD computation. In SIMD mode, the instructions are broadcast by a control unit (**CU**) and executed synchronously by all the enabled PEs using data contained in their local memories. There exists a mechanism for the CU to globally disable PEs and for PEs to transmit the results of tests on local data to the CU.

In MIMD mode, PEs execute instructions from their local memory, asynchronously with respect to each other. To change from MIMD parallelism to SIMD parallelism, PEs begin to fetch instructions from a logical address space corresponding to a queue of instructions sent by the CU. Any instruction fetches other than from this queue return the program to the MIMD mode of parallelism. Thus, any branch statement can change the mode of computation. The CU can also broadcast data to the PEs via the SIMD queue allowing them to use values computed dynamically by the CU. More details of the PASM prototype are discussed in the text as needed.

The goal of this research is to study the inherent properties of mixed-mode parallel architectures and their computational characteristics. The PASM prototype is used as a tool in the experiments. The results obtained from the experiments are evaluated and analyzed in terms of the properties of the modes of parallelism so the conclusions will be as general and implementation detail independent as possible.

Experimentation has been performed on a PASM prototype submachine with four PEs. Unless otherwise noted, all the instructions that would be executed in a larger system are executed in the four PE configuration. Because the execution time of the parallel EGT algorithm is related to the data set size in each PE and not the number of PEs, the results obtained on a larger system would be similar to those presented.

Several algorithm studies have been performed using PASM demonstrating the potential of a mixed-mode architecture. Examples include parallel FFTs [7], bitonic sequence sorting [13], and matrix multiplication [12]. Although the previous studies have discovered several important concepts about mixed-mode computing, the work in this paper differs in that the previous studies have concentrated on algorithms that typically would be a single phase in a larger task. In this research, the mapping of multiple phase algorithms to mixed-mode architectures is explored. The result of the juxtaposition of phase implementations that are optimal in isolation is the focus of this study. General discussions of different aspects of the trade-offs among SIMD mode, MIMD mode, and mixed-mode parallelism can be found in [6, 15, 16].

3. The EGT Algorithm
3.1. Overview of Algorithm

To study the mapping of algorithms to mixed-mode parallel processors, a detailed discussion of an image processing algorithm, the EGT algorithm [26] is presented. The paper does not attempt to argue the quality of the EGT algorithm from an image processing perspective (it is irrelevant), but instead uses the algorithm as a vehicle to examine computational characteristics that might arise in a variety of applications.

The purpose of the EGT algorithm is to determine the optimal threshold to be used for **segmentation** [21] (requantizing the image from **pixels** (picture elements) with 256 gray levels to one of two levels (zeros and ones)). Any pixel with a value less than the threshold level has a zero in the corresponding position of the segmented image. Any pixel with a value greater than or equal to the threshold level has a corresponding one. The EGT algorithm is designed for processing images of size $M \times M$ pixels consisting of an object immersed in a background. Ideally, but not in practice, the object is all one gray level and the background is another gray level (forming a "two level" image). In practice, the image may have variable lighting so that the object, as well as the background, may have slowly varying gray levels. Such a situation arises, for example, in the automated inspection of printed circuit boards.

The goal of the EGT algorithm is to determine the *best* threshold level that will differentiate the object from the background after segmentation, i.e., the object will be all ones and the background all zeros, or vice-versa. The **figure of merit** (FOM) for determining the *best* threshold level is how well the edges of the object in the segmented image match the areas of high gradient in the original image, i.e., areas with a large rate of change in pixel values among neighboring pixels. An **edge** in the segmented image is defined as a zero-to-one or one-to-zero transition in pixel values. Thus, a pixel in the segmented image is considered an **edge pixel** if it has at least one neighboring pixel of opposite value. The gradient at each pixel in the original image is found by applying the Sobel operator [11]. The FOM is given by the average gradient (Sobel value) per edge pixel generated by the given threshold. A threshold value with a large FOM generates edges that highly correspond (on the average) to areas with large gradients in the original image. Thus, the threshold with the largest FOM is the one used to segment the image. The EGT algorithm consists of four conceptual steps: 1) use the Sobel operator to calculate the gradient for each pixel, 2) determine the FOM of each possible threshold (1 to 255), 3) determine the threshold level with the largest FOM, 4) segment the image based on the threshold determined in step 3. The following subsection discusses these steps in more detail.

3.2 The Basic EGT Algorithm

This subsection describes the basic algorithm from [18] that achieves the desired effect presented in the above conceptual outline. Although a serial algorithm is described, only the slight modifications described in the next subsection are required to create the parallel implementation. The algorithm assumes that the $M \times M$ image to be processed is contained in the two-dimensional pixel array $\mathbf{PX(i,j)}$, where $0 \le i,j \le M-1$. After the optimum threshold is found, it is used to segment the image and the result is placed in the two-dimensional array $\mathbf{SEG\_PX(i,j)}$, where once again $0 \le i,j \le M-1$.

To determine the optimum threshold, several quantities are computed for each pixel in the image. Applying the Sobel operator to a given pixel, $PX(i,j)$, involves a calculation based on the pixel and its eight nearest neighbors. The Sobel operation first calculates two components of the gradient (S_x and S_y), using the equations given in Fig. 3.1. The two components of the gradient can be combined into a single value for each pixel and stored in **GR** by the equation given in Fig. 3.1. This equation can be implemented with integer arithmetic (as opposed to floating point) and thus was chosen for this study.

As each pixel is processed, statistics are updated for each possible threshold, T ($1 \leq T \leq 255$). The statistics are stored in two arrays **EDGE_PX_COUNT(T)** and **EDGE_GR_SUM(T)**. The number of edge pixels that would be produced by segmenting the image with a threshold value of T is stored in EDGE_PX_COUNT(T), and the sum of the gradients corresponding to those edge pixels are stored in EDGE_GR_SUM(T). After all the pixels have been processed, the statistics are used to calculate the FOM, for each T.

The complete algorithm, given in Fig. 3.2, consists of four separate loops. The first loop initializes the threshold statistic arrays. Next, the doubly-nested main loop performs the following four steps for each PX(i,j), $0 \leq i,j \leq M-1$. First, the two components of the gradient are calculated and then combined by the equations given in Fig. 3.1. Second and third, the maximum pixel value, **L_MAX**, and minimum pixel value, **L_MIN**, among the 3×3 window of pixels around the PX(i,j) are determined, i.e., the minimum and maximum among PX(i,j) and its eight nearest neighbors. Only for threshold values between L_MIN+1 and L_MAX would some pixels in the window be changed to zeros and some to ones, creating an edge (a zero-one or one-zero transition) involving PX(i,j). The final step in the main loop updates the threshold statistic arrays, EDGE_PX_COUNT(T) and EDGE_GR_SUM(T), for L_MIN+1 $\leq$ T $\leq$ L_MAX (the loop "for T=L_MIN+1 to L_MAX" is not executed if L_MIN=L_MAX)

Once all the pixels are processed, the statistics collected for the threshold values are used to calculate the FOM, for each possible threshold value. The FOM is computed as the sum of the edge pixel gradients divided by the number of edge pixels produced from the threshold, giving the average gradient per edge pixel. As the FOM are computed, the threshold with the largest FOM is determined, and stored in the variable **BEST_THRS**. That threshold is used in the final loop to segment the image.

For sake of efficiency, the two-dimensional arrays PX(i,j) and SEG_PX(i,j) are implemented as one-dimensional arrays. Pointers are used to access the array.

3.3. Parallel Implementation of EGT

The algorithm described above can be easily implemented on a parallel processing system with **N** PEs. The N PEs are configured logically as a $\sqrt{N} \times \sqrt{N}$ grid. The $M \times M$ image is divided into N sub-images, each of size $L \times L$ pixels, where $L = M/\sqrt{N}$. Each sub-image is loaded into the local memory of a PE (assuming a distributed memory system), and each PE processes its own local sub-image. The algorithm in Fig. 3.2 must be altered by reducing the bounds of the index variables i and j so that $0 \leq i,j \leq L-1$. The combined segmented sub-images will match the segmented image produced by a serial system that generates an individual threshold for each sub-image.

Implementation on a distributed memory system requires an additional step to resolve the problem that arises when processing the pixels on the sub-image borders. Because the Sobel, local minimum, and local maximum operations are based on a 3×3 window around PX(i,j), some pixels needed when processing the sub-image border points are contained in an adjacent PE's local memory. Inter-PE communication must take place to transfer these sub-image border pixels. Also, each corner point must be transferred to the PE containing the corresponding catercorner sub-image. In general, $4(L+1)$ pixels must be transferred for each PE. It is assumed that all PEs may transfer corresponding sub-image border pixels simultaneously. If all these transfers are performed prior to the main EGT loop, the data will be available when needed.

4. Phase Evaluation

This section presents several implementations for each phase of the EGT algorithm on a mixed-mode architecture. The phases are: transfer of sub-image border points, Sobel calculation, local minimum/maximum determination, threshold statistics update, FOM calculation, and segmentation. The goal of this discussion is to optimize the implementation of each phase in isolation. Particular attention is paid to the selection of parallelism modes and its effect on performance.

4.1. Transfer of Sub-image Border Points

The first phase of the parallel EGT algorithm transfers the sub-image border points to PEs containing the neighboring sub-images through messages passed via the interconnection network. The interconnection network is assumed to be circuit-switched. Although it is possible for communication paths to be blocked, it is assumed that the interconnection network is robust enough to support the permutations needed for a logical $\sqrt{N} \times \sqrt{N}$ grid of PEs, as is the case for the extra-stage cube network in PASM.

Three possible implementations of this phase are the PEs can be in SIMD mode, MIMD mode using a polling scheme, or MIMD mode with explicit synchronization between transfers. The key difference between these three implementations is the way in that the PEs know when data is available to be read from the network.

The SIMD algorithm is very efficient because the PEs are implicitly synchronized. After the source PE sends data into the network, only a small deterministic delay is needed before the destination PE can read the data. When this phase is implemented in MIMD, PEs are not necessarily synchronized even though they are executing the same sequence of instructions. Therefore, each PE must poll the network to verify that data has arrived before reading its network port. Furthermore, in general, SIMD allows the CU to execute certain instructions (e.g., increment and test loop indices) concurrently with the PEs' computations [17]. This **CU overlap** can result in significantly improved performance over MIMD, where the PEs execute all the instructions.

A third alternative is to execute the phase in MIMD, but instead of polling the network, the PEs explicitly synchronize after the data has been written into the network. The performance of this approach depends upon the synchronization overhead. One method to explicitly synchronize when the PEs are in MIMD is to produce a barrier synchronization [20] among the processors. A **barrier synchronization** can be implemented in PASM by accessing a single word from the logical address space corresponding to the CU instruction queue [9]. This method of synchronization is very efficient and incurs very little overhead, i.e., the time for a single reference to the CU instruction queue.

The performance of the SIMD and two MIMD implementations of this phase was measured on the PASM prototype with various sub-image sizes. Due to the overhead of polling and the lack of CU overlap, the MIMD version with polling takes approximately twice as long to execute compared to the SIMD version. Although the barrier synchronized MIMD version is better than the polling MIMD version, it is 25% slower than the SIMD version due to the CU overlap in SIMD. See [5] for a graph of this data.

4.2. Main EGT Loop

The next three phases, Sobel calculation, local minimum/maximum, and the threshold statistics update, are the central portion of the EGT algorithm. These phases are executed once for each pixel in the sub-image, and therefore have a time complexity of $O(L^2)$.

4.2.1. Sobel Calculation

The Sobel phase has four major components, as seen in lines 4 through 6 in Fig. 3.2 and Fig. 3.1. First, the loop control must be performed for both the i and j loops. Second, the addresses of the appropriate elements in the PX array must be computed. These addresses are used in the third step to calculate S_x and S_y. Finally, S_x and S_y are combined to form the gradient.

Each of these four basic operations can take place in either SIMD or MIMD mode. For example, the loop control can be performed on the PEs in MIMD mode or overlapped on the CU in SIMD mode. The computation of the PX addresses is independent of the data on each PE and can also be executed on the CU in SIMD mode. The computation of S_x and S_y uses only simple mathematical operations, independent of the actual data values. Thus, once the addresses are calculated for this portion there is no difference in execution time when implemented in either parallelism mode. On the contrary, when combining the two components into a single gradient data conditional statements are needed to form the absolute values of S_x and S_y. In general, data conditional statements execute differently in each mode. Forming the absolute value in MIMD mode is straightforward using a simple *if-then-else* statement.

Typically, data conditional statements are less efficient in SIMD. In this study, three implementations of data conditional statements were evaluated for each pertinent phase of the EGT algorithm. First, the present PASM scheme where PEs inform the CU of value of the local conditional (global masking) was measured. Second, an improved scheme where PEs disable themselves, as used in the MPP [3, 4], MasPar [8], and the Connection Machine [27] (local PE masking), was emulated [22]. Finally, a special boolean function was used to remove the need for masking, while increasing the instruction count (an example of this type of boolean function is given in Subsection 4.2.3). In this paper, only the SIMD technique which performed the best in each situation is reported.

As shown in Fig. 4.1, the benefit of CU overlap in SIMD (using emulated local PE masking) outweighs its inefficiency in executing conditionals. A mixed-mode approach, consisting of the loop control and Sobel operation in SIMD, and the absolute value in MIMD, has an execution time between the uni-mode versions. This is influenced by the small, but noticeable, overhead of switching modes introduced by the current implementation of C used for these experiments.

4.2.2. Local Minimum/Maximum Determination

The calculation of the local minimum and maximum involves determining the smallest and largest values of the nine pixels in a 3×3 window. This phase requires eight data conditional statements for each of the operations.

Measurements on the PASM prototype of L_MIN and L_MAX in lines 7 through 12 in Fig. 3.2 executed L^2 times show that a MIMD implementation achieves the best performance. This is because in MIMD the PEs are independent, allowing each PE to execute the "then" statement or not, depending on the data in its local memory. Those PEs that do not execute the "then" may continue on to the next "if" statement without waiting for the other PEs to finish their execution of the "then." The best SIMD implementation (using emulated local PE masking) is less efficient when executing a sequence of conditionals because when a PE does not need to execute the "then" statement, it must remain idle until the other PEs have completed it. The time for a "then" statement is incurred if at least one PE needs the instructions. More detailed information is given in [5].

4.2.3. Threshold Statistics Update

The threshold statistics update (**TSU**) routine is the most important phase of the EGT algorithm because it contains the inner-most loop and may execute for more time than the other phases (see Fig. 3.2). Because its flow of control depends upon the PE data, it will be shown that this phase is best implemented in MIMD.

During each invocation of the TSU phase, the EDGE_GR_SUM(T) and EDGE_PX_COUNT(T) arrays must

be updated for each T (L_MIN $<$ T $\leq$ L_MAX). This requires each PE to loop through the arrays with the index beginning at the local minimum plus one and finishing at its local maximum.

The MIMD implementation is given in lines 13 through 15 of Fig. 3.2. Because the PEs are independent, each PE is able to begin and end its loop based on its local data.

This phase is not well suited to SIMD because each PE may need to perform a different number of iterations based on its own L_MIN and L_MAX values. The CU must issue a single instruction stream that will allow each PE to iterate the proper number of times.

Let $D_P(i,j) = (L\_MAX - L\_MIN)$ when processing PX(i,j) on PE P. In MIMD mode, PE P performs $D_P(i,j)$ iterations of the TSU loop and then continues processing the next pixel. In SIMD mode, the CU must broadcast the loop body at least $\max_P[D_P(i,j)]$ times, $0 \leq P < N$, resulting in two performance penalties. First, some PEs may be idle waiting for other PEs to execute this required number of iterations. Second, the CU must somehow determine $\max_P[D_P(i,j)]$, $0 \leq P < N$.

One way to implement this in SIMD is to have the CU loop until all the PEs are done, as shown in Fig. 4.2. Each PE contains its own local counter, T, that is initialized to the value of L_MIN. During each iteration, if a PE's local counter is less than its L_MAX, the corresponding threshold's statistics are updated. Otherwise, the PE becomes disabled and will not execute the remaining iterations. The CU continues to broadcast the loop body until no more PEs are enabled. Performing the data conditional statement with a PE local enable mask or boolean function mechanism will not be an advantage over a global mask in this situation because the CU must receive the condition code bits for the "while" statement before broadcasting the loop body.

An alternative SIMD implementation, shown in Fig. 4.3, removes the need to transfer the condition code bits from the PEs to the CU. Instead, the CU always iterates 255 times, i.e., the maximum number of iterations. The PEs can then use the boolean function to update the threshold statistics appropriately. The boolean function returns a zero if its argument evaluates to false and minus one (all bits one) if the argument evaluates true. In this algorithm, on PEs where L_MIN $<$ T $\leq$ L_MAX the boolean function sets $B = 11...111 = -1$, so that with two's complement arithmetic, $-B = +1$. On PEs where T is outside this range, B is set to zero. The third SIMD implementation, shown in Fig. 4.4, uses the interconnection network to find the maximum iteration count among the PEs. This iteration count is transferred to the CU, that uses it as the loop index bound. The PEs use the boolean function to update their threshold statistics, similar to the previous SIMD implementation described. This method is potentially beneficial because only a single data transfer is needed from the PEs and CU, and only the number of iterations needed are executed. A penalty is incurred by the need to determine the maximum iteration count among the PEs via the interconnection network. Using a recursive-doubling [25] scheme, this requires $\log_2 N$ inter-PE data transfers (for N PEs) that are not needed in the other SIMD implementations.

The graph in Fig. 4.5 shows the performance of each of the implementations discussed above. The execution time of each version is plotted versus $D = D_P(i,j)$. This data was collected by executing the TSU phase for a single pixel with $D_P(i,j) = D$ in all PEs P. As predicted, the MIMD implementation performs best for all values of D. The SIMD implementation that always iterates 255 times requires a constant amount of time, independent of the data. Compared to this version, the implementation with the "while" statement in SIMD mode attains a better performance for $D < 120$. As the value of D increases, the overhead of testing the condition codes in the "while" implementation is larger than the time to perform the additional $255 - D$ iterations executed in the constant loop of 255 iterations.

The recursive doubling version is much slower because of the PASM prototype's inefficient mechanism used to transfer data from the PEs to the CU. The dotted curve in Fig. 4.5 labeled "SIMD(rec doub)*" corresponds to the recursive doubling implementation without the approximately 65,000 clock cycles needed for the PE to CU data transfer and is included to attempt to circumvent this artifacts of the prototype implementation details. Because the execution time of the recursive doubling function is $O(\log_2 N)$, the $N = 4$ curve representing this scheme would be raised for larger N, while the other curves remain constant. Each doubling in the number of PEs requires an additional step in the recursive doubling algorithm. To account for the extra network setup and data transfers, an additional 2000 clock cycles (approximately) are required for each doubling of the PE count above four. For example, the curve for a 1024 PE configuration would be raised by 16000 clock cycles (independent of D).

In summary, the TSU phase is best implemented in MIMD. The juxtaposition of this phase with the remaining program is analyzed in detail in Section 5.

4.3. Figure of Merit Calculation

The FOM phase determines the threshold that achieved the highest average gradient per edge pixel. Each PE calculates a BEST_THRS for its sub-image. This is a simple routine that contains a loop with 255 iterations independent of the image size or number of PEs (see Fig. 3.2). This phase has asymptotic time complexity of $O(1)$ and does not greatly influence the overall execution time of the EGT algorithm and therefore will not be discussed in detail.

4.4. Segmentation

The segmentation phase involves setting each pixel in the segmented image to either zero or one, depending on value of the corresponding pixel in the original image in comparison to the threshold level. The PEs can perform this operation on their sub-images concurrently, so the asymptotic time complexity is $O(L^2)$. The straightforward MIMD implementation is given in lines 23 trough 27 of Fig. 3.2.

The MIMD algorithm can be implemented in SIMD, but the performance would be poor because the if-then-else statement would be serialized. A better implementation, given in Fig. 4.6, removes the conditional statement. A segmentation table, SEG_TABLE(V), is created that has an entry for each possible pixel value V, $0 \leq V \leq 255$. The entry for V in the segmentation table is the corresponding segmentation value of a pixel with value V, i.e., if V < BEST_THRS then SEG_TABLE(V) = 0, otherwise SEG_TABLE(V) = 1. The table is first created using the boolean function, and then each pixel in the image is segmented by indexing into the table (a very quick operation).

The relative performance of the straightforward algorithm implemented in MIMD, and the table method implemented in MIMD and SIMD is influenced by three factors. First, there is $O(1)$ overhead to initialize the table when using the table method. Second, the number of instructions in the body of the segmentation loop is slightly less in the table method, causing the straightforward MIMD implementation to be slower than the MIMD table method. Third, in SIMD the loop control can be overlapped on the CU. As the sub-image size grows, the table initialization becomes negligible, and the SIMD version clearly outperforms either MIMD version.

5. Evaluation of Mode Selections
5.1. Phase Optimized Implementation

This section discusses several implementations of the complete EGT algorithm. A phase optimized version based on the results in Section 4 is compared to a pure MIMD and several pure SIMD implementations.

The results discussed in Section 4 determine the mode selection for each phase of the EGT algorithm that produces the best performance of that phase in isolation. This phase optimized implementation has the transfer of sub-image border points, Sobel, and segmentation phases in SIMD. The local minimum/maximum determination, TSU, and FOM phases are implemented in MIMD. Because the loop control for the main EGT loop is considered part of the Sobel phase, it is executed on the CU in SIMD and can be overlapped with the PEs' execution of the loop body (i.e., the Sobel, local minimum/maximum, and TSU phases).

5.2. Experimental Data Sets

The following subsections present experimental data obtained by measuring the execution time of the different EGT algorithms on the PASM prototype. Two test images were selected to provide a variety of testing conditions that can be precisely analyzed.

Test Image I, shown in Fig. 5.1, consists of a square object immersed in a background. The difference between the gray levels of the object and the background is given by the parameter D(edge) $(0 \leq D(edge) \leq 255)$. For an $M \times M$ image, the object has dimensions $M/2 \times M/2$ and is centered within the background. The data is distributed such that each PE processes $L \times L$ pixels, including a $L/2 \times L/2$ corner of the object.

Test Image II, shown in Fig. 5.2, consists of a series of stripes. Each stripe is two pixels wide, and has a gray level of D(edge), and the background has a gray level of zero. Each sub-image has a series of $L/16$ stripes separated by 14 rows of zero-valued pixels. There is a one-pixel wide border of zero-valued pixels on all sides of each sub-image. The stripes are organized such that when divided into four sub-images they are skewed by four pixels so that no two sub-images contain a stripe in the same relative position. If the four sub-images were to be laid one another, no stripes would overlap and two rows of zero-valued pixels would separate each stripe. Test Image II was selected to study the effect of the parameter $D_P(i,j)$, the TSU iteration count defined in Section 4. For each pixel position i, j in the sub-image, exactly one sub-image has a non-zero $D_P(i,j)$.

Tests were performed using four sizes of each test image. Image sizes of 64×64, 128×128, 256×256, and 512×512 pixels were partitioned into 32×32, 64×64, 128×128, and 256×256 sub-images for four PEs, respectively.

5.3. Phase Optimized Mixed-Mode vs. Pure MIMD
5.3.1. Analysis of Test Image I Performance

The graph in Fig. 5.3 plots the execution times of two complete EGT algorithm implementations when processing various sizes of Test Image I. This subsection compares and analyzes the performance of the phase optimized and pure MIMD implementations. As seen from the graph, the phase optimized version Test Image I is slower than the pure MIMD version for all sub-images sizes tested. For reasons discussed below, as the image size increases the margin between the execution times of the two implementations diminishes.

There are several opposing factors that influence the relative performance of the programs. Fig. 5.4 shows the breakdown by phases for Test Image I of size $M = 64$ ($L = 32$). The CU overlaps the pixel pointer incrementing in the phase optimized version, while that time is included in the Sobel section of the pure MIMD version. The transferring of sub-image border points, Sobel, and segmentation phases are shown to be more efficient in SIMD, an advantage for the phase optimized program. This advantage does not outweigh the relatively large amount of time lost during the TSU phase. The TSU phase is implemented in MIMD mode in both programs, so why does its execution appear to take longer in the phase optimized version? The answer lies in what happens after the TSU phase.

The phase optimized program executes the TSU phase for a given PX(i,j) in MIMD mode and then returns to SIMD mode to execute the Sobel phase for the next pixel. As discussed in Section 4, when PE P processes PX(i,j) in MIMD mode, the time to execute the TSU phase is proportional to $D_P(i,j)$, the number of iterations PE P executes in the TSU loop when processing PX(i,j). Because entering SIMD mode implies synchronization among all of the PEs, when $D_P(i,j)$ differs among the PEs for a given pixel position, some PEs remain idle waiting for the PE with the largest $D_P(i,j)$ to finish the TSU phase. The effective iteration count for all the PEs becomes $\max_P[D_P(i,j)]$, $0 \le P < N$.

For a given PX(i,j), the TSU phase consists of incrementing two arrays of variables $D_P(i,j)$ times. Let $\mathbf{T_{inc}}$ be the time to increment a single variable. In general, when processing the $L \times L$ sub-images using the phase optimized implementation, the total time spent executing the TSU phase, $\mathbf{T_{PO}(TSU)}$, is given as:

$$T_{PO}(TSU) = \sum_{i=0}^{L-1} \sum_{j=0}^{L-1} [\max_P[2T_{inc} \times D_P(i,j)]], \; 0 \le P < N.$$

If the Sobel phase is executed in MIMD, as in the pure MIMD implementation, once a PE has completed the TSU phase, it can immediately proceed with the execution of the Sobel phase, i.e., it does not need to synchronize with other PEs. In general, the total execution time for the TSU phase in the pure MIMD implementation, $\mathbf{T_M(TSU)}$, is the maximum among the PEs to independently process the entire sub-image, or:

$$T_M(TSU) = \max_P[\sum_{i=0}^{L-1} \sum_{j=0}^{L-1} [2T_{inc} \times D_P(i,j)]], \; 0 \le P < N.$$

$T_{PO}(TSU)$ is said to have a "sum of max's" execution time for the TSU phase, whereas the $T_M(TSU)$ has a "max of sum's" execution time. The "sum of max's" is always greater than or equal to the "max of sum's," meaning $T_{PO}(TSU) \ge T_M(TSU)$. $T_{PO}(TSU)$ equals $T_M(TSU)$ only when, for each pixel, all PEs execute the identical number of iterations in the TSU phase, i.e., for any fixed i and j, $D_P(i,j) = D_0(i,j), 0 \le P < N$.

This phenomenon is analogous to the situation analyzed in [12] when the execution time for a sequence of SIMD instructions depends upon the data, e.g., the PASM prototype's integer multiply instruction. Because synchronization occurs after every instruction, similar to the synchronization after the TSU phase, a "sum of max's" execution time occurs for the sequence of SIMD instructions.

Most pixels in Test Image I have neighbors of the same value, i.e., $D_P(i,j) = 0$. The exception occurs when a PE is processing one of the $L+1$ outside edge pixels of the object or one of the $L-1$ inside edge pixels. These 2L pixels in each sub-image have D(edge) = 80, the value of $D_P(i,j)$ where PX(i,j) is on an edge of the object. In Test Image I, D(edge) is the same for all edge pixels and all PEs. When the PEs are executing the phase optimized program, they are all concurrently processing the same pixel position. For Test Image I, the edge pixels are processed in pairs. When a PE is processing an edge pixel at least one other PE is also processing an edge pixel. For most pixel positions only two sub-images contain an edge.

In Test Image I, there is a difference in $D_P(i,j)$ among the PEs for all pixel positions on the edges of the object, except for the four pixels surrounding the corner in each sub-image. When a PE is processing one of these four pixels, the other PEs are processing a corresponding corner pixel in their sub-images. There are a total of 4L pixel positions where $D_P(i,j) = D(edge)$ for at least one PE, i.e., $\max_P[D_P(i,j)] = D(edge)$, $0 \le P < N$, for 4L values of i and j. The other pixel positions have $D_P(i,j) = 0$ for all P. Recall from above that for any single PE only 2L pixels have $D_P(i,j) = D(edge)$, and the rest have $D_P(i,j) = 0$. Therefore $[\sum_{i=0}^{L-1} \sum_{j=0}^{L-1} [2T_{inc} \times D_P(i,j)]] = (2L) \times D_P(edge)$ for all PEs.

Combining the above information with the equations for $T_{PO}(TSU)$ and $T_M(TSU)$, the execution time for their entire TSU phase can be given as:

$$T_{PO}(TSU) = 4L \times 2T_{inc} \times D(edge) = 8L \times T_{inc} \times D(edge)$$

$$T_M(TSU) = (2L) \times 2T_{inc} \times D(edge) = (4L) \times T_{inc} \times D(edge).$$

The pure MIMD program executes the TSU phase for Test Image I approximately twice as fast as the phase optimized version. From the breakdown of the phase execution times in Fig. 5.4, there are other phases that execute faster in the phase optimized program. That particular figure has D(edge) = 80 and therefore shows the TSU phase dominating the execution time. But if D(edge) is decreased, the TSU phase has less effect on the total execution time. Fig. 5.5 shows the effect of different D(edge) values on the total execution time for Test Image I of size L = 32. For large values of D(edge), the MIMD implementation executes substantially faster because the TSU phase dominates the execution time. But, for small values of D(edge), i.e., 1 or 2, the phase optimized version performs best. With smaller values of D(edge), the TSU phase has less impact on the total execution time. As the other phases begin to dominate, the relative performance of the phase optimized program increases.

The sub-image size also changes the relative performance. The number of edge pixels in Test Image I is O(L), so that as the sub-image size grows at $O(L^2)$, the proportion of edge pixels to non-edge pixels decreases. As the image size increases, the other phases begin to dominate over the TSU phase in determining the total execution time. This causes the phase optimized performance to approach that of the MIMD implementation as the image size increases, as shown in Fig. 5.3.

5.3.2. Analysis of Test Image II Performance

This subsection applies the analysis developed in the previous subsection to Test Image II. Every sub-image of Test Image II consists of L/16 stripes, each two pixels wide. As with Test Image I, $D_P(i,j) = D(edge)$ for all pixels where $D_P(i,j) > 0$, however, for Test Image II D(edge) = 255. Each stripe has $4L + 4$ edge pixels, giving a total of $(L/16)(4L+4) = (L^2 + L)/4$ pixels in each sub-image where $D_P(i,j) = D(edge)$.

The positioning of the stripes in Test Image II is a "worst case" for the phase optimized implementation because all L^2 pixel positions have an edge pixel in only one sub-image. When the phase optimized program processes a given pixel position, only one PE executes D(edge) iterations in the TSU phase. The other PEs remain idle, effectively serializing this phase across the PEs. The total amount of time spent executing the TSU phase is the same as a serial program. The pure MIMD version does not encounter this serialization effect resulting in an algorithm approximately three times as fast as the phase optimized implementation, as seen in Fig. 5.6.

5.4. Phase Optimized Mixed-Mode vs. Pure SIMD

Two pure SIMD implementations have been created and tested on the PASM prototype. The two versions have the best SIMD implementation of each phase with different TSU phases. One version has the TSU phase implemented with the constant 255 iteration loop given in Fig. 4.3. The other version contains the "while" version given in Fig. 4.2. For this test image with D(edge) = 255, MAX_D = 255 for all pixels, and the Fig. 4.4 version would always be longer than the Fig. 4.3 version.

As seen in Fig. 5.3 and Fig. 5.6, the phase optimized implementation performs better than either of the pure SIMD versions for both test images. These results occur because, as shown in Subsection 4.2.3, the TSU phase is more efficiently implemented in MIMD. The SIMD versions also suffer from the "sum of max's" phenomenon in that, for a given PX(i,j), the CU must broadcast the TSU loop body $\max_P[D_P(i,j)]$, $0 \le P < N$, times for the Fig. 4.2 version and all 255 times for the Fig. 4.3 version.

5.5. Analysis

This section has shown that the relative performance of the phase optimized and pure MIMD implementations of the EGT algorithm is data dependent. Both versions have their disadvantages. The main weakness of the pure MIMD version is its lack of CU overlap in the Sobel phase. Although the phase optimized version exploits CU overlap, this requires the PEs to enter SIMD mode, forcing implicit synchronization. Synchronizing after every invocation of the data dependent TSU phase results in the "sums of max's" phenomenon. This phenomenon has less impact as the variation of $D_P(i,j)$ among the PEs decreases. The results were based on the use of both experiments and algorithm complexity analyses.

A hybrid of these two versions can be created by implementing the transfer of border points and segmentation phases in SIMD mode, and all three phases in the main loop, as well as the FOM loop, in MIMD mode. This version is essentially the pure MIMD version with the transfers and segmentation in SIMD. Because these phases only play a minor role in the total execution time of the algorithm, the new hybrid program performs less than 1% better than the pure MIMD version for both of the test images.

On the PASM prototype, fetching SIMD instructions is slightly faster than fetching MIMD instructions due to differences in the implementation technology of the PE memory and the instruction queue for PEs that is in the CU. This has only limited impact because the instruction execution and data fetching occur at the same speed in both modes. Furthermore, the fetching of the next instruction is often partially overlapped with execution of the current instruction. Because the results presented in this section show that a pure MIMD implementation is better than phase optimized approach, taking into account the difference in instruction fetch time would only substantiate this conclusion.

Many other options of implementing the entire algorithm were considered, but performance was not improved compared to the phase optimized and MIMD approaches. For example, separating the Sobel, Local Min/Max, and TSU phases into their own loops results in lowered performance due to the overhead of storing and retrieving the data multiple times.

If the TSU phase were such that processors were required to communicate or synchronize between each successive pair of iterations, the choice of modes would be different. This inter-iteration data independence of the TSU is a key feature of this algorithm. Without this independence, mixed-mode performance would most likely be best.

6. Summary and Conclusions

While experiments involving other computational situations have shown the advantages of the mixed-mode capability (e.g., [13]), in the case of the EGT algorithm it provides little improvement. The results presented here demonstrate the attention needed to the problem of mapping algorithms to mixed-mode architectures. Clearly, there exists an inherent problem of using the phase optimization approach. When each phase of the EGT algorithm was optimized in isolation, an implementation of the entire algorithm was created that was less than optimal. This can potentially be true for any algorithm whose execution times are data dependent and subtasks can be performed on different processors without synchronization. In these cases, the most efficient mapping can only be determined by applying a more global approach that considers the impact of the temporal juxtaposition of code executed in different modes of execution. In addition, statistical information about the characteristics of the data may need to be incorporated into the mapping decision process.

As stated earlier, the purpose of this paper is to contribute to the understanding of mixed-mode parallelism. This involves exploring both the strengths and weaknesses of this approach to massively parallel computation. The long term goal of these studies is to develop an overall theory of mixed-mode computation that can be used to determine when and how mixed-mode parallelism can best be exploited.

Acknowledgements: The authors thank Prof. Thomas L. Casavant of the University of Iowa, Wayne G. Nation of Purdue University, and the reviewers for their many useful comments.

References

[1] M. Auguin, F. Boeri, "The OPSILA computer," in *Parallel Languages and Architectures*, M. Consard, ed., Elsevier Science Publishers, Holland, 1986, pp. 143-153.

[2] M. Auguin, F. Boeri, "Experiments on a parallel SIMD/SPMD architecture and its programming," *France-Japan Artificial Intelligence and Computer Science Symp. 87*, Nov. 1987, pp. 385-411.

[3] K. E. Batcher, "Design of a massively parallel processor," *IEEE Trans. Computers*, v. C-29, Sep. 1980, pp. 836-844.

[4] K. E. Batcher, "Bit serial parallel processing systems," *IEEE Trans. Computers*, v. C-31, May 1982, pp. 377-384.

[5] T. B. Berg, S. D. Kim, H. J. Siegel, "Limitations imposed on mixed-mode performance of optimized phases due to temporal juxtaposition," *Tech. Report*, E.E. School, Purdue, 1991.

[6] T. B. Berg, H. J. Siegel, "Instruction execution trade-offs for SIMD vs. MIMD vs. mixed-mode parallelism," *5th Int'l Parallel Processing Symp.*, May 1991.

[7] E. C. Bronson, T. L. Casavant, L. H. Jamieson, "Experimental application-driven architecture analysis of an SIMD/MIMD parallel processing system," *IEEE Trans. Parallel & Distributed Systems*, v. 1, Apr. 1990, pp. 195-205.

[8] P. Christy, "Software to support massively parallel computing on the MasPar MP-1," *IEEE Compcon*, Feb. 1990, pp. 29-33.

[9] H. G. Dietz, T. Schwederski, M. T. O'Keefe, A. Zaafrani, "Static synchronization beyond VLIW," *Supercomputing '89*, Nov. 1989, pp. 416-425.

[10] P. Duclos, F. Boeri, M. Auguin, G. Giraudon, "Image processing on a SIMD/SPMD architecture: OPSILA," *9th Int'l Conf. on Pattern Recognition*, Nov. 1988, pp. 430-433.

[11] R. O. Duda, P. E. Hart, *Pattern Classification and Scene Analysis*, John Wiley and Sons, New York, NY, 1973.

[12] S. A. Fineberg, T. L. Casavant, T. Schwederski, H. J. Siegel, "Non-deterministic instruction time experiments on the PASM system prototype," *1988 Int'l Conf. on Parallel Processing*, Aug. 1988, pp. 444-451.

[13] S. A. Fineberg, T. L. Casavant, H. J. Siegel, Experimental analysis of a mixed-mode parallel architecture using bitonic sequence sorting. *J. Parallel & Distributed Computing*, v. 11, Mar. 1991, pp. 239-251.

[14] M. J. Flynn, "Very high-speed computing systems," *Proc. IEEE*, v. 54, Dec. 1966, pp. 1901-1909.

[15] R. F. Freund, "Optimal selection theory for superconcurrency," *Supercomputing '89*, Nov. 1989, pp. 699-703.

[16] L. H. Jamieson, "Characterizing parallel algorithms," in *The Characteristics of Parallel Algorithms*, L. H. Jamieson, D. B. Gannon, R. J. Douglass, eds., MIT Press, Cambridge, MA, 1987, pp. 65-100.

[17] S. D. Kim, M. A. Nichols, H. J. Siegel, Modeling overlapped operation between the control unit and processing elements in an SIMD machine. *J. Parallel & Distributed Computing*, Special Issue on Modeling of Parallel Computers, Aug. 1991.

[18] J. T. Kuehn, H. J. Siegel, D. L. Tuomenoksa, G. B. Adams III, "The use and design of PASM," in *Integrated Technology for Parallel Image Processing*, S. Levialdi, ed., Academic Press, San Diego, CA, 1985, pp. 133-152.

[19] G. J. Lipovski, M. Malek, *Parallel Computing: Theory and Comparisons*, John Wiley and Sons, Inc., New York, NY, 1987.

[20] S. F. Lundstrom, G. H. Barnes, "A controllable MIMD architecture," *1980 Int'l Conf. on Parallel Processing*, Aug. 1980, pp. 165-173.

[21] O. R. Mitchell, A. P. Reeves, K.-S. Fu, "Shape and texture measurements for automated cartography," *1981 IEEE CS Conf. on Pattern Recognition and Image Processing*, Aug. 1981, pg. 367.

[22] W. G. Nation, S. A. Fineberg, M. D. Allemang, T. Schwederski, T. L. Casavant, H. J. Siegel, "Efficient masking techniques for large-scale SIMD architectures," *Frontiers '90; The 3rd Symp. on the Frontiers of Massively Paral-*

lel Computation, Oct. 1990, pp. 259-264.

[23] H. J. Siegel, L. J. Siegel, F. C. Kemmerer, P. T. Mueller, Jr., H. E. Smalley, Jr., S. D. Smith, "PASM: a partitionable SIMD/MIMD system for image processing and pattern recognition," *IEEE Trans. Computers,* v. C-30, Dec. 1981, pp. 934-947.

[24] H. J. Siegel, T. Schwederski, J. T. Kuehn, N. J. Davis IV, "An overview of the PASM parallel processing system," in *Computer Architecture,* D. D. Gajski, V. M. Milutinovic, H. J. Siegel, B. P. Furht, eds., IEEE CS Press, Wash., DC, 1987, pp. 387-407.

[25] H. S. Stone, "Parallel computers," in *Introduction to Computer Architecture,* H. S. Stone, ed., Science Research Associates, Inc., Chicago, IL, 1975.

[26] R. E. Suciu, A. P. Reeves, "A comparison of differential and moment based edge detectors," *1982 IEEE CS Conf. on Pattern Recognition and Image Processing,* June 1982, pp. 97-102.

[27] L. W. Tucker, G. G. Robertson, "Architecture and applications of the Connection Machine," *Computer,* v. 21, Aug. 1988, pp. 26-38.

$$S_x = (1/4)[(PX(i-1,j-1) + 2(PX(i,j-1)) + PX(i+1,j-1)) - (PX(i-1,j+1) + 2(PX(i,j+1)) + PX(i+1,j+1))]$$
$$S_y = (1/4)[(PX(i-1,j-1) + 2(PX(i-1,j)) + PX(i-1,j+1)) - (PX(i+1,j-1) + 2(PX(i+1,j)) + PX(i+1,j+1))]$$
$$GR = |S_x| + |S_y|$$

Fig. 3.1: Sobel Operation for PX(i,j)

```
1: for T = 1 to 255 {  /* Initialize threshold statistics */
2:   EDGE_GR_SUM(T) = 0
3:   EDGE_PX_COUNT(T) = 0 }

4: for i = 0 to M−1 {  /* Main EGT loop */
5:   for j = 0 to M−1 {
6:     GR = Sobel(i,j)  /* Use equations in Fig. 3.1 */
7:     L_MIN = min[ PX(i−1,j−1), PX(i−1,j), PX(i−1,j+1),
8:                 PX(i,j−1), PX(i,j), PX(i,j+1),
9:                 PX(i+1,j−1), PX(i+1,j), PX(i+1,j+1)]
10:    L_MAX = max[ PX(i−1,j−1), PX(i−1,j), PX(i−1,j+1),
11:                 PX(i,j−1), PX(i,j), PX(i,j+1),
12:                 PX(i+1,j−1), PX(i+1,j), PX(i+1,j+1)]
13:    for T = L_MIN+1 to L_MAX {  /* TSU */
14:      EDGE_GR_SUM(T) += GR /* += means increment */
15:      EDGE_PX_COUNT(T) += 1 }}}
16: BEST_THRS = MAX_FOM = 0
17: for T = 1 to 255 {  /*calculate FOM */
18:   if EDGE_PX_COUNT(T) > 0 {  /* Check divide by zero */
19:     FOM = EDGE_GR_SUM(T) / EDGE_PX_COUNT(T)
20:     if FOM > MAX_FOM {  /* Find largest FOM*/
21:       MAX_FOM = FOM
22:       BEST_THRS = T }}}

23: for i = 0 to M−1 {  /* Segmentation */
24:   for j = 0 to M−1 {
25:     if PX(i,j) < EST_THRS
26:       then SEG_PX(i,j) = 0
27:       else SEG_PX(i,j) = 1 }}
```

Fig. 3.2: Basic EGT Algorithm

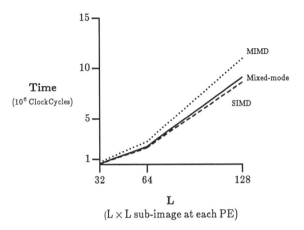

Time
(10^6 ClockCycles)

MIMD
Mixed-mode
SIMD

L
(L × L sub-image at each PE)

Fig. 4.1: Execution Time for Sobel Phase

```
LOCAL_T = L_MIN
while(no_more_PEs_enabled) {
    LOCAL_T = LOCAL_T + 1
    if LOCAL_T ≤ L_MAX
      then  EDGE_GR_SUM(LOCAL_T) += GR
            EDGE_PX_COUNT(LOCAL_T) += 1
    else  disable_PE }
```

Fig. 4.2: SIMD Algorithm with Condition Codes for TSU

```
for T = 1 to 255  {
    Broadcast_from_CU_to_PEs(T)
    B = boolean((L_MIN < T) and (T ≤ L_MAX))
    EDGE_GR_SUM(T) += (GR and B)
    EDGE_PX_COUNT(T) += B  }
```

Fig. 4.3: SIMD Algorithm Without Condition Codes for TSU

```
LOCAL_D = L_MAX − L_MIN
MAX_D = max_among_PEs(LOCAL_D)  /* recursive doubling */
transfer_from_PEs_to_CU(MAX_D)
LOCAL_T = L_MIN+1
for T = 1 to MAX_D {
    B = boolean(LOCAL_T ≤ L_MAX))
    EDGE_GR_SUM(LOCAL_T) += (GR and B)
    EDGE_PX_COUNT(LOCAL_T) += −B
    LOCAL_T += 1  }
```

Fig. 4.4: SIMD Algorithm with Recursive Doubling for TSU

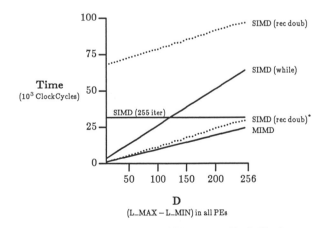

Time
(10^3 ClockCycles)

SIMD (rec doub)
SIMD (while)
SIMD (255 iter)
SIMD (rec doub)*
MIMD

D
(L_MAX − L_MIN) in all PEs

Fig. 4.5: Execution Time for TSU Phase for a Single Pixel

```
for V = 0 to 255
    SEG_TABLE(V) = − boolean(V ≥  BEST_THRS)
for i = 0 to L−1 {
    for j= 0 to L−1 {
        SEG_PX(i,j) = SEG_TABLE(PX(i,j))  } }
```

Fig. 4.6: Table Method of Segmentation

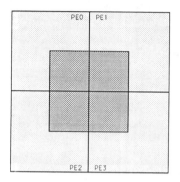

Fig. 5.1: Test Image I

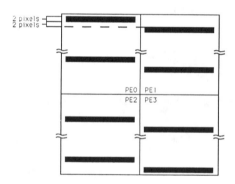

Fig. 5.2: Test Image II

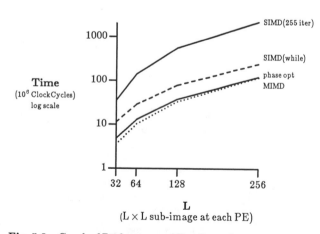

Fig. 5.3: Graph of Performance of Test Image I

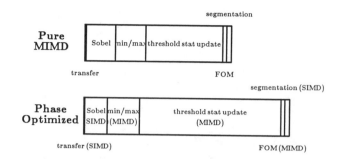

Fig. 5.4: Breakdown of Execution Times for Phase Optimized and MIMD Versions Processing Test Image I (Subimage Size 32×32)

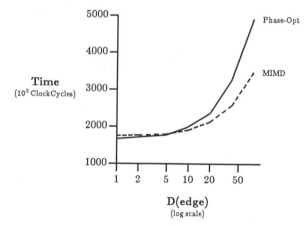

Fig. 5.5: Execution Time vs. D(edge) for Test Image I (L=32)

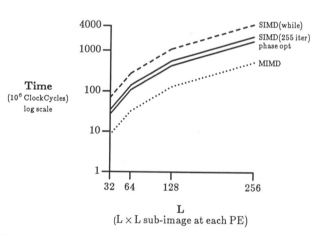

Fig. 5.6: Graph of Performance of Test Image II

MANIPULATION OF PARALLEL ALGORITHMS TO IMPROVE PERFORMANCE

Linda F. Wilson and Mario J. Gonzalez
Computer Engineering Research Center
Department of Electrical and Computer Engineering
The University of Texas at Austin
Austin, Texas 78712

Abstract

Due to the increased use of parallelism in computer systems, designers need tools to evaluate job performance and match problems with candidate architectures. This paper investigates the manipulation of parallel algorithms to increase efficiency on a multiprocessor system with a given number of processors. Algorithms are modeled in this paper using job profiles, which are geometric models that can be used to determine how a job's structure limits its performance in terms of efficiency and speedup on multiple processor systems. A heuristic to restructure parallel algorithms to improve performance is presented, and estimates of increased performance due to the heuristic are given based on extensive simulation.

Keywords: multiprocessing, parallel algorithms, performance evaluation, program structure, algorithm design

Introduction

Multiprocessor systems can enhance performance by devoting multiple processors to a *single program*. To effectively utilize multiple processors, the software must be designed to take advantage of the available resources. A *parallel program* consists of two or more processes, each of which is an independent sequential program that uses hardware and software resources. Independent processes execute concurrently, pausing only when interaction between processes is necessary.

Due to the interaction between processors and the competition for shared resources, the structure of the software written for multiprocessor systems has a large impact on performance. Processor utilization is "limited by the lack of a practical methodology for designing application programs to run on such systems" [5]. In response to the need for a methodology, this paper investigates the manipulation of parallel algorithms to increase efficiency on a multiprocessor system with a given number of processors.

Background

The Job Profile

In 1971, T. C. Chen introduced the concept of using a space-time diagram to represent the processing of a job [1]. "Space-time" referred to the occupation of equipment (in this case, processors) through time. Chen's "standard job" had two parts: one strictly sequential and the other purely parallel. Chen used the term "width" to represent the parallelism of the job at a particular time. Figure 1 shows the space-time diagram of a standard job where the sequential part is modeled by a width of one with duration t_1 and the parallel part is represented by a width of W and duration t_2. Among other things, Chen showed that only a relatively small amount of sequential processing in a problem that otherwise has a high degree of parallelism can dramatically reduce system efficiency and the effective number of processors.

Specifically, Chen showed that if ST_j is the area covered by the space-time diagram of a job, and if ST_p is the area covered by multiple processors moving in time across the job, then three common performance indices—efficiency (η), completion time (τ), and speedup (S)—can be expressed as shown in equations (1)–(3).

$$\eta = \frac{\text{space-time area of job } (ST_j)}{\text{space-time area covered by processors } (ST_p)} \quad (1)$$

$$\tau = \frac{\text{space-time area covered by processors } (ST_p)}{\text{number of processors executing job } (n)} \quad (2)$$

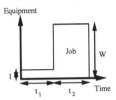

Figure 1: The Space-time Diagram of a Standard Job

$$S = \frac{\text{execution time with one processor}}{\text{execution time with n processors}} = \frac{ST_j}{\frac{ST_p}{n}} = \eta\, n \quad (3)$$

Originally, Chen applied the concept of a profile to a pipelined execution of instructions. He then proceeded to generalize the profile concept to the application of multiple processors across the space-time diagram of a job.

Job Profiles

As stated in [10], job profiles are designed to convey information about a problem using simple geometric techniques. A directed graph and its job profile representation are given in Figure 2. In this representation we assume that a single subtask can be processed in one unit of time. The precedence implied by the directed edges is preserved in the job profile. Subtasks in each column can be executed in parallel, and any given column cannot start execution until all the subtasks in the preceding column are completely processed. The "width" of each column represents the number of processors needed to execute the column in one time unit. Processing of a job is represented by a block of n processors sweeping across the job profile.

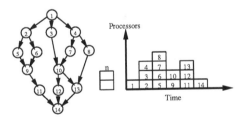

Figure 2: Precedence Graph and Discrete Job Profile Representations of a Job

Using the job profile model developed by Ullah and Gonzalez [7], the space-time area of a job (ST_j) is computed as

$$ST_j = \sum_{i=1}^{C} W_i, \quad (4)$$

where C equals the number of columns in the profile and W_i is the height of the ith column in the job profile (assuming each column is one unit in length). The space-time area covered by the processors (ST_p) is determined by the scheduling discipline and the number of processors. Under the *processor sharing* scheduling discipline, processors continually switch between tasks when $W_i > n$. Processor sharing assumes that task switching time and communication time are both zero, and the space-time area (ST_p) is defined by:

$$ST_p = \sum_{i=1}^{C} \max(n, W_i). \tag{5}$$

With the *processor sweeping* scheduling discipline, each processor is dedicated to a subtask until completion, and no preemption occurs between subtasks. Given n processors, W_i tasks of unit length can be completed in $\lceil \frac{W_i}{n} \rceil$ units of time. The space-time area (ST_p) for processor sweeping is therefore defined by:

$$ST_p = n \left(\sum_{i=1}^{C} \left\lceil \frac{W_i}{n} \right\rceil \right). \tag{6}$$

Processor sweeping is a realistic scheduling discipline, whereas processor sharing is more of a theoretical one. Due to space limitations, only processor sweeping will be discussed in this paper.

Using the above equations, Ullah and Gonzalez developed upper and lower bounds for efficiency, execution time, and speedup for a job represented by a profile of the form shown in Figure 2 and processed using a work-conserving scheduling policy (i.e., one that never leaves a processor idle if there is a task available for execution). In deriving the bounds, Ullah and Gonzalez assumed the ideal case (i.e. the effects of task switching time and communication time were assumed to be negligible).

Special Job Profiles

In [8], Weems introduced the eager and lazy job profiles. The *eager job profile* for a particular algorithm is generated by scheduling each task at the earliest possible time which will not violate any precedence constraints. The length of the eager job profile, τ_∞, is the execution time of the job using an infinite number of identical processors. The *lazy job profile* is created by scheduling each task at its latest possible starting time such that the total length of the profile is τ_∞. Figure 3 shows a simple precedence graph and its corresponding eager and lazy job profiles.

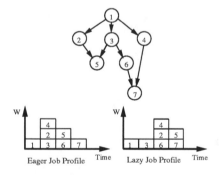

Figure 3: Simple Precedence Graph with Eager and Lazy Profiles

Although Weems proved that the eager and lazy job profiles for a particular precedence graph are unique, he also showed that an algorithm may have many other job profile representations. To represent and categorize the wide range of possible profiles for a particular job, Weems introduced the concept of delayability. The *task delayability*, $\delta(T_i)$, of a task T_i is defined as the amount of time the task can be delayed from its current initiation time without increasing the minimum overall execution time, τ_∞.

$$\delta(T_i) = t_\ell(T_i) - \chi(T_i), \tag{7}$$

where $\delta(T_i)$ is the task delayability, $t_\ell(T_i)$ is the latest possible initiation time for the task (corresponding to the lazy profile), and $\chi(T_i)$ is the current starting time of the task. Notice that the maximum task delayability, $\delta_{max}(T_i)$, is computed by:

$$\delta_{max}(T_i) = t_\ell(T_i) - t_e(T_i), \tag{8}$$

where $t_e(T_i)$ is the earliest possible starting time (corresponding to the eager profile).

The *delayability* (d) of a job profile is defined as the sum of the task delayabilities.

$$d = \sum_{i=1}^{p} \delta(T_i) = \sum_{i=1}^{p} \left[t_\ell(T_i) - \chi(T_i) \right], \tag{9}$$

where p is the number of tasks in the profile. Observe that the eager profile has maximum delayability while the lazy profile has zero delayability.

Factors Affecting Job Profile Manipulation

Movability

Although task delayability provides a measure of the potential for moving a task from one column to another, it does not indicate if the task can move without violating precedence constraints. Thus, a new job profile characteristic must be defined.

The *movability*, m, of a task is the ability to move the task from its current starting time to a later initiation time, assuming no other task moves, without violating precedence constraints or increasing τ_∞. If $m(T_i) > 0$, task T_i can move $m(T_i)$ units of time without violating any precedence constraints. The *initial movability*, m_e, of a task is the movability of the task when all tasks are arranged according to the eager job profile. The *successor set*, $\mathcal{S}_i$, of a task T_i is the set of all tasks T_j such that $T_i \prec T_j$ (i.e. T_j is an immediate successor of T_i). The *predecessor set*, $\mathcal{P}_i$, of a task T_i is the set of all tasks T_k such that $T_k \prec T_i$ (i.e. T_k is an immediate predecessor of T_i).

Task movability is a measure of how far a task may be moved from its present column *even if no other task moves*. This is in contrast with task delayability, which measures the maximum amount a task may move *if* all of its succeeding tasks have moved to their latest possible starting times. Movability and delayability are related in that the maximum potential movability of a task is equal to its task delayability. Thus, task delayability serves as an upper bound for task movability.

The movability of a task is dependent upon the current location of the task's successors. Therefore, task movability can be computed using the following equation:

$$m(T_i) = \begin{cases} \min_{T_j \in \mathcal{S}_i} \{ \chi(T_j) - [\chi(T_i) + \mu(T_i)] \}, & \text{if } \mathcal{S}_i \neq \emptyset \\ t_\ell(T_i) - \chi(T_i), & \text{if } \mathcal{S}_i = \emptyset \end{cases} \tag{10}$$

where $\chi(T_i)$ is the current starting time of task T_i and $\mu(T_i)$ is the time required to execute T_i. The initial movability of a task is computed by using equation (10) with $t_e(T_i)$ as the initiation time of task T_i. For ease of discussion, it is assumed in this paper that all tasks require one unit of time for execution.

Observe that if one task, T_k, is moved from its current initiation time to a time x units later, its movability decreases x units, and the movability of each of T_k's immediate predecessors *may* increase by x units. Task movability, therefore, changes through time and must be recomputed after any task is moved. Naturally, the movability of a particular task cannot increase if none of its immediate successors moves to a later initiation time.

Bounds on ST_p

Given a fixed number of processors, efficiency and speedup for a particular job can be increased by reducing ST_p, the space-time area of the job swept by the processors. A *critical task* is one in which $t_e(T_i) = t_\ell(T_i)$ and $\delta(T_i) = 0$. Since the critical tasks of an algorithm cannot be moved without increasing τ_∞, the location of the critical tasks significantly affects the minimum possible ST_p that can be achieved by job profile manipulation.

The *critical job profile* is the job profile consisting of only the critical tasks in the precedence graph. The width of column i in the critical job profile is denoted by $W_{crit\ i}$. The *critical* ST_p, $ST_{p\ crit}$, is the space-time area swept by the processors while executing the critical job profile. For processor sweeping,

$$ST_{p\ crit} = n \sum_{i=1}^{C} \left\lceil \frac{W_{crit\ i}}{n} \right\rceil = nC + n \sum_{i=1}^{C} \left\lceil \frac{W_{crit\ i} - n}{n} \right\rceil , \qquad (11)$$

where $W_{crit\ i}$ is the width of the critical task profile for column i.

In [9], Wilson derived the following lower bound for ST_p under processor sweeping:

$$ST_p \geq \max \left(ST_{p\ crit}, n \left\lceil \frac{ST_j}{n} \right\rceil \right) = ST_{p\ min}. \qquad (12)$$

The *theoretical minimum* ST_p, $ST_{p\ min}$, of a job is the absolute minimum ST_p as bounded by the number of tasks in the job and the critical task profile. Due to precedence constraints, the theoretical minimum ST_p may not be achievable. However, $ST_{p\ min}$ provides a measure as to how "good" a particular job profile is with respect to ST_p. In the next section, $ST_{p\ min}$ will be used to evaluate a job manipulation heuristic to improve ST_p for the processor sweeping scheduling discipline.

Job Profile Manipulation Under Processor Sweeping

The Processor Sweeping Heuristic

When the processor sweeping scheduling discipline is used, ST_p can be calculated using the following equation, which was derived in [9]:

$$ST_p = nC + n \sum_{i=1}^{C} \left\lceil \frac{W_i - n}{n} \right\rceil . \qquad (13)$$

Note that ST_p is minimized when the summation is minimized. By equation (13), a single task move can decrease ST_p if a move from column i to column k decreases ST_{p_i}, while leaving ST_{p_k} unchanged, where ST_{p_i} is the area in column i that is swept by the processors. Notice that moving one task from column i will decrease ST_{p_i} if and only if $ST_{p_i} = rn + 1$ before the move and $ST'_{p_i} = rn$ after it, where r is a non-negative integer. In other words, ST_{p_i} will decrease if it is left with a "full" column (i.e. W_i is an integer multiple of n). Because a sweep processes n tasks at a time, changes in ST_p and ST_{p_i} occur n units at a time.

A task move from column i to column k leaves ST_p unchanged if the move has no effect on ST_{p_i} and ST_{p_k} or if ST_{p_i} decreases while ST_{p_k} increases. Furthermore, ST_p can increase only if moving a task from i to k leaves ST_{p_i} unchanged and forces ST_{p_k} to increase by adding one task to an already full column. In the processor sweeping heuristic, a task will be moved as long as the move does not increase ST_p.

Processor Sweeping Job Manipulation Heuristic:

1. Using critical path methods (CPM) from graph theory, compute the initial movability of each task in the precedence graph. If no tasks have positive (non-zero) movability, stop.

2. Build the eager profile. Compute $ST_{p\ crit}$, $ST_{p\ min}$, and $ST_{p\ eager}$. If $ST_{p\ eager} = ST_{p\ min}$, stop.

3. For each column k, where k varies from $C - 1$ to 1, do the following:

   (a) Choose a task i with positive movability $m(i)$ from all of the movable tasks in column k according to one of the procedures below:

      i. Simple: Choose a task at random.

      ii. Maxfirst: Choose a task at random from among those with the highest movability.

      iii. Maxadd: Choose a task at random from among those with the most movable predecessors that can gain movability.

   (b) Evaluate each of the candidate columns c, where $c = k + 1$ to $k + m(i)$ (i.e. column c is within $m(i)$ units of column k). If at least one of the candidate columns is not full (i.e. $W_c \neq rn$, where r is a non-negative integer), move task i to the candidate column with $W_c \neq rn$ that is farthest from column k.

   (c) If a task i was moved, update the width of both of the two columns involved in the move, and recompute the movability of each of i's predecessors.

   (d) Compute the current ST_p. If $ST_p = ST_{p\ min}$, stop.

   (e) If column k contains another movable task that has not been examined, goto Step 3(a) and choose another task to evaluate. Otherwise, $k = k - 1$.

   (f) If $k = 0$, stop. Otherwise, goto Step 3(a) and repeat for the new column k.

The goal of the processor sweeping heuristic is to fill as many columns as possible (i.e. to make W_i equal to a multiple of n) without violating precedence constraints. Notice that filling columns includes moving tasks to fill the area bounded by $W = n$.

An important feature of the processor sweeping heuristic is the fact that each column must be processed only once. According to the heuristic, a task is moved as far as possible without increasing ST_p or violating precedence constraints. In this manner, tasks with several units of movability fill columns that are far away so that tasks with little movability can fill nearer columns. Once a task is placed, it cannot be moved again. Furthermore, a movable task remains in its initial column if and only if all later columns within reach are full. Since tasks in later columns are moved before those in earlier ones, a second pass of the heuristic cannot possibly change ST_p since no additional moves can be made.

Since there could be several movable tasks in a particular column, several methods for selecting the next task to place should be considered to determine if the method used has considerable impact on the value obtained for ST_p. Step 3(a) gives three possible task selection methods: Simple, Maxfirst, and Maxadd. Under Simple, tasks are selected at random from among all of the tasks in the column with positive movability. With Maxfirst, the task with the maximum movability is selected. If two or more tasks have the same movability, $m(i)$, and $m(i)$ is the largest movability in the column, the next task to move is selected at random from among those with movability $m(i)$. The Maxadd method chooses the task whose move could add the most movability to its predecessors. When a task T_i is considered, all of its predecessors are examined to determine how many of them *could* gain movability as a result of T_i's move. The task selected for placement is the one with the most predecessors that could gain movability. If there is a tie, a task is chosen at random from among the tied tasks.

After the processor sweeping heuristic is used to reshape a profile, some columns may contain several movable tasks that could not move because all of the reachable columns were already full. Because those tasks with positive movability could not move, some tasks in earlier columns lost the opportunity to move. In fact, one task that is left in its early profile column could prevent many more tasks from moving. Thus, it is important to move as many tasks as possible without increasing ST_p.

If a column that has already been processed by the heuristic still contains at least n tasks with positive movability, a block of n tasks can be moved without affecting ST_p. Therefore, Steps 3(e) and 3(f) of the sweeping heuristic can be modified as follows:

3. (e) If column k still contains at least n tasks with positive movability, move a block of n tasks to the farthest possible column. Keep moving blocks of n tasks until column k contains fewer than n tasks with positive movability.

   (f) Set $k = k - 1$. If $k = 0$, stop. Otherwise, goto Step 3(a) and repeat for the new column k.

Under this modified version of the heuristic, a second pass of the heuristic over the entire profile *may* improve ST_p. If a block of n

tasks is moved to column i, there may be enough movable tasks in the combined set of tasks to move a block of n tasks from column i to column k. To avoid examining all possible consequences of a block move, a second pass of the heuristic may be used to move additional blocks of n tasks. In the second pass, only Steps 3(e) and 3(f) of the modified heuristic are required.

Although the processor sweeping heuristic cannot guarantee that the minimum achievable ST_p will be reached, it can be shown that the heuristic will never make ST_p worse than the ST_p provided by the eager job profile [9]. The simulation discussed in the next section will demonstrate that other task placement methods can make ST_p worse than $ST_{p\ eager}$.

Simulation of the Processor Sweeping Heuristic

To test the effectiveness of the processor sweeping heuristic, computer simulation was used to determine average results for various types and sizes of precedence graphs. Both sparse and dense precedence graphs were used in the simulation, and 1000 graphs of each type were generated for each of the following size categories: (1) 50-100 tasks, (2) 100-150 tasks, and (3) 150-200 tasks.

To determine the role of task selection in the heuristic, five different versions of the sweeping heuristic were examined. The first three versions used the original heuristic (without block moves) and were named according to the task selection method used: Simple, Maxfirst, and Maxadd. The last two versions consisted of one and two passes, respectively, of the modified heuristic (with block moves). The modified versions of the heuristic used the Maxadd method to select tasks for moving; thus, they will be referred to as Maxadd+n (1) for the first pass and Maxadd+n (2) for the second pass of the heuristic. For comparison with the different versions of the heuristic, three other tasks placement methods were simulated: (1) scheduling tasks according to the eager job profile; (2) placing tasks according to the lazy profile; and (3) moving tasks at random without violating precedence constraints.

Figure 4 contains information about the percentage of sparse and dense graphs with 100 to 150 tasks that achieved the optimal $ST_{p\ min}$ as a result of the various task placement methods. Due to space limitations, only the results for the Eager, Lazy, Random, and Simple placement schemes are given in Figure 4. Although the average results for the Maxadd+n heuristic were the best overall, Simple is much easier to implement, and the results for the various versions of the heuristic were comparable to those for Simple. Using the processor sweeping heuristic, 52 to 94% of the 1000 sparse graphs were able to reach the guaranteed optimal ST_p, while 22 to 100% of the dense graphs achieved the theoretical minimum ST_p as a result of the heuristic. Under the other task placement methods (eager, lazy, and random), very few of the sparse graphs reached $ST_{p\ min}$, and most of the dense graphs achieved $ST_{p\ min}$ only when n was large. Since dense graphs typically have fewer movable tasks than sparse graphs, the percentages for dense graphs are generally higher than those for sparse graphs for each task placement method.

Summary

The results from the simulation of the processor sweeping job manipulation heuristic can be summarized by the following set of design guidelines for job profile manipulation:

Guideline 1:	If processor sweeping is used and the highest possible performance must be achieved at all cost, the Maxadd+n modified version of the processor sweeping heuristic should be used with one or two passes.

Guideline 2:	If processor sweeping is used and high performance needs to be obtained at little cost, the Simple version of the processor sweeping heuristic should be used.

Guideline 3:	If the processor sweeping discipline is used to execute an algorithm and the processor sweeping heuristic cannot be used, the eager job profile should be used since it provides better average results than the lazy profile.

These guidelines can be used by designers to determine the appropriate method to restructure algorithms to match existing computer architectures.

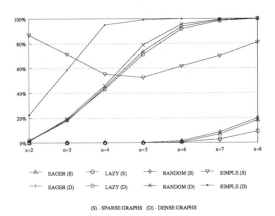

Figure 4: Percentage of 1000 Sparse and Dense Graphs with 100 to 150 Tasks in which $ST_{p\ final} = ST_{p\ min}$

References

[1] T. C. Chen. "Parallelism, pipelining and computer efficiency". *Computer Design*, pages 69–74, January 1971.

[2] S. Fernbach. "Parallelism in computing". *Proceedings of the 1981 International Conference on Parallel Processing*, pages 1–4, 1981.

[3] M. J. Flynn and J. L. Hennessy. "Parallelism and representation problems in distributed systems". *IEEE Transactions on Computers*, C-29(12):1080–1086, December 1980.

[4] L. S. Haynes, R. L. Lau, D. P. Siewiorek, and D. W. Mizell. "A survey of highly parallel computing". *Computer*, pages 9–24, January 1982.

[5] K. Hwang and F. A. Briggs. *Computer Architecture and Parallel Processing*. McGraw-Hill, New York, 1984.

[6] A. Silberschatz and J. L. Peterson. *Operating System Concepts*. Addison-Wesley, Reading, Massachusetts, alternate edition, 1988.

[7] N. Ullah and M. J. Gonzalez, Jr. "Analysis of multiple processor performance using job profiles". *Proceedings of the First Annual IEEE Symposium on Parallel and Distributed Processing*, pages 82–89, 1989.

[8] M. L. Weems. "Continuous job profiles: A model for performance analysis of parallel algorithms". Master's thesis, The University of Texas at Austin, Austin, Texas, May 1990.

[9] L. F. Wilson. "Design guidelines for parallel algorithms using job profiles". Master's thesis, The University of Texas at Austin, Austin, Texas, December 1990.

[10] L. F. Wilson and M. J. Gonzalez. "Design guidelines for parallel algorithms using continuous job profiles". *Proceedings of the Fifth International Parallel Processing Symposium*, April 1991.

Using Simulated Annealing for Mapping Algorithms onto Data Driven Arrays [1]

Bilha Mendelson
IBM Israel
Science & Technology
Haifa 32000, Israel

Israel Koren
Dept. of Electrical and Computer Engineering
University of Massachusetts
Amherst, MA 01003

Abstract Control-driven arrays provide high levels of parallelism and pipelining for inherently regular computations. Data-driven arrays can provide the same for algorithms with no internal regularity. The purpose of this paper is to establish a method for mapping any given algorithm onto data driven array. The method, based on the simulated annealing algorithm, aims to find an efficient mapping that minimizes the area, and maximizes the performance of the given algorithm or find a tradeoff between the two.

1 Introduction

A variety of topologies and architectural designs of processor arrays have been proposed and many computational algorithms for these arrays have been devised [4]. These include globally synchronous systolic arrays and globally asynchronous wavefront arrays [3],[6]. Most existing algorithms for these arrays were developed for problems with inherent regularity. These computations proceed in the processor array in a predetermined manner and achieve high performance through parallelism and pipelining.

There are however, many computationally demanding problems which do not exhibit high regularity. The dataflow mode of computation seems to be an appropriate approach to follow. We adopt therefore, the approach proposed in [5] where the algorithm is first represented in the form of a data flow graph (DFG) and then mapped onto a data driven processor array. The processors in this array will execute the operations included in the corresponding nodes (or subsets of nodes) of the DFG, while regular interconnections of these elements will serve as edges of the graph. The data driven processor array is a programmable and homogeneous array which is composed of an hexagonal mesh of identical processing elements (PEs). The hexagonal topology of the array serves only as an example of a regular structure.

A data driven PE is capable of performing arithmetic, logical, routing and synchronization operations, and can contain more than one operation. In particular, the routing operations can be performed in parallel to the arithmetic and logic operations.

A single connection link between adjacent PEs in the array was shown to be a critical limitation [11, 1]. It affects both the communication between nodes in different PEs and the capability to include more operations in the same PE. We follow an alternative approach: two communication links connect every two adjacent PEs. Since each input and output operand is held in an internal register, the number of nodes assigned to a PE is limited by the total number of registers in the PE. In our design each PE contains 16 internal registers (12 for communicating with its six neighbors and 4 scratchpad registers).

Figure 1 shows a data–flow graph representing the computation of the two coefficients A and B in the solution $Y(t) = A\cos wt + B\sin wt$ of a spring-mass system with an external force $F(t) = F_0\cos wt$. Figure 2 depicts a possible mapping of the graph in Figure 1 onto a regular processor array.

A method for mapping a given algorithm onto a data driven array to achieve a better performance has been developed. The mapping involves assigning every node of the DFG to a data driven PE in a data driven processor array. The goal of an efficient mapping is to minimize the

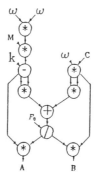

Figure 1:
A data flow graph.

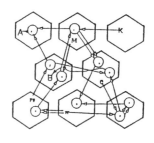

Figure 2: The mapping of the graph in Figure 1 onto a hexagonally connected array.

area (i.e., the number of PEs used) as well as optimize the performance (pipeline period and latency) of the given algorithm, or find a tradeoff between the area, pipeline period and latency criteria. The search for an optimal mapping is a combinatorial optimization problem.

Simulated annealing is known to be a powerful algorithm for solving combinatorial optimization problems [7]. Section 2 describes the use of the simulated annealing for mapping DFGs. The simulated annealing procedure may become very time consuming when applied to large problems. Section 3 shows how to deal with large programs by dividing the given algorithm into several blocks and mapping each of them separately. Later, these blocks will be merged together in order to put them as close to each other as possible. Conclusions and future research are presented in the last section.

[1]This work was done while B. Mendelson was with the University of Massachusetts at Amherst and was supported in part by SRC under contract 90-DJ-125.

2 Simulated Annealing and the Mapping Problem

Simulated annealing (SA) was shown to be a powerful algorithm for solving combinatorial optimization problems [7]. The SA algorithm is used in VLSI design [12] (for placement, floorplan design, channel routing, etc.) and in the design of multicomputer systems. Johnson et al. [2] have compared the performance of the SA algorithm and other heuristic algorithms for the graph bisection problem. Their conclusion is that if the given problem has a regular structure, it is likely that a good heuristic algorithm will yield good results. However, when there is no such structure it is better to use SA.

Finding a mapping for an algorithm that minimizes area and optimizes performance is an NP-complete problem. The DFG for an arbitrary algorithm has no particular structure and it is very difficult to find a good heuristic algorithm for generating an optimal mapping. Therefore, we choose to use the SA algorithm. This method aims to find an efficient mapping that minimizes the area, and maximizes the performance of the given algorithm or find a tradeoff between the two.

To use the SA algorithm for the mapping process we have to define some basic parameters that the process needs. These include the initial configuration [2], acceptable moves and a cost function. The initial configuration is some fea-

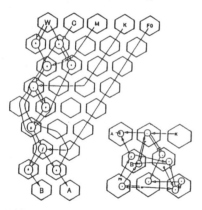

(a) Initial configuration   (b) Final configuration

Figure 3: Mapping using the simulated annealing process.

sible mapping where every DFG node is assigned to a PE (the simple mapping procedure presented in [5] can be used to generate such an initial configuration).

Acceptable moves that change a configuration are: move a node from one PE to another and exchange nodes between PEs. The first move allows the nodes to move from one PE to another PE which has not used all its capacity, while the purpose of the second move is to allow reassignment of nodes even for PEs that used all of their capacity.

A result of such a mapping is shown in Figure 3. Figure 3(a) shows the initial configuration and Figure 3(b) the result after performing the SA process. In this example the goal was to minimize the area. Notice that the area has decreased by 60%, which is a substantial improvement over the initial configuration.

---

[2]A certain mapping of the DFG onto the hexagonal array is called a *configuration*.

In most applications, there is a need to combine area minimization and performance maximization. We suggest therefore, three criteria for evaluating a mapping:

Area - number of PEs utilized by the mapped DFG.
Pipeline period - mean time between successive results.
Latency - time elapsed from entering the input operands until the output is produced.

A good mapping is one that optimizes all three criteria according to the weight coefficients that are supplied to the SA algorithm. The cost function, which is used to evaluate the new configuration is given by $\lambda_a A + \lambda_p P + \lambda_l L$ where A is the area (number of PEs) being used, P is the estimated pipeline period, L is the estimated latency and λ_i's are weight coefficients ($\Sigma \lambda_i = 1$, $i = a, p, l$).

Minimizing the area is achieved by compressing several DFG nodes into the same PE. The nodes that are ready to be executed and are assigned to the same PE are executed sequentially. The pipeline period of the mapped algorithm is therefore,

$$P = \max_E(\sum_{i=1}^{n_i} EX(node_i))$$

where $EX(node_i)$ is the execution time of node i, n_i is the number of nodes assigned to PE_i and E is the set of all PEs that some node has been assigned to them. The pipeline period of the mapped algorithm may be affected by the way the compression is done. It can increase if in the assignment process the total execution time of nodes assigned to the same PE exceeds the estimated pipeline period. The chosen weight coefficients determine the kind of optimization that is performed.

The temperature in the SA procedure controls the probability of accepting a new configuration, which not necessarily reduces the cost function. This temperature is reduced geometrically by the factor r, called cooling rate. Figure 4 shows the average number of iterations needed by the SA algorithm to obtain the final mapping for four different values of $(\lambda_a, \lambda_p, \lambda_l)$. As the cooling rate increases, the temperature reduces slower. Therefore, as was expected, in all the cases the number of iterations increases as the cooling rate increases. We can observe from the figure that the number of iterations is sensitive to the value of λ_l. The average number of iterations increased by 71.1% when λ_l was changed from 0 to 0.3 (because additional routing nodes are needed when a node is moved from one PE to another). Adding routing nodes affects the latency more than the pipeline period and the area, and therefore, the algorithm needs more iterations to converge.

Using a performance estimation method for evaluating the potential performance of algorithms for data driven machines [9] (and the execution times from [1]), we know that for the example in Figure 1 the optimal pipeline period and latency are 16 and 50, respectively. By applying an exhaustive search for the best mapping, we found out that the minimum number of PEs needed for the optimal mapping is 7. We denote the probability to get the best achievable mapping by p. SA is a random algorithm and therefore, there is a need to perform several runs and choose the best mapping. The probability to achieve a good mapping in n runs is:

$$P_n = 1 - (1 - p)^n$$

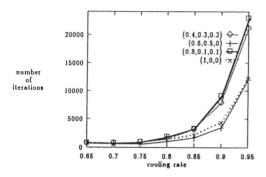

Figure 4: Average number of iterations needed by the SA to obtain a mapping of the algorithm in Figure 1.

From this expression we obtain the minimum number of runs of the SA algorithm to ensure a good mapping with probability P_n.

Table 1 shows the minimum number of runs needed to get a good mapping for $P_n = 0.95$ and different values of $(\lambda_a, \lambda_p, \lambda_l)$. In order to obtain the entries in this table, we have run the SA on the example many times. The mappings with area and performance no worse than the desired area and performance were counted. (A value of ∞ indicates that a good mapping was never found.) The probability to get a good mapping in a single run, p, was calculated by dividing the total number of good mappings by the total number of simulated annealing runs.

We used the execution times from [1] (i.e., divide takes 16 clock cycles, multiply 8 clock cycles and the rest take one clock cycle). Column 3 of the table gives the number of runs that are needed in order to achieve the global minimum mentioned above which uses only 7 PEs and pipeline period and latency of 16 and 50, respectively. Column 4 (5) shows this number for a mapping which uses at most 8 (9) PEs, pipeline period of 17 (19) and latency of 55 (60).

It can be seen that there is a need for few runs (2-5) of the SA algorithm, in order to achieve the optimum area and performance, if we select a cooling rate of 0.95. If the use of 8 PEs and reduction of 10% in the performance is allowed, only one run in most of the cases is needed.

When only one run of the SA algorithm is allowed, there is a need to know the average performance of this method. Figure 5 shows the mean value of the area, pipeline period and latency obtained for the algorithm in Figure 1.

Table 1: Number of runs needed to get a mapping with $P_n = 0.95$ for the algorithm in Figure 1.

$(\lambda_a, \lambda_p, \lambda_l)$	Cooling rate	15 PEs pip. period=8 latency=15	18 PEs pip. period=10 latency=18
(0.4,0.3,0.3)	0.90	∞	∞
	0.95	∞	273152.5
(0.5,0.5,0)	0.90	∞	∞
	0.95	∞	130709.8
(0.8,0.1,0.1)	0.90	∞	∞
	0.95	∞	373269.6
(1,0,0)	0.90	∞	∞
	0.95	334527.2	53956.0

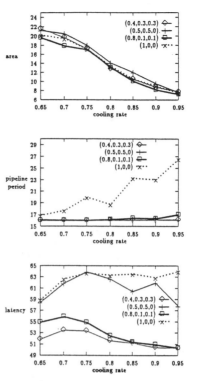

Figure 5: Average area and performance achieved when mapping the example in Figure 1.

We have mapped several algorithms onto data driven arrays and although our experience with the SA algorithm is limited, we can still recommend ways to choose the parameters of the SA. The best performance and area minimization is achieved when $r = 0.95$. It is true when considering the average behavior of SA as well as when calculating the number of iterations needed to get the best achievable mapping. In some cases, when $r < 0.95$, the SA uses a larger number of iterations without improving the mapping. Therefore, a cooling rate of 0.95 is advised for mapping graphs.

As we increase λ_a and λ_p, the effect of the change on the pipeline period is not significant. However, it has a significant impact on the area. Since in our examples the pipeline period is almost insensitive to the value of λ_p, area minimization can still be achieved without adversely affecting the pipeline period by slightly reducing the pipeline period coefficient and slightly increasing the area coefficient.

In summary, SA proved to be an efficient procedure for mapping algorithms onto homogeneous arrays. If $r = 0.95$ is chosen, only a few runs of the SA algorithm are needed to obtain a satisfactory mapping. It was also proved that area minimization can be obtained in addition to performance maximization by choosing the right weight coefficients.

2.1 Improvements of the Initial Mapping

In the previous section, we presented the result of the SA algorithm adapted to the mapping problem. These results have been obtained by applying the SA algorithm to an initial mapping that was done using the method presented in [5] (e.g., assigning a single node to a PE). We have found that if we try to put several nodes together in the same PE after the initial mapping has been done, the quality of the SA results improves.

This compaction phase is done by shifting the DFG nodes in six directions (down-right, down-left, left, up-left, up-right and right), as will be explained in the next section. The idea is to minimize the number of unutilized PEs in the mapped array.

3 Mapping large algorithms

When the number of nodes of the DFG is sufficiently small the SA algorithm gives satisfactory results. However, when the number of nodes increases, an optimal solution is not reached. But still the results of the mapping process are good and close to the optimum. Besides the degradation in the quality of the result, the number of iterations increases dramatically with the increase in the number of DFG nodes.

To this end, we suggest to represent a large algorithm not as a flat DFG but as a hierarchical tree whose nodes are sub-graphs of the DFG. The leaves of the hierarchical tree will then be mapped, using SA, and placed on the PE array.

3.1 Dividing the Graph to Sub-graphs

An algorithm consists of three basic structures: arithmetic (logic), it-then-else, and loop structures. We can use these structures to represent the algorithm as an hierarchical tree. Each node in the hierarchical tree will represent a structure of the algorithm. The leaves of the hierarchical tree will be composed of structures that can not be further separated.

A conditional structure is separated into three siblings: *Then*, *Else* and *Cond*. Each one of the parts will be further separated if it contains a separable structure. A loop structure node will be decomposed into two parts: *Body* and *control*. The body can be replicated several times to improve performance [8]. The leaves of the hierarchical tree are arithmetic/logic structures that can not be further separated.

3.2 Mapping the hierarchical tree

The mapping process follows basically a bottom up approach. It starts from the leaves, maps them side by side and climbs up the tree until it reaches the root. The algorithm for mapping, using the hierarchical approach, is shown in Figure 6. The algorithm forms clusters in the hierarchical tree. It starts with the leaves at the lowest level and then tries to place all the nodes of the sub-graphs that belong to the same parent node together. Only in a later stage of the algorithm, leaves that are not connected to the same parent node in the hierarchy tree will be placed.

Each stage of the algorithm starts by classifying the leaf nodes at the lowest level of the hierarchy tree into distinguishable sets (S_i) according to the leaves' parent nodes. Then, for each set S_i it selects the node in the hierarchy tree that has the most links to other nodes and maps it using SA. This node is then removed from the set S_i and added to the set of mapped nodes, M_i. If there was more than one leaf in S_i, then the next node is chosen from the set S_i using the same criteria. This node is then mapped, but this time, in order to increase the array utilization, an additional parameter is added to the SA process, namely the needed aspect ratio for the resulting mapping. The aspect ratio is determined according to the nodes that had been already mapped. All the links that connect the already mapped nodes in M_i to S_i are inspected. The side where

```
l:=lowest level of the hierarchical tree
repeat
    for i:=1 to k      {k is the number of nodes in level (l-1)}
        S_i:={ all the leaves in level l of the hierarchical tree that
                are children of node_i in level (l − 1) }
    for i:=1 to k
        M_i := ∅
    for i:=1 to k
        begin
            repeat
                choose a leaf from S_i with the highest connectivity, s
                map s onto the hexagonal array
                S_i := S_i ∩ s
                M_i := M_i ∩ s
            until (S_i = ∅)
            for i:=1 to k
                compact mappings in M_i
        end
    l := l − 1
until (l = 0)        reach the root
```

Figure 6: The algorithm for mapping large algorithms.

the majority of them lie, is chosen to be the side where the new node will be placed, termed *chosen side*. The SA will get this parameter and attempt to achieve the desired aspect ratio. This can be done by slightly changing the area weight function by adding a term which is a function of how far is the current aspect ratio from the desired one. After the leaf had been mapped, it is added to M_i. This process continues until S_i is empty (i.e., all the leaves of a mutual parent have been mapped). Then, a compaction phase is applied to the mapped nodes of M_i. The compaction phase tries to reduce the unutilized PEs in the mapped array, (e.g., tries to reduce the size of the rectangular bounded by the first and the last rows of the mapped algorithm and the first and the last columns of it). The compaction phase goes around the rectangular surrounding the mapped algorithm and tries to migrate DFG nodes from one PE to a neighbor PE in the compaction directions, to reduce the rectangular size. This process is performed on all the S_i sets at the lowest level. Then, each set M_i replaces the parent node of the leaves that have been added to it and becomes a leaf at level $(l − 1)$. In the next iteration, only the leaves that were not mapped earlier will be mapped using SA. The others will be placed in the array without any modification (only the compaction phase will be performed on them). The hierarchical mapping terminates when the root of the hierarchy tree is reached (i.e., $l = 0$).

3.3 Example

We demonstrate the way large algorithms are mapped through the example in Figure 7. The outer multiply node is replicated 3 times, as well as the body part, to improve the performance [10].

The algorithm is divided into four major parts: the body replicated 3 times and the control part. Each body part is further divided hierarchically until the leaves' level is reached. Since the number of DFG nodes in the leaves is small, the whole body part has been chosen as a leaf of the tree.

Since all the body parts are identical and do not feed data one to each other, they were placed one besides the other and were combined together to create the complete mapping of the given algorithm.

```
sum:=0
for i=1 to 1000
    if a=b
        then if c=d
                then r= g⋆(h[i]+q[i])
                else r=j[i]+k[i]
            endif
        else r=e[i]⋆f[i]
    endif
    sum:=sum+r
endfor
```

Figure 7: A nested if–then–else in a loop structure.

The compaction algorithm was then applied to the resulting mapping and by migrating nodes from one PE to another, the total area of the mapping has been reduced. The final mapping is shown in Figure 8. The latency of the mapped algorithm is 68 clock units and the pipeline period is 4 which are not far from the estimations (60 and 3, respectively).

4   Conclusions

A simulated annealing algorithm has been developed to map a given DFG onto an hexagonal homogeneous data driven array. It has been shown that it is efficient for small to moderate size algorithms. Another method has been presented for dealing with mapping of large graphs.

The above mapping method assumed an unbounded array. This is not a realistic assumption. Therefore, there is a need to solve the problem of assigning DFG nodes to PEs that are on several chips. The connections available among

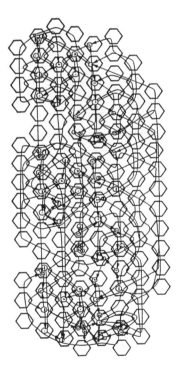

Figure 8: The mapping of the example in Figure 7 after compaction.

chips have to be provided besides the PE characteristics. Once those are given, one can map the algorithm onto an array consisting of several chips using a method similar to the one presented above with additional restrictions.

References

[1] R. Conrad and I. Koren. MPPE: A multiple port processing element for data-driven arrays. Tech. Rep. TR-90-CSE-4, University of Massachusetts, Amherst, 1990.

[2] David S. Johnson et al. Optimization by simulated annealing: An experimental evaluation, part I. *Oper. Res.*, 37:865–892, Nov.-Dec. 1989.

[3] A.L. Fisher, H.T. Kung, et al. Design of the PSC: A programmable systolic chip. In *Proc. Third Caltech Conf. on VLSI*, pp. 287–302, March 1983.

[4] J.A.B. Fortes and B.W. Wah. Special issue on systolic arrays–from concept to implementation. *IEEE Computer*, 20(7), July 1987.

[5] I. Koren, B. Mendelson, I. Peled, and G. M. Silberman. A data-driven VLSI array for arbitrary algorithms. *IEEE Computer*, pp. 30–43, October 1988.

[6] S.Y. Kung et al. Wavefront array processors – concept to implementation. *IEEE Computer*, 20(7):18–33, July 1987.

[7] P.J.M. Van Laarhoven and E.H.L Aats. *Simulated Annealing: Theory & Applications.* D.Reidel Publishing Company, 1987.

[8] B. Mendelson and I. Koren. Estimating the potential parallelism and pipelining of algorithms for data flow machines. Tech. Rep. TR-90-CSE-5, University of Massachusetts, Amherst, 1990.

[9] B. Mendelson, B. Patel, and I. Koren. *Designing Special-purpose Co-processors Using the Data Flow Paradigm* in *Advanced Topics in Data Flow Computing*, chapter 21, pp. 547–570. Prentice-Hall, 1991.

[10] B. Mendlson. *Mapping of Algorithms onto Programmalbe Data Driven Arrays.* PhD thesis, ECE Dept. University of Massachusetts, Amherst, 1990.

[11] S. Weiss, I. Spillinger, and G. Silberman. Techniques for mapping data flow programs onto vlsi processing arrays. In *Proc. Int'l Conf. on Fundamental Programming Languages and Computer Architecture*, 1989.

[12] D.F. Wong, H.W. Leong, and C.L. Liu. *Simulated Annealing for VLSI Design.* Kluwer Academic Publishers, 1988.

MICROPROCESSOR ARCHITECTURE
WITH MULTI-BIT SCOREBOARD CONCURRENCY CONTROL*

Thang Tran and Chuan-lin Wu
Department of Electrical and Computer Engineering
The University of Texas at Austin
Austin, Texas 78712

Abstract -- This paper describes a superscalar microprocessor architecture with an innovative multi-bit scoreboard technique for handling data dependencies. The multi-bit scoreboard prevents the pipeline from stalling with any dependency except a true data dependency (a read-after-write contention). Another innovative feature is the connection of the instruction buffers and decoding units in a First-In-First-Out shifting structure that permits proper accessing of the register file, setting the bits of the multi-bit scoreboard, and feeding of instructions to the functional units. The architecture has multiple decoding and functional units which issue instructions for out-of-order execution and completion. To maintain the precision of the machine in case of interrupts and exceptions, a novel instruction-retire buffer is implemented to keep track of the program counters and other processor state. The model architecture includes a branch prediction unit, a large register file, a temporary register set, instruction buffers, an instruction cache, and a data cache. Detailed simulation studies of this superscalar microprocessor with 4 decoding units show an improvement in execution speed of 4X over a single-pipeline processor. The benchmark-program simulation also derives optimal parameters for this architecture.

Introduction

As technology progresses, the power of microprocessors increases at an incredible pace. More functions can now be implemented on the same chip, and clock speed is being greatly increased. One important performance measurement for microprocessors is the number of clock cycles required for the execution of one instruction, or, inversely, the number of instructions executed per clock cycle. For a processor with a single functional unit, the performance limit is one instruction per clock cycle. However, much greater speeds can be achieved when more decoding and functional units are placed on a chip so that instructions can be executed in parallel, as is the case with the superscalar microprocessor.

The major problems with the parallel execution of a sequential-instruction stream are flow control and the handling of data dependencies. Because the flow of sequential instructions is frequently disrupted by branch instructions, many branch-prediction techniques have been devised to minimize this effect, such as the branch target cache [1], the folding branch [2], indicator bits embedded into branch instructions [1,3], and the branch-history table [4,5,6]. To handle data dependencies with single or multiple pipelines, a scoreboard [7,8], reservation stations [9,10,11], a variation of reservation stations [12,13,14,15], and compiler techniques [16] have previously been used. The software solution complements the hardware implementation. The advantages of various types of hardware solutions are discussed in [13]: (1) dynamic dependencies, such as those determined by conditional branches, are not known at compile time; (2) the architecture is evaluated independently of the quality of the compiler; and (3) the available execution resources are handled more effectively. The Intel860 [*] and RS6000 [*] have multiple functional units but they avoid some dependency problems by having separate register files and decoding units for floating point and integer ALU operations. These processors have only a single path for integer operations and are targeted more toward scientific applications.

In processors with multiple functional units, instructions may be dispatched and completed out-of-order. Three types of data dependency must be handled by any superscalar architecture: true dependency (read-after-write contention), anti-dependency (write-after-read contention), and output dependency (write-after-write contention). Only true dependency should stop the flow. The register file is used for quick local references of data. For the parallel execution of instructions from one program, the correct sequence of instructions must be retained in reference to the data in the register file. In processors with only one pipeline, a single-bit scoreboard is sufficient to detect any contention in read/write and to halt the flow until the contention has been cleared.

The major goal for this project was the construction of a superscalar microprocessor architecture that can average two instructions per clock cycle for non-scientific applications. To help achieve this, three types of status bit are used in the scoreboard: read, write, and temporary. As will be

---

* This work is supported in part by the National Science Foundation under grant MIP-8819652, by Advanced Micro Devices Incorporated and by Cray Research Incorporated.

shown, this type of scoreboard in conjunction with a set of temporary-result registers will keep the flow moving through the pipeline, except for the true dependencies (the read-after-write contentions). The FIFO structure of the instruction buffers and decoding units continuously supplies instructions for decoding and retains the sequence of instructions to access the register file and to set the scoreboard correctly. Because the instructions can complete out-of-order, the instruction-retire buffer keeps track and updates the processor states for proper operation of interrupts.

The second section of this paper describes the multi-bit scoreboard architecture, along with the FIFO of the decoding units for dispatching and executing instructions out-of-order, and the instruction-retire buffer for properly updating the processor states. The third section demonstrates how to set the status bits of the scoreboard. The fourth section discusses evaluation methods and benchmark simulation. Section five discusses the simulation results and the optimal model of this architecture with respect to cost and performance. The conclusion contains some comparisons of this architecture's performance with the currently available RISC microprocessors.

3—Bit Scoreboard Architecture

This superscalar architecture handles three types of data dependency: true, anti-, and output dependency. This architecture provides the necessary hardware for issuing instructions out-of-order while retaining sequential program state for the handling of traps/interrupts through the instruction-retire buffers.

Model

As a concept to speed up instruction execution, pipelined architecture has been widely implemented in most of today's microprocessors and computers. Pipelined architecture achieves close to the ideal case of 1 instruction/cycle if there is no branch and if all instructions move through the pipe at the same speed. In our model, the pipeline consists of 4 stages: instruction fetch, instruction decode, execution, and write-back. Figure 1 presents the architecture's block diagram, which contains the following blocks: an instruction cache, instruction buffers, multiple-decoding units, multiple-execution units, a register file with multi-bit scoreboard, temporary registers, a data cache, and a branch-prediction unit. Instructions are fetched from either instruction cache or external memory to the instruction buffers and shifted into the decoding units. The status bits of the scoreboard are checked in decoding; if there is no true dependency, the bits are set, instruction is dispatched, and operands are fetched from the register file to the queue buffer of the functional unit. At completion of execution, the

bits are reset, and the register file is updated if necessary. The branch prediction unit predicts the next stream of instruction from the current branch instruction in decoding. Data cache handles data needed by the load/store instructions.

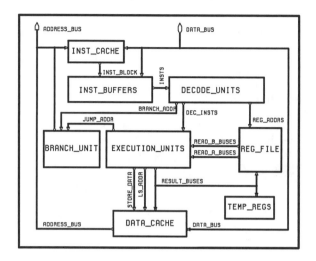

Figure 1 Top-Level Block Diagram.

To access external memory, two different buses are used for addresses and instruction/data. Instruction and data caches share the buses; the data bus can be 64-bit where 2 instructions/data can be fetched in one cycle. On-chip instruction and data caches are necessary to improve the performance of the processor and to reduce its external-memory traffic [17]. This is more essential to superscalar architecture where multiple instructions are executed in each cycle.

The data cache is accessed by the load/store execution unit which handles load, store. Because stores and load-misses update the cache and/or the external memory, these instructions must wait to be completed in order.

Each functional unit handles a different class of operations: load/store, branch, integer ALU, shifter, and floating-point ALU. Each contains a queue buffer, and executes instructions taken from its queue. The queue buffer holds instructions for more than 1 instruction dispatched from decoding units, and for the functional unit being busy of executing of previous instruction. While there are separate pipes for FP add, multiply and divide, they share a common queue buffer. The branch functional unit has a bus routed directly to the branch-history table for updating.

The simplest and oldest branch-prediction technique is the Branch-History Table (BHT). Research shows that, for a BHT size of 8 entries, the correctness ratio is greater than 90% [5]. While branch instructions might disrupt the stream of instructions, they fortunately have locality, often being used in loops or nested loops. Therefore, in multiple-instruction execution, the performance improvement brought about by branch prediction can be expected to be 30-40% as shown later in evaluation. In this model the BHT is a fully associative cache. It contains the address of the branch instruction, the target address of the branch and control bits for previous outcomes of the branch. Because not-taken branches are sequential, the BHT keeps only taken branches. If the branch instruction is not in the BHT, the result is assumed to be not-taken. In decoding, the branch-prediction unit decides the next instruction address. If predicted taken, all instructions after the branch are invalidated and a new sequence will be fetched. All predicted instructions after the branch will be tagged as conditional. When the branch executes, the outcome will update the BHT and all instructions tagged as conditional will be cleared from the pipe if the BHT has mispredicted; otherwise just the branch tags are reset.

Instruction Issuing

Instruction buffers are used to absorb discontinuities in instruction fetching caused by predicted-taken branches. Instruction buffers receive multiple instructions from either the instruction cache or the external memory, and forward these instructions to the decoding units. When the buffer has space to load the next block of instructions, the next block will be fetched automatically. If the next block is in the instruction cache it will be loaded into the instruction buffer, or else the next instructions are fetched from external memory.

The new shifting structure allows the sequence of instructions to be retained in the decoding units for accessing the register file and setting the scoreboard correctly. A FIFO implementation is required for the instruction buffers and the decoding units. Each instruction buffer and decoding unit has a valid bit. If the operands of an instruction in a decoding unit can be accessed from the register file without conflict, and if the read buses are available to transfer data from the register file to the functional unit, then the decoding unit is ready to accept another instruction in the next cycle; otherwise this decoding unit is busy. In the next cycle, the decoding units which are busy will be moved to the top of the FIFO, instructions from the instruction buffers will move into the available decoding units, and instructions from the instruction cache or external memory will be fetched into instruction buffers. The decoded instruction and its operand data are routed directly to the specific functional unit for execution. Figure 2 is a block diagram of the pipe movement from instruction fetching to execution. All decoding units set the

status bits of the scoreboard at the same time. Because the top decoding units have a higher priority in the setting of the status bits, the next decoding unit must check both the status bits and any set signal from the top decoding units. The set signal is treated the same as the status bit itself.

When the execution of the instruction has been completed, the bits of the scoreboard must be reset. First, the read bits must be reset. The branch tag from branch prediction and the read-and-write bits of the scoreboard must be checked for data to be written to the register file or to temporary registers.

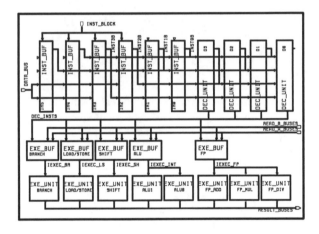

Figure 2 Pipeline Movement of Instructions.

Temporary Registers

The temporary register has its own control logic for writing data back to the register file. An allocate-bit denotes that this register is already in use (reserved). A valid-bit denotes that valid data are written into the temporary register. A branch-bit is produced by instructions after branch prediction. A register pointer points to the location in the register file. In every cycle, if there are valid data in the temporary register and the resulting buses are not busy, the temporary register will try to write data back to the register file if the write-and-read bits of the scoreboard are not set for that register pointer.

Instruction-Retire Buffer and Interrupt/Trap Handling

A instruction-retire buffer is implemented to handle interrupts, traps, exceptions, serial instructions, branch misprediction and updating-processor states. Each instruction-retire buffer entry consists of a tag and instruction status information. The instruction-retire buffer also consists of a set of program counters (PCs) pointing to the first instruction in the pipe and taken branch targets. When an instruction is in decode, a tag indicating the next available buffer location is assigned to it; this tag is passed along with the instruction to execution. After the execution of the instruction, the entry in the instruction–retire buffer is updated as valid. Valid instructions are retired by popping them off of the instruction–retire buffer in sequence and updating the processor states (pc, condition codes), as long as there is no exception. If an exception condition exists for an instruction that is about to retire, a trap is taken. This ensures that the earliest exception in the instruction sequence is recognized first. The pc's of incomplete instructions are frozen until the return from the trap has been executed. When the instructions are fetched from this set of PCs some instructions may be executed again. The trap routine should not change the source operands of the trapped instruction, at least not in a way that is visible to the users. If the instruction can be executed, then it is free of dependencies and can update the register file directly.

The instruction–retire buffer also handles branch misprediction, serial, and load/store instructions. In the case of misprediction, the instruction–retire buffer keeps track of instructions after the branch. When these instructions are retired, the result data are removed from the temporary register and the corresponding temporary tags in the register file are reset. Serial instructions are those which update special registers and processor states. A serial instruction is executed when it is on top of the instruction–retire buffer and when no other instruction is in the pipe; the decoding units are stalled after dispatching the serial instruction. The store and load–miss instructions, which update the data cache and/or the memory, must be executed in order because of a possible exception. They are executed when reaching the top of the instruction–retire buffer.

Setting of Scoreboard Bits

For instructions which can not be dispatched, the decoding units must remember current settings of the status bits for the next cycle to avoid redundancy. The statuses of instruction in the decoding unit are: READ0, READ1, and WRITE indicate that the read and result operands need to be checked, RD0_SET, RD1_SET indicating that the read status needs to be incremented, RD_CON indicates a write–after–read contention, and WR_TEMP indicates that the temporary bit is set for the result operand. Figure 3 contains the flow chart for setting the read bits. The read count is increased by RD_SET, and, as the execution of the instruction is completed, the read count is decreased. If either the temporary or write bit is

set, the instruction must wait in decoding; this is true dependency. If the temporary bit is not set, the read bit should be set for anti–dependency checking of subsequent instructions. Figure 4 has the flow chart for setting the write or temporary bits from write–operands. The result operand is used to set the write bit in decoding and reset the write bit after execution. In decoding, if both the read and write bit of the resulting operand are not set, then the write bit is set, or else the temporary bit is set. As the temporary bit is set, a temporary register is allocated and a tag is attached to the instruction going to execution. This process allows the instruction to be executed despite the write–after–write or write–after–read contention. If the temporary bit is set, then the instruction must wait in decoding until the next cycle. If an instruction in decoding needs the data that is being written back to the register file, the data is forwarded directly to the read bus. For instruction that must wait in decoding until next cycle, if the RD_CON or WR_TEMP is set, the read and write status bits are checked again. The temporary register is only used if there is contention.

The example in Figure 5 shows the dependency checking, the instruction dispatching in decoding units, and the setting/clearing of bits in the scoreboard. The arrangement of the write and read operands in the instruction sequence covers all cases of conflicts. After the first cycle, instructions 1, 3, and 4 can be dispatched while instruction 2 is held back in the decoder and shifted into decoder location 0. Instruction 2 has a true dependency with instruction 1, instruction 3 has both output– and anti–dependencies with instruction 1, and instruction 4 has an output dependency with instruction 2. With the use of the temporary bit, both instructions 3 and 4 are allowed to dispatch. Even though instruction 2 cannot be dispatched, all the bits in the scoreboard for this instruction are updated in this cycle. In the next cycle, instructions 1 and 3 complete execution; instruction 1 updates the register file, while instruction 3 puts the result into a temporary register. Instruction 2 needs only to check the write bit of R1, which is being reset by the write back of instruction 1, for dispatching. The result is forwarded from the result bus to the read bus. Instruction 6 has only anti–dependency which can be dispatched. Since R1 and R3 of the register file have the temporary bits set, instructions 5 and 7 cannot be dispatched. The read bits for R1 and R3 cannot be set by instruction 5 or 7 until the temporary bits are clear. In the next cycle, instructions 2 and 4 write back to the register file, and all bits of R3 and the read bit of R1 are cleared. As the read bit of R1 is cleared, the result operand in the temporary register must be written to R1 and the temporary bit cleared. At this time instructions 5 and 7 can set the read bits and be dispatched.

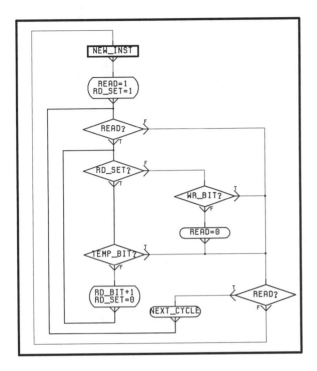

Figure 3 Flow Chart of Setting Read–Status Bit.

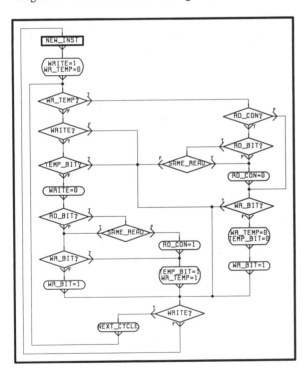

Figure 4 Flow Chart of Setting Write–Status Bit.

INST#	DISPATCH	WR_REG	RD0_REG	RD1_REG	WRITE	RD_CON	WR_TEMP	READ0	RD0_SET	READ1	RD1_SET
1	Y	R1	R1	R2	1	0	0	1	1	1	1
2	N	R3	R1	R2	1	0	0	1	1	1	1
3	Y	R1	R0	R2	1	0	0	1	1	1	1
4	Y	R3	R0	-	1	0	0	1	1	1	1

```
1 - OP  R1,R2 -> R1
2 - OP  R1,R2 -> R3
3 - OP  R0,R2 -> R1
4 - OP  R0    -> R3
5 - OP  R4,R1 -> R4
6 - OP  R0    -> R2
7 - OP  R0,R3 -> R2
```

REG	RD_BIT	SET	CLR	WR_BIT	SET	CLR	TP_BIT	SET	CLR
R0	0	11		0			0		
R1	0	11		0	1		0	1	
R2	0	111		0			0		
R3	0			0	1		0	1	
R4	0			0			0		

INST#	DISPATCH	WR_REG	RD0_REG	RD1_REG	WRITE	RD_CON	WR_TEMP	READ0	RD0_SET	READ1	RD1_SET
2	Y	R3	R1	R2	0	0	0	1	0	0	0
5	N	R4	R4	R1	1	0	0	1	1	1	1
6	Y	R2	R0	-	1	0	0	1	1	0	0
7	N	R2	R0	R3	1	0	0	1	1	1	1

*WRITE BACK INSTRUCTIONS 1 AND 3
*INSTRUCTION 3 PUT R3 INTO TEMPORARY REGISTER

REG	RD_BIT	SET	CLR	WR_BIT	SET	CLR	TP_BIT	SET	CLR
R0	2	1	1	0			0		
R1	2	1	1	1	1	1	1		
R2	3		11	0			0	1	
R3	0			1			1		
R4	0	1		0	1		0		

INST#	DISPATCH	WR_REG	RD0_REG	RD1_REG	WRITE	RD_CON	WR_TEMP	READ0	RD0_SET	READ1	RD1_SET
5	Y	R4	R4	R1	0	0	0	0	0	1	1
7	Y	R2	R0	R3	1	0	3	0	0	1	1

*WRITE BACK INSTRUCTIONS 2 AND 4
*TEMPORARY REGISTER UPDATES R1

REG	RD_BIT	SET	CLR	WR_BIT	SET	CLR	TP_BIT	SET	CLR
R0	2		1	0			0		
R1	1	1	1	0			1		1
R2	1		1	0	1		1		1
R3	0	1		1		1	1		1
R4	1			1			0		

Figure 5 Example of Setting Multi–Bit Scoreboard for Dispatching Instructions.

Evaluation Method

To evaluate for benchmark applications, we use a trace–driven simulation system based on the AM29000 [*] RISC processor. The trace provides the opcodes, physical addresses of register file, and calculated memory addresses for both instructions and data. The HighC29K [18] compiler is used to produce the assembly–language codes for the trace system. The trace is generated for each benchmark used in simulation. The benchmark programs are:

(1) Stanford: includes 8 small–integer benchmark programs which computes permutations, solves Towers of Hanoi, solves the 8–queens problem 50 times, multiplies two 40x40–integer matrices, solves Soma Cube–type problems, quick–sorts 5000 elements, bubble–sorts 500 elements, and binary–tree–sorts 5000 elements.

(2) Diff: compares 2 text files and reports all differences.

(3) Grep: reports all occurrences of a given string in a text files.

(4) Dhrystone1.1: a popular benchmark which attempts to replicate the "average" use of various "C" programming constructs.

(5) Stanfp: is single–precision floating point benchmark from Stanford.

(6) Linpack: double–precision benchmark program.

The execution time for the functional units is based on the AM29050 [*] RISC processor timing.

For the integer ALU, the shifter, the branch unit, and the load/store unit the execution time is 1 clock-cycle. The simulation of the floating point functional unit includes issue-rate and latency. The issue-rate is how soon the next instruction may enter execution, and the latency is the time of execution for an instructions until completion. The issue-rate and latency times in number of clock cycles are:

	Issue-rate	Latency
Floating-point add	1	3
Integer multiply	1	3
Single precision multiply	1	3
Double precision multiply	3	5
Single precision divide	10	10
Double precision divide	17	17
Square root	55	55

In the implementation of any computer architecture, the trade-off between cost and performance is an important issue. In this project, the method of finding the model with the optimum balance between these two factors is to start with the maximum-performance model, with unlimited large components. Each component is then varied in size to reach the optimal level of utilization and cost. Number of decoding units, read buses, result buses, ALU functional units, the cache size, and branch-history-table size are parameters whose optimum values are determined by this method.

Discussion of Simulation Results

The processor throughput with the maximum-performance model for floating-point and integer benchmarks versus the number of decoding units are shown in Table 1 and Table 2. To evaluate the effect of the multi-bit scoreboard and branch-history-table, we simulated the maximum model with a single-bit scoreboard, and without branch prediction. The simulation results are shown in Table 2. The last two columns of the maximum-model section show the incremental and overall gains for increasing the number of decoding units, and the last column of the other two sections show the percentage gains of the maximum performance model over the other two models. For the multi-bit scoreboard with 4 decoding units, the performance gain over the single-bit scoreboard is 26.4%. For a model with branch prediction and 4 decoding units, the performance gain is 39.1%. These features can be implemented with this superscalar architecture.

TABLE 1 Floating-Point Throughput of Maximum Model.

Feature Constraints	Decoding Units	Throughput (inst/cycle)			Incremental % change	Overall % change
		Linpack	Stanfp	Ave.		
Maximum Model	1	0.627	0.8	0.714		
	2	0.708	0.885	0.797	11.6 %	11.6 %
	4	0.75	0.987	0.869	9.0 %	21.7 %
	8	0.75	0.987	0.869	0.0 %	21.7 %
	16	0.751	0.987	0.869	0.0 %	21.7 %

TABLE 2 Effect of Multi-Bit Scoreboard and Branch Prediction on Performance.

Feature Constraints	Decoding Units	Throughput (inst/cycle)					Incremental % change	Overall % change
		Diff	Grep	Stanf	Urhyl	Ave.		
Maximum Model	1	0.999	0.997	0.999	0.899	0.974		
	2	1.554	1.373	1.486	1.238	1.411	44.9 %	44.9 %
	4	1.923	1.768	1.808	1.468	1.748	23.9 %	78.6 %
	8	2.325	1.916	1.935	1.554	1.933	11.1 %	98.5 %
	16	2.341	1.916	1.936	1.595	1.947	0.7 %	99.9 %
							Feature Gain (%)	
Single-bit Scoreboard	1	0.916	0.998	0.966	0.876	0.937	3.9 %	
	2	1.156	1.349	1.284	1.116	1.221	15.6 %	
	4	1.261	1.557	1.448	1.243	1.377	26.4 %	
	8	1.268	1.621	1.512	1.282	1.421	36.8 %	
No Branch Prediction	1	0.997	0.997	0.998	0.894	0.972	0.2 %	
	2	1.172	1.123	1.321	1.040	1.164	21.2 %	
	4	1.308	1.183	1.405	1.109	1.251	39.1 %	
	8	1.308	1.183	1.405	1.127	1.256	53.9 %	

To further evaluate the performance of caches, we simulated other architecture models. The zero model has single-bit scoreboard, no branch prediction, and no cache. The instruction-cache model is the zero model with instruction cache, and the ID-cache model is the zero model with both the instruction and data caches. The simulation results are shown in Table 3. For the zero model, the performance gain with increasing number of decoding units is neglegible. However, the last column of the instruction-cache model shows a greater than 50% gain with added instruction cache, and the last column of the ID-cache model shows a gain of greater than 30% with added data cache over the instruction-cache model.

TABLE 3 Effect of Caches on Performance of Zero Model.

Feature Constraints	Decoding Units/	Cache Size (Kbytes)	Throughput (inst/cycle)					Feature Gain (%)
			Diff	Grep	Stanf	Urhyl	Ave.	
Zero Model	1	0	0.411	0.421	0.434	0.432	0.425	
	2	0	0.426	0.426	0.453	0.434	0.435	
	4	0	0.426	0.429	0.452	0.437	0.436	
Instruction Cache	1	1	0.629	0.690	0.757	0.572	0.662	55.9 %
	1	2	0.641	0.703	0.757	0.654	0.689	62.3 %
	1	4	0.643	0.703	0.757	0.687	0.670	64.3 %
	2	1	0.654	0.739	0.805	0.588	0.697	60.1 %
	2	2	0.654	0.739	0.805	0.675	0.718	65.1 %
ID-cache (Instruction and Data Caches)	1	1	0.875	0.975	0.966	0.712	0.882	33.2 %
	1	2	0.903	0.989	0.967	0.800	0.915	32.8 %
	1	4	0.902	0.989	0.967	0.848	0.927	32.4 %
	2	1	0.962	1.069	1.153	0.745	0.982	41.8 %
	2	2	0.996	1.097	1.153	0.868	1.029	43.2 %

The first step in the process of obtaining optimal model parameters is to vary the number of decode units. Second, the number of read and result buses is reduced to a point sufficient to maintain the same performance level, while the number of integer–ALU units is determined by observation of the utilization of each unit. Third, cache size (currently the data and instruction caches are the same size) is varied to observe the instruction– and data–cache hit ratios, and the throughputs. Fourth, from simulation, the BHT is reduced to a reasonable size. Integer benchmarks are used for trade–off evaluation. In reference to Table 2 (incremental % change), when 4 instead of 8 decoding units are selected, the penalty in performance is 11% for integer benchmarks. Four decoding units should be chosen for the optimal model.

The first two sections of Table 4 show the throughputs versus the number of read buses and result buses for integer benchmarks. From these figures, we may conclude that six read buses and three result buses are sufficient to maintain performance. The number of buses should be greater than two for double–precision floating point operation. The utilization of more than two integer–ALU units is 4% or less for all integer benchmarks with 4 decoding units. The design should be implemented with 2 integer–ALU units.

Cache–size simulations for throughputs, instruction cache hit rates, and data cache hit rates are shown in the third, fifth, and sixth sections of Table 4, respectively. The cache organization is 2–way associative, 4 words per block. A cache size of 4K bytes is selected for the optimized model. While a cache size of 4K bytes is shown to be optimal for this set of benchmarks, larger cache sizes may be required to support the same level of performance for real–world applications. However, no aspect of this architecture precludes the use of larger caches.

Since the BHT is fully associative, its size should be small to keep the access time short. The fourth and last sections of Table 4 shows the throughput and the correct–prediction ratio versus BHT size, respectively. The optimal BHT size for the programs of interest is 16 entries. The number of temporary registers is less than 10 for all benchmarks, and the number of instruction–retire buffer entries is kept at 16. The average performance of this optimal model is 1.66 instructions per cycle which is about 4 times the performance of the zero model in Table 3.

TABLE 4 Integer throughput, Cache Hit Ratio, and Correct–Prediction Ratio.

Feature Constraints		Throughput (inst/cycle)					Incremental % change	Overall % change
		Diff	Grep	Stanf	Drhy1	Ave.		
Number of Read Buses	4	1.866	1.588	1.69	1.37	1.629		
	6	1.92	1.758	1.805	1.461	1.736	6.6	6.6
	8	1.923	1.76	1.808	1.468	1.74	0.6	6.8
Number of Result Buses	2	1.76	1.737	1.792	1.433	1.681		
	3	1.916	1.758	1.805	1.458	1.734	3.2	3.2
	4	1.92	1.758	1.805	1.461	1.736	0.1	3.3
Cache Size (Kbytes)	1	1.797	1.691	1.79	1.057	1.584		
	2	1.875	1.736	1.791	1.273	1.669	5.4	5.4
	4	1.889	1.737	1.791	1.384	1.7	1.9	7.3
	8	1.893	1.737	1.791	1.415	1.709	0.5	7.9
BHT Size	4	1.805	1.587	1.692	1.177	1.545		
	8	1.805	1.53	1.766	1.18	1.57	1.6	1.6
	16	1.87	1.785	1.791	1.267	1.658	5.6	7.3
	32	1.887	1.737	1.791	1.31	1.681	1.4	8.8
	64	1.888	1.737	1.791	1.374	1.698	1.0	9.9
Cache Hit Ratio (%)								
Instruction Cache Size (Kbytes)	1	93.6	94.6	97.1	80.5	91.5		
	2	94.5	95.2	94.3	90.5	94.3	3.1	3.1
	4	97.4	95.2	97.1	93.1	95.7	1.5	4.6
	8	97.4	95.2	97.1	93.2	95.7	0.0	4.6
Data Cache Size (Kbytes)	1	96.3	98.9	99.6	94.0	95.0		
	2	96.8	99.6	99.7	94.3	97.6	2.7	2.7
	4	97.1	99.7	99.7	94.3	97.7	0.1	2.8
	8	97.3	99.7	99.7	94.5	97.8	0.1	2.9
Correct Prediction Ratio (%)								
BHT Size	4	89.9	44.5	36.2	36.6	51.8		
	8	89.9	50.6	60.3	37.6	59.6	15.1	15.1
	16	95.6	77.8	67.3	53.7	73.6	23.5	42.1
	32	98.4	81.7	67.3	65.3	78.2	6.3	51.0
	64	98.6	81.7	67.3	83.2	82.7	5.8	59.7

Conclusion

This paper describes a new superscalar architecture –a faster implementable microprocessor which can average close to 2 instructions per cycle of execution. Given the high integration level of current technologies, this combination of features can be justified in terms of cost–effectiveness, as shown by the simulation results. To enhance its performance, the processor is equipped with the following features: a branch–history table, an instruction cache, a data cache, simple temporary registers to handle branch and some read/write contentions, multiple decoding units, instruction buffers, a instruction–retire buffer to keep track of the sequential state, a large register file, and many functional units. The multi–bit scoreboard effectively handles data dependencies to reduce pipe stalls; a simple branch–prediction technique improves architecture performance close to 40%; and the caches supply the necessary instruction and data bandwidth.

The optimal model consists of the following features: 4 decoding units, 6 read buses, 3 result buses, 4K byte, two–way associative caches, a fully associative branch–history table with 16 entries, and 2 integer–ALU units. The architecture's highest performance is 97 MIPS and 203,000 Dhrystones per second at 50 MHZ, while the optimal performance is 83 MIPS and 165,000 Dhrystones per second. The optimal model performs at 87% of the theoretical maximum and 400% of the zero model. The benchmarks show that the performance is a significant improvement over that of the currently available RISC processors and superscalar microprocessors. Table 5 presents a comparison between this superscalar's performance and other available RISC processors. The bottom half contain all performances normalized to 25MHZ.

TABLE 5 Performance Comparison

Processor	Reported Frequency (MHZ)	MIPS	Dhry1.1 (Dhrys/sec)	Linpack DP (MFLOPS)
SPARC[*]	25	15.0	19,000	1.50
AM29K[*]	25	17.0	32,000	1.71
R3000[*]	25	17.0	35,600	3.80
M88000[*]	25	20.0	34,000	-
Intel-960CA[*]	33	28.0	41,600	-
Intel-860[*]	40	27.0	83,000	10.00
This Processor	40	66.0	132,000	29.50
SPARC[*]	25	15.0	19,000	1.50
AM29K[*]	25	17.0	32,000	1.71
R3000[*]	25	17.0	35,600	3.80
M88000[*]	25	20.0	34,000	-
Intel-960CA[*]	25	21.0	31,000	-
Intel-860[*]	25	17.0	51,900	6.25
This Processor	25	41.5	82,500	18.40

Future research could investigate complementary post-compiler optimization to maximize this superscalar performance. The current compiler is designed for single-pipelined architecture, but can be modified for parallel execution of instructions.

References

[1] K. F. Lee and A. J. Smith, "Branch-Prediction Strategies and Branch Target Buffer Design," *IEEE Computer*, pp. 6–22, Jan. 1984.

[2] D. R. Ditzel and H. R. McLellan, "Branch Folding in the CRISP Microprocessor: Reduce Branch Delay to Zero," *Proc. of the 14th Annual Symp. on Comp. Arch.*, pp. 2–9, 1987.

[3] P. G. Emma, et. al., "Comprehensive Branch Prediction Mechanism for BC," *IBM Technical Disclosure Bulletin*, Vol. 28, No. 5, pp. 2255–62, Oct. 1985.

[4] A. G. Liles, Jr. and B. E. Willner, "Branch Prediction Mechanism," *IBM Technical Disclosure Bulletin*, Vol. 22, No. 7, pp. 3013–16, Dec. 1979.

[5] R. N. Rechtschaffer and F. J. Sparacio, "Comprehensive Branch- Prediction Mechanism for BC," *IBM Technical Disclosure Bull.*, Vol. 28, No. 5, pp. 2255–62, Oct. 1985.

[6] J. E. Smith, "A Study of Branch Prediction Strategies," *IEEE Proc. Eighth Symp. Computer Architecture*, pp. 135–148, May 1981.

[7] J. E. Thornton, *Design of a Computer — The Control Data 6600*, Glenview, IL: Scott, Foresman and Co., 1970.

[8] R. M. Russel, "The CRAY-1 Computer System," *Comm. ACM*, Vol. 21, pp. 63–72, Jan. 1978.

[9] R. M. Tomasulo, "An Efficient Algorithm for Exploiting Multiple Arithmetic Units," *IBM Journal*, Vol. 11, pp. 25–33, Jan. 1967.

[10] R. M. Keller, "Look-Ahead Processors," *ACM Computing Surveys*, Vol. 7, No. 4, pp. 177–195, Dec. 1976.

[11] S. Weiss and J. E. Smith , "Instruction Issue Logic in Pipelined Super- computers," *IEEE Trans. Comput.*, Vol. C-33, No. 11, pp. 110–118, Nov. 1984.

[12] K. Murakami, et. al., "SIMP (Single Instruction Stream/Multiple Instruction Pipelining): a Novel High-Speed Single Processor Architecture," *Proc. of the 16th Annual Symp. on Comp. Arch.*, pp. 78–85, 1989.

[13] R. D. Acosta, J. Kjelstrup, and H. C. Torng, "An Instruction Issuing Approach to Enhancing Performance in Multiple Functional Unit Processors," *IEEE Trans. Comput.*, Vol. C-35, No. 9, pp. 815–828, Sept. 1986.

[14] M. D. Smith, M. Johnson, and M. A. Horowitz, "Limits on Multiple Instruction Issue," *Proceedings, 16th Annual Int. Sym. on Computer Architecture*, pp. 290–302, Jun. 1989.

[15] G. S. Sohi and S. Vajapeyam, "Instruction Issue Logic for High-Performance, Interruptable Pipelined Processors," *Proceedings, 14th Annual Int. Sym. on Computer Architecture*, pp. 27–34, Jun. 1987.

[16] L. Wang and C. L. Wu, "Distributed Instruction Set Computer," *Proc. 88 Int. Conf. Parallel Processing*, St. Charles, Illinois, Aug. 1988.

[17] F. Brigge, "Cache Memory Performance in a Unix Environment," *Comp. Arch. News*, Vol. 14, No. 3, June 1986.

[18] HighC29K compiler, Meta Ware, Inc., and Advanced Micro Devices, Inc., 1989.

[*] AM29000 and AM29050 are trademarks of Advanced Micro Devices, Inc.
SPARC is the trademark of Sun Micro Systems, Inc.
M88000 is the trademark of Motorola, Inc.
Intel 960CA, 960KA, and 860 are trademarks of Intel, Inc.
RS6000 is the trademark of IBM, Inc.

PERFORMANCE ANALYSIS OF AN
ADDRESS GENERATION COPROCESSOR

Paul T. Hulina   Lee D. Coraor   Shih-Wei Sun

Department of Electrical and Computer Engineering
121 EEE
The Pennsylvania State University
University Park, PA 16802

ABSTRACT -- Most applications of general purpose VLSI processors are developed using high level languages. Information is generally handled in structured form; that is, procedures and data structures. Considerable processing time is spent performing address calculations required to access the data structures. In this paper, an alternative to software address generation, a hardware memory reconfiguring unit (MRU), or address coprocessor is evaluated. Assembly language programs that implement convolution, correlation, FFT, and matrix multiplication algorithms are evaluated for performance for a system which uses the MRU and a system which does not. The MRU system exhibits a speed-up of between approximately 1.5 and 2.5.

BACKGROUND

The design of a general purpose VLSI processor is aimed at high average performance for a wide range of applications. However, the efficient solution of some tasks demands the enhancement of the processing capabilities of the processor (e.g. floating point computations). In spite of the advances of microelectronic technology, the addition of functional units to the VLSI processor is restricted by the limited amount of real estate available.

A viable alternative for increasing computer system performance is based on the use of special purpose units which take some of the computation burden from the main processor. Floating-point processors, CRT controllers, disk controllers, and DMA controllers, are but a few examples of currently available devices. In some cases, these units are designed to perform certain restricted data processing tasks in a very efficient manner. In other cases, they help in the management of the computer systems resources. For example, memory management units (MMU's) assist the main processor in controlling the elements of the memory hierarchy of a computer system, which consists of a cache, primary storage, and secondary storage. Efficient memory management is crucial in virtual memory computer systems.

Most applications are currently developed using structured high level languages (HLL's) such as Pascal, Ada, etc. In an HLL, information is generally handled in a structured form, e.g. as procedures and data structures. In general, HLL compilers will generate a considerable amount of code just to navigate through the data structures. This fact aggravates the problem of the access overhead. Efficient address generation for data structures is therefore a primary goal in order to obtain high performance in a computer system. This is particularly true in a RISC environment where address calculations are almost exclusively performed in software since the available addressing modes in a RISC machine are limited. Another area where address computation is important is that of digital signal processing (DSP). Significant effort has been focused on the design of special purpose processors [1,2]. More recently, Nwachukwu [3] has proposed an array indexing unit to be used for address generation in an array processor to produce a system which is more versatile than conventional FFT array processors. Although these machines each have their own advantages, each is also designed for a particular set of tasks. The success of the current special purpose machines is strongly determined by the intended task domains. Hence the machines tend to be well suited for one particular application, and usually inflexible.

Instead of building special purpose architectures, some designers have concentrated their efforts on using general purpose processors to achieve versatility at the expense of having to generate addresses in software. One example is the fast Fourier transform algorithm and its many variations [4,5]. Although these algorithms improve the efficiency of the various DSP operations, a great deal of time is wasted in address calculations since a general purpose processor is used.

OBJECTIVES

This paper evaluates an alternative to software address generation which consists of a hardware unit that is tailored to function more efficiently with data structures associated with DSP applications. This device can in fact be viewed as a memory reconfiguring unit (MRU) or a coprocessor for accomplishing address transformations.

The objective was to design an MRU which was easy to interface between host and memory and which would not require any modifications either to the host processor or the memory. Special opcodes have not been introduced into the instruction repertoire nor does the design presume any modifications to the existing operating system. The function of a unit such as the MRU is to provide the CPU with a set of specialized addressing modes as defined by algorithms which are frequently used in several major applications.

To demonstrate that the MRU can effectively generate different primitive address sequences, i.e., sequential addressing, sequential with offset addressing, shuffled addressing, bit-reversed addressing, and reflected addressing [6][7], some frequently used algorithms; convolution, correlation, fast Fourier transform, and matrix equation; are studied in this research. These algorithms are selected because they utilize different address sequences. The performance of a computer system with the MRU and one without the MRU will be evaluated respectively to illustrate the speedup of the MRU.

DESIGN APPROACH

The approach which has been used here is to classify a desired address sequence into a composition of a series of "primitive" address sequences. A different address sequence may then be generated by specifying a different composition of these primitive address sequences. The selection of the primitive address sequences is based on a survey of frequently used algorithms for the solution of real-time DSP problems. They are summarized in Table 1. Based on an addressing sequence of length N, if A_i is address i contained in the range $0 \le i \le N-1$, the primitive addressing sequences are defined as follows:

(1) Sequential addressing
$A_0, A_1, A_2, ..., A_{N-1}$.
(2) Sequential with offset (k) addressing
$A_{0+k}, A_{1+k}, A_{2+k}, ..., A_{N-1+k}$.
(3) Shuffled addressing (base r, N/r=p)
$A_0, A_p, A_{2p}, ..., A_1, A_{p+1}, A_{2p+1}, ..., A_2, A_{p+2}, A_{2p+2}, ..., A_{N-1}$.
(4) Bit-reversed addressing (eg. N=8)
$A_0, A_4, A_2, A_6, A_1, A_5, A_3, A_7$.
(5) Reflected addressing
$A_0, A_{N-1}, A_1, A_{N-2}, ..., A_i, A_{N-i}, ..., A_{N/2-1}, A_{N/2}$.

The performance analysis in this paper will emphasize the evaluation of algorithms which employ these five different addressing sequences.

PROPOSED MRU SYSTEM

As shown in Figure 1, the signal flow between the processor and memory is through the MRU and dictates that the MRU be transparent to normal signal flow i.e., no appreciable delay should be introduced when instructions or unmapped data are to be fetched from memory. The system functions in one of three modes, depending on the address sent out by the CPU. (The term mapped data refers to data sequences which need to be accessed in specified addressing patterns while unmapped data refers to data which need not).

The MRU is a memory mapped device in which each user accessible register is assigned a unique memory address. When the CPU writes to any of these MRU registers the MRU is placed in the "INIT" mode for one instruction cycle. The CPU thus initializes the MRU registers to set up the desired primitive address sequences.

If the host processor accesses data which is not to be mapped or if it fetches an instruction, the MRU is in the "PASS" mode and lets the signals, data and address, through to the memory/processor unaltered. Only a decode delay is introduced by the MRU in order to determine the memory space being addressed, whether mapped or unmapped.

When the host wants to access mapped data it sends out a base address which is recognized by the MRU as one of the base addresses specified during initialization. The MRU enters the "MAP" mode for one instruction cycle. Each time a particular primitive address sequence is referenced the CPU will send out the same base address. The base address thus indicates the address sequence to be generated by the MRU and corresponds to the location at which the first piece of mapped data is stored for this pattern.

Thus by specifying an addressing pattern in a short initialization routine (with the MRU in the INIT mode), and by specifying a base address whenever the address pattern is to be used (with the MRU in the MAP mode), the MRU makes the memory appear reconfigurable. The MRU provides a set of specialized hardware addressing modes so that primitive addressing sequences may be generated.

FUNCTIONAL HARDWARE DESCRIPTION

Primitive Generation Units (PGUs) are the basic component in the MRU and are used to generate a specific pattern. Several PGUs are combined with additional logic into an Offset Generation Unit (OGU) which selects the appropriate pattern to be used and controls the PGUs. A Decode and Control Logic (DCL) unit determines the system mode and is also used to trap/regenerate signals to/from the CPU and to/from memory. A functional representation of the memory reconfiguring unit is shown in Figure 2.

Decode and Control Logic (DCL) Unit

The DCL determines the MRU operating mode and provides for control of signal flow between the host processor and the memory. Address registers are also included in this unit, so that the OGU may use the supplied base addresses to generate the final mapped addresses.

The mode selection and data switching portion of the Decode Logic Unit (DLU), as shown in Figure 3, determines the system mode by decoding the address lines sent by the host processor. Figure 4 is the functional diagram of the Signal Flow Control (SFC) unit which is responsible for all signal flow. Notice that this unit basically performs a multiplexing function. For example, the addresses to memory come from either the CPU or the OGU depending on whether or not address mapping is to occur. The operation of DCL can be described as follows [6]:

PASS mode. When an address associated with either an instruction or unmapped data is sent by the host processor, the DLU decodes the address in order to determine that the MRU is in the PASS mode. Consequently the SFC allows the address, data, and signals to pass through with only a decode delay.

INIT mode. If the address sent by the host processor indicates the INIT mode, the DLU will send the subsequent data to the OGU in order to initialize the appropriate registers and counters. Typically, the data includes the length of sequence, the type of sequence, constant offset, and the other control words. During the INIT mode, base address registers are loaded with base addresses by the SFC. The address indicating the INIT mode is sent out repeatedly until all data required have been written.

MAP mode. A mappable base address is decoded by the DLU which compares the coming address with previously stored base address. If a match is found, the SFC intercepts the R/W signal and holds the data bus until the mapped address is generated. After the mapped address generated by the OGU is sent to the SFC, it enables the R/W signal and the data flow. Every time the base address is received, the DCL activates the OGU and the appropriate PGU to generate the requested addressing sequence.

Offset Generation Unit (OGU)

Three control word registers in the OGU contain the control information necessary to activate each Primitive Generation Unit (PGU). The control word registers are loaded by the host processor while the MRU is in the INIT mode. In the adder, the base address from the DCL will be added with the element of the requested addressing sequence from the PGU to provide the final mapped address. The final mapped address is then routed back to the SFC in the DCL, which then generates the R/W signal and puts the address on the address bus and passes the contents of the data bus to/from the memory.

The PGU is an essential part of the OGU. It consists of four different hardware units which are responsible for generating the primitive addressing sequences [6]:

(1) Sequential addressing sequences can be generated by a simple parallel-input parallel-output up/down counter.
(2) Sequential with offset (k) addressing sequences can be generated by an adder/subtractor unit. One of the two operands is obtained from a sequential counter and the other is obtained from the constant offset k stored in a register which has been previously loaded while the MRU was in the INIT mode.
(3) A counter-multiplexer combination unit as shown in Figure 5 can provide shuffled and bit-reversed addressing sequences. To generate shuffled addressing for a sequence of $N=2^n$, the output of an n-bit counter is input to n, n-to-1 multiplexers. Each multiplexer is controlled by $\log_2 n$ control bits corresponding to the different base of the required shuffled addressing sequence and thus selects one of n inputs for generating an n-bit word. For the bit-reversed addressing sequence, the control bits in the unit are simply set to select the counter output bits in reversed order, i.e., counter outputs $C_{n-1}, C_{n-2}, ..., C_1, C_0$ are selected in the order $C_0, C_1, C_2, ..., C_{n-2}, C_{n-1}$.
(4) Reflected addressing sequences may be generated by a hardware unit as shown in Figure 6. In the reflected address generation unit, the counter is incremented every other clock cycle. For a counter value "i", the ith element A_i is fetched and on the next cycle, "i" is subtracted from the constant (N-1) to fetch the element A_{N-1-i}.

The final offset is selected by a 4-to-1 multiplexer corresponding to the requested addressing pattern. The functional diagram of PGU is shown in Figure 7.

ALGORITHM ANALYSIS

To demonstrate that the primitive addressing sequences can be efficiently generated within the MRU, several frequently used digital signal processing algorithms such as convolution, correlation, fast Fourier transform, and matrix manipulation were studied. These algorithms were implemented on an HP/UNIX system using assembly language. The time required for program execution in assembly language was calculated and an analysis of the performance will be discussed in the next section.

The assumption made for all the algorithms to be implemented is that information such as the length of the sequence, the specific primitive addressing sequence, and data type have already been stored in the host processor. Therefore, in the initialization of MRU system, this information is sent to the proper registers in the MRU.

Convolution

If $A(x)$ and $B(x)$ are functions defined on $(-\infty, \infty)$, the convolution, $C(X)$ of A and B, is the function defined by [8]:

$$C(X) = \lim_{T \to \infty} \int_{-T}^{T} A(t)B(x-t)dt$$

For convenient numerical calculation, the discrete convolution is typically used. The discrete convolution of two sequences A and B with equal length N is the set of numbers C_i, $i=0, \ldots, N-1$, defined by [9] as :

$$C_i = \sum_{j=0}^{i} A_j B_{i-j}$$

The algorithm can be written as:

```
for        (i = 0 to N-1) do
           C_i = 0;
    for        (j = 0 to i) do
               C_i = C_i + A_j B_{i-j};
```

Actually there are only two simple addressing sequences, sequential addressing and sequential with offset addressing, used for sequences A, B, and C in the above algorithm. But the addressing sequence for A and B can be treated as reflected addressing, i.e., when the sequences of A and B are stored in consecutive memory locations, the address of the first item of A is loaded as the base address. Then for every iteration, the value $(N+i+1)$ is used to reset the reflected address generation unit as the length of that sequence. Thus the reflected addressing sequence with at most a length of N is performed. For example, when N=3:

(1) In the first iteration (i=0), the reflected addressing sequence is 0, 3. The actual data fetched are A0, B0.
(2) In the second iteration (i=1), the reflected addressing sequence is 0, 4, 1, 3. The actual data fetched are A_0, B_1, A_1, B_0.
(3) In the third iteration (i=2), the reflected addressing sequence is 0, 5, 1, 4, 2, 3. The actual data fetched are A_0, B_2, A_1, B_1, A_2, B_0.

This implementation requires the use of only two base address registers and two OGUs.

Correlation

The correlation of two functions with infinite lengths is defined by [10] as:

$$C(X) = \lim_{T \to \infty} \int_{-T}^{T} A(t)B(x+t)dt$$

Similarly, the discrete correlation of two sequences A and B with equal length N is the set of numbers C_i, $i=0, \ldots, N-1$, defined by [10] as:

$$C_i = \sum_{j=0}^{N-i-1} A_j B_{i+j}$$

The algorithm can be written as:

```
for        (i = 0 to N-1) do
           C_i = 0;
    for        (j = 0 to N - i - j) do
               C_i = C_i + A_j B_{i+j};
```

Both addressing sequences of A and C are sequential addressing, and the sequence of B is sequential with offset (i) addressing.

Fast Fourier Transform

The Fourier transform is a masterpiece of applied mathematics which has been used in almost every branch of science and engineering. The fast Fourier transform (FFT) is a method for efficiently computing the discrete Fourier transform (DFT) of a series of data samples . The discrete Fourier transform (DFT) is defined by [11] as:

$$A_r = \sum_{k=0}^{N-1} X_k e^{-j(2\pi rk/N)} \qquad r = 0, 1, 2, \ldots, N-1$$

where A_r is the rth coefficient of the DFT, X_k denotes the kth sample of the time series which consists of N samples, and $j=(-1)^{1/2}$. Both A_r and X_k could be complex numbers.

Suppose a time series having N samples is divided into two functions [11], Y_k and Z_k, each of which has only half as many points, N/2. Y_k is composed of the even-numbered points (X_0, X_2, X_4, ...), and Z_k is composed of the odd-numbered points (X_1, X_3, X_5, ...). They have a DFT defined by

$$B_r = \sum_{k=0}^{N/2-1} Y_k e^{-j(4\pi rk/N)}$$

$$C_r = \sum_{k=0}^{N/2-1} Y_k e^{-j(4\pi rk/N)} \qquad r = 0, 1, 2, \ldots, N/2-1$$

but the DFT wanted is A_r, where for r = 0,1,...., N-1 we have:

$$A_r = \sum_{k=0}^{N/2-1} [Y_k e^{-j(4\pi rk/N)} + Z_k e^{-j(2\pi r(2k+1)/N)}]$$

and can be written as:

$$A_r = \sum_{k=0}^{N/2-1} Y_k e^{-j(4\pi rk/N)} + e^{-j(2\pi r/N)} \sum_{k=0}^{N/2-1} Z_k e^{-j(4\pi rk/N)}$$

$$= B_r + e^{-j(2\pi r/N)} C_r \qquad 0 \le r \le N/2$$

and further:

$$A_{r+N/2} = B_r + e^{-j(2\pi (r+N/2)/N)} C_r$$

$$= B_r - e^{-j(2\pi r/N)} C_r \qquad 0 \le r \le N/2$$

The general procedure for solving an FFT with an N-sample series (where N is a power of 2) is:

(1) Decompose the N-point DFT into two N/2-point DFTs and determine the corresponding recomposition equations.
(2) Decompose each N/2-point DFT into two N/4-point DFTs and determine the corresponding equations.
(3) Continue the process until N/2 two-point DFTs result. This terminates the process.

The result is presented in Figure 8 for an 8-point DFT. In the diagram, it can be observed that the left nodes are in bit-reversed sequence from top to bottom. For an input series of N numbers, the indices of these points can be written in binary number form [10]:

$$I = b_{M-1} 2^{M-1} + b_{M-2} 2^{M-2} + ... + b_1 2^1 + b_0 2^0$$

where $M = \log_2 N$ and b_i, $i = 0, 1, 2, ..., M-1$, are either 0 or 1. The bit-reversed version of the indices can be simply written as:

$$I_R = b_0 2^{M-1} + b_1 2^{M-2} + ... + b_{M-2} 2^1 + b_{M-1} 2^0$$

The algorithm for accomplishing the bit reversal can be described as follows [10]

```
for        (k = 0 to N - 1) do
           newadr = 0
           maddr = k;
           for        (i=0 to M - 1) do
                      lrmndr = mod(maddr,2);
                      newadr = newadr + lrmndr 2^{M-1-i};
                      maddr = maddr/2;
                      X(newadr) = X temp(k);
```

After each index is reversed, the data item corresponding to the index will be selected in increasing order. The final algorithm of the FFT can be given as follows with the series A which has the elements positioned in bit-reversed sequence [12]:

```
for        (i = 1 to log N) do
           W = e^{2nj/2i};
           W_p = 1;
           for        (p = 0 to 2^{i-1} - 1) do
                      for        (k = p to N - 1 step 2^i) do
                                 t = W_p A(k + 2^{i-1});
                                 u = A(k);
                                 A(k) = u + t;
                                 A(k + 2^{i-1}) = u - t;
                      W_p = W_p W;
```

Series A is fetched by a shuffled addressing sequence with different bases in every iteration. For example, when N=8, we have:

(1) In the first iteration, the sequence of index is 0, 1, 2, 3, 4, 5, 6, 7. (base=8)
(2) In the second iteration, the sequence of index is 0, 2, 4, 6, 1, 3, 5, 7. (base=4)
(3) In the third iteration, the sequence of index is 0, 4, 1, 5, 2, 6, 3, 7. (base=2)

Matrix Manipulation

It is often required in practice to solve a system of n linear equations. The nth-order set of simultaneous, linear algebraic equations in the n unknowns x_1, x_2, x_3, ..., x_N can conveniently be represented by the matrix equation form:

$$a_{11}x_1 + a_{12}x_2 + a_{13}x_3 + ... + a_{1n}x_n = b_1$$
$$a_{21}x_1 + a_{22}x_2 + a_{23}x_3 + ... + a_{2n}x_n = b_2$$
$$\bullet \bullet \bullet \qquad\qquad \bullet \bullet \bullet$$
$$a_{n1}x_1 + a_{n2}x_2 + a_{n3}x_3 + ... + a_{nn}x_n = b_n$$

or more simply by

$$AX = B$$

where A is an n-square matrix of coefficients having elements a_{ij}, X and B are n by 1 column vectors with elements x_i and b_i respectively.

One method of solving the system of simultaneous equations which is represented by a matrix equation is to calculate the inverse of the coefficient matrix [13]. For the set defined in above, the matrix equation is

$$X = A^{-1}B$$

where A^{-1} is the inverse of A. The method for finding the inverse of a matrix is described in [13] as the following algorithm. The only limitation is that a_{11} must not be zero. If it is, rearrange the equations so that the element in the first row and first column is not zero. This method can be expressed in algorithmic form as follows:

```
a_{1,n+1};
for        (i = 2 to n) do
           a_{i,n+1} = 0;
for        (m = 1 to n) do
           for        (j = 1 to n) do
                      a_{n+1,j} = a_{1,j+1}/a_{j,1};
           for        (i = 1 to n) do
                      for        (j = 1 to n) do
                                 a*_{i,j} = a_{i,j+1} - a_{i,1} a_{n+1,j};
           for        (i = 1 to n-1) do
                      for        (j = 1 to n) do
                                 a_{i,j} = a*_{i+1,j};
           for        (j = 1 to n) do
                      a_{n,j} = a_{n+1,j};
```

When A is an m x n matrix other than an n-square matrix, the equation still can be solved. In this case the n x m transpose matrix A^T is multiplied on both sides of the matrix equation as follows:

$$A^T A X = A^T B$$

or it can be written as:

$$PX = Q$$

where P is an n-square matrix. This equation can be solved using the method of inversion.

The program solving this matrix equation requires matrix addition, multiplication, and transposition, and will employ sequential addressing and sequential with offset addressing.

RESULTS

By evaluating the assembly language programs, the execution time can be estimated to evaluate the performance of the computer system employing the MRU. The program execution time is obtained by calculating the number of clock cycles taken by the instructions in the program. The MC68020 microprocessor is used as the host processor and the MC68881 floating-point coprocessor is employed when needed in both operations with the MRU and without the MRU. Calculating the execution time is based on the MC68020 and the MC68881 instruction execution times in terms of external clock cycles. Because the MC68020 and the MC68881 are used, there are three values given for each instruction [14][15][16]:

(1) Best Case - The best case reflects the time when the instruction is in the cache and benefits from the maximum overlap due to other instructions.
(2) Cache Case - When the instruction is in the cache but has no overlap, the value of cache-only-case is employed.
(3) Worst Case - The worst case reflects the time when the instruction is not in cache or the cache is disabled and there is no instruction overlap.

T_{MRU} and T_{NMRU} represent the execution time of the algorithms, in clock cycles, with the MRU and without the MRU respectively. Table 2 gives equations which specify the execution time for the four algorithms to be compared. The equations in

Table 2 consist of N^3, MN^2, N^2, MN, $N\log_2 N$, N, $\log_2 N$, and constant terms. The N^3, MN^2, N^2, MN, and $N\log_2 N$ terms are due to nested loops with different boundary values. The N and $\log_2 N$ terms result from single loops. The equations in the table were formulated under the following assumptions:

(1) The length of the data sequences in the convolution, correlation, and FFT programs is N. The dimension of the coefficient matrix in the matrix equation program is M x N, with M not necessarily equal to N.
(2) In the initialization of the MRU operation, the host processor loads the base address to MRUBARX, selects the required primitive addressing sequence in MRUCWRX0, loads the length of the sequence to MRUCWRX1, and determines the type of the sequence element. That is, the number of bytes occupied by each element in MRUCWRX2, where X = A, B, or C.
(3) The host processor can restart the sequence output at any time during the program processing with MRUCWRX3. The offset used in sequential with offset addressing sequence and the base required in shuffled addressing sequence are loaded to MRUCWRX4.
(4) MRUBARX, MRUCWRX0, MRUCWRX1, MRUCWRX2, MRUCWRX3, and MRUCWRX4 are considered to function in the same manner as memory locations.
(5) The values ss, so, sh, br, and rs represent sequential, sequential with offset, shuffled, bit-reversed and reflected addressing sequence respectively. They are treated the same as immediate data values.
(6) The MRU is considered sufficiently fast so that the unmapped address delay through the MRU is negligible and does not cause the host processor to insert wait states. In the PASS mode the only delay is a two gate delay. In the MAPPING mode the addresses have been precalculated by the MRU and only a two gate delay results.

Provided in Appendix A are the programs for implementing the convolution algorithm. Shown in Appendix A.1 and A.2 are assembly language programs with and without the MRU respectively. In a similar fashion programs for the other three algorithms were developed but for brevity, they are not included here.

Figure 9 through Figure 12 represent the graphs of speed factor versus length N and were obtained by plotting the equations given in Table 2. The speed factor is defined as T_{NMRU} divided by T_{MRU} and is a speedup measure of the MRU system. As shown in Table 3, the speedup varies between approximately 1.5 and 2.4 for large values of N. The following observations are drawn by analyzing the results presented in the four figures:

(1) The convolution and correlation programs deal with fixed-point arithmetic operations. They have similar algorithms but employ different addressing sequences. The correlation program obtains higher speed factors than the convolution program does, i.e., selecting sequential and sequential with offset sequences saves more time than selecting reflected sequence.
(2) The FFT and matrix equation programs operate on floating-point data. The speed factors also demonstrate that including the MRU increases the system performance with a speed factor of at least 1.5 when N is greater than 10.
(3) The computer system with the MRU always achieves higher performance than the system without the MRU. In the range between 2 and 1000 the larger the value of N is, the higher the speed factor is. As N approaches infinity, the speed factor is almost the same as that obtained when N is equal to 1000.
(4) The matrix equation program manipulates 2-dimensional data, whereas the convolution, correlation, and FFT programs employ 1-dimensional data. It can be expected that the 2-dimensional convolution, correlation, and FFT computations with the MRU system will obtain high performance.
(5) The programs included in this study are not the only way to implement the algorithms. Therefore the values of execution time obtained are not absolutely exact. The factors affecting

the execution time of a program also include the length of the data sequence, the type of the sequence element, the required primitive addressing sequence, and if the instructions are in cache or not.

CONCLUSION

A computer system employing the MRU can speed up some frequently used digital signal processing computations by factors of as much as 2.41 compared with a system without the MRU. The result is achieved by comparing the execution time of the system with and without the MRU when performing the convolution, correlation, FFT, and matrix equation computations. The higher performance is obtained by the MRU hardware realization which reduces the overhead associated with repetitive calculation of addresses in software. The time savings depends on factors such as the length of the data sequence, the type of sequence element, and the required primitive addressing sequence.

The MRU system can be modified in hardware to adapt to unique environments found in digital signal processing applications and also to the more general demands required in general purpose computing. The MRU system can be easily expanded to incorporate new features without any modification to either host processor or memory.

ACKNOWLEDGEMENT

This work was supported in part by the National Science Foundation under grant number MIP-8912455.

REFERENCES

[1] Jonathan Allen,"Computer Architecture for Signal Processing", Proceedings of the IEEE, Vol.63, No.4, April 1978.
[2] B. A. Bowen and W. R. Brown, VLSI Systems Design for Digital Signal Processing, Prentice-Hall, NJ, 1985.
[3] E. O. Nwachukwu, "Address Generation in an Array Processor", IEEE Transactions on Computers, Vol C-34 No 2, February 1985.
[4] Richard E. Blahut, Fast Algorithm for Digital Signal Processing, Addison-Wesley, Mass., 1985.
[5] C. S. Burrus and T. W. Parks, Topics in Digital Signal Processing, John Wiley, UK, 1985.
[6] Vinita Signhal, Paul T. Hulina, and Lee D. Coraor, "A Memory Reconfiguring Unit for Versatile Work-Stations", 1986 IEEE Workstation Technology and Systems Conference.
[7] Paul T. Hulina and Lee D. Coraor, "Coprocessor Architecture for Efficient Address Computation and Memory Accessing", Computer Systems Science and Engineering, Vol.5, No.3, July 1990.
[8] David Kahaner, Cleve Moler, and Stephen Nash, Numerical Methods and Software, Prentice-Hall, Englewood Cliffs, N.J., 1989.
[9] Chi-Tsong Chen, System and Signal Analysis, Holt, Rinehart and Winston, Inc., New York, N.Y., 1989.
[10] Robert D. Strum and Donald E. Kirk, First Principle of Discrete Systems and Digital Signal Processing, Addison-Wesley, Mass., 1988.
[11] G-AE Subcommittee on Measurement Concepts, "What is the Fast Fourier Transform?", IEEE Transaction on Audio and Electroacoustics, Vol.AU-15, No.2, June 1967.
[12] Aho, Hopcroft, and Ullman, The Design and Analysis of Computer Algorithms, Addison-Wesley, Mass., 1974.
[13] Salvatore DeAngelo and Paul Jorgensen, Mathematics for Data Processing, McGraw-Hill, New York, N.Y., 1970.
[14] Motorola Inc., MC68020 32-Bit Microprocessor User's Manual, Second Edition, Prentice-Hall, Englewood Cliffs, N.J., 1985.
[15] Motorola Inc., MC68881 Floating-Point Coprocessor User's Manual, First Edition, Prentice-Hall, Englewood Cliffs, N.J., 1985.
[16] Thomas L. Harman, The Motorola MC68020 and MC68030 Microprocessor Family, Prentice-Hall, Englewood Cliffs, N.J., 1989.

Appendix A.1

```
# # # # # # # # # # # # # # # # # # # # # # # # # # # # # # # # # # # # # #
#                    Convolution  (With MRU)                             #
#                                                                        #
#     Assumptions:                                                       #
#     (1) Data are stored in consecutive memory locations starting at    #
#         address PARAM.                                                  #
#     (2) The first item stored is the length n of array a and b, which   #
#         is represented by 2 byte integer.                              #
#     (3) The data items of array a and b are represented by 2 byte       #
#         integers and the numbers of array c are 4 byte integers,       #
#     (4) Treat array a and b as one sequence.                           #
# # # # # # # # # # # # # # # # # # # # # # # # # # # # # # # # # # # # # #

         text
Start:   lea     PARAM, %a0        # Initialize parameters.

         mov.w   &rs, MRUCWRA0     # Select reflected sequence for the
                                   # sequence which includes a and b.
         mov.w   &ss, MRUCWRC0     # Select sequential sequence for c.
         clr.l   %d2
         mov.w   (%a0), %d2
         mov.w   %d2, MRUCWRC1     # Load the length of array c.
         mov.w   &2, MRUCWRA2      # Load the type of the data element.
         mov.w   &4, MRUCWRC2
         mulu    &2, %d2           # Calculate the bytes occupied
                                   # by array a or b.
         mova    %a0, %a1          # Determine the location of array a.
         adda    &2, %a1
         mova    %a1, MRUBARA      # Initialize base address register.
         mova    %a1, %a3          # Determine the location of array c
         adda    %d2, %a3
         adda    %d2, %a3
         mova    %a3, MRUBARC      # Initialize base address register.

         clr.l   %d5               # Set i to zero.
         mov.w   &0, MRUCWRC3      # Start the sequence from the
                                   # beginning.
Loop1:   clr.l   %d0               # Set c[i] to zero.
         mov.w   &0, MRUCWRA3      # Start the sequence from the
                                   # beginning.
         clr.l   %d6               # Set j to zero.
         clr.l   %d7               # Calculate (n+i+1).
         mov.w   %d5, %d7
         add.w   (%a0), %d7
         addi.w  &1, %d7
         mov.w   %d7, MRUCWRA1     # Load the length of the sequence
                                   # which consists of a and b.
Loop2:   clr.l   %d1
         mov.w   MRUMAA, %d1       # Fetch a[j].
         addi.w  &1, %d6           # Update j to get the next mapped
                                   # address.
         muls    MRUMAA, %d1       # Calculate a[j]*b[i-j].
         add.l   %d1, %d0          # Update c[i].
         addi.w  &1, %d6           # Update j and check bound.
         mov.w   %d5, %d2
         addi.w  &1, %d2
         mulu    &2, %d2
         cmp.w   %d6, %d2
         blt     Loop2

         mov.l   %d0, MRUMAC       # Save c[i].
         addi.w  &1, %d5           # Update i and check bound.
         cmp.w   %d5, (%a0)
         blt     Loop1

End:     illegal
```

Appendix A.2

```
# # # # # # # # # # # # # # # # # # # # # # # # # # # # # # # # # # # # # #
#                    Convolution  (Without MRU)                          #
#                                                                        #
#     Assumptions:                                                       #
#     (1) Data are stored in consecutive memory locations starting at    #
#         address PARAM.                                                  #
#     (2) The first item stored is the length n of array a and b, which   #
#         is represented by 2 byte integer.                              #
#     (3) The data items of array a and b are represented by 2 byte       #
#         integers and the numbers of array c are 4 byte integers,       #
# # # # # # # # # # # # # # # # # # # # # # # # # # # # # # # # # # # # # #

         text
Start:   lea     PARAM, %a0        # Initialize parameters.
         clr.l   %d2
         mov.w   (%a0), %d2        # Load the length of the array.
         mulu    &2, %d2           # Calculate the bytes occupied
                                   # by array a or b.
         mova    %a0, %a1          # Determine the location of array a.
         adda    &2, %a1
         mova    %a1, %a2          # Determine the location of array b.
         adda    %d2, %a2
         mova    %a2, %a3          # Determine the location of array c.
         adda    %d2, %a3
```

```
Loop1:   clr.l   %d5               # Set i to zero.
         clr.l   %d0               # Set c[i] to zero.
         clr.l   %d6               # Set j to zero.
Loop2:   clr.l   %d7               # Set k to zero.
         mov.w   %d5, %d7
         sub.w   %d6, %d7          # k=i-j.
         mov.w   %d6, %d2          # Calculate the offset of a[j].
         mulu    &2, %d2
         clr.l   %d1
         mov.w   (%a1, %d2), %d1   # Fetch a[j].
         mov.w   %d7, %d2          # Calculate the offset of b[k].
         mulu    &2, %d2
         muls    (%a2, %d2), %d1   # Calculate a[j]*b[k].
         add.l   %d1, %d0          # Update c[i].
         addi.w  &1, %d6           # Update j and check bound.
         cmp.w   %d6, %d5
         ble     Loop2

         mov.w   %d5, %d2          # Calculate the offset of c[i].
         mulu    &4, %d2
         mov.l   %d0, (%a3, %d2)   # Save c[i].
         addi.w  &1, %d5           # Update i and check bound.
         cmp.w   %d5, (%a0)
         blt     Loop1
End:     illegal
```

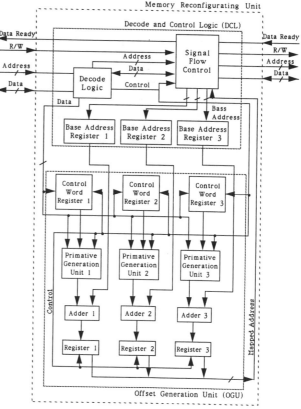

Figure 1   Signal Flow Between the Host Processor and Memory via the MRU

Figure 2   Overall Functional Diagram of the MRU

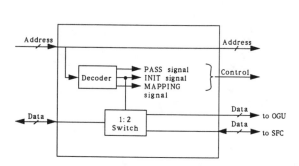

Figure 3   Functional Diagram of the Decode Section of the DLU

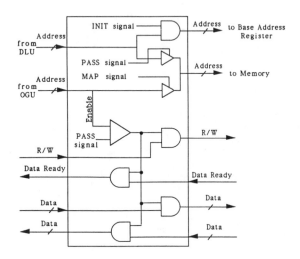

Figure 4   Functional Diagram of Signal Flow Control (SFC) Unit

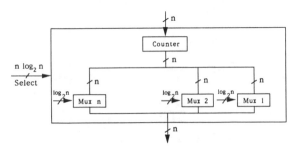

Figure 5   Counter-Multiplexer Unit for Shuffled and Bit-Reversed Address Generation

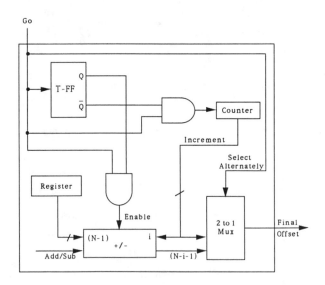

Figure 6   Reflected Address Generation Unit

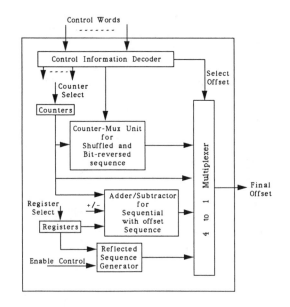

Figure 7   Functional Diagram of the PGU

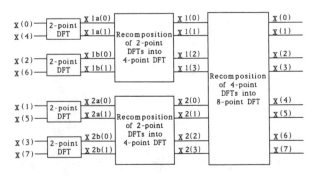

Figure 8   Calculation of DFT by Successive Recomposition of 2-point DFTs.

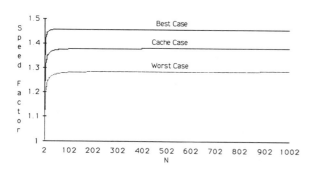

Figure 9   Speed Factor vs. Length N in Convolution Program

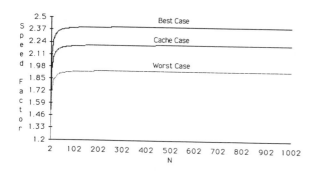

Figure 10   Speed Factor vs. Length N in Correlation Program

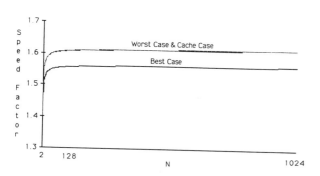

Figure 11   Speed Factor vs. Length N in FFT Program

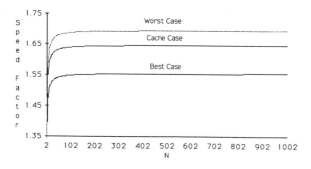

Figure 12   Speed Factor vs. Length N in Matrix Equation Program

Table 1

Summary of the Common Computations and Primitive Addressing Sequences

Computation	Addressing Sequence(s)
Matrix operations. (Multiplication, Addition, Adjoint, Transpose, etc.)	Sequential, Sequential with offset,
Fast Fourier transform.	Shuffled, Sequential, Bit-reversed.
2 dimensional fast Fourier transform.	Shuffled, Sequential, Bit-reversed.
Convolution.	Reflected, Sequential.
2 dimensional convolution.	Reflected, Matrix addressing.
Correlation.	Sequential, Sequential with offset.
2 dimensional correlation.	Sequential with offset, Matrix addressing.

Table 2   The Execution Time Required for Different Computations

	Best Case	Cache Case	Worst Case
T_{NMRU} (Convolution)	$43N^2+78N+50$	$63.5N^2+121.5N+81$	$74N^2+135N+102$
T_{MRU} (Convolution)	$29.5N^2+53.5N+91$	$46N^2+100N+137$	$57.5N^2+132.5N+171$
T_{NMRU} (Correlation)	$44.5N^2+79.5N+51$	$67.5N^2+125.5N+81$	$76.5N^2+147.5N+102$
T_{MRU} (Correlation)	$18.5N^2+44.5N+116$	$30.5N^2+78.5N+173$	$39.5N^2+103.5N+219$
T_{NMRU} (FFT)	$2009.5N\log_2 N+320N+364\log_2 N+84$	$2098N\log_2 N+352N+398\log_2 N+121$	$2111.5N\log_2 N+365N+410\log_2 N+145$
T_{MRU} (FFT)	$1279.5N\log_2 N+270N+439\log_2 N+139$	$1293N\log_2 N+282N+498\log_2 N+201$	$1301N\log_2 N+288N+514\log_2 N+244$
T_{NMRU} (Matrix Eq.)	$690N^3+298MN^2+848N^2+483MN+456N+456$ or $690N^3+898N^2+456N+453$	$764N^3+327MN^2+953N^2+537MN+520N+571$ or $764N^3+1011N^2+492N+567$	$807N^3+343MN^2+1274N^2+567MN+558N+613$ or $807N^3+1075N^2+516N+607$
T_{MRU} (Matrix Eq.)	$444N^3+198MN^2+767N^2+333MN+619N+795$ or $444N^3+757N^2+581N+722$	$464N^3+207MN^2+833N^2+351MN+751N+1001$ or $464N^3+819N^2+692N+921$	$476N^3+213MN^2+880N^2+363MN+877N+1183$ or $476N^3+857N^2+804N+1064$

Table 3   Speed Factors as N Approaches Infinity for the Computations

	Best Case	Cache Case	Worst Case
Convolution	1.46	1.38	1.29
Correlation	2.41	2.21	1.94
FFT	1.57	1.62	1.62
Matrix Equation	1.55	1.65	1.70

A Percolation Based VLIW Architecture

Arthur Abnous, Roni Potasman, Nader Bagherzadeh,
*and Alex Nicolau**

Department of Electrical and Computer Engineering
University of California, Irvine
Irvine, CA 92717

Abstract

This paper describes a Very Long Instruction Word (VLIW) architecture that is based on a Percolation Scheduling (PS) compiler. The architecture exploits fine-grain (instruction-level) parallelism in order to speed up program execution. In order to take advantage of the capabilities of the PS compiler, we have devised a novel multiway branching scheme that improves the overall performance of the architecture and allows speculative execution of the operations in the branch and branch delay slots. Also, we describe a modified pipeline structure that reduces the frequency of pipeline hazards in a VLIW processor.

1   Introduction

The VLIW architecture is considered to be one of the promising methods of increasing performance beyond standard RISC architectures. While RISC architectures take advantage of only temporal parallelism (by using pipelined functional units), VLIW architectures can also take advantage of spatial parallelism by using multiple functional units to execute several operations concurrently. Similar to superscalar architectures, the VLIW architecture can reduce the Clock Per Instruction (CPI) factor by executing several operations concurrently. Superscalar processors schedule the execution order of the operations at run-time. They demand more hardware support in order to manage synchronization among concurrent operations. Since VLIW machines schedule operations at compile-time (allowing global optimizations), they tend to have relatively simple control paths– following the RISC methodology.

Although the idea of VLIW architecture had been known for years, there were no real implementations of such architectures mainly due to the lack of compilation tools for such machines. Recently, however, with rapid advances in compiler optimization techniques [1], this has changed. These new fine-grain compilation techniques compute a static parallel schedule from an originally sequential program. Percolation Scheduling is one of these promising techniques. In recent years, there have been several efforts to design and develop VLIW architectures [2], [3], [4].

2   The Compiler

Exploiting fine-grain parallelism is a critical part of exploiting all of the parallelism available in a given program, mainly because highly irregular forms of parallelism are often not visible at coarser levels. Our goal is to extend the potential parallelism by compacting across basic block boundaries while still preserving program correctness.

Our compiler system is based on Percolation scheduling. PS is a system of semantics-preserving transformations that convert an original program graph into a more parallel one. Its core consists of four transformations (*Move-op, Move-cj, Delete, and Unify*), which are defined in terms of adjacent nodes in the program graph. The transformations are atomic and thus can be combined with a variety of guidance rules (heuristics) to direct

*This work is supported in part by a grant from the University of California MICRO Project in cooperation with Northrop Corporation.

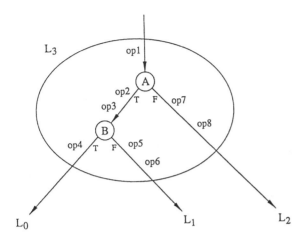

Figure 1: A node in our model

the optimization process. In Reference [1] it was shown that the core transformations are *complete* with respect to the set of all possible local, dependency-preserving transformations on programs.

2.1 The Machine Model

The input program is represented by a control and data flow graph (CDFG). The vertices (nodes) correspond to operations executed in each step. The edges represent flow of control from one node to its successor. Initially all nodes contain a single operation. Making a program "more parallel" involves compaction of several operations into one node while preserving the semantics of the sequential code. The architecture allows the presence of multiple conditional jumps (as well as other operations) per node. In the model we use, operations form a tree-like structure in the node. One (and only one) execution path is selected from the entry point down to the unique successor of the node. The actual path is selected according to the condition codes in the tree. For example, if we assumed that condition A is true and condition B is false in Figure 1, then $op1$, $op2$, $op3$, $op5$, and $op6$ are to be executed and the successor of this node is L_1. The execution of the node can be conceptualized as three steps:

1. Operands (for all operations) and condition code registers are read.

2. All operations are executed, condition codes are evaluated and a path to the unique successor instruction is chosen.

3. The results of operations on the path chosen are written back to register file/memory.

2.2 The Scheduling Process

The process of compaction involves the motion of operations from a node to its predecessor, thus trying to move all operations in the program as high as possible while maintaining program correctness. In this process, where it is assumed that there are no machine resource constraints (yet), only data dependency and flow dependency restrict the parallelism (i.e. a node may include more operation than physically executable by our machine).

After having the unlimited-resources schedule we begin the Resource Constraint Scheduling (RCS) process, where we generate a schedule given the resource limitations of the target machine. Since the core transformations of PS are general, any additional heuristics (e.g. the global resource constrained parallelism technique [5]) can be applied. The (compacted) program graph is scanned from the first node down to all nodes of the program. For each node we compare the number of existing operations to the number of operations executable by the machine. If there are no resource constraint violations, we proceed to the successor nodes of the current node. However, if there is a resource constraint violation, we have to defer some of the operations. The deferring function is the *mobility* of the operation, or in other words, the 'freedom' we have in scheduling this operation in later cycles without stretching the total execution time. This process continues until there are no resource constraint violations in the current node. Eventually, none of the nodes in the program graph violate any of the resource constraints.

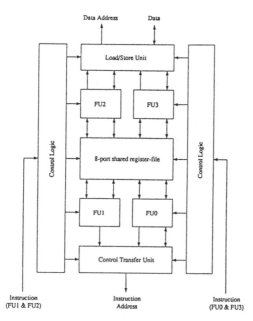

Figure 2: Block diagram of the processor

3   The Processor Architecture

In this section, we present an overview of the architectural features of the processor under development at our group. Two key aspects are stressed: an efficient instruction execution pipeline designed to reduce the frequency of pipeline stalls due to data hazards in the context of a VLIW processor with pipelined functional units, and support for multiway branch operations. The architecture has been designed to reflect the execution model assumed by our PS compiler.

3.1   Overview

Since our goal was to build a VLSI microprocessor, the key technological constraints that have influenced the architecture are the ones that arise in VLSI system design. Also, the average degree of parallelism extracted by our compiler was another factor that we have taken into consideration. Feasibility and efficiency considerations for a VLSI implementation of the processor architecture led to the current configuration, which includes four functional units and provides an engine that

can take advantage of the parallelizing capabilities of our PS compiler.

The organization of the processor was developed to facilitate an efficient VLSI implementation. A key aspect was an emphasis on locality of communication between different blocks of the processor. This is reflected in Figure 2 which illustrates the block diagram of our processor. The processor contains four integer functional units connected through a shared multiport register file. The processor is pipelined and can issue four operations packed into a long instruction word in each clock cycle. The Control Transfer Unit controls the instruction fetching sequence and can execute multiway branch operations. The Load/Store Unit establishes a connection to a data cache subsystem. The processor has some of the typical attributes of a pipelined RISC processor. It has a simple instruction set that is designed with efficient pipelining and decoding in mind. The instruction set follows the register-to-register execution model. All instructions have a fixed size, and there are only a few instruction formats. Arithmetic, logic, and shift operations are executed by all of the functional units. Load/Store operations are executed by FU2 and FU3, which are connected to the data cache subsystem through the Load/Store Unit. In our architecture, Load/Store operations use the register indirect addressing mode instead of the more typical displacement addressing mode. Branch operations are executed by FU0 and FU1, which are connected to the Control Transfer Unit. The processor can execute three-way branch operations. Also, to increase execution throughput, the processor performs speculative execution of the operations in the branch and branch delay slots.

3.2   Pipeline Structure

Figure 3 demonstrates the pipeline structure of a typical RISC processor [6]. An instruction is executed in five stages: IF (Instruction Fetch), ID (Instruction Decode), EX (EXecute), MEM (MEMory access), and WB (Write Back). This pipeline structure provides for the efficient execution of instructions in

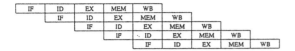

Figure 3: The pipeline structure of a typical RISC processor

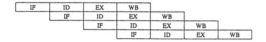

Figure 4: The pipeline structure of our VLIW processor

RISC machines.

In the above pipeline, the delay between the RD and WB stages is two cycles. This means that the result generated by an operation will not be written back to the register file in time for the next two operations in the pipeline to read it. To alleviate the possible read-after-write (RAW) hazards caused by this delay, RISC processors use internal bypassing of result operands [6]. The cost of bypassing in a RISC with one functional unit is minor compared to the number of cycles that it saves. It requires $2d$ comparators (where d is the delay between RD and WB) and the necessary pathways from the pipeline registers to the inputs of the ALU. If the bypassing circuitry is carefully designed, it will not lengthen the clock cycle of the processor. However, in a VLIW with n functional units, bypassing becomes a costly function to perform. The number of required comparators alone makes bypassing from each functional unit to every other functional unit quite expensive; we will need $2dn^2$ comparators. The required busses pose an even bigger problem.

Thus, bypassing cannot be used at a global scale to eliminate RAW hazards. Frequent RAW hazards can be avoided by reducing the delay between RD and WB from two cycles to

one cycle. This means that there will not be a MEM stage which implies the use of the register indirect addressing mode instead of the displacement addressing mode. This suggests the pipeline in Figure 4 which is the one used in our VLIW processor. In this pipeline structure, the EX stage is used to access the data cache for Load/Store operations. For other operations, it is used to perform an ALU operation.

3.3 Branching and Speculative Execution

The processor provides for the execution of three-way branch instructions. This was motivated by the execution model assumed in our compiler. Three-way branching is implemented by branch operations that have a pair of conditions instead of the usual single condition. Branch instructions have the following form:

$$BRc_1c_2 \quad cc_1, cc_2, offset$$

c_1 and c_2 represent conditions and are either $T(True)$ or $F(False)$. cc_1 and cc_2 are condition code registers. c_1 corresponds to cc_1, and c_2 corresponds to cc_2. A pair of such branch operations, encoded according to the condition codes in the control tree of the instruction node are issued to FU0 and FU1, which are connected to the Control Transfer Unit and can perform three-way branch operations.

The instruction node in Figure 1 has three successor nodes. This means that there are three possible target addresses: T_0 is $PC + offset_0$, T_1 is $PC + offset_1$, and T_2 is $PC + 1$. The code generator will generate a one-to-one mapping from $\{L_0, L_1, L_2\}$ to $\{T_0, T_1, T_2\}$. For example, if L_0, L_1, and L_2 are mapped into T_0, T_1, and T_2 respectively, the following branch operations will be scheduled for execution:

$$BRTT \quad A, B, offset_0 \qquad \text{executed by FU0}$$
$$BRTF \quad A, B, offset_1 \qquad \text{executed by FU1}$$

The above operations instruct the processor to branch to T_0 ($PC + offset_0$) if A is true and B is true, branch to T_1 ($PC + offset_1$) if A is true and B is false, or to fetch the instruction at T_2 ($PC + 1$), otherwise.

Benchmark	Unlimited Resources	Limited Resources
Livermore L1	7.4	2.1
Livermore L3	5.0	3.3
Livermore L4	2.6	2.2
Livermore L5	4.0	2.4
Livermore L9	3.4	1.9
Livermore L11	5.0	3.3
Livermore L12	5.0	3.3
Livermore L13	2.7	2.1
Livermore L14	3.7	2.6
Dijkstra	2.3	2.0

Table 1: Speed-up factors

To increase throughput, the processor performs speculative execution in the branch and branch delay slots. The operations in the branch and branch delay slots are assigned tags depending on which edge of the control tree of the instruction node they reside on. All of them are issued but only the ones with the tags corresponding to the path chosen at run-time will complete and write to the register file. While executing the branch operations, the Control Transfer Unit also computes a completion condition flag based on the outcome of the branch operations. The operations with a tag matching to the completion condition flag are allowed to complete execution and write their results to the register file.

4   Simulation Results

In this section, we present the results of our simulations. The main objective is to compare the speed-up factor of a machine with unlimited resources to that of our processor. The hardware limitations are due mainly to the total number of functional units available, the number of load/store and branch operations that can be issued and executed in each cycle, and the operand bypassing mechanism. The speed-up factors for various benchmarks are presented in Table 1. As expected, the speed-up factors drop when resource constraints are taken into account.

5   Conclusion

We presented a new VLIW architecture that exploits fine-grain parallelism extracted by Percolation Scheduling. The architecture contains multiple pipelined functional units that run in parallel. Included in the architecture is support for multiway branch operations. Given the context of a pipelined VLIW processor, we have employed a simplified pipeline structure refined to reduce the frequency of pipeline stalls due to data hazards. We also presented the results of our simulations comparing the performance of our processor to that of a machine with unlimited resources.

References

[1] A. S. Aiken, Compaction-Based Parallelization, Ph.D. Dissertation, Cornell University, August 1988.

[2] K. Ebcioglu, Some Design Ideas for a VLIW Architecture for Sequential-Natured Software, *Proceedings IFIP*, 1988.

[3] R. P. Colwell, R. P. Nix, J. J. Odonnell, D. P., Papworth, and P. K. Rodman, A VLIW Architecture for a Trace Scheduling Compiler, *IEEE Transactions on Computers*, Vol. 37, 1988, pp. 967-979.

[4] J. Labrousse and G. A. Slavenburg, CREATE-LIFE: A Modular Design Approach for High Performance ASIC's, *COMPCON* 1990.

[5] K. Ebcioglu, and A. Nicolau, A *global* resource-constrained parallelization technique, *Proc. ACM SIGARCH ICS-89: International Conference on Supercomputing*, Crete, Greece June 2-9 1989.

[6] P. Chow, and M. Horowitz, Architectural Tradeoffs in the Design of MIPS-X, *Proceedings of the 14th Annual International Symposium on Computer Architecture*, pp. 300-308, Pittsburg, Pennsylvania, June 1987.

THE TMS320C40 AND ITS
APPLICATION DEVELOPMENT ENVIRONMENT:
A DSP FOR PARALLEL PROCESSING

Ray Simar Jr.

Texas Instruments, Inc.
P.O. Box 1443 MS 701
Houston, Texas USA 77251-1443

ABSTRACT -- The TMS320C40 is the first parallel-processing general-purpose DSP. The TMS320C40 architecture combines six communication ports for direct processor to processor communication, a six channel DMA coprocessor for concurrent I/O, a high performance DSP CPU, memory, program cache, 32-bit global and local memory buses, two timers and an on-chip analysis module for parallel debug. This level of monolithic integration allows the TMS320C40 to achieve 275 MOPS (Million Operations Per Second) and 320 Mbytes per seconds of I/O. Special attention has been paid to the creation of a suitable environment for application development. This includes a parallel run-time support library and a unique debugger for debugging multiple TMS320C40s.

INTRODUCTION

As digital signal processing technology continues to advance, more and more applications require immense levels of floating-point performance. More and more it has been found that parallel processing is the only way to meet the performance requirements of these applications.

As is well understood, DSP applications lend themselves extremely well to decompositions across multiple digital signal processors. This is a natural result of the fact that DSP algorithms map well onto signal flow-graphs and signal flow-graphs are a natural way to express the underlying parallelism in an algorithm or application. Furthermore, this suitability for parallelization is found in simple algorithms, such as filters, and more complex algorithms such as Fast Fourier Transforms and high-speed modems.

It is also important to note that a whole host of other computationally intensive problems are well suited to parallel processing. These include problems in the areas of robotics, 3D graphics, neural networks, and finite element analysis.

It is this increasing demand in a growing number of applications for ever higher levels of performance that has driven the development of the TMS320C40.

TMS320C40 ARCHITECTURE

The goal of the TMS320C40 is to allow the system designer to provide any desired level of performance through parallel processing. Rather than having system performance limited by a single processor, multiple processors, operating in parallel, can be used to reach whatever performance level is needed. The TMS320C40 achieves this goal by providing six high-speed bidirectional communication ports, a six channel DMA coprocessor for concurrent I/O, and a high performance

DSP CPU compatible with the TMS320C30. The TMS320C40's block diagram is shown in Figure 1.

Communication Ports

Each communication port of the TMS320C40 (Figure 2) is a 20 Mbytes/sec direct processor to processor connection. The physical interface consists of an 8-bit wide data bus and four control lines. All six communication ports provide an aggregate bandwidth of 120 Mbytes/sec.

The connection between TMS320C40s is a direct connection requiring no external components. Each communication port contains a 64-byte input FIFO buffer and a 64-byte output FIFO buffer. All handshaking and synchronization between processors is done in dedicated hardware. Transfers between processors are performed simply by reading and writing the communication ports at memory mapped addresses.

These six communication ports are essential to the support of parallel processing. The ports allow a wide variety of topologies to be implemented. Figure 3 illustrates a pipelined linear array, bidirectional ring, tree, and 2D array. Any number of additional configurations are possible including hexagonal grids, 3D grids, and 6 dimensional hypercubes.

DMA Coprocessor

The six channel DMA coprocessor provides concurrent I/O to maximize sustained CPU performance. The DMA coprocessor can move data to and from off-chip memory, on-chip memory, and the six communication ports while the CPU processes data.

The parallel operation of the DMA coprocessor and the CPU has significant performance implications. The CPU is able to continuously process data at its maximum rate and the DMA is able to absorb any delays due to system I/O bottlenecks.

Each DMA channel is self programming. Whenever a DMA channel is finished a series of data transfers it can automatically reinitialize its control registers and begin another series of data transfers. This self-programming or autoinitialization capability is important for providing a very high level of CPU and I/O concurrency. Not only does the CPU not have to stop to do I/O, it also does not have to stop to program the DMA.

In a single cycle, the DMA coprocessor can perform up to three operations per cycle: a 32-bit data transfer, update an internal address register, and update a transfer counter. This gives the DMA coprocessor a peak operational rate of 75 million operations per second.

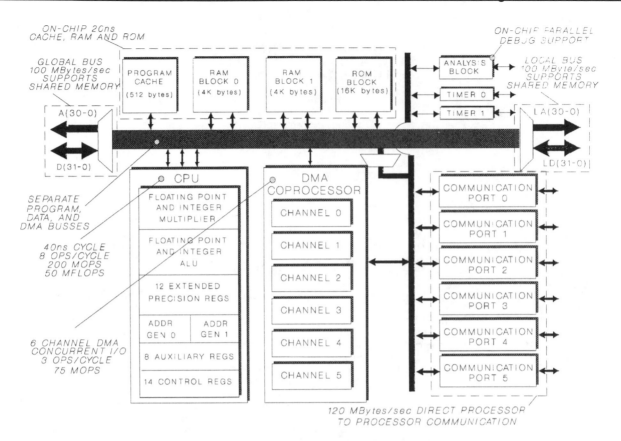

Figure 1. TMS320C40 block diagram. The on-chip communication ports support direct processor-to-processor communication. The DMA coprocessor provides concurrent I/O. The on-chip analysis block supports parallel debugging.

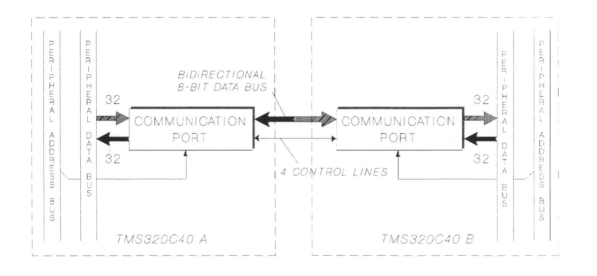

Figure 2. Communication port and direct processor to processor connection.

TMS320C40 CPU

The TMS320C40 CPU is optimized for computationally intensive DSP applications. It is capable of 200 million operations per second when operating with a 40 nanosecond cycle time. This high operational count is the result of the ability of the CPU to perform eight operations per cycle:
- Floating-point or integer multiply.
- Floating-point or integer addition or subtraction.
- Two data accesses.
- Two address register updates.
- Zero-time branch and loop-counter update.

Special support is also provided for floating-point division, floating-point square-root, byte and half word manipulation, and IEEE floating-point format conversions. The TMS320C40 instruction set is an assembly-language compatible superset of the TMS320C30.

External Memory Interfaces

In additional to the six communication ports and their 8-bit data buses, the TMS320C40 has two 32-bit wide external interfaces supporting up to 16 Gbytes of memory for maximum system flexibility. These are the global memory interface and the local memory interface. Each memory interface is capable of running at up to 100 Mbytes/sec.

Both the global memory interface and the local memory interface have been specially designed to simplify the construction of shared bus and shared memory systems. Processor status and bus-lock pins are available to determine the type of memory access (program, data, and DMA) for intelligent bus arbitration in shared memory systems. Separate asynchronous address and data bus enable pins are available to instantaneously tristate the respective buses for high speed bus arbitration in shared memory systems. Finally, two independent memory strobes and ready signals on each bus support simple interfaces to high and low-speed memories.

DEVELOPMENT TOOLS FOR PARALLEL DSP

Development tools for parallel DSPs are critical to the development of sophisticated DSP applications. For the TMS320C40, development tools have been developed to support code generation, simulation, and hardware design and verification for parallel systems.

Code Generation and Simulation Tools

The TMS320C40 is supported with an optimizing ANSI C compiler and a parallel processing run-time support library for interprocessor message passing.

The parallel processing run-time support library provides a comprehensive set of functions supporting message passing, DMA transfers, semaphores, and timing. The message passing functions support synchronized transfers and asynchronous transfers over the communication ports.

The DMA functions support all of the capabilities of the TMS320C40 DMA coprocessor. This includes simple unsynchronized block moves, block moves synchronized with external interrupts and transfers of complex data. The latter is especially suited for Fast Fourier Transform operations.

Semaphore based *lock()* and *unlock()* functions support accesses to shared global memory. A *myid()* function allows processors to identify themselves and timer functions support the measurement of elapsed time and putting the CPU to sleep for a specified period of time.

The industry standard SPOX DSP operating system provides parallel processing support and DMA coprocessor and communication port drivers.

The assembler, linker, and loader support the mapping of program and data to specific processors in a multiprocessor system.

Simulation tools are available in the form of hardware verification models and full functional models from Logic Automation Inc., and a state accurate single processor simulator from Texas Instruments.

Debugging Parallel Processors

A parallel debugger, the XDS 510, has been developed for the TMS320C40 (Figure 4). The XDS 510 supports global starting, stopping and stepping of multiple TMS320C40s in the target system. Key to this ability is the on-chip analysis module and the JTAG interface of the TMS320C40 (Figure 1 and Figure 4).

The on-chip analysis module of the TMS320C40 supports setting of program, data, and DMA breakpoints, a program counter trace-back buffer, and a dedicated timer for profiling. The XDS 510 communicates with the analysis module via the serial JTAG test port. The JTAG test port can be connected to any number of TMS320C40s and, in concert with the analysis module, and under the control of the XDS 510, supports the global starting and stopping of all processors in the system and global single-stepping of the system. Also, if a particular TMS320C40 hits a breakpoint, this breakpoint can halt all the TMS320C40s in the system.

The user controls all debug activity through a windowed user interface. A window can be opened for each processor in a system. Each window is then used to debug the associated individual processor in C or assembly language. Special windows are also available for controlling the global starting, stopping, stepping, and breakpoint capability.

SUMMARY

The TMS320C40 is the first parallel processing general-purpose DSP. The six high-speed communication ports and six channel DMA-coprocessor make the construction of parallel DSP systems easy. At 275 MOPS and 320 Mbytes/second the TMS320C40 supports the construction of systems with a wide variety of architectures and the highest levels of performance. Specialized development tools make the construction of parallel processing systems easy. In summary, the TMS320C40 is designed to meet the device requirements and development tool requirements for parallel DSP.

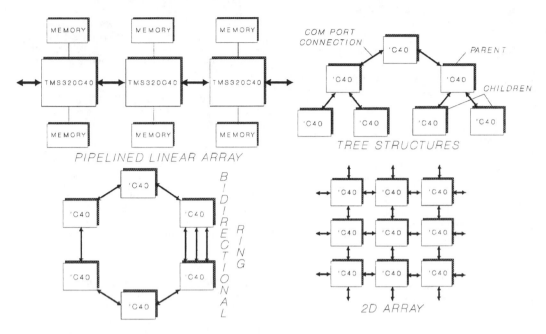

Figure 3. Examples of a few of the many parallel processor networks possible with the TMS320C40.

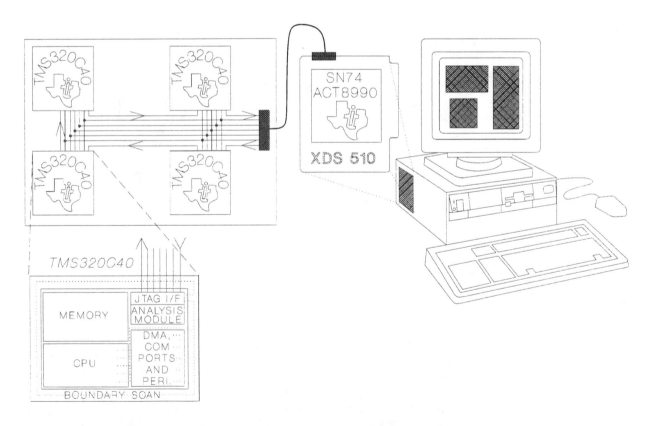

Figure 4. The XDS 510 parallel debugger. The on-chip analysis module, JTAG interface, and XDS 510 support the debug of multiple TMS320C40s in the target system. The XDS 510 may be PC based or workstation based.

EFFICIENT SYNCHRONIZATION SCHEMES FOR LARGE-SCALE SHARED-MEMORY MULTIPROCESSORS

Kanad Ghose and Der-Chung Cheng
Department of Computer Science
State University of New York, Binghamton, NY 13902-6000
ghose@bingvaxu.cc.binghamton.edu

ABSTRACT: Efficient synchronization mechanisms are critical for the success of large-scale shared-memory multiprocessors. In this paper, we present highly efficient synchronizers for barrier and critical section synchronization that do not introduce any spin traffic in the network. These synchronizers are structurally simple, scale well with the size of the system and implement immediately-reusable and time-constrained barriers and starvation-free critical section synchronization. We also present a novel way to implement software combining based on our barrier synchronizer that outperforms published algorithms for software combining and renders software combining as a viable alternative to expensive combining hardware for distributing hot-spot traffic in large scale systems. Trace-driven simulations confirm the usefulness of the proposed synchronizers.

1. INTRODUCTION

In conventional-style parallel processing paradigms, parallel components have to be synchronized using synchronization primitives. The commonly used synchronization primitives are related to:

1. Implementation of barriers, where processors are forced to wait at a certain common point in their code till all other processors have caught up. Barrier synchronization is extremely useful in executing loop iterations in parallel.

2. Implementation of reporting, where a processor merely indicates that it has crossed a certain point in its code.

3. Implementation of mutual exclusion, where only one process is allowed at a time into the critical section involving updates to shared variables.

Processor synchronization is a critical issue in shared memory multiprocessors. Without the benefit of an efficient synchronization mechanism, synchronization traffic can create hot-spots due to repeated accesses of synchronization variables in the process of spinning [16], [21]. Efficient synchronization primitives are also needed to use the limited memory/network bandwidths effectively and to allow the exploitation of concurrency with the minimum possible synchronization overhead. Reducing the synchronization overhead is all too critical in applications for shared memory systems, where inter-process interactions are likely to be more frequent.

Several techniques have been proposed in the existing literature to implement synchronization on shared variables in an efficient manner [2-4], [8-14], [17-26].

In this paper, we will propose the design of dedicated synchronization hardware for large-scale shared memory multiprocessors that allow highly efficient synchronization. We particularly target large-scale shared-memory systems involving 32 (or 64) or more processors, since they are relatively more prone to the ill effects of spinning on synchronization variables. (We explain and substantiate this later in the paper.) Specifically, we propose two sets of hardware synchronizers -- one set for implementing barrier synchronization and another set for implementing mutual exclusion. We will also present a novel software combining scheme based on our proposed barrier synchronizer that allows low-latency combining and provides better performance than any existing technique for software combining (which also require the use of other synchronization primitives), such as [8] and [24]. We will evaluate the performance of our synchronizers through simulation using synthesized traces and then indicate the potential advantages of our synchronizers over existing synchronizers.

This paper is organized as follows. Section 2 provides an overview of existing synchronization mechanisms. Section 3 introduces two versions of an immediately-reusable barrier synchronizer and shows

some of their not-so-conventional applications. Section 4 presents a simple starvation-free critical section synchronizer that supports spin locking without generating any network traffic. Section 5 represents an assessment of our synchronizers; Section 6 embodies the conclusions.

2. EXISTING HIGH-PERFORMANCE SYNCHRONIZERS

In this section, we briefly describe the existing hardware proposed for high-performance synchronization.

The synchronization support in the Sequent Balance multiprocessor is within a dedicated chip (the SLIC chip [3]) in the form of 64-bit registers holding bit locks that are kept coherent using a broadcast bus. Spins on the lock bits are thus localized within each chip. The dedicated broadcast bus limits the applicability of this mechanism to large-scale systems.

The barrier synchronizer proposed for the Burrough's FMP [18], is in the form of a tree of AND gates. Processors are connected to the inputs of the AND gates at the leaf level and raise these lines high on reaching the barrier; the processors then spin till the AND tree outputs a one - this happens when all processors have reached the barrier. The PAX system [12] provides support for barrier synchronization in a similar manner.

These AND-based barrier hardware are not reusable as such - till all processors have lowered their signal lines, a processor cannot reuse the barrier. The AND-ing mechanism provides no indication as to when it is safe to reuse the barrier. To signal when all processors have lowered their signal lines after reaching the barrier, all processors observe the output of a gate that ORs these signal lines. When the output of the OR-gate falls to zero *after* the output of the AND gate became a one and fell to zero, it is safe to reuse the barrier. Thus, the barrier hardware is not *immediately* reusable.

The barrier hardware proposed by Beckmann and Polychronopoulos [4] uses separate flags for reporting the arrival of a processor at the barrier and for recording the status of the barrier. Processors report their arrival at the barrier by resetting their respective report latches. A NOR-logic is used to detect the resetting of all report latches and thus detect the crossing of the barrier. To avoid races, the last participating processor enables the barrier firing logic (by resetting the "BREN" latch [4]). This requires each processor to have a connection to its report latch, as well the BREN latch. Moreover, if the number of participating processors are not known in advance, it is impossible to determine which processor resets BREN! Because of this, Beckmann's and Polychronopoulos' barrier hardware cannot be used to implement software combining with low latencies, as we can with our barrier (Sec. 3.3). Also, the logic used to make barriers immediately reusable is relatively complex -- requiring additional latches, and self-clearing logic.

Software combining [25] represents a way of addressing locks in a distributed manner, by allowing access to a (primary) lock through a series of secondary locks. By distributing these locks over memory modules, hot-spot traffic due to spinning over locks are drastically reduced. A tree of locks ('combining tree') is essentially formed, with the primary lock at the root, the secondary locks at all internal nodes and the processors at the leaves. A processor grabs the primary lock by "moving" up this tree, level-by-level. Distributed synchronizers [12] embed the combining tree within the switches of a MIN like the Omega network, where the tree interconnection is available intrinsically. It uses counters within each switch to implement the secondary locks. Although distributed synchronizers provide integrated support for mutual exclusion, barriers and reporting, it has some apparent drawbacks. These include the need to access the counter in each switch indivisibly, the fact that synchronization is still done primarily in software (and thus slow), the fact that the basic design is incapable of handling several multiple synchronizations simultaneously without complicating the switch hardware. Finally, this approach is applicable only to systems employing MINs like the Omega network and it also requires normal requests through the MIN to be distinguishable from synchronization requests.

The Fetch-and-operate synchronization primitives represent the best approach for reducing hot-spot traffic due to synchronization. The Fetch-and-Add (F&A) and other similar primitives are combinable - F&A requests, as well as normal loads and stores directed at the same memory address can be combined within the interconnection network into a single request. A considerable amount of hardware is needed within each switching element [1] for combining -- this increases the switch latencies considerable and makes a combining network impractical. Studies [16] have shown that three-way combining is more effective than two-way combining, thus dictating a more complex switch design. In a MIN with combining logic, the utilization of the combining logic on the input side, where the possibilities of combining are substantially lower. Finally, coding reusable barriers using F&As can be somewhat tricky [1] and hot-spot traffic can still result from concurrent read requests (from spins) that fail to get combined.

Other hardware-based synchronization primitives include Thomson's parallel adder-based synchronizer [23], the Compare-and-Swap instructions of the S/370 extended architecture [23], the full-empty bit based synchronization mechanism of the Denelcor HEP [13], the barrier synchronizers proposed by O'Keefe and Dietz [19, 20], the Cedar synchronization instructions [26], the hardware proposed by Sohi et al [22] to support combining of 'restricted' Fetch-and-Ops and the F&Op-based synchronizer proposed in [17]. These synchronizers have been discussed and compared in some depth in [7] and lack of space does not permit a detailed discussion in this paper.

In [2], Anderson proposed several software techniques to reduce the performance degradation due to spin locks. One technique (called 'backoff') requires the introduction of delays in the spin loop to reduce the frequency of testing the locks. Variations are possible depending on where the delay is inserted - between every testing of the lock, between every memory reference and after discovering (through a read) that a lock is available and so on. The other technique proposed by Anderson requires each processor to be put on a queue in the shared memory where they spin on *dedicated* locks that indicate whose turn it is to grab the lock. Both of these techniques lead to substantial improvements compared to several existing synchronization techniques involving spin locks [2]. The QOSB primitive proposed by Goodman et al [8] follows the same philosophy as Anderson's approach and is amenable to implementation in systems employing coherent caches.

3. A SCALABLE HARDWARE BARRIER

In this section we will propose a simple hardware for implementing *immediately re-usable* barriers. This hardware is iterative in nature and can be quickly scaled to the system size. Immediate reusability of the barrier will ensure that 'eager' processors do not get hung up after successfully crossing the barrier, waiting for slower processors to acknowledge the crossing of the barrier. (As an example, in the PAX system (Section 2), processors acknowledge the crossing of the barrier by lowering their signal lines.) The potential applications of immediately reusable barriers are indicated later. We will also introduce what we call a time-constrained barrier to allow processors to stay out of barriers selectively. Finally we will indicate some not-so-conventional applications of these barrier synchronizers. (Conventional applications of barrier synchronizers are well-known!)

3.1 A Simple Immediately Reusable Barrier

Immediate re-usability of a barrier will be useful in the following contexts:

- Applications where the concurrent units go through a sequence of several barriers and where the execution time between reporting at two consecutive barriers differ widely from one concurrent unit to the other. This scenario is possible for barriers used to synchronize loop iterations (as in cycle shrinking, self-scheduling

(guided or chunked) and so on) with conditionals within the loop body.

- Any other application where a fuzzy barrier [10] can be used.

- Any application, in general, that requires a very large number of barrier synchronization.

The logical structure of our basic hardware for implementing barriers is depicted in Figure 1(a). Every processor has an associated 1-bit latch, which is initially reset. The output of these latches are AND-ed and the output of the AND gate is used to simultaneously clear all the latches. Each processor indicates its arrival at the barrier by setting the latch associated with it. After setting this latch, each processor waits, possibly in a busy-waiting loop, till the contents of its associated latch falls back to zero. This happens when all processors have reached the barrier - a situation commonly termed as 'barrier crossing' or 'barrier firing'. Since the latches are reset without any explicit acknowledgement from each processor, the barrier hardware is usable immediately. Automatic resetting of the latches also ensures freedom from potential races that may otherwise be present if processors have to reset the latches explicitly after detecting the crossing of the barrier. This is a potential problem in the barrier logic presented in [4], when the numbers of processors participating in a barrier are not known in advance.

Figure 1(b) depicts a cascadable (iterative) version of this simple barrier logic.

Another advantage of the immediately reusable barrier hardware, albeit minor, is that each processor reporting at the barrier is freed from the mundane task of checking that all other participating processors have seen the crossing of the barrier (usually by busy waiting). It is noteworthy that the software implementation of reusable barriers can be somewhat tricky [1] or can be somewhat complex and slow [24].

3.2 Implementing Partial Barriers - Time-Constrained Barriers

Time-constrained barriers are useful when processors participating in the barrier can elect to stay out. The usefulness of partial barriers have been alluded to in existing literature, such as [8, 19, 24]. When a processor decides to stay out, it must not prevent the barrier from firing. One way to do this is to force this processor to issue a one (i.e., report) to all barriers in which it is not participating. This is, quite clearly, an infeasible solution for several reasons:

1. A race condition will be introduced when a processor tries to participate in a barrier from which it stayed off.

2. The processor has to set its reporting flag back for all barriers that it is not participating in when these barriers fire.

The time constrained barrier (TCB) allows a processor to register its intent to participate in the barrier within a timeout period following the firing of the barrier. If a processor does not register its intent to participate within this timeout period, the TCB hardware automatically generates the reporting signal (to allow the barrier to function properly). This timeout mechanism also guarantees freedom from race conditions.

Figure 2 depicts the basic logic associated with the TCB. The basic cascaded logic of the simple barrier of the last section is augmented to include the timeout-based masking logic. The timeout signal is used as depicted in Figure 2(a) -- processors register their intention to participate in the barrier by taking the line labeled 'Mask' high when the timeout signal is active (i.e., high). This line can be held high as long the processors wants to participate in the barrier -- there is no need to take it high every time the barrier fires. If the 'Mask' line is low when the timeout period expires, the NOR-logic automatically generates a one to enable the propagation of the 'reporting' signals from the other processors in the group and in other groups to the left. When the barrier fires (i.e., all participating processors report), the rightmost AND gate generates the reset signal for the status/report latches (Ls) and simultaneously triggers a one-shot like logic to generate the timeout signal. The duration of the timeout pulse can be prespecified (using some static timing analysis, for example) or set by a dedicated control processor dynamically, as shown in Figure 2(b).

3.3 Using the Time-Constrained Barrier for Software Combining

Software combining [8], [24] is useful in relatively large shared memory systems as a means of distributing hot-spot traffic and as a means to allow programs to still make use of combinable synchronization primitives. This appears to be a viable approach, since combining hardware is very complex and expensive.

The basic TCB can be used very effectively to combine simultaneous Fetch-and-Op operations [9] directed at a common memory location in software (i.e., perform software combining as described in [8], [24]) and provide low latencies for this combining. The novel approach presented here does not require the use of spin locks (as used in [8], [24]) - it also allows Fetch-and-Ops to be combined using a software pipeline. The proposed approach also does not require complicated and complex hardware like a combining network [9] or [22].

Software combining, as exemplified in [25], essentially requires all processors to wait till the slowest processor has turned in a request for combining. When combinable requests from *all* the processors are available, the actual request combining can terminate and send responses back to the processors. What this mode of software combining does is to increase the effective latency of the software combining process - a lazy processor can be late in submitting its Fetch-and-Op request and thus hold up others.

One way to decrease the overall latency of the software combining process is to have successive time windows and combine only the requests that have been submitted within the same time window. The duration of this time window is optimally chosen: making it too small would reduce the number of requests that can be combined; making it too large will increase the latency of the combining operation. We use the TCB hardware to implement combining in this manner as follows. In the following, we describe how the Fetch-and-Op operation targeting a common shared variable V is implemented.

The basic idea here is to combine requests, then combine these combined requests and so on till a single request emerges. This is done by using a software combining tree. The timeout signal associated with the TCB is used to time out the submission of combinable Fetch-and-Op requests from the processors. Processors intending to submit a Fetch-and-Op operation must raise their 'Mask' line before the expiration of the timeout period. Processors are statically assigned to memory modules where they have to write the arguments to the Fetch-and-Op operation. Requests are combined two-at-a-time and one of the processors turning in a request has the responsibility of propagating the combined request to the next "upper" level in the combining tree. If a processor P1 discovers that its statically assigned peer processor P2 has not turned in any request, then P1 is responsible for propagating its own request to the next "upper" level in the combining tree, where its request is likely to get combined with the request from its peer at that level. The combining operation proceeds as follows:

[1]. The combining operation associated with a time window starts at the "leaf" level of the combining tree when processors desiring to submit a combinable request in this window raise their 'Mask' lines.

[2] Each participating processor writes its argument for the Fetch-and-Op into a preassigned memory location to be read by its peer. These locations are initially assigned to null values (-1 is used in the code given later.)

[3] Participating processors arrive at the barrier. The barrier fires when every participating processor has written its argument in the location to be seen by its peer.

[4]. Participating processors then determine if their peers are participating or not -- the preassigned locations remain set at the null value if a peer is absent. If the peer is absent, the (other) processor propagates its own request to the next higher level in the combining tree by repeating step [1]. If both peers have submitted a request, the processor from the right subtree relinquishes the responsibility of propagating the combined request to the next higher level to the processor from the next subtree. Again, the process of step [1] repeats. Combined requests thus propagate up the tree in a lock-step fashion. Processors not elected

to go up the tree simply wait at the barrier.

[5] A processor going up the combining tree can figure out (from the number of executions of step [3]), when it has crossed the root of the combining tree and have been elected to perform the memory operations for the combining, targeting the variable V.

[6] Once the combined Fetch-and-Op operation is performed on the shared variable, steps similar to [2] through [5] are performed to generate responses for the participating processors. Again, locations used to register requests from peers are used to generate responses on the backward trip.

The code that is executed by each processor to perform combining in this manner, using the TCB hardware is given in the Appendix.

Several aspects of this combining algorithm are noteworthy:

- The same TCB hardware can be used to assist the combining of requests from different time windows on the same shared variable.

- The **same** TCB barrier can be used to implement Fetch-and-Op combining on **different** shared variables if the assignment of the combining nodes over the memory modules are done appropriately.

- In effect, the combining algorithm sets up a **software pipeline**: when requests combined in a time window are going up from level L to level (L+1) in the software combining tree, requests in the following time window are going up from level (L-1) to L.

- Note that software combining, as done here, avoids the need to synchronize progression through levels in the combining tree using software, by using the TCB instead. This drastically cuts down the number of accesses to shared variables.

- Note that the static assignment of memory modules to peer processors are specific to the processor ids and the level in the tree at which combining takes place. Also, the memory assignment is done in a way to distribute combining-related memory traffic uniformly over the memory modules and thus avoid hot spots.

- A potential memory conflict occurs when two processors, supposed to be peer at a level L "meet" each other, with one attempting to go up the tree and the other on its way down. (This happens when these processors submitted requests for combining in two different time windows). This conflict is in the form of simultaneous read attempts, and thus requires no synchronization.

Note that in the scheme we have suggested, the latency of the combined Fetch-and-Op operation is of the order of $n = \log_2 N$, where N is the number of processors. A simple analysis of our combining algorithm indicates that in the best case, the total number of concurrent memory accesses needed is 8.N, i.e., O(N). We assume that instructions and local data are either cached or kept in the local memory. For the software combining algorithms suggested by Goodman et al [8] and Tang and Yew [24], the best case number of concurrent memory accesses is $O(N.\log_2 N)$, again assuming that code and private variables are held locally. The worst-case number of memory accesses in all these schemes, including ours is $O(N.\log_2 N)$. This corresponds to the case when requests cannot be combined. The proposed software combining algorithm is thus expected to perform considerably better than the existing software combining algorithms. This advantage is a consequence of using the very modest barrier hardware - note, however, that some form of synchronization primitives are also required in [8] and [24].

3.4 Other uses of the Barrier Hardware

The basic barrier hardware of Section 3.1 can be used to implement the 'Max' (and 'Min') operations, whose purpose is to find out the maximum (or minimum) of N values, one held by each of the N processors. We need a single line running through the entire system, which will be used to implement Max and Min in a bit-serial fashion. Each processor has two s-bit registers, DATA and MAX/MIN; DATA holds the argument to the Max (or Min) operation and MAX/MIN is the register in which the global maximum (or minimum) is constructed, bit by bit.

I-155

The Max function is implemented as follows, with each participating processor executing the following code:

```
FOR j := (s-1) DOWN TO 0 DO /* (s-1) is the msb */
    BEGIN
        Use the barrier logic to meet at a common point;
        IF (DATA[j] = 1) THEN raise common line;
        Use the barrier logic to wait till others catch up;
        Read common line into MAX/MIN[j];
        IF (DATA[j] <> MAX/MIN[j])
        THEN clear DATA[j-1] through DATA[0];
        Lower common line
    END;
```

The maximum value is in MAX/MIN at the end of this step. This algorithm can be easily adapted to handle signed numbers.

'Min' can be implemented in a similar fashion. Note that it is quite possible to incorporate this logic into the barrier hardware. The operations Max and Min require $O(s)$ time to complete and, in theory, independent of the number of processors. The bit-serial concurrent evaluation of Max and Min is useful in many parallel algorithms, such as parallel sorting and searching, parallel determination of spanning trees (such as described in [5]) and so on.

4. HARDWARE SUPPORT FOR MUTUAL EXCLUSION

Our hardware for implementing mutual exclusion essentially implements a software combining tree in hardware very much like distributed synchronizers. However, there are significant differences between our design and that of distributed synchronizers:

- The *entire* mutual exclusion and scheduling logic is built into a surprisingly nominal amount of hardware. Synchronization and its associated scheduling does not require any software assistance.

- Our technique does not insist on using a MIN -- it can be used, in principle, with *any* kind of interconnection.

- The node hardware in our tree is simpler and does not use counters, leading to lower delays within a node.

- Our approach allows a processor to submit a request for entry into the critical section and then cancel it prematurely.

Our critical section synchronizer hardware is in the form of a tree with a gating logic at each node in this tree. Each node selects, using a rotating priority, a request submitted to it from a lower level and passes it up to the next upper level, as explained later. hereafter called the DMS (Distributed Mutex Synchronizer), since it implements the mutual exclusion operation in a distributed fashion.

4.1 Operation of the DMS

Each node in the DMS (Figure 3) can accommodate 'm' sets of requests for entry or exit from the critical section. For all internal nodes, these requests come from the nodes connected to it at the immediately lower level in the tree; for nodes at the leaf of the tree, these requests originate in the $N = m^k$ processors. Each node, in addition, has one line going to each of the 'm' requesters. These lines are used to inform a requester when its request for entry into the critical section is granted.

Figure 4 depicts how the DMS is used in a shared-memory multiprocessor with 8 processors and 2-input nodes in the DMS. A node in the DMS is depicted as a "DMS node" in this diagram. For convenience, each set of lines between nodes or between the processors and the leaf nodes in the DMS, are shown as bold lines in the diagram. Each processor, attempting to enter the critical section, submits a request to enter the critical section to the DMS and waits, possibly, but not necessarily, in a busy-wait loop till it receives an acknowledgement to enter the critical section from the DMS. This acknowledgement arrives on one of the 3 links between a processor and the leaf node element associated with it. On receiving the acknowledgement, the processor uses the interconnection network of the shared-memory multiprocessor to access data in the critical section. **Hot spot traffic generated due to lock access and spinning on the**

locks are thus entirely absent in the interconnection network. On exit from the critical section, the processor signals the DMS to select another request to enter the critical section. The DMS hardware implements a *rotating priority scheduling algorithm* within each node to guarantee freedom from starvation.

4.2 Functional Characteristics of a Node in the DMS

Figure 3 depicts the logic within a node in the DMS, where each node can handle 4 requests (i.e., m=4). As long as a request for entry into the critical section is present on an input to a node, the corresponding "Request" line is held high. This line is to be held in this state until the selected processor exits from the critical section or when it cancels its request to enter the critical section. Each node uses a rotating priority to select one of the requests present on its input and propagate the selected request upwards to the next level in the tree through the "Request Output" line.

When a request propagated all the way up to the root level is finally selected by the root node, a signal indicating the requesting processor to enter the critical section is passed down from the root by using the "Ack" lines associated with the nodes. This is accomplished by tying the "Request Output" line of the root node to its "Ack" input, as depicted in Figure 4. The logic that selects a request at each node is also responsible for steering the "Ack" signal to the selected requester.

A requester can cancel request it submitted or indicate its exit from the critical section by strobing the "Unlock/Cancel" line to its associated node in the DMS. When a node receives an "Unlock/Cancel" signal corresponding to a request that it selected (and propagated), it does two things:

1. It propagates the "Unlock/Cancel" signal to the next upper level, if any.

2. It selects the next pending request to the right of the canceled/exiting request, by rotation, and propagates that request upwards.

If the "Unlock/Cancel" signal does not correspond to the request that was selected by the node, then the cancel/request request is not propagated upwards.

The control logic associated with a switch selects a pending request on one of its 'm' "Request" inputs by using a rotating mask and passes it upwards in the tree. This mask is a pattern of m-bits with exactly one bit set and all other bits reset. The latches depicted in Figure 3 contain this mask. For convenience, we will refer to the aggregate of these 'm' latches as a selection mask register, SM. When a request, originating on the j-th input to the node (say) terminates, a request, if present, on "Request i" is selected next, if and only if there are no requests present on the lines numbered (j+1 mod m), (j+2 mod m), ..(i-1 mod m). As an example for m=4, when the request on input #3 terminates, the request on input #1 will be selected only when a request is absent on input #0. The request selection logic described here, thus implements a right-rotating priority and is realized as shown in Figure 3. *This rotating priority logic ensures that starvation is absent, as shown in the next section.*

4.3 Starvation-Free Property of the DMS

The mutual exclusion implemented by the DMS is starvation-free A process that has submitted a request to enter its critical section is guaranteed that it will be admitted by a certain time. To see this, consider the implications of the scheduling-by-rotation at the root node. The root node has exactly 'm' subtrees with at most N/m pending requests in each subtree, where 'N' is the number of processors in the system. After the root node has selected a request for entry into the critical section from a subtree, it will select one request from each of the other (m-1) subtrees before it schedules a second request from the same subtree. In the worst case, a request in one of these subtrees will get scheduled after the (N/m-1) other requests in the same subtree. Thus, the request submitted by a processor will be scheduled after at most (N/m-1).(m-1) other requests.

One should note here that implementing such a starvation-free solution in software would have been very inefficient, since exits from the critical section would have required the invocation of a scheduler.

5. AN ASSESSMENT

In this section, we first present simulation results that indicate speedups possible using our critical section and barrier synchronizers. Admittedly, we are comparing the performance of our synchronizers against conventional implementation based on the very simple-minded Test-and-Set primitive. We have done this deliberately to show what difference in performance an efficient synchronization scheme can bring about. A comparison of the performance of our scheme with some of the synchronizers already proposed is quite beyond the scope of this paper. Instead, we will indicate which of the existing synchronizers should fare as well as the synchronizers we have proposed and compare them qualitatively with ours; we will also comment on the limitations of our synchronizers.

5.1 Simulation Results

The performance of our barrier synchronizer and the critical section synchronizers were evaluated through simulation. The simulator assumes that a packet-switched Omega network using 2×2 switching elements is used. In the case of conflicting requests through the two input ports of a switching element, these ports were favored alternately. The simulator takes in as input the address traces generated by processors as they ran the programs we selected as benchmarks (a PDE solver, **doall** loops and a simple program involving critical section access). These address traces were derived from the simulated execution of hand-compiled versions of the sources. *We had to generate address traces in this manner since we had no access to actual address traces for a large-scale shared-memory system, involving up to 256 processors.*

Figure 5 (a) depicts the speedup possible using our barrier synchronizer on a 2-D Poisson solver against a conventional implementation using TEST&SETs for the synchronization. For a N-processor system, the size of the grid space is assumed as P×P, where P is the highest integer such that $P^2 \leq N$. (The code for the Poisson solver is really a **doacross** loop.) (We define speedup as the ratio of the execution time using TEST&SET to the execution time using one of our synchronizers.)

Figure 5(b) depicts the speedup possible using our barrier synchronizer for pre-scheduled loops with a barrier synchronization at the end of an iteration. The trace for this graph assumes that the total number of iterations equal to twice the number of processors. The 'long' and 'short' loop benchmarks have 104 and 14 instructions in the TEST&SET versions within the loop body, respectively, and are coded as an outer **do**-loop that executes twice with a **doall** loop as its body. In both cases, the loop bodies include accesses to data specific to an iteration, but kept in the shared memory modules.

Figure 5 (c) depicts the speedup possible with our critical section synchronizer w.r.t to an implementation that uses TEST&SET, against the number of processors in the system. The four graphs correspond to four program fragments with different number of instructions within the critical section and different total number of instructions within the program fragment.

The simulation results depicted in figures 5 (a) through 5 (c) corroborate the fact that speedups possible with our scheme increase as the number of processors increase, thus allowing shared memory systems to be more scalable. In MIN-based shared-memory multiprocessors, the memory-access latencies increase with the number of processors, and thus conventional synchronization schemes that induce memory traffic will take a heavier toll on the performance for systems that use a relatively large number of processors.

5.2 A Comparison With Existing Synchronization Mechanisms

A common characteristic of our barrier and critical section synchronizers is their ability to keep synchronization traffic out of the interconnection network, and thus reduce spin overhead to zero. In systems using our synchronizers, spinning is confined "locally"; many

proposed synchronizers, such as the SLIC chip [3], [4], [8], the PAX barrier [11], [18, 19], Thomsons barrier [21] have the same ability. Although, software techniques for reducing spin overheads do exist [2], they may still affect other shared memory access. Anderson reports result for systems with up to 20 processors in [2]; In fact, the impact of Anderson's scheme in large-scale MIN-based systems (such as the RP3) employing several more processor remains to be studied!

Efficient barrier synchronizers, such as the PAX barrier [11], the Burrough's FMP barrier [18], O'Keefe and Dietz's barriers [19, 20], the barrier proposed by Beckmann and Polychronopoulos [4], the fuzzy barrier [10] should perform almost as well as our barrier hardware for the benchmark applications we considered. However, our barrier seems to have the following advantages:

1. The hardware needed is much simpler compared to many of these schemes, such as [4], [10], [18], [19, 20], and the Thomson barrier [23]. The only barrier hardware that is simpler is the PAX barrier [11] - but, then, it does not have some of the abilities of our barrier, as listed below.

2. Processors do not need to wait to make sure that other processors have observed the crossing of the barrier. The facilities that make this possible also allow the our barrier to be immediately reusable. (The potential uses of a reusable barrier were listed earlier in Section 3.) The barrier hardware of [4] provides the same ability, but requires the last processor to report at the barrier to enable the barrier. This, in turn, requires an additional overhead (to figure out who was the last processor to report at the barrier; in [4], a fetch-and-add primitive is used to do this).

3. The time-constrained barrier introduced in Section 3.3 makes possible a very efficient implementation of software combining, that is considerably superior to the other published approaches to software combining [8], [24]. Further, only one copy of the time-constrained barrier is needed to implement the software combining of independent Fetch-and-Ops, including Fetch-and-Ops performed in different time windows. The time-constrained barrier also allows such operations as MAX and MIN (Sec. 3.4) to be performed in constant time.

4. The barrier synchronizers we have proposed can be scaled very easily with the size of the system. Many efficient barrier schemes, such as [4], [10], [19-20] and the Thomson barrier [23] are not easily amenable to such scaling. System-size scalability is an important requirement, since efficient synchronizers are more useful in large-scale systems.

Although, our barrier does provide a limited form of partial barrier synchronization, it is still not as versatile in this respect as [10], [19, 20]. These three schemes, however, rely on a sophisticated compiler to pre-schedule the barriers.

The critical section synchronizer that we proposed reduces spin overhead to zero and improves the performance dramatically, as shown from the simulation results. Only, the SLIC chip [3] can provide a comparable performance. However, the SLIC chip has two drawbacks: First, unlike our design, it does not provide any facility for fair scheduling of the deferred request. Second, unlike our scheme, it requires a broadcast capability to keep local copies of lock states consistent - a fact that may inhibit its use in large-scale systems. The critical section synchronizer proposed in [12] has the ability to schedule requests in a fair manner, albeit using a more complex scheme and hardware assistance. Goodman et al's synchronization primitive [8], as well as Anderson's queueing technique for reducing spins also have the same abilities, but these schemes require some -if not none - software overhead for queue management.

It is worth noting that our critical section synchronizer has a delay of $O(\log_m N)$ *gate delays* (if m-input DMS nodes are used) -- this is at most a few *processor cycles* in most systems and thus significantly faster than conventional (e.g., TEST&SET-based or Fetch-and-Add based) synchronizers for MIN-based which have a best-case delay of $O(\log_2 N)$ cycles. Our critical section synchronizer is also considerably faster than the SLIC synchronizer, which uses a message-passing dedicated bus to keep copies of lock bits coherent (Sec. 2.1).

I-157

The limitation of our critical section synchronizer, as such, appears to be in terms of handling multiple critical section accesses at the same time - we require one critical section synchronizer for each simultaneously active critical section. We do have a hardware scheme to essentially time-multiplex one copy of the critical section synchronizer across several groups of synchronizing processes, but we feel that such a scheme may not be necessary in real applications. In most cases, the Operating System/Compiler should be able to "schedule" the critical section hardware among the few groups that exist -- this, in fact, is the manner in which the SLIC chip was used.

6. CONCLUSIONS

It is well-known that synchronization primitives involving spinning on synchronization variables have a very adverse effect on the performance of large-scale shared-memory systems, which tend to use a MIN. We presented the design and application of simple but highly efficient synchronizers that implement re-usable and partial barriers and starvation-free critical section synchronization.

The proposed barrier synchronizers have some desirable characteristics:

1. Their performance rivals that of the best existing techniques for high-performance synchronization.

2. They offer a variety of additional capabilities that most of the best existing synchronizers do not have. These capabilities include the ability to scale easily with the size of the system, to allow barriers to be immediately re-usable, and to allow the ability to implement partial barriers.

We also introduced a novel application of our very simple time-constrained barrier in the form of implementing low-latency software combining. The resulting algorithm outperforms all published algorithms for software combining (namely, those in [8] and [24]). We feel that the ability to do low-latency software combining is important in large-scale systems since it is a more viable alternative to the vastly complicated and under-utilized combining network/hardware. Low-latency software combining allows the powerful combinable primitives to be used and yet avoid hot spots.

The proposed hardware for critical section synchronization is considerably simpler and faster than the two existing schemes that come close in terms of functional capabilities [3], [12]. It is essentially a gating logic based on concept similar to software combining trees and allow starvation-free, high-speed synchronization. It is also scalable easily with the size of the system.

Two common characteristics of our barrier and critical section synchronizers is in their ability to keep synchronization traffic out of the processor-memory interconnection and to allow a very simple interface between the processors and the synchronization hardware. We tested the performance of our barriers using a simulator for a Omega network based system, using representative address traces (Section 5.1). The results indicate our synchronizers to be extremely effective in enhancing the performance, especially for systems using 32 or more processors.

REFERENCES

[1] G. Almasi and A. Gottlieb, *Highly Parallel Computing*, Benjamin-Cummings, 1989.

[2] T. E. Anderson, "The Performance of Spin Lock Alternatives for Shared-Memory Multiprocessors", IEEE Trans. on Parallel and Distributed Systems, Vol. 1, No. 1 (Jan. 1990), pp. 6-16.

[3] B. Beck and B. Kasten, "VLSI Assist in Building a Multiprocessor Unix System", Proc. Usenix Conf., 1985, pp. 255-275.

[4] C. Beckman and C. Polychronopoulos, "Fast Barrier Synchronization Hardware", CSRD Report No. 986, Univ. of Illinois, 1990

[5] G. E. Blleloch, "Scans as Primitive Parallel Operators", IEEE Trans. on Computers, Vol. C-38, No. 11 (Nov. 1989), pp. 1526-1538.

[6] D. Cheng, "Efficient Synchronization Techniques for Large-Scale Shared Memory Multiprocessors", Masters thesis, Dept. of Computer Science, State Univ. of New York at Binghamton, May 1990.

[7] K. Ghose and D. Cheng, "Synchronization Schemes for Large-Scale Shared Memory Multiprocessors", Tech. Rep. CS-TR-90-51,

Dept. of CS, SUNY-Binghamton, 1990.

[8] J. R. Goodman, M. K. Vernon and P. J. Woest, "Efficient Synchronization Primitives for Large-Scale Cache-Coherent Multiprocessors", Proc. Architectural support for Programming Languages and Operating Systems III, 1989, pp. 64-75.

[9] A. Gottlieb, R. Grishman, C. P. Kruskal, K. P. McAuliffe, L. Rudolph, and M. Snir, "The NYU Ultracomputer - Designing an MIMD Shared Memory Parallel Computer", IEEE Trans. on Computers, Vol C-32, No. 2 (Feb. 1983), pp. 175-189.

[10] R. Gupta and M. Epstein, "High Speed Synchronization of Processors Using Fuzzy Barriers", Int'l. Jrnl. of Parallel Programming, Vol. 19, No. 1, 1990, pp. 53-73.

[11] T. Hoshino, *PAX Computer: High-Speed Parallel processing and Scientific Computing*, Addison-Wesley, 1989.

[12] D. N. Jayasimha, "Distributed Synchronizers", Proc. International Conference on Parallel Processing, 1988, pp. 23-27.

[13] J. S. Kowalik, editor, *Parallel MIMD computation: HEP Supercomputer & Its Applications*, MIT Press, 1985.

[14] R. Lee, "On Hot Spot Contention", Computer Architecture News, Vol. 13, 1985. pp 15-20.

[15] C. Lee, C. P. Kruskal, and D. J. Kuck, "The effectiveness of Combining in shared Memory Parallel Computers in the Presence of Hot Spots", In Proc. of the Intl. Conf. on Parallel Processing, 1986, pp. 35-45.

[16] G. Lee, "A Performance Bound on Multistage Combining Network", IEEE Trans. on Computers, Vol C-38, No 10 (Oct. 1989), pp. 1387-1395.

[17] G. J. Lipovski and P. Vaughan, "A Fetch-and-Op Implementation for Parallel Computers", Proc. 15th Ann. Sym. on Computer Architecture, 1988, pp. 384-392.

[18] S. F. Lundstrom, "Application Considerations in the System Design of Highly Concurrent Multiprocessors", IEEE Trans. on Computers, Vol. C-36, No. 11 (Nov. 1987), pp. 1292-1309.

[19] M. T. O'Keefe and H. Dietz, "Hardware Barrier Synchronizer: Static Barrier MIMD (SBM)", Proc. Int'l. Conf. on Parallel Proc., 1990, Vol. I, pp. 35-42.

[20] M. T. O'Keefe and H. Dietz, "Hardware Barrier Synchronizer: Dynamic Barrier MIMD (DBM)", Proc. Int'l. Conf. on Parallel Proc., 1990, Vol. I, pp. 43-46.

[21] G. F. Pfister and V. A. Norton, "Hot Spot Contention and Combining in Multistage interconnection Networks", IEEE Trans. on Computers, Vol. C-34, No.10 (Oct. 1985), pp. 943-948.

[22] G. S. Sohi, J. E. Smith and J. R. Goodman, "Restricted Fetch-and-Φ Operations in Parallel Processing", Proc. 3rd Int'l. Conf. on Supercomputing, 1989, pp. 410-416.

[23] H. S. Stone, IEEE Video Notes: *Communication and Synchronization in Parallel Processing Systems*, 1989, IEEE CS Press No. 1991N.

[24] P. Tang and P-C. Yew, "Software Combining Algorithms for Distributed Hot-Spot Addressing", Jrnl. of Parallel and Distributed Computing, Vol. 10, No. 2 (Oct. 1990), pp. 130-139.

[25] P. Yew, N. Tzeng, and D. H. Lawrie, "Distributing Hot-spot addressing in large-scale multiprocessors", Proc. Intl. Conf. on Parallel Processing, 1986, pp. 51-58.

[26] C. Zhu and P. Yew, "A Synchronization Scheme and its Applications for Large Multiprocessor Systems", Proc. 4-th. Int'l. Conf. on Distributed Computing Systems, 1984, pp. 486-491.

APPENDIX

Software Combinable Fetch_And_Op Using the TCB

The combinable Fetch-and-Op operation, which uses the time-constrained barrier of Section 3 is of the form:

Fetch-and-Op (V, e)

where V is a shared variable and e is a local supplied as argument. The operation 'Op' can be a dyadic arithmetic operation (such as +) or boolean operations. We assume that the locations used in the memory modules for storing arguments and peer information are all initialized to a null value, say -1, i.e.,

L[*].module_no := -1, and R[*].module no := -1.

We use a Pascal/C-like notation to express the Fetch-and-Op operation and the various procedures it uses.

```
const H := log N;  /* L = height of comb. tree */

function Fetch_and_Op(V: address; argument: integer)
var local_response, height: integer; direction: boolean;
{    height := 1; /* start from the leaf level*/
     Combine;    /* invoke combining procedure*/
     return local_response
}/* Fetch_and_Op */

function Module_number(Processor_ID, height: integer): integer;
var I, module_no: integer;
     /* Although shown here as a FOR loop, the following code
     is either kept in local table or computed using a fast
     shift-add sequence */
{    module_no := 0;
     for I := (H+1) downto height do
          {module_no := module_no + (N DIV **(H + 1 - I));
          Module_number := module_no;}
}/* Module_number */

procedure Combine;
var T:       record of
               module_no: integer;
               arg: integer;
             end;
{    Set Mask;   /* indicate interest in participation*/
     /* "butterfly" communication among peers*/
     address := Module_number(Processor_ID, height);
     direction := Processor_ID MOD (2**height);
     /* 'direction' decides which of two peers can go up */
     T.module_no := Processor_ID;
     T.arg := argument;
     if (direction = 0) then R[address] := T else L[address] := T;
B1:  if (not(barrier)) then goto B1;
     if (direction = 0) then T := L[address] else T :=R[address];
     /* determine which processor should proceed to the next
     level and also restore initial value in the module holding
     peer info for later re-use */
     if (direction = 0) then {
          if (T.module <> -1) then
               argument := argument Op T.argument;
          /* 'Op' is the operation in F&Op*/
          L[address].module_no := -1;}
     if (direction = 1) then {
          R[address].module_no := -1;
          if (T.module = -1) then direction := 0
          else /* not elected to go up, wait for response from peer*/
               for I := 1 TO 2*(H - height ) do
                 { Set Mask;
B2:                if (not(barrier)) then goto B2;}
     Generate_Response;
     return; }
     if (height = H) then {
          /* root of tree has been reached*/
          local_response := V; V := local_response + argument;
          Generate_Response;}
     else {height := height + 1; Combine;}
     /* Continue to the next level of combining*/
}/* Combine */

procedure Generate_Response;
{    Set Mask;    /* indicate interest in participation*/
     address := Module_number(Processor_ID, height);
     if (direction = 0) Response[address] := local_response;
     /* processor selected to go up writes response */
B3:  if (not (barrier)) then goto B3;
     if (direction = 1)
     /* processor not selected to go up reads response */
          { local_response := Response[address]; direction := 0;
          local_response := local_response Op T.argument;}
          /*'Op' is the operation in F&Op*/
     if (height > 1) then
          {height := height - 1; Generate_Response;}
} /* Generate_Response */
```

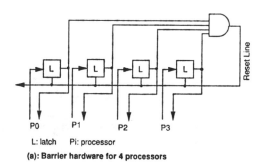

L: latch   Pi: processor

(a): Barrier hardware for 4 processors

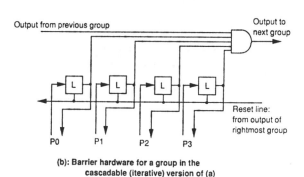

Output from previous group

Output to next group

Reset line: from output of rightmost group

(b): Barrier hardware for a group in the cascadable (iterative) version of (a)

Figure 1. Immediately Reusable Barriers

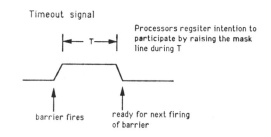

Timeout signal

Processors regsiter intention to participate by raising the mask line during T

barrier fires   ready for next firing of barrier

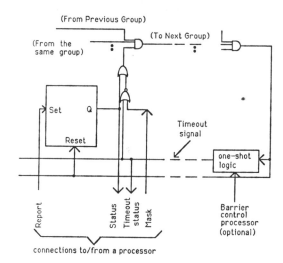

(From Previous Group)

(To Next Group)

(From the same group)

Set   Q

Reset

Timeout signal

one-shot logic

Report

Status   Timeout   Mask
         status

Barrier control processor (optional)

connections to/from a processor

Figure 2. Time–Constrained Barriers

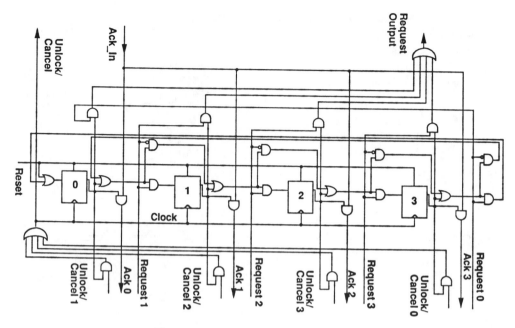

Figure 3. Internal Logic of a DMS Node

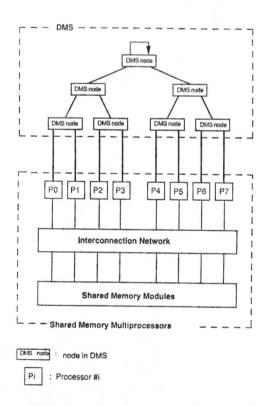

Figure 4. The Distributed Mutex Synchronizer in Use

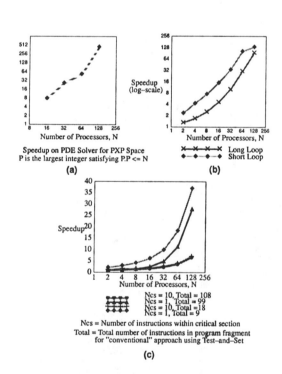

Figure 5. Simulation Results

MISC: a Mechanism for Integrated Synchronization and Communication using Snoop Caches

Takashi MATSUMOTO   Tomoyuki TANAKA   Takao MORIYAMA   Shigeru UZUHARA

IBM Research, Tokyo Research Laboratory
5-19, Sanbancho, Chiyoda-ku, Tokyo 102, Japan*

Abstract

We propose a mechanism for efficiently performing inter-processor synchronization and communication on a shared-memory, shared-bus multiprocessor system. A typical use of this mechanism is for communication among fine-grain statically-scheduled parallel instruction streams produced by a parallelizer for a DOACROSS loop. Every word in shared memory is augmented with a *synchronization bit*, which indicates whether data has been written into the word. This bit is cached together with the word. A processor attempting to read from an unwritten word is blocked and later awoken when another processor writes into it. To make this scheme efficient, we introduce new cache coherency protocols called *all-write* and *all-read/write* that are to be specified for pages containing communication variables, and the inter-cache snoop control mechanism. We consider the effective uses of the mechanism, including ways of quickly reusing the words used as communication variables. The proposed mechanism provides a natural solution to the producer-consumer type synchronization problem by integrating the communication of a data and notification of its arrival into the operations of snoop caches.

1   Introduction

The advances in VLSI technology have resulted in the development of a wide variety of multiprocessor systems, and the problem of multiprocessor synchronization has become an important issue. In dataflow and message-passing machines [5, 24, 23, 6] the communication of the data and synchronization are integrated as an indivisible operation. However, on shared-memory machines, if the communication is to be done through shared-memory, an additional synchronization mechanism must be provided. In this paper, we propose a mechanism for intergrated synchronization and communication for shared-memory machines using snoop cache. We call our system "MISC" (a Mechanism for Integrated Synchronization and Communication).

We designed the mechanism to be used on a homogeneous shared-memory multiprocessor as follows. The target application program is first statically analyzed and divided into groups of fine-grain instruction streams that communicate frequently. An example of such a group is a set of instruction streams produced by a parallelizer for a single DOACROSS loop, or a sequential loop. These instruction streams are scheduled to processors and preempted as a group by the operating system. The operating system decides the processors on which these instruction streams should run, and specifies them as a "processor group" (see Section 2.3). We assume that the instruction streams are statically scheduled so that most writes take place before the corresponding reads; even if the reader is blocked to wait for the corresponding write, the write is expected to happen soon enough so that a context-switch is not necessary.

Types of synchronization required for multiprocessors include producer-consumer, mutual exclusion, barriers, fork-join, those for control dependency, etc. Of these, producer-consumer type synchronization is most common, and therefore most important, in closely cooperating, fine-grain parallel processing. MISC provides an architectural solution to the producer-consumer type synchronization, and aims to reduce the cost of compile-time analysis of parallelizers. This mechanism should be used in conjunction with other synchronization mechanisms for other types of synchronization (see Section 4).

In Section 2, we describe the mechanism being proposed. After giving the outline of the basic idea, we describe the new snoop cache protocols proposed for MISC, and the inter-cache snoop control mechanism required for the protocols. We then describe how to reuse the communication variables using synchronization polarity bits, and consider an optimization using word valid bit for the case the cache block size and the word size a different. In Section 3, We consider the effective use the mechanism. In Section 4, we compare our mech with other proposals providing architectural supp multiprocessor synchronization. Lastly in Secti state the conclusions.

2   MISC: a Mechanism for Int Synchronization and Com

2.1   The basic idea

A synchronization bit is added to memory (Figure 1). This bit sp been written to the word alrea now read it. This bit has the

*e-mail: matumoto@trl.vnet.ibm.com

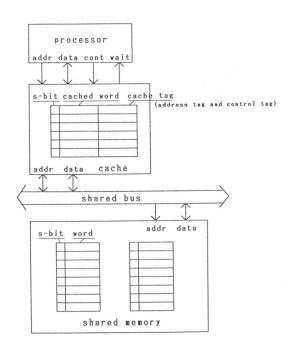

s-bit:synchronization bit

Figure 1: The memory system augmented with synchronization bits

is written to the word, and has the value FULL after it.[1] All synchronization bits are initialized to EMPTY.

The basic idea of MISC is as follows. When reading from a word, the synchronization bit associated with the word is first checked, and if it is EMPTY, the processor is blocked and waits until the bit becomes FULL before reading the word. When writing to a word, three actions are taken: (1) the synchronization bit associated with the word set to FULL; (2) the data is written to the word; and (3) all the processors waiting on the word are awoken.

The idea of providing each memory cell with a FULL/EMPTY status bit to solve the read-before-write race problem is not new. The idea appeared earlier in HEP and I-Structure storage. What we are proposing using synchronization bits is a way to provide an integrated synchronization and communication mechanism for shared-memory machines with processors of conventional hardware.

Variants of I-Structure storage [5] is widely used in dataflow and message-passing machines [23, 6, 11, 27]. In this mechanism, when a read is attempted on a unwritten word, the "processor context" of the token attempting the read is appended to the list of deferred read requests associated with the word, and waits for the write request

---

[1] For the moment the reader may assume that the value ~~pty~~ is 0, and full is 1. In Section 2.4 we show how this ~~gnment~~ is reversed in order to quickly reuse communica~~~~ variables. We say "set" to mean making this bit full, ~~"clear"~~ to mean making it empty.

for the word.

However, we assume a multiprocessor system consisting of more conventional processors with registers for which context switch is expensive, and that the fine-grain tasks are statically scheduled to the instruction streams to a certain extent, so that the corresponding write expected to be performed before the time it takes to do a context switch. Therefore, we only add the status bit to a memory word, and to block a processor attempting a read on a unwritten word.

Denelcor HEP multiprocessor [12], which predates I-structure storage, maintains multiple number of processor contexts within a processor. Processor contexts that issued unsatisfied read requests are enqueued in a separate unit, and read requests are repeatedly reissued until they are satisfied. Thus processors are not blocked, and effectively busy-wait.

We assume a conventional processor with a single context, and opt to block the processor in such a case, and have the synchronization bit to be loaded in the caches (as if it were a part of the word) in order for this blocking to be performed locally (Figure 1).

In the figure, as well as in the following description, we shall assume that one cache block consists of only one word for the ease of discussion. See Section 2.5 for the case when they differ.

2.2 The *all-write* and *all-read/write* snoop protocols for MISC

A straightforward implementation of the scheme just described would be terribly inefficient, for there would be a contention of bus accesses. To reduce bus traffic in shared-memory multiprocessor systems various snoop cache protocols have been proposed [3]. Write-back protocols, which shows higher performance than the simple write-through protocol, is classified into two groups: invalidate and update protocols. In *invalidate* (or *write-invalidate*) protocols, the corresponding blocks in other caches are invalidated. *Write once*, *Symmetry*, *Illinois*, and *Berkeley* protocols are of this type. In *update* (or *write-broadcast*) protocols, written data is broadcasted to other caches, and the corresponding blocks are updated. *Firefly* and *Dragon* protocols are of this type.

Neither of these two protocol types are ideal for MISC. If we use an invalidate protocol, a write to a communication variable (a variable used for inter-processor communication) would cause the corresponding blocks in other caches to be invalidated, requiring additional bus accesses when other processors try to read the variable. Similarly with an update protocol, even if the communication variable is written to before it is read by other processors, the broadcast will not result in the expected update in case the other caches do not contain the corresponding block, again requiring additional bus accesses. Therefore, we propose new protocols called *all-write* and *all-read/write* to be used for pages containing all communication variables. In Section 2.3 we show how to specify and switch snoop protocols on a per-page, per-bus-access basis.

When a processor does a write to a word specified to use the *all-write* protocol, the access is output to the shared-bus, and all the caches that can load a copy of the word without causing a write-back operation load copies of this word. This action is referred to as an "all-write operation".

In case there are multiple reader processors for a variable, if one processor reads a word, it is likely that the other processors will soon read the same variable. The suitable protocol in this case is the *all-read/write* protocol. Using this protocol, a write causes an all-write operation as above, and in addition, upon a read-miss (which causes an access to be output on the shared-bus), all the caches that can load a copy of the word without causing a write-back operation load copies of this word.[2]

Note that all-write and all-read/write protocols are broad protocols types, as are invalidate and update protocols, and we only specify the key actions; the rest of the specification in a particular protocol of an all-write or all-read/write protocol family are to be determined according to the hardware features of the target system.

The proposed mechanism in more detail, including the snoop actions of the caches, is as follows (see Figure 1).

When reading from a word, a copy of the word (together with the associated synchronization bit) is first loaded in the cache if it is not already included. (In case of a read-miss, this word is broadcasted to other caches when *all-read/write* protocol is specified.) Then the cache checks the synchronization bit. If it is FULL, the cache returns the contents of the word to the processor. If it is EMPTY, the cache puts the processor in a WAIT state using the WAIT signal line.[3] The cache continues snooping the bus line even while the processor is in the WAIT state. When a write to this word is performed by another processor, the snooping mechanism of the cache updates the word and the synchronization bit. Thereafter the processor is restored from the WAIT state and reads the contents of the word.

When writing to a word, first the cache associated with the processor that issued the write sets the associated synchronization bit, and the word is output on the shared-bus. Then all the caches that can load a copy of the word without causing a write-back operation load copies of this word. If the cache that has loaded a copy finds that the processor is waiting on this word, the cache restores the processor from the WAIT state, the processor reads the word and resumes execution.

A multiprocessor system with hierarchical caches

Even with processor group specification units (described in Section 2.3), all-read and all-write operations cause many accesses to the caches, and increases the chances of access collisions to the caches from the processors and the shared memory. Constructing the caches with dual-port memory would alleviate this problem, but is expensive. An alternative approach is to introduce hi-

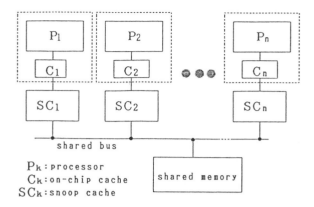

Figure 2: A multiprocessor system with hierarchical caches

erarchical caches (Figure 2). In this organization each processor is associated with two caches: a smaller cache, which is fast enough to be comparable to the processor speed, and a cache for snooping the bus, which must be large enough to lower the access frequency. The bus collision problem would be alleviated, for a processor's accesses to the snoop cache are made more infrequent, and all-read and all-write operations may cause an update only in the snoop cache.

2.3 Inter-cache snoop control mechanism

We presented two snoop cache protocols that are suitable for communication variables, but these protocols are not suited for all variables. For example, local variables and other data that are to be used only by one processor for a certain length of time should be specified to use an invalidate protocol, whereas semaphore keys, synchronization variables, and other data that are used for inter-processor communication should be specified to use an update, all-read, all-write, or all-read/write protocol. The inter-cache snoop control mechanism [14, 17] solves this problem.

Per-page protocol switch

It is desirable to control the actions of the snoop caches to use the suitable protocol for the variable being used, but it is too expensive to provide specification of the protocol to each word, so this data (the protocol type bits T_0, T_1, and T_2) is encoded in the unused bits of the page entry (Figure 3).

Every time the processor (or the memory management unit) makes an access to the shared-memory, these bits (cached in the TLB) are output to the protocol type bus (Figure 4).[4] All the caches snoop the signals on this protocol type bus, in addition to the shared-bus, and acts accordingly. Thus it is possible to specify and dynamically switch the protocol to be used on a per-bus-access

---

[2]These protocols are variants of the *all-read* protocol [14], which was suggested by Dr. Norihisa SUZUKI (August 1988).

[3]The processor must be able to accept interrupts even in a WAIT state to enable process preemption by the operating system.

[4]There must be output lines for these unused bits in the processor (such as Am29000 [1]), or alternatively the most significant bits of the physical address can be used to specify the protocol type.

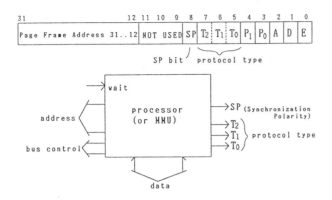

Figure 3: A page entry and a processor showing the bits and lines required for MISC

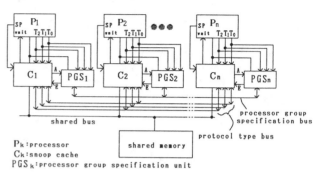

Figure 4: The organization of MISC

basis. This additional snoop action occurs simultaneously with the snooping of the shared-bus, and no additional cycles are required to determine the protocol type.

There are microprocessors which provide means of specifying actions of the cache (write-back or write-through, cachable or non-cachable, etc.) on a per-page basis, such as Motorola's MC68040 [21] and MC88200 [20]. The crucial difference between these and the inter-cache snoop control mechanism is that in the latter a bus action by one cache (a read or write) directs the snoop actions in other caches by way of protocol type bits and the protocol type bus.

The idea of providing special instructions that direct the snoop cache operations has also been proposed [25, 8]. We assume a processor of more conventional organization, and in addition, a comparable effect of instruction-level protocol switching can be achieved by mapping several logical pages to the same physical page, each specifying different protocols, and accessing by different logical addresses.

Processor group specification unit

A processor group specification unit (PGSU) is added to each cache in order to restrict the all-read and all-write operations to the related processors and thereby increasing the cache hit rate (Figure 4). Before an application is run, each processor (or PGSU) is assigned an identification number for the processor group it is in. (This would be done by the operating system.) Whenever a processor initiates an all-read or all-write operation, the associated PGSU outputs the assigned identification number to the special bus, called the processor group specification bus. Other caches observes this bus and initiates a snoop action only when the identification number matches the identification number assigned to that PGSU. The bus width required for the processor group specification bus is $log_2 (n + 2) - 1$ (or, to be more exact, $\lceil log_2 \lfloor (n+2)/2 \rfloor \rceil$) where n is the number of processors in the system.

2.4 Synchronization polarity bit

We now present a mechanism for quickly reusing such words used as communication variables. As shown in Figure 3 one of the used bits in the page entry is used as the the *synchronization polarity* bit (or "SP bit"). If the SP bit is 1, the synchronization bit of 1 means FULL, and 0 means EMPTY. If the SP bit is 0, these conditions are reversed. This is specified to a snoop cache by the synchronization polarity signal line from the processor (Figure 4).[5]

When all the communication variables in a page are used up, they can be made ready for reuse only by reversing the SP bit in the corresponding page entry. This assumes that the processor has the ability to efficiently maintain consistency among the TLBs in the processors (such as Inter-Agent Communication mechanism of i80960 [22]).

The above method is suitable for reclaiming a large set of communication variables at one time, for example, when reclaiming an entire array. For reclamation of variables in smaller units (e.g. a fixed set of variables specified by the parallelizer as communication variables, rather than an entire array), it is also possible to reuse variables in a page without actually reversing the SP bit in the page entry at run-time. In Figure 5 two logical pages with logical addresses addr1 and addr2 are assigned to the physical page that includes variable X in a program, and the SP bit for these addresses are reversed in the page entries. For the first use of X, reference to the variable is compiled as offset_X[addr1], then for the second use of X, it is compiled as offset_X[addr2], and so on. Alternating the logical addresses in this manner achieves the effect of reversing the SP bit in the page entry, and is suitable for reusing variables in units smaller than a page. (This can be done for entire arrays also.)

Whether a set of words are all used up should be detected through a separate barrier-type synchronization mechanism, such as Elastic Barrier[6] [14, 16, 17, 18] or Fuzzy Barrier [9].

---

[5]We assume that this signal line is available, as for the protocol type bits.

[6]The basic mechanism of Elastic Barrier is given in the Appendix.

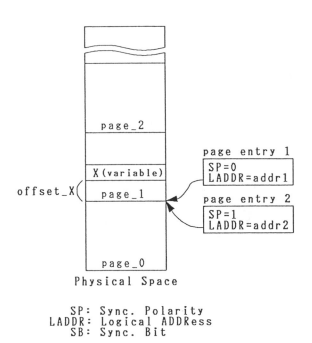

Physical Space

```
SP:    Sync. Polarity
LADDR: Logical ADDRess
SB:    Sync. Bit
```

Figure 5: Reusing a communication variable quickly by mapping two logical pages to the same physical page

2.5 Considerations for when the block size and the word size differ

In the preceding description, we have assumed that one cache block consists of only one word for the ease of discussion, but in most present-day cache systems a cache block usually consists of several (4, 8, etc.) words. A cache tag (address tag and control tag) is provided for each block, but a synchronization bit must be provided for each word. In conventional snoop cache systems, data are always handled one block at a time; in case of a write-miss, first the whole block is loaded in the cache, then the word is written in. Thereafter the block is broadcasted to other caches or to shared-memory if appropriate. Here we consider a way to avoid reading the entire word from shared-memory for a write-miss when *all-write* or *all-read/write* protocol is in use.

Assuming that it is possible to broadcast on the shared-bus the one word which has just been written, we provide a per-word valid bit (WV bit) to each word in the cache. The WV bit is used as follows: On a write-miss, a block is created in the cache, the word is written in, and only the corresponding WV bit is set. (And that word is broadcasted to other appropriate caches.) On a read-miss, the entire block is read in through the shared-bus, and all the WV bits are set. On a write-hit, the corresponding WV bit is set. On a read-hit, the word is simply read if the corresponding WV bit is set, otherwise, the entire block is read in through the shared-bus, and all the WV bits are set.

```
(a)       DO 5 i = 4,n
             X(i) = X(i) + X(i+2) + X(i-3)
          5 CONTINUE

(b)       C(1) = X(1)
          C(2) = X(2)
          C(3) = X(3)
          C(n+1) = X(n+1)
          C(n+2) = X(n+2)
          DO 5 i = 4,n
             C(i) = X(i) + X(i+2) + C(i-3)
          5 CONTINUE
```

Figure 6: Conversion for a simple DOACROSS loop.

3 Applications of MISC

To use MISC, a simple DOACROSS loop wherein substitution is performed linearly (Figure 6(a)) is converted to use newly allocated communication variables for those variables that have dependency across iterations (Figure 6(b)). Following the program fragment (b), array C replaces array X, which can be reused.

If processes are to compete to acquire the loop iteration count at run-time, the number of processors to execute this kind of loop can be changed at run-time, since MISC guarantees synchronization for each array element. To ensure that preemption occurs only at the end of an iteration, the operating system must employ a preemption with advance notification and each process must check the preemption request before fetching the iteration count. Alternatively the user-level scheduler must ensure that the execution is continued until the end of the iteration [15, 19].

For a more complex loop wherein whether the assignment occurs is determined at run-time (Figure 7(a)), the above conversion (Figure 7(b)) is unsatisfactory, for if the assignment to a variable does not occur, a process attempting to read this variable may be suspended forever. It is necessary to make sure that assignment always occur (Figure 7(c)).

Another typical use for MISC is when producer and consumer processors communicate data through a buffer. In this situation the communication can be done with minimal overhead; synchronization for each data is automatically guaranteed.

4 Related work

Earlier we contrasted our mechanism with HEP and I-Structure storage. We now compare the mechanism with more recent work.

Other proposed mechanisms with the similar goal as ours (producer-consumer synchronization in fine-grain, closely-cooperating, and statically-scheduled parallel processes) are (1) high speed static synchronization, (2) Elastic Barrier, and (3) register channels.

High speed static synchronization [4, 26] aims to ensure all the precedence relations specified. It ensures precedence relations only, whereas in MISC the data transfer is incorporated with the precedence (producer-consumer) synchronization. This mechanism requires

```
(a)      DO 5 i = 4,n
           IF (Y(i) .EQ. 0) GO TO 5
           X(i) = X(i) + X(i+2) + X(i-3)
      5 CONTINUE

(b)      C(1) = X(1)
         C(2) = X(2)
         C(3) = X(3)
         C(n+1) = X(n+1)
         C(n+2) = X(n+2)
         DO 5 i = 4,n
           IF (Y(i) .EQ. 0) GO TO 5
           C(i) = X(i) + X(i+2) + C(i-3)
      5 CONTINUE

(c)      C(1) = X(1)
         C(2) = X(2)
         C(3) = X(3)
         C(n+1) = X(n+1)
         C(n+2) = X(n+2)
         DO 5 i = 4,n
           IF (Y(i) .EQ. 0) GO TO 6
           C(i) = X(i) + X(i+2) + C(i-3)
           GO TO 5
      6    C(i) = X(i)
      5 CONTINUE
```

Figure 7: Conversion for a more complex DOACROSS loop

more compile-time dependency analysis than MISC.

Elastic Barrier [14, 16, 17, 18] is a lightweight and general barrier synchronization mechanism. What has just been stated for high speed static synchronization applies to Elastic Barrier also. MISC is not intended for barrier type synchronization, and is to be used with a barrier mechanism, such as Elastic Barrier to complement each other.

While the goal of the the **register channel** mechanism [10] is similar to ours, the implementation is completely different. Register channels are shared by processing elements (PEs) within a processor chip, and are used for communication between the PEs. MISC is intended for a shared-bus connected multiprocessor system, where processors, not PEs within a processor, communicate through shared-memory, shared bus, and snoop caches. Register channel requires a complex connection where a PE is connected to every other PE within a chip, in contrast to a simple shared-bus connection of MISC.

Other mechanisms that address the multiprocessor synchronization problem using snoop caches include the **QOSB** primitive[7], **lock-based cache scheme** [13], and **Cache with Synchronization Mechanism** [2]. These schemes are directed toward the lock and mutual exclusion problems; while efficient for cases where processes may run in any order, these are not as efficient for the goal of MISC: fine-grain producer-consumer synchronization for statically-scheduled parallel fine-grain instruction streams.

5 Conclusions

We have proposed a mechanism for efficiently performing inter-processor synchronization and communication on a shared-memory, shared-bus multiprocessor system. We have shown how the mechanism using a memory word augmented with a *synchronization bit* provides a natural and efficient solution to the producer-consumer type synchronization problem by integrating the communication of a data and notification of its arrival into the operations of snoop caches. We also proposed new cache coherency protocols called *all-write* and *all-read/write* that are to be specified for pages containing communication variables, and the inter-cache snoop control mechanism. We considered the effective uses of the mechanism, including ways of quickly reusing the words used as communication variables.

We are planning to evaluate this mechanism in more detail, through simulation experiments to evaluate the effectiveness of the mechanism in various typical situations.

Acknowledgements

We are grateful to Prof. Hidehiko TANAKA and others at Tanaka Lab. (the University of Tokyo), where Matsumoto is a postgraduate research student. We would also like to thank our managers, especially Dr. Tsutomu KAMIMURA and Dr. Norihisa SUZUKI, for supporting our work.

References

[1] Advanced Micro Devices, Inc. *Am29000 32-Bit Streamlined Instruction Processor Users Manual.* Advanced Micro Devices, Inc., 1988.

[2] Amano, H., Terasawa, T., and Kudoh, T. Cache with Synchronization Mechanism. INFORMATION PROCESSING 89. IFIP, 1989, pp. 1001–1006.

[3] Archbald, J. and Baer, J. Cache Coherency Protocols: Evaluation Using a Multiprocessor Simulation Model. *ACM Transactions on Computer Systems.* 4, 4 (November 1986), pp. 273–298.

[4] Arita, T., Takagi, H., Kawamura, T., and Sowa, M., Flow Control Method of the PN Processor. In *Proceedings of the 39th Semiannual IPSJ Convention.* October 1989, pp. 1896–1897 (in Japanese).

[5] Arvind and Iannucci, R.A. A Critique of Multiprocessing von Neumann Style. In *Proceedings of 10th International Symposium on Computer Architecture.* June 1983, pp. 426–436.

[6] Dally, W.J., Chien, A., Fiske, S., Horwat, W., Keen, J., Larivee, M., Lethin, R., Nuth, P., Wills, S., Carrick, P., and Fyler, G. The J-MACHINE: A Fine-Grain Concurrent Computer. INFORMATION PROCESSING 89. IFIP, 1989, pp. 1147–1153.

[7] Goodman, J.R., Vernon, M.K., and Woest, P.J. Efficient Synchronization Primitives for Large-Scale Cache-Coherent Multiprocessors. In *Proceedings of*

Third International Conference on Architectural Support for Programming Languages and Operating Systems. April 1989, pp. 64–75.

[8] Goto, A., Shinogi, T., Chikayama, T., Kumon, K., and Hattori, A., Processor Element Architecture for a Parallel Inference Machine, PIM/p. *IPSJ Journal of Information Processing.* 13, 2 (1990), pp. 174–182.

[9] Gupta, R. The Fuzzy Barrier: A Mechanism for High Speed Synchronization of Processors. In *Proceedings of Third International Conference on Architectural Support for Programming Languages and Operating Systems.* April 1989, pp. 54–63.

[10] Gupta, R. Employing Register Channels for the Exploitation of Instruction Level Parallelism. In *Proceedings of Principle and Practice of Parallel Processing.* March 1990, pp. 118–127.

[11] Iannucci, R.A. Toward a Dataflow/von Neumann Hybrid Architecture. In *Proceedings of 15th International Symposium on Computer Architecture.* June 1988, pp. 131–140.

[12] Jordan, H.F. Performance Measurement on HEP— A Pipelined MIMD Computer. In *Proceedings of 10th International Symposium on Computer Architecture.* June 1983, pp. 207–212.

[13] Lee, J. and Ramachandran, U. Synchronization with Multiprocessor Caches. In *Proceedings of 17th International Symposium on Computer Architecture.* 1990, pp. 27–37.

[14] Matsumoto, T. Fine Grain Support Mechanisms. *IPSJ SIG Reports on Computer Architecture*, 89–ARC–77, July 1989, pp. 91–98 (in Japanese).

[15] Matsumoto, T. Synchronization and Processor Scheduling Mechanisms for Multiprocessors. *IPSJ SIG Reports on Computer Architecture*, 89–ARC–79, November 1989, pp. 1–8 (in Japanese).

[16] Matsumoto, T. A Generalized Barrier-Type Synchronization Mechanism. In *Proceedings of Joint Symposium on Parallel Processing '90.* May 1990, IPSJ/IEICE/JSSST, pp. 49–56 (in Japanese).

[17] Matsumoto, T. A Study of FGMP: Fine Grain Multi-Processor. *Transactions of Information Processing Society of Japan.* 31, 12 (December 1990), pp. 1840–1851 (in Japanese).

[18] Matsumoto, T. Elastic Barrier: A Generalized Barrier-Type Synchronization Mechanism. To appear in *Transactions of Information Processing Society of Japan.* (in Japanese).

[19] Moriyama, T., Negishi, Y., Uzuhara, S., and Matsumoto, T. A Multiprocessor Resource Management Scheme which Considers Program Grain Size. *IPSJ SIG Reports on Computer Architecture*, 90–ARC–83, July 1990, pp. 103–108 (in Japanese).

[20] Motorola Inc. *Motorola Semiconductor Technical Data MC88200.* BR589/D, 1988.

[21] Motorola Inc. *Motorola Semiconductor Technical Data MC6840.* MC68040/D Rev1 (Replace NP455), 1990.

[22] Myers, G.J. and Budde, D.L. *The 80960 Microprocessor Architecture.* John Wiley & Sons, Inc., New York, 1988.

[23] Papadopoulos, G.M. Implementation of a General-Purpose Dataflow Multiprocessor. MIT/LCS/TR-432, MIT, 1988. Also in *Proceedings of 17th International Symposium on Computer Architecture.* 1990.

[24] Shimada, T., et al. An Architecture of a Data Flow Machine and Its Evaluation. In *Proceedings of COMPCON84 Spring.* IEEE, 1984, pp. 486–490.

[25] Shinogi, T., Matsumoto, A., Chikayama, T., Goto, A., and Hattori, A. The Architecture of a Processor Element of PIM/p. In *Proceedings of the 37th Semiannual IPSJ Convention.* September 1988, pp. 137–138 (in Japanese).

[26] Takagi, H., Kawamura, T., Arita, T., and Sowa, M. High Speed Static Synchronization without any Additional Waiting. In *Proceedings of Joint Symposium on Parallel Processing '90.* May 1990, IPSJ, pp. 57–64 (in Japanese).

[27] Toda, K., Nishida, K., Uchibori, Y., and Shimada. T., An Architecture of the Macro-dataflow Computer CODA In *Proceedings of Joint Symposium on Parallel Processing '90.* May 1990, IPSJ, pp. 185–192 (in Japanese).

(An earlier version of this paper was presented as: Matsumoto T., Tanaka T., Moriyama T., and Uzuhara S. A Mechanism for Integrating Communication and Synchronization using Snoopy Cache. "RYUKYU" Summer Workshop on Parallel Processing (Okinawa, Japan, July 18–20). 1990. In IEICE Technical Report (CPSY-90-42). Institute of Electronics, Information and Communication Engineers, 1990, pp. 25–30 (in Japanese).)

APPENDIX
Elastic Barrier: A Generalized Barrier-Type Synchronization Mechanism

This appendix describes the basic mechanism of Elastic Barrier, intended to be used with MISC. In the following text, fine grain processing units (fundamental processing units) are called *tasks*, instruction streams which consist of tasks are called *shreds*, and a group of communicating shreds is called a *process*. At processor scheduling time, the operating system allocates the specified number (the number of shreds in a process) of real processors to a process. All shreds in a process are scheduled to processors and preempted simultaneously (as a group) by the operating system.

We invented Elastic Barrier [14, 16, 18] to make efficient execution of fine grain concurrency possible. It is easy to design a barrier-type light synchronization mechanism to rendezvous all of the shreds in a process to a point. We first describe this barrier-type mechanism and then expand it to cover general light synchronization.

In Figure 1 the whole system diagram including the synchronization mechanism is illustrated. We here opt for a shared-bus method as an inter-processor connection for communication. To reduce contentions of data communication we prepare another communication bus for synchronization information called the *synchronization bus*. The number of lines in this synchronization bus is equal to the number of the processors with each line corresponding to a specific processor. Each processor has its own *synchronization controller*, which detects completion of inter-processor synchronization using the synchronization bus. When each processor reaches a barrier synchronization point, the processor activates the corresponding line of the synchronization bus through its own controller. Each controller has a *group register* to enroll processors which synchronize; all the allocated processors in a process are enrolled there. Each controller constantly monitors the synchronization bus by referring to the register, to check whether the lines corresponding to all of the processors in a process are active or not; i.e. the completion of barrier synchronization. Processors do not execute further instructions until the completion of the preceding synchronization is detected.

Information on inter-processor (inter-shred) synchronization is put into instruction streams (shreds) in such a way that we establish a new field or tag in a processor's instruction code or new prefix instructions for synchronization. Barrier synchronization information requires only 1 bit to express a synchronization request for rendezvous just before (or after) execution of the instruction to which this bit is attached. To eliminate overhead of the procedure which checks flags or variables for rendezvous, we stop processors temporarily with hardware mechanisms (just like wait-state in slow memory accesses) until the synchronization completes. To make OS's preemption of processor resources always possible, processors must have interrupt facilities which are valid during wait-state.

To this point the explanation has only been for a barrier-type synchronization mechanism. In general syn-

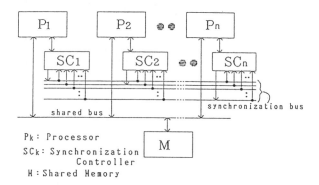

Pk: Processor
SCk: Synchronization Controller
M: Shared Memory

Figure 1: System with the Synchronization Mechanism

chronization, the combination of processors which need to synchronize varies according to the synchronization point, even if the processors (shreds) in the same process are considered. If a system holds information on the combination at each synchronization point, it synchronizes only the shreds which need synchronization at each rendezvous point. However, that method requires a rather large amount of hardware because each processor must know about all of the others' combinations or synchronization points. The number of lines in the synchronization bus must be at least number of processors squared if it is adopted. Making a tradeoff, we adopt a dummy synchronization request and insert it at a point in a shred where synchronization is not needed, but where other shreds in the same process will synchronize nearby. Utilizing this approach, all shreds in a process rendezvous at each synchronization, and no additional hardware is needed.

Calculating the points to insert dummy requests is heavy and difficult. Moreover, even if they are calculated as precisely as possible, overhead may occur owing to causes which are impossible to predict before execution; such as latencies of bus contention or cache miss.

So far we have discussed only the type of synchronization where processors wait until all of them reach a certain point. However, in practice, most synchronization among fine grain tasks is for producer-consumer relationships, and keeps the execution order of tasks intact. Since producer tasks do not need to wait for starts of consumer tasks, unnecessary overhead is created if they wait.

With only slight modifications of the mechanism, we avoid these cases of overhead. We extend the synchronization information from 1 bit to 2 bits (thereby giving 4 directions for synchronization), and provide the synchronization controller with three new counters. Each processor outputs the information to the corresponding synchronization controller just before execution of the instruction to which it is attached.

Figure 2 shows the block diagram of the controller which connects to the corresponding processor through data lines and three control lines: SSIG0 (Synchronization SIGnal 0) and SSIG1 are inputs for synchroniza-

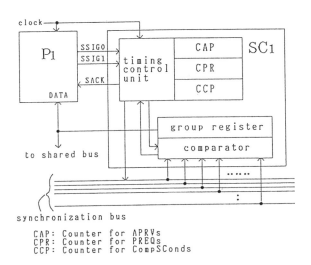

Figure 2: Synchronization Controller (SC)

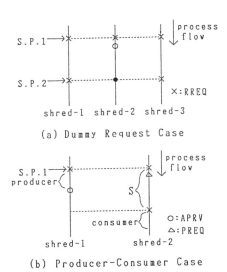

(a) Dummy Request Case

(b) Producer-Consumer Case

Figure 3: Overhead Reduction using APRV and PREQ

tion information, SACK (Synchronization ACKnowledge) gives the processor permission to leave the wait-state.

The rough meaning of the four types of the synchronization information are as follows:

NONE The processor does nothing for synchronization.

RREQ After outputting this to the controller, the processor stays in wait-state until it receives a SACK signal from the controller. (Real REQuest)

APRV After outputting this, the processor continues execution of instructions without waiting for any signals from the controller. This lets the controller give approval for the completion of later synchronization. (APpRoVal)

PREQ After outputting this, the processor continues execution without waiting for any signals. This is used to check for completion of synchronization in advance of the actual synchronization point. (PreREQest)

Note: RREQ and NONE correspond to 1 bit of synchronization information before expansion.

Instead of describing the roles of three new counters separately, we explain the usages and advantages of 4 types of synchronization information, and introduce the counters within the context of the explanations. 'CompSCond' (Completion of Synchronization Condition) occurs when all lines on the synchronization bus according to processors enrolled in the group register become active.

Since our mechanism is based on barrier synchronization, one CompSCond is required at each synchronization point. However, we have circuits for synchronization which are separate from the processors; namely, synchronization controllers. We can eliminate unnecessary processor waits by working the controllers as independently

as possible. For this purpose, we have extended the synchronization information.

Figure 3a shows a means to eliminate the overhead which accompanies a dummy synchronization request. In Figure 3a each small cross marks the place where a RREQ is attached, and shred-2 needs a dummy request near S.P.2 (Synchronization Point 2). Following the method before extension of the mechanism, we calculate the place on the shred-2 corresponding to S.P.2, and then insert a RREQ as a dummy there (solid dot). It may, however, be shifted from the real S.P.2 in execution by unpredictable causes, resulting in unnecessary waits. If we use an APRV instead of the RREQ and insert it at the hollow dot (which hereafter denotes an APRV), we can avoid the overhead. The controller of shred-2 activates its line for S.P.2 when it receives the APRV signal just after the completion of synchronization at S.P.1. On the other hand, the processor corresponding to shred-2 continued to execute succeeding instructions. To prepare the case of continuous APRVs on the same shred, we need *a counter* for APRVs before their CompSCond. We call it CAP. CAP is incremented by 1 when the controller receives an APRV, and decremented when the controller activates its line. If CAP's content is equal to zero, the controller does not cause further activations.

Figure 3b shows another overhead elimination which accompanies a producer-consumer relationship. An APRV (hollow dot) on shred-1 corresponds to the producer and the second RREQ (cross) on shred-2 represents the consumer. We ought to use APRVs for the production completion points, since producers have no reason to wait for starts of corresponding consumers. If we do not use PREQs and the producer on shred-1 finishes its task early, the CompSCond corresponding to the synchronization takes place just at the point of the shred-2 second RREQ. However, we can allow the CompSCond to occur anywhere in section S, thinking about characteristics of producer-consumer synchronization. Moreover, the earlier CompSConds take place, the more the

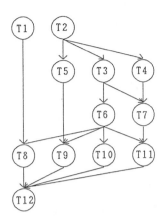

Figure 4: Task Graph of the Example

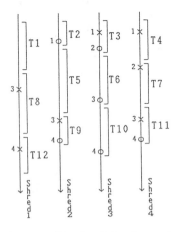

Figure 5: Shreds of the Example

possibility of overhead reduction increases. Therefore we insert a PREQ at a small triangle (which hereafter presents a PREQ), and allow the CompSCond to happen anywhere between the PREQ and the corresponding RREQ. The synchronization controller for shred-2 activates its line when it receives the PREQ signal, while the processor continued the execution just like in the case of APRV. When the processor reaches the RREQ, it asks the controller whether it is able to go further. If the CompSCond has taken place, the controller outputs SACK at once to let the processor go. If not, the controller makes the processor stay in wait-state until the CompSCond takes place. We make further extension on this facility; if a shred has continuous RREQs, we allow the corresponding PREQs all to be inserted before the RREQs. To implement this facility, two counters are needed in the controller; one is *a counter* (CPR) for PREQs before their CompSCond, and the other is *a counter* (CCP) for CompSConds before their RREQs. CPR works just like CAP, for PREQs instead of APRVs. CCP is incremented by 1 when the controller detects an CompSCond by PREQs, and decremented when the controller outputs SACK to the processor for the corresponding RREQ. If CCP's content is equal to zero, the controller does not output SACK. CAP is given a priority in the activation actions. While RREQs are allowed not to have PREQs, each PREQ always has a corresponding RREQ on the same shred, and between them there are no RREQs that do not have their PREQs.

Before using this Elastic Barrier mechanism, compilers arrange the order of the synchronization in each basic block of a process. Then the mechanism does the lightest synchronization within the limit of keeping the pre-decided order intact. The overhead of each synchronization point is at most an electric signal delay (1–2 clocks) from one processor's SSIG to others' SACK.

Figures 4, 5, and 6 show an example for insertions of synchronization information. Figure 4 shows the task graph of the example where each node represents a fine grain task and each directed edge corresponds to dependency between two tasks. Figure 5 illustrates a task al-

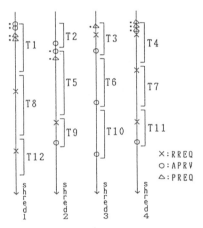

Figure 6: Positions where Synchronization Information is Attached

location in which the procedure is divided into 4 shreds. The length of a task indicates the number of instructions, and also the task process time with no processor wait. In the figure, the combination of the cross and the hollow dot which have the same index number corresponds to an edge in Figure 4, and the instruction at the cross must not be processed earlier than the instruction at the hollow dot owing to dependency. Figure 6 displays the positions where synchronization information is inserted for this example. Hollow dots indicate APRVs, triangles represent PREQs and crosses correspond RREQs. NONEs are adopted where nothing is marked, if all instructions must have one of 4 types each.

Finally, we will briefly compare Elastic Barrier with the 'Fuzzy Barrier' [9]. While Elastic Barrier has aspects which are similar to the 'Fuzzy Barrier', we devised ours quite independently in 1988. Owing to the adoption of APRV information and counters, ours is more general for producer-consumer type synchronization, more convenient for the insertion of dummy requests and is superior with regards to the elimination of overhead. Our mechanism can also realize 'fuzzy barriers' easily [16].

Wired-NOR Barrier Synchronization for Designing
Large Shared-Memory Multiprocessors

Kai Hwang and Shisheng Shang

Department of EE–Systems, University of Southern California

Los Angeles, CA 90089–0781

Abstract: We propose the use of wired-NOR logic to support barrier synchronization in large, shared-memory multiprocessors. Through electrical analysis, we prove that these wires can easily synchronize 525 processors in 540 ns using off-the-shelf bipolar transistors. The required sync wires and the multiprogramming degree are closely related. The hardware synchronization is shown 100 to 1000 times faster than the software barrier synchronization. Operational issues are addressed on synchronizing large, multiprogrammed, multiprocessor systems using the barrier wires.

1 Introduction

Usually, synchronization mechanisms are most realized in software [1, 3]. Some other systems use combining networks [2, 5]. Recently, O'Keefe and Dietz [6, 7] have proposed the use of dedicated synchronization hardware in MIMD multiprocessors. The software approach requires longer time to complete a synchronization, but is more flexible to use. However, in a large multiprocessor, the time to complete a synchronization increases quadratically with respect to the number of processors. The high cost of the combining network and its limited speed make it inefficient to build very large multiprocessors. The method by O'Keefe and Dietz works only for a monoprogrammed multiprocessor, where the barrier patterns must be predicted at compile time. Another drawback is that the associated control logic may become very complex, as the number of processors becomes very large.

Different from the above approaches, we introduce a fast hardware mechanism, which integrates scalable hardware with system software to support barrier synchronization. The environment we considered is a tightly-coupled, shared-memory multiprocessor system. We emphasize on scalable, large-scale systems, which support multiprogramming at the process level. In this paper, we consider a multiprocessor system using a modified *run-to-completion* static scheduling [10]. That is, each process once initiated is allowed to continue execution uninterrupted until completion. When there are two or more processes allocated to the same processor, the processes are scheduled locally by the processor in a round-robin fashion.

In Section 2, we describe the wired-NOR design of the synchronization bus. We discuss the scalability issues and present some timing analysis of the synchronization bus in Section 3. Then we demonstrate in Section 4 how to use the proposed hardware to accomplish fast barrier synchronization. Section 5 presents a scheme to combine the barrier synchronization hardware with software semaphores in shared memory. The effect of bus width on the degree of multiprogramming is discussed in Section 6. The contributions of this research work are summarized in Section 7.

2 Wired-NOR Barrier Synchronization

This research was supported by USA/NSF Grant No. 89-04172. For all correspondence, contact Prof. Kai Hwang via Email kaihwang@panda.usc.edu, or FAX (213) 740-4449, or Dept. of EE-Systems, USC, Los Angeles, CA 90089-0781, USA.

We propose to use m synchronization lines for an n-processor multiprocessor, as shown in Fig. 1. We make use of wired-NOR design and broadcast capability of the bus to reduce the hardware complexity. Each processor uses two binary vectors $X = (X_0, X_1, \cdots, X_{m-1})$ and $Y = (Y_0, Y_1, \cdots, Y_{m-1})$ for distributed synchronization control. These two vectors are memory mapped and also accessible by other processors. Using distributed control, this scheme has the advantage of reduced network traffic caused by extensive synchronization.

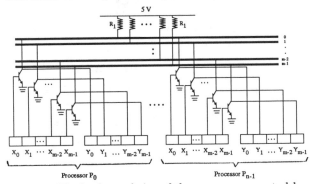

Figure 1: The hardware design of the concurrency control bus with m synchronization lines for an n-processor system. (P_i: processor i, X_i: the ith bit in control vector X, and Y_i: the ith bit in control vector Y)

Each control vector X represents the location of a *set/reset circuit*, which affects voltage levels on all synchronization lines. Each X_i bit is the input to an NPN bipolar transistor. The vector Y represents the location of a *probing circuit* for sensoring the voltage on each line. Each Y_i bit is the output of the transistor and is connected to one synchronization line. The vectors X and Y are program accessible. The vector X can be read and written, but the vector Y is read only.

Each synchronization line is wired-NOR by n NPN bipolar transistors and is connected to n probing circuits. For example, when any X_i bit is high, the transistor is closed and it pulls down the current. The voltage level of line i ($0 \le i \le m-1$) becomes low. When all X_i are low, all transistors connected to line i are open. Thus, the voltage level of line i becomes high. We will show later how to use this electrical property to accomplish barrier synchronization.

Next, we want to decide the number, m, of synchronization lines required in the concurrency control bus design, given the following system parameters:

k The degree of multiprogramming supported by the multiprocessor.

ℓ_i The number of processes created in a program i.

P_i The number of processors allocated for a program i.

b_i The number of active synchronization points demanded in a program i.

S_i The number of synchronization lines required for a program i.

Note that $\ell_i \geq P_i$ for $i = 1, 2, \cdots, k$, and $\sum_{j=1}^{k} P_j \leq n$, where n is the total number of processors in the system. An active synchronization point is initialized but not reached the rendezvous point of a program. Based on the line allocation policy to be discussed in Section 4, we can estimate that

$$S_i = b_i \lceil \ell_i / P_i \rceil + 1 \qquad (1)$$

Let S be the total number of required synchronization lines, when the maximum degree of multiprogramming is k. That is, $S = \sum_{i=1}^{k} S_i$ becomes a lower bound on m. Thus, the degree of multiprogramming is restricted by the number of synchronization lines and by the application programs run on the system. After describing our synchronization schemes, we will discuss the effect of the synchronization bus width on the degree of the multiprogramming in Section 6.

3 Scalability and Timing Analysis

In the this section, we discuss two hardware design issues: (1) how many transistors (gates) can be connected to each synchronization line, and (2) how fast the hardware can respond, when a synchronization point is reached. To answer the first question, we have to consider the DC characteristics of the circuit at logic 0 or 1 levels. Using TTL family, the equivalent circuit of a synchronization line at logic 0 is shown in Fig. 2, where Q_1 is an NPN transistor, Q_2 and R_3 form the input stage of a probing circuit, R_1 is a pullup resistor, and R_2 corresponds to the output stage of a set/reset circuit.

In the worst case, there is only one transistor pulling down the synchronization line. The total pulldown current should not exceed the maximum current, that is allowed by an NPN transistor. The Q_1 transistor must operate in saturation to lock the line

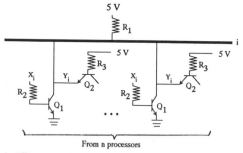

Figure 2: The equivalent circuit of a synchronization line i at the logic-0 level. $(0 \leq i \leq m - 1)$

in a logic 0 level. In this case, Q_2 is also saturated. Similarly, Fig. 3 illustrates the equivalent circuit when a synchronization line is at the logic 1 level. In this case, the Q_2 must be operated in an inverted mode, and Q_3 and Q_4 are saturated. One constraint is that the pullup circuit should be capable of driving all the probing circuits to keep m Q_2 in the inverted mode.

The delay time of a synchronization line is determined below. The delay time t_d consists of the switching time (t_s) of a transistor plus the rise time (t_r). The rise time is modeled as an RC circuit. It is determined that $t_r = 2.2RC = 2.2nR_1C_i$. Therefore, we have:

$$t_d = t_s + 2.2nR_1C_i \qquad (2)$$

Typical values of the above electrical parameters [9] used are listed in Table 1 based on using TTL devices. After some derivations, the delay time becomes $t_d = 35 + 55nR_1$ (R_1 in kΩ). There is a tradeoff between delay time and power consumption. That is, the smaller the delay time we have (for smaller R_1), the larger the power it will be consumed.

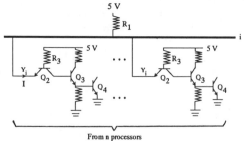

Figure 3: The equivalent circuit of a synchronization line i at the logic-1 level. $(1 \leq i \leq m)$

Table 1: Typical values for electrical parameters used in the bus design with TTL devices.

$I_{C_1(max)}$	800 mA	h_{FEI}	0.5
$V_{BE(sat)}$	0.8 V	R_2	100 Ω
$V_{BC(inv)}$	0.7 V	R_3	4 kΩ
$V_{CE(sat)}$	0.2 V	t_s	35 ns
$V_{EC(inv)}$	0.3 V	C_i	25 pF
h_{FE}	100		

The above analysis points out that this design using standard TTL logic can be scaled up to support a system with more than 500 processors. The delay times and the numbers of processors for different values of R_1 are illustrated in Fig. 4. Using higher speed and higher power bipolar transistors, one can further reduce the delay time and thus increase the scalability of the wired-NOR synchronization bus to support systems with more than a thousand processors.

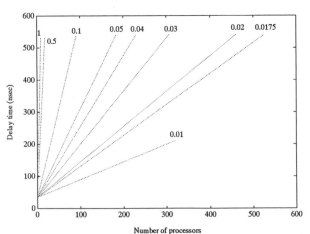

Figure 4: Delay times vs. the number of processors for different values of R_1 (in kΩ).

4 Barrier Synchronization Operations

We show next how fast barrier synchronization can be implemented with the proposed wired-NOR bus lines. We first specify the syntax of a parallel C program construct that can be supported by the wired-NOR bus, as follows.

$fork$(func[(arg1, arg2,$\cdots$)], nproc);
void (*func)();
unsigned int nproc;
$sync$();
$init\_barrier$(bp, func, nproc);
barrier bp;
void (*func)();
unsigned int nproc;
$wait\_barrier$(bp);
barrier bp;

Multiple child processes are created by a *fork* primitive here. The number of processes (*nproc*) created is determined either statically at compile time or dynamically at run time. Once child processes are created, the *fork* initiates them to execute a subprogram *func* with a given set of arguments.

The code of the function need not to be duplicated for each child process, instead it is shared by all child processes. In fact, only the process control block and local variables are needed in each child process. The *sync* primitive causes the parent process to spin, until all child processes finish their tasks. The parent process can perform other computations between the *fork* and the *sync* primitives, concurrently with the execution of child processes.

Besides the *fork* and *sync* primitives, we use an *init_barrier* to initialize a barrier as a rendezvous point of *nproc* child processes, and it should reside in the parent process. We define a special data type, called *barrier*, to specify different barrier points. A *wait_barrier* is used to delay the child process to a busy wait, until all child processes have reached the same barrier. At that moment, all child processes stop spinning. A barrier can be used many times by the child processes. To see how to use the synchronization bus to implement the parallel primitives, we use an example to illustrate the synchronization operations using the *fork* and *sync* primitives.

Example 1: The synchronization of 7 processes on 4 processors using 3 wired-NOR lines.

The forking starts in Step 1. Seven processes are distributed to 4 processors with 1 or 2 processes resident in each processor. The system call assigns $\lceil\frac{7}{4}\rceil+1$ lines, say line 0, 1, and 2, for synchronization.

Besides one line is used for protection, each process has allocated with an exclusive synchronization line. For instance, line 0 is allocated to process 0 and line 1 is assigned to process 4, and so on. If processor i has 2 processes, the system call sets bits X_0, X_1, and X_2. If processor i has only 1 process, the system call sets bit X_0 and bit X_2.

Step 1: Forking (use of 3 synchronization lines)

Step 2: Process 0, Process 1, Process 4, and Process 6 reach the synchronization point

Step 3: All child processes reaching the synchronization point

Step 4: Resetting the line for protection (deallocation of 3 synchronization lines)

Suppose processes 0, 1, 4, and 6 reach the synchronization point in Step 2. Each of them resets its corresponding bit in X. For process 0, it resets bit X_0, for process 1, it resets bit X_0, for process 4, it resets bit X_1, and for process 6, it resets bit X_1. Then, these processes start to test the values of bit Y_0 and Y_1. They do not stop reading the 2 bits, until both become 1, when all processes arrive at the synchronization point (see Step 3).

In Step 4, all child processes reset bit X_3, and the synchronization process is complete. When the *sync* in the parent process detects that the Y_3 bit becomes high, it returns 3 allocated lines back to the operating system.

The use of 2 synchronization lines for each synchronization point is especially important when the system is required to support multiprogramming. If only one line is used for synchronization, the parent process cannot tell whether all child processes catch the voltage change on the line. Therefore, the allocated synchronization line can not be released and returned to the operation system. Only when all child processes are complete, the line becomes available. In this case, to deallocate the synchronization line, the operating system has to either monitor the termination of all child processes or delay the deallocation until the completion of the user program. The first approach imposes a lot of overhead on the operating system, while the second approach results in poor utilization of synchronization lines.

Previous studies [7, 8] neglected this phenomenon by assuming that either all processes execute at the same speed or no reuse of the lines, which are totally impractical. Our method alleviates this problem by using a dedicated protection line to ensure that all processes have reached the synchronization point and have acknowledged. When the parent process detects the fact that the protection line becomes high (i.e. all child processes have acknowledged), it releases the allocated lines and return them back to the operating system for other programs. As a result, the synchronization lines can be highly utilized.

The functions of *init_barrier* or *wait_barrier* are similar to those of *fork* and *sync*, except an additional reinitialization step is required. Interested readers may consult the pseudo codes for *init_barrier* and *wait_barrier* in [4]. We obtain that, when ℓ child processes with b active synchronization points are executed on P processors ($\ell \geq P$), $b\lceil\ell/P\rceil + 1$ synchronization lines are needed.

5 Using Barriers With Semaphores

One can combine the use of wired-NOR synchronization bus and with counting semaphores in shared memory. The purpose is to handle cases limited by insufficient number of synchronization lines. We use two sets of primitives for combined use of hardware barrier and software semaphores, as specified below:

m_fork(func[(arg1, arg2, ···)], nproc);
void (*func)();
unsigned int nproc;
m_sync();
m_init_barrier(bp, func, nproc);
barrier bp;
void (*func)();
unsigned int nproc;
m_wait_barrier(bp);
barrier bp;

The system calls employ one software semaphore among the all the child processes assigned to the same processor. The use of semaphores implies better support for medium to large-grain parallelism. We assume that the operating system provides kernels, such as *wait* and *signal*, to facilitate the suspension and resumption of processes using these semaphores. In *Example 2*, 5 child processes are executed on 3 processors using this combined approach. The letter in parenthesis distinguishes *active* (A) from *blocked* (B) process.

Example 2: Multiple synchronizations of 5 processes on 3 processors using 3 barrier wires and 6 counting semaphores.

The initialization of the *barrier* is performed in Step 1. The number of processes assigned to a processor is determined. The ith synchronization line is allocated for the *barrier*. In addition, 3 semaphores (S1) are demanded for each of the allocated processors. The initial value of each S1 is equal to the number of processes to be executed on the corresponding processor.

Step 1: Initializing the barrier
(use of 1 synchronization line and 3 semaphores)

Step 2: Forking (use of 2 synchronization lines and 3 semaphores)

Step 3: All child processes arriving at the barrier

Step 2 is for process forking. The allocation and initialization of 3 semaphores (S2) are performed in the *m_fork* system call. The values of semaphores S2 are the same as their counterparts initialized in Step 1, which is equal to the number of child processes accessing it. Only two synchronization lines ($i+1$ and $i+2$) are allocated to each occurrence of *m_fork*: one of the line is used for probing, and the other signals the completion of synchronization. The semaphores are distributed to distinct

cache blocks to avoid unnecessary cache invalidations due to false sharing.

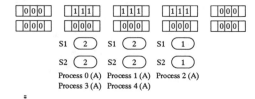

Step 4: Reinitialization of the barrier

Step (k-1): All child processes reaching the synchronization point

Step k: Resetting the line for protection
(deallocation of 3 lines and 6 semaphores)

When reaching the barrier point, all processes, except the last one in each processor, are blocked by the *wait* kernel. Only active processes consume the CPU time in this case. When all child processes reach the *barrier* (see Step 3), the values of the semaphores S1 all become 0, all X_i bits are reset by the active child processes, and the synchronization line i becomes high again.

Step 4 reinitializes the *barrier*. Each active child process sets X_i bit to 1 again, reloads the value of the S1 to its initial value. The blocked process in each processor are activated by the *signal* kernel. The Steps 3 and 4 may be repeated as many times as the *barrier* is used.

When all child processes arrive at the same synchronization point, the values of S2 becomes all 0. The active child processes reset X_{i+1} bits, and the Y_{i+1} bits become high (see Step S_{k-1}). In Step S_k, each active child process resets bit X_{i+2}, and wakes up those blocked processes. When the *m_sync* in the parent process detects that the Y_{i+1} bit becomes high, it releases 3 lines to the operating system and destroys 6 semaphores used.

The *m_init_barrier* and *m_wait_barrier* operate similarly as *m_fork* and *m_sync*, except an additional reinitialization step is required for the semaphores. We obtain that, when ℓ child processes with b active synchronization points are executed on P processors ($\ell > P$), $b + 1$ synchronization lines and bP counting semaphores are demanded. The pseudo codes for these primitives were specified in [4]. Details are skipped here.

6 Effect on Multiprogramming

We want to determine the relation between the width of the synchronization bus and the degree of multiprogramming in a shared-memory multiprocessor. Assume that $b_i = b$ and $\ell_i = \ell$, where $i = 1, 2, \cdots, k$. Consider a machine with n processors and m synchronization lines. Assume that $P_i = \min(n/k, \ell)$ processors are allocated to each program i. By Eq. 1, the required number of synchronization lines, S, is approximated by:

$$S \approx \sum_{i=1}^{k}(b\lceil \ell \cdot k/n\rceil + 1) \quad (3)$$
$$\approx b\ell k^2/n + k$$

Since $S \le m$, we obtain the degree of multiprogramming as follows:

$$k \le \frac{-n + \sqrt{n^2 + 4b\ell mn}}{2b\ell} \quad (4)$$

The value of $b\ell$ depends on the characteristic of the parallel programs executed on the system, and it can be used to approximate the system workload. For fixed values of $b\ell$ and n, the degree of multiprogramming k increases with the order of $\sqrt{m}$. Figure 5 illustrates the effects of multiprogramming with respect to the width of the synchronization bus for a 32-processor and a 128-processor systems respectively.

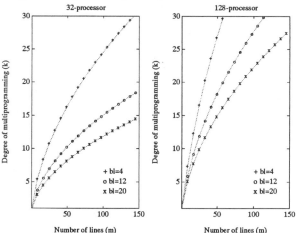

Figure 5: The degree (k) of multiprogramming supported versus the number (m) of synchronization lines used under the workload ($b\ell$) variation from 4 to 20.

Next, we analyze the synchronization line allocation policy used in Section 5, when the semaphores are jointly used. Let S_i' be the number of synchronization lines demanded by program i, and S' be the total number of lines required, when the degree of multiprogramming is k. Then, we have:

$$S_i' = b + 1 \quad (5)$$

And $S' = \sum_{i=1}^{k} S_i' = k(b+1)$. Since $S' \le m$, the degree of multiprogramming becomes:

$$k \le \frac{m}{b+1} \quad (6)$$

The degree of multiprogramming increases in proportional with the number of synchronization lines used, and inversely with the number of synchronization points demanded in each program.

For a fixed m, using semaphores can support a higher degree of multiprogramming than that without. The above analysis gives a static snapshot of the multiprogramming capability. By estimating the degree of multiprogramming and analyzing the workload in the target benchmarks, a computer architect can determine the appropriate synchronization bus width, i.e. m, in designing a shared-memory multiprocessor.

7 Conclusions

We have proposed a wired-NOR bus for fast barrier synchronization in shared-memory multiprocessors. The distributed synchronization control reduces the network traffic, while the processors are waiting for synchronization. Through electrical analysis, we have shown that the synchronization bus can easily support up to 500 processors using off-the-shelf TTL components. We have defined two sets of *fork* and *join* primitives, which facilitate data and program partitionings commonly demanded in parallel programming. One set supports fine-grain parallelism, while the other supports medium to large-grain parallelism. Through examples, we have demonstrated that the synchronization bus is capable of supporting these parallel primitives.

The synchronization bus width directly affects the degree of multiprogramming. It provides useful guideline for computer architects to design efficient multiprocessors. The proposed scheme is technologically feasible in constructing a very large, scalable, shared-memory multiprocessor. The research work is being continued to remove the restriction requiring the system operating in a *run-to-completion* environment.

References

[1] N.S. Arenstorf and H.F. Jordan. Comparing Barrier Algorithms. *Parallel Computing*, 12:157–170, 1989.

[2] A. Gottlieb, R. Grishman, C.P. Kruskal, K.P. McAuliffe, L. Rudolph, and M. Snir. The NYU Ultracomputer — Designing an MIMD Shared Memory Parallel Computer. *IEEE Transaction on Computers*, C-32(2):175–189, Feb 1983.

[3] G. Graunke and S. Thakkar. Synchronization Algorithms for Shared-Memory Multiprocessors. *Computer*, pages 60–69, Jun 1990.

[4] K. Hwang and S. Shang. Fast Synchronization of Large Multiprocessors Using Wired-NOR Barriers and Counting Semaphores. *Technical Report* CENG: 91–06, Department of EE–Systems, University of Southern California, Los Angeles, CA 90089, Mar 1991.

[5] G.J. Lipovski and P. Vaughan. A Fetch-And-Op Implementation for Parallel Computers. In *Proc. of The 15th Annual International Symposium on Computer Architecture*, pages 384–392, 1988.

[6] M.T. O'Keefe and H.G. Dietz. Hardware Barrier Synchronization: Dynamic Barrier MIMD (DBM). In *Proc. of 1990 International Conference on Parallel Processing*, pages I43–I46, 1990.

[7] M.T. O'Keefe and H.G. Dietz. Hardware Barrier Synchronization: Static Barrier MIMD (SBM). In *Proc. of 1990 International Conference on Parallel Processing*, pages I35–I42, 1990.

[8] C.D. Polychronopoulos. Compiler Optimizations for Enhancing Parallelism and Their Impact on Architecture Design. *IEEE Transaction on Computers*, C-37(8):991–1004, Aug 1988.

[9] Texas Instruments, Inc., Dallas, Texas. *The TTL Logic Data Book*, Mar 1988.

[10] J. Zahorjan and C. McCann. Processor Scheduling in Shared Memory Multiprocessors. In *Proc. of ACM SIGMETRICS 1990 Conference on Measurement and Modeling of Computer Systems*, pages 214–225, May 1990.

A New Synchronization Mechanism

Tong Chen Chuan-Qi Zhu
Computer Center
Fudan University
Shanghai 200433 P.R. China

Abstract -- As the grain size becomes smaller, more efficient synchronization operations are needed. In this paper, we give the design of a new synchronization architecture and some new primitives based on the hardware to support process-oriented scheme and barrier. Unlike traditional architecture, there will be no busy waiting and even no synchronization variables in memory. The partition of processors is also supported in this architecture machine. Any subset of processors is able to execute barrier in several clock ticks. Advantages of the discussed architecture are showed in some examples.

I. Introduction

The development of VLSI and RISC technology makes it possible to implement 64-bit floating point processor on a single chip, such as Intel 80860 and 80960. Multiprocessor systems built with such kind of chips are attractive in cost and performance. Due to the limitation of the technology at present and in the near future, the system BJ-1 which we are building and many other existing systems are middle-scale systems. Our research, supported by National Science Foundation of China, aims to design an efficient and practicable synchronization architecture for BJ-1. The new architecture presented in this paper can be applied to other middle-scale shared-memory multiprocessor systems as well.

To speed up the execution of programs, parallelism should be exploited. The main parallelism of a program is in loops. However, data dependence must be enforced. It brings about the problem of synchronization among processors. Most synchronizations are invoked by the barriers of doall loop and the data dependence in doacross loops. Some special hardware, such as the concurrency bus in Alliant FX series [1], has been used to implement barrier efficiently. As for data dependence in doacross loop, process-oriented scheme [2] is promising. We put forward a architecture of synchronization with some special hardware to support process-oriented scheme and to execute barrier efficiently. Synchronization in our system costs very little. There will be no extra dependence imposed by the way of synchronizing, and no busy waiting which often degrades the performance greatly.

The underlying system also supports the partition of processors. Sometimes doacross loop, constrained by data dependence, can not make much benefit from a large number of processors. A few processors are enough. This is one of the reasons to partition processors into groups. Which group a processor belongs to is determined by its group number given at run time. As a result, our system is able to partition the processors into groups arbitrarily and move processors from one group to another easily. So it is possible to schedule the processors dynamically and keep processors balanced among groups.

II. Process-Oriented Scheme

A do loop may be treated as dosequencial loop, doacross loop, or doall loop[3] according to the circumstance. Iterations of a dosequencial loop are executed sequentially just as a regular FORTRAN or Pascal loop does. Iterations of a doall loop are data independent and can be executed simultaneously. Only barrier is needed to synchronize the execution of all iterations. But the execution of a doacross loop is a bit complicated. Iterations of a doacross loop can be executed simultaneously while the data dependence among the iterations has to be enforced. There are three types of data synchronization schemes for doacross loops: the data-oriented scheme [4], the statement-oriented scheme and the process-oriented scheme. The details of these scheme can be found in [2]. This paper focuses on the process-oriented scheme only.

In process-oriented scheme a process corresponds to a iteration of a loop, which may contain some inner loops, and a processor can execute one process at a time. The scheduling strategy ensures that process[i1] should be started before process[i2] when it precedes process[i2] in the precedent graph.

Suppose there are two statements S1 and S2. If S1 depends on S2, S1 is called a sink and S2 is called a source. A statement may be a source and/or a sink or neither. In each process, a dedicated integer synchronization variable step is used to record the execution phase of sources. The step is initialized to 0 at the beginning of execution of the process and is increased by 1 after a source is completed. Process-oriented scheme uses step for synchronization. When a sink is about to be executed, the step of its source's iteration will be tested to find out if the source has been completed. The procedure is demonstrated in example 2.1.

example 2.1

		the value of step
	DOACROSS I=1,N	0
S1	A[I+1]=I*I	1
S2	B[I+3]=A[I-1]+I	2
S3	C[I]=B[I-1]+2	2
	ENDDO	

This work is supported by National Science foundation of China.

S2 depends on S1 with dependent distance 2 and S3 depends on S2 with dependent distance 4. So S1 is a source, S2 is a sink for S1 and a source for S3 and S3 is a sink. Let process[i] represent the process of the i'st iteration and step[i] represent the step of process[i]. The step is initial to be zero when the iteration is to be executed. After the execution of source statement S1 and S2, the step is incremented. The step remains to be 2 after execution of sink statement S3 . The value of step after the execution of each statement is listed in example 2.1. Because of the data dependence process[i] has to test whether step[i-2]>0 before executing statement S2. If the condition is true, processor[i] can execute S2. Otherwise it has to wait until step[i-2]>0. Similarly, S3 of process[i] has to wait for (step[i-4]>1).

In this paper, step is also used in barrier synchronization. Processor which encounters a barrier sets its step to be 11...1, the largest possible step in the system, and tests if the steps of all the processors in the barrier are all 11...1 (or in other words, steps>11...10). If so, processors waiting at the barrier can go. Otherwise they have to wait until the condition becomes true. Hardware of the underlying system is simplified significantly because the barrier and data dependence are handled in almost the same way.

III. Architecture

Brief View

A synchronization hardware mechanism is presented here. For the sake of easy understanding, how the new type of synchronization works is described briefly first . The data synchronization, based on the process-oriented scheme, is realized in the kind of inquire-reply approach. The processor issues an inquiry when it needs to synchronize with other processors, and then decide whether it has to wait or not according to the answer from the relevant processors. The processor which replied is responsible for informing the waiting processor to resume executing when the waited step is completed.

The structure concerned with the synchronization is shown in Fig.3.1. There are two parts, a concurrency control bus and some special synchronization hardware added to each processor. The bus is the path to transmit inquiry and reply. The synchronization hardware, including bus controller, synchronization registers, and waiting counter array, takes control of synchronization

Special Synchronization Hardware

1.Synchronization Registers. There is a set of synchronization registers, PG, PI and PS, containing respectively the group number, the iteration being executed and the current step of the iteration. They maintain the necessary information for synchronization.

2.Waiting Counter Array. The waiting counter array keeps the traces of inquiries. There are N-1 counters in waiting counter array when there are N processors in the system. The waiting counter array of a processor has a counter for every processor except itself. Each counter may be one of the two states, enabled and disabled. When processor[i] rejects processor[j]'s inquiry (i,j are iteration numbers), the difference between the step of processor[i] and the step of the inquiry is written into counter[j] of processor[i]'s waiting counter array and counter[j] is enabled. All enabled counters in waiting counter array are decreased by one when processor[i] executes a source. That counter[j] is decreased to 0 means that the step processor[j] is waiting for has been completed and processor[j] should be told to go on. It is implemented in the following way. Whenever it is 0, an enabled counter raises its own INFORM to 1 for one cycle and then disables itself in the next cycle. The signal GO[j] is the logic OR of all INFORM[i,j]. So GO[j] is able to tell processor[j] to resume when possible.

3. Concurrency Control bus. The concurrency control bus is used for interprocessor communication. It is a synchronized bus. Let's assume that the bus works just as an ordinary synchronized bus does. Details about requesting of the control of the bus, waiting for acknowledgement and putting data on the bus are abbreviated to one action, putting data on the bus, in the following description.

The bus consists of BS,BG,BP,BI,REQUIRE and REJECT. The REQUIRE and REJECT are one-bit signal line and the bits of BS, BG ,BP and BI are sufficient. BG, BI, BS and BP are used to transmit the group number,the iteration number, the step and processor ID respectively. BG and BI make up the target address of the inquiry and BP points out the source address. BS contains the required step. INQUIRE, which is designed for the implementation of barrier synchronization, is also used to point out the target processors of an inquiry. Asserted INQUIRE implies that the current inquiry goes to all the processors in the same group of the inquiring processor. REJECT is used for processors to reply the inquiry. Asserted REJECT gives a negative reply to the inquiry and makes the inquiring processor wait.

It should be noted that the REJECT conveys the answer of inquiry issued in the previous bus cycle. The require and reply can be overlapped to enhance the capability of the bus. For the convenience of description, REJECT, if not specified, refers to the one in next bus cycle.

4. Bus controller. The bus controller keeps monitoring the communications on the bus and taking correct reaction. The bus controller and the PE work concurrently. The PE is not disturbed by inquiries. The bus controller is responsible for replying the inquiry. First the bus controller must determine whether the inquiry is to it or not. When (BG=PG and BI=PI) or (BG=PG AND INQUIRE is asserted) holds, the bus con-

troller knows it is the target of the inquiry. Then it tests whether the step in inquiry has been executed. If BS is greater than PS, the bus controller give a negative answer, asserting REJECT, and the difference of the steps, BS-PS, is loaded into the counter[BP] of the waiting counter array. Otherwise there is no reply.

When a processor receives an inquiry of barrier of the same processor group and REJECT is not asserted, obviously all processors in its group are now ready for barrier with PS=1...1, and therefore the bus controller of the processor will assert its own RESUME to inform the PE to surpass the barrier. The above description can be abbreviated to when (BG=PG and INQUIRE is asserted and REJECT is not asserted), RESUME is asserted.

IV. Synchronization Primitives

Definition

We provide three synchronization primitives based on the hardware introduced above. They are WAIT (ITERATION, STEP), ADVANCE and BARRIER, which are defined as follows.

```
WAIT(ITERATION,STEP)
{
Put (PG, ITERATION, STEP, and  processor
ID) on (BG, BI, BS, and BP);
If REJECT is asserted, wait until its GO
rises to 1.
}

ADVANCE
{
Increase the register PS by 1;
Decrease every enabled waiting counters
 by 1;
}

BARRIER
{
   Load PS with 11...1;
   Put  PG on BG and assert INQUIRE;
   If REJECT is asserted, wait;
   Exception:
   { Whenever RESUME is asserted, return }
}
```

Detailed Explanation

WAIT(I,K). WAIT(I,K) is placed before a sink which depends on step k of iteration I to ensure that the step K of iteration I has been executed. When REJECT is tested during the execution of WAIT(I,K), it may be either asserted or not. Asserted REJECT means that the iteration I is executed but the step K has not been finished, and the inquiring processor has to wait for its GO signal rising to 1, which occurs only when step K is completed. When REJECT is not asserted, there are two cases. First there is a processor executing the iteration. Second there is no processor executing the iteration. In the former case, the bus controller confirms that the PS is not less than BS. In the latter case, scheduling strategy implies that iteration I has been completed. Either case means that the step K of iteration I has been done. So the sink

does not need to wait.

When a sink has several sources, several WAIT()s are needed. Their order makes little difference. Because the number of related sources is not large in most cases, the primitives of WAIT() can be arranged in any order. Some optimization can be made to reduce the number of WAIT() needed, details can be seen in [5].

ADVANCE. ADVANCE is placed after every source. It increases the STEP of the process stored in PS by one and decreases each enabled counters by one. It means that they are one step closer to the sources waited. When the source that a processor waits for is reached, the processor's waiting counter becomes 0 and the related INFORM signal will tell the waiting processor to resume its execution.

BARRIER. BARRIER intends to implement barrier synchronization among the processors which bear the same group number. It is taken for granted that the processors waiting at the same barrier are in the same group. When it encounters a barrier, the processor loads its PS with 11..1,the largest step number, and then broadcast the inquiry through the concurrency bus and makes sure that all processors in its group are ready for the barrier synchronization. The broadcasting is indicated by asserted INQUIRE. If REJECT is asserted, there must be at least one processor is busy in executing some iteration. Otherwise all processor are ready. Not all processors have to issue inquiry because some may intercept the communication on bus and get to know that it can pass the barrier. RESUME, asserted by the bus controller,is the signal to inform the related processor to pass the barrier.

Comments

Several advantages can be seen from the explanation above. The primitives are fully supported by hardware . All operations except the waiting in the primitives can be completed in one or two cycles. The waiting time, imposed by the dependence, is inevitable. The overhead for synchronization in this architecture is low. Another and more important reason for efficiency is that there is no shared variables in memory, no access to memory and no busy waiting in the primitives. Some serious problems usually caused by synchronization, such as surplus access to memory which degrade the performance of system greatly, are avoided. It is reasonable that synchronization variables are in synchronization registers which are attached to every processor because only the process being executed needs synchronization in process-oriented scheme.

V. Examples

Example 5.1 simple loop.

```
       DOACROSS I=1,N
    s1     A[I]=I*3
           ADVANCE;
           WAIT(I-2,1);
           WAIT(I-3,2);
```

```
s2      B[I]=A[I-2]+B[I-3];
        ADVANCE;
        . . .
            get-next-iteration;
        ENDDO

    get-next-iteration
    IF (there is an iteration which has not
        been assign to any processor)
        execute the iteration;
    ELSE
        BARRIER
        further scheduling
    ENDIF
    END
```

It is straightforward to apply the primitives to **parallel** programs. Insert proper WAITs be**fore each** sink and ADVANCE after source and **BARRIER** at the end of iterations.

However, it is recommended not to use BARRIER directly when iterations of a doacross loop are executed by fewer processors. It will be more efficient that not all but only those iterations after whose completion there is no iteration has not been assigned to processor execute BARRIER. BARRIER is invoked by scheduling scheme and a lot of barriers can be deleted. Details of scheduling scheme can be found in [6].

Example 5.2 Multiply-nested loop:

```
        DOACROSS I=1,N1
            DOACROSS J=N2
s1              A[I,J]=.....
                ADVANCE;
            . . .
                WAIT((I-2)||(J-3),1);
s2              ...=A[I-2,J-3]
            . . .
            ENDDO
        ENDDO
```

Example 5.2 shows that there is little difference between the way of handling the simple loop and the multiply--nested loop in Our system. A bit connection operator || is used to denote the multi-dimension iterations. So it is not forced to linearize the iteration. Moreover, there is no test for boundary and no coalesced dependence caused by boundary problem.

Example 5.3 IF statement

```
        DO ...
            IF ... THEN
                BLOCK1
            ELSE
                BLOCK2
                ADVANCE
            ENDIF
        ENDDO
```

As for IF statement, each branch is treated separately. When the two branches reunite,the branch which has fewer sources has to execute some extra ADVANCEs to ensure that the PS is the same at the end of either branch. In example 5.3, there are one more sources in BLOCK1 than in BLOCK2. Therefore an ADVANCE is appended to BLOCK2.

VI. Summery

Certain synchronization hardware mechanism proposed for the process-oriented synchronization scheme is introduced in this paper. Several synchronization primitives, based on the synchronization hardware design, are also suggested. Some examples of parallel program show the advantages of the proposed architecture. The demand of busy waiting in synchronizations is eliminated by hardware to reduce the communication on the concurrency control bus. The underlying system is effective and efficient for data synchronizations in doacross loop and barrier synchronization.

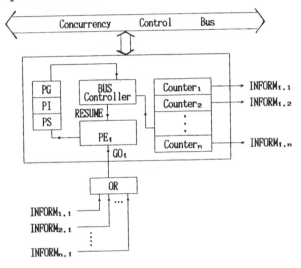

Fig. 3.1 Architecture

References

[1] "FX/series Architecture Manual" Alliant Computer Systems Corp., Jan 1986

[2] Hong-Men Su and Pen-Chung Yew. "On Data Synchronizations for Multiprocessor Systems" proc. of 16th Annual Intl. Symp. on Computer Architecture, Jerusalem, pp. 416-423

[3] M. J.Wolfe, "Optimizing Supercompilers for Supercomputers" Ph. D. Thesis, University of Illinois at Urbana-Champaign.

[4] Chuan-Qi Zhu and Pen-Chung Yew. "A Scheme to Enforce Data Dependence on Large Multiprocessor Systems" IEEE Trans. on Software Eng., VOL. SE-13, No.6, pp. 726-739, June 1987

[5] ZhiYuan Li and Walid Abu-Sufah. "On Reducing Data Synchronization in Multiprocessed Loops". IEEE Trans. on Computers, VOL C36, No.1, pp. 105-109, Jan. 1987

[6] Zhixi Fang, Peiyi Tang,Pen-Chung Yew and Chuan-Qi Zhu. "Dynamic Processor Self-Scheduling for General Parallel Nested Loops" IEEE Trans. on Computers, VOL C37, No.7,pp. 919-929, July 1990

Scalar-Vector Memory Interference in Vector Computers[1]

Ram Raghavan and John P. Hayes

Advanced Computer Architecture Laboratory
Department of Electrical Engineering and Computer Science
University of Michigan
Ann Arbor, MI 48109-2122.

ABSTRACT

Memory interference occurs when two or more concurrent data requests are addressed to the same main memory bank. In vector supercomputers, such memory contention can significantly reduce memory bandwidth and overall system performance. Though numerous studies have been made of the interference among vector accesses, only a few results have been reported in the literature on memory contention between scalar and vector accesses. In this paper, we analyze scalar-vector interference in conventional interleaved memories, when the memory activity is both high and low. We then present some simulation results that validate our analytical results. We also obtain scaling relations which show how the degree of memory interleaving, the bank cycle time, and the number of memory ports must change relative to one another if a fixed level of memory system efficiency is to be maintained.

1 Introduction

Vector supercomputers require a very high CPU-memory bandwidth to achieve good performance. A traditional way of obtaining this high bandwidth is to use an interleaved memory, in which the physical memory addresses are distributed among several memory banks, each of which is independently accessible (Figure 1). Successive memory addresses in an m-way interleaved memory are assigned to consecutive banks, modulo m, and the low-order $\log_2 m$ bits of the memory address indicate the address of the bank in which the requested datum resides.

In spite of features such as interleaved main memory, benchmarks run on vector computers have shown that the actual performance of these machines is much less than their maximum achievable performance [Dong 87]. One important reason for this degradation is memory interference that occurs when two or more concurrent memory accesses are addressed to the same memory bank. Two main types of memory access conflicts can occur with interleaved memories. *Memory bank conflicts* occur when a busy bank is accessed (busy bank conflict) or when two or more accesses are made to the same bank simultaneously (simultaneous bank conflict). The second type, called a *memory path conflict*, occurs when two memory accesses addressed to different banks interfere with each other in the network connecting the CPU(s) and the memory banks. Both types of memory conflicts often limit the *effective memory bandwidth*, that is, the average

[1]This research was supported by the Office of Naval Research under Contract No. N00014 85 K 0531.

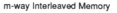

m-way Interleaved Memory

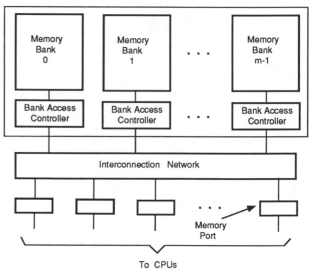

Figure 1: General structure of an interleaved memory.

instantaneous bandwidth achieved, to only a small fraction of the maximum memory bandwidth required for best performance.

Most of the earlier studies, such as [Bail 87, Cheu 86, Harp 89, Oed 85, Ragh 90, Wijs 89] to mention only a few, deal with the interaction among two or more vector accesses, a problem referred to here as *vector-vector interference*. However, in many commercial shared-memory supercomputers such as the Cray X-MP/48, more than one CPU may access the main memory concurrently. Since both scalar and vector data can be concurrently accessed, it is important to study the effect of interference between vector and scalar accesses, called *scalar-vector interference*, on the effective memory bandwidth.

Note that in vector machines with scalar caches, cache misses initiate block transfers between the cache and the main memory. Since cache blocks are contiguous sets of words, they can be considered as vectors, and hence any memory interference they cause falls into the category of vector-vector interference. In this paper, we address the scalar-vector interference problem of the first type that occurs in vector machines without a scalar cache. Often below we will characterize the performance of a memory system by its *efficiency*, which is informally defined as the ratio of the number of data elements delivered by the memory to the number requested in any large time interval, assuming the memory access rate is the highest possible. When the efficiency is 100%, data are being transferred with the maximum effective bandwidth.

The result of scalar-vector interaction in a conventional memory can be a severe drop in the effective memory bandwidth. In the rest of the paper, we will use m for the degree of memory interleaving, n_c for the memory bank cycle time (measured in processor clock cycles), and k for the number of concurrent vector accesses. As an example of the effects of scalar-vector interference, consider an interleaved memory with $m = 16$ and $n_c = 2$. Let $k = 8$ vector streams A, B, C, ..., H be accessed concurrently from the memory resulting in 100% efficiency. Figure 2(a) shows the state-time diagram of such a system under steady-state conditions, where all streams are assumed to have been started at time $t = 0$. The X-axis corresponds to the memory banks and the Y-axis to time (clock cycles) increasing downward. The symbol A in column j of row t means that bank j is busy at time t accessing an element of vector A.

Let now a scalar access be made to bank 0 at, say, $t = 1025$, after the steady state has been reached. As a result, the vector stream F is blocked for two clock cycles and is serviced by the memory bank only at time $t = 1027$. Figure 2(b) shows the state-time diagram of the new steady state reached following the scalar access, assuming there are no bank buffers to hold unserviced requests. The new efficiency is only about 63%. This example demonstrates that scalar-vector interference can introduce conditions that reduce the effective memory bandwidth significantly, if buffers are not used. In fact, our simulations show that a large drop in efficiency occurs about 99% of the time for the example considered here, and is independent of the time and the bank at which the scalar reference is made, under the steady-state condition shown in Figure 2(a).

In spite of the seriousness of this problem, however, not much work on scalar-vector interference has been reported. We know of only one study concerned with the vector access delay in the presence of scalar accesses. Calahan [Cala 89] uses experimental results to obtain an expression for the delay incurred by a scalar reference that interferes with vector accesses:

$$D_{\text{sca,avg}} = A + \frac{n_c^2 r_a}{2m} \qquad (1)$$

where A is a constant that accounts for the delay due to memory refreshing in dynamic random access memories, r_a is the average access rate from all ports, measured in accesses per clock cycle, m is the number of banks, and n_c the memory cycle time in units of processor cycle time. He then extends this model to vector accesses to find the delay incurred by a vector access in the presence of other accesses:

$$D_{\text{vec,avg}} = \frac{n_c^2 r_a}{2m} \left(1 + \frac{L_{\text{test}} - 1}{L_{\text{load}}} \right) \qquad (2)$$

where L_{test} is the length of the test vector whose delay is $D_{\text{vec,avg}}$, and L_{load} is the length of the vectors accessed concurrently from other memory ports. In this paper, we derive more accurate expressions for the delay that account for different levels of activity in the memory system. *High memory activity* is said to exist

Time	\multicolumn{16}{c	}{Banks}
	0	1	2	3	4	5	6	7	8	9	10	11	12	13	14	15
1025	F	G	C	B	H	E	E	H	B	C	G	F	D	A	A	D
1026	F	H	C	G	H	F	E	E	B	D	G	C	D	B	A	A
1027	A	H	B	G	C	F	D	E	E	D	F	C	G	B	H	A
1028	A	A	B	D	C	G	D	B	E	E	F	H	G	C	H	F
1029	D	A	A	D	F	G	C	B	H	E	E	H	B	C	G	F
1030	D	B	A	A	F	H	C	G	H	F	E	E	B	D	G	C
1031	G	B	H	A	A	H	B	G	C	F	D	E	E	D	F	C
1032	G	C	H	F	A	A	B	D	C	G	D	B	E	E	F	H
1033	B	C	G	F	D	A	A	D	F	G	C	B	H	E	E	H
1034	B	D	G	C	D	B	A	A	F	H	C	G	H	F	E	E
1035	E	D	F	C	G	B	H	A	A	H	B	G	C	F	D	E

(a)

Time	\multicolumn{16}{c	}{Banks}
	0	1	2	3	4	5	6	7	8	9	10	11	12	13	14	15
1025	*	G	C	B	H	E	E	H	B	C	G	F	D	A	A	D
1026	*	H	C	G	H	.	E	E	B	D	G	C	D	B	A	A
1027	F	H	B	G	C	.	D	E	E	D	.	C	G	B	H	A
1028	F	.	B	D	C	g	D	B	E	E	.	H	G	C	H	.
1029	a	.	.	D	.	g	C	B	H	E	E	H	B	C	G	.
1030	a	a	.	.	.	f	C	G	H	.	E	E	B	.	G	C
1031	d	a	A	.	.	f	.	G	C	.	F	E	E	.	.	C
1032	d	b	A	A	.	H	.	.	C	.	F	.	E	e	.	F
1033	G	b	H	A	a	H	B	.	.	.	.	.	.	e	E	F
1034	G	C	H	.	a	A	B	.	.	G	.	B	.	D	E	e
1035	e	C	G	.	F	A	A	.	.	G	d	B	.	D	.	e

(b)

Figure 2: Accessing eight streams in an interleaved memory with $m = 16$ and $n_c = 2$: (a) conflict-free, and (b) with conflicts (marked by *) due to a single scalar access.

when a majority of the banks are busy most of the time. This is possible, for example, when a large number of vectors are accessed concurrently. Similarly, *low memory activity* exists when only a few banks are busy, which occurs when a majority of the memory ports are inactive.

In section 2, we determine the maximum reduction in efficiency of interleaved memory systems due to a single scalar access. This analysis also yields the average-case efficiency of interleaved memories during high memory activity. The results obtained here are also validated through simulations. In section 3, we calculate the average delay in accessing a vector in an interleaved memory due to scalar-vector interaction during low memory activity. Section 4 discusses how the degree of memory interleaving m must be scaled as the bank cycle time and the number of memory ports increase.

2 High Memory Activity

Scalar-vector interference is highly likely when memory accesses are made from several ports. In the 4-processor Cray X-MP/48, for instance, such interference is possible between requests emanating from different CPUs. In this section, we derive an approximate expression for the worst-case efficiency of an interleaved memory that results from initiating a single scalar access when several vector accesses are in progress.

Any analysis of scalar-vector interference is made difficult by the fact that scalar references tend to be random, while vector accesses are highly regular. Therefore, we must make some simplifying assumptions about either the vector or the scalar accesses so that both are amenable to the same type of analysis. Deterministic analysis is feasible when only a few vector streams are in progress. Probabilistic analysis, on the other hand, is useful when the memory activity is high, because the interaction among a large number of vector streams leads to severe memory interference, which has the effect of randomizing all the memory references [Bail 87, Buch 90]. Note that this assumption of random references is approximate, and is only justified when the number of vector streams is large.

To analyze scalar-vector interference under high activity conditions, we define a Markov chain model of the memory. Figure 3 lists the major assumptions made in our model. To study the steady-state behavior, we assume that the streams are infinitely long. In reality, vectors are of finite length and thus our analysis can be expected to give an asymptotic limit on the effects of scalar-vector interference.

The main reason for restricting the number of vector streams entering into a simultaneous bank conflict to two is to keep the number of possible states of a memory port within manageable limits. However, our simulation results indicate that this assumption is generally valid with high memory activity. For instance, we simulated the memory system considered in Figure 2(b) and assumed that it operated at 100% efficiency until a scalar access was introduced at bank 0 at time $t = 1025$. All streams that accessed a busy bank were considered blocked until the bank became free. In case of a simultaneous bank conflict, one of the streams was chosen for service while the others were blocked. Figure 4 shows the percentage of the time that one or two vector streams were blocked at each bank; the rest of the time no stream was blocked at these banks. It can be seen from the figure that simultaneous bank conflicts between two streams occured about one percent of the time, and similar types of conflicts among three vector streams almost never occured.

We will find the following theorem useful in formulating the Markov transition matrix in the next section. The theorem limits to a large extent the types of conflicts that can occur in an interleaved memory under the stated conditions.

Theorem 1 *Let a memory system with m banks and bank cycle time n_c be in a conflict-free steady state. If the memory system is operating at 100% efficiency due to m/n_c concurrent vector accesses, then a single scalar access can result only in simultaneous bank conflicts in the new steady state.*

1. The memory system consists of m banks with a bank cycle time n_c (in units of processor cycle time) and is accessed from p memory ports. For simplicity, we assume n_c is a divisor of m.

2. The memory ports are connected to the memory banks through a $p \times m$ crossbar interconnection network. Memory path conflicts in the network are ignored.

3. No bank buffers are present.

4. At any instant, $k = m/n_c \leq p$ vector streams are in progress. This ensures high memory activity.

5. All vector streams are initiated simultaneously at time $t = 0$ and are infinitely long.

6. All streams visit all the banks; that is, all vector strides are coprime with m. Hence it is assumed that a stream can access any idle bank at any instant with equal probability.

7. If the scalar access and a vector access contend for service, the scalar access is given preference.

8. In the new steady state after the scalar access, no more than two vector streams can enter into a simultaneous bank conflict at any instant at a particular bank.

9. The number k of concurrent vector streams is large and the distribution of states of the ports is stationary. In the new steady state, the fraction x of banks that are busy at any instant is a constant.

Figure 3: Main assumptions made in the Markov chain model for scalar-vector interference in a conventional memory system.

Proof Before the scalar access is made, each bank is assumed to be busy all the time. Therefore, when a scalar access is chosen for service, some stream must lose its turn and be blocked for n_c clock cycles. When the delayed vector stream is later serviced, a second vector stream that arrives just at that instant at the same bank must be blocked for n_c clock cycles. As a result, the first vector stream will arrive at other banks n_c clock periods later than in the case where no scalar access is made, and thus enter into simultaneous bank conflicts with other vector streams. By extending this argument, it is easy to see that each delayed vector stream arrives exactly $l n_c$ clock cycles late, for some integer l, at succeeding memory banks. A busy bank conflict cannot occur since that would mean that some bank completes service to a memory access in fewer than n_c clock cycles, which is clearly impossible. Thus, only simultaneous bank conflicts can take place in the memory system. ∎

Following Weiss [Weis 89] we define *efficiency* as a function of the number of banks, the bank cycle time, and the size of the buffers at the banks, and use it as a measure of the steady-state performance of the memory system. The expected efficiency E_1 for a single vector stream over all possible strides d, $1 \leq d \leq m$, is defined as

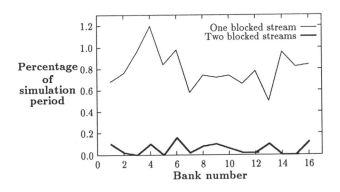

Figure 4: Percentage of the time for which one or two vector streams (out of eight streams) are blocked at each bank in an interleaved memory with $m = 16$, $n_c = 2$.

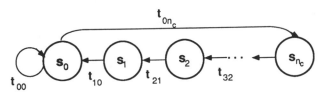

Figure 5: Markov chain model of a memory port in an highly active interleaved memory.

$$E_1 = \sum_{d=1}^{m} N_{\exp}(d) \bigg/ \sum_{d=1}^{m} N_{\max}(d) \qquad (3)$$

where $N_{\max}(d)$ is the maximum possible number of completed memory accesses in a selected time interval for a vector with stride d, and $N_{\exp}(d)$ is the expected number of memory accesses of the same vector stream serviced within the same time interval. Equivalently, we can define the expected efficiency as

$$E_1 = \sum_{d=1}^{m} T_{\min}(d) \bigg/ \sum_{d=1}^{m} T_{\exp}(d) \qquad (4)$$

where $T_{\min}(d)$ is the minimum possible time needed to access a fixed number of elements of a vector with stride d, and $T_{\exp}(d)$ is the expected time to access the same number of vector elements in the steady state.

The Markov chain model presented here closely resembles that used by Bailey [Bail 87] in his analysis of vector-vector interference. The main difference between the two models is that Theorem 1 reduces the types of conflicts to one, while Bailey's model allows for all possible busy bank conflicts. In our model, we assume that the steady state of a port accessing a vector from the memory can be approximately modeled by a Markov chain of $n_c + 1$ states, $s_0, s_1, \ldots, s_{n_c}$, as shown in Figure 5. The state s_0 at time j denotes that the vector associated with the port accesses an idle bank at time j, while any other state s_i denotes that the vector stream is blocked at a busy bank, and i clock periods remain before the bank can service the request from the

blocked stream. Since in the new steady state, a fraction x of the banks are busy at any instant, mx/n_c banks complete service and become available for other vector streams.

Let T denote the Markov transition matrix for a port, where the entry t_{ij} is the probability that a port in state s_i will move into state s_j during the next clock cycle. As a result of Theorem 1, $t_{0j} = 0$, for $1 \leq j \leq n_c - 1$. Assuming that a stream can access, with equal probability, any of the banks that become idle, the probability of a stream entering into a simultaneous bank conflict with another stream can be approximated by

$$t_{0n_c} \approx \left(\frac{mx}{n_c} - 1 \right) \bigg/ m(1 - x)$$

This assumes that all other streams have accessed distinct banks, an assumption justified by the data presented in Figure 4. Since, among the two competing streams, one is chosen arbitrarily and the other blocked, we get

$$t_{0n_c} = \frac{1}{2} \frac{mx - n_c}{mn_c(1 - x)} \qquad (5)$$

Then an element t_{ij} in row i and column j of the transition matrix T is given by

$$t_{ij} = \begin{cases} \dfrac{1 - mx + n_c}{2mn_c(1 - x)} & \text{if } i = j = 1 \\[2mm] \dfrac{mx - n_c}{2mn_c(1 - x)} & \text{if } i = 1 \text{ and } j = n_c + 1 \\[2mm] 1 & \text{if } 1 \leq i \leq n_c \text{ and } 0 \leq j \leq n_c - 1 \\[2mm] 0 & \text{otherwise} \end{cases}$$

Let $p_0, p_1, \ldots, p_{n_c}$ denote a priori probabilities of the port being in the $n_c + 1$ states. The Markov chain described by the transition matrix T is an ergodic process, and therefore the a priori probability of the port being at any state at any given instant is the limiting frequency of appearance of that state. Hence, the probability vector $\mathbf{p} = (p_0, p_1, \ldots, p_{n_c})$ can be determined by solving the set of linear equations $\mathbf{p}T = \mathbf{p}$, viz.,

$$\begin{aligned} p_0 \left(1 - \frac{mx - n_c}{2mn_c(1 - x)} \right) + p_1 &= p_0 \\ p_1 = p_2 = \cdots = p_{n_c-1} &= p_{n_c} \\ p_{n_c} &= p_0 \frac{mx - n_c}{2mn_c(1 - x)} \end{aligned}$$

Since $\sum_{i=0}^{n_c} p_i = 1$, we get

$$p_0 + n_c p_0 \frac{mx - n_c}{2mn_c(1 - x)} = 1 \qquad (6)$$

As the fraction x of banks busy at any instant in the new steady state is assumed to be constant, the expected number of banks that are newly accessed at any instant must equal the expected number of banks that become idle at that instant. In other words,

$$kp_0 = mx/n_c \qquad (7)$$

Substituting $p_0 = mx/kn_c$ in equation (6), we solve for x obtaining

$$x = \frac{2m + n_c(2k-1) - \sqrt{(2m + n_c(2k-1))^2 - 8mkn_c}}{2m} \quad (8)$$

The preceding analysis leads us to the following theorem.

Theorem 2 *In a memory system with degree of interleaving m and bank cycle time n_c operating at 100% efficiency due to $k = m/n_c$ concurrent vector accesses, a single scalar access may, on the average, reduce the k-stream efficiency to*

$$E_k = \frac{2m + n_c(2k-1) - \sqrt{(2m + n_c(2k-1))^2 - 8mkn_c}}{2kn_c} \quad (9)$$

Proof The efficiency E_k for k streams can be defined similarly to the single stream efficiency E_1 in (3) as follows. If there are no conflicts, k data items must be delivered by the memory system every instant. However, in the presence of conflicts, only mx/n_c data items are delivered. Therefore, the k-stream efficiency is

$$E_k = mx/kn_c \quad (10)$$

Substituting for x from equation (8) yields equation (9). ∎

Although the expression for efficiency in Theorem 2 is complex, it simplifies considerably in the case of high memory activity (large k), as the following corollary shows.

Corollary 1 *In a memory system that operates at 100% efficiency while servicing a large number of vector accesses, a single scalar access can be expected to reduce its efficiency by about 40%, making the maximum expected efficiency about 60%.*

Proof When $2k \gg 1$, equation (9) can be approximated by

$$E_k \approx \frac{kn_c - \left(\sqrt{m^2 + k^2 n_c^2} - m\right)}{kn_c}$$

$$\approx 1 - \left(\sqrt{1 + \left(\frac{m}{kn_c}\right)^2} - \frac{m}{kn_c}\right) \quad (11)$$

Since E_k is smallest when $k = m/n_c$, the largest reduction in efficiency occurs when $m = kn_c$:

$$E_k = 1 - (\sqrt{2} - 1) = 2 - \sqrt{2} = 0.586 \quad (12)$$

Thus, in the new steady state, only about 60% of the banks can be expected to be busy at any clock cycle. This result implies an expected loss of about 40% in the effective memory bandwidth due to the introduction of a single scalar access. ∎

The efficiency value above can also be considered to be the average-case efficiency in the presence of memory interference, if we assume that all vector streams are initiated simultaneously. Note that when $n_c \ll 4m$, the average efficiency is independent of n_c, m and k. The estimate above for the efficiency is only approximate due to the following assumptions: (1) vectors are of infinite length, (2) the memory is operating at 100% efficiency with the maximum number of streams in progress, and (3) as a result of scalar-vector interference, a vector stream can access any idle bank with equal probability. This E_k can be taken as an average-case efficiency due to the approximately random interference among the several vector accesses.

When several scalar memory accesses are made in the new

steady state, we now prove that the bandwidth cannot be reduced much below that given in equation (12). Since, in the new steady state, only about half the banks are busy at any given time, this is equivalent to each bank being busy only about 50% of the time. As a result, no new cascade of bank conflicts is possible, which implies that the bandwidth cannot be affected significantly by any new scalar references.

Finally, we point out that, when bank buffers are provided, 100% memory efficiency can be maintained despite sporadic scalar accesses. In this case, since the bank buffers hold the unserviced memory requests, no stream is stalled, and the cascading memory conflicts are avoided.

To test the validity of the efficiency obtained for the case of high memory activity, we wrote a cycle-by-cycle simulator for interleaved memories [Ragh 91]. The simulation model assumes that the memory banks are connected to the memory ports through a crossbar interconnection network. Each memory port is capable of initiating a memory access in each clock cycle. Since the theory presented earlier considers only memory bank conflicts, the simulation model assumes that no memory path conflicts occur within the interconnection network. During a simultaneous memory bank conflict, only one of the requests is chosen for service according to a random priority scheme. Finally, a port stalls when its request is not immediately serviced and resubmits its request during the next clock cycle.

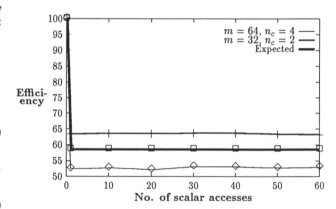

Figure 6: Measured and expected efficiencies in the presence of scalar-vector interaction.

In the first series of experiments, we assumed that the degree of interleaving $m = 32$, the bank cycle time $n_c = 2$, and the number of concurrent vector streams $k = 16$. The aim of our simulation experiments was to measure the drop in efficiency due to one or more scalar accesses, and hence we chose the start banks and strides for the vector streams to achieve 100% efficiency initially. Then the scalar accesses were introduced, each after the previous one had been serviced, at arbitrary banks during arbitrary clock cycles, and the final steady-state efficiency was measured. The second series of experiments was the same as before except that $m = 64$, $n_c = 4$, and $k = 16$. Figure 6 shows the steady-state efficiency in the two cases.

The small discrepancy between the measured and expected efficiencies occurs because our Markov chain model assumes that the scalar-vector interference transforms the highly regular vector accesses into random scalar accesses. This assumption only approximately models the vector access activity. Our experiments also show that the average-case efficiency is about 63% when $m = 32$ and about 53% when $m = 64$, and is independent of the start banks of the streams. Hence equation (12) can be taken as a good approximation to the average-case efficiency.

The low efficiencies that result from scalar-vector interference and depicted in Figure 6 are due to the fact that buffers to hold unserviced requests are absent. The inefficiency introduced by a single scalar access can be eliminated if, say, a buffer of size one is present at the input of each memory bank.

So far we have restricted our attention to the case when the memory activity is high, that is, the number of concurrent vector streams is the maximum, or $k = m/n_c$. Similar analysis can also be done for low memory activity. However, since probabilistic techniques such as Markov chain analysis cannot be used when the memory activity is not stationary, low memory activity is studied using deterministic analysis in the next section.

3 Low Memory Activity

To approximate low memory activity, we modify the model used earlier (Figure 3) and assume that a single infinitely long vector with stride d is accessed, and that the memory system has already reached a steady state. This ensures that no more than n_c banks are busy at any instant. As before, we let a single scalar access be made to a randomly chosen bank and assume that no bank buffers are present. Since efficiency will not be affected dramatically by a single scalar access in this case, we use the expected delay caused by the scalar access on the vector access as a measure of the severity of interference.

The expected delay D_{exp} encountered by the vector access depends on the bank cycle time n_c, the degree of interleaving m, and the number of clock cycles that elapse between two successive visits to the same bank, called the return number r [Oed 85]. Clearly, at any time, only $\min(n_c, r)$ banks will be active while the others remain idle. Since scalar access can be delayed only if it accesses a busy bank, we have

$$D_{\text{exp}} = \frac{1}{m} \sum_{i=1}^{\min(n_c, r)} \Delta_i \qquad (13)$$

where Δ_i is the delay incurred by accessing the ith busy bank.

The following theorem gives an upper bound on the expected delay experienced by the vector access.

Theorem 3 Let $m = 2^\beta$, for some β, be the degree of memory intreleaving, and n_c the bank cycle time such that n_c divides m. Then the expected delay D_{exp} for a vector access due to interference from a single scalar access is

$$D_{\text{exp}} \leq \frac{n_c(n_c + 1)}{2m} \qquad (14)$$

Proof There are three cases to consider: (1) $r < n_c$, (2) $n_c \leq r < 2n_c$, and (3) $r \geq 2n_c$. In the first case, only r banks are busy at any instant. In fact, these banks are continuously busy, and therefore if the scalar access is addressed to any of these banks, the vector access will be delayed by exactly n_c clock cycles. But,

since r divides m, n_c divides m, and $r < n_c$, it must be true that $r \leq n_c/2$. The probability of accessing any of these r banks is r/m, therefore

$$D_{\text{exp}} = \frac{rn_c}{m} < \frac{n_c(n_c + 1)}{2m}$$

If $n_c \leq r < 2n_c$ and the scalar access is issued to a busy bank, then the vector access incurs a constant delay of $2n_c - r$ at the same bank. On the other hand, if the scalar access enters into a simultaneous bank conflict with the vector access, or arrives at an idle bank that belongs to the access set of the vector stream, then the vector stream incurs a delay of $n_c - j$ clock cycles, where $0 \leq j \leq r - n_c$. Thus the sum of all possible delays experienced by the vector access is given by

$$D_{\text{total}} = \sum_{i=1}^{n_c-1} (2n_c - r) + \sum_{j=0}^{r-n_c} (n_c - j) \qquad (15)$$

Simplifying (15), we obtain the expected delay as

$$D_{\text{exp}} = \frac{1}{m} \times D_{\text{total}} = \frac{n_c^2 + (2r - 1)n_c - r(r - 1)}{2m} \qquad (16)$$

Since $r < 2n_c$, we can rewrite (16) to get

$$D_{\text{exp}} < \frac{n_c^2 + (4n_c - 1)n_c - 2n_c(2n_c - 1)}{2m} = \frac{n_c(n_c + 1)}{2m}$$

Lastly, if $r \geq 2n_c$, then the scalar access can delay the vector access for j clock cycles, $1 \leq j \leq n_c$. Thus the total delay for the vector access is

$$D_{\text{total}} = \sum_{j=1}^{n_c} j = \frac{n_c(n_c + 1)}{2} \qquad (17)$$

Hence the expected delay is

$$D_{\text{exp}} = \frac{1}{m} \times D_t = \frac{n_c(n_c + 1)}{2m} \qquad (18)$$

The theorem follows from combining the results for the three cases above. ∎

Comparing the bound (14) with Calahan's empirical expression (1), we see that when the overall access rate $r_a = 1$ (single stream), we have an additional factor that is linearly dependent on the bank cycle time n_c, and independent of the type of memory (static or dynamic) used. Thus the bound (14) is more general than the expression for the delay found in (1). We can also extend equation (14) to find the following upper bound for the average expected delay due to k scalars or k concurrent vectors, where $k \ll m/n_c$:

$$D_{\text{k,avg}} \leq \frac{kn_c(n_c + 1)}{2m} \qquad (19)$$

4 Scaling Relations for Memory Designs

In some supercomputers, such as the Cray X-MP/48, the degree of memory interleaving m has been chosen to make $m = k_{\max}n_c$, where n_c is the bank cycle time and $k_{\max}$ is the maximum number of concurrent memory references. This linear relationship often leads to large inefficiencies in practical systems since it ignores memory interference entirely. Using the expressions for efficiency derived earlier, we draw some conclusions in this section about how m should be scaled when n_c and k are increased. Scaling relations for conventional memory systems have been derived by Bailey [Bail 87], Tang and Mendez [Tang 89], and Bucher and Calahan [Buch 90], but there is some discrepancy among their conclusions. For example, Tang and Mendez do not consider the bank cycle time n_c in their analysis and yet arrive at the same conclusions as Bucher and Calahan, whose analysis does account for n_c. In this section, we also compare our scaling relations with those obtained by others.

The expressions for efficiency and delay derived in the last two sections can also be considered as scaling relations. Thus, when the memory activity is high, equation (11) implies that if the same efficiency E_k must be maintained, then the number of banks m must be increased linearly with the bank cycle time n_c and the number of memory ports k. If the memory activity is low, however, then in order to keep the same efficiency, the bound (14) implies that m must increase approximately quadratically with n_c, and linearly with k.

The difference between the scaling relations for high and low memory activity can be explained qualitatively as follows. The memory interference is high when the number of active ports is high. Therefore, the resultant low efficiency can be easily maintained by increasing m linearly with n_c. On the other hand, a high efficiency during low memory activity can be preserved only by increasing m at a much faster rate than n_c.

We now compare our scaling relations with those of others. Bailey [Bail 87] analyzes the effect of bank contention among k concurrent vector accesses on the efficiency of the memory systems used in vector computers, assuming that no bank buffers are present. He approximates a vector access by a sequence of random scalar accesses, and uses a Markov chain for his analysis, similar to the one described in the previous section. The major difference between our Markov model and Bailey's is that Bailey's model allows any type of busy bank conflict, while our model assumes only simultaneous bank conflicts occur.

Assuming that each request can be addressed to any bank with equal probability, Bailey arrives at the following fraction x of banks that are busy at any time when k vector streams are in progress:

$$x = \frac{\sqrt{1 + 2kq^2 n_c(n_c + 1)/m} - 1}{q(n_c + 1)} \qquad (20)$$

Substituting equation (20) in the expression (10), and taking $q = 1$ for vector accesses, we get

$$E_k = \frac{m}{kn_c} \frac{\sqrt{1 + 2kn_c(n_c + 1)/m} - 1}{(n_c + 1)} \qquad (21)$$

From equation (20), it is easy to see that if the same level of efficiency is to be maintained, then any increase in the number of ports k by a factor i must be matched by a similar increase in the number of banks m. However, when the bank cycle time increases by a factor j, the number of banks must be increased by a factor j^2. The quadratic relationship between m and n_c is due to the fact that $n_c + 1$ types of conflicts are possible in Bailey's model, compared to only one (simultaneous bank conflicts) in our model.

Tang and Mendez [Tang 89] also consider the memory in a vector processor that is concurrently accessible from several ports. Like Bailey, they model a vector access by a sequence of scalar accesses, but use combinatorial analysis to derive an expression for the efficiency. They assume that all ports are active, which implies high memory activity. They conclude that to preserve a constant level of memory efficiency, the number of banks should increase in proportion to the bank cycle time and the number of memory ports. While these linear relationships agree with those we obtain, their analysis entirely ignores the effect of bank cycle time on the efficiency. Though they found a discrepancy of about 25% between the measured efficiency and that predicted by their model, they attribute it, without much justification, to section conflicts, which are ignored in their model.

Bucher and Calahan [Buch 90] use queueing models to analyze the performance of the memory systems in vector computers. They model vector streams as sets of uniformly distributed scalar accesses and the memory banks as independent single servers. Each server is assumed to be accessed at a rate of $\alpha = k/mn_c$ requests per clock cycle, where $k \leq m$ is the number of ports. The authors define the *utilization u* of the memory system as the fraction of the time that a bank is kept busy.

Bucher and Calahan then use an open queueing model for low memory activity and a closed queueing model for high memory activity in interleaved memories. In the open queueing model, the memory requests are assumed to leave the system after being serviced. When the system is in equilibrium, the authors show that the memory utilization is $u = \alpha n_c$, and the average delay for a single request is

$$D_{\mathbf{avg}} = \alpha \frac{n_c^2}{2(1 - \alpha n_c)} \qquad (22)$$

They conclude after some approximation that the delay is proportional to n_c^2. However, since $u < 1$ because of conflicts, $1/(1 - \alpha n_c)$ can be expanded into a power series yielding

$$D_{\mathbf{avg}} = \frac{\alpha n_c^2}{2} \sum_{i \geq 0} \alpha^i n_c^i \qquad (23)$$

This implies that the delay due to interference is, in fact, dependent on all powers of n_c.

For analyzing the average-case behavior when a large number of ports are active concurrently, Bucher and Calahan employ a closed queueing model, in which an old memory request is replaced after being serviced by a new one, after an arbitrary interval of time and is resubmitted to a randomly chosen bank. The authors then show that the average delay for a single request is

$$D_{\mathbf{avg}} = \frac{t_a(u' - 1)}{2} + t_a\sqrt{\frac{(u' - 1)^2}{4} + \frac{u' n_c}{2t_a}} \qquad (24)$$

where t_a is the average interval between requests issued by a port and u' is the memory utilization in the absence of queueing delays. Bucher and Calahan assume that $u' \gg 1$ under "extremely heavy traffic" conditions, and approximate equation (24) by $D_{\mathbf{avg}} = kn_c/m$. This approximation assumes that a bank can be busy more than 100% of the time, which is not possible. In summary, the conclusions drawn from their open and closed queue models agree with our scaling relations for low and high memory activity conditions, respectively, but the validity of some of the approximations they use is unclear.

Finally, we note that our results, while confirming the general scaling relations obtained by others, also identify explicit relationships among the degree of interleaving, the bank cycle time, and the number of active memory ports.

5 Conclusions

We have analyzed the interference among vector and scalar accesses that can occur in the conventional interleaved memories of a vector computer and showed that performance can be reduced by more than 40% when a few scalar accesses interfere with a large number of vector accesses already in progress. We also derived an upper bound on the delay that a vector stream can be expected to experience due to scalar-vector interaction during low memory activity. The resulting expressions for efficiency provide several scaling relations among the number of memory banks, the number of memory ports, and the memory bank cycle time for different levels of memory activity.

References

[Bail 87] D.H. Bailey, "Vector computer memory bank contention," *IEEE Transactions on Computers,* Vol.C-36, pp.293–298, March 1987.

[Buch 90] I.Y. Bucher and D.A. Calahan, "Access conflicts in multiprocessor memories: Queueing models and simulation studies," *Proceedings of Supercomputing, '90,* pp.428–438, November 1990.

[Cala 89] D.A. Calahan, "Some results in memory conflict analysis," *Proceedings of Supercomputing '89,* pp.775–778, November 1989.

[Cheu 86] T. Cheung and J.E. Smith, "A simulation study of the Cray X-MP memory system," *IEEE Transactions on Computers,* Vol.C-35, pp.613–622, July 1986.

[Dong 87] J. Dongarra, "Performance of various computers using standard linear equations software in a Fortran environment," in W.J. Karplus, Ed., *Multiprocessors and Array Processors.* San Diego,CA: Simulation Councils Inc., pp.15–33, January 1987.

[Harp 89] D.T. Harper III, "Address transformations to increase memory performance," *Proceedings of the 1989 International Conference on Parallel Processing,* pp.237–241, August 1989.

[Oed 85] W. Oed and O. Lange, "On the effective bandwidth of interleaved memories in vector processor systems," *IEEE Transactions on Computers,* Vol.C-34, pp.949–957, October 1985.

[Ragh 90] R. Raghavan and J.P. Hayes, "On randomly interleaved memories," *Proceedings of Supercomputing '90,* pp.48-57, November 1990.

[Ragh 91] R. Raghavan, *Memory Architectures for Vector Processing.* Ph.D. Dissertation, The University of Michigan, 1991.

[Tang 89] P. Tang and R.H. Mendez, "Memory conflicts and machine performance," *Proceedings of Supercomputing '89,* pp.826–831, November 1989.

[Weis 89] S. Weiss, "An aperiodic storage scheme to reduce memory conflicts in vector processors," *Proceedings of the 16th International Symposium on Computer Architecture,* pp.380–385, June 1989.

[Wijs 89] H.A.G. Wijshoff, *Data Organization in Parallel Computers.* Boston, MA: Kluwer, 1989.

DESIGN AND EVALUATION OF FAULT-TOLERANT
INTERLEAVED MEMORY SYSTEMS

Sy-Yen Kuo
Department of Electrical Engineering
National Taiwan University
Taipei, Taiwan
R.O.C.

Ahmed Louri
Department of Electrical
and Computer Engineering
University of Arizona
Tucson, Arizona

Sheng-Chiech Liang
Department of Electrical
and Computer Engineering
University of Arizona
Tucson, Arizona

Abstract- A highly reliable interleaved memory system for uniprocessor and multiprocessor computer architectures is presented. The memory system is divided into *groups*. Each group consists of several *banks* and furthermore, each bank has several *memory units*. The error model is defined at the low-level memory units. A memory unit is faulty if any single or multiple fault results in the loss of the entire unit. Spare memory units as well as spare banks are incorporated in the system to enhance reliability. A faulty memory unit is replaced by a spare unit within a bank first, and if the bank has no redundancy remaining for the faulty unit, the whole bank will be replaced by a spare bank at the next higher level. The structure of the reconfigurable memory system is designed in a way such that the replacement of faulty units (banks) by spare units (banks) will not disturb memory references if each bank (group) has at most two spare units (banks). If there are more than two spare units in a bank, a second-level address translator is designed which can prohibit references to faulty memory units by address remapping. Detailed analysis is performed to establish the reliability models. Reliability figures are derived to evaluate systems with various amounts of redundancy. The results show that the system reliability can be significantly improved with little overhead. The property of user transparency in memory access is retained. Both the hardware and the time overhead cost are lower than those of other previous approaches.

I. INTRODUCTION

Due to the increasing requirement of high speed computation in scientific applications, various advanced parallel architectures have appeared in recent years [1,2,3,4]. An architecture with powerful processors and highly concurrent topology is essential in designing a high performance computer. However, this is not enough. Usually a proper memory system with high bandwidth is also required [5]. In tightly coupled multiprocessing environment, main memory is a primary system resource shared by all the processors. Memory contention caused by two or more processors simultaneously attempting to access the same bank of a memory system can degrade the system performance significantly.

The memory bandwidth can be increased by organizing memory chips in banks to read or write multiple words at a time rather than a single word. For example, a 16-MB main memory takes 512 memory chips of 256k×1 bits and can be organized into 16 banks of 32 memory chips. Various memory interleaving techniques which resolve some of the contention by allowing concurrent access to more than one memory bank have been widely used to avoid severe performance degradation. The skewed interleaved memory systems in [6,7] and the general ordered interleaved systems which include high-order, low-order, and hybrid interleaved systems [4] are some examples. The analysis of bandwidth and access interference was investigated in [8,9] for the skewed interleaved system, and in [5,10] for the ordered interleaved system.

Since a faulty memory bank will either degrade the system performance significantly (for a high-order interleaved memory) or the entire system will be out of order (for a low-order or skewed interleaved memory), two approaches have been proposed in the literature to exclude the faulty banks from normal operation. The approach in [11] is for multiprocessors based on multi-stage interconnection networks, which excludes the faulty banks by task remapping (reprogramming the interconnection networks). The other approach in [12] is for uniprocessors with low-order interleaved memory systems, which excludes the faulty memory banks by a second-level address remapping technique.

In this paper, we present a novel fault-tolerant memory system based on the general ordered interleaved memory system and the results can be extended to other types of interleaved systems. This memory system is composed of groups of *banks* and the access to each bank uses the low-order interleaving technique. Each bank contains several *memory units* (MUs). Each *MU* is usually a memory chip. Spare banks and spare *MUs* are added in each group and each bank, respectively. This technique results in a reconfigurable interleaved memory system with two-level redundancy. The redundancy strategy at the low level is similar to that in [13]. However, at the high level, the reconfiguration scheme is quite different. The major difference is that no complex switches are used in our approach in order to replace a faulty bank by a spare bank. Faulty *MUs* are replaced by spare *MUs* within a bank first, and the replacement goes to the bank level if the faulty bank (a bank with faulty *MUs*) runs out of redundancy.

Analysis is performed to determine the amount of redundancy required in each level according to the application requirements. The system performance is preserved until the memory system runs out of redundancy. Time delays introduced by fault tolerance are negligible if each bank has at most two spare *MUs*. For the general reconfiguration scheme with more than two faulty *MUs* in a bank, a new second-level address remapping controller is presented. The approach is much simpler in hardware complexity and more efficient in terms of time overhead than the *SLAT* (second-level address translator) in [12].

II. SYSTEM STRUCTURE OF
THE INTERLEAVED MEMORY

In this section some background on the interleaved memory system is introduced first. The structure and the addressing scheme for the *fault−tolerant interleaved memory* system (*FTIM*) are then described. Comparisons between the nonredundant ordered interleaved system and the fault-tolerant interleaved memory system are also discussed.

A. Interleaved Memory System

In general, in an N-way high-order or low-order interleaving, we have N banks in an interleaved memory system where N is a power of 2, i.e., $N=2^n$ and n is an integer. Assume that there is a total of $M=2^m$ words in the memory system and therefore, a physical address has m bits. In such a system, n bits of the address suffice to select a bank and the remaining $m−n$ bits are used to select a word within a bank. If the n bits are the high-order bits of the address space, the scheme is a *high−order interleaving* scheme whereas a *low−order interleaving* scheme results if the low order n bits are used to select a bank. A low-order interleaved memory generally has a higher bandwidth than a high-order interleaved memory, and is frequently used to reduce memory interference [14]. Low-order interleaving is preferred if the memory interference problem is the main concern. However, the major drawback is that it is not modular. A failure in a single bank affects the entire address space and will almost certainly be catastrophic to the whole system [14]. The high-order interleaved memory provides better system reliability than the low-order interleaved memory since a failed bank affects only a localized area of the address space and therefore, provides a graceful performance degradation [14]. This is because the failed bank can be logically isolated from the system and the memory can be informed so that no process address space is mapped onto the failed bank.

B. Fault-Tolerant Interleaved Memory System with Two-Level Redundancy

The memory structure employed is a two-level memory system similar to that in [12] which has an instruction cache and a main memory as shown in Fig. 1. The main memory with a total of $M=2^m$ words is interleaved with the N-way low-order interleaving scheme where $N=2^n$. Moreover, these N independent memory banks are divided into $L(=2^l)$ groups with each group consisting of K banks in the fault-tolerant interleaved memory ($FTIM$) system. By using an address latch in each group, a group of $K=2^{n-l}$ banks ($l \leq n$) which are fully interleaved can be multiplexed on an internal memory bus. This approach is called the $C-access$ method [14]. In the $C-access$ method, K banks are accessed *concurrently* which increases the available bandwidth of the memory system to K times the bandwidth of a single bank. Depending on the memory access time, K can be equal to N so that the system contains only one group or equal to one so that the system has N groups and only one bank can be accessed at a time.

In addition, each bank can have 4, 8 or more MUs. The address format of a $FTIM$ system is shown in Fig. 2(b). Compared with the original address format for the low-order interleaved memory as shown in Fig. 2(a), we can see the following differences: (1) the N memory

banks are grouped into L groups with the higher order l bits of the n bits to select a group and the lower order $n-l$ bits to select a bank within a group; (2) depending on its size, a bank can have $Q=2^q$ independent MUs. Thus, the high order q bits of the $m-n$ bits are used to select an MU and the remaining low order bits of the $m-n$ bits are used to select a word within an MU.

With this organization, we can have more cost effective fault tolerance approaches. Instead of having spare banks at the high level only, we can have a two-level redundancy scheme by introducing spare MUs in the low level. After the faulty MUs have been identified, spare MUs are used to replace them first, and then a faulty bank is replaced by a spare bank if the faulty bank runs out of spare MUs. An example of the hierarchical modular memory system is shown in Fig. 3. In this example, four banks are grouped together to form a group. Each bank has four normal MUs and a spare MU. Each normal MU is designated as $MU(i,j)$ where i and j are the row number and the bank number of the MU, respectively. An MU denoted by a R is a spare MU.

Depending on the application requirements, which will be discussed in Section III, spare MUs can be added at the bottom or the top of each bank. These spare MUs are provided as local redundancy within a bank, and can be used to replace faulty MUs in the same bank. Similarly, spare banks can be included in each group. In Fig. 3, one spare MU is added to each bank and a spare bank is attached to each group.

III. DESIGN OF THE FAULT-TOLERANT INTERLEAVED MEMORY SYSTEM

In this section, we will present the high-level structures of two reconfigurable memory systems with different redundancy strategies and the design methodology of the two corresponding reconfiguration controllers. In order to evaluate the fault tolerance technique and to describe the effects of faults on a memory system, the specific fault model will be defined first.

A. Fault Model

We only consider faults which will cause the loss of a complete MU. Not every fault in the MU will lead to the loss of an entire MU (the cause of a failure in an MU or how to diagnose a faulty MU is not

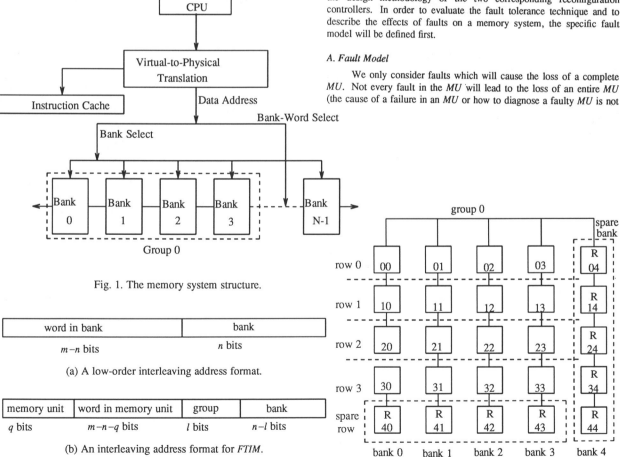

Fig. 1. The memory system structure.

word in bank	bank
$m-n$ bits	n bits

(a) A low-order interleaving address format.

memory unit	word in memory unit	group	bank
q bits	$m-n-q$ bits	l bits	$n-l$ bits

(b) An interleaving address format for $FTIM$.

Fig. 2. Addressing formats of interleaved memory systems.

Fig. 3. An interleaved memory system with two-level redundancy.

the subject matter of this paper) For example, most memory chips are implemented with error correcting code[15] such that any single bit error in a word can be automatically corrected. In this situation, faulty bits in the memory will not always cause the entire faulty *MU* to be discarded. This fault model is more effective than the one presented in [12] where a fault will cause the loss of an entire memory bank.

If a faulty *MU* is detected in a low-order interleaved memory system which does not have fault tolerance capability, the entire system will not function. Therefore, it is desirable to maintain spare *MU*s in each bank and/or spare banks in each group to replace faulty *MU*s or banks. The fault tolerance technique is implemented by assuming that there is an external mechanism to detect and locate the presence of a faulty *MU*. This assumption can be easily accomplished by an error detecting code. When faults occur in the memory system, program execution will be halted and the system management will be informed about the faulty *MU*. Correct information will be recovered from a backup storage and stored in the designated spare *MU*.

B. Easily Reconfigurable System Structure with Two-Level Redundancy

A simple and easily reconfigurable *FTIM* system which can tolerate up to two faulty *MU*s in each bank and two faulty banks in a group will be presented first. We will also discuss how the faulty *MU*s in a bank can be replaced by spare *MU*s in the same bank without reducing the system performance. An example system with two spare rows and a spare bank(column) is used to illustrate the design methodology of the reconfigurable *FTIM* and its reconfiguration controller. This approach can then be used to design a reconfigurable *FTIM* system with two spare rows and two spare columns. However, for applications with the requirements of very high reliability and continuous operation for a long time, more than two spare rows or columns may be necessary. In this case, the reconfiguration controller will be more complicated than that of a *FTIM* system and will be discussed in the next subsection.

In a reconfigurable *FTIM* system with two spare rows, the two spare *MU*s in each bank are placed with one at the bottom and the other at the top of the bank (Fig. 4(a)). A switching element is associated with each *MU* (it can be attached to the "enable" line of each *MU*) to control the selection of the unit. The input and output connections of a switching element are shown in Fig. 4(b) which are controlled by two reconfiguration control registers (*CRA* and *CRB*). The selection line $s=0$ means that the corresponding switch is not activated by the reconfiguration controller and thus the *MU* selected by the *MU* selection line is enabled. Otherwise, if $s=1$, the switch is activated so that it will skip the selected *MU* and replace it with its adjacent top or bottom *MU*. The selection of top or bottom adjacent *MU* depends on the control register that is used to activate it. Each field in a reconfiguration control register controls a bank. The number of bits in each field is a function of the number of *MU*s in the corresponding bank. A bank with l *MU*s (normal and spare) will have r bits in the corresponding field of both *CRA* and *CRB* where r is the smallest integer such that $2^r \geq l-2$. Memory units in a bank are numbered from 0 to l. The bit-patterns in *CRA* and *CRB* determine whether a faulty *MU* will be replaced by its adjacent fault-free *MU* either at the top or the bottom. Table I illustrates the meaning of a field in *CRA* and *CRB* for the example in Fig. 4. For instance, if the field for bank i in *CRA* has the pattern 01, switches of $MU(0,i)$ and $MU(1,i)$ are activated and therefore, the top spare *MU* in bank i will be used to replace a faulty *MU* in that bank.

A faulty *MU* ($MU(3,1)$) in bank 1 is shown in Fig. 4(a). The corresponding switch setting to replace the faulty *MU* in bank 1 is completed by setting the contents in the second field of *CRA* and *CRB* to 00 and 10, respectively. Therefore, switches of $MU(3,1)$, $MU(4,1)$, and $MU(5,1)$ in this bank are activated with *CRB*=10 and switches of $MU(0,1)$, $MU(1,1)$, and $MU(2,1)$ are not activated with *CRA*=00. If there are two faulty *MU*s in a bank, for example, $MU(1,2)$ and $MU(3,2)$ of bank 2 in Fig. 4(a), the bits in the third field of *CRA* and *CRB* are set to 01 and 10, respectively, such that faulty $MU(1,2)$ is replaced by the spare *MU* at the top and faulty $MU(3,2)$ by the spare *MU* at the bottom. The two spare *MU*s should not be all at the top or all at the bottom of a bank, since from this example we can see that an *MU* can only be replaced by one of its adjacent neighbors due to the simple structure of the switch. As a result, this structure can tolerate up to two faulty *MU*s in each bank. However, it has the advantages of regular and easily reconfigurable structure as well as simple switches. Its hardware overhead is low (only simple switches and two reconfiguration control registers are needed). Moreover, time overhead for memory reference introduced by fault-tolerance scheme is negligible since after the system has been reconfigured, only a two-gate delay is introduced in switch selection.

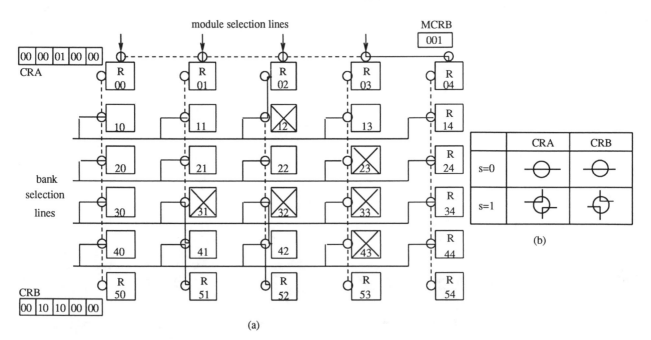

(a)

(b)

Fig. 4. An example *FTIM*.

If only one spare *MU* is added to the bottom (top) of each bank, then *CRA* (*CRB*) can be removed and the number of bits in each field for a bank with l *MUs* is equal to the smallest r such that $2^r \geq l$. The output connection of a switch will be either in the normal state or in the reconfiguration state and the switch has only one selection line from *CRB*(*CRA*).

The replacement of a faulty bank by a spare bank can be implemented with an approach similar to that for the *MU* level. Only a switching element is needed at each bank selection, and if it is activated the entire bank is replaced by its adjacent neighbor bank. A *bank control register* (*BCR*) is used to select the bank to be switched out if there is only one spare bank. Two control registers (*BCRA* and

Table I. *MUs* activated by *CRA* and *CRB*

$l=6$ $(r=2)$		
	CRA	*CRB*
00	normal	normal
01	memory units 0-1	memory units 4-5
10	memory units 0-2	memory units 3-5
11	memory units 0-3	memory units 2-5

BCRB) are needed if one spare bank is added before the first bank and one spare bank is added after the last bank. The bit-patterns of *BCRA*(*BCRB*) are similar to those of the *CRA*(*CRB*) with the exception that the *BCRA* and the *BCRB* have only one field. An example reconfiguration is shown in Fig. 4(a) in which the faulty fourth bank is replaced by a spare bank. Three bits ($r=3$) are required in the *BCR*, since the value 2^r should be greater than $l=5$. However, the reconfiguration time to replace a faulty bank will be significant since memory bank contents have to be copied between banks.

C. Design of a General System Reconfiguration Controller

In some applications, it may not be feasible to add a switch to each *MU* or the system may be required to have extremely high reliability and availability. Therefore, more than two spare rows and two spare columns may be necessary to achieve the reliability and availability requirements. The above simple and easily reconfigurable *FTIM* system can then be modified to form a more general reconfigurable *FTIM* system. Two approaches to designing a general reconfigurable *FTIM* systems are presented here. These two designs have two or more spare rows and a reconfiguration controller to tolerate multiple faulty *MUs* in each bank without any switch attached to an *MU*. The first design, the *general reconfigurable fault—tolerant interleaved memory* (*GRFTIM*) system, can have a flexible number of spare *MUs* in each bank. A faulty *MU* in the *GRFTIM* system will not be accessed by memory

references and will be replaced by the first fault-free *MU* in the same bank. The second design is the *RHFTIM* system which represents the *fault—tolerant interleaved memory system with reduced hardware cost*. Although the *RHFTIM* system still has a flexible number of spare *MUs* as in the *GRFTIM* system, a faulty *MU* now will cause the entire row containing the faulty *MU* isolated. It is easy to see that the *GRFTIM* system is more efficient in reconfiguration and spare utilization than the *RHFTIM* system, however, it requires more extra hardware and introduces larger time delays as discussed in the following paragraphs.

Let the memory module have $Q(=2^q)$ *MUs* in each bank and R $(=2^r, r \geq 1)$ spare *MUs* at the bottom of each bank. In the design of the *GRFTIM* system, an additional address translator (Fig. 5) is used in the system. We call this address translator the *second—level address remapping controller* (*SLARC*). A *multiplexer* (*MUX*) is used to reduce time delays incurred by the second-level address remapping in the absence of faults. As shown in Fig. 5, if there are faulty *MUs* in any of the banks, output from the *memory unit error indicator* (*MUEI*) is set to one and the physical address output from the *MUX* is the output of the *SLARC*. Otherwise, output from the *MUEI* is zero and the physical address from the multiplexer (*MUX*) is the output from the virtual-to-physical translation directly without being further processed by the *SLARC*. The *MUEI* is a one-bit flag used to enable both the *SLARC* and the *MUX*. The *MUEI* is set when a faulty *MU* is detected.

In addition to the *MUEI*, there are two other inputs to the *SLARC*. One is the first-level physical address after the virtual-to-physical translation. This physical address has $q+n$ bits. The lowest n bits are used to select a memory bank. In the low-order interleaved memory system, the highest q bits of the physical address are the same as the highest q bits of the virtual address and are used to select a *MU* in a bank. These highest q bits are extracted from the output of the first-level address translation and remapped so that the output addresses from the *SLARC* will always point to a fault-free *MU*. The other input to the *SLARC* is a vector, the *memory unit status indicator* (*MUSI*), with $2^{q+r}(=$ the number of *MUs* in a bank) bits. Each bit in an *MUSI* vector indicates whether the corresponding *MU* is faulty or not. The 2^r bits in the *MUSI* vector represent the R spare *MUs* in each bank. Since the *MUSI* information is associated with each bank (*i.e.*, each bank has its own *MUSI* vector), a pile of registers is required to maintain the information for all the banks. The use of a cache memory may be a cost effective way if the number of banks in a system is large (for example, greater than 32). In an N-way low-order interleaved memory system, selection of a specific *MUSI* information will then depend on the bank selection lines in the lowest n bits ($N=2^n$) of the output address from the virtual-to-physical address translation. For example, assume that $MU(2,3)$ of bank 3 is selected after the first-level virtual-to-physical translation. The bank selection lines should first be used to select the *MUSI* vector associated with bank 3. Then bit 2 of the

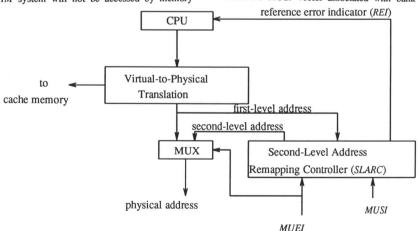

Fig. 5. Two-level address translation of *GRFTIM* with the *SLARC*.

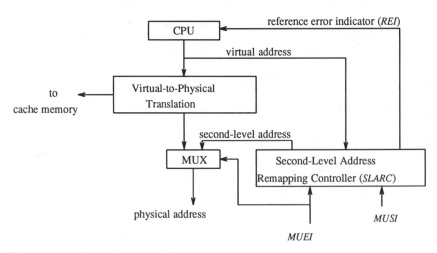

Fig. 6. Address translation of *RHFTIM* with the *SLARC*.

selected *MUSI* is used to identify whether the selected *MU* is fault-free (if it is set to "0") or faulty (if it is set to "1"). The *MUSI* vector is updated every time a faulty *MU* is detected and located.

From the above discussion, we know that the *SLARC* can introduce a significant time delay since an additional cache memory reference cycle is required. Although the use of a register file can reduce the reference time, the time overhead of the *GRFTIM* is still larger than that of the simple *FTIM*. The time delay introduced by the second-level address remapping may be intolerable in some applications. In these applications, the *RHFTIM* system can be used. For the *RHFTIM* system, only one set of the *MUSI* information is stored in a register with 2^{q+r} bits and no additional memory reference will be needed to select a *MUSI* information. Each bit in the *MUSI* now represents the status of an entire row instead of a single *MU* in a bank. For example, if the mth bit of the *MUSI* is set to one, it indicates that there is one or more faulty *MU*s in the mth row. Therefore, a single faulty *MU* in a row will cause the entire row to be treated as faulty. A faulty row is then bypassed and replaced by its nearest available fault-free row.

The major advantage of this *RHFTIM* system is that the extra time delay incurred by the *SLARC* is approximately zero since no additional memory reference is introduced in this design. The second-level address remapping operation in the *RHFTIM* (as shown in Fig. 6) can be performed in parallel with the first-level virtual-to-physical address translation. This is due to the fact that the high order q bits which represent the offset address to select a word in a *MU* can be extracted directly from the virtual address without being translated by the first-level address translation and directly input to the *SLARC*. The *SLARC* can then start the address remapping without waiting for the bank selections (from the output of the first-level virtual-to-physical translation) to select an *MUSI* information of a designated bank. The contents of the *MUSI* are now stored in a single register only.

However, the *RHFTIM* system is inefficient in memory space utilization and the allocation of spare *MU*s. For example, if there is only one faulty *MU* in a certain row, the entire row space is isolated after the second-level address remapping. The situation is even worse when there is a large number of banks in a group, because this modified approach will waste many fault-free *MU*s. However, if there are no more than eight banks in a group or the number of *MU*s in a bank is much bigger than the number of banks in a group, then this modified approach may be more efficient than the *GRFTIM* system since it uses less hardware and introduces less time delay as will be discussed in section IV.

There are two outputs from the *SLARC* in both the *GRFTIM* and the *RHFTIM* systems. One is the physical address after the second-level remapping which will always point to a fault-free *MU*. The other output is the *reference error indicator(REI)*. Eventually, after the

system has run out of spares, the *SLARC* will set the *REI* to indicate that an error access is made to access a faulty *MU* which can not be replaced by a spare *MU*.

The structure of the *SLARC* is shown in Fig. 7. The *SLARC* consists of a decoder and a *MU selection circuit(MUSC)*. The inputs to the *MUSC* consist of a 2^{q+r}-bit *MUSI* vector (which contains the locations of the faulty *MU*s in the banks), and the 2^q outputs from the decoder. The outputs of the *MUSC* are 2^{q+r} physical *MU* select signals. If the mth output line from *MUSC* is set, the mth physical *MU* is selected after the second-level address remapping. Therefore, a faulty *MU* will never be selected by the *MUSC*.

In a fault-free bank, the high order q bits of the physical address are decoded to determine the *MU* to be selected. However, if faulty *MU*s exist in the memory bank, the actual *MU* selected will be remapped by the *MUSC* onto a physical *MU* in the following way. If there is one faulty *MU* in a bank, say physical *MU* m, then the output line $m+i$ ($i{\geq}0$) from decoder is remapped to $m+i+1$. For multiple faults, the remapping mechanism is extended in a manner that the next available fault-free physical *MU* is selected. The logic of the *MUSC* can be realized by switches or multiplexers. Realization of the entire logic of the *SLARC* can be completed by using the *switching logical unit(SLU)* in [12].

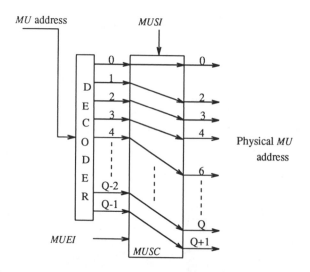

Fig. 7. The second-level address remapping controller.

I-192

The example shown in Fig. 7 indicates that the physical *MU* 1 and *MU* 5 of a bank are faulty. Due to the faulty *MU* 1, the *MU* selection line 1 is remapped to select *MU* 2 and line 2 remapped to select *MU* 3. Another faulty *MU* (*MU* 5) will cause the unit selection line 4 remapped to select *MU* 6 and so on. Although the *SLARC* in the above discussion is designed for the *GRFTIM* system, it can also be used in the *RHFTIM* system without any modification.

IV. RELIABILITY AND COST ANALYSIS

Reliability models to evaluate the above fault tolerance techniques are presented in subsection A. Numerical results are then derived to demonstrate how much improvement can be achieved on the system reliability and how a proper amount of redundancy can be determined according to the system requirements. Effect on hardware overheads and extra time delays in memory references introduced by the fault tolerance techniques is discussed in subsection B.

A. Reliability Analysis

The reliability $R(t)$ of a system is the conditional probability that the system operates correctly throughout the interval $[t_0, t]$ given that it was operating correctly at time t_0 [16]. In order to facilitate the reliability evaluation, we make the following assumptions: (1) the *MUs* become faulty randomly and independently, with a constant failure rate λ, (2) each *MU* has the capability of single error correction using an error correcting code so that λ is very small [15], *e.g.*, $\lambda=10^{-2}$/unit time (a unit time can be 10^6 hours). Therefore, the reliability of a *MU* is represented by $R=e^{-\lambda \cdot t}$.

Let n represent the number of normal *MUs* in each bank and r represent the number of spare *MUs*. The reliability R_b of a bank in either the *FTIM* or the *GRFTIM* system is then given by

$$R_b = \sum_{k=0}^{r} P(k) \times \binom{n+r}{k} \times (e^{-\lambda \cdot t})^{n+r-k} \times (1-e^{-\lambda \cdot t})^k,$$

where $P(k)$ is the probability that a memory bank with k faulty *MUs* can be successfully reconfigured.

In a memory system with n_b normal banks and r_b spare banks, the system reliability R_s of the *FTIM* or the *GRFTIM* is then given by

$$R_s = \sum_{k=0}^{r_b} P_b(k) \times \binom{n_b+r_b}{k} \times (R_b)^{n_b+r_b-k} \times (1-R_b)^k,$$

since faulty *MUs* in the simple *FTIM* system and the complex *GRFTIM* system can be replaced by spare *MUs* in the low-level and then goes to the high-level banks if a faulty bank is out of spares.

Since the reconfiguration controllers of the *FTIM* and the *GRFTIM* systems are always able to completely replace the faulty *MUs* with the spare *MUs* if the number of faulty *MUs* is less than the number of spare *MUs*, we have $P(k)=1$ when $k \le r$ and $P(k)=0$ when $k>r$. Similarly, we have $P_b(k)=1$ when $k \le r_b$ and $P_b(k)=0$ when $k>r_b$.

The reliabilities of an example system (which has 16 banks with 16 *MUs* in each bank) are evaluated for several values of λ and various amounts of redundancy. The notation *mrnc* in Fig. 8 and Fig. 9 is used to indicate that the memory system has m spare rows and n spare columns (*i.e.*, the system has $(16+m) \times (16+n)$ *MUs*). In Fig. 8, we show the system reliabilities with $\lambda=0.1$. The system reliability drops to zero very fast if no redundancy is included and it is improved significantly when two spare rows and two spare columns are incorporated in the system. In addition, the spare rows contribute more in increasing the reliability than the spare columns. For example, the curve $2r0c$ in Fig. 8 is above the curve $1r1c$. The reason is that the reconfiguration controller is more efficient in the low-level replacement (the replacement of faulty *MUs*) than in the high-level replacement (the replacement of faulty banks). From this result, the *GRFTIM* system can be provided with redundant rows only and without spare banks in each group. In addition, the hardware cost of the reconfiguration controller (which is similar to the *MU* level reconfiguration controller) can also be reduced.

Fig. 9 shows that if $\lambda=0.01$, the system reliability is improved significantly even only one spare row is added to the system. It also shows that the system reliability is approximately equal to 1 when more than one spare row and one spare column are added to the system. In the case of adding only one spare row to the system, from the $1r0c$ curves in Fig. 8 and Fig. 9, we can see that the system reliability is improved more effectively with $\lambda=0.01$ than with $\lambda=0.1$.

Fig. 10 shows the reliabilities of the system with two spare rows and two spare columns and with various values for λ. From this figure we can see that if the failure rate is higher than one per unit time, the system reliability will be very low even there are two spare rows and two columns. However, since the value of λ is very small in practical applications, the use of two spare rows and two spare columns in the *FTIM* system seems appropriate.

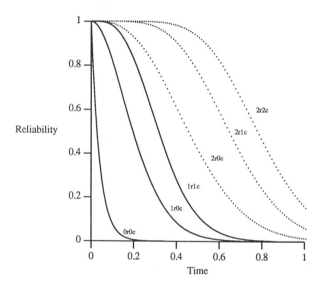

Fig. 8. System reliability with $\lambda=0.1$.

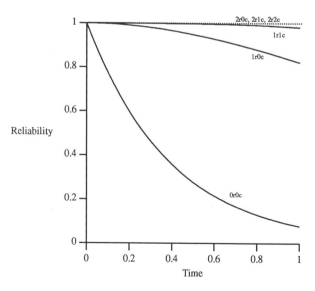

Fig. 9. System reliability with $\lambda=0.01$.

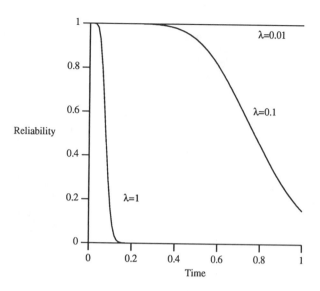

Fig. 10. System reliabilities of a $2r\,2c$ *FTIM* with various λ values.

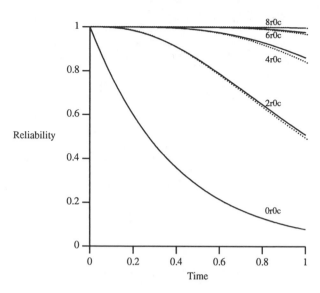

Fig. 11. System reliabilities of the generalized *FTIM*.

In the *RHFTIM* system, the simplified *SLARC* will treat the entire row as faulty even there may be only one faulty *MU* in that row. The reliability in this case will then be quite different from the above reliabilities. Let the number of banks in a group be represented by n_b, the number of *MU*s in a bank be represented by n_r, and the number of spare *MU*s in each bank by r_r. The reliability of a row is then given by

$$R_m = (e^{-\lambda \cdot t})^{n_b},$$

and the system reliability R_s is therefore

$$R_s = \sum_{k=0}^{r_r} P_m(k) \times \begin{bmatrix} n_r + r_r \\ k \end{bmatrix} \times (R_m)^{n_r + r_r - k} \times (1 - R_m)^k.$$

Two example systems, 16×16 and 32×8, are considered with $\lambda = 0.01$ to evaluate the reliability enhancement with various number of spare rows added. The results are shown in Fig. 11, where the dotted curves represent the results of a 16×16 memory array and the solid curves

represent the results of a 32×8 array (32 *MU*s × 8 banks). From this figure, we can see that the reliabilities of these two systems are very close with the same amount of redundancy. This is due to the fact that the reliability is much more dominated by the number of *MU*s in the array than by the topology configuration. Also by comparing with Fig. 9, we find that this fault tolerance technique is not as efficient as the simple *FTIM* system. In a *FTIM* system with two spare rows, the reliability can be improved to approximately one. In a *RHFTIM* system, eight spare rows are required to achieve the same reliability level. Both the *GRFTIM* and the *RHFTIM* systems have reliabilities approaching one if two and eight spare rows are included, respectively.

B. Overhead Analysis

Let the memory system originally have $N (=2^n)$ banks in each group and $Q (=2^q)$ *MU*s in each bank. In the simple *FTIM* system design, two spare rows and two spare columns are included. Each *MU* has a switch and a redundant link for fault tolerance. In addition, two reconfiguration control registers are used to control the functions of each switch. Since the function of a switch is very simple (Fig. 4(b)), the hardware overhead introduced by the switches and reconfiguration control registers is very small compared with the cost of the whole memory array. After the system has been restructured, extra time delay incurred due to the fault tolerance scheme is the time delay of a switch. This time delay is a two-gate delay since a switch can be implemented with a two-level logic.

The hardware overhead in the *RHFTIM* is caused by the 2^{q+r}-bit *MUSI* register and the *SLARC*. Although the cost of an *SLARC* is

higher than the cost of a reconfiguration controller for the *FTIM*, it is much less than that of the *second-level address translator* (*SLAT*) in [12]. Since the *SLARC* can be implemented by the same technique as the *SLU* circuit in *SLAT* and the cost of the decoder and the reference error generator is less than that of other circuits in *SLAT* (the *SLAT* has four more logical circuits and each of them is as complex as the *SLU*). Therefore, the cost of the *SLARC* is less than that of the *SLAT*. In addition, in our structure, only little time delay is caused by the *MUX* if there are no faults. All the reconfiguration circuits can be implemented by combinational circuits so that the time overhead is not significant compared with the memory reference cycle. In addition, as shown in Fig. 6, the *SLARC* can be executed in parallel with the virtual-to-physical address translation which makes the time delay negligible.

An additional memory reference cycle is introduced in the *GRFTIM* system in order to fetch the *MUSI* information of each bank. A register file or a cache memory can be used to store the *MUSI* information. In a system with N banks and $Q + R$ *MU*s in each bank, the *GRFTIM* system will require N registers. Each register has 2^{q+r} bits to store an *MUSI* information. Although the *GRFTIM* has a larger register file and a bigger time delay than the *RHFTIM* system, it is more effective in improving the system reliability. From Fig. 9 and Fig. 11, we see that it takes four times more redundancy in the *RHFTIM* system than in the *GRFTIM* to achieve the same reliability.

From the above analysis, we know that the *FTIM* has the advantages of simple reconfiguration controller and high efficiency in spare utilization. If the value of λ is less than 0.01, the *FTIM* will be the best choice among all approaches. The proposed approach here can be generalized to designing fault-tolerant interleaved memory systems based on skewed or prime factor. Unlike the technique in [12], the way of interleaving is not modified by *SLARC* and every time there is a fault in the memory system, only the corresponding faulty *MU* is switched out. No bank is discarded and therefore, this technique can be used in other interleaved memory systems.

V. SUMMARY

A fault-tolerant interleaved memory system for highly reliable computer architectures has been presented. Fault tolerance is achieved by including hardware redundancy in the system to enchance the reliability. This system has two levels of redundancy. In the lower level, spare *MU*s are included in each bank to tolerate faulty *MU*s in the same bank. In the higher level, spare banks are used to replace faulty banks which do not have spare *MU*s left to replace faulty *MU*s. Based on our analysis, the system reliability of each of the three fault-tolerant interleaved memory systems, the *FTIM*, the *GRFTIM*, and the *RHFTIM*, is much higher than that of a nonredundant system. A simple and efficient reconfiguration controller designed specifically for a system with at most two spare rows and two spare columns is given. For memory system with more than two redundant rows, a second-level address remapping controller is presented. These techniques are very simple and have less hardware overhead cost and time delays than other previous techniques.

REFERENCES

[1]   K. Hwang, "Advanced parallel processing with supercomputer architectures," *Proceedings of the IEEE*, vol. 75, pp. 1348-1379, Oct. 1987.

[2]   D. K. Pradhan and etc. and D. K. Pradhan and etc., and J. P. Hayes, *Computer architecture and organization.* Englewood Cliffs, NJ: McGraw Hill, 1984.

[3]   R. Duncan, "A survey of parallel computer architectures," *IEEE Computer*, pp. 5-16, Feb. 1990.

[4]   H. S. Stone, *High-performance computer architecture* Reading, MA: Addison-Wesely, 1990.

[5]   D. P. Bhandarkar , "Analysis of memory interference in multiprocessors," *IEEE Trans. Comput.*, vol. c-24, pp. 897-908, Sept. 1975.

[6]   D. J. Kuck and R. A. Stokes, "The Burroughs scientific processor (BSP)," *IEEE Trans. Comput.*, vol. c-31, pp. 363-376, May 1982.

[7]   D. T. Harper III and J. R. Jump, "Vector access performance in parallel memories using a skewed storage scheme," *IEEE Trans. Comput.*, vol. c-36, pp. 1440-1449, Dec. 1987.

[8]   D. H. Lawrie and C. R. Vora, "The prime memory system for array access," *IEEE Trans. Comput.*, vol. c-31, pp. 435-442, May 1982.

[9]   C. -L. Chen and C. -K. Liao, "Analysis of vector access performance on skewed interleaved memory," *Proc. International Conference on Parallel Processing*, vol. 3, pp. 387-394, 1989.

[10]   G. Burnett and E. G. Coffman , "A study of interleaved memory systems ," *Proc. AFIPS 1970 Spring Joint Comput. Conf.*, pp. 467-474, 1970.

[11]   M. S. Algudady, C. R. Das, and W. Lin, "Fault-tolerant task mapping algorithms for MIN-based multiprocessors," *Proc. International Conference on Parallel Processing*, 1990.

[12]   K. -C. Cheung, G. S. Sohi, K. -K. Saluja, and D. -K. Pradhan, "Design and analysis of a gracefully degrading interleaved memory system," *IEEE Trans. Comput.*, vol. c-39, pp. 63-71, Jan. 1990.

[13]   M. Wang, M. Cutler, and S. Y. H. Su, "Reconfiguration of VLSI/WSI mesh array processors with two-level redundancy," *IEEE Trans. Comput.*, vol. c-38, pp. 547-554, April 1989.

[14]   K. Hwang and F. A. Briggs, *Computer architecture and parallel processing.* McGraw Hill, 1984.

[15]   T. R. N. Rao and E. Fujiwara, *Error-control coding for computer systems.* Englewood Cliffs, NJ: Prentice-Hall, 1989.

[16]   B. W. Johnson, *Design and analysis of fault tolerant digital systems.* Reading, MA: Addison Wesley, 1989.

MEMORY SYSTEM FOR A STATICALLY SCHEDULED SUPERCOMPUTER

Chandra S. Joshi

Silicon Graphics Inc.,[1]
2011 North Shoreline Boulevard,
Mountainview, CA 94039.

Brad A. Reger

Silicon Graphics Inc.,[1]
2011 North Shoreline Boulevard,
Mountainview, CA 94039.

John R. Feehrer

Optoelectronic Computing
systems Center,[1]
University of Colorado,
Box 425, Boulder, CO 80309.

Abstract

Memory system design is crucial to both the performance and the cost of computer systems, and even more so in supercomputers. VLIW supercomputer memory systems share some important characteristics with more conventional systems, but they also differ in some important ways. The key influences on the design of VLIW memory systems are their underlying compiler technologies, the CPU's cycle time, and the target cost. The design trade-offs between architecture and implementation of such memory systems can not be derived from first principles, are not obvious, yet have profound effects on the system performance and cost. Memory systems augmented with advanced compiler techniques like memory disambiguation can reduce the design complexities without compromising performance. Hence such system designs can balance the performance, cost and capacity more effectively than those achieved by more conventional supercomputers.

In this paper we discuss the architecture and implementation trade-offs in the design of memory system for high performance VLIW machines, with specific details about the memory system design of Multiflow's Trace /500 supercomputer.

Introduction

Recent years have seen a rise in the development of machines whose architectures are heavily influenced by their targeted compiler technology[2], [4], [14], [15]. Multiflow's trace scheduling compiler played a major role in the design of the company's TRACE series of VLIW machines [1], [2]. Multiflow's VLIW machines use wide instruction words, resembling horizontal microcode, that directly control the operations of multiple register files, pipelined memory units, branch units, integer ALU's, and pipelined floating point units. All operations take a statically-predictable number of machine cycles to complete. Pipelining is exposed at the instruction set level, and all machine resources are managed by the compiler. The compiler performs static analysis of memory references and has complete knowledge about the memory

system parameters - total number of memory references that can be issued per cycle, the memory latency, the use of the system buses, total number of memory banks in the machine and the bank busy time. Since the Multiflow machines were primarily targeted for compute-intensive scientific and engineering markets, the memory system designs of these machines have been biased towards providing a sustained high bandwidth with large capacity. In light of the design choice of not having a data cache, special attention was paid to hide the effects of long memory latencies such as more registers, integrated register files for functional units and aggressive implementation techniques.

The continuing disparity between the CPU cycle time and the main memory cycle time makes the trade-offs in memory system design of high performance systems a very complex task. As an example, CRAY-2 design offered a less costly and large memory capacity at the expense of long bank busy times and long latencies, greater than 50 clock cycles. CRAY X-MP, on the other hand, used fast and expensive SRAMs, which optimized performance but not the cost and the total capacity. The memory subsystem supporting a VLIW supercomputer requires performance and features similar to those of a traditional supercomputer. However, such memory system designs can take advantage of the advanced compiler techniques in optimizing the performance and the cost.

This paper compares the merits and deficiencies of compile time vs. run time supported memory systems. Then we discuss what deficiencies or performance anomalies were encountered in the memory system design of the first commercial statically scheduled minisupercomputer TRACE/300 developed by Multiflow, and how these were addressed in the second generation TRACE /500 VLIW supercomputer. Besides the compiler-related architectural improvements, we also discuss the trade-offs/enhancements made relating to disparity between the CPU speed and the main memory, how they affected the scalar and the vector performance, the capacity and the cost. Reference [1], [2] describe in detail the architecture and the implementation of these machines.

---

1. This work was done between 1987 and March 1990, while the authors were at Multiflow Computer.

Memory Disambiguation and hardware support

Statically scheduled machines, like Multiflow's TRACE series VLIW machines, rely heavily on their compiler's capability of analyzing conflicts involving multiple memory references. Multiflow's Trace scheduling compiler used the *memory disambiguator [2], [13],* which passed judgement on the feasibility of simultaneous memory references. It performed two types of analysis - one that would answer the question, when a loop is unrolled, for example, whether a store to A(I) and a load from A(I+J) can be reordered. That is, it tries to answer the data dependency between the loads and the stores of two different basic blocks or between two different unrollings of a given loop. The second type of analysis the compiler performed was relative *bank disambiguation,* whether the two memory references could have a conflict, modulo the number of banks. The *disambiguator* returns one of the following three answers in either type of analysis.

i) No - The compiler then schedules those references as tightly as it can and the operations happen at full memory bandwidth.

ii) Yes - The compiler does not schedule these memory operations simultaneously and it tries to find some other useful work to fill these slots.

iii) Maybe - These situations arise where indirect addressing is used - references of the form A(B(I)), common in sparse matrix computations. These situations are also common in the case of references to two arrays passed in as arguments to a subroutine (so that their base addresses are unknown).

For memory reference patterns which fall in the first category, both a static and a dynamic scheduler will perform identically as there are no conflicts. A good description of dynamic scheduling is presented in [8]. For memory access patterns falling in the second category, when there is an actual conflict between two references, a dynamic scheduler can not work its way around the conflicts. A compiler, however, can look beyond the basic schedule and find something useful to fill these slots created due to the spreading of these memory operations. In the worst case scenario, with no additional work to do, the compiler puts mnops (multi-cycle nops) in these slots and the performance degrades to a level similar to that of a dynamic scheduler.

For situations falling in the third category, when the *disambiguator returns MAYBE,* a dynamic scheduler has an advantage as it only sees the real conflicts. In the case of disambiguation involving a LOAD and a STORE, the static scheduler has no choice but to schedule these references

such that no memory hazards exist. In Multiflow's TRACE / 300, this problem was aggravated in the wider machines as they could do a maximum of two 64-bit stores per cycle and only two of the four clusters in the TRACE 28/300 could perform stores. To achieve full memory bandwidth, the compiler had to fill half of the available bandwidth with loads along with two stores- requiring disambiguation of these loads with stores. Memory disambiguation of these loads with stores was not always possible as normally the loads and the stores were accessing different arrays. We also found that these unbalanced and asymmetric hardware resource constraints made a negative impact on the quality of the compiled code and even if we achieved the optimum code quality, it required considerable amount of effort and time. In essence, since the compiler already does an enormous amount of work for these statically scheduled machines, making the hardware simple and symmetrical helps compiler in producing better quality code.

In analysis involving relative bank disambiguation with the disambiguator returning its answer as potential bank conflict involving two memory references, a certain amount of hardware support is needed as the performance penalty could be very high if the bank busy times are long. Again, the dynamic scheduler has a slight advantage here as it sees the real conflicts and not the potential. However, the compiler can use bank stall heuristics and trade off lengthening of the schedule with less memory bank conflict. The Multiflow's TRACE /300 had hardware support for bank stall but not card stall. That is, the machine stalled when a reference was made to an already busy bank. However, the hardware could not support multiple memory references to the same card in a single clock cycle (*card conflict*). The bank stall mechanism alone was sufficient hardware support for the "MAYBE" answer of the disambiguator for the TRACE 7/300, which could issue only one memory operation per clock cycle. So, if the compiler "rolled its dice" and scheduled these operations anyway, if conflict occurred, the machine stalled for (bank busy time -1) cycles, but the program functioned correctly. With the lack of the hardware support for card conflicts, the compiler was forced to schedule in wider TRACE 28/300 machines, four undisambiguated load or store operations in four clock cycles instead of a potential of one. On a statistical average, the memory efficiency in such situations reduced by 50% compared to having hardware support that could resolve card conflicts at run time. This hardware support would stall the machine for (total number of requests to the same card -1) cycles.

We fixed both of these performance anomalies of TRACE / 300 in TRACE /500. Use of advanced VLSI technology and denser packaging technology allowed for the implementation of four 64-bit stores per cycle in the TRACE /500. Design of the card-conflict resolving hardware is discussed later in the implementation section. Having the hardware support for a true multiported memory also allowed us to implement a unique capability of splitting a TRACE 28/500 into two 14 -wide processors and then later rejoining them in a single VLIW machine, under program control.

Memory reference batching

For vector operations, by loading the vectors in batches, a statically scheduled machine can achieve performance levels similar to or better than those achieved by dynamically scheduled machines. In a decoupled machine which has their address generation hardware decoupled from their functional units, an operation involving two vectors (a simple vector add operation as described in Equation 1) might proceed in the following manner.

$$C(I) = A(I) + B(I) \qquad \text{(EQ 1)}$$

Issue Load of vector A(I) and Load of vector B(I) simultaneously in a decoupled machine

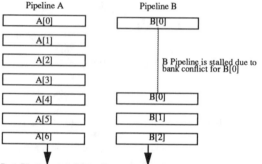

For a statically scheduled machines, use of load batching to reduce bank conflict

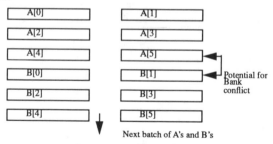

FIGURE 1. Static Vs. Dynamic Scheduling

For a decoupled vector machine like the CRAY X-MP[9], the two vector loads of A[I] and B[I] are issued simultaneously followed by a vector add operation. If there is a memory conflict between the two vector accesses, the decoupled design of the hardware stalls one of the address generation pipelines until the time that the conflict is resolved. In the pipeline diagram, Figure 1, load of vector B[I] is stalled for four clock cycles and from then on both vectors A[I] and B[I] proceed at the rate of one vector ele-

ment every clock cycle. In this mode, hardware only stalls once for the whole vector operation. The decoupled hardware staggers the two vector LOAD operation at run time such that their bank access patterns do not cause any conflicts. This is achieved at the expense of designing complex decoupled address and execution units which synchronize via some scorboarding mechanism. A similar level of performance can be achieved in a statically scheduled environment by performing the LOADs of the two vectors in batches as shown in the Figure 1. The whole vector operation is divided into batches such that instead of performing a LOAD of A[0] followed by a LOAD of B[0], a number of LOADs of A[I] are done before any LOADs of B[I] are performed. There is a potential of bank conflict every time we switch from vector A to vector B but it does not happen for every element of the vector. This type of scheduling however requires a large number of registers that can hold the values of A[I], while the first element of B[I] arrives. Also, there is a latency penalty as the ADD operation can not proceed until both A[I] and B[I] are loaded in the register file. But, the throughput can come close to that achieved by a decoupled architecture. If the compiler can successfully *bank disambiguate these two load operations,* it can stagger these loads to avoid the bank conflict altogether. Again, for situations when bank disambiguation is not possible, it can "roll the dice" and use bank stall heuristics to reduce the performance penalty due to the conflicts. In the TRACE 14/500, the floating register file had 128 64 bit registers so that the compiler could use this technique to alleviate performance loss due to bank conflicts.

Design of the Parallel memory for a faster CPU

The worsening disparity between supercomputer CPU speeds and the main memory makes the memory design for a balanced system even harder. The memory subsystem has to not only meet the performance needs of the faster CPUs, but provide a large capacity as well. It is common knowledge that longer memory latency will directly affect the scalar performance, recent studies [3] have shown that vector performance could also degrade substantially due to long bank busy times. The study in [3] reported that the *memory efficiency for vector type of operations, is directly proportional to the number of interleaved banks in the machine and inversely proportional to the square of the bank busy times and also inversely proportional to the number of processors.* For most vector type of operations in statically compiled environments, the bank conflicts come into play when the compiler switches from issuing LOADs for one vector to the issuing of LOADs for another vector, typical of the code generated by the Multiflow compiler. The following simple statistical analysis predicts the memory stalls in this scenario and it matches the analysis of [3]. The analysis is valid for a 28-wide VLIW that could issue four memory load/store operations per clock cycle.

Vector Performance in a static scheduling environment

Assuming a machine that has n functional units and each of these functional units can perform one memory reference per cycle. The machine has m memory banks, with a bank busy time of b cycles. Considering a typical code sequence generated by the Multiflow compiler, a batch of references to one unit-stride vector A followed by a batch of references to another unit-stride vector B. If the batches are large enough, we need only model the probability of a stall at the transition. Figure 2 shows the reference patterns for a four cluster machine, TRACE -28 VLIW. Assuming a code sequence that is issuing 4 memory references per cycle to sequential banks with vector A[0] is aligned with memory bank 0.

Cycle	Cluster 0 references	Cluster 1 references	Cluster 2 references	Cluster 3 references
0	A[0]	A[1]	A[2]	A[3]
1	A[4]	A[5]	A[6]	A[7]
2	A[8]	A[9]	A[10]	A[11]
3	A[12]	A[13]	A[14]	A[15]
4	B[0]	B[1]	B[2]	B[3]

FIGURE 2. Static scheduling of two vectors

After cycle 3, we switch over to generating references for vector B, whose memory bank alignment is unknown. The worst possible stall in this case is $(b-1)$ cycles. This will occur if any of {B[0], B[1], B[2], B[3]} coincides with any of {A[12,13,14,15]}. This is the same as the condition that B[0] coincides with any of{A[9,10,11,12,13,14,15]}.

In general, there will be $(2n - 1)$ possible choices, assuming n functional units, for B which will cause a $(b-1)$ cycle stall. This has the probability of $(2n - 1)/m$. Similarly, a $(b-2)$ cycle stall will occur if B coincides with any of the n possibilities {A[5,6,7,8]}. This has the probability of n/m.

There is the same probability n/m for a $(b-3), (b-4),...,1$ cycle stall. Thus the total expected stall caused by a transition from A to B is

$$ES = ((2n-1)/m)*(b-1) + (n/m) * ((b-2) + (b-3) +.... + 1)$$

$$ES = ((b-1)/m) * (2n - 1 + (n/2)*(b-2)).$$

In this scenario also, as reported in [3], *memory efficiency drops down proportional to the square of the bank busy time.*

Although there have been a number of studies done [5], [6], [7], [17] which use more complex bank interleaving techniques to improve the memory efficiency for vector operations of arbitrary strides, these schemes require fairly complex address generation logic. Trying to adopt one of these schemes would have made the design not only more complex but more importantly add at least 1 -2 clock cycles to the already relatively long latency. We opted for not worsening our scalar performance by 10-20% to gain performance for vector operations that involved strides of more than 16 or 32. Also, most of these interleaved schemes, like Cydra's Pseudo-randomly interleaved memory [16], even out performance losses on powers of 2 strides for single memory reference streams. However, for multiple independent vector streams, simple modulo schemes can work better than some of the other interleaving schemes[12]. We used the relationship described in [3] and our simple analysis described above to increase the total number of interleaved banks to 256, to meet the memory performance needs of the faster CPU.

Scalar Performance

Memory latency directly affects the scalar performance of a computer system. Cache memories provide a low latency and can be very effective for scalar codes. However the caches fall short when the data sizes are large compared to the size of the cache, which is the case for majority of codes that are being run on supercomputers. Although statically scheduled VLIW's are probably one of the best architectures at hiding the memory latency, they also perform at a level completely dictated by the memory latency when executing certain type of scalar codes, e.g. traversing a linked list. Since the compiler does an enormous amount of work for statically scheduled machines, a shorter memory latency becomes an important design parameter for achieving good compile speeds. In light of our decision to not incorporate a data cache and to not use static RAMS for the main memory (for cost reasons), we looked at other places in the machine to enhance the architecture so as to minimize the effects of a longer memory latency.

To offset some of the affects of a relatively long memory latency, we worked hard at having more registers in both the integer and the floating register files. The partitioning and implementation allowed double the amount of integrated registers for two functional units in one place. Large number of registers, as discussed earlier, were provided to buffer up the values of a long vector due to the long memory latency. Secondly, large number of registers helped in keeping the CSE'd (common subexpression) values around. Additionally, the need to spill the registers is reduced for situations where the compiler becomes greedy for LOADS to take advantage of high levels of scalar parallelism. An integrated register file helped reduce the data movement between the functional units and between the functional units and the main memory. All these improvements in other parts of machine helped offset the effects of a longer memory latency.

There were couple of other small architectural enhancements that eliminated some of the side effects of long memory pipeline. A common strategy employed by the compiler is to schedule operations that exhibit long latencies, such as memory accesses, early in code sequences so that other useful work can be done in the meantime. When long loops are winding down, this often means that some memory loads were initiated that are discovered to be unnecessary shortly thereafter (after a conditional branch is taken). If there were a way to "crush" such memory operations, the machine resources that were reserved by the compiler at the time of initiation would become free again. We provided such an operation in the /500 for this purpose.

A simple addition in machines with multiple functional units would be to have the capability of providing a broadcast mechanism. The memory system provided a mechanism to do broadcast loads. A single load operation can load a value from memory into both floating register files, or alternately into all integer register files. This helps to avoid unnecessary copying of address pointers around the clusters.

TRACE /500 memory system - Implementation

The design parameters that influenced Multiflow's TRACE /500 memory system were the desired architectural enhancements discussed earlier in the paper, high bandwidth, low latency, low bank busy time, greater than a gigabyte of physical memory, low cost, air cooling and schedule. It became almost impossible to optimize all of the above parameters. The three parameters large capacity, low cost and air cooling precluded us from using static RAMs (SRAMs) as our main memory choice for RAMs - the traditional supercomputer choice for main memory. To help offset the risk associated with the clock rate, uncertainty in DRAM prices and our performance simulation results, we designed the memory system to use either 40ns access 1 Megabit TTL Bicmos DRAMs, or 60 ns access 1 Megabit TTL DRAMs. Figure3 shows a block diagram of the TRACE /500 system [1].

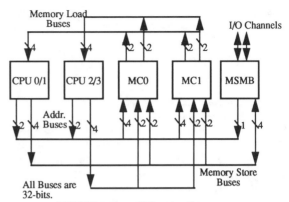

FIGURE 3. Trace /500 system Buses

Physical Organization

The memory system is interleaved across a maximum of 256 32-bit banks with a total main memory capacity of one Gigabyte. Each of the two memory controllers in the system can host up to 16 daughter cards. Each daughter card has eight interleaved banks, implemented with one of the two kinds of RAMs mentioned above. The memory system has four CPU address ports with each port capable of doing a 64-bit or 32-bit memory reference. The peak memory bandwidth for CPU accesses is two Gbytes/sec. and one Gbyte/ sec. for I/O and cache miss transfers. The memory system is organized as a set of 8 interleaved quadrants, each carrying a set of interleaved daughter cards with eight banks. Interleaving is on 32-bit boundaries, same as the earlier Trace / 300; the memory system ignores the BIW (byte in word) part of the address. The low order three bits of the address, after ignoring the byte in word, specify one of the eight quadrants, the next two bits specify one of the 4 the daughter cards in the quadrant and the next three bits specify the bank on a daughter card. A 64-bit memory reference requires access from both the even and the odd quadrant. The even quadrants 0, 2, 4,6 are housed in the even memory controller and the odd quadrants 1,3,5,7 are physically located on the odd memory controller.

Eight different memory configurations are supported in the / 500 with a minimum memory capacity of 32 banks with 32 Megabytes of memory, and a maximum storage of 1 Gigabyte of physical storage implemented across 256 banks. The system was designed to incorporate 4M DRAM devices, boosting the total capacity to 4 Gigabytes.

The minimum number of quadrants (0, 1, 2, 3) are physically installed across two memory controllers. The even quadrants are in the even memory controller and the odd quadrants are in the odd memory controller. Consequently, both memory controllers are required in a minimally-configured system. Figure 4 shows a block diagram of the memory controller.

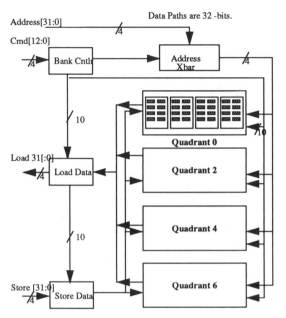

FIGURE 4. Even Memory controller

The Bank controller chip controls all the other chips (Load Data, Store Data and the Address crossbar) and also provides the controls for the daughter cards.

Latency and Bank-busy times

Due to the fast cycle time of the CPU, the memory latency and bank-busy time in clock ticks became worse in the TRACE /500 compared to the TRACE /300. As a result, the memory latency of the machine slipped to 12 cycles with 40ns access DRAMs and to 14 cycles with 60 ns access DRAMs. Since we opted for not having a data cache, we worked hard in other areas of the machine to keep the performance degradation due to the longer memory latency to a minimum. Integrated register files with more registers, broadcast load capabilities etc., as discussed earlier, helped in hiding the relatively long latency. Besides these architectural enhancements, we spent a great deal of design effort on optimizing the physical implementation of the memory pipeline. Use of latches instead of flops at couple of pipeline stages and about 2000 tuned wires and 3 months of routing effort, saved two cycles from the memory latency.

The other important memory design parameter, bank-busy time slipped to 6 cycles with 40ns access DRAMs and to 11 cycles with 60ns access DRAMs. TRACE /300 had a bank busy time of 4 cycles. As discussed earlier, to compensate for these relatively higher bank-busy times for keeping the

memory efficiency comparable to TRACE /300, we increased the total number of banks in the machine to 256 (TRACE /300 had 64 banks). Although, we had increased the maximum interleaving to 256 banks, with 1M x 1 DRAMs the memory offering would have been a Gbyte - an expensive alternative for users who could use the performance advantage of high interleaving without necessarily needing a Gbyte of memory. To make our memory offering more flexible, we decided to use 256K X 4 DRAMs and developed an SEC-DED and up to four-bit-nibble error-detect code for these DRAMs. The code we developed is a small subset of the code described in [11]. Some of the properties of this code are that it requires eight check bits for 32 bits of data; each check bit generation requires 12 data bits and each data bit is used in generating exactly 3 check bits. This worked out well for our implementation inside our gate arrays, since 12 bits can be gated with only two levels of 4-input XOR gates, about 2 ns. Adding 2 ns of ECC time to the 12-beat memory latency was a worthwhile trade-off for avoiding the design difficulties of dealing with ECC stalls of TRACE /300. In TRACE /300, ECC was done in parallel with the forwarding of the un-corrected data to the CPU's. If the data word had a single bit error, the CPU was stalled before it could use the data. The memory system then resent the corrected data to the CPU a couple of cycles later and then unstalled the machine. Also, once the machine went into the ECC stall, all the memory requests that were at different stages of the memory pipeline were allowed to drain. Making all of this logic work required a considerable amount of effort and having a 2 ns ECC path made the trade-off in favor of having flow-through ECC very easy.

A memory reference to a busy bank causes a bank stall and the performance is degraded proportionally to the number of cycles the bank remains busy before it can start the new request. The faster clock rate of the /500 makes the machine skid a couple of cycles before the memory controller's stall can take effect. Hence, all input address/command ports have a three-deep queue to hold all the requests due to this skid. Although this command buffering at the input ports is not as good as having it at the banks themselves, it did help our bank stalls as the two memory controllers work on their queues independently and thus avoid future memory stalls.

Card conflict resolution

The major architectural improvement to the /500 memory system was the ability to resolve, in hardware, multiple requests to the same quadrant (same as a memory controller on TRACE 14/300, a card). This ability was also key in making the machine to able to split and run as two independent 14-wide processors. The card conflict problem in the TRACE 14/300 arose because of two hardware constraints: first no two memory references could be directed to the same memory controller and, secondly, the memory controllers also shared the output buses and the compiler statically allocated all the resources including the use of all the system buses. The compiler had to serialize multiple memory references, if either the references were directed to the same controllers or they were targeted for different controllers who shared the output bus. Under worst case scenario the performance degraded to the level of a 7-wide machine instead of a full 28-wide VLIW.

Our choice of ECL (Emitter coupled logic) family as our implementation technology required that we do point to point buses which eliminated the constraint posed by bus conflicts. The idea of how to solve card conflict appeared when we were trying to solve the case of multiple requests that were targeted to different quadrants but to banks which were busy for a different number of cycles. Instead of waiting for all the banks to become free before starting any of the requests in a given cycle, to avoid future stalls we wanted to start a address port's request as soon as its bank became free. In order to execute the requests in order, we did not want to start another request from the same input address port until all the requests that came in a given cycle had been initiated. In the input logic we designed an address port disable logic such that a port was disabled, after its request had been initiated, until all the other port's requests that arrived in the same cycle had been started. In order to keep the load chips synchronized with the input logic, as every quadrant's request was started, a tag was sent to the load chips which carried this information in a pipelined fashion and used it to direct the data returning from the quadrants to the right set of CPU LOAD buses.

Once we had a solution for handling multiple memory requests to the same quadrant, we extended the memory system to support split 14-wide multiprocessor system [1]. The four address ports were split into two sets of two ports from each of the 14-wide CPU by having a separate stall network for each set. In multiprocessor mode, the input buffers and the LOAD buses associated with a 14-wide CPU were controlled independently and the memory stalls only affected the CPU whose requests were blocked. In this mode of operation, the hardware pushed two vector streams apart by stalling one of the processors, if they accessed the same bank, such that all future conflicts could be avoided - just like a decoupled architecture, as shown in top half of Figure 1. Scan based configuration could assign a higher priority to either of the two processors. In the default mode, the priority of the two processors toggled every cycle.

Cost

Early in the design process, we ruled out the use of ECL SRAMs (static RAMs) as our main memory choice for

RAMS, for several reasons. Within the realms of the packaging technology that we were considering, we would have been able to provide a maximum of 256 Mbytes of ECL static memory at a factory cost of approximately $250,000. Use of SRAMs would have brought us closer to the use of liquid cooling too. In contrast, for the same amount of money we could offer a Gbyte of memory with 256 interleaved banks with 40ns access Bicmos DRAMs. Our other possible choice for DRAMS, 60ns access 1 Mx1, DRAMs would have costed about $140,000 for a Gbyte of memory.

Conclusions

Design of a supercomputer memory subsystem, like any other design process, requires a very balanced approach. All the different parameters of a high performance memory system, the architecture it supports, the machine cycle time, latency, bandwidth, scalar performance, vector performance, total capacity and cost- all of them need to be evaluated simultaneously to design an effective memory subsystem. The Trace /500 memory system with large interleaving provides sustained high memory bandwidth for vector codes. The use of DRAMs for main memory enabled us to provide a large capacity at a low cost. Architectural improvements due to integrated register files with large number of registers helped in minimizing the effects of a relatively long memory latency of DRAMs. Use of static scheduling techniques like memory disambiguation and memory reference batching can be used very effectively in achieving high levels of performance. In summary, the balanced design of the Trace /500's memory system supports a statically scheduled VLIW that matches the performance of a single CPU CRAY Y-MP[1], offers very high capacity and at a fraction of the cost of the conventional supercomputers.

The major drawback of a completely statically scheduled machines is their lack of object code compatibility from one generation to the next, largely because memory technology does not scale (in terms of speed) at the same rate as the CPUs. With the product design cycle of a new computer system approaching two years, lack of object code compatibility becomes a big obstacle in the acceptance of any new computer system. The static scheduling techniques should be augmented with some interlocking techniques like register scorboarding to keep the object code compatibility across few generations of machines.

Due to the worsening disparity between the CPU speed and the main memory cycle time, bank busy time [3] becomes the key parameter that determines the vector performance of a computer. The adverse effects of long bank busy times can be offset by increasing the number of banks. However, increasing the number of independent banks in a machine beyond what we implemented (256), will be difficult for future memory system designs. CDRAMS (Cached DRAMs) is a potential technology that could provide virtually thousands of banks by implementing caches inside DRAMs. The access out of the cache is on the order of 10-20 ns, and these caches act like SRAMs, such that they are only busy for the time that they are being accessed, no precharge time penalty as encountered in standard DRAMs. The caches inside the DRAMs have large line sizes and it only requires one DRAM cycle to read the entire cache line. Even with a 2 ns CPU, the bank busy time will be on the order of 5 -10 clock cycles, not in hundreds- a better match to the CPU speeds. Although different interleaving schemes[12],[16] even out the performance for different strides and perform better for non unit-stride vector codes, techniques like use of CDRAMS and maybe multiple level memory hierarchies need be investigated to provide high throughput memory subsystems for multiple processors each with multiple memory reference streams.

With the advent of so called "killer micros", supercomputers are losing ground on scalar codes and use of data caches in supercomputer environment would be interesting to explore. The behavior of data caches for large data sets has not been extensively modeled until recently. Some of the recent studies[13] can provide some insights for the compilers into how to manage data caches for large applications. Here also, static scheduling techniques can be effectively used to improve the performance of a program[13]. A combined integrated approach, both hardware and software, as used in the design of the memory system for TRACE /500, should be considered for designing future generations of high performance memory systems.

Acknowledgments

We would specially like to thank Bob Colwell who provided innumerable suggestions for the earlier drafts of the paper. Shannon Hill along with the authors implemented the Multiflow memory system. Various discussions with Paul Rodman, Dave Papworth and Eric Hall helped crystallize some of the design parameters. Woody Lichtenstein provided the bank stall analysis. Greg Smith provided the cost analysis. We would also like to thank James Tornes who provided numerous simulation tools and verified the multiported operation of the memory system.

References

1. Robert P. Colwell, W. Eric Hall, Chandra S. Joshi, David B. Papworth, Paul K. Rodman, James E. Tornes, "Architecture and Implementation of a VLIW supercomputer," Supercomputing' 90, Nov. 1990, New York city, New York.

2. Robert P. Colwell, Robert P. Nix, John J. O'Donnell, David B. Papworth, Paul K. Rodman, "A VLIW Architecture for a Trace Scheduling Compiler," IEEE Trans. on Comp., V. 37, N.8, Aug. 1988.

3. David Bailey," Vector Computer Memory Bank contention," IEEE Trans. on Comp., V. 36, N. 3, March 1987.

4. J. A. Fisher, "Very Long instruction word architectures and the ELI-512," Proc. 10th Symp. Comput. Architecture, IEEE, June 1983, pp 140-150.

5. Shlomo Weiss, "An aperiodic storage scheme to reduce memory conflicts in Vector Processors," Proc. of 16th Annual Symp. on Computer Arch., pp 380, May 1989.

6. G.S. Sohi, High-Bandwidth Interleaved memories for Vector Processors - a simulation Study, Tech. Report 790, Computer Sciences Dept., University of Wisconsin-Madison, Sept. 1988.

7. D.J. Kuck and R.A. Stokes, "The Burroughs Scientific Processor (BSP)," IEEE Trans. on Comp. V. C-31, N. 5. May 1982.

8. James E.Smith. "Dynamic Instruction Scheduling and the Astronautics ZS-1," IEEE Computer, July 1989.

9.Tony Cheung and James E Smith. "A simulation study of the CRAY X-MP memory system," IEEE Trans. on Comp. V. C-35., N. 7, July 1986

10. J.R. Ellis, Bulldog, A compiler for VLIW Architectures, Cambridge, MA MIT Press,1986.

11.Shigeo Kaneda, "A class of odd-weight-column SEC-DED-SbED codes for Memory system Applications," IEEE Trans. on Comp., V. 33, N.8, August 1984.

12. Ram Raghvan and John P. Hayes, "On Randomly Interleaved Memories," Supercomputing'90, Nov. 1990,pp 49.

13. David Callahan, Allan Porterfield, "Data Cache Performance of supercomputer Applications," Supercomputing'90, Nov. 1990, pp 49.

14. J.L. Hennessy, N. Jouppi, F. Baskett, and J. Gill, "MIPS: A VLSI processor architecture," Proc. CMU Conf. VLSI Syst. Computat., Oct. 1981, pp 337-346.

15. B.R. Rau, D.W.L. Yen, W. Yen and R.A. Towle, "The Cydra 5 Departmental Supercomputer: Design Philosophies, Decisions and Trade-offs," Computer, Vol. 22 No. 1, January 1989.

16. B.R. Rau, M.S. Schlansker and D.W.L. Yen, "The Cydra 5 stride-insensitive memory system," Proc. International Conference on Parallel Processing, pp. 242-246, 1989.

17 David T. Harper III, J. Robert Jump, "Vector Access Performance in Parallel Memories Using a Skewed Storage Scheme," IEEE Trans. on Comp., V. C-36, N. 12, Dec.1987.

Message Vectorization for Converting Multicomputer Programs to Shared-Memory Multiprocessors*

Dhabaleswar K. Panda and Kai Hwang
University of Southern California
Department of EE-Systems
Los Angeles, CA 90089-0781

Abstract: We present a new concept of *message vectorization* as a replacement to the traditional mailbox-type scalar communication. A mapping methodology is provided to convert *send* and *receive* message operations into equivalent memory *write* and *read* access steps. Similar to the concept of vectorizing computational steps, our proposed methodology allows communication vectorization for memory-based message passing and leads to significant reduction in communication overhead. Three different multiprocessor configurations, single bus-based, crossbar-connected, and orthogonally-connected, are evaluated in their capabilities to support message vectorization through shared-memory. The results lay foundation for efficient and systematic conversion of message-passing programs written for multicomputers into multiprocessors using the shared memory.

1. Introduction

It is well known that a multicomputer program can be executed on a shared-memory system by some mapping methodology, where computing nodes of a multicomputer are mapped to processors of the shared-memory system and message-passing operations are carried out by using memory-based mailboxes [13]. However, no work has been done in investigating the effect of architectural parameters of a shared-memory system like interconnection network, memory access time, and data contention on the efficiency of such a mapping [2, 3]. Another challenge is whether this mailbox type of *scalar communication* is the best possible solution to the mapping problem at hand.

The crust of the problem depends on efficient conversion of *send* and *receive* operations in a multicomputer program to appropriate *write* and *read* memory operations on a shared-memory system. The mailbox type of scalar communication works well for programs involving only *permutation* type of communication. However, most parallel algorithms in scientific and numerical applications involve *broadcast, multicast,* and *personalized multicast* operations [6, 7]. The scalar communication, due to network and memory-access conflicts [9], limits the efficiency of mapping *broadcast* and *multicast* communication patterns onto a shared-memory system.

Interleaved memory organizations provide vector-oriented memory accesses and support vector/matrix data structures at the hardware level. These interleaved memory organizations associated with a smart *on-the-fly data manipulator* [11] have been demonstrated to support fast data manipulation [12] in shared-memory multiprocessors. In this paper, we take this vector-oriented approach in developing a general framework for mapping message-passing operations of any multicomputer program, irrespective of its underlying architecture, onto various shared-memory multiprocessor systems.

We consider a partitioned multicomputer program and define a set of *primitive communication patterns*. For these patterns, the mapping scheme emphasizes on the extraction of as many vector memory operations as possible, subject to the hardware operational constraints of a given shared-memory system. Using memory-based communication, we transform each primitive communication pattern to respective memory-based *send* and *receive* operations. We propose a two-step scheme for message vectorization. Using a sample program, we illustrate our concept for three different shared-memory multiprocessor configurations.

2. A Communication Model for Multicomputer Programs

Consider a multicomputer system consisting of a set of computer modules and a set of communication links. Each computing node is equipped with its own processing unit, local memory, and hardware for message communication over the message-passing network [1].

* All rights reserved by the authors on April 15, 1990. This research is supported by NSF Grant No. MIPS 89-04172 at the University of Southern California from 1989 to 1991. For all correspondence, contact Prof. Kai Hwang via Email kaihwang@panda.usc.edu, or Phone (213) 740-4470.

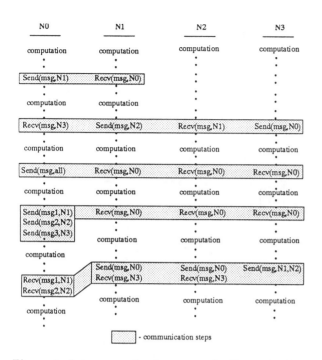

Figure 1: Inter-procedural communication steps in a sample program for a 4-node multicomputer (N0-N3: computing nodes).

We consider a partitioned multicomputer program [8, 13]. The procedures in the program are assumed to communicate using *send* and *recv* communication constructs. The procedures are assumed to be statically allocated to multicomputer nodes without any dynamic load balancing. We assume the message passing scheme to be *unblocked send* and *blocked recv*. The program is assumed to be deadlock free.

We consider converting a m-node multicomputer program onto an n-processor shared-memory system, where $m \geq n$. Though our methodology works for any value of m, our illustrations and examples in this paper are based on $m = n = 4$ for easy understanding. Figure 1 shows a sample program for a 4-node multicomputer. Computation and communication steps alternate in each procedure. With exchange of messages (through appropriate *send* and *recv* constructs), the procedures get synchronized and the program execution proceeds in a wave like manner.

The communication steps in Fig. 1 demonstrate different communication patterns. For an example, the first step is an *one-to-one (personalized)* message exchange. The second step is a *many-to-many (personalized)* message exchange. The third and fourth steps implement *one-to-all* broadcast and *all-to-one* operations, respectively. The last step combines *multicast* (broadcasting a message to selected others) and *many-to-one* operations.

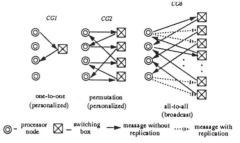

Figure 2: Architecture independent *communication graphs* (CGs) for selective primitive communication patterns.

We categorize the communication steps of any multicomputer program into various patterns. Table 1 lists all possible communication patterns together with their characteristics identifiers, C1–C12. Use the notation (Nx,Ny) to represent a single message transmission from node Nx to node Ny and ((Nx1,Ny1), (Nx2,Ny2)) to represent transmission of multiple messages from a set of source nodes, Nx1 and Nx2, to a set of destination nodes, Ny1 and Ny2. The communication steps can now be represented as a sequence of communication operations. For an example, the multicomputer program in Fig. 1 is represented by following program-level communication operations:

Step 1: $C1(N0, N1)$
Step 2: $C2((N1, N2), (N3, N0))$
Step 3: $C3(N0, N1, N2, N3)$
Step 4: $C4((N0, N1), (N0, N2), (N0, N3))$
Step 5: $C10((N1, N0), (N2, N0))$;
$\quad\quad C8(N3, N1, N2)$

It can be seen that some communication patterns are subsets of others. For an example, *many-to-many* pattern can be expressed as a subset of *all-to-all* with selective masking. Hence, we identify a set of patterns which are mutually independent, maximal in nature, and can be used for representing all other patterns. Such a set is identified as *minimal communication set*, $CS = \{C1, C2, C3, C4, C5, C6, C7\}$, where C1 to C7 are primitive communication patterns characterized in Table 1.

Each of these primitive patterns, independent of the underlying multicomputer architecture, can be represented as a *communication graph* (CG). Figure 2 shows unidirectional communication graphs CG1, CG2, and CG6. An appropriate number of switch boxes, equal to total number of messages, are used to provide required message transfers. The *indegree* and *outdegree* of each switch box is one. The communication graphs incorporate both *broadcast* and *personalized* type of message communication.

Table 1: Possible Multicomputer Communication Patterns in a m-node Multicomputer ($sr=$ number of sources, $ds=$ number of destinations, $1 < sr, ds < m-1$).

Identifier	Communication pattern	Type	No. of source nodes	No. of destn nodes	No. of distinct messages	Total no. of messages	Degree of replication
C1	one-to-one	personalized	1	1	1	1	1
C2	permutation	personalized	m	m	m	m	1
C3	one-to-all	broadcast	1	m-1	1	m-1	m-1
C4		personalized	1	m-1	m-1	m-1	1
C5	all-to-one	personalized	m-1	1	m-1	m-1	1
C6	all-to-all	broadcast	m	m-1	m	m(m-1)	m-1
C7		personalized	m	m-1	m(m-1)	m(m-1)	1
C8	one-to-many	broadcast	1	ds	1	ds	ds
C9		personalized	1	ds	ds	ds	1
C10	many-to-one	personalized	sr	1	sr	sr	1
C11	many-to-many	broadcast	sr	ds	sr	sr.ds	ds
C12		personalized	sr	ds	sr.ds	sr.ds	1

3. Vectorized Access in Shared-Memory Multiprocessors

In this section, three different multiprocessor configurations using interleaved shared memories are characterized. A data manipulator to carry out efficient vectorized memory access is emphasized. Concepts of structural connectivity graphs and the associated operational subgraphs are introduced to determine valid scalar and vector-oriented operations.

3.1 Interleaved Memory Organizations

Figure 3 illustrates three shared-memory configurations. Consider an n processor system with n memory modules as shown in Fig. 3a. The memory modules, $M_0, M_1, \ldots, M_{n-1}$, are connected to a single bus B_0. The processors, $P_0, P_1, \ldots, P_{n-1}$, are connected to the bus through n identical *memory controllers*, $MC_0, MC_1, \ldots, MC_{n-1}$. The memory controller supports *interleaved-read* and *write* accesses to the memory modules. Memories are n-way one-dimensionally interleaved. Each processor also has scalar access capability to any of the memory modules. The system provides fully shared-memory capability. An optional interprocessor interrupt bus is used to enable fast synchronization among the processors.

Figure 3b represents an n processor system with n^2 memory modules, $M_{0,0}, M_{0,1}, \ldots, M_{n-1,n-1}$. These memory modules are connected to n different buses. The processors are connected to the n buses through a system interconnect. Without any loss of generality, we assume a crossbar-connected interconnect for our analysis. This configuration supports one-dimensional memory interleaving and provides fully shared-memory capability.

Figure 3c shows an orthogonally-connected multiprocessor configuration with n processors, $2n$ buses, two-dimensional memory interleaving. This architecture concept was originally reported in [5]. A design implementation of this orthogonal multiprocessor was reported in [4]. A group of n row buses, $RB_0, RB_1, \ldots, RB_{n-1}$, are directly connected to the processors in the horizontal dimension. The remaining n column buses, $CB_0, CB_1, \ldots, CB_{n-1}$, are distributed across the two-dimensional memory organization in an orthogonal way. Consider indices in the range $0 \leq i, j, k \leq n-1$. A memory module $M_{i,j}$ is connected to two buses, RB_i and CB_j. Both RB_i and CB_i are controlled by the same memory controller MC_i. Both row and column buses support n-way interleaving across memory modules connected to them. This configuration allows a memory module $M_{i,j}$ to be shared between two processors P_i and P_j. But, only one of these two processors can access the memory module at a given time. This memory organization provides conflict-free memory access and makes the multiprocessor a partially shared-memory system.

3.2. Scalar and Vector Memory accesses

Consider a conflict-free scalar memory access by a processor P_i to a memory module. As shown in Fig. 4a, there is a fixed time overhead α to initiate this memory access. This overhead includes time to activate the memory controller, the memory controller putting required address on the bus, address propagation delay on the bus, and memory access time. For a *read* operation, data element from the memory module is read out of the bus with a cycle time of τ. Similarly for a *write* operation, the data is written to the bus with a cycle time of τ. Thus, the time overhead for a single scalar memory access is $t_s = (\alpha + \tau)$.

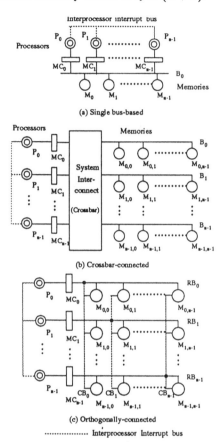

Figure 3: Three shared-memory multiprocessor configurations using interleaved memory organizations: (a) Single bus-based multiprocessor, (b) Crossbar-connected multiprocessor, and (c) Orthogonally-connected multiprocessor(P_i: Processor, M_i or $M_{i,j}$: Memory module, B_i: Interleaved access Bus, MC_i: Memory Controller).

Consider an *interleaved (vector) read/write* operation as shown in Fig. 4b. For a *read* operation, data elements from the memory modules are streamed out of the bus in every minor cycle with a pipelined cycle time of τ. Similarly for a *write* operation, data elements are streamed in via the bus first and then a *parallel-write operation* is activated. Thus, the overall time to perform a single *vector operation* is $(\alpha + n\tau)$.

Figure 4c shows the protocol for *block interleaved read/write* operations. In this protocol, a set of vectors are read(written) to(from) consecutive words of the memory modules in consecutive interleaved cycles. We assume an overhead of β to increment subse-

quent addresses of the memory modules. For a block with l vectors, this block read/write operation takes $(\alpha + \beta(l-1) + n\tau l)$ time. Using the current bus technology, one can have $\tau = 50$ nsec, $\alpha = 800 - 1000$ nsec and $\beta = 100$ nsec for $n \le 32$.

3.3 Vector Memory Access Using a Data Buffer Manipulator

Consider pipelined data transfer operations during an *interleaved-read* access. The n elements are loaded from n memory modules to a set of n data buffers associated with the processor. Similarly, during a *store* operation, the data elements are read from different buffers and written to the memory modules. Let e, $0 \le e \le n - 1$, be the indices of these buffers. Consider any permutation or mapping ρ with the set of indices $E = \{0, 1, \ldots, n - 1\}$.

As the data elements are transmitted (or received) to (or from) the interleaved bus in a pipelined manner, the source (or destination) buffers can be selected based on ρ. We define ρ as an *index set* and the operation of selecting appropriate buffer during a load/store operation as *indexing*. We provide a novel index manipulator hardware [12], where index sets are generated off-line during compile time as specified by the programmer and stored in fast *index memories* local to the processors. During each memory access, a desired index set is selected from the index memory to implement required data buffer manipulation.

Figure 5a shows the functional organization of an example data buffer manipulator. The interleaved bus is assumed to have 4 memory modules. The processor has 4 data buffers. Figure 5b shows the data buffer manipulation scheme during an *interleaved-read* operation. As the data elements are read from the pipelined bus in sequence, they are written into the buffers 0,3,1, and 2 respectively under the control of an appropriate index set. Similarly, during an *interleaved write* operation, the elements are read from the buffers 1,1,2, and 0 respectively and written to the memory modules as shown in Fig. 4c. This feature provides a versatile mechanism for on-the-fly message replication and provides uniform overhead for implementing memory-based *broadcast* and *multicast* operations.

3.4 Structural and Operational Connectivity

We take a graph-theoretic approach to analyze various conflict-free scalar and vector operations in a given shared-memory multiprocessor. Consider a single bus-based system as shown in Fig. 3a. The connectivity between the set of n processors, the set

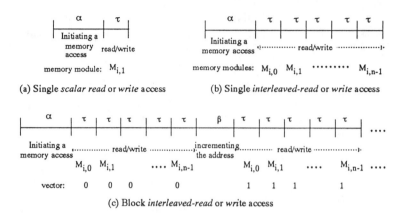

(a) Single *scalar read* or *write* access

(b) Single *interleaved-read* or *write* access

(c) Block *interleaved-read* or *write* access

Figure 4: Different protocols for implementing scalar and vector memory access operations.

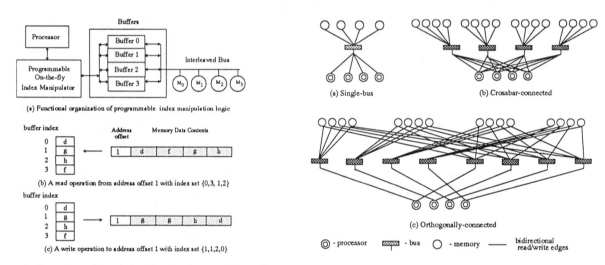

(a) Functional organization of programmable index manipulation logic

(b) A read operation from address offset 1 with index set {0,3, 1,2}

(c) A write operation to address offset 1 with index set {1,1,2,0}

Figure 5: Functional organization and operating principle of an example data buffer manipulator with four memory modules on an interleaved bus.

(a) Single-bus

(b) Crossbar-connected

(c) Orthogonally-connected

⊚ - processor   ▥ - bus   ◯ - memory   —— bidirectional read/write edges

Figure 6: Structural connectivity graphs of three shared-memory multiprocessor configurations.

of n memory modules, and the bus B_0 can be expressed by a connectivity graph as shown in Fig. 6a. The bidirectional nature of the edges reflects both read and write capabilities. We define such graphs as *structural connectivity graphs*. Figure 6b and 6c show structural connectivity graphs for crossbar-connected and orthogonally-connected multiprocessor configurations, respectively.

These graphs do not reflect operational characteristics of the system. To analyze conflict-free memory operations, we introduce a concept of *operational subgraphs* for these graphs. Consider a single bus-based system as shown in Fig. 3a. Four different memory access operations: *scalar read, scalar write, interleaved (vector) read*, and *interleaved (vector) write* are possible on this configuration. An electrical connectivity is established between a processor, the bus, and the respective memory modules to implement each of these

operations.

For an example, if processor P_0 writes data to memory module M_1 during a cycle, there is a connectivity from P_0 to M_1 through the bus B_0. This connectivity constitutes a subgraph of the structural connectivity graph shown in Fig. 6a. We associate such subgraphs with *valid* (conflict-free) or *invalid* (conflict) operations in a system. The significance of these subgraphs is that each of the corresponding memory operation can be implemented in a single step. We denote these subgraphs as *operational subgraphs*. Figure 7 illustrates some operational subgraphs for a single bus-based system.

Valid operational subgraphs depend on the hardware operational constraints of a system. These constraints differ for scalar and vector accesses. Table 2 reflects the hardware operational constraints in terms of the *indegree* and *outdegree* for processor, bus, and memory nodes in an operational subgraph.

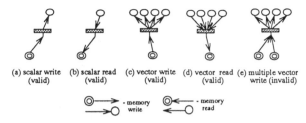

Figure 7: Operational subgraphs (valid and invalid) of a single bus-based system.

Definition 1 *A valid operational subgraph representing a scalar or vector memory access operation on a shared-memory multiprocessor X is a subgraph of its structural connectivity graph satisfying the constraints as shown in Table 2.*

Table 2: Graph-theoritic Constraints for Nodes in an Operational Subgraph (indeg= indegree, outdeg = outdegree).

Type of Nodes	Scalar model		Vector model	
	indeg	outdeg	indeg	outdeg
Processor	1	1	1	1
Bus	1	1	$> 1, \leq n$	$> 1, \leq n$
Memory	1	1	1	1

4. Mapping Communication Operations

A formal methodology is presented for mapping communication operations of a multicomputer onto shared-memory multiprocessors. The principle of message vectorization is emphasized by transforming the sample program onto three shared-memory multiprocessors.

4.1 Memory-based Scalar Communication

Consider the communication graphs in Fig. 2. Each switching box can be replaced by a shared-memory location. The sender processor writes the message to this memory location and the receiver processor reads it back. These two operations are scalar operations and take two time steps of t_s duration each. For a single bus-based system as shown in Fig. 3a, memory module M_j is used for passing messages from P_i to P_j, $1 \leq i, j \leq n$. For other two architectures, memory module $M_{i,j}$ is used for the same purpose.

These two *send* and *receive* operations can be represented graph-theoretically as *memory-based communication graphs* (MCGs). Each primitive communication pattern Ci can be converted into equivalent memory-based communication graphs $MCGi_{send}$ and $MCGi_{recv}$, respectively. Figure 8 illustrates the graphs for communication patterns C1 and C5.

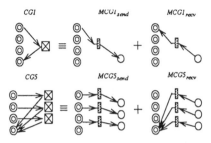

Figure 8: Transformation of communication graphs (CG) to memory-based communication graphs (MCG).

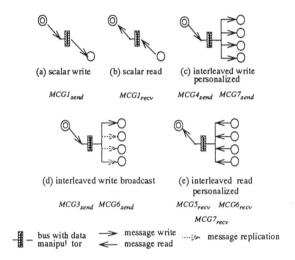

Figure 9: Functionality of data manipulator to support single step memory access and functionally equivalent memory-based communication graphs (MCGs).

4.2 Vectorized Communication

Consider the memory-based communication graph $MCG5_{recv}$ as shown in Fig. 8. If all memory modules are connected to a single bus, the destination processor can perform an *interleaved read* operation to the memory modules and receive all messages in one vectorized memory access step. In the absence of vectorized memory access, this step is performed by multiple (as many as three for this example) scalar memory access steps. Thus, vectorized memory access provides efficiency in implementing communication-intensive patterns (C3–C7) on shared-memory multiprocessors. As we discussed in section 3.3, the data manipulator hardware provides uniform overhead for both *broadcast* and *multicast* operations. Figure 9 shows five different functionality of the data manipulator to support single step memory access. Functionally equivalent memory-based communication graphs are also identified.

4.3 The Mapping Methodology

The objective of message vectorization is to reduce overall communication overhead. This demands that maximum number of communication operations should be implemented in vectorized manner. The following 4-step methodology is proposed to convert a multicomputer program onto a target shared-memory architecture X.

1. Reducing communication operations of a multicomputer program in terms of primitive communication patterns in the communication set CS.

2. Determining a subset of the primitive patterns that can be implemented on X using only scalar access. Determining another subset which demonstrate potential for being implemented using vector access.

3. Transforming communication operations of a program into equivalent memory-based communication steps by using optimal combination of scalar and vector accesses.

4. Restructuring memory communication steps with appropriate synchronization primitives.

We emphasize on step 2 due to its crucial nature. We introduce *memory-based communication sets* (MCSs). For a given shared-memory architecture X, we define two sets, MCS_X^{scalar} and MCS_X^{vector}. MCS_X^{scalar} consists of *send* and *receive* operations of primitive patterns which can be implemented on the architecture X using single-step scalar access only. MCS_X^{vector} encapsulates *send* and *recv* operations which can be implemented using single-step vectorized access.

We assume the use of data manipulator hardware to support vectorized memory access. The operations demonstrating vectorization capability are $MCG3_{send}$, $MCG4_{send}$, $MCG5_{recv}$, $MCG6_{send}$, $MCG6_{recv}$, $MCG7_{send}$, and $MCG7_{recv}$. The remaining operations are implemented as scalar access. The communication graphs for operations demonstrating vectorization are replaced with equivalent graphs as given in Fig. 9. These equivalent graphs are checked for their operational subgraph properties in architecture X. This leads to:

1. $MCS_{bus}^{scalar} = \{C1_{send}, C1_{recv}\}$

2. $MCS_{bus}^{vector} = \{C3_{send}, C4_{send}\}$

3. $MCS_{crossbar}^{scalar} = \{C1_{send}, C1_{recv}, C2_{send}, C2_{recv}\}$

4. $MCS_{crossbar}^{vector} = \{C3_{send}, C4_{send}, C6_{send}, C7_{send}\}$

5. $MCS_{orthogonal}^{scalar} = \{C1_{send}, C1_{recv}, C2_{send}, C2_{recv}, C3_{recv}, C4_{recv}, C5_{send}\}$

6. $MCS_{orthogonal}^{vector} = \{C3_{send}, C4_{send}, C5_{recv}, C6_{send}, C6_{recv}, C7_{send}, C7_{recv}\}$

4.4 Transforming Communication Operations

Consider the sample program presented in Fig. 1. First the communication patterns C8 and C10 are reduced to primitive patterns C3 and C5, respectively. Now we transform this communication model with 5 primitive communication patterns (C1-C5) into different multiprocessors.

A. Single Bus Multiprocessor System:

For a single bus-based system, $C1_{send}$ and $C1_{recv}$ operations can only be implemented in scalar mode. However, $C3_{send}$ and $C4_{send}$ operations demonstrate potential for vector access. That leaves all other operations, $C2_{send}, C2_{recv}, C3_{recv}, C4_{recv}, C5_{send}$, and $C5_{recv}$, to be implemented as multiple $C1$ operations. Based on these constraints, communication operations of the sample program are transformed into memory-based communication steps of a single bus-based system as follows [3]:

Step 1: $C1_{send}(P_0 \rightarrow M_1)$, $C1_{recv}(M_1 \rightarrow P_1)$
Step 2: $C1_{send}(P_1 \rightarrow M_2)$, $C1_{send}(P_3 \rightarrow M_0)$,
$\qquad C1_{recv}(M_2 \rightarrow P_2)$, $C1_{recv}(M_0 \rightarrow P_0)$
Step 3: $C3_{send}(P_0 \rightarrow M_1, M_2, M_3)$, $C1_{recv}(M_1 \rightarrow P_1)$,
$\qquad C1_{recv}(M_2 \rightarrow P_2)$, $C1_{recv}(M_3 \rightarrow P_3)$
Step 4: $C4_{send}(P_0 \rightarrow M_1, M_2, M_3)$, $C1_{recv}(M_1 \rightarrow P_1)$,
$\qquad C1_{recv}(M_2 \rightarrow P_2)$, $C1_{recv}(M_3 \rightarrow P_3)$
Step 5: $C1_{send}(P_1 \rightarrow M_0)$, $C1_{send}(P_2 \rightarrow M_0)$,
$\qquad C1_{recv}(M_0 \rightarrow P_0)$, $C1_{recv}(M_0 \rightarrow P_0)$;
$\qquad C3_{send}(P_3 \rightarrow M_1, M_2)$, $C1_{recv}(M_1 \rightarrow P_1)$,
$\qquad C1_{recv}(M_2 \rightarrow P_2)$

B. Crossbar-connected Multiprocessor System:

Step 1: $C1_{send}(P_0 \rightarrow M_{0,1})$, $C1_{recv}(M_{0,1} \rightarrow P_1)$
Step 2: $C2_{send}((P_1 \rightarrow M_{1,2}), (P_3 \rightarrow M_{3,0}))$,
$\qquad C2_{recv}((M_{1,2} \rightarrow P_2), (M_{3,0} \rightarrow P_0))$
Step 3: $C3_{send}(P_0 \rightarrow M_{0,1}, M_{0,2}, M_{0,3})$,
$\qquad C1_{recv}(M_{0,1} \rightarrow P_1)$,
$\qquad C1_{recv}(M_{0,2} \rightarrow P_2)$, $C1_{recv}(M_{0,3} \rightarrow P_3)$
Step 4: $C4_{send}(P_0 \rightarrow M_{0,1}, M_{0,2}, M_{0,3})$,
$\qquad C1_{recv}(M_{0,1} \rightarrow P_1)$,
$\qquad C1_{recv}(M_{0,2} \rightarrow P_2)$, $C1_{recv}(M_{0,3} \rightarrow P_3)$
Step 5: $C1_{send}(P_1 \rightarrow M_{1,0})$, $C1_{send}(P_2 \rightarrow M_{2,0})$,
$\qquad C1_{recv}(M_{1,0} \rightarrow P_0)$, $C1_{recv}(M_{2,0} \rightarrow P_0)$,
$\qquad C3_{send}(P_3 \rightarrow M_{3,1}, M_{3,2})$,
$\qquad C1_{recv}(M_{3,1} \rightarrow P_1)$, $C1_{recv}(M_{3,2} \rightarrow P_2)$

C. Orthogonal Multiprocessor System:

---

[3] We use the following notations in our memory-based message communication: $(Px \rightarrow M_y)$ for *scalar write* operation from processor Px to memory module M_y, $(Px \rightarrow M_y, M_z, ..., M_w)$ for *vector write* operation, $(M_y \rightarrow Px)$ for *scalar read* operation by processor Px from memory module M_y, and $(M_y, M_z, ..., M_w \rightarrow Px)$ for *vector read* operation.

Step 1: $C1_{send}(P_0 \to M_{0,1})$, $C1_{recv}(M_{0,1} \to P_1)$

Step 2: $C2_{send}((P_1 \to M_{1,2}), (P_3 \to M_{3,0}))$,
$\qquad C2_{recv}((M_{1,2} \to P_2), (M_{3,0} \to P_0)$

Step 3: $C3_{send}(P_0 \to M_{0,1}, M_{0,2}, M_{0,3})$,
$\qquad C3_{recv}((M_{0,1} \to P_1), (M_{0,2} \to P_2), (M_{0,3} \to P_3))$,

Step 4: $C4_{send}(P_0 \to M_{0,1}, M_{0,2}, M_{0,3})$,
$\qquad C4_{recv}((M_{0,1} \to P_1), (M_{0,2} \to P_2), (M_{0,3} \to P_3))$

Step 5: $C5_{send}((P_1 \to M_{1,0}), (P_2 \to M_{2,0}))$,
$\qquad C5_{recv}(M_{1,0}, M_{2,0} \to P_0)$; $C3_{send}(P_3 \to M_{3,1}, M_{3,2})$,
$\qquad C3_{recv}((M_{3,1} \to P_1), (M_{3,2} \to P_2)$

Each of the shared-memory communication steps above gets implemented in a producer-consumer manner with explicit synchronization between them. We assume a static barrier MIMD hardware synchronization scheme [10] on the interprocessor interrupt bus. Each receiver processor executes a *wait-sync* operation on the sender processor before a *read* operation. Similarly, every sender processor executes an *activate-sync* operation after each *write* operation. Since we assume our original multicomputer program to be deadlock-free, this synchronization scheme ensures no deadlock while being converted to a shared-memory environment. These synchronization primitives are incorporated to the code appropriately before execution.

5. Conclusions

The major contribution of this paper lies in developing a systematic methodology to convert message-passing operations in a multicomputer program into vectorized memory access operations in a shared-memory multiprocessor. The mapping methodology widens the scope of shared-memory multiprocessors to run programs written for both multiprocessors as well as multicomputers. It lays foundation for efficient and economic program conversion from multicomputers to multiprocessors. The key concept of the mapping method is presented here. The readers are referred to two related papers [11, 12] for data manipulator hardware design and the scope of interleaved access for fast data movement.

In recent years, the development of scalable shared-memory multiprocessors with distributed model of memory is approaching to resemble message passing architectures in several ways. Similarly, memory-based interconnection schemes are being proposed to be used in a multicomputer environment due to less software overhead associated with memory-based message passing. Our results emphasizes on bridging the gap between these two models of parallel computing and opens up new research directions for architecture-independent and communication-independent program development.

The proposed mapping methodology greatly enhances the program portability from multicomputers to multiprocessors. Continued research efforts are encouraged to achieve the reverse mapping from multiprocessors to multicomputers.

References

[1] W. C. Athas and C. L. Seitz. Multicomputer: Message-passing Concurrent Computers. *IEEE Computer*, pages 9–25, August 1988.

[2] A. Baratz and K. McAuliffe. A Perspective on Shared-memory and Message-memory Architectures. In J. L. C. Sanz, editor, *Opportunities and Constraints of Parallel Processing*, pages 9–10. Springer-Verlag, 1989.

[3] K. Gharachorloo. Towards More Flexible Architectures. In J. L. C. Sanz, editor, *Opportunities and Constraints of Parallel Processing*, pages 49–53. Springer-Verlag, 1989.

[4] K. Hwang, M. Dubois, D. K. Panda, and et.al, OMP: A RISC-based Multiprocessor using Orthogonal-Access Memories and Multiple Spanning Buses. *ACM Supercomputing Conference, Amsterdam*, pages 7–22, June 11-15 1990.

[5] K. Hwang, P.S. Tseng, and D. Kim. An Orthogonal Multiprocessor for Parallel Scientific Computations. *IEEE TC*, C–38(1):47–61, Jan 1989.

[6] S. Lee and K. G. Shin. Interleaved All-to-all Reliable Broadcast On Meshes and Hypercubes. In *ICPP90*, pages III: 110–113.

[7] X. Lin and L. M. Ni. Multicast Communication in Multicomputer Networks. In *ICPP90*, pages III:114–118.

[8] B. Lint and T. Agerwala. Communication Issues in the Design and Analysis of Parallel Algorithms. *IEEE TSE*, SE-7(2):174–188, Mar 1981.

[9] C. Maples. A High-Performance, Memory-Based Interconnection System for Multicomputer Environments. *Supercomputing '90, New York*, pages 295–304.

[10] M. T. O'Keefe and H. G. Dietz. Hardware Barrier Synchronization: Static Barrier MIMD (SBM). *ICPP90*, pages I: 35–42.

[11] D. K. Panda and K. Hwang. Reconfigurable Vector Register Windows for Fast Matrix Manipulation on the Orthogonal Multiprocessor. *ASAP Conference, Princeton*, pages 202–213, Sep 5-7, 1990.

[12] D. K. Panda and K. Hwang. Fast Data Manipulation in Multiprocessors Using Parallel Pipelined Memories. *JPDC, Special Issue on Shared-Memory Systems*, June 1991.

[13] J. X. Zhou. A Parallel Computer Model Supporting Procedure-Based Communication. *Supercomputing '90, New York*, pages 286–294.

Allocation and Communication in Distributed Memory
Multiprocessors for Periodic Real-time Applications

Shridhar B. Shukla

Code EC/Sh, Dept. of Elect. & Computer Engineering

Naval Postgraduate School

Monterey, CA 93943-5000

and

Dharma P. Agrawal

Computer Systems Laboratory, Dept. of Elect. & Computer Engineering

North Carolina State University

Raleigh, NC 27695-7911

Abstract: This paper investigates task allocation and intertask communication in distributed memory multiprocessors to support task-level parallelism in periodic real-time applications. *Wormhole* routing, used in the second generation of these machines, makes the communication time a function of channel contention in the network rather than the distance between communicating tasks. We show that the oblivious contention resolution policy of wormhole routing leads to *output inconsistency* in which a constant output period is not guaranteed. To eliminate this behavior, we present a framework for *scheduled* routing in which communication processors provide explicit flow-control by independently executing switching schedules computed at compile-time. Schedule computation requires that channel utilization in the network be less than unity. We outline contention-based task allocation and path selection strategies to minimize such utilization and compare its performance with conventional strategies. Our approach is based on the communication requirements of an application expressed as a set of time bounds on intertask messages. Simulation results are presented to show that the proposed mapping procedure is more suitable for periodic real-time applications than the conventional approaches.

Key words: task allocation, routing, scheduling, multicomputers, real-time

1   Introduction

Mapping an application on distributed memory multiprocessors (also referred to as multicomputers) involves problem partitioning, task allocation, node scheduling, and message routing. These steps are interdependent. For example, the sources and destinations of messages, which affect their possible paths, are determined by task allocation. Similarly, task scheduling at a node affects the instants at which messages will be ready for transmission and, therefore, determines the contention for channels. In spite of such interdependence, researchers have typically investigated these steps separately, notwithstanding their combined effect on the final system performance. For example, routing algorithms are typically analyzed to optimize the network latency and not the performance seen at the application level [Dal87, Nga89]. Moreover, the traffic generated by individual nodes is assumed to be identical for all nodes and each node is assumed to communicate equally with all other nodes. While many scheduling and allocation strategies make use of deterministic models of computation, the routing algorithms are studied using statistically generated traffic patterns. Although such separate considerations may be justified for general applicability and to make the algorithms amenable to analysis, they are unjustified in application-driven real-time situations where the system level performance is of primary importance.

When input data arrives periodically and a sequence of processing steps must be carried out on each input set, as in artificial vision, task-level pipelining presents itself as an effective means of parallel programming [AJ82]. It refers to periodic invocation of a set of tasks. For its successful implementation, the mapping procedure must ensure that the multicomputer supports a processing rate equal to the input arrival rate. Second generation multicomputers, such as the Intel iPSC/2 and Ametek Series 2010 use *wormhole* routing [Dal87] for sending messages between nodes. This blocking variant of *cut-through* routing makes the time to send a message between two nodes relatively insensitive to the distance between communicating nodes. On the other hand, it depends primarily upon the path setup time, which is a direct function of the contention in communication channels. We show that the oblivious contention resolution of *wormhole* routing flow-control makes it unsuitable for periodic real-time applications. Moreover, when compile-time task allocation is used, the network communication pattern, and therefore, the channel contention is determined by task allocation. Therefore, we propose an integrated mapping strategy which incorporates the task system specification into both task allocation and message routing to enhance and guarantee system throughput. The organization of this paper is as follows. Task-level pipelining and the effect of *wormhole* routing on throughput along with the conditions that lead to *output inconsistency* are described in Section 2. Communication requirements for pipelining a set of tasks and scheduled routing are presented in Section 3. Task allocation and path selection to support scheduled routing are presented in Section 4. The reduction in channel contention by our approach over the conventional mapping approach is presented therein. We conclude with remarks on our approach and suggestions for further research.

2   Task-level Pipelining

A task-flow graph (TFG) is a directed acyclic graph whose vertices represent tasks and the directed edges represent messages between tasks. Each task is a set of operations executed sequentially. A message consists of data or information transferred between tasks. A TFG is specified as a 2-tuple of sets $\{S_T, S_M\}$. $S_T = \{T_1, T_2, \ldots, T_{N_t}\}$ is the set of N_t tasks. Each T_i is associated with c_i, the num-

Shridhar Shukla is supported in part by the Research Initiation Program of NPS and *Dharma Agrawal* is supported in part by NSF grant No. MIP-8912767.

ber of operations to be performed to execute T_i. $S_M = \{M_1, M_2, \ldots, M_{N_m}\}$ is the set of N_m messages between tasks. Message M_i contains m_i bytes and is sent from task T_{is} to task T_{id}. T_{id} cannot be started before M_i reaches it. S_M defines a precedence relationship over S_T such that $T_j \prec T_k$ if there is a path from T_j to T_k. This model treats identical messages from a task as distinct at the application level if they are destined for different tasks. In task-level pipelining, the TFG is executed periodically and a task sends messages to other tasks at the end, and not during, its execution. TFG execution begins when tasks with no preceding tasks, referred to as *input* tasks, begin upon receipt of an external signal such as input arrival. It is completed when all tasks that do not precede any other, called *output* tasks, complete. The start of a TFG execution by input arrival is called as an *invocation*. Depending upon the interval between the start of successive invocations, the task sizes, and precedence relationships among tasks, different tasks execute in different invocations at the same time.

Assume a link bandwidth B and processing speed s_i for task T_i. Let τ_c be the processing time for the longest task and τ_m be the transmission time for the longest message. In general, partitioning techniques attempt to minimize the communication overhead [AJ88]. In large-grain parallelism as represented by TFG's, it is reasonable to assume that the longest task is longer than the longest message. Maximum throughput is obtained when the input arrival period $\tau_{in} = \tau_c$. The minimum latency for an invocation can be computed in terms of the critical path length. Consider a $S_{CP} = \{T_{i_1}, M_{j_1}, T_{i_2}, M_{j_2}, \ldots, T_{i_k}, M_{j_k}, T_{i_{k+1}}, \ldots, T_{i_L}\}$ of tasks and messages such that T_{i_1} is an input task, T_{i_L} is an output task, and $T_{i_k} \prec T_{i_{k+1}}$, $1 \leq k \leq (L-1)$. The sum, Λ, of the task execution and message transfer times in S_{CP} is called the critical path length of the TFG if it is maximum among all such sequences and S_{CP} is called the critical path. Note that there may be several critical paths in a TFG. The minimum time for completion of an invocation is Λ. Let $t_f^j(.)$ denote the time at completion, in j^{th} TFG invocation, of a task or message transmission that appears in the parenthesis. Similarly, let $t_s^j(.)$ denote the start of such events in the jth invocation. Since all input tasks start executing simultaneously, the latency for invocation j, λ^j, is computed as $max(t_f^j(T_i)) - t_s^j(T_k)$ where T_k is any input task and T_i is any output task. If τ_{out}^j is the interval between completions of the jth and $(j-1)^{th}$ invocations, then

$$\tau_{out}^j = \tau_{in} + \lambda^j - \lambda^{j-1}$$

If invocation period is τ_{in}, the TFG is pipelined successfully if

$$\tau_{out}^j = \tau_{in} \ \forall \ j = 1, 2, 3, \ldots \tag{1}$$

Obviously, $\tau_{in} \geq \tau_c$; otherwise, there will be an infinite accumulation at the input of the slowest task. Eq. (1) states that the interval between arrival of two inputs must be the same as that between the generation of two outputs for any pair of successive invocations. If this condition is violated, the TFG pipelining is defined to have **output inconsistency (OI)**.

Task-level pipelining is illustrated in Fig. 1 by the example of Fast Fourier Transform (FFT) which is a highly regular, computationally intensive algorithm used in a variety of signal and image processing applications. We obtain the TFG for a 2-D FFT in terms of that of a 1-D FFT. 1-D FFT of an N point vector of inputs can be computed in $\log N$ stages of *butterfly* operations with $N/2$ *butterflies* per stage. Each stage is divided into p parallel tasks, with

$N/2p$ butterflies per task. When the tasks in stage i finish computations, they send their outputs to tasks in stage $(i+1)$. The TFG for a 16-point, 1-D FFT with $p = 4$ is shown in Fig. 1(a). The number of operations and bytes transferred are computed assuming 6 arithmetic operations per *butterfly* and 2 bytes per data item. The TFG for 2-D FFT uses a $2 \log N$ stages to transform a $N \times N$ matrix of inputs. N 1-D FFT's are computed for rows followed by another N 1-D FFT's for columns. Tasks in the first $\log N$ stages perform 1-D FFT's on all N rows with each task performing $(N^2/2p)$ butterflies. Tasks in stage $\log N$ send data to tasks in stage $\log N + 1$ in such a way that the second set of $\log N$ stages perform N column transforms. A TFG for a 16×16 2-D FFT with $p = 4$ is shown in Fig. 1(b).

2.1 Inconsistency in Output Generation for Wormhole Routing

Wormhole routing is a blocking variant of *cut-through* routing and imposes deterministic path selection via its routing function to guarantee freedom from deadlocks. When two messages require a link simultaneously, its flow-control hardware resolves contention using a *first-come-first-served* (FCFS) policy [Dal87]. For task-level pipelining, the latency for each invocation is determined by the delays encountered by messages in that invocation. Given a task allocation, the predetermined routing function determines the shared links and associated messages. Since messages of different invocations co-exist in task-level pipelining, this service policy results in delays that are unequal over successive invocations, causing OI. It results if the conditions given below are true with respect to the routing function, TFG messages, and the invocation period.

Claim: *Let M_1 from T_{1s} to T_{1d} and M_2 from T_{2s} to T_{2d} be two messages such that $T_{1d} \prec T_{2s}$ and all four tasks are in the critical path. If the paths assigned to M_1 and M_2 have a link in common and τ_{in} is such that $t_s^0(M_2) < t_s^0(M_1) + \tau_{in} < t_f^0(M_2)$, then output inconsistency is present.*
Proof: Refer to [SA91].

Although we have considered only a pair of interacting messages above, when a complex TFG is mapped on a large network, several messages may interact to cause OI. Even when path selection is sensitive to the network load and makes use of the multiple equivalent paths in the network, as in adaptive cut-through routing [Nga89], OI may result. For example, consider three messages M_1 from T_{1s} to T_{1d}, M_2 from $T_{2s}(= T_{1s})$ to T_{2d}, and M_3 from T_{3s} to T_{3d} such that $T_{2d} \prec T_{3s}$, and T_{2d}, T_{3s}, T_{3d} are in the critical path. Assume that task allocation makes messages M_1 and M_3 between adjacent nodes and M_2 between non-adjacent nodes. If message M_1 blocks the first link in one of the paths of M_2 and the adaptive flow-control commits it to a path that contains a link used by M_3, an argument similar to the one above can be given to show that OI results. Thus, messages which are less important to the completion of the current invocation may receive priority over others in such routing schemes because path selection and flow-control are *oblivious* to the application requirements.

3 Scheduled Routing

Scheduled routing removes this gap between application requirements and routing policies by explicit flow-control resulting from execution of node switching schedules (described later in this section) to guarantee constant throughputs [Shu90]. In this technique, path selection and flow-control are based on the communication requirements of a TFG derived by using Eq. (1) as follows. Consider $\tau_{in} = \tau_c$, giving the maximum throughput. Since each task is executed once every τ_c time units, it must receive messages from its preceding tasks every τ_c time units. To prevent an infinite accumulation in message buffers, each message must reach its destination within τ_c after its generation. For an *invocation* period of τ_{in}, each message must be delivered *once* every τ_{in} time units. Based on these observations, time bounds can be assigned over a single interval $[0, \tau_{in}]$ as follows. Consider a single TFG *invocation* starting at $t = t_0$. Since each task begins execution as soon as messages from all preceding tasks are received, M_i is available for transmission at $r_i^0 = t_f^0(T_{is})$ and must complete transmission by $d_i^0 = r_i^0 + \tau_c$. Thus, transmission of a message can proceed over an interval as long as the longest task execution. For each $[r_i^0, d_i^0]$ pair, assign two time instants, a release-time r_i and a deadline d_i, such that

$$r_i = r_i^0 \bmod \tau_{in} \tag{2}$$

$$d_i = d_i^0 \bmod \tau_{in} \tag{3}$$

M_i must be transmitted in the interval $[r_i, d_i]$ if $d_i > r_i$ or in the intervals $[0, d_i]$ and $[r_i, \tau_{in}]$ when $d_i \leq r_i$, so that its deadline can be guaranteed when the TFG is pipelined with period $\tau_{in} \geq \tau_c$. Scheduled routing guarantees these time-bounds by incorporating them in path selection and flow-control which is implemented as communication schedules executed by communication processors (CP's) at the multicomputer nodes independently.

We present a comparison between performance of *wormhole* and scheduled routing in Fig. 2 for the example of a 256×256 FFT on a binary 6-cube. (For other examples and performance comparisons, the reader is referred to [Shu90].) Our objective is to observe the OI due to *wormhole* routing for various values of load on the network and at the same time illustrate that scheduled routing is indeed feasible. We define normalized load as the ratio of the *minimum possible* input arrival period and the required input arrival period, *i. e.*, $\frac{\tau_c}{\tau_{in}}$. Thus, the maximum load corresponds to unity when the TFG is invoked at the rate $\frac{1}{\tau_c}$. Normalized throughput is defined to be the ratio $\frac{\tau_{out}}{\tau_{in}}$ where τ_{out} is the output generation interval. Processing speeds of AP's of the multicomputer have been selected in such a way that $\frac{\tau_m}{\tau_c} = 1$. Twelve different values of the input period (τ_{in}) are selected between its minimum value of τ_c and $5\tau_c$. Very large values of the input period are not interesting because messages from different invocations do not contend with each other. If wormhole routing results in OI, we indicate it by plotting an up-down *spike* for the corresponding load value to indicate that the throughput is not constant for all invocations at that input period. For scheduled routing, if a feasible schedule is obtained for a load value, $\tau_{out} = \tau_{in}$, giving a normalized throughput of unity. The spikes for wormhole routing indicate OI for the corresponding input periods. On the other hand, a feasible flow-control schedule exists for all the load points as indicated by a horizontal line for unity throughput. When a schedule exists, the throughput equals unity and the latency remains constant. This is the principal benefit of scheduled routing. In the rest of this paper, we formulate the mathematical programming problems resulting from scheduled routing and present task allocation and message communication strategies to support such routing.

3.1 Node Switching Schedules

In scheduled routing, messages are transmitted only when there is a clear path between the source and destination nodes. M_i can be transmitted completely if such paths are set up for a total duration of $(\frac{m_i}{B})$ time units during $[0, \tau_{in}]$. Since each message has an unobstructed path, the run-time flow-control overhead evaporates and deadlocks become a non-issue. Such transmission is made possible by determining a path for M_i not only in terms of its source and destination addresses, but also in terms of its deadline and release-time at compile-time. Consider a communication schedule, Ω, that specifies a switching schedule, ω_i, for each node N_i of the multicomputer. ω_i specifies the connections between input channels and output channels to be setup in the CP at node N_i over $[0, \tau_{in}]$. The node switching schedules are such that execution of ω_i at N_i results in a clear path for each message for the required duration within its timing constraints.

In such explicit flow-control, the communications processor (CP) at each node is required to store and execute its switching schedule ω_i. In Fig. 3, we show a functional block diagram of the CP. The controller sets up the cross-bar and multiplexer according to the schedule being executed. If the topology has degree n, each CP has $n + 1$ input and $n + 1$ output connections, n of which are to and from adjacent nodes. It also connects to the input/output buffers the application processor (AP) at that node. We assume that the CP has separate buffers for each input and output channel from the adjacent nodes. This allows the CP to simultaneously transmit and receive messages on separate channels for its resident AP. A message coming from a neighboring CP, however, can be sent on only one outgoing channel. Each ω_i is a set of commands that directs the CP at node N_i to set up such connections. By simultaneously executing these schedules at all CP's, clear paths are established for delivering each message within its timing constraints. Such simultaneous execution requires a certain degree of synchronization between adjacent CP's. Such synchronization can be easily achieved by way of periodically exchanged synchronizing messages [Shu90].

When the time available for transmission of a message is greater than its length, the message has a *slack*. Presence of *slack* can be used in computing ω_i's so that all message deadlines are met and no link capacity is exceeded in any interval over the input period. Thus, a message must be allocated to different intervals of $[0, \tau_{in}]$ based upon the links used and the timing constraints. We refer to this problem of partitioning a message among intervals of $[0, \tau_{in}]$ as the **message-interval allocation** problem. In addition, simultaneous availability of all the links for a message must be ensured by the node switching schedules of nodes in the message path so that unobstructed paths are set up. Therefore, all the message allocations to a single interval must be scheduled explicitly for transmission within the interval. We refer to this problem of ensuring simultaneous availability of links as the **interval scheduling** problem.

The distinct r_i and d_i values of all messages divide $[0, \tau_{in}]$ into non-overlapping intervals. Let $t_0 = 0 \leq t_1 \leq t_2 \leq t_3 \leq \ldots \leq t_K = \tau_{in}$ be the K distinct endpoints of intervals in which messages are available for transmission. Interval $[t_{k-1}, t_k]$ is denoted by Δ_k and a message is *active* in Δ_k if it is available for transmission during $[t_{k-1}, t_k]$. Let $A = [a_{ik}]$ be an $N_m \times K$ matrix, called the *message activity* matrix, such that $a_{ik} = 1$ if M_i is active in Δ_k and 0 otherwise. If there are N_l links in the topology, define an $N_m \times N_l$ *path assignment* matrix as $B = [b_{ij}]$ such that $b_{ij} = 1$ if message M_i uses link L_j and $b_{ij} = 0$ otherwise. For each message,

$$\sum_{k=0}^{K-1} a_{ik}(t_{k+1} - t_k) \geq \frac{m_i}{B} \qquad (4)$$

An equality in (4) indicates that M_i has no slack. A message with no slack utilizes the links in its path fully in its *active* intervals and no other message can use these links during these intervals without exceeding link capacity or missing a deadline. On the other hand, if (4) is an inequality, links used by M_i have room for carrying other messages during its *active* intervals. While it is possible to determine how much more data can be carried, the exact intervals in which other messages can be transmitted on these links cannot be determined until message-interval allocation is performed.

3.2 Message-Interval Allocation

If two messages share a link and are active in the same interval they must be considered together to determine their allocation to intervals. Also, two messages that share links and intervals independently with a third message must also be considered together. Therefore, if the message *activity* matrix A and the path assignment matrix B are given, the messages of a TFG can be partitioned into disjoint subsets based on the following definitions.

Definition: Two messages M_p and M_q are related if and only if either $\exists$ a link L_i and an interval Δ_k such that $b_{pi} = b_{qi} = 1$ and $a_{pk} = a_{qk} = 1$ or $\exists$ a message M_r which is related to both M_p and M_q.

Definition: A message set, S_R, is maximal if and only if $\forall M_i, M_j \in S_R$, M_i and M_j are related and $\not\exists M_k \notin S_R$, such that M_k and M_i are related.

The *transitivity* and *closure* of the relation between messages defined above partitions S_M into disjoint subsets. Let S_R^i denote the ith *maximal* subset. The maximality of each subset ensures that $S_R^i \cap S_R^j = \emptyset$ for $i \neq j$. Such subsets of S_M can be obtained by performing row operations on an identity matrix and the matrices A and B. Thus, the message-interval allocation problem can now be solved separately for all the maximal subsets of S_M.

Consider a maximal subset S_R with $Q_n = \{n_1, n_2, \ldots\}$ as the indices of messages in it. Also, let $Q_l = \{l_1, l_2, \ldots\}$ and $Q_k = \{k_1, k_2, \ldots\}$ be the indices of the links and intervals used by messages in S_R. Let x_{hj} denote the time for which M_{n_h} is transmitted in Δ_{k_j}, i. e., between $[t_{(k_j-1)}, t_{k_j}]$. A feasible allocation of each message to the intervals in which it is active is obtained by assigning values to x_{hj} such that

$$\sum_{k_j \in Q_k} a_{n_h k_j} x_{hj} = \frac{m_{n_h}}{B} \ \forall n_h \in Q_n \qquad (5)$$

$$\sum_{n_h \in Q_n} a_{n_h k_j} b_{n_h l_i} x_{hj} \leq t_{k_j} - t_{k_j-1} \ \forall l_i \in Q_l, k_j \in Q_k \quad (6)$$

Constraints (5) require that the total allocation for each message equals its duration at the given bandwidth. Constraints (6) require that the total allocation of all messages to a link in each interval does not exceed the interval length. This allocation problem is analogous to the one formulated in [LM81] to schedule periodic tasks on multiple processors. For each task, first the intervals (generated by the release-times and deadlines) are identified in which it is available for processing. The tasks are then allocated to processors in appropriate intervals so that the processing performed by each processor is less than its capacity in an interval. The similarity between the two formulations is

that the messages constitute periodic tasks and links constitute processors. The difference is that a message may be required to be processed by more than one link at a time, but a task must be processed by exactly one processor at any time. If the set of constraints (5)-(6) is feasible, a suitable message-interval allocation can be obtained. The values of x_{hj} given by a feasible solution are then used to initialize an $N_m \times K$ message-interval allocation matrix, $P = [p_{ik}]$, where p_{ik} gives the time for which M_i is transmitted in Δ_k.

3.3 Interval Scheduling

The outcome of message-interval allocation is information about how much of each message is transmitted in an interval. The solution guarantees that no link exceeds its capacity in any interval. We now recall that explicit flow-control requires all the links assigned to a message to be available simultaneously so that a clear path from the source to destination is available. Therefore, interval scheduling must now be performed for each interval used by each maximal subset. The objective of interval scheduling is to construct a schedule for each interval so that the links that are used by messages with non-zero allocations in that interval are indeed available simultaneously. This problem is one of preemptively scheduling tasks on a set of identical processors where a task may require simultaneous use of more than one processors. The set of links used in an interval could be considered as the set of processors and messages with non-zero allocations could be considered as the tasks. An integer programming formulation for this scheduling problem has been described in [BDW86] and is adapted to this application as described below.

Let $Q_n^k = \{n_1, n_2, \ldots\}$ be the indices of messages in a maximal subset that have non-zero allocations in Δ_k. Let $Q_l^k = \{l_1, l_2, \ldots\}$ be the indices of links used by messages in Q_n^k. We first define a *link feasible* set of messages.

Definition: A message set $Q_{lfs} = \{n_1, n_2, \ldots\}$ is link feasible if $\not\exists n_i, n_j \in Q_{lfs}$ such that M_{n_i} and M_{n_j} use the same link.

Clearly, messages in a link feasible set can be transmitted simultaneously. Let N_{lfs} be the number of link feasible sets possible for Q_n^k. Associate a variable y_j with j^{th} link feasible set representing the time for which all the messages in it are transmitted simultaneously. Let Q_i be the set of indices of link feasible sets of messages in Q_n^k which contain message M_i. Consider the integer programming problem of minimizing

$$\sum_{j=1}^{N_{lfs}} y_j \ s.t. \ \sum_{j \in Q_i} y_j = p_{ik} \ \forall i \in Q_n^k$$

If $\sum_{j=1}^{N_{lfs}} y_j \leq \Delta_k$, then the interval is schedulable; otherwise, the messages with nonzero allocations in Δ_k require more time than the interval length. Once again, this problem may be solved as a standard integer programming problem. The number of constraints is proportional to the number of messages in the set but the number of variables is $O(N_{lfs})$. Depending upon the particular path assignment for the messages in Q_n^k, the number of variables may be $O(2^{N_k})$ where N_k is the number of messages in Q_n^k.

4 Contention-based Mapping

For a given task allocation and path selection, the above integer programming problems must have a feasible solution for a feasible set of ω_i's (i. e., a feasible Ω). An optimal mapping in this case is one that leads to a feasible Ω if one exists. A feasible Ω is one that does not require any link of the network to exceed its capacity and ensures that no link is required by two messages at the same time over $[0, \tau_{in}]$. Thus, the mapping algorithms used should be driven by the objective to make the integer programs have feasible solutions. In this section, we define an appropriate objective function and present task allocation and path selection algorithms based on it.

In the presence of deadlines and release-times for individual messages, two different link utilizations may be defined.

Definition: Link Utilization, U_j^l, for link L_j is defined as the ratio of the sum of transmission times of all messages carried by L_j and the sum of lengths of all intervals in which at least one message is active on L_j. Thus, if Q_j is the set of values of index k such that $\exists$ a message M_i with $a_{ik}b_{ij} \neq 0$,

$$U_j^l = \frac{\sum_{i=1}^{N_m} b_{ij} m_i}{\sum_{k \in Q_j} (t_{k+1} - t_k)}$$

$U_j^l \leq 1$ indicates that the total transmission requirement of all the messages using link L_j does not exceed the total time in which it is used. Now consider three messages using a link L_j. Let two of the messages be without slack and require use of L_j in Δ_k. The lengths and timing constraints of all three messages together may be such that $U_j^l \leq 1$. But, the two messages without slack demand more capacity than the link can offer in Δ_k. This combination of no-slack messages using a link in the same interval gives a hot-spot on the link even though the third message has enough slack to make $U_j^l < 1$. A mapping must eliminate such hot-spots in the network. Therefore, *spot* utilization may be defined as below.

Definition: Spot utilization, U_{jk}^s, is defined as the number of no-slack messages using L_j in the interval $[t_{k-1}, t_k]$.
Thus, $U_{jk}^s = \sum_{i=1}^{N_m} a_{ik} b_{ij} \ \forall \ M_i$ for which (4) is an equality. Consideration of link and spot utilizations separately allows us to concentrate on the hot-spots created by no-slack messages for a given path assignment. A *spot* refers to a link-interval pair. Only one no-slack message can use a link in a given interval if its endpoints are the same as the deadline and release-time of the message. Therefore, the path assignment must be such that $U_{jk}^s \leq 1$ for all the spots. It should be noted that spot utilization is computed only from messages that have no slack. The utilization of a link in an interval by a message that has slack cannot be determined until the message-interval allocation is performed. However, U_j^l can be computed before such allocation. Therefore, an assignment can be rejected if the link utilization is greater than unity. It is not guaranteed that $U_i^l \leq 1 \ \forall \ i$ leads to a feasible message-interval allocation. Non-existence of a feasible message-interval allocation is detected if the integer program given in the next section is not feasible. Thus, the net utilization over $[0, \tau_{in}]$ of links used by messages with slack can be determined, but their utilization profile cannot be determined unless message-interval allocation is done.

To increase the probability of finding a feasible Ω, we look for assignments that spread the message traffic as evenly as possible by minimizing the peak values of utilizations. Thus, the quantity to be minimized by path assignment becomes:

$$U = \max(U_j^l, U_{jk}^s) \tag{7}$$

where the maximum is taken over all possible values of j and k. If $U \leq 1$, scheduled routing can be attempted; otherwise, the communication requirements exceed the network capacity.

The message sources and destinations are determined by task allocation. Conventional task allocation algorithms optimize various measures of communication cost such as cardinality [Bok82], dilation [SC90], and weighted network distance [Lo88]. All these measures attempt to place communicating tasks as close as possible. If the multicomputer uses a circuit-switched technique such as *wormhole* or *scheduled* routing, the communication cost is more sensitive to the path setup time rather than the path length. Therefore, the allocation must minimize the contention in the network.

Typically, after task allocation minimizes some measure of communication cost, several messages must travel more than one hop. Multicomputer networks provide multiple, equivalent paths between non-adjacent nodes. *Wormhole* routing imposes selection of a fixed path between a non-adjacent pair of nodes to guarantee freedom from deadlocks in its blocking path setup policy. In scheduled routing, deadlocks are impossible due to compile-time allocation of links. Therefore, a message can select a path from multiple alternatives to reduce channel contention further. Thus, the task allocation and subsequent path selection strategies minimize the peak utilization, U, as defined in Eq. (7) so that message-interval allocation and interval scheduling can be attempted.

4.1 Heuristics for Task Allocation and Path Assignment

The algorithm AllocateTasks of Fig. 4 proceeds by obtaining a constructive initial solution which is iteratively improved by pairwise interchange [HK72] with respect to the objective function Eq. (7). The initial solution is obtained by allocating tasks one-by-one to unoccupied nodes so that the resulting peak utilization is the lowest. This assignment is then improved incrementally by attempting pairwise exchange of task positions so that the peak is lowered. If the current allocation cannot be improved by such pairwise interchange, its peak is compared with that of the best allocation obtained so far. If it is smaller, the best allocation is updated. A new, random allocation is then generated and improved iteratively. If the peak utilization of the current is greater than the best one so far, AllocateTasks terminates. For each allocation, the *link* and *spot* utilizations are computed using the deadlock-free E-cube path assignment for intertask messages [Dal87].

Once AllocateTasks terminates, algorithm AssignPaths of Fig. 5 assigns paths for each multi-hop message so that the peak utilization is lowered further. The initial E-cube path assignment is improved in the following manner. If the peak is due to a *spot*, say (L_j, Δ_k), all the messages that are active in the Δ_k and use L_j are identified. When the peak is due to a link all the messages that use it over any of the K intervals are identified. For each multi-hop message among these, all the alternative paths are considered and the one that causes the largest drop in utilization is identified. Then, the message-path combination that causes the largest drop among all multi-hop messages is selected and the message is reassigned to its new path. If there are no messages with alternative paths using the link or spot with the peak, the assignment cannot be improved any further.

It may not be possible to change the peak utilization by reassigning any of the messages on (L_i, Δ_k) except to make a peak of the same value appear on a different link and/or spot. In this case, the reassignment that causes this change is made so that the algorithm moves to a different point in the link-interval space. AssignPaths reassigns a message even if the peak remains the same but appears on a different link or spot. This helps in reducing the peak when there are multiple appearances of the same peak value in different links and/or spots. Both AssignPaths and AllocateTasks *repeatedly* disturb the current solution randomly to escape the local minima. This technique has been shown to be effective in obtaining good solutions to computationally complex problems without using exhaustive search [Bok82].

4.2 Performance Results

We have obtained values of U of Eq. (7) using task allocation that minimizes the communication cost measured by the sum of message distances weighted by message sizes (Quadratic Assignment Problem [HK72]) followed by deadlock free E-cube path selection of *wormhole* routing and compared them with those obtained by using Allocate-Tasks followed by AssignPaths. The application selected is 256×256 FFT on a binary 6-cube and a 64-node 2-D torus. Scheduled routing treats the links of a network as processors, and therefore, networks with larger number of links tend to perform better. Since a 2-D torus has fewer links than a binary 6-cube, we select a link bandwidth, B, of 64 bytes/μsec ($\frac{\tau_m}{\tau_c} = 1.0$) for the latter and of 128 bytes/μsec ($\frac{\tau_m}{\tau_c} = 0.5$) for the former. As a result, the lowest value of the peak link utilization for the 6-cube is 1.0 and for the torus it is 0.5. This bandwidth value makes the longest message of the same duration as the longest task, and therefore, the minimum value of Eq. (7) becomes unity. In Fig. 6, we plot the peak utilization for QAP followed by E-cube mapping and AllocateTasks followed by AssignPaths mapping. As shown in Fig. 6(a), the proposed mapping strategy is able to reduce the peak to the minimum for all values of the load, whereas the QAP-E-cube mapping gives a peak utilization value that exceeds the link capacity in several cases. For the 2-D torus, our approach leads to utilizations lower than unity in all cases as shown in Fig. 6(b), whereas the QAP-E-cube mapping leads to hot-spots for several period values. Thus, scheduled routing, which requires that the peak utilization be unity or less, is supported by AllocateTasks and Assign-Paths while the conventional task allocation and routing strategies lead to hot-spots in the network. In Fig. 7, we plot the peak utilization for B = 96 bytes/μsec ($\frac{\tau_m}{\tau_c} = 0.67$) for binary 6-cube and 2-D torus. Or approach is able to eliminate the hotspots for all but three load points in case of the torus. It is possible that the contention-based mapping does not yield a feasible Ω if the required throughput is very high. This indicates that the network bandwidth is insufficient for the given TFG and that either the throughput requirements should be relaxed or problem partitioning should be refined to yield a TFG with lower communication requirements. It should be noted that the time-bounds used by scheduled routing are obtained in Eqns. (2) and (3) by making message transmissions spread over an interval equal to the longest task. Thus, this technique provides throughput guarantees by sacrificing the latency for individual data sets. This, however, is easily justified for periodic real-time applications.

5 Concluding Remarks

Wormhole routing, shown to be *output inconsistent* in this paper, can be refined by using virtual-channel flow control which allows a local, priority-based contention resolution at each node for contending packets [Dal90b]. Although the network throughput is enhanced, it cannot guarantee that a packet with the highest priority among contending packets at one node will be the highest priority packet at the next node. Such flow control may reduce the instances of OI, but unlike scheduled routing, does not guarantee a constant throughput. By using compile-time scheduling of links and distributed execution of node switching schedules, the proposed technique eliminates flow-control overhead. However, if run-time link scheduling is to be used, some level of flow-control overhead is necessary.

The routing technique proposed here is similar to that of input regulation to reduce contention [SLCG89] coupled with spatial time-division multiple access (TDMA) transmission [NK85]. An important difference is that a multicomputer network is not a broadcast communication medium. Therefore, complete knowledge in terms of computation and communication of the application is assumed to solve the resulting scheduling problems. Extension of the approach to cases where only stochastic knowledge of the application is available is a challenging area for further research.

We have presented a contention-based mapping strategy in which, *a priori* knowledge about the application requirements is exploited to compute a task allocation as well as path selection. In [SA91], such knowledge is used only for path selection. While both approaches are effective, it remains to be investigated at which stage of the mapping procedure such knowledge is used most effectively.

Since all the CP's execute their schedules independently, this technique is scalable to a multicomputer of any size if a set of feasible communication schedules can be computed. Scheduled routing takes the view that the links of the network are *message processors* and imposes constraints on the allocation and scheduling of these to obtain clear paths for messages. Therefore, the more the number of links in the network for a given number of nodes, greater is the possibility of obtaining a feasible Ω for a given bandwidth. The feasibility of a schedule also depends upon the effectiveness of task allocation and path assignment that determine the constraints set up for message-interval allocation and interval scheduling problems.

A necessary condition for existence of feasible solutions to these problems is absence of hot-spots, *i. e.*, the communication requirements of the application must not require any of the links in the network to exceed their capacity during any interval equal to the invocation period. Based on this requirement, we have proposed task allocation and path selection strategies that reduce hot-spots in the network. The performance of these algorithms clearly shows that they are more suitable for periodic real-time applications than conventional mapping schemes. We are currently investigating the applicability of this scheme to larger networks and applications in which complete knowledge of computation and communication requirements is not available.

References

[AJ82] D. P. Agrawal and R. Jain. A pipelined pseudoparallel system architecture for real-time dynamic scene analysis. *IEEE Transactions on Computers*, 31(10):952–962, 1982.

[AJ88] R. Agrawal and H. V. Jagadish. Partitioning techniques for large-grained parallelism. *IEEE Transactions on Computers*, 37(12):1627–1634, December 1988.

[BDW86] J. Blazewicz, M. Drabowski, and J. Weglarz. Scheduling multiprocessor tasks to minimize schedule length. *IEEE Transactions on Computers*, pages 389–393, May 1986.

[Bok82] S. H. Bokhari. On the mapping problem. *IEEE Transactions on Computers*, C-30(3):207–214, March 1982.

[Dal87] W. J. Dally. *A VLSI Architecture for Concurrent Data Structures*. Kluwer Academic Publisher, 1987.

[Dal90a] W. J. Dally. Performance analysis of k-ary n-cube interconnection networks. *IEEE Transactions on Computers*, C-39(6):775–785, 1990.

[Dal90b] W. J. Dally. Virtual-channel flow control. In *17th Annual International Symposium on Computer Architecture*, pages 60–68, 1990.

[DS87] W. J. Dally and C. L. Seitz. Deadlock-free message routing in multiprocessor interconnection networks. *IEEE Transactions on Computers*, C-36(5):547–553, 1987.

[HK72] M. Hannan and J. M. Kurtzberg. A review of the placement and quadratic assignment problems. *SIAM Review*, 14:324–342, April 1972.

[LM81] E. L. Lawler and C. U. Martel. Scheduling periodically occurring tasks of multiple processors. *Information Processing Letters*, 12(1), 1981.

[Lo88] V. M. Lo. Heuristic algorithms for task assignment in distributed systems. *IEEE Transactions on Computers*, pages 1384–1397, November 1988.

[Nga89] John Y. Ngai. *A Framework for Adaptive Routing in Multicomputer Networks*. PhD thesis, California Institute of Technology, 1989.

[NK85] R. Nelson and L. Kleinrock. Spatial TDMA: A collision-free multihop channel access protocol. *IEEE Transactions on Communications*, pages 934–944, September 1985.

[SA91] Shridhar B. Shukla and Dharma P. Agrawal. Scheduling pipelined communication in distributed memory multiprocessors for real-time applications. In *Proceedings of the 18th Annual International Symposium on Computer Architecture*, May 1991.

[SC90] K. G. Shin and M.-S. Chen. On the number of acceptable task assignments in distributed computing systems. *IEEE Transactions on Computers*, 39(1):99–110, January 1990.

[Shu90] Shridhar B. Shukla. *On Parallel Processing for Real-time Artificial Vision*. PhD thesis, North Carolina State University, Raleigh, NC 27695, 1990.

[SLCG89] M. Sidi, W.-Z. Liu, I. Cidon, and I. Gopal. Congestion control through input rate regulation. In *IEEE Global Telecommunications Conference and Exhibition*, pages 1764–1768, November 1989.

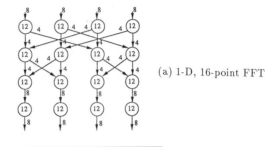

(a) 1-D, 16-point FFT

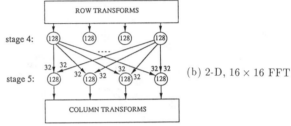

(b) 2-D, 16 × 16 FFT

Figure 1: Task flow graph for FFT, $p = 4$

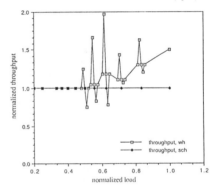

Figure 2: 2-D FFT on binary 6-cube, B = 64 bytes/μsec

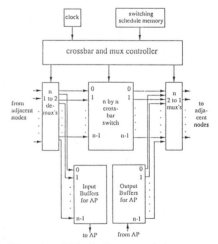

Figure 3: Functional block diagram of the communication processor

```
/*AllocateTasks: Allocate tasks to minimize U
assuming E-cube path selection*/
     compute initial allocation;
     /*by constructive assignment*/
     assign initial to best & current allocations;
     AFLAG = false; /*termination flag*/
     while not(AFLAG)
       begin
         do /*loop for iterative improvement*/
           begin
             IFLAG = false; /*false when cost of current*/
                 /*cannot be reduced*/
             for each task
               begin
                 /*selected in decreasing order of contribution*/
                 /*to peak of current allocation*/
                 examine task moves that reduce U of current;
                 /*pairwise exchange*/
                 if there is a peak reducing move
                   begin
                     update current to cause maximum reduction
                       by reallocating one task;
                     IFLAG = true;
                   end
               end
           end
         while (IFLAG);
         if peak(current) < peak(best)
           begin
             update best allocation;
             /*probabilistic jump to escape local minima*/
             /*randomly selected pairs of tasks are swapped*/
             shuffle current allocation randomly;
           end
         else AFLAG = true;
       end
```

Figure 4: Task allocation assuming E-cube path selection

```
/*AssignPaths: to minimize U*/
     compute E-cube assignment initial;
     set best & current to initial;
     AFLAG = false; /*termination flag*/
     while not(AFLAG)
       begin
         do /*iterative improvement*/
           begin
             IFLAG = false; /*false when current*/
                 /*cannot be improved */
             find L_i/(L_i, Δ_k) in current with peak U;
             find reroutable messages on L_i/(L_i, Δ_k);
             for each such message
               select a path with largest peak
                 reduction/peak repositioning;
               if ∃ a reroute to reduce peak
                 select the reroute with largest reduction;
               else select a reroute to reposition peak;
               if ∃ a reroute that changes peak(current)
                 select it to update current;
                 IFLAG = true;
           end
         while (IFLAG);
         if ((peak(current) < peak(best)) or
             (position(current) ≠ position(best)))
           begin
             set best to current;
             /*attempt to escape local minima*/
             assign paths randomly to each message;
           end
         else AFLAG = true;
       end
```

Figure 5: Path assignment heuristic

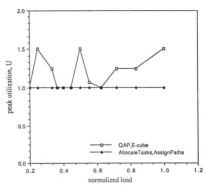

(a) Binary 6-cube, B= 64 bytes/μsec

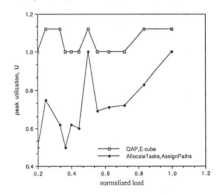

(b) 2-D Torus, B= 128 bytes/μsec

Figure 6: Maximum Link/Spot Utilizations

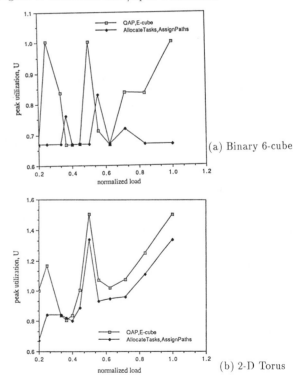

(a) Binary 6-cube

(b) 2-D Torus

Figure 7: Maximum Link/Spot Utilizations at B = 96 bytes/μs

I-219

Broadcast Networks for Fast Synchronization

Carl J. Beckmann
Constantine D. Polychronopoulos
Center for Supercomputing Research and Development
and Dept. of Electrical and Computer Engineering
University of Illinois at Urbana–Champaign
Urbana, Illinois, 61801

Abstract — This paper proposes a modification to packet–switched multistage interconnection networks (MINs) to allow unrestricted multicasting. We show how such a *broadcast network* can be used for efficient synchronization by eliminating hot spots due to spin-waiting. Simulation results show that hardware broadcasting not only decreases synchronization overhead, but also increases network performance for non-synchronization traffic. The hardware complexity of broadcast networks can be significantly less than *combining networks*, making them a more practical alternative for real systems. Their possible use in supporting distributed cache coherence is also discussed.

1. Introduction

The feature that busses possess – which allows relatively straightforward hardware cache coherence schemes – is the ability to broadcast writes to those caches with a copy of a particular memory location. This paper proposes a hardware enhancement to MIN switches which allows efficient multicasting to arbitrary sets of processors, and can be used for efficient synchronization by eliminating spin–waits over the network.

The broadcast network proposed in this paper is shown to reduce the adverse effects of spin–waiting in MIN–based machines: If memory writes signalling synchronization events can be broadcast, then processors can spin–wait on locally cached variables instead of repeatedly polling shared memory. This not only decreases overhead by immediately detecting synchronization events as they occur, but also frees useful bandwidth by eliminating spin–wait traffic.

The broadcast capability of the network described below has several important features. It provides a *full multicast* capability, i.e. any processor can broadcast to any arbitrary subset of processors. This is in contrast to the broadcast capability described in [Sieg85] and [SiMc81] which allows only certain subsets of processors as destinations, or full broadcasting as suggested for cache–based synchronization in [Berk88]. Furthermore, it is a *demand–based* strategy in the sense that processors must request those broadcasts they wish to receive. Finally, the source of a broadcast need not know the destinations, since packet routing is handled entirely by the network hardware. These features guarantee that broadcasting has minimal adverse impact on software, and more importantly, it does not require the operating system to explicitly partition the machine in order to take advantage of the broadcast capability.

2. Broadcast Network Operation

When a processor begins a synchronization, it performs a special memory read operation which establishes a virtual circuit between the memory module and the processor. This circuit is represented by routing bits in each switch, which are set if broadcasts are to be routed to the corresponding switch output, as shown in Figure

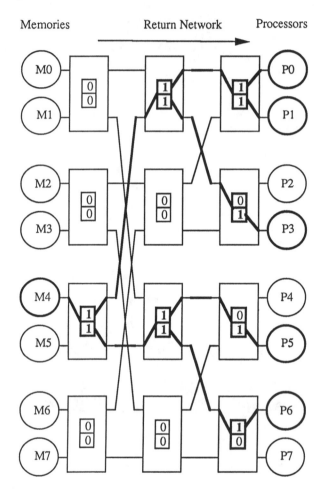

Figure 1. Routing Tree in Return Network

1. As processors enter a synchronization phase, they dynamically create these virtual circuits, building up *routing trees* in the memory–to–processor return network.

When a processor detects the occurrence of a synchronization, it notifies the waiting processors by performing a special memory write which triggers a broadcast. After leaving the memory, the broadcast follows the routing bits in the return network switches back to the waiting processors. These processors are thus notified immediately of the occurrence of the synchronization event, without having to continuously poll shared memory. The broadcast operation may also clear the routing bits on its way back to the processors, if the routing tree is not to be re–used.

Note that since the routing information is set up in the network by the waiting processors, the processor performing the broadcast need not know which processors it is broadcasting to. This information may be difficult to obtain in a dynamic scheduling environment, and also difficult to encode efficiently in a network packet.

Multiple Routing Trees— In order to differentiate between routing trees for different synchronization variables, a *broadcast tag* is associated with each variable. The broadcast tag is the address of the routing tree bits in the switches' routing tree memories or *routing tables*. Any broadcast operation must specify a broadcast tag in addition to the memory address of the corresponding synchronization variable.

Since the number of unique tags is equal to the number of memory locations in each switch routing table and the size of these tables is limited by hardware technology, broadcast tags are a limited system resource which must be allocated by the run–time system to ensure that multiple concurrent jobs do not use the same tags at the same time. In [BePo90] it is shown that in a P–processor system, up to $P-1$ barriers may be concurrently active, hence $P-1$ broadcast tags would be needed to implement barrier synchronization.

The Broadcast Switch— The switches in the return network can be implemented either with a queued–output design (Figure 2), or a queued–input design (Figure 3) more suited to wider switches. No modification is need to the forward network switches.

If the number of broadcast tags provided by the hardware is $O(P)$, then the amount of memory required for a routing table is comparable to that required for the switch buffers in an ordinary switch. For example, a 256–location table for a degree–two switch requires 512 bits of storage. This is the same amount of storage required, e.g., for two four–word, 64–bit FIFO buffers.

By contrast, *combining network* switches [GGKM83] [PPBGH85] [PfNo85] require complex combining queues, an ALU for computing intermediate combined results (e.g., for combining fetch–and–adds), and storage for combined messages. Also, due to the complexity of the combining queues, wide combining switches (usually with degree greater than four) are difficult to implement. Furthermore, the forward and return networks cannot be kept separate in combining networks. We therefore believe the hardware complexity and cost of broadcast networks to be significantly lower than those of combining networks.

3. Synchronization with a Broadcast Network

This section shows how a broadcast network can be used for efficient synchronization, and presents simulation results for both barrier synchronization (both linear and logarithmic) and mutual exclusion. Other types of synchronization are also discussed in [BePo91]. Two performance measures were of primary interest:

• The improvement of synchronization time, and

• the improvement of throughput–delay characteristics of simultaneous non–synchronization traffic.

The simulations assumed a dance–hall architecture with 64 processors and 64 memories interconnected via an omega network with degree four (4x4) switches. We assumed 32 bit wide data paths for the switches and memories. Each message requires up to four clock cycles on each link: 32 bits of control (including an 8–bit broadcast tag), 32 bits of address and up to 64 bits of data.

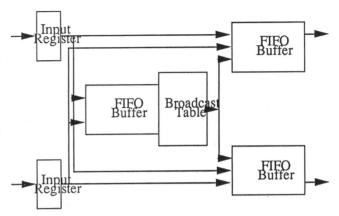

Figure 2. Queued–output Switch

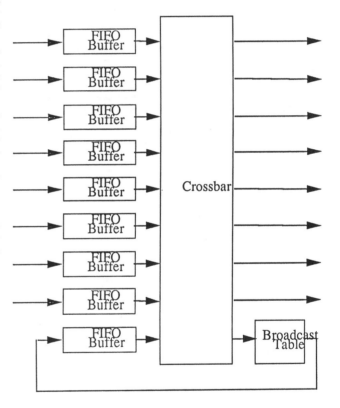

Figure 3. Queued–input Switch

The network was packet–switched with *cut–through* pipelining [ReFu87] for efficiency. Routing table access time was conservatively modelled as three clock cycles. The switches use the queued–output design of Figure 2. Memory cycle time was optimistically modelled as four clock cycles (the time required to send or receive a message on a network link), which tends to favor the non–broadcast network. The simulators were built using the tools developed as part of CSRD's CHIEF simulation environment [BCVY90] [Beck91].

In the experiments conducted, a given number of processors participated in synchronization while the remaining processors generated random background traffic. At each clock cycle, each of the background processors generated a memory read to a random memory module with a given probability. The probabilites 0.01, 0.02, 0.05, 0.1 and 0.2 were used to generate the five different background traffic rates for most of the results presented below.

In the algorithms presented in this section the prefix `req_bc:` indicates that the following memory operation adds this processor to the corresponding memory location's routing tree. The clause `broadcast:` indicates that the following memory operation is a broadcast, and `bc_clear:` indicates a broadcast which also clears the corresponding routing tree. The call `wait_bc`(*variable, value*) repeatedly receives broadcasts to *variable*, returning control when *value* is received.

3.1. Linear Barrier Synchronization

In synchronizing at a counter–based barrier such as the one shown in Algorithm 1, each processor first checks into the barrier by performing a fetch–and–add on the barrier variable. During the fetch–and–add the processor also appends itself to the routing tree rooted at that barrier variable. It then waits for a broadcast of the final write to the barrier, indicating the last processor has synchronized. The last processor to perform the fetch–and–add clears the barrier and broadcasts the event to all participating processors, simultaneously erasing the routing tree to avoid unnecessary future broadcasts. In Algorithm 1, the variable `num_proc` holds the number of processors synchronizing at the barrier.

Algorithm 1. Broadcast Network Barrier Algorithm

```
my_barrier = fetch_and_add(
               req_bc: barrier, 1)
if (my_barrier .lt. num_proc)
     wait_bc(barrier, 0)
else
     bc_clear: barrier = 0
endif
```

Figure 4 shows simulation results comparing a broadcast network using Algorithm 1 with an ordinary network using a comparable non–broadcast linear barrier algorithm [ArJo87]. The curves show *barrier depth*, defined as the time between a processor's arrival at the barrier and when it is notified of its completion, averaged over all processors. The solid curves correspond to barriers without broadcasting, the dotted curves with broadcasting. The five curves in each set correspond to the five different background traffic rates.

Both algorithms show a roughly linear relationship between the number of processors and the barrier depth. However, the constant of proportionality (slope) is about three times higher in the non–broadcasting case. With broadcasting, the average slope is about four clock cycles per processor, which is equal to the memory cycle time. Without broadcasting it is about 13 clock cycles, or $3\frac{1}{3}$ memory operations: In the presence of broadcasting, each processor only performs a single fetch–and–add to increment the counter, while in the non–broadcasting case each processor apparently polls the synchronization variable an average of $2\frac{1}{3}$ times before synchronizing.

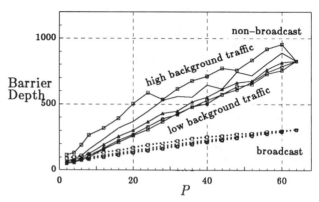

Figure 4. Linear Barrier Depth

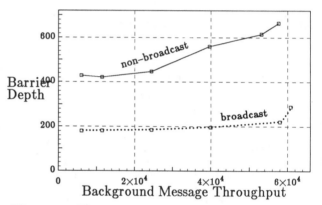

Figure 5. Barrier Depth versus Background Traffic

Effect of Background Traffic on Barrier Performance— Figure 5 illustrates the sensitivity of a 32–processor barrier on background traffic in more detail. Barrier depth is plotted against the actual throughput achieved by the background traffic. The throughput is computed as the total number of memory operations performed by the 32 background processors during a 10,000 clock cycle simulation. For comparison, the theoretical maximum throughput would be 80,000 operations (32 processors * 10,000 cycles / 4 cycles per packet). Not only is the barrier depth lower with broadcasting, but it also increases less rapidly with increasing background traffic.

Effect of Barrier Synchronization on Background Traffic— Figure 6 illustrates the effect of barrier synchronization on non–synchronization traffic. Again, 32 processors performed barrier synchronization while the other 32 generated random background traffic. The two curves plot the average network latency for a background memory operation, versus the total throughput achieved by the background traffic. The six points on each curve correspond to issuing probabilities of 0.01, 0.02, 0.05, 0.1, 0.2 and 0.3. At a generation probability of 0.2, the network achieved roughly 10% better throughput with 20% less latency with broadcasting. This figure clearly shows that background traffic is less affected by synchronization with a broadcast network.

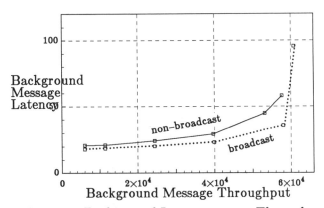

Figure 6. Background Latency versus Throughput

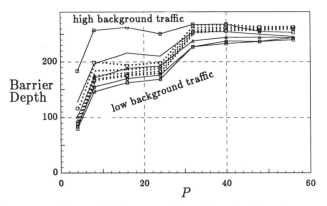

Figure 7. Degree 5 Tree Barrier Depth

3.2. Tree Barriers

In a *degree K* tree barrier, processors first synchronize in groups of K using linear barriers. The last processor to synchronize in each group participates in the next higher level barrier, and so on, until all processors have synchronized. A degree K tree barrier requires $\lceil \log_K P \rceil$ levels to synchronize P processors.

Figures 7 and 8 show barrier depth versus the number of processors synchronizing, for tree barrier synchronization, for $K{=}5$ and $K{=}7$, respectively. For low background traffic rates the ordinary network outperforms the broadcast network in both of these cases. This is attributed to the higher latency in the return network when setting up routing trees, which becomes more apparent for small linear barriers (see Figure 4). As the background traffic is increased the broadcast network outperforms the ordinary network, illustrating its superior robustness to background network traffic.

3.3. Mutual Exclusion

Efficient mutual exclusion can be performed with a broadcast network, using the *bakery algorithm* shown in Algorithm 2. Each processor takes a number using fetch–and–add, then waits its turn until the processor with the previous number is done.

Figures 9 and 10 show simulation results for mutual exclusion using Algorithm 2. Note that unlike test–and–set based algorithms, Algorithm 2 both is fair and starvation–free (even without hardware broadcasting). In these experiments, each participating processor coninuously executed a critical section containing two reads and one write to shared memory.

Algorithm 2. Mutual Exclusion
with a Broadcast Network

```
my_number = fetch_and_add(next)
temp = req_bc: serving
if (temp .ne. my_number)
    wait_bc(serving, my_number)
endif
```
critical section code
```
...
broadcast: serving=my_number+1
```

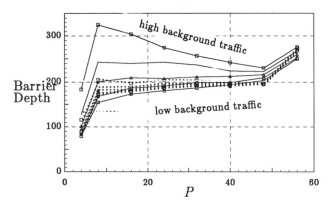

Figure 8. Degree 7 Tree Barrier Depth

Figure 9 shows the average critical section depth seen by each processor, i.e., the average time required for a processor to acquire, execute and release the critical section. The dotted curves, representing the broadcast network's performance, are straight lines indicating that the critical section depth is $O(P)$. The non–broadcast network shows super–linear depth, perhaps $O(P^2)$ (although this tapers off again as P becomes large). We again notice that the sensitivity of the broadcast network to background traffic is less severe.

Figure 10 shows the overall critical section throughput, i.e. the total number of times the critical section is executed by all processors, versus the number of processors synchronizing. With the broadcast network the throughput is roughly constant, independent of the number of processors. Without broadcasting the critical section throughput *decreases* as P increases; this is due to the network and memory contention from the additional processors.

This behavior is easy to explain. With broadcasting, each execution of the critical section requires a fixed number of memory accesses over the network. Thus, the amount of network traffic does not depend on the number of processors P, but only on the rate at which critical sections are executed. The critical section throughput should thus be constant, and the depth seen by each processor $O(P)$. Without broadcasting, $P{-}1$ processors spin–wait on the lock variable, so the average time for the owner of the critical section to hand it off to its successor is no longer $O(1)$ but $O(P)$. Thus, broadcasting can improve mutual exclusion by a factor of P for high–contention locks.

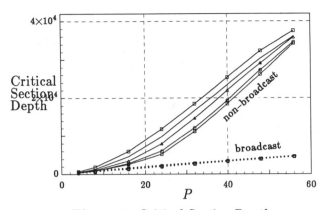

Figure 9. Critical Section Depth

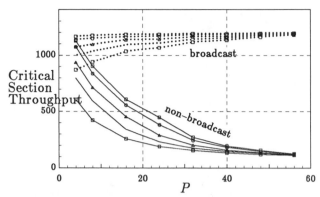

Figure 10. Critical Section Throughput

4. Other Uses for Broadcast Networks

In addition to synchronization, broadcasting could be used in parallel numerical algorithms in which data are shared by many processors, but only updated by one processor at any given time. Each processor initially caches a copy of the data in its local memory simultaneously adding itself to the data's routing tree. Whenever a processor updates any of the data, the updates are broadcast to all processors in the routing tree. Synchronization would still be required to guarantee the proper sequence of updates (and could be performed using broadcasting!), but redundant reads to shared memory after each update would be completely eliminated.

Taking the idea of controlled broadcasting further, one could envision using broadcast networks to implement general hardware cache coherence in MIN–based systems. Whenever a shared–writable atom is fetched from memory, a processor could automatically add itself to the routing tree for that atom. If write–through caches were used at the processors and the network treated all writes to shared atoms as broadcasts, then these writes would be received by all processors with a copy of the atom in their cache.

5. Conclusions

We suggest that broadcast networks offer a practical and cost–effective alternative to combining networks for reducing performance degradation due to hot–spots. Three features make them more amenable to hardware implementation:

- The forward and return networks can be kept separate, with no changes to the forward network.
- Wider switches are practical, resulting in shallower networks with low latency.
- A reasonable number of broadcast tags can be accomodated without an order of magnitude increase in the number of register/memory bits already required in each switch.

A more in–depth comparison can be found in [BePo91].

Broadcast networks have the potential of allowing hardware–enforced cache coherence to be implemented in MIN–based systems. However, the implementation and performance issues associated with this need to be explored in greater depth before the viability of this approach can be fully assessed.

6. Acknowledgements

This work was funded in part by Department of Energy grant DOE DE–FG02–85ER25001, National Science Foundation grants NSF MIP–8410110 and NSF–CCR–89–57310, and AT&T grant AT&T–AFFL–71–POLYCH, and by a Shell Doctoral Fellowship. Their support is gratefully acknowledged.

REFERENCES

[ArJo87] N. S. Arenstorf, H. F. Jordan, "Comparing Barrier Algorithms", ICASE Report No. 87–65, NASA contractor Report 178377, September 1987.

[BCVY90] J. Bruner, H. Cheong, A. Veidenbaum, P.–C. Yew, "Chief: A Parallel Simulation Environment for Parallel Systems," CSRD Report No. 1050, CSRD, University of Illinois, November 1990.

[Beck89] C. Beckmann, "Reducing Synchronization and Scheduling Overhead in Parallel Loops," M.S. Thesis, University of Illinois at Urbana–Champaign, August 1989.

[Beck91] C. Beckmann, "CARL: An Architecture Simulation Language," CSRD Report No. 1066, CSRD, University of Illinois, January 1991.

[Berk88] Wayne Berke, "A Cache Technique for Synchronization Variables in Highly Parallel, Shared Memory Systems," Ultracomputer Note #151, NYU, December 1988.

[BePo90] C. Beckmann, C. Polychronopoulos, "Fast Barrier Synchronization Hardware," *Proceedings Supercomputing '90*, pp. 180–189, New York, NY, November 1990.

[BePo91] C. Beckmann, C. Polychronopoulos, "Broadcast Networks for Fast Synchronization," CSRD Report No. 1070, Center for Supercomputing Research and Development, University of Illinois at Urbana–Champaign, January 1991.

[GGKM83] A. Gottlieb, R. Grishman, C. P. Kruskal, K. P. McAuliffe, L. Rudolph, M. Snir, "The NYU Ultracomputer — Designing a MIMD, Shared–Mmeory Parallel Machine (Extended Abstract)," *IEEE Transactions on Computers*, Vol. C–32, No. 2, February 1983, pp. 175–189.

[PBGH85] G. Pfister, W. Brantley, D. George, S. Harvey, W. Kleinfelder, K. McAuliffe, E. Melton, V. Norton and J. Weiss, "The IBM Research Parallel Processor Prototype (RP3): Introduction and Architecture," *Proceedings of the 1985 International Conference on Parallel Processing*, August 1985, pp. 764–771.

[PfNo85] G. Pfister and V. Norton, "'Hot Spot' Contention and Combining in Multistage Interconnection Networks," *Proceedings of the 1985 International Conference on Parallel Processing*, August 1985, pp. 790–797.

[ReFu87] D.A. Reed, R.M. Fujimoto, *Multicomputer Networks: Message–Based Parallel Processing*, MIT Press, 1987, p. 82.

[Sieg85] H. J. Siegel, *Interconnection Networks for Large–Scale Parallel Processing: Theory and Case Studies*, Lexington Books, 1985.

[SiMc81] H. J. Siegel, R. J. McMillen, "The Multistage Cube: a Versatile Interconnection Network," *Computer* 14, December 1981, pp. 65–76.

[YeTL87] P.–C. Yew, N.–F. Tzeng and D. H. Lawrie, "Distributing Hot–Spot Addressing in Large–Scale Multiprocessors," *IEEE Transactions on Computers*, Vol. C–36, No. 4, April 1987.

Arc-Disjoint Spanning Trees on Cube-Connected Cycles Networks

Pierre Fraigniaud
LIP - IMAG
Ecole Normale Supérieure de Lyon
69364 Lyon Cedex 07, France
pfraign@lip.ens-lyon.fr

Ching-Tien Ho
IBM Almaden Research Center
San Jose, CA 95120
ho@ibm.com

Abstract   We give an explicit construction of three arc-disjoint spanning trees in the cube-connected cycles. These disjoint trees allow to realize optimal broadcasting algorithms or fault-tolerant broadcasting algorithms. The speed-up of the broadcasting algorithm based on the arc-disjoint spanning trees over the naive one-spanning-tree algorithm is about a factor of three. The heights of these disjoint trees are optimal within a factor of two.

1   Introduction

Cube-connected cycle (CCC) network was introduced by Preparata and Vuillemin [14] as an attractive interconnection for general-purpose multiprocessors. They proved that these networks can perform the ascend/descend type algorithms with the same time complexity as the hypercube networks. While the hypercube topology has the logarithmic growth of the degree with respect to the number of the processors and is hard to scale up, the CCC topology has a bounded degree 3 and requires much less hardware and layout area. Thus, CCC appeared as a good candidate to the construction of large interconnection networks.

Parallel algorithms implemented on a distributed system often require intensive communications among the processors. One of the most frequently used communication primitives is *broadcasting* where a processor sends a message to all other processors in the network [9]. This paper mainly deals with solutions of the broadcasting problem on CCC.

Finding edge-disjoint spanning trees rooted at a given vertex allows to define efficient broadcasting algorithms ([9,11,3]) as well as fault-tolerant *one-to-all* or *all-to-all* broadcasting ([6,15]). Indeed, n arc-disjoint spanning trees (ADST's) allows to partition the original message into n packets of the same size and to broadcast them independently, using the disjoint spanning trees. For reliable communication, the original message is duplicated and sent through the different trees. Thus, at least one copy of the message is received by all processors even if up to $n - 1$ messages are lost due to link failure [6,15]. In this paper, we find three ADST's for the family of *cube-connected cycle* graphs, which is the maximum possible number.

The paper is organized as follows. Notations and definitions are described in the next section. We mainly recall the notion of compounded cycles into graphs. In Section 3, we present general properties of such a structure. Section 4 presents our main result, that is the construction of three ADST's of small depth in any CCC. We then present communication complexity of broadcasting in CCC based on these spanning trees in Section 5. In Section 6, we simply summarize our related results including the general CCC (of cycle length greater than the dimension) and folded CCC. Finally, Section 7 concludes the paper.

2   Definitions and Notations

Throughout the paper, Q_n denotes an n-dimensional Boolean cube or n-cube. The n-dimensional cube-connected cycle [14], denoted $CCC(n)$, can be viewed as an n-cube with each node being replaced by a cycle of n nodes and each node corresponding to one cube dimension.

A point-to-point interconnection network can be modeled as a graph G, where the vertex set $V(G)$ corresponds to the processors with local memories and the edge set $E(G)$ to the communication links. We assume that the communication links are full-duplex (i.e., bidirectional). Let G be an undirected graph, we shall denote G^* the corresponding symmetric digraph, i.e., each undirected edge (i, j) in G corresponds to two arcs (directed edges) in G^*. We will describe *arc-disjoint* spanning trees in the context of symmetric digraphs.

We denote $d(u)$ the degree of the vertex u. G is a d-regular graph if and only if $d(u) = d$ for all $u \in V(G)$. We also denote κ (resp. λ) the vertex (resp. edge) connectivity of G, that is the minimum number of vertices (resp. edges) need to be removed in order to disconnect the graph or reduce the graph to one vertex. Clearly $\kappa \leq \lambda \leq d_{min}$ where d_{min} is the minimum degree of G.

We recall the notion of *compound graphs* which is used, for instance, to construct large interconnection networks ([2]), or to obtain graphs which have good broadcasting time ([4]).

Definition 1 Let G be any graph. Replace each vertex u of G by a copy B_u of another graph B and join the copies using the edges of G (if there is an edge between u and v in G, then put an edge between B_u and B_v). The associated graph noted $B[G]$ is called the *compound* of B in G.

Note that compound graph is different from product graph in that the former has less edges than the latter (unless one of the graph has only one vertex). In the following, we shall consider only *compound of cycles* into graphs. Thus we can give a less general definition, specifying the way to interconnect the compounded cycles.

Definition 2 Let G be any graph. The associated compound of cycles into G, denoted by $C[G]$, is constructed as follows: Each vertex u of G is replaced by a cycle C_u of length $d(u)$: $C_u = \{(u, 0), (u, 1), \cdots, (u, d(u) - 1)\}$. Each vertex (u, i) of a cycle is connected in $C[G]$ to $(u, (i - 1) \bmod d(u))$ and $(u, (i + 1) \bmod d(u))$ on the cycle. For each edge (u, v) in G, we put an edge $((u, i), (v, j))$ in $C[G]$ such that each vertex of $C[G]$ has at most 3 neighbors: two on the cycle and one on another cycle. An edge on a cycle is called cycle-edge and an edge between two cycles is called G-edge.

Note that, for a given graph G, there are many ways of compounding cycles into it. For instance, consider the n-cube Q_n. The cube connected cycle $CCC(n)$ is obtained by replacing each vertex u of Q_n by a cycle C_u of length n. Moreover, if u and v are two vertices of Q_n which differ by only one bit i, then put an edge between C_u and C_v attaching this edge, respectively, to (u, i) and (v, i). Then $CCC(n) = C[Q_n]$.

We denote $C[G]^*$ the symmetric digraph associated with $C[G]$. In all that follows, a cycle arc of the digraph $C[G]^*$ is said *clockwise* if it is an arc of form $((u, i), (u, (i+1) \bmod d(u)))$, and *counterclockwise* otherwise.

3 Properties of compounded cycles

Clearly, the number of nodes in $C[G]$ is $\sum_{u \in V(G)} d(u)$, which is reduced to dN if G has N vertices and degree d. If the minimum degree of G is greater than two then the degree of any $C[G]$ is 3. However, the edge connectivity (and hence the node connectivity) in this case may be less than 3. Nevertheless, we can prove the following lemma based on the node connectivity of G. (see [7] for the proof.)

Lemma 1 Let G be a regular graph of degree $d \geq 3$. If the vertex connectivity of G is d, then the vertex connectivity of any corresponding $C[G]$ is 3. $\square$

Since the edge-connectivity of $C[G]$ is not greater than 3, it is not possible to construct two (undirected) edge-disjoint spanning trees in $C[G]$ (simply by counting the number of edges). However, it is possible to find three ADST's rooted at any given vertex $x \in C[G]^*$. We recall a weak version of Edmonds' Theorem [5].

Theorem 1 (*Edmonds*) *Let G be a directed graph with arc connectivity λ, and let u be any node in G. There exist λ arc-disjoint spanning trees in G all rooted at node u.* $\square$

Corollary 1 *Let G be a regular graph of degree $d \geq 3$ and vertex connectivity d. Let x be any node in $C[G]^*$. Then, there exist 3 ADST's in $C[G]^*$ rooted at node x.* $\square$

4 Arc-disjoint spanning trees in CCC's

In this section, we give constructions of three ADST's of $CCC(n)$ based on the ADST's defined by Johnsson and Ho [9] in the hypercubes. We will call G-edges the *cube-edges*.

In [16], Shiloach gives an algorithm to construct ADST's in any digraph. This algorithm is general and does not allow to minimize the heights of the trees. To generate ADST's for graphs that are in the same family but different sizes (such as Q_3 and Q_4) using Shiloach's Algorithm, it has to be executed for each size. Moreover, the time complexity of the algorithm is quite high: $O(k^2 N(m + N))$ where k is the number of spanning trees, and N and m are the number of nodes and arcs, respectively, in the digraph.

We now present the main result of this paper. We only give a sketch of the proof. See [7] for a complete proof.

Theorem 2 *There exist 3 ADST's of any $CCC(n)^*$ for $n \geq 3$, and the depths of these trees are at most $4n$.*

Remark: In comparison with the diameter of the $CCC(n)$ stated in [12], which is $2n + \lfloor \frac{n}{2} \rfloor - 2$, the maximum depth of the trees is close to the optimal.

Sketch of the proof : The existence of these 3 trees is given by Corollary 1. We give the construction rules of the 3 trees step by step.

Step 0: Tools.

Johnsson and Ho give in [9] the definition of the $nESBT$ (that is n Edge-disjoint Spanning Binomial Tree) in an n-cube. The $nESBT$ is composed of n trees T_j, $0 \leq j \leq n - 1$ described using a *Father* and *Children* functions for any node $u = (u_{n-1}, \ldots, u_0)$ in the n-cube. They proved that these n trees are arc-disjoint in Q_n^*. We shall use these trees to build 3 ADST's in $CCC(n)^*$. We give a construction of three ADST's rooted at $(0, 0) \in CCC(n)^*$, that is the vertex 0 of the cycle C_0. General constructions dedicated to vertices (u, i) follow by vertex transitivity.

Step 1: Initialization.

Let us define in Q_n^*, $S_0 = T_0$, $S_1 = T_1$ and $S_2 = \cup_{i \geq 2} T_i$. We also call S_0, S_1 and S_2 the corresponding sets of cube-arcs in $CCC(n)^*$. For convenience, we assume the cube-arcs of S_i are marked with color (i), $i = 0, 1$. Assume the cube-arcs of S_2 corresponding to the tree T_s, $s \geq 2$ are marked color $(2.s)$, where s forms a set of different "shades" of the same color (2). (The notion of shades will be used later.) We will give a coloring of some remaining cycle-arcs and modify coloring of certain cube-arcs to make up S_0, S_1 and S_2 as 3 ADST's rooted at $(0, 0) \in CCC(n)^*$.

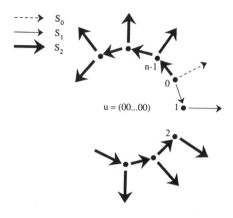

Figure 1: Coloring of the source at step 2.

Step 2: Connections on cycles.

It is possible to mark the cycle-arcs such that the incoming cube-arcs and the outgoing cube-arcs of the same color on each cycle are connected. On cycle C_0, the clockwise arc from $(0, 0)$ is marked with color (1). All the counterclockwise arcs, but those incoming to $(0, 0)$ and $(0, 1)$, are marked with color (2) (see Figure 1). Thus the connections are realized on C_0. On all other cycles $C_u, u \neq 0$, there are two incoming cube-arcs marked with colors (0) and (1), respectively, and $n - 2$ incoming cube-arcs marked with color (2). The distribution of the outgoing cube-arcs is different. But, following the functions *Father*(u, j) and *Children*(u, j) defined in [9], if the father of a vertex $u \in Q_n$ in the j-th tree of nESBT is connected through dimension k, then the children of u (if they exist) are the neighbors of u obtained by complementing bit position x of u where x (cyclically) starts from bit k to bit j and from right to left. Any other tree having its father connected through any dimension between k and j does not have any children. Thus we can color all intermediate clockwise arcs without conflict. We mark all the clockwise arcs of C_u with color (0) (resp.(1)) from the incoming cube-arcs of color (0) (resp.(1)) to the outgoing cube-arcs of the same color. For $i = 2, \ldots, n - 2$, we mark all the clockwise cycle-arcs with color (2) insuring the connection between incoming cube-arcs of color $(2.s)$ and the outgoing cube-arcs of the same color and shade.

For example, if $u = (0101010) \in Q_7$ (see Figure 2(a)), then we do not mark cycle-arcs with color (0) since u has no children in the tree T_0 of the nESBT of Q_n. The father of u in the first spanning tree is connected through dimension 5, and its children through dimensions 6, 0 and 1. Thus the cycle-arcs $((u, 5), (u, 6))$, $((u, 6), (u, 0))$ and $((u, 0), (u, 1))$ are marked with color (1). The father of u in the second, fourth and sixth spanning trees of Q_n^* are respectively through dimensions 2, 4 and 6, and u does not have children in these trees. The father of u in the third spanning tree is connected through dimension 1 and its children through dimensions 2 and 3. Thus, the cycle-arcs $((u, 1), (u, 2))$ and $((u, 2), (u, 3))$ are marked with color (2). Similarly, the cycle-arcs $((u, 3), (u, 4))$ and $((u, 4), (u, 5))$ are marked with color (2).

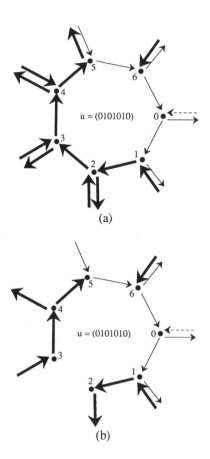

(a)

(b)

Figure 2: Node (0101010) after step 2 (a) and after step 3 (b).

Step 3: Removing redundant incoming arcs.

The sets S_0 and S_1 are two trees, but S_2 is not a tree. Indeed, some vertices can have two different incoming arcs marked with color (2). Here we remove the incoming cube-arcs in C_u of color $(2.s)$, satisfying: 1) u has no children in the s-th tree of the nESBT and 2) its symmetric cube-arc is marked with another shade of color (2). More precisely, we remove the cube-arc of color (2) incoming to vertices (u, i) if u has a child through dimension i in another tree $T_j, j \geq 2, j \neq i$.

For instance, if $u = (0101010) \in Q_7$ (see Figure 2(b)), then we remove the incoming cube-arcs to $(u, 2)$ and $(u, 4)$. Note that this modification also implies that the outgoing cube-arc from $(u, 3)$ is removed by applying these rules to cycle C_v where $v = (0100010)$ is the neighbor of u through dimension 3. Similar rules are applied to the outgoing cube-arcs from $(u, 5)$.

Next, we remove each clockwise cycle-arc of color (2), if it is incident to a node whose incoming cube-arc is also of color (2). For instance, if $u = (0101010) \in Q_7$ (see Figure 2(b)), then we remove the cycle-arc $((u, 2), (u, 3))$. At this point S_2 is a tree since it is a connected graph whose vertices have only one incoming arc.

Step 4: Making S_2 a spanning tree.

On C_0, we mark with color (2) the cycle-arc $((u, 2), (u, 1))$. On $C_u, u \neq 0$, at most 2 nodes have no incoming arcs of color (2), i.e., these 2 nodes have one incoming cube-arc colored with either (0) or (1). It is easy to give a strategy of coloring cycle-arcs such that all nodes have an incoming arc of color (2) and the height of S_2 is increased by at most one.

Step 5: Making S_0 and S_1 spanning trees.

We denote $|u|$ the parity of u, that is $|u| = 0$ (resp. 1) if the number of 1-bits in u is even (resp. odd).

Step 5.1: It is possible to color cycle-arcs such that:

- all the nodes of an even-parity cycle have exactly one incoming arc of color (1) and another incoming arc of color (2);

- all the nodes of an odd-parity cycle have exactly one incoming arc of color (0) and another incoming arc of color (2).

On C_0, we mark all the clockwise cycle-arcs with color (1) except for the one incoming to $(u, 0)$. We mark color (2) the cycle-arc $((u, 2), (u, 1))$. (see Figure 1). For nodes $u \neq 0$ we mark the cycle-arcs depending on the value of $|u| = 0$ and those of u_1 and u_0.

After this step, each node has at least 2 incoming arcs marked with two different colors. Moreover, our construction insures that the directed subgraphs S_0 and S_1 are connected.

Step 5.2: Coloring uncolored cube-arcs.

We recall that if a node $u \in Q_n$ has an even (resp. odd) parity then all its neighbors have an odd (resp. even) parity. All the nodes of C_0, but $(0,0)$, are only colored (1) and (2). But all the incoming cube-arcs are not colored yet and are all originated from cycles in which each node has one incoming arc of color (0). whose nodes are colored (0). Then we can mark these incoming cube-arcs (but the one of $(0,0)$) with color (0). Thus, all nodes in C_0 are colored with three colors.

We apply the same rules to all other cycles, using the complementary color on their neighbors. Note that for a given node in these cycles, either an incoming cube-arc or an incoming-cycle is not yet colored, if there is one. In either case, we mark this uncolored arc with the third color. Thus, each node but the root has its three incoming arcs colored with three different colors. It is possible to show that the directed subgraph S_0 (resp. S_1) is connected [7]. Thus we have obtained three ADST's of CCC(n).

Heights of the trees:

We denote $h(S_i)$ the height of the tree S_i. It can be derived (see [7]) that

$$h(S_0) \leq 4n - 1;$$
$$h(S_1) \leq 4n;$$
$$h(S_2) \leq 3n. \qquad \square$$

5 Broadcasting on CCC

We first define our communication model. We follow the notations in [9] where τ is the overhead (start-up) associated with each communication, t_c is the data transmission time per byte and M is the message size in bytes. Thus, the communication cost of a message transfer from one processor to one of its neighbors is $T = Mt_c + \tau$. We assume that communication can take place on all ports of each processor concurrently.

Corollary 2 *Broadcasting a message of length M can be done in* $(\sqrt{\frac{Mt_c}{3}} + \sqrt{(4n-1)\tau})^2$.

Proof: We use the 3 ADST's (of heights $\leq 4n$). The original message is partitioned into 3 equal parts, and each part is pipelined in each trees (see [9]). $\square$

Note that the lower bound is $(\lceil M/3 \rceil + d - 1)t_c + d\tau$ where d is the diameter of CCC(n) ($d \approx 2.5n$). Note also that when communication is restricted to one port at a time, it is possible to use these 3 ADST's in broadcasting M elements with complexity $(\sqrt{Mt_c} + \sqrt{O(n)\tau})^2$. In this case, the lower bound is $(M + d - 1)t_c + d\tau$.

6 Our related results

In this section, we remark on some related results in attaining optimal or nearly optimal number of ADST's for a few related cases. All the ADST's obtained in the following do not use the expensive algorithm of Shiloach [16].

First, for a general n-dimensional CCC of size $(n + c)2^n$ with $c > 0$, i.e., each cycle is of length $n + c$, we are able to construct 2 ADST's in it, which is optimal because some nodes have degree two. They are obtained using any 2 arcs-disjoint spanning trees of Q_n^* and turning clockwise on the cycles to map the first tree, and counterclockwise to map the second tree. This can be generalized as follows: given any graph G and at least two ADST's in G, we are able to construct two ADST's in $C[G]^*$.

Second, our approach in defining the three ADST's in CCC can be adapted to the n-dimensional folded CCC in certain cases. The folded hypercube F_n is a hypercube Q_n enhanced with one extra link per node connecting to the node farthest away [1,13,10]. This allows mainly to decrease the diameter to about 50% as well as to attain better fault tolerance. The n-dimensional folded cube-connected cycle with $n + 1$ nodes per cycle is constructed from the folded hypercube as CCC from the hypercube. Ho has shown in [8] that it is possible to generalize the construction of n ADST's in Q_n to obtain $n+1$ ADST's in F_n. Based on the ADST's in F_n and the technique described in this paper, we are able to construct three ADST's in the n-dimensional folded CCC when n is odd. When n is even, F_n is not bipartite. Certain argument in the proof of Theorem 2 cannot be similarly applied. In this case, we are only able to construct 2 ADST's.

7 Conclusion

The cube-connected cycles (CCC) are compounded cycles into hypercubes. We have given an explicit construction in CCC for three arc-disjoint spanning trees of heights optimal within a factor of two. They are constructed based on the knowledge of arc-disjoint spanning trees on hypercubes [9], and the techniques described in the paper.

Many problems remain open. For instance, is it possible to decrease the heights of these trees to be as close as possible to the diameter? How to efficiently construct three arc-disjoint spanning trees in the folded CCC when the dimension is even? Is there a general construction of three arc-disjoint spanning trees in a $C[G]^*$ knowing existing ones in G^*? More theoretically, is there a general bound on the heights of disjoint spanning trees?

References

[1] G.B. Adams and H.J. Siegel. The extra stage cube: a fault-tolerant interconnection network for supersystems. *IEEE Trans. Computers*, 31(5):443–454, May 1982.

[2] J.-C. Bermond, C. Delorme, and J-J. Quisquater. Strategies for interconnection networks: some methods from graph theory. *JPDC*, 3():433–449, 1986.

[3] J.-C. Bermond and P. Fraigniaud. Broadcasting and gossiping in de Bruijn networks. Research report LIP, ENS Lyon, France, 1990.

[4] J.-C. Bermond, P. Hell, A.L. Liestman, and J.G. Peters. Broadcasting in bounded degree graphs. To appear in SIAM Journal of Computing.

[5] J. Edmonds. Edges-disjoint branchings, combinatorial algorithms. In R.Rustin, editor, *Combinatorial Algorithms 91-96*, Algorithmics press, New York, 1972.

[6] P. Fraigniaud. Fault-tolerant gossiping on hypercube multicomputer. To appear in the proceedings of EDMCC2, Munchen FRG, 1991.

[7] P. Fraigniaud and C.-T. Ho. *Arc-disjoint spanning trees on cube-connected cycles networks*. RJ 7931 (72914), IBM, January 1991.

[8] C.-T. Ho. Full bandwidth communications on folded hypercubes. In *1990 International Conf. on Parallel Processing, Vol. I*, pages 276–280, Penn State, 1990.

[9] S.L. Johnsson and C.-T. Ho. Optimum broadcasting and personalized communication in hypercubes. *IEEE Trans. Comp.*, 38(9):1249–1268, 1989.

[10] S. Latifi and A. El-Amawy. On folded hypercubes. In *1989 International Conf. on Parallel Processing, Vol. I*, pages 180–187, Penn State, 1989.

[11] D. MacKenzie and S. Seidel. *Broadcasting on three multiprocessor interconnection topologies*. CS-TR- 89-01, Michigan Technical University, 1989.

[12] D.S. Meliksetian and C.Y.R. Chen. Communication aspects of the Cube-Connected Cycles. In *1990 International Conf. on Parallel Processing, Vol. I*, pages 579–580, Penn State, 1990.

[13] F.J. Meyer and D.K. Pradhan. Flip-trees: fault-tolerant graphs with wide containers. *IEEE Trans. Computers*, 37(4):472–478, April 1988.

[14] F. Preparata and J. Vuillemin. The cube connected cycles: a versatile network for parallel computation. *Communications of the ACM*, 24(5):300–309, 1981.

[15] P. Ramanathan and K.G. Shin. Reliable broadcast in hypercube multicomputers. *IEEE Transactions on Computers*, 37(12):1654–1657, 1988.

[16] Y. Shiloach. Edge-disjoint branching in directed multigraphs. *IPL*, 8(1):24–27, 1979.

CACHE COHERENCE ON A SLOTTED RING

Luiz A. Barroso and Michel Dubois
EE-Systems Department
University of Southern California
Los Angeles, CA 90089-1115

Abstract -- The Express Ring is a new architecture under investigation at the University of Southern California. Its main goal is to demonstrate that a slotted unidirectional ring with very fast point-to-point interconnections can be at least ten times faster than a shared bus, using the same technology, and may be the topology of choice for future shared-memory multiprocessors. In this paper we introduce the Express Ring architecture and present a snooping cache coherence protocol for this machine. This protocol shows how consistency of shared memory accesses can be efficiently maintained in a ring-connected multiprocessor. We analyze the proposed protocol and compare it to other more usual alternatives for point-to-point connected machines, such as the SCI cache coherence protocol and directory based protocols.

1. Introduction

It is a well known fact that bus-based architectures, which are the most popular topologies in current commercial systems, have serious electrical problems that have kept them from reaching higher bandwidths, thus limiting their scalability [4]. Moreover, processors clocked at 40 MHz are already available on the market, and we believe that even a small number of these could generate enough traffic to saturate current interboard buses, in a caching shared-memory multiprocessor. On the other hand, point-to-point unidirectional interconnections can be made much faster than bus interconnections since they do not require arbitration cycles, and lack most of the electrical problems with shared buses, namely signal propagation delays and signal reflection/attenuation. Current IEEE standard proposals are based on 500 Mbps, 16 bits wide unidirectional point-to-point interconnects [17]. The *Express Ring* is a cache-coherent shared-memory multiprocessor with the simplest type of point-to-point interconnection network: a unidirectional ring. We believe that such a ring can be made at least ten times faster than state-of-the-art buses, therefore showing much better scalability.

The use of local caches in a shared-memory multiprocessor greatly enhances overall memory access performance, since a large fraction of processor references is likely to be present in the cache, and is resolved at processor speed. Private caches also affect performance by decreasing the traffic in the interconnect, thus avoiding conflicts and

This work is funded by NSF under Grant No. MIP-8904172. Luiz A. Barroso is also supported by CAPES/MEC, Brazil, under Grant No. 1547/89-2.

queueing delays. However, in order to enforce a consistent view of multiple cached copies of shared writable data, one requires a *cache coherence protocol*. In this paper we develop a cache coherence protocol for a slotted unidirectional ring multiprocessor. We believe that the proposed protocol is a simple and efficient solution for the coherence problem in a ring architecture, and we compare it with other protocol options, such as directory based schemes [18][3] and the SCI cache coherence protocol [12].

Snooping protocols [8][13] are very popular in bus-based systems, mainly because of their simplicity and low cost of implementation. Unfortunately this class of protocols relies heavily on the broadcasting of information, and is inadequate for most point-to-point interconnected systems, where broadcasting is generally very expensive. Several cache coherence protocols for point-to-point interconnected machines have recently been proposed in the literature [18], most of them making use of directories to keep track of the copies for every memory block. These schemes are relatively complex to implement. Because of the natural broadcasting structure of the Express Ring, we believe that an adapted version of a snooping cache coherence protocol performs better than a directory based mechanism, with the advantages of being simpler to implement and scale to larger configurations.

In the next Section we briefly describe the Express Ring architecture, its pipeline features and memory organization. In Section 3 we present a cache coherence protocol for this architecture, including some implementation issues. The performance characteristics of the proposed protocol are presented in Section 4, in which we also compare it with other alternatives in the context of the Express Ring architecture. Final conclusions are drawn in Section 5.

2. The Express Ring Architecture

The Express Ring intends to demonstrate that a high-speed slotted unidirectional ring is not only feasible but may be superior to a shared bus as the interconnection structure for shared-memory multiprocessors. Our design philosophy was to keep the interconnection structure very simple, in order to be able to clock it as fast as current technology allows. We also wanted to avoid any centralized arbitration, as well as a complex interconnection access control mechanism.

An impo_____ Ring architecture is that _____ erned, the system behavi__ _____ cture, since all ring transa_____. This characteristic, tog_____ large private caches, mak_____ eneral purpose computing.

The Expr__ _____ h is schematically show_____ high- performance proces_____ otted ring structure. Each _____ igure 2, consists of 2 proc_____-port cache, a fraction of _____ pace (including main mer_____ the ring interface. Transf_____ the same clock, but the _____ ors operation.

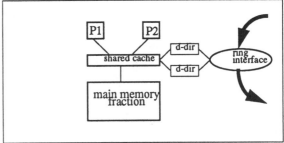

Processor Cluster diagram

Figure 2.

centralized arbitration is required for the packet slots: a processor cluster may use the first empty slot that passes through it to send a message. Such a scheme may lead to starvation of a cluster, specially when the ring traffic is relatively heavy, and the utilization of the slots approaches 100%. A simple and effective mechanism to prevent starvation, without requiring any central arbiter or reservation of slots, is discussed in Section 3.

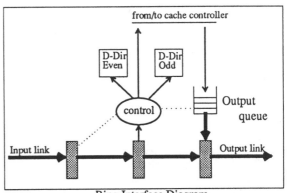

Ring Interface Diagram

Figure 3.

Routing in this architecture is very straightforward. An incoming message is either consumed by the cluster or forwarded to the next cluster in the ring order. The ring interface has a very fast logic that scans the first bits of each message and determines whether it has to be forwarded or removed from the ring. A block diagram of the ring interface is shown in Figure 3. It is important to note that the speed of the forwarding logic is critical in the Express Ring design, and is likely to become the bottleneck on the pipeline clock frequency.

There is a single physical address space shared by all processors in the system. The memory is distributed among the clusters in the ring and the location of a memory block is determined by the higher address bits. The processor cluster to which a block is mapped is called the block's *home cluster*. Swapping space is provided by disks attached to some of the clusters. A single *dirty* bit, is associated with each cache block in main memory. The dirty bit is used to indicate wether the current main memory copy of a block is up-to-date, and it is critical to the performance of the

Using a ring t_____g elements is not a new _____ en done in the design and a_____ns for local area networks _____] and the Cambridge sl_____g interconnections have a____ _____ ___ ___ context of more tightly-coupled systems, as in the CDC CYBERPLUS system [7] and the MIT Concert Multiprocessor [9]. However, we are not aware of the use of a slotted ring network as the interconnection of a shared-memory cache coherent multiprocessor system. The main benefit of the slotted ring is that several messages can be transmitted at the same time, which increases the communication bandwidth.

The Express Ring interconnection network can be viewed as a circular pipeline, with 3*P stages, where P is the number of processor clusters (or nodes) in the system, i.e., each processor cluster accounts for 3 pipeline stages, as shown in Figure 3. At each ring clock cycle, 64-bit wide packet slots are transferred from a stage of the pipeline to the next one. Messages are inserted in the ring by filing packet slots with useful information, and removed by marking a packet slot as empty. There are 3*P packet slots circulating in the ring. The pipeline clock can be extremely fast, since no proposed coherence protocol.

A dual-directory serves coherence requests coming from the ring; it contains a copy of the local cache directory (tags + state information), so that snooping activity does not interfere with normal processor accesses to the cache. The dual directory is 2-way (even/odd) interleaved to be able to keep up with the rate at which coherence messages arrive from the ring. Caches may contain blocks that map to the local cluster's physical space or to any other cluster's memory. Both cache misses and invalidations may generate messages to the ring in order to satisfy local accesses and maintain global system coherence.

3. The Express Ring Cache Coherence Protocol

3.1 Description

In this Section we present a snooping cache coherence protocol for the Express Ring architecture. We propose a write-back/write-invalidate protocol with five cache block states and two memory block states. Three of the cache block states, *Invalid* (INV), *Read-Shared* (RS) and *Write-Exclusive* (WE), are taken from the basic write-back/ write-invalidate snooping protocol. A cache block in INV state is assumed to be not present in the cache. A cache block in RS state can be accessed for reads but not for writes; several RS cached copies of a block may exist at the same time. A cache block in WE state can be accessed for both reads and writes, and the protocol ensures that when a block is in WE no other cache has a valid (WE or RS) copy of it.

Transitions among these three states in a bus-based system are virtually conflict free, since the bus arbitration mechanism itself takes care of imposing an order among global coherence events. However, the situation in the Express Ring is not so simple. In this system several messages can be traversing the ring at any given time, and two or more clusters may be attempting to access the same block concurrently, causing a conflict. Conflicts may have to be solved by forcing one or more clusters to abort their state transitions. In our protocol we represent state transitions that are subject to conflicts by two extra cache states: *Read-Pending* (RP) and *Write-Pending* (WP). We identify such transitions with specific cache states because it simplifies the protocol description, but it is important to note that pending states do not have to be coded explicitly in the cache directory. Registers in the snooper could keep track of blocks to which transactions are pending in the ring. In order to support lockup-free caches it is necessary to allow more than one cache block to be in a pending state in a cluster at the same time [5].

A cache block in RP (WP) state is in transition to a RS (WE) state but has not yet committed, i.e. has not yet received the corresponding acknowledgment. An acknowledgment for a miss is a message that provides the missing block, and the acknowledgment for an invalidation is a signal that indicates that all system caches successfully invalidated their copies of a block. A cache block in RP (WP) state may also abort its transition to RS (WE) state if some kind of negative acknowledgment is received. We discuss the acknowledgment mechanisms in detail later in this section. The diagram in Figure 4 shows the possible states of a block in a given cache (i). W(i) and R(i) denote respectively write and read operations generated at the local cluster (i), and (j) denotes any cluster other than (i).

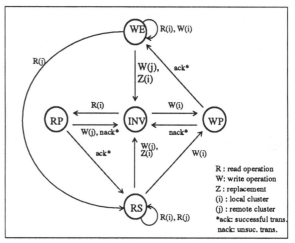

State transition diagram of a block in cache (i)

Figure 4.

The two memory states indicate wether the current main memory copy of a block is valid or not, and they are implemented by a dirty bit, as described in the previous section. The memory copy of a block is not valid when some system cache has a copy of that block in WE state. We note that when a block's memory state is valid, there may or may not be cached copies of it. The two memory states are named *modified/unmodified*.

Coherence actions are procedures that have to be enforced by the coherence protocol in order to preserve an overall consistent view of the shared memory. Coherence actions are triggered by one of the following events: read miss, write miss, invalidation and replacement. Below we explain the protocol behavior in further detail by describing the state transitions and data movement involved in every coherence action.

Read Miss: A read miss immediately changes the cache state of the missing block from INV to RP. The requesting cluster's cache state changes to RS when the missing block is received. If the memory block state is *unmodified* the home cluster provides the block to the requesting cluster, and the memory block state remains *unmodified* after the transition is completed. Otherwise the *dirty cluster* (a cluster with the WE copy) sends the valid copy to the requesting cluster and to the home cluster, so that the memory copy is also updated. The cache state in the dirty cluster changes to RS and the memory state changes to *unmodified*.

Write Miss: The local cache state of the missing block goes immediately to WP, and changes to WE when a valid copy of the block is received. In a write miss, not only a copy of the block has to be fetched but also all other existing copies have to be invalidated. The block is provided by the home cluster if the memory state is *unmodified* or by the dirty cluster, if the memory state is *modified*. The final cache state in all clusters other than the requesting one is INV, and the final memory state is *modified*.

Invalidation: Generated when a processor attempts to modify a block that is present in its cluster's cache in RS state. An invalidation action ensures that the cluster first acquires an unique copy of the block before it is allowed to modify it. The requesting cluster cache state goes immediately to WP, and changes to WE when an acknowledgment is received. One should note that invalidations only happen when the memory state is *unmodified*. The final cache state in all clusters other than the requesting one is INV, and the memory state changes to *modified*.

Replacement: Generated when a cache needs the block frame that is presently occupied by a block in WE state (i.e., a block that has been altered). The block has to be *flushed* to the home node, so that the updates are not lost. The requesting cache state goes immediately to INV, and the memory state changes to *unmodified*. Blocks in any other cache state (WP, RP, RS and INV) do not have to be written back since they have not been altered.

3.2 Implementation

We now discuss some implementation issues of our protocol, also detailing its behavior in conflicting situations. The protocol is implemented as a set of *coherence messages* exchanged through the slotted ring in order to implement the coherence actions described earlier. A miss in a cache block that is mapped to the local cluster's physical space (i.e., the requesting cluster and the home cluster are the same) is called a local miss. A local read miss does not generate a coherence message to the ring if the block's memory state is *unmodified*. In all other situations misses, invalidations and replacements cause coherence messages to be sent to the ring. A summary of the protocol coherence messages is shown in Table 1.

3.2.1 Coherence Messages

Responses to a *Read-Block* message (generated when a read miss occurs) vary depending on the current memory state of the block. If the block's memory state is *unmodified*, the home cluster contains a valid copy and provides it to the requesting cluster by replying with a *Send-Block* message. However if the block's memory state is *modified*, the copy has to be provided by the dirty cluster. In this situation, the dirty cluster replies with a *Send-Block-Update* message that

Table 1.
Protocol Commands/Responses

Event	Command	Response
read miss	Read-Block	Send-Block Send-Block-Update
write miss	Read-Exclusive	Send-Block
invalidation	Invalidate	(none)
replacement	Send-Block	(none)

contains the requested block. The requesting cluster forwards a *Send-Block* message to the block's home node when it receives the *Send-Block-Update* message, so that the main memory can be updated.

A *Read-Exclusive* message is generated when a write miss occurs. Either the dirty cluster (if the block is *modified*) or the home cluster (if the block is *unmodified*) replies to a *Read-Exclusive* message by providing a valid copy of the block. In this case, the reply is always a *Send-Block* message, even if the block's memory state is *modified*. In other words, we do not update the main memory copy of a block when a WE copy simply migrates from one cluster to another. A Read-Exclusive message also invalidates every other copy of that block in the ring.

An *Invalidate* message is generated when a cluster attempts to write to a block in RS state in its local cache. In this case there is no need to fetch a copy of the block, but it is important to invalidate all other cached copies of it.

3.2.2 Message Formats

There are two message formats on the ring. *Read-Block*, *Read-Exclusive*, and *Invalidate* are very short messages, named *probe* messages, with the following format:

|<message type>|<requester addr>|<block addr>|<ack>| = 44 bits
    4   6   32   2

Send-Block and *Send-Block-Update* messages (called *block* messages) are larger since they carry a cache block. Here we consider a cache block size to be eight 32-bit words (256 bits). The format of the block messages is:

|<message type>|<requester addr>|<block addr>|<block>| = 298 bits
    4   6   32   256

Probe messages fit in a single 64-bit packet, and occupy only one pipeline stage (a *probe slot*) in the Express Ring, while block messages occupy 5 consecutive pipeline stages (a *block slot*). Probe slots and block slots are arranged in *frames*, that circulate through the slotted ring. Frames are composed of one probe slot for blocks with even addresses (even probe slot), one probe slot for blocks with odd addresses (odd probe slot), one block message slot (which have an even or odd address), and an extra slot used for intercluster interrupt signals. The frame format is shown in Figure 5.

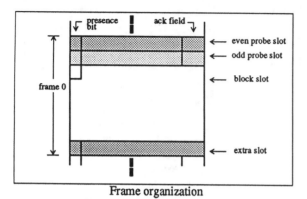

Frame organization

Figure 5.

We say that a given slot is empty if no valid information is currently being transmitted through it. A cluster wanting to transmit a probe has to wait until it detects an empty probe slot of the same parity. A block message may be transmitted using any empty block slot. There are three basic motivations to arrange probe and block slots in frames:

(1) Enforce a minimum interarrival time for probes with same parity, giving enough time for the interleaved dual-directory to respond. Using frames, the minimum interarrival time for probes with same parity will be eight ring clock periods, which gives 40 ns. (ring clock freq. of 200 MHz) for the dual-directory to respond, instead of 5 ns.

(2) Eliminate the problem of having to deal with two different sizes of messages on the ring, which simplifies arbitration. If any pipeline slot could be used for probe and block messages a cluster waiting to transmit a block message would not be able to determine whether there are enough consecutive empty slots to transmit its message when it sees an empty slot. Transmitting block messages in non-consecutive pipeline slots, or temporarily removing messages from the ring to accommodate a block message, would make the ring interface much more complex, and that would certainly slow down the ring clock.

(3) Provide an efficient acknowledgment mechanism for probes. The need for such mechanism will become clear in Section 3.2.3.

A probe message traverses the whole ring exactly once, being consumed by the cluster that issued it. This is necessary because all the clusters' directories have to be consulted on a miss or an invalidation. However, a block message does not circulate through the whole ring, but only through the path between its source and destination. Assuming balanced access patterns, a block message traverses one half of the ring on the average, and this is the reason why we have chosen to put two probe slots and one block slot per frame. The sender of a probe and the destination of a block message are responsible for removing them from the ring. A message is removed by resetting its presence bit (see Figure 5).

When a probe message passes through a cluster it generates a dual-directory lookup but does not wait for the result of the lookup; rather, it is forwarded to the next cluster without delay. Any acknowledgment for a probe uses the <ack> field of the next probe of same parity. This mechanism is included to resolve conflicting situations that may arise when probes for the same block are issued by distinct clusters in a very short time interval. It also permits a cluster to reject a probe.

3.2.3 Conflict Resolution

Conflicting situations occur when more than one cluster issue a probe for a given block in a very short time interval. The protocol has to be able to resolve these situations not only keeping the global state of the block consistent, but also avoiding deadlocks, livelocks and starvation of a cluster. To be able to deal with such situations there is an <ack> field in every probe slot. Acknowledgments for an even (odd) probe in frame i will be put in the <ack> field of the even (odd) probe of the subsequent frame (frame i+1). The reason why acknowledgments for a probe are not "piggybacked" in the same probe slot is to allow the ring interface logic enough time to respond. The <ack> field consists of two bits. Those bits are defined as:

a-bit : acknowledgment bit. Set by the home cluster or the dirty cluster of the block.
ns-bit: no-snoop bit. Set by a cluster that was not able to consult its dual-directory in time to acknowledge a probe.

Both the a-bit and the ns-bit are reset by the requesting cluster when a probe is sent. The a-bit is set by the cluster that has the valid copy of a block being requested by a *Read-Block* or a *Read-Exclusive* probe (which can be either the home cluster or the dirty cluster), signaling that the request was received and that the requested block will be provided. When the requesting cluster sees the probe returning, it removes it from the ring and waits for the acknowledgment that comes in the next frame. If the a-bit is set it waits for the requested block to arrive (still in a pending state) to complete the transition, otherwise it re-issues the probe in the next available slot. When a *Read-Exclusive* or an *Invalidate* probe for a given block passes through a cluster that has that block in RP state, that block's transition is immediately aborted.

Another hazardous situation occurs when multiple *Invalidate* or *Read-Exclusive* probes are issued concurrently. In this case, the cluster with the valid copy serves as the *arbitrator cluster*. The probe that reaches the arbitrator cluster first has its a-bit set. The cluster that receives the probe with the a-bit set "wins" the arbitration and is allowed to write in the block (i.e., change from WP to WE state). The remaining clusters receive the a-bit reset and have to abort the request. Note that a cluster that aborts an *Invalidate* request must retry with a *Read-Exclusive* probe, since it no longer can assume that its previous RS copy of the block is valid.

The ns-bit is set by a cluster that has its dual-directory busy, and cannot accept a given probe. This bit may not be needed if the engineering of the ring interface always allows enough time to snoop at any request. The behavior of the protocol when the ns-bit is necessary is well explained in [1].

While we do not think that fairness will be a serious problem in the Express Ring interconnection, it is possible that a given cluster has to wait an arbitrary long period of time until it finds an empty slot to send a message. If that turns out to be a problem the following policy can be adopted. We simply enforce that every time a cluster removes a message from the ring it does not use that slot but passes it to the next cluster, even if it has messages to transmit. This mechanism works like a token passing scheme, where the token is actually the empty slot. The upper bound on the time to acquire an empty slot of any kind will be a small multiple of the total ring traversal time, since there are multiple frames in the slotted ring.

4. Protocol Evaluation

In this section we evaluate the expected latencies to satisfy coherence messages in the Express Ring cache coherence protocol, as well as in a directory based protocol and in the SCI cache coherence protocol. We compare the three protocols in the context of the Express Ring architecture. Brief descriptions of centralized directory and the SCI protocols are given below.

4.1 Centralized Directory Protocols

In this class of protocols [14] every coherence message is directed to the block's home cluster, which keeps track of the state and location of all cached copies of the block. In directory protocols, the home cluster allows sharing of a block for read-only copies but enforces exclusive access for writable copies. The home cluster satisfies all misses if there is no writable (dirty) copy of the block, forwarding the request to the dirty cluster otherwise. When receiving an invalidation, the home cluster sends invalidation messages selectively only to the clusters sharing a particular block, and replies to the requesting cluster once all copies have been invalidated. The main difference between directory and snooping schemes is that directory-based protocols do not rely on the broadcasting of coherence information.

4.2 SCI Cache Coherence Protocol

The SCI standard, which is being defined by IEEE [17], proposes a type of distributed directory protocol in which a linked list of clusters is maintained for each cached block, called the *sharing list*. The list is implemented by forward and backward pointers in each cache block frame. Coherence requests are sent to the home cluster which satisfies them if the block is currently not cached. If the block is being cached, the home cluster forwards the request to the cluster at the head of the sharing list for that block, which takes the appropriate coherence actions and responds to the requester. In case of a read miss, the missing cluster is inserted at the head of the list either by the home cluster or the previous head. In case of a write miss or an invalidation, all clusters are sequentially removed from the sharing list until the writing cluster is the only one left.

4.3 Latency Analysis

In analyzing the latencies we differentiate miss latencies from invalidation latencies. The latency to satisfy write misses, which can be seen as both a miss and an invalidation, can be approximated by whatever action takes longer, e.g., fetch the missing block or invalidate current cached copies. We will further assume that directory and SCI protocols make use of message formats similar to our probes and block messages, so that the latencies being considered here are actually probe message round-trip latencies.

Miss Latency

In our snooping protocol, only one ring traversal is required to satisfy a miss. A probe is inserted and removed from the ring by the requesting (missing) cluster. Note that even though all clusters in the system snoop on the probe, the time for each cluster to snoop does not add up to the miss latency because a probe message does not wait for the result of the snoop, but is immediately forwarded to the next cluster. In other words, the snooping time overlaps with the probe round-trip time. Thus, the total miss latency is roughly the time for a probe message to traverse the whole ring plus the time to fetch a block and insert it in the ring. This is true because as soon as the probe passes through the cluster with the valid copy, the block is fetched and the block message is sent. We call the time for a probe to complete a round-trip through the ring a *cycle*, and we use it as the basic unit to compare latencies, since the time to fetch a block and insert it in the ring is independent of the protocol used.

It is interesting to note that the miss latency, in the snooping protocol, is independent of the position of the cluster with the valid copy, which is why we claim that the slotted ring is an uniform access interconnection under this protocol, similar to a shared bus. This feature is not present in the other two protocols.

In a directory scheme [14] the miss latency depends on the global state of the missing block. If the block is not dirty, the latency is equal to one cycle, since in this case the home cluster provides a valid copy of the block. However, if the block is dirty, the position of the dirty cluster in relation to the home cluster significantly affects the miss latency. If the dirty cluster is in the path from the home cluster to the requesting cluster, the miss will be satisfied in one cycle (but with two directory lookups), otherwise it will take two cycles for the missing block to reach the requesting cluster.

In the SCI protocol, the home cluster satisfies a miss only if the block is not currently being cached, i.e., the sharing list is empty. In this case the miss latency is one cycle. If the block is being cached, the head of the list is responsible for providing a copy of the block to the missing cluster, and then the latency of satisfying the miss depends on the relative position of the cluster at the head of the sharing list. If the cluster at the head of the list is in the path from the requesting cluster to the home cluster, the latency will be two cycles.

Invalidation Latency

Invalidation messages in the snooping protocol also require only one cycle, unless a given cluster rejects the invalidation probe (by setting the ns-bit). We will not consider this possibility since such rejections would compromise the performance of the other protocols as well.

In the directory protocol, invalidations require more than one cycle if the clusters to be invalidated are not in the path from the home node to the requesting node, but no more than three cycles are required in any situation.

Invalidations in the SCI protocol may require 0 to P cycles (where P is the number of clusters in the ring). If the cluster that generates the invalidation is the only one in the sharing list, no message is generated and the write operation takes place with no further delay. However, if there are N clusters in the sharing list, the invalidation may involve $N-1$ cycles if the order in which the clusters appear in the sharing list is exactly inverted with respect to their order in the ring direction.

4.4 Discussion

The above analysis shows that the proposed cache coherence protocol outperforms the centralized directory protocol in terms of latency of coherence accesses. The only situation in which the SCI protocol may outperform the snooping protocol is when a cluster attempts to write to a RS block and it has the only cached copy of that block. In this case the snooping protocol sends an invalidation because the cluster does not know that it has the only cached copy and that there are no other copies to invalidate. In the SCI protocol, if the cluster at the head of the list sees itself as the only one in the sharing list, it modifies the block without delay. However, in all remaining situations, the SCI protocol exhibits much longer latencies than the snooping protocol, which we believe would lead to worse average performance figures.

It is also quite straightforward to extend our snooping protocol in order to avoid invalidation messages when a

cluster has a private RS copy. This can be accomplished by including a *Read-Exclusive* cache state, as it has been done in the Illinois protocol [15]. We chose to omit the *Read-Exclusive* state from the previous snooping protocol description for the sake of simplicity, but its inclusion is a minor modification in the overall protocol structure.

The snooping protocol is also likely to generate less traffic than the other protocols considered, due to the fact that probes will never travel more than once through the ring (under conflict free transactions). There is no waste of bandwidth in broadcasting miss or invalidation probes in this architecture.

By arranging probes into frames and interleaving the dual-directory we hope to be able to avoid probe rejections completely, so that the ns-bit can be removed from the design.

To give an estimate of the processing power of the Express Ring with our coherence protocol consider a configuration with 16 clusters (32 processors) and a 200 MHz pipeline clock rate (the slotted ring has 48 stages which accommodates 6 frames). On the average, a miss occupies a probe slot for a full ring traversal and a block slot for half a ring traversal. Based on this, we can say that two misses can be satisfied as a frame makes a full ring traversal, and the expression for the maximum system miss rate that can be supported by the Express Ring under these assumptions is

$$\text{Maximum Miss Rate} = 6/(48*tc) = 25 \text{ M misses/sec.,}$$

Where tc is the ring clock cycle (5 ns). For cache hit rates of 98%, and an average memory access per instruction rate of 1.2, such an Express Ring is able to support a peak performance of

$$25 \times 10^6/(1.2*0.02) = 1.04 \text{ GIPS}$$

The latency of a probe request in such a system is less than 250 ns, which is very competitive with the latency of current interboard buses. A configuration with 2 pipeline stages per cluster and a 500 MHz clock could support 64 clusters (128 processors) with a probe round-trip latency under 300 ns.

5. Final Remarks

The objective of this study was to demonstrate that cache coherence can be efficiently maintained in a slotted ring architecture, by means of a relatively simple cache coherence protocol which is based on the classic bus snooping protocol. We believe that the Express Ring protocol demonstrates that point, and is an attractive alternative to maintain coherence in a shared-memory multiprocessor. The proposed protocol not only outperforms directory-based protocols and the SCI protocol, but is also less complex than both alternatives, in a ring-based architecture.

Another important characteristic of our protocol is that, regardless of the relative positions of clusters in the ring, all ring transactions have the same latency, making the slotted ring a uniform access media. As a result, at the programmer level the Express Ring behaves as a very fast bus-connected multiprocessor, which makes it appropriate for general purpose computing.

The main limitation of the Express Ring architecture is that latency of accesses increases linearly with the number of clusters in the ring. Because of this, configurations with hundreds of clusters would experience very large delays to satisfy coherence requests. To be able to deal with that, we plan to experiment with latency tolerant techniques such as lockup-free caches [5] and delayed consistency [6]. Such techniques allow computation to be overlapped with communication, thus reducing the penalty for ring transactions. We are also interested in investigating how our protocol can be extended to allow configurations that include multiple Express Rings connected together in various manners. The Aquarius Multi-Multi protocol [2] may be an interesting scheme to adopt.

6. Acknowledgments

The authors wish to thank Dr. Kai Hwang for his support and Dr. Alvin Despain for valuable discussions about implementation issues. Mr. Pong Fong also helped us with some insight in the behavior of the SCI protocol.

7. References

[1] L. Barroso and M. Dubois, A Snooping Cache Coherence Protocol for a Ring Connected Multiprocessor, USC Tech. Report CENG 91-03, January 1991.

[2] M. Carlton and A. Despain, "Multiple-Bus Shared Memory System", IEEE Computer, Vol. 23, No. 6, June 1990, pp. 80-83.

[3] D. Chaiken, et al., "Directory-Based Cache Coherence in Large-Scale Multiprocessors", IEEE Computer, Vol. 23, No. 6, June 1990, pp. 49-59.

[4] D. Del Corso, M. Kirrman, and J. Nicoud, Microcomputer Buses and Links, Academic Press, 1986.

[5] M. Dubois and C. Scheurich, "Lockup-Free Caches in High-Performance Multiprocessors", The Journal of Parallel and Distributed Computing, January 1991, pp. 25-36.

[6] M. Dubois et al., Delayed Consistency and Its Effects On The Miss Rate of Parallel Programs, USC Technical Report, CENG 91-14, April 1991.

[7] M. Ferrante, "CYBERPLUS and MAP V Interprocessor Communications for Parallel and Array Processor Systems", Multiprocessors and Array Processors, W. J. Karplus editor, The Society for Computer Simulations, 1987, pp. 45-54.

[8] J. Goodman, "Using Cache Memory to Reduce Processor/Memory Traffic", Proc. of the 10th Int. Symp. on Computer Architecture, June 1983, pp. 124-131.

[9] R. Halstead Jr. et al., "Concert: Design of a Multiprocessor Development System", Proc. of the 13th Int. Symp. on Computer Architecture, June 1986, pp. 40-48.

[10] A. Hooper, R. Needham, "The Cambridge Fast Ring Networking System," IEEE Trans. on Computers, Vol. 37, No. 10, October 1988, pp. 1214-1224.

[11] D. Farber and K. Larson, "The System Architecture of the Distributed Computer System - the Communication System", Symp. on Computer Networks, Polytechnic Institute of Brooklyn, April 1972.

[12] D. James, "SCI (Scalable Coherent Interface) Cache Coherence", Cache and Interconnect Architectures In Multiprocessors, M. Dubois and S. Thakkar editors, Kluwer Academic Publishers, Massachusetts, 1990, pp. 189-208.

[13] R. Katz et al., "Implementing a Cache Consistency Protocol", Proc. of the 12th Int. Symp. on Computer Architecture, June 1985, pp. 276-283.

[14] D. Lenoski et al., "The Directory-Based Cache Coherence Protocol for the DASH Multiprocessor", Proc. of the 17th Int. Symp. on Computer Architecture, June 1990, pp. 148-160.

[15] M. Papamarcos and J. Patel, "A Low Overhead Coherence Solution for Multiprocessors with Private Cache Memories", Proc. of the 11th Int. Symp. on Computer Architecture, New York, 1986, pp. 414-423.

[16] J. Pierce, "How Far Can Data Loops Go?", IEEE Trans. on Communications, Vol COM-20, June 1972, pp. 527-530.

[17] SCI (Scalable Coherent Interface): An Overview, IEEE P1596: Part I, doc171-i, Draft 0.59, February 1990.

[18] P. Stenstrom, "A Survey of Cache Coherence Schemes for Multiprocessors", IEEE Computer, Vol. 23, No. 6, June 1990, pp. 12-25.

A SOLUTION OF CACHE PING-PONG PROBLEM IN RISC BASED PARALLEL PROCESSING SYSTEMS

Jesse Fang
CONVEX Computer Corporation

Mi Lu*
TEXAS A & M University

Abstract -- Cache ping-pong problem arises very often in RISC based parallel processing systems where each processor has its own local cache and employs a copy-back protocol for the cache coherence. To solve the problem that large amount of data moving back and forth between the caches in different processors, techniques associated with parallel compiler need to be developed.

Based on the concept proposed in [1] about the relations between array element accesses and enclosed loop indices in nested parallel loops, we present an algorithm in this paper to reduce the unnecessary data movement between the caches for parallel loops with multiple array subscript expressions. By analyzing the array subscript expressions in the nested parallel loop constructs, the compilers can use the algorithm to prepare information at compile time and let the processor execute the corresponding iterations of parallel loops in terms of the data in its cache at execution time. It benefits the parallel programs in which parallel loops are enclosed by a sequential loop and have multiple different subscript expressions for the same array, whose elements are repeatedly used in the different iterations of the outmost sequential loop.

1. Introduction

An important task of a parallel compiler is to identify the parallel nature contained in a sequential program, and to generate parallel code implemented on a parallel architecture. Recently, parallel compilers such as Illinois PARAFRASE compiler [2]–[4] and Rice Parallel Fortran Converter [5]–[7] have been developed incorporating with new theory and advanced technology.

It is assumed in most of the compiler systems that shared memory architectures are provided, and a large memory block is directly addressable by all the processors in equal time intervals. However, hierarchical memory system is widely applied in today's RISC based parallel systems. RISC processors, such as MIPS, IBM RIOS, SPARC and Intel 860, are new products in computer industry which combine semiconductor and compiler back-end techniques. RISC architectures have a private cache associate with each processor. To increase the memory bandwidth, even more than one level of cache can be used, and the size of the cache may be very large. It should be noted that the access time to a private cache is much faster than to the global memory or to the caches of other processors.

Poor cache hit ratios in such hierarchical memory multiprocessor systems are due to the following two reasons: the data requested by a processor are in the global memory and the data requested by a processor are in the caches of other processors. The first reason is the same as the one for the traditional cache utilization on the uniprocessor systems, in which frequently used data are desired to be kept as much as possible in cache or local memory. One of the hardware solutions to obtain high cache hit ratios is to provide large-size cache. However, when executing parallel code, the frequently used data may be shared by multiple processors, which may run the multiple threads for a parallel program at the same time. Therefore, increasing cache size cannot improve the cache hit ratios in parallel program execution. The large cache size may even result in severe inefficiencies when a parallel code requires data moving back and forth between the caches in different processors. This phenomena is called "cache ping-pong phenomena" in shared memory multiprocessor systems in [1]. The time needed for a processor to access the caches of other processors is close to the access time needed to the global memory, since both of them have to go through the data bus or interconnection network. Furthermore, it can be even worse than the global memory access because of the increase of the traffic on data bus or interconnection network and hence the degradation of the bus or network bandwidth.

Past research in this area has been focused on the data locality improving by the program restructuring, which may enhance the cache hit ratio on both uniprocessor and multiprocessor systems. Similar phenomena has been studied for virtual memory systems. W. Abu-Sufah, D. Kuck, and D. Lawrie presented some source program transformation techniques to improve the paging behavior of the programs [8]. These transformations, referred to as "loop-block", include breaking iterative loops into smaller loops (strip-mining) and then recombining and reindexing these smaller loops (loop-fusing and loop-interchange). Since then, a number of loop-blocking algorithms have been developed for different computer architectures such as "loop-tiling" [9] and "loop-jam" [10]. These algorithms exploited and took advantage of the high degree of data reuse for the computation within a block. However, for most of the parallel code with complicated program constructs, the benefit of the blocking algorithms is very limited. For instance, a parallel loop nest is enclosed by a serial loop and there is loop-carried data dependence in the outermost serial loop. If the data are repeatedly used in the different iterations of the serial loop, the blocking technique does not avoid the cache ping-pong problem for the data with dependence in the outerloop and independence in the enclosed parallel loop.

The recent research on cache or local memory ping-pong problem by Z. Fang [1] presented an overview of mathematical concepts for the problem. The concept defines the relation between the array element accesses and the enclosed loop indices in nested parallel constructs. The relation determined by an array subscript expression can be used to partition the iteration space into equivalence classes. All vectors in an equivalence class may access some common array elements at the execution time. However, the method in [1] to calculate the next vector in an equivalence class in terms of the previous vectors is based on an assumption, which only allows one subscript expression for an array variable in the nested loop. This assumption limits the results in [1] to be applied in real application programs. In this paper, we present an algorithm to solve the "cache ping-pong" problem for moving general nested parallel loops, in which an array variable may have more than one subscript expression in the same or different statements of the loop body. The algorithm is executed in a on-line fashion, finding for a linear integer system the next vector from a sequence of the stored vectors.

The rest of the paper is organized as follows. In section 2, the cache or local memory ping-pong problem on our simple machine model is introduced. In addition, the preliminary concepts and the overall approach to solve the cache ping-pong problem on our simple program model are presented. We describe in section 3 the main results in a simple case which has only single array subscrip expression. These results are then extended to the more complicated case which involves multiple subscript expressions or multiple array variables. Algorithms for eliminating the unnecessary data movement between the caches are presented. Section 4 shows the experimental results in a parallel compiler prototype. Parallel code is executed with or with out the proposed compiler strategy and the results are compared. Finally, the paper concludes with Section 5.

*This research is partially supported by the National Science Foundation under Grant No. MIP-8809328.

2. Background

2.1 Machine Model

In a RISC based shared memory multiprocessor system, a number of RISC processors and global memory modules are connected by data-bus or interconnection network. The concurrent execution of multiple threads in parallel programming are ensured by a set of primitives, provided by the system, including fetch/increment or semaphore instructions.

The cache design in most of the RISC processor design has the following characteristics: 1) local to a processor, 2) its size is large enough, 3) it uses copy-back and coherence strategy, 4) its line size is more than one word. In order to simplify the presentation, in this paper we assume that the cache memory is of one level, and the line size is one word. The results can be extended to more complicated machine models considering more levels of local memories and more words in the cache line size.

2.2 Cache Pin-Pong Problem

In a parallel program, a **thread** is referred to as the execution of a piece of code specified by parallel constructs [11]. It can be viewed as a unit of programmer-defined or parallel-compiler-specified work. As in a common parallel construct, a thread in a parallel-loop is the execution of an iteration (or a chunk of iterations if we use strip-mining or other techniques) of the loop, and the threads spawned on entering the parallel-loop merge at the end of the loop. The order in which the iterations of the loop are performed is arbitrary, and the processor on which the parallel-loop is entered is not necessarily the same on which the code following the parallel-loop is executed.

In addition, parallel-loops may be nested with sequential constructs when executed on multiprocessor systems, and some frequently used data may be repeatedly used and modified by different threads. If the threads accessing the same data are not assigned to the same processor, the set of data may be unnecessarily moved back and forth between the caches in the systems. This phenomena is called **cache ping-pong phenomena** in shared memory multiprocessor systems.

The following example in [1] shows the problem in a nested parallel construct:

```
DIMENSION A(1000, 1000)

I=0
WHILE DO 10, 100
I=I+1
    PDO 20 J = 1, 100
        DO 30 K = 1, 100
S:          A(I+2*J+5*K, I+J+3*K) =
            A(I+2*J+5*K, I+J+3*K) + ...
30      CONTINUE
20      CONTINUE
10  CONTINUE
```

In this example the statement S does not have data dependence in the DO J loop. If there is no other loop-carried dependence between statements of the loop body, the J loop can be parallelized. There are a total of 10,000 threads $T_{I,J}$ in the execution of the parallel loop (both I and J from 1 to 100). Each thread requests 100 elements of array A. Many of the array elements are repeatedly accessed in these threads.

For instance, thread $T_{1,1}$ requests data $A(8,5)$, $A(13,8)$, $A(18,11)$, $A(23,14)$, ..., $A(498,299)$, $A(503,302)$ for the innermost serial loop index K from 1, ..., 100 respectively. Meanwhile, thread $T_{2,3}$ requests data $A(13,8)$, $A(18,11)$, $A(23,14)$, ..., $A(498,299)$, $A(503,302)$, $A(508,305)$ and thread $T_{3,5}$ requests data $A(18,11)$, $A(23,14)$, ..., $A(503,302)$, $A(508,305)$, $A(513,308)$. It can be observed that there exists a list of threads: $T_{1,1}, T_{2,3}, T_{3,5}, T_{4,7}, ..., T_{49,99}$, which reuse most of the array elements accessed in the previous thread. If the threads of the list are assigned to different processors, the data of array A are unnecessarily moved back and forth between caches in the system.

For instance, when $I = 1$, thread $T_{1,1}$ is assigned into processor 2. Note that loop I is serial. After the first iteration of loop I is completed, the processors need to be reassigned for the threads of the second iteration of loop I that contains a parallel loop J. If thread $T_{2,3}$ is assigned to processor 4, 99 array elements need to be moved from cache in processor 2 to the cache in processor 4. This unnecessary data movement not only slows down the execution, but also degrades the bus or network bandwidth because it tremendously increases bus traffic. If thread $T_{3,5}$ is assigned to processor 2 in the third iteration of loop I, these data need to be moved back from the cache in processor 4 to the cache in processor 2.

In general, loops are the largest resource for parallelization in application programs. Parallel loops are the most common parallel program constructs either defined by user directives or detected by automatic parallel compilers. The cache ping-pong phenomena shown in the above example are very common in parallel code for scientific computation.

In order to gather evidence of the array access patterns in a wide range of applications, we studied a large number of benchmark programs. They include Linpack benchmark, Perfect Club benchmark, Application programs in mechanical CAE, Computational chemistry, Image and Signal Processing and Petroleum applications. These benchmarks were analyzed by PROFILE to recognize the most expensive routines and loop nests at execution time, which were chosen for the future research. We found that almost all of the most time-consuming loop nests contain at least three level loops. About 60% of these loop nests contain at least one level parallel loop [12]. 95% of the parallel loops can be moved from the innermost loop after using loop-interchange technique. Only 6% of the parallelable nested loops have the parallel outermost loop. 94% parallel loops are enclosed by serial loops, that includes the loop nests in which a parallel loop appears in the outermost loop level in a subroutine, but the subroutine is called by a call-statement contained by a serial loop. Most of these loop nests are not perfect-nested. 53% of the nested loops involve only one major array, which usually is two-dimensional or three-dimensional with a small size in the third dimension. Most of the loop bounds are passed by parameters, while there are a few simple triangular nested loops. 83% of the array variables in the parallel program constructs have more than one subscript expression either in the same parallel loop or in the different parallel loops enclosed by the same serial loop. This paper intends to solve the cache ping-pong problem in such program models in which parallel loops are enclosed by a single serial loop and may contain other loops with multiple array subscript expressions in the loop body. This model can cover almost half parallel loop nests in our study for wide range benchmarks in applications. If more than one serial loop encloses the parallel loops, only the inner serial loop needs to be concerned in the program model. If more than one level loops either parallel or serial are enclosed by the parallel loops, only the outer loop needs to be in our model. The model allows multiple subscript expressions in the parallel loops, but only uses single array variable. The multiple array variables with multiple subscript expressions are not included by the model discussed in this paper. More research work is required to create a general role to handle the cache ping-pong problem in more general program models.

2.3 Preliminaries

In this section, the preliminary concepts relevant to the iteration space and data dependence analysis are reviewed and the notations used in this paper are introduced.

Standard definitions are used to analyze the array accesses [2,3, 4 5, 6, 7]. Considering a nested parallel construct of k loops of the form

```
DO I₁ = L₁, U₁
    DO I₂ = L₂, U₂
        DO Iₖ = Lₖ, Uₖ
S₁:         A(h(I₁,I₂,...,Iₖ) + a) = ...
S₂:         ... = A(g(I₁,I₂,...,Iₖ) + b) +...
30      CONTINUE
```

```
20    CONTINUE
      .....
10  CONTINUE
```

where the array A is of dimension d, both **a** and **b** are offset vectors in Z^d. This loop is not necessary to be perfect-nested. The loop bounds are not required to be constants. The function **h** and **g** are linear:

$$\mathbf{h}, \mathbf{g} : \ Z^k \ \rightarrow \ Z^d.$$

The **iteration space** denoted as **C** is defined by the product $\prod_{j=1}^{k} N_j$, where N_j is the range of the j-th index, $[\ L_j : U_j\ ]$. The **domain space** denoted as **D** is defined by the product $\prod_{i=1}^{d} M_i$, where M_i is the size of array A in the ith dimension. Any array subscript expressions in statements of a parallel nested loop can be more precisely defined by:

$$\mathbf{h}, \mathbf{g} : \ \mathbf{C} \ \rightarrow \ \mathbf{D}$$

or we say that the array subscript expressions define maps:

$$\mathbf{h}, \mathbf{g} : \ \prod_{j=1}^{k} N_j \ \rightarrow \ \prod_{i=1}^{d} M_i.$$

There exists a total order in the iteration space **C** that is defined by the point in time at which the element is executed. If we say a vector **t** is greater than a vector **s**, where

$$\mathbf{t} \ = \ (t_1, t_2, ..., t_k)$$

and

$$\mathbf{s} \ = \ (s_1, s_2, ..., s_k),$$

then there is a point m, which is from 1 to k, such that $t_i = s_i$ for $i < m$ and $t_m > s_m$.

The standard data dependence definition in [13-15] is given as follows. If two statements access the same memory location, we say that there is a data dependence between them.

In general, if a dependence is inside of an iteration of a loop, the dependence is called loop-independent dependence. If a dependence is across iterations of a loop, it is called loop-carried dependence.

Loop-independent dependence does not cause the cache or local memory ping-pong problem if an iteration of a parallel loop must be performed by one thread. **Distance vector** [4-6] in dependence analysis shows the distance between two iterations that reference the same memory location. If the distance is t, a loop-carried flow dependence from S_1 to S_2 within a DO I loop has at least one variable which is computed in S_1 and referenced in S_2 after t iterations. The DO I loop should be executed sequentially or synchronized by some additional synchronizer to keep the execution order of the statements in the dependence.

Dependence analysis associated with a distance vector is a good approach to describe the data reference relationships between iterations in a loop. However, the cache or local memory ping-pong problem, which is involved in multi-level loops in a nested parallel construct, is more complicated. Furthermore, some loops do not have an explicit distance vector such as the example shown in section 2.2, but they can be parallelized by the Banerjee test. The dependence analysis approach is not enough to describe the nature of the cache ping-pong problem.

The overall nature of the cache or local memory ping-pong problem in the simple program model in section 2.2 is described below. If the outermost loop is parallel in a nested parallel construct, there is no cache ping-pong problem, because the different parallel loop iterations never access the same memory locations. If the outermost loop is serial, and encloses parallel loops, the dependences carried by the serial loop may cause the data moving back and forth between threads that execute the iterations of the parallel loops in the different iterations of the outer serial loop. Some array elements may be reused in the different iterations of the outermost serial loop due to the loop-carried dependences in the parallel loops. Meanwhile, these array elements need to be moved in the caches between processors in each iteration of the outermost serial loop due to the parallel loops enclosed by the serial loop.

To help develop our algorithm to solve the cache ping-pong problem, we have two major constraints on our program model.

The first constraint is that this paper concentrates on parallel nested constructs in which only one-level loops are parallel or there may exist multi-level parallel loops but only one-level loops are parallelized. This constraint is reasonable to match our simple machine model described in section 2.1, which does not have the hardware processor cluster concept.

The second constraint is that we assume all dependences in the nested parallel construct use a unique iteration space. Usually, the assumption is not acceptable in application programs. If two parallel loops have different loop bounds, the iteration spaces must be different for the dependences carried by the loops. However, the iteration space mapping and transferring will make the main results too complicated. It will also make the proofs tedious. The results presented in this paper can be extended to the general program model without the constraint by adding the space transformation technique between the iteration spaces, which is similar to, but much more difficult than, the linear space transformation.

To simplify our discussion in the paper, we have the following assumptions in the program model to describe the main results in the next sections:

1. All functions representing array subscript expressions are linear mapping: $\mathbf{C} \rightarrow \mathbf{D}$.

2. There are only one level parallel loops in the nested parallel construct, which are enclosed by an outermost serial loop.

3. There may exist one-level serial loop enclosed by the parallel loops.

4. All data dependences in the nested parallel construct use the same iteration space.

The program model on which we develop the main results may have three loop levels: the outermost serial loop, the middle parallel loop level, and the innermost loop level either serial or parallel. They are not necessary to be perfect nested. There may exist more than one parallel loops in the middle level enclosed by the single outermost serial loop. Loop bounds can be any variables. These loops contain only one array variables with multiple subscript expressions. The program structures are not important as long as the data dependence uses a unique iteration space.

3. Main Results

3.1 Mathematical Concepts

The most definitions and lemmas in this section are extended from the definitions and theorems in [1] by adding multiple array subscript expressions instead of only one expression in [1].

Definition 1.   Reduced Iteration Space is a subspace in the iteration space for the parallel constructs described in Section 2.3 by removing the dimension of the innermost loop index.

A linear function **h** defined in the program model that is specified by an array subscript expression, is a map from the **reduced iteration space**, $N \times M$, to the set of subsets of the **domain space**.

$$\mathbf{h} : \ N \ \times \ M \ \rightarrow \ 2^{D_1 \times D_2}$$

where the upper bounds of the outermost serial loop and the middle parallel loop are N and M respectively. The dimensions of the array is D_1

The following example illustrates the typical parallel program constructs, which contain three level loops and r different array subscript expressions.

```
DO i = 1, N
   .....
   PDO j = 1, M
      .....
      DO k = 1, L
         .....
         A(a_{1,1}i + b_{1,1}j + c_{1,1}k + d_{1,1}, a_{1,2}i + b_{1,2}j +
         c_{1,2}k + d_{1,2}) = ....
```

$...= A(a_{2,1}i + b_{2,1}j + c_{2,1}k + d_{2,1}, a_{2,2}i + b_{2,2}j + c_{2,2}k + d_{2,2}) + ...$

$A(a_{r,1}i + b_{r,1}j + c_{r,1}k + d_{r,1}, a_{r,2}i + b_{r,2}j + c_{r,2}k + d_{r,2}) = ...$

30   CONTINUE

20   CONTINUE

10 CONTINUE

the linear function $\mathbf{h}_m$ is

$$f_m(i,j,k) = a_{m,1}i + b_{m,1}j + c_{m,1}k + d_{m,1}$$

and

$$g_m(i,j,k) = a_{m,2}i + b_{m,2}j + c_{m,2}k + d_{m,2},$$

where m is from 1 to r, assuming there are r different array subscript expressions in the parallel construct.

To collect the sets of vectors in the reduced iteration space, which may access common memory locations within the corresponding threads, we define a set of elements of array A, which are accessed within thread T_{i_0,j_0} by linear function $\mathbf{h}_m(i,j,k)$ defined by the m-th subscript expression as follows:

Definition 2. For a given pair i_0 and j_0, the set of elements $A(f_m(i_0, j_0, k), g_m(i_0, j_0, k))$ of array A, which are accessed within thread T_{i_0,j_0} by statement subscripted by the linear function $\mathbf{h}_m(i,j,k)$, is denoted by $A_{i_0,j_0}^{(m)}$, where $1 \leq k \leq L$ and $1 \leq m$

$$A_{i_0,j_0}^{(m)} = \{A(f_m(i_0, j_0, k), j_0, \text{ where } k \in [1,L]\}.$$

Since both f_m and g_m are linear in terms of i, j, and k, it is obvious to have the following lemma, which is useful in the rest of this section.

Lemma 1. In a program construct described above, if there exist two vectors in iteration space, (i,j,k) and (i',j',k'), such that

$$f_m(i,j,k) = f_m(i',j',k')$$

and

$$g_m(i,j,k) = g_m(i',j',k'),$$

then for any constant n_0, we have a series of vectors in the space, $(i,j,k+n_0)$ and $(i',j',k'+n_0)$ satisfying the following equations:

$$f_m(i,j,k+n_0) = f_m(i',j',k'+n_0)$$

and

$$g_m(i,j,k+n_0) = g_m(i',j',k'+n_0)$$

where $1 \leq k'+n_0 \leq L$ and $1 \leq k+n_0 \leq L$.

It is clear from Lemma 1 that if

$$A_{i_1,j_1}^{(m)} \cap A_{i_2,j_2}^{(m)} \neq \Phi,$$

then threads T_{i_1,j_1} and T_{i_2,j_2} should be assigned into a same processor, because they may access some common elements of array A subscripted by the linear function $\mathbf{h}_m$.

Lemma 2 In the program model described in Section 2.3., we have two vectors (i,j,k) and (i',j',k') holding the equations

$$f_m(i,j,k) = f_m(i',j',k')$$

and

$$g_m(i,j,k) = g_m(i',j',k'),$$

if they satisfy the following condition:

$$i' - i = \alpha_m = b_{m,1} c_{m,2} - b_{m,2} c_{m,1}$$

$$j - j' = \beta_m = a_{m,1} c_{m,2} - a_{m,2} c_{m,1}$$

$$k - k' = \gamma_m = a_{m,2} b_{m,1} - a_{m,1} b_{m,2}.$$

Definition 3. For a given pair i_0 and j_0, the data set of elements of array A denoted by A_{i_0,j_0}, which may be accessed within thread T_{i_0,j_0} by statements with the reference to the array variable A, is the union of the sets $A_{i_0,j_0}^{(m)}$, where m is from 1 to r.

It is clear from the above description,

$$A_{i_1,j_1} \cap A_{i_2,j_2} \neq \Phi,$$

then threads T_{i_1,j_1} and T_{i_2,j_2} should be assigned into a same processor, because they may access some common elements of array A at the execution of the parallel construct.

Definition 4. In a program model described in Section 2.3, for a given pair i_0 and j_0, U_{i_0,j_0} denotes a set of vectors (i,j) in the reduced iteration space of size $N \times M$, which satisfy the following condition:

$$U_{i_0,j_0} = \{(i,j) \mid A_{i_0,j_0} \cap A_{i,j} \neq \Phi\}$$

By Lemma 1, the definition can be described as:

$$S_{i_0,j_0} = \{(i,j) \mid \exists k_0 \text{ and } k, m_0 \text{ and } m \text{ such that}$$

$$f_{m_0}(i_0,j_0,k_0) = f_m(i,j,k) \text{ and } g_{m_0}(i_0,j_0,k_0) = g_m(i,j,k)\}$$

What we need to do is is to prove that the set $\mathbf{S}_{i_0,j_0}$ is an equivalence class, then to find a way to calculate j_2 from the current i_2 and the previous vector (i_1, j_1) in the equivalence class.

In order to develop an approach to compute the vector series in the set $\mathbf{S}_{i_0,j_0}$, we need to introduce some necessary notations. In the above example, for the linear function $\mathbf{h}_m$, where $1 \leq m \leq r$, let us denote:

$$\alpha_m = b_{m,1} c_{m,2} - b_{m,2} c_{m,1}$$
$$\beta_m = a_{m,1} c_{m,2} - a_{m,2} c_{m,1}$$
$$\gamma_m = a_{m,2} b_{m,1} - a_{m,1} b_{m,2}.$$

3.2 Partition of Iteration Space

As shown in Section 3.1, each linear function $\mathbf{h}_m$, where $1 \leq m \leq r$, gives a particular value of α_m, β_m, and γ_m.

Section 3.1 gave the definition of a set of vectors (i,j) in the reduced iteration space, U_{i_0,j_0}, in which each vector may access some elements of array A that are referenced in thread T_{i_0,j_0} for the given pair (i_0, j_0).

Lemma 3. The set of vectors (i,j) in Definition 4, U_{i_0,j_0}, is the same set shown below, if there is only one linear function in the parallel construct, indicated by the superscript (1).

$$S_{i_0,j_0}^{(1)} = \{(i,j) \mid i = i_0 + p \times \alpha \text{ and}$$

$$j = j_0 + p \times \beta \text{ for } p \in Z\}$$

Now we extend the definition from one linear function to r linear functions in the parallel loop. As shown in the program model in Section 3.1, there are r pairs of subscript expressions for an array variable. We have a list of r triples: $(\alpha_1, \beta_1, \gamma_1)$, $(\alpha_2, \beta_2, \gamma_2)$, ..., $(\alpha_m, \beta_m, \gamma_m)$, ..., $(\alpha_r, \beta_r, \gamma_r)$.

If the relationship defined by the following definition can partition the reduced iteration space, then the corresponding threads may access some common elements of the array that are stored in the local cache by thread T_{i_0,j_0} for given (i_0, j_0).

Definition 5. In reduced iteration space of loop i and loop j, if there are r different pairs of subscript expressions for an array variable, we define the set of the pairs of i and j so that the corresponding threads may access some common elements of the array subscripted by these linear functions, which are accessed in the cache or local memory by thread T_{i_0,j_0}.

$$S_{i_0,j_0}^{(r)} = \{(i,j) \mid i = i_0 + \sum_{m=1}^{r} p_m \times \alpha_m \text{ and}$$

$$j = j_0 + \sum_{m=1}^{r} p_m \times \beta_m \text{ for } p_1, ..., p_r \in Z\}.$$

Theorem 1. The set of vectors in a reduced iteration space of loop i and loop j, $S_{i_0,j_0}^{(r)}$ defined in Definition 5, is a subset of the set of vectors, $\mathbf{U}_{i_0,j_0}$ defined in Definition 4.

The proof is straightforward following Lemma 2 and Lemma 3. All threads corresponding to the vectors in $S_{i_0,j_0}^{(r)}$ may access some common array elements at execution of the parallel construct. Since Definition 5 is only a subset of Definition 4, the approach described in the paper is not the optimal solution for the cache ping-pong problem, but it can significantly reduce the cache ping-pong phenomena at the execution time. The following theorem shows that the relation defined in the above definition can partition the reduced iteration space. Therefore, we can reduce the unnecessary data moving between processors and improve the system performance.

Theorem 2. $S_{i_0,j_0}^{(r)}$ is an equivalence class in the reduced iteration space of loop i and loop j.

Proof: If we have

$$S_{i_0,j_0}^{(r)} \cap S_{i_1,j_1}^{(r)} \neq \Phi,$$

then there exists (i,j) belonging both $S_{i_0,j_0}^{(r)}$ and $S_{i_1,j_1}^{(r)}$.

Therefore there exist $p_1, p_2, ..., p_r$ such that

$$i = i_0 + \sum_{m=1}^{r} p_m \times \alpha_m$$

$$j = j_0 + \sum_{m=1}^{r} p_m \times \beta_m.$$

Meanwhile, there exist $p_1', p_2', ..., p_r'$ such that

$$i = i_1 + \sum_{m=1}^{r} p_m' \times \alpha_m$$

$$j = j_1 + \sum_{m=1}^{r} p_m' \times \beta_m.$$

So, we have

$$i_1 = i - \sum_{m=1}^{r} p_m' \times \alpha_m = i_0 + \sum_{m=1}^{r} (p_m - p_m') \times \alpha_m$$

$$j_1 = j - \sum_{m=1}^{r} p_m' \times \beta_m = j_0 + \sum_{m=1}^{r} (p_m - p_m') \times \beta_m.$$

Therefore,

$$(i_1, j_1) \in S_{i_0,j_0}^{(r)}.$$

In the same way, we have:

$$(i_0, j_0) \in S_{i_1,j_1}^{(r)}.$$

Finally, we have

$$S_{i_0,j_0}^{(r)} = S_{i_1,j_1}^{(r)}.$$

By Theorem 2 and Definition 5, it is obvious that there is less memory access from one processor to caches or local memories of the other processor, if all these threads in the same equivalence class are assigned to one processor. We describe the fact in Theorem 3 and omit the proof.

Theorem 3. Every tread $T_{i,j} \in S_{i_0,j_0}^{(r)}$ may reuse some data in the other threads belonging to the same equivalence class $S_{i_0,j_0}^{(r)}$, which can significantly reduce the access of the data referenced by the threads belonging to the other equivalence classes.

3.3 Computing Vectors in An Equivalence Class

To assign all threads belonging to the same equivalence class to the same processor at execution time, an iterative algorithm is designed to calculate the next vector in $S_{i_0,j_0}^{(r)}$ in terms of the current vector. The algorithm can be used to compute the current loop index of the middle parallel loop from the current outermost serial loop index and the previous serial and parallel loop indices.

The initial vector for each equivalence class, $S_{i_0,j_0}^{(r)}$, or the initial value of $(p_1, ..., p_r)$ needs to be prepared at compilation time. The lower loop bound of the outermost serial loop is the initial value for i_0. All the other indexes of the outermost loop, whose value is less than $\min(\alpha_1,....,\alpha_r)$, are initial too. For the given initial value of i_0, by Definition 5, it seems to be required to find all possible $(p_1, ..., p_r)$ such that $\sum_{m=1}^{r} p_m \times \alpha_m = 0$ to calculate the initial value of j_0, but it is not necessary. In the program model in section 2.2, Loop j is parallel. For any fixed i_0, there do not exist j_1 and j_2 such that

$$\mathbf{A}_{i_0,j_1} \cap \mathbf{A}_{i_0,j_2} \neq \Phi.$$

By this assumption, a string $(p_1, ..., p_r)$ satisfying the above equation must satisfy another equation $\sum_{m=1}^{r} p_m \times \beta_m = 0$.

Therefore the algorithm to compute the initial vector for each equivalence class, $S_{i_0,j_0}^{(r)}$ is straightforward. First let i_0 equal to the lower bound of Loop I, we have M initial vectors (i_0, j), where j is from 1 to M in our program model, for M equivalence classes. The list of $(p_1, ..., p_r)$ satisfying the equation $\sum_{m=1}^{r} p_m \times \alpha_m = 0$ is computed, which is useful to find the initial value of j_0 as well as the next vector in the equivalence class at the execution time. Then increasing value of i_0, calculate the list of $(p_1, ..., p_r)$ in the same way as in the first step, until the value of i_0 is equal to $\min(\alpha_1,....,\alpha_r)$. This computation needs to solve the integer linear equation $\sum_{m=1}^{r} p_m \times \beta_m = 0$. However, it is at the compilation time, so our approach does not affect the execution performance although the algorithm to solve the integer linear equation has high time complexity.

From Definition 5, we need to find the next vector in an equivalence class in the increasing order of component i. Assuming the initial vector is given, we are to compute component j step by step. Let an integer linear system $L : \{\mathbf{p}, \alpha, I\}$ be defined as follows:

$$\mathbf{p} = (p_1, p_2, \cdots, p_r)^T$$
$$\alpha = (\alpha_1, \alpha_2, \cdots, \alpha_r)$$
$$I = \alpha \cdot \mathbf{p}$$

where $1 \leq I \leq N$ with N being positive integer, and α, $\mathbf{p}$ are all r-dimensional positive integer vectors.

Consider the following problem: given initial $\mathbf{p}^0$, find $\mathbf{p}^1$ such that no $\mathbf{p}' \in Z^{+r}$ satisfying $\alpha \cdot \mathbf{p}^0 < \alpha \cdot \mathbf{p}' < \alpha \cdot \mathbf{p}^1$. Furthermore, given $\mathbf{p}^i$, find $\mathbf{p}^{i+1}$ such that there exist no $\mathbf{p}'$ satisfying $\alpha \cdot \mathbf{p}^i < \alpha \cdot \mathbf{p}' < \alpha \cdot \mathbf{p}^{i+1}$. We first give a graph representation of the system.

Definition 6. A labeled digraph $G = (V, A, W)$ for the above integer system is defined as follows:

- $\{I^1, I^2, \cdots, I^n\}$ where $I^1 < I^2 < \cdots < I^n$

- $A = \{(I^l, I^{l'})|0 \leq l < l' \leq n$ and $\exists \alpha_m$ such that $I^{l'} = I^l + \alpha_m\}$

- $\{w(I^l, I^{l'}) = \alpha_m| I^{l'} = I^l + \alpha_m$ for $(I^l, I^{l'}) \in A\}$.

According to the definition, each arc $(I^l, I^{l'})$ in G is associated with a label, say α_m. If $\mathbf{u}(I^l, I^{l'})$ is an r-dimensional mth unit vector, then $\alpha_m = \alpha \cdot \mathbf{u}(I^l, I^{l'})$ for $(I^l, I^{l'}) \in A$.

The following lemma describes the path which can be found in the constructed graph in terms of unit vectors.

Lemma 4. Let I^s and I^t be the two nodes in the digraph defined by Definition 6, and $I^s = \alpha \cdot \mathbf{p}^s$ and $I^t = \alpha \cdot \mathbf{p}^t$. There is a path π from I^s to I^t such that $\pi = I^{l_1}, I^{l_2}, \cdots, I^{l_p}$, where $I^{l_1} = I^s$ and $I^{l_p} = I^t$, if and only if

$$\mathbf{p}^t = \mathbf{p}^s + \sum_{k=1}^{p-1} \mathbf{u}(I^{l_k}, I^{l_{k+1}}).$$

Proof: From Definition 6, we have

$$
\begin{aligned}
I^t &= I^{l_{p-1}} + w(I^{l_{p-1}}, I^t) \\
I^{l_{p-1}} &= I^{l_{p-2}} + w(I^{l_{p-2}}, I^{l_{p-1}}) \\
&\cdots \\
I^{l_2} &= I^s + w(I^{l_1}, I^{l_2}) \\
I^t &= I^s + \sum_{k=1}^{p-1} w(I^{l_k}, I^{l_{k+1}}).
\end{aligned}
$$

Notice that $w(I^{l_k}, I^{l_{k+1}}) = \alpha \cdot \mathbf{u}(I^{l_k}, I^{l_{k+1}})$ and $I = \alpha \cdot \mathbf{p}$, we have

$$
I^t = I^s + \sum_{k=1}^{p-1} \alpha \cdot \mathbf{u}(I^{l_k}, I^{l_{k+1}})
$$

and

$$
\mathbf{p}^t = \mathbf{p}^s + \sum_{k=1}^{p-1} \mathbf{u}(I^{l_k}, I^{l_{k+1}}).
$$

This exploited the relationship of $\mathbf{p}^s$ and $\mathbf{p}^t$ such that there exists a path from I^s to I^t. The definition below defines the concepts of accessible node and addressable digraph.

Definition 7. Let G be a digraph described in Definition 6. A node $I^l \in V$ is *accessible* if and only if there exists a path from I^1 to I^l.

If all the nodes in the digraph are accessible, the digraph is referred to as *addressable* digraph.

As an example, the addressable digraph for $I = 3p_1 + 7p_2$ is shown in Figure 1.

Corollary 1. In an addressable digraph, for any node $I^{l'}$ with $l' > 1$, there exists another node I^l and coefficient α_m such that the corresponding $\mathbf{p}^l$ and $\mathbf{p}^{l'}$ satisfy $\mathbf{p}^{l'} = \mathbf{p}^l + \mathbf{u}(\alpha_m)$.

Proof: In fact, $\pi' = I^{l_1}, I^{l_2}, \cdots, I^{l_{p-1}}$ is a subpath of path $\pi = I^{l_1}, I^{l_2}, \cdots, I^{l_p}$ defined in Lemma 4. If we let $I^l = I^{l_{p-1}}$ and $I^{l'} = I^{l_p}$, then applying Lemma 4 to the subpath π', we have

$$
\mathbf{p}^{l_{p-1}} = \mathbf{p}^{l_1} + \sum_{k=1}^{p-2} \mathbf{u}(I^{l_k}, I^{l_{k+1}}).
$$

Comparing it with

$$
\mathbf{p}^{l_p} = \mathbf{p}^{l_1} + \sum_{k=1}^{p-1} \mathbf{u}(I^{l_k}, I^{l_{k+1}}),
$$

we have $\mathbf{p}^{l_p} = \mathbf{p}^{l_{p-1}} + \mathbf{u}(I^{l_{p-1}}, I^{l_p})$, that is, $\mathbf{p}^{l'} = \mathbf{p}^l + \mathbf{u}(\alpha_m)$.

This means, in an addressable digraph, $I^{l'}$ can be always found from I^l by tracing the arc $A(I^l, I^{l'})$. Correspondingly, given the vector $\mathbf{p}^{l'}$, we can always find $\mathbf{p}^l$ by applying the unit vector $\mathbf{u}(\alpha_m)$, and visa versa.

Let $I^1, I^2, \cdots, I^l, \cdots, I^q$ be a sorted sequence of position integers for an integer linear system $I = \alpha \cdot \mathbf{p}$ satisfying the condition that there exists no $\mathbf{p} \in Z^{+r}$ such that $I^l < \alpha \cdot \mathbf{p} < I^{l+1}$ for $l = 1, 2, \cdots, q-1$. We consider the following problem.

Given the initial vector $\mathbf{p}^0$, find $\mathbf{p}'$ such that there is no $\mathbf{p}'$ satisfying $\alpha \cdot \mathbf{p}^0 < \alpha \cdot \mathbf{p}' < \alpha \cdot \mathbf{p}^1$. Interactively, given $\mathbf{p}^l$, find $\mathbf{p}^{l+1}$ such that there is no $\mathbf{p}'$ satisfying $\alpha \cdot \mathbf{p}^l < \alpha \cdot \mathbf{p}' < \alpha \cdot \mathbf{p}^{l+1}$. This problem is to be solved in an on-line fashion.

According to Lemma 4 and the Corollary 1 described previously, all the I^{l+1} in the above sequence can be found by $I^k + \alpha_m$ for some k. Meanwhile, $\mathbf{p}^{l+1}$ can be found by $\mathbf{p}^k + \mathbf{u}(\alpha_m)$ correspondingly. Note that α_m may not exist for some $m = 1, 2, \cdots, r$.

Consider an addressable digraph representing the above system. We first define the relationship between the nodes I^l and I^{l+1}, and the relationship between the nodes I^{l+1} and I^k as are described.

Definition 8. Given an addressable digraph as defined in definition 7 representing an integer linear system. Node I^{l+1} is referred to as *successive node of node I^l* if there is no $\mathbf{p} \in Z^{+r}$ such that $I^l < \alpha \cdot \mathbf{p} < I^{l+1}$, for $l = 1, 2, \cdots, q-1$, and denoted as *successive*(I^l). Node I^k and node $I^{k'}$ are *adjacent node* if they are connected by an arc pointed from I^k to $I^{k'}$. I^k is the *start node* of $I^{k'}$ associated with α_m denoted as *start-node*$_{\alpha_m}(I^k)$, and $I^{k'}$ is the *end node of I^k associated with α_m*, denoted as *end-node*$_{\alpha_m}(I^k)$.

If $I^l = \alpha \cdot \mathbf{p}^l$ and $I^l = \alpha \cdot \mathbf{p}^{l+1}$ then $\mathbf{p}^{l+1}$ is referred to as the *next vector of $\mathbf{p}^l$*.

Note that the out-degree for each node I^l, $l \le q - max\{\alpha_1, \cdots, \alpha_r\}$, is r where r is the dimension of the system. However, the in-degree of a node could be less than r in any case.

Definition 9. If the in-degree of a node given in the digraph defined in Definition 8 is equal to r with r being the dimension of the vector $\mathbf{p}$, then the node is referred to as a *full* node.

Lemma 5. Let I^j and I^k be the start nodes of I^l associated with α_j and α_k respectively. Assume the successive node of I^j and I^k are $I^{j'}$ and $I^{k'}$ respectively. In other words, $I^j = $ start-node$_{\alpha_j}(I^l)$, $I^k = $ start-node$_{\alpha_k}(I^l)$, successive$(I^j) = I^{j'}$ and successive$(I^k) = I^{k'}$ (see Figure 2). If $I^{j'} - I^j < I^{k'} - I^k$, then end-node$_{\alpha_j} - I^l < $ end-node$_{\alpha_k} - I^k$.

Proof: Since

$$
\begin{aligned}
\text{end-node}_{\alpha_j}(I^{j'}) &= I^{j'} + \alpha_j \\
\text{end-node}_{\alpha_k}(I^{k'}) &= I^{k'} + \alpha_k,
\end{aligned}
$$

$$
\begin{aligned}
\text{end-node}_{\alpha_j}(I^{j'}) - I^l &= I^{j'} + \alpha_j - I^l \\
&= I^{j'} - I^j
\end{aligned}
$$

and

$$
\text{end-node}_{\alpha_k}(I^{k'}) - I^l = I^{k'} - I^k
$$

for the same reason. Thus $I^{j'} - I^j < I^{k'} - I^k$ results in end-node$_{\alpha_j} - I^l < $ end-node$_{\alpha_k} - I^k$.

Theorem 4. If node I^l is a full node and $\mathbf{p}^l$ is corresponding to I^l. Consider all the start-nodes$_{\alpha_m}(I^l)$, for $m = 1, 2, \cdots, r$, and successive(start-nodes$_{\alpha_m}(I^l)$). If $m = m_0$ such that

$$
\text{successive(start-nodes}_{\alpha_{m_0}}(I^l)) - \text{start-nodes}_{\alpha_{m_0}}(I^l) \quad (1)
$$

is minimum, say Δ_{min}, then successive$(I^l) = I^l + \Delta_{min}$.

If successive(start-nodes$_{\alpha_{m_0}}(I^l)) = I^{j_0}$ and the corresponding input of the system is $\mathbf{p}^j$, then the next vector of I^l is $\mathbf{p}^{j_0} + \mathbf{u}(\alpha_{m_0})$.

This theorem can be proved by applying Lemma 5 inductively.

Corollary 2: If I^l is not a full node, and start-nodes$_{\alpha_{m'}}(I^l)$ does not exist, then $I^l - \alpha_{m'}$ should be used to substitute it in (1), and successive(start-nodes$_{\alpha_{m'}}(I^l)$) should be node I^j such that $I^j - (I^l - \alpha_{m'}) > 0$ and $I^j - (I^l - \alpha_{m'}) > 0$ is the minimum, for all $0 \le j \le q$.

If we consider $I^l - \alpha_{m'}$ as a virtual node, similar proof as for Lemma 5 can be used for this corollary.

Given a vector $\mathbf{p}^l$ as described in Definition 7 and Definition 8, the next vector of $\mathbf{p}^l$, $\mathbf{p}^{l+1}$ can be found by the following algorithm.

Algorithm NEXT

input: a sequence of vectors $\mathbf{p}^1, \mathbf{p}^2, \cdots, \mathbf{p}^l$ {to the integer linear system }

output: $\mathbf{p}^{l+1}$ { next vector of $\mathbf{p}^l$ }

1. $I^l = \alpha \cdot \mathbf{p}^l$ { as is executed on line, $I^{l-1}, I^{l-2}, \cdots$ should be already calculated and stored in a sorted order }

2. **for** $m = 1$ to r **do**
   begin

3.   start-nodes$_{\alpha_m}(I^l) = I^l - \alpha_m$

4.   **if** $P_m \neq 0$ **do**
            $I^j = $ successive(start-nodes$_{\alpha_m}(I^l)$)

5.   **else**
            find node I^j such that $I^{j-1} < $ start-nodes$_{\alpha_m}(I^l) < I^j$

6. $\Delta_m = I^j - $ start-nodes$_{\alpha_m}(I^l)$

7.   **if** $\Delta_m < \Delta_{m-1}$ **do**
        begin

8. $\Delta_{min} = \Delta_m$

9. $m_0 = m$
        end {of **if** }

10. $\mathbf{p}^{l+1} = \mathbf{p}^j + \mathbf{u}(\alpha_{m_0})$

11. **end** {of **for** }

Theorem 5. Algorithm NEXT finds the next vector $\mathbf{p}^{l+1}$ of vector $\mathbf{p}^l$ correctly.

Proof: First, it is easy to see that I^i found in step 4 or step 5 is a node in the addressable digraph. $\mathbf{p}^j$ is a vector such that $\alpha \cdot \mathbf{p}^j = I^j$. According to the definition of successive node and the statement in step 5, $I^j > $ start-nodes$_{\alpha_m}(I^l)$. Since start-nodes$_{\alpha_m}(I^l) = I^l - \alpha_m$, and as specified in step 2, $I^j > I^l - \alpha_m$ and $I^j + \alpha_m > I^l$ for any α_m. Here $I^j + \alpha_m$ is a node in the addressable digraph, and $\mathbf{p}^j + \mathbf{u}(\alpha_m)$ is the corresponding vector satisfying $\alpha \cdot (\mathbf{p}^j + \mathbf{u}(\alpha_m)) = I^j + \alpha_m$ according to Corollary 2.

Since $m_0 \in \{m | m = 1, 2, \cdots, r\}$ (step 9) and $\mathbf{p}^{l+1} = \mathbf{p}^j + \mathbf{u}(\alpha_{m_0})$ (step 10), we have proved that $I^j + \alpha_{m_0}$ is a node, $I^j + \alpha_{m_0} > I^l$, and $\mathbf{p}^{l+1} = \mathbf{p}^j + \mathbf{u}(\alpha_{m_0})$ satisfies $\alpha \cdot \mathbf{p}^{l+1} = I^j + \alpha_{m_0}$.

Now we prove that there is no vector $\mathbf{p}'$ existing such that $I^l < \alpha \cdot \mathbf{p}' < \alpha \cdot \mathbf{p}^{l+1}$.

i) I' can not be accessed by α_{m_0}, otherwise $I' - \alpha_{m_0}$ is a node. This is because $I^j < I' < I^{l+1}$, and $I^l + \alpha_{m_0} < I' - \alpha_{m_0} = I^j$, which means that there is a node existing between $I^l + \alpha_{m_0}$ and I^j, thus contradicts the generating of I^j stated in step 4 or step 5.

ii) I' can not be accessed by $\alpha_{m'}$, for $m = 1, 2, \cdots, r$ other than m_0. If I' can be accessed by $\alpha_{m'}$, for $m' \in \{m | m = 1, 2, \cdots, r$ and $m \neq m_0\}$, then $I' - \alpha_{m'}$ is a node. After executing the **for** loop in step 2 for $m = m'$, $(I' - \alpha_{m'}) - (I^l - \alpha_{m'})$ will be stored in Δ_m, and which is equal to $I' - I^l$.

After executing the **for** loop in step 2 for $m = m_0$, $(I' - \alpha_{m_0}) - (I^l - \alpha_{m_0})$ will be stored in Δ_m and which is equal to $I^{l+1} - I^l$. Now, if $I^l < I' < I^{l+1}$, then $I' - I^l < I^{l+1} - I^l$. $\Delta_m|_{m=m'} < \Delta_m|_{m=m_0}$, which contradicts the execution of step 7 to step 9. Thus I' is not possibly existing such that $I^l < I' < I^{l+1}$.

In fact, by executing Algorithm NEXT, we examined the possibility to access I^{l+1} from the found I's by different α_m's, for $m = 1, 2, \cdots, r$. For each specific α_m, we noticed that if I^l and I^{l+1} are closest, $I^l - \alpha_m$ and $I^{l+1} - \alpha_m$ are also closest. So, instead of finding the node closest to I^l, we search for the node closest to $I^l - \alpha_m$, for all the $m = 1, 2, \cdots, r$, utilizing the known information of the found vectors and found nodes.

The on-line feature of the algorithm limited the memory used and thus reduced the executing time.

Let the largest entry in α be α_{max}, for $m = 1, 2, \cdots, r$. To exam the range of start-nodes$_{\alpha_m}(I^l)$ notice that the smallest start-nodes$_{\alpha_m}(I^l) = I^l - \alpha_{max}$. Therefore, we have $I^l - \alpha_{max} < $ start-nodes$_{\alpha_m}(I^l) < I^l$.

In the integer system, there are at most α_{max} nodes existing in the interval between $I^l - \alpha_{max}$ and I^l. To compute a new output, say I^l, it is sufficient to store the previous results $I^j, \cdots, I^{l-2}, I^{l-1}$ and the corresponding vectors $\mathbf{p}^j, \cdots, \mathbf{p}^{l-2}, \mathbf{p}^{l-1}$ such that $j = I^l - \alpha_{max}$.

Step 2 shows that to find one next vector, the algorithm needs to execute r times for a vector with r dimensions. Step 6 involves a binary search on a sequence of α_{max} data items. $O(log\alpha_{max})$ time is hence needed. Other steps take only constant time. So the time complexity of Algorithm NEXT is $O(rlog\alpha_{max})$.

4. Experimental Results

We have implemented the results in this paper in a parallel compiler prototype, which performs the dependence analysis and parallel transformations for FORTRAN programs. The prototype computes the α, β, γ in Theorem 1, and the initial vectors at compiler time, then uses the information for dynamic scheduling as in [16] at execution time. The parallel code generated by the prototype is running on a shared memory multiprocessor simulator based on a commercial MIPS-based system simulator. The system can simulate 4 to 16 processors, various sizes of cache memories, various cache coherence protocols, various cache line sizes and various memory bandwidth of data bus, crossbar or interconnection network. The processor and the cache in the simulation system are based on MIPS R3000 and R4000. The prototype simulates different scheduling strategies also. They include static scheduling, self-scheduling, and guided-self-scheduling. The experimental results show that these scheduling approaches are slightly different on the execution performance. But the technique presented in this paper to reduce the cache ping-pong problem made a significant improvement for the execution performance, no matter which scheduling approach was used in the experiments. We employ the self-scheduling scheme in [16] in our prototype and simulator, because it is the most simple scheme in the implementation.

In examining the experimental results, the reader should be aware that some of the improvements cited may have been achieved because the huge cost of the cache is missing in RISC architecture. We compared the parallel code execution with or without the compiler strategy presented in this paper for the cache ping-pong problem and found significant enhancement by eliminating the unnecessary data moving back and forth between processors. In the experiment, we assume that the memory and crossbar bandwidth is proportionally improved when the number of processors gets increased.

<u>Gaussian Elimination</u> Gaussian Elimination is a basic matrix operation that is used in many application programs. We use a 1K by 1K array in the experimental benchmark. The following is the kernel code of the program:

```
DO k = 1, N
........
    PDO j = k+1, N
    DO i = k+1, N
        ........
        A(i,j) = A(i,j) - A(i,k)*A(k,j)/A(k,k)
        ........
    CONTINUE
    CONTINUE
CONTINUE
```

Number processor	Original Serial Code	Parallel Code with Cache Ping-Pong	Parallel Code without Cache Ping-Pong	Speed Up
4	285.0s	163.5s	102.2s	1.6
8	285.3s	109.8s	59.9s	1.7
16	286.2s	83.6s	39.8s	2.1

Linpack Benchmark Linpack benchmark is vectorized/parallelized code. We chose the loops containing SAXPY and SMXPY subroutine calls, and inlined these routines in the loops, most of which have three level loops: serial, parallel, serial. As the table below shows, our approach achieved better performance from the original parallel code that doesn't have any consideration for the cache Ping-Pong problem. To make this measurement, we use 1K by 1K Linpack benchmark.

Number processor	Original Serial Code	Parallel Code with Cache Ping-Pong	Parallel Code without Cache Ping-Pong	Speed Up
4	514.7s	327.5s	234.0s	1.4
8	514.3s	226.9s	151.3s	1.5
16	515.2s	161.0s	94.7s	1.7

A Complete Application We also performed the test on a complete application program benchmark, a computational chemistry application program. The kernel of the most frequently used routine has the following form:

```
      K = 0
100   K = K + 1
      ...........
      PARALLEL DO I = 1, M
      ...........
        DO 200 J = 1, M
          X(I+K,J+I+K) = ......
          ...........
          .... = X(I+K, J+I+K) * .......
200     CONTINUE
      END_PARALLEL_DO
      ...........
      IF (K .LT. MAX_BOUND) GO TO 100
```

The dimension of the array X is 3K by 1k. The approach presented in the paper needs to be modified slightly, because the outermost serial loop is an IF_loop, which doesn't have explicit loop bound. The table below shows the results when the modified approach is applied to eliminate cache Ping-Pong.

Number processor	Original Serial Code	Parallel Code with Cache Ping-Pong	Parallel Code without Cache Ping-Pong	Speed Up
4	1732.0s	1415.6s	832.7s	1.7
8	1734.1s	843.2s	481.8s	1.75
16	1735.7s	557.8s	309.9s	1.8

5. Conclusions

A compiler technique to solve the "cache ping-pong" problem which is very interesting and very important in parallel processing has been developed. In particular, a very common parallel program construct has been studied in which parallel loops are enclosed by a serial loop and the array elements are reused in the parallel loops in different iterations of the enclosed serial loop. Efforts have been made to reduce the data movement between the caches for parallel programs. Methods of calculating the appropriate parallel loop indices for each processor in terms of the data stored in its cache have been used. Although researches in this area are not completed yet, our efforts are significant in the sense that it introduced the mathematical concepts, analyzed the array subscript expressions, and provided the effective approach as a base solution for further cache ping-pong eliminating tasks in parallel processing systems.

Reference

[1] Fang, Z, "Cache or Local Memory Thrashing and Compiler Strategy in Parallel Processing Systems", Proc. of the International Conference on Parallel Processing, Aug. 1990, pp. II-271 - II-275.

[2] Kuck, D. J., The Structure of Computer and Computations. Vol. 1, John Wiley and Sons, New York, 1978.

[3] Wolfe, M. J., Techniques for Improving the Parallelism in Programs, Report 78-929, Dept. of Computer Science, Univ. of Illinois at Urbana-Champaign, July 1978.

[4] Kuck, D. J., Kuhn, R. H., Leasure, B., and Wolfe, M.,"The Structure of an Advanced Vectorizer for Pipeline Processor", Proc. IEEE Computer Society Fourth International Computer Software and Applications Conf., Oct. 1980.

[5] Allen, J. R., and Kennedy, K., PFC: A Program to Convert Fortran to Parallel Form, Rep. MASC-TR82-6, Rice Univ. Mar. 1982.

[6] Allen, J. R. and Kennedy, K., "Automatic Loop Interchange", Proc. of the ACM SIGPLAN 84 Symposium on Compiler Construction, June 1984, pp. 233-246.

[7] Callahan, D., Cooper, K. D., Kennedy. K., and Torczon. L.,"Interprocedural Constant Propagation", Proc. of SIGPLAN 86 Symposium on Compiler Construction, July, 1986, pp. 59-67.

[8] Abu-Sufah, W., Kuck, D. and Lawrie, D., "On the Performance Enhancement of Paging Systems Through Program Analysis and Transformations", IEEE Transactions on Computers, Vol. C-30, No. 5, May 1981.

[9] Wolfe, M., "Iteration Space Tiling for Memory Hierarchies", Proc. of the Third SIAM Conf. on Parallel Processing, Los Angeles, CA, Dec. 1-4, 1987.

[10] Callahan, D., Carr, S., and Kennedy, K., "Improving Register Allocation for Subscripted Variables", Proc. of the ACM SIGPLAN'90 Conference on Programming Language Design and Implementation, White Plains, NY, June 20-22, 1990.

[11] Leasure, B., et. al. "PCF Fortran: Language Definition by the Parallel Computing Forum", Proc. International Conferences on Parallel Processing, Aug. 1988.

[12] Jesse Fang and Mi Lu, An Iteration Partition Approach for Cache or Local Memory Thrashing on Parallel Processing, Technical Report TAMU-ECE 91-02, Department of Electrical Engineering, Texas A&M University.

[13] Kennedy, K., Automatic Translation of Fortran Programs to Vector Form, Tech. Rep. pp. 476-029-4, Rice University, Houston TX, Oct. 1980.

[14] Kuck, D., Kuhn, R., Leasure, B., Padua, D., and Wolfe, M., "Dependence Graph and Compiler Optimizations", Conf. Record of 8th ACM Symposium on Principles of Programming Languages, Jan. 1981.

[15] Padua, D. and Kuck, D., "High-speed Multiprocessors and Compilation Techniques", IEEE Trans. Comput., C-29, Sept. 1980, pp. 763-776.

[16] Fang, Z., Yew, C., Tang, T. and Zhu, C.,"Dynamic Processor Self-scheduling for General Parallel Nested Loops", IEEE. Trans. Comput., Vol. 39, No. 7, July 1990, pp. 919-929.

A Lockup-free Multiprocessor Cache Design

Per Stenström, Fredrik Dahlgren, and Lars Lundberg

Department of Computer Engineering, Lund University
P.O. Box 118, S-221 00 LUND, Sweden

Abstract

Shared memory latency is an important performance limitation of large-scale, shared-memory multiprocessors. Although it can be reduced by using private caches and pipelined networks, pipelining is restrictive due to access order requirements as imposed by the programming model.

By letting the compiler (or user) provide access order information to the memory system, considerable performance improvements can be obtained. In this paper, we report on the design of a processing element memory subsystem, in essence a lockup-free cache, that exploits this and controls pipelining and cache coherence for a general class of pipelined networks.

1   Introduction

Large-scale, shared-memory multiprocessors rely on private caches to reduce performance degradation due to contention and network latency [10]. Even if caches are infinitely large, communication between the caches are needed in order to maintain *cache consistency*.

In order to hide the network latency, one can use pipelined networks and exploit *memory access pipelining*. The buffered MIN in IBM RP3 [8] is one example. The correctness of parallel programs relies on a well-defined *access order model*. Sequential consistency [7] is the strictest model and assumes that the result of the execution is the same as if memory accesses are performed in program order. In general, pipelining must be avoided which degrades performance.

Other models restrict access order to synchronization points. They have been formulated differently in terms of pipelining requirements such as the *weakly ordered model* originally proposed by Dubois et al. [2], the DRF0-model [1], and *release consistency* [4]. By letting the compiler (or user) provide access ordering information, we can formulate sufficient conditions that can be exploited by the memory system.

This results in performance improvements due to less restrictive pipelining requirements.

In this paper, we formulate information that is extracted from the parallel program, and sufficient conditions needed to maintain various access order models. This information is exploited by a lockup-free cache controller in the global memory request pipelining control. We present a complete VLSI-design of the lockup-free cache controller and outline the requirements on other parts of the memory system.

Related work on this topic is the uniprocessor, lockup-free cache proposal by Kroft [6]. Scheurich and Dubois proposed a multiprocessor lockup-free cache which supports the weakly ordered model and is aimed at bus-oriented architectures [9]. Important objectives of our design were to support general access order models and make it applicable to a general class of networks.

2   Access order information

In this section, we consider three access order models; sequential consistency, weak ordering, and release consistency. We use a classification scheme for variables and define a relation that constrains access order for the correctness of parallel programs under these models.

2.1   Access order relations

The access order relation is based on the notion of "performing a memory access" specified by Dubois et al.[2]. In Section 3, we use our classification scheme and relation to formulate sufficient pipelining requirements in terms of memory access performance.

Consider an *access sequence* $S = (a_1, a_2, \ldots, a_n)$ as specified by the program for a single processor, where $a_i \in C$ which is a set of *consistency properties*. The elements of $C = \{weak$ (w), *ordinary* (o), *strong* (s), *fence* (f), *gather* (g)$\}$. The parallel program is correct if accesses are performed in the order specified by S.

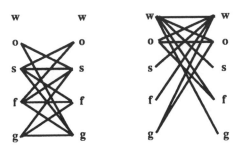

Figure 1: The R_{ao} (left) and $\overline{R_{ao}}$ (right) relations.

The *access order* relation $a_i \, R_{ao} \, a_{i+1}$ is a binary relation on C that constrains access order between two consecutive accesses: a_i must be **performed** before a_{i+1}. Since access order can be relaxed, $R_{ao} \subset C \times C$. In Figure 1 (to the left), we show all $(a_i, a_{i+1}) \in R_{ao}$ as solid lines.

The relation $\overline{R_{ao}}$ (to the right in Figure 1) is defined such that $\overline{R_{ao}} \cap R_{ao} = \emptyset$ and $\overline{R_{ao}} \cup R_{ao} = C \times C$. $\overline{R_{ao}}$ thus specfies when access order may be relaxed: If $(a_i, a_{i+1}) \in \overline{R_{ao}}$, then a_{i+1} may be performed before a_i without violating parallel program correctness. That is, given an access sequence S, all permutations of S that are constructed by successively applying $\overline{R_{ao}}$ to S are correct.

2.2 Access order models

R_{ao} does not constrain access order for sequential program correctness. For the multiprocessor organization we consider, we will show that this is maintained. In the following we show how R_{ao} maintains parallel program correctness under various access order models.

Sequential consistency (SC): SC was formulated by Lamport [7]. Dubois et al. [2] stated a sufficient condition to maintain SC in terms of memory access performance:

> If all memory accesses are performed in program order, SC is maintained.

For the subset $C_1 = \{weak \ (w), strong \ (s)\} \subset C$, the condition above is satisfied by R_{ao} (on C_1) for all $(a_i, a_{i+1}) = (s, s)$. In other words, if shared, read-write variables are denoted *strong*, SC is maintained. Access order relaxation can be obtained for read-only data. Such data are denoted *weak*.

Weak ordering: We now consider the subset $C_2 = \{weak \ (w), ordinary \ (o), strong \ (s)\} \subset C$. The rationale behind *weak ordering* [2] is to restrict access order to explicit synchronization points. By denoting synchronization variables *strong*, and other shared,

read-write variables *ordinary*, access order relaxation between *ordinary* accesses is obtained. For instance, $((o, o) \in \overline{R_{ao}})$. However, $\{(o, s), (s, o)\} \subset R_{ao}$. In this case, R_{ao} and $\overline{R_{ao}}$ are binary relations on C_2.

Release consistency: Programming paradigms that use critical sections to access shared variables rely on some synchronization primitives such as Acquire to enter a c.s. and Release to exit. Gharachorloo et al. [4] formulated an access order relaxation, denoted *release consistency*, for which we associate the consistency property *fence* (f) with Acquire and *gather* (g) with Release, instead of associating strong with both of them. This results in the following relaxations: $\{(o, f), (g, o)\} \subset \overline{R_{ao}}$, that is, an Acquire operation may be performed before a previously issued *ordinary* access. However, *ordinary* accesses must be performed before a subsequently issued Release operation is performed.

Correctness of the parallel program below relies on release consistency. All variables (A,B,C, and D) are *ordinary*. Assume that processor P0 enters the critical section before P1. This implies that D will be assigned the value of A (that is 1) due to access order constraints. Access order may be relaxed as demonstrated:

P0:		P1:
A:=1;	w_1, o_2	...
Acquire(X);	w_3, f_4	Acquire(X);
B:=A;	w_5, o_6, o_7	D:=C
C:=B;	w_8, o_9, o_{10}	Release(X);
Release(X);	w_{11}, g_{12}	...

P0 specifies the access sequence $S = (w_1, o_2, w_3, f_4, w_5, o_6, o_7, w_8, o_9, o_{10}, w_{11}, g_{12})$. Here, instruction fetches are *weak*, data accesses are *ordinary*, and the primitives Release and Acquire are *fence* and *gather*, respectively. R_{ao} states that all permutations of *weak* and *ordinary* accesses are correct as long as f_4 is performed before o_i, $i \geq 6$ and all o_i, $i \leq 10$ are performed before g_{12}. Finally, R_{ao} constrains f_4 to be performed before g_{12}.

3 Memory system requirements

Figure 2 shows the multiprocessor organization considered. Each processing element (PE), which consists of a processor and a lockup-free cache, is connected to the shared memory modules via a pipelined interconnection network with the following characteristics:
(1) There is exactly one bidirectional path between a PE and each memory module. (2) Each path is a

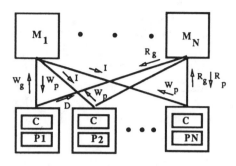

Figure 2: Multiprocessor organization.

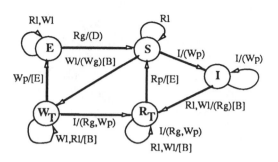

Figure 3: Cache coherence state transitions.

memory access pipeline such that two consecutively issued accesses cannot pass each other.

In order to exploit pipelining, each cache is lockup-free, meaning that it can handle multiple global accesses. Two important implications on the cache controller to be described in Section 4 are pipelining control and cache coherence maintenance. We describe how these issues are solved separately below.

3.1 Pipelining control

We assume that all processors issue memory accesses in program order. From the viewpoint of a sequential program for a processor, data and instructions may reside in its cache or in any memory module.

Although pipelining may result in that two consecutively issued memory accesses from the same processor are performed at memory or in the cache out of program order, sequential program correctness is maintained by the following reason:

Consider an access sequence $S = (W_{i-1}, R_i, W_{i+1})$ to the **same** variable, where W denotes WRITE accesses and R denotes READ. Sequential program correctness, in terms of access order, is assured iff R_i returns the value written by W_{i-1}. This is maintained by network assumption (1) above. Thus, pipelining is allowed from the viewpoint of sequential program correctness.

For a parallel program, data may be replicated at several caches. A read and write access may be local (satisfied by the private cache) or global. In order to notify the cache controller when a global memory access has been performed, we associate a response transaction PERFORMED with each of them. The pipelining condition can now be stated:

Condition: (Pipelining): Given an access sequence $S = (a_1, a_2, \ldots, a_n)$. Then for any two consecutive accesses (a_i, a_{i+1}) a_{i+1} may not be issued until a_i is performed iff $(a_i, a_{i+1}) \subset R_{ao}$. Conversely, a_{i+1} may be issued before a_i is performed iff $(a_i, a_{i+1}) \subset \overline{R_{ao}}$.

The lockup-free cache controller to be presented controls pipelining according to this condition by the means of a set of five *access order states* $P = \{S_n, S_o, S_s, S_f, S_g\}$. S_n means that there is no outstanding global access. The other states S_i mean that one or several accesses of consistency property i are outstanding. Consequently, the cache controller can maintain R_{ao} by keeping track of the type of outstanding accesses. In Section 4, we will describe how these states are supported by the cache controller.

3.2 The cache coherence protocol

Previously published cache coherence protocols [10] do not support lockup-free caches for pipelined networks. The write-invalidate protocol to be described attacks this problem and is applicable to multiprocessor organizations according to Figure 2.

The global actions are as follows (see Figure 2). A write access by P_1 to a replicated block results in a global write access W_g to be sent to the corresponding memory module that stores the block (M_1). M_1 broadcasts invalidations (I) to those caches having a copy of the block. These respond by sending W_p to notify M_1 that the block is invalidated. Finally, C_1 is notified that the write access is performed by receiving W_p from the memory module. If two processors issue global writes to the same block simultaneously, one will receive W_p while the other receives I.

A read miss from processor P_N results in a global read access R_g to be sent to the memory module that stores the block (M_N). M_N keeps track of where the block resides and retransmits the access to that cache (C_1). C_1 responds with the block (D) which is routed to P_N as R_p. However, if there is a pending invalidation for the block, M_N will respond with I. These actions assure that a read and write access are performed when the cache receives R_p and W_p, respectively.

The actions taken by the cache controller are spec-

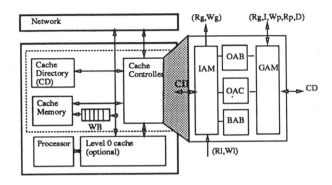

Figure 4: PE and cache controller organization.

ified by the Mealy state transition graph in Figure 3. Each cache block is in any of the states (E)xclusive, (S)hared, (I)nvalid, Writing (W_T), or Reading (R_T). Cache controller actions are a result of a *stimulus* which can be: processor reads and writes (R_l, W_l) or global actions or responses as described above. In the graph, we denote a cache controller action: stimulus/(global actions),[internal actions].

If the block is in the cache (state S or E) a read can be carried out locally. If a read miss is encountered (state I), transition to R_T is done and a global read access is issued. The read is also buffered (denoted [B]). Likewise, if a write to a shared block (state S) is encountered, transition to state W_T is done and a global write access is issued. The write is buffered also in this case. R_T and W_T are *transient* states and will be left when the global access is performed. If the processor issues reads and writes to a block in a transient state, these accesses will be buffered. When the global access is performed, all buffered accesses are carried out in issuance order (denoted [E]).

4 Lockup-free cache implementation

Figure 4 shows the organization of each PE. Besides the optional level 0 cache (to be discussed later), the lockup-free cache consists of a cache controller, a cache directory (CD) that stores the state of each cache block, the cache memory, and a write buffer (WB).

4.1 Functional units

The functional units of the cache controller (see Figure 4) are: Internal Access Mechanism (IAM), Global Access Mechanism (GAM), Outstanding Access Buffer (OAB), Outstanding Access Counter (OAC), and the Blocked Access Buffer (BAB). This functional decomposition was found to achieve as high degree of parallelism as possible.

Each internal access (R_l or W_l) consists of an address, consistency property, a sequence number (to keep track of issuance order) and data (in case of W_l).

The IAM processes internal accesses as follows: First, the consistency property is checked against the access order state (P). If R_{ao} is violated, the access cannot be issued and is buffered in BAB. Second, the state of the block is checked in the cache directory and actions are taken according to the state transition graph from Figure 3. If the action [B] is taken (as a result of a read miss or global write access), the internal access is buffered in OAB.

The GAM processes global accesses and responses as follows: The state of the block is checked in the cache directory and cache coherence actions are taken according to Figure 3. When a transient state is left, the action [E] is taken. It results in a sequential search for all entries in the OAB that match the address. These accesses will be handled in issuance order (as depicted by the sequence number) by the IAM.

The access order state is kept track of by IAM. The access order relation specifies that exactly one outstanding access of type *strong*, *fence*, or *gather* is allowed. However, several outstanding accesses of type *ordinary* might be allowed. The OAC (a counter), which is incremented whenever an ordinary global access is issued and decremented when the response is received, keeps track of these.

4.2 VLSI-design characteristics

The cache controller has been successfully implemented using Genesil silicon compilation system [5], which is based on parameterized module generators. The IAM and GAM are implemented by finite state machines consisting of about 50 states each.

The OAB is implemented by data paths, a finite-state machine (100 states), and an 8 entry buffer, each storing the address (32 bits), consistency property (3 bits), type (read, write) (1 bit) and sequence number (5 bits). The BAB is a FIFO which is implemented by a data path, a RAM, and a finite state machine (50 states). Data associated with buffered accesses are stored in a write buffer outside the chip (WB in Figure 4) in order to limit the chip area and number of pins. If any of the buffers (OAB or BAB) are full, the cache will block.

The chip size is 9.5×11.0 mm^2 using a 1.5 μ CMOS process (see Figure 5). It contains about 32600 transistors. As can be seen, the buffers (OAB and BAB) constitute the major part of the chip (40% of the core area). The chip contains 141 pins distributed as fol-

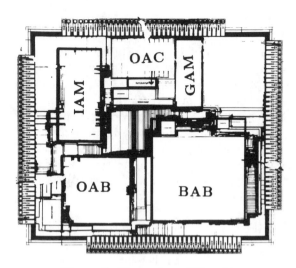

Figure 5: Floor-plan of the chip.

lows: Two access interfaces (local and global) consisting of 42 bits each, two cache directory interfaces (10 bits each) to facilitate duplication of the cache directory to avoid contention. The other 37 pins are used for memory, network, and processor (or level 0 cache) handshaking.

The chip is controlled by a two-phase clock. The most common case is to handle a cache hit which takes a clock cycle. The cache directory check and cache memory access constitute two stages in the chip pipeline. Circuit simulations have shown that the chip can be operated at a peak speed of 10 MHz. We will discuss the implications on this in the next section.

5 Discussion

A VLSI-implementation of a lockup-free cache controller which controls pipelining and cache coherence under various access order models has been presented. The cache controller is applicable to a general class of multiprocessors with pipelined networks.

Release consistency can result in considerable performance improvements due to less restrictive pipelining requirements [3]. On the other hand, less strict models may outrule the use of multiprocessors for applications relying on stricter models. Our cache design can support various models by letting the compiler (or user) classifying variables into consistency properties.

Our lockup-free cache design improves over the proposal by Scheurich and Dubois [9] in the following respects: It implements a cache coherence protocol for a general class of pipelined networks and it can support cache blocks with arbitrary block sizes. Furthermore, we have made a complete realization which gave the following important insights.

A major part of our present design is the buffer space. A critical design parameter is to find out which size is optimal for the majority of applications. Due to the complex protocols and buffer management algorithms implemented by the cache controller, the cache access time is limited to 100 ns/cache access for the technology we used. For cache sizes relevant for multiprocessors (several mega bytes), this will match the access time of RAM-memory in the same technology. However, it necessitates the use of a lockup-free cache hierarchy.

Figure 4 shows an optional level 0 cache. Surprisingly enough, this cache can be lockup-free without buffers and complex management and can thus be faster. The reason is that cache misses propagates to the level 1 cache and are book-kept there. In order not to violate cache coherence or access order, it must propagate all writes to the level 1 cache. The access order states kept track of by the level 1 cache must also be visible to the level 0 cache.

References

[1] S. V. Adve and M. D. Hill. Weak Ordering — A New Definition. In *Proc. of the 17th Annual Int. Symposium on Computer Architecture*, pages 2–14, 1990.

[2] M. Dubois, C. Scheurich, and F. Briggs. Memory Access Buffering in Multiprocessors. In *Proc of 13th International Symposium on Computer Architecture*, pages 434–442, 1986.

[3] K. Gharachorloo, A. Gupta, and J. Hennessy. Performance evaluation of memory consistency models for shared-memory multiprocessors. In *ASPLOS IV*, April 1991.

[4] K. Gharachorloo, D. Lenoski, J. Laudon, P. Gibbons, A. Gupta, and J. Hennessy. Memory Consistency and Event Ordering in Scalable Shared-Memory Multiprocessors. In *Proc. of 17th International Symposium on Computer Architecture*, pages 15–26, May 1990.

[5] S.C. Johnson. Silicon compiler lets system makers design their own chips. *Electronic Design*, (Oct 4):167–181, 1984.

[6] D. Kroft. Lockup-free Instruction Fetch/Prefetch Cache Organization. In *Proc of 8th Int. Symp. on Computer Architecture*, 1981.

[7] L. Lamport. How to Make a Multiprocessor Computer That Correctly Executes Multiprocess Programs. *IEEE Transactions on Computers*, C-28:690–691, 1979.

[8] G. F. Pfister, W. C. Brantley, D. A. George, S. L. Harvey, W. J. Kleinfelder, K. P. McAuliffe, E. A. Melton, V. A. Norton, and J. Weiss. The IBM Research Parallel Processor Prototype (RP3): Introduction and Architecture. In *Proceedings of the 1985 international conference on parallel processing*, 1985.

[9] C. Scheurich and M. Dubois. Concurrent Miss Resolution in Multiprocessor Caches. In *Proc of Int. Conf. on Parallel Processing*, pages 118–125, 1988.

[10] P. Stenström. A Survey of Cache Coherence Schemes for Multiprocessors. *IEEE Computer*, 23(6), June 1990.

STREAMLINE: CACHE-BASED MESSAGE PASSING IN SCALABLE MULTIPROCESSORS

Gregory T. Byrd[*] and Bruce A. Delagi[†]
Dept. of Electrical Engineering
Stanford University

Abstract. Traditional shared-memory multiprocessors are designed to efficiently support the shared variable model of computation. Accordingly, they offer an efficient means of demand-driven communication, where data is *fetched* when it is needed by the consuming process. On the other hand, data-driven communication, in which values are *sent* proactively when they are produced, is not so well supported. Our research is an attempt to quantify the effect of providing hardware support for data-driven communication to scalable shared-memory multiprocessors.

In order to investigate this idea, we have developed the *StreamLine* mechanism. StreamLine is intended to extend a multiprocessor cache coherence mechanism, in order to provide proactive, cache-to-cache data transfers, as well as assist in the management of message queues and the scheduling of data-driven computations. This paper describes the basics of StreamLine, some ideas toward a potential implementation, and some preliminary simulation results.

Introduction

Large-scale MIMD parallel computers can be divided into two broad classes of architecture — *shared-memory* machines (often called multiprocessors) and *message-passing* machines (often called multicomputers or distributed memory machines) [7]. Each of these classes of machines is based on an underlying model of computation and communication.

In shared-memory machines, communication between processors is accomplished by reading and writing data in agreed-upon shared locations (i.e., *shared variables*). Shared variable communication naturally supports a *demand-driven* style of computation — a consumer retrieves data from global memory as and when it is needed. This can be very efficient, especially if each application deals with only a small piece of a large, shared data structure.

In contrast, message-passing machines are designed to support a *data-driven* computational model — computation is initiated by sending data directly to a consumer process. The consumer need not fetch the data from the producer, and process synchronization is implicit in the sending and receiving of messages. A message-handling coprocessor is often provided to allow messages to be routed and received without interrupting the applications processor [9].

Programmers want to use the model that is most efficient for a particular application, be it memory-oriented or message-oriented, regardless of the underlying hardware architecture. Either model can, of course, be implemented on either class of machine, but there are significant performance penalties imposed by trying to use the "unsupported" model on a given architecture.

Consider, for instance, an implementation of message passing on a shared-memory multiprocessor with invalidate-based cache coherence. As a process produces data, it is written into the *producer's* cache, not the consumer's. Thus when the consumer wants the data, it must retrieve it, involving a round trip through the network. Also, the synchronization involved with signalling the availability of the data is a separate operation from the actual writing (and reading) of the data.

Message-passing machines, on the other hand, typically involve too much overhead with message delivery and interpretation to efficiently provide fine-grained access to shared data.

We believe that it is important to provide a single architecture to support both demand-driven and data-driven computation, so that parallel programs can be tailored to the needs of the application, rather than the needs of the underlying machine. In particular, we are investigating the potential of adding support for data-driven computation to scalable shared-memory multiprocessors.

As part of the investigation, we propose a communication mechanism called *StreamLine*. StreamLine is intended to extend a multiprocessor cache coherence mechanism, in order to provide proactive, cache-to-cache data transfers, as well as assist in the management of message queues and the scheduling of data-driven computations. This paper described the basic mechanism, describes a possible implementation, and presents some preliminary performance data, based on simulation. Plans for future studies will also be discussed.

The Mechanism

StreamLine is the name of a cache-based mechanism for message passing in multiprocessors with a global address space. The mechanism combines the abstraction of a stream with cache and memory management techniques to provide for proactive data transfer between caches. This section provides background on the stream abstraction and a behavioral description of the StreamLine mechanism.

Streams

A *stream* is a data structure which represents a sequence of values communicated between threads of computation [1, 5]. The values may be self-referential entities (such as integers, strings, or arrays), pointers to global data, or pointers to other streams. The sequence itself is potentially infinite, meaning that there is no inherent limit to the number of values which may be passed over time.

[*] North Carolina Supercomputing Center, P.O. Box 12889, Research Triangle Park, NC 27709
[†] Digital Equipment Corp., 654 High Street, Palo Alto, CA 94301

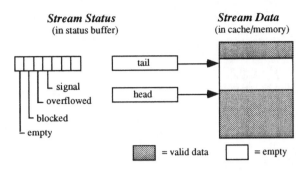

Figure 1: Stream status and data.

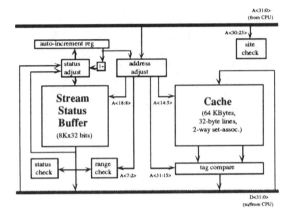

Figure 2: StreamLine cache organization.

Stream values are produced and consumed incrementally, and operations on streams are *non-strict* — that is, the operations can be applied to the stream before the entire sequence of values has been computed. Producers write to a stream; this causes a value to be placed at the end of the sequence. Writes from multiple producers are merged non-deterministically. Consumers read from a stream, which returns the first value in the sequence and (optionally) removes it from the sequence.

In addition to communication, a stream also provides a means of synchronization among processes. Upon reading an empty stream, a consumer may become blocked and may later be resumed when a value is written to the stream.

Streams as Memory Objects

StreamLine implements the stream abstraction by associating each stream with a region of memory, which is used to hold a subset of the values in the sequence repesented by the stream. This memory is managed as a circular buffer — items are added to the sequence by writing to the location at the *tail* of the buffer, and items are removed by reading from the *head* of the buffer. (See figure 1.) The locations between the head and tail contain the stream values which have been produced but not yet consumed.

Status information for each stream must be maintained. A minimal set of status includes:

- **head:** pointer to the beginning of valid data;
- **tail:** pointer to the first location following valid data;
- **empty:** a flag which is set when there is no valid data on the the

- **blocked:** a flag which is set when there is a thread waiting for data to be written on the (empty) stream;
- **signal:** a flag which indicates whether an exception signal should be generated upon reading an empty stream.

Stream status is used to constrain and augment memory accesses to the stream data buffer. For instance, a read operation to a location outside the head and tail is an invalid operation. Additionally, a write to an empty stream may cause some other activity, such as activating a computation which is waiting for data.

A Proposed Implementation

This section sketches a possible implementation of the StreamLine mechanism. We describe the implementation in terms of a scalable multiprocessor architecture, in the range of a hundred to a few thousand processors, with a distributed global address space. We assume off-the-shelf high-performance processors with on-chip cache and virtual memory hardware, a torus or mesh interconnection network utilizing cut-through routing [3], and a directory-based hardware cache coherence mechanism [2]. Each processing site contains a processor, some second-level cache, a portion of the global memory, and an interface to the interconnection network.

We propose to implement StreamLine as part of the multiprocessor's memory system. Since nearly all stream operations require knowledge of a stream's status, we provide a special *stream status buffer* (SSB) to allow the checking of status information in parallel with data access. We also implement special cache operations, through control registers and accesses to reserved memory locations, for accessing and manipulating stream status and for requesting message transmission between streams.

Cache Organization

The organization of the second-level cache is shown in figure 2. The right half of the figure shows a traditional cache organization: 64K-byte cache, 2-way set-associative, 32-byte lines. We assume that the physical address can be examined to determine whether an access is for a stream data word, a stream status word, or a non-stream data word.

The remainder of the figure shows the StreamLine-specific part of the implementation, which is accessed in parallel with the normal cache. If the access is to a non-stream address, then the StreamLine hardware has no affect. The addition of the stream hardware should not slow down the non-stream accesses appreciably. Furthermore, simple stream operations, such as reads, should be able to proceed at the same rate as non-stream operations.

Cache Operation

On every stream access, the *stream number* field of the address is used to look up a status word in the SSB. The status word feeds into a *status check* module, which confirms that the status word is valid and that the current access is allowable, and a *range check* module, which verifies that the address of the access is in the range between the current head and tail.

Stream Reads. If a read access is to an address in the valid range, the data is returned immediately. If the address is not in the valid range, the cache may stall the processor, meaning that no data is returned and the stall timer is started. If the stall timer overflows before data is written into the location, then a *timeout* exception is sent to the processor. Alternatively, if the stream is empty and the **signal** bit is set in the status word. then an *empty-stream* exception is sent to the processor.

Stream Writes. Write accesses to streams are not checked against the head and tail. A write to any location in the stream buffer is accepted. It is up to software to prevent the overwriting of valid data. This allows, for instance, the message at the head of a stream to be mutated and resent without copying the entire message to some other stream.

Status Accesses. If the *status bit* of a stream address is set, then the access is targetted toward the stream's status word, not a data word. Several special-purpose accesses are required for the stream status word, such as reading or writing a specific status field, initializing the entire status word, or reading and setting the write lock. Unused bits in the status address may be used to encode the access type.

Auto-Increment Writes. One of the special encodings of the status address is used to support "auto-increment" writes, in which data is written to the tail of the stream and the tail is post-incremented. This form of write is used extensively by the network. With the hardware presented above, each auto-increment write would require at least two SSB access cycles, since the tail must be read, incremented, and written back into the SSB. This is not desirable, especially for the network, since we would like to remove message from the network as quickly as possible. Thus, in order to provide single-cycle auto-increment writes, we utilize *auto-increment registers*.

Data Transfer. To initiate a cache-to-cache data transfer, the address of an *output stream* is written into the **stream-output** control register. This causes the cache to send the message which is found at the head of the output stream. The message contains both the targets of the transmission and the data to be transmitted.

The first word of a message indicates the number of targets. In the simplest case, this is a positive integer and is followed by the addresses of the target streams. A negative offset may be used, followed by a pointer to another stream which contains the actual targets.

Following the targets is a count of the number of data words in the message. Again, if this count is positive, then the data itself follows. Otherwise, the next word is a pointer to a *data stream*, where the actual count and data are written.

The sending cache writes the message targets and data to the outgoing network port. If the cache experiences a miss on any of part of the message, it should request the cache line from memory and then forward it directly to the network, rather than writing it into the cache, since it will not likely be accessed again soon.

On the receiving end, the network port first acquires the write-lock for the target stream—the target stream status is retrieved in the process. The data is written onto the target stream as it comes over the network, using auto-increment writes.

Performance

In this section, we present some preliminary results on the performance of our proposed implementation of the StreamLine mechanism. The results are based on simulation studies, so we first briefly describe the simulation environment. We then present measurements for a single application experiment, based on Cholesky factorization of a sparse matrix. Future experiments will also be described.

Parameter	Value
Processor cycle time	25 ns
Second-level cache size	64 Kbytes
Cache cycle time	25 ns
Cycles per cache read/write	2
Cache line size	4 words
Cache coherence	invalidate, fully-mapped [2]
Network cycle time	25 ns
Network channel width	32 bits
Memory access time	80 ns

Table 1: Simulated machine parameters.

network cycles	stream time (μsec)	non-stream time (μsec)	stream speedup
1	377.73	492.65	1.30
2	382.18	626.90	1.64
4	399.13	980.25	2.46

Table 2: Sparse Cholesky factorization, 1138BUS, 64 processors.

Simulation Enviroment

In order to evaluate the performance of our proposed implementation, we have created a set of component models using the SIMPLE simulation environment [4]. SIMPLE is an event-driven behavioral simulator, written in object-oriented LISP. SIMPLE provides an extensive, flexible instrumentation system and an application-level interface for driving simulation models. Component models were built to describe the base multiprocessor architecture, described above, as well as the proposed StreamLine extensions. Application programs, written in extended LISP, drive the models directly, rather than address traces.

An Experiment—Sparse Cholesky Factorization

We chose Cholesky factorization[a] of a sparse matrix as the initial experiment for understanding the performance of the StreamLine mechanism.

We implemented two versions of the Cholesky factorization algorithm, one with streams and one without. Both implementations are data-driven—a column's data is explicitly copied to the columns which need it, rather than the demand-driven approach of reading the data as needed.

We simulated the applications on an 8-by-8 mesh of processors. The machine parameters used in the simulations are shown in table 1. One instruction is issued per processor cycle.

The simulation results are shown in table 2. The dataset is the "1138BUS" matrix from the Harwell-Boeing collection [6]— a symmetric positive-definite matrix with 1138 columns and 2596 non-zero elements. Execution times for the stream and non-stream implementations are shown in the first and second columns, respectively. The rightmost column shows the ratio of the non-stream execution time to the stream execution time.

The first row of the table shows that the StreamLine implementation results in a 30% speedup in execution time for the

---

[a]Only the numerical factorization stage is considered here; symbolic factorization and other data manipulation routines are assumed already complete.

default machine parameters. This is primarily due to the directed nature of the data transmission in the non-stream case, even though the sender of the data writes it directly into the receiver's buffer, the data actually resides in the *sender's* cache until the receiver reads it. At that time, one or more read read requests must be sent over the network, and the data must be returned, resulting in a substantial waiting time for the reader regardless of when the data was actually written.

The second and third rows of the table show what happens when network performance is decreased, relative to the processor cycle. StreamLine is much less dependent on the speed of the network, since multiple request-reply transactions are not required for data transfer and synchronization. In fact, increasing the network cycles from one to two caused an execution time increase of only 1%, compared to a 27% increase in the non-stream case, resulting in a speedup of 1.64. If we make the network four times slower than the processor, the stream-based implementation shows a speedup of 2.46.

The one-cycle network was chosen as the default, mainly because of the current match between processor cycle times and state-of-the-art cut-through routing hardware [8]. This ratio may change, however, if processor cycle times outpace the ability to drive signals between chips.

Further Experiments

We realize that an experiment based on a single application does not provide a basis for general conclusions. Further experiments are in progress with other applications, including Gaussian elimination with pivoting (dense matrices), a sparse triangular solver, and an explicit PDE solver (with an SOR-like communication pattern).

Also, the results reported here assume that there are no misses in the first-level cache, i.e., no misses due to instruction fetch or references to private (non-shared) data. We have implemented a stochastic simulation model of such misses and will investigate the effects of varying private miss rates.

We also plan to repeat the experiments with further variations of the simulated machine parameters, such as the cache service time, in order to determine how the StreamLine mechanism performs under different system environments.

Conclusions

This work was motivated by two observations. First, a data-driven computational model is sometimes superior to a demand-driven one, both in terms of performance and ease of programming. Second, the shared-memory machine model, especially when coupled with an invalidate-based coherence scheme, is heavily oriented toward the demand-driven model of computation.

Our goal was to develop a mechanism that integrates support for data-driven computation with strengths of the reference-oriented shared memory model, resulting in an architecture that better supports a wide range of applications and programming paradigms.

The StreamLine mechanism augments a hardware-based cache coherence mechanism to provide directed cache-to-cache transfers of data. It provides support for the efficient management of message queues, and for the data-driven scheduling of computational threads.

Though it is too early to say much about the performance benefits of StreamLine, the results so far indicate that there are some situations which show a substantial improvement. In fact, in our early experiments, the performance gained by streams is much more than the gains due to a faster or larger cache. Our remaining task is to explore the space of applications and system parameters to determine the cases where streams are beneficial, and to quantify that benefit relative to cost.

Acknowledgements

The work described in this paper has been supported by Digital Equipment Corporation, by DARPA Contract F30602-85-C-0012, by NASA Ames Contract NCC 2-220-S1, and by Boeing Contract W266875. G. Byrd was also supported by an NSF Graduate Fellowship and by a DARPA/NASA Assistantship in Parallel Processing.

The authors would like to thank the users and developers of the SIMPLE/CARE simulation system, especially Nakul Saraiya, Sayuri Nishimura, Max Hailperin, Russell Nakano, and Manu Thapar; the administration and staff of Stanford's Knowledge System Laboratory, especially Rich Acuff and James Rice; and Matt Reilly, for reviewing an early version of the paper.

References

[1] H. Abelson and G. J. Sussman. *Structure and Interpretation of Computer Programs.* The MIT Press, Cambridge, MA, 1985.

[2] J. K. Archibald. *The Cache Coherence Problem in Shared-Memory Multiprocessors.* PhD thesis, Univ. of Washington, Feb. 1987.

[3] W. J. Dally and C. L. Seitz. Deadlock-free message routing in multiprocessor interconnection networks. *IEEE Trans. Comput.*, C-36(5):517–553, May 1987.

[4] B. A. Delagi et al. Instrumented architectural simulation. In *Proc. Third Intl. Conf. on Supercomputing*, volume 1, pages 8–11. Intl. Supercomputing Inst., Inc., 1988.

[5] B. A. Delagi et al. LAMINA: CARE applications interface. In *Proc. Third Intl. Conf. on Supercomputing*, volume 1, pages 12–21. Intl. Supercomputing Inst., Inc., 1988.

[6] I. S. Duff, R. G. Grimes, and J. G. Lewis. User's guide for the Harwell-Boeing Sparse Matrix Collection. Feb. 1988.

[7] R. Duncan. A survey of parallel computer architectures. *Computer*, 23(2), Feb. 1990.

[8] C. L. Seitz, K. M. Chandy, and A. J. Martin. Submicron systems architecture: Semiannual technical report. Technical Report Caltech-CS-TR-89-12, Dept. of Computer Science, California Inst. of Technology, 1989.

[9] C. L. Seitz et al. The architecture and programming of the Ametek Series 2010 multicomputer. In *Third Conf. on Hypercube Concurrent Comp. and Appl.*, volume 1, pages 33–36. ACM Press, 1988.

DECOMPOSITION OF PERFECT SHUFFLE NETWORKS[a]

Kenneth E. Batcher
Dept. of Mathematics and Computer Science
Kent State University
Kent OH 44242-0001

email: batcher@cs.kent.edu

Abstract--A perfect shuffle network interconnects 2^n processors by passing 2^n items through one
stage of 2^{n-1} switches n times. A large perfect shuffle network in a massively parallel machine can be
decomposed into modules of various sizes so the machine can be efficiently spread across integrated-cir-
cuit chips, printed-circuit boards, and racks of
boards, etc. Three different decomposition methods
are discussed here.

KEY WORDS: perfect shuffle, decomposition, massive
parallelism, shuffle-exchange, interconnection
networks

1. Introduction

In 1971 Stone described perfect shuffle networks
[1,2]. Perfect shuffle networks use less hardware
than the popular multi-stage (shuffle-exchange) networks like the Omega network and the Butterfly
network [3,4]; rather than pass 2^n items through n
stages of switches one time, a perfect shuffle network passes the 2^n items through one stage of
switches n times.

Perfect shuffle networks can emulate the routing of
items on a hypercube [2]; each pass through the
switches emulates the routing of items along one of
the n dimensions of an n-cube. Rather than n links
per processor in an n-cube, a perfect shuffle network has a constant number of links per processor.
Theoretically, an architecture based on the perfect
shuffle should be more scalable than a hypercube
architecture.

Despite these advantages a perfect shuffle network
has never been used in any implemented machine
whereas multi-stage networks and hypercubes are
very popular. If one is building a massively parallel
machine with several thousand processors one is
concerned with several practical problems. One
problem is how to decompose the machine into
modules of various sizes so it can be efficiently
spread across chips, boards, and racks, etc. We have
been examining this problem with respect to the
perfect shuffle network. This paper shows the results of some initial investigations. Section 2 defines
a perfect shuffle network, sections 3, 4, and 5 describe three different ways of decomposing the networks, and section 6 presents some conclusions.

2. Definition

Perfect shuffle networks are usually described as a
set of 2 x 2 switches interconnected with a shuffle
connection. For some positive integer n, the network
has 2^n processors (indexed 0 through 2^n-1) and
2^{n-1} switches (indexed 0 through 2^{n-1}-1). Switch
$S(i)$ feeds processors $P(2i)$ and $P(2i-1)$ for $0 \leq i <
2^{n-1}$. One input of switch $S(i)$ is fed by processor
$P(i)$ and the other input is fed by processor
$P(i+2^{n-1})$. Figure 1 shows an eight-processor, four-
switch network (n = 3).

(a) This material is based upon work supported by
the National Science Foundation under Grant No.
MIP-9004127. The Government has certain rights in
this material.

Any opinions, findings, and conclusions or recommendations expressed in this material are those of
the author and do not necessarily reflect the views
of the National Science Foundation.

There is an alternate description. For $0 \leq i < 2^{n-1}$ the operation of switch $S(i)$ can be performed by a link between processors $P(2i)$ and $P(2i+1)$ to exchange data between the processors when necessary. The shuffle connections go directly from processor outputs to processor inputs. Figure 2 shows the alternate description of the network for $n = 3$. With the alternate description a perfect shuffle network has 2^n processors, $P(0)$, $P(1)$, ..., $P(2^n-1)$, with three links per processor. Processor $P(j)$ is connected to its exchange mate, $P(j\pm 1)$, through an exchange link, to processor $P(2j \bmod 2^n-1)$ through a shuffle-out link, and to processor $P(2^{n-1} j \bmod 2^n-1)$ through a shuffle-in link. The shuffle-in links for processors $P(0)$ and $P(2^n-1)$ are connected to the shuffle-out links of the same processors. All links are bi-directional; an inverse shuffle is performed when all processors send data out on their shuffle-in links and accept data on their shuffle-out links.

For large n it is hard to show the interconnection of 2^n processors so we use a simpler model. The index of each processor is written as an n-bit binary vector. The index of the exchange mate of a processor is found by complementing the right-most bit of the processor index. During a shuffle the destination index of an item is found by shifting its source index left end-around one position. Conversely, during an inverse-shuffle the destination index of an item is found by shifting its source index right end-around one position. Figure 3 shows the source and destination indices for the exchange, shuffle, and inverse-shuffle data movements.

3. 2 x 2 Decompositions

There are many ways that an end-around shift of an n-bit index can be decomposed into end-around shifts on subsets of the bits. Each shift decomposition leads to a decomposition of a perfect shuffle network of 2^n processors.

3.1 - A 32-processor example

As an example consider a 32-processor network. Processor indices are 5-bit vectors, $(b4, b3, b2, b1, b0)$. A 5-bit vector can be shifted left end-around in two steps. First, the left-most three bits are shifted left end-around to obtain the 5-bit vector $(b3, b2, b4, b1, b0)$. Then, the right-most three bits of this vector are shifted left end-around to obtain the vector $(b3, b2, b1, b0, b4)$. The resultant vector is a left end-around shift of the original vector by one position. This shift decomposition leads to a decomposition of a 32-processor network into four modules with eight processors per module. The first shift step (involving the left-most index bits) results in a major shuffle of data between the four modules while the second shift step (involving the right-most index bits) results in a minor shuffle of data within each module as shown in figure 4. For clarity we omit the exchange links between the processors and the return paths from the right side of the figure back to the left side.

Note in figure 4 how the major shuffle treats the 32 links as eight bundles with four links per bundle. Each eight-processor module has two bundles of shuffle-in links on its left side and two bundles of shuffle-out links on its right side. The links within each bundle run in parallel from a module to a module. We call this a 2 x 2 decomposition because each module has two shuffle-in bundles and two shuffle-out bundles. The reason each module has two shuffle-in bundles and two shuffle-out bundles is that there was one common index bit between the two steps of the shift decomposition; the center bit of the 5-bit indices was involved in both the first (major-shuffle) step and the second (minor-shuffle) step.

3.2 - General Case

Theorem 1: A 2^n-processor perfect shuffle network can be decomposed into 2^{n-k} identical modules with 2^k processors in each module for $2 \leq k < n$. Each module has two shuffle-in bundles and two shuffle-out bundles. Each bundle has 2^{k-1} links.

Proof: Express each processor index as an n-bit binary vector, $(b_{n-1}, b_{n-2}, ..., b_k, b_{k-1}, b_{k-2}, ..., b_1, b_0)$. A left end-around shift of the leftmost n-k+1 bits of this vector produces $(b_{n-2}, b_{n-3}, ..., b_{k-1}, b_{n-1}, b_{k-2}, ..., b_1, b_0)$. This is a shuffle of 2^{n-k+1} bundles with 2^{k-1} links in each bundle.

A left end-around shift of the rightmost k bits of $(b_{n-2}, b_{n-3}, ..., b_{k-1}, b_{n-1}, b_{k-2}, ..., b_1, b_0)$ produces $(b_{n-2}, b_{n-3}, ..., b_{k-1}, b_{k-2}, ..., b_1, b_0, b_{n-1})$. The second step is a shuffle of 2^k links from two bundles and the resultant vector is a left end-around shift of the original n-bit binary vector. The second step can be implemented within a module containing 2^k processors, two shuffle-in bundles, and two shuffle-out bundles. QED.

3.3 - A 65536-processor example

A large perfect shuffle network can be spread out across racks, cards, and chips, etc., by repeated 2 x 2 decompositions. As an example we spread a 65536-processor network across 16 racks with 32 printed-circuit cards in each rack, 32 chips on each card, and 4 processors in each chip. Each processor has a 16-bit index; the leftmost four index bits select the rack, the next five index bits select the card within the rack, the next five index bits select the chip on the card, and the rightmost two index bits select the processor in the chip.

First we decompose a shift of the 16-bit processor indices into a shift of the leftmost 5 bits followed by a shift of the rightmost 12 bits (n = 16, k = 12). Each rack has 4096 processors, two shuffle-in bundles and two shuffle-out bundles (each bundle has 2048 links).

Then we decompose the shift of the 12 rightmost processor index bits into a shift of the leftmost 6 bits followed by a shift of the rightmost 7 bits (n = 12, k = 7). Each rack has 32 cards. Each card has 128 processors, two shuffle-in bundles and two shuffle-out bundles (each bundle has 64 links).

Finally we decompose the shift of the 7 rightmost processor index bits into a shift of the leftmost 6 bits followed by a shift of the rightmost 2 bits (n = 7, k = 2). Each card has 32 chips. Each chip has 4 processors, two shuffle-in bundles and two shuffle-out bundles (each bundle has 2 links).

3.4 - Implementing large bundles

Decomposition of a perfect shuffle network leads to a number of bundles with several shuffle links in each bundle. Each bundle runs from one module to one module. At the micro-level (e.g., between chips on a board) the number of links in a bundle is small and the bundling can be implemented by running all links of a bundle with parallel routes. At the macro-level (e.g., between racks of the system) the number of links in a bundle is large. Since all links in the bundle go from one place to one place one can easily multiplex the links over one or a few high bandwidth paths such as fibre optic cables.

4. 1 x 1 x 1 Decompositions

Note the similarities between figures 1 and 4. In figure 1 each 2 x 2 switch (and its associated pair of processors) is replaced by an 8-processor module in figure 4. Each link of figure 1 is replaced by a 4-link bundle in figure 4. The routing of links in figure 1 is the same as the routing of bundles in figure 4. In general, any 2 x 2 decomposition gives us a figure similar to a network with 2 x 2 switches. Section 2 showed an alternate description for a perfect shuffle network using processors with three links. This suggests that there should be a way of decomposing a network into modules with three bundles per module. We call these 1 x 1 x 1 decompositions.

4.1 - A 32-processor example

As an example consider a 32-processor network. Processor indices are 5-bit vectors, $(b4, b3, b2, b1, b0)$. A 5-bit vector can be shifted left end-around in three steps. First, the left-most three bits are shifted left end-around to obtain the 5-bit vector $(b3,$

b2, b4, b1, b0). Then, the third and fourth bits, b4 and b1, are swapped to obtain (b3, b2, b1, b4, b0). Finally, the right-most two bits of this vector are shifted left end-around to obtain the vector (b3, b2, b1, b0, b4). The resultant vector is a left end-around shift of the original vector by one position. This shift decomposition leads to a decomposition of a 32-processor network into eight modules with four processors per module. The first shift step (involving the left-most index bits) results in a major shuffle of data between the eight modules. The intermediate swap step results in a data exchange between pairs of modules. The final shift step (involving the right-most index bits) results in a minor shuffle of data within each module as shown in figure 5.

Note the similarities between figures 2 and 5. Each processor in figure 2 is replaced by a four-processor module in figure 5. Each shuffle link of figure 2 is replaced by a 4-link bundle in figure 5. Each exchange link in figure 2 is replaced by a 4-link bundle connecting two modules of a pair in figure 5.

4.2 - General Case

Theorem 2: A 2^n-processor perfect shuffle network can be decomposed into 2^{n-k} identical modules with 2^k processors in each module for $2 \le k \le n-2$. Each module has a shuffle-in bundle, a shuffle-out bundle, and an exchange bundle. Each bundle has 2^k links.

Proof: Express each processor index as an n-bit binary vector, $(b_{n-1}, b_{n-2}, \ldots, b_{k+1}, b_k, b_{k-1}, b_{k-2}, b_{k-3}, \ldots, b_1, b_0)$. A left end-around shift of the left-most n-k bits of this vector produces $(b_{n-2}, b_{n-3}, \ldots, b_{k+1}, b_k, b_{n-1}, b_{k-1}, b_{k-2}, b_{k-3}, \ldots, b_1, b_0)$. This is a shuffle of 2^{n-k} bundles with 2^k links in each bundle.

A swap of bits b_{n-1} and b_{k-1} in $(b_{n-2}, b_{n-3}, \ldots, b_{k+1}, b_k, b_{n-1}, b_{k-1}, b_{k-2}, b_{k-3}, \ldots, b_1, b_0)$ produces $(b_{n-2}, b_{n-3}, \ldots, b_{k+1}, b_k, b_{k-1}, b_{n-1}, b_{k-2}, b_{k-3}, \ldots, b_1, b_0)$. This is an exchange of data between the

higher halves (where $b_{k-1} = 1$) of even modules (where $b_{n-1} = 0$) with the lower halves (where $b_{k-1} = 0$) of odd modules (where $b_{n-1} = 1$).

A left end-around shift of the rightmost k bits of $(b_{n-2}, b_{n-3}, \ldots, b_{k+1}, b_k, b_{k-1}, b_{n-1}, b_{k-2}, b_{k-3}, \ldots, b_1, b_0)$ produces $(b_{n-2}, b_{n-3}, \ldots, b_{k+1}, b_k, b_{k-1}, b_{k-2}, b_{k-3}, \ldots, b_1, b_0, b_{n-1})$. This is a shuffle of 2^k links and the resultant vector is a left end-around shift of the original n-bit binary vector. The last step can be implemented within a module containing 2^k processors, a shuffle-in bundle, an exchange bundle, and a shuffle-out bundle. QED.

5. Cycle Structure Decompositions

The decompositions described in sections 3 and 4 are based on decompositions of end-around shifts of processor indices. They always decompose a network into identical modules where the number of modules and the number of processors per module are both integral powers of two. In this section we describe decompositions where the number of modules and the number of processors per module are not powers of two. The modules are not identical.

5.1 - A 16-processor example

End-around shifting any 4-bit vector four places returns the original vector. Vector (0000) returns to itself after one shift and vector (1111) also returns to itself after one shift. Vectors (0101) and (1010) are in a cycle with a period of two. Vectors (0001), (0010), (0100), and (1000) are in one cycle with a period of four, vectors (0011), (0110), (1100), and (1001) are in another cycle with a period of four, and vectors (0111), (1110), (1101), and (1011) are in a third cycle with a period of four. The set of 16 four-bit vectors is divided into six cycles: two cycles containing one vector each, one cycle containing two vectors, and three cycles containing four vectors each.

We define the *ones-count* of a vector to be the number of 1-bits in the vector. Note that end-around shifting a vector does not change the number of 1's in the vector. We define the *ones-count* of a cycle to be the ones-count of any of its vectors.

We can decompose a 16-processor network into six modules corresponding to these six cycles. Exchange links go between processors with index vectors differing only in the right-most position; a processor in a module with a ones-count of k exchanges data with a processor in a module with a ones-count of k+1 or k-1. Figure 6 shows the decomposition of a 16-processor network; processor indices are shown as decimal numbers, shuffle links as thick lines, and exchange links as thin lines.

5.2 - General Case

The advantage of a cycle structure decomposition is that the processors in each module are linked together with shuffle links in a simple cycle easing the task of building the modules. The only links between modules are exchange links. Usually the exchange links show no simple structure.

The number of processors in a cycle must be a divisor of n. Processor 0 is always in its own 1-processor cycle and processor 2^n-1 is always in its own 1-processor cycle. If n is prime then the remaining 2^n-2 processors are in $(2^n-2)/n$ identical cycles with n processors in each cycle.

A cycle structure decomposition can be applied to the alternate description of any network. Since the routing of link bundles in a 1 x 1 x 1 decomposition mimics the routing of links in the alternate description of a smaller network, one can use the cycle structure decomposition to group the modules of the 1 x 1 x 1 decomposition to minimize the lengths of the shuffle bundles (at the expense of the exchange bundles).

6. Conclusions

The 2 x 2 decompositions described in section 3 decompose a perfect shuffle network into a number of modules where the connections to each module can be grouped into two shuffle-in bundles and two shuffle-out bundles. The routing of the link bundles between modules mimics the routing of links between the 2 x 2 switches of a smaller perfect shuffle network.

The 1 x 1 x 1 decompositions described in section 4 decompose a perfect shuffle network into a number of modules where the connections to each module can be grouped into a shuffle-in bundle, a shuffle-out bundle, and an exchange bundle. The routing of the link bundles between modules mimics the routing of links between the processors in the alternate description of a smaller perfect shuffle network.

The 2 x 2 and 1 x 1 x 1 decompositions can be applied recursively to decompose a large perfect shuffle network into racks, cards, chips, etc. The links of large bundles can be multiplexed over a few high bandwidth paths to simplify the construction.

The cycle structure decompositions described in section 5 can be used to group the modules of a 1 x 1 x 1 decomposition to minimize the physical lengths of shuffle bundles. The lengths of exchange bundles may grow.

A large perfect shuffle network can be decomposed into modules of various sizes so it can be efficiently spread across chips, boards, and racks, etc. Before a perfect shuffle network can be used to interconnect the processors of a massively parallel machine we must also solve the problem of making the network fault tolerant; this problem needs further research.

References

[1] Stone, H. S., "Parallel processing with the perfect shuffle", *IEEE Transactions on Computers*, vol C-20, pp 153-161, 1971.

[2] Stone, H. S., *High-Performance Computer Architecture*, Second Edition, Addison-Wesley, Reading MA (1990), 459 pp.

[3] Siegel, H. J., *Interconnection Networks for Large-Scale Parallel Processing: Theory and Case Studies*, Second Edition, McGraw-Hill, (1990),

[4] Crowther, W., *et al.*, "Performance Measurements on a 128-node Butterfly Parallel Processor", *Proc. of the 1985 Int'l. Conf. on Parallel Processing*, pp. 531-540, 1985.

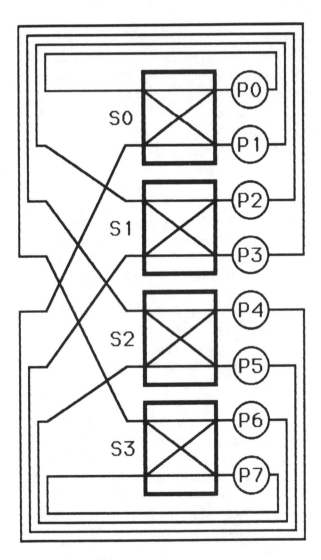

Fig. 1 - Perfect Shuffle Network with eight Processors (P0-P7) and four 2x2 Switches (S0-S3).

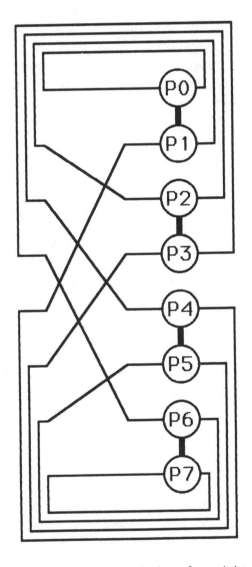

Fig. 2 - Alternate description of an eight-processor perfect shuffle network.

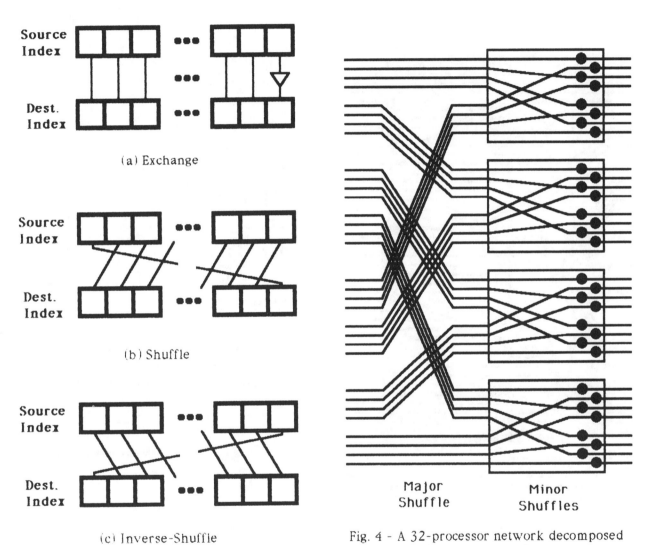

(a) Exchange

(b) Shuffle

(c) Inverse-Shuffle

Fig. 3 - Simple model of Data Movements

Major Shuffle

Minor Shuffles

Fig. 4 - A 32-processor network decomposed into four 8-processor modules.

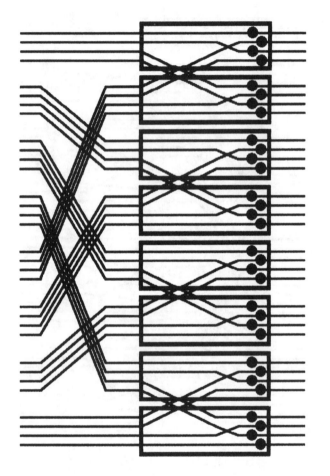

Fig. 5 - A 32-processor network decomposed
into eight 4-processor modules.

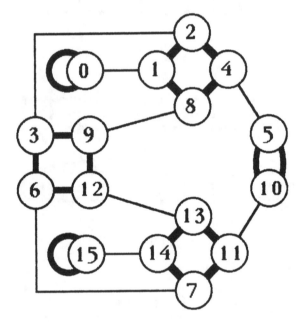

Fig. 6 - A Cycle Structure Decomposition
of a 16-processor network

FAST SELF-ROUTING PERMUTATION NETWORKS *

Chingyuh Jan and A. Yavuz Oruç

Electrical Engineering Department and Institute of Advanced Computer Studies

University of Maryland, College Park, MD 20740

ABSTRACT

This paper presents two new self-routing permutation networks using radix sorting techniques. For n inputs, both networks have $O(n \lg n)$ cost; the first network has $O(\lg^2 n)$ routing time and the second network has $O(\lg^3 n)$ routing time[1]. Both these networks have better cost than the well-known Batcher sorters that can be used as self-routing permutation networks. The only other networks that match the cost of the networks given in the paper are the Benes and Waksman networks, but neither of these is self-routing.

Key Words: concentrator, expander, permutation network, radix permuter, radix sorting, self-routing network.

1 INTRODUCTION

A permutation network, or a permuter is a device which can connect a set of n terminals onto another set of terminals equal in size in any one of $n!$ ways, and finds applications in processor communications in parallel computers [Fe81,Th78]. One aspect of permutation networks which lacks a satisfactory solution to date concerns their routing. Here *routing* refers to deciding the states of the switches in a permutation network in order to achieve an arbitrarily specified permutation. The literature contains several routing algorithms for permutation networks which are either too slow or too expensive. For example, the well–known *looping algorithm* [Wa68] takes $O(n \lg n)$ routing time for the Benes permutation network [Be65] which has only $O(\lg n)$ depth, and thus is too slow. There exist parallel algorithms which can route in $O(\lg^2 n)$ steps, but they require n processors and $O(n^2)$ connections [LPV81,NS81].

The main problem with most permutation networks reported in the literature, including Clos designs [Cl53], is that they were designed with the objective of realizing all permutations with as few switches as possible, but not specifically with a view to have a simple routing algorithm. To alleviate this problem with the known permutation networks, sorting networks can be used as they are inherently self-routing and there exist sorting networks with $O(n \lg^2 n)$ cost and $O(\lg^2 n)$ routing time [Ba68]. Another approach is to intersperse the routing hardware with the actual network as was done in [KO89b]. Using this approach, one can design a self-routing permutation network with $O(n \lg^2 n)$ cost and $O(\lg^2 n)$ routing time[2].

In this paper, we use radix sorting techniques [Kn73] to improve upon these results. More specifically, we present two new self-routing permutation networks. For n inputs, both networks have $O(n \lg n)$ cost; the first network has $O(\lg^2 n)$ routing time and the second network has $O(\lg^3 n)$ routing time. Here, the cost is measured in terms of elementary switching devices, simple binary adders and comparators. The routing time is proportional to the number of such components along the longest path(s) between an input and an output. The first network attains the same order of routing time as a Batcher's sorter takes but at a cost that is less than that of such a sorter by a factor of $\lg n$. The only other networks which match the cost of the networks given in this paper are the Benes and Waksman networks, but neither of these networks is self-routing.

The remainder of the paper is organized as follows. In Section 2, we give a brief overview of radix sorting. Section 3 describes the first network and a linear size self-routing concentrator which is used in the construction of this network. The second network is described in Section 4 and the paper is concluded in Section 5.

2 RADIX SORTING

Radix sorting works as follows. Given a set A of binary numbers, each with $b = \lg n$ bits, examine their most significant bits and divide them into two sets: those whose most significant bits are zero, and those whose most significant bits are 1. Iterate this process for the second bit, third bit, and finally the least significant bit, each time dividing the numbers in each set into two other sets. Call the two sets which are created by examining the most significant bits of the numbers, A_0 and A_1; those created by examining the next most significant bits, $A_{00}, A_{01}, A_{10}, A_{11}$, and finally those created by examining the least significant bits $A_{00...0}, A_{00...1}, \ldots, A_{11...1}$. Then, the number whose binary representation is $a_{b-1} \ldots a_1 a_0$ will belong to set $A_{a_{b-1} \ldots a_1 a_0}$, and the numbers in A will be sorted in ascending order into sets $A_{00...0}, A_{00...1}, \ldots, A_{11...1}$. If A contains each of the numbers $0, 1, \ldots, n-1$ exactly once, then each of these sets will contain exactly one number whose binary representation matches the index of that set. Therefore, the numbers in A are permuted into these indexed sets according to their binary representations. We call these sets the *index sets* of A. In particular, set $A_{a_{b-1}a_{b-2}...a_{b-s}}$ where $a_{b-1}, a_{b-2}, \ldots, a_{b-s} \in \{0,1\}$ and $1 \le s \le b$, is called an index set of degree s. If $A = \{0, 1, \ldots, n-1\}$ then the index set of A of degree s contains exactly 2^{b-s} elements.

---

*This work is supported in part by the National Science Foundation under Grant No: CCR-8708864, and in part by the Minta Martin Fund of the School of Engineering at the University of Maryland.

[1]All logarithms are in base 2 unless otherwise stated.

[2]These complexities are for the number switches, and they were multiplied by $\lg n$ in [KO89b] to express them at the bit level.

The most direct application of radix sorting to routing in permutation networks is obtained by demultiplexing each input onto all n outputs through a binary tree of $n-1$ vertices. This can be realized by connecting the leaf nodes of a set of n binary trees to the leaf nodes of another set of n binary trees such that each binary tree in the first set is connected to a binary tree in the second set exactly once. The roots of the n binary trees in the first set correspond to the inputs of the network, and the roots of the n binary trees in the second set corresponds to its outputs. Accordingly, routing a permutation through this network amounts to decoding the destination bits of each of the inputs over the binary trees which, effectively, is equivalent to the radix sorting method outlined above. This is graphically illustrated in Figure 1 for $n = 4$. Notice that only the first set of binary trees are used in decoding the destination addressses of the inputs. Once the inputs reach the leaves of the first half of trees, i.e., the index sets $A_{00}, A_{01}, A_{10}, A_{11}$, then the second half of trees concentrate these to the appropriate outputs.

3   NETWORK DESIGN 1

The main idea of the radix sorting scheme outlined in the preceding section is to keep separating the inputs into disjoint index sets until all the inputs find their destinations in the index sets $A_{00...0}, A_{00...1}, \ldots, A_{11...1}$. In separating the inputs into disjoint index sets, a path is made available between each input and all such sets to which it may potentially belong at a given stage. However, since there are only n inputs propagating through the network at any given stage in time, some of these paths are wasted. For example, in the network of Figure 1, each input has a path to two index sets at the first stage and to four index sets at the second stage even though an input can belong to only one of the index sets at each stage. This suggests that we may reduce the cost of the network construction given in Figure 1, by somehow concentrating the "live" inputs before they reach their final destinations. We formalize this idea in the form of a recursive network construction which is shown in Figure 2.

Call this construction a *radix permuter*, and denote it by $RP_1(n)$ where n is the number of inputs and outputs to the network. With no loss of generality assume that n is a power of 2. Let $f_i = 2^{s_i}; 1 \le s_i \le \lg n$ be a positive integer which, obviously, divides n. $RP_1(n)$ is recursively defined by using a 3-stage network structure. The first stage encompasses a network with n inputs and $f_i n$ outputs, called an

$(n, f_i n)$-*distributor*, and denoted $DIS(n, f_i n)$. The second stage consists of f_i concentrators, each with n inputs, and n/f_i outputs, denoted $CON(n, n/f_i)$. A concentrator with n inputs and n/f_i outputs is a switching device which can map any subset of n inputs onto some outputs equal in number on a one-to-one basis. The third stage of $RP_1(n)$ consists of f_i radix permuters with n/f_i inputs, denoted $RP(n/f_i)$. The parameter f_i is called the *fanout factor* of $RP_1(n)$.

The outputs of $DIS(n, f_i n)$ are partitioned into f_i sets of n outputs. Each of these sets of outputs represents an index set of degree s_i, and each index set is identified by a distinct s_i-bit code which is its degree as defined in the previous section. The $DIS(n, f_i n)$ connects an input to an output in an index set if and only if the leftmost s_i bits in the destination address of that input match the degree of that index set. It should be obvious that the distributor connects a total of n/f_i inputs to an equal number of outputs in each index set, and the remaining outputs in each index set just float. Those inputs that occupy the outputs in each index set are called the *live* inputs. The concentrators in the second stage are used to concentrate the live inputs in each index set onto the inputs of the permuters in the third stage. Once the live inputs reach the inputs of the permuters, they can then be permuted onto their destinations. Given these facts, the following statement is immediate.

Theorem 1: For all $n = 2^r; r \ge 1$, $RP_1(n)$ can realize all $n!$ permutations between its inputs and outputs. ||

To complete the design of $RP_1(n)$, we need to give constructions for its distributor and concentrator components. For the distributors, we use the simple recursive construction shown in Figure 3. When fully decomposed, the distributor consists of 1×2 demultiplexers as shown in Figure 4(a) for $n = 4$ and $f_1 = f_2 = 4$. Each input to the distributor is given a $\lg n$-bit binary address which specifies its destination on the output side. The inputs are routed to their destinations through the distributor in a self-routing fashion by decoding the leftmost $s_i = \lg f_i$ bits from left to right as illustrated in Figure 4(b) for $n = 4$ and $f_1 = f_2 = 4$.

Let $C_{DIS}(n, f_i n)$ denote the number of 1×2 demultiplexers in this distributor. Then

$$C_{DIS}(n, f_i n) = n + 2C_{DIS}(n, \frac{f_i}{2} n). \qquad (1)$$

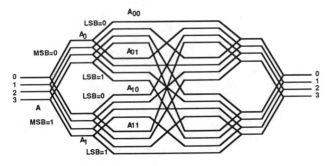

Figure 1: A 4-input permutation network using radix sorting.

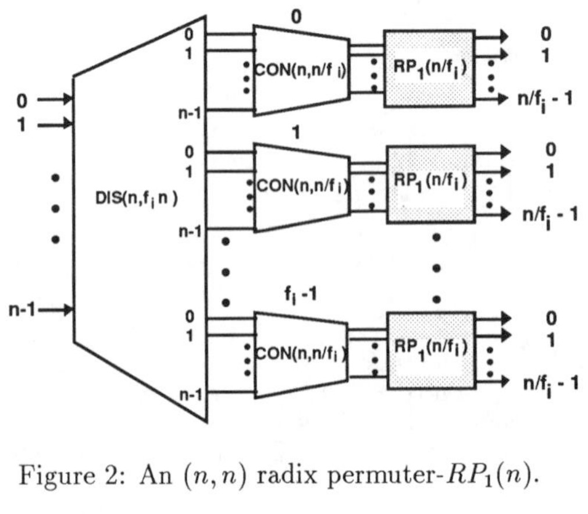

Figure 2: An (n, n) radix permuter-$RP_1(n)$.

The solution of this recurrence with the boundary condition $C_{DIS}(n, 2n) = n$ yields

$$C_{DIS}(n, f_i n) = n(f_i - 1). \quad (2)$$

Let $D_{DIS}(n, f_i n)$ denote the delay through this network. If each 1×2 demultiplexer is assigned a unit delay, then

$$D_{DIS}(n, f_i n) = 1 + D_{DIS}(n, \frac{f_i}{2} n) \quad (3)$$

or,

$$D_{DIS}(n, f_i n) = \lg f_i = s_i. \quad (4)$$

As for the concentrators in Figure 2, we shall use a construction that is obtained by superposing a cube network onto a binary tree as shown in Figure 6 for $n = 16$. The cube network is attached to the binary tree via its nodes at level $d; 0 \leq d \leq \lg n$, where the root of the tree is at level 0, and its leaves are at level $\lg n$. The binary tree portion of the concentrator uses $2n - 1$ nodes that are simple arithmetic units to perform comparison and addition operations. The cube network has 2^d inputs and as many outputs, and encompasses $2^{d-1} d$ 2×2 switches. Assuming that each node in the binary tree and each 2×2 switch in the cube network has unit cost, the total cost of this concentrator for n inputs is $2n - 1 + 2^{d-1} d$.

The concentrator receives m live inputs through its n leaf nodes at level $\lg n$, and concentrates these live inputs upon its leftmost m outputs, where $1 \leq m \leq n$. The concentration takes place in two phases: First, the binary tree portion of the construction is used in a prefix circuit fashion as in [KO89b] to rank the live inputs, and store these ranks in the leaf nodes. This is illustrated in Figure 5, and can be done in a self-routing fashion in $O(\lg n)$ time. In the second phase, the ranks of the live inputs are used to route these inputs to the leftmost m leaf nodes. To do this in a self-routing fashion, each node in the binary tree is given a binary address. These addresses correspond to the index sets described in the earlier section. This correspondence is illustrated in Figure 6 for $n = 16$, where $-$ entries signify "don't care" bit positions. The routing is then carried out by a simple decoding technique that compares each rank with the binary address of the current node. If the rank does not match the binary address then the input with that rank is propagated up. This continues until either a match occurs or the input

in question reaches a node at level d. If the matching takes place, then the input is propagated from that node down to its destination by using the rightmost i bits of its rank as an address, where $\lg n - i$ is the level at which the matching occurs. On the other hand, if an input reaches a node at level d and does not match the address of that node then its destination must reside in another subtree rooted at level d, and consequently it enters the cube network in order to be routed to the root node of the subtree to which its destination belongs. Once it reaches that root node then it is propagated down to bring it to its destination in that subtree. Both of these cases are illustrated in Figure 6.

The second phase of the routing scheme that is just described can be performed in several ways. Using a completely serial scheme, each live input can reach its destination in no more than $2(\lg n - d) + d = 2 \lg n - d$ steps; d steps through the cube network, and $2(\lg n - d)$ to climb up and down the subtrees. Considering that there may be as many as n live inputs, this procedure takes a total of $n(2 \lg n - d)$ steps. Recalling that the overal cost of the concentrator is $2n - 1 + 2^{d-1} d$, it follows that, for $d = \lg(n/\lg n)$, this procedure provides a self-routing concentrator with $O(n)$ cost and $O(n \lg n)$ routing time.

To reduce the routing time, some parallelism can be introduced. One approach is to allow exactly one input from each of the 2^d subtrees to start its trip towards its destination with

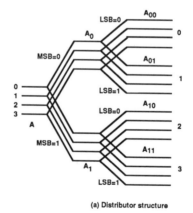

(a) Distributor structure

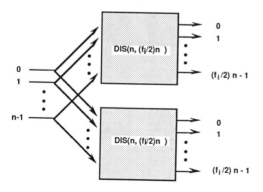

Figure 3: An $(n, f_i n)$ distributor.

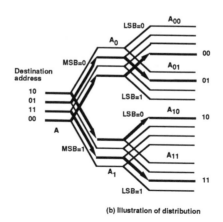

(b) Illustration of distribution

Figure 4: A $(4, 16)$ distributor when fully decomposed.

the provision that nodes in different substrees can start in parallel. This will ensure that no conflicts can arise within each subtree, but there may be conflicts once and if the inputs reach the root nodes of the subtrees. This is because, the destinations of some inputs may belong to the same subtree. This conflict problem may be formalized as follows. We are given a set of 2^d numbers, each containing d bits. These numbers identify the 2^d subtrees whose roots are connected to the inputs of the cube network. Some of these numbers may be identical, and the goal is to route these numbers through the cube network as fast as possible.

At this point, we need to establish some facts about the patterns of ranks that may occur in our concentrator and the routing properties of the cube network.

Definition 1: A rank pattern over the leaf nodes of a binary tree is an assignment of consecutive integer values $0, 1, 2, \ldots$ to the live leaf nodes such that the left most live leaf node is assigned rank 0, the next left most live leaf node is assigned 1, etc. ||

Definition 2: An *induced* rank pattern of degree d is a rank pattern whose entries are obtained by retaining the leftmost d bits in the corresponding ranks. For example, for the rank pattern shown in Figure 5, the induced rank pattern of degree 2 is 0000111122. ||

Definition 3: A *sample* rank pattern of degree d is an ordered set of ranks obtained by selecting one rank from a induced rank pattern in such a way that at most one rank is selected from each subtree rooted at level d. If exactly one rank is selected from each subtree then the sample rank pattern is called *complete*, and is otherwise called *incomplete*. For example, for the induced rank pattern 0000111122, $0, 0, 0, 1$, $0, 0, 1, 2$, and $0, 0, 0, 2$ are all complete sample rank patterns, and $-, 0, 1, 2$ is an incomplete sample rank pattern. ||

Proposition 1: For any rank pattern, the number of ranks in any subtree rooted at level d is no more than $n/2^d$.
Proof: It is obvious. ||

Proposition 2: For any rank pattern, the ranks in any sample rank pattern of degree d that reach level d are in nondecreasing order for all $d = 0, 1, \ldots, \lg n$.
Proof: It is obvious. ||

Theorem 3: Any sample rank pattern of degree d drawn from a induced rank pattern over the leaf nodes of a binary tree of height $\lg n$ and in which ranks at adjacent roots differ by at most one can be self-routed through a 2^d-input cube network in $O(d + n/2^d)$ time.
Proof: The proof is given by constructing a self-routing procedure as follows. Let R_u and R_l be the two ranks that enter an arbitrary but fixed switch S_i in stage $i; 0 \leq i \leq d - 1$, in the cube network at its upper and lower inputs, respectively. Let $r_{u,i}$ and $r_{l,i}$ denote the ith bits of R_u and R_l, respectively. We distinguish between a "live" rank and a "dummy" rank in that when obtaining a sample rank pattern from an induced rank pattern, some subtrees may not have any live ranks, and this will result in incomplete sample rank patterns in which case the empty roots are filled by so called dummy ranks. Thus, there are four cases to consider. If both R_u and R_l are dummy then the switch is set arbitrarily to either the identity or transpose state. If R_u is dummy then

the switch is set by $r_{l,i}$ and if R_l is dummy then it is set by $r_{u,i}$. The last case is that neither R_u and R_l is dummy. In this case, if $r_{u,i} = 0$, and $r_{l,i} = 1$ then the switch is set to the identity, and if $r_{u,i} = 1$, and $r_{l,i} = 0$ then the switch is set to the transpose state. We further show that the other two cases, i.e., $r_{u,i} = r_{l,i} = 0$, or $r_{u,i} = r_{l,i} = 1$, are impossible except when $R_u = R_l$. First, for $i = 0$, by the hypothesis, the adjacent ranks differ by at most one, and hence they must either be identical, or they must differ in their rightmost bit positions. Suppose, for some $i; 1 \leq i \leq d-1$, some two ranks R_x and R_y enter a switch S_i at stage i. As argued in [KO89b], this implies that R_x and R_y must have the same values in the right most i bits, i.e., bits $0, 1, \ldots, i-1$. Furthermore, they must have originated from some two inputs that are tied to a cube network of 2^{i+1} inputs that is contained in the original 2^d-input cube network. Given that the ranks differ by at most one, this implies that $|R_x - R_y| \leq 2^{i+1} - 1$. Thus, R_x and R_y cannot differ in bit positions $d-1, d-2, \ldots, i+1$, and therefore either $R_x = R_y$, or they differ in bit position i. This proves that the only way that a conflict can occur at a switch is when both ranks are identical. Given this, it is easy to see that a rank in a sample rank pattern satisfying the hypothesis can be in conflict with no more than the number of ranks of the same value. But this is bounded by the number of leaf nodes in a subtree rooted at level d, which is $n/2^d$. We conclude that any such sample rank pattern can be self-routed through the cube network in $O(d+n/2^d)$ steps. ||

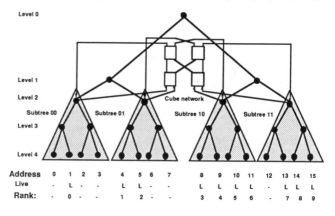

Figure 5: A self-routing concentrator.

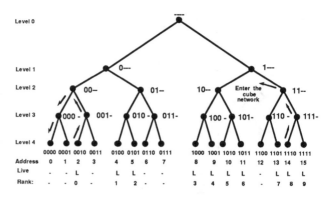

Figure 6: Illustration of routing on the concentrator.

Given this fact, the only task that remains is to establish that any induced rank pattern of degree d can be decomposed into a set of sample rank patterns such that the adjacent ranks in each sample rank pattern differ by at most one. The existence of such a decomposition is facilitated by the following results.

Proposition 3: Let F_i be the set of live inputs whose ranks are destined to subtree $i; 0 \leq i \leq 2^d - 1$, and let m be the number of live inputs. Then $|F_i| = n/2^d, 0 \leq i \leq \lfloor \frac{m}{n/2^d} \rfloor - 1$, and $0 < |F_i| < n/2^d$ for $i = \lfloor \frac{m}{n/2^d} \rfloor$.
Proof: It is obvious. ||

Proposition 4: The induced ranks of the live leaf nodes in any k adjacent subtrees rooted at level $d; 0 \leq d \leq \lg n - 1$ can differ by at most k, for $2 \leq k \leq 2^d$.
Proof: It follows from Proposition 3. ||

Theorem 4: In any induced rank pattern of degree d, the ranks associated with the leftmost live leaf nodes in any two adjacent subtrees rooted at level d differ by at most one.
Proof: By Proposition 4, any two induced ranks in any two adjacent subtrees T_i and T_{i+1} rooted at level d can at most differ by two. If they differ by zero or one then there is nothing to prove. On the other hand, if they differ by two then it is easy to see that the two subtrees must have three consecutive induced ranks $R_x, R_x + 1, R_x + 2$. However, Proposition 3 implies that R_x must belong to T_i and $R_x + 1$ must belong to T_{i+1}, and the statement follows. ||

Proposition 5: Suppose that two adjacent subtrees T_i and T_{i+1} rooted at level d have three consecutive induced ranks $R_x, R_x + 1$, and $R_x + 2$. Let N_x, N_{x+1}, and N_{x+2} be the numbers of ranks $R_x, R_x + 1$ and $R_x + 2$, in that order, contained in subtrees T_i and T_{i+1}. Then $N_x + N_{x+2} \leq N_{x+1}$.
Proof: Since $R_x + 1$ is the middle rank, by Proposition 3, $N_{x+1} = n/2^d$. Since the two subtrees combined together can at most have $2n/2^d$ live inputs, $N_x + N_{x+2} \leq 2n/2^d - N_{x+1} = n/2^d = N_{x+1}$. ||

Theorem 5: Given any induced rank pattern of degree d associated with the subtrees rooted at level d. If we remove the leftmost rank from each of the subtrees, then the remaining ranks form an induced rank pattern in which the ranks of the leftmost live inputs in any two adjacent subtrees differ by at most one.
Proof: It follows from Proposition 5. ||

These results establish that any induced rank pattern of degree d can be decomposed into a sequence of sample rank patterns in which the adjacent ranks differ by at most one. By Theorem 3, each of these sample patterns can be self-routed through the cube network superposed with binary tree as shown in Figure 5 in $O(d + n/2^d)$ steps. Furthermore, each such sample pattern can be self-routed through the binary tree in $\lg n - d$ steps to the roots of the subtrees rooted at level d, by using a leftchild priority routing scheme, i.e., if two live inputs reach the same node the one on the left proceeds to the next level, thereby ensuring that the leftmost live input reaches the root. It will take additional $\lg n - d$ steps to route each sample pattern back through the binary trees so as to have the live inputs reach their final destinations. This can also be done on a self-routing basis, this time

using the rightmost $\lg n - d$ bits of the ranks in the sample rank pattern. Since each subtree has at most $n/2^d$ leaf nodes, the number of sample rank patterns in decomposition of any induced rank pattern is at most $n/2^d$. Thus, combining the ranking and concentration steps we have established

Theorem 6: The network in Figure 5 can self-concentrate any pattern of live inputs in $O(\frac{n}{2^d}(2(\lg n - d) + d + n/2^d) + \lg n)$ steps. ||

In particular,

Corollary 6.1: When $d = \lg(n/\lg n)$, the network given in Figure 5 has $O(n)$ cost and it can concentrate, on a self-routing basis, any pattern of live inputs in $O(\lg^2 n)$ steps. ||

The routing time can still be reduced to $O(\lg n)$ by noting that all the inputs in any subtree rooted at level $\lg(n/\lg n)$ can reach the roots of these subtrees in $O(\lg n)$ time if pipelining is used. The formal proof of this claim can be found in [Ja91].

We have established that both the distributor and concentrator sections of the radix permuter in Figure 2 can be realized by self-routing connectors. Combining this fact with the overall structure of the network we conclude:

Theorem 7: For all $n = 2^r; r \geq 1$, $RP_1(n)$ can realize all $n!$ permutations between its inputs and outputs in a self-routing fashion. ||

The cost and routing time of this network can be determined as follows.

Theorem 8: $RP_1(n)$ has $O(kn^{1+1/k})$ cost and $O(k \lg n)$ delay for each $k; 1 \leq k \leq \lg n$.

Proof: Let $C_{RP_1}(n)$ denote the cost of $RP_1(n)$. From the construction of $RP_1(n)$ we have

$$C_{RP_1}(n) = C_{DIS}(n, f_1 n) + f_1 C_{CON}(n, n/f_1) + f_1 C_{RP_1}(\frac{n}{f_1})$$

where $C_{DIS}(n, f_1 n)$ denotes the cost of the distributor network in the first stage, and $C_{CON}(n, n/f_1)$ denotes the cost of each of the concentrators in the second stage. Upon replacing $C_{DIS}(n, f_1 n)$ with the right hand side of Equation 2, and letting $C_{CON}(n, n/f_1) = \alpha n$ where $\alpha \leq 3$ is some positive constant, we have

$$C_{RP_1}(n) = (f_1 - 1)n + \alpha f_1 n + f_1 C_{RP_1}(\frac{n}{f_1}).$$

It is easy to show that the solution of this recurrence with fanout factors $f_1, f_2, \ldots, f_k$, and the boundary condition $n = f_1 f_2 \ldots f_k$ yields

$$C_{RP_1}(n) \leq \beta n (f_1 + f_2 + \ldots + f_k) \tag{5}$$

where $\beta = \alpha + 1$. Minimizing this over $f_1, f_2 \ldots, f_k$ subject to the side condition $n = f_1 f_2 \ldots f_k$ gives $f_1 = f_2 = \ldots = f_k = n^{\frac{1}{k}}$, and therefore we have

$$C_{RP_1}(n) \leq \beta k n^{1 + \frac{1}{k}}. \tag{6}$$

Now, let $D_{RP_1}(n)$ denote the delay through $RP_1(n)$. From the construction of $RP_1(n)$ we have

$$D_{RP_1}(n) = D_{DIS}(n, f_1 n) + D_{CON}(n, n/f_1) + D_{RP_1}\left(\frac{n}{f_1}\right)$$

where $D_{DIS}(n, f_1 n)$ denotes the depth of the distributor network in the first stage, and $D_{CON}(n, n/f_1)$ denotes the depth of each of the concentrators in the second stage. Upon replacing $D_{DIS}(n, f_1 n)$ with the right hand side of Equation 4, we have

$$
\begin{aligned}
D_{RP_1}(n) &\leq s_1 + \alpha \lg n + D_{RP_1}\left(\frac{n}{f_1}\right) \\
&\leq s_1 + \ldots + s_k + \alpha \lg n + \alpha \lg \frac{n}{f_1} \\
&\quad + \ldots + \alpha \lg \frac{n}{f_1 \cdot \ldots \cdot f_k}
\end{aligned}
\tag{7}
$$

where α is a positive constant. It is easy to show that the solution of this recurrence with the boundary condition $n = f_1 f_2 \ldots f_k$ and $\lg n = s_1 + s_2 + \ldots + s_k$ yields

$$
\begin{aligned}
D_{RP_1}(n) &\leq \lg n + k\alpha \lg n = (k\alpha + 1)\lg n \\
&\leq k(\alpha + 1)\lg n = O(k \lg n)
\end{aligned}
\tag{8}
$$

It is immediate from the above results that, when $k = \lg n$, this network has $O(n \lg n)$ cost and $O(\lg^2 n)$ delay, or equivalently $O(\lg^2 n)$ routing time since the network is self-routing.

4 NETWORK DESIGN 2

In this section, we present a second network construction that gives the same asymptotic cost. We also show that the routing time of this network is $O(\lg^3 n)$ by using a conservative estimate. This construction is obtained by replacing the distributors in Figure 2 by another kind of distributors, and the concentrators by certain kind of bipartite graphs, called *expanders* [Pi73,Bas81]. The schematic of the construction is shown in Figure 7.

The distributor to be used in this case is called a *copier*, and denoted $CP(2n, 2f_i n)$. It is a bipartite graph with $2n$ inputs and $2f_i n$ outputs followed by a set of masking elements as shown in Figure 8. As before, the outputs are partitioned into index sets of degree s_i, and each input is connected to exactly one output in each index set. The mask elements that are connected to the outputs in each index set mask out the inputs that do not belong to that index set. This is illustrated in Figure 9. It is obvious that this distributor is self-routing. Its cost is $4nf_i$, and its routing time is $O(1)$. As for the expanders, we need the following facts adopted from [Or91].

Definition 4: (Bassalygo) Let $E(qt, pt)$ denote a bipartite graph with qt inputs and pt outputs where q, p, t are positive integers. $E(qt, pt)$ is called (α, β)-*expanding* if any i of the qt inputs $(1 \leq i \leq \lfloor \alpha qt \rfloor)$ are connected to at least $\lfloor \beta i \rfloor$ of the pt outputs[3]. ‖
We write $E(\alpha, \beta, qt, pt)$ to denote a bipartite graph with qt inputs, pt outputs that is (α, β)-expanding. ‖

---

[3]A subset of inputs, A is said to be connected to a subset of outputs, B, if each output in B is connected to at least one input in A.

Theorem 9: There exists an $(1/4, 1, 2t, t)$ expander with $10t$ edges.
Proof: See [Or91] ‖

Theorem 10: There exists a bipartite graph $E(2t, t)$ with $10t$ edges in which every $t/2$ inputs are connected to some $t/2$ distinct outputs by nonoverlapping edges.
Proof: See [Or91] . ‖

With these results in place, the following result is immediate.

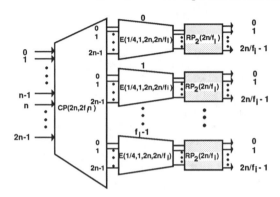

Figure 7: An (n, n) radix permuter–$RP_2(n)$.

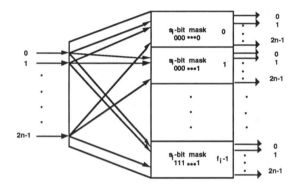

Figure 8: An $(n, 2f_i n)$-copier–$CP(2n, 2f_i n)$.

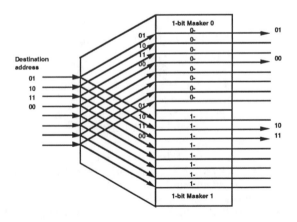

Figure 9: Illustration of the copying scheme on $CP(2 \times 4, 2 \times 2 \times 4)$.

Theorem 11: For all $n = 2^r; r \geq 1$, $RP_2(n)$ can realize all $n!$ permutations between its inputs and outputs in a self-routing fashion. ||

Remark: We note that, on the input side, only the first n of the $2n$ inputs are used. The reason that we use $2n$ inputs, rather than n inputs stems from the fact that the expanders need to have more inputs than outputs. This makes the overall design to have $2n$ inputs and an equal number of outputs. Nonetheless, it can be shown that when the $RP_2(2n/f_i)$ permuters in the third stage are recursively decomposed, the inputs will be mapped to outputs in such a way that each input is connected to either of a pair of outputs where the pairs are formed from top to bottom on the output side. Using this fact, we can then collect the n inputs from the pairs of outputs by using simple 2×1 multiplexers. ||

Let $C_{RP_2}(n)$ and $D_{RP_2}(n)$ denote the cost and delay of this network in that order, we have

$$C_{RP_2}(n) = \alpha_1 f_1 n + \alpha_2 f_1 n + f_1 C_{RP_2}(\frac{n}{f_1}) \qquad (9)$$

and

$$D_{RP_2}(n) = \beta + D_{RP_2}(\frac{n}{f_1}) \qquad (10)$$

where α_1, α_2 and β are some positive constants. The solutions of these recurrences with $f_1 = f_2 = \ldots = f_k = n^{\frac{1}{k}}$ yield,

$$C_{RP_2}(n) = O(kn^{1+\frac{1}{k}}) \qquad (11)$$

and

$$D_{RP_2}(n) = O(k). \qquad (12)$$

It follows that $RP_2(n)$ attains the bound given in [KO89a]. Furthermore, for $k = \lg n$, this second network has $O(n \lg n)$ cost and $O(\lg n)$ delay. However, the routing time of this network construction cannot immediately be concluded from its total delay, as the expanders used in the network cannot be set in $O(1)$ time. Realizing one-to-one assignment on an expander is no more difficult than finding a maximal cardinality matching in a bipartite graph. In [MVV87], it was shown that this can be done in $O(\lg^2 n)$ time. This then implies that the routing time of $RP_2(n)$ is $O(\lg^3 n)$.

5 CONCLUDING REMARKS

We have presented two self-routing permutation networks. Both networks have $O(n \lg n)$ cost and polylogarithmic routing time. However, it remains open whether or not a self-routing permutation network with $O(n \lg n)$ cost and $O(\lg n)$ routing time can be constructed.

REFERENCES

[Bas81] L. A. Bassalygo, "Asymptotically optimal switching circuits," *Problems of Information Transmission,* Vol. 17, No. 3, July-Sep. 1981, pp. 206-211.

[Ba68] K.E. Batcher, "Sorting networks and their applications," *in Proceedings of 1968 Spring Joint Computer Conference,* pp. 307–314.

[Cl53] C. Clos, "A study of non–blocking switching networks," *The Bell System Technical Journal,* Mar. 1953, pp. 406–424.

[Be65] V.E. Beneš, *Mathematical Theory of Connecting Networks and Telephone Traffic,* Academic Press Pub. Company, New York, 1965.

[Fe81] T–Y Feng, "A survey of interconnection networks," *IEEE Computer,* Dec. 1981, pp. 12–27.

[Ja91] Chingyuh Jan, "Optimal connector designs for parallel processors," *Ph.D. Diss.,* In preparation.

[Kn73] D.E. Knuth, *The Art of Computer Programming,* Vol. 3 Reading, MA, Addison–Wesley, 1973.

[KO89a] D. Koppelman, and A. Yavuz Oruç, "Time-space tradeoffs in parallel communications," *in Proceedings of Allerton Conference,* 1989, Urbana, IL.

[KO89b] D. Koppelman, and A. Yavuz Oruç, "A self-routing permutation network," *in Journal of Parallel and Distributed Computing,* No. 10, 1990 pp. 140-151.

[Or91] A. Yavuz Oruç, "Permuters for parallel processors," *submitted to IEEE-TPDS.*

[Ma77] M.J. Marcus, "The theory of connecting networks and their complexity: a review," *Proceedings of the IEEE,* Vol. 65, No. 9, Sep. 1977, pp. 1263–1271.

[MVV87] K. Mulmuley, U. Vazirani, and V. Vazirani, "Matching is as easy as matrix inversion," *Proceedings of ACM Symposium of Theory of Computing,* 1987, pp. 345-354.

[NS82] D. Nassimi and S. Sahni, "Parallel algorithms to set up the Benes permutation network," *IEEE Transactions on Computers,* Vol. C-31, No. 2, Feb. 1982, pp. 148–154.

[Pi73] M. S. Pinsker, "On the complexity of a concentrator," *In Proc. 7th Int. Teletraffic Congress,* Stockholm 1973, pp. 318/1-318/4.

[Pip82] N. Pippenger, "Telephone switching networks," *in Proc. of Symposia of Applied Mathematics,* Vol. 26, May 1982, pp. 101–113.

[LPV81] G. F. Lev, N. Pippenger, and L. G. Valiant "A fast parallel algorithm for routing in permutation networks," *IEEE Transactions on Computers,* Vol. C–30, No. 2, Feb. 1981, pp. 93–100.

[Wa68] A. Waksman, "A permutation network," *Journal of the ACM,* Vol. 15, No. 1, Jan. 1968, pp. 159–163.

[Th78] C. D. Thompson, "Generalized interconnection networks for parallel processor intercommunication," *IEEE Transactions on Computers,* Vol. C-27, Dec. 1978, pp. 1119-1125.

ON SELF-ROUTABLE PERMUTATIONS IN BENES NETWORK

Nabanita Das, Krishnendu Mukhopadhyaya and Jayasree Dattagupta
Electronics Unit
Indian Statistical Institute
Calcutta - 700 035, India

Abs..act: In this paper, we define an equivalence relation on the set of all possible permutations, namely the 'input group transformation', that enables us to enhance the set of permutations realizable by self-routing techniques in Benes network to a great extent. It shows that the classes of permutations defined as LE (linear equivalent) and IΩE (inverse omega equivalent) are all routable in Benes network by the same self-routing algorithm developed for the routing of LC (linear complement) class of permutations only [7]. The analysis of a typical subset of LE, namely the BPE (bit-permute equivalent) class shows that $|BPE| = n! \, 2^{N-1}$ compared to $|LC| < 2^{n(n+1)}$, $n = \log_2 N$. An $O(n)$ time algorithm is also presented here, to determine whether a permutation P belongs to BPE class or not.

I. Introduction

In multi-computer or multi-processor systems, to establish communication paths dynamically among different modules (processing elements and/or memory modules) multi-stage interconnection networks (MIN's) are used. The communication from a set of N inputs to a set of N outputs may be represented by an NxN permutation. A rearrangeable interconnection network is one which can realize any arbitrary permutation in a single pass. Benes network is a well known rearrangeable network, with a recursive structure using 2x2 switching elements as shown in fig. 1. In an NxN Benes network, determining the switch settings to realize an arbitrary permutation takes $O(N.n)$ time on a uniprocessor system [1] in contrast to its propagation delay $O(n)$, where $n = \log_2 N$.

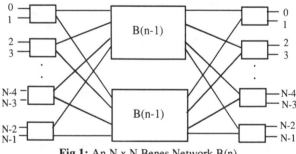

Fig.1: An N x N Benes Network B(n)

Many attempts have been made to develop fast self-routing algorithms for many useful classes of permutations required in parallel computers. Lenfant developed a parallel control algorithm for the 'FUB' permutations [2], that includes only

$O(2^{2n})$ out of $2^n!$ all possible permutations. Nassimi and Sahni proposed a simpler algorithm for self-routing a class of permutations that includes the BPC (bit-permute complement) class, the IΩ (inverse-omega) class etc. [3]. Boppana and Raghavendra [7] developed a new self-routing technique for LC (linear complement) class of permutations, which also routes the IΩ class of permutations as well.

In this paper, the idea of 'input group transformation' [8] on permutations, that partitions the whole set of permutations into a set of equivalent classes, is applied to identify a much larger set of permutations realizable in Benes network with a complexity $O(n)$. Here we show that if any permutation is routable by the algorithm of [7], any other permutation P' belonging to the same equivalent partition as P is also routable by [7]. Hence it reveals that all permutations belonging to the same partition as any LC permutation or any IΩ permutation, defined as LE (linear equivalent) and IΩE (inverse omega equivalent) class respectively is routable by [7]. The exact enumeration of the whole set of permutations routable by [7] is difficult. However, the study of the BPE (bit permute equivalent) class of permutations generated from BP class, which is a typical subset of LC, shows that $|BPE| = n! 2^{N-1}$, compared to $|BP| = n!$ and $|LC| < 2^{n(n+1)}$. It actually gives us some idea about the degree of extension of the set of permutations realizable by self-routing algorithm using our notion of 'input group transformation'. We also develop here an algorithm to determine whether a permutation belongs to the BPE class or not. The algorithm is of complexity $O(n)$.

II. Input Group Transfomations & Their Properties

Definition 1: For an NxN Benes network, the inputs are grouped in different levels, as shown in fig. 2.

An **input group** in level i, $0 \le i \le n$, is represented as G(i,x), where $G(i,x) \equiv \{x, x+1,...,x+2^i-1\}$, $x = p.2^i$ for some p, $0 \le p < 2^{n-i}$.

Two input groups G(i,x) and G(i,y), $0 \le i < n$, are said to be **adjacent** if $x = p.2^{i+1}$, for some p, $0 \le p < 2^{n-i-1}$ and $y = x+2^i$.

Fig. 2 shows that the set of inputs $\{0,1,2,3\}$ is the input group G(2,0); G(2,0) , G(2,4) are two adjacent groups at level 2.

Definition 2: An **input group interchange** tI(j:x), applied on a permutation P interchanges inputs of two adjacent groups G(j,x) and G(j,x+2^j), $0 \le j < n$, following a rule $k \longleftrightarrow k + 2^j$, $x \le k < x + 2^j$. This process generates another permutation P' and is denoted by tI(j:x)[P] $\longrightarrow$ P'.

Example 1 : Given a permutation P: 1 7 2 3 6 5 0 4, (the permutation is represented as the sequence of outputs corresponding to the sequence of inputs 0 to N-1, i.e for P, the input-output mappings are $0 \to 1, 1 \to 7, 2 \to 2, 3 \to 3, 4 \to 6, 5 \to 5, 6 \to 0, 7 \to 4$ respectively), let us apply an input group interchange $tI(1:4)$ on P. We get another permutation P', i.e., $tI(1:4)[P] \to P'$, where P' : 1 7 2 3 0 4 6 5.

Definition 6 : An **input group transformation** T is an ordered sequence of input group interchanges, without repetition.

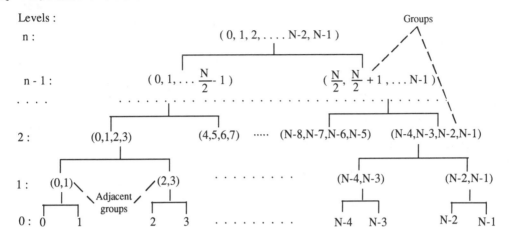

Fig. 2 : The input groups at different levels

Definition 3: Given a permutation P, an **output cluster** $C_o(j,x)$ defines the set of outputs corresponding to the inputs of the group $G(j,x)$.

The **set of output clusters** at level j, $\{C_o(j,.)\}$ is the set of all output clusters at level j.

Example 2: For P and P' of example 1, $C_o(2,0) = \{1,7,2,3\}$ and $\{C_o(1,.)\} = \{\{1,7\}, \{2,3\}, \{0,4\}, \{6,5\}\}$.

Lemma 1: Any sequence of input group interchanges keeps the set of output clusters at any level unchanged.

Proof : By input group interchange at level j, two output clusters $C_o(j,x)$ and $C_o(j,x+2^j)$ are actually interchanging their positions with respect to inputs, i.e., now $C_o(j,x) \longleftrightarrow C_o(j,x+2^j)$. But if we consider $\{C_o(j,.)\}$, we find that nothing has been changed, which is true for all levels.

In example 1, $\{C_o(j,.)\}$ of P and P' are same at all levels j.

Definition 4: A sequence of group interchanges $\{ tI(l_k:y_k) \; tI(l_{k-1}:y_{k-1}) \dots tI(l_1:y_1)\}$ is said to be **ordered** if $l_1 \geq l_2 \geq \dots \geq l_k$ and for $j > i$, if $l_i = l_j$ then $y_i \geq y_j$.

Remark: The ordering of $tI(l_i:y_i)$'s according to y_i (when l_i's are equal) is immaterial since they are independent of each other. It is included in definition 4 only to ensure the uniqueness of a sequence of group interchanges.

Definition 5: Two sequences of input group interchanges R_1 and R_2 are **equivalent** if for every permutation P, $R_1[P] \equiv R_2[P]$.

Lemma 2 : An ordered sequence of input group interchanges with repetition is equivalent to a shorter ordered sequence of group interchanges.

Proof: As the sequence is ordered, any repetition would be consecutive. We also note that for any permutation P, $tI(x:y)[tI(x:y)[P]] \equiv P$.

Remark: Input group transformation defines an equivalence partition on the set of all permutations. If a permutation P' is derivable from another permutation P by the application of some T, i.e., $T[P] \equiv P'$, we will say that $P' \sim P$ (P' is related to P) and it is easy to see that '$\sim$' is an equivalence relation.

Theorem 1 : Let T and T' be two different input group transformations. For any permutation P, $T[P] \neq T'[P]$.

Proof: Let $T = tI(x_i:y_i) \dots tI(x_1:y_1)$ and $T' = tI(w_j:z_j) \dots tI(w_1:z_1)$. Since T and T' are different, for some k, $(x_k:y_k) \neq (w_k:z_k)$. Without loss of generality, we may assume $(x_1:y_1) \neq (w_1:z_1)$.

Case I : $x_1 \neq w_1$. Without loss of generality, let $w_1 < x_1$. Since, in T' all interchanges are in levels lower than x_1, the output clusters of T'[P] at level x_1 would not only be the same as that of P, but they would be in the same positions also. But in T[P], because of $tI(x_1:y_1)$ their positions would change.

Case II : $x_1 = w_1$. So $y_1 \neq z_1$. Without loss of generality, let $z_1 < y_1$. Here also, since all the group interchanges in T' at level x_1 are from z_1 or lower, the output cluster of T'[P] of size 2^{x_1} and starting from y_1 remain undisturbed, whereas in T[P] it gets changed.

Theorem 2 : For any sequence of input group interchanges, $R = tI(x_k:y_k) \dots tI(x_1:y_1)$, there exists an equivalent input group transformation $T = tI(w_m:z_m) \dots tI(w_1:z_1)$.

Proof: We shall prove the result by induction on k. The result is trivially true for k=1. Let it be true for some k-1.

Take $R = tI(x_k:y_k) \dots tI(x_1:y_1) = R_1 \, tI(x_1:y_1)$ where $R_1 = tI(x_k:y_k) \dots tI(x_2:y_2)$.

By induction hypothesis, R_1 can be replaced by an equivalent input group transformation $T_1 = tI(w_m:z_m) \dots tI(w_1:z_1)$.

Now, if $x_1 > w_1$, $T_1 tI(x_1,y_1)$ gives the equivalent input group transformation.

Let $x_1 \leq w_1$. Take $i = \min \{j \mid w_j \leq x_1\}$. Let $R_2 = tI(w_{i-1}:z_{i-1})$ ... $tI(w_1:z_1)$. The output z corresponding to the input y_1, in the permutation $R_2[I]$ (where I is the identity permutation), tells us the starting point of the output cluster to which the original $tI(x_1:y_1)$ was applied.

So, $tI(w_m:z_m)...tI(w_i:z_i)tI(x_1:z)R_2$ is equivalent to R. Still $tI(x_1:z)$ may not be in the correct position. But as ordering within the same level of group interchanges does not affect the end transformation, we place $tI(x_1:z)$ in the correct position (removing both in case of duplication) to get the required input group transformation.

Definition 7 : Given a permutation P, let S(P) denote the set of all permutations derivable from P by application of input group transformations. Then S(P) is said to be the equivalent set of P.

Lemma 3 : Given any permutation P, the cardinality of its equivalent set S(P) is 2^{N-1}.

Proof: For an NxN system, at any level j we can apply input group interchanges independently on $2^{n-(j+1)}$ groups. Hence the total number of distinct sequences of input group interchanges at that level is $2^{2^{n-(j+1)}}$. By lemma 2 and theorem 1, we get the total number of distinct input group transformations for this system as $\sum_{j=0}^{n-1} 2^{2^{n-(j+1)}} = 2^{N-1}$.

By theorem 1, we get that from any permutation P, by applying all possible input group transformations we can generate a set of 2^{N-1} permutations, which is the equivalent set of P.

III. Self Routing in Benes Network

Here we will use the self routing technique described in algorithm 1 of [7]. We will refer to it as algorithm BR. For the first (n-1) stages, a switch at stage i ($1 \leq i < n$) will determine the smaller one of the two destination tags attached to its two inputs and will be set up according to its i-th bit. For the next n stages, a switch at stage j, ($n \leq j < 2n$) is set up by the (2n-j)th bit of the destination tag of each input.

Example 3: The routing in an 8x8 Benes network for a permutation P : 6 4 0 2 3 1 7 5 given by algorithm BR is shown in fig.3.

In [7] it has been mentioned that this self-routing technique with complexity $O(n)$ is applicable to LC class of permutations as well as $I\Omega$ permutations. But note that P of example 3 is neither a LC permutation nor an $I\Omega$ permutation.

Theorem 3 : If any permutation P is routable by the routing algorithm BR, then any permutation $P'\varepsilon S(P)$ is also routable by algorithm BR, where S(P) is the equivalent set of P.

Proof: We route both P and P' by algorithm BR in the first n-1 stages. Now from lemma 1, the set of output clusters $\{C_0(j,.)\}$ of P and P' are the same, for $1 \leq j \leq n$. Firstly let us start with stage 1. Since $\{C_0(1,.)\}$ of P and P' are same, the input pairs of the switches at stage 1 are the same for both. Now our routing algorithm routes the inputs depending on the two inputs of any switch at any stage. Hence for both P and P', we will have the same set of inputs at the upper half and the lower half of stage 2, i.e. $C_0(n-1,x)|_P = C_0(n-1,x)|_{P'}$ for all possible values of x. Besides this, the routing ensures that at the input of stage 2 also $\{C_0(j,.)\}$ will be the same for P and P' for $1 \leq j \leq n$.

In this way, the routing in each successive stage k, $1 \leq k < n$, will rearrange the inputs of stage (k+1), such that $C_0(n-k,x)|_P = C_0(n-k,x)|_{P'}$ for all possible values of x, keeping the sets $\{C_0(j,.)\}$ same for both P and P' for $1 \leq j \leq n$.

Thus after (n-1) stages, we will get at the input of stage n, $C_0(1,x)|_P = C_0(1,x)|_{P'}$ for all possible values of x, i.e., each switch at stage n will have same pairs of inputs for P and P'.This proves that if P is routable by BR, so also is P'.

Definition 8 : Let C represent any class of permutations, then CE is defined as the union of all equivalent classes generated by the permutations $P \varepsilon C$, and is denoted by, $CE = \bigcup_{P \varepsilon C} S(P)$.

Corollary1: Any permutation $P \varepsilon CE$ where $C = L \cup I\Omega$, is routable by algorithm BR.

The enumeration of the number of permutations included in CE is quite difficult. Instead, we attempt to calculate the cardinality of BPE only, which is a subset of CE.

IIIA. BPE (Bit-permute equivalent) Class of Permutations

BP (bit-permute) class of permutations [3] is a typical subset of L [7]. For any permutation $P \varepsilon BP$, for all input I (binary representation $I_n I_{n-1} ... I_1$) and output O (binary representation $O_n O_{n-1} ... O_1$), the non-singular binary matrix Q_{nxn} that satisfies $O^T = Q \times I^T$, is essentially the identity matrix, with some row interchanges. The effect is that in P, for all input-output pairs I and O, $O_j = I_k$, for $1 \leq j,k \leq n$ and $O_i \neq O_j$ for $i \neq j$.

Remark: Since BP is a subset of L, BPE (bit-permute equivalent) class of permutations is a subset of CE.

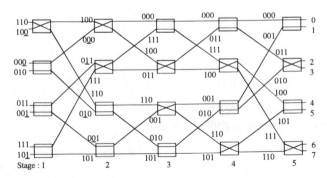

Fig. 3: Routing of permutation P of example 3 in B(3). (At any stage i, $1 \leq i \leq 2$, the underlined input bit determines the routing)

Example 4 : Given a permutation P^* ε BP, determined by the rule, $I_3 I_2 I_1 \rightarrow I_2 I_1 I_3$, i.e., P^* : 0 2 4 6 1 3 5 7, generate P by the application of T $\equiv \{tI(0:0), tI(0:4), tI(0:6), tI(1:0)\}$ on P^*, i.e., $T[P^*] \rightarrow P$. Hence P : 6 4 0 2 3 1 7 5, whose routing is shown in fig. 3. Note that , P ε BPE.

Lemma 4 : Let P_1 and P_2 denote two distinct BP-permutations. Then there does not exist any input group transformation T such that $T[P_1] \equiv P_2$.
Proof: Let $P_1 : I_n I_{n-1}...I_1 \rightarrow O_{n,1} O_{n-1,1}...O_{1,1}$ and $P_2 : I_n I_{n-1}...I_1 \rightarrow O_{n,2} O_{n-1,2}...O_{1,2}$ where $O_{i,k} = I_j$ for $1 \leq i,j < n$ and $k=1,2$.
Now say $O_{i,1} = O_{i,2} = I_j$ for $1 \leq j < m$, $1 \leq i < n$. Then it is easy to see that both P_1 and P_2 will have the same sets of output clusters at all levels $l \leq m$. But, at level (m+1) the output clusters are different and by lemma 1, no input group transformation can make them equal. Hence follows the lemma.

Lemma 5 : The cardinality of the set BPE is n! 2^{N-1}.
Proof: Follows from lemma 4 and lemma 3.

IV. Recognition of Permutations in BPE

Given any permutation P, we present an algorithm BPE-ID, that will decide whether P ε BPE or not and will output the BP that generates P by input group transformation.
Our algorithm starts with a given permutation P and tries to apply a sequence of group interchanges that will transform P into a BP, if P ε BPE.

Lemma 6: A permutation P is a BP permutation iff:

a) the outputs O(x) and $O(x+2^i)$ corresponding to the least inputs x and $x+2^i$ of two adjacent input groups G(i,x) and $G(i,x+2^i)$ differ in a unique bit position for all possible values of x and it is true for all i, $1 \leq i < n$ and

b) the output O(x) corresponding to the least input x of G(i,x) is always the least element of $C_0(i,x)$ for all possible values of i and x.
Proof: Since a BP permutation is defined by a definite bit-permute rule, any BP will satisfy the above conditions.
Again if any permutation P satisfies the above rules, we can easily construct a unique BP-rule to describe P.

Given any P, our algorithm checks P for condition b) of lemma 6, first at level 1 and rearranges the output clusters $C_0(1,x)$, if necessary, by applying tI(0:x) and for all clusters it also checks condition a). If it succeeds, it proceeds to higher levels. Ultimately it results a BP at level n or fails at an intermediate level.

4.1 Algorithm BPE-ID: The permutation P is stored in the form of an array OUT(.), where OUT(i) stores the output corresponding to the input i.

1. **For** k=1 **to** n **do**
    1.1 d=|OUT(0) – OUT(2^{k-1})|
        If d$\neq 2^j$, 0$\leq$j<n, **output** "P is not in BPE" and **terminate**.
    1.2 **For** p=0 to 2^{n-k}-1 **do**
        1.2.1 **If** OUT(p.2^k) > OUT(p.2^k+2^{k-1}) **then** apply tI(k-1: p.2^k).
        1.2.2 **If** OUT(p.2^k+2^{k-1}) – OUT(p.2^k)$\neq$ d, **then output** "P is not in BPE" and **terminate**.
        1.2.3 **Next** p.

    1.3 **Next** k
2. **Output** the string OUT that represents the generator BP and **terminate**.

4.2 Complexity analysis of the algorithm BPE-ID:

At any stage k, $1 \leq k \leq n$, our algorithm compares two outputs OUT(x) and OUT(x+2^{k-1}) and if necessary, it applies tI(k-1:x), where x=p.2^k for $0 \leq p \leq 2^{n-k}$-1.For each p, an extra comparison is needed to check condition a) of lemma 6.Therefore, at stage k, it will require 2^{n-k+1} comparisons and N/2 input interchanges in the worst case.

The total number of comparisons involved is $\sum\limits_{k=1}^{n} (2^{n-k+1}+N/2)$.

Hence the complexity of the algorithm is $O(N \log N)$ on a uniprocessor system. But the comparisons and interchanges at each level can be performed in parallel using a system with N/2 PE's. This will reduce the complexity of the algorithm to $O(\log N)$ only.

V. Conclusion

The application of input group transformation on permutations, extends the applicability of the self-routing technique [7] to a large set of permutations for Benes network. However this idea can be generalized for any (2n-1)-stage network, such as (2n-1)-stage shuffle-exchange network, where the transformation rule should be determined by the specific interconnection structure of the network. Moreover, it is interesting to note that besides the LE and IΩE classes of permutations, there exist some other classes of permutations as well, which can be routed by some self-routing techniques other than [7].

References

1.A.Waksman, "A Permutation network", *J. Assoc. Comput. Mach.*, vol. 15, No. 1, pp. 159-163, 1968.
2.J.Lenfant, "Parallel permutations of Data : A Benes Network Control Algorithm for Frequently Used Permutation", *IEEE Trans. Comput.*, pp. 637-647, July 1978.
3.D.Nassimi and S.Sahni, "A self-routing Benes Network and Parallel Permutation Algorithms", *IEEE Trans. Comput.*, pp. 332-340, May 1981.
4.—, "Parallel Algorithm to Set up the Benes Permutation Network", *IEEE Trans. Comput.*, pp. 148-154, Feb. 1982.
5.K.Y.Lee, "On the Rearrangeability of 2($\log_2$N) - 1-stage Permutation Networks", *IEEE Trans. Comput.*, pp. 412-425, May 1985.
6.D.Nassimi, "A Fault-tolerant Routing Algorithm for BPC Permutations on Multistage Interconnection Networks", *Proc. of 1989 International Conference on Parallel Processing*, pp. I278 - I287, 1989.
7.R.Boppana and C.S.Raghavendra, "On Self-Routing in Benes and Shuffle Exchange Networks", *Proc. of Int. Conf. on Parallel Processing*, Aug.1988, pp.196-200.
8.N.Das, B.B.Bhattacharya and J.Dattagupta, "Analysis of Conflict Graphs in Multistage Interconnection Networks", *Proc. of the 1990 IEEE Region 10 Conf. on Computer and Communication Systems (TENCON-90)*, Sept. 1990,Hongkong,,pp.175 -179.

REALIZING FREQUENTLY USED PERMUTATIONS ON SYNCUBE

Zhiyong Liu and Jia-Huai You

Department of Computing Science, University of Alberta

Edmonton, Alberta, Canada T6G 2H1

Abstract

We present a new implementation scheme of hypercube, called *Syncube*. Communication is faster through syncube than that through a traditional hypercube with the same hardware complexity. We show that the set of permutations realizable in one pass ($\log_2 N$ steps) through syncube without precomputation is largely greater than that through an MIN, such as Ω network. We present a new algorithm to realize BPC-permutations on hypercube as well as on syncube. We give schemes which can make full use of the connections among all the nodes in every dimension, and can realize $\log_2 N$ independent permutations simultaneously. In addition, syncube has a merit over MINs in that it can benefit localized communications.

Key words: Interconnection network, hypercube, permutations, communication bandwidth, localized communication.

1 Introduction

Hypercube is a promising topology for parallel and distributed processing systems on which various algorithms can naturally be implemented [3].

Let N be the number of nodes in the system, and $n = \log_2 N$. Although hypercube topology has merits over some multistage interconnection networks (MINs), the topology itself does not specify how the $\log_2 N$ links are switched out of a node in a physical implementation, nor does it say how the $\log_2 N$ links are selected in their utilization. The conventional design of hypercube networks and their routing algorithms have some limitations that prevent full exploitation of the strong potentiality of the hypercube topology.

In previously proposed schemes, the time length of a "cycle" (or "step") increases with $O(\log_2 N)$, or more hardware is needed and thus the hardware complexity is no longer $O(N \log_2 N)$; it will become $O(N(\log_2 N)^2)$.

It has been noticed by McCrosky [10] that the time length to move all messages one step in Hillis's algorithm increases with $O(\log_2 N)$. In Borodin-Hopcroft's algorithm [1], if $\log_2 N$ packets are to be sent from $\log_2 N$ input ports to $\log_2 N$ corresponding output ports at a time, full connectivity among these input and output ports is necessary. As a result, the hardware complexity of the entire system will be $O(N(\log_2 N)^2)$. If the packets are to be sent out one at a time, then a $\log_2 N$ times longer "cycle" is needed.

A hypercube structure, called generalized folding cube (GFC), with routing algorithm has been proposed in [2]. Hardware complexity of GFC is $O(N(\log_2 N)^2 2^{2p})$, where 2^p permutations can be realized simultaneously for an integer p.

We have developed a new structure, *Syncube*, and corresponding algorithms for hypercube connected topology in this paper. Using techniques developed here, n permutations can be realized simultaneously while the hardware complexity of syncube remains to be $O(N \log_2 N)$, and the time length of a "cycle" will not increase as any function of n. A key technique in syncube is a way by which all the links on every dimension are kept busy all the time, so bandwidth is fully increased. For these reasons, communication in syncube is more efficient than the previously proposed schemes for hypercube.

2 The Structure of Syncube

In a hypercube, if node A is connected by the ith dimensional link to node B, we say A is the ith *imagery* node of B, and denote it as $IMG^{(i)}(A) = B$.

A *Syncube (Synchronized Hypercube)* is an implementation of hypercube topology in which a node consists of a *PE* and a *router*. The function of a router is to route messages. The conceptual structure of a router in node S is shown in Figure 1, where $R_i(S)$ denotes the ith register in node S, and is connected to $R_j(S)$ and $R_j(IMG^{(i)}(S))$, where $j = i + 1 \bmod n$; by the symmetry of the definition, $R_i(IMG^{(j)}(S))$ is also connected to $R_j(S)$.

Obviously, the hardware complexity of the system is $O(N \log_2 N)$.

The key idea in our approach is to avoid conflicts in message transmissions. We say a conflict occurs if a register is to hold more than one message according to the routing algorithm. Our goal is to develop routing algorithms which will not cause any conflict in the routing procedure, so that buffer is not necessary and the "cycle" will depend only on hardware technology.

A *subscript mapping scheme* $M(n-1, n-2, \cdots, 0) = (v_{n-1}, v_{n-2}, \cdots, v_0)$ is a mapping such that $(v_{n-1}, \cdots, v_0)$ is a permutation of $(n-1, n-2, \cdots, 0)$. A subscript mapping scheme is called *shift mapping* if $v_i = k \pm i \bmod n$, where $0 \le k \le (n-1)$. A subscript mapping scheme gives an order in which the links on the n dimensions are used.

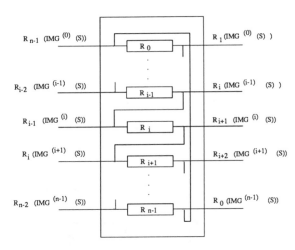

Figure 1: The Structure of a Router in Node S

We are now ready to give a routing scheme for syncube, which will be called *SMVDT-Routing* (Shift Mapping Virtual Destination Tag Routing).

Given a shift subscript mapping scheme $M(n - 1, n - 2, \cdots, 0) = (v_{n-1}, v_{n-2}, \cdots, v_0)$, where $v_i = k \pm i \bmod n$, SMVDT-Routing works as follows: At the beginning, the message on node S, $0 \leq S \leq N - 1$, is put in R_k. Then, in the routing procedure, every message is transmitted along the same dimension at a step, starting from dimension v_0 upwards to dimension v_{n-1}. At step i, if the v_ith bit of the address of the current node is not the same as the v_ith bit of the destination address of the packet, then transmit the packet from R_{v_i} of the current node to $R_{v_{i+1}}$ in its v_ith imagery node; else transmit the packet from R_{v_i} of the current node to $R_{v_{i+1}}$ of the same node. By doing this, the destination address is matched with the source address one bit in each step, dimension by dimension. Formal description of the algorithm is omitted here.

In a hypercube, each time the links on one dimension is eliminated, the hypercube is split into two subcubes. Let $subcube(d_1, d_2, \cdots, d_k)$, $(1 \leq k \leq n)$ of a hypercube be the set of the hypercubes obtained by eliminating all the links on the $d_1th, d_2th, \cdots, d_kth$ dimensions, where $0 \leq d_1, d_2, \cdots, d_k \leq n - 1$.

We quote the definition of *Complete Residue System modulo m* ($CRS \pmod m$) from [6]: A *Complete Residue System modulo m* ($CRS \pmod m$) is a set of integers which contains exactly one representative of each residue class $\pmod m$.

We define *Virtual Module Operation* $VMOD^{(l)}_{(k_1, \cdots, k_l)}(B)$ for an integer B. The result of virtual modulo operation is an integer obtained from dropping the bits on dimensions $k_1, k_2, \cdots, k_l$ in the binary representation of B. For example, with $B = 100111$, $VMOD^{(1)}_{(1)}(B) = 10011$, and $VMOD^{(2)}_{(1,3)}(B) = 1011$.

Let $sub \in subcube(d_1, d_2, \cdots, d_l)$, and $Numbset(sub)$ be the set that consists of all the numbers (the destination addresses of the messages) residing on sub. Let $(i : j)$

denote the set of integers from i to j. We have

Theorem 1 *Given a permutation, if there exists a shift mapping scheme* $M(n-1, n-2, \cdots, 0) = (v_{n-1}, v_{n-2}, \cdots, v_0)$ *where* $v_i = k \pm i \bmod n$, *such that*

$$\forall \; sub \in subcube(v_{n-1} : v_{n-m}), \; 1 \leq m \leq n - 1,$$

we have

$$\left\{ VMOD^{(m)}_{(v_{n-1} : v_{n-m})}(D) \mid D \in Numbset(sub) \right\}$$

is a $CRS \pmod{2^{n-m}}$, *then the permutation can be realized by SMVDT-Routing in n steps on syncube.*

This condition is a special case of Theorem 1 in [8] in that the subscript mapping there can be arbitrary, not necessarily a shift mapping; we omit the proof here.

Now let's choose n shift mapping schemes $M_0 = (n - 1, n - 2, \cdots, 0), M_1 = (n - 2, \cdots, 0, n - 1), \cdots\cdots, M_{n-1} = (0, n - 1, \cdots, 1)$. Because of the structure of the routers, it can be shown that n permutations can be passed independently. Any permutation satisfying the condition given in Theorem 1 can be passed in n steps. Therefore, n permutations satisfying the condition given in Theorem 1 can be realized simultaneously in n steps without conflict [9].

3 Passability of Frequently Used Permutations on Syncube

3.1 Permutations Satisfying the Passable Conditions

It can be shown [8] that any permutation passable on Ω, Ω^{-1} and indirect n-Cube are passable on syncube.

However, there do exist some permutations which are not passable on the above MINs but are passable on syncube. For example, the four permutations $\delta(x)$, $\lambda(x)$, $\delta^{-1}(x)$, and $\lambda^{-1}(x)$ which are needed in a skewing scheme are not passable on indirect n-Cube [5], but they are passable on syncube.

Among the five families of frequently used permutations given in [7] which can be realized on Benes network in $2 log_2 N - 1$ steps, two families, i.e., $\lambda^{(n)}_{j,k}$ and $\delta^{(n)}_{j,k}$, satisfy the condition given in Theorem 1, and thus can be realized in $log_2 N$ steps on syncube.

The remaining three families in [7] are subsets of BPC-permutations (Bit Permute Compliment permutations) researched in [11, 12, 13, 4]. BPC-permutations can be realized in $2 log_2 N - 1$ or $2 log_2 N$ steps on MINs. We will develop a new algorithm to realize BPC-permutations on syncube.

3.2 BPC-permutations

Let $D = d_{n-1} d_{n-2} \cdots d_0$ be a destination address, $S = s_{n-1} s_{n-2} \cdots s_0$ be its source address. Let $(f_{n-1}, f_{n-2}, \cdots, f_0)$

be a permutation of $(n-1, n-2, \cdots, 0)$, a BPC-permutation is such that $d_i = s_{f_i}$ or $d_i = \bar{s}_{f_i}$.

Any permutation can be represented by "cycles". A cycle of length t $(1 \le t \le n)$ is a set $(v_1, v_2, \cdots, v_t)$ where $f_{v_1} = v_2$, $f_{v_2} = v_3$, $\cdots$, and $f_{v_t} = v_1$.

In a cycle $(v_1, v_2, \cdots, v_t)$, if $v_i = \max\{v_1, v_2, \cdots, v_t\}$, $1 \le i \le t$, we say that v_i is the *top dimension* of the cycle. We use $TOPSET$ to denote the set of the top dimensions of all the cycles in a given permutation.

Our idea for the realization of BPC-permutations is as follows: first arrange any BPC-permutation into a permutation passable on hypercube, this takes at most $n-1$ steps; and then use SMVDT-Routing given in Section 2 to realize it in n steps. We have developed an algorithm, called *BPC-Partition*, to arrange any BPC-permutation into a passable one. The algorithm transmits the packet on any node based solely on one bit of the destination address; in particular, it does not need to compare the bit with the address of the current node. In the following algorithm, D is the destination address of the packet residing on node S in *any step*.

Algorithm BPC-Partition;
    BEGIN
        /* The N nodes do the following parallelly */
        FOR $j := 0$ TO $n-2$ DO
            IF $(j \notin TOPSET)$ and $(d_j = 1)$
                THEN send the packet to $IMG^{(j)}(S)$;
    END;

Note that this algorithm should be viewed as a generic description of routing for BPC-permutations in that once a BPC-permutation is given, the $TOPSET$ can be cooperated into a specific form of the above algorithm so that no computation is needed to decide whether a dimension belongs to $TOPSET$.

The proofs of the following theorems are omitted here and can be found in [9].

Let $S^{(i)}$ denote the node on which the number D resides at step i (so, $S^{(0)}$ is the source address of the packet whose destination address is D).

Theorem 2 *Let D and D' be the two numbers residing on the two nodes connected by any jth dimensional link at step j, we have:*

$$\begin{cases} d_j = d'_j & \text{if } j \notin TOPSET, \\ d_j \ne d'_j & \text{if } j \in TOPSET. \end{cases}$$

Theorem 3 *In the transmission process of Algorithm BPC-Partition, no conflict will occur at any step.*

Theorem 4 *Let D and D' be the two numbers residing on node $S = s_{n-1}^{(j)} s_{n-2}^{(j)} \cdots s_0^{(j)}$ and node $S' = s_{n-1}'^{(j)} s_{n-2}'^{(j)} \cdots s_0'^{(j)}$ respectively at the jth step, S and S' be two distinct nodes in any subcube in subcube$(0, 1, \cdots, j-1)$. If $d_{n-1} \cdots d_{j+1} = d'_{n-1} \cdots d'_{j+1}$, and $d_j = d'_j$, then we have*

$$\begin{cases} s_j^{(j)} = s_j'^{(j)} & \text{if } j \notin TOPSET, \\ s_j^{(j)} \ne s_j'^{(j)} & \text{if } j \in TOPSET. \end{cases}$$

Theorem 5 *After the transmission procedure of Algorithm BPC-Partition, any BPC-permutation will become a permutation passable on hypercube under the subscript mapping scheme $M(n-1, n-2, \cdots, 0) = (0, 1, \cdots, n-1)$.*

Because of Theorem 5, we can use Algorithm BPC-Partition to arrange any BPC-permutation into a permutation passable on hypercube in at most $\log_2 N - 1$ steps, and

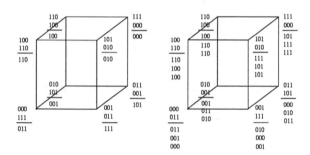

Figure 2: The Realization of a BPC-permutation

then use SMVDT-Routing to realize the permutation in at most $\log_2 N$ steps. In such a way, any BPC-permutation can be realized in at most $2 \log_2 N - 1$ steps. One can show, for example, a perfect shuffle permutation will be realized in $2 \log_2 N - 1$ steps, a bit reversal permutation will be realized in $2 \log_2 N - \lfloor \log_2 N \rfloor - 1$ steps, and a vector reversal permutation will be realized in $\log_2 N$ steps.

An example is given in Figure 2, where $N = 8$, and the permutation is a bit reversal with every bit being complemented: if $S = s_2 s_1 s_0$, then $D = \bar{s}_0 \bar{s}_1 \bar{s}_2$, and $TOPSET = \{1, 2\}$ since there are two cycles (1) and (0, 2). In the figure, a column of numbers is shown on each node; the first number in each column is the source (the serial number of the node), the second one is the destination of the message, and the rest shows the sequence of the numbers residing on the node step by step. The left part of the figure shows how the permutation has been arranged into a passable one by algorithm BPC-Partition (with one step), and the right part shows the realization of the permutation by algorithm SMVDT-Routing (with three steps).

It is interesting to see that Algorithm BPC-Partition can work under any shift subscript mapping scheme for any BPC-permutation. However, under different subscript mapping scheme, a BPC-permutation may have different dimensions in $TOPSET$.

Theorem 6 *Under any shift subscript mapping scheme, any BPC-permutation can be arranged into a permutation passable on syncube under the same mapping scheme.*

According to Theorem 6, n BPC-permutations can be realized simultaneously on syncube in at most $2n-1$ steps by choosing n appropriate subscript mappings. The precise description of the algorithm can be found in [9].

4 Benefiting Locality

Comparing with MINs, syncube has another advantage, i.e., it can benefit localized communications. Let's consider again the frequently used permutations given in [7]. A large number of permutations in the five families are permutations *within segments*. They are realized in the same number of steps on MINs as the permutations which are not within segments. When a permutation is within segments, it is obvious that there is locality in the communication. Because the syncube structure retains the flexible partitionality of hypercube topology, permutations within segments of different sizes can be realized in different steps. This feature can make the average time spent on communication even shorter.

As an example, it is not difficult to see that, a cyclic shift within segments of size 2^{n-j} can be realized in $n-j$ steps on syncube. In a parallel computation, if the cyclic shifts occur within $2^1, 2^2, \cdots, 2^n$ evenly, the average delay for the communication will be $\frac{1}{2} \times \log_2 N$ steps.

5 Conclusion

We have presented the syncube structure for implementation of hypercube and routing algorithms. Hardware complexity of syncube remains to be $O(N \log_2 N)$ while the time length of a "step" does not increase with any function of the size of a network. By keeping all the links busy all the time, n permutations can be realized on this structure simultaneously. This makes the bandwidth of communication fully increased. The structure has merits over MINs in that the number of permutations passable on it is largely greater than that on MINs, and it can benefit communications with locality so that the delay time for communications is shorter than on MINs.

Finally, we remark that with limited additional links syncube may realize all kinds of the frequently used permutations in $\log_2 N$ steps. This issue is under investigation.

Acknowledgement: This research is supported by the NSERC of Canada under the Grant OGP9225.

References

[1] A. Borodin and J. E. Hopcroft. Routing, merging and sorting on parallel models of computation. In *Proc. of the 14th Annual Symposium on Theory of Computing*, pages 338–344, May 1982.

[2] S. B. Choi and A. K. Somani. The generalized folding-cube. In *Proc. of the 1990 International Conference on Parallel Processing*, volume I, pages 372–375, Aug. 1990.

[3] M. T. Heath, editor. *Hypercube Multiprocessors 1987*. SIAM, Philadelphia, 1987.

[4] S. T. Huang and S. K. Tripathi. Self-routing technique in perfect-shuffle networks using control tags. *IEEE Trans. on Computers*, C-37(2):251–256, Feb. 1988.

[5] D. Lee. On access and alignment of data in a parallel processor. *Information Processing Letters*, 33(1):11–14, Oct. 1989.

[6] K.Y. Lee. On the rearrangeability of $2(\log_2 N) - 1$ stage permutation networks. *IEEE Trans. on Computers*, C-34(5):412–425, May 1985.

[7] J. Lenfant. Parallel permutations of data: A Benes network control algorithm for frequently used permutations. *IEEE Trans. on Computers*, C-27(7):637–647, July 1978.

[8] Z. Liu and J. You. Routing algorithms for hypercube. Technical Report 90-26, Department of Computing Science, University of Alberta, Edmonton, Alberta, Canada T6G 2H1, 1990.

[9] Z. Liu and J. You. A new implementation for hypercube connection. Technical Report 91-09, Department of Computing Science, University of Alberta, Edmonton, Alberta, Canada T6G 2H1, 1991.

[10] C. McCrosky. Message passing in synchronous hypercubes. Technical Report 86-4, Department of Computational Science, University of Saskatchewan, Saskatoon, Canada, 1986, (to appear in Computer System Science and Engineering).

[11] D. Nassimi and S. Sahni. A self-routing Benes network and parallel permutation algorithms. *IEEE Trans. on Computers*, C-30(5):332–340, May 1981.

[12] D. Nassimi and S. Sahni. Optimal BPC permutations on a cube connected SIMD computer. *IEEE Trans. on Computers*, C-31(4):338–341, April 1982.

[13] P. C. Yew and D. H. Lawrie. An easily controlled network for frequently used permutations. *IEEE Trans. on Computers*, C-30(4):296–298, April 1981.

Inter-Section Locality of Shared Data in Parallel Programs

Jih-Kwon Peir[1], Kimming So[2], Ju-Ho Tang[1]

[1]Computer Science Department
IBM T. J. Watson Research Center
Yorktown Heights, New York 10598

[2]IBM Austin
Advanced Workstation Division
Austin, Texas 78758

Abstract

Numerous software schemes have been proposed to enforce cache coherence among the private caches in a highly parallel multiprocessor system. In this study the concept of inter-section locality is introduced to estimate the effectiveness of a software coherence scheme and to provide a way for better understanding of the behavior of the shared data in a parallel program. Most of the software schemes rely on how well they can keep the shared data in cache for reuse across program sections or synchronization barriers. Therefore, the effectiveness of these schemes depends on the reuse hit ratio from one section to another. Using traces collected from several multiprocessor applications, our trace-driven simulation with a spectrum of cache sizes shows that this locality is in general quite low. The major reason is the working set of shared data across the barriers is too large for any cache of reasonable size.

It is also found that, because the overall cache performance is normally dominated by the intra-section locality but not the inter-section locality, even if the inter-section locality can be as high as 100%, the overall performance difference between any two software schemes can be marginal. The results suggest that the sophisticated software coherence schemes may not be cost effective.

Keywords: Cacheability controls, Cache coherence scheme, cache-based highly parallel system, Inter-section locality, Intra-section locality, Parallel program characteristics, Trace-driven simulation.

1. Introduction

A cache storage is a critical component in a multiprocessor (MP) system. It can keep each processor executing at its full speed by substantially reducing the average memory access time for individual processor as well as the memory contention in the entire MP system. However, in designing a shared memory MP system where each processor is equipped with a cache memory, it is necessary to maintain a data sharing coherence among multiple caches such that a memory access for a piece of data from a processor is guaranteed to receive the latest version of the data in the system. Traditionally, cache coherence is enforced by hardware either through a shared snoopy bus [Swe&Smi86] or through a centralized control [IBM82]. The general approach is to maintain a single consistent copy of every piece of data in the system; so when an update to a cache entry, called a *line*, of a particular cache is made, a hardware scheme guarantees that the same line in other cache(s) get either invalidated or updated accordingly. These two hardware schemes are known as *invalidation* and *broadcast* schemes, respectively.

The hardware solution is effective only when there are a small number of processors in an MP system. For a highly parallel cache-based shared memory system with multistage interconnection network (MIN), the complexity of coherence circuitry and communication penalty across the entire system limit the number of processors to the order of 10 only [Pfi + 85]. A highly parallel system that does not guarantee hardware coherence normally requires a proper software coherence control mechanism.

A basic SPMD (Single-Program-Multiple-Data) parallel programming model is used in this study. Under this model, a program is partitioned into a sequence of parallel and serial sections. Synchronization points (or barriers) are inserted at the end of each section to maintain proper data dependence. A hardware coherence scheme under the model may create heavy but unnecessary inter-cache traffic even for a Doall loop in which all the pieces of work are data independent. This is due to the fact that a cache line usually contains a multiple pieces of data which can independently be used by several processors during program execution. In its worst case for the invalidation scheme mentioned above, a *ping-ponging* phenomenon may be encountered when more than one processor keep updating different parts of the same line in about the same time. This is known as *false-sharing* [Egg89]. In the sequel, the part of program data that has the potential of being read or written by more than one processor at runtime is called the *shared data* of the parallel program.

Recently, there have been numerous proposals of software-based cache coherence schemes [Vei86] [Che&Vei88][Cyt + 88] [Che&Vei89][Min&Bae89]. Such *software coherence schemes* usually allow inconsistency lines temporarily coexist among multiple caches as long as the processors are guaranteed not to access any stale data, i.e., data modified in a remote cache. The unnecessary coherence control traffic due to false-sharing can be eliminated. The simplest software coherence scheme is to invalidate the entire cache [Vei86] [Lee + 87] or an entire page [Smi85] whenever there is a potential of accessing a piece of stale data. Apparently, the indiscriminate invalidation may cause unnecessary cache misses. The best possible way is to invalidate only those lines that have been updated by other processors. However, without special hardware supports the scheme is not practical for it requires a search of the entire cache per invalidation.

Several schemes have been proposed to approximate this best invalidation scheme. For example, a scheme is described in [Che&Vei88] to continue use of unmodified shared data across the sections. The scheme uses the compiler to identify all the reads to a variable before a write as *cache-reads*. A piece of cache-read data can continue be used even after it has been marked invalid between sections. Two similar schemes proposed in [Min&Bae89] [Che&Vei89] use a complicated tagging scheme to detect the stale data. By using a counter for each word in the cache and careful manipulation of this counter during each cache access, it is shown that explicit invalidation is not needed and even a write-shared variable in a cache can be continuously accessed as long as its counter indicates so. By assuming certain static scheduling mechanism and an infinite cache with unity line size, this software scheme yields miss ratios comparable to those of the directory-based hardware scheme [Min&Bae90].

Our work started off with a set of applications, a parallel program tracing tool PSIMUL [So + 88], and a trace-driven cache model to evaluate the coherence schemes. There are two simple built-in coherence schemes inside PSIMUL:

1. *Optimistic scheme* - there is no cache invalidation between sections.

2. *Pessimistic scheme* - the whole cache content is invalidate at the end of every section. This scheme is the same as the one studied in [Vei86].

These two schemes provide bounds for any reasonable software coherence scheme. There is another coherence scheme, called *Invalidate-On-Modified*, which is not provided by PSIMUL but is included in our study. Under the scheme all modified data are invalidated at the end of a section. The performance results of this scheme provide us with some idea of those software schemes which have to invalidate the modified shared data in order to maintain data coherence. Finite cache sizes ranging from 16KB (kilobytes) to 2MB (megabytes) with two different line sizes: 16 and 128 bytes have been used for the study.

Surprisingly, our results indicated that there is no substantial difference in cache misses between the optimistic and the pessimistic schemes. In particular, it is found that the usual performance metrics such as overall cache miss ratio and the frequency of moving

data among caches are not enough to estimate the actual performance gain of a software coherence scheme. The major reason is the overall cache performance is very likely dominated by the locality of data accesses within the parallel sections of a program. In addition, it is found that with reasonably cache sizes, not only the shared data have a poor cache hit ratio, they also cause most of the cache misses.

The effectiveness of these software coherence schemes depends very much on the behaviors of parallel programs and in particular the access patterns of the shared data. In order to estimate the performance gains of a software coherence scheme, it is necessary to quantify the benefit of keeping a piece of shared data in a cache at the end of a section, i.e., if the shared data ever be reused by the processor in a subsequent section. To this regard, there are two types of locality of shared data references in a program: the *intra-section* locality is measured within each section and the *inter-section* locality measured from one section to another. The applications selected in the study are Simple, Weather code, Tomcat, and FFT; they are in the area of scientific computation. The applications have been parallelized for SPMD computation and are traced by PSIMUL at the instruction level. A brief description is give in Appendix 1.

Our trace-driven simulation of these applications shows that the inter-section locality is generally very low; and it is much lower for software schemes that invalidate modified cache lines at the end of each section. For example, for a 1MB cache with the invalidate-on-modified scheme. the inter-section cache hit ratios are mostly below 10% for both Simple and Weather code except for the case of a 64-way MP. Application Tomcat contains numerous small sections. Because the working set of each of these sections is small, the application has somewhat higher reuse hit ratios of 35-45% with the optimistic scheme. However, the reuse hit ratio drops to the 10% range under the invalidate-on-modified scheme. Maximum inter-section locality is achievable when the cache has a capacity large enough to retain almost all of the accessed shared data. For example, application FFT has a shared data size of 2 MB; its inter-section hit ratios can approach 100% when a cache of about 2MB is used. Unfortunately, because all shared data are modified in FFT computation, there is no inter-section locality at all for FFT for any software scheme that invalidates modified cache lines between sections.

For the rest of this paper, Section 2 contains a background of SPMD computation model and the tracing technique used in this study. Inter-section locality and its measurement are introduced in section 3. Section 4 contains the performance results and their implication. Our concluding remarks are in Section 5. A brief description of the four applications is included in Appendix 1.

2. Background and Methodology

This section describes how SPMD model is applied to software coherence schemes and technique in trace-driven simulation.

2.1 SPMD computational model and software coherence schemes

In SPMD computation [Pfi84] of a parallel program, the same program is sent to all of the participating processors and individual processor executes different pieces of work by operating on different pieces of data. A parallel program is laid out as a sequence of sections of code. There are two types of sections. A parallel sections contains a multiple pieces of work, each of which can be assigned independently to a processor at program execution time. A serial section contains one piece of work which can only be executed by one processor. Data dependency between two sections is likely resolved by a barrier which forces all the executing processors to wait for each other at the end of the first section.[1] The shared data used to implement a critical section, or in a barrier are specifically called *synchronization variables*; the rest will simply be referred to as shared variables or data.

The goal of any software scheme is the same as any hardware coherence scheme, i.e., to maintain the coherence of data sharing among the processors and at the same time to maximize the use of local cache during the execution of a parallel program. The synchronization variables usually are not cached at execution time. A software coherence scheme is especially effective to cope with Doall loops in which all the pieces of work are independent of each other. In this case, all the shared variables used in a parallel section are cacheable and will be flushed to the memory, if necessary, at the end of the section. Therefore a software coherence scheme is different from another by how it balances between the following two factors:

- the frequency and scope of flushing shared data from a local cache to maintain coherence

- the way to keep the shared data in a local cache to be reused by the processor from one section to another in order to maximize the cache hit ratio.

For a software coherence, it is easier to keep in a cache the shared read-only data than the read-modify data for reuse, where the read-modify data are the data modified by a processor during the execution of a section. Therefore the amount of shared read-modify data in a section is important and will be part of our study.

2.2 Tracing Technique

Trace-driven simulation is used in our study. The execution of a parallel program is traced by PSIMUL and the processor traces are used to drive a cache model to produce the performance data of various coherence schemes.

Generation of MP traces and trace-driven MP cache simulation are known to be time consuming and complicated. For the purpose of performance estimation of software schemes for SPMD computations, however, a way is derived to substantially reduce the effort. First, it is noticed that the amount of computation and results of these applications are independent of the number of processors being used. Second, under the assumption that a static scheduling algorithm is being used in the execution of a parallel program, it is possible to focus our study of shared data accesses to only one processor. To this regard, it suffices to use a built-in scheduler in PSIMUL to produce a trace to drive the local cache of the processor. This technique that allows us to examine one processor trace at a time greatly facilitates our study. If necessary, the complete MP cache performance can be obtained by either averaging or sampling over all or part of the individual's.

As an option of EPEX/Fortran preprocessor, Preface [Ber&So88] calls to a marker routine with different parameters are inserted in the parallel program to reflect different synchronization events of the program, e.g., barriers, beginning of a parallel section, end of a unit of work, etc. Recognizing the invocations of the marker routine, PSIMUL understands the attributes of the units of work in a parallel program as well as part of their inter-unit dependency during tracing time. This allows us to implement a scheduling algorithm within PSIMUL and produce MP traces which reflect the resulting schedule.

For the purpose of this study, PSIMUL adopts the static self-scheduling algorithm. For example, in a parallel section processor i always executes the units of work of the: i th, $(i+n\_pe)$th, $(i+2n\_pe)$th ..., where n_pe represents the number of processors in an MP system. Here a unit of work can be one iteration or a chunk of iterations in a parallel do loop. A trace containing only these pieces of work is then used to drive a cache model for our study. In order to further reduce our effort for studying inter-section locality, a fully associative cache model is used in PSIMUL to produce for every section two sequences of distinct cache lines, i.e., line tags: the sequence of lines that initially filled up the cache and the sequence of lines that were left in the cache at the end of the section. Each record associated with a line also contains information of how the line was used in the section, e.g., read-only or read-modify. The usage of these sequences is described in the next section.

Since the applications being studied are highly parallelizable and computationally intensive, the total amount of computation is about the same disregarding whether they are executed on a uniprocessor (UP) system or an multiprocessor (MP) system. In this paper, a trace generated by one simulated processor executing the entire parallel program will be called a *UP trace* and a trace generated by one of the simulated processors in a parallel execution, either 16 or 64 processors, will be called a *MP trace*. In general it is expected that a UP trace has a larger working set but has a stronger overall memory reference locality. The inter-section locality of these traces will be compared in the study.

---

[1] Note that the general SPMD model allows data dependency between pieces of work in a parallel section. The execution of such parallel section requires proper synchronization and coherence instructions to maintain the data dependency. The four parallel applications in this study do not utilize this property and thus make our model of computation a little simpler.

3. Inter-Section Locality

In order to measure the inter-section locality of a parallel application, it suffices to observe the cache activities during the transition between sections, i.e., one needs to take a snapshot of the cache content at the end of a section and to measure how it is reused by the processor at the beginning of the next section. This approach allows us to filter out the cache activities within each section, which are for the interest of the intra-section locality.

Let us start off with a few terminologies. During the execution of a section, a line is *first-referenced* when the first time an access to the line is made. Similarly, a line is *last-referenced* when the last time an access to the line is made. For the *i-th* section in a parallel program, there are two referencing sequences defined below. The *first-sequence*, F^i, which is denoted by $F_1^i, F_2^i, F_3^i, ..., F_n^i$, is the sequence of lines ordered according to their first-references in the *i-th* section. The length of this sequence n can be as large as the total number of shared lines of the program. The second sequence of lines is called the *last-sequence*, L^i, and is denoted by $L_n^i, L_{n-1}^i, ..., L_2^i L_1^i$. They are ordered from LRU (least recently used) to MRU (most recently used) lines. Since these two sequences are produced by PSIMUL at the end of every section, their lengths are also limited by the number of lines in the cache of PSIMUL.

These two sequences are used to study inter-section locality of the shared data for a particular processor. Basically, the last-sequence defines what are stayed in the cache at the end of the section from the MRU (most-recent-used) position to the LRU (least-recent-used) position. Whether these cache lines are reused before they are replaced will depend on the first-sequence of the next section. In our simulation, the cache content is first initiated by L^1. Then, F^2 is treated as a series of requests to a fully-associate cache with LRU replacement scheme. The simulation of F^2 requests continues until the number of simulated requests is equal to the size of the cache. At this point, all the cache lines left from the previous sections are either referenced or replaced. The process of defining cache content by L^{i-1} and simulating cache hits and misses of F^i continues until the last-sequence of the last section. The total hits and misses of all the first-sequences are accumulated for measuring the inter-section locality:

$$reuse\ hit\ ratio = \frac{total\ cache\ hits}{total\ cache\ references}$$

For a given line size, various cache sizes can be simulated by the same set of sequences as long as they are smaller than the cache size used in PSIMUL where these sequences were generated. A maximum of 2MB were used in this study. Because a fully associative cache was used for simplicity, the results tend to provide an optimistic measurement of inter-section locality. One should note that however, the technique presented can as well be applied to other different cache structures, more sophisticated software coherence schemes, and scheduling algorithms.

In Figure 1, we show an example with seven lines in the *last-sequence* and the *first-sequence* of two adjacent sections. Assume the cache has a size of four lines.

From the last-sequence, we can determine the lines stayed in the cache at the end of section *i-1*. Because of the LRU replacement, the four cache blocks from the MRU to the LRU position are: 103, 106, 101 and 104. This four blocks are the same as the last four blocks in L^{i-1}. To see if those lines are *reused* in section *i*, first it is noticed that although there are 3 lines (101, 103, 104) common to the last 4 lines in L^i and the 4 first lines in F^i, due to the LRU replacement in the cache simulation, only lines 101 and 103 are cache hits. The processor actually misses lines 102 and 104. Consequently the reuse hit ratio between these two sections is at best 50% under the assumptions. This example also demonstrates the importance of using reasonable cache parameters and operations.

```
Last-referenced sequence of section i-1
    102 105 107 104 101 106 103

First-referenced sequence of section i
    101 102 103 104 105 106 107
```

First-Sequence	Hit/Miss	Cache Content MRU			LRU
\<end of i-1\>		103	106	101	104
101	Hit	101	103	106	104
102	Miss	102	101	103	106
103	Hit	103	102	101	106
104	Miss	104	103	102	101

Figure 1. A Inter-Section Locality Example

It is assumed in this example the cache size is smaller than the total number of referenced lines in each section. The section size varies in real applications. It is possible that some sections have only a few referenced lines. At the end of such sections, the cache may consist of the data lines left over from earlier sections. Therefore, the locality crossing multiple sections can also be captured in the study.

In addition to the cache size, the inter-section cache reuse hit ratio depends largely on the sequences of the first and the last references in every section. In Figure 2, we demonstrate two extreme cases. In the best case, the inter-section locality is fully captured despite the cache size. In the worst case, however, the inter-section locality is completely lost unless a larger cache is provided.

```
-------------------------------------
Best Case:
  L_{i-1} = 107 106 105 104 103 102 101
  F_i     = 101 102 103 104 105 106 107
  Sizes   =  1   2   3   4   5   6   7
  Ratios  = 1.0 1.0 1.0 1.0 1.0 1.0 1.0
-------------------------------------
Worst Case:
  L_{i-1} = 101 102 103 104 105 106 107
  F_i     = 101 102 103 104 105 106 107
  Sizes   =  1   2   3   4   5   6   7
  Ratios  = 0.0 0.0 0.0 0.0 0.0 0.0 1.0
-------------------------------------
```

Figure 2. Two Extreme Cases in Captured The Inter-Section Locality

4. Performance Analysis

The section begins with some observations on the differences between private and shared data in the four applications described in Section 2. Then our measurements of inter-section locality of the applications are presented and the results are interpreted based on the source codes of parallel applications. In the performance study, the cache size varies from 16KB to 2 MB; the line size is fixed at 128 bytes and only the LRU replacement algorithm is used. Except in Section 4.1, our inter-section locality study is confined to shared data only.

4.1 Shared and Private Data in Parallel Programs

Besides a small number of synchronization variables, the sizes of the shared data are 10.0 MB, 11.4 MB, 7.8 MB, and 2MB in Simple, Weather Code, Tomcat, and FFT, respectively. In general, they dominate the data set size of the applications but they contribute only a small portion of the memory references of these programs. To compare the intra-section locality of shared data with that of private data, a full processor trace, instead of only the first and last sequences, is used to drive a cache model. The cache model accepts all the private and shared memory references and produces overall cache misses for all the sections.

The results in Figure 3, are obtained with a cache size of 256KB and a line size of 128 bytes. The cache is flushed at the end of every section; thereby, no effect of inter-section locality is shown. As indicated in the Figure, the miss ratios of the private data are insignificantly low but in contrast, those of the shared data are extremely high for the cache sizes being studied. Moreover, the cache misses are mostly caused by the shared data accesses. For example in Simple shared data access is 14.6% of the total memory access which including both instruction and data accessed, shared data misses contribute to 99.6% of the total misses, the miss ratio of shared data access is 9.1%, and the miss ratio of private data access is 0.006%. However, one should notice that the intra-section hit ratios of the shared data are still much higher than the inter-section hit ratios to be presented in the next subsection.

The results in Figure 3 indicate that the cache misses are dominated by those of the shared data, it suffices

to confine our study of inter-section locality to only the shared data. The results in the next subsection were obtained by applying the first and last sequences of shared data accesses to our cache model.

4.2 Inter-Section Locality Measurement

The results of measuring the inter-section locality of the four applications are presented in this subsection. As described in Section 3, the UP and MP traces of each application are used to drive our cache model. For the MP, only the simulation results from the second processor trace are presented; in general the variations among the processors are found to be small. The simulation using UP and MP traces will respectively be called an *UP* and an *MP environment* below.

There are two simple software coherence schemes being considered: the optimistic scheme and the invalidate-on-modify scheme. These schemes are described in Section 1. In general our simulation results show that the inter-section cache reuse hit ratios are very low in all four applications. Because some of the observations are application dependent, the inter-section locality of each applications will be presented separately in the sequel.

4.2.1 Simple

There are a few interesting observations that can be made out of the results of Simple in Figure 4:

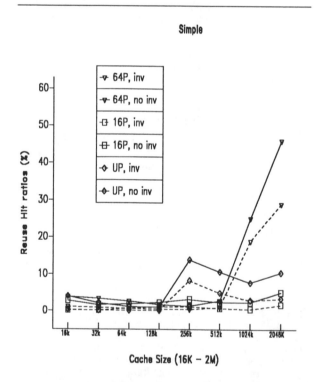

Figure 4. Inter-section cache hit ratios for Simple

- For the optimistic scheme in the UP environment, the average reuse hit ratios is only 6.4% for all the cache sizes being considered. Even for large caches of 1MB and 2MB, the ratios are only 7.7% and 10.5%, respectively.

	Simple	Weather	Tomcat	FFT
Shared data (%):				
reference	14.6	15.2	40.0	18.2
misses	99.6	98.2	99.7	99.7
miss ratio	9.1	21.0	2.2	26.6
Private data (%):				
miss ratio	0.006	0.072	0.301	0.016

Figure 3. Intra-section Locality

There are two major reasons as indicated from the source code of the application. First, Simple requires very large cache to capture inter-section locality. There are 20 shared arrays in Simple, each is half MB in size. Most of the major loops in Simple access 4 or more of these arrays. As a result, even for a 2MB cache, only a part of the shared arrays can be retained in the cache after every section. This factor of a large working set is compound by the fact that a processor does not necessarily reference the same set of shared arrays in consecutive sections. Therefore, an even larger cache is required to hold the shared arrays across multiple sections to capture inter-section locality.

The second reason is the effect of the order of memory references among the shared arrays or within a shared array. It is observed that in most of the Doall loops in Simple, several shared arrays are normally accessed at the same time. Without a large enough cache, they are contenting the limited cache space and eventually only a portion of each of these arrays can be left at the end of a section. On the other hand, the accessing order within each shared array tends to follow a same direction. For example, as a consequence that loop iterations are normally indexed from small to high, all references to the shared arrays tend to start from the lower addresses to the higher addresses of an array. Under this circumstance, even a shared array is used in consecutive sections, the portion of the referenced array (high addresses) left in the cache at the end of the first section not necessarily get reused at the beginning of the second section. If the cache is not large enough they may be replaced before being accessed again. This seems to match the worst case scenario shown in Figure 2.

- Like any other observation of the behavior of scientific computation, the reuse hit ratios can have abrupt jump or drop due to the change of a parameter in the experiment. For example, a jump of reuse hit ratios occurred when the cache is doubled from 128KB to 256KB in the UP environment. This incident owes to the boundary computation in Simple. There are 4 parallel sections that only use the boundary elements of several two dimensional shared arrays in the computation and each of these sections is followed by a section which accesses all the elements within these arrays and can have high reuse rate. Because the working set of each of these 4 sections is 256 KB, a 256KB cache is enough to hold the entire working set to be reused in the next section. In this case, further increase in the cache size can hardly improve the reuse hit ratios for these 4 sections.

- The reuse hit ratio do not always improve with the increase of the cache size. It is shown from the figure that 512KB and 1MB caches have lower reuse hit ratios comparing with the 256KB cache in the UP environment. This is due to the fact that both the numerator and the denominator in the hit ratio are changing at the same time. For instance, when the cache size is doubled the total references accounted for the inter-section locality are almost doubled; but the reuse hits may not increase at the same rate. For the cases of 256KB and 512KB

caches, we found that the inter-section hits are 5374, and 7647, but the reuse hit ratios are 13.9% and 10.7% respectively.

- In an MP environment, increase of the number of the processors or the cache size tends to hold the working set better, and it normally results in a higher reuse hit ration. This is quite obvious from Figure 4 for the case of 64 processors with one or two MB caches; the reuse hit ratios reach as high as 25.1% and 46.1% respectively.

Surprisingly however, the MP performance can be worse than the UP's in some cases. The reason again is due to the changes of working set in the MP environment. For example, almost all the loops in Simple are row major. The parallelization is normally performed at the outer loop to reduce the number of barrier synchronizations. Such partition mechanism produces a stride-N access pattern at each processor. Since each cache line consists of 16 double-words and each processor executes one out of every 16 iterations of the outer loop (i.e. the chunk size is equal to 1), the working set size of each processor in a 16-way MP system remains almost the same as in the UP environment. Furthermore, the jump of the reuse hit ratio that is observed in UP environment between 128KB and 256KB caches is much less eminent because each processor now only has to execute part of the boundary computation.

In a 64-way system, the working set size reduces to about one quarter. As a result, the inter-section locality starts to take advantage of large caches. This actually shows a poor use of cache space in highly parallel system. For instance, with a 1MB cache in each processor, the total cache size in the system is equal to 64MB - 6.4 times of the total shared variable size in Simple. The above results suggest that a proper use of domain decomposition technique can be effective to reduce the shared data space in each processor during a parallel computation.

- The results above are from the optimistic scheme, in order to maintain cache coherence, some software schemes need to invalidate part or all the modified lines at the end of each section. Our study of the invalidate-on-modify scheme shows that the reuse hit ratios decreases substantially. For example, for a 1MB cache, the reuse hit ratios are only 2.9% and 18.9% in the UP and 64-way MP environments respectively. This is due to the fact that the chance of a shared variable is read-only throughout a section is not very high.

4.2.2 Weather code

Both Simple and Weather code require a very large cache to capture inter-section locality, but the latter produces more uniform results than Simple. According to Figure 5, the reuse hit ratio of the Weather code increases monotonically with larger cache sizes and number of processors. For example, the inter-section reuse hit ratios for UP, 16-way, and 64-way are 6.7%, 12.3%, and 35.4% for 1MB cache without invalidation. The ratios drop to 1.8%, 2.7%, and 17.7% with invalidation.

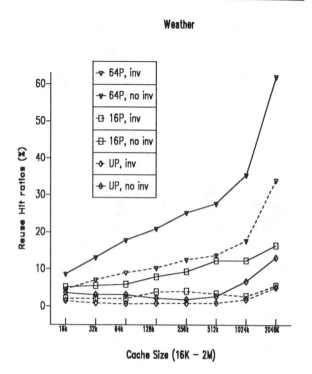

Figure 5.   Inter-Section cache hit ratios for Weather Code

4.2.3 Tomcat

Refer to Figure 6, the results of Tomcat are very different from those of Simple and Weather. The reuse hit ratios are considerably higher and the inter-section locality seems not very sensitive to the cache size. After a careful examination of the source code, it is found that the main execution of the program is over a nested loop of a relatively small Doall loop inside an iterative loop. Its execution therefore creates a large amount of small sections; each one only touches a small number of cache lines. Loop interchanging is possible to reduce the number of sections and synchronization points. However, in doing so, many stride-1 accesses will become stride-n accesses. Therefore, it is decided to maintain the original parallel program for the sake of overall cache performance.

As an illustration, the numbers of references and hits of each section are plotted in Figure 7. During the execution of the program of 399 sections, 384 of them are the parallel sections within the iterative loop and only 223 cache lines are referenced in each of these sections. This results in an inter-section hit ratios of 40-50% for the cache size of 1 MB. Furthermore, because of these small sections, the inter-section hit ratios are not sensitive to the cache size. Nonetheless, the reuse hit ratios drop to around 10% when modified lines are invalidated at the end of every section for the invalidate-on-modify scheme.

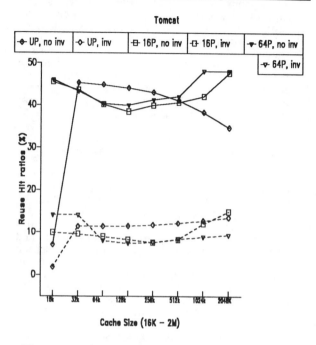

Figure 6.   Inter-Section cache hit ratios for Tomcat

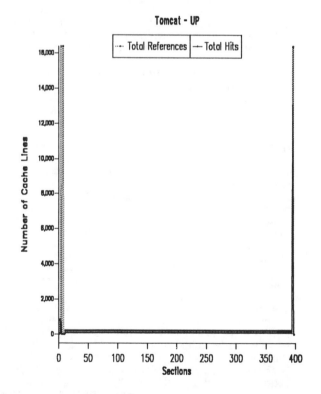

Figure 7.   Inter-Section cache references and hits for Tomcat

4.2.4 FFT

FFT is a much smaller program with a shared data of about 2 MB. Figure 8 shows that the reuse hit ratios jump higher when the cache is larger than 512 KB. The reuse hit ratios are approaching 100% when no invalidation is applied and almost the entire shared data fit into a 2MB cache. On the other hand, because all the shared variables in each section are modified in FFT, regardless of the chosen cache size, the program has zero inter-section locality for invalidate-on modified scheme.

One noticeable anomaly happens when the cache size is 1MB the reuse hit ratio (11.7%) for the 16-way MP is much worse than that (29.0%) in the UP environment. There are two reasons found in the source code. First, each processor of a 16-way MP has almost the same working set as that of the UP environment like the Simple workload. Second, after partitioning, the referencing sequences in each processor of a 16-way system is much closer to the worst case scenario shown in Figure 2 than the UP's.

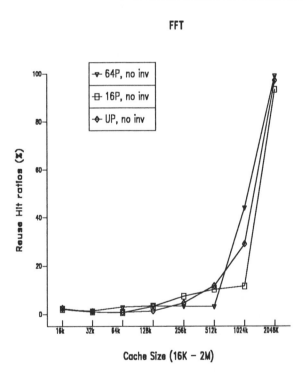

FFT

Legend:
- 64P, no inv
- 16P, no inv
- UP, no inv

Y-axis: Reuse Hit ratios (%)
X-axis: Cache Size (16K – 2M)

Figure 8. Inter-Section cache hit ratios for FFT

5. Concluding Remarks

The concept of inter-section locality of a parallel program and the reuse hit ratio to measure it are introduced in this study. They can be used to estimate the performance gains of a software coherence scheme and to have a better understanding of the behavior of the shared data in parallel programs.

In a shared memory parallel system where each processor has a local cache, a software coherence has a chance to be effective under the following two conditions:

1. The memory accesses to the shared data during inter-section transitions contribute to a significant portion of the overall memory accesses.

2. The reuse hit ratio for the shared data should be reasonably high.

However, these factors were found to be negative in this study. First, for reasonable cache structures and for programs with sufficiently large parallel sections, the overall cache performance is likely dominated by the intra-section locality. Second, the reuse hit ratios are very low because the total shared data accessed in a section are in general larger than the cache space. A piece of data left in the cache at the end of a section usually gets replaced at the beginning of the next section before ever has a chance to be reused. Therefore, with reasonable cache structures, the performance difference among the software coherence schemes are found to be small, and it may not be cost effective to apply the sophisticated software schemes on such programs.

This study also suggested that it is very important to optimize the use of local cache in a highly parallel system. By proper partition the shared data among processors, the effective cache size in each processor can be much larger. To this regard, it is very difficult to rely completely on a parallelizing compiler to optimize the cache usage. Parallel algorithms with domain decomposition techniques may be the right direction.

We would like to continue the work in the following directions:

1. The static self-scheduling algorithm used in this study seems to be most natural and appropriate. It is important to see if there are factors in a scheduling algorithm that might significantly affect the performance.

2. This study realized the inter-section locality with the use of a cache. It is necessary to try more applications and more cache parameters in order to provide a full spectrum of comparison.

3. Our results may also depends on how the applications were parallelized. A very simple parallelizing technique - Doall loops, is used throughout all four applications. It might be interesting to compare the inter-section locality for different parallel versions of the same application, such as coarse-grain and fine-grain parallelism.

6. References

[Alm&Got89] G.S. Almasi and A. Gottlieb, *Highly Parallel Computing*, The Benjamin/ Cummings Publishing Co., 1989.

[Axe+83] T.S. Axelrod, P.F. Dubois and P.G. Eltgroth, "A Simulation for MIMD Performance Prediction - Application to the S-1 MkIIa Multiprocessor", Proc. of the 1983 Intern'l Conf. on Parallel Processing, Aug. 1983, pp350-358.

[Ber&So88]   D. Bernstein and K. So, "PREFACE - A Fortran Preprocessor for Parallel Workstation Systems," IBM IBM Research RC 136000, Mar. 1988.

[Che&Vei88] H. Cheong and A. V. Veidenbaum, "A cache coherence scheme with fast selective invalidation," Proc. of the 15th Annu. Int. Sym. on Computer Architecture, p. 299-307, June 1988.

[Che&Vei89] H. Cheong and A.V. Veidenbaum, "A Version Control to Cache Coherence," Proceedings of 1989 Int'l Conf. on Supercomputing, p 322-330, June 1989.

[Cyt+88]   R. Cytron, S. Karlovsky, and K. McAuliffe, "Automatic Management of Programmable Caches," Proc. of the 1988 Int. Conf. on Parallel Processing, Aug. 1988.

[Egg89]   S. J. Eggers, "Simulation Analysis of Data Sharing in Shared Memory Multiprocessors," Ph.D. thesis, University of California, Berkeley, Feb. 1989.

[Egg&Kat88] S. J. Eggers and R. H. Katz, "A Characterization of Sharing in Parallel Programs and Its Application to Coherency Protocol Evaluation," Proceedings 15th Ann. Int'l Symp. on Computer Architecture, p 373-382, May 1988.

[Gen+88]   W. Gentzsch, F. Szelenyi, and V. Zecca, "Use of Parallel FORTRAN for some Engineering Problems on the IBM 3090 VF Multiprocessor," IBM European Center for Scientific and Engineering Computing, ICE-0023, July 1988.

[IBM82]   IBM Corp., *IBM 3081 Functional Characteristics*, GA22-7076 Poughkeepsie, New York, 1982.

[Lee+87]   R. L. Lee, P. C. Yew, and D. H. Lawrie, "Multiprocessor Cache Design Considerations," Proc. of the 14th Annu. Int. Sym. on Computer Architecture, p. 253-262, June 1987.

[Min&Bae89] S. L. Min and J.-L Baer, "A Timestamp-based Cache Coherence Scheme," Proceedings of 1989 Int'l Conf. on Parallel Processing, p23-32, Aug. 1989.

[Min&Bae90] S. L. Min and J.-L. Baer, "A Performance Comparison of Directory-based and Timestamp-based Cache Coherence Schemes," Proc. of the 1990 Int. Conf. on Parallel Processing, pp 305-311, Aug. 1989.

[Pfi84]   As far as we know, the term SPMD was coined by Greg Pfister during the early stage of the RP3 project at IBM T.J. Watson Research Center.

[Pfi+85]   G.F. Pfister, W.C. Brantley, D.A. George, S.L. Harvey, W.J. Kleinfelder, K.P. McAuliffe, E.A. Melton, V.A. Norton, and J. Weiss, "The IBM Research Parallel Processor Prototype (RP3): Introduction and Architecture, " Proceedings, 1985 Int'l Conf. on Parallel Processing, pp 764-771, Aug. 1985.

[Smi+85]   A. J. Smith, "CPU Cache Consistency with Software Support and Using 'One Time Identifiers'," Proc. of the Pacific Computer Communications '85 pp. 153-161, Oct. 1985.

[So+88]   K. So, F. Darema, D. George, A. Norton, and G. Pfister, "PSIMUL- A System for Parallel Simulation of Parallel Programs," in *Performance Evaluation of Supercomputers*, published by North-Holland, June 1988.

[Sto+85]   J.M. Stone, V.A. Norton, F. Darema-Rogers, E. Melton, and G. Pfister, "The VM/EPEX Fortran Preprocessor Reference," IBM Research RC 11408, Sept. 1985.

[Swe&Smi86] P. Sweazey and A.J. Smith, "A Classes of Compatible Cache Consistency Protocols and their Support by IEEE Futurebus," Proceedings, 13th Ann. Int'l Symp. on Computer Architecture, 1986, pp 414-423.

[Vei86]   A. V. Veidenbaum, "A Compiler-assisted Cache Coherence Solution for Multiprocessors," Proc. 1986 Int'l Conf. on Parallel Processing, Aug. 1986.

Appendix 1: Applications

There are four parallel applications used in this study. These are popular scientific applications which have been parallelized for SPMD computation. Specifically they are parallelized in EPEX/Fortran [Sto+85] and have been executed in real MP systems before.

1. Simple [Axe+83] - An application that models the hydrodynamic and thermal behavior of fluids in two dimensions. The problem size is 256 and the data are in double precision.

2. Weather code [Alm&Got89, Chapter 2] - A weather forecasting code which predicts from one state of the atmosphere (3D, surface and altitude) to another. the NASA GLAS/GISS fourth order general circulation model for The grid size is 108x9x72 and the data are double precision.

3. Tomcat [Gen+88] - A mesh generation routine that uses Poison equations. The mesh size is 384x384. Its sequential version is a member of the Spec benchmarks.

4. FFT - A two dimensional Fast Fourier Transform that uses radix-2 algorithm. The data used in the program are single precision and the problem size is 512x512.

Analytical Modeling for Finite Cache Effects

Jin Chin Wang, Michel Dubois*, Faye' A. Briggs

Tandem Computers Incorporated
M.P. Division
14231 Tandem Boulevard
Austin, Texas 78728-6699
(512)244-8397
jcwang%maui%halley.uucp@cs.utexas.edu

*Department of Electrical Engineering - Systems
University of Southern California
Los Angeles, CA 90089-0781
(213)740-4475
Dubois@priam.usc.edu

Abstract

In this paper, finite cache effects are studied. First, we extend the *access burst program model* to include the case of finite cache systems. Replacements are assumed to be uniformly distributed throughout the whole execution. Then, we apply the model to a cache coherence protocol and the finite cache systems are modeled by a discrete Markov chain. An approximate solution is found for each component of the coherence overhead. The accuracy of the model is verified by comparing the model predictions with execution-driven simulation results for six parallel algorithms.

1. Introduction

The three approaches generally used to evaluate the performance of multiprocessor systems are measurements, trace-driven simulations, and analytical program models. These approaches vary in terms of cost, flexibility, and accuracy. Measurements have the highest credibility; however, they are limited to existing machines. Some hardware parameters, such as cache block size, are not adjustable. Trace-driven simulations of real programs are usually time-consuming, generally restricting the systems being simulated to small processor configurations. Moreover, measurements and trace-driven simulations fail to explain observed performance levels [2] and the results are only valid for specific benchmarks run on the specific architecture. Therefore, in order to analyze performance on large, proposed architectures, analytic modelling is needed.

A good analytical program model to evaluate multiprocessor systems can be particularly useful because the analytical program model can be used to quickly analyze new architecture ideas. Such a model has theoretical foundations and can be used to explain experimental data. An objective of this research is to find a representative analytical program model which captures program behavior in terms of shared block accesses, reaches the same level of accuracy as that of trace-driven simulations, and has much higher flexibility. Of course, to be applicable, an analytical program model must be validated against measurements or trace-driven simulations. The major contribution of this paper is to define an analytical program model, and validate it against execution-driven simulations.

Recently, several researchers studied cache coherence by assuming the cache sizes are infinite [1][3][6]. Although the infinite cache model provides a convenient environment for analyzing the execution of a wide class of problems on multiprocessors, in reality, cache size is always finite in the sense that one can find important problems whose working set size cannot fit into multiprocessor caches. In such a case, the infinite cache model fails to represent the real-world model. Clearly, finite cache effects should be explored.

In the next section, we apply the *access burst program model* [3] to a finite cache multiprocessor system. The system is modeled by a discrete Markov chain. An approximate solution is found for each component of the coherence overhead. In Section 3, we verify the accuracy of the program model by comparing the model predictions with simulations results for six parallel algorithms. Section 4 summarizes the paper results and outlines future research directions.

2. The Access Burst Program Model for Finite Caches

In [3][10], we introduced a program model called the *access burst program model* to analyze the inherent behavior of shared writable blocks in cache-based multiprocessor systems. While a processor accesses a shared writable datum, no other processor is able to interleave accesses to the same datum. Accesses to a shared writable datum by one processor therefore occur in uninterrupted bursts. When a cache block contains more than one data element, the locality of accesses also contributes to access bursts. In the model, we defined five parameters to characterize shared writable blocks, which are q_s, the fraction of references to shared writable blocks, J, the average number of processors accessing shared writable blocks, W, the fraction of Write bursts, l the average length of bursts and f, the fraction of access bursts such that the first access is a Write. We applied the model to study the shared block contention in infinite cache systems [3].

In finite cache systems, we assume replacements are uniformly distributed throughout the whole execution and no replacement occurs in the processor which generates the current burst for shared writable block i. We define r_i as the fraction of number of replacements to number of accesses for the shared writable block i. Note that for infinite cache systems, the value of r_i is always zero since no replacement can occur.

The *global state* of the shared writable block i is described by the number of caches possessing a copy of the block and by the status RO or RW of the block. The global states are denoted by MEM, 1_RW, 1_RO, 2_RO,..., J_RO, where MEM is the state in which no cache has a copy of the shared writable block i. A state transition occurs at the time when a burst is completed by a processor or at the time when a copy of the shared writable block is replaced. The average number of accesses in between two state transitions, b_i, for the shared writable block i can be computed as follows:

$$b_i = \frac{l_i}{r_i \times l_i + 1} \qquad (1)$$

where l_i is the total number of accesses per access burst, and $r_i \times l_i + 1$ is the total number of state transitions in a burst which is the sum of the number of state transitions caused by replacements $(r_i \times l_i)$ and an additional transition which ends the burst. The probability of starting a state transition is therefore $q_s \times p_i / b_i$, where p_i is the fraction of shared writable block accesses that are for the particular shared writable block i. The probability that a state transition is caused by a replacement for the shared writable block i, R_i, is therefore

$$R_i = \frac{r_i \times l_i}{r_i \times l_i + 1} = r_i \times b_i \qquad (2)$$

The Markov chain for the state transitions of the shared writable block i is shown in Figure 1. The transition probabilities from state k_RO, $1 < k < J_i$, are found as follows.

• From state k_RO to state $(k-1)\_RO$: The probability of this transition is the probability that a copy of the block is replaced, which is R_i.

• From state k_RO to state $(k+1)\_RO$: The probability of this transition is the product of the probability that the transition is generated by the end of a burst, $(1 - R_i)$, the probability that the next burst contains only Read accesses, $(1 - W_i)$, and the probability that the access burst is made in one of the $J_i - k$ other caches, $(J_i - k)/J_i$.

• From state k_RO to state k_RO: This is the case when the state transition is generated because of the end of a burst, and the next access burst contains only Read accesses in one of the k caches. The transition probability is $(1 - W_i) \times (1 - R_i) \times k/J_i$.

• From state k_RO to state 1_RW: This is the case when the state transition is generated because of the end of a burst, and the next access burst modifies the block. The transition probability is $W_i \times (1 - R_i)$.

The transition probabilities from states MEM, 1_RO, 1_RW and J_i_RO are derived from similar arguments.

From the definitions of cache coherence events listed in [3], we can compute the probability of occurrence of each coherence event. When no copy of the shared writable block i is present in the system, a miss occurs at the beginning of a state transition, (i.e., a new access burst). If there are k copies in the system, a miss occurs at the beginning of a new access burst when the next processor starting the access burst is one of the $(J_i - k)$ processors without a copy in their caches. Therefore, the fraction of references to the shared writable block i which miss in the cache is equal to the fraction of state transitions causing a miss divided by the average number of accesses in between two state transitions.

$$Pr(M_{Ii}) = \frac{1}{b_i} \times (Pr(MEM) + Pr(1\_RW) \times \frac{(J_i - 1)}{J_i} \times (1 - R_i)$$
$$+ \sum_{k=1}^{J_i - 1} \frac{(J_i - k)}{J_i} \times (1 - R_i) \times Pr(k\_RO) \bigg) \qquad (3)$$

In order to find $Pr(M_{Ii})$ analytically, we first compute XS_i, the average number of copies of the shared writable block i in steady state.

$$XS_i = Pr(1RW) + \sum_{k=1}^{J_i} k \times Pr(k\_RO) \qquad (4)$$

From Equations (3) and (4), we have

$$b_i \times Pr(M_{Ii}) = (1 - R_i) \times \left(1 - \frac{XS_i}{J_i}\right) + R_i \times Pr(MEM)$$

In steady state, the number of misses is equal to the number of copies which are either replaced or invalidated; this yields

$$b_i \times Pr(M_{Ii}) = R_i + \frac{(J_i - 1)}{J_i} \times XS_i \times W_i \times (1 - R_i) \qquad (5)$$

From the above two equations, we have

$$XS_i = \frac{J_i \times (1 - 2 \times R_i + R_i \times Pr(MEM))}{(1 - R_i) \times (1 + (J_i - 1) \times W_i)} \qquad (6)$$

Substituting in Equation (5), we have

$$Pr(M_{Ii}) = r_i + \frac{1}{b_i}\left(\frac{(J_i - 1)W_i(1 - 2R_i + R_i Pr(MEM))}{1 + (J_i - 1) \times W_i}\right)(7)$$

If we lump the states 1_RW, and k_RO ($k = 1,...,J_i$) into a state called CHE, the Markov diagram in Figure 1 can be reduced to the Markov diagram in Figure 2. From the reduced discrete Markov diagram, the global flow balance equation can be written as

$$Pr(MEM) \times (1 - R_i) = (1 - Pr(MEM)) \times R_i \times \delta \qquad (8)$$

where δ is equal to $R_i \times \frac{Pr(1\_RO) + Pr(1\_RW)}{1 - Pr(MEM)}$ and $0 < \delta < 1$. Therefore,

$$R_i \geq Pr(MEM) \qquad (9)$$

This yields

$$Pr(M_{Ii}) \leq r_i + \frac{1}{b_i} \times \left(\frac{(J_i-1) \times W_i \times (1-R_i)^2}{1+(J_i-1) \times W_i} \right) \quad (10)$$

That is,

$$Pr(M_{Ii}) \leq r_i + \frac{1}{l_i} \times \left(\frac{(J_i-1) \times W_i \times (1-R_i)}{1+(J_i-1) \times W_i} \right) \quad (11)$$

Similarly, the rest of the coherence events can be equated and solved analytically, which are listed as follows:

The fraction of accesses to the shared writable block i invalidating RO copies in other caches is given by

$$Pr(INRO_i) = \frac{1}{b_i} \times \left(Pr(1RW) \times W_i \frac{(J_i-1)}{J_i} \times (1-f_i)(1-R_i) \right. \quad (12)$$

$$\left. + \sum_{i=1}^{J_i} Pr(iRO) \times W_i \times (1-R_i) \right)$$

$$\geq \frac{1}{b_i} \times \frac{W_i(1-R_i)^2(J_i(1-W_if_i)-(1-W_i)(1-R_i)-W_i(1-f_i))}{J_i-(1-W_i) \times (1-R_i)}$$

$$\geq \frac{1}{l_i} \times \frac{W_i(1-R_i)(J_i(1-W_if_i)-(1-W_i)(1-R_i)-W_i(1-f_i))}{J_i-(1-W_i) \times (1-R_i)}$$

The fraction of references to the shared writable block i changing the state from RW to RO is

$$Pr(CSRW_i) = \frac{1}{b_i} \left(Pr(1RW) \frac{J_i-1}{J_i}(1-R_i) \right)((1-W_i)+W_i(1-f_i))$$

$$= \frac{1}{b_i} \times \left(\frac{(J_i-1) \times W_i \times (1-W_i \times f_i) \times (1-R_i)^2}{J_i-(1-W_i) \times (1-R_i)} \right)$$

$$= \frac{1}{l_i} \times \left(\frac{(J_i-1) \times W_i \times (1-W_i \times f_i) \times (1-R_i)}{J_i-(1-W_i) \times (1-R_i)} \right) \quad (13)$$

The fraction of references to the shared writable block i, such that a write miss occurs and a remote cache has a RW copy of the block i, is therefore

$$Pr(INRW_i) = \frac{1}{b_i} \times Pr(1RW) \times W_i \times \frac{J_i-1}{J_i} \times f_i \times (1-R_i)$$

$$= \frac{1}{b_i} \times \left(\frac{(J_i-1) \times W_i^2 \times f_i \times (1-R_i)^2}{J_i-(1-W_i) \times (1-R_i)} \right)$$

$$= \frac{1}{l_i} \times \left(\frac{(J_i-1) \times W_i^2 \times f_i \times (1-R_i)}{J_i-(1-W_i) \times (1-R_i)} \right) \quad (14)$$

The second term on the right hand side of inequality (11) is an upper bound for the invalidation miss rate, $Pr(M_{Ii})$. The difference between this term and the miss rate in the infinite cache model is $(1-R_i)$ which appears in the numerator of inequality (11). Since R_i is a proper fraction (i.e., $0<R_i<1$) the above inequalities and equalities support theorems [11], which are in synchronous trace-driven simulations, the system with infinite caches always has a larger number of M_I, IN_RO, CS_RW and IN_RW than the system with finite caches.

Unlike the infinite cache system, in the finite cache system can obtain only approximate system effects. This is due to the assumption that shared writable blocks are analyzed in isolation, by summing the individual contributions of all shared writable blocks. In the finite cache model, when a system is in steady state, the invalidation miss rate generated by shared writable blocks is:

$$Pr(M_{Ii}) = q_s \times \sum_{i=1}^{N_s} p_i \times Pr(M_{Ii}) \quad (15)$$

where N_s is the total number of shared writable blocks. The average system coherence penalty generated by shared writable blocks is:

$$\lambda_{coherence} = q_s \times \sum_{i=1}^{N_s} P_i \times \lambda_i \quad (16)$$

where $\lambda_i = Pr(M_{Ii}) \times \lambda_M + Pr(IN\_RO_i) \times \lambda_{IN\_RO} + Pr(CS\_RW_i) \times \lambda_{CS\_RW} + Pr(IN\_RW_i) \times \lambda_{IN\_RW}$.

3. Multitasked Algorithms

In this section, the accuracy of the analytical model for finite cache systems is verified by comparing the model predictions with the execution-driven simulation results of six parallel algorithms. The six algorithms are the Jacobi iterative [12], the S.O.R. iterative [12], the quicksort [9], the bitonic merge sort [7], the FFT [4], and the image component labeling algorithms [8]. The simulation methodology described in [10] was applied. In this methodology, a parallel algorithm is actually executed on a uniprocessor and the multiprocessing effect is obtained by executing the process of each simulated processor in turn. The simulator *switches* from one simulated processor to another on each data access and synchronization primitive execution.

We express all penalties in units of the penalty of transferring a single word between a cache and main memory, that is, $\lambda_{word} = 1$. If the penalty to read a word from main memory is the same as the penalty to write a word to main memory, then we can estimate the performance improvement due to the caching of shared writable blocks as $1-\lambda_{coherence}$. In particular, if $\lambda_{coherence} > 1$, then caching shared writable blocks is not productive. The penalties of different coherence events in the Basic cache coherence protocol are: $\lambda_M = \lambda_{CS\_RW} = \lambda_{IN\_RW} = 0.75+0.25 \times B$, and $\lambda_{IN\_RO} = 0.5$ [3].

For simplicity, we study only the direct mapping cache which has no replacement policy associated with the cache organization. A probe is inserted in the execution-driven simulator to derive the values of parameter r for the six parallel algorithms, which are illustrated in [10]. With the values of parameters J, W, l, and f discussed in [3][10], the system invalidation miss ratio and the system coherence penalty for the six parallel algorithms are computed and shown in Figures 3 to 14.

From these figures, it can be seen that larger cache sizes result in higher invalidation miss ratios and a bigger coherence penalty. This is because the average life time of a shared writable block in the large cache is longer. As a result, the coherence penalty for the shared writable block can be expected to be higher. The model predictions and simulation results are close for the Jacobi iterative, the S.O.R. iterative, the quicksort and the image component labeling algorithms. The discrepancy observed between model predictions and simulation results for the FFT and the bitonic merge sort algorithms can be explained as follows: In the bitonic merge sort algorithms, the number of bursts is equal to the number of synchronization points, which is usually less than ten. However there are thousands of replacements in the simulations when the cache size is small. This violates the assumption that no replacement occurs in the processor which generates the current burst. For the FFT algorithm, shared writable blocks will definitely be replaced before the next burst begins when the cache size is less than N/B, where N is the problem size and B is the cache block size. That is, the coherence penalty for shared writable blocks is equal to zero. From equation (2) we can see that R never takes the value of one. This implies the coherence penalty is never zero for the FFT algorithm; therefore, the model predictions are far from accurate for the three algorithms. The model need to be further refined to take care of such inaccuracies for finite cache systems.

4. Concluding Remarks and Future Suggestion Work

In this paper, a simple program model, namely the *access burst program model*, used to capture the program behavior in terms of shared block accesses in cache-based multiprocessor systems was applied to finite cache multiprocessor systems. The replacements are assumed to be uniformly distributed throughout the whole execution. The finite cache systems are modeled by a discrete Markov chain. An approximate solution was found for each component of the coherence overhead. The accuracy of the model predictions was verified by comparing the model predictions with execution-driven simulation results for six parallel algorithms. The model, based upon stochastic processes, appears to be a good approximation to shared block contention effects in multiprocessor systems.

In the model, all processors are assumed to have the same probability of starting the next access burst on the same block, after a burst is completed. It is clear that this may not be a good approximation for some parallel algorithms such as the FFT algorithm. Rather, it seems that a processor which has just finished an access burst has a higher probability of starting the next one. The model could therefore be refined by introducing a new parameter which captures this *affinity*.

The research presented in this paper provides a basis for further investigations. Additional algorithms should be studied in order to understand the program behavior in detail. Different types of shared blocks have to be identified and classified based on the model parameter characteristic. Since different types of shared blocks may have very distinct reference patterns, different optimization strategies should be adopted. This information certainly helps parallel compilers to improve system performance.

5. References

[1] A. Agarwal, R. Simoni and J. Hennessy, An Evaluation of Directory Schemes for Cache Coherence, In *Proceedings of 15th Annual International Symposium on Computer Architecture*, pages 280-289, June 1988.

[2] P. Bitar, A Critique of Trace-Driven Simulation for Shared-Memory Multiprocessors, In M. Dubois and S. S. Thakkar, editors, *Cache and Interconnect Architectures in Multiprocessors*, pages 37-52, Kluwer Academic Publishers, 1990.

[3] M. Dubois and J.C. Wang, Shared Data Contention in a Cache Coherence Protocol, In *Proceedings of International Conference on Parallel Processing*, pages 146-155, August 1988.

[4] Digital Signal Processing Committee, editor, *Programs for Digital Signal Processing*, IEEE Press, 1979.

[5] IBM Corporation, Special Issue on IBM 3081, *IBM Journal of Research and Development*, 26(1):2-19, January 1982.

[6] S. J. Eggers and R. H. Katz, A Characterization of Sharing in Parallel Programs and Its Application to Coherency Protocol Evaluation, In *Proceedings of 15th Annual International Symposium on Computer Architecture*, pages 373-382, June 1988.

[7] M. J. Quinn, *Design Efficient Algorithm for Parallel Computers*, McGraw-Hill, 1987.

[8] A. Rosenfield and A. Kak, *Digital Picture Processing*, Volume 1-2, Academic Press, 1982.

[9] R. Sedgewick, *Quicksort*, New York: Garland Publishing, Inc., 1980.

[10] J.C. Wang, *Analytical Modeling for Shared Block Contention in Cache Coherence Protocols*, Ph.D. thesis, University of Southern California, December 1990.

[11] J.C. Wang and M. Dubois, Correlation between cache size and coherence protocol overhead, In *Proceedings of the International Computer Symposium 1990*, pages 811-814, December 1990.

[12] D. Young, *Iterative Solution of Large Linear Systems*, Academic Press, New York, 1971.

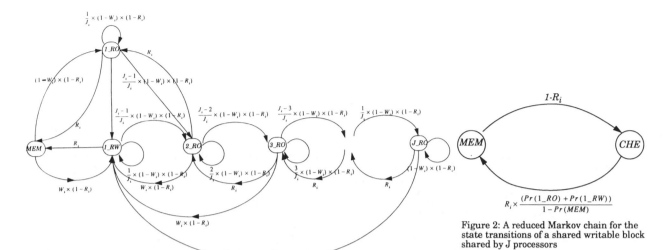

Figure 2: A reduced Markov chain for the state transitions of a shared writable block shared by J processors

Figure 1: Markov chain for the state transitions of a shared writable block shared by J processors

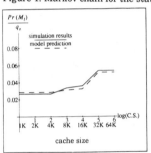

Figure 3: The system invalidation miss ratio for the Jacobi iterative algorithm

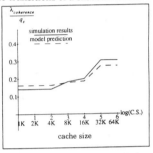

Figure 4: The system coherence penalty for the Jacobi iterative algorithm

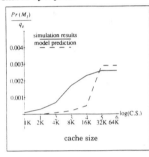

Figure 9: The system invalidation miss ratio for the bitonic merge sort

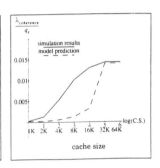

Figure 10: The system coherence penalty for the bitonic merge sort

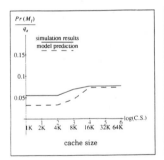

Figure 5: The system invalidation miss ratio for the S.O.R. algorithm

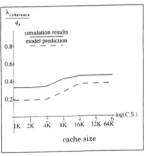

Figure 6: The system coherence penalty for the S.O.R. algorithm

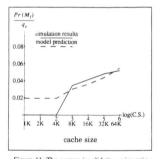

Figure 11: The system invalidation miss ratio for the FFT

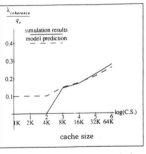

Figure 12: The system coherence penalty for the FFT

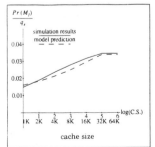

Figure 7: The system invalidation miss ratio for the quicksort

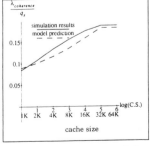

Figure 8: The system coherence penalty for the quicksort

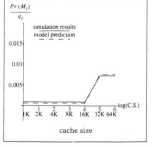

Figure 13: The system invalidation miss ratio for the image component labeling algorithm

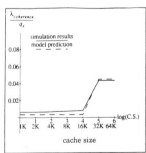

Figure 14: The system coherence penalty for the image component labeling algorithm

THE EFFECTS OF NETWORK DELAYS ON THE PERFORMANCE OF MIN-BASED CACHE COHERENCE PROTOCOLS

Sandra Johnson Baylor and Yarsun Hsu

IBM T. J. Watson Research Center
Yorktown, Heights, NY 10598

Abstract -- The effects of network delays on the performance of distributed directory-based cache coherence protocols are evaluated. A tightly-coupled multiprocessor architecture including data and instruction caches and a multistage interconnection network is assumed. The effects of network delays on system performance is evaluated for a particular cache coherence protocol. The evaluation methodology used is trace-driven simulation. The results show that a very large fraction of all requests traversing the network (averaging 90%) is attributed to shared data misses and coherence related requests. Also, the time required to process these requests are very large, ranging from 25 to 95 cycles. Varying the network delay compounds this problem as quantitatively demonstrated by the results.

I. Introduction

The use of private caches in a shared-memory multiprocessor may result in the well-known cache coherence problem [1]. Hardware-based solutions to this problem are generally designed for bus-based or multi-stage interconnection network-(MIN) based systems. MIN-based systems generally use directory-based protocols [2] and may include an order of magnitude more processors than bus-based systems. The central medium used to enforce cache coherence is located on the memory side of the MIN. Therefore, network delay in enforcing coherence can be a significant factor in degrading the performance of these systems.

This paper presents the results of a study conducted to quantify the multiprocessor performance degradation that results from MIN delay when incorporating a cache coherence protocol in the system. A distributed-directory-based protocol is used. The MIN delay is varied to determine its effect on system performance.

The following section (Section II) discusses the motivations for this work. Section III presents descriptions of the cache coherence protocol used and the MIN network evaluated. Section IV briefly discusses the simulation methodology and presents the results of the evaluations. Section V discusses some conclusions and suggestions for further study.

II. Motivation

Numerous distributed-directory protocols have been proposed and a significant number of MINs have been evaluated. However, most of the performance evaluation studies of MINs consists of models assuming independent and identically distributed network request rates. We model the MINs using realistic requests originating from the processor and its associated private cache. We also study the various types of requests traversing the MINs. Also, we are not aware of any detailed study that quantitatively considers the impact of MIN network delay on the performance of multiprocessors implementing such protocols. Our objective was to quantify the impact of this MIN delay on system performance.

Agarwal, et. al., evaluated the performance of directory-based and snoopy-based protocols for a 4 processor multiprocessor. Min and Baer [3] conducted a performance comparison of Censier and Feautrier's [4] directory-based hardware protocol and their timestamp-based software protocol [5] for MIN-based multiprocessors with private data caches, Chaiken, et. al. [6], compared the performance of several non-broadcast directory schemes.[1] None of these studies quantifies the effects of MIN delay on system performance.

III. System Organization and Cache Coherence Protocol

A. System Organization

Illustrated in Figure 1 is the architecture assumed in this study. It consists of P processors with associated data and instruction caches, a packet-switched multistage interconnection network (MIN) and M memory modules ($P = M$ in this study). The MIN actually consists of two networks: a request network, which routes requests from the processors to the memory modules; and a response network, which routes requests from the memory modules to the processors. Associated with each memory module is the *distributed global directory* (DGD) used in enforcing cache coherence.

The request and response networks used are modified omega networks [7] as illustrated in Figure 2a for eight inputs and outputs. Both the request and response networks consist of $\log_2 P$ stages of $\frac{P}{2}$ 2x2 switches. The traditional omega network routing methodology is assumed, *i.e.*, the low-order bits of the address are used to determine the routing of messages through the network. Illustrated in Figure 2b is the basic building block of the network, the switch. This 2x2 switch includes single word registers for both inputs (I and J) and outputs (P and Q) to buffer a single packet. The switch is also composed of four FIFO queues used to buffer messages traversing the network. Two output queues, one each for outputs P and Q, are associated with each input. Although not shown in the figure, the FIFO queues are bypassed if they are empty and the destination output register is empty. A message crossing either of the networks consists of a header packet and a variable number of data packets, dependent upon the type of request. Each packet consists of one 32-bit word. Blocking networks are used, *i.e.*, all packets of a message are transferred sequentially so that no other packet will be able to advance to an output register until all packets composing a message have advanced. More information on the network is found in [8].

Inputs to the request network include cache miss processing requests, write-back data and coherence protocol requests. Depending on the type of request, the request network routes the request to a memory module and/or the DGD associated with the module. Inputs to the response network include line or word transfers and invalidation and change-state coherence protocol requests (see Section III.B). These requests are routed directly to the cache.

Each memory module has a six cycle access time. However, decoding of network requests and data access is assumed to be pipelined. Consecutive cache lines are interleaved across the memory modules in order to reduce contention. Requests to the memory module originate from the processors, the private caches and/or the distributed global directory.

B. Coherence Protocol

The coherence protocol used in this study is a directory-based protocol described in [9]. Coherence state information is located in the private caches (local information) and the DGD/memory modules (global information). Each line present in a cache is in one of four possible

---

[1] Schemes that do not require broadcast invalidations.

non-transient states: exclusive read-only (EX), exclusive read-write (RW), read-only (RO) and invalid (INV). A line in the exclusive state is present in only one cache and has not been modified. An exclusive read-write line is present in one cache and modified. Read-only lines are unmodified and may be present in more than one cache. Invalid lines hold data that is no longer valid; however, space has been allocated for the line in the cache. As a result, no replacement algorithm is needed for future accesses to this line.

Each line present in the DGD is also in one of four possible states: exclusive read-only (GEX), exclusive read-write (GRW), read-only (GRO) and not present in any cache (NP). Along with the state of the line, the DGD includes a processor identifier (PID) to determine the cache owning the exclusive copy of the line (if the global state is GEX or GRW). If the global state is GRO then the PID becomes a counter, denoting the number of caches having a copy of the line [10]. Read misses, write misses and write hits signal the operation of the protocol by submitting coherence requests to the global directory.

IV. Simulation Methodology and Results

The trace-driven simulator used to evaluate the coherence protocols explicitly models each functional block of the system shown in Figure 1. Instructions are assumed to be non-self-modifying. Also, an infinite instruction cache is assumed and is not modelled in this study. Moreover, the processor is not modelled in this study; however, each reference issued by the processor is assumed to be sent to the cache. In reality, the presence of processor registers reduces the number of requests issued to the cache. Therefore, the data collected in this study presents a lower bound on system performance. A processor cycle is simulated by utilizing the period of time during which one reference, if any, is issued by each processor in the system, i.e., the time in which each processor services the next element in its reference stream. A discussion of the trace collection methodology and the applications evaluated is found in [9]. Table 1 lists the various simulation parameters for each subsystem and their associated values. While simulating a large number of processors is desirable, limited resources prohibited the evaluation of more than 32 processors.

We concentrate on the various types of cache miss requests that traverse the request network and their corresponding replies. A quantification of the performance degradation that occurs as a result of the network delays and contention when processing these requests is determined. We divide cache misses into four categories; shared load misses, shared store misses, modify requests and private misses (loads and stores). While modify requests are not misses, they are included because processing them results in the same performance degradation as misses. Shared load and store misses result in greater processing penalties than private misses. This is because processing these misses may result in two round-trip passes across the MINs. For example, if a processor requests a load to a shared line that is not in its cache but present and modified in another cache, the processor initially issues a line transfer request. When the DGD receives this request it sends a change state signal to the cache owning the line (first round-trip pass). This cache then writes back the modified data. When the modified data is received, the entire line is transferred to the cache of the processor requesting the load (second round-trip pass). Similar scenarios occur for shared store misses.

Of the total misses occurring in the system, the percentages of misses in which the processor requests shared loads, shared stores, modifies and private data are calculated. The processing times for each type of miss is determined by measuring the time taken to present the request to the network from the cache and then deliver the data to the processor. Also, the average network delay for both MINs is calculated as well as the total algorithm execution time and the average effective CPU access time. Although three applications were evaluated in this study, most of the results presented are for one application, hydrodynamics. This is because the behavior of the results was similar for all of the applications evaluated. However, in cases where an application deviated from the general trend, they are discussed in detail.

Enumerated in Table 2 is the memory reference statistics for the applications evaluated. The values shown for the total number of references are for a 32 processor system, whereas the remaining data is independent of the number of processors. This data includes the percentage of references that are instructions and data, the percentage of references that are private and shared, the percentage of data that is private and shared and the percentage of shared data references that are reads and writes. The percentage of references attributable to data ranges from 36% to 43%. Also, the percentage of data references that are shared range from 9% for hydrodynamics to 23% for the MVS FFT application. Moreover, the percentage of shared data assignable to reads ranges from 32% to 35%.

Presented in Figure 3 is the total algorithm execution time versus the cache line size, the network switching time per stage and the number of processors for both hydrodynamics and the MVS FFT. For a given number of processors and network switching time, this parameter generally increases as the line size increases. This is because the larger line sizes result in longer line transfer times and more contention in the response network. For a given number of processors and line size, the execution time intuitively increases as the network switching time increases. However, the rate of increase varies for the two application. The execution time increases at a larger rate for the MVS FFT application. This is largely attributed to the fact that a greater percentage of shared data misses and modify requests are processed across the network for this application. This results in longer average miss processing times and therefore longer execution times.

Illustrated in Figure 4 is the average effective CPU access time versus the cache line size, the network switching time and the number of processors for hydrodynamics. This parameter generally increases as the line size and the network switching time increase. Observe that with a range of 1.6 to 2.5 cycles, these times are somewhat large. To understand the reason for these large values, we calculated the percentage of requests traversing the MINs that are shared load misses, shared store misses, modify requests and private misses. These percentages are shown in Figure 5 versus the cache line size for 16 processors and hydrodynamics. (This parameter is independent of the network switching time per stage). The overwhelming majority of the requests traversing the network are for shared data misses and modify requests.

Also, as the line size increases the percentages of shared load misses, private misses and modify requests decreases while the shared store miss percentages increase. This is because a shared store request issued by the processor will more than likely result in a shared store request (as opposed to a modify request) for the larger line sizes. This is a direct result of the false sharing that occurs as line sizes increase [11].

Presented in Figure 6 is the average network delay versus the cache line size and the network switching time per stage for both the request and response networks. This parameter generally increases as the line size and the network switching time increase. However, the rate of increase (versus the line size) is higher for the response network. This is because entire lines are transferred across this network causing longer transfer times for the larger line sizes. With the exception of transferring the modified words in a line on a write-back, only single packet messages (not included the message header) traverse the request network.

Shown in Figure 7 is the time required to process a shared load miss, a shared store miss, modify requests and private misses versus the cache line size and the network switching time per stage for 16 processors (hydrodynamics). The time generally increases as the line size and the network switching time increases. Observe that these processing times are very large, ranging from 25 to 95 cycles. These large processing times (even when the network switching time is 0 cycles) explain the large average effective CPU access times illustrated in Figure 4. This shows that maintaining hardware coherence can be very costly (even when only a small fraction of all references are shared) when false sharing occurs.

IV. Conclusions

Presented in this paper is an evaluation of the effects of network delay on the performance of a shared-memory multiprocessor implementing a directory-based cache coherence protocol. The results of trace-driven simulations show that the overwhelming majority of the requests traversing the network are for shared data and coherence related activity. Processing these requests require 25 to 95 cycles, depending on the network switching time, the number of processors the cache line size and the level and type of sharing. These processing times result in large average effective CPU times ranging from 1.6 to 2.5 cycles.

It is preferable to simulate more than 32 processors; however, our limited resources prevented such a study. We did not use trace compaction or other techniques that may be used to evaluate larger numbers of processors and such procedures may by used in future work. Also, more work is needed to evaluate other types of applications such as data dependent applications. Also, a detailed study of the miss processing times for applications exhibiting small amounts of false sharing is needed to determine the scalability of the coherence protocol evaluated as well as the effects of network delay and bandwidth on the applications.

Acknowledgements

We would like to thank Frederica Darema and Shing-Chong Chang for supplying us with their application codes and Caroline Benveniste for supplying us with her trace scheduling program. Also, Kevin McAuliffe and Bharat Rathi for participating in the development and performance evaluation of the coherence protocol.

REFERENCES

1. M. Dubois and F. A. Briggs, "Synchronization, Coherence and Ordering of Events in Multiprocessors." *Computer*, vol. 21, no. 2, pp. 9-21, February 1988.

2. A. Agarwal, R. Simoni, J. Hennessy, and M. Horowitz, "An Evaluation of Directory Schemes for Cache Coherence," *Proc. of the Int. Symp. on Computer Architecture*, pp.280-289, 1988.

3. S. L. Min and J. L. Baer, "A Performance Comparison of Directory-Based and Timestamp-Based Cache Coherence Schemes," *Proc. of the Int. Conf. on Parallel Processing*, vol. 1, pp. 10-20, August 1990.

4. L. M. Censier and P. Fequtrier, "A New Solution to Choerence Proglems in Multicache Systems," *IEEE Trans. on Computers*, vol. C-27, no. 12, pp. 1112-1118, December 1978.

5. S. L. Min and J. L. Baer, "A Timestamp-based Cache Coherence Scheme," *Proc. of the Int. Conf. on Parallel Processing*, vol. 1, pp. 23-32, August 1989.

6. D. Chaiken, C. Fields, K. Kurihara and A. Agarwal, "Directory-Based Cache Coherence in Large-Scale Multiprocessors," *Computer*, vol 23, no. 6, pp. 48-58, June 1990.

7. D. H. Lawrie, "Access and Alignment of Data in an Array Processor," *IEEE Trans. on Computers*, vol. C-24, pp. 1145-1155, December 1975.

8. Y. Hsu, C. Benveniste, J. Ruedinger, and C. J. Tan, "A Large Interconnection Network for the High Parallel Systems," *IBM Research Report RC 15036*, pp. 1-18, October 1988.

9. S. J. Baylor, K. P. McAuliffe, B. D. Rathi, "An Evaluation of Cache Coherence Protocols for MIN-Based Multiprocessors," *Int. Symp. on Shared Memory Multiprocessing*, pp.1-32, April 1991.

10. S. J. Baylor, K. P. McAuliffe, B. D. Rathi, "The Effects of Directory PIDs on the Performance of MIN-Based Cache Coherence Protocols," *IBM Research Report RC 16304*, pp. 1-22, November 1990.

11. S. J. Eggers and R. H. Katz, "The Effect of Sharing on the Cache and Bus Performance of Parallel Programs," *ASPLOS III, pp. 257-270, April 1989.*

Table 1. Simulation Parameters, PID-Processor Identifier

SUBSYSTEM	PARAMETER	VALUE
Processor	number	16-32
Cache	size	64 Kbytes
	line size	16-64 bytes
	placement policy	2-way set-associative
	update policy	write-back
	access time	1 CPU cycle
Network	switch size	2x2
	packet size	32 bits
	message size	variable
	data path	32 bits
	blocking?	yes
	contention free switching time	2 CPU cycles
Distributed Global Directory (DGD)	size	16K bytes
	placement policy	direct-mapped
	number of PIDs	1
	access time	1 CPU cycle
Memory Module	access time	6 CPU cycles

Table 2. Memory Reference Statistics, 32 Processors. P-Private, S-Shared.

CODE	TOTAL REFS	REFS		REFS		DATA REFS		Shared DATA	
		I	D	P	S	P	S	R	W
MVS FFT	2,498,601	59	41	91	9	77	23	68	32
Sort	6,672,295	64	36	97	3	92	8	65	35
Hydro	1,469,046	57	43	96	4	91	9	65	35

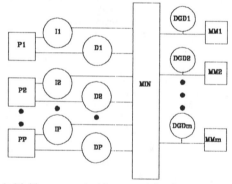

Figure 1. Multiprocessor Architecture. P-Processors, I-Instruction Cache, D-Data Cache, DGD-Distributed Global Directory, MM-Memory Module.

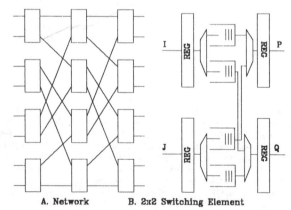

A. Network   **B. 2x2 Switching Element**

Figure 2. Interconnection Network and Switch.

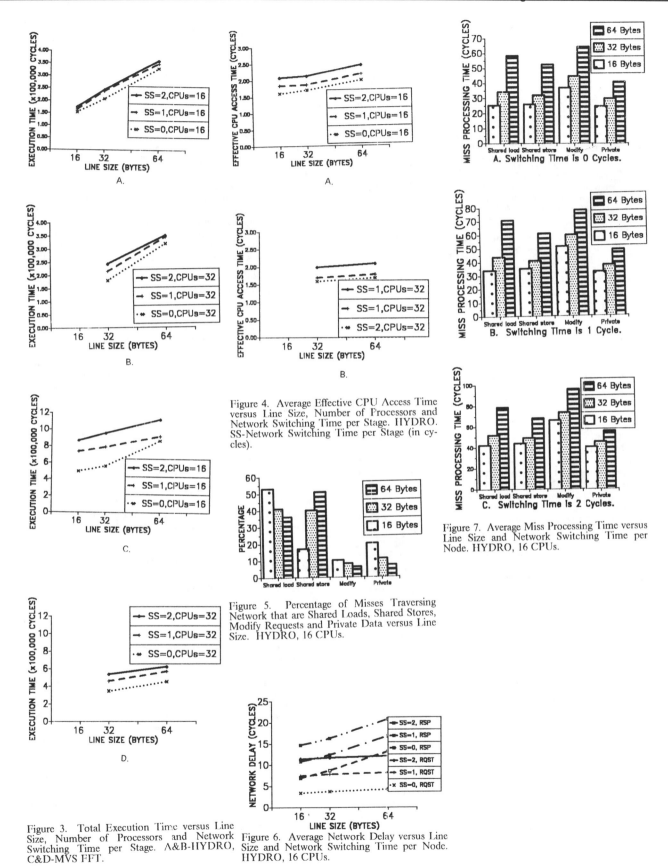

Figure 4. Average Effective CPU Access Time versus Line Size, Number of Processors and Network Switching Time per Stage. HYDRO. SS-Network Switching Time per Stage (in cycles).

Figure 5. Percentage of Misses Traversing Network that are Shared Loads, Shared Stores, Modify Requests and Private Data versus Line Size. HYDRO, 16 CPUs.

Figure 7. Average Miss Processing Time versus Line Size and Network Switching Time per Node. HYDRO, 16 CPUs.

Figure 3. Total Execution Time versus Line Size, Number of Processors and Network Switching Time per Stage. A&B-HYDRO, C&D-MVS FFT.

Figure 6. Average Network Delay versus Line Size and Network Switching Time per Node. HYDRO, 16 CPUs.

A CACHE COHERENCY MECHANISM WITH LIMITED COMBINING CAPABILITIES FOR MIN–BASED MULTIPROCESSORS

Kanad Ghose and Sreenivas Simhadri
Department of Computer Science
State University of New York, Binghamton, NY 13902–6000
ghose@bingvaxu.cc.binghamton.edu

ABSTRACT: Large-scale shared memory multiprocessors employing multi-stage interconnection networks (MINs) usually have to employ private caches to avoid the detrimental effect of long network and memory latencies. When shared-writable data is cached in these private caches, incoherency problems result. In this paper, we propose a new technique for implementing cache coherency in MIN-based systems that alleviates the drawbacks of the existing techniques and provides a limited form of request combining within the MIN. We accomplish this by maintaining relatively small *directories* at intermediate levels in the MIN. A write to a block is broadcast to keep cached copies coherent. Information maintained in the switch directories is used to filter out these broadcasts from caches that do not have a copy of the block being written. Use of the switch directories also allows cache-to-cache transfers on a miss, keeping the transfer traffic confined to the processor end of the MIN.

1. Introduction

In large scale shared memory systems, caching provides an important mechanism for reducing delays due to resource contention. In shared memory systems that employ multistage interconnection networks(MINs), caching also provides a way of avoiding the long network and memory access latencies. The use of a cache in a shared memory system has a hidden cost in the form of the overhead for keeping these caches consistent in the face of reads and writes to shared, cached variables.

To date, several hardware-based cache coherency mechanisms have been proposed for shared memory multiprocessors employing a MIN. One of these is an adaptation of the basic Censier and Feautrier scheme, [4], requiring presence bits in the entry for each block in memory directories. These presence bits identify the caches that hold copies of the block. Stenstrom's scheme, [15], improves the scheme in [4] by allowing cache-to-cache transfers and uses memory directories and presence bit vectors in each cache. The scheme proposed in [2] by Bhuyan, Liu and Ahmed relies on the use of a single bus for exchanging coherence control information among the processor caches. A time stamp based mechanism proposed in [10] by Min and Baer that requires data structures to be time stamped at the time of the modification and additional hardware at the processor end to check the validity of a read or write operation based on time stamps. Extensive compile time marking is needed by this scheme for judiciously associating time stamp with the structures. Mizrahi *et al*,[11], proposed a technique which uses storage within the switching elements of a MIN to reduce the effective memory latency and avoids incoherence by maintaining a unique copy of the data. This in itself requires a mechanism akin to cache coherence hardware. Other Schemes proposed for cache coherence in MIN based systems are [1], [3]. These are the enhancements of [2] and [4], respectively.

The techniques stated above, for maintaining coherent caches in MIN-based systems suffer from the following problems. First, there are bottlenecks in the form of centralized directories [4] or coherence control busses [2]. Second, many of these schemes are hardware-intensive (e.g., large directory entries in [15], [4]; complex logic within the caches for generation and use of coherency control information in [2]; time stamp registers and associated logic in [10]). Third, many of these techniques make inefficient use of the network bandwidth (e.g., by requiring change of ownership before a write [15]; by disallowing cache-to-cache transfers [4]; by generating traffic to keep a single copy current [11] etc.). Finally, the protocol overhead can be substantial in many of these schemes.

In [5], Cheong and Viedenbaum proposed a compiler directed approach for ensuring cache coherence, similar to the approach taken in [10]. [9]. [16], are also techniques that involve a compiler directed solution. The general problem with the software-directed approaches are in the form of higher miss ratios due to overly conservative invalidations (or broadcasts) and in their inability to cope with dynamic sharing.

In this paper we will introduce a new cache-coherence scheme for shared-memory multiprocessors to alleviate the limitations of existing mechanisms. Our scheme keeps cache coherence related traffic to a minimum by employing distributed directories within some switching elements of the MIN and incorporates cache-to-cache transfers and a limited form of request combining to reduce network traffic further and speed up the service of cache misses. The salient features of our proposed scheme are:

(1) Cache coherence traffic within the MIN is kept to a minimum by filtering out unnecessary traffic.

(2) Bottlenecks due to centralized directories are avoided by using distributed directories within switching elements of the MIN.

(3) All cached copies of a block are maintained consistent, thus obviating the concept of ownership and the associated overhead. This also reduces misses on recently-invalidated blocks.

(4) Our scheme supports a limited form of request combining, thus further reducing network traffic and allowing better performance.

The proposed scheme is not as hardware-intensive as it appears: this is because directories are present only in regularly-spaced stages of the MIN and the combining logic is simple.

This paper is organized as follows: In Section 2, we will define the various terms used in this paper. The hardware required for the scheme is described in section 3. The protocol is described in detail in Section 4. In the subsequent sections the hardware and implementation issues and the performance evaluation of this scheme are discussed.

2. Definitions

Figure 1 depicts a MIN-based system (with a Omega-like MIN) employing the proposed cache coherence hardware. Each processor has a private cache (P-Cache), where shared variables, as well as private variables are cached. A memory directory (MD) is associated with each memory module, with an entry for each shared block cached from that module. The MDs are finite in size and retain only the most recently used entries. The older entries are written back to the memory module itself and can be retrieved. Switch directories (SDs) exist within the switches at regularly-spaced intervals within the MIN. The SDs are used to selectively direct write updates and requests for missing blocks to only the minimum necessary number of P-caches and thus serve as filters in constraining the network traffic to the minimum. The SDs and MDs hold entries for shared blocks only - a tag bit associated with every block, set as a result of compile time marking prevents entries for private blocks to be made in the SDs and MDs. The SDs, like the MDs, are used like a cache and retain entries for the most recently used blocks.

The access paths to a particular memory module from processors in a MIN, such as the commonly used Omega network and its variants [6], define a tree rooted at the memory module with the processors at the leaves. The tree so formed is a binary tree if 2×2 switching elements are used. Assume that the switch directories in our proposed system are located every k levels, for the purpose of explaining the cache coherency mechanism, the paths from processors to a particular memory module can be reckoned as a K-ary tree, with interior nodes corresponding to SDs. If 2×2, switching elements are used, $K = 2^k$, where k is the separation between adjacent levels in the MIN that have SDs. We now provide the definitions of some terms that we will use in the rest of this paper:

(1) *Locality of sharing:* The locality of sharing for a node, henceforth referred to as locality, is defined as the leaves of the subtree rooted at this node. The leaves are the P-Caches in the MIN. The locality of a P-Cache is itself, and the locality of a SD entry is all the processors reachable through it.

(2) *Nearest Common Ancestor(NCA):* The nearest common ancestor of two P-Caches P1 and P2 with respect to a common block B they share, is defined as the switch node which is the root of the smallest subtree with these P-Caches as leaves. NCA(B,P1,P2) indicates the NCA of these P-Caches sharing the block B.

(3) *Current NCA:* The current NCA of a block B is defined as the NCA of *all* P-Caches that currently have a copy of B.

We will also use the qualifier *above* to allude to the SD entries of a block that are located at switch levels (i.e., in switching stages) closer to the memory side than the current NCA of the block. The qualifier *below* is used to refer to SD entries located at switching stages closer to the processor side than the current NCA. These qualifiers will be used in Section 3.2 and in the description of the protocol in Section 4.

3. Switch Directories

The purpose and the usage of SDs are described in this section. In section 3.1, we describe the methods used for filtering unnecessary broadcasts and invalidates. In the subsequent sections we describe the format of the SD entries and its functional characteristics.

3.1 Filtering out redundant MIN traffic: One way to reduce redundant network traffic is the scheme proposed by Stenstrom [15] that uses presence bit vectors in every processor cache to direct broadcasts/invalidations. The drawback of this scheme is in the form of rather big presence bit vectors, each with 1 bit/processor and each to be placed in every cache holding the corresponding block. This can also be wasteful, since a the sharing of a block can be confined to a few processors.

In our cache coherence mechanism, we use SD entries as filters to confine network traffic to the smallest subtree necessary. In contrast to Strenstrom's scheme, only K-bit presence vectors are needed in each SD entry of a block (where $K=2^k$, k being the separation of consecutive stages of the MIN with SDs) to locate the block within one of the K subtrees rooted at the SD. (Note that K is expected to be substantially smaller than the number of processors.)

Each network packet has an operation field indicating the type of the operation to be performed on the memory location. The SD-logic uses this information, in conjunction with the information in the SD entry to process a packet appropriately.

3.2 Nature of SD and MD Entries: Each entry in a SD has the following fields:

(1) **Mode Field:** It keeps track of the NCA and takes the values *above*, *below* or *NCA* depending on the status of the NCA.

(2) **Presence Bit Vector(PBV)[0..K-1]:** This vector indicates the presence of a block in one or more of the K ($=2^k$) subtrees.

(3) **Wait Bit Vector(WBV)[0..K]:** This vector keeps track of the pending requests for a memory block. When the block is received by this SD, it is broadcast to all the subtrees with the corresponding WBV bit set. The Kth bit indicates that the parent is waiting for this block.

(4) **Memory address tag for the block:** This is the usual tag present in a set associative cache's entry.

(5) **Usage information:** Some LRU approximation information is used for replacing the SD entries.

Although the SD seems to be big, it is not so for two reasons. First, the entries in the SD correspond only to the shared blocks and second, the size of each entry is (2.K+3+T+U), where T and U are the sizes of the memory tag and the LRU information.

Each memory directory(MD) associated with a memory module retains entries for the recently used shared blocks cached from that module. The MD entries have the following fields: the memory address tag, a field to indicate the current NCA of the block, an inhibit flag (which when set inhibits the memory from supplying the block) and a usage information field.

4. The Cache Coherency Protocol

Missing blocks are always supplied from the nearest P-Cache that has a copy as the protocol guides a miss request using information in the SD entries. Writes proceed in two phases: first, a write request propagates upto the NCA, possibly competing with similar write requests. The updated value supplied by the write is sent simultaneously to all P-Caches possessing the block being written. The SD entries are again used to direct the updated value to only the P-Caches. Our protocol implements the notion of sequential consistency as described in [13], [17]. We now describe our protocol in detail ommiting the scenario associated with a read hit as a read hit proceeds without any delay.

4.1 Reads for a uncached block When a block B is cached for the first time in processor P1, the request is satisfied by the memory module where B is mapped. When this request goes up the MIN a trail of entries will be set up. In these entries the WBV bit corresponding to P1's subtree would be set. When the request proceeds to the Memory Directory of B(MD(B)), the NCA field in B's entry in the MD(B) is set to P1 to nominate P1 as the current NCA and the cached bit for this entry is set to the mode "cached" to inhibit the memory from supplying the block. The MD(B) then sends a packet to P1 with the missing block. As this block proceeds towards P1, it updates the SD entries it had created earlier on its way up in the following manner:

• The WBV bit for P1's subtree is reset and the mode fields upto the current NCA (in this case P1) are set to *above*

• the PBV bit corresponding to the current NCA's subtree is set

• When the Current NCA is reached, its mode field is set to *NCA* and the block is deposited in the P-Cache of P1

4.2. Reads for a previously cached block: When a processor P2, has a miss on the block B previously cached in one or more P-Caches, it sends a request towards the memory module. This requests sets up a trail of SD entries with corresponding WBV bits set, exactly as before. The requests could be either from out of the current locality or from within the locality, each are handled as described below.

4.2.1. Requests from Out of Locality: Two situations arise here, a previous trail could be encountered or the previous trail has been wiped out by the replacement of corresponding SD entries. **(a). The request progresses all the way to the memory.** Assume that the entry for B in MD(B) indicates SD(k) as the current NCA. When the request arrives at the memory controller the NCA field in MD(B) entry is set to the new NCA. MD(B), simultaneously, redirects the request from P2 to SD(k). As this redirected request proceeds down to SD(k), it re-establishes the trail of SD entries for B from the memory module to SD(k) by recreating entries if needed. The entries in the SD's up to the new NCA are set with mode field *above*. A request of supply and NCA update trickles down towards SD(k).

When SD(k) receives the supply request for B, it sends a copy to P2 via the new NCA. The WBVs corresponding to B in this path are reset and the mode set to *below* as the missing block progresses to P2.

The entry created at the new NCA has its WBV bit set and the mode bit set to *NCA*. The bit corresponding to SD(k)'s and P2's subtrees are also set in the PBV of the SD entry at new NCA to reflect the new locality of sharing. As the redirected request proceeds down to SD(k) from the new NCA, it modifies SD entries for B as follows -- The WBV bits are reset and corresponding PBV bits set and the mode field is set to *below*. The block is supplied to all subtrees with WBV bits set.

(b). The request encounters the trail of SD entries from MD(B) to the current NCA (say SD(k)) at a SD(say SD(j)), with B's entry in the *above* mode. This indicates that the current NCA entry in MD(B) for the block B has to be updated to NCA(B,SD(k),P2). This prompts the following actions:

A request to update the NCA entry in the MD(B) from SD(j) is sent up; this would re-establish a trail of SD entries from SD(j) to MD(B) in case the trail was broken due to replacement of entries in the *above* mode. Simultaneously, SD(j) redirects the miss request to the old NCA, which then routes it to a P-Cache, as in Case (a). The setting and resetting of the WBV and PBVs are similar to the previous cases.

4.2.2. Requests from within the Current Locality of Sharing: The request encounters the trail of SD entries from MD(B) to the current NCA (say SD(k)) at a SD(say SD(j)), with B's entry in the *below* mode. This indicates that there is no need to update the NCA field in B's entry in MD(B). In this case SD(j) redirects the copy to any subtree that has a copy of B by consulting the PBV in B's entry. As a consequence some P-Cache in the subtree rooted at SD(j) supplies the missing block to P2 via SD(j). The WBVs in the SD entries along the path from P2 to SD(j) are set and reset accordingly. Also, as the missing block progresses down from SD(j) to P2, the PBV of the SD entries in this path are set to indicate the presence of the block in P2's subtree.

4.2.3. Combining simultaneous reads: Combining of two or more read miss requests is done when a request R1 meets an SD entry at SD(j) (say) for B with one or more bits in the WBV set. This indicates that there is at least one other request pending for B that will be serviced through this SD. Combining essentially involves setting the WBV bit for R1's subtree in B's entry at SD(j). The redirection or propagation of R1 is stopped. When the missing block is supplied through SD(j), it will redirect it to all the subtrees whose WBV bits are set and update the PBV appropriately. Any updates of the NCA would be handled as in case (c) above (depending on the mode of the SD entry).

4.3 Write Misses: Write Misses are handled essentially in the same way as read misses, however there are the following fundamental differences.

• The value being written is sent up the tree along with the block address.

• When the old NCA receives the updated value as part of the redirected write miss request, it broadcasts the update to all subtrees that indicate a presence of the block, as evident from the PBV.

Write requests are combined in a different manner; a write request which comes in when another write request is pending is ignored. This essentially amounts to dropping one of several simultaneous writes, as in [8]. A write request for a block B which meets a SD entry in SD(j) say, with B's entry in the *below* mode waits at SD(j) till the WBV bit in B's entry are all reset. Therefore, in this case we do not combine a write with pending read requests. When a write request meets the current NCA or meets a trail of SD entries in the "above" mode and eventually gets redirected to the current NCA it can be combined with pending reads if the NCA entry for B has bits in the WBV set. This is done as follows. The NCA broadcasts the value in the write miss request to all subtrees that indicate the presence of B. As this updated value progresses towards the P-Caches that have a copy of B, WBV vectors on the trail are reset. Notice that the

block value supplied in response to the read misses simply gets terminated on its way up to the P-Caches where the read misses occurred, since the WBV bits would be reset.

4.4 Write Hits: Write hits are always confined within the locality of the current NCA as the request comes from a P-Cache which already has the block. The write hit request carrying the updated value for a block B is propagated to the current NCA for B which broadcasts the updated value to all subtrees that indicate a presence of the block B. The write hit proceed in two phases. The request is first directed towards the NCA and in the second phase the data is broadcast to all the P-Caches having the block. Simultaneous write requests from more than one P-Cache for the same block are arbitrated at every SD level. The surviving request proceeds towards the NCA, thus imposing a serializing constraint on the write requests. Also note that the requests for writes can come from the top end of the NCA which are essentially write misses. These write misses involve longer latencies than the write hits themselves and thus can be given a static priority over write hits. As in the case of write misses, a write supercedes the block value to be supplied in response to an earlier pending read miss, essentially combining a write with reads. This is done to permit supply of the updated value to the P-Cache requesting it.

4.5. Replacement Policies for SD and MD Entries: The replacement policy for SD entries will guarantee that the SD entries in *below* mode are not replaced if possible. In the dire case when it is imperative to replace an entry in the *below* mode, all entries in the subtree rooted at the victim's SD are invalidated. If this entry were the NCA then the memory copy is updated. The selection of an *above* entry as victim for replacement wipes out all SD entries in the path from the victim's SD to the root. To prevent the replacement of an SD entry corresponding to a block that is used frequently the *usage counts are propagated periodically up the tree* (piggy-backed on other packets) to SDs at higher levels. A SD entry marked as the *NCA* is never replaced.

Note that when a block entry to be replaced is the last one in the entire set of SDs, then a memory write takes place to update the memory copy and the status of the MD entry for the block. The SD adjacent to the memory module for the block (i.e., at the top most level) is responsible for these actions.

The MD also uses the usage information propagated up the tree for replacements. In the extreme case when the entry for a block B has to be replaced, the MD induces a write back of the block to get a consistent memory copy and invalidates copies in all the P-Caches that have cached the block. In the process of invalidating the P-Caches, the SD entries for B are also removed. Note also that an entry for a block in the MD gets replaced automatically with the replacement of a SD entry at a higher level, as described earlier

4.6. Handling Replacements in the P-Caches: Replacements in the P-Caches are done in a manner similar to the replacement of SD entries. The replacement of a block in a P-Cache triggers the replacement of a SD entry if the block was the only one in the subtree rooted at that SD. If this is not the case then only the presence bit for the block subtree is reset. Again a write back to the memory takes place if the top most SD discovers the block being replaced as the sole cached copy. It is worth noting here that write backs are thus deferred as much as possible again contributing positively to the performance of the protocol.

5. SD Hardware and Implementation Issues

The hardware associated with a switch cache is similar with the hardware associated with combining switches, as in the one proposed for the NYU Ultracomputer [8], *albeit considerably simpler*. This simplicity results from the fact that we do not have combining queues (and an associated matching logic).

Figure 2 depicts the hardware incorporated within switches that have a directory. All incoming requests are buffered in an input queue IQ, from where they are picked up, one-at-a-time, put in the latch L and compared with entries in a set-associative directory (consisting of D1, D2, L, C1 and C2). On a match (i.e., a hit), the matching entry is steered to the function unit F, where it is processed in a manner specific to the input request, and finally put in the output queue, OQ. On a miss, the replacement logic R first performs the replacement and forwards the request to F. Note again, that the matching logic is associated with C1 and C2 -- and not with every entry in IQ or OQ.

The logic within F and R is somewhat more complicated than the corresponding logic associated with an ordinary cache; however, the SD functions are still simple enough to make the implementation of F and R simple (and viable). Processing within F and R mainly involves bit-level operations. In any case, *the SD operations are pipelined with the operation of the packet-based MIN, in effect hiding delays within a switching element with a SD.*

6. Performance Evaluation

A performance evaluation for our scheme should take into account the effect of combining and filtering. It is thus not easy to analyze the performance of our protocol theoretically along the lines of [1] or [2]. A better approach would be to use trace-driven simulation, which is likely to be very computing-intensive for large systems (64+ processors). We are not aware of the open availability of realistic traces for large shared-memory systems. In what follows we do a very approximate performance analysis for our coherence mechanism to evaluate its overall latency on a read miss or a write.

As a first order approximation we will assume that a block can be found with probability 'q' in any of the SDs located in the current locality (i.e., the subtree rooted at the current NCA). The penalty on a read miss or a write in terms of the number of levels that have to be traversed to service these requests from within the locality is:

$$X1 = \sum_{i=1}^{N_s} 2i.P(i)$$

where N_s is the height of the NCA from the P-Caches and p(i) is the probability of finding the missing block at a height 'i' from the P-Caches. Clearly, $P(1) = q$ and $P(i) = (1- P(i-1)).q$, for $i > 1$.

The contribution X1 can thus be estimated based on q and N_s. To find the contribution to the read miss or write penalty outside the region of locality, assume that these requests meet a trail in a level above the current NCA with probability 'r'. The average height at which these requests will meet the trail outside the locality is $h_{avg} = (L + N_s)/2$. The contribution to the penalty due to requests outside the locality is thus

$$X2 = r.4.h_{avg} + (1 - r).(2.L + 2.h_{avg})$$

where L is the total number of levels in the switch.

$$X2 = r.4.h_{avg} + (1 - r).(2.L + 2.h_{avg})$$

where L is the total number of levels in the switch.

The total delay (in terms of the number of switch levels in the k-ary tree) that have to be traversed on a read miss or a write is thus

$$(\alpha.X1+(1 - \alpha).X2).$$

where α is the Fraction of reads to writes

The equations given above are only for the coherence penalities imposed due to read misses and writes. Figure 3. depicts how these penalties vary with various values of 'q', 'r' and N_s for representative values of α. As expected, the penalties are lower when most of the accesses are confined within the locality of sharing. The overall performance depends on, of course, such things as the P-cache hit ratio, fractions of reads/writes etc.

To be fair, one should also note that the switch latencies in a system using our protocol are likely to be higher than that in MINs without any directory/logic within the switches. Thus the number of levels traversed is not a true indicator of the actual delays. However, one should also note that our technique offers a number of other substantial advantages (as detailed in Section 7), that can quite possibly

offset the effect of longer switch latencies. We expect the result of trace-driven simulations to confirm this in the near future.

7. Assessment and Conclusions

The proposed cache coherence mechanism maintains sequential consistency as introduced in [13] and has at least the following potential advantages:

(1) It does not appear to suffer from the limitations of existing schemes (Sec. 1.2).

(2) It provides a cheap way of combining coherence related traffic within the SDs. This feature is absent in existing cache coherence mechanisms for MIN-based systems, and can offer a significant performance edge for two reasons:

(3) Combining avoids a *strict* serialization of the coherence traffic, like the schemes suggested in [15], [11] and capitalizes on the fact that shared reads and writes occur in bursts.

The combining logic needed is fairly simple, as outlined in Section 5 and does not increase switch latencies beyond what is needed to support the protocol without combining features. The combining logic is also considerably simpler than that employed in [8].

(4) In shared memory multiprocessors with MINs, hot-spot/tree saturation effects cause a drop in overall system throughput by blocking network traffic significantly -- this traffic congestion is confined to the upper part of the MIN. Our cache coherence mechanism, in contrast, confines the bursty network traffic to hot spots to the lower part of the MIN since each SD effectively allows only one out of K (= 2^k) simultaneous requests to a hot module to progress to the next switching level. This alleviates the tree saturation effect.

(5) The coherence mechanism keeps all shared blocks consistent -- this is done using a broadcast mechanism, that can have other applications, as well (e.g., implementing barriers). This reduces the miss rate on blocks that otherwise would have to be invalidated. In a MIN based system without a bus-like broadcast facility invalidation overheads are about the same as write broadcasts, thus our scheme is not made inefficient when writes are broadcasted.

(6) Write backs from the P-Caches are deferred till the last cached copy of a block gets replaced. This, in effect, is done by filtering the write backs, again ensuring a conservative use of the limited network bandwidth.

Our scheme is compared with the schemes proposed in [15] and [4] in Figure 4 to show how coherence related traffic flows through the MIN. We can easily see the potential advantages of our scheme from this figure. The results of the very rudimentary analysis, as presented in Figure 3 seem to corroborate this. We intend to collect real traces from a large-scale shared memory system to evaluate the performance of our scheme in the near future.

References

[1] Algudady, M.S., Das, C.R. and Thazuthaveetil, M.J. "A write update cache coherency protocol for MIN based multiprocessors with accessibility based split caches", in Proc. Supercomputing 1990.
[2] Bhuyan, L.N., Liu, B.C. and Ahmed,I. "Analysis of MIN Based Multiprocessors With Private Cache Memory," Proc. 1989 Int. Conf. on Parallel Processing, p.I51-I58.
[3] Brooks, E.D. and Hoag, J.E. "A scalable coherent cache system with incomplete directory state", In Proc. 1990 Int. Conf. on Parallel Processing, Vol I Architecture, Aug 1990.
[4] Censier, L.M. and Feautrier, P. " A New Solution to Coherence Problem In Multicache Systems." IEEE TC C-27, 12(DEC. 1978), 1112-1118.
[5] Cheong, H. and Veidenbaum, A.V., "Stale Data detection and coherence enforcement using flow analysis", In Proc. 1988 Int. Conf. on Parallel Processing, Vol I Architecture, p.138-145, Aug 1988.
[6] Feng, T.Y. "A survey of Interconnection networks", IEEE Computer, p. 12-27, Dec 1981.
[7] Gornish, E.H., Granston, E.D. and Veidenbaum, A.V. " Compiler-

directed data Prefetching in Multiprocessors with memory hierarchies"

[8] Gottlieb, A., Grishman, R., Kruksal, C.P., McAuliffe, L., Rudolph, L. and Snir, M. "The NYU Ultracomputer - Designing an MIMD shared memory parallel computer", IEEE Trans. on Computers, Vol. C.32., No.2.,p. 175-189, Feb 1983.

[9] McAuliffe, K.P. "Analysis of Cache Memories in Highly Parallel Systems." PhD Thesis, NYU, May 1986.

[10] Min, S.L., and Baer, J.L. "A Time-stamp-based cache Coherence Scheme." In Proc. 1989 Int. Conf. on Parallel Processing, p.23-32, August 1989.

[11] Mizrahi, H.E., Baer, J.L., Lazowska, E.D. and Zahorjan, J. "Extending the memory hierarchy into Multiprocessor Interconnection Networks: A performance analysis", In Proc. 1989 Int. Conf. on Parallel Processing, p.41-50.

[12] Pfister, G.F. and Norton,V.A. "Hot Spot Contention and Combining in Multistage Interconnection Networks," IEEE Trans. on Computer, Vol. C-343, No.10, p.943-948, Oct. 1985.

[13] Scheurich, C. and Dubois, M. "Correct Memory Operation of Cache-Based Multiprocessors." Proc. of the Int. Symp. on Computer Architecture, p.234-243, 1987.

[14] Stenstrom, P. "A survey of cache coherence schemes for multiprocessors", IEEE Computer June 1990, p.12-24.

[15] Stenstrom, P. "Reducing contention in shared memory multiprocessors", Computer, Nov 1988, p. 26-37.

[16] Veidenbaum, A.V. "A Compiler-Assisted Cache Coherence Solution for Multiprocessors." In Proc. 1986 Int. Conf. on Parallel Processing. p.1029-1036, August 1986.

[17] Ghose, K and Simhadri, S.N. "A New Cache Coherency Mechanism with Limited Combining Capabilities for MIN-based Multiprocessors", Technical Report CS-TR-90-52, Dept. of Computer Science, State University of New York at Binghamton, Sep 1990.

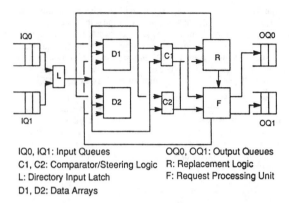

IQ0, IQ1: Input Queues   OQ0, OQ1: Output Queues
C1, C2: Comparator/Steering Logic   R: Replacement Logic
L: Directory Input Latch   F: Request Processing Unit
D1, D2: Data Arrays

Figure 2. 2–way Set Associative Directory Within 2X2 Switch

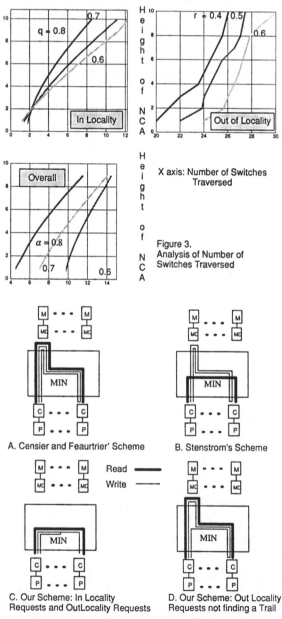

X axis: Number of Switches Traversed

Figure 3.
Analysis of Number of Switches Traversed

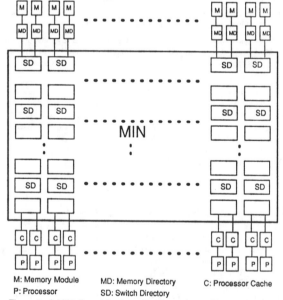

M: Memory Module   MD: Memory Directory   C: Processor Cache
P: Processor   SD: Switch Directory

Figure 1 : A MIN–Based System with Directories at Alternate Levels

A. Censier and Feaurtrier' Scheme   B. Stenstrom's Scheme

Read ——
Write ——

C. Our Scheme: In Locality Requests and OutLocality Requests   D. Our Scheme: Out Locality Requests not finding a Trail

Figure 4. Comparision of Delays in various schemes referred

Effects of Program Parallelization and Stripmining Transformation on Cache Performance in a Multiprocessor*

Manish Gupta
Department of Computer Science
University of Illinois at Urbana-Champaign
Urbana, IL 61801

David A. Padua
Center for Supercomputing Research and Development
University of Illinois at Urbana-Champaign
Urbana, IL 61801

Abstract

The results of studies of cache performance on traditional uniprocessor systems cannot be assumed to extend directly to multiprocessor systems. In this paper, we study the change in memory referencing behavior of sequential numeric programs when they are parallelized. Through trace-driven simulation, we show that cache performance tends to degrade as we increase the number of processors in the system. The performance degradation is particularly severe when larger cache blocks, which worked well for the uniprocessor case, are used. However, it is possible to improve the behavior of the program if proper attention is paid to preserving spatial locality during the parallelization step. We show that applying stripmining transformation to various parallel loops in the programs leads to considerable improvements in cache performance and restores some of the performance benefits of using larger cache blocks.

1 Introduction

Cache memories have been widely used in conventional high performance computers to reduce the effective memory access time. A number of multiprocessor systems with shared memory have also been designed to use caches [2, 11]. In most of these systems, each processor has its own private cache. Besides providing fast access to frequently used data, cache memories are also expected to help reduce contention on the shared bus, or reduce traffic in the interconnection network.

Recent studies have shown that multiprocessors tend to have higher cache miss rates compared to uniprocessors [5, 8]. The most important issue that arises in the context of multiprocessors is that of cache coherence. Several schemes have been proposed for ensuring cache coherence, and studies have been done [1, 4] on the actual performance degradation caused by the cache coherence overhead. This overhead may be the result of *true sharing* or *false sharing* [12] of data. True sharing refers to the sharing of the same memory word by different processors. False sharing is exhibited by caches with multi-word blocks when different processors access different words that happen to be in the same block.

In this paper, we study the changes in memory referencing behavior of sequential programs when they are parallelized, and the implications of these changes for cache performance. The block (line) size of a cache is one of the most important factors affecting its performance. Hence, we focus a lot of our attention on the effect of block size on the miss ratio and how this relationship is modified when we move from a uniprocessor system to a multiprocessor system. We show that given the same cache parameters, the performance of cache tends to degrade as we increase the number of processors in the system. However, through transformations on the source code, it is possible to improve considerably the spatial locality of the parallel program. We simulate the execution of programs under one such simple transformation, the stripmining transformation [9], and study the extent to which it improves the performance of cache.

*This work was supported by the U.S. Department of Energy under grant no. DOE DE-FG02-85ER25001.

2 Methodology

Our analysis of the memory referencing behavior of parallel programs is based on extensive trace-driven simulation, as explained below.

2.1 Multiprocessor Model

The simulations were designed to model a multiprocessor system consisting of one to eight processors connected to global memory through a shared bus. A number of popular multiprocessors have such a shared-bus architecture, such as the Encore Multimax [2] and the Sequent Balance [11] systems. We assume that each processor maintains its own private cache and that it does not have any other local memory. The system uses a write-through policy for caches and a snoopy cache coherence protocol with write-invalidates [1].

2.2 Simulation System

The simulation system consists of two components – a process trace generator, and a trace-driven simulator. A program trace gives the sequence of addresses accessed by a program. The only memory references indicated by our trace generator are the data references; instruction fetches are ignored. The long trace generated contains information about each memory reference, such as the processor that generates that reference, whether the operation is a read or a write, and the memory address. The trace also indicates which processor executes each iteration of a parallel loop. This allows us to determine the iteration to which each memory reference made inside a parallel loop corresponds. We use that information to simulate static scheduling of *doall* loops, with successive iterations of the loop going to different processors.

The simulator is driven by such a program trace. The cache organization is modeled by the simulation system as being set-associative. The number of blocks per set and the number of sets are parameters to the simulator. By varying these parameters, we can also simulate a direct-mapped cache or a fully associative cache. For each memory read, we search the cache on that processor to see if the reference results in a cache hit or miss. In the case of a miss, we bring the block containing that data into the cache, replacing another block from the set into which it is brought if that set is already full. Each cache uses the LRU replacement strategy for blocks. Because a write-through policy is assumed for caches, each memory write operation is treated as a cache miss. The data is assumed to be written into the main memory, and if that block is present in any other processor's cache, the entry for that block is invalidated.

2.3 Application Programs

Six different scientific application programs from the Linpack and Eispack libraries and the Perfect Benchmarks[TM] [3] were used for this study. The sequential programs were parallelized using the KAP source-to-source restructurer [7]. For subroutines from Linpack and Eispack libraries, we wrote driver routines that generate their own test matrices and invoke those subroutines. The program corresponding to each subroutine was traced for two different data sizes, namely, matrices of sizes 64 x 64 and 128 x 128. For the programs from Perfect Benchmarks, we modified some constants governing the array sizes, so that we could run these programs with smaller data sizes.

The first three programs use routines from the Linpack library. The first program uses *sgefa*, which factors a real matrix by gaussian elimination. *Sqrdc* uses householder transformations to compute the QR factorization of a real matrix, and *ssvdc* reduces a real rectangular matrix by orthogonal transformations to its diagonal form. The fourth program uses the routine *minfit* from Eispack, which determines the singular value decomposition of a double precision real rectangular matrix. The remaining two programs are from the Perfect Benchmarks. The program *flo52* is a two-dimensional code providing an analysis of the transonic inviscid flow past an airfoil by solving the unsteady Euler equations, and *trfd* simulates the computational aspects of a two-electron integral transformation. These programs were chosen because they exhibit a fairly high degree of parallelism.

3 Memory Referencing Behavior on Multiprocessors

For this study, we concentrate on programs exhibiting parallelism only at the loop level. These computationally intensive numerical programs spend most of their time in loops, and hence that is the level at which almost all present parallelizing compilers and program restructurers attempt to expose parallelism.

3.1 Effect of Parallelization

Examining how the transformation of a sequential *do* loop into a parallel *doall* loop changes the memory referencing patterns of the loop code gives us some insight into how the behavior of a program changes when it is executed on a multiprocessor. For example, consider the following simple loop:

$$
\begin{aligned}
&\text{doall } i = 1, n \\
&\quad y(i) = y(i) + a * x(i) \qquad\qquad (1) \\
&\text{enddo}
\end{aligned}
$$

When this loop is executed using self-scheduling, we have different processors executing successive iterations; for example, we may have processor 1 accessing $x(1)$ and $y(1)$, processor 2 accessing $x(2)$ and $y(2)$, and so on. The use of a large cache block size causes pollution of cache space. When a processor accesses an array element, the entire block containing that element is brought into the cache, and the processor may use only one of those elements. Another problem caused by the use of multi-word cache blocks is false sharing, which leads to unnecessary invalidation of cache blocks. When processor 1 writes into $y(1)$, the block containing $y(1)$ is invalidated on the caches of all other processors, even though none of them is using the value of $y(1)$.

These problems do not arise if the caches use single-word blocks. However, using single-word cache blocks is not a very satisfactory solution.

3.2 The Stripmining Transformation

A better approach to these problems seems to be to make sure that each processor works on as large a part of the block of elements it accesses, as possible. This can be achieved by transforming the loops of the program with stripmining, and ensuring that the base of each array resides on a cache block boundary. In the following discussion, we assume that the memory allocator takes care of aligning the base

of each array appropriately. The stripmining transformation adds another level of nesting to the loop so that the outer loop steps through the index set of the original loop in blocks of some size. The original loop now steps through each block. By choosing the block size in stripmining to be the same as the block size (in words) of the cache, we try to ensure that a set of iterations in which array elements residing in a single block are referenced is assigned as a chunk to a single processor. For example, the loop (1) shown above would be transformed as follows:

$$
\begin{aligned}
&\text{doall } i1 = 1, n, blk\_size \\
&\quad \text{do } i = i1, min(n, i1 + blk\_size - 1) \\
&\quad\quad y(i) = y(i) + a * x(i) \\
&\quad \text{enddo} \\
&\text{enddo}
\end{aligned}
$$

It is clear that regardless of how various iterations of the doall loop are scheduled, each processor accesses all the elements of arrays x and y that reside in a single block, and no block is shared among processors. For loops in which successive iterations have references to successive elements of various arrays, such a transformation is expected to eliminate false sharing, or reduce it considerably.

We now briefly describe how we "simulate" the stripmining transformation. Making changes in the source program would require a new set of traces to be produced for each transformed program. It is easier to modify the simulator so that each processor is assigned a suitably sized chunk of iterations rather than a single iteration at a time. This is possible because the trace indicates which processor executes each iteration of the *doall* loop. We assign all the references corresponding to a *doall* iteration to the processor that would have executed that iteration in the transformed program. Thus we achieve the same results as those from a normal simulation driven by the trace of the stripmined program.

4 Simulation Results

The simulations were run for cache sizes varying from 8 Kbytes to 128 Kbytes. The caches of size 64 and 128 Kbytes behave virtually as infinite caches for the programs we ran because we used small data sizes. Qualitatively, the inferences drawn from these simulations remain almost the same for all cache sizes in the range of 8 to 128 Kbytes. We used small data sizes for all programs in order to get better accuracy, given that our trace sizes had to be limited to 250,000 references per processor. Correspondingly, the figures we show in this section are those obtained for cache size of 8 Kbytes.

4.1 Effect of Number of Processors

As we increase the number of processors, the amount of cache memory per processor is kept the same. Hence the total amount of available cache memory increases. This factor would contribute to a decrease in the miss ratio. However, since global variables are cached on reference by all the processors, some space is now used up to hold repeated information. The invalidation of blocks to ensure cache coherence also contributes to an increase in the miss ratio.

Figure 1 shows the variation of miss ratio with change in the number of processors for two different block sizes. Most of the programs show a fairly significant increase in the miss ratio with increase in the number of processors. This rise in the miss ratio becomes more pronounced as the cache blocks are made bigger. For many programs, the miss ratio increases sharply as we move from a uniprocessor to a two-processor system. Similar results of worsening of the miss ratio have been reported in [8]. These results point to a decrease in the effectiveness of cache with regard to data references on larger multi-

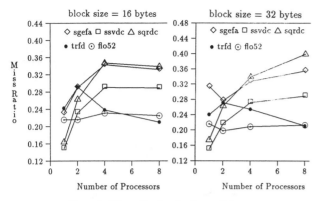

Figure 1: Miss ratios for simple parallel programs

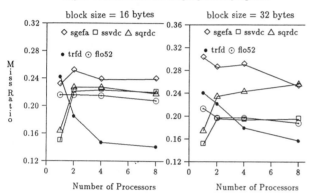

Figure 2: Miss Ratios for stripmined parallel programs

processor systems. The figures also seem to confirm that the effects of larger block size, such as pollution of cache space and false sharing, come into play in a significant manner in multiprocessor systems.

Figure 2 shows what happens when the programs are transformed through stripmining of loops. There are significant improvements in the miss ratio for larger numbers of processors. For programs where the miss ratio increased sharply on increasing the number of processors from one to two, the magnitude of increase is much less. In most programs, the miss ratio now shows a decline as the number of processors is increased, as opposed to the increasing trend shown for programs without stripmining. This may be attributed to the reduction in false sharing brought about by stripmining. These results demonstrate the effectiveness of stripmining transformation in improving the performance of cache for multiprocessor systems using multi-word cache blocks.

4.2 Effect of Block Size

The plots in Figure 3 show the effect of block size on miss ratio for various programs. We are particularly interested in studying whether and how the optimal (with respect to miss ratio) block size changes as we move from a uniprocessor system to a multiprocessor system. For each program, we show the variation of miss ratio with block size for three cases – the sequential program running on a uniprocessor, the "simple" parallel program (the version produced by KAP) running on a multiprocessor with eight processors, and the stripmined parallel program running on the multiprocessor.

As expected, all the sequential programs show an initial decline in miss ratio as the block size is increased. The simple parallelized version of most programs, however, shows an increase in miss ratio. In fact, this rise in miss ratio is particularly sharp as the block size is increased beyond one word. Similar results have been reported by

[8], and some other researchers [6] have also argued in favor of using single-word blocks in multiprocessors. However, it can be seen that the stripmining transformation on parallel programs makes the use of higher sized blocks much more attractive. For the *sgefa* and *trfd* programs, the optimal block size changes from 4 bytes to 16 bytes, and the miss ratios at all higher sized blocks are much lower than those for the original parallel program. In case of *minfit* and *flo52*, the miss ratios steadily decrease with increasing block size even for the simple parallel program, and get still lower for the stripmined parallel program. The figures for *sqrdc* and *ssvdc*, on the other hand, show a single word block as being optimal even for the stripmined program. However, even in these cases, a comparison with the miss ratios exhibited by the simple parallel program shows that stripmining does considerably improve the memory referencing behavior of the programs.

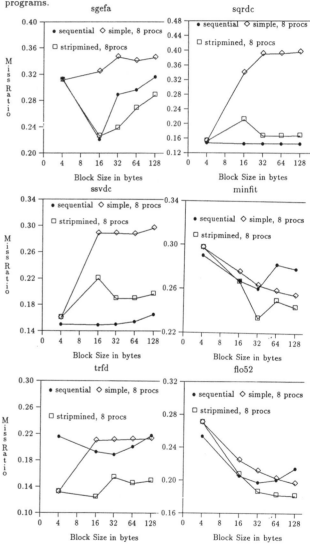

Figure 3: Miss Ratios v/s Block Size

Another interesting fact one can observe from these plots is that a parallel program transformed with stripmining shows a memory behavior quite similar to that of the underlying sequential program. In all cases, the particular block size leading to the minimum miss ratio for a sequential program also yields good performance, in terms of miss ratio, for the stripmined parallel program. While straightforward parallelization of a program leads to changes in its memory referencing behavior so that smaller cache blocks become more attractive, the stripmining transformation seems to undo some of those changes so that bigger blocks once again yield higher hit ratios, in spite of the problem of cache coherence posed by the write operations. All of the parallel programs we used have a significant number of write operations, ranging from 12% to 20% of all the memory references. Thus, with suitable transformations on the source code, we can bring the memory referencing behavior of a parallel program closer to that of the underlying sequential program.

It is worth noting that we have considered only data caches. If the processor architecture uses the same cache for instructions and data, the actual cache performance is expected to be even better at higher block sizes, because instructions usually show much better locality than data.

4.3   Number of Blocks Invalidated

Figure 4 shows how the number of cache blocks invalidated due to write operations on other processors varies with the block size used. The results shown by simulations are hardly surprising but extremely significant. Without the stripmining transformation, the number of block invalidations increases steeply as the block size is increased beyond one word, then stabilizes at a high value for larger blocks. This increase can be attributed to false sharing. Programs transformed using stripmining behave much better, and the number of block invalidations is considerably lower at higher block sizes. Because the behavior shown by all programs is very similar, we have chosen two representative samples to show the plots.

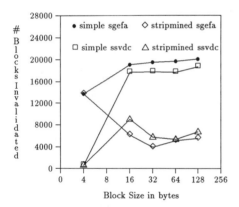

Figure 4: Number of blocks invalidated v/s block size

5   Conclusions

We simulated the performance of processor caches in a bus-based shared memory multiprocessor. We used the parallel memory traces of several numeric programs, each having a significant number of write operations. A simple analysis of a typical parallel program resulting from the parallelization of loops in the original sequential program shows a decline in the spatial locality of the parallel program. This is confirmed by simulations which show a deterioration in the performance of caches as the number of processors in the system is increased. The performance degradation is particularly severe for caches with larger blocks. As a result, we usually get the smallest miss ratios on using smaller cache blocks containing just a single word each in multiprocessors, as opposed to the uniprocessor systems, where block sizes in the range of 4 to 16 words give better results.

However, the use of stripmining transformation on the parallel programs changes the scenario completely. The spatial locality of the parallel programs improves considerably, and we often get the smallest miss ratios on using cache blocks of sizes 4 to 8 words each. To quite an extent, the parallel programs show memory referencing behavior similar to that of the corresponding sequential programs.

From this study, we conclude that cache performance tends to degrade as we move from a uniprocessor to a multiprocessor system. However, a lot of this degradation results from how parallel programs are written or produced. If proper attention is paid to preserving the property of locality of the program, fairly good cache performance can also be obtained in a multiprocessor system.

6   Acknowledgements

We thank Sanjoy Ghosh for writing the process trace generator that we used to generate traces for this study. We also wish to thank the referees for their valuable suggestions.

References

[1] J. Archibald and J.-L. Baer. Cache coherence protocols: Evaluation using a multriprocessor simulation model. *ACM Transactions on Computer Systems*, 4(4):273–298, November 1986.

[2] C. G. Bell. Multis : A new class of multiprocessor computer. *Science*, 228:462–467, 1985.

[3] The Perfect Club. The perfect club benchmarks: Effective performance evaluation of supercomputers. *International Journal of Supercomputing Applications*, 3(3):5–40, Fall 1989.

[4] M. Dubois and F. A. Briggs. Effects of cache coherency in multiprocessors. *IEEE Transactions on Computers*, C-31(11):1083–1099, November 1982.

[5] S. J. Eggers and R. H. Katz. Evaluating the performance of four snooping cache coherency protocols. In *Proc. 16th Annual Symposium on Computer Architecture*, pages 2–15, 1989.

[6] J. R. Goodman. Using cache memory to reduce processor-memory traffic. In *10th Annual Symposium on Computer Architecture*, pages 124–131, June 1983.

[7] Kuck & Associates, Inc. *KAP User's Guide, Version 6*. Champaign, IL, 1988.

[8] D. J. Lilja, D. M. Marcovitz, and P. C. Yew. Memory referencing behavior and a cache performance metric in a shared memory multiprocessor. Technical Report CSRD 836, University of Illinois, June 1989.

[9] D. A. Padua and M. J. Wolfe. Advanced compiler optimizations for supercomputers. *Communications of the ACM*, 29(12):1184–1201, December 1986.

[10] S. Thakkar, P. Gifford, and G. Fielland. The balance multiprocessor system. *IEEE Micro*, pages 57–69, February 1988.

[11] J. Torrellas, M. S. Lam, and J. L. Hennessy. Shared data placement optimizations to reduce multiprocessor cache miss rates. In *Proc. 1990 International Conference on Parallel Processing*, pages 266–270, August 1990.

FAULT-TOLERANT ROUTING ON A CLASS OF REARRANGEABLE NETWORKS

Yao-ming Yeh and Tse-yun Feng

Department of Electrical and Computer Engineering
The Pennsylvania State University
University Park, PA 16802

Abstract — Existing fault-tolerant routing schemes for rearrangeable networks consider switch control faults on the Benes network only. In this paper, we propose a general fault-tolerant routing scheme for a class of $2log_2N$ stage rearrangeable networks. Networks in this class can be either symmetric or asymmetric in their structure, regular or irregular in their inter-stage connections, and with $2log_2N$ or $2log_2N - 1$ stages. Our scheme considers both switch and link faults in the network. For the failure of control line in the switches, a fault-free routing scheme is proposed to tolerate multiple faults on each stage of the network. For the failure of switch data lines and inter-stage links, a graceful degradation routing scheme is developed to configure the network so that the loss is ensured to be minimal for weighted permutation requests.

Key words — *The Benes networks, Fault-free routing schme, Fault-tolerant routing scheme, Graceful degradation routing scheme, Rearrangeable networks*

1. INTRODUCTION

Multistage interconnection networks (MINs) are vulnerable to the failure of switches and links in the network. Two approaches can enhance MINs' fault-tolerant capability: to use extra hardware (such as extra switches or links [1][2][3][9][13]) to provide alternate paths and/or use extra software (such as extra routing passes [15] or special routing operation [8][10]) to bypass the faults. In [18], we proposed a class of $2log_bN$ stage rearrangeable networks, for any b. An interconnection network is rearrangeable [4] if it can realize arbitrary permutation between input and output terminals. This class of networks have multiple paths for every input-output pair, therefore, pure software approach of fault-tolerant routing is possible for these networks. Comparing to the cost of hardware approach, to develop fault-tolerant routing scheme with pure software enhancement is more promising than the hardware approach.

The existing fault-tolerant schemes only apply to the Benes network with very limited capability. Agrawal [3] used an extra switch attached to either side of the network to compensate a faulty switch in the network. Agrawal's scheme can only tolerate single fault in the network. Huang [7][8] also provided a fault-tolerant scheme on the Benes network, which is extended from his routing scheme on the finite state model. Nassimi [10] applied fault-tolerant self-routing to bypass the fault. Nassimi's scheme can tolerate single fault on each stage and no fault on center stage. Wysham and Feng [17] extended the looping algorithm to tolerate the switch control faults on the Benes networks. Above fault-tolerant schemes only consider switch control faults with limited fault-tolerant capability for the Benes network.

In this paper, we develop the fault-tolerant routing scheme for all possible faults in the class of the rearrangeable networks using (2×2) switches. We consider both switch faults and link faults in the network. Our scheme can provide fault-free routing for switch control faults with multiple faults presented in one stage of the network. We

also develop a graceful degradation routing scheme to ensure minimum loss for switch path faults and link faults. Our scheme is extended from our general routing scheme previously proposed in [18].

Feng et.al. [5][6] proposed fault diagnosis schemes for MIN's to pinpoint the faulty spot and distinguish the type of fault. Here, we assume fault diagnosis to be available as needed with respect to the MIN's studied and do not discuss it further.

The remaining sections of this paper are organized as follows: Section 2 describes the notations and concepts in the switch labelling scheme and the original routing algorithm which are related to the fault-tolerant routing scheme. Section 3 introduces the symmetric properties of the network. These properties provide the fault-tolerant capability in the routing scheme. Section 4 describes five algorithms in our fault-tolerant routing scheme. Fault-free routing for switch control faults are given in Algorithms $4 \sim 6$. Minimum-loss routing for switch path faults and link faults are given in Algorithms 7 and 8. Conclusions are given in Section 5.

2. PRELIMINARIES

In this section, a switch labelling scheme and the original routing scheme are briefly described. They are commonly applicable to a class of rearrangeable networks which are equivalent to the Benes network. These equivalent networks can have either regular or irregular inter-stage connections, either symmetry or asymmetry structure, and even, either $2log_bN$ or $2log_bN - 1$ stages.

2.1 Switch labelling scheme

The switch labelling scheme consists of two steps: outside-in decomposition, and inside-out decomposition. These two steps can generate an outside-in code and an inside-out code on every switch in the network.

A. Outside-in decomposition

In a $2log_bN$ stage network, the levels of a network are labelled from 1 to n stage-by-stage from outside-in. A switch is denoted as $S_{ia,j}$ for switch located in input half of the network, or $S_{oa,j}$ for switch in output half. Where the first subscript is the stage index; the second subscript is the switch index within a stage, as shown in Figure 1.

The idea of outside-in decomposition is to label the switches based on the subnetwork decomposition level by level from outside-in till the center stages. To support this mechanism, we can send 2N signals simultaneously from both input and output terminals to propagate through stage by stage till they meet in the center of the network. Initially, each signal carry a label with no value. When a signal traverses through one stage, a value i, $0 \leq i \leq b-1$, is attached behind its label according to following rule:

R1: *attach i when the signal goes through the ith port of the switch (i.e., consider the output ports for input half of the network; and input port for output half).*

The label of each switch is derived from the label of all its incoming signals. This label is defined as *outside-in-code* as below.

Definition 1: In a $2log_b N$ stage network, an *outside-in-code*, OC, of a switch in level m (i.e, either switch $S_{im,j}$ or $S_{om,j}$, for $0 \leq j \leq N-1$), is defined as an m digits base b number: $OC[1:m]$, where $0 \leq OC[k] \leq b-1$ for $1 \leq k \leq m-1$. The value of each digit $OC[k]$ is derived from rule R1.

The collection of the switches with the same outside-in-code form a set of *complete residue system module N/b* (i.e., CRS) [18]. These switch sets are defined as below.

Definition 2: In a $2log_b N$ stage network, a *CRS switch set* is the set of switches that have similar outside-in-code. This outside-in-code is called the *code* of this CRS switch set. A CRS switch set with code x is denoted as CRS_x. The *size* of a CRS switch set is the number of switches in the set. It is denoted as $|CRS_x|$.

Lemma 1: In a $2log_b N$ stage network, the size of each CRS switch sets on level a is $2*b^{n-a}$, for $1 \leq a \leq log_b N$.

Proof: Refer [18].

B. Inside-out decomposition

The operation of inside-out decomposition is like the reflection of outside-in decomposition. The mechanism of inside-out decomposition is to label the switches level-by-level from center stage to both input and output terminals. Similar procedure is used to derive the *inside-out-code* of each switch except the rule is modified as:

R2: *attach i when the signal goes through the ith port of the switch (i.e., consider the input ports for input half of the network; and output port for output half).*

Definition 3: In a $2log_b N$ stage network, an *inside-out-code* of a switch in level m (i.e, either switch $S_{im,j}$ or $S_{om,j}$, for $0 \leq j \leq N/b$), is defined as an $n-m$ digits base b number: $IC[1:m]$, where $0 \leq IC[k] \leq b-1$ for $1 \leq k \leq n-m$. The value of each digit $IC[k]$ is derived from rule R2.

Lemma 2: In a $2log_b N$ stage network, a CRS switch set with size k contains the switches with unique inside-out-codes values from 0 to $(k/2)-1$ in base b representation on each half of the network.

Proof: Refer [18].

From Lemma 1, labelling the outside-in-code on every switch of a $2log_b N$ stage network reflects the subnetwork construction in the network; furthermore, from Lemma 2, labelling the inside-out-code on the switches indicate their relative location to construct a subnetwork. Figures 1 and 2 show the result of outside-in and inside-out decompositions on baseline-baseline^{-1} and omega-omega^{-1} networks respectively. The outside-in-code and inside-out-code of each switch is indicated within the switch. Two $(N/b) \times (2log_b N)$ matrices are defined to store these codes: *outside-in matrix* and *inside-out matrix*.

2.2 A GENERAL ROUTING SCHEME

With the help of our switch labelling scheme, a general routing scheme is developed for the whole class of $2log_b N$ stage networks.

Two isomorphic trees are used in our routing algorithm. A network tree describes the network configuration, which is constructed from outside-in and inside-out matrices. A permutation tree describes how a permutation request is decomposed to fit in a network. They are defined as below.

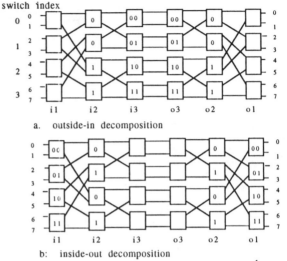

switch index

a. outside-in decomposition

b: inside-out decomposition

Figure 1: switch labelling on a baseline-baseline^{-1} network

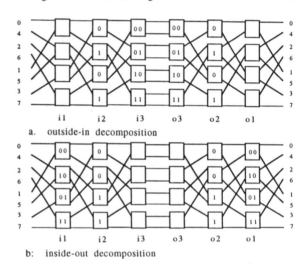

a. outside-in decomposition

b: inside-out decomposition

Figure 2 : switch labelling on an omega-omega^{-1} network

Definition 4: The CRS switch sets in a $2log_b N$ stage network form a *network tree* of depth n and degree b, where $n = log_b N$. A network tree is denoted as $\Phi(n)$, where n is the depth of the tree. A node of the network tree is a CRS switch set. Each node is identified by the code of its CRS switch set. A subtree of the network tree describes a subnetwork of this $2log_b N$ stage network. A subtree is denoted as $\Phi_a(d)$, where subscript a is its root's CRS switch set code and d is its depth.

Definition 5: A $(N \times N)$ permutation request for a $2log_b N$ stage network forms a *permutation tree* of depth n and degree b, where $n = log_b N$. Its root is the permutation request itself. A permutation tree is denoted as $\Psi(n)$, where n is the depth of the tree. A node of the permutation tree is a *reducible sub-permutation* which can be fully assigned to a number of switches. The siblings of a node are the equally decomposed into b sub-permutations from that node.

The outline of the general routing scheme is as below.

Algorithm 1: routing on a $(N \times N)$ network using $(b \times b)$
switches
begin
1: contruct the network tree from outside-in and inside-out matrices;
2: construct the permutation tree by input-output pair permutation;
3: map the permutation tree to the network tree;
4: assign the switch settings in the network tree;
end.

The detail routing procedure for a network using (2×2) switches is shown in Algorithms 2 and 3. In Algorithm 2, a queue (first-in-first-out list) is used to implement breadth-first mechanism in constructing the permutation tree. A procedure $queue(p)$ is used to insert a node to the end of queue and a procedure $dequeue(p)$ to delete a node from the head of the queue. Since the network tree is constructed when a network is specified, Step 1 of Algorithm 1 is not necessary when the same network is used to realize a new permutation request.

In Algorithm 3, the procedure $decompose(P, P_0, P_1)$ decomposes a node in the permutation tree, in other word, it decomposes a reducible subpermutation into two siblings. In this algorithm, the decomposition on a subpermutation is performed on a relation table. In a relation table, the *input buddy* of an input-output pair is the other input-output pair in the same column, meanwhile, its *output buddy* is the other input-output pair in the same row. This algorithm uses $inputbuddy(w)$ and $outputbuddy(w)$ to denote the input and output buddies of an input-output pair w respectively. Since the code assigned to an input-output pair is either 0 or 1, the binary complement operation is applied on variable c.

For example, in an omega-omega^{-1} network, a permutation request is given as following:

$$P = \begin{pmatrix} 0 & 1 & 2 & 3 & 4 & 5 & 6 & 7 \\ 5 & 1 & 2 & 7 & 6 & 3 & 0 & 4 \end{pmatrix}$$

Figure 3 shows the whole routing procedure. The switches are mapped to each column and row of every relation table from their IC and OC as indicated. The switch setting of each switch is derived from the code assigned to its first input-output pair: 0 as "through" and 1 as "cross". This routing scheme needs $O(Nlog_2 N)$ in sequential time and $O((log_2 N)^2)$ in parallel time.

3. ALTERNATE PATHS AND SYMMETRIC PROPERTIES

It is well known that a $log_2 N$ stage interconnection network provides an unique path for every input-output pair. To add an extra stage to the $log_2 N$ stage network will create two alternate paths for every input-output pair. If we keep adding stages to the network, an extra stage doubles the alternate paths in the network. Eventually, a $2log_2 N$ stage interconnection network has N alternate paths for every input-output pair. In this section, we will explore the property of these alternate paths and identify the symmetric property in $2log_2 N$ stage networks

3.1 Redundant switches in $2log_2 N$ stage networks

Waksman [16] pointed out the existence of the redundant switches in a Benes network. It is stemmed from the alternate paths of the Benes network. We can combine his idea with the concept of CRS switch set to identify the redundant switches in every $2log_2 N$ stage network. It is described in the following theorem.

Theorem 1: In a $2log_2 N$ stage network, at least one redundant switch is presented in each CRS switch set.

Proof: refer [19].

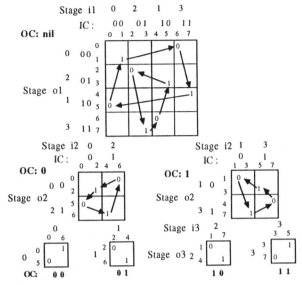

a. relation table

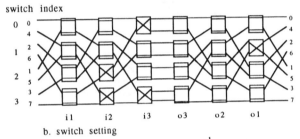

b. switch setting

Figure 3 : Routing on an omega-omega^{-1} network

Algorithm 2: routing on a $(N \times N)$ network using (2×2)
switches
begin
  $queue(P)$;
  while queue not empty **do**
    $dequeue(pptr)$;
    $decompose(pptr, p_0, p_1)$;
    $switchsetting(pptr)$;
    if p_0 decompose **then** $queue(p_0)$;
    if p_1 decomposible **then** $queue(p_1)$;
  endwhile
end.

Algorithm 3: input-output pair decomposition on a $(m \times m)$ sub-permutation P

procedure $decompose(P, P_0, P_1)$;
begin
1: **for** $i = 0$ to m **do** { initialization }
    mark "x" on the relation table for $P(i)$;
  endfor
2: map the switches to each row and column of the relation table;
3: **while** some "x"s remain in the relation table **do** { code assignment }
    select a "x" as the start point w (i.e., select the first "x" in a column);
    $c \leftarrow 0$; { set code 0 }
    $P(w) \leftarrow c$; { assign code to input-output pair w }
    while $inputbuddy(w)$ is unassigned **do**
      $w \leftarrow inputbuddy(w)$; $c \leftarrow \bar{c}$
      $P(w) \leftarrow c$;
      if $outputbuddy(w)$ is unassigned **then**
        $w \leftarrow outputbuddy(w)$; $c \leftarrow \bar{c}$
        $P(w) \leftarrow c$;
      endif;
    endwhile;
  endwhile;
4: **for** $i = 0$ to m **do** { collect P_0 and P_1 }
    if $P(i) = 0$ **then** $P_0(j_0) \leftarrow P(i)$; $j_0 \leftarrow j_0 + 1$;
    if $P(i) = 1$ **then** $P_1(j_1) \leftarrow P(i)$; $j_1 \leftarrow j_1 + 1$;
  endfor
end.

Note that the redundant switch can be any switch of a CRS switch set. These redundant switches are the resources to provide alternate paths for a given permutation requests. We can achieve fault-tolerant routing with the help of these redundant switches when switch faults and link faults are presented in a network.

For a $2log_2N$ stage network, since the center level has $N/2$ switch sets, all the switches in one of the center stage can be regarded as redundant switches. Therefore a $2log_2N - 1$ stage network is a $2log_2N$ stage network without the redundant center stage.

3.2 Subnetwork symmetry

Here, we will show that different mapping on both network tree and permutation tree creates alternate paths for the same permutation. In addition, different mapping also affects the switch setting of the subnetworks. In order to describe the relationship among them, first, we define the switch settings of a network (and its subnetwork) with a binary matrix as below.

Definition 6: The control matrix of a $2log_2N$ stage network is a $(2log_2N) \times (N/2)$ binary matrix and is denoted as $M_{\Phi(n)}$, where $\Phi(n)$ is the network tree which represents this network. Every element in $M_{\Phi(n)}$ represents a switch control bit which is either 0 (through) or 1 (cross). We use a triple-vector to represent the control matrix of a network: $M_{\Phi(n)} = (M_{CRS}, M_{\Phi_0(n-1)}, M_{\Phi_1(n-1)})$, where M_{CRS} is a $2 \times (N/2)$ binary matrix represents the control settings of the switches in its root node $CRS(N,k)$. $M_{\Phi_0(n-1)}$ and $M_{\Phi_1(n-1)}$ are the control matrices of its subnetworks $\Phi_0(n-1)$ and $\Phi_1(n-1)$ respectively.

Above triple-vector defined can also represent the control matrix of a subnetwork $\Phi_a(m)$ by the representation: $M_{\Phi_a(m)} = (M_{CRS_a}, M_{\Phi_{a0}(n-1)}, M_{\Phi_{a1}(n-1)})$, where a is a binary number which represents the index of the subnetwork, and $0 < m < n$.

Theorem 2: Subnetwork symmetry

In a $2log_2N$ stage network $\Phi(n)$, if a permutation request P is realized by a control matrix $M_{\Phi(n)}$, where $M_{\Phi(n)} = (M_{CRS}, M_{\Phi_0(n-1)}, M_{\Phi_1(n-1)})$, then another control matrix $M'_{\Phi(n)}$ also realize P, where

$$M'_{\Phi(n)} = (\overline{M_{CRS}}, M_{\Phi_1(n-1)}, M_{\Phi_0(n-1)})$$

Proof: refer [19].

Theorem 2 can be applied to any subnetwork $\Phi_a(m)$ within $\Phi(n)$. The property of subnetwork symmetry reveals the concept that each subnetwork within a $2log_2N$ stage network is an independent unit. Once a fault is presented in the network, we need only change the routing of the smallest subnetwork which contains this faulty switch. In addition, this subnetwork symmetric property is independent of the permutation request, it only depends on the network configuration.

With the subnetwork symmetric property, a network with n level of subnetworks provides 2^n alternate paths for every input-output pair. We can derive these alternate paths by applying the operation defined below.

Definition 7: The operation of *subnetwork-complement-and-swap*, $NCAS(M_{\Phi_a(m)})$, to a subnetwork $\Phi_a(m)$ is defined as
$$NCAS(M_{\Phi_a(m)}) = NCAS(M_{CRS_a}, M_{\Phi_{a0}(m-1)}, M_{\Phi_{a1}(m-1)})$$
$$= (\overline{M_{CRS_a}}, M_{\Phi_{a1}(m-1)}, M_{\Phi_{a0}(m-1)})$$

Consider the code assignment shown in Figure 4, to perform $NCAS(M_{\Phi(n)})$ is equivalent to replace all the assigned indices by applying a mapping $(0 \rightarrow 1 ; 1 \rightarrow 0)$ to CRS table, the decomposed sub-permutations are the same except that they swap the subnetworks assigned to them. Originally, the decomposed sub-permutations are as follows

$$P_0 = \begin{pmatrix} 0 & 2 & 4 & 6 \\ 5 & 2 & 6 & 0 \end{pmatrix} \quad P_1 = \begin{pmatrix} 1 & 3 & 5 & 7 \\ 1 & 7 & 3 & 4 \end{pmatrix}$$

P_0 and P_1 are assigned to Φ_0 and Φ_1 respectively. After applying $NCAS(M_{\Phi(n)})$, P_0 is assigned to Φ_1 and P_1 to Φ_0. For more specific, to perform $NCAS(M_{\Phi_a(m)})$ generates two effects: First, the switch settings of CRS_a table's corresponding switches (include input and output sides) are complemented. Second, the control settings of its subnetworks Φ_{a0} and Φ_{a1} are swapped.

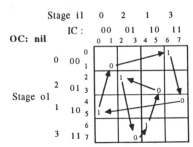

Figure 4: Subnetwork symmetry: complement all codes

3.3 Loop symmetry

Beside subnetwork symmetry, we will explore the symmetric property which depends on permutation requests rather than network configuration.

Opferman and Tsao-Wu [11][12] defined the loop and discussed its properties in their looping algorithm. A loop is defined as follows.

Definition 8: A *loop*, L, is a set of input-output pairs, in which if an input-output pair $\begin{bmatrix} x \\ y \end{bmatrix} \in L$, then its input buddy $\begin{bmatrix} \hat{x} \\ y' \end{bmatrix} \in L$, and also its output buddy $\begin{bmatrix} x' \\ \hat{y} \end{bmatrix} \in L$. Where $\hat{x}$ and $\hat{y}$ are the buddies of x and y respectively, and $x \neq x'$, $y \neq y'$. The number of input-output pairs in a loop is called the *order* of the loop, which is always even.

Lemma 3: A CRS switch set with size $2m$ forms at least one loop and at most m loops.

Proof: Refer [19].

Similar to subnetwork symmetry, the effect to apply a mapping $(0 \rightarrow 1; 1 \rightarrow 0)$ to each loop individually in a multiple-loop relation table will form a different decomposed sub-permutation pair for its lower level, as shown in Figure 5.

Theorem 3: Loop symmetry

In a CRS switch set, if m loops are formed from a permutation P, then 2^{m-1} distinct decomposed sub-permutation pairs can be formed from this switch set.

Proof: refer [19].

Loop symmetry depends on the permutation request given, whereas, it is independent of the network configuration. If two different networks are applied with the same permutation request, they have the same loop property. For a multiple-loop CRS switch set, we can derive alternate paths in a loop by applying the operation defined below.

Definition 9: The operation of *loop-complement-and-swap*, $LCAS(M_{\Phi_{La}(m)})$, to a subset $\Phi_{La}(m)$ of subnetwork $\Phi_a(m)$ is defined as

$$LCAS(M_{\Phi_{La}(m)}) = LCAS(M_{CRS_{La}}, M_{\Phi_{La0}(m-1)}, M_{\Phi_{La1}(m-1)})$$

$$= (\overline{M_{CRS_{La}}}, M_{\Phi_{La1}(i-1)}, M_{\Phi_{La0}(i-1)})$$

Since loop symmetry is similar to network symmetry, we can also apply a redundant switch to a loop as in Theorem 1.

Theorem 4: To realize a permutation P in a $2log_2 N$ stage network, one redundant switch is presented in every loop of every reducible sub-permutation.

Proof: refer [19]. □

Both subnetwork and loop symmetric properties can provide us to enumerate all the possible switch settings for a given permutation. Investigating these possible cases, we can select the best one to tolerate the faults.

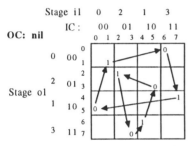

Figure 5: Loop symmetry: complement a loop

4. FAULT-TOLERANT ROUTING

In this section, the fault-tolerant routing scheme for all possible faults in the class of $2log_2 N$ stage networks is described. We consider both switch faults and link faults in the network. Our scheme can provide fault-free routing for switch control faults with multiple faults presented in one stage of the network. We also develop a graceful degradation routing scheme to ensure minimum loss for switch path faults and link faults.

4.1 Fault model

Feng and Wu [5] suggested a fault model for the purpose of fault diagnosis on MIN's. In this model, a (2×2) switch is considered as a (2×2) crossbar circuit. A (2×2) crossbar has four cross-points. Two possible cases (i.e. either connected or not connected) are assumed for each crosspoint, therefore, 16 possible states are derived for a (2×2) switch. These possible states consist of 2 valid states (i.e. "through" connection when control signal is 0; "cross" connection when control signal is 1) and 14 faulty states.

To meet our need for fault-tolerant routing, the faults in a switch are classified as *switch control faults* and *switch path faults*. A switch control fault is a switch configuration fault in which a switch is unable to respond to the control signal. The switch control faults are further classified as *stuck-at-0* and *stuck-at-1* as shown in Figure 6.a. A switch path fault is a transmission path fault in which a switch is unable to transmit the data from its input ports to its output ports. Considering the number of paths which can be passed through a (2×2) switch, switch path faults are further classified as *stuck-one-path* and *stuck-two-path* as shown in Figure 6.b.

Beside the switch faults, link faults may also present

in an interconnection network. A link fault is that the link can not transmit the data, which can be simulated by the stuck-one-path faults on the two stages of switches to which the faulty link connects. It is shown in Figure 6.c. Above classification covers all possible cases in Feng and Wu's fault model and it will simplify our fault-tolerant routing scheme.

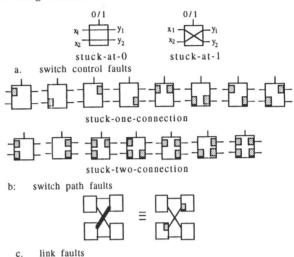

a. switch control faults

b: switch path faults

c. link faults

Figure 6 : Fault model

4.2 Fault-tolerant routing for switch control faults

A switch control fault makes the faulty switch as fixed configuration: stuck-at-0 makes "through" connections; and stuck-at-1 makes "cross" connections. The faulty switch can still transmit the data. No path is stuck when switch control fault occurs. However, undesired permutation will happen when the desired control configuration of the faulty switch conflicts with the stuck configuration. Our goal here is to overcome this situation and develop a fault-free configuration with the presence of switch control faults in the class of $2log_2 N$ stage network.

From Theorem 1, a redundant switch in a CRS switch set can be selected from arbitrary switch in this set. The function of a faulty switch with switch control fault is just the same as what a redundant switch has. Therefore, the idea of our fault-tolerant routing is to select the faulty switch as the redundant switch. To achieve this mechanism, we should modify two places in our routing scheme. First, locate the faulty switch in the network tree. This can be easily determined from the outside-in-code and inside-out-code of the faulty switch: outside-in-code decides the node (i.e., CRS switch set); inside-out-code decides the position within the node. Through switch mapping, we can locate the input-output pairs which pass through the faulty switch. Second, apply fault-free input-output pair decomposition on the node which contains the faulty switch. The code assignment operation is modified so that the switch setting of the faulty switch is fixed and set at the stuck-at-fault. In other words, the input-output pair on the upper port of the faulty switch (i.e., the first entry on the faulty switch) is set as 0 when stuck-at-0 or set as 1 when stuck-at-1.

The procedure to configure fault free switch settings is shown in Algorithms 4 and 5. From Algorithm 4, a faulty switch only affects the routing on the node which contains this faulty switch. In Algorithm 5, the procedure $faultdecompose(P, P_0, P_1)$ is modified from procedure $decompose(P, P_0, P_1)$. Only the modified Step 3 is shown in this algorithm. To adjust the code assignment so that the switch settings do not conflict with the faults, we start the code assignment from the input-output pair on the faulty switch. This modified operation has the same effect as selecting the faulty switch as the redundant switch. This scheme just adjusts the start point of code assignment according to the location of the faulty switch. There is no extra overhead involved, therefore, its time and space complexity is the same as the original routing scheme.

Algorithm 4: fault-tolerant routing on a $(N \times N)$ network using (2×2) switches

```
begin
  queue(P);
  while queue not empty do
    dequeue(pptr);
    if sub-permutation pptr contains faulty switch
      then faultdecompose(pptr, p_0, p_1);
      else decompose(pptr, p_0, p_1);
    endif;
    switchsetting(pptr);
    if p_0 decomposable then queue(p_0);
    if p_1 decomposable then queue(p_1);
  endwhile
end.
```

For example, fault-free routing for stuck-at-1 faults located at $S_{i1,3}$, $S_{i2,1}$, and $S_{o2,0}$ is shown in Figure 7. This algorithm can tolerate one switch control fault in a loop. From Lemma 3, a CRS switch set with size $2m$ can tolerate at least one fault and at most m faults. If we consider its fault-tolerant capability independent of the permutation request, the number of faults can be tolerated is equal to the number of CRS switch sets in the network if only the faults are evenly distributed to every CRS switch set in the network. There are $N-1$ CRS switch sets for an $(N \times N)$ network, therefore, $N-1$ switch control faults can be tolerated in the network.

Algorithm 5: procedure $faultdecompose(P, P_0, P_1)$

```
3:  while some "x"s remain in the relation table do { code assignment }
      if unassigned faulty switch in the relation table
        then select the first "x" of the faulty switch as w;
          if stuck-at-0 then c ← 0;
          if stuck-at-1 then c ← 1;
        else select the first "x" in the column as w;
          c ← 0; { set code 0 }
      endif
      P(w) ← c; { assign code to input-output pair w }
      while inputbuddy(w) is unassigned do
        w ← inputbuddy(w); c ← c̄
        P(w) ← c;
        if outputbuddy(w) is unassigned then
          w ← outputbuddy(w); c ← c̄
          P(w) ← c;
        endif
      endwhile
    endwhile;
```

4.2.1 Fault-free routing for $2log_2 N - 1$ stage networks: center stage fault-tolerance

Algorithms 4 and 5 use the redundant switches in the network to tolerate the switch control fault. This scheme provides the maximum fault-tolerant capacity on the center stages of the $2log_2 N$ stage network, which can tolerate $N/2$ faults on the center stages. However, a $2log_2 N - 1$ stage network, such as the Benes network, does not have the redundant center stage. Thus, this scheme can not provide any fault-tolerant capability to the fault in its single center stage. We develop a different procedure to tolerate the fault located in the center stage of a $2log_2 N - 1$ stage network. This procedure use $NCAS$ operation to relocate the switch settings when a conflict occurred on the faulty switch. In Algorithm 6, the switches on the center stage are numbered as a $n - 1$ bits binary code from

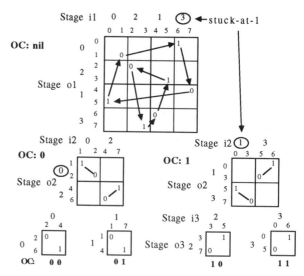

a. relation table

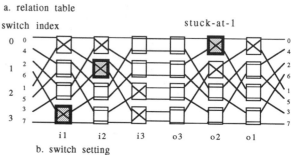

b. switch setting

Figure 7: Fault-free routing for switch control faults

$00 \ldots 0$ to $11 \ldots 1$ with the switch on top as $00 \ldots 0$. Assume the faulty switch is located at $F[1 : n-1]$. A binary operation $\oplus$ is used in the algorithm, which is bit-wise exclusive-OR on two binary numbers.

Algorithm 6: fault-tolerant routing on the center stage of a $2log_2 N - 1$ stage network

```
begin
  derive the switch setting of whole network by original routing scheme;
  if the faulty switch, F[1 : n−1], conflicts with its control setting then
    select a substitute switch, S[1 : n−1], in the center stage;
    D[1 : n−1] = F[1 : n−1] ⊕ S[1 : n−1];
    for i = 1 to n−1 do
      if D[i] = 1 then NCAS(M_{Φ_{S[1:i-1]}(n-i+1)});
    endfor
  endif
end.
```

For the switch setting in Figure 3.b (i.e., eliminate the redundant stage o3), if the faulty switch is stuck-at-0 at $S_{3,0}$, then $F[1 : n - 1]$ is 00. The candidates for the substitute switch are $S_{3,1}$ and $S_{3,2}$. If we select the nearest to the faulty switch as the substitute switch, then the number of switches to be relocated is minimum. It is equivalent to find the candidate switch with the minimum distance code to the faulty switch. $S_{3,1}$ is selected as the substitute switch by this principle, as a result, $S[1 : n - 1]$ is 01. The distance code is $01 = 00 \oplus 01$. Thus, we should perform $NCAS(M_{\Phi_0(1)})$. The result fault-free control setting is shown in Figure 8.

The shortage of this algorithm is that it fails when we cannot find a substitute switch on the center stage from the initial switch setting. This situation happens when the switch settings of all switches in the center stage are the same and also conflict with the fault.

For the faults in non-center stages of a $2log_2N - 1$ stage network, we can use the procedure as described in previous section to derive the fault-free routing.

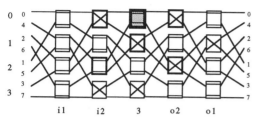

Figure 8: Fault-free routing for fault on center stage

4.3 Fault-tolerant routing for switch path faults

Some input-output pairs will be stuck if a switch path fault happens in the network. Therefore, it is impossible to provide fault-free routing for switch path faults and link faults. However, for weighted connection requests, we can find the best switch settings such that the network can achieve minimum-loss configuration. We define a set of weighted connection as below:

Definition 10: A set of weighted connection requests assigned to a network is described as a weighted permutation:

$$P = \begin{pmatrix} x_1 & x_2 & \dots & x_N \\ y_1 & y_2 & \dots & y_N \\ w(x_1) & w(x_2) & \dots & w(x_N) \end{pmatrix}$$

where $w(x_i)$ is a positive value represents the weight assigned to the input-output pair $\begin{bmatrix} x_i \\ y_i \end{bmatrix}$.

For example, a weighted permutation is shown below

$$P = \begin{pmatrix} 0 & 1 & 2 & 3 & 4 & 5 & 6 & 7 \\ 5 & 1 & 2 & 7 & 6 & 3 & 0 & 4 \\ 5 & 7 & 3 & 1 & 2 & 4 & 6 & 8 \end{pmatrix}$$

In this weighted permutation, a larger weight indicates the higher priority for that input-output pair.

In order to investigate the input-output pairs which are affected by switch path faults, we define the fault projection set as below:

Definition 11: A *fault projection set* of a faulty switch is the set of input-output pairs which are possible to pass through the stuck path of this switch. If the faulty switch is located in input half, the fault is projected to input terminals; and vice versa. Assume the inside-out-code of a faulty switch $S_{a,j}$ is denoted as $IC[1:m]$, where $1 \leq m \leq n-1$. The fault projection set of this faulty switch is denoted as: $FPS(S_{ia,j}) = \{ \begin{pmatrix} x \\ y \end{pmatrix} \mid \forall \ x[1:n]$ such that $x[1:m] = IC[1:m]\}$ or $FPS(S_{oa,j}) = \{ \begin{pmatrix} x \\ y \end{pmatrix} \mid \forall \ y[1:n]$ such that $y[1:m] = IC[1:m]\}$

Above definition consider a faulty switch as an unit, thus, it can only describe the fault projection set of a stuck-two-path fault. To describe the fault projection set of a stuck-one-path fault, we can project the faulty spot to a switch in its outer level and derive the fault projection set of this switch. Figure 9 shows the fault projection set of both stuck-one-path and stuck-two-path faults. On the input half, a stuck-two-path fault is located at switch $S_{i2,2}$. $S_{i2,2}$ has inside-out-code $IC[1] = 1$ and outside-in-code $OC[1] = 0$. Its fault projection set is $1 \times \times$, which

covers input terminals $4 \sim 7$. And on the output half, a stuck-one-path is located at switch $S_{o2,1}$. Its projected switch on the outer level is $S_{o1,0}$ which has inside-out-code $IC[1:2] = 00$, therefore, its fault projection set is $00\times$, which covers output terminals $0 \sim 1$.

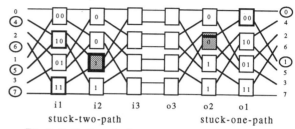

Figure 9: Fault projection set

4.3.1 Routing with the minimum-loss configuration

The criteria to tolerate the switch path faults is to reduce the total weight passing through the stuck paths to minimum. Different scheme should be considered for stuck-one-path and stuck-two-path faults. They are described in Algorithms 7 and 8 respectively.

Algorithm 7: fault-tolerant routing for stuck-one-path faults
begin
  derive the permutation tree;
  derive fault projection set $FPS(one\ path\ of\ S_{a,j})$;
  stuck_path ← *minimum weight in FPS(one path of $S_{a,j}$)*;
  locate the permutation node which contains *stuck_path*;
  map this node to the network node contains the faulty switch;
  map other nodes accordingly;
  determine the switch settings of all nodes;
end.

Algorithm 8: fault-tolerant routing for stuck-two-path faults
begin
  derive fault projection set $FPS(S_{a,j})$;
  while *tree_level* $\leq a$ **do**
    decompose(pptr);
    for all loops covered by $FPS(S_{a,j})$ **do**
      apply $LCAS$ on node pptr to form all sub-permutation pairs $\{P_0, P_1\}$;
    endfor
    if faulty switch $S_{a,j}$ in sub-permutation pptr **then**
      locate *stuck_path*;
      calculate *weight(stuck_path)*;
      candidate_set ← *candidate_set* ∪ *stuck_path*
    endif
  endwhile
  stuck_pairs ← *minimum weight in candidate_set*;
  locate the permutation node which contains *stuck_pair*;
  map this node to the network node contains the faulty switch;
  map other nodes accordingly;
  eliminate the unneeded permutation nodes;
  derive the remaining levels of the permutation tree;
  determine the switch settings of all nodes;
end.

In Algorithms 7 and 8, instead of setting the switches of each network node immediately after its corresponding permutation node is decomposed, their switch setting is determined after the whole permutation tree is formed. This is because the mapping between these two trees can not be determined before the node containing the faulty switch is reached. Since the permutation node involved in Algorithm 7 is the same as the original routing algorithm, the extra overhead is small without increasing the time complexity of the routing scheme. Algorithm 8 is a brute force scheme, which evaluates all the possible branches affected by the fault projection set. However, by enumerating all the possible cases covered by the fault projection set, we can always find the minimum-loss configuration when a stuck-two-path fault is presented in the network. The extra overhead involved in Algorithm 8 depends on the location of the faulty switch and the number of loops presented in its upper levels. The more loops and the inner the faulty switch located, the more extra overhead is needed to configure the minimum-loss routing.

5. CONCLUSIONS

The alternate paths and symmetric properties on the class of $2log_2N$ stage rearrangeable networks are introduced. Based on these properties, a general fault-tolerant routing scheme for both switch and link faults for every network in the class is proposed. For switch control faults, a fault-free routing scheme is devised. With the same time and space complexity as the original routing algorithm, the proposed algorithm can tolerate multiple faults on each stage of the network. A fault-free routing scheme for the fault in the center stage of a $2log_2N - 1$ stage network is also developed. For switch path faults and link faults, a minimum-loss routing scheme is introduced. Our scheme considers stuck-one-path and stuck-two-path faults separately. The proposed algorithms select the best configuration among all possible cases so that the loss due to stuck paths is minimal. The extra overhead for this graceful degradation scheme depends on the location of the stuck path. It has more flexibility for the the stuck paths in the inner levels of the network, however, acquiring more extra overhead to derive minimum-loss configuration.

REFERENCES

[1] G.B. Adams, D.P. Agrawal and H.J. Siegel, "A survey and comparison of fault-tolerant multistage interconnection networks," *IEEE Computer Magazine*, June 1987, pp. 14-27.

[2] G.B. Adams, and H.J. Siegel, "The extra stage cube: a fault-tolerant interconnection network for supercomputer," *IEEE Trans. Computers*, vol. C-31, no. 5, May 1982, pp. 443-454.

[3] D.P. Agrawal, "Testing and fault tolerance of multistage interconnection networks," *IEEE Computer*, Apr. 1982, pp. 41-53.

[4] Tse-yun Feng, "A survey of interconnection networks," *IEEE Computer*, Vol. 14, Dec. 1981, pp. 12-27,

[5] T.Y. Feng and C.W. Wu, "Fault diagnosis for a class of multistage interconnection networks," *IEEE Trans. Computers*, vol. C-30, no. 10, Oct. 1981, pp. 743-758.

[6] T.Y. Feng and W. Young, "Fault diagnosis for a class of rearrangeable networks," *J. Paral. Distr. Comp.*, vol. 3, 1986, pp. 23-47.

[7] Shing-tsaan Huang and Satish K. Tripathi, "Finite state model and compatibility theorey: new analysis tools for permutation networks," *IEEE Trans. Computers*, vol. C-35, no. 7, July 1986, pp. 591-601.

[8] Shing-tsaan Huang and Chin-hsiang Tung, "On fault-tolerant routing of Benes network," *Journal of Information Science and Engineering*, vol. 4, July 1988, pp. 1-13.

[9] V.P. Kumar and S.M. Reddy, "Augmented shuffle-exchange multistage interconnection networks," *IEEE Computer Magazine*, June 1987, pp. 30-40.

[10] David Nassimi, "A Fault-tolerant routing algorithm for BPC permutation on multistage interconnection networks," *Proceedings, 1989 International Conference on Parallel Processing*, 1989, pp. I.278-I.287.

[11] D.C. Opferman and N.T. Tsao-Wu, "On a class of rearrangeable switching networks-part I: control algorithm," *Bell System Technical Journal*, Vol. 50, No.5, 1971, pp. 1579-1600.

[12] D.C. Opferman and N.T. Tsao-Wu, "On a class of rearrangeable switching networks-part II: enumeration studies and fault diagnosis," *Bell System Technical Journal*, Vol. 50, No.5, 1971, pp. 1601-1618.

[13] C.S. Raghavendra and A. Varma, "Fault-tolerant multiprocessors with redundant-path interconnection networks," *IEEE Trans. Computers*, vol. C-35, no. 4, April 1986, pp. 307-316.

[14] John Paul Shen and John P. Hayes, "Fault-tolerant of dynamic-full-access interconnection networks," *IEEE Trans. Computers*, vol. C-33, no. 3, March 1984, pp. 241-248.

[15] A. Varma and C.S. Raghavendra, " Fault-tolerant routing in multistage interconnection networks," *IEEE Trans. Computers*, vol. C-38, no. 3, March 1989, pp. 385-393.

[16] A. Waksman, "A permutation network," *J. Ass. Comput. Mach.*, vol. 15, Jan. 1968, pp. 159-163.

[17] S. Wysham and Tse-yun Feng, " On routing a faulty Benes network," *Proceedings, 1986 International Conference on Parallel Processing*, vol. I, 1990, pp. 351-354.

[18] Yao-ming Yeh and Tse-yun Feng, " On a class of rearrangeable networks," Submitted to *IEEE Trans. Computers*.

[19] Yao-ming Yeh, " Fault-tolerance and generalization On a class of multistage interconnection networks," Ph. D. Dissertation, Dept. Electrical & Computer Engineering, The Pennsylvania State University, Aug. 1991.

Fault Side-Effects in Fault Tolerant Multistage Interconnection Networks

T. Schwederski   E. Bernath   G. Roos
schwederski@mikroelektronik.uni-stuttgart.dbp.de
Institute for Microelectronics Stuttgart
Allmandring 30
W-7000 Stuttgart 80
F.R. Germany

W. G. Nation   H. J. Siegel
hj@ecn.purdue.edu
Parallel Processing Laboratory
School of Electrical Engineering
Purdue University
West Lafayette, IN 47907-0501, USA

Abstract

Many fault tolerant (FT) multistage interconnection networks have been devised that tolerate one or more faults. It is shown that the effects of some realistic faults can propagate through multiple stages and cause non-faulty components to behave erroneously; such faults with *side effects* can cause "FT" networks to lose their fault tolerance capabilities even due to a single fault. The PASM prototype interconnection network is examined as a case study of realistic faults with side effects. A fault model that includes side effects is introduced. An enhanced Extra Stage Cube network is proposed that tolerates faults with or without side effects.

1.   Introduction

Design of fault tolerant networks for parallel machines has been widely considered. The *multistage interconnection networks* are one important class, and many fault tolerant networks of this class have been devised [1]. Generally, multistage interconnection networks consist of two or more *stages* of *switches* that connect sources and destinations, e.g., [2,3]. *Fault tolerant (FT) networks* enhance the basic structure by adding redundancy to bypass faulty links and/or switches e.g., [1,4-7].

The fault models that are employed in these networks either assume that a faulty component cannot be used at all, or they assume a partial failure (e.g., a switch sends the correct data to the wrong destination) [1]. In all of these fault models, the faults are local and have no effect on non-faulty links or stages. For many realistic failures, this assumption is not valid. For example, in a study on fault identification of non-fault-tolerant multistage cube networks, Davis et al. [8] introduce a more general fault model that also includes stuck paths that might span the network. Thus, a single fault may have *fault side effects (FSEs)* on network components without hardware faults.

In many cases, the impact of such side effects will be detrimental. Stuck paths, for example, may prohibit processors from sending to or receiving from the network. This will be shown in a study of an actual FT network with realistic faults. To examine FSEs, a general model is presented, and the properties of FSE-tolerant networks are discussed. FT networks may fail in the presence of FSEs [9]; they can be modified by adding appropriate hardware such that they are FSE-tolerant. An enhanced version of the Extra Stage Cube FT network is proposed that achieves FSE-tolerance with little added hardware complexity. To illustrate the required overhead, the enhancements are discussed in the context of an actual prototype interconnection network.

This work was sponsored in part by the EUREKA research program PROMETHEUS, subprogram PROCHIP, by the Bundesministerium für Forschung und Technologie under contract TV 8926.3, and by the Office of Naval Research under grant number N00014-90-J-1483.

2.   Case Study of the PASM ESC Network

The processing elements of the protoype of the PASM (Partitionable SIMD/MIMD machine) [10] parallel processing system system are connected by an Extra Stage Cube (*ESC*) network with 16 inputs and 16 outputs [1,11]. The ESC is designed to tolerate any single fault and is robust in the presence of multiple faults. Figure 1 shows an ESC network for N=8. The PASM prototype network is constructed from 2-by-2 interchange boxes and bypass switches; it is circuit switched, has a data path width of 16 bits with two parity bits, and is capable of broadcasts. Paths through the network are established and released by a path request (*PRQ*) / path grant (*PGRT*) mechanism. Sources and destinations will be referred to as PEs.

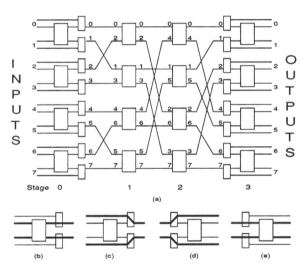

Figure 1: (a) Extra Stage Cube Network for N=8; (b) Input stage enabled; (c) Input stage disabled; (c) Output stage disabled; (e) Output stage enabled

Consider a stuck path. If any PRQ signal becomes stuck while in the asserted state, the path can no longer be dropped. Thus, at least one destination becomes permanently unreachable. Consider why the ESC cannot tolerate such a fault. During normal operation, the input stage is enabled and the output stage is disabled. To clarify the discussion, the ESC from Figure 1 is used as an example. Let the top box of stage 2 be faulty in such a way that it has established a path from its upper output (link 0) to network output 0, and can no longer drop this path. This is indicated in Figure 2a. While the output stage is enabled, PE 0 cannot be reached because the stuck path blocks all data transfers to PE 0 from both the upper and the lower input of the output stage box. Disabling the output stage does not alleviate the problem. When the output stage is disabled, PE

Figure 2: (a) Stuck path to PE 0 with bypass disabled; (b) Stuck path from PE 0 with bypass enabled

0 can only be reached from the top output of the faulty box that is blocked due to the stuck path (Figure 2b). Thus, PE 0 cannot be reached due to a single fault, and this single fault will therefore cause the ESC to lose its FT capabilities.

Stuck paths are an example of forward fault propagation, i.e., from the fault in the direction of the network outputs. Consider an example of backward propagation (i.e., from fault to network inputs). Assume that a PGRT signal is stuck at link K. Then, any path request that uses link K will immediately receive a PGRT, whether or not the path can be established to the destination. As a consequence, the source will change the data on the output link from the routing tag to the first data item. Because the path will, in general, not yet have been routed to the destination, an incorrect data item will be used for routing, an incorrect destination will be reached, and data will be lost.

3.   FSE Network Models and Properties
3.1   Notation and Definitions

The previous discussion has shown that it is insufficient to model faults as localized entities because faults can propagate. The fault models that are commonly used for the design and analysis of FT networks must therefore be modified. To complement the fault model that describes the physical failures in the system, a *fault propagation (FP) model* is introduced.

The fault model provides an abstraction of a physical failure. A standard model is used that assumes that any network component can deviate from its specified behavior due to a single static hardware failure; transient faults are not considered. Examples of faults that fit this model are stuck-at-faults, open connections, etc. The immediate effect of such a fault could be data and control corruption or non-usability of a component (e.g., a link is broken and no data can pass the link). This could manifest itself in data not being transmitted, incorrect data being transmitted, data being transmitted to an incorrect destination, and spurious data generation. Spurious data generation would be especially problematic in packet-switched networks because it could cause portions of the network to saturate due to hot spot contention [12].

To simplify the following discussion, link faults are treated as switch faults, similar to the approach in [13]. Link faults need therefore not be considered because they can be accounted for by switch faults.

Now consider the FP model. To develop such a model, FSEs are classified; the terminology is defined first.

One of the most important properties of FSEs is their *reach*, i.e., to what extent a fault propagates through the network. Definition 1 formalizes this concept.

Definition 1:   Let a fault occur at a switch at stage K. Let the fault have effects at a switch at stage F in the direction of the destination and at a stage G in the direction of the source. The *forward reach* R_F of an FSE is then given by $|F-K|$, and the *backward reach* R_B is given by $|K-G|$.

As shown in the case study in Section 2, FSEs may influence all network stages, and the maximum value of R_F or R_B may span all of the network (i.e., be M−1 for an M-stage network). The hardware implementation of switches, links, and protocols has a significant impact on R_F and R_B. One primary design goal of a network that is to tolerate FSEs is to minimize the maximum values of R_F and R_B, ideally to 0 (no side effects).

A second factor that determines the consequences of FSEs is their span, i.e., the number of switches at each stage that are influenced by the fault, as introduced in Definition 2.

Definition 2:   The *span* S_L of an FSE at stage L is given by the number of switches at stage L that are affected by the fault.

Many FSEs will be restricted to a single forward or backward path as discussed in Section 2. For these, the span is one for all stages. A stuck broadcast path originating at the input stage could span the complete network at the output stage. Clearly, the span of FSEs should also be kept to a minimum.

If a network is to tolerate FSEs, the propagation of faults must be stopped by some appropriate hardware. The required properties of such hardware are given in Definition 3:

Definition 3:   An *isolation point* is a network component that is constructed such that it passes data and control signals freely if set to the operating state, and stops data, control, and fault propagation if set to the disconnecting state.

Isolation points can be located inside the interconnection network, where they might separate links between switches, or they can be located at sources and destinations, where they stop faults from affecting processors or memories. Isolation points may be constructed from hardware devices (e.g., buffers that can be disabled), or they may be conceptual only. In the DR network [14], for example, spare PEs are provided. By not using spare PEs, these are effectively acting as isolation points, so that such a PE can be modelled as being separated from the network by an isolation point, even though such a separation does not exist in the physical sense. For simplification, explicit isolation points will be used throughout the discussion.

Because isolation points may be hardware devices that are set by some control logic, faults of isolation points must be considered. It is assumed that both isolation points and their control lines may fail. Thus, an isolation point can stop fault propagation only if the isolation point and the control logic are fault free, and the control logic is not affected by fault propagation.

3.2   Fault Side-Effect Model

To determine the effects of fault propagation, appropriate models must be developed. These may be very general and permit propagation with large reach and span, or may be restricted. In this paper, a rather general propagation model is assumed. Other fault propagation models are discussed in [9].

Fault Propagation (FP) Model: A fault has side effects that have maximum reach and maximum span, both in forward and backward directions, along all paths that can be reached from the fault until a source, a destination, or a disconnecting isolation point is encountered. It is assumed that all elements that may be affected by faults do not operate correctly.

Most failures will not have such significant effects, but this model is well suited to illustrate the cababilities of an enhanced ESC to tolerate FSEs.

3.3. Properties of FSE-Tolerant Networks

Before a network can be examined for its capability to tolerate FSEs, the meaning of such a property must be defined.

Definition 4: Let an FT interconnection network maintain property P in the presence of a single fault without side effects. This property could be full access, dynamic full access, etc. [1]. If the network maintains property P under the assumption that faults propagate according to fault propagation model FP, it can tolerate FSEs under the FP propagation model.

In the presence of a fault, a network can always be modelled such that the network is partitioned into a faulty network portion *FNP* that contains the fault and all its propagation paths, and the operational, fault-free portion *ONP*. In the best possible case, FNP would contain only the faulty switch. Then, no fault propagation occurs, and conventional fault modelling suffices. In the worst case, the network could contain, for example, all network output ports so that no messages could be passed through the network (e.g., stuck broadcast path at the input stage of a network without isolation points).

Because FNP contains all fault propagation paths, and fault propagation stops at isolation points only, all connections between ONP and FNP are by definition separated by isolation points. Paths from FNP and ONP can reach sources and destinations, and these connections may be isolation points or not. Theorem 1 states under what conditions a network can tolerate FSEs.

Theorem 1: A fault-tolerant interconnection network with fault-tolerance property P can tolerate FSEs under the FP model if and only if the fault-free network portion ONP provides property P, and all paths connecting FNP to sources or destinations are separated by disconnecting isolation points (see Figure 3).

Proof:

If: Because all fault propagation paths entering or leaving FNP are isolated, ONP is not affected by faults, and no operational PE is affected by a fault. Because ONP provides property P, the network can tolerate FSEs under the FP model. q.e.d.
Only-If: (a) If the fault-free portion ONP does not provide property P, the network loses the fault tolerance capabilities for which it was designed.
(b) Recall from Section 3.1 that all spare PEs are considered to be isolated from the network. It is assumed that all used PEs are required for property P. If any path that connects FNP to sources or destinations is not isolated, faults will propagate from FNP to a used PE, and this affected PE will malfunction. Because all PEs not needed for property P are isolated, at least one PE required for property P malfunctions. Thus, property P is not maintained. q.e.d.

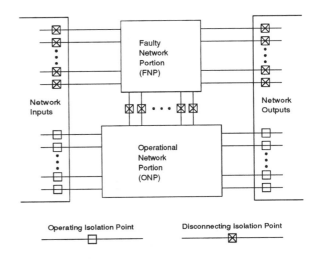

Figure 3: Conceptual view of network partitioning required for fault propagation tolerance

4. Enhancing the Extra Stage Cube Network

In this section, enhancements to the Extra Stage Cube network are discussed to illustrate techniques that can be employed to achieve FSE tolerance. Minor modifications to the ESC are suggested that take advantage of its inherent architectural properties. Let the Enhanced ESC (*EESC*) have the same basic structure as the ESC (i.e., each PE has two links into the network as shown in Figure 1) with the following differences.
(a) All PEs can disconnect themselves from any bypass or interchange box input/output, independently of the bypass state. Such an isolation can be accomplished, for example, by disabling appropriate tristate buffers or ignoring port values.
(b) An enabled bypass isolates the bypass-to-interchange box connection, and a disabled bypass isolates the bypass-to-PE connection.
(c) The output K of an input stage interchange box and the input K of an output stage interchange box can operate as isolation points controlled by PE K.
(d) The bypass and stage enable-lines can operate as disconnecting isolation points so that a fault in a bypass can be stopped from propagating into the PEs.
The model of a 2-by-2 input stage interchange box is depicted in Figure 4. All isolation points required by the above modifications are indicated in the figure. The necessary control lines from PEs to bypasses and interchange boxes are also shown. In Section 5, simple hardware will be proposed that fits these requirements.

Theorem 2: The EESC can tolerate FSEs under the FP model.

Proof: Link faults are treated as PE, interchange box, or bypass faults. PE failures need not be considered because the ESC is not designed to operate under such faults. Thus, only interchange box and bypass failures must be discussed. Five cases must be distinguished: (1) faults in input stage interchange boxes, (2) faults in output stage interchange boxes, (3) faults in interior interchange boxes, (4) faults in input stage bypasses, and (5) faults in output stage bypasses.

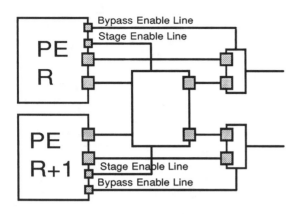

Figure 4: Enhanced ESC input stage and bypass circuits. Shaded squares are isolation points

(1) Faulty input stage interchange box (Figure 5a). By enabling the input stage bypass and isolating the PE outputs that connect to the input stage interchange box, the fault cannot propagate to PEs or into the network. Modifications (a) and (b) are utilized. With respect to Theorem 1, FNP contains the faulty interchange box, and ONP contains the remaining network. All connections between ONP and FNP, and to the PEs connected to FNP, are isolated. ONP provides property P because the fault with propagation paths is equivalent to a single input stage box fault under which the ESC assumptions are valid.

(2) Output stage interchange box fault. Similar to (1).

(3) Faulty interior interchange box (Figure 5b). Without loss of generality, assume that the fault occurred at an interchange box with even numbered inputs. By enabling the input stage and the output stage, and isolating all even numbered input stage outputs and all even numbered output stage inputs, all even numbered interior boxes are disconnected (upper middle rectangle labelled "E" in Figure 5c). Enhancements (b) and (c) are required. Thus, FNP contains all even numbered interior boxes and all even numbered PE bypasses. ONP contains the input stage interchange boxes, the output stage interchange boxes, odd numbered PE bypass circuits, and all interior interchange boxes with odd numbered input links. FNP is isolated from ONP and from the PEs by construction. ONP provides property P because a PE can first send its data to the odd numbered input stage interchange box output, and then route to its destination through the rest of the network. (This is analogous to the way the original ESC avoids a single non-propagating interior interchange box fault.) Thus, the complete path stays within ONP, and Theorem 1 holds true.

(4) Faulty input stage bypass (Figure 5c). Assume that the fault occurs at an even-numbered bypass. Then, the fault is isolated by enabling input and output stages, isolating the bypass connections of all even-numbered PEs (data and enable lines), isolating all even-numbered input stage interchange box outputs, and isolating all even-numbered output stage interchange box inputs. Clearly, the partitioning in this case is identical to that in the case of an interior fault as illustrated by Figure 5c. The only difference is the need to isolate the PE outputs that connect to the even-numbered bypasses. Thus, the same proof as for case (3) applies.

(5) Faulty output stage bypass. Similar to (4). q.e.d.

In contrast to the original ESC fault assumptions, the EESC under the FP model is no longer robust in the presence of multiple faults. As a simple example, consider two faulty interior boxes, one with even, one with odd link numbers. In this case, all even numbered interior boxes belong to FNP because of the first fault, and all odd ones belong to FNP due to the second fault. Thus, no interior interchange box is left in ONP, and thus no paths through ONP are possible.

The total available network bandwidth as compared to the original ESC assumptions is reduced. This is due to the necessity of utilizing only half of the network in case of a single error, which halves the available bandwidth. In the original ESC, only those paths that would have to use the faulty interchange are rerouted, so that only a small loss in bandwidth occurs due to a single fault.

5. EESC Implementation in the PASM Prototype

To illustrate the hardware overhead that is required to enhance the ESC, the PASM prototype interconnection network is examined as an example to demonstrate one approach. The enhancements discussed in Section 4 do not require modification of interior boxes. Only the bypasses, input and output stages, and the PE interfaces need to be discussed. Consider the

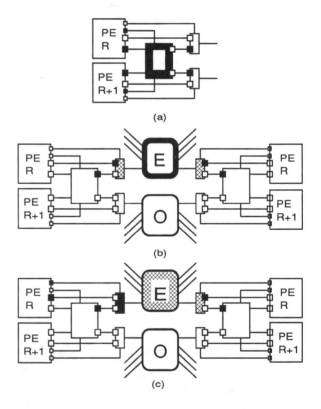

Figure 5: (a) Isolating an input stage interchange box fault; (b) Isolating an interior stage interchange box fault; (c) Isolating an input stage bypass fault. Rectangles labelled "E" represent the subnetwork of all interior interchnage boxes whose links have even numbers. "O" is the same for those with odd numbers

input stage and input bypass structure as shown in Figure 5. In the input stage, four sets of buffers (drivers) (B_UU, B_LU, B_UL, and B_LL) implement the required 2-by-2 crossbar switch. By providing two box enable lines UP_En and LW_En that connect to the box control logic, the buffers can be disabled so that they serve as isolation points; this feature is required to isolate the input stage interchange box from bypass failures. Each bypass contains two buffers that can be enabled through a bypass enable line. Thus, input stage faults can be isolated, and all requirements that are needed to make the ESC FSE-tolerant are met.

This hardware structure serves to illustrate two points. First, little overhead is sufficient in many cases to avoid the fault propagation problem; the hardware of Figure 6 has negligible overhead as compared to a solution that does not consider fault propagation. Secondly, it shows that fault propagation must be carefully considered during the design phase. For example, if no fault propagation were considered, the bypass buffers BYBU and BYBL could be merged with the input stage box buffers. While this does not affect functionality, the input stage would lose its capability to tolerate fault propagation [9].

6. Conclusion

The concept of fault side effects was introduced. It was shown that these side effects may render "fault tolerant" networks inoperable, even as a result of a single fault. A fault model that includes fault side effects was presented and properties of FSE-tolerant networks were described. The capability of some networks for handling fault side effects can be enhanced significantly with little hardware overhead. It is important during the network design and implementation to consider features required to eliminate fault propagation. The model and implementation example given here will aid the network architects and designers to judge the fault tolerance of a particular network in its capabilities in the presence of fault side effects.

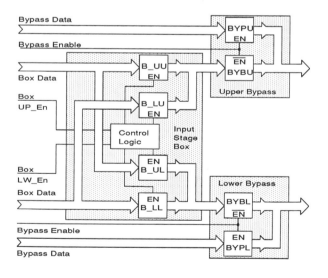

Figure 6: Implementation detail of an enhanced ESC input stage and bypass in the PASM prototype

References

[1] G. B. Adams III, D. P. Agrawal, and H. J. Siegel, "A survey and comparison of fault-tolerant multistage interconnection networks," *Computer*, Vol. 20, No. 6, June 1987, pp. 14-27.

[2] G. J. Lipovski and M. Malek, *Parallel Computing: Theory and Comparisons*, John Wiley and Sons, Inc., New York, NY, 1987.

[3] H. J. Siegel, W. G. Nation, C. P. Kruskal, and L. M. Napolitano, "Using the multistage cube network topology in parallel supercomputers," *Proceedings of the IEEE*, Vol. 77, No. 12, December 1989, pp. 1932-1953.

[4] K. Padmanabhan and D. H. Lawrie, "A class of redundant path multistage interconnection networks," *IEEE Transactions on Computers*, Vol. C-32, No. 12, December 1983, pp. 1099-1108.

[5] D. S. Parker and C. S. Raghavendra, "The gamma network," *IEEE Transactions on Computers*, Vol. C-33, No. 4, April 1984, pp. 367-373.

[6] S. M. Reddy and V. P. Kumar, "On fault-tolerant multistage interconnection networks," *1984 International Conference on Parallel Processing*, August 1984, pp. 155-164.

[7] J. P. Shen and J. P. Hayes, "Fault-tolerance of dynamic-full-access interconnection networks," *IEEE Transactions on Computers*, Vol. C-33, No. 3, March 1984, pp. 241-248.

[8] N. J. Davis IV, W. T.-Y. Hsu, and H. J. Siegel, "Fault location techniques for distributed control interconnection networks," *IEEE Transactions on Computers*, Vol. C-34, No. 10, October 1985, pp. 902-910.

[9] T. Schwederski, E. Bernath, G. Roos, W. G. Nation, and H. J. Siegel, *An analysis of fault side-effects in fault tolerant multistage interconnection networks*, IMS Technical Report TB01-91, May 1991.

[10] H. J. Siegel, T. Schwederski, J. T. Kuehn, and N. J. Davis IV, "An overview of the PASM parallel processing system" in *Computer Architecture*, D. D. Gajski, V. M. Milutinovic, H. J. Siegel, and B. P. Furht, eds., IEEE Computer Society Press, Washington, DC, 1987, pp. 387-407.

[11] G. B. Adams III and H. J. Siegel, "The extra stage cube: a fault-tolerant interconnection network for supersystems," *IEEE Transactions on Computers*, Vol. C-31, No. 5, May 1982, pp. 443-454.

[12] G. F. Pfister and V. A. Norton, "'Hot spot' contention and combining in multistage interconnection networks," *IEEE Transactions on Computers*, Vol. C-34, No. 10, October 1985, pp. 933-938.

[13] A. Varma and C. S. Raghavendra, "Reliability analysis of redundant-path interconnection networks," *IEEE Transactions on Reliability*, Vol. 38, April 1989, pp. 130-137.

[14] M. Jeng and H. J. Siegel, "Design and analysis of dynamic redundancy networks," *IEEE Transactions on Computers*, Vol. C-37, No. 9, September 1988, pp. 1019-1029.

NEW BOUNDS ON THE RELIABILITY
OF TWO AUGMENTED SHUFFLE-EXCHANGE NETWORKS

Bernard Menezes   *Umesh Bakhru*   *Randolph Sergent*

Dept. of Elec. Eng. and Inst. for Advanced Computer Studies

University of Maryland

College Park, MD 20742, U.S.A.

Abstract: This paper examines the reliability resulting from two forms of redundancy, spatial and temporal, in multistage interconnection networks. An example of the former is the extra-stage shuffle-exchange network (SEN+) and of the latter is a SEN with wraparound connections (i.e. each output is connected back to its corresponding input.) We first develop upper and lower bounds on reliability of the SEN+ which are an improvement over existing ones. The new bounds are developed by examining which switches in the SEN+ are affected by failed switches, and which switches must remain operational in order to guarantee full access (i.e. connection between every input (processor) and output (memory) pair). We then examine the SEN with wrap-around connections. This provides full access under a variety of fault conditions though communication between some processor-memory pairs may require several passes through the network. Upper and lower bounds on reliability are obtained by deriving necessary and sufficient conditions for full access.

1. Introduction

An important class of networks proposed for connecting multiple processing elements (PE's) and multiple memory elements (ME's) is the multistage shuffle-exchange network (SEN). This has $N = f^n$ inputs and N outputs and is made up of $n = log_f N$ stages[1], each stage comprised of N/f switching elements of size $f \times f$. A property of the SEN is the existence of a unique path between each source-destination pair. Consequently, the failure of even a single switch will destroy the *full access* property (i.e. each PE will no longer be able to communicate with every ME).

This paper is concerned with estimating the reliability of two augmented SEN's. The first of these is the extra-stage SEN (or SEN+). This network provides two paths between each PE-ME pair which are disjoint except in the first and last stages. The other network provides wrap-around connections between each ME and its corresponding PE. In this case, a packet, unable to reach its destination in one pass through the network, may have to make several passes before reaching its final destination (if at all). No redundant switches are added in the latter case but fault tolerance is provided using temporal redundancy. This property of the network whereby full access is facilitated by multiple passes through the network is termed *Dynamic Full Access* (DFA) [SHEN84].

The goal of this paper is to obtain improved upper and lower bounds on reliability, R, as a function of switch reliability, r in the above two schemes employing 2×2 switches ($f = 2$). We define r as the probability that a switch is operational and R as the probability of full access Also, a switch is assumed to

be either operational or completely failed (i.e. unable to pass information).

The paper is organized as follows. In Section 2, we investigate reliability of the SEN+. After surveying some of the earlier work on obtaining reliability bounds for the SEN+, we lay the theoretical foundation for the new bounds. The expressions for the lower bounds are derived and the upper bounds are developed. Our results are presented and a comparison with the old bounds is made. In Section 3, necessary and sufficient conditions for network connectivity in the SEN with wrap-around connections are obtained. These are used to derive upper and lower bounds on reliability. Finally, we present our results and a comparison between the two schemes.

2. Bounds on Reliability of the SEN+

We first discuss work done in obtaining existing reliability bounds for the SEN+ and then investigate how these may be improved.

2.1 Existing Work

The SEN (or Omega network) [LAWR75] is one of a family of networks, which include the banyan [GOKE73], baseline [WU80], multistage cube [SIEG81], etc. that are topologically equivalent. To understand the bounds developed in [BLAK89] and in this paper, the SEN+ is redrawn as an extra-stage banyan (Figs. 2.1(a), (b).) An $N \times N$ extra-stage banyan can be redrawn as two $N/2$ banyans linked by a begining and ending stage. The two $N/2$ banyans will be refered to as Subnet 1 and Subnet 2 (Fig. 2.1(c)).

To obtain the lower bound in [BLAK89], note that no faulty switch in the first and last stages and at least one *completely* operational subnet guarantees full access. Using the series-parallel reliability model, we obtain, after simplification

$$R > 2r^{N/4(logN + 3)} - r^{N/2(logN + 1)} \qquad (1)$$

To obtain the upper bound in [BLAK89], note that no two switches that occupy *corresponding* positions in the two subnets can be simultaneously faulty. (For example, switches 1 and 3 in stage 1 of Fig. 2.1(b).) Since there are many combinations of failed switches which will cause the network to be failed other than such pairs, the system reliability will be overestimated. Thus,

$$R < r^N[1 - (1 - r)^2]^{\frac{N}{4}(logN - 1)} \qquad (2)$$

While the results from these estimates of system reliability were close to the actual values for the small networks examined in [BLAK89], they diverge considerably for larger networks. Thus, an attempt was made to find better bounds for the reliability of the SEN+.

---

[1] Unless otherwise stated, all logarithms in this paper are to base 2

2.2 Our Approach to Obtaining Improved Bounds

The reliability of an $N \times N$ SEN+ can be expressed as

$$R = r^N \sum_{i=0}^{2N'} c_i r^{N'-i}(1-r)^i \tag{3}$$

where $N' = (N/4)(\log N - 1)$ is the number of switches in each subnet and c_i is the number of ways in which i switches in the subnets may fail without affecting full access.

To find c_i, we first introduce the following definition.

Definition: A *switch tree* rooted at switch j, denoted $SW[j]$, is the set of all switches that have a direct path from or to j. The nodes in $SW[j]$ are said to be covered by j.

When a switch, j, fails in one subnet, then all paths through j will have failed. For the whole network to continue operating, none of the nodes corresponding to $SW[j]$ in the other subnet can have failed. In the case of multiple node failures, f_1, f_2, . . . f_j, all nodes in the other subnet corresponding to $SW[f_1] \cup SW[f_2] \cup \ldots SW[f_i]$ must be fault-free if full access is to be guaranteed.

Example: Figure 2.2 shows a subnet of a 32×32 SEN+. The switch trees rooted at switch x and y are shown in two different hatchings. If both x and y are failed, all nodes in the other subnet corresponding to hatched nodes in Figure 2.2 should be working for reliable operation. □

To obtain the coefficients, c_i, in Equation 3, we must examine the number of ways in which k and $i - k$ failures can occur in subnets 1 and 2, $0 \le k \le i$, without destroying connectivity.

Obtaining an exact expression requires computing the number of switches covered by each k-node failure set in Subnet 1 which is very time-consuming. Instead, we obtain bounds on c_i by computing the minimal and maximal unions of k switch trees. These are used respectively to obtain lower and upper bounds on reliability as below.

Let M_k and m_k be respectively the maximum and minimum number of switches covered by k faults in an $N/2 \times N/2$ SEN.

For Lower Bound: Use Eqn. 3 with

$$c_i = \sum_{j=0}^{i} max \left\{ \binom{N'}{j}\binom{N'-M_j}{i-j}, \binom{N'}{i-j}\binom{N'-M_{i-j}}{j} \right\} \tag{4}$$

For Upper Bound: Use Eqn. 3 with

$$c_i = \sum_{j=0}^{i} min \left\{ \binom{N'}{j}\binom{N'-m_j}{i-j}, \binom{N'}{i-j}\binom{N'-m_{i-j}}{j} \right\} \tag{5}$$

Thus, our approach to obtaining lower and upper bounds on the reliability of the SEN+ reduces to obtaining the size of the maximal and minimal union of k switch trees. These problems are rather involved and cannot be presented here due to space limitations. We instead summarize our main results on this subject.

First, we introduce the address string of a switch as a convenient notation. Using this, a switch in an $N \times N$ SEN is represented by a $2(n-1)$-bit string of symbols from the set $\{0, 1, x, *\}$. (x is a don't care symbol and may be used to denote sets of switches.) The switch in row i of stage j is denoted by $(*)^j$ concatenated with the binary representation for i concatenated with $(*)^{n-j-1}$. For example, switch 2 in stage 1 of a 32×32 SEN is denoted $*0010(*)^3$ and switch 5 in stage 4 is denoted $(*)^4 0101$. A property of this labeling scheme that will find repeated application is presented in the following theorem.

Theorem 2.1: Two switches, $(*)^i a_{n-2}...a_0 (*)^{n-i-1}$ and $(*)^j b_{n-2}...b_0 (*)^{n-j-1}$ where $0 \le i - j \le n - 1$, are connected by a path if $a_{k+i-j} = b_k$, $n - i + j - 2 \le k \le 0$.

Proof: Switch $(*)^j b_{n-2}...b_0 (*)^{n-j-1}$ in stage j is connected to switches $(*)^{j+1} b_{n-3}...b_0 0 (*)^{n-j-2}$ and $(*)^{j+1} b_{n-3}...b_0 1 (*)^{n-j-2}$ in stage $j + 1$ since the connection between adjacent stages in the SEN is the perfect shuffle. Thus, the given switch is connected by a direct path to all switches of the form $(*)^i b_{n-2-i+j}...b_0 x^{i-j}(*)^{n-i-1}$ or to switches of the form $(*)^i a_{n-2}...a_{i-j} x^{i-j}(*)^{n-i-1}$ and the theorem follows. □

Example: The address strings of switch 30 in stage 2 and switch 25 in stage 4 of a 64×64 SEN are shown below

$$* \ * \ 1 \ 1 \ 1 \ 1 \ 0 \ * \ * \ * \quad \text{and}$$
$$* \ * \ * \ * \ 1 \ 1 \ 0 \ 0 \ 1 \ *$$

Since the two address strings have identical overlapping substrings, there exists a path between the two switches from Theorem 2.1. □

2.3 Maximal Union of Switch Trees

Having motivated the problem, we first summarize the procedure used in obtaining the maximal union of k switch trees in an $N \times N$ SEN, $k \le (N/2)\log N$ (refered to as a maximal k-union). We employ a greedy strategy, i.e., the maximal k-union is obtained by augmenting the maximal $(k-1)$-union with the switch tree containing the largest number of uncovered nodes. For this purpose, we introduce a list of switches, S, so that the union of switch trees rooted at the first k switches in S is maximal. This strategy has been shown to result in a maximal union in the detailed version of our paper. We reproduce the main ideas and an example here.

A failure in one of the end stages (stages 0 and $n-1$) has a switch tree of size $N - 1$ while a failure in the middle stage of the SEN has a switch tree of size $O(\sqrt{N})$. In fact, we have shown that maximal coverage can be obtained by considering faults rooted at stages 0 and $n - 1$ exclusively. The address strings of faults in S that cover the 64×64 SEN are shown below.

$$* * * * * 0 0 0 0 0$$
$$0 0 0 0 0 * * * * *$$
$$* * * * * 1 0 0 0 0$$
$$0 0 0 0 1 * * * * *$$
$$* * * * * 0 1 0 0 0$$
$$* * * * * 1 1 0 0 0$$
$$0 0 0 1 0 * * * * *$$
$$0 0 0 1 1 * * * * *$$

Note that the fault set is comprised of clusters of 2^i nodes, alternating between stages 0 and $n - 1$. Also, note that their addresses differ in bits closest to the center of the address string. The maximal coverage for i faults is plotted in Fig. 2.3 for $N = 64$. Note that the entire network of 192 switches is covered by only 8 faults.

2.4 Minimal Union of Switch Trees

Obtaining the minimal union of k switch trees is more complicated than the maximal unions. The approach used here is quite different from that used in obtaining the maximal union of k switch trees. In particular, we show that the greedy strategy used there cannot be employed here. To see why, consider the 16×16 SEN (Figure 2.4). Let Σ_i denote a minimal covering set of i nodes and let m_i be its cardinality. It is easily verified that

- $\Sigma_5 = \{a, b, c, d, e\}$ and there exist Σ_1, Σ_2, Σ_3 and Σ_4 such that $\Sigma_1 \subset \Sigma_2 \subset \Sigma_3 \subset \Sigma_4 \subset \Sigma_5$.
- However, there exist no Σ_5, Σ_6 such that $\Sigma_6 \supset \Sigma_5$. For example, $\Sigma_6 = \{a, b, c, d, f, g\}$.

Note that the greedy strategy works until $k = 5$ but fails for $k = 6$. While the greedy strategy does not work universally i.e. Σ_i is not a subset of Σ_j for all i and all $j > i$, it is nevertheless true for specific values of i.

Our primary goal here is to obtain the plot m_i (the minimum number of switches covered by i faults) versus i. Our strategy can be summarized as follows: First, we obtain isolated points on the plot. For example, we have shown that a $k \times k$ banyan in the center of the network is a minimal covering fault set of size $(k/2) log k$. A SEN at the center is one that is equally distant (to within at most one stage) from stages 0 and $n - 1$. For example, there is no set of 12 faults that covers fewer switches than the 8×8 banyan shown in the box in Fig. 2.4. Points corresponding to banyans of size 8×8, and 16×16 are so indicated in the plot of m_i versus i for $N = 32$ (Fig. 2.5).

Another characteristic of these plots is that all points beyond $i = (N/4)(log N - 1)$ can be obtained from those for which $0 < i < (N/4)(log N - 1)$. Moreover the segment between $i = (N/4)(0.5 log N - 1)$ and $i = (N/4)(log N - 1)$ is identical to that between
$i = (N/4)(log N - 1)$ and $i = (N/4)(1.5 log N - 2)$ save for a vertical shift. Finally, some segments of the plot can be obtained from the corresponding plot for the $(N/2) \times (N/2)$ network. These facts are exemplified in Figure 2.5 for the 32×32 and 64×64 networks.

2.5 Results on Reliability Bounds for the SEN+

The maximal and minimal unions developed in the preceeding sections were used to calulate the reliability of the SEN+ network, using Equations (4) and (5). Fig. 2.6(a) shows R versus r from 0.990 to 0.999 for a 256-input SEN+ (above this range of r, R tends rapidly to 1 and below .99, it tends to 0). Included for comparison, are the old bounds. The distance between the new bounds is about half to one-third of the distance between the old bounds.

Fig. 2.6(b) shows R versus N for $r = .99$ and .999. Note that for a given R, the improvement in the bounds increases for larger networks (and larger r.) Thus, we expect the improvement in the bounds to be greater for larger networks.

3. Reliability Bounds for the SEN with Wrap-around Connections

In this section, we examine necessary and sufficient conditions for DFA in the SEN with wrap-around connections. These are then used to obtain upper and lower bounds on reliability.

3.1 Upper Bounds on Reliability of the SEN with Wrap-around Connections

A necessary condition for DFA in this network is that the switches in the first and last stages be operational. This provides a simple upper bound on network reliability viz. $R \leq r^N$. A better upper bound is obtained from the following two theorems.

Theorem 3.1 Let S_a be the set of 2^i switches in stage i, $0 \leq i \leq \lceil (n-1)/2 \rceil$, defined by $S_a = \{(*)^i a_{n-i-2} \ldots a_0 x^i (*)^{n-i-1}\}$ where $a_{n-i-2} \ldots a_0$ is the binary representation of a, $0 \leq a \leq 2^{n-i-1} - 1$. Then, at least one switch in each S_a must be operational for DFA.

Proof: Let Q_a be the set of 2^i switches in stage 0 defined by $Q_a = \{x^i a_{n-i-2} \ldots a_0 (*)^{n-1}\}$. Consider a PE adjacent to a switch in Q_a. From Theorem 2.1, a path from that PE to any ME must pass through a switch in S_a and the theorem follows. □

Thus the switches in stage i may be partitioned into 2^{n-i-1} sets, 2^i switches per set, such that at least one switch in each set must be operational as a necessary condition for not disconnecting any PE. A similar partitioning of stage i switches, $\lceil n/2 \rceil \leq i \leq n-1$, is used in defining a necessary condition for access to MEs. This and Theorem 3.1 imply that all switches in stages 0 and $n-1$, at least one-half of the switches in stages 1 and $n-2$, at least one-fourth of the switches in stages 2 and $n-3$, etc., must be operational for DFA. This yields an improved upper bound on reliability viz.

$$R \leq r^N \prod_{i=1}^{n/2} [1 - (1 - r)^{2^i}]^{(N/2^i)}, \quad n \text{ even}$$

$$\leq r^N [1 - (1 - r)^{\sqrt{N/2}}]^{\sqrt{N/2}} \prod_{i=1}^{\lfloor n/2 \rfloor} [1 - (1 - r)^{2^i}]^{(N/2^i)}, \quad n \text{ odd} \quad (6)$$

We sum the minimum number of working switches necessary for DFA in each stage to obtain

Corollary 3.1.1 A lower bound on the number of working switches required to ensure that any PE-ME pair can communicate in a finite number of passes through the network is $2(N - \sqrt{N})$ for even n and $2N - 3\sqrt{N/2}$ for odd n.

3.2 Lower Bounds on Reliability of the SEN with Wrap-around Connections

To obtain lower bounds on reliability we seek sufficient conditions for DFA. Specifically, we ask, "What is the minimal set of operational switches that guarantees DFA ?". We next define a subgraph and show that it provides DFA and is also minimal.

Definition: Subgraph **A** of the SEN with wrap-around connections is

$$\bigcup_{a_{(n-3)/2} \ldots a_0 = 0 \ldots 0}^{1 \ldots 1} SW[(*)^{(n-1)/2} a_{(n-3)/2} \ldots a_0 a_{(n-3)/2} \ldots a_0 (*)^{(n-1)/2}] \quad \text{for odd } n$$

$$\bigcup_{a_{n/2-1} \ldots a_0 = 0 \ldots 0}^{1 \ldots 1} SW[(*)^{n/2} a_{n/2-1} \ldots a_1 a_{n/2-1} \ldots a_0 (*)^{n/2-1}] \quad \text{for even } n$$

Because of space limitations, all theorems will only be proved for the case of odd n.

Theorem 3.2: Subgraph **A** guarantees full access in at most two passes.

Proof: Let $S = s_{n-2} \ldots s_0 (*)^{n-1}$ be the switch connected to the source PE and let $D = (*)^{n-1} d_{n-2} \ldots d_0$ be the switch connected to the destination ME. Let

$$(*)^{(n-1)/2} s_{(n-3)/2} \dots s_0 s_{(n-3)/2} \dots s_0 (*)^{(n-1)/2},$$

$$(*)^{n-1} s_{(n-3)/2} \dots s_0 d_{n-2} \dots d_{(n-1)/2}, \text{ and}$$

$$(*)^{(n-1)/2} d_{n-2} \dots d_{(n-1)/2} d_{n-2} \dots d_{(n-1)/2} (*)^{(n-1)/2}$$

be the address strings of X, D' and X' respectively. From Theorem 2.1, there exists a path from S to X and from X to D', and both of these paths are contained in the switch tree rooted at X. Since $SW[X] \subset \mathbf{A}$, there exists a path between S and D' in $\mathbf{A}$. Also, D' is connected to stage 0 switch, $s_{(n-3)/2} \dots s_0 d_{n-2} \dots d_{(n-1)/2}(*)^{n-1}$ via the wrap-around connection. Again, from Theorem 2.1, there exists a path from this node to X' and from X' to D and these paths are contained in $SW[X']$. Since $SW[X'] \subset \mathbf{A}$, there also exists a path between $s_{(n-3)/2} \dots s_0 d_{n-2} \dots d_{(n-1)/2}(*)^{n-1}$ and D and the theorem follows. □

Theorem 3.3 $\mathbf{A}$ is a minimally connected subgraph of the network.

Proof: For odd n, $\mathbf{A}$ is made up of $2^{(n-3)/2+1} = \sqrt{N/2}$ switch trees rooted at the centermost stage. Each of these switch trees is made up of $1 + 2(2 + 4 \dots + 2^{(n-1)/2})$ switches. Thus, the total number of nodes in $\mathbf{A}$ is at most $2N - 3\sqrt{N/2}$. Since $\mathbf{A}$ guarantees DFA, it must have at least that many nodes from Corollary 3.1.1. Hence, $\mathbf{A}$ has precisely $2N - 3\sqrt{N/2}$ nodes and is minimal in respect to connectivity. □

Example: Subgraph $\mathbf{A}$ for $N=16$ is shown in Figure 3.1 (dark edges). It has a total of 24 operational switches. □

The survival of $\mathbf{A}$ is a sufficient condition for DFA which leads to the following lower bound for reliability.

$$R < r^{2(N-\sqrt{N})} \quad for\ even\ n$$

$$R < r^{2N-3\sqrt{N/2}} \quad for\ odd\ n \qquad (7)$$

3.3 Results on Reliability for the SEN with Wraps

Fig. 3.2 shows R versus r for the SEN with wrap-around connections for $N=256$. The two upper bounds are r^N and Eqn. 6. The first lower bound is Eqn. 7. In the full paper, we have derived a much better lower bound (lbound2 in Fig. 3.2) obtained by considering two maximally disjoint minimal subgraphs. Finally, we have also performed simulations which indicate that R for the SEN+ is closer to the lower bound while for the SEN with wraps, R is closer to the upper bound. (For example, at $r=.999$, R is .735 for the SEN+ and .78 for the SEN with wraps.) We are currently investigating the extra-stage cube [ADAM82] which provides fault tolerance in the first and last stages. Preliminary results indicate that its reliability is considerably higher than the two schemes considered here.

REFERENCES

[ADAM82] Adams III and H.J. Siegel, "The Extra Stage Cube: A Fault-Tolerant Interconnection Network for Supersystems" *IEEE Trans. Comput.*, vol. c-31, May 1982, pp. 443-454.

[BLAK89] J.T.Blake and K.S.Trivedi, "Multistage Interconnection Network Reliability", *IEEE Trans. Comput.*, C-38, Nov. 1989, pp. 1600-1604.

[GOKE73] L.R. Goke and G.J. Lipovski, "Banyan Networks for Partitioning Multiprocessor Systems" *Proc. Intl. Symp. Comput. Architecture*, 1973, pp. 21-28.

[LAWR75] D.H.Lawrie, "Access and Alignment of Data in an Array Processor," *IEEE Trans. on Computers*, Vol. C-24, Dec. 1975, pp. 1145-1155.

[SHEN84] J.P. Shen and J.P. Hayes, "Fault-Tolerance of Dynamic-Full-Access Interconnection Networks", *IEEE Trans. Comput.* vol. C-33, no.3, March 1984, pp. 241-248.

[SIEG81] H.J.Siegel and R.J.McMillen, "The Multistage Cube: A Versatile Interconnection Network," *Computer*, Vol 14, Dec. 1981, pp. 65-76.

[WU80] C.Wu and T.Feng, "On a Class of Multistage Interconnection Networks", *IEEE Transactions on Computers*, C-28, Aug. 1980, pp. 694-702.

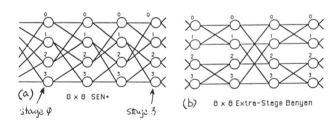

(a) 8 x 8 SEN+   stage φ   stage 3

(b) 8 x 8 Extra-Stage Banyan

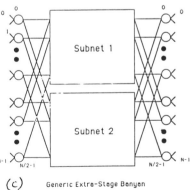

(c) Generic Extra-Stage Banyan

Fig. 2.1 The SEN and Banyan extra-stage versions

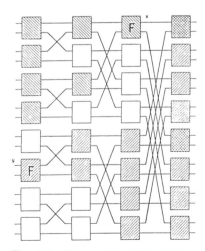

Fig. 2.2 Subnet of a 32x32 SEN+

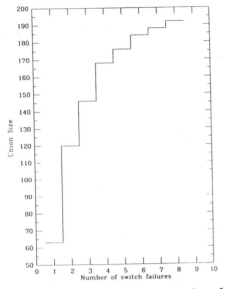

Fig. 2.3 Maximal union vs. number of

switch failures for a 64x64 SEN

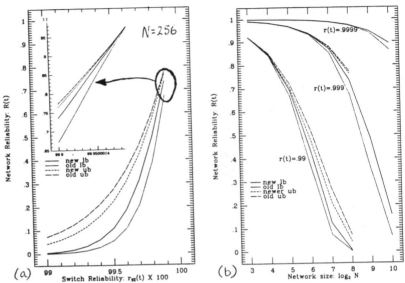

Fig. 2.6 Upper and lower bounds on the SEN+ reliability

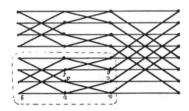

Fig. 2.4 16x16 SEN with five

and six switch failures

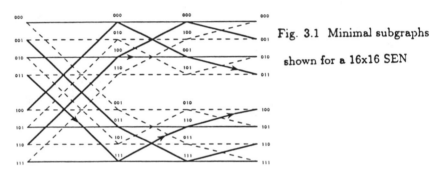

Fig. 3.1 Minimal subgraphs

shown for a 16x16 SEN

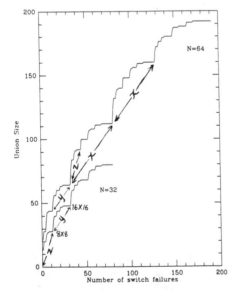

Fig. 2.5 Minimal union vs. number

of switch failures

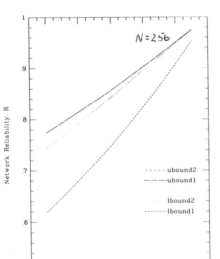

Fig. 3.2 Upper and lower bounds on

reliability of the wrap-around SEN

A RELIABLE BUTTERFLY NETWORK FOR
DISTRIBUTED-MEMORY MULTIPROCESSORS

Nian-Feng Tzeng

The Center for Advanced Computer Studies
University of Southwestern Louisiana
Lafayette, LA 70504

Abstract —— Interconnecting processors in accordance with the butterfly topology to construct a distributed-memory multiprocessor is advantageous, because every butterfly node has a fixed degree. In this paper, we examine a reliable butterfly-based multiprocessor which preserves its full rigid butterfly configuration even in the presence of faults. The proposed butterfly system can tolerate any single and many multiple node/link failures, giving rise to significantly improved reliability. Reconfiguration in response to an arising fault in our design is easy and may be performed in a distributed manner. A system after reconfiguration is ensured to provide the same high performance. Reliability results show that our design compares favorably with an earlier design.

1. Introduction

The butterfly network has the distinct property that every constituent node requires a fixed degree, independent of the system size. The property of a constant degree is extremely advantageous, because it makes possible the use of a single type of building block for constructing an arbitrarily large distributed-memory multiprocessor based on the butterfly topology. The cost of systems so constructed would be lowered as the building block approach becomes popular.

An essential design consideration of a large network system is its reliability. When the system size grows, the probability of having all system components fault-free during a given operation period falls quickly and could soon reach an unacceptably low point. It is thus necessary to incorporate redundancy in the system design, especially for a large system, to ensure proper continuing operation even after some components fail, improving reliability. As soon as a failure arises and is detected, a fault-tolerant system reconfigures itself so as to isolate the failed components. It is often desirable that a reconfigured system retains its rigid original structure so as to support the service level the system is designed to aim at, particularly for a high-performance system, like the butterfly-based multiprocessor.

We propose a method for significantly enhancing the reliability of butterfly-based distributed-memory multiprocessors, where each butterfly node is not simply a switch only, but includes, in addition to storage, computation and communication logics. The reconfiguration process in response to an arising failure in this design involves only several nodes, and can be carried out in a distributed manner. A reconfigured system is assured to deliver the same high performance since it maintains the full rigid butterfly topology.

A **b**utterfly-based **d**istributed-memory **m**ultiprocessor (called the BDM for short) of size $N = L \times (\log_2 N + 1)$ is arranged as L levels of $(\log_2 L + 1)$ stages, with nodes in consecutive stages interconnected by the butterfly pattern. Fig-ure 1 illustrates the BDM with 32 nodes (i.e., $L = 8$). Every link between two nodes can be used for bidirectional data transmission. Stages in the BDM are numbered rightwards in a sequence from 0 to $\log_2 L$, with 0 for the leftmost stage. Similarly, the nodes in every stage are labeled downwards from 0 to $L - 1$ (referred to as their *level* numbers), with 0 for the top node. A node at the lth level of the gth stage in the BDM can be uniquely denoted by (g, l).

Every node has degree 4, with two connected to nodes in the preceding stage and the other two connected to nodes in the next stage. A node in the two end stages utilizes its two links for connecting other nodes, and the remaining links may be used for the I/O purpose. Let the links of a node used for connecting other nodes in the same level be called the *straight* (or S for short) links, and the rest be the *cross* (or C for short) links, then the BDM interconnection topology can be formally described below: (i) nodes (g, l) and $(g+1, l)$ are connected through an S link, and (ii) nodes (g, l) and $(g+1, l+(-1)^{b_g} \times 2^g)$ are connected through a C link, where $0 \le g < \log_2 L$, $0 \le l < L$, and b_g is the bit with weight 2^g in the binary representation of l ($= b_{\log_2 L - 1} \cdots b_1 b_0$).

2. Proposed Reliable BDM Design

In order to ensure that a BDM maintains its full strict structure in the presence of a fault, we may (i) augment the connections between levels in a way that if a direct link exists between nodes (g_1, l_1) and (g_2, l_2), then an extra link is placed between nodes (g_1, l_1) and (g_2+1, l_2), and (ii) add an extra node to the right end of every level, as depicted in Fig. 2, where the added node in level l is numbered (s, l). For example, a direct link is added between $(0, 0)$ and $(2, 1)$, since there is a direct connection existing between $(0, 0)$ and $(1, 1)$ in the original BDM given in Fig. 1. Similarly, an extra link is added between node $(2, 4)$ and extra node $(s, 0)$, as a cross connection exists between $(2, 4)$ and $(3, 0)$.

When node $(1, 1)$ fails and is bypassed, for example, a pair of direct connections still exists between levels 0 and 1, with one between $(1, 0)$ and $(0, 1)$ and the other between $(0, 0)$ and $(2, 1)$, effectively making node $(2, 1)$ replace the role of the failed node (see Fig. 3(a) for the system after reconfiguration). With the extra links added, a pair of direct connections is also present between levels 1 and 3 after the said fault occurs, with one between $(3, 1)$ and $(1, 3)$ and the other between $(2, 1)$ and $(2, 3)$, as illustrated in Fig. 3(a). The link between $(2, 1)$ and $(2, 3)$ is realized by adding a control switch to the intersection of cross links between $(1, 1)$ & $(2, 3)$ and $(2, 1)$ & $(1, 3)$, and setting the switch to the "V" state (see Fig. 4(a)). Node $(3, 1)$ therefore replaces the role of $(2, 1)$ (which replaced the failed node) successfully. The spare node $(s, 1)$ then replaces the role of $(3, 1)$, and a pair of direct connections exists between levels 1 and 5.

This work was supported by the NSF under Grant MIP-8807761.

The interconnection of each pair of cross links in the BDM is equipped with a control switch, which can be set at either the "V" state or the "X" state, as shown in Fig. 4(a). A pair of additional links added to the BDM is referred to as an *added cross connection pair,* and one control switch is placed at the intersection of each added cross connection pair, as depicted in Fig. 4(b). A BDM with added cross connection pairs and spare nodes is referred to as the RBDM (which stands for the reliable BDM).

The control switch used is normally much simpler [1] than a node, and can be made more reliable as pointed out by earlier studies [2]. The use of control switches at regular cross connection pairs would somewhat reduce the reliability of individual cross connection pairs. However, if the switch is made very reliable (we can afford to do so, since the total number of such switches in a BDM with N nodes is less than $N/2$), reliability degradation could be negligibly small. Furthermore, should such a switch or a cross connection fail, the failure can be treated as the node failure(s) and is tolerable. This design thus results in enhanced reliability if a node is relatively more complex than a switch (a common case), as will be seen in the next section.

Four switches are added to each node so that a failed or unused node can be bypassed, as shown in Fig. 4(c). The spare node is connected to the external for the I/O purpose, and two parallel links are present between a spare node and its immediate left neighbor (see Fig. 2) so that when the spare node is unused and bypassed, I/O data can directly reach the immediate left neighbor. A single failure in any one of the two links is tolerable by utilizing the spare node to replace the role of its (healthy) left neighboring node (which is bypassed), as shown in Fig. 3(b).

Formally, a RBDM design with L $(= 2^n)$ levels can be defined as a collection of $L \times (n+2)$ nodes, organized in $n+2$ stages (where the stage of spare nodes is numbered stage $n+1$). Every node now has two additional links for making extra connections (called E links for short), as depicted in Fig. 4(c). The interconnection style through C and S links is the same as that given in the previous section for a regular BDM, except that two parallel connections are added between (n, l) and $(n+1, l)$ through C and S links, for all $0 \leq l < L$. The interconnection through E links is as follow: nodes (g, l) and $(g+2, l+(-1)^{b_g} \times 2^g)$ are connected through an E link, where $0 \leq g < n$, $0 \leq l < L$, and b_g is the bit with weight 2^g in the binary representation of l. A complete RBDM with 8 levels is illustrated in Fig. 2.

The proposed RBDM has L spare nodes and $L \times (n+2)$ extra links (2 in the expression accounts for the two parallel connections present between a spare and its left neighbor), when compared with its regular counterpart. As a result, the node overhead ratio is $1/(n+1)$, and the link overhead ratio equals $(n+2)/2n$.

Any single node failure is tolerable in the RBDM. When a fault happens to an active node, the failed node is bypassed and its role is replaced by its physical successor (to its right in the same level), which in turn is replaced by its successor, etc., until the spare in the level is brought into the system. Each level can tolerate up to one node failure.

Likewise, the RBDM can tolerate any single link failure or cross connection pair failure. When a fault happens to a C link, it is treated as the failure of either one of the two nodes connected by the link, and thus causes reconfiguration in the

same way as a node failure. If an S link fails, slightly different reconfiguration takes place. When the link between $(1, 7)$ and $(2, 7)$ in Fig. 3(b) becomes faulty, for example, node $(2, 7)$, although healthy, is bypassed, and its role is replaced by $(3, 7)$. The two cross connections between levels 7 and 5 exist between $(1, 7)$ & $(2, 5)$ and between $(3, 7)$ & $(1, 5)$ (through the bypassed node $(2, 7)$). The direct connection between two logically adjacent nodes in level 7, $(1, 7)$ and $(3, 7)$, is realized by the added cross connection pair between levels 7 and 5, when the associated control switch is set to the "V" state (see Fig. 4(b)). The spare node in level 7 is brought in to replace the role of node $(3, 7)$, and a pair of cross connections between levels 7 and 3 are established between $(3, 7)$ & $(3, 3)$ and between $(4, 7)$ & $(2, 3)$, as illustrated in Fig. 3(b). This design assures its full strict topology after a link fails.

The failure of a cross connection pair, say between level l and level $l_2 = l + \gamma_g$, can be tolerated as follows: it is treated as the failures of both $(g+1, l)$ and $(g+1, l_2)$, and evokes reconfiguration accordingly. This is made possible because we simply substitute the failed pair using the added cross connection pair present between the two levels.

Reconfiguration Process

At the beginning of operation when no fault exists, all the control switches on C links are set to the "X" state to implement the butterfly interconnection pattern, and all the E links are disabled. After a fault arises and is detected, certain control switches on C links change their settings, and also some E links are activated with the control switches on them set appropriately. This is accomplished locally, with only several nodes involved.

To simplify subsequent explanation, the C link from (g, l) to $(g-1, l_1)$ is denoted as $C_{g-1}^{(g,l)}$, and the C link to $(g+1, l_2)$ is $C_{g+1}^{(g,l)}$. Similarly, the E links from (g, l) respectively to $(g-2, l_3)$ and $(g+2, l_4)$ are referred to as $E_{g-2}^{(g,l)}$ and $E_{g+2}^{(g,l)}$. The procedure in response to a node failure is addressed first. Assuming that a node fails at stage g in level l of the RBDM with L $(= 2^n)$ levels. Reconfiguration due to this failure requires the participation of nodes at stage k in level l, for all $g \leq k \leq n+1$. Specifically, the failed node is bypassed, and the nodes to its right carry out the following steps:

(1) For $k \geq 2$:
   a) If node (k, l) has active $E_{k-2}^{(k,l)}$, reconfiguration fails; else the node activates $E_{k-2}^{(k,l)}$, with the switch on it set to the "X" state.
   b) If the node at the other end of this activated link, namely, $(k-2, l+\gamma_{k-2})$, is faulty or bypassed, reconfiguration fails; else the node is informed of this change so that it would enable $E_k^{(k-2, l+\gamma_{k-2})}$ and disable $C_{k-1}^{(k-2, l+\gamma_{k-2})}$ (with the switch on it kept unchanged) for communicating with level l;
(2) For $k \leq n$:
   a) If the switch on $C_{k-1}^{(k,l)}$ of node (k, l) is at the "X" state, the switch is changed to the "V" state; else reconfiguration fails.
   b) If the node now connected directly to (k, l) is faulty, reconfiguration fails as well;
(3) For $k < n$:
   Node (k, l) deactivates $C_{k+1}^{(k,l)}$ (with the switch on it kept unchanged).
(4) One link between nodes (n, l) and $(n+1, l)$ is disabled.

The last two steps are for disabling unused links. When a fault occurs at stage 1 in level 1 of Fig. 2, for example, nodes $(1, 1)$, $(2, 1)$, $(3, 1)$, and $(4, 1)$ carry out this procedure, and the affected links upon reconfiguration are darkened in Fig. 3(a). After reconfiguration, every node in the intermediate stages has exactly four active links, and the full strict butterfly topology is preserved. It is clear from the above procedure that an arising fault at stage g causes no more than $3 \times (n + 1 - g)$ nodes involved in the reconfiguration procedure. A failure at a later stage leads to fewer participators. This procedure carries out appropriate reconfiguration in response to a tolerable fault pattern, and provides an unsuccessful signal if a fault pattern is intolerable.

The fault at a C link (or the switch on it) can be treated as a node failure, and evokes the same procedure as above. Reconfiguration in response to an S link failure is slightly different. Consider the failure of S link between (g, l) and $(g+1, l)$. This fault results in (i) node (g, l) activating $E_{g+2}^{(g,l)}$ with the switch on it set to the "V" state, (ii) node $(g+1, l)$ connecting $C_g^{(g+1,l)}$ directly to its S link to the next stage (effectively bypassing this node), and (iii) node $(g+2, l)$ enabling $E_g^{(g+2,l)}$ and changing the switch on $C_{g+1}^{(g+2,l)}$ from the "X" state to the "V" state. Additionally, each node to the right of node $(g+2, l)$ in level l evokes the procedure given above for a node failure. The affected links due to reconfiguration for two link failures are darkened in Fig. 3(b). The number of nodes involved in reconfiguration for the failed S link is no more than $3 \times (n - g)$, which again is stage-dependent. Note that the fault at any link between a spare node and its left neighbor is treated as a fault at the left neighbor and is reconfigured accordingly, as shown in Fig. 3(b). In this case, three nodes participate in reconfiguration.

3. Reliability Analysis

The RBDM is considered failed when the number and the locations of the faults prevent it from reconfiguring back to its corresponding BDM. Many multiple link failures are tolerable, and in practice a link tends to be far more reliable than a node. For simplicity, our reliability analysis does not consider the failure of individual links, with the failure rate of a cross connection pair included in that of the nodes which are connected by the pair. The failures of added cross connection pairs are taken into account separately.

The reliability for all nodes (including spares) is assumed to be equal and exponentially distributed, with a constant failure rate λ (i.e., $R = e^{-\lambda t}$). An added cross connection pair plus its control switch is assumed to have reliability $R_{cs} = e^{-\lambda_{cs} t}$. It can be shown that an arising node failure at level l in an RBDM with $L = 2^n$ levels causes at most n added cross connections (plus their switches) to be activated during reconfiguration, and fewer such connections are activated when the fault arises at a later stage. In the presence of tolerable multiple faults, say f faults, no more than fn added cross connections are enabled and have to be healthy. The reliability of the RBDM can be expressed as

$$R_{RBDM}(t) > \beta_0 R^N + \beta_1 R^{N-1}(1-R)R_{cs}^n + \beta_2 R^{N-2}(1-R)^2 R_{cs}^{2 \cdot n}$$

$$+ \cdots + \beta_i R^{N-i}(1-R)^i R_{cs}^{i \cdot n} + \cdots + \beta_N (1-R)^N R_{cs}^{N \cdot n},$$

where $N = L \times (n + 2)$ is the total number of nodes (including spares) in the system, and β_i is the number of ways in which i tolerable faults can occur in the system. β_i equals 0 for i exceeding the number of maximum tolerable faults, L.

In order to evaluate RBDM reliability, it is necessary to calculate the number of possible locations for the ith tolerable faulty node to occur, given that the system has already reconfigured itself successfully in response to $i - 1$ faults, $i \geq 2$. This number is dependent on the positions of those existing $i - 1$ faults. The two initial β values are $\beta_0 = 1$ and $\beta_1 = N$, and a lower bound on the number of possible locations for the ith fault can be derived as follows.

After a node, say (g, l), failed and the system is reconfigured, a subsequent node failure may not be tolerable even when it is at a level other than l. In Fig. 3(a), for example, a subsequent failure at $(0, 0)$ is intolerable because such a failure requires $(1, 0)$ to substitute the role of the failed node during reconfiguration, and reconfiguration is never successful as a direct connection between $(1, 0)$ and $(2, 1)$ cannot be established. Node $(0, 0)$ is disallowed to fail subsequently and becomes a *critical* node. In general, the failure of (g, l) in the RBDM introduces g critical nodes at level $l + \gamma_{g-1}$, $g + 2$ critical nodes at level $l + \gamma_g$, $g + 3$ critical nodes at level $l + \gamma_{g+1}$, etc., where $\gamma_t = (-1)^{b_t} \times 2^t$ with b_t being the bit with weight 2^t in the binary representation of l. Totally, there are $\delta = g + (g+2) + (g+3) + \cdots + n + (n+1)$ critical nodes created in all the levels other than level l. The maximum value of δ is $\delta_{max} = n(n+3)/2$. A subsequent fault is tolerable, provided that it is not any one of the δ critical nodes, nor is in the same level that involves the failed node. The second fault can thus be at no fewer than $N - \Delta$ (where Δ is $\delta_{max} + (n+1)$) possible locations for any given first failure, implying that $\beta_2 \geq N(N - \Delta)/2$. Any existing two tolerable faults introduce no more than 2δ critical nodes, and a subsequent fault can be at no less than $N - 2\delta - 2(n+1)$ locations. There are β_2 distinct fault patterns in the presence of two failures, and each pattern allows a third fault to be at no less than $N - 2\Delta$ locations, yielding $\beta_3 \geq N(N - \Delta)(N - 2\Delta)/(2 \times 3)$.

Along the same line, it is easy to see that in general, for any given fault pattern with $i - 1$ failures, the ith tolerable fault can be at no fewer than $N - (i-1)\Delta$ possible locations, as $i - 1$ failures introduce no more than $(i-1)\delta$ critical nodes. Since there are no less than $\beta_{i-1} = N(N - \Delta)(N - 2\Delta) \cdots (N - (i-2)\Delta)/(i-1)!$ distinct fault patterns in the presence of $i - 1$ failures, we have

$$\beta_i \geq \beta_{i-1}(N - (i-1)\Delta)/i = N(N - \Delta)(N - 2\Delta) \cdots (N - (i-1)\Delta)/i!,$$

for all $i \geq 2$. Note that the expression on the right-hand side equals 0 if $N \leq (i-1)\Delta$. With β_i lower bounds available, we may calculate the RBDM reliability lower bound using the above inequality for $R_{RBDM}(t)$.

The reliability lower bounds on the RBDM with $L = 2^4$ and 2^8 levels are depicted in Fig. 5, where the node failure rate λ is assumed to be 1.0 per unit time (e.g., 10^6 hours), and the failure rate of a cross connection plus its switch, λ_{cs}, is 0.1 per unit time. The exact reliabilities of the corresponding BDM and RECBAR [3] are also provided in this figure under the same node failure rate (i.e., 1.0 per unit time) but a zero link failure rate (i.e., totally fault-free links, an optimistic assumption). Notice that the assumption of fault-free links makes it simple to compare the RECBAR with the BDM (and also the RBDM), since they have different types and amounts of links. The RBDM has a significantly improved reliability over its BDM counterpart and is more reliable than the RECBAR. The reliability gap between the RBDM and the BDM (or the RECBAR) increases as the system size grows, and in reality, the gap is expected to be larger than that shown in the figure.

4. Conclusions

We have presented a reliable butterfly network suitable for the distributed-memory multiprocessor construction. Our design goal is to maintain the full rigid butterfly interconnection style even in the presence of faults (not just to assure that all workable nodes are connected), so that a reconfigured system can deliver the same high performance. The proposed design tolerates any single and many multiple node/link failures, significantly enhancing system reliability. Reconfiguration in response to an arising fault is simple, involves only a small fraction of system nodes, and can be carried out in a distributed fashion. It is shown that our design is more reliable than a prior reconfigurable design.

References

[1] M. Blatt, "Effects of Switch Failure on Soft-Configurable WSI Yield," *Proc. 1990 Int'l Conf. Wafer Scale Integration*, Jan. 1990, pp. 152-159.

[2] I. Koren and M. A. Breuer, "On Area and Yield Considerations for Fault-Tolerant VLSI Processor Arrays," *IEEE Trans. Computers*, vol. C-33, pp. 21-27, Jan. 1984.

[3] V. Balasubramanian and P. Banerjee, "A Fault Tolerant Massively Parallel Processing Architecture," *J. Parallel and Distributed Computing*, 4, pp. 363-383, 1987.

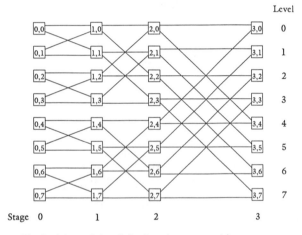

Fig. 1. A butterfly-based distributed-memory multiprocessor.

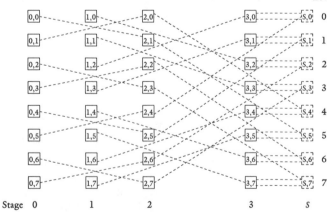

Fig. 2. The reliable BDM obtained from Fig. 1 by adding redundancy. (Extra links and spares are denoted by dashed lines. Regular links are omitted intentionally for clarity.)

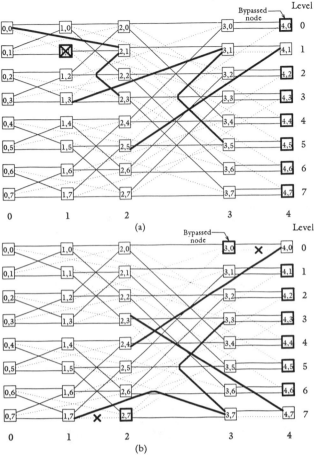

Fig. 3. A reconfigured system in the presence of (a) node (1, 1) failure, and (b) two link failures as marked. (Inactive links are denoted by dotted lines. Bold squares denote bypassed nodes.)

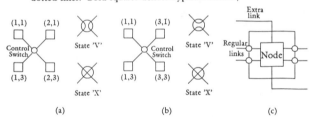

Fig. 4. (a) A pair of cross links with a control switch. (b) An added cross connection pair. (c) A node with four switches to allow "bypassing."

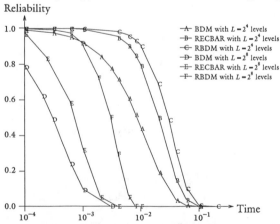

Fig. 5. Reliability comparison among BDM, RECBAR, and RBDM under $\lambda = 1.0$ and $\lambda_{cs} = 0.1$ per unit time.

MULTIPLE-FAULT TOLERANT CUBE-CONNECTED CYCLES NETWORKS[1]

C. Jimmy Shih and Kenneth E. Batcher
Department of Mathematics and Computer Science
Kent State University
Kent, Ohio 44242-0001

Abstract

Most fault-tolerance schemes for Cube-Connected Cycles (CCC) networks are limited to single-fault tolerance in the criterion of permutation connections. We are therefore proposing our 2 and 3-fault tolerance schemes. In this paper we first present the proposed structure which picks many appropriate connections and adds these connections and their associated PEs to an original network. As faults occur, the faulty connections and PEs will be removed and remaining connections will form a substitute network called physical network. Since the physical network is no longer the same as the original network, a physical-to-logical mapping mechanism is required so that the physical network can be used in the same way as before. We prove that the mapping function is an isomorphism, therefore, the reconfiguration process we use is correct.

Introduction

The CCC network is essentially based on the Hypercube topology. It replaces each processor of the Hypercube with many processors connected together in a cycle. Preparata and Vuillemin precisely defined a general CCC network as the following [Prep81]:

A CCC is a network of identical PEs, where each PE has three ports. Each link which connects two PEs is bi-directional. The CCC has $N \leq 2^s$ PEs, where $1 \leq s \leq r+2^r$ and $r \geq 1$. Each PE can be addressed as m, where $0 \leq m < N$. A PE with an address m can be alternatively represented as a tuple (ℓ, p), such that $m = \ell * 2^r + p$, $0 \leq p < 2^r$, and $0 \leq \ell < 2^{s-r}$. ℓ and p have (s-r) bits and r bits respectively. Each PE has F, B, and L ports, thus F port of a PE(ℓ, p) is F(ℓ, p), etc. The inter-connections are as follows: F(ℓ, p) is connected to B$(\ell, (p+1) \bmod 2^r)$; B(ℓ, p) is connected to F$(\ell, (p-1) \bmod 2^r)$; L(ℓ, p) is connected to L$(\ell+e2^p, p)$; where $e = 1 - 2 \operatorname{bit}_p(\ell)$ and $\operatorname{bit}_p(\ell)$ is the pth bit of ℓ.

There are many fault-tolerance schemes available to CCC networks. However most of them are single-fault tolerant in the criterion of permutation connections [Adam87] [Bene86] [Bene88] [Tzen90]. Therefore we are proposing our scheme to enhance fault-tolerant capability for CCC networks with respect to the following specifications:

---

[1]   This material is based upon work supported by the National Science Foundation under Grant MIP-9004127.

1. Fault Model	Include all network components; unusable if faulty
2. Criterion	Cube realizable permutations
3. Capability	More than single-fault
4. Routing complexity	The same as the original complexity

Table 1 Proposed Fault-tolerance Specifications

This paper is organized in the following way. In the second section we define terms, and in the third section we present our proposed structure which locates appropriate links for redundancy. In section four we prove the mapping function is isomorphic. The isomorphism is then used in the reconfiguration process. In section five we carry out the reconfiguration process with an example and finally we conclude the paper with further study.

Definitions

As the n-CCC was defined, n was used to specify the dimensions of CCC networks. The redundant connections (include their related PEs) which will be added to the network to enhance fault-tolerance are denoted as k. So the combination of n and k, denoted (n,k), means that k redundancies (connections and PEs) are added to an n-dimensional network. Usually we define the first dimension connection as the link whose two ends are different in the least significant bit of their addresses. However, these k connections are chosen from the link whose two ends are different in more than one bit. For the purpose of identifying different connections, we shall define it in such a way that k connections can be easily distinguished.

Definition 1: The m-connection is a set of links; each link has address differences $(d_0 \ d_1 \ ... \ d_{n-1})$ between its two ends (cycles) such that
$$m = 2^0 d_0 + 2^1 d_1 + ... + 2^{n-1} d_{n-1}.$$

Thus, the first dimension of a 3-CCC is 1-connection whose address differences are (1 0 0). The second dimension is 2-connection whose address differences are (0 1 0). The third dimension is 4-connection whose address differences are (0 0 1). These vectors can be combined together to form a matrix to represent the connections of a 3-CCC. This introduces the idea of the connection matrix.

Definition 2: A connection matrix, C, is a matrix which consists of more than one m-connections; each one of them is represented in a transposed vector of the address difference.

Thus, these three transposed vectors, $(1\ 0\ 0)^T$, $(0\ 1\ 0)^T$, and $(0\ 0\ 1)^T$, can be put together and form the connection matrix, C, of a 3-CCC network.

$$C = \begin{bmatrix} 1 & 0 & 0 \\ 0 & 1 & 0 \\ 0 & 0 & 1 \end{bmatrix}$$

If an extra connection is added to a 3-CCC network, we shall reconstruct a new connection matrix to represent a connection matrix with redundancy. For example, the 7-connection is added to the 3-CCC network, the transposed vector of the address difference of the 7-connection, $(1\ 1\ 1)^T$, should be appended to the previous C connection matrix (if there are more than one redundant m-connections they can be combined together and form a redundancy matrix. denoted R_m). This new connection matrix with redundancy is denoted C_r.

$$C_r = \begin{bmatrix} 1 & 0 & 0 & 1 \\ 0 & 1 & 0 & 1 \\ 0 & 0 & 1 & 1 \end{bmatrix}$$

Since C_r contains redundant connections, faulty m-connections (1-connection for example) are removed from C_r when faults occur. The remaining connection matrix becomes the physical connection matrix, denoted C_p, which will be used for reconfiguration.

$$C_p = \begin{bmatrix} 0 & 0 & 1 \\ 1 & 0 & 1 \\ 0 & 1 & 1 \end{bmatrix}$$

The connection matrix having the original m-connections, such as C in the previous example, is treated as the logical connection matrix, denoted C_t, because application programs have a logical view of the network as C_t. When the physical connections are no longer the same as the logical connections a mapping mechanism is required to transform the physical back to the logical so that programs will work the same as before. This brings in the definition of the mapping matrix, M:

Definition 3: The mapping matrix, M, is an n x n matrix which satisfies the following conditions:
1) $M \cdot C_t = C_p$ and
2) $C_t = M^{-1} C_p$ where C_t and C_p are n x n matrices. The matrix operations are defined over GF(2).

For verification of the second condition we have to detect whether M is invertible. One simple way of detecting an invertible matrix is to find out the determinant of the matrix. If the determinant of the matrix is not zero, the matrix is invertible. These arguments induce the following theorem to simplify definition 3:

Theorem 1: Given a square matrix M, a logical connection matrix C_t, and a physical connection matrix C_p, all defined over GF(2) such that $M \cdot C_t = C_p$. If det M is not zero then M is the mapping matrix between C_t and C_p.
Proof: Omitted due to simple linear algebra fact.

Proposed Structure

The structure which we are proposing has two layers. In the first layer, k appropriate m-connections are added to an original network among

cycles. In the second layer, a PE of every added m-connections and a bypass mechanism for these PEs have to be added to every cycle. So when a faulty PE occurs, data which approaches the faulty PE can be bypassed and redirected to a functioning PE. Since the second layer structure has been well studied and designed [Batc79] [Adam82], we will concentrate on the first layer structure.

The way to find out k appropriate m-connections basically follows theorem 1. We want to choose redundant m-connections in a way that after moving out f columns (faulty m-connections) from (n+k)-column C_r, the resultant (n+k-f)-column C_p' contains an invertible n-column M matrix.

As described earlier C_r has two parts, the logical connections submatrix, C_t, and the redundancy submatrix, R_m. C_t is always an identity matrix, therefore, we shall generate many candidate R_m first. The second step is to verify whether the resultant C_p' contains an invertible M matrix. The following discussions follow these two steps.

Theorem 2: There are two necessary conditions to form a redundancy matrix, R_m, so that the C_p' may contain an invertible M matrix.
i.   Each row of R_m must has at least f Hamming weight.
ii.   Each row of R_m must has at least f-1 Hamming distance with other rows.
Proof: Omitted due to limited space.

Based on theorem 2 we should be able to write a program to generate many k-column candidate R_m. For example, given f=2 and k=3 we should be able to get many candidate R_m; one of them is [(0 1 1)(1 0 1)(1 1 0)(1 1 1)] -- or (14- 13- 11-) if we pack every column into a decimal and suffix each decimal with a dash to stand for an m-connection.

We now turn to the second step to verify that C_p' contains an invertible M matrix. The verification process is carried out by an ad hoc program which only fits a particular (n,k) scheme. As mentioned before, C_p' has (n+k-f) columns and k is usually larger than f, the C_p' is more than n columns. In other words, there are many n-column C_p if we randomly pick n columns from C_p'. We want to verify that at least one of them (C_p) is invertible. The following program is a short example which does verification for (4,3) scheme and f=2 case:

Program 1: Verification of a R_m in (4,3) f=2
Input: A 4 by 4 logical connection matrix, C_t, and a 4 by 3 redundancy matrix, R_m.
Output: "good pattern" which says that the R_m is able to correct all cases of 2 faults or "bad pattern" which does not.

```
1. Success ← 0; det C_p ← 0; total ← 0;
   nk ← n+k; nk_f ← nk-f;
2. Form 4 x 7 C_r matrix by appending R_m to C_t
3. For ichi = 1 to nk-4 step 1
4.    For ni = ichi+1 to nk-3 step 1
5.      For son = ni+1 to nk-2 step 1
6.        For si = son+1 to nk-1 step 1
7.          For go = si+1 to nk step 1
8.            Form 4 x 5 matrix, C_p', by appending the
               following columns together C_r[*,ichi],
               C_r[*,ni], C_r[*,son], C_r[*,si], C_r[*,go].
```

```
9.         For one = 1 to nk_f-3 step 1
10.          For two = one+1 to nk_f-2 step 1
11.            For three = two+1 to nk_f-1 step 1
12.              For four = three+1 to nk_f step 1
13.                Form 4 x 4 matrix ,C_p, by appending
                   the following columns
                   together:C_p'[*,one],    C_p'[*,two],
                   C_p'[*,three], C_p'[*,four].
14.                det C_p ← finDeterminant (C_p);
15.                If (det C_p = 1)
16.                  then (success ← success + 1;
17.                  det C_p ← 0;
18.                  jump out of the loops which form C_p;)
19.              endFor_four;
20.            endFor_three;
21.          endFor_two;
22.        endFor_one;
23.        total ← total + 1;
24.      endFor_go;
25.     endFor_si;
26.    endFor_son;
27.   endFor_ni;
28. endFor_ichi;
29. If (success = total)
30. then (print "good pattern")
31. else (print "bad pattern");
```

The following are more good patterns we have found for two fault and three fault cases (shown in table 2, 3 respectively). For example, the second row of table 2 is a good pattern which consists of 14-connection, 13-connection, and 11-connection, for a 4-CCC network.

(n,k)	Good Pattern:
(4,3)	(14- 13- 11-)
(11,4)	(1479- 1753- 1898- 1972-)
(26,5)	(67104769- 66588641- 59215645- 40266971- 22370999-)

Table 2 Examples of Good Pattern for f=2

(n,k)	Good Pattern:
(4,4)	(7- 11- 13- 14-)
(11,5)	(1207- 1371- 1645- 1934- 2032)
(26,6)	(67104769- 66588641- 59215645- 40266971- 22370999- 55359854)

Table 3 Examples of Good Pattern for f=3

Since we only provide patterns for the largest n which k redundancy can cover -- for example, in the case f=3 as k=5 n is 11; or as k=6 then n becomes 26 -- the question arises about other (n,k) which are in between, for example, (11,5) and (26,6) in the case f=3. Our answer is that with a same number of redundancy, as long as an n-CCC has a good pattern, (n-1)-CCC should have a good pattern as well. The following theorem supports our answer:

Theorem 3: If (n,k) scheme corrects all cases of f faults then (n-1,k) scheme corrects all cases of f faults.
Proof: Omitted due to limited space.

Isomorphism of the Mapping Function

In this section, we want to show that the mapping between the logical address and the physical address is an isomorphism (two sets are of the same form). In our case the set of logical addresses and the set of physical addresses after reconfiguring transformation are isomorphic. Let us start with the following well-known concept of algebra: Let $T:V \rightarrow W$. If the mapping is both one-to-one and onto, then T is an isomorphism.

For our case, we define x_ℓ as an n by 2^n matrix which contains 2^n logical addresses. Similarly, x_p is an n by 2^n matrix which contains 2^n physical addresses. The M, an n by n mapping matrix which maps x_ℓ to x_p, i.e., $M \cdot x_\ell = x_p$.

In terms of algebraic expression, we shall define a linear operator, T_M, such that $T_M(x_\ell) = M \cdot x_\ell$. And we want to prove that T_M is a one-to-one and onto mapping. The next three theorems are arranged in the way that the first two provide stepping stones and the final one concludes the proof.

Theorem 4: Let $T:V \rightarrow W$ be a linear transformation between two finite-dimensional vector spaces, V and W. If T is invertible, then T is an isomorphism.
Proof: Omitted due to simple linear algebra fact.

Theorem 5: Let T_A be the linear operator on R^n defined by $T_A(x) = A \cdot x$, if A is an invertible n x n matrix then T_A is also invertible.
Proof: Omitted due to simple linear algebra fact.

Theorem 6: Let M be an invertible n x n matrix, x_ℓ and x_p are n x 2^n matrices, such that $M \cdot x_\ell = x_p$. Also, let T_M be the linear operator, defined by $T_M(x_\ell) = M \cdot x_\ell$. Then, T_M is an isomorphism.
Proof: Omitted due to limited space

Analogously, we may prove its inverse transformation, $T'_{M^{-1}}$ is an isomorphism as well.

Theorem 7: Let M be an invertible n x n matrix. x_ℓ, x_p are n x 2^n matrices such that $x_\ell = M^{-1}x_p$. Let $T'_{M^{-1}}$ be the linear operation defined by $T'_{M^{-1}}(x_p) = M^{-1}x_p$. Then $T'_{M^{-1}}$ is an isomorphism.
Proof: Omitted due to similarity to theorem 6.

How theorem 6 applies to our case is that the set of logical addresses (x_ℓ) is isomorphic to the set of physical addresses (x_p) through the logical-to-physical transformation T_M. Theorem 7 applies to our case in the reverse way. The set of physical addresses is isomorphic to the set of logical addresses through the physical-to-logical transformation, $T'_{M^{-1}}$. Namely, we have the following two legitimate equations to transform from the physical to the logical and vice versa:
$$M \cdot x_\ell = x_p; \quad x_\ell = M^{-1} \cdot x_p.$$

Reconfiguration Process

When faults occur the reconfiguration program shall step in to carry out the reconfiguration process. As the (ℓ,p) tuple suggested there are two layers of the reconfiguration process. In the first layer it maps logical cycles to physical cycles (ℓ part) and in second layer it maps logical

PEs to physical PEs (p part). Let us begin with the first layer.

It first locates the substituted m-connections from the remaining connections. Once these m-connections are found it shall form the mapping matrix, M. With M matrix it then performs the logical-to-physical mapping, namely, $M \cdot x_l = x_p$. The mapping result should be available to the system which in turn directs communications among cycles as routings are requested.

Let us take the previous example -- a (4, 3) scheme has (14- 13- 11-) redundancy and tolerate 2 faults -- to go through the entire reconfiguration process. Since the example is able to tolerate two faults we shall randomly pick two faulty m-connections, say 2- and 4-connections. Consequently the second and third columns in C_r matrix should be crossed out. The resultant C_p' is therefore a 4 x 5 matrix which has five m-connections. The program 1 has to find out four m-connections which are able to reconfigure the system, i.e., to locate a 4 by 4 invertible matrix, M, from C_p'. In our case M happens to be 1-, 8-, 14-, and 13-connections. The logical-to-physical mapping, $M \cdot x_l = x_p$, is performed as follows:

$$\begin{bmatrix} 1 & 0 & 0 & 1 \\ 0 & 0 & 1 & 0 \\ 0 & 0 & 1 & 1 \\ 0 & 1 & 1 & 1 \end{bmatrix} \begin{bmatrix} 0 & 1 & 0 & 1 & 0 & 1 & 0 & 1 & 0 & 1 & 0 & 1 & 0 & 1 & 0 & 1 \\ 0 & 0 & 1 & 1 & 0 & 0 & 1 & 1 & 0 & 0 & 1 & 1 & 0 & 0 & 1 & 1 \\ 0 & 0 & 0 & 0 & 1 & 1 & 1 & 1 & 0 & 0 & 0 & 0 & 1 & 1 & 1 & 1 \\ 0 & 0 & 0 & 0 & 0 & 0 & 0 & 0 & 1 & 1 & 1 & 1 & 1 & 1 & 1 & 1 \end{bmatrix}$$

$$= \begin{bmatrix} 0 & 1 & 0 & 1 & 0 & 1 & 0 & 1 & 1 & 0 & 1 & 0 & 1 & 0 & 1 & 0 \\ 0 & 0 & 0 & 0 & 1 & 1 & 1 & 1 & 0 & 0 & 0 & 0 & 1 & 1 & 1 & 1 \\ 0 & 0 & 0 & 0 & 1 & 1 & 1 & 1 & 1 & 1 & 1 & 0 & 0 & 0 & 0 \\ 0 & 0 & 1 & 1 & 1 & 1 & 0 & 0 & 1 & 1 & 0 & 0 & 0 & 0 & 1 & 1 \end{bmatrix}$$

or $(0\ 1\ 2\ 3\ 4\ 5\ 6\ 7\ 8\ 9\ 10\ 11\ 12\ 13\ 14\ 15)_{logical\ cycle} \rightarrow$
$(0\ 1\ 8\ 9\ 14\ 15\ 6\ 7\ 13\ 12\ 5\ 4\ 3\ 2\ 11\ 10)_{physical\ cycle}$
The m-connection mapping is:
$(1-\ 2-\ 4-\ 8-)_{logical\ conn} \rightarrow (1-\ 8-\ 14-\ 13-)_{physical\ conn}$.
The second layer mapping which maps PE within a cycle is: $(0\ 1\ 2\ 3)_{logical\ PE} \rightarrow (0\ 3\ 4\ 5)_{physical\ PE}$.

From now on data which is supposed to move to (2,3) before faults occur now has to go to (8,5) and uses 13-connection instead of 8-connection, and etc. These changes will be directed by the system, therefore, application programs remain unchanged. Restoration of programs and data back to the system right after the reconfiguration is done should also follow described procedures.

Conclusions

This research presented connections of CCC networks in the concept of vector space and solved the multiple-fault tolerance issue. Our next effort will try to solve the decomposition issue in the manner that both optimal redundancy and uniform layout in terms of overall hardware cost can be achieved.

Bibliography

[Adam82] G. B. Adams III and H. J. Siegel, "The Extra Stage Cube: A Fault-Tolerant Interconnection Network for Supersystems," IEEE Trans. Comput., May 1982, pp. 443-454.

[Adam87] G. B. Adams III, D. P. Agrawal, and H. J. Siegel, "A survey and Comparison of Fault-Tolerant Multistage Interconnection Networks," IEEE Computer, June 1987, pp. 14-27.

[Bane86] P. Banerjee, S. Y. Kuo and W. K. Fuchs, "Reconfigurable Cube-Connected Cycles Architectures," Proc. 16th Int'l Symp. Fault-tolerant Computing, July 1986, pp. 286-291.

[Bane88] P. Banerjee, "The Cubical Ring Connected Cycles: A Fault-Tolerant Parallel Computation Network," IEEE Trans. Comput., Vol. 37, 5, May 1988, pp. 632-636.

[Batc79] K. E. Batcher, "MPP - A Massively Parallel Processor," in Proc. Int. Conf. Parallel Processing, Aug. 1979, pp. 249.

[Batc80] K. E. Batcher, "Architecture of a Massively Parallel Processor," Seventh Annual Symposium on Computer Architecture, May 1980, pp. 168-173.

[Batc80] K. E. Batcher, "Design of a Massively Parallel Processor," IEEE Trans. Comput., Vol. C-29, no. 9, Sept. 1980, pp. 836-840.

[Batc82] K. E. Batcher, "Bit-Serial Parallel Processing Systems," IEEE Trans. Comput., Vol. C-31, no. 5, May 1982, pp. 377-384.

[Chen90] S. K. Chen, "n^+-Cube: The Extra Dimensional n-Cube," Proceeding ICPP, Aug. 1990, I583-584.

[Lati90] S. Latifi, "The efficiency of the Folded Hypercube in Subcube Allocation," proceeding ICPP, Aug. 1990, I218-221.

[Prep81] F. P. Preparata and J. Vuillemin, "The Cube-Connected Cycles: A Versatile Network for Parallel Computation," Comm. ACM 24, 5, May 1981, pp. 300-309.

[Sieg87] H. J. Siegel, W. T. Hsu, and M. Jeng, "An Introduction to the Multistage Cube Family of Interconnection Networks," The Journal of Supercomputing, 1, 1987, pp. 13-42.

[Tzen90] N. F. Tzeng, S. Bhattacharga, and P. J. Chuang, "Fault-Tolerant Cube-Connected Cycles Structures Through Dimensional Substitution," proceeding ICPP, Aug. 1990, I433-440.

AN ENHANCED DATA-DRIVEN ARCHITECTURE

Hallo Ahmed   Lars-Erik Thorelli   Johan Wennlund

Dept. of Telecommunication and Computer Systems
The Royal Institute of Technology (KTH)
100 44 Stockholm, Sweden

Abstract Conventional dataflow architectures suffer from a number inefficient mechanisms. The most serious of these deficiencies are: control and handling of run-time parallelism, execution of sequential instructions, and storage and processing of arrays of data. The enhanced data-driven architecture attempts to solve some of these problems by providing a dynamic queueing mechanisms for supporting multiple activations of vector nodes without the need for tagging. It supports efficient processing and handling of arrays of data by vectorising loops instead of unfolding them. And it achieves efficient processing of sequential nodes by providing each node with the address of one of its successors and using this information for prefetching nodes, thus bypassing parts of the scheduling pipe of the machine. The architecture has been simulated and a prototype processing element has been implemented.

1. Introduction

In the dataflow model of computation [1, 2], a program is represented by a directed graph in which the nodes denote operations and the arcs connecting them represent data dependencies. Nodes become enabled when their operands arrive, thus computations are *data-driven*. This model supports concurrent execution of enabled node; firing of nodes is asynchronous because of its data-driven nature; and computations are free from side-effects because there is no notion of shared data storage and intermediate results are conveyed directly by means of *tokens*.

These properties imply that multiprocessor architectures based on the dataflow model need not suffer from the synchronisation and coordination overheads incurred in von Neumann architectures. Unfortunately, apart from one or two exceptions (such as the SIGMA-1 [3]), dataflow machines have not yet achieved the performance dictated by their underlying model of computation.

We believe, along with others [3] - [7], that inefficiencies in existing dataflow architectures tend to outweigh any performance gains resulting from the data-driven computational model. The most serious of these inefficiencies are the handling and control of parallelism; the storage and processing of structured data, such as arrays; and the large overheads associated with processing fine grain nodes.

In this paper we present a novel dataflow architecture with a number of extensions aimed at improving overall performance and achieving supercomputer performance in the field of numerical computations. The main objectives of our architecture are:

1. To control run-time parallelism more efficiently by providing a dynamic queueing mechanism [8]. This scheme facilitates multiple activations of nodes without the need for tagging and its associated overheads.
2. To achieve high performance by supporting an efficient storage scheme for structured data and providing a vector processing capability similar to conventional pipelined vector architectures. The efficiency of this approach has also been demonstrated by others [3, 5, 6].
3. To accomplish efficient processing of sequential nodes and to support chaining of vector operations. This is achieved by providing a prefetching mechanism and a number of registers [8, 9].

The architecture presented in this paper has been simulated in detail and is currently in the final stages of its implementation. Simulation results as well as preliminary tests on the hardware confirm the validity of our design decisions.

The remainder of this paper is organised as follows: in the next section the drawbacks of existing dataflow architectures are discussed in more detail. In § 3 the enhanced data-driven model is presented. In § 4 the implemented machine is described. Simulation results and some concluding remarks are presented in § 5 and 6, respectively.

2. Problem Definition

Conventional dataflow architectures, both static [10, 11] and dynamic [12, 13], suffer from a number of deficiencies. In this section we will describe the most serious of these deficiencies and their effect on performance.

2.1. Exploiting Parallelism

In static architectures multiple activations of nodes are prohibited, thus limiting the amount of parallelism exploited by these machines to that detected at compile time. As a result, graphs with small and medium degrees of parallelism do not achieve high performance. This feature is noted in the Hughes Data-Flow Multiprocessor [14], for example.

On the other hand, dynamic architectures support multiple activations of nodes at run-time. This scheme has the following drawbacks:

a. The generation of multiple instances of nodes requires additional overhead instructions for manipulating tags and controlling parallelism. According to Arvind et. al. [15], to allow maximal parallelism while preserving determinacy, dynamic dataflow graphs require as much as 150% overhead instructions.

b. Since the generation of new activations of nodes, functions, or loop iterations is data-driven, highly parallel programs may generate large numbers of partially enabled nodes – diadic nodes which have received only one operand. This is also noted in [16] based on experience from the Manchester dataflow computer. Deadlock may result if the number of such nodes exceeds the resources of the machine (in this case the size of its matching store). However, although deadlock can be prevented by throttling techniques [6, 16, 17] at the cost of some overhead, the generation of large numbers of partially enabled nodes can only be reduced by throttling and not suppressed. Each partially enabled node causes a mismatch in the matching store, consequently producing a bubble (lost pipeline cycles) in the functional unit and leading to performance degradation.

2.2. Handling and Processing Structured Data

Structure storage schemes both in dynamic and static architectures generally [a] support element-wise access to data structures. In static architectures, structured data are stored in the heap as trees having one or more root nodes with array elements at the leaves [20]. Elements are reached over a directed path from a root node. Drawbacks of this scheme are the excessive storage space required for representing structured data (in spite of structure sharing) and the delays associated with accessing them [4, 21].

In dynamic architectures, in order to improve parallelism, storage is allocated to structured data before expressions producing them are invoked. This scheme is implemented by associating a presence bit with every element in the store. An attempt to read an empty location is deferred until an element is written in that location. This description applies to both I-structures [22] and the structure store [16]. Disadvantages of this scheme are:

a. Deferred read requests must be processed when the elements finally arrive. As noted by Gajski et. al. [4], checking for deferred reads on every write slows down access of I-structures in comparison

with the simple von Neumann model.

b. The number of transferred packets in the system increases because each read operation requires a request packet and a value packet returned from the structure store. Moreover, overhead instructions are required for calculating the address of every array element [3, 4, 6].

2.3. Fine Granularity

Conventional dataflow models, static and dynamic, do not take advantage of the simple data dependence relation between successive instructions. Every enabled diadic node must be produced either by matching or updating operations, i.e. tokens *must* enter the scheduling pipe of the machine. It is possible, however, to avoid these overheads if some intermediate results could somehow *bypass* parts of the scheduling pipe. In the von Neumann model, sequential threads are efficiently processed; a program counter selects the successor instruction and a register set is used for passing intermediate results.

To appreciate the gravity of this problem in dataflow architectures, we draw the reader's attention to the fact that in dynamic models, the arrival of the first operand of a diadic node when both operand are variables, *always* results in a mismatch and consequently produces a bubble. This problem is clearly stated in [23] where it is shown that for a certain non-numeric program nearly 29% of the ALU's cycles are consumed by bubbles. However, we expect this number to increase significantly for problems involving more dynamic parallelism.

3. The Model

The main characteristics of our model may be summarised in the following points.

1. The nodes of the dataflow graph can be either *scalar* or *vector* nodes. A scalar node may have up to two scalar type operands (Figure 1). However, a vector node must have at least one vector type operand and at most one additional scalar or vector operand (Figure 2). A vector type operand is a pointer to a vector structure.

2. Input arcs of scalar nodes, called scalar arcs, carry at most one token. A scalar node fires when both its operands are available (for a diadic node) and a flag, called the *current node bit*, is set to one (Figure 1).

3. Vector nodes may carry, on their vector arcs, any number of vector pointers which are queued on these arc in a first come - first served (FCFS) manner, as depicted in Figure 2. A vector node fires when both its operands are available, consuming the first operands at the head of its input queues. Immediately after firing, the vector node could become enabled again if both its queues are not empty. If a resource (a functional unit) is available, the node

---

[a] Implementations of a few instructions that can fetch more than one element from structure storage are reported in [18, 19]. However, the large grain nature of these instructions leads to long successions of unsatisfied matching operations caused by the fine grain nature of the architecture, and in particular its matching store.

can fire again while its first instance is still being processed. Thus, multiple activations of a vector instruction can be processed in parallel. The only constraint is that the results of these parallel vector operations must also be handled (stored, communicated, and processed) in an FCFS manner, which can be achieved with a suitable result buffering scheme.

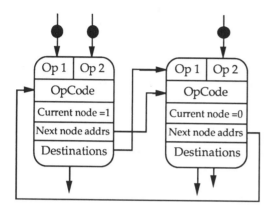

FIGURE 1: Structure of scalar nodes.

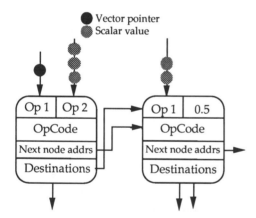

FIGURE 2: Structure of vector nodes.

4. Each node in the dataflow graph carries the address of one of its successors, as shown in figures 1 and 2. This *next node address* is used for checking the status of a successor while the node itself is being processed. If the next node already has one operand then it is *prefetched* and gets the other operand *directly* from the node currently being processed. This scheme facilitates bypassing parts of the scheduling pipe and allows processing two or more sequential nodes by saving intermediate results in registers and supplying them directly to successor nodes. In addition, vector nodes utilise the next node address information for chaining vector operations.

5. In our structure storage scheme, structured data are stored as one dimensional arrays of a finite length. Vector operations on these arrays are atomic, i.e. arrays are produced or consumed as a whole

without partial updates. A vector node writes its result *once* to structure memory, however, consumer nodes are allowed unlimited *read* access to the structure. This scheme leads to efficient memory utilisation (only one copy per vector structure) and less memory contention (only one write access per vector structure).

6. A number of sequential scalar nodes can be grouped into a *task* structure, as depicted in Figure 1. Only one node in a task (initially the first) is enabled at a time. When a node is processed it resets a single bit flag (the current node bit) and transfers control to its successor by setting its flag using the next node address. This flag in conjunction with the next node address functions as a simple program counter transferring control from one node to its successor. This scheme guarantees that nodes inside a task have at most one token on their input arcs. However, inputs to a task structure are synchronised explicitly using synchronisation nodes whenever necessary.

Vectorisation of dataflow graphs along with the vector queueing scheme aim at solving the problem of handling and control of parallelism. In our model, loops are vectorised instead of being unfolded. A few tens of iterations of a loop in the dynamic model – with all its tag management instructions, structure memory transactions, mismatches in the matching store and its associated bubbles – can be replaced with a few vector and scalar operations in our model. It has been reported by Hiraki et. al. [3] that vectorising loops on the SIGMA-1, which is a dynamic vector dataflow computer, results in three to four times speed up compared to unfolding techniques.

At a higher level, our model supports concurrent activations of vector nodes. This level of parallelism is supported by the vector queueing scheme, without the need for tags or overhead instructions.

The problem of handling and processing structured data is solved by providing a vector processing capability and an efficient structure storage scheme. And finally, the complications arising from fine granularity are solved by giving each node the address of one of its successors and providing a prefetching mechanism and a number of registers.

4. Architecture

We have implemented one processing element using off-the-shelf components and Programmable Logic Devices (PLDs). The prototype processing element, illustrated in Figure 3, consists of a six stage pipe (scalar nodes use only five stages) with a cycle time of 100 nS per stage or a maximum performance of 10 million floating point operations per second (10 MFLOPS). A short description of each stage along with its function is given below.

Stage 1, fetch and control unit: This unit selects a node for processing. The address of which comes either from the ready queue, which contains pointers to all enabled nodes in graph memory, or from the next node address register (located in the fetch and control unit). At the end of each cycle, this register is updated with the address of the successor of the node currently being processed. The fetch and control unit obtains this address by reading the next node address field of the current node in graph memory (see also Figures 1 and 2). The content of this register is selected in the next cycle if a signal received from stage 2 indicates that the successor already has one operand. Otherwise, an address is taken from the head of the ready queue. The selected address is forwarded to graph memory.

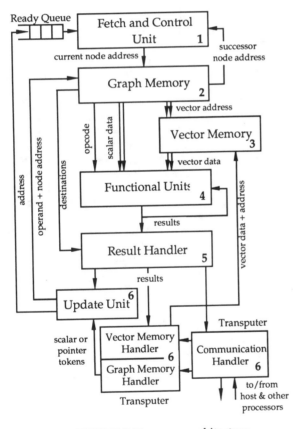

FIGURE 3: Processor architecture.

Stage 2, graph memory: The dataflow graph along with all scalar operands (including vector pointers) are stored in graph memory which is partitioned at the hardware level to facilitate independently addressed fields. The following access operations take place concurrently during one cycle.

1. If node type is vector, its operands (only if they are vector pointers) are written to two address registers in vector memory.

2. If node type is scalar, operand values are written to operand registers in the functional unit. In this case pipeline stage 3 is bypassed.

3. Status bits of the current node are modified: presence bits of variable operands and current node bits are cleared.

4. Destination addresses (where to send the results) of the node to be processed are written to first in-first out queues in the result handler, where they wait until the corresponding data is computed. Up to 3 destinations are allowed.

5. The operation code is written to a register in the functional unit.

6. The next node address field of the selected node (the current node) is written to a register in graph memory and a copy of it is fed back to stage 1. The status of the successor (its presence and current node bits) is checked and a signal is sent to the fetch and control unit indicating if the successor node is executable.

Stage 3, vector memory: This memory is used for storing structured data in the form of arrays with a maximum length of 64×64 bits. It is a multiported memory allowing overlapped read/write access in the same cycle. Vector operands are accessed using two vector address registers. The first element of each vector is written to an operand register in the functional unit. To access consecutive vector elements, the vector address registers are incremented by one after every read cycle. Note that this stage is bypassed when scalar nodes are processed.

Stage 4, functional unit: This stage consists of three functional units, an integer ALU (AM29C332), an integer multiplier (AM29C323) and a floating point unit (AM29C325). The operands received from stage 2 (scalars) or stage 3 (vectors) are processed according to the operation code delivered by stage 2, and results are written into the functional unit's output register. If a result is an input to a successor node, its value is taken directly from the output register instead of circulating it through the pipe.

Stage 5, result handler: This unit consists of three first in-first out queues. Destination addresses delivered to this stage from graph memory are buffered in these queues where they are combined with the results delivered from the functional unit. Each queue is capable of buffering 128 floating point scalars (or two floating point vectors). Local (internal) scalar results are written to the first queue (one instance of data per destination). Local vector results are written to the second queue (only one copy along with all its destinations). Finally, all external results destined to other processing elements, scalar and vector, are written to the third queue (one copy per result along with addresses).

Stage 6, result distribution: This stage consists of three parallel units, each taking its input from one of the queues of stage 5.

1. The update unit: Reads a scalar result along with its address from the first queue, and updates the corresponding operand and its presence bit in graph memory. Upon updating, the node is checked to see if it is enabled, in which case its address is appended to the ready queue. This unit also receives pointer and scalar operands from the graph memory handler for updating vector and scalar nodes. The result of this unit – an enabled node address – is fed back to stage 1 forming a circular pipe.

2. Vector memory and graph memory handler: This unit consists of one T801 transputer and some additional logic and memory. The vector memory handler is responsible for memory allocation to vectors read from the second queue of stage 5, and external vectors received from the communications handler. The graph memory handler is responsible for managing vector pointer queues on the arcs of vector nodes, and supplying the update unit with vector pointers (after the vector memory handler writes them to vector memory) as well as external scalar tokens received by the communication handler from other processors.

The transputer runs at 25 MHz, therefore this stage is asynchronous with the rest of the pipe. In addition to its speed, a transputer was chosen because it provides a flexible – programmable – environment for testing various allocation, garbage collection, and queue management schemes for vector arcs. Moreover, it comes at an affordable price.

3. Communication handler: The main element in this unit is also a T801 transputer which is responsible for communication with the host computer and other processors utilising the transputer links. Downloading data and code from the host also takes place through this unit. The communication handler reads the contents of the third queue in stage 5 and forms scalar or vector packets to be sent to the host or to other processors. This unit also receives external scalar tokens which are forwarded to the graph memory handler, and vector structures which are forwarded to the vector memory handler. This stage is also asynchronous relative to the main pipe.

Our motivations for implementing a prototype were primarily to verify the feasibility of realising the enhanced data-driven architecture in a reasonable way and to obtain realistic parameters for our simulator. Our goal was not the implementation of a state-of-the-art machine which would compete in terms of performance with existing computers.

Because of these considerations, several properties of the model were not implemented in this prototype, such as the ability to support multiple activations of vector nodes and vector chaining. We also

did not use pipelined functional units, because of the relatively slow pipeline period for each stage (100 nS). Instead, fast non-pipelined functional units were used. In addition we limited the size of the data to 32 bits.

5. Performance

The prototype processing element is not fully operational at the time of writing this paper (April 1991). It is currently being tested and debugged. The architecture was simulated in detail by writing a hardware level simulator ($\approx$ 2500 lines) in the process oriented and C based language CSIM [24]. Input to the simulator consists of hand-compiled dataflow graphs.

Our benchmarks include partial differential equations (PDEs), vector summation and matrix multiplication programs which are executed on a simulated enhanced dataflow multiprocessor with a torus interconnection network. The effect of different job mixes on performance by executing several programs in a multiprogramming environment were also studied.

The effect of the prefetching mechanism, facilitated by providing each node with the address of one of its successors, is depicted in Table 1. The benchmark is a vectorised Laplace PDE for a mesh of 16x64 points. The program is run for 40 iterations on one processing element. Row A in Table 1 represents results associated with a dataflow graph exploiting the next node address information, while row B shows results for the case where such information is not used.

The first column shows the time taken to execute the graph. The second two columns represent functional unit utilisation and number of nodes executed by this unit. The next two columns are fetch and control unit statistics. The first entry is the average length of the ready queue which contains addresses of enabled nodes. The other column shows the number of addresses read from this queue. The number of nodes updated by the update unit is represented by the next column.

Finally, the last three entries are related to vector memory. The average length of the queue of vectors to be written to vector memory is shown in column Q Length. The next entry is the average response time of vector memory, i.e. the average time required to store a vector in memory, including waiting time. And the last column is the number of write operations to vector memory.

The results presented in Table 1 demonstrates the significance of prefetching and its associated register transfer mechanism in improving performance. Most of the results shown in the table are self explanatory, therefore we will describe the few entries which are not.

		Functional Unit		Fetch & Control Unit		Update Unit	Vector Memory		
	Time μS	Util.	# Nodes	Q Length	Q Fetches	# Updates	Q Length	Res. Time	# VMW
A	31 078	0.96	10 880	28.4	4 480	7 600	2.59	25.2 μS	3200
B	41 868	0.81	10 880	5.7	8 320	11 440	23.66	154.8 μS	6400

TABLE 1: Effect of the next node address information and its associated prefetching mechanism on performance when the program is executed on one processor.

The average length of the ready queue is higher in the case where prefetching is performed (row A) and the number of fetch operations is lower. The reason is that most of the time the fetch and control unit is *prefetching* successor nodes which belong to the same thread of instructions rather than fetching nodes from other threads. Another observation is that in case B, vector memory becomes a bottle neck. The average number of vectors waiting to be stored (Q Length) is 23.66, while in case A it is much lower.

In order to investigate the combined effect of eliminating prefetching, i.e. reducing node granularity, and decreasing vector length, i.e. reducing data granularity, we performed a test involving a fine grain version of the Laplace program (without the next node information) for a mesh of 16×64 elements. Vector length was reduced by half after each run and the corresponding processing time was measured. The program was executed on eight processing elements so that the effect of communication on performance is also observed. The program was mapped uniformly to the processors such that each one was responsible for evaluating two adjacent columns of the mesh.

Note that vector length was reduced to a minimum of 2 elements. The reason why it was not reduced further is that our algorithm becomes meaningless for a one dimensional mesh.

The effect of varying both data and node granularity on performance can be seen very clearly from the graph of Figure 4. Reducing data granularity has, as expected, profound influence on performance. The value of t_2 represents processing time for the fine grain graph when vector length is 64, while t_1 represents the time when prefetching is utilised.

The relationship between processing time and the number of processors is depicted in Figure 4.5. Processing time decreases linearly as the load is distributed to more processors, however, this linear behaviour ceases in this example when the number of processors is more than 8. The reason is the variations in the degree of available parallelism in a processor. We observed that when the program was executed by up to 8 processors the degree of parallelism was greater than 1. On the other hand,

when the program was allocated to 16 processors this value droped to 0.123, i.e. the processor is idle most of the time since not enough enabled nodes are available. In addition, this implies that most of the communication will not overlap with computations.

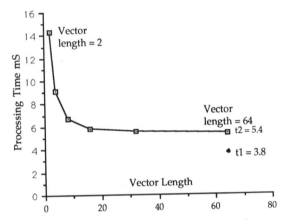

FIGURE 4: Effect of node granularity and vector length on processing time when a 16x64 mesh is executed on 8 processors.

6. Conclusions

The basic data-driven model has many drawbacks, such as: excessive communication and matching demands due to its fine granularity, inefficient handling and processing of structured data, inefficient execution of sequential nodes, and the absence of an intermediate temporary storage scheme, such as registers.

The enhanced data-driven architecture presented in this paper attempts to solve some of these problems by introducing new mechanisms and improving existing ones. In this scheme, nodes of the dataflow graph can handle vector, as well as scalar data, to achieve high-performance in numerical computations. A vector queueing mechanism is used to control run-time parallelism and synchronise the flow of vector operands. Multiple threads of

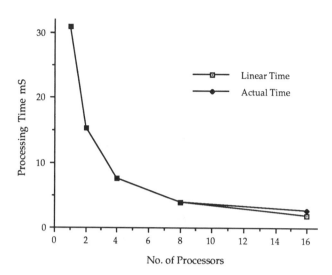

FIGURE 5: Processing time of the Laplace Equation for a 16x64 grid and 40 iterations.

execution are supported and results within a thread are passed by means of register transfer instead of tokens, thus optimising execution of sequential nodes. Finally, a degree of control is allowed to override the data-driven nature of computations in order to eliminate node-level synchronisation in scalar computations.

Behaviour of this architecture was simulated and results indicate the effectiveness of its mechanisms. We have also implemented one processing element to demonstrate the feasibility of realising the various mechanisms of the enhanced data-driven architecture.

References

[1]   R.M. Karp, R.E. Miller, _Properties of a model for parallel conventions: determinacy, termination, queueing_, RC-1285, IBM Research Center, Yorktown Heights, New York, (1964).

[2]   J.E. Rodriguez, _A Graph Model for Parallel Computation_, Ph.D. Thesis, MIT, Cambridge, Mass., (Sep. 1967).

[3]   K. Hiraki, S. Sekiguchi, T. Shimada, "Efficient Vector Processing on a Dataflow Supercomputer SIGMA-1", _Proceedings of the Int. Conf. on Supercomputing_, (1988).

[4]   D.D. Gajski, D.A. Padua, D.J. Kuck, R.H. Kuhn, "A Second Opinion on Data Flow Machines and Languages". _IEEE Computer_, (Feb, 1982).

[5]   W. Najjar, J.-L. Gaudiot, "A Hierarchical Data-Driven Model For Multi-Grid Problem Solving", _High Performance Computer Systems_, Editor, E. Gelenbe. Elsevier Science Publishers B. V. (1988).

[6]   H. Yamana, T. Marushima, T. Hagiwara, Y. Muraoka, "System Architecture of Parallel Processing System-Harray-", _Proceedings of the Int. Conf. on Supercomputing_, (1988).

[7]   S. Sakai, Y. Yamaguchi, K. Hiraki, Y. Kodama, T. Yuba, "An Architecture of a Dataflow Single Chip Processor", _Proceedings of the 16th Annual International Symp. on Computer Architecture_, (1989).

[8]   H. Ahmed, "A Dataflow Model Based on a Vector Queueing Scheme", _Proceedings of EUROMICRO-90, Part 2_, (1990).

[9]   H. Ahmed, _An Enhanced Data-Driven Architecture for Efficient Scalar and Vector Processing_, Ph.D. Thesis, CSALAB, The Royal Inst. of Technology, Stockholm, Sweden, TRITA-TCS-9103, (1991).

[10]   J.B. Dennis, "Static Data Flow Computation", _Control Flow and Data Flow: Concepts of Distributed Programming_, Editor M. Broy, Vol. F14. Springer-Verlag Berlin, (1985).

[11]   J.B. Dennis, "Dataflow Computation for Artificial Intelligence", _Parallel Processing for Supercomputers & Artificial Intelligence_, McGraw-Hill, (1988).

[12]   Arvind, K.P. Gostelow, "The U-Interpreter", _IEEE Computer_, (Feb. 1982).

[13]   J. Gurd, C.C. Kirkham, I. Watson, "The Manchester Prototype Dataflow Computer", _Comm. of the ACM_, (1, 1985).

[14]   J.-L. Gaudiot, et. al., "A Distributed VLSI Architecture for Efficient Signal and Data Processing", _IEEE Transactions on Computers_, (Dec, 1985).

[15]   Arvind, D.E. Culler, G.K. Maa, "Assessing the Benefits of Fine-Grain Parallelism in Dataflow Programs", _Proceedings of Supercomputing '88_, (1988).

[16]   P. Barahona, W. Boehm, J. Gurd, C. Kirkham, A. Parker, J. Sargeant, I. Watson, _The Dataflow Approach to Parallel Computation_, Dep. of Computer Science, University of Manchester, England, Technical report, (1986).

[17]   Arvind, D.E. Culler, _Managing Resources in a Parallel Machine_, Laboratory for Computer Science, MIT, Cambridge, Mass. CSG Memo 257, (1985).

[18]   T. Shimada, et. al., "Evaluation of a Prototype Dataflow Processor of the SIGMA-1 for Scientific Computations", _Proceedings of the 13th ACM Symposium on Computer Architecture_, (1986).

[19]   W. Böhm, J. Gurd, Y.M. Teo, "The Effect of Iterative Instructions in Dataflow Computers", _Proceedings of the International Conference on Parallel Processing_, (1989).

[20]   J.B. Dennis, _First Version of a Data Flow Procedure Language_, Laboratory for Computer Science, MIT, May 1975.

[21]   D.P. Misunas, _A Computer Architecture for Data-Flow Computation_, Cambrige, Mass., M.Sc. Thesis, MIT/LCS/TM-100, (March, 1978).

[22]   Arvind, R.E. Thomas, _I-Structure: An Effective Data Structure for Functional Languages_, Laboratory for Computer Science, MIT, Cambridge, Mass., LCS-TM178, (1978).

[23]   G.M. Papadopoulos, D.E. Culler, "Monsoon: an Explicit Token-Store Architecture", _Proceedings of the International Symp. on Computer Arch_, (1990).

[24]   H.D. Schwetman, _CSIM Reference Manual_ (Revision 11), Microelectronics and Computer Technology Corporation, Austin, TX, ACA-ST-252-87, (October, 1987).

A FINE GRAIN MIMD SYSTEM WITH HYBRID EVENT-DRIVEN/DATAFLOW
SYNCHRONIZATION FOR BIT-ORIENTED COMPUTATION

Mark R. Thistle
Thomas Lawrence Sterling

Supercomputing Research Center
Institute for Defense Analyses
17100 Science Drive
Bowie, Maryland 20715

Abstract --The relationship between event driven and data driven synchronization is explored in the framework of an operational system and real world application programs. Hardware support for fine grain event driven synchronization is coupled with software supported data driven synchronization to produce a hybrid synchronization model that permits efficient imperative program execution. The Zycad System Development Engine provides the event-based parallel execution engine on which is structured a static dataflow programming model. Example problems on an actual operational hybrid system reveal that high efficiency is possible by exploiting event-driven execution through a data driven model with low average temporal overhead.

Keywords: fine grain MIMD, programming methodology, static dataflow, digital logic simulation, event-driven synchronization

INTRODUCTION

MIMD fine grain parallel processing requires a large synchronization name space and hardware support for synchronization mechanisms to achieve efficient processing and high scalability.[1] To be useful, such systems also require a parallel programming methodology that is imperative in nature, reveals fine grain parallelism of the user applications, and is intuitive to the programmer.[2] Many different semantic schema for fine grain parallel processing have been considered and some implemented. Two of particular interest that are closely associated are event-driven synchronization and data-driven synchronization.[3][4] The former has received substantial attention in the domain-specific area of simulation.[5][6] The latter is the basis for the dataflow models of computation.[7]

Event-driven synchronization invokes an operation whenever an argument state change occurs. Data-driven synchronization elicits a compute operation only when all the necessary argument values have been derived. The two possess many similarities. They are asynchronous, support parallel execution, permit a wide range of task granularity, and imply message passing mechanisms. They are distinguished by their semantics as they relate to computation control flow. Event-driven semantics are more flexible in that they may be applied to achieve diverse flow control relationships. They are also more primitive in that the base semantics do not themselves establish a complete imperative flow control discipline. Data-driven semantics are more restrictive in the choice and method of coordinating parallel computation but are complete as an imperative model of computation.

The importance of event-driven synchronization is that it demands the support only of locally atomic mechanisms. A single event message incident upon some operation within the physically distributed name space invokes an action completely local to that location. In comparison, some number of separate data or signal messages must accrue, distributed in time, before a basic operation is performed under data-driven synchronization. Some intermediate state must be preserved to conduct the nonatomic data-driven synchronization protocol. Ultimately the data-driven synchronization mechanism may be assembled from of more primitive event-driven synchronization mechanisms.

This paper investigates the relationship between these two models through the implementation of an operational parallel processing system. The hardware support was acquired in the form of a Zycad System Development Engine (SDE), a 64-processor multiprocessor optimized for digital gate-level logic simulation, which may be more generally viewed as bit-level computation.[8] Special purpose hardware in the SDE synchronizes execution of an operation upon receipt of result packets. Because of the restricted value space of operation results (binary or ternary data only), a new result is issued only if its value is different from the value of its previously issued result value. Thus total communication and computation may be significantly reduced. On top of execution model is constructed software support for a parallel programming methodology exploiting the static dataflow programming paradigm. Efficiency is examined from the perspective of overhead and effectiveness is evaluated through scalability measurements. Examples demonstrating the implications of this merger of semantics and mechanisms include application programs for bit-level processing algorithms. It is shown that high efficiency can be achieved through this hybrid architecture in spite of the processing of independently schedulable fine grain tasks.

The flow control semantics that are imposed on the event driven synchronization mechanisms of the SDE are derived from the static dataflow model of computation. While somewhat restrictive because of the lack of reentrant subroutines, the model is entirely satisfactory for our study of the relations between the event driven and data driven execution models and sufficient for our example application programs. In itself, the SDE does not provide programmer defined flow control constructs.[9] To achieve an imperative flow control discipline, conditional branching and iteration are asserted by software. The acyclic blocks of parallel computation contained within such constructs are implemented directly with the event driven primitives. The static dataflow model establishes the semantic constructs of a flow control discipline while retaining access to the fine grain program parallelism and direct support for event-driven flow control in acyclic program segments.

The remainder of this paper, after providing additional details about the Zycad SDE, focuses on the method by which the hardware event driven synchronization mechanism is augmented with the software dataflow flow control semantics. Examples illustrate the application of this hybrid execution system to real world problems. Quantitative results demonstrate the overhead and scalability characteristics of this system architecture. We close with consideration of future work.

ARCHITECTURE OF THE ZYCAD SDE

The event-driven synchronization paradigm is most often associated with discrete event simulation. It is generally implemented in software on top of traditional sequential execution hardware platforms and is therefore not usually considered as a viable execution paradigm for application areas other than simulation. The Zycad SDE, however, is a special purpose engine that provides direct hardware support for event-driven synchronization. Although the SDE's intended use is the execution of digital logic simulations, it provides a unique opportunity to study event-driven synchronization as a model of computation applicable to a more general class of application than digital circuit simulation. The architecture of the SDE is described briefly in this section to facilitate a better understanding of the opportunities and constraints encountered in pursuit of the objective of this work.

The SDE is a message-driven multiprocessor designed specifically for discrete time simulation of gate-level digital systems. It consists of 64 gate evaluation processors, operating under a 92ns clock and interconnected by a three-level bus hierarchy. There are additional hardware modules for memory model and bus arbitration support. The various hardware modules are contained within G-modules, each of which is a self sufficient unit of execution. An overview of the architecture is shown in Figure 1. The primitive operations executed by the SDE hardware include simple 3-input, 1-output logical functions (gates), D-type flip-flops and latches, nodes for combining multiple drivers, and user-configurable hardware-assisted memory. Every logic element may be specified to perform its operation with a either a zero or unit simulated time delay. Gates are statically mapped at load or compile time onto the processors, which perform the evaluation and state update for their local gates. Each of the 64 processors has a capacity of 16,384 gates, for a total capacity of 1 million gates in the largest configuration. In the broader context of general bit-oriented computation, a "gate firing" is equivalent to an instruction execution.

The processors and other hardware modules work in concert to implement an event-driven simulation algorithm, which is controlled by a single controller module, the Multifunction (M) Module. When an event occurs at a logic element (the output of the logic element changes state), messages are sent to all other processors containing the logic elements to which the source element output is logically connected. These destination processors evaluate the gates through table lookup to determine whether their output state is modified. The key observation is that a logic element is executed (evaluated) every time any one of its inputs change state. (Not all inputs on memory and latch primitives trigger evaluation.) If a logic element produces a new output state different from its previous state, then the state transition is considered an "event". Events are queued to be distributed to all affected gates (fanout) either during the same simulation clock period (for a zero delay gate) or during the next simulation clock period (for a unit delay gate). All modules in the system process the events scheduled for the current simulation time independently. When there are no more events in any module to process at the current time, all modules are synchronously advanced to the next time step by the M module, which controls the simulation, and begin processing again.

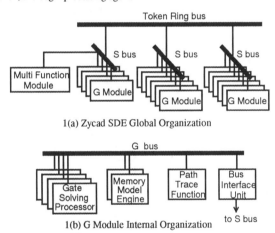

1(a) Zycad SDE Global Organization

1(b) G Module Internal Organization

Figure 1. Zycad SDE Architecture

RELATIONSHIP BETWEEN DATAFLOW AND SDE EXECUTION

The following is a close examination of the relationship between the Zycad SDE machine-level behavior reflecting its event-driven synchronization mechanisms and the static dataflow computing model employing data-driven flow control. From these observations it will be seen that, while the SDE permits efficient fine grain operations, it does not directly support an imperative programming methodology.

The static dataflow model of computation is an alternative to the classical von Neumann architecture for the exploitation of fine-grained parallelism to achieve substantial performance gains. Synchronization and firing of operations are performed independently and limited only by the data precedence constraints of the application algorithm. The dataflow model is value-oriented and functional in that there are no global side-effects. Synchronization is data driven. An operation is performed when the necessary operands are available and its result can be accepted by its receptor operations. The static dataflow architecture determines the placement of its operations among the parallel processing elements at compile time but determines when they are to be performed and in which order at run time. This permits a significant degree of adaptability to run-time-varying conditions resulting from data-dependent conditional program flow.[10]

In the SDE, the primitive operators are value-oriented, precluding the possibility of global side-effects. (The latches and flip-flops provide only local storage acting like buffers and gates. Even the memory model

supported in the SDE is accessible from a single set of statements, preventing global side-effects.) This functional aspect is the same as that for reduction and dataflow models.[11] Each primitive operation is independently schedulable on the SDE. This provides the fine grain tasks also found in dataflow computers. The control state of a program at any instant is distributed among all of the operations and not established by a small number of global registers. At compile time, operations to be performed are assigned to specific processing elements. However, the actual times of their execution are determined at run time. This static mapping and dynamic scheduling distinguishes the SDE from SIMD (Single Instruction Stream Multiple Data Stream) and systolic models, but strengthens the similarity with static dataflow computers.[12][13]

The native instruction set of the SDE does not support procedural specification of algorithms. Programs (simulation models) on the SDE are non-reentrant. There is never more than one instantiation of a given operator at one time, although an operator may be invoked many times in succession. Again, static dataflow programs, as well as systolic arrays, have the same property. More importantly, however, there is no flow control discipline to govern the stages of execution. It is difficult to determine the order in which operations will execute within a virtual time step as well as the precise virtual time step at which a particular value will be valid. This problem occurs for non-uniform program structures whose flow is dependent on intermediate values. Examples include case statement-like structures and iterative structures. The difficulty results from the event-driven execution mechanism of the SDE. The SDE evaluates an operation whenever any of the input arguments change state. In some cases, when a new input value is received, no new output will result from the operation because the new input arguments generate the same output value. This is an important optimization incorporated in the SDE for simulation purposes but it removes the program synchronization information from the event. At other times, operations will be performed producing result values that are temporarily wrong because they represent the effects of change in only one of the operation input arguments and not a complete new set of input arguments. Thus, for a new set of input argument values, several executions of the same operation could be performed. Hardware mechanisms in the SDE eliminate some of these redundant and sometimes erroneous values but with the effect of losing the program synchronization information. Therefore, the event-driven mechanism is distinguished from the data-driven model of computation, as well as the demand-driven and lazy evaluation schemes.[14][15][16]

The conclusion from these observations is that the Zycad SDE does not directly support the execution model required for the static dataflow model.

The SDE does support, however, correct execution of an important subset of dataflow program segments, those that exhibit an acyclic structure. Exploiting the zero-delay attribute of operations in the SDE, it is known that when, in the virtual-time sense, all of a zero-delay code segment's input values are valid, then all of the result values of that code segment are valid one time step later. For many applications, most program operations occur within such code segments. The uncertainty of timing arises from cyclic flow and hierarchical structures with conditional control. The SDE does not incorporate a uniform discipline for building such structures; there does not exist a set of primitive operators which support a model of flow control. A model of computation must be imposed by fabricating the necessary flow control primitives from the native operations of the SDE.

IMPLEMENTATION OF THE DATAFLOW MODEL ON THE SDE

From the previous section it is clear that the Zycad SDE instruction set does not conforms to the dataflow model of computation. In order to study the relationship of the data driven and event driven execution paradigms on the SDE hardware platform, the dataflow model was imposed on the SDE. It was found that the acyclic code segments of zero delay operations for the Zycad map directly to equivalent code segments of static dataflow operators. But the overriding requirement is that the input arguments to such functional structures be valid prior to execution. This implies a synchronization infrastructure that coordinates the execution of the separate acyclic code segments which make up an application program. This section describes the elements of such an infrastructure, built from the primitive operators of the SDE instruction set and compatible with the static dataflow paradigm.

Value Passing Synchronization

The objective of the control flow infrastructure is to assert discipline into the passing of values between interconnected code segments. This means that there is an explicit indication that a program value is available. An ancillary set of data is required to supply such indications. These additional data will be referred to as *synchronization signals*. The control infrastructure is constructed of such signals and the flow control discipline is enforced through the use of these signals. The static dataflow model of computation determines the conditions, known as the *firing rules*, required for any given operation to be executed. For *strict* operators, which include the vast majority of operations performed in most programs, the firing rules require that all input argument values be valid and that all receptor operators be prepared to accept the next result value. These criteria can be satisfied directly by dedicated signals or indirectly by the overall constraints imposed by the flow graph topology.

Figure 2 shows in abstract form the relationship between two connected code segments where it is assumed that a data value produced by the left-hand code segment, **A**, is consumed by the right-hand code segment, **B**. This value becomes valid when the *data signal* is active. The line directed from code segment **B** to code segment **A** represents the *acknowledge signal* indicating that **B** is prepared to accept a new result value from **A**. The forward and backward synchronization signals move data values across the interface between code segments **A** and **B** in a disciplined manner, permitting correct programs to be constructed from such interconnected segments according to the precepts of the dataflow model.

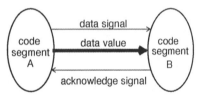

Figure 2. Inter Code Segment Synchronization

A simple ensemble of Zycad SDE primitive operators is assembled to support this synchronization protocol. This arrangement, shown in Figure 3, is referred to as a *flow control cell*. It buffers the data value and maintains a flag that indicates the status of the data value between the source and destination operators. A flow control cell is placed between every pair of operators that must synchronize the passing of data. The flow control logic is complemented by additional logic associated with each operator to enforce the firing rule. The firing rule for a strict operator requires that all input arguments be available and output destinations be ready to accept new values. The firing rule for an operator can be realized by a n-way Boolean AND function, where n is the number of input arguments and result destinations. All of the flow control cells associated with inputs to an operator must provide active *data signals* for the operator to fire and those flow control cells receiving the result from the operator must provide active *acknowledge signals*. Figure 4 illustrates the SDE primitive realization of a strict operator with three inputs and two destinations for the output.

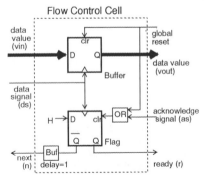

Figure 3. Synchronization Flow Control Cell

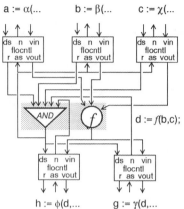

Figure 4. Flow Control for Strict Operators

The execution of an operator involves several data value transfers. Its input argument values are transferred from the argument source operators and its result data value is transferred to some number of descendant operators that are to consume its result. For each of these transfers, an operator is associated with one side or the other of a flow control cell. For each of its input arguments, the operator is associated with the consumer side of the respective flow control cells. For each of its descendant actors that accept its result value, the operator is associated with the source side of the respective flow control cells. In general, a dataflow graph (program) is mapped onto the SDE by instantiating a flow control cell for every arc in the graph where the full flow control discipline must be enforced and an SDE primitive or set of primitives for each operator function, accompanied by the appropriate firing rule logic.

The communication of data between a producer and consumer operator involves the transfer of data from the producer to the intermediate flow control cell and from the flow control cell to the consumer operator. The data value is held in the *buffer* flip-flop in the control cell. The *data signal* (the firing signal) from the source code segment is also latched by the *flag* flip-flop. The *data signal* triggers the storing of the data value in the *buffer*, assuming that the *acknowledge signal* is inactive. The *data signal* also causes an active (high) value to be stored into the *flag* flip-flop. The output of the *flag* is used to produce two output signals, which are used in a manner somewhat different from that suggested by Figure 2 but to the same effect. Instead of the *data signal* being forwarded to the consuming code segment directly, this ephemeral signal is trapped by the *flag* and kept constant. A value named *ready* is the output of the flag and is forwarded to the consuming code segment. As long as *ready* is active, its associated data value is current and available for use by code segment **B**.

The *data signal* from the source code segment to the producer side of a flow control cell is active for one virtual time step. This signal follows, by one simulated time step, the time when the data value output of the producer operator first becomes valid. This delay is due to the unit delay boolean AND function which implements the operator firing rule. It ensures that the data inputs to the flow control flip-flops are valid when the *data signal* latches their values. The one cycle duration is due to the interaction of the flow control cells and the operator and will be discussed later. Thus, the data value is transferred to the consumer code segment from the source code segment with a one virtual time step delay. The BUFFER gate on the *data signal* input is required to ensure that the rising edge of the clock inputs to the flip-flops occurs after the *acknowledge signal* has gone inactive.

The *acknowledge signal* from the consuming code segment is also an ephemeral value that is active for only one virtual time step. It is produced by the firing logic of the consuming operator. Instead of passing it directly to the source code segment, it too is processed by the flow control cell flag. Upon receipt of the *acknowledge signal*, the *flag* clears its contents (makes them inactive). The output of the *flag* is complemented and sent to the source code segment as a value called *next* with a virtual time delay of one cycle. When *next* is active, the source code segment is alerted that its consumer code segment is prepared to accept its next result value. At the same time, the *ready* value is inactive (being the complement of *next*) indicating that the data value is not available yet. The flow control cell *flag* thus regulates the flow of the data value across the interface between the two code segments. A timing diagram for the flow control cell is shown in Figure 5. With the *next* value active, the source code segment is free to begin the data value transfer. When the data value is valid, the *data signal* is asserted for one time step by the producer code segment. Some time later, after the consuming code segment has employed the data value and is ready for a new one, it asserts an active *acknowledge* signal for a duration of one time step.

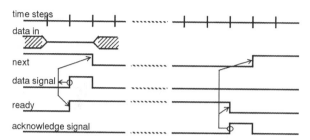

Figure 5. Timing Diagram for Flow Control Cell

The firing rule logic is implemented by an n-way AND gate is a unit-delay function which, in conjunction with the flow control cell logic, creates an active firing signal one simulated tick after the firing conditions have been met. The Zycad SDE primitive operators allow up to three input arguments. For n greater than three, an appropriate tree structure of operators is assembled to perform the n-way function. The flow control cells for the input arguments to the operator must provide active *ready* signals indicating that the input argument values are ready. The flow control cells of the descendant operators must provide active *next* signals indicating that these operators are ready to receive the next set of result values from the operator. The firing signal is active for a single simulated time step. Since the inputs to the firing logic AND function, the *ready* signals from the input flow control cells and the *next* signals from the output flow control cells, are made inactive by the firing signal (output of the AND function), the firing signal will cause its own inputs to change such that its output will again be inactive one time step later.

The operator function is implemented directly using SDE primitive operations. The input values to the function come from the data value outputs of the input side flow control cells. The result value of the function is applied to the data value inputs of the output side flow control cells. In Figure 4, the operator function is represented by *f*, to depict any of the possible SDE primitive operators. In fact, *f* need not be just a single operator but may be any acyclic graph code segment of basic operators. The number of input cells (at the top of the figure) can be made as large as necessary to accommodate the number of input values employed by *f*. The output cells can be organized in multiple groups, each group associated with a distinct output value from the *f* graph segment. Only the operators at the end of the code segment need synchronize with the set of inputs to the segment. No explicit synchronization is necessary for operators within the acyclic code segment. This use of aggregate synchronization is what allows an efficient implementation of the dataflow synchronization while retaining the advantages of the event-driven execution mechanism.

Conditional Operators

Most programs of sufficient complexity to be of interest exhibit structure with conditional flow control. Entry and/or exit from iterative code blocks is usually conditioned on the value of some intermediate data element which either is computed within the loop or is an element of an input data stream used by the loop. The dataflow model supports conditional flow control through the use of special non-strict operators that direct/consume tokens to/from one or more arcs conditioned on the Boolean value of one of the input arguments. One possible set of conditional operators that provides the building blocks for all conditional control flow structures within the static dataflow paradigm consists of the *switch* and *select* operators.[17] These operators are non-strict in that to fire, they do not require all source cells to be ready and all consumer cells to provide active next signals. Nor do they necessarily apply acknowledge signals to all of their source cells or data signals to all of their consumer cells. This section describes the Zycad SDE realization of the *switch* and *select* operators.

The dataflow *switch* operator delivers an input argument X to one of two sets of descendent actors. The choice of which of the two sets of actors to send X is determined by the Boolean value of the *switch* operator's

conditioning input argument S. The *switch* operator constructed with SDE primitives is shown in Figure 6 along with the flow control cells on the input and output arcs to and from the operator, respectively, for a configuration with two destinations for each output. The *switch* operator requires both input arguments to be available for the operator to fire as indicated by active *ready* values from each of the input cells. Upon execution, both cells receive acknowledge signals. This is consistent with strict firing rules. Where operation of the switch operator becomes non-strict is in its method of dealing with its two sets of descendent actors and respective output cells. Since only one of the two sets receives the input X value, only the cells of that set are required to be prepared to accept the next result value of the operator. Therefore, only that group of cells need have asserted active *next* values. Also, upon execution of the operator, only these cells receive *data signals*. The other set of cells, not having been designated by the Boolean value of S, has no effect on the flow control of the operator or is affected by its execution.

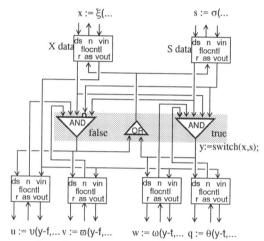

Figure 6. Flow Control of the Switch Operator

The *select* operator is the second of the two conditional operators used and it too employs a non-strict firing rule. The *select* operator has three rather than two input arguments. It conditionally selects between its two input values, X and Y, to produce its result value dependent on the Boolean value of its third input argument, S. If S is *0*, then the *select* operator's result value will be that of its X input. If S is *1*, the result value is that of its Y input. The *select* operator is assembled from Zycad SDE primitive instructions as shown in Figure 7. The firing rule of the *select* operator is non-strict in that only the S input value and the one it selects from its choice of X and Y need be available for it to fire, assuming the output cells are prepared to accept the next result value from the operator. Only the selected input, X or Y, along with the conditional S cell is sent *acknowledge signals* upon execution of the operator. In either case, the output cells receive *data signals* upon operator firing. The SDE primitive operation that produces the result value for the *select* operator is a two-to-one multiplexor, selecting the values from the X or Y cell buffers conditioned by the value of the S cell buffer.

Programing Environment

The initial implementation of a programming environment for the SDE uses the Logic Description Generator (LDG) for program specification.[18] LDG is a Lisp-based tool developed at the IDA Supercomputing Research Center for efficient textual specification of logic designs and was extended to support the semantics of a dataflow programming language. A Zycad Dataflow Translator (ZDT) was developed to synthesize the necessary dataflow operator and flow control logic from the dataflow program specification according to the translation methodology previously outlined.[19] ZDT generates a description of the dataflow program in Zycad's hardware description language ZILOS, including the necessary initialization patterns and control statements.[20] The ZILOS program is then assembled, loaded and executed by the Zycad simulation environment. LDG and ZDT support the specification of placement information for each dataflow primitive operator. The individual primitive operators (gates) constructed to implement a given dataflow primitive operator are atomically mapped to the same physical processor. Additional tools have been developed to collect and analyze performance statistics from program execution on the SDE.[21] Gate event and evaluation counts, message traffic profiles and hardware utilization statistics are collected and displayed graphically to evaluate the performance of the SDE on an application.

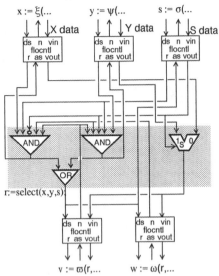

Figure 7. Flow Control for the Select Operator

EXAMPLE

An example program is provided to illustrate the methodology presented here. The program is a median filter, a simple algorithm borrowed from the field of signal processing, which compares two bit streams with varying offsets. Performance measurements for the median filter, are compared with those on other architectures. Also, scalability and mapping experiment results are presented.

Median Filter

The second example is a median filter program, which is a simple signal processing algorithm that was implemented on the SDE in order to demonstrate the ability to exploit the event-driven execution mechanism of the SDE through the dataflow programming model.[22] Specifically, this algorithm demonstrates the ability to eliminate the flow control overhead from acyclic code segments and exploit the native execution speed of the machine.

The fundamental computation of the algorithm is determining the median value of W B-bit unsigned integers. A stream of such values is produced as an N-long input stream is passed by a W wide window within which the median value is computed. The implementation on the SDE consists of a sorting network, from which is taken the median value of the numbers falling within a W-long window produced by the array of input operators. The program structure is illustrated in Figure 8. The sorting network provides a large acyclic program segment. The sorting network is a bitonic sorter, which is configured as a multistage network of 2x2 B-bit switches.[23][24] To first order, the number of primitive (bit-level) dataflow operators in the median filter program is determined by the contribution from the switch, which for a window width of W and input values containing B bits is $O(BW\log_2^2 W)$.

Flow Control Overhead Reduction

One measure of the cost of imposing the dataflow model on the SDE is the spacial cost in terms of the number of instructions required to implement the flow control discipline. To evaluate this cost, two versions of the median filter algorithm were developed. The first imposes a full flow control discipline between every source/destination pair of operators in the program. Since primitive operators are all bit-level, the spacial overhead due to the creation of flow control cells between each pair of operators is extremely large. Accurate counts were developed for the numbers of dataflow operators, arcs in the dataflow graph (which cause flow control cells to be generated), and machine level operations generated to implement the

program on the SDE. These counts are functions of the parameters of the algorithm: the width of the window (W), the number of bits in each data word (B), and the number of elements in the input stream (N). It should be observed that N has only a second order effect, determining the number of instructions attributed to I/O operators. The spacial overhead consists of the number of instructions used to implement the flow control discipline required by the dataflow computing model, i.e. the operations which form the flow control cells controlling communication between dataflow operators and those performing the firing rule logic. An instruction count analysis of the program showed that overhead instructions dominate the program, comprising 90% of the total code size of the program. Figure 9 shows a plot of total instruction count as a function of window size for the program (for fixed word width B=16) and also the portion of the total instruction count attributed to overhead.

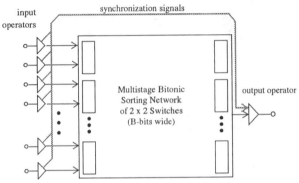

Figure 8. Optimized Median Filter Structure

An optimized version of the program was created by specifying at the source code level the removal of the flow control cells within the acyclic sorting network. The sorting network then becomes a zero-delay combinational logic circuit which executes with no additional overhead imposed by the programming model. In order to remove the flow control within the sorting network, while retaining overall program flow control, the block of acyclic sorting code was bracketed by synchronization signals from all the input operators to the output operator to condition the consumption of the median value output from the network with the arrival of all the inputs. Since the sorting network to first order determines the program size, elimination of the overhead in the sorting network reduces the total program size by up to a factor of 10 from its original value. This reduction is evident in Figure 9. This optimization results in a program that contains only about 10% overhead. Since program size bears directly on the temporal cost of program execution in this event driven architecture, this spacial reduction in the portion of the program that is overhead suggests an equivalent reduction in the temporal execution overhead leading to a performance improvement of an order of magnitude.

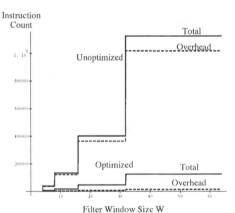

Figure 9. Program Instruction Counts for the Unoptimized and Optimized Median Filter Programs (B=16,N=1000).

Performance

To test the hypothesis that significant efficiency improvement would be achieved by the removal of flow control overhead, execution times were collected for both the unoptimized and the optimized median filter programs. The execution times for several different size problems and a given length random number input stream are shown in Table 1 for both the optimized and unoptimized versions. (It was found that the input data could significantly affect execution time on the SDE, due to the event-driven execution mechanisms in the SDE. All-zero data resulted in a factor of 4 decrease in execution time. Random data does not necessarily result in worst case performance, but provides a reasonable point of comparison.) Reductions of factors from 6.5 to 8.5 in the machine level instruction count for the optimized program were observed. Since fewer operations are used in the optimized program and each processor in the SDE has a finite storage capacity of 16384 operations, the optimized program requires fewer processors for the minimal configuration of processors. In order to compare the performance of the SDE for the two implementations, the application specific metric of "filter input words per second per processor" is calculated for the SDE. For example, with parameters N=1000 input words, W=15 long window, and B=16 bit words, the unoptimized programs required 135047 machine-level instructions, corresponding to a minimum of nine active processors, and executed in 34.49 seconds. The optimized version required a minimum of only two processors, with 17926 machine-level instructions, and executed in 13.8 seconds. The normalized performance measures are 3.22 words/sec/proc for the unoptimized program and 36.23 words/sec/proc for the optimized version, a factor of nine performance improvement when the unnecessary flow control overhead is eliminated. The data in Table 1 show that the optimized programs consistently outperformed the unoptimized programs of the same size by approximately an order of magnitude. Thus a factor of 10 reduction in program size due to overhead removal resulted in approximately a factor of 10 increase in performance. The optimization technique was successful in reducing the overhead required to impose the flow control discipline to a tolerable level.

Table 1. Median Filter Performance on the Zycad SDE for N=1000 Long Input Stream

Program Size		Unoptimized Program			
Window Size(W)	Bits per word(B)	# Insts	Min # Procs	Time(s)	Words/sec/proc
7	16	41095	3	10.28	32.42
15	8	66279	5	14.11	14.17
15	16	135047	9	34.49	3.22
31	16	402825	25	85.98	0.46

Program Size		Optimized Program			
Window Size(W)	Bits per word(B)	# Insts	Min # Procs	Time(s)	Words/sec/proc
7	16	6574	1	3.15	317.46
15	8	8882	1	7.98	125.40
15	16	17926	2	13.80	36.23
31	16	48202	3	87.06	3.83

The normalized performance metric also allows comparison with other architectures. The projected peak performance of the SDE can be calculated by assuming that a given input stream could be segmented into independent portions, each of which could be executed by a separate copy of the program using only the minimum number of processors necessary for each program copy. The number of copies of the program that could be executed concurrently is determined by dividing the total number of processors in the SDE (64) by the number of processors required for each copy, which is just the size of the program divided by the instruction capacity of a processor. The SDE's peak performance in input words per second could then be compared to the performance of other machines for a median filter algorithm using the same metric. (This performance measure of the SDE is optimistic in that it assumes linear a speedup characteristic. Such linear speedup is generally not achievable, even with independent programs, due to resource contention.) For the purpose of comparison, a sequential version of the median filter algorithm was written in C. Performance measurements

were taken on a 20MHz Sparcstation-1. The sequential program uses a bubble sort algorithm on a linked list data structure with no optimization. In addition, benchmarks from a (unoptimized) median filter implementation, also using a bubble sort technique, on a 16K processor Connection Machine 2 were obtained [25]. Comparison of the performance of the SDE (optimized program) with the Sparcstation 1 and the CM-2 is shown in Table 2.

Table 2. Median Filter Performance Comparison Using the Application Specific Metric of Words–per–Second Processed by the Median Filter for N = 65536 Long Input Stream

Program Size		Filter Words/second			Performance Ratio	
W	B	SDE(peak)	Sparc1	CM-2(16K)	Sparc1/SDE	CM-2/SDE
7	16	25773	12850	9062573	0.5	352
15	16	2642	12365	2255604	4.7	854
31	16	261	11108	520706	42	1995

The timings in Table 2 provide a coarse indication of the effectiveness of the event-driven synchronization architecture compared to both a sequential RISC uniprocessor and an SIMD architecture for this application. It is difficult to compare the timings since they represent different formulations of the algorithm, each tailored for a particular architecture. To make a gross comparison, however, it can be assumed that effectively the same amount of work is done in all three implementations. A "back of the envelope" comparison of peak bit operation rates reveals that the Sparc, a 32-bit processor executing at 20 MHz, is capable of 640 million bit operations per second, approximately the same peak rate as the SDE, which has 64 processors operating at 10.87 MHz for a peak rate of 696 million bits operations per second. The peak execution rate of a 16k-processor CM-2, for which the processor clock rate is 7 MHz, is approximately 115 billion bit operations per second, or 165 times greater than the SDE.

The measured performance comparisons in Table 2 indicate that of performance of the SDE relative to the other architectures deteriorates as program size increases. For the smallest program size listed in the table, the peak SDE performance actually exceeds that of the Sparcstation and it is within a factor of two of the expected performance relative to the CM-2. However, for each doubling of the filter window size, the SDE loses an order of magnitude in performance relative to the Sparcstation and a factor of 2.4 relative to the CM-2.

Program Mapping

The mapping strategy applied in obtaining the program execution times on SDE was the default mapping for the programming tools. Just as for the dataflow model of computing and any statically mapped architecture, the allocation of operations to processors is critical for performance of the event driven architecture. The mapping functions available for the SDE are somewhat limited, however. They are all based on the parsing order of the program source file. Consequently the program specification order rather than the data dependence graph determines the placement of operations. The justification for this mapping heuristic is that elements of a program which are "nearby" in the program description are more likely to interact than elements which are distant in the program description, an appropriate heuristic for logic simulation. In order to reduce communication requirements, since the bus interconnect is a likely bottleneck, an attempt is made to place such neighboring elements in the same or neighboring processor.

The default mapping function fills each processor to its capacity with machine level instructions before allocating instructions to the next processor. Processors are assigned instructions in increasing processor number order. Therefore, the minimum number of processors are active in the default mapping. The problem with this approach is that the last processor may contain only a few instructions, creating a load imbalance. A second mapping function allows the program to be evenly distributed across p processors. This eliminates the static load balance problem. In addition to these machine-instruction-level mapping functions, several mapping functions with the granularity of a dataflow operator were developed. With these functions, all machine-level instructions which comprise a dataflow operator and its associated input arc synchronization control cells are mapped atomically to a processor. The semantics of the dataflow model support the specification of placement for each operator. Three general dataflow-level mapping functions were developed. The first is a segmented scheme similar to the default machine-level instruction mapping function where contiguous sets of dataflow operators are mapped to processors, evenly distributing the operators across p processors. The second function interleaves dataflow operators such that operator i is mapped to processor $(i \mod p)$. The third scheme randomly distributes dataflow operators across p processors. Table 3 shows a comparison of execution times for four mapping functions applied across 64 processors for the given size median filter program.

Table 3. SDE Execution Times for Different Mappings of an Unoptimized Median Filter Program of size {N=1000,W=15, B=16} onto 64 Processors

Mapping Function	Execution Time(sec)
Machine-level Segmented	25.40
Dataflow-level Segmented	7.87
Dataflow-level Interleaved	10.84
Dataflow-level Random	11.90

Slower by more than a factor of two than any of the dataflow level mappings is the machine-level instruction mapping scheme. Because the functional information indicating which gates interact within a dataflow operator is ignored, greater communication is required across processor boundaries. The most effective allocation scheme is the dataflow level segmented mapping. This scheme best exploits the locality reflected in the program description by keeping neighboring instructions as close as possible. The remaining two mappings result in slightly longer execution times, still only half that of the machine level scheme. These times confirm that, like any other statically mapped architecture, the allocation of operations to resources has a significant impact on the performance of this event-driven machine. It is possible that data dependence graph based allocation schemes may result in even greater improvement in performance.[26]

Program Scalability

Another property of the SDE architecture that was studied in the context of the median filter program was scalability. The ability to make effective use of available parallel computing resources is critical to any parallel architecture. Table 4 shows speedup for an optimized median filter program of size {N=1000, W=15, B=16}. The most effective mapping strategy, the Dataflow-level Segmented scheme, was employed. The poor scalability observed reflects the inability of the program to make effective use of the processing resources as indicated by the imbalanced distribution of processor utilizations typified in Figure 10 for an 8 processor mapping.

Table 4. Median Filter Speedup for N=1000,W=15,B=16

# procs	Speedup
2	1.28
4	1.76
8	2.64
16	3.75
32	4.56
64	5.15

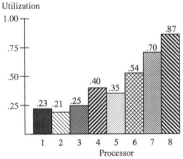

Figure 10. Processor Utilization for Optimized Median Filter of Size N=1000, W=15, B=16 on 8 Processors

To understand the source of this observed processor starvation, a more detailed examination of the execution dynamics was undertaken. The program generates a result every three simulated ticks; a one tick delay to produce the inputs to the sorting network from the input operators, and a two tick backward acknowledge propagation delay from the output operator to the input operators. Therefore, since the majority of the program consists of operations in the sorting network, processor utilization is limited to about 33%. Program scalability is poor due to the starvation of processors. Because of the fixed capacity of processors, it is not always possible to statically assign enough operations per processor such that there are sufficient numbers of active operations in each simulation cycle to achieve high utilization. Therefore utilization for even the minimum number of processors is limited, causing speedup to be poor as more processors are added. For this application, the maximum utilization achievable is constrained by the extra synchronization requirements imposed on the problem to enforce dataflow synchronization. This is a fundamental constraint for this problem since there is no other computation which can be scheduled to hide the synchronization latency. However, the implementation of the imperative programming methodology imposes no fundamental limitation given enough parallel activities to hide the small latency imposed by the model.

CONCLUSIONS

We have observed the similarities between the event-driven and data-driven synchronization paradigms and explored their relationships in realizing a fine grain parallel computing system directed by an imperative parallel programming methodology. It was determined that each brings important capability but is incomplete. A complementing association of the two, merging the semantics of the data-driven model with the physical implementation of the event-driven mechanisms, yielded a fully operational parallel computing system. An imperative programming methodology based on the static dataflow model of computation was developed for the Zycad SDE, an event driven, bit-oriented gate-level simulator. This methodology has enabled the mapping of procedural formulations of bit manipulating algorithms to the SDE instruction set and the execution of those algorithms by the SDE's parallel processing facilities. The work reported here demonstrates the validity of this approach.

An important question to be addressed is that of program mapping. Operators are statically mapped to processors at load time. Allocation among the 64 possible processors is complicated by the trade-offs between locality avoiding communication costs and distribution making greatest use of parallel resources. The event driven nature aggravates the allocation process because operator firing criteria include argument values for which availability is not predictable at compile time. It was shown that the speedup curve is very nearly flat for the dataflow segmented mapping strategy. The best resource utilization is achieved by using close to the minimum number of processors required for the number of instructions in the program. With this mapping, multiple copies of a program can be executed concurrently.

In terms of absolute performance, the SDE did not perform as well as predicted by peak computing rate calculations relative to the two other machines examined, a Sparcstation 1 and a 16k Connection Machine-2. Additional study is required in order to identify the sources of loss that contribute to this performance degradation. One possibility is that the event-driven synchronization model is much less effective for computation in which there is structure that can be exploited in other architectures, such as vector or data parallelism. In applications in which there is little or no identifiable structure (digital simulation is a good example of this) the even-driven architecture may well have an advantage.

Two interesting consequences of this work may be in the areas of unanticipated applications. One possible important application may be the field of circuit simulation. The SDE is used as a single level simulation vehicle, i.e., representation of the target system must all be at the gate-level. But hierarchical system descriptions incorporating functional specification of subsystems greatly facilitate system design and accelerate the system development cycle. The imperative programming methodology described here may be employed to develop functional descriptions and still execute them on the SDE in conjunction with gate level circuit characterizations. A second more speculative application of the technique relates to chip design. The method takes a procedural description of an algorithm and derives a gate-level description to perform the same function. This may be viewed as a silicon compilation tool for dataflow graphs. Such a tool could prove valuable for rapid prototyping of complicated algorithms on special purpose IC's.

ACKNOWLEDGEMENTS

The authors wish to thank Al Schwartz for the implementation of the Zycad Dataflow Translator. Thanks to the Zycad Corporation for their assistance in understanding the SDE architecture.

REFERENCES

[1]Dennis, J.B., "Programming Generality, Parallelism and Computer Architecture," *Information Processing '68*, North Holland, Amsterdam, 1969, pp. 484-492.

[2]Ackerman, W.B., "Data Flow Languages", *AFIPS Conference Proceedings, Vol. 48: Proceedings of the 1979 National Computer Conference*, pp. 1087-1095.

[3]Reuveni, A., "The Event Based Language and its Multiple Processor Implementations," MIT/LCS/TR-226, Laboratory for Computer Science, M.I.T., January, 1980.

[4]Dennis, J.B. and Misunas, D.P., "A Preliminary Architecture for a Basic Data-Flow Processor," *Proceedings of the Second Annual Symposium on Computer Architecture*, Dec. 1974, pp. 126-132.

[5]Ziegler, B. P., Theory of Modeling and Simulation, John Wiley, New York, 1976.

[6]Kobayashi, H., Modeling and Analysis: An Introduction to System Performance Evaluation Methodology, Addison-Wesley, Menlo Park, CA, 1978.

[7]Arvind, Culler, David E., "Dataflow Architectures," *Annual Review of Computer Science, 1986*, pp. 225-53.

[8]Hansen, Steve, "System Development Engine," *Proceedings of the Third International Zylink User Conference*, April 1988, pp. 53-73.

[9]Zycad Corp., "System Development Engine, Zycad Intermediate Form Reference Manual," Pub. 2400-1032-02A, St. Paul, MN, October 1987.

[10]Arvind, Culler, David E., Maa, Gino K., "Assessing the Benefits of Fine-Grain Parallelism in Dataflow Programs," *International Journal of Supercomputer Applications*, Vol. 2, No. 3, 1988, pp. 10-36.

[11]Mago, G. A., "A Network of Microprocessors to Execute Reduction Languages, Part 1," *International Journal of Computer and Information Sciences*, Vol. 8, No. 5, October 1979, pp. 349-385.

[12]Batcher, Kenneth E., "Design of a Massively Parallel Processor," *IEEE Transactions on Computers*, Vol C-29, No. 9, September 1980, pp. 836-840.

[13]Kung, H.T., "Why Systolic Architectures?" *Computer*, Vol. 15, No. 1, Jan. 1982, pp. 37-46.

[14]Treleaven, P.L. and Hopkins, R.P., "Decentralized Computation," *Eighth Symposium on Computer Architecture*, May 1981, pp. 279-290.

[15]Junes, Simon L. Payton, The Implementation of Functional Programming Languages, Prentice Hall, New York, 1987.

[16]Pingali, Keshav, "Lazy Evaluation and the Logic Variable," Research Topics in Functional Programming, David A. Turner, ed., Addison-Wesley, New York, 1990, pp. 171-198.

[17]Sharp, John A., Data Flow Computing, Ellis Horwood Limited, New York, 1985.

[18]Gokhale, M.B., "The Logic Description Generator: A Functional Hardware Description Language", SRC-TR-89-008, IDA Supercomputing Research Center, Bowie, MD,. September 1989.

[19]Thistle, Mark R., "The Zycad SDE Dataflow Translator," SRC-TN-90-303, IDA Supercomputing Research Center, Bowie, MD, October 1990.

[20]Zycad Corp., *ZILOS Reference Manual*, Menlo Park, CA, 1989.

[21]Thistle, Mark R., "A Performance Analysis Toolkit for the Zycad SDE," SRC-TN-90-304, IDA Supercomputing Research Center, Bowie, MD, October 1990.

[22]Gallagher, N.C., and Wise, G.L., "A Theoretical Analysis of the Properties of Median Filters," IEEE Transactions on Acoustics, Speech, and Signal Processing, Vol ASSP-29, pp. 1136-1141, December 1981.

[23]Batcher, K. E., "Sorting Networks and Their Applications," *1968 Spring Joint Comput. Conf., AFIPS Conf.* Vol. 32, Washington, D.C.: Thompson, 1968, pp. 307-314.

[24]Stone, H.S., "Parallel Processing with the Perfect Shuffle," *IEEE Trans. Comput.*, C-20, pp. 153-161, Feb. 1971.

[25]Reese, Craig F., "An Introductory Look at 1D Signal Processing on the Connection Machine," unpublished report, Supercomputing Research Center, May, 1989.

[26]Bokhari, Shahid H., Assignment Problems in Parallel and Distributed Computing, Kluwer Academic Publishers, Boston, 1987.

Liger: A Hybrid Dataflow Architecture Exploiting Data/Control Locality

Tzi-cker Chiueh

Department of Electrical Engineering and Computer Science
University of California, Berkeley, CA 94720

Abstract

Dataflow machines use inherent parallelism in the programs to mask the synchronization overhead and long memory access delays; whereas von Neumann machines take advantage of the program deterministicness by pipelining the execution. We believe that both approaches are essential to truly high-performance machines. In this paper, a new hybrid architecture called *Liger* is developed that aims at integrating these two paradigms in a coherent fashion, thus making use of parallelism of every possible granularity. Among the novel features of *Liger* are providing a natural framework for merging vector pipelining and dataflow parallelization, a source-broadcasting tag matching mechanism, and the decoupling of synchronization and data storage. Examples are presented to illustrate the power of this architecture.

1. Introduction

One of the key ideas underlying high performance computation machines is to *overlap* the execution of computation tasks. Different computation models use different techniques for overlapping computation. In von Neumann model, the order of instruction execution is to a large extent pre-determined at compile time, and is manifested as a program counter in the hardware. Furthermore, the communication among instructions is through some sort of shared storage such as registers or memory. Because of the statically-determined execution order and the communication-through-storage paradigm, *pipelining* of sequential instructions can be made very efficient, and thus becomes the predominant overlapping technique for boosting the performance of von Neumann machines. RISC architecture pushes this model even further by streamlining the instruction set so that each instruction takes approximately the same number of operations and each operation takes approximately the same amount of time. Moreover, by letting the compiler takes more and more control over the execution dynamics, runtime execution becomes more predictable and the pipeline efficiency is further improved. However, it is the same predictability which facilitates pipelined execution that prevents von Neumann machines from tolerating long-latency operations such as synchronization and global memory access, because of the difficulty of exploiting parallelism and switching large context states. As a result, the von Neumann model in its purest form is generally considered very difficult to scale up for very large scale computation structures.

In contrast, every instruction in dataflow model is viewed as an independent computation task. From the very beginning, there is no such thing as total ordering of instruction execution in dataflow machines, only the partial ordering as dictated by the data dependency relationships among instructions (or *actors*). A single unified synchronization mechanism is provided at the hardware level that handles all the inter-actor synchronization. An actor is allowed to be fired only when all of its operands are available, and it is up to the *producers* of the data operands to tell the *consumers* that they are available. Because of no superficial imposition of instruction execution ordering beyond what data dependency dictates, the parallelism intrinsic in the programs is exposed to the dataflow model in a natural way. Moreover, dataflow programming languages typically take a *functional* view towards computation, which further eliminates artificial instruction ordering constraints due to limited storage. Typically the communication among actors is through explicit flow of *tokens* , which carry both synchronization information and data operands. Since synchronization is performed on an actor-by-actor basis, there is no way to pipeline the execution of actors directly related by data dependency relationships, because by definition the dependents can start only after the dependees are completed. However, it is possible to pipeline the execution of the actors that are *not* related in a partial ordering, i.e., when they are independent. Consequently *parallelism* becomes the major tactics in dataflow machines to overlap the computation. More to the point, parallelism is used to mask long-latency operations in large-

scale computation machines. Therefore, the dataflow model is potentially more scalable provided that the applications have sufficient intrinsic parallelism.

In state-of-the-art machines, some kind of mixture from dataflow and von Neumann paradigms are emerging on the horizon. For example, the computation tasks in the message-passing architecture of von Neumann school communicate with each other through sending messages as opposed to through shared storage. Another example, some dataflow machines try to minimize the synchronization overhead by aggregating several actors into a larger-granularity unit and following the sequential instruction ordering within these units. Several researchers [IANN88] [EKAN86] have recognized that as a computation model for very large scale machines, neither is suitable; only a careful blend of the two can do the job. The question then is what is the optimum point in the spectrum between dataflow and von Neumann models. In this paper, we investigate the weakness and strength of each model and propose a new architecture that integrates the advantages of both models in a coherent fashion. Our approach is to start with the dataflow model and to correct the associated deficiency, which we are to discuss in detail next.

Because dataflow machines perform synchronization at the hardware level for each fired actor, the synchronization overhead stands out as the major bottleneck towards efficient implementation. In *dynamic* dataflow machines, tag matching for the "partner tokens" is required to exploit the parallelism among different invocations of the same code segments; in static dataflow machines, acknowledgement feedback from the consumers of the tokens is needed to enforce the one-token-per-arc semantics of the dataflow program graph. In either case, additional machinery needs to be built into the hardware to implement the abstract dataflow model. Since dynamic dataflow model can expose more parallelism than its static counterpart, we will focus on the former from now on.

The synchronization overhead associated with dynamic dataflow machines manifests itself in two forms: excessive token traffic and ignorance of locality. Because synchronization is performed on an actor-by-actor basis, the tokens that carry synchronization information typically have to run through the wait-matching store *twice* for a two-input-operand actor to fire. This puts serious burden over the inter-processor networks as well as the already hard-pressed tag matching unit. Moreover, the architecture doesn't guarantee that instructions that are related logically through data dependency relationship will execute physically in temporal proximity, therefore contiguous instructions in the hardware pipeline may well come from vastly different code-segments. Program locality is not as apparent in dataflow machines as in von Neumann machines. Consequently the effectiveness of caching degrades.

The solution to these two problems is to bring some order to the otherwise completely non-deterministic instruction execution mechanism. Based on the observation that a substantial amount of inter-actor synchronization requirement could have been satisfied by enforcing static execution ordering within subsets of actors, the idea is to provide support at the instruction-set and hardware level for "gluing" a set of actors into an indivisible unit and only perform synchronization across the units. We call such units the *macro-actors*. The actors within a macro-actor are restricted to execute sequentially and controlled by a program counter.

Intuitively, this aggregation technique could improve *control locality* by allowing pipelining the execution of actors within a macro-actor, as well as *data locality* because software techniques can be used to ensure that actors of the same block touch related data items.

If partitioning of programs into macro-actors is created along the data dependency relationships, the macro-actors correspond to sequential code segments. On the other hand, when partitioning is done by clustering the set of actors that perform the same operation on different values, each macro-actor thus formed corresponds to a vector instruction in von Neumann machines. A hybrid architecture called *Liger* is developed in this paper, which embodies the idea of macro-actor in a novel way, and combines the advantages of efficient pipelining of von Neumanns and those of tolerating long-latency operations through parallelism as in dataflows.

The rest of the paper is organized as follows. In section two, the computation model of *Liger* is presented to show where it is in the dataflow-von Neumann spectrum. The hardware architecture and the software support of *Liger* is discussed in section three. Section four concludes this paper with planned future works.

2. The Computation Model of *Liger*

In Liger, programs are first partitioned into macro-actors, which are aggregates of actors. Associated with each macro-actor is a *run-length count,* the number of actors encapsulated within the macro-actor. At any time, each macro-actor has a *current actor,* which denotes the next actor to execute in the macro-actor. Because all the actors within a macro-actor is supposed to execute sequentially, there will be one and only one current actor associated with each active macro-actor at any time.

A macro-actor can be activated when the input operands of the associated current actor are available. Once activated, the actors within the macro-actor are executed sequentially as in von Neumann machines. By arranging the actors in the instruction memory in a physical contiguous order, a simple program-counter-based mechanism is sufficient for scheduling the actors within a macro-actor. Every time when an actor is completed, the associated run-length counter is decremented. Macro-actors are automatically terminated when the run-length count reaches zero. Macro-actors can be suspended and restarted during its life time when encountering long-latency operations or when some of the input operands of the current actor are not available.

Although the basic scheduling unit is a macro-actor, the basic unit for storage allocation is a code block that generally contains one or more macro-actors. Typically a code block is either a procedure call or a loop invocation. By storage, we are referring to data storage as well as synchronization storage. In *Liger*, the data operands do not flow among actors. They are kept in shared storage waiting to be operated upon. However, the synchronization information do flow among actors, serving the purpose of notifying waiting macro-actors. Because of this distinction, data storage and synchronization storage are separated in the architecture. Furthermore, unlike dataflow models where token storage are allocated/deallocated on an actor-by-actor basis, *Liger* allocates (deallocates) required storage when the code block is invoked (completed). Macro-actors can not span code block boundaries.

The instruction set of *Liger* looks exactly like modern RISC instructions. In particular, the operands to be operated upon are kept in registers. There is no flowing of tokens among actors associated with the same macro-actor. They communicate through registers and synchronization is implicit in the instruction sequencing. This is important from implementation viewpoint because bypassing or short-circuiting techniques can be used to improve the pipeline efficiency. There are, however, token flowing among macro-actors. However, these tokens only carry synchronization information. Each actor in a macro-actor can *selectively* choose to send out tokens to other macro-actors. Through proper partitioning techniques, the interaction among macro-actors in the form of token exchange can be reduced. As a result of large-granularity synchronization, the total token traffic is significantly cut down as compared to dynamic dataflow machines.

The macro-actor can also contain a single von Neumann vector instruction, in which case the component operations can be pipelined efficiently as in conventional vector processors. This is one of the novel features of *Liger* that renders a natural marriage between vector pipelining and dataflow parallelization. Consequently *Liger* can exploit both structured and unstructured parallelism. This feature also distinguishes *Liger* from previous proposals on combining von Neumann and dataflow computation models. In the next section, we will describe in detail the architecture to implement the computation model described above, and explains how von Neumann and dataflow paradigms can fit into a unified hardware framework.

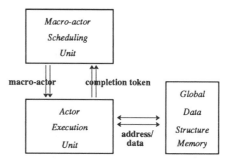

Figure 1 The General Architecture of *Liger*

3. The *Liger* Machine

3.1. The Hardware Architecture

As shown in Figure 1, *Liger* consists of three components: a *macro-actor scheduling unit* (MSU), an *actor execution unit* (AEU), and a global data structure memory. MSU is responsible for updating the synchronization status of the macro-actors and issuing fireable macro-actors when AEU makes a request. AEU carries out the computation work and selectively sends back synchronization control information called *completion tokens*. The data structure memory implements the I-structure semantics and serves as a global data repository. However, the emphasis of this paper will be only on the design of AEU and MSU.

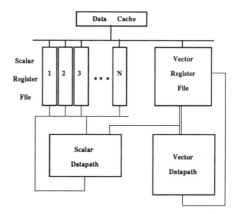

Figure 2   The Data Path of Actor Execution Unit

3.1.1. Actor Execution Unit

As shown in Figure 2, AEU has a similar datapath organization as modern von Neumann machines. It has both a scalar and a vector section. Each code block is allocated a scalar register set upon invocation. One novel feature of this organization is that the vector register file is visible to all currently active code blocks. Furthermore, each register in the vector register file is individually addressable; therefore vector data can be operated directly by the scalar section.

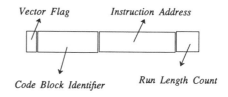

Figure 3   The Structure of A Macro-Actor

The structure of a macro-actor is shown in Figure 3. It comprises four fields: vector flag, code block identifier, instruction address, and run length count. The vector flag of a macro-actor specifies whether the macro-actor is a vector instruction. When the macro-actor is a vector instruction, the run-length count field represents the vector length. When AEU receives a macro-actor, the run-length count field is loaded into the run-length counter if the macro-actor is not a vector instruction, and into the vector control register otherwise. This encoding dictates that a macro-actor is either a single vector instruction or a sequential code segment.

The architecture allows concurrent operation of the scalar and the vector sections. That is, there can be one macro-actor in the scalar unit and one in the vector unit simultaneously. The sequencing of actors in the execution of a macro-actor is based on a program counter. After an actor is fetched from the instruction memory, the program counter is incremented and the run-length counter is decremented. When the run-length counter reaches zero, a new macro-actor is brought into the AEU from MSU in the next cycle. In other words, while the last actor of the previous macro-actor is working towards completion, the new macro-actor can start its computation, thus making efficient use of the hardware pipeline. A macro-actor can be suspended when it encounters a long-latency operation such as global memory access, or when some of the input register operands for its current actor are not yet available. Upon suspension, the current actor is recirculated back to MSU, together with the current instruction address and the run-length counter, and the execution of all the following actors is aborted as well. Finally a new macro-actor from the ready queue of MSU is brought in. The suspension operation takes only two to three cycles because the associated register set does not need to be swapped. The new macro-actor operates on its corresponding register set without interference.

Like modern RISC machines, AEU assumes a five-stage pipeline: instruction fetch, decode, execute, memory access, and write back. Modern implementation techniques used in high-performance microprocessors such as instruction/data caching and bypassing can be readily employed in AEU. Each register has a ready bit, which is set when a data item is written, and can be reset, *selectively* by the instructions that read it. In the decode stage, the *readiness* of the register operands is checked, and the actor in question is suspended when some of the input operands are not ready. Because of the single-assignment semantics of dataflow languages, this register model is sufficiently general to handle all the synchronization requirements of the programs.

The generic format of the instructions is shown in Figure 4. It is a typical RISC instruction format augmented with several flags directing the synchronization control operations. The *synchronize* flag

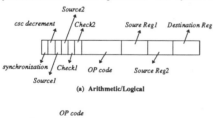

(a) **Arithmetic/Logical**

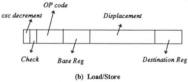

(b) **Load/Store**

Figure 4   The Instruction Format of *Liger*

indicates whether at the completion of this instruction a completion token should be send back to MSU. The functionality of the *CSC-decrement* flag will be discussed when we explain the working of MSU. The *source1* and *source2* flags indicate whether the readiness flag of the respective register operands should be turned off after the reads. The *check1* and *check2* flags specify whether the ready-bit check of the input operands should be checked during the decode stage. By turning off check bits in the instructions, it is possible to bypass

checking in the case of constants and loop invariants. As for global memory access, *Liger* supports only one addressing mode: displacement plus base. This memory access will first go to data cache. Load/store instructions are completed when the requested data are in the cache and are suspended when a cache miss occurs. The suspension of memory access instructions is different from the other cases in that one of the source instruction addresses is now loaded with the self instruction address in order to get re-activated when the memory responds back. Meanwhile, a packet consisting of the memory address and the tag of the instruction is sent to the data structure memory. The memory controller will then send back the requested data item, using the given tag as a destination address, and activate the memory access instruction again. The structure of the instruction tag and the matching mechanism is explained in the next subsection.

3.1.2. Macro-actor Scheduling Unit

The scheduling model of *Liger* uses a tag matching mechanism. Since instruction tags are just names for distinguishing different invocations of an actor, it really doesn't matter how these names are created as long as they are different We choose a hierarchical naming scheme in which the tag of an instruction comprises three parts: a physical processor number (PE), which is the identity of the processor that carries out the execution of the instruction, a code block identifier (CBI) and the instruction address (IA). All instructions within a code block share the same CBI. The instruction addresses are determined at compile time while the PE and CBI fields are created at run time. CBI is actually the base address of the activation frame for the corresponding code block. By associating the invocation names with the addresses of the allocated storage, there is no need for other tag generation mechanisms. Furthermore, as soon as the activation frame is allocated/deallocated, so is the corresponding code block identifier created/deleted.

As far as synchronization is concerned, there are three fundamental differences that distinguishes *Liger* from most previous dynamic dataflow architectures. First, the notion of synchronization points and data operands are decoupled. Therefore data values do not circulate as in conventional dynamic dataflow machines. In *Liger*, the data operands are stored in the registers of AEU, while the synchronization points are explicitly represented in MSU. As a result, the allocation of hardware resources to synchronization points and data operands can be optimized according to each individual's situation.

Second, the set of synchronization points and registers used in a code block are allocated and deallocated as a whole when the code block is activated and completed respectively. This makes allocation/deallocation of storage more efficient, and allows application of certain compile-time storage optimization such as graph-coloring register allocation. Upon initiation of a code block, the runtime system allocates the needed resources and releases them when the associated code block is completed. The set of synchronization points, called the *synchronization frame*, together with the corresponding register set, form the *context* of the code block.

Third, tag matching in *Liger* adopts an unusual *source broadcasting* mechanism. In previous dataflow machines, the producers of data operands explicitly send out a token to each of the consumers of the operand. Consequently, there are as many tokens as there are consumers for each data object produced. This tag matching paradigm is awkward because excessive token traffic is generated, consumers names must be stored explicitly in the instructions, and dummy "repeat" instructions are needed to duplicate the tokens and send to the consumers if each instruction can only hold a fixed number of consumer identities. In contrast to this point-to-point notification approach, each *Liger* actor broadcasts the resultant data object along with its own identity when it exits the pipeline of AEU. Each entry of the synchronization frame contains the identities of the source instructions that produce the input operands. Each synchronization frame entry compares the broadcasted identity with those stored, and marks the associated operand ready when a match occurs.

As shown in Figure 5, MSU comprises a set of synchronization frames, each of which in turn consists of a code block identifier register (CBIR), a completion status counter (CSC), a set of synchronization points, one for each instruction in a code block, and a message data queue. Associated with each synchronization frame is a register set in AEU. Upon a procedure call or a loop invocation, a new code block identifier, which is stored in CBIR, a synchronization frame and a register set is allocated. In order to detect the completion of a code block, the completion status counter is built into each synchronization frame. The way CSC works is as follows. When the programs are compiled, each code block is partitioned into one or more major-actors. When a code block is invoked, the CSC is initialized to be the number of major-actors in it. The last instruction of each major-actor is required to decrement CSC when it exits the pipeline of AEU and sends back a completion token. When the completion status count

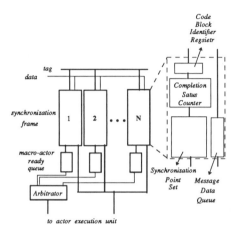

Figure 5   The Organization of Macro-actor Scheduling Unit

reaches zero, the code block is known to be completed. Then both the synchronization frames and register file, and the code block identifier can be recycled. The *CSC-decrement* flag of each instruction controls the operation of CSC when the completion token flows through the synchronization frame.

Synchronization Within A Code Block

Each synchronization point is composed of five fields: self instruction address, left source instruction address, right instruction address, synchronization status, and run-length. When an instruction exits the AEU pipeline, whether a completion token is sent back to MSU is controlled by the associated synchronization flag, which in turn is based on compile-time analysis. When it chooses to broadcast a completion token, the hardware forms the token by concatenating the instruction's tag with the associated CSC-decrement flag. The tag matching process in MSU actually proceeds in a pipelined fashion. The PE field of the tag determines to which processor the token should be sent. The CBI field of the tag is compared against the contents of all the CBIR's in that processor. Only matched synchronization frames can proceed. The CSC-decrement flag dictates whether CSC needs to be decremented when the token passes through the completion status counter. Finally, the IA field of the tag is compared against the contents of the left source and right source instruction addresses of all the synchronization points in the synchronization frame whose CBI matches the token's CBI. The synchronization status field of those matched synchronization points will be modified according to the following rule:

[1]   When the synchronization status field is originally 0, then it is changed into 1.

[2]   When the synchronization status field is originally 1, then it is changed into 0 and the associated instruction address and run-length fields are down loaded into a ready macro-actor queue for firing.

There are comparison logic built into the self instruction address, and two source instruction address fileds of each synchronization point. As can be seen, the tag matching process is broken down into stages (first CBI, then IA) to mitigate the scale of fully associative comparison, thus eliminating potential bottleneck. Because of this source-broadcasting approach to synchronization, as opposed to destination-designation approach used in conventional dynamic dataflows, an actor whose output is consumed by multiple recipients can notify them in *one* token passing. On the other hand, for those actor whose output is used by one or two consumers can bypass this notification step entirely through the sequential scheduling mechanism in a macro-actor. As a result, the token traffic is significantly reduced without sacrificing parallelism.

During the execution in AEU, a macro-actor can be suspended due to the unavailability of certain input operands, or to a long-latency memory access. When suspended, the macro-actor simply re-circulates the CBI-PE tag and the run-length count back to the MSU. Using the same tag matching mechanism, the associated run-length count is latched into the corresponding field of the synchronization point that matches the tag. Only in this case, it is the self instruction address field against which the broadcasted tag is compared.

Synchronization Across Code Blocks

Previous description of tag matching only applies to the case where the producer actors and the consumer actors belong to the same code block. When they are in different code blocks, the previous method won't work because the consumers have a different CBI than the producer's, and therefore never be able to match the broadcasted tag from the producer. To remedy this, we have to examine the kinds of inter-codeblock communication that need to be supported. As mentioned earlier, the code blocks in *Liger* correspond to either a loop invocation or a procedure call. As a result, there are only two kinds of inter-codeblock communication: arguments passed among procedures, and loop invariants passed from the enclosing context to the inner loop invocations. We shall explain how these two cases are handled in the following.

For inter-procedure argument passing, the received procedure has to make up several dummy instructions, one for each received argument. All the consumer actors of an input argument will use the addresses of these dummy instructions as their source instruction addresses for synchronization purpose. On the sending procedure side, the IA field of the tag of these dummy instruction is known at compile time, while the PE and CBI fields must be communicated at run-time, as described below. When procedure A calls procedure B, the PE and CBI of procedure B is known to procedure A because it is procedure A that requests the allocation of procedure B's context. The PE and CBI of procedure A, however, must be communicated to procedure B through explicit message sending.

The messages exchanged among code blocks look like conventional tagged tokens and are composed of two components: a PE-CBI tag for the destination code block, and a data object to be communicated. Based on the PE-CBI tag, the system can route the message to the corresponding code block, and the data objects will be put into a hardware message data queue of the synchronization frame associated with that code block. An explicit *receive* instruction is needed to bring the data object from the message queue to the register set for further processing. Because *receive* instructions take only one operand, the firing order of these instructions is identical to the arrival order of the messages, thus making it possible to use a simple queue mechanism for storing incoming data objects. This inter-codeblock communication mechanism can be used for exchanging the PE-CBI tags of code blocks, for passing procedure arguments and for returning memory data.

For communication among the enclosing code block and the inner loop invocations, *Liger* makes it a policy that all loop invariants must be available before the first loop iteration begins. As a result, all the loop invariants are passed from the enclosing code block to the inner loop iteration exactly the same way as procedure argument passing, at the time that the loop iteration is initiated. Although this policy may diminish parallelism in some cases, it is crucial for the correct operation of the source-broadcasting synchronization mechanism. The fundamental assumption of source broadcasting is that *all* consumer actors are existing in the synchronization frames, waiting for the producer to complete. When the producers and the consumers belong to the same code block, this doesn't pose any problem since the synchronization points for producers and consumers are allocated at the same time. However, When producers and consumers are in different code blocks, especially in the context that enclosing contexts pass values to the inner loop invocations, it is NOT always the case that the synchronization points for the consumer actors are already allocated. The source broadcasting mechanism might fail if no special machinery is developed to solve this problem. Our approach is to let the compiler ensure the readiness of loop invariants before any inner loop iteration is invoked, and therefore skip the synchronization step altogether because once a loop iteration is started, the loop invariants are by construction ready.

In summary, inter-codeblock communication is performed with destination-designation naming, as opposed to the source-broadcasting approach used in intra-codeblock communication. Since the amount of inter-codeblock communication is typically much smaller than that of intra-codeblock communication, we believe this hybrid synchronization naming scheme strikes a balance between reducing the token traffic and keeping down the comparison hardware required. The macro-actor ready queue of each code block is maintained separately. In order to exploit the data locality, the arbitrator that decides which macro-actor goes first when multiple fireable macro-actors are available, tends to fire the macro-actors from the same code block contiguously. In other words, the arbitrator will switch to another code block only when the current code block does not have other fireable macro-actors. Among code blocks, some kind of priority mechanism is possible to enforce certain compile-time scheduling decision. The next macro-actor to be fired typically exists well before the AEU sends a request. That is, the operations of AEU and MSU are decoupled and can run concurrently.

3.2. The Software Support

The major software challenge for *Liger* is how to partition the programs in such a way that the data/control locality is exploited without sacrificing the parallelism. The minimum requirement of the partitioning algorithm is *deadlock-free*. The non-sequentiality of programming languages dictates that the order of instruction execution may depend on inputs and therefore can't be completely determined at compile time. Instruction pairs whose execution order depends on inputs are said to be *non-sequential*. Because the execution order of the tokens within a macro-actor is strictly sequential, the partitioning algorithm must avoid deadlock by refraining non-sequential instruction pairs from being kept in the same macro-actor. Iannucci's method of dependency sets [IANN88] is proposed exactly to fit this role. Beyond this constraint, one can choose appropriate partitioning parameters to find the optimal tradeoff between parallelism and locality: exploit just enough parallelism to mask synchronization and memory-access latency, and keep the pipelines full by taking advantage of data/control locality.

Because the actor execution unit provides vector as well as scalar sections, the programs can also be vectorized to make use of the vector facility. The interaction between vectorization and partitioning is minimal because the scalar and vector sections can run in parallel. As a result, a typical strategy is to vectorize the vectorizable part and leave the rest for partitioned scheduling. In order to make available the results of vector instructions, the contents of the vector register files is visible to all other code blocks in the same AEU. That is, the vector register file in the vector section is shared by all the code blocks in the corresponding scalar section. Consequently, the vector register file is globally allocated across procedures and code blocks.

The contents of the synchronization frame associated with each code block are compiled as part of the code. Theoretically, each instruction in the code block needs a synchronization point. In practice, because the synchronization within a macro-actor is implicit in the sequential scheduling, the number of synchronization points is typically slightly more than that of macro-actors in that code block. Upon initialization, a free synchronization frame is selected and the pre-compiled contents are loaded into the frame before starting anything. *Liger* uses customized logic (e.g., comparison) for synchronization, which requires fixing the size of the synchronization frame at the hardware level. When a code block's synchronization frame is too big for a hardware frame, two hardware frames with the same code block identifier are used. The completion status count for each frame in this case is set to the number of macro-actors in the entire code block as before. By the same token, the number of hardware synchronization frames in each MSU or in the entire system is fixed. This implies that certain parallelism-throttling mechanism [RUGG87] or K-bounded loop technique [CULL88] are needed to make sure that parallelism is properly managed to match the system resource bounds.

4. Conclusion

In pursuit of the optimum point in the spectrum of possibilities between von Neumann and dataflow, one must carefully examine the strength and weakness of various architectural features. Furthermore, one needs to distinguish the fundamental bottlenecks associated with the computation models from the implementation choices. It should be clear from the previous presentation that *Liger* is really a synthesis of existing architectural ideas as opposed to a brand new invention. However, the work presented here is original in the sense that it puts as

much emphasis on locality as on parallelism. This simple shift of conceptual paradigm results in an unique architecture that can make use of parallelism of every granularity.

The novelty of *Liger* consists in the support for compile-time and run-time exploitation of locality, without unnecessarily restricting the parallelism. The idea of macro-actor encompasses both sequential code sequences and vector instructions, aiming to benefit from the pipeline implementation knowledge originating from von Neumann model. Synchronizing at a larger granularity implies less synchronization overhead. Moreover, the notion of source-broadcasting tag matching, together with codeblock-based hardware synchronization frame, further reduces the number of synchronization tokens needed. Separating data storage from synchronization point storage also promotes the autonomy between the macro-actor scheduling unit and the actor execution unit, which in turn enhances the parallelism among the two. Due to the hardware support for synchronization among macro-actors and among code blocks, the compiler has tremendous flexibility in finding an optimal tradeoff between parallelism and pipeline efficiency. As future work, a simulation model is under construction to quantitatively evaluate the strength and weakness of *Liger*, as well as more research to gain a deeper understanding of the resource management issues in this particular architecture.

REFERENCE

[AHME89]   H. Ahmed, "A Vector Dataflow Architecture" TRIAC-TCS-8905, Department of Telecommunication Systems-Computer Systems, Stockholm, Sweden, September 1989.

[BOHM89]   A. Bohm, J. Gurd, Y. Teo, "The Effect of Iterative Instructions in Dataflow Computers," 1989 International Conference on Parallel Processing, pp 201-208, Chicago, August 1989.

[CULL88]   D. Culler, Arvind, "Resource Requirements of Dataflow Programs," 1988 International Symposium on Computer architecture, pp 141-150, Honolulu, Hawaii, May 1988.

[DENN88]   J. Dennis, G. Gao, "An Efficient Pipelined Dataflow Processor Architecture," Supercomputing 88', pp 368-373, Orlando, Florida, November 1988.

[EKAN86]   K. Ekanadham, "Multi-tasking on a Dataflow-like Architecture," RC 12307, IBM T.J. Watson Research Laboratory, Yorktown Heights, NY, November 1986.

[GOTT89]   I. Gottlieb, "Dynamic Structured Data Flow: Preserving the Advantages of Sequential Processing in a Data Driven Environment," 1988 International Conference on Parallel Processing, pp 240-243, Chicago, Illinois, August 1988.

[GRAF90]   V. Grafe, J. Hoch, "The Epsilon-2 Hybrid Dataflow Architecture" 1990 International Conference on Parallel Processing, pp 88-93, Chicago, August 1990.

[IANN88]   R. Iannucci "Towards A Dataflow/von Neumann Hybrid Architecture," 15th International Symposium on Computer Architecture, pp 131-140, Hawaii, May 1988.

[IRAN88]   K. Irani, K. Luc, "Elimination of Bottlenecks in Dynamic Dataflow Processors," Supercomputing 88', pp 80-86, Orlando, Florida, November 1988.

[PAPA90]   G. Papadopoulos, D. Culler, "Monsoon: an Explicit Token-Store Architecture," 17th International Symposium on Computer Architecture, pp 82-91, Seattle, May 1990.

[RUGG87]   C. Ruggiero, Throttle Mechanisms for the Manchester Dataflow Machines. PhD Thesis University of Manchester, Manchester, England, July 1987.

[STER88]   T. Sterling, D. Wills, E. Chan, "Tokenless Static Dataflow Using Associative Templates," Supercomputing 88', pp 70-79, Orlando, Florida, November 1988.

[UCHI88]   K. Uchida, T. Temma, "A Pipelined Dataflow Processor Architecture Based on a Variable Length Token Concept," 1988 International Conference on Parallel Processing, pp 209-216, Chicago, Illinois, August 1988.

DATA FLOW MODEL FOR A HYPERCUBE
MULTIPROCESSING NETWORK

ADNAN SHAOUT and DANIEL SMYTH
THE ELECTRICAL AND COMPUTER
ENGINEERING DEPARTMENT
THE UNIVERSITY OF MICHIGAN - DEARBORN
DEARBORN, MICHIGAN 48128

ABSTRACT: This paper develops a useful theoretical model for the transfer of data inside a hypercube multiprocessing network. This model defines a statistical parameter, λ, which can be interpreted both as an average distance travelled by data within the network and as a measurement of how much memory sharing occurs within the network during the execution of a program. Also, a simple model for the node in a distributed memory hypercube is derived, in order to show data flow. In addition, methods of estimating time required for data transfer are developed. Finally a simulation for the data flow model is done.

1.   INTRODUCTION:

The subject of randomized data transfer within a hypercube network has been analyzed recently in at least two papers, one by Mitra and Cieslak [2] and another by Abraham and Padmanabhan [1].

Mitra and Cieslak model the hypercube interconnection structure as an indirect extended omega network in order to analyze message passing algorithms. This model, while useful for future architectural modelling, is not exactly what is needed for the modelling of a direct hypercube network. Abraham and Padmanabhan do model a direct hypercube network, but use the hypothesis that each node is equally likely to send a message to any other node in the entire array. This is a very good assumption when analyzing *system* software, however, it is less good when modelling *applications* software. The model presented here is specifically adapted for the analysis of applications software to be run on distributed-memory, direct binary n-cube multiprocessors.

A main feature of this model is that it is homogeneous, that each node is identical to every other node, and that the probabilities involved depend only upon relative positions within the structure.

2.PROBABILISTIC DATA MODEL:

POSTULATE 1:   The probability that a given processor requires the use of a data word from another processor memory k levels away is $a_0 p^{n-k} q^k$, where p is a measure of the locality of memory use, $q = 1 - p$, and a_0 is a normalizing constant.

Given a base processor, P_0, with an arbitrarily assigned address $A_0 = 000...00$, we can observe that the set of processors $\{P_k\}$, k levels away from P_0, have addresses with k ones and n-k zeros. Thus there are $n!/((n-k)!\, k!)$ processors in $\{P_k\}$.

Note that $\sum_{k=0}^{n} n!/((n-k)!\, k!) = (1+1)^n = 2^n$.

LEMMA 1:   Since there are $n!/(n-k)!\, k!$ processors k levels away, the probability, P_{MT}, that a processor is using any memory equals a_0.
$$P_{MT} = a_0.$$

DEFINITION 1: λ is defined as the mean distance that a word must travel between its source and destination nodes.

LEMMA 2: $\lambda = nq$.

For example, using Lemma 2, if $q = 0$, then each processor operates independently of the remaining node processors and memories. As expected, $\lambda = 0$, indicating that no bus transfer of data takes place. On the other extreme, if $p = q = 0.5$, then each node memory is shared equally by all node processors, and so on average each word must travel half way through the structure to reach the node processor which requires it. This is confirmed by calculating $\lambda = nq = n/2$. Therefore q (or λ) can be seen as a measurement of memory sharing as well as distance.

3.HYPERCUBE BUS NETWORK:

A processor at level k with respect to a base processor has k busses leading toward the base node and n-k leading away from it. Therefore, the closer a data word gets to its final destination, the fewer busses are available to transfer it. In other words, it must be *funnelled* through the bus structure. This process is shown in the diagram below, showing the number of nodes in either direction with which a node at a given level can communicate.

Level:0   1   2   k   n
    o →n 1← o →n-1 2← o →n-2 k← o →n-k   n← o

Suppose a processor, P_m, is at level m with respect to a final destination processor P_0 at level 0. The subset of processors that can send data directly through P_m to P_0 is a 2^{n-m} subcube of processors, with each processor having an address containing all of the ones in A_m. For $k \geq m$, there are $(n-m)!/((k-m)!(n-k)!)$ nodes in level k of this subcube. Each node can send its data through any of $k!/((k-m)!m!)$ nodes at level m.

LEMMA 3: The probability that P_m is used for data transfer to P_0 is $p_{m,0}$, where:

$$p_{m,0} = a_0 (m!(n-m)!/n!) \sum_{k=m}^{n} (n!/(k!(n-k)!)) p^{n-k}q^k$$

LEMMA 4: The probability that P_m is used to transfer data to any processor m levels away equals $P_{m,0}$, where:

$$P_{m,0} = a_0 \sum_{k=m}^{n} (n!/(k!(n-k)!)) p^{n-k}q^k .$$

LEMMA 5: The probability that a processor is used to transfer data to any node, including itself is P_{DT}, where $P_{DT} = a_0(1+\lambda)$.

LEMMA 6: The probability that P_m is used for data transfer, except from itself, to P_0 is $\overline{p_{m,0}}$, where:

$$\overline{p_{m,0}} = a_0 (m!(n-m)!/n!) \sum_{k=m+1}^{n} (n!/(k!(n-k)!)) p^{n-k}q^k .$$

LEMMA 7: The probability that P_m is used to transfer data to any processor m levels away, except form itself, is $\overline{P_{m,0}}$, where

$$\overline{P_{m,0}} = a_0 \sum_{k=m+1}^{n} (n!/(k!(n-k)!)) p^{n-k}q^k .$$

LEMMA 8: The probability that a processor is used to transfer data to any node, excluding itself is $\overline{P_{DT}}$, where:

$$P_{DT} = a_0(\lambda - (1-p^n)).$$

Once again the significance of the mean data transfer length, λ, as a system performance parameter can be seen. Transforming these probabilities into data flow rates will show this significance more clearly. In order to do so, a simple model of the individual memory and CPU node processor will be needed.

4. SIMPLIFIED HYPERCUBE NODE STRUCTURE:

It is necessary to distinguish between memory operations and CPU operations.

Data transfer between nodes in a hypercube network is generally serial rather than parallel in nature. This is done to reduce the number of interconnection bus lines required. To prevent slowing down the CPU, it is necessary to interpose some type of buffer between the bus and the CPU that can be read in the same way as a memory. This produces the simplest model for the node: one where the input bus lines are considered to enter the node via the memory. The basic node structure is shown in Figure 1:

4.1 DATA FLOW THROUGH THE NODE MODEL:

In this section, certain quantities will be referred to as data flow rates, but use expressions previously defined as probabilities. This is not a contradiction. The probability of a certain type of data transfer (or flow) multiplied by the total flow rate of all types of data will yield the flow rate of the given type. Therefore, the probability expressions can be used

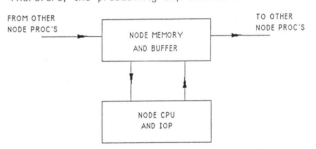

Figure 1: Simple Node Structure

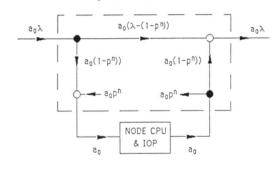

Figure 2: Data Flow inside Node CPU/Memory

as relative flow rates. The reason for doing so is because flow is easier to visualize in a structure. The simple node structure is drawn in Figure 2 below indicating data flow rates.

To demonstrate that the arrows in Figure 2 are in the proper direction, the following inequality is analyzed:

LEMMA 9: $1 - p^n \leq \lambda$.

This implies that $a_0(\lambda - (1-p^n)) \geq 0$, as expected, since negative data flow rates are not possible.

4.2 ESTIMATING λ FROM DATA FLOW:

Although the data transfers within node memory are complicated and, in practice, may be difficult to define, those transfers to and from a node memory are well defined and can be used to estimate λ.

DEFINITION 2: The external-to-internal data flow ratio, L, is defined as:

L=(Number of data transfers along external busses)/(Number of data transfers between node memory and CPU (or IOP)).

$$L = (0.5)(2a_0\lambda) / (2a_0) = \lambda/2$$

(The 0.5 term comes from the fact that each bus transfer is shared by two node memories).

Thus:

$\lambda = 2\times$(# of data transfers along external busses) / (# of data transfers between node memory and CPU (or IOP))

λ is very dependent upon the algorithm in use, in fact, almost totally so, and can be used to rate algorithms in order of the amount of data transfer time required. A smaller value of λ will indicate a faster algorithm.

4.3 ESTIMATING MEAN DATA TRANSFER TIME:

DEFINITION 3: Cycle time, t_C, is the time required to send a data word along a bus between adjacent nodes, given that the two node memories involved have completed their handshaking arrangements.

DEFINITION 4: The mean data transfer time, T_{DT}, is defined as the number of data transfer cycles the average message takes between the time it is generated by its source node and the time its destination node finally receives it, using semi-random data transfer.

To find expressions for T_{DT}, several assumptions must be made. The first is that data flow is unbiased, that is, there is no way to indicate priority; each memory accepting data randomly selects the bus from which it will receive data. This is obviously a worst-case assumption, but it is in accord with the probabilistic and homogeneous nature of the model. The second assumption is that handshaking occurs in parallel with data transfer. Therefore, by the time one data transfer is completed, the processors involved have already decided upon which data they will send or receive next.

Let the probability that a processor is available for receiving input be P_{acc}.

DEFINITION 5: then $P = P_{acc}/n$ is the probability that the adjacent processor will gain access during a given cycle, where n is the number of near neighboring processors.

DEFINITION 6: $Q = 1 - P$ is the probability that the adjacent processor will not gain access during a given cycle. The mean time for the data word to reach the final destination is

$$t_{1,0} = t_c \sum_{k=1}^{\alpha} kPQ^{k-1} = t_c/P = t_c/(1-Q).$$

Since the probability that a word k levels away from its final destination will not be transfered during a given cycle is Q^k,

$$t_{k,k-1} = t_c/(1-Q^k).$$

Therefore,

$$t_{m,0} = t_c \sum_{k=1}^{m} 1/(1-Q^k).$$

LEMMA 9: The mean data transfer time, T_{DT}, is:

$$T_{DT} = t_c nq + t_c \sum_{k=1}^{\infty} Q^k (1-(p + qQ^k)^n) / (1-Q^k)$$

The first term is to be expected, since it shows that transmission time is proportional to the mean distance it must travel through the network. The second term is due mainly to the fact that data must compete for bus access, and reflects the assumption that there is no special priority given to the data closest to their final destination. Again, it would be of use if an expression for T_{DT} could be found, independent of n, for those cases when n is difficult to determine.

$$(p + qQ^k)^n = (1 - q(1 - Q^k))^n = (1 - (\lambda(1 - Q^k)/n))^n$$

$$\lim_{n\to\infty} (p + qQ^k)^n = \exp(-\lambda(1 - Q^k))$$

$$T_{DT} \cong t_c\lambda + t_c \sum_{k=1}^{\infty} Q^k (1-\exp(-\lambda(1 - Q^k))) / (1-Q^k)$$

The single index summations shown above have the unfortunate property of being very slow to converge as $P\to0$. For small n, the following algorithm produces results more quickly and accurately.

Let $S(Q,0) = 0$, and $S(Q,k) = S(Q,k-1) + Q^k/(1-Q^k)$.

Let $f(\lambda,0) = (1-(\lambda/n))^n$, and

$$f(\lambda,k) = f(\lambda,k-1)\times(((n+1)/k)-1)/((n/\lambda)-1), \text{ then}$$

$$T_{DT} = t_c(\lambda + \sum_{k=1}^{n} S(Q,k)\times f(\lambda,k)).$$

This algorithm for calculating T_{DT} is independent of n, which is the desired result.

4.4 ESTIMATING P and Q:

The probability that a given bus is given access to transfer data to or from a node processor, P, is mainly dependent upon how many lines are leading into that processor. A good first order estimate of P is:

$\lambda/(2n(1+\lambda)) < P < 1/2n$,

since there are n data lines leading into a node and only input or output can be done at one time. The λ-dependent term can be used when internal node operations take up an appreciable portion of the processor's time, or when data transfer into the node occurs in parallel. The estimate for Q is then, obviously,

$$Q = 1 - P.$$

The relatively large values for Q cause T_{DT} to become large. For example, if $\lambda = 0.5$, $n = 4$, $P = 0.125$, $Q = 0.875$, then

$T_{DT} = t_C (0.5 + 3.169) = 7.338 (0.5 \, t_C)$.

If the time were dependent only upon the distance travelled inside the structure, T_{DT} would equal $0.5 \, t_C$. This shows that funnelling can produce a considerable slowdown in transferring data.

5. APPLICATION:

An example of an application for this method is to analyze data routing in a program to evaluate Laplace's Equation in two dimensions using the relaxation method. Assume that voltage V is to be calculated over a square region of $64 \times 64 = 4096$ points. The computer is to be a directly-connected hypercube of $N = 16$, $n = 4$. During each cycle, new values of V must be calculated for $4096 / 16 = 256$ points within each node processor. This calculation is done using the previous values for these points, and for the $4 \times 16 = 64$ points immediately bordering the points contained in one node processor. This data must be brought in from other nodes. From this, we can compute:

$\lambda = 64 / (64 + 256)$ $= 0.20$,

$q = \lambda / n = 0.20 / 4$ $= 0.05$,

$p = 1 - q = 1 - 0.05$ $= 0.95$,

$P \cong 1 / 2n = 1 / 8$ $= 0.125$ (est.),

$Q = 1 - P \cong 1 - 0.125$ $= 0.875$ (est.),

$a_0 = 64 + 256 = 320$ words/cycle,

$a_0 p^n = 320 \times 0.95^4 = 260.6$ words/cycle,

$a_0 (1 - p^n) = 59.4$ words/cycle.

From these calculations, a data flow diagram is constructed in Figure 3 below:

From the data calculated above, T_{DT}, assuming random routing of data, can be calculated.

$$T_{DT} = t_C n q + t_C \sum_{k=1}^{\infty} Q^k (1 - (p + q Q^k)^n) / (1 - Q^k)$$

$T_{DT} = t_C (0.20 + 1.345) = 1.545 \, t_C$.

The proportion of time required due to

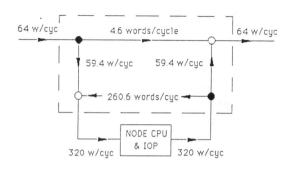

Figure 3: Data Flow Diagram Showing Sample Calculation (Assuming Random Data Flow)

funnelling is

$1.345 / (0.20 + 1.345) = 0.871$. This implies that tagging the data with its destination and letting it work its way through the network semi-randomly is very inefficient and wasteful compared to software synchronized methods which explicitly command the processors when to communicate with each other.

7. CONCLUSIONS:

This model shows the importance, when writing software for this type of system, of keeping the ratio of bus data transfers to internal node transfers to a minimum. This will insure a minimum value for λ, which will insure the small average data transfer path and the shortest possible mean data transfer time. Also, the expressions for T_{DT} demonstrate the pitfall of allowing data to be transferred without giving proper priority to those closest to their final destinations.

8. REFERENCES:

[1] Abraham, S. and Padmanabhan, K., "Performance of Direct Binary n-Cube Network for Multiprocessors," Proc. 1986 International Conf. on Parallel Processing

[2] Mitra, D. and Cieslak, R., "Randomized Parallel Communications," Proceedings 1986 International Conf. on Parallel Processing.

Two Techniques to Enhance the Performance of Memory Consistency Models

Kourosh Gharachorloo, Anoop Gupta, and John Hennessy

Computer Systems Laboratory
Stanford University, CA 94305

Abstract

The memory consistency model supported by a multiprocessor directly affects its performance. Thus, several attempts have been made to relax the consistency models to allow for more buffering and pipelining of memory accesses. Unfortunately, the potential increase in performance afforded by relaxing the consistency model is accompanied by a more complex programming model. This paper introduces two general implementation techniques that provide higher performance for all the models. The first technique involves *prefetching* values for accesses that are delayed due to consistency model constraints. The second technique employs *speculative execution* to allow the processor to proceed even though the consistency model requires the memory accesses to be delayed. When combined, the above techniques alleviate the limitations imposed by a consistency model on buffering and pipelining of memory accesses, thus significantly reducing the impact of the memory consistency model on performance.

1 Introduction

Buffering and pipelining are attractive techniques for hiding the latency of memory accesses in large scale shared-memory multiprocessors. However, the unconstrained use of these techniques can result in an intractable programming model for the machine. Consistency models provide more tractable programming models by introducing various restrictions on the amount of buffering and pipelining allowed.

Several memory consistency models have been proposed in the literature. The strictest model is *sequential consistency* (SC) [15], which requires the execution of a parallel program to appear as some interleaving of the execution of the parallel processes on a sequential machine. Sequential consistency imposes severe restrictions on buffering and pipelining of memory accesses. One of the least strict models is *release consistency* (RC) [8], which allows significant overlap of memory accesses given synchronization accesses are identified and classified into acquires and releases. Other relaxed models that have been discussed in the literature are *processor consistency* (PC) [8, 9], *weak consistency* (WC) [4, 5], and *data-race-free-0* (DRF0) [2]. These models fall between sequential and release consistency models in terms of strictness.

The constraints imposed by a consistency model can be satisfied by placing restrictions on the order of completion for memory accesses from each process. To guarantee sequential consis-

tency, for example, it is sufficient to delay each access until the previous access completes. The more relaxed models require fewer such delay constraints. However, the presence of delay constraints still limits performance.

This paper presents two novel implementation techniques that alleviate the limitations imposed on performance through such delay constraints. The first technique is *hardware-controlled non-binding prefetch*. It provides pipelining for large latency accesses that would otherwise be delayed due to consistency constraints. The second technique is *speculative execution for load accesses*. This allows the processor to proceed with load accesses that would otherwise be delayed due to earlier pending accesses. The combination of these two techniques provides significant opportunity to buffer and pipeline accesses regardless of the consistency model supported. Consequently, the performance of all consistency models, even the most relaxed models, is enhanced. More importantly, the performance of different consistency models is equalized, thus reducing the impact of the consistency model on performance. This latter result is noteworthy in light of the fact that relaxed models are accompanied by a more complex programming model.

The next section provides background information on the various consistency models. Section 3 discusses the prefetch scheme in detail. The speculative execution technique is described in Section 4. A general discussion of the proposed techniques and the related work is given in Sections 5 and 6. Finally, we conclude in Section 7.

2 Background on Consistency Models

A consistency model imposes restrictions on the order of shared memory accesses initiated by each process. The strictest model, originally proposed by Lamport [15], is sequential consistency (SC). Sequential consistency requires the execution of a parallel program to appear as some interleaving of the execution of the parallel processes on a sequential machine. Processor consistency (PC) was proposed by Goodman [9] to relax some of the restrictions imposed by sequential consistency. Processor consistency requires that writes issued from a processor may not be observed in any order other than that in which they were issued. However, the order in which writes from two processors occur, as observed by themselves or a third processor, need not be identical. Sufficient constraints to satisfy processor consistency are specified formally in [8].

A more relaxed consistency model can be derived by relating memory request ordering to synchronization points in the program. The weak consistency model (WC) proposed by Dubois et al. [4, 5] is based on the above idea and guarantees a consistent view of memory only at synchronization points. As an example, consider a process updating a data structure within a critical section. Under SC, every access within the critical section is delayed until the previous access completes. But such delays are unnecessary if the programmer has already made sure that no other process can rely on the data structure to be consistent until the critical section is exited. Weak consistency exploits this by allowing accesses within the critical section to be pipelined. Correctness is achieved by guaranteeing that all previous accesses are performed before entering or exiting each critical section.

Release consistency (RC) [8] is an extension of weak consistency that exploits further information about synchronization by classifying them into acquire and release accesses. An *acquire* synchronization access (e.g., a lock operation or a process spinning for a flag to be set) is performed to gain access to a set of shared locations. A *release* synchronization access (e.g., an unlock operation or a process setting a flag) grants this permission. An acquire is accomplished by reading a shared location until an appropriate value is read. Thus, an acquire is always associated with a read synchronization access (see [8] for discussion of read-modify-write accesses). Similarly, a release is always associated with a write synchronization access. In contrast to WC, RC does not require accesses following a release to be delayed for the release to complete; the purpose of the release is to signal that previous accesses are complete, and it does not have anything to say about the ordering of the accesses following it. Similarly, RC does not require an acquire to be delayed for its previous accesses. The data-race-free-0 (DRF0) [2] model as proposed by Adve and Hill is similar to release consistency, although it does not distinguish between acquire and release accesses. We do not discuss this model any further, however, because of its similarity to RC.

The ordering restrictions imposed by a consistency model can be presented in terms of when an access is allowed to perform. A read is considered *performed* when the return value is bound and can not be modified by other write operations. Similarly, a write is considered *performed* when the value written by the write operation is visible to all processors. For simplicity, we assume a write is made visible to all other processors at the same time. The techniques described in the paper can be easily extended to allow for the case when a write is made visible to other processors at different times. The notion of being performed and having completed will be used interchangeably in the rest of the paper.

Figure 1 shows the restrictions imposed by each of the consistency models on memory accesses from the same process.[1] As shown, sequential consistency can be guaranteed by requiring shared accesses to perform in program order. Processor consistency allows more flexibility over SC by allowing read operations to bypass previous write operations. Weak consistency and release consistency differ from SC and PC in that they exploit information about synchronization accesses. Both WC and RC allow accesses between two synchronization operations to be pipelined, as shown in Figure 1. The numbers on the blocks denote the order in which the accesses occur in program order. The figure shows that RC provides further flexibility by exploiting information about the type of synchronization.

The delay arcs shown in Figure 1 need to be observed for correctness. The conventional way to achieve this is by actually delaying each access until a required (possibly empty) set of

---

[1]The weak consistency and release consistency models shown are the WCsc and RCpc models, respectively, in the terminology presented in [8].

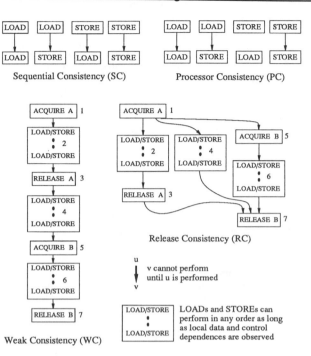

Figure 1: Ordering restrictions on memory accesses.

accesses have performed. An interesting alternative is to allow the access to partially or fully proceed even though the delay arcs demand that the access be delayed and to simply detect and remedy the cases in which the early access would result in incorrect behavior. The key observation is that for the majority of accesses, the execution would be correct (according to the consistency model) even if the delay arcs are not enforced. Such accesses are allowed to proceed without any delay. The few accesses that require the delay for correctness can be properly handled by detecting the problem and correcting it through reissuing the access to the memory system. Thus, the common case is handled with maximum speed while still preserving correctness.

The prefetch and speculative execution techniques described in the following two sections are based on the above observations. For brevity, we will present the techniques only in the context of SC and RC since they represent the two extremes in the spectrum of consistency models. Extension of the techniques for other models should be straightforward.

3 Prefetching

The previous section described how the delay constraints imposed by a consistency model limit the amount of buffering and pipelining among memory accesses. Prefetching provides one method for increasing performance by partially proceeding with an access that is delayed due to consistency model constraints. The following subsections describe the prefetch scheme proposed and provide insight into the strengths and weaknesses of the technique.

3.1 Description

Prefetching can be classified based on whether it is *binding* or *non-binding*, and whether it is controlled by *hardware* or *soft-*

ware. With a binding prefetch, the value of a later reference (e.g., a register load) is bound at the time the prefetch completes. This places restrictions on when a binding prefetch can be issued, since the value will become stale if another processor modifies the location during the interval between prefetch and reference. Hardware cache-coherent architectures, such as the Stanford DASH multiprocessor [18], can provide prefetching that is non-binding. With a non-binding prefetch, the data is brought close to the processor (e.g., into the cache) and is kept coherent until the processor actually reads the value. Thus, non-binding prefetching does not affect correctness for any of the consistency models and can be used as simply a performance boosting technique. The technique described in this section assumes *hardware-controlled non-binding prefetch.* We contrast this technique with previously proposed prefetch techniques in Section 6.

Prefetching can enhance performance by partially servicing large latency accesses that are delayed due to consistency model constraints. For a read operation, a *read prefetch* can be used to bring the data into the cache in a read-shared state while the operation is delayed due to consistency constraints. Since the prefetch is non-binding, we are guaranteed that the read operation will return a correct value once it is allowed to perform, regardless of when the prefetch completed. In the majority of cases, we expect the result returned by the prefetch to be the correct result. The only time the result may be different is if the location is written to between the time the prefetch returns the value and the time the read is allowed to perform. In this case, the prefetched location would either get invalidated or updated, depending on the coherence scheme. If invalidated, the read operation will miss in the cache and access the new value from the memory system, as if the prefetch never occurred. In the case of an update protocol, the location is kept up-to-date, thus providing the new value to the read operation.

For a write operation, a *read-exclusive prefetch* can be used to acquire exclusive ownership of the line, enabling the write to that location to complete quickly once it is allowed to perform. A read-exclusive prefetch is only possible if the coherence scheme is invalidation-based. Similar to the read prefetch case, the line is invalidated if another processor writes to the location between the time the read-exclusive prefetch completes and the actual write operation is allowed to proceed. In addition, exclusive ownership is surrendered if another processor reads the location during that time.

3.2 Implementation

This subsection discusses the requirements that the prefetch technique imposes on a multiprocessor architecture. Let us first consider how the proposed prefetch technique can be incorporated into the processor environment. Assume the general case where the processor has a load and a store buffer. The usual way to enforce a consistency model is to delay the issue of accesses in the buffer until certain previous accesses complete. Prefetching can be incorporated in this framework by having the hardware automatically issue a prefetch (read prefetch for reads and read-exclusive prefetch for writes and atomic read-modify-writes) for accesses that are in the load or store buffer, but are delayed due to consistency constraints. A prefetch buffer may be used to buffer multiple prefetch requests. Prefetches can be retired from this buffer as fast as the cache and memory system allow.

A prefetch request first checks the cache to see whether the line is already present. If so, the prefetch is discarded. Other-

wise, the prefetch is issued to the memory system. When the prefetch response returns to the processor, it is placed in the cache. If a processor references a location it has prefetched before the result has returned, the reference request is combined with the prefetch request so that a duplicate request is not sent out and the reference completes as soon as the prefetch result returns.

The prefetch technique discussed imposes several requirements on the memory system. Most importantly, the architecture requires hardware coherent caches. In addition, the location to be prefetched needs to be cachable. Also, to be effective for writes, prefetching requires an invalidation-based coherence scheme. In update-based schemes, it is difficult to partially service a write operation without making the new value available to other processors, which results in the write being performed.

For prefetching to be beneficial, the architecture needs a high-bandwidth pipelined memory system, including lockup-free caches [14, 21], to sustain several outstanding requests at a time. The cache will also be more busy since memory references that are prefetched access the cache twice, once for the prefetch and another time for the actual reference. As previously mentioned, accessing the cache for the prefetch request is desirable for avoiding extraneous traffic. We do not believe that the double access will be a major issue since prefetch requests are generated only when normal accesses are being delayed due to consistency constraints, and by definition, there are no other requests to the cache during that time.

Lookahead in the instruction stream is also beneficial for hardware-controlled prefetch schemes. Processors with dynamic instruction scheduling, whereby the decoding of an instruction is decoupled from the execution of the instruction, help in providing such lookahead. Branch prediction techniques that allow execution of instructions past unresolved conditional branches further enhance this. Aggressive lookahead provides the hardware with several memory requests that are being delayed in the load and store buffers due to consistency constraints (especially for the stricter memory consistency models) and gives prefetching the opportunity to pipeline such accesses. The strengths and weaknesses of hardware-controlled non-binding prefetching are discussed in the next subsection.

3.3 Strengths and Weaknesses

This subsection presents two example code segments to provide intuition for the circumstances where prefetching boosts performance and where prefetching fails. Figure 2 shows the two code segments. All accesses are to read-write shared locations. We assume a processor with non-blocking reads and branch prediction machinery. The cache coherence scheme is assumed to be invalidation-based, with cache hit latency of 1 cycle and cache miss latency of 100 cycles. Assume the memory system can accept an access on every cycle (e.g., caches are lockup-free). We also assume no other processes are writing to the locations used in the examples and that the lock synchronizations succeed (i.e., the lock is free).

Let us first consider the code segment on the left side of Figure 2. This code segment resembles a producer process updating the values of two memory locations. Given a system with sequential consistency, each access is delayed for the previous access to be performed. The first three accesses miss in the cache, while the unlock access hits due to the fact that exclusive ownership was gained by the previous lock access. Therefore, the four accesses take a total of 301 cycles to perform. In a system with release consistency, the write accesses are delayed until the

lock L	(miss)
write A	(miss)
write B	(miss)
unlock L	(hit)

Example 1

lock L	(miss)
read C	(miss)
read D	(hit)
read E[D]	(miss)
unlock L	(hit)

Example 2

Figure 2: Example code segments.

lock access is performed, and the unlock access is delayed for the write accesses to perform. However, the write accesses are pipelined. Therefore, the accesses take 202 cycles.

The prefetch technique described in this section boosts the performance of both the sequential and release consistent systems. Concerning the loop that would be used to implement the lock synchronization, we assume the branch predictor takes the path that assumes the lock synchronization succeeds. Thus, the lookahead into the instruction stream allows locations A and B to be prefetched in read-exclusive mode. Regardless of the consistency model, the lock access is serviced in parallel with prefetch for the two write accesses. Once the result for the lock access returns, the two write accesses will be satisfied quickly since the locations are prefetched into the cache. Therefore, with prefetching, the accesses complete in 103 cycles for both SC and RC. For this example, prefetching boosts the performance of both SC and RC and also equalizes the performance of the two models.

We now consider the second code segment on the right side of Figure 2. This code segment resembles a consumer process reading several memory locations. There are three read accesses within the critical section. As shown, the read to location D is assumed to hit in the cache, and the read of array E depends on the value of D to access the appropriate element. For simplicity, we will ignore the delay due to address calculation for accessing the array element. Under SC, the accesses take 302 cycles to perform. Under RC, they take 203 cycles. With the prefetch technique, the accesses take 203 cycles under SC and 202 cycles under RC. Although the performance of both SC and RC are enhanced by prefetching, the maximum performance is not achieved for either model. The reason is simply because the address of the read access to array E depends on the value of D and although the read access to D is a cache hit, this access is not allowed to perform (i.e., the value can not be used by the processor) until the read of C completes (under SC) or until the lock access completes (under RC). Thus, while prefetching can boost performance by pipelining several accesses that are delayed due to consistency constraints, it fails to remedy the cases where out-of-order consumption of return values is important to allow the processor to proceed efficiently.

In summary, prefetching is an effective technique for pipelining large latency references even though the consistency model disallows it. However, prefetching fails to boost performance when out-of-order consumption of prefetched values is important. Such cases occur in many applications, where accesses that hit in the cache are dispersed among accesses that miss, and the out-of-order use of the values returned by cache hits is critical for achieving the highest performance. The next section describes a speculative technique that remedies this shortcoming by allowing the processor to consume return values out-of-order regardless of the consistency constraints. The combination of prefetching for stores and the speculative execution technique for loads will be shown to be effective in regaining opportunity for maximum pipelining and buffering.

4 Speculative Execution

This section describes the speculative execution technique for load accesses. An example implementation is presented in the latter part of the section. As will be seen, this technique is particularly applicable to superscalar designs that are being proposed for next generation microprocessors. Finally, the last subsection shows the execution of a simple code segment with speculative loads.

4.1 Description

The idea behind speculative execution is simple. Assume u and v are two accesses in program order, with u being any large latency access and v being a load access. In addition, assume that the consistency model requires the completion of v to be delayed until u completes. *Speculative execution for load accesses* works as follows. The processor obtains or assumes a return value for access v before u completes and proceeds. At the time u completes, if the return value for v used by the processor is the same as the current value of v, then the speculation is successful. Clearly, the computation is correct since even if v was delayed, the value the access returns would have been the same. However, if the current value of v is different from what was speculated by the processor, then the computation is incorrect. In this case, we need to throw out the computation that depended on the value of v and repeat that computation. The implementation of such a scheme requires a *speculation mechanism* to obtain a speculated value for the access, a *detection mechanism* to determine whether the speculation succeeded, and a *correction mechanism* to repeat the computation if the speculation was unsuccessful.

Let us consider the speculation mechanism first. The most reasonable thing to do is to perform the access and use the returned value. In case the access is a cache hit, the value will be obtained quickly. In the case of a cache miss, although the return value will not be obtained quickly, the access is effectively pipelined with previous accesses in a way similar to prefetching. In general, guessing on the value of the access is not beneficial unless the value is known to be constrained to a small set (e.g., lock accesses).

Regarding the detection mechanism, a naive way to detect an incorrect speculated value is to repeat the access when the consistency model would have allowed it to proceed under non-speculative circumstances and to check the return value with the speculated value. However, if the speculation mechanism performs the speculative access and keeps the location in the cache, it is possible to determine whether the speculated value is correct by simply monitoring the coherence transactions on that location. Thus, the speculative execution technique can be implemented such that the cache is accessed only once per access versus the two times required by the prefetch technique. Let us refer back to accesses u and v, where the consistency model requires the completion of load access v to be delayed until u completes. The speculative technique allows access v to be issued and the processor is allowed to proceed with the return value. The detection mechanism is as follows. An invalidation or update message for location v before u has completed indicates that the value of the access may be incorrect.[2] In addition, the lack of invalidation and update messages indicates

---

[2]There are two cases where the speculated value remains correct. The first is if the invalidation or update occurs due to false sharing, that is, for another location in the same cache line. The second is if the new value written is the same as the speculated value. We conservatively assume the speculated value is incorrect in either case.

that the speculated value is correct. Cache replacements need to be handled properly also. If location v is replaced from the cache before u completes, then invalidation and update messages may no longer reach the cache. The speculated value for v is assumed stale in such a case (unless one is willing to repeat the access once u completes and to check the current value with the speculated value). The next subsection provides further implementation details for this mechanism.

Once the speculated value is determined to be wrong, the correction mechanism involves discarding the computation that depended on the speculated value and repeating the access and computation. This mechanism is almost the same as the correction mechanism used in processors with branch prediction machinery and the ability to execute instructions past unresolved branches. With branch prediction, if the prediction is determined to be incorrect, the instructions and computation following the branch are discarded and the new target instructions are fetched. In a similar way, if a speculated value is determined to be incorrect, the load access and the computation following it can be discarded and the instructions can be fetched and executed again to achieve correctness.

The speculative technique overcomes the shortcoming of the prefetch technique by allowing out-of-order consumption of speculated values. Referring back to the second example in Figure 2, let us consider how well the speculative technique performs. We still assume that no other processes are writing to the locations. Speculative execution achieves the same level of pipelining achieved by prefetching. In addition, the read access to D no longer hinders the performance since its return value is allowed to be consumed while previous accesses are outstanding. Thus, both SC and RC complete the accesses in 104 cycles.

Given speculative execution, load accesses can be issued as soon as the address for the access is known, regardless of the consistency model supported. Similar to the prefetch technique, the speculative execution technique imposes several requirements on the memory system. Hardware-coherent caches are required for providing an efficient detection mechanism. In addition, a high-bandwidth pipelined memory system with lockup-free caches [14, 21] is necessary to sustain multiple outstanding requests.

4.2 Example Implementation

This subsection provides an example implementation of the speculative technique. We use a processor that has dynamic scheduling and branch prediction capability. As with prefetching, the speculative technique also benefits from the lookahead in the instruction stream provided by such processors. In addition, the correction mechanism for the branch prediction machinery can easily be extended to handle correction for speculative load accesses. Although such processors are complex, incorporating speculative execution for load accesses into the design is simple and does not significantly add to the complexity. This subsection begins with a description of a dynamically scheduled processor that we chose as the base for our example implementation. Next, the details of implementing speculative execution for load accesses are discussed.

We obtained the organization for the base processor directly from a study by Johnson [11] (the organization is also described by Smith et al. [23] as the MATCH architecture). Figure 3 shows the overall structure of the processor. Only a brief description of the processor is given; the interested reader is referred to Johnson's thesis [11] for more detail. The processor

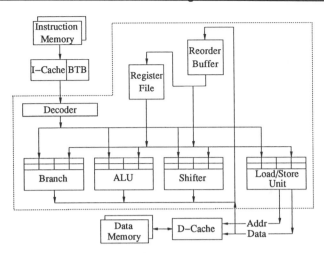

Figure 3: Overall structure of Johnson's dynamically scheduled processor.

consists of several independent function units. Each functional unit has a *reservation station* [25]. The reservation stations are instruction buffers that decouple instruction decoding from the instruction execution and allow for dynamic scheduling of instructions. Thus, the processor can execute instructions out of order though the instructions are fetched and decoded in program order. In addition, the processor allows execution of instructions past unresolved conditional branches. A branch target buffer (BTB) [16] is incorporated into the instruction cache to provide conditional branch prediction.

The *reorder buffer* [22] used in the architecture is responsible for several functions. The first function is to eliminate storage conflicts through register renaming [12]. The buffer provides the extra storage necessary to implement register renaming. Each instruction that is decoded is dynamically allocated a location in the reorder buffer and a tag is associated with its result register. The tag is updated to the actual result value once the instruction completes. When a later instruction attempts to read the register, correct execution is achieved by providing the value (or tag) in the reorder buffer instead of the value in the register file. Unresolved operand tags in the reservation stations are also updated with the appropriate value when the instruction with the corresponding result register completes.

The second function of the reorder buffer is to allow the processor to execute instructions past unresolved conditional branches by providing storage for the uncommitted results. The reorder buffer functions as a FIFO queue for instructions that have not been committed. When an instruction at the head of the queue completes, the location belonging to it is deallocated and the result value is written to the register file. Since the processor decodes and allocates instructions in program order, the updates to the register file take place in program order. Since instructions are kept in FIFO order, instructions in the reorder buffer that are ahead of a branch do not depend on the branch, while the instructions after the branch are control dependent on it. Thus, the results of instructions that depend on the branch are not committed to the register file until the branch completes. In addition, memory stores that are control dependent on the conditional branch are held back until the branch completes. If the branch is mispredicted, all instructions that are after the branch are invalidated from the reorder buffer, the reservation stations and buffers are appropriately cleared, and decoding and execution is started from the correct branch target.

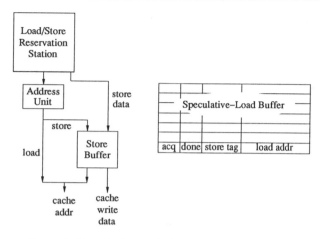

	Speculative–Load Buffer			
acq	done	store tag	load addr	

Figure 4: Organization of the load/store functional unit.

The mechanism provided in the reorder buffer for handling branches is also used to provide precise interrupts. Precise interrupts are provided to allow the processor to restart quickly without need to save and restore a lot of state. We will discuss the effect of requiring precise interrupts on the implementation of consistency models later in the section. Thus the reorder buffer plays an important role by eliminating storage conflicts through register renaming, allowing conditional branches to be bypassed, and providing precise interrupts.

To implement the speculative load technique, only the load/store (memory) unit of the base processor needs to be modified and the rest of the components remain virtually unchanged. Figure 4 shows the components of the memory unit. We first describe the components shown on the left side of the figure. These components are present regardless of whether speculative loads are supported. The only new component that is required for supporting speculative loads is the *speculative-load buffer* that will be described later.

The *load/store reservation station* holds decoded load and store instructions in program order. These instructions are retired to the address unit in a FIFO manner. Since the effective address for the memory instruction may depend on an unresolved operand, it is possible that the address for the instruction at the head of the reservation station can not be computed. The retiring of instructions is stalled until the effective address for the instruction at the head can be computed. The *address unit* is responsible for computing the effective address and doing the virtual to physical translation. Once the physical address is obtained, the address and data for store operations are placed into the *store buffer*. The retiring of stores from the store buffer is done in a FIFO manner and is controlled by the reorder buffer to assure precise interrupts (the mechanism is explained in the next paragraph). Load operations are allowed to bypass the store buffer and dependence checking is done on the store buffer to assure a correct return value for the load. Although the above implementation is sufficient for a uniprocessor, we need to add mechanisms to enforce consistency constraints for a multiprocessor.

Let us first consider how access order can be guaranteed for sequential consistency. The conventional method for achieving this is to delay the completion of each access until its previous access is complete. We first consider how the store operations are delayed appropriately. In general, a store operation may need to be delayed until certain previous load or store operations are completed. The mechanism for delaying the store is aided by the

fact that stores are already withheld to provide precise interrupts. The mechanism is as follows. All uncommitted instructions are allocated a location in the reorder buffer and are retired in program order. Except for a store instruction, an instruction at the head of the reorder buffer is retired when it completes. For store instructions, the store is retired from the reorder buffer as soon as the address translation is done. The reorder buffer controls the store buffer by signaling when it is safe to issue a store to the memory system. This signal is given when the store reaches the head of the reorder buffer. Consequently, a store is not issued until all previous loads and computation are complete. This mechanism satisfies the requirements placed by the SC model on a store with respect to previous loads. To make the implementation simpler for SC, we change the policy for retiring stores such that the store at the head of the reorder buffer is not retired until it completes also (for RC, however, the store at the head is still retired as soon as address translation is done). Thus, under SC, the store is also delayed for previous stores to complete and the store buffer ends up issuing stores one-at-a-time.

We now turn to how the restrictions on load accesses are satisfied. First, we discuss the requirements assuming the speculative load mechanism is not used. For SC, it is sufficient to delay a load until previous loads and stores have completed. This can be done by stalling the load/store reservation station at loads until the previous load is performed and the store buffer empties.

For speculative execution of load accesses, the mechanism for satisfying the restrictions on load accesses is changed. The major component for supporting the mechanism is the speculative-load buffer. The reservation station is no longer responsible for delaying certain load accesses to satisfy consistency constraints. A load is issued as soon as its effective address is computed. The speculation mechanism comprises of issuing the load as soon as possible and using the speculated result when it returns.

The speculative-load buffer provides the detection mechanism by signaling when the speculated result is incorrect. The buffer works as follows. Loads that are retired from the reservation station are put into the buffer in addition to being issued to the memory system. There are four fields per entry (as shown in Figure 4): load address, acq, done, and store tag. The load address field holds the physical address for the load. The acq field is set if the load is considered an acquire access. For SC, all loads are treated as acquires. The done field is set when the load is performed. If the consistency constraints require the load to be delayed for a previous store, the store tag uniquely identifies that store. A null store tag specifies that the load depends on no previous stores. When a store completes, its corresponding tag in the speculative-load buffer is nullified if present. Entries are retired in a FIFO manner. Two conditions need to be satisfied before an entry at the head of the buffer is retired. First, the store tag field should equal null. Second, the done field should be set if the acq field is set. Therefore, for SC, an entry remains in the buffer until all previous load and store accesses complete and the load access it refers to completes. Appendix A describes how an atomic read-modify-write can be incorporated in the above implementation.

We now describe the detection mechanism. The following coherence transactions are monitored by the speculative-load buffer: invalidations (or ownership requests), updates, and replacements.[3] The load addresses in the buffer are associatively

---
[3]A replacement is required if the processor accesses an address that maps onto a cache line with valid data for a different address. To avoid deadlock, a replacement request to a line with an outstanding access needs to be delayed until the access completes.

checked for a match with the address of such transactions.[4] Multiple matches are possible. We assume the match closest to the head of the buffer is reported. A match in the buffer for an address that is being invalidated or updated signals the possibility of an incorrect speculation. A match for an address that is being replaced signifies that future coherence transactions for that address will not be sent to the processor. In either case, the speculated value for the load is assumed to be incorrect.

Guaranteeing the constraints for release consistency can be done in a similar way to SC. The conventional way to provide RC is to delay a release access until its previous accesses complete and to delay accesses following an acquire until the acquire completes. Let us first consider delays for stores. The mechanism that provides precise interrupts by holding back store accesses in the store buffer is sufficient for guaranteeing that stores are delayed for the previous acquire. Although the mechanism described is stricter than what RC requires, the conservative implementation is required for providing precise interrupts. The same mechanism also guarantees that a release (which is simply a special store access) is delayed for previous load accesses. To guarantee a release is also delayed for previous store accesses, the store buffer delays the issue of the release operation until all previously issued stores are complete. In contrast to SC, however, ordinary stores are issued in a pipelined manner.

Let us consider the restriction on load accesses under RC. The conventional method involves delaying a load access until the previous acquire access is complete. This can be done by stalling the load/store reservation station after an acquire access until the acquire completes. However, the reservation station need not be stalled if we use the speculative load technique. Similar to the implementation of SC, loads are issued as soon as the address is known and the speculative-load buffer is responsible for detecting incorrect values. The speculative-load buffer description given for SC applies for RC. The only difference is that the acq field is only set for accesses that are considered acquire accesses under RC. Therefore, for RC, an entry remains in the speculative-load buffer until all previous acquires are completed. Furthermore, an acquire entry remains in the buffer until it completes also. The detection mechanism described for SC remains unchanged.

When the speculative-load buffer signals an incorrect speculated value, all computation that depends on that value needs to be discarded and repeated. There are two cases to consider. The first case is that the coherence transaction (invalidation, update, or replacement) arrives after the speculative load has completed (i.e., done field is set). In this case, the speculated value may have been used by the instructions following the load. We conservatively assume that all instructions past the load instruction depend on the value of the load and the mechanism for handling branch misprediction is used to treat the load instruction as "mispredicted". Thus, the reorder buffer discards the load and the instructions following it and the instructions are fetched and executed again. The second case occurs if the coherence transaction arrives before the speculative load has completed (i.e., done field is not set). In this case, only the speculative load needs to be reissued, since the instructions following it have not yet used an incorrect value. This can be done by simply reissuing the load access and does not require the instructions following the load to be discarded.[5] The next subsection further

illustrates speculative loads by stepping through the execution of a simple code segment.

4.3 Illustrative Example

In this subsection, we step through the execution of the code segment shown at the top of Figure 5. The sequential consistency model is assumed. Both the speculative technique for loads and the prefetch technique for stores are employed. Figure 5 also shows the contents of several of the buffers during the execution. We show the detection and correction mechanism in action by assuming that the speculated value for location D (originally in the cache) is later invalidated.

The instructions are assumed to be decoded and placed in the reorder buffer. In addition, it is assumed that the load/store reservation station has issued the operations. The first event shows that both the loads and the exclusive prefetches for the stores have been issued. The store buffer is buffering the two store operations and the speculative-load buffer has entries for the three loads. Note that the speculated value for load D has already been consumed by the processor. The second event occurs when ownership arrives for location B. The completion of store B is delayed by the reorder buffer, however, since there is an uncommitted instruction ahead of the store (this observes precise interrupts). Event 3 signifies the arrival of the value for location A. The entry for load A is removed from the reorder buffer and the speculative-load buffer since the access has completed. Once load A completes, the store buffer is signaled by the reorder buffer to allow store B to proceed. Since location B is now cached in exclusive mode, store B completes quickly (event 4). Thus, store C reaches the head of the reorder buffer. The store buffer is signaled in turn to allow store C to issue and the access is merged with the previous exclusive prefetch request for the location.

At this point, we assume an invalidation arrives for location D. Since there is a match for this location in the speculation buffer and since the speculated value is used, the load D instruction and the following load instruction are discarded (event 5). Event 6 shows that these two instructions are fetched again and a speculative load is issued to location D. The load is still speculative since the previous store (store C) has not completed yet. Event 7 shows the arrival of the new value for location D. Since the value for D is known now, the load access E[D] can be issued. Note that although the access to D has completed, the entry remains in the reorder buffer since store C is still pending. Once the ownership for location C arrives (event 8), store C completes and is retired form the reorder buffer and the store buffer. In addition, load D is no longer considered a speculative load and is retired from both the reorder and the speculative-load buffers. The execution completes when the value for E[D] arrives (event 9).

5 Discussion

The two implementation techniques presented in this paper provide a greater opportunity for buffering and pipelining of accesses than any of the previously proposed techniques. This section discusses some implications associated with the two techniques.

---

[4]It is possible to ignore the entry at the head of the buffer if the store tag is null. The null store tag for the head entry signifies that all previous accesses that are required to complete have completed and the consistency model constraints would have allowed the access to perform at such a time.

[5]To handle this case properly, we need to tag return values to distinguish between the initial return value, which has not yet reached the processor and

needs to be discarded, and the return value from the repeated access, which is the one to be used. In addition, in case there are multiple matches for the address in the speculative-load buffer, we have to guarantee that initial return values are not used by any of the corresponding loads.

Example code segment:

```
read  A    (miss)
write B    (miss)
write C    (miss)
read  D    (hit)
read  E[D] (miss)
```

Event	Reorder Buffer	Store Buffer	acq	done	st tag	ld addr	Cache Contents
1 reads are issued and writes are prefetched	ld E[D] ld D st C st B ld A	st C st B	✓ ✓ ✓	 ✓	 st C	ld E[D] ld D ld A	A: ld pending B: ex–prf pending C: ex–prf pending D: valid E[D]: ld pending
2 ownership for B arrives	ld E[D] ld D st C st B ld A	st C st B	✓ ✓ ✓	 ✓	 st C	ld E[D] ld D ld A	A: ld pending B: valid exclusive C: ex–prf pending D: valid E[D]: ld pending
3 value for A arrives	ld E[D] ld D st C st B	st C st B	✓ ✓	 ✓	 st C	ld E[D] ld D	A,D: valid B: valid exclusive C: ex–prf pending E[D]: ld pending
4 write to B completes	ld E[D] ld D st C	st C	✓ ✓	 ✓	 st C	ld E[D] ld D	A,D: valid B: valid exclusive C: ex–prf pending E[D]: ld pending
5 invalidation for D arrives	st C	st C					A: valid B: valid exclusive C: ex–prf pending D: invalid
6 read of D is reissued	ld E[D] ld D st C	st C	✓		st C	ld D	A: valid B: valid exclusive C: ex–prf pending D: ld pending
7 value for D arrives; read of E[D] is reissued	ld E[D] ld D st C	st C	✓ ✓	 ✓	 st C	ld E[D] ld D	A,D: valid B: valid exclusive C: ex–prf pending E[D]: ld pending
8 ownership for C arrives	ld E[D]		✓			ld E[D]	A,D: valid B,C: valid exclusive E[D]: ld pending
9 value for E[D] arrives							A,D,E[D]: valid B,C: valid exclusive

Figure 5: Illustration of buffers during execution of the code segment.

The main idea behind the prefetch and speculative load techniques is to service accesses as soon as possible, regardless of the constraints imposed by the consistency model. Of course, since correctness needs to be maintained, the early service of the access is not always helpful. For these techniques to provide performance benefits, the probability that a prefetched or speculated value is invalidated must be small. There are several reasons why we expect such invalidations to be infrequent. Whether prefetched or speculated locations are invalidated loosely depends on whether it is critical to delay such accesses to obtain a correct execution. If the supported consistency model is a relaxed model such as RC, delays are imposed only at synchronization points. In many applications, the time at which one process releases a synchronization is long before the time another process tries to acquire the synchronization. This implies that no other process is simultaneously accessing the locations protected by the synchronization. Correctness can be achieved in this case without the need to delay the accesses following an acquire until the acquire completes or to delay a release until its previous accesses complete. For cases where the supported consistency model is strict, such as SC, the strict delays imposed on accesses are also rarely necessary for correctness. We have observed that most programs do not have data races and provide sequentially consistent results even when they are executed on a release consistent architecture (see [8] for a formal definition of programs with such a property). Therefore, most delays imposed on accesses by a typical SC implementation are superfluous for achieving correctness. It is important to substantiate the above observations in the future with extensive simulation experiments.

A major implication of the techniques proposed is that the performance of different consistency models is equalized once these techniques are employed. If one is willing to implement these techniques, the choice of the consistency model to be supported in hardware becomes less important. Of course, the choice of the model for software can still affect optimizations, such as register allocation and loop transformation, that the compiler can perform on the program. However, this choice for software is orthogonal to the choice of the consistency model to support in hardware.

6   Related Work

In this section, we discuss previous work regarding prefetching and speculative execution. Next, we consider other proposed techniques for providing more efficient implementations of consistency models.

The main advantage of the prefetch scheme described in this study is that it is non-binding. Hardware-controlled binding prefetching has been studied by Lee [17]. Gornish, Granston, and Viedenbaum [10] have evaluated software-controlled binding prefetching. However, binding prefetching is quite limited in its ability to enhance the performance of consistency models. For example, in the SC implementation described, a binding prefetch can not be issued any earlier than the actual access is allowed to be issued.

Non-binding prefetching is possible if hardware cache-coherence is provided. Software-controlled non-binding prefetching has been studied by Porterfield [20], Mowry and Gupta [19], and Gharachorloo et al. [7]. Porterfield studied the technique in the context of uniprocessors while the work by Mowry and Gharachorloo studies the effect of prefetching for multiprocessors. In [7], we provide simulation results, for processors with blocking reads, that show software-controlled prefetching boosts the performance of consistency models and diminishes the performance difference among models if both read and read-exclusive prefetches are employed. Unfortunately, software-controlled prefetching requires the programmer or the compiler to anticipate accesses that may benefit from prefetching and to add the appropriate instructions in the application to issue a prefetch for the location. The advantage of hardware-controlled prefetching is that it does not require software help, neither does it consume instructions or processor cycles to do the prefetch. The disadvantage of hardware-controlled prefetching is that the prefetching window is limited to the size of the instruction lookahead buffer, while theoretically, software-controlled non-binding prefetching has an arbitrarily large window. In general, it should be possible to combine hardware-controlled and software-controlled non-binding prefetching such that they complement one another.

Speculative execution based on possibly incorrect data values has been previously described by Tom Knight [13] in the context of providing dynamic parallelism for sequential Lisp programs. The compiler is assumed to transform the program into sequences of instructions called transaction blocks. These blocks are executed in parallel and instructions that side-effect main memory are delayed until the block is committed. The sequential order of blocks determines the order in which the blocks can commit their side-effects. The mechanism described by Knight checks to see whether a later block used a value that is changed by the side-effecting instructions of the current block, and in case of a conflict, aborts the execution of the later block.

This is safe since the side-effecting instructions of later blocks are delayed, making the blocks fully restartable. Although this scheme is loosely similar to the speculative load scheme discussed in this paper, our scheme is unique in the way speculation is used to enhance the performance of multiprocessors given a set of consistency constraints.

Adve and Hill [1] have proposed an implementation for sequential consistency that is more efficient than conventional implementations. Their scheme requires an invalidation-based cache coherence protocol. At points where a conventional implementation stalls for the full latency of pending writes, their implementation stalls only until ownership is gained. To make the implementation satisfy sequential consistency, the new value written is not made visible to other processors until all previous writes by this processor have completed. The gains from this are expected to be limited, however, since the latency of obtaining ownership is often only slightly smaller than the latency for the write to complete. In addition, the proposed scheme has no provision for hiding the latency of read accesses. Since the visibility-control mechanism reduces the stall time for writes only slightly and does not affect the stall time for reads, we do not expect it to perform much better than conventional implementations. In contrast, the prefetch and speculative load techniques provide much greater opportunity for buffering and pipelining of read and write accesses.

Stenstrom [24] has proposed a mechanism for guaranteeing access order at the memory instead of at the processor. Each request contains a processor identification and a sequence number. Consecutive requests from the same processor get consecutive sequence numbers. Each memory module has access to a common data structure called *next sequence-number table* (NST). The NST contains P entries, one entry per processor. Each entry contains the sequence number of the next request to be performed by the corresponding processor. This allows the mechanism to guarantee that accesses from each processor are kept in program order. Theoretically, this scheme can enhance the performance of sequential consistency. However, the major disadvantage is that caches are not allowed. This can severely hinder the performance when compared to implementations that allow shared locations to be cached. Furthermore, the technique is not scalable to a large number of processors since the *increment network* used to update the different NST's grows quadratically in connections as the number of processors increases.

The detection mechanism described in Section 4 is interesting since it can be extended to detect violations of sequential consistency in architectures that implement more relaxed models such as release consistency. Release consistent architectures are guaranteed to provide sequential consistency for programs that are free of data races [8]. However, determining whether a program is free of data races is undecidable [3] and is left up to the programmer. In [6], we present an extension of the detection mechanism which, for every execution of the program, determines *either* that the execution is sequentially consistent *or* that the program has data races and may result in sequentially inconsistent executions. Since it is not desirable to conservatively detect violations of SC for programs that are free of data races and due to the absence of a correction mechanism, the detection technique used in [6] is less conservative than the one described here. In addition, the extended technique needs to check for violations of SC arising from performing either a read or a write access out of order.

7 Concluding Remarks

To achieve higher performance, a number of relaxed memory consistency models have been proposed for shared memory multiprocessors. Unfortunately, the relaxed models present a more complex programming model. In this paper, we have proposed two techniques, that of prefetching and speculative execution, that boost the performance under all consistency models. The techniques are also noteworthy in that they allow the strict models, such as sequential consistency, to achieve performance close to that of relaxed models like release consistency. The cost, of course, is the extra hardware complexity associated with the implementation of the techniques. While the prefetch technique is simple to incorporate into cache-coherent multiprocessors, the speculative execution technique requires more sophisticated hardware support. However, the mechanisms required to implement speculative execution are similar to those employed in several next generation superscalar processors. In particular, we showed how the speculative technique could be incorporated into one such processor design with minimal additional hardware.

8 Acknowledgments

We greatly appreciate the insight provided by Ruby Lee that led to the prefetch technique described in this paper. We thank Mike Smith and Mike Johnson for helping us with the details of out-of-order issue processors. Sarita Adve, Phillip Gibbons, Todd Mowry, and Per Stenstrom provided useful comments on an earlier version of this paper. This research was supported by DARPA contract N00014-87-K-0828. Kourosh Gharachorloo is partly supported by Texas Instruments. Anoop Gupta is partly supported by a NSF Presidential Young Investigator Award with matching funds from Sumitomo, Tandem, and TRW.

References

[1] Sarita Adve and Mark Hill. Implementing sequential consistency in cache-based systems. In *Proceedings of the 1990 International Conference on Parallel Processing*, pages I: 47–50, August 1990.

[2] Sarita Adve and Mark Hill. Weak ordering - A new definition. In *Proceedings of the 17th Annual International Symposium on Computer Architecture*, pages 2–14, May 1990.

[3] A. J. Bernstein. Analysis of programs for parallel processing. *IEEE Transactions on Electronic Computers*, EC-15(5):757–763, October 1966.

[4] Michel Dubois and Christoph Scheurich. Memory access dependencies in shared-memory multiprocessors. *IEEE Transactions on Software Engineering*, 16(6):660–673, June 1990.

[5] Michel Dubois, Christoph Scheurich, and Fayé Briggs. Memory access buffering in multiprocessors. In *Proceedings of the 13th Annual International Symposium on Computer Architecture*, pages 434–442, June 1986.

[6] Kourosh Gharachorloo and Phillip Gibbons. Detecting violations of sequential consistency. In *Symposium on Parallel Algorithms and Architectures*, July 1991.

[7] Kourosh Gharachorloo, Anoop Gupta, and John Hennessy. Performance evaluation of memory consistency models

for shared-memory multiprocessors. In *Fourth International Conference on Architectural Support for Programming Languages and Operating Systems*, pages 245–257, April 1991.

[8] Kourosh Gharachorloo, Dan Lenoski, James·Laudon, Phillip Gibbons, Anoop Gupta, and John Hennessy. Memory consistency and event ordering in scalable shared-memory multiprocessors. In *Proceedings of the 17th Annual International Symposium on Computer Architecture*, pages 15–26, May 1990.

[9] James R. Goodman. Cache consistency and sequential consistency. Technical Report no. 61, SCI Committee, March 1989.

[10] E. Gornish, E. Granston, and A. Veidenbaum. Compiler-directed data prefetching in multiprocessors with memory hierarchies. In *International Conference on Supercomputing*, pages 354–368, September 1990.

[11] William M. Johnson. *Super-Scalar Processor Design*. PhD thesis, Stanford University, June 1989.

[12] R. M. Keller. Look-ahead processors. *Computing Surveys*, 7(4):177–195, 1975.

[13] Tom Knight. An architecture for mostly functional languages. In *ACM Conference on Lisp and Functional Programming*, 1986.

[14] D. Kroft. Lockup-free instruction fetch/prefetch cache organization. In *Proceedings of the 8th Annual International Symposium on Computer Architecture*, pages 81–85, 1981.

[15] Leslie Lamport. How to make a multiprocessor computer that correctly executes multiprocess programs. *IEEE Transactions on Computers*, C-28(9):241–248, September 1979.

[16] J. K. F. Lee and A. J. Smith. Branch prediction strategies and branch target buffer design. *IEEE Computer*, 17:6–22, 1984.

[17] Roland Lun Lee. *The Effectiveness of Caches and Data Prefetch Buffers in Large-Scale Shared Memory Multiprocessors*. PhD thesis, University of Illinois at Urbana-Champaign, May 1987.

[18] Dan Lenoski, James Laudon, Kourosh Gharachorloo, Anoop Gupta, and John Hennessy. The directory-based cache coherence protocol for the DASH multiprocessor. In *Proceedings of the 17th Annual International Symposium on Computer Architecture*, pages 148–159, May 1990.

[19] Todd Mowry and Anoop Gupta. Tolerating latency through software-controlled prefetching in shared-memory multiprocessors. *Journal of Parallel and Distributed Computing*, June 1991.

[20] Allan K. Porterfield. *Software Methods for Improvement of Cache Performance on Supercomputer Applications*. PhD thesis, Department of Computer Science, Rice University, May 1989.

[21] Christoph Scheurich and Michel Dubois. Concurrent miss resolution in multiprocessor caches. In *Proceedings of the 1988 International Conference on Parallel Processing*, pages I: 118–125, August 1988.

[22] J. E. Smith and A. R. Pleszkun. Implementation of precise interrupts in pipelined processors. In *Proceedings of the 12th Annual International Symposium on Computer Architecture*, pages 36–44, June 1985.

[23] Michael Smith, Monica Lam, and Mark Horowitz. Boosting beyond static scheduling in a superscalar processor. In *Proceedings of the 17th Annual International Symposium on Computer Architecture*, pages 344–354, May 1990.

[24] Per Stenstrom. A latency-hiding access ordering scheme for multiprocessors with buffered multistage networks. Technical Report Department of Computer Engineering, Lund University, Sweden, November 1990.

[25] R. M. Tomasulo. An efficient hardware algorithm for exploiting multiple arithmetic units. *IBM Journal*, 11:25–33, 1967.

Appendix A: Read-Modify-Write Accesses

Atomic read-modify-write accesses are treated differently from other accesses as far as speculative execution is concerned. Some read-modify-write locations may not be cached. The simplest way to handle such locations is to delay the access until previous accesses that are required to complete by the consistency model have completed. Thus, there is no speculative load for non-cached read-modify-write accesses. In the following, we will describe the treatment of cachable read-modify-write accesses under the SC model. Extensions for the RC model are straightforward.

The reorder buffer treats the read-modify-writes as both a read and a write. Therefore, as is the case for normal reads, the read-modify-write retires from the head of the reorder buffer only when the access completes. In addition, similar to a normal write, the reorder buffer signals the store buffer when the read-modify-write reaches the head. The load/store reservation station services a read-modify-write by splitting it into two operations, a speculative load that results in a read-exclusive request and the actual atomic read-modify-write. The speculative load is issued to the memory system and is also placed in the speculative-load buffer. The read-modify-write is simply placed in the store buffer. The store tag for the speculative load entry is set to the tag for the read-modify-write in the store buffer. The done field is set when exclusive ownership is attained for the location and the return value is sent to the processor. The actual read-modify-write occurs when the read-modify-write is issued by the store buffer. The entry corresponding to the speculative load is guaranteed to be at the head of the speculative-load buffer when the actual read-modify-write is issued. In the case of a match on the speculative load entry before the read-modify-write is issued, the read-modify-write and the computations following it are discarded from all buffers and are repeated. The return result of the speculative read-exclusive request is ignored if it reaches the processor after the read-modify-write is issued by the store buffer (the processor simply waits for the return result of the read-modify-write). If a match on the speculative load entry occurs after the read-modify-write is issued, only the computation following the read-modify-write is discarded and repeated. In this case, the value used for the read-modify-write will be the return value of the issued atomic access. The speculative load entry is retired from the speculative buffer when the read-modify-write completes.

Efficient Storage Schemes for Arbitrary Size Square Matrices in Parallel Processors with Shuffle-Exchange Networks

Rajendra V. Boppana and C. S. Raghavendra

Department of Electrical Engineering-Systems

University of Southern California, Los Angeles, CA 90089-0781

Abstract. We present storage schemes based on linear permutations to store and access $N \times N$, for any integer N, data arrays in parallel processors with shuffle-exchange type interconnection networks. For parallel access of the most important templates, namely, row, column, main diagonal, and square block, simple criteria are given based on the linear permutations involved. This is the first such solution to provide access of the important templates of arbitrary size square matrices without memory and network conflicts using minimum number of memory modules and a shuffle-exchange type network.

Key words: data alignment, linear permutations, matrix storage, parallel memory systems, scrambled skewed storage, shuffle-exchange networks.

1   Introduction

In parallel processors, access to shared data in the memory modules plays an important role in the overall performance. For specific problems, knowing the data access patterns by processors, one can allocate data to memory modules initially so that high parallelism can be achieved in data access at run time. By the parallel access of data, we mean that the access of data is free of memory conflicts (that is, no memory module contains more than one element of the data to be accessed) and interconnection network conflicts (that is, one pass through the interconnection network is sufficient to transfer data from processors to memory modules or vice versa). In this paper we address the problem of storing an $m \times m$ matrix, $m \geq N$, in the N memory modules such that various N-(element)subsets (also called, *templates*) of the matrix are retrieved from memory modules without memory and interconnection network conflicts.

Previously known results [3, 4, 5, 7, 9, 10, 11, 12] either assume that N is a power of 2 and provide template access without memory and network conflicts or, for arbitrary N, discuss only the issue of template access without memory conflicts. In this paper, we present a class of scrambled schemes, called *clip* schemes, to store an $N \times N$ matrix for any integer N. These schemes are based on the class of *composite linear permutations*. As a main result of this paper, we specifically show some storage schemes in the proposed class which provide conflict free access of rows, columns, main and back diagonals, and square blocks (when N is square of an integer) of data arrays using the generalized shuffle-exchange networks.

2   Preliminaries

We assume that there are N processors and N memory units in the parallel processor system. Let $N = p_1^{r_1} \times \cdots \times p_k^{r_k}$, where $p_1, \cdots, p_k$ are distinct and primes, $k \geq 1$, $r_1, \ldots, r_k \geq 1$, and $n = r_1 + \cdots + r_k$. Let d_x, $0 \leq x < n$, be the xth digit of the sequence $p_1, \ldots, p_1 (r_1 \text{ times}), p_2, \ldots, p_2 (r_2 \text{ times}), \ldots, p_k$. Let $w_0 = 1$ and $w_x = \prod_{y=1}^{x} d_y$, $1 \leq x < n$. Each processor (memory module) is given a unique and distinct index i, $0 \leq i < N$, which can be represented in mixed radix form as $i_{n-1} \ldots i_0$ with d_x and w_x as the radix and weight of the xth digit, i_x. However, it is treated as an n-digit column vector $(i_0, \ldots, i_{n-1})^T$ in the matrix-vector computations defined below. Similarly, given an n-digit column vector $(j_0, \ldots, j_{n-1})^T$, its value is computed as $\sum_{x=1}^{x=k} j_x w_x$.

We use modulo arithmetic in the matrix-vector computations involving vectors and matrices defined in mixed radix form. Specifically, if two digits of radix p are added (or multiplied) then the result is given in modulo p. For example, the modulo addition and multiplication of the two digits 2 and 3 of radix 5 are 0 and 1, respectively. We never perform modulo arithmetic between digits of different radices. We use $\oplus$ to represent the modulo addition and juxtaposition to indicate multiplication.

Subsets of a matrix. A *template* T of an $N \times N$ data matrix is a set of N element positions $\{(\lambda_x, \mu_x) | 0 \leq x < N\}$, with $(\lambda_0, \mu_0) = (0, 0)$ [11]. By the access of a template T, we mean accessing of the N elements of the matrix whose positions are given by the template. Row (T_r), column (T_c), diagonal (T_d), and square block (T_s) templates are the four important templates reported in the literature [4, 9]. These templates are defined as follows.

Definition 1 $T_r = \{(0, j) | 0 \leq j < N\}$, $T_c = \{(i, 0) | 0 \leq i < N\}$, $T_d = \{(i, i) | 0 \leq i < N\}$, and $T_s = \{(i, j) | 0 \leq i, j < \sqrt{N}\}$.

It is clear that T_r, T_c, T_d, and T_s correspond to the positions given by, respectively, row 0, column 0, main diagonal, and the square block $S_{0,0}$ of the data matrix. The square block $S_{i,j}$ of an $N \times N$ matrix, N square of some integer, is the $\sqrt{N} \times \sqrt{N}$ submatrix with (i, j) in the top left position. Given a template $T = \{(0, 0), (\lambda_x, \mu_x) | 1 \leq x < N\}$, the set $\{(a, b), (a \oplus \lambda_x, b \oplus \mu_x) | 1 \leq x < N\}$, defines the affine template $T(a, b)$ of T. For example, ith row is the affine row template, $T_r(i, 0)$. The back diagonal of a matrix is the diagonal starting at its upper right corner and ending at its lower left corner; it is an affine

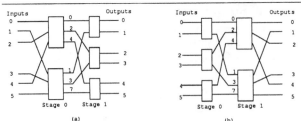

Figure 1: Examples of 6×6 Omega and inverse Omega networks. Addresses of lines before each stage are as indicated.

template of some template, for example, for $N = 2^n$, $T_d(0, N-1)$.

Processor-memory interconnection networks. We assume that $N \times N$ Omega or generalized shuffle-exchange networks [8, 1], which are less expensive compared to the alternatives such as the crossbar, are used for processor-memory interconnection. An important property of these networks is the ability to setup paths from inputs to outputs using only the output address with simple digit-controlled routing (at each stage, a particular digit of the address is used) [8]. An $N \times N$ Omega network [8] consists of n stages, numbered $0, \ldots, n-1$ left to right. Stage $(n-1-x)$, $0 \le x < n$, contains N/d_x switches of size $d_x \times d_x$ which are interconnected to the preceding stage with a d_x-shuffle. In a p shuffle, i, $0 \le i < N-1$, is mapped to $pi \bmod (N-1)$ and $(N-1)$ is mapped to itself. The Omega network, for $N = 2 \times 3$, and its inverse are shown in Figures 1(a) and 1(b).

Composite linear permutations. Using the concept of linear transformations [6], we define composite linear transformations on the set obtained by the cartesian product of vector spaces. For primes p, q and $r_1, r_2 > 0$, let $_pV_{r_1}$ and $_qV_{r_2}$ be two vector spaces over $GF(p)$ and $GF(q)$ respectively. Let $V = {_pV_{r_1}} \times {_qV_{r_2}}$. An element $i \in V$ is a $(r_1 + r_2)$-tuple with first r_1 elements (compactly denoted $i_{0:r_1-1}$) in $GF(p)$ and the remaining elements in $GF(q)$. Under the composite linear transformation, any $i \in V$ is mapped to some $j \in V$ given by the following matrix equation.

$$\left(\frac{j_{0:r_1-1}}{j_{r_1:r_1+r_2-1}} \right) = \left(\begin{array}{c|c} {_pQ_{r_1}} & 0 \\ \hline 0 & {_qQ_{r_2}} \end{array} \right) \left(\frac{i_{0:r_1-1}}{i_{r_1:r_1+r_2-1}} \right) \quad (1)$$

The matrix, denoted Q, in the above equation is called the characteristic matrix of the composite linear transformation; entries of its submatrix $_pQ_{r_1}$ are in $GF(p)$ and those of $_qQ_{r_2}$ are in $GF(q)$. A characteristic matrix is said to be nonsingular if and only if each of its submatrices corresponding to a particular radix is nonsingular. In that case, the mapping is called 'composite linear permutation'.

A composite linear permutation on a set of N ele-

ments has its $n \times n$ characteristic matrix as follows.

$$\begin{pmatrix} p_1 Q_{r_1} & 0 & \cdots & 0 \\ 0 & p_2 Q_{r_2} & \cdots & 0 \\ \vdots & \vdots & & \vdots \\ 0 & 0 & \cdots & p_k Q_{r_k} \end{pmatrix}$$

If the columns of Q are represented as $q_0, \ldots, q_{n-1}$, then $Q = (q_0, \ldots, q_{n-1})$ and $Qi = i_0 q_0 \oplus \cdots \oplus i_{n-1} q_{n-1}$. A composite affine linear permutations is of the form $j = Qi \oplus c$ for some $c \in V$. The composition of two composite affine linear permutations is again a composite affine linear permutation [2].

3 Inverse Omega passable composite linear permutations

Only a subset of the class of composite linear permutations are realized by a generalized Shuffle-Exchange network in one pass. Therefore, we characterize this subset of composite linear permutations so that network conflicts are eliminated in the data access by using the storage schemes based on such permutations.

Definition 2 *For a matrix $Q = (q_{i,j})_{n \times n}$, the "leading" principal submatrix $Q(x)$, $0 \le x < n$, of Q is defined as follows.*

$$\begin{pmatrix} q_{0,0} & \cdots & q_{0,x} \\ \vdots & & \\ q_{x,x} & \cdots & q_{x,x} \end{pmatrix}$$

Theorem 1 *A composite linear permutation, π, with characteristic matrix Q on the set of N numbers is realized by an $N \times N$ inverse Omega network if only if all the leading principal submatrices of Q are nonsingular.*

For proof, see [2]. It is easy to show that if a composite linear permutation is passed by the inverse Omega network, then any affine composite linear permutation is passed by it. The composite linear permutations passable by an Omega network can be characterized in an analogous manner [2].

4 Clip schemes

The clip storage scheme: Element (i, j) of the data matrix is stored in location i of the memory unit with index $i \oplus \pi(j)$, where π is some composite linear permutation.

The most important characteristic of a clip scheme is the composite linear permutation (hence, the corresponding 'Q' matrix) used. This composite linear permutation gives the complete information about the storage scheme. Therefore, a clip scheme, which defines an $N \times N$ mapping matrix, can be compactly represented using a much smaller $n \times n$ characteristic matrix, Q. In Figure 2(a), we give the mapping matrix for storing a 9×9 data matrix in 9 memory units, as specified by the clip scheme using the composite linear permutation with the characteristic matrix given

0	6	3	1	7	4	2	8	5
1	7	4	2	8	5	0	6	3
2	8	5	0	6	3	1	7	4
3	0	6	4	1	7	5	2	8
4	1	7	5	2	8	3	0	6
5	2	8	3	0	6	4	1	7
6	3	0	7	4	1	8	5	2
7	4	1	8	5	2	6	3	0
8	5	2	6	3	0	7	4	1

$$\begin{pmatrix} 0 & 1 \\ 2 & 0 \end{pmatrix}$$
(b)

(a)

Figure 2: The mapping and characteristic matrices of a clip scheme to store a 9×9 data matrix.

in Figure 2(b). With clip schemes, address computations can be done on-the-fly, since the arithmetic to compute the address of a data element is performed on individual digits of its indices. There is no propagation of carry from one digit to another during an address computation. When N is a power of 2, the address generation circuit consists of only exclusive-or gates [3]. In the general case, simple digit arithmetic circuits can be used [2].

Conflict-free access of templates of interest. For each template, T, of interest, we would like to know whether the template access is free of memory conflicts. If it is so, then we want to determine the data transfer function, $f(T)$, to be realized by the interconnection network in the access of that template or any of its affine template. For the row, column, diagonal, and square block templates, which are of interest to us here, this is typically done by showing that the index of the memory unit that a given processor has to access is given by some matrix-vector computation that is in the form of equation 1. Whenever the matrix appearing in the computation is non-singular, the template access is free of memory conflicts and the corresponding data transfer function is a composite affine linear permutation. From this matrix, we can also determine whether the data transfer function is realized by an Omega or inverse Omega network. The theorems below state the criteria for providing conflict-free access of rows, columns, main and back diagonals, and square blocks. Proofs of these results are given in [2].

Theorem 2 *For any clip scheme, T_r and T_c are free of memory conflicts and $f(T_r)$ and $f(T_c)$ are composite linear permutations, given by the matrices Q and I (identity matrix), respectively.*

Theorem 3 *For a clip scheme with characteristic matrix Q, T_d is free of memory conflicts if and only if the following matrix is non-singular.*

$$Q'' = Q \oplus I$$

Furthermore, whenever Q'' is non-singular, $f(T_d)$ is a composite linear permutation given by Q''.

Theorem 4 *For a clip scheme with characteristic matrix Q, the back diagonal is accessible free of memory conflicts if and only if the matrix $(I - Q)$ is nonsin-*

gular. Furthermore, when it is nonsingular, the data transfer function is a composite linear permutation.

For the access of square block templates, we assume that N is square of an integer, that is, $r_1, \ldots, r_k$ are all even. The mixed radix form considered thus far does not use all the radices in representing values in the set $\{0, \ldots, \sqrt{N} - 1\}$. Therefore, it does not facilitate a systematic study of the square block templates, since both indices of an element of the template are less than $\sqrt{N}$. So, we slightly modify the representation of i, $0 \leq i < N$. In the modified form, d_x, $0 \leq x < n/2$, is the xth digit of the sequence $p_1, \ldots, p_1, p_2, \ldots, p_2, \ldots, p_k, \ldots, p_k$, where each p_y, $1 \leq y \leq k$, is repeated $r_y/2$ times; for $n/2 \leq x < n$, $d_x = d_{x-n/2}$. Weights w_x are computed as products of d_x as defined before. Then, i is represented in mixed-radix form $i_{n-1} \ldots i_0$ such that i_x, $0 \leq x < n$, is of radix d_x and of weight w_x. For example, for $N = 2^4 \times 3^2 \times 5^6$, 100 is represented in the earlier notation as $(000000)_5(20)_3(0100)_2$, here the subscript indicates the radix of the digits within the parentheses, and in the modified notation given above as $(000)_5(0)_3(00)_2(013)_5(1)_3(00)_2$. Similarly, rows and columns of a characteristic matrix expressed in the earlier notation are rearranged to obtain the corresponding characteristic matrix in the modified notation. Note that the modified notation does not affect the results for the conflict-free access of the various templates given above. Let I_n, or simply I when there is no confusion, is the $n \times n$ identity matrix; xth column of I is denoted by e_x.

Theorem 5 *For a clip scheme with characteristic matrix Q, T_s is free of memory conflicts if and only if the characteristic matrix Q' (obtained from Q such that, for $0 \leq i < n/2$, $q'_i = q_i$, and, for $n/2 \leq i < n$, $q'_i = e_{i-n/2}$) is nonsingular. Furthermore, whenever Q' is non-singular, $f(T_s)$ is a composite linear permutation given by Q'.*

All the above results obtained for various templates are applicable for any of the affine templates derived from these templates [2]. For example, if the transfer function for the access of column 0 (column template, T_c) is a composite affine linear permutation, then the transfer function for the access of any other column (an affine template of T_c) is also a composite affine linear permutation.

5 Some practical clip schemes

Thus far, we have described the criteria for the access of various templates when a clip scheme is used. For each template access of interest, we showed that the template is free of memory conflicts if and only if the characteristic matrix of the storage scheme or some matrix derived from it is nonsingular. Furthermore, conflict-free access of row and column templates is always guaranteed. It still remains to be shown whether there are any characteristic matrices which satisfy the constraints given in theorems 3, 4 and 5. In such a case, to realize the data transfer functions of these templates without network conflicts, the characteristic matrix of a storage scheme and its derivatives should satisfy Theorem 1.

For any N, a square number, we use the mixed radix form discussed in the context of access of the square block templates. Let $L_{n/2}$ $(U_{n/2})$ be an $\frac{n}{2} \times \frac{n}{2}$ lower (upper) triangular characteristic matrix such that an entry on the diagonal has value $\lceil p/2 \rceil$ where p is the radix of that entry. Consider a characteristic matrix of the following form.

$$\Gamma_n = \left(\begin{array}{c|c} L_{n/2} & I_{n/2} \\ \hline L_{n/2} & 0 \end{array} \right) \qquad (2)$$

We may omit the subscript n if there is no confusion.

Theorem 6 *For any N not a multiple of 2 or 3, Γ defines a clip storage scheme for which row, column, main diagonal, back diagonal, and square block templates are free of memory conflicts. Furthermore, the data transfer functions for the access of all these templates are realized by an $N \times N$ inverse Omega network.*

This is easily proved [2] by showing that the characteristic matrix corresponding to the data transfer function of each template access is nonsingular and satisfies Theorem 1. With a little experimentation, many schemes satisfying Theorem 6 can be constructed. For example, the characteristic matrices with $L_{n/2}$ replaced by $U_{n/2}$ in (2) can be used for clip schemes with the same results. Also a diagonal entry of $L_{n/2}$ or $U_{n/2}$ could be any value other than 0,1 or $p-1$, where p is the radix of that entry, in the above example characteristic matrices without affecting the validity of the results.

For the case when N is a multiple of 2 or 3, the above schemes provide access of the templates without memory conflicts; however, to avoid network conflicts in the access of templates, an inverse Omega network is required for row, column, and square block accesses, and an Omega network is required for the main and back diagonal accesses. This is because the smallest leading submatrix, which is simply a 1×1 matrix, of the characteristic matrix can not take three non-singular values for when the radix is 3 or 2. See [2] for more details. Clip schemes for conflict-free access of templates in parallel processors with Omega networks can be developed similarly.

6 Summary

In this paper, we considered a class of scrambled skewing schemes, called the clip schemes. These schemes have the following advantages: (a) use of simple and popular Omega type interconnection networks to realize the data transfer functions of a template, (b) compact representation of the storage scheme, and (c) efficient address generation methods. These aspects are crucial for the use of a storage scheme. The proposed schemes are flexible and could be used in storing multidimensional and larger size matrices [2]. The main contribution of the paper is the development of some special clip schemes for use in parallel processors with shuffle-exchange type interconnection networks.

Further work in this direction could be in developing systematic procedures to obtain schemes that allow conflict free access of the templates of interest. One can analyze the transfer functions for subsets that are regular but not affine templates and investigate methods to realize them using an Omega or a similar network. Also, further work is warranted in realizing composite linear permutations by Omega networks with binary switches.

References

[1] L. N. Bhuyan and D. P. Agrawal. Design and performance of generalized interconnection networks. *IEEE Trans. on Computers*, C-32(12):1081–1090, December 1983.

[2] R. V. Boppana and C. S. Raghavendra. Data access and alignment of arbitrary size square matrices in parallel processors with multistage dynamic and static interconnection networks. Technical report, Dept. of EE-Systems, Univ. of Southern Cal., Univ. Park, Los Angeles, CA 90089-0781, 1991. In preparation.

[3] R. V. Boppana and C. S. Raghavendra. Generalized schemes for access and alignment of data in parallel processors with self-routing interconnection networks. *J. of Parallel and Distributed Computing*, 11:97–111, 1991.

[4] P. Budnik and D. J. Kuck. The organization and use of parallel memories. *IEEE Trans. on Computers*, c-20(12):1566–1569, 1971.

[5] J. M. Frailong, W. Jalby, and J. Lenfant. XOR-Schemes: A flexible data organization in parallel memories. In *Proc. International Conf. on Parallel Processing*, pages 276–283, 1985.

[6] K. Hoffman and R. Kunze. *Linear Algebra*. Prentice-Hall, second edition, 1971.

[7] K. Kim and V. K. Prasanna Kumar. Perfect latin squares and parallel array access. In *Proc. International Symp. on Comput. Arch.*, pages 372–379, 1989.

[8] D. H. Lawrie. *Memory-Processor Connection Networks*. PhD thesis, University of Illinois, Urbana, 1973.

[9] D. H. Lawrie. Access and Alignement of Data in an Array Processor. *IEEE Trans. on Computers*, c-24(12), 1975.

[10] D.-L. Lee. Scrambled storage for parallel memory systems. In *Proc. International Symp. on Comput. Arch.*, pages 232–239, 1988.

[11] H. D. Shapiro. Theoretical limitations on the efficient use of parallel memories. *IEEE Trans. on Computers*, c-27(5):421–428, 1978.

[12] H. A. G. Wijshoff and J. V. Leeuwen. The structure of periodic storage schemes for parallel memories. *IEEE Trans. on Computers*, c-34(6):501–505, 1985.

Recent results on the parallel access to tree-like data structures - the isotropic approach, I

Reiner Creutzburg

University of Karlsruhe
Faculty of Informatics
Institute of Algorithms
and Cognitive Systems
P. O. Box 6980
D-7500 Karlsruhe 1
(phone: +49-721-608 4325)
(fax: +49-721-696893)

Lutz Andrews

Deutsches Textilforschungszentrum NW
Frankenring 2
D-4150 Krefeld 1

Abstract

This note deals with recent results of a series of papers on the problem of parallel access to trees and is motivated by the research work on parallel memory design for trees. Known results on parallel access to complete q-ary subtrees and extended q-ary subtrees are summarized. Then, new results on the parallel access to other tree-like data structures are presented.

1 Introduction

One of the most significant challenges in parallel computers is the parallel memory organization. In order to obtain great performance, data in parallel memory have to be accessed at the highest throughput. Therefore data must be processed in parallel.

Classical SIMD computers perform vector processing [3-4,15,19-21,25,30]. A lot of research work has been done in designing parallel memories to access arrays [2-4,10-11,13-15,19-26,28,30].

In general parallel memories are designed for obtaining conflict-free access to arrangements of cells that belong to a specified set of data templates. In case of a single template T the smallest number of memory modules needed obviously is $N = |T|$ (the size of T), but more modules may be required. In general the problem arises of finding the smallest number of memory modules for storing data so as to have conflict-free access for a set of templates of interest.

Trees are another important data structure in computer science [18]. It is an interesting problem to design parallel memories to access trees [1,2,5-9,11-13,15,22,23,27,29,30]. We choose the terminology given in [17] to describe a tree. Consider a labelled q-ary tree such that the immediate successors of the node x are $qx + 1, qx + 2, \ldots, qx + q$, and the label of the root is 0. The level of a node is defined by initially letting the root be at level 1. The level of every node is one more than the level of its immediate predecessor. The height t of a tree is defined as the maximum level of any node in the tree.

Parallel access to a subtree means the conflict-free access to all the nodes of the subtree. Whenever a set of data items must be retrieved but is not available from distinct memory modules and hence cannot be retrieved in one cycle we speak of a "conflict".

We will always consider the case of **isotropic** storage of the nodes of a tree-like data structure. We remark that there also exists semi-isotropic and non-isotropic storage schemes [12-15].

In recent papers the parallel access to
- complete binary subtrees [1,5,6,11-13,15,22,23,27,29,30],
- complete q-ary subtrees [1,5,6,15,30],
was investigated.

For possible application in parallel logic programming the complete extended binary subtrees are of particular interest. We studied
- complete left-extended binary trees,
- complete right-extended binary trees,
- complete general-extended binary trees,
in detail [8,9]. Then, the results were generalized in [1,7] for the case of

- complete left-extended q-ary trees,
- complete right-extended q-ary trees,
- complete general-extended q-ary trees,
and are described in detail in [1].

The aim of this paper is to summarize and classify known results and to present new results for other tree-like data structures.

2 Module assignment functions

If we speak about parallel storage of general data structures and parallel access the module assignment functions and address functions play an important role. In [15] module assignment functions and address functions for general data structures are described. We will concentrate here only on module assignment functions. We consider now module assignment functions for q-ary trees.

A module assignment function S for q-ary trees is a mapping from the set of the labels of a complete q-ary tree to the N memory modules. We denote the set of indices of memory modules by

$$E_N = \{0, 1, \ldots, N - 1\}.$$

In the following we will always consider the case of **isotropic** storage [15] to tree-like data structures. This assumption means if two arbitrary nodes x_1 and x_2 are stored in the same memory module

$$S(x_1) = S(x_2) \bmod N$$

then their successor nodes

$$qx_1 + 1, qx_1 + 2, \ldots, qx_1 + q; \quad qx_2 + 1, qx_2 + 2, \ldots, qx_2 + q,$$

respectively, are stored in the same memory modules

$$S(qx_1 + 1) = S(qx_2 + 1); S(qx_1 + 2) = S(qx_2 + 2); \ldots;$$
$$S(qx_1 + q) = S(qx_2 + q) \bmod N,$$

respectively.

A module assignment function is called **conflict-free** with respect to a tree, if for two nodes x_1 and x_2 $(x_1 \neq x_2)$

$$S(x_1) \neq S(x_2) \bmod N.$$

2.1 Recursively linear module assignment functions

A special recursively linear module assignment function S for binary trees was introduced in [27] by the following recurrence equations

$$\left.\begin{array}{rcl} S(0) & = & 0 \\ S(2x + 1) & = & aS(x) + 1 \\ S(2x + 2) & = & aS(x) + 2 \end{array}\right\} \bmod N \qquad (1)$$

where a is an integer with $0 < a < N$.

A generalized form of (2.1) for q-ary trees was given in [5,6]

$$\left.\begin{array}{lll} S(0) & = & 0 \\ S(qx+1) & = & aS(x)+1 \\ S(qx+2) & = & aS(x)+2 \\ \cdots & & \cdots \\ S(qx+q) & = & aS(x)+q \end{array}\right\} \bmod N \qquad (2)$$

where a is an integer with $0 < a < N$.

The general recursively linear module assignment function S was introduced in [5,6,15] and has the following form:

$$\left.\begin{array}{lll} S(0) & = & 0 \\ S(qx+1) & = & a_1 S(x)+b_1 \\ S(qx+2) & = & a_2 S(x)+b_2 \\ \cdots & & \cdots \\ S(qx+q) & = & a_q S(x)+b_q \end{array}\right\} \bmod N \qquad (3)$$

where a_i and b_i are integers with $0 < a_i, b_i < N$ for $i = 1,...,q$.

3   Known results

In [1,5-7,15] the following results for isotropic subtree access are given and proved:

Theorem 1 *A parallel conflict-free access to all the nodes of an arbitrary complete q-ary subtree of height t of a q-ary tree $(q > 1, t > 1)$ with a special recursively linear module assignment function S is possible, if the number of memory modules N is equal to*

$$N = q^{t-2}(q(t-1)+1).$$

A minimization of N according to (3) is possible. Several examples are given in [1,15].

Theorem 2 *The minimal number N of needed memory modules for the parallel conflict-free access to general extended q-ary subtrees of height t is equal to*

$$N = qt.$$

The minimal number N of needed memory modules for the parallel conflict-free access to complete left- or right extended q-ary subtrees is equal to

$$N = (q-1)t+1.$$

The related module assignment functions are given in [1,7,15].

Theorem 3 *A parallel conflict-free access to all t layers of a complete q-ary subtree of height t $(t > 0)$ of a complete q-ary $(q > 0)$ tree is possible with*

$$N = q^{t-1} \qquad (4)$$

memory modules using a special recursively linear module assignment function S. The number N given in (4) is minimal.

Theorem 4 *A parallel conflict-free access to all the nodes of an arbitrary layer of the length l $(l > 0)$ and to a layer of length smaller than l in each level h $(h > 0)$ of a complete q-ary subtree $(q > 0)$ is possible with*

$$N = l$$

memory modules using a special recursively linear module assignment function S.

4   New results

In several previous papers [1,5-9,11-13,15] we considered the parallel access to complete q-ary subtrees of height t and to complete general extended q-ary subtrees of height t of a complete q-ary tree.

Now we deal with the simultaneous parallel conflict-free access to complete q-ary subtrees and to complete left-, right- or general extended q-ary subtrees of a complete q-ary tree with N memory modules by using the same module assignment function [1].

Furthermore, we deal with the access to complete q-ary subtrees of height $t - 1$ and to the last level t of a complete q-ary subtree of height t.

4.1   Simultaneous parallel access to complete q-ary subtrees and to complete-extended q-ary subtrees

We use the following notations in this chapter
  t_t height of a complete q-ary subtree,
  t_c height of a complete general extended q-ary subtree,
  t_{lc} height of a complete left extended q-ary subtree,
as shown in the figure at the end of this paper.

The following example is illustrative.

Example 1 *We consider the access to a complete binary tree of height $t_t = 3$ consisting of 7 nodes with $N = 10$ memory modules. We use the memory module assignment function given already in [27]:*

$$\left.\begin{array}{lll} S(0) & = & 0 \\ S(2x+1) & = & 6S(x)+1 \\ S(2x+2) & = & 6S(x)+2 \end{array}\right\} \bmod 10. \qquad (5)$$

From Theorem 1 it is known that a conflict-free access to the 7 nodes of all 10 subtrees of height $t_t = 3$ is possible. Furthermore, this function allows the conflict-free parallel access to all the 7 nodes of an arbitrary left extended binary subtree of height $t_{lc} = 4$. The relating 10 subtrees are shown in the following figure

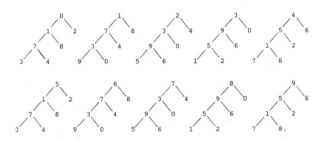

Now we generalize the results of Theorems 1-2 in the following Theorem [1].

Theorem 5 *A parallel conflict-free access to all the $(q^{t_t} - 1)/(q-1)$ nodes of an arbitrary complete q-ary subtree of height t_t of a complete q-ary tree $(q > 0, t_t > 1)$, to all the*
  $q(t_c - 1)+1$ nodes of an arbitrary complete general extended q-ary subtree of height $t_c = t_t + p + 1$
and to all the
  $q(t_{lc} - 1)+1$ nodes of an arbitrary complete left extended q-ary subtree of height $t_{lc} = q(t_t + p - 1) + 2$
with a special recursively linear module assignment function S according to (2) is possible, if the number of memory modules N in (2) is equal to

$$N_p = q^{t_t-2}(q(t_t+p)+1) \qquad (6)$$

with $p \in \{-1, 0, 1, 2, 3, ...\}$ and if the parameter a in (2) satisfies the condition

$$a = \begin{cases} 1 \quad \bmod \quad q(2+p)+1 & \text{in the case } t_t = 2, \\ 0 \quad \bmod \quad q & \\ & \text{in the case } t_t = 3, \\ 1 \quad \bmod \quad q(3+p)+1+1 & \\ q \quad \bmod \quad q^{t_t-2} & \\ & \text{in the case } t_t > 3. \\ 1 \quad \bmod \quad q(t_t+p)+1 & \end{cases} \qquad (7)$$

Remark 1 *For $p = -1$ we obtain from (6) that*

$$N_{-1} = q^{t_t-2}(q(t_t-1)+1).$$

This case is the evidence of Theorem 1 and we have $t_c = t_t$ and $t_{lc} = q(t_t - 2) + 2$.

The **proof** is given in [1].

The Table 1 lists a lot of special cases of Theorem 5 concerning parallel access to an arbitrary complete q-ary subtree of height t_t and to an arbitrary complete left- or general extended q-ary subtree of height t_{lc} and t_c, resp., of a complete q-ary tree [1].

4.2 Parallel access to a complete q-ary subtree of height t-1 and to the last layer of level t of a q-ary subtree of height t

Now we consider the parallel (linear) access to all nodes of a complete q-ary subtree of height $t - 1$ and to all nodes of the last layer t of a complete q-ary subtree of height t. Note that in this case we can access all nodes of an arbitrary q-ary subtree of height t in two steps. The number of needed memory modules for the conflict-free parallel access is (in general) smaller than the number of memory modules which we need in the case that we conflict-free access to all nodes of a complete q-ary subtree of height t in only one step. We mention that we can conflict-free access to all these nodes of a complete q-ary subtree of height t in two steps (see above) by using the same module assignment function.

We deal with the following example:

Example 2 *Consider the access to the 7 nodes of the complete binary subtree of height $t = 3$ of the complete binary tree. We can conflict-free access to the 3 nodes of the complete binary subtree of height $t = 2$ and to the 4 nodes of the 3rd layer of the complete binary subtree of height $t = 3$ with $N = 6$ memory modules if we use the following module assignment function S:*

$$\left.\begin{array}{rcl} S(0) &=& 0 \\ S(2x+1) &=& 4S(x) + 1 \\ S(2x+2) &=& 4S(x) + 2 \end{array}\right\} \bmod 6. \quad (8)$$

The 6 relating subtrees are the following:

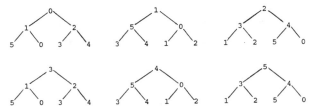

Remark 2 *It is easy to show that the number $N = 6$ of needed memory modules for the conflict-free access to the 7 nodes of a complete binary subtree of height $t = 3$ in two steps (see above) is the minimal number of necessary memory modules not only for linear module assignment functions.*

Theorem 6 *In a complete q-ary subtree of height t $(t > 2)$ of a complete q-ary tree $(q > 1)$ the parallel access to all the $(q^{t-1} - 1)/(q - 1)$ nodes of a complete q-ary subtree of height $t - 1$ of the complete q-ary tree and to all the q^{t-1} nodes of the last layer t is possible with*

$$N = q^{t-2}(q(t-2)+1) \quad (9)$$

memory modules using the recursively linear module assignment function S:

$$\left.\begin{array}{rcl} S(0) &=& 0 \\ S(qx+1) &=& aS(x) + 1 \\ S(qx+2) &=& aS(x) + 2 \\ \dots && \dots \\ S(qx+q) &=& aS(x) + q \end{array}\right\} \bmod q^{t-2}(q(t-2)+1) \quad (10)$$

with

$$a = \left\{\begin{array}{lll} 0 & \bmod & q \\ 1 & \bmod & q+1 \\ q & \bmod & q^{t-2} \\ 1 & \bmod & q(t-2)+1 \end{array}\right. \begin{array}{l} \text{\textit{in the case } } t = 3, \\ \\ \text{\textit{in the case } } t > 3. \end{array} \quad (11)$$

The **proof** is given in [1].

The Table 2. lists some special cases concerning parallel access that can be derived from Theorem 6.

t	1-ary tree	binary tree	ternary tree	quad-tree	...	7-ary tree	oct-tree
3	3 nodes $N = 2$ modules	7 nodes $N = 6$ modules	13 nodes $N = 12$ modules	21 nodes $N = 20$ modules	...	57 nodes $N = 56$ modules	73 nodes $N = 72$ modules
4	4 nodes $N = 3$ modules	15 nodes $N = 20$ modules	40 nodes $N = 63$ modules	85 nodes $N = 144$ modules	...	400 nodes $N = 735$ modules	585 nodes $N = 1088$ modules
5	5 nodes $N = 4$ modules	31 nodes $N = 56$ modules	121 nodes $N = 270$ modules	341 nodes $N = 832$ modules		2801 nodes $N = 7546$ modules	
6	6 nodes $N = 5$ modules	63 nodes $N = 144$ modules	364 nodes $N = 1053$ modules		...		

Table 2. List of necessary numbers N of memory modules for parallel, recursively linear, complete q-ary subtree access of height $t - 1$ of a complete q-ary tree and to all nodes of the last level t of the complete q-ary subtree of height t of a complete q-ary tree that can be derived from Theorem 6.

With the help of a computer we found the minimal number of needed memory modules in some cases for binary trees. They are presented in the following Table 3. Note that these numbers of memory modules are minimal if we use recursively linear module assignment functions of the type (3) for binary trees.

t	3	4	5	6
$q = 2$	7 nodes $N = 6$ modules	15 nodes $N = 20$ modules	31 nodes $N = 42$ modules	63 nodes $N = 108$ modules

Table 3.

References

[1] ANDREWS, L.: *Beiträge zur parallelen Speicherung von Bäumen.* Diploma work. Berlin 1988 (in German).

[2] ANISIFOROV, A.: *Parallel subtree access in dyadic memories.* Diss. (Leningrad 1989).

[3] BATCHER, K. E.: *STARAN parallel processor system hardware.* In: Proc. Fall Joint Computer Conf. AFIPS Conf., AFIPS Press, **43**, 1974, pp. 405-410.

[4] BUDNIK, - KUCK, D. J.: *The organization and use of parallel memories.* IEEE Trans. Comput., **C-20**, 1971, pp. 1566-1569.

[5] CREUTZBURG, R.: *Parallel optimal subtree access with recursively linear memory function.* In: PARCELLA'86 (Eds.: T. Legendi, D. Parkinson, R. Vollmar, G. Wolf) Akademie-Verlag: Berlin 1986, pp. 203-209.

[6] CREUTZBURG, R.: *Parallel linear conflict-free subtree access.* Proc. Internat. Workshop Parallel Algorithms Architectures (Suhl 1987), (Eds.: A.Albrecht, H. Jung, K. Mehlhorn) Akademie-Verlag: Berlin 1987, pp. 89-96.

[7] CREUTZBURG, R.: *Parallel conflict-free optimal access to complete extended q-ary trees.* In: PARCELLA'88 (Eds.: G. Wolf, T. Legendi, U. Schendel) Akademie-Verlag: Berlin 1988, pp. 248-255.

[8] CREUTZBURG, R.: *Parallel conflict-free access to extended binary trees.* Proc. IMYCS (Smolenice 1988), pp. 115 - 121.

[9] CREUTZBURG, R. - ANDREWS, L.: *Optimal parallel conflict-free access to extended binary trees.* In: Recent Issues in Pattern Analysis and Recognition (Eds.: V. Cantoni, R. Creutzburg, S. Levialdi, G. Wolf) LNCS **399**, Springer: New York 1989, pp. 214-225.

[10] GÖSSEL, M. - REBEL, B.: *Parallel memory with recursive address computation.* In: Parallel Computing (Ed.: M. Feilmeier) Elsevier: London 1984, pp. 515-520.

[11] GÖSSEL, M. - REBEL, B.: *Data structures and parallel memory.* In: PARCELLA'86 (Eds.: T. Legendi, D. Parkinson, R. Vollmar, G. Wolf) Akademie-Verlag: Berlin 1986, pp. 49-60.

[12] GÖSSEL, M. - REBEL, B.: *Memories for parallel subtree access.* In: Parallel Algorithms and Architectures (Eds.: A. Albrecht, H. Jung, K. Mehlhorn) (Suhl 1987) Akademie-Verlag: Berlin 1987, pp. 122-130.

[13] GÖSSEL, M. - REBEL, B.: *Parallel memories for picture processing (in Russian).* Isvest. Akad. Nauk., Techniceskaja Kibernetika. No. 2 (1988), pp. 198-206.

[14] GÖSSEL, M. - REBEL, B.: *Parallel access to rectangles.* In: Recent Issues in Pattern Analysis and Recognition (Eds.: V. Cantoni, R. Creutzburg, S. Levialdi, G. Wolf) LNCS **399**, Springer: New York 1989, pp. 201-213.

[15] GÖSSEL, M., REBEL, B., CREUTZBURG, R.: *Speicherarchitektur und Parallelzugriff.* Akademie-Verlag: Berlin 1989., also: *Memory Architecture and Parallel Access.* (English version in print)

[16] HOCKNEY, R. W. - JESSHOPE, C. R.: *Parallel Computers.* Hilger: Bristol, 1981.

[17] HOROWITZ, E. - SAHNI, S.: *Fundamentals of Data Structures.* Computer Science Press: Woodland Hills (Ca.) 1976.

[18] KNUTH, D. E.: *The Art of Computer Programming, Fundamental Algorithms.* Addison-Wesley: Reading (MA) 1968.

[19] KUCK, D. J. - STOKES, R. A.: *The BURROUGHS scientific processor.* IEEE Trans. Comput., **C-31**, 1982, pp. 363-376.

[20] LAWRIE, D. H.: *Access and alignment, in an array processor.* IEEE Trans. Comput., **C-24**, 1975, pp. 1145-1155.

[21] LAWRIE, D. H. - VORA, C. R.: *The prime memory system for array access.* IEEE Trans. Comput., **C-31**, 1982, pp. 435-442.

[22] METLITZKI, E. A.: *Data structures and parallel memory organization based on dyadic storage schemes.* In: Recent Issues in Pattern Analysis and Recognition (Eds.: V. Cantoni, R. Creutzburg, S. Levialdi, G. Wolf) LNCS **399**, Springer: New York 1989, pp. 189-200.

[23] METLITZKI, E. A. - KAVERSNEV, V. V.: *Parallel Memory Systems.* Leningrad 1989 (in Russian, in print)

[24] REBEL, B. - GÖSSEL, M.: *Ein paralleler Speicher.* Research Report, Central Institute Cybernetics and Information Processes. Berlin, November 1982.

[25] REBEL, B. - GÖSSEL, M.: *On conflict-free parallel memory access.* Computers and Artificial Intelligence, **2**, 1983, pp. 395-401.

[26] SHAPIRO, H. D.: *Theoretical limitations on the use of parallel memories.* Univ. Illinois, Dept. Comp. Sci., Report No. 75-776, December 1975.

[27] SHIRAKAWA, H.: *On a parallel memory to access trees.* In: Memoirs of Research Institute of Science and Engineering, Ritsumeikan Univ. Kyoto, Japan No.46 (1987) pp. 57 - 62. (as report 1984)

[28] WIJSHOFF, H. A. G. - VAN LEEUWEN, J.: *The structure of periodic storage schemes for parallel memories.* IEEE Trans. Comput. **C-34** (1985), pp. 501-505

[29] WIJSHOFF, H. A. G.: *Storing trees into parallel memories.* In: Parallel Computing 85 (Eds.: M. Feilmeier, J. Joubert, U. Schendel) Elsevier: Amsterdam 1986, pp. 253-261.

[30] WIJSHOFF, H. A. G.: *Data Organization In Parallel Computers.* Kluwer: Amsterdam 1989

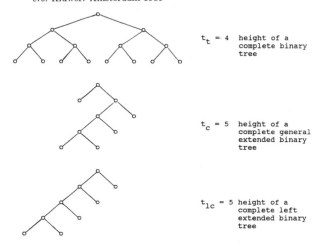

$t_t = 4$ height of a complete binary tree

$t_c = 5$ height of a complete general extended binary tree

$t_{lc} = 5$ height of a complete left extended binary tree

t_t	q=2	q=3	q=4	q=5	q=6
3	7 nodes $N_{-1}=10$ $t_{lc}=4;t_c=3$	13 nodes $N_{-1}=21$ $t_{lc}=5;t_c=3$	21 nodes $N_{-1}=36$ $t_{lc}=6;t_c=3$	31 nodes $N_{-1}=55$ $t_{lc}=7;t_c=3$	43 nodes $N_{-1}=78$ $t_{lc}=8;t_c=3$
	$N_0=14$ $t_{lc}=6;t_c=4$	$N_0=30$ $t_{lc}=8;t_c=4$	$N_0=52$ $t_{lc}=10;t_c=4$	$N_0=80$ $t_{lc}=12;t_c=4$	$N_0=114$ $t_{lc}=14;t_c=4$
	$N_1=18$ $t_{lc}=8;t_c=5$	$N_1=39$ $t_{lc}=11;t_c=5$	$N_1=68$ $t_{lc}=14;t_c=5$	$N_1=105$ $t_{lc}=17;t_c=5$	$N_1=150$ $t_{lc}=20;t_c=5$
4	15 nodes $N_{-1}=28$ $t_{lc}=6;t_c=4$	40 nodes $N_{-1}=90$ $t_{lc}=8;t_c=4$	85 nodes $N_{-1}=208$ $t_{lc}=10;t_c=4$	156 nodes $N_{-1}=400$ $t_{lc}=12;t_c=4$	259 nodes $N_{-1}=684$ $t_{lc}=14;t_c=4$
	$N_0=36$ $t_{lc}=8;t_c=5$	$N_0=117$ $t_{lc}=11;t_c=5$	$N_0=272$ $t_{lc}=14;t_c=5$	$N_0=525$ $t_{lc}=17;t_c=5$	$N_0=900$ $t_{lc}=20;t_c=5$
	$N_1=44$ $t_{lc}=10;t_c=6$	$N_1=144$ $t_{lc}=14;t_c=6$	$N_1=336$ $t_{lc}=18;t_c=6$	$N_1=650$ $t_{lc}=22;t_c=6$	$N_1=1116$ $t_{lc}=26;t_c=6$
5	31 nodes $N_{-1}=72$ $t_{lc}=8;t_c=5$	121 nodes $N_{-1}=351$ $t_{lc}=11;t_c=5$	341 nodes $N_{-1}=1088$ $t_{lc}=14;t_c=5$	781 nodes $N_{-1}=2625$ $t_{lc}=17;t_c=5$	1555 nodes $N_{-1}=5400$ $t_{lc}=20;t_c=5$
	$N_0=88$ $t_{lc}=10;t_c=6$	$N_0=432$ $t_{lc}=14;t_c=6$	$N_0=1344$ $t_{lc}=18;t_c=6$	$N_0=3250$ $t_{lc}=22;t_c=6$	$N_0=6696$ $t_{lc}=26;t_c=6$
	$N_1=104$ $t_{lc}=12;t_c=7$	$N_1=513$ $t_{lc}=17;t_c=7$	$N_1=1600$ $t_{lc}=22;t_c=7$	$N_1=3875$ $t_{lc}=27;t_c=7$	$N_1=7992$ $t_{lc}=32;t_c=7$

Table 1. Some special cases concerning parallel access to an arbitrary complete q-ary subtree of height t_t and to an arbitrary complete left- or general extended q-ary subtree of heights t_{lc} and t_c, resp., of a complete q-ary tree that can be derived from Theorem 5.

Multiple-Port Memory Access in Decoupled Architecture Processors

Shlomo Weiss

Department of Computer Science
University of Maryland
Baltimore, MD 21228

Abstract

We introduce a decoupled computer that supports multiple memory transfers per cycle. A detailed design is presented that addresses the problems of multiple instruction issue in the Access Unit, the order in which operands arrive from memory to the Execute Unit, and memory hazards. The result is a decoupled computer in which the memory transfer rate is sufficiently high to often keep the Execute Unit pipeline full, leading to a performance level limited by the processing rate of the arithmetic pipeline.

1   Introduction

Superscalar computers [1, 2, 3] are distinguished by the ability to issue multiple instructions per cycle. In particular, memory operations can be overlapped with the execution of arithmetic or logic instructions. A decoupled access/execute computer [3] is a superscalar machine that consists of two primary computational units: an Access Unit (AU) and an Execute Unit (EU). Instructions are split between the two units according to their type. Memory reference instructions are directed to the Access Unit; the Execute Unit handles arithmetic instructions.

This organization has two important properties: (a) operands may be loaded by the AU prior to their use by the EU, thus hiding memory access delays, and (b) more than one instruction may be issued in each cycle, surpassing the instruction issue bottleneck. However, within each unit instructions are still decoded and issued at the rate of one instruction per cycle. Since, in a scalar processor, one memory instruction must be issued for each memory reference, and this is done by the AU at the maximum rate of one-per-cycle, the maximum rate at which operands may be fetched from memory is also one-per-cycle. This is true even if the memory system itself is designed to support a higher data transfer rate.

The AU issue rate of one instruction per cycle is often the performance bottleneck in a decoupled computer. In this paper, we present a design of a decoupled computer that eliminates this bottleneck by supporting multiple memory transfers per cycle. In the proposed design a number of operand streams may be simultaneously active between the memory and the EU. This requires multiple memory instructions to be decoded and issued in each cycle, and a mechanism to maintain the order in the operand streams arriving from memory to the EU, without serializing memory references. Also, memory hazards caused by loads that bypass store instructions must be detected and handled. The implementation of this design is transparent to the programmer and requires no changes in the instruction set of the decoupled computer.

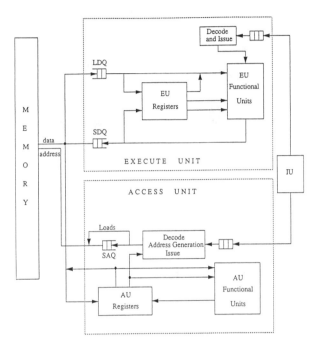

Figure 1: Decoupled computer organization

2   Decoupled Architecture Overview

The basic structure of a decoupled computer is shown in Figure 1. The Instruction Unit (IU) dispatches up to two instructions per cycle, one instruction to the Access Unit (AU) and the other to the Execute Unit (EU). The EU processes arithmetic and logic instructions. The primary floating-point functional unit in the EU is a multiply-add pipeline, which performs the operation F1 ← F2 + F3*F4. When the pipeline is full, it produces one multiply-add result in every cycle, and the maximum performance of the EU is increased to two floating-point operations per cycle. The EU register file provides three outputs, for three reads in a cycle to support the multiply-add primitive, and two inputs to allow simultaneous writes from a functional unit and from memory.

The AU handles memory instructions. Store instructions are decoded, the effective address generated, and then transferred to the Store Address Queue (SAQ). Results to be stored in memory arrive from the EU functional units to the SDQ. When the SDQ is not empty, the operand at its head is matched with the address at the head of the SAQ and the store operation is performed. Often results are not available yet when the corresponding store is decoded, and therefore a number of store instructions may be pending in the SAQ.

Load instructions are decoded and the effective address is generated and checked against pending stores in the SAQ. A match causes a hold issue condition in the AU until the conflict goes away. When there are no conflicts, load instructions bypass stores. Once the load is performed and the memory accessed, operands for arithmetic and logic instructions arrive from memory to the LDQ. From the LDQ, operands can be input directly to the EU functional units.

3 AU Instruction Issuing Bottleneck

In a scalar processor, performance is limited by the rate at which instructions are issued. Within each of the two units of a decoupled computer, the AU and the EU, instructions pass one at a time through the issue stage of the pipeline. We define the *issue bound* to be the number of cycles taken by a loop iteration assuming that the only limitation is set by the maximum rate at which instructions are decoded and issued. The issue bound can be reached given favorable data dependency conditions.

Operation	AU issue bound cycles/iteration	EU issue bound cycles/iteration	Bottleneck
V = V + S	2	1	AU
V = V + V	3	1	AU
V = V + S*V	3	1	AU
V = V*V + V	4	1	AU
V = S*V + S*V	3	2	AU
V = V*V + V*V	5	2	AU

Table 1: AU and EU issue bound for selected vector and vector/scalar operations. Assumes multiply-add functional unit is available in the EU.

Table 1 shows selected vector and vector/scalar operations. As an example, the vector/scalar addition V = V + S requires one load and one store instruction per loop iteration (the scalar is kept in a register). The AU issues these two memory operations in two cycles. The EU decodes and issues the single floating-point add instruction in one cycle. The table indicates that the AU issuing rate is the bottleneck in all the examples shown.

4 Decoupled Architecture with Multiple Memory Ports

We have seen that the memory transfer rate of one operand per cycle is often the performance bottleneck in a decoupled architecture. This rate is limited by two factors: (a) the instruction issue bandwidth of the AU, and (b) the single memory port structure of the decoupled computer. Because of these two restrictions, a decoupled computer cannot use efficiently a memory system that supports a transfer rate of multiple words per cycle.

In the rest of this paper we present the design of a decoupled computer with multiple memory ports. Figure 2 shows the AU and the EU and their interface to an interleaved memory system. Also shown the IU and the instruction dispatch buses; for clarity, we have omitted the instruction reload bus between the IU and the memory.

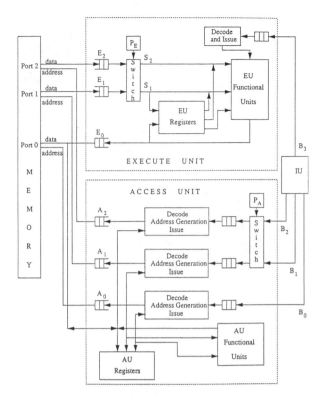

Figure 2: Decoupled computer with three memory ports

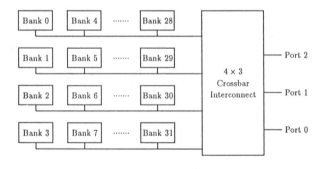

Figure 3: Interleaved memory system with three ports

The interleaved memory is shown in Figure 3. It consists four sections, each section being connected by its own address bus and data bus to a 4 × 3 crossbar interconnect. Through the crossbar, the three address buses are connected to three address ports in the AU (A_0, A_1, and A_2), and the three data buses are connected to three data ports in the EU (E_0, E_1, and E_2). Each of these ports has a queue (a FIFO buffer) to queue addresses in the AU or data in the EU. Store addresses are sent by the AU to port A_0; the store can be performed when the data is available in port E_0. The other ports are used for load addresses (A_1 and A_2) and load data (E_1 and E_2). One data bus (from port 0) is connected also to the AU and is used to load and store A registers.

Each of the three AU ports decodes its own instruction, generates the effective address, and, if necessary, updates an index register. Hence each port must have an integer adder. This requirement is not different than in the CRAY X-MP, where the address of each element in a vector stream is computed by adding a fixed stride to the previous address. In the case of the decoupled computer, however, there is the additional requirement that three index registers must be accessed in a single cycle. Therefore, the AU register file has to support three register reads and three register writes per cycle (while most machines do not require such a high-bandwidth to access registers, in the RS/6000, for example, the general purpose register file supports three reads and two writes in a cycle).

The interleaved memory in Figure 3 is similar to the CRAY X-MP memory system. In the CRAY X-MP the three memory ports available to the CPU can be simultaneously active in vector mode. After decoding and issuing a vector instruction, the control hardware takes over and performs the memory transfer (up to 64 operands). Multiple vector streams may be active even though instructions are issued at the rate of one-per-cycle.

A decoupled computer is a scalar machine and one instruction must be decoded for each memory reference. A multiple-port memory system cannot be efficiently used without increasing the instruction issue bandwidth of the AU. Another and more subtle problem in a decoupled architecture has to do with the way in which the AU and EU cooperate in performing memory operations. In a decoupled computer the AU decodes and issues memory instructions; the EU accepts and uses the operands loaded from memory or supplies the results to be stored. The matching of store addresses to results to be stored is done based on their order. Similarly, in the EU, the matching of an operand arriving from memory to the correct instruction is based on the operand arrival order.

This works in the decoupled computer of Figure 1 because the EU has a single input port (the LDQ) and a single output port (the SDQ). In a decoupled computer with multiple memory ports a mechanism is needed to maintain the order of operands loaded from memory.

5   Order of Operands Loaded from Memory

Referring to Figure 2, in each cycle, the IU dispatches up to three instructions (one store and two loads) to the AU. Therefore, three buses (B_0, B_1, and B_2), are needed between the IU and AU. The single store instruction (if there is one to be dispatched in that cycle) is sent on bus B_0. Buses B_1 and B_2 are used for loads, with the convention that the load instruction on B_1 precedes the one on B_2 in the instruction stream. If a single load is dispatched, that instruction is sent on bus B_1, while B_2 is not used in that cycle.

As they arrive at the AU on bus B_0, stores are directed to port A_0. Load instructions are assigned to ports A_1 and A_2 in a round-robin fashion. To that end, a one-bit pointer (P_A), initially set to 0, is maintained. P_A is incremented modulo 2 by the number of loads that arrive in each cycle. Its value controls a switch that connects buses B_1, B_2 to ports A_1, A_2 according to the following table.

P_A	Bus B_1	Bus B_2
0	A_1	A_2
1	A_2	A_1

A load instruction that was performed by the AU through port A_1 returns an operand to port E_1 in the EU. A similar correspondence is maintained between ports A_2 and E_2. The order in which the two input ports E_1 and E_2 are accessed by the EU is maintained by a one bit pointer P_E, initially set to 0. In each cycle, P_E is incremented modulo 2 by the number of operands retrieved from the input ports. Its value determines the connection between ports E_1, E_2 and switch outputs S_1, S_2 according to the following table.

P_E	Switch output S_1	Switch output S_2
0	E_1	E_2
1	E_2	E_1

Register loads are performed through switch output S_1 only. In a single cycle, two operands may be transferred from ports E_1 and E_2 directly to a functional unit through switch outputs S_1 and S_2. If a single operand is to be transferred to a functional unit, it is always taken from switch output S_1. Whether that operand is taken from E_1 or E_2 is determined by the above table.

Instruction issue may be held in either port A_1 or A_2 because of memory conflicts, memory hazards (see section 6), or some other resource conflict. Within each port instructions are issued strictly in order. A hold issue condition in one of the ports does not block the other port. Therefore, the two streams of load instructions may get out of order with respect to one another.

As a result, the two streams of operands arriving to ports E_1 and E_2 of the EU may "slip" relative to each other. This order is restored in the EU by the round-robin discipline in which operands are retrieved from ports E_1 and E_2. Since the issue of load instructions in the AU is alternated between ports A_1 and A_2, and the retrieval of operands by the EU is alternated between ports E_1 and E_2, it is guaranteed that the EU accepts operands exactly in the same order as if the load instructions were processed in a serial fashion.

6   Memory Hazards

In the typical operation of a decoupled computer, the AU runs ahead and loads operands that are used by the EU. This allows memory delays to be hidden and tends to reduce gaps in the pipeline. Therefore, it is essential to permit load instructions to bypass stores, since stores are delayed waiting for results. This can be safely done only if there are no memory hazards and loads do not conflict with pending stores.

In the decoupled computer of Figure 1 this is done by comparing each load instruction, before it is issued, to stores queued in the SAQ. In the decoupled computer with multiple memory ports (Figure 2) the detection of memory hazards is somewhat more complex due to two factors: (a) two loads may be issued in each cycle, and both must be compared against pending stores, and (b) loads and stores dispatched to the AU in the same cycle must also be hazard free.

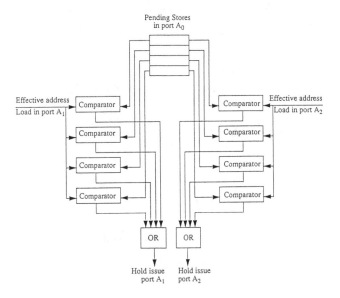

Figure 4: Memory hazard detection

Operation	AU issue bound cycles/iteration	EU issue bound cycles/iteration	Bottleneck
V = V + S	1	1	—
V = V + V	1	1	—
V = V + S*V	1	1	—
V = V*V + V	2	1	AU
V = S*V + S*V	1	2	EU
V = V*V + V*V	2	2	—

Table 2: AU and EU issue bound – multiple-port decoupled architecture of Figure 2. Assumes multiply-add functional unit is available in the EU.

We address item (b) by imposing two rules on the way instructions are dispatched from the IU to the AU: (1) at most one store is dispatched in each cycle, and (2) the store is the last instruction in the order in which these instructions (dispatched in the same cycle) appear in the instruction stream. Hazards between memory instructions dispatched in the same cycle are prevented since the loads always precede the single store instruction.

As for item (a) above, Figure 4 shows the block diagram of a circuit that detects memory hazards for load instructions prior to issue. Two comparators (their size equal to the number of bits in the effective address) are required for each slot in the pending store queue in port A_0. The addresses of the pending stores are simultaneously compared with both load addresses; this is done after the two load instructions have been decoded and their effective addresses generated, but prior to issue. A match produces a hold issue signal for that port.

7   Discussion and Conclusions

In a decoupled computer, the AU instruction issue bandwidth limits the maximum memory transfer rate to one word per cycle. This often becomes the performance bottleneck, as one arithmetic pipeline can accept two operands (or three operands in case of a multiply-add pipeline) and produce one result per cycle. Hence even a single arithmetic pipeline is kept underutilized.

We have introduced a decoupled computer that supports multiple memory transfers per cycle. The EU pipeline is often kept full, as shown in Table 2 for a number of vector and vector/scalar operations. The issue bound (a lower bound on the number of cycles required to issue the instructions in each iteration, see section 3), is shown for the AU and the EU. The AU is the bottleneck in one case, the operation V = V*V + V, which is processed in a single cycle per iteration by the multiply-add pipeline, but requires four memory references. Otherwise, the

EU pipeline is kept saturated and the peak performance level, limited by the processing rate of the multiply-add pipeline, is sustained.

One of the primary characteristics of a decoupled architecture is the use of architectural queues for memory operations. The task of the programmer (or the compiler writer) would be somewhat more complex if there would be a need to distinguish between all the different queues in the multiple-port decoupled architecture. Fortunately, this is not necessary. Queues E_1 and E_2 may be viewed logically as the LDQ; this "LDQ" may supply two operands in a single cycle. Similarly, E_0 is the SDQ. As for the AU, ports A_0, A_1, and A_2 are not visible to the programmer. The format of a load instruction is simply "LDQ ← address" and the format for a store instruction is "address ← SDQ", as in the single-port decoupled architecture.

Therefore, the design introduced in this paper enhances the memory transfer rate but requires no changes in the instruction set of the decoupled computer. The multiple memory ports are an implementation issue and the programmer does not need to be aware of their existence. The programmer can continue to maintain the same logical view of the decoupled architecture as in Figure 1, in which there is a single stream of addresses in the AU and a single stream of operands in the EU.

References

[1] International Business Machines Corporation, Austin, TX. *IBM RISC System/6000 Technology*, 1990. SA23–2619.

[2] B. R. Rau, D. W. L. Yen, W. Yen, and R. A. Towle. The Cydra 5 departmental supercomputer: design philosophies, decisions, and trade-offs. *Computer*, 22:12–35, Jan. 1989.

[3] J. E. Smith. Decoupled access/execute computer architectures. *ACM Transactions on Computer Systems*, 2:289–308, Nov. 1984.

ELIMINATING FALSE SHARING

Susan J. Eggers and Tor E. Jeremiassen
Department of Computer Science, FR-35
University of Washington
Seattle, WA 98195

Abstract – The performance and scalability of bus-based, shared memory multiprocessors is limited by the amount of bus traffic. Previous studies have shown that for machines with large caches and a write-invalidate coherency protocol, most of the bus traffic stems from coherency overhead. Some of this overhead is unavoidable, a direct consequence of the true sharing activity in the program. However, for caches with multi-word cache blocks, coherency overhead can also be caused by multiple processors accessing different words in the same block, known as false sharing.

This paper measures the amount of false sharing in several coarse-grained parallel applications. We present two simple transformations for restructuring shared data, and apply them to the shared objects responsible for most of the false sharing. Simulations of traces from the modified programs indicate that the transformations reduce the miss rates significantly. Shared misses were reduced by an order of magnitude more than in a previous study which applied a different type of transformation.

Introduction

Shared-memory multiprocessors are considered by many to be an efficient way of obtaining the high performance of parallel machines, while still keeping within the traditional programming paradigm. Many of these machines are based on a single bus and have private caches for each processor, giving rise to the cache coherency problem. From a programming perspective the simplest coherence solution lies in snooping cache coherency protocols [2].

One of the major performance problems of parallel programs executing on these machines is the overhead of maintaining cache coherency, i.e. the additional bus traffic generated by the protocols. For workloads with good processor locality [1], write-invalidate protocols incur less coherency overhead than write-broadcast protocols [7]. Current parallel programs show a high level of processor locality [1, 7], and it is reasonable to believe that it will continue to be the case to an even higher degree as programmers become more educated about the performance implications of data sharing.

Coherency-related bus traffic for write-invalidate protocols has two sources: one is invalidation signals which guarantee the writing processor the only cached copy of a shared address; the other is cache misses for rereferenced, invalidated data, called *invalidation misses*. Some of the coherency overhead is unavoidable, because multiple processors are updating the same shared addresses, and the system needs to provide each reading processor with an address's most current value. For unit-word cache blocks, this is the only source of coherency overhead and has been denoted *true sharing* [13]. However, for multi-word cache blocks overhead is also caused by multiple processors accessing different words within the same cache block. Since this overhead is not caused by the actual sharing of data, it has been denoted *false sharing* [13]. Eliminating the type of overhead induced by false sharing holds promise for performance improvements: execution time is reduced, and the decrease in coherency-related bus traffic enables the system to support more processors.

Since false sharing is a product of multi-word cache blocks, one would expect that it would increase as the block size increases. Previous studies [4, 6] have implied that this may be the case. While large block sizes benefit uniprocessors, false sharing may reduce or even reverse this effect on multiprocessors. By being able to significantly reduce the number of misses induced by false sharing, one should be able to utilize the benefits of larger cache blocks on multiprocessors.

Our study explicitly documents the presence of false sharing in three coarse grained parallel programs. We found that false sharing misses increase, at times substantially, as block size increases. For two applications the amount of false sharing misses was sufficient to adversely affect performance to a significant degree. In these applications we found that the number of false sharing misses accounts for up to 40% of all cache misses. We were able to pinpoint which shared objects were the main sources of false sharing and present two simple program transformations that reduce false sharing misses without resorting to padding the data structures. The results of these transformations are reductions in the false sharing miss rate of 40% to 75% and a reduction in the total miss rate of 20% to 30%.

The remainder of the paper is organized as follows: section 2 describes the methodology and workload for the studies; section 3 analyzes false sharing in the untransformed applications; the transformations are presented in section 4; section 5 analyzes the applications after applying the transformations; section 6 discusses related work; and section 7 concludes.

Methodology and Workload

The results in this study were obtained through trace-driven simulations. The traces were generated on a 20-processor Sequent Symmetry Model B [11] via a software tracing system for parallel programs called MPtrace [8].

Cache behavior was analyzed using a multiprocessor simulator that emulates a simple, shared memory architecture with up to 12 processors. The processors are assumed to be RISC-like, with each instruction taking one cycle. The block size was varied from 4 to 256 bytes. For simplicity, all simulations were run assuming an infinite cache. All traces were simulated for two million references for each processor after a cold start interval.

For each simulation, data on the total number of references and misses are collected. Total misses are broken down into misses on shared data, invalidation misses, and false sharing misses (false sharing misses are detected whenever an invalidation miss occurs, and the referenced word has not been accessed by any other processor since the block was invalidated). Maps of the distribution of false sharing misses by cache block and by word are produced and are used to identify the shared structures that incur false sharing misses.

Three coarse-grained parallel applications were used in the study. They are written in C and generate parallel processes that are permanently scheduled on a particular processor. Topopt [5] performs topological optimization on VLSI circuits using a parallel simulated annealing algorithm. Pverify [12] determines whether two boolean circuits are functionally identical. Psplice [10] is a circuit simulator that combines the original direct method solution with waveform relaxation.

The C compiler used was the Dynix C compiler with default options.

Analyzing False Sharing

This section discusses the results of simulating the original, unmodified programs. The results are presented as graphs of the various miss rates as the block size is varied from 4 to 256 bytes.

False sharing in Topopt (see Figure 1) and Pverify (see Figure 2) increases significantly as block size increases, and accounts for a large proportion of total misses, even for moderate size cache blocks. For Pverify the false sharing miss rate actually determines the trend of the total miss rate, because (1) it increases sharply as block size increases, (2) it comprises a substantial proportion of the total miss rate and (3) the other components of the total miss rate are seemingly unaffected by changes in block size. The false sharing increases are direct consequences of the rising number of independent data objects being allocated to the same cache block (as block size increases).

The miss rates for Psplice are shown in Figure 3. The total miss rate shows a declining trend, with only a small amount of false sharing evident relative to the total miss rate, and virtually none for cache block sizes smaller than 32 bytes.

In summary, false sharing exhibited bimodal behavior. When shared data was structured against the pattern of cross-processor accessing, sizable amounts of false sharing occurred. When shared data was allocated so that each processor's portion was fairly independent, false sharing levels were acceptable.

Eliminating False Sharing

As part of this study we want to demonstrate that for those programs that exhibit a large amount of false sharing, simple shared data restructuring can significantly improve cache performance. We look at shared data reference patterns and data object allocation, rather than using an in-depth knowledge of the program semantics. It is our belief that this analysis ultimately can be included as part of an optimizing compiler; however the implementation of such a compiler is outside the scope of this paper.

We are considering neither padding data structures or array elements nor aligning objects on cache block boundaries, as was done in [13]. The former may waste cache locations on empty structures; and the latter requires that the compiler be cognizant of the cache configuration which would reduce portability.

Identifying Data Structures

The programs used in this study were modified to output the address and size of each data object that was dynamically allocated in shared space. By using this profiling information, the symbol table and the false sharing miss count, we were able to attribute the false sharing misses to specific data objects.

In Topopt, we found that there were two sets of objects that were the main sources of false sharing misses. The first are vectors that describe elements of the gate array input to the program, the other is a collection of fixed vectors used for interprocess communication and synchronization. Both sets of vectors are allocated across processors, and are used in a producer-consumer fashion between the slave-processors and the master-processor, with each element only being written by one processor.

In Pverify a data structure used to represent wires in the circuit is responsible for most false sharing misses. The data structure contains a vector that is indexed by the processor number. Each element of this vector is accessed only by one processor, and thus the structure is plagued with false sharing misses.

Transforming the Data Structures

False sharing occurs in Topopt and Pverify because data that is not shared or only shared in a limited way is allocated together according to similar semantic properties rather than sharing usage. The transformations attempt to combine data objects with similar sharing properties. This involves collecting objects that are almost exclusively written by one processor, so that few other processors will access the data. For fixed-sized data structures the easiest way to do this is to first collect the vectors with similar properties and then transpose them; the structures are transformed from n vectors with p elements each into one vector with p elements, each element being a record with n fields.

For structures that are allocated and replicated according to the size of their input, this approach cannot be used without modification. Instead we use another technique, called *indirection*, in which blocks of memory are allocated to each processor and each element of the original vector contains a pointer to a value in a block instead of the value itself. The values are then written into the processor specific blocks. The benefits should be significant for all block sizes, since the transformation is not tailored to one in particular. The cost is modest in space (the size of a pointer for each object) compared to other techniques such as padding; however, the transformation does result in one added memory access every time an element of the original vectors is accessed. This gave rise to 5% more memory references in Topopt and about 15% in Pverify. Nevertheless, the transformation reduces false sharing on these vectors, making it more likely that these pointers, as well as the elements they point to, will be in the cache. Hence, the cost of each memory reference is likely to be very small. Furthermore, since this transformation, by reducing the false sharing, reduces coherency traffic on the bus, the cost of accessing the bus will be smaller.

Automating the Transformations

In automating the transformations the primary task is to identify the shared data structures, or elements thereof, that are accessed on a per-processor basis, and that by being allocated adjacent to each other, will tend to cause false sharing.

There are several foreseeable problems in implementing the analysis for automating these transformations. The first problem is that the transformations change data structures rather than code. While code transformations allow for separate compilation, any transformation of the global data structures will require recompilation of all dependent modules.

The second problem is the aliasing problem that can occur with the use of pointers in C. This problem is not unique to applying our transformations. Techniques have been proposed to perform static analysis in the presence of pointers [9]. Should they fail the accuracy of the analysis will deteriorate and the transformations may no longer be safe.

The third problem is detecting the portions of the code in which the parallelism occurs. This of course depends heavily on the model of parallel programming that is being used. The programs we analyzed are coarse-grained, i.e. the parallelism is on the process level.

The fourth problem is to perform the analysis to determine which variables in the private space (1) have different values across all the processes and (2) are used to index expressions of shared data structures. Once these shared data structures have been identified, dependence between different accesses can be detected by using, for instance, Banerjee's equations [3]. Transformations can then be performed on those data structures where elements are used on a per-processor basis.

In a simple model, which both Topopt and Pverify are instances of, paralellism is easily detected since it is created through the execution of a *fork* statement. *Fork* duplicates the private address space of the parent process in the child process. The fork statement is placed inside a loop, and each child process executes the main processing part of the program with distinct set of values for the induction variables. By performing interprocedural symbolic constant propagation on the induction variables and a general interprocedural side-effect analysis, shared data structures indexed by linear expression of the induction variables will be exposed. Once these shared data structures have been identified, dependence between different accesses can be solved as mentioned above, and the transformations can be performed on the data structures that are accessed on a per-processor basis. It is clear that this simple approach to the analysis will not be able to discover all possible opportunities to apply the transformations. However, in many cases it will be sufficient.

Analyzing Optimizations

Figures 4 and 6 show the miss rates for the modified versions of Topopt and Pverify respectively. The false sharing miss rate has declined an average of 46% for Topopt and and 75% for Pverify over the original programs. For Topopt, for the largest two block sizes, 128 and 256 blocks, the benefits of eliminating false sharing were outweighed by a loss in spatial locality for shared data. Until this point the total miss rate of the transformed program was 12% to 24% lower than the original.

For Pverify, on the other hand, the benefits of eliminating false sharing is seen across all block sizes. Not only has the miss rates been decreased, as seen by an average decrease in the total miss rate of 24%, but the sharply increasing trends of the miss rates have been moderated significantly.

Bus utilization results are presented to demonstrate the effects of the shared data transformations on total machine performance. This particular metric is useful, because the system bus is the bottleneck to better performance in bus-based, shared memory multiprocessors.

Bus utilization for the modified and the original versions of Topopt and Pverify are shown in Figures 5 and 7 respectively. For Topopt, bus utilization has been decreased by up to 23%, with an average decrease of almost 10%, compared to the original. Considering only the 5 smallest block sizes[a], the average decrease in bus utilization is about 14%. For Pverify, the average reduction in bus utilization is 13%, 24% if only the four smallest block sizes are counted. The coherency related bus traffic has been reduced considerably, with an average reduction of 30% for Topopt and 58% for Pverify across all block sizes.

Related Work

Evidence of false sharing is supported in two other studies [4, 6] These studies measured cache misses and

---

[a]One would not design a cache configuration that resulted in bus utilization greater than 85% when there are better alternatives.

bus traffic in coarse-grained parallel programs, and found that in many of the applications both metrics increased as block size increased. The authors hypothesize that unrelated objects in shared memory reside in the same cache block; and that contention among the processors for these cache blocks is responsible for the increase in the metrics.

The studies in [13] measure false sharing directly. Their studies revealed amounts and patterns of false sharing similar to the results in this work. They suggested five techniques to reduce false sharing misses, of which scalar and record expansion (to a specific cache block boundary) were most effective. These two optimizations eliminated on average 3% of the shared misses with 16 processors and 4% with 32 processors. They point out that these transformations may reduce the effectiveness of prefetching, increase the working set of a program and increase the conflict misses in a finite cache. Our work differs from theirs in the type of transformations applied (collect and transpose or allocate and indirection without cache block padding versus padding) and the benefit from the transformations (on average we were able to eliminate 30% to 40% of the shared data misses).

Conclusion

We have traced three parallel applications for bus based shared memory multiprocessors, and have run simulations for each of these traces with infinite caches and cache block sizes ranging from 4 to 256 bytes. The results of the simulations show that the amount of false sharing varies significantly across the selected programs. In some programs the false sharing is negligible, while in others it is significant enough to determine the trend of the total miss rate. For these programs we found that the false sharing miss rate increases substantially as the block size increases. Therefore there exists a real potential for improving their performance by reducing the amount of false sharing.

We found that the majority of false sharing misses could be traced to a set of structures that exhibited the same usage and sharing patterns. We presented two data transformation techniques aimed at reducing the amount of false sharing misses. After applying the transformations, the false sharing miss rate was reduced typically by 40% to 75%, and the toatal miss rate by 20% to 30%. The transformations decreased shared misses by an order of magnitude more than those used in a previous study. Total bus utilization was reduced by an average of 11%, while the coherency related bus traffic was reduced be an average of 45%. The improvements in performance not only reduced execution time, but the lower bus traffic allows for of using more processors in the computers (up to 5 more for 16 byte cache blocks in the best case for Topopt and Pverify).

It is our belief that these transformations and the recognition of the usage and reference patterns of these structures are simple and general enough so that these techniques can be applied automatically during compilation.

Acknowledgements

We owe many thanks to those who provided our application traces and the tracing software:. David Keppel and Eric Koldinger (MPtrace), Srinivas Devadas (Topopt), and Hi-Keung Tony Ma (Pverify). We also thank David Wood, David Keppel and Eric Koldinger for providing helpful comments on an earlier draft of this paper. This research was supported by IBM Contract No. 18830046 and the National Science Foundation under a Presidential Young Investigator Award (MIP-9058439).

*

References

[1] Anant Agarwal and Anoop Gupta. Memory-reference characteristics of multiprocessor applications under mach. In *Proceedings of the 1988 ACM SIGMETRICS Conference on Measurement and Modeling of Computer Systems*, pages 215–225, 1988.

[2] J. Archibald and J-L. Baer. An evaluation of cache coherence solutions in shared-bus multiprocessors. *ACM Transactions on Computer Systems*, 4(4):273–298, November 1986.

[3] U. Banerjee. Data dependence in ordinary programs. Master's thesis, University of Illinois Urbana-Champaign, Department of Computer Science, November 1976.

[4] David R. Cheriton, Anoop Gupta, Patric D. Boyle, and Hendrik A. Goosen. The vmp multiprocessor: Initial exerience, refinements and performance evaluation. In *Proceedings of the 15th Annual International Symposium on Computer Architecture*, pages 410–421, 1988.

[5] S. Devadas and A. R. Newton. Topological optimization of multiple level array logic. In *IEEE Transactions on Computer-Aided Design*, November 1987.

[6] S. J. Eggers and R. H. Katz. The effect of sharing on the cache and bus performance of parallel programs. In *Proceedings of the 3rd International Conference on Architectural Support for Programming Languages and Operating Systems*, pages 257–270, April 1989.

[7] S. J. Eggers and R. H. Katz. Evaluation of the performance of four snooping cache coherency protocols. In *Proceedings of the 16th Annual International Symposium on Computer Architecture*, pages 2–15, May 1989.

[8] S. J. Eggers, D. R. Keppel, E. J. Koldinger, and H. M. Levy. Techniques for efficient inline tracing on a shared-memory multiprocessor. In *Proceedings of the International Conference on Measurement and Modeling of Computer Systems*, May 1990.

[9] Vincent A. Guarna Jr. A technique for analyzing pointer and structure references in parallel restructuring compilers. In *Proceedings of the 1988 International Conference on Parallel Processing*, pages 212–220, August 1988.

[10] G.K. Jacob, A.R. Newton, and D.O. Pederson. Parallel linear-equation solution in direct-method circuit simulators. In *IEEE International Symposium on Circuits and Systems*, volume 3, pages 1056–1059, Philadelphia PA, May 1987.

[11] R. Lovett and S. Thakkar. The symmetry multiprocessor system. In *Proceedings of the 1988 International Conference on Parallel Processing*, pages 303–310, August 1988.

[12] H-K. T. Ma, S. Devadas, R. Wei, and A. Sangiovanni-Vincentelli. Logic verification algorithms and their parallel implementation. In *Proceedings of the 24th Design Automation Conference*, pages 283–290, July 1987.

[13] J. Torrellas, M. S. Lam, and J. L. Hennessy. Shared data placement optimizations to reduce multiprocessor cache miss rates. In *Proceedings of the 1990 International Conference on Parallel Processing*, volume II, pages 266–270, August 1990.

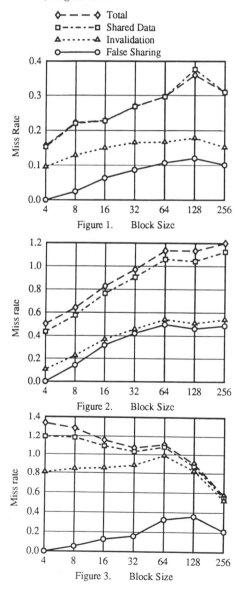

Figure 1.   Block Size

Figure 2.   Block Size

Figure 3.   Block Size

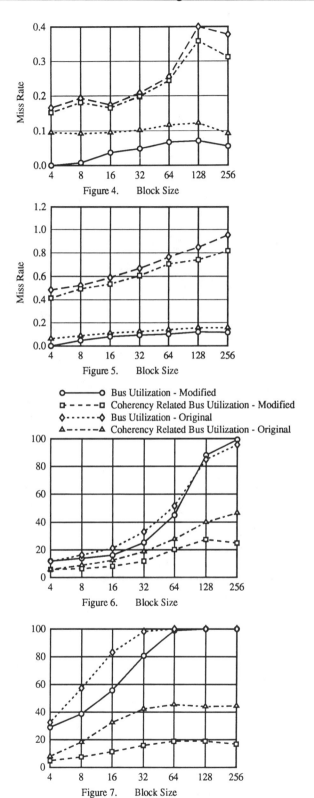

Figure 4.   Block Size

Figure 5.   Block Size

Figure 6.   Block Size

Figure 7.   Block Size

Effective Load and Resource Sharing in Parallel Protocol-Processing Systems

T. V. Lakshman and Dipak Ghosal
Bell Communications Research
331 Newman Springs Road
Red Bank, NJ 07701.

Yennun Huang
AT&T Bell Labs
600 Mountain Avenue
Murray Hill, NJ 07974.

Satish K. Tripathi
Dept. of Computer Science
University of Maryland
College Park, MD 20770.

Abstract

We study architectural and performance issues concerning the effective use of parallel computing in protocol processing and switching systems. The motivation for this work is the expected large processing capacities needed in future switching systems, such as ISDN switching systems, which have to provide connection management services for tens of thousands of users. Such switching systems are high-performance protocol-processors in which the workload is a large ensemble of short (processing times of a few milliseconds) tasks. A key performance issue in parallel protocol-processing systems, is the interaction between load-sharing policies (needed in parallel systems for effective capacity utilization) and call-record management schemes (used to access and update the shared call context information). In a non-shared memory parallel system, the cost of call-record access and update increases non-linearly with the number of processors where a call record can possibly reside (for example, if a call-record dynamically moves around in the system then the search cost grows non-linearly with the number of processors). This overhead makes the task service time increase with system size. By constraining call-records to reside (or to be replicated) only in a few processors the access overhead can be reduced. However, reducing the call-record management overhead in this manner also constrains the size of the set of processors to which an arriving task may be assigned (by the load-sharing policy) for processing. Since increasing the choice of processors available to the load-sharing policy increases the capacity utilization (and potentially performance), there is a conflict between the requirements for load-sharing and the requirements for resource management. We study the effect of these two competing factors on system performance and determine the optimal set size for load-sharing and resource management. This set size is an important parameter in the design of efficient algorithms and architectures for parallel protocol-processing systems.

1 Introduction

This paper studies architectural and performance issues concerning the effective use of parallel computing in telecommunications switching systems. The context for this paper is the future telecommunications network resulting from the on-going evolution of the network from a transport mechanism to a software-intensive information network. In such a network, the service paradigm of the voice call will be replaced by a new paradigm involving the transport and management of information units under various constraints of time, space, accuracy, and variability. This increase in network functionality in conjunction with the rapid rise in transmission bandwidth and protocol complexity necessitates corresponding increases in the processing capacities of switching systems which have to provide connection management services to tens of thousands of users.

The necessary large processing capacities can be cost effectively achieved by extensive use of parallel computing. The workload, in this switching application, is a large collection of short tasks. Each task is a protocol message handler which typically needs only a few milliseconds of processing. These tasks are not completely independent because many messages may belong to the same connection (call) in which case these messages share a common call context. This common context information consisting of the call state and other connection parameters are stored in call records. Processing of a task involves the access and update of the call record of the call to which the task belongs. In a parallel system, this access and update causes overhead. The amount of overhead depends on the call-record management scheme. If the parallel system had a common global memory then all call records could be stored in this common memory. The call record access and update overhead would then be determined primarily by memory contention. However, the delay due to memory contention rises rapidly with the number of processors and this makes large-scale parallelism using a global shared-memory architecture impractical. Shared memory can only be used to share data amongst small groups of processors.

In a non-shared memory architecture, a call-record can be stored in the local memories of some subset of the processors in the system. This subset can either be a priori determined (as for example, by applying a hash function to the call identifier) or can change dynamically depending on the system workload. If the call-record location is not pre-determined, then before processing each task a search has to be performed to locate the call record. This introduces overhead which grows non-linearly with the number of processors that the call record can possibly reside in (the system size in the worst case). Similarly, the update cost increases with the number of call record copies. By con-

This project is supported in part by NSF grant CCR-9002351.

straining call-records to reside (or to be replicated) only in a few processors the access overhead can be reduced. However, reducing the call-record management overhead in this manner also constrains the size of the set of processors to which an arriving task may be assigned (by the load-sharing policy) for processing. In distributed systems, the probability of capacity waste due to nodes being idle when tasks are waiting at other nodes is large and grows rapidly with system size [1]. Load-sharing reduces this capacity waste, and increasing the choice of processors available to the load-sharing policy increases the capacity utilization (and potentially performance). Clearly, in this parallel protocol-processing application, there is an inherent interaction and conflict between the requirements of load-sharing policies and the requirements of efficient call-record management schemes. To our knowledge, this interaction between load-sharing and resource management has not been previously addressed in the context of an important practical application.

We study the effect of these two competing factors on system performance. Specifically, we study the following question: If we are given N processors, how should the system be organized so as to maximize the benefits of load-sharing and minimize the overheads? If load-sharing policies did not affect call-record management overhead then for load-sharing purposes it would be best to be able to assign an incoming task to any processor. However, for this application, such an organization need not give the best performance since the call-record management overhead can offset the performance gains from load-sharing. It may be better to organize the system as interconnected groups of processors with static load-sharing policies (such as random routing) used to assign tasks to groups and dynamic load-sharing policies used within groups to assign tasks to processors. Then, an important parameter in the design of efficient parallel protocol-processing systems is the group size which maximizes performance[1]. We use a multi-class priority queueing model to determine this optimal group size (see [2], [3], [4] for related work).

The next section gives an overview of protocol-processing using an ISDN system as an example. Section 3 describes some typical call record management schemes that were implemented in an experimental parallel protocol-processing system. This section also discusses the impact of different call-record management schemes on load-sharing policies. Section 4 presents a performance model for a system using replicated call-records and analyzes load-sharing policies to be used for such systems. Our analysis shows that the effectiveness of load-sharing is very sensitive to the call-record management overhead. Hence the system architecture and call-record management scheme must be chosen to minimize the call-record management overhead. These and other re-

sults are discussed in section 5. The sensitivity of the results to various modeling approximations are also discussed in this section. Concluding remarks are in section 6.

1 Overview of Protocol Processing

This section presents a brief summary of distributed protocol-processing using the ISDN (Integrated Services Digital Network) layer 3 protocol as an example. The ISDN layer 3 protocol is used for establishing, maintaining, and clearing ISDN connections (calls) between ISDN users. Protocol entities at the user's terminal equipment (TE) and the switching central-office (CO) execute the required protocol actions. Four layer 3 protocol entities are involved in call-management: two CO protocol entities managing each end of a call, and two TE protocol entities corresponding to the originating and destination TEs. CCITT Recommendation Q.931 [5] [6] specifies the layer 3 protocol entities as communicating finite-state machines. A call-management function, such as a call set-up, is performed through a sequence of finite-state machine transitions with some call-processing action being performed at each state. State transitions are triggered either by peer-to-peer layer 3 messages or by internal events such as time-outs, local connect requests, etc. Peer-to-peer layer 3 messages are sent using ISDN layer 2 (CCITT Recommendation Q.921 [7]) which provides reliable, sequenced links between layer 3 entities.

A single call set-up and disconnect may require as many as 70 layer 3 and layer 2 messages with many of these messages requiring the execution of about 1500 instructions on an MC68020 processor (the number of instructions can be much more for complex messages exercising the various permitted ISDN options).

When user A wants to call user B, A instructs the layer 3 at TEA to send a Setup message to the CO. This is done by first passing the message to layer 2 at TEA. If there is no established link between the layer 2 at TEA and a layer 2 process at the CO, then this link is first established using a sequence of layer 2 control messages (Set Asynchronous Balanced Mode Extended, Unnumbered Acknowledgment, etc). This link is then used to send the Setup message from the TEA layer 3 to the CO layer 3. The CO layer 3 performs various actions on receipt of a message. Some typical actions are listed below:

1. Parse received message to establish syntactic validity of messages and to extract the required information fields.

2. Determine the protocol finite-state machine state and other context information (such as timer values) for the connection to which the message being processed belongs. Except for the parsing of a few message fields, all processing requires retrieval of the call record (creation of call record for set-up messages) since the needed context information is stored in call records. Since the task allocator may not assign messages to

---

[1] Note that organizing the system in groups does not preclude scaling of the system. Groups only imply that dynamic load-sharing is restricted to processors within a group and that call-records need be replicated only within a group. Static load-sharing is used to assign traffic to groups. Also, note that there are other factors such as availability which also impact the architecture. However, here we only consider the effects of load-sharing.

processors where its call-record is stored, retrieval of call records may need a distributed search. Before reading the call record, synchronization with other layer 3 entities may be needed since the call record may contain variables shared by both ends of a connection. For a distributed implementation, this may also involve waiting for previous updates to complete.

3. Timer starts/stops.

4. Call processing functions and protocol actions such as communication with a switch controller and computing the next call state.

5. Composition of new messages (often requiring the filling in of several fields from templates).

6. Update of call record. For a distributed implementation, this again needs synchronization and possibly writes of multiple copies of call-records.

The call goes into the active state after a sequence of message exchanges involving the protocol (layer 3 and layer 2) entities at TEA (originating side), TEB (destination side), and the corresponding protocol entities at the CO. A typical layer 2 and layer 3 message sequence for setting up an ISDN call is shown in figure 1.

On receipt of a Setup message, the CO can acknowledge with a Setup Ack message. As can be seen from figure 1, the layer 3 at the originating TE sends a Setup message by passing this message as a Data message to its layer 2. Since at this point a layer 2 to layer 2 connection is assumed not to exist between the TE and the CO, a Sabme and an Ua message are sent to establish the connection. The Setup message is then transmitted over the layer 2 link as an Info message which is acknowledged by an Rr message at the layer 2 level. The layer 2 at the CO then sends this message as a Data message to the layer 3 at the CO responsible for handling TEA. This message exchange then continues at the destination end of the connection. Note that there is a considerable interchange of messages (and correspondingly processing) for setting up just one connection. Since the CO can typically serve requests from tens of thousands of such customers, the workload at the CO is a large collection of connection management requests arriving over the various ISDN lines.

2 Protocol Processing using Parallel Architectures

When using parallel processors for protocol processing, the key performance issues are low-overhead access to the shared call-record information and effective load-sharing to reduce response time. For moderate amounts of parallelism, both these requirements are easily met by using a common global memory to store both the call-record information and the task queue. Since the task queue is in common memory, processors cannot be idle when there are tasks waiting unlike a non-shared memory parallel system. Load-sharing is simple

and this system gives a bound on the performance of load-sharing policies in a non-shared memory system. In the next section, we analyze a system with a common memory task-queue. With common memory, call record management overhead is limited to the concurrency control overhead and this overhead (ignoring memory contention) is much lower than the corresponding overhead for non-shared memory architectures. However, memory contention precludes large-scale parallelism using a shared memory architecture.

With non-shared memory architectures, the call-record management overhead varies widely with the record management scheme used. Some typical call-record management schemes are:

1. Static assignment of call-records to processors: A hash function is applied to the call-record key (a combination of the call identifier, terminal identifier, etc) and the hash value is used to permanently assign call-records to processors. When choosing a processor to process a message, it is not required that the processor holding the call-record for the message be chosen (though that might be a good heuristic). When a call record is needed, the layer 3 process at the processor where the message is being processed sends a call-record request to its local call-record management process. This process computes the hash function to determine the processor where the call-record is stored, communicates with the remote call-record management process which sends back a copy of the call-record after performing any necessary synchronization (for example, some other process may be using the call record). The static assignment scheme has the least overhead (especially if all messages for a call are processed at the processor where the call-record is stored). However, the static nature of the assignment prevents adaptation to traffic fluctuations. Call-records exist in the system for relatively long periods and load-sharing at the call level will essentially be a static load sharing scheme which does not take the task arrival instant system state into account in making task assignment decisions.

2. Static assignment with replication: Here the hash function maps the call-record key to a set of processors. Each processor in this set holds a copy of the call-record. A majority consensus algorithm is used for access and update of copies. Apart from fault-tolerance, replication can aid load-sharing since an arriving task can be assigned to any processor which has a copy of the call-record (note that complete replication is a special case of this scheme).

3. Dynamic storage: The call record is initially stored at the processor where it is created (usually the processor chosen to process the set-up message for the call). Subsequently the call-record moves to different processors based on requests. To access a call record, its current location has to be determined by distributed search and this search cost is proportional to the number of processors that the call record can possibly be

in. With this scheme, the load-sharing policy has no constraint on the processors to which it can assign tasks and this increases the effectiveness of load sharing (at the expense of increased call record access cost). Note that the same is true for full replication of call records except that, in this case, the call-record overhead is in updating the multiple copies and not in the search.

4. Functional partitioning: Here the set of processors is partitioned into two disjoint sets. The processors in one partition only execute call-record management functions and processors in the other partition only execute the layer 3 protocol procedures. Though this scheme has architectural appeal because of the separation of data management and switching functions, there is also more potential for wasted capacity since it is possible that some processors in one partition are idle while all processors in the other partition are busy and have tasks waiting to be processed. Also, improper choice of partition size will quickly degrade performance since one of the partitions will become the bottleneck.

These call-record management schemes were implemented on an 18 processor bus-based parallel system. The system was used to process layer 3 messages using different call record management schemes with simulated traffic. Experimentally it was seen that the overheads of call-record management grow with group size and can offset the load-sharing gains. In the next section, we quantitatively assess this interaction between load-sharing and resource management by analyzing different parallel protocol-processing schemes and comparing their performance.

3 Performance Model and Analysis

For the analysis in this section, we use a multi-class performance model with call-record management requests and call-processing requests forming separate classes. The cost of record management in non-shared memory architectures is taken into account in the model and its impact on load-sharing policies is analyzed. First we analyze an ideal load-sharing policy where the load-sharing is perfect because the task queue is assumed to be in (hypothetical) global memory. In section 4.2, we consider the distributed memory case in which each processor has its own queue. This corresponds to no load-sharing and is a special case of the common-memory system (case when group size is 1). In Section 4.3, we analyze a realistic and practical load-sharing policy — load-sharing using a random location policy and a threshold transfer policy [2]. Our consideration of this policy is motivated by the results in [2] that even simple policies perform almost as well as complex policies which use a lot of state information. We believe that this result would hold even more strongly for protocol-processing applications since the task processing times are small (in comparison to typical

computer systems applications) and the overhead of complex policies could well be equal to the task service time.

3.1 Common Memory Queue

Figure 2 shows the performance model for a system with the task queue in common shared memory. The system is partitioned into groups with each group having a shared memory where tasks waiting for service at that group are stored. As can be seen from figure 1, there is a long sequence of message exchanges between the CO and the TE to set up a connection. This sequence can be viewed as a single task either exiting the CO or reentering the CO (in the form of a response from the TE) after a suitable delay and this interchange can be modeled by introducing a feedback path from the servers to their inputs. A delay center is introduced into the feedback path to model the delay at the TE.

Let N be the number of processors and let g be the number of groups. Then the number of processors per group, denoted by c, is $\frac{N}{g}$. We assume that tasks (messages) arrive from a Poisson point source with rate λ. Each task after service may either exit the system with probability p_{out} or may be routed back through the feedback path with probability p_{in}. The service time at the delay center is taken to be exponentially distributed with mean γ.

Each call-processing request is serviced in two phases. The first phase, which is the processing phase, requires carrying out typical message-processing computations (outlined in section 2). We assume that this time is exponentially distributed with mean S. At the end of the message-processing computations, the call record must be updated to reflect the new protocol state (and to reflect changes in context variables). The second phase consists of broadcasting a message to all the other processors in the group to update copies of the modified call record in the local memories of processors in the group. These update requests are modeled as a separate update request class and these are assumed to have preemptive priority over the call-processing request class. The service time for the update request class jobs is assumed to be exponentially distributed with mean $\frac{1}{\mu_1}$.

It is assumed that the model queueing network is a Jackson network [8]. This is an approximation because update requests have preemptive priority over call-processing requests. Based on the assumptions, each multiple server behaves like an independent $M/M/c$ queue and the delay center like an $M/M/\infty$ queue. Let $\bar{Q}_c$ and $\bar{R}_c$ denote the mean queue length and the mean response time respectively of the $M/M/c$ queue. The total arrival rate into the delay center, denoted by λ_{dc}, is given by

$$\lambda_{dc} = \frac{\lambda}{1 - p_{in}} p_{in} \qquad (1)$$

Let $\bar{Q}_t$ denote the mean number of requests in the system (including the number in the delay center and the multiple server queues). $\bar{Q}_t$ can be expressed as

$$\bar{Q}_t = g \times \bar{Q}_c + \lambda_{dc}\gamma$$

$$= gQ_c + \lambda\gamma\frac{\imath\cdots}{1 - p_{in}}$$

The total response time for a call-processing request $\bar{R}_t$ can be obtained using Little's Law, i.e., $\bar{R}_t = \frac{Q_t}{\lambda}$. Using the previous equation for $\bar{Q}_t$ and noting that

$$\bar{R}_c = \frac{\bar{Q}_c}{\frac{\lambda}{(1 - p_{in})g}} \qquad (2)$$

it can be shown that

$$\bar{R}_T = \frac{p_{in}}{1 - p_{in}}(\bar{R}_c + \gamma) + \bar{R}_c \qquad (3)$$

To determine $\bar{R}_c$ we need to know the effective service rate (accounting for preemptions from higher priority requests) of call-processing requests. We use the shadow server approximation [9] to determine this effective service rate. Let λ_1 be the mean arrival rate of the high-priority update traffic. We assume that these arrivals are from a Poisson point source. Since μ_1 is the service rate, the utilization of the server due to high-priority traffic is

$$\rho_1 = \frac{\lambda_1}{\mu_1} \qquad (4)$$

Based on the shadow server approximation, we inflate the service time of low-priority traffic by $1 - \rho_1$ and we get the effective service time of call-processing requests to be

$$S_{eff} = \frac{S}{1 - \rho_1} \qquad (5)$$

The arrival rate of call-record update requests depends on the arrival rate of call-processing requests. Let λ_{in} denote the total arrival rate of call-processing requests to a processor group. Then this rate is given by

$$\lambda_{in} = \frac{\lambda}{(1 - p_{in})g} \qquad (6)$$

Each call-processing request generates $c - 1$ update requests. So the total rate of generation of update requests is $\lambda_{in}(c - 1)$. Since this total is equally divided amongst the c servers, the arrival rate of update traffic (high-priority traffic) to each server is

$$\lambda_1 = \frac{\lambda_{in}(c - 1)}{c} \qquad (7)$$

Note that λ_1 increases as g decreases. Using Eq.(6) in Eq.(4) we get

$$S_{eff} = \frac{S}{1 - \frac{\lambda_{in}(c-1)}{c\mu_1}} \qquad (8)$$

Note that if μ_1 is very large (the time to perform updates is very small in comparison to the processing time S), then $S_{eff} = S$. Also, note that if λ_{in} increases (by increase of either λ or p_{in}) the effective service time of the call-processing requests increases due to the higher update traffic (and hence preemptions).

Assuming S_{eff} to be the service time of each server in the multiple server queue and further assuming that this service time is exponentially distributed, one can obtain the mean queue length $\bar{Q}_c$ and consequently the mean response time $\bar{R}_c$ using standard techniques [8].

3.2 Distributed Memory Without Load Sharing

This is a special case (group size is 1) of the analysis presented in the previous section. In this scheme, there is no replication of data and each processor has its own queue. Since there is no replication of the call record, there are no update requests and no associated overhead of updating the replicated copies. Each processor is a simple $M/M/1$ queue with arrival rate $\frac{\lambda}{N}$ and exponentially distributed service time S. The mean call setup time can be obtained from using Eq.(3) by setting the number of groups g equal to N.

3.3 Distributed Memory with Load Sharing

As before, the system consists of N nodes divided into g groups each of size c. Call-records are replicated only in the processors within a group. Furthermore, each processor has its own queue for storing tasks. The call-processing requests are assumed to be generated from a Poisson point source with rate λ. A request after being processed is routed into the feedback path with probability p_{in} in which case it joins the externally arriving call request stream after an exponentially distributed delay with mean γ.

When a request arrives at a node, it is processed in two phases. The first phase is the load-sharing phase. The load sharing scheme that we consider is a simple policy based on a random location policy and a threshold transfer policy [2]. When a request arrives at a node, if the queue length is less than a threshold, T, the local node processes the request. Otherwise, the request is transferred to some other node selected at random. To prevent too many task transfers, there is a limit, L, on the number of times that a task can be transferred. Note that with this load sharing policy, there is no need to exchange state information in deciding where to transfer a newly arrived request that cannot be locally processed. Let C denote the mean time required for transferring a call request from one node to another node.

The second phase is the processing phase in which a call request is actually processed. There are two steps in the servicing of a request by a node. As in the previous section, in the first phase the protocol specific message-processing computations are performed. This usually changes the protocol state and the values of context variables (for example, timer variables) and hence the results have to be written back into the call record (and all its copies). We assume that the service time for the first phase is exponentially distributed with mean S. The second phase of processing consists of broadcasting the updates to all the processors. We assume that the scheduling discipline at each processor is such that servicing these broadcast update messages have preemptive priority for processing over all other processing tasks. The update service time is assumed to be exponentially distributed with mean μ_1.

Our analysis of the above model follows the same reasoning as that adopted in [2]. However, we cannot directly use the expressions in [2] since we must account for the additional complexity in our model due to the high priority update traffic. We do this by using the shadow server approximation [9] that we used in the analysis of the common memory queue system. Let λ_1 be the mean rate of the Poisson point source characterizing the arrival process of the high priority update traffic. Since μ_1 is the service rate, the utilization of the server due to the high priority traffic is $\rho_1 = \frac{\lambda_1}{\mu_1}$. Based on the shadow server approximation, we inflate the service time of the low priority traffic by the factor $1 - \rho_1$. Thus the effective service time of call-processing requests is given by

$$S_{eff} = \frac{S}{1 - \rho_1} \qquad (9)$$

and the effective service time for transferring a call-request is given by:

$$C_{eff} = \frac{C}{1 - \rho_1} \qquad (10)$$

Now λ_1 can be determined from the fact that every arriving request generates $c - 1$ high priority update requests and that these are equally divided among the c processors in a group. Then, λ_1, the arrival rate of high priority update traffic per processor is given by

$$\lambda_1 = \lambda_{in}\left(\frac{c - 1}{c}\right) \qquad (11)$$

where $\lambda_{in} = \frac{\lambda}{(1 - p_{in})g}$ where λ is the external arrival rate and g is the group size.

To obtain the response time of the each processor we use the method used in [2]. Here, we only point out the key step in the analysis. For a detailed derivation, see [2]. The important parameter that needs to be derived is the rate at which transferred tasks arrive at a processor. Since a random location policy is used, the arrival rate of transferred tasks is independent of the server state (the number of call requests in the node). As before, the threshold queue length is denoted by T and the transfer limit by L. Since tasks are transferred when the queue length is greater than or equal to T and since each task is transferred at most L times, the arrival rate of transferred tasks is given by

$$\lambda_t = \lambda\left[\sum_{l=1}^{L} p_T(1 - p_{T+}) + p_{T+}\right]^l \qquad (12)$$

where p_T is the conditional probability that a node is in state T given that it is in the processing state and p_{T+} is the unconditional probability that the node is in the transferring state.

The probability of a node being in the transfer state is equal to the server utilization due to performing task transfers and due to the processing of tasks that cannot be transferred because of the transfer limit. From this it can be shown that [2]

$$p_{T+} = \frac{p_T S_{eff}(\lambda + \lambda_t) - (S_{eff} - C_{eff})\lambda_t}{1 - (1 - p_T)S_{eff}(\lambda + \lambda_t)} \qquad (13)$$

The state transitions of the system until state T form a simple birth-death model with arrival rate $\lambda + \lambda_t$ and service

time S_{eff}. By solving this model, we get

$$p_T = \frac{\rho^T(1 - \rho)}{1 - \rho^{T+1}} \qquad (14)$$

where $\rho = (\lambda + \lambda_t)S_{eff}$. Using Eqs.(10)-(12) one can obtain a non-linear equation with a single unknown λ_t which can be solved using standard numerical techniques.

Once λ_t is known, the probabilities p_T and p_{T+} are known using Eqs. (11) and (12) respectively. Once these are known the mean queue length when the system is in the processing state can be found and by solving the M/M/1 preemptive priority HOL queue we can obtain the system response-time R_T [2].

4   Results and Discussions

In this section, we discuss results of our analysis (these results agree with our preliminary simulation results). As pointed out before, one of the key performance issues to be considered in this application is the conflict between the larger group sizes needed for effective load-sharing and the smaller group sizes needed to keep the call-record management overhead low. Because of this conflict, there is an optimal group size for a given set of system parameters. In the following discussion, $N = 16$, $p_{in} = .5$, $\gamma = .2$, and $S = .1$.

Figure 3 shows the relationship between response time and group size, for different arrival rates λ, when the common memory queue organization (perfect load-sharing) is used. With this perfect load-sharing scheme the response time should decrease monotonically with group size, if there were no call-record management overhead. Instead the response time first decreases with group size due to the effect of load-sharing (though the rising service time due to update-overhead reduces the rate of decrease) until an optimal group size is reached. Beyond this size any further decrease in response time with group size is subsumed by the increased update overhead. The following observations from the figure are noteworthy:

- The response time is the highest for $g = 1$ for all arrival rates. Note that this corresponds to the no load-sharing case.

- At low loads, for group sizes greater than 1, the response time increases with group size. This is because at low loads the benefits of load-sharing are marginal and the rising call-record management overhead due to increases in group size also increases the response time.

- The optimal group size increases with the load. This is because the gain due to load-sharing is more pronounced at higher loads and hence a higher call-record management overhead can be tolerated. However, the rising call-record management overhead eventually subsumes the benefits of load-sharing.

An important parameter determining the group size (and efficiency of load-sharing) is the ratio of the service times for update-requests and the service time for call-processing requests. We call this ratio the overhead factor(OF).

Figure 4 shows the normalized response time[2] as a function of the group size for different overhead factors and for a fixed arrival rate $\lambda = 20$. The results show that for very high overhead factors and for large group sizes, the response time of even the perfect load-sharing scheme can be worse than that of the no load-sharing scheme. Our analysis confirms our experimental observation that keeping the call-record management overhead as low as possible is crucial to obtaining good performance. Using small group sizes (preferably with shared memory) keeps the overhead low and even small groups give significant load-sharing gains[3].

The performance of the distributed memory organization using the threshold transfer policy and random location policy was also studied with respect to group size and overhead factors. The performance was very similar to that of the common memory queue organization except that since the load-sharing is not perfect the gains due to load-sharing are smaller. Also, in this case there is the additional overhead of task transfers. Because of these factors, the optimal group size for this case is smaller than for the perfect load-sharing case.

Figure 5 compares the performance of various schemes. The response times (for different group sizes and for low overhead factor) of the various distributed memory schemes is plotted along with that of the no load-sharing system. It can be seen that, for group size equal to 16, the distributed memory schemes perform worse than the no load-sharing scheme which perform worse than the perfect load-sharing scheme. This difference in performance becomes even more when the overhead factor is increased and for a distributed memory system with these parameters the group size of 16 is a wrong choice.

Figure 6 compares the other policies. From the figure it is clear that:

- The optimal group size for the distributed memory organization is 2.

- For almost all group sizes, the common-memory queue system almost always performs better than the corresponding distributed memory system.

- The best group size for the common memory queue system is 4.

Again from the above results, it is clear that the call-record management overhead must be kept low. This is even more so when the effect of various modeling assumptions is considered. Below we discuss the effect of these assumptions.

- Bus contention: This was ignored in the model and taking the contention into account will tend to decrease the optimal group size.

---

[2]The response time is normalized with the response time of the limiting case with group size 1

[3]If the burstiness of task arrivals increases the group size would also have to increase

- The shadow server approximation: This approximation underestimates the actual response time. However, since we use the same approximation for all schemes we expect the same relative error in all cases.

5 Concluding Remarks

We analyzed load-sharing and resource management issues in the context of an important practical application — the design of parallel systems for providing connection management services to a very large population of users (such as in ISDN switching systems). A key performance issue in such systems is the close interaction between the requirements for effective load-sharing and the requirements for effective resource management (resource management being access and update of a call-record database). Using a multi-class performance model, we analyzed the impact of this interaction on the choice of parallel architectures for this protocol-processing application. Our analysis shows that the effectiveness of load-sharing (necessary in any large parallel system) is very much dependent on keeping call-record management overhead low. This overhead can be kept low by architectural and algorithmic means. One possible architecture is one in which processors are partitioned into groups with the call-record database replicated within a group, with dynamic load-sharing restricted to processors within a group, and with static load-sharing being used to assign tasks to groups. The group size is determined by two competing factors: the need to have large group sizes to increase the benefits of load-sharing and the need to have small group sizes to keep the resource-management overhead low. We analytically determined the optimal group size for various load-sharing policies. For the system parameters that we used in the analysis, small groups performed well and for such sizes it is feasible to used shared memory for storage of call-records within groups. However, availability constraints will still necessitate replication (or checkpoints) of call-records and cause update-overhead even in a shared-memory architecture. An extension to this work would be to consider the impact of a more efficient update-algorithm – such as the $O(\sqrt{N})$ algorithm in [10] – on performance and availability. Another extension is to consider the effects of real-time requirements of the application and try to determine the response-time distribution.

References

[1] M. Livny and M. Melman, "Load balancing in homogeneous broadcast distributed systems," in *Proceedings of the ACM Computer Network Performance Symposium*, pp. 47–55, ACM, 1982.

[2] D. Eager, E. Lazowska, and J. Zahorjan, "Dynamic load sharing in homogeneous distributed systems," *IEEE Trans. Software Engineering*, vol. SE-12, pp. 662–675, May 1986.

[3] D. M. Diaz, P. S. Yu, and B. T. Bennett, "On centralized versus geographically distributed database systems," in *Proceedings of the 7th International Conference on Distributed Computing Systems*, pp. 64–71, ieee, September 1987.

[4] S. Zhou, "A trace-driven simulation study of dynamic load balancing," *IEEE Traction on Software Engineering*, vol. SE-14, pp. 1327–1341, September 1988.

[5] CCITT COM XI-R 174, "Recommendations Q.930(I.450) ISDN user-network interface layer 3 - general aspects and Q.931 ISDN user-network interface layer 3 specification for basic call control," in *Series I Recommendations of the CCITT Blue Book*, CCITT, 1988.

[6] BELLCORE, "ISDN access call control switching and signaling requirements." Technical Advisory TA-TSY-000268, 1988. Issue 3.

[7] CCITT COM XI-R 43-E, "Recommendations Q.920(I.440) ISDN user-network interface data link layer - general aspects and Q.921(I.441) ISDN user-network interface data link layer," in *Series I Recommendations of the CCITT Blue Book*, CCITT, 1988.

[8] L. Kleinrock, *Queueing Systems*, vol. 1. New York: John Wiley and Sons, 1975.

[9] K. C. Sevcik, "Priority scheduling disciplines in queueing network models for computer systems," in *Proc. IFIP Congress 77*, (Amsterdam), North-Holland Publishing Co., 1977.

[10] T. V. Lakshman and D. Ghosal, "A new symmetric $O(\sqrt{N})$ multiple-copy update-algorithm and its performance evaluation." Submitted to the 1991 International Conference on Data Engineering, July 1990.

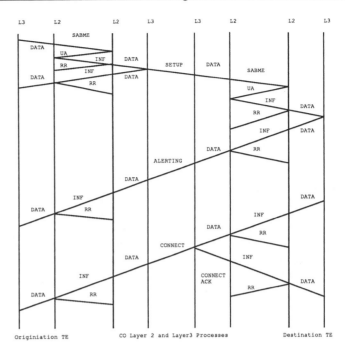

Figure 1: Call Setup Message Sequence

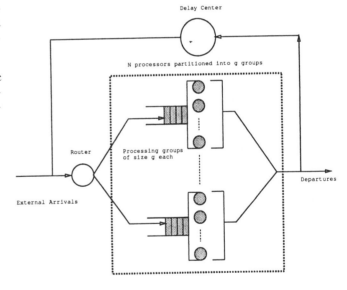

Figure 2: System Model with Common Memory Task Queue

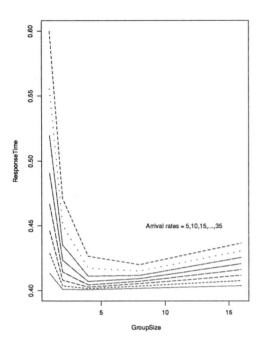

Figure 3: Response Time vs. Group Size for different Arrival Rates (OF=.05)

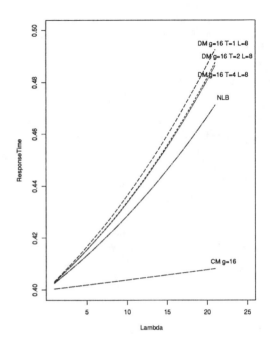

Figure 5: Comparison of load-sharing schemes for group size = 16 (DM:distributed memory, CM:common memory,T:threshold,L:transfer limit)

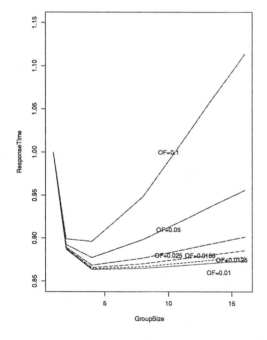

Figure 4: Response Time vs. Group Size for different overhead factors(OF) and arrival rate = 20

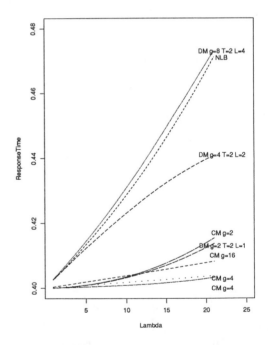

Figure 6: Comparison of load-sharing schemes for different group sizes (DM:distributed memory, CM:common memory,T:threshold,L:transfer limit)

A Distributed Token-Driven Technique for Parallel Zero-Delay Logic Simulation on Massively Parallel Machines

Moon Jung Chung   *Yunmo Chung*

Department of Computer Science
Michigan State University
East Lansing, MI 48824

Abstract

This paper proposes a new efficient distributed token-driven technique for parallel zero-delay logic simulation on massively parallel machines. Each message (or signal) is associated with a token. Tokens are used to synchronize all messages driven from the same input vector. A gate executes messages based on the readiness of tokens or comparison of token identifiers. A higher degree of parallelism can be achieved than with any other compiled-code simulation techniques. According to experimental results on the Connection Machine, a massively parallel SIMD machine, the distributed token-driven technique outperforms both distributed event-driven simulation and compiled-code simulation for some benchmark digital circuits.

1   Introduction

Distributed event-driven simulation techniques have been studied for parallel logic simulation because they have the ability to easily handle asynchronous designs and timing analysis in parallel processing environments. Recently, however, as the demands for maintainability, testability, portability, and high speed increase, synchronous designs have been preferred for VLSI applications [1].

Several compiled-code simulation techniques have been proposed for synchronous designs. In compiled-code simulation, the circuit description is translated into Assembler or Pascal code. The code is compiled and then run simultaneously [2]. Each process is evaluated once for every input vector during each simulation cycle independent of the input patterns. Compiled-code simulation provides significant increases in performance over event-driven simulation by means of the high speed of execution of compiled-code. Since this technique does not require event queue management, it is extremely efficient.

As efficient approaches to compiled-code simulation, several levelized compiled-code simulation (LCC) techniques have been proposed. In these techniques, initially a logic levelization (or scheduling) procedure is required which specifies the sequence in which the logical functions are processed from primary inputs or data output of storage elements [1]. As an attempt to improve the LCC simulation, LECSIM, a levelized event-driven compiled logic simulator, has been proposed which integrates the advantages of event-driven simulation and levelized compiled simulation [3]. LECSIM runs faster than traditional unit-delay event-driven simulators, but slower than traditional levelized compiled simulators. In [4], both a PC-set (set of potential change times) method and a parallel technique have been presented for unit-delay compiled simulation. The degree of parallelism may be limited, i.e very low, when we consider those techniques in parallel processing environments.

As discussed in [3], intermediate events in unit-delay simulation are used for detecting timing anomalies like hazards and race conditions. As the complexity of synchronous circuits increases, however, testing the functional behavior of a circuit is done before performing timing analysis for the simplicity of simulation. In this case, zero-delay logic simulation is usually used for such functional testing.

This paper presents a cost-effective parallel zero-delay logic simulation technique suitable for massively parallel SIMD (Single Instruction Multiple Data) processing environments. The technique has the following advantages: (1) high parallelism is achieved, (2) compared with distributed event-driven simulation, output control as well as unnecessary intermediate messages are not needed, and (3) a levelization procedure is not required. A gate processes messages based on the readiness of tokens arrived at each input port of the gate. Tokens are used to synchronize all the messages from the same input vector. Therefore, unnecessary immediate messages (or events) are not generated. When a massively parallel SIMD machine is used as a target machine, the number of simulation cycles [5] required can be predicted and is smaller than any other distributed event-driven simulation techniques.

The architecture of massively parallel SIMD machines offers significant advantages over MIMD (Multiple Instruction Stream Multiple Data Stream) machines in parallel logic simulation because logic simulation requires small-grain operations such as NOT, AND, and OR. In addition, partitioning and load balancing problems in logic simulation are not serious for circuits with a large number of gates in massively parallel SIMD environments since in general they have a large number of available processors. For example, as currently available massively parallel SIMD machines, the Connection Machine and the MasPar have up to 64K and 16K processing elements, respectively. The distributed token-driven simulation technique proposed in this paper can be used on the above parallel machines.

2   Token-Driven Simulation

A circuit being simulated can be considered a directed circuit graph in which nodes represent processes and arcs indicate event communications between the nodes. In this paper, we assume that each process is assigned to one processor on a massively parallel SIMD machine. In logic simulation, processes correspond to gates while arcs represent links between gates.

In conventional distributed event-driven simulation, each simulation output from a process is carried by an *event*, which is time-stamped with a simulation time to be executed. The term *token* used in this paper is conceptually different from an event. In distributed token-driven simulation, a token carries an output signal from a process without being timestamped. An identifier may be assigned to each token, if necessary, for synchronizing messages accordingly for correct simulation.

In distributed event-driven simulation on a massively parallel SIMD machine, each process repeats the same procedure which consists of *event selection*, *event evaluation*, *new event propagation* and *queue manipulation*. The time period to perform such a procedure is called a *simulation cycle* [5]. Each simulation cycle is synchronized in SIMD environments. In the same way, the simulation cycle can be defined in the distributed token-drive simulation by substituting *event* with *token* in the above procedure.

2.1 Two Approaches

Depending on the digital circuits to be simulated, the distributed token-driven technique for zero-delay logic simulation can be divided into two schemes: *without identifiers* and *with identifiers*. In the other words, one scheme uses tokens, each with a signal only, while the other uses tokens, each with its token identification number as well as a signal.

Using Tokens Without Identifiers

The scheme without token identifiers is efficient for combinational circuits. No token is labeled with a number. An *active port* at a gate is defined to be a port which is supposed to receive tokens during simulation. Each active port has its own queue to store tokens. During every simulation cycle, the following steps will be performed:

1. For all gates, a gate is chosen as an active gate if each active port has at least one token.

2. Each active gate produces an output signal from input signals of the tokens.

3. Each active gate propagates a token with the output signal to all the successors of the gate.

4. Each gate which receives new tokens puts them into its corresponding queues.

Since simulation is performed for a combinational circuit, the simulation terminates only if all the active ports in the simulation network have empty queues. There are no intermediate output signals produced from simulation. Hence, it is very easy to recognize the resulting vectors without output control.

Using Tokens With Identifiers

The scheme using tokens with identifiers which can simulate sequential circuits, is more general than the above scheme. During every simulation cycle, the following steps will be performed:

1. For all gates, a gate is chosen as an active gate if each active port has at least one token.

2. Each active gate produces an output signal from input signals of tokens with the smallest identifier.

3. Each active gate propagates a token with both the output signal and a new identifier to all the successors of the gate.

4. Each gate which receives new tokens puts them into its corresponding queues.

Before starting simulation, it is very important to determine how an identifier is assigned to a token which has a signal. Numbers used as token identifiers must be non-decreasing. For example, timestamps of input vectors can be used as token identifiers in distributed token-driven simulation.

A flip-flop can be considered a gate in logic simulation. In step 3, the assignemnt of a new identifier is based on the function of the active gate. For example, a *normal* gate like AND or OR gives the same (smallest) identifier (considered in step 2) as a new identifier, while a new identifier at a D flip-flop is set to the smallest identifier plus the (input vector) clock interval. Asynchronous designs need more careful generation and assignment of identifiers than synchronous designs since asynchronous circuits are complicated.

According to the above procedure, simulation may not end since step 1 is not satisfied (or processed) and then there might be some unprocessed tokens forever. To flush the unprocessed tokens at the end of the simulation, a token with infinity as its identifier is added to the end of input at each input gate. In this case, even though a queue with only a token with infinity as its identifier is empty, the token is used to satisfy step 1.

2.2 Additional Considerations

The second approach can be used instead of the first by giving the same identifier for all tokens from an input vector. Additional factors which affect the performance of the proposed token-driven simulation will be discussed.

Token Size

In most of the distributed event-driven simulation techniques, events, each with a digital signal, traverse a simulation network model for a circuit under simulation. On a parallel machine which has massively parallel processors and slow communication time, the communication overhead may cause significant degradation of the simulation performance. To cope with the overhead in the distributed token-driven simulation technique, consecutive output signals occurring at a gate are grouped into a token and then the token is propagated into all the successors of the gate. In this case, both the number of simulation cycles and the simulation time can be significantly reduced. Distributed token-driven simulation without identifiers can efficiently use tokens, each having more than one signal.

As the number of input vectors increases, good performance can be obtained if the token size increases accordingly. However, the performance goes down if the token size exceeds a certain limit since there is a trade-off between communication time reduction and evaluation delay. The optimal token size for the best performance depends on both the critical path length of a circuit and the number of input vectors.

Simultaneous Signal Evaluations

In distributed event-driven simulation on massively parallel machines, each process has a function table which contains each output signal for all possible input signal combinations according to the function of the process [5]. Only signals carried by events with the same timestamp can be evaluated at a process during each simulation cycle since only one signal is associated with an event. In other words, signals for at most one input vector are involved with signal evaluation.

In the proposed distributed token-driven simulation technique, however, each active gate can simultaneously evaluate as many signals as carried by a token. In other words, signals produced from some input vectors can be simultaneously evaluated at a process due to the following two reasons: a token carries more than one signal; each active process produces output signals independently of a new output signal.

A function table is modified for simultaneous signal evaluation in our distributed token-driven simulation technique. Let m and n be the number of input ports at a process and the token size in bits, respectively. Since each input port is associated with a circular queue to store tokens, a process needs m queues, $Q_0, Q_1, \ldots Q_{m-1}$. Let t_{ij} be the j-th token in Q_i. In addition, b_{pqr} is defined as the r-th input signal in token t_{pq}. In this case, the first tokens t_{i0} ($i = 0 \ldots m-1$) in each queue contain the following input signals:

$$t_{00} (b_{000}, b_{001}, b_{002}, \ldots, b_{00,n-1}),$$
$$t_{10} (b_{100}, b_{101}, b_{102}, \ldots, b_{10,n-1}),$$
$$\ldots\ldots\ldots\ldots$$
$$t_{m-1,0} (b_{m-1,00}, b_{m-1,01}, \ldots, b_{m-1,0,n-1}).$$

Here, let s be the number of input vectors which can be involved with simultaneous signal evaluation. In others words, all signals derived from s consecutive input vectors can be simultaneously evaluated at a process if each input port of the process has a token. All signals in the first s signals of the first tokens in each queue are involved with simultaneous signal evaluation using a function table lookup operation. A function table contains s output signals in each element. An index into the function table for the table lookup operation is defined to be the concatenation of the following bits:

$$b_{m-1,0,s-1}, b_{m-2,0,s-1}, \ldots, b_{00,s-1},$$
$$b_{m-1,0,s-2}, b_{m-2,0,s-2}, \ldots, b_{00,s-2},$$
$$\ldots\ldots\ldots\ldots$$
$$b_{m-1,00}, b_{m-2,00}, \ldots, b_{000}.$$

We now discuss the data structure for a function table, and how each output signal of an element of the function table must be computed. The function table is a two-dimension array. An element of the table is defined to be a binary value f_{ij}, ($i = 0, 1, 2, \ldots, (2^{(m*s)}-1)$ and $j = s-1, \ldots, 1, 0$). For an index i of a function table, s elements (to be stored), $s_{i0}, s_{i1}, \ldots, s_{i,s-1}$ are computed as follows:

1. index i is converted into a $(m*s)$-bit binary number, say

$$a_{m-1,s-1}, a_{m-2,s-1}, \ldots, a_{0,s-1},$$
$$\ldots\ldots\ldots\ldots$$
$$a_{m-1,1}, a_{m-2,1}, \ldots, a_{0,1},$$
$$a_{m-1,0}, a_{m-2,0}, \ldots, a_{0,0}.$$

2. f_{ij} is computed by applying the gate function with bits $a_{m-1,j}, a_{m-2,j}, \ldots, a_{0,j}$, where the gate function indicates a digital function of the gate like AND, OR, etc.

Given an index i in a table lookup operation, output signals f_{ij} ($j = s-1, s-2, \ldots, 1, 0$) come out from the table.

Figure 1 shows the data structures for queues and a function table in case of $s = 3$ and $m = 3$ at a process. For good performance, the token size is a multiple of s so that each signal evaluation is involved with s input vectors. The simultaneous signal evaluation strategy can be optionally used depending on a target machine. The strategy is recommended for a parallel machine whose element access time from a large array used as a function table is not slow.

Feedback Loops

The proposed distributed token-driven simulation technique can be applied to any circuit with feedback loops if it contains at least one flip-flop. The feedback loop can then

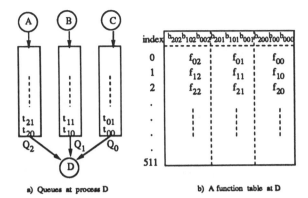

a) Queues at process D   b) A function table at D

Figure 1: An example for simultaneous evaluation

be broken at the flip-flops by treating flip-flop inputs as primary outputs and their outputs as primary inputs, as used in unit-delay compiled simulation [4]. In the distributed token-driven technique with identifiers, a simulation ends when a predetermined identifier limit is reached.

3 Performance Evaluation

The scheme for logic simulation proposed in [5] was used to implement our distributed token-driven simulation technique. In this paper, the following data structures are considered. First, each gate in a circuit being simulated is assigned to a processor in a target machine. Second, each input port of a gate is associated with a simple FIFO (first-in first-out) queue to store tokens sent to the port since each port receives tokens in processing order. All queues belonging to a process must be contained in a processor. Finally, signal evaluation is performed based on a table lookup operation as in [5].

The distributed token-driven simulation technique without identifiers was implemented on the Connection Machine with 32K processors for parallel zero-delay logic simulation. A 32 × 32 array multiplier and ISCAS85 [6] are used as benchmark circuits for the performance evaluation of the proposed technique.

One of the advantages of our distributed token-driven simulation is that the number of simulation cycles can be predicted. Let T and N be the critical path length(in nodes) of a testbed and the number of input vectors for the simulation, respectively. In this case, the number of simulation cycles for simulation is exactly

$$T + N - 1.$$

The above number can be obtained in case of token size 1 since no unnecessary messages (or events) are involved in simulation at all. When a token contains more than one signal, the number of simulation cycles is computed as

$$T + \lceil \tfrac{N}{S} \rceil - 1$$

where S is a token size in signals.

According to the comparison of the experimental results of our distributed token-driven simulation technique with those of distributed event-driven simulation given in [5], our technique runs much faster than traditional event-driven simulation techniques. There are several reasons for the good performance. First, the number of simulation cycles is much less than any other distributed event-driven simulation techniques. Second, less communication costs for sending tokens

Test Circuits	LECSIM (SUN3/260)	PC-set	Parallel	Token-driven
C1355	309	84.9	9.8	3.9
C1908	500	162.7	54.3	4.1
C6288	1487	1757.3	369.3	5.2

Table 1: Performance Comparison (in seconds)

No. input vectors	2100	4200	8400	10000
C6288	3.59	4.99	7.77	8.84
Mul-32	3.38	4.70	7.16	8.10

Table 2: Execution Times (in seconds)

is required than the cost for sending events since the size of a token is smaller than that of an event. In addition, grouping consecutive output signals into a token is allowed to reduce communication costs. Third, queue manipulation for tokens is much simpler than that for events. There are no unnecessary events during simulation. Finally, a simulation cycle is very simple.

Since there are no parallel zero-delay simulation schemes in existence for performance comparison, the experimental results of compiled-code simulation are considered for the comparison. Table 1 shows the comparison of the simulation techniques for some ISCAS85 benchmark circuits with 5,000 randomly generated vectors. As a unit of the measurement in the distributed token-driven simulation technique, *CM time* is used which is the time it takes to execute parallel instructions on the sequencer of the Connection Machine. In the measurement, the time period for only simulation is considered, i.e., the figures listed do not include the time required for reading vectors, printing output, or translating circuit description. LECSIM was measured on SUN 3/260 [3]. The experimental results of both the PC-set and the parallel technique in [4] are shown in the table. The experimental results for our token-driven simulation were obtained based on the token size of 21, i.e. each token contains 21 signals. According to the experimental results in the table, distributed token-driven simulation runs much faster than any other compiled logic simulation techniques. Since the execution of distributed token-driven simulation is a pipelined style, the execution times are proportional to the number of input vectors rather than the size of test circuits for a large number of input vectors. The execution time of token-driven simulation with up to 10,000 randomly generated input vectors is given in Table 2. In the measurements, a 32-bit array multiplier with 8256 gates and ISCAS85/C6288 are considered as test circuits.

Depending on the token size, the execution time may vary since the size affects the number of simulation cycles, communication costs, and signal evaluation time. According to experimental results, better performance can be achieved if the token size increases up to a certain point. The reason is that we can achieve significant reduction in both the communication cost and the number of simulation cycles.

We now compare the proposed distributed token-driven simulation with traditional distributed event-driven simulation. Output control is not required in our distributed token-driven simulation since no unnecessary messages occur. Dead-

lock never occurs during simulation since each active gate unconditionally propagates a token to all the successors of the gate. Determining active gates based on the selective-trace approach in our technique does not cause any rollback and state-saving overhead as required in Time Warp. The computation of a global clock is not necessary because each gate works based on the readiness of tokens in its queues independently of processing of other gates. One of great advantages of our distributed token-driven simulation technique is that queue manipulation is simple. In addition, the size of each queue is relatively small due to smaller token sizes compared with event sizes.

4 Conclusions

The distributed token-driven simulation presented herein is a new efficient parallel zero-delay logic simulation technique in massively parallel processing environments, which integrates the advantages of distributed event-driven simulation and compiled-code simulation. The experimental results demonstrate that the presented technique in this paper achieves high parallelism and outperforms both traditional distributed event-driven simulation and compiled-code simulation when we consider only functional simulation rather than timing simulation.

We are currently investigating the optimal token size for a given critical path and the number of input vectors, and measuring the performance of the distributed token-driven simulation technique with identifiers for ISCAS89 benchmark circuits. We aim to implement the proposed technique in other massively parallel environments. Further studies will consider the extension of the proposed distributed token-driven simulation technique to circuit designs with multi-delay.

ACKNOWLEDGMENTS

The authors would like to acknowledge the National Center for Supercomputing Applications, University of Illinois at Urbana-Champaign, for allowing the use of their resources.

References

[1] L. Wang *et al.*, "SSIM: A software levelized compiled-code simulator," in *Proceedings of the 24th Design Automation Conference*, pp. 2–8, 1987.

[2] A. Ruehli and G. Ditlow, "Circuit analysis, logic simulation, and design verification for VLSI," *Proceedings of the IEEE*, vol. 71, pp. 34–48, January 1983.

[3] Z. Wang and P. Maurer, "LECSIM: A levelized event driven compiled logic simulator," in *Proceedings of the 27th Design Automation Conference*, pp. 491–496, 1990.

[4] P. Maurer and Z. Wang, "Techniques for unit-delay compiled simulation," in *Proceedings of the 27th Design Automation Conference*, pp. 480–484, 1990.

[5] M. Chung and Y. Chung, "Efficient parallel logic simulation techniques for the Connection Machine," in *Proceedings of the Supercomputing '90*, pp. 606–614, November 1990.

[6] F. Brglez, P. Pownall, and R. Hum, "Accelerated ATPG and fault grading via testability analysis," in *Proceedings of the IEEE International Symposium on Circuits and Systems*, June 1985.

Parallel Hough Transform for Image Processing on a Pyramid Architecture

Alireza Kavianpour and Nader Bagherzadeh[*]
Department of Electrical and Computer Engineering
University of California, Irvine
Irvine, CA 92717

Abstract

This paper considers the problem of detecting lines in images using a pyramid architecture. The approach is based on the Hough Transform calculation. A pyramid architecture of size n is a fine-grain architecture with a mesh base of size $\sqrt{n} * \sqrt{n}$ processors each holding a single pixel of the image. The pyramid operates in an SIMD mode. Two algorithms for computing the Hough Transform are explained. The first algorithm initially uses different angles, θ_j's, and its complexity is $O(k + \log n)$ with $O(m)$ storage requirement. The second algorithm computes the Hough Transform in a pipeline fashion for each angle θ_j at a time. This method produces results in $O(k * \log n)$ time with $O(1)$ storage, where k is the number of θ_j angles used and n is the number of pixels.

Key Words: Hough transform, image processing, parallel processing, pyramid architecture.

1   Introduction

Parallel computing for image processing has recently received considerable attentions [1, 3, 10, 13]. Technological advances have made the design of fine-grain architectures possible and many practical algorithms have been implemented. The Hough Transform [7] is a powerful tool in shape analysis. It extracts global features from images; however, because of its computational complexity, it is not easily implementable in real-time for some applications. One approach for achieving real-time implementation of the Hough Transform is by parallel processing. Various configurations have been suggested for parallel implementation of the Hough Transform. A Hough Transform algorithm on a fine-grain SIMD machine with broadcasting capability has been proposed by Ibrahim [6]. Recently several efforts have been made to speed up the computation of the Hough Transform by utilizing parallelism. Systolic arrays for regular and modified

Hough Transform have been proposed by Chuang and Li [2]. An $O(k\sqrt{n})$ time Hough Transform algorithm for n pixels is given by Silberberg [12] where k is the number of angles used for the calculation. A number of more efficient Hough Transform algorithms with $O(k + \sqrt{n})$ time complexity have been suggested [3]. Algorithms on mesh connected arrays have been described by Kannan and Chuang [8]. An algorithm for the Hough Transform calculation on a linear array has been considered by Fisher and Highnam [5], but because of the communication latency between remote processors it takes as much as the serial version of the algorithm to complete. Blanford explains a generalized Hough Transform [1]; the method is similar to the standard Hough Transform except that the image is divided into circular, parabolic, or elliptical bands, as opposed to straight lines. Little et al. [9] described a possible implementation of the Hough Transform on the Connection Machine.

2   Pyramid Architecture

A pyramid computer with the base size of n is an SIMD machine that can be viewed as being constructed from $(1/2) \log n + 1$ levels of a two-dimensional mesh connected processor array, where the Lth level, $0 \leq L \leq (1/2) \log n$, is a two-dimensional mesh connected processor array of size $n/(4^L)$. A mesh connected computer of size n is a collection of n processing elements arranged in a $\sqrt{n} * \sqrt{n}$ grid, where each processing element except for those along the border, is connected to its four neighbors. Each processing element at level L is connected to its neighbors at level L and four children at level $L - 1$ ($L > 0$) and a parent at level $(L+1)$, $(L < (1/2) \log n)$. Thus, each internal processing element has nine connections. Figure 1 illustrates a pyramid with a $4 - by - 4$ base configuration. Circles represent processing elements and lines represent communication paths. All of the processing elements in the pyramid operate in a strict SIMD mode under the direct control of a single node. Each processing element has its own memory and all of the communication links are bidirectional. The total number of processing elements is given by: $\sum_{L=0}^{(1/2) \log n} n/(4^L) = (4n/3) - 1/3$.

The pyramid topology has been proposed as an architecture for high-speed image processing where its simple geometry adapts naturally to many types of

[*]Acknowledgement: The work presented was supported in part by the UC MICRO program under Grant No. 89-108 in cooperation with Rockwell International. Additional funds were supplied by Irvine Faculty Research Fellowship.

problems [3]. Pyramids are more attractive than meshes because they provide the potential for solving problems with logarithmic time complexity.

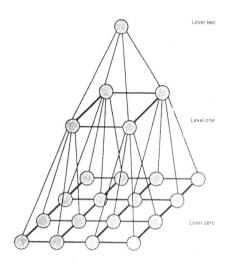

Figure 1: A 4-by-4 base pyramid

3 Hough Transform

The detection of lines and curves in an image is a fundamental problem in image processing. The problem is often solved by the Hough Transform. The Hough Transform and its generalization have frequently been used to detect object boundaries in image analysis. In the simplest case, the picture contains a number of discrete black figure points lying on a white background. The problem is to detect the presence of groups of collinear or almost collinear figure points. The Hough Transform was first introduced as a method for detecting complex patterns of points in binary image data. This method has many desirable features. Since, each image point can be treated independently, the method can be implemented using parallel processing.

The Hough Transform algorithm was introduced by Paul Hough in a patent filed in 1962 [7]. Rosenfeld et al. showed that it can be used to detect curves [11]. Later Duda and Hart [4] suggested that straight lines can be parameterized by the length ρ and orientation angle θ of the normal vector to the line from the image origin. This relationship is given by $\rho = x * \cos\theta + y * \sin\theta$.

A straight line in the plane is uniquely specified by two parameters such as (a, b) in the Cartesian space (a and b represent slope and intercept respectively), or (ρ, θ) in the polar(parameter) space where ρ is the length of the line segment from the origin per-

pendicular to the line, and θ is the angle that the line makes with the positive x axis, measured clockwise. The most commonly used coordinate system in the Hough Transform implementation is (ρ, θ). With n image points the ranges of (ρ, θ) can be taken to be ρ: $[-M, M]$ and θ: $[-\pi, \pi]$ where M is $\sqrt{(2n)}$ (in pixel units). If we restrict θ to be in $[0, \pi]$ range, then normal parameters for the line will be unique. With this restriction, every line in the $x - y$ plane corresponds to a unique point in $\rho - \theta$ plane. Given a set of n figure points, $(x_0, y_0), (x_1, y_1), ..., (x_{n-1}, y_{n-1})$, it is possible to find a set of straight lines that connect some of these points together. For each point (x_i, y_i) the equation $\rho = x_i * \cos\theta + y_i * \sin\theta$ creates a set of sinusoidal curves on the (ρ, θ) plane by varying θ from 0 to π. Curves corresponding to the set of collinear figure points will have a common point of intersection which defines the line connecting them in the Cartesian coordinates. A point in the Cartesian plane corresponds to a sinusoidal curve in the parameter plane, and a point in the parameter plane corresponds to a straight line in the Cartesian plane.

Given n pixels $(x_0, y_0), (x_1, y_1), ..., (x_{n-1}, y_{n-1})$, (x_i, y_i) is transferred into a sinusoidal curve in $\rho - \theta$ plane defined by $\rho = x_i * \cos\theta + y_i * \sin\theta$. The common point of intersection in $\rho - \theta$ plane, represented as (ρ_i, θ_i), defines a line intersecting at the collinear points. When it is not necessary to determine the lines exactly, we can specify the acceptable error resolution ρ_{res} and θ_{res}. The parameter space ρ and θ can be quantized into m ρ values $\rho_1, \rho_2, ..., \rho_m$ and k θ angles, $\theta_1, \theta_2, ..., \theta_k$ with an accumulator array of size $k * m$ each entry corresponds to a point defined by a pair of (ρ_i, θ_i). A region is treated as a two-dimensional array of accumulator cells. For each point (x_i, y_i) in the Cartesian plane the (ρ, θ) of the cell selected by the curve with the relation $\rho = x_i * \cos\theta + y_i * \sin\theta$ is incremented. A given cell in the two-dimensional accumulator array eventually records the total number of curves intersecting at the point represented by a given (ρ, θ). After all the pixels have been treated, the parameter array is inspected to find those cells with the largest counts. If the count value in a given cell (ρ_i, θ_i) is k, then precisely k pixels lie (to within the quantization error) along the line whose parameters are (ρ_i, θ_i). Because of the importance of this problem, and its complexity of $O(kn)$ on a single processor computer, where n is the number of edge pixels and k is the number of quantized values (ρ or θ), a lot of attention have been devoted to the development of efficient fine-grain SIMD parallel algorithm.

4 An Efficient Parallel Hough Transform

A processor in a $\sqrt{n} - by - \sqrt{n}$ base pyramid architecture is addressed using a triple index scheme

(L, j, i). In this notation L is the level number ($0 \leq L \leq (1/2) \log n$) and j is the group number ($1 \leq j \leq n/(4^{L+1})$ for all L such that $0 \leq L < (1/2) \log n$). The group number is designated clockwise using Hamiltonian path starting from the upper left group (a group is a 4-connected set of processors; for root processor, group number is zero). Finally, i is the index of each processor within a group numbered clockwise ($0 \leq i \leq 3$). Group refers to those processors that have the same parent. We assume that processors at level zero (base) store a whole image at once. Let's consider ρ_{res} as the resolution or maximum increment between two consecutive ρ values. This value is directly related to the size of the accumulator array, $h(\theta_j, \rho_{jp})$. If there are m ρ values, then the size of the accumulator array $h(\theta_j, \rho_{jp})$ for a particular angle θ_j is m, i.e., $h(\theta_j, \rho_{jp})$, $1 \leq p \leq m$: array$[1, 2, ..., m]$ of integers. θ_{res} is the resolution or maximum increment between two consecutive θ values, i.e., $\theta_{res} = 180/k$, where k is the number of angles considered for the Hough Transform calculation. Given a pyramid of a specified base size, it is always possible to choose ρ_{res} and θ_{res} so that the accumulator array will meet the memory requirement. For example, if only 16 θ angles are available then $\theta_{res} = 180/16 = 11.25$ degree. The procedure for calculating the Hough Transform has two steps. In step one, the accumulator array values are computed. In step two, maximum value in the array is determined. In the following algorithms, first we describe how accumulator array is formed, then we present a method for finding a maximum of the array.

4.1 Algorithm One

We assume the number of angles is equal to $n/4$ (n is the number of base level processors). Also, each processor at level zero (base) stores the following information in its local memory:

x_{ji} represents the x coordinate of the pixel processed by $p(0, j, i)$. y_{ji} represents the y coordinate of the pixel processed by $p(0, j, i)$. ρ_{res} represents the resolution or maximum increment between two consecutive ρ_{jp} values. Processor at level zero in group j stores $\cos \theta_j$ and $\sin \theta_j$ values. We assume each value of $x_{ji}, y_{ji}, \rho_{res}, \cos \theta_j$, and $\sin \theta_j$ occupies $b = \lceil \log \sqrt{n} \rceil$ bits of local memory.

At level one parent processor of group j stores $h(\theta_j, \rho_{jp})$, for all values of ρ_{jp} such that $1 \leq p \leq m$. This is an $array[1, 2, ..., m]$ of integers, initially all entries of $h(\theta_j, \rho_{jp})$, $1 \leq p \leq m$, are equal to zero. This array occupies $m * b$ bits of local memory. Algorithm One consists of two parts: Sum and Maximum.

4.1.1 Algorithm Sum

There are three phases:

1) In this phase, each processor $p(0, j, i)$ in the group j computes $\rho_{jp} = \lfloor (x_{ji} * \cos \theta_j + y_{ji} * \sin \theta_j)/\rho_{res} \rfloor$.

2) In phase two, ρ_{jp}'s from level zero are transferred to parent processors at level one, and the count corresponding to $\rho_{jp}th$ position of $h(\theta_j, \rho_{jp})$, $1 \leq p \leq m$ for θ_j is incremented, $(h[\rho_{jp}] = h[\rho_{jp}] + 1)$.

3) Coordinates of four pixels in group j located at the base level (x_{ji} and y_{ji}, $i = 0, 1, 2, 3$), are transferred to neighbor processors at the group $j mod(n/4) + 1$ using clockwise Hamiltonian path over all $n/4$ groups.

Phases one, two, and three are repeated $n/4$ times. At the end of this algorithm, for each θ_j, parent processor of group j, $p(1, j, i)$ has the accumulated sum of $h(\theta_j, \rho_{jp})$, $1 \leq p \leq m$, for all of the pixels in the image.

4.1.2 Algorithm Maximum

Each processor at level one will send the first count (corresponding to ρ_{j1}) of $h(\theta_j, \rho_{jp})$, $1 \leq p \leq m$, to its parent processor. Parent processors select maximum value of the first count of $h(\theta_j, \rho_{jp})$, $1 \leq p \leq m$, received from their four children and send this value to parents at higher levels. Moreover, parent processors store the computed maximum value for the next iteration. This process is repeated for other entries of $h(\theta_j, \rho_{jp})$, $1 \leq p \leq m$. At the end of this process, the apex processor (processor at level $(1/2) \log n$) has the global maximum count for all values of ρ_{jp} and θ_j in the parameter space.

The total number of operations for all three phases is twelve. For $n/4$ angles, the total number of operations is $12 * n/4$. In general for k angles the time complexity is $O(k)$. At the end of this algorithm processors $p(1, j, i)$ will have the corresponding values of $h(\theta_j, \rho_{jp})$, $1 \leq p \leq m$, for k angles given by $\theta_1, \theta_2, ..., \theta_j, ..., \theta_k$. For each count of $h(\theta_j, \rho_{jp})$, $1 \leq p \leq m$, there is one transfer and three comparison operations. Since for other values of count in $h(\theta_j, \rho_{jp})$, $1 \leq p \leq m$ calculation can be completed in a binary tree fashion, then the maximum time is proportional to the number of levels in the pyramid. Thus, the time complexity for the Algorithm Maximum is $O(\log n)$. The total time complexity of Algorithm One is given by $O(k + \log n)$. Each processing element at the base needs $6 * b$ bit storage for storing $x_{ji}, y_{ji}, \rho_{res}, \rho_{jp}, \cos \theta_j$, and $\sin \theta_j$. Each processing element at level one needs m*b bits for storing the $h - array$. The amount of memory needed is proportional to $h - array$ with m ρ_{jp} entries resulting in an $O(m)$ storage for each processing element.

4.2 Algorithm Two

In this algorithm ρ_{jp} is computed for one value of θ_j at a time. The accumulated sum and the value of maximum are calculated in parallel. We assume:

x_{ji} represents the x coordinate of the pixel processed by $p(0, j, i)$. y_{ji} represents the y coordinate of the pixel processed by $p(0, j, i)$. ρ_{res} represents the resolution or maximum increment between two consecutive ρ_{jp} values.

We assume each processor at level one stores the following information:

ρ_{jp} = The computed value ρ for a given angle θ_j and c_{jp} = The total count for ρ_{jp}. Initially ρ_{jp} and c_{jp}, $1 \le p \le m$, are equal to zero.

The four phases of the algorithm are described below:

1) In this phase, $\cos \theta_j$ and $\sin \theta_j$ values are sent to each processor at the base level, then $\rho_{jp} = \lfloor (x_{ji} * \cos \theta_j + y_{ji} * \sin \theta_j)/\rho_{res} \rfloor$ is calculated at each processor.

2) In this phase, each processor transfers the value of ρ_{jp} to its parent processor.

3) In this phase, m processors at level one collect ρ_{jp} values that are computed by the base processors and the corresponding counter cip will be incremented. The computed value of ρ_{jp}'s for a given angle θ_j is transferred to the corresponding processor at level one using the shortest path. This implies that the tree connection in the pyramid or Hamiltonian path in the first level is used. The selection of the path depends on the position of the ρ_{jp} with respect to the corresponding level one processor. At the end of this phase m processors at level one will have the accumulated sum ρ_{jp} for a particular angle θ_j (m entries of $h - array$ are stored in m different processors).

4) Maximum of c_{jp}, $1 \le p \le m$, will be calculated using an algorithm similar to Algorithm Maximum as described before.

Phases one through four are repeated for all values of θ_j, $j = 1, ..., k$.

Thus, the total time complexity for k angles is $O(k + k * \log n + \log n) = O(k * \log n)$. Each processor requires a constant amount of storage, and the amount of memory needed is $O(1)$.

5 Summary

In this paper two new and efficient parallel algorithms for calculating the Hough Transform on a pyramid architecture were described. In these algorithms both mesh and tree connections of a pyramid were exploited.

References

[1] R. P. Blanford, "Dynamically Quantized Pyramids for Hough Vote Collections, " *Proc. IEEE Workshop Computer Architecture Pattern Anal. Machine Intell.*, pp. 145-152, Oct. 1987.

[2] H. Y. H. Chuang and C. C. Li, "A systolic array processor for straight line detection by modified Hough transform," *Proc. IEEE Workshop Computer Architecture Pattern Anal. Image Database Manag.*, pp. 300-304, Nov. 1985.

[3] R. E. Cypher, J. L. C. Sanz, and L. Snyder, "The Hough Transform has O(N) complexity on SIMD N*N mesh array architecture," *Proc. IEEE Workshop Computer Architecture Pattern Anal. Machine Intel.*, pp. 115-121, Oct. 1987.

[4] R. O. Duda and P. E. Hart, "Use of the Hough transformation to detect lines and curves in pictures," *CACM*, 15, (1), pp. 11-15, 1972.

[5] A. L. Fisher and P. T. Highnam, "Computing the Hough Transforms on a Scan Line Array Processor," *IEEE Trans. on PAMI*, vol. 11, no. 3, pp.262-265, March 1989.

[6] H. A. H. Ibrahim, J. R. Kender, and D. E. Shaw, "The Analysis and Performance of Two Middle-level Vision Tasks on a Fine-grained SIMD Tree Machine," *Proceedings IEEE Conf. Comput. Vis. Pattern Recog.* pp. 248-256, June 1985.

[7] P. V. C. Hough, "A method and means for recognizing complex patterns," U.S. Patent 3,069,654, 1962.

[8] C. S. Kannan and H. Y. H. Chuang, "Fast Hough Transform on a Mesh Connected Processor Array," *Information Processing Letters* vol. 33, pp. 243-248, 1990.

[9] J. J. Little, G. Bleccoch, and T. Cass, "Parallel Algorithms for Computer Vision on the connection Machine," *The First IEEE Inter. Conf. on Computer Vision* pp. 587-591, 1987.

[10] Yi Pan and Henry Y.H. Chuang, "Parallel Hough Transform Algorithms on SIMD Hypercube Array," 1990 *International Conference on Parallel processing*, pp. 83-86, Aug. 1990.

[11] A. Rosenfeld, Picture processing by computer, Academic Press, 1969.

[12] T. M. Silberberg, "The Hough Transform on the geometric arithmetic parallel processor," *Proc. IEEE Workshop Computer Architecture Pattern Anal. Machine Intel.*, pp. 387-393, Oct. 1985.

[13] S. L. Tanimoto, T. J. Ligocki, and R. Ling, "A Prototype of pyramid machine for hierarchical cellular logic," *Parallel Hierarchical Computer Vision*, L. Uhr, Ed. London, 1987.

UWGSP4: MERGING PARALLEL AND PIPELINED ARCHITECTURE
FOR IMAGING AND GRAPHICS

H.W. Park, Ph.D., **T. Alexander,** Ph.D., **K.S. Eo,** Ph.D., and **Y. Kim,** Ph.D.

Image Computing Systems Laboratory
Department of Electrical Engineering, FT-10
University of Washington
Seattle, WA 98195

Abstract

In view of the proliferation of imaging and graphics in various application areas, special computer architectures optimized to either imaging or graphics are being emphasized. In particular, application areas requiring both imaging and graphics are gradually increasing. In order to satisfy the increasing number of these applications, and to achieve high performance for both imaging and graphics, we have designed and are implementing a system called UWGSP4 (University of Washington Graphics System Processor #4).

UWGSP4 is configured with a parallel architecture for image processing and a pipelined architecture for computer graphics. A high bandwidth central shared memory with a sophisticated interconnection network is provided to enable the parallel and pipelined computational architecture to yield high sustained performance (over 50% of the peak performance in some image processing applications). The system is being implemented using four different ASIC (application specific integrated circuit) chips and three different printed-circuit boards.

1 Introduction

The disciplines of image processing and computer graphics are becoming more important, and have widespread applications. Imaging, or image processing, includes image enhancement, image restoration, image reconstruction, image compression and decompression, etc., and generates a new image from an original image. Computer graphics, on the other hand, generates images from a data base of objects such as polygons and lines.

Several kinds of special-purpose workstations have been designed and commercialized for either imaging or graphics. Among them, the Ardent Titan Graphics Supercomputer [1], the Silicon Graphics Superworkstation [2] and the AT&T Pixel Machine [3] provide high polygon throughput by using dedicated graphics hardware; however, they cannot provide high imaging performance. On the other hand, the MPP [4], the Connection Machine [5], and the CMU Warp [6] can achieve high image computing rates, but not high graphics performance.

The UWGSP (University of Washington Graphics System Processor) series of imaging and graphics workstations have been developed at the University of Washington since 1986. UWGSP1, developed in 1986, was a first generation fixed point imaging and graphics subsystem interfaced to an IBM PC/AT host. UWGSP1 consisted of a TMS34010 graphics system processor (GSP) closely coupled with a TMS32020 digital signal processor (DSP), and provided a flexible, low-cost (less than $7000 total cost) and medium performance (it takes about 3 seconds to perform 3 x 3 convolution on a 512 x 512 image) imaging platform. Since a fixed point 16-bit arithmetic is not sufficient for many image processing applications, we developed a second generation (UWGSP2) in 1988, which used a floating point processor (ACT8837). The imaging performance of UWGSP2 was about twice as fast as UWGSP1. UWGSP3

[7], developed in the beginning of 1990, utilized a multiprocessor configuration for imaging and graphics, and consisted of a TMS34020 GSP and four TMS34082 floating point graphics coprocessors that can be configured into a pipelined or SIMD (single instruction, multiple data streams) mode depending on the algorithm. The peak performance of UWGSP3 is 160 MFLOPS, with a sustained performance of about 30 to 40 MFLOPS.

None of the commercial workstations and the UWGSP series workstations mentioned above can implement both imaging and graphics processing efficiently in a single platform. UWGSP4, which is the main topic of this paper, is designed to provide high performance both in imaging and graphics. In particular, our architecture focuses on achieving a high sustained computational rate, over 50% of the peak performance for both imaging and graphics.

The following two sections of this paper describe the architecture of UWGSP4, and emphasize the specific architectural features which are optimized for imaging and graphics. In the last section, some simulated performance figures for several imaging and graphics algorithms are given.

2 System Architecture of UWGSP4

As shown in Fig. 1, UWGSP4 consists of four parts: a parallel vector processor, a shared memory, an interconnection network, and a graphics subsystem. Since most image processing operations, such as image transforms and convolution, can be computed by parallel processors at very high speeds, our parallel vector processor consists of up to 16 vector processing units which provide a peak of 1,280 million floating point operations per second (MFLOPS) of computing power. Further, they are optimized to yield a high sustained computational rate

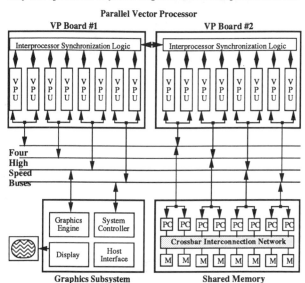

Parallel Vector Processor

Fig. 1 Overall block diagram of UWGSP4

in image processing. Each of the 16 vector processing units can operate independently or in synchronism with the others.

All communications between processing units are performed through the shared memory. The interconnection network between the processing units and the shared memory consists of a crossbar network connected to four high speed buses. The shared memory and the interconnection network are essential for the realization of high computing performance, since most of the limitations in parallel computing systems arise from insufficient data transfer bandwidth. The shared memory uses a 32-way interleaving scheme in order to achieve a 1,280 Mbytes/sec memory access bandwidth with standard DRAMs.

Each high speed bus runs at a speed of 80 MHz, and consists of a 32-bit path for address and data, together with some control signals. Since 16 vector processing units are connected to the shared memory via four high speed buses, each high speed bus provides four vector processing units with a communication path to the shared memory. An ECL (emitter-coupled logic) bus interface is used to sustain the 80 MHz clock rate, and the high speed bus is time-shared between the four vector processing units. The sustained data transfer rate of all four high speed buses is 1,280 Mbytes/sec, which is matched to the memory bandwidth of the shared memory.

A crossbar network is one of the best interconnection networks from the viewpoint of bandwidth and availability [8]. Our interconnection network is organized as an 8 x 8 x 40-bit crossbar using 64 crosspoint switches. Since the depth of the crossbar is four bytes for data plus some control signals (a total of 40 bits) and the cycle time of the network is 40 MHz, the total data transfer rate of our crossbar network is 1,280 Mbytes/sec.

Finally, the graphics subsystem supports the host interface, the graphics processing, and an image display. The graphics processing section consists of a parallel-pipelined architecture capable of providing a high graphics performance (about 200,000 Gouraud shaded polygons/sec). The image display section supports 24-bit full color images, and double frame buffers are incorporated to support smooth animation. The host interface provides a 20 Mbytes/sec data transfer rate between UWGSP4 and a host computer.

The following subsections describe each part in detail.

2.1 Parallel Vector Processor

The parallel vector processor is the primary computation engine in the system, and is used mainly for imaging and general mathematical computations. Figure 2 shows the block diagram of a single vector processing unit (VPU); as shown in Fig.1, the complete parallel vector processor consists of 16 such VPUs over two boards. Each VPU consists of two floating point units (FPUs), a set of scalar and vector register files, an ASIC chip for control and instruction issue, a pixel formatter unit (PFU) for pixel handling, a unified instruction and data cache, and a bus interface unit (BIU) for interface to the high speed bus.

The BIU provides the signal conversion between the standard 40 MHz TTL-level interface to the VPUs and the 80 MHz ECL-level interface to the high speed bus. Four VPUs are connected to a single high speed bus, in two pairs, as shown in Fig.1. Each pair of VPUs shares the high speed bus in an alternating fashion. Each BIU has a bus arbiter that controls the arbitration between the VPUs which communicate on the same bus phase.

Two FPUs, implemented using 74ACT8847 CMOS floating point processor chips from Texas Instruments, are used for each VPU. The FPUs operate in an alternating fashion with

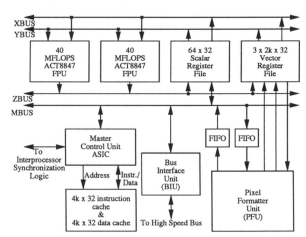

Fig. 2 Block diagram of a single vector processing unit

a 20 MHz two-phase clock. Each floating point processor possesses a full set of arithmetic and logic instructions, and can handle IEEE 754 standard single- and double-precision floating point operands as well as 32-bit integer data values. The arithmetic logic unit (ALU) and the multiplier within the floating point processor can operate independently or be used simultaneously when performing pipelined multiply-accumulates. The peak performance of a single floating point processor is 40 MFLOPS, so that one VPU provides a peak computing rate of 80 MFLOPS.

Control of the VPU, including instruction fetch/issue and data cache handling, is the domain of the master control unit ASIC. It fetches and interprets instructions, and controls the FPUs so as to execute the desired arithmetic and logical operations. The ASIC is implemented using a 30,000 gate CMOS standard-cell custom IC. It also performs all the control and sequencing necessary for proper operation of the VPU.

A set of vector and scalar register files facilitate the fast execution of vector instructions and scalar (bookkeeping) calculations. Sixty four four-ported 32-bit scalar registers are provided. During scalar execution, the control ASIC manipulates the four ports to move data to and from the two FPUs and the data cache. In addition, a set of three 2048-word vector register files are also provided to each VPU. Each vector register file has a separate read and write port. In operation, the control ASIC loads two of the vector register files with arrays of input operands, and then causes the FPUs to perform the desired computations on the arrays to generate a third array, which is stored into the third register file.

In general, image pixel data consists of 8-bit or 16-bit unsigned integer values, whereas image processing is performed in floating point for accuracy. In UWGSP4, data conversion between integer and floating-point value is carried out by a special pixel formatter unit (PFU) which is implemented with a field programmable gate array (FPGA). Thanks to the PFU, each VPU can provide a high computing rate without the burden of having to convert integer into floating-point value, and vice versa.

2.2 Shared Memory and Interconnection Network

In parallel computers with central shared memory, the sustained performance of the system is usually limited by the interconnection network and the memory bandwidth of the shared memory. Therefore, the configuration of the shared memory and interconnection network should be designed to minimize data access conflicts. In order to increase memory

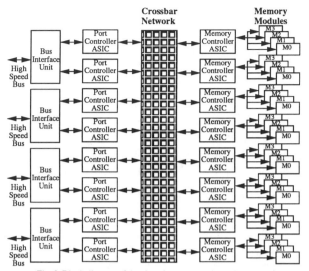

Fig. 3 Block diagram of the shared memory and crossbar network

bandwidth and to match the relatively slow memory to the processor speed (the processor cycle time is generally 4 to 10 times faster than the memory cycle time), a memory interleaving scheme has been used [9]. In UWGSP4, 32-way memory interleaving is used to achieve a 1,280 Mbytes/sec memory bandwidth.

Figure 3 shows the block diagram of the shared memory and crossbar network. The shared memory and interconnection network consist of four high-speed buses, four bus interface units (BIUs), eight port controllers (PCs), an 8 x 8 x 40-bit crossbar, eight memory controllers (MCs), and eight four-way (total 32-way) interleaved memory modules.

For the interconnection network between the parallel vector processor and the shared memory, a combination of multiple high-speed buses and a crossbar network was adopted. The BIUs interface the high-speed buses to the PCs; each BIU interfaces one high speed bus to two PCs. All shared memory components, except the high speed buses and BIUs, operate with a 40 MHz clock cycle, whereas the high speed buses use an 80 MHz clock cycle. Therefore, two PCs can communicate with one high speed bus in an alternating fashion without any conflict. The BIUs interface the TTL-level logic in the shared memory to the ECL bus.

The PCs were designed using 12,000 gate CMOS gate-array ASICs. The PCs translate memory access commands from the VPUs into simple commands which can be executed by the MCs, and control the crosspoint switches in conjunction with the MCs. The VPUs can access a scalar data item, a column or row vector, or 2-D array data with a single command.

The crossbar network provides a separate path between any PC and any MC at all times, so that eight PCs can communicate with eight MCs simultaneously. The crossbar network is implemented with discrete Advanced Schottky TTL transceiver chips.

The MC generates the physical address of each data word from the command received from the PC, and was designed as an 8,000 gate CMOS gate-array ASIC. Each MC controls four interleaved DRAM memory modules, and accesses vector data from the memory modules at an access rate of 160 Mbytes/sec. Also, DRAM refresh is carried out by the MCs. In order to accommodate multiple memory chips of different capacities, the memory controller ASIC was designed to be able to use 1 Mbit, 4 Mbit, or 16 Mbit DRAMs.

The four memory modules connected to each MC provide a 25 nsec access time in the case of row vector data, which can utilize the four-way interleaving and page-mode access capability of the DRAM modules. Memory depth is 32 bits. All memory accesses are word-based, but any VPU can selectively write to any specific byte within a word by using byte mask. Since the shared memory has eight MCs, each of which controls four interleaved memory modules, a total of 32 interleaved memory modules are supported, which provides a maximum memory space of 1 Gbyte (256 Mwords).

2.3 Graphics Subsystem

The graphics subsystem is the primary agent for maintaining and drawing the image, and also for generating realistically-shaded three-dimensional images from scene descriptions. As shown in Fig. 4, the graphics subsystem consists of two independent polygon processing pipelines, four bit-blit interpolator (BBI) ASICs, a Z-buffer and a double-buffered frame buffer, a TMS34020-based graphics processor and system controller, an overlay buffer, a cursor generator, RAMDACs and a host interface.

The two polygon processing pipelines are responsible for the front-end processing for three-dimensional computer image synthesis. The operations provided by the pipelines include geometric transformation, back-face culling, illumination, clipping, projection and slope calculation. The polygon processing pipelines consist of nine Intel i80860 CPUs operating in a parallel-pipelined fashion. One of the nine CPUs, called the head processor, communicates with the shared memory to extract the actual polygons by traversing an object hierarchy, and distributes polygon processing jobs to the two pipelines. Each pipeline is composed of four CPUs (corresponding to four pipeline stages). The head processor dynamically assigns polygon rendering jobs to each pipeline stage in such a way as to balance the load between the eight CPUs of two pipelines. The computation results of these pipelines are fed to four BBI ASICs, where scan conversion is performed in conjunction with hidden surface removal.

The BBI ASICs carry out image drawing to the screen (frame buffer), using the span data computed by the polygon processing pipelines. The four ASICs, each of which is a 20,000 gate CMOS standard-cell custom IC, are capable of

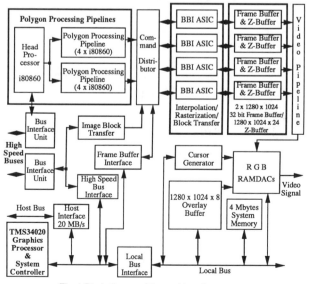

Fig. 4 Block diagram of the graphics subsystem

drawing and filling image pixels at the rate of 40 million pixels per second. They also control the memory in the frame buffer and Z-buffer, and carry out screen refresh functions under the supervision of the system controller. Another feature incorporated into the BBI chips is the ability to support a multi-windowed display with minimal overhead, based on an extended Z-buffer concept. The combination of the polygon processing pipelines and four BBIs is capable of delivering over 200,000 100-pixel polygons per second with Gouraud shading. Other special functions which are provided by the BBIs are transparency, antialiasing, texture mapping, and Z-buffer handling.

The double frame buffer consists of two 1280 x 1024 pixel buffers built using video RAMs (VRAMs) with each pixel being 32 bits deep to support 24-bit true color and realistic animation. The other eight bits in the 32 bits of frame buffer are used for supporting transparency, antialiasing and alpha blending. The Z-buffer, whose size is 1280 x 1024 x 24 bits, is implemented in VRAMs as well to significantly speed up three-dimensional rendering algorithms. The image data stored in the frame buffer is serially transferred to the RAMDACs to be displayed on the monitor.

The system controller of the TMS34020 maintains the host interface, acts as a central controller for the UWGSP4 system, and controls the 1280 x 1024 x 8 bit overlay buffer. A 4 Mbyte local memory is provided for storing programs and data associated with the system controller.

All communications between UWGSP4 and the host computer are performed through the host interface in the graphics subsystem; the data transfer rate is 20 Mbytes/sec. The graphics subsystem occupies two high speed buses to store and load data to and from the shared memory. One interface with a high speed bus is used to load graphics data from the shared memory to the polygon processing pipelines. The other interface is used to transfer the image data, program and any control information between the shared memory and the host or the graphics subsystem.

3 Performance Enhancements

UWGSP4 is a special purpose computing system intended mainly for imaging and graphics. Even though general operations are also supported, the full performance of UWGSP4 cannot be obtained unless the parallel vector architecture is completely utilized. This section describes in detail several hardware features which contribute to the high computing rate in imaging and graphics.

3.1 Parallel Processing

Most image processing operations can be performed in a parallel fashion by the vector processors. For example, the 2-D fast Fourier transform (FFT), which is a popular transformation in image processing, is performed by taking 1-D FFT's of the image rows followed by 1-D FFT's of the image columns. If the image size is N x N (N is a power of 2), up to N processors can perform the 1-D FFT of N rows and N columns, independently of each other. Therefore, the computing speed is N times as fast as that of a single processor system (if the memory bandwidth does not limit the computing power). Also, two-dimensional convolutions can be done by the multiple processors at very high speeds.

In UWGSP4, 16 vector processing units can perform arithmetic or logical computations independently or in synchronism. In order to efficiently utilize all of the 16 processing units in image processing, a simple interprocessor synchronization scheme and fast parallel data access are primary requirements. A com-

bination of a token-ring scheme implemented in hardware (the interprocessor synchronization logic shown in Fig.1), and software test-and-set or fetch-and-add operations, is provided for interprocessor synchronization. Each processing unit has two lines (named *token_in* and *token_out*), which are used to pass a token around the 16 processing units. Each processing unit passes a token received on its *token_in* line onto the *token_out* line (which is connected with the *token_in* line of the next processing unit) unless it wants to access any shared variable in the shared memory for synchronization. The shared variable can only be accessed by the processing unit which currently possesses the token, so that only one processing unit among the 16 can access the shared variable.

3.2 Vector Addressing

In general, image processing involves large arrays of data and hence lends itself well to vector processing. Vector operations in UWGSP4 are implemented with microcode. For efficient vector operations, the vector processing units contain vector data addressing hardware which allows the vector register file to be accessed with different addressing patterns. The vector data addressing hardware permits most imaging algorithms to be performed in a single vector operation. Figure 5 shows the vector addressing hardware, which consists of several counters and increment registers. The three vector register files have their own vector addressing hardware. The registers and counters are pre-set according to the algorithm before the vector operation.

For example, let us assume that a vector register file contains a 32 x 32 block (1 kword) of an image in row-major order and an image convolution with a 5 x 5 kernel is to be performed. The registers and counters contain the values as shown in Fig. 6. Other vector addressing schemes can be supported by altering the values in the registers and counters; most imaging operations, including the FFT, can be supported. In the case of the FFT, the last stage of the FFT produces the results in bit-reversed order. The third register file, which stores the computation result from the FPUs, provides bit-reverse hardware in the vector address generator to unscramble the result data.

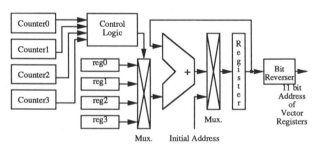

Fig. 5 Addressing hardware for vector registers

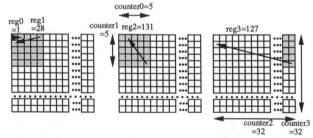

Fig. 6 Register and counter values set up in vector addressing hardware for 5 x 5 convolution

3.3 Pixel Formatter

Most image pixel data stored in the shared memory consists of one-byte or two-byte packed integer values. However, most image computation should be done in floating point for accuracy. In order to perform image processing, the image data loaded from the shared memory should be converted to floating point before the computations. Also, computation results which are in floating point should be converted to one-byte or two-byte packed integer values before transfer to the shared memory. The data conversion between floating point and integer values is referred to as pixel formatting. Pixel formatting is performed by a special hardware unit, consisting of a multilevel pipelined architecture. The conversion rate of this unit is 40 million pixels per second, i.e., one pixel integer value can be converted to a 32-bit floating-point value in a 25 nsec cycle time, and vice versa. In this case, the image pixel value can be 1, 4, 8, or 16-bit unsigned integer value, all of which can be handled by the pixel formatter hardware. The other function of this unit is to transfer data stored in the third vector register file to the other vector register files so that the results can be used as source operands if necessary.

3.4 Pipelined Graphics Engine

Graphics operations based on polygon data are different from image processing operations. Most graphics operations are 3 x 3 or 4 x 4 matrix operations, or scalar operations, whereas image processing consists mostly of larger matrix operations or transformations. Therefore, general three-dimensional graphics operations cannot fully utilize the parallel vector processor. Also, the scan-line algorithm with hardware Z buffer is more suited to a pipelined architecture rather than a parallel architecture [10]. For the high graphics performance, UWGSP4 uses nine i80860 CPUs for one head processor and two four-stage polygon processing pipelines, and four BBI ASIC chips for scan conversion. Load balancing between the pipeline stages is very important for highest performance. The combination of the two four-stage pipelines of polygon processing and four BBI ASICs was designed for global optimization of load balancing of the graphics engine.

4 Simulation Results of Performance

It is estimated that image processing performance, for images of reasonable size (e.g., 512 x 512 pixels) will be real time for simple operations like contrast enhancement and convolution, or near real time for more complex imaging operations such as FFT and image warping. Initial simulations show that a 5 x 5 convolution, for example, can be performed on a 512 x 512 image in under 18 msec, which corresponds to about 700 MFLOPS sustained performance. The simulations also predict that UWGSP4 will be capable of carrying out a 2-D FFT at rates of over 600 MFLOPS, or, equivalently, under 33 msec for a 512 x 512 image. For comparison, these computing rates for 5 x 5 convolutions and 2-D FFT are about 15 times and 75 times faster than the CMU Warp computer, respectively: a 10-cell Warp can perform a 5 x 5 convolution in 284 msec and a 2-D FFT in 2.5 sec on 512 x 512 images [6].

The graphics performance is simulated to be over 200,000 Z-buffered, Gouraud shaded 100-pixel polygons rendered per second to the display screen. In addition, the system supports pixel transfer rates of up to 40 million pixels per second from the shared memory to the frame buffer (display screen). Therefore, 25 frames of 1k x 1k images can be displayed in a second. When drawing lines, the system is expected to reach speed of up to 250,000 3-D shaded 100-pixel lines per second. In comparison, the Silicon Graphics 4D/240GTX system [2] provides about 100,000 Z-buffered, lighted quadrilaterials per second and the Titan Supercomputer [1] can render up to 200,000 Gouraud shaded triangles/sec.

5 Conclusions

UWGSP4 was designed to provide near real-time image processing and graphics performance. Our goal is to achieve a peak imaging performance of 1,280 MFLOPS and a graphics rate of 200,000 polygons/sec with Gouraud shading, and also to support a high sustained performance of more than 50% of the peak performance. This sustained rate is nearly impossible in general purpose parallel computing systems.

As mentioned above, the disciplines of imaging and graphics will merge even more closely in the 1990s. For example, advanced medical applications require high speed image processing such as contrast enhancement, arithmetic and logical operations, rotation, zoom, window and level, image compression and decompression, cine display, and image analysis, together with compute-intensive graphics operations such as 3-D reconstruction and volume rendering. Also, military applications need advanced image processing and analysis as well as graphics operations such as vector map handling and terrain modeling. As a final example, the rapidly growing field of scientific visualization simultaneously demands very high floating-point performance and real-time graphics rendering. We expect that the UWGSP4 imaging and graphics system will be an integrated platform with very high performance to be used in the application areas of scientific visualizations, high performance military and medical diagnostic workstations, CAD/CAM, realistic animation, and so on.

References

[1] B.S. Borden, "Graphics Processing on a Graphics Supercomputer," IEEE Computer Graphics & Applications, Vol. 9(4), pp. 56-62, 1989.

[2] K. Akeley, "The Silicon Graphics 4D/240GTX Superworkstation," IEEE Computer Graphics & Applications, Vol. 9(4), pp. 71-83, 1989.

[3] M. Potmesil and E.M. Hoffert, "The Pixel Machine: A Parallel Image Computer," ACM Computer Graphics, Vol. 23(3), pp. 69-78, 1989.

[4] K.E. Batcher, "Design of a Massively Parallel Processor," IEEE Trans. on Computer, Vol. C-29(9), pp. 836-840, 1980.

[5] L.W. Tucker and G.G. Robertson, "Architecture and Applications of the Connection Machine," IEEE Computer, Vol. 21(8), pp. 26-38, 1988.

[6] M. Annaratone, E. Arnould, T. Gross, H.T. Kung, M. Lam, O. Menzilcioglu, and J.A. Webb, "The Warp Computer: Architecture, Implementation, and Performance," IEEE Trans. on Computer, Vol. C-36(12), pp. 1523-1538, 1987.

[7] K.S. Mills, G.K. Wong, and Y. Kim, "A High Performance Floating-Point Image Computing Workstations for Medical Applications," SPIE Medical Imaging IV, Vol. 1232, pp. 246-256, 1990.

[8] K. Hwang and F.A. Briggs, Computer Architecture and Parallel Processing, McGraw-Hill, Inc., 1984.

[9] T. Cheung and J.E. Smith, "A Simulation Study of the CRAY X-MP Memory System," IEEE Trans. on Computer, Vol. C-35(7), pp. 613-622, 1986.

[10] J.D. Foley, A. van Dam, S.K. Feiner, J.F. Hughes, Computer Graphics: Principles and Practice (2nd ed.), Addison-Wesley Publishing Co., 1990.

APPLICATION OF NEURAL NETWORKS IN HANDLING LARGE INCOMPLETE DATABASES: VLSI Design and Performance Analysis

B. Jin, S.H. Pakzad, and A.R. Hurson
Department of Electrical and Computer Engineering
The Pennsylvania State University
University Park, PA 16802

Abstract

Maybe algebra operations have been proposed for handling databases containing incomplete information. The major difficulties that hinder the use of such a set of operations are its tendency to return irrelevant data and low quality results, which offers low physical performance. As a solution, an artificial neural network-based decision support system is proposed which learns and constructs a knowledge base to filter out the low quality and irrelevant resultant data. The proposed filter constructs its knowledge base according to the data semantics of the underlying database and the user's query. This paper discusses the implementation of the decision-making network based on the reconfigurable VLSI design of a Basic Neural Unit (BNU) chip. Simulation results are studied for the performance analysis.

1. Introduction

In a database system, missing information can be interpreted as the "value at present unknown". The inclusion of missing information provides unknown data for the attributes in a data model. As the relational database model has been matured, researchers have studied the issue of how to handle missing information [1,2,4,8]. This study is based on Codd's maybe algebra [2] that adopts a three-valued logic. Such an extended capability allows the user to probe the database for potential data relationships that might be useful but cannot be retrieved by true relational algebra [1]. While bringing a better information utilization, this set of maybe operations also has its drawbacks. For example, maybe join operator has the potential to return irrelevant data relationships which are not defined according to the semantics of the underlying database. Moreover, it might generate extremely large result relations, which could cause a drastic degradation in performance if most of the resultant tuples are not of interest to the user. [4] shows that the loss in physical performance is paralleled by a loss in logical performance. Therefore, it is desired to develop a dynamic filter with a learning capability to adjust itself, according to the specific characteristics of the underlying database and the user's query. To fulfill the requirements of such a dynamic mechanism, a neural network-based decision support system is proposed. In a neural network, learning algorithm sets the connection strengths to their appropriate values based on valid input/output pairs (i.e. training pairs) [9]. In this way, the network learns and constructs a knowledge base to filter out the useless information. The dynamic nature of the databases and the user's query requires expandability and reconfigurability at the underlying decision-making network. Such a requirement creates a major problem for the traditional neuron-centered design techniques. In this study, a weight-centered design principle is used instead, which yields a high parallelism and reconfigurability for the constructed networks. A Basic Neural

Unit (BNU), based on the weight-centered approach, is designed to be used as a basic building block in constructing an expandable and reconfigurable decision-making network.

This paper is organized as follows: Section 2 overviews the concept of the neural network-based decision support system. The VLSI design of BNU and the construction of a decision-making network using BNU's are discussed in Section 3. Section 4 presents the simulation results in order to analyze the performance of the proposed decision support system. Finally, Section 5 concludes the paper.

2. Architecture of a Neural Network-Based Decision Support System

A neural network-based decision support system is shown in Figure 1. It is composed of four modules: i) adaptive learning and decision-making network; ii) knowledge acquisition; iii) decision-controlled buffer; and iv) user-system interface.

Adaptive Learning and Decision-Making Network

This module is a three layer neural network, an input layer, a hidden layer, and an output layer (Figure 1). The decision-making network is characterized by a set of neurons, pattern of interconnection, propagation rule, output function, and learning rule. It is a fully connected network. Network input is a binary bit pattern representing the resultant tuples generated by maybe operations. Each input unit can take a value of either "1" (meaning known attribute value) or "0" (meaning unknown or null attribute value). The number of input units in the decision-making network varies according to the number of attributes in the resultant tuples. Hidden neuron-units compose a layer of abstraction, pulling features from the input pattern [6,7,9]. The decision responses are shown at the output layer. They represent the classifications associated with each input pattern. For example, the leftmost output unit represents the "keep" decision, which means that the associated input tuple should be passed to the user for its high quality information; while the rightmost unit represents the "drop" decision due to the low quality and/or erroneous information. Besides these two categories, we can define certain intermediate categories representing some degrees of data quality. For example, resultant tuples that fall into "keep(75%)" category are to be passed to the user along with appropriate explanations provided by the user-system interface module. A tuple may also fall into a category of "keep(50%)", in which case the system cannot decide to keep it or drop it but still pass it to the user.

The decision-making network functions in either of the two operational modes, learning mode or decision-making mode [3]. Based on a supervised learning algorithm of the "generalized delta rule", the network learns from a set of training data generated by the knowledge acquisition module.

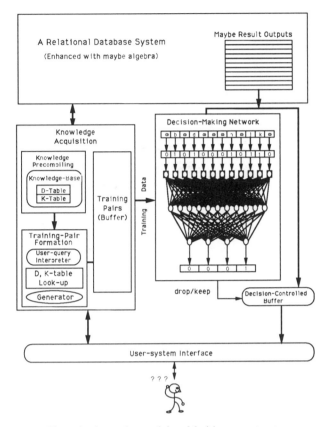

Figure 1. A neural network-based decision support system

Once trained, the network is switched to the decision-making mode in which it receives input patterns generated by maybe operations and responds to the closest classification (e.g. "keep", "drop", etc.). Generalization is considered to be successful if the network responds correctly, with high probability, to input patterns that were not included in the training set [6]. The decision made by the network is transferred to the decision-controlled buffer as a control signal.

Knowledge Acquisition

The knowledge acquisition module generates a set of training pairs for the decision-making network. In a database environment, it is required for decision-making process to understand the semantics of the underlying database as well as the requirements of the user's query. We proposed a two-level hierarchical knowledge acquisition module, as shown in Figure 1, for the decision support system [5].

At the first level, knowledge is acquired by extracting relationships that are implicitly/explicitly defined in a database system, independently of the user's query. This can be done by a knowledge precompiling unit. Two concepts in a relational database system form the theoretical foundation for knowledge precompiling: *data dependencies* and *key attributes* . The knowledge base consists of a D-table and a K-table. The D-table stores the knowledge captured by an analysis of the data dependency relationships. The K-table is constructed to store the knowledge extracted from various kinds of key attributes, i.e., *candidate keys* (including *primary keys*, *alternate keys*) and *foreign keys* [10]. Hence the knowledge precompiling provides a long-term knowledge with a strong portability.

In order to form the training pairs, the decision support system should also acquire proper information from the user's query. This knowledge can be extracted by a user-query interpreter unit at the second level of knowledge acquisition. At this level, knowledge acquisition takes place after a query is submitted. The acquired knowledge, referred to as short-term knowledge, differs from query to query and cannot be shared. A user-query interpreter is designed to analyze and recognize the set of query-dependent information. Consequently, the long term and short term knowledge is used, along with a set of heuristic criteria, to form the training pairs. This set of heuristic rules is determined by a careful analysis of the user's requirements and the semantics of the database [5]. It should be enforced by the system designer. Moreover, it should be flexible enough to satisfy the requirements of the ever changing environment.

Decision-Controlled Buffer

Decision-controlled buffer is a delay device which simply holds the resultant tuples and waits for the output from the decision-making network. When a decision arrives, the buffer will either pass a resultant tuple to the user (if a "keep" decision was made) or filter out the tuple (if a "drop" decision was made). The delay time would depend on the execution speed of the decision-making network.

User-System Interface

It functions as a dialogue interface between the user and the decision support system, providing a communication environment for both. It explains the result to the user, and more importantly, links the user to the knowledge acquisition module through a dialogue system.

3. A Reconfigurable VLSI Design for Decision-making Network

Software simulations have been conducted to demonstrate the feasibility of the proposed decision support system [5]. While the simulation results have shown the feasibility of the proposed scheme, they did not give any insight regarding the parallel capability of the proposed decision-making network. The effectiveness of the parallel processing in decision support can only be exploited by implementing the decision-making network in hardware.

In VLSI implementation, the requirement for additional connections complicates the design strategy where strong modularity, expandability, and reconfigurability are necessary. In neuron-centered design, Figure 2(a), once a PE or a set of PE's is implemented, it is usually difficult to expand the size of the network with the increasing number of inputs. Due to this restriction, we proposed a new approach, which we shall name weight-centered design [3]. Figure 2(b) shows a 4-input, 1-output weight-centered processing element. It consists of 4 weight units, 1 accumulator, and a nonlinear function unit. It is functionally equivalent to a 4-input, 1-output neuron-centered PE. In the proposed weight-centered PE, two ports are left open for the purpose of expansion and reconfiguration. Horizontally, inputs can be sent to more weight units to increase the number of output neurons. Vertically, weight-centered PE's can be stacked up to increase the number of input neurons.

Based on the concept of weight-centered design, a Basic Neural Unit (BNU) was designed [3]. Besides the expandability and reconfigurability, embedding more weight units into a single silicon chip also serves the main objective in designing the BNU. Due to the fabrication process available to us, a BNU that consists of 30 weight units, 10 inputs and 3 outputs, was fabricated by the IBM. A photomicrograph of the

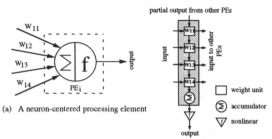

(a) A neuron-centered processing element

(b) A weight-centered processing element

Figure 2. Neuron-centered design v.s. weight-centered design

BNU chip is shown in Figure 3(a). BNU is packed in a 120-pin package.

The varying physical characteristics of databases and the dynamic nature of the user's queries require varying number of neurons on input, output, and hidden layers. As a result, the supporting decision-making network should have the ability to be reconfigured to satisfy the requirements of the user's query. This can be accomplished by using BNU chips as basic building blocks in constructing a decision-making network. On one hand, a BNU itself forms a 10-input, 3-output, two layer network. On the other hand, it can be connected to other BNU's to construct a larger network. In expanding a multilayer network, a column of BNU's can be added to the input-to-hidden layer board to increase the number of input neurons by 10, and the number of hidden neurons can be increased by 9 after adding a row of BNU's. Moreover, adding one more layer board will increase one layer in the decision-making network. Figure 3(b) shows a 30-input, 18-output, two layer network constructed by BNU chips.

4. Performance Analysis

To study the performance of the proposed decision support system, a number of simulation runs have been conducted. Two main relations, a student relation and a company relation, were generated.

STUDENT (stud_ssn, last_name, first_name, class_standing, major, GPA, area_of_interest)
COMPANY (code, name, required_major, minimum_GPA, area_of_interest, location)

A large volume of raw data was generated for the relations in the underlying database. The generated database represents about 600 students, 80 companies, 138 majors, and 100 areas of interest. The GPA ranges from 1.0 to 4.0. The simulation was run based on four classes of queries which can cover a variety of maybe join operations, such as joins over single and multiple attributes, θ-join and attribute maybe join [8]. These query classes are shown as following:

Query class 1 --- The first class of queries was performed to represent a maybe join operation on the company's hiring area of interest and a student's area of interest. The null value in the student relation simply meant that the student was not sure about what his/her area of interest was.

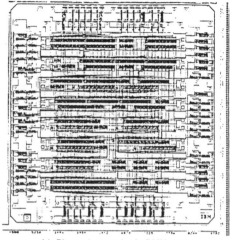

(a) Photomicrograph of a BNU chip

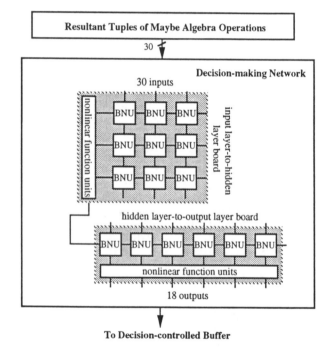

(b) A decision-making network constructed by 18 BNU chips
(30 inputs , 18 outputs, three layer)

Figure 3. Constructing a decision-making network using BNUs

Query class 2 --- The second type of queries represents a θ-join operation based on the minimum GPA requirement for hiring, i.e., a company's minimum GPA requirement.

Query class 3 --- The third class was performed to simulate maybe join operation on multiple attributes using area of interest, major, and the GPA as the join attributes. Even if a student doesn't know his/her area of interest, the student's GPA and major might satisfy a company's criteria and therefore should be considered.

Query class 4 --- The fourth class was run to evaluate the attribute maybe-join on major and area of interest.

Having been processed, these queries generated total of 608, 412, 337, and 153 tuples in the resultant relations. For the purpose of performance comparison, the degree of the resultant relation was 9 for every query. Hence the decision-making network had an input layer of 9 units which received output data from the 9 attributes of the resultant relations. It had a hidden layer of a flexible number of units. The network had an output layer of 5 units representing 5 categories, i.e., "keep", "keep with 75% degree of certainty", "uncertain (50% certainty)", "drop with 75% degree of certainty", and "drop".

For each class of queries, the training pairs were automatically generated by the knowledge acquisition module according to the semantics of the underlying database. There were four sets of training data used in the simulation, each having 5, 10, 15, and 20 training pairs, respectively. In every simulation run, the decision-making network was first trained by one set of training data. Then, the result of the queries (i.e., unknown data to the network) were passed to the network where proper decisions were made.

The simulations focused on two issues: i) the relationship between training speed   the number of training pairs and the number of hidden units in the network. ii) relative performance improvement of the hardware implementation of the network against the emulated network.

For query class 1, four cases were analyzed based on the number of training pairs (i.e. 5, 10, 15, and 20 training pairs, respectively). In each case the number of hidden neurons was varied and the corresponding training speeds were observed. Figure 4 shows the simulation results. As we can conclude:

• The training time increases as the number of training pairs increases.
• When a small number of hidden units (e.g. < 10) was used, the decision-making network was either not trained or trained using a long training time. For each training set, there was a lower limit on the number of hidden units that showed the start of a good training performance.
• The training time can be improved by choosing a configuration with relatively large number of hidden units (e.g. ≥ 10). However, after a certain point, an increase in the number of hidden units did not improve the training time. Moreover, our simulation results showed that for each case there existed an upper limit on the number of hidden units over which the training time increased abruptly and was irregular.

In a separate case, simulation was run to investigate the hardware effectiveness of the proposed BNU chip against the software emulation of the decision-making network. For this analysis, a configuration consisting of a three layer network was used. For each query class, the execution time of the trained network was observed by varying the number of hidden units. Approximately 1550 tuples were involved in the test data. Note that the software simulator was developed in a uniprocessor environment, rather than a parallel processing system. Figure 5 depicts the results of the operation. As one can conclude, the execution time of the hardware network was constant and independent of the number of hidden units. This is due to the inherent parallelism in the operation of the neuron network chip. As a matter of fact, the response time is a function of the number of layers in the network. In contrast, performance of the emulated network is a function of the number of hidden units in the network.

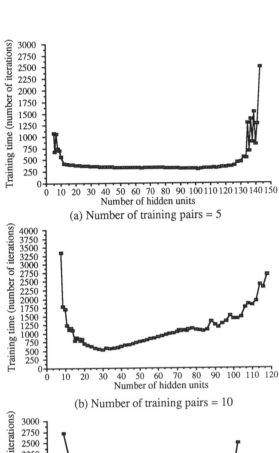

(a) Number of training pairs = 5

(b) Number of training pairs = 10

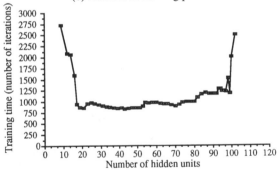

(c) Number of training pairs = 15

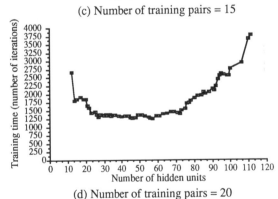

(d) Number of training pairs = 20

Figure 4. Decision-making network training time as a function of the number of training pairs and the number of hidden units

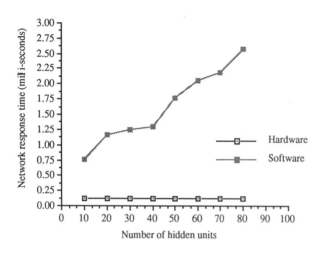

Figure 5. Decision-making network response time (Software simulation v.s. BNU-based hardware implementation)

5. Conclusion

A neural network-based decision support system for handling large databases containing incomplete information is proposed. The network is used to filter out low-quality and erroneous tuples among the set of maybe join results. In addition, the VLSI implementation of the decision-making network based on the concept of weight-centered approach was also addressed. The BNU chip is used as a basic building block to implement decision-making network. It offers a high degree of modularity, expandability and reconfigurability. A simulator was developed to analyze the validity of the proposed model for a variety of maybe join operations.

Our simulation results clearly demonstrated the feasibility of the proposed decision support system. However, in some cases the execution time of the maybe operations (e.g. maybe join) and the cardinality of the resultant relation might be drastic to justify the application of maybe operations. It is interesting to investigate the prefiltering ability of the proposed decision-making network. This allows the input relations to go through the filter in order to eliminate any input tuples that might have the potential of generating either erroneous and/or low quality resultant tuples. This can also be extended to enforce the proposed attribute maybe operations [4,8] through the decision-making network.

References

[1] Biskup, J., "A Foundation of Codd's Relational Maybe-Operations," ACM Transaction Database Systems, Vol. 8, No. 4, 1983, pp. 608-636.

[2] Codd, E.F., "Extending the Database Relational Model to Capture Meaning," ACM Transactions on Database Systems, Vol. 4, No. 4, 1979, pp. 397-434.

[3] Hurson, A.R., Jin, B., and Pakzad, S.H., "Neural Network-based Decision Making for Large Incomplete Databases," Proceedings of Parallel Architectures and Language Europe, PARLE91, June, 1991.

[4] Hurson, A.R., Miller, L.L., and Pakzad, S.H., "Incomplete Information and the Join Operation in Database Machines," Proceedings of Fall Joint Computer Conference, 1987, pp. 436-443.

[5] Jin, B., Hurson, A.R., and Miller, L.L., "Neural Network-Based Decision Support For Incomplete Database Systems: Knowledge Acquisition and Performance Analysis," Proceedings of Analysis of Neural Net Applications Conference, ANNA-91, May 1991.

[6] Kohonen, T., "State of the Art in Neural Computing," In Proceedings, IEEE First International Conference on Neural Networks, June 1987, pp. I79-I90.

[7] Lippermann, R.P., "An Introduction to Computing with Neural Nets," IEEE ASSP Magazine, April 1987, pp. 4-22.

[8] Pakzad, S.H., Hurson, A.R., and Miller, L.L., "Maybe Algebra and Incomplete Data in Database Machine ASLM," Journal of Database Technology, 1990.

[9] Rumelhart, D.E., Hinton, G.E., and Williams, R.J., "Learning Internal Representations by Error Propagation," in Parallel Distributed Processing (PDP): Explorations in the Microstructure of Cognition, Vol. I: Foundations, MIT Press, Cambridge, Massachusetts, 1986, pp. 318-362.

[10] Ullman, J.D., Principles of Database Systems, 2nd Edition, Computer Science Press, Rockville, MD 1982.

BRANCH-AND-COMBINE CLOCKING OF ARBITRARILY
LARGE COMPUTING NETWORKS

Ahmed El-Amawy

Electrical and Computer Engineering Department
Louisiana State University, Baton Rouge, LA 70803

Abstract -- We describe a novel scheme for global synchronization of arbitrarily large computing structures such that clock skew between any two communicating cells is bounded above by a constant. The new clocking scheme is simple to implement and does not rely on distribution trees, phase-locked loops or handshake protocols. Instead, it utilizes clock nodes which perform simple processing on clock signals to maintain a constant skew bound irrespective of the size of the computing structure. Among the salient features of the new clocking scheme is the interdependence between network topology, skew upper bound, and maximum clocking rates achieveable.

We use a 2-D mesh framework to present the concepts, to introduce three network designs and to prove some basic results. For each network we establish the (constant) upper bound on clock skew between any two communicating processors, and show its independence of network size. Besides theoretical proofs, simulations have been carried out to verify correctness and to check the workability of the designs, and a 4×4 network has been built and successfully tested for stability. Other topics such as node design, clocking of nonplanar structures (such as hypercubes), and the new concept of fuse programmed clock networks are also discussed.

1. Introduction

The simplest way of controlling pipelined and concurrent computing structures is the use of a global clock to connect the sequence of computation with time [1]. This design discipline is usually referred to as the synchronous discipline. Synchronous systems are by far the best known and most widely used in VLSI implementations and could also be attractive for multi-chip/ multi-board implementations.

The clock skew problem has long been recognized as a major obstacle in implementing very large synchronous systems (possibly) employing thousands or millions of processing cells [2]-[9]. The term clock skew refers to the variation in effective arrival times of a clocking event at different clocked cells. Ideally, a clocking event should arrive at (and affect) all the cells at the same time. Due to a combination of several effects: threshold voltages, variation in signal propagation delays on wires, and different buffer delays, the effective clock arrival times might well vary from cell to cell. Clock skew can be a problem even within a single chip [1].

The problem clock skew introduces is that it directly relates to clocking rate [9], [1], [2], [5]. Larger skews would necessarily impose lower clock rates if the synchronous system is to operate correctly. A simplified condition that generally holds states that [5] the clock period should at least equal the sum of 1) the time to distribute the clock 2) the time to perform the computation; and 3) maximum clock skew between any two communicating cells. However, with all current clock distribution schemes clock skews grow with system size. S-Y Kung states [2]: "For very large systems, the clock skew incurred in global clock distribution is a nontrivial factor causing unnecessary slow down in clock rate." Some designers try to minimize the effects of clock skew by equalization of wire lengths, careful screening of off-the-shelf parts, symmetric design of the distribution network, and design guidelines to reduce skew due to process variations [10]-[12].

Fisher and Kung [5] investigated clock skew problems in large 1-D, and 2-D arrays clocked by an H-Tree. The H-Tree guarantees that every computing cell is at the same physical distance from the root. Using two models in which all sources of skew are treated together as a distance metric, they have shown that if small delay variations are not ignored (summation model) clock skew tends to grow with system size. We directly quote Fisher and Kung [5]: "The paper shows that under this assumption (summation model, 2-D case) it is impossible to run a clock such that the maximum skew between two communicating cells will be bounded by a constant as systems grow" ..... "Therefore, unless operating at possibly unacceptable speeds, very large systems controlled by global clocks can be difficult to implement because of the inevitable problem of clock skews and delays".

Some researchers investigated alternative methods for controlling very large computing structures [1], [2], [5], [7]. The methods are either based on the asynchronous mode or on a hybrid synchronous/asynchronous approach. S-Y Kung and colleagues [2], [8], [9] introduced the concept of wavefront arrays, where Processing Elements (PEs) work asynchronously but information transfer between a PE and its immediate neighbors is by mutual convenience.

Seitz [1], [7] described an alternative discipline known as "self-timing" for controlling large systems. Self-timed systems are interconnections of parts called elements. Elements can be thought of as performing computational steps whose initiation is caused by signal events at their inputs and whose completion is indicated by signal events at their outputs. Self-timed elements can be designed as synchronous systems with internal clock that can be stopped synchronously and restarted asynchronously [1], [7]. A similar scheme was described by Fisher and Kung [5] for systolic array control.

This paper describes a new clock distribution scheme that is both simple and efficient [13]. The scheme makes it possible to guarantee tight synchronization between any two communicating processors regardless of the size of the computing network. Unlike all previous alternative methods a single clock source is needed for the whole network. This is very attractive for VLSI/WSI implementations since the single crystal can be located external to the chip/wafer. Moreover, the new approach makes it possible to extend the best known and most popular timing discipline (the synchronous discipline) to cases where it was previously considered impractical and/or too inefficient. Because no handshaking is needed, the new scheme allows the use of very fast clocking rates. Further, the new scheme eliminates the possibility of metastable failure associated with most alternatives of the synchronous discipline. Thus, the designer of a large computing structure is being offered a new alternative that is simple, inexpensive, and efficient.

In Section 2 we use a 2-D mesh framework to introduce the concepts, to establish some basic results, and to present three alternative *valid* network designs. For each of these networks the (constant) skew bound is established and shown to be independent of network size.

In Section 3 we discuss such relevant issues as node and network design, and clocking nonplanar computing networks (such as hypercubes). We also discuss the new concept of fuse programmed clock networks. Section 4 is a conclusion.

2. Basics of The Clocking Scheme

The new clocking scheme we now describe is based on the premise of clock processing [13]-[15]. Clock signals are distributed, combined and processed much the same way data signals are. Simple timing constraints, that are easily realizable, are imposed on the timing of clock signals. To distinguish between a clock processing entity and a data processing entity we shall hereafter reserve the term "clock node" or "node" for the former and "data call", "cell", or "processor" for the latter. An important function of a clock node is to enforce the adherence of clock signals to the timing constraints. A node also creates a local reference for neighboring nodes in the clock network.

Here we shall utilize the 2-D mesh framework to introduce the underlying concepts and prove some basic results. Three clock network designs with different cost/performance characteristics are described. More precisely the three networks use clock nodes with degree = 2, 3 and 4. Clocking processor networks with higher dimensionalities or different topologies will be discussed in Section 3.

Consider a 2-D computing network in which cells are connected in a mesh fashion such that each cell is connected to its four neighbors with the exception of boundary cells which have fewer neighbors. As a preliminary we associate a clock node to each data cell, such that each node clocks its associated cell as illustrated in Figure 1. This assumption will be relaxed in Section 3 to allow a node to clock several cells.

The principle underlying the new scheme can be termed "Branch-and-Combine" (or BaC for short). Starting from a single clock source, the clock signal is distributed to neighboring nodes which in turn act as sources to neighboring nodes down the network. Each node receives at least two inputs which are then combined into one signal in accordance with the function and timing constraints enforced by the node. Copies of this signal are then sent to neighboring nodes and used to clock the associated cell. The process repeats until all cells are clocked.

2.1 Preliminaries

For clarity we omit data cells from the figures to follow, unless otherwise noted, keeping in mind the node per cell assumption. We presume thus that clock nodes are located at mesh points and are identified by their coordinates. It is also presumed, for each network discussed, that nodes are of the same type and that the degree = fan-in = fan-out = F. Although we assume positive edge triggering, throughout this paper, other triggering methods are equally applicable, with minor changes. For this section we assume node to cell delay negligible. In Section 3 we will show how variations in node to cell delay can be accounted for.

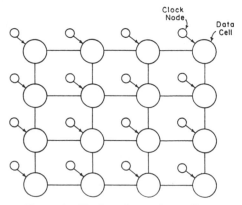

Figure 1: Clock node per data cell.

2.1.1 Notation

The following notation will be used in the remainder of this paper:

t^+ = time of $0 \rightarrow 1$ transition

$N(i,j)$ = node (i,j)

$ND(i,j) = g \pm \delta$ = delay through $N(i,j)$

$LD(i,j,k,l) = m \pm \varepsilon$ = link delay from $N(i,j)$ to $N(k,l)$.

$t_{i,j}^+(I_x)$ = time at which input I_x of $N(i,j)$ makes a $0 \rightarrow 1$ transition.

$\overset{\vee}{t}_{i,j}^+ = Min\{t_{i,j}^+(I_x), t_{i,j}^+(I_y), \cdots\}$, where $I_x, I_y, \cdots$ are inputs to $N(i,j)$

$t_{i,j}^+(o)$ = time at which $N(i,j)$ outputs a $0 \rightarrow 1$ transitions. Generally, if θ is a variable, then $\overset{\vee}{\theta}$ will be used to denote minimum value (lower-bound) on θ, and $\overset{\wedge}{\theta}$ will denote the maximum value (upper-bound) on θ. A $*$ superscript (e.g. θ^*) will be used to denote a realization (actual value for some case). Also the following assumptions will be used to simplify expressions:

$$(node + link) \ delay = g \pm \delta + m \pm \varepsilon = \alpha \pm \beta$$
$$\alpha = g + m, \ \beta = \delta + \varepsilon, \ \Delta = \alpha + \beta$$

2.2 Network Analysis

As mentioned earlier three clock network designs will

be introduced. But since the basic principles are identical in all three cases, we arbitrarily elected to detail the analysis of the network shown in Figure 2 (F=3). Deriving results for the other two networks (F=2 and F=4) becomes straightforward after the analysis of the Figure 2 network is completed. It will thus suffice to summarize the results for the other two networks.

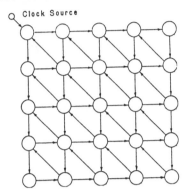

Figure 2: Clock network with F = 3.

Consider the clock network in Figure 2. Inputs to N(i,j) will be labeled I_x, I_y, and I_z, for the x,y, and diagonal directions respectively, as illustrated below. Because of the cyclic nature of the network, the reader may suspect that the network could oscillate. Indeed this could happen if appropriate timing constraints are not imposed on network signals. In fact, the very cyclic nature of the network is the reason, skew can be bounded by a constant!! [15].

An attempt to analyze the behavior of the final network design could easily lead to circular arguments. Therefore we follow an alternative approach in which we start with a very simple (and unrealistic) model to establish some facts and derive some results. We then modify the model and show that results based on the first model still hold true under the new set of conditions. We then present the final (practical) model and prove that results still hold.

2.2.1 Model 1

The model consists of the following conditions. Those conditions marked with * will be progressively relaxed.

1- Network is reset for all t < 0.

2-* Clock source S produces a step signal at t = 0.

3- a) The clock distribution network topology is that shown in Figure 2.
   b) All outputs of a given node are copies of the same signal.

4- $\alpha \geq 2\beta$. This means that (node + link) delays are allowed to vary by 50%, but not more. For analysis purposes we notice here that since $\beta = \delta + \varepsilon$, this condition implies $\alpha > 2\varepsilon$ and also $\Delta = \alpha + \beta > 2\beta$.

Although practically feasible the condition is merely intended to simplify following the proofs but is not necessary otherwise.

5- Network delays are relatively time invariant. This means delays can vary slowly over time within the specified bounds such that changes within a short interval (such as a few clock periods) are so small that they can be considered constants. The condition is important for subsequent models but not this one, however.

6-* Each node functions as follows: In response to the first arriving $0 \rightarrow 1$ transition on any of its inputs, it produces a step output after one node delay such that $t_{i,j}^{+}(o) = \overset{y}{t}_{i,j}^{+} + g \pm \delta$; $\forall$ i, j.

Conditions 6 suggests a node design such as that in Figure 3. Notice that the node cannot make a $0 \rightarrow 1$ output transition unless triggered. Notice also that clock skew between two communicating data cells is nothing but the difference between $t^{+}(o)$ of their corresponding nodes.

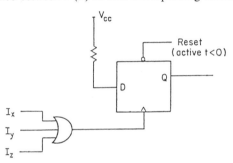

Figure 3: Node structure for Model 1.

Lemma 1: Under model 1, the clock network will not oscillate and every node will produce a single $0 \rightarrow 1$ output transition.

Proof: Since each node can only make a single $0 \rightarrow 1$ transition, and since each node has at least one input connected to S via a network path, each node is guaranteed a trigger condition. Once all nodes have been triggered, all node outputs remain high indefinitely. Thus the network will not oscillate and every node will output exactly a single $0 \rightarrow 1$ transition. $\square$

Theorem 1: The clock network described by Model 1 guarantees clock skew between any two communicating cells to be bounded above by 2Δ (constant), regardless of network (mesh) size.

Proof: Consider any arbitrary section of an arbitrarily large network adhering to Model 1, such as that in Figure 4. Sections containing top or left boundary nodes can be included by assuming any unused input to be permanently grounded. For bottom and right boundary nodes, simply ignore the unused outputs. Because of symmetry we only need to consider any two adjacent nodes (differing in only one coordinate). Thus without loss of generality, consider nodes $N(i+1, j+1)$ and $N(i+1,j+2)$ of Figure 4. Two cases of interest can be distinguished. Either node could receive its trigger before the other. The case when both nodes are triggered simultaneously is trivial since skew $\leq 2\delta < 2\Delta$.

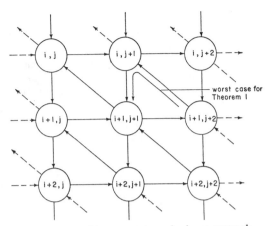

Figure 4: An arbitrary section of a large network.

Case 1: $\overset{\vee}{t}{}^+_{i+1,j+1} < \overset{\vee}{t}{}^+_{i+1,j+2}$

In this case,

$$\overset{\vee}{t}{}^+_{i+1,j+2} = \overset{\vee}{t}{}^+_{i+1,j+1} + ND\,(i+1,j+1) +$$
$$LD\,(i+1,j+1,i+1,j+2)$$
$$= t^+_{i+1,j+1}(o) + LD\,(i+1,j+1,i+1,j+2)$$

Thus,

$$t^+_{i+1,j+2}\,(o) \le t^+_{i+1,j+1}\,(o) + \overset{\wedge}{LD}\,(i+1,j+1,i+1,j+2) +$$
$$\overset{\wedge}{ND}\,(i+1,j+2);\ or$$
$$t^+_{i+1,j+2}(o) - t^+_{i+1,j+1}\,(o) \le m + \varepsilon + g + \delta$$

Hence for this case maximum skew is $\le \Delta$

Case 2: $\overset{\vee}{t}{}^+_{i+1,j+1} > \overset{\vee}{t}_{i+1,j+2}$

In the worst case $N(i+1,j+2)$ produces an output $0 \to 1$ transition that reaches $N(i,j+1)$ before the other two inputs; and $N(i,j+1)$ produces an output that reaches $N(i+1,j+1)$ also before the two other inputs (see illustration in Figure 3). To determine maximum skew we only need to consider this worst case scenario.

$$\overset{\vee}{t}{}^+_{i,j+1} = t^+_{i,j+1}(I_z) \le t^+_{i+1,j+2}(o) + \overset{\wedge}{LD}\,(i+1,j+2,i,j+1)$$

By simple extension we have

$$t^+_{i+1,j+1}(o) \le\ t^+_{i+1,j+2}(o) + \overset{\wedge}{LD}\,(i+1,j+2,i,j+1) +$$
$$\overset{\wedge}{ND}\,(i,j+1) + \overset{\wedge}{LD}\,(i,j+1,i+1,j+1) +$$
$$+ \overset{\wedge}{ND}\,(i+1,j+1)$$

Therefore maximum skew is

$$t^+_{i+1,j+1}(o) - t^+_{i+1,j+2}(o) \le 2(m + \varepsilon + g + \delta) = 2\Delta$$

Hence in all cases, skew between any two communicating cells is $\le 2\Delta$. $\square$

The following result is very useful for future developments.

Theorem 2: Under Model 1, maximum skew between any two inputs to the same node is bounded above by 3Δ.

Proof: Define a hop as signal propagation through a node and one of its output links. Clearly a hop delay $= \alpha \pm \beta$. The proof is very simple since an input to a certain node can be reached from any of the two other inputs in at most

three hops. The only exception is top and left boundary cells, where unused inputs are grounded. But those inputs do not trigger their respective nodes and can be ignored. Arbitrarily let

$$\overset{\vee}{t}{}^+_{i,j} = t^+_{i,j}(I_x) \text{ and } \hat{t}{}^+_{i,j} = t^+_{i,j}(I_z)$$

It is clear that in the worst case

$$t^+_{i,j}(I_z) \le t^+_{i,j}(I_x) + \overset{\wedge}{ND}\,(i,j) + \overset{\wedge}{LD}\,(i,j,i,j+1)$$
$$+ \overset{\wedge}{ND}\,(i,j+1) + \overset{\wedge}{LD}\,(i,j+1,i+1,j+1)$$
$$+ \overset{\wedge}{ND}\,(i+1,j+1) + \overset{\wedge}{LD}\,(i+1,j+1,i,j)$$
$$\le t^+_{i,j}(I_x) + 3\Delta$$

Hence maximum skew between $I_x(i,j)$ and $I_z(i,j)$ is 3Δ.

For the other cases, we observe the similarity by tracing signal path from the first arriving input to the last arriving input to the same node, which takes at most 3 hops. Figure 5 illustrates all six possible cases. Observe the worst case assumption, that the signal sourced at $N(i,j)$ arrives before other inputs to any node in the path, in every case. $\square$

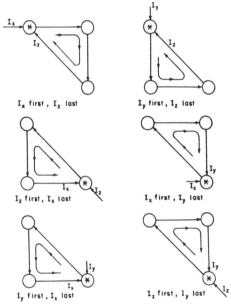

Figure 5: Illustration of proof to Theorem 2.

2.2.2 Model 2

We change two of Model 1 conditions to arrive at a slightly more realistic model. Model 2 differs from Model 1 in conditions 2 and 6. The new conditions are:

2'- The clock source S produces a single clock pulse at $t=0$, whose duration T_h (high portion time) is bounded as follows $3\Delta + \mu \le T_h \le 3\Delta + \mu'$ where $\mu' > \mu > 0$.

6'- Each node functions as follows:

a) When triggered by a $0 \to 1$ input transition, the node produces a pulse whose duration T_h is as defined in condition 2'.

b) any $0 \to 1$ input transition arriving within $\overset{\vee}{T}_h$ units of time after the node is triggered will be ignored. However any two input transitions separated by $\overset{\vee}{T}_h$

or more may trigger the node twice, thus producing two pulses (which is undesirable). It will definitely produce two pulses if inputs are separated by $\geq T_h^*$ (the actual duration of the first arriving input). Condition 6′ suggests a node whose structure is exemplified in Figures 6.

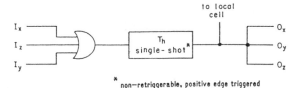

Figure 6: Assumed node structure for Model 2.

Thus Model 2 consists of conditions 1,2′,3,4,5, and 6′. We observe here, that (unlike Model 1) one could suspect the network to oscillate under this model.

Theorem 3: Under Model 2, the clock network will not oscillate and each node will produce exactly one pulse.

Proof Outline: Due to the lengthy nature of the proof and the limited space, we only provide a brief outline here. Interested readers can find the full proof in [14].

We use induction on network size to prove the theorem.

Induction hypothesis: If a network of size $i \times i$ does not oscillate and each of its nodes produces exactly one pulse, then a network of size $(i+1) \times (i+1)$ will not oscillate and each of its nodes will produce exactly one pulse.
Basis: 2×2 network.

We consider a 2×2 network with the only feedback connection snipped open at $I_z(1,1)$. We prove that this network will not oscillate and that every node will produce exactly one pulse. We then prove that by connecting $I_z(1,1)$, the network will still not oscillate and that every node will produce exactly one pulse. This is done by dividing time into three intervals $J_1: t \leq 0$, $J_2: 0 \leq t \leq t^*$, $J_3: t^* < t \leq \infty$, where t^* is the exact time at which the first $0 \rightarrow 1$ transition arrives at $I_z(1,1)$ while open. We show sequentially that connecting the feedback link during J_1, J_2 and J_3 will not alter network behavior.

Induction Step: Assume true for $i \times i$ and show true for $(i+1) \times (i+1)$.

To prove the above we disconnect all feedback links from row $(i+1)$ and column $(i+1)$ to the $i \times i$ network at the receiving nodes. We then prove that with these open connections, $(i+1) \times (i+1)$ network will not oscillate and that every node will produce exactly one pulse. We then iteratively connect the feedback links one after the other starting with $I_z(1,i)$ (and because of similarity $I_z(i,1)$). We show in every case, using a method similar to that in the 2×2 case, that network behavior will not change. This is repeated until all: $I_z(k,1)$ (and $I_z(i,k)$); $1 \leq k \leq i$ are connected. □

Lemma 2: The clock network described by Model 2 guarantees the maximum skew between any two communicating cells to be bounded above by 2Δ, where Δ is max-imum (node + link) delay, regardless of network size.

Proof: Since every clock node produces exactly one pulse, we can consider the skew to be the time difference between arrival times of $0 \rightarrow 1$ transitions at the two data cells in concern. The rest of the proof is hence identical to the proof of Theorem 1. □

Lemma 3: Under Model 2, maximum skew between any two inputs to the same node is bounded above by 3Δ, where Δ is maximum (node + link) delay.

Proof: The proof is identical to the proof of Theorem 2. □

2.2.3 Model 3

This final model must permit a free running clock, to be practical. We observe that under Model 2 when S produces a single pulse so will every node in the network. After the output pulse expires, a node outputs logic 0 permanently. If we wait until all nodes in the network become idle (all inputs and outputs at logic 0) and then apply a second pulse at S, the nodes will produce pulses (a second pulse each) and then turn idle in a manner similar to the response to the first pulse. However, to apply a second pulse, we do not have to wait until all nodes turn idle. All we have to consider is worst case timing at any node in the network, and determine a safe minimum separation time between pulses such that clocking events resulting from consecutive S pulses would not intersect. If the same node structure associated with Model 2 is assumed, we must allow the OR gate output to drop for at least one singleshot setup time before we apply the next clocking event to the node. Worst case timing for consecutive clocking of a node is illustrated in Figure 7.

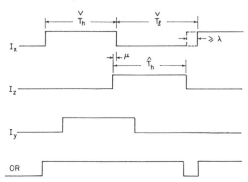

Figure 7: Worst case timing for consecutive clocking of a node.

We must also establish a lower bound on the duration of the low part of a pulse T_l. This is given by

$$T_l \geq \hat{T}_h + 3\Delta + \lambda - \check{T}_h = 3\Delta + \mu' - \mu + \lambda$$

To arrive at Model 3 we only have to modify conditions 2′ and 6′ to become:

2″. The clock source S produces a continuous stream of pulses starting at t=0. The timing of these pulses must adhere to the following bounds.

$$3\Delta + \mu \leq T_h \leq 3\Delta + \mu' \text{ and } T_l > 3\Delta + \mu' + \lambda - \mu$$

where $\mu' > \mu > 0$ and λ is the setup time for the singleshot.

6". Each node functions as follows: Whenever triggered, the node will produce an output pulse that adheres to the timing constraint defined in 2". All input $0 \rightarrow 1$ transitions separated in time by $< \overset{\vee}{T_h}$ will be ignored except the first. Any two input transitions separated in time by $\geq \overset{\wedge}{T_h}$ will trigger the node twice. We can hence assume the node structure in Figure 6.

Thus Model 3 consists of conditions 1,2",3,4,5, and 6". Observe the implied assumption that the minimum clock period $= 6\Delta + \mu' + \lambda$.

Because delays are assumed relatively time-invariant and because of conditions 2" and 6", clocking events will propagate through the network in a manner analogous to the propagation of electromagnetic waves and will not intersect. Thus we have.

Theorem 5: Under Model 3, clocking events will not intersect. $\square$

Due to Theorem 5, skew results for the previous models apply to this model as well. We thus have

Theorem 6: Under Model 3, the maximum clock skew between any two communication cells is bounded above by 2Δ, where Δ is the maximum (node + link) delay. $\square$

Theorem 7: Under Model 3, maximum skew between any two inputs to the same node is bounded above by 3Δ, where Δ is maximum (node + link) delay. $\square$

2.2.4 F=4 and F=2 Networks

Consider the (F=4) clock network shown in Figure 8. It can be observed that any pair of nodes associated with two communicating cells is on a cycle of length 2. Hence,

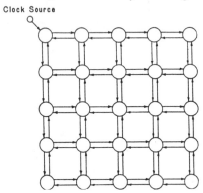

Figure 8: A valid clock network with F = 4.

the skew between the two inputs to a node is $\leq 2\Delta$. This can be verified by noticing that any node input can be reached from the other input in two hops. Also since any node can be a local source to a neighboring node with one hop delay, maximum skew between communicating cells is Δ. Model 3 can be modified to accommodate the F=4 network (assuming similar node structure) by specifying: $\mu + 2\Delta \leq T_h \leq 2\Delta + \mu'$, and $T_l > 2\Delta + \mu' + \lambda - \mu$. Hence minimum clock period $= 4\Delta + \mu' + \lambda$. Clearly this network allows for a faster clock and provides lower skew bound compared to the F=3 network, but requires 33%

additional connections.

The (F=2) network shown in Figure 9 is also valid and incurs less link cost than the two other networks considered so far. Every pair of nodes associated with two communicating cells is on a cycle of length 4. It can thus be easily verified that maximum skew between any pair of communicating cells is 3Δ and maximum skew between any two inputs to a node is 4Δ. It can similarly be verified that the minimum clock period in this case (assuming an adapted Model 3) is $8\Delta + \mu' + \lambda$ [14].

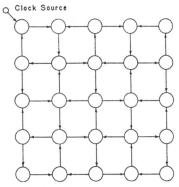

Figure 9: A valid clock network with F = 2.

From the above it is clear that a trade-off between hardware complexity and skew bound exists. For any of the three networks presented, one can state:

a) maximum skew between communicating cells = $(5-F)\Delta$; $2 \leq F \leq 4$,

b) maximum skew between any two node inputs = $(6-F)\Delta$; $2 \leq F \leq 4$.

2.2.5 Invalid Networks

As will be discussed in the next section not every network topology is valid for BaC clocking. For example the network is Figure 10 is invalid and will not bound the skew by a constant.

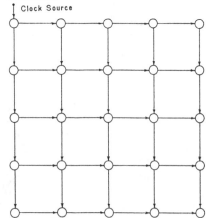

Figure 10: An invalid clock distribution network.

2.2.6 Simulation and Test Results

We have simulated the three valid networks presented above (F = 2,3,4) on Encore's Multimax using Turbo C. We have assumed a 500×500 network in each case. Node and link delays have been lumped together and assigned as (output) link delays. A random link delay ranging from 7 to 15 (corresponding to $\alpha - \beta$ and $\alpha + \beta$) units of time has been assigned to different links using a random number generator. (The values 7-15 and their units are actually irrelevant here). For each network we have checked for:

1- Lack of oscillation.

2- Maximum skew between node inputs.

3- Maximum skew between neighboring nodes

In each case simulation results completely matched theoretical expectations. We have also simulated the network in Figure 10 up to size 500×500 under similar assumptions. In that case skew grew without apparent limit as network size grew larger. Also a 4 x 4, F=3 network was built of SSI chips (Schottky TTL) using the node design shown in Figure 11. The network was tested successfully at 8 MHz for lack of oscillation. The waveforms at different node outputs were observed on an oscilloscope and found synchronized, clean and undistorted.

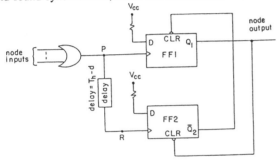

Figure 11: A more efficient node design.

3. Network and Node Design Issues

In this section we discuss some issues related to node and network design for BaC clocking. The new concept of fuse programmed clock networks (which is a spinoff of BaC clocking) is also discussed. Further, clocking data networks with different topologies including nonplanar structures is addressed.

3.1 Node Design

The design in Figure 6 is indeed not unique, and may not be the best. To exemplify we present two alternative node designs. All node designs described in this paper exhibit approximately two gate delays.

It is tempting to consider a purely combinational logic design such as that in Figure 12. The OR performs the combining function. The limiter logic is utilized to limit the length of output pulse width to T_h. In the figure we use an inverter and a cascade of buffers to invert signal P and provide the necessary delay ($= T_h$). However the design in Figure 12, requires a minimum clock cycle $> 3\,T_h$ because the signal P(R) could remain high (low) for approximately $2T_h$ which rules out another input pulse for at least that long ($2\,T_h$) after the earliest input drops. Therefore although this design is very simple, it is not as efficient as the one in Figure 6 in terms of clocking rate.

To avoid the problem pointed out above consider the

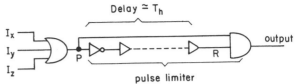

Figure 12: Purely combinational logic node design.

slightly more complex node design in Figure 11. For simplicity, assume flip-flop switching time = d. A $0 \rightarrow 1$ transition on P at t=0 sets FF1 at $t = d$. At $t = T_h{-}d$, FF2 is triggered. A $t = T_h$, Q_2 drops causing FF1 to clear at $T_h{+}d$. Thus the output Q_1 is high for T_h as desired. This design allows the low portion of the clock signal to be as short as that achievable with the circuit in Figure 6.

3.2 Fuse Programmed Networks

It is evident that combining inputs at a clock node necessitates imposing a lower bound on the low (0) portion of clock signals. ORing function at the input of a node has been assumed so far to implement combining, which necessiates a pulse limiter after the OR. This latter function can be eliminated if the order in which input signals arrive at the node is known. This requires 1) the order be time invariant; 2) logic at each node to detect the arrival order of inputs and; 3) all node inputs, except the first arriving, to be disconnected from the node by fusing them out.

The first requirement is practical to assume since variations in delays with time are usually small, and take place relatively slowly. Moreover, in VLSI/WSI delays tend to increase or decrease uniformly over the network. The third requirement is feasible given the ready existence of the technology such as in programmable logic arrays PLA's. This leaves the second requirement of input order detection to discuss.

On one side, the existence of such logic will eliminate the need for any processing by the node. Also the logic would only be used to program the network (fuse burning), and will not interfere with normal operation. On the other side, the cost of input order detection logic is a direct function of F(node degree). An RS latch can be utilized to decide on the relative arrival times of a pair of inputs. For $F > 2$ a tree like structure of latches may be needed. Another option is to use multiplexers to select inputs to a single RS latch at each node. Input connections may then be fused based on multiplexer select values as well as the state of the RS latch.

Once F-1 inputs are disconnected from the node, a single input would still be connected. An OR gate may be needed, however, due to the uncertainty at fabrication time as to which input would remain connected. Since after programming the connections, the potential for lengthening clock pulse duration no longer exists. Thus, in theory, the timing constraints can be dropped altogether. Further, with this method, the OR naturally serves the buffering function often employed in other clocking schemes.

3.3 Properties of BaC Networks

As we pointed out a above not every topology is valid. In [15], using graph models, the author has formally defined the necessary and sufficient conditions for a network to be valid. In this subsection, due to space limitations we summarize the main points of the theoretical work in [15].

(i) For a network topology to be valid for BaC clocking such that skew upper bound is constant, irrespective of network size it must contain cycles such that:

Each pair of communicating data processors must be clocked by [13], [15]:

    a) the same clock node; or

    b) a pair of clock nodes which are linked together in at least one cycle of finite length; or

    c) a pair of nodes to which there are finite paths from a common node or form two other nodes linked together in a finite cycle.

(ii) Timing constraints on clock signal (and accordingly on node function) are determied by the topology and delays of the chosen network. The maximum skew ϕ between any two inputs to the same node (over the entire network) determines the timing constraints as follows [13], [15]

$$\mu + \phi \leq T_h \leq \phi + \mu'; \text{ and } T_l \geq \phi + \mu' + \lambda - \mu$$

where μ, μ', and λ are as defined earlier.

(iii) Let k be the length of the longest minimum length cycle containing any pair of nodes satisfying (i)-c above. Also let l be the length of the longest minimum length path satisfying the same condition. Then the skew bound between any pair of communicating processors (cells) is [13], [15]

$$= (k-1)\Delta + 2l\beta$$

where β and Δ are as defined earlier.

(iv) Because of the cyclic nature of BaC networks, the clock source may be connected to any one of several nodes. In cases satisfying condition (i)-b above, (for every pair of directly communicating cells), the global source can be connected to any node without affecting network behavior. This is the case with the $F = 2,3,4$ networks presented earlier [13], [15]. We have tested that result with the 4×4, $F = 3$ network we built and found the above to be true. This may be an attractive feature from a fault-tolerance viewpoint, particularly in WSI applications.

(v) The clock network need not only contain nodes. Buffers can also be used. In such cases data processors can be clocked from nodes or buffers provided that condition (i)-c above is satisfied [13], [15].

Figure (13) shows a valid BaC network containing cycles of different forms and spacings. Figure (14) shows the utilization of BaC clocking in buffered networks. This latter design may prove to be of practical significance due to its relative low cost.

3.4 Nonplanar BaC Networks

BaC clocking can be utilized to control data processing networks of nonplanar topologies such as hypercubes and m-D meshes. One straightforward way is to assign a

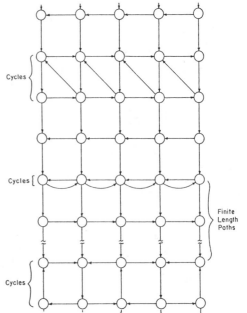

Figure 13: A valid Bac network with different cycles.

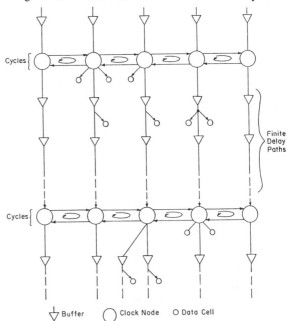

Figure 14: Valid BaC network with buffers.

clock node to each processor and connect the nodes in the same way data processors are connected. Alternatively, the node can be placed within the processor and two extra wires are added to the data links to carry clock signals back and forth. This method may be relatively costly but guarantees skew upper bound to be $\leq \Delta$. We now provide an example showing the applicability of BaC clocking to hypercube processor networks with selectable cost/performance criterion. The example also shows clocking several processors from a single clock node safely.

3.4.1 A Hypercube Example

An n-dimensional hypercube consists of 2^n cells labeled 0 through 2^{n-1}. Two cells are directly connected iff their labels differ in only one bit. An n-dimensional hypercube can be partitioned into 2^{n-k} subcubes of dimension k each; $k \leq n$. In such a case each subcube would contain 2^k cells. One method of partitioning is to consider the binary address of a cell, $A = a_{n-1}a_{n-2} \cdots a_k a_{k-1} \cdots a_1 a_0$ as having two parts. Bits 0 through a_{k-1} represent the in-subcube label, whereas bits k through $n-1$ represent the subcube label.

For this example we assign a single clock node to each subcube, and give it the same label as the associated subcube. Thus we employ 2^{n-k} clock nodes labeled from 0 to $2^{n-k}-1$. Two clock nodes are connected together in a 2-cycle (i.e. there is a link form each node to the other) if and only if their labels differ in only one bit. The clock node associated with a subcube is responsible for clocking all the cells in that subcube. The network containing the clock nodes is itself an $(n-k)$ hypercube. Figure 15 illustrates a 4-cube partitioned into 4 subcubes of dimension 2, and the connections among the clock nodes. The figure also illustrates clock distribution within subcube 10, which contains nodes 1000, 1001, 1010, 1011.

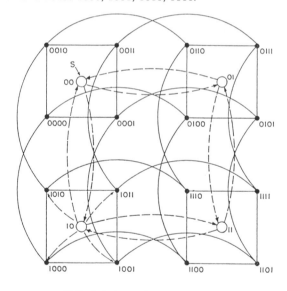

○ Clock Node ● Data Cell

Figure 15: BaC clocking of hypercube networks.

The six conditions of Model 3 (as applied to the $F=4$ network) are imposed on this network. In addition, we assume that the delay from a clock node to a cell in its associated subcube is $v \pm \eta$. It can be easily verified that the upper skew bound is $\Delta \pm 2\eta$ in this case. Note that η depends on k implicitly.

4. Conclusion

We have presented a novel clocking scheme which enables global synchronization of large computing structures with unlimited size such that skew is bounded by a constant. We have shown theoretically that, if clock signals follow a set of simple constraints, the proposed clock networks will not oscillate despite the cyclic nature of the networks. Three "valid" clock networks for 2-D mesh-based topologies have been proposed. Implementation issues such as network and node design, and fuse programming BaC-type networks have been addressed.

References

[1] C.L. Seitz, "System Timings," in C. Mead and L. Conway, Introduction to VLSI Systems, Ch. 7, **Addison Wesley**, Reading, MA, 1980.

[2] S-Y. Kung, "On supercomputing with systolic/wavefront arrays," **Proc. IEEE**, Vol. 72, No. 7, July 1984.

[3] D. Wann and M. Franklin, "Asynchronous and clocked control structures for VLSI based interconnection networks," **IEEE Trans. Comput.**, Vol. C-32, No. 3, March 1983, pp. 284-293.

[4] S-Y. and R.J. Gal-Ezer, "Synchronous vs. asynchronous computation in VLSI array processors, in **Proc. SPIE Conf.**, Arlington, VA, 1982.

[5] A. Fisher and H.T. Kung, "Synchronizing large VLSI processor arrays," **Proc. 10th Annual Intl. Symp. Comp. Archit.**, pp. 54-58, 1983. Also in the **IEEE Trans. Comp.**, Vol. C-34, No. 8, Aug. 1985, pp. 734-740.

[6] C.L. Seitz, "Concurrent VLSI architectures, **IEEE Trans., Comp.** vol., C-33, No. 12, Dec. 1984, pp. 1247-1263.

[7] C.L. Seitz, "Self-timed VLSI systems," **Proc. Caltech Conf. on VLSI**, Jan. 1979, pp 345-355.

[8] S.Y. Kung, K.S. Arun, R.J. Gal-Ezer, and D.V. Bhaskar Rao, "Wavefront array processor: language, architecture and application," **IEEE Trans. Comput.**, Vol. C-31, No. 11, Nov. 1982, pp. 1054-1066.

[9] S.Y. Kung, S.C. Lo, S.N. Jean and J.N. Hwang, "Wavefront array processors- concept to implementation," **Computer**, Vol. 20, No. 7, July 1987, pp. 18-33.

[10] J.P. Fishburn, "Clock Skew Optimization," **IEEE Trans. Comput.**, Vol. C-39, No. 7, July 1990, pp. 945-951.

[11] P.M. Kogge, "The Architecture of Pipelined Computers, **McGraw Hill**, New York, 1991, pp. 21-39.

[12] M. Shoji, "Elimination of process-dependent Clock Skew in CMOS VLSI," **IEEE J. Solid-State Circuits**, Vol. SC-21, No. 5, Oct. 1988, pp. 875-880.

[13] A. El-Amawy, "Arbitrarily large clock networks with constant skew bound," **U.S. Patent**, Pending.

[14] A. El-Amawy, "Clocking arbitrarily large computing structures under constant skew bound," **ECE Dept., Louisiana State University, Tech. Report #09-90C-AE**, Sept. 1990.

[15] A. El-Amawy, "Properties of Branch-and-Combine Clock Networks," **ECE Dept., Louisiana State University, Tech. Report #01-91C-AE**, Jan. 1991.

EMULATION OF A PRAM
ON LEVELED NETWORKS

Michael A. Palis and *Sanguthevar Rajasekaran*
Department of Computer and Information Science
University of Pennsylvania
Philadelphia, PA 19104-6389

David S. L. Wei
Department of Computer Science
Radford University
Radford, VA 24142

Abstract: We present an optimal algorithm for emulating a CRCW PRAM on leveled networks. As corollaries, we show that one step of a CRCW PRAM can be emulated optimally by the star graph and the n-way shuffle network, which have sub-logarithmic diameters. We also present a $4n + o(n)$ step emulation algorithm for an $n \times n$ mesh, which is faster than the emulation implied by Ranade's algorithm.

1 Introduction

We consider the problem of emulating a PRAM on a processor interconnection network (ICN). In an ICN, each processor has its own private memory and can communicate with other processors only by explicit packet routing. We identify a class of *leveled networks* which can be used to model a variety of ICN's including the n-cube, n-way shuffle [13], star graph [1], and the mesh. We show that one step of a concurrent read, concurrent write (CRCW) PRAM can be emulated in $O(\ell)$ steps on a leveled network with diameter ℓ, which is optimal. As corollaries, we obtain optimal emulations of the CRCW PRAM on the n-way shuffle and star graph, which have sub-logarithmic diameter. Our result generalizes Ranade's emulation algorithm [9], who showed that one step of a CRCW PRAM can be emulated in $O(\log N)$ time on an N-node butterfly.

We also present an efficient emulation of the CRCW PRAM on a two-dimensional mesh. Although Ranade's emulation technique can be applied to an $n \times n$ mesh, the underlying constant in the $O(n)$ time bound is roughly 100, which makes it impractical. In this paper we provide a faster emulation algorithm whose time bound is only $4n + o(n)$. This algorithm also has some nice 'locality' properties. In particular, if each request for memory access originates within a distance d of the location of the memory, then the algorithm terminates in $6d + o(d)$ steps. Moreover, each link in the network only requires a queue size of $O(1)$.

2 Leveled Networks

An (N, ℓ) *leveled network* consists of $\ell + 1$ groups of nodes such that each group has N nodes and these groups form a sequence of $\ell+1$ columns, say $c_1, c_2, \ldots, c_{\ell+1}$. Column c_1 and column $c_{\ell+1}$ are identified; thus, although there are $\ell+1$ columns of N nodes each, the total number of nodes is ℓN. The only links in the network are between nodes in c_i and nodes in either c_{i-1} or c_{i+1} (provided these columns exist). Every node in each column has at most d incoming and outgoing links where d is the degree of the network. For each node in the first column, there exists a unique path of length ℓ connecting it to any node in the last column. Clearly, the diameter of the network is ℓ.

A leveled network is called **nonrepeating** if it satisfies the following property: if any two distinct paths from the first column to the last column share some links and then diverge, these two paths will never share a link again.

Leveled networks are interesting because the problem of packet routing in various (single-stage) ICNs (such as the n-cube) can be reduced to an equivalent packet routing problem on a leveled network. Given an N-node single-stage network $\mathcal{N}$ with diameter D, the first step is to select a path of length at most D for every $\langle source, destination \rangle$ pair. The collection $\mathcal{C}$ of all such paths is then represented as an (N, D) leveled network whose links are defined as follows: (1) there is a link from node u of column c_i to node v of column c_{i+1} if and only if there is some path $p \in C$ whose i-th edge connects nodes u and v in $\mathcal{N}$; (2) for every node u, there is a link from node u in column c_i to node u in column c_{i+1}, $1 \leq i \leq D$. The set of links defined by (2) take care of paths in $\mathcal{C}$ which are less than D in length. For such a path p, the corresponding path in the leveled network will follow the same sequence of nodes (in increasing columns), and then extended to the last column by following the links in (2).

The n-cube, n-way shuffle [13], star graph [1], mesh, and a host of other single-stage networks can all be represented as nonrepeating leveled networks. For instance, the leveled network representation of the n-cube is the butterfly, which is easily seen to be nonrepeating. For lack of space, we refer the reader to [14] for the leveled network representations of the n-way shuffle, star graph, and other single-stage networks.

3 Routing on a Leveled Network

Suppose we are given a leveled network and a set of packets, each of which specifies a $\langle source, destination \rangle$ pair. Initially, the packets are placed in their *source* processors in the first column of the leveled network. We consider the problem of routing the packets through the network such that: (1) at most one packet passes through any link of the network at any time, and (2) at the end of the routing, the packets are in their destination processors in the first column (or the last column, since these two columns coincide). A paradigmatic case of general routing is **permutation routing** in which initially there is exactly one packet in each node of the first column and the destinations form some permutation of the sources. A more general case is **partial h-relations routing** in which there are at most h packets in every node and there are no more than h packets with the same destination. In this short paper, we focus on permutation routing; however, the results extend to other routing problems as well.

The routing algorithm is an adoption of the two-phase randomized routing algorithm invented by Valiant [13]. During phase 1, each packet from the first column of the leveled network is sent to a random node in the last column by traversing a random link at each level. During phase 2, each packet is sent from the random node to its true destination along the unique path. The queueing discipline is first-in first out (FIFO): packets contending for the same outgoing link are queued up, and the packet which is in the queue the longest time is sent out.

The following theorem shows that with high probability (abbreviated as w.h.p. hereafter) the algorithm routes all packets in time linear in the network diameter. (By 'high probability' we mean a probability of $\geq (1 - N^{-\alpha})$, for any constant $\alpha \geq 1$, N being the number of nodes in a single column of the leveled network.)

Theorem 3.1 For an (N, ℓ) nonrepeating leveled network, any permutation routing of N packets can be completed in $O(\ell)$ steps w.h.p., provided that $d \geq 2$ (where d is the degree of the network) and $\ell = \Omega(\log N/(\log \log N))$. For each link, the queue size is $O(\ell)$ w.h.p.

Proof Sketch: Let π be an arbitrary packet in the network. We compute an upper bound on the delay this packet suffers as follows. Let δ_i stand for the number of packets that meet π's path for the first time at level i (for $1 \leq i \leq \ell$). Since the leveled network is nonrepeating, we can use the *queue line lemma* [13] to infer that an upper bound for the delay π suffers is $\sum_{i=1}^{\ell} \delta_i$.

The generating function for Prob.$[\delta_i = k]$ is calculated as $G_i(x) = e^{x/d^2}$, d being the degree of the network. The generating function (call it $G(x)$) for $\sum_{i=1}^{\ell} \delta_i$ is the product of the G_i's. $G(x)$ simplifies to $e^{x\ell/d^2}$. This immediately implies that

$$Prob.[\sum_{i=1}^{\ell} \delta_i \geq q] \leq \sum_{x=q}^{\infty} \left(\frac{\ell}{d^2}\right)^x \frac{1}{x!}.$$

If q is chosen to be $c\alpha\ell$ (for some appropriate constant c), the above probability can be shown to be no more than $N^{-\alpha}$ $\square$.

Note: The above proof assumes that there is only one packet to start with at any node in the first column of the leveled network. A similar proof can be given to show that even if there are ℓ packets to start with in every node the above algorithm still terminates in $O(\ell)$ steps w.h.p. (For details see [14]).

4 PRAM Emulation on any ICN

We consider the problem of emulating a PRAM with N processors and shared memory of size M on an N-node ICN. For simplicity, we assume that the PRAM is exclusive read, exclusive write (EREW); the emulation result can be extended to the more general concurrent read, concurrent write (CRCW) PRAM using 'message combining' (see [14]).

Our emulation algorithm is based on Karlin and Upfal's technique called *parallel hashing* [3]. The idea is to map the M shared memory cells of the PRAM onto the local memory modules of the N processors of the ICN. The mapping is obtained by randomly choosing a hash function h from the following class of hash functions:

$$H = \{h | h(x) = ((\sum_{0 \leq i < \rho} a_i x^i) \bmod P) \bmod N\}$$

where P is a prime, $P \geq M$, $a_i \in Z_P$, and ρ is a positive constant.

The above class of hash functions has the following interesting property:

Fact 4.1 [3] If N items are mapped into $N/2^i$ buckets using a random hash function (from the class defined before), the maximum number (call it Y_i) of items mapped into a single bucket satisfies:

$$Prob.[Y_i \geq j] \leq N \left(\frac{2^i}{j - \rho}\right)^{\rho}.$$

Consider the case of distributing N packets among N processors using the above hashing scheme. Fact 4.1 implies that if ρ is chosen to be some constant multiple of $\log N/(\log \log N)$, then each processor will get $O(\rho)$ packets.

Given the above address mapping, the memory requests (read or write) of the PRAM processors can be simulated on the ICN as follows. Suppose that PRAM processor i wants to access shared memory location j. On the ICN, this step is accomplished in two phases: (1) processor i sends a packet (encoding the request) to processor $h(j)$; and (2) if the packet was a 'read' request, processor $h(j)$ sends the contents of memory cell j back to processor i. Each of these two phases corresponds to a routing task. In the first phase, there is at the most one packet originating from any node and there are $O(\log N/(\log \log N))$ packets destined for any node w.h.p. Whereas in the second phase there are $O(\log N/(\log \log N))$ packets starting from any node and at the most one packet destined for any node.

5 PRAM Emulation on a Leveled Network

We make use of the address mapping given in the previous section. A single step of the PRAM is simulated as follows. Let processor i want to access memory cell j. On the leveled network: 1) processor i in the first column sends a packet (encoding the request) to processor $h(j)$ in the last column (which, recall, coincides with the first column); 2) if the request was a 'read', $h(j)$ sends the contents of cell j back to processor i.

From Theorem 3.1 and Fact 4.1, we can prove the following:

Theorem 5.1 Each step of an EREW PRAM with N processors and shared memory of size M can be emulated by an (N, ℓ) leveled network with degree d, in $O(\ell)$ steps w.h.p.

6 Emulating a PRAM on the Mesh

Mesh Connected Computers (MCCs) are increasingly being accepted as a feasible model for building machines for many reasons including their simple interconnection, linear scalability etc. A MCC is nothing but an $n \times n$ square grid in which each grid point corresponds to a processing element and each edge corresponds to a communication link. We assume that in a single step each processor can perform a local computation (like a comparison) and also communicate with all its 4 (or less) neighbors. This version of the MCC is called MIMD and has been assumed in previous works (see e.g.,[4, 5, 6, 8, 13]).

Although an asymptotically optimal algorithm for emulating a PRAM on the mesh is implied by Ranade's algorithm [9], the underlying constant in the time bound is very high (> 100). In this section, we present an algorithm for emulating a single step of an EREW PRAM on an $n \times n$ mesh in time $4n + o(n)$. Moreover, the algorithm has some nice 'locality' properties: if each memory request originates within a distance d of the location of the memory, the algorithm terminates in time $6d + o(d)$ steps with high probability. Our algorithm needs a queue size of only $O(1)$ with high probability.

As before, the emulation algorithm consists of mapping the shared memory locations of the PRAM onto the n^2 memory modules of the mesh using a random hash function (see Section 4). The following fact will be used in our analysis.

Fact 6.1 (1) If N items are mapped into N buckets, then w.h.p. each bucket will get $O\left(\frac{\log N}{\log \log N}\right)$ items.

(2) If $N = n^2$ items are mapped into βn buckets, the maximum number of items mapped into any bucket will be $\frac{n}{\beta} + O(n^{3/4})$ w.h.p.

(3) If N items are mapped into N buckets, and if S is a collection of $\log N$ buckets, the number of items mapped into S will be $O(\log N)$ w.h.p.

6.1 The Routing Algorithm

As explained in Section 4, after the address mapping, the emulation problem reduces to two phases of routing (one phase to route the packets to their destinations, and another phase, to send back the contents of memory locations requested by 'read' packets). A large number of articles have been written on the problem of solving permutation routing on the MCC (see e.g., [4, 5, 6, 8, 13]).

In both the phases we make use of the same routing algorithm. The algorithm is the same as the one given in [4], with a different analysis. Each phase will be finished in $2n + o(n)$ steps w.h.p.

We first present a routing algorithm with the promised time bound but which needs a queue size of $O(\log N)$. Later we will describe how to bring down the queue size to $O(1)$. Partition the $n \times n$ mesh into horizontal slices with ϵn rows in each slice (for some ϵ to be fixed later). There are three stages in the algorithm. Contention for edges are resolved by 'furthest destination first' queueing discipline. Let π be a packet that originates in node (i, j) (i.e., in row i and column j) and whose destination is (k, l).

stage1

π chooses a random node (call it (i', j)) in the same column and slice of its origin and traverses to that node along column j.

stage2

π traverses along row i' to the node (i', l).

stage3

Finally the packet π traverses along column l to the node (k, l).

6.2 Analysis of the Routing Algorithm

Consider the following routing problem on a linear array of size m. There are k_i packets to start with at node i (for $1 \le i \le m$) such that $\sum_{i=1}^{m} k_i = m'$. Each packet chooses a random node in the linear array as its destination. How fast can this routing be performed (assuming the 'furthest destination first' priority scheme)?

The answer is $m' + o(m)$ for the following reason. Let i be the origin of a packet π and let j be its destination. W.l.o.g. assume j is to the right of i. Since all the links are bidirectional, for the worst case analysis we can assume all the packets are traversing from left to right. The number of packets that will have a higher priority than π is given by the binomial $B(m', \frac{m-i}{m})$. Using Chernoff bounds, this number is no more than $\frac{(m-j)m'}{m} +$

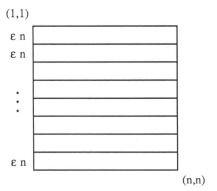

(1,1)

ϵn

ϵn

$\vdots$

ϵn

(n,n)

Partitioning of the Mesh

$o(m)$ w.h.p. Applying the queue line lemma [13], the time needed for π to reach its destination is no more than $(j - i) + \frac{(m-j)m'}{m} + o(m)$ w.h.p. In the worst case this time bound is $n' + o(n)$.

If we apply the above fact to the first stage of our routing algorithm we see that the time bound for stage 1 is no more than $\epsilon n + o(n)$. (Realize that the number of packets originating from any column slice is no more than $\epsilon n + o(n)$ w.h.p. (see Fact 6.1-(2)). If we fix ϵ to be $1/(\log n)$, the time needed for stage 1 is $o(n)$ w.h.p.

In the second stage of the routing algorithm, consider any row j. How many packets will there be in row j at the beginning of stage 2? Using Fact 6.1-(2) and Chernoff bounds one can readily see that this number is no more than $n + o(n)$ w.h.p. Given this fact, we can use arguments similar to the one given for linear array to prove that both the second and the third phases will be completed in $n + o(n)$ each w.h.p.

Using Fact 6.1-(1), we can also prove a queue size of $O(\log n)$. This proves the following:

Theorem 6.1 *The routing algorithm described terminates in $2n + o(n)$ steps w.h.p. The queue size is $O(\log n)$.*

The above theorem together with the emulation algorithm described before will yield the following:

Theorem 6.2 *Each instruction of an EREW PRAM can be emulated on the MCC in $4n + o(n)$ steps w.h.p. The queue size of the processors is $O(\log n)$.*

We can reduce the queue size of the above algorithm to $O(1)$ making use of Fact 6.1-(3). The improvement will parallel the $2n + O(\log n)$ time routing time algorithm presented in [4], with a slightly different analysis. In similar lines we can also prove the following:

Theorem 6.3 *If each memory request originates within a distance of d of the location of the memory, the above emulation algorithm terminates in $6d + o(d)$ steps w.h.p.*

7 Conclusions

In this paper we have presented optimal algorithms for emulating a CRCW PRAM on leveled networks, which model a variety of (single-stage) interconnection networks, including the n-way shuffle and the star graph. We also presented a $4n + o(n)$ step emulation algorithm for an $n \times n$ mesh, which is faster than the emulation implied by Ranade's algorithm.

References

[1] Akers, S., Harel, D. and Krishnamurthy, B., 'The Star Graph: An Alttractive Alternative to the n-Cube,' Proc. 1987 International Conference of Parallel Processing, pp. 393-400.

[2] Alt, H., Hagerup, T., Mehlhorn, K., and Preparata, F., 'Deterministic Simulation of Idealized Parallel Computers on More Realistic Ones,' SIAM Journal on Computing, vol. 16(5), 1987, pp. 808-835.

[3] Karlin, A., and Upfal, E., 'Parallel Hashing–An Efficient Implementation of Shared Memory,' Proc. ACM Symposium on Theory Of Computing, 1986, pp. 160-168.

[4] Krizanc, D., Rajasekaran, S., and Tsantilas, T., 'Optimal Routing Algorithms for Mesh-Connected Processor Arrays,' Proc. AWOC 1988, Springer-Verlag Lecture Notes in Computer Science 319, pp. 411-422. Also submitted to Algorithmica, 1989.

[5] Kunde, M., 'Routing and Sorting on Mesh-Connected Arrays,' VLSI Algorithms and Architectures: Proc. AWOC 1988, Springer-Verlag Lecture Notes in Computer Science 319, pp. 423-433.

[6] Leighton, T., Makedon, F., and Tollis, I.G., 'A $2n - 2$ Step Algorithm for Routing in an $n \times n$ Array With Constant Size Queues,' Proc. 1989 ACM Symposium on Parallel Algorithms and Architectures, pp. 328-335.

[7] Palis, M., Rajasekaran, S., and Wei, D., 'General Routing Algorithms for Star graphs,' in Proc. Fourth International Parallel Processing Symposium, April, 1990, pp. 597-611.

[8] Rajasekaran, S., and Overholt, R., 'Constant Queue Routing on a Mesh,' Proc. Symposium on Theoretical Aspects of Computer Science, Hamburg, Germany, Feb. 1991.

[9] Ranade, A.G., 'How to Emulate Shared Memory,' Proc. IEEE Symposium on Foundations Of Computer Science, 1987, pp. 185-194.

[10] Snyder, L., 'Type architectures, shared memory, and the corollary of modest potential,' Annu. Rev. Comput. Sci. 1, 1986, pp. 289-317.

[11] Upfal, E., and Wigderson, A., 'How to Share Memory in a Distributed System,' IEEE Symposium on Foundations Of Computer Science, 1984, pp. 171-180.

[12] Valiant, L.G., 'A bridging model for Parallel Computation,' CACM, August 1990, Vol.33, No 8, pp. 103-111.

[13] Valiant, L.G., and Brebner, G.J., 'Universal Schemes for Parallel Communication,' Proc. ACM Symposium on Theory Of Computing, 1981, pp. 263-277.

[14] Wei, D.S.L., 'Fast Parallel Routing and Computation on Interconnection Networks,' Ph.D. Thesis, Univ. of Pennsylvania, Jan. 1991.

OPTIMAL NONBLOCKING NETWORKS WITH PIN CONSTRAINTS [*]

Minze V. Chien and A. Yavuz Oruç

Electrical Engineering Department and Institute of Advanced Computer Studies

University of Maryland, College Park, MD 20740

ABSTRACT

This paper considers techniques to partition nonblocking networks into identical modules of switches in such a way that each module is contained in a single chip and that the cost of the overall network is kept as small as possible. It is shown that the well-known rearrangeable Clos networks can be partitioned into identical modules with an arbitrarily specified number of inputs such that the cost of the overall network is asymptotically minimum. An extension of this result is also given for nonblocking Clos networks.

1 Introduction

In many parallel computer systems connection networks are used to provide communication paths among processors [8,12]. Earlier connection networks were designed and optimized primarily based on cost, depth, and set-up (or routing) time. For example, the Benes network provides an asymptotically minimum cost solution to the permutation problem with a logarithmic depth [4]. Batcher's odd-even merge, and bitonic sorting networks provide near minimum cost solutions to the same problem with a self-routing mechanism and logarithmic delay if pipelining is used [3].

While cost, delay and set-up time are critical factors in designing connection networks, another key factor that must be considered is the modularity of the design. Current VLSI technologies can accomodate a large number of switches in a single chip, but the mere fact that one chip cannot have too many pins precludes the possibility of implementing a large connection network on a single chip. It was noted in [9] that the pin limitations rather than chip area are what constraint the design of large networks. For example, a single chip implementation of a connection network for a parallel computer with 1024 processors will need at least 1024 pins even if the inputs and outputs are multiplexed, and data are processed in a bit-sliced fashion. For larger size networks the problem gets much worse, and one must resort to partitioning the problem into smaller sizes.

In this paper, we propose a partitioning approach which insures layout uniformity among the modules and yet avoids wasting any chip space. Our approach is based on a simple idea that the switches in a connection network can be adjusted to any size without increasing the overall cost of the network. In earlier designs reported in the literature, minimum cost connection networks could only be obtained by using constant size switches [6,10,12,14]. More specifically, one obtains such a network by starting out with a Clos network whose outer stages contain 2×2 switches, and recursively decomposing the middle stage switches. This makes it difficult to layout large size Clos networks without wast-

ing chip space. We propose two schemes which preserve the minimum cost attribute of these networks at any switch size. The first scheme relies on the fact that when the switches of a Clos network are replaced by Benes networks then the resulting network also has minimum cost regardless of the switch sizes used. In particular, this approach yields a mininum cost n-input network encompassing $\sqrt{n} \times \sqrt{n}$ switches. This provides a good solution for moderately large values of n, for example, when $n = 10000$, each switch can be laid out in a chip with about 200 pins without multiplexing inputs and outputs. For larger values of n, we introduce a second approach which also results in minimum cost networks encompassing identical size switches. This approach relies on the same principle but with recursive techniques added. Basically, we decompose the switches until they all have the same size, and then replace them with Benes networks. In addition to being minimum cost, all of our designs yield logarithmic or polylogarithmic depth as well.

2 Clos Networks

A Clos network [7] is a cascade of three stages of crossbar switches (or other smaller size Clos networks) and is typically defined with three parameters, n, m, and q, where n is the number of inputs to the network, m is the number of inputs (outputs), and q is the number of outputs (inputs) of each switch in the first (third) stage. A graphical representation of such a network is depicted in Figure 1. It is well-known that the Clos network is rearrangeable when $q \geq m$, and strictly nonblocking when $q \geq 2m - 1$.

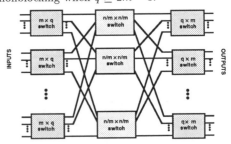

Figure 1: A 3-stage Clos network.

Given the recursive structure of Clos networks, we shall express the cost and depth of a Clos network in terms of recurrences. The cost and depth of a Clos network with a constant number of inputs will both be assumed to be a constant. In particular, it will be assumed that a 2×2 switch has unit cost and unit depth. We shall denote the cost and depth of an n-input Clos network with parameters m and q, by $C(n; m; q)$ and $D(n; m; q)$, respectively. The cost and depth of a switch with m inputs (outputs) and q outputs (inputs) will be denoted by $C(m, q)$ and $D(m, q)$ in that order. We then have directly from Figure 1,

$$C(n; m; q) = qC(\frac{n}{m}, \frac{n}{m}) + 2\frac{n}{m}C(m, q), \qquad (1)$$

[*]This work is supported in part by the National Science Foundation under Grant No: CCR-8708864, and in part by the Minta Martin Fund of the School of Engineering at the University of Maryland .

All logarithms in this paper are in base 2, and $\lg n$ denotes $\log_2 n$.

$$D(n;m;q) = D(\frac{n}{m},\frac{n}{m}) + 2D(m,q). \tag{2}$$

For rearrangeable Clos networks, $m = q$ so that these equations reduce to

$$C(n;m) = mC(\frac{n}{m}) + 2\frac{n}{m}C(m), \tag{3}$$

$$D(n;m) = D(\frac{n}{m}) + 2D(m) \tag{4}$$

where $C(m) = C(m,m), C(\frac{n}{m}) = C(\frac{n}{m},\frac{n}{m})$ and $D(\frac{n}{m}) = D(\frac{n}{m},\frac{n}{m}), D(m) = D(m,m)$.
For nonblocking Clos networks, $q = 2m - 1$ so that

$$C(n;m) = (2m-1)C(\frac{n}{m}) + 2\frac{n}{m}C(m,2m-1), \tag{5}$$

$$D(n;m) = D(\frac{n}{m}) + 2D(m,2m-1). \tag{6}$$

For rearrangeable networks, one can minimize Equation 3 to obtain a network with $O(n\lg n)$ cost and $O(\lg n)$ depth [4]. For nonblocking networks, Equation 5 can be used to obtain a network with $O(n^{1+1/\lg^{\frac{1}{2}}n})$ cost and $O(n^{1/\lg^{\frac{1}{2}}n})$ depth [6]. Because of the reasons mentioned in the previous section, these networks may not result in the best VLSI implementation of a Clos network. In fact, it was shown in [11] that a single chip implementation of a nonblocking Clos network takes as much, or even more space for some Clos networks than a single crossbar. In the following sections we present partitioning techniques that preserve the minimum cost feature of these networks.

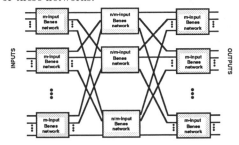

Figure 2: A 3-stage Clos network with minimum cost switches.

3 First Approach

Our first approach relies on replacing the switches in a rearrangeable Clos network by minimum-cost networks such as those given by Benes [4] and Waksman [13], as shown in Figure 2. This replacement obviously leaves the rearrangeability of the original network intact. Let $C_B(n;m)$ and $D_B(n;m)$ denote the cost and depth of this network. An m-input Benes network uses $-\frac{m}{2} + m\lg m$ 2×2 switches, and has $-1 + 2\lg m$ depth. Likewise, an $\frac{n}{m}$-input Benes network uses $-\frac{n}{2m} + \frac{n}{m}\lg\frac{n}{m}$ 2×2 switches, and has $-1 + 2\lg\frac{n}{m}$ depth. Substituting these in Equations 3 and 4, we have

$$C_B(n;m) = -\frac{3n}{2} + n\lg\frac{n}{m} + 2n\lg m, \tag{7}$$

$$D_B(n;m) = -3 + 2\lg\frac{n}{m} + 4\lg m. \tag{8}$$

It is easily verified that, for all $m, 1 \le m \le n$, $C_B(n;m) = O(n\lg n)$, and $D_B(n;m) = O(\lg n)$. Therefore,

Proposition 1: For all $m, 1 \le m \le n$, the network in Figure 2 has asymptotically minimum cost and logarithmic

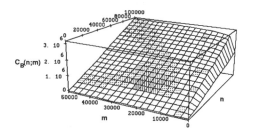

Figure 3: The cost of the network using Approach 1 as a function of n and m; $1 \le n \le 10^5, 1 \le m \le 5\cdot 10^4$.

depth. ||

Figure 3 depicts a 3-D plot of $C_B(n;m)$ for $1 \le n \le 10^5$ and $1 \le m \le 5\cdot 10^4$ obtained by using Mathematica. The figure illustrates that the cost increases with increasing m, but it is asymptotically minimum over all values of m; $1 \le m \le 5\cdot 10^4$. In particular, for $n = 10^5$, $n\lg n \simeq 1.6\cdot 10^6$ and $C_B(n;m)$ falls between $2\cdot 10^6$ and $3\cdot 10^6$ for, $n = 10^5$, and all $m, 1 \le m \le 50000$.

This result ensures an asymptotically minimum cost, but it requires two types of switches; those with m inputs, and those with $\frac{n}{m}$ inputs. To obtain a network encompassing a single type of switch, we let $m = \frac{n}{m}$. This results in a network consisting of only $\sqrt{n}\times\sqrt{n}$ switches whose cost and depth are given by

$$C_B(n;\sqrt{n}) = \frac{3n}{2}(\lg n - 1), \tag{9}$$

$$D_B(n;\sqrt{n}) = -3 + 3\lg n. \tag{10}$$

As expected, $C_B(n;\sqrt{n})$ is asymptotically minimum, and $D_B(n;\sqrt{n})$ is logarithmic in n. In particular, for $n = 10^5$, this design yields a network with about $15\cdot 10^5$ 2×2 switches. In terms of Benes networks, this design uses $3\sqrt{n} = 3\cdot\sqrt{10^5} \simeq 960$ such networks, each with $\sqrt{n} = \sqrt{10^5} \simeq 320$ inputs and consisting of about $\sqrt{10^5}\lg\sqrt{10^5} - \sqrt{10^5}/2 \simeq 2400$ 2×2 switches. In VLSI, this amounts to a chip with about 320 pins and about 24000 transistors assuming that inputs and outputs are multiplexed and a 2×2 switch is realized by about 10 transistors.

4 Second Approach

Even though the partitioning scheme described in the previous section results in minimum cost designs with a single type of switch, it cannot be used for large n as the number of inputs to the switches will become prohibitive; for $n = 10^6$, the switches will have 1000 inputs and outputs, and surely no current technology can produce a chip with 1000 pins. Even for, $n = 10^5$, the preceding computation shows that a chip with 300 pins will be needed.

We can modify our first approach to obtain networks with any switch size m which satisfy the identity $m = n^{\frac{1}{r+1}}$ where r is a nonnegative integer. Without loss of generality, suppose that $\frac{n}{m} \ge m$ in Equation 3. That is, the switches in the center stage are larger in size than the switches in the outer stages. We can then decompose the switches in the center stage iteratively until the switches at the bottom most level of the iteration have m inputs. This will ensure that all switches have the same number of inputs. In algebraic terms, this amounts to recursively decomposing the first terms in Equations 3 and 4 by substituting $\frac{n}{m}$ for n until their first terms reduce to $C(m)$ and $D(m)$, respectively. Suppose that this happens after r iterations. Then it can be shown that

$$C(n;m) = m^r C(\frac{n}{m^r}) + 2\frac{rn}{m} C(m). \qquad (11)$$

Now assuming that the arguments of both recurring terms are equal, i.e., $\frac{n}{m^r} = m$, we find $r = \frac{\lg n}{\lg m} - 1$, and so the above equation becomes

$$C(n;m) = (2\frac{n}{m}\frac{\lg n}{\lg m} - \frac{n}{m})C(m) \qquad (12)$$

Similarly, it can be shown that the depth recurrence in Equation 4 gives, for the same value of r,

$$D(n;m) = (2\frac{\lg n}{\lg m} - 1)D(m). \qquad (13)$$

Now, if the switches with m inputs are replaced with m-input Benes networks then Equations 12 and 13 become

$$C_{RB}(n;m) = (2\frac{n}{m}\frac{\lg n}{\lg m} - \frac{n}{m})(m \lg m - \frac{m}{2}) \qquad (14)$$

$$D_{RB}(n;m) = (2\frac{\lg n}{\lg m} - 1)(2\lg m - 1) \qquad (15)$$

or,

$$C_{RB}(n;m) = 2n \lg n - n \lg m - n\frac{\lg n}{\lg m} - n/2 \qquad (16)$$

$$D_{RB}(n;m) = 4\lg n - 2\frac{\lg n}{\lg m} - 2\lg m + 1 \qquad (17)$$

where the subscript RB is used to emphasize that these formulas are obtained by replacing the m-switches by Benes networks after recursively decomposing the switches in the center stage. It is easily verified that, for all $m, 1 \le m \le \sqrt{n}$, $C_{RB}(n;m) = O(n \lg n)$, and $D_{RB}(n;m) = O(\lg n)$. Therefore,

Proposition 2: For all $m, 1 \le m \le \sqrt{n}$, if the switches in the center stage of the network in Figure 2 are first recursively decomposed until each switch in the center has m inputs, and then all the switches in the network are replaced by m-input Benes networks, then the resulting network has asymptotically minimum cost and logarithmic depth. ||.

Figure 4 shows a 3-D plot of $C_{RB}(n;m)$ for $1 \le n \le 8 \cdot 10^5$ and $1 \le m \le 1000$ obtained by using Mathematica. As in the earlier case $C_{RB}(n;m)$ is asymptotically minimum over n; $1 \le n \le 8 \cdot 10^5$, for all values of m, $1 \le m \le 1000$.

In particular, for $n = 10^5$, this design yields a network with about $24 \cdot 10^5$ 2×2 switches. In terms of Benes networks, this design requires 10^4 50-input Benes networks, each consisting of about $50 \lg 50 - 50/2 \simeq 250$ 2×2 switches. In VLSI terms this amounts to a chip with about 50 pins and about 2500 transistors assuming that inputs and outputs are multiplexed and a 2×2 switch is realized by about 10 transistors. Comparing these with the earlier design, it is seen that for the same number of inputs, the total number of switches remains roughly about 10^6 while the number of chips has increased by tenfold with the number of pins on each chip reduced by a factor of six.

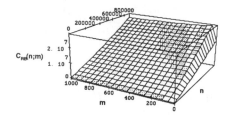

Figure 4: The cost of the network using Approach 2 as a function of n and m; $1 \le n \le 8 \cdot 10^5, 1 \le m \le 1000$.

5 Extension to Nonblocking Networks

The results given in the earlier sections can be extended to nonblocking networks as well. Our starting point is the recurrences which are given in Equations 5 and 6. Recall that these represent the cost and depth of a 3-stage nonblocking Clos network. As stated in the introduction section, a 3-stage nonblocking Clos network can be decomposed to obtain nonblocking networks with $O(n^{1+\lg^{1/2} n})$ switches. Rather than doing this, we shall use a nonblocking network construction due to Cantor [6]. For n inputs and as many outputs, this network uses $8n \lg^2 n$ 2×2 switches. Even though nonblocking networks with $O(n \lg n)$ switches exist [1], a Cantor network is easier to construct than such networks, and therefore will be used in our designs. However, before we can use it, we replace each of the $m \times 2m - 1$ switches in outer stages by two $m \times m$ switches as was suggested in [6]. This is done so that all switches in the network have an equal number of inputs and outputs, and thus can be replaced by a Cantor network. It is easy to see that the resulting network remains nonblocking after this modification.

With these facts, we can now replace $C(m, 2m-1)$ by $2C(m)$ and $D(m, 2m - 1)$ by $D(m)$ in which case, Equations 5 and 6 become

$$C(n;m) = 2mC(\frac{n}{m}) + 4\frac{n}{m}C(m), \qquad (18)$$

$$D(n;m) = D(\frac{n}{m}) + 2D(m). \qquad (19)$$

Let $C_A(n;m)$ and $D_A(n;m)$ denote the cost and depth of the network obtained by replacing each of the $m \times m$ switches in the outer stages of the modified 3-stage nonblocking network by an m-input Cantor network, and each of the $\frac{n}{m} \times \frac{n}{m}$ switches in the center stage by an $\frac{n}{m}$-input Cantor network. We then have from Equations 18 and 19

$$C_A(n;m) = 16n \lg^2 \frac{n}{m} + 32n \lg^2 m \qquad (20)$$

$$D_A(n;m) = \lg^2 \frac{n}{m} + 2\lg^2 m. \qquad (21)$$

Again, it is clear that, for all $m, 1 \le m \le n$, $C_A(n;m) = O(n \lg^2 n)$, and $D_A(n;m) = O(\lg^2 n)$. Therefore,

Proposition 3: For all $m, 1 \le m \le \sqrt{n}$, if the switches in the modified nonblocking Clos network are replaced by Cantor networks then the resulting network has $O(n \lg^2 n)$ cost and $O(\lg^2 n)$ depth. ||.

As in Section 3, we can obtain nonblocking networks with a single type of switch by letting $\frac{n}{m} = m$. This yields a nonblocking network with $12n \lg^2 n$ cost and $\frac{3}{4} \lg^2 n$ depth.

Finally, using the second approach, we can avoid having large size switches by decomposing the center stage switches as in rearrangeable networks. Using the recurrences given in Equations 18 and 19, we can obtain networks with any switch size m which satisfy the identity $m = n^{\frac{1}{r+1}}$ where r is a nonnegative integer as follows. Without loss of generality, suppose that $\frac{n}{m} \geq m$ as before. We can then decompose the switches in the center stage iteratively until the switches at the bottom most level of the iteration have m inputs. In algebraic terms, this amounts to recursively decomposing the first terms in Equations 18 and 19 by subisuting $\frac{n}{m}$ for n until their first terms reduce to $C(m)$ and $D(m)$, respectively. Suppose that this happens after r iterations. It can be shown that

$$C(n;m) = (2m)^r C\left(\frac{n}{m^r}\right) + (2^r - 1)\frac{4n}{m}C(m). \quad (22)$$

Now assuming that the arguments of both recurring terms are equal, i.e., $\frac{n}{m^r} = m$, we find $r = \frac{\lg n}{\lg m} - 1$, and $m^r = \frac{n}{m}$ so the above equation becomes

$$C(n;m) = 2^r m^r C(m) + (2^r - 1)\frac{4n}{m}C(m) \quad (23)$$

$$= (2^r \frac{5n}{m} - \frac{4n}{m})C(m) \leq 2^{\frac{\lg n}{\lg m}-1}\frac{5n}{m}C(m) \quad (24)$$

$$\leq \frac{5n^{1+1/\lg m}}{m}C(m), \quad (25)$$

Similarly, it can be shown that

$$D(n;m) = (2\frac{\lg n}{\lg m} - 1)D(m). \quad (26)$$

Upon replacing $C(m)$ by $8m\lg^2 m$, $D(m)$ by $\lg^2 m$, and letting $C_{RA}(n;m)$ and $D_{RA}(n;m)$ denote the cost and depth of the resulting network, we obtain

$$C_{RA}(n;m) \leq 40n^{1+1/\lg m}\lg^2 m, \quad (27)$$
$$D_{RA}(n;m) \approx 2\lg n\lg m. \quad (28)$$

These equations show that, for $m = O(n)$ $C_{RA}(n;m) = O(n\lg^2 n)$, and $D_{RA}(n;m) = O(\lg^2 n)$. For $m = O(1)$, these equations imply that $C_{RA}(n;m) = O(n^2)$, and $D_{RA}(n;m) = O(\lg n)$. Finally, if $m = O(\lg n)$, $C_{RA}(n;m) = O(n^{\frac{1+1}{\lg\lg n}}\lg^2\lg n)$, and $D_{RA}(n;m) = O((\lg n)\lg\lg n)$. These results indicate that not all values of m give $O(n\lg^2 n)$ cost nonblocking networks even though this appraoch can still be used to control the chip size and number of pins for moderately large values of n, e.g., 10^6 or less which is for all practical purposes, an upper bound on the number of processors that may be contained in a single parallel computer system.

6 Summary and Concluding Remarks

We have presented two techniques for partitioning 3-stage rearrangeable Clos networks into modules of Benes networks with an arbitrarily specified number of inputs in such a way that the total cost of the partitioned network is asymptotically minimum. The first technique can be used to realize asymptotically minimum cost rearrangeable networks on VLSI chips with $O(\sqrt{n})$ pins, while the second technique can accomodate VLSI chips with any number of pins desired without increasing cost beyond $O(n\lg n)$. These partitioning techniques also work for nonblocking networks to a certain

extent if Benes networks are replaced by Cantor networks. In this case, the first technique can be used to realize nonblocking networks on VLSI chips with $O(\sqrt{n})$ pins and $O(n\lg^2 n)$ cost, while the second technique can accomodate VLSI chips with any number of pins desired, but at a higher cost. This can be alleviated by using $O(n\lg n)$ cost nonblocking networks such as those given by Bassalygo and Pinsker [1,2]. However, the main problem with such an approach is that the construction of these nonblocking networks relies on certain types of bipartite graphs, called expanders, which are difficult to construct explicitly.

References

[1] L. A. Bassalygo and M. S. Pinsker, "Complexity of optimum nonblocking switching without reconnections," *Problems of Information Transmission*, Vol. 9, No. 1, Jan.-Mar. 1974, pp. 64-66.

[2] L. A. Bassalygo, "Asymptotically optimal switching circuits," *Problems of Information Transmission*, Vol. 17, No. 3, July-Sep. 1981, pp. 206-211.

[3] K. E. Batcher, "Sorting networks and their applications," *Proc. AFIPS 1968 Summer Joint Computer Conference*, Vol. 32, pp. 307-314.

[4] V. E. Beneš, "Optimal rearrangeable connecting networks," *Bell Systems Technical Journal*, Vol. 43, July 1964, pp. 1641-1656.

[5] V. E. Beneš, *Mathematical Theory of Connecting Networks and Telephone Traffic*, Academic Press Pub. Company, New York, 1965.

[6] D. G. Cantor "On non-blocking switching networks," *Networks*, Vol. 1, 1972, pp. 367-377.

[7] C. Clos "A study of non-blocking switching networks," *BSTJ*, Vol. 32, 1953, pp. 406-425.

[8] T-Y Feng, "A survey of interconnection networks," *IEEE Computer*, Dec. 1981, pp. 12–27.

[9] M. A. Franklin, D. F. Wann, and W. J. Thomas, "Pin limitations and partitioning of VLSI interconnection networks," *IEEE Trans. Comp.*, C-31, Nov. 1982, pp. 1109-1116.

[10] D. S. Parker, Jr., "Notes on shuffle/exchange-type switching networks," *IEEE Trans. Comput.*, Vol. C-29, Mar. 1980, pp. 213-222.

[11] W. Payne, F. Makedon and W. Daasch, "High-speed interconnection using the Clos network," *First Int. Conf. on Supercomputing*, Lecture notes in Comp. Sci., Vol. 297, June 1987, pp. 96-111.

[12] H. S. Stone, "Parallel processing with the perfect shuffle," *IEEE Trans. Comput.*, Vol. C-20, Feb. 1971, pp. 153-161.

[13] W. A. Waksman, "A permutation network," *JACM*, May 1968, pp. 159-163.

[14] C. L. Wu, and T. Feng, "On a class of multistage interconnection networks," *IEEE Trans. Comput.*, Vol. C-29, Aug. 1980, pp. 694-702.

DESIGN AND IMPLEMENTATION OF A VERSATILE INTERCONNECTION NETWORK IN THE EM-4

Shuichi SAKAI   Yuetsu KODAMA   Yoshinori YAMAGUCHI

Computer Science Division, Electrotechnical Laboratory
1-1-4 Umezono, Tsukuba-shi, Ibaraki 305, JAPAN

Abstract This paper presents the design and prototype implementation of an interconnection network in a highly parallel dataflow computer **EM-4**, that will have more than a thousand processing elements. As a first step, a single chip processing element **EMC-R** was designed and fabricated in 1989, and an **EM-4** prototype system with 80 **EMC-R**s has been fully operational since April 1990. The peak performance of this prototype is 1 GIPS.

The interconnection network of the **EM-4** prototype adopts circular omega topology. This paper first examines the features of this topology, then proposes a node grouping method, a node addressing method and a self-routing algorithm on each node. Then store-and-forward deadlock prevention mechanisms are proposed, and automatic load distribution mechanisms attached to this network are presented. Next, network implementation in the **EM-4** prototype system is described. The network consists of distributedly controlled switching units and connection lines, and actually performs 14.63 GB/s. The switching unit was implemented as a unit of the **EMC-R** and performs packet communication concurrently with and independently of the other functional units on the chip.

1   Introduction

The main role of interconnection networks [1] is to realize data communication between processors or between a processor and a memory unit, and to do so with little delay and throughput high enough for the execution of application programs. So the interconnection networks must provide communication that enables the maximally available performance of processing elements (PEs). To create the most appropriate network, every designer must consider (1) target execution performance of the computer being developed and (2) properties of the programs which will be executed on the computer.

Current VLSI technologies have enabled large-scale networks, so research into interconnection networks can be conducted in a real parallel computer, as well as in theory and simulations. Interconnection networks which connect tens of PEs, hundreds of PEs, or tens of thousands of PEs have already been implemented [2] [3] [4] [5]. Researchers face problems such as optimization of physical connections and hotspot contentions.

This paper presents design and implementation of an interconnection network in the **EM-4** [6] [7] [8], an advanced dataflow machine that will have more than a thousand PEs. A circular omega topology is adopted for an **EM-4** network and a new node-grouping methods, a routing

method, a store-and-forward deadlock (SFD) prevention method, and an automatic load balancing method are proposed and implemented in this machine. Section 2 will describe the adoption of a circular omega topology. A method of grouping PEs, a routing algorithm will be proposed in this section. Section 3 will propose a method of preventing SFDs in the network and will also make a suggestion for the automatic load distribution mechanisms embedded in the network. Section 4 will describe design and implementation of the switching unit in the **EM-4**, which is a component of the network. It will also show the network implementation in the **EM-4** prototype which consists of 80 PEs. The prototype system was designed in 1988-1989, and it has been fully operational since April 1990. Its processing performance is 1 GIPS and the communication performance is 14.63 GB/s.

2   Circular Omega Network

2.1   Network Topology

The parallelism that a highly parallel computer exploits can be classified into three levels: functional level, block level and instruction level. For extracting each level parallelism in order to realize the most efficient execution, the whole system should be constructed in a layered structure. In the **EM-4**, a two-layered structure is considered: every function is allocated to a group of PEs, each block including individual instructions is allocated to a certain PE. Therefore, a lower level is a group and a higher layer is a whole system.

We used a circular omega topology (Figure 1) for the interconnection network of the **EM-4** prototype system. A circular omega network consists of switching units (SUs) and interconnection lines which connect them into the topology of an omega network [12]. Each SU is connected to a PE and makes data switching and data transfer independently of and concurrently with the PE. As the word "circular" implies, input lines of the first stage SUs are connected to output lines of the last stage SUs. In this paper, each connection path is assumed to consist of unidirectional lines (directions of arrows in Figure 1); there are no bidirectional communications on each path. In addition, we here assume that all the communications on the network are realized as a packet transfer.

On these assumptions, the circular omega network satisfies the following conditions: (1) hardware is O(N), where N is the number of PEs; (2) the distance between PEs is less than or equal to O(logN); (3) uniform structure; (4)

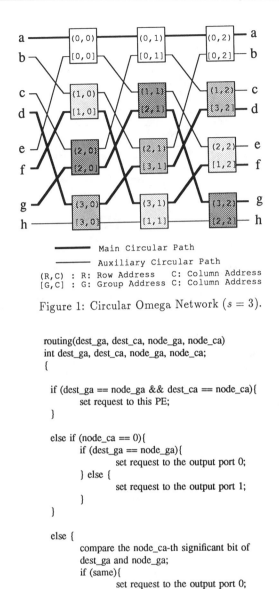

Main Circular Path
Auxiliary Circular Path
(R,C) : R: Row Address   C: Column Address
[G,C] : G: Group Address C: Column Address

Figure 1: Circular Omega Network ($s = 3$).

```
routing(dest_ga, dest_ca, node_ga, node_ca)
int dest_ga, dest_ca, node_ga, node_ca;
{

    if (dest_ga == node_ga && dest_ca == node_ca){
        set request to this PE;
    }

    else if (node_ca == 0){
        if (dest_ga == node_ga){
            set request to the output port 0;
        } else {
            set request to the output port 1;
        }
    }

    else {
        compare the node_ca-th significant bit of
        dest_ga and node_ga;
        if (same){
            set request to the output port 0;
        } else {
            set request to the output port 1;
        }
    }

}
```

dest_ga, dest_ca:destination group address and
column address, respectively;
node_ga, node_ca: node group address and column
address, respectively
output port 0:output link to the same group node
output port 1:output link to another group node

Figure 2: Routing Algorithm in each SU.

self-routing is possible (see 2.3); and (5) store-and-forward deadlocks(SFDs) [9] can be prevented with small cost (see Section 3). Robustness of the network will be examined in another paper.

2.2 Grouping and Addressing

In a circular omega network, we can set a closed loop called a circular path. Suppose that there is a circular omega network with s stages and suppose that the set of nodes of this network is named N_s which is addressed by a doublet (R, C) (R: raw number, C: column number). Then there is a division of N_s that is expressed as follows.

$$J_s = \{ \{(R^0, 0), (R^1, 1), ..., (R^{s-1}, s-1)\}| \qquad (1)$$
$$(R^x, x) \text{ and } (R^{x+1}, x+1) \text{ are directly}$$
$$\text{connected by the network path }\}$$

Proof of the existence of J_s will be shown in another paper. Here J_s is called a *circular-path division* of the circular omega network and each element of the J_s is called a *circular path*. The circular omega topology has two circular-path divisions. One is illustrated as thick lines in Figure 1 and the other is illustrated as thin lines in Figure 1. The circular-path division which contains a circular path $(0,0)$, $(0,1)$, ..., $(0, s-1)$ is called a *main circular-path division* $J_s^{(m)}$ (thick lines) and the other circular-path division is called an *auxiliary circular-path division* $J_s^{(a)}$ (thin lines). In the same way, the element of the former is called a *main circular path* and that of the latter is called an *auxiliary circular path*.

In the **EM-4**, the main circular-path division is used for grouping, that is, every element of the main circular-path division is defined as a group in the **EM-4** system. By this, data communication inside the group becomes local communication inside a main circular path. The group address of each node is different from that of any other node in the same auxiliary path to which it belongs.

When this grouping is adopted, the node address should be represented as $[G, C]$ rather than (R, C), where G is a group number and C is a column number. In the following, we adopt $[G, C]$-representation of nodes, where two addressing systems indicate the same node as the same address if C is equal to 0 (Figure 1).

2.3 Routing Algorithm

Figure 2 shows the self-routing algorithm in each node in the circular omega network. In this paper, self-routing is defined as a procedure of each node to decide the destination output port for each packet in real time. This self-routing algorithm was obtained from the shuffle connection pattern of the omega network and the conversion rule from (R, C)-system to $[G, C]$-system.

By this self-routing algorithm, one hop makes one certain bit of the destination group address equal to that of the current node group address except at the first column nodes (Note that each SU can change two bits of the group address, but the less significant bit will again be changed by the next stage SU). This means that in one lap of the circular omega network, each packet reaches the node whose group address is the same as that of the destination node. In another lap, this packet will reach the destination, since the routing inside the group will be adopted.

In this way, it is proved that all the packets reach their destination within two laps of a circular omega network.

3 Special Facilities

3.1 Store-and-Forward Deadlock Prevention

When one uses a circular omega topology with packet buffers, there is a possibility of store-and-forward deadlocks (SFDs) [9].

To prevent SFDs, we could make a connection structure of buffers in the network loop-free. Structured Buffer Pool (SBP) method [9] is a typical method to realize this. However, the SBP method needs as many buffers in each node as the longest path length (diameter) of the network. In a real highly parallel machine, the adoption of this method is impossible because of its hardware cost.

In a circular omega network, the SFD problem is solved in the followings way. Set three buffer banks at each node. At first, each packet must use the first buffer bank. When it goes across the border between a first-stage node and a second-stage node, it will use the second buffer bank. When a second-buffer-bank packet goes across the border, it will use the third buffer bank. As was shown in 2.3, no packet can go round the network twice, so the packet stored in the third bank buffer will reach the destination without reaching the second stage. Therefore loops do not occur in the network and SFDs are completely prevented.

Dally [11] proposes similar SFD prevention mechanisms in the cube, CCC, and so on. Our work was done independently [13] and showed that SFDs can be avoided with three buffer banks and three logical channels in the circular omega topology, while CCC needs four buffer banks and four logical channels.

3.2 Automatic Load Distribution Facilities

Function level dynamic load balancing is essential in a parallel computer. To perform this, it is necessary to monitor the whole system, to know which PE is the least loaded, to decide to which PE a new task will be allocated, and to allocate the task before execution of a new task. The central control of dynamic load balancing is almost impossible in a highly parallel computer since its time overhead is so large. The distributedly-controlled load-balancing method is advantageous for its quick response, but it is difficult to monitor the global load situation.

Recently it was proposed that by attaching dynamic load distribution facilities to each network SU, global load balancing is carried out efficiently in a distributed way [13] [14]. Topologies of networks with such facilities were restricted to tree or multistage topology like omega. In the **EM-4**, we proposed, for the first time, automatic load balancing mechanisms attached to a circular omega topology.

To achieve dynamic load balancing, auxiliary circular paths are used. Suppose that a PE will call a new function. Then this PE delivers a special packet, *an MLPE packet* (Minimum-Load PE packet), an output port connected to the auxiliary circular path. The MLPE packet always holds the minimum load value and the least loaded PE number which it goes through in the following way: at the starting point, the MLPE packet holds its sender's load value and its PE number; when it goes through a certain SU in the auxiliary circular path, the SU compares the load value of the PE connected to it, and if the value is less than

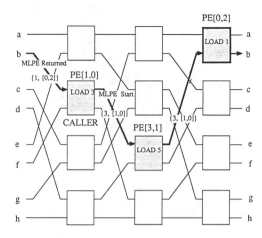

{L, [G, C]} : Content of the MLPE Packet Data
When MLPE returned back, L becomes the minimum load which PE[G,C] has.

Figure 3: Automatic Load Balancing with an Auxiliary Circular Path.

that of the packet, data of the MLPE packet will be automatically rewritten to the current PE's value; otherwise the MLPE packet keeps its value and goes to the next SU. By this mechanism, when the MLPE packet comes back to the starting point, it holds the least loaded PE number and its load value. Figure 3 shows this. In this figure, PE[1,0] generates an MLPE packet and, after the circulation, it obtains the least loaded PE number [0, 2] and its load 1.

By this method, called the auxiliary-path load balancing, each MLPE packet scans s different groups , where s is the number of network stages (see 2.2). When the total number of the PEs increase, coverage of PEs by this load balancing method becomes relatively small. The efficacy of this method will be reported in another paper.

4 Implementation

4.1 EM-4 Prototype System

Before implementing an **EM-4** system with 1,024 PEs, a prototype system with 80 PEs was designed and implemented [8]. The **EM-4** prototype system is a dataflow machine with 80 **EMC-R**s, each of which is a single chip gate-array processor containing 45,788 gates. The **EM-4** prototype system consists of sixteen PE group boards each of which includes five **EMC-R**s, a global network, a packet interface switch, a packet interface processor and a host computer.

4.2 Switching Unit

In the **EM-4** prototype, data communication is performed with a fixed size (two word) packet. Figure 4 shows the switching unit (SU) organization of the prototype. The SU contains six ports: two network input ports (IP0 and IP1), one input port from the processing part of the **EMC-R**, two network output ports (OP0 and OP1), and one output port to the processing part of the **EMC-R**. Each network input port has a three-bank buffer, which provides SFD-free packet switching. Routing is carried out in the Rout-

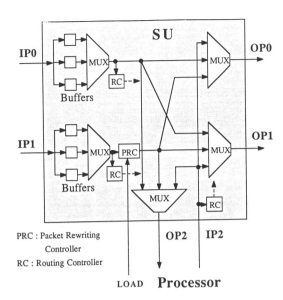

Figure 4: SU Organization of the **EMC-R**.

Table 1: Hardware Profile of a Switching Unit

Switching Method	pipelined packet switching
Transfer Mode	fully synchronous
Clock	single-phase, 80 ns
Routing	self-routing
Packet	39 bits/word * 2 words
Communication Path	39 data lines 2 request lines 3 ready lines
Signal Lines/Chip	176 (= 44 * 4)
Buffer	2 words/bank * 3 banks/port * 2 network ports (total 468 bits)
Special Mechanism	automatic load balancing
Data Transfer Delay	80 ns/word (including routing and arbitration.)
Data Transfer Rate	60.9 MB/s/port
Number of Gates	9,179

ing Controller (RC in the Figure) at the latter part of each input port. Transfer requests are arbitrated in an arbiter of each output port, if collision of the destination occurs. After the arbitration, the output port selects and sends data to the next stage SU or to the processing part. Routing, arbitration and sending are carried out distributedly and independently in each port.

The SU is included in the **EMC-R** and performs all its actions independently of and concurrently with the processing part of the chip.

The routing algorithm realized in each routing controller (RC) is shown in Figure 2 plus routing toward the packet interface switch.

Furthermore, the SU contains automatic load balancing facilities whose mechanisms are described in 3.2. The PRCs (Packet Rewriting Controllers) in Figure 4 are the load-balancing circuits. The PRC rewrites an MLPE packet according to the load value of the packet and that of the processor connected to the SU. Even if the rewriting occurs, the transfer speed of the MLPE packet is the same as that of normal data packets.

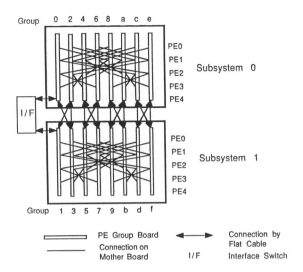

Figure 5: **EM-4** Network Implementation.

Figure 6: **EM-4** Prototype.

In the **EM-4** prototype, load value is generated automatically by hard-wired circuits from the number of waiting packets, the size of usable frame area and the size of usable structure memory area. The SU realizes a real-time transfer of a packet, one word per clock, including routing, arbitration, transfer and buffering.

Table 1 shows the hardware profile of the SU. The word size is 39 bits, the buffer bank size is two packet words, and there are 9,179 gates. The throughput of each port is 60.9 MB/s and the transfer delay is 80ns/word.

4.3 Whole Network Implementation

In the **EM-4** prototype all the data switching facilities are implemented on **EMC-R** chips; there are no additional switching circuits outside the chips. The data communication circuits consist of drivers and connection lines beside the **EMC-Rs**.

Figure 5 illustrates the physical connection patterns of the prototype interconnection network. Each intra-group network, or a main circular path, is realized in the connection paths and SUs on each PE group board. There are no additional drivers for it, since each connection path is shorter than 30 cm. Inter-group connections are implemented in the following way. The prototype is divided into two parts, an upper subsystem and a lower subsystem each of which have 40 PEs. In each subsystem, eight PE group boards are connected to a mother board where connections inside the subsystem are realized (thick lines in Figure 5). Inter-part connection is by cables (arrows in Figure 5). On each PE group board, there are four connectors for this connection, two for input and two for output. Connection between the prototype system and the packet interface processor is by cables.

From Figure 5, it can be stated that both of the connection patterns on each mother board and the connection patterns of cables are regular and the length of each path is short for the whole size of the rack (1.5m long). These properties made the network implementation easy.

The prototype system was implemented and has been fully operational since April 1990. The physical size of it is $60\,cm \times 92\,cm \times 140\,cm$ (Figure 6).

5 Conclusion

This paper first stated that a circular omega topology is highly advantageous for the highly parallel computer **EM-4** , and proposed the node-grouping method, the routing algorithm, the SFD prevention method, and the automatic load distribution mechanisms attached to the network. It also showed the implementation methods of the network on the **EM-4** prototype and the performance evaluation where the real machine was used. The automatic load balancing mechanisms have been evaluated and will be reported in another paper and close estimations of the prototype system performance are now being carried out.

In the next phase, the **EM-4** with 1,024 PEs will be implemented and evaluated with its network.

Acknowledgment

The authors wish to thank Dr. Toshitsugu Yuba, Director of the Computer Science Division, and Mr. Toshio Shimada, Chief of the Computer Architecture Section for supporting this research, and the staff of the Computer Architecture Section for their fruitful discussions.

References

[1] Wu, C.L. and Feng, T., Eds.: Tutorial: Interconnection Networks for Parallel and Distributed Processing, IEEE Computer Society (1984).

[2] Pfister, G.F., Brantley, W.C., George, D.A., Harvey, S.L., Kleinfelder, W.J., McAuliffe, K.P., Melton, E.A., Norton, V.A. and Weiss, J.: The IBM Research Parallel Processor Prototype (RP3): Introduction Architecture, Proc. of ICPP85, pp.764-771 (1985).

[3] Tucker, L.W. and Robertson, G.G.: Architecture and Applications of the Connection Machine, IEEE Computer, Vol.21, No.12, pp.26-38 (1988).

[4] Gustafson, J. L., Hawkinson, S. and Scott, K.: The Architecture of a Homogeneous Vector Supercomputer, Proc. of ICPP86, pp.649-652 (1986).

[5] Hiraki, K., Sekiguchi, S. and Shimada, T.: System Architecture of a Dataflow Supercomputer, TENCON87, pp.1044-1049 (1987).

[6] Sakai, S., Yamaguchi, Y., Hiraki, K., Kodama, Y. and Yuba, T.: An Architecture of a Dataflow Single Chip Processor, Proc. of ISCA89, pp.46-53 (1989).

[7] Yamaguchi, Y., Sakai, S., Hiraki, K., Kodama, Y., and Yuba, T.: An Architectural Design of a Highly Parallel Dataflow Machine, Proc. of IFIP89, pp.1155-1160 (1989).

[8] Kodama, Y., Sakai, S. and Yamaguchi, Y.: A Prototype of a Highly Parallel Dataflow Machine EM-4 and its Preliminary Evaluation, Proc. of InfoJapan90, pp.291-298 (1990).

[9] Raubold, E. and Haenle, J.: A Method of Deadlock-Free Resource Allocation and Flow Control in Packet Networks, Proc. of ICCC 1976, pp.483-487 (1976).

[10] Preparata, F.P., Vuillemin, J.: The Cube-Connected Cycles: A Versatile Network for Parallel Computation, Commun. ACM, pp.300-309 (1981).

[11] Dally, W.J. and Seitz, C.L.: Deadlock-Free Message Routing in Multiprocessor Interconnection Networks, IEEE Trans. Comput., Vol.C-36, No.5, pp.547-553 (1987).

[12] Lawrie, D.H.: Access and Alignment of Data in an Array Processor, IEEE Trans. Comput., Vol.C-24, No.12, pp.1145-1155 (1975).

[13] Sakai, S.: Interconnection Networks in a Highly Parallel MIMD Computer, Doctoral Dissertation, University of Tokyo (1985).

[14] Watson, I., Woods, V., Watson, P., Banach, R., Greenberg, M. and Sargeant, J.: Flagship: A Parallel Architecture for Declarative Programming, Proc. of ISCA88, pp.124-130 (1988).

A Massively Parallel Processing Unit with a Reconfigurable Bus System RIPU

Wen-Tsuen Chen[†], Chia-Cheng Liu[‡] and Ming-Yi Fang[†]

[†]Department of Computer Science
National Tsing Hua University
Hsinchu, Taiwan 300, Republic of China

[‡]Department of Information Engineering
Feng Chia University
Taichung, Taiwan 400, Republic of China

Abstract

This paper describes the design of a massively parallel processing unit with a reconfigurable bus system (RIPU). The design of the RIPU is motivated by the architectural needs to support fine grained reconfiguration for a broad range of applications. It is a two-dimensional mesh-connected parallel processing array with a grid-shaped reconfigurable bus system, and operated in an SIMD fashion. We also provide a high-level parallel programming language and environment for conveniently developing parallel programs on the RIPU array. Preliminary results show the RIPU system can dynamically match the computational structures of a wide variety of applications.

1. Introduction

Advances in VLSI technology have made possible the construction of highly parallel processors consisting of more than several hundred processor elements. One of the most attractive massively parallel architectures is the mesh due to its simplicity and low wiring complexity [1], [4]. However, because of its local connectivity nature, a long communication delay will be incurred when data have to be moved over a long distance. To overcome its communication inefficiency, many architecture enhancements were added to the mesh. These include mesh with broadcasting buses [11], reconfigurable mesh [8], and polymorphic torus [6]-[7].

Both reconfigurable mesh and polymorphic torus can be categorized as reconfigurable SIMD parallel processors, which have been proved to be very powerful computing models [13]. One of its most distinguished features is the utilization of the reconfigurability of its interconnection network to dynamically match the communication needs of the algorithms. In addition, the reconfigurability of the network can be used to bypass faulty processors from the system so as to achieve fault-tolerant capability. Since the reconfigurable SIMD parallel architectures can dynamically change its network topology on per instruction basis, the performances of many algorithms developed on such a model can approach those on an ideal PRAM model. In fact, it is shown that many problems can be solved on reconfigurable SIMD parallel architectures with constant time complexity [6], [13], [14].

The design of a massively parallel processing system with reconfigurable integrated parallel processing units (RIPUs) is described in this paper. The motivation for building the RIPU stems from the need to support fine-grained reconfiguration for a broad range of applications. For examples, in vision or image computations, more than one network topology are usually needed at different stages of the computation. It is thus desirable to incorporate the reconfiguration into the system. The RIPU system is a two-dimensional mesh-connected parallel processing array with a grid-shaped reconfigurable bus system, operated in an SIMD fashion. It is an extension and enhancement of our previous system, IPU, which had been implemented in Tsing Hua

University [2]-[3]. A RIPU chip consists of 4 processing elements (PEs), each of which has a locally controllable bus switch built to adjust the configuration of the bus system. Each PE contains a 4-bit ALU, a 64 * 4-bit RAM, several 4-bit registers, some multiplexers, and a programmable bus switch combined with a control register to store the switch configuration settings. The die size is about 360×360 mil². The RIPU Development System that acts as a bridge between the host computer and the RIPU array is designed as well. We also provide a high level parallel programming environment for conveniently designing parallel programs on the RIPU array.

There have been several related works incorporating reconfiguration into a mesh [12] or a torus, such as reconfigurable mesh [8] and YUPPIE [7]. The RIPU is similar to the YUPPIE in functionality, but aims toward more general-purpose computations. It is designed to provide an environment for experimentation with a wide variety of algorithms that explore reconfiguration computation. Moreover, the RIPU involves more switch settings than those allowed by the previous architectures, and hence more interconnection patterns can be realized on the RIPU. Consequently, it is more flexible for the RIPU to configure the bus system than the above two architectures.

In the following we shall describe the design of the RIPU system, including the hardware architecture of the RIPU array, RIPU chip, and its programming environment.

2. System Architecture

Overview

The RIPU system consists of a large-scale single instruction stream multiple data stream (SIMD) RIPU array. The system block diagram is shown in Figure 1. Basically, there are two kinds of interconnection networks in the RIPU array, the PNET (physical network) and the INET (internal network). The INET is a reconfigurable bus whose interconnection patterns are programmable. The PNET is a mesh interconnection with its boundary connected in either torus mode or spiral mode, which is programmable. We call this connection scheme the **Dynamic Torus connection** [7]. The

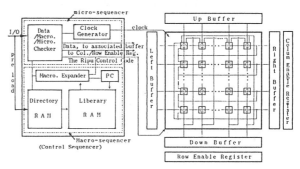

Figure 1. The block diagram of the RIPU system.

RIPU array is a passive back end unit rather than a stand-alone one. It is attached to a host computer. Four buffers (Up, Down, Left, Right) located at the boundary of the RIPU array are used to latch the data between the RIPU array and the host computer. The RIPU array and INET switch setting are controlled by the **controlling sequencer** which serves as an interface for communication between the host and the RIPU array. Thus, the entire system is primarily made up of three parts: the RIPU array, the controlling sequencer, and the host.

Programming in the RIPU system

For easy programming the RIPU system, we develop a parallel programming language which is essentially an extended version of the normal C language. Some extended parallel language structures are embedded in the normal C language to form a new language, called the **extended parallel C language**. This language is intrinsically a high-level one so that the programming efforts can be reduced to a minimum. At the same time, a **preprocessor** is designed to translate extended parallel processing statements and the RIPU assembly language into the RIPU control codes.

Operation of the RIPU System

The operation of the RIPU system is briefly described as follows. An application program in extended parallel C language is first prepossessed and compiled. The part of normal C program is compiled and executed in the host computer. Whereas the part of program in extended parallel statements is translated into a sequence of the RIPU instruction codes. The host computer then sends these RIPU instruction codes to the RIPU array via the controlling sequencer. Finally, when the RIPU array completes the task, the host computer collects the results from the RIPU array via controlling sequencer.

To reduce the communication overhead between the host and the RIPU array, we provide a set of macro-codes (i.e. library functions) each of which is composed of a sequence of micro-codes for some primitively and frequently used functions. Thus, the controlling sequencer consists of macro expansion mechanism called macro sequencer and micro controlling sequencer. There are two RAMs in the controlling sequencer: one is used to store all of the macro programs called macro library RAM, and the other stores their starting addresses called directory RAM. These data can be downloaded prior to the initialization of the system.

Hence, when the host computer sends a macro-code and parameters to the sequencer, Macro Expander performs macro expansion (see Figure 1) to generate the corresponding micro-codes by consulting the directory RAM and the macro library RAM. These micro-codes are then sent directly to the RIPU array from the micro controlling sequencer. If there were no macro instructions then a sequence of micro-codes should be sent from the host computer to the RIPU array one by one, which is more time-consuming.

Furthermore, the host can create a "window" (mask) in the RIPU array. The "window" is masked on/off according to the contents of Row Enable Register and Column Enable Register of the controlling sequencer in every instruction cycle. Each PE in this window can execute the current instruction but those outside the window do not. In other words, a PE can be masked on/off under the control of the host. By using the "window operation", we can do more than one operation on the same data in RIPU array asynchronously.

3. The RIPU Processor Architecture

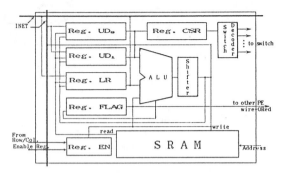

Figure 2. The RIPU processing element.

The RIPU Processor Element

The functional diagram of a PE is composed of several major blocks as depicted in Figure 2. Each PE can be viewed as a completely independent 4-bit processor with the following features:

- Each PE contains four 4-bit registers: Registers UD_1/UD_0 and LR used as operands for ALU, Register UD_0/UD_1 used for data movement and connected with the bus, and Register CSR used for storing the configuration settings of the bus switch. Register UD_0 is used as the target of data movement operations while Register UD_1 is used as an operand of ALU operations and vice versa.

- One flag register FLAG is used to store the status (overflow, zero, sign and carry) of a PE. The status output of a PE is wire-ORed with the other PEs in the chip.

- One enable register EN is used to control the execution of each PE. Once it has been reset, the states of the PE (i.e., the register and memory contents) cannot be updated. In this way it is possible to selectively turn off some PEs and reactivate them when needed.

- There is a 64*4-bit static RAM built in a PE, which serves as the working storage for each PE.

- The ALU performs 4-bit arithmetic and logical operations, such as addition with or without carry, subtraction with or without carry, decreasing Register UD, bitwise OR, bitwise XOR, bitwise AND and logical NOT of Register UD. The result of the ALU is directly fed into the shifter when needed. The shifter performs rotation left/right with carry. The purpose for including the shift operations is mainly to provide the multiplication and division operations.

- A PE can be selectively enabled/disabled by the host computer externally or by the condition code within Register FLAG internally.

- The INET switch can be internally configured by a PE or externally configured by the host. Externally configuration capability is useful when the PE is faulty. By adjusting the settings of the bus switches properly, a number of network topology, such as mesh, tree, ring, mesh, pyramid, and hypercube, can be efficiently realized on the RIPU array.

The Switch Design

Reconfigurability has been incorporated in several related works [1], [4]-[11]. Our work retains the useful features while providing more dynamic control and flexibility. We attempt to reconfigure the topology at instruction level to support fine-grained reconfigurability. A new network configuration is set by the current instruction, but takes effect at next instruction cycle. That is, we attempt to overlay in part the next configuration setting time with the current instruction execution time. Consequently, we can shorten the time for setting the configuration for the next instruction.

The INET is totally programmable. At the P(i,j), there is an INET switch S(i,j) which is a complete graph for four directions (U, D, L, and R). By "complete" we mean that each direction can be connected to all other directions.

The INET, whose data paths is shown in Figure 3, is a switch able to drive the Register UD or Register LR to the physical links (U, D, L, and R) as well as to short-circuit these physical links. When the bus is activated, the PE can receive the data currently passing through the bus if the PE is not masked off. As shown in Figure 4, fifteen "control patterns" are supported, and only one of them can be implemented in each node at one time.

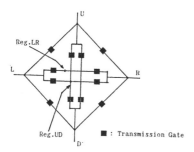

Figure 3. The INET data paths.

A Switch Function to Support Connection Network

The SWITCHSET function is provided as a function for supporting the INET control. The format is SWITCHSET(i,j,s_k) or SWITCHSET(condition, s_k), where i, j are the coordinates of a PE, "condition" is the internal status of a PE, and s_k is a number indicating one of the "control pattern" of the switch setting. The former is used to externally set the configuration setting while the latter determines the configuration by the internal status of a PE. The control patterns are dependent on the switch design and the number of registers to be connected to the INET. By proper selecting of the switch control patterns, a number of network topologies can be efficiently realized.

The Instruction Set and The Assembly Language

Multiple fields instruction word (long instruction word) is used in RIPU design to achieve massive parallelism. There are eleven fields in one instruction, including Condition, UD_0-source, UD_1-source, LR-source, FLAG-source, UD_0/UD_1, RAM-read/write-source, ALU Operation-Code, Shifter Operation-Code, CSR-Setting-on/off and RAM-Address. One RIPU assembly statement, ending with the symbol ";", can contain many mnemonic names of different fields. As an example, a program in the RIPU assembly language, that performs the addition of two 1-nibble numbers, is illustrated in the following.

ripu$add1(a, b, c)

Figure 4. The switch control patterns.

```
/*  @C = @A + @B
    RAM[c]←RAM[a] + RAM[b] */
#define  MESH  0
(1)\%UD0=MESH, %A=UD1, %SW=ON;
    /*  SWITCHSET( ,MESH) is used in the
        following instruction cycle */
(2)\%UD0=0, %LR=0;
    /*  No %SW=ON in the instruction means
        previous configuration is retained*/
(3)\%A=%UD0, %F=%A, %FLAG=%FS;
(4)\%UD0=%RAM[a];
(5)\%LR =%RAM[b];
(6)\%A=%UD0, %F=%A+%LR+%C, %SM=%F,
    %RAM[c]=%SM, %FLAG=%FS;
```

In the above program, statement (1) sets the configuration to the mesh topology for the following operations. %A=UD1 and %SW=ON mean that Register UD_1 is used as the operand to the ALU and that the data of Register UD_0 is latched by Register CSR. Statement (2) clears the contents of Registers UD_0 and LR. Statement (3) is purely used for clearing the carry flag. Statement (4) and (5) read the memory contents from locations a and b, and store them into Registers UD_0 and LR, respectively. Statement (6) adds UD_0 and LR with carry, whose result, %SM, is written into memory location c, and then the operation status is saved in Register FLAG.

4. Programming the RIPU

In this section, we shall describe the programming environment of the RIPU system. The overall environment consists of three parts: the extended parallel C language, the RIPU assembly language and the preprocessor.

Extended C language

In the following, we shall give a program in the extended parallel C language to demonstrate the programming of the RIPU. The problem we want to solve is matrix multiplication, which can be stated as follows: Given two $n * n$ matrices A = (a_{ij}) and B = (b_{ij}), the product D = (d_{ij}) is computed as

$$d_{ij} = \sum_{k=0}^{n} a_{ik} b_{kj} \quad \text{where } 0 \le i, j \le n\text{-}1.$$

It is clear that it can be easily solved by the column/row broadcasting method in RIPU. For example, consider the case of $n = 4$. Initially, all elements of matrices A and B are both located at the host. At step 1, a_{00} ... a_{30} are broadcasted by the host along each row and b_{00}...b_{03} along each column at the same time. In general, each PE has three data elements @A, @B and @D, containing a_{ij}, b_{ij} and d_{ij} respectively. Step 2 computes @A * @B and adds the result to @D. After 8 steps, broadcasting and computation are completed and @D in each PE_{ij} contains d_{ij} of matrix D.

The complete program for the matrix multiplication problem by using the column/row broadcasting method is depicted as follows :

```
#include "ripu$main_define.sys";
#include <stdio.h>
main()
/*   4*4 Matrix Multiplication D = A * B */
{
int   leftbuffer[4], upbuffer[4], downbuffer[4];
int   ha[4][4], hb[4][4], hd[4][4], i, j, k;
\RIPU_MAIN_BEGIN;
\RIPU_DECLARE_BEGIN;
\NIBBLE * 2  @A, @B, @D;  /* 8-bit signed integer */
\RIPU_DECLARE_END;

    .
    .    /*   Read data for matrices A and B, which is
    .         written in normal C language. */
\CELL[0:3,0:3];
    /* Select active PEs, active rows 0:3, active columns 0:3 */
\@A = 0;
\@B = 0;       /*  initialization  */
\@D = 0;
for (i = 0; i <= 3; i++)
{   \H_V_LUbroadcast(leftbuffer, 4, @A, upbuffer, 4, @B);
    /*  The array "leftbuffer" with length 4 is sent to @A of
        each row and the array "upbuffer" with length 4 is
        sent to @B of each column. */
    \@D = @D + @A * @B;
}
for (i = 0; i <= 3; i++)
{   \HFMOVELEFT(leftbuffer,4,@D,@D);
    /*  The host receives data from the RIPU. */
    for (j = 0; j <= 3; j++)
        hd[j][i] = leftbuffer[j];
}

    .
    .    /*  Print matrix D, written in normal C language. */
    .
\RIPU_MAIN_END;
}
```

In the above program, if we do not have bus reconfigurability, we need 14 steps due to the input data skew on the mesh-connected SIMD machine using the systolic method for matrices multiplication [2]-[3]. In the program, each statement preceded by the symbol "\" is an extended RIPU statement. These statements will be executed by the RIPU processor. The leading symbol "\" is used to distinguish it from normal C statements. We had built macros for all of the extended statements except for the declaration statements. The name of each variable in the RIPU declaration statement is preceded by the symbol "@". After we declare a RIPU variable, say @A, each RIPU PE allocates a memory location for @A. The memory allocation problem is handled by the Preprocessor.

The Preprocessor

We have designed a preprocessor to translate the extended parallel processing statements into normal C statements before invoking the C compiler. This preprocessor also translate a program written in the RIPU assembly language into the corresponding micro-codes.

In fact, the preprocessor is a translator which translates RIPU parallel processing statements into corresponding function calls. We has built a RIPU library for these function calls, which are written in the RIPU assembly language. Each library function corresponds to a primitive RIPU operation such as arithmetic operation, logical operation and I/O operation, etc. The functions in RIPU library are prepossessed and compiled into RIPU instructions in advance. When we want to build an executable task, the RIPU codes of suitable functions in RIPU library would be linked in.

5. Summary

With the advance in VLSI technologies, it is possible to build a massive array of PEs to speed up the computation time for various applications. In this paper, we have described the design of a massively parallel processing array called the Reconfigurable Integrated Parallel Processing Unit. We also describe a parallel programming language for writing the RIPU programs. It is basically an extended C language enhanced with a set of statements for parallel programming. A RIPU program is compiled in the host and a sequence of micro-codes is generated. The micro-codes are then downloaded to the controlling sequencer for controlling the execution of each of the PE in the RIPU array. Reconfigurability has been incorporated into the RIPU array. By changing the configuration of bus switch built into each PE, we can easily realize a number of topologies. Preliminary results show the RIPU system can dynamically match the communication needs for a wide variety of applications.

References

[1]   K. E. Batcher, "Design of a massively parallel processor," IEEE Trans. on Computers, Vol. C-29, pp. 836-840, Sept. 1980.

[2]   W. T. Chen, et. al, "Design of a Parallel Processing Unit with Systolic VLSI Chips", Journal of information Science and Engineering, Vol. 7, no. 2, 1991.

[3]   W. T. Chen, et. al, "Design of a Systolic VLSI chip for Massively Parallel Processing," Technical Report, Institute of Computer Science, National Tsing Hua University, Hsinchu, Taiwan, R. O. C., Dec. 1988.

[4]   M. J. B. Duff, "CLIP4: A Large scale integrated circuit array parallel processor," in Proc. IJCPR, 1976, pp728-733.

[5]   W. D. Hillis, The Connection Machine. Cambridge, MA: MIT Press, 1985.

[6]   H. Li, and M. Maresca, "Polymorphic-Torus Network" IEEE Trans. on Computers, Vol. C-38, pp. 1345-1351, Sept. 1989.

[7]   M. Maresca and H. Li, "Connection autonomy in SIMD computers: A VLSI implementation," J. Parallel Distrib. Computing 7, pp. 302-320, 1989.

[8]   R. Miller, V.K.P. Kumar, D. Reisis, and Q. Stout, "Meshes with reconfigurable buses," in Proc. 5th MIT conf. Advanced Res. VLSI, Mar. 1988.

[9]   S. F. Reddaway, "DAP-A flexible number cruncher," in Proc. LASL Workshop Vector Parallel Processors, 1978, pp. 233-234.

[10] L. Snyder, "Introduction to the configurable highly parallel computer," IEEE Computer, Vol. 15, pp.47-56, Jan, 1982.

[11] Q. Stout, "Mesh-connected computer with broadcasting," IEEE Trans. on Computers, Vol. C-32, pp. 826-830, Sept. 1983.

[12] T. Fountain, "Processor Arrays: architectures and applications," London, Academic, 1987

[13] B. F. Wang and G. H. Chen, "Two-dimensional processor array with a reconfigurable bus system is at least as powerful as CRCW model", Inform. Process. Lett. 36, pp.31-36, 1990.

[14] B. F. Wang and G. H. Chen, "Constant time algorithm for the transitive closure and some related graph problems on processor arrays with reconfigurable bus systems", IEEE Trans. on Parallel and Distrb. System, Vol.1 No.4, Oct. 1990.

Performance Evaluation of Multicast Wormhole Routing in 2D-Mesh Multicomputers*

Xiaola Lin, Philip K. McKinley, and Lionel M. Ni

Department of Computer Science
Michigan State University
East Lansing, MI 48824-1027

Abstract

Multicast communication services, in which the same message is delivered from a source node to an arbitrary number of destination nodes, are being provided in new generation multicomputers. The nCUBE-2 is the first multicomputer that directly supports broadcast using wormhole routing. This paper shows that the broadcast wormhole routing adopted in the nCUBE-2 is not deadlock-free, a property that is essential to wormhole routed networks.

Four multicast wormhole routing strategies for 2D-mesh multicomputers are proposed and studied. All of the algorithms are shown to be deadlock-free. These are the first deadlock-free multicast wormhole routing algorithms ever proposed. A simulation study has been conducted that compares the performance of these multicast algorithms under dynamic network traffic conditions in an 8×8 mesh. The results indicate that a dual-path routing algorithm offers better performance than any of a tree-based, multiple-path, or fixed-path algorithm.

1 Introduction

Efficient routing of messages is critical to the performance of multicomputers. Historically, commercial multicomputers have supported only single-destination, or *unicast*, message passing. More recently, multicomputers have begun to provide *multicast* communication services, in which the same message is delivered from a source node to an arbitrary number of destination nodes [1] [2]. Multicast communication is finding increasing demand in many parallel programs, including simulation of computer networks and electronic circuits, particle dynamics calculations, and image processing.

Providing support for multicast communication involves several, often conflicting, requirements. First, it is desirable that the message delay from the source to each of the destinations be as small as possible. Second, the amount of network traffic must be minimized. Third, the routing algorithm must not be computationally complex. Finally, the multicast communication protocol must be deadlock-free. Multicast deadlock refers to the situation in which two or more messages are blocked forever due to the interaction of their acquiring, waiting for, and releasing network resources such as communication channels and message buffers.

Designing multicast protocols and routing algorithms to meet these requirements naturally depends on the topology of the network used to interconnect nodes in the multicomputer. In this paper, we restrict our discussion to the *two-dimensional mesh* (2D-mesh) topology. A $x \times y$ 2D-mesh comprises $x \cdot y$ nodes interconnected in a grid fashion, with each node having at most four neighbors. The 2D-mesh topology is represented with a mesh graph $M(V, E)$, in which each node in $V(M)$ corresponds to a processor and each edge in $E(M)$ corresponds to a communication channel.

The 2D-mesh topology has become increasingly important to multicomputer design due to its many desirable properties, including regularity, low cross-section bandwidth, and the fixed degree of nodes. At least two new generation multicomputers, the *Symult 2010* [3] and the Intel *Touchstone* project, have adopted the 2D-mesh topology.

The design and resulting performance of multicast communication protocols also depend on the underlying switching mechanism used in the multicomputer. Early multicomputers used store-and-forward message passing. In order to decrease the amount of time spent transmitting data, the *virtual cut-through* method [4] was introduced. In virtual cut-through, a message is stored at an intermediate node only if the next channel is busy. In order to avoid buffering messages altogether, *wormhole routing* [5] has been adopted in advanced multicomputers. In wormhole routing, a message consists of a sequence of *flits* (flow control digits). The header flit(s) of the message governs the route, and the remaining flits of the message follow in a pipeline fashion. Unlike virtual cut-through, in which messages can be removed temporarily from the network, blocked messages remain in the network. Therefore, the routing criteria and the deadlock properties of wormhole routing are quite different from those of store-and-forward or virtual cut-through. Because of its low network latency and the small amount of dedicated buffer space required at each node, wormhole routing has become the most promising switching technology and has been adopted in Symult 2010 [3], nCUBE-2 [1], iWARP [6], and Intel's Touchstone project. Wormhole routing has also been used in fine-grained multicomputers, such as MIT's J-machine [7] and Caltech's MOSAIC [8].

Multicast routing for multicomputers has been studied previously in [9, 10], in which various graph models and multicast routing algorithms were proposed, but the deadlock problem was not addressed. In [11], three multicast protocols were pre-

---

*This work was supported in part by the NSF grants ECS-8814027 and MIP-8811815.

sented to deal with the deadlock problem, however, the switching mechanism employed was based on virtual cut-through, and no routing algorithms were proposed in the paper.

In this paper, we study deadlock-free multicast communication in 2D-mesh multicomputer networks using wormhole routing. In Section 2, we discuss why extensions of one-to-one deadlock-free routing algorithms are not directly applicable to multicast communication. Section 3 describes a deadlock-free, tree-like routing scheme that uses two channels instead of one for each unidirectional channel. In Section 4, heuristic multicast routing algorithms based on a path-like model are presented; these methods are also deadlock-free, but do not require double channels. In order to compare these routing algorithms, we have simulated their operation under a variety of network conditions. Results of these simulations are presented in Section 5. Section 6 contains our concluding remarks.

2 Multicast Routing and Deadlock

In multicomputer networks, communication channels and message buffers constitute the set of *permanent reusable resources*. Messages are the entities that compete for these resources. Deadlock refers to the situation in which a set of messages each is blocked forever because each message in the set holds some resources also needed by the other.

A practical multicast routing algorithm must be deadlock-free and should transmit the source message to each destination node using as little time and as few communication channels as possible. For a given set of destination nodes, the multicast routing algorithm might deliver the message along a common path as far as possible, then branch in order to deliver the message to each destination node. Essentially, the message follows a tree-like route to the destinations. One possible approach to implementing such a multicast tree is to extend deadlock-free unicast routing algorithms to handle multicast traffic.

For example, the *e*-cube algorithm has been proved to be deadlock-free for one-to-one communication in *n*-cubes [12]. The nCUBE-2 [1], which is the first hypercube multicomputer to use wormhole routing, supports broadcast and a special form of multicast in which the destination nodes form a subcube. A tree is created in which each path from the source to a destination uses *e*-cube routing. In order to avoid buffering, when the tree branches and a node must send the message on multiple channels, all of the required channels must be available before transmission on any of them may take place. Hence, the branches of the tree proceed forward in lockstep. Blockage of the tree in this manner can prevent completion of the message transfer even to those destination nodes to which paths have been successfully created. Moreover, deadlock can occur using such a routing scheme. Figure 1 shows a deadlock configuration on a 3-cube. Suppose that nodes 000 and 001 simultaneously attempt to transmit broadcast messages, $M0$ and $M1$, respectively. Figures 1a and 1b, respectively, depict the would-be trees for these broadcasts. The broadcast originating at node 000, $M0$, has acquired channels [000, 001], [000, 010], and [000, 100]. The header flit has been duplicated at node 001 and is waiting on channels [001, 011] and [001, 101]. The $M1$ broadcast has already acquired channels [001, 011], [001, 101] and [001, 000], but is waiting on channels [000, 001] and [000, 010]. The two broadcasts will block forever.

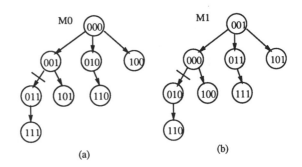

Figure 1. Deadlock in a 3-cube multicomputer.

In a similar manner, we may attempt to extend unicast routing on a 2D-mesh to encompass multicast. One deadlock-free unicast routing algorithm requires that messages be sent first in the X-direction and then in the Y-direction. An extension of this routing method to include multicast is shown in Figure 2. As in the hypercube, all branches of the tree must proceed in lockstep. When progress of a tree is blocked, all messages requiring any segment of the tree are also blocked.

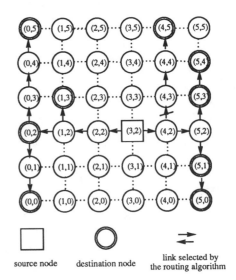

Figure 2. An X-first multicast routing pattern.

Moreover, the above routing algorithm can lead to deadlock. Figure 3 shows a deadlock configuration in which two multicasts, $M0$ and $M1$, have the following communication patterns:

```
M0: source: (1,1); destinations: (0,2), (3,1)
    links acquired: [(1,1),(0,1)], [(1,1),(2,1)];
    required link:  [(2,1),(3,1)];

M1: source: (2,1); destinations: (0,1), (3,0);
    links acquired: [(2,1),(1,1)], [(2,1),(3,1)];
    required link:  [(1,1), (0,1)];
```

In Figure 3a, $M0$ has acquired channel [(1,1),(0,1)] and requires channel [(2,1),(3,1)], while $M1$ has acquired channel

$[(2,1),(3,1)]$ and requires channel $[(1,1),(0,1)]$. Figure 3b shows the details of the situation for the nodes $(0,1)$, $(1,1)$, $(2,1)$, and $(3,1)$. Because neither node $(1,1)$ nor node $(2,1)$ buffers the blocked message, channels $[(1,1),(0,1)]$ and $[(1,1),(0,1)]$ cannot be released. Deadlock has occurred.

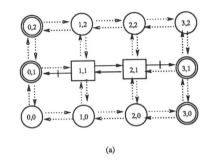

(a)

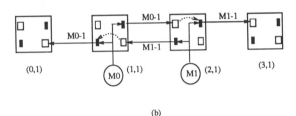

(0,1)   (1,1)   (2,1)   (3,1)

(b)

Figure 3. A deadlock situation in a 3×4 mesh.

Our objective is to develop and evaluate deadlock-free multicast routing algorithms in order to support reliable and efficient parallel computations in multicomputers. In the next two sections, we present deadlock-free multicast wormhole routing schemes based on the concept of network partitioning. The basic idea is to divide a multicast into several sub-multicasts, each routed in a different acyclic subnetwork. Because the subnetworks are disjoint and acyclic, no cyclic resource dependency can develop. Thus the routing schemes are deadlock-free.

3 Double-Channel Multicast Routing

Our first deadlock-free multicast wormhole method uses a modification of X-first routing algorithm discussed earlier and shown to be susceptible to deadlock. In order to break the cycle of channel dependencies, we first double each channel on the 2D-mesh, then partition the network into four subnetworks, $N_{+X,+Y}, N_{-X,+Y}, N_{-X,+Y}$ and $N_{+X,-Y}$. Subnetwork $N_{+X,+Y}$ contains the unidirectional channels with address $[(i,j),(i+1,j)]$ and $[(i,j),(i,j+1)]$, subnetwork $N_{+X,-Y}$ contains channels ith address $[(i,j),(i+1,j)]$ and $[(i,j),(i,j-1)]$, and so on. Figure 4 shows the partitioning of a 3×4 mesh into the four subnetworks.

For a given multicast, the destination node set D will be divided into at most four subsets, $D_{+X,+Y}, D_{-X,+Y}, D_{-X,+Y}$ and $D_{+X,-Y}$, according to the relative positions of the destination nodes and the source node u_0. Set $D_{+X,+Y}$ contains the destination nodes to the upper right of u_0, $D_{-X,+Y}$ contains the destinations to the upper left of u_0, and so on. The multicast will be divided into at most four submulticasts from u_0 to

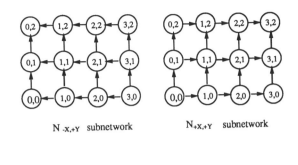

$N_{-X,+Y}$ subnetwork $N_{+X,+Y}$ subnetwork

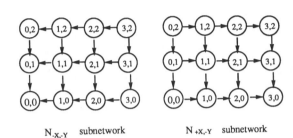

$N_{-X,-Y}$ subnetwork $N_{+X,-Y}$ subnetwork

Figure 4. Network partitioning for 3×4 mesh.

each of $D_{+X,+Y}, D_{-X,+Y}, D_{-X,+Y}$, and $D_{+X,-Y}$. The submulticast to $D_{+X,+Y}$ will be implemented in subnetwork $N_{+X,+Y}$ in X-first manner, $D_{-X,+Y}$ in subnetwork $N_{-X,+Y}$, and so on. Algorithms 1 and 2 below give the message preparation and routing algorithm, respectively.

Algorithm 1. Message preparation for the double-channel X-first tree routing algorithm

Input: Destination set D, source node address $u_0, u_0 = (x_0, y_0)$;

Output: (at most) Four lists of destination nodes: $D_{+X,+Y} D_{-X,+Y}, D_{-X,-Y}$, and $D_{+X,-Y}$.

Procedure:

1. Divide D into four sets $D_{+X,+Y}, D_{-X,+Y}, D_{+X,-Y}$, and $D_{-X,-Y}$ such that
$$D_{+X,+Y} = \{ (x,y) | x > x_0, y \geq y_0 \}$$
$$D_{-X,+Y} = \{ (x,y) | x \geq x_0, y > y_0 \}$$
$$D_{+X,-Y} = \{ (x,y) | x < x_0, y \leq y_0 \}$$
$$D_{-X,-Y} = \{ (x,y) | x \geq x_0, y < y_0 \}.$$

Algorithm 2. The double-channel X-first routing algorithm for subnetwork $N_{+X,+Y}$.

Input: Destination set D, local address $u, u = (x, y)$;

Procedure:

1. If $x < Min\{x_i | (x_i, y_i) \in D\}$, send message and D to $(x+1, y)$, stop.

2. If $(x, y) \in D$, then $D \leftarrow D - \{(x,y)\}$ and a copy of the message is sent to the local node.

3. Let $D_Y = \{(x_i, y_i) | x_i = x, (x_i, y_i) \in D\}$. If $D_Y \neq \emptyset$, send message and D_Y to $(x, y+1)$; If $D - D_Y \neq \emptyset$, send message and $D - D_Y$ to $(x+1, y)$.

As an example, consider the 6×6 mesh in Figure 2. At the source node, $(3,2)$, the destination set D is partitioned into four subsets as follows.

$$D = \{(0,0), (0,2), (0,5), (1,3), (4,5), (5,0), (5,1), (5,3), (5,4)\}$$

$$
\begin{aligned}
D_{+X,+Y} &= \{(4,5), (5,3), (5,4)\} \\
D_{-X,+Y} &= \{(0,5), (1,3)\} \\
D_{-X,-Y} &= \{(0,0), (0,2)\} \\
D_{+X,-Y} &= \{(5,0), (5,1)\}
\end{aligned}
$$

The message will be sent to the destination nodes in $D_{+X,+Y}$ through subnetwork $N_{+X,+Y}$, to the nodes in $D_{-X,+Y}$ using subnetwork $N_{-X,+Y}$, and so on. The routing pattern is shown in Figure 5. It is straightforward to show that this algorithm avoids deadlocks [13].

Assertion 1 *The double-channel X-first routing algorithm is deadlock-free.*

While the multicast tree approach avoids deadlock, a major disadvantage is the need for double channels. It may be possible implement double channels with virtual channels [12], however, early analysis shows that the signaling for multicast communication may be quite complex; we are continuing to study this problem. We next introduce multicast routing algorithms that are deadlock-free and which do not require additional channels (either physical or virtual). We can see that deadlock situations arise in multicast trees when copies are created and diverge at intermediate nodes. The algorithms in the next section avoid deadlock by enforcing the rule that a message, once in the network, never be copied.

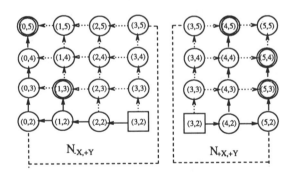

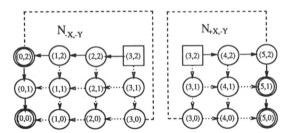

Figure 5. The routing pattern of the tree algorithm.

4 Path-Like Multicast Routing

We first present a network partitioning strategy based on Hamilton paths that is fundamental to the deadlock-free routing schemes. A *Hamilton path* visits every node in a graph once and only once. It is not difficult to see that a 2D-mesh has many Hamilton paths.

In our algorithms, each node i in a multicomputer is assigned a label, $\ell(i)$. In a network with N nodes, the assignment of the label to a node is based on the order of that node in a Hamilton path, where the first node in the path is labeled 0 and the last node in the path is labeled $N - 1$. Each node is represented by its integer coordinate (x,y). The label assignment function ℓ for a $m \times n$ mesh can be expressed as

$$
\ell(x,y) = \begin{cases} y*n + x & \text{if } y \text{ is even} \\ y*n + n - x - 1 & \text{if } y \text{ is odd} \end{cases}
$$

Figure 6(a) shows such a labeling in a 4×3 mesh. The labeling effectively divides the network into two subnetworks. The *high-channel subnetwork* contains all of the channels from lower labeled nodes to higher labeled nodes, and the *low-channel network* contains all of the channels from higher labeled nodes to lower labeled nodes.

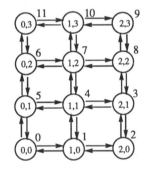

(a) physical network

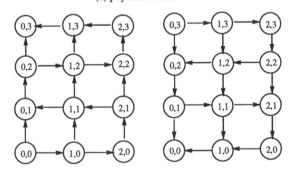

(b) high-channel network   (c) low-channel network

Figure 6. The labeling of a 4×3 mesh.

The first step in finding a deadlock-free multicast algorithm for the 2D-mesh is to define a routing function that uses these two subnetworks in such a way as to avoid channel cycles. Let N be the node set of the 2D-mesh. One routing function $R : N \times N \to N$, is defined as $R(u,v) = w$, such that w is adjacent to u. If $\ell(u) < \ell(v)$, then w is the node adjacent to u such that $\ell(w)$ is maximum among all such nodes. If

$\ell(u) > \ell(v)$, then w is the node adjacent to u such that $\ell(w)$ is minimum among all such nodes.

For two arbitrary nodes r and t, a path from r to t can be selected by the routing function R as $(v_1, v_2, \ldots, v_k)$, where $v_1 = r$, $v_i = R(v_{i-1}, t)$, $1 < i \leq k$, and $v_k = t$. The path is a partial order preserved for the assignment ℓ if $\ell(v_i) < \ell(v_{i+1})$ for $1 \leq i < k$ (that is, in the high-channel network), or $\ell(v_i) > \ell(v_{i+1})$ for $1 \leq i < k$ (that is, in the low-channel network). It has been proved in [14] that, for two arbitrary nodes r and t in a 2D-mesh, the path selected by the routing function R is a shortest path from r to t, and it is a partial order preserved for the label assignment function ℓ. We are now ready to define two multicast routing heuristics that use the routing function R.

Dual-Path Multicast Routing

The first heuristic routing algorithm divides the destination node set D into two subsets, D_H and D_L, where every node in D_H has a higher label than that of the source node u_0, and every node in D_L has lower label than that of u_0. Multicast messages from u_0 will be sent to the destination nodes in D_H using the high-channel network and to the destination nodes in D_L using the low-channel network. Algorithm 3 gives the message preparation part of the dual-path routing algorithm.

Algorithm 3. Message preparation for the dual-path routing algorithm.
Input: Destination set D, local address u_0, and node label assignment function ℓ;
Output: Two sorted lists of destination nodes: D_H and D_L placed in message header.
Procedure:

1. Divide D into two sets D_H and D_L such that D_H contains all the destination nodes with higher ℓ value than $\ell(u_0)$ and D_L the node with lower ℓ value than $\ell(u_0)$.

2. Sort the destination nodes in D_H and D_L using the ℓ value as the key, respectively. Put these sorted nodes in list D_H in ascending order and D_L in descending order, respectively.

3. Construct two messages, one containing D_H as part of the header and the other containing D_L as part of the header.

The source node divides the destination node set into two subsets: D_H and D_L, which are then sorted in ascending order and descending order, respectively, with the label of each node used as its key for sorting. Algorithm 4, the dual-path routing algorithm, uses a distributed routing method in which the routing decision is made at each intermediate node. Thus the dual-path routing algorithm is executed at each intermediate node. Upon receiving the message, each node first determines whether its address is the first destination node in the message header. If so, the address is removed from the message header and the message is delivered to the host node. At this point, if the address field of the message header is not empty, the message is forwarded toward the first destination node in the message header using the routing function R.

Algorithm 4. The dual-path routing algorithm
Input: A message with sorted destination list $D_M = \{d_1, \ldots, d_k\}$, a local address w and node label assignment ℓ;
Procedure:

1. If $w = d_1$, then $D'_M = D_M - \{d_1\}$ and a copy of the message is copied to the local node; otherwise, $D'_M = D_M$.

2. If $D'_M = \emptyset$, then terminate the message transmission.

3. Let d be the first node in D'_M, and let $w' = R(w, d)$.

4. The message is sent to node w' with address destination list D'_M in its header.

In order to show that the dual-path routing algorithm is deadlock-free, we need only prove that there is no cycle dependency among the channels, because the cycle dependency of the resource is a necessary condition of the deadlock situation. By the definition of channel partitioning scheme, each of the high-channel and low-channel subnetworks comprises a separate sets of channels in the multicomputer. There are no cycles within each subnetwork, and hence no cyclic dependency can be created among the channels.

Assertion 2 *Dual-path multicast routing is deadlock-free.*

Multi-Path Multicast Routing

The performance of the dual-path routing algorithm is dependent on the location distribution of destination nodes. Consider the example shown in Figure 7 for a 6×6 mesh topology. The total number of channels used to deliver the message is 33 (18 in the high-channel network and 15 in the low-channel network). The maximum distance from the source to a destination is 18 hops. In order to reduce the average length of multicast paths and the number of the channels used for a multicast, an alternative is to use a *multi*-path multicast routing algorithm, in which the restriction of having at most two paths is relaxed.

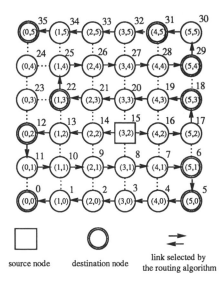

source node   destination node   link selected by the routing algorithm

Figure 7. An example of dual-path routing in a 6×6 mesh.

In a 2D-mesh, most nodes have outgoing degree 4, so up to 4 paths can be used to deliver a message, depending on the location of the source node. The only difference between multi-path routing and dual-path routing concerns the message preparation algorithm at the source node. Algorithm 5 shows the message preparation of the multi-path routing algorithm, in which we further partition the destination sets D_H and D_L of the dual-path algorithm for multiple multicast paths. Basically, D_H is divided into two sets, one containing the nodes with their x-coordinates greater than or equal to that of u_0 and the other containing the rest of the nodes in D_H. D_L is divided in the same way.

Algorithm 5. Message preparation for the multi-path routing
Input: Destination set D, local address $u_0 = (x_0, y_0)$, and node assignment ℓ;
Output: Sorted destination node lists $D_{H1}, D_{H2}, D_{L1}, D_{L2}$ for 4 multicast paths.
Procedure:

1. Divide D into two sets D_H and D_L such that D_H contains all the destination nodes with higher ℓ value than $\ell(u_0)$ and D_L the nodes with lower ℓ value than $\ell(u_0)$.

2. Sort the destination nodes in D_H and D_L using the ℓ value as the key. Place these sorted nodes in list D_H in ascending order and D_L in descending order, respectively.

3. (Assume that $v_1 = (x_1, y_1)$ and $v_2 = (x_2, y_2)$ are the two neighboring nodes to u_0 with higher labels than that of u_0.)
   Divide D_H in to two sets, D_{H1} and D_{H2} as follows:
   $D_{H1} = \{(x,y)|x \le x_1 \text{ if } x_1 < x_2, \, x \ge x_1 \text{ if } x_1 > x_2\}$ and
   $D_{H2} = \{(x,y)|x \le x_2 \text{ if } x_2 < x_1, \, x \ge x_1 \text{ if } x_2 > x_1\}$.
   Construct two messages, one containing D_{H1} as part of the header and sent the message to v_1, and the other containing D_{H1} as part of the header and sent the message to v_2.

4. Similarly, divide D_L into D_{L1} and D_{L2}.

For the multicast example shown in Figure 7, the destination set is first divided into two sets D_H and D_L at source node $(3,2)$, with $D_H = \{(5,3),(1,3),(5,4),(4,5),(0,5)\}$ and $D_L = \{(0,2),(5,1),(5,0),(0,0)\}$. D_H is further divided into two subsets D_{H1} and D_{H2} at Step 3, with $D_{H1} = \{(5,3),(5,4),(4,5)\}$ and $D_{H2} = \{(1,3),(0,5)\}$. D_L is also divided into $D_{L1} = \{(5,1),(5,0)\}$ and $D_{L2} = \{(0,2),(0,0)\}$. The multicast will be performed using four multicast paths, as shown in Figure 8. Note that multi-path routing requires only 20 channels in the example, and the maximum distance from the source to destination is 6 hops. Hence, this example shows that multi-path routing can offer significant advantage over dual-path routing in terms of generated traffic and the distance between the source and destination nodes.

Assertion 3 *Multi-path multicast routing is deadlock-free.*

Fixed-Path Multicast Routing

For purposes of comparison, we describe a third multicast algorithm called *fixed-path* routing. Fixed-path routing is similar to dual path routing except that each path traverses all possible horizontal links before traversing a single vertical link. The upper path visits all nodes in increasing order until the last destination is reached. Similarly, the lower path visits all nodes in decreasing order until the last destination is reached. For the source and destinations shown in Figure 7, the total number of channels required to deliver the message using fixed-path routing would be 35 (20 in the high-channel network and 15 in the low-channel network). The maximum distance from the source to a destination is 20 hops. Clearly, fixed-path routing is not as efficient as the other two approaches. However it is very simple to implement. We will discuss fixed-path routing further in the next section.

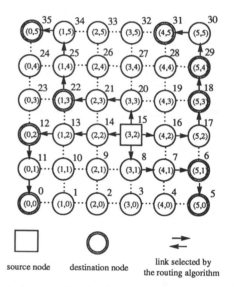

Figure 8. An example of multi-path routing in a 6×6 mesh.

5 Simulation Study

The performance of a multicast routing algorithm not only is measured in terms of the delay and traffic resulting from single multicast message, but also depends on the interaction of the multicast message with other network traffic. In order to study these effects on the performance of the proposed multicast routing algorithms, we have written a simulation program to model the multicast communication in 8×8 mesh networks. The simulation program used to model multicast communication in 2D-mesh networks is written in C and uses an event-driven simulation package, CSIM [15]. Although they are not shown in the figures, all simulations were executed until the confidence interval was smaller than 5 percent of the mean, using 95 percent confidence intervals.

In order to compare the tree-like and path-like algorithms fairly, we first simulated each on a network that contained double channels. Figure 9 plots the average network latency for various network loads. The average number of destinations for a multicast is 10, and the message size is 128 bytes. The speed of each channel is 20 Mbytes/second. All three algorithms display good performance at low loads. As the load is increased, the path algorithms are not negatively affected as soon as the tree-like algorithm. This result is explained by the fact that in tree-like routing, when one branch is blocked, the entire tree is

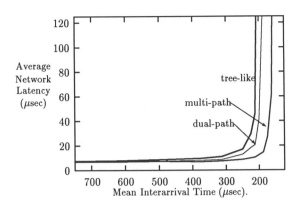

Figure 9. Varying load on a double-channel mesh.

blocked. This type of dependency does not exist in path-like routing. Multi-path routing outperforms dual-path routing because, as we have shown earlier, paths tend to be shorter and less traffic is generated. Hence, the network will not saturate as quickly.

The disadvantage of tree-like routing increases with the number of destinations. Figure 10 compares the three algorithms, again using double channels. The average number of destinations varies from 1 to 45. In this set of tests, every node generates a multicast messages with an average time between messages of 300 μsec. The other parameters are the same as for the previous figure. With larger sets of destinations, the dependencies among branches of the tree become more critical to performance and causes the delay to increase rapidly. The path algorithms still perform well, however. Notice that the dual-path algorithm outperforms the multipath algorithm for large destination sets. This result will be explained shortly. Our conclusion from Figures 9 and 10 is that tree-like routing is not particularly well suited for 2D-mesh networks. First, it requires double channels in order to be deadlock-free. Second, its performance is worse than that for path-based schemes. The remainder of the simulation results concern only single link, path-based approaches.

Figure 11 compares, for various numbers of destinations, the amount of "additional" traffic resulting from multi- and dual-path routing. Additional traffic is found by subtracting the number of destinations from the number of channels involved in the multicast. This is a static measurement and does not depend on network traffic conditions, but gives an indication of the efficiency of the algorithm. As shown by example in Figures 7 and 8, multi-path routing usually requires fewer channels than dual-path routing. Because the destinations are divided into four sets rather than two, they are reached more efficiently from the source, which is approximately centrally located among the sets.

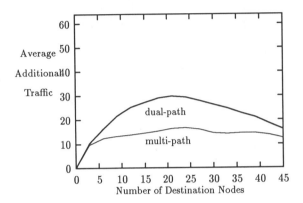

Figure 11. Average additional traffic.

Figure 12 plots the average network latency time for various network loads for multi-path and dual-path routing. Only single channels are used, the average number of destinations for a multicast is 10, and the message size is 128 bytes. The speed of each channel is 20M bytes/second. Both algorithms display good performance at low loads. As the load is increased, multipath routing offers slight improvement over dual-path routing. This is likely due to the fact that multipath routing introduces less traffic to the network.

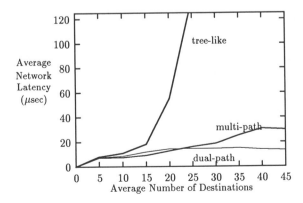

Figure 10. Varying destinations on double-channel mesh.

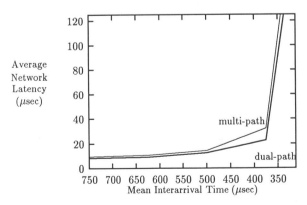

Figure 12. Varying load for path routing.

One may conclude from the results given thus far that multipath routing is superior to dual-path routing. However, a major disadvantage of multipath routing is not revealed until both the load and number of destinations are relatively high. Figure 13 shows that under these conditions, dual-path routing performs much better than multi-path routing. The reason is somewhat subtle. When multi-path routing is used to reach a relatively large set of destinations, the source node will likely send on all of its outgoing channels. Until this multicast transmission is complete, any flit from another multicast or unicast message that routes through that source node will be blocked at that point. In essence, the source node becomes "hot spot." In fact, every node currently sending a multicast message is likely to be a hot spot. As the load increases, these hot spots will reduce system throughput and increase message latency.

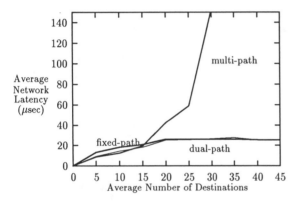

Figure 13. Varying destination number for path routing.

Hot spots are much less likely to occur in dual-path routing, and this fact accounts for its stable behavior under high loads with large destination sets. Although all of the outgoing channels at a node can be simultaneously busy, this can only result from two or more messages routing through that node. Figure 13 also compares fixed-path routing to multi-path routing or dual-path routing. For a small number of destinations, fixed path routing traverses many unnecessary channels, creating more traffic and needlessly blocking more messages than multi- or dual-path routing. For a large enough number of destinations, however, dual- and fixed-path routing effectively become identical performance. Because fixed-path routing is much simpler than dual-path routing, it may be the best choice for messages with large destination sets.

6   Concluding Remarks

In this paper, we have studied deadlock-free multicast wormhole routing for 2D-mesh multicomputers. A tree-based algorithm was presented that uses double channels in order to avoid deadlock. Three deadlock-free path-based algorithms, which do not require additional channels, were also presented and evaluated. Among these, the dual-path routing algorithm offered the best overall performance for different traffic loads and destination set sizes. The major disadvantage of multipath routing is that hot spots may occur under certain conditions, significantly degrading communication performance. A simpler fixed-path routing algorithm offers performance equal to the dual-path algorithm for large destination sets.

References

[1] NCUBE Company, *NCUBE 6400 Processor Manual*, 1990.

[2] C. L. Seitz, J. Seizovic, and W.-K. Su, "The C programmer's abbreviated guide to multicomputer programming," Tech. Rep. Caltech-CS-TR-88-1, Department of Computer Science, California Institute of Technology, Jan. 1988.

[3] C. L. Seitz, W. C. Athas, C. M. Flaig, A. J. Martin, J. Seizovic, C. S. Steele, and W.-K. Su, "The architecture and programming of the Ametek Series 2010 multicomputer," in *Proceedings of the Third Conference on Hypercube Concurrent Computers and Applications, Volume I*, (Pasadena, CA), pp. 33–36, Association for Computing Machinery, Jan. 1988.

[4] P. Kermani and L. Kleinrock, "Virtual cut-through: A new computer communication switching technique," *Computer Networks*, vol. 3, no. 4, pp. 267–286, 1979.

[5] W. J. Dally and C. L. Seitz, "The torus routing chip," *Journal of Distributed Computing*, vol. 1, no. 3, pp. 187–196, 1986.

[6] S. B. Borkar, R. Cohn, G. Cox, S. Gleason, T. Gross, H. T. Kung, M. Lam, B. Moore, C. Peterson, J. Pieper, L. Rankin, P. S. Tseng, J. Sutton, J. Urbanski, and J. Webb, "iWarp: An integrated solution to high-speed parallel computing," in *Proceedings of Supercomputing'88*, pp. 330–339, Nov. 1988.

[7] W. J. Dally, "The J-machine: System support for Actors," in *Actors: Knowledge-Based Concurrent Computing* (Hewitt and Agha, eds.), MIT Press, 1989.

[8] W. C. Athas and C. L. Seitz, "Multicomputers: Message-passing concurrent computers," *IEEE Computer*, pp. 9–25, Aug. 1988.

[9] Y. Lan, A. H. Esfahanian, and L. M. Ni, "Multicast in hypercube multiprocessors," *Journal of Parallel and Distributed Computing*, pp. 30–41, Jan. 1990.

[10] X. Lin and L. M. Ni, "Multicast communication in multicomputers networks," in *Proceedings of the 1990 International Conference on Parallel Processing*, pp. III–114–III–118, Aug. 1990.

[11] G. T. Byrd, N. P. Saraiya, and B. A. Delagi, "Multicast communication in multiprocessor systems," in *Proceedings of the 1989 International Conference on Parallel Processing*, pp. I–196 – I–200, 1989.

[12] W. J. Dally and C. L. Seitz, "Deadlock-free message routing in multiprocessor interconnection networks," *IEEE Transactions on Computers*, vol. C-36, pp. 547–553, May 1987.

[13] X. Lin, P. McKinley, and L. M. Ni, "Deadlock-free multicast wormhole routing in in 2d-mesh multicomputers," Tech. Rep. MSU-CPS-ACS-38, Department of Computer Science, Michigan State University, Apr. 1991.

[14] X. Lin and L. M. Ni, "Deadlock-free multicast wormhole routing in in multicomputer networks," in *Proceedings of the 18th Annual International Symposium on Computer Architecture*, May 1991.

[15] H. D. Schwetman, "CSIM: A C-based, process-oriented simulation language," Technical Report PP-080-85, Microelectronics and Computer Technology Corporation, 1985.

O(LOGN/LOGLOGN) RANDOMIZED ROUTING IN DEGREE-LOGN "HYPERMESHES"

Ted Szymanski

Department of Electrical Engineering and
Center for Telecommunications Research
Columbia University
email: teds@apollo.ctr.columbia.edu

Abstract: Comparing N node hypercubes and the recently proposed "hypermeshes" [17], each built with degree $logN$ crossbars, the hypermesh uses fewer degree-$logN$ crossbars and directed links by a factor $O(loglogN)$ and has a smaller diameter by a factor $O(loglogN)$, which has both practical and theoretical significance; For $N = 64K$ the reduction in degree-$logN$ crossbar cost and diameter over a hypercube is a factor of four. The hypermesh can realize all hypercube nearest neighbor permutations and all Omega, Omega-inverse, ASCEND and DESCEND permutations, each in minimum distance. Therefore a hypermesh based machine can execute the majority of parallel algorithms designed for hypercube and Omega based machines, including Bitonic sort, FFT, and those dealing with matrices, without modification. The following theoretical properties are proved for degree-$logN$ hypermeshes (and degree-$logN$ multistage networks); (1) any set of h permutations can be realized within $O(h \cdot logN / loglogN)$ unit-route steps with overwhelming probability by using randomized routing, (2) any set of h permutations can be realized in $O(h \cdot logN / loglogN)$ deterministic steps when path precomputation is allowed, and (3) the expected delay is $O(logN / loglogN)$ and the bandwidth is $\Theta(N)$ when packets arrive continuously at the inputs. The only other known degree-$logN$ networks that share property (1) are the d-way shuffles considered by Valiant, however they do not share (2) and cannot share (3). Equally importantly, hypermeshes can be recursively partitioned such that all partitions obey (1)-(3). These performance bounds are optimal in that no other degree-$logN$ network can be faster than a hypermesh by more than a constant factor (however d-way shuffles meet (1) with somewhat lower hardware cost).

1. Introduction

Most large multiple processor computers are based on one of two distinct interconnection network architectures; multistage interconnection networks such as the Omega or SW-banyan networks are popular with general purpose MIMD (Multiple Instruction stream, Multiple Data stream) machines such as the Butterfly/Monarch machines from BBN. Point-to-point networks such as the binary hypercube are popular in highly specialized SIMD (Single Instruction stream, Single Data stream) machines such as the CM series from Thinking Machines.

A "hypermesh" interconnection network was recently proposed in [17]. This network is architecturally distinct from the two main classes of networks used in current large scale machines, namely multistage networks and point-to-point networks. The "hypermesh" can be modelled as a hypergraph with N nodes arranged in n-dimensional hyperspace, where all nodes whose addresses differ in exactly one base-b digit belong to a net, and where every "net" can realize permutations of data between all its members. The hypermesh can be built with conventional electronic crossbars without requiring advanced technology, and in this paper we refer to an electronic hypermesh.

Practical advantages of hypermeshes over banyans and hypercubes were identified in [17]; The hypermesh can realize all

This research was supported by CTR through NSF Grant CDR-88-11111.

Omega, Omega Inverse, DESCEND and ASCEND permutations in one pass and in minimum distance. The majority parallel algorithms, such as the Bitonic sort, the FFT, finite element analysis and matrix algorithms, use these permutations. The hypermesh, like the hypercube, can execute all of these algorithms while imposing the minimum number of data transfer steps. Equally importantly the software algorithms do not require structural modification.

It is possible to embed any "radix-2 mesh/toroid" (i.e., a toroid with a power of 2 nodes along each dimension) into a hypercube such that nearest neighbors in the mesh/toroid are also nearest neighbors in the hypercube. Highly specialized SIMD machines often execute "continuum models" obeying partial differential equations (i.e., in the simulation of wind tunnels) and discrete particle models, each of which requires data transfers over nearest neighbors in a mesh/toroid [16].

The ability to embed meshes/toroids is extremely important: Current supercomputer performance is measured in deliverable MegaFlops per Million dollars. A highly specialized SIMD architecture which cannot embed meshes/toroids in unit dilation will suffer excessive delays when executing these algorithms, which will translate into a lack of generality and real dollar losses. Using degree-$logN$ crossbars, both the hypercube and the hypermesh can embed radix-2 meshes/toroids in unit dilation; however the hypermesh uses fewer degree-$logN$ crossbars and fewer directed links by a factor of $O(loglogN)$ and has a smaller diameter by a factor of $O(loglogN)$. For $N = 64K$ the reduction over a hypercube in degree-$logN$ crossbars and diameter is a factor of 4, and the reduction in directed links is a factor of 2, which should represent a significant decrease in the complexity of the backplane wireability.

The previous arguments indicate the advantages of the hypermesh over the hypercube. The hypermesh also has advantages over multistage networks such as the Omega network. It was shown in [17] that one hypermesh dimension performs the function of one stage in an Omega network. In an Omega network a packet must traverse every stage from input to output. In a hypermesh a packet only traverses necessary dimensions; therefore the hypermesh can be thought of as an Omega network where packets can bypass stages if those stages did not change the packet's route. By forcing every packet to traverse every dimension in a hypermesh, the hypermesh can therefore simulate identically the Omega (or Omega-Inverse) multistage networks.

One potentially severe drawback of banyans (including the Omega and Omega Inverse networks) is the unique-path-property; there exists a single path between every input and every output. This property results in potential traffic imbalances; certain permutations (i.e., bit-reversal) can cause some internal links to carry the traffic of $\sqrt{N}$ connections. The hypermesh has numerous minimum distance paths between sources and destinations, reducing the imbalance problem significantly. Due to the unique path property banyans are also very susceptible to failures; the failure of any link (or node) will disable all communications between processors whose unique paths included that link. Hypermeshes are immune to such failures since there exist many minimum distance paths between every source/destination pair.

Finally, it is well known that the Omega network is functionally partitionable, i.e., the states of the first stage of nodes can be deterministically set, so that an $N \times N$ Omega network simulates two smaller independent $N/2 \times N/2$ Omega networks (this reasoning applies recursively). In this manner a large parallel processor can be reconfigured into two or more smaller independent parallel processors. Note however that every packet must still traverse every stage in the initial multistage network. Hence there is no decrease in message latency by partitioning the Omega network; In contrast, hypermeshes are also partitionable (recursively), yet a packet always traverses minimum distance paths.

Improvements by large factors usually occur along with fundamental architectural changes (in this case the use of hypergraphs rather than graphs and multistage networks) rather than incremental technological improvements. Since the hypermesh is a novel class of network, it is desirable to examine theoretical aspects of hypermeshes and compare the results with hypercubes, each built with degree-$logN$ nodes.

The following theoretical properties are proven for degree-$logN$ hypermeshes (and degree-$logN$ multistage networks); (1) any set of h permutations can be realized within $O(h \cdot logN / loglogN)$ unit-route steps with overwhelming probability by using randomized routing, (2) any set of h permutations can be realized in $O(h \cdot logN / loglogN)$ deterministic steps when path precomputation is allowed, and (3) the expected delay is $O(logN / loglogN)$ and the bandwidth is $\Theta(N)$ when packets arrive continuously at the inputs. The only other known degree-$logN$ networks that share property (1) are the d-way shuffles considered by Valiant et. al. [19], however they do not share (2) and cannot share (3) since they lack the hardware necessary for pipelining. Equally importantly, hypermeshes can be recursively partitioned such that all partitions obey (1)-(3). (It has never been established if d-way shuffles are partitionable, and it is known that 2-way shuffles are not partitionable.) These performance bounds are optimal in that no other degree-$logN$ network can be faster than a hypermesh by more than a constant factor.

In other words, the advantages of the hypermesh over the hypercube or multistage networks are based in theory. The future direction in architectures for the "continuum model" could favor either nearest-neighbor connections or perfect-shuffle connections (i.e., the Omega network) according to Stone [16]. Dally has shown that the delay in 2D meshes is significantly lower than hypercubes if (1) each network is implemented completely in VLSI and has comparable area, and (2) the delay through a node is negligible, so that the delay in passing through $\sqrt{N}$ nodes is negligible [7]. Both assumptions are not necessarily met in a "discrete component environment", where networks consist of small crossbars, usually of bounded degree and bounded IO bandwidth, interconnected in various manners. Two recent papers has shown that hypercubes can outperform 2D meshes in a discrete component environment [2][6].

This paper is organized as follows. Section 2 introduces the "hypermesh" topology and compares practical implementations of Omega networks, hypermeshes and hypercubes. Section 3 presents some relevant properties of hypermeshes and properties of banyans. Section 4 presents the theoretical results. Section 5 contains some concluding remarks.

2. HyperMesh Topology

An n dimensional hypermesh can be characterized by the number of nodes aligned along each of the n dimensions and can be represented in a mixed radix scheme by a series of factors

$$\left\{ d_{n-1}, d_{n-1}, \cdots, d_0 \right\}$$

where

$$N = d_{n-1} \cdot d_{n-2} \cdots d_0$$

If the hypermesh is "regular" then all factors are identical and denote this topology as a n dimensional radix-b hypermesh, or simply a d^n hypermesh; each node occupies one point in an n-dimensional space, with d nodes aligned (or interconnected) along any dimension. To uniquely specify a node, n base d digits are required, where each digit specifies the node's position in one dimension. Assuming that the dimensions are labeled $n-1,...,0$, then each node's unique integer address can be viewed as n-digits in base d, i.e.

$$digit\ vector = x_{n-1}, x_{n-2}, \cdots x_0 \rightarrow$$
$$integer\ address = x_{n-1} \cdot d^{n-1} + x_{n-2} \cdot d^{n-2} + \cdots x_0 \cdot d^0$$

A binary Omega network with 4 stages (i.e., $\Omega_{2,4}$) is shown in fig. 1a, a radix 4 Omega network with two stages (i.e., $\Omega_{4,2}$) is shown in fig. 1b, and a regular 4^3 hypermesh is shown in fig. 1c. Each horizontal, vertical or diagonal line is a hypergraph net which represents a 4×4 crossbar that interconnects the 4 nodes connected to the net. The size of a hypermesh, the number of links and the number of switches can be adjusted by using a mixed radix system. A mixed radix system can also be used to adjust the relative bandwidth over different hyperspace dimensions; It may be desirable to have much higher bandwidth over the lower dimensions to exploit locality; note that an Omega network cannot be "tuned" in this manner. A mixed radix system is particularly useful since the number of nodes grows extremely rapidly in a regular hypermesh. To realize ASCEND/DESCEND permutations (which are defined over a power of 2 elements), the radices n_i must be expressed as powers of 2; denote a mixed radix hypermesh as a series of radices, one for each dimension and in order $n-1,...,0$. Hence a (4,8,8) hypermesh has 3 dimensions, with 4 nodes aligned in dimension 2 and 8 nodes aligned along dimensions 1 and 0.

The routing algorithm in a hypermesh is similar to the routing algorithm in a torus. Assume the hypermesh is packet switched and that each packet carries an n-digit destination tag containing the address of the destination, as usual. Each node compares its own address with the packet's destination address; Suppose they differ in the $j-th$ digit. The packet can be sent out over the switch in the $j-th$ dimension, to the node with the same $j-th$ digit as the destination. If the node's address and the destination tag differ in m digits, then the packet can exit over either of m crossbars and still follow a minimum distance path. In fig. 1c node 0's address in base 4 is $0,0,0_4$ and node 15's address is $0,3,3_4$. There are 2 minimum distance paths between 0 and 15, given by $0 \rightarrow 12 \rightarrow 15$ and $0 \rightarrow 3 \rightarrow 15$ Note that the routing algorithm also applies to mixed radix hypermeshes.

The maximum distance between any 2 nodes in an n dimensional hypermesh (or equivalently the diameter of the hypermesh) is n. This definition assumes that a transmission from a node, over a hypergraph net (crossbar switch) to the destination node, is counted as one hop, as is usually the case for hypergraphs. There are $n!$ minimum distance paths between any two Processors separated by a distance of n. Note that if all the radices are precisely 2 then the hypermesh degenerates into a hypercube (with two nodes aligned along each dimension and with $logN$ dimensions) and hence to achieve better performance than a hypercube switches of degree > 2 must be used in each dimension.

2.1. Practical Implementation Details

In this section the practical implementation of three networks, the Omega network, the hypermesh and the hypercube, is examined. Networks can be packet switched or circuit switched. Assume an MIMD environment, where the paths of packets are not precomputed in advance to avoid link contention, as in an SIMD environment. In a packet switched network data is fragmented into relatively small packets, which are transferred from node to node in a "store-and-forward" manner until they reach their destination. To avoid "loosing" packets a packet from node A to node B is transmitted forward only if B has sufficient buffer space, which requires a flow control signal to be transmitted from B to A. In a circuit switched network a "connection" is established between two nodes A and B, which requires the exclusive use of every directed link that it traverses. To acknowledge a successful connection establishment in hardware a signal must flow from B to A, requiring transmissions "backwards" along the same path.

The links into the first stage of an Omega network and the links leaving all stages can be unidirectional; this implementation does not allow "backwards" flow control signals which requires the use of higher level protocols between processors to acknowledge communications and hence will probably render the scheme useless in a high performance parallel computer. Therefore the Omega network implementation should use bi-directional links between stages. A bi-directional link requires drivers and terminators at each end. The use of bi-directional links may also allow data (as well as flow control) to flow in the reverse direction.

In a hypermesh all nodes with addresses that differ in exactly one hyperspace dimension are interconnected with a switch. Whether the switch is one-sided or two-sided then the interconnection of b nodes requires b bidirectional links or $2b$ directed links (b links into the crossbar and b links exiting the crossbar). Similarly the links between nodes in a hypercube are bidirectional.

A b^n hypermesh requires $b^{n-1} \cdot n$ crossbar switches of degree b and $b^n \cdot 2n$ directed links. A b^n Omega network requires $b^{n-1} \cdot n$ crossbar switches of degree b and $b^n \cdot 2n$ directed links. A $2^{nlogb} = b^n$ hypercube requires 2^{nlogb} crossbar switches of degree $nlogb$ and $2^{nlogb} \cdot nlogb$ directed links. (The $2N$ links into and out of each node in each network have not been counted.) Various properties of radix 16 hypermeshes are shown in table 1. The b^n hypermeshes and the b^n Omega networks have identical costs, which are $O(loglogN)$ lower than binary hypercubes of similar sizes.

3. Functional Equivalence of the Hypermesh to the Ω, Ω^{-1} Networks

Let $\Omega_{d,n}$ denote a n-stage Ω network with d^n input/output terminals, built with $d \times d$ crossbar switches in each stage, with a "d-way shuffle" pattern before each stage. The "d-way shuffle" is defined here as ρ^{logd}, where ρ is the well known perfect shuffle

$$\rho\,[x_{n-1}\,x_{n-2}\,\cdots\,x_1\,x_0\,] = x_{n-2}\,x_{n-3}\,\cdots\,x_0\,x_{n-1}$$

where x_j denotes a bit. Note that ρ^{logd} is also equivalent to a single left-wise rotation of $x_{n-1}\,\cdots\,x_0$ where the x_i are interpreted as base-d digits rather than bits.

Let E^y denote the permutations satisfying

$$\text{if}\quad E[x_{n-1}\,\cdots\,x_y\,\cdots\,x_0] \rightarrow E[x_{n-1}\,\cdots\,\overline{x_y}\,\cdots\,x_0]$$
$$\text{then}\ \ E[x_{n-1}\,\cdots\,\overline{x_y}\,\cdots\,x_0] \rightarrow E[x_{n-1}\,\cdots\,x_y\,\cdots\,x_0]$$

Properties (1) & (2) are well known (i.e., see [15]). Properties (3)-(7) were proven in [17]

Property 1:

$$\Omega_{2,logN} = E^{n-1} \cdots E^0 \tag{1}$$

Property 2:

$$\Omega_{2,logN}^{-1} = E^0 \cdots E^{n-1} \tag{2}$$

Property 3: Let $N = d^n$ and where $d = 2^j$ for some $j \geq 1$;

$$\Omega_{d,n}^{-1} \supset E^0 \cdots E^{n-1} \tag{3}$$

Property 3 is illustrated in fig. 1b, which shows how an $\Omega_{4,2}$ network contains an embedded $\Omega_{2,4}$ network; although the binary topology has been transformed from that in fig. 1a, the two binary networks have the same functionality since each implements $E^3 \cdot E^2 \cdot E^1 \cdot E^0$.

Property 4: A d^n hypermesh can simulate $\Omega_{d,n}$ (and $\Omega_{d,n}^{-1}$), with all routes traversing only minimum distance paths.

(1) represents the well known "ASCEND" permutations and (2) the "DESCEND" permutations. The majority of SIMD algorithms make use of ascend or descend permutations, including the FFT, the Bitonic sort, and most matrix operations (i.e., see [12][15][16]). Hence the hypermesh can execute all these algorithms; furthermore it only traverses dimensions when necessary and hence it can be significantly faster than multistage networks.

Property 5: The n-dimensional base-d hypermesh is recursively partionable into independent hypermeshes, in size increments of d^{n-1}.

Property 6: Define a dual b^n hypermesh as a n dimensional base-b hypermesh where each link can support 2 connections simultaneously (rather than just 1). The dual b^n hypermesh can simulate the $2n - 1$ stage Benes network (built with nodes of degree b) and hence it is rearrangeably nonblocking.

Property 7: Define a dual-hypercube as a hypercube where each link can support 2 connections simultaneously. The dual hypercube is rearrangeable.

The following two properties of banyans are well known and repeated here for completeness. By definition every banyan has a unique path between each input port and each output port. As a consequence of the definition every $b^n \times b^n$ banyan has two features:

Property 8: Every node in stage 1 is the root of a "fan out tree"; from stage 1 the fan-out tree can reach exactly b^{s-1} nodes in stage s, and it can reach all b^{n-1} nodes in stage n (traversing links from left to right).

Property 9: Every node in stage n is the root of a "fan-in tree"; from stage n the fan-in tree can reach exactly b^{n-s} nodes in stage s, and it can reach all b^{n-1} nodes in stage $s = 1$ (traversing links from right to left).

3.1. 2D Meshs, Hypermeshes and Hypercubes

An attempt is made in this section to compare 3 networks in a discrete component environment. Assume that (1) a crossbar of bounded degree and bounded IO bandwidth can reside on a single IC, (2) crossbars are interconnected with twisted differential pair or perhaps fibers; in either case the signal velocity is about $2/3\ c$ (c is the speed of light). Similar assumptions are commonly made in the literature when comparing different networks.

Currently 64×64 GaAs crossbars which fit on a single IC are commercially available for about 1,200 dollars. Each crossbar link has a bandwidth of 200 Mbit/sec, and each crossbar requires a separate application specific IC to perform application specific routing functions. The IC does not contain buffering and thus a separate IC is needed for that function. When such a crossbar is used as a $k \times k$ node (for $k \leq 64$) assume that each "node link" is composed of $64/k$ "crossbar links" arranged in parallel.

To avoid processor intervention when routing messages in a 2D mesh assume that a crossbar based node handles these functions independently. In a 2D mesh each Processor has a degree 5 node; one connection to each nearest neighbor and one to the processor itself. A 4K Processor mesh would require 4K nodes; using a GaAs IC for each node, each node link would use $64/5 = 12.8$ crossbar links for an inter-node link bandwidth of 2.56 Gbit/sec. Therefore the time required to transmit a 128-bit packet between adjacent nodes is 50 *nanosec*.

In a 4K Processor hypercube each processor requires a degree 13 node. Using a GaAs IC for each node, each node link would use $64/13 = 4.92$ crossbar links for an inter-node link bandwidth of .985 Gbit/sec. Therefore the time to transmit a 128-bit packet between two neighboring nodes is 130 *nanosec*. Note that the propagation delay in the interconnect between nodes will be negligible if the nodes are within a few meters apart since the signal velocity is about $2/3 \; c$.

A number of choices exist for the hypermesh; a 8^4, 16^3 and 64^2 hypermesh can all interconnect 4K Processors. Consider a 2D 64^2 hypermesh with 64 rows and 64 columns, with a switch (or hypergraph net) in each row and each column, for a total of 128 switches. To use the same number of 64×64 GaAs ICs as the 2D mesh and the hypercube, assume that each hypermesh switch uses 32 GaAs ICs in parallel. The inter-node link bandwidth is then 6.4 Gbit/sec and the time to transmit a 128-bit packet between 2 nodes in the same row or column is then 20 *nanosec*. (In an MIMD machine each Processor requires a degree-2 switch which is an added overhead. In an SIMD machine each node never receives 2 packets simultaneously; the degree-2 switch can be implemented by having the processor read/write to 2 separate registers, one for the row transfer and one for the column transfer. In general, in an SIMD machine the switch of degree $loglogN$ associated with each Processor is not necessary and can be replaced by $loglogN$ registers, one register for each net/dimension.)

Consider implementing a Bitonic sort (or odd-even merge sort) of $4K$ elements on each architecture. Thompson and Kung's implementation of the odd-even merge algorithm on 2D meshes requires $6 \sqrt{N}$ data transfer steps (which is essentially optimal); In a 2D mesh the total data transfer time is then $1.92 \cdot 10^4$ *nanosec*. The hypermesh and hypercube algorithms are identical; each requires 78 data transfers and 78 floating point compare-exchange operations. The total data transfer time in a hypercube is then $1.01 \cdot 10^4$ *nanosec* and the total data transfer time in a hypermesh is $1.56 \cdot 10^3$ *nanosec*. The hypercube is faster than a 2D mesh by a factor of $1.92/1.01 = 1.9$. The hypermesh is faster than a 2D mesh by a a factor of $19.2/1.56 = 12.3$. The hypermesh is faster than a hypercube by a factor of $10.1/1.56 = 6.47$.

The hypermesh and hypercube algorithms as given exploit the bidirectional links between nodes as follows: to implement a compare-exchange operation between neighbors A and B, A sends its number to B, B simultaneously sends its number to A, and after the transfers are complete, each simultaneously compares the original number and the one just received, selects the appropriate one and discards the other. The 2D mesh algorithm as given does not exploit the bidirectional links between nodes, which are "hardwired" and cannot be made to act as one unidirectional link with twice the bandwidth, without added overhead. However even if the 2D mesh bidirectional links could be made to operate as unidirectional links with twice the bandwidth the hypermesh would be faster than a 2D mesh by a factor of roughly 6 for the parallel sorting algorithm.

When executing an algorithm which requires data transfers only to the four nearest neighbors (N,S,E,W), the hypermesh is faster than a 2D mesh by a factor of 2.5 or 1.25, depending on whether or not a bidirectional internodal link can act as a unidirectional link with twice the bandwidth and whether the algorithm can exploit such a feature. Note that ultimately propagation delays over some physically long hyperspace dimensions may limit the hypermesh performance, but this effect can be reduced with optical interconnect for those dimensions. Finally these figures should not be taken as absolutes, rather they represent an estimate of transmission delays for certain algorithms using some reasonable models.

4. Randomized Routing: A Bound on the Delivery Time In Buffered Hypermeshes

Define a "partial h-relation" as a set of Nh packets to be delivered; each source has at most h packets initially and each destination appears on at most h packets. The delivery time is defined as the time after which all packets have been delivered.

Theorem 1: Consider any $b^n \times b^n$ banyan followed by any other $b^n \times b^n$ banyan. Phase 1 performs in the first network and phase two performs in the second inverse. The time to complete either phase of a randomized routing algorithm is

(1) For any $k > nh$ where $n = \lceil \log_b N \rceil$ for any radix $b \geq 2$ and $\beta = (b-1)/b$;

$$Pr[time \; complete \; phase \; 1 \; \geq \; n + h + k] \; \leq \; Nh \cdot \left[\frac{enh\beta}{k} \right]^k \cdot e^{-nh\beta}$$

(2) For bounded b, each phase in any partial h-relation is completed in time $O(logN)$ with overwhelming probability; In particular, for any $Z \geq 2 \cdot e$;

$$Pr[time \; complete \; each \; phase \; \geq \; n + h + Zhn] \; \leq \; hN^{-(Zh-1)}$$

Proof: During phase 1, every packet randomly and independently selects a phase 1 destination over the range $0, \cdots, N-1$ inclusive. Let the stages in each banyan be labeled from 1 to n. The phase 1 destination can be viewed as a vector of n base-b digits, with digit i $(1 < i \leq n)$ controlling the routing decision for stage i. The phase 1 destination can be selected by generating n random digits from the range $0,...,b-1$ inclusive.

Since the first half of the network is a banyan, then every packet must follow a unique path from its input port to its phase 1 destination. Consider any three packets labeled X, Y and Z. Since all phase 1 destinations are selected randomly and independently, then the event that X and Y intersect during phase one is independent from the event that X and Z intersect during phase 1. It is possible for the paths of two (or more) packets to intersect during phase 1. However, if the paths diverge then they will never again intersect during phase 1 (as a consequence of the fan-in tree property). Thus the first half of the network is "non-overtaking", i.e., suppose the paths of packets X and Y intersect and share one or more common links (called the "common subpath"); then if Y is ahead of X at some point while traversing the common subpath then Y will always remain ahead of X while traversing the common subpath. The "nonovertaking" property leads to the following fact:

Fact 1: All packets whose paths have intersected the path of X prior to stage s cannot contribute to Xs queueing delays at stage s.

Consider the case where some packet labeled X will be the last packet to be delivered during phase 1. The worst case expected delay of packet X enter the 1st stage is $(h-1) < h$, since each source has at most h packets and X could be the last to enter.

Having entered a switch s^* in the first stage, by fact 1 X's queueing delay while in the first stage (which does not count its own deterministic transmission time) is due to packets arriving from the $b-1$ other sources connected to s^*. Each of these sources has h packets, and each packet is equally likely to select any of the b outgoing links. Hence the expected number of these which must traverse the same link as X is $h \cdot (b-1)/b$. Let $\beta \equiv (b-1)/b$. Hence X's queueing delay in stage 1 is $< h$.

Consider the switch s^* in stage i of the banyan which X visits during phase 1. By the definition of a banyan and property 9 it follows that switch s^* is the root of a fan-in tree which spans b^s sources, with b^{s-1} sources reachable traveling backwards from any link entering s^*. Hence the number of sources whose packets may for the first time intersect with X in stage s is $(b-1) \cdot b^{s-1}$. By the definition of a banyan and property 8 every source can reach b^s links leaving stage s. By the random nature of the routing decisions made in each stage, the path of a packet from any source is equally likely to intersect any of these b^s links leaving stage s. Therefore the expected number of new packets whose paths will intersect with Xs in stage s is

$$h \cdot (b-1) \cdot b^{s-1} \cdot \frac{1}{b^s} = h\beta$$

Therefore over all n stages the expected number of packets whose paths will intersect with that of X is $nh\beta$. The $Nh-1$ packets can be viewed as forming a set of $Nh-1$ Bernoulli trials; for convenience inflate the number of trials to Nh by adding a dummy event with probability 0. The Nh trials are then Poisson trials, and the expected number of successes in the Nh Poisson trials is $nh\beta$. Thus the probability of success in any one trial is $n\beta/N$.

Fact 2 (Hoeffding [8]): If T is the number of successes in N independent Poisson trials with probabilities $p_1, ..., p_N$, then if $\sum p_i = Np$ and $m \ge Np + 1$ is an integer then

$$Pr(T \ge m) \le B(m, N, p)$$

where $B(m, M, p)$ is the probability of at least k successes in M Bernoulli trials, with the probability of success in any trial equal to p.

Packet $X's$ queueing delay in phase 1 is at most equal to the number of packets that intersect its route in Phase 1. Therefore for $k > nh\beta$

$$Pr[X's \text{ queueing delay in Phase 1} \ge k]$$
$$\le B\left[k, Nh, n\beta/N\right]$$

Since there are at most Nh packets to consider the probability that at least one of them experiences a delay of at least k is upper bounded by

$$Nh \cdot B\left[k, Nh, n\beta/N\right]$$

(This upper bound is pessimistic [19][10] but it does yield a simple closed form expression. The bound is true even if the trials are dependent.)

Fact 3: Valiant's bound on Chernoff's bound states that for any $m > Mp$,

$$B\left[m, M, p\right] \le \left[\frac{eMp}{m}\right]^m \cdot e^{-Mp}$$

Therefore for any $k > nh\beta$,

$$B\left[k, Nh, n\beta/N\right] \le \left[\frac{ehn\beta}{k}\right]^k \cdot e^{-nh\beta}$$

Therefore for any $k > nh\beta$

$$Pr[\text{time complete phase } 1 \ge n + h + k] \le Nh \cdot \left[\frac{ehn\beta}{k}\right]^k \cdot e^{-nh\beta}$$

By symmetry phase two can be viewed by running phase 1 "backwards" and hence the same bound applies for phase 2. Hence, part (1) is proved. (Note that for partial h-relations, add dummy trials with zero probability of success; the same arguments yield the same conclusion.)

For arbitrary bounded b, let $Z \ge b \cdot e$ and $k \ge Zhn$. Each phase is completed in time $O(logN)$ with overwhelming probability;

$$Pr[\text{time complete each phase } \ge n+h+Zhn] = Nh \cdot \left[\frac{ehn\beta}{Zhn}\right]^{Zhn} \cdot e^{-nh\beta}$$
$$< hN \cdot b^{-Zhn} \cdot e^{-nh\beta} < hN \cdot N^{-Zh}$$

(note that $\beta^{Zhn} \ll 1$) so that for bounded h

$$Pr[\text{time complete each phase } \ge h + O(logN)] \le hN^{-(Zh-1)}$$

Hence part (2) is proved. $\square$

Consider a $2n-1$ stage network formed by the concatenation of any two n stage binary banyan networks ($b = 2$; last stage of the 1st banyan is redundant if the second topology is the inverse of the first). By application of theorem 1, the time to complete either phase of a randomized routing algorithm can be bounded as follows.

The delay to complete either phase in any degree-2 banyan ($b = 2$, $n = logN$ and $\beta = 1/2$) is:
(1) For any $k > nh/2$ where $n = \log_2 N$;

$$Pr[\text{time complete phase } 1 \ge n+h+k] \le Nh \cdot \left[\frac{ehn}{2k}\right]^k \cdot e^{-nh/2}$$

(2) Each phase in any partial h-relation is completed in time $O(logN)$ with overwhelming probability; In particular, for any $Z \ge e$

$$Pr[\text{time complete phase } 1 \ge n+h+Zhn] \le Nh \cdot \left[\frac{ehn}{2Zhn}\right]^{Zhn} \cdot e^{-nh/2}$$
$$< hN \cdot 2^{-Zhn} \cdot e^{-nh/2} < hN \cdot N^{-Zh}$$

so that for h bounded

$$Pr[\text{time complete each phase } \ge h + O(logN)] \le hN^{-(Zh-1)}$$

This proof of Theorem 1 exploits symmetry arguments and hence the theorem applies to arbitrary banyans of arbitrary degree (unlike results in [10]).

Claim: The bounds in Theorem 1 are applicable to a b^n hypermesh.

Proof: By property (4) the b^n hypermesh can simulate a $b^n \times b^n$ Omega or a $b^n \times b^n$ Omega inverse network, by constraining the order in which dimensions are traversed. In phase 1 let the hypermesh simulate a $b^n \times b^n$ Omega network acting as the randomization network. In phase 2 let the hypermesh simulate a $b^n \times b^n$ Omega inverse network acting as the routing network. (Assume that every packet must traverse every hyperspace dimension in the hypermesh, so that the "bypassing" of stages is disabled.) It follows that theorem 1 must be valid for hypermeshes. $\square$

Surprisingly, without modification Theorem 1 does not bound the delivery time to $O(logN/loglogN)$ for degree $b = logN$; From theorem 1, the time to complete either phase of the randomized routing algorithm in a b^n hypermesh (or dual banyan), where

$N \le b^n$, $b = \log_2 N$ and $n = \lceil logN / loglogN \rceil$, is for any $k > nh\beta$

$$Pr[time\ complete\ phase\ 1 \ge n + h + k] \le Nh \cdot \left[\frac{ehn\beta}{k}\right]^k \cdot e^{-nh\beta}$$

For fixed $Z \ge \beta C$ where fixed $C > e$

$$Pr[time\ complete\ each\ phase \ge n + h + Zhn] \le Nh \cdot \left[\frac{ehn\beta}{\beta Cnh}\right]^{\beta Chn} \cdot e^{-nh\beta}$$

$$\le Nh \cdot e^{\beta Cnh} \cdot C^{-\beta Cnh} \cdot e^{-nh\beta}$$

For fixed C the probability is not bounded since (for some fixed K)

$$\lim_{n \to \infty} Nh \cdot e^{-nh\beta(ClogC - C)} = \lim_{n \to \infty} b^n \cdot h \cdot e^{-nK} \to \infty$$

Note that the difficulty with the previous method is that it "charges" packet X with a unit delay for every other packet whose path has intersected X's, even if the packet in question does not actually delay X, i.e., if a packet Y uses a link while X is using some other link then Y does not actually delay X yet the worst case analysis still charges X a unit delay. To prove $O(logN / loglogN)$ delivery time for randomized routing as $N \to \infty$, make the restriction that each node on the outer stages of the Benes network has just one active source, which requires extra hardware. (Alternatively we can assume that there are $O(1)$ active sources in each first stage node, or that there are $O(b)$ copies of the network between active sources and destinations.)

Theorem 2: Consider any $b^n \times b^n$ banyan followed by any other $b^n \times b^n$ banyan. Each node in the first stage has b input terminals; let a single active source (with $\le h$ packets) be connected to each node in the 1st stage; the remaining $b - 1$ input terminals are inactive (i.e., they have 0 packets with probability one). Similarly, a single destination is associated with each node in the last stage of the second banyan. Phase one destinations are still selected from all reachable outgoing links of the first banyan.

Let N denote the number of active sources in the banyan. For $N = b^{n'}$, the banyan requires $n' + 1$ stages of $b \times b$ nodes, with $b^{n'}$ nodes per stage, for a total of $b^{n'} \cdot (n' + 1)$ nodes. Let $n = n' + 1$. As before, phase 1 performs in the first banyan and phase two performs in the second. The time to complete either phase of a randomized routing algorithm is
(1) For any $k > nh$

$$Pr[time\ complete\ phase\ 1 \ge n + h + k] \le Nh \cdot \left[\frac{enh}{bk}\right]^k \cdot e^{-nh/b}$$

(2) For $b = O(logN)$ and for $Z > e$ and bounded h

$$Pr[time\ complete\ phase\ 1 \ge h + O(logN / loglogN)]$$
$$\le h \cdot N^{-(Zh - 1)}$$

Proof: The proof of theorem 2 is similar to that of theorem 1. By assumption each node in stage 1 is fed by exactly one source with $\le h$ packets. Consider some packet labelled X. X's queueing delay to enter the first stage 1 is $< h$.

Consider the switch s^* in stage i of the banyan which X visits during phase 1. By the definition of a banyan and property 9 and the fact that each first stage node has one active source, it follows that switch s^* is the root of a fan-in tree which spans b^s input terminals and b^{s-1} active sources, with b^{s-2} active sources reachable by traveling backwards from any link entering s^* (for $s \ge 2$). Hence the number of sources whose packets may for the first time intersect with X in stage s ($s \ge 2$) is $(b-1) \cdot b^{s-2}$. By the

definition of a banyan and property 8 every source can reach b^s links leaving stage s. By the random nature of the routing decisions made in each stage, the path of a packet from any source is equally likely to intersect any of these b^s links leaving stage s. Therefore the expected number of new packets whose paths will intersect with Xs in stage s ($s \ge 2$) is

$$h \cdot (b-1) \cdot b^{s-2} \cdot \frac{1}{b^s} = \frac{h\beta}{b}$$

The expected number of packets whoses paths will intersect with that of X on links leaving stages $2...n$ is

$$\frac{(n-1)h\beta}{b} < \frac{nh}{b}$$

After adding a dummy trial as before, the Nh packets can be viewed as forming a set of Nh Poisson trials where the expected number of successes in the Nh Poisson trials is $\le nh/b$. Thus the probability of success in any one trial is $nh/(bNh)$. Packet X's queueing delay in phase 1 is at most equal to the number of packets that intersect its route in Phase 1. Applying Hoeffding's result it follows that for $k > nh/b$

$$Pr[X's\ queueing\ delay\ in\ Phase\ 1 \ge k]$$
$$\le B\left[k, Nh, nh/(bNh)\right]$$

Since there are at most Nh packets to consider the probability that at least one of them experiences a delay of at least k is upper bounded by

$$Nh \cdot B\left[k, Nh, nh/(bNh)\right]$$

Therefore for any $k > nh/b$

$$Pr[time\ complete\ phase\ 1 \ge n + h + k]$$
$$\le Nh \cdot \left[\frac{ehn}{bk}\right]^k \cdot e^{-nh/b}$$

By symmetry phase two can be viewed by running phase 1 "backwards" and hence the same bound applies for phase 2. Hence part (1) is proved.

Let $N = b^{n'}$ where $b = \log_2 N$; $n' = \lceil loglogN \rceil$ and $n = n' + 1$, thus the node degree is $O(logN)$ and the diameter is $O(logN / loglogN)$. For $Z \ge e$ and for $Zhn > nh/b$ it follows that

$$Pr[time\ complete\ phase\ 1 \ge n + h + Zhn] \le Nh \cdot \left[\frac{ehn}{bZhn}\right]^{Zhn} \cdot e^{-nh/b}$$

$$\le Nh \cdot N^{-Zh} \le h \cdot N^{-(Zh-1)}$$

For bounded h the delay is $O(logN / loglogN)$, and hence part (2) is proved. $\square$

For theorem 2 to apply to partitions, the hypermesh must be partitioned so that each partition obeys the active source restriction (i.e., the outermost stage of the Benes network being simulated should have one active source). After partitioning, b grows relative to the diameter. Furthermore, since inactive stages are bypassed when simulating the Benes network, the partitions actually become faster.

Note that proofs regarding randomized routing are usually of theoretical interest; Valiant's proof of $O(logN)$ delivery time in hypercubes was the first rigorous proof that the hypercube could self-route an arbitrary permutation in $O(logN)$ time. It was shown that the hypermesh can match and exceed the performance of a hypercube in this framework. Practically we would not recommend using randomized routing as suggested by Valiant since it is quite inefficient; in practice there are alternative schemes such as "hot potato" or "deflection" routing which perform remarkably well [6]. Note however that it has never been rigorously proven that hot-potato routing delivers a permutation in $O(logN)$ time.

4.1. Rearrangeability in Dual Hypermeshes

Define a "dual" hypermesh as either of the following; (1) two hypermeshes, with the output of one being fed into the other, or (2) a hypermesh where each link can support two connections simultaneously, for example by use of Time Division Multiplexing (TDM) or Space Division Multiplexing (SDM).

In [17] it was proved that the dual hypermesh (and similarly defined dual hypercube) was rearrangeable. The proof follows simply from Properties 1 and 3, which together show that a b^n hypermesh can simulate a binary Omega network. It is well known that the concatenation of a strict buddy binary banyan and its inverse is rearrangeable. By Properties 1-3 a dual b^n hypermesh can simulate the concatenation of a binary Omega and binary Omega inverse network and hence must be rearrangeable. Let the node degree be $logN$. The number of stages in a radix-$logN$ Omega network is exactly $\lceil logN/loglogN \rceil$; the corresponding hypermesh has the same number of dimensions. Each pass through a dual hypermesh requires $2 \cdot \lceil logN/loglogN \rceil$ unit-route steps. Without pipelining a set of h permutations can be performed in deterministic time $2 \cdot h \cdot \lceil logN/loglogN \rceil$. With pipelining (i.e., having a different permutation moving along in each stage simultaneously) a distinct permutation can be realized in each step.

4.2. A Routing Algorithm for Base 2^t Hypermeshes

The routing of any permutation through the radix 2^t Omega+Omega Inverse network can be found using Andresen's routing algorithm. The routing for a radix 2^t dual hypermesh can be computed from the routing through a radix 2^t Omega+Omega-Inverse network; There is a correspondence between positions in the hypermesh and positions in the Omega network. The average path distance in the hypermesh can be reduced as follows (when pipelining is not used). If some connection labeled X appears at some node s^* in the first half of the dual hypermesh and if it reappears at the same node in the last half of the dual hypermesh, then this loop need not be traversed in the hypermesh (if the degree n node allows bypassing of all intermediate dimensions.)

4.3. Performance Analysis, Renewal Case

The previous theorem and bounds apply for realizations of h-relations in a "nonregenerative" traffic model, where traffic is not continually appearing at the inputs to be delivered to the destinations. Define a "partial λ-relation" as traffic specification such that the packet arrivals at each source form a stationary Poisson process, such that the intensity (or average rate) of each source is no more than λ and the intensity at each destination is no more than λ. Let the links have exponentially distributed service times. Each node has b output queues of infinite capacity and the network is asynchronous and buffered.

Theorem 3; In the multistage network formed by the concatenation of two b^n banyans, the exact delay over any and all λ-relations is given by

$$D = \frac{2 \cdot log_b N - 1}{1 - \lambda}$$

Proof: (see [18] for a proof.)

By setting $b = logN$ it follows immediately that the bandwidth in the dual hypermesh is $\Theta(N)$ and the expected delay is $O(logN/loglogN)$.

5. Conclusions

The "hypermesh" is a novel type of interconnection network with some very interesting properties. It can realize all permutations over nearest neighbors in a hypercube and it can realize all Omega and Omega-Inverse passable permutations, each in minimum distance. Most parallel algorithms use precisely these permutations. As a result a hypermesh based machine can execute most algorithms designed for either a hypercube or Omega based machine without structural modification and without added delays.

Meshes/toroids which can be embedded into a hypercube with unit dilation can also be embedded into a hypermesh in unit dilation. The hypermesh can execute algorithms on these structures with no loss of efficiency. Any target graph which can be embedded into a hypercube with dilation k can be embedded into a hypermesh with dilation $\leq k$ (since nodes are never further apart in a hypermesh than a hypercube).

Using switches of degree-$logN$, the hypermesh has a smaller diameter than a hypercube by a factor of $O(loglogN)$ and its uses fewer degree-$logN$ switches and directed links by a factor of $O(loglogN)$. The factor of $O(loglogN)$ improvement in cost and diameter over a hypercube has both practical and theoretical significance. For large networks the decrease in backplane wires alone can be significant. The improvement of the hypermesh over both the hypercube and multistage networks such as the Omega were verified in the theoretical properties as well.

6. References

[1] S. Andresen, "The Looping Algorithm Extended to Base 2^t Rearrangeable Switching Networks", IEEE Trans. Comm., Oct. 1977, pp. 1057-1063

[2] S. Abraham and K. Padmanabhan, "Constraint Based Evaluation of Multicomputer Networks", Int. Conf. Parallel Processing, 1991, pp. 521-525

[3] V.E. Benes, "The Mathematical Theory of Connecting Networks and Telephone Traffic", Academic Press, New York, 1965

[4] C. Berge, "Graphs and Hypergraphs", North-Holland, 1974

[5] J.C. Bermond, J. Bond and J.F. Scale, "Large Hypergraphs of Diameter 1", in Graph Theory and Combinatorics, Ed. B. Bollobas, Academic Press, 1984

[6] C. Fang and T.H. Szymanski, "An Analysis of Deflection Routing in Multidimensional Regular Mesh Networks", IEEE Infocom'91, April 1991, pp. 859-868

[7] W.J. Dally, "Performance Analysis of k-ary n-cube Interconnection Networks", IEEE Trans. Comput., June 1990, pp. 775-785

[8] W. Hoeffding, "On the Distribution of the Number of Successes in Independent Trials", Ann. Math. Statistics 27 (1956), pp. 713-721

[9] D.H. Lawrie, "Access and Alignment of Data in an Array Processor", IEEE Trans. Comput., Dec. 1975, pp. 1145-1155

[10] D. Mitra and A. Cielsak, "Randomized Parallel Communications on an Extension of the Omega Network" JACM, Vol. 34, No. 4, Oct. 1987, pp. 802-824

[11] M.C. Pease, "The Indirect Binary n-Cube Microprocessor Array", IEEE Trans. Comput., Vol. C-26, May 1977, pp. 458-473

[12] F.P. Preparata and J. Vuillemen, "The Cube-Connected Cycles: A Versatile Network for Parallel Computation", CACM, May 1981, pp. 300-309

[13] I.D. Scherson and S. Sen, "Parallel Sorting in Two-Dimensional VLSI Models of Computation", IEEE Trans. Comput., Feb. 1989, pp. 238-249

[14] H.J. Siegel, "The Theory Underlying the Partitioning of Permutation Networks", IEEE Trans. Comput., Vol. C-29, No. 9, Sept. 1980, pp. 791-780

[15] H.J. Siegel, "Interconnection Networks for Large Scale Parallel Processing; Theory and Case Studies", Lexington Books, 1985

[16] H.S. Stone, "High Performance Computer Architecture", Addison-Wesley, 1987

[17] T.H. Szymanski, "A Fiber-Optic "Hypermesh" for SIMD/MIMD Machines", IEEE Supercomputing 90, Nov. 1990, pp. 103-110

[18] T.H. Szymanski and C. Fang, "Randomized Routing of Virtual Connections in Extended Banyans", Int. Conf. Parallel Processing, 1991

[19] L.G. Valiant and G.J. Brebner, "Universal Schemes for Parallel Communications", Proc. 13th Annual ACM Symp. on Theory of Computing, 1981, pp. 263-277

Table 1a: Maximum distance				
size ->	16^3	16^4	16^5	16^6
binary hypercube	12	16	20	24
d-ary n-cube	24	32	40	48
hyper-mesh	3	4	5	6

Table 1b: # Nodes 1 hop away				
size ->	16^3	16^4	16^5	16^6
binary hypercube	12	16	20	24
d-ary n-cube	6	8	10	12
hyper-mesh	45	60	75	90

Table 1c: Cost Comparison			
logN->	8	8	8
type	Omega	hmesh	hcube
links	3072	3072	4608
crossbars	192(8)	192(8)	512(9)

Table 1d: Cost Comparison			
logN->	16	16	16
type	Omega	hmesh	hcube
links	524,288	524,288	1,048,576
crossbars	16,384	16,384	65,536

Table 1: (c-d): Cost comparison between logN-degree hypermeshes and hypercubes (value in brackets is node degree).

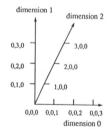

node addresses (in base-4)

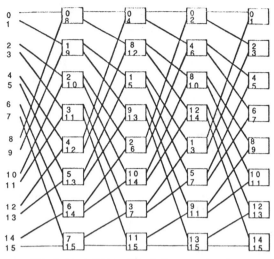

Figure 1a: A 4 stage base 2 Omega network.

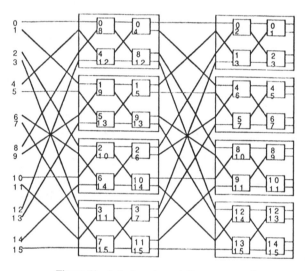

Figure 1b: A 2 stage base 4 Omega network.

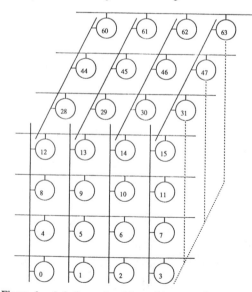

Figure 1c: A 3 dimensional base 4 hypermesh.

RANDOMIZED ROUTING OF VIRTUAL CONNECTIONS
IN EXTENDED BANYANS

Ted Szymanski & Chien Fang

Department of Electrical Engineering and
Center for Telecommunications Research
Columbia University
New York, NY, 10027
email:teds@apollo.ctr.columbia.edu

Abstract: The following properties are proven for the concatenation of two self-routing banyans; With internal buffering within the nodes and a packet switched mode of operation; (1) the network's carried load equals the offered load λ for $\lambda < 1$, and (2) the expected delay is $O(\log_b N)$ (where b is the node degree) regardless of the traffic pattern, provided that no output is overloaded. Therefore the concatenation of any two buffered banyans yields a robust self-routing switch with $\Theta(N)$ bandwidth. The above approach requires re-sequencing queues at the output ports, and internal buffering may lead to slower implementations than internally unbuffered switches. Without internal buffering within the nodes and a circuit switched mode of operation, the following is proven; (1) the expected blocking probability of an individual connection request is $O(\log_b N (k/d)^d)$, where d is the link "dilation" factor and k is a constant. (A d-dilated link can carry d connections simultaneously.) By increasing the dilation the blocking probability can be made arbitrarily small, regardless of the traffic patterns, provided that no output is overloaded. The unbuffered switch uses an extremely simple routing algorithm, is easily implemented in hardware, does not require any internal "backtracking" or "path hunting", and may be useful in applications requiring very fast circuit establishment.

1. Introduction

An ideal $N \times N$ switch has the following properties: (1) it requires $O(\log N)$ stages and $O(N \log N)$ crosspoints, (2) it is self-routing in time $O(\log N)$ and (3) it is deterministically nonblocking regardless of the traffic pattern (provided that there are no destination conflicts). The only network proven to meet (1) - (3) is the AKS sorting network [1] (for permutation traffic), which is of theoretical interest only. The only practical self-routing network proven to have $\Theta(N)$ bandwidth is the "Batcher-Banyan" [5] and related sorting/permutation networks, with $\Theta(N (\log N)^2)$ crosspoints. In fact finding an efficient explicit construction for an ideal $N \times N$ switch is a fundamental problem in parallel computation. However, if these constraints are relaxed slightly some near optimal solutions with practical uses can be achieved.

An $N \times N$ crossbar switch is internally nonblocking regardless of the traffic pattern, provided that no output port is overloaded. The switch therefore exhibits an optimal $\Theta(N)$ bandwidth but its cost of $O(N^2)$ crosspoints is prohibitive for large N. There are known switch architectures which have $\Theta(N)$ bandwidth, $O(N \log N)$ nodes and $O(\log N)$ depth, i.e., the well-known Benes rearrangeable network [3]. However, this switch relies on a global routing algorithm and on connection rearrangement, making it impractical for many applications. The well-known 3 stage Clos network can be made deterministically nonblocking, without the need for connection rearrangement. However the cost grows to $O(N^{3/2})$ crosspoints and a global routing (or path hunt) algorithm is still required.

Randomization is a powerful technique used often in theoretical computer science. In a pioneering paper, Valiant et. al. [15]

proposed randomized routing as a self-routing scheme which can provably deliver a single set of Nh packets within time $O(h \log N)$ in a hypercube with "overwhelming probability", regardless of the traffic pattern, provided that no destination was overloaded. However the algorithm cannot be pipelined in a hypercube to allow for the continuous delivery of packets, thus the hardware has a suboptimal steady state bandwidth of $O(hN/h \log N)$. Mitra et. al. [10] extended Valiant's work and considered randomized routing in the "extra stage Omega network"; it was proven that the "fully extended (binary) Omega network" (essentially the concatenation of two binary Omega networks) has $\Theta(N)$ bandwidth in the steady state. However, their proof was complicated and lacked generality. Furthermore, the extended Omega networks have no efficient 2D or 3D implementations. Our first result proves in a simpler manner that the concatenation of a buffered banyan and its inverse has $\Theta(N)$ bandwidth when using a randomized algorithm, provided that no output is overloaded. The first banyan routes each packet from its input to a random destination; the second banyan routes the packet to its final destination. Such a multistage network also has efficient 2D and 3D implementations. Fig. 1 shows a 3D construction for two concatenated banyans.

There have been an enormous number of papers recently in the communications literature, presenting designs for so called "fast packet switches". However, it is rarely proven that such switches are robust in the face of arbitrary traffic patterns; very often their cost is many times that of two buffered banyans, and often their control is significantly more complicated. For example, the "Batcher-Banyan" switch [5] being constructed at Bell Communications Research is one such switch; while it does have $\Theta(N)$ bandwidth, its cost is a factor of $O(\log N)$ more than one or two buffered banyans, and its control is considerably more complicated.

If the traffic is randomly distributed, even internally nonblocking switches (i.e. crossbars or the Batcher-Banyan) exhibit a blocking probability of roughly e^{-1}, which is due solely to destination conflicts. A simple solution to this problem is to "speed up" the switch relative to the inputs/outputs so that it can deliver multiple packets to an output port simultaneously. Note that this approach, if implemented in hardware with parallel switches, is never crosspoint effective, and in very large scale switches the $\Omega(N^2)$ or $\Omega(N (\log N)^2)$ crosspoint cost will dominate all others.

2. Topology and Routing algorithms for Multistage Banyans

The basic building block is a b-input port b-output port crossbar, which has been modified to allow up to d "virtual connections" to share any one port. These building blocks are then used in a multi-stage switch architecture. Assume each input port has $h < d$ active sources generating connection requests, and each output port can receive up to d connections simultaneously.

By definition every banyan has a unique path between each input port and each output port . As a consequence every $b^n \times b^n$ banyan, with b^n input ports and b^n output ports, has two features:

This research was supported through CTR by NSF Grant CDR-88-11111.

Property 1: Assume stages are labeled 1 to n from left to right. Every node in stage 1 is the root of a "fan out tree"; from stage 1 the fan-out tree can reach exactly b^{s-1} nodes in stage s, and it can reach all b^{n-1} nodes in stage n.

Property 2: Every node in stage n is the root of a "fan-in tree"; from stage n the fan-in tree can reach exactly b^{n-s} nodes in stage s, and it can reach all b^{n-1} nodes in stage $s=1$.

Any strict buddy $b^n \times b^n$ banyan network can be topologically rearranged into a 2 stage network, with "factors" of size $b^{\lceil n/2 \rceil} \times b^{\lceil n/2 \rceil}$ in one stage, and "factors" of size $b^{\lfloor n/2 \rfloor} \times b^{\lfloor n/2 \rfloor}$ in the other stage. A $k \times k$ factor is either a $k \times k$ crossbar or a k-input k-output banyan. A radix-b strict buddy banyan is defined as one with the following property; in every stage except the last, each node has $b-1$ "buddy" nodes that is connected to the same successor node in the next stage.

3. Traffic Imbalances In a Banyan

Let $l(i,j)$ denote link j in stage i ($0 \leq j \leq N-1$) and $1 \leq i \leq \log_b N \equiv n$ in a $b^n \times b^n$ banyan. Define a "partial h-relation" as a traffic specification (i.e., a set of packets) such that each source has at most h packets and each destination appears on at most h packets. Partial h-relations will be used when analysing a synchronous version of the proposed architecture where the switch attempts to deliver all packets in a partial h-relation in each time slot. Define a "partial λ-relation" as traffic specification such that the packets at each source form stationary renewal processes, such that the intensity (or average rate) of each source is no more than λ and the intensity at each destination is no more than λ. Partial λ-relations will be used when analysing an asynchronous version of the switch architecture.

Define the "traffic imbalance" of a link $l(i,j)$, denoted $t(h',i,j)$, as the number of packets whose paths must traverse that link given a partial h-relation denoted h' and an assignment of paths such that all links are as evenly loaded as possible. Define $T(i,j)$ as the maximum of $t(h',i,j)$ over all i, all j and all h-relations, for $h = 1$. In other words $T(i,j)$ represents the maximum number of connections which are forced to use some link in the network, over all possible permutations. The following properties apply to any $b^n \times b^n$ banyan and are not difficult to prove.

Property 3: $\max t(h',i,j)_{all\ h',\ all\ j} = \min(b^i, b^{n-i}) \cdot h$

Property 4: For $h = 1$, $T(i,j) = b^{\lfloor n/2 \rfloor} = O(\sqrt{N})$

Property 5: The number of $h=1$-relations (i.e., permutations) which result in a traffic intensity of i or more on a link leaving stage s is given by

$$\begin{bmatrix} b^s \\ i \end{bmatrix} \cdot \begin{bmatrix} b^{n-s} \\ i \end{bmatrix} \cdot i! \cdot (N-i)!$$

Property 6: Assuming $h = 1$ and that all permutations are equally likely, the probability a link leaving stage s has a traffic intensity of i or more is given by

$$\begin{bmatrix} b^s \\ i \end{bmatrix} \begin{bmatrix} b^{n-s} \\ i \end{bmatrix} / \begin{bmatrix} N \\ i \end{bmatrix}$$

Properties 3-6 indicates that the traffic imbalance problem in a multistage network can be so severe that a single banyan cannot be used as a robust general purpose switch. In a parallel processing application data skewing algorithms can ensure that traffic is reasonably distributed. Our main results of $\Theta(N)$ bandwidth are proven for single banyans with random traffic; in telecommunication applications a randomization network (another banyan) may be required.

4. Randomized Routing

By definition "complete randomization" is possible when each input port of the distribution network can reach every input port to the routing network. Let $r(n)$ be the number of stages in the randomization (routing) networks. For complete randomization set $r = n$. However if the wires joining both halves perform no permutations, then the last stage of the distribution network is redundant and can be removed, yielding $r = n-1$ stages for complete randomization. Phase 1 of the routing algorithm applies to the randomization network and phase 2 applies to the routing network. During phase 1 every packet randomly and independently selects a phase 1 destination over the range $0, \cdots, N-1$ inclusive. The phase 1 destination can be viewed as a vector of n base-b digits, with digit i ($1 < i \leq n$) controlling the routing decision for stage i. Therefore the phase 1 destination can also be selected by generating n random digits from the range $0,...,b-1$ inclusive. In either case the outcome of a packet's routing decision at each stage is totally random and independent from the routing decision made by any other packets. Therefore at any stage a packet is randomly distributed over the output links of that stage. Fig. 2 shows the concatenation of two $2^3 \times 2^3$ banyans for complete randomization.

Property 7: In a multistage network formed by the concatenation of any radix b strict buddy banyan and its inverse, then

$$T(i,j) = 1$$

Proof: It can be shown that such a network is rearrangeable by establishing a topological equivalence to the radix b Benes network. Therefore it is not possible to select any permutation which forces any link to carry more than one connection. □

4.1. Incomplete Randomization

It may be undesirable to use complete randomization when $n > 2$, let $r < n-1$. A reasonable option is $r = n/2$. When incomplete randomization is used, every packet selects a phase one destination randomly from all those that it can reach. Hence the route taken by each packet at each stage of the randomization network is still totally random.

Property 8: Assuming 1-relations

$$T(i,j) = b^{\lfloor n/4 \rfloor} = O(N^{1/4})$$

Proof: First, assume that each $\sqrt{N} \times \sqrt{N}$ factor in fig. 3 is a crossbar switch. In this case the network is a 3 stage Clos network and by property 7 the traffic intensity over the inter-factor links is unity. The inter-factor links need not be considered any further. Now view each factor as a 2 stage $\sqrt{N} \times \sqrt{N}$ banyan built with $N^{1/4} \times N^{1/4}$ nodes. By property 4 the maximum traffic imbalance in each factor is $O(N^{1/4})$. □

5. Performance Analysis, Renewal Case

Let the sources be fed by Poisson processes with intensity λ and let the link transmission time be exponentially distributed. Each node has b output queues of infinite capacity and the network is asynchronous and buffered. Let $T(i,j)$ now denote the maximum intensity on $l(i,j)$ for all λ-relations, all i and all j.

Theorem 1: For full λ-relations (let $n = \log_b N$);

(1) In a fully extended banyan ($r = n-1$),

$$T(i,j) = \lambda \quad \text{for } 1 \leq i \leq 2n-1 \text{ and } 0 \leq j \leq N-1$$

(2) In a partially extended banyan ($r = n/2$, n even);

$$T(i,j) \leq \lambda \cdot b^{n/4}$$

Proof: For case (1); Consider the links leaving stage i of the randomization network. Transform the banyan into a 2 stage network with factors of size $b^i \times b^i$ in the first stage and factors of size $b^{n-i} \times b^{n-i}$ in the second. Let Θ denote the set of sources feeding one particular factor in the first stage, let Φ denote the set of links leaving that factor, and let l^* and l^{**} denote two distinct members of Φ. Let $\lambda(s,l)$ denote the steady state intensity contribution from source s over link l. By the nature of the routing algorithm, within every factor every output link is equally likely and hence

$$\sum_{s \in \Theta} \lambda(s,l^*) = \sum_{s \in \Theta} \lambda(s,l^{**})$$

and by conservation of flow

$$\sum_{l \in \Phi} \sum_{s \in \Theta} \lambda(s,l) = b^i \cdot \lambda$$

which together imply that

$$\sum_{s \in \Theta} \lambda(s,l^*) = \lambda \quad \text{for} \quad all \quad l^* \in \Phi$$

Hence part (1) holds for every stage in the randomization network. By symmetry the traffic intensities in phase two are identical and hence (1) applies for every stage in the routing network. Hence part (1) is proved.

For $r = n/2$ (and n even), an argument similar to the proof of property 8 indicates that all inter-factor links have intensity λ and the maximum imbalance occurs within the factors. Since $O(N^{1/4})$ source-destination mappings must traverse a single link within a factor in the worst λ-relation (from property 8), then part (2) follows immediately. $\square$

Theorem 2: (1) For the concatenation of any strict buddy $b^n \times b^n$ banyan and its inverse, with the innermost two stages merged into 1, i.e., $r = n-1$,

$$D = \frac{2 \cdot \log_b N - 1}{1 - \lambda}$$

(2) For a partially extended b^n banyan ($r = n/2$ and n even)

$$D = \frac{n+1}{1 - \lambda} + \frac{1}{1 - \lambda \cdot b^{n/4}} + 2 \cdot \sum_{i=1}^{n/4-1} \frac{1}{1 - \lambda \cdot b^i}$$

Proof: (1) follows from theorem (1) part (1) and the fact that the network considered is a system of feed-forward queues with exponential link transmission times fed by independent Poisson Processes. The result follows from Burke's theorem (see [6]). For part (2), view the network as a three stage network, with factors of size $b^{n/2} = \sqrt{N}$ in each stage. By theorem (1) all inter-factor links have intensity λ; By the nature of the routing algorithm all queues in the randomization network and those in the second stage of $\sqrt{N} \times \sqrt{N}$ factors in the routing network have intensity λ. Thus all $2(n/2)$ stages of links (except those in the routing banyan's first stage of $\sqrt{N} \times \sqrt{N}$ factors) have intensity λ. Within the first stage of factors in the routing banyan, property 3 indicates that link intensities start at b, keeps increasing by a factor of b for each stage, reach a maximum of $b^{n/4}$ and then keeps decreasing by a factor of b per stage, until the last stage (where the intensity is λ). $\square$

The previous results illustrate the relief from traffic imbalances for the two particular extended banyans considered in this paper. Theorem 2 establishes rigorously that the cancatenation of a banyan and its inverse yields a robust switch with $\Theta(N)$ bandwidth, with an optimal $O(N\log N)$ cost and with an extremely simple routing algorithm. Mitra et. al. [10] first derived a similar bound, which was applicable only to extended Omega networks. While the theorem assumes exponentially distributed link transmission times and infinite buffering within the nodes, in practice the link transmission time is not exponentially distributed and the buffering will be finite. While it is considered intractable to estab-

lish exact analytic results for such a multistage switch useful approximate models, along with some crosspoint-efficient queueing schemes within the nodes, can be found in [14].

One drawback of the above architecture is the need for re-sequencing queues. The following theorem examines an internally unbuffered switch.

6. A Bound on the Blocking Probability in Banyans

Consider the concatenation of any two $b^n \times b^n$ dilated banyans (where d is the dilation factor). The first banyan performs the randomization phase and the second performs the routing phase. (If the second topology is the inverse of the first then the last stage in the first banyan is redundant and can be removed, so that $r = n-1$. However, it is retained in following proof.) The following two facts are useful.

Fact 1 (Hoeffding [4]): If T is the number of successes in N independent Poisson trials with probabilities p_1, ..., p_N, then if $\sum p_i = Np$ and $m \geq Np + 1$ is an integer then

$$Pr(T \geq m) \leq B(m,N,p)$$

where $B(m,M,p)$ is the probability of at least k successes in M Bernoulli trials, with the probability of success in any trial equal to p.

Fact 2: Valiant's bound of Chernoff's bound [15] states that for any $m > Mp$,

$$B\left[m,M,p\right] \leq \left[\frac{eMp}{m}\right]^m \cdot e^{-Mp}$$

Theorem 3: The conditional blocking probability pb, defined below, is upper bounded by

$$1 - \left[1 - \left[\frac{eh}{d}\right]^d \cdot e^{-h}\right]^{2\log_b N}$$

Proof: Define a state of a d-dilated $b^n \times b^n$ banyan with N input ports, and with h connection requests per port, as a configuration with the following: (1) an assignment of destinations to the connection requests (which corresponds to an assignment of routes through the banyan for each connection request), (2) in the flow of connection requests from left to right, when $> d$ requests arrive at one dilated link, d are selected randomly and propagated and the remaining requests are blocked. There are $(Nh)^N$ possible assignments of destinations and for each assignment there are one or more different states, each reflecting a different choice of requests to be blocked (when a choice exists).

Define A_s as the event that a request offered to the first stage is not blocked when attempting to exit a stage s switch, and pa as the end-to-end probability that a connection is not blocked given that it was offered;

$$pa \equiv Pr[A_1] \cdot Pr[A_2|A_1] \cdot Pr[A_3|A_1 \cap A_2] \cdots Pr[A_n | \bigcap_{i=1}^{n-1} A_i]$$

Finally define $pb \equiv 1 - pa$.

We will first establish a lower bound on $Pr[A_1]$. The $b \times b$ nodes in the first stage are statistically independent and identical. The number of input ports that can reach any output link leaving stage 1 is b^1; these ports submit exactly $h \cdot b^1$ connection requests. By symmetry these requests are evenly distributed over the b^1 links leaving each node in stage 1.

The connection requests may appear either synchronously or asynchronously. Whether the requests arrive synchronously or asynchronously, they can be veiwed as being routed serially, i.e., one at a time. Consider the jth request to be serviced. The probability that the jth request encounters a saturated link is given by

$$B(d,j-1,1/N') \leq B(d,N'h,1/N')$$

where $N' = b$ and where the inequality is obtained by inflating the number of trials to $N'h$ by the addition of $N'h-j+1$ dummy trials with zero probability of success and then applying Fact 1. (Note that when dummy trials with zero probability of success are added the inequality still applies.) The probability the last (out of $N'h$) request to be serviced encounters a saturated link is

$$B(d,N'h-1,1/N') \leq B(d,N'h,1/N') \leq \left[\frac{eh}{d}\right]^d \cdot e^{-h} \qquad (1)$$

where the last inequality is obtained by inflating the number of trials to $N'h$ by the addition of one dummy trial with zero probability of success, and then applying Fact 1 and 2. A request that encounters a saturated link is blocked.

If the requests arrive simultaneously rather than individually the bound still applies to each and every request since the number of blocked requests in stage 1 depends only on the destination assignment rather than the service order.

Since over all assignments the nodes in the first stage are identical and independent then

$$Pr[A_1] \geq \left[1 - \left[\frac{eh}{d}\right]^d \cdot e^{-h}\right]$$

$Pr[A_s \mid A_1 \cap A_2 \cap \cdots \cap A_{s-1}]$ represents the probability that a request is not blocked when leaving stage s given that it did not block in all previous stages. To establish a lower bound for this probability, first factor the banyan into a 2-stage network with factors of size $b^s \times b^s$ in the first stage and factors of size $b^{n-s} \times b^{n-s}$ in the second stage. The first stage factors are independent and identical, and therefore we only need to consider one factor. Within a factor all input ports are statistically indistinguishable and all output ports are also statistically indistinguishable. Within a factor route an arbitrary connection request, X, from any input of a stage 1 node to any input of a stage s node. Consider all possible states in which this route holds (i.e. all states in which X's route has been established up to the input of a particular stage s node). Since the number of input/output ports in a factor is given by $N' = b^s$, and there are exactly h requests at each input, the total number of connection requests that can compete with X for outgoing links at the stage s node is $N'h-1$. Given that X has reached stage s, no information can be infered about the destinations of these $N'h-1$ requests, therefore they can be considered as independent Bernoulli trials. Define a successful outcome of a trial as the event that a request selects the same link to exit as X at the stage s node. The probability of success in each trial is $1/N'$ since by the random routing assumption each request selects any one of the N' output ports of the factor with equal probability. Some of the requests will be blocked when more than $d-1$ requests attempt to traverse the same link leading into the stage s node as X. The blocked requests can be considered as dummy trials with zero probability of success and Fact 1 ensures that the bound on blocking probability still holds if we assume that they reach stage s and compete for output links. Similar reasoning applies to all requests within the same factor.

Consider routing the requests through the stage s node; assume serial routing is used, as before. However, given that X's route has already been established up to the input of a stage s node we can assume it is the last request to be routed through the stage s node in order to obtain an upper bound on the blocking probability. Therefore all the $N'h-1$ requests are routed serially through stage s before X is routed from the input of stage s to the output of stage s. Hence, the probability that X encounters a saturated

link and blocks given that it did not block in all previous stages is upper bounded by eq. (1), where the number of trials is inflated to $N'h$ by the addition of a dummy trial with zero probability of success, and then applying Fact 1 and 2. Therefore, for $2 \leq s \leq n$

$$Pr[A_s \mid \bigcap_{i=1}^{s-1} A_i] \geq \left[1 - \left[\frac{eh}{d}\right]^d \cdot e^{-h}\right]$$

In all $r = n$ stages, pa, is then lower bounded by

$$\left[1 - \left[\frac{eh}{d}\right]^d \cdot e^{-h}\right]^n$$

and pb is upper bounded by

$$1 - \left[1 - \left[\frac{eh}{d}\right]^d \cdot e^{-h}\right]^n$$

Note that the preceding bounds are valid for the first half of the switch (the distribution network). In the second half of the switch the traffic is not randomly distributed (due to the h-relation constraint). The following arguments indicate that a worst case can assume every connection is equally likely to appear at every node in the same stage (in absense of blocking in previous stages), and that the connection is equally likely to exit over any of the b outgoing links from a node. Given that i connections arrive at a node in stage m ($1 \leq m \leq n$) of the routing network, the exact probability that j chosen requests select a particular output link leading to b^{n-m} outputs (in a 1-relation) is given by (see [12])

$$\left[\begin{matrix} b^{n-m} \\ j \end{matrix}\right] / \left[\begin{matrix} b^{n-m+1} \\ j \end{matrix}\right] < \left[\frac{1}{b}\right]^j$$

The exact probability that $i-j$ remaining requests do not select a particular output link leading to b^{n-m} outputs, given that i requests arrived and that j chosen requests have already selected that output link is given by (see [12])

$$\left[\begin{matrix} (b-1)b^{n-m} \\ i-j \end{matrix}\right] / \left[\begin{matrix} b^{n-m+1}-j \\ i-j \end{matrix}\right] > \left[1 - \frac{1}{b}\right]^{i-j}$$

Hence, given that i requests arrive at a node in stage m of the routing network a worst case can assume that the connections are evenly distributed over the output links, yielding a higher probability of blocking than there is in reality. The same reasoning indicates that a connection is less likely to arrive at a node given that other connections have already arrived there. However, once again a worst case can assume connections are equally likely to arrive at any node in every stage. Therefore the bound stated in theorem 3 follows. Note that due to the h-relation constraint the last stage in the routing network cannot block any requests. If $r+n = n-1 + n = 2log_bN-1$, then the blocking probability is upper bounded by

$$1 - \left[1 - \left[\frac{eh}{d}\right]^d \cdot e^{-h}\right]^{2log_bN-2} \qquad \square$$

Finally these bounds also apply for randomly distributed traffic (see Fig. 4).

6.1. Asymptotic Performance

Letting $x = (eh/d)^d \cdot e^{-h}$ and for $x \ll 1$ then pb of a $logN$ stage banyan can be simplified by applying the binomial expansion;

$$pb \leq logN \cdot \left[\frac{eh}{d}\right]^d \cdot e^{-h} + O\left[\left[(eh/d)^d e^{-h}\right]^2\right]$$

This bound applies to a dilated banyan even if the traffic is randomly distributed (rather than just conforming to permutations or

h-relations). For small eh/d this approximation is usually good to many digits.

Observe that if the dilation is bounded then asymptotically $pb \to 1$. However if the dilation is $O(\log\log N)$ then it can be shown that the blocking probability approaches zero as $N \to \infty$, yielding a nearly optimal self-routing, unbuffered switch (the only suboptimal aspect in the theoretical sense is the $\log\log N$ growth in the dilation.) Fig. 5 shows the upper bound and exact analysis of the blocking probability of asymptotically large banyans with $O(\log\log N)$ dilation. While the switch is not deterministically non-blocking, we point out that every switch architecture must have $pb > 0$ when the traffic is randomly distributed. In practice setting $h=d$ maximizes throughput (see [13]).

It has been brought to our attention that Arora, Leighton and Maggs have extended [9] to yield a self-routing switch.

7. References

[1] M. Ajtai, J. Komlos and E. Szemeredi, "An O(NlogN) Sorting Network", Proc. 15th ACM Symp. Theory of Computation, pp. 1-9, 1983
[2] K.E. Batcher, "Sorting Networks and Their Applications", Proc. 1968 Spring Joint Computer Conf.
[3] V.E. Benes, "The Mathematical Theory of Connecting Networks and Telephone Traffic", Academic Press, New York, 1965
[4] W. Hoeffding, "On the Distribution of the Number of Successes in Independent Trials", Ann. Math. Statistics 27 (1956), pp. 713-721
[5] A. Huang and S. Knauer, "Starlite: A Wideband Digital Switch", in proc. Globecom, Dec. 1988
[6] L. Kleinrock, "Queueing Systems, Vol. 2", 1976, Wiley
[7] R.R. Koch, "Increasing the Size of a network by a Constant Factor can Increase Performance by More than a Constant Factor", Proc. 29th Ann. Symp. Foundations of Computer Science, pp. 221-230, Oct. 1988
[8] C.P. Kruskal and M. Snir, "The Performance of Multistage Interconnection Networks for Multiprocessors", IEEE Trans. Comput., Vol C-32, pp. 1091-1098, Dec. 1983
[9] T. Leighton and B. Maggs, "Expanders Might be Practical: Fast Algorithms for Routing Around Faults in MultiButterflies", Proc. 1989 Symp. Foundations of Computer Science, pp. 384-390, 1989
[10] D. Mitra and A. Cieslak, "Randomized Parallel Communications on an Extension of the Omega Network", JACM, Vol. 34, No. 4, Oct. 1987, pp. 802-824
[11] T.H. Szymanski and V.C. Hamacher, "On the Permutation capability of Multistage Interconnection Networks", IEEE Trans. Comput., Vol. C-36, No. 7, pp. 810-822, July 1987
[12] T.H. Szymanski and V.C. Hamacher, "On the Universality of Multipath Multistage Interconnection Networks", Journal of Parallel and Dist. Computing, No. 7, pp. 541-569, 1989
[13] T.H. Szymanski and C. Fang, "An Asymptotically Nonblocking $O(\log N)$-Depth, $O(\log\log N)$-Degree Randomized Routing Switch", submitted for publication.
[14] T.H. Szymanski and C. Fang, "Design and Analysis of Buffered Crossbars and Banyans with Cut-Through Switching", Proc. IEEE Supercomputing 90, pp. 264-273
[15] L.G. Valiant and G.J. Brebner, "Universal Schemes for Parallel Communications", Proc. 13th Annual ACM Symp. on Theory of Computing, 1981, pp. 263-277

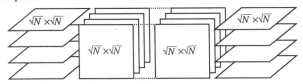

Fig. 1 A 3D construction of two concatenated banyans

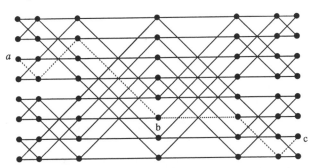

Fig. 2 A fully extended $2^3 \times 2^3$ banyan with randomized routing. The dotted line shows a request at a destined for c is first randomly routed to b before it is finally routed to c.

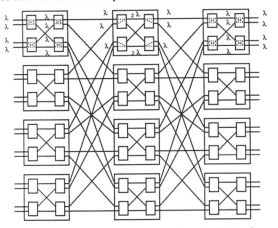

Fig. 3 A partially extended banyan, $n=4$, $r=2$, illustrating worst case intensity.

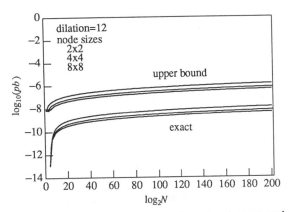

Fig. 4 Blocking probability: upper bound and exact analysis (from [11]) for banyan networks with fixed link dilation.

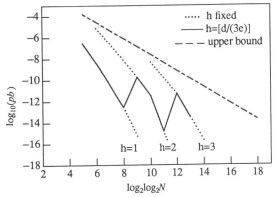

Fig. 5 Blocking probability: upper bound and exact analysis (from [11]) for banyan networks with link dilation = $\lceil 2\log_2\log_2 N \rceil$.

FAULT-TOLERANT MESSAGE ROUTING AND ERROR DETECTION SCHEMES FOR THE EXTENDED HYPERCUBE

J. Mohan Kumar[*], and L. M. Patnaik[*,**]

*Microprocessor Applications Laboratory
** Department of Computer Science and Automation &
Supercomputer Education and Research Center
Indian Institute of Science
BANGALORE 560 012 INDIA
email: lalit@vigyan.ernet.in

ABSTRACT

We discuss fault-tolerant aspects of the extended hypercube architecture. The network controllers of the extended hypercube which are primarily used to handle intermodule communication, have been effectively used for error detection and message routing. Error detection algorithms which run concurrently with typical application examples, viz., matrix multiplication and multinode broadcast algorithm have been discussed. The network controllers facilitate fault-tolerant message routing.

Keywords: Extended Hypercube, Multinode Broadcast, Fault-tolerance, Message Routing, Error Detection.

I. INTRODUCTION

The Extended Hypercube(EH), comprising of dedicated communication processors in addition to the processor elements, is found to be efficient in executing communication-intensive parallel algorithms [1]. In this paper we discuss fault-tolerant message routing techniques and algorithm-based error detection mechanisms for the EH. In general, for a large number of processors it is necessary to cope up with the problem of faulty processors and links. The EH consists of two types of nodes, viz., the processor elements (PEs) and the network controllers (NCs). The basic building block of an EH consists of a k-cube of PEs and an NC. The nodes are arranged in hierarchical levels, with PEs at the lowermost level and NCs at the higher levels. At every level of hierarchy of the EH, there are one or more hypercubes. The presence of extended links and network controllers in the EH makes it possible to have a highly robust topology inspite of the low degree of connectivity. The topological and architectural features of the EH discussed in [2] are summarized in the second section.

A network is said to be fault-tolerant as long as all the active processors are connected by fault free links and processors [3]. Another definition says, a network is fault-tolerant if under failure of one or more processors, a number of active processors remain connected together according to the original topology of the network for which it was designed [4]. A network is said to be robust if its performance does not decrease significantly in case of topology changes. Connectivity is a measure of the robustness of the network [5]. High degree of connectivity, as in the case of the binary hypercube, prevents non-faulty processors from being disconnected. However, a network with a high degree of connectivity is very expensive as the cost of the network is directly proportional to the degree of connectivity [5]. The degree of connectivity in an EH is lower than that of a hypercube of comparable dimension, and yet the EH is found to have a robust topology.

In [4] Banerjee et. al discuss algorithm-based fault detection schemes for detection of faults on a hypercube topology while concurrently executing algorithms such as matrix multiplication, gaussian elimination and fast fourier transforms. In our experimental set up, the EH is being used for implementing multinode broadcast and total exchange algorithms to simulate Artificial Neural Networks (ANNs). In the following sections, we discuss techniques to detect faults in the EH while concurrently executing algorithms related to the implementation of matrix multiplication and multinode broadcast algorithm (used in the simulation of artificial neural networks (ANNs)), and our studies indicate a very high percentage of error detection and low overheads. Fault-tolerant message routing techniques and reconfiguration techniques are also discussed. It is found that the EH is robust to link faults, since messages can be passed on one of the alternate paths via NCs.

The paper is organized as follows. In the second section, we discuss the architectural and topological features of the EH. In the third section, we discuss algorithm-based error detection schemes for detecting faults. Fault-tolerant message routing techniques and reconfiguration schemes are presented in section four. The fifth section discusses the experimental studies carried out on a multitransputer system. Finally, the sixth section concludes the paper.

II. EXTENDED HYPERCUBE

In this section, a formal introduction to the EH architecture is presented. The EH architecture discussed earlier [6], [2], is suited for hierarchical expansion of multiprocessor systems. The basic module of the EH consists of a k-cube and an additional node for handling communication - the Network Controller (NC). There are a total of $[2^k*(k/2+1)]$ links in the basic module, consisting of the $[(2^k*k)/2]$ links of the hypercube and 2^k links connecting individual processor elements to the NC as indicated in fig. 1.

An EH consisting of a k-cube and one NC will be referred to as the basic module or EH(k,1) in the rest of this paper. The EH(k,1) has two levels of hierarchy : a k-cube at the zeroth level and an NC at the first level. The hypercube consisting of the PEs is referred to as the HC(k,0). An EH(k,2) has 2^k number of k-cubes of PEs at the zeroth level, a k-cube of NCs at the first level and one NC at the second level. In general, an EH(k,l)(l is the degree of the EH), consists of a k-cube of eight NCs/PEs at the (l -1)st level and one NC at the l th level. The NCs/PEs at the (l-1)st level of hierarchy form a k-cube. We refer to this cube as HC(k, l -1). The NCs at the (l-2)nd level of hierarchy form 2^k distinct k-cubes which will be called HC(k, l -2). The HC(k,0)s are all computation HCs and the HC(k,l)s (for l > 0) are all communication HCs. The basic module of the EH denoted by EH(k,1) is a constant pre-definable building block and the node configuration remains the same regardless of the dimension of the EH. The EH architecture can easily be extended by interconnecting appropriate number of basic modules. For example, we can interconnect eight EH(3,1)s (**basic modules**)to get a 64-node EH - an EH(3,2), shown in fig.2. The topology formed by the 3-cube of NCs at the first level, and the controller at the second level is identical to that of the basic module. Thus we have a hierarchical structure consisting of 64 PEs at the zeroth level (lowest level), eight NCs at the first level, and one NC at the second level. The nodes at level j $(0 \leqslant j < l)$ are called as the children of the node at level (j + 1). The EH has two types of links, viz., hypercube links which form the k-cubes and the EH links which connect the nodes at jth level to the nodes at (j + 1)st level. Further, the path between any two PEs of the EH via one or more NCs is called an extended link.

The nodes of the EH are addressed as described below. The address of an arbitrary node at the zeroth level is written as $D_l D_{l-1} ... D_0$, where D_i $(0 \leqslant i \leqslant l)$ is a k-bit mod 2^k number, e.g., in an EH(3,l), D_i is 3-bit mod 8 number. In an EH(k,l) a node at the zeroth level has an (l +1)-digit address, a node at the first level has an l -digit address ..., the solitary node at the l th (topmost)level has a single digit address. The NC at the topmost level retransmits messages to/from the host system from/to the PEs via the hierarchical network of NCs. In an EH(k,1), message passing between neighboring nodes of the hypercube is via the direct communication links among them whereas communication between any two non-neighboring nodes of the hypercube is via the NC. Message passing operation in local communication (within a k-cube) involves 1)the source PE, 2)upto (k-1) PEs within the k- cube, and 3) the destination PE, whereas, message passing operation in global communication involves 1)the source PE, 2)upto 2(l -1) NCs, and 3) the destination PE.

III. ERROR DETECTION MECHANISMS

A fault is a physical defect in the system whereas an error is the manifestation of the fault at the logical level[7]. The mechanism employed for the detection of errors in multiprocessor systems is a very important issue in the design of fault-tolerant parallel computers. Several error detection techniques have been reported in the literature. There are two types of faults, viz., permanent (or static) faults and transient (or intermittent) faults. Errors due to permanent faults can be detected by conventional approaches, such as the off-line testing suggested by Armstrong et. al [8]. However, in multiprocessor networks, transient faults are more frequent than permanent faults. Most of the on-line error detection algorithms affect the performance of the system. There is a need to have fault detection mechanisms which run concurrently with the application algorithms. In [4] Banerjee et. al present such algorithm-based error detection schemes in which the fault detection mechanisms are tailored to the application algorithms such as matrix multiplication, gaussian elimination and fast fourier transforms. The algorithm based fault detection mechanisms are implemented on a hypercube topology and the results of the analysis have been validated by the experimental tests.

In this section we present algorithm-based fault detection mechanisms for the EH while executing algorithms related to matrix multiplication and ANN implementation [9]. The presence of NCs in EHs facilitates execution of fault detection algorithms without affecting the performance of the multiprocessor system. In our studies, it is assumed that the node at the topmost level, which is connected to the host system is fault-tolerant. This is a reasonable assumption, as all multiprocessor systems have some kind of a host which monitors all the nodes. In our scheme, the nodes at the lower most layer, the PEs of the HC(k,0)s are checked by the NCs at the first level. Then the NCs of the HC(k,1)s are checked by the NCs at the second layer. Likewise, the fault detection algorithm is performed in a hierarchical fashion and finally the NC at the lth level checks the NCs of the HC(k,l -1).

A. Matrix Multiplication

The execution of an error detection algorithm which runs concurrently with the matrix multiplication algorithm on a hypercube topology is described by Banerjee et. al in [4]. We describe execution of an error detection algorithm which runs concurrently with the matrix multiplication algorithm on an EH(k,l) topology. However, in our scheme the error detection algorithm is executed in a hierarchical fashion and since NCs are employed for error detection, there is no overhead on the PEs.

We perform the matrix multiplication,

$$C = A \times B$$

where C, A, and B are full matrices of sizes p x r, p x q, and q x r respectively. We perform the matrix multiplication in a manner similar to the one discussed in [4], however the partitioning of matrix A is performed by the NCs in a distributed fashion. It is assumed that the sizes of the matrices are multiples of 2^{k*l}, i.e p, q, and r are multiples of 2^{k*l}. The NC at the topmost level partitions the matrix A into 2^k number of rectangular strips and sends one strip each to the nodes at the (l -1)th level. The nodes at the (l -1)th level may be NCs or PEs, if they are NCs again each NC subpartitions its strip into 2^k number of smaller strips. This kind of subpartitioning is continued by NCs at all levels j > 1. Finally each PE at level 0 receives a strip of matrix A, we shall denote this strip as A1 0, indicating the strip received at level 0. Likewise a strip of A received at level j is denoted by Al j and comprises of $p/[2^{k*l}] * 2^{k*j} = p * 2^{k*(j-l)}$) rows of A. The matrix B is sent to all PEs, and each PE performs the submatrix multiplication, $C^{l\,0} = A^{l\,0} \times B$ and sends the submatrix of the result matrix to the NC at level 1. The NCs at level 1 gather all the $C^{l\,1}$s from their respectve children and form $C^{l\,1}$s. Similarly, $C^{l\,j}$s are formed at all levels, and finally the NC at the topmost level forms the result matrix $C^{l\,l} = C$.

The above algorithm is modified to perform error detection checks. Each NC at level 1 performs a column check on the PEs directly below it. Thus the error detection is performed in a distributed manner as each of the $2^{k*(l-1)}$ number of NCs at level 1 check 2^k number of PEs. The error detection procedure for every NC at level 1 is described in the following steps, 1) Receive subpartition $A^{l\,1}$ from the NC at level 2 or the host (if it is an EH(k,1)), partition $A^{l\,1}$ into 2k number of $A^{l\,0}$s and distribute them to 2^k PEs. 2) Compute $C^{l\,1}_y = A^{l\,1} \times B_y$, where $C^{l\,1}_y$ is the yth column of $C^{l\,1}$ and B_y is the yth column of B ($0 < y < r$). 3) Gather $C^{l\,0}$s computed at PEs and form $C^{l\,1}$. 4) Compare $C^{l\,1}_y$ with the yth column of $C^{l\,1}$. The two should be equal if there are no faulty PEs.

Steps #1 to #4 are executed by all NCs at level 1 simultaneously. To validate the result of step #4, we perform column check on two columns, y_1 and y_2, where $0 \leqslant y_1, y_2 < r$. If the result of the comparison is 'fail' in both cases, then one of the PEs corresponding to the row in which the yth column of $C^{l\,1}$ and $C^{l\,1}_y$ differ is faulty. Since each row element of $C^{l\,1}$ is contributed by a different PE, the NC at level 1 can identify the faulty PE. In the above algorithm the NCs at level 1 are assumed to be non-faulty. However, the NCs at level 1 can be checked for faults by the NCs at level 2 by performing steps #1 to #4 on $C^{l\,2}$ at level 2. In general the above procedure can be executed at all levels for $1 \leqslant j < l$ and the NCs/PEs tested in a bottom up fashion.

The experimental results show that the percent error coverage is very high as indicated in table 1, and the results are comparable to those obtained by Banerjee et. al in [4]. However, in our scheme the implementation of the error detection scheme does not cause any execution time overhead on the PEs. In [4], the processors perform additional computation to determine the column checksum and moreover, every processor executes the error detection algorithm with two different mates. Hence, the execution time overhead is as high as 35% for a 4-cube performing a 64 X 64 matrix multiplication and reduces as the size of the matrix increases. In our scheme the PEs perform those computations which are related to actual matrix multiplication only. Additional computations, and comparisons related to error detection are performed by the NCs. Further, the error detection overhead does not affect the performance of the NCs since the error detection computations are performed by the NCs while waiting for the result submatrices from the PEs.

B. Multinode Broadcast

In a multinode broadcast algorithm, each node broadcasts the same message to all other nodes of the multiprocessor system. The performance aspects of the binary hypercube and the EH in executing the multinode broadcast algorithm is discussed in [1]. Consider the EH(3,1) shown in fig.1, consisting of an HC(3,0) at the zeroth level and an NC at the first level. The EH(3,1) is one basic module of an EH(3,l) comprising of l levels of hierarchy. The nodes $PE_0, PE_1, ..., PE_7$ broadcast messages $a_0, a_1, a_2, ..., a_7$ respectively. At the completion of the multinode broadcast cycle, each PE_i (i =0...7) will have $a_0, a_1, a_2, ..., a_7$ in its memory. In general, the fault detection algorithm is executed on an HC(k,l) as follows, 1) On each PE_i, $A_i = [\Sigma \, a_i] / N$ ...(1) is computed. 2) All PEs send their respective A_i to NC at the next higher level. 3) The NC compares A_i's computed at at all PE_i's. The A_i's should all be identical since identical inputs are used at every PE_i. If the input from any PE_i is different from those at the rest, then processor PE_i is faulty. The NC identifies the faulty processor and a message indicating the address of the faulty PE is sent to the other PEs in the k-cube. In our application, the PEs perform arithmetic operations on the data received at the end of each multinode broadcast, and the computation of A_i is the error detection overhead.

IV. FAULT-TOLERANT MESSAGE ROUTING AND RECONFIGURATION TECHNIQUES

In a multiprocessor network, there can be two types of faults, transient and permanent. If a transient failure repeats more than once for the same set of inputs, then it is classified as permanent fault. In case of loosely coupled multiprocessor systems, the permanent fault may be a link fault or a processor fault. In case of a link fault, the faulty link has to be isolated and the processors at the two ends of the link should be connected via an alternate path. If a processor is faulty, it should be isolated and the network reconfigured to perform with graceful degradation or a spare processor should take the place of the faulty processor.

The Extended Hypercube topology is very robust against link faults, as the extended links of the EH can be effectively used in place of faulty links. This results in a saving in the cost of the network since the use of redundant links increases the node connectivity. However, to make the EH fault-tolerant to processor faults, it is necessary to use spare processors. In the following analysis we discuss the reconfiguration techniques employed to tackle faults in an EH.

A. Link Faults

In [10], Chen et. al have discussed a Depth-First Search approach for fault-tolerant routing in binary hypercubes. We present an algorithm for fault-tolerant routing of messages on the EH. The EH has two types of links, viz., the hypercube links and the EH links. The hypercube links of level j, form one or more k-cubes. As described in section II, a node at level j, represented by $D_l D_{l-1} ..., D_i, ..., D_j$, where $D_i (j \leqslant i \leqslant l)$ is a k-bit mod 2^k number. This node at level j is connected to a node at level (j + 1), viz., $D_l D_{l-1} ... D_{j+1}$, which in turn is connected to the node $D_l D_{l-1} ... D_{j+2}$ at level (j +2). There may be one or more k-cubes at all levels j, for $0 \leqslant j < l$. Messages between nodes of the EH may be of the following types, 1) message between two nodes of the same k-cube, 2) message from a node at level j to a node at level k (j>k), 3) message from a node at level j to a node at level k (j<k), 4) message between two nodes belonging to different k-cubes of level j, (for $0 < j < l-1$).

A.1 Fault-tolerant Message Routing

Messages between two nodes of the same k-cube can be routed through one of the (k + 1) disjoint paths. EH(3,1) shown in fig. 1 has hypercube links, 00--01,00-- 02,00--04,01--03,03--02,02--06,06--07,07--05,05--04,04--06,01--05, and 03--07 which form HC(3,0), and EH links 00--0,01--0,02--0, 03--0,04--0,05--0,06--0,and 07--0 which connect the PEs with the NC. A path 0i--0--0j ($0 < i,j < 7$) is an extended link of the EH. Suppose one of the hypercube links say 03--07, which connects nodes 03 and 07, is faulty, then messages from 03 to 07 can take the path 03---0---07, via the NC. In general if any link connecting nodes 0i and 0j is faulty, the path 0i--0--0j can be used. Suppose one of the EH links, 0i--0 is faulty, then messages from 0i to 0 can be rerouted as 0i--0j--0 (0j is a neighbor of 0i). In effect each k-cube of the EH is (k + 1) fault tolerant.

Procedure for fault-tolerant message routing

{If $D^s_l D^s_{l-1} ... D^s_0 = D^d_l D^d_{l-1} ... D^d_0$

  {destination is source, terminate} else

    {x:: = 0;

    for i = 1 to l do

      { if $D^s_{l-i} = D^d_{l-i}$ then

        { x:: = l-i

ftutran from $D^s_1 D^s_{1-1} \dots D^s_0$ to $D^s_1 D^s_{1-1} \dots D^s_x$

fthctran from $D^s_1 D^s_{1-1} \dots D^d_x$ to $D^d_1 D^d_{-1} \dots D^d_x$

ftdtran from $D^d_1 D^d_{1-1} \dots D^d_x$ to $D^d_1 D^d_{1-1} \dots D^d_0$ } }

}

}

end fault-tolerant message routing

ftutran / Fault-tolerant message routing from a node at a lower level to a node at a higher level/

Procedure ftutran

{ for p = 0 to x do

    { if $D^s_1 D^s_{1-1} \dots D^s_p D^s_{p+1}$ is not faulty then

        {transfer from $D^s_1 D^s_{1-1} \dots D^s_p$ to $D^s_1 D^s_{1-1} \dots D^s_x$}

    else

        {transfer from $D^s_1 D^s_{1-1} \dots D^s_p$ to $D^s_1 D^s_{1-1} \dots D^s_{p'}$

        transfer from $D^s_1 D^s_{1-1} \dots D^s_{p'}$ to $D^s_1 D^s_{1-1} \dots D^s_{p+1}$}

    }

}

endftutran

fthctran / $D^s_1 D^s_{1-1} \dots D^s_{x+1} = D^d_1 D^d_{1-1} \dots D^d_{x+1}$/

/ Fault-tloerant messge routing between two nodes of the same k-cube /

Procedure fthctran

{ if all links $D^s_1 D^s_{1-1} \dots D^s_p - D^s_{p'}$ are faulty then

    { transfer from $D^s_1 D^s_{1-1} \dots D^s_x$ to $D^s_1 D^s_{1-1} \dots D^s_{x+1}$

    and $D^s_1 D^s_{1-1} \dots D^s_{x+1}$ to $D^d_1 D^d_{1-1} \dots D^d_x$}   else

    {transfer from $D^s_1 D^s_{1-1} \dots D^s_x$ to $D^d_1 D^d_{1-1} \dots D^d_x$ via $D^s_1 D^s_{1-1} \dots$

                             $D^s_p - D^s_{p+1}$}

}

end fthctran

ftdtran / Fault-tolerant message routing from a node at ahigher level to a node at a lower level/

Procedure ftdtran

{ for p = x down to 1

    if $D^d_1 D^d_{1-1} \dots D^d_p - D^d_{p-1}$ is not faulty t hen

    { transfer from $D^d_1 D^d_{1-1} \dots D^d_p$ to $D^d_1 D^d_{1-1} \dots D^d_{p-1}$}   else

    {transfer from $D^d_1 D^d_{1-1} \dots D^s_p$ to $D^d_1 D^d_{1-1} \dots D^d_{(p-1)'}$

    transfer from $D^d_1 D^d_{1-1} \dots D^d_{(p-1)'}$ to $D^d_1 D^d_{1-1} \dots D^d_{(p-1)}$ }

}

endftdtran

In the above algorithms $D^y_1 D^y_{1-1} \dots D^y_p - D^y_{p'}$ represents a link connecting $D^y_1 D^y_{1-1} \dots D^y_p$ to $D^y_1 D^y_{1-1} \dots D^y_{p'}$.

B. Processor Faults

The EH has two types of nodes, viz., the PEs and NCs. We shall consider PE faults and NC faults. In fig. 3a, is shown the diagram of an augmented EH(3,1). The augmented EH(3,1) consists of one additional PE, 0x for the 3-cube and a spare NC. There is a link between the NC and 0x. 0x is used if any processor of the k-cube is found to be faulty. If PE, 0y (y = 0, ..., 7) is faulty, then 0y is isolated and all tasks originally meant for 0y are assigned to 0x. The direct links of the faulty processor 0y are replaced by extended links and messages from the former neighbors of 0y are routed to 0x via the NC. The scheme is illustrated in the fig.3b. Since the EH has a hierarchical recursive structure the idea of having one spare PE and one spare NC for every EH(k,1) can be extended to an EH(k,l). The method suggested above is cost-effective as the degree of connectivity of the PEs is not altered while the degree of node of the NC increases by one. The above scheme has been implemented on transputer-based

EH(k,1)s. Link faults and processor faults were simulated and the performance of the augmented EH under fault conditions was studied. The experimental results obtained while executing multinode broadcast algorithm under fault conditions on augmented EH(k,1)s are presented in table 2.

V. Experimental Studies

The experimental studies are carried out on a testbed based on a reconfigurable multitransputer system. Each PE of the system consists of a T800 transputer and 256 K bytes of memory and each NC comprises of a T800 transputer, a C004 crossbar switch and 2 M bytes of memory. The T800 has an in-built floating point unit and four links for serial communication. The transputers are connected to one another by the direct links or via a programmable crossbar switch. The network comprises of T800 transputers, whose links are connected to a crossbar link switch. The network can be configured to the desired topology by suitably programming the switch. The programs for fault simulation and error detection are written in OCCAM 2. The master transputer communicates with the host system, an IBM PC 386 system via a link adapter. Errors in the links during transfer of messages among transputers are checked by in-built mechanisms.

Various types of transient and permanent errors as suggested by Balasubramanian et. al [11], have been simulated. The percent error coverage for the EH while executing matrix multiplication and multinode broadcast are shown in table 1. There is **no execution time overhead** on the PEs for error detection while performing matrix multiplication. The execution time overhead on the NC at level 1 due to the matrix multiplication error detection algorithm is shown in fig.4a. Figs.4b and 4c show the overheads on the PEs and the NCs (at level 1) respectively, for error detection while executing multinode broadcast.. An EH whose links are faulty is called an injured EH [10]. We make use of the fault-tolerant message routing algorithm described earlier to route messages via one of several alternative paths available in the EH. EH link faults and hypercube link faults are simulated and time taken by injured EHs of various dimensions to execute multinode broadcast algorithm is tabulated in table II.

VI. CONCLUSIONS

The use of NCs in the EH facilitates implementation of error detection algorithms which run concurrently with the application programs. We find that the percent error coverage is very high and yet there is no error detection overhead for matrix multiplication, and very small overhead for multinode broadcast. In case of link faults, the additional links among PEs and NCs are effectively used to achieve fault-tolerant message routing. Further, we have suggested methods to use spare PEs which can be used in the event of processor faults.

REFERENCES

1. J. Mohan Kumar, L.M. Patnaik, and Divya K. Prasad, "A Transputer-Based Extended Hypercube," Microprocessing and Microprogramming, vol. 29, Nov. 1990, pp. 225-236.

2. J. Mohan Kumar, and L.M. Patnaik, " Extended Hypercube: A Hierarchical Network of Hypercubes," to be published in IEEE Trans. on Parallel and Distributed Systems.

3. D. K. Pradhan, "Fault-tolerant Multiprocessor Link and Bus Network Architectures," IEEE Trans. on Computers, pp. 33-45, Jan. 1985.

4. P. Banerjee, J.T. Rahmeh, C. Stunkel, V.S. Nair, K. Roy, V. Balasubramanian, and J.A. Abraham, "Algorithm-based Fault Tolerance on a Hypercube Multiprocessor," IEEE Trans. on Computers, Vol. 39, No. 9, Sept. 1990, pp. 1132-1145.

5. B. Becker, and H-U. Simon, " How Robust is the n-Cube?," Technical Report 05/1986 SFB 124 -B1 Universitat des Saarlandes,Germany.

6. N. Jagadish, J. Mohan Kumar, and L.M. Patnaik, " An Efficient Interconnection Scheme Using Dual-Ported RAMS,' IEEE MICRO, Oct.1989, pp. 10-19.

7. J.-C. Laprie, "Dependable Computing and Fault-Tolerance: Concepts and Terminology," Proc. 15th Annu.Symp. Fault-Tolerant Comput., June 1985, pp. 2-11.

8. J.R. Armstrong, and F.G. Gray, "Fault Diagnosis in a Boolean n-cube Array of Multiprocessors," IEEE Trans. on Comput., Vol. C- 30, pp. 587-590, Aug. 1981.

9. H. Yoon, J.H. Nang, and S.R. Maeng, " Parallel Simulation of Multilayered Neural Networks on Distributed-Memory Multiprocessors, " Microprocessing and Microprogramming 29 (1990) pp. 185-195.

10. M-S. Chen, and K.G. Shin, " Depth-First Search Approach for Fault-Tolerant Routing in Hypercube Multicomputers," IEEE Trans. on Parallel and Distributed Computing, Vol.1, No.2, Apr.1990, pp. 152-159.

11. V. Balasubramanian, and P. Banerjee, "Tradeoffs in the Design of Efficient Algorithm-Based Error Detection Schemes for Hypercube Multiprocessors," IEEE Trans. on Software Engineering, Vol.16, No. 2, Feb. 1990, pp. 183-196.

Type of Error	Matrix Multiplication			Multinode Broadcast		
	EH(2,1)	EH(3,1)	EH(4,1)	EH(2,1)	EH(3,1)	EH(4,1)
Transient Bit Error	87.40	86.20	85.00	89.60	88.30	86.10
Transient Word Error	100.00	100.00	100.00	100.00	100.00	100.00
Permanent Bit Error	90.90	90.60	90.10	92.70	92.20	91.60
Permanent Word Error	100.00	100.00	100.00	100.00	100.00	100.00

Table 1

Percent Error Coverage for Matrix Multiplication (of size 64X64 and 64-bit Precision Numbers) and Multinode Broadcast

EH size	Execution time				
	With No Faults	1 Link Fault	2 Link Faults	3 Link Faults	1 Processor Fault
EH(2,1)	1.00	1.14	1.32	1.56	1.20
EH(3,1)	1.00	1.00	1.16	1.22	1.24
EH(4,1)	1.00	1.00	1.10	1.14	1.32

Table 2

Execution Time for Multinode Broadcast Under Fault Conditions

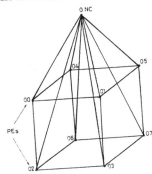

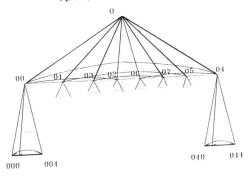

Fig.1 Basic Module of the Extended Hypercube (EH) e.g. EH(3,1)

Fig. 2 EH(3,2)

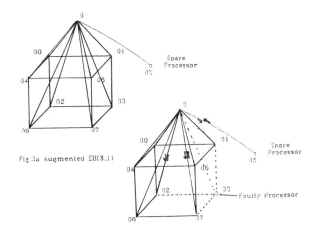

Fig.3a Augmented EH(3,1)

Fig. 3b Message Routing in an Augmented EH(3,1) with Processor Fault

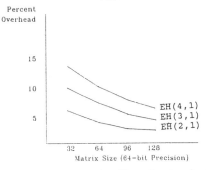

Fig. 4a Execution Time Overhead on NC for Matrix Multiplication

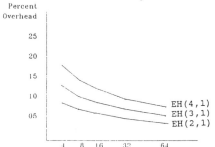

Fig.4b Execution Time Overhead on PEs for Multinode Broadcast

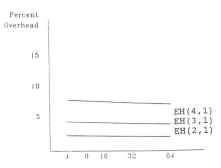

Fig. 4c Execution Time Overhead on NC for Multinode Broadcast

TWO ALGORITHMS FOR MUTUAL EXCLUSION IN A DISTRIBUTED SYSTEM

Kia Makki
Paul Banta
Ken Been
Roy Ogawa

Department Of Computer Science
University Of Nevada at Las Vegas
Las Vegas, Nevada 89154

Abstract

This paper presents two algorithms which achieve mutual exclusion in a distributed system using token passing. The performance of both algorithms improves to a constant number of messages needed to enter the critical section as the number of sites requesting access to the critical section increases.

The proposed algorithms make use of a "Token Queue" which is a part of the token that contains a list of all sites which are requesting the token. The token and queue are sent to the first site on the token queue followed by the second and the third etc. Also, each site maintains a record of which site is the last site on the token queue so that a single token request can be made in order for a site to be placed on the token queue.

Key Terms

Distributed System
Mutual Exclusion
Token Passing
Token Queue

Introduction

Given a distributed system, in which sites communicate with each other only by sending messages and do not share a common memory or clock, it is necessary to guarantee the integrity of shared resources by restricting the use of such resources to one user at a time. This is known as the mutual exclusion problem in a distributed system.

In this paper we present a preliminary algorithm followed by two specific algorithms for solving the mutual exclusion problem in a distributed system. Our algorithms are "Token Based", whereby a token is passed among sites, and only the site which possesses the token is allowed to enter its critical section. Attached to the token is a queue which lists the sites scheduled to receive the token in the future. When a site is finished executing its critical section it must send the token to the next site on the queue. A site requests the token (puts itself on the token queue) by sending a single token request to a site which it believes will be using the token in the near future. Additional messages must be sent in order to keep sites informed about which site is a good one to request the token from.

In an environment with few requests being made for the token, both of our algorithms require N messages to be transferred per critical section execution, where N is the number of sites in the system. As the number of token requests increase, the efficiency of both algorithms increase to the point of requiring C messages to be transferred per critical section, where C is a constant which can be as small as 2.

Previous Work

Mukesh Singhal [4,5] has delineated two classes of solutions to the mutual exclusion problem for distributed system; "assertion driven" solutions and "token based" solutions. In the assertion driven algorithms, a site may enter its critical section only when an assertion becomes true at that site. The assertion is devised so as to become true at only one site at a time. Most of these algorithms require that a site which wants to execute its critical section send request messages to some subset of the sites in the system. The requesting site may enter its critical section when it receives a reply to each of its requests. The algorithm of Ricart and Agrawala [2], which improves on an earlier algorithm from Lamport, requires the requesting site to receive permission from all other sites in the system. Therefore, 2*(N-1) messages are sent per critical section execution. Singhal [4] proposes a "dynamic information structure" in which the set of sites from which a particular site must request permission "evolves with time as sites learn about the state of the system". Although this algorithm requires 2*(N-1) messages per critical section execution in the worst case, it adjusts itself nicely to non-uniform traffic environments, maximizing performance when some sites request critical section execution much more than others.

In the token based algorithms, the token is a unique and singular message which is passed among the sites. Only the site which has the token is allowed to enter its critical section. Ricart and Agrawala [3] propose a token based scheme in which the requesting site sends requests to all other sites. The site which has the token then sends it to the requester. Suzuki and Kasami [6] have proposed a similar strategy. Both algorithms require N messages per critical section execution.

Description Of The Preliminary Algorithm

We present two versions of our algorithm in this paper as well as a "Preliminary Algorithm". The preliminary algorithm is not free from starvation and is intended to demonstrate the main concept of how we use the token queue. It is also intended to demonstrate the simplicity and elegance of our algorithms without bogging down in the details of how to avoid starvation. The two modifications to the preliminary algorithm are presented later in the paper. They focus primarily on how to avoid starvation.

In our algorithms, we make use of a "Token Queue". The token queue contains a list of all the sites which have requested the token. The token and token queue are passed to the first site on the token queue, and then to the next, and so on. Each time the token is passed to a new site, that site is removed from the token queue. For example, if sites S_8, S_2, and S_4 have requested the token, the token and queue would look like figure 1.

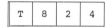

Figure 1 - Token With Token Queue

Sites which are not on the token queue are sent messages informing them of the last site on the token queue. If a sites becomes interested in the critical section, it would send a single request to the last site on the token queue. The last site on the token queue would hold the request until the token arrives, and then place the request at the end of the token queue. Sites only need to be updated about the last site on the token queue when the last site on the token queue changes. For example, if the token queue contains sites S_8, S_2, and S_4, and the token is sent to sites S_8 followed by S_2 without any new sites being added to the token queue then the last site on the token queue is still S_4. In this situation, no sites need to be informed that the last site on the token queue is still S_4. If, however, when the token arrived at S_4 three new sites were added to the token queue, S_5,

S_3, and S_1, then all sites other than S_4, S_5, S_3, and S_1 would need to be informed that S_1 was the new last site on the token queue. S_4, S_5, S_3, and S_1 will each update themselves that S_1 is the last site on the token queue when they receive the token.

In addition to storing the last site on the token queue, each local site also keeps track of its own state and maintains a local queue of token requests that it has received. A site may be in one of four states. They are "N"ot requesting the token, "R"equesting the token, "E"xecuting the critical section, and "H"olding the token. These states are summarized in table 1.

State	Symbol	Wants Token	Has Token
Not req.	N	N	N
Requesting	R	Y	N
Executing	E	Y	Y
Holding	H	N	Y

Table 1 - States Of A Site

Each site must also maintain a local queue of token requests. For example, lets assume that site S_4 is "R"equesting the token and also happens to be the last site on the token queue. All other sites in the system will send their request for the token to S_4. Since S_4 knows that the token will eventually be sent to it, it keeps the token requests in its local queue. When the token does arrive at S_4, S_4 adds its local queue to the token queue.

Initialization

In order to initialize the system, the token must be created with an empty token queue, every site must be "N"ot requesting, every site must know the location of the token, and every site must have an empty local queue. This is to say that there have been no requests made for the token, and every site knows the location of the token.

We can arbitrarily say that site S_1 is the site that will create the token with an empty token queue, and it must be in the "H"olding state. All other sites will then be in the "N"ot Requesting state and know that site S_1 is a good site to ask for the token.

The Preliminary Algorithm

The preliminary algorithm is given as a set of procedures. The procedures represent the necessary responses to make in given situations. It is assumed that all data structures are available to each procedure including the token and token queue.

The preliminary algorithm is not free from starvation. We will present two methods for correcting later in the paper.

```
procedure request_token ;

begin

    site_state := "R" ;
    send ( token_request, good_site ) ;

end .
```

This procedure is called when a site wishes to enter its critical section. The state of the local site is changed to "R"equesting the token. A request for the token is sent to a "Good Site" to ask for the token. The good site to ask is the last known site on the token queue.

```
procedure receive_good_site_update ;

begin

    receive ( good_site ) ;

end .
```

This procedure is called when a site receives an update about the last site on the token queue. The local variable "good_site" is updated to reflect that the last site on the token queue is a good site to send a token request to.

```
procedure receive_request ;

begin

    if ( ( site_state = "R" ) or
          ( site_state = "E" ) or
          ( site_state = "H" )     ) then

      begin
        enqueue( token_request,
                    local_queue ) ;
        if ( site_state = "H" ) then
          call send_token ;
      else
        send ( token_request, good_site ) ;
      end if ;

end .
```

This procedure is called when a site receives a request for the token. If the site is requesting the token, executing its critical section, or is holding the token, the request is stored in the local queue. If the site is not requesting the token, then the token request is forwarded to the last known site on the token queue.

If the site is in the "H"olding state this implies that the token queue is empty and the local site is holding the token until it receives some information on where to send it. In this event, a call is also made to send the token to the next site. The send_token routine will add the local queue to the token queue before sending the token.

```
procedure receive_token ;

begin

    dequeue ( first item, token queue ) ;
    site_state := "E" ;
    call critical_section ;
    call send_token ;

end .
```

This procedure is called when the token is received. The first item is removed from the token queue. This item should be the local site since the token is sent to the first site on the token queue. Next, the state of the local site is changed to "E"xecuting the critical section and the critical section is entered. Finally, a call to send the token to the next site on the token queue is made.

```
procedure send_token ;

begin

   if ( local_queue not empty ) then
   begin
      enqueue ( local_queue,
                token_queue ) ;
      for ( i = each site not
               on token_queue ) do
         send ( token_queue[ last ],
                          site[ i ] ) ;
   end if ;

   if ( token_queue not empty ) then
   begin
      good_site := token_queue[ last ] ;
      send ( token,
           site[ token_queue[ first ] ] ) ;
      site_state := "N" ;
   else
      site_state := "H" ;
   end if ;

end .
```

This procedure is called to send the token to the next site on the token queue. First, if the local queue has requests on it, these requests are added to the token queue. Also, all sites not on the token queue are updated about the last site on the token queue. This is done only if the local queue was not empty which implies that the last site on the token queue has changed. Finally, if the token queue is not empty, the token is sent to the first site on the token queue and the site state is changed to "N"ot requesting. If the token queue happens to be empty, then the token is not sent and the site state is changed to "H"olding which signifies that the local site is waiting for a request to be received in order to send the token.

Explanation

The basic idea behind the preliminary algorithm is that the token has a queue of sites that must be served. If a site, S_i, knows the last site on the token queue, S_j, it can send its request to that site. If the token queue is long enough, the request from S_i will reach S_j before the token does. Site S_j, being in the "R"equesting state, knows that the token will eventually be sent to it. Site S_j, therefore stores the token request from S_i in it's own local queue. When the token reaches site S_j, S_i's token request is placed at the end of the token queue.

In the event that the token request from site S_i reaches site S_j after the token has come and gone, S_j will forward S_i's request to the new last site on the token queue.

All the sites that are not on the token queue must be kept informed about the last site on the token queue. Therefore, each time the last site on the token queue changes, each site not on the token queue must be updated. These updates do not necessarily occur in a timely manner. An information packet is sent to each site and the token is then passed on. There is no guarantee that any of the information packets sent will reach their destinations before the token reaches the next site or the next or the next etc.

Correctness Of The Algorithm

We claim that the preliminary algorithm presented in section 2, achieves mutual exclusion and is free from deadlock. We note, however, that it is not free from starvation. Two modifications to the preliminary algorithm which prevent starvation are discussed later in the paper.

Mutual Exclusion

In order to achieve mutual exclusion, we must show that at most only one site will be in its critical section at a time. This is trivial due to the nature of token passing. We claim that there is only one token and that a site may enter its critical section only if it possesses the token. Possession of the token by a site, S_i, guarantees that no other site has the token and therefore that no other site is in its critical section.

Deadlock

Deadlock would occur if a cycle of requests was formed. For example, sites S_i and S_j both become interested in the critical section simultaneously. Site S_i thinks that site S_j is the last site on the token queue and sends its request to site S_j. Site S_j thinks that site S_i is the last site on the token queue and sends its request to site S_i. Both sites would be in the "R"equesting state when they received the other ones request. Therefore, they would each place the other's request on their own local queue and continue to wait...forever.

We claim that this sequence of events can never happen and examine several situations to prove this fact.

If sites S_i and S_j have never requested the token in the past, then there is no way that either one would be considered as the last site on the token queue. Therefore, neither site would send its request to the other.

If only one of the sites, say S_i, has never

requested the token in the past, then there is no way that site S_j would consider site S_i as the last site on the token queue. Therefore, site S_j would not send its request to site S_i.

We can now limit our investigation to the situation where both sites S_i and S_j have requested the token in the past. Since mutual exclusion guarantees that sites S_i and S_j did not posses the token at the same time, we can say that one site, say S_j, possessed the token after the other site, S_i. Therefore, when site S_j passed the token to the next site, it saved the last site on the token queue as a good site to request the token from. Since site S_i had already been serviced, site S_i could not have been the last site on the token queue. Therefore, there is no way that site S_j could consider site S_i as the last site on the token queue.

Starvation

We claim that the preliminary algorithm, as presented in section 2, is not free from starvation. In this section we show the starvation problem and discuss two modifications which will correct the problem.

Starvation occurs when a site S_i issues a token request to site S_j. By the time the token request is received by site S_j, the token has already come to site S_j and been sent to the next site, site S_k. Site S_j knows this because it is in the "N"ot requesting state. Site S_j therefore, forwards S_i's token request to site S_k. By the time the token request is received by site S_k, the token has again come and gone. As you can see, there is no guarantee that this type of a cycle will ever end. Essentially, site S_i's token request may never catch up to the token in order to be put on the token queue and therefore, site S_i will never receive the token and will starve.

Modification 1

In the first modification to the preliminary algorithm, all token requests are numbered. When site S_i makes a token request it gives the token request a unique number (unique to site S_i). For example, a site could number its requests 0, 1, 2, 3, etc. When site S_j receives S_i's request, it not only forwards the token request to site S_k, but stores the request in its own local queue along with the unique request number. Likewise, when site S_k receives S_i's request it does the same.

The basic idea behind this scheme is that the number of sites that the token can be sent to while still eluding S_i's request is reduced each time S_i's request is received and forwarded. As site S_i's request is placed on more and more local

queues, the token has fewer and fewer sites that it can travel to which are not aware of S_i's request. In the worst case, S_i's request will be sent to N-1 sites before it is put on the token queue.

The reason that requests are given unique numbers is to avoid having requests placed on the token queue multiple times. Since multiple sites record copies of a single request, we must ensure that a particular request be placed on the token queue only once. In order to achieve this, the token must also have a history queue. The history queue records the most recently serviced request for each site in the distributed system. For example, site S_i has made four requests numbered 0, 1, 2, and 3. Each one of these has been serviced. Therefore, the ith entry in the history queue contains a 3 indicating that S_i's request number 3 has been satisfied. However, site S_j may not be aware of this information, and may have S_i's request number 2 in its local queue. If the token were sent to site S_j in this particular situation, site S_j would compare its own local queue with the history queue of the token. Since site S_i's request number 3 has been satisfied, all previous requests by site S_i must also have been satisfied, including request number 2. Therefore, site S_j knows not to place request number 2 on the token queue.

Modification 2

The second modification to the preliminary algorithm involves artificially slowing the token down in order to allow requests to "catch up" and thereby eliminate starvation. In order to apply this modification, we must be guaranteed that there is some maximum amount of time that it takes for a message to be sent from any node to any other node. We will call this time T_{max}. For purposes of discussing the algorithm we will assign T_{max} the value 1. That is to say that it takes at most 1 unit of time, or 1 time period for a message to be sent from any given node to any other given node.

This modification also associates a delay flag with exactly one site on the token queue. The site which is flagged as the "delay site" uses the token to enter its critical section, adds all sites on its local queue to the token queue, updates all sites not on the token queue about the last site on the token queue, and then waits 2 time periods until it can send the token to the next site on the token queue.

The selection of the delay site is made by the previous delay site. Say site S_i is the current delay site. When it receives the token, it uses it, adds its local queue to the token queue, sends the necessary updates, and then selects the last site on the token queue as the delay site. The reason for selecting the last site is to minimize the number of delays. We want the delay to occur at the latest possible time. This is achieved by selecting the last site on the token queue.

We claim that 2 units of time are necessary and sufficient to prevent starvation. In order to show this, we look at the worst case for all relevant message traffic during the 2 units of time. We assume that site S_i is the delay site. One consequence of this is that S_i is also the site that all other sites think is a good site to request the token from. It will take 1 unit of time for site S_i to send an update to each site not on the token queue. In the worst case, the update will reach site S_j immediately after S_j has sent its request for the token to site S_i. It will take another 1 unit of time for S_j's request to reach S_i. This guarantees that site S_j's request will be received by site S_i before the token has left and therefore will be added to the token queue.

Initially, the last site on the token queue, say S_i, is picked as the delay site. When the token is finally sent to S_i, S_i waits the 2 time units. At this point, there is no delay site on the token queue since S_i has just been serviced. Before sending the token, site S_i must assign a new delay site. In order to reduce the number of delays to a minimum, the last site on the token queue, say S_j, is chosen.

Performance Analysis

We look at the performance analysis of the preliminary algorithm in an extremely light token request environment and an extremely heavy token request environment. We then make some general comments about the performance.

Light Token Request Analysis

In an environment with very light token requests, N messages are required per critical section execution where N is the number of sites. Lets assume that the token is waiting at some site, S_i, with no sites on the token queue. Another site, S_j, becomes interested in its critical section and sends its request to site S_i. Site S_i places S_j's request on the token queue and informs all other sites that the last site on the token queue is site S_j. The token is then sent to site S_j. In this example, there was 1 token request, N-2 last site update messages, and 1 token transfer. This adds up to exactly N messages.

Heavy Token Request Analysis

In an environment with very heavy token requests, C messages are required per critical section execution where C is a constant which can be as small as 2. Lets assume that the token queue has been built up to the point where all N sites are on it. The token is sent to the first site on the queue, S_i. It is impossible for S_i to have any sites in its local queue since all the sites are on the token queue. Therefore, site S_i cannot add a last site onto the token queue. Since the last site on the token queue has not changed, no last site update messages are required to be sent. The token is then passed to the next site, S_j. While S_j is using the token, site S_i again becomes interested in the token. Since site S_i just had the token, it is aware of the last site on the token queue, S_z. Site S_i sends its token request to site S_z. When site S_j is done with the token, it has no sites on its local queue, and does not change the last site on the token queue. Therefore, no last site update messages are required to be sent. Site S_j sends the token to the next site on the queue, S_k. While S_k is using the token, site S_j becomes interested in the token. Since S_j just had the token, it is aware that S_z is the last site on the token queue, and sends its token request to S_z. As you can see, all token requests will be sent to site S_z where they will be stored in the local queue until the token arrives. In this situation there will be N token transfers while passing the token to each site on the token queue and N token requests being sent to S_z. When the cycle is complete, the token queue will once again have all N sites on it and N token requests will have been satisfied. This yields 2 messages per critical section execution.

Comments On Performance

As we have shown, in light token request environments, our algorithm has no performance improvement over other algorithms. However, as token requests increase, more and more sites are placed on the token queue and subsequently do not need to be sent last site update messages. Our algorithm is designed to perform extremely well with very heavy token requests.

References

[1] K. Raymond, "A Distributed Algorithm For Multiple Entries To A Critical Section", *Information Processing Letters*, pp. 189-193, 27 February 1989

[2] G. Ricart and A. K. Agrawala, "An Optimal Algorithm For Mutual Exclusion In Computer Networks", *Communications Of The ACM*, pp. 9-17, January 1981

[3] G. Ricart and A.K. Agrawala, "On Mutual Exclusion In Computer Networks", *Communications Of The ACM*, pp. 146-148, February 1983

[4] M. Singhal, "A Dynamic Information-Structure Mutual Exclusion Algorithm For Distributed Systems", *IEEE*, 1989

[5] M. Singhal, "A Heuristically-Aided Algorithm For Mutual Exclusion In Distributed Systems", *IEEE Transactions On Computers*, pp. 651-662, May 1989

[6] I. Suzuki and T. Kasami, "A Distributed Mutual Exclusion Algorithm", *ACM Transactions On Computer Systems*, pp. 344-349, November 1985

A Modular High-Speed Switching Network
for Integration of
Heterogeneous Computing Resources[†]

Wen-Shyen E. Chen, Young Man Kim, and Ming T. Liu

Department of Computer and Information Science

The Ohio State University

2036 Neil Avenue

Columbus, OH 43210-1277

Abstract

A high-bandwidth, low-latency network is essential to realize high-performance computing over the network [1]. The Nectar system [1] is a small-scale example of the high-speed network backplane developed at Carnegie Mellon University for heterogeneous systems.

For the large-scale switching networks that have been proposed in the literature to achieve a maximum throughput of 100%, either internal speed-up or high hardware complexity is needed [2]. As an example, the ATOM switch [3] requires large internal speed-up; the Knockout switch [4] has $O(N^2)$ crosspoints, where N is the network size.

In this paper, we propose a modular switching network based on the parallel-in-parallel-out (PIPO) queueing mechanism. Performance analysis shows that the PIPO switching network is capable of delivering the same maximum throughput without internal speed-up. The switching network has hardware complexity of $O(N \log N)$ and has a regular and uniform structure. Therefore, it has the advantages of (a) being highly scalable due to the modular structure, (b) ease of VLSI implementation, and (c) uniformity of switching elements used. With the employment of packet distributor, the PIPO switching network utilizes efficiently the buffers in the network. As a result, it is a good candidate as large-scale high-speed networks for heterogeneous systems.

1 Introduction

We have witnessed the trend in the evolution of high-performance computing systems moving from homogeneous parallel machines toward heterogeneous multicomputers over networks [1]. (Fig. 1 illustrates such a general computing environment with high-performance computing resources interconnected by a high-speed network.) The driving forces for such a trend are the needs to (a) integrate parallel machines in a general computing environment, (b) increase the accessibility of the high-performance computing resources, and (c) accommodate applications requiring heterogeneous computing resources. A high-bandwidth (up to several Gb/s), low-latency (less than 100 μs) network is essential to make

[†]Research reported herein was supported by U.S. Army CECOM, Ft. Monmouth, NJ, under contract No. DAAB07-88-K-A003. The views, opinions, and/or findings contained in this paper are those of the authors and should not be construed as an official Department of the Army position, policy or decision.

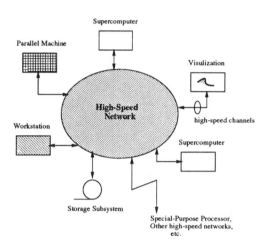

Figure 1: A general computing environment.

high-performance computing over the network feasible [1]. The high-bandwidth can sustain the requirements of high-speed hosts and the low latency, allowing concurrent processing of small-grain computations. The Nectar system [1] is a small-scale example of the high-speed network backplane developed at Carnegie Mellon University for heterogeneous systems.

Several switching networks suitable for large-scale systems have been proposed in the literature [5, 6, 9]. Based on the placement of buffers, these switching networks can be classified into input queueing or output queueing, among others [10]. An example of switching networks employing an input queueing mechanism is the crossbar with FIFO buffers placed on each of the input nodes to the network. Due to head-of-line (HOL) packet blocking, this type of networks has a maximum throughput of about 0.586, for large network size [10]. Switching networks that employ an output queueing mechanism have been proved to achieve the best delay/throughput performance [10]. The ATOM switch [3] is an output queueing network at the cost of N-time internal speedup, where N is the size of the switching network. One of the first output queueing networks without N-time internal speed-up is the Knockout switch [4]. However, large Knockout switches suffer from high hardware costs due to high hardware complexity ($O(N^2)$ crosspoints) [9]. The switching networks proposed in [11] are another effort to implement the output queueing mechanism without incurring internal speed-up. Nevertheless, $O(N^2 \log N)$ switching elements (SEs) are used in these networks. Recently, Kim and Leon-Garcia [2] proposed a self-routing multistage switching network consisting of modified (2×2) SEs, distributors, and buffers located between stages and before the output nodes. The SEs and distributors are elegantly connected as a multi-stage network, and packets are allowed to be queued in the intermediate stages. (Therefore, it uses neither the input nor the output queueing mechanism.) It needs internal speed-up of two and $O(N \log^2 N)$ SEs, and approaches a maximum throughput of 100%, as achieved by the output queueing networks. However, since SEs and distributors of different configurations are used alternatively from the input stage to the output stage, the configuration is not uniform. In addition, internal speed-up of two might not be desirable from a hardware implementation point of view [6].

In this paper, we propose a high performance packet switching network with hardware complexity of $O(N \log N)$ and without internal speed-up. The new switching network consists of (4×4) packet switches (PSs) with a connection pattern similar to that of a banyan network [12]. There are $\log_2 N$ stages of PSs; each stage has $\frac{N}{2}$ PSs. The PS employs a parallel-in-parallel-out (PIPO) queueing mechanism and is nonblocking. (We use "packet switch" here to denote the SE employing a PIPO queueing mechanism.) This switching network, denoted as the PIPO switching network, preserves the self-routing property of banyan networks and achieves the maximum throughput of 100% if buffer modules of proper sizes are used for the PSs in the network. Since PSs of the same configuration are used uniformly throughout the stages and since the network has a regular and uniform structure, it has the advantages of being highly scalable due to the modular structure and ease of VLSI implementation.

The rest of the paper is organized as follows. Section 2 describes the structure and the operation of the PIPO switching network. Section 3 analyzes the performance of the PIPO switching network and discusses the results. Section 4 concludes this paper.

2 The PIPO Switching Network

As discussed in Section 1, implementing the output queueing mechanism at the switching network level requires either high hardware complexity or internal speed-up. The multistage switching network reported in [2] provides an alternative with moderate hardware complexity and speed-up to match the maximum performance of the output queueing network.

Another approach, as will be described in this section, is to implement PIPO queueing mechanism at the switching element level. The packet switches are then connected as a self-routing banyan network. The PIPO queueing mechanism is described as follows.

2.1 PIPO Queueing Mechanism

The major performance bottleneck of the conventional (2×2) SE, such as the (2×2) crossbar, is internal packet contention. One remedy is to speedup the SEs in the network by two and use additional packet distribution switches as in [2]. Another alternative is employing the output queueing mechanism at this SE level without internal speed-up*. As a result, two packets destined for

---

*Note that the Knockout switch implements the output queueing in the *network* level. However, it is much simpler to implement

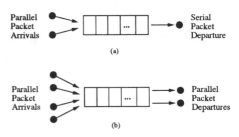

Figure 2: (a) The PISO queueing mechanism. (b)The PIPO queueing mechanism.

the same output port of an SE can be placed in the corresponding output buffer simultaneously, and the major bottleneck can be removed. This configuration is denoted as a 2-parallel-in-1-serial-out (PISO) queue as depicted in Fig. 2(a). In an SE of such configuration, incoming packets can be dropped only when there is no space available in the output buffer. Since during each cycle two packets can arrive at an output buffer and only one packet can be delivered out of the buffer, the buffer has higher probability to be full, resulting in higher packet loss probability and lower throughput. One simple solution is to introduce more spaces into the output buffers, thereby increasing the delays experienced by packets.

A better approach with reduced packet delay is to allow two packets to be sent out from the output buffer simultaneously. In this case, at most four, instead of two, packets should be able to enter one output buffer, resulting in a 4-parallel-in-2-parallel-out (PIPO) queue in the buffer. Such a queue is depicted in Fig. 2(b). Probabilistic analysis shows that the latter configuration with smaller buffer size will result in much smaller probability for buffer overflow. Consequently, the same amount of loss probability can be achieved with a smaller buffer size compared to that of the former configuration. Since delay is proportional, in general, to the size of the buffer, less delay in the PIPO switching network is expected (as will be validated in Section 3).

Using the terms defined in this section, the Knockout switch [4] uses K-parallel-in-1-serial-out (PISO) queueing mechanism (in network level), in that K incoming packets destined for the same output node can be put into the output queue simultaneously. As a result, the internal speed-up is not needed to achieve near 100% throughput performance. However, as discussed in Section 1, the disadvantage of the Knockout switch is the hardware complexity of the packet concentrator $(O(N^2))$. Since this "single-step" packet concentrator results in high complexity, a natural extension of the idea behind the Knockout switch is to employ "multi-step" packet concentrators, in that the functionality of the packet concentrator in the Knockout switch is distributed to multiple stages. The network proposed by Kim and Leon-Garcia [2] implements this idea. Nevertheless, the latter network resolves packet contention in individual SEs by internal speed-up of two and by using expensive distributors. As a result, the hardware complexity of the network is $O(N \log^2 N)$, instead of more moderate $O(N \log N)$. In addition, the configurations of the distributors used are not the same between any two stages. As will be clear in the coming sections, the requirement of internal speed-up can be removed by,

---

the output queueing mechanism in (2×2) SEs, compared to implementing the same queueing mechanism at the $(N \times N)$ network level.

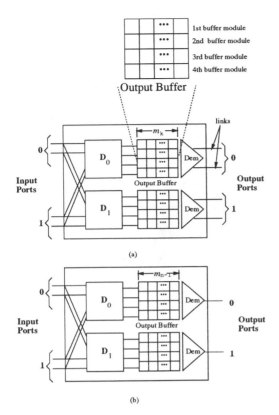

D_i : (4 x 4) Packet Distributor

Figure 3: The packet switch used in the PIPO switching network: (a) The packet switch used at all but the last stages, (b) The packet switch used at the last stage

for example, our PISO or PIPO queueing mechanism implemented at the SE level. Because more delay may be experienced by packets in the PISO switching network, we will concentrate on the PIPO switching network in this paper.

2.2   The Packet Switch (PS)

The PS in the PIPO switching network implements the PIPO queueing mechanism as discussed in the previous subsection. It is a nonblocking switch consisting of two (4×4) *packet distributors*, two sets of buffer modules, and two 4-to-2 demultiplexers (two 4-to-1 demultiplexers are used in the PSs at the last stage). The upper input (output) port is labeled as "0" and the lower input (output) port is labeled as "1." The label is used in routing packets. There are two links associated with each of the input/output ports. The structure of the PS is illustrated in Fig. 3. Note that the packet distributor associated with the 0 (1) output port is labeled as D_0 (D_1). Each output buffer consists of four buffer modules; each buffer module is of size m_k. As shown in Fig. 3(b), the PSs in the last stage use 4-to-1 demultiplexers since each network output node can accept at most one packet per stage cycle.

When a packet enters a PS, it is switched to packet distributor D_0 or D_1, depending on the output port requested. The packet distributor receives packets from its

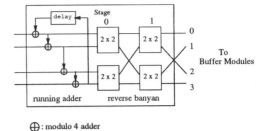

Figure 4: The packet distributor.

four input links and distributes all the packets in a cyclic fashion to the four buffer modules in the output buffer without contention. (The implementation of the packet distributor will be described in detail later.) The 4-to-2 (4-to-1) demultiplexer then chooses two (one) packets from the four buffer modules *in FIFO order* and sends the packets to the output port of the PS. The two output links associated with the output port can deliver two packets to the next stage during each stage cycle. Since in the PS, arriving packet destined for one output port does not interfere with (i.e., conflict or delay) the others destined for the same output port and multiple links are provided for outgoing packets from an output buffer, the PS implements the PIPO queueing mechanism. Moreover, no internal speed-up is needed.

2.3   The Packet Distributor

The packet distributor is used to distribute packets at its inputs evenly without blocking to four buffer modules connected to it, so that the buffers can be used and shared more fairly. This is done by providing an exclusive path for a packet to be placed in one of the four buffer modules without contention. A (4×4) packet distributor consists of four conventional SEs connected as a reverse banyan network, a mirror image of banyan network. The structure of the packet distributor, depicted in Fig. 4, is very similar to that of the distributor proposed in [2]. It has a running adder that helps route packets to a buffer module through the (4×4) reverse banyan network. The running adder computes the addresses of the buffer modules to which the packets at the input links are to be routed. The result from the running adder is then fed back to the first link at the beginning of the next stage cycle, as shown in Fig. 4. Since exclusive-OR gates are major components to compute modulo 4 additions, the running adder is not complex to implement.

The four buffer modules connected to a packet distributor are labeled from 0 to 3. Let the counter C be the address of the output buffer module to which the next arriving packet is to be placed. Initially, when all the buffer modules are empty, the counter C is set to 0. After that, C is set to T (mod 4), where T is the total number of the active packets coming into the packet distributor since the network started to operate. Therefore, C contains the address of the next buffer module that is ready to accept the packet. The value of the updated C for use at the next stage cycle can be easily obtained from the running adder at the lowest line as will be explained below.

Assume that the format of the packets to be delivered to the PIPO switching network is as shown in Fig. 5. Each packet has an activity bit, a destination address, and an information field. The activity bit is set to 1 (0) if the

activity bit	destination address	information

Figure 5: The format of a packet.

packet is full (empty). The packets at the input links of the packet distributor are padded with the addresses of the buffer modules that they have been assigned. The addresses of the buffer modules are obtained as follows. At the beginning of each stage cycle, $\mathcal{C}$ is sent to the top link of the the packet distributor. If the link has an active packet to send, $\mathcal{C}$ becomes the padded address of the packet. $\mathcal{C}$ is then added by the activity bit and propagated to the link just below. If the activity bit is 0, then the value of $\mathcal{C}$ will be propagated down to the link below directly. This process continues to the last link of the packet distributor. This operation forces the active packets to have the buffer module address increased by 1 in modulo 4 from $\mathcal{C}$ to $\mathcal{C} + j$, where j is the number of active packets at the input links of the packet distributor in the current stage cycle.

After the buffer module addresses have been padded into the active packets, the reverse banyan network then routes the packets to proper buffer modules without blocking. A packet can be routed to the proper buffer module by examining the padded address $d_0 d_1$: If d_{1-k} of the padded address is 0 (1), then the packet should be routed through the 0 (1) output at stage k, as shown in Fig. 4. The nonblocking property of the reverse banyan network is the result of the consecutive buffer module addresses (modulo 4) padded to the active packets at the input links of the packet distributor. This property is proved in [2, 13] and is omitted here. After the packets reach the buffer modules, the padded addresses are taken away from the packets.

2.4 The Structure and Operation of the PIPO Switching Network

The $(N \times N)$ PIPO switching network consists of $\log_2 N$ $(=n)$ stages of PSs. The linking pattern of the PSs is the same as that of banyan networks, except that there are two links, instead of one link, connecting an output port of a PS and an input port of another PS at the next stage. A $(2^4 \times 2^4)$ PIPO switching network is shown in Fig. 6. Note that the bold lines in Fig. 6 represent double-links and the thin lines, connecting the PSs at the last stages to the output nodes, represent single-links. (Although there are two links between an input node and an input port of a PS at the first stage, only one links will be used. Another link is left open.) In a PIPO switching network of size $N (= 2^n)$, the stages are labeled in a sequence from 0 to $n-1$ starting from the input (leftmost) stage. Each input (output) port is numbered by n bits, $p_0 p_1 \ldots p_{n-2} p_{n-1}$, where p_0 and p_{n-1} are the most and the least significant bits, respectively. At each stage, a PS is identified by $n-1$ bits, $p_0 p_1 \ldots p_{n-2}$, which is the binary representation of its location within the stage from the top. The connection pattern of the banyan network provides a unique path between any pair of input/output nodes.

The routing of a packet is done by examining its destination tag, $d_0 d_1 \ldots d_{n-1}$, which is a part of each packet. At stage x, the routing depends only on d_x; if d_x is 0 (1), the packet accepted by the PS is first switched to the packet distributor D_0 (D_1). The packet is then assigned a buffer module connected to the packet distributor, and is routed to that buffer module by the packet distributor. As a re-

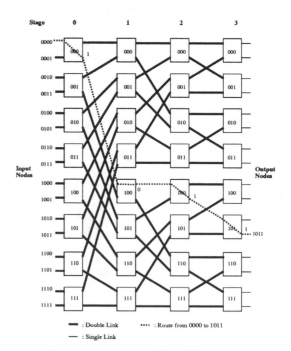

Figure 6: The structure of the $(2^4 \times 2^4)$ PIPO switching network.

sult of the nonblocking property of the reverse banyan in the packet distributor, no contention occurs. The packet will be placed in the assigned buffer module if space is available. The demultiplexer will then choose two (one if in the last stage) packets in the front of the buffer modules and route them to the correct output port. An example of routing a packet from input node 0000 to output node 1011 in a $(2^4 \times 2^4)$ PIPO switching network is illustrated in Fig. 6, with the route plotted in dotted line. Since each of the PSs can be set up in a distributed fashion and a packet can be routed to its destination without involvement of a central controller, the PIPO switching network is a self-routing network.

3 Performance Analysis

3.1 Performance Model

We analyze the PIPO switching network under the uniform traffic model as in [14, 15], i.e., the following assumptions are made.

- Packets are generated at each network input node with equal probability.

- Packets are of a fixed size and are directed uniformly over all of the network output nodes.

- The pattern of packet distribution is identical in every input node and is statistically independent of SEs of any switching stage in the network.

Two performance measures of merit for networks are throughput and delay. The *throughput* of a network is defined as the average number of packets passed by one output node per stage cycle. The *delay* is the average number of stage cycles required for a packet to pass through the network. In the PIPO switching network, packets can always be advanced to the next stage. However, the acceptance of the packets by the next stage depends on the availability of the buffer space at that stage. Therefore, *loss probability* is also an important performance measure.

We first define the following variables in a similar manner as in [16], and then derive a set of state equations relating these variables.

$N =$ The number of input/output nodes, i.e., the size of the PIPO switching network.

$n =$ The number of stages in the PIPO switching network. ($n = \log_2 N$.)

$\lambda_{IN} =$ The network input rate.

$m_k =$ The size of a buffer module at stage k.

$b_k =$ The total number of the buffer spaces in a output buffer consisting of several buffer modules at stage k. (In this paper, $b_k = 4 \cdot m_k$.)

$p_j(k) =$ The probability that an output buffer of a PS at the k-th stage has j packets in it. ($0 \leq k \leq n-1, 0 \leq j \leq 4$.) Note that $\sum_{j=0}^{b_k} p_j(k) = 1.0$. ($p_0, \overline{p_0}, p_{b_k}$, and $\overline{p_{b_k}}$ are the probabilities that the corresponding buffer module is empty, not empty, full, and not full, respectively.)

$q_j(k) =$ The probability that j packets are *coming to* a buffer module of a PS at stage k. ($0 \leq k \leq n-1, 0 \leq j \leq 4$.)

$r_j(k) =$ The probability that j packets in a buffer module of an SE at stage k are able to move forward to the next stage. (If $0 \leq k \leq n-2$, then $0 \leq j \leq 2$ since in the intermediate stages, at most two packets can move forward in one stage cycle. If $k = n-1$, then $0 \leq j \leq 1$ since in the last stage, at most one packet can move to the corresponding output node.)

$S =$ The throughput.

$D =$ The delay.

$L_k =$ The packet loss probability at stage k, $0 \leq k \leq n-1$.

$L =$ The network packet loss probability.

3.2 Performance Evaluation

In this subsection, we follow the model defined in [15] to derive the relations among these variables, except that the analysis in this paper relates the variables in their steady states. The derivation of the following equations for $q_j(k)$'s, $p_j(k)$'s and $r_j(k)$'s assumes the independence of packets in different buffer modules, i.e., uniformity in destinations of these packets, regardless of the contents in the other buffer modules, as in [16]. Although the performance analysis done in this paper is only for the PIPO switching networks with $b_k = 4m_k$, with simple modification the model can be applied to other PIPO switching networks with different configurations.

The Intermediate stages ($0 \leq k \leq n-2$)

Variables $q_j(0)$'s, which are the probabilities that j packets are coming to the buffer module of an PS in the first stage, can be expressed in terms of the network input rate λ_{IN} as follows.

$$q_0(0) = (1 - 0.5\lambda_{IN})^2 \qquad (1)$$
$$q_1(0) = \lambda_{IN}(1 - 0.5\lambda_{IN}) \qquad (2)$$
$$q_2(0) = 0.25(\lambda_{IN})^2 \qquad (3)$$
$$q_3(0) = q_4(0) = 0.0 \qquad (4)$$

From the transition diagram in Fig. 7, the following equations can be derived. These equations can be applied to a buffer module at any PS in the intermediate stages.

$$p_0(k) = p_0(k) \cdot q_0(k) + p_1(k) \cdot q_0(k)$$
$$+ p_2(k) \cdot q_0(k) \qquad (5)$$
$$p_1(k) = p_0(k) \cdot q_1(k) + p_1(k) \cdot q_1(k)$$
$$+ p_2(k) \cdot q_1(k) + p_3(k) \cdot q_0(k) \qquad (6)$$
$$p_2(k) = [p_0(k) + p_1(k) + p_2(k)] \cdot q_2(k)$$
$$+ p_3(k) \cdot q_1(k) + p_4(k) \cdot q_0(k) \qquad (7)$$
$$p_j(k) = [p_0(k) + p_1(k) + p_2(k)] \cdot q_j(k)$$
$$+ \sum_{l=3}^{j+2} p_l(k) \cdot q_{j+2-l}(k), \qquad (8)$$
$$3 \leq j \leq 4$$
$$p_j(k) = \sum_{l=j-2}^{j+2} p_l(k) \cdot q_{j+2-l}(k), \qquad (9)$$
$$5 \leq j \leq b_k - 2.$$
$$p_{b_k-1}(k) = \sum_{l=b_k-3}^{b_k} p_l(k) \cdot q_{b_k+1-l}(k) \qquad (10)$$
$$p_{b_k}(k) = 1.0 - \sum_{j=0}^{b_k-1} p_j(k) \qquad (11)$$

Depending on the number of packets in a buffer module, one, two or none of the packets can advance to the next stage. If there are more than two packets in the buffer module, then two packet at the heads of the buffers can move forward. This scenario is captured by the following equations.

$$r_0(k) = p_0(k) \qquad (12)$$
$$r_1(k) = p_1(k) \qquad (13)$$
$$r_2(k) = 1 - [p_0(k) + p_1(k)] \qquad (14)$$

To simplify the presentation, let

$$h_0(k+1) = r_0(k) + 0.5r_1(k) + 0.25r_2(k) \qquad (15)$$
$$h_1(k+1) = 0.5r_1(k) + 0.5r_2(k) \qquad (16)$$
$$h_2(k+1) = 0.25r_2(k) \qquad (17)$$

The following equations hold:

$$q_0(k+1) = [h_0(k+1)]^2 \qquad (18)$$

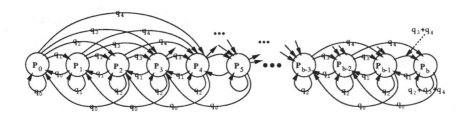

Figure 7: The transition diagram of an m_k-output-buffer module in the intermediate stages. (Note that $b_k = 4 \cdot m_k$. We use b here to denote b_k.)

$$
\begin{aligned}
q_1(k+1) &= 2h_0(k+1) \cdot h_1(k+1) & (19) \\
q_2(k+1) &= 2h_0(k+1) \cdot h_2(k+1) \\
&\quad + [h_1(k+1)]^2 & (20) \\
q_3(k+1) &= 2h_1(k+1) \cdot h_2(k+1) & (21) \\
q_4(k+1) &= [h_2(k+1)]^2 & (22)
\end{aligned}
$$

The Last Stage ($k = n-1$)

According to the transition diagram in Fig. 8, we have:

$$
\begin{aligned}
p_0(n-1) &= p_0(n-1) \cdot q_0(n-1) \\
&\quad + p_1(n-1) \cdot q_0(n-1) & (23) \\
p_1(n-1) &= p_0(n-1) \cdot q_1(n-1) \\
&\quad + p_1(n-1) \cdot q_1(n-1) \\
&\quad + p_2(n-1) \cdot q_0(n-1) & (24) \\
p_j(n-1) &= p_0(n-1) \cdot q_j(n-1) \\
&\quad + \sum_{l=1}^{j+1} p_l(n-1) \cdot q_{j+1-l}(n-1) & (25) \\
&\quad 2 \le j \le 4 \\
p_j(n-1) &= \sum_{l=0}^{4} p_{j-3+l}(n-1) \cdot q_{4-l}(n-1) & (26) \\
&\quad 5 \le j \le b_{n-1}-1 \\
p_{b_{n-1}}(n-1) &= 1.0 - \sum_{j=0}^{b_{n-1}-1} p_j(n-1) & (27)
\end{aligned}
$$

Since during any stage cycle, the PSs in the last stage can deliver at most one packet, the following equations can be derived.

$$
\begin{aligned}
r_0(n-1) &= p_0(n-1) & (28) \\
r_1(n-1) &= 1.0 - p_0(n-1) & (29)
\end{aligned}
$$

The throughput (S), delay (D), and packet loss probability (L) can be expressed as follows.

$$
S = [1.0 - p_0(n-1)] \tag{30}
$$

$$
D = \left(\sum_{k=0}^{n-2} \frac{1}{\sum_{j=1}^{b_k} \frac{p_j(k)}{\lfloor \frac{j+1}{2} \rfloor \cdot \overline{p_0}(k)}} \right.
$$

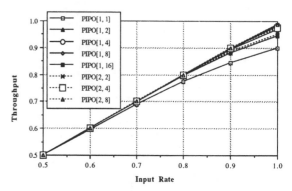

Figure 9: The throughput of some ($2^{10} \times 2^{10}$) PIPO switching networks.

$$
\left. + \frac{1}{\sum_{j=1}^{b_{n-1}} \frac{p_j(n-1)}{j \cdot \overline{p_0}(n-1)}} \right) \tag{31}
$$

$$
Q_k = \sum_{j=0}^{4} j \cdot q_j(k) \tag{32}
$$
$$
0 \le k \le n-1
$$

$$
\begin{aligned}
L_k &= \{ p_{b_k-1}(k) \cdot q_4(k) + p_{b_k}(k) \\
&\quad \cdot [2q_4(k) + q_3(k)] \} / Q_k, & (33) \\
&\quad 0 \le k \le n-2 \\
L_{n-1} &= \{ p_{b_{n-1}-2}(n-1) \cdot q_4(n-1) + p_{b_{n-1}-1} \\
&\quad \cdot [2q_4(n-1) + q_3(n-1)] + p_{b_{n-1}} \\
&\quad \cdot [3q_4(n-1) + 2q_3(n-1) \\
&\quad + q_2(n-1)] \} / Q_{n-1} & (34) \\
L &= 1.0 - \prod_{k=0}^{n-1} (1 - L_k) = \frac{\lambda_{IN} - S}{\lambda_{IN}} & (35)
\end{aligned}
$$

3.3 Numerical Results

In this subsection, we present the results of performance analysis for the PIPO switching network. We use PIPO[Y, X] to denote the PIPO switching network with $m_k = $Y, $0 \le k \le n-2$ and $m_{n-1} = $X, i.e., we assume that the sizes of the buffer modules in the intermediate stages are equal. The rationale is that all the PSs in the intermediate stages are the same, and they all employ the 4-parallel-in-2-parallel-out queueing mechanism.

Fig. 9 shows the throughput of some ($2^{10} \times 2^{10}$) PIPO switching networks. Note that Fig. 9 illustrates only the

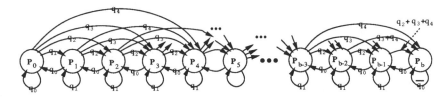

Figure 8: The transition diagram of an m_{n-1}-output-buffer module in the last stages. (Note that $b_{n-1} = 4 \cdot m_{n-1}$. We use b here to denote b_{n-1}.)

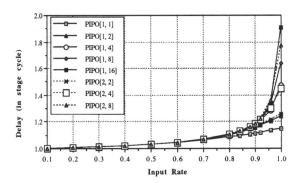

Figure 10: The delay of some $(2^{10} \times 2^{10})$ PIPO switching networks.

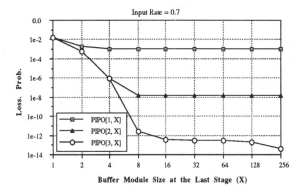

Figure 11: The packet loss probabilities of some $(2^{10} \times 2^{10})$ PIPO switching networks under network input rate 0.7.

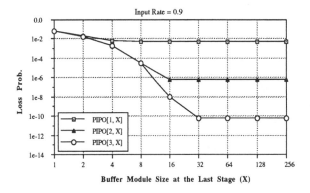

Figure 12: The packet loss probabilities of some $(2^{10} \times 2^{10})$ PIPO switching networks under network input rate 0.9.

throughput of the PIPO switching network with input rates ≥ 0.5 since the analytic results showed that when input rates are less than 0.5, throughput is almost the same as the input rate. From this figure, we note that the throughput of the network increases significantly from PIPO[1, 1] to PIPO[1, 2]. The trend of throughput improvement continues until the size of the buffer module at the last stage reaches 8 (PIPO[1, 8]). In addition, the throughput of the PIPO switching network approaches 100% in PIPO[1,16]. Since there is only one link connected to each output node, the network is unable to deliver the packets in the buffer module at the last stage as effectively as the intermediate stages do. Consequently, buffer modules of larger sizes are needed in the last stage to reduce packet loss probability and to provide higher throughput. As shown in this figure, increasing the size of the buffer modules (from 1 to 2) in the intermediate stages has little effect in the improvement of the throughput of the PIPO switching network.

The delay of the same PIPO switching networks is depicted in Fig. 10. For network input rate $(\lambda_{IN}) \leq 0.9$, the delay experienced by a packet in any of the network configurations in this figure is moderate. However, for inputs rate greater than 0.9, the delay of a packet in the PIPO switching network grows rapidly, especially in networks with larger m_{n-1}. This is due to the increment in the the total buffer size $\sum_{k=0}^{n-1} b_k$ and the accumulation of the undelivered packets.

Figs. 11 and 12 show the packet loss probabilities versus X ($= m_{n-1}$), the size of the buffer module at the last stage, when the network input rates are 0.7 and 0.9, respectively. We observe that the packet loss probability reaches a minimum and no further improvement is obtained by increasing X. The packet loss probability of the PIPO[2, 16] is in the order of 10^{-8} with the input rate 0.7, and is in the order of 10^{-6} with the input rate 0.9.

By increasing Y, the packet loss probability can be further reduced (down to the order of 10^{-10} for PIPO[3, 8].)

From this result, we know that the output buffer, which employs the 4-parallel-in-1-serial-out queueing mechanism, at the last stage is the bottleneck for the throughput of the network, and most of the packet-dropping is done in the last stage if X is not much larger than Y. This bottleneck at the last stage is common for the switching networks in the literature [2, 3, 4, 11] since incoming packets which have the same destination will have to meet each other eventually at that destination. However, only one packet at a time can be delivered to the destined output node.

Note that high-performance packet switching systems which employ the PISO queueing mechanism somewhere in the network will have performance bottlenecks in those parts that use the PISO queueing mechanism. For example, the Knockout switch [4] has performance bottle-necks in the PISO queues following the concentrators that implement the knockout mechanism. The output buffers placed between the network and output nodes of the networks proposed in [2] also use the PISO queueing mechanism and thus become the performance bottlenecks in the networks.

4 Conclusion

In this paper, we have proposed a new self-routing high-speed switching network that can achieve the maximum performance of 100% throughput with moderate hardware complexity and without internal speed-up. This switching network consists of $O(N \log N)$ (4×4) packet switches that implement a parallel-in-parallel-out queueing mechanism. Each packet switch has two links associated with each input (output) port, two packet distributors, two sets of output buffer modules, and two demultiplexers. The connection pattern of the packet switches is the same as that of banyan networks. The performance analysis results of the PIPO switching network show that the network will approximate 100% throughput performance with a uniform input traffic pattern as the sizes of the buffer modules increase. The delay incurred is small since only a smaller number of buffers ($m_k \leq 2$) are used at the intermediate stages. The packet loss probability of the PIPO switching network can be in the order of 10^{-8} to 10^{-10} if buffer modules of proper sizes are used.

The proposed PIPO switching network is modular, highly scalable, and very suitable for simple VLSI implementation. With the employment of packet distributors, the PIPO switching network utilizes efficiently the buffers in the network. As a result, it is a good candidate for the high-speed network for large-scale heterogeneous systems.

Acknowledgment

The authors would like to thank Professor Hyong S. Kim at the Department of Electrical and Computer Engineering, Carnegie Mellon University, for his helpful comments.

References

[1] H. T. Kung, "Moving into the Mainstream." Keynote Speech in the 19th International Conference on Parallel Processing, August 1990.

[2] H. S. Kim and A. Leon-Garcia, "A Self-Routing Multistage Switching Network for Broadband ISDN," *IEEE Journal on Selected Areas in Communications*, vol. SAC-8, pp. 459–466, April 1990.

[3] H. Suzuki, H. Nagano, T. Suzuki, T. Takeuchi, and S. Iwasaki, "Output-Buffer Switch Architecture for Asynchronous Trasfer Mode," in *Proceedings of ICC '89*, pp. 99–103, 1989.

[4] Y. S. Yeh, M. G. Hluchyj, and A. S. Acampora, "The Knockout Switch: A Simple, Modular Architecture for High-Performance Packet Switching," *IEEE Journal on Selected Areas in Communications*, vol. SAC-5, pp. 1274–1283, October 1987.

[5] H. Ahmadi and W. E. Denzel, "A Survey of Modern High-Performance Switching Techniques," *IEEE Journal on Selected Areas in Communications*, vol. SAC-7, pp. 1091–1103, September 1989.

[6] F. A. Tobagi, "Fast Packet Switch Architectures for Broadband Integrated Services Digital Network," *Proceedings of IEEE*, vol. 78, pp. 133–167, January 1990.

[7] W.-S. E. Chen, K. Y. Lee, Y.-W. Yao, and M. T. Liu, "FB-Banyans and FB-Delta Networks: Fault-Tolerant Networks for Broadband Packet Switching," *International Journal of Digital and Analog Cabled Systems*, vol. 2, pp. 327–341, October 1989.

[8] W.-S. E. Chen, Y. M. Kim, Y.-W. Yao, and M. T. Liu, "FDB: A High Performance Fault-Tolerant Switching Fabric for ATM Switching," in *Proceedings of the 10th IEEE International Phoenix Conferecne on Computers and Communications*, pp. 703–709, March 1991.

[9] J. J. Degan, G. W. R. Lunderer, and A. K. Vaidya, "Fast Packet Technology for Future Switches," *AT&T Technical Journal*, vol. 68, pp. 36–50, March/April 1989.

[10] M. J. Karol, M. G. Hluchyj, and S. P. Morgan, "Input Versus Output Queueing on a Space-Division Packet Switch," *IEEE Transactions on Communications*, vol. COM-35, pp. 1347–1356, December 1987.

[11] H. Imagawa, S. Urushidani, and K. Hagishima, "A New Self-Routing Switch Driven with Input-to-Output Address Difference," in *Proceedings of IEEE GLOBECOM '88*, pp. 1607–1611, December 1988.

[12] L. R. Goke and G. J. Lipovski, "Banyan Networks for Partitioning Multiprocessing Systems," in *Proceedings of the 1st Annual Symposium on Computer Archchitectures*, pp. 21–28, December 1973.

[13] H. S. Kim and A. Leon-Garcia, "Non-blocking Property of Reverse Banyan Networks." Submitted for publication.

[14] J. H. Patel, "Performance of Processor-Memory Interconnections for Multiprocessors," *IEEE Transactions on Computers*, vol. C-30, pp. 771–780, October 1981.

[15] H. Yoon, K. Y. Lee, and M. T. Liu, "Performance Analysis of Multibuffered Packet-Switching Networks in Multiprocessor Systems," *IEEE Transactions on Computers*, vol. C-39, pp. 319–327, March 1990.

[16] Y. C. Jenq, "Performance Analysis of a Packet Switch Based on Single-Bufferd Banyan Network," *IEEE Journal on Selected Areas in Communications*, vol. SAC-1, pp. 1014–1021, December 1983.

DYNAMICALLY RECONFIGURABLE ARCHITECTURE
OF A TRANSPUTER-BASED MULTICOMPUTER SYSTEM[a]

Lan Jin* Lan Yang** Chane Fullmer* Brian Olson*

* Department of Computer Science
California State University - Fresno
Fresno, CA 93740-0109
email: jin@csufres.csufresno.edu

** Department of Computer Science
California State Polytechnic University
Pomona, CA 91768
email: lyang@csupomona.edu

Abstract -- The performance of problem solving on a multicomputer system heavily depends on the degree of match between system architecture and parallel algorithm. Thus dynamic reconfiguration has been the basic requirement to designing the architecture of our multicomputer system. In this paper, we proposed three approaches to dynamic reconfiguration based on different strategies: a fixed topology, multiple topologies, and random topology. The issues of hardware organization and software control of dynamic reconfiguration applicable to each approach are addressed. A dynamically reconfigurable multicomputer system together with its control software was designed and implemented using transputers and occam language. Its reconfiguration capability allows any random topology to be created in run time of a program. Test problems were solved on our transputer-based multicomputer system, proving its effectiveness and improved performance in distributed problem solving.

1. Introduction

In a distributed multicomputer system, message-passing style of communication across point-to-point links between computer nodes may cause significant overhead that degrades system performance. Therefore, in distributed problem solving, it is important to take advantage of proximity of data by acquiring a perfect match between system architecture and parallel algorithm. Many approaches can be taken to achieve this goal. On a fixed topology of processors, one has to develop special parallel algorithms to fit the architecture or to design special-purpose architectures to fit the problem. For a general-purpose multicomputer system, it is more reasonable to design a reconfigurable architecture in which the interconnection topology can be varied and optimized to best match any required communication pattern of the problem. In the ideal case, a dynamically reconfigurable architecture would enable the programmer to choose different, always optimal interconnection topologies at different stages of a problem solving process so that the effective performance can be significantly improved. Our research has been motivated by seeking and applying different methods of dynamic reconfiguration in the design and implementation of a highly reconfigurable multicomputer system.

The remainder of this paper is organized as follows. In the next section, we will propose three approaches to dynamic reconfiguration of multicomputer architecture based on different strategies. Then hardware organization and software control of dynamic reconfiguration applicable to these approaches will be addressed in sections 3 and 4 respectively. In section 5, we will describe the design, implementation and testing of our reconfigurable transputer-based multicomputer system. Finally, conclusion will be given.

(a) This work has been supported in part by a California State University Research, Scholarship, and Creative Activity Award. Additional support was provided by a research grant from California State University - Fresno and affirmative Action Faculty Development Program from California State Polytechnic University, Pomona.

2. Approaches to Dynamic Reconfiguration

We have investigated and proposed three approaches to dynamic reconfiguration based on different strategies.

The first approach is based on a fixed topology. Its idea is to choose a static topology as the basis which is fixed but has a good reconfigurability. This basic topology realized on a direct interconnection network with passive and dedicated point-to-point links requires no special software control. The programmer, however, has to map the problem on it for matching the required specific topology. As analyzed in [1], hypercube can be used for this purpose. To embed a ring (Figure 1b) or a multidimensional mesh (Figure 1c) in a hypercube, it is sufficient to align the nodes in each dimension of the mesh or ring in a Gray-code sequence of node numbers. Embedding an n-tree in an n-cube can be done by two methods: either adding a diagonal link to the root node (Figure 1d) or using dual nodes as the root of the tree (Figure 1e). The advantages of dynamic reconfiguration based on a fixed topology are simplicity of hardware, high link bandwidth, ease of reconfiguration, and no overhead of software control. However, the flexibility of reconfiguration is limited.

The second approach is based on multiple topologies. Several useful topologies are designed for coexistence in the same system and selected using switches. If these topologies

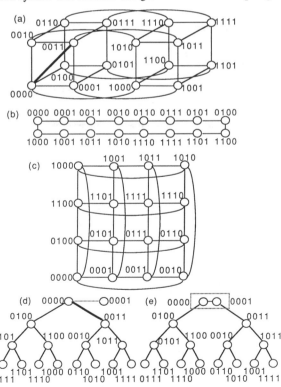

Figure 1. Reconfiguration of a hypercube (a)
into a ring (b), a mesh (c) and a tree (d,e)

have many common links, we may use a small number of switches to change only a part of their links. Switch-setting control statements are included into the problem solving program for accomplishing the run-time reconfiguration task. Compared with the first approach, this method has potentially more flexibility of reconfiguration, though at some sacrifice of hardware complexity, link bandwidth, and software control.

The third approach is intended to realize a random topology by means of dynamic reconfiguration. By random topology we mean any arbitrary interconnection topology, from simple loop to fully connected graph, from regular graph to irregular graph, from a single system-wide topology to a set of multiple local topologies. In fact, it represents a dynamically established fully-connected graph. It is characterized by an extraordinary property that when any two nodes are going to communicate with each other, a neighboring relationship can be established between them. This new possibility of dynamic reconfiguration can be realized by introducing a complete dynamic interconnection network into the system. The most important advantage of this approach is its incomparably high reconfiguration power. It enables every node not only to be connected to every other node, but also to communicate directly with as many nodes as it requires, not limited by the maximum number of physical links each node provides. The only problem is that it may not be allowed to talk different dialogues to all the nodes at the same time. If necessary, the communication events must be properly scheduled. The reconfiguration software will play a significant role in this regard. Thus the software complexity and overhead for random topology will be higher than the second approach due to the large variation and randomness of its control.

3. Hardware Organization of Dynamic Reconfiguration

The above-mentioned approaches to dynamic reconfiguration require hardware support of different complexities.

For the first approach, all the processor nodes are connected in a fixed topology. For the sake of flexibility, it is preferred to hardwire two links from each node into a simple loop, while leaving the remaining links free by leading them to the jumpers or edge connectors of the board, as shown in Figure 2a (assuming 8 nodes each with 4 links). Depending on the specific topology chosen as the basis of the hardware organization, these free links can be connected together additionally. Note that the choice of loop as the hardwired topology is reasonable because, on one hand, the loop best fits the execution of algorithmic parallelism on a pipeline of processors, and, on the other hand, it is the minimum configuration which can be found in many other topologies, as can be seen in Figure 1.

The hardware organization applicable to the second approach is shown in Figure 2b, in which all the free links in Figure 2a are connected to a switching network. These links are thus put under the control of reconfiguration software. Loop remains the hardwired topology of the organization because it is, most probably, one of the multiple topologies we should consider in this approach. To gain the maximum reconfigurability, we may use crossbar switch as the switching network. A special control processor CP should be provided to execute the control software for performing reconfiguration.

The third approach to dynamic reconfiguration requires the hardware organization to offer maximum interconnection power. Therefore, all the links of all the processor nodes are connected to a dynamic interconnection network under the software control from a control processor, as shown in Figure 2c. For expandability of multicomputer system, additional links connected with outside the board must be considered. They are taken to the edge connectors of the board and through

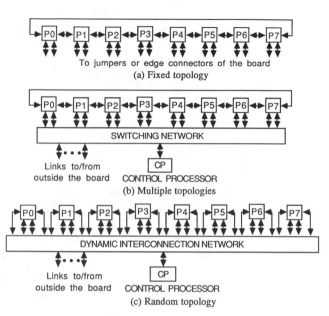

To jumpers or edge connectors of the board
(a) Fixed topology

SWITCHING NETWORK

Links to/from outside the board   CONTROL PROCESSOR
(b) Multiple topologies

DYNAMIC INTERCONNECTION NETWORK

Links to/from outside the board   CONTROL PROCESSOR
(c) Random topology

Figure 2. Hardware Organization for dynamic reconfiguration

the edge connectors further connected to the interconnection networks of other boards. The system can thus be expanded simply through interboard connections. Furthermore, the control processors on all the participating boards should be connected together for close cooperation of their actions.

4. Software Control of Dynamic Reconfiguration

Dynamic reconfiguration needs strong software support. In this paper, we use occam language and its environment for this support. Based on CSP (Communicating Sequential Processes) and message passing paradigms, occam can express any arbitrary degree of interconnections between cooperating and concurrently executing processes, that is, an arbitrary number of concurrent occam processes may mutually communicate by means of messages transmitted along an arbitrary number of occam channels. However, direct implementation of occam programs in a multicomputer system may be constrained by the number of physical links available on each processor node. We should apply software means to reduce the gap between static occam program structure and its real implementation. Three techniques of software control by occam are proposed here to support the reconfiguration of the architecture.

The first technique is to use message bypass or message through-routing. If two processes are located on two distant processors, messages between them can be routed via a series of bypass occam processes, located successively on a series of physically interposed processor nodes. Messages can thus be transported from any source process to any distant destination process within the multicomputer network. In this case, the issue of a message and the receipt of that message between bypass processes are synchronized events in the occam model.

Channel multiplexing [2] is a method which allows several communication channels to share a single physical link by passing their messages into a multiplexing process for transmission, thus overcoming the physical link limitation between neighboring processors. The receiving processes accepts messages from respective senders via a demultiplexing process. The multiplexing and demultiplexing processes reside at each end of a single physical link, allowing multiple pairs of processes to communicate in multiplexed fashion. In this case, each message is tagged by its source or destination processor number as its mark. Either discriminated channel or sequential channel can be used to pass the tagged message.

The third technique is a very effective way of supporting the three approaches to dynamic reconfiguration proposed in this paper. With this possibility of run-time allocation of link connections between processors, theoretically speaking, all the possible parallelism in any occam program can be exploited.

For dynamic reconfiguration based on a fixed topology, the software support is simple. Since the new topology is an embedded graph of the current topology, what we need to do is to disable some existing links and enable other ones. Thus, under the new topology, occam channels will be placed only onto these newly enabled free links.

For dynamic reconfiguration based on multiple topologies, we have introduced an occam process as a switch controller, which accepts reconfiguration requests, waits for current communications on all affected channels to be finished, and then changes the switch setting of the network.

For dynamic reconfiguration based on random topology, any two processors may request a direct link at any time under any configuration. This can be implemented by centrally controlled link resource management or distributed link scheduling system. Here we describe a centrally controlled link resource management. In this scheme, if two processes reside on two distinct processors that are not connected by a direct link at the moment when a communication is ready to take place, a link request must be made first to the link resource manager, asking for a direct link to be established between them. The link resource manager receives the request, puts it into a request queue, and serves all the requests according to some priority, combining the first-come first-serve strategy and the availability of the requested link. The link resource manager has to check the statuses of the links, wait for communications on them to finish, and then establish new connections according to the request.

Despite the obvious advantages of random topology, reconfiguration control and link resource scheduling are complicated. To avoid reconfiguration overhead, we are working on the combination of dynamic reconfiguration based on multiple topologies and random topology, that is to use a small set of common topologies as the basis and to supplement them occasionally with random connections.

5. Implementation and Testing of the System

Among commercially available VLSI microprocessor chips, we chose transputer for implementation of our multicomputer system. We use specifically INMOS T800 which integrates a 32-bit 15-MIPS central processor, a 64-bit floating point processor, 4 Kbytes of fast RAM plus an interface of external memory, and four 20-Mbits/sec bidirectional communication links onto a single chip. The transputer family was specially designed for construction of multicomputer systems. It provides not only convenience of interconnection and simplicity of communication between chips, but also ease of executing concurrent programs. The system composed of transputers can be programmed in occam, which was designed in combination with the transputer architecture for taking full advantage of the special features of transputers.

The schematic diagram of our transputer-based multicomputer system is shown in Figure 3. It is composed of three major units: an 8-node reconfigurable transputer network, a master processor T800, and a host machine SUN-3/260. The network is organized as a VMEbus slave board IMS B014, which consists of eight transputer modules, each equipped with a T800 transputer, 1 Mbyte RAM, and 4 bidirectional links connected to two 32X32 crossbar switches C004. A configuration processor T212 is connected to the configuration

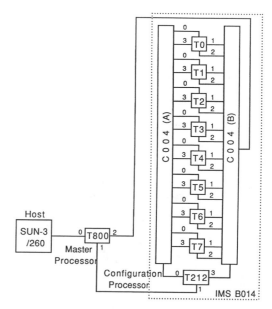

Figure 3. Schematic diagram of a reconfigurable multi-transputer system

links of the two C004's for issuing reconfiguration commands based on the program it received from the master processor. The master processor can communicate with the network via an additional link 3 connected to the crossbar switch, and through the switch to any one of the network transputers.

The reconfiguration software of our multicomputer system was developed in occam 2 and employs a hybrid adaptation of dynamic reconfiguration from a set of known topologies and random reconfiguration. It allows any random topology to be created.

The reconfiguration software consists of two modules. The first module, *setdvt.occ*, is the actual procedure that handles the request from the user program on the master processor. It is included in the user program via the #USE construct in occam 2. As an input parameter, the setdvt() procedure requires a table of bytes describing the interconnections requested by the problem. The user program makes a simple procedure call passing the table to setdvt(). The table used may be created in advance and put in #include files, or it may be created by the user process at run time. The table format was designed for the user at as high a conceptual level as possible and can be thought of as an array of bytes in rows of four columns. Each row in the table represents an interconnection between two transputers as follows:

[byte1,byte2,byte3,byte4]

where byte1 and byte4 are transputer module numbers (from 0 to 7), and byte2 and byte3 are link numbers (from 0 to 3) to be connected. Because of the hardware configuration shown in Figure 3, only links 0 and 3, and links 1 and 2 may be connected together through the C004 switches. Since the crossbar switches do not provide a way to determine if there is activity on the links passing through them, the master processor must be aware that the network links being connected are finished with any prior communications before the request for a reconfiguration can be made.

The second module, *reconfig.occ*, resides on the configuration processor T212. The module constantly monitors the link from the master processor -- waiting to receive a configuration table, one byte at a time, from setdvt(). A lookup table of actual interconnection codes is used to translate the configuration table into the appropriate commands which the crossbar switches require. If an illegal link request is detected, the module sets an error flag to return to setdvt() and bypasses all further interconnections for the current reconfiguration request. Otherwise the connections are set and an "OK" flag is returned to setdvt().

To validate the operation of the reconfiguration software, we have measured the reconfiguration overhead and solved several test problems on our system. The cost of doing any given reconfiguration is 0.256 ms for the first connection and 0.176 ms for each additional connection. Compared with the total execution time of the problem, this overhead is usually small. Three sample problems are given below: data broadcast [3], matrix multiply, and implementation of algorithms in functional language FP [4].

The data broadcast problem was solved on a variable topology of N processors in $\log_2 N$ steps, as shown in Figure 4 for N = 8. At each step the system doubles the number of processors and establishes a new topology among them. This algorithm has shown the distinct advantage of dynamic reconfiguration because it takes the same broadcast time as hypercube, but can be generalized to any number of N with a maximum of only 4 links per node independent of N.

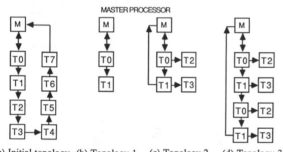

(a) Initial topology (b) Topology 1 (c) Topology 2 (d) Topology 3

Figure 4. Dynamic reconfiguration for data broadcast problem

The Matrix Multiply problem was solved on matrices ranging in size from 16X16 to 160X160 using dynamic reconfiguration in two phases (see Figure 5). In the data load phase, a dynamic star topology was established by locating the master processor M at the center. In the computation phase, a mesh topology was used and the results were accumulated at transputer T6. Therefore, before the multiply started, the master processor was switched to T6. When the operation was complete, the results at T6 can be output directly to the master processor. The same problem was solved on fixed mesh topology with T0 always connected to the master processor. A comparison of the results has showed a 18.6--20.7% speed improvement in favor of dynamic reconfiguration.

The FP problem combines two functions PRODUCTS and FACTORIAL for calculating first the product of a list of integers, and then the factorial of this product. Using a compiler specially developed for our system, the FP program was converted to occam processes, which can be represented in two connected dataflow graphs, one for each function. Then each dataflow graph was partitioned into sections (see Figure 6) to be mapped onto the transputer system. The nature of the generated dataflow graphs is such that they do not easily

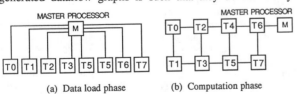

(a) Data load phase   (b) Computation phase

Figure 5. Dynamic reconfiguration
for matrix multiply problem

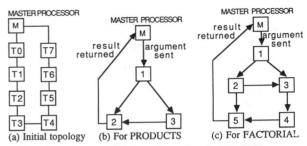

(a) Initial topology   (b) For PRODUCTS   (c) For FACTORIAL

Figure 6. Dynamic reconfiguration for FP problem

conform to a fixed topology. Furthermore, if they were made to conform to the same fixed topology, communication efficiency would drop. Therefore, this problem provided a good example for testing our dynamic reconfiguration software. Dynamic reconfiguration of topologies took place as computation flowed from one dataflow graph to the next dataflow graph. In our case, we had an initial topology which was configured as a loop so that the boot worm can load all the transputer nodes with their executable codes. The average of 25 runs gave the total execution time and reconfiguration time to be 61.76 ms and 3.008 ms respectively. With problems of moderate size, reconfiguration time accounts for only 4.87% of the total execution time.

6. Conclusion

We have proposed and discussed three approaches to designing a highly reconfigurable transputer-based multicomputer system. The approach based on a fixed topology requires least software support and provides, to a certain extent, the capability of reconfiguration. The approach based on multiple topologies requires moderate hardware and software support and can be used in many systems to provide a number of most commonly used topologies applicable to various classes of parallel algorithms. The approach of random topology leads to the most powerful and flexible system, in which dynamic reconfiguration makes it possible to implement a fully connected topology, whose size is not restricted by the maximum number of physical links per processor. It requires, however, intensive software support. The reconfiguration software developed for our multi-transputer system takes advantage of occam 2 and allows any random topology to be created in run time of a program. Its effectiveness has been preliminarily validated by test programs solved on the system. We are still working on other sample problems and also experimenting on new methods of implementation of dynamic reconfiguration.

References

[1] B. Jin and L. Jin, "A New Approach to Hypercube Network Analysis," Proceedings of the 9th Int'l Conf. on Distributed Computing Systems, (1989), pp.263-268.

[2] P. Jones, "The Implementation of a Run-Time Link-Switching Environment for Multi-Transputer Machines," Proc. of the 2nd Conf of The North American Transputer User Groups (NATUG2), (1989), pp.107-122.

[3] S.G. Akl, The Design and Analysis of Parallel Algorithms, Prentice Hall, Englewood Cliffs, NJ, (1989), pp.45-47.

[4] L. Yang and L. Jin, "A Dataflow Implementation of Functional Language FP on Transputers," Proc of the 4th Conf of The North American Transputer User Groups (NATUG4), (1990), pp.153-163.

A Cell-Based Data Partitioning Strategy for Efficient Load Balancing in A Distributed Memory Multicomputer Database System

Kien A. Hua
Dept. of Computer Science
University of Central Florida
Orlando, FL 32816, USA

Chiang Lee
Inst. of Information Engineering
National Cheng-Kung University
Tainan, Taiwan, ROC

Honesty C. Young
IBM Research Division
Almaden Research Center
San Jose, CA 95120, USA

Abstract – This paper deals with load balancing in distributed memory parallel database computers. In such an environment, a relation is typically partitioned and distributed across a set of processing nodes. An efficient load balancing strategy, therefore, is the determining factor for the performance improvement of such systems. We devised a greedy algorithm for the initial data distribution, and an iterative method for subsequent rebalancing. We also developed an analytical model to compare our scheme to two known methods. The results showed that the proposed technique exhibits significant rebalancing costs reduction (in terms of response time and throughput) over existing approaches.

1 Introduction

There are a number of limitations and bottlenecks in "conventional" (von Neumann-type) computers for non-numeric processing in general and database management in particular. The limitations and bottlenecks are in the areas of I/O bandwidth, I/O rate, the aggregate processing power, and physical memory size. Recognizing the limitations of "conventional" computers, applying multiple processors to database problems has been an active area of research for many years. In the past decade, the computer industry has made tremendous progress in microprocessor, memory, and communication technology. As the speed of microprocessors approaches mainframe performance and their costs decrease rapidly, there has been an increasing interest in developing microprocessor-based parallel database machines. Several research prototypes have been built [1, 3, 7, 9] and have drawn great attention from the computer industry. In recent years, a few commercial products have been introduced in the market [4, 10]. We can expect the cost of microprocessors and communication hardware to continue to drop at a drastic rate, and this approach to database management will become even more attractive in the 90's.

Larger relations are horizontally partitioned over a set of *processing nodes* (PNs) in distributed memory parallel database servers, such as the one we described in [6]. Data partitioning and repartitioning are critical to achieve good system performance in terms of *scaleup* and *speedup*. By good scaleup, we mean that if the processing capacity and the problem size are increased by the same proportion, the response time remains the same. By good speedup, we mean the response time of a given problem can be reduced inversely proportional to the increased processing capacity. In this paper, we try to tackle the data partitioning and repartitioning problem on the distributed memory database servers. The same algorithm can also be used to balance the data among several disk drives uniformly accessible from a set of PNs.

We use a well known heuristics for the *bin packing problem* as a subroutine in our algorithms and introduce the notion of *cell*, which is a unit of data partition and movement. That is, all records belonging to the same cell are stored on the same PN. During tuple rebalancing phase, tuples belonging to the same cell are moved together to the same destination PN. A cell is a logical unit and its size is allowed to be changed during run time.

In this paper, we discuss the tuple balancing based on the static information. That is, our criteria is to make sure that each of the PN assigned to store a given relation has about the same amount of tuples from a given relation. The dynamic rebalancing scheme is more involved than the static one, because it depends on the characteristics of the queries, such as the host variables. Two queries running at the same time may have completely different characteristics.

The rest of this paper is organized as follows. Section 2 explains the data partitioning problem and offers our solution. Section 3 has an analytical model and the performance comparison with two known repartitioning methods. Our conclusions are in section 4.

2 Storage Structure and Load Balancing

In this section, we describe the load balancing problem and provide our solution for the initial data distribution and subsequent data rebalancing. In a distributed memory message-passing database server, most large relations are partitioned into data fragments and distributed across a set of PNs. In order to achieve good *speedup* for data intensive queries and good *scaleup* for high throughput transactions, tuples of a given relation must be evenly partitioned among all the domain PNs assigned to store that relation. As we mentioned before, we concentrate on the static data balancing. That is, our goal is to keep the number of records stored on each PN the same, regardless of their usage pattern. To simplify the discussion, we assume all records of a single relation have the same size. The extension to accommodate variable record size is straightforward and will not be discussed here.

In our approach, a relation is partitioned into many *cells*. A cell is a logical unit and the number of cells of a relation is much larger than the number of PNs used to store that relation. To facilitate data balancing, we use one level of indirection. Tuples are assigned to cells, and cells are mapped to PNs. A directory is used to keep the mapping information and to map a cell identifier to a PN during runtime. As far as data balancing (data reorganization) is concerned, a cell is a unit of data movement. That is, tuples within the same cell are assigned to the same PN. As the cell sizes may vary over time, we also describe a mechanism to split/merge cells.

A count is associated with each cell that has the number of tuples in the cell, and is updated when a record is inserted into or deleted from a cell.

When a cell is moved from one PN to another, local indices must be adjusted accordingly. The amount of local index adjustment is proportional to the amount of date moved across the node boundary. The introduction of cell does not complicate the global index. The concept of using cell as a logical level in the storage hierarchy does not seem to change the complexity of keeping the global indices consistent.

These notation will be used in the following sections:

- n: number of PNs in the system
- c: number of cells in the system for a given relation
  (In general, c is much large than n, and it is determined on a per relation basis.)

- $||R||$: relation size in tuples
- $||PN_i||$: number of tuples that have been allocated to PN_i
- $||C_i||$: number of tuples in cell i of a cell list
- $||C_{i,j}||$: number of tuple in the j-th largest cell in PN_i

In the subsequent subsections, we will discuss the initial data distribution algorithm and the data rebalancing algorithm.

2.1 Initial Distribution Algorithm

A traditional horizontal partitioning strategy declusters a relation into n partitions, and assigns each partition to a distinct PN. A serious disadvantage of this scheme is the high cost for performing load rebalancing. We will discuss the performance issues in Section 3. A greedy algorithm to achieve this goal is given in the following:

Initial Distribution Algorithm

1. Partition the relation into c cells, $c \gg n$.
2. Compute $||C_i||$ for $1 \leq i \leq c$.
3. Sort the list of $||C_i||$ where $1 \leq i \leq c$ into descending order. Let LC be the sorted list.
4. Let $||PN_i|| = \emptyset$, $1 \leq i \leq n$.
5. Repeat until LC $= \emptyset$ /* until LC is empty */

   - Find a PN_i where $||PN_i|| = \min_{1 \leq j \leq n}(||PN_j||)$;
     /* If there is more than one PN that satisfies this condition, select one arbitrarily */
   - Assign the first (*i.e.*, largest) cell, C_1 from LC, to PN_i and update the directory information;
   - $||PN_i|| = ||PN_i|| + ||C_1||$; /* allocate a cell to PN_i */
   - LC $=$ LC $-\{C_1\}$; /* remove C_1 from LC */
   - For each $||C_i||$ in LC, rename it to $||C_{i-1}||$;

Conceptually, the allocation algorithm is executed by first sorting in descending order the cell sizes. The cells are then assigned to the PNs in the sorted order. The assignment is done by allocating the currently largest cell in the sorted list to the currently smallest PN. The cell is then removed from the list. This process is repeated until the sorted list is exhausted. The initial distribution algorithm is similar to the "best fit decreasing" algorithm described in [5].

2.2 Data Rebalancing Algorithm

As insertions and deletions occur to the local data in each PN, the size of local files may change so that gradually, a non-uniform distribution of file sizes will appear. This causes unbalanced disk load (*i.e.*, data skew) to the PNs and degrades overall system performance. In this situation, a redistribution of data is necessary to resume good system performance. However, the cost of rebalancing may even be higher than the cost of processing data in a skewed circumstance. For example in Bubba's design [2], if the system estimates that the rebalancing cost is higher than the cost of processing data under the skewed situation, the redistribution will not be performed. In other words, the system will tolerate a possibly high cost on processing skewed data simply because redistribution of the entire relation would cost even more. The high cost for load rebalancing in these traditional horizontal partitioning schemes is due to the fact that these techniques do not take advantage of the already balanced portion of the data but reshuffle the data space entirely. The proposed strategy addresses this problem, and the detailed algorithm is given in the following:

Data Rebalancing Algorithm

Phase 1:

1. Each PN sorts their cell sizes into descending order. Let L_i be the sorted list generated by PN_i: $L_i = \{||C_{i,1}||, ..., ||C_{i,k}||\}$
2. $\forall i$, PN_i sends L_i to a designated coordinator

Phase 2:

1. Determine the average number of tuples each PN should store in a balanced system: $Ave = (||R||/n)$.
2. For i = 1 to n do
   - $L_i' = \emptyset$; /* L_i' is the retaining list */
   - $sum_i = 0$;
   - repeat until $((L_i = \emptyset)$ or $((sum_i + \{C_{i,1}\}) > Ave))$
     - $sum_i = sum_i + ||C_{i,1}||$;
     - $L_i' = L_i' \cup \{C_{i,1}\}$; /* $C_{i,1}$ is retained */
     - $L_i = L_i - \{C_{i,1}\}$ /* remove $C_{i,1}$ from L_i */
     - For each $||C_{i,k}|| \in L_i$, rename it to $||C_{i,(k-1)}||$;

Phase 3: /* L_i has the cells need to be moved from PN_i */

1. Merge $L_i, 1 \leq i \leq n$, into a single sorted list LC;
2. $||PN_i|| = sum_i$, for $1 \leq i \leq n$;
3. Apply step 5 of the Initial Distribution Algorithm to LC;

Except during the final data movement period, this algorithm is executed at a chosen coordinator that manipulates only the directory information. The main objective of this algorithm is to rebalance the PN loads at as little cost as possible. Basically, the algorithm first sorts the cell sizes in each PN into sorted lists. Then, the average number of tuples a PN should retain after the balancing is computed. We retain larger cells first, because it is more effective to rebalance smaller cells in phase 3. Every PN retains its local cells up to the average number of tuples. The remaining cells are then merged into a single sorted list and they are allocated to the PNs using the initial distribution algorithm. We see that this algorithm tries to have as many larger cells at their current host PN as possible. Only those smaller cells are moved between PNs. It is easier to balance large number of small cells than to balance small number of large cells during phase 3 of the rebalancing algorithm. An analytical model will be given and the performance issues will be treated in more detail in Section 3.

This algorithm can also handle the case when an additional PN (called it PN_{new}) is added to the set of PN to store the given relation R. At the beginning of the algorithm, PN_{new} does not have any cells of R. At end of the rebalancing, PN_{new} has its fair share of cells of R. With careful adjustment of boundary conditions, this algorithm can also be used to do the rebalancing when a PN used to store some cells of R is removed from the set of PN used to store R.

2.3 Environment Adaptation

A relation can be partitioned using evenly spaced key ranges. This simple scheme may encounter problems if an extremely non-uniform distribution of the data occurs in the data space. When only very few of the cells are much denser than the remaining cells, the data rebalancing algorithm as described in the last section may fail to correct the severely skewed environment. This problem can be resolved by splitting those denser cells into many smaller cells so that they can be allocated to many more PNs by the data rebalancing algorithm. The split process thus can be considered as a preprocessing procedure that complements the data rebalancing algorithm in the case of extremely non-uniform data distribution. Permitting different interval sizes for the key ranges allows us to handle the non-uniformity in the distribution of data in the data space. In order to handle frequent values nicely, one should allow a key value (for range partitioning) or a bucket (for hash partitioning) be mapped to a set of cells.

In practice, the split process should be needed rarely. If the number of cells is sufficiently large, the majority of non-uniform data distribution cases would involve many cells and the data placement strategy described in the last two sections should be adequate to handle most situations.

As the data distribution changes, an earlier dense area in the data space may become sparse. For efficiency, cells need to be merged when an areas becomes sparse. Since the coordinator has all the cell directory information during the data repartitioning phase, merging two cells stored on two different PNs is as easy as merging two cells stored on the same PN. One way to determine the merge criteria is to keep the size of each cell large enough for efficient I/O performance.

3 Performance Analysis

In this section we derive the cost function, in terms of response time, of three data rebalancing schemes. We also compare the amount of inter PN data transmission, but do not give the detailed communication costs due to space limitations. Since the granularity of load rebalancing performed in the proposed algorithm is at the cell (instead of tuple) level, it is abbreviated as CB (Cell-Based) algorithm hereafter for the sake of convenience. We also give two more traditional schemes, Range Readjust (RR) and Rehash (RH) algorithms, and derive their cost functions. Finally, we compare these strategies and give our remarks.

3.1 Evaluation Model

In this evaluation, we compare the cost of balancing a relation in a system with N PNs. For simplicity, we assume that each PN has the same number of cells (c/n), and PN_1 contains $\delta \times 100\%$ more tuples than each of the remaining PNs in the system. δ is called the degree of skew. Tuples within each PN are still uniformly distributed among the local cells. We make this assumption just to simplify the analytical model, but the amount of inter-PN data movement does not change because of this assumption.

A set of parameters is designed for cost evaluation. The parameters are similar to those used in [8]

- Workload Parameters:
    - $||R||$: the relation size in tuples
    - r: the size of a tuple in bytes
    - $||C_s||$: the size in tuples of the skewed cell
    - $||C_u||$: the size in tuples of the un-skewed (smaller) cell
    - σ: the degree of data skew, which is defined as:
      $$\sigma = \frac{||C_s|| - ||C_u||}{||C_s||} = 1 - \frac{||C_u||}{||C_s||}$$
- System Parameters:
    - μ: the CPU processing rate in MIPS
    - w_{io}: the total I/O bandwidth for each PN
    - w_{comm}: the communication bandwidth among PNs
    - I_{cpu}: The CPU path length for processing a tuple in any step
- Measurement Parameters
    - T_{io}: the time cost in seconds for disk accesses
    - T_{comm}: the time cost in seconds for transferring data between PNs
    - T_{cpu}: the CPU time in seconds for processing tuples
    - C(CB): the total response time of the CB algorithm
    - C(RR): the total response time of the RR algorithm
    - C(RH): the total response time of the RH algorithm

The relation size is assumed to be 1 million tuples. The size of each tuple is $r = 200$ bytes. The processing rate of each PN is $\mu = 20$ MIPS. The bandwidth of disk I/O is $w_{io} = 5$ MBytes/sec, and that of communication among PNs is $w_{comm} = 40$ MBytes/sec. The CPU pathlength for processing a tuple in any step costs $I_{cpu} = 500$ instructions. The number of PNs in the system is $n = 32$, and the number of cells for the corresponding relation is $c = 640$. Finally, we note that in the cost computation, the split and merge costs will not be considered in any algorithm since they are not expected to be used frequently in "real life" applications.

3.2 Cost of CB Algorithm

Recall that each PN retains as much of its own data as possible during Phase 2, and the leftover data will be distributed evenly to all PNs during Phase 3. Under the skew assumption in our model, the number of cells retained by PN_1 at the end of Phase 2 of the algorithm is:

$$x_1 = \frac{(n-1)(c/n)||C_u|| + (c/n)||C_s||}{n||C_s||} = \frac{c}{n^2}(n - \sigma(n-1))$$

During Phase 3, these extra cells are assigned to the appropriate PNs according to step 5 of the initial distribution algorithm. The amount of cells being moved from PN_1 to other PNs are:

$$x_2 = \frac{c}{n} - x_1 = \frac{c}{n^2}(\sigma(n-1)) = \frac{c}{n}\frac{\sigma(n-1)}{n}$$

We note that the computation of the CB algorithm is performed on the cell directory. There is no access to the data records. Its cost is negligible comparing to the disk I/O's and data communication costs involved during the actual load balancing (*i.e.*, , data redistribution) process. For data intensive applications, the data transfer block is large and the data are clustered on the disk. Therefore, the seek time and rotational latency are very small, and the total response time is dominated by the data transfer time. We also use large block size for communication, and the communication time is also dominated by the data transfer time. The cost for load balancing using CB algorithm is thus:

$$C(CB) = \max\left(\frac{x_2||C_s||r}{w_{io}}, \frac{x_2||C_s||r}{w_{comm}}\right)$$

The first term is the cost to load the tuples to be transferred, and the second term is the communication for transferring those tuples to their target PNs. Since these two operations can be performed in parallel, the maximum of them is counted. Note that the cost to store tuples back to disks in the receiving PNs is not in the equation because it is identical to the cost of loading them in the sending PN, and these two actions are performed in parallel.

3.3 Cost of RR Algorithm

In this environment, the relation is declustered into n partitions based on key ranges. To balance the data load for each PN, the RR algorithm surveys the currently unbalanced tuple distribution in each PN and readjusts the range of each PN properly. Tuples no longer belonging to the newly defined range are transferred to the proper destination PNs. In other words, the rebalance is accomplished by shifting the boundaries of the ranges so that the new ranges are of equal size. The range readjust algorithm is given below:

Range Readjust Algorithm:

1. Each PN builds in parallel a local tuple distribution table which describes the distribution of the data in its own range (*i.e.*, the number of tuples located within each smaller predetermined subranges);

2. Each PN sends its tuple distribution table to a designated coordinator;

3. After receiving the local tables from all PNs, the coordinator finds the new ranges that each PN should have in order to balance the data loads;

4. These new key ranges are broadcast to all PNs;

5. Using the newly defined key ranges, each PN checks in parallel if any tuple need to be transfer to their new ranges in other PNs. If so, transfer them to the appropriate PNs.

There are two main tasks in this algorithm. The first task is to read tuples from disk to build the distribution table in each PN. The second task is to read the tuples that need to be transferred and to send these tuples to the proper PNs. We thus have

$$C(RR) = \frac{||PN_1||r}{w_{io}} +$$
$$\max\left(\frac{(||PN_1|| - (||R||/n))r}{w_{io}}, \frac{(||PN_1|| - (||R||/n))r}{w_{comm}}\right)$$

where

$$||PN_1|| = \frac{||R||}{1 + (n-1)(1-\sigma)}$$

The first term in $C(RR)$ represents the cost for reading tuples from disk to build the distribution tables; the second term is the cost of reading the affected tuples and sending them to the proper PNs. The CPU time is negligible in this expression. The communication time for transferring the distribution table and broadcasting the new key range information by the coordinator is also assumed to be negligible. The derivation of the expression is based on the fact that C(RR) is dictated by the bottleneck PN_1 which has skewed data. We will also show the amount of inter-PN data movement in a later section.

3.4 Cost of RH Algorithm

In this subsection, we describe a naive way to rebalancing the load on a hash partitioning environment. The cell-based algorithm we proposed in this paper is also applicable to hash partitioning environment by using many hash buckets and treating each bucket as a cell.

In this environment, data partitioning is performed by hashing the relation into n partitions using a predetermined hash function. Each partition is then assigned to each of the PNs in the system. To balance the data load in the case of data skew, the RH Algorithm rehashes tuples in each PN to their new location using a new hash function. More specifically, when a rebalance of load needs to be performed, each PN will load its data and hash them to n buckets using a new hash function. The number of buckets is identical to the number of PNs in the system. We assume that the hash function is perfect so that bucket sizes are equal within each PN. After the hashing, buckets are transferred to their corresponding PNs and the load of each PN is in perfect balance. Since hashing techniques are often seen in the literature, we will not give the detailed algorithm here.

The cost of this algorithm is derived as follows.

$$C(RH) = \max(T_{io}, T_{cpu}, T_{comm})$$

$$T_{io} = \frac{||PN_1||r}{w_{io}}; \quad T_{cpu} = \frac{||PN_1||I_{cpu}}{\mu};$$
$$T_{comm} = \left(\frac{n-1}{n}\right)\frac{||R||r}{w_{comm}}.$$

Similar to the previous algorithms, these equations are obtained by assuming that all PNs perform hash in parallel. Thus PN_1, the PN that has data skew, causes the bottleneck and dominates the cost of the algorithm. In the equation, T_{io} is the time cost to load tuples from disk for hashing. We have assumed that hashed data are stored in a set of communication blocks, each corresponding to a PN. During the hash process, once a communication block is full, it is immediately transferred to the corresponding PN, not written back to disk. T_{cpu} is the cost of building hash buckets in memory. T_{comm} is the cost to transfer

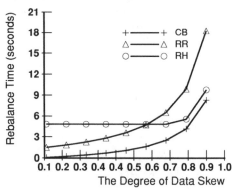

Figure 1: Response Times of the Rebalancing Algorithms.

buckets. Since we assume that this hash is a perfect randomization process, every tuple will have $1/n$ possibility to be kept in the original PN; or $(n-1)/n$ possibility to be hashed to those buckets that will eventually be transferred to other PNs. In total, therefore, there are $((n-1)/n)||R||$ tuples transferred from their original PN to other PNs.

3.5 Performance Comparisons

In this subsection, we compare the performance of our cell-based algorithm with other two (RR and RH) traditional methods mentioned above. Two measurements are used. One is the response time based on the analytical models just described, the other is the amount of inter-PN data movement. The latter one is a good measurement of the system throughput. The formulas to determine the amount of data movement across the PN boundary are relatively simple, and they are not discussed in this paper.

The response time (in second) of performing various rebalancing algorithms for various degrees of data skew are plotted in Figure 1.

The result shows that the CB algorithm is consistently the one of least cost among various algorithms. The RH curve shows a constant cost within a wide range of σ, this is due to the fact that T_{comm} dominates C(RH) for medium and small σ's, and T_{comm} does not vary with σ. Only when the degree of skew is extremely large will the T_{io} dominate T_{comm}. The RR algorithm performs better than RH when σ is not too high. This is because both T_{io} and T_{comm} in C(RR) are low for small σ. When σ is high, the cost for building the distribution table in the skewed PN swamps the algorithm's performance.

Finally, all three curves share a common characteristic, the slops are much steeper for large σ. This observation suggests that it is advantageous to perform load balancing frequently to avoid the high cost of rebalancing seriously skewed environment. Depending on the rate of the skew effect (*i.e.,* how quickly the data becoming skew), the derivative of the load rebalancing cost function can be used to determine how often one should perform the load balancing for his/her system

For some operations, such as the directory update and the global index update, the amount of work is proportional to the amount of inter-PN data movement. The data moved across the PN boundary also uses the communication bandwidth. Therefore, the amount of inter-PN data movement is a good indicator of system throughput. The percentages of data moved across PN boundary are plotted in Figure 2.

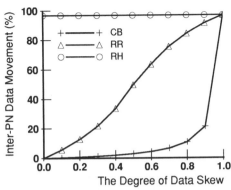

Figure 2: Percentages of Data Moved Across PN Boundary of the Rebalancing Algorithms.

As we can see that the amount of data movement for RH method is independent of the degree of skew. The amount of data movement for RR method grows almost linearly with the degree of skew. The amount of data movement for CB algorithm is very low for a large range of the degree of skew, and the knee of the curve occurs when the degree of the skew (σ) is 0.8 or higher. Thus, it is relatively cheap to use the cell-based rebalancing algorithm when the data is not highly skewed. Since it is cheap to use the CB algorithm, we can afford to run it often and keep the system from being severely skewed.

4 Conclusions

The advent of microprocessor, memory and communication technology has made the microprocessor-based parallel processing a very attractive approach for managing very large databases. Fully utilizing the performance potential of this technology, however, requires a careful design of the distributed file structure and an efficient load balancing strategy. In [6] we described the general approach of designing and implementing an efficient distributed memory database server. In this paper, the emphasis of our discussion is on the storage structure and the load balancing issues of the design. Since effective storage structure is an important lever for load balancing, it is normally the determining factor for the performance of a multiprocessor system. Some serious flaws in today's designs have placed a limit on the performance improvement of existing parallel database computers. In this paper, we proposed a cell-based file structure and an accompanying load balancing strategy that ensemble a very efficient combination for supporting parallel database processing in a distributed memory environment.

Hash partitioning may have better self-balancing characteristics (*i.e.*, the degree of skew does not get too large in a short period of time). However, hash partitioning may not be the most desirable partitioning strategy, because it does not support range query well. Further more, the amount of inter-PN data movement for the rehash scheme is independent of the degree of skew. Thus, it is expensive to rebalance a hash partitioned system. Even if hash partition is desirable in certain context, the idea of cells should still be used and use the directory information to retain as many local cells as possible.

The two primary advantages of the proposed cell-based rebalancing scheme are summarized in the following:

- The granularity for load balancing is at the cell level as opposed to the tuple level in the traditional approaches. Since the number of items we need to examine in the pro-

posed scheme is significantly reduced, the computation overhead associated with a rebalancing process is minimized.

- Instead of reshuffling the entire data space as in the traditional load rebalancing methods, the proposed technique takes advantage of the already balanced portions of the data. Data movement is kept to minimal during redistribution allowing us to devise a load rebalancing strategy with little communication overhead.

We develop an analytical model to compare our schemes with two existing more traditional load balancing strategies. The results show that the proposed technique significantly reduce rebalancing cost (in terms of either response time or the amount of inter-PN data movement) over the other methods.

Finally, due to the high cost of load rebalancing in the traditional structures, most today's systems rather tolerate a possibly high cost on processing skewed data than perform load rebalancing if the redistribution of the entire relation is estimated to cost even higher. The proposed scheme is aimed to overcome this problem. It can be shown that our technique can always provide some savings regardless of the degree of skew. In this sense, the proposed load rebalancing algorithm can be used as an optional pre-processing phase for those data intensive database operations.

References

[1] H. Boral, W. Alexander, L. Clay, G. Copeland, S. Danforth, M. Franklin, B. Hart, M. Smith, and P. Valduriez. Prototyping Bubba, a highly parallel database system. *IEEE Transactions on Knowledge and Data Engineering*, 2(1):4–24, March 1990.

[2] G. Copeland, W. Alexander, E. Boughter, and T. Keller. Data placement in bubba. In *Proceedings of the 1988 ACM SIGMOD International Conference on Management of Data*, pages 99–108. ACM, June 1988.

[3] D. J. DeWitt, S. Ghandeharizadeh, D. Schneider, A. Bricker, H.-I. Hsiao, and R. Rasmussen. The Gamma database machine project. *IEEE Transactions on Knowledge and Data Engineering*, 2(1):44–62, March 1990.

[4] S. Englert, J. Gray, T. Kocher, and P. Shah. A benchmark of NonStop SQL release 2 demonstrating near-linear speedup and scaleup on large databases. Technical Report 89.4, Tandem Computer Inc., May 1989.

[5] M. R. Garey and D. S. Johnson. *Computers and Intractability, A Guide to the Theory of NP-Completeness*. W. H. Freeman and Company, New York, San Francisco, 1979.

[6] K. A. Hua and H. C. Young. Designing a highly parallel database server using off-the-shelf compnents. In *Proceedings, International Computer Symposium*, December 1990.

[7] M. Kitsuregawa, H. Tanaka, and T. Moto-oka. Application of hash to data base machine and its architecture. *New Generation Computing*, 1(1):63–74, 1983.

[8] M. S. Lakshmi and P. S. Yu. Limiting join factors of join performance on parallel processors. In *Proceedings of the 5th International Conference on Data Engineering*, pages 488–496, February 1989.

[9] R. Lorie, J.-J. Daudenarde, G. Hallmark, J. Stamos, and H. Young. Adding intra-transaction parallelism to an existing DBMS: Early experience. *IEEE Data Engineering Bulletin*, 12(1):2–8, March 1989.

[10] Teradata. *DBC/1012 Data Base Computer Concepts and Facilities*. Teradata Corporation, Los Angeles, CA., 1988. Teradata Document C02-0001-05.

Area Efficient Computing Structures for Concurrent Error Detection in Systolic Architectures [†]

by

M. O. Esonu , A. J. Al-Khalili and S. Hariri [*]

Department of Electrical and Computer Engineering, Concordia University,
1455 De Maisonneuve Blvd. West, Montreal, Quebec, Canada, H3G 1M8 .
* Electrical and Computer Engineering Department, Syracuse University,
113 Link, Syracuse, N.Y. 13244 U.S.A.

Abstract

A method of designing testable systolic architectures is proposed in this paper. In our approach, redundant computations are introduced at the algorithmic level by deriving two versions of a given algorithm. The transformed dependency matrix (TDM) of the first version is a valid transformation matrix while the second version is obtained by rotating the first TDM by 180 degrees about any of the indices that represent the spatial component of the TDM. Concurrent error detection (CED) systolic array is constructed by merging the corresponding systolic array of the two versions of the algorithm. The merging method attempts to obtain the self testing systolic array at minimal cost in terms of area and speed. It is based on rescheduling input data, rearranging data flow, and increasing the utilization of the array cells. The resulting design can detect all single permanent and temporary faults and the majority of the multiple fault patterns with high probability. The design method is applied to an algorithm for matrix multiplication in order to demonstrate the generality and novelty of the approach to design testable VLSI systolic architectures.

Index words - **Concurrent Error Detection, data flow computations, VLSI algorithms, space - time mapping, VLSI architectures, systolic arrays, systolic architectures, parallel processing, real-time systems.**

I Introduction

Systolic architectures have received a great deal of attention as an effective means of exploiting VLSI technology to produce high - performance, special - purpose devices for compute - bound problems [1,2]. The uniformity of individual modules and their regularity with localized interconnection make it attractive to consider wafer - scale integration of such arrays to achieve small size, high reliability and performance, low power consumption and cost - effectiveness [3]. Such special - purpose VLSI architectures often have dense circuits. Since the susceptibility of dense VLSI circuits to transient faults is increasing, in systolic array structures typical of VLSI algorithm implementations, a failure, either transient or permanent, in even one processing element can produce unreliable results. Therefore the use of Concurrent Error Detection (CED) capability scheme to ensure the reliability of the computed results is essential.

Several CED schemes for systolic arrays have been proposed in the literature [4-11]. Gulati and Reddy [6] proposed a CED scheme called Comparison with Concurrent Redundant Computation (CCRC), in which each computation and its redundant counter part are performed in two adjacent cells simultaneously. This approach can be applied to a class of 1-D or 2-D systolic arrays in which data as well as the (sub) results keep moving from cell to cell during computation. Wu [7] proposed a similar approach as in [6] which is applicable to unidirectional data flow linear systolic arrays. Cosentino [8] proposed a concurrent error correction scheme which is also based on concurrent redundant computation (CRC) approach. The scheme is restricted to a class of systolic arrays in which the partial results must stay in the cells. Patel and Fung [9], Cheng and Patel [10] presented CED schemes which use the approach of RESO (recomputation with shifted operands) to test a systolic array by repeating every computation with shifted operands. Gupta and Bayoumi [11] proposed a scheme termed as LOED (Logarithm based On-line Error Detection) which is based on the use of logarithmic coding (the addition theorem of logarithms) to test systolic cells. Also the relationship between concurrent error detection via CRC and space-time transformation T of a given systolic array design has been established in [12].

Some of the disadvantages suffered by most of the existing schemes [4-10] include : The reduction of the throughput of a unidirectional array by 50%. The incapability of detecting all transient faults without incurring major overheads in hardware. The detection of only single faults is allowed, multiple fault detection requires high hardware overhead. These techniques are limited to only those systolic implementations in which the data as well as (sub) results keep moving. They cannot be applied to systolic arrays where data are stored in the cells.

The scheme developed in [11] overcomes some of the above disadvantages of the other scheme however it requires a high hardware overhead with complex circuitry to do so.

This paper proposes an interesting method for designing testable systolic architectures. Our scheme is based on CRC approach and space-time mapping of algorithms into systolic arrays. In our approach, redundant computations are introduced in the VLSI algorithms, such that when these algorithms are mapped into specific VLSI systolic architectures, certain degree of observability and controllability are inherent in these architectures that allow concurrent error detection in the systolic architectures. We obtain two transformed dependency matrices (TDM's) that represent the two different versions of the given algorithm. The first TDM is obtained by selecting a valid transformation matrix which transforms the dependency matrix of the algorithm into the new transformed dependency matrix. On the other hand, the second one is obtained by rotating the systolic array corresponding to the first TDM by 180 degrees about any of the indices that represent the spatial component of the TDM. These TDM's are mapped into respective systolic arrays. Concurrent error detection (CED) systolic array is

† This work has been supported by a grant from the Natural Sciences and Engineering Research Council of Canada

constructed by merging the corresponding systolic array of the two versions of the algorithm.

A careful observation of the data flow characteristics of the CED systolic array reveals that, two identical sets of output results (one for each version) may need to be computed at the same computational site and at the same time. Our objective is to identify such computations so that by adding extra hardware only at those computational sites at which they will be computed and by re-scheduling the input data (such that the original input data scheduling is satisfied), the interaction between the two wavefronts can be isolated. Two corresponding sets of computed output results are produced at the same time by the CED systolic array. Concurrent error detection is achieved by comparing these output results using totally self-checking circuits.

This approach offers the following advantages: It is area efficient, there is no need to replicate all the hardware in the CED systolic array. Redundant hardware is introduced only at those computational site where it is needed. All the single transient and permanent faults in the CED array can be detected. In addition to detecting all single faults, this method has the capability of detecting multiple fault patterns with high probability of coverage. By rotating the systolic array of the first version of the algorithm by 180 degrees about one of the indices, it becomes possible to schedule the two independent computations so as to start at the same time with both using the same computational space but at different times. Consequently, there is no high-delay cost incurred. It can also be applied to any systolic array implementation, whether the case in which the data as well as (sub) results keep moving or the case in which the partial results stay in the cells. There is no reduction in throughput. This method provides an effective means of designing CED systolic array architectures and the proposed scheme overcomes the major disadvantages of the other CED schemes.

The design method is applied to an algorithm for matrix multiplication in order to demonstrate the generality and novelty of the approach to design testable VLSI systolic architectures. The cost of this error detection capability is minimal and this includes, the time required to flush out the computations of the two versions of the algorithm, additional hardware (in worst case, the area is increased by less than 1% only, for large values of the matrix dimension N) in some of the processing elements in the CED systolic array and the totally self-checking circuits to compare the two output results of the generated output variables for equality.

II Preliminaries

An important problem associated with designing a systolic array is the mapping of algorithms into systolic array architectures. An interesting approach to develop a mapping is the one called Space - Time representation of computation structures [13-18]. Slight variations exist among the methodologies proposed in the literature based on the space - time mapping of algorithms into systolic arrays. However, the most important problem is how to derive a valid transformation. Moldovan and Fortes presented a six step procedure to find a valid transformation matrix of a given algorithm [13,14,15,17]. Their procedure is an exhaustive search to find a solution that minimizes the parallel execution time. A heuristic approach to derive a valid transformation in polynomial time, based on restricted row operations and restricted normal form of a matrix, is presented in [15].

In general, the space - time mapping approach involves (i) The formulation of the algorithm as a set of linear recurrence equations. (ii) The process of pipelining to eliminate all data broadcasting which is costly in VLSI implementations. To pipeline an algorithm, all algorithm's variables are supplied with any indices which they are deficient in so that they become pipelined with respect to those indices. (iii) Then a linear transformation T is applied to the dependency matrix D to obtain the transformed dependency matrix Δ whose first row is slightly negative while the other elements are 0, 1 or -1 .

For the purpose of illustrating this approach, we will consider three - index nested loop algorithms. An example of such algorithms is the matrix multiplication algorithm, and one which had been used in [14] is as follows:

Algorithm :

```
DO 10  I,J,K = 1,N
   B(I,J,K) = B(I,J-1,K)
   C(I,J,K) = C(I-1,J,K)
   A(I,J,K) = A(I,J,K-1) + B(I,J,K) * C(I,J,K)
10 CONTINUE
```

The corresponding constant dependency matrix of the (matrix multiplication) algorithm is

$$D = \begin{matrix} i \\ j \\ k \end{matrix} \begin{bmatrix} 0 & 0 & -1 \\ 0 & -1 & 0 \\ -1 & 0 & 0 \end{bmatrix} \qquad (2.1)$$

If we choose the transformation T [14] ,

$$T = \begin{bmatrix} 1 & 1 & 1 \\ 0 & 1 & 1 \\ 0 & 0 & 1 \end{bmatrix} \quad \text{then,} \quad \Delta = TD = \begin{bmatrix} -1 & -1 & -1 \\ -1 & -1 & 0 \\ -1 & 0 & 0 \end{bmatrix} \qquad (2.2)$$

The mapping of the index set (i,j,k) into new index set $(\hat{i},\hat{j},\hat{k})$ using the transformation T is given as follows [13,14,15] :

$$(\hat{i},\hat{j},\hat{k}) = T(i,j,k)$$
$$(\hat{i},\hat{j},\hat{k})^t = T(i,j,k)^t \qquad (2.3)$$

Thus, using the T given in Eq.(2.2),

$$\begin{bmatrix} \hat{i} \\ \hat{j} \\ \hat{k} \end{bmatrix} = \begin{bmatrix} 1 & 1 & 1 \\ 0 & 1 & 1 \\ 0 & 0 & 1 \end{bmatrix} \begin{bmatrix} i \\ j \\ k \end{bmatrix} = \begin{bmatrix} i + j + k \\ j + k \\ k \end{bmatrix} \qquad (2.4)$$

The corresponding VLSI systolic array that implements this matrix multiplication algorithm is shown in Fig.1 . All the cells in the array are identical, and the structure of a cell results from the computations required by the algorithm as well as the timing and data communication dictated by the transformed data dependencies (Δ). Figure 2 depicts the structure of the cell. It consists of an adder, multiplier, registers for storing data of each variable and no delay elements.

III Concurrent Error Detection

3.1 Fault Model

In this paper, we assume that faults can occur in the refined level of the processing element such as the computational unit, input/output latch registers, communication links and switches. We will assume that at most one of these modules or parts of the PE is faulty at any given time. Also, it is assumed that both temporary and permanent faults can occur in the array. It is further assumed that the outputs of the faulty cells may assume any logical values independent of the inputs.

3.2 The Proposed Scheme

The developed testing scheme achieves concurrent error detection (CED) through concurrent redundant computation (CRC). It is based on the observation that at any given time the data flow computational activity is located in only some cells in the CED systolic array. That is, at any given time instant, some of the cells in the CED systolic array are performing meaningful computations of the results of the data variables while some are not. Hence, there is inherent spatial redundancy in the array which could be exploited to perform concurrent redundant computations. Two independent computations can be launched into the CED systolic array in such a way that they are performed on different but partly overlapping regions of the array. If two corresponding computations of the independent wavefronts have to share the same computational resource at the same time, then spatial redundancy can be added only at this resource to ensure that the two computations are simultaneously but separately computed on the different computational resources. Thus at the time instant when the computational wavefront of the required computation reaches the faulty cell, its redundant counter part would have been confined to a fault-free region of the array. Consequently, a comparison of the corresponding results would lead to the detection of the fault and this is true because of our single fault assumption.

In this approach, redundant computations are introduced in the VLSI algorithms. We obtain two transformed dependency matrices (TDM's) that represent the two different versions of the given algorithm. The first TDM is obtained by selecting a valid transformation matrix which transforms the dependency matrix of the algorithm into the new transformed dependency matrix. However, the second one is obtained by rotating the systolic array corresponding to the first TDM by 180 degrees about any of the indices that represent the spatial component of the TDM. These TDM's are mapped into respective systolic arrays. Concurrent error detection (CED) systolic array is constructed by merging the corresponding systolic array of the two versions of the algorithm.

Since the systolic array corresponding to the second version of the algorithm is derived from that of the first version, therefore, both arrays have the same number of processing elements and similar interconnection patterns. However, the direction of propagation of the two data flows may be different. Hence, in this method, a merge of the two systolic arrays is equivalent to superimposing the corresponding computational sites in the two arrays. For example, let S represent the first systolic array with the following computational sites: $\{(\hat{j}_1, \hat{k}_1), (\hat{j}_2, \hat{k}_2),..., (\hat{j}_n, \hat{k}_n)\}$ and let S' represent the second systolic array with the computational sites: $\{(\hat{j}_1', \hat{k}_1'), (\hat{j}_2', \hat{k}_2'),..., (\hat{j}_n', \hat{k}_n')\}$. Therefore, a merge of the two arrays represented by $C = S \cup S'$ results in a systolic array which has the following computational sites: $\{((\hat{j}_1, \hat{k}_1), (\hat{j}_1', \hat{k}_1'))$, $((\hat{j}_2, \hat{k}_2), (\hat{j}_2', \hat{k}_2'))$,..., $((\hat{j}_n, \hat{k}_n), (\hat{j}_n', \hat{k}_n'))\}$. If the computational sites $(\hat{j}_i, \hat{k}_i)$ and $(\hat{j}_i', \hat{k}_i')$, (for $i = 1,2,...,n$), are the same then the two cells are the same and hence can be represented by one cell. However, it is important to note that this resultant cell must satisfy the communication requirements of the two original cells.

Although the data flow characteristics of the two versions are different, two identical sets of output results (one for each version) may need to be computed at the same computational site and at the same time. Therefore our goal

is to identify such computations so that by adding extra hardware only at those computational sites at which they will be computed and by re-scheduling the input data (such that the original input data scheduling is satisfied), the interaction between the two wavefronts can be isolated. In other words, by introducing redundant hardware in only some of the CED systolic array cells, the two independent computations can be separately computed. There is no need to duplicate all the hardware in the cells of the CED array. Also due to the input data rescheduling, all the interconnection links need not be replicated. Two corresponding sets of computed output results are produced at the same time by the CED systolic array. Concurrent error detection is achieved by comparing these output results using totally self-checking circuits.

As mentioned above, we derive the second version of the algorithm by rotating the systolic array of the first version by 180° about any of the principal axis (horizontal or vertical axis). This has the advantages in that it makes it easier to merge the two systolic arrays just by superimposing their respective computational sites. Thus, minimizing the number of cells in the CED systolic array and producing a regular structure. Also, rotation by 180° makes it possible to launch the two independent computations into the CED systolic array at the same time, with both wavefronts propagating towards the center of the array. Then redundant hardware is added only where these two wavefronts meet at the same time. Therefore, because of these advantages, rotation by 180° will lead to better implementation. The mathematical analysis and a procedure for the derivation of the second version of the algorithm is given in [19].

3.3 Application of the Proposed Scheme

In order to illustrate our design scheme, we will consider the matrix multiplication algorithm whose dependency matrix is given in Eq.(2.1). As an example of the transformed dependency matrix (TDM) of one version of the algorithm, we will choose the TDM given in Eq.(2.2). The corresponding VLSI implementation of this TDM and the cell structure are as shown in Figs. 1 and 2. To obtain the TDM corresponding to the second version of the algorithm we *rotate*, the systolic array corresponding to the TDM of the first version, by 180 degrees about the horizontal-axis or the vertical-axis (for a 2-D array). Thus, if we rotate the corresponding VLSI implementation of the TDM in Eq.(2.2) by 180° about the vertical-axis and using the results derived in [19], we obtain the second transformation matrix T' and hence Δ',

$$T' = \begin{bmatrix} 1 & 1 & 1 \\ 0 & -1 & 1 \\ 0 & 0 & 1 \end{bmatrix} \text{ and, } \Delta' = \Delta_{rot-vert} = T'D = \begin{bmatrix} -1 & -1 & -1 \\ -1 & 1 & 0 \\ -1 & 0 & 0 \end{bmatrix} (3.16)$$

We have proven [19] that this TDM generated by rotation preserves the order of computation of the given algorithm and also, all other dependency constraints are not violated. The mapping of the index set (i, j, k) into new index set $(\hat{i}', \hat{j}', \hat{k}')$ using the transformation T' is given by ,

$$\begin{bmatrix} \hat{i}' \\ \hat{j}' \\ \hat{k}' \end{bmatrix} = \begin{bmatrix} 1 & 1 & 1 \\ 0 & -1 & 1 \\ 0 & 0 & 1 \end{bmatrix} \begin{bmatrix} i \\ j \\ k \end{bmatrix} + \begin{bmatrix} 0 \\ C \\ 0 \end{bmatrix} = \begin{bmatrix} i + j + k \\ -j + k + C \\ k \end{bmatrix} (3.17)$$

Figure 3 consists of the mapping of the index set into VLSI arrays using the transformation matrices T and T' (for N=3). In this figure, $\hat{i}' = \hat{i}$.

i	j	k	$\hat{i}$	$\hat{j}$	$\hat{k}$	$\hat{j}'$	$\hat{k}'$
			time	processor		processor	
1	1	1	3	2	1	4	1
1	1	2	4	3	2	5	2
1	1	3	5	4	3	6	3
1	2	1	4	3	1	3	1
1	2	2	5	4	2	4	2
1	2	3	6	5	3	5	3
1	3	1	5	4	1	2	1
1	3	2	6	5	2	3	2
1	3	3	7	6	3	4	3
2	1	1	4	2	1	4	1
2	1	2	5	3	2	5	2
2	1	3	6	4	3	6	3
2	2	1	5	3	1	3	1
2	2	2	6	4	2	4	2
2	2	3	7	5	3	5	3
2	3	1	6	4	1	2	1
2	3	2	7	5	2	3	2
2	3	3	8	6	3	4	3
3	1	1	5	2	1	4	1
3	1	2	6	3	2	5	2
3	1	3	7	4	3	6	3
3	2	1	6	3	1	3	1
3	2	2	7	4	2	4	2
3	2	3	8	5	3	5	3
3	3	1	7	4	1	2	1
3	3	2	8	5	2	3	2
3	3	3	9	6	3	4	3

Fig.3 : Mapping of index set into VLSI arrays using transformation matrices T and T' (for N=3).

The VLSI systolic array structure for the TDM in Eq.(3.16) is shown in Fig.4 (for N=3). The array structure is the same as that in Fig.1, with the difference being that the data variable B in Fig.4 propagates in the opposite direction to that in Fig.1. Also the cell structure of Fig.4 is similar to that depicted in Fig.2. The CED systolic array, shown in Fig.5, is constructed by merging the systolic arrays of Figs. 1 and 4, using the merging techniques described in section 3.2.

In our proposed approach of achieving CED in Fig.5, rather than duplicating all the functional blocks in each cell, additional hardware are added only to those cells where the corresponding results of the two independent computations have to be computed at the same time. As seen from Figs.3 and 5, the results of the coefficients a_{i2} (i=1,2,3) for the two versions need to be computed in cells (3,1), (4,2) and (5,3) (at t=4,5,6 , t=5,6,7 and t=6,7,8 for the respective coefficients). If the same functional blocks in these cells are to be used to compute the two independent results and if any of these blocks fails, this fault might have the same effect on the two output results and hence will not be detected. Consequently, the functional blocks in cells (3,1), (4,2) and (5,3) are duplicated in order to compute the corresponding results separately. The input data for variable A are re-scheduled so as to satisfy the original data scheduling. In order to synchronize the data propagation of each of the computations, additional delay elements are introduce in the interconnection lines for variable B. The input data for variable C are stored in the cells.

Figure 6 depicts the CED systolic array (for N=3) designed using the described technique. The array consists of two types of cell structures. In the first type, the functional blocks are not duplicated, while the functional blocks are duplicated in the second type. The respective structure of the two types of cells are shown in Fig.7(a) and Fig.7(b). Figure 6 also contains 6 latches which are used as delay elements to synchronize the data propagation of the two independent computations. Also, as observed in this figure, the corresponding results of the two versions arrive at the output of the array at the same time. Therefore additional circuitry is not required to synchronize the arrival of the output results. The output results of the two independent computations are compared for discrepancies using a totally self-checking (TSC) comparator as shown in Fig.6. It is interesting to know that with this technique, CED can be achieved at very low cost.

Figure 8 shows the CED systolic array implementation of the matrix multiplication algorithm for N=4. All the cells in the array are identical. Since no two identical computations need to be computed using the same set of cells, hence there is no duplication of any functional blocks in the cells. The only added hardware are the latches (three in each interconnection line for variable B) which are used to synchronize the propagation of the corresponding data elements for each computation, into the systolic array. As in Fig.6, the input data for variable A are re-scheduled to satisfy the original input data scheduling for this variable. The corresponding output results appear at the output of the array at the same time and these are compared for discrepancies using the TSC error detecting circuits. The purpose of comparing Figs.6 and 8 is to point out the differences between the two CED systolic arrays. For instance, in Fig.6, the functional blocks of some of the cells are duplicated because they are required to compute the identical output results of the two independent computations. Consequently, Fig.6 consists of two types of cell structure. However in Fig.8, the functional blocks of the cells are not duplicated, hence all the cells are identical and CED is obtained almost at no cost.

3.4 Procedure for Designing Area Efficient CED Systolic Architectures

In this subsection, we present the following five-step procedure to design a CED systolic array architecture using the proposed scheme :
1. The TDM of a given algorithm is determined and then mapped into a systolic architecture.
2. A second version of the algorithm is determined by rotating the systolic array in step 1 by 180 degrees about any of the principal axis (horizontal, vertical or diagonal axis).
3. The two systolic arrays are merged to construct the CED systolic array.
4. Add extra hardware only in those cells of the CED array that would be computing the corresponding output results for the two independent computations.
5. The input data of some of the variables in the algorithm are re-scheduled to satisfy their original input data scheduling. Extra time is required for the completion of the computations and this corresponds to the maximum number of extra delay elements introduced in the interconnection path of any of the data variables.

IV Analysis of the Fault Coverage of the Proposed Scheme

For the proposed scheme, a functional fault model has been assumed. A thorough analysis of this CED technique shows that all single faults in the CED systolic array or a faulty component in the error detecting logic itself would be detected. For instance, in Fig.6, the output result of one of the a_{11} is calculated in cells (2,1), (3,2) and (4,3) while the alternate result is calculated in cells (4,1), (5,2) and (6,3). A discrepancy in the two results indicates a fault or malfunction (temporary or permanent) either in any of the functional blocks of any of these cells or in any of the interconnection lines in the systolic array or in any of the added delay elements. Similarly, the results of, for instance, a_{12} are calculated with the duplicated functional blocks in cells (3,1), (4,2) and (5,3). In this case also, a fault either in any of the functional blocks in these cells or in any interconnection line in the array will produce a discrepancy in the two answers. Since the error detecting circuits are totally self-checking, hence any single fault in the CED systolic array or a faulty component in the error detecting logic itself would be detected. Therefore all single faults in the systolic array of Fig.6 are detectable. Similarly, in Fig.8, consider a single functional module failure of, for instance, the multiplier in the cell (3,2). This fault will cause the partial results of a_{i1} and a_{i4} ($i=1,2,..,4$) computed in this cell to be erroneous. Since all the partial results of these outputs are not computed by only one cell, therefore, the error propagates to the other cells until it reaches the output. The erroneous results computed in this cell will corrupt the other partial results computed in cells (4,3) and (5,4). Consequently, the final results of the output a_{i1} and a_{i4} will be unreliable. The redundant results of these outputs are computed using cells (5,1), (6,2), (7,3) and (8,4) which are fault-free. As a result, the redundant results will be fault-free, and a comparison of the original output results with their redundant counterparts will produce a discrepancy. Hence, this fault is detectable. In addition, most of the multiple fault patterns can be detected as long as the corresponding output results of the two independent computations are not similar. For example, a multiple fault pattern which comprises of single faults in cells (3,1), (4,2), (5,3), (4,1), (5,2) and (6,3) in Fig.6 produces a detectable error pattern, since no corresponding output results of a_{ij} would be affected by the fault pattern. Although few examples have been given, however, it can be easily shown that all single (permanent and temporary) faults and most multiple fault patterns in the CED arrays of Figs. 6 and 8 are detectable.

In Figs.6 and 8, a single fault in either register A, registers C and C', TSC circuit, an adder or multiplier in any of the cells, will manifest itself as an error in the results of some outputs a_{ij} of one version of the algorithm. The consequence is that, the occurrence of any of these faults will activate the error output of only one TSC circuit. However, a fault in either register B, register B' or any interconnection line in the array will be detected by both TSC circuits. Consequently, though we cannot be able to locate the individual faulty modules, but we can be able to determine which modules have probably failed.

V Area and Time Overhead of this scheme

The silicon overhead includes the additional functional blocks introduced only in those cells that are required to compute the corresponding output results of the two independent computations. It also includes the delay elements used to synchronize the flow of input data into the CED systolic array and the error detection circuitry required at the output of the array. As regards to the time, since the two independent computations are launched into the CED systolic array at the same time, there comes a time when a computational resource would be requested by the two different computations at the same time. In order to resolve this conflict, extra time is introduced by delaying one computation (using the delay elements) until the other has com-

pletely utilized the computational resource. Therefore in this respect, we can say that our scheme involves overhead in time. For instance, in Fig.6, one extra clock cycle is required to complete the computation of the two independent results. The number of the extra clock cycles required corresponds to the maximum number of the delay elements introduced in the path of the interconnection line of any of the variables in the algorithm.

In Fig.6, from the way by which the input data for variable A are re-scheduled, it appears that one computation is launched after the other. Consequently, at most twice the number of clock cycles would be required to complete the computation of the two independent results. The systolic array of Fig.6 requires 7 clock cycles to complete the computation of one version of the algorithm. However, the CED systolic array of Fig.6 requires only 8 clock cycles to complete the two computations. Hence, given the nature of the data flow into the CED systolic array after re-scheduling the input data, we can conclude that the proposed scheme does not involve any overhead in time. There is , of course, no loss in throughput. One of the advantages of the scheme is that though the input data is re-scheduled, the corresponding output results of the two computations arrive at the output of the array at the same time. This facilitates the comparison of the two answers without the introduction of any additional control or synchronization circuitry.

For N equal to an odd number (as in Fig.6), the area overhead is N while the time redundancy is $(N-2)$. The complexity of the non-redundant systolic array is N^2 for the number of processors and $(3N-2)$ clock cycles are required to complete the execution of the matrix multiplication algorithm. Therefore, the area overhead ratio is of $O(1/N)$, and the time overhead ratio is $O((N-2)/(3N-2))$. For large values of N ($N = 100$), the percentage hardware and time redundancy ratios are 1% and 33% respectively. If N is even, there is almost no area overhead (0%), and the time overhead ratio is of $O((N-1)/(3N-2))$ (33%).

VI Comparison of our CED scheme with other CED schemes

Several CED schemes have been proposed which achieve fault detection by duplication of operation, either of individual cells or the whole array. We briefly compare our design against other relevant schemes. For the schemes proposed in [6,7,8], besides requiring one additional row of cells , the error detection logic is required in each cell. The schemes in [6,7] are limited to only those systolic implementations in which the data as well as (sub) results keep moving. While the scheme in [8] is restricted to a class of systolic arrays in which the partial results may stay in the cells. There is a reduction in throughput of a unidirectional array by 50% in [6,7]. This is highly undesirable in situations where a high throughput is a crucial requirement. These schemes [6,7,8] cannot detect all transient faults. They allow only single faults, multiple fault coverage requires high hardware overhead. Our scheme requires one additional row of cells which is as a result of the duplication of the functional blocks in those cells that compute the corresponding output results of the two independent computations. Although it requires extra delay elements, however it does not require the error detecting logic in every cell in the array. There is no reduction in throughput. The scheme can be applied to any systolic array implementation. It can detect all single (both permanent and temporary) faults and most multiple fault patterns.

In the application of RESO [9,10], besides requiring one additional row, the scheme needs three additional cells in every row. Also, the method is effective only under a single faulty cell assumption. When our approach is applied to the same array requires only one additional row of cells, some delay elements and yet the fault model allows one full row of cells to be faulty at a time. The scheme developed in [11] overcomes some of the short falls of the other schemes [4-10]. It has extra hardware that allows a cell to test itself thereby allowing multiple faulty cells. The same self-testing systolic cell can then be used to build any array structure. Although each cell can achieve self-testing in the scheme proposed in [11], however the hardware required to do so makes the structure of the cell to be very complex. In addition to the functional modules of the cells, the error detection logic incorporated in each cell, consists of a counter, shift register, a comparator and a memory. Alternatively, it may consist of a ROM, memory address register and a memory buffer register. All of these increase the complexity of the cell and make it susceptible to more failures. In our scheme, the structure of the cells is simple although the cells are not self-testing. However our scheme detects all single permanent and temporary faults and most multiple fault patterns. The LOED scheme [11] detects all permanent faults but only most of the transient faults. Since the cells are self-testing, LOED has the ability to detect some of the multiple fault patterns.

VII Conclusion

In this paper, we have presented an area efficient CED scheme for VLSI systolic array architectures. This scheme offers solutions to several problems of the other known CED schemes. It is applicable to a wide class of VLSI implementation of algorithms. The technique is not limited to only those systolic implementations where the data as well as (sub) results keep moving. It can also be applied to those implementations where either the data or the (sub) results are stored. It has a feature of exploiting the advantages of the interleaved computations without any reduction in the throughput of the array structure. The scheme can detect all single permanent and temporary faults and a majority of the multiple fault patterns. This scheme can be applied to any systolic array structure (e.g. 2D, tree - connected etc.).

References

[1] H. T. Kung, "Why Systolic Architectures?," *IEEE Computer*, Vol. C-31, pp. 37-46, Jan. 1982.

[2] H. T. Kung and C. E. Leiserson, "Systolic Arrays (for VLSI), *In I. S. Duff and G. W. Stewart (editor), Sparse Matrix Proceedings*, 1978, pp. 256-282, SIAM 1979.

[3] F. T. Leighton and C. E. Leiserson, "Wafer-Scale Integration of Systolic Arrays," *IEEE Trans. Computers*, Vol. C-34, pp. 448-461, May 1985.

[4] S-W Chan and C-L Wey, "The Design of Concurrent Error Diagnosable Systolic Arrays for Band Matrix Multiplication," *Proc. IEEE Trans. on Computer-Aided Design*, Vol.7, No.1, pp. 21-37, January 1988.

[5] J. Abraham, P. Banerjee, C-Y. Chen, W. Fuchs, S-Y Kuo and A. Reddy, "Fault Tolerance Techniques for Systolic Arrays," *IEEE Computer*, pp. 65-74, 1987.

[6] R. K. Gulati and S. M. Reddy, "Concurrent Error Detection in VLSI Array Structures," *Proc. IEEE Intl. Conf. on Computer Design*, pp. 488-491, 1986.

[7] C-C Wu and T-S Wu, "Concurrent Error Correction in Unidirectional Linear Arithmetic Arrays," *Proc. 17-th Intl. Symp. on Fault-Tolerant Computing*, pp. 136-141, 1987.

[8] R. J. Cosentino, "Concurrent Error Correction in Systolic Architectures," *Proc. IEEE Trans. on Computer-Aided Design*, Vol. 7, No. 1, pp. 117-125, January 1988.

[9] J. H. Patel and L. Y. Fung, "Concurrent Error Detection in ALU's by Recomputing with Shifted Operands," *IEEE Trans. Comput.*, Vol C-31, pp. 589-595, 1982.

10] W-T Cheng and J. H. Patel, "Concurrent Error Detection in Iterative Logic Arrays," *FTCS*, pp. 10-15, June 1984.

[11] S. R. Gupta and M. A. Bayoumi, "Concurrent Error Detection In Systolic Arrays For Real-Time DSP Applications," *VLSI Signal Processing III*, edited by Robert W. Brodersen and Howard S. Moscovitz, IEEE Press, 1988.

[12] H. F. Li, C. N. Zhang and R. Jayakumar, "Latency of Computational Data Flow and Concurrent Error Detection in Systolic Arrays," Tech. Report in Dept of Compt Sc., Concordia Univ., Montreal Que., Canada.

[13] D. I. Moldovan, "On the Design of Algorithms for VLSI Systolic Arrays," *Proc. IEEE,* Vol. 71, No. 1, January 1983.

[14] W. L. Miranker and A. Winkler, "Spacetime Representations of Computational Structures," *Journal of Computing*, Vol. 32, 1984.

[15] C. N. Zhang and H. F. Li, "Mapping Data Flow Computation into Universal Systolic Arrays," submitted to Journal of Parallel and Distributed Computing, 1989.

[16] M. O. Esonu, A. J. Al-Khalili and S. Hariri, "Design of Optimal Systolic Arrays: A Systematic Approach," *Proc. IEEE Symposium on Parallel and Distributed Processing*, pp. 166-173, Dallas, Texas, Dec. 9-13, 1990.

[17] D. I. Moldovan and J. A. B. Fortes, "Partitioning and Mapping Algorithms into Fixed Size Systolic Arrays'" *IEEE Trans. Comput.*, Vol. C-35, No. 1, pp. 1-12, January 1986.

[18] M. O. Esonu, A. J. Al-Khalili and S. Hariri, "On the Design of Optimal Fault-Tolerant Systolic Array Architectures," accepted for presentation at the *5-th Int'l Parallel Processing Symp.*, Anaheim, California, April 30 - May 2nd, 1991.

[19] M. O. Esonu, A. J. Al-Khalili and S. Hariri, "Area Efficient Architectures for Concurrent Error Detection in Systolic Arrays," submitted to the *IEEE Trans. on Computers*, 1990.

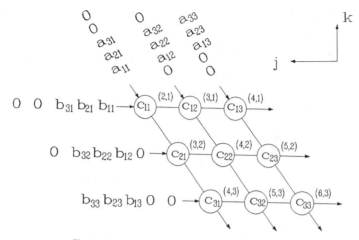

Fig.1 VLSI array structure that implements the matrix multiplication algorithm (for $N=3$).

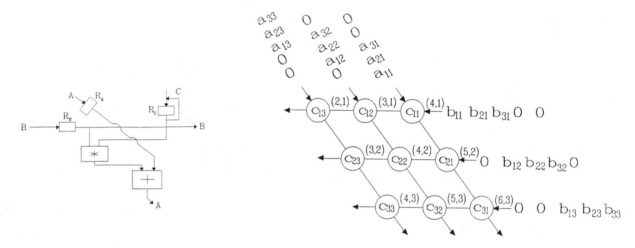

Fig.2 The structure of the cell in Fig.1.

Fig.4 The VLSI array structure resulting from rotation of Fig.1 by 180° about the vertical axis.

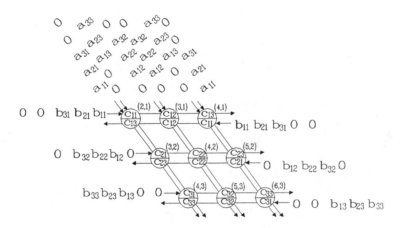

Fig.5 The VLSI array structure resulting from merging the systolic arrays of Figs 1 and 4 (for $N=3$).

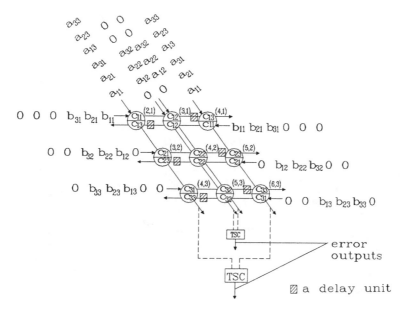

Fig.6 CED systolic array for matrix multiplication (for $N=3$).

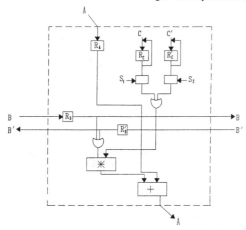

Fig.7(a) The cell structure of type I cell in Fig.6 .

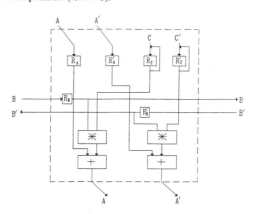

Fig.7(b) The cell structure of type II cell in Fig.6 .

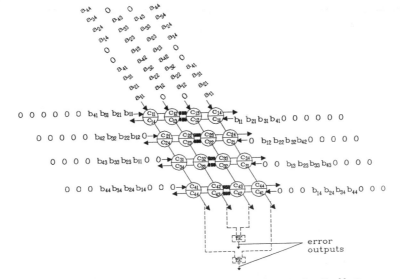

Fig.8 CED systolic array structure for matrix multiplication (for $N=4$).

Channel Multiplexing in Modular Fault Tolerant Multiprocessors †

M. Sultan Alam
Rami G. Melhem

Department of Computer Science
University of Pittsburgh
Pittsburgh, PA 15260

ABSTRACT

A systematic approach to the design of general fault tolerant multiprocessors is described. This design approach utilizes the concept of multiplexing different channels of a link to reconfigure a fault tolerant system in the presence of faults. Various fault tolerant design schemes for specific multiprocessors have been previously proposed. Only a few of these schemes generalize to other architectures and a few can implement arbitrary degrees of redundancy. The design approach presented in this paper is simple and can be applied to any modular architecture with any degree of redundancy. This approach also incorporates useful design criteria such as incremental design and efficient reconfigurability. We apply this approach to the construction of fault-tolerant binary trees and fault-tolerant hypercubes.

1. Introduction

Many fault tolerant designs have been proposed in the literature for specific multiprocessor architectures. For example, fault tolerant designs for hypercubes have been proposed by Rennels, Chau et al. and Banarjee et al.[4,6,13]. Hayes presents a fault tolerant design for non-homogeneous trees[10], a fault tolerant design for homogeneous trees has been proposed in[11]. Fault tolerant designs for binary trees have been proposed by Hasan et al. [9], Adit Singh[14] and Raghavendra et al.[12]. Banarjee et al. proposed two fault-tolerant design schemes that tolerate one faulty cycle or one faulty node per group in the cube-connected-cycles architecture[3]. None of these designs are presented in a general form that applies to a large class of architectures and only a few can implement an arbitrary degree of redundancy. A fault tolerant design method has been proposed by Chakravarty et al. for the pyramid, the mesh, the hypercube and the binary tree architecture[5]. Although this design method can be used for different architectures and can implement an arbitrary degree of redundancy. The Diogenes method[7] realizes a k-FT version of a graph by decomposing it into m outerplanar graphs using k spare nodes. This method can however be very expensive in terms of the total layout area and wire requirement when $m > 1$. An automorphic approach to the design of fault-tolerant multiprocessors have been proposed by Dutt and Hayes[8]. It can implement any degree of redundancy in regular multiprocessor graphs that are circulant. For non-circulant multiprocessor graphs the edge-supergraph of the multiprocessor graph needs to be constructed efficiently in order to implement fault-tolerance. For some graphs the degree of a node of the edge-supergraph is approximately twice the degree of the node in the original multiprocessor graph.

The following aspects should be taken into consideration when a reconfigurable fault tolerant multiprocessor is designed: (a) the ease of applicability to different architectures, (b) incremental design, which is the ability to built a larger system from smaller systems, (c) simple reconfiguration algorithms, (d) fast reconfiguration, which implies the ability to locally reconfigure around any new set of faults without undoing any earlier reconfiguration, (e) efficient use of redundant hardware. In this paper we present a new design method that addresses the above issues. Although this design approach can be applied to any multiprocessor system (with any redundancy), in this paper we consider systems with nodes of the same degree. This is because such systems are easy to describe and also because most of the multiprocessors used today usually have nodes with the same degree. For example, binary hypercubes, binary trees(except the root and the leaves), torus, CCC's, rings etc. have nodes of the same degree.

In order to achieve fault tolerance without degrading the system performance spares need to be added to the system to replace faulty nodes. If any spare node in a system can replace any faulty primary node, then the redundancy scheme is called *global redundancy*. Usually fault tolerance is required in large systems because the probability of node failure in such systems is high. But implementing *global redundancy* in large systems may be prohibitively expensive in terms of hardware overhead. The alternative is to use *local redundancy*. Systems that employ *local redundancy* may be built from modules. Each module consists of a set of spares and a set of primary nodes, where the spare nodes in a module can only replace the primary nodes of that module. The use of such modules to build fault tolerant systems has several advantages: 1) local and fast fault detection and reconfiguration; 2) ease of system construction, scalability and modular replacement and 3) simple reconfiguration algorithms. The design approach discussed in this paper is particularly attractive for systems that employ modular *local redundancy*, but can also be used for systems with *global redundancy*.

The rest of the paper is organized as follows: In Section 2, we describe our design approach and show how to apply it to multiprocessor systems in general. The design of fault tolerant multiprocessor systems from fault tolerant modules is described in Section 3. The channel requirement of the links are described in Section 4. In Section 5, we apply this design approach to binary hypercube systems. Finally, Section 6 contains the conclusion.

2. Modular Fault Tolerant Systems

A fault tolerant multiprocessor system can be viewed as a set of primary nodes and a set of spare nodes that can replace them. A primary node in the system is assumed to have m primary node neighbors. The m neighbors of a primary node P_i are denoted by $N_j(P_i)$, for $j = 1,2,...,m$. The link between $N_j(P_i)$ and P_i is called the jth link of P_i. The set of spare nodes that can replace a primary node P_i is called the *spare set* of P_i which is denoted by $SS(P_i)$. For example in Figure 1(a), P_i is a primary node and the spare nodes that can replace it are in the set $SS(P_i)$. Also the spare sets of each primary node $N_j(P_i)$ are denoted by the set $SS(N_j(P_i))$. Note that $SS(P_i)$ and $SS(N_j(P_i))$ may or may not be mutually exclusive. We assume that the systems considered in this paper have some kind of fault detection mechanism. When a fault is detected in a primary node P_i, reconfiguration is initiated by replacing P_i by a spare node S_r from the set $SS(P_i)$. After reconfiguration the adjacency relationship among the *live* nodes should be preserved, where the *live nodes* are defined to be the union of the set of all non-faulty primary nodes and the set of spare nodes that have replaced all faulty primary nodes. That is, after reconfiguration S_r should become a neighbor of $N_j(P_i)$, for $j=1,2,\cdots,m$, and also if a node $N_j(P_i)$ was faulty and was replaced by the spare $S_g \in SS(N_j(P_i))$, then S_r should become a neighbor of S_g. In this context we consider two nodes to be neighbors if there exists a link or a dedicated connection between the two nodes, where a dedicated connection between two nodes is defined to be a connection through at least one intermediate node which is similar to a permanent *circuit switched* connection. The store and forward method is not used at the intermediate nodes to establish such connections since it causes considerable delay in message propagation.

Figure 1 and the figures in the rest of the paper can be explained as follows: Larger circles denote primary nodes and the smaller ones denote spare nodes. A spare node that replaces a primary node is shown by an arrow between the two nodes, where the tail of the arrow represents the faulty node (crossed out) and the head arrow represents the spare node. For example in Figure 1(b), node P_i fails and S_r replaces it. In the same figure the connections established between node S_r and node $N_j(P_i)$ are through node P_i, where $j=1,\cdots,m$. In this context node P_i is used as a switch for the dedicated connections passing through it. In order to become a neighbor of a node Y, the spare S_r should initiate a process for establishing a dedicated connection to node Y unless there exists a link between them. The following assumptions are necessary for the above reconfiguration strategy to function properly:

A1. No two nodes fail at the same instant.

A2. No node fails during reconfiguration.

A3. A faulty node can switch non-faulty links through it.

Note that assumption A3 does not necessarily imply that the router of the faulty node has to be fail-safe. An extra switch may be added to a node (where the probability of the switch failure is assumed to be negligible compared to the probability of a node failure) to switch all its links when the node fails. The reconfiguration procedure can be initiated either

† This work has been partially supported by NSF Grant MIP-89‑‑‑‑‑

locally or by a global controller depending on the system design. If local reconfiguration is desired then the spare node that replaces a faulty node can initiate the reconfiguration as will be shown later. A fault tolerant architecture must have the following architectural features to guarantee proper reconfiguration:

F1. There must exist either a link or some mechanism to establish a dedicated connection from any spare in $SS(P_i)$ to the primary node $N_j(P_i)$, for $j = 1,2,...,m$.

F2. There must exist either a link or some mechanism to establish a dedicated connection from any spare in $SS(P_i)$ to any spare in $SS(N_j(P_i))$, for $j = 1,2,...,m$.

If a failed primary node P is replaced by a spare node S then to preserve the adjacency relationship among the *live* nodes using direct connections, S should have direct links to all the neighbors of P. This kind of interconnection is called a *single hop* or a *direct* interconnection. The implementation of the *single hop* interconnection scheme may be prohibitively expensive for large systems since each node may require a large number of links (communication ports). If S is permitted to establish dedicated connections through at most x nodes, where $x \geq 1$, then the interconnection is called *(x+1)_hop* interconnection. In order to keep the reconfiguration time within allowable limits, the value of x will be limited to one. Therefore, in this paper, a *2-hop* interconnection design approach will be used to build fault tolerant systems.

In this paragraph we define some terms that will be used in the rest of this paper. The links between two primary nodes are called *primary links* and the links between a spare node and a primary node are called *interface links*. *Spare links* are the links between two spare nodes. A *non-primary* link is either a *spare link* or an *interface link*. The primary set, $PS(S)$, of a spare node S is the set of primary nodes that S can replace. The flexibility, $flex(S)$, of a spare node S is the cardinality of $PS(S)$. For a primary node P, the degree of coverage, $deg(P)$, of P is the cardinality of the spare set $SS(P)$. In a given system, if $flex(S)$ is constant for each S and $deg(P)$ is constant for each P, then, these constants are referred to by $flex$ and deg, respectively. In terms of system implementation, a constant degree of coverage and a constant flexibility means that the spare nodes and the primary nodes may be homogeneous. In this paper, we only consider systems with constant deg and $flex$.

A Fault Tolerant Basic Block, FTBB, is defined as a minimal set of nodes such that, for any primary node P and any spare node S in FTBB, the nodes of $SS(P)$ and $PS(S)$ are in FTBB. In other words, spare nodes inside an FTBB can only replace primary nodes from that FTBB. The size of a FTBB is given by a tuple (M,K), where M and K are the number of primary and spare nodes, respectively. If each spare node in an FTBB can replace any of the primary nodes in that FTBB then that FTBB is said to have *full spare utilization* (FSU) (for further explanation see[2]). Note that the size of an FTBB with *full spare utilization* is $(flex,deg)$. Two FTBBs are considered to be neighbors if a primary node from one FTBB is a neighbor of a primary node in the other FTBB. A neighbor of FTBB F is denoted by $N(F)$. The number of neighboring FTBBs of an FTBB is denoted by F_N, which is usually the same for all FTBBs in a homogeneous systems. In the next section we describe how features F1 and F2 can be incorporated in fault tolerant systems built from identical FTBBs with *full spare utilization*. In such systems, if the spares in FTBB F are labelled $S_1,\cdots,S_{deg}$ and the spare nodes in FTBB F', where $F'=N(F)$, are labelled $S_1',\ldots,S_{deg}'$, then the mirror image of S_i in FTBB F' is defined to be the spare node S_i', for $i=1,2,\cdots,deg$.

3. Channel Multiplexing in the 2-Hop Reconfiguration Scheme

In this scheme the links of a failed primary node are not used for reconfiguration. Also, the state of the faulty node is transferred to the node that replaces it without transferring the states of any other node. This decreases the reconfiguration time. Since primary links cannot be used for reconfiguration, extra links need to be added to the system to achieve fault tolerance. In this interconnection scheme the primary nodes are interconnected according to the required topology. The interconnection of the non-primary links is given below.

Conditions for Constructing 2-hop Reconfigurable Systems

(C1). There should be a direct link between a spare node S and a primary node P if $S \in SS(P)$.

(C2). The spare nodes within an FTBB should be fully interconnected. That is, each spare node within an FTBB has direct links to $deg-1$ other spare nodes within that FTBB.

(C3). There should be a direct link from a spare to its mirror images in neighboring FTBBs.

We use Figure 2 to describe the interconnection. The interconnection among two neighboring FTBBs F and F' are shown in Figure 2. In this figure the interconnection is described in terms of the non-primary links. The interconnection among the primary nodes are architecture dependent and therefore are not shown. The rest of the interconnection is not architecture dependent and can be described as follows: A spare node S_r, where $S_r \in$ FTBB F, has links to all the nodes in F and also has a link to each of its mirror image in neighboring FTBBs. If any primary node P_i fails in FTBB F and is replaced by the spare S_r, then in order to ensure 2-hop dedicated connections we use algorithm RECON which is given below.

Algorithm RECON

1. Replace P_i by spare node S_r and disable all the links of P_i

2. For j = 1 to m do { i.e., consider each of P_i's neighbors }

   2.0 Let $N_j(P_i) \in$ FTBB F' and S_r' be the mirror image of S_r in FTBB F', if $F' \neq F$.

   2.1 if $N_j(P_i)$ is not faulty then

      2.1.1 if $F = F'$ then use the link between S_r and $N_j(P_i)$ as the jth link of S_r

      2.1.2 else establish a dedicated connection from S_r to $N_j(P_i)$ via S_r' to be used as the jth link of S_r

   2.2 else { $N_j(P_i)$ is faulty and say S_k' replaced it }

      2.2.1 if $F = F'$ then use the link between S_r and S_k' as the jth link of S_r

      2.2.2 else establish a dedicated connection from S_r to S_k' via S_r' to be used as the jth link of S_r

Note that in step 2.2.2 of the above algorithm, if $S_k' = S_r'$, then a direct connection can be established from S_r to S_k'. Given that a system is made of FTBBs with FSU and conditions C1, C2 and C3 are met, it can be easily verified that after reconfiguration all the dedicated connections have at most 2-hops. Note that in RECON the steps 2.1.1 and 2.2.1 are only executed if P_i and $N_j(P_i)$ belong to the same FTBB. In this case conditions C1 and C2 ensure a direct connection from S_r to $N_j(P_i)$ or a direct connection from S_r to S_k'. Conditions C1 and C3 ensures that step 2.1.2 can establish a 2-hop dedicated connection. Similarly, C2 and C3 ensures the execution of step 2.2.2.

We use a fault tolerant tree architecture to describe how algorithm RECON works. In the fault tolerant tree architecture shown in Figure 3, the thick lines represent the active links and the thin lines represent the unused links. A FTBB in Figure 3 (shown in a rectangular box) consists of three primary nodes and a spare node. A spare in a FTBB can only replace any primary node within that FTBB. Note that the system in Figure 3 satisfies conditions C1, C2 and C3. Figure 4 is used to describe how algorithm RECON works on the fault tolerant tree system in Figure 3. In this figure P_i is one of the faulty nodes and S_r is the spare that replaces it. Since there exists a direct link between S_r and $N_j(P_i)$, where $j=$**3** and 2, a direct connection is established between S_r and $N_j(P_i)$. However, there is no direct link between S_r and $N_1(P_i)$, therefore S_r establishes a dedicated connection between itself and $N_1(P_i)$ through the spare node S_r'. The reconfiguration process in $FTBB_3$ can be similarly explained.

In this scheme each primary node requires $m+deg$ links and each spare requires $flex+deg+F_N-1$ links but each of the non-primary links of this scheme may require more than one channel to support 2-hop reconfiguration. That is, even though there may be a spare available to replace a faulty node in an FTBB, the reconfiguration may fail due to insufficient channels. For example in Figure 4, if all the non-primary links had one channel, then the failure of two primary node P_i and P_j cannot be tolerated even though there are two spares available to replace them. This is because the non-primary link between the nodes $N_1(P_i)$ and S_r' needs two channels to preserve the adjacency relationship among the live nodes after reconfiguration. This observation leads us to the two lemmas in the next section. In these lemmas we derive the channel requirement of the non-primary links in order to utilize all the available spares in an FTBB and to support 2-hop reconfiguration.

4. Channel Requirement of the 2-hop Scheme

The primary links in this scheme require at most one channel since they are not used for reconfiguration. We use L to denote the number of primary node neighbors of P within FTBB F, where $P \in$ FTBB F. We assume that L and m for each primary node is the same, where m is the number of primary node neighbors of P. If that is not the case then these lemmas can easily be extended by incorporating different values of L's and $m's$ for different primary nodes. For systems made from FTBBs with FSU we state the following two lemmas without proof. These lemmas address the channel requirement of the non-primary links and their proofs are given in [1].

Lemma 1: Each interface link in a system constructed from FTBBs with FSU requires at most $m-L+1$ channels in order to fully utilize the spare nodes.

The following definitions are used in the next lemma. For a primary node P, where $P \in$ FTBB F, we use $J_{P,F'}$ to denote the number of primary node neighbors that P has in a neighboring FTBB F'. Also, $Q_{F,F'}$ denotes the maximum value of $J_{P,F'}$ for any primary node P in FTBB F.

Lemma 2: In a system constructed from FTBBs with FSU, the spare links across two FTBBs, F and F', require at most $\min\{2flex, Q_{F,F'} + Q_{F',F}\}$ channels each and the spare links within FTBBs require at most $2(m-L)+1$ channels each in order to fully utilize the spare nodes.

The above two lemmas specify the number of channels required per link for the worst case scenario. The probability that the worst case fault configuration occurs is very low. In the next section, we show that 2-hop reconfiguration can be achieved with very high probability if a constant number of channels are used in the spare and interface links.

5. Application of the Channel Multiplexing Scheme to Binary Hypercubes

In this section, we show how the design approach discussed in section 3 can be implemented in binary hypercubes. The reconfiguration algorithm in this section is executed locally and distributedly. That is, there is no need for a global controller to execute the reconfiguration algorithm. Therefore, the switches added to the spare nodes should be able to switch one channel to another based on some reconfiguration information sent in a message. The concept of reconfiguration by sending messages will be clear in algorithm RECON_CUBE. The fault tolerant binary hypercube architecture used in this case has 50% redundancy with $flex=4$ and $deg=2$ (i.e. FTBB size (4,2)). The dimension of the cube is denoted by n. In this architecture the 2^{n-1} spare nodes are interconnected as an $n-1$-dimensional *spare cube* and each of the 2 spares in an FTBB is connected to 4 primary nodes in that FTBB. In general, FTBB's may be constructed across any two dimensions, we chose dimension 1 and 2. Each primary node has an n-bit address $P_n \cdots P_1$ and the address of an FTBB is denoted by $n-2$ most significant address bits of any primary node that belongs to it. Two spares are added to each FTBB and are connected to all the primary nodes within that FTBB. In Figure 5, the labelled nodes belong to one FTBB and four such FTBB's are used to construct a 4-dimensional fault-tolerant hypercube. Note that this fault tolerant architecture satisfies conditions C1, C2 and C3. The two spares in an FTBB are neighbors and are referred to as twins. Formally, the twin of a spare node S is $twin(S) = N_2(S)$. A link between two primary nodes across dimension i is called the ith link. When a primary node fails and a dedicated connection is established to replace the ith link, that dedicated connection is referred to as the ith l-link (i^{th} logical link). Since there are two spare nodes adjacent to each primary node, this architecture supports the failure of any two nodes in each FTBB.

When a failure of a primary node P is detected the following algorithm is used for reconfiguration:

Algorithm RECON_CUBE

(1) Replace the faulty node P with one of the available spare nodes S in $SS(P)$. If no such spare exists declare system failure.

(2) Disable all the links(channels) of P, and let S execute the following procedures.
   (a). Reconfigure_link_inside_FTBB(1)
   (b). Reconfigure_link_inside_FTBB(2)

(3) S sends the address of P to all its spare neighbors except its twin.

(4) Each spare (say S') that receives a message sent by S in step 3 executes the following:

   if S' receives the message from S via the jth link then Reconfigure_link_outside_FTBB(j)

Even though this algorithm reconfigures the system when a primary node P fails, it can also be used to handle the failure of spare nodes. If a spare node s fails and s did not replace any other primary node then no reconfiguration is required. But if s replaced a primary node P and failed later on, then to perform a proper reconfiguration, the spare node S, where $S=twin(s)$, must not be faulty or have replaced any other primary node. In this case the above reconfiguration algorithm can be executed such that the faulty node P is being replaced by S.

The procedure Reconfigure_link_inside_FTBB(i) enables a connection between S and $N_i(P)$ to be used as the ith l-link. If $N_i(P)$ is faulty, then it enables a channel from the link between S and the spare that replaced $N_i(P)$ to be used as the ith l-link. Reconfigure_link_outside_FTBB(i) essentially does the same thing; If $N_i(P)$ is not faulty then it connects S to $N_i(P)$ through an intermediate spare node, otherwise, it connects S to the spare that replaces $N_i(P)$. For more details see[1].

We use Figure 6 to explain how the algorithm RECON_CUBE works. In Figure 6, primary node A fails and spare node D' replaces it. Using step 2 of the reconfiguration algorithm all the links of the faulty processor are disabled and the procedure Reconfigure_link_inside_FTBB() is executed by spare node D' once with the parameter 1 and then with the parameter 2. Executing the procedure with the parameter 1 results in the use of a channel from $D'F$ as the 1st l-link. Similarly, Reconfigure_link_inside_FTBB(2) enables a channel from link $D'D$ to be used as the 2nd l-link. In step 3, D' sends messages to spare node C' and I' with the address of the failed primary node A. In step 4, spare node C' connects a channel from link $D'C'$ to a channel from link $C'B$, and the connection $D'C'B$ is used as the 3rd l-link. Similarly, in step 4 spare node I' switches a channel from link $D'I'$ to a channel from $I'J$ and the connection $D'I'J$ is used as the 4th l-link.

Next, we show that the above reconfiguration will succeed with high probability if only a few channels per non-primary link are only used. Theoretically, hypercubes built from FTBBs with FSU need $m-L+1$ channels for the interface links, $2(m-L)+1$ channels for the spare links within FTBBs and 2 channels for the spare links across FTBBs in order to utilize all the spare nodes. If we restrict the number of channels in each of these links then the 2-hop reconfiguration may fail due to the lack of channels for some fault configurations. In order to get a better estimate we resort to simulation. The systems simulated for this experiment were built from FTBBs identical to the ones in Figure 5.

A simulation software tool was designed to obtain results for fault tolerant binary hypercubes of dimension j, where $4 \leq j \leq 10$. We denote the reliability of a node by $r = e^{-\lambda t}$, where λ and t denotes the failure rate and time respectively. Typical values of r range from 1.0 to 0.905, for $\lambda t = 0.0$ to 0.1. Also, i is used to denote the number of channels per non-primary link, where $i \geq 1$. For a given pair of j and r the simulation software generates a set of faults FS such that the system is *alive*, i.e. no more that 2 nodes are faulty in any FTBB (note that there are only two spares per FTBB). Faulty nodes in FS are then chosen in random order to perform reconfiguration using RECON_CUBE. The reconfiguration process may either complete successfully or fail due to insufficient channels (i.e. the number of channels required in a non-primary link for reconfiguration is greater than i). For a given set of j, r and i the simulation software repeats the procedure 1 million times and keeps track of the number of times the reconfiguration procedure failed due to insufficient channels. This information is then used to calculate the reconfiguration success rate for valid fault configurations, which is denoted by RSR.

Figure 7(a) and Figure 7(b) uses 2 and 3 channels per non-primary link respectively to perform reconfiguration after a node fails. The curves labelled j_d in this figure represent the RSR's of fault tolerant hypercubes of dimension j, where $4 \leq j \leq 10$. The reliability of a fault tolerant hypercube with i channels per non-primary link is calculated as follows: For a given value of λt we find the reliability of the system with unlimited number of channels per non-primary link, say R_∞, and then we find RSR for the same value of λt with i channels in each non-primary link. The product of R_∞ and RSR gives the reliability of the system with i channels per non-primary link. $R_\infty = \left[r^6 + C_1^6\, r^5\,(1-r) + C_2^6\, r^4\,(1-r)^2 \right]^N$, where N is the number of modules (FTBBs) in the system. From Figure 8 it is clear that 3 channels per non-primary link can achieve almost 100% RSR hypercubes of dimension 7 and 8.

In Figure 8 the RSR and the reliability of fault tolerant hypercubes of dimensions 4 to 9 are plotted with respect to the number of channels in each non-primary link. The RSR and the reliability in this figure are given for $\lambda t = 0.5$. For example, if one channel in each non-primary link is used then the RSR is about 2% in an 8-d cube. The RSR increases from 2% to 78% for an additional channel. The RSR can be further improved to about 100% by adding a third channel in each non-primary link (except the spare links across FTBBs). But the reliability improvement by adding a fourth channel to a 8-d cube with 3 channels per non-primary link is almost negligible. Actually the point of diminishing return for the 8-d cube starts after 2 channels per non-primary link. Note that from lemma 2 and lemma 1, a 8-d cube requires 7 and 13 channels per interface link and per spare link across FTBBs respectively.

6. Conclusion

In this paper we have presented a general design approach for constructing modular fault tolerant multiprocessors. This design approach allows the implementation of an arbitrary degree of redundancy. It is flexible, easy to apply to different architectures and, ensures fast reconfiguration by using at most 2-hop interconnections. Further more, this scheme has the ability to locally reconfigure around any new set of faults without undoing any earlier reconfiguration. In this scheme, channel multiplexing on non-primary links help utilize the redundant hardware efficiently. Finally, we note that this scheme may be applied to modular systems that are constructed from FTBBs without *full spare utilization* [2].

References

1. M. Sultan Alam and Rami G. Melhem, ''Channel Multiplexing in Modular Fault Tolerant Multiprocessors,'' *Technical Report TR 91-4, Department of Computer Science*, University of Pittsburgh, Jan 16, 1991.

2. M. S. Alam and R. G. Melhem, ''An Efficient Modular Spare Allocation Scheme and its Application to Fault Tolerant Binary Hypercubes,'' *IEEE Trans. on Parallel and Distributed Systems*, vol. 2, no. 1, pp. 117-125, Jan 1991.

3. Prithviraj Banarjee, S. Y. Kuo, and W. K. Fuchs, ''Reconfigurable Cube-Connected Cycles Architectures,'' *Proc. 16th Int. Symp. Fault Tolerant Computing*, pp. 286-291, June 1986.

4. P. Banarjee, J. T. Rahmeh, C. B. Stunkel, V. S. Nair, K. Roy, and J. A. Abraham, ''An Evaluation of System Level Fault Tolerance on the Intel Hypercube Multiprocessors ,'' *Proc. 18th Int. Symp. on Fault-Tolerant Computing*, pp. 362-367, June 1988.

5. S. Chakravarty and S. J. Upadhyaya, ''A Unified Approach to Designing Fault-Tolerant Processor Ensembles,'' *Proc. Int. Conf. on Parallel Processing*, Aug 1988.

6. Siu-Cheung Chau and Arthur L. Liestman, ''A Proposal for a Fault-Tolerant Binary Hypercube Architecture,'' *Proc. of the Int. Symp. on Fault-Tolerant Computing*, pp. 323-330, 21-23rd June 1989.

7. F. R. K. Chung, F. T. Leighton, and A. L. Rosenberg, ''Diogenes: A Methodology for Designing Fault-Tolerant VLSI Processor Arrays,'' *Proc. 13th Int. Symp. Fault Tolerant Computing*, pp. 26-31, June 1983.

8. Shantanu Dutt and John P. Hayes, ''An Automorphic Approach to the Design of Fault Tolerant Multiprocessors,'' *Proc. 19th Int. Symp. Fault Tolerant Computing*, pp. 496-503, June 1989.

9. A. S. Mahmudul Hasan and Vinod K. Agrawal, ''A Fault Tolerant Modular Architecture for Binary Trees,'' *IEEE Trans. on Computers*, vol. c-35, no. 4, pp. 356-361, April 1986.

10. J. P. Hayes, ''A Graph Model for Fault Tolerant Computing Systems,'' *IEEE Trans. on Computers*, vol. c-25, pp. 875-883, Sept 1976.

11. M. B. Lowrie and W. K. Fuchs, ''Reconfigurable Tree Architectures using Subtree Oriented Fault Tolerance,'' *IEEE Trans. on Computers*, vol. C-36, pp. 1172-1182, Oct 1987.

12. C. S. Raghavendra, A. Avizienis, and M. Ercegovac, ''Fault-Tolerance in Binary Tree Architectures,'' *IEEE Trans. on Computers*, vol. C-33, pp. 568-572, June 1984.

13. D. A. Rennels, ''On Implementing Fault-Tolerance in Binary Hypercubes,'' *Proc. IEEE Fault Tolerant Computing*, pp. 344-349, 1985.

14. Adit Singh, ''A Reconfigurable Modular Fault Tolerant Binary Tree Architecture,'' *Proc. 17th Int. Symp. Fault Tolerant Computing*, pp. 298-304, June 1987.

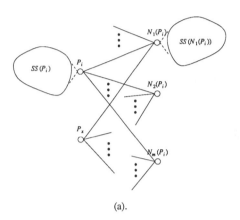

(a).

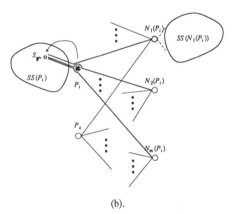

(b).

Figure 1. Graph model of a fault tolerant multiprocessor system

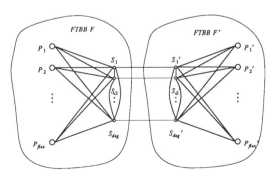

Figure 2. Two neighboring FTBBs in the 2 hop interconnection scheme

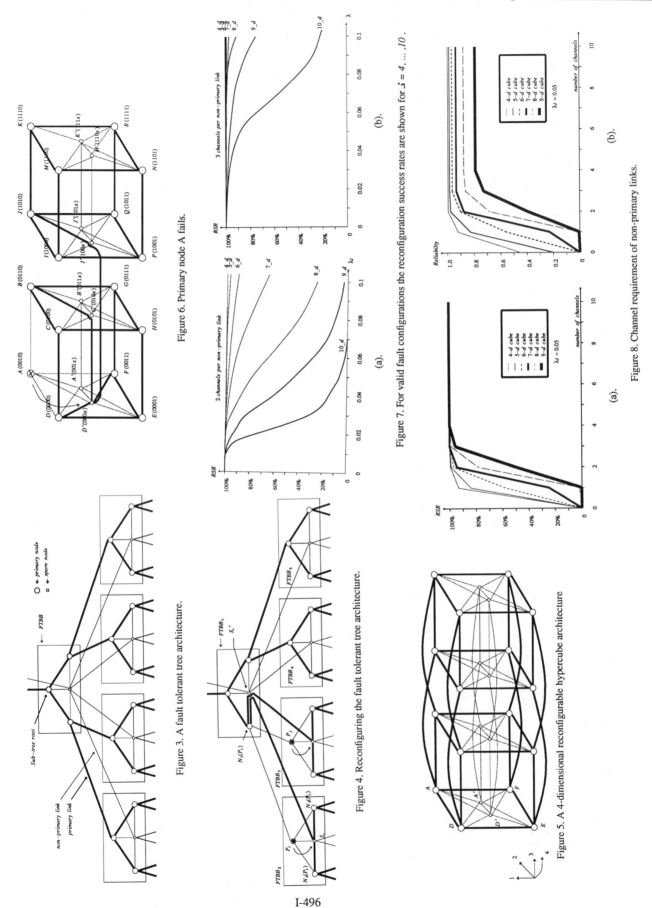

Figure 6. Primary node A fails.

(a).

(b).

Figure 7. For valid fault configurations the reconfiguration success rates are shown for $j = 4, ..., 10$.

Figure 8. Channel requirement of non-primary links.

(a).

(b).

Figure 3. A fault tolerant tree architecture.

Figure 4. Reconfiguring the fault tolerant tree architecture.

Figure 5. A 4-dimensional reconfigurable hypercube architecture

A CACHE-BASED CHECKPOINTING
SCHEME FOR MIN-BASED MULTIPROCESSORS*

Mazin S. Algudady, Chita R. Das and Matthew J. Thazhuthaveetil

Department of Electrical and Computer Engineering
The Pennsylvania State University
University Park, PA 16802

ABSTRACT

A checkpointing and error recovery scheme for MIN (multistage interconnection network)-based multiprocessors with private caches is presented. This scheme is integrated in a cache coherence protocol designed for MIN-based multiprocessors. In this scheme, checkpoints are initiated at two instants: modified blocks replacement and remote processors' read/write misses. The latter instant takes care of the *Domino* effect. At a checkpoint, processor's internal registers are saved in a recovery stack, recently modified blocks are marked as *unwritable* and modified private blocks are flushed into the recovery stack. System performance, in terms of processing power and coherence bus utilization, is analyzed using a comprehensive evaluation methodology.

1. INTRODUCTION

Hardware complexity of MIN-based shared-memory multiprocessors makes such systems prone to failure. This complexity with the demand for continuously operational systems make reliability assurance an important design issue. In order to achieve high reliability, such systems should recover from *transient* errors without overall system restart since these errors occur most frequently. One technique to achieve this, referred to as *checkpointing and error recovery*, is by periodically saving the execution signature of a process. The purpose of including checkpoints in a system is to provide a return path for a process to a state known to be *correct* and *uncorrupt*.

Early proposed error recovery schemes were for distributed systems that utilize message passing as their communication technique [1]. These are generally not applicable to shared-memory multiprocessors due to the sharing of variables in a global memory. In multiprocessors with private caches, cache coherence protocols are used to maintain coherency among the nodes [2]. It is possible to modify these protocols to take care of transient processor errors. A *cache-aided rollback error recovery* (CARER) scheme was proposed for uniprocessors and was extended later to bus-based shared-memory multiprocessors [3]. The scheme was integrated in the *Dragon* write-broadcast coherency protocol. A second error recovery scheme for shared-memory multiprocessors was proposed and integrated in the *Illinois* coherence protocol [4].

Multiple rollback propagation, referred to as the *Domino* effect, is a problem in which the rollback of one process propagates to other processes, and hence necessitates rollback to early checkpoints. This problem must be dealt with when developing a checkpointing scheme. Otherwise, a process might require resumption of execution at a very early checkpoint [5].

In this paper, we propose a checkpointing and error recovery scheme for MIN-based shared-memory multiprocessors with private caches. This recovery scheme is integrated in the cache coherence protocol that we developed

* This work is sponsored in part by the National Science Foundation under grants CCR-8810131 & MIP-8710753.

for such systems [6]. A cache coherence scheme takes care of shared memory blocks; private blocks are taken care of by flushing modified private blocks into memory or a recovery stack. The implementation of this checkpoint and error recovery scheme requires the use of counters to distinguish between modified blocks that belong to a checkpoint session and those that do not, as explained in [4].

Checkpoints are initiated by hardware at the following two instants: (i) When an updated shared block is replaced. (ii) When an updated copy of a shared block is to be read/written by a remote processor. At both instants, the block is written back to main memory. The second instant takes care of the *Domino* effect. At a checkpoint session, in addition to satisfying a processor request, the following activities take place: All cache blocks modified since the previous checkpoint session are marked *unwritable*, the processor's internal registers are saved, and all modified private blocks are saved in the stack [4].

The architecture for which our scheme is suitable is a shared-memory multiprocessor with N processor nodes connected through a MIN, as shown in Figure 1. Each processor node includes a private cache and a local memory module; the local memory module is globally accessible. The shared memory is thus distributed among the nodes of the multiprocessor as in the BBN Butterfly. All the cache controllers are connected to a coherence control bus. Each cache controller snoops at all bus transactions. Also connected to this bus is a central table for shared blocks status information. We use a *recovery stack* to store part of a checkpoint information [4].

Our performance study comprises of three phases: (i) computing shared blocks steady-state probabilities of the states imposed by the protocol, (ii) MIN delay queueing model, and (iii) the single node queueing model.

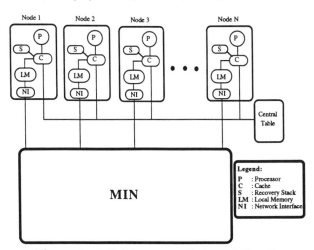

Figure 1: A Multiprocessor System of Size N

The rest of the paper is organized as follows: In section 2, we touch the modified cache coherence protocol. Section 3 presents the performance study. This is followed by results and discussion in section 4. Finally, section 5 concludes this work.

2. THE CACHE COHERENCE PROTOCOL

The protocol uses four global states: *Modified-Transient, Modified-Permanent, UnModified-Transient,* and *UnModified-Permanent*. *Transient* states indicate that the block is currently in transit in response to a block transfer request. A block in a *Transient* state cannot be accessed in cache until the data arrives and its state changes to *Permanent*. A block in a *Permanent* state is available in cache. In addition, three local state flags are present per cache block in each cache controller: *modified, valid* and *unwritable* bits. A block in state *unwritable* is modified; it is referred to as *unwritable* because it must be written back to the recovery stack before it can be updated. However, a block in state *modified* (*unwritable* flag is reset) is writable in the sense that it can be updated directly without any checkpoint activity required. Possible state transitions for a shared data block in a cache are shown in Figure 2; the protocol is not described here due to space limit. Detailed description of the protocol can be found in [6].

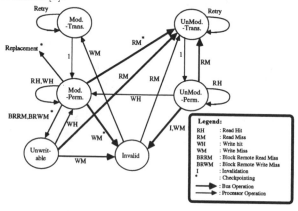

Figure 2: The Protocol Transition Diagram

3. PERFORMANCE ANALYSIS

3.1 Workload Model

We assume that all processor nodes are identical, with each processor having a cache of size ψ blocks. A processor follows a series of "think" and "wait" (after making a request to its cache) periods. The thinking time is assumed to follow an exponential distribution with an average value z. We also assume a *least recently used* (LRU) replacement policy and uniform memory referencing.

It is necessary to introduce the following notations.

Notation

f_r : probability of a read operation.
f_w : probability of a write operation; $f_r + f_w = 1$.
h : system cache hit ratio under coherence.
md : probability that a private block is modified.
N : number of nodes in the system (system size).
N_{sh} : number of shared blocks in the system.
q_p/q_s : probability of requesting a private/shared block; $q_p + q_s = 1$.
S_b : coherence control bus service time.
S_c : cache service time.
S_s : stack service time.
U_p : processor utilization.

3.2 Shared Blocks Steady-State Probabilities

The states of a shared block are modelled as a *discrete time Markov chain*, as shown in Figure 3, which is later solved for its steady-state probabilities. For mathematical simplicity, we assume that all state transition times are equal. A *Not-Present* state is included for evaluation purposes. A block is in state *Invalid* if it occupies a cache frame but is invalid due to the cache coherence protocol. A block is in state *Not-Present* if it does not occupy a cache frame. In Figure 3, the following notation is used: $r(= U_p q_s f_r / z N_{sh})$ represents the probability of a shared block read request, $w(= U_p q_s f_w / z N_{sh})$ is the probability of a shared block write request, $f(= U_p(1-h)/z\psi)$ represents the probability of replacement of a shared block, $O_w(= 1 - (1-w)^{N-1})$ is the probability of one remote write request, $O_r(= (1-O_w)-(1-w-r)^{N-1})$ is the probability of one remote read request with no simultaneous remote write request, and $O(= 1-[(1-w-r)^{N-1}]+f)$ is the probability of at least one remote read or write request or replacement of a shared block.

The steady-state probabilities of the states in the *Markov* model are defined as: *Not-Present* π_0, *UnModified-Transient* π_1, *Modified-Transient* π_2, *UnModified-Permanent* π_3, *Modified-Permanent* π_4, *Invalid* π_5, and *Unwritable* π_6. Due to space limit, we only give equations for the steady-state probabilities, π_i, as follows:

$$\pi_0 = 1/\{1 + \theta[\frac{1}{\alpha\gamma-r}(1 + r + w + \frac{r}{\gamma}) + 1 + \frac{1}{\gamma}] + w[1 + \frac{\alpha}{f\beta}(1+O)]\}$$

$$\pi_1 = [1 + r/(\alpha\gamma - r)]\theta\pi_0$$
$$\pi_2 = [1 + \theta/(\alpha\gamma - r)]w\pi_0$$
$$\pi_3 = [1 + r/(\alpha\gamma - r)]\theta/\gamma\pi_0$$
$$\pi_4 = w\alpha/f\beta\pi_0$$
$$\pi_5 = \theta\pi_0/(\alpha\gamma - r)$$
$$\pi_6 = wO\alpha\pi_0/f\beta$$

where $\alpha = f + w + r$, $\beta = f + w + O + O_w + O_r$, $\gamma = f + r + O_w$, and $\theta = r + \frac{\alpha w O_r}{f\beta}(1 + O)$.

3.3 A System-Level Performance Model

In order to compute system-level performance measures, we first separate the MIN from the system model and analyze it independently for determining its effect on

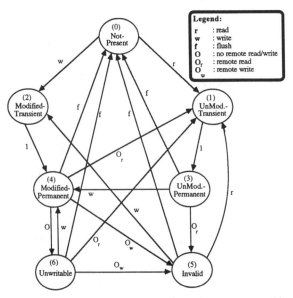

Figure 3: Markov Model for Shared Block States

system parameters. Since the MIN is used only for accessing remote memory modules, it determines the delay incurred in fetching/writing back a block from/to a local memory module and a remote memory module depending on the traffic in the MIN. Let these average delays be denoted as S_l and S_g, respectively. We then use S_l and S_g as simple delay centers in the high-level single node queueing model. Detailed description of this model is in [6].

We assume that all the nodes involved in the execution of a job have identical task assignment. This implies that all the nodes are identical; therefore, analysis of any single node is sufficient for determining system behavior. The single node model expresses the flow of a processor request exactly in terms of the request being hit or miss, private or shared, etc. with all relevant probabilities, as shown in Figure 4. It is mixed in the sense that it includes both open and closed class customers [7].

The single node model assumes that a processor *thinks* for a certain amount of time, z, then requests a block from its cache (represented as a queueing center) with a rate $\lambda_p (= U_p/z)$. A request is hit with probability P_h; the processor resumes thinking after a delay related to the cache service time, S_c. Otherwise, a request is either a miss (private or shared) or requires extra operations such as broadcasting invalidation signals due to a write request on an unmodified block, or writing an *unwritable* block into the stack; they constitute a probability given by $P_m (= (1-h) + f_w q_s (\pi_3 + \pi_6))$. Private misses, given by the probability $P_{pr} (= q_p (1-h))$, are serviced either locally with a probability $P_{pl} (= \frac{1}{N} P_{pr})$, or globally from a remote memory module with a probability $P_{pg} (= \frac{N-1}{N} P_{pr})$.

Shared misses and requests requiring broadcasting invalidations need to use the coherence control bus, which is represented as a queueing center with service time S_b. Writes on unmodified blocks, given by the probability $P_{iv} (= f_w q_s \pi_3)$, require broadcasting invalidations through the bus and updating the central table. Write

requests to *unwritable* blocks, given by the probability $P_s (= f_w q_s \pi_6)$, require writing the blocks into the recovery stack before the writes can proceed. This is represented by a delay center with service time S_s. Shared block misses require central table consultation through the coherence control bus for the status of the block. They are serviced from a memory module either directly or after writing back a block's modified copy from a remote cache. These are given by the probabilities $P_{sm} (= q_s (1-h)(1 - \pi_4 - \pi_6)^{N-1})$ and $P_{sr} (= q_s (1-h)(1 - (1 - \pi_4 - \pi_6)^{N-1}))$, respectively. Requests distribution to local and global memory modules is similar to that for private blocks. When a shared block miss is serviced directly, it is with probability $P_{sml} (= \frac{1}{N} P_{sm})$ if the fetch is local. Otherwise, if the fetch is global from a remote memory module, it is with probability $P_{smg} (= \frac{N-1}{N} P_{sm})$. Writing back a block to a memory module and receiving acknowledgment incurs a delay of S_l (if the write back is local) and S_g (if the write back is remote). The probability for the former case is given as $P_{srl} (= \frac{1}{N} P_{sr})$, and as $P_{srg} (= \frac{N-1}{N} P_{sr})$ for the latter case. However, after a remote write back, block fetch could be either local or remote, given by the probabilities $P_{srgl} (= \frac{1}{N} P_{srg})$ and $P_{srgg} (= \frac{N-2}{N} P_{srg})$, respectively. After each shared block fetch, the block's entry in the central table is updated through the bus.

The single node model also takes care of checkpointing, write backs, invalidation signals and block fetches, which are represented as open class customers. Such customers inflate the service time of various centers since the model is mixed. Traffic due to write backs, P_{wb}, includes private and shared block replacements and remote requests to write back modified blocks. This is given by

$$P_{wb} = (1-h)\left\{ \frac{[\psi - N_{sh}(1-\pi_0)]md + N_{sh}(1-\pi_0)(\pi_4 + \pi_6)}{\psi} + q_s[1 - (1 - \pi_4 - \pi_6)^{N-1}] \right\}.$$

Write backs combine two parts: one due to remote write backs through the MIN which increases the MIN delay and is represented by $P_{wbg} (= \frac{N-1}{N} P_{wb})$, and another due to local write backs which inflate the local delay and is given by $P_{wbl} (= \frac{1}{N} P_{wb})$.

Coherence bus traffic due to the remaining $N-1$ processors, given by the probability P_b, consists of traffic due to invalidations $q_s f_w \pi_3$, central table consulting due to a shared block miss $q_s (1-h)$, updating after a block fetch $q_s (1-h)(1 - \pi_4 - \pi_6)^{N-1}$, a write back $(1-h)(\frac{N_{sh}(1-\pi_0)(\pi_4 + \pi_6)}{\psi})$ and finally updating and acknowledgment after writing back a modified block from a remote cache, $2q_s (1-h)(1 - (1 - \pi_4 - \pi_6)^{N-1})$. After simplification, P_b is expressed as

$$P_b = 1 - \left\{ 1 - [q_s f_w \pi_3 + q_s(1-h) + (1-h)(\frac{N_{sh}(1-\pi_0)\theta}{\psi}) + \delta] \right\}^{(N-1)},$$

where $\theta = \pi_4 + \pi_6$, and $\delta = q_s(1-h)[2 - (1-\theta)^{N-1}]$.

The traffic that keeps the cache busy, given by the probability P_l, consists of traffic due to storing a block in a cache due to a miss (private or shared), requests from the central table to write back a modified block and checkpointing.

$$P_l = (1-h)\left\{ 1 + \frac{q_s}{N_{sh}}(\pi_4 + \pi_6) + \frac{(1-\pi_0)N_{sh}(\pi_4 + \pi_6)}{\psi} + q_s[1 - (1 - \pi_4 - \pi_6)^{N-1}] \right\}.$$

The traffic that keeps the recovery stack busy include traffic due to storing *unwritable* blocks from the cache after replacement and write hits on such blocks, and saving modified private blocks at checkpointing. These are given by the probability P_{os}.

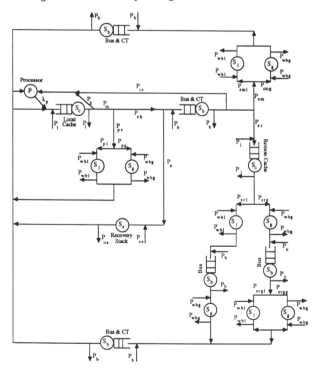

Figure 4: Single Node Queueing Model

$$P_{os} = \frac{(1-h)(1-\pi_0)N_{sh}\pi_6}{\psi} + q_s f_w \pi_6 + [\psi - (1-\pi_0)N_{sh}]$$
$$md\{\frac{(1-h)(1-\pi_0)N_{sh}\pi_4}{\psi} + q_s[1-(1-\pi_4-\pi_6)^{N-1}]\}.$$

The single node queueing model is analyzed using the mixed model MVA technique [7] to compute processor utilization and coherence bus utilization. The model is solved assuming the presence of one customer, which depicts the case that a processor generates a request and waits for its response.

4. RESULTS & DISCUSSION

The analytical model discussed in the previous section is used to quantify the effect of including checkpoints in the protocol. The MIN delay and single node queueing models were solved for a *BBN Butterfly* like multiprocessor system with a MIN made up of 4×4 switching elements. It is assumed that $h = 0.9$, the service times of the cache and the stack are 1 cycle each. Coherence bus service time S_b, varies depending on the number of nodes in the system N: $S_b = 1$ *cycle* when $N = 16$, $S_b = 2$ *cycles* when $N = 64$ and $S_b = 4$ *cycles* when $N = 256$.

Tables 1 and 2 show the system processing power $(U_p \times N)$ and coherence bus utilization for various system sizes and degrees of sharing. Results shown include both cases: without and with checkpoints. Without checkpoint implies that the *unwritable* state in the protocol described in section 2 is not required. Table 1 also gives the percentage degradation of system processing power due to the inclusion of checkpoints. Similarly, Table 2 shows the percentage decrease in coherence bus utilization. Inclusion of checkpoints reduces bus utilization compared to checkpoint exclusion, since decrease in processor utilization causes less requests to be generated to the local cache and thereby, less use of the bus. From Table 1, it is clear that inclusion of checkpoints degrades system processing power by about 10%. We have done extensive *discrete-event simulation* in [6] to verify the accuracy of our analytical model; it is not presented here due to space limit.

System processing power and coherence bus utilization are also computed when checkpoints are initiated at one instant only; read/write misses from remote processors. The single node queueing model is the same (Figure 4) with the same set of equations described in section 3.3; only terms related to checkpointing at replacement in P_l and P_{os} are removed. As expected, the amount of degradation is less compared to the degradation associated with the two-instant checkpointing scheme since the number of checkpoints in a certain time frame is less. (Relevant processing power and coherence bus utilization Tables are not presented here due to space limit.) However, this reduc-

tion in number of checkpoints results in longer rollback when a fault occurs. Therefore, it is quite obvious that there is a trade off between the instants when checkpoints are initiated and the length of the rollback period.

5. CONCLUSIONS

A cache-based checkpointing & error recovery scheme for MIN-based multiprocessors to take care of *transient* processor errors is presented in this paper. At checkpoint instants, a process's signature is saved. System performance measures in terms of processing power and coherence bus utilization are evaluated using a comprehensive evaluation methodology. It is observed that the inclusion of checkpoints and error recovery scheme in a system degrades system performance measures compared to those of a fault free system.

REFERENCES

[1] R. Koo and S. Toueg, "Checkpointing and Rollback-Recovery for Distributed Systems," *IEEE Transactions on Software Engineering*, vol. C-34, pp. 56-65, January 1985.

[2] M. Dubois and F. A. Briggs, "Effects of Cache Coherency in Multiprocessors," *IEEE Transactions on Computers*, vol. 31, no. 11, pp. 1083-1099, November 1982.

[3] R. E. Ahmed, R. C. Frazier and P. N. Marinos, "Cache-Aided Rollback Error Recovery (CARER) Algorithm for Shared-Memory Multiprocessor Systems," in *Proceedings of the 20th International Symposium on Fault-Tolerant Computing*, pp. 82-88, June 1990.

[4] K. WU, W. K. Fuchs and J. H. Patel, "Error Recovery in Shared-Memory Multiprocessors Using Private Caches," *IEEE Transactions on Parallel and Distributed Systems*, vol. 1, no. 2, pp. 231-240, April 1990.

[5] Y. Lee and K. G. Shin, "Design and Evaluation of a Fault-Tolerant Multiprocessor Using Hardware Recovery Blocks," *IEEE Transactions on Computers*, vol. C-33, no. 2, pp. 113-124, February 1984.

[6] Mazin Yousif, *Effective Use of Caches in MIN-Based Multiprocessors*, Ph.D. Dissertation, The Pennsylvania State University, August 1991.

[7] E. Lazowska, J. Zahorjan, G. S. Graham and K. C. Sevcik, *Quantitative System Performance*, Prentice-Hall, Inc., 1984.

Table 1: System Processing Power for Different System Sizes and Degrees of Sharing with Two Checkpointing Instants

q_s \ N	16	64	256	16	64	256	16	64	256
	Without Recovery			With Recovery			% of Degradation		
0.05	11.61	44.73	149.50	11.16	42.48	141.32	3.86	5.02	5.47
0.08	11.29	42.11	141.06	10.54	39.84	132.54	6.61	5.39	6.03
0.10	11.10	41.05	135.91	10.35	37.98	127.10	6.76	7.48	6.48
0.12	10.94	40.04	131.07	10.13	37.58	122.43	7.46	6.15	6.59
0.15	10.689	38.59	124.45	9.86	35.22	115.65	7.78	8.75	7.07
0.18	10.45	37.22	118.35	9.43	33.87	109.33	9.77	9.00	7.63
0.2	10.19	36.35	109.90	9.18	33.01	99.90	9.87	9.189	9.10
0.25	9.92	34.40	97.79	8.81	31.03	92.98	11.20	9.79	4.92

Legend:
$\psi = 512$, $md = 0.3$, $f_r = 0.75$, $f_w = 0.25$, $S_c = 1$ cycle, $S_s = 1$ cycle,
$S_b = 1, 2,$ & 4 cycles for $N = 16, 64,$ & 256 respectively, $h = 0.9$, $N_{sh} = 2048$

Table 2: Coherence Bus Utilization for Different System Sizes and Degrees of Sharing with Two Checkpointing Instants

q_s \ N	16	64	256	16	64	256	16	64	256
	Without Recovery			With Recovery			% of Degradation		
0.05	2.17	16.35	71.93	2.15	15.98	67.25	0.92	2.26	6.51
0.08	2.89	20.69	76.90	2.85	20.08	71.45	1.14	2.95	7.09
0.10	3.33	23.45	78.10	3.28	22.43	72.65	1.50	4.33	6.98
0.12	3.80	25.88	80.50	3.71	24.63	74.68	2.37	4.83	7.23
0.15	4.40	28.91	86.30	4.25	27.45	78.65	3.50	5.05	8.86
0.18	4.90	31.20	93.40	4.65	29.43	83.12	5.02	5.67	11.01
0.2	5.23	32.36	96.23	4.91	30.15	86.54	6.10	6.83	10.07
0.25	6.01	34.24	99.99	5.54	31.24	98.36	7.80	8.76	1.63

Legend:
$\psi = 512$, $md = 0.3$, $f_r = 0.75$, $f_w = 0.25$, $S_c = 1$ cycle, $S_s = 1$ cycle,
$S_b = 1, 2,$ & 4 cycles for $N = 16, 64,$ & 256 respectively, $h = 0.9$, $N_{sh} = 2048$

CRAFT : COMPILER-ASSISTED ALGORITHM-BASED
FAULT TOLERANCE IN DISTRIBUTED MEMORY MULTIPROCESSORS[*]

V. Balasubramanian and P. Banerjee

Center for Reliable and High Performance Computing
Coordinated Science Laboratory
University of Illinois

ABSTRACT

In this paper, we discuss the the implementation of a source-to-source restructuring compiler, CRAFT (CompileR assisted Algorithm-based Fault Tolerance),for the synthesis of low-cost system-level checks for general numerical Fortran programs using the notion of algorithm-based checking for distributed memory multiprocessors.

1. INTRODUCTION

Algorithm-based fault tolerance deals with low-cost system-level techniques for fault detection in multiprocessors by modifying actual application algorithms running on the system to perform the on-line checking. This technique had been originally proposed for special purpose array processors for specific applications [1-3]. Recently, there has been some results reported on the design of algorithm-based checking mechanisms for various parallel applications on hypercube multiprocessors, in the presence of finite precision arithmetic [4].

All the above strategies proposed were algorithm-specific; they had been designed by exploiting some particular features of the algorithms. Recently, Balasubramaniam and Banerjee have proposed a theoretical basis for synthesizing algorithm-based checking techniques for general applications at the compiler level [5]. The basic approach outlined by them was to identify linear transformations in Fortran DO loops, restructuring program statements to convert non-linear transformations to linear ones, and to insert system-level checks based on this property. In this paper, we discuss the the actual implementation of a source-to-source restructuring compiler, CRAFT (CompileR assisted Algorithm-based Fault Tolerance), for the synthesis of low-cost system-level checks for general numerical Fortran programs, based on the above approach.

2. LINEAR TRANSFORMATIONS IN LOOPS

2.1. Linearity Property

Let us consider indexed program statements of the form $S(I^1, I^2, ..., I^k)$, where k denotes the number of distinct loop surrounding S and I^j is the index of the jth loop. We are interested in statements S which are linear transformations [5]. The general form of a potentially linear statement having two loop levels, I and J (can be generalized to more than two levels) can be written as

$$D[I,J] = f(I, J, A[I], B[J], C[I,J])$$

where f is some real-valued function; $D[I,J]$ denotes an array variable whose subscripts are real functions of I and J; $A[I, J]$ denotes a set of such variables, $\{A_1[I, J], A_2[I, J], ... \}$. $A[I]$ and $B[J]$ denote sets of array variables $A[I]$ and $B[J]$, respectively.

A brute force application of the linearity test described in [5] is not feasible since it involves complex algebraic manipulations and consumes valuable time. We have addressed the problem at a *symbolic (structural) level*. Let exp_1 and exp_2 be two expressions conforming to the general form of the R.H.S. expression of an assignment. Using the commutative and associative properties of addition, and the distributivity of multiplication over addition, one

can translate the linearity test to the following three *recursive linearity rules*:

(R1) $exp_1 \pm exp_2$ is linear in $I(J)$ if and only if both exp_1 *and* exp_2 are linear in $I(J)$.

(R2) $exp_1 * exp_2$ is linear in $I(J)$ if and only if either exp_1 is linear in $I(J)$ and exp_2 does not involve $I(J)$, or vice versa.

(R3) exp_1 / exp_2 is linear in $I(J)$ if and only if exp_1 is linear in $I(J)$ and exp_2 does not involve $I(J)$.

In addition, we have the following four fairly obvious linearity rules for simple expressions and functions:

(R4) A constant is not linear in any variable.

(R5) A simple variable is linear in itself.

(R6) An array variable is linear in its subscript variables.

(R7) A function of an expression is linear in the variables the expression is linear in, if and only if the function itself is linear. Examples of non-linear functions are trigonometric functions, square roots, etc.

2.2. Symbolic Linearity and Program Linearity

The successful application of the linearity property of a statement is heavily dependent on the program context of the statement. For example, in the double loop structure of Fig. 1, statement $S1$ is linear in I, and statement $S2$ is linear in both variables K and J as per the recursive linearity rules of Section 2.1. However, K varies with I (=$2I+1$), which is indicated by its presence in the outset of the I-loop. Thus, statement $S1$ is not strictly linear in I.

We introduce the concept of program linearity of a statement to address such situations. A *symbolically linear* statement i.e, one satisfies the linearity conditions of Section 2.1, is said to be *program linear* in the context of a particular program, if this property can be directly exploited to develop a suitable checking scheme (to be discussed in Section 4) for on-line error detection. Statement $S2$ in the figure is program linear in both K and J.

```
       K = 1
       DO I = 1, M
           L = I * 2
           DO J = 1, N
S1             A(I, J) = A(I, J) + K * C(I)
S2             A(K, J) = B(K, J) + C(J) * D(K, L)
           ENDDO
           K = K + 2
       ENDDO
```

Fig. 1. *Illustration of program linearity*

2.3. CRAFT Implementation of Symbolic and Program Linearity Detection

We have used Parafrase-2 [6], an existing source-to-source vectorizing package, to aid us in the CRAFT project. The LIN-TEST pass of our compiler performs automatic detection of linear statements in a given program followed by linearization of other statements. LINTEST is executed in two phases or sub-passes. The first of these is the GETLIN phase, which goes through the entire

[*] This research was supported by the Office of Naval Research under Contract N00014-91J-1096.

program and identifies symbolically linear statements. The second phase uses the above information and calls one of the following two procedures, depending on the nature of the statement currently being processed: (1) the PROGLIN procedure; (2) the MAKELIN procedure. The GETLIN phase performs step-by-step detection of symbolic linearity for all assignment statements (inside loops). The recursive nature of the expression tree corresponding to a particular statement suggests a direct application of the recursive linearity rules of Section 2.1.

The key step in the identification of program linearity of a symbolically linear statement is to determine whether the input variables to the statement are functions of the iteration variables (the loop indices) associated with loops that the statement is nested in. This requires a detailed flow analysis of the entire loop body for each of these loops. Parafrase-2, although targeted to achieve this objective, currently does not implement all the detailed flow analysis. However, we observe that, in general, output variables of assignment statements inside a particular loop do depend on that loop index, unless they are assigned to a constant expression; this is a relatively rare occurrence. Based on this observation, any variable occurring in the outset of a particular loop is assumed to be dependent on the iteration variable of that loop.

2.4. Compiler Transformations to Introduce Linearity

Many assignment statements encountered in Fortran programs violate linearity rules $R1$–$R7$ for both I and J, and, hence, are not linear. However, it is possible to restructure the statement such that it satisfies these rules for at least one loop index which then qualifies it as a linear transformation. Again, this restructuring needs to be done at a symbolic level. In [5], we had listed some very simple restructuring rules. We now provide a more general set of rules which are also easy to implement. The following four simple techniques can be used by the compiler to introduce linearity into statements

(T1) If an expression of the form $exp_1 + exp_2$ is such that one component (say, scalar or array variable) does not contain the loop index in question, the variable can be expanded to accommodate that index.

(T2) If an expression of the form $exp_1 * exp_2$ is such that both exp_1 and exp_2 contain the index in question, variable substitution can be performed by substituting this expression by a single variable containing the relevant loop indices. An additional statement needs to be placed before the restructured statement that simply assigns the multiple variable term to the single variable.

(T3) If an expression of the form exp_1 / exp_2 violates rule $R3$, then variable substitution can be performed similar to the one described above, which removes variables containing the loop index in question from the denominator.

(T4) Violation of rule $R7$ can be similarly taken care of by substituting the non-linear expression by a simple array variable.

The MAKELIN procedure, within the CRAFT compiler, operates actively on all assignment statements which are not symbolically linear, except those that satisfy two conditions simultaneously: (1) their R.H.S. expressions are non-linear, and (2) they write to scalars or non-linear expressions. In other words, if the L.H.S. of an assignment statement is a scalar or an array expression involving only constants (e.g. A(2)), then it will be ignored, unless its R.H.S. expression is linear in some variable; in such a case, the output variable is expanded to accommodate this *linearizing variable*. If the output variable is, however, a linear array expression, the R.H.S. expression is linearized with respect to the linearity variable. If multiple linearity variables exist in either of these cases, then the one associated with the innermost surrounding loop of the parent statement is chosen.

3. LINEARITY-BASED SYSTEM-LEVEL CHECKING

Once a statement is identified to be a linear transformation (or is restructured to satisfy the linearity property), it is possible to use this very information to develop system-level checks for detecting errors during computations using a checksum encoding [5].

The ADDCHECK pass of the CRAFT compiler operates on the linearized version of the original program and generates suitable check code for providing on-line detection of errors that may occur during actual execution of the program. The check code for each program linear (or linearized) statement, includes three sets of statements: (1) *summation statements*, each denoting the summation of the various values of an array variable (or scalar) involving the linearity variable, for different values of this iteration variable, (2) *check statements*, that denote the modified computation of the output using these summations, and the comparison that follows, and (3) *initialization statements*, that denote the proper initialization of the *summation variables* before they are made to store the summation values. We now describe the generation of each set.

3.1. Summation Statements

The generation of summation statements first involves identification of the leaves of the statement expression tree and selection of only those expressions that involve the linearity variable. For each such expression (represented generically by *exp*), a summation statement is built and included in the program, as follows. Assume for the moment that *exp* is an array. First, a suitable operand list for the summation variable is created. This consists of all variables that occur along with the linearity variable in *exp*, but which are associated only with loops inside the check loop. Next, an expression corresponding to the new summation variable, say *sum*, is built. Using the *exp* and *sum* expressions, a summation statement of the form *sum* = *sum* + *exp* is built. Finally, the summation statement is inserted just before or immediately after the linear statement in question, depending on whether *exp* corresponds to an input variable or an output variable, respectively. Such an insertion guarantees that the linear statement share all its parent conditional statements with the summation statement, so that there is a one-to-one correspondence between the executions of the linear and summation statement.

Fig. 2 shows the check code inserted by the ADDCHECK pass for statement $S2$ in the loop structure of Fig. 1.

3.2. Check and Initialization Statements

The first of the check statements denotes the alternate computation of the output summation using the input summations. The structure of this statement is similar to that of the original linear statement, but for the replacement of each input array (or scalar) involving the checking variable with the corresponding summation variable, and the presence of a new summation variable at the output. This is shown in Fig. 2 where check statement $C1$ is obtained from $S1$ by replacing $B(K,J)$ and $C(J)$ with $SUM1$ and $SUM2$,

respectively. The next check statement is actually a subroutine call for comparison of the two output summations to check for possible mismatches. This is indicated as $C2$ in the figure. It is possible that the check loop is not the innermost loop in a multiply nested loop structure, in which case, all loops inside the check loop that surround the linear statement, need to be duplicated and placed around the check statements to ensure the validity of the comparisons.

The generation of initialization statements is fairly simple. This involves a duplication of the expressions corresponding to the various summation variables so that they may be used in the construction of assignment statements having their R.H.S. expressions set to the constant, zero. These statements are placed before the

```
        DO I = 1, M
        . . .

I1          SUM0 = 0
I2          SUM1 = 0
I3          SUM2 = 0
            DO J = 1, N

            . . .

SM1             SUM1 = SUM1 + B(K, J)
SM2             SUM2 = SUM2 + C(J)
S2              A(K, J) = B(K, J) + C(J) * D(K, L)
SM3             SUM0 = SUM0 + A(K, J)
            ENDDO
C1          SUM3 = SUM1 + SUM2 * D(K, L)
C2          CALL COMPARE(SUM3, SUM0)

        . . .

        ENDDO
```

Fig. 2. *Illustration of ADDCHECK pass on example loop of Fig. 1.*

check loop and are indicated as $I1$–$I3$ in the figure. As in the case of check statements, all loops occurring inside the check loop that affect the linear statement must be duplicated and placed around the initialization statements.

4. APPLICATION TO REAL PROGRAMS

This section presents the results of application of the compiler to real scientific Fortran programs. Some of these routines are sequential programs. We also consider parallel versions of two of the above programs developed for a hypercube multiprocessor, by running the compiler on the corresponding node programs. The characteristics of a program with regard to the amount of linearity available in key loops remain more or less unchanged when one shifts from the sequential to the parallel domain.

Table 1 presents the results of our runs on six application programs of varying complexity. Six sample routines from the LIN-PACK library, (namely, DGEFA and DGESL) and EISPACK library (namely TRED2 and TQL2), and two programs from the Perfect Club Benchmark Suite (namely TRFD and MDG and their subroutines) [7], are considered first.

The next six columns of the table show the results obtained at the end of the LINTEST pass for each of the above six programs along with their subroutines, if present. For a particular program: (1) *Exec* indicates the total number of executable statements inside the program. (2) *Assign* denotes the number of assignment statements present within loops. (3) *Lin* indicates the number of assignments (out of (2)) that were detected to be symbolically linear, and is, naturally, the outcome of the GETLIN phase. (4) *Pglin* pertains to candidate statements for the PROGLIN procedure, and shows the number of symbolically linear statements (out of (3)) that are program linear as well. (5) *Mklin* represents the results of the MAKELIN procedure, and shows the number of statements from (2) (but not in (3)) that were transformed to be symbolically linear and were program linear as well. (6) *Pslin* denotes the number of statements (out of (5)) that are attributed to variable expansions. Such expansions can be avoided because of the way in which the summations are carried out during actual checking. Hence, these statements represent a special class of non-linear statements that are amenable to checking without the necessity for any restructuring, and are called *pseudo-linear*. (7) The term *lin-checkable* is used to denote all such statements which lend themselves to linearity-based checking. Note that statements that are not lin-checkable can be checked by the simple duplication-and-comparison approach. The total number of lin-checkable statements per row is nothing but the

Table 1. *Linearity characteristics of six application programs*

| Program | Number of statements | | | | | | % lin-checkable | |
	Exec	Assign	Lin	Pglin	Mklin	Pslin	Static	Dynamic
DGEFA	43	16	4	4	5	5	56.3	98.0
DGESL	49	18	3	3	9	9	66.7	74.2
TRED2	92	46	12	12	16	16	60.9	66.4
TQL2	84	47	3	2	14	14	34.0	95.0
TRFD*	67	23	3	2	1	0	12.5	57.4
INTGRL	28	11	0	0	2	2	18.2	38.1
OLDA	100	35	6	6	13	13	54.3	97.1
MDG*	100	6	2	2	2	2	66.7	99.9
MDMAIN	61	11	3	3	0	0	27.3	30.0
BNDRY	13	6	0	0	6	6	100.0	100.0
INITIA	97	34	0	0	12	12	35.3	41.4
CORREC	10	4	0	0	2	2	50.0	50.0
INTRAF	59	47	4	4	5	4	18.8	29.7
INTERF	108	81	9	9	2	2	13.6	11.0
POTENG	69	43	0	0	2	2	4.7	10.9

Table 2. *Linearity characteristics of two hypercube node programs*

| Program | Number of statements | | | | | | % lin-checkable | | | |
| | Exec | Assign | Lin | Pglin | Mklin | Pslin | Static | Dynamic | | |
								P=4	P=8	P=16
PDGEFA	72	25	5	5	10	10	60.0	93.3	89.1	84.1
PTRED2	362	138	8	8	32	32	29.0	58.9	56.5	52.4

total number of program linear and linearized statements ((3) + (4)) for the same row. Each number in the *static* column is obtained by dividing this number by the total number of assignments statements that were present initially or created during substitutions ((2) + (5)-(4)). The dynamic results present a more realistic picture as they are an estimate of the total number of executions of all lin-checkable statements with respect to the total number of executions of all assignment statements, for a given input data size.

What can one infer from these values? Since, any error that occurs during the execution of a lin-checkable statement is detected by the corresponding check code with a high probability, one can expect programs having reasonably high dynamic counts (the first five programs) to give high error coverages. Correspondingly, programs with very low dynamic counts such as some subroutines in the MDG program cannot be expected to give high error coverages, if linearity-based checking were the *sole* concurrent error detection scheme to be used. The coverages for on-line error detection of such programs can be boosted up by the intelligent use of duplication techniques. Note that the results for DGEFA, DGESL and TQL2 were estimated using an input matrix size of 100, whereas an input size of 1000 was used for the TRED2 routine. The parameter values for the TRFD and MDG routines were defaults provided by the Benchmark Suite.

Finally we report on results results for parallel versions of the DGEFA and TRED2 routines in Table 2. Since DGEFA is a highly structured program as mentioned in the previous section, the static count for PDGEFA remains more or less the same indicating the fact that the linearity of the program as a whole has not been affected by parallelization. The dynamic count drops down slightly as expected. Since the program parallelization causes the number of iterations of the parallel loop to decrease by a factor equal to the number of processors, the relative number of executions of lin-checkable statements inside the parallel loop also goes down by the

same factor. Parallelization of the TRED2 routine, however, introduces lots of non-linear statements because of the somewhat loosely structured nature of the original program. Hence, the static percent is decreased quite a bit as is evident from the Tables 1 and 2. The drop in the dynamic count is comparatively lower, as many of these non-linear statements occur outside the most frequently executed loops in the program.

4. CONCLUSIONS

In recent years, the area of *Algorithm-based fault tolerance* has gained importance and popularity since it provides a low cost system level technique for fault detection in multiprocessors without any hardware modifications. The difficult task of synthesizing algorithm-based checking techniques for general applications was recently addressed in [5]. This paper has discussed the implementation of an actual compiler that synthesizes suitable low-cost checks based on the above approach. It should be noted that parallelizing compilers for distributed memory machines such as the hypercube, are currently under development [8,9]. We are currently working on attaching our compiler as a back end to such a parallelizing compiler.

REFERENCES

[1]   K. H. Huang and J. A. Abraham, ''Algorithm-Based Fault Tolerance for Matrix Operations,'' *IEEE Trans. Comput.*, vol. C-33, pp. 518-528, Jun. 1984.

[2]   J. Y. Jou and J. A. Abraham, ''Fault Tolerant FFT Networks,'' in *Proc. 15th Int. Symp. on Fault-Tolerant Computing*. Ann Arbor, Michigan, Jun. 1985.

[3]   A. L. N. Reddy and P. Banerjee, ''Algorithm-based Fault Detection Techniques in Signal Processing Applications,'' *IEEE Trans. Comput.*, vol. 39, pp. 1304-1308, Oct. 1990.

[4]   P. Banerjee, J. T. Rahmeh, C. Stunkel, V. S. Nair, K. Roy, V. Balasubramanian, and J. A. Abraham, ''Algorithm-Based Fault Tolerance on a Hypercube Multiprocessor,'' *IEEE Trans. Comput.*, vol. 39, pp. 1132-1145, Sept. 1990.

[5]   V. Balasubramanian and P. Banerjee, ''Compiler Assisted Synthesis of Algorithm-Based Checking in Multiprocessors,'' *IEEE Trans. Computers*, vol. 39, pp. 436-447, Apr. 1990.

[6]   C. D. Polychronopoulos, M. Girkar, M. R. Hagighat, C. L. Lee, B. Leung, and D. Schouten, ''Parafrase-2: An Environment for Parallelizing, Partitioning, Synchronizing, and Scheduling Programs on Multiprocessors,'' in *Proc. 18th Int. Conf. on Parallel Proc.*, St. Charles, IL, pp. 39-48, Aug. 1989.

[7]   The Perfect Club, ''The Perfect Club Benchmarks: Effective Performance Evaluation of Supercomputers,'' *Int. Journal of Supercomputing Applications*, vol. 3, pp. 5-40, Fall 1989.

[8]   D. Callahan and K. Kennedy, ''Compiling Programs for Distributed-Memory Multiprocessors,'' *The Journal of Supercomputing*, vol. 2, pp. 151-169, Oct. 1988.

[9]   H. Zima, H. Bast, and H. Gerndt, ''SUPERB: A Tool for Semi-automatic MIMD/SIMD Parallelization,'' *Parallel Computing*, vol. 6, pp. 1-18, 1988.

EXPERIMENTAL EVALUATION OF MULTIPROCESSOR CACHE-BASED ERROR RECOVERY

Bob Janssens and W. Kent Fuchs

Center for Reliable and High-Performance Computing
Coordinated Science Laboratory
University of Illinois
Urbana, IL 61801

ABSTRACT

Several variations of cache-based checkpointing for roll-back error recovery in shared-memory multiprocessors have been recently developed. By modifying the cache replacement policy, these techniques use the inherent redundancy in the memory hierarchy to periodically checkpoint the computation state. Three schemes, different in the manner in which they avoid rollback propagation, are evaluated in this paper. Employing full address traces from five different parallel applications running on the Encore Multimax shared-memory multiprocessor, we simulate the performance effects of integrating the recovery schemes in the cache coherence protocol. This is the first paper to experimentally evaluate multiprocessor cache-based checkpointing techniques using address traces. Our results indicate that the cache-based schemes have the potential of facilitating checkpointing at a reasonable cost, provided that the problem of variability in checkpoint frequency can be resolved.

1 INTRODUCTION

Checkpointing and rollback are often used in fault-tolerant computer systems to allow recovery from errors without a global restart of computation. Checkpointing degrades overall system performance during normal execution. To reduce the cost incurred by implementing error recovery capability, it is important that checkpointing overhead be kept small. One approach to user transparent checkpointing is the cache-based design developed for recovery from transient processor errors [5]. It differs from conventional recovery cache techniques in that it uses the inherent redundancy in the memory hierarchy for maintaining checkpoints. The technique has been extended to shared-memory multiprocessors by integrating the checkpointing and recovery algorithms with cache coherence protocols [1, 11].

This paper presents the first experimental evaluation of the impact on system performance for several variations of cache-based error recovery in bus-based multiprocessors using full multiprocessor address traces. Under analytical models with estimated workload and system parameters, it has been shown that cache-based error recovery results in slight performance degradation [11]. However, the derived performance measures vary widely with the assumed values of the parameters. In this paper, we use actual address traces from the execution of five parallel programs on the Encore Multimax shared-memory parallel processor to more accurately evaluate the performance of cache-based error recovery designs. An extended version of this paper can be found in reference [6].

This research was supported in part by the National Aeronautics and Space Administration (NASA) under Grant NASA NAG 1-613, in cooperation with the Illinois Computer Laboratory for Aerospace Systems and Software (ICLASS), and by SDIO/IST and managed by the Office of Naval Research under contract N00014-88-K-0656.

2 CACHE-BASED CHECKPOINTING

2.1 Description of Algorithm

In the CARER cache-based checkpointing proposal [5], a checkpoint is taken every time a dirty cache line is replaced and written back to main memory. In a set associative cache, the replacement policy is modified to avoid a checkpoint whenever possible by selecting a clean line for replacement. A checkpoint is also taken when an external event such as I/O or an interrupt occurs. At a checkpoint, the internal processor registers are saved and all dirty lines in the cache are marked *unchangeable*. Unchangeable lines may be read, but have to be written back to memory before they are modified. The checkpoint state consists of all the unchangeable lines in the cache and the contents of main memory. When a rollback is necessary, all cache lines except the unchangeable lines are invalidated and the saved processor registers are restored, restarting the computation at the last checkpoint. Since the checkpoint state is stored partially in the cache and partially in main memory, both should be highly reliable or protected by their own error detection and recovery mechanisms.

2.2 Extension to Multiprocessors

Multiprocessor systems present the additional problem of rollback propagation. A rollback of one process may necessitate a rollback of other processes that may have received corrupted data from the originating process. These extra rollbacks in turn may cause more rollbacks, resulting in a domino effect. To come to a consistent state several processes may have to roll back multiple steps, requiring that multiple checkpoints be kept for every processor.

The simplest way to avoid the domino effect is through fully synchronized checkpointing and rollback [1]. A checkpoint on one processor immediately causes a checkpoint on all other processors. Similarly, a rollback on one processor causes all others to roll back to the last checkpoint. This method requires that the originating processor can communicate a checkpoint to all other processors in time for them to establish their checkpoints immediately after the completion of the currently executing instruction. An enhancement to this simple scheme is flagged synchronized checkpointing and rollback. In this scheme every processor keeps a flag for every other processor, indicating whether it has communicated with that processor since the last checkpoint [1]. In a shared-memory system, interprocessor communication occurs whenever one processor reads data modified by another processor. In the flagged scheme, a checkpoint on one processor propagates only to those processors for which the flag is raised. Similarly, a rollback on one processor causes rollbacks only on the processors with the flag for the originating processor raised.

Table 1: Characteristics of address traces.

program	description	input	tot. num. of references	data reads	data writes	code reads
gravsim	N-body sim.	two colliding 128-body clusters	92,178,814	33,266,880	6,392,078	52,519,859
tgen	test generator	ISCAS benchmark s208 w. 20 faults	101,264,382	32,613,809	4,461,889	64,188,674
fsim	fault simulator	ISCAS benchmark s208 w. 80 vectors	149,918,375	50,950,933	3,958,919	95,008,523
pace	circuit extractor	two bit slices of an ALU	87,861,165	23,266,576	7,842,338	56,752,251
phigure	global router	469 cell circuit	132,998,231	38,244,233	11,530,981	83,223,027

Communication-induced checkpointing is an alternative to the synchronized methods. The source processor takes a checkpoint immediately following the communication of data between two processors [11]. The source processor is now prevented from modifying the sent data upon a subsequent checkpoint. The destination processor is allowed to roll back past the processor interaction. As long as synchronization variables are cached, correct access ordering is ensured [11].

2.3 Integration into Coherence Protocol

In most cases, the checkpointing schemes described above can be integrated into existing cache coherence protocols by adding an unwritable state for every dirty state. The synchronized schemes can be extended only to coherence protocols that write back dirty lines only when room is needed in the cache. In protocols where a dirty shared line is written back to main memory whenever there is a read miss on that line in another processor, a checkpoint has to be taken on every such miss. But this is exactly the condition for taking additional checkpoints in the communication-induced checkpointing scheme. Thus for these protocols, the checkpoints taken in the synchronized schemes are a superset of those taken in the communication-induced scheme.

The Berkeley write-invalidate coherence protocol [7] was used in this study. To integrate the checkpointing schemes into the Berkeley protocol, an *unwritable (owned)* and a *shared unwritable* state have to be added to the four original states of *invalid, valid, dirty (owned),* and *shared dirty*. In the communication-induced checkpointing scheme, a bus read miss on a dirty line will always cause a checkpoint, forcing the state into *shared unwritable*. Therefore, a *shared dirty* state is not needed.

3 SIMULATION EXPERIMENTS

3.1 Methodology and Workload

The address traces used in this study were generated by a version of the TRAPEDS address trace generator written for the Encore Multimax shared-memory multiprocessor [6, 9]. A discussion of alternative multiprocessor address tracing techniques can be found in reference [10].

The traces used were from five parallel programs written in C using Encore's standard parallel library. Gravsim is a gravitational N-body simulator using the Barnes and Hut algorithm [3]. Tgen is a test pattern generator for sequential and combinational circuits using a parallel search algorithm [8]. Fsim is a parallel fault simulator for digital circuits [8]. Pace is a parallel circuit extractor [2]. Phigure is global router using structured hierarchical decomposition of independent tasks [4].

All traces were generated on seven processors. Table 1 describes the characteristics of the generated traces. All applications were traced from beginning to end, resulting in a trace of more than 10 million memory references per processor in every case.

3.2 Experimental Results

Simulations were performed using the Berkeley cache coherence protocol for the fully synchronized, flagged synchronized, and communication-induced checkpoint schemes. Cache sizes were varied from 4K to 128K bytes. Line size was held fixed at 16 bytes. Direct mapped and two-way set associative caches were simulated. Cache miss ratios, writeback rates, and checkpointing rates were calculated on a per-processor basis and then averaged to arrive at a single number for each application.

In set associative caches, the modified replacement policy used in the cache-based error recovery schemes has only a slight effect on the performance. The modification of the LRU replacement policy affects the miss ratio and the base bus traffic (traffic resulting from writebacks needed for the computation). The average miss ratio for the five programs is shown in Figure 1. The average bus traffic in terms of the number of writebacks per 100,000 references is shown in Figure 2. In the direct mapped cache, the miss ratio and base bus traffic remain the same, since the replacement policy is not modified. In the two-way set associative cache, the miss ratio increases by an insignificant amount and the base bus traffic actually decreases slightly with the modified replacement policy of the recovery schemes. The total number of writebacks consists of those needed for checkpointing in addition to the base writebacks. The frequency of these additional writebacks depends heavily on the frequency of checkpointing and will therefore be discussed after introducing the checkpointing frequency data.

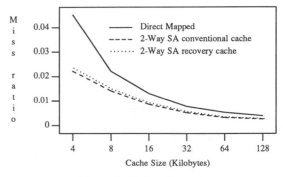

Figure 1: Cache miss ratios.

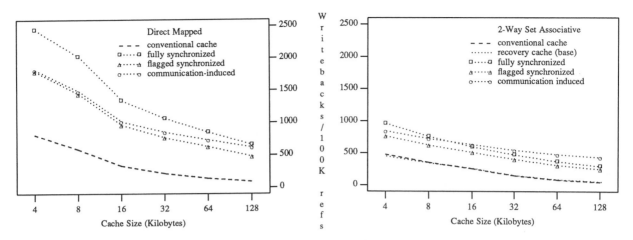

Figure 2: Writeback traffic.

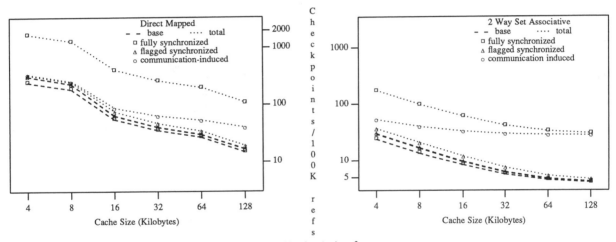

Figure 3: Checkpointing frequency.

Figure 3 contains the average checkpointing frequencies for the three recovery schemes. The base checkpointing frequency includes all checkpoints that are taken due to the necessity of writing back a dirty cache line. The total checkpointing frequency also includes the checkpoints induced in other processors. The base checkpointing frequency is about twice as high in the direct mapped cache as in the set associative cache. It reduces with increased cache size since a large cache has fewer writebacks of dirty lines. The extra induced checkpointing frequency varies with the recovery scheme used.

In the fully synchronized scheme, a checkpoint on one processor causes concurrent checkpoints on all other processors. Therefore, the total checkpointing frequency is approximately seven times the base frequency. The large number of induced checkpoints actually reduces the number of base checkpoints by clearing the cache frequently of dirty lines.

The flagged synchronized scheme drastically reduces the induced checkpointing frequency, since checkpoints are induced only if communication has occurred with the originating processor. Even in programs with large amounts of communication, the flagged synchronized scheme performs much better than the fully synchronized scheme. In programs with little communication the increase over the base checkpointing frequency is negligible.

The communication-induced scheme is even more sensitive to interprocessor communication. When there is a high rate of communication, the total number of checkpoints taken exceeds that of the fully synchronized scheme and does not improve with higher cache sizes. In the absence of a large amount of communication, the scheme performs as well as the flagged synchronized scheme.

In all three schemes, even if the average checkpointing frequency is low, there is no guaranteed minimum checkpointing interval, which means that errors with long latencies might not be detected. Most of the checkpoint intervals in the fully synchronized schemes are small. Increasing the cache size, especially in the set associative cache, will introduce longer checkpoint intervals, but in all cases almost 50% of the intervals will stay below 100 references. In the larger caches, the proportion of small intervals actually increases, even though their absolute number decreases. Under the communication-induced checkpointing scheme, there are proportionally fewer small checkpoint intervals. But increasing the cache size shifts the distribution only marginally towards larger intervals. The flagged synchronized scheme causes the fewest small checkpoint intervals. However, even in the largest cache there are still checkpoint intervals of fewer than ten accesses.

The total writeback traffic is influenced by the checkpoint frequency. At any time, a certain percentage of the modified cache lines will be unwritable instead of dirty. On a write hit,

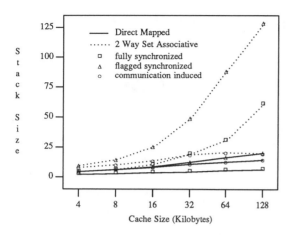

Figure 4: Rec. stack size to capture 90% of the extra writebacks.

an unwritable line has to be written back to main memory before it can be updated. The chance of such an extra writeback is determined by the percentage of unwritable lines present in the cache. The more frequently checkpointing occurs, the more likely it is for a modified block to be in the unwritable rather than dirty state. A comparison of Figures 2 and 3 shows that the number of extra writebacks indeed increases with increased checkpoint frequency. The extra writeback frequency of the synchronized schemes shows variability with the cache size, while that of the communication-induced scheme does not, just as in the case of checkpoint frequency. The extra writebacks can easily double the total traffic. In large caches, almost all of the traffic is due to extra writebacks.

The correlation of extra writeback traffic and number of checkpoints has a hidden side effect that can be used to increase performance with the addition of a small amount of hardware. The data written back due to modification or invalidation of a cache line is checkpoint data; from the computation's point of view it has been overwritten. It is possible to store this data in a small recovery stack local to every cache, eliminating the extra bus traffic [11]. The recovery stack is cleared at every checkpoint. If the recovery stack is full, the writeback can always proceed to the shared memory. Figure 4 shows the size of the recovery stack needed so that, in the average program, 90% of the extra writebacks will not have to go to shared memory. Because of its success at keeping the checkpointing frequency low, the flagged synchronized recovery scheme needs the largest recovery stack.

4. CONCLUSIONS

Trace-driven simulation was used to measure the performance of cache-based error recovery in a shared-memory multiprocessor. The results answer many questions that remained unresolved after analytical performance studies of the cache-based recovery method. The cache miss ratio is not degraded by the addition of the recovery scheme to the cache protocol. The writeback traffic increases significantly. However, a small recovery stack at every cache can intercept most of the extra writebacks caused by the recovery schemes, avoiding the necessity of extra bus bandwidth. The checkpointing frequency of the schemes is high, especially in smaller caches. The communication-induced checkpointing scheme suffers from a high number of induced checkpoints, regardless of cache size. The induced checkpointing frequency of

the flagged synchronous scheme is lower and it decreases with stack size. In all of the schemes, however, small checkpoint intervals cannot be completely eliminated.

Cache-based checkpointing allows the implementation of error recovery capability at a lower cost, both in extra hardware and performance, than traditional recovery caches. Its major disadvantages are the uncontrollability and variability of the checkpoint frequency. The initiation of a checkpoint is controlled by the cache parameters and the communication pattern of the application program. Future research should be directed at finding low-cost methods that eliminate these disadvantages.

ACKNOWLEDGEMENTS

We would like to thank Mark Bellon, Randy Brouwer, Srinivas Patil, and Krishna Prasad Belkhale for supplying the applications that were traced for this study. Also, we very much appreciate Craig Stunkel's help in adapting his TRAPEDS methodology to the Encore Multimax.

REFERENCES

[1] R. E. Ahmed, R. C. Frazier, and P. N. Marinos, "Cache-Aided Rollback Error Recovery (CARER) Algorithms for Shared-Memory Multiprocessor Systems," *Proc. 20th Int. Symp. on Fault-Tolerant Computing*, 1990, pp. 82–88.

[2] K. P. Belkhale, *Parallel Algorithms for Computer-Aided Design with Applications to Circuit Extraction*, Ph. D. Thesis, Univ. of Illinois, Urbana, IL, Nov. 1990, Tech. Report CRHC-90-15.

[3] M. Bellon, "Parallelizing Gravitational N-body Algorithms on MIMD Architectures," Motorola Urbana Design Center, Urbana, IL.

[4] R. J. Brouwer and P. Banerjee, "PHIGURE: A Parallel Hierarchical Global Router," *Proc. 27th Design Automation Conf.*, 1990, pp. 650–653.

[5] D. B. Hunt and P. N. Marinos, "A General Purpose Cache-Aided Error Recovery (CARER) Technique," *Proc. 17th Int. Symp. on Fault-Tolerant Computing*, 1987, pp. 170–175.

[6] B. L. Janssens, "Generation of Multiprocessor Address Traces and their Use in the Performance Analysis of Cache-based Error Recovery Methods," M. S. Thesis, Univ. of Illinois, Urbana, IL, March 1991, Tech. Report CRHC-91-10.

[7] R. H. Katz et al., "Implementing a Cache Consistency Protocol," *Proc. 12th Int. Symp. on Computer Architecture*, 1985, pp. 276–283.

[8] S. Patil, *Parallel Algorithms for Test Generation and Fault Simulation*, Ph. D. Thesis, Univ. of Illinois, Urbana, IL, Sep. 1990, Tech. Report CRHC-90-12.

[9] C. B. Stunkel and W. K. Fuchs, "TRAPEDS: Producing Traces for Multicomputers via Execution Driven Simulation," *Proc. ACM SIGMETRICS and Performance Int. Conf. on Measurement and Modeling of Computer Systems*, 1989, pp. 70–78.

[10] C. B. Stunkel, B. Janssens, and W. K. Fuchs, "Address Tracing for Parallel Machines," *Computer*, Vol. 24, No. 1, Jan. 1991, pp. 31–38.

[11] K.-L. Wu, W. K. Fuchs, and J. H. Patel, "Error Recovery in Shared Memory Multiprocessors Using Private Caches," *IEEE Trans. on Parallel and Distributed Systems*, Vol. 1, No. 2, Apr. 1990, pp. 231–240.

Base-m n-cube : High Performance Interconnection Networks for Highly Parallel Computer PRODIGY

Noboru TANABE, Takashi SUZUOKA, Sadao NAKAMURA,
Yasushi KAWAKURA and Shigeru OYANAGI
TOSHIBA R&D Center
1, Komukai Toshiba-cho, Saiwai-ku, Kawasaki 210, Japan
tanabe@isl.rdc.toshiba.co.jp

Abstract

A new interconnection network, named "base-m n-cube", is proposed for highly parallel machines. Base-m n-cube is a natural extension of binary k-cube, and can decrease the node degree without sacrificing performance and generality. Base-m n-cube is also superior to binary k-cube with various properties such as distance, latency, complexity of logic, wiring ease, embedding and reliability. The authors have developed a prototype base-m n-cube machine, named PRODIGY, which realizes faster communication speed than any commercial hypercube machines. Base-m n-cube is evaluated by PRODIGY and simulation. The result shows that base-m n-cube is a promising approach for powerful highly parallel machines.

1.Introduction

Interconnection networks have been an important research area for highly parallel machines over the years. Recently, binary k-cubes[1][2] are receiving much attention for their generality, small diameter, and routing ease. A number of highly parallel binary k-cube machines are commercialized, such as Connection Machine[3][4], NCUBE[5][6], and iPSC[7]. However, these machines have limited communication power, owing to the bit serial links used. The major reason for bit serial links is the wiring difficulty. Binary k-cube requires the node degree of log2 N, which is a serious restriction for wiring to provide parallel links for highly parallel machines.

This paper proposes a new interconnection network, named "base-m n-cube", for highly parallel machines. It is a natural extension of binary k-cube, and can decrease the node degree without sacrificing performance and generality. The authors have also developed a prototype base-m n-cube machine, named PRODIGY[14][15], and evaluated the achieved practical communication performance. The result shows that using base-m n-cube is a promising approach for powerful interconnection network for highly parallel machines.

This paper describes the definition, various properties of base-m n-cube, and its evaluation, based on the practical implementation and simulation.

2.Base-m n-cube interconnection networks

A.Definition

In this section, the base-m n-cube network is defined, in contrast to binary n-cube.

The binary n-cube is defined by the statement that two nodes, whose binary k-bit addresses differ in exactly one bit position, are directly connected. Figure 1 shows binary 4-cube(16 nodes).

In the same way, the base-m n-cube is defined by the statement that two nodes, whose m-ary (m>=2) n-digit addresses differ in exactly one digit, are indirectly connected

through m * m crossbar switches. Figure 2 shows base-4 2-cube(16 nodes). Each rectangle in Fig.2 represents a crossbar switch.

There are some other interconnection networks which are similar to base-m n-cube. The neighboring nodes for base-m n-cubes are the same as for some alpha networks(generalized hypercubes)[8][9][10]. Figure 3 shows alpha network(16 nodes) which has the same neighboring nodes as base-4 2-cube. The difference between base-m n-cube and alpha networks is the degree of nodes. If a crossbar switch is implemented in an LSI, base-m n-cubes are easier to implement than alpha networks, because many wires can be integrated within LSI.

Base-m n-cubes and k-ary n-cubes[11][12] are similar in name, but different in topology. Figure 4 shows 4-ary 2-cube(16 nodes). K-ary n-cube is organized as the n-dimensional ring of length k, whereas base-m n-cube is organized as the n-dimensional crossbar networks of m*m. K-ary n-cube is a hypercube, only when k=2. Base-m n-cubes are hypercubes in all m.

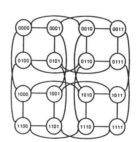

Fig.1 Binary 4-cube(16nodes)

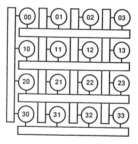

Fig.2 Base-4 2-cube(16nodes)

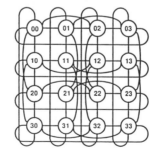

Fig.3 Alpha network(16nodes)

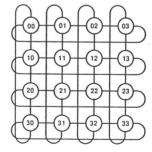

Fig.4 4-ary 2-cube(16nodes)

B.Routing

Deadlock in the interconnection networks occurs when no message can advance toward its destination, because of the loop in the requesting resources.

Deadlock-free self-routing algorithms[2] for binary n-cube have already been known. For example, message can be routed without deadlock by sending to the leftmost different bit direction in the message's destination address and correcting the bit.

The deadlock-free self-routing algorithm for base-m n-cube can be easily generated by expanding the above algorithm. A node in base-m n-cube can be identified by n-digit m-ary address. For example, a message can be routed without deadlock by sending to the leftmost different digit direction in the message's destination address and correcting the digit.

Wormhole routing can be realized on base-m n-cube. Namely, the router selects the direction to the leftmost different digit, and a crossbar switch connects the correspondent path and corrects the digit.

3. Properties of the base-m n-cube

In this section, several properties of the base-m n-cube are discussed in comparison with binary k-cube.

A. Distance and latency

The distance is important for achieving the desired performance, when the routing algorithm is store and forward type. The **diameters** of binary k-cube and base-m n-cube are as follows.

$$\text{binary k-cube} : D2 = k = \log_2 N$$
$$\text{base-m n-cube} : Dm = n = \log_m N$$

where N : total node number

Because m>=2, the diameter of base-m n-cube is shorter than that of binary k-cube.

The **average distances** of these networks are as follows.

$$\text{binary k-cube} : AD2 = k/2 = (\log_2 N)/2$$
$$\text{base-m n-cube} : ADm = n * (m-1)/m$$
$$= \{(m-1) * \log_m N\}/m$$

Figure 5 shows the average distance of various combinations of total node number:N, and crossbar switch size:m. From the distance standpoint, using big crossbar switches is better.

The network latency is deeply related to distance, especially in the case of store and forward (S&F) routing[11].

Under the assumption that the latencies of router and crossbar switch are the same as 1 cycle time of a channel, the binary k-cube and base-m n-cube **latencies for S&F routing** are as follows.

$$\text{binary k-cube} : Ts2 = Tr * (d * L / W)$$
$$= d * Tc * L / W$$
$$\text{base-m n-cube} : Tsm = Tx * d + Tr * (d * L / W)$$
$$\doteq d * Tc * L / W$$

where
L : message length,
W : channel width,
d : distance from source node to destination node,
Tx : latency through crossbar switch,
Tr : latency through router,
Tc : channel cycle time for W bit data forwards to the next.

Since the time through a crossbar switch is negligible compared with the time required for storing the entire message, the base-m n-cube latency is represented as the same approximate equation of binary k-cube. Because the

base-m n-cube distance is shorter than that for the binary k-cube, the S&F routing latency on base-m n-cube is shorter than that on binary k-cube.

The binary k-cube and base-m n-cube **latencies for wormhole(WH) routing**[11] are as follows.

$$\text{binary k-cube} : Tw2 = Tr * (d + L / W)$$
$$= Tc * (d + L / W)$$
$$\doteq Tc * L / W$$
$$\text{base-m n-cube} : Twm = Tx * d + Tr * (d + L / W)$$
$$= Tc * (2d + L / W)$$
$$\doteq Tc * L / W$$

Since the time through router and crossbar switch is negligible compared with the time for storing the entire message, the WH routing latency on the base-m n-cube is represented as the same approximate equation as that on the binary k-cube. The WH routing latencies don't depend on the distance, in the case without contention.

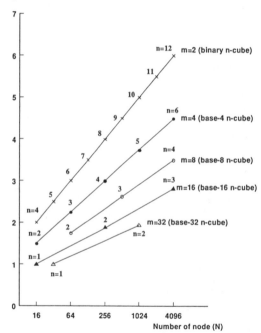

Fig.5 Average distance

B. Logic complexity

The logic complexity can be represented by the number of crosspoints. Here, the numbers of crosspoints for the router and crossbar switch are calculated for WH routing, because the router complexity for WH routing is higher than that for S&F routing.

Since router needs 1 more port for memory access in addition to routing ports, the **numbers of crosspoints per chip** for binary k-cube and base-m n-cube are as follows.

$$\text{router for binary k-cube} : sr2 = (k + 1)^2$$
$$= (\log_2 N + 1)^2$$
$$\text{router for base-m n-cube} : srm = (n + 1)^2$$
$$= (\log_m N + 1)^2$$
$$\text{crossbar for base-m n-cube} : sx = m^2$$

If x^2 crosspoints can be integrated per chip, the maximum

number of nodes for each network can be calculated by setting $k = x - 1$, $m = x$, and $n = x - 1$. Then, the maximum configuration for binary k-cube is only $N = 2^{(x-1)}$ nodes, but the maximum configuration for base-m n-cube is $N = x^{(x-1)}$ nodes. Because base-m n-cube can distribute logic to a router and a crossbar switch, it can reduce the complexity of each chip, and can connect more nodes than the binary k-cube.

The **numbers of LSIs for router and crossbar switch** are as follows.

router for binary k-cube : $Cr2 = N = 2^k$
router for base-m n-cube : $Crm = N = m^n$
crossbar for base-m n-cube : $Cx = n * m^{n-1}$
$$= (\log_m N) * m^{n-1}$$

If $m \ge n$, then $Crm \ge Cx$. This means that base-m n-cube increases the number in chips, but the difference of chip numbers between these networks is smaller than N in many cases. It is not so critical.

The **total amounts of crosspoints** for binary k-cube and base-m n-cube are as follows.

binary k-cube : $S2 = Cr2 * sr2$
$$= N * (\log_2 N + 1)^2$$
$$= 2^k * (k + 1)^2$$
base-m n-cube : $Sm = Crm * srm + Cx * sx$
$$= N * (\log_m N + 1)^2 + (\log_m N) * m^{n+1}$$
$$= m^n * (n + 1)^2 + n * m^{n+1}$$

Figure 6 shows an example of the number of total crosspoints in various m and n combination. We can see that there are some cases that base-m n-cube is superior to binary k-cube, both in regard to the diameter and to the total amount of crosspoints.

Number of total crosspoints

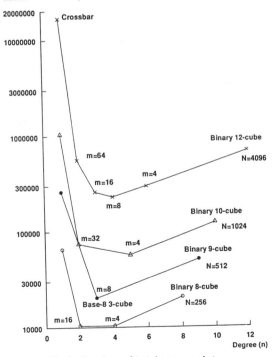

Fig.6 Number of total crosspoints

C.Wiring

The most critical point in interconnection networks for highly parallel systems is the wiring limit at least now. Highly parallel binary k-cube machines, for example Connection Machine and NCUBE, are running into the wall of inter-board connection. In this section, the limits of LSI's pin and inter-board connection are discussed under the assumption of normal implementation.

1)LSI pin limit

If a router is integrated in an LSI, the network degrees are closely related to the LSI's pin. When total node numbers are the same($N = 2^k = m^n$), the degree of binary k-cube is $\log_2 m$ times larger than base-m n-cube. Therefore, the router used for the base-m n-cube is easier to implement than that for the binary k-cube.

The **LSI pin counts** for binary k-cube and base-m n-cube are shown as follows.

router for binary k-cube : $Pr2 = k * l + c$
router for base-m n-cube : $Prm = n * l + c$
crossbar for base-m n-cube : $Pxm = m * l$

where l : pin count for a bidirectional link,
c : pin count for memory bus, etc.

The pin count for a state of the art LSI is under 300. If $Pc = 80$(address,data:32bits, etc), $l = 20[2*(8bits data + handshake)]$, then the maximum numbers of nodes are calculated as follows.

binary k-cube : $300 \ge 2 * k * 10 + 80$ $\quad k \le 11$
$\quad N \le 2048$
base-m n-cube : $300 \ge 2 * n * 10 + 80$ $\quad n \le 11$
$\quad 300 \ge 2 * m * 10$ $\quad m \le 15$
$\quad N \le 15^{11}$

The maximum number of nodes by base-m n-cube is markedly larger than that by binary k-cube. Even if a parallel channel is used, the LSI pin limit is not so critical, under the assumption that one LSI contains one node.

2)Inter-board connection limit

Even if the LSI pin limit is not so critical, there are no binary k-cube machines whose links are parallel channels. The inter-board connection limit is too critical to construct a highly parallel binary k-cube with parallel channels. The limit is that only up to four 200-pin connectors can be attached to one side of a 60 cm board.

Both the degrees and the numbers of nodes implemented in a board are related to the inter-board connection difficulty. Under the assumption that the number of node per board is $Nb = 2^b = m^i$ (Nodes are connected with binary b-cube or base-m i-cube within a board.), the **number of links for inter-board connection per node** is shown as follows.

binary k-cube : $O2 = k - b$
base-m n-cube : $Om = n - i$
$$= n - \log_m Nb$$
$$= k/\log_2 m - b * \log_m 2$$
$$= (k - b)/\log_2 m$$
$O2 / Om = \log_2 m$

This means that the binary k-cube requires ($\log_2$ m)-times more links for inter-board connection than base-m n-cube.

The **numbers of lines for inter-board connection per board** for binary k-cube and base-m n-cube are shown as follows.

$$\text{binary k-cube : L2} = O2 * l * Nb + h$$
$$= (k - b) * l * Nb + h$$
$$\text{base-m n-cube : Lm} = Om * l * Nb + h$$
$$= (k - b)/\log_2 m * l * Nb + h$$
$$<= L2$$

where l : number of lines for a bidirectional link,
h : number of lines between host and nodes in a board

Therefore, base-m n-cube is always easier to wire than binary k-cube. The majority of power consumed by a machine is used to drive signals over a long wire. Because base-m n-cube has fewer lines relating to inter-board connection than binary k-cube, base-m n-cube can save power dissipation and chips for driving lines.

Figure 7 shows the maximum node number limited by the wiring bottleneck between board and backplane(L<=800). The maximum node number for highly parallel binary k-cube machine is 4096. However highly parallel base-m n-cube can satisfy the restriction with parallel channels. If many nodes are implemented on a board, like NCUBE, markedly larger nodes can satisfy the restriction with serial base-m n-cube than with serial binary k-cube.

If a few nodes are implemented on a board, like iPSC(1 node per board), the wiring bottleneck between board and backplane is not an important problem, but the wiring bottleneck between backplanes is critical.

Number of nodes (N)

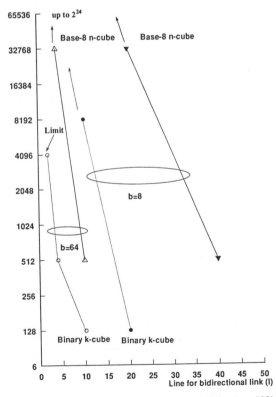

Fig.7 Maximam node number (wiring restriction:L<=800)

D.Embedding

The optimal topologies for interconnection networks depend on the communication patterns used in their applications. There are many parallel algorithms for typical network topologies (e.g. mesh, tree, binary k-cube). Therefore, the embedding capability is one of the most important measures for the generality of the topologies.

The number of neighboring nodes is related to the embedding capability. The **numbers of neighboring nodes** in binary k-cube and base-m n-cube are shown as follows.

$$\text{binary k-cube : E2} = k = n * \log_2 m$$
$$\text{base-m n-cube : Em} = n * (m-1)$$
$$Em/E2 = (m-1)/\log_2 m$$
$$d/dm(Em/E2) = (\ln m + 1/m - 1)/\{(\ln 2) * (\log_2 m)^2\} > 0$$

As m>=2, base-m n-cubes have a larger numbers of neighboring nodes than binary k-cubes. The larger m is, the greater the difference in number for neighboring nodes between binary k-cube and base-m n-cube.

If $m = 2^i$ (i is a natural number), the neighboring nodes for binary k-cube on every node are the subsets of those for base-m n-cube. The reason is that every pair of binary node numbers which are exactly 1 bit different, are also 1 digit different in m-ary representation.

In order to show the embedding capability of base-m n-cube, some examples of topologies will be given, which base-8 3-cube(512 nodes, 21 neighbors) can embed with dilation 1.

* ring   (up to 512 nodes)
* binary tree   (up to 256 nodes)
* 2D mesh and torus   (up to 512 nodes)
* 3D mesh and torus   (up to 512 nodes)
* k-ary n-cube   (up to 512 nodes)
* binary k-cube   (up to 512 nodes)
* cube connected cycles   (up to 64 nodes)

E.Fault tolerance and reliability

Similar to the binary k-cube, there are many paths to connect two nodes in base-m n-cube. If these paths are selected randomly in the system by wormhole routing, deadlock may occur. However, it is possible to avoid deadlock by storing messages few times on wormhole routing.

From the previous section, it is easy to see that the number of neighboring nodes in the base-m n-cube is n*(m-1). This implies that up to n*(m-1)-1 nodes can fail without disrupting the network. Because the number of neighboring nodes in the base-m n-cube is larger than in the binary k-cube, base-m n-cube has less probability that the operational nodes cannot communicate due to a failure of neighboring nodes than the binary k-cube.

The number of total links between chips is related to reliability. Here we assume that each router and each crossbar switch is integrated in a chip. The total number of links between chips for binary k-cube and base-m n-cube are shown as follows.

$$\text{binary k-cube : R2} = k * 2^{k-1}$$
$$\text{base-m n-cube : Rm} = n * m^n = n * 2^k$$
$$R2/Rm = k / (2 * n) = n * \log_2 m / (2 * n) = \log_2 m / 2$$

When m > 4, the base-m n-cube has fewer links between chips than the binary k-cube. High dimensional crossbar and low dimensional router make the base-m n-cube more reliable, because the reliability of intra-chip communication is usually higher than inter-chip, due to noise.

The reliability of communication between two nodes of base-m n-cube is also higher than that of binary k-cube, because of the short diameter and the average distance.

4.A highly parallel base-m n-cube machine : PRODIGY[15]

The authors have developed a prototype base-m n-cube machine, named PRODIGY. One purpose of the PRODIGY development is to measure the actual performance of new interconnection network base-m n-cube. Results of several experiments, using PRODIGY, are given in the next section. In this section, an overview of PRODIGY and some lessons from the implementation are shown.

A.System overview

PRODIGY has 512 processing elements(PEs) connected by base-8 3-cube. The base-8 3-cube for PRODIGY is constructed with two kinds of LSIs(router LSI, crossbar LSI).

Figure 8 shows base-8 3-cube. In Fig.8, a circle represents a PE with a router LSI, and a rectangle represents an 8 * 8 crossbar switch. The number in the circle shows node identifier with 8-ary representation.

Tables 1 and 2 give specifications for these LSIs. These LSIs have full-duplex 8 bit parallel communication channels with hand-shake control. The communication speed for each channel is 5Mbyte/s(10MHz) in each direction.

Figure 9 shows the PE organization. Each PE consists of a 16 bit microprocessor(SCC68070,10MHz), 2 Mbyte DRAM and a router LSI. 8 PEs are connected via a crossbar LSI on a board. 8 boards are connected with 8 crossbar LSIs through backplane. 8 backplanes are connected with 64 crossbar LSIs through cables. Figure 10 shows the PRODIGY with 128 PEs in operation. Because 6 ports of each 8 * 8 crossbar switch for inter-backplane connections are remained for the extension, it is easy to extend up to 512 PEs.

input ports	4
output ports	4
data width(for link)	8 bits/port
data width(for memory)	16 bits
communication speed	5 MB/s/port
routing algorithm	wormhole
frequency	10 MHz
device technology	CMOS gate array
package	179 pin PGA
number of gate	17K gates

Table.1   Specifications of router LSI

input ports	8
output ports	8
data width(for link)	8 bits/port
swiching elements	8 × 8
communication speed	5 MB/s/port
broadcasting	available
frequency	10 MHz
device technology	CMOS gate array
package	223 pin PGA
number of gate	15K gates

Table.2   Specifications of crossbar switch LSI

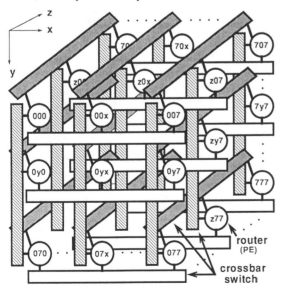

Fig.8 Base-8 3-cube

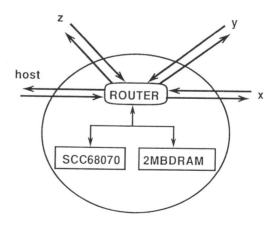

Fig.9 PE organization

Fig.10 PRODIGY in operation

B.Lessons from the implementation

There are some changes in thinking through the implementation of PRODIGY. Some of them have been applied to the discussion in Section 3 in this paper. Some of them about the hardware amount are described as follows.

1) The number of gate is not so big a problem for the integration of the router and the 8 * 8 crossbar switch.

2) Because the router and the 8 * 8 crossbar switch can be integrated in LSI, there are not so many of these chips in the total system. The majority of chips are line drivers and RAMs. A large amount of power is dissipated in lines for inter-board connections.

3) The LSI pin bottleneck is not as serious a problem as that of inter-board connection. The most critical point on highly parallel interconnection networks with parallel channels is inter-backplane connection.

4) If channel data width is 8bits on base-8 n-cubes, n = 2 is very easy to implement. However, the larger n is, the more difficult it is to connect between backplanes.

5.Network performance evaluation

The properties of base-m n-cube have already been shown in Section 3. However the base-m n-cube performance is not clear merely from those discussions. Some simulations[14] have been made to evaluate the base-m n-cube performance. In this section, not only the evaluation by simulations, but also the actual performance of base-m n-cube and binary n-cube on PRODIGY is shown.

A.Communication speed with no contention on PRODIGY

The maximum communication speed per channel of PRODIGY is 5Mbytes/sec. Figure 11 shows the maximum communication speed for highly parallel hypercube machines[3]-[7]. PRODIGY has the fastest communication channels in these machines.

From Fig.11, it can be seen that the communication channel performance on PRODIGY is much better than the commercial binary n-cube machines. The reason is that PRODIGY is implemented with 8 bit parallel channels for

base-m n-cube.

Moreover, PRODIGY retains the possibility to improve communication speed by using higher clock and lower overhead protocol.

Figure 12 shows the actual communication speed with no contention on PRODIGY. There are several kinds of software overhead for setting DMA controller, and for managing message buffer. These are independent from message length. Because of the overhead, the shorter the message length is, the slower the communication speed is.

It takes 77 micro-seconds for 2 bytes message transmission on PRODIGY. This time is shorter than times for iPSC/2(900 µs)[13] or NCUBE(384 µs)[13].

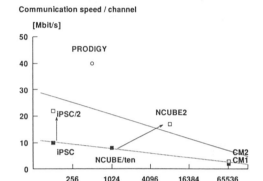

Fig.11 Maximam communication speeds
for highly parallel hypercube machines

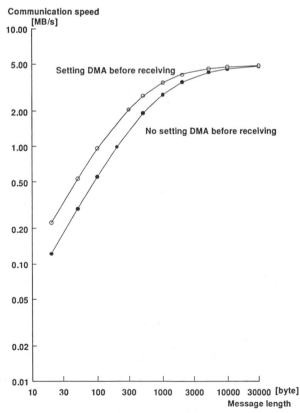

Fig.12 Communication speed of PRODIGY(no contention)

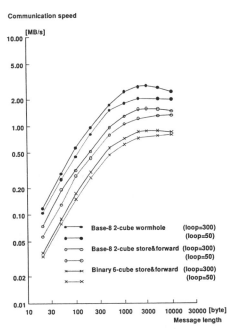

Fig.13 Random communication speed of PRODIGY
(message num. = 64)

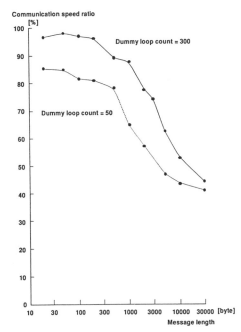

Fig.14 Effect of contention on communication speed of PRODIGY

B.Communication speed with contention on PRODIGY

In order to evaluate the interconnection network performance, it is necessary to discuss not only static performance but also dynamic performance. The dynamic performance closely depends on the communication pattern for each application. Here, dynamic performance of random communication patterns is discussed for general problems.

The present PRODIGY has only 128 PEs connected with incomplete base-8 3-cube. Therefore, the actual communication speed with contention for base-8 2-cube can be measured instead of base-8 3-cube, on PRODIGY.

The upper two lines in Fig.13 show the actual communication speed with contention for base-8 2-cube on PRODIGY, using WH routing.

These results were measured in the following manner. Each PE has one message to be sent initially. A PE having some message sends to a randomly selected PE after executing dummy loop. In this experiment, the CPU load is varied by the dummy loop count. Each PE measures the time required for communication using a timer and counts up the message sending instances.

Peak speed(2.8MB/s, loop=300) can be attained, when message length is 3Kbytes. This point has little influence from software overhead, as shown in Fig.11.

Figure 14 shows the communication speed ratio between Fig. 12(no contention) and Fig.13(with contention). It can be seen that over 90% performance is attained when the message length is shorter than 1KB, as shown in Fig.14. It can also be seen that the network bandwidth tends to hold at 40%, even in the worst-case situation.

C.Wormhole routing vs. store & forward routing

As PRODIGY can accomplish WH routing, S&F routing can also be handled by adding a few bytes of routing information at the top of the message. By using this ability,

the WH routing performance is compared with that for S&F routing on PRODIGY.

The middle two lines in Fig.13 show the actual communication speed with contention for base-8 2-cube on PRODIGY, using S&F routing.

The tendency shown in the graph for S&F routing is almost the same as that for WH routing. The WH routing peak speed is 1.8 times faster than that of S&F routing. As the diameter of base-8 2-cube is only 2, the WH routing effect is not so large.

We can not perform experiment for larger network on PRODIGY. So, simulation was carried out to determine the network size effect tendency.

Figure 15 shows the network size effect on the processing throughput (number of messages generated in a unit time) measured by simulation. The simulation model is similar to that used in the experiments on PRODIGY. The difference between WH routing and S&F routing increases as the networks size becomes large.

D.Base-m n-cube vs. binary n-cube

As already discussed in Section 3, PRODIGY can emulate binary n-cubes. By using this ability, the actual base-m n-cube performance is compared with that for binary n-cube on PRODIGY.

The lower two lines in Fig.13 show the actual communication speed with contention for binary 6-cube on PRODIGY using S&F routing.

These results are measured in almost the same way as was used in the experiment for base-8 2-cube S&F routing. The difference in these experiments is the routing information contents added to the top of the message at source PE.

The tendency shown in the graph for binary 6-cube S&F routing is almost the same as that for base-8 2-cube. The peak speed for base-8 2-cube S&F routing is 1.8 times faster than that for binary 6-cube S&F routing. As the average distance of binary 6-cube is only 3, the effect of distance is not so marked.

In Fig.15, the throughput is normalized by that for binary k-cube. The result shows that the difference between base-m n-cube S&F routing and binary n-cube S&F routing slightly increases as the network size grows.

We can not perform the experiment for binary n-cube WH routing on PRODIGY, because the router for the binary k-cube WH routing requires more bypass paths for routing than that for the base-m n-cube. Simulation was accomplished to determine the difference in performance between binary k-cube WH routing and base-m n-cube WH routing.

The upper two lines in Fig.15 show that binary k-cube WH routing is always better than base-8 n-cube. However the difference in performance is very small, if the channel bit widths are the same.

It has already been discussed in section 3 that a highly parallel binary k-cube with parallel channels is very difficult to construct. However, a highly parallel base-m n-cube with parallel channels is not so difficult to construct. Therefore, higher performance can be actually attained by the base-m n-cube than by the binary k-cube.

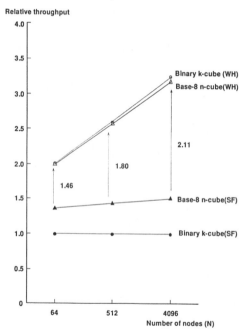

Fig.15 Size effect by size on processing throughput

(by simulation, message length=100, message num.=N, processing time=300, communication pattern=random)

6.Conclusion

This paper has proposed new interconnection networks, base-m n-cubes, and compared their basic properties with those for the binary k-cubes. The base-m n-cube interconnection networks contain binary k-cubes. In addition, the base-m n-cube has been shown to be superior to the binary k-cubes in regard to many properties. Shorter distance, shorter latency, lower logic complexity, wiring ease, larger number of neighbors, better embedding, better fault tolerance, and better reliability make this new structure very promising.

In addition, the authors have implemented a prototype base-8 3-cube machine, named PRODIGY, which has the fastest communication channels in hypercube machines without high clock rate.

The actual random communication performance for three routing algorithms was measured on PRODIGY. By using base-8 2-cube wormhole routing, the performance drop by contention is under 10%, in the case that the message length is shorter than 1KB. This is one evidence that PRODIGY can be used with small attention paid to the network topology. From our observation, base-8 2-cube wormhole routing is 1.8 times faster than base-8 2-cube store & forward routing, and base-8 2-cube store & forward routing is 1.8 times faster than binary 6-cube store & forward routing.

By simulation, the authors have shown that these differences become larger as the network size grows, and base-8 n-cube wormhole routing is as fast as binary k-cube wormhole routing. Because a highly parallel base-m n-cube can be implemented with parallel channels, like PRODIGY, results indicate that much higher performance can be actually attained by base-m n-cube than by binary k-cube.

References

[1] C.L.Seitz,"The Cosmic Cube", CACM, Vol.28, pp.22-23, 1985

[2] S.Abraham and K.Padmanabhan,"Performance of Direct Binary n-Cube Network for Multiprocessors", IEEE Trans. on Computers, Vol.38, No.7, pp.1000-1011, 1989

[3] W.D.Hillis,"The Connection Machine", The MIT Press, 1985

[4] L.W.Tucker and G.G.Robertson,"Architecture and Application of Connection Machine", IEEE Computer, Vol.21, No.8, pp.29-38, 1988

[5] J.P.Hayes, T.Mudge, Q.F.Stout,"A Microprocessor-based Hypercube Supercomputer", IEEE MICRO, 1986

[6] NCUBE Corp.,"NCUBE 2 6400 Series Supercomputer Technical Overview", 1989

[7] Intel Scientific Computers,"The INTEL iPSC/2 System product-information", Intel Scientific Computers, 1987

[8] L.N.Bhuyan and D.P.Agrawal,"A General Class of Processor Interconnection Strategies", Proc. 9th Ann. Symp. Computer Architecture, pp.90-98, 1982

[9] L.N.Bhuyan and D.P.Agrawal,"Generized Hypercube and Hyperbus Structures for a Computer Network", IEEE Trans. on Computers, Vol.33, No.4, pp.323-333, 1984

[10] D.P.Agrawal, V.K.Janakiram, G.C.Pathak,"Evaluating the Performance of Multicomputer Configurations", IEEE Computer, Vol.19, No.5, pp.23-37, 1986

[11] W.J.Dally,"Performance Analysis of k-ary n-cube Interconnection Networks", IEEE Trans. on Computers, Vol.38, No.7, pp.1000-1011, 1990

[12] W.J.Dally,"A VLSI Architecture for Concurrent Data Structures", Kluwer Academic Publishers, 1987

[13] M.Annaratone, C.Pommerell and R.Ruhl, "Interprocessor Communication Speed and Performance in Distributed-memory Parallel Processors", Proc. 16th Ann. Symp. Computer architecture, pp.315-324, 1989

[14] T.Suzuoka, N.Tanabe, S.Nakamura, S.Fujita, S.Oyanagi, "Network Performance of the Prodigy Parallel AI Machine", Systems and Computers in Japan, Vol.20, No.9, 1989. Translated from Trans. IEICE Vol.J71-D, No.8, pp.1496-1501, 1988

[15] N.Tanabe, T.Suzuoka, S.Nakamura, S.Oyanagi, "Hardware of Prodigy Parallel AI Machine", IEICE SIG. CSPY89-51, pp.39-44, (in Japanese), 1989

FAULT-TOLERANT RESOURCE PLACEMENT
IN HYPERCUBE COMPUTERS

Hsing-Lung Chen

Department of Electronic Engineering
National Taiwan Institute of Technology
Taipei, Taiwan, R. O. C.

Nian-Feng Tzeng

Center for Advanced Computer Studies
University of Southwestern Louisiana
Lafayette, LA 70504

Abstract —— Multiple copies of a certain resource often exist in the hypercube computer to reduce the potential delay in accessing the resource. After one resource copy becomes unavailable due to a fault, potential performance degradation is kept minimal if the resource copies are placed in a way that access to the resource is guaranteed to take the same number of hops from every node even in the presence of faults. Such a resource placement achieves *fault tolerance* and is of our concern.

We first consider placing multiple copies of a certain resource such that every cube node without the resource is adjacent to a specified number of resource copies. The use of our developed perfect and quasi-perfect codes makes it possible to arrive at solutions to this placement problem in a simple and systematic manner for an arbitrary hypercube computer. We then deal with the generalized resource placement in the hypercube such that every node can reach no less than a specified number of resource copies in no more than certain hops, using as few resource copies as possible. A fault-tolerant placement we get also leads to reduced potential contention in accessing a shared resource copy, particularly suitable for large-scale hypercubes.

1. Introduction

A certain resource in the hypercube computer might have to be shared by processing nodes because (1) it is too expensive to equip every node with a dedicated copy of the resource, or (2) it might be unnecessary to install at every node a copy of the resource which is not used very frequently. Resources range from hardware devices (like I/O processors, disks, special-purpose units, etc.) to software modules (like library routines, compilers, data files, etc.). The access of a shared resource tends to involve delay attributed by possible contention with other access requests and by communication latency from a node without such a resource to reach a node with the resource. The first delay component (due to contention) can be reduced by lowering the number of nodes which share a copy, whereas the second component is kept small by minimizing the mean number of hops taken to arrive at a node with the resource.

It is often that multiple copies of each type of resource exist in a hypercube to reduce the delay of shared resource access, thereby improving system performance. The use of multiple copies also enhances reliability because the loss of one copy would not cause one type of resource totally unavailable. Distributing resource copies in a hypercube with an attempt to optim-

ize system performance measures of interest has been investigated recently [1], [2], [3]. Livingston and Stout have studied [1] the minimum number of resource copies needed to meet certain specified requirements when distributing the resource in a hypercube computer. A method has been proposed in [2] for mapping the I/O processors onto a hypercube such that each cube node is adjacent to at least one I/O processor. More recently, efficient algorithms [3] for allocating a given number of resource copies to a hypercube system in an effort to optimize a defined performance measure have been developed.

All the prior articles aimed to optimize specified measures under the situation that every cube node is connected with only one copy of the resource. The lose of one node (and its associated resources) would make several nodes redirect their requests destined for the lost node to other remote nodes. Redirected requests not only take more hops to reach their destination nodes, but also increase contention at those nodes, possibly making the access delay of shared resources significantly larger. As a result, the hypercube after reconfiguration in response to an arising fault could suffer from considerable performance degradation simply because of the increase in access delay. While an extension to the basic method of [2] can yield an embedding with every node adjacent to two I/O processors so that the failure of an I/O processor would not increase the number of hops any node has to travel in order to get access to one I/O processor, the extended method seems applicable only to a few cube sizes and becomes increasingly complicated when the degree of required adjacency grows.

Excessive sharing of one resource copy is likely to degrade overall system performance because of unnecessarily long access delay. In a large-sized system, it might require that every node be in connection with multiple copies of each resource to prevent any copy from becoming congested or a bottleneck, ensuring good performance. In this paper, we consider placing copies of a certain resource in a way that every node is adjacent to a specified number of resource copies, and then that every node can reach a specified number of resource copies in no more than certain hops. Placing a resource this way could avoid a reconfigured system from experiencing significant performance degradation and achieves fault tolerance in the sense that access to the resource can be guaranteed within the same hops from every node even in the presence of faults. Fault-tolerant placement results also reduce the potential access contention of any shared copy and are particularly desired for large-scale hypercubes from the reliability and performance standpoints.

The proposed resource placement strategy is based on our developed linear block codes, which can easily be applied to get a placement with any degree of adjacency in any sized hypercube. In Section 2, necessary notations and useful background are given. Section 3 deals with the resource placement in a way that every cube node without the resource is adjacent to multiple copies of the resource. The generalized resource placement is treated in Section 4.

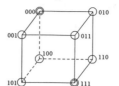

Fig. 1. A 3-dimensional hypercube.

Fig. 3. A 2-adjacency perfect placement in Q_2.

(Nodes with double circles have one resource copy each.)

2. Preliminaries

2.1. Notations and Background

An n—dimensional hypercube computer, denoted by Q_n, consists of 2^n nodes, each addressed by a unique n—bit identification number. A link exists between two nodes of Q_n if and only if their node addresses differ in exactly one bit position. A link is said to be along dimension i if it connects two nodes whose addresses differ in the ith bit (where the least significant bit is referred to as the 0th bit). Q_3 is illustrated in Fig. 1. Two nodes in a hypercube are said to be *adjacent* if there is a link present between them. The *distance* between any two cube nodes is the number of bits differing in their addresses. The number of *hops* needed to reach a node from another node equals the distance between the two nodes. A d—dimensional subcube in Q_n involves 2^d nodes whose addresses belong to a sequence of n symbols $\{0, 1, *\}$ in which exactly d of them are of the symbol $*$ (i.e., the *don't care* symbol whose value can be 0 or 1).

Our resource placement strategy is based on the *linear block code*, which is briefly reviewed subsequently. For more detailed discussion, please refer [4, Ch. 3], [5, Ch. 3], or any text book on coding theory. A (binary) block code of size M over bit symbols $\{0, 1\}$ is a set of M binary sequences of length n called *codewords*. It is often that M equals 2^k for some integer k, and we are interested in this case only, referred to as an (n, k) code. An (n, k) block code, denoted by $\Psi(n, k)$, can be described concisely using a k by n matrix $\mathbf{G}$ called the *generator matrix*. For a linear block code, any codeword is a linear combination (performed by modulo-2 additions over corresponding bits) of the rows of its associated generator matrix $\mathbf{G}$. Consider a simple example of a binary linear code $\Psi(3, 1)$ with the generator matrix:

$$\mathbf{G} = \begin{bmatrix} 1 & 1 & 1 \end{bmatrix}. \tag{1}$$

This linear code involves totally 2^1 codewords (as $k = 1$), which are the possible linear combinations of the row of the generator matrix $\mathbf{G}$ given above, i.e., 000 and 111.

Let W be a subspace, then the set of vectors orthogonal to W is called the *orthogonal complement* of W. If W, a subspace of a vector space of n-tuples, has dimension k (i.e., consisting of k basis vectors), then the orthogonal complement of W has dimension $n-k$ [4, pp. 41]. The subspace formed by code $\Psi(n, k)$ has dimension k, so its orthogonal complement has dimension $n-k$. Whether an n-tuple $\mathbf{c}$ is a codeword of code $\Psi(n, k)$ or not can be checked by using an $(n-k)$ by n matrix $\mathbf{H}$, called a *parity-check matrix* of the code. $\mathbf{c}$ is a codeword if and only if it is orthogonal to every row vector of $\mathbf{H}$, namely, $\mathbf{c} \cdot \mathbf{H}^T = \mathbf{0}$, where $\mathbf{H}^T$ is the transpose of matrix $\mathbf{H}$. Matrix $\mathbf{H}$ can be constructed as follows: every row of $\mathbf{H}$ is a basis vector of the orthogonal complement of $\Psi(n, k)$. For example, a parity-check matrix $\mathbf{H}$ of the binary code $\Psi(3, 1)$ with the generator matrix as Eq. (1) is

$$\mathbf{H} = \begin{bmatrix} 1 & 0 & 1 \\ 0 & 1 & 1 \end{bmatrix}. \tag{2}$$

Notice that the choice of $\mathbf{H}$ is not unique, and any chosen $\mathbf{H}$ satisfies $\mathbf{G} \cdot \mathbf{H}^T = \mathbf{0}$. This equality is essential and is used frequently throughout this paper.

If a parity-check matrix $\mathbf{H}$ for a binary code has r rows, then each column is a r-bit binary number and there are at most $2^r - 1$ independent columns. The "largest" parity-check matrix $\mathbf{H}$ that can be obtained is r by $2^r - 1$. In other words, the columns of this "largest" parity-check matrix $\mathbf{H}$ consist of all the possible nonzero r—tuples (totally, $2^r - 1$ of them). The matrix given in Eq. (2) is such a parity-check matrix. This "largest" parity-check matrix $\mathbf{H}$ is special and, in fact, is the *parity-check matrix of a Hamming code*. The generator matrix of a Hamming code is $2^r - 1 - r$ by $2^r - 1$. The Hamming code is a special linear code $\Psi(n, k)$ such that n equals $2^r - 1$ and k is equal to $2^r - 1 - r$. Like linear code $\Psi(3, 1)$, codes $\Psi(7, 4)$, $\Psi(15, 11)$, and $\Psi(31, 26)$ are examples of the Hamming codes. One possible parity-check matrix $\mathbf{H}$ of the Hamming code $\Psi(7, 4)$ is

$$\mathbf{H} = \begin{bmatrix} 1 & 1 & 0 & 1 & 1 & 0 & 0 \\ 1 & 0 & 1 & 1 & 0 & 1 & 0 \\ 0 & 1 & 1 & 1 & 0 & 0 & 1 \end{bmatrix}. \tag{3}$$

Any linear block code possesses the following basic properties [5, pp. 64].

Property 1: Let $\mathbf{H}$ be a parity-check matrix of a linear code. If no $d-1$ or fewer columns of $\mathbf{H}$ add to $\mathbf{0}$, the code has a minimum weight (i.e., minimum distance) of no less than d.

Property 2: Let $\mathbf{H}$ be a parity-check matrix of a linear code. The minimum weight (i.e., minimum distance) of the code is equal to the smallest number of columns of $\mathbf{H}$ that sum to $\mathbf{0}$.

For the parity-check matrix of Hamming code $\Psi(7, 4)$ given in Eq. (3), as an example, the minimum weight (i.e., distance) of the code is no less than three (necessary to achieve single-error correction) because no two or fewer columns could sum to $\mathbf{0}$, as all the columns of $\mathbf{H}$ are nonzero and no two of them are alike.

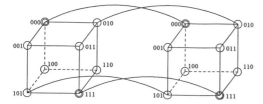

Fig. 2. A possible placement of 2^2 resource copies in Q_4.
(Nodes with double circles have one resource copy each. Some links are omitted intentionally for clarity.)

2.2. Resource Placement Overview

I/O processors allocation

Reddy *et al.* [2] considered allocating I/O processors (the resource) to cube nodes in a way that every node is adjacent to at least one I/O processor. A perfect allocation results if and only if every cube node is adjacent to exactly one I/O processor. A perfect allocation is shown to exist in only some cubes, namely, only for Q_n with $n = 2^r - 1$. Since the Hamming codes are known to be perfect codes [4, 5], a perfect allocation is obtained by placing the I/O processors at Q_n nodes with addresses being the codewords of Hamming code $\Psi(n, k)$. An example perfect allocation for Q_3 is illustrated in Fig. 1, where the two resource copies are placed at the codewords of Hamming code $\Psi(3, 1)$ defined by Eq. (1).

The allocation of I/O processors in Q_n for $n \neq 2^r - 1$ is also considered in [2]. Let n' be the largest value that is less than n and is equal to $2^r - 1$. The allocation for Q_n is obtained by viewing Q_n as a union of $2^{n-n'}$ n'-dimensional subcubes and by arriving at a perfect allocation in each of the subcubes individually. For example, an allocation for Q_4 is derived by finding a perfect allocation for each of the two 3-dimensional subcubes, as depicted in Fig. 2.

Resource allocation

Efficient algorithms for optimal or near-optimal resource allocation in hypercubes have been developed by Chiu and Raghavendra recently [3]. The performance measure of resource allocation to be optimized is the *resource diameter*, defined as the maximum resource distance among all the cube nodes, where the resource distance of a node is the minimum number of hops to a node equipped with a copy of the resource. An allocation with the resource diameter minimized is highly desirable.

The hierarchical allocation strategy proposed in [3] for placing 2^k copies of a resource in Q_n is equivalent to finding an optimal partition of the given system parameters n and k. Specifically, the strategy searches for a partition of n into l components, $n_1, n_2, \cdots, n_l$ (i.e., $n = n_1 + n_2 + \cdots + n_l$), and a partition of k into l components, $k_1, k_2, \cdots, k_l$ (i.e., $k = k_1 + k_2 + \cdots + k_l$), such that the summation of all the resource diameters d_i, $1 \leq i \leq l$, is minimized, where d_i is the resource diameter of the allocation of the ith partitioned component pair (n_i, k_i) according to a perfect code (like the Hamming code and the Golay code) or a basic strategy below.

The basic strategy of placing 2^k resource copies in Q_n is as follows [3]: Q_n is first partitioned into 2^{k-1} subcubes, each with dimension $n-k+1$; and then two copies of the resource are placed in each $(n-k+1)$-dimensional subcube at any two opposite corners, say at locations $(00 \cdots 0)$ and $(11 \cdots 1)$. Figure 2 illustrates a possible placement of 2^2 resource copies in Q_4, following the basic strategy. It is shown [3] that a placement resulted from the basic strategy has the resource diameter of $\left\lceil (n_i - k_i + 1)/2 \right\rceil$.

3. Multiple-Adjacency Resource Placement

All aforementioned resource allocation methods deal mainly with placing multiple copies of a resource in a way that every cube node is guaranteed to have access to one and only one resource copy within a resource diameter. For a large system, it might be desirable to consider the placement of a resource in a way that every node is assured to get access to multiple resource copies (rather than one single copy) within a specified number of hops. Such a resource placement is fault-tolerant in the sense that the resource diameter of the placement remains unchanged even after a fault arises. In the following, we discuss the linear block code-based resource placement strategy which achieves the desired goal.

3.1. Perfect Placement

Definition 1: A hypercube is said to have a $j-adjacency$ *perfect placement* if it is possible to place the resource copies in the hypercube such that (1) each node without the resource is adjacent to exactly j copies of the resource, and (2) no resource copies are placed at neighboring nodes.

Condition (2) above is to ensure that the highest adjacency is achieved for a given number of resource copies. The necessary condition for hypercube Q_n to have a $j-adjacency$ perfect placement is provided next.

Lemma 1: Q_n has a $j-adjacency$ perfect placement only if n is equal to $j \times (2^r - 1)$ for an integer r.

Proof: Each node in Q_n is adjacent to n nodes, one along each dimension. If every node without the resource is adjacent to exactly j copies of the resource and no resource copies are placed at neighboring nodes, then each resource copy is shared, on an average, by (n/j) neighboring nodes. This implies that each copy serves totally $(n/j)+1$ nodes, including the node where the copy resides. Therefore, when a $j-adjacency$ perfect placement exists, $(n/j)+1$ should divide 2^n, the total number of nodes in Q_n, or equivalently, $2^n = i \times (n/j + 1)$ for an integer i. It is readily shown from the preceding equation that i has to be a power of 2, so does $(n/j + 1)$. As a result, $n/j + 1$ equals 2^r for some integer r, yielding $n = j \times (2^r - 1)$.

According to Lemma 1, a 2–adjacency perfect placement exists in Q_n for $n = 2, 6, 14, 30$, and so on. A 2–adjacency perfect placement for Q_2 is shown in Fig. 3. Naturally, one would wonder if a j–adjacency perfect placement always exists in Q_n, for any $n = j \times (2^r - 1)$. In fact, the condition of $n = j \times (2^r - 1)$ is also sufficient for Q_n to have a j–adjacency perfect placement, and a systematic procedure is obtained next to arrive at such a placement.

A linear block code satisfying the property that each non-codeword is adjacent to exactly j codewords serves as the basis of our placement procedure. The codewords of this code specify the locations where the resource copies should be placed. Let this block code be referred to as a j–*adjacency perfect code*, then the parity-check matrix $\mathbf{H}$ of the j–adjacency perfect code for Q_n with $n = j \times (2^r - 1)$ is composed of j cascade copies of all the possible nonzero r–tuples. Since the parity-check matrix of Hamming code $\Psi(2^r - 1, 2^r - 1 - r)$ comprises (one copy of) all the possible nonzero r–tuples (totally, $2^r - 1$ columns, as described in Section 2), the parity-check matrix $\mathbf{H}$ of the j–adjacency perfect code can be represented as

$$\mathbf{H} = \begin{bmatrix} \mathbf{H}_a & \mathbf{H}_a & \cdots & \mathbf{H}_a \end{bmatrix}, \quad (4)$$

where $\mathbf{H}_a$ is the parity-check matrix of Hamming code $\Psi(2^r - 1, 2^r - 1 - r)$. With $\mathbf{H}$ obtained, we can derive the generator matrix of the desired code, $\mathbf{G}$, using the equality $\mathbf{G} \cdot \mathbf{H}^T = \mathbf{0}$. $\mathbf{H}$ is an r by n matrix, and $\mathbf{G}$ is an $(n - r)$ by n matrix. In fact, the j–adjacency perfect code is the linear block code $\Psi(n, n - r)$.

To give an example, consider the case of a 2–adjacency perfect code for Q_n, $n = 6$. The parity-check matrix of the perfect code, according to Eq. (4), can be put in the form

$$\mathbf{H} = \begin{bmatrix} 1 & 0 & 1 & 1 & 0 & 1 \\ 0 & 1 & 1 & 0 & 1 & 1 \end{bmatrix}, \quad (5)$$

where the first three columns are all the possible nonzero 2–tuples (and constitute the parity-check matrix of Hamming code $\Psi(3, 1)$, see Eq. (2)), and the next three columns are another copy of all nonzero 2–tuples. The generator matrix of the code with parity-check matrix $\mathbf{H}$ is

$$\mathbf{G} = \begin{bmatrix} 1 & 1 & 1 & 0 & 0 & 0 \\ 1 & 0 & 0 & 1 & 0 & 0 \\ 0 & 1 & 0 & 0 & 1 & 0 \\ 1 & 1 & 0 & 0 & 0 & 1 \end{bmatrix},$$

a $(6 - 2)$ by 6 matrix. Since all the codewords of this code are obtained from linear combinations of the four rows of $\mathbf{G}$, we have the locations: 000000, 111000, 100100, 010010, 110001, 011100, 101010, 001001, 110110, 010101, 100011, 001110, 000111, 011011, 101101, and 111111, at which resource copies should be placed, as depicted in Fig. 4. This placement result satisfies that every node without the resource is adjacent to exactly two copies of the resource.

The linear code so constructed remains to be shown that each non-codeword is adjacent to exactly j code-

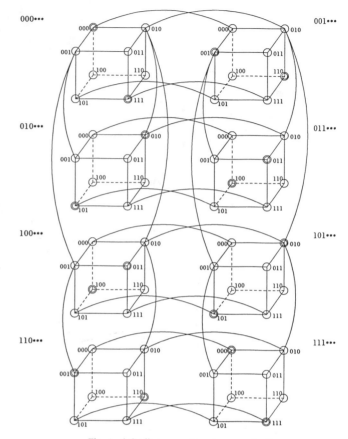

Fig. 4. A 2-adjacency perfect placement in Q_6.

(Nodes with double circles have one resource copy each. Some links are omitted intentionally for clarity.)

words and no two codewords are adjacent. The following lemma reveals this fact.

Lemma 2: For any linear block code with parity-check matrix $\mathbf{H}$, each non-codeword is adjacent to exactly j codewords and no two codewords are adjacent.

Proof: Let $\mathbf{e}_x$ denote an n–tuple with all positions being 0 except the xth position (note that the first position is the leftmost one). For example, $\mathbf{e}_2 = [0\,1\,0\,0\,\cdots\,0\,0]$. If codeword $\mathbf{u}_1$ is adjacent to a non-codeword $\mathbf{v}$, then, there exists $\mathbf{e}_{i_1}$ (where i_1 is the position of the bit differing between $\mathbf{u}_1$ and $\mathbf{v}$) such that $\mathbf{u}_1 + \mathbf{e}_{i_1} = \mathbf{v}$ (recall that the addition is in mod 2). Since $\mathbf{u}_1$ is a codeword, satisfying $\mathbf{u}_1 \cdot \mathbf{H}^T = \mathbf{0}$, we have

$$\mathbf{v} \cdot \mathbf{H}^T = (\mathbf{u}_1 + \mathbf{e}_{i_1}) \cdot \mathbf{H}^T = \mathbf{u}_1 \cdot \mathbf{H}^T + \mathbf{e}_{i_1} \cdot \mathbf{H}^T = \mathbf{e}_{i_1} \cdot \mathbf{H}^T.$$

It is easy to show that $\mathbf{e}_{i_1} \cdot \mathbf{H}^T$ is nothing but the i_1^{th} column of $\mathbf{H}$. According to the method for constructing $\mathbf{H}$, exactly j columns in $\mathbf{H}$ are identical. Suppose that the i_1^{th}, i_2^{th}, $\cdots$, and i_j^{th} columns of $\mathbf{H}$ are identical columns, then $\mathbf{e}_{i_1} \cdot \mathbf{H}^T$, $\mathbf{e}_{i_2} \cdot \mathbf{H}^T$, $\cdots$, and $\mathbf{e}_{i_j} \cdot \mathbf{H}^T$, are all equal, since they are nothing but the i_1^{th} column, the i_2^{th} column, $\cdots$, and the i_j^{th} column of $\mathbf{H}$, respectively. Thus, we have

$$\mathbf{e}_{i_1} \cdot \mathbf{H}^T = \mathbf{e}_{i_2} \cdot \mathbf{H}^T = \cdots = \mathbf{e}_{i_j} \cdot \mathbf{H}^T = \mathbf{v} \cdot \mathbf{H}^T.$$

Adding $\mathbf{v} \cdot \mathbf{H}^T$ to each above expression, we get

$$(\mathbf{v} + \mathbf{e}_{i_1}) \cdot \mathbf{H}^T = (\mathbf{v} + \mathbf{e}_{i_2}) \cdot \mathbf{H}^T = \cdots = (\mathbf{v} + \mathbf{e}_{i_j}) \cdot \mathbf{H}^T$$
$$= (\mathbf{v} + \mathbf{v}) \cdot \mathbf{H}^T,$$

which is equal to $\mathbf{0}$, as $(\mathbf{v} + \mathbf{v})$ yields an all-zero n-tuple. From the above equation, it is clear that there are at least j codewords, namely, $\mathbf{v} + \mathbf{e}_{i_1}$, $\mathbf{v} + \mathbf{e}_{i_2}$, $\cdots$, and $\mathbf{v} + \mathbf{e}_{i_j}$, which are adjacent to any non-codeword $\mathbf{v}$.

Let us assume that a non-codeword $\mathbf{v}$ is adjacent to $j+1$ or more codewords. That is, there exist $j+1$ or more codewords $\mathbf{u}_1, \mathbf{u}_2, \cdots, \mathbf{u}_w, w > j$, such that

$$\mathbf{v} = \mathbf{u}_1 + \mathbf{e}_{i_1} = \mathbf{u}_2 + \mathbf{e}_{i_2} = \cdots = \mathbf{u}_w + \mathbf{e}_{i_w}, \quad (6)$$

$$i_1 \neq i_2 \neq \cdots \neq i_w.$$

Since $\mathbf{u}_1, \mathbf{u}_2, \cdots,$ and $\mathbf{u}_w$ are all codewords, they satisfy $\mathbf{u}_1 \cdot \mathbf{H}^T = \mathbf{u}_2 \cdot \mathbf{H}^T = \cdots = \mathbf{u}_w \cdot \mathbf{H}^T = \mathbf{0}$. From the following result

$$(\mathbf{u}_1 + \mathbf{e}_{i_1}) \cdot \mathbf{H}^T = \mathbf{u}_1 \cdot \mathbf{H}^T + \mathbf{e}_{i_1} \cdot \mathbf{H}^T = \mathbf{e}_{i_1} \cdot \mathbf{H}^T$$

and Eq. (6), we have

$$\mathbf{e}_{i_1} \cdot \mathbf{H}^T = \mathbf{e}_{i_2} \cdot \mathbf{H}^T = \cdots = \mathbf{e}_{i_w} \cdot \mathbf{H}^T.$$

This equation implies that there are $w > j$ equivalent columns in $\mathbf{H}$ (as $\mathbf{e}_{i_m} \cdot \mathbf{H}^T$ is the i_m^{th} column of $\mathbf{H}$, for all $1 \leq m \leq w$). According to the construction of $\mathbf{H}$, however, there are only j equivalent columns in $\mathbf{H}$. This contradiction results from our assumption that a non-codeword is adjacent to $j+1$ or more codewords.

The j-adjacency perfect code is a linear block code, and one can show from Properties 1 and 2 given in Section 2 that the minimum distance of a j-adjacency perfect code, $j \geq 2$, constructed from Eq. (4) is exactly 2, as follows. Since all the columns of $\mathbf{H}$ are nonzero, the code has a minimum distance of no less than two (from Property 1). For $j \geq 2$, there are two columns alike in $\mathbf{H}$ and they sum to $\mathbf{0}$, implying that the minimum distance of a j-adjacency perfect code, $j \geq 2$, is exactly 2 (from Property 2). Thus, resource copies are never placed at two neighboring nodes following a j-adjacency perfect code. The Lemma thus follows.

$\Box$

From Lemmas 1 and 2, we immediately arrive at the next theorem.

Theorem 1: A j-adjacency perfect placement exists in Q_n if and only if n is equal to $j \times (2^r - 1)$ for an integer r.

The number of resource copies needed to achieve a j-adjacency perfect placement in Q_n with $n = j \times (2^r - 1)$ is $\xi = 2^{n-r}$ (because the linear code dictating such a placement is $\Psi(n, n-r)$), and ξ cannot be reduced, leading to an optimal result. The ratio of the number of nodes with the resource to the number of nodes without the resource in Q_n equals $1 : (2^r - 1)$.

A perfect placement, when it exists, is optimal, but it is present only for a few cube sizes. For most of the

cases, perfect placements cannot be obtained. Fortunately, we may get a *quasi-perfect* placement systematically for any cube size.

3.2. Quasi-Perfect Placement

Definition 2: A hypercube is said to have a j-*adjacency quasi-perfect placement* if it is possible to place the resource copies in the hypercube such that (1) each node without the resource is adjacent to at least j copies, but no more than $j+1$ copies, of the resource, and (2) no resource copies are placed at neighboring nodes.

Again, a linear block code is to be derived whose codewords specify the locations where the resource copies are placed. The parity-check matrix $\mathbf{H}_q$ of a code for a j-adjacency quasi-perfect placement can be constructed from the parity-check matrix $\mathbf{H}$ of a code for a j-adjacency perfect placement described earlier. Specifically, $\mathbf{H}_q$ of a code for a j-adjacency quasi-perfect placement in Q_n, $n \neq j \times (2^r - 1)$, is derived from $\mathbf{H}$ of a code for a j-adjacency perfect placement in Q_{n_p}, such that n_p is the largest integer less than n and satisfies $n_p = j \times (2^r - 1)$ for an integer r. After $\mathbf{H}$ is constructed according to Eq. (4), we get $\mathbf{H}_q$ as follows:

$$\mathbf{H}_q = \begin{bmatrix} \mathbf{H} & \mathbf{C}_1 & \mathbf{C}_2 & \cdots & \mathbf{C}_{n-n_p} \end{bmatrix}, \quad (7)$$

where $\mathbf{C}_i$, $1 \leq i \leq n - n_p$, is the ith column of $\mathbf{H}$. Because $\mathbf{H}$ is an r by n_p matrix, $\mathbf{H}_q$ is an r by n matrix. The generator matrix $\mathbf{G}_q$ corresponding to $\mathbf{H}_q$ above can be solved from the relationship $\mathbf{G}_q \cdot \mathbf{H}_q^T = \mathbf{0}$, giving rise to an $(n-r)$ by n matrix. The code for a quasi-perfect placement in Q_n with $n \neq j \times (2^r - 1)$ is again $\Psi(n, n-r)$, which contains 2^{n-r} codewords.

As an example, let us derive $\mathbf{H}_q$ for a 2-adjacency quasi-perfect placement in Q_7. In this case, n_p equals 6, and $\mathbf{H}$ for Q_6 is given in Eq. (5), leading to

$$\mathbf{H}_q = \begin{bmatrix} 1 & 0 & 1 & 1 & 0 & 1 & 1 \\ 0 & 1 & 1 & 0 & 1 & 1 & 0 \end{bmatrix},$$

where the last column is a repetition of the first column. Its generator matrix is

$$\mathbf{G}_q = \begin{bmatrix} 1 & 1 & 1 & 0 & 0 & 0 & 0 \\ 1 & 0 & 0 & 1 & 0 & 0 & 0 \\ 0 & 1 & 0 & 0 & 1 & 0 & 0 \\ 1 & 1 & 0 & 0 & 0 & 1 & 0 \\ 1 & 0 & 0 & 0 & 0 & 0 & 1 \end{bmatrix}.$$

This code includes 32 codewords, which determine the locations where resource copies are placed in Q_7. Among those 96 cube nodes without involving the resource, 64 of them are adjacent to 2 resource copies each, and the rest are adjacent to 3 resource copies each. Interestingly, the ratio of the number of nodes with the resource to the number of nodes without the resource equals 1:3, which is the same as the ratio of the 2-adjacency perfect placement in Q_6 discussed above. This is because the code for Q_n is always $\Psi(n, n-r)$, no matter whether n equals $j \times (2^r - 1)$ or not.

It is to be noted that the placement so constructed for Q_7 requires the same number of resource copies (32 copies) as does that obtained by Reddy *et al.* [2] (where the resource is the I/O processor) using a somewhat complicated enumeration method or an overlapping method. The two prior methods, however, are not generally applicable as our systematic technique in two aspects: (1) they can apply to Q_n with n limited only to $2^r - 1$ for an integer r, and (2) they become more complex as the degree of resource adjacency increases. In contrast, our technique can universally be applied to get a placement with any degree of adjacency in any sized hypercube.

The following theorem indicates that a placement in Q_n according to the linear code $\Psi(n, n-r)$ with parity-check matrix given by Eq. (7) meets the j—adjacency quasi-perfect requirement. A proof of this theorem is provided in [6].

Theorem 2: For any linear block code with parity-check matrix $\mathbf{H}_q$, each non-codeword is adjacent to at least j codewords, but no more than $j+1$ codewords, and no two codewords are adjacent.

4. Generalized Resource Placement

This section investigates the generalized resource placement where every node in Q_n can reach at least j copies, $j \geq 2$, of a resource in no more than h hops, $h \geq 1$. A placement of this sort is fault-tolerant in that the resource diameter remains unchanged even after $(j-1)$ faults arise and up to $(j-1)$ resource copies become unavailable. Our goal is to find such a placement that uses as few resource copies as possible. This is accomplished by employing the j—adjacency perfect code developed earlier.

We may envision Q_n as a "bulky hypercube" with dimension α (denoted by BQ_α) such that every node of the "bulky hypercube", called a supernode, is a subcube with dimension $n-\alpha$ and every link between two neighboring supernodes in BQ_α represents $2^{n-\alpha}$ parallel links between the corresponding nodes in the two supernodes. For example, Q_3 shown in Fig. 1 can be envisioned as BQ_1, whose every node is a 2—dimensional subcube and every link represents four parallel links, as depicted in Fig. 5. To obtain a generalized resource placement, we apply the j—adjacency perfect code to BQ_α. For the j—adjacency perfect code to exist, α should be equal to $j \times (2^r - 1)$ for some integer r (from Theorem 1).

A certain number of resource copies are placed inside each BQ_α supernode whose address is a codeword of the chosen j—adjacency perfect code. Let the supernode involving resource copies be denoted by $SN^\supset$, and the supernode without resource copies be denoted by $SN^\neg$. Since perfect code $\Psi(\alpha, \alpha-r)$ is used for placing the resource copies in BQ_α, totally there are $2^{\alpha-r}$ $SN^\supset$'s. The way of placing resource copies is identical in each supernode $SN^\supset$, and different placement techniques require different numbers of resource copies. We choose to place resource copies in every $SN^\supset$ in a hierarchical fashion, as explained below.

The supernode $SN^\supset$ (which is a cube with dimension $n-\alpha$) is viewed as composed of 2^{δ_1} subcubes with dimension $n-\alpha-\delta_1$, and each of the subcubes is a supernode

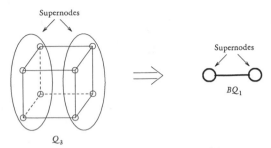

Fig. 5. A "bulky hypercube," in which a supernode is a Q_2.

with respect to $SN^\supset$, called SN_1. Similarly, an SN_1 in turn is considered to be composed of 2^{δ_2} subcubes with dimension $n-\alpha-\delta_1-\delta_2$, and each of the subcubes is a supernode SN_2 with respect to this SN_1, and so on. More specifically, $SN^\supset$ is divided into l levels, $l \geq 1$, by partitioning the number $n-\alpha$ into l positive integers, δ_1, δ_2, $\cdots$, δ_l, such that $n-\alpha = \delta_1 + \delta_2 + \cdots + \delta_l$. At any level y, $1 < y \leq l$, there are 2^{δ_y} supernodes SN_y's within a supernode SN_{y-1}. If each supernode SN_y is treated as an "entity" (i.e., a node), these 2^{δ_y} supernodes form a δ_y—dimensional cube at level y. Resource copies are placed in the δ_y—dimensional cube according to a Hamming code strategy or a basic strategy proposed by Chiu and Raghavendra [3] (explained in Section 2). To allow the use of a Hamming code strategy at level y, the partition of $n-\alpha$ has to satisfy $\delta_y = 2^r - 1$, for an integer r. Figure 6 illustrates an example where $SN^\supset$ is divided into 2 levels, with a Hamming code strategy and a basic strategy followed respectively to place resource copies in the first level and the second level.

The resource diameter is one (meaning that each node can have access to a resource copy in no more than one hop) at the level to which a Hamming code strategy is applied. On the other hand, if a basic strategy is applied to level y, the resource diameter at that level is not fixed and is dependent upon the number of locations at which the resource copies are placed, namely, the resource diameter $= \lfloor (\delta_y - c_y + 1)/2 \rfloor$, where the total number of locations involving resource copies is 2^{c_y}. Let d_y represent the resource diameter at level y, then a partition on $n-\alpha$ is said *feasible* if $d_1 + d_2 + \cdots + d_l$ equals $h-1$. Only the feasible partition is of our concern (as it tends to involve fewer resource copies than other partitions).

In the following, we examine the nodes in $SN^\neg$'s and the nodes in $SN^\supset$'s separately to see if they meet the requirement that each of them gains access to at least j resource copies in no more than h hops, for a feasible partition in the supernode $SN^\supset$. Let us consider a node inside any supernode $SN^\neg$ first. The node can reach j $SN^\supset$'s in exactly one hop (as a j—adjacency perfect code is used). Since a feasible partition is followed in every supernode $SN^\supset$, every node inside a supernode $SN^\supset$ can get access to at least one resource copy in no more than $h-1$ hops (as at least one copy is placed in an $SN^\supset$). As a result, every node inside $SN^\neg$ can reach at least j resource copies in no more than h hops.

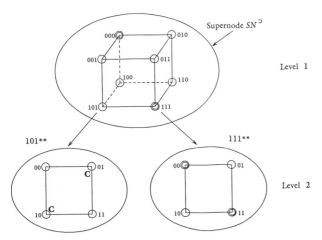

Fig. 6. Following a Hamming code strategy and a basic strategy to place resource copies respectively in level 1 and level 2 of the partitioned $SN^\supset$.

(A node with resource is denoted by a double cycle. A critical node has symbol "**C**" next to it.)

Next, a node inside any supernode $SN^\supset$ is considered. The resource diameter of $SN^\supset$ is $h-1$ (for only the feasible partition of $SN^\supset$ is concerned). There is at least one node $X \in SN^\supset$ such that X takes $h-1$ hops to reach a resource copy inside $SN^\supset$. A node of this kind is referred to as a *critical node*. In Fig. 6, for example, critical nodes are marked and each of them takes two ($=h-1$) hops to reach one resource copy. A critical node X in one supernode $SN^\supset$ cannot reach any resource copy outside this supernode in h hops or less because (1) resource copies are placed in every supernode $SN^\supset$ in the same way, and (2) X takes $h-1$ hops to reach a resource copy inside this supernode, which is not adjacent to any other supernode $SN^\supset$ (recall that supernode $SN^\supset$ addresses are the codewords of a j-adjacency perfect code, in which the distance of two codewords is exactly 2). As a result, to evaluate the number of resource copies reachable from X in no more than h hops, we need to consider the resource copies only *inside* the supernode $SN^\supset$ where X resides. On the other hand, a non-critical node may reach a resource copy inside another supernode $SN^\supset$ in h hops or less. For example, non-critical node (10100) in Fig. 6 can reach the resource copy at node (11100) inside another $SN^\supset$ (whose distance is 2 from the supernode $SN^\supset$ shown in Fig. 6; such a supernode always exists following the property of the j-adjacency perfect code) in 3 ($=h$) hops. It is readily shown that a non-critical node O can get access to at least j resource copies in h hops or less, as follows. Node O may reach at least one resource copy (say, at node O') inside $SN^\supset$ in no more than $h-2$ hops, and from node O', $j-1$ other nodes with resource (i.e., the corresponding nodes in the $j-1$ $SN^\supset$'s which are adjacent to the supernode $SN^\supset$ where nodes O and O' reside) can be reached in two hops (as a result of the j-adjacency perfect code). Therefore, O can have access to at least $1+(j-1)$ resource copies in no more than $(h-2)+2$ hops. Non-critical nodes thus meet our placement requirement. What remains to be considered is whether critical nodes

satisfy the placement requirement. The number of resource copies reachable from a critical node in no more than h hops will be derived subsequently. To meet our placement goal, we search for the feasible partition which requires a minimum number of resource copies to satisfy that a critical node reaches no less than j resource copies in no more than h hops. Such a partition could lead to a desirable placement.

As mentioned earlier, a partition produces δ_y-dimensional cubes at level y. Suppose that each Q_{δ_y} has 2^{c_y} locations placed with the resource following either a Hamming code or a basic strategy, giving rise to the resource diameter of d_y. A critical node $X \in$ supernode $SN^\supset$ takes $d_1, d_2, \cdots,$ and d_l hops (which amount to $h-1$ hops) respectively in level 1, level 2, $\cdots,$ and level l to reach a resource copy. Let $SN_y(X)$ denote the node SN_y which contains critical node X in Q_{δ_y}. In Fig. 6, for example, $SN_1(X=10101)$ represents node (101) in level 1. Let $f_y(X)$ and $g_y(X)$ be the numbers of such nodes SN_y's $\in Q_{\delta_y}$ that involve resource copies and are reachable from $SN_y(X)$, for a critical node X, in exactly d_y hops and d_y+1 hops, respectively. It is observed that $f_y(X)$ and $g_y(X)$ are identical for any critical node $X \in Q_{\delta_y}$ (because either a Hamming code or a basic strategy is followed in level y). For simplicity, $f_y(X)$ and $g_y(X)$ are denoted respectively by f_y and g_y. The expressions for f_y and g_y are given in the next lemmas, whose proofs can be found in [6]. These results allow us to determine if a feasible partition satisfies our placement goal.

Lemma 3: If a placement in Q_{δ_y} is obtained following a Hamming code strategy, $f_y = 1$ and $g_y = (\delta_y-1)/2$.

Lemma 4: When a basic strategy is followed in Q_{δ_y} to have 2^{c_y} locations placed with the resource, then $f_y = 1$ and $g_y = c_y$, if $\delta_y - c_y + 1$ is odd; otherwise, $f_y = 2$ and $g_y = 2(c_y-1)$.

Let f and g be the numbers of such nodes ($\in SN^\supset$) that involve resource copies and are reachable from any critical node X in exactly $h-1$ ($= \sum_{y=1}^{l} d_y$) hops and in exactly h hops, respectively. The next theorem follows immediately from Lemmas 3 and 4.

Theorem 3: $f = \prod_{y=1}^{l} f_y$ and $g = \sum_{y=1}^{l} g_y \left(\prod_{i=1;\, i \neq y}^{l} f_i \right).$

This theorem reveals that a critical node in $SN^\supset$ can reach exactly f (or g) resource copies in $h-1$ (or h) hops, for a given placement in the supernode $SN^\supset$. In Fig. 6, for instance, $f_1 = 1$, $g_1 = 1$, $f_2 = 2$, and $g_2 = 0$, yielding $f = 2$ and $g = 2$. The two ($=f$) resource copies reachable from a critical node, say (10101), in two hops are those at nodes (11100) and (11111). Likewise, the two ($=g$) resource copies reachable from the same critical node in three hops are those at nodes (00000) and (00011). In order to meet our placement requirement, any feasible partition of $n-\alpha$ which results in $f+g$ less than j is discarded. The desirable placement is obtained by searching possible α's and partitions of $n - \alpha$ for the result that requires the fewest resource copies. Formally, our generalized placement strategy is given below:

(a) For a given set of n, j, and h, select a j—adjacency perfect code to determine the locations of supernodes $SN^\supset$'s.

(b) A supernode $SN^\supset$ is partitioned into several levels, with resource copies placed in each subcube formed at a level following a Hamming code or a basic strategy. Only feasible partitions are considered.

(c) Find the partition using the fewest resource copies.

(d) Repeat (a)-(c) above for every possible j—adjacency perfect code to search for the desirable placement.

Since the perfect codes (i.e., the j—adjacency perfect codes and the Hamming codes) exist only in a few instances for practical values of n and j, the number of possible placements is quite limited and the search is fairly efficient. The desirable placement is not necessarily unique. Consider the case of $n = 16$, $j = 6$, and $h = 3$ as an example. The only available 6—adjacency perfect code is $\Psi(6, 5)$ in this case (as the possible $\alpha = 6(2^r - 1)$, $\alpha \leq n$, is solely 6), and each supernode $SN^\supset$ is a 10—dimensional cube. We have the following possible placements in every $SN^\supset$:

(1) Hamming code $\Psi(3, 1)$ for each subcube at the first level, and Hamming code $\Psi(7, 4)$ for each subcube at the second level;

(2) Hamming code $\Psi(3, 1)$ for each subcube at the first level, and a basic strategy using 2^γ resource copies for each subcube Q_7 at the second level, where γ satisfies $\lfloor (7 - \gamma + 1)/2 \rfloor = 2 - 1$ (since only feasible partitions are of our concern and d_1 is 1).

(3) Hamming code $\Psi(7, 4)$ for each subcube at the first level, and a basic strategy using 2^γ resource copies for each subcube Q_3 at the second level, where γ satisfies $\lfloor (3 - \gamma + 1)/2 \rfloor = 2 - 1$.

(4) a basic strategy using 2^γ resource copies for each Q_{10}, where γ satisfies $\lfloor (10 - \gamma + 1)/2 \rfloor = 2$.

It is clear that placement (1) needs 2^{1+4} resource copies per $SN^\supset$ and yields $f + g = 1 + (1 + 3)$; placement (2) needs $2^{1+\gamma}$ resource copies per $SN^\supset$ with the smallest $\gamma = 5$; placement (3) needs $2^{4+\gamma}$ resource copies per $SN^\supset$ with the smallest $\gamma = 1$; and placement (4) needs 2^γ resource copies per $SN^\supset$ with the smallest $\gamma = 6$. Placement (1) should be discarded, as $f + g = 5 < j$. Placement (2) for $\gamma = 5$ is a desirable placement, as it takes 2^6 resource copies per $SN^\supset$ and leads to $f + g = 1 + (1 + 5) > j$. In this case, placement (3) for $\gamma = 2$ and placement (4) for $\gamma = 6$ are also desirable placements.

In general, a Hamming code strategy utilizes resource copies more efficient than a basic strategy, so a placement with as many levels following the Hamming code strategy as possible tends to require fewer resource copies; this sort of placement is a desirable placement, provided that it yields $f + g$ no less than j. It should be noted that any placement obtained from interchanging two levels of a desirable placement is also a desirable placement. As an instance, placing resource copies in every $SN^\supset$ of the above example as follows is a desirable placement: a basic strategy using 2^5 resource copies for each subcube Q_7 at the first level and Hamming code $\Psi(3, 1)$ for each subcube at the second level, as it is attained by interchanging the two levels of placement (2).

5. Conclusions

We have investigated the placement of resource copies in a hypercube computer such that the resource diameter remains unchanged even after cube nodes fail and their associated resource copies become unavailable, achieving fault-tolerance. The use of j—adjacency perfect codes enables the placement of resource copies in a simple and systematic way that ensures every cube node without the resource to be adjacent to exactly j resource copies. Multiple-adjacency perfect placements exist in hypercubes with a few selected sizes. For other cube sizes where perfect placements are unattainable, j—adjacency quasi-perfect codes are developed to help arrive at the resource placement satisfying that every cube node without the resource is adjacent to at least j, but no more than $j + 1$, resource copies.

The j—adjacency perfect code is applied to a generalized resource placement problem where every cube node gets access to at least j resource copies in no more than a specified number of hops. This generalized placement problem is solved by searching for the most efficient placement that meets our placement requirement. The resource placements we obtained appear interesting and could be useful for a large-scale hypercube computer to enhance its reliability and performance.

References

[1] M. Livingston and Q. F. Stout, "Distributing Resources in Hypercube Computers," *Proc. 3rd Conf. Hypercube Concurrent Computers and Applications,* Jan. 1988, pp. 222-231.

[2] A. L. N. Reddy, P. Banerjee, and S. G. Abraham, "I/O Embedding in Hypercubes," *Proc. 1988 Int'l Conf. Parallel Processing,* Aug. 1988, pp. 331-338.

[3] G.-M. Chiu and C. S. Raghavendra, "Resource Allocation in Hypercube Systems," *Proc. 5th Distributed Memory Computing Conf.,* Apr. 1990, pp. 894-902.

[4] R. E. Blahut, *Theory and Practice of Error Control Codes,* Reading, MA: Addison-Wesley Publishing Co., 1983.

[5] S. Lin and D. J. Costello, *Error Control Coding Fundamentals and Applications,* Englewood Cliffs, NJ: Prentice-Hall, 1983.

[6] N.-F. Tzeng and H.-L. Chen, "Fault-Tolerant Resource Placement in Hypercube Computers," Tech. Rep. *TR 91-8-3,* CACS, Univ. of SW Louisiana, Lafayette, LA 70504.

Multiphase Complete Exchange on a Circuit Switched Hypercube

Shahid H. Bokhari

Department of Electrical Engineering, University of Engineering & Technology,
Lahore, Pakistan

Abstract—For a circuit switched hypercube of dimension d ($n = 2^d$), two algorithms for achieving complete exchange are known. These are (1) the Standard Exchange approach that employs d transmissions of size 2^{d-1} blocks each and is useful for small block sizes, and (2) the Optimal Circuit Switched algorithm that employs $2^d - 1$ transmissions of 1 block each and is best for large block sizes.

A unified multiphase algorithm that includes these two algorithms as special cases is described. The complete exchange on a hypercube of dimension d and block size m is achieved by carrying out k partial exchanges on subcubes of dimension d_i, $\Sigma_{i=1}^{k} d_i = d$ and effective block size $m_i = m 2^{d-d_i}$. When $k = d$ and all $d_i = 1$, this corresponds to algorithm (1) above. For the case of $k = 1$ and $d_1 = d$, this becomes the circuit switched algorithm (2). Changing the subcube dimensions d_i varies the effective block size and permits a compromise between the data permutation and block transmission overhead of (1) and the startup overhead of (2).

For a hypercube of dimension d, the number of possible combinations of subcubes is $p(d)$, the number of partitions of the integer d. This is an exponential but very slowly growing function (e.g. $p(7) = 15, p(10) = 42$) and it is feasible to enumerate over these partitions to discover the best combination for a given message size. The multiphase approach has been analyzed for, and implemented on, the Intel iPSC-860 hypercube.

Introduction

On a distributed memory parallel computer system, the complete exchange communication pattern requires each of n processors to send a different m byte block of data to each of the remaining $n - 1$ processors. This communication pattern arises in many important applications such as matrix transpose, matrix-vector multiply, 2-dimensional FFTs, distributed table look-ups etc. It is also important in its own right since, being equivalent to a complete directed graph, it is the densest communication requirement that can be imposed on an interconnection network. The time required to carry out the complete exchange is an important measure of the power of a distributed memory parallel computer system.

There are two algorithms for complete exchange on circuit switched hypercubes like the Intel iPSC-2, Intel iPSC-

860, and Ncube-2. The first is the Standard Exchange algorithm: for an d-dimensional hypercube, this algorithm uses d transmissions of 2^{d-1} blocks each. On circuit switched machines this algorithm is useful for small block sizes ($< \approx 200$ bytes, on the iPSC-860). The second is the Optimal Circuit Switched algorithm, that uses $2^d - 1$ transmissions of 1 block each: the transmissions are carefully scheduled to avoid link contention.

We describe a unified multiphase algorithm that carries out the complete exchange on a hypercube of dimension d as set of k "partial" exchanges on subcubes of dimensions d_i, $\Sigma_{i=1}^{k} d_i = d$ with effective block size $m_i = m 2^{d-d_i}$. The motivation here is to reduce the time required for the complete exchange by compromising between the data permutation and block transmission overhead of the Standard Exchange algorithm and the startup overhead of the Optimal algorithm.

For a hypercube of dimension d there are $p(d)$ possible generalized algorithms, where $p(d)$ is the number of partitions of the integer d. Although $p(d)$ is an exponential function, it grows very slowly. For example, $p(7) = 15$, $p(10) = 42$, and $p(20) = 672$. It is thus quite feasible to enumerate over all partitions to find the algorithm best suited for a given block size.

For the case where $k = d$ and each $d_i = 1$, the unified algorithm degenerates into the Standard Exchange algorithm. When $k = 1$ and $d_1 = d$, it becomes the Optimal algorithm. The unified algorithm thus includes the two known algorithms as special (although extreme) cases.

Measurements on the Intel iPSC-860 show that the multiphase approach can substantially improve performance for block sizes in the 0–160 byte range. This range corresponds to 0–40 floating point numbers and is commonly encountered in practical numeric applications. While our measurements are for the Intel iPSC-860, our techniques are applicable to all circuit switched hypercubes that use the common 'e-cube' routing strategy. The Intel iPSC-2 and the Ncube-2 are examples of such machines.

Circuit Switched Hypercubes

Circuit-switched communications distinguish the new hypercubes, (Intel iPSC-2, -860, and Ncube-2) from older machines. In these machines, a dedicated path is set up between two processors when communication is desired. Messages then flow through this path without involving intervening processors. The path between source and destination is determined by repeatedly applying the the 'e-cube' routing rule until destination is reached.

Research supported by the National Aeronautics and Space Administration under NASA contract NAS1-18605 while the author was in residence at the Institute for Computer Applications in Science & Engineering, Mail Stop 132C, NASA Langley Research Center, Hampton, VA 23665-5225.

e-cube rule: Starting with the right hand side of the binary label of the source, move to the processor whose label more closely matches the label of the destination.

The user has no control over *how* a message is routed between two processors. The fixed routing algorithm completely determines this path. Because of this, we can encounter *edge* and *node* contention. Edge contention is the sharing of an edge (i.e. a communication link) by two or more paths. Similarly, node contention is the sharing of a node. Measurements on the iPSC-860 [1] reveal that edge contention has a disastrous impact on communication time, while node contention has no measurable effect.

Since circuit switching provides very fast communications, it is generally felt that it eliminates most, if not all, of the inefficiencies caused by communication overhead. In particular, it is a common belief that programmers can ignore the details of the interconnection network, since communication overhead is negligible. This is false: as we shall see in this paper, very careful consideration of the interconnection network is necessary if the full power of the machine is to be utilized.

Algorithms for Complete Exchange

Of the two algorithms for complete exchange that are currently in use, Standard Exchange[5] is well known, while the Optimal Circuit Switched algorithm[7, 9] is a recent development. The former requires only $\log n$ transmissions of $n/2$ blocks each and has better performance for small block sizes. The latter uses $n-1$ transmissions of 1 block each and has better performance for large block sizes. Both algorithms completely avoid edge contention.

The Standard Exchange Algorithm

The Standard Exchange algorithm [5] uses $\log n$ transmissions of size $n/2$ blocks each. All transmissions are along paths of length 1, thus there is no contention. This algorithm incurs overhead because of the shuffling of blocks and because it transmits $\frac{n}{2} \log n$ blocks instead of the optimal number, $n-1$. It is, nevertheless, competitive for small block sizes. This is because there are only $\log n$ transmissions (as opposed to $n-1$ for the algorithm discussed below) and thus the overhead of starting up a message is not incurred as frequently.

```
procedure Standard_Exchange;
    for j = d - 1 downto 0 do
        if (bit j of mynumber = 0) then
            message = blocks n/2 to n - 1
        else
            message = blocks 0 to n/2 - 1;
        send_message_to_processor((mynumber) ⊕ (2^j));
        shuffle blocks;
    end;
end;
```

The Optimal Circuit Switched Algorithm

For complete exchange, each processor must send its ith block to processor i, but is free to schedule its transmissions to avoid edge contention. Of the many possible schedules[2] that completely avoid contention, we use the one developed by Schmiermund and Seidel[7]. This has the property that the entire communication pattern is decomposed into a sequence of pairwise exchanges. This is very useful on the Intel iPSC-2 and iPSC-860 because of certain idiosyncrasies of their communication hardware.

```
procedure Optimal_Circuit_Switched;
    for i = 1 to n - 1 do
        send_block_to_processor((mynumber) ⊕ (i));
    end;
```

Analysis of Run Times

Let us define the following performance parameters of our hypercube. Transmission: τ μsec. per byte, data permutation: ρ μsec. per byte, startup(latency): λ μsec., distance impact: δ μsec. per dimension. The time taken by a message of size m bytes to cross d dimensions is $\lambda + \tau m + \delta d$; the time to shuffle m bytes of data within memory is ρm.

In Standard Exchange, d transmissions of $m2^{d-1}$ bytes each over dimension 1, take $d(\lambda + \tau m2^{d-1} + \delta)$ seconds. There are d shuffles on 2^d blocks of m bytes each, taking $d(\rho m2^d)$ seconds. The total time is thus $d(\lambda + (\tau + 2\rho)m2^{d-1} + \delta)$.

For the Optimal Circuit Switched algorithm there are $2^d - 1$ transmissions of blocks of m bytes. At each step, all pairs of processors are at identical distances from each other. Thus the overall distance impact equals the average path length in a hypercube, which is $d2^{d-1}/(2^d - 1)$ The total time is $(2^d - 1)(\lambda + \tau m + \delta \frac{d2^{d-1}}{2^d - 1})$.

Standard is better than Optimal whenever

$$m < \frac{(2^d - d - 1)\lambda + d(2^{d-1} - 1)\delta}{(d2^{d-1} - 2^d + 1)\tau + d2^d\rho}.$$

For a hypothetical machine of dimension 6 with $\tau = \rho = 1, \lambda = 200$ and $\delta = 20$, Standard Exchange is better for $m < 30$.

The Multiphase Approach

In our multiphase approach the complete exchange is carried out as a set of two or more "partial" exchanges. This permits us to use the Circuit Switched algorithm for block sizes for which it is ordinarily inefficient and provides significant performance gains. In fact our multiphase approach is a unified algorithm that *includes* the Standard Exchange and Circuit Switched algorithms as special cases.

Motivation and Example

Given that the Standard Exchange algorithm is competitive for small message sizes and the Circuit Switched algorithm performs best at large message sizes, it is possible to combine these algorithms to obtain performance better than either. Recall that we have $n = 2^d$ nodes on our hypercube. Ordinary complete exchange is based on the exchange of sets of n blocks per processor. We can envisage a "partial" exchange that is carried out simultaneously on all subcubes of dimension $d_1 < d$ but based on $n = 2^d$ blocks (not 2^{d_1} blocks) per processor. By carefully permuting our data blocks, we can then execute another partial exchange on all subcubes of dimension $d_2 = d - d_1$, again with 2^d blocks and not 2^{d_2} blocks. The end result will be that a complete exchange on the hypercube of dimension d is carried out in two phases, using messages that are longer than the messages that would have been used if a single phase approach had been employed. What we have achieved here is an effective "lengthening" of messages that lets us take advantage of the Circuit Switched algorithm for message sizes for which it is normally unsuited. The price paid is the overhead of data permutation, which is required to align blocks to that they finish up in the correct position. Figure 1 illustrates this approach for a dimension 3 hypercube.

An example. Suppose we have to carry out the complete exchange of block size 24 on our hypothetical 6-d hypercube with $\tau = \rho = 1, \lambda = 200$ and $\delta = 20$. We have seen that Standard Exchange is best on this machine for blocksizes of less than 30 bytes. For 24 bytes the Standard algorithm takes $15144\mu sec$. Let us see what happens if we carry out this exchange in two phases of dimension 2 and 4 respectively. The first phase on dimension 2 subcubes with an effective block size of $24 \times 2^{6-2} = 384$ bytes takes $1832\mu sec$. using the Circuit Switched algorithm. The next exchange on dimension 4 subcubes with effective block size $24 \times 2^{6-4} = 160$ bytes takes $6040\mu sec$., again using the Circuit Switched algorithm.

To this must be added the overhead of shuffling data, which is $\rho m 2^d$ per phase. This totals $3072\mu sec$. The total time for the two phase approach is thus $10944\mu sec$., which is substantially faster than the Standard algorithm.

General Algorithm

A complete exchange on a hypercube of dimension d with $n = 2^d$ processors and block size m is done using a set of partial exchanges $\mathcal{D} = \{d_1, d_2, \cdots, d_k\}$, where each d_i specifies a subcube dimension. Obviously $|\mathcal{D}| = k, 1 \leq k$, and $\Sigma_{i=1}^k d_i = d$. The jth partial exchange is done on the set of subcubes determined by bits $\Sigma_{i=1}^j d_i - d_j$ to $\Sigma_{i=1}^j d_i$ of the hypercube node labels.

In a partial exchange 2^d blocks of size m each are exchanged, regardless of cube dimension. Hence the time required for the ith phase is obtained from (1) or (2) with m replaced by $m2^{d-d_i}$ bytes (the *effective block size*).

```
procedure Multiphase;
{ d:     dimension of the hypercube
  k:     number of phases (subcubes) in partition D
  d_i:   dimension of the ith subcube in partition D
  start: starting bit of subcube label
  stop:  ending bit of subcube label }
```

```
start = d - 1;
for i = 1 to k do {Partial exchange}
    stop = start - d_i + 1;
    compute effective blocksize;
    for j = 1 to (2^{start-stop+1} - 1) do
        send_effective_block_to_processor
            ((mynumber) ⊕ (j2^{stop}));
    shuffle blocks d_i times;
    start = stop - 1;
end;
end;
```

When $k = d$, all d_is are 1 and the outer i loop is executed k times with $start = stop = d - 1, d - 2, \cdots, 1, 0$. The inner j loop is executed only once for each i and Multiphase degenerates into Standard Exchange. When $k = 1$ and thus $d_1 = d$, the outer loop is executed only once. $stop$ always equals 0, j takes on the values $1, 2, \cdots, 2^d - 1$ and Multiphase becomes Optimal Circuit Switched.

Minimizing the Execution Time

Given a hypercube of dimension d, there are many different combinations of subcube dimensions and algorithms that can be used to obtain a multiphase algorithm. The sequence of dimensions is unimportant, as long as the shuffles are carried out correctly. The optimal set can be obtained by enumerating over all the partitions of d. For each partition $\mathcal{D} = \{d_1, d_2, \cdots, d_k\}$ we select the best algorithm at each phase. This procedure is not as expensive as it appears at first sight, since we are enumerating over the partitions of hypercube *dimension* and not *size*. It is a classical result [4] that the number of partitions of an integer d is $p(d) \sim \frac{1}{4\sqrt{3}d} e^{\pi\sqrt{2/3}\sqrt{d}}$. Exact values can easily be calculated. Some values of interest are: $p(5) = 7, p(7) = 15, p(10) = 42, p(15) = 176, p(20) = 627$. Thus for a thousand node hypercube (the largest commercially available in 1990) we need to enumerate only 42 partitions—a trivial number.

Implementation on the iPSC-860

We have implemented the Multiphase algorithm on the Intel iPSC-860 hypercube. Complete details of our implementation may be found in [3].

For efficiency, we use FORCED message types and employ the *Pairwise Synchronized Exchange* technique of [6, 7, 8]. We use FORCED types for "pairwise synchronisation" messages as well as for the actual data transfers. We post all receives for all messages before a global synchronisation.

Measured Performance Characteristics

As discussed above, the time for a message of size m bytes to cross d dimensions is $\lambda + \tau m + \delta d$. When messages of the FORCED type are used and all receives are posted before transmission begins, the values λ and τ are $95.0\mu sec$. and $0.394\mu sec./byte$, respectively. The value of δ is $10.3\mu sec$.

per dimension. The λ for a zero byte message is significantly better, being 82.5μsec. When using these measured parameters to predict the time required by the multiphase algorithm, we must remember that each pairwise exchange is preceded by an exchange of zero byte synchronization messages. Thus we have the *effective* values of $\lambda = 177.5\mu$sec. and $\delta = 20.6\mu$sec./dimension.

The time for global synchronization on a cube of dimension d has been measured at 150$d\mu$sec. The time for data permutation (shuffling) is $\rho = 0.54\mu$sec./byte. This is slower than the time to transmit data because of the overhead of computing the permutation. This occurs because we have implemented our algorithm in C using a compiler that does not take many of the powerful features of the iPSC-860 into account. It should be possible to significantly improve this figure by using assembly language and/or an optimizing compiler. This will change our final measured timings somewhat, but will not affect our overall approach, which is valid even if the cost of permutation is zero.

The time for a partial exchange on a subcube of dimension d_i within a hypercube of dimension d is thus $(2^{d_i-1} - 1) \cdot (177.5 + 0.394m + 20.6\frac{4 \cdot 2^{d_i-1}}{2^{d_i}-1} + 0.54 \cdot 2^d m) + 150d$. When $d_i = d$, the shuffling can be omitted, since d-shuffles of 2^d blocks are equivalent to the identity permutation.

Evaluation of Multiphase Algorithm

Measured timings for the Multiphase algorithm on an Intel iPSC-860 hypercube of dimension 7 are shown in Figure 2.[2] Each combination is denoted by its set of subcubes. In particular, the Standard Exchange algorithm is denoted by $\{1, 1, 1, 1, 1, 1, 1\}$ and the Optimal Circuit Switched Algorithm by $\{7\}$. For dimension 7, the number of combinations are 15. Although we have measured the performance of all combinations, to avoid congested plots we show only those combinations that form the hull of optimality (i.e. only the best combination for every blocksize). The only exception is the Standard Exchange Algorithm ($\{1, 1, \cdots\}$), which is shown for purposes of comparison, even though it is never optimal on the iPSC-860 for dimension 7. Dashed lines on our plots indicate predicted values and solid lines show actual measurements.

As expected, the Optimal Circuit Switched algorithm is always optimal for large enough block size. There are three optimal combinations in all: $\{2, 2, 3\}, \{3, 4\}$ and $\{7\}$. $\{7\}$ (Optimal Circuit Switched) is optimal beyond 160 bytes and $\{2, 2, 3\}$ optimal for 0 to 12 bytes. The combination $\{3, 4\}$ leads to a factor of two improvement over both the Standard Exchange and Optimal Circuit Switched Algorithms at blocks of 40 bytes. There is good agreement between the predicted and observed run times.

Conclusions

Circuit switched machines have only recently made an appearance as commercial products. These machines provide powerful communication mechanisms but, as the results of this paper show, very careful algorithm design is required to optimize performance.

We have addressed the implementation of complete exchange and have described a multiphase algorithm that unifies two previously known algorithms and yields performance better than either over some ranges of message sizes. Similar techniques can be applied to other communication patterns. Since complete exchange is the most demanding pattern, the time taken by our multiphase algorithm is an upper bound on *any* communication requirement.

Acknowledgements— I am indebted to my colleagues at ICASE, NASA-Langley and NASA-Ames for their generous assistance.

References

[1] S. H. Bokhari. Communication overheads on the Intel iPSC-860 hypercube. ICASE Interim Report 10, May 1990.

[2] S. H. Bokhari. Complete exchange on the Intel iPSC-860 hypercube. Technical Report 91-4, ICASE, January 1991.

[3] S. H. Bokhari. Multiphase complete exchange on a circuit switched hypercube. Technical Report 91-5, ICASE, January 1991.

[4] Emil Grosswald. *Topics from the Theory of numbers*. Birkhäuser, Boston, 1984.

[5] S. Lennart Johnsson and Ching-Tien Ho. Matrix transposition on boolean n-cube configured ensemble architectures. *SIAM J. Matrix Anal. Appl.*, 9(3):419–454, July 1988.

[6] Ming-Horng Lee and Steve R. Seidel. Concurrent communication on the Intel iPSC/2. Technical Report CS-TR 9003, Dept. of Computer Science, Michigan Tech. Univ., July 1990.

[7] Thomas Schmiermund and Steve R. Seidel. A communication model for the Intel iPSC/2. Technical Report CS-TR 9002, Dept. of Computer Science, Michigan Tech. Univ., April 1990.

[8] Steve Seidel, Ming-Horng Lee, and Shivi Fotedar. Concurrent bidirectional communication on the Intel iPSC/860 and iPSC/2. In *Proceedings of the Sixth Distributed Memory Computing Conference*, April 1991.

[9] Steve R. Seidel. Circuit switched vs. store-and-forward solutions to symmetric communication problems. In *Proc. 4th. Conf. Hypercube Concurrent Computers and Applications*, volume 1, pages 253–255, 1989.

---

[2]Timings for dimension 5 & 6 are reported in [3].

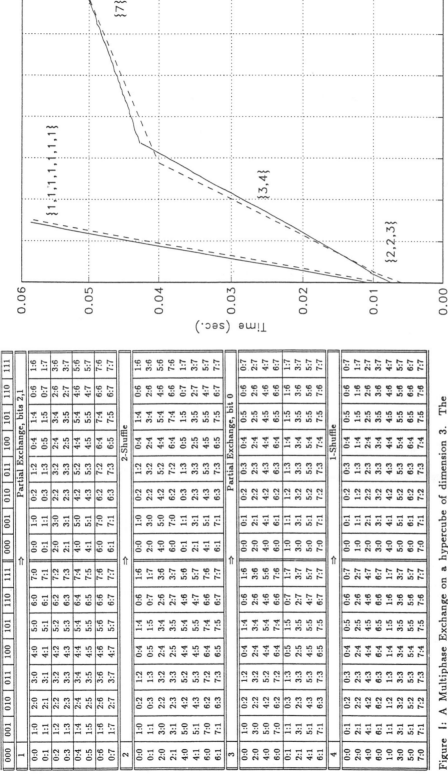

Figure 1: A Multiphase Exchange on a hypercube of dimension 3. The first row gives the binary labels of processors. Data blocks are arranged in columns. The first partial exchange is on the 2 subcubes of dimension 2 determined by bits 2 and 1; data are moved in superblocks of size 2. This is followed by a 2-shuffle. The second partial exchange is on the 4 subcubes of dimension 1 determined by bit 0; data are moved in superblocks of size 4.

Figure 2: Performance of the Multiphase algorithm for a 128 node ($d = 7$) Intel iPSC-860. Three partitions are optimal in this case: {2,2,3}, {3,4} and {7}. For block size 40 bytes, the time taken by the Standard Algorithm {1,1,1,1,1,1,1} equals the time taken by the Optimal Circuit Switched Algorithm {7} and is 0.037 sec. The time taken by the Multiphase algorithm {3,4} is 0.016 sec., which is more than twice as fast.

HIERARCHICAL UNI-DIRECTIONAL HYPERCUBES

Chih-Hsiang Chou
David H.C. Du

Department of Computer Science
University of Minnesota
Minneapolis, Minnesota 55455

Abstract -- We propose in this paper a new class of hierarchical interconnection topologies. Each topology has a uni-directional hypercube at the lower level and the upper level topology varies from a simple ring to trivalent directed graphs. These topologies have fewer degree than the uni-directional hypercubes while retain many of their advantages. Characteristics of twelve different topologies within this class are compared.

Introduction

Optical fiber has been long the choice of medium for high speed, high bandwidth and long range communication. To take its advantages and cope with its inherent properties, many interconnection topologies with uni-directional links have been proposed in the past [2,3,4,6,7,10,11]. Among them, the uni-directional hypercubes (UHC_2s in particular) have shown many desirable characteristics such as short diameter and efficient routing which are most applicable to both hypercube based Metropolitan Area Networks (MANs) and massively parallel systems. Although the degree of each node in a UHC_2 with degree n (denoted by n-UHC_2) and $N = 2^n$ nodes is $\log N$ only, it is still considerably large when compared to other topologies as N increases. We propose in this paper a new class of topologies, the Hierarchical Uni-directional Hypercubes (HUHCs) which not only alleviates the degree growth problem of the UHC_2s, but also retains many of their advantages. In addition, deflection routing [8] with efficient routing decision can also be applied to HUHCs with degree of four to achieve high throughput performance.

Hierachical Uni-Directional Hypercubes

An HUHC with degree n (the n-HUHC) has two levels of topology in hierarchy. At the lower level is an n-UHC_2. We will refer to a node in the n-UHC_2 by its address A where $0 \le A < 2^n$. To construct the HUHC, each node A is decomposed into n nodes first, denoted by (A, i) where $0 \le i < n$, such that the link connected to or from the port i of A is connected to or from (A, i) respectively. Then all the (A, i)'s are interconnected by a low degree upper level topology.

Several properties of an n-HUHC are readily observable:

(1) An n-HUHC has $n 2^n$ nodes.

(2) The degree of each node is only one more than the degree of the upper level topology, independent of n.

(3) Node (A, i) in an n-HUHC is always connected to or from node $(A \text{ xor } 2^i, i)$.

In the following discussion we categorize different HUHCs according to the degree of their upper level topologies.

Upper Level Topologies with Degree Two

The simplest upper level topology of an n-HUHC is a uni-directional ring. We are particularly interested in comparing the following three ring schemes as they are more regular than the others.

(a) Each node (A, i) connects to node $(A, i \oplus 1)$.

(b) Each node (A, i) connects to node $(A, i \ominus 1)$.

(c) (A, i)'s are connected as in (a)/(b) if A is even/odd.

$\oplus$ and $\ominus$ are addition and subtraction modulus n respectively, i.e. $x \oplus y = (x+y) \bmod n$ and $x \ominus y = (n+x-y) \bmod n$. Figures 1 and 2 show a 3-UHC_2 and a 3-HUHC with scheme (a). From the following theorem [3], we can easily derive that schemes (a) and (b) have the same diameter and average distance.

Theorem 1: For each shortest path in an n-HUHC with scheme (a), there is a unique corresponding shortest path in an n-HUHC with scheme (b), and both shortest paths have the same length.

Upper level topologies with degree two other than the rings are possible. For example:

(d) Each node (A, i) connects to its two neighboring nodes $(A, i \oplus 1)$ and $(A, i \ominus 1)$ if $Polarity(A, i)$ [2,3] is negative.

For this scheme to work properly, n must be even. Figure 3 shows a 4-HUHC with this scheme. We note that all the schemes degenerate to a special case of the Cube-connected Cycles (CCC) proposed in [9] if we replace all the links by bi-directional links. In the table 1, we compare the diameter and average distance of the CCC with schemes (a) to (d). The minimum among the four schemes are displayed in boldface.

As we can see, when $n \ge 6$ both schemes (a) and (b) give the best results of diameter and average distance. This is desirable because both schemes are more regular than the others and the size of an MAN usually demands n to be larger than 6. The diameter is $4n-1$ if n is even, or $4n+1$ if n is odd and $n \ge 7$. When $6 \le n \le 16$ the diameter is less than $1+3.2\log N$. On the other hand, the average distances are comparable to the diameters of the corresponding CCCs, which are about $2.5n$ for large n. Scheme (c) has good results when $n < 6$, however the number of nodes it has is usually insufficient for large networks like MANs. The only case where scheme (d) performs better is when $n = 4$. From the discussion above, we conclude that either scheme (a) or (b) is more suitable HUHC upper level topology with degree two.

Upper Level Topologies with Degree Three

More upper level topologies with degree three are conceivable. In general, only n-HUHCs with even n will be regular if the degree of the upper level topology is odd. Therefore we assume n is even in the following discussion of n-HUHC.

Many regular bi-directional topologies with degree three, the so called *trivalent* or *cubic* graphs, have been proposed [1,5,9] with diameters ranging from $1.5\log N$ to $O(\sqrt{N})$. Although much has been studied about the trivalent graphs, little is known about the regular uni-directional topologies with degree three. Furthermore, only the Cube-connected Cycle has a simple and distributed routing algorithm. Nevertheless, there is a simple way to construct upper level topologies with degree three based on the four schemes proposed above. The idea is to connect one extra link between pairs of nodes in the same upper level topology as in the chordal ring [1]. For scheme (a), the most natural connection between a pair would be from node (A,i) to node $(A,i \oplus w)$ where w is an odd integer constant and $Polarity(A,i)$ is negative. Since w is odd, the polarity of $(A,i \oplus w)$ is necessarily positive and every node has two incoming and two outgoing links. Hence the n-HUHC constructed is always regular. Other regular two-connected n-HUHCs can be derived from schemes (b) to (d) in a similar way. The formal definition of each scheme is given as below. For the value of w, we need to consider $w = 1, 3, ..., n-1$ only since $i \oplus (w+n) = i \oplus w$.

(e) As in scheme (a), in addition a link is connected from (A,i) to $(A,i \oplus w)$ if $Polarity(A,i)$ is negative.

(f) As in scheme (b), in addition a link is connected from (A,i) to $(A,i \ominus w)$ if $Polarity(A,i)$ is negative.

(g) (A,i)'s are connected as in (e) if A is even, otherwise connected as in (f).

(h) As in scheme (d), in addition a link is connected from (A,i) to $(A,i \oplus w)$ if $Polarity(A,i)$ is positive.

Form Theorem 1, an n-HUHC with either scheme (e) or (f) and the same w will have the same diameter and average distance. On the other hand, an n-HUHC with scheme (h) and $w = W$ or $w = n-W$ will also have the same diameter and average distance because of symmetry. Table 2 below shows the diameters and average distances under different schemes for different values of n and w. Overall, for any given n, schemes (e) and (f) have the minimal diameter and average distance among all schemes. This again is desirable. The optimal w for schemes (e) and (f) when $n = 4, 6, 8$ and 10 is $3, 3, 3$ and 7 respectively.

Other Two-Connected Topologies Based on the UHC$_2$

In addition to the four schemes above, four other schemes of two-connected topology based on the UHC$_2$ are possible. They are constructed by simply merging the two nodes (A,i) and $(A,i \text{ xor } 1)$ together into one node in the n-HUHCs (n must be even) with schemes (a) to (d) and are named schemes (i) through (l). Figure 4 shows the scheme (i) based on a 4-HUHC with scheme (a). Strictly speaking, none of the schemes from (i) to (l) can be classified as an HUHC. But since they are also hierarchical topologies based on the UHC$_2$ in the general sense, we will denote them as HUHC$_2$s to distinguish them from the previous HUHCs (or HUHC$_1$s

explicitly). By the same convention, an HUHC$_2$ based on an n-UHC$_2$ will be denoted by n-HUHC$_2$. We compare in table 3 the diameters and average distances of the four schemes of an n-HUHC$_2$. We see again that the first two schemes have the shortest diameter ($= 2n-1$) and average distance among all the schemes for any given n.

Discussions

Diameter of Schemes (a) and (b)

We observe that the shortest path from node (A,a) to node (B,b) can be in general decomposed into two parts. The first part consists of links belonged to the upper level topologies and the second part has links from the lower level UHC$_2$ topology. If we concatenate links of each part together, then we have two paths. One traverses in the upper level topology from node a to node b with length $\delta_{ab} = kn + (b \ominus a) + \delta_{AB}$ where k is an integer. The other traverses in the UHC$_2$ from node A to B with length δ_{AB}. Through detail analysis [3], we have the following results of diameters of schemes (a),(b):

(1) When n is even the diameter is $4n-1$.

(2) When n is odd and $n \leq 5$ the diameter is $3n+7$.

(3) When n is odd and $n > 5$ the diameter is $4n+1$.

The low degree, logarithmical diameter and simple distributed routing of schemes (a) and (b) make them suitable for optical fiber based MANs.

Diameter of Schemes (i) and (j)

Similarly, for scheme (i), the length of the shortest path from (A,a) to (B,b) (assume a and b are even) is

$$\frac{kn}{2} + \frac{b \ominus a}{2} + \delta_{AB} + 2c$$

where c is an integer ≥ 0. And the diameter is $2n-1$.

Comparisons of Different Two-Connected Topologies

The two-connected topologies have such advantages as low and regular degree. In addition, deflection routing can be easily implemented on some two-connected topologies which requires less routing decision time and queueing buffers. These together make the two-connected topologies the very promising topologies for the MANs implemented with point-to-point connected optical fibers. However, there are also many characteristics like diameter, average distance, multiple paths, geographical layout, etc. that make each two-connected topology outstanding over others.

We will first compare the diameter and average distance of the MSX, MSN with the proposed schemes (e), (f), (i) and (j) as shown in Table 4. For a given number of nodes N, the diameter and average distance of each topology are chosen from the optimal configuration of that topology. For example, in the case of MSN, the topology has equal number of rows and columns. If there is no optimal configuration with N nodes available, then the diameter and average distance are chosen from the optimal configuration which has the maximal number of nodes less than N. For example, the MSN configuration chosen for $N = 2048$ has 44 rows, 46 columns and 2024 nodes. Diameter and average distance of the approximated configuration are denoted in italics in the table.

Next, for the same configurations chosen above, we compare the multiple paths characteristic among them. The comparison is judged in terms of the average number of shortest path, α. First, a distance d ($1 \le d \le$ diameter) is given. Then for each pair of nodes with distance d, we find the number of shortest paths (each has length d) between them. Finally, α is evaluated as the total number of all such shortest paths divided by the total number of such pairs. The results of α vs. d when $N = 1024$ is shown in figure 5.

As we can see from the table, for any given number of nodes N, the MSX always has the shortest diameter and average distance among all two-connected topologies. Its diameter is equal to $\log N$ only and the average distance is even less than that. However, it lacks multiple paths between any two nodes. The α of MSX is always 1.00 regardless of the value of d, which means on the average there is only one shortest path between any two nodes. For this reason, the reliability and average throughput of MSX are the lowest among all. Also, as mentioned before, the MSX doesn't have a feasible mapping from its topology to the geographical layout.

The diameter of MSN grows in $\sqrt{N}$ order which is reasonably small when N is less than 100. Beyond that, however, it grows rapidly and is the highest among all. The average distance of MSN also grows rapidly in the same order, and is about half of the diameter plus one. MSN's long diameter and average distance shortcomings are compensated by the deflection routing and its adequate number of multiple paths between nodes with short distance. For nodes with long distance, for example $d > 12$ in figure 5, the MSN has less multiple paths (hence smaller α) than both schemes (i) and (j) have. Note that the latter two have shorter diameter than the MSN, so there is no cases of d greater than their diameter to compare.

The schemes (i) and (j) of an n-HUHC$_2$ have a diameter of $2n-1$ where $N = n 2^{n-1}$. Since $\log N = \log n + n - 1$, the diameter is less than $2\log N$, but still grows in $2\log N$ order. Moreover, their average distance is about $\log N$ only, therefore is quite comparable to the best result of MSX when N is large. On the other hand, HUHC$_2$ also has the highest α among all for large d and N. In theses cases, it has more flexible routing decision to make and can achieve higher throughput than MSX and MSN. The main disadvantage of the HUHC$_2$ is their limited number of exact configuration of N nodes. To overcome this, we suggest using schemes (e) and (f) of the HUHC which have N in between two of the exact configurations of the HUHC$_2$. Their diameter and average distance also grow in the $2\log N$ and $\log N$ order respectively, only with a higher coefficient than the HUHC$_2$.

Conclusion

Twelve hierarchical uni-directional hypercube topologies have been proposed in this paper. They share the same advantages as hypercubes like short diameter, short average distance and distributed routing, and are applicable to metropolitan area networks using optical fibers as transmission media. In particular, a class of two-connected topologies are compared with MSX and MSN which have comparable diameter and average distance with respect to the former and the same deflection routing with respect to the latter. The network expansion problem inherent in the hypercube topologies is solved by adding nodes to the upper level topology as well.

References

[1] B. Arden and H. Lee, "Analysis of Chordal Ring Network," *Proceedings of the Workshop on Interconnection Networks for Parallel and Distributed Processing*, (4, 1980), pp. 93-100

[2] C.H. Chou and D. H.C. Du, "Uni-Directional Hypercubes," *Supercomputing '90*, (11, 1990), pp. 254-263

[3] C.H. Chou, *Uni-Directional Hypercubes*, Ph.D. Thesis, University of Minnesota, (11, 1990)

[4] "Proposed Standard Distributed Queue Dual Bus (DQDB) Metropolitan Area Network (MAN)," *IEEE*, P802.6/D6, (11, 1988)

[5] W. Leland and M. Solomon, "Dense Trivalent Graphs for Processor Interconnection," *IEEE Transactions on Computers*, (3, 1982), pp. 219-222

[6] I. Limb and C. Flores, "Description of Fasnet - A Unidirectional Local Area Communication Network," *AT&T Tech. J.*, (9, 1982), pp. 1413-1440

[7] N. Maxemchuk, "Regular and Mesh Topologies in Local and Metropolitan Area Networks," *AT&T Tech. J.*, (9, 1985), pp. 1659-1686

[8] N. Maxemchuk, "Comparison of Deflection and Store-and-Forward Techniques in the Manhattan Street and Shuffle-Exchange Networks," *IEEE INFOCOM*, (3, 1989), pp. 800-809

[9] F.P. Preparata and J. Vuillemin, "The Cube-connected Cycles: A Versatile Network for Parallel Computation," *CACM*, (5, 1981), pp. 300-309

[10] F. Ross, "An Overview of FDDI: The Fiber Distributed Data Interface," *IEEE Journal on Selected Areas in Communications*, (9, 1989)

[11] F. Tobagi, F. Borgonovo, and L. Fratta, "Expressnet: A High-Performance Integrated-Services Local Area Network," *IEEE Journal on Selected Areas in Communications*, (11, 1983)

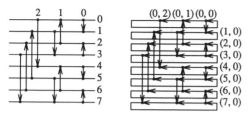

Figure 1. 3-UHC$_2$ Figure 2. Scheme (a) of 3-HUHC

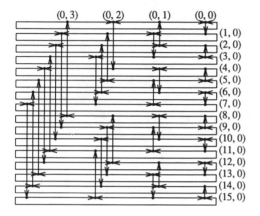

Figure 3. Scheme (d) of 4-HUHC

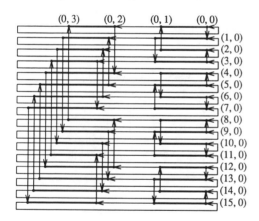

Figure 4. Scheme (i) of 4–HUHC$_2$

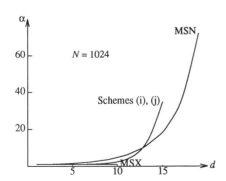

Figure 5. Comparisons of α when $N = 1024$

Table 1. Comparisons between CCC and HUHCs with different schemes

n	$N=n2^n$	CCC		scheme (a),(b)		scheme (c)		scheme (d)	
		dia	avg	dia	avg	dia	avg	dia	avg
2	8	4	2.29	**7**	**3.57**	**7**	**3.57**	7	4.00
3	24	6	3.22	16	6.35	**14**	**6.17**		
4	64	8	4.70	15	8.03	15	8.13	**13**	**7.43**
5	160	10	5.99	22	11.04	**20**	**11.01**		
6	384	13	7.56	**23**	**13.25**	25	13.50	25	13.84
7	896	15	8.99	**29**	**16.45**	30	16.51		
8	2048	18	10.60	**31**	**18.84**	35	19.28	33	20.66
9	4608	20	12.10	**37**	**22.19**	40	22.35		
10	10240	23	13.73	**39**	**24.66**	45	25.31	41	28.07

Table 2. Comparisons among schemes (e) to (h) with different n and w

n	$N=n2^n$	w	scheme (e),(f)		scheme (g)		scheme (h)	
			dia	avg	dia	avg	dia	avg
2	8	1	7	3.57	7	3.57	7	3.57
4	64	1	15	8.03	15	8.13	**11**	**6.11**
		3	**11**	**6.11**	11	6.11	11	6.11
6	384	1	23	13.25	25	13.50	17	9.85
		3	**15**	**9.25**	15	9.37	17	9.62
		5	17	9.85	17	9.86	17	9.85
8	2048	1	31	18.84	35	19.28	22	13.97
		3	**19**	**12.36**	21	12.56	21	13.18
		5	23	13.15	23	13.30	21	13.18
		7	22	13.97	23	14.09	22	13.97
10	10240	1	39	24.66	45	25.31	27	18.23
		3	24	15.82	27	16.09	27	16.82
		5	24	15.72	27	16.11	27	16.47
		7	24	**15.44**	27	15.66	27	16.82
		9	27	18.23	29	18.50	27	18.23

Table 3. Comparisons of HUHC$_2$s with different schemes

n	$N=n2^{n-1}$	scheme (i),(j)		scheme (k)		scheme (l)	
		dia	avg	dia	avg	dia	avg
2	4	**3**	**2.00**	3	2.00	3	2.00
4	32	**7**	**4.19**	7	4.19	7	4.19
6	192	**11**	**6.85**	12	6.97	11	7.00
8	1024	**15**	**9.65**	17	9.94	17	10.37
10	5120	**19**	**12.42**	22	12.90	20	13.49

Table 4. Comparisons of different two-connected topologies

N	MSX		MSN		schemes (e),(f)		schemes (i),(j)	
	dia	avg	dia	avg	dia	avg	dia	avg
4	2	1.33	2	1.33			3	2.00
8	3	1.96	3	2.00	7	3.57		
32	5	3.58	5	3.30			7	4.19
64	6	4.49	9	5.02	11	6.11		
192	7	5.44	13	7.45			11	6.85
384	8	6.40	19	10.46	15	9.25		
1024	10	8.37	33	17.01			15	9.65
2048	11	9.37	45	23.49	19	12.36		
5120	12	10.36	71	36.49			19	12.42
10240	13	11.36	101	51.49	24	15.44		

Consecutive Requests Traffic Model in Multistage Interconnection Networks *

Yann-Hang Lee†,   Sandra E. Cheung†   and Jih-Kwon Peir‡

†Dept of Computer and Information Sciences
University of Florida
Gainesville, FL 32611

‡IBM T.J. Watson Research Center
Yorktown Heights, NY 10598

Abstract

The Consecutive Request (CR) traffic pattern in a shared memory multiprocessor system is formed when each processor issues a sequence of requests to successive memory modules, This type of dependent traffic pattern is studied in multistage interconnection networks. A CR traffic may appear in the parallel access of vectors or cache lines. We show that a MIN constructed with conventional switches has a degraded performance under a CR traffic due to the high degree of conflict and low link utilization throughout the whole network. To remedy this problem we propose two solutions: Bit-reverse mapping and Dynamic Priority scheme. These two schemes and their combination not only improve the performance of the CR traffic and achieve the same performance for the uniform traffic, but can also cost effectively be incorporated in the conventional switch design. A simulation model using a Delta network is used to confirm the performance of the proposed schemes.

1   Introduction

The principal issue in the design of a shared memory multiprocessor system is to provide an efficient means of communication between the processing elements (PEs) on one hand, and the shared memory modules on the other. An attractive approach to this, which at the same time is the best compromise between cost and efficiency, is to use a multistage interconnection network (MIN) [3]. MINs have been widely used in massively parallel systems such as the Illinois Cedar [5] and the NYU Ultracomputer [4].

A MIN can be built with a complexity of $O(N \log_2 N)$ in contrast to the $O(N^2)$ complexity of a fully connected crossbar network. MINs, in their basic form, consist of $\log_2 N$ stages of switches with links connecting them. The basic requirement of a MIN is to provide a path between any

source and destination. Feng [3] classifies this type of MINs as "blocking" networks, since a memory request may be blocked when two paths share the same link. A topological equivalence relationship was established for this class of networks [15] which means that they can perform the same set of permutation functions if the destinations are rearranged appropriately. Buffers can be added to each switch element for temporary storage of blocked messages [1]. This results in a better utilization of switch elements.

The network performance is dependent on the rate of requests as well as their traffic patterns. In a centralized-control SIMD environment, the network performance can attain its optimum when certain permutations appear in the memory access traffic [10]. Unfortunately these special patterns cannot always be guaranteed in an environment with asynchronous executions on PEs. In most network performance analyses the common assumption is that all PEs issue their requests "independently" and "uniformly" to all memory modules. However, when the memory access traffic does not follow this assumption, the performance is often overestimated.

In order to access the memory modules uniformly, certain data allocation schemes, such as an interleaved pattern, may be used. However, data allocation alone cannot eliminate nonuniform traffic completely, since applications in the real world do not generally generate uniform traffic accesses. The "hot spot" traffic pattern where traffic to specific memory modules is much higher than to other modules, is an example of this, degrading the network performance dramatically as observed in [13]. Various solutions have been proposed to this problem [13, 6, 2, 12]. Other types of nonuniform traffic are also considered in [9].

Even though the memory accesses could be distributed uniformly, the independence assumption might not be valid as in the case when consecutive memory access in which a request targeted to memory location i is followed by a request to location $i+1$. This kind of access pattern may

---
*This research was supported in part by a grant from Florida High Technology and Industry Council

occur when processors access a *stride-1* vector data or load program codes. This consecutive request pattern can also be caused by the transfer of a cache block between cache and shared memory if each processor is associated with a primary cache. When the interleaved memory allocation scheme is used (assuming low-order interleaving), these consecutive memory accesses will be targeted to contiguous memory modules. In this case, the network receives sequences of dependent requests which may experience repeated network conflicts as observed in [14]. We shall refer to this class of network traffic as the Consecutive Requests traffic pattern.

In this paper, we will study the impacts of the Consecutive Requests (CR) traffic pattern in MINs. We will show that, under the CR traffic, the network composed of the conventional switch elements will experience heavy conflicts in the initial stages. We propose a mapping scheme, the Bit-Reverse address mapping, on every destination address with a proper rearrangement of the memory modules, which reduces the number of conflicts in the initial stages by reordering the destinations of continuous requests. We also propose to give priority to a particular input port when a continuous arrival of input messages are destined to the same output port, which will increase the overlap of the uses of the two links at each switch.

Vector accesses are usually preceded and followed by periods of uniform traffic and synchronization. We consider vector accesses both under these circumstances as well as under an idle network. Non-CR messages can obstruct the smooth passage of the CR traffic. Other factors which might affect the performance of our proposed schemes are the starting destination of the request sequences and the time at which they were issued. We study all these cases and evaluate the performance of the proposed solutions through simulation.

The evaluations show that, under the CR traffic, the performance of the original switch design is badly degraded and the solutions proposed significantly mitigate this degradation. Evaluations of the new designs are also done under the uniform traffic pattern to fulfill the nature of non-deterministic memory accesses in multiprocessor systems. These results show that the performance of the proposed schemes is the same as that of the primitive design for uniform traffic.

The paper is divided into five sections. Section 2 presents the model of MINs and discusses the CR traffic pattern. It also describes how a MIN performance can be deteriorated by this traffic patterns if conventional switch elements are used. In Section 3, we propose the Bit Reverse transformation method and Dynamic Priority scheme and show how these approaches can alleviate the network congestion problems. The simulation results are presented in Section 4 to

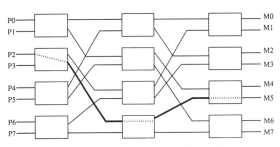

Figure 1: An 8 × 8 Delta Network

confirm the performance of our proposals. A conclusion is followed in Section 5.

2 Consecutive Request Traffic Pattern

2.1 Network Model

The MIN is known to be a cost-effective way of interconnecting processing elements (PEs) and memory modules (MMs) in a shared-memory system. An $N \times N$ Delta Network[11] constructed with 2×2 crossbar switches is shown in Figure 1, where N is equal to 8. The network is composed of $log_2 N$ stages with $N/2$ switches each. The routing controls are distributed to each of the switches by the corresponding bit of the destination address. Take the path from *PE2* (010) to *MM5* (101), in Figure 1, for example. The first bit "1" of the destination address is used by the switch in the first stage to route the message to the "lower" port. The second and third bits "0" and "1" are used in the same way to route the message to the "upper" and "lower" ports of the switches in the corresponding stages. This routing method is also known as the destination tag routing.

A conventional 2×2 switching element with single buffers is shown in Figure 2. Each input/output port has an associated register for receiving/sending messages between stages. A buffer is attached to each output register and serves as a waiting queue when the output register is busy. The network is operated with a synchronous clock. In the first part of the clock cycle, a message in an output register can be transferred to an input register in the next stage. Then, in the second part of the clock cycle, switches will route the messages in its input ports to the output buffers or output registers provided no conflicts occur in the routes and spaces are available. Meanwhile, an acknowledge signal will be passed back to the preceding stage to indicate whether a message can be received in the next clock or not.

When two messages from both input registers of a switch are destined to the same output port at the same time, an insertion conflict occurs. One of the messages has to be delayed by one cycle, and the victim is usually chosen in a

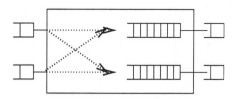

Figure 2: A Buffered 2×2 Switch

round robin fashion. This conflict may propagate back to the preceding stage and prohibit a successive message from transmission.

In order to reduce the insertion conflict, a split queue design was proposed[8], with two queues at each output port and an input register for each queue. It reduces the insertion conflicts by allowing two input messages to be successfully transferred to the queues in one cycle. The conflict at the single output port is resolved using a round robin priority scheme.

The performance of multistage interconnection networks has been evaluated extensively [11, 7, 1, 8]. These evaluations were established under the assumption that the MIN was driven by the independent and uniform traffic which can be characterized as the following:

(1) Messages are generated at each source node by independent and identical random processes.

(2) Each source node generates with the same probability at each cycle a message, and sends it with equal probability to each node.

The above assumptions seem true when memory elements are interleaved in a low-order fashion, where in the absence of hot spots, the requests to a portion of memory space will be spread over all memory modules. However, it is questionable that two successive requests from a processor are independent. If they were dependent, the question is whether switching elements will experience an unbalanced traffic and correlated conflicts.

2.2 CR Traffic Model

To investigate the impact of dependent requests, we consider the traffic pattern called "Consecutive Requests" (CR) which has the following characteristics.

1. A sequence of requests are generated at each source node and are destined to consecutive sink nodes, for example, the requests generated at processor *PE1* to memory module *MM4* is followed by the requests to memory modules *MM5*, *MM6* and *MM7*.

2. This sequence of requests are generated continuously without interference from other requests.

One place where this access pattern occurs is in vector accessing. A large number of numeric arrays and numerous iterations of computations involving array accesses are used in engineering and scientific applications. These data are distributed to many memory modules to prevent memory contention and increase computational parallelism. While executing these, very often multiple processors make use of instruction pipelining. Consequently, each processor will continuously request the data from memory modules with only one cycle in between.

The same consecutive pattern occurs in cache block fetches and program loading. If a cache block is allocated to consecutive memory modules with interleaved memory, the controller would need to address successive words and issue consecutive requests.

2.3 The Effect of CR Pattern

In this section we will describe the effect of the CR traffic in a network with conventional switches. Consider the activities in a switch in the first stage when the same CR sequence arrives simultaneously at each of its input ports. If the requests start with destination 000 in an 8×8 system, the first half of the CR messages are addressed to the first half of destinations. The sequence of the first bits, used to control routing, remains the same; we define the duration in which routings remain unchanged as *Duration of Identical Routings* (DIR). We can see that the possibility of insertion conflicts is high during this half of the CR sequence since the DIRs of both inputs are the same and long, i.e. both of the inputs try to route messages to the same output port for many cycles. The same thing occurs during the second half of the CR sequence which is directed to the lower half of memory.

In the second and the subsequent stages, conflicting messages will not only affect one of the messages at the input ports but also messages at the stages before it as well. We examined the network behavior and found that the CR traffic can severely deteriorate overall network performance for the following reasons.

1. The usages of the two links connected to the two output ports of a switch are not fully overlapped, i.e. resources are not fully utilized. As in the above example, at the output ports of a first-stage switch, the lower link is idle while the upper link is busy during the first half of the input CR sequences. The reverse occurs in the latter half.

2. The DIR is still long in the latter stages. The length of the DIR has an adverse effect on the performance of the CR traffic, because it determines the period of underutilization of the switches. If the round robin

scheme is used the average message delay of any processor is at least twice the distance from processor to memory, since the messages will be blocked at every stage. The DIR becomes larger as the network size increases.

If we can find a way that can prevent or reduce the above two phenomena, we can mitigate the performance degradation caused by the CR traffic. For simplicity, we shall call the above two effects created by the CR traffic as Effect(1), and Effect(2) in the subsequent sections.

The congestion problem caused by the CR traffic pattern cannot be avoided using the split queue design [8] if the round robin priority scheme, because the design can only prevent the insertion conflicts of the output queues and does not offer too much relief for the two effects of CR traffic. Based on this thought, and by using conventional switches, we will attempt to provide an efficient way for reducing the effects of CR traffic.

3 Dynamic Priority and Bit-Reverse Scheme

From the above, we can see that the network performance will degrade due to poor link utilization and long durations of identical routings. In this section, we propose two approaches to remedy these deficiencies. They are easily incorporated in conventional switches and have the same performance as the conventional switch design under an independent and uniform traffic pattern.

3.1 Bit-Reverse Mapping

We consider the conventional switch design which uses the round robin priority scheme to resolve insertion conflicts. When a processor issues consecutive requests starting from the same memory module or approximately the same memory module, the destination addresses of these requests vary in low-order bits. Since the routing of messages in the first stage depends on the first bit of the destination address, the routes of successive requests will be overlapped in the first stages of the network. One way to split the routes of successive requests and to reduce their overlap is to adopt a bit-reverse mapping in the destination addresses and to rearrange the memory modules accordingly. In the interface between the processor and the network, we reverse the order of the destination addresses, for example, in an 8×8 MIN system, 3 (011) is mapped to 6 (110). The consecutive requests 0,1,2,3,4,5,6,7 are then mapped to 0,4,2,6,1,5,3,7. Correspondingly, the memory locations are also distributed according to the bit-reverse order in memory modules. By doing this, we have not changed the network performance of independent requests. The performance is the same as that without address mapping on uniform traffic since the

destination addresses are generated randomly and are distributed to each destination with equal probability. No matter with which mapping function the addresses are mapped, they are still randomly distributed.

The Bit-Reverse mapping scheme can reduce the effect of CR traffic pattern because of the reduction of Effect(1) and Effect(2) in the first stage. Any two successive requests from a processor will have unoverlapped routes, except that they are received from the same input port. Also, the DIR is short in the first stage (only one cycle). If sequences arrive synchronously at the two input ports, a routing conflict arises. No conflict other than the initial one can occur in the first stage, after which both the output ports are kept busy. The usage of both links of the same switch in the first stage is therefore well overlapped.

The main drawback to the Bit-Reverse mapping scheme is that Effect(1) becomes serious as the DIR increases in the latter stages. Since the round robin priority scheme is used to resolve insertion conflicts, all the messages that route to the same destination will cluster together at the stages approaching the destination. Despite this deficiency, we expect that the advantages of the Bit-Reverse address mapping will lead to improvements in the network performance. This will be confirmed in the next section.

3.2 Dynamic Priority Scheme

From the previous sections, we observed that when the round robin priority scheme is used to select which of the two conflicting messages is routed to the output port, Effect(1) and Effect(2) of the CR traffic could not be avoided. In order to reduce these effects we propose to use a dynamic priority scheme to resolve the insertion conflicts.

Adhering to the conventional switches with no address mapping mechanism, we will now give preference to a message arriving at a particular input port in the case when successive input messages are trying to route to the same output port.

In Dynamic Priority any input port can be initially given the priority and it will keep this priority as long as it continues having messages in subsequent cycles. The priority circuit indicates the register which has the *current* priority, and once the priority register has no message but the other one does, the priority circuit will revert to reflect this change of traffic.

Dynamic Priority causes the CR sequence of the priority port to pass rapidly, allowing the overlap of the link usages from the same switch to be higher and Effect(1) is reduced. Moreover, since the sequence of messages to the second and subsequent stages remains in a consecutive order (i.e. not mixed with the sequence issued by other processors), the DIR length is halved in these stages. So, Effect(2) can be reduced in the second stage and subsequent stages.

An intuitive drawback of this approach, is that the unfair message-passing would cause uneven responses to the processors, and for them to perceive unfair services. However, if all the processors in the network are performing vector accesses, this unfairness within each switch element does not reflect on the performance of the overall system. Also, in uniform traffic, this unfairness is not present mainly because the DIR of any stage in this type of traffic is on the average two.

3.3 The Combined Approach

The hybrid model which combines both the Dynamic Priority scheme as well as the Bit Reverse mapping offers the same benefits as the ones described above, but with one added benefit. Observe that the Dynamic Priority allows for rapid passage of messages in the early stages but in the latter stages it is still almost round robin since the DIR in this stage is not long. In Bit Reverse mapping however, the DIR becomes larger in the last few stages and the Dynamic Priority could alleviate this in the same manner as described above. We therefore expect this scheme to have the advantages of both.

4 Performance Evaluation

4.1 Simulation Model

The simulation model used is a 64 × 64 buffered Delta network. A larger network has been used in additional studies and have yielded similar results. The total message capacity of the queues at the output ports is varied from four, six to eight messages each. The reason the size of eight was chosen is that in [7] it was shown that a queue size of eight could model infinite queue size, so that blockage caused by a full queue can be almost eliminated. A processor has at most one outstanding request and stops issuing a request when the corresponding input register of the first stage of the MIN is busy. Once the processor starts issuing CR messages, it will generate one every clock cycle, if it can insert it into the network. This dependency of subsequent is the main characteristic of the CR traffic. The time at which the processors start their vector access is also varied, from synchronous to asynchronous.

The location of the vector is an important factor in the performance of the CR traffic. Processors could be performing simultaneous operations on the same vector. This would generate the CR traffic in which all processors make the same sequence of requests to memory modules, we shall refer to this as the Type A access. In some cases, each processor works on its private vector. This may result in an access sequence with differing initial locations, we shall refer to this as the Type B access.

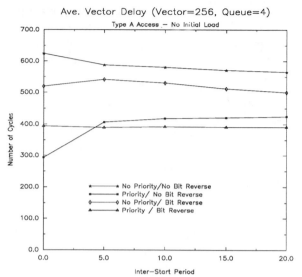

Figure 3: Type A Access - No Initial Load

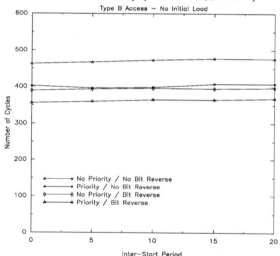

Figure 4: Type B Access - No Initial Load

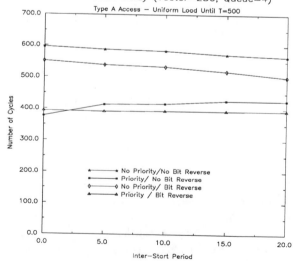

Figure 5: Type A Access - Uniform Load for 500 Cycles

Finally, our model studies vector accesses under an idle network before the access, and also under an initial load of uniform accesses. In the following subsections we shall discuss in more detail the issues brought up and the effect of these on the performance of the CR traffic.

4.2 Discussion of Results

Our simulation has primarily focused on the relative performance of the conventional switch which uses the round robin strategy and our proposed schemes, the Dynamic Priority, Bit Reverse mapping, and their combination. In the discussion which follows we will refer to the round robin strategy with no memory mapping as the conventional design. The measurements we took for the CR traffic model are the average message delay, the throughput (number of messages which reach memory per cycle), and the average sequence (vector) delay, which is defined as the average number of cycles it takes for all the processors to complete their vector access. In our simulations, we have collected these measurements for the conventional design, the Dynamic Priority, Bit-Reverse, and the hybrid scheme. The time that each processors begins to issue their CR requests is distributed within the inter-start period.

The simplest scenario is the one in which there is no initial load on the system and the processors issue a Type A access. Figure 3 shows the average vector delay for these four schemes under these assumptions. From this figure we can conclude that the Dynamic Priority scheme gives the best results if the processors start synchronously. In the asynchronous case this scheme deteriorates but the combined scheme performs slightly better than the Dynamic Priority alone. It is interesting to observe that the combined approach has a rather stable behavior. In all cases our proposed schemes yield better performance than the conventional design.

Type B access causes a less number of conflicts and its performance can be seen in Figure 4. In this case, the hybrid scheme yields the best results while the individual schemes perform alike. Still all of them perform better than the conventional switch design.

It is of importance in which sequence the allocated vector is being accessed since this determines the sequence of requests issued by a processor and also the number of conflicts. Although Type B access yields a far better performance, a mapping function is required in order to map the vectors in different memory modules. Our simulation has concentrated on Type A access and for completeness we have shown how our schemes would perform under Type B access.

Processors usually do a series of random accesses before and after performing vector accesses. The results of the four schemes in a Type A access where the CR traffic is preceded

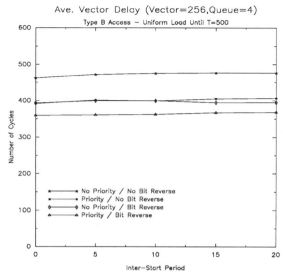

Figure 6: Type B Access - Uniform Load for 500 Cycles

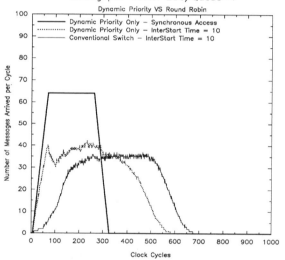

Figure 7: Throughput for Dynamic Priority and Round Robin

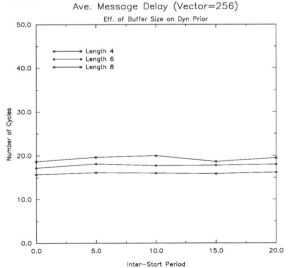

Figure 8: Effect of Buffer Size on Message Delay

by a uniform traffic period of 500 cycles at a request rate of 0.5 is shown in Figure 5. The initial load has a larger effect on the performance of the synchronous Dynamic Priority scheme, since this scheme is more vulnerable to the network being idle before the access. The hybrid scheme shows the most stability, as can be observed from Figure 3 and 5 that there is no change in the curve. We also obtained the case in which the same initial load precedes a Type B access. The results are presented in Figure 6. Remarkably, the initial load has no effect on the average vector delay in the case of a Type B access.

Another factor which we measured is the throughput under the CR traffic. From Figure 7 it can be seen that the throughput obtained under synchronous access using the Dynamic Priority scheme is the highest.

Figures 7 and 9 show an interesting property of the Dynamic Priority scheme under synchronous and Type A access, that irrespective of the length of the vector, after an initial setup time (different for each processor), each processor will deliver messages with the same message delay. Once the first request reaches memory, they keep arriving at a constant rate of one per cycle. This indicates that after the bottleneck is resolved in the early stages, the pipeline of requests is operating to its fullest capacity.

The first message arrives at memory in a total of $s + 1$ cycles, where s is the number of stages of the network. At each subsequent cycle there is one more processor whose first request arrives. If the vector is at least as long as the number of memory modules, this constant will increment continuously until all processors have one arrival per cycle. It starts decrementing at a constant rate of one processor per cycle once the first processor finishes its vector. If the vector is shorter than the number of memory modules, the curve still has the same shape but some processors will have finished their entire vector even before others have begun. Figure 9 depicts the case in which a vector larger than the number of memory modules is accessed in a network of 64×64 processors. From this the average vector delay is given by:

$$AvVectorDelay = (\frac{1}{N})(\sum_{i=1}^{N}(s + v + i - 1))$$
$$= s + v - 1 + (\frac{N+1}{2})$$

- s = number of stages in the MIN
- N = number of processors in the network
- v = number of elements in the vector

From the simulation results it is clear that the CR traffic has a lot to gain from the Dynamic Priority Scheme and suffers a lot under the conventional switch design. This scheme

gives continual priority to one port until that is exhausted. Thus the utilization of the links can be overlapped and the overall turnaround time for all the processors is reduced compared to that of the the round robin strategy.

The size of the buffers inside of the switches does not have a great impact on the overall performance of Dynamic Priority scheme. As Figure 8 shows, for a vector of length 256 the average message delay degrades slightly as the buffer space increases. The improvement in the average vector delay only becomes significant in the asynchronous case (as can be seen in Figure 10). However, even with a small amount of buffer space inside a switch, the proposed schemes perform well.

To be complete, we provide the graph which shows that the uniform traffic suffers the same message delay in all four schemes. This can be seen in Figure 11.

5 Conclusion

The Consecutive Requests traffic model is a type of dependent traffic which can be encountered both in vector processors as well as in conventional processors accessing cache lines in parallel. When a MIN with the primitive buffered switch design is used to interconnect processors and memory modules, the CR traffic suffers a great number of conflicts in the initial stages, especially if the vector accesses all start from the same memory module. Two schemes are proposed; the Dynamic Priority scheme is a modification of the conventional round robin scheme and the Bit Reverse scheme is a memory mapping scheme used to distribute the accesses and reduce the number of conflicts in the early stages of the MIN.

Through simulations, we have shown that the Dynamic Priority, Bit Reverse mapping, and their combination can significantly improve the network performance under CR. The schemes do not deteriorate the performance under uniform traffic and can be implemented rather cost effectively. Specially, the combined scheme provides a quite stable vector delay no matter whether the requests are issued synchronously or asynchronously, or whether the network has queued messages at the time the vector accesses are initiated.

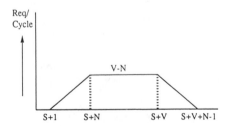

Figure 9: Throughput of Dynamic Priority

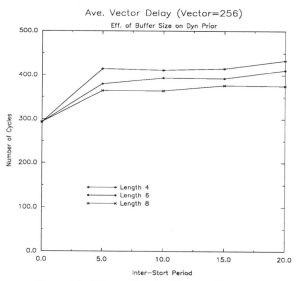

Figure 10: Effect of Buffer Size on Vector Delay

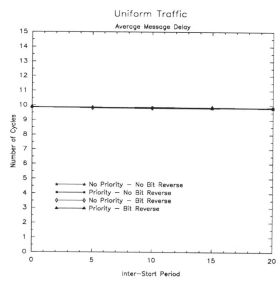

Figure 11: Uniform Traffic Under Proposed Schemes

References

[1] D.M. Dias, and J.R. Jump, "Analysis and simulation of buffered delta networks," *IEEE Trans. on Comp.* C-30, (Apr, 1981), pp. 273-282.

[2] D.M. Dias, and M. Kumar, "Preventing Congestion in Multistage Networks in the Presence of Hotspots," *Proc. 1989 Int. Conf. on Par. Proc.* (1989), pp. I.9-13.

[3] T.-Y. Feng, and C.-L. Wu, "Fault-Diagnosis for a Class of Multistage Interconnection Networks," *IEEE Trans. on Comp.* C-30, (Oct, 1981), pp. 743-757.

[4] A. Gottlieb, et al., "The NYU Ultracomputer - Designing an MIMD Shared Memory Parallel Computer," *IEEE Trans. on Comp.* C-32, (Feb, 1983), pp. 175-189.

[5] D.D. Gajski, D.J. Kuck, D.H. Lawrie, and A.H. Sameh, "Cedar - A Large Scale Multiprocessor," *Proc. 1983 Int. Conf. on Par. Proc.* (1983), pp. 524-529.

[6] W.S. Ho, and D.L. Eager, "A Novel Strategy for Controlling Hot Spot Congestion," *Proc. 1989 Int. Conf. on Par. Proc.* (1989), pp. I.14-18.

[7] C.P. Kruskal, and M. Snir, "The Performance of Multistage Interconnection Networks for Multiprocessors," *IEEE Trans. on Comp.* C-33 (Dec, 1983), pp. 1091-1098.

[8] M. Kumar, and J.R. Jump, "Performance Enhancement in Buffered Delta Networks Using Crossbar Switches and Multiple Links," *Journal of Par. and Dist. Comp.* Vol. 1, No. 1, 1984, pp. 81-103.

[9] T. Lang, and L. Kurisaki, "Nonuniform Traffic Spots (NUTS) in Multistage Interconnection Networks," *Proc. 1988 Int. Conf. on Par. Proc.* (1988), pp 191-195.

[10] D.H. Lawrie, "Access and Alignment of Data in an Array Processor," *IEEE Trans. on Comp.* C-24, (Dec, 1975) pp. 1145-1155.

[11] J.H. Patel, "Performance of Processor-Memory Interconnections for Multiprocessors," *IEEE Trans. on Comp.* C-30, (Oct, 1981), pp. 771-780.

[12] J.-K. Peir, and Y.-H. Lee, "Look-Ahead Routing Switches for Multistage Interconnection Networks," to appear in *Journal of Par. and Dist. Comp.*

[13] G.F. Pfister, and V.A. Norton, "'Hot-Spot' Contention and Combining in Multistage Interconnection Networks," *IEEE Trans. on Comp.* C-34 (Oct, 1985), pp. 943-948.

[14] K.-H. Shen, "Preventing Performance Degradation of Multistage Interconnection Networks Under Two Traffic Patterns," *Master Thesis, CIS Dept., University of Florida*, Fall, 1989.

[15] C.-L. Wu, and T.-Y. Feng, "On a Class of Multistage Interconnection Networks," *IEEE Trans. on Comp.* C-29, (Aug, 1980) pp. 694-703.

AN APPROACH TO THE PERFORMANCE IMPROVEMENT
OF MULTISTAGE INTERCONNECTION NETWORKS
WITH NONUNIFORM TRAFFIC SPOTS

Nian-Feng Tzeng

The Center for Advanced Computer Studies
University of Southwestern Louisiana
Lafayette, LA 70504

Abstract —— When nonuniform reference patterns produce unproportionally heavy traffic at certain switching elements in a multistage interconnection network (MIN), network performance is degraded due to tree saturation, which is also known to cause hot-spot contention in the MIN. In this paper, we consider an approach to MIN performance improvement in the presence of nonuniform traffic spots by means of supporting a considerably increased traffic flow using multiple queues. Simulation studies have been carried out to evaluate the performance of MINs with nonuniform traffic spots, and the results show that the proposed design indeed can improve network performance significantly.

1. Introduction

Processors and memory modules of a large-scale multiprocessor system may be interconnected by a multistage interconnection network (MIN), whose efficiency tends to critically dictate overall system performance. The performance of a MIN could degrade significantly when one or several memory modules get heavy access from a set of processors, possibly causing hot-spot contention due to tree saturation [1]. A memory module which receives excessive access is called a hot spot. The tree saturation phenomenon caused by hot-spot access leads to drastic network performance degradation, if a certain memory module gets a request rate larger than its service rate. Solutions have been proposed to eliminate or alleviate the adverse impact of hot-spot contention.

A different type of nonuniform traffic, called *nonuniform traffic spots*, may exist in a MIN and could also lead to serious network performance degradation, as explored by Lang and Kurisaki [2]. Unlike the hot-spot case, this type of nonuniform traffic makes no heavy access to any memory module, but it causes unproportionally high traffic at certain switching elements in the MIN. When the arrival rate of a switching element exceeds its service rate, tree saturation results, and the specific switching element becomes a nonuniform traffic spot (NUTS). In the presence of NUTS, all the buffers along the tree rooted at NUTS and extended all the way back to every involved processor are totally full. This tree saturation phenomenon, like that in the hot-spot case, can degrade network performance remarkably. It is clear that the use of combining switches cannot deal with this type of nonuniform traffic, simply because the requests contributing to NUTS do not necessarily head for the same destination.

Lang and Kurisaki [2] suggest two solutions for overcoming nonuniform traffic that results in NUTS. These solutions depend entirely on rerouting contention requests over paths which get around the NUTS, thus lessening their impact successfully. In this paper, we investigate an effective approach capable of improving network performance effectively when nonuniform traffic exists. The main idea of this approach is to incorporate multiple queues in each switch input and in each network primary input. As a result, regular requests which are not involved in NUTS can travel through the network, even in the presence of tree saturation, giving rise to improved performance. The proposed approach can profitably be applied to deal with any kind of nonuniform traffic. We investigate the performance of our design in the presence of NUTS caused by different kinds of nonuniform traffic patterns through extensive simulation.

Nonuniform Traffic Spots

Every switch in the same stage of a MIN carries an equal share of traffic under the uniform reference patterns, and requests flow over the MIN smoothly. When a nonuniform reference pattern arises, however, a MIN may suffer from serious contention, giving rise to substantial performance degradation. In particular, if any link in a MIN receives a request rate larger than its service rate, tree saturation develops and many processors get affected. In Fig. 1, for example, link l_1 receives an unproportionally high request rate, if $P0$, $P4$, $P8$, and $P12$ have higher access preference to $M0$, $M1$, $M2$, and $M3$. This might happen in a hierarchical requesting fashion where a set of processors has a larger reference rate to a cluster of memory modules, because those processors communicate with one another through the cluster of memory modules more frequently. Suppose a processor generates requests at a rate of r, out of which $c\%$ is directed to the cluster of preferred memory modules and the rest is uniformly distributed across all memory modules, including those preferred memory modules. For a size N network in which $f\%$ processors are involved in referring the preferred memory modules, tree saturation arises if $Nfrc + r(1-c) > 1$. A saturation tree developed in Fig. 1 is rooted at the top switch of stage 1 and affects all the four processors involved in preference access.

In this study, we are interested in the worst case scenario that the hierarchical requesting fashion can possibly result. We classify the degree of communication among processors into a two-level hierarchy, with one level for those "intimate" processors and the other level for the remaining processors. Specifically, processors in a size N system are partitioned into g equal groups, and a processor has a higher communication rate with processors in the same group than with processors outside the group. Memory modules are also divided into g equal clusters, and all of the N/g processors in one group share a common cluster of preferred memory modules.

To identify the worst possible result for an Omega network under this kind of nonuniform traffic, groups of processors are interleaved in a way that processors 0, g, $2g$, $\cdots$ belong to one group; processors 1, $g+1$, $2g+1$, $\cdots$ are in another group, and so on; whereas contiguous memory modules are assigned to clusters, namely, modules 0, 1, 2, $\cdots$, $g-1$ belong to one cluster; modules g, $g+1$, $g+2$, $\cdots$, $2g-1$ are in another group, etc. Every group of processors have a distinct cluster of preferred memory modules. In this situation, more than one internal link may get the request

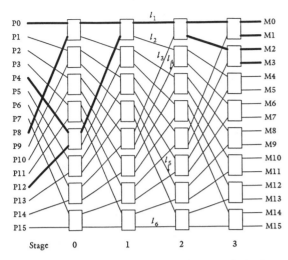

Fig. 1. A nonuniform traffic spot at the top switch of stage 1.

rate high enough to cause tree saturation, if $g > 1$, making multiple saturation trees present in the network. Note that Type II nonuniform traffic discussed in [2] is a special case of this two-level requesting hierarchy for which g equals 2.

A different kind of nonuniform traffic occurs when processors references follow a bit-reversal permutation pattern. It is known that the bit-reversal permutation, like the perfect shuffle permutation, arises frequently in parallel processing [3]. Unfortunately, a bit-reversal permutation produce serious contention in an Omega network, as was addressed in [2]. It is readily shown that a bit-reversal permutation over an Omega network of size 16 (see Fig. 1) gives rise to high request rates on four links between stages 1 and 2 (namely, l_1, l_4, l_5, and l_6 as marked in Fig. 1). Nonuniform traffic results from realizing such a permutation and could produce NUTS.

An actual program execution is hardly the case that processors issue requests following the bit-reversal permutation for their whole lifetime. Instead, all processors may generate such a permutation once in several cycles, and in the remaining cycles, they issue requests randomly or according to other reference patterns. In this study, we assume that every processor is involved in generating the bit-reversal permutation with a given probability, and for the rest of the time, it generates random requests.

2. Proposed Configuration

The buffer design dictates switch complexity and the network behavior. Previously, Tamir and Frazier [4] carried out a comprehensive discussion on the design of high-performance VLSI communication switches. They explored several different multi-queue buffer designs, including the statically allocated multi-queue (SAMQ) and the dynamically allocated multi-queue (DAMQ). The multi-queue buffer makes it possible to allow more requests to pass through every switch at a time than a conventional single-queue buffer and is thus preferable. It is concluded in their study [4] that the DAMQ buffer performs the best among all design alternatives considered, achieving a clear performance gain. Though it can handle uniform traffic very well, the DAMQ design fails to sustain acceptable bandwidth in the presence of nonuniform traffic, since the DAMQ design places no limit on the number of slots a single queue can take. Experimental results in the next section will illustrate that, in the presence of NUTS, the DAMQ

design suffers from low saturated bandwidth because queues at ports along the way from processors to the NUTS location(s) "hog" the majority of available storage, hindering all other communication through those ports.

In the following, we propose a switch buffer design which overcomes the hogging problem while ensuring high switch storage utilization. Multiple queues are employed at each switch input, with one separate queue for requests to be directed to one output port. Each queue is assigned one slot permanently and also shares a common storage pool with other queues in the whole switch, as shown in Fig. 2. Requests arriving at a queue make use of its permanent slot first and then get additional slots from the shared pool whenever necessary. Hardware circuits are responsible for maintaining all the queues as well as the storage pool. The advantage of assigning a permanent slot to each queue is two-fold. First, this can prevent any single queue from exhausting all storage and, thus, assures that communication over all the ports continues even in the presence of saturation trees caused by nonuniform traffic. Second, the needed bandwidth for the shared pool could be reduced, because the permanent slot of a queue, once empty, is chosen first to accommodate a request entering that queue, leading to a reduced mean number of requests contending for the shared pool in each cycle.

Because of an efficient use of storage, the capacity of shared storage need not be large to attain good performance under any traffic distribution. A more efficient use of switch storage can reduce total buffer slots in a switch considerably. The price paid is a more complex storage management in comparison with the SAMQ design. However, we believe that chip area reduction due to the storage savings in our design is expected to be much larger than the extra chip area required for more complex control circuitry.

Overall Configuration

A MIN constructed from the proposed new switches can prevent a saturation tree from blocking traffic *inside* the MIN, enabling fluent regular traffic. However, this MIN still fails to sustain high bandwidth simply because saturation trees tend to inhibit the majority of requests from *entering* the MIN, since the network primary input involves only a single queue. All requests behind the single request at the queue head which is bound for NUTS are blocked, even though, once admitted to the MIN, they may proceed through the MIN despite the presence of saturation trees. This problem can be overcome by employing a multi-queue at each primary input so that no single request would block all other requests at the same primary input, possibly allowing requests to enter the network even when saturation trees exist. The provision of having a multi-queue at each primary input is critical and indispensable. Different switch designs have been shown to yield the same low saturated bandwidth under nonuniform traffic when only a single queue exists at each primary input [4, 5].

With a multi-queue at each primary input, a request, after it is issued, is placed at a queue determined by the destination of the request, waiting to enter the network. Here, we consider that $\eta = 4$ queues are present at each primary input (as 4×4 switches are our MIN building blocks). The most significant two tag (i.e., destination) bits of a request are used to determine the queue for holding the request. This process is *not a part of routing*; it is only for partitioning requests into classes based on their destinations and placing requests of the same class at one queue. As a result, any requests bound for NUTS located at one class of destinations would not inhibit

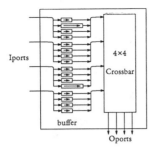

Fig. 2. The new 4×4 switch design.

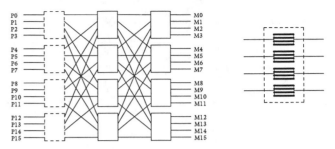

Fig. 3. The overall configuration of the proposed MIN with size 16, where each dashed square is as shown in the right.

requests heading for the destinations of all other classes from entering the network, ensuring a larger bandwidth. After a request is admitted to the network, a destination-tag routing procedure [6] is followed to set up a path for the request.

The multi-queue is assumed to be implemented at a separate entity external to a processor so that (1) several multi-queues can be put together and form a unit to achieve a better use of storage by sharing space and to save the chip count, and (2) the processor I/O interface is left unchanged. Every unit so constructed requires similar functions as what are needed for the switching element employed to construct the MIN (although a switching element is more complex because it needs, in addition to these functions, the switching capability). A 4×4 building block for the MIN may thus be in place of a unit at network primary inputs so as to reduce the types of chips needed for the overall configuration, even though a switching element contains an unnecessary crossbar.

The overall configuration with size 16 is illustrated in Fig. 3, where every four multi-queues at the primary inputs are housed in a unit and the constituent switching element is of size 4. As a result of this added column in front of the MIN, every request incurs an extra delay when passing through the network. The extra delay at this column is assumed to be the same as that at any MIN stage. Our configuration offers a good bandwidth even in the presence of saturation trees, since it maintains fluent regular traffic.

3. Experimental Study

Experimental Model

A size 256 Omega network composed of 4×4 switching elements is simulated. The network transmits requests from processors to memory modules, operating in the packet-switched mode. A request forms one packet, which takes a slot. A switch input port during a network cycle may receive at most one request, and a received request is placed at an appropriate queue after being admitted to the input port. When the request at any queue head cannot advance to the next stage in a cycle due to conflicts with other requests or to the lack of buffer space, the request stays at its original position and reattempts in the subsequent cycle; no request is dropped inside the network. Thus, there is no need to retransmit any request admitted to the network and sequential consistency of request transmission is preserved. During a cycle, a memory module can serve one request at the network primary output to which the module is connected.

Let the storage size at each switching element be Q. Since the 4×4 switch is the building block of the simulated network, according to our design, each input port contains

four queues, and every queue has one slot assigned permanently and also shares available storage with the other fifteen queues inside the switch (see Fig. 2). As a result, sixteen out of the Q slots at a switch are dedicated, while the remaining $(Q-16)$ slots are shared by all the sixteen queues. As for the DAMQ and conventional single-queue 4×4 switches, each input port has $Q/4$ fixed slots; whereas each queue in the SAMQ switch has $Q/16$ fixed slots (since an SAMQ switch involves 16 static queues, with four for each input port).

Requests are generated as follows. For the hierarchical requesting case, processors are partitioned into $g \, (> 1)$ equal groups, so are memory modules. The way of partitioning processors and memory modules is described earlier. Each group of processors is assigned a cluster of preferred memory modules; no two groups share the same cluster. A processor has $c\%$ generated requests destined for any one of the preferred $256/g$ memory modules with an identical probability, and the rest requests are randomly distributed across all memory modules. For the bit-reversal permutation case, all processors are simultaneously involved in generating a bit-reversal permutation once in, on an average, $1/(r \cdot p)$ cycles, while in other cycles, every processor issues random requests according to the independent Poisson process, where r is the request generation rate per processor per cycle. In other words, a fraction p of the requests issued by any processor are contributed to bit-reversal permutations, with the rest being uniformly distributed.

Simulation Results

The measures of merit in our study include mean latency and normalized bandwidth. Mean latency indicates the average time (in terms of cycles) of a request from its generation to its exit from the network, and normalized bandwidth (called bandwidth for simplicity) is the average number of requests arriving at one memory module per network cycle. Note that bandwidth is chosen as one axis for plotting network latency-bandwidth characteristics in Figs. 4 and 5, although it is a collected measure under various request generation rates (r).

Latency versus bandwidth under hierarchical requesting with $c = 0.2$ is depicted in Fig. 4. For either g value shown, our configuration can significantly alleviate the impact of NUTS on performance when compared with a conventional network (which is constructed from single-queue switches), giving rise to much higher saturated bandwidth. While it starts with slightly larger latency (due to the added column in the network input side), our network eventually enjoys less latency when the conventional network is close to saturation. It is interesting to observe from the figure that a network

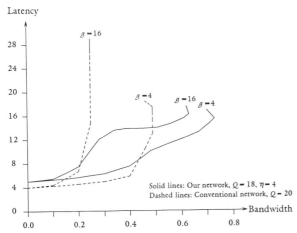

Fig. 4. Latency versus bandwidth under hierarchical requesting with $c = 0.2$.

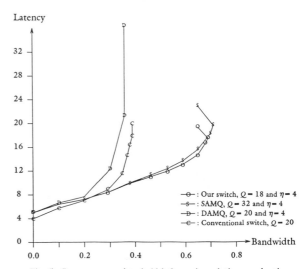

Fig. 5. Latency versus bandwidth for various designs under the bit-reversal permutation pattern with $p = 0.1$.

possesses better latency-bandwidth characteristics for $g = 4$ than for $g = 16$. The reason is as follows. For $g = 16$, there are only 16 links (out of 256 links) used for delivering preferred memory requests from stage 1 to stage 2 (in the middle of the MIN), each responsible for carrying the preferred memory requests originating from all 16 processors in a group (an analogous situation can be seen in Fig. 1, where link l_1 is used to transmit preferred memory requests from 4 processors in one group). On the other hand, when g is 4, preferred memory references from one group are carried by multiple links (rather than a single link) between stages 1 and 2, and no link is as heavily loaded as in the prior case. In general, the most serious tree saturation (and thus the worst performance) in a size N network would result if g equals $\sqrt{N}$, because all preferred memory references from one group are then transmitted by only one single link in the middle of the network. Notice that the curve of our network for $g = 16$ stays relatively flat when bandwidth is between 0.33 and 0.5 due to the reason that as bandwidth grows, both regular and hierarchical requests experience consistently increasing delays but more regular requests then contribute to bandwidth (since saturation trees are present) and thus hold the average delay down (for they have much lower latency than the hierarchical requests). When bandwidth goes beyond 0.5, regular requests have large enough delays that cause the mean latency of total requests to swell gradually.

The latency-bandwidth characteristics of networks with NUTS caused by bit-reversal permutation patterns are given in Fig. 5. Our configuration possesses far better characteristics after saturation trees (due to NUTS) develop. The characteristics of networks constructed from SAMQ switches and from DAMQ switches are also provided in Fig. 5. Each primary input of the networks is equipped with four queues ($\eta = 4$). Our network has virtually the same peak bandwidth as the SAMQ-based network, even though storage in our switch is $Q = 18$, considerably less than $Q = 32$ in an SAMQ switch. This confirms that our switch design could achieve a given performance level using significantly less storage space than a switch without sharing storage. The DAMQ-based network has the worst characteristics mainly because of the severe hogging problem when traffic becomes heavy.

4. Conclusions

An approach has been investigated and proved effective in tackling nonuniform traffic, enhancing network performance in the presence of NUTS. The main idea of this approach lies in the use of multiple queues at each switch input and at each network primary input. As a result, regular requests are allowed to pass through the network fluently even when saturation trees exist. This configuration is expected to have reasonably low hardware complexity and comes with two other advantages. First, requests issued from a processor to a memory location are always delivered by our configuration in sequence, preserving sequential consistency. Unlike the solutions addressed in [2], where requests may get out-of-sequence and resequencing logics are needed at network primary outputs, our configuration suffers from no buffering and resequencing overhead at primary outputs. Second, our approach is also applicable to multiprocessors based on the message-passing architecture (such as hypercubes) and fast packet-switching fabrics in the telecommunication network so as to lessen any possible detrimental impact caused by nonuniform traffic. Our configuration thus appears interesting and useful.

References

[1] G. F. Pfister and V. A. Norton, "'Hot-Spot' Contention and Combining in Multistage Interconnection Networks," *IEEE Trans. Computers*, vol. C-34, pp. 943-948, Oct. 1985.

[2] T. Lang and L. Kurisaki, "Nonuniform Traffic Spots (NUTS) in Multistage Interconnection Networks," *Journal of Parallel and Distributed Computing*, **10**, pp. 55-67 (1990).

[3] J. Lenfant, "Parallel Permutation of Data: A Benes Network Control Algorithm for Frequently Used Permutations," *IEEE Trans. Computers*, vol. C-27, pp. 637-647, July 1978.

[4] Y. Tamir and G. L. Frazier, "High-Performance Multi-Queue Buffers for VLSI Communication Switches," *Proc. 15th Int'l Symp. Computer Architecture*, May 1988, pp. 343-354.

[5] N.-F. Tzeng, "Alleviating the Impact of Tree Saturation on Multistage Interconnection Network Performance," to appear in *Journal of Parallel and Distributed Computing*, June 1991.

[6] D. H. Lawrie, "Access and Alignment of Data in an Array Processor," *IEEE Trans. Computers*, vol. C-24, pp. 1145-1155, Dec. 1975.

Modeling and Analysis of "Hot Spots" in an Asynchronous $N \times N$ Crossbar Switch

Eugene Pinsky[1] and Paul Stirpe[2]

Computer Science Department

Boston University

Boston, MA 02215

e-mail: pinsky@cs.bu.edu stirpe@cs.bu.edu

ABSTRACT

One of the most promising design approaches to high speed networks and large-scale multiprocessors is the use of optical interconnections. The basic component of such a system is a switch that has the capacity of interconnecting a large number of inputs to outputs. In this paper we analyze an asynchronous crossbar switch model for lightwave circuit-switched networks that incorporates "hot spots" (non-uniform access patterns). We develop efficient computational algorithms to compute the performance measures and quantify the effect of imbalance on the system performance.

Key Words: performance analysis, modeling, crossbar interconnection networks, hot spots, asynchronous.

INTRODUCTION

One of the most promising approaches to building high speed networks and distributed multiprocessors is the use of optical interconnections. The basic component of such a system is a switch (interconnection network) that has a capacity of interconnecting a large number of inputs to outputs. In this paper we present an analytical model of an $N \times N$ crossbar switch operating in asynchronous mode.

The crossbar switch, due to its excellent performance characteristics, is a basic building block in the design of many switching systems. The crossbar is internally non-blocking and all electrical paths experience unit delay when propagating through the switch. However, the circuit complexity of an $N \times N$ crossbar is $O(N^2)$, prohibiting the implementation of large (1024×1024) electrical switches. As a result of the severe limitations in size of electrical crossbar switches, researchers proposed the Multistage Interconnection Network (MIN), consisting of small crossbar switches (eg. 2×2) as basic building blocks [1, 5].

Optical crossbar switches have the potential to overcome the circuit complexity limitations of electrical crossbars, and thus provide a means of designing large crossbar switches. Several design approaches, such as beam stearing and beam spreading/masking have been proposed [2]. To date, technological limitations, such as the switching speed of spatial light modulators, inhibits implementation.

We believe that this technology will eventually become feasible and that the performance modeling of such switching systems is important. Since lightwave networks have tremendous bandwidth and light cannot be buffered, we believe such networks will be comprised of asynchronous non-buffered intermediate network switching nodes (e.g. optical crossbars). In such a system, a request to establish a route will contain the information encoded in its header that would specify which output to use at each intermediate switch in the interconnection network for the subsequent data transfer. The system operates in circuit-switching mode where all routing decisions are shifted to the periphery of the interconnection network.

The main contribution of this paper is the incorporation of "hot spots": non-uniform access. Most models of interconnection networks assume that outputs requests are uniformly distributed. However, hot spots could have a tremendous impact on performance [6]. It is desirable, therefore, to analyze the relationship between achievable throughput, concurrency and the access pattern for switches of large size.

In this paper we present an analytic model of hot spots phenomena for an asynchronous $N \times N$ crossbar switch. We derive efficient algorithms to compute the bandwidth and the non-blocking probabilities, and compare the performance measures of the asynchronous and synchronous models. For the asynchronous model considered, we will show that hot spots *do not* have a significant impact on the performance.

THE MODEL

We consider an $N \times N$ crossbar switch. The requests for connection for any input i arrive according to a Poisson distribution with rate $\tilde{\lambda}$. We assume there exist non-uniform access patterns (*hot spots*). Requests are divided into two classes: "hot spot" requests and "ordinary" requests. Hot spot requests, arriving at input i always require the connection to the same output j_i. Assume that "hot spot" requests arrive at any input i according to a Poisson distribution with rate $\tilde{\lambda}_1$. Ordinary requests have a uniform traffic pattern: an ordinary request is equally likely to be addressed to any of the N outputs. Assume that ordinary requests arrive at any input i according to a Poisson distribution with rate $\tilde{\lambda}_2$ (thus $\tilde{\lambda} = \tilde{\lambda}_1 + \tilde{\lambda}_2$). The rate $\tilde{\lambda}_1$ measures the "skewness" of traffic access pattern: $\tilde{\lambda}_1 = 0, \tilde{\lambda}_2 > 0$ corresponds to a uniform traffic pattern (there are no hot spots) whereas the case $\tilde{\lambda}_1 > 0, \tilde{\lambda}_2 = 0$ corresponds to the trivial case of N dedicated connections for hot spot requests. It is of interest to consider the case $\tilde{\lambda}_2 \neq 0$ and to establish the effect of non-uniform access on the asymptotic behavior of the system as $N \mapsto \infty$.

We assume that the time of a "hot spot" and of an ordinary connection is distributed exponentially with mean $1/\mu_1$ and $1/\mu_2$ respectively. We define $\lambda_1 = \tilde{\lambda}_1$, $\lambda_2 = \tilde{\lambda}_2/N$ and loads $\tilde{\rho}_1 = \tilde{\lambda}_1/\mu_1$, $\tilde{\rho}_2 = \tilde{\lambda}_2/\mu_2$, $\rho_1 = \lambda_1/\mu_1$, $\rho_2 = \lambda_2/\mu_2$, $\tilde{\rho} = \tilde{\rho}_1 + \tilde{\rho}_2$. We assume that there are no store-and-forward

---

[1]Supported in part by the National Science Foundation Grant #CCR-9015717

[2]The author is employed and supported by IBM Corp., Endicott, N.Y.

buffers at the internal switches and that a blocked request is cleared from the system.

The state-space $\Gamma(N)$ of the system can be described by (n_1, n_2), the number of configurations with n_1 hot spot and n_2 ordinary requests. Therefore, if we define $\overline{n}(t) = (n_1(t), n_2(t))$ to be the state of the system at time t, then the underlying stochastic process $\overline{n}(t)$ is Markov, and it has a unique steady-state probability distribution of the form:

$$\pi(n_1, n_2) = \frac{1}{Z(N)} \cdot \frac{N!(N - n_1)!}{[(N - (n_1 + n_2))!]^2} \cdot \frac{\rho_1^{n_1}}{n_1!} \cdot \frac{\rho_2^{n_2}}{n_2!}$$

where $Z(N)$ is the normalization constant. It can be shown that we can replace the exponential service distribution by any distribution with the same mean, due to the insensitivity of the underlying stochastic process [3].

PERFORMANCE MEASURES

Let us now derive the expressions for average performance measures: average number of connections, throughput and non-blocking probability for both hot spot and ordinary requests, and quantify their dependence on the access pattern.

The average number of hot spot connections is

$$E_1(N) = \frac{\rho_1}{Z(N)} \cdot \frac{\partial Z(N)}{\partial \rho_1} = N\rho_1 \frac{Z(N-1)}{Z(N)}$$

The average number of ordinary connections is

$$E_2(N) = N^2 \rho_2 \frac{Z(N-1)}{Z(N)} - N(N-1)\rho_1\rho_2 \frac{Z(N-2)}{Z(N)}$$

In the absence of hot spots ($\rho_1 = 0$), the concurrency would have been $N^2\rho_2 Z(N-1)/Z(N)$. The amount $N(N-1)\rho_1\rho_2 Z(N-2)/Z(N)$ is lost due to contention with hot spot requests.

The non-blocking probability that a hot spot request is not blocked is therefore,

$$B_1(N) = \frac{Z(N-1)}{Z(N)}$$

By conservation principles, the load $N\tilde{\lambda}_2 B_2(N) = \mu_2 E_2(N)$. Therefore, the non-blocking probability that an ordinary request is accepted is

$$B_2(N) = \frac{Z(N-1)}{Z(N)} - \frac{(N-1)}{N}\rho_1 \frac{Z(N-2)}{Z(N)}$$
$$= B_1(N)\left[1 - \frac{E_1(N-1)}{N}\right]$$

COMPUTATIONAL ALGORITHMS

Despite these simple expressions for the averages, to compute them exactly requires the exact computation of the partition function. However, a direct evaluation is difficult due to the factorial terms in the partition function. We proceed by developing an efficient recursive algorithm to compute $Z(N)$ and thus the performance measures.

We proceed as follows. First, from the definition of Laguerre polynomials [7],

$$L_N(x) = \sum_{i=0}^{N}(-1)^i \binom{N}{N-i}\frac{x^i}{i!}$$

we can write $Z(N)$ as

$$Z(N) = N!\sum_{n_1=0}^{N}\frac{\rho_1^{n_1}}{n_1!}\rho_2^{N-n_1}L_{N-n_1}\left(-\frac{1}{\rho_2}\right)$$

From the generating function of Laguerre polynomials [7], we compute the exponential generating function of $Z(N)$ as follows:

$$G(t) = \sum_{N=0}^{\infty}\frac{Z(N)}{N!}t^N = \frac{1}{1-\rho_2 t}\exp\left(\rho_1 t + \frac{t}{1-\rho_2 t}\right)$$

Differentiating with respect to t we have:

$$\frac{\partial G(t)}{\partial t} = \sum_{N=0}^{\infty}\left[\frac{\rho_1 Z(N)}{N!} + \sum_{k=0}^{N}(\rho_2 + k + 1)\rho_2^k\frac{Z(N-k)}{(N-k)!}\right]t^N$$

On the other hand,

$$\frac{\partial G(t)}{\partial t} = \sum_{N=0}^{\infty}(N+1)\frac{Z(N+1)}{(N+1)!}t^N$$

Equating the corresponding coefficients to obtain a recurrence relation, we obtain the following algorithm:

> **Algorithm 1:**
> $Z(0) = 1$
> for $n = 1$ till $N - 1$
>
> $$Z(n+1) = \rho_1 Z(n) + \sum_{k=0}^{n}\left\{(\rho_2 + k + 1) \times \left[\prod_{j=1}^{k}(n-k+j)\rho_2\right]Z(n-k)\right\} \quad (1)$$
>
> Compute $B_1(N), B_2(N), E_1(N), E_2(N)$
> **Finished !**

Let us analyze the complexity of Algorithm 1. There are N iterations. For a typical step, in the computation of $Z(n)$ we sum n terms where term k requires $O(k)$ operations. However, we note that

$$\prod_{j=1}^{k}(n-k+j)\rho_2 = (n-k+1)\rho_2\prod_{j=1}^{k-1}(n-(k-1)+j)\rho_2$$

and therefore, term k can be computed from term $k-1$ and takes $O(1)$. It follows then that each $Z(n)$ can be computed in $O(n)$ operations. Since $1 + 2 + \cdots + N = N(N+1)/2$, it follows that the complexity of the algorithm is $O(N^2)$.

We now present a mean-value type of algorithm to compute the blocking probabilities. This algorithm is cast directly in terms of blocking probabilities. The main advantage of this approach is numerical stability, which is obtained at increased computational cost.

To derive the algorithm, we divide both sides of equation 1 by $Z(N+1)$ to obtain

$$1 = \frac{\rho_1 Z(N)}{Z(N+1)} + \sum_{k=0}^{N}(\rho_2 + k + 1)\rho_2^k N(N-1)\cdots(N-k+1)\frac{Z(N-k)}{Z(N+1)}$$

Note that

$$\frac{Z(N-k)}{Z(N+1)} = \frac{Z(N-k)}{Z(N-k+1)}\frac{Z(N-k+1)}{Z(N-k+2)}\cdots\frac{Z(N)}{Z(N+1)}$$

$$= B_1(N+1)\prod_{j=1}^{k}B_1(N-k+j)$$

We can, therefore, use the above equations to obtain the following algorithm:

Algorithm 2:
$B_1(0) = 0$
for $n = 1$ till $N - 1$

$$B_1(n+1) = \left[\rho_1 + \sum_{k=0}^{n}\left\{(\rho_2 + k + 1) \times \right.\right.$$
$$\left.\left.\prod_{j=1}^{k}\left[(n-k+j)\rho_2 B_1(n-k+j)\right]\right\}\right]^{-1}$$

Compute $B_2(N)$, $E_1(N)$ and $E_2(N)$
Finished !

Let us analyze the complexity of Algorithm 2. There are N iterations. In a typical step during the computation of $B_1(n + 1)$ there are n terms with term k requiring $O(k)$ multiplications. Therefore, the complexity of computing each $B_1(n)$ is of $O(n^2)$. It follows then that the complexity of the algorithm is $O(N^3)$ (recall that $1^2 + 2^2 + \cdots N^2 = N(N+1)(2N+1)/6$). This is a factor of N larger than the complexity of the algorithm 1.

SOME NUMERICAL EXAMPLES

Let us first consider the case $\tilde{\lambda}_1 = 0$ (no hot spots). We compare our results to some previous results for the analysis of a synchronous crossbar model. For the synchronous model it is assumed that at each input i, random independent, uniformly distributed requests are generated over all inputs with probability p, the average number of requests generated per cycle [5]. Blocked requests are discarded such that requests between clock cycles are independent. Interference arises only over contention at the switch outputs j.

In the limiting case for very large N, the bandwidth $E(N, p)$ and probability of acceptance $P_A(N, p)$ (non-blocking probability) for the synchronous model is approximated by [5]:

$$E(N, p) \simeq N(1 - \exp(-p)), \quad P_A(N, p) \simeq \frac{(1 - \exp(-p))}{p}$$

Note that asymptotically, P_A becomes independent of N. In the case of heavy traffic $p = 1$ and $N \mapsto \infty$, the classical asymptote for the acceptance probability is $1 - e^{-1} = 0.638$.

To compare this synchronous case with our model, we note that $\tilde{\rho}$ "corresponds" to p. Note that in the asynchronous case, switch interference can arise both from concurrent requests to the same input i or the same output j. Moreover, contention may also occur at any time during an active i-to-j connection. Therefore, the non-blocking probability is lower compared to the synchronous case.

In figure 1, we compute the values of the acceptance probability for both synchronous and asynchronous models.

As expected, this probability is lower for the asynchronous case. In figure 2 we fix the rate of ordinary requests ($\tilde{\rho}_2$) and vary the rate of incoming hot spot requests (from $\tilde{\rho}_1 = 0$, corresponding to a uniform access pattern, to $\tilde{\rho}_1 = \tilde{\rho}_2$). In figure 3 we fix the overall $\tilde{\rho}$ and vary $\tilde{\rho}_2$ (so that $\tilde{\rho}_1 + \tilde{\rho}_2$ is fixed). We considered 8 cases where $\tilde{\rho}_1$ varied from 0% of $\tilde{\rho}$ to 60% of $\tilde{\rho}$. For all of these cases, the blocking probability of ordinary requests appears to be insensitive to the relative values of $\tilde{\rho}_1/\tilde{\rho}$ and depends only on the overall traffic per input, namely $\tilde{\rho}$.

It is interesting to compare our conclusions on the effect of hot spots with some previous results, primarily with those presented in [6] for multistage interconnection networks. In that work it is shown by simulation that as little as %0.125 imbalance in a 1,000-node network could limit the network performance to less than 50% of its maximum value. In the analysis of a large asynchronous crossbar with hot spots, considered in this paper, we do not observe similar dramatic effects for such slight traffic imbalances.

Finally, in figure 4 we considered the efficiency (concurrency per input) for hot spot and ordinary connections. We fix $N = 100$, total traffic $\tilde{\rho} = 0.0023$ and increase the traffic of hot spot requests $\tilde{\rho}_1$. This increase in $\tilde{\rho}_1$ decreases the traffic of ordinary requests $\tilde{\rho}_2$ since $\tilde{\rho} = \tilde{\rho}_1 + \tilde{\rho}_2$ remains fixed. For higher rates of hot spot requests, the efficiency $E_1(N)$ increases, whereas the efficiency $E_2(N)/N$ of ordinary requests decreases. As we increase $\tilde{\rho}_1$, ordinary requests become "starved": most of the connections correspond to "hot spot" requests.

CONCLUSION

In this paper we presented a model of hot spots in an asynchronous crossbar switch. The model could be applicable to the performance analysis of lightwave circuit-switched networks. We derived a simple recurrence relation to exactly compute the bandwidth and non-blocking probability. In the future we hope to derive and rigorously prove the asymptotic results for this model, extend this analysis to asynchronous lightwave multistage networks, analyze the model under bursty traffic conditions and analyze other asynchronous network switches.

REFERENCES

[1] L. Bhuyan and C. Lee. An interference analysis of interconnection networks. In *Proc. International Conference on Parallel Processing*, 1982.

[2] A. Hartmann and S. Redfield. Design sketches for optical crossbar switches intended for large-scale parallel processing applications. *Optical Engineering*, 28(4):315–327, April 1989.

[3] F. Kelly. *Reversibility and Stochastic Networks*. J. Wiley and Sons, New York, 1979.

[4] F. Kelly. The optimization of queueing and loss networks. In O. Boxma and R. Syski, editors, *Queueing Theory and Its Applications*, pages 375–392. North-Holland, 1988.

[5] J. Patel. Performance of processor-memory interconnections for multiprocessors. *IEEE Trans. Computers*, 30(10):771–780, October 1981.

[6] G. Pfister and V. Norton. Hot spot contention and combining in multistage interconnection networks. *IEEE Trans. Computers*, 34(10):943–948, October 1985.

[7] G. Szego. *Orthogonal Polynomials*. Amer. Math. Soc., Providence, R.I., 1937.

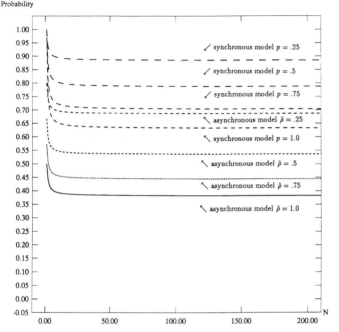

Figure 1. The Acceptance (Non-Blocking) Probability

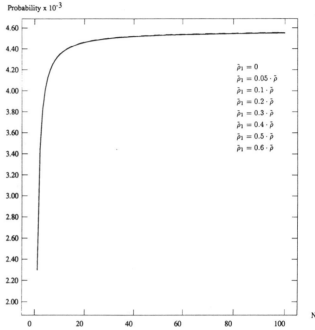

Figure 3. Effect of Hot Spots on Blocking Probability of Ordinary Requests

$(\tilde{p} = 0.0023)$

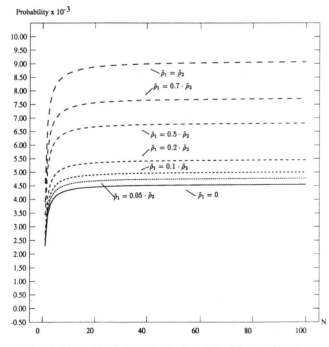

Figure 2. Effects of Hot Spots on Blocking Probability of Ordinary Requests

$(\tilde{p}_2 = 0.0023)$

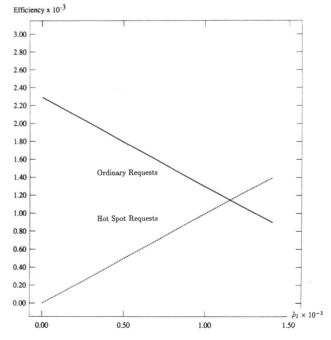

Figure 4. The Starvation Phenomenon

PERFORMANCE OF MULTISTAGE COMBINING NETWORKS[†]

Byung-Chang Kang, GyungHo Lee[††] and Richard Kain

Department of Electrical Engineering
University of Minnesota
Minneapolis, MN 55455

Abstract -- The performance of shared memory MIMD computer systems can be seriously degraded when many processors access a shared variable at the same time. A *combining network,* in which multiple requests can be combined into a single request, can mitigate this problem. Despite the benefit, the performance of the combining network has not been much investigated. Most previous studies of combining networks rely on either simulations that require long execution times or analytical methods based on unrealistic assumptions. In this paper we present a new analytic model with more practical assumptions and some analytic results concerning the performance of multistage combining networks.

1. INTRODUCTION

In shared memory parallel systems, many processors may access the same memory location at the same time. These concurrent requests create "hot-spot contention," which can create a serious performance bottleneck on the interconnection network between processors and the memory [1]. Since only one request arriving at a switch node can pass through one output port, if multiple requests desire the same output port, the other requests must wait at a buffer within the switch element (or other requests are rejected at the switch inputs--see [2]). Thus, the requests are serialized. Serialization increases network latency, which in turn limits processor utilization. When traffic is heavy, the switch buffer can be saturated, so requests from the previous stage are blocked. The blocked buffer then saturates the buffer at the previous stage, and so on. Eventually, the processors at stage 0 can be blocked from issuing further requests. This phenomenon is known as "tree saturation" [1]. Saturation reduces the effective network bandwidth, which limits the number of processors one can employ in a parallel system.

To avoid the performance limitation due to contention, *combining* has been suggested [1] [3]. With combining, multiple requests directed toward the same output port may be combined into a single request so that network traffic is reduced without reducing its throughput. Although other approaches, such as making the traffic more or less uniform throughout the system [2] [4] [6], can reduce saturation, they maintain the total traffic on the network and introduce artificial network delays. Combining seems to be a better choice because it does not degrade the throughput.

Despite the potential benefits of combining, the performance of a *multistage combining network*, a multistage interconnection network composed of combining switches, has not been fully investigated. Pfister and Norton [1] claimed that combining improved the performance, in terms of network latency, based on the simulation of 64x64 multistage network. Network delays under uniform and nonuniform traffic are analyzed in [5] but the analysis does not consider combining. Lee [7] analyzed the performance of an idealized combining network, in which an infinite number of combinable requests in a queue can be combined into a single request at a switch. In this paper, we analyze the performance of combining networks with more practical assumptions and compare the network bandwidth and delays with and without combining.

2. NETWORK AND TRAFFIC MODEL

We model the NYU Ultracomputer's combining network [1], which is a buffered rectangular omega network. We assume the following: The network is constructed with 2x2 switches. So there are n $(= \log_2 N$, N=the system size) stages in the network. The network is synchronous and packet switched. The network cycle time and the service time are both 1. The enqueueing and dequeueing processes are overlapped. Each request is a single packet. Each queue can accept up to two distinct requests, one from each input port, during each cycle. If during some cycle a queue has only one empty space and two requests that cannot be combined are directed to it, only one request is accepted by the queue and the other remains on the queue at the previous stage. At most two input requests can be combined to form a queued (combined) request. Processors have infinite buffers at their output ports and memory modules have infinite buffers at their input ports. Even if the combining queue is full, a combinable request can enter the queue if the request can be combined with an entry already in the queue.

We use the "single hot-spot traffic model [1]," in which each memory request has a fixed probability q of accessing the same variable. The requests fall into two groups: the *non-combinables*, which access each memory module with equal probability, and the *combinables*, which access the single shared variable. Let r_c be the rate of combinables and r_n be the rate of non-combinables. If r is the request rate $(=r_n + r_c)$ issued by a processor, then $r_c = qr$ and $r_n = (1 - q)r$. Let r_i and $r_{c,i}$ denote the request rate and the rate of combinables at stage i, respectively. Note that r_n is the same for all stages because non-combinables are uniformly distributed to all memory modules. The network stages are numbered starting from the processor side. Without combining, the request rate at a switch at stage i along a path to the hot-spot will be

$$r_i = r_n + 2r_{c,i-1} = r_n + 2^i r_c . \qquad (1)$$

Thus r_i can easily exceed 1 after a few stages even if q is very small. With a request arrival rate greater than 1, a finite queue is easily saturated, and requests from the previous stage will be blocked. Thus tree saturation occurs, slowing down the system. In a combining network, however, this phenomenon can be mitigated by two mechanisms: (1) Combining reduces the effective request rate at a switch so that the blocking probability at the switch can be greatly relieved. It mitigates the saturation effect. (2) Since a combined request at the final stage represents many requests that have been combined within the network, the network bandwidth can be greatly improved. So combining mitigates the serialization effect.

[†] This work was supported in part by the Graduate School at the University of Minnesota and SamSung Electronic Co. Ltd.
[††] The second author is currently with Samsung Electronics-Computers in Seoul, Korea, on leave of absence from the University of Minnesota.

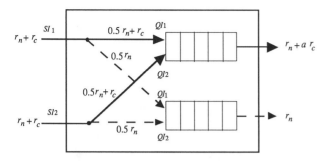

Fig. 1. Request rates at a 2x2 combining switch. Solid lines show the hot spot path. SI_i and QI_i (i=1,2) represent the input port of the switch and the combining queue, respectively. The non-combinable rate at QI_i is a half of the one at SI_i.

---

3. ANALYSIS OF COMBINING NETWORK

We model the combining queue with a Markov process. The following terms will be used in the rest of the paper:

R : The state that a combining queue in a switch is *ready to combine*. This state occurs when the combining queue contains a combinable request that can be used for combining.

e_i: The *potential combining probability*. -- The probability that the combining queue of stage i is in the R state.

l: The *queue size* -- The maximum number of requests a queue can have.

j: The *queue length* -- The number of requests in the queue.

$S_{i,j}$: The state that a queue at stage i has length j.

$p_{i,j}$: The probability that a combining queue at stage i is in state $S_{i,j}$.

b_i : The blocking probability at the queue in stage i.

t_i^y: The transition probability from state $S_{i,j}$ to state $S_{i,j+y}$ ($-1 \leq y \leq +2$).

As shown in Fig. 1, the non-combinable rate at each input port of the combining queue (QI_i) is a half of the non-combinable rate at each input port of the switch (SI_i). Let $r_m = 0.5r_n$. Table 1 shows the probability of possible combinations of request arrivals at input ports and the queue size changes that depend on the queue's state. From the table we find the probability of each possible queue size transition:

$$t_i^{+2} = b_{i+1}[r_m^2 + 2r_m r_{c,i-1}(1 - e_i)]$$

$$t_i^{+1} = b_{i+1}[2r_m(1 - r_m - r_{c,i-1}) + 2r_{c,i-1}(1 - r_m - r_{c,i})(1 - e_i)$$
$$+ (2r_m r_{c,i-1}e_i + r_{c,i-1}^2)] + (1 - b_{i+1})[r_m^2 + 2r_m r_{c,i-1}(1 - e_i)]$$

$$t_i^{-1} = (1 - b_{i+1})[2r_{c,i-1}(1 - r_m - r_{c,i-1})e_i + (1 - r_m - r_{c,i-1})^2] \quad (2)$$

where $b_i = p_{i,l-1}t_i^{+2} + p_{i,l}(t_i^{+1} + t_i^{+2})$

When the network is in equilibrium, the probability of each state is:

$$p_{i,j} = \begin{cases} \sigma_{i,0}p_{i,0} & \text{if } j = 1 \\ \sigma_{i,1}p_{i,j-2} + \sigma_{i,0}p_{i,j-1} & \text{if } 2 \leq j \leq l \end{cases}$$

and $\displaystyle\sum_{j=0}^{l} p_{i,j} = 1$ for any i. $\quad (3)$

where $\quad \sigma_{i,0} = \dfrac{t_i^{+1} + t_i^{+2}}{t_i^{-1}}, \quad \sigma_{i,1} = \dfrac{t_i^{+2}}{t_i^{-1}}$

In the NYU Ultracomputer's combining scheme, a combining queue is in the R state only if the number of combinables is odd. Also, a combinable at the head of the queue cannot be used for combining because of the overlap of the enqueueing process with the dequeueing process. Thus the potential combining probability e_i is:

$$e_i = \sum_{j=1}^{l-1} p_{i,j} \left[\sum_{k=1}^{m} \binom{j}{2k-1} r_{c,i-1}^{2k-1}(1 - r_{c,i-1})^{j-(2k-1)} \right] \quad (4)$$

where $\quad m = \begin{cases} \dfrac{j}{2} & \text{if } j \text{ is even} \\ \dfrac{j+1}{2} & \text{if } j \text{ is odd} \end{cases}$.

A combinable entering an input port of a queue can be combined with a request in the queue (if the queue is in the R state) or a combinable at the other input port. Therefore, the probability of combining is:

$$p_{comb,i} = [2r_m r_{c,i-1} + 2(1 - r_m - r_{c,i-1})r_{c,i-1}] e_i + r_{c,i-1}^2$$
$$= [2(1 - r_{c,i-1}) e_i + r_{c,i-1}] r_{c,i-1} \quad (5)$$

A combinable is removed from the traffic for each combining occurring at the switch, so the output combinable request rate at a switch is the total input combinable request rate reduced by the probability of combining at the switch:

$$r_{c,i} = 2r_{c,i-1} - p_{comb,i} \quad (6)$$

Network latency is an important performance measure. Let the *delay* of a request be the number of cycles between the request's departure and arrival at a switch. Without combining, two components affect the average delay: d_{conf}, the delay due to conflicts between requests at input ports and d_{queue}, the queueing delay. A conflict delay arises if two input requests are directed to a single queue; one of the requests has to spend one more cycle in the queue than the other. The average delay is nonzero even if the queue is empty. The average value of d_{conf} is one half of the request rate at an input port [8], which is $r_m + r_{c,i-1}$ from Fig. 1. Queueing delay arises if the queue is not empty; d_{queue} is simply the queue length($=j$).

Table 1. The changes the number of requests($=y$) in a combining queue, where:

X: No request at the input port
N: A non-combinable at the input port
C: A combinable at the input port.
R: The combining queue is in ready to combine state.
$\neg R$: The combining queue is not in state R.

Requests at		Probability of	Not blocked		blocked	
QI_1	QI_2	input combination	$\neg R$	R	$\neg R$	R
X	X	$(1 - r_m - r_c)^2$	-1	-1	0	0
X	N	$r_m(1 - r_m - r_c)$	0	0	+1	+1
X	C	$r_c(1 - r_m - r_c)$	0	-1	+1	0
N	X	$r_m(1 - r_m - r_c)$	0	0	+1	+1
N	N	r_m^2	+1	+1	+2	+2
N	C	$r_m r_c$	+1	0	+2	+1
C	X	$r_c(1 - r_m - r_c)$	0	-1	+1	0
C	N	$r_m r_c$	+1	0	+2	+1
C	C	r_c^2	0	0	+1	+1

With combining, the analysis of the delay becomes more complicated because non-combinables and combinables have different delays that depend on the queue's state. First, we consider the delay of non-combinables. Since combining reduces the queue length by 1 for each pair of combinables, d_{queue} with combining is the queue length reduced by the number of combinable request pairs: $d_{queue} = j - \lfloor$ number of combinables $/2 \rfloor$. When the queue is in the $\neg R$ state, combining cannot occur. Consequently, d_{conf} is the same as the one without combining; i.e., $d_{conf} = (r_m + r_{c,i-1})/2$. When the queue is in the R state, a combinable entering an input port will be removed through combining and will not conflict with other input requests. Therefore $d_{conf} = r_m/2$. The average delay of non-combinables ($d_{n,i}$) is:

$$d_{n,i} = \sum_{j=0}^{l} p_{i,j} \left[\sum_{k=1}^{m_1} \left(j - k + \frac{r_m}{2} \right) \binom{j}{2k-1} r_{c,i-1}^{2k-1} (1-r_{c,i-1})^{j-(2k-1)} \right.$$
$$\left. + \sum_{k=0}^{m_2} \left(j - k + \frac{r_m + r_{c,i-1}}{2} \right) \binom{j}{2k} r_{c,i-1}^{2k} (1-r_{c,i-1})^{j-2k} \right] \quad (7)$$

where

$$m_1 = \begin{cases} \dfrac{j}{2} & \text{if } j \text{ is even} \\ \dfrac{j+1}{2} & \text{if } j \text{ is odd} \end{cases} \quad \text{and} \quad m_2 = \begin{cases} \dfrac{j}{2} & \text{if } j \text{ is even} \\ \dfrac{j-1}{2} & \text{if } j \text{ is odd} \end{cases}$$

Now consider the delay of combinables. Suppose the queue is in the R state. A combinable entering the queue can jump into the middle of the queue through combining. The queueing delay of combinables depends on the distribution of the combinables in the queue. An incoming combinable must be combined with the last ($= n_c^{th}$) combinable in the queue because all other combinables have been combined already. The delay can be obtained from the distance of the n_c^{th} combinable from the head of the queue. Assuming combinables are evenly distributed among requests in the queue, the average distance of the n_c^{th} combinable from the head of the queue is:

$$\text{Average distance of } n_c^{th} \text{ request} = \frac{n_c (j+1)}{n_c + 1} - 1 \quad (8)$$

If the queue is not blocked from the next stage, this is the average delay of combinables when the queue is in the R state.

Suppose the queue is in the $\neg R$ state. Since requests cannot jump into the middle of the queue, d_{queue} of combinables is the same as d_{queue} of non-combinables. A combinable at an input port cannot conflict with another combinable at the other port of the queue. So, $d_{conf} = r_m/2$. Therefore, the average delay of combinables becomes:

$$d_{c,i} = \sum_{j=0}^{l} p_{i,j} \left[\sum_{k=1}^{m_1} \left(\frac{(2k-1)j-1}{2k} \right) \binom{j}{2k-1} r_{c,i-1}^{2k-1} (1-r_{c,i-1})^{j-(2k-1)} \right.$$
$$\left. + \sum_{k=0}^{m_2} \left(j - k + \frac{r_m}{2} \right) \binom{j}{2k} r_{c,i-1}^{2k} (1-r_{c,i-1})^{j-2k} \right] \quad (9)$$

The average delay for the traffic at stage i is:

$$d_i = \frac{r_m d_{n,i} + r_{c,i} d_{c,i}}{r_m + r_{c,i}} \quad (10)$$

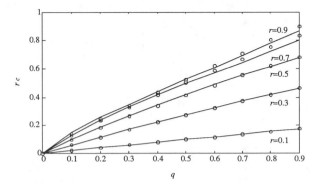

Fig. 2. The output combinable request rate at the final stage of multistage network.

4. EXPERIMENTS AND RESULTS

Consider the final stage of a multistage network. From the assumptions, requests at the final stage are never blocked from the next stage (memory). So, $t_i^{+2}=0$, $\sigma_{i,1}=0$ and $\sigma_{i,1}$ can be expressed as a function of e_i. Then Eq. (3) can be simplified as:

$$p_{i,j} = \frac{\sigma_{i,0}^j (1-\sigma_{i,0})}{1-\sigma_{i,0}^{l+1}} \quad (11)$$

From Eqs. (4) and (11), we can express e_i as a function of $\sigma_{i,1}$. Since e_i and $\sigma_{i,1}$ are functions of each other, we solve by iteration: Initially, an arbitrary value is given to e_i ($0 \le e_i \le 1$). Then calculate $\sigma_{i,0}$. From the calculated $\sigma_{i,0}$, obtain a new e_i. From the new e_i, calculate $\sigma_{i,0}$. This process repeats until the error (Δe_i) between the new e_i and the old e_i becomes negligible ($\Delta e_i < \varepsilon$). The final e_i is plugged into Eqs. (5)-(10) to obtain $r_{c,i}$ and d_i. The results obtained with $\varepsilon = 10^{-4}$ and $l=4$ are shown in Fig. 2 and Fig. 3. The figures demonstrate that our analysis (solid lines) is well matched with the simulation results (circles) for a wide range of r and q.

At the other stages b_i and t_i^{+2} ($0 \le i \le l$) may not be zero. Although it is hard to solve Eqs. (2)-(6) directly, it is possible to obtain a reasonable solution if we impose an extra iteration level to obtain the b_i values. Initially, an arbitrary value is given to b_i ($0 \le b_i \le 1$). Then we calculate e_i and $r_{c,i}$, from stage 1 to the final stage sequentially, by the same iteration method used in the final stage case. Then we have a new set of b_i ($0 \le i \le n$). Using the new set of b_i, we calculate e_i and $r_{c,i}$ again for all stages. This process repeats until the error (Δb_i) between the new b_i and the old b_i enters a tolerance range for all i ($1 \le i \le n$). To evaluate $p_{i,j}$ ($0 \le j \le l$) from the values of e_i, $\sigma_{i,0}$, b_{i+1} and $\sigma_{i,1}$, we solve Eq. (2) numerically.

The results of the analysis for a 10-stage network with $l=4$ are compared with simulation results in Fig. 4. With $\Delta b_i = 10^{-3}$ and $\Delta e_1 = 10^{-4}$, the results show that our analysis accurately explains the behavior of the multistage combining network for most practical values of r and q. For large r and q, the analysis results are not so accurate as with low r and q: $r_{c,i}$ from the analysis is higher than $r_{c,i}$ from simulations. This can be explained by the fact that the request waiting time (and the traffic rate) at later stages has a large variance [6] [9] while our analysis does not consider the variance. This large variance

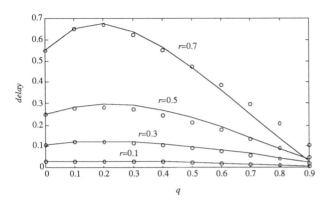

Fig. 3. The network delay at at the final stage of multistage network.

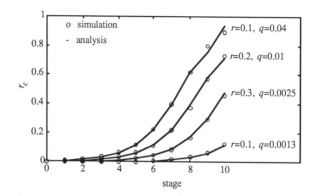

Fig. 4. The output combinable request rate at a 10-stage network.

increases the probability of combining because the probability of combining includes the second moment of $r_{c,i}$ as shown in Eq. (5).

5. COMBINING NETWORK PERFORMANCE

The network's performance can be measured by its bandwidth and delay. With a fixed physical configuration, one can increase network utilization and bandwidth by increasing the average request rate. The network delay of a non-combining network exponentially increases as the request rate approaches one [9]. Therefore, there is a trade off between bandwidth and delay. In combining networks, on the other hand, bandwidth and delay may not be mutually exclusive because a higher request rate can provide a higher combining probability that reduces delay.

From the equations in Section 4, we can obtain the bandwidth of combining networks by finding out the maximum r_0 that does not cause network saturation. The results show that the bandwidth of a combining network is larger than the bandwidth of a non-combining network. For example, a combining network has approximately 5 times larger bandwidth than a non-combining network when the system has 1024 processors, $l=4$ and $q=0.01$. If q is 0.1, the bandwidth gain becomes approximately 20. This illustrates that a larger q provides a larger bandwidth gain. The results also show that the bandwidth gain is scalable with the system size, so combining is more efficient in large systems.

From the results in [1] and our analysis, we can compare the delay of a combining network with a non-combining network. The results show that the delay in a combining network is indeed much shorter than the delay in non-combining network. As an example, the delay in a combining network is 7 times shorter than one in a non-combining network when the system has 1024 processors, $l=4$, $r=0.05$ and $q=0.005$. When q is larger, the delay difference between a combining network and a non-combining network becomes much larger because network saturation occuring in the non-combining network causes the delay to increase without bound whereas a combining network has a finite delay. When q is small, combining reduces the network delay even if network saturation does not occur on the non-combining network.

6. CONCLUSION

We have developed an analytical model that can explain the behavior of the combining network used in the NYU Ultracomputer. Simulations show that the analysis is accurate for most practical values of request rate and combinable rate.

From the model, we were able to obtain the combining probability, request rate, and delay at each stage, the network bandwidth, and the network delay. The results from the analysis show that combining can improve the network bandwidth and the delay considerably. The trend of performance improvement against network size shows that a larger system can have a better performance gain. Although combining networks are criticized because of their hardware cost, which has been estimated to be between 6 and 32 times more than that of "normal" networks [1], this performance improvement might justify their hardware cost. Since the cost-effectiveness of combining networks is scalable to the system size, our results suggest that designers of a large system aiming at cost-effective high performance should consider employing combining networks.

References
[1] G.F. Pfister and V.A. Norton, "Hot-spot contention and combining in multistage interconnection networks," Proc. 1985 Int. Conf. on Parallel Processing, pp. 790-797.
[2] R.J. Thomas, "Behavior of the Butterfly parallel processor in the presence of memory hot-spots," Proc. 1986 Int. Conf. on Parallel Processing, pp 46-50.
[3] Allan Gottlieb, Ralph Grishman, Clyde P. Kruskal and Kevin P. Mcauliffe, "The NYU ultracomputer- designing an MIMD shared memory parallel computer," IEEE tran. computer 1983, Feb. 1983, pp. 175-189.
[4] Wing S. Ho and Derek L. Eager, "A novel strategy for controlling hot-spot congestion," Proc. 1989 Int. Conf. on Parallel Processing, vol. I. pp. 14-18.
[5] Clyde P. Kruskal, Marc Snir, and Alan Weiss, "The distribution of waiting times in clocked multistage interconnection networks," Proc. 1986 Int. Conf. on Parallel Processing, pp. 12-19.
[6] D. Dias and M. Kumar, "Preventing congestion in multistage networks in the presence of hotspots," Proc. 1989 Int. Conf. on Parallel Processing, vol. I. pp. 9-13.
[7] GyungHo Lee, "A performance bound of multistage combining networks," IEEE trans. computers, Oct. 1989, pp. 1387-1395.
[8] Nian-Feng Tzeng, "Design of a novel combining structure for shared-memory multiprocessors," 1989 Int. Conf. on Parallel Processing. vol. I. pp. 1-8.
[9] Clyde Kruskal and Marc Snir, "The performance of multistage interconnection networks for multiprocessors," IEEE trans. computers, Dec. 1983. pp. 1091-1098.

EFFECT OF HOT SPOTS ON MULTIPROCESSOR SYSTEMS USING CIRCUIT SWITCHED INTERCONNECTION NETWORKS

Lizyamma Kurian and Matthew J. Thazhuthaveetil

Department of Electrical and Computer Engineering

The Pennsylvania State University

University Park, PA 16802

Abstract: It has been pointed out by Pfister and Norton that nonuniform traffic created by *hot spots* causes drastic bandwidth reduction in buffered multistage interconnection networks (MINs), due to tree saturation. Tree saturation is a congestion effect; a hot spot would cause congestion effects in any kind of interconnection network. We study hot spots in circuit-switched crossbars, multiple-buses, generalized MINs and extra-stage networks. We observe that the relative bandwidth degradation due to hot spots is higher in crossbars than in MINs. We also suggest a more meaningful measure for hot spots than reference percentage.

1. Introduction

The contention for shared lock and synchronization data in highly parallel shared-memory multiprocessors gives rise to non-uniform traffic and serious performance impacts. The effect of such a non-uniform traffic pattern consisting of a single hot spot of higher access rate superimposed on a background of uniform traffic was studied by Pfister and Norton [7]. R.Lee [4] studied the effect of varying the sources contributing to the hot spot. Pfister and Norton proposed hardware combining to solve the problem and Yew et.al [8] proposed software combining. Pfister and Norton stated that the effect is independent of network topology, switching mode (packet or circuit switched), or whether the network is used for memory access or message passing. They inferred that it requires only a multistage interconnection network (MIN) with distributed routing and a non-uniform traffic pattern to cause performance degradation due to hot spots. In this paper we show that it does not necessarily require a MIN or tree saturation to cause degradation. Any interconnection network other than a single shared bus suffers from performance degradation in presence of non-uniform traffic and as will be shown from our simulation results, under appreciable hot spots, crossbars, MINs and multiple-bus systems provide the same performance.

In this paper we present an evaluation of the impact of hot spots on the performance of various systems as measured by the effective memory bandwidth (EMBW) of the interconnection network. For all systems studied in this paper, it is assumed that all the processors are synchronized. In the presence of a hot spot of $h\%$, $h\%$ of all memory requests go to one memory location and the rest of the memory requests are uniformly distributed among all memory locations. The cycle time of all memory modules is assumed to be the same and constant and propagation delays and arbitration times associated with the interconnection are included in the memory cycle. It is also assumed that a processor issues a new request in the cycle after receiving memory service with probability r and that rejected requests are resubmitted.

The inherent memory and network conflicts in any multiprocessor system cause severe bottlenecks degrading the performance far below the ideal dotted line curve in Fig. 4 of [7] (reproduced as Fig. 1 here). We concur with [7] and [4] that hot spots cause performance degradation, but if the bandwidth attainable by a MIN under uniform referencing conditions (0% hot spot) is considered, we find that the performance degradation is not as severe as portrayed in this figure.

2. Hot Spots in Crossbars

It is popularly assumed that the main cause of performance degradation due to hot spots is that the cold traffic (traffic directed at the non hot memory modules) is affected by the hot traffic, which may be the case for packet switched buffered MINs. In crossbar systems, there is no network contention and cold traffic can be directed to the corresponding memory module as soon as it is generated. Since cold traffic is not affected by hot traffic, it may seem that a crossbar system will not be affected by hot spots. But on detailed analysis and simulation, it becomes evident that circuit-switched crossbars are severely affected by hot spots.

We may derive the bandwidth of a system with a crossbar network, in presence of a hot spot of percentage h as follows. All the assumptions regarding the system stated in the introductory section apply here. If the system has p processors and m memories, and a request rate of r, the number of processors referencing the hot memory module in the first cycle is

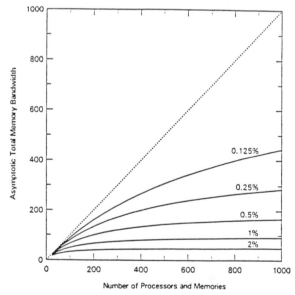

Fig. 4.   Asymptotically maximum total network bandwidth as a function of the number of processors for various fractions of the network traffic aimed at a single hot spot.

Fig.1 Results from [7]

$$prh + \frac{pr(1-h)}{m} \qquad (2.1)$$

A maximum of one of these requests is satisfied per cycle. Hence, n_1, the minimum number of processors that queue up for the hot memory module after 1 cycle is

$$n_1 = prh + \frac{pr(1-h)}{m} - 1 \qquad (2.2)$$

In the second cycle, all these blocked processors resubmit their requests. In addition, processors from the group that requested other modules in the first cycle (or those that requested none), may reference the hot memory module depending, on the request rate and hot spot percentage. Therefore, the number of processors requesting the hot memory module in the second cycle is

$$n_1 + (p - n_1)rh + \frac{(p - n_1)r(1 - h)}{m} \qquad (2.3)$$

and at most one of them is satisfied. Hence, n_2, the number of processors that queue up at the end of the second cycle is

$$n_2 = n_1 + (p - n_1)rh + \frac{(p - n_1)r(1 - h)}{m} - 1 \qquad (2.4)$$

Proceeding in a similar fashion, n_j, the number of processors that are in the queue after the jth cycle is

$$n_j = n_{j-1} + (p - n_{j-1})rh + \frac{(p - n_{j-1})r(1 - h)}{m} - 1 \quad (2.5)$$

Now we can derive the steady state minimum blocking probability,

$$PB_{ss} = \frac{n_j|_{j \to \infty}}{pr} \qquad (2.6)$$

The upper bound on steady state acceptance probability is then given by

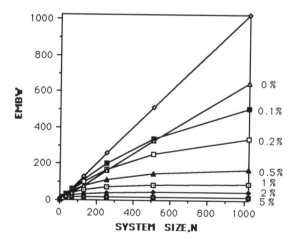

Fig.2 Crossbar asymptotic limits calculated from eq. 2.8, along with crossbar bandwidth for uniform referencing superimposed on it.

$$PA_{ss} = 1 - \frac{n_j|_{j \to \infty}}{pr} \qquad (2.7)$$

Hence, the upper bound on effective memory bandwidth is

$$EMBW = PA_{ss}pr = pr - n_j|_{j \to \infty} \qquad (2.8)$$

This is an upper bound since contention for cold memory modules is not considered. Under appreciable hot spots, memory contention among the cold modules will be negligible and hence ignoring this factor does not completely invalidate the expression.

The asymptotic limits calculated from equation 2.8 are plotted in Fig. 2 along with bandwidth from a crossbar for 0% hot spots. It is interesting to observe that equation 2.8 yields the same results as the asymptotic limits in [7], which emphasizes the fact that the inherent nature of hot spots is sufficient to create serious performance degradation in spite of a network that offers no contention. Simulation was performed to confirm the analytical results. It is observed that the upper bound in equation 2.8 approaches the exact value under appreciable hot spots. Under uniform referencing or negligible hot spots the expression deviates from the exact value significantly, since contention among cold memory modules is ignored. Under uniform referencing conditions, expressions from [6] could be used to obtain the bandwidth. The effective memory bandwidth of crossbar systems with number of processors and memories ranging from 8 to 256, obtained from simulation with request rate 1.0 appear in Fig. 3. In this figure, we are also comparing the crossbar with MINs and extra-stage MINs as will be explained in the forthcoming sections. Hot spots of 1%, 2% and 5% are considered here.

3. Hot Spots in Bus Based Systems

Multiple-bus systems do suffer from performance degradation due to hot spots, but the relative degradation is less than that of a crossbar. In Fig. 4, we present simulation results of a 16×16 system and 12×12 system with request rate 1.0. We have plotted results for bus numbers N, $\frac{N}{2} + 1$, $\frac{N}{4}$ and the single bus case, where N is the number of processors and memory modules. Hot spots of 5%, 10%, 32% and the uniform referencing case are considered. It may be observed that as the intensity of the hot spot increases, the relative difference in bandwidth between systems with fewer numbers of buses and the corresponding crossbar tends to zero. Hence it is safe to conclude that the crossbar is even more underutilized than what was mentioned in [3] and [2]. This finding also stresses the fact shown by them that multiple bus interconnection has better characteristics than that of a crossbar.

One may also observe from Fig. 4 that a single shared bus interconnection is not affected by hot spots. This result is to be expected since regardless of the access request distribution, a single bus connection could only connect one memory module to a processor per cycle.

4. Hot Spots in MINs

Previous research on hotspots concentrated on hot spots in packet-switched MINs. We conducted simulation studies on the impact of hot spots on circuit-switched MINs. Our results confirm Pfister and Norton's statement that the degradation does not depend on whether the network is circuit-switched or packet-switched. The degradation experienced by circuit-switched networks emphasizes that tree saturation, the cause of degradation in packet switched networks need not necessarily be present in or-

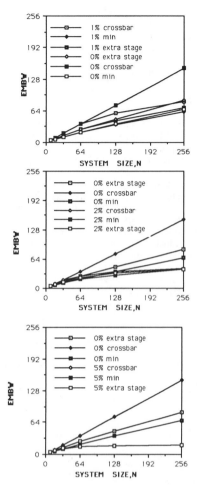

Fig.3 Effective memory bandwidth of crossbar, MIN and extra-stage networks, as a function of the number of processors, for hot spot percentages of 1%, 2% and 5%.

der to cripple networks. The effective memory bandwidths of MINs of sizes ranging from 8 to 256 are presented in Fig. 3, which also provides a comparison between crossbars and MINs. We considered MINs constructed with 2×2 switching elements.

5. Performance of Extra-stage MINs

It has been suggested that extra-stage MINs would yield higher bandwidth than ordinary MINs, though little performance evaluation effort has gone in this direction. Since the extra-stage network was proposed primarily for reliability concerns, most of the work in that area has been concerned with evaluating parameters related to fault tolerance. We study the performance of extra-stage MINs in the presence of hot spots, as well as assuming uniform referencing.

The extra-stage MINs considered here are derived from generalized MINs or delta networks by the addition of one stage of switching elements at the input side of the network. This extra-stage provides the MIN with multiple distinct paths between each source and destination. These networks are different from those in [1], in that no multiplexers or demultiplexers are used and that the extra-stage is always operational.

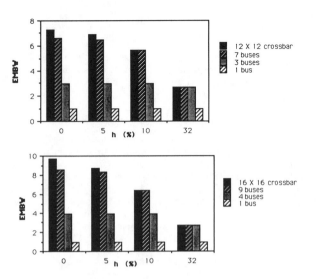

Fig.4 Comparison of effective memory bandwidth of crossbar and multiple-bus systems under various hot spot percentages, for system sizes 16 X 16 and 12 X 12.

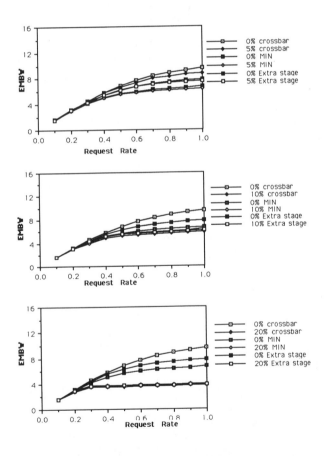

Fig.5 Effective memory bandwidth vs request rate for 16 X 16 crossbar, MIN and extra-stage MIN systems under various hot spot percentages.

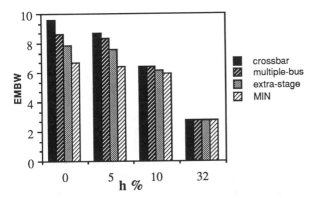

Fig.6 Comparison of 16 X 16 systems with crossbar, multiple-bus system with 9 buses, MIN and extra-stage MIN interconnection, for various hot spot percentages.

The bandwidth obtained by these extra-stage MINs is plotted in Fig. 3 for various hot spot percentages. Since crossbar and MIN bandwidths are also plotted in the same figure, one may obtain a direct comparison between the different systems. As expected, the extra-stage MIN provides a bandwidth in between that of the crossbar and the general MIN under uniform referencing conditions and mild hot spots. The difference in bandwidth reduces as the hot spot intensity increases and as seen from Fig. 3, at 5% hot spot, the curves for crossbars, MIN and extra-stage are clustered together. For smaller systems like 8×8 and 16×16, there is still difference in performance, but this is not visible in the scales chosen for this figure. In order to provide a detailed picture of a smaller system, performance of a 16×16 system is plotted in Fig. 5 for various request rates from 0.1 to 1.0, for various hot spot percentages. The relative difference in bandwidth breaks down as the hot spot intensity approaches 20%.

6. Measuring Hot Spots in Degrees

We observe that when the non-uniformity of referencing due to the hot spot does not exceed twice the normal frequency of referencing, the degradation effect is not very adverse. For example, in a 16×16 system with uniform referencing, each memory is referenced with the frequency 5.88%. Hot spots of 1% or 2% do not critically affect this system, but as the hot spot intensity approaches 5%, the system starts to show signs of 'fatigue'. For an $N \times N$ system, the threshold hotspot frequency that starts to affect the system critically may be identified as $\frac{100}{(N+1)}$ %. Since the severity of effect of an h% hot spot clearly depends on the size of the system, a size dependent description may be used instead of describing hot spots in terms of reference percentages. When the hot memory module is referenced with twice the frequency of others, we describe the hot spot as a one 'degree' hot spot; if it is referenced with thrice the frequency, the intensity of the hot spot is two 'degrees'. The hot spot percentage corresponding to d degrees is thus calculated as $\frac{100d}{(N+d)}$ %. Our simulation results show that the percentage bandwidth degradation caused by a d degree hot spot is the roughly the same for corresponding systems of any size. For instance, a 1 degree hotspot causes a degradation of roughly 10% in a crossbar, whereas for MINs, multiple-buses and extra stage MINs the degradation is about 5% only. Quantifying hotspots

in degrees also enables one to say that in presence of a hotspot of 4 degrees, crossbars, MINs, multiple-buses etc provide the same bandwidth. One may observe this very clearly from Fig. 6 where some results from Fig. 3 and Fig. 5 are combined to give a comprehensive comparison of all systems studied in this paper.

7. Conclusion

Tree saturation is typical of buffered packet-switched MINs, but performance degradation due to hot spots is observed in all networks. In fact, it is only nominally dependent on the specific nature of the network. The extent by which the bandwidth degrades is greater for crossbars than for any other network. Crossbars are thus more vulnerable to hot spot degradation than MINs and they are heavily underutilized in the presence of high hot spots. If large hot spots are common, we may conclude that crossbars are unacceptable not only for cost reasons, but also for performance reasons. In general, we observe that a network with better performance in the absence of hot spots suffers higher relative degradation in the presence of hot spots.

References

1. G.B.Adams III and H.J.Siegel, "The Extra-stage Cube : A Fault Tolerant Interconnection for Supersystems", IEEE Transactions on Computers, May 1982.

2. C.R.Das and L.N.Bhuyan, "Bandwidth availability of multiple bus multiprocessors", IEEE Transactions on Computers, vol.C-34, No.10, October 1985.

3. T.Lang, M.Valero and I.Alegre, "Bandwidth of crossbar and multiple bus connections for multiprocessors", IEEE Transactions on Computers, Vol. C-31, No.12, pp 1227-1234, Dec 1982

4. R.Lee, "On "hot spot" contention", ACM SIG Computer Architecture, vol.13, pp. 15-20, Dec 1985.

5. Y.C.Liu and C.J.Jou,"Effective Memory Bandwidth and Processor Blocking Probability in Multiple-Bus Systems", IEEE Transactions on Computers, Vol. C-36, No.6, pp. 761-764, June 1987.

6. J.H.Patel, "Performance of Processor-Memory Interconnections for Multiprocessors", IEEE Transactions on Computers, Vol C-30, No.10, pp 771-780, October 1981.

7. G.F.Pfister and V.A.Norton, ""Hot Spot" contention and combining in multistage interconnection networks", IEEE Transactions on Computers, Vol. C-34, No.10, pp 943-948, Oct 1985.

8. P.C.Yew, N.F.Tzeng and D.H.Lawrie, "Distributing Hot-Spot Addressing in Large-Scale Multiprocessors", IEEE Transactions on Computers, Vol. C-36, No.4, pp 388-395, April 1987.

9. G.B.Adams III, D.P.Agrawal and H.J.Siegel, " A Survey and Comparison of Fault-Tolerant Multistage Interconnection Network", IEEE Computer, June 1987.

10. V.P.Kumar and S.M.Reddy, "Augmented Shuffle Exchange Multistage Interconnection Networks", IEEE Computer, June 1987, pp 30-40.

11. T.N.Mudge, J.P.Hayes, G.D.Buzzard and D.C.Winsor, "Analysis of Multiple Bus Interconnection Networks", Proceedings of the 1984 Conf. on Parallel Processing, pp. 228-232.

RECONFIGURATION OF COMPUTATIONAL ARRAYS WITH MULTIPLE REDUNDANCY[*]

Rami Melhem
and
John Ramirez

Department of Computer Science
The University of Pittsburgh
Pittsburgh, PA 15260

ABSTRACT

Multiple redundancy schemes for fault detection and correction in computational arrays are proposed and analyzed. The basic idea is to embed a logical array of nodes onto a processor/switch array such that d processors, $1 \leq d \leq 4$, are dedicated to the computation associated with each node. The input to a node is directed to the d processors constituting that node, and the output of the node is computed by taking a majority vote among the outputs of the d processors. The proposed processor/switch array is versatile in the sense that it may be configured as a non-redundant system or as a system which supports double, triple or quadruple redundancy. It also allows for spares to be distributed in the processor/switch array in a way that permits spare sharing among nodes, thus enhancing the overall system reliability.

If some nodes or switches in the array are initially defective or become faulty at run time, a reconfiguration algorithm may be used to remap the array and restore its specified functionality. We study the reconfiguration problem for arrays with multiple redundancy and we explore many possible reconfiguration algorithms that apply to linear and two dimensional arrays. The performance of the algorithms are analyzed analytically using Markov techniques, using probability arguments, and via simulation.

1. INTRODUCTION

Most research in fault tolerant computational arrays has concentrated on defect avoidance and fault coverage (see for e.g. [1, 2, 8, 13, 14, 16]) while only few research efforts have been directed toward fault detection and correction in such arrays. The roving spare technique [15] and the weighted checksum coding [3, 6] are two approaches that may be used to detect (and in some cases correct) transient or permanent run-time faults. However, in these approaches, a latency period may elapse before faulty processors are detected. If the processor array is subject to severe recovery time constraints, or if the production of faulty results may be disastrous, then the use of modular redundancy seems appropriate.

Active modular redundancy has been used in highly reliable computing systems (see e.g. [4, 5, 10, 17]) to detect and dynamically mask faults. Recently, Kiskis and Shin [7] suggested a technique for embedding triple modular redundancy into hypercubes by assigning each task to three processors in the hypercube. Their goal is to mask any single fault and yet, retain the logical hypercube connection. Using the same principle, modular redundancy in the context of computational arrays may be achieved by replicating each node in a logical array d times. The input to a node is directed to its d replicas, and the output of the node is computed by taking a majority vote among the outputs of the d replicas. In Section 2 of this paper, we introduce a versatile architecture which is based on interleaved arrays of processors and switches, and which permits the implementation of different degrees of redundancy. Specifically, logical arrays may be embedded onto this architecture with $d=1$ for no fault tolerance, $d=2$ for fault detection, $d=3$ for fault correction/masking and $d=4$ if additional sparing is required.

The flexibility of the suggested architecture also allows for reconfiguration strategies that may be applied at fabrication time, compile time or at run time. At fabrication or compile time, reconfiguration may be applied to embed a specified logical array onto the architecture in a way that avoids defects or faults, respectively. At run time, reconfiguration may be used to repair the array after faults or to improve the quality of the embedding, thus improving the reliability of the system [11]. The fault model applied in this paper assumes that processors and switches may fault, but that a faulty switch may be used as a short circuit connection [9].

In Section 3, we introduce run-time reconfiguration strategies for linear arrays, and we analyze them using Markov chain techniques. In Section 4, we present two-dimensional versions of the reconfiguration algorithms. The Markov analysis of the two-dimensional versions leads to an explosion in the number of states. For this reason, we use a probabilistic approach to analyze that case, and we support our analysis by providing simulation results. We conclude in Section 5 with some remarks.

2. A VERSATILE ARCHITECTURE FOR LINEAR REDUNDANT ARRAYS

Consider the Processor Switch/Voter Array (PSVA) shown in Figure 1(a). In this figure, each circle represents a processor and each square represents a switch/voter element that we will refer to as "switch". In general, a linear PSVA which contains $2(n_s+1)$ processors, $proc_{q,i}$, $q=0,1$, $i=0,...,n_s$, and n_s switches, $switch_i$, $i=1,...,n_s$, may be constructed by connecting each $switch_i$ to the four processors, $proc_{q,i-1}$ and $proc_{q,i}$, $q=0,1$. In order to obtain a versatile architecture that is able to support multiple redundancy, we assume that each $switch_i$, is a flexible device which may hold active connec-

---

[*] This work is, in part, supported under NSF grant MIP-8911303.

tions to a subset C_i of the four processors connected to it. Depending on c_i, the cardinality of C_i, $switch_i$ may be in one of the four following modes that are described in conjunction with Figure 1(b).

1) If $c_i = 0$, then $switch_i$ is in a *pass* mode in which I_i is connected to O_i,

2) If $c_i = 1$, then $switch_i$ is in a *straight* mode in which data received on the input I_i is sent to the processor in C_i, and the output of that processor is sent to O_i.

3) If $c_i = 2$, then $switch_i$ is in a *checking* mode in which data received on the input I_i is sent to the two processors in C_i. The outputs of these two processors are compared, and if the two outputs agree, the result is sent to O_i, otherwise, an error flag is set.

4) If $c_i > 2$, then $switch_i$ is in a *voting* mode in which data received on the input I_i is sent to all the processors in C_i. The results of a majority vote on the outputs of the processors in C_i is sent to O_i. If one of the outputs disagrees with the majority, the corresponding processor is removed from C_i.

In the above discussion, it was assumed that the PSVA will be used to embed linear arrays in which data flow is uni-directional. Arrays with bi-directional data flow may be supported if the four above switch modes are redefined in terms of bidirectional flow.

The embedding of an n_a-node logical array onto an n_s-switch PSVA may be specified in terms of a mapping function, *map*, which specifies for each node i, $1 \le i \le n_a$, the switch associated with that node. The processors associated with node i are the ones in $C_{map(i)}$. That is the ones that are actively connected to $switch_{map(i)}$. For example, if each node is to consist of two processors and a switch in a checking mode, then $map_2(i) = i$ and $C_i = \{proc_{0,i-1}, proc_{1,i-1}\}$. For triply redundant systems in which a node consists of three processors and a switch in a *voting* mode, the embedding is specified by

$$map_3(i) = i + \lfloor \frac{i-1}{2} \rfloor \qquad i = 1, \cdots, n_{3,max} = n_s - \lfloor \frac{n_s}{3} \rfloor$$

and $C_{map_3(i)} = \{proc_{0,map_3(i)-1}, proc_{1,map_3(i)}, proc_{q,map_3(i)-q}\}$, where $q = 0$ or 1 if i is even or odd, respectively. Finally, for quadruple redundancy, the mapping is

$$map_4(i) = 2i - 1 \qquad i = 1, \cdots, n_{4,max} = n_s - \lfloor \frac{n_s}{2} \rfloor \quad (1)$$

and $C_{map_4(i)} = \{proc_{q,map_4(i)-k} ; q = 0,1 \text{ and } k = 0,1\}$. For map_3 and map_4, any switch that is not used to embed a node is set to the *pass* mode. For example, Figure 1(c) shows the embedding of a 4-node array into the PSVA of Figure 1(a) using map_4. In this figure, as well as in the rest of this paper, bold lines indicate active connections and thin lines indicate non-active connections.

The reliability analysis for PSVA arrays is straight forward. For instance, consider the mapping described by equation (1) and assume that the PSVA fails when at least one node loses its fault correction capability. That is, when for some i, either $switch_{map_4(i)}$ fails or two processors connected to that switch fail. The system's reliability in this case is

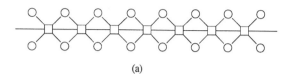

(a)

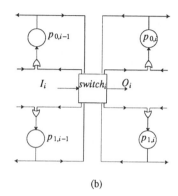

(b)

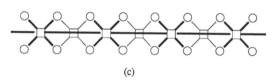

(c)

Figure 1 - A linear PSVA with $n_s = 7$

$$R_{map_4} = (r^4 + 4 r^3 (1-r))^{n_{4,max}} r_s^{n_{4,max}}$$

where r is the reliability of each processor and r_s is the reliability of each switch. That is, r_s is the probability that a switch cannot function properly in any mode except the *pass* mode.

In addition to the flexibility of choosing the required degree of redundancy in PSVA's that are initially fault free, it is also possible to embed redundant arrays onto defective PSVA's. Specifically, given a PSVA and assuming that some processors and switches in the array are defective, then it is possible to embed a logical linear array with a given redundancy onto the defective PSVA if defective switches can operate correctly when set to the *pass* mode. For example, in Figure 2, we show the embedding of a triply redundant 4-node array onto a defective 10-switch PSVA. In this figure, an X indicates a fault.

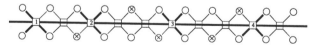

Figure 2 - Embedding a triply redundant array onto a 10-switch PSVA using Greedy

One way of embedding a maximal d-redundant logical array, $d \le 4$, onto a defective PSVA with n_s switches is to apply the following greedy algorithm (referred to as Greedy):

1) Let $node = 0$ and mark all defective/faulty processors in the PSVA as $unavailable$;

2) For $k = 1, \cdots, n_s$ do

If $switch_k$ is not defective/faulty then

   2.1) Let H_k = set of processors that are not marked and are connected to $switch_k$;

   2.2) If $|H_k| \geq d$ then

      2.2.1) Let C_k be a subset of H_k that contains exactly d processors. Give the processors $proc_{0,k-1}$ and $proc_{1,k-1}$ a higher priority for inclusion in C_k ;

      2.2.2) Mark the processors in C_k $unavailable$;

      2.2.3) Let $node = node + 1$ and $map_g(node) = k$;

3) $n_g = node$; /* the size of the embedded array */

Although greedy algorithms are not usually optimal, it may be shown [12] that the above greedy strategy embeds a maximal logical linear array. That is, no other algorithm can embed an array with more than n_g nodes in the PSVA. The yield, $Y_{Greedy}(d, n_a, n_s)$, resulting from applying Greedy to embed a n_a-node logical array, for a given n_a, onto a n_s-switch PSVA is the probability of successfully completing the embedding. This yield may be computed using a layered Markov chain [12].

A problem different from embedding a maximal size array onto a defective PSVA is the embedding of a fixed size array onto a defective PSVA so as to maximize the reliability of the resulting system. For example, consider a mapping $map()$ which embeds a n_a-node logical array onto a n_s-switch PSVA such that each node, i, $i=1, \cdots, n_a$, is mapped to a switch, $switch_{map(i)}$, with $c_{map(i)} \geq 3$. That is a switch which is actively connected to at least three non-defective processors. Clearly, if $n_{a,3}$ and $n_{a,4}$ are the numbers of nodes mapped to switches with $c_{map(i)} = 3$ and 4, respectively, then $n_{a,3} + n_{a,4} = n_a$. Now if system failure is declared when the array is not capable of correcting errors, then the reliability of the PSVA system is

$$R_{map,3} = (r^3)^{n_{a,3}} \ (r^4 + 4r^3(1-r))^{n_{a,4}} \ (r_s)^{n_a}$$
$$= (r^3 r_s)^{n_a} \ (4 - 3r)^{n_{a,4}}$$

In order to obtain a mapping which maximizes the reliability, we need to find the mapping that maximizes $n_{a,4}$. For the details of a quadratic algorithm that finds such a mapping we refer to 12]. However, both the quadratic algorithm and Greedy are centralized reconfiguration algorithms suitable for off line defect tolerance. In the next section, we will show that, in addition to tolerating initial defects, it is possible to apply run time reconfiguration to increase the life expectancy of PSVA's.

3. RUN TIME RECONFIGURATION OF PSVA

Run-time faults may be dealt with in PSVA's by providing enough redundancy at each node. For example, if the mapping (1) is used to map a logical array onto a non-defective PSVA, then four processors will constitute each node; three processors may be used in a TMR mode and the fourth may be used as a spare to replace any of the other three. This idea of using sparing along with TMR has been used in the design of FTMP [5], where a pool of spares is made available to replace any faulty unit in a triply redundant module. In PSVA's, however, the use of spares is restricted in the sense that a spare can only replace one of three specific processors. This restriction greatly reduces the survivability of the system.

In order to improve the survivability of the PSVA system, a run-time reconfiguration algorithm may be applied when a node becomes defective, that is, when two of the four processors constituting that node fail. This algorithm should remap the logical array onto the PSVA such that at least three processors constitute each logical node. Greedy (with $d=3$) may be applied for the reconfiguration. However, this may require extensive remapping of nodes, and thus extensive relocation of processors, which may be prohibited at run time.

3.1. A Shift_1 reconfiguration algorithm

Ideally, a run-time reconfiguration algorithm should be a simple algorithm which may be executed distributively and which only requires the remapping of defective nodes (constant time complexity). If the initial mapping is given by (1), then alternate switches are used to map consecutive nodes, and a simple remapping strategy may be applied. Namely, if node i is defective and node $i-1$ is mapped to a switch connected to four non-faulty processors, then shift node i one switch to the left. In other words, let node i borrow one processor from node $i-1$. If such a shift fails, then a shift to the right is attempted if one processor can be borrowed from node $i+1$. Otherwise, the remapping is declared unsuccessful. An implicit assumption is that no two processors become faulty simultaneously, and that no fault occurs during reconfiguration. The above reconfiguration policy implies the following:

1) A node can borrow at most one processor from one of its neighbors.

2) A node can borrow a processor from one of its neighbors only if that neighbor has not been shifted previously due to other faults.

Specifically, if node $i-1$ has not been previously remapped, then it cannot lend node i more than one processor, while if, due to previous faults, node $i-1$ has been shifted to the left, then it did so because more than one of the processors connected to $switch_{map(i-1)}$ was faulty. Given that $i-1$ cannot borrow more than one processor from node $i-2$, then it cannot leave any processor to node i. The same applies to right shifts. Note that the second implication does not mean that two neighboring nodes i and $i-1$ cannot be both shifted in a given configuration. For instance, consider the following case: node i becomes defective and node $i-1$ is not shifted. Thus, i is shifted to the left. Later, $i-1$ becomes defective and $i-2$ is not shifted. Thus, $i-1$ is shifted to the left. The result is a configuration in which both i and $i-1$ are shifted.

The reconfiguration policy described above may be executed distributively and requires that only the defective nodes be shifted. With such a local reconfiguration algorithm, the ability of the PSVA to survive a given number of faults, nf, depends on the order at which the faults occur. The worst case is when the order of faults prohibits configurations that have two shifted neighboring nodes. Hence, given a certain distribution of nf faults, the worst case probability that the distributed algorithm will survive the nf faults is equal to the probability of success of a global reconfiguration algorithm which examines the defective nodes in any order, and tries to remap those nodes such that the resulting configuration does not have shifted neighboring nodes. Such a global algorithm, which we call Shift_1, is described next for the case where the initial mapping is $map(i)=2i-1$. Although the algorithm is simple, we will use a precise formal notation to describe it. This will simplify formal performance analysis.

Given a defective PSVA, Shift_1 finds (if possible) a new mapping, $remap(i) = map(i) + \Delta(i)$, in which no node has less than three processors. The algorithm considers the logical nodes in the order $i=1,...,n_a$, and for each i, sets $\Delta(i) = 0$, -1 or 1. The function $R(i)$ is used to keep track of the number of non-faulty processors that node i makes available for possible use by node $i+1$, with $R(i)=-1$ meaning that node i had borrowed a processor from node $i+1$. The number of non-faulty processors connected to $switch_k$ is indicated by h_k. Also, $OK(switch_k)$ and $OK(proc_{u,k})$ are predicates that are true if $switch_k$ and $proc_{u,k}$ are not faulty, respectively, and $OK(x, y)$ is used to indicate that both $OK(x)$ and $OK(y)$ are true.

For $i = 1, \cdots, n_a$ do

1) IF $OK(switch_{map(i)})$ and $R(i-1) \neq -1$ and $h_{map(i)}=4$, THEN
    1.1) set $\Delta(i) = 0$
    1.2) if $OK(switch_{map(i)+1})$ then set $R(i)=1$ else set $R(i)=0$,

2) ELSE IF $OK(switch_{map(i)})$ and [$R(i-1) \neq -1$ and $h_{map(i)} = 3$]
    or [$R(i-1) = -1$ and $h_{map(i)} = 4$]
    THEN set $\Delta(i) = 0$ and $R(i)=0$,

3) ELSE IF $i>1$ and $OK(proc_{0,map(i)-1}, proc_{1,map(i)-1})$
    and $R(i-1) = 1$,
    THEN set $\Delta(i) = -1$ and $R(i) = 0$,

4) ELSE IF $i < n_a$ and $OK(switch_{map(i)+1})$ and $R(i-1) \neq -1$,
    and $OK(proc_{0,map(i)+1}, proc_{1,map(i)+1})$
    THEN set $\Delta(i) = 1$ and $R(i) = -1$,

5) ELSE declare the remapping not successful

More descriptively, for a given logical node i, the algorithm first tries $remap(i) = map(i)$ (steps 1 and 2). This is possible if $switch_{map(i)}$ is not faulty and is connected to at least three non faulty processors ($h_{map(i)} \geq 3$). Moreover, if node $i-1$ had borrowed one of the processors connected to $switch_{map(i)}$, then $h_{map(i)}$ should be equal to 4. Note that, in step 1, we set $R(i)=1$ only if $switch_{map(i)+1}$ is not faulty because, otherwise, node $i+1$ cannot be shifted to the left, and thus, effectively, $R(i)=0$. If steps 1 and 2 fail, then a shift to the left is possible (step 3) if node $i-1$ had a extra processor (indicated by $R(i-1) = 1$). Finally, in step 4, a shift to the right is attempted. However, such a shift is only attempted if node $i-1$ had not been shifted to the right ($R(i-1) \neq -1$). This condition is added to prohibit configurations in which two neighboring nodes are shifted.

Let p and p_s be the probabilities that a processor and a switch, respectively, is faulty. We may find the probability that Shift_1 will successfully complete the remapping by using a Markov chain. The chain consists of four states, s_0, s_1, s_{-1} and s_f, corresponding to the result of executing an

iteration in Shift_1 (see Figure 3). After a certain iteration i, the Markov process is in state s_f if Shift_1 fails to map node i (step 5). Otherwise, the process is in state s_0, s_1 or s_{-1} if $R(i) = 0$, 1 or -1 respectively. If after iteration $i-1$, the process is in state s_u for some $u =0$, 1, -1 or f, then the state of the process after iteration i will depends on the probabilities of $OK(switch_{map(i)})$ and $OK(switch_{map(i)+1})$ (each is equal to p_s), and on the fault probabilities of the four processors connected to $switch_{map(i)}$. To simplify the notation, we use $q_s=(1-p_s)$ and we use $\pi_4 = p^4$ for the probability that the four processors connected to $switch_{map(i)}$ are not faulty. Similarly, we use $\pi_3=p^3(1-p)$, $\pi_2=p^2(1-p)^2$ and $\pi_f=4p(1-p)^3+(1-p)^4$.

Let σ_u^i be the probability of being in state s_u after the ith transition. That is, for $u=0,1,-1$, σ_u^i is the probability that $R(i)=u$. Let also σ^i be the vector containing σ_u^i, for $u=0,1,-1$ and f, in that order. With this, the Markov process is described by

$$\sigma^i = A \ \sigma^{i-1}$$

where, the transition matrix, A, may be found directly from the algorithm to be:

$$
\begin{bmatrix}
\pi_4 p_s q_s + 4\pi_3 p_s & \pi_4 q_s + 2\pi_3(1+p_s)+\pi_2 & \pi_4 p_s & 0 \\
\pi_4 p_s^2 & \pi_4 p_s & 0 & 0 \\
\pi_4 q_s p_s + 2\pi_3 p_s p_s + \pi_2 p_s & 2\pi_3 q_s p_s + \pi_2 p_s & 0 & 0 \\
\pi_4 q_s^2 + \pi_3(2q_s+2q_s^2)+5\pi_2+\pi_2 q_s+\pi_f & 4\pi_3 q_s^2+4\pi_2+\pi_2 q_s+\pi_f & 1-\pi_4 p_s & 1
\end{bmatrix}
$$

Clearly, the first node cannot borrow a processor from its left neighbor (it does not have one). Hence, the Markov process is initiated at state s_0. Also, the last node, n_a, cannot be shifted to the right. That is, the embedding is successful only if, after n_a iterations, the Markov process is in state s_0 or s_1. Hence, the probability that Shift_1 is successful is:

$$P_{Shift\_1}(n_a) = Pr(remap \text{ is not } defective)$$
$$= (1 \ 1 \ 0 \ 0) \ A^{n_a} \ (1 \ 0 \ 0 \ 0)^T$$

Note that P_{shift_1} includes the probability that the original mapping, map, was not defective. In order to isolate the effect of restructuring, we compute the probability that the original mapping, $map(i) = 2i-1$ is not defective. This is given by

$$P_{map}(n_a) = Pr(map \text{ is not } defective) = (p_s \ (\pi_4 + 4\pi_3))^{n_a}$$

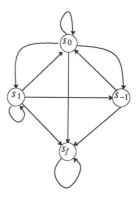

Figure 3 - Markov chains for the Shift_1 algorithm

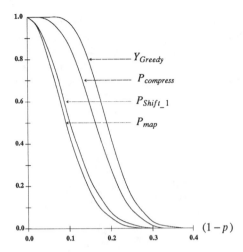

Figure 4 - comparative performance of different restructuring algorithms ($n_a = 15, n_s = 29$).

Both P_{shift} and P_{map} are shown in Figure 4 for a 15-node array mapped to a 29-switch PSVA. In that figure, we also plot $Y_{Greedy}(3,15,29)$ because it is the upper bound on the performance of any remapping algorithm that tries to reduce the number of defective logical nodes to zero. The fourth curve, that of $P_{compress}$, shows the performance of a slightly more complicated run-time restructuring algorithm that we discuss next.

3.2. A Shift_1/compress reconfiguration algorithm

In order to improve the probability of successful remapping, we may allow the run-time algorithm to cause, if necessary, successive shifts of neighboring nodes. Specifically, if node i is defective and it cannot be remapped to $switch_{map(i)-1}$ because this will make node $i-1$ defective, then we try to remap node $i-1$ to $switch_{map(i-1)-1}$, and we repeat this process inductively. The process stops successfully if no more shifting is needed, and unsuccessfully if the shift is not possible because of the unavailability of non-faulty processors. If the shift to the left with compression fails, then a similar shift is attempted to the right before reconfiguration is declared unsuccessful. Compression is expected to improve the performance of the run time restructuring, and still allows for distributed restructuring. However, the run time complexity of the distributed restructuring increases from constant to $O(n_a)$. Given that each node may not be shifted by more than one switch, we call this algorithm Shift_1/compress.

With compression allowed during reconfiguration, the probability of the PSVA to survive a given number of faults, nf, does not depend on the order at which the faults occur. Hence, this probability may be computed by studying a global algorithm, which, given an initial mapping, $map()$, and a distribution of nf faults, considers the logical nodes in the order $i=1,...,n_a$, and for each node finds $remap(i) = map(i) + \Delta(i)$. As in Shift_1, the value of $\Delta(i)$ is restricted to 0, 1 or -1. But unlike Shift_1, the algorithm emulates compression by trying the case $\Delta(i) = -1$ first. This strategy, however, will compress the nodes to the left, even when such a compression is not needed. Unnecessary compression may be compensated for

by adding a second phase to the global algorithm which shifts back (to the right) any node that has been shifted unnecessarily to the left. This second phase does not affect the probability of successfully reconfiguring the PSVA, and thus will not be discussed here.

Because a node may cause one of its neighbors to be shifted, restructuring with compression allows a node to borrow up to two processors from one of its neighbors. Hence, in the following description of the global Shift_1/compress algorithm, we will use $-2 \leq R(i) \leq 2$ to indicate the number of processors that node i left for possible use by node $i+1$, with $R(i) = -1$ and -2 indicating that i had been shifted to the right and borrowed 1 or 2 processors, respectively, from node $i+1$. As in Shift_1, we will start from $map(i) = 2i-1$ but we will partition h_k, the number of non-faulty processors connected to $switch_k$, into $h_k = h_k^+ + h_k^-$, where h_k^- is the number of non-faulty processors in $\{proc_{0,k}, proc_{1,k}\}$ and h_k^+ is the number of non-faulty processors in $\{proc_{0,k+1}, proc_{1,k+1}\}$. That is, we will distinguish between the processors that are to the left and to the right of $switch_k$. Following is the algorithm.

For $i = 1, \cdots, n_a$ do

1) IF $i > 1$ and [$R(i-1)=1$ and $h_{map(i)}^-=2$] or [$R(i-1)=2$ and $h_{map(i)}^->0$], THEN
   1.1) set $\Delta(i)=-1$,
   1.2) if $OK(switch_{map(i)+1})$ then set $R(i) = h_{map(i)}^+$,
   else set $R(i) = 0$.

2) ELSEIF $OK(switch_{map(i)})$ and $R(i-1)=0$ and $h_{map(i)}=4$, THEN
   2.1) set $\Delta(i)=0$,
   2.2) if $OK(switch_{map(i)+1})$ then set $R(i) = 1$ else set $R(i) = 0$

3) ELSEIF $OK(switch_{map(i)})$ and [$R(i-1) \geq 0$ and $h_{map(i)} = 3$] or [$R(i-1) = -1$ and $h_{map(i)} = 4$],
   THEN set $\Delta(i) = 0$ and $R(i) = 0$,

4) ELSEIF $i < n_a$ and $OK(switch_{map(i)+1})$ and $h_{map(i)}^+ + h_{map(i+1)}^- \geq 3$,
   THEN set $\Delta(i) = 1$ and $R(i) = h_{map(i)}^+ - 3$.

5) ELSE declare the remapping not successful

In order to analyze this algorithm, we construct a Markov chain similar to the one used for Shift_1 but with six states. One state, s_f to indicate the failure of iteration i, and five states, s_u, $u = 0,1,2,-1,-2$ corresponding to $R(i) = u$ after iteration i. The results of the analysis are plotted in Figure 4. As expected, compression improves the performance of Shift_1 considerably.

4. PSVA FOR TWO-DIMENSIONAL ARRAYS

Linear PSVA's may be generalized to two-dimensional arrays by connecting an $n_s \times n_s$ array of switches, $switch_{i,j}$, $i,j=1,...,n_s$, to an $(n_s+1) \times (n_s+1)$ array of processors $proc_{i,j}$, $i,j=0,...,n_s$, such that $switch_{i,j}$ is connected to $proc_{i-1,j-1}$, $proc_{i-1,j}$, $proc_{i,j-1}$ and $proc_{i,j}$. As in the linear case, switches should accommodate pass, straight, checking and voting modes. The three latter modes are defined as before except that each processor connected to $switch_{i,j}$ receives four inputs from that switch and sends four outputs back to the switch. In order to retain the reconfiguration flexibility of the PSVA, the four pass modes shown in Figure 5 should be accommodated. These modes are similar to the ones described in [9].

Figure 5 - The *pass* modes of a switch in 2-dimensional PSVA

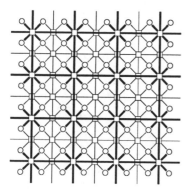

(a) A 4×4 array ($d = 4$)

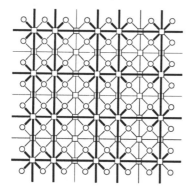

(b) A 4×5 array ($d = 3$)

Figure 6 - Embedding 2-dimensional arrays onto a 7×7 PSVA.

The embedding of an ($n_x \times n_y$)-node logical array onto a PSVA may be specified in terms of a mapping function $map(i,j) = (map_x(i,j), map_y(i,j))$, which maps each node (i,j) to $switch_{map(i,j)}$, and by the sets $C_{map(i,j)}$ that specify the set of non-faulty processors actively connected to each $switch_{map(i,j)}$ to constitute logical node (i,j). For example, the function $map(i,j) = (2i-1, 2j-1)$ specifies an embedding in which node (i,j) is mapped to $switch_{2i-1,2j-1}$, thus allowing quadruple redundancy (see Figure 6(a)). Similarly for triple redundancy we may use $map(i,j) = (2i-1, j+\lfloor \frac{j-1}{2} \rfloor)$ (see Figure 6(b)). In these two cases the specification of the sets $C_{map(i,j)}$ is straight forward.

Besides offering a choice of the degree of redundancy, the flexible *pass* modes of the switches offer the capability of constructing redundant arrays in the presence of defective processors and switches and of reconfiguring the array at run time to improve its future reliability. Unlike the linear case, however, there are some conditions that should be satisfied for a mapping function to represent a valid embedding of a logical array onto a two-dimensional PSVA. Specifically, if the four ports used to connect a given switch to the other switches are

labeled by "north", "south", "east" and "west", then the switch array should allow for the north, south, east and west ports of each $switch_{map(i,j)}$ to be connected, respectively, to the south port of $switch_{map(i-1,j)}$, the north port of $switch_{map(i+1,j)}$, the west port of $switch_{map(i,j+1)}$ and the east port of $switch_{map(i,j-1)}$. Necessary and sufficient conditions that guarantee valid embeddings are presented and analyzed in [12]. For the simple run time reconfiguration algorithms presented here, the fact that the resulting mappings are valid is straight forward to verify.

The *Shift_1* reconfiguration algorithm of Section 3 may be generalized to two dimensional arrays if shifting is allowed in any of the four normal directions. Specifically, starting from a given fixed mapping, map, $Shift\_1$ finds a new mapping, $remap = map + \Delta$, where the addition operation is extended to tuples such that $(u,v)+(u',v') = (u+u',v+v')$. The value of Δ is restricted to (0,0), (1,0), (0,1), (-1,0) or (0,-1). In the remainder of this section, we study this strategy assuming that the initial fixed mapping is $map(i,j) = (2i-1, 2j-1)$. We consider the problem of mapping a $n_a \times n_a$ logical array onto a $n_s \times n_s$ PSVA with faulty/defective processors such that each node is at least triply redundant. To simplify the algorithms and the analysis we will assume perfect switches. However, as was shown in linear PSVA's, faulty switches may be incorporated in the analysis.

As discussed in Section 3, the Shift_1 reconfiguration algorithm allows a node to borrow at most one processor from a neighbor. Also, to analyze the worst case performance of Shift_1, we do not allow the configuration resulting from $remap$ to have two neighboring nodes that are shifted. With this restriction, $remap$ is guaranteed to be valid.

In the following sequential form of Shift_1, the logical nodes are considered in the order $(1,1), (1,2), \cdots,$ $(1,n_a), (2,1), (2,2), \cdots,$ and an attempt is made to remap each node with $\Delta = (0,0), (-1,0), (0,-1), (0,1)$ or $(1,0)$, in that order. Given that node (i,j) is remapped before nodes $(i,j+1)$ and $(i+1,j)$, a function $R(i,j)$ is used to indicate the number of processors that node (i,j) makes available for possible use by $(i,j+1)$ and $(i+1,j)$. As in the linear case, $R(i,j) = 0, 1$ or -1, with the convention that if $R(i,j) = 1$, then the processor $proc_{map(i,j)}$ is the one that is made available to be used by node $(i,j+1)$ if $(i,j+1)$ is shifted to the left, or by node $(i+1,j)$ if $(i+1,j)$ is shifted upward. The case $R(i,j) = -1$, indicates that (i,j) had borrowed one processor from either $(i+1,j)$ or $(i,j+1)$. In other words, $R(i,j) = -1$ implies that $\Delta(i,j)$ is either (0,1) or (1,0). As before, h_k is the number of non-faulty processors connected to $switch_k$.

For $i = 1, \cdots, n_a$ do
For $j = 1, \cdots, n_a$ do

1) IF $h_{map(i,j)} = 4$ and $\Delta(i-1,j) \neq (1,0)$ and $\Delta(i,j-1) \neq (0,1)$
THEN set $\Delta(i,j) = (0,0)$ and $R(i,j) = 1$,

2) ELSE IF $[h_{map(i,j)} = 4$ and $\Delta(i-1,j) \neq (1,0)]$
or $[h_{map(i,j)} = 4$ and $\Delta(i,j-1) \neq (0,1)]$
or $[h_{map(i,j)} = 3$ and $\Delta(i-1,j) \neq (1,0)$ and $\Delta(i,j-1) \neq (0,1)]$
THEN set $\Delta(i,j) = (0,0)$ and $R(i,j) = 0$,

3) ELSE IF $j > 1$ and $OK(proc_{map(i,j)-(1,1)}, proc_{map(i,j)-(0,1)})$ and
$\Delta(i-1,j) \neq (1,0)$ and $R(i,j-1) = 1$,
THEN set $\Delta(i,j) = (0,-1)$ and $R(i,j) = 0$,

4) ELSE IF $i > 1$ and $OK(proc_{map(i,j)-(1,1)}, proc_{map(i,j)-(1,0)})$ and
$\Delta(i,j-1) \neq (0,1)$ and $R(i-1,j) = 1$ and $\Delta(i-1,j+1) \neq (0,-1)$
THEN set $\Delta(i,j) = (-1,0)$ and $R(i,j) = 0$,

5) ELSE IF $j < n_a$ and $OK(proc_{map(i,j)+(0,0)}, proc_{map(i,j)-(1,0)})$ and
$\Delta(i,j-1) \neq (0,1)$ and $\Delta(i-1,j) \neq (1,0)$
THEN set $\Delta(i,j) = (0,1)$ and $R(i,j) = -1$,

6) ELSE IF $i < n_a$ and $OK(proc_{map(i,j)+(0,0)}, proc_{map(i,j)-(0,1)})$ and
$\Delta(i,j-1) \neq (0,1)$ and $\Delta(i-1,j) \neq (1,0)$
THEN set $\Delta(i,j) = (1,0)$ and $R(i,j) = -1$,

7) ELSE declare the remapping not successful

Note that steps 1 and 2 correspond to $remap(i,j) = map(i,j)$, and that steps 3, 4, 5 and 6 correspond to shifting (i,j) leftward, upward, rightward and downward, respectively. Node (i,j) may be shifted leftward only if node $(i-1,j)$ has not been shifted downward and node $(i,j-1)$ has a processor to give away. Similarly, (i,j) may be shifted upward only if node $(i,j-1)$ has not been shifted rightward and $(i-1,j)$ has a processor to give away and that processor has not been taken by $(i-1,j+1)$. Finally, node (i,j) may be shifted rightward or downward only if neither $(i-1,j)$ nor $(i,j-1)$ has been shifted toward it.

In order to analyze the probability of success of Shift_1, we define seven possible states, s_u, $1 \leq u \leq 7$, for a node (i,j) after its mapping. Namely, a node (i,j) is in state s_u if at the $(i,j)^{th}$ iteration of algorithm Shift_1, step u is applied. If a Markov process technique is used to compute the probabilities of being in one of the above states, then a chain with a very large number of states is needed because the result of mapping node (i,j) depends on the results of mapping each of the nodes $(i-1,j)$, $(i-1,j+1)$ and $(i,j-1)$. An alternative technique is to assume that $P_u(i,j)$ is the probability that node (i,j) is in state s_u after its mapping, and then write the equations relating $P(i,j)$ with $P(i-1,j)$, $P(i-1,j+1)$ and $P(i,j-1)$, where $P(i,j)$ is a vector containing the seven probabilities $P_u(i,j)$ in the order $u = 1, \cdots, 7$. By solving the resulting system of difference equations, we may find the probability that all the nodes are mapped successfully. The probability equations are derived from Shift_1 in a straight forward way. For example, if π_u, $u = 2,3,4$ are as defined in Section 3, then the conditions in step 1 are satisfied when $h_{map(i,j)} = 4$ (with probability π_4), node $(i-1,j)$ is not shifted downward (with probability $1 - P_6(i-1,j)$) and node $(i,j-1)$ is not shifted rightward (with probability $1 - P_5(i,j-1)$). Hence,

$$P_1(i,j) = \pi_4(1 - P_6(i-1,j))(1 - P_5(i,j-1))$$

Similarly, from steps 2-6 in Shift_1, we obtain:

$$P_2(i,j) = 4\pi_3(1 - P_6(i-1,j))(1 - P_5(i,j-1))$$
$$+ \pi_4 P_6(i-1,j)(1 - P_5(i,j-1)) + \pi_4(1 - P_6(i-1,j))P_5(i,j-1)$$

$$P_3(i,j) = \pi_2(1 - P_6(i-1,j))P_1(i,j-1)$$

$$P_4(i,j) = \pi_2 P_1(i-1,j)(1 - P_5(i,j-1))(1 - P_3(i-1,j+1))$$

$$P_5(i,j) = \pi_2(1 - P_6(i-1,j))(1 - P_5(i,j-1))$$

$$P_6(i,j) = \pi_2(1 - P_6(i-1,j))(1 - P_5(i,j-1))$$

$$P_7(i,j) = 1 - P_1(i,j) - P_2(i,j) - P_3(i,j) - P_4(i,j) - P_5(i,j) - P_6(i,j)$$

Now, by setting $P(0,j) = P(i,0) = (0\,1\,0\,0\,0\,0\,0)^T$ for any i and j, we may iteratively compute $P(i,j)$, $i,j = 1,...,n_a$. The mapping is successful if no node ends up in state s_7, no node at the right boundary is shifted rightward, and no node at the bottom boundary is shifted downward. That is, the probability of success of Shift_1 is given by:

$$P_{Shift\_1}(n_a \times n_a) = [\prod_{i=1}^{n_a-1} \prod_{j=1}^{n_a-1}(1 - P_7(i,j))]$$

$$* [\prod_{j=1}^{n_a-1}(1 - P_7(n_a,j) - P_6(n_a,j)]$$

$$* [\prod_{i=1}^{n_a-1}(1 - P_7(i,n_a) - P_5(i,n_a)]$$

$$* [1 - P_7(n_a,n_a) - P_6(n_a,n_a) - P_5(n_a,n_a)]$$

The probability $P_{Shift\_1}(8,8)$ is plotted in Figure 7 for the embedding of an 8×8 logical array onto a 15×15-switch PSVA. From this figure it is clear that the probability of success of Shift_1 is an improvement over $P_{map}(8 \times 8) = (\pi_4 + 4\pi_3)^{64}$, the probability of success without any remapping.

It is possible to improve the performance of Shift_1 by allowing compression in any of the four normal directions, thus allowing a node to borrow up to two processors from any of its neighbors. As in the linear case, Shift_1/compress allows a node to borrow a processor from a shifted node. However, some restrictions should be imposed on the shifting in order to obtain a valid mapping. Specifically, a node (i,j) may be shifted to the left only if $node(i-1,j)$ is not shifted upward or downward. Similar conditions should be imposed when shifting (i,j) in the other directions.

The above restrictions complicates the analysis of Shift_1/compress because the state of node (i,j) depends on the state of $(i-1,j-1)$ in addition to the states of $(i-1,j)$, $(i,j-1)$ and $(i-1,j+1)$. However, a lower bound on the performance of the two dimension Shift_1/compress algorithm may be obtained if the one dimension Shift_1/compress algorithm is applied row-wise to the two dimensional array. In this case, the probability of successful embedding is $P_{row\_wise/compress}(n_a \times n_a) = (P_{compress}(n_a))^{n_a}$. As is clear

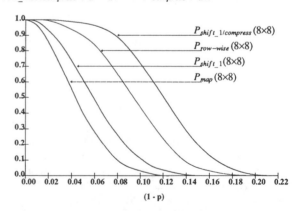

Figure 7 - The probability of successful reconfiguration in 2-dimensional PSVA.

from Figure 7, the row wise compression performs much better than the two dimensional Shift_1 without compression.

Finally, in order to gain more confidence in our analysis and to study the performance of the two-dimensional Shift_1/compress algorithm, we wrote a simulation package that simulates the three remapping algorithms; Shift_1, row_wise/compress, and Shift_1/compress. The simulation results for the first two algorithms agree closely with the analytical results, and the simulation results for the third algorithm are shown in Figure 7. These results show that by allowing compression in the four directions, we obtain a much more robust reconfiguration algorithm.

5. CONCLUDING REMARKS

We have presented a versatile architecture for implementing multiple redundancy in computational arrays, and we have studied reconfiguration algorithms that may be applied to this architecture. The algorithms may be applied either to tolerate initial defects or to cover for run time faults. We distinguished, however, between defect tolerance algorithms and run time algorithms according to the amount of global restructuring that is required. This distinction is natural because the algorithms that require global restructuring may outperform those that are restricted to local restructuring, but are less suitable for run time reconfiguration.

Given the flexibility of the PSVA architecture, we emphasized the strategies underlying the reconfiguration algorithms. These strategies may be applied to different instances of the restructuring problem. For example, the Shift_1 and Shift_1/compress algorithms discussed in Sections 3 may be applied if we start from a fixed mapping which is sparser or denser than $map(i) = 2i - 1$. If we start from a sparser mapping (e.g. $map(i) = 2i + \lfloor \frac{i}{2} \rfloor$), we improve the probability of completing run time reconfiguration successfully, thus increasing the reliability, while if we start from a denser mapping (e.g. $map(i) = i + \lfloor \frac{i-1}{2} \rfloor$), we lower that probability and decrease the reliability. In both cases, the reconfiguration strategies and the analysis techniques discussed in Section 3 apply.

Although it is assumed that switch/voting elements may fail, those elements are not used in a redundant mode, and thus should be self-checking elements. This also implies that errors in switches cannot be masked. The reliability of a switch/voter may be improved and the masking of voting errors may be accomplished if direct connections are provided among the four processors connected to each switch. This allows the results of each of the d processors constituting a node to be transmitted to the other $d-1$ processors, thus allowing d independent voting to determine the correct result distributively [7]. By relieving the switch/voter elements from the voting tasks, the function of those elements reduces to that of switching among ports, thus allowing for more reliable elements.

References

1. J. Abraham, P. Banerjee, C. Chen, W. Fuchs, S.Y. Kuo, and A. Reddy, ''Fault Tolerance Techniques for Systolic Arrays,'' *IEEE Computer*, pp. 65-74, July 1987.

2. M. Alam and R. Melhem, ''An Efficient Spare Allocation Scheme and its Application to Fault Tolerant Binary Hypercubes,'' *IEEE Trans. on Parallel and Distributed Systems*, vol. 2, no. 1, pp. 117-126.

3. C. Anfinson and F. Luk, ''A Linear Algebraic Model of Algorithm-based Fault Tolerance,'' *IEEE Trans. on Computers*, vol. 37, no. 12, pp. 1599-1604, 1988.

4. R. Harper, J. Lala, and J Deyst, ''Fault Tolerant Parallel Processor Architecture Overview,'' *Proc. of FTCS 18*, pp. 252-257, 1988.

5. A. Hopkins, ''FTMP - A Highly Reliable Fault Tolerant Multiprcessor for Aircraft,'' *Proc. of IEEE*, vol. 66, no. 10, pp. 1221-1239, 1978.

6. K. Huang and J. Abraham, ''Algorithm-based Fault-tolerance for Matrix Operations,'' *IEEE Trans. on Computers*, vol. C-36, no. 6, pp. 518-528, June 1984.

7. D. Kiskis and K. Shin, ''Embedding Triple-Modular Redundancy into a Hypercube Architecture,'' *Proc. of the Third Conf. on Hypercube Concurrent Computers and Applications*, pp. 337-245, 1988..

8. I. Koren, ''A Reconfigurable and Fault-tolerant VLSI Multiprocessor Array,'' *Proc. of the 8th Symposium on Computer Architecture*, pp. 425-442, 1981.

9. S.Y. Kung, S.N. Jean, and C.W. Chang, ''Fault-Toelrant Array Processors Using Single Track Switches,'' *IEEE Trans. on Computers*, vol. 38, no. 4, pp. 501- 514, 1989.

10. C. Kwan and S. Toida, ''Optimal Fault-Tolerant Realizations of Some Classes of Hierarchical Tree Systems,'' *Proc. of FTCS 11*, pp. 176-178, 1981.

11. R. Melhem, ''Bi-level Reconfigurations of Fault Tolerant Arrays in Bi-modal Computational Environments,'' *Int. Symposium on Fault Tolerant Computing*, 1989. (To appear in IEEE Trans. on Computers)

12. R. Melhem and J. Ramirez, ''Reconfiguration of Multiply Redundant Computational Arrays,'' Technical Report 90-2, Dept. of Computer Science, University of Pittsburgh, 1990.

13. R. Negrini, M. Sami, and R. Stefanelli, ''Fault Tolerance Techniques for Array Structures used in Supercomputing,'' *IEEE Computer*, pp. 78-87, Feb. 1986.

14. A. Rosenberg, ''The Diogenes Approach to Testable Fault-Tolerant Array Processors,'' *IEEE Trans. on Computers*, vol. C-32, no. 10, pp. 902-910, 1983.

15. L. Shombert and D. Siewiorek, ''Using Redundancy for Concurrent Testing and Repairing Systolic Arrays,'' *The Seventeenth Int. Symp. on Fault-Tolerant Computing*, pp. 244-249, 1987.

16. A. Singh and H. Youn, ''An Efficient Restructuring Approach for Wafer Scale Processor Arrays,'' *Proc. of the Int. Workshop on Defect and Fault in VLSI Systems*, 1988.

17. N. Theuretzbacher, ''VOTRICS: Voting Triple Modular Computer Systems,'' *Proc. of FTCS 16*, pp. 144-150, 1986.

RECONFIGURABLE MULTIPIPELINES

Yoon-Hwa Choi

Department of Computer Science
University of Minnesota
Minneapolis, MN 55455

Abstract

Pipelining has been widely used in advanced computer architectures for enhancing their performance. These pipelines typically consist of several functionally different stages. One of the problems in implementing such a system on a VLSI chip or wafer is that the probability of having a fault is not negligible. Consequently, it is desirable to have some means of achieving fault tolerance. In this paper, we present a reconfiguration algorithm for multipipelines in the presence of faults. Unlike other approaches, the algorithm can cover faults in switching elements as well as faults in stages of multipipelines.

1. Introduction

In supercomputers, multipipelines are often used to perform vector operations. For the implementation of such an architecture on a single chip or wafer, fault tolerance has been a critical design issue. Due to the simplicity of the structure, spare PE's and switching elements can be incorporated without considerable circuit overhead to improve the yield or reliability.

The problem of recovering fault-free linear pipelines in the presence of faulty stages has been investigated in [1]. An optimal distributed algorithm in the sense that the maximum number of pipelines under any fault pattern can be recovered has been proposed. However, it is based on the assumption that switching elements are fault-free. Switch programming is done internally by each switching elment, and thus requires considerable circuit overhead. The testing circuit added to each PE has been lightly treated, and its correct function is implicitly assumed.

It is a fact that the increasing complexity of digital integrated systems causes severe problems with respect to testing of these systems. Cost effectiveness and low overhead are of the utmost importance since the extra circuits added for testing and reconfiguration might outweigh the intended benefits. Boundary scan design [2], which provides controllability and observability of primary inputs and outputs of integrated circuits, leads us to develop more practical solutions to testing or reconfiguring pipelines. By employing the boundary scan concept, we can test both the PE's and switching elements [5][6].

This paper presents an efficient reconfiguration algorithm for a multipipeline architecture. The main contribution of this paper is the proposed reconfiguration algorithm that can also cover faults in switching elements, typically assumed to be fault-free.

2. Reconfigurable Multipipelines

A common approach to achieving defect tolerance involves spare PE's and flexible interconnection structure [3][4]. In our reconfigurable multipipelines, we add only one switching element between two adjacent stages of each pipeline as shown in Fig. 1 to minimize the circuit overhead. The connection between PE's in the jth stage and PE's in the $(j+1)$th stage can be realized by programming the switching elements $\{S_{i,j}, 1 \leq i \leq m\}$.

The structure of a switching element is shown in Fig. 2, where each switching element is able to realize the four states. In the design all switch control registers in the same column are connected into a shift register for testing and reconfiguration purposes. Once they are tested fault-free, each switching element can receive correct switch control information. Three switch control bits (instead of two) are used to select a switch state, and thus no decoding circuit is needed.

In the design of multipipelines, boundary scan concept is employed so that the input and output registers in each PE can be connected into a shift register chain in the test mode. All boundary scan paths in the same column can also be connected into a longer path through which test vectors can be scanned in or responses can be scanned out. The PE's and switching elements can be tested through boundary scan paths [5][6].

The main issue in this paper is how to reconfigure multipipelines in the event of faults in the system. In the case that all switching elements are fault-free, a simple algorithm in [1] can find an optimal number of pipelines. It finds the uppermost pipeline, second uppermost pipeline, and so on until it reaches the bottom.

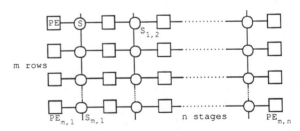

Fig. 1. Reconfigurable multipipeline

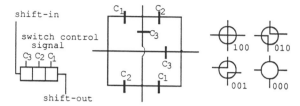

Fig. 2. Switching element and its four states.

The problem we are dealing with here is different since failures in switching elements are taken into account and faulty switching elements will be utilized in pipeline configuration. If a fault in a switching element is regarded as a total failure, i.e., the switching element is unusable, the performance (the number of recoverable pipelines) will degrade considerably. If a faulty switching element is to be utilized and the desired configuration (switch state) cannot be realized due to a fault in it, an alternative configuration needs to be chosen. Since each switching element has ten possible configurations (see Fig. 4), it is difficult or might be impossible to determine an optimal configuration using its local information.

Our reconfiguration algorithm is based on a greedy strategy and generates three switch control bits for each switching element based on local optimization criterion to be addressed later. For each switching element $S_{i,j}$ the following information will be used as the inputs.
(1) Fault status (faulty/fault-free) of its left and right PE's, denoted by $F_{i,j}$ and $F_{i,j+1}$.
(2) Configuration of the switching element $S_{i-1,j}$ (see Fig. 4).
(3) Fault status (type of fault) of the switching element $S_{i,j}$ itself (to be described in the next section).

In the case that there is a fault in a multipipeline system, a 100% utilization of fault-free PE's is highly unlikely. As a result, some fault-free PE's cannot be included in the working pipelines. Those fault-free PE's left out are called *dormant PE's* in this paper, and treated as a PE with a *fictitious fault*. Adding more channels between stages may reduce the number of dormant PE's. However, it requires extra circuit overhead which cannot be acceptable in practical design.

3. Fault Model

In order to utilize faulty switching elements in pipeline configuration, we need to identify the functional capability of a switching element in the presence of a fault. Since a PE is the smallest replaceable unit, once a fault is identified in a PE, it will be removed from consideration. However, a switching element is the critical component in recovering the multipipelines, its utilization is extremely important. In pipeline configuration, we use the following fault model: all single stuck-at-α or stuck-open faults on a data link, all single stuck-at-α faults on a control line and single bridging faults between data links, a single fault in a switching element, and multiple faults as long as each faulty switching element has a single fault in it.

A switching element and its corresponding graph model is given in Fig. 3. The nodes in the graph correspond to all the nets connecting transistors, and the edges represent the

transistors. A short (or a bridging fault) between two data links in a switching element can be modeled as a fault on a single edge (a control stuck-at fault) in Fig. 3(b)). A stuck-open fault on l_1 or l_2 can be treated as a PE fault.

A fault in a switching element does not imply that it cannot be used in pipeline configuration. Since a switching element can have four states as shown in Fig. 2, it can be utilized as long as the required state can be realized. Moreover, in many cases only one of the two paths available for a given state is needed in pipeline configuration. Once we distinguish switch states from the required configurations, each switching element can realize ten configurations shown in Fig. 4. We use $conf_i$ to represent the configuration (i) in Fig. 4. In the figure, $conf_{3-8}$ show a partial utilization of a switching element.

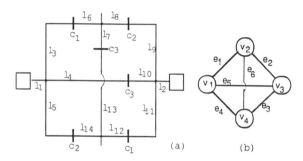

Fig. 3. Graph model for a switching element.

Prior to presenting the reconfiguration algorithm we need to identify which configurations can still be realized for each type of fault.
(1) A stuck-at-α fault at node v_1 or v_3 does not allow any connection passing through the node. Thus it can be treated as a PE fault. A stuck-at-α fault at v_2 or v_4 does not allow any vertical connection through it.
(2) A stuck-open fault may occur when there is a broken link. Thus it can be readily seen which configuration cannot be realized for the fault.
(3) A short (or a bridging fault) on an edge in Fig. 3(b) does not cause an error if the edge is part of the connection to be established. A short on e_1 or e_3 does not allow the switching element to be in either 010 or 100 state. However, 001 state is still valid. Similarly, a short on e_2 or e_4 allows the switching element to be in 010 state. A short on e_5 or e_6 allows none of 010 and 001 states. But the switching element can be in 100 state. If only one of the two paths available for a given state is used in pipeline configuration, a short does not cause any problem as long as the PE driving the unused path

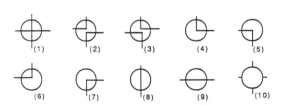

Fig. 4. Ten configurations for a switching element.

is in a high impedance state. In other words, $conf_{4-9}$ are valid even if there is a bridging fault.

4. Reconfiguration Algorithm

In the reconfigurable multipipelines, different stages may perform different operations and thus the corresponding PE's are different. A faulty PE in a stage can only be replaced by a fault-free PE in the same stage. Switching elements between stages are used to realize the replacement. For ease of description, we first define two terms, *desired configuration* and *mobility*, as follows.

Definition: A *desired configuration* for a switching element in reconfigurable multipipelines is said to be the configuration to be realized by the switching element when there is no switch fault.

Definition: Let a multipipeline architecture consist of m rows and n columns (or stages) of PE's. Then the *mobility* $z_{i,j}$ of $PE_{i,j}$ is defined to be the number of fault-free PE's below it in the jth column of PE's. That is, $z_{i,j} = \sum_{k=i+1}^{m} F_{k,j}$. Here $F_{k,j}$ (=0/1) indicates that $PE_{k,j}$ is faulty/fault-free.

We also define $z_{0,j}$ to be the number of fault-free PE's in the jth column of PE's. That is, $z_{0,j} = \sum_{k=1}^{m} F_{k,j}$. Then the number of fault-free PE's (in each stage) that can be utilized in pipeline configuration is bounded above by $Min \{z_{0,j}, j=1,2,..,n\}$.

Our pipeline reconfiguration algorithm visits each switching element row by row (from the first row), from left to right. For each switching element visited, its desired configuration is identified first. If the configuration cannot be realized due to a fault in the switching element, an alternative configuration will be selected. An alternative configuration is determined based on the configuration of $S_{i-1,j}$, the type of fault in $S_{i,j}$, and the mobilities of $PE_{i,j}$ and $PE_{i,j+1}$. The decision process is detailed in Fig. 5-6. Since the configuration of $S_{i,j}$ is determined assuming that the subsequent switching elements can comply with its request, backtracking is sometimes needed if the request cannot be satisfied by the subsequent switching elements due to a fault. This local optimal strategy will lead to a good result although it is not guaranteed to find an optimal solution.

We represent the switch by a circle. The lines inside the circle represent the connections between four terminals of the switching element. PE's are denoted by rectangular boxes. A faulty PE is marked by an X in the corresponding box. A dormant PE, i.e., a PE with a fictitious fault, created by the reconfiguration algorithm is indicated by a + in the corresponding box. The role of a switching element is to connect two PE's in adjacent stages to form a pipeline. Thus the configuration of switching element(s) above $S_{i,j}$ indicates whether $S_{i-1,j}$ requests a left PE ($PE_{i,j}$), a right PE ($PE_{i,j+1}$), or no PE to $S_{i,j}$. If $S_{i-1,j}$ is in $conf_4$, for example, no request is made from $S_{i-1,j}$. If $S_{i-1,j}$ is in $conf_3$, however, $S_{i-1,j}$ is looking for a right PE (i.e., $PE_{i,j+1}$). A request can propagate down in the case that the switching element $S_{i,j}$ can-

not comply with the request. The request for a lower left (right) PE is indicated by a * in the lower left (lower right) corner.

Each switching element $S_{i,j}$ in the multipipeline will be in one of the following twelve situations. We classify the twelve cases into three groups ($G1, G2, G3$) according to the type of request from $S_{i-1,j}$.

$G1$ (no request for a PE): Four possible cases are shown in Fig. 5. The second column shows the desired configuration. The third column shows the reaction in case the requirement cannot be satisfied. In Case 1, if the switching element cannot realize $conf_9$ due to a fault, one of the two alternatives, (a)(b), needs to be selected. Both configurations create a dormant PE, which is unavoidable. In selecting one of them for $S_{i,j}$, the mobilities $z_{i,j}$ and $z_{i,j+1}$ will be used. If $z_{i,j} \leq z_{i,j+1}$, then $PE_{i,j+1}$ will be made unusable and a request for a right PE will be generated (Case 1(a)). Otherwise we create a fictitious fault in $PE_{i,j}$ and generate a request for a left PE (Case 1(b)). If a fictitious fault is created in $PE_{i,j}$, then $S_{i,j-1}$ needs to be reprogrammed. Of course, this reprogramming might create another fictitious fault. Cases 2-4 are self-explanatory.

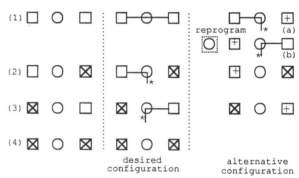

Fig. 5. Switch programming when no request for a PE.

$G2$ (request for a right PE): Four possible cases are shown in Fig. 6. The first case, Case 5, requires 010 state if the switching element is fault-free. When the switching element is faulty, the following alternative configurations need to be considered depending on the fault type.

a) If there is a short on e_5 or e_6, $conf_1$ (100 state) will be selected since both fault-free PE's can still be utilized.

b) If there is a short on e_1 or e_3, then $conf_4$ (010 state) will be chosen. The right PE should be included to respond to the request for a right PE. Thus a fictitious fault is created in its left PE and no request for a PE will be generated. The same is true for a stuck-at fault at v_4 or a stuck-open fault on l_5 or l_{14} (or e_4).

c) If there is a stuck-open fault on l_8 or l_9 (or e_2), it is impossible to accept the request for a right PE. Hence we ignore the request and choose $conf_9$ (100 state). As a result, the switching element $S_{r,j}$ ($r < i$) first generated the request for a right PE cannot be included in pipeline configuration. Thus a fictitious fault is created in its left PE (i.e., $PE_{r,j}$) and the switching elements between $S_{r,j-1}$ and $S_{i,j-1}$ in the $(j-1)$th column need to be reprogrammed. Here we should note that a stuck-at fault at v_2 in $S_{i,j}$ is the same as a stuck-at fault at v_4 in $S_{i-1,j}$. Thus a request for a PE from $S_{i-1,j}$ cannot be generated if there is a stuck-at fault at v_2 in $S_{i,j}$.

In Case 6, $conf_8$ is desired. If the configuration cannot be realized, it is impossible to propagate the request for a right PE generated from above. Thus we ignore the request for a right PE and handle the problem in the same way as in Case 2 in Fig. 5. Of course the actions taken in Case 5(c), creating a fictitious fault and reprogramming switching elements, also need to be done here. In Case 7, if $conf_4$ cannot be realized due to a stuck-open or control stuck-at-0 fault on e_2, the request needs to be either ignored (Case 7(a)) or propagated down to the next row (Case 7(b)). In selecting one of them, mobility will be used. If $z_{i,j} \leq z_{i,j+1}$, then $conf_8$ will be chosen (Case 7(b)). Otherwise, $conf_7$ will be selected. Case 8 can be handled trivially, and so no explanation is needed.

$G3$ (request for a left PE): Four possible cases need to be considered. This is similar to $G2$, and thus no explanation is necessary.

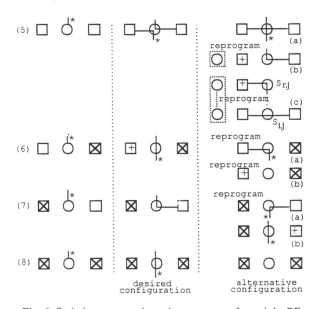

Fig. 6. Switch programming when a request for a right PE.

Due to faults in switching elements and PE's, dormant PE's have been created and a request for a PE generated from above cannot be satisfied. As a result, backtracking is sometimes needed. However, the number of faulty switching elements is expected to be small and a faulty switching element can still realize some configurations. Thus the probability that reprogramming will call another reprogramming is low. The simulation results in the next section will show that the additional time needed to reprogram switching elements is manageably small.

The reconfiguration algorithm is illustrated by an example given in Fig. 7. Fig. 7(a)-(c) show the switch programming of the first and second rows, respectively. Due to a stuck-at fault at v_4 in $S_{1,3}$, a fictitious fault is created in

$PE_{1,3}$, and $S_{1,2}$ is reprogrammed (see Fig. 7(b)). Reprogramming due to a short on e_1 in $S_{5,3}$ is shown in Fig. 7(e).

The reconfiguration algorithm has been presented assuming that each PE can be in a high impedance state if it is not included in working pipelines. However, any failure in the mechanism can be handled similarly since we can easily identify which path cannot be used. The proposed algorithm is to run on an external computer and the resulting switch setting information is shifted in through the chain of switch control registers. Thus the circuit overhead required for achieving fault tolerance has been minimized. We point out that most distributed reconfiguration algorithms presented in literature are not practically useful due to considerable circuit overhead.

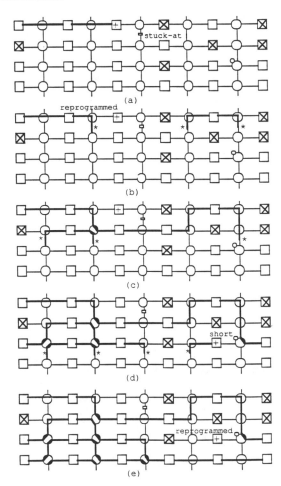

Fig. 7. An illustration of pipeline configuration.

5. Performance Evaluation

Monte Carlo simulation is performed to estimate the number of recoverable pipelines for 1000 fault patterns with both PE and switch faults. In the simulation fault patterns are generated based on independent PE and switch failures. Let r denote the ratio of the PE and the switch failure rates. In other words, $r=(1-Yield(PE))/(1-Yield(S))$. The simulation is performed for various values of PE yield and r. Fig. 8 shows the average number of recoverable pipelines m_r for

$m=10$ and $n=10$. The results reveal how the yield is degraded by switch failures as compared to perfect switch for our algorithm.

In the figure, we can find that m_{10} (i.e., $r=10$) is almost equal to m_{oo}. The uppermost lines indicate the least number of good PE's in any column, i.e., $z_0=\min\{z_{0,j}, 1\leq j\leq n\}$. The gap between z_0 and m_r represents the performance degradation due to the lack of channels between stages. Even if a single channel is employed in our design, we can obtain reasonably high performance. For example, m_{10} is close to m_0 if PE yield is 0.9. The figure also shows that the average number of recoverable pipelines when no switching elements are employed is unacceptably low even for high PE yields.

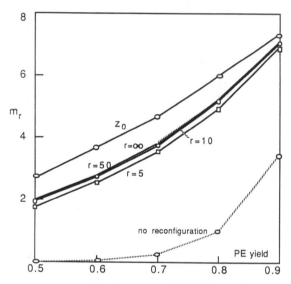

Fig. 8. m_r versus PE yield.

We have also obtained the recovery rate $R_r=m_r/m_{oo}$ for m and n ranging from 4 to 10. The results are shown in Fig. 9. For a given value of n, the recovery rate decreases as m increases. For a given number of rows, the recovery rate also decreases as the number of stages increases. It is well known that the speedup that a linear pipeline can provide is always bounded above by the number of stages in the pipeline. In actual practice, pipelines rarely exceed 10 to 20 stages. Thus we do not consider the asymptotic behavior of the reconfiguration algorithm.

The proposed reconfiguration algorithm requires reprogramming in the case that a fictitious fault is created or a request for a PE (generated by a switching element) can not

be satified. That is, a switching element might need to be programmed more than once. We ran the simulation program to evaluate how much additional time is needed for switch programming. The results (in Fig. 10) have shown that in wide range of PE yields, from 0.5 to 0.9, the optimal performance can be obtained by calling reprogramming fewer than three times even for a relatively low switch yield ($r=10$).

The average number of recoverable pipelines m_{10}						
PE yield	$q=1$	$q=2$	$q=3$	$q=4$	$q=5$	$q=10$
0.5	1.794	1.929	1.938	1.938	1.938	1.938
0.6	2.566	2.695	2.708	2.711	2.711	2.711
0.7	3.497	3.684	3.711	3.711	3.711	3.711
0.8	4.994	5.133	5.164	5.164	5.164	5.164
0.9	6.869	6.976	6.995	6.995	6.995	6.995

Fig. 10. m_{10} versus the number of iterations q.

6. Conclusion

A fault-tolerant design of multipipelines for supercomputing has been presented. Fault tolerance has been achieved by programming switching elements. Only one track of switching elements between stages has been used to minimize hardware overhead. The primary value of our results is to fill a common but crucial void in research on reconfigurable architectures: the problem of how to handle faults in switching elements. The utilization of faulty switching elements in pipeline configuration has considerably improved the number of recoverable pipelines.

References

[1] R. Gupta, A. Zorat, and I.V. Ramakrishnan, "Reconfigurable multipipelines for vector supercomputers" IEEE Trans. Computers, Vol. 38, No. 9, pp. 1297-1307, Sept. 1989.

[2] JTAG, "JTAG boundary-scan architecture standard proposal," version 2.0, Mar. 30, 1988.

[3] F. Lombardi, R. Negrini, M.G. Sami, and R. Stefanelli, "Reconfiguration of VLSI arrays: A covering approach," Proc. IEEE Int. Symp. Fault-Tolerant Computing, pp. 251-256, 1987.

[4] M. Chean, and J.A.B. Fortes, "A taxonomy of reconfiguration techniques for fault-tolerant processor arrays," IEEE Computer, pp. 55-69, Jan. 1990.

[5] A. Hassan, V.K. Agarwal, J. Rajski, and B.N. Dostie, "Testing of glue logic interconnects using boundary scan architecture," Proc. IEEE Int. Test Conf., pp. 700-711, Sept. 1989.

[6] R.P. Van Riessen, H.G. Kerkhoff, A. Kloppenburg, "Designing and implementing an architecture with boundary scan," IEEE Design & Test of Computers, pp. 9-25, Feb. 1990.

Recovery rate R_{10}							
	PE yield=0.8				PE yield=0.9		
m	$n=4$	$n=6$	$n=8$	$n=10$	$n=4$	$n=6$	$n=8$	$n=10$
4	0.984	0.977	0.963	0.963	0.995	0.993	0.992	0.988
6	0.982	0.957	0.922	0.927	0.989	0.984	0.982	0.988
8	0.965	0.936	0.894	0.898	0.988	0.985	0.974	0.966
10	0.947	0.915	0.862	0.854	0.985	0.978	0.963	0.950

Fig. 9. $R_{10}(=m_{10}/m_{oo})$ for various values of m and n.

Embedding of Multidimensional Meshes on to Faulty Hypercubes *

Pei-Ji Yang,[†] Sing-Ban Tien, C.S. Raghavendra

Department of Electrical Engineering-Systems
University of Southern California
Los Angeles, CA 90089-0781

Abstract

A scheme for embedding multidimensional meshes on to faulty hypercubes is presented. In an n-dimensional hypercube, given any $1 \leq f \leq n$ faulty nodes, this scheme can embed a mesh with dilation 2 of size at least $2^{m_1} * \cdots * 2^{m_i-1} * (2^{m_i} - k) * 2^{m_i+1} * \cdots * 2^{m_d}$ where $k = \min(f, \lceil 2^{m_i}/2^{n-f+1} \rceil)$, $1 \leq i \leq d$ and $n = m_1 + m_2 + \cdots + m_d$. The processor utilization of this embedding scheme is high. In particular, (1) for one-dimensional mesh (linear array or ring), the processor utilization of embedding is 100%; (2) for $1 \leq f \leq n - m_i + 1$, a mesh with the optimal size of $2^{m_1} * \cdots * 2^{m_i-1} * (2^{m_i} - 1) * 2^{m_i+1} * \cdots * 2^{m_d}$ is obtained. The embedding scheme can be constructed with time complexity O(n) if $f < n$ and O($f^2 + n$) if $f \geq n$.

1 Introduction

Due to its regularity and nice structural properties [6], hypercube has become a popular multiprocessor network. Most importantly, several versions of hypercube architectures are now commercially available, which makes experimental implementation of parallel algorithms possible. To execute these algorithms on hypercubes, we need to embed their underlying graph structures onto hypercube; for example, we embed mesh network on to hypercube for executing multigrid algorithms [2].

Embedding a given graph G on to a host graph H means mapping nodes of G to nodes of H such that edges of G also map to edges of H. A relaxed condition is to allow paths rather than edges between mapped nodes [3]. Two costs associated with graph embedding are: 1) *dilation*, which is the maximum distance in H between the images of adjacent nodes of G. 2) and *utilization*, which is the ratio of total number of the nodes in G to total number of fault-free nodes that are allocated for G. For some task graphs (e.g. chain, ring, tree, mesh, etc.), embeddings in hypercubes have been studied by many researchers [2, 4]. However, embedding tasks on to in the faulty hypercubes have been considered to a much lesser extent.

In this paper, embeddings of multidimensional meshes on to faulty hypercube will be studied. Our problem is related to finding the maximal size of task graphs in the residual hypercubes in the presence of faulty processor nodes. We focus on finding efficient embedding schemes which can tolerate number of faulty nodes on the order of the dimensions n.

Some techniques propose to embed a smaller size fault-free cube in the faulty hypercube, as in [1, 3]. Techniques for embedding task graphs in fault-free hypercubes are then applied. However, in general, embedding of a larger size task graph can be obtained if the embedding is attempted in the residual hypercube directly.

In [8], a 2-D mesh of size $2^{m_1} * 2^{m_2}$ is used as an example and an exhaustive search method is employed to choose the combinations of m_1 dimensions among the n dimensions of hypercube for determining if the required size mesh exists in a faulty hypercube.

In this paper, a dilation 2 mesh embedding is presented which obtains a mesh of size at least $2^{m_1} * \cdots * 2^{m_i-1} * (2^{m_i} - k) * 2^{m_i+1} * \cdots * 2^{m_d}$ where $k = \min(f, \lceil 2^{m_i}/2^{n-f+1} \rceil)$, $1 \leq i \leq d$ and $n = m_1 + m_2 + \cdots + m_d$. The processor utilization of this embedding scheme is high. In particular, (1) for one-dimensional mesh (linear array or ring), the processor utilization of embedding is 100%; (2) for $1 \leq f \leq n - m_i + 1$, a mesh with the optimal size of $2^{m_1} * \cdots * 2^{m_i-1} * (2^{m_i} - 1) * 2^{m_i+1} * \cdots * 2^{m_d}$ is obtained. A construction procedure with time complexity O(n) if $f < n$ and O($f^2 + n$) if $f \geq n$ is also described.

In our technique, we assume that the faulty nodes information is globally available. Faulty nodes information can be gathered very quickly in hypercube. Further, fault occurrence is a rare phenomenon.

The main idea of our approach is in using the faulty nodes information to assign proper dimensions of the cube to some direction of mesh such that faults in that direction are isolated and can be easily bypassed with dilation 2. The strategy of assigning dimensions is based on a concept called *free dimension*. A dimension i is a free dimension if there is no pair of faulty nodes that are neighbors along dimension i.

2 Preliminaries

An n-dimensional hypercube, denoted as Q_n, consists of $N (= 2^n)$ nodes which are labeled from 0 to $2^n - 1$. The labels are represented in the form of binary strings, $b_n \ldots b_1$, $b_i = 0, 1$, where $1 \leq i \leq n$. Two nodes are adjacent if and only if their labels differ in exactly one bit.

Q_n can be partitioned into subcubes of various sizes. In this paper, Q_n is always partitioned in such a way that all the subcubes are of the same size and spanned by the same set of dimensions, i.e., in their representations *'s are in the same bit positions. A partitioned Q_n with k-subcubes can be viewed as Q_{n-k} whose nodes are k-subcubes, called supernodes. The nodes of Q_n are called collapsed (or squeezed) along these k dimensions to form Q_{n-k}.

A d-dimensional mesh can be represented by $p_1 * p_2 * \cdots * p_d$. In this paper, we assume that the mesh size in each direction is a power of 2, i.e., $p_i = 2^{m_i}$ for $1 \leq i \leq d$.

There are two fault models defined in [3]. The first model assumes that, in a faulty node, the computational function of the node is lost while the communication function remains intact; this is the partial fault model. The second model assumes the communication function is lost, therefore, the computational function is rendered useless regardless it is faulty or not; this is the total fault model. We assume the partial fault model.

For the Intel iPSC/2 Direct Connect Communication technology [5], there is a switching circuit, called Direct Connect Router, between each processor and the connection links. Thus, the failure of processor node does not imply the failures of the links that are connected to this processor. As the switching circuitry is far less prone to failure than full capability processor element, the partial fault model is more suitable for the iPSC/2 type hypercube computer.

3 Free Dimension Concept

In this section, a key concept – *free dimension* – underlying our techniques is introduced. We analyze the worst case scenarios and show that when number of faulty nodes $f \leq n$, there exists at least one free dimension in Q_n. The concept is then used to partition Q_n into subcubes such that each subcube contains at most one faulty node.

Definition 1 *Two adjacent faulty nodes x_i and x_j are said to*

*This research is supported by the NSF Grant no. 8452003.

[†]This author's research is supported by a fellowship from CSIST, Taiwan, R.O.C.

occupy *a dimension k of Q_n, if x_i differs from x_j in the kth bit of their labels. A dimension i is* **free** *if there is no such pair of faulty nodes occupying dimension i.*

Let F denote the set of faulty nodes, and let $f = |F|$. F induces a graph $G_F = (F, E)$, where the vertex set is the set of faulty nodes and for $x_1, x_2 \in F$, there is an edge between x_1 and x_2 iff x_1 and x_2 are adjacent in Q_n. An edge of G_F is labeled with the dimension occupied by the two end faulty nodes of this edge. We show in the following that when $f \leq n$, there is at least one free dimension in Q_n. First, we consider that G_F is connected.

Lemma 1 *A connected subgraph induced by F can occupy at most $f - 1$ dimensions.*

Proof: Let T be a spanning tree of G_F. It is clear that every edge of T occupies a dimension of Q_n. For an edge e in G_F but not in T, there exists a path P in T between two end nodes of edge e, and $P + e$ forms a cycle. Since G_F is a subgraph of Q_n which contains no cycle of odd length, the label of an edge of a cycle in G_F must appear even number of times. Thus, in path P, there must exist an edge that also occupies the same dimension as e, or equivalently, the dimension occupied by e has already been counted in the dimensions occupied by path P (in T). Since there are $f - 1$ edges in T, G_F can occupy at most $f - 1$ dimensions. $\square$

Now, consider the case that G_F is not connected.

Theorem 1 *Given a set of faulty nodes F in Q_n with $f \leq n$, the number of dimensions occupied by F is at most $f - m$, where m is the number of connected components in G_F.*

As $m \geq 1$, therefore, the number of dimensions occupied by F is at most $f - 1$ when G_F forms a single connected component.

Corollary 1 *Given a set of faulty nodes F, if $f \leq n$, then there exist at least $n - f + 1$ free dimensions in Q_n.*

The worst case situation in which $n + 1$ faulty nodes render no existence of free dimension in Q_n can happen when $n + 1$ faulty nodes form some specific structure, e.g., a star. In practice, this rarely happens since the faults tend to scatter throughout the entire hypercube. Therefore, much more faults can be allowed, when free dimensions still exist.

Example 1: Consider $F = \{0000, 0001, 0010, 0100, 1000\}$ in Q_4. 5 faulty nodes occupy each dimension of Q_4. However, consider $F' = \{0010, 0101, 1001, 1100, 1111\}$ in Q_4. 5 faulty nodes do not occupy any dimension of Q_4.

In the following, we will show how the free dimension concept can be used to partition the cube into subcubes such that at most one faulty node is contained in each subcube.

Theorem 2 *Given any $f \leq n$ faulty nodes, Q_n can be partitioned into $(n - f + 1)$-dimensional subcubes such that each subcube spanned by the same set of dimensions contains at most one fault.*

Proof: From Corollary 1, a free dimension, say i, can be found since $f \leq n$. Squeeze Q_n along dim i to Q_{n-1} whose node is a 1-dimensional subcube spanned along i. Since Q_n is collapsed along a free dimension i, where no pair of adjacent faulty nodes lies, Q_{n-1} has the same number of faulty supernodes f.

The above *squeezing process* can be repeated $n - f + 1$ times so that Q_n becomes Q_{f-1}, and in this case, Corollary 1 cannot be applied to find another free dimension, since there are f faulty supernodes. Note that in the process of collapsing nodes along free dimensions, the number of faulty supernodes is always f. In Q_{f-1}, each supernode is an $(n - f + 1)$-dimensional subcube. Since, in the squeezing process, no two faulty nodes can be collapsed into a supernode by the definition of free dimension, each supernode of size 2^{n-f+1} contains no more than one fault. $\square$

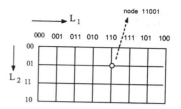

Figure 1: Two dimensional Gray code for an $8*4$ grid embedded in a 5-cube.

4 Embedding Meshes onto Faulty Hypercubes

A d-directional mesh of size $2^{m_1} * 2^{m_2} * \cdots * 2^{m_d}$, where $n = m_1 + m_2 + \cdots + m_d$, (called a full size mesh) can always be embedded in an n-cube by finding d Gray code sequences $L_1, L_2, \cdots, L_d$ for d directions of mesh, respectively, and representing the required mesh by the d-dimensional Gray code (e.g. Binary Reflected Gray Code) ordering by $L_1, L_2, \cdots, L_d$ [6]. For example, an $8 * 4$ mesh can be embedded in a 5-cube as shown in Fig. 1.

For mesh embedding in a fault-free hypercube, the assignment of the n dimensions in Q_n to $m_1, m_2, \cdots, m_d$ dimensions for d directions of mesh can be arbitrary. However, for embedding mesh on to a faulty hypercube, the assignments of dimensions to different directions must be done carefully. If we assign dimensions of cube to directions of mesh in a proper way then we may obtain a larger size of mesh in the residual hypercube. Therefore, one has to find the right assignment and this is the main difference between embeddings in fault-free and faulty hypercubes.

We are assuming the partial fault model as explained in Section 2. We use faulty nodes information to assign the dimensions to directions of mesh such that faulty nodes can be easily bypassed with dilation 2. The free dimension concept is used to partition the cube into subcubes such that each subcube contains at most one faulty node. Linear arrays are embedded in subcubes with dilation at most 2. These linear arrays are then lined up to form the entire mesh. Our technique of assigning the dimensions carefully leads to the following theorem.

Theorem 3 *Given any $f \leq n$ faulty nodes, there exists a mesh of size at least $2^{m_1} * \cdots * 2^{m_{i-1}} * (2^{m_i} - k) * 2^{m_{i+1}} * \cdots * 2^{m_d}$ with dilation 2 in the residual hypercube, where i is some direction of mesh, $k = \min(f, \lceil 2^{m_i}/2^{n-f+1} \rceil)$, and $n = m_1 + m_2 + \cdots + m_d$.*

Proof: Without loss of generality, assume $i = 1$. We show how we can choose a combination of m_1 dimensions among n dimensions of Q_n such that the faulty nodes can be "isolated" and bypassed with dilation 2.

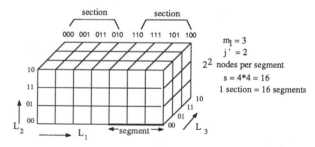

Figure 2: Notations example for an $8 * 4 * 4$ mesh embedded in a 7-cube.

Given $f \leq n$ faults, we can recursively find $j \geq n - f + 1$ free dimensions by the squeezing process as stated in Theorem 2. Choose the first m_1 free dimensions if $j > m_1$ and let $j' = \min(j, m_1)$. Assign these j' to the first j' of the m_1 dimensions in the first direction of mesh. All other $n - j'$ dimensions of

Q_n are arbitrarily assigned to other directions of mesh. Define a *segment* as a subcube of size $2^{j'}$ nodes spanned by these j' free dimensions. Thus, there are $2^{m_1-j'}$ segments in L_1 and there is at most one faulty node per *segment* by Theorem 2. If we view a $2^{m_1} * 2^{m_2} * \cdots * 2^{m_d}$ mesh from the first direction, there are $s(= 2^{m_2} * 2^{m_3} * \cdots * 2^{m_d})$ segments that have the same embedding scheme in L_1. We define these s segments with the same embedding scheme as a *section*. We show our notations by an example of $8 * 4 * 4$ mesh in Fig. 2. In this figure, we assume that only two free dimensions can be found ($j' = 2$). Thus, the size of a *segment* is 2^2 nodes, and the size of a *section* is $4 * 4$ segments.

We then define the *marking procedure* as follows. *Marking procedure* is executed *per segment* basis. Two cases are considered. The first case is when there exists at least one faulty node in the *section* where the *segment* is located. In this case, if there is a faulty node in this *segment*, then mark this faulty node. If not, then mark the last node of this *segment* as a pseudo-faulty node. The second case is when there is no faulty node in the *section* where the *segment* is located. In this case, no node is marked, i.e., all nodes in the *segment* can be used. After this marking procedure, a linear array of size $2^{j'}-1$ or $2^{j'}$ nodes can be embedded with dilation 2 inside each *segment* of L_1. Every *segment* of a *section* either marks one node or zero node. As a consequence, two neighboring mesh points along direction 2, $\cdots$, or direction d have the same or difference 1 Gray code values of L_1, and thus, the dilation is no more than 2.

Next, we investigate the dilation between two mesh points that connect two *segments* along the first direction. As shown

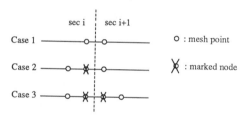

Figure 3: Three cases between two mesh points which connect two segments along the first direction of mesh.

in Fig. 3, there are three cases: 1) there is no marked node (faulty or pseudo-faulty node) between these two mesh points, and, thus, the dilation is one; 2) there is one marked node between two mesh points and the dilation is two; 3) there are two marked nodes between two mesh points, i.e., the last node of *segment* i and the first node of *segment* $i+1$ are both marked. In the third case, due to the "reflected property" of Binary Reflected Gray Code, the dilation is one between these two mesh points.

From the above discussion, the dilations in all cases are no more than two. The size of mesh obtained can be calculated as follows. Each *segment* can mark one node depending on whether there exist faulty nodes in that *section*. Since there are f faulty nodes in Q_n, at most f *sections* are marked. Thus, total marked out nodes is at most $s * \min(f, 2^{m_1-j'})$, which is no more than $s * \min(f, \lceil 2^{m_1}/2^{n-f+1} \rceil)$. Hence, the size of mesh obtained is no less than $(2^{m_1} - \min(f, \lceil 2^{m_1}/2^{n-f+1} \rceil)) * 2^{m_2} * \cdots * 2^{m_d}$. $\square$

In the above theorem, only one direction of mesh is degraded in size and this direction can be any direction of mesh. For our convenience, we will call this direction as *degraded direction*. Note that the size of mesh stated in this theorem is a lower bound. Also, the number of tolerated faulty nodes can be greater than n as long as there exists a free dimension in Q_n.

To obtain the **largest size** of mesh, we should choose the direction with the largest m_i as the *degraded direction*, as stated in the following theorem without proof.

Theorem 4 *Let S_i be the size of mesh obtained by choosing the ith direction of mesh as the degraded direction. If $m_i = \max_{k=1}^d m_k$, then $S_i = \max_{k=1}^d S_k$.*

4.1 Discussion and Examples

Any single faulty node will render the size of embedded mesh degrade from $2^{m_1} * \cdots * 2^{m_i} * \cdots * 2^{m_d}$ to $2^{m_1} * \cdots * 2^{m_{i-1}} * (2^{m_i} - 1) * 2^{m_{i+1}} * \cdots * 2^{m_d}$, $1 \le i \le d$, $n = m_1 + m_2 + \cdots + m_d$. Thus, with one fault, to embed a mesh onto hypercube, $2^{m_1} * \cdots * 2^{m_{i-1}} * (2^{m_i} - 1) * 2^{m_{i+1}} * \cdots * 2^{m_d}$ is the **optimal size** we can achieve. When there are more faults, from the above discussion, we can get the optimal size of mesh with any given set of $1 \le f \le n - m_i + 1$ faulty nodes; this is true because the largest mesh obtained in this case has $2^{m_i} - 1$ nodes along the ith direction of mesh. For $f > n - m_i + 1$, by choosing the largest m_i as the *degraded direction*, we get the largest possible mesh with our scheme. This may or may not be the optimal size mesh.

Now, let us discuss the processor utilization of mesh embedded in faulty hypercube. From Theorem 3, it is easy to verify that the processor utilization of the embedding is at least

$$\frac{(2^{\sum_{j \ne i} m_j}) * (2^{m_i} - k)}{2^n - f} \qquad (1)$$

where i is some direction of mesh, $k = \min(f, \lceil 2^{m_i}/2^{n-f+1} \rceil)$, and $1 \le f \le n$. For $f < n - 2$, it is easy to verify that the utilization ratios are close to one.

Theorem 3 can also be applied in several scenarios:

1. The required mesh is not a power of 2 in some direction j. (For example, a mesh of size $8 * 8 * 6$.) In this case, we can assign direction j as the *degraded direction* and our embedding scheme can be used to check if the required mesh can be embedded.

2. The required mesh is allowed to have some performance degradation in some direction. Again, we assign this direction as the *degraded direction*. In this case, our embedding scheme can be used to check how much degradation of mesh occurs in the faulty hypercube.

3. The largest size of mesh in the residual hypercube is required. In this case, the direction with the largest m_i is chosen to be the *degraded direction* in order to obtain the largest size of mesh.

In the following, we give examples to show how our scheme works.

In Example 2, we show how the free dimension concept helps us to assign dimensions of Q_n to directions of mesh such that faults are distributed in such a way that they can be easily bypassed with dilation 2.

Example 2: Let $d = 2$, $m_1 = 3$, $m_2 = 2$, and the given fault set $F = \{01011, 01010, 01110, 01111, 10010, 10110\}$ in Q_5. If we assign dimension 1, 2, 3 of Q_5 to m_1, and dimension 4, 5 of Q_5 to m_2, faults are distributed in mesh as shown in Fig. 4, and mesh size we can embed is only $6 * 3$. From Theorem 2, we can recursively squeeze twice and find two free dimensions, i.e. dimension 2 and dimension 5. If we assign dimension 2, 5 to m_2, faults are shown in mesh as in Fig. 5. After marking out two good nodes, we get an $8 * 3$ mesh with dilation 2. From this example, we also show that our scheme can tolerate number of faults more than n as long as we can find free dimensions. The processor utilization of this example is $(8 * 3)/(32 - 6) = 0.923$.

Example 3: Let $d = 2$, $m_1 = 4$, $m_2 = 3$, $f = 3$, $n = 7$. From Theorem 2, we can recursively squeeze $n - f + 1 = 5$ times and find 5 free dimensions. However, as both m_1 and m_2 are less than 5, we can only assign 4 or 3 free dimensions to m_1 or m_2.

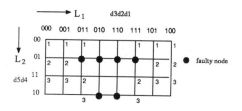

Figure 4: Arbitrary 6 faults in a 5-cube.

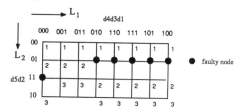

Figure 5: Fault pattern after dimension reassignment.

respectively. If we assign 4 free dimensions to m_1, we get a mesh of size at least $15 * 8$; whereas if we assign 3 free dimensions to m_2, we get a mesh of size at least $16 * 7$. Thus, from this example, we should assign free dimensions to the largest m_i in order to get the largest size of mesh. However, if the user requires a mesh of size $16 * 6$, we must assign free dimensions to m_2.

4.2 1-D Mesh – Linear Array or Ring Case

Note that, in our scheme, we mark out the last fault-free node of *segment* only if there exist faults in that *section*. For one dimensional mesh, i.e., linear array (or ring) case, there is only one *segment* per *section*. Since we don't need to mark out fault-free node if there is no fault in a *segment*, all available fault-free nodes are used and the processor utilization is 100%. In other words, using our scheme, a **maximum length** linear array or ring with dilation 2 is obtained if there exists a free dimension. Hence, the following corollary is true.

Corollary 2 *Given any $f \leq n$ faulty nodes, there exists a linear array (or ring) of length $2^n - f$, with dilation 2, in Q_n.*

Consider the Gray code sequence $12131214\cdots$, which embeds a linear array or ring in a fault-free Q_n; the dimension 1 appears in alternate positions in this sequence. Such a sequence can be obtained with any dimension j in the alternating position to embed a ring. Now, if three consecutive nodes are faulty in this sequence, then j cannot be a free dimension. Since we consider $f \leq n$, there exists a free dimension which can be used in this alternating sequence. With the partial fault model, it can be verified that either the faulty nodes are bypassed (when two consecutive faulty nodes exist) or we can route through a faulty node. We use all the fault-free nodes and thus obtain the optimum size linear array or ring.

4.3 A Construction Algorithm

In this section, we describe a construction algorithm that can embed the mesh of size as stated in Theorem 3 as long as free dimensions exist in Q_n. Details can be found in [9]. Following the constructive proof as stated in Theorem 3, this construction algorithm can be divided into three parts.

The first part is to find free dimensions with a given set of f faulty nodes. Two methods are provided in [9]: one is executed in a distributed manner, and can identify the free dimensions in $O(n)$ steps. This method can be applied to $f < n$ cases. On the other hand, the second method, executed in a central processor, takes $O(f^2 + n)$ steps, and can be used with arbitrary number of faulty nodes.

In the second part, the free dimension concept is used to partition Q_n into *segments*. The partitioning is a recursive squeezing process as stated in the proof of Theorem 2. We then

assign the free dimensions found in the squeezing process to the largest direction of mesh and arbitrarily assign other dimensions of Q_n to other directions of mesh. After these dimension assignments, each *segment*, as defined in the proof of Theorem 3, contains at most one faulty node, and each node can find the labels of its physical neighbors via the conversion between Gray code and binary code, which can be accomplished in $O(n)$ steps [4].

In the third part, in each *segment*, the *marking procedure* as stated in the proof of Theorem 3 is executed. By broadcasting the faulty node label in each *segment* and *section*, every node knows if there is a faulty node in its *segment* or *section* in constant time. The faulty node or the last node (in Gray code sequence) of *segment* is marked if there exist faulty nodes inside the *section*. After the *marking procedure*, each mesh point can calculate all its logical neighbors in mesh. Thus, the mesh is formed. Except broadcasting, all other steps in the second and third part can be executed in a distributed manner involving only local operations. Thus, the time complexities of the second and third part are $O(n)$.

From the above discussions, the time complexity of our construction algorithm is dominated by the first part and is $O(n)$ if $f < n$ and $O(f^2 + n)$ if $f \geq n$.

5 Conclusions

In this paper, a key concept, called *free dimension*, is introduced. The free dimension concept can be used to partition the cube into subcubes such that each subcube contains at most one faulty node. Based on the free dimension concept, we presented a dilation 2 mesh embedding scheme which has high processor utilization in faulty hypercubes. Especially for $1 \leq f \leq n - m_i + 1$, an optimal size of mesh is obtained. In addition, for one dimensional mesh case, i.e., linear array or ring, the processor utilization of the embedding is 100%. Our embedding scheme can also be applied to $f > n$ cases as long as there exist free dimensions. Furthermore, all embedding schemes can be applied to torus (wraparound mesh).

As we assume partial fault model in this paper, dilation 2 has the same meaning as distance 2. If the total fault model is used, the distance-2 neighbors may have dilation of at most 4, if number of faulty nodes is less than n; and may have dilation of at most 6, if number of faulty nodes is less than $2n - 2$ [7].

References

[1] B. Becker and H. Simon, "How robust is the n-cube ?" Information and Computation, 1988, pp.162-178.

[2] T.F. Chan and Y. Saad, "Multigrid algorithms on the hypercube multiprocessor," IEEE Trans. on Comput., Nov. 1986, pp.969-977.

[3] J. Hastad, T. Leighton and M. Newman, "Reconfiguring a hypercube in the presence of faults," ACM Theory of Computing, 1987, pp. 274-284.

[4] S.L. Johnson, "Communication efficient basic linear algebra computations on hypercube architectures," J. Parallel and Distributed Computing, 1987, pp.133-172.

[5] S.F. Nugent, "The iPSC/2 Direct-Connect communications technology," Hypercube Conf. 1988.

[6] Y. Saad and M.H. Schultz, "Topological properties of Hypercubes," IEEE Trans. Comput., July 1988, pp.867-872.

[7] S.B. Tien and C.S. Raghavendra, "Algorithms and bounds for shortest paths and diameter in faulty hypercubes," 28th Allerton Conf. on Communication, Control, and Computing, Univ. of Illinois at Urbana-Champaign, Oct. 1990.

[8] J. Wang and Fusun Özgüner, "Embeddings, communication and performance of algorithms in faulty hypercubes," Proc. 5th Distributed Memory Computing Conf., April 1990, pp.1455-1464.

[9] P.J. Yang, S.B. Tien, and C.S. Raghavendra, "Embedding of multidimensional meshes on to faulty hypercubes," Technical Report, Dept. of Elec. Eng, Univ. of Southern Calif.

A WAY OF DERIVING LINEAR SYSTOLIC ARRAYS FROM A MATHEMATICAL ALGORITHM DESCRIPTION: CASE OF THE WARSHALL-FLOYD ALGORITHM

Jean Frédéric MYOUPO

L.R.I. Bat. 490, URA 410 CNRS, Université Paris-Sud, 91405 Orsay, France

Abstract: We Generalize the geometric considerations introduced in [2] for matrix product to the transitive closure and shortest path problems. We use it to derive a modular version of the linear algorithm in [18]. This modular version generalizes the array in [12]. Each cell of the array has a local memory of constant size k. If n>k, we need $3n^2(1+2/k)+O(n)$ time steps to solve the transitive closure or shortest path problems, on an array of $n^2/k+O(n)$ processors. Otherwise $3n^2+O(n)$ time steps are sufficient to solve the same problems on an array of 2n-1 processors. In the first case, these complexity results asymptotically tend to $3n^2+O(n)$ respectively when k grows, which is a better improvement of the complexity results of such systolic algorithms. Two new modular linear algorithms for matrix product are also obtained from this method: The first one is partially-pipelined running in time $n^2(1+2/k)+O(n)$ on an array of $n^2/k+O(n)$ processors if n>k, otherwise $n^2+O(n)$ time steps are sufficient to solve the same problem on an array of 2n-1 processors . For k>1, these complexity results are better than the previously known ones for equivalent models [16]. The second one is also partially-pipelined with an output operation: it runs in time $(1+(k+1)/k)n^2+O(n)$ on an array of $n^2/k+O(n)$ processors if n>k. Then its time of execution tends to $2n^2+O(n)$ when k grows, which is a better improvement of the complexity results of such systolic algorithms. Some results presented here appeared partially in [13].

Key Words : Design of systolic Algorithms, Modular Linear Systolic Arrays, Parallel Algorithms, Complexity, Warshall-Floyd Algorithm.

1. Introduction

Solving the transitive closure or shortest path problem with a systolic array of processors (referred to as cells) has been studied intensively in the past. An impressive number of arrays has been proposed in the literature; among others [1, 4-13, 18-20, 22-25]. In a typical application, such arrays would be attached as a peripheral device to a host computer which would insert input values and extract output values from them. For the transitive closure (shortest path), rectangular arrays have been studied in [4, 6, 7, 10, 11, 19], hexagonal arrays in [8] and linear arrays in [9, 12, 13, 18]. These linear arrays are either non modular [12, 18] or are not efficient and very hard to handle [5, 9]. Our goal is to design efficient modular linear arrays easy to handle with non complicated controls. The algorithm proposed in [12] requires $3n^2+O(n)$ time steps and each processor has a local memory of size n, where n is the order of the adjacency matrix of the considered graph. It runs on an array of 2n-1 processors. No time delay between any two consecutive cells depends on n as in [18], consequently it is not modularly extensible as well as those in [9, 18].

In this paper we discuss an other way of designing modular linear systolic arrays for transitive closure, which can also be applied to shortest path problems. Each token travelling in the array through a channel encounters a time delay (between two consecutive cells) which does not depend on n. Each cell has k storages, where k is fixed. This algorithm, which is a three-pass one as in [4, 17, 18], runs in $3n^2(1+2/k)+O(n)$ time steps on $n^2/k+O(n)$ cells if n>k. For n≤k we have the array in [12]. With a slight modification we can derive a partially-pipelined linear algorithm for matrix product running in time $n^2(1+2/k)+O(n)$ on an array with the same size. The geometric method we use can be applied to linearize a class of sequential algorithms: The Algebraic Path Problem in general and Dynamic Programming.

2. The Warshall-Floyd Algorithm

Given the input matrix A, the sequential Warshall-Floyd algorithm to find A* is well known [see 1 or 8 for details]:

$$a_{ij}^{(p+1)}=a_{ij}^{(p)}+a_{ip}^{(p)}*a_{pj}^{(p)}; \quad 1\leq i, j, p\leq n \qquad (1)$$

This algorithm solves both the transitive closure and the shortest path problems. The difference between the the two problems is the following:

i) For the transitive closure problem, the input matrix A is the adjacency matrix. The operation + is the Boolean OR operation; the operation * is the Boolean AND operation. The output A* is the transitive closure matrix.

ii) For the shortest path problem, the input A is the distance matrix. The operation + is the minimum operation; the operation * is the addition operation. The output A* is the shortest path matrix. Readers can find examples in [8] (pp. 606-607) which shows how the Warshall-Floyd algorithm works.

3. Linear Array Model

We compute the recurrence (1) on a linear array of 2(n)n-1 cells if n≤k. For n>k, we compute (1) on a linear array of (n+k-1)q(n)+k(q(n)+e(n)-1)+n+r-1 cells if r(n)≠0; or on a linear array of (n+k-1)q(n)+k(q(n)+e(n)-1) cells otherwise; where q(n) and r(n) are respectively the quotient and the rest of the Euclidean division of n by k; that is n=kq(n)+r(n), 0≤r(n)<k, e(n)=1 if r(n)≠0; e(n)=0 otherwise. The technique used here derives essentially from [2, 15, 16, 22]. Our array contains two data channels BA, BA', four control channels BC1, BC2, BC'1, BC'2 and an address channel BAD as shown in Fig. 1.

We use two copies of matrix A, say A and A'; let a_{ij} denote the (i, j)[th] entry in A and a'_{ij} denote the same entry in A'; a_{ij} travels in BA and a'_{ij} travels in BA'. Each token travelling in BA, BC1, BC2 and BAD encounters a delay of 1 clock cycle between any two consecutive cells. Those travelling in BA', BC'1 and BC'2 encounter a delay of 2 clock cycles. The control tokens C'2 and C2 are integer controls travelling in BC'2 and BC2 respectively. The integers C1 and C'1 are control tokens travelling in BC1 and BC'1 respectively.

A token enters a cell from the left (its input) at the beginning of a cycle and emerges from the right (its output) at the end of a cycle (possibly updated). Each cell has a local memory of size k, where k is fixed. Let c_{ij} be the token whose address is in the channel BAD of the input of the appropriate cell; then

$$c_{ij}=c_{ij}+x*y \qquad (2)$$

where x and y are the cell's inputs in BA and BA' respectively. The operation of a cell in any clock cycle is the following:

Step 1. -If C2-C'2=0 then (2) is performed and the result is stored at the location indicated in the channel BAD of its input.

Step 2.-If there is a control token in its input control channel BC1, then it changes the contents of its input token of the channel BA' to the updated value of c_{ij}.

Step 3.-If there is a control token in its input control channel BC'1, then it changes the contents of its input token of the channel BA to the updated value of c_{ij} .

Although initially a_{ij} and a'_{ij} are the same, these values change as the algorithm progresses. The assignment (2) is performed if and only if C2=C'2; there is no computation otherwise.

4. Geometric Approach

The operations of the systolic array we design here is the result of the projection of the space-time diagram of equation (1) on the space axis. This projection must be such that the trajectories of the elements a_{ij} and a'_{ij} become simple lines of defined slopes, in the plane whose axes are cell'axis and time axis. Then the resulting timing function, which gives for any c_{ij} its partial computation dates, and the task allocation function, which assigns to a cell of the array the computation of c_{ij}, can be defined easily: just determine $t_0(a_{ij})$ and $t_0(a'_{ij})$, 1≤i, j≤n, the dates at which a_{ij} and a'_{ij} respectively are inserted in the array. If we know the slopes of the trajectories (which are lines) of a_{ij} and a'_{ij}. Then the dates and the cells of the different intersections of the trajectories of the elements a_{ip} and a'_{pj} give the timing and the task allocation functions of c_{ij}.

Concretly, consider the plane whose axes are cell's axis and time axis. The trajectory of a token travelling in the array through a channel with a time delay μ between 2 consecutive cells , is a line of slope μ. For example, let $t_0(a_{ij})$ be the time at which a_{ij} is inserted in the array, $t(a_{ij})$ the time at which a_{ij} reaches the cell numbered s and μ the time delay through the channel in which a_{ij} is travelling. Then the equation of the trajectory of a_{ij} in the plane defined above is $t(a_{ij})=\mu s+t_0(a_{ij})$. Then we have to define slopes which can give efficient arrays. Our method consists of drawing all the trajectories of the elements a_{ij} and a'_{ij} in this plane. Suitable controls are used to avoid the points where the accumulation $c_{ij}=a_{ij}+a_{ip}*a'_{pj}$ with p≠p' could occur. At the intersection of the trajectory of a_{ip} and the trajectory of a'_{pj} the partial computation $c_{ij}=c_{ij}+a_{ij}*a'_{pj}$ of c_{ij} is performed. This partial computation will be noted i.p.j or simply ipj . We believe that this method, which was introduced in [2] just for Matrix Product, can help to linearize and derive better complexity results for a class of algorithms: Gauss-Jordan Elimination, that is the Algebraic Path Problem in general. It has been used successfully in [14] to get faster linear systolic arrays for Dynamic Programming.

Remark 1. The effects of the controls C2 and C'2 are the following: C2 and C'2 are set to the same value. That is C2=C'2. For all p, $1 \leq p \leq n$, C2 is inserted when a_{ip} is inserted in the array and C'2 is inserted when a'_{pj} is inserted in the array. More generally, let C2(a_{ip}) and C'2(a'_{pj}) be the values of C2 and C'2 accompanying a_{ip} and a'_{pj} respectively. Then C2(a_{ip}) =C'2(a'_{pj}) for all p such that $1 \leq p \leq n$, and C2(a_{ip})$\neq$ C'2($a'_{p'j}$) for $p \neq p'$.

4.1. Case $n \leq k$

Remark 2. For $n \leq k$ we obtain the array in [12] ; each cell has a local memory of size n. a_{ij} travels with an address which indicates in what location of the memory the result of the computation in which a_{ij} is involved must be stored. This address in this case is i.The matrices A and A' have to be stored between passes. This is done by the host computer which extracts the elements a_{ij} and a'_{ij} at the end of the pass z at precise times and inserts them in the array at the beginning of the pass z+1 at precise times, $1 \leq z < 3$.

We just need to insert the tokens in the array as in [12] :

1- Begin the first pass at T_1, the second pass at $T_2=T_1+T(n)$ and the third pass at $T_3=T_2+2T(n)$. Where $T(n)=(n+4)(n-1)$

2- In every pass z $(1 \leq z \leq 3)$:

a)- Insert a_{ij} in BA at time $(j-1)n+n+i-2+T_z$.

b)- Insert a'_{ij} in BA' at time $(i-1)n+n-j+T_z$.

c)- Insert a control token in BC1 when a_{ij} is inserted on BA.

d)- Insert a control token in BC'1 when a'_{ii} is inserted on BA'.

e)- Set C2=C'2. Insert C2 in BC2 when a_{ip} is inserted. Insert C'2 in BC'2 when a'_{pj} is inserted. This common value of C2 and C'2 must be different from the one accompanying $a_{ip'}$ and $a'_{p'j}$ respectively with $p \neq p'$.

f)- Insert the address i in BAD when a_{ij} is inserted on BA.

g)- Extract a_{ij} from cell 2n-1 at time $(j-1)n+3n+i-3+T_z$.

h)- Extract a'_{ij} from cell 2n-1 at time $(i-1)n+n-j+2(2n-1)+T_z$.

Definition 1: D_p is the embedding of the set of the computation points i.p.j (where the elements a'_{pj} are involved) in pass 1 in the plane whose axes are cell's axis and time axis. i.p.j or ipj represents the point where $c_{ij}=c_{ij}+a_{ip}*a'_{pj}$.

Example 1: For n=4, we have D_1={ 1.1.1, 1.1.2, 1.1.3, 1.1.4, 2.1.1, 2.1.2, 2.1.3, 2.1.4, 3.1.1, 3.1.2, 3.1.3, 3.1.4, 4.1.1, 4.1.2, 4.1.3, 4.1.4 }; D_2={ 1.2.1, 1.2.2, 1.2.3, 1.2.4, 2.2.1, 2.2.2, 2.2.3, 2.2.4, 3.2.1, 3.2.2, 3.2.3, 3.2.4, 4.2.1, 4.2.2, 4.2.3, 4.2.4 }; D_3={ 1.3.1, 1.3.2, 1.3.3, 1.3.4, 2.3.1, 2.3.2, 2.3.3, 2.3.4, 3.3.1, 3.3.2, 3.3.3, 3.3.4, 4.3.1, 4.3.2, 4.3.3, 4.3.4, } and D_4={ 1.4.1, 1.4.2, 1.4.3, 1.4.4, 2.4.1, 2.4.2, 2.4.3, 2.4.4, 3.4.1, 3.4.2, 3.4.3, 3.4.4, 4.4.1, 4.4.2, 4.4.3, 4.4.4 }.

The controls C2 and C'2 are used to avoid the points where the accumulation $c_{ij}=c_{ij}+a_{ip}*a'_{p'j}$ with $p \neq p'$ could occur. Fig. 2. shows an example of this array for n=4. At the intersection of the trajectory of a_{ip} and the trajectory of a'_{pj} the partial computation $c_{ij}=c_{ij}+a_{ip}*a'_{pj}$ of c_{ij} is performed.

Remark 3 : The elements a_{ip} and a'_{pj} involved in the computation in D_p are travelling with C2(a_{ip}) and C'2(a'_{pj}) respectively such that C2(a_{ip}) =C'2(a'_{pj}). Set C2(D_p) this common value of C2 and C'2 in D_p. Then C2(D_p)$\neq$C2($D_{p'}$) for $p \neq p'$. The geometric configuration of the array in pass z is the same in pass z+1, $1 \leq z < 3$.

4.1.1 Construction of the Array

To construct the array , proceed as follows:

1)---Construct D_1: to do this, insert $a_{11}, a_{21}, \ldots, a_{n1}$ in this order every clock cycle in the array through the first cell; the elements a'_{1j} are inserted consequently; a'_{11} being inserted at the same time with a_{11} (see Figure 2).

2)--- D_p, p>1 is obtained from D_{p-1} by a translation of the vector (0, n) i.e if F is this translation, then $D_p=F(D_{p-1})$. This completes the description of the algorithm. At the end of the three passes, $c_{ij}=1$ if and

only if the transitive closure of the matrix A has a 1 in that position.

Proposition 1.(task allocation and timing functions): a_{ip} and a'_{pj} reach cell i+j-2 at time pn+2i+j-4+T_z. Then c_{ij} is computed in cell i+j-2

Proof: It is obvious from the equations of the trajectories of a_{ip} and a'_{pj}.

Theorem 1: T(n) is the time during which all the computations for the pass z are completed in the cells. Then in this case the complexity in time of our algorithm is $3n^2 +O(n)$.

Proof. a_{nn} is inserted at time $t_0(a_{nn})=n^2+n-2+T_z$. It spends t'=2n-2 time steps to reach the last cell of the array where it is involved in the last computation of the zth pass; consequently the time during which all the computations are completed in the cells is $t_0(a_{nn})+t'$; then it is easy to see that $T(n)=t_0(a_{nn})+t'$. Since the array is a three-pass one, the time of execution of the algorithm is $3n^2 +O(n)$.

4.2. Case $n > k$

Remark 4. For n>k and i=k, some cells of the array have already used their k storages so that they cannot store any new c_{ij} (see Figure 3.a n=4, k=1). Then the problem is, how to insert a_{ij}, i>k so that c_{ij} should be stored in a cell whose k memory locations are still empty .

Example 2. All the geometric constructions we are going to carry out now derive essentially from the case $n \leq k$. To illustrate the idea of this paper, let us look at an example for n=4 and k<n. Then each cell of the array can store only k elements c_{ij}. Consider D_1 of the figure 2. Each cell of D_1 has a local memory of size 4 . The method consists of splitting up D_1 into q(n) subsets if r(n)=0 or into q(n)+1 subsets otherwise, n=q(n)k+r(n) (i.e q(n) and r(n) are respectively the quotient and the rest of the Euclidean division of n by k), such that each cell covered by one of these subsets can store only k elements c_{ij} and two of them (subsets) have no cell in common:

- a) n=4 and k=1. Then q(n)=4. Split up D_1 of the case $n \leq k$ into 4 subsets $D_1^{\;1}, D_1^{\;2}, D_1^{\;3}$ and $D_1^{\;4}$. $D_1^{\;kv}$ is a subset of D_1 whose elements i1j are such that (v-1)k<i$\leq$kv, $1 \leq v \leq q(n)$. $D_1^{\;1}$={1.1.1, 1.1.2, 1.1.3, 1.1.4}, $D_1^{\;2}$={2.1.1, 2.1.2, 2.1.3, 2.1.4}, $D_1^{\;3}$={3.1.1, 3.1.2, 3.1.3, 3.1.4}, $D_1^{\;4}$={4.1.1, 4.1.2, 4.1.3, 4.1.4}. $D_1=D_1^{\;1} \cup D_1^{\;2} \cup D_1^{\;3} \cup D_1^{\;4}$ with $D_1^{\;kv} \cap D_1^{\;kv'} \neq \emptyset$, $v \neq v'$, $1 \leq v$, $v' \leq q(n)$, where $\cup$ and $\cap$ are respectively the operations *Union* and *Intersection* of sets. D_p is obtained from D_{p-1} by a translation of the vector (0, 4) (see Fig. 3. a.).

- b) n=4 and k=2. Then q(n)=2. Split up D_1 of the case $n \leq k$ into 2 subsets $D_1^{\;2}$ and $D_1^{\;4}$: $D_1^{\;2}$={1.1.1, 1.1.2, 1.1.3, 1.1.4, 2.1.1, 2.1.2, 2.1.3, 2.1.4}, $D_1^{\;4}$={3.1.1, 3.1.2, 3.1.3, 3.1.4, 4.1.1, 4.1.2, 4.1.3, 4.1.4}; $D_1=D_1^{\;2} \cup D_1^{\;4}$ and $D_1^{\;2} \cap D_1^{\;4} \neq \emptyset$. D_p is obtained from D_{p-1} by a translation of the vector (0, 5) (see Fig. 3. b.).

- c) n=4 and k=3. Then q(n)=1 and r(n)$\neq$0. Split up D_1 of the case $n \leq k$ into 2 subsets $D_1^{\;3}$ and $D_1^{\;4}$: $D_1^{\;3}$={1.1.1, 1.1.2, 1.1.3, 1.1.4, 2.1.1, 2.1.2, 2.1.3, 2.1.4, 3.1.1, 3.1.2, 3.1.3, 3.1.4} and $D_1^{\;4}$={4.1.1, 4.1.2, 4.1.3, 4.1.4}. $D_1=D_1^{\;3} \cup D_1^{\;4}$ and $D_1^{\;3} \cap D_1^{\;4} \neq \emptyset$. D_p is obtained from D_{p-1} by a translation of the vector (0, 6).

In the general case n>k, set n=kq(n)+r(n) and consider D_1 of the case $n \leq k$. Each cell in D_1 must store n elements c_{ij}. The idea consists of splitting up this D_1 into q(n) subsets if r(n)=0; into q(n)+1 otherwise. These subsets must not have cells in common and the cells covered by each of them subsets) can store only k elements c_{ij}.

Definition 2: $D_p^{\;kv}$ is a subset of D_p whose elements i.p.j are such that k(v-1)<i$\leq$kv. Set i=kq(i)+r(i) where q(i) and r(i) are respectively the quotient and the rest of the Euclidean division of i by k. Define e(i) =1 if r(i)$\neq$0; e(i)=0 otherwise. Note also that $D_p=D_p^{\;k} \cup D_p^{\;2k} \cup ... \cup$

$D_p^{Kq(n)+r(n)}$ and $D_1^{kv} \cap D_1^{kv'} \neq \emptyset$ or $D_1^{kv} \cap$ $D_1^{kq(n)+r(n)} \neq \emptyset$. $1 \leq p \leq n$, $v \neq v'$ and $1 \leq v$, $v' \leq q(n)$. Where $\cup$ and $\cap$ are respectively the operations *Union* and *Intersection* of sets. $D_p^{Kq(n)+r(n)}$ is a subset of D_p whose elements i.p.j are such that $kq(n) < i \leq kq(n) + r(n)$.

4.2.1 Construction of the Array

To construct the array in this case, proceed as follows:

a- Set D_1^k : to do this, insert a_{11}, a_{21},..., a_{k1} in this order every clock cycle in the array through the first cell numbered 0; the a'_{1j} 's are inserted consequently; a'_{11} being inserted at the same time with a_{11}. (see Fig. 3. a. , Fig. 3. b., n=4).

b- Construction of D_1^k: Follow the trajectory of a'_{11}, traverse all the cells with no empty storage, and stop in the cell numbered m=n+2k-1 at time tm=2n+4k-2+t_0(a'_{11}). Insert $a_{k+1, 1}$ in the array such that k+1.1.1 is stored in m. The points k+1.1.j, j>1, are stored in the cells following m. This gives the subset of D_1^k. We achieve the construction of D_1^k as in section a) . In the same way, construct D_1^{2k}, ... , $D_1^{kq(n)+r(n)}$.

c- Construction of D_{p+1}^k: Consider the point (m-1, tm). Draw the trajectory of $a_{2k, p+1}$. It must be the first line (of slope 1) below the point (m-1, tm), which does not overlap the trajectory of $a_{k+1, p}$. Precisely, this trajectory is at 2 time steps below the point (m-1, tm). That is the trajectory of $a_{2k, p+1}$ goes through the ponit (m-1, tm-2)=(n+2k-2, 2n+4k-4+t_0(a'_{11})).

Proposition 2:

a)- The number of cells in a D_1^{kv} is n+k-1, $1 \leq v \leq q(n)$.

b)- The number of cells in $D_p^{kq(n)+r(n)}$ (r≠0) is n+r(n)-1.

Consequently the number of cells in the array is (n+k-1)q(n)+k(q(n)+e(n)-1), (r(n)=0) or (n+k-1)q(n)+k(q(n)+e(n)-1)+n+r(n)-1, (r(n)≠0).

c)- There are n+2k-1 time steps from the trajectory of $a_{i,p}$ to the trajectory of $a_{i+k,p}$.

That is, D_1^{kv} is obtained from $D_1^{k(v-1)}$ by a translation of the vector (n+2k-1, 2n+4k-2).

d)- There are n+k-1 time steps from the trajectory of $a_{i,p}$ to the trajectory of $a_{i,p+1}$. That is D_p is obtained from D_{p-1} by a translation of the vector (0, n+k-1).

Proof. a)- Without loss of generality, we find the number of cells in D_1^k; a_{11} is inserted in the array by the cell 0. Its trajectory is a line of slope 1. Since a_{11} is involved in the computation in every cell, it needs n steps to achieve the n computations in which it participates. Then a_{11} has run over n cells. Each trajectory of an element a_{i1} gains one more cell (at its last computation point) with respect to the trajectory of $a_{i-1,1}$; there are k-1 such elements a_{i1} inserted after a_{11}; then it is easy to see that n+k-1 is the number of cells in D_1^k.

b)- Using the same argument as in a)-, $a_{kq(n)+1,p}$ runs over n cells; there are (r-1) elements a_{ip}, i>kq(n)+1, which are inserted after it; consequently $D_p^{kq(n)+r(n)}$ covers n+r-1 cells.

Then it is easy to see that the number of cells of the array is (n+k-1)q(n)+k(q(n)-1+e(n)) or (n+k-1)q(n)+k(q(n)-1+e(n))+n+r(n)-1.

c)- Let $t(a'_{ij})$ be the time at which a'_{ij} reaches the cell numbered s. Then $t(a'_{ij})$ =2s+t_0(a'_{ij}) is the equation of its trajectory in the plane defined previously; t_0(a'_{ij}) is the time at which a'_{ij} is inserted in the array. Without loss of generality, set i=1, p=1, j=1 and t_0(a'_{11})=0. So, $t(a'_{ij})$ =2s; according to the construction of D_1^{2k}, a'_{11} reaches cell numbered m=n+2k-1 (i.e s=m) at time 2n+4k-2 where k+1.1.1 is set, i.e t($a_{k+1,1}$) in cell m is 2n+4n-2; $t(a_{k+1,1})$=s+ t_0($a_{k+1,1}$) is the equation of the trajectory of $a_{k+1,1}$. Consequently, at cell m, we have n+2k-1+t_0($a_{k+1,1}$)=2n+4n-2 i.e t_0($a_{k+1,1}$)=n+2k-1. a_{11} and a'_{11} are inserted at

time 0; so, the trajectory of $a_{k+1,1}$ is at n+2k-1 time steps from the trajectory nof a_{11}.

d)-Without loss of the generality (because of the symmetry of the construction), we prove this part setting i=1 and p=1. From the construction of D_{p+1}^k, the trajectory of $a_{2k, p+1}$ goes through the ponit (m-1, tm-2)=(n+2k-2, 2n+4k-4). Then t_0($a_{k, 2}$)=n+2k-2 and t_0(a_{12})=n+2k-2-(k-1)=n+k-1. Then there are n+k-1 time steps from the trajectory of a_{11} to the trajectory of a_{12}.

Remark 5. a_{ip} travels with an address which indicates where the computation in which it is involved must be stored; it is defined as follows: i=kq(i)+r(i) then the location (address) of c_{ij} is r(i) if r(i). This address is inserted in the array at the same time with a_{ip}.

As in [15-17, 22] our linear array is a three pass one:

1- Begin the first pass at T_1, the second pass at T_2=T_1+T(n) and the third pass at T_3=T_2+2T(n). Where T(n)=n^2+n(2q(n)+e(n)+k-2)+k(4q(n)+3e(n)-4)-1-q(n)+r(n) if r(n)=0 or T(n)=n^2+n(2q(n)+e(n)+k-1)+k(4q(n)+3e(n)-4)+r(n)-q(n)-3+r(n) otherwise. e(n)=1 if r(n)≠0; e(n)=0 otherwise.

2- In every pass z (1≤z≤3):

a)- Insert a_{ij} on BA at time (j-1)(n+k-1)+r(i)+n-2+(q(i)+e(i)-1)(n+2k-1)+T_z; where i=kq(i)+r(i), e(i)=1 if r(i)≠0; e(i)=0 otherwise (this follows from c) and d) of proposition 2).

b)- Insert a'_{ij} on BA' at time (i-1)(n+k-1)+n-j+T_z.

c)- Insert a control token on BC1 when a_{ij} is inserted on BA.

d)- Insert a control token on BC'1 when a'_{ii} is inserted on BA'.

e)- Set C2=C'2. Insert C2 in BC2 when a_{ip} is inserted. Insert C'2 in BC'2 when a'_{pj} is inserted. This common value of C2 and C'2 must be different from the one accompanying a_{ip}' and $a'_{p'j}$ for p≠p'.

f)- Insert the address r(i) in BAD when a_{ij} is inserted on BA.

g) Extract a_{ij} from the last cell numbered m_{last} at time (j-1)(n+k-1)+r(i)+n-2+(t+e(i)-1)(n+2k-1)+m_{last}+T_z.

h) Extract a'_{ij} from the last cell numbered m_{last} at time (i-1)(n+k-1)+n-j+2m_{last}+T_z.

Where m_{last}=(n+k-1)q(n)+k(q(n)-1+e(n))+n+r(n)-1 if r(n)≠0; m_{last}=(n+k-1)q(n)+k(q(n)-1+e(n)) otherwise.

This completes the description of the algorithm. At the end of the three passes, c_{ij}=1 if and only if the transitive closure of the matrix A has a 1 in that position.

Proposition 3. (task allocation and timing functions): a_{ip} and a'_{pj} reach cell r(i)+j-2+(q(i)+e(i)-1)(n+2k-1) at time (p-1)(n+k-1)+2r(i)+j+n-4+2(q(i)+e(i)-1)(n+2k-1). Then c_{ij} is computed in cell r(i)+j-2+(q(i)+e(i)-1)(n+2k-1).

Proof: It is obvious from the equations of the trajectories of a_{ip} and a'_{pj}.

Theorem 2: T(n) is the time during which all the computations are completed in the cells at a pass z. Then the complexity in time of this algorithm is $3n^2$(1+2/k)+O(n). It runs on an array of n^2/k+O(n) cells.

Proof. a_{nn} is inserted at time t_0(a_{nn})=(n-1)(n+k-1)+n-2+(q(n)+e(n)-1)(n+2k-1)+r(n)+T_z. It spends t'=(n+k-1)q(n)+n+r(n)-2+k(q(n)-1+e(n)) or t'=(n+k-1)q(n)-1+k(q(n)-1+e(n)) time steps to reach the last cell of the array where the last point n.n.n is computed; consequently the time during which all the computations are completed in the cells is t_0(a_{nn})+t'; then it is easy to see that T(n)=t_0(a_{nn})+t'. Since q(n)=(n-r(n))/k, T(n)=n^2(1+2/k)+O(n); for the three passes the complexity in time of our algorithm is $3n^2$(1+2/k)+O(n). It requires (n+k-1)q(n) +n+r-1+k(q(n)-1) or (n+2k-1)q(n)+k(q(n)-1); i.e n^2/k+O(n) cells.

5. Matrix Product

5.1 Partially-Pipelined Linear Algorithm

Since the transitive closure and the matrix product are computationally equivalent [1], matrix product A*B can run on an array which is equivalent to the one constructed above for transitive closure and shortest path problems : just replace + and * by real or complex addition and multiplication respectively; replace A' by B and suppress the controls C1 and C'1. In this case we only need one pass. If n≤k, it runs on an array of 2n-1 cells in n^2+O(n) time steps. For n>k, n^2/k+O(n) cells and n^2(1+2/k)+O(n) time steps are sufficient to solve the problem. For k>2, these complexity results are better than all the previously known ones for equivalent models [16]. The $3n^2$-n-1 time steps required by the " Run Over Technique" in [16] is practically just a particular case of our result when k=1. The complexity results in [17] are also a particular case of ours for k≤n

5.2 Partially-Pipelined Linear Algorithm with an output Operation

We want to find a matrix C which is the product of two matrices A and B i.e C=A*B.

Set n>k. This section brings slight modifications of the previous partially-pipelined linear algorithm: Consider the channels BA, BC2, BC'2, BAD, the control tokens C2, C'2 and the address r(i) or k as defined previously. Replace BA' by BB. Introduce two new channels BC, BC1 and BC3. The tokens travelling in BC, BC1 and BC3 encounter a delay between two consecutive cells of 1, k+1 and 1 clock cycles respectively. C1 is an integer travelling in BC1. Each element of the matrix C is transferred in BC when it is completely computed. The elements b_{ij} of the matrix B travels in BB. C3 is an integer travelling in BC3. It is a control token used to set the k locations of the local memory of a cell to 0 initially for all i, j such that 1≤i, j≤n. Let c_{ij} be the token whose address is in the channel BAD of the input of the appropriate cell; then

$$c_{ij}=c_{ij}+x*y \qquad (3)$$

where x and y are the cell's inputs in BA and BB respectively. The operation of a cell in any clock cycle is the following:

Step 1.- If C3=0, the it sets the k locations of its local memory to 0.

Step 2.- If C2-C'2=0 then (3) is performed and the result is stored at the location indicated in the channel BAD of its input.

Step 3.-If C1>0, the updated value of c_{ij} is transferred in BC.

The partially -pipelined linear algorithm with input and output operations is the following:

a)- Insert a_{ij} on BA at time (j-1)(n+k-1)+r(i)+n-2+(q(i)+e(i)-1)(n+2k-1); where, e(i)=1 if r(i)≠0; e(i)=0 otherwise.

b)- Insert b_{ij} on BB at time (i-1)(n+k-1)+n-j.

c)- Set C1 to 1 and insert it in BC1 when a_{in} is inserted in BA for all i, 1≤i≤n. Set it to -1 and insert it in BC1 when a_{ij} is inserted in BA for all j≠n.

d)- Set C2=C'2. Insert C2 in BC2 when a_{ip} is inserted, insert C'2 in BC'2 when b_{pj} is inserted. This common value of C2 and C'2 must be different from the one accompanying $a_{ip'}$ and $a'_{p'j}$ for p≠p'.

e)- Insert the address r(i) in BAD when a_{ij} is inserted on BA.

f)- Set C3 to 0 and insert it at time 0. Set it to 1 and insert it at all time different from 0.

This completes the description of the algorithm that we suppose to start at time O.

Theorem 3: For n>k, the product of two n-square matrices can be computed, including input and output operations, in time $(1+(k+1)/k)n^2+O(n)$ on an array of $n^2/k+O(n)$ cells.

Proof: It is easy to see that a_{1n} and b_{n1} meet in cell O at time t(1)=(n-1)(n+k-1)+n-1, where C2=C'2. c_{11} is completely computed when they meet. Since C1 is 1, c_{11} is transferred in BC. The array has n^2/k +O(n) cells. It then travels in BC till it reaches the last cell of the array numbered m_{last}, where it is the last element of the matrix C to be extracted. Let t(2) be this travel time . Then t(2)=(k+1)m_{last} =(k+1)n^2/k+O(n) (the time delay through BC is k+1). Then all the elements of the matrix C are already extracted at time t(1)+t(2)=(n-1)(n+k-1)+n-1+(k+1)n^2/k+O(n)=$(1+(k+1)/k)n^2$+O(n).

Comparatively with the arrays in [25], ours are modular and more easy to handle, which is a great advantage. We could have obtained better time complexity results by setting the time delay between two consecutive cells through BC less than k+1. Unfortunately, in that case, we have overlappings of some trajectories of some elements c_{ij}. This situation can be seen on an example: Consider the figure 3.b, n=4 and k=2 and suppose that the time delay between two consecutive cells through BC is 2. Then c_{12} is ready in cell 1 at time 19 and it is transferred in BC at the same time. It travels in BC for 2 clock cycles and arrive in cell 2 at time 21. But at the same time in cell 2, c_{22} is transferred in BC.

6. Conclusion

In this paper we have used simple geometric considerations to design modular linear systolic arrays for the transitive closure or shortest path and matrix product. For n≤k, the transitive closure or shortest path is solved in $3n^2$+O(n) time steps on an array of 2n-1 cells.

Otherwise, $3n^2(1+2/k)+O(n)$ time steps are sufficient to solve the same problem on an array of $n^2/k+O(n)$ cells. This time complexity tends asymptotically to $3n^2+O(n)$ when k grows. For n>k the array has the property that the number of cells of the array decreases when the size of the memory of each cell increases and vice-versa [17]. The proof of correctness of the linear algorithm presented here for transitive closure carries out essentially the same operations as the three-pass algorithm of Guibas et al., which is described in [4, 12, 18, 22].

For k≥2 the partially-pipelined linear algorithm proposed here, for the matrix product, is faster and requires less cells than those in [16, 20]. The partially-pipelined linear algorithm with an output operation we obtain runs in time $(1+(k+1)/k)n^2+O(n)$ on an array of $n^2/k+O(n)$ processors if n>k. Then its time of execution tends to $2n^2+O(n)$ when k grows. Which is a better improvement of the complexity results of such systolic algorithms.

Acknowledgements: I am grateful to Professor M. Tchuente for introducing me in the area of parallel algorithms, specially in systolic array algorithms.

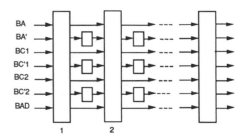

Fig. 1.

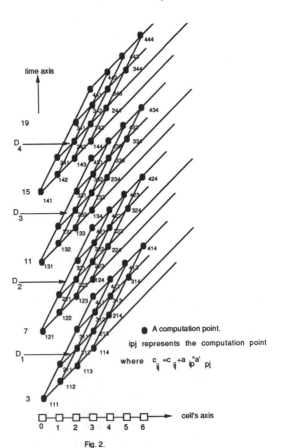

Fig. 2.

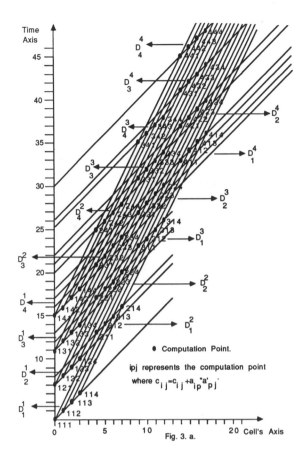

Fig. 3. a.

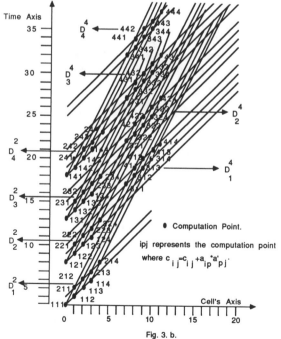

Fig. 3. b.

References

[1] A. V. Aho, J. Hopchoft and J. D. Ullman: *The design and analysis of computer algorithms* Addison-Wesley (1974).

[2] A. Benaini and M. Tchuente: Matrix product on linear systolic arrays, in Parallel and Distributed Algorithms, M. Cosnard and al. eds, North Holland, 1989.

[3] R. Dewangan and C. Pandu Rangan: A simple imple-mentation of Warshall's algorithm on a systolic chip. in Graph-Theoretic Concepts in Computer Science-*International Workshop WG'86*, Bernried(1986), Lecture notes in Compu-ter Science **246.**

[4] L. J. Guibas, H. T. Kung and C. D. Thompson : Direct VLSI Implementation of combinatorial algorithms; *Proc. Conference on very large Scale Integration: Architecture, Design, Fabrication*; California Institute of Technology (1979).

[5] C. H. Huang and C. Langauer: An incremental mechanical deve-lopment of systolic solutions to the algebraic path problem. *Acta Informatica 27*, (1989), 97-124.

[6] S. Y. Kung, P. S. Lewis and S. C. Lo: On optimal mapping algorithms to systolic arrays with application to the transitive closure problem, in *Proc. 1986 IEEE Int. Symp. Circuits Syst..*

[7] S. Y. Kung and S. C. Lo: A spiral systolic architecture algorithm for transitive closure problems, in *Proc. IEEE Int. Conf. Comput. Design*, 1985.

[8] S. Y. Kung, S. C. Lo and P. S. Lewis: Optimal systolic design for transitive closure and the shortest path problems; IEEE *Trans. Comput. C-36*, No. 5, 1987, PP. 603-614.

[9] P. Lee and Z.Kedem: Synthesizing linear array algorithms from nested for loop algorithms.*IEEE Trans. Comput., C-37* (1988).

[10] P. S. Lewis and S. Y. Kung: Dependence graph based design of systolic arrays for the algebraic path problem . *Proc. 12th ann. Asilomar Conf. Signals Syst., Comput.*, 1986.

[11] F. C. Lin and Wu: Systolic arrays for transitive closure algorithms in *Proc. Int. Symp.VLSI Syst. Designs*, Taipei May 1985.

[12] J. F. Myoupo : A linear systolic array for transitive closure problems, *Proc. Int. Conf. Parallel Process.*, 1990. Vol. I.

[13] J. F. Myoupo : Transitive closure and matrix product on modular linear systolic arrays, *Proc. SCCC Int. Conf. Comput. Science*. Santiago de Chile, July 1990.

[14] J. F. Myoupo:" Dynamic programming on linear systolic arrays", to appear in the *Proc. 28th Annu. Allerton Conf. Contr., Commun., Comput.* Oct. 1990.

[15] V. K. Prasanna Kumar and Y. C. Tsai: Designing linear systolic arrays. *J. parallel distrib. Comput.* **7** (1989), 441-463.

[16] S. K. Prasanna Kumar and Y. C. Tsai: On mapping algorithms to linear and fault-tolerant systolic srrays, IEEE *Trans. Comput.* , vol. **38,** No. 3 PP. 470-478, 1989.

[17] I. V. Ramakrishnan and P. J. Varman: Synthesis of an optimal family of matrix multiplication algorithms on linear arrays; Tech. Rep. Univ. of Maryland Dept. Comput. Sci. Proc. *ICPP*, 1985.

[18] I. V. Ramakrishnan, P. J. Varman : Dynamic programming and transitive closure on linear pipelines; *Proc. Int. Conf. on Parallel Processing*, 1984.

[19] Y. Robert and D. Trystram: An orthogonal systolic array for the algebraic path problem, *Computing*, 39,PP. 1987.

[20] G. Rote: A systolic array algorithm for the algebraic path problem; *Computing;* Vol. **34**, PP. 192-219, 1985.

[21] U. Schwiegelshohn and L. Thiele: Linear systolic arrays for matrix computation. *J. parallel distrib. Comput.* **7** (1989).

[22] J. D. Ullman *Computational aspects of VLSI*, Computer Science Press (1984).

[23] J. L. A. Van de Snepscheut: A derivation of a distributed implementation of Warshall's algorithm. *Science of Computer Programming* **7** (1986), 55-60.

[24] S. W. Warman Jr.: A modification of Warshall's algo-rithm for the transitive closure of binary relations; *CACM*. Vol. **18** , 1975.

[25] S. Warshall: A theorem on boolean matrices; *JACM*, Vol. **9**,1972.

B-SYS: A 470-Processor Programmable Systolic Array[†]

Richard Hughey and Daniel P. Lopresti
Department of Computer Science
Brown University
Box 1910
Providence, RI 02912

Abstract

This paper presents an architecture for programmable systolic arrays that provides simple and efficient systolic communication. The Brown Systolic Array is a linear implementation of this Systolic Shared Register architecture; a working 470-processor prototype system performs 108 MOPS. A 32-chip, 1504-processor implementation could provide 5 GOPS of systolic co-processing power on a single board.

Keywords: systolic array, parallel processing, VLSI, SIMD, sequence comparison.

Introduction

The systolic array is perhaps the most significant architectural development inspired by the revolution in VLSI fabrication technology. By pumping data through a regular network of hundreds or thousands of simple processing elements, computationally intensive problems can be solved many times faster on systolic arrays than on traditional machines. Since the introduction of the systolic paradigm, systolic co-processors have been proposed to solve a wide variety of problems. However, new VLSI single-purpose processing elements can require many months to design, simulate, fabricate, and test. Because of this great effort required to construct new arrays, research has turned to programmable systolic arrays. In this paper, we present both a general architecture for programmable systolic arrays and a working implementation based upon it. This *Systolic Shared Register* (SSR) architecture preserves the simple communication of single-purpose systolic arrays while providing the user with a fully programmable systolic co-processor.

To evaluate the SSR model, the Brown Systolic Array (B-SYS) has been designed, simulated, fabricated, tested, and used. B-SYS is particularly well suited to combinatorial problems: those requiring only integers and simple operations. These problems present large potential gains from systolic processing since small processing elements are sufficient, and the smaller each individual processing element, the easier it is to construct massively parallel machines. Many combinatorial problems (including data compression, dynamic programming, graph, image processing, and other problems) can efficiently use hundreds of thousands or even millions of processing elements; to achieve such high processor densities in a reasonable amount space, small and simple processing elements must be used.

A B-SYS prototype system has been running since the early fall of 1990. Ten B-SYS chips are linked together to form a small one-board, 470-processor linear systolic array which has been used to run various combinatorial algorithms. This prototype performs 108 million 8-bit operations per second (MOPS). We estimate that an easily expandable, low-cost single-board system of 1504 processing elements could perform five billion 8-bit additions each second (5 GOPS). The two major sections of this paper are devoted to the Systolic Shared Register paradigm and the Brown Systolic Array, respectively.

Systolic Shared Register Architectures

Perhaps the most critical issue in the design of programmable systolic arrays is that of systolic communication; the simple methods of single-purpose systolic systems should be maintained, but the architecture must be fully programmable. To address this issue (and others), the Systolic Shared Register architecture features a regular topology, broadcast instructions, systolic shared registers, and a stream model of data flow. The first two items follow directly from the algorithmic requirements of the class of systolic algorithms. The third provides efficient systolic communication and more flexible dispositions of computation and memory throughout the VLSI chip. Finally, the stream model of data flow is an elegant method for programming systolic co-processors.

Regular Topology

To support systolic algorithms, all of which use a regular topology by definition, the systolic co-processor itself should have a regular topology [1]. A fixed-degree regular network with nearest-neighbor connections (such as a line, square mesh, or hexagonal mesh) provides for the orderly and extensible mapping of algorithms to the systolic array.

Broadcast Instructions

Per-processor programmability currently requires too much area for inclusion in high-density systolic arrays, a result of both the size of the program store and the complexity of the instruction sequencer. To form arrays with hundreds of thousands of processing elements, there should be many processing elements on each chip; otherwise, the physical space used by the co-processor would be excessively large. Typical independently programmed systolic processing elements, such as iWarp, only have one processing element per chip [2]. Additionally, when large numbers of MIMD processors (hundreds or thousands) are used for systolic algorithms, the programs stored in each processor tend to be similar: by definition, systolic algorithms require only a small number of cell programs [1]. Thus, the inherent power of a MIMD machine is often not used for systolic algorithms.

Shared Registers for Communication

In a programmable systolic array, data cannot be moved automatically since different applications require different

---

[†]This research was supported in part by NSF grants MIP-8710745 and MIP-9020570 and by ONR and DARPA under contract N00014-83-K-0146 and ARPA order no. 6320, Amendment 1.

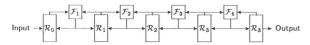

Figure 1: Linear SSR processor array with data stream.

types of data movement; data movement must be explicitly specified. Several machines provide this ability with special instructions and special queues or registers for data movement [2, 3]. In an SSR architecture, however, data moves through the array as a natural result of the user's program.

Figure 1 shows a segment of a linear SSR array. Each functional unit ($\mathcal{F}_i$) is adjacent to two register banks. Each functional unit can access data values from the register banks directly to its west and east for both input and output. Input to and output from the array can take place at either end. Systolic communication in an SSR architecture uses these shared register banks. There are no internal registers in the functional units, though a small number of flag bits may be present. This separation of computation and storage is a hallmark of the SSR architecture.

The use of broadcast instructions and relative addressing of register banks prevents contention; dual-ported memory is not required since no two functional units attempt to access the same register bank simultaneously. Simple switches between functional units and register banks can control the flow of information.

With Systolic Shared Registers, data flow takes place concurrently with computation. For example, the instruction

$$E_2 \leftarrow \min(W_0, W_1)$$

not only performs a minimization but also moves data through the array from west to east.

The SSR paradigm also does not arbitrarily limit the number of communication channels between processors; the user may decide to use the first few registers of each register bank for local storage, the next for transmission of data from west to east, the next for another data stream, and so on. In summary, shared registers provide a small, fast, flexible, and elegant implementation of systolic communication.

Stream Programming

Finally, SSR machines are programmed using a data stream paradigm, allowing the inputs and outputs at both sides of a linear array to be linked to file streams or functions by the host program. In mesh architectures, all edge processing elements, singly or in groups, may be linked to data streams. For example, a data stream of n-element vectors forming an $n \times n$ matrix could be linked to the entire west side of a mesh. The use of a *logical* data stream is crucial. It is possible that a single stream of data may use different physical registers at different times, changing register meanings to produce a more efficient program. This stream model of data flow allows the programmer to ignore such detail.

Other SSR Networks

The Systolic Shared Register paradigm is not restricted to linear arrays. Figure 2 shows abstract designs for hexagonal and octagonal mesh SSR machines. In the octagonal mesh, each functional unit can communicate with eight neighboring functional units through its four adjacent register banks.

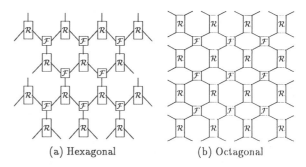

(a) Hexagonal   (b) Octagonal

Figure 2: SSR networks.

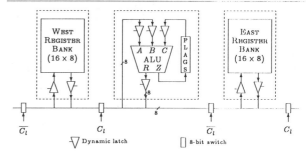

Figure 3: Segment of the B-SYS array.

Although there are two boundary edges on corner memory nodes in the hexagonal mesh and three in the octagonal mesh, just one I/O interface for each boundary register bank is needed since only one value is written to any specific register bank during the execution of an instruction. This is a result of using SIMD broadcast instructions and relative addressing of the register banks.

The Brown Systolic Array

As mentioned above, the Brown Systolic Array is a working implementation of a linear SSR machine. Its original purpose was to provide a programmable string comparison coprocessor suitable for the Human Genome Project, which is analyzing the three billion nucleotides of human DNA [4]. The need for inexpensive yet powerful hardware for this project is clear, and B-SYS is well suited to the task of performing many different comparison algorithms for the computational biologist. As a side advantage, B-SYS (being programmable) is able to perform any combinatorial problem suitable for a linear systolic array.

The architecture of an array segment is shown in Figure 3. As can be seen, each 8-bit functional unit has local flags (8) and buffers as well as access to two 16-word register banks via complementary 8-bit switches. A context flag can control conditional instruction execution. The B-SIM behavioral simulator (with both TeX snapshots and array animation using the Tango system [5]) proved crucial for algorithm development and early evaluation of this architecture.

The fully custom Brown Systolic Array chip was designed using Berkeley's Magic design suite and the COSMOS simulator [6, 7]. The 86-pin, $6.9 \, \text{mm} \times 6.8 \, \text{mm}$ chip was fabricated by MOSIS using a $2 \, \mu\text{m}$ CMOS process in May, 1990.

As is illustrated by the floor-plan (Figure 4), the B-SYS

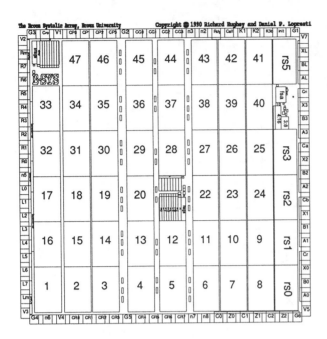

Figure 4: Floor-plan of the Brown Systolic Array.

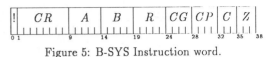

Figure 5: B-SYS Instruction word.

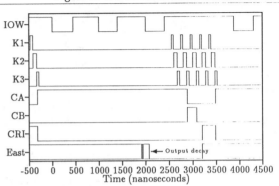

Figure 6: B-SYS clocking for sequence comparison

chip contains 47 functional units, each layed out with its west register bank (identical to the structure shown in the expanded processor) and one additional register bank. Each row of the chip has its own control circuitry, including a finite state automata (FSA) for control signal generation and two decoders, one for the 16-element register banks and one for the eight flags. Also visible are row buffers for the 16 bits of function information associated with each instruction.

Because of the simplicity of the SSR architecture, the 1730 transistors of the functional unit and register bank pair are densely packed in a 999 μm × 607 μm cell. The modestly sized chip with 47 functional units has 85 000 transistors. A larger die with a smaller technology could fit well over 500 functional units and register banks on a single chip.

Because sending information on and off chip is relatively slow, it is not practical for end functional units to read operands from adjacent chips. For this reason, the register bank shared by the rightmost processing element of one chip (block 47 in Figure 4) and the leftmost of the next (block 1) is duplicated. This performs the function of a write-through cache so that chip I/O does not take place during the loading of instruction operands. Whenever a result is written to the right, the value is sent both to the shadow register bank (above the B-SYS logo) and off-chip. At the same time, an input is received (either from an adjacent chip or the board) for the leftmost register bank (part of block 1 in Figure 4).

The B-SYS ALU is loosely based on the OM1 ALU described in Mead and Conway [8]. Logic for calculating arbitrary functions computes the propagate and generate signals which control the Manchester carry chain. The carry chain requires approximately 4 ns to evaluate.

The 38-bit B-SYS instruction word is shown in Figure 5. Eight bits specify the 3:1 function of the result logic block (CR) and four bits specify 2:1 functions for each of the carry logic blocks (CG, CP). Register bank addresses require five bits to specify a register bank (east or west) in addition to the register number (A, B, R), while flag addresses require three bits (C, Z). A single bit indicates whether or not the context flag should be obeyed (!).

B-SYS' 84 pins are grouped into five categories: clocking (4), instruction (38), data (9 for each side, including a control line for masking), diagnostic output (6), and source voltages (12). The remaining 6 pins are unused. B-SYS is controlled with three clocks (K1–K3) and an initialization signal (init). The clocks correspond to decoding the memory address and precharging memory (or evaluating the carry chain), accessing the register bank, and clearing the register selection. Figure 6, a transcription of logic analyzer data, illustrates clocking on the prototype. Lowering the init signal sends the chip through the three phases of reading two operands (CA and CB) and writing the result (CRI). The majority of time is spent copying the instruction to the board, as indicated by the IOW line. The decay of B-SYS output from the previous instruction in 2 ms can also be seen.

Timing simulation produced the conservative estimate that an entire instruction evaluation could be accomplished within 320 ns (20 ns clocks with various times between pulses). The chips were tested with a Hewlett Packard 16500A logic analysis system. The simplicity of the SSR architecture led to a fast testing scheme. Of the twenty-four chips delivered, ten performed satisfactorily at 400 ns per instruction (and could be run at faster speeds), though the complete array is run at a slower speed (Figure 6). The remaining chips suffered assorted fabrication defects which were individually pinpointed to specific rows, processing elements, and bits.

A wire-wrap ISA board buffers instructions between the B-SYS chips and the Intel 80386 (25 MHz) host. Each instruction requires three 16-bit ISA I/O bus cycles (Figure 6); this instruction transfer time dominates the prototype's 108 MOPS performance. Simple board modifications could increase the array speed to 200 MOPS, while a

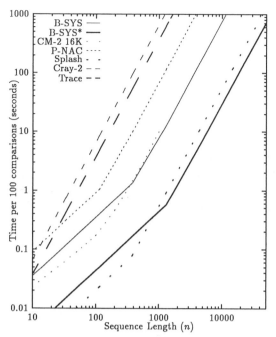

Figure 7: Many-against-one sequence comparison.

truly sophisticated board with its own controller could run the B-SYS array at full speed, increasing performance to 400 MOPS, or perhaps even closer to the chip's maximum of 3.3 MOPS per functional unit (1.8 GOPS for a 470-element array).

B-SYS was evaluated with a one-against-many sequence comparison algorithm appropriate to the Human Genome Project. The B-SYS prototype performs 470-character one-against-many sequence comparison 81 times faster than its 80386 host. Figure 7 compares B-SYS to several systolic co-processors and supercomputers. P-NAC, Splash, Cray-2, and Trace timings are extrapolated from experimental 100×100 data [9]. CM-2 timings are from experimental data. P-NAC is a single-purpose nMOS co-processor. Splash is a 2-board linear array of field-programmable gate arrays. B-SYS* is a proposed 32-chip (1504-element) board with its own controller. The Splash programmable hardware system is twice as fast as B-SYS* when $n < 400$, but the higher processor density on the single-board B-SYS* system lets it perform 1.8 times faster than Splash when $n > 1500$. With a redesigned 500-processor B-SYS chip, a 32-chip (16000-element) B-SYS system would perform 2500 times faster than P-NAC and 20 times faster than Splash when $n > 16000$.

Splash is the only programmable competitor to B-SYS* but it does not feature full or simple programmability; even slight algorithmic changes require complete reconfiguration, often lowering the systolic cell density. For example, B-SYS performance will not change for 7-bit ASCII characters, while Splash cell density will drop by a third, moving Splash's bend between $O(n)$ and $O(n^2)$ performance to $n = 250$ and increasing the performance factor difference to 2.7 when $n > 1500$. Other variations, such as the affine

cost functions used by computational biologists, will produce even starker contrasts.

Ease of programmability is also a concern. For this problem, P-NAC requires no lines of code, B-SYS requires tens of lines of code, the Connection Machine and other supercomputers require slightly more lines, and the Splash system requires nearly 3000 lines of code. With the development of programming tools for the automatic support of systolic streams, the programming and testing of new B-SYS applications requires little additional time beyond that required for selection of an appropriate systolic algorithm [10].

Conclusions

The Brown Systolic Array features regular interconnections for systolic data flow; SIMD broadcast instructions for control; shared registers for communication and computation; and data streams for the ease of programming. The working 470-processor prototype provides over 100 MOPS of systolic co-processing power.

Programs for several more B-SYS applications have been generated, including data compression, sequence comparison variations, and graph algorithms [10]. The power of B-SYS is its simple and efficient programmability and high concentration of processing elements. A more sophisticated co-processor board board could increase the Brown Systolic Array's performance to a limiting 5 GOPS for a 32-chip board, or even 50 GOPS with a redesigned, 500-element B-SYS chip. The Systolic Shared Register architecture forms a strong basis for the construction of small and powerful systolic co-processors.

References

[1] M. J. Foster and H. T. Kung, "The design of special-purpose VLSI chips," *Computer*, pp. 26–40, Jan. 1980.

[2] S. Borkar *et al.*, "iWarp: An integrated solution to high-speed parallel computing," in *Proc. Supercomputing '88*, pp. 330–339, IEEE, Nov. 1988.

[3] L. W. Tucker and G. G. Robertson, "Architecture and applications of the Connection Machine," *Computer*, pp. 26–38, Aug. 1988.

[4] J. D. Watson, "The Human Genome Project: Past, present, and future," *Science*, pp. 44–48, 6 Apr. 1990.

[5] J. T. Stasko, "Tango: A framework and system for algorithm animation," *Computer*, pp. 27–39, Sept. 1990.

[6] R. N. Mayo *et al.*, "1990 DECWRL/Livermore Magic Release," Research Report 90/7, Digital Western Research Laboratory, Palo Alto, CA, Sept. 1990.

[7] D. Beatty *et al.*, *User's Guide to COSMOS version 1.2.* Carnegie-Mellon University, Oct. 1988.

[8] C. A. Mead and L. A. Conway, *Introduction to VLSI Systems.* Reading, MA: Addison-Wesley, 1980.

[9] M. Gokhale *et al.*, "Building and using a highly parallel programmable logic array," *Computer*, pp. 81–89, Jan. 1991.

[10] R. Hughey, *Programmable Systolic Arrays.* PhD thesis, Department of Computer Science, Brown University, Providence, RI, 1991.

GENERAL ANALYTIC MODELS FOR THE PERFORMANCE ANALYSIS OF UNIQUE AND REDUNDANT PATH INTERCONNECTION NETWORKS

Mahib Rahman
VLSI Technology Inc.
2200 Park Central North, Suite 600
Pompano Beach, FL 33064

David G. Meyer
School of Electrical Engineering
Purdue University
West Lafayette, IN 47907

Abstract — General analytic models for the performance analysis of unique and redundant path unbuffered interconnection networks are presented. By simply defining a particular term in these models, a given type of memory request pattern may be taken into account. Memory request patterns employed to illustrate the usage of these models include random, permutation, and a generalized memory request pattern which is presented and referred to as the *preferred* request pattern. It is shown that the random and permutation memory request patterns are special cases of the preferred memory request pattern. The results of these analyses are presented with respect to the expected bandwidth and cost-effectiveness.

Introduction

Various analyses have been performed for *unique path* and *redundant path* interconnection networks in a circuit switching environment. Patel [11] has presented an analytical model for the performance analysis of unique path networks under random requests. In addition, Kruskal and Snir [7] have proposed a model for analyzing redundant path banyan networks under random requests. More recently, Szymanski and Hamacher [14] have extended these analytical models for analyzing unique and redundant path banyan networks under the permutation memory request pattern assumption. Also, analyses of unique path and multiple-bus networks under favorite requests have been performed by Bhuyan [2].

In this paper, we present general analytic models for the performance analysis of particular unique and redundant path networks. By simply defining a basic term in these models, a given type of memory request pattern may be taken into account. Memory request patterns employed to illustrate the utility of these models include random, permutation, and a generalized memory requesting model referred to as the *preferred* request pattern. Unique path networks considered in the analyses include crossbars and the general class of $a^n \times a^n$ banyans. In addition, redundant path networks analyzed include d-dilated and r-replicated $a^n \times a^n$ banyans, gamma, and IADM (Inverse Augmented Data Manipulator) networks.

Basic Assumptions

In this section we define some fundamental assumptions used in our analyses. We first define the most fundamental set of assumptions used in all of our analyses (independent of the type of memory request pattern employed).

Assumptions A: Given any type of memory request pattern, the following are assumed true: 1) at the beginning of every network access cycle, each processor generates a new memory request with probability $p\_req(0)$ [11]; 2) if two or more requests select the same output link of a switch in a given stage, only one of these requests is randomly selected and accepted by the network, with the others being *blocked*; and 3) the requests blocked by the network in a given cycle are ignored and not resubmitted as new requests in the next cycle [11].

The last assumption listed above is used to simplify the analyses. In reality, of course, rejected requests are expected to be resubmitted as new requests in the next cycle. Extensive simulations [11, 16] have, however, shown the effect of removing this data dependency assumption on the performance to be only slightly different from that if the third assumption is omitted. In addition to the above fundamental assumptions, specific assumptions must be established for the case in which a random memory request pattern is presumed.

Assumptions B: Random request pattern assumptions: 1) the patterns of request arrivals at the inputs of the same switch are independent; 2) requests arriving at the inputs of a switch are uniformly distributed over the outputs of that switch; and 3) for each stage in the network, the pattern of request arrivals at the inputs of that stage have the same distribution [7].

Under permutation and preferred memory request patterns, however, the independence assumptions of Assumptions B are no longer valid. Hence, new assumptions for these types of request patterns must also be introduced.

Assumptions C: Permutation request pattern assumption: in a set of parallel memory requests, each memory module is selected by at most one processor [14].

We now introduce the preferred memory request pattern. This is a more general memory requesting model for which the random and permutation memory request patterns are special cases. It is based upon the principle of the favorite memory requesting model where processors have favorite memory modules they access most of the time [2]. However, some of the basic assumptions employed by the preferred request pattern differ from those of the favorite request pattern model.

Assumptions D: Preferred request pattern assumptions: 1) each processor has a preferred memory which it attempts to access most of the time; 2) two or more processors may not have the same preferred memory; 3) we have prior knowledge from trace-driven analyses of a factor p_pref that specifies the probability with which each processor selects its preferred memory; 4) the probability that a processor issues a preferred request is independent of whether any other processor issues a preferred request; and 5) if a preferred type of request is not issued by a requesting processor, it is equally likely to access any one of N memory modules in an N×N processor-memory network.

This type of request pattern may occur in MIMD (Multiple-Instruction-Multiple-Data) systems where a given processor accesses its localized data at its preferred memory module most of the time, except when an interprocessor communication is desired. The preferred memory request pattern assumptions which differ from that of the favorite requesting model [2] are Assumptions 2) and 5). Our former assumption is realistic since processors selecting *distinctively* preferred memory modules will minimize the occurrence of memory conflicts. In addition, Assumption 5) is practical because it specifies that if a processor does not access its critical data block in its preferred memory module, it is equally likely to access another data block in any one of the N memory modules *including* the preferred module.

In the next section, we generalize the model of [14] into a form which will enable us to analyze crossbar and unique path multistage networks under various memory request patterns by simply modifying an explicit term.

Generalized Model

We will first derive the analytical model for an N×N crossbar switch, and then extend it for multistage networks employing crossbars as their basic switching element. Let $p\_req(0)$ be the probability with which each processor issues a request at the input of a crossbar network. Hence, assuming that the events occurring at the inputs of an N×N crossbar switch are independent, the probability with which i requests $(0 \leq i \leq N)$ are issued at N input ports of the switch is given by a binomial distribution,

$$\binom{N}{i} (p\_req(0))^i (1 - p\_req(0))^{N-i} .$$

We now introduce an important term in the model which may be modified to account for a particular type of memory request pattern. This term is denoted as $p\_sel(q,r)$. It is defined as the probability with which an input port (processor) selects a particular output port (memory device) of the crossbar network given that q requests have already selected this output port and r requests have selected *other* output ports.

Given that i memory requests are issued, the probability that j $(j \leq i)$ of these i requests select a particular output port is given by

$$\prod_{q=0}^{j-1} p\_sel(q,0) .$$

In addition, the probability that the remaining $i - j$ requests do not select this output port is defined as

$$\prod_{r=0}^{i-j-1} (1-p\_sel(j,r)) .$$

Thus, by combining the previous two results, we may define the probability with which at least one of the given i requests selects a particular output port as the following conditional distribution:

$$\sum_{j=1}^{i} \binom{i}{j} \prod_{q=0}^{j-1} p\_sel(q,0) \prod_{r=0}^{i-j-1} (1-p\_sel(j,r)) .$$

Henceforth, by conditioning the number of requests between one and N, we may specify the probability of acceptance at a particular output port of an N×N crossbar switch as

$$p\_acc(1) = \sum_{i=1}^{N} \binom{N}{i} (p\_req(0))^i (1-p\_req(0))^{N-i}$$

$$\cdot \sum_{j=1}^{i} \binom{i}{j} \prod_{q=0}^{j-1} p\_sel(q,0) \prod_{r=0}^{i-j-1} (1-p\_sel(j,r)) . \quad (1)$$

Using the same arguments as Patel [11], we can extend the above probability of acceptance expression to that of an n-stage unique path multistage network composed of $a \times a$ crossbar switches. By noting that the outputs of crossbar switches in stage $m-1$ become inputs to crossbar switches in stage m, we may modify Eq. (1) as follows:

$$p\_acc(m) = \sum_{i=1}^{a} \binom{a}{i} (p\_acc(m-1))^i (1-p\_acc(m-1))^{a-i}$$

$$\cdot \sum_{j=1}^{i} \binom{i}{j} \prod_{q=0}^{j-1} p\_sel(m,q,0) \prod_{r=0}^{i-j-1} (1-p\_sel(m,j,r)),$$

$$p\_acc(0) = p\_req(0), \quad 1 \leq m \leq n \quad (2)$$

where $p\_acc(m)$ is the probability (or rate) at which an output link of a switch in stage m accepts a request. Also, $p\_sel(m,q,r)$ is the probability with which a request at the input of a switch in stage m selects a particular output link of the switch given that q requests at the switch input have also selected this link and r requests at the switch input have chosen other output links. The initial condition in the recurrence relation of Eq. (2) is $p\_acc(0)$. This term is synonymous to $p\_req(0)$, the probability with which a processor issues a request. The probability of acceptance at an output port in the last stage of an n-stage network may be computed by evaluating $p\_acc(n)$ in Eq. (2). From this result, the bandwidth of an N×N network is determined by simply computing:

$$BW = N \cdot p\_acc(n) . \quad (3)$$

Crossbar Network

In this section we will illustrate the use of Eq. (1) to determine the acceptance rate of N×N crossbar networks under random, permutation, and preferred memory request patterns. This is achieved by simply modifying $p\_sel(q,r)$ to account for a given type of request pattern.

Analysis Under Random Requests.
The random memory request pattern assumes that every processor is equally likely to select any one of the N memory modules and the requests are statistically independent. Hence, by Assumptions B, the expression for $p\_sel(q,r)$ for an N×N crossbar network is independent of q and r:

$$p\_sel(q,r) = \frac{1}{N} , \quad 0 \leq q \leq N-1, \; 0 \leq r \leq N-2 . \quad (4)$$

The probability of acceptance for an N×N crossbar network under random requests may thus be computed by employing this result in Eq. (1). Performing this substitution, Eq. (1), under random requests, reduces to the well known result of Patel [11],

$$p-acc(1) = 1 - (1 - \frac{p\_req(0)}{N})^N .$$

Analysis Under Permutation Requests.
By Assumptions C, the probability with which a given processor selects a particular memory module is dependent upon whether any other request at the input of the crossbar network selects this output module. If no other processor issues a request for the indicated memory module, the probability with which the given processor selects this output module of the N×N crossbar network is 1/N, assuming that the processor is unbiased. On the other hand, if any other request selects the specified module, this probability is zero under the permutation memory request pattern assumption. In summary,

$$p\_sel(q,r) = \begin{cases} \dfrac{1}{N} , & q=0, \; 0 \leq r \leq N-2 \\ 0 , & otherwise. \end{cases} \quad (5)$$

By substituting this result in Eq. (1), $p\_acc(1)$ simplifies to $p\_req(0)$. Of course, since there are no conflicts in the crossbar network under permutation requests, this result is quite obvious; however, we again illustrate the advantage of the analytical model by adjusting $p\_sel(q,r)$

to account for a particular type of memory request pattern.

Analysis Under Preferred Requests. By the second assumption in Assumptions D, a processor will select a particular memory module due to a preferred request only if *none* of the other q requests that have already selected this module issue a preferred request. The probability of the latter event occurring is given by

$$(1-p\_pref)^q$$

where p_pref indicates the probability with which a processor issues a preferred request. Hence, given that a processor selects a particular memory module and that q requests have already selected this module, the probability that the indicated processor issues a preferred request is given by

$$p\_pref \cdot (1-p\_pref)^q \ .$$

If we are also given that r requests have selected other memory modules, the probability that k of these r requests have issued preferred requests is given by

$$\binom{r}{k} (p\_pref)^k (1-p\_pref)^{r-k} \ , \quad 0 \le k \le r \ .$$

Given that a preferred request is issued and that k processors have already selected k other memories due to preferred requests, the probability that a particular memory module is selected is $1/(N-k)$. Thus, by conditioning k between zero and r, we may derive the probability with which a particular memory is selected, given that a preferred request is issued and r requests have selected other memories:

$$\sum_{k=0}^{r} \binom{r}{k} (p\_pref)^k (1-p\_pref)^{r-k} \cdot \frac{1}{N-k} \ .$$

By combining the above results, the probability with which a particular memory module is selected due to a preferred request, given q and r, is defined as

$$p\_pref \cdot (1-p\_pref)^q \cdot \sum_{k=0}^{r} \binom{r}{k} \frac{(p\_pref)^k \cdot (1-p\_pref)^{r-k}}{N-k} \ .$$

In the case for which a not preferred request is issued, the processor is equally likely to choose any one of N memory modules. This probability is given simply by

$$(1-p\_pref) \cdot \frac{1}{N} \ .$$

By conditioning upon the mutually exclusive events of a preferred and not preferred request being issued, the probability with which a particular memory module is selected, given q and r, is defined as

$$p\_sel(q,r) = p\_pref \cdot (1-p\_pref)^q \cdot$$

$$\sum_{k=0}^{r} \binom{r}{k} (p\_pref)^k (1-p\_pref)^{r-k} \cdot \frac{1}{N-k}$$

$$+ (1-p\_pref) \cdot \frac{1}{N} \ , \quad 0 \le q \le N-1, \ 0 \le r \le N-2 \ . \quad (6)$$

It may be apparent to the reader that for $p\_pref = 0$, Eq. (6) reduces to that for random requests (Eq. (4)). Similarly, in the case of $p\_pref = 1$, Eq. (6) reduces to that for permutation requests (Eq. (5)). Hence, the random and permutation request patterns are *special* cases of the preferred request pattern proposed in this paper.

$a^n \times a^n$ Banyan Networks

We now illustrate the usage of the analytical model defined by the recurrence relation of Eq. (2) for unique path multistage networks. The type of networks which will be used to exemplify this are $a^n \times a^n$ banyan networks. The banyan network, defined initially by Goke and Lipovski [6], is a general network encompassing a broad class of multistage networks. Informally, an $a^n \times a^n$ banyan network has been defined to be an n-stage network composed of $a \times a$ crossbar switches, where each input of the network has a unique path to each output of the network. Illustrations of $2^3 \times 2^3$ unique path banyans may be found in [7, 14].

Analysis Under Random Requests. Each output of a switch in this network leads to a unique set of memory modules. Hence, the probability with which a request to an $a \times a$ crossbar switch in stage m of the network under this type of requests is independent of m, q, and r:

$$p\_sel(m,q,r) = \frac{1}{a},$$

$$1 \le m \le n, \ 0 \le q \le a-1, \ 0 \le r \le a-2. \quad (7)$$

This result may be used in the recurrence relation of Eq. (2) to compute $p\_acc(n)$, the acceptance rate at the output stage of the network. It should again be understood that this substitution reduces Eq. (2), under random results, to that of Patel [11]

$$p\_acc(n) = 1 - (1 - \frac{p\_req(0)}{a})^a \ .$$

Analysis Under Permutation Requests. The output link selection probability for unique path banyans, under permutation requests, have been derived by Szymanski and Hamacher [14]. Hence, we will not present this derivation but rather define their result explicitly as

$$p\_sel(m,q,r) = \quad (8)$$

$$\begin{cases} \dfrac{a^{n-m}-q}{a^{n-m+1}-q-r} \ , & \text{for } q < a^{n-m}, \ (q+r) < a^{n-m+1}, \ 1 \le m \le n \\ 0 \ , & \text{otherwise.} \end{cases}$$

Analysis Under Preferred Requests. The fourth assumption in Assumptions D specifies that a processor issues a preferred request independent of whether any other processor issues such a request. Hence, given that q requests have selected a particular output link of a switch in stage m, the probability that s of these q requests are preferred requests is given by

$$\binom{q}{s} (p\_pref)^s (1-p\_pref)^{q-s} \ , \quad 0 \le s \le q \ .$$

In addition, if we are also given that r requests have selected other output links of this switch in stage m, the probability with which t of these r requests are preferred requests is defined as

$$\binom{r}{t} (p\_pref)^t (1-p\_pref)^{r-t} \ , \quad 0 \le t \le r \ .$$

Let the probability with which a request selects a particular output link of stage m, given that s requests have already selected this link due to preferred requests and t requests have selected other links due to the same type of requests, be denoted as $p\_sel\_perm(m,s,t)$. The second assumption in Assumptions D specifies that two or more preferred requests may not be directed to the same memory module. Hence, some careful investigation will reveal that $p\_sel\_perm(m,s,t)$ is simply $p\_sel(m,s,t)$ due to permutation requests (see Eq. (8)). Hence,

$p\_sel\_perm(m,s,t) =$

$$
\begin{cases}
\dfrac{a^{n-m}-s}{a^{n-m+1}-s-t} \;, & \text{for } s<a^{n-m},\; (s+t)<a^{n-m+1},\; 1\le m \le n \\[2mm]
0 \;, & \text{othewise.}
\end{cases}
$$

By conditioning s and t between their respective limits, we may define the probability with which a request selects a particular output link of stage m, given that it is a preferred type of request, as

$$
\sum_{s=0}^{q} \binom{q}{s} (p\_pref)^q (1-p\_pref)^{q-s}
$$

$$
\cdot \left(\sum_{t=0}^{r} \binom{r}{t} (p\_pref)^r (1-p\_pref)^{r-t} \cdot p\_sel\_perm(m,s,t) \right).
$$

Thus, the probability that a particular output link of stage m is selected due to a preferred request, given q and r, is defined as

$$
p\_pref \cdot \sum_{s=0}^{q} \binom{q}{s} (p\_pref)^q (1-p\_pref)^{q-s}
$$

$$
\cdot \left(\sum_{t=0}^{r} \binom{r}{t} (p\_pref)^t (1-p\_pref)^{r-t} \cdot p\_sel\_perm(m,s,t) \right).
$$

Using the fifth assumption in Assumptions D, a not preferred type of request is equally likely to select any one of a^n memories. Hence, the probability that a request selects a particular output link of an $a\times a$ switch in stage m due to a not preferred type of request is

$$
(1-p\_pref) \cdot \frac{1}{a} \;.
$$

By conditioning upon the mutually exclusive events in which preferred and not preferred requests are issued, respectively, the probability with which a particular output link of an $a\times a$ switch in stage m is selected, given q and r, is

$$
p\_sel(m,q,r) = p\_pref \cdot \sum_{s=0}^{q} \binom{q}{s} (p\_pref)^q (1-p\_pref)^{q-s}
$$

$$
\cdot \left(\sum_{t=0}^{r} \binom{r}{t} (p\_pref)^t (1-p\_pref)^r \right.
$$

$$
\left. \cdot p\_sel\_perm(m,s,t)\right) + (1-p\_pref) \cdot \frac{1}{a} \;. \tag{9}
$$

It is worthwhile to note that this result reduces to that under random requests (Eq. (7)), and that under permutation requests (Eq. (8)) for $p\_pref=0$ and $p\_pref=1$, respectively. This again illustrates the generality of the preferred memory requesting model presented in this paper.

Performance Analysis of Redundant Path Networks

In this section, we present general analytical models for the performance analysis of various redundant path multistage networks. First, we will extend the general model for unique path networks to that for redundant path banyans in the following subsection.

d-Dilated, r-Replicated $a^n\times a^n$ Banyan Networks

Two types of redundant path banyan networks have been proposed in the literature. The first type is the d-

dilated banyan network [6, 7]. The d-dilation of a banyan network has been defined to be the network obtained by replacing each link of this network by d distinct links. In other words, a request entering a switch at stage m, $0 \le m \le n-1$, of a d-dilated $a^n \times a^n$ banyan network may exit using any of d edges connected to a successor switch at the next stage. An illustration of a 2-dilated $2^2 \times 2^2$ banyan network may be found in [7, 14].

The second type of redundant path banyan proposed is the r-replicated banyan. Here, the r-replication of a network has been defined to be the network consisting of r identical copies of the original network [6, 7]. An illustration of a 2-replicated $2^2 \times 2^2$ banyan network may also be found in [7, 14]. It follows that a d-dilated, r-replicated $a^n \times a^n$ banyan network is derived from r copies of an d-dilated $a^n \times a^n$ banyan.

Kruskal and Snir [7] have analyzed the performance of d-dilated, r-replicated $a^n \times a^n$ banyan networks under random requests. Szymanski and Hamacher [14] have extended these results to analyze these networks under permutation requests. In this Section, we propose a more simple and general model than that presented in [14] to analyze such banyan networks under a given type of memory request pattern.

Assumptions specified in [7] for d-dilated $a^n \times a^n$ banyan networks are as follows: 1) if $j \le d$ requests to a switch are competing for d redundant output links of the switch which lead to the same successor switch at the next stage, all of these j requests are accepted; and 2) if $j > d$ requests are competing at a switch for the indicated d output links, only d of j input requests are chosen at random and forwarded with the remaining ones being ignored. Based upon these two assumptions, we introduce two new assumptions which are appropriate for our particular analyses: 3) if $j \le d$ requests at the input of a switch are issued for the d output links of the switch which lead to the same successor switch in the next stage, then the specified j requests are uniformly distributed among the d output links; and 4) if $j > d$ requests are competing for the indicated d output links, then only d of these j requests are chosen at random and are uniformly distributed among the d output links.

The final assumption in our analysis is related to r-replicated $a^n \times a^n$ banyan networks: 5) if a processor at the input of the network issues a request, it randomly chooses one of r copies of an $a^n \times a^n$ banyan network.

In a d-dilated, r-replicated banyan, assumptions 3) through 5) imply that a processor at the input of stage 0 is equally likely to select any one of $r \cdot d$ output links of a $1{:}(r \cdot d)$ demultiplexer in this stage of the network. The rate at which an output link of stage 0 issues a request is thus given by

$$
p\_acc(0) = p\_req(0)/(r\cdot d) \tag{10a}
$$

where $p\_req(0)$ is the probability that a processor issues a request.

Switches in stages 1 through n of d-dilated, r-replicated $a^n \times a^n$ banyan networks are composed of $a\cdot d \times a \cdot d$ crossbar switches. Assuming that the events occurring at the inputs of such a switch in stage m are independent, the probability that i requests are issued at the $a\cdot d$ inputs of this switch is given by

$$
\binom{a\cdot d}{i} (p\_acc(m-1))^i (1-p\_acc(m-1))^{a\cdot d-i} ,
$$

$$
1\le m \le n, \;\; 1\le i \le a\cdot d \;.
$$

Also, the probability that at least one of these i requests select a particular set of d output links which lead to same successor switch in the next stage is defined as

$$\sum_{j=1}^{i} \binom{i}{j} \prod_{q=0}^{j-1} p\_sel(m,q,0) \prod_{r=0}^{i-j-1} (1-p\_sel(m,j,r))$$

where $p\_sel(m,q,r)$ is the probability that the indicated set of d output links in stage m is selected given that q requests have already selected this set, and r requests have chosen other such sets.

Given that j requests have selected a particular set of d outputs links which lead to the same successor switch in the next stage, the probability that a specific one of these d output links accepts a request is dependent upon the magnitude of j. Using assumptions 3) and 4) defined earlier in this section, we may define this probability as

$$p\_sel\_dil(j) = \begin{cases} j/d, & for\ j < d \\ 1, & otherwise. \end{cases}$$

Thus, the probability that a particular d-dilated output link is chosen, given that i requests are issued, is defined as

$$\sum_{j=1}^{i} \binom{i}{j} \prod_{q=0}^{j-1} p\_sel(m,q,0) \prod_{r=0}^{i-j-1} (1-p\_sel(m,j,r)) \cdot p\_sel\_dil(j).$$

By conditioning the number of requests at the input of an $a \cdot d \times a \cdot d$ switch in stage m $(1 \leq m \leq n)$ between one and $a \cdot d$, we may derive the rate at which a particular output link of this switch is selected as

$$p\_acc(m) = \sum_{i=1}^{a \cdot d} \binom{a \cdot d}{i} (p\_acc(m-1))^i (1-p\_acc(m-1))^{a \cdot d - i}$$

$$\cdot \sum_{j=1}^{i} \binom{i}{j} \prod_{q=0}^{j-1} p\_sel(m,q,0) \qquad (10b)$$

$$\cdot \prod_{r=0}^{i-j-1} (1-p\_sel(m,j,r)) \cdot p\_sel\_dil(j), \quad 1 \leq m \leq n.$$

In the output stage of the network (stage $n+1$), there are $r \cdot d$ paths to each memory module. Hence, $r \cdot d$:1 multiplexer switches are employed at this stage to accept data from only one of these paths. The probability that there is at least one request at the inputs of such switches at stage $n+1$ is given by

$$p\_acc(n+1) = \sum_{i=1}^{r \cdot d} \binom{r \cdot d}{i} (p\_acc(n))^i (1-p\_acc(n))^{r \cdot d - i}.$$

This result may be simplified further to yield the network acceptance rate as

$$p\_acc(n+1) = 1 - (1-p\_acc)^{r \cdot d}. \qquad (10c)$$

Thus, our model for evaluating the acceptance rate of d-dilated, r-replicated $a^n \times a^n$ is defined by Eqs. (10a) through (10c).

There exists a one-to-many equivalence mapping of a distinct output link in stage m of a unique path $a^n \times a^n$ banyan to a particular set of d redundant output links in stage m of an d-dilated, r-replicated $a^n \times a^n$ banyan. Hence, some careful consideration will show that $p\_sel(m,q,r)$ as defined in this section is equivalent to that for unique path banyan networks. The definitions for $p\_sel(m,q,r)$ under random, permutation, and preferred memory request patterns defined earlier for unique path banyans are therefore also applicable to this analysis of redundant path banyans.

Gamma Networks

Another type of redundant path network is the gamma network [10]. This network is in the Plus-Minus 2^i (PM2I) class of interconnection schemes [5] between stages of a network. Varma and Raghavendra [15] have analyzed the gamma network for two special cases in which particular data routing algorithms are employed. Also, Yoon *et al.* [17] have used the analytical model of Patel [11] to analyze these networks under random requests. The problem with extending this model (used correctly by Patel for analyzing unique path networks) to that for redundant path networks such as the gamma network is that the independence assumption associated with the outputs of a given switch are no longer valid. This is due to the dependence relation among the redundant output links of a switch in the gamma network. For this reason, Kruskal and Snir [7] have introduced new assumptions for analyzing redundant path networks (see Assumptions 1) and 2) in the previous section). These assumptions are therefore also used in this section.

An $N \times N$ gamma network is composed of $n+1$ stages $(n = \log_2 N)$ of switches where the stages are assumed numbered from 0 to n from the input stage to output stage. The processors are connected to the network at the input of stage 0 which consists of 1:3 demultiplexer switches. Stages 1 through $n-1$ employ 3×3 crossbar switches. In addition, the memory modules are linked at the output of stage n which is composed of 3:1 multiplexer switches. The straight output link of a switch in stage m leads to a distinct set of memory modules, unlike its $\pm 2^m$ redundant output links where each leads to the same set of memory ports. It is due to this characteristic (the unequal redundancy factor in outputs of a switch in the gamma network) that prevents us from directly applying the general model for redundant path banyans to gamma networks. An illustration of an 8×8 gamma network may be found in [10, 15].

The probability that a request is accepted at an output link of an 1:3 demultiplexer switch in stage 0 is given by

$$p\_acc(0) = p\_req(0)/3 \qquad (11a)$$

where $p\_req(0)$ is the probability with which a processor issues a request. The factor of $1/3$ indicates the *average* rate at which a given request selects a particular output link of the demultiplexer in stage 0. The effect of this agrees with the assignment of a probability of $1/2$ to the straight output link, and $1/4$ to each of the two redundant $\pm 2^m$ output links. Hence, the above result is consistent with the assumptions of Kruskal and Snir [7] defined previously.

The 3×3 crossbar switches in stages 1 through $n-1$ have different acceptance rates at their straight and $\pm 2^m$ output links, respectively, due to the redundancy property of the $\pm 2^m$ output links. Thus, we may take this redundancy into account by defining the *average* rate of acceptance at a given output link in stage m as

$$p\_acc(m) = \frac{1}{3}(p\_acc\_st(m) + 2 \cdot p\_acc\_pm(m)),$$

$$1 \leq m \leq n-1. \qquad (11b)$$

where $p\_acc\_st(m)$ is the acceptance rate at the straight output link, and $2 \cdot p\_acc\_pm(m)$ is the sum of the acceptance rates at the pair of PM2I (or $\pm 2^m$) redundant output links.

The straight output link of a switch in the gamma network leads to a unique set of memory modules analogous to any output link in unique path banyans. Hence, Eq. (2) may be modified to reflect the acceptance rate at the straight output link of a 3×3 crossbar switch in the gamma network as

$$p\_acc\_st(m) = \sum_{i=1}^{3} \binom{3}{i} (p\_acc(m-1))^i (1-p\_acc(m-1))^{3-i}$$

$$\cdot \sum_{j=1}^{i} \binom{i}{j} \prod_{q=0}^{j-1} p\_sel(m,q,0)$$

$$\cdot \prod_{r=0}^{i-j-1} (1-p\_sel(m,j,r)), \quad 1 \le m \le n-1 . \quad (11c)$$

Using assumptions defined earlier for redundant path banyans, the probability with which a request selects a particular $\pm 2^m$ output link of a switch in stage m ($1 \le m \le n-1$), given that there are j requests to the indicated pair of redundant PM2I output links, is defined as

$$p\_sel\_pm(j) = \begin{cases} j/2, & \text{for } j < 2 \\ 1, & \text{otherwise.} \end{cases}$$

Hence, the acceptance rate at a particular $\pm 2^m$ output link may be defined by modifying Eq. (11c) as

$$p\_acc\_pm(m) = \sum_{i=1}^{3} \binom{3}{i} (p\_acc(m-1))^i (1-p\_acc(m-1))^{3-i}$$

$$\cdot \sum_{j=1}^{i} \binom{i}{j} \prod_{q=0}^{j-1} p\_sel(m,q,0)$$

$$\cdot \prod_{r=0}^{i-j-1} (1-p\_sel(m,j,r))$$

$$\cdot p\_sel\_pm(j), \quad 1 \le m \le n-1 . \quad (11d)$$

The acceptance rate at an output of a 3:1 multiplexer switch in the last stage is given by the probability that there is at least one request at the three inputs of this switch. The indicated rate may be given by

$$p\_acc(n) = 1 - (1-p\_acc(n-1))^3 . \quad (11e)$$

Thus, our analytical model for the gamma network is defined by the recurrence relations of Eqs. (11a) through (11e). To exemplify the usage of this model, the term $p\_sel(m,q,r)$ need only be modified to take into account a particular type of memory request pattern. This term may be derived under random, permutation, and preferred memory request patterns in a manner similar to that described earlier for unique path banyans.

IADM Networks

The IADM network is topologically equivalent to the gamma network, the difference between these networks being the type of switches employed. The IADM network uses multiplexer-demultiplexer types of switches as opposed to crossbar type switches used in the gamma network. Due to the size constraints of this paper, we are unable to include our general model for the IADM network as well as its exemplifications under the various types of request patterns. The reader may, however, refer to [12] for such data.

Network Costs

In this section we will present some rather simplistic *costs* for the various types of networks analyzed in this paper. These cost figures will be used to determine the *cost-effectiveness* of the indicated networks. The cost-effectiveness of a given network will be expressed as the ratio of its expected bandwidth to its network cost. Also, the cost of a network is assumed to be directly proportional to the sum of the gate complexities of switches within the network. Hence, this cost figure ignores gate complexities associated with the control of the network, layout complex-

ity, and signal buffering due to fan-in/fan-out limitations. However, for the purpose of comparing the various networks with respect to the cost-effectiveness, we believe that the indicated cost model is reasonably accurate.

The basic building blocks of a network switch are N:1 multiplexers and 1:N demultiplexers. For large N, the number of gates in both such elements is proportional to N where, in this case, a gate is defined to be an $(\log_2 N + 1)$-input NAND logic gate. Hence, for the purpose of simplicity, the cost or gate complexity of an N:1 multiplexer or an 1:N demultiplexer will be assumed to be N hereafter. For instance, an N×N crossbar switch may be viewed as being composed of N N:1 multiplexers. Thus, its cost will be defined to be N^2. Also, a unique-path $a^n \times a^n$ banyan network is composed simply of $\log_2 N$ stages of $N/2$ $a \times a$ crossbar switches. Assuming that the cost of a crossbar switch is a^2, the cost of this type of network can thus be given by $(a^2 N \log_2 N)/2$. Similar cost analyses may be performed to derive the costs of other networks discussed in this paper. A summary of the results of such analyses is presented in Table I.

Discussion of Results

We first define the conditions in which the following graphs are obtained. To be consistent with processor request rates used in previous analyses, it is assumed that every processor always issues a request during each cycle, or $p\_req(0) = 1$. Also, under preferred requests, we assume that a processor issues a preferred memory request, given that there exists a request in a given access cycle, at a **70 percent** rate, or $p\_pref = 0.7$. In reality, the indicated rate is, however, quite dependent upon the quality of the memory allocation algorithm employed by the multiprocessor system.

Note that for notational convenience in graphs presented henceforth, $a^n \times a^n$ banyan networks are denoted as banyan-a.

Figs. 1 and 2 illustrate our bandwidth results under random, permutation and preferred request patterns for crossbar and $2^n \times 2^n$ banyans respectively. We note that, under permutation requests, the crossbar network performs considerably better than that under random requests. This is to be expected since all requests are accepted by this non-blocking network under the former type of requests. In the case of the $2^n \times 2^n$ banyan, however, we observe a much smaller performance improvement under permutation requests. The reason is that, under the specified type of requests, blocking may occur in all stages of the banyan network except for the output stage. In addition, both these figures illustrate that under preferred requests where p_pref is 0.7, the networks perform better than under random requests but worse than under permutation requests. This is to be expected, since the network acceptance rate due to preferred requests consists of components due to each of the latter types of request patterns. It should be noted that for other multistage networks analyzed in this paper, similar such bandwidth results were observed under the specified types of request patterns.

Figs. 3 through 5 compare the expected bandwidths of various N×N networks under random, permutation, and preferred memory request patterns. As may be observed, the relative positions of the bandwidth curves remain unchanged with respect to each other as the request pattern is altered. However, a number of observations are in order. In addition to our results for higher dilated and replicated banyans, we conclude that d-dilated $a^n \times a^n$ banyans perform better than d-replicated $a^n \times a^n$ banyans. This is to be expected, due to the higher complexity switches employed by the former type of banyans. Finally, we note that although the gamma and IADM networks are topologically equivalent, the crossbar switches employed in the gamma network causes it to perform much better than the IADM network (which uses multiplexer-demultiplexer types of switches).

Fig. 6 illustrates the effect of varying the probability with which a processor issues a preferred request on the expected bandwidth of various types of 1024×1024 multistage networks. We, however, note only a small improvement in performance as p_pref is varied from zero to unity. The reason is that, in a blocking multistage network, the network blockage rate is dominated by all stages except for the output stage.

Finally, Fig. 7 illustrates the cost-effectiveness of the various $N \times N$ networks under the preferred request pattern. As defined earlier, the cost-effectiveness of a given network is defined as the ratio of the expected network bandwidth to the network cost. For large N, due to its high gate complexity, the crossbar network is the least cost-efficient. Also, the $2^n \times 2^n$ path banyan is observed to be the most cost-efficient of the networks considered for similar network sizes.

Summary

General analytic models for the performance analysis of unique path and several redundant path interconnection networks have been presented. By defining a particular term in these models to reflect the data routing behavior under a specific type of memory request pattern, we were able to model network performance. These models were used to analyze the performance of the various networks under random, permutation, and preferred memory request patterns. The random and permutation request patterns were found to be special cases of the more general preferred memory requesting model presented in this paper. Finally, the results of our analyses were presented with respect to the expected bandwidth and cost-effectiveness.

References

[1] G. B. Adams, D. P. Agrawal, H. J. Siegel, "Fault-Tolerant Multistage Interconnection Networks," *Computer*, vol. 20, pp. 14-27, June 1987.

[2] L. N. Bhuyan, "An Analysis of processor-memory interconnection networks," *IEEE Trans. Comp.*, vol. C-34, pp. 279-283, March 1985.

[3] S. Cheemalavagu and M. Malek, "Analysis and Simulation of banyan interconnection networks with 2×2, 4×4, and 8×8 switching elements," *1982 Real-Time Systems Symposium*, pp. 83-89, Dec. 82.

[4] C. Clos, "A study of nonblocking switching networks," *Bell Syst. Tech. J.*, pp. 406-424, Mar. 1953.

[5] T. Y. Feng, "Data manipulating functions in parallel processors and their implementations," *IEEE Trans. Comp.*, vol. C-23, pp. 309-318, March 1974.

[6] G. R. Goke and G. J. Lipovski, "Banyan networks for partitioning multiprocessor systems," in *Proc. 1st Annu. Symp. Comp. Arch.*, 1973, pp. 21-28.

[7] C. P. Kruskal and M. Snir, "The performance of multistage interconnection networks for multiprocessors," *IEEE Trans. Comp.*, vol. C-32, pp. 1091-1097, December 1983.

[8] D. H. Lawrie, "Access and alignment of data in an array processor," *IEEE Trans. Comp.*, vol. C-24, pp. 1145-1155, Dec. 1975.

[9] R. J. McMillen and H. J. Siegel, "Routing schemes for the augmented data manipulator network in an MIMD system," *IEEE Trans. Comp.*, vol. C-31, pp. 1202-1214, Dec. 1982.

[10] D. S. Parker and C. S. Raghavendra, "The gamma network," *IEEE Trans. Comp.*, vol. C-33, pp. 367-373, Apr. 1984.

[11] J. H. Patel, "Performance of processor-memory interconnections for multiprocessors," *IEEE Trans. Comp.*, vol. C-30, pp. 301-310, October 1981.

[12] M. Rahman and D. G. Meyer, *General analytic models for performance analysis of unique and redundant path interconnection networks*, Purdue University TR-EE 90-52, September 1990.

[13] H. J. Siegel, *Interconnection networks for large-scale parallel processing: theory and case studies*, Lexington Books, D. C. Heath and Co., Lexington, MA, 1985.

[14] T. H. Szymanski and V. C. Hamacher, "On the permutation capability of multistage interconnection networks," *IEEE Trans. Comp.*, vol. C-36, pp. 810-822, July 1987.

[15] A. Varma, and C. S. Raghavendra, "Performance Analysis of a redundant-path interconnection network," in *Int. Conf. Parallel processing*, 1985, pp. 474-479.

[16] D. W. L. Yen, J. H. Patel, and E. S. Davidson, "Memory interference in synchronous multiprocessors systems," *IEEE Trans. Comp.*, vol. C-31, pp. 1116-1121, November 1982.

[17] H. Yoon, K. Y. Lee, and M. T. Liu, "Performance analysis and comparison of packet switching networks," *Proc. 1987 Int. Conf. Parallel Proc.*, Aug. 1987, pp. 542-545.

Table I. Cost summary of $N \times N$ networks.

Network	Cost
Crossbar	N^2
Unique path $a^n \times a^n$ banyan	$\frac{1}{2} a^2 N \log_2 N$
d-dilated, r-replicated $a^n \times a^n$ banyan	$\frac{1}{2} (ad)^2 r N \log_2 N + 2 dr N$
IADM	$6 N \log_2 N$
Gamma	$3N(3\log_2 N - 1)$

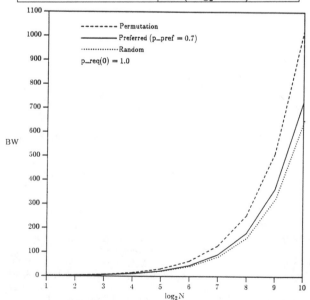

Fig. 1. Expected bandwidth, $N \times N$ crossbar networks.

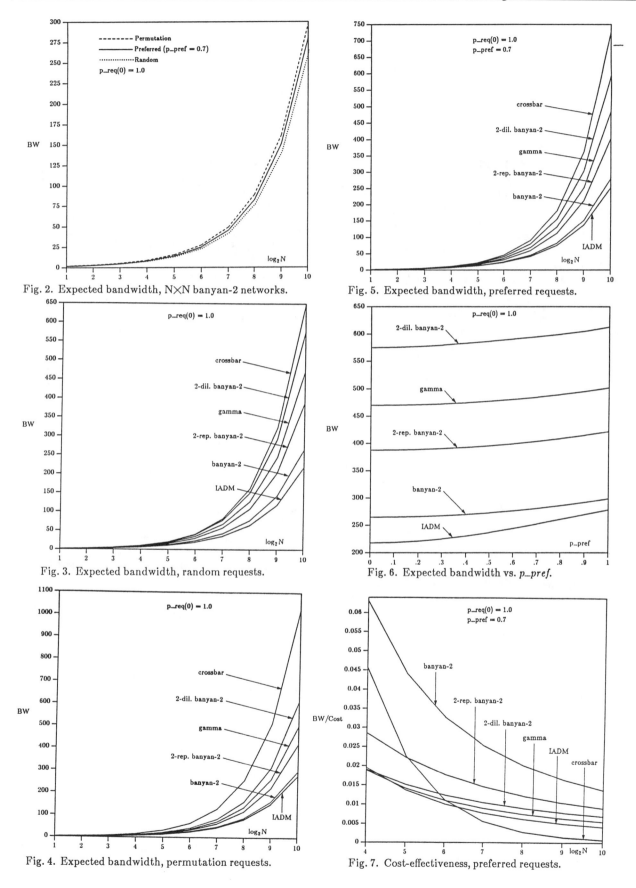

Fig. 2. Expected bandwidth, N×N banyan-2 networks.

Fig. 5. Expected bandwidth, preferred requests.

Fig. 3. Expected bandwidth, random requests.

Fig. 6. Expected bandwidth vs. *p_pref*.

Fig. 4. Expected bandwidth, permutation requests.

Fig. 7. Cost-effectiveness, preferred requests.

Performance Evaluation of Multistage Interconnection Networks with Finite Buffers*

Jianxun Ding and Laxmi N. Bhuyan
Department of Computer Science
Texas A &M University
College Station, TX 77843, USA

ABSTRACT

In this paper, we propose an approach to analyzing a class of multiprocessor systems that uses Multistage Interconnection Networks (MINs) as the communication mechanisms between the processors and the main memory modules. The MINs have finite buffers at the input side of their switch elements and operate in a synchronous packet-switching mode. We first examine the important issue of different clock periods in the synchronous MIN analysis. Then we extend the analysis to split buffered MINs that show a large improvement in performance. Finally we propose a hybrid method based on a probabilistic model and a discrete M/D/1 model to incorporate the processor waiting time for the memory access in a multiprocessor environment. The performance of multiprocessor systems with different types of finite buffered MINs are compared.

1 Introduction

In a multiprocessor system, processors communicate with each other through accessing shared memory modules or by passing messages over an interconnection network (IN). Among various INs, the Multistage Interconnection Networks (MINs) are extremely suitable for large scale shared memory multiprocessors [1, 2, 3, 4]. A MIN can be operated synchronously or asynchronously; it also can be circuit switched or packet switched [4]. There have been a lot of research on the synchronous circuit switched MINs [5, 6, 7, 8]. The analytical models for the circuit switched MINs are relatively simpler than those of packet switched MINs, where switches with finite buffers are involved. However, the synchronous buffered MINs are simple to implement and have the following advantages. The packets pass through the network stages in a pipelined way, thus resulting in high throughput. In case of a conflict, a blocked packet is stored in an intermediate switch, and does not occupy the whole path. Most of today's experimental multiprocessor systems use this class of MIN's [1, 2, 3].

The analysis of MIN's with finite buffers usually brings in the use of complex Markov chains. For example, very complex equations of single buffered MIN's were derived by Dias and Jump [9] when they modeled the operation of a switch by Timed Petri Net. For multibuffered MIN's (buffer size in a switch port is greater than one), they used simulation approach. To simplify the analysis, researchers usually change the finite buffer model to the infinite buffer model [9, 10, 11]. Jiang, Bhuyan, and Muppala [11] used the Mean Value Analysis (MVA) algorithm [12] to evaluate the behavior of the asynchronous packet switched MINs with infinite buffers. Since the

finite buffered MINs can not have accurate closed form formula, Jenq [13] used a probabilistic model and by iteratively evaluating a set of probabilistic formulas, he was able to get the latency and throughput for single buffered MINs . Yoon, Lee, and Liu [14] extended Jenq's model to cases where the number of units of a buffer at the MIN switch is more than one. Kim and Garcia extended Jenq's analysis to non-uniform traffic patterns [15].

However, these studies have made one or more assumptions that are not close to the real systems: (1) assuming that the buffers in the switch have infinite units [9, 10, 11]; (2) assuming that buffers are associated with each output port in stead of with each input port [10, 11]; (3) assuming that the basic synchronous clocks (cycles) are big enough to let control signals pass from the last stage to the first stage [13, 14, 15]; and (4) analyzing the packet switched MINs alone in stead of analyzing them in a multiprocessor environment [9, 10, 13, 14, 15].

The assumption (1) described above is obviously an approximate approach. Only limited buffer units can be built in a single switch. With infinite buffer analysis, the approximate results may not be good under different workloads. For the assumption (2), it has been pointed out [10, 16] that putting buffers at the output side is very costly, because each output buffer must be able to accept more than one packet within a single cycle. To implement the output buffering, either the switches must operate as fast as the sum of their input ports' speeds (which is not possible since input ports themselves are mostly connected to the preceding switches), or a buffer must have as many write ports as the number of input ports, to be able to handle simultaneous packet arrivals. In practice, many existing switches have their buffers at the input side [17]. For the assumption (3), the basic synchronous clocks should be small and just let the control signals pass one stage to ensure the packets passing through stages in a pipelined way. We shall discuss this issue in section 2 . The assumption (4) does not consider the effects of the other components outside the MINs. For instance, all the studies, except [11], assumed that the input rate of requests at the first stage of the MINs is fixed and independent. In a multiprocessor system, however, a processor has to wait until the previous memory request gets satisfied. The waiting time includes both memory access time and MIN latency.

In this paper, we present an analysis that considers all the above issues. We propose a hybrid method to evaluate the performance of the multiprocessor systems that have synchronous packet switched MINs with finite buffers. In section 2, a probabilistic MIN analysis considering one "small" cycle per stage is presented. In section 3, the probabilistic model is extended to the split buffered MINs that result in increased performance. Section 4 moves the analysis of sections 2 and 3 to a multiprocessor environment. Finally section 5 concludes the paper.

---
*This research is supported by the NSF grant MIP-9002353

2 A Probabilistic Model for Finite Buffered MINs

2.1 Clock Design

In this section we first try to understand the importance of the clock length in a synchronous network analysis. An 8×8 MIN with multiple buffers in a switch is illustrated in Fig. 1.

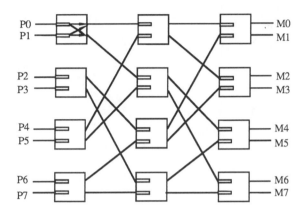

Figure 1: An 8×8 Delta Network with Finite Buffers

Jenq [13] examined the probabilistic properties between the input, the output, and the state of buffers in a switch and derived a set of equations for single buffered MINs. Through iteratively evaluating this set of formulas, the latency and throughput of a MIN can be obtained. Kim and Garcia extended Jenq's model to non-uniform traffic patterns [15]. Yoon, Lee, and Liu [14] extended the model to the finite multibuffered MINs. Thus Jenq's model has become somewhat standard to the analysis of synchronous packet switched MINs.

However, the clock synchronization mechanism in these models is not practical to MIN implementations. They all assume that a MIN operates in a "big" cycle format. Each cycle τ has two phases, $\tau = \tau_1 + \tau_2$, as shown in Fig. 2(a). In the first phase τ_1, control signals are passed across the network from the "last stage toward the first stage", so that every packet knows whether it can go to a next stage's input buffer or should stay in the current buffer. In the second phase τ_2, selected packets may move forward by one stage.

In a MIN, a packet can move forward only if it is selected among the competing packets by the routing logic of the switch, In Jenq's model, a packet can move when *either* the buffer of the switch to which it is destined in the next stage is not full *or* a packet in the next stage buffer will move forward, creating a space in the buffer. The control mechanisms for this kind of synchronization are very complicated because they must recognize whether the current buffer is full or not. If it is full, the control mechanism must find out whether the packet from the current buffer can be sent to the next stage during τ_1. If there is another packet for the same destination in the other buffer of the switch (see Fig. 1), the conflict has to be solved. Only then can the control circuit tell the preceding switch about the current buffer's status: *full* or *not full*. The main problem is that the control signals in one stage cannot be decided until the succeeding stage's control signals arrive. If we consider that the control signals passing across one stage costs λ time

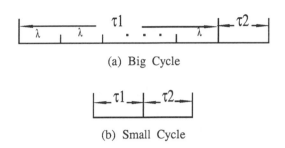

(a) Big Cycle

(b) Small Cycle

Figure 2: Comparison of Two Different Types of Cycles

units, then in a $2^n \times 2^n$ network using 2×2 switches, τ_1 equals to $n \times \lambda$ time (see Fig. 2(a)). Though the packets are passed across stages in a pipelined way, the decision, whether packets can pass through a stage or not, is made sequentially. The length of every τ_1 is then in the order of $O(n)$. So different sized networks will have different τ_1's, and that would require different clock periods, posing extreme difficulties in system design and expansion. It also brings difficulties in the performance analysis because it is not justifiable to compare the latencies of different sized networks just in terms of the number of cycles.

In our analysis, we assume that all the switches operate in a synchronous mode, but our clock cycles have a fixed length for any network size. During a cycle, control information is exchanged only between two successive stages that are required for the data transfer. Every cycle consists of two phases, τ_1 and τ_2, as shown in Fig. 2(b). In τ_1, every input port in a switch informs its connected preceding switch about the buffer status, *full* or *not full*. If the succeeding switch's input port has a *not full* buffer, a packet destined to this output is selected and it can pass in the next phase τ_2. Otherwise, no packet can go through this output port. It may be realized that when the first packet of a *full* buffer in a switch is able to move forward in the second phase τ_2, we lose one cycle in our mechanism. It is because the preceding switch's output port does not allow any packet to pass in the next phase τ_2, due to the following switch's *full* condition obtained in the first phase τ_1. In the worst case, if all the buffers on a path from the first stage to the last stage are all *full*, we may lose n small cycles. Taking a closer look, however, we will find (see Fig. 2(b)) that we gain a lot of time on an average by changing from the "big" cycles ($O(n)$) to "small" cycles ($O(1)$). We will compare the two mechanisms quantitatively in the subsection 2.3 and see that the small cycle mechanism gives better performance. Moreover, generating control signals becomes much simpler because the switch control circuit does not have to solve the conflicts first and then tell the preceding switch about the buffer status, etc. All it needs to do is to specify the status of the buffers : *full* or *not full*.

2.2 Analysis

To analyze our model of synchronous finite buffered MINs, we use a similar technique as Jenq [13], and Yoon, Lee, and Liu's [14]. Consider a general class of $b^n \times b^n$ MINs — delta networks [5, 9, 14] with n stages ($N = b^n$) and every $b \times b$ switch having an FIFO buffer of l units at each input port. We assume that the input packets are directed uniformly over all the output ports. Therefore, the packet distribution patterns are statistically identical for all the switches in a stage and we can just analyze a single switch in stead of analyzing all the b^{n-1} switches in a stage. The following variables are defined.

- $p_i(k,t)$: probability that a buffer at the stage k has i packets at the beginning of the tth cycle. We define $\overline{x}$ as $1-x$ for any variable x, i.e. $p_i(k,t) + \overline{p_i}(k,t) = 1$.

- $q(k,t)$: probability that a packet has come out of a stage $k-1$ switch and appeared on one of the b input ports of the stage k switch during the tth cycle. Note that $q(1,t)$ represents the input rate. If in the tth cycle a packet comes to a *full* first stage input port, it will be rejected. In the next cycle, the arrival probability of a packet $q(1,t+1)$ still equals to $q(1,t)$.

- $r'(k,t)$: probability that a packet in a buffer at the stage k is selected to be a candidate to pass that switch by winning the contention among the contending packets at the tth cycle, given that there is a packet in that buffer.

- $r(k,t)$: probability that a packet in a buffer at the stage k is able to move forward during the tth cycle, given that there is a packet in that buffer.

- S : normalized throughput (throughput normalized with respect to the network size N, i.e. average number of packets passed by the network per output link per cycle).

- d : normalized delay (delay normalized with respect to the number of cycles, i.e. average number of cycles taken for a packet to pass a single stage).

The probability that there are j non-empty buffers of a stage k switch in the tth cycle is $C_b^j\, \overline{p_0}(k,t)^j p_0(k,t)^{b-j}$, where C_b^j denotes the combination of j out of b. The probability that there is at least one packet from these j non-empty buffers destined to a specific output port is $[1-(1-1/b)^j]$. So the probability that a packet is ready to come to a stage $k+1$ input port during the tth cycle is equal to the probability that there is one packet coming out from the connecting output port of a k stage switch, when there is at least one packet in the corresponding buffers.

$$q(k+1,t) = \sum_{j=1}^{b} C_b^j\, \overline{p_0}(k,t)^j p_0(k,t)^{b-j}[1-(1-1/b)^j]$$
$$= 1 - [1-\overline{p_0}(k,t)/b]^b, \qquad 2 \le k \le n. \quad (1)$$

A packet must compete with other j $(0 \le j \le b-1)$ packets to pass the current switch and the chances to be chosen out are $1/(j+1)$. Also, it may be destined to any one of b output ports.

$$r'(k,t) = \sum_{j=0}^{b-1} C_{b-1}^j\, \overline{p_0}(k,t)^j p_0(k,t)^{b-1-j} \times$$
$$[1-(1-1/b)]^{j+1}\frac{b}{j+1}$$
$$= \frac{q(k+1,t)}{\overline{p_0}(k,t)}, \qquad 1 \le k \le n. \quad (2)$$

In our model, a selected packet can enter a buffer only when it is *not full* (in contrast to Jenq's model, we do not care whether the first packet of the next stage buffer is able to move forward or not).

$$r(k,t) = r'(k,t)\,\overline{p_l}(k+1,t), \qquad 1 \le k \le n-1. \quad (3)$$

Once a packet is available at a network output at the beginning of a cycle, it is removed during that cycle.

$$r(n,t) = r'(n,t) = \frac{q(n+1,t)}{\overline{p_0}(n,t)}. \quad (4)$$

We can draw a similar state transition diagram of a buffer as in [14] and get our probability formulas for the buffer status as follows:

$$p_0(k,t+1) = \overline{q}(k,t)[p_0(k,t)+p_1(k,t)r(k,t)], \quad (5)$$
$$p_1(k,t+1) = [q(k,t)]\,[p_0(k,t)+p_1(k,t)r(k,t)]$$
$$+\overline{q}(k,t)[p_1(k,t)\overline{r}(k,t)+p_2(k,t)r(k,t)], \quad (6)$$
$$p_i(k,t+1) = [q(k,t)]\,[p_{i-1}(k,t)\overline{r}(k,t)+p_i(k,t)r(k,t)]$$
$$+\overline{q}(k,t)[p_i(k,t)\overline{r}(k,t)+p_{i+1}(k,t)r(k,t)],$$
$$2 \le i \le l-2. \quad (7)$$
$$p_{l-1}(k,t+1) = q(k,t)\,[p_{l-2}(k,t)\overline{r}(k,t)+p_{l-1}(k,t)r(k,t)]$$
$$+\overline{q}(k,t)p_{l-1}(k,t)\overline{r}(k,t)+p_l(k,t)r(k,t), \quad (8)$$
$$p_l(k,t+1) = q(k,t)\,p_{l-1}(k,t)\overline{r}(k,t)+p_l(k,t)\overline{r}(k,t), \quad (9)$$

where all the k's range from 1 to n. It can be verified that $\sum_{i=0}^{l} p_i(k,t) = 1$. Note that when the buffer length l equals to one or two, we will have slightly different formulas.

With the steady input load $q(1,t)$, we can solve the above equations iteratively and get the time-independent steady-state values, $r(k)$ and $p_i(k)$, where $1 \le k \le n$ and $1 \le i \le l$. The normalized throughput — the average number of packets coming out of the network per output link per cycle, can be obtained by $p_0(n)$.

$$S = 1 - [1-\overline{p_0}(n)/b]^b. \quad (10)$$

Let $R(k)$ denote the average probability that a packet in a buffer in stage k is able to move forward, then the normalized delay — the average number of cycles for a packet to pass a stage, can be derived by the following formula [13, 14],

$$d = \frac{1}{n}\sum_{k=1}^{n}\frac{1}{R(k)}, \text{ where } R(k) = r(k)\sum_{i=1}^{l}[p_i(k)/\overline{p_0}(k)]\frac{1}{i}. \quad (11)$$

2.3 Results

Some interesting results, obtained from the above analysis, are shown in Table 1 and Table 2. In order to compare the results of the two models, we let both models have a base time unit — the small τ in our revised model. For an n staged MIN, Jenq's τ equals $(n+1)/2$ of our τ if we assume that the interval for the control signals to pass across a stage equals to the interval for a packet to pass across a stage ($\tau_1 = \tau_2$).

In Table 1, the normalized throughput is the average number of packets passed by the network per output link per "big" cycle (in Jenq's model) or per "small" cycle (in our model). The absolute throughput is the average number of packets passed by the network per output link in terms of the base cycle — a small cycle. The ratio shows the advantage of our model over that of Jenq's. We can see that though we have smaller normalized throughput than that of Jenq's model, we have much higher

Table. 1 Throughput comparisons of "big" and "small"
cycled synchronization mechanisms
(1024×1024 MIN with buffer length = 1)

Input Rate	Normalized Throughput			Absolute Throughput		
	Jenq's	Ours	ratio	Jenq's	Ours	ratio
0.2	0.197	0.157	0.798	0.036	0.157	4.387
0.4	0.366	0.221	0.603	0.067	0.221	3.317
0.6	0.439	0.229	0.522	0.080	0.229	2.872
0.8	0.450	0.231	0.512	0.082	0.231	2.819
1.0	0.453	0.231	0.510	0.082	0.231	2.807

Table. 2 Delay comparisons of "big" and "small" cycled
synchronization mechanisms
(1024×1024 MIN with buffer length = 1)

Input Rate	Normalized Delay			Absolute Delay		
	Jenq's	Ours	ratio	Jenq's	Ours	ratio
0.2	1.069	1.297	1.214	58.773	12.972	0.221
0.4	1.206	1.717	1.424	66.305	17.167	0.259
0.6	1.403	1.930	1.376	77.166	19.299	0.250
0.8	1.502	2.013	1.340	82.626	20.133	0.244
1.0	1.550	2.058	1.327	85.269	20.581	0.241

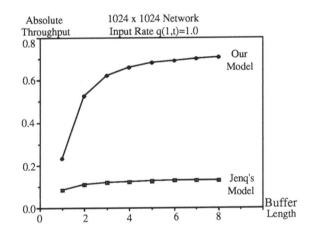

Figure 3: Absolute Throughput versus Buffer Length

(approximately 2.8 to 4.4 times higher) absolute throughput.

In Table 2, the normalized delay is the average number of "big" cycles (Jenq's model) or "small" cycles (our model) taken for a packet to pass a single stage. The absolute delay is the average number of small cycles taken for a packet to pass the whole network. Table 2 shows that although it takes more number of cycles (about 1.4 times higher than Jenq's model) for a packet to pass a stage in our revised model, it takes much less time (only about one fourth the delay time of Jenq's model) for a packet to pass a whole network in the same base time unit, because our cycle period is much smaller than that of Jenq's.

Fig. 3 plots the absolute throughput versus the buffer length for the two models. It also shows that our model outperforms Jenq's model under the multibuffer conditions.

3 A Probabilistic Model for Split Buffered MINs

In the previous section, we analyzed the synchronous, packet switched MINs with finite buffered switches. In those switches, every switch has a single FIFO buffer at each input port. We call this type of switches the conventional buffered switches and the MINs built by this type of switches the Conventional buffered MINs (CMINs). There is one drawback in this type of CMINS. If the first packets of input buffers are destined to the same output port in a switch, a conflict occurs and only one "lucky" packet can be selected to pass. The other "unlucky" packets, being at the head of their FIFO buffers, will block the succeeding packets in these buffers, even if these succeeding packets are destined to other idle output ports.

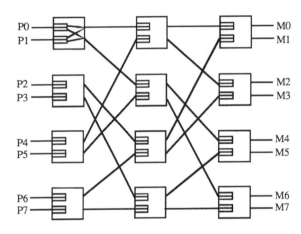

Figure 4: An 8×8 Split Buffered Delta Network (b=2)

One natural way to solve this problem is to split every input buffer into small buffers as shown in Fig. 4, where each small buffer connects to one output port. Let the network operate synchronously with each cycle τ still consisting of two phases, $\tau = \tau_1 + \tau_2$. In the first phase τ_1, every input port at the stage j informs its connected preceding stage $j-1$'s output port about their split buffers' status, *full* or *not full*. If all the split buffers are *full*, the connected preceding stage switch's output port is blocked. Otherwise, the routing logic randomly selects one packet from the stage $j-1$ switch whose stage j address matches one of the *not full* split buffers in the stage j switch's input port. In the second phase τ_2, this selected packet can move forward one stage.

It is obvious that the Split buffered MINs (SMIN) will have better performance than CMINs. However, no formal analysis for SMINs exists in literature. Tamir and Frazier [16] used Markov chain to analyze a single 2-by-2 split buffered switch. For the whole network of switches, [16] and [18] used simulation methods to show the superiority of split buffered networks.

As in section 2, we develop a probabilistic model for SMINs. The variables defined in the subsection 2.2 need not be changed, except that all the meanings relating to a single buffer there must now be repeated for all split buffers.

The following equations can be derived for the SMINs:

$$q(k+1,t) = \sum_{j=1}^{b} C_b^j \, \overline{p_0}(k,t)^j p_0(k,t)^{b-j}$$

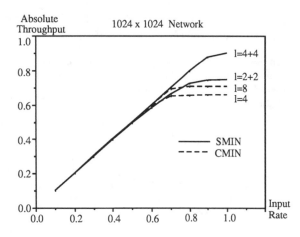

Figure 5: Absolute Throughput versus Input Rate for CMIN and SMIN (b=2)

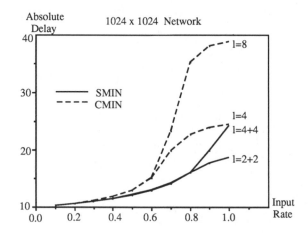

Figure 6: Absolute Delay versus Input Rate for CMIN and SMIN (b=2)

$$
\begin{aligned}
&= 1 - p_0(k,t)^b, \qquad\qquad 2 \le k \le n. \\
r'(k,t) &= \sum_{j=0}^{b-1} C_{b-1}^j\, \overline{p_0}(k,t)^j p_0(k,t)^{b-1-j} \frac{1}{j+1} \\
&= \frac{1}{b\,\overline{p_0}(k,t)} q(k+1,t), \qquad 1 \le k \le n. \\
r(k,t) &= r'(k,t)\,\overline{p_l}(k+1,t), \qquad 1 \le k \le n-1. \\
r(n,t) &= r'(n,t) = \frac{1 - p_0(n,t)^b}{b\,\overline{p_0}(n,t)}, \\
p_0(k,t+1) &= [1 - q(k,t)/b][p_0(k,t) + p_1(k,t)r(k,t)], \\
p_1(k,t+1) &= [q(k,t)/b][p_0(k,t) + p_1(k,t)r(k,t)] + \\
&\quad [1 - q(k,t)/b][p_1(k,t)\overline{r}(k,t) + p_2(k,t)r(k,t)], \\
p_i(k,t+1) &= [q(k,t)/b][p_{i-1}(k,t)\overline{r}(k,t) + p_i(k,t)r(k,t)] + \\
&\quad [1 - q(k,t)/b][p_i(k,t)\overline{r}(k,t) + \\
&\quad p_{i+1}(k,t)r(k,t)], \qquad 2 \le i \le l-2. \\
p_{l-1}(k,t+1) &= [q(k,t)/b][p_{l-2}(k,t)\overline{r}(k,t) + p_l(k,t)r(k,t)] + \\
&\quad [1 - q(k,t)/b]p_{l-1}(k,t)\overline{r}(k,t) + \\
&\quad p_l(k,t)r(k,t), \\
p_l(k,t+1) &= [q(k,t)/b]\, p_{l-1}(k,t)\overline{r}(k,t) + p_l(k,t)\overline{r}(k,t).
\end{aligned}
$$

Solving the above equations iteratively, we can get the time-independent steady-state values $p_0(k)$ and $r(k)$ ($1 \le k \le n$). Then the absolute throughput S — the average number of packets passed by the network per output link per cycle can be expressed as

$$ S = 1 - p_0(n)^b, $$

and the absolute delay D — the average number of cycles taken for a packet to pass a whole network can be expressed as

$$ D = \sum_{k=1}^{n} \frac{1}{R(k)}, \quad \text{where } R(k) = r(k) \sum_{i=1}^{l} [p_i(k)/\overline{p_0}(k)]\frac{1}{i}. $$

Fig. 5 plots the absolute throughput versus the input rate for the 1024 × 1024 CMINs and SMINs with total buffer sizes of 4 and 8. The input rate $q(1,t)$ is 1.0 and a packet is rejected if the corresponding first stage buffer is full. It shows that when the traffic load is high ($q(1,t) > 0.7$), the absolute throughput of SMINs outperforms that of CMINs.

Fig. 6 plots the absolute delay versus the input rate for the 1024×1024 CMINs and SMINs with total buffer sizes of 4 and 8. It shows that the effects of the buffer size on the absolute delay become apparent after the input load reaches 0.6 for CMIN's, in contrast to an input load of 0.8 for SMINs. The absolute delay of the CMINs on an average is also much high compared with that of the SMINs.

4 Analysis of a Multiprocessor System

Suppose we have a multiprocessor system with N ($N = 2^n$) processors and N memory modules. We use two $N \times N$ n-stage MINs as the communication media (see Fig. 7) — one for forwarding memory requests from the processor side to the memory side (we call it FN, *the Forward Network*), and the other one for returning results or acknowledgements (we call it RN, *the Return Network*).

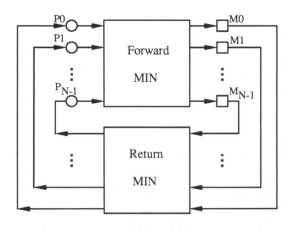

Figure 7: A Multiprocessor System

Once a processor issues a memory request, the request packet goes through the FN, joins the memory waiting queue and accesses the memory. The result packet waits in the RN's waiting queue, goes through the RN, and finally returns to the processor. In the mean time, the processor just waits until the

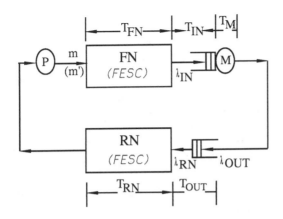

Figure 8: The Path of the Analytical Model

result/acknowledge packet comes back. The complete path of a processor's request is shown in Fig. 8.

When a packet traverses its own path, it will interfere with other packets. Therefore, we cannot analysis each request independently. However, under the uniform traffic load, all the paths are statistically the same for any packets. So we can regard the FN and the RN as the *flow equivalent service centers (FESC)* [12] and can use the probabilistic model to calculate their delays and throughputs. After we obtain the delay and the throughput, we can incorporate them with the other components on the path. For the memory modules, we can always assume that they have enough buffer units to store the incoming and outgoing packets. There are two buffer queues in a memory module, one on the input side, and the other on the output side. The probability of an arrival packet to each queue in a cycle is fixed in the steady state. Also the events whether there is an arrival packet in distinct cycles are statistically independent. For the memory input queue, the corresponding memory module can be considered as a server. Since a packet comes out from the FN with the probability of λ_{IN} (which equals to the absolute throughput of the FN) in a cycle and the memory has fixed access time, the memory input queue and the memory together can be approximately considered as an discrete M/D/1 system. For a similar reason, the memory output queue and the return MIN can be also considered as an discrete M/D/1 system because if we regard the return MIN as an FESC, it also has fixed service time (the absolute delay we get from the probabilistic model).

We assume that the whole system operates in a synchronous mode. The system clock period equals to the cycle period of the MINs as discussed in section 2. All processors are synchronized with the cycles. This means a processor can only issue a memory request at the beginning of a cycle. Once a memory request is issued, the processor will wait until it receives the result/acknowledgement. Then the processor becomes active with the probability of m per cycle to generate memory request. Under the uniform traffic load, all the requests are independently and uniformly distributed over all the memory modules. Thus, all the processors are statistically the same and we just need to evaluate the behavior of a single processor. Let the average delay of the FN be T_{FN}, the average waiting time in the memory input queue be T_{IN}, the memory access time be T_M, the average waiting time in the memory output

queue be T_{OUT}, and the average delay of the RN be T_{RN}, as shown in Fig. 8.

At the first step, we use m as the initial $q(1,t)$ (input rate to FN) for the probabilistic model to get the T_{FN} and the λ_{IN}, according to the discrete M/D/1 model, we can get

$$T_{IN} = \frac{T_M \lambda_{IN}}{2(1 - T_M \lambda_{IN})}(T_M - 1).$$

For the T_{OUT}, we must first get T_{RN}. We use the following procedure:

1. Let the initial input rate λ_{RN} to RN equal to λ_{OUT} ($= \lambda_{IN}$).

2. Perform the probabilistic modeling described in section 2.2 to get the average delay T_{RN} and the probability that a first stage's buffer is *full*, $p_l(1)$.

3. Let $\lambda_{RN} = \lambda_{OUT}/(1 - p_l(1))$.

4. Go to 2 and iterate.

After the procedure converges, we obtain T_{RN}. Since we have T_{RN} and λ_{OUT}, we can get T_{OUT} using the formula similar to T_{IN}.

As mentioned above, once a processor issues a memory request, it must wait until the result/acknowledgement packet returns. Therefore, the effective memory request rate will be

$$m' = \frac{m}{1 + m(T_{FN} + T_{IN} + T_M + T_{OUT} + T_{RN})}.$$

We substitute m' for m in the first step and iteratively evaluate the above procedure until the m' converges. Then the utilization of a processor is given by [6],

$$U = \frac{1}{1 + m(T_{FN} + T_{IN} + T_M + T_{OUT} + T_{RN})}.$$

The system response time T (processors' average waiting time) is

$$T = T_{FN} + T_{IN} + T_M + T_{OUT} + T_{RN}.$$

We apply the analytical method to the CMINs and SMINs. Fig. 9 plots the utilization versus the memory request probability in a 1024×1024 CMIN with two different memory access time, $Tm = 2$ cycles and $Tm = 8$ cycles. The buffer size is 4. The figure shows that the higher the memory access time, the smaller the processor utilization. However, the difference between the curves is not large because of the large network latency. It also shows that when the memory request probability grows from 0.0 to 0.1, the processor utilization drops dramatically from 1.0 to around 0.25.

An interesting phenomenon is that if we plot the utilization of the corresponding split buffered MINs with $2+2$ buffer queues in a switch, we get the same curves as in Fig. 9. This is true though the analysis of CMINs and SMINs shows that the 4-buffered CMIN has lower throughput and higher delay than those of $2+2$ buffered SMINs (see Fig. 5 and Fig. 6).

Fig. 10 plots the processor waiting time versus network size for CMINs with memory access time of 2 and 8. The memory request rate is 0.1 and the buffer size equals to 2. It shows that the longer the memory access time or the larger the network size, the longer the processor waiting time. If we plot the

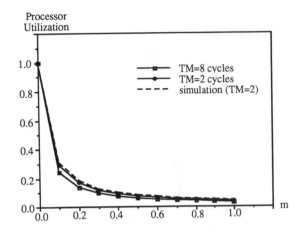

Figure 9: Processor Utilization versus Memory Request Rate for a 1024×1024 Multiprocessor System (Buffer Size = 4)

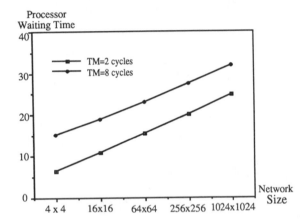

Figure 10: Processor Waiting Time versus Network Size (Memory Request Rate = 0.1, Buffer Size = 2)

CMINs or SMINs with different buffer sizes, we again get the same curves as those of CMINs regardless of the buffer sizes. In fact, no matter how large the buffer we have or how big the average performance improvement we achieve from the SMINs, the processor utilizations and the waiting time are mainly determined by the T_M and the network size. Our simulation results confirm these results.

The reasons for the above phenomena are understandable. In a multiprocessor system, once a processor issues a memory request packet, it will be waiting until the acknowledgement/results packet is received. So, the packet population in the whole network is very low (the maximum number equals to the number of the processors). Therefore, the effective memory request rates for the MINs are very small. If they are less than 0.1, the network performance are always the same for the CMINs and SMINs, regardless of their buffer sizes (see Fig. 5 and 6). The utilization of the MINs can be improved by providing a cluster of processors at an input switch as is done in the Cedar [3] and the IBM RP3 System [2]. As a result, the other processors in the cluster can issue their memory requests while one processor is waiting for the acknowledgement/result packet. Therefore, with the same number of processors, the

network size will be much smaller and utilizations of various system resources can be improved. However, the individual processor's utilization does not improve by using a cluster of processors. The processor utilization can be improved by applying private cache memories so that the processor remains busy for a long period of time before sending out request for a block, or by providing multiprogramming mechanism (context switching). Thus the request rate m can be kept less than 0.1, resulting in high processor utilization.

5 Conclusions

In this paper, we first examined some important issues in the synchronous packet switched MIN analysis. This class of MINs has always been analyzed by assuming some unrealistic conditions (infinite buffer queues, big synchronous clock, and isolated analyzing, etc.). We presented a new scheme based on a "small" synchronous clock. Analysis showed that our model has better performance behavior than the previous model. We then analyzed the split buffered MINs that show a large performance improvement over those of conventional MINs. Finally, we proposed a new approach to analyzing the whole multiprocessor system with packet switched MINs. The new approach is a hybrid method based on a probabilistic modeling, the discrete M/D/1 modeling, and the flow equivalent technique. Our results showed that the effect of increasing the buffer size or using split buffered technique is insignificant, with regard to the multiprocessor efficiency, in a system with fixed number processors of which one processor connecting to one network port. In order to improve the system performance, we have to apply other techniques like clustering more processors in a network node, or using cache memories with the processors.

References

[1] "Butterfly GP1000 Overview," *BBN Advanced Computers, Inc.*, Nov. 1988.

[2] G. F. Pfister, W. C. Brantley, D. A. George, S. L. Harvey, W. J. Kleinfelder, K. P. McAuliffe, E. A. Melton, V. A. Norton, and J. Weiss, "The IBM research parallel processor prototype (RP3): introduction and architecture," *Proc. 1985 Int'l Conf. on Parallel Processing*, pp. 764–771, 1985.

[3] D. Gajski, D. Kuck, and A. Sameh, "Cedar — A large scale multiprocessor," *Proc. 1983 Int'l Conf. on Parallel Processing*, pp. 524–529, 1983.

[4] L. N. Bhuyan, Q. Yang, and D. P. Agrawal, "Performance of multiprocessor interconnection networks," *Computer*, vol. 22, no. 2, pp. 25–37, Feb. 1989.

[5] J. H. Patel, "Performance of processor-memory interconnections for multiprocessors," *IEEE Trans. on Comput.*, vol. C-30, no. 12, pp. 771–780, Dec. 1981.

[6] J. H. Patel, "Analysis of multiprocessors with private cache memories," *IEEE Trans. on Comput.*, vol. C-31, no. 4, pp. 296–304, Apr. 1982.

[7] M. Lee and C. L. Wu, "Performance analysis of circuit switching baseline interconnection networks," *Proc. 11th annu. Comp. Architecture Conf.*, pp. 82–90, 1984.

[8] L. N. Bhuyan, "An analysis of processor-memory interconnection networks," *IEEE Trans. on Comput.*, vol. C-34, no. 3, pp. 279–283, Mar. 1985.

[9] D. M. Dias and J. R. Jump, "Analysis and simulation of buffered delta networks," *IEEE Trans. on Comput.*, vol. C-30, no. 4, pp. 272–282, Apr. 1981.

[10] C. P. Kruskal and M. Snir, "The performance of multistage interconnection networks for multiprocessors," *IEEE Trans. Comput.*, vol. C-32, pp. 1091–1098, Dec. 1983.

[11] H. Jiang, L. N. Bhuyan, and J. K. Muppala, "MVAMIN: Mean value analysis algorithms for multistage interconnection networks," *TR 88-3-3*. Center for Advanced Computer Studies, University of Southwestern Louisiana, 1988. Also to appear *J. Parallel and Distrib. Comput.*

[12] E. D. Lazowska, J. Zahorjan, G. S. Graham, and K. C. Sevcik, *Quantitative System Performance*, Prentice-Hall, Inc., 1984.

[13] Y. C. Jenq, "Performance of a packet switch based on single-buffered banyan network," *IEEE J. Select. Areas Commun.*, vol. SAC-3, no. 6, pp. 1014–1021, Dec. 1983.

[14] H. Yoon, K. Y. Lee, and M. T. Liu, "Performance analysis of multibuffered packet-switching networks in multiprocessor systems," *IEEE Trans. on Comput.*, vol. C-39, no. 3, pp. 319–327, Mar. 1990.

[15] H. S. Kim and Alberto-Garcia, "Performance analysis of buffered banyan networks under nonuniform traffic patterns," *IEEE INFOCOMM'88*, pp. 4A.4.1–4A.4.10, 1988.

[16] Y. Tamir and L. Frazier, "High-performance multi-queue buffers for VLSI communication switches," *Proc. 15th annu. Comp. Architecture Conf.*, pp. 343–354, 1988.

[17] M. Snir and J. Solworth, "The Ultraswitch — a VLSI network node for parallel processing," *Ultracomputer Note 39*, 1982. Courant Institute, NYU.

[18] W. J. Dally, "Virtual-channel flow control," *Proc. 17th annu. Comp. Architecture Conf.*, pp. 60–68, 1990.

Effects of Arbitration Protocols on the Performance of Multiple-bus Multiprocessors

Qing Yang [1]

Dept. of Electrical Engineering

The University of Rhode Island

Kingston,RI 02881

ABSTRACT

Earlier performance studies on multiple-bus multiprocessor systems [1, 2, 3, 4] assume a random selection of competing requests for bus assignment and ignore the effects of realistic bus arbitration schemes on the performance of such systems. This paper presents analytical models for multiple-bus systems by taking into account bus arbitration protocols. Three analytical models are derived for rotating priority, round-robin priority and queueing priority, respectively. Simple probabilistic and queueing analysis is used to study the rotating and queueing priority protocols. For round robin protocol, we use the concepts of renewal theory by reversing the *states* and *parameters* in the definition of *renewal counting process*. Each of our analyses models exactly the behavior of the corresponding priority protocol with little computation cost. The analytical models are used to carry out performance analysis and comparison of different priority protocols. Numerical results show that the round-robin protocol gives the lowest bus access delay among the three protocols.

1 Introduction

The multiple-bus multiprocessor considered in this paper is shown in Figure 1, which consists of N processors, M memory modules and B system buses. The buses are connected to all the processors and memory modules. The key element in such a system is bus arbiter that resolves bus conflicts and memory contentions [3]. Different arbitration protocols affect significantly the performance of the multiple-bus systems. It is therefore both practically useful and theoretically important to study the multiple-bus system based on realistic arbitration schemes.

A number of arbiter protocols for the shared bus system stand out in the literature [5, 6, 7, 8, 9, 10]. Among these protocols are equal priority, rotating priority, round robin and FIFO (queueing priority). An *equal priority* protocol can be considered as the bus allocation scheme in which all the processors have equal chance of getting access to the system buses at each bus cycle. In most of the bus-based shared memory multiprocessors, all processors in the system are often deemed basically equal due to the uniform access time of the global memory. As a result, the scheduling of multiple tasks to processors can be done dynamically and be greatly simplified. Therefore, the hardware design of such systems should support "fair allocation" of resources to the processors.

In practice, it is often impossible to have a strictly equal priority. A close approximation to the equal priority is to ensure the access of a bus by a requesting device within N bus cycles for an N-device system. This consideration leads to a *rotating priority* protocol in which each device is given a time-varying priority. After each bus cycle, the priority is incremented until it reaches the maximum, after which the priority rolls over to the minimum and repeats the cycle. Consider Figure 1 as an example. Suppose that the priorities of processor 1 through processor N at a given cycle are N, N-1, ...2, and 1, respectively. At the next bus cycle, priority of all processors are incremented by 1 except that the priority of processor 1 becomes 1 and processor 2 assumes the highest priority N. Another cycle moves the highest priority N to processor 3 and so on. Thus the highest priority circulates around all the devices. No matter how high the system load is, it is guaranteed for any device to get access to a bus within $N - B$ cycles for a B-bus system.

[1]This research is supported by National Science Foundation under grant No. CCR-8909672

There is still a potential unfairness with the rotating priority scheme, particularly under unevenly distributed workload. Consider a situation in a 2-bus system in which processors $i - 1, i$, and $i+1$ are most active and other processors are inactive during certain period of time. Suppose that the highest priority rotates along devices 1, 2, ... N, 1, and so on. During $N - 2$ out of N cycles processors $i - 1$ and i have higher priority than processor $i+1$ and hence get buses. Processor $i + 1$ can get bus only when the highest priority reaches processor i or processor $i + 1$. As a result, the protocol discriminates against processor $i + 1$ for most of the time. The *round robin priority* protocol overcomes the unfairness problem of the above rotating priority scheme. In the round robin protocol, each device also has a time varying priority. However, instead of increasing priority level by 1 after every cycle, the priority changes dynamically depending on the system load. After each bus cycle, the device immediately following the last one that got a bus assumes the highest priority. The priority of a device is therefore incremented by a variable after each cycle depending on the system load. This protocol can adapt to the system load quite well and hence achieve relative fairness.

The *queueing priority* protocol assigns buses to requests according to the times of their arrivals. This type of protocol is widely used in packet communications. The highest priority is given to the request that came earliest while the buses were busy. As a result, the protocol ensures uniform and fair delay among the different request packets.

We concentrate on performance analysis of the following three priority protocols. (1) Rotating priority protocol; (2) Round robin priority protocol and (3) Queueing priority protocol. Notice the the difference between the rotating priority protocol and round robin priority protocol is that the former shifts the highest priority exactly one device per cycle irrespective of the system load, whereas the latter moves the highest priority a random number of devices depending on the system load.

2 Performance Analysis

Our analyses are based on the following assumptions.

1. The system buses are packet-switched [11]. All devices (N processors and M memory modules) connected to the buses are assumed to be independent and capable of receiving packets from B buses simultaneously. Also each device has enough buffer size to store the incoming packets.

2. All processors and memory modules are synchronized at bus cycle which is fixed in length. At the beginning of a bus cycle, a processor issues a memory request with probability $m1$, and similarly, with probability $m2$ a memory module may want to request for a bus to send back a response packet to the requesting processor. For simplicity, we assume $m1$ and $m2$ are equal to the same value represented by m.

3. A request issued at the current cycle is assumed to be independent of the request in the previous cycle. This *request independence* assumption has been widely accepted in performance evaluation of multiprocessor systems [4] though it is unrealistic. Errors introduced by this assumption are negligible over a wide range of work load. It is also a reasonable assumption here since our main purpose is to study relatively the performance of the multiple-bus interconnections based on different priority protocols.

4. The transmission time of a request packet or a response packet is assumed to be equal to one bus cycle.

The performance measure we are interested in here is *bus access delay*. It is defined as the average number of bus cycles needed for a device to finish a bus operation after it issues a bus request. We denote the bus access delay by D_p, where the subscript p can be one of the symbols in (ro, rr, q) corresponding to rotating, round robin and queueing priority protocols, respectively.

2.1 Rotating Priority Protocol

With rotating priority, the device that has the highest priority in the current cycle will have the lowest priority at the following cycle. If device h, for instance, has the highest priority, then we assume that device $h+1$ has the second highest priority followed by $h+2$, and so on. Device $h-1$ is the one with the lowest priority. Thus, the priority rotates along the chain of devices. Let us consider the $[(h+n)mod(M+N)]$th device in the chain. There are n devices that have higher priority than the device in consideration. If less than B of these n devices request for bus at the current cycle, then the device will get a bus. If the number of devices requesting for bus among the n devices is greater than B, then the device in consideration has to wait. The probability that the device will get a bus, assuming that the highest priority is fixed at device h, is given by

$$P_n^F = Prob.\ \{No.\ of\ devices\ requesting\ for\ bus\ among$$
$$the\ n\ devices\ < B\}$$
$$= \begin{cases} 1, & if\ n \le B \\ \sum_{j=0}^{B-1} \binom{n-1}{j} m^j (1-m)^{n-1-j}, & otherwise. \end{cases} \quad (1)$$

Let $P_i^{ro}(k)$ be the probability that device i can get a bus at the kth cycle after it has issued a request with rotating priority. If i is equal to h (Device h is assumed to have the highest priority) or i is greater than h but $i - h \le B - 1$, then device i will get a bus at the requesting cycle. Otherwise, the ith device may or may not get a bus depending on the distance (in terms of the number of devices) between the ith device and the hth device and the number of devices that are requesting for bus between h and i. The probability that the ith device gets a bus in this case can be specified by equation (1) by substituting n in (1) with the distance between device i and device h. Since the priority rotates after every bus cycle among $M + N$ devices, the probability that device i belong to any particular priority level is $\frac{1}{N+M}$ at a random cycle. Therefore, $P_i^{ro}(1)$ is given by

$$P_i^{ro}(1) = \frac{B}{N+M} + \frac{1}{N+M} \sum_{j=1}^{N+M-B} P_{j+B-1}^F. \quad (2)$$

The first term in the above equation gives the probability that the highest priority is at one of the devices that are less than or equal to B devices apart before device i. The second term is the probability that device i gets a bus even though it is more than B devices apart from device h. If device i did not get access to a bus, it would be one of the N+M-B devices that are far away from device h. However, at the next cycle, the highest priority moves one device closer to device i. Hence, if device i were B+1 apart from device h and did not get a bus at the previous cycle (with probability $\frac{1-P_B^F}{N+M-B}$), it would get a bus at the current cycle with probability 1. Similarly, if the device i were $B + l$ apart from device h and did not get a bus (with probability $\frac{1-P_{B+l-1}^F}{N+M-B}$), the probability of getting a bus now is P_{B+l-2}^F. Therefore, $P_i^{ro}(2)$ is given by

$$P_i^{ro}(2) = (1 - \frac{B}{N+M}) \cdot [\frac{1-P_B^F}{N+M-B} \cdot 1 + \frac{1-P_{B+1}^F}{N+M-B} \cdot P_B^F$$
$$+ \frac{1-P_{B+2}^F}{N+M-B} \cdot P_{B+1}^F + \cdots + \frac{1-P_{B+l-1}^F}{N+M-B} \cdot P_{B+l-2}^F \cdots].$$

Similar reasoning yields

$$P_i^{ro}(3) = (1 - \frac{B}{N+M}) \cdot (1 - \frac{1}{N+M-B}) \cdot$$
$$[\frac{(1-P_{B+1}^F)(1-P_B^F)}{N+M-B-1} \cdot 1 + \frac{(1-P_{B+2}^F)(1-P_{B+1}^F)}{N+M-B-1} \cdot P_B^F$$
$$+ \frac{(1-P_{B+3}^F)(1-P_{B+2}^F)}{N+M-B-1} \cdot P_{B+1}^F + \cdots].$$

And in general

$$P_i^{ro}(k) = (1 - \frac{B}{N+M}) \left[\prod_{l=1}^{k-2}(1 - \frac{1}{N+M-B-l+1})\right] \cdot$$
$$\frac{\sum_{j=1}^{N+M-B-k+2} \prod_{n=1}^{k-1}(1 - P_{B+j+n-2}^F) \cdot P_{B+j-2}^F}{N+M-B-k+2}, \quad (3)$$

for $1 < k < N + M - B$. From equation (3), it can be verified that the probability of a device getting a bus within $N + M - B$ cycles is 1.

The bus access delay is given by

$$D_{ro} = \sum_{k=1}^{N+M-B} k P_i^{ro}(k). \quad (4)$$

2.2 Round Robin Priority Protocol

The analysis of round robin priority protocol can be done in the similar way as for rotating priority protocol. The major difference of the round robin priority protocol from the rotating priority protocol is that the highest priority shifts random number of devices after every cycle as compared to 1 device per cycle for rotating priority protocol. The number of devices that the highest priority skips after each cycle depends on the request rate of devices. If the number of bus requests generated at a cycle is greater than B, the number of skipped devices can be considered as the number of trials until the Bth success in a sequence of Bernoulli trials. Therefore, the number has a Negative Binomial distribution. The distribution function is given by

$$P_s(n) = \binom{n-1}{B-1} m^B (1-m)^{n-B} \quad (5)$$

Consider a requesting device that is d devices apart from the highest priority h currently. Suppose that the device did not get a bus at the requesting cycle. Let s_1, s_2, $\cdots$, and s_r represent the number of devices the highest priority h moves after the first cycle, the second cycle, and the rth cycle following the requesting cycle, respectively. The total number of devices that the highest priority h has passed after the rth cycle, σ_r, is given by

$$\sigma_r = s_1 + s_2 + \cdots + s_r.$$

Let d be the distance between device h that has the highest priority and the requesting device. Our primary interest here is to derive, for a given distance (d), the number of cycles that are needed for the highest priority h to move to a position close enough to the requesting device so that the device gets a bus. In other words, we are interested in the number of occurrences of s_i's within d. This observation lends itself to the idea of the renewal theory since s_i's are independent and identically distributed. However, the random process $\{s_i | i = 1, 2, \cdots\}$ is not defined over time. Rather, the random variable s_i is interpreted as the number of devices between the occurrence of the ith renewal and the $(i-1)$th renewal. The renewals correspond to the assignment of the highest priority after each bus cycle. Let $N_s(d)$ be the number of renewals or the number of movements that h makes within d devices. Clearly, $N_s(d) = k$ if and only if $\sigma_k \le d < \sigma_{k+1}$. Therefore, the probability distribution of $N_s(d)$ is given by

$$Pr(N_s(d) = k) = Pr(\sigma_k \le d) - Pr(\sigma_{k+1} \le d)$$
$$= \sum_{i=1}^{d} Pr(\sigma_k = i) - \sum_{i=1}^{d} Pr(\sigma_{k+1} = i).$$

Notice that h moves at least B devices after every cycle due to the protocol. Thus, s_i must be greater than or equal to B and $Pr(\sigma_k = i) = 0$ for all $i \leq kB$ for the system with B buses. Therefore

$$Pr(N_s(d) = k) = \sum_{i=kB}^{d} Pr(\sigma_k = i) - \sum_{i=(K+1)B}^{d} Pr(\sigma_{k+1} = i). \quad (6)$$

But

$$Pr(\sigma_k = i) = P_s^k(i),$$

where $P_s^k(i)$ denotes the k-fold convolution of $P_s(i)$ with itself. Let $G_{s_i}(z)$ and $G_{\sigma_k}(z)$ be the probability generating functions of s_i and σ_k, respectively. Then

$$G_{\sigma_k}(z) = G_{s_1}(z) \cdot G_{s_2}(z) \cdots G_{s_k}(z).$$

Since s_i has a negative binomial distribution with generating function

$$G_{s_i}(z) = \left(\frac{mz}{1 - (1-m)z}\right)^B,$$

and s_i's are identically distributed, we have

$$G_{\sigma_k}(z) = \left(\frac{mz}{1 - (1-m)z}\right)^{kB}.$$

Therefore σ_k also has a negative binomial distribution with parameters m and kB.

Let $P_i^{rr}(j)$ be the probability that device i obtains a bus at cycle j. If the number of occurrences of s_i within d is k, then the requesting device will get access to a bus at cycle $k+1$. This reasoning leads to

$$
\begin{aligned}
P_i^{rr}(k+1) &= \frac{1}{N+M} \sum_{d=1}^{N+M-1} Pr(N_s(d) = k) \\
&= \frac{1}{N+M} \sum_{d=1}^{N+M-1} \left[\sum_{i=kB}^{d} P_s^k(i) - \sum_{i=(K+1)B}^{d} P_s^{k+1}(i) \right],
\end{aligned}
$$

where $P_s^k(i)$ is given by

$$P_s^k(i) = \binom{i-1}{kB-1} m^{kB}(1-m)^{i-kB}.$$

Here, we have assumed that i can be any value between 1 and $N+M$ equally likely at a random cycle. $P_i^{rr}(1)$, is given by

$$P_i^{rr}(1) = 1 - \sum_{k=2}^{\lceil \frac{N+M}{B} \rceil} P_i^{rr}(k) \quad (7)$$

where $\lceil y \rceil$ is the minimum integer number that is greater than or equal to y. The bus access delay is therefore given by

$$D_{rr} = \sum_{k=1}^{\lceil \frac{(N+M)}{B} \rceil} k P_i^{rr}(k). \quad (8)$$

2.3 Queueing Protocol

The queueing protocol can be analyzed by means of queueing theory analysis. We use a general service center (center 0) with a load dependent service rate to represent the bus system and a infinite-server service center (center 1) to represent requesting devices. It is assumed that the queue network is separable in the sense that it satisfies the conditions for a product form solution. It has been shown in the previous studies [12] that the technique can provide a fairly accurate estimation of the system performance of real multiprocessors

Let v_0 and v_1 be the relative throughput of service centers 0 and 1, respectively. Let μ be the average service rate of a bus. The relative utilizations of the two centers, ρ_0 and ρ_1, are

$$\rho_0 = \frac{v_0}{m}, \quad \rho_1 = \frac{v_1}{\mu}.$$

The joint distribution of k_0 customers in center 0 and k_1 customers in center 1 is given by [13]

$$P(k_0, k_1) = \frac{1}{C(N+M)} \frac{\rho_0^{k_0}}{k_0!} \cdot \frac{\rho_1^{k_1}}{\beta(k_1)}, \quad (9)$$

for $k_0 + k_1 = N + M$, where

$$\beta(k_1) = \begin{cases} k_1!, & for\ k_1 < B; \\ B \cdot B^{k_1 - B}, & for\ k_1 \geq B \end{cases} \quad (10)$$

The quantity $C(N+M)$ in equation (11) is a normalization constant that makes the probabilities sum to 1. The computation of $C(N+M)$ can be performed using the convolution algorithm [13]. Since the queueing model for the system is a single chain network with two centers and all the customers coming out of one center will join the other center, we have $v_0 = v_1$. For numerical stability, we choose $v_0 = B$ in computing $C(N+M)$.

Once the normalization constant is obtained, the queue length distribution of the bus queue is readily derived based on the above joint distribution, and hence the mean queue length Q_L. According to the throughput theorem, the throughput of the bus queue, T_B, is given by

$$T_B = v_1 \frac{C(N+M-1)}{C(N+M)}.$$

The average bus access delay D_q can then be found from Little's formula, applied to the bus queue system

$$D_q = \frac{Q_L}{T_B}.$$

One should notice that the above analysis assumes a closed queue network. Once a device generates a bus request, it waits for the response. This is not consistent with the *request independence* assumption used in the analyses of the other four protocols. In order to make a fair performance comparison among the three protocols, we should adjust the request rate of the queueing analysis so that the comparison is based on the same system load.

For the analyses of the previous four protocols, the *request independence* assumption implies that a device generates a bus request with a probability m in a cycle independently from the previous cycle. The bus requests that did not get a bus in the previous cycle are discarded in the current cycle. In other words, there are on average $m(N+M)$ bus requests that are generated every cycle. The queueing analysis presented in this section, on the other hand, assumes that the failed requests wait in the queue. Thus the total number of bus requests in the current cycle is the sum of newly generated bus requests and all the pending requests that have not got a bus yet. Denote the adjusted request rate for the queueing protocol by p_a. Let

$$m \cdot (N+M) = Q_L + p_a \cdot (N+M-Q_L),$$

we have

$$m = \frac{Q_L + p_a \cdot (N+M-Q_L)}{N+M}.$$

For a given adjusted request rate p_a, our queueing model can derive the bus access delay D_q corresponding to it. This delay should be compared with the bus access delays of the other protocols corresponding to the request rate m of the above equation. The performance comparison is therefore based on the same system load.

3 Results And Conclusions

The bus access delay determines, in major part, the memory latency time in a shared memory multiple-bus multiprocessor system. Increasing the number of buses decreases the delay but the hardware cost increases. Thus, a system designer should have a clear performance estimation to guide his/her parameter selection in designing a multiple-bus multiprocessor system. Based on the analytical models presented in the previous section, we present some numerical results to show the performance of the different arbiter designs.

Performance comparison of different priority protocols is shown in Figure 2 for a single-bus system of 32 devices. It is interesting to notice that at low request rate queueing priority protocol outperforms the rotating priority protocol. When the request rate exceeds 0.4, the rotating priority protocol gives better performance than the queueing protocol. This is because the rotating priority has bounded delay no matter how high the system load is. The situation changes when one more bus is added to the system as shown in Figure 3 which plots the bus access delay for a 2-bus system. In this case, the rotating priority has higher delay than the queueing priority for all request rates from 0.1 through 0.9. The round robin priority protocol performs the best among the three protocols in both cases.

Figure 4 shows performance comparison by varying the system size from 2 to 64 devices for a fixed request rate of 0.2 and 2 buses. With the small request rate (0.2), the delays of the queueing priority, and round robin priority are very close for the size range considered. This small difference in bus access delay at low load suggests that a system designer select one of the two protocols freely. The attention can be focused on the simplicity of the hardware implementation and minimization of propagation delay.

Figures 5 and 6 plot the bus access delay vs. system sizes for higher request rate (0.8) for systems with 2 buses and 4 buses, respectively. For the 2-bus system, the rotating priority protocol shows the similar performance as the queueing priority protocol for small system sizes. As shown in Figure 5, the performance differences among the three priority protocols increase as the system size increases. The performance differences are even larger for the 4-bus system as shown in Figure 6. The rotating priority suffers from more delay than the other two and the round robin priority protocol shows the lowest delay among the three protocols.

We conclude that the round robin priority performs the best among the three protocols due to its adaptive nature. The performance difference among the different priority protocols at high load increases as the number of buses increases.

Acknowledgement

The author would like to thank Mr. R. Ravi for his help in simulation and some numerical computations.

References

[1] T. N. Mudge, J. P. Hayes, G. D. Buzzard, and D. C. Winsor, "Analysis of multiple-bus interconnection networks," *Journal of Parallel and Distributed Computing, 3(3)*, Sept. 1986.

[2] K. B. Irani and I. H. Onyuksel, "A closed-form solution for the performance analysis of multiple-bus multiprocessor system," *IEEE Tran. on Computer*, vol. C-33, pp. 1004–1012, Nov. 1984.

[3] Q. Yang and L. Bhuyan, "Performance of multiple-bus interconnections for multiprocessors," *Journal of Parallel and Distributed Computing (8)*, pp. 267–273, March 1990.

[4] L. N. Bhuyan, Q. Yang, and D. P. Agrawal, "Performance of multiprocessor interconnection networks," *IEEE Computer*, pp. 25–37, Feb. 1989.

[5] L. N. Bhuyan, "Analysis of interconnection networks with different arbiter designs," *J.P.D.C.*, pp. 384–403, 1987.

[6] T. Lang and M. Valero, "M-users B-servers arbiter for multiple-bus multiprocessors," in *Microprocessing and Microprogramming*, pp. 11–18, 1982.

[7] T. N. Mudge, J. P. Hayes, and D. C. Winsor, "Multiple-bus architectures," *IEEE Computer, Special Issue on Interconnection Networks*, pp. 42–48, June 1987.

[8] Q. Yang and R. Ravi, "Design and analysis of multiple-bus arbiters with different priority schemes," in *Proc. PARBASE-90*, March 1990.

[9] S. M. Mahmud and M. Showkat-UI-Alam, "A new arbitration circuit for synchronous multiple bus multiprocessor systems," in *Proc. of the 90 Int. Conf. on Sys. Integration*, April 1990.

[10] M. K. Vernon and U. Manber, "Distributed round-robin and first-come first-serve protocols and their application to multiprocessor bus arbitration," in *Proc. 15th Int. Symp. on Comput. Arch.*, pp. 269–277, 1988.

[11] Q. Yang and L. N. Bhuyan, "Analysis of packet-switched multiple-bus multiprocessor systems," *IEEE Transactions on Computers*, March 1991.

[12] Q. Yang, L. Bhuyan, and B.-C. Liu, "Analysis and comparison of cache coherence protocols for a packet-switched multiprocessor," *IEEE Trans. on Comput..*, vol. 38, pp. 1143–1153, August 1989. Special Issue on Distributed Computer Systems.

[13] K. S. Trivedi, *Probability and Statistics with Reliability, Queueing, and Computer Science Applications.* Prentice Hall, 1982.

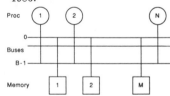

Fig 1. A Multiple-bus Multiprocessor System

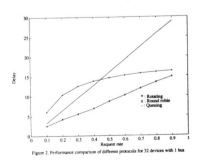

Figure 2. Performance comparison of different protocols for 32 devices with 1 bus

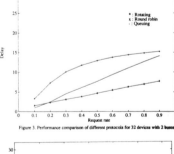

Figure 3. Performance comparison of different protocols for 32 devices with 2 buses

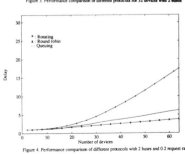

Figure 4. Performance comparison of different protocols with 2 buses and 0.2 request rate

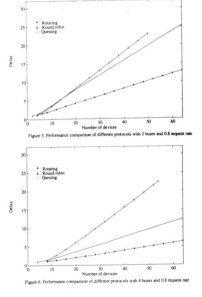

Figure 5. Performance comparison of different protocols with 2 buses and 0.8 request rate

Figure 6. Performance comparison of different protocols with 4 buses and 0.8 request rate

PERFORMANCE EVALUATION OF HARDWARE SUPPORT
FOR MESSAGE PASSING IN DISTRIBUTED MEMORY MULTICOMPUTERS *

Jiun-Ming Hsu

IBM Corporation
T. J. Watson Research Center
P.O. Box 218
Yorktown Heights, NY 10598

Prithviraj Banerjee

Center for Reliable and High-Performance Computing
Coordinated Science Laboratory
University of Illinois at Urbana-Champaign
Urbana, IL 61801

Abstract

This paper presents the performance evaluation of some hardware support for message passing schemes in distributed memory multicomputers. We evaluate the performance of a *message passing coprocessor* (MPC) and a *virtual channel router* (VCR) under different communication workload in both hypercube and mesh topologies. We show that the software and hardware overhead of communication can be reduced significantly by these two processors. In this paper, we also propose several optimal-adaptive routing strategies which can take advantage of the virtual channels and cached circuits in the VCR to obtain optimal performance. Combining optimal-adaptive routing with virtual channels and cached circuits, a set of most often used circuits could be maintained in the network, i.e. the network can *adapt* to the applications.

1. Introduction

Distributed memory multicomputers have recently offered a cost-effective and feasible approach to supercomputing by connecting a large number of low-cost processors with direct links. This type of architecture is more readily scaled up to large numbers of processors than multiprocessor designs based on globally shared memory. In distributed memory multicomputers, synchronization and data sharing are achieved by explicit message passing. Hence, the speed and efficiency of communication are very important in the overall performance of such machines.

The goal of our research is to reduce the communication overhead by supporting message passing in hardware. We have designed two devices: a *message passing coprocessor* (MPC) [1] and a *virtual channel router* (VCR) [2], to perform message passing in hardware. Our objective is to minimize the communication overhead as much as possible so that the message transmission rate is close to the actual bandwidth of the communication links. The detailed architecture of the MPC and VCR was presented in [1, 2]. We will give a brief overview of these two devices in Section 2.

This paper reports the performance evaluation of the MPC and VCR in both hypercube and mesh networks. We performed simulation using synthetic benchmarks to evaluate different hardware support configurations (with caching enabled and disabled, different numbers of virtual channels per link, and etc.), under a variety of communication workload (light/heavy traffic, different spatial and temporal locality, message sizes, and etc.), for both hypercube and 2-D/3-D mesh networks.

Adaptive routing techniques are known to reduce the message routing time of multicomputers by avoiding congested links [3-6]. This paper describes some existing adaptive routing techniques for both hypercube and mesh topologies. We also propose several

optimal-adaptive routing strategies which can take advantage of the virtual channels and cached circuits in the VCR to provide deadlock-free, low-latency message routing. The performance evaluation results are also presented.

2. Hardware Support for Message Passing

The MPC and VCR can be employed in any type of processor networks like the hypercube or 2-D/3-D meshes, with serial or parallel communication links. The block diagram of a multicomputer node processor with the MPC and VCR is shown in Figure 1. The MPC is a microprogrammable processor with its own local memory and local bus. Its functions include message passing, process blocking and unblocking, message buffer management, and buffer copying. It supports software caching by performing pre-processing and buffer pre-allocation for the expected (*cached*) messages.

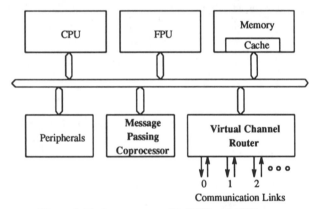

Figure 1. Node processor with MPC and VCR

The processor nodes are connected to form a hypercube (or mesh) topology by the bi-directional links in the VCR. In case of a mesh, the links are 8 bits wide; in case of a hypercube, they are 1 bit wide. The VCR supports fixed path, circuit-switched routing. It keeps the circuit (from source to destination) connected even after the message transmission is completed to exploit the locality in message destination. This reserved circuit is called *cached circuit* [7,8]. When a new message generated at a node has the same destination as the last message (a destination *hit*), the new message can be sent immediately using the cached circuit, thereby saving the circuit set-up time (hardware overhead) in the routers. In order to increase the network connectivity and thereby increasing the hit rate, each physical communication link is divided into multiple *virtual channels* [9]. Performance evaluation indicated that two virtual channels per link can provide high message hit rate for most of the applications [1,2]. Therefore, in our implementation, we used two virtual channels per link. Each node can keep two outgoing and two incoming cached circuits at the same time.

* This research was supported in part by the National Science Foundation Presidential Young Investigator Award under Grant NSF MIP 86-57563 PYI, and in part by an equipment grant NSF CCR 87-05240.

Table 1 summarizes the timing of the VCR for hypercube and mesh networks. Table 2 summarizes the total message latency and software overhead of the hypercube and mesh networks with the MPC and VCR, for message sizes of 100 and 1K bytes. The timing for the missed message includes both destination and buffer misses. The readers are referred to [10] for a detailed timing analysis.

Table 1. VCR timing information

	Hypercube	3-D Mesh
VCR clock cycle	35 nSec	25 nSec
Link width	1 bit	8 bits
Link bandwidth	28M bits/sec	20M bytes/sec
Data packet latency	5 cycles (175 nSec)	3 cycles (75 nSec)
Routing probe latency	27 cycles (945 nSec)	11 cycles (275 nSec)
Path delink	13 cycles (455 nSec)	4 cycles (100 nSec)

Table 2. Communication latency of MPC and VCR

(a) For hypercube network

Message Size (Bytes)	Total latency (μSec)		Software time (μSec)	
	Hit	[Miss]	Hit	[Miss]
100	55.2	[68.8]	8.2	[15.7]
1K	397.2	[410.8]	8.2	[15.7]

(b) For 2-D mesh network

Message Size (Bytes)	Total latency (μSec)		Software time (μSec)	
	Hit	[Miss]	Hit	[Miss]
100	10.7	[15.5]	4.1	[7.9]
1K	56.9	[61.7]	4.1	[7.9]

3. Routing Strategies in Multicomputers

3.1. Fixed and Adaptive Routing in Hypercubes

In the hypercube, routing a message from one node, whether it is the originating or intermediate node, to another involves resolving the *address difference* (exclusive-or) between these two nodes. Each non-zero digit in the address difference must be resolved once and only once, for shortest-path routing. For example, to route a message from node 0000 to 1111 in a four dimensional hypercube, the routing algorithm needs to search for a path (or circuit) which resolves the address difference $A=1111$. This is equivalent to a tree search in the search space. The following routing algorithms define how the tree is searched. The non-optimal algorithms were proposed by previous researchers; we propose the optimal ones which will be shown to work better with the VCR.

Fixed-path, E-cube routing: In *E-cube* routing, the address difference is resolved from higher to lower dimension. If there is no free or cached channel in the link, the message is blocked, it does not backtrack. In the 0000 to 1111 example, the message is routed through nodes 1000, 1100, and 1110.

Random routing: In *Random* routing, the message is randomly routed to any non-zero link in A which has free or cached channels. If all the non-zero links in A are busy, the message backtracks. It is used for comparison purpose only.

Exhaustive routing: In *Exhaustive* routing, the router performs a depth-first search [3] in the search space. The router will look at the non-zero links in A *sequentially* and choose the *first* link that has a free or cached channel. It will backtrack when all the links are blocked.

Optimal-exhaustive routing: In *Optimal-exhaustive* routing, a depth-first search is also performed, but unlike the *Exhaustive* routing, all the non-zero links in A are examined *simultaneously*. The router chooses the link that has the least number of active channels, and it tries to avoid tearing down the cached circuits. Hence, this algorithm tends to avoid the traffic in the network by greedily choosing the least congested route. It also backtracks when all the link are blocked.

Hyperswitch routing: *Hyperswitch* routing [5, 6] is similar to the *Exhaustive* routing, except that a *history flag* is used to eliminate the search space of some possible routes. The history flag is zero in the originating node (the root of the search tree), and passed to the branches of the search tree, together with the routing probe. Whenever a search fails in a particular link, the bit corresponding to that link is set. This history flag is passed down the tree only; nodes use the history of their ancestors. In any stage of the routing, if the bit corresponding to a link is set, that link will not be used to search for a free channel, thus limiting the number of nodes visited in a search.

Optimal-hyperswitch routing: The *Optimal-hyperswitch* routing is similar to the *Optimal-exhaustive* routing, except that the *history flag* is also used to eliminate some search space. The non-zero links in A will be examined only if the corresponding bits in the history flag are not set.

Spare-dimension routing: The *Spare-dimension routing* also performs depth-first search on the search tree, like the *Exhaustive* routing. But when all the non-zero links in A are blocked, the message will be routed to the first zero-link in A which has a free or cached channel; i.e. the message will go an extra dimension. The message is allowed to detour only once, and not allowed to go back to the link that it just came from.

Optimal-spare routing: This is similar to the *Optimal-exhaustive* routing except that the message is allowed to detour once when all the non-zero links in A are blocked. When searching for a detour link, the optimal strategy is also used.

3.2. Fixed and Adaptive Routing Techniques in Meshes

We extend the above adaptive routing algorithms to the mesh topologies. Only the optimal, shortest-path algorithms are described and evaluated. In a 2-D mesh network, to route a message from node 20 to 03, the routing algorithm needs to search for a path which resolves the address difference $A=(-2, 3)$. Therefore the message needs to go two hops in the *down* direction, and three hops in the *right* direction. This is also equivalent to a tree search in the search space.

Fixed-path, E-cube routing: The address difference is resolved from higher to lower dimension.

Random routing: The message is randomly routed to any non-zero link in A which has free or cached channels.

Optimal-exhaustive routing: The router performs a depth-first search. All the non-zero links in A are examined *simultaneously*. The router chooses the link that has the least number of active channels, and it tries to avoid tearing down the cached circuits.

Optimal-zig-zag routing: The *Zig-zag* routing algorithm [4] requires the message move towards the diagonal in order to keep the remaining search space as large as possible, thus decreasing the probability of being blocked. In the 20 to 03 example, the message is routed through nodes 21 and 12. The *Optimal-zig-zag* routing also follows the *Zig-zag* path, but when there are choices, it will follow the optimal policy. That is, it puts *Zig-zag* policy before optimal policy.

All the above routing algorithms, with the exception of the *E-cube* routing, are adaptive. The deadlock problem is avoided by backtracking, i.e. they do not hold the routing circuit and wait for a congested link. These algorithms are also greedy (or distributed); they only look at local information. All of the algorithms use only shortest paths, except the *Spare-dimension* and *Optimal-spare* routing in the hypercube, which are allowed one detour. The VCR needs one or two extra clock cycles to perform the adaptive search. The backtracking time is 245 nSec for the hypercube and 100 nSec for meshes.

4. Performance Evaluation of the Hardware Support

This section summarizes the performance evaluation of the hardware support for message passing and adaptive routing strategies. A more detailed description of the simulation methodology and results can be found in [10]. The timing parameters used in the simulation were reported in Section 2. The inputs to the simulator, the synthetic communication benchmarks, were chosen to be similar to the characteristics of real application programs. An LRU stack size of two and hit rates of 0.8 and 0.1 were used to model the temporal locality of message destination. Both uniform and localized communication were used. When not specified, the VCR has the configuration of two virtual channels per link, with circuit caching enabled. The results collected includes the average message latency (includes both software and hardware times), average hardware latency (includes the circuit setup time and the actual message transmission time), average circuit setup time, and average destination hit rate, for all the messages.

Table 3 shows the hardware time, circuit setup time, and destination hit rate as a function of message rate for three kinds of VCR configurations: 1, 2, and 4 virtual channels per link. The circuit hit rate increases as the number of virtual channels increases, hence the average circuit set-up time decreases. However, because of the additional overhead for more virtual channels, the hardware time does not improve too much after two virtual channels. Therefore, two virtual channels per link are adequate in maintaining a reasonable hit rate and keeping the overhead down.

Figure 2 compares the message latency (including the software time) and the circuit setup time of the *cache-2VC* and *no-cache* cases, under localized communication. As the traffic becomes heavier, the software and hardware time savings of the *cache-2VC* also increases. This shows that the hardware support enables the applications to exploit lower level parallelism. Overall, caching schemes are more useful when the average message size is small, and when there exists high temporal locality in the communication pattern. The virtual channels provide link sharing and increase the connectivity of the network, thereby reducing the circuit set-up time. Simulation results on the mesh networks also confirmed the above observations.

5. Performance Evaluation of Routing Techniques

Table 4 shows the simulation results of hypercube routing techniques on message hardware latency, circuit setup time, and destination hit rate, under different message traffic. In general, the optimal algorithms perform better than the non-optimal adaptive ones, which in turn perform better than the *E-cube* routing. Under heavy traffic, the optimal algorithms are significantly better. *Random* routing has shorter latency than the *Hyperswitch* and *Spare-dimension* under heavy traffic. This indicates that it is undesirable to use a history flag to eliminate some search space, nor is it desirable to go an extra dimension to search for routes.

Figure 3 plots the message destination hit rate and hardware time versus message rate for the three algorithms: *E-cube, Exhaustive,* and *Optimal-exhaustive.* For the *E-cube,* the hit rate decreases when traffic increases due to the contention in the network. The *Exhaustive* adaptive routing avoids contention, thereby increasing the destination hit rate and reducing message latency. The *Optimal-exhaustive* routing further avoids the traffic in the network, hence it has the lowest message latency and highest hit rate. Combining cached circuits and virtual channels with *Optimal-exhaustive* routing, it is possible that a set of most often used circuits can be maintained in the network, i.e. the network can *adapt* to the applications. In mesh networks, the *Optimal-exhaustive* routing also outperformed other algorithms (Table 5).

6. Conclusion

We have presented the performance evaluation of hardware support for message passing in distributed memory multicomputers. Simulation using synthetic benchmarks confirmed the gains in the use of virtual channels, cached circuits, and microprogrammable message passing coprocessor in the hypercube and mesh networks. Virtual channels increase the connectivity of the network, thereby reducing the hardware latency of message transmission. Software/hardware caching schemes exploit locality in messages and further reduce overhead. The reduction in message latency enables the multicomputers to exploit more lower level parallelism in the application programs. The optimal-adaptive routing strategies can further take advantage of the virtual channels and cached circuits in the VCR to obtain optimal performance.

References

[1] J.-M. Hsu and P. Banerjee, "A Message Passing Coprocessor for Distributed Memory Multicomputers," in *Proc. Supercomputing '90*, New York, NY, pp. 720-729, Nov. 1990.

[2] J.-M. Hsu and P. Banerjee, "Hardware Support for Message Routing in a Distributed Memory Multicomputer," in *Proc. 1990 Int'l Conf. on Parallel Processing*, vol. 1, St. Charles, IL, pp. 508-515, Aug. 1990.

[3] M.-S. Chen and K. G. Shin, "Depth-First Search Approach for Fault-Tolerant Routing in Hypercube Multicomputers," *IEEE Trans. on Parallel and Distributed Systems*, vol. 1, April 1990.

[4] H. G. Badr and S. Podar, "An Optimal Shortest-Path Routing Policy for Network Computers with Regular Mesh-Connected Topologies," *IEEE Trans. on Computers*, vol. 38, No. 10, pp. 1362-1371, October 1989.

[5] E. Chow, H. Madan, J. Peterson, D. Grunwald, and D. Reed, "Hyperswitch Network for the Hypercube Computer," *Proc. 15th Int'l Symp. Computer Architecture*, May 1988.

[6] D. C. Grunwald and D. A. Reed, "Networks for Parallel Processors: Measurements and Prognostications," in *Proc. 3rd Conf. on Hypercube Concurrent Computers and Applications*, Pasadena, CA, pp. 610-619, Jan. 1988.

[7] H. G. Badr, D. Gelernter, and S. Podar, "An Adaptive Communications Protocol for Network Computers," *Performance Evaluation*, vol. 6, pp. 35-51, 1986.

[8] D. A. Reed, P. K. McKinley, and M. F. Barr, "Performance Analysis of Switching Strategies," *Proc. 1987 Symp. of the Simulation of Computer Networks*, pp. 130-141, Aug. 1987.

[9] W. J. Dally, "Virtual-Channel Flow Control," in *Proc. 17th Int'l Symp. on Computer Architecture*, Seattle, WA, pp. 60-68, May 1990.

[10] J.-M. Hsu, "Performance Measurement and Hardware Support for Message Passing in Distributed Memory Multicomputers," Technical Report CRHC-91-5, Coordinated Science Lab., Univ. of Illinois, Urbana, IL, Jan. 1991.

Table 3. Comparison of different number of virtual channels (hypercube)

(a) Hardware time and setup time vs. message rate

Hardware (Setup) Time in μSec	Message interarrival time (μSec)				
	40	50	60	70	100
1 VC/link	26.3 (11.73)	24.3 (9.75)	22.4 (7.86)	21.5 (6.95)	19.8 (5.20)
2 VC/link	23.6 (4.04)	21.1 (2.75)	19.9 (2.22)	19.5 (2.05)	18.3 (1.76)
4 VC/link	22.0 (0.93)	19.8 (0.64)	19.2 (0.56)	18.5 (0.54)	17.6 (0.52)

(b) Hit rate vs. message rate

Hit Rate	Message interarrival time (μSec)				
	40	50	60	70	100
1 VC/link	0.22	0.22	0.22	0.22	0.22
2 VC/link	0.61	0.63	0.63	0.63	0.64
4 VC/link	0.83	0.86	0.88	0.88	0.88

Hypercube size: 64 nodes   Message size: 48 bytes   LRU stack size: 2   Spatial locality: uniform
Messages per node: 500   Average interval: 40 – 100 μSec   LRU stack hit rate: 0.8, 0.1   Non-blocking communication

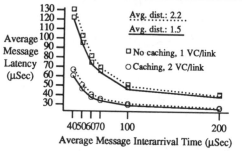

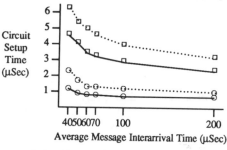

Hypercube size: 64 nodes
Messages per node: 500
Message size: 48 bytes
Average interval: 40 – 200 μSec
LRU stack size: 2
LRU stack hit rate: 0.8, 0.1
Spatial locality: localized
Non-blocking communication

Figure 2. Message latency and circuit setup time vs. message rate (hypercube)

Table 4. Comparison of adaptive routing techniques (hypercube)

(a) Hardware time and setup time vs. message rate

Hardware (Setup) Time in μSec	Message interarrival time (μSec)				
	7	15	30	40	60
E-cube	33.7 (8.81)	33.7 (8.88)	29.2 (6.78)	23.9 (4.28)	20.1 (2.29)
Random	31.8 (6.98)	30.1 (6.14)	25.5 (4.08)	22.5 (2.81)	20.8 (2.17)
Hyperswitch	35.6 (10.24)	33.7 (9.25)	27.4 (5.95)	23.4 (4.05)	20.1 (2.50)
Spare-dim.	33.3 (7.88)	31.3 (7.00)	25.5 (4.53)	22.0 (3.00)	19.8 (2.09)
Exhaustive	30.7 (6.94)	29.6 (6.26)	24.5 (4.15)	21.3 (2.77)	19.7 (2.11)
Opt-hypersw.	23.5 (4.21)	22.8 (3.89)	20.8 (2.85)	19.4 (2.23)	18.4 (1.78)
Opt-spare	23.7 (4.25)	22.3 (3.50)	20.9 (2.86)	19.5 (2.24)	18.4 (1.81)
Opt-exhaust.	23.4 (4.22)	22.3 (3.50)	20.6 (2.75)	19.2 (2.01)	18.4 (1.77)

(b) Hit rate vs. message rate

Hit Rate	Message interarrival time (μSec)				
	7	15	30	40	60
E-cube	0.41	0.41	0.51	0.61	0.64
Random	0.44	0.45	0.56	0.64	0.62
Hyperswitch	0.39	0.41	0.52	0.61	0.63
Spare-dim.	0.43	0.45	0.57	0.64	0.64
Exhaustive	0.47	0.47	0.58	0.65	0.65
Opt-hypersw.	0.59	0.59	0.65	0.68	0.67
Opt-spare	0.59	0.60	0.65	0.68	0.66
Opt-exhaust.	0.60	0.60	0.65	0.68	0.67

Hypercube size: 64 nodes   Message size: 48 bytes   LRU stack size: 2   Spatial locality: uniform
Messages per node: 500   Average interval: 7 – 60 μSec   LRU stack hit rate: 0.8, 0.1   Non-blocking communication

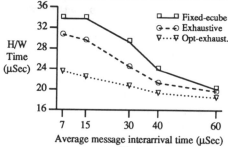

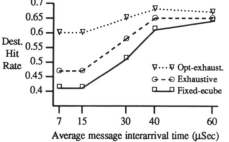

Hypercube size: 64 nodes
Messages per node: 500
Message size: 48 bytes
Average interval: 7 – 60 μSec
LRU stack size: 2
LRU stack hit rate: 0.8, 0.1
Spatial locality: uniform
Non-blocking communication

Figure 3. Destination hit rate and hardware time vs. message rate (hypercube)

Table 5. Comparison of adaptive routing techniques (meshes)

(a) 3D mesh: hardware time and setup time vs. message rate

Hardware (Setup) Time in μSec	Message interarrival time (μSec)				
	3	7	15	30	60
E-cube	9.09 (2.25)	8.70 (1.98)	8.08 (1.68)	6.75 (1.01)	6.04 (0.88)
Random	8.31 (1.42)	8.18 (1.33)	8.07 (1.31)	7.20 (1.08)	6.56 (1.07)
Opt. zig-zag	7.85 (1.27)	7.56 (1.07)	7.42 (1.03)	6.73 (0.93)	6.10 (0.91)
Opt-exhaust.	7.25 (1.10)	7.05 (0.97)	6.86 (0.92)	6.30 (0.82)	5.80 (0.81)

(b) 3D mesh: hit rate vs. message rate

Hit Rate	Message interarrival time (μSec)				
	3	7	15	30	60
E-cube	0.50	0.52	0.54	0.54	0.54
Random	0.52	0.53	0.53	0.53	0.53
Opt. zig-zag	0.57	0.58	0.58	0.58	0.58
Opt-exhaust.	0.61	0.63	0.63	0.63	0.63

Mesh size: 4^3, 8^2 nodes   Message size: 96 bytes   LRU stack size: 2   Spatial locality: uniform
Messages per node: 500   Average interval: 3 – 60 μSec   LRU stack hit rate: 0.8, 0.1   Non-blocking communication

PARALLEL ITERATIVE REFINING A* SEARCH

Guo-jie Li
National Research Center
for Intelligent Computing Systems
P. O. Box 2704
Beijing, P. R. China

Benjamin W. Wah
Coordinated Science Laboratory
University of Illinois
1101 W. Springfield Avenue
Urbana, IL 61801
wah@aquinas.csl.uiuc.edu

ABSTRACT

In this paper, we study a new search scheme called Iterative Refining A* (IRA*). The scheme solves a class of combinatorial optimization problems whose lower-bound values and heuristic solutions of partial problems are easy to compute. An example of problems in this class is the symmetric traveling-salesman problem for a set of fully-connected cities. The scheme consists of a sequence of guided depth-first searches, each searching for a more accurate solution than the previous one. It is an extension from Korf's IDA* search, except that it prunes by approximation rather than by thresholding. In IRA*, global upper bound (or incumbents), rather than lower-bound information, is passed between the successive depth-first searches. The time required by IRA* for finding optimal solutions to problems in this class is very close to that required by a best-first search, while the memory space required is the same as that of a depth-first search. We derive the performance bounds of parallel IRA* and found that it is more robust and has much better worst-case behavior than a parallel depth-first search.

INDEX TERMS: Approximation, combinatorial optimization, iterative-deepening A*, iterative-refining A* (IRA*), parallel processing, performance bounds.

1. INTRODUCTION

NP-hard combinatorial optimization problems (COPs) require an exponential amount of computation time and memory space in the worst case in solving them. It has been proved that the A* algorithm with an admissible lower-bound function expands the minimum number of nodes in finding the optimal solution or a solution with a guaranteed deviation from the optimal solution. *Our objective in this paper is to find an efficient and robust parallel search scheme for solving COPs so that the memory space required is bounded, and that the ratio of the total number of nodes expanded in the best sequential (or A*) algorithm to that of the proposed (parallel) search scheme is*

This research was supported in part by National Aeronautics and Space Administration under grant NCC 2-481 and in part by National Science Foundation under grant MIP 88-10584.

This research was undertaken when the G. J. Li was a visiting scholar in the Coordinated Science Laboratory, University of Illinois, Urbana-Champaign.

International Conference on Parallel Processing, St. Charles, IL, August 13-15, 1991.

maximum.

There are a number of recent results on parallel combinatorial searches [1, 3]. Most of them are focused on load balancing, variable sharing, and reducing communication overheads. A number of studies were devoted to finding the best speedup of a parallel algorithm as compared to a sequential algorithm using the *same* search strategy. For instance, near-linear or superlinear speedups have been observed for a parallel depth-first search as compared to a sequential depth-first search [1]. Similar behavior has been observed for parallel IDA* search as compared to a sequential one [5]. These issues are important in parallel processing; however, they do not indicate the actual speedup achieved in solving the given problem. *The speedup achieved by parallel processing must be compared against the best sequential algorithm for solving the given problem and not the sequential version of the parallel algorithm.* Using the number of nodes expanded as a measure of performance, the best sequential performance is achieved by the best-first strategy. In this paper we characterize the aggregate speedup achieved as a function of the speedup achieved by parallel processing (using the same search strategy) and the slowdown resulted due to the use of a more space-efficient algorithm (such as depth-first search) than the best-first search.

In general, combinatorial optimization problems can be roughly divided into two classes according to whether feasible solutions can be found easily. In the first class, finding feasible solutions may be as hard as finding optimal ones. Examples of problems in this class include integer programming, solving 15-puzzle problems, and finding vertex covers. These problems can be solved efficiently by IDA* [2]. In the second class, it is relatively easy to find feasible solutions and hence upper-bound values to partially evaluated problem instances. Examples of problems include the symmetric traveling-salesman problem on fully-connected cities and many unconstrained scheduling problems. In this paper, we address problems in the second class. Without loss of generality, we discuss our results with respect to minimization problems.

In developing an efficient parallel search strategy, it is necessary to decompose the search into tasks so that the processing resources are scheduled to search independent subtrees. It is important that memory and time limitations be observed, that anomalies in parallelism be addressed, and that the scheme be applicable to a wide variety of problems. The Iterative Refining A* (in short, IRA*) search proposed in this paper are developed with these objectives in mind.

In general, parallelism of COPs can be exploited in two levels: intra-node and inter-node levels. Current techniques in parallel processing, such as pipelining and array processing, are suitable for intra-node parallelism. In contrast, inter-node parallelism can be broadly classified into four types.

(a) In *tree-splitting* or *tree-decomposition*, each processor has a local list for storing active nodes, each constituting a disjoint part of the search tree. When a processor finishes evaluating all nodes in the local list, it gets an unsearched node from another processor. The major issues here are in finding efficient load balancing and variable-sharing (or information communication) strategies and in coping with anomalies in parallelism.

(b) In *parallel window searches*, each processor expands the nodes whose lower bounds fall into a given window, namely, a lower-bound interval. This approach is suitable for bottom-up searches, such as game-tree search, but suffers from the high overhead of node distribution found in top-down searches.

(c) In *parallel iterative searches*, the tree searched is expanded in successive iterations and may overlap with each other. In the original IDA* algorithm [2], thresholds are defined so all nodes with values less than a given threshold are searched in an iteration. Trees in successive iterations are searched sequentially. These trees can be searched in parallel. Parallel iterative search has the benefit of obtaining a desirable approximate solution early in the process and in communicating these solutions to all trees searched in parallel.

(d) The last type of inter-node parallelism is based on exploiting multiple heuristics and has not been addressed seriously in the literature. In this form, multiple heuristics are run in parallel. This takes advantage of the large fluctuations in computation times of various heuristics in generating feasible solutions.

The parallel IRA* search scheme studied in this paper utilizes the four types of inter-node parallelism discussed above and addresses three important issues: memory-space limitation, detrimental anomalies in parallel search [4], and achieving a performance as close as possible to the theoretical minimum number of nodes expanded in a best-first search. It performs better than a pure depth-first search or a guided depth-first search, which only address the issue on memory limitation. We derive bounds on performance of parallel IRA* searches. These bounds show that the IRA* algorithms are more robust than pure depth-first searches with respect to parallel processing and approximate computations.

In this paper, we evaluate our results using the traveling-salesman problem. These problems are generated by calling the random number generator *rand()*, which randomly determines the location of cities on a 100-by-100 Cartesian plane. All cities are assumed fully connected, and distances between cities are symmetric; that is, the triangular inequality is satisfied. The lower-bound function used here is computed by the spanning-tree method, and the upper bounds by a hill-climbing heuristic. Since the method for computing the upper bound generates a feasible solution, the global minimum upper bound is the same as the incumbent.

2. A BASIC PRINCIPLE OF PARALLEL PROCESSING

In general, two criteria can be used to measure the efficiency of parallel solution of a COP.

Type I   Minimize $T(P)$

          Subject to $\alpha \geq \alpha_r$ and $M(P) \leq M_{max}$,

Type II: Maximize α

          Subject to $T(P) \leq T_{max}$ and $M(P) \leq M_{max}$,

where $T(P)$ and $M(P)$ are the computation time and memory space spent by the algorithm using P processors; α and α_r are, respectively, the achieved and desired accuracies of the solution; and T_{max} and M_{max} are the time and space constraints. In this paper, the criterion used is of Type I.

Let $T_{best}(1)$ be the computation time required by the best sequential algorithm for solving a given problem. Since T_{best} is fixed, Type I criterion is equivalent to

maximize $\dfrac{T_{best}(1)}{T(P)}$ subject $\alpha \geq \alpha_r$ and $M(P) \leq M_{max}$

Let T(1) be the underlying sequential algorithm corresponding to the parallel algorithm to be measured. Then,

$$\frac{T_{best}(1)}{T(P)} = \frac{T_{best}(1)}{T(1)} \cdot \frac{T(1)}{T(P)} = R \cdot S \qquad (2.1)$$

where $R = \dfrac{T_{best}(1)}{T(1)}$ and $S = \dfrac{T(1)}{T(P)}$. It is clear that R characterizes the efficiency of the underlying sequential algorithm, and S characterizes the efficiency of the parallel implementation of the underlying sequential algorithm. A desirable objective of a parallel search is that it achieves the maximum of $R \cdot S$ rather than only R or S individually, and that it satisfies the constrains listed in Type I or II criterion.

A parallel best-first search may achieve the maximum $R \cdot S$. However, it violates the memory-space constraints unless the COP solved is very small. For the original IDA* algorithm, the average S achieved for different problem instances may be close to P, although anomalous behavior occurs, as reported by Rao, Kumar, and Ramesh [5]. R achieved, however, is small for solving the second class of COPs, such as TSPs. For the conventional depth-first search, R achieved is quite small, and S is unpredictable due to its anomalous behavior: acceleration anomalies may result in superlinear speedups, and detrimental anomalies are frequent. *The basic idea in this paper is, therefore, on obtaining both R and S as large as possible and on making S predictable under the restriction of memory space.*

3. EFFECTS OF INITIAL UPPER-BOUNDS

Most search algorithms including A* can be formulated as a version of a branch-and-bound algorithm (B&B). In the B&B computation for solving a minimization problem, the global minimum upper bound, or *incumbent*, is initially set to a given value u_0 (or infinity) and monotonically decreases until it achieves f_o, the optimal solution, which remains unchanged until the end of computation. In the following sections, the incumbent initially obtained before starting the computation of an iteration is called the *initial upper bound* of the iteration.

In general, a combinatorial search process can be split into two phases:

     (a) computation to obtain an optimal solution

Table 1. Comparison of T_f and T_v of guided depth-first searches for 5 symmetric 20-city TSPs.

	Initial Seeds				
	1	2	3	4	5
T_f	54806	44143	181299	146335	106305
T_v	12981	13483	42658	10602	29547
T_b	16235	23491	77292	13867	48062

(b) computation to verify its optimality.

To analyze the performance independent of the details of the underlying computer architecture, we use the number of nodes expanded as the performance measure throughout this paper. Let T_f and T_v be the numbers of nodes expanded before and after the first optimal solution is found. Let T_b be the number of nodes expanded in an A* search. Obviously, the total number of nodes expanded in any search algorithm, T, satisfies

$$T = T_f + T_v \geq T_b$$

In an A* algorithm, the best-first search strategy is adopted, so $T_f = T_b$ and $T_v = 0$. This implies that the initial upper bound has no impact on the A* algorithm. However, in depth-first searches, T_f is usually much larger than T_v and T_b. Table 1 shows T_f, T_v, and T_b for 5 symmetric TSPs of 20-cities. These examples show that in most cases, T_f is the dominant part of search time for depth-first searches. To reduce the computation time of depth-first searches, the most important concern is to make T_f as small as possible.

For general heuristic searches, T_f depends mainly on the accuracy of the guidance function, while T_v depends uniquely on the accuracy of the lower-bound function. The only way to reduce T_v is to design more accurate lower-bound functions, which is difficult since it is problem dependent. In a depth-first search, since no heuristic function is used to guide the selection of nodes, the upper-bound function has a strong influence on T_f, especially when approximations are applied.

So far, a lot of work has been carried out on improving the guidance and lower-bound functions, but the impact of the upper-bound function has not been studied seriously. In this paper, we do not attempt to compare different upper-bound functions, but rather study the influence of the initial upper bound on the performance of a search.

We have extensively studied the computational behavior of guided depth-first searches[1] for solving TSPs and knapsack problems. It is found that a decrease of the initial upper bound results in a significant decrease in the computation time in most cases. When an optimal solution is sought, a smaller initial upper bound will always have a positive effect, i.e., reducing T_f for the depth-first search. Due to space limitation, this claim is stated without proof in Theorem 1.

Theorem 1. Suppose that $A(u_1)$ and $A(u_2)$ are B&B algorithms using the different initial upper bounds u_1 and u_2 and that $u_1 > u_2$. Assume that dominance relation is consistent with the lower-bound function[2]. Let $T(u_1)$ and $T(u_2)$ be the computation time to find the optimal solution by $A(u_1)$ and $A(u_2)$, respectively. Then $T(u_1) \geq T(u_2)$.

Theorem 1 holds for any search strategies. However, a smaller initial upper bound does not always guarantee a reduction in computation time when approximate B&B algorithms are employed. Anomalies may happen, although they do not happen frequently.

4. ITERATIVE REFINING A* SEARCH

Having realized that the initial upper bound has a vital influence on T_f in depth-first searches, we need to answer the next question: how to obtain a small initial upper bound with the minimum overhead. This can be found by an efficient heuristic algorithm, which finds a feasible solution to the problem. This approach is not taken in this paper as it is problem dependent. Our goal is to develop a general problem-independent scheme for finding initial upper bounds efficiently.

Note that the initial upper bound cannot be assigned arbitrarily. If the assigned initial upper bound is less than f_o, then it is not a true upper bound but a threshold. Thresholds have been used in IDA* search [2] and can be selected optimally to minimize the total search overhead [7]. However, the criterion for their selection is not based on minimizing the overhead for finding an initial upper bound.

An approximate B&B algorithm can be used to find suboptimal solutions of COPs. During an approximate B&B procedure, a node P_i generated will be terminated if

$$g(P_i) \geq \frac{u(P_i)}{1+\varepsilon} \qquad \varepsilon \geq 0, \qquad (4.1)$$

where $u(P_i)$ is a global upper bound for P_i and $g(P_i)$ is its lower bound. u^f, the final suboptimal solution obtained by the approximate B&B algorithm, deviates from f_o, the optimal solution value, by

$$\frac{u^f}{1+\varepsilon} \leq f_o \leq u^f \qquad \varepsilon \geq 0. \qquad (4.2)$$

Usually, ε_f ($= (u^f - f_o)/f_o$), the final error achieved, is much smaller than the allowed error ε. Since f_o is unknown in advance, g^f, the global minimum lower bound, instead of f_o, is often used to get ε_g, the upper bound of the error of the suboptimal solution obtained. Note that $\varepsilon_g = (u^f - g^f)/g^f$. If α, the worst-case accuracy of the solution, is defined as f_o/u^f, then $\alpha = 1/(1 + \varepsilon)$.

Approximations significantly reduce the amount of time needed to arrive at suboptimal solutions. Let $A(\varepsilon_1)$ and $A(\varepsilon_2)$ be B&B algorithms with ε_1 and ε_2, respectively, and $\varepsilon_1 < \varepsilon_2$. If a suboptimal solution u^f is found by both algorithms, then in most cases, $A(\varepsilon_2)$ finds u^f earlier than $A(\varepsilon_1)$ does, though anomalous behavior may appear when a depth-first search is employed.

The above claim has been verified experimentally and is demonstrated in Figure 1. A symmetric TSP is solved by a

---

1. In a guided depth-first search, the children of a node are searched in monotonic order of lower bounds.

2. A dominance relation D is said to be consistent with the lower-bound function if $P_i \, D \, P_j$ implies that $g(P_i) \leq g(P_j)$, where $g(P_i)$ and $g(P_j)$ are nodes in the B&B search tree. More detailed discussion on dominance relations can be found in the reference [4]

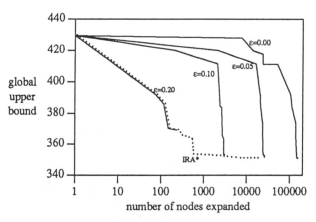

Figure 1. The effect of error allowance on global upper bound for guided depth-first searches. (TSP of 20-cities, initial seed = 4)

guided depth-first search with four different error allowances, 0.2, 0.1, 0.05 and 0. The computational behaviors are measured in terms of the rates of decrease of the global upper bounds. Figure 1 shows that the larger the error allowance is, the faster the global upper bound decreases. For instance, when $\varepsilon = 0.2$, 240 nodes are expanded to reach the global upper-bound value 369.2, while 138,000 nodes are expanded in order to reach the same global upper-bound value when $\varepsilon = 0$.

Comparing the four curves and considering the effect of the initial upper bounds, we can arrange the depth-first searches in such a way that the approximation search with a large ε is implemented first; the suboptimal solution obtained is then transferred to the next iteration as the initial upper bound, in which a smaller ε is used. As a result, the time for solving a COP can be reduced significantly!

For the above example, the same problem can be solved in four iterations with the error allowances decreasing from 0.2 to 0. It takes 241 node expansions to finish the first iteration; the second iteration with $\varepsilon = 0.1$ and initial incumbent 369.2 takes 491 node expansions; the third iteration with $\varepsilon = 0.05$ and initial incumbent 353.3 takes 2,419 node expansions; finally, the fourth iteration with initial incumbent 351.8 takes 14,229 node expansions to find the optimal solution 351.0 and complete the search. The computational behavior of this implementation is reflected by the dotted curve in Figure 1, which shows an order-of-magnitude improvement in performance (see Case 4 in Tables 2 and 3). Since the suboptimal solution is refined iteratively, we call this scheme an *Iterative Refining A** (in short, IRA*) search.

The IRA* search scheme is outlined in Figure 2. The stopping criterion can be the number of iterations, the final error allowance, or other conditions. An upper-bound function can be used to compute u_1^0, the initial incumbent of the root. The key of the IRA* search scheme is in *refining*, *i.e.*, choosing the proper strategy to reduce the error allowance. The goal of refining the computation except the last iteration is to find an optimal or desired suboptimal solution as early as possible. Once the desired solution is found, the remaining time is only used to verify optimality.

```
INPUT:     initial error allowance ε₁;
           initial incumbent u₁⁰ if available;
           problem parameters;
OUTPUT:    Optimal or suboptimal solution;
BEGIN
    WHILE (stopping-criterion is not satisfied) {
        DO approximate depth-first search from the root of the
            tree with εᵢ and initial incumbent uᵢ⁰
        UNTIL no active node remains in the stack or the time
            constraint is exceeded;
        IF (time constrain is exceeded) THEN
            report the achieved solution and the error bound εg;
        ELSE {
            (a) record the suboptimal solution and εg;
            (b) set new error allowance εᵢ₊₁ < εᵢ;
            (c) set uᵢ₊₁⁰, the initial incumbent of the next iteration,
                equal to uᵢᶠ, the final incumbent in this iteration;
        }
    }
END
```

Figure 2. The IRA* search scheme

Table 2. Performance comparison of various search algorithms for 5 20-cities TSPs. (T_g: time for guided depth-first search; T_r: time for IRA* search; and T_b: time for best-first search)

	Initial Seeds				
	1	2	3	4	5
T_g	67787	57626	223957	156937	135852
T_r	20828	51043	88864	17380	54889
T_b	16235	23491	77292	13867	48062
T_r/T_b	1.28	2.17	1.15	1.25	1.14

Table 3. Time of each iteration in IRA* for 5 20-cities TSPs. $T(\varepsilon)$ is the time needed in an iteration of IRA* to achieved an approximation degree ε.

Approx. Searches	Initial Seeds				
	1	2	3	4	5
$T(0.2)$	769	210	407	241	303
$T(0.1)$	1071	2291	2329	491	1279
$T(0.05)$	1886	9055	8836	2419	5245
$T(0)$	17102	39487	77292	14229	48062
$T(0)/T(0.05)$	9.17	4.36	8.75	5.89	9.16

The sequence of refinements must be chosen properly. If the difference in error allowances between successive iterations is too large, then the final upper bound found in each iteration may have little influence on the next iteration. In contrast, if the difference is too small, then the global upper bound may remain unchanged in some iterations; hence, these iterations do not contribute to improving the solution. From our experience, four iterations are sufficient for the problems we have tested, and either the optimal or a good suboptimal solution is found in the iteration with $\varepsilon = 0.05$.

Table 2 compares the performance among a guided depth-first search, IRA*, and a best-first search[3] for 5 20-city symmetric TSPs.[4] Due to space limitations, other results which exhibit similar behavior are not shown. The results show that the performance of IRA* is very nice. In all instances except when the initial seed is 2, the search times required by IRA* are only 14 to 28 percent more than that for the best-first search, and the space required is linear. To our knowledge, no other general search scheme has achieved this high performance.

We now explain the reason why IRA* is efficient and analyze its performance. Note that TSP is a strict NP-hard problem, whose computation time grows exponentially when the accuracy of the solution is increased. Table 3 demonstrates this claim. The time required in an approximation search with $\varepsilon = 0.05$ is only about 20 percent of T_b. On the other hand, since the lower-bound function for TSP based on the spanning-tree method is not very tight, it is quite possible that a node provides a new incumbent but its lower bound is relatively loose. Thus, an optimal or a good suboptimal solution is usually found before the last iteration, and the overhand of IRA* is quite small.

The computational behavior of the approximate depth-first searches are related to the sequences of incumbents, which are problem-instance dependent. Therefore, a theoretical analysis of the performance of IRA* is very difficult. The following theorem shows the minimum time required by IRA*.

Throughout this paper, all analysis are based on the assumption that *the number of nodes whose lower bound is less than g increases exponentially with g*. Empirically, we have verified this assumption for strict NP-hard COPs, such as symmetric TSPs for fully connected cities [6]. This assumption holds true in a large range of lower-bound values, from the minimum lower-bound value to some value larger than the optimal solution. Let n_0 and B_0 be, respectively, the numbers of nodes whose lower bounds are minimum and are less than the optimal-solution value. n_0 is one for most COPs, *i.e.*, only the root of the B&B search tree has the minimum lower bound.

Lemma 1. Assume that n(g), the number of nodes whose lower bounds are less than g in the B&B tree, increases exponentially with g within the range $g_r < g < L$, and that $n(g_r) = n_0$, then

$$n(g) = n_0 \left\lceil \frac{B_0}{n_0} \right\rceil^{\left\lceil \frac{g - g_r}{f_o - g_r} \right\rceil}, \qquad (4.3)$$

where f_o and g_r are the optimal-solution and minimum lower-bound values, respectively, and $L > f_o$.

Proof. According to the assumption, we can denote n(g) as an exponential function.

$$n(g) = M \cdot A^g, \qquad (4.4)$$

where M and A are constants. From the following conditions,

$$M \cdot A^{g_r} = n_0 \qquad (4.5)$$

$$M \cdot A^{f_o} = B_0, \qquad (4.6)$$

we get $A = \left\lceil \frac{B_0}{n_0} \right\rceil^{\left\lceil \frac{1}{f_o - g_r} \right\rceil}$ and $M = n_0 \left\lceil \frac{B_0}{n_0} \right\rceil^{\left\lceil \frac{-g_r}{f_o - g_r} \right\rceil}$ □

In terms of the above result, we can analyze the best-case behavior of IRA*.

Theorem 2. Let ε_i be a series of error allowances, $\varepsilon_i > \varepsilon_{i+1}$, $1 \leq i < k$, and $\varepsilon_k \geq 0$. Let T_r be the computation time required by IRA* for solving a COP and $n_0 = 1$. Then $T_r \geq \sum_{i=1}^{k} B_0^{r_i}$, where r_i satisfies $\frac{f_o}{1+\varepsilon_i} - g_r = r_i(f_o - g_r)$.

Proof. Consider iteration i in which the error allowance is ε_i. From Eq. (4.1), all nodes whose lower bound is less than $\frac{f_o}{1+\varepsilon_i}$ have to be expanded; thus, at least $n\left\lceil \frac{f_o}{1+\varepsilon_i} \right\rceil$ nodes are expanded in iteration i. From Lemma 1,

$$n\left\lceil \frac{f_o}{1+\varepsilon_i} \right\rceil = B_0^{\left\lceil \frac{\frac{f_o}{1+\varepsilon_i} - g_r}{f_o - g_r} \right\rceil} \qquad (4.7)$$

Note that, in the above equation, $n(\cdot)$ is a function notation. Let $\frac{f_o}{1+\varepsilon_i} - g_r = r_i(f_o - g_r)$. Eq. (4.7) shows that at least $B_0^{r_i}$ nodes have to be expanded in iteration i. The theorem is proved by considering k iterations. □

Only a loose upper bound of T_r can be derived due to its anomalous behavior. However, our results in Table 3 demonstrate that most of the overhead of IRA* is incurred in the last iteration. If the optimal solution is found before this, then T_r is usually close to T_b, since all nodes expanded in the last iteration have to be expanded in a best-first search as well. In general, this is not the worst case. We now discuss the worst-case behavior of IRA*.

Theorem 3. Let k be the number of iterations of an IRA*, ε_k be 0, and the optimal solution be found in the last iteration. Let t_j be the computation time required in the j-th iteration, and n_0 be 1. Then $t_k < B_0^{r_k+1}$, and $t_{k-1} < B_0^{r_{k-1}+1}$, where $r_k = \frac{\varepsilon_{k-1} f_o}{f_o - g_r}$ and $r_{k-1} = \frac{(\varepsilon_{k-2} - \varepsilon_{k-1}) \cdot f_o}{(1+\varepsilon_{k-1})(f_o - g_r)}$.

Proof. Since the optimal solution is found in the last iteration and using the approximate lower-bound test defined in Eq. (4.1), we can claim that $u_{k-1}^0 = u_{k-2}^f < (1+\varepsilon_{k-2})f_o$; otherwise, the optimal solution must be found before the $(k-1)$-th iteration. Similarly, $u_{k-1}^f < (1+\varepsilon_{k-1}) \cdot f_o$, and $g_{k-1}^f < u_{k-1}^0 / (1+\varepsilon)$. The worst case happens when the incumbent is unchanged until the last node is expanded in the $(k-1)$-th iteration. This implies that

$$t_{k-1} < n\left\lceil \left\lceil \frac{1+\varepsilon_{k-2}}{1+\varepsilon_{k-1}} \right\rceil \cdot f_o \right\rceil$$

From Eq. 4.4, we have

$$t_{k-1} < M \cdot A^{\left\lceil \frac{(1+\varepsilon_{k-2}) \cdot f_o}{1+\varepsilon_{k-1}} \right\rceil}$$

3. T_b is obtained in the last iteration of IRA*, since $u_k^0 = f_o$. It is very difficult to solve a large symmetric TSP by a best-first search due to the space constraint.

4. For asymmetric TSPs and TSPs on incompletely connected graphs, problems with several hundred cities can be solved. The solvable size of TSPs on symmetric and fully connected graphs is quite small. Even for 20 cities, the number of nodes expanded in the search tree may exceeds one million.

Hence,

$$t_{k-1} < B_0^{\left[1 + \frac{(\varepsilon_{k-2} - \varepsilon_{k-1}) \cdot f_o}{(1 + \varepsilon_{k-1})(f_o - g_r)}\right]} \qquad (4.8)$$

Applying the above analysis to the k-th iteration, we get

$$t_k < B_0^{\left[1 + \frac{\varepsilon_{k-1} f_o}{f_o - g_r}\right]} \qquad (4.9)$$

The theorem follows from Eq's (4.8) and (4.9) and the definition of r_i, $i = k-1, k$. $\quad\square$

Theorems 2 and 3 are of importance only in verifying the performance of IRA* and not in predicting it, as B_0 is problem-instance dependent and is difficult to estimate. Some parameters in the above theorems can be estimated. For example, g_r is easy to obtain by abstracting the information of the root node. From our experience, f_o is close to the mean of the root's lower-bound and upper-bound values. More accurate estimate of f_o can be obtained by using statistical extreme-value distribution theorem. According to this theorem, TSP can be viewed as $(n-1)!/2$ samples and each upper bound as a local minimum, and the optimal solution satisfies the Weibull distribution.

To verify this upper bound, we examine a 20-city TSP with $u_k^0 = 398.792$, $g_r = 303.165$, $f_o = 395.792$, $T_b = 552101$, and $t_k = 756385$. Since B_0 is not available, we use T_b instead of B_0 in order to derive a loose upper bound of t_k. In this example, $u_{k-1}^f = u_k^0 = 398.792$; hence, the achieved ε_{k-1} is $(u_k^0 - f_o)/f_o = 3/f_o$, and r_k is 0.033. From Theorem 3, we get $t_k = 756385 < T_b^{1+r_k} = 854093$.

The sequence of ε used in IRA* is called a *refinement schedule*. A good refinement schedule should be determined in terms of the distribution of nodes by lower bounds and the accuracy of the initial incumbent. The design of an optimal refinement schedule is open at this time. In this paper, a static refinement schedule is adopted. We have examined a number of dynamic refinement schedules and found that the search performance of IRA* for solving TSPs is not sensitive to changes in the refinement schedule, since the computation time of the last iteration is dominant and is close to T_b in most cases.

5. PARALLEL ITERATIVE REFINING A*

In studying parallel IRA*, we focus on improving its speedup with respect to the *best* sequential algorithm. To simplify the analysis, we assume that the overheads of load balancing and communication are negligible. We assume that the computational model is a loosely coupled multicomputer, and that the B&B tree is split into a number of subtrees that are searched in parallel. We found earlier that when multiple lists are used, a parallel search may risk searching a large number of extra nodes [4]. Our objective here is to reduce this risk. We further assume that the parallel computation is synchronized by iterations. Note that the search trees in different iterations have different sizes.

Before presenting the theoretical analysis, we need to explain the representation of the search tree. The search trees used in the literature are usually depicted by depth. For the sake of understanding IRA*, the search trees used here are depicted by cost (see Figure 3). Note that nodes in the same cost level do not mean that they are located in the same depth of the original tree.

In iteration i, the global upper bound decreases from u_i^0 to u_i^f (it is possible that $u_i^f = u_i^0$). When a parallel IRA* is implemented, the speedup is related to the interval $\frac{u_i^0 - u_i^f}{1 + \varepsilon_i}$. Similar to the discussion in Section 3, we define r_i as $\frac{u_i^0 - u_i^f}{(1 + \varepsilon_i)(f_o - g_r)}$. Since $(u_i^0 - u_i^f) < (f_o - g_r)$, we have $0 \le r_i \le r_{max} \le 1$. r_{max} reflects the maximum threshold interval among all iterations. Define

$$C_i = \frac{n \left\lceil \frac{u_i^0}{1 + \varepsilon_i} \right\rceil}{n \left\lceil \frac{u_i^f}{1 + \varepsilon_i} \right\rceil}.$$ From Lemma 1, we have $C_i = A^{r_i(f_o - g_r)}$. Let

$C_{max} = \max_i(C_i)$, namely, $C_{max} = A^{r_{max}(f_o - g_r)}$. C_{max} reflects the ratio between the numbers of nodes expanded with different thresholds and is an important parameter for characterizing the behavior of IRA* searches. In a conventional depth-first search, $r_{max} = 1$; hence, $C_{max} = B_0$ if $n_0 = 1$.

In the following discussion, notations (1) and (P) are used to denote the parameters used in serial and in parallel computing.

Theorem 4. Assume that the search tree is decomposed into multiple subtrees, that the processors are synchronized in node expansions, that $u_i^0(P) = u_i^0(1)$, and that $u_i^f(P) = u_i^f(1)$. Let $T_r(1)$ and $T_r(P)$ be the computation times required for the IRA* algorithms using one and P processors, respectively. The speedup, $\frac{T_r(1)}{T_r(P)}$, are limited within the following bounds.

$$\max\left[1, \frac{P}{C_{max}}\right] \le \frac{T_r(1)}{T_r(P)} \le C_{max} \cdot P \qquad (5.1)$$

Proof: The proof is in two parts. Due to space limitation, we show only the part for finding the lower-bound speedup; the part for finding the upper-bound speedup is similar.

In finding the lower-bound speedup, since the parallel IRA* algorithm is synchronized by iterations, $T_r(P) = \sum_{i=1}^{k} t_i(P)$. For the i-th iteration in both serial and parallel cases, the nodes whose lower bounds are less than $\frac{u_i^f}{1 + \varepsilon_i}$ must be expanded. These nodes are called *essential nodes*, while other nodes expanded are called *nonessential nodes*. In the worst case, no nonessential node is expanded in the serial case but all nonessential nodes are expanded in the parallel case. This situation is possible when u_i^f, the final incumbent of iteration i, is not found until $\frac{1}{P} \cdot n \left\lceil \frac{u_i^0}{1 + \varepsilon_i} \right\rceil$ essential nodes have been expanded in the serial case, and all these nodes are assigned to processor 1 in the parallel case if $P > C_i$. In addition to essential nodes, all nonessential nodes are expanded in the parallel case, and these nonessential nodes are assigned to processor 2 to processor P. Figure 3 shows the worst case for parallel IRA* in which $P > C_i$ and

$$t_i(1) = n \left\lceil \frac{u_i^f}{1 + \varepsilon_i} \right\rceil, \qquad t_i(P) = \frac{1}{P} \cdot n \left\lceil \frac{u_i^0}{1 + \varepsilon_i} \right\rceil$$

Without loss of generality, the initial steps for generating P nodes in the parallel IRA* search and the ceiling function is

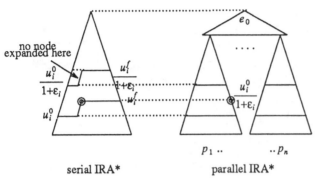

serial IRA*   parallel IRA*

Figure 3. The worst case behavior of parallel IRA*.

---

omitted in this proof. This means that the overhead of synchronization is ignored. To derive the upper bound of minimum speedup, *i.e.*, the maximum speedup in the worst case, we neglect the overhead of load balancing. Accordingly,

$$\frac{t_i(1)}{t_i(P)} = \frac{n\left\lceil \dfrac{u_i^f}{1+\varepsilon_i} \right\rceil}{\left\lceil \dfrac{1}{P} \right\rceil \cdot \left\lceil n\left\lceil \dfrac{u_i^0}{1+\varepsilon_i} \right\rceil \right\rceil} \quad \text{if } P > C_i \qquad (5.2)$$

From Eq.(5.2) and the definition of C_i, we have

$$\frac{P}{C_i} \le \frac{t_i(1)}{t_i(P)} \quad \text{if } P > C_i . \qquad (5.3)$$

When $P \le C_i$, it is possible that in the parallel case processor 1 expands the same nodes as expanded in the serial case, and all nodes expanded by other processors are unnecessary. This implies that

$$1 \le \frac{t_i(1)}{t_i(P)} \quad \text{if } P \le C_i. \qquad (5.4)$$

The lower bound is proved by combining Eq's (5.3) and (5.4), and using the definition of $C_{\max}$. □

Note that the speedup is dominated by the last iteration in which the threshold interval is restricted within $\varepsilon_{k-1} f_o$, and oftentimes $u_k^0 = u_k^f$. This implies that C_k is very small and is very likely equal to 1. Consequently, in practice, the bounds of speedup are tighter than those given in Theorem 4, especially when load balancing is applied.

The above theorem shows the robustness of parallel IRA*. Since $C_{\max}$ is small in IRA*, the condition that $P > C_{\max}$ is generally satisfied. Parallel IRA* also avoids the detrimental anomalies and poor slowdown often found in parallel depth-first searches; that is, at least $\dfrac{P}{C_{\max}}$ times speedup can be gained. The cost of this gain is in the loss of large super-linear speedups, although super-linear speedup less than $C_{\max}P$ is still achievable. In contrast, the possible speedups in conventional parallel depth-first search may range from 1 to $B_0 \cdot P$.

Table 4. Simulation results on parallel IRA* and PGDFS for solving 5 20-city TSPs using 16 processors.

Init. Seed	1	2	3	4	5
Sequential Best-First Search					
T_b	16235	23491	77292	13867	48062
IRA* and Parallel IRA*					
T_r	20828	51043	88864	17380	54889
T_r^1	26879	48404	87739	177783	55025
$T_r(16)$	1697	3043	5501	1127	3457
R_r	0.78	0.46	0.87	0.80	0.88
H_r	0.77	1.05	1.01	0.98	1.00
U_r	15.84	15.91	15.95	15.78	15.92
S_r	12.27	16.77	16.15	15.42	15.88
SP_r	9.57	7.72	14.05	12.30	13.90
GDFS and Parallel GDFS					
T_g	67787	57626	223957	156937	135852
T_g^1	122035	49340	219463	42032	85028
$T_g(16)$	7643	3099	13732	2643	5340
R_g	0.24	0.41	0.35	0.09	0.35
H_g	0.56	1.17	1.02	3.73	1.60
U_g	15.97	15.92	15.98	15.90	15.92
S_g	8.87	18.60	16.31	59.38	25.44
SP_g	2.12	7.58	5.63	5.25	9.00

Table 5. Summary of various speedups for 10 TSP instances using 16 processors and random seeds ranging from 1 to 10. (T_b: time for best-first search; T_g: time for guided depth-first search; T_r: time for IRA* search.)

average speedup	$\dfrac{T_b(1)}{T_r(16)}$	$\dfrac{T_b(1)}{T_g(16)}$	$\dfrac{T_g(16)}{T_r(16)}$	$\dfrac{T_g(1)}{T_g(16)}$	$\dfrac{T_r(1)}{T_r(16)}$	$\dfrac{T_g(1)}{T_r(16)}$
Method 1	11.42	5.06	2.26	14.34	14.69	32.38
Method 2	11.37	5.67	3.37	20.26	14.96	64.45

Method 1 computes the speedup as the ratio of the total computation times of the 10 problem instances using different search strategies. Method 2 computes the speedup as the average speedup of individual problem instances.

---

6. SIMULATION RESULTS OF PARALLEL IRA*

To verify the theoretical results derived above, we have simulated the performance of parallel IRA* (PIRA*) and parallel guided depth-first searches (PGDFS) using 16 processors. Results for 5 symmetric TSPs of 20 fully-connected cities are shown here. Other results are not shown due to space limitation.

The parallel search is implemented as follows. Initially, a serial best-first search is carried out to produce 16 active nodes, which are distributed into 16 separated lists. A parallel search is then carried out with the 16 nodes as the roots of subtrees. In each step, a processor expands one node, provided there are 16 or more active nodes among the lists. If a list is empty, it gets a node from another list. Since we are focusing on the speedup (SP) of the parallel search, we use the number of steps as our performance measure. The refinement schedule adopted in the parallel IRA* search is the same as that in the serial one.

The simulation results for PIRA* and PGDFS are shown in Table 4. The sequential search times, T_b, T_r, and T_g, are taken from Table 2. $T_g(P)$ and $T_r(P)$ are the times required by PGDFS and PIRA*, respectively. Since the sets of the nodes expanded in a serial search and the corresponding parallel search are not identical, we denote the total number of nodes expanded in all processors in PGDFS (resp. PIRA*) by T_g^1 (resp. T_r^1).

Let H be the ratio of the number of the nodes expanded in a serial search and that in the corresponding parallel search. H reflects the amount of extra work or savings in a parallel search as compared to the serial one. During a parallel search, especially in the first and last few steps, some processors may be idle as there are insufficient active nodes. The parameter U characterizes the speedup of T_r^1 (resp. T_g^1) over $T_r(P)$ (resp. $T_g(P)$). The simulation results show that in almost every step all processors are busy. That is, U is very close to 16. Using the above definitions and recalling R and S defined in Section 2, we have the following relations among these parameters.

$$SP = R \times S = R \times (H \times U) \qquad (6.1)$$

Tables 4 demonstrates that PIRA* is much more efficient and robust than PGDFS. For the 5 TSP instances, the speedups (SP) range from 7.72 to 14.05 in PIRA*, with an average speedup[5] of 12.03. For PGDFS, the speedups range from 2.12 to 9.00, with an average speedup of 5.51. In PIRA*, the number of nodes expanded is close to that of the serial IRA*. As a result, poor slowdown in performance is avoided, and S, the speedups with respect to the same search scheme, are close to the number of processors (ranging from 12.27 to 16.77). Obviously, all these speedups obtained by simulations are within the bounds derived in Section 5. In PGDFS, the values of parameter H is more diverse than those in PIRA*; hence, anomalies occur more frequently (S ranges from 8.87 to 59.38). Even in the case when S equals 59.38, the real speedup (SP) is only 5.25, since R is too small.

Table 5 compares different methods for computing speedups. We focus in this paper on $T_b(1)/T_r(P)$, namely, the speedup of the parallel algorithm as compared to the best sequential algorithm. This is shown in column 2 in Table 5. We see that PIRA* has a speedup of around 11 as compared to a sequential best-first search.

For research on parallel algorithms, what is important is the speedup based on the same search strategy. This is illustrated by the results in columns 5 and 6 in Table 5, which show that the speedups are closer to linear and can even be superlinear. This linear or superlinear speedup has been observed frequently by researchers in this area. However, these speedups do not represent the merits of parallel processing in actually solving the original problem.

The speedups achieved by PIRA* as compared to traditional sequential methods based on guided depth-first search are more dramatic. These are shown in the last column in Table 5, which show that a speedup of over 32 can be achieved.

7. CAVEATS

The effectiveness of the IRA* algorithm depends on the complexity of the COP to be solved. For problems with fully polynomial-time approximation algorithms, such as the knapsack problem, the IRA* algorithm may not be effective. A necessary condition for IRA* to be effective is that the computation time of the problem instance to be solved must be reduced substantially by approximate depth-first searches. Second, since IRA* is based on depth-first searches, it acquires the drawbacks and anomalous behavior of approximate depth-first searches. Overall, anomalies happen much less frequently in IRA* than in depth-first searches because we are performing a sequence of depth-first searches, each of which may compensate the anomalous behavior of another.

ACKNOWLEDGEMENTS

We would like to thank Mr. Lon-chan Chu for his helpful discussions and for providing WISE, an integrated Search Environment which implements a B&B search of the traveling-salesman problem. We are also indebted to Dr. Akiko Aizawa for her technical assistance.

REFERENCES

[1] V. Kumar, et al. (ed.), in *Parallel Algorithm for Machine Intelligence and Vision*, Springer-Verlag, New York, 1990.

[2] R. E. Korf, "Depth-First Iterative Deepening: An Optimal Admissible Tree Search," *Artificial Intelligence*, vol. 27, pp. 97-109, North-Holland, 1985.

[3] V. Kumar, K. Ramesh, and V. N. Rao, "Parallel Best-First Search of State-Space Graphs: A Summary of Result," *Proc. National Conf. Artificial Intelligence*, pp. 122-127, AAAI, 1988.

[4] G. J. Li and B. W. Wah, "Computational Efficiency of Combinatorial OR-Tree Searches," *Trans. on Software Engineering*, vol. 16, no. 1, pp. 13-31, IEEE, Jan. 1990.

[5] V. N. Rao, V. Kumar, and K. Ramesh, "A Parallel Implementation of Iterative-Deepening-A*," *Proc. of National Conf. Artificial Intelligence*, pp. 878-882, AAAI, 1987.

[6] B. W. Wah and L.-C. Chu, "TCA*--A Time-Constrained Approximate A* Search Algorithm," *Proc. Int'l Workshop on Tools for Artificial Intelligence*, pp. 314-320, IEEE, Nov. 1990.

[7] B. W. Wah, *MIDA*: An IDA* Search with Dynamic Control*, Research Report CRHC-91-09, Center for Reliable and High Performance Computing, Coordinated Science Laboratory, Univ. of Illinois, Urbana, IL 61801, March 1991.

---

5. The average speedup is computed by dividing the summation of T_b's by the summation of $T_r(16)$'s (or $T_g(16)$'s) for the 5 TSPs.

Scheduling in Parallel Database Systems

S. Dandamudi and C. Y. Chow

School of Computer Science, Carleton University
Ottawa, Ontario K1S 5B6, Canada

Abstract - Scheduling in parallel processing systems is an important performance issue. In this paper, we study the impact of scheduling on performance of parallel database systems. We do not model a particular database system in our study. Rather, we use an abstract model of a shared-nothing type of architecture. Several database machines have been built/proposed with such an architecture. Our results, obtained via simulation, indicate that, in I/O intensive applications, I/O scheduling policy is the most important factor in determining overall system performance. Processor scheduling policies play insignificant role in determining the overall system performance. The results also indicate that the desired locking granularity is not influenced by the type of scheduling algorithm used.

1. INTRODUCTION

The performance of parallel processing systems is affected by several factors. Scheduling in such systems is an important factor affecting the overall system performance. In this paper, we study the impact of scheduling in parallel database systems.

A parallel database system can be built by connecting together many processors, memory units, and disk devices through a global interconnection network. Depending on how these components are interconnected there can be a variety of possible architectures. Here we consider the shared-nothing architecture in which each processor has its own (private) disk and memory units. Shared-nothing database systems are attractive from the standpoint of scalability and reliability [9]. Several commercial and experimental systems such as the Non-Stop SQL from Tandem, DBC/1012 from Teradata, the Gamma machine at the University of Wisconsin [4], and the Bubba system at MCC [6] belong to this type of architecture.

Locking is one of the common techniques used in database systems to achieve concurrency control. The systems mentioned in the last paragraph use lock-based concurrency control mechanisms. Database systems that use physical locks support the notion of granule [7,8]. A granule is a lockable unit in the entire database and typically all granules in the database have the same size. Before accessing the database entities, a transaction must lock the required granules. However, if the required granules are locked by another transaction in the system, the transaction will be blocked. The granule can be a column of a relation, a single record, a disk block, a relation or the entire database. Locking granularity refers to the size of lockable granule. Fine granularity improves system performance by increasing concurrency level but it also increases lock management cost. Coarse granularity, on the other hand, sacrifices system performance but lowers the lock management cost. Locking granularity issues are discussed in [2].

Here we study the impact of scheduling algorithms on the performance of shared-nothing database systems. We use simulation as the principal means of performance evaluation. The remainder of the paper is organized as follows. In Section 2, a brief description of the simulation model is presented. Section 3 describes the scheduling algorithms considered in this study. The results are presented in Section 4 and conclusions are given in Section 5.

2. THE SIMULATION MODEL

The simulation model, which is an extension of the model used by Ries and Stonebraker [7,8], used in this study is shown in Figure 1. This is a closed model with a fixed number of transactions $ntrans$ in the system at any instance. The transactions cycle continuously through the system. Every time a transaction completes execution, a new transaction with a new set of parameter values is created so that the system always sees $ntrans$ transactions.

All relations are assumed to be horizontally partitioned across all disk units in the system using the round robin strategy [3]. We assume a *fork and join* type of transaction structure. In this model, a transaction is assumed to be split into several sub-transactions each working on fragments of the relations involved. For example, if the operation is the *selection* operation, then each processor will be wor-

king on a smaller relation to retrieve partial response to the selection operation. These partial responses are then combined to derive the final response. Thus, a transaction is not considered complete unless all the sub-transactions are finished. It is also assumed that each sub-transaction consists of two phases: an I/O phase and a CPU phase.

A transaction goes through the following stages.

(1) The transaction at the head of the pending queue is removed and the required locks are requested. If the locks are denied, the transaction is placed at the end of the blocked queue. System will record the blocked transaction. This record is used to help release blocked transactions after the blocking transactions finished their processing and released their locks. (The lock conflict computation and the blocked transaction release process are described in [3].)

(2) After a transaction receives all its locks, it is split into $npros$ (the total number of processors in the system) sub-transactions and placed at the end of the I/O queue of its assigned processor.

(3) After receiving the required I/O and CPU service, the sub-transaction waits in the waiting area for other sub-transactions of its parent transaction to complete. A transaction is considered complete when all of its sub-transactions are completed. A completed transaction releases all its locks and those transactions blocked by it. The releasing process of blocked transactions is described in [3].

Note that the transactions request all needed locks before using the I/O and CPU resources. Thus deadlock is impossible. In addition, it is assumed that the cost to request and set locks includes the cost of releasing those locks. The cost is assumed to be the same even if the locks are denied. In addition, we assume that processors share the work for locking mechanism. This is justified because relations are equally distributed among the system resources. Moreover, lock processing has priority over the running transaction for the CPU and I/O resources.

The following input parameters are used to drive the simulation.

$npros$	=	the number of processors in the system.
$dbsize$	=	the number of accessible entities in the entire database.
$ltot$	=	the number of locks in the database system.
$ntrans$	=	the number of transactions in the system.
$maxtransize$	=	the maximum transaction size in the system.
$cputime$	=	CPU time required by a transaction to process one entity of the database.
$iotime$	=	I/O time required by a transaction to process one entity of the database.
$lcputime$	=	CPU time required to request and set one lock.
$liotime$	=	I/O time required to request and set one lock.
$tmax$	=	the number of time units to run the simulation.

The following statistics are recorded for each simulation run.

$totcom$	=	number of transactions completed at $tmax$.
$throughput$	=	$totcom/tmax$ = the number of completed transactions per unit time.
$aveRT$	=	average response time (i.e., from the time the transaction enters the pending queue till the time it completes all processing and releases locks).
$aveBT$	=	average blocking time (blocking time is the time spent by a transaction in the blocked queue awaiting the release of the needed locks).
$aveST$	=	average synchronization time (synchronization time is defined as the time between the completion of the first sub-transaction to the completion of the last sub-transaction of the same transaction).

For details on transaction decomposition and characterization, and on lock conflict computation, see [3].

3. SUB-TRANSACTION SCHEDULING POLICIES

There are several scheduling policies that are applicable to processor scheduling. These policies can be broadly divided into two classes: policies that are independent of the transaction characteristics and policies that need certain information on transactions. We consider two policies belonging to each category.

Transaction independent policies

- *First-Come-First-Served (FCFS):* This is the simplest scheduling policy that does not require any knowledge about transaction characteristics. The sub-transactions are placed in the order of their arrival to the queue. When a resource (either CPU or I/O) becomes available the sub-transaction at the head of the waiting queue is assigned this resource.

- *Round Robin (RR):* This is a preemptive policy in which each sub-transaction is given a fixed quantum of service and if the sub-transaction is not completed it is preempted and returned to the tail of the associated queue. Note that this scheduling policy is useful only for processor scheduling, and not for I/O scheduling.

Transaction dependent policies

- *Shortest Job First (SJF):* This policy requires service time requirements of sub-transactions. If this knowledge is known *a priori*, this policy schedules that sub-transaction which has the smallest service time requirement among the sub-transactions that are waiting to be scheduled. In the context of databases, an estimate of this information can be obtained. For example, if the transaction involves a join operation, then selectivity characteristics of the relations involved can be used to obtain a good estimate of the complexity involved in processing the transaction.

- *Highest Priority First (HPF):* A potential disadvantage with the above strategy is that it is possible (depending on transaction characteristics) that some sub-transactions of a given transaction receive quick service (because their service requirements are smaller) while other sub-transactions of the same transaction, which have larger service demand, may be waiting for service at other processors (or I/O servers). This effectively increases transaction completion time. (Recall that a transaction is not considered complete unless all the sub-transactions of that transaction are finished.) In this policy, each transaction is given a priority that is inversely proportional to the total size of the transaction. When a transaction is divided into sub-transactions, they carry this priority information. The scheduling decisions are made based on this information. Thus all sub-transactions receive the same treatment in all queues irrespective of their actual service demand.

Note that for processor scheduling, all four policies could be applied. However, for I/O scheduling, preemptive policies are not applicable. Therefore, we study the impact of the remaining three scheduling policies for I/O scheduling. We have implemented all four policies for processor scheduling. However, as shown in Section 4.5, processor scheduling policies have insignificant role in determining the overall system performance. Therefore, in the following, we consider the same scheduling policy for both processor and I/O scheduling (for example, if we use the HPF policy for I/O scheduling then the same policy is used for processor scheduling).

4. RESULTS AND DISCUSSION

This section presents the results obtained from the simulation experiments. We use independent replication strategy to obtain confidence intervals. This strategy provided 95% confidence intervals that are less than 5% of the mean values (in most cases) reported in the following sections. Unless otherwise indicated, the results correspond to the input parameter values given in Table 1. Also, the transaction size is uniformly distributed over the range 1 and *maxtransize*.

Table 1 Input parameters used in the simulation experiments

parameters	values	parameters	values
dbsize	5000	*iotime*	0.20
npros	20	*lcputime*	0.01
tmax	12000	*liotime*	0.20
cputime	0.05	*maxtransize*	500

4.1. Impact of Number of Transactions

By fixing the input parameters as in Table 1, several experiments were conducted by varying the number of transactions (*ntrans*) from 1 through 300. For these experiments, the number of locks (*ltot*) is fixed at 100. The results are reported in Figure 2. These results show that FCFS policy is more sensitive to the number of transactions in the system. This is an expected result. The reason is that a sub-transaction requiring large I/O service time could keep the server busy and hence introduce inordinate delays for other smaller sub-transactions. This increased delay has multiplicative effect because this delay not only affects the associated transaction but also permeates the effect on the entire system. This is because transactions will not release locks until all their sub-transactions are finished. This increases lock hold time resulting in more transactions being placed in the blocked queue. As can be seen from Figure 2 the blocking time is the main component that causes the increase in total response time. Also, a single large sub-transaction can delay completion of all other transactions. This results in an increase in synchronization time. However, the synchronization time is substantially smaller (less than 10%) than the blocking time (not shown in Figure 2).

The performance of SJF is marginally better than HPF. While the difference in performance of these two policies increases with the number of transactions, it is still less than 10% for the input parameter values considered here. As with FCFS, it is the blocking time that contributes to the increase in response time. The effect of synchronization time (not shown) is relatively minor.

The system throughput in Figure 2 shows that as the number of transactions increases the system throughput increases initially (up to about 25 transactions in this case[1]) and then starts falling off. Note that, during this initial phase, all three scheduling policies have roughly the same throughput. The initial increase is due to the increased load while lock denial is not a problem yet. But as the number of transactions increases, contention for data increases causing more and more lock denials. The drop in throughput after this point is due to the fact that each arriving transaction makes a lock request for a set of locks and each such lock request takes away useful processor and I/O resources from actual transaction processing to lock processing (which is an overhead). Thus the CPU and I/O resources are preforming more overhead activity as the number of transactions increases. Note that lock processing has higher priority than sub-transaction processing. The drop in throughput is more severe with FCFS. This is because, with more transactions in the system, there will be more lock requests to be processed and this simply makes a long sub-transaction even longer. As shown in the next section, this lock processing overhead increases with the number of locks in the system and has a serious impact on system performance.

4.2. Effect of Number of Locks

In this section we discuss the effect of number of locks. With the input parameters set as in Table 1, several simulation experiments were run by varying the number of locks (*ltot*) from 1 to *dbsize* = 5000 for each run. The number of transactions is fixed at *ntrans* = 200 for this set of simulation experiments. The results are presented in Figure 3.

The response time curves in Figure 3 show the tradeoff between the locking overhead and the allowed degree of concurrency. That is, increasing the number of locks initially decreases the response time by allowing concurrent execution of more transactions until a certain point after which the overhead for locking predominates without increasing the concurrency level and the response time increases. Note that as the number of locks increases the associated overhead to process a transaction lock request increases proportionately. For reasons discussed in the previous section, FCFS is more sensitive to the number of locks.

As discussed in Section 4.1, it is the blocking time that is the major component of the response time. Unlike in Section 4.1, the synchronization time (not shown) increases substantially when the number of locks is large. While all the scheduling policies exhibit increased sensitivities to larger number of locks, synchronization

---

[1] Note that this number as well as the actual peak throughput increases as the transaction size decreases - see Section 4.3.

time of FCFS increases much more quickly. The reason is that the overhead of lock processing increases in direct proportion to the number of locks.

An important observation is that the type of scheduling policy has negligible impact on the desired locking granularity (desired size of lockable entities). While all three policies support coarse granularity (dividing the entire database of 5000 entities into 10 to 100 lockable entities is sufficient to ensure best performance), the penalty for not maintaining the optimum number of locks is more for FCFS compared to that for the other two policies.

From Figures 3b and 2b we note that large number of locks and large number of transactions tend to decrease system throughput (correspondingly results in an increase in response time). The combined effect of large number of transactions with large number of locks will result in substantial increase in lock processing overhead because of the multiplicative effect (i.e., the overhead is proportional to the product of the number of transactions and the number of locks). This indicates that implementing a higher-level scheduler that essentially controls the admission rate of transactions would be very useful. This scheduler effectively rejects transactions when there are too many in the system. Thus, by rejecting these transactions, the lock processing overhead can be reduced thereby increasing system throughput (see [3] for more details). This is analogous to the "backoff" strategy in Ethernet. Similar strategies have successfully been used in multistage interconnection networks to reduce hot-spot contention [5] and for thread management in shared-memory multiprocessors [1]. Note that, if such a higher-level scheduler is employed, all three sub-transaction scheduling policies will have similar performance as discussed in Section 4.1.

4.3. Effect of Transaction Size

In this section the impact of the average transaction size is presented. Note that the number of entities required by a transaction is determined by the input parameter *maxtransize* (maximum transaction size). The transaction size is uniformly distributed over the range 1 and *maxtransize*. Thus the average transaction size is approximately *maxtransize/2*. In this set of experiments, all the input parameters are fixed as in Table 1 except for *maxtransize* which is set at 100 and 2500. This corresponds to an average transaction size of 1% and 25% of the database, respectively.

Figure 4 shows the system throughput as a function of number of transactions. These results were obtained with the number of locks set at *ltot* = 100. We make three important observations from this figure. First, the throughput is less sensitive to the number of transactions when the transaction sizes are small. This is a direct consequence of reduced lock processing overhead. Note that, for a fixed number of locks, the lock processing overhead is directly proportional to the transactions size. Second, the performance difference among the scheduling policies diminishes with increasing transaction size. When the average transaction "touches" 25% of the database, this difference tends to be marginal. The reason is that each transaction requires more locks and hence only few transactions can successfully get the required locks and proceed to the processing stage. Third, for small transactions, SJF and HPF maintain the throughput within 10% of the peak as the number of transactions increases. In contrast, FCFS is sensitive to the number of transactions even for small transactions.

Figure 5 shows the throughput as a function of number of locks. The number of transactions is fixed at *ntrans* = 200 for this set of experiments. Several observations can be made form the data presented in Figure 5. First, the observations made about the data in Figure 4 are generally valid even when the number of locks is varied. That is, for large transactions, the difference among the three scheduling algorithms is marginal. For small transactions, there is substantial improvement in system throughput associated with HPF and SJF. Note that, from Figure 4a, if the number of transactions is reduced to about 25, the difference is again marginal. Second, the optimal point (at which the throughput peaks) shifts toward the right (i.e., toward larger number of locks or finer granularity) with decreasing transaction size. For systems with smaller transaction sizes, increasing the number of locks reduces unnecessary lock request failures. For systems with larger transaction sizes, increasing the number of locks does not improve the degree of concurrency

(because more database entities are required by each transaction) whilst increasing the lock overhead substantially. However, all these optimal points have a number of locks that is less than 100. This reinforces the observation of Section 4.2 that coarse granularity is sufficient. This conclusion is valid even when there is a large variance in the transaction size (the results are not presented here).

4.4. Impact of Lock I/O Time

In previous experiments the lock I/O time (*liotime*) was set equal to the transaction I/O time (*iotime*). The next set of simulation experiments were run with *liotime* set equal to 0. Note that *liotime* = 0 models the situation in which the concurrency control mechanism maintains the lock table in main memory.

The impact of lock I/O time as a function of number of locks can be seen by comparing Figures 6a and 3b. It can be observed that when the lock table is kept in main memory, there is only a marginal difference among the three scheduling algorithms. Also, as the lock I/O time decreases, the throughput and the optimum number of locks increases. This is due to the reduced penalty associated with lock processing. In particular, when the lock table is kept in main memory, the optimum number of locks increases to 1000, implying that fine granularity improves performance marginally. Note, however, even in the case of zero lock I/O time, 10 locks are still good. Thus, coarse granularity works fine in this case as well.

The impact of number of transactions is shown in Figure 6b (also see Figure 2b). It can be noted that the above observations are generally valid. In conclusion it appears that even when the lock table is kept in main memory coarse granularity is sufficient.

4.5. Effect of Processor Scheduling Policies

This section investigates the impact of processor scheduling policies and processor service time variance on overall system performance. Note that all four scheduling policies described in Section 3 are applicable to processor scheduling. We have conducted several simulation experiments to study the impact of processor scheduling and selected results, which are typical, are presented in Table 2. The notation "I/O scheduling policy/Processor scheduling policy" is used to identify a combination of scheduling policies. Thus in Table 2 the I/O scheduling policy is fixed and the processor scheduling policy is varied. All system parameters are fixed as in Table 1. Unlike in previous sections, these results are obtained by using exponentially distributed processor service times (previous sections have used uniform distribution in the range mean ±5%). It can be seen from these results that the actual processor scheduling policy used has insignificant impact on overall system performance. This observa-

Table 2 Throughput comparison of processor scheduling policies

#locks	FCFS/FCFS	FCFS/RR	FCFS/SJF	FCFS/HPF
1	0.0682	0.0682	0.0682	0.0682
10	0.2364	0.2361	0.2364	0.2372
100	0.1254	0.1251	0.1258	0.1255
1000	0.0200	0.0198	0.0198	0.0199

#transactions	FCFS/FCFS	FCFS/RR	FCFS/SJF	FCFS/HPF
1	0.1003	0.1003	0.1003	0.1003
10	0.2718	0.2723	0.2717	0.2713
100	0.1903	0.1895	0.1901	0.1900
200	0.1254	0.1251	0.1258	0.1255

Table 3 Throughput comparison of processor service time distributions

#locks	FCFS/FCFS		FCFS/RR	
	Exponential	Uniform	Exponential	Uniform
1	0.0682	0.0684	0.0682	0.0686
10	0.2364	0.2359	0.2361	0.2368
100	0.1254	0.1251	0.1251	0.1257
1000	0.0200	0.0198	0.0198	0.0199

#trans-actions	FCFS/FCFS		FCFS/RR	
	Exponential	Uniform	Exponential	Uniform
1	0.1003	0.1013	0.1003	0.1013
10	0.2718	0.2723	0.2723	0.2732
100	0.1903	0.1893	0.1895	0.1893
200	0.1254	0.1251	0.1251	0.1251

tion is valid for other scheduling policy combinations as well as other service time variances. Table 2 show the impact of changing processor service time distribution from uniform distribution to exponential distribution. All system parameters are fixed as in Table 1. The data suggest that the processor service time variance also has insignificant impact on overall system performance.

5. CONCLUSIONS

We have analyzed the effects of scheduling algorithms on the performance of shared-nothing parallel database systems. The results show that, in I/O intensive applications, I/O scheduling policy is the most important factor in determining the overall system performance. Processor scheduling policies have insignificant effect on overall system performance.

Our results also indicate that coarse locking granularity is sufficient and this is not influenced by the type of scheduling algorithm used in the system. This is mainly due to lock processing overhead that increases in direct proportion to the number of locks in the database. This overhead increases with increasing number of transactions in the system. This suggests that a higher-level transaction scheduler that effectively "chokes" the incoming transactions depending on the system load would be effective. If such a transaction-level scheduler is employed, all three sub-transaction scheduling policies will have similar performance.

ACKNOWLEDGEMENT

We gratefully acknowledge the financial support provided by the Natural Sciences and Engineering Research Council of Canada and Carleton University.

REFERENCES

[1] T. E. Anderson, E. D. Lazowska, and H. M. Levy, "The Performance Implications of Thread Management Alternatives for Shared-Memory Multiprocessors" *IEEE Trans. Computers*, Vol. 38, No. 12, December 1989, pp. 1631-1644.

[2] S. P. Dandamudi and S. L. Au, "Locking Granularity in Multiprocessor Database Systems," *Proc. Data Eng. Conf.*, Kobe, Japan, April 1991.

[3] S. P. Dandamudi and C. Y. Chow, "Performance of Transaction Scheduling Policies for Parallel Database Systems," *Proc. Int. Conf. Distributed Computing Systems*, Arlington, Texas, May 1991

[4] D. DeWitt et al. , "GAMMA - A High Performance Backend Database Machine," *Proc. 12th VLDB Conf.*, Kyoto, Japan, 1986.

[5] W. S. Ho and D. L. Eager, "A Novel Strategy for Controlling Hot Spot Congestion," *Proc. 1989 Int. Conf. Parallel Processing*, 1989, vol. I, pp. 14-18.

[6] B. C. Jenq, B. C. Twichell, and T. Keller, "Locking Performance in a Shared Noting Parallel Database Machine," *IEEE Trans. Knowledge and Data Eng.*, Vol. 1, No. 4, December 1989, pp. 530-543.

[7] D. R. Ries and M. Stonebraker, "Effects of Locking Granularity in a Database Management Systems," *ACM Trans. Database Syst.*, Vol. 2, No. 3, September 1977, pp. 233-246.

[8] D. R. Ries and M. Stonebraker, "Locking Granularity Revisited," *ACM Trans. Database Syst.*, Vol. 4, No. 2, June 1979, pp. 210-227.

[9] M. Stonebraker, "The Case for Shared Nothing," *Database Eng.*, Vol. 9, No. 1, March 1986, pp. 4-9.

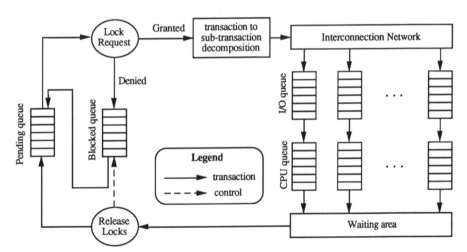

Figure 1 Simulation model

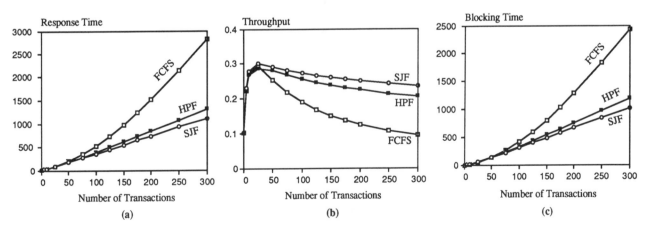

Figure 2 Performance of scheduling algorithms as a function of number of transactions

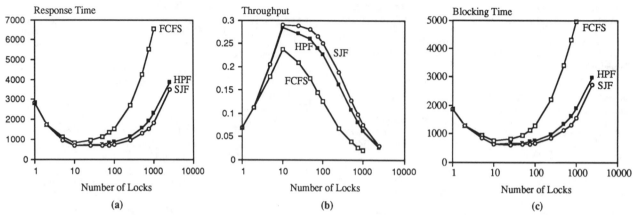

Figure 3 Performance of scheduling algorithms as a function of number of locks

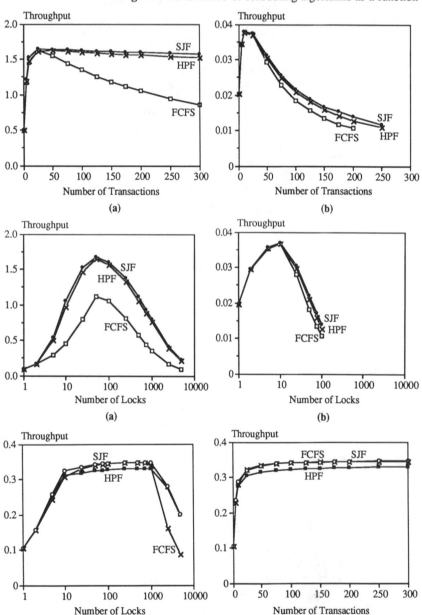

Figure 4 Performance of scheduling algorithms as a function of number of transactions (a) Small transactions (average transaction size = 1% of database) (b) Large transactions (average transaction size = 25% of database)

Figure 5 Performance of scheduling algorithms as a function of number of locks (a) Small transactions (average transaction size = 1% of database) (b) Large transactions (average transaction size = 25% of database)

Figure 6 System throughput with *liotime* = 0

ALLOCATING PARTITIONS TO TASK PRECEDENCE GRAPHS

B. Narahari and Hyeong-Ah Choi

Dept. of Electrical Engineering & Computer Science
The George Washington University
Washington,DC 20052

Abstract

This paper discusses the problem of allocating partitions of processors to nodes of a task precedence graph, where each node represents a subtask. Each subtask can be executed, in parallel, on one of a number of distinct partitions and must be assigned to one of these choices of partitions. We are given the execution time on a partition and interpartition communication time between any allocatable pair of partitions of related subtasks. The objective is to select a partition for each subtask such that the overall completion time of the entire task is minimized. This paper formulates this problem, and shows that it is NP-hard even if the maximum degree of the precedence graph is at most 3 and the number of choices for each subtask is at most 2. Polynomial time algorithms are discussed for the cases where the precedence graph is a tree, and when the maximum degree of the precedence graph is 2, thus closing the gap between the P and NP cases in terms of the degree of the graph.

1. Introduction

A partitionable parallel processing system consists of a set of processors, and common system resources, and can be configured into a number of simultaneous partitions each of which executes a task in parallel. Examples of such systems include PASM [9], the NCube system [7]. Partitionable parallel architectures are highly suitable for applications where a task may be decomposed into several parallelized subtasks, and each subtask has different processing requirements and can be performed concurrently. This property is found in many image understanding tasks. In the case where there are precedence constraints between the sibtasks, the overall completion time for the entire task depends on the partitions allocated to the subtasks, the interpartition communication time between dependent subtasks, and also on the precedence constraints between the subtasks.

This paper discusses the following partition allocation problem. We are given a task represented by a precedence graph over n subtasks, where each node of the precedence graph denotes a subtask. An edge from subtask T_i to T_j denotes that T_j cannot start execution until T_i has completed. Each subtask T_i must be assigned to one of q_i partitions, where q_i is the total number of allocatable partitions for this subtask. We assume that

any two sets of choices q_i and q_j are disjoint, *i.e.*, the partitions are disjoint and thus there is no restriction on the number of processors. This is justified in a situation where the parallel processor has allready been partitioned into Q partitions, where $Q = \sum_i q_i$ is the total number of partitions. As part of the input to the problem, we are given the number of choices for each subtask, the execution time for each of the choices and the data movement time, *i.e.*, the interpartition communication time to transfer data, between any pair of partitions of related subtasks. This time is referred to by the acronym IPC time. The objective is to choose a partition for each subtask such that the overall completion time is minimized. The completion time includes the execution times and the IPC time required to move data between partitions of related tasks. From the definition of our problem, it follows that removing the assumption/condition that the choices are disjoint reduces our problem to the standard, intractable, multiprocessor (with communication time) scheduling problem [6].

The fundamental problem of allocating and scheduling partitions to a set of n independent tasks, so as to minimize the overall completion time for all tasks, was studied in depth by Krishnamurti and Ma [8]. They showed that their problem, with the condition that the number of simultaneously scheduled tasks is bounded, is NP-hard, and they discuss an approximation algorithm In their problem, each task can be executed on a number of partition sizes; different sizes can be viewed as different choices. The problem of partitioning an interconnection networks and scheduling independent tasks has been studied in depth (for example [2]). There are some fundamental differences in the partition allocation problem studied in this paper and those studied in the literature. Specifically, they do not consider task precence graphs and they do not consider data movement time between choices of partitions. In our problem we are only focusing on determining what choices to select so as to minimize the objective function, and thus is a partition allocation/assignment problem as opposed to a partition scheduling problem.

In this paper we formulate the partition allocation problem and investigate its complexity. We show that the problem is NP-hard even if the total degree of the graph is at most 3 and the number of choices for each task is at most 2. We give a polynomial time algorithm when the degree is 2, thus closing the gap between the P and NP cases in terms of the degree of the graph. We describe a polynomial time algorithm when the precedence graph is a tree. In the following section we formulate our problem and Section 3 discusses the complexity of the problem and presents polynomial time algorithms for special cases. Section 4 provides a concluding remarks and ongoing endeavors.

2. Problem Formulation

As an input to the partition allocation problem we are given a task precedence graph, choices of partitions for each subtask, execution times on each choice of partition, and the IPC times between partition choices of dependent subtasks. Let $G_0 = (V(G_0), E(G_0))$ be the task precedence graph, where the set of vertices $V(G_0) = \{v_1, v_2, \ldots, v_n\}$ denotes the n subtasks and $E(G_0)$ is the set of directed edges denoting the precedence constraints. If $(v_i, v_j) \in E(G_0)$ then subtask v_j cannot start until v_i has completed and in addition, the edge indicates that data must be transfered from v_i to v_j. The set of choices of partitions, for each subtask, and the execution and data movement times associated with the choices can be represented by a weighted communication graph G_1. Each subtask T_i, denoted by node v_i in the precedence graph, has q_i choices of partitions that can be allocated to it. The vertex set of G_1 is defined as

$$V(G_1) = \bigcup_{1 \leq i \leq n} V_i, \quad \text{where} \quad V_i = \{v_i^1, v_i^2, \ldots, v_i^{q_i}\}$$

The vertex subset V_i corresponds to subtask T_i where v_i^p represents the p-th choice (of partition) for T_i. In other words, each node v_i in the precedence graph is replaced by q_i nodes, where each of the q_i nodes represents a choice of a partition for the subtask T_i. The edge set of G_1 is defined such that for any two sets V_i and V_j, $1 \leq i, j \leq n$, either

$E(V_i, V_j) = \emptyset$ if $(v_i, v_j) \notin E(G_0)$ or
$E(V_i, V_j) = \{(v_i^p, v_j^q) \mid 1 \leq p \leq q_i \text{ and } 1 \leq q \leq q_j\}$ if $(v_i, v_j) \in E(G_0)$,
where $E(V_i, V_j)$ represents the set of edges whose tails are in V_i and whose heads are in V_j.

For each vertex $v_i^p \in V(G_1)$ a positive number $w(v_i^p)$ represents the execution time on the p-th choice for subtask T_i. Similarly, to each edge $e = (v_i^p, v_j^q) \in E(G_1)$, there is assigned a positive number $w(e)$ representing the IPC time required to send the output of the p-th choice for T_i to the the q-th choice for T_j. In other words, the time to migrate data from partition p allocated to subtask T_i to the partition q allocated to subtask T_j. Figure 1 shows a precedence graph and the corresponding communication graph with $q_i = 2$ for all i.

Given any communication graph G_1, we define the upper bound on the number of choices in the graph, α,

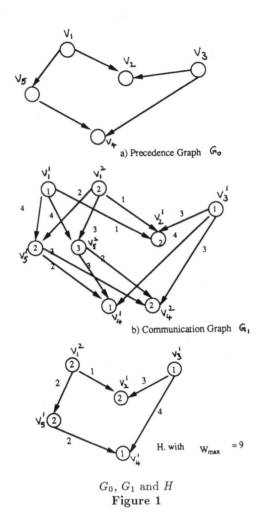

a) Precedence Graph G_0

b) Communication Graph G_1

H. with $W_{max} = 9$

G_0, G_1 and H
Figure 1

as: $\alpha = \max_{1 \leq i \leq n}\{q_i\}$. The degree of a node v, $\deg(v)$ is the number of incoming and outgoing edges at that node, and and the maximum degree of the graph G, denoted $\Delta(G)$, is defined as: $\Delta(G) = \max_{v \in V(G)}\{\deg(v)\}$.

A subgraph $H(V, E)$ of $G(V, E)$ is called an *induced subgraph* of $G(V, E)$ if $V(H) \subseteq V(G)$ and $E(H) = \{(u, v) \mid (u, v) \in E(G), u, v \in V(H)\}$ [5]. The constraint placed in the partition allocation problem is to select one choice (among the q_i partitions) for each subtask T_i, $1 \leq i \leq n$, i.e., from each node set V_i in G_1 select one node v_i^p. The set of selected nodes and the edges connecting these nodes would form an induced subgraph of G_1 that is isomorphic to the precedence graph G_0. We call such a graph an *allocation graph*. Therefore the constraint can be simply stated as: from the communication graph G_1 select an allocation graph H. The nodes and edges in H would denote the choices that are selected and their weights dictate the resulting execution and data movement times. Since the sets of choices are disjoint and can be allocated independently, an allocation of any partition v_i^p, $1 \leq i \leq n, 1 \leq p \leq q_i$, for subtask T_i will not affect the set of allowable choices for any other subtask T_l. While this is not always a realistic assumption, it does reduce the complexity; studying the problem with the removal of this assumption is part of our ongoing research. In what follows we define the completion time resulting from an allocation graph H.

For any allocation graph $H \subseteq G_1$, let P_H be a directed path in H. The sum of the weights of the vetices in the path, denoted by $W(V(P_H))$, and the sum of the weights of the edges in the path, denoted by $W(E(P_H))$, can be defined as:

$$W(V(P_H)) = \sum_{v \in V(P_H)} w(v),$$

$$W(E(P_H)) = \sum_{e \in E(P_H)} w(e).$$

The weight of the path P_H, denoted $W(P_H)$, can now be defined as:
$W(P_H) = W(V(P_H)) + W(E(P_H))$.

The completion time of the task is the weight of the "heaviest" path, i.e., the path with the largest weight denoted $W_{\max}(H)$.

$$W_{\max}(H) = \max\{W(P_H) \mid P_H \text{ is a path in } H\}$$

We can now define the partition allocation problem with the objective of minimizing overall completion time of the task, which we denote as the PA-MAX Problem, as: "Find an allocation graph H of G_1 such that $W_{\max}(H)$ is as small as possible." The decision version of this problem can be stated as:
INSTANCE: A precedence graph G_0, the communication graph G_1 and deadline time D
QUESTION: Is there an allocation graph H such that $W_{\max}(H) \leq D$.

The variable $W_{\max}(H)$ denotes the completion time of the task resulting from an allocation corresponding to the vertices in H. If we define the objective to be the sum of the vertex and edge weights in H, then our problem with this objective reduces [3] to the task assignment problem in distributed systems [4]. The allocation graph H gives the set of partitions that must be selected and scheduled to give us minimum completion time or minimum total system time, respectively, on the parallel architecture. Figure 1 gives an allocation graph for the communication graph in the example. We note that in our problem we are only focusing on determining what choices to select,i.e., determine H, so as to minimize objective function and thus can be viewed as an assignment problem as opposed to a scheduling problem.

3. Problem Complexity

This section discusses the complexity of PA-MAX and we start with an intractability result. We will show that the problem is NP-hard even for a very restricted

case where the degree of the precedence graph is at most 3 and the maximum number of choices is 2. The proof of NP-hardness follows from a reduction to the 3-SAT problem [6].

Theorem 3.1: *The PA-MAX problem is NP-hard even if $\alpha = 2$, $\Delta(G_0) = 3$ and G_0 is bipartite.*
Proof. In [3]. ∎

3.1 Polynomial Time Special Cases

Although the PA-MAX problem is NP-hard, we show that some useful special cases can be solved in polynomial time. We first consider the case when the precedence graph is a tree. We then consider the case when the maximum degree of the precedence graph is 2. Due to the constraints of space we shall briefly outline our algorithms and their complexities. For the case where the underlying graph of the precedence graph is a partial k-tree, we have shown an $O(\alpha^{k+2}n)$ algorithm in [3].

Theorem 3.2: *Given a precedence graph G_0 represented by a tree (i.e., G_0 does not have any cycle) and a communication graph G_1, there exists an $O(|E(G_1)|)$ algorithm to find an allocating graph H of G_1 such that $W_{\max}(H)$ is as small as possible.*
Proof. When G_0 is a tree, the children of a node are its predecessors and the root is the terminal node. Before we describe our algorithm, we define a function $f(v_i^j)$, for each vertex in G_1, that denotes the earliest time that j-th choice for subtask v_i can complete.

Initially, we set $f(v_i^j) = w(v_i^j)$ for all $v_i^j \in V(G_1)$ such that the corresponding vertex v_i in G_0 is a leaf. For each nonleaf node $v_i \in V(G_0)$, the value of $f(v_i^j)$ is then defined as follows. For $v_i \in V(G_0)$ let $\{v_{i_1}, v_{i_2}, ..., v_{i_p}\}$ be the set of predecessors of v_i in G_0. Note that vertex $v_i \in V(G_0)$ corresponds to subset $V_i \subseteq V(G_1)$ and vertex $v_{i_l} \in V(G_0)$ to subset $V_{i_l} \subseteq V(G_1)$ for all l, $1 \leq l \leq p$. For each $v_i^j \in V_i$, we define $f(v_i^j)$ as

$$f(v_i^j) = \max_{1 \leq l \leq p} \{ \min_{1 \leq j' \leq q_{i_l}} \{w(v_{i_l}^{j'}, v_i^j) + w(v_i^j) + f(v_{i_l}^{j'})\}\}.$$

This can be done by first computing $f(v_i^j)$ for all $v_i^j \in V(G_1)$ such that the predecessors of $v_i \in V(G_0)$ are all leaves. We then recursively compute the values of all the remaining vertices until the values of the vertices in $V(G_1)$ corresponding to the root (or terminal) vertex in $V(G_0)$ are computed. After we have completed the computation of $f(v_i^j)$ for all $v_i^j \in V(G_1)$, we have $W_{\max}(H) = \min\{f(v_r^j)|1 \leq j \leq q_r\}$, where $v_r \in V(G_0)$ is the root of G_0. In each step of the above computation, if there exists a tie, we can arbitrarily select one among them. The time complexity of our algorithm is $O(|E(G_1)|) = (\alpha^2 n)$ since each edge (u, v) is considered exactly once to compute $f(v)$. When the number of choices of partitions for each subtask is bound by a constant c, the number of edges in G_1 is $O(c^2 n)$, and thus the complexity of the algorithm is $O(n)$ where n is the number of subtasks. ∎

We next show that an optimal allocation graph of G_1 can be found in polynomial time if $\Delta(G_0) \leq 2$. This result, together with the NP-hardness result in Theorem 3.1 completely closes the gap between the polynomially solvable cases and the NP-hardness cases in terms of the maximum degree of the precedence graph.

Theorem 3.3: *Given a precedence graph G_0 with $\Delta(G_0) \leq 2$ and a communication graph G_1, there exists an $O(\alpha^3 n) = O(\alpha \mid E(G_1) \mid)$ time algorithm to find an allocating graph H of G_1 such that and $W_{\max}(H)$ is as small as possible.*

Proof. Note that if $\Delta(G_0) \leq 2$, the underlying graph of G_0 (i.e., the graph without considering the directions of edges) is either a simple path or a cycle. As the former case can be solved using the algorithm for the latter case, we assume that the underlying graph of G_0 is a cycle. Let $V_{in} = \{v \in V(G_0) \mid \text{indeg}_{G_0}(v) = 0\}$ and $V_{out} = \{v \in V(G_0) \mid \text{outdeg}_{G_0}(v) = 0\}$. Before we describe the procedure, note that each vertex $z \in V(G_0 - (V_{in} \bigcup V_{out})$ belongs to exactly one directed path in G_0 connecting from a vertex in V_{in} to a vertex in V_{out}.

> Step 1: Construct a directed graph G_2 from G_1 as follows. Let $V(G_2) = \bigcup_{1 \leq i \leq n} V_i$, where $V_i \subseteq V(G_1)$ corresponds to $v_i \in V(G_0)$ such that $v_i \in V_{in}$ or $v_i \in V_{out}$ (note that $V_i = \{v_i^1, .., v_i^{q_i}\}$). The edge set is defined such that $E(G_2) = \{(v_i^a, v_j^b) \mid 1 \leq a \leq q_i$ and $1 \leq b \leq q_j$, where there exists a directed path in G_0 from v_i to v_j . For each edge $e = (v_i^a, v_j^b) \in E(G_2)$, set $w(e)$ to be the length of a shortest path (i.e., the sum of the weights of the vertices and the edges in the path) in G_1 from v_i^a to v_j^b.
>
> Step 2: Find a cycle C_2 in the underlying graph of G_2, such that $\mid V(C_2) \bigcap V_i \mid = 1$ for each $v_i \in V_{in} \cup V_{out}$ and $\max\{w(e) \mid e \in E(C_2)\}$ is as small as possible.
>
> Step 3: From C_2 obtained in Step 2, construct the subgraph C_1 of G_1 by replacing each edge in C_2 by the corresponding path in G_1.

The subgraph C_1 constructed from C_2 establishes the desired allocation graph H of G_1. To analyze the time complexity of the above steps, we first note that Step 1 can be done in $O(\alpha \mid E(G_1) \mid)$ time. This follows from the following discussion. Let $P_0(i, j)$ be a directed path in G_0 from $v_i \in V_{in}$ to $v_j \in V_{out}$ and let $P_1(i, j)$ be a subgraph of G_1 induced on $\{v_{l_1}, v_{l_2}, \ldots, v_{l_{q_l}} \mid v_l \in V(P_0(i, j))\}$. Then compute the length of a shortest path from v_{i_a}, for $1 \leq a \leq q_i$, to each of v_{j_b} for all b, $1 \leq b \leq q_j$, applying the algorithm in the proof of Theorem 3.2 q_i times. This takes $O(q_i \mid E(P_1(i, j)) \mid) = O(\alpha \mid E(P_1(i, j)) \mid)$ time. Since each edge $e \in E(G_1)$ belongs to only one of such path $P_1(i, j)$, we can do Step 1 in $O(\alpha \mid E(G_1) \mid)$ time.

To find the cycle in Step 2, select an arbitrary vertex $v_i \in V(G_2)$ and compute the value of cycle through v_i, i.e, $\min_{1 \leq p \leq q_i}\{\max\{w(e) \mid e$ belongs to a cycle going through $v_i^p\}\}$. This can be done again using the algorithm described in the proof of Theorem 3.2 and it takes $O(q_i \mid E(G_2) \mid) = O(\alpha \mid E(G_1) \mid)$.

Finally, Step 3 can be done in $O(\mid E(G_1) \mid$ time which gives the desired time complexity. ∎

In addition to the two cases mentioned in this section, we have shown in [3], a $O(\alpha^{k+2} n)$ time algorithm for the case where the underlying graph of the precedence graph is a k-tree. k-trees, which are a subclass of chordal graphs, have been well studied and include a large family of graphs such as series parallel graphs and trees [1].

4. Conclusions

This paper formulated the and discussed the problem of allocating partitions, of processors, to nodes of a task precedence graph. We showed that the problem remains NP-complete even when the total degree of each node is less than or equal to 3 and when the number of choices, for each subtask is at most 2. Although these are discouraging results, the polynomial time special cases, for trees and degree 2 graphs, form a useful class of precedence graphs. Future efforts include investigation of approximation and heuristic algorithms and studying the problem when we need to allocate disjoint partitions but the set of choices is not disjoint.

References

[1] S. Arnborg, D.G. Corneil, and A. Proskurowski, "Complexity of finding Embeddings in a k-tree", *SIAM J. Alg. and Disc. Methods,* Vol.8, No.2, pp. 277–284, 1987.

[2] M. Chen, and K.G. Shin, "Processor Allocation in an N-Cube Multiprocessor Using Gray Codes," *IEEE Transactions on Computers,* Vol.C-36,No.12, pp. 1396–1407, December 1987.

[3] Hyeong-Ah Choi and B. Narahari, "Allocating Partitions to Task Precedence Graphs" *Technical Report.*

[4] W.W. Chu, L. J. Holloway, M. Lan, and K. Efe, "Task Allocation in Distributed Data Processing", *Computer,* pp. 57–69, November 1980.

[5] S. Even, Graph Algorithms, Comp. Science Press, 1979.

[6] M. R. Garey and D.S. Johnson, Computers and Intractability: A Guide to the Theory of NP Completeness, W.H. Freeman, San Francisco, 1979.

[7] J.P. Hayes, T.N. Mudge, Q.F. Stout, and S. Colley, "Architecture of a Hypercube Supercomputer" *Proceedings of the 1986 International Conference on Parallel Processing.*

[8] R. Krishnamurti and Y.E. Ma, "The Processor Partitioning Problem in Special-Purpose Partitionable Systems", *Proc. 1988 International Conference on Parallel Processing,* Vol. 1, pp. 434–443.

[9] H. J. Siegel, L. J. Siegel, F.C. Kemmerer, P.T. Mueller,Jr., H.E. Smalley, and S.D. Smith, "PASM : A Partitionable SIMD/MIMD System for Image Processing and Pattern Recognition," *IEEE Transactions on Computers,* Vol.C-30, No.12, December 1981.

ALGORITHMS FOR MAPPING AND PARTITIONING
CHAIN STRUCTURED PARALLEL COMPUTATIONS

Hyeong-Ah Choi and B. Narahari

Dept. of Electrical Engineering and Computer Science
The George Washington University
Washington, DC 20052

Abstract

This paper presents efficient algorithms for optimally mapping and partitioning chain structured parallel computations onto linear arrays and partitionable linear arrays. A chain structured computation is a program whose modules are configured as a chain, in other words the task graph is a linear array. We first present an $O(mp)$ algorithm for optimally mapping a chain of m modules onto p processors, thereby providing an improvement over the recent best known $O(mp \log m)$ algorithm. We then extend our technique, and provide an $O(mnp)$ algorithm to map n independent tasks, where each is a chain of m modules, to a partitionable linear array of p processors.

1. Introduction

A parallel, pipelined or distributed computation can be represented by a task graph where the nodes represent modules of the program and edges denote the communication needed between modules. The weights of the nodes denotes the computational load and the edge weight denotes the communication load. In parallel computing it is important to map a task graph onto a parallel computer such that the execution time is minimized. The execution time the program is dictated by the processor assigned the largest computational load, which also includes the time it spends on communication. The general mapping problem, also known as the task allocation problem, is known to be intractable[1]. The objective in this paper is to develop algorithms to construct exact optimal solutions for special instances of the problem where the task graph is a *chain structured computation* (CSC), *i.e.*, the task graph is a linear array.

In recent work [2], Bokhari has shown that by placing some simple constraints, optimal solutions (that satisfy the constraints) can be found in polynomial time for chain structured pipelined or parallel computations. While chain structured computations are simple and seem restrictive, they are found frequently in applications such as image processing, signal processing and scientific computing [2][8][9]. In addition, a simple solution to many problems is to pipeline the sequential modules thus resulting in a chain structured computation where the module with the slowest rate of execution dictates the speed of the system.

In this paper we present efficient polynomial time algorithms for mapping CSCs onto two models of parallel architectures supported by a linear array interconnection network. The first problem we study is that of mapping a single CSC with m modules to a parallel architecture with p processors connected as a linear array. The best known algorithm is an $O(mp \log m)$ algorithm for homogeneous processors developed by Iqbal and Bokhari [5], and in this paper we describe a carefully crafted $O(mp)$ algorithm. We then consider the problem of mapping n independent tasks, each task being a CSC of m modules, onto a multitasking partitionable linear array of p processors. We show how our earlier technique can be used to provide an $O(mnp)$ algorithm for this problem.

Our solution technique consists of defining an *assignment graph* and the optimal mapping can then be defined in terms of paths in this assignment graph. We then use a dynamic programming approach to develop efficient algorithms to find the optimum mapping. Since a linear array can be embedded with unit dilation cost in networks such as meshes and hypercubes [3], our algorithms can be used to map CSCs to these networks by labelling the nodes of these networks according to that specifed by the embedding function. The next section formulates the problem addressed in this paper.

2. Problem Formulation

A chain structured computation (CSC) is represented by a weighted graph $G = (V, E)$ where $V = \{v_1, v_2, \ldots, v_m\}$ and $E = \{(v_i, v_{i+1}) | 1 \leq i \leq m - 1\}$. The set of vertices V represents the modules, and the edges E represent communication between modules.

Each node v_i in G is assigned a weight w_i which denotes the execution time of the module on a single processor. Each edge (v_i, v_{i+1}) is assigned a weight c_i denoting the amount of communication on the edge if the two modules are mapped to adjacent processors. We assume that $c_0 = c_m = 0$ and we assume that the communication time between modules mapped to the same processor is negligible. This paper assumes that the weights of the nodes, i.e., the execution times of the modules, are non-uniform since equal weights would reduce the problem to an embedding problem. A *subchain* of $G = (V, E)$ is any set of contiguous modules, and the subchain consisting of modules $\{v_i, v_{i+1}, \ldots, v_j\}$ is denoted $< i, j >$.

A linear array of p processors is represented by a graph $G_p = (V_p, E_p)$ where $V_p = \{P_1, P_2, \ldots, P_p\}$ denotes processors and $E_p = \{(P_i, P_{i+1}) | 1 \leq i \leq p - 1\}$ denotes the communication links. We assume that all processors and all communication links are homogeneous, i.e., a module will have the same execution time on any processor. We assume that a processor cannot overlap computation and communication and communication must be synchronized. We further assume that *all* processors must be utilized.

For a task graph $G = (V, E)$ and an architecture graph $G_p = (V_p, E_p)$, a mapping π is a function from V to V_p. To reduce the solution space and thus make the problem tractable, we use the *contiguity constraint*, used first by Bokhari in [2] (and in [8] and [5]). The *contiguity constraint* placed on our mappings can be stated as: for any two modules v_i and v_{i+1}, $\pi(v_i) = \pi(v_{i+1})$ or if $\pi(v_i) = P_l$ then $\pi(v_{i+1}) = P_{l+1}$.

From the contiguity constraint above, the mapping π always maps a subchain $< i_k, j_k >$ to a processor k. Therefore the time taken by processor T_k, which is assigned the subchain

$< i_k, j_k >$, is the computational load assigned to this processor. This can be defined as: $T_k = (\sum_{l=i_k}^{j_k} w_l) + c_{i_k - 1} + c_{j_k}$. This computation load represents the 'cost' of the processor under the mapping and an efficient mapping must minimize the *bottleneck cost* which is the maximum among all processor costs. The processor with the bottleneck cost is called the *bottleneck processor* [2]. Therefore the cost $C(\pi)$ of a mapping can be defined as: $C(\pi) = \max_{k \in V_p} \{ T_k \}$. The objective in our problem is to find the optimal mapping (i.e., minimize the bottleneck cost). Since any mapping assigns modules to processors, we shall also use the word assignment to mean a mapping.

3. Mapping CSC to Linear Array

The problem of finding an optimal mapping of a CSC, which represents one task, to a linear array of processors was first studied by Bokhari [1] and an $O(m^3 p)$ algorithm was presented. In recent work an $O(m^2 p)$ algorithm [8] and an $O(mp \log m)$ algorithm [5] were presented. The algorithm in [5] uses a probe function that given a cost ω searches for a feasible mapping with a bottleneck cost ω. Our approach differs significantly in the sense that we use a dynamic programming approach to construct the mapping with the optimal bottleneck cost. In order to do this we define an *assignment graph* G^* for the problem, such that a path in G^* will correspond to a mapping and the cost of a path corresponds to the cost of the mapping.

Given a chain structured computation $G = (V, E)$, where $V = \{v_1, \ldots, v_m\}$, the assignment graph $G^* = (V^*, E^*)$ is an edge weighted graph (i.e., only edges are assigned weights) where $V^* = \{v_0\} \cup V$ and $E^* = \{(v_i, v_j) | 0 \le i < j \le m\}$. The weight of each edge is defined such that for each $e = (v_i, v_j) \in E^*$, $w(e) = (\sum_{l=i+1}^{j} w_l) + c_i + c_j$. Recall, from the problem definitions, that this is equal to the cost of a processor T_k when subchain $< i+1, j >$ is mapped to it. Figure 1 shows an example of a chain structured task and the corresponding assignment graph constructed as defined above.

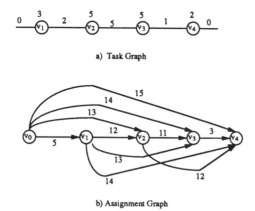

a) Task Graph

b) Assignment Graph

Assignment Graph G^*
Figure 1

Now consider a path P in G^* from v_0 to v_m using exactly p edges. This path corresponds to a mapping of nodes in V (which are the modules of the chain) to nodes in V_p (the processors), such that if (v_i, v_j) is the k-th edge in the path then

modules $v_{i+1}, v_{i+2}, \ldots, v_j$ are mapped to processor P_k. Note that this is a subchain, i.e., the subchain $< i+1, j >$ above, and the contiguity constraint is satisfied. Define the cost of the path P, denoted $C(P)$, to be maximum edge weight over all edges in P, this is also called the *bottleneck cost* in the path. This can be defined formally as:

$$C(P) = \max_{i_k} \left\{ w\big((v_{i_k}, v_{i_{k+1}})\big) \right\}.$$

Note that this is nothing but the bottleneck cost defined earlier. Now observe that to find the optimum mapping, we need to find a path using exactly p edges, from v_0 to v_m, with minimum cost (such a path is also called the bottleneck path). An example of a bottleneck path of length 3 with cost 12, in Figure 1, is the path (v_0, v_1, v_2, v_4). We now use the concepts provided by the assignment graph to develop our dynamic programming algorithm.

Let $d^k(0, i)$ denote the bottleneck cost of an optimal path, using exactly k edges, from v_0 to v_i in G^*. Therefore $d^p(0, m)$ denotes the cost $C(\pi)$ of the optimal mapping. From the definitions, $d^1(j, i)$, $j < i$, is the weight of the edge (v_j, v_i) in G^*. Any path, in G^*, using exactly k edges (i.e., of length k) consists of a path using $k - 1$ edges, from v_0 to v_j for some j, and an edge from v_j to v_m. The path from v_0 to v_j must be the optimal path using $k - 1$ edges with the least bottleneck cost among all paths using $k - 1$ edges from v_0 to v_j. Therefore from the principal of optimality we have the following relation for $d^k(0, i)$,

$$d^k(0, i) = \min_{k-1 \le j < i} \left\{ \max \left\{ d^{k-1}(0, j), d^1(j, i) \right\} \right\} \quad (1)$$

The simple dynamic programming technique defined by relation (1), can be used to construct a straightforward algorithm to compute $d^k(0, i)$ for $1 \le k \le p$, and $k \le i \le m$, and thus obtain the optimal mapping (corresponding to the solution for $d^p(0, m)$). Computing the value of $d^k(0, i)$ using the above equation clearly takes $O(m)$ steps for each i, and thus the algorithm takes $O(m^2 p)$ time. Note that initially only $d^1(0, i)$, $1 \le i \le m$, need to computed and this takes $O(m)$ time since $d^1(0, i) = d^1(0, i+1) - w_{i+1} + c_i - c_{i+1}$. Once these values are computed, each $d^1(j, i)$ can be computed only when needed in constant time by using the relation: $d^1(j, i) = d^1(0, i) - d^1(0, j) + 2c_j$. By exploiting the properties of CSCs, when they are mapped to homogeneous processors, Iqbal and Bokhari [5] provided an $O(mp \log m)$ algorithm. In what follows, we recall and use some of their observations and provide a carefully crafted algorithm that runs in $O(mp)$ time. This speedup is achieved by proving that at each iteration on k (to compute $d^k(0, i)$), we can determine all values of $d^k(0, i)$ for all i, $k \le i \le m$, by searching only $O(m)$ values in $O(m)$ time.

3.1 Improved Algorithm

A faster algorithm, based on the concept of transforming any chain to a monotonic chain, is outlined in what follows.
Def.: A chain is called a *monotonic chain* if for all i, $1 \le i \le m$, $c_i < w_{i+1} + c_{i+1}$ and $c_i < w_i + c_{i-1}$.

In [5] it is proved that any non-monotonic chain can be *condensed* to a monotonic chain of smaller size, by merging/condensing modules that violate the monotonicity condition, *without* affecting the cost of the optimal mapping. An $O(m)$ condensation procedure that condenses any chain of m modules to a monotonic chain in $O(m)$ time is described in [5]. Given the condensation theorem and the above procedure, we can therefore assume without loss of generality that *all* CSCs studied in this paper are monotonic chains. (Note that if they are not, then they can be transformed to a monotonic chain in $O(m)$ time). In what follows we first state some properties of monotonic chains, whose proofs are shown in [4], and then use these properties to modify our algorithm such that it runs in $O(mp)$ time.

Lemma 3.1: *For any monotonic chain* $G = (V, E)$ $d^1(j, i) < d^1(j - 1, i)$, *for* $1 \le j \le i - 1$ *and* $1 \le i \le m$.

Lemma 3.2: *For any monotonic chain* $G = (V, E)$, $d^1(j, i + 1) > d^1(j, i)$ *where* $0 \le j < i$ *and* $1 \le i \le m - 1$.

Lemma 3.3: *For any monotonic chain* $G = (V, E)$, $d^k(0, i) \le d^k(0, i + 1)$ *where* $1 \le k \le p$ *and* $k \le i \le m$.

The following properties, of $d^1(j, i)$ and $d^{k-1}(0, j)$, follow as a direct consequence of the above lemmas.

- $d^{k-1}(0, j)$ is a non-decreasing function of j,
- $d^1(j, i)$ for fixed i is a decreasing function of j, and
- If $i_1 > i_2$ then, for all j, the function $d^1(j, i_1)$ always lies above (i.e., is greater than) the function $d^1(j, i_2)$.

From the above we can conclude the following:

- If $d^k(0, i_1) = \max[d^{k-1}(0, j_1), d^1(j_1, i_1)]$ for some $j_1 < i_1$, and $d^k(0, i_2) = \max[d^{k-1}(0, j_2), d^1(j_2, i_2)]$ for some $j_2 < i_2$, then $j_2 \ge j_1$ if $i_2 > i_1$.

The interpretation of the above observation is that the minimum value of $d^k(0, i)$ cannot occur at a value of j less than j_1 if the minimum value of $d^k(0, i - 1)$ occured at j_1. Therefore, when we search for optimal $d^k(0, i + 1)$ we need not search for all possible assignments of j, and instead we start by $j = j_{\text{old}}$ where j_{old} (the previous value of j) was the value of j at which the minimum for $d^k(0, i)$ occured. For the total number of steps, to compute $d^k(0, i)$ for all i, to be $O(m)$ we must make sure that at each value of i we do not search *all* remaining values of j.

From definition $d^k(0, i)$ is the minimum value of the function $h(j) = \max\{f(j), g(j)\}$, where $f(j) = d^{k-1}(0, j)$ and $g(j) = d^1(j, i)$. It was shown that $f(j)$ is a non-decreasing function and $g(j)$ is a decreasing function and thus we have three cases. Firstly, if $f(j)$ and $g(j)$ intersect then the function h cannot have more than one minimum, and during the search for the minimum we need only check the one value of $h(j)$ which follows the minimum. The second case is when f is entirely above g in which case the minimum value is the first value of j (namely, $j = k - 1$). For both these cases, we can stop searching once the value of $d^k(0, i)$ starts increasing from the previous minimum value $d^k_{\text{old}}(0, i)$ (with the value of j_{old}). Finally, we can have the case and when g is entirely above f then the minimum value is the last value of j (namely, $j = i - 1$). Thus, based on the properties proved above, we conclude that at each iteration of k, all values of $d^k(0, i)$, for $k \le i \le m$, can be found in $O(m)$ steps, and hence the algorithm has complexity $O(mp)$.

4. Mapping CSCs to Partitionable Array

A partitionable parallel architecture can simultaneously execute many tasks with different computational requirements and different degrees of parallelism. The architecture can be partitioned into subsets of processors where each subset operates independently of others and computes a task. Examples of partitionable architecures include the PASM system [10], and hypercube based systems such as intel's iPSC [6]. The problem of allocating and scheduling partitions, of processors,

to minimize the earliest completion time of all tasks has been well studied and has been shown to be NP-complete [7].

In this section we consider the allocation of n chain structured tasks to a partitionable linear array parallel architecture with p processors. By placing the constraint that all tasks must be assigned at least one processor, i.e., no task is queued, we can solve the problem in $O(mnp)$ time (this constraint forces the assumption that $n \le p$). The solution to this partitioning problem must determine both: a partition of processors to be allocated to each task and the mapping of the modules of each task to processors in its partition. Since each task must be assigned an independent partition, we cannot assign modules of different tasks to the same processor and thereby cannot reduce the problem to one of mapping a single program over p processors. The objective is to minimize the time of the bottleneck processor.

A set of n tasks is defined by a set of task graphs $G_c = \{G^1 \cup G^2, \ldots, \cup G^n\}$. Each task i is represented by the graph $G^i = (V^i, E^i)$ where G^i is a CSC with m modules, such that $V^i = \{v_1^i, v_2^i, \ldots, v_m^i\}$ and $E^i = \{(v_j^i, v_{j+1}^i) | 1 \le j \le m - 1\}$. The weight of a module j of task i is denoted as w_j^i, and the weight of edge (v_j^i, v_{j+1}^i), the communication time, is denoted c_j^i. The set of chains is therefore $G_c = (V_c, E_c)$ where $V_c = \bigcup_{1 \le i \le n} V^i$ and $E_c = \cup_{1 \le i \le n} E^i$. The superscript denotes the task/chain to which a module or edge belongs.

A *contiguous partition* of k processors is defined as any set of processors $\{P_{i+1}, \ldots, P_{i+k}\}$. A mapping π must map modules of all tasks to the p processors and is defined as: $\pi : V_c \mapsto V_p$. The mapping π must satisfy two constraints: (1) the contiguity constraint and (2) the partition constraint. The *partition constraint* can be defined as: for any $\pi : V_c \mapsto V_p$, if $\pi(v_k^i) = \pi(v_l^j)$, $1 \le k, l \le m$, then $i = j$ for some $1 \le i, j \le n$.

The cost $C(\pi)$ of the mapping is defined, as before to be the bottleneck processor. Since the architecture is a linear array of homogeneous processors, we can assume without loss of generality that the partitions are adjacent, *i.e.*, the partition for G^i consists of the contiguous subchain of processors $\{P_{l+1}, \ldots, P_{l+k_i}\}$, where $l = \sum_{j=1}^{j=i-1} k_j$ (the number of contiguous processors assigned to tasks 1 through $i - 1$). The justification, whose proof is trivial, lies in the fact that given any n partitions each of size k_i we can always rearrange them such that the above property is satisfied.

4.1 The Assignment Graph and the Algorithm

Given a set of n tasks, represented by the graph $G_c = (V_c, E_c)$ we first relabel the vertices in V_c such that the node v_j^i is relabelled $u_{(i-1) \cdot m + j}$ and, letting $k = (i - 1) \cdot m + j$, the weight of u_k, denoted w_k, is the weight w_j^i of v_j^i. Thus, any node labelled $u_k \in V_c$ represents the node v_j^i (the j-th module of task i) where $j = ((k - 1) \mod m) + 1$ and $i = \lceil k/m \rceil$. In an identical manner we can relabel the edges (and edge weights) such that c_j^i is replaced by c_l. To avoid this notational confusion, one could simply view each of the n chains as consisting of nodes $\{u_{(i-1) \cdot m + 1}, \ldots, u_{i \cdot m}\}$. Figure 2(a) illustrates an example of two tasks G^1 and G^2 where the relabelling of their nodes is shown below each node; for example v_2^2 is labelled u_5 and u_3 denotes node v_3^1.

Given a set of chains G_c, with $V_c = \{u_1, \ldots, u_{mn}\}$, we define the assignment graph for this mapping problem to be an edge weighted graph $H^* = (V^*, E^*)$ where: $V^* = \{u_0\} \bigcup V_c$ and

$$E^* = \left\{ (u_i, u_j) \Big| km \le i < j \le (k + 1)m, 0 \le k \le n - 1 \right\}$$

Each edge $e = (u_i, u_j) \in E^*$ in the graph is assigned a weight $w(e)$ where $w(e)$ is defined identical to the weights defined for the assignment graph G^* in the previous problem. There is however, one major difference between the two assignment graphs; since two modules from the same task cannot be mapped to the same processor, there is no edge between nodes u_i and u_j if they are not in the same task unless u_i is the last (m-th) module of the previous task. This constraint is captured by the condition $k \cdot m \leq i < j \leq (k+1)m$ where $0 \leq k \leq n-1$, and thus if $i < km$ and $j > km$ there is no edge between u_i and u_j.

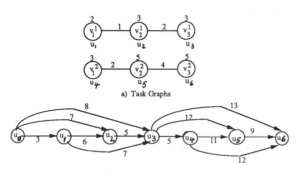

a) Task Graphs

(b) The Assignment Graph

Assignment Graph H^*

Figure 2

Now consider a path P in H^* from u_0 to u_{mn} using exactly p edges. This path corresponds to a mapping of nodes in V_c to processors in V_p such that, if (u_i, u_j) is the l-th edge in the path then subchain $< i + 1, j >$ is mapped to P_l (this situation is identical to paths defined in G^*). Note that this is a contiguous subchain. Secondly, since there is no edge in H^* between u_i and u_j if they are not in the same task, modules from different tasks cannot be mapped to the same processor thereby satisfying the constraints of the mapping. Therefore, every path from u_0 to u_{mn} corresponds to a valid mapping. In a manner identical to the earlier problem, define the cost of the path $C(P)$ to be the bottleneck cost:

To find an optimal mapping we need to find a minimum bottleneck cost path using exactly p edges from u_0 to u_{mn}; this is identical to that defined in in the previous section and we can use relation (1) to determine $d^p(0, mn)$ which corresponds to the optimal mapping. In using relation (1), we need to observe that if there is an edge (u_j, u_i) in H^* then $d^1(j, i)$ is the weight of the edge, and if there is no such edge, i.e., if $j < km$ and $i > km$ for some $1 \leq k \leq n-1$, then $d^1(j, i) = \infty$. Note that by setting a cost equal to infinity we are not allowing that assignment of modules to processors.

Directly implementing the the relation for $d^k(0, i)$ results in a $O(m^2 n^2 p)$ algorithm. An immediate improvement, in the algorithm complexity, can be made by observing that to determine $d^k(0, i)$ we need only check for the values of j where $d^1(j, i)$ is defined; this is equivalent to checking only those $O(m)$ values of j which are the indices of modules that belong to the same task or j is the index of the last module of

the previous task. Thus, it follows that the improved algorithm will only need to check for $O(m)$ values of j for each of the $O(mn)$ values of i. This implies an $O(m^2 np)$ complexity. But observe that since all chains are monotonic all m values of i which fall within a task must satisfy the monotonic properties proved in the last section. Using the search technique outlined in the improved algorithm in the last section we can find m values of $d^k(0, i)$, where $(k-1)m < i \leq km$ for $0 \leq k \leq n$, in $O(m)$ time and thus find all $O(mn)$ values of $d^k(0, i)$ in $O(mn)$ time at each iteration of k. Therefore our algorithm runs in $O(mnp)$ time, and in terms of the total size N of the input, where $N = mn$, our algorithm runs in $O(Np)$ time.

5. Conclusions

This paper presented efficient algorithms for the problem of optimally mapping chain structured parallel computations onto different models of parallel architectures which have a linear array interconnection network. We first presented an $O(mp)$ algorithm for mapping a chain of m modules onto p processors thereby providing improvements over the best known $O(mp \log m)$ algorithms. We were able to generalize the technique to map n independent tasks to a partitionable linear array of p processors in $O mnp$ time. In [4] we discuss algorithms to map chain structured computations to parallel architectures with a powerful host, where part of a task can be assigned to the host. Further research may be directed towards generalizing our technique to other classes of parallel computations.

6. References

[1] S.H. Bokhari, "On the Mapping Problem", *IEEE Transactions on Computers,* Vol.C-30, pp.207–214, March 1981.

[2] Shahid H. Bokhari, "Partitioning Problems in Parallel, Pipelined and Distributed Computing", *IEEE Trans Comp,* Vol. 37, No.1, January 1988, pp. 48–57.

[3] Tony F. Chan and Youcef Saad, "Multigrid Algorithms on the Hypercube Multiprocessor," *IEEE Transactions on Computing,* Vol C-35, No 11, 1986.

[4] H.-A. Choi and B. Narahari, "Efficient Algorithms for Mapping and Partitioning a Class of Parallel Computations", Technical Report, GWU-IIST, March 1991.

[5] M.A. Iqbal and S.H. Bokhari, "Efficient Algorithms for a class of Partitioning Problems", ICASE Report No.90-49, July 1990, NASA Contractor Report 182073.

[6] Intel Corp., *Intel iPSC System Overview,* 1986.

[7] R. Krishnamurti and Y.E. Ma, "The Processor Partitioning Problem in Special-Purpose Partitionable Systems", *Proc. 1988 International Conference on Parallel Processing,* Vol. 1, pp. 434–443.

[8] D.M. Nicol and D.R. O'Hallaron, "Improved Algorithms for Mapping Pipelined and Parallel Computations", to appear in *IEEE Trans. Computers,* 1991.

[9] D.A. Reed, L.M. Adams, and M.L. Patrick, "Stencils and Problems Partitionings: Their influences on the performance of multiple processor systems", *IEEE Trans. Comp.,* C-36, No. 7, July 1987, pp. 845–858.

[10] H. J. Siegel, L. J. Siegel, F.C. Kemmerer, P.T. Mueller, Jr., H.E. Smalley, and S.D. Smith, "PASM : A Partitionable SIMD/MIMD System for Image Processing and Pattern Recognition," *IEEE Transactions on Computers,* Vol.C-30, No.12, December 1981.

Mapping Task Trees onto a Linear Array

Dipak Ghosal [1,2]   *Amar Mukherjee* [1,3,6]   *Ramakrishna Thurimella* [4]

Yaacov Yesha [4,5]

Abstract

Assume that a bounded degree tree of n nodes is given where the nodes represent computational tasks and the tree structure represents the dependencies among tasks. We show how to map such a tree onto a linear array of processors. The size of the array depends on the height h of the tree. The parallel time and the size of the processor array used are $O(\sqrt{n}\log n)$ and $O(\sqrt{n})$ respectively for $h \leq \sqrt{n}$, and they are $O(h \log n)$ and $O(n/h)$ respectively for $h > \sqrt{n}$. It is proved that the time achieved and the number of processors used are both optimal up to a factor of $O(\log n)$.

1 Introduction

An important consideration in mapping the computational structure of a parallel algorithm onto a parallel or distributed architecture is to keep the communication overhead at a minimum while maintaining at the same time a good load balance. Most practical parallel processing architectures do not provide direct links between each pair of processors because the cost of such interconnection schemes becomes prohibitively large. Given the precedence graph of a parallel algorithm and a processor architecture, the problem of finding an efficient mapping that minimizes the computation and communication time is, in general, a very difficult problem and has been the subject of intense investigation in recent years.

In this paper, we present a class of efficient mapping algorithms for the computational structures represented by directed bounded-degree trees onto linear arrays. Trees play important roles as computational structures for several applications such as search algorithms, parsing and divide-and-conquer strategies.

The linear array processor architecture forms an impor-

tant class of regular bounded-degree interconnection structures that are suitable for VLSI or Wafer Scale Integration(WSI). In addition to the regular structure, the simplicity of the linear array interconnection provides several practical advantages over higher dimensional multiprocessor arrays. Linear arrays are easily implementable and being modular they can be expanded to accommodate increased problem size via WSI or off chip communication. More importantly, linear arrays can be easily synchronized by a clock whose rate is independent of the size of the array [FK83]. Linear arrays can be reconfigured for fault-tolerance and the number of input/output pins can be bounded independent of the size of the arrays [Ku80]. A well known example of a linear array is the WARP machine [Ku84]. The primary disadvantage of the linear array architecture is that the diameter is directly proportional to the number of processors. This implies that the interprocessor communication grows linearly with the the number of processors. As a result, designing efficient algorithms for linear arrays is difficult.

In this paper we prove that a bounded-degree tree of n nodes and height h (the maximum nubmer of nodes in any path from a leaf node to the root of the tree is h) can be mapped onto a linear array of size $O(\min \sqrt{n}, n/h)$ such that the total execution time, i.e. the time for computation and for communication, is $O(\max\{h, \sqrt{n}\} \log n)$. We also prove a time lower bound of $\Omega(\max\{h, \sqrt{n}\})$, thus establishing an optimality of our upper bound on the time up to a $\log n$ factor. We remark that our lower bound is proved for a model that is more general than the one used for the upper bound in the sense that it allows a task to be mapped to more than one processor. Note that in T time units, m processors can execute, *independent of the underlying architecture*, at most mT tasks. Therefore, $\Omega(\min\{\sqrt{n}, n/h\}/\log n)$ is a lower bound on the number of processors needed to achieve our time upper bound. Hence the number of processors used by our scheme is optimal up to a $\log n$ factor.

The formulation of the mapping problem described above is new and there are only a few prior similar results. Perhaps the work that comes closest to ours is that of embedding arbitrary *binary* trees with constant dilation on to a hypercube. See for example [Wa89], [Bo81],[BI85]. There are important differences however. The first difference is that the number of processors used by any mapping scheme that is one-to-one from tasks to processors uses $O(n)$ processors which is asymptotically inferior to $O(n/h)$. The second difference is that our algorithm gives a mapping for a bounded degree architecture as opposed to the $\log n$ degree network

---

[1]This work was initiated while these authors were with Institute for Advanced Computer Studies, The University of Maryland at College Park, College Park, MD 20742.

[2]Bell Communications Research, Red Bank, NJ 07701.

[3]Department of Computer Science, The University of Central Florida, Orlando, FL.

[4]Institute for Advanced Computer Studies, The University of Maryland at College Park, College Park, MD 20742.

[5]Also, Department of Computer Science, The University of Maryland Baltimore County, Baltimore, MD 21228.

[6]Also on sabbatical at Johns Hopkins University, Baltimore, MD.

that is usually assumed for the embedding results.

We note that in the special case that the task tree is an expression-evaluation tree with associative operations, the problem is popularly known by the name "tree contraction". Making assumptions about the operations in an expression lets one restructure the task tree. In other words, the dependencies are "less rigid" because one can form an expression with a few variables by substituting variables for the values of nodes whose values are unknown at the time of evaluation. Consequently, there is a vast amount of literature on this problem, including several parallel algorithms that achieve $O(\log n)$ parallel time. Our problem, however is different, in the sense that we do not assume that the task tree is an expression evaluation tree.

Others [Ch88], [HMR83] have considered embedding various topologies on the hypercube. Papadimitriou and Ullman [PU84] consider a computation represented by an $n \times n$ 'diamond' DAG and prove a lower bound of $\Omega(n^3)$ on the product of communication and computation time. The model considered in [PU84] allows a task to be mapped to more than one processor.

2   Preliminaries

A bounded degree tree of n nodes and height h is given. Figure 1a shows an example tree with $n = 80$, $h = 9$ and maximum in-degree 4. The nodes represent computational tasks, and the tree structure represents dependencies among tasks.

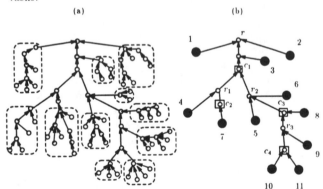

(a)   (b)

Figure 1 : a) A task tree T with $n = 80$ tasks, height 9 and maximum degree 4, and b) its intermediate representation where each subtree in the dashed box of the left figure is replaced by a single gray node. The subtree corresponding to the gray node numbered i, $1 \leq i \leq 11$, has $\langle$size, height$\rangle$ respectively $\langle 8, 5 \rangle$, $\langle 9, 4 \rangle$, $\langle 5, 3 \rangle$, $\langle 7, 4 \rangle$, $\langle 8, 4 \rangle$, $\langle 3, 2 \rangle$, $\langle 9, 4 \rangle$, $\langle 6, 4 \rangle$, $\langle 7, 3 \rangle$, $\langle 4, 2 \rangle$, $\langle 5, 2 \rangle$.

Tasks are mapped to processors of a linear array. For a given mapping, the processors perform computation subject to the following rules:

1. Computation is synchronized.

2. In one unit of time, each processor can either execute a task that is mapped to it, or transmit one value that is available to it to one of its neighbors.

3. No more than one value may be transmitted over the same link at the same time unit.

4. After a task is executed, its value is available to the processor to which it is mapped.

5. If a value is transmitted to a processor, it becomes available to it.

and an additional rule that reflects the dependency structure represented by the tree:

6. A non-leaf task cannot be executed before the values of all its children are available to the processor to which that task is mapped.

An extension of the above model allows mapping each task to more than one processor. If a task is mapped to a processor, we say that that processor contains a copy of that task. The above rules are modified as follows: A processor may execute a copy of a task that it contains (rather than simply execute a task that is mapped to it). Also Rule 6 should be changed to: A copy of a non leaf task cannot be executed before the values of copies of all its children are available to the processor that contains it. The time of the schedule is redefined to be the number of time units that pass until at least one copy of each task is executed.

For a given schedule of computation and communication as above, the time of the schedule T is defined as the number of time units needed to execute all the tasks. We aim to achieve

Objective 1: time T that is close to optimal, and

Objective 2: number of processors P close to the minimum required to achieve time T.

More formally, the above objectives can be stated as follows. A time T is called optimal if it is *minimum* (up to a constant factor) over all mappings and schedules. For a given time T that is needed to evaluate a tree, a schedule S is called *processor-optimal* with respect to T if the number of processors used is *minimum* (up to a constant factor), over all mapping and schedules that achieve time T.

Note that the mappings in the literature that use n processors could not achieve Objective 2 within $\log n$ factor, unless $n = O((n/T)\log n)$, which means $T = O(\log n)$. So if a tree has height, say n^ϵ ($\epsilon > 0$) then n processors is far from optimal.

In case of a complete binary tree of n nodes and height $\log n$, there appear to be several ways to obtain a mapping to a linear array such that the above mentioned objectives can be met, i.e. the tree can be mapped on to an array of size $\sqrt{n}$ such that a time of $O(\sqrt{n})$ can be achieved. For example, note that there are $\sqrt{n}$ subtrees rooted at level $\lceil h/2 \rceil$ each of size $O(\sqrt{n})$. By mapping the each subtree to one processor, we can evaluate the whole tree in $O(\sqrt{n})$ time. It turns out (as proved in Section 4.2) that this is the best time possible for the linear array architecture. Intuitively, the reason is that to achieve a better time we need more processors, but increasing the number of processors on a linear array increases the diameter of the array making it inefficient for the processors to communicate. That is, the best balance is achieved between the computation time and the communication time when the size of the array is $\sqrt{n}$. In general, as shown in the subsequent sections a number of processors equal to $\min(n/h, \lceil \sqrt{n} \rceil)$ gives the best total time.

Henceforth, for a tree of height h, let $\max(h, \lceil \sqrt{n} \rceil)$ be denoted by h'. Assume that by a path (resp., tree), we mean a directed path (resp., directed tree with edges directed towards the root). It is easy to show that every tree T with n nodes has a node v whose removal disconnects the tree into smaller trees such that the number of nodes in each smaller tree is $\leq n/2$. v is referred to as a *centroid* of the tree. Consider the following argument that shows the existence of a centroid. Start from the root r. If r is a centroid, then we are done. Otherwise, there is a child of r whose subtree contains $> n/2$ nodes; move to that child. Now, either the current node is a centroid or it has a child whose subtree contains more than $n/2$ nodes; if the current node is not a centroid, move to that child. Repeat this search until a centroid is found. As the tree has a finite depth, we are sure to find a node that satisfies the requirements for a centroid. It is easy to show that there can be more than one vertex that satisfies the requirements for a centroid, i.e. there need not be a unique centroid for a tree.

3 Mapping and Execution

3.1 Partitioning the Task Tree

Using a centroid, the task tree is recursively partitioned to obtain an efficient mapping. In the following, we define a height-compressed tree T_{short} based on these partitions such that T_{short} can also be viewed as a task tree but one whose height is only $O(\log n)$ *.

We define T_{short} in terms of another tree which we refer to as T_{temp}.

T_{temp} is defined recursively below. In this tree each node represents either a subtree of T or a path in T:

1. The *root* node r represents T.
2. A node v representing a subtree T' is a *leaf* if the number of nodes of T' is $\leq h'$ (recall that $h' = \max(h, \lceil \sqrt{n} \rceil)$). Otherwise, let a be a centroid of T'. The node v has two types of children : One corresponding to each of

---

*All logarithms are to the base 2 unless explicitly stated otherwise.

the subtrees resulting from the removal of the path $\mathcal{P}$ from a to the root of T', and one for $\mathcal{P}$.

Figure 2 shows the tree T_{temp} and T_{short} for the tree T shown in Figure 1.

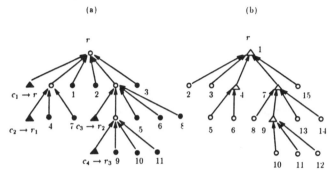

Figure 2 : a) T_{temp}, and b) T_{short} for T of Figure 1. These trees are obtained by using the centroids c_1, c_2, c_3 and c_4 marked in square boxes in Figure 1. The labels next to the nodes of T_{short} correspond to the labels of a preorder traversal. Each node of T_{short} corresponds to either a path or a subtree of T. The node of T_{short} with preorder label i, $1 \leq i \leq 15$, has respectively $3, 8, 9, 2, 7, 9, 2, 8, 2, 7, 4, 5, 6, 3, 5$ tasks.

For convenience, the nodes of T_{temp} that represent subtrees (paths) are marked as circles (triangles). Several observations about T_{temp} are in order at this point. The height of T_{temp} is at most $O(\log(n/h'))$. Each circular node represents tasks that are *fully evaluatable*, i.e., they do not depend on any tasks other than those in the circle. Each leaf, whether triangular or circular, represents no more than h' tasks. Note that the number of components resulting from the removal of a vertex v is equal to the degree of v in T. Therefore, the maximum degree of T_{temp} is one more than the maximum degree of T.

Define a partial order on the leaves of T_{temp} as follows. Leaf a is less than leaf b if there exists a' and b' such that a' (resp. b') belongs to the set of nodes of T represented by a (resp. b) and there is a directed path in T from a' to b'. It is easy to see that this partial order is well defined and that it can be represented by a directed tree of height $\leq \lceil \log(n/h) \rceil$. This tree is T_{short}. Refer to Figure 2. It can be easily seen that T_{short} can be easily obtained from T_{temp} by replacing each light circular node x by a triangular node and removing the triangular child of x. Hence, the maximum degree of T_{short} is the same as the maximum degree of T. T_{short} is an important structure that we use heavily in the rest of the paper.

It is easily verified that the leaves (internal nodes) of T_{short} are the circular (triangular) leaves of T_{temp}. The tasks represented by the leaves of T_{temp} (the triangular and circular leaves together) are a partition of the tasks of T.

3.2 Mapping

Assume that the processors are numbered $P_1, P_2, \ldots, P_k$ in sequence in the left to right order where $k = \lceil n/h' \rceil$. Each processor is assigned less than $2h'$ tasks as shown below. We will say that a processor is *available* if it has been assigned less than h' tasks. A preorder traversal is performed on T_{short} and the tasks represented by a node of T_{short} are assigned to that processor which has the least index among the available processors. As each node of T_{short} represents no more than h' tasks, each processor gets less than $2h'$ tasks. Using the above scheme, the assignment of nodes of T_{short} in Figure 2 on linear array of size 9 is shown in Figure 3.

3.3 Execution

Evaluating a node of T_{short} is equivalent to evaluating a subtree of T. In particular, when the root of T_{short} is evaluated, T is completely evaluated. Notice that each node of T_{short} is either a path of T or a fully evaluatable subtree of T. Therefore, when a node v of T_{short} is being evaluated there is a unique task of T which can be evaluated only at the end. The value of the node v of T_{short} is defined as the value of that unique task.

The execution proceeds as shown below. Let l be the height of T_{short}. Recall that $l \leq \lceil \log(n/h') \rceil$. In the following, a node of T_{short} is said to be at level j if it's distance to the root of T_{short} is $j - 1$. Note that the root is at level 1.

For $j = l$ *down to 1* **do**

    1. **In Parallel** *evaluate all nodes of T_{short} at level j.*

    2. *For each evaluated node v, let w be the parent of v in T_{short}:*

        Route the result of evaluating v to the processor containing w.

We give below the trace of the execution for the task tree of Figure 1. The routing algorithm for communicating values follows basically pipelining and is explained in detail at the beginning of Section 4.2.

Figure 3 : The Execution Trace. Note that the numbers in the box represent the label of the nodes in T_{short} as shown in Figure 2. Boxes from left to right represent processors $P_1, P_2, \ldots, P_9$ respectively.

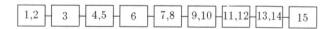

Step 1 : Evaluate Nodes 10, 11 and 12.

Computation 9 time units.

Communication 2 time units for sending 11 and 12 from P_7, P_7 to P_6, P_6 respectively.

Step 2 : Evaluate Nodes 5,6,8,9,13 and 14

Computation 9 time units.

Communication 4 time units for sending 6, 13 and 14 from P_4, P_8, P_8 to P_3, P_5, P_5 respectively.

Step 3 : Evaluate Nodes 2,3,4,7 and 15.

Computation 9 time units.

Communication 8 time units for sending 3, 4 and 15 from P_2, P_3, P_9 to P_1, P_1, P_1 respectively.

Step 4 : Evaluate Node 1.

Computation 3 time units.

Communication none.

The total execution time consists of 30 units of computation and 14 units of communication.

4 Analysis

4.1 Computation

Each processor contains $\leq 2h'$ tasks. Hence each level of T_{short} can be evaluated in $\leq 2h'$ time. As there are at most $\lceil \log(n/h') \rceil$ levels in T_{short}, the computation time is $\leq 2h' \log n$.

4.2 Communication

An upper bound on the total communication time is stated in Theorem 1. The proof of Theorem 1 is based on certain properties which characterize the communication between the processors. These properties are formally stated in the following lemmas. The proofs of all the subsequent lemmas and theorems are omitted due to lack of space (see [GMTY91] for details).

Lemma 1 (Pipelining) *Routing a total of p values to the same destination from various sources over a processor distance of $\leq q$ on a linear array can be done in $\leq p + q$ steps.*

Lemma 2 (Link-disjointness) *If the destinations for two values are u and v, where u is neither an ancestor nor a descendant of v in T_{short}, then the paths taken by these values are link-disjoint.*

Lemma 3 *(Routing distances increase geometrically)*
The processor-distance traveled by a value during iteration j, where j ranges from the height of T_{short} down to 1, in Step 2 of the For loop is $\leq \lceil (n/2^{j-1})/h' \rceil$.

Theorem 1 (Upper bound on total communication) *The total time for communication is $O(h' \cdot d \cdot \log(n/h') + n/h')$ where d is the maximum degree of T.*

4.3 Bounds on Total Time and Number of Processors

For each choice of $h' \geq h$, the communication plus the computation time is seen to be $O(h' \cdot d \log(n) + (n/h'))$ while we use $\lceil n/h' \rceil = O(\min(\sqrt{n}, n/h))$ processors. Since $h' = \max(h, \lceil \sqrt{n} \rceil)$, the expression for total execution time is $O(\max(h, \sqrt{n}) \log n)$. In the next section we will prove that the total time achieved by our mapping and schedule of execution is optimal up to a $\log n$ factor.

As shown in Section 1, the number of processors is optimal up to a $\log n$ factor.

4.4 Optimality up to $\log(n)$ Factor (Lower Bound)

Theorem 2 *Any mapping requires time $\Omega(\max(\sqrt{n}, h))$ even when a task is allowed to be mapped to more than one processor.*

5 Conclusion

In this paper we presented a scheme for mapping task trees onto a linear array. The size of the linear array used depends on the height h of the tree. When $h \leq \sqrt{n}$, our mapping scheme achieves a parallel time $O(\sqrt{n} \log n)$ using $O(\sqrt{n})$ processors and when $h > \sqrt{n}$ the parallel time is $O(h \log n)$ using $O(n/h)$ processors. We proved that the time achieved and the number of processors used are both optimal up to a factor of $O(\log n)$.

There are a number of interesting generalizations to this problem. One of the problem of the linear array structure is that the diameter is directly proportional to the number of processors. This restricts the number of processors that can be used. For a fixed k-dimensional array the diameter is $O(n^{1/k})$. As a result one could use a larger number of processors to achieve a better execution time. The mapping mentioned in Section 2 for a complete binary tree of n nodes and height $\log n$ can be modified for higher dimensional arrays such that the time and the processor-time product are both optimal. But this method breaks down if the task tree to be evaluated is not complete binary tree. It is interesting to see if there are efficient ways to map an arbitrary task

tree on to k-dimensional arrays for $k > 1$. Generalizations of the task tree such as non-uniform execution time of the nodes and weighted edge reflecting non-uniform communication requirement should also be addressed.

References

[BI85]   S.N. Bhatt and I.C. Ipsen, How to embed trees into hypercubes, Tech. Rep. YALEU/DCS/RR-443, Yale University, Dec. 1985.

[Bo81]   S.H. Bokhari, On the mapping problem, *IEEE Trans. on Computers*, 30, 1981, pp. 550-557.

[Ch88]   M. Chan, Dilation-2 embedding of grids into Hypercubes, *Proc. Int. Conf. on Parallel Processing*, 1988, pp. 295-298.

[FK83]   A.L.Fisher and H.T.Kung, Synchronizing Large VLSI Array Processor, Proc. of *10th Annual IEEE/ACM Symposium on Computer Architecture*,1983, pp.54-58.

[GMTY91]   D. Ghosal, A. Mukherjee, R. Thurimella, and Y. Yesha, Mapping Task Trees onto a Linear Array, *Bellcore TM-NWT-018949*, Bell Communications Research, Red Bank, NJ 07701, 1991.

[HMR83]   J. Hong, K. Melhorn, and A.L. Rosenberg, Cost trade-offs in graph embeddings, with applications, *JACM*, 30, 1983, pp. 709-728.

[Ku80]   H.T. Kung, Why Systolic Architectures, *IEEE Computer*, 15(1), 1980 (Jan.),pp.37-46.

[Ku84]   H. T. Kung, Systolic algorithms for the CMU WARP processor, Proc. *7th Int. Conf. on Pattern Recognition*, July 1984, pp. 570-577.

[PU84]   C.H. Papadimitriou and J.D. Ullman, A communication time tradeoff, *SIAM J. of Computing*, **16** (4), 1987, pp. 639-646.

[Ul84]   J.D. Ullman, *Computational Aspects of VLSI*, Computer Science Press, 1984.

[Wa89]   A. Wagner, Embedding arbitrary binary trees in a Hypercube, *J. of Parallel and Distributed Computing*, 7, 1989, pp. 503-520.

AP1000 ARCHITECTURE AND PERFORMANCE OF LU DECOMPOSITION

Takeshi Horie, Hiroaki Ishihata, Toshiyuki Shimizu,
Sadayuki Kato, Satoshi Inano, and Morio Ikesaka
Fujitsu Laboratories LTD.
1015 Kamikodanaka, Nakahara-ku,Kawasaki 211, Japan

Abstract

The AP1000 is a highly parallel computer with distributed memory. The design goal for the AP1000 is to execute fine-grain data-parallel programs efficiently. In this paper, we present the AP1000 architecture and the performance of LU decomposition to show that our architecture is suitable for fine-grain data-parallelism.

1   Introduction

The AP1000 is a distributed-memory, message-passing, parallel computer. The design goal for the AP1000 is to execute fine-grain data-parallel programs efficiently. The design concepts of the AP1000 are as follows:

Low latency of communication The delay of point-to-point communication and 1-N (broadcast) communication must be reduced to execute data-parallel processing. This time must be less than 10 to 20 microseconds. It is noted that we must reduce not only the latency of the network but also message handling overhead both at sending and receiving. To minimize this overhead, iWarp [1] supports systolic communication. However, with respect to the flexibility of a cell's programs, we consider that memory communication is better than systolic communication and that the overhead for memory communication must be minimized.

High throughput communication The network must be able to handle a lot of messages per unit time. When the network is congested, network performance must not be affected. Some new routing schemes have been proposed such as wormhole [2] and cut-through routing [3]. However, these schemes have poor throughput of communications.

Fast barrier synchronization A barrier synchronization mechanism is quite effective for implementing the synchronous algorithm in which the application iterates the segments. However, several message-passing machines such as iPSC [4] and nCUBE [5] have no special hardware for barrier synchronization.

Efficient data distribution and collection In many parallel programs, data must be distributed from the host computer to cells before calculation. After calculation, the results dispersed in cells must be collected by the host computer. In both cases, the host computer must interact with many cells.

The remainder of this paper describes the AP1000 architecture and the performance of the LU decomposition. Section 2

gives an overview of the AP1000. Section 3 compares the performance of the LU decomposition on the AP1000 with that of a conventional message-passing machine.

2   AP1000 Architecture

2.1   System Configuration

The AP1000 consists of 16 to 1024 processing elements, called cells, connected by three independent networks called the torus network (T-Net), the broadcast network (B-Net), and the synchronization network (S-Net). A message controller (MSC) interfaces a processing element and networks.

Figure 1 shows the AP1000 architecture.

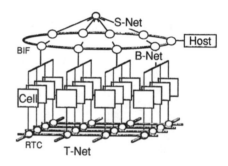

Figure 1: AP1000 Configuration

2.2   MSC

The MSC minimizes message transfer setup overhead for both sending and receiving. The MSC has the following functions:

Line sending Line sending sends the messages in cache memory to the networks. Even if there is no specified data in cache memory, direct memory access will send them to the networks. The cpu can inject messages into the network using the write operations. This approach to message sending has two advantages. First, there is no memory transfer to compose messages in memory. Second, there is no delay to set up DMA control registers.

Buffer Receipt The message buffers are implemented as circular buffers. The MSC allocates buffer space and transfers messages into that area. This approach has no interruption when the message is received, and a program can execute it in parallel with message reception.

2.3 T-Net

The T-Net has a two-dimensional torus topology. We proposed a new routing scheme for the T-Net which incorporates the structured buffer pool algorithm into the wormhole routing. This scheme provides low latency, high throughput communication, and deadlock avoidance. The routing controller(RTC) handles the flow control and the routing to perform communications between any cells. In addition to point-to-point communications, the RTC has a broadcast function. Each channel of the RTC is 16 bits and has a transfer rate of 25 Mbytes/s. The delay at the intermediate node is 80 ns.

2.4 B-Net and S-Net

The topology of the B-Net is a compound network including hierarchical common buses and a ring structure network to build a system with up to 1024 cells. The B-Net consists of a 32-bit bus with a transfer rate of 50 Mbytes/s. The B-Net is used for 1-N communication such as data broadcast. The B-Net also performs data distribution and collections.

The S-Net has a tree topology, and is used for barrier synchronization and status checking. There are 40 independent events for each. The delay of the barrier synchronization is 1.6 microseconds.

2.5 Cell Configuration

Each cell consists of the SPARC IU and FPU, cache memory, the MSC, DRAM, and the network devices. The SPARC operates at 25 MHz. There is 128 KB of cache memory. The BIF has been developed to implement the S-Net and B-Net. There are 16 Mbytes of DRAM. The MSC, RTC, and DRAM are connected by a 32-bit internal bus which is synchronous.

3 LU Decomposition and Performance

3.1 LU Decomposition

We examine the LU decomposition method for obtaining the solution of linear equations, $Ax = b$. The solution consists of the following three stages: (1) LU decomposition with partial pivoting, (2) forward substitution, (3) backward substitution. We chose a scattered distribution for data. This is because a scattered distribution is suitable for minimizing load imbalance and the hot spot of communications.

3.2 Performance

We have already developed a system with 256 (8 × 8) cells. We compared the performance on the AP1000 with that using a conventional architecture.

The AP1000 has special hardware for message communications. The line sending and buffer receipt of the MSC are used to transfer messages between cells. For both the broadcast and the point-to-point communications the T-Net is used. To identify the messages to be received, a tag is attached to each message. When the IU receives the messages, the IU searches the circular buffers for the specified tag.

A conventional message-passing machine has no special hardware for message handling. When the message reaches the destination cell, the message interrupts the IU and the IU sets up the DMA to copy the message to memory. When the IU sends the message, the IU copies the message into the message buffer and sets up the DMA to transfer the message into the network. These are implemented on the AP1000 using the T-Net and ordinary DMA functions of the MSC.

Figure 2 shows the the performance of parallel LU decomposition on the AP1000 and on a conventional parallel machine. For any matrices the performance of the AP1000 is better than that of conventional machine. As the number of cells increases, the difference of performance becomes larger. This is because the amount of calculation becomes smaller and high communication overhead degrades the performance. For applications such as matrix operations, the latency of communications is the most important factor for parallel machines.

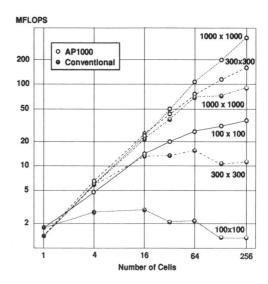

Figure 2: Performance on AP1000

4 Conclusion

We presented the architecture of the AP1000 in detail. We developed the parallel LU decomposition program and compared its performance on the AP1000 with that on a conventional machine. The result showed that the AP1000 can achieve higher performance at fine-grain parallelism than the conventional message-passing computer.

References

[1] S.Borkar, et al., *Supporting Systolic and Memory Communication in iWarp*, 17th ISCA, 1990, pp.70-81.

[2] W. J. Dally, et al., *Deadlock free message routing in multiprocessor interconnection network*, IEEE Trans. Comp., 1987, C-36, (5), pp.547-553.

[3] P.Kermani, et al., *Virtual Cut-Through: A New Computer Communication Switching Technique*, Computer Networks 3, 1979, pp. 267-286.

[4] D.S.Scott, et al., *The Scalability of Intel Concurrent supercomputers*, 3rd ICS, 1988, pp.121-124.

[5] J.F.Palmer, *The NCUBE Family of Parallel Supercomputers*, ICCD-86, 1986, p.107.

PERFORMANCE MODELING OF QUARTET MULTIPROCESSOR KERNELS WITH FAILURES

Nikolaos G. Bourbakis,IBM,S.Jose,CA
Fotios N. Barlos, GMU,VA

ABSTRACT

In this paper the performance modeling of two multiprocessor quartet kernels used on the HERMES system architecture is presented briefly. HERMES is a multiprocessor, heterogeneous vision system architecture consisted of $(N/2)x(N/2)$ processor nodes in a 2-D array configuration. Since the overall configuration of the HERMES system is based on the repetition of the quartet (M-M, B-B) kernels, the evaluation of the kernels' performance is important for the overall evaluation of HERMES syste. Special emphasis is given on the performance of these two kernels when failures occur on the processor nodes.

This work is a part of the HERMES research project

1. INTRODUCTION

Performance analysis has always been an important part for the design and implementation of multiprocessor system architectures. Various methods have been used to evaluate the performance of multiprocessor structures. One of them is the analytic modeling (deterministic, probabilistic) [1,2,3].
In this paper, we discuss the performance evaluation of two multiprocessor kernels (Mememory-to-Memory "M-M" and Bus-to-Bus "B-B") by using the probabilistic modeling. The M-M and B-B kernels are used for the design of a multiprocessor vision system architecture called HERMES. HERMES is, a multibit, multiprocessor system architecture consisted of $(N/2^i)x(N/2^i)$ processor-nodes in a 2-D array configuration, where NxN is the size of the input image and "i" is a resolution parameter. Since the HERMES system functions in a parallel-hierarchical (bottom-up and top-down) manner in various levels, we study the performance of these two kernels by using three consecutive functional levels. In particular, we consider that the main M-M or B-B kernel consists of four adjacent nodes, where one of them "plays" also the role of the father-node for the others.
In our study we consider also one more node, which 'plays' the role of the grant-father. Thus, the examined kernel includes nodes which functions in three consecutive levels. The choice of this type of kernel's functionality models the operation of HERMES nodes at any functional level. In this paper special emphasis is given on the performance evaluation of these two multiprocessor kernels when failures occur on the processor nodes.
This paper is organized into five main parts. The second part provides a brief description of the HERMES system. The third part deals with the performance evaluation schemes of the examined kernel. The fourth part discusses the performance of kernels with failures and the last part concludes the overall presentation.

2. THE "HERMES" SYSTEM

In this section we provide a brief presentation of the HERMES system assisting the reader to understand the concepts discussed in later sections. The HERMES system is an autonomous, heterogeneous, multibit, multiprocessor architecture [4,5]. The general configuration of HERMES is shown in fig.1. The HERMES hybrid structure includes two main components: a 2-D photoarray of NxN cells and the main 2-D array of processors. HERMES receives the images directly from the environment, using the photoarray, and processes them in a parallel-hierarchical (top-down and bottom-up) asynchronous manner. In particular, "orders" go down and "abstracted" image information goes up along the HERMES hierarchy. Each processor node receives directly the image pixels from its own dedicated

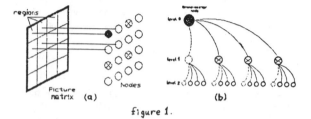

figure 1.

square region, see fig.1, processes them and provides the results upwards to its own father-node. The father node selects all the information provided by its son-nodes, integrates it and provides the results to its own father-node (grant-father-node). The same process is repeated until the full master node receives the accumulated information from its own son-nodes.
Since the overall structural and functional design of the HERMES system depends on the structure and operation of the quartet kernel, we present here the two interconnected kernels as shown in fig.2. In particular, in the B-B scheme each processor-son-node communicates with its own father-node through a common bus. In the M-M scheme each son-node communicates with its father-node through its own dedicated bus and its dedicated memory portion in the father-node's storage memory. The M-M scheme allows communication of the three son-nodes with their father-node simultaneously.

3. TRANSMISSION TYPES ON THE QUARTET KERNELS

The study of the performance analysis of the two interconnected kernels requires the expanding of the quartet kernel into a new kernel of five nodes (a grant-father-node,"GFN",a father-node,"FN",and three son-nodes "SNi",i=1,2,3). The new kernel is called "expanded kernel". The operations performed on the expanded kernel are schematically depicted in the table 1. Where the "arrow" (--->) represents transmission of "commands" from a node which functions at an upper level to a node which functions at the immediate lower level. The "arrow" (<---) represents transmission of abstracted image information upwards.

TABLE 1
=============

Transmission type (i=1,2,3)	Communication Schemes	
	M-M	B-B
GFN --> FN --> SNi	-	I
GFN --> FN <-- SNi	M,I	B,I
GFN <-- FN --> SNi	-	-
GFN <-- FN <-- SNi	M	B

M : Performance degradation due to a full memory portion.
B : Performance degratation due to an occupied bus.
I : Performance degratation due to an interrupt issued from GFN to FN.
- : Future performance degratation.

In our study we estimate the average number of active sons (SNi,i=1,2,3) during a transmission process. This is called "processing power". We also calculate the average inactive time of a son node.

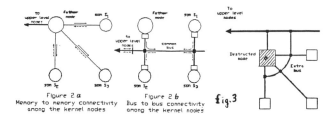

Figure 2.a
Memory to memory connectivity
among the kernel nodes

Figure 2.b
Bus to bus connectivity
among the kernel nodes

fig.3

4. PERFORMANCE ON KERNELS WITH FAILURES

In a previous work, we had study the performance of the multiprocessor
M-M and B-B kernels under the assumption that their operation was regular without failures or any other type of communication problems [5].
In this section we study the performance of the same kernels under the
occurance of failures on them.

Definition: A failure (f) on a multiprocessor kernel (K) is defined as
the inability of a number of processor nodes (XNs), XN∈{K},
to interact appropriately with the rest nodes YNs ∈ {{K}-{XNs}} under a
predefined protocol (p).

where, XN,YN ∈ {GFN, FN, SNi}, {K} represents the nodes of the kernel K.

Two different type of failures are discussed here:
(i) Son node failure f(SNi) and (ii) Father node failure f(FN)
For each of these two types of failure we answer the following three
questions: (1) What is the loss of the image information? (2) How the
failure is recognized by the system? (3) Can the system recover from
failure and how?

4.1. FAILURE OF SON NODE (SNi)
The failure f(SNi) is the simplest case between the types of failures.
If a failure f(SNi) occurs, this means that a processor node SNi is
unable to communicate with any other node in the kernel configuration.
Since each node SNi does not have any successor node, this type of
failure does not affect seriously the operation of the other nodes.
The loss of image information in this case is 25% of the information
carried in the quartet kernel. The failure can be sensed by the father
node, when it does not receive any more replies from the destructed
node. The system can recover the loss of the image information after
an appropriate process by using the rest available information either
at this particular kernel or at a higher level of operation. From
hardware point of view there is not automatic recover. The only reprimand is the change of the collapsed processor node with a new one if
possible.

4.2. FAILURE OF FATHER NODE (FN)
The failure f(FN) means that a father node FN is unable to communicate
with the rest nodes of the expanded kernel. Since there are two different interconnected schemes (M-M, B-B), the failure of the father node
generate probably different software and hardware problems. In particular, in the B-B scheme the loss of information is 100% for this kernel
if there is not software recovery process. The failure can be sensed by
the grantfather node. If however, there is a recovery process then the
the grantfather node will receive the information of the three grantsons (SNi). In this case the loss of information is 25% of the quartet
kernel. The recovery process requires extra control of the common bus
by the grantfather node. In case that the scheme is M-M then the loss
of information is 100% if there is not recovery (hardware and software)
process. If however there is a recovery, which means that there is an
extra bus line which crosses the sons' bus lines, then the loss of
information is only 25% of the quartet kernel, as shown in fig. 3.
In this particular case both GFN and SNi nodes sense the failure and
GFN receives the information of SNi.

4.3. MULTIPLE FAILURES ON HERMES
In case however that many failures occur at the same time on the HERMES
structure, then always the node "GFNs" or "FNs" untertake the recovery
process by defining the appropriate information flow. In case that the

full master node fails, then the operating system "sitting" on the four
top master nodes will define the appropriate schedule for the information flow. More details about failures on the HERMES structure are given
in [6].

4.4. PERFORMANCE
In this paragraph we present the performance of the multiprocessor
kernel when failures occur.
A. Failure of a Son Node (SNi)

The formula (1) represents the processing power of the son nodes for the
first type of transmission with no failures:
$$P = [3*(1+R1)/(1+R1+R2*r1)] \qquad (1)$$
where, R1=λ1/μ1, R2=λ1/μ2, r1=λ2/μ1.
The formula (2) represents the processing power of the son nodes for the
second type of transmition with no failures:
$$P = [3-N+3*(R3/N)] \qquad (2)$$
where, N=λ1[(1/μ1)+(r1/μ2)+(1/μ3)+(r2/μ1)], R3=λ1/μ3, r2=λ2/μ3.
For the third type of transmission the model is the same with the second
type except that the ratios R1,R2,R3,r1,r2 are different. In the fourth
type of transmission there is no degradation. Thus, figure 4 shows that
if a son failure f(SNi) occurs in a kernel then the processing power is
higher since there are fewer "customers" requiring service in the
queuing process.
B. Failure of a Father node (FN)

In this case we use the formulas above appropriately and the results are
provided graphically in figure 5. Thus, if a father node failure occurs
the processing power is lower, since more nodes require service from
the grantfather node.

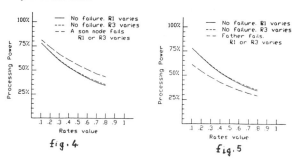

fig.4   fig.5

5. CONCLUSIONS

In this paper we examined the performance of two multiprocessor quartet
kernels, used for the design and evaluation of the HERMES system, when
failures occur on the processor nodes. The conclusion was that if son-
nodes fail the processing power increases, while if father nodes fail
the processing power decreases. The occurance of multiple failures on
the HERMES structure is discussed in a future work [6].

REFERENCES

[1] Kleinrock L.,"Queuing Systems",vol.I&II,J.Wiley & sons, 1975-76
[2] Lazowsska   ,et.al."Quantitative system performance",Prentice
    Hall 1984.
[3] Marsan M.A.,et.al."Performance models of multiprocessor systems",
    MIT Press, 1986.
[4] N.Bourbakis,"Design of real-time supercomputing vision architectures", Proc.IEEE Conf. on Supercomputing,vol.I,pp.392-98,May 1987,CA
[5] F.Barlos,"Functional models, Performance evaluation and Failures
    recovery of the HERMES structure",Master's Thesis,Spring 1989,GMU.
[6] N.Bourbakis, F.Barlos and C-H.Chien,"Performance analysis and
    failures recoveries mechanism on multiprocessor structures"

AN ARCHITECTURE FOR THE ROLLBACK MACHINE

Jonathan R. Agre and Peter A. Tinker

Rockwell International Science Center, Thousand Oaks, CA 91360

Conventional program parallelization relies on predictions of dynamic program behavior based on analysis of the static program text. While this technique works well for many important problems, it cannot extract sufficient parallelism from some programs characterized by very dynamic data and control structures found, for example, in artificial intelligence, discrete simulation, and computer vision. This problem is addressed by using an *optimistic* strategy that allows concurrent execution of fragments in the hope that most potential mutual dependences will be found to be illusory at run time. Optimistic scheduling results in temporary data dependence violations when dependent fragments execute out of order. When this occurs, a locally consistent state is restored by retracting the effects of out-of-order fragments using *rollback*.

Other related work in support of rollback-based parallel execution of monolithic general-purpose programs can be found in [4]. An approach that combines hardware assistance with programmer-supplied tasking and ordering annotations is described by Fujimoto [2]. His *Virtual Time Machine* is intended for discrete event simulation programs and does not support arbitrary programs, but is the closest relative of the work described here. This paper presents a summary of an architecture for efficient implementation of the *Rollback Machine* (RBM) [1], based on an execution model reported in [3]. We concentrate on the memory system, which is responsible for recording sufficient information to enable (1) performing read and write operations on potentially shared memory locations; (2) detecting data dependence violations and initiating rollback; and (3) providing efficient reclamation of obsolete memory areas.

The basic RBM architecture is assumed to be an MIMD set of processors and a set of memory modules, connected by a communication medium capable of passing data among processors and memories. We assume that the medium supports multicast. Each processor has a private memory for code and local data, and is connected to a shared memory with a uniform virtual address space. A program for the RBM is divided at compile time or run time into code fragments called *tasks* that are scheduled dynamically at run time. The partitioning of a program may be arbitrary — in particular, data dependences may exist among concurrently executing tasks. This partitioning can be expected to affect performance, but will not affect correctness. Parallelism is exploited by initiating execution of multiple tasks as they become available.

We assume that (1) the machine runs a single program at a time; (2) no interrupts can alter the program flow of control; (3) tasks communicate only through shared memory; and (4) a task contains no branches to addresses outside the task. A task executes instructions, and in particular it performs memory read and write operations. Each processor has a *task stack* in its local memory. Tasks that are awaiting execution or rollback are stored on the task stack along with linkage information that controls the progression of task execution. New tasks become available as a result of one task performing a *spawn* operation that terminates the task and generates zero, one, or two new tasks representing further work. The sequential execution of the program can be represented as an *execution tree*, which is a binary tree of the tasks. The *program order* of task executions (the order of their execution on a sequential machine) is generated by traversing the tree in a preorder sequence. Each of the nodes of the execution tree can be uniquely labeled with a *timestamp* that allows a lexicographic ordering of the nodes according to the program order (e.g., Dewey Decimal). Given two tree node timestamps one can determine their relative position in this ordering.

In general, each RBM processor proceeds independently, repeatedly executing the earliest timestamped task on its task stack. If a task generates other tasks, these are added to the task stack. When a processor's task stack contains no tasks, it becomes idle and migrates a task from another processor. When no tasks remain in the system, the program terminates. When a task reads a shared variable, it receives the value written with the largest timestamp earlier than the timestamp of the reader. This value is obtained by searching among the timestamped values associated with that variable. When a task writes a new value to a variable, this value is paired with the writer's timestamp and added to the set of values for that variable. Data dependence violations are detected when a write operation is performed. Any tasks that read the variable's previous value but should have read the new value are rolled back. When a task is rolled back, all memory operations performed by the task are retracted, and any descendants must be eliminated. Normally, a rolled-back task is once more made available for execution; the descendant tasks of a rolled-back task are not re-executed. Retraction of a write operation during rollback may cause further rollback:

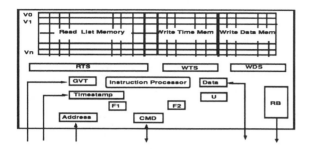

Figure 1: Time Indexed Memory System

tasks that already read the retracted value did so in error and must be rolled back themselves. This "cascade" effect is an inherent danger of the RBM and must be carefully controlled as in our scheduling strategy [3].

The time-indexed shared memory is organized into a set of memory banks with each bank containing a disjoint set of shared variables. Each memory bank consists of a bank controller (BC), that is interfaced to the communication switch, and a storage area (SA) for the shared variables (see Figure 1). The memory is accessed by a variable address that corresponds to a column of memory containing all the time-based history of the variable. It stores three lists for each of its shared variables in a column: 1) the read time list (RT) - the time stamps of the tasks which read the variable, 2) the write time list (WT) - the time stamps of the tasks which wrote the variable, and 3) the write data (WD) list - the data corresponding to each time-stamped write of the variable. The memory is organized as an array of size $V \times 4N$ words where V is the number of shared variables, $2N$ is the size of the RT and N is the size of the WT and WD. When a read command is given the address is placed in the memory address register (MAR) and the entire column corresponding to the address is transferred to the memory buffer register (MBR) (i.e., 4N words). The entire MBR is accessible (in a parallel transfer mode outside the SA to the BC. It is the function of the BC to maintain the three data structures kept in the SA in sorted order of increasing time stamp. The BC responds to memory commands from the processors that arrive via the communication switch. The possible commands are as follows: **Init** - Initialize memory bank; **SetGVT p,t** - Set current minimum (GVT) value to time stamp t (from processor p); **Read x,t** - Read the variable x at time stamp t; **Write x,t,d** - Write the data d at variable x at time stamp t; **RetractRead x,t** - Undo the read of variable x at time stamp t; **RetractWrite x,t** - Undo the write of variable x at time stamp t.

It is also possible for the BC to initiate a multicast transfer on the communication switch to the processors consisting of the following command: **Rollback** $t_1, \cdots, t_n$. The effect of this is to rollback the tasks corresponding to the time stamps given. The SA is directly connected to three shift register units called the read time shifter (RTS), the write time shifter (WTS), and the write data shifter (WDS). A memory access will parallel transfer the appro-

priate lists from the MBR to the corresponding shifter unit. The RB structure is a stack that is employed to temporarily hold the time stamps of the tasks to be rolled back. All of the necessary manipulations of the shared variable data can be accomplished through simple operations on the shifter units. Each shifter unit consists of $N + 1$ shift registers, an additional buffer register B, plus a control unit. Each register is capable of a word parallel transfer from the SA and of executing simple circular shift commands. Two important operations are incorporated into the memory commands as by products of the specified operations: a) fossil collection- the removal of data and time stamps that are older than GVT and no longer useful, and b) flow control - causing a rollback of the furthest ahead tasks when the RT or WT are full.

The **Read x,t** command causes the acccess of the column of data corresponding to the variable's address x. Time stamp t will be entered in the RT list in the correct order. The WT list will be searched to find the largest timestamp t' less than or equal to t and the data corresponding to t' will be returned. Initially, the shifter units are loaded in parallel and then the two functions **InsertRead** and **GetData** are executed in parallel. Following this, **SendData** transmits the data to the requesting process. Lastly, **SendRollback** notifies the task to be rolled back if the RT list was full (i.e., flow control). The write command initially loads the registers and then calls the functions **InsertWrite** and **RollbackReads** in parallel. However, these two functions must synchronize with Register U on the upper bound for tasks to be rolled back. Lastly, **SendRollback** communicates the tasks to be rolled back due to invalid reads.

Performance analysis of the architecture is currently under investigation using a simulation model that runs a synthetic program model that is parametricly specified and probabilistic executed. Preliminary simulation results have achieved speedups of up to 2.5 using five machines

References

[1] Jonathan R. Agre and Peter A. Tinker. An architecture for the Rollback Machine. Technical Report TR 90-06, Rockwell International Science Center, Thousand Oaks, CA, January 1990.

[2] R.M. Fujimoto. The Virtual Time Machine. Technical Report UUCS-88-019, University of Utah, Salt Lake City, UT 84112, December 1988.

[3] Pete Tinker. Task scheduling for general rollback computing. In *Proceedings of the International Conference on Parallel Processing*, volume II, pages 180–183, St. Charles, IL, August 1989. University of Pennsylvania.

[4] Pete Tinker and Morry Katz. Parallel execution of sequential Scheme with ParaTran. In *Proceedings of the Conference on Lisp and Functional Programming*, pages 28–39, Snowbird, UT, July 1988. ACM.

THE ARCHITECTURE OF A MULTILAYER DYNAMICALLY RECONFIGURABLE TRANSPUTER SYSTEM

Marek Tudruj   Miroslaw Thor

Institute of Computer Science, Polish Academy of Sciences,

P.O.Box 22, 00-901 Warsaw, Poland

Abstract -- This paper presents main features of the dynamically reconfigurable multilayer transputer system DIRECT. The system is built of three transputer layers interconnected by means of crossbar switches. Dynamic on-demand link reconfiguring and dynamic loading of transputers are applied. Non-standard control exchange, external to the links, has been extensively used in the system to improve its efficiency.

Introduction

Transputer has become one of the most popular processors for multiprocessor system design. Systems with fixed planar interconnection networks based on neighbor to neighbor connections were its basic application area. The INMOS C004 crossbar switch [1] made possible construction of reconfigurable transputer networks such as ParSiFal [2], MultiCluster [3], SuperNode [4]. In these networks the dynamic reconfigurability of link connections is possible in hardware, but the systems have basically single-level structures. This paper outlines the architecture of a multilayer transputer system with novel structural and functional features which enable the execution of highly parallel OCCAM programs in really dynamic multiprocessor environment. The name of the system is DIRECT which is an acronym for Directly Interconnected Reconfigurable External Control Transputers. In the DIRECT system the on-demand dynamically configured interprocessor communication is based on: OCCAM channel access multiplexing to transputer links, dynamic reconfiguring of link connections and new solutions for control exchange between processor layers.

The proposed methodology preserves the synchronous communication model of the OCCAM language.

Architecture of the DIRECT system

The system has the following general features:
- hierarchical structure built of three processor layers dedicated to different kinds of tasks,
- a dedicated communication structure in each processor layer,
- on-demand, user program driven reconfigurable processor interconnections,
- external to links control exchange for interlayer processor communication,
- the use of transputers as processors and transputer compatible crossbar switches,
- OCCAM as the user programming language,

The first processor layer in the system is composed of working transputers which execute elementary user processes in OCCAM programs. The second processor layer consists of control transputers which execute global control constructs of OCCAM programs, perform control over crossbar switches and provide supervise control. The third is a system layer whose transputers perform global system functions, user process allocation algorithms and the loading of transputers. Working transputer layer is composed of four 8-transputer clusters. Transputers in each cluster can communicate locally, each using any of its three links connected through a local crossbar switch LSwi. The 4-th link of each working transputer is used to connect transputers from different clusters through the global switch GSw. LSwi and GSw are (32 x 32) INMOS C004 switches.

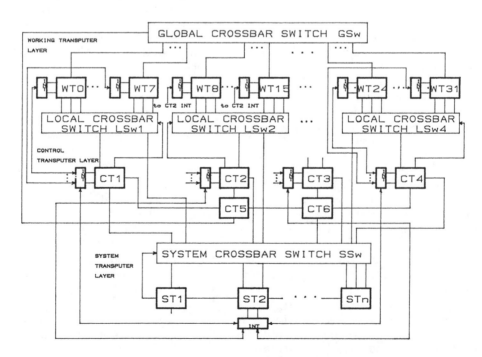

Fig.1 The general structure of the DIRECT system

In a multilayer system such as DIRECT, the implementation of inter-process communication demands intensive exchange of control. Hence, in the DIRECT system, additional control signals are exchanged outside transputer links to improve its efficiency. These signals include:
i) Connection Requests from working to control transputers and from control to system transputers,
ii) Event Request signals from control to working transputers and system to control transputers,
iii) Analyse, Reset and BootFromROM signals from control to working transputers and from system to all remaining transputers in the system.
Issuing of external control signals is program driven through the execution of transputer store instructions to the reserved external memory locations. The signals are generated in decoders in special control interfaces connected to transputers, (see the INT blocks in Fig.1), through a logical combination of reserved address values and memory control signals.

External communication in the Direct system

External communications are executed between communicating processes allocated to separate transputers. Two types of external communications are possible: local and global ones. They require working transputer link connections through local or global crossbar switches, respectively. Each transputer in the system is supplied with four operating system processes, called Communication Servers (COMSi). They multiplex process accesses to links and are organized as looping high priority processes starting with ALT constructs. The COMSi processes use tables which store the crossbar type, links and transputers for each communication.

The organization of external communications between processes in working transputers includes three phases:
- initial control transmission from working to control transputers to transfer configuration requests,
- synchronization of reconfiguration requests in control transputers,
- configuration of a crossbar switch and the transmission through thus connected links.
In the initial phase a working transputer sends a Connection Request to a local control transputer (through their control interfaces). The control transputer identifies the requesting working transputer and its link ready for communication. Then it configures the local switch LSw to get connection with the indicated working transputer link to get control information specifying the reason of the request. In the synchronization phase working transputer configuration request activates special WAIT processes in the control transputer responsible for configuring the corresponding crossbar switch, which suspend on their mutual internal communication. The communication of WAIT processes synchronise the requests for configuring. In local communications only one control transputer is used in the synchronization phase. In the global communications several control transputers are used. In the configuration phase the control transputer in which the synchronization took place issues a configurating message to the crossbar switch. The sending process should not output before links are connected. It is assured by a delay of the acknowledgement sent to the outputing process after the synchronization, during which link connections are configured.

Dynamic task loading to working and control transputers

Dynamic task loading is performed while there are still some programs to be run in the transputer memory. Such a loading has to be supported with thoroughful memory space management, with operating on active process queues, with channel access multiplexing to transputer links and with some other supervisory actions which can not be done while programming exclusively in OCCAM language. Thus programming at transputer assembler level is necessary. In the DIRECT system dynamic task loading is performed from the system layer transputers to working and control transputers. It is executed by the resident operating system processes. There are two initiators possible in the system for dynamic task loading: a working transputer or a subset of system transputers. The general algorithms of loading organization in these two cases can be found in [6].

The use of the proposed dynamic mechanisms

The dynamic link connection reconfigurability and dynamic task loading mechanisms, included in the DIRECT system, can be used for different program execution paradigms. One of such possible paradigms can be an optimized parallel execution of OCCAM programs. An optimization strategy for the user OCCAM process allocation to working transputers, developed for the DIRECT system, assumes dynamic, step by step task loading which will assure the balanced load of working transputers. Thus, based on heuristics, the total execution time of parallel programs can be reduced. The strategy applied is based on the analysis of graph representations of OCCAM programs [5].

Conclusions

The working and the system transputer layers are dynamically on-demand reconfigurable. The three transputer layers operate simultaneously, and thus, they execute in parallel the elementary user processes with global control and systems functions. In the DIRECT system the total throughput of the link connection establishing sub-system is increased as compared to systems with a single large crossbar switch. Special control interfaces improve the efficiency of control exchange. The system can be fairly easily built of commercially available components and subsystems.

References

[1] INMOS Ltd., Transputer Ref.Man., Prent. Hall, 1988
[2] Capon, P.C. et all, ParSiFal – A Parallel Simulation Facility, IEEE Colloq. Digest 91, 1986.
[3] Gosch J., A New Transputer Design from West German Startup, Electronics, March, 1988.
[4] Muntean T., SUPERNODE, Architecture Parallele, Dynamiquement Reconfigurable de Transputers, 11-emes Journ. sur l'Inform., Nancy, Jan. 1989.
[5] Szturmowicz M., Tudruj M., A Multi-Layer Transputer Network for Efficient Parallel Execution of OCCAM Programs, 15th EUROMICRO Symp. Short Notes Proceedings, Cologne, Sept., 1989.
[6] M.Tudruj, M. Thor, A Multilayer Dynamically Reconfigurable Transputer System, Parcella '90 Workshop Proceedings, Berlin, Sept., 1990.

CONCURRENT AUTOMATA AND PARALLEL ARCHITECTURE

T. Y. Lin

Mathematics and Computer Science
San Jose State University, San Jose, CA 95192

INTRODUCTION

This paper is motivated by our recent negative solution to Peterson conjecture [1] and Tanenbaum's comments on the organization of the stack of IBM PC [2,pp.226]. We are exploring the "high level" computer architecture problems from the automata theory. Our approach is unconventional. However, we do believe that it is a reasonable intellectual adventure and that it will be useful, especially in the real time systems. To us, architecture is "the abstraction of the computer as seen by algorithm designer", "abstraction of the hardware relationships".

CHURCH'S HYPOTHESIS

A computer is a computing machine, it defines (or abstracts to) a computational model. Such a model is a representation of the computing hardware. Church's hypothesis asserts that the Turing machine model is equivalent to a computer. However, we maps hardware differently. Church maps the maximal capability of a hardware to an automaton, while our computation model represents the "normal" or "natural" capability only. For example, IBM PC, by Church hypothesis, is equivalent to the Turing machine. In practice, IBM PC , due to its defect, is utilized up to the level of finite automata (FA) only. In our view, IBM PC in normal programming enviroment is equivalent to FA only. We are exploring the computation models of hardwares from practical point of view. Our computation model is the normal or natural capability of a hardware, NOT the maximal capability. Such computation model will be called **natural computation model**. This new map is an "engineer's representation" of a computing machine.

NATURAL COMPUTATION MODELS (NCM)

It is well known that the difference between a Turing machine (TM) and a Finite Automaton (FA) is that the former has the ability to rewrite symbols on its input tape [3, pp. 184]. To be more precise, it rewrites its input tape in a "non-finitary pattern" (i.e., in a pattern which can not simulated by an FA). Since a microprocessor has no devices to store or to rewrite its input, it is equivalent to FA only.

FA-Hypothesis: The natural computation model of a bare microprocessor is a finite automata.

In his book, Tanenbaum stated that "Intel 8088's memory management architecture is very primitive. ..... does not even detect stack overflow, a defect that has a major implications..." [2,pp.226]. In general, if a machine has no stack overflow detection, software designers are forced to design their software based on the assumption that there are only a finite length stack.(The software can overcome this difficulty by polling, whcih is wasetful, unnatural and inpratcial. A computation model based on bounded stack is Not a "genuine" pushdown automata. By observing that a bounded stack can be simulated by finite states, we have

Proposition 1. A pushdown automaton with bounded stack is a finite automaton.

By the proposition, we can conclude that a microprocessor which has no stack overflow exception is "naturally" a FA. On the other hand, using the strategy as stated by Tannebaum, computers equipped with an overflow exception can be organized naturally in such a way that they appear to the programmers as machines with indefinite length stack.

PDA-Hypothesis: A Microprocessor with Stack Overflow Exception (hardware gives warning when stack is overflowed) is a Genuine PushDown Automata (GPDA).

CONCURRENT AUTOMATA

Concurrent automata are ordinary Petri nets augmented with any kind of automata, such as finite automata, pushdown automata, linear bounded automata or Turing machines. Roughly, the augmentation proceeds as follows. At each transition, one place an automaton, say Pushdown Automaton (PDA) to process the color of a token (color is a string of input alphabet). The PDA process the prefix of the color and send the token with new color(=remaining string) to output place. [1]

These concurrent automata are extensions of CLASSICAL automata and Petri nets. In particular, "Concurrent PushDown Automata (CPDA)" lie properly between ordinary Petri nets and Turing machines. The existence of such CPDA's provide counter examples to Peterson's conjecture.

Intuitively each individual automaton represents an individual machine. A Petri net represents a concurrent control system. So the concurrent automata are the natural computation models for parallel systems, and any other tightly coupled multimachines.

The concurrent pushdown automata that solve the Peterson's conjecture confirmed the intuitive belief that the computational power of the total system can be strictly greater than that of each individual system in an order of **magnitude**. This leads us to a new type of parallel architecture.

PARALLEL ARCHITECTURES

Computing Power of Parallel Systems

"Superlinear speedup" has been the central concern of parallel processing. Here, we propose a similar but more **qualitative** question: Is there a "superlinear increase in modeling (computing) power?". Here "modeling power" means the capability in modeling and solving problems. Using the automata theory, it is the modeling power. For example, an FA can solve only problems which are equivalent to some regular languages. An FA can NOT parse a "genuine" context free language.

EPN Parallel Architecture and its NCM

Here we propose a parallel architecture whose natural computation model (NCM) is the concurrent pushdown automata (CPDA). Our architecture is similar to the connection machine, but more powerful. The Petri net control will be done by the conventional CPU, as in the connection machine. In connection machine, the PE (processing element) is a small FA. For us, we require each PE to be a genuine PDA that is a microprocessor with stack overflow exception. We will call this organization EPN-architecture (Extended Petri Net Architecture). The machine built by this architecture is called EPN-machine. The computing power of an EPN-machine is strictly greater than that of Pushdown automata and Petri nets. In other words, the EPN-machine can solve any problems which are solvable by the "join" of pushdown automata and Petri nets. More precisely, EPN-machine can solve problem which are equivalent to the extended Petri net language.

NCM of Connection Machine.

The PE of connection machine is simple, its NCM is an FA. Therefore the NCM of the connection machine is a collection of FA's which is controlled by a programmable communication network. The NCM of communication network is a Petri net. So, using our terminology the NCM of a connection machine is Concurrent FA. As an automaton, a Concurrent FA is equivalent to a Petri net. Therefore, a connection machine can only solve PROBLEMS, which are equivalent to a Petri net language. In computing power, the connection is less powerful than our EPN-machine. However,we should remak here that, since the Host of connection machine is a conventional machine, one can develop a communication network which is (not very naturally) equivalent to the Turing machine. So the (maximal) computation model of Connection machine is the Turing machine (implied by Church's hypothesis).

CONCLUSIONS

In this paper we introduce natural computation models for hardwares and Concurrent Automata for parallel systems. One should not that most of concurrent automata are equivalent to some classical automata, except the concurrent pushdown automata which is discovered in [1].

Most of research on parallel systems focus only on their computing **speed**, so the modeling capability of their problem domains is usually smaller. Many of scientific problems are equivalent to regular language. A parallel system whose modeling capability is equivalent to a finite automata is sufficient. However, as the applications are extending to AI area. The parallel systems should begin to focus on their modeling power. This paper is a small contribution to that direction.

Finally, we would like to stress again that we model the hardware from engineering point of view. We model the practical ("normal" or "natural") computing power of a hardware, not the ideal maximal capability of the hardware-the famous Church's hypothesis.

References

[1] T. Y. Lin, Extended Petri Nets and Peterson's Conjecture, Proc. of Fourth Annual Parallel Symposium, April, 1990.

[2] A. Tannenbaum, Operating Systems: design and implementation, Prentice-Hall, 1987.

[3] J. Hopfcroft and J.Ullman, Intr. Automata theory, Languages and Computations, Addison-Wesly, 1976.

TOWARDS A GENERAL-PURPOSE PARALLEL SYSTEM
FOR IMAGING OPERATIONS

Hamid R. Arabnia

Department of Computer Science
The University of Georgia
415 Graduate Studies Research Center
Athens, Georgia 30602, U.S.A.

Abstract — This short paper introduces a pipeline of processor rings that can be (logically) reconfigured so as to accommodate all possible numerically balanced rings.

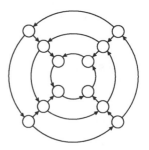

Figure 1. A multi-ring network for imaging operations

Introduction

In general, imaging operations can be parallelized using one of the strategies that follow.

(a) *Image-parallelism* — The image space is divided into a number of sub-images where each is allocated to a processor for processing.

(b) *Data-parallelism* — The graphics primitives (such as lines/vectors, planes, circles, ...) that define an image are divided into smaller portions of data where each is assigned to a processor for processing.

(c) *Process-parallelism* — This is the decomposition of a process into a number of subprocesses, and allocating each subprocess to a processor for execution.

Almost all parallel algorithms in imaging operations use one or another of these strategies. We believe that the processor network proposed in this paper has the potential to encourage the design of imaging algorithms which can incorporate more than one of the above strategies. This approach, whereby two or three of the strategies are combined, we have termed *image-data-process-parallelism*, or *IDP-parallelism*.

A Processor Network For Imaging Operations

The example network shown in figure 1 (processors are represented by circles and the arrows denote channels/links) supports all major parallelization approaches used in imaging operations. The network is made up of a number of rings of processors (three rings in this example). The maximum number of channels for any number and size of rings required for this set up is two input and two output channels. A problem whose process is decomposed into P subprocesses and whose data is decomposed into D smaller portions can exploit a network made up of P rings where each ring has D processors. In the network shown in figure 1, $D = 4$ and $P = 3$. Processor networks based on ring topology, as shown in the figure, support the IDP-parallelism approach.

Ring Reconfigurability

A network of processors having P rings each containing D processors where P and D are constants, is too restrictive for many imaging operations. For example, a problem may have two or more phases: in one phase it may require more (or fewer) rings than another phase. The network used in the two phases cannot be transformed into each other without changing the inter-processor communication connections manually. This manual change in connections makes the network too restrictive for real-time imaging applications. Therefore, the problem is to find a reconfigurable network of processors that supports the multi-ring topology where each processor has only two input and two output channels. We have devised such a network which is introduced here.

Consider a ring of n (a power of 2 for simplicity) processors where $P_0 \rightarrow P_1$, $P_1 \rightarrow P_2$, ... , $P_{n-1} \rightarrow P_0$ (the arrows show the direction of the links). This network contains within it (as its subsets), one ring of n processors and n rings of one processor each. In order to construct R rings of k processors each ($R \times k = n$), the following two extra channels for each processor, P_i , need to be added to the network:

$$P_i \rightarrow P_{(i+R)\,mod\,n}$$
$$P_i \leftarrow P_{(i+n-R)\,mod\,n}$$ (1)

As an example, consider the 16 processor ring shown in figure 2(a). In order to construct a pipeline of 2 rings of 8 processors (*i.e.*, $R=2$, $n=16$, and $k=8$), we find (by applying the two equations, (1), to each processor) that the following channels need to be added to the network:

$P_0 \rightarrow P_2$, $P_0 \leftarrow P_{14}$, $P_1 \rightarrow P_3$, $P_1 \leftarrow P_{15}$, $P_2 \rightarrow P_4$, $P_2 \leftarrow P_0$, $P_3 \rightarrow P_5$, $P_3 \leftarrow P_1$, $P_4 \rightarrow P_6$, $P_4 \leftarrow P_2$,

$P_5 \rightarrow P_7$, $P_5 \leftarrow P_3$, $P_6 \rightarrow P_8$, $P_6 \leftarrow P_4$, $P_7 \rightarrow P_9$, $P_7 \leftarrow P_5$, $P_8 \rightarrow P_{10}$, $P_8 \leftarrow P_6$, $P_9 \rightarrow P_{11}$, $P_9 \leftarrow P_7$, $P_{10} \rightarrow P_{12}$, $P_{10} \leftarrow P_8$, $P_{11} \rightarrow P_{13}$, $P_{11} \leftarrow P_9$, $P_{12} \rightarrow P_{14}$, $P_{12} \leftarrow P_{10}$, $P_{13} \rightarrow P_{15}$, $P_{13} \leftarrow P_{11}$, $P_{14} \rightarrow P_0$, $P_{14} \leftarrow P_{12}$, $P_{15} \rightarrow P_1$, and $P_{15} \leftarrow P_{13}$. This will result the network shown in figure 2(b) which consists of a pipeline of 2 rings of 8 processors. The two rings within figure 2(b) are shown explicitly in figure 3. In addition, each processor is connected to the corresponding processor in the other ring to form a two-level pipeline (*i.e.*, $P_0 \rightarrow P_1$, $P_2 \rightarrow P_3$, and so on). Other ring configurations can be constructed in a similar way.

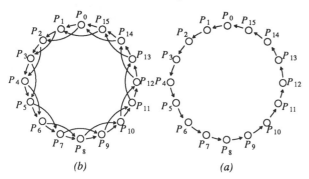

Figure 2. Construction of pipeline of rings

A reconfigurable network that will accommodate all possible numerically balanced rings can now be constructed. In order to see how, suppose each processor in the network is connected to a switch as shown in figure 4 (each black circle denotes a switch with identification S_i). Each dashed arrow is an input channel and is connected to ANY switch in front of it. If a switch is replaced with another one, then this channel will become disconnected from the original switch and will become connected to the new one. Each solid arrow next to a dashed arrow is an output channel and is ALWAYS connected to the switch that has the same subscript as its processor. Having the set up shown in figure 4, all possible ring configurations are now possible. To construct R balanced rings (*i.e.*, R rings of k where $k \times R = n$; n being the number of processors), each switch is moved by R steps counterclockwise round the original ring. For example, to construct 2 balanced rings (*i.e.*, $R = 2$) where there are 16 processors (refer to figure 4), the switches are moved by 2 steps round the ring, resulting the set up shown in figure 5. The resultant two rings within figure 5 is shown explicitly in figure 3. In addition, to form the pipeline, each processor is connected to the corresponding processor in the other ring (*i.e.*, $P_0 \rightarrow P_1$, $P_2 \rightarrow P_3$, ...). It is important to note that the technique shown here is a general method; that is, it is applicable to any number of processors (not just 16).

The ring network of processors introduced here can be reconfigured by software into all possible balanced configurations. The network can be built using the existing switching technology; this is the subject of our next paper.

Because of fewer communication channels, the proposed network would be significantly less expensive

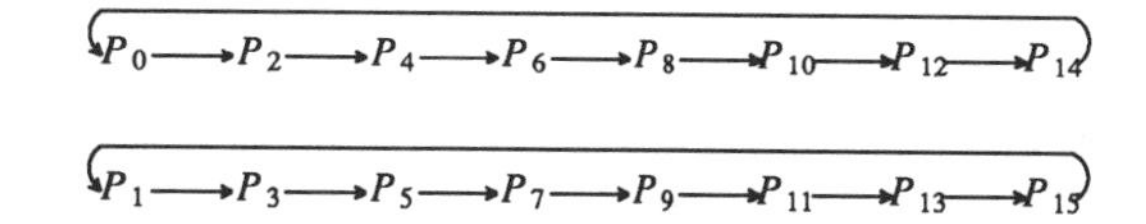

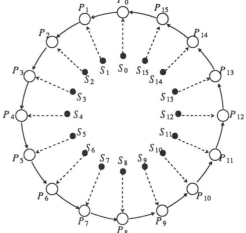

Figure 3. Two rings of processors within figures 2(b) and 5

(economically) to build than hypercube-based architectures (with dimensions larger than 4).

Acknowledgements

This work was partly supported by The University of Georgia Research Foundation via Faculty Research Grant. We acknowledge the help of The National Science Foundation (NSF) via The Computer and Information Science and Engineering (CISE) Research Instrumentation Program (Grant numbers: CCR-8717033 and CDA-8820544). I am indebted to Professors Rod Canfield and Jeff Smith for many stimulating discussions about this subject.

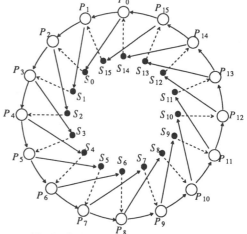

Figure 4. A 16 processor ring with switches

Figure 5. Moving switches by 2 steps

A Fast 1024-Point FFT Architecture

Tom Chen   Li Zhu

Department of Electrical Engineering

Colorado State University

Fort Collins, CO 80523

Abstract

This paper presents a 1024-point, column FFT design architecture. The entire system contains 256 radix-4 butterfly processors. These butterfly processors are connected using circuit switching networks. Assuming the data wordlength is 16 bits, the proposed 1024-point, column FFT engine is capable of processing 1024 samples in 559 clock cycles. If the clock frequency is 20MHz, it will take 28 μs to complete a 1024-point FFT. The architecture is inherently defect and fault tolerant, therefore, is suitable for implementation on a single ULSI chip.

1 Introduction

Several single-chip FFT processors have been produced [1, 2, 3] in the past. These chips are designed for specific transform lengths. The FFT architecture discussed in this paper employs 256 radix-4 butterfly processors and is capable of processing 1024 samples simultaneously. We use a column FFT algorithm to maximize the utilization of each butterfly processor as well to maximize parallelism. The incorporation of a large switch network together with the butterfly processors allows easy and efficient implementation of such a flexible interconnection scheme. Bit-serial arithmetic is used for the butterfly processor for its simple communication and control, efficient computation, and better $area \times time^2$ performance [4].

The remaining part of the paper is divided into four sections. The fast Fourier transform in general is discussed in Section 2. Section 3 discusses the architecture of the column FFT engine using the column FFT algorithm. Simulation results of the column FFT engine will be presented in Section 4. Finally, we summarize our approach and give several concluding remarks in Section 5.

2 The Fast Fourier Transform

Fourier transform plays a very important role in spectral analysis. It converts sampled data from the time domain to the frequency domain, or vice versa. To compute N-point DFT in a straightforward manner, $N \times N$ operations are needed. For real-world applications, N ranges from several hundreds to several thousands. Hence, the number of required operations can be prohibitively large. FFT [5, 6] speeds up the Fourier transform dramatically by utilizing the DFT's symmetrical and periodical properties. It reduces the number of required operations from $N \times N$ to $N log_2 N$. Among various variations of FFT algorithms, Cooley-Tukey algorithm is the most widely used algorithm and is often referred to as the *FFT*.

To explain column FFT algorithm, we use a 64-point Cooley-Tukey FFT algorithm as an example. Figure 1 shows the flow diagram of a 64-point, radix-4, FFT. Each square in the diagram represents a radix-4 butterfly processor. Each radix-4 butterfly performs three complex multiplications and

eight complex additions. To transform 64 samples from the time domain to the frequency domain, these samples have to be processed through three columns of the butterfly operations. According to the flow diagram, there are 16 butterfly processors in a column for a 64-point transform.

The concept of column FFT processing involves trying to use one column of the butterfly processors to carry out the entire transformation repeatedly. For example, in Figure 1, a butterfly processor column is used to process the first column of the transform. Then the outputs are properly reordered and are presented again to the butterfly processor column to process the second column of the transform, and the process continues to subsequent columns. A column FFT algorithm called the "Constant Geometry Algorithm" (CGA) was presented in [7], which provides a constant topology for transferring the output data after each column process. However, the disadvantage of such an implementation is its lack of flexibility. Once the wiring is fixed, it is not possible to expand the system. Architectural expandability is an important issue in designing high performance processors, because the silicon processing technology limits the maximum silicon area being used to implement logic. Hence it is not practical to integrate an FFT system with a very long transform length on to one piece of silicon. Therefore, it is widely desired for any FFT systems with shorter transform lengths to have expandability in order to easily and arbitrarily form a new FFT system with a longer transform length using the available smaller FFT systems. Bearing this in mind, we choose the FFT algorithm shown in Figure 1 as the target algorithm to be implemented. As will be shown in the next section, this algorithm allows us to efficiently include the architectural expandability.

3 The 1024-point Column FFT architecture

3.1 Architectural Features and System Operations

Figure 2 shows the target architecture for the 1024-point column FFT engine. There are a total of 256 radix-4 butterfly processors in the architecture. Each input and output of a butterfly processor goes through a shift register. The length of the shift register is fixed to the wordlength of input data. In our case, it is 16 bits. A butterfly processor and the 16 associated shift registers (note that the input data are complex numbers) form a complete processing element referred to here as the *butterfly processing element*, or BPE.

The 256 BPEs are grouped into 16 groups (B00 to B15). Each BPE in a group is connected to all the other BPEs in the same group through a crossbar switch matrix. Figure 3 shows the schematics of the B block containing 16 BPEs as a group. 16 radix-4 BPEs have 64 complex inputs and 64 complex outputs. Therefore, the crossbar switch matrix is organized as two 64-row by 128-column arrays. One is taking care of the real part of the data transfer and the other is taking care of the imaginary part of the data transfer. The switching structure allows programmable assignment of the horizontal rows to the data signals in the system with flexible routing strategies (multiple choices for any point-to-point connection). The BPEs can be placed on the top or the bottom of the matrix. Each input and output of the BPEs is connected to a column of the switch matrix. The switch cell is implemented using a CMOS transmission gate, and hence, is bidirectional. The switch setting can easily be programmed using the row and column load lines, which in turn are controlled by the row and column decoders. Four B blocks are grouped into a bank of B blocks. There are four banks of B blocks. They are labeled as G0-G3.

The eight SA blocks (SA0-SA7) in Figure 2 provide data transfer channels for the sampled input data to reach their destinations ready for the first column processing. Each SA block has 256 inputs and 512 outputs. Figure 4 shows the schematics of the 'SA' block. According to Figure 1, for the first column processing, the first four input data will be sent to the first BPE in the BPE column in Figure 1. Assume that the entire column of BPEs in the architecture shown in Figure 2 is labeled in increasing order from block 'B00' to 'B15'. Therefore, the first 64 input data will be sent to 'B00'; the next 64 input data will be sent to 'B01'; the third group of 64 input data will pass through 'SA00' and 'SA01' and be sent to 'B02'; and so on. SA blocks are also used to gather the final results to the output registers after the entire FFT is finished. In this case, the results from 'B00' and 'B01' will be sent to the output registers through 'SA0' and 'SA1', and the results in 'B02' and 'B03' will be sent to the output registers through 'SA1' only.

The eight SB blocks (SB0-SB7) in Figure 2 provide data transfer channels to separate the 128 incoming data from the corresponding SA blocks and route them to two separate adjacent 'B' blocks accordingly. For example, 'SB0' is responsible for separating data for 'B00' and 'B01'. Figure 5 shows the structure of the SB blocks. The SA and SB blocks together will provide communication between different B blocks during the column processing.

For a 1024-point FFT, there are five columns in total. According to Figure 1, during the first and the second column processing, output data from each BPE are sent to the nearest 16 BPEs. In other words, communication after the first and the second column processing is restricted within each B block. Therefore, data transfer after these two columns can be carried out simultaneously throughout the network, The time required to perform data transfer for the first two columns is the wordlength of the input data times the clock cycle period. The data transfer after the third column processing is within the nearest 64 BPEs, and therefore, it is within a G block. To accomplish the data transfer after the third column processing, four 64 × 128 switch matrices in each G block are linked via two SB blocks to form a 64 × 512 switch matrix. The newly formed switch matrix in each G block can transfer 64 data at a time. Hence, to carry out data transfer within each G block, four time phases, each of which has the length of the wordlength times the clock cycle period, are needed. Each phase allows one B block to transfer its output data to all the other B blocks within the same

G block. Data transfer after the fourth column processing requires connections between the G blocks. This is achieved by linking all the SA and SB blocks in a row together (there are two rows of SA and SB blocks). Two rows of SA and SB blocks provide a total of 256 inter-bank communication channels. These 256 channels allow one G block to send its output data to all the other G blocks. Therefore, four G blocks will send out their output data in turn, requiring four timing phases to finish the entire data transfer.

The switch matrix programming is overlapped with butterfly computation. Once the output data from a column processing are shifted into the input shift registers of the BPEs, the butterfly computation starts. At the same time, the process of programming the switch matrices for next column data transfer also starts. This programming process will finish before the output data is available in the BPEs' output shift registers.

3.2 Expandability and Fault Tolerance

The column FFT architecture discussed in this paper can be easily expanded to form a new column FFT with a longer transform length. Let us take a 4096-point column FFT engine as an example. Figure 6 shows how four 1024-point column FFT engines can be connected together to form a 4096-point column FFT engine. Data transfer during the first five columns is the same as the transfers performed in the 1024-point column FFT. Data transfer after the fifth column requires connections between 1024-point FFT engines. This is achieved with two additional SA blocks and four time phases. During each time phase, one 1024-point FFT engine takes control of the entire switch network and transmits its output data to other 1024-point FFT engines. The expansion topology is somewhat similar to the binary tree structure.

Fault-tolerance is another important issue in the design of ultra large scale systems. Figure 7 shows how redundant BPEs can be included at the top and the bottom of the switch matrix. A faulty BPE can be replaced by a working redundant BPE with proper programming of the switch matrix. The switch matrix itself is inherently fault tolerant. When a faulty switch cell is found, the original communication channel through the faulty switch cell (indicated by the dashed line) can be replaced by an alternative channel (indicated by the solid line) in the switch matrix.

4 Simulation Results

The entire 1024-point FFT system is implemented and simulated using VHDL [8]. Timing verification at the logic level during data transfer through switch matrices is also carried out. The results from the simulations show that the entire system functions as expected. The entire 1024-point can be processed in 559 clock cycles. Assuming the system clock runs at 20MHz, it takes 28 μsec to complete a 1024-point FFT.

5 Concluding Remarks

We have presented a 1024-point column FFT architecture. The employment of the crossbar switch matrix in the system allows us to efficiently address issues such as fault-tolerance and system expandability. With the advances in VLSI tech-

A Parallel Architecture for Image Processing

Wen-Kuang Chou and *David Y.Y. Yun*

Lab. of Intelligent & Parallel Systems (LIPS)
University of Hawaii at Manoa
and PICHTR
Holmes Hall #493, 2540 Dole Street
Honolulu, HI 96822

1. Introduction and Summary

Modern space and earth exploratory technologies have enabled rapid gathering of scientific data. As the collection of image data accumulates, it becomes increasingly important to compress them to save storage and improve transmission. The primary concern in image compression is that of converting an analog or digital picture to the smallest sequence of bits such that it (often after transmission) can be used to reconstruct a high quality replica of the original image. Basically, there are three major classes of image encoding schemes: predictive coding, transform coding, and hybrid coding [3]. It is well known that the transform coding can achieve better compression ratio than predictive or hybrid coding and can be less sensitive to errors than predictive and hybrid coding [3, 7].

However, the transform coding suffers from the often excessive complexity for implementation. As a result, most of the past research has aimed at improving the single transform algorithm such as Fast Fourier Transform (FFT) [5] and Fast Cosine Transform (FCT) [4] but not at combining two or more popular transforms in a common architecture. Latest developments on FFT and FCT are primarily concerned with variations of the basic algorithms in [5] and [1], which can be better adapted to parallel processing in some special purpose machine. However, if a special machine has to be developed for every different transform and executed separately, not only the hardware cost will be prohibitive but it also will be impossible to run applications needing to use more than one transform. Thus, it is more important to consider architectures capable of processing more than one popular transforms. In this way, the hardware cost can be reduced while the same computational efficiency is maintained. In two-dimensional image processing, it usually employs one-dimensional transform twice; once for horizontal transform and once for vertical transform. It has been suggested that different types of transform can be used in the horizontal and vertical transforms [3] such as FFT for horizontal transform and FCT for vertical transform. Our primary objective is to achieve a uniform representation for more than one transform so that a single parallel architecture can support heterogeneous 2-D transforms.

The initial work of the Fast Fourier Transform for parallel processing in a special purpose perfect-shuffle network [2, 8] is proposed by Pease [8], which is concentrated on single algorithm again. In this paper, a regular and uniform equation for both FFT and FCT which consists of a sequence of elementary operations is presented. Based upon the equational regularity, a common parallel architecture which is composed of a single-stage perfect shuffle network and a (*log* N - 1)-stage sparse connection network is proposed to realize both FFT and FCT, where N is the number of input points. The significance of this result lies in the use of a common network architecture to handle both important transforms for image processing. This can result in reduced hardware cost through increased architecture applicability.

2. Notations of FFT and FCT

Given N equally spaced data sequence $\{x_k; k = 0, 1, ..., N-1\}$, where N is assumed to be a power of 2, the discrete Fourier transform of these points is given by the following equation:

$$f_j = \frac{1}{N} \sum_{k=0}^{N-1} x_k exp\left(-\frac{2\pi i}{N}jk\right) \quad \text{for } j = 0, 1, ..., N-1, \quad (1)$$

where $i = \sqrt{-1}$; the discrete Cosine transform of these points is given by the equation as follows:

$$g_j = \frac{2\varepsilon_j}{N} \sum_{k=0}^{N-1} x_k cos\left[\frac{\pi(2k+1)j}{2N}\right] \quad \text{for } j = 0, 1, ..., N-1, \quad (2)$$

where $\varepsilon_j = 1/\sqrt{2}$ if j =0; otherwise, it is 1.

Let $\mathbf{x}$, $\mathbf{f}$, and $\mathbf{g}$ be the column vectors of $\{x_k\}$, $\{f_k\}$, and $\{g_k\}$ respectively, $0 \le k \le N-1$. Since ε_j in eq. (2) affects only the amplitude of g_0, we shall take ε_j as unity with g_0 scaled up by $\sqrt{2}$. Let's define the matrix $\mathbf{F}$, whose elements are $F_{jk} = exp(-2\pi ijk/N)$, and the matrix $\mathbf{C}$, whose elements are $C_{jk} = cos(\pi(2k+1)j/2N)$. Then the discrete Fourier transform defined by eq. (1) and the discrete Cosine transform defined by eq. (2) can be written as $\mathbf{f} = \mathbf{F}\,\mathbf{x}/N$, and $\mathbf{g} = 2\mathbf{C}\,\mathbf{x}/N$, respectively. For clarity, the normalization factor 1/N and 2/N will be neglected until the end.

Let $\mathbf{F}(n)$ and $\mathbf{C}(n)$ be the n × n matrices. The key to the FFT algorithm and FCT algorithm lies in the recursive representation. Let ω be the principal Nth root of unity, then the elements of $\mathbf{F}$ are given by $F_{jk} = \omega^{jk}$. In order to represent $\mathbf{F}$ in recursive form, the rows of $\mathbf{F}$ should be permuted in *bit-reversal* order and let the result be $\hat{\mathbf{F}}$. For FFT, $\hat{\mathbf{F}}(N)$ can be represented by recursively as shown in [8],

$$\hat{\mathbf{F}}(N) = \begin{bmatrix} \hat{\mathbf{F}}(N/2) & \hat{\mathbf{F}}(N/2) \\ \hat{\mathbf{F}}(N/2)\mathbf{K} & -\hat{\mathbf{F}}(N/2)\mathbf{K} \end{bmatrix} = \begin{bmatrix} \hat{\mathbf{F}}(N/2) & \phi \\ \phi & \hat{\mathbf{F}}(N/2) \end{bmatrix} \begin{bmatrix} \mathbf{I} & \phi \\ \phi & \mathbf{K} \end{bmatrix} \begin{bmatrix} \mathbf{I} & \mathbf{I} \\ \mathbf{I} & -\mathbf{I} \end{bmatrix} \quad (3)$$

where $\mathbf{K} = diag(\omega^0\ \omega^1\ \omega^2\ \omega^3\ \cdots)$ and ϕ indicates the null matrix with appropriate dimension. $\mathbf{I}$ is an identity matrix with appropriate dimension.

Let $\hat{\mathbf{C}}$ be the resulting matrix after the rows of the matrix $\mathbf{C}$ are permuted in bit-reversal order first and then the columns are permuted in such a way that the first half consists of even indexed columns in ascending order and the second half consists of odd indexed columns in descending order. The recursive representation of $\hat{\mathbf{C}}$ as shown in [6] is given by

$$\hat{\mathbf{C}}(N) = \begin{bmatrix} \hat{\mathbf{C}}(N/2) & \hat{\mathbf{C}}(N/2) \\ \mathbf{U}\hat{\mathbf{C}}(N/2)\mathbf{K} & -\mathbf{U}\hat{\mathbf{C}}(N/2)\mathbf{K} \end{bmatrix}.$$

Then, it can be factored as follows:

$$\hat{\mathbf{C}}(N) = \begin{bmatrix} \mathbf{I} & \phi \\ \phi & \mathbf{U} \end{bmatrix} \begin{bmatrix} \hat{\mathbf{C}}(N/2) & \phi \\ \phi & \hat{\mathbf{C}}(N/2) \end{bmatrix} \begin{bmatrix} \mathbf{I} & \phi \\ \phi & \mathbf{K} \end{bmatrix} \begin{bmatrix} \mathbf{I} & \mathbf{I} \\ \mathbf{I} & -\mathbf{I} \end{bmatrix} \quad (4)$$

where $\mathbf{U} = \mathbf{RLR}$, and $\mathbf{K} = diag(cos\ \phi_0, cos\ \phi_1, ...)$, $\phi_j = (j + 1/4)(2\pi/N)$. The matrix $\mathbf{R}$ is the permutation matrix for performing the bit-reversal arrangement. The matrix $\mathbf{L}$ is a lower triangle matrix defined as

$$\mathbf{L} = \begin{bmatrix} 1 & 0 & 0 & 0 & \cdots & 0 \\ -1 & 2 & 0 & 0 & \cdots & 0 \\ 1 & -2 & 2 & 0 & \cdots & 0 \\ -1 & 2 & -2 & 2 & \cdots & 0 \\ \cdot & \cdot & \cdot & \cdot & & \cdot \\ \cdot & \cdot & \cdot & \cdot & & \cdot \\ \cdot & \cdot & \cdot & \cdot & & \cdot \\ -1 & 2 & -2 & 2 & \cdots & 2 \end{bmatrix}.$$

3. Uniform Representation of FFT and FCT

The *Kronecker product* of matrices $\mathbf{A}$ and $\mathbf{B}$ is defined as

$$\mathbf{A} \times \mathbf{B} = (\mathbf{A}b_{ij}). \quad (5)$$

Let $\mathbf{P}$ be the permutation matrix performing *perfect shuffle*, and $\mathbf{P}^{-1}$ be the permutation matrix performing *inverse perfect shuffle*.

The following lemmas are related to the permutation matrices $\mathbf{P}$ and $\mathbf{P}^{-1}$, which are important to prove Theorem 2 and Theorem 3.

Lemma 1: Let $m = log\ N$. Then, $\mathbf{P}^m = \mathbf{P}^{-m} = \mathbf{I}$, and $\mathbf{P}^{-(m-1)} = \mathbf{P}$.

Lemma 2: Suppose that M is a power of 2, then $\mathbf{P}(A(M) \times I(2)) = (I(2) \times A(M))\mathbf{P}$, or, $\mathbf{P}(A(M) \times I(2))\ \mathbf{P}^{-1} = I(2) \times A(M)$.

By iteratively applying Lemma 2, Theorem 1 can be derived as follows:

Theorem 1:

$$I(N/2) \times B(2) = \mathbf{P}^{m-1}(B(2) \times I(N/2))\mathbf{P}^{-(m-1)},\ \text{where}\ m = log\ N.$$

From the definition of Kronecker product, eq. (3) can be expressed by

$$\hat{\mathbf{F}}(N) = (\hat{\mathbf{F}}(N/2) \times I(2))\ D(N)\ (I(N/2) \times \hat{\mathbf{F}}(2)), \quad (6)$$

where $D(N)$ is the $N \times N$ *quasi-diagonal* matrix, quasidiag($\mathbf{IK}$). Let $E(N)$ be quasidiag($\mathbf{IU}$). Then, eq. (4) can be translated to

$$\hat{\mathbf{C}}(N) = E(N)\ (\hat{\mathbf{C}}(N/2) \times I(2))\ D(N)\ (I(N/2) \times \hat{\mathbf{C}}(2)). \quad (7)$$

The following two theorems show that the transform matrices $\hat{\mathbf{F}}$ for FFT and $\hat{\mathbf{C}}$ for FCT can be expressed by a sequence of elementary operations.

Theorem 2: [8]

$$\hat{\mathbf{F}}(N) = AQ_1PAQ_2PAQ_3 \cdots AQ_{m-1}PAP, \quad (8)$$

where $A = \hat{\mathbf{F}}(2) \times I(N/2)$, Q_j's are diagonal matrices, $1 \leq j \leq m-1$, and $m = log\ N$.

Theorem 3:

$$\hat{\mathbf{C}}(N) = \hat{E}_{m-1}\hat{E}_{m-2} \cdots \hat{E}_1\ BY_1PBY_2PBY_3 \cdots BY_{m-1}PBP, \quad (9)$$

where $B = \hat{\mathbf{C}}(2) \times I(N/2)$, $\hat{E}_j = E(N/2^{m-j-1}) \times I(2) \times \cdots \times I(2)$, which is a quasi-diagonal matrix, $Y_j = \mathbf{P}^{-(j-1)}\hat{D}_j\mathbf{P}^{(j-1)}$, which is a diagonal matrix, and $\hat{D}_j = D(N/2^{m-j-1}) \times I(2) \times \cdots \times I(2)$, $1 \leq j \leq m-1$.

Investigating the eq. (8) and eq. (9), we find that the right half of eq. (9) is much similar to eq. (8). If we neglect the common factor $1/\sqrt{2}$ in $\hat{\mathbf{C}}(2)$, $\hat{\mathbf{C}}(2)$ is the same as $\hat{\mathbf{F}}(2)$. As a result, A is equal to B. Both Q_j and Y_j are diagonal matrices but with different diagonals, $1 \leq j \leq m-1$. The diagonals of Q_j are the elements in the set $\{\omega^0, \omega^1, ..., \omega^{N-1}\}$, while the diagonals of Y_j are the elements in the set $\{cos\phi_0, cos\phi_1, ..., cos\phi_{N-1}\}$, where $\phi_k = (k + 1/4)(2\pi/N)$. A uniform representation for FFT and FCT is given as follows:

$$\hat{\mathbf{T}}(N) = H_{m-1}H_{m-2} \cdots H_1\ AW_1PAW_2PAW_3 \cdots AW_{m-1}PAP \quad (10)$$

If $\hat{\mathbf{T}}(N)$ is used to perform FFT, H_j is equal to $I(N)$, and W_j is equal to Q_j. If $\hat{\mathbf{T}}(N)$ is for FCT, H_j is equal to $\hat{E}_j$, and W_j is equal to Y_j.

4. A Common Parallel Architecture for FFT and FCT

Based upon the uniform representation in eq. (10), a common parallel architecture is proposed to realize both FFT and FCT. The architecture of this common machine is shown in Figure 1(a), which consists of a single-stage perfect shuffle network to compute the right half of eq. (10) (i.e., $AW_1PAW_2PAW_3 \cdots AW_{m-1}PAP$) and a ($log\ N$ -1)-stage parallel network to execute the left half of eq. (10) (i.e., $H_{m-1}H_{m-2} \cdots H_1$). The single-stage perfect shuffle network consists of N storage nodes and N/2 multiply-add modules. The function of multiply-add module is shown in the Figure 1(b). The ($log\ N$ -1)-stage parallel network is composed of $log\ N$ - 1 sparsely connected networks.

Given the input vector $\mathbf{x}$, it is preprocessed as a bit-reversal order, denoted as $\hat{\mathbf{x}}$, for computing FFT. However, for computing FCT, $\mathbf{x}$ should be partitioned into one half of even subscripts (even half) and the other half of odd subscripts (odd half). Then, the even half is reordered in subscripts with ascending order and the odd half is reordered in subscripts with descending order. The concatenation of these two halves is denoted as $\hat{\mathbf{x}}$. The equation $AW_1PAW_2PAW_3 \cdots AW_{m-1}PAP\hat{\mathbf{x}}$ can be executed in the single-stage shuffle network in m iterations by following:

[1] the equation is executed from right to left direction;

[2] P means performing perfect shuffle permutation;

[3] A is executed by multiply-add module;

[4] W_j are the connection weights of the links into multiply-add modules.

Given appropriate input $\hat{\mathbf{x}}$ and W_j (or weights), the results of FFT (i.e., $\mathbf{f}$) and the semi-results of FCT are obtained after m iterations by performing perfect shuffle permutation and doing some simple arithmetic operations in the multiply-add modules. The semi-results of FCT are scaled with $1/(\sqrt{2})^m$, denoted as $\hat{\mathbf{g}}$. Then $\hat{\mathbf{T}}(N)\hat{\mathbf{x}}$ for FCT in eq. (10) can be expressed by $\hat{E}_{m-1}\hat{E}_{m-2} \cdots \hat{E}_1\hat{\mathbf{g}}$. This part can be realized by a (m-1)-stage sparsely connected network since each quasi-diagonal matrix $\hat{E}_j$ can be expressed by a sparse connection. So, $\hat{\mathbf{g}}$ is fed into a fixed connection network with m-1 stages. The final results of FCT (i.e., $\mathbf{g}$) with bit-reversal order are obtained at the most right side. The time for computing FFT is $log\ N$ and the time for computing FCT is $2log\ N$ - 1.

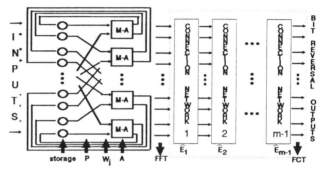

(a) Common network structure for FFT and FCT

$$Y_0 = W_0X_0 + W_1X_1$$
$$Y_1 = W_0X_0 - W_1X_1$$

(b) Mutiply-Add module

Figure 1. (a) A common parallel architecture for FFT and FCT; (b) M-A module.

References

[1] N. Ahmed, T. Natarajan, and K.R. Rao, "Discrete Cosine Transform", *IEEE Trans. on Computers*, C-23(1), Jan. 1974, pp. 90-93.

[2] K.E. Batcher, "Sorting Networks and Their Applications", *AFIPS Conference Proceedings*, 1968 SJCC, **32**, Washington, DC., 1968, pp. 307-314.

[3] R.J. Clarke, **Transform Coding of Images**, Academic Press, New York, 1985.

[4] W.H. Chen, C.H. Smith, and S.C. Fralick, "A Fast Computational Algorithm for the Discrete Cosine Transform", *IEEE Trans. on Commun.*, **COM-25(9)**, Sep. 1977, pp. 1004-1008.

[5] J.W. Cooley and J.W. Tukey, "An Algorithm for the Machine Calculation of Complex Fourier Series", *Math. Comput.*, **19(2)**, April 1965, pp. 297-301.

[6] H.S. Hou, "A Fast Recursive Algorithm for Computing the Discrete Cosine Transform", *IEEE Trans. on ASSP*, ASSP-35(10), Oct. 1987, pp. 1455-1461.

[7] A. Leger, J.L. Mitchell and Y. Yamazaki, "Still Picture Compression Algorithms Evaluated for International Standarisation", *IEEE Globcom*, 1988, pp. 1028-1032.

[8] M.C. Pease, "An Adaptation of the Fast Fourier Transform for Parallel Processing", *JACM*, **15(2)**, April 1968, pp. 252-264.

Improving The Efficiency of Multi-rate Signal Processing Architectures *

Sanjay R. Deshpande[†] & Prem P. Jain[‡]

Abstract

We observe that for *multi-rate* algorithms, such as decimation/interpolation used in real time signal processing, VLSI implementations with single time-base designs (such as systolic) are fundamentally inefficient in terms of hardware utilization. Improvement in efficiency can be achieved only by relaxing the "single clock" requirement and sharing of components. This is demonstrated here via two, more efficient, schemes suggested for an illustrative FIR filtering required during decimation.

1 Introduction

Real time signal processing requires performing computations at different sampling rates, e.g., decimation filtering, which can be formally characterized as being *multi-rate* (defined herein). The process of decimation involves a change in the sampling frequency of a signal from a higher value f_0 to a lower value f_k. It has been shown that the computational complexity of a decimation filter is greatly reduced if the decimation is carried out in more than one stages [2]. So the overall filter for decimating by, say, 2^l is better implemented in multiple stages, say k, with the FIR stages numbered 1 through k+1. Each stage i of the first k FIR stages is followed by a decimation block of order D_i. The arrangement is shown in Figure 1. Note that $\prod_{i=1}^{k} D_i = 2^l$. (For the reasons of simplicity of discussion, decimation factors D_i are assumed to be powers of 2.)

We start with some formal ideas which lead to the conclusion that multi-rate computations require the relaxation of single-clock or systolic style of design to ensure efficiency. The discussion motivates a new, multi-clock, style of design which is illustrated and generalized via the analysis of the above decimation computation. The result is: two possible schemes that display improved architectural efficiency over a single-clock design.

2 Multi-Rate Functions

A function $\mathbf{F}$ from $A \longrightarrow B$ for which $\forall (a, b) \in \mathbf{F}$, $|a| = m$, $|b| = n$. is denoted as $\mathbf{F}^{m;n}$. A function $\mathbf{F}^{m;n}$ is *p-input-unique* if $\forall a_i, a_j \in \text{Domain}(\mathbf{F}^{m;n})$, $|a_i - \bigcup_{j \neq i} a_j| = p$, $p \geq 1$. A function $\mathbf{F}^{m;n}$ is *q-output-unique* if $\forall b_i, b_j \in \text{Range}(\mathbf{F}^{m;n})$, $|b_i - \bigcup_{j \neq i} b_j| = q$, $q \geq 1$. A p-input-unique, q-output-unique $\mathbf{F}^{m;n}$ is denoted by $\mathbf{F}_{p;q}^{m;n}$. We will consider functions which have m=p and n=q, which, for brevity, are denoted as $\mathbf{F}_{p;q}$.

To define multi-rate functions, the concept of time must be introduced into the last definition. This is done by asserting the following: For $\mathbf{F}_{p;q}$, **each element of input set a_i is accessed at distinct time instant; the same is true for elements of the output set b_j.**

It will be convenient to assume that consecutive time instants are separated by a fixed *time-interval*. Thus, for $\mathbf{F}_{p;q}$, the input set occupies p time-intervals, and the output set occupies q time-intervals. Single rate and multi-rate functions can now be formally defined as:

Definition 1 *An* $\mathbf{F}_{p;q}$, *where* $p = q$, *or* $p \neq q$, *is called a multi-rate function.*

A function in which p = q, is called *rate-invariant*, in which p = q = 1, is called *single rate*, and in which $p \neq q$, is called *rate-variant*. An algorithm in which all functions are single-rate is called a *single-rate algorithm*; otherwise it is called *multi-rate*. Generalization to multiple input and output domains is straight-forward [3].

It has been shown in [3] that for a multi-rate algorithm, a hardware implementation that uses a single clock will result in inefficient utilization of hardware resources. A simple example can be offered here to illustrate this fact:

Consider a two stage decimation filter (similar to Figure 1, but inluding only the two leading FIR pre- and post-filtering blocks and the intervening decimation block). Assume that the two FIR blocks are of orders K1 and K2 respectively, and that the decimation factor is 2, which implies that only every other value output by FIR_1 is passed on to FIR_2. For convenience, let K2 be an even number.

An FIR filter of order K is described by the following recurrence relation: $y_n = \sum_{i=0}^{K-1} a_i x_{n-i}$. Using systolic cells similar to the one described in [5], its efficient systolic implementation would require a chain of K cells, executing at the sampling rate. The two stage decimation filter, if implemented as a single systolic array, would therefore require K1+K2 such cells. Notice that the model of a systolic array as a Moore Machine ([6]) forces the use of a single clock basis for both sections.

However, because of the intervening decimation block, the latter FIR does useful computation only 50% of the time, and is unutilized the other half of the time (actually, computes on null data), thus leading to architectural inefficiency. (Note that if the decimation factor is greater than 2, the efficiency is worse.)

3 Efficient Architectures For FIR Filters For Decimation

If we relax the constraint of a single clock, we can obtain an architecture that is lower in cost. We will use three clocks C1, C2 and C3 such that C1 has twice the frequency as C2 and

Figure 1: Arrangement of FIR filters for decimation by 2^l.

C3, and C3 has the phase opposite to C2. The K1 order FIR is implemented, as before, using only the C1 clock. We can implement the K2 order FIR as shown in Figure 2. During alternate cycles, the cells compute different halves of two different sums which are eventually combined to produce the correct output. To match the speeds of the x and y inputs of each cell so that the cells collect appropriate terms, an extra register has to be added per cell in the y path (lower path).

Assuming, A, M, R, and m to be the costs of an adder, a multiplier, a register, and a multiplexor, and c to be the number of registers per systolic cell, the cost advantage of the latter scheme is given by, $\frac{K2}{2}(A + M + (c - 2)R - m) - A - 2R$, which is almost always positive if we assume that K2 > 4 and $c \geq 3$.

To specify the architectural requirements for a general decimation example of the kind shown in Figure 1, let us consider

*This work was done when both authors were at the University of Texas at Austin, Austin, TX 78712. The research was supported in parts by DARPA under grant N00039-86-C-0167 and by ONR-URI under grant N00014-86-K-0763. †International Business Machines, AWD – Future Systems Technology, Austin TX 78758. ‡CAE Plus, Austin, TX 78759.

each of clocks, FIFO buffer and multiplexors for the general case of filter section order of K following a decimation factor of 2^i. For the sake of simplicity, we will assume also that the filter order K is expressible as 2^j, $j > i$. (The mapping scheme is independent of this assumption, although hardware utilization is not.)

Clocks: Assuming that a cell is capable of executing at the rate of clock C_0, to implement a filter of order K following a decimation of factor 2^i, we need 2^i other clocks, C_1 through C_{2^i}, each with a clock rate $2^{i\,th}$ that of C_0. Every two consecutive members of these other clocks are separated in phase by $\frac{2\pi}{2^i}$ radians, where T_1 is the period of C_0.

Adders and FIFO buffers: Per cell, the total number of registers required in series to slow the rate of y data, with respect to the x data, for the decimation factor of 2^i is 2^i+1.

During each cycle of C_0, a $2^{i\,th}$ part of the computation corresponding to a y_i is output from the array. These individual parts of the computation have to be accumulated to form the final output. In general, for a decimation factor of 2^i, for each of the clocks, C_1 through C_{2^i-1}, we need $\frac{K}{2^i}$ deep FIFOs which are connected through an adder network. The clock C_{2^i} is used to clock the output register. The number of adders required is 2^i-1. However, since the different adders are not active simultaneously, we could further reduce their number to one if suitable switching hardware.

Multiplexors: For a decimation factor of 2^i, the number of coefficients stored per cell is 2^i. Each cell accesses its coefficients in the order of increasing subscript value. (See Figure 3.) However, the cycle of C_0 during which a cell accesses its lowest-subscript coefficient varies with the cell and is a function of the decimation factor. The C_0 cycle during which cell j+1 accesses its lowest-subscript coefficient lags that for cell j by 2^i+1 cycles.

An Architecture More Suitable For VLSI

In the next scheme we use a constant sized array, time multiplexed over multiple related computations, without paying any penalty in computational rate. This new scheme utilizes a filter design heuristic, based on an empirical relationship described in [2] (not discussed in this paper), that allows one to make each of the FIR blocks of the same order.

Cell With Memory: The cells used in the first scheme have multiplexors to allow them to access different coefficients. Instead, a similar coefficient access capability can be provided using a memory. (Such cells have been used in actual systolic architectures such as Warp [1].)

Bidirectional Cell: The new cells are designed to handle data flowing in both directions. This is achieved by providing a couple of crossbar switches over the x and y data paths of the original cell as shown in Figure 4. At appropriate times the control settings of the crossbars are assigned values to provide a straight-through or a cross-over connection changing the direction of the data-flow in the array.

Mapping and Scheduling If we assume that the cells are capable of computing at a rate twice that of the first FIR stage (or the input sampling rate), the same hardware can be used by all the stages of the decimation process.

Recalling that the filter design heuristic results in the same order, K, of the FIR filter for each of the decimation stages, instead of folding an order K filter over a smaller array, we will use K cells for each section. But with K cells, the array could, in effect, be exercised at $\frac{1}{D}$ of the original rate, or equivalently, only once every D cycles of the original clock.

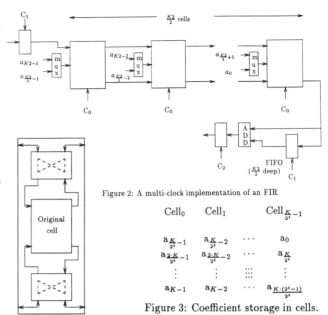

Figure 2: A multi-clock implementation of an FIR

Cell$_0$	Cell$_1$		Cell$_{\frac{K}{2^i}-1}$
$a_{\frac{K}{2^i}-1}$	$a_{\frac{K}{2^i}-2}$	$\cdots$	a_0
$a_{\frac{2\cdot K}{2^i}-1}$	$a_{\frac{2\cdot K}{2^i}-2}$	$\cdots$	$a_{\frac{K}{2^i}}$
$\vdots$	$\vdots$	$\vdots\vdots$	$\vdots$
a_{K-1}	a_{K-2}	$\cdots$	$a_{\frac{K\cdot(2^i-1)}{2^i}}$

Figure 3: Coefficient storage in cells.

Figure 4: A bidirectional cell.

If we now denote the *cumulative* decimation factors at each of the decimation stages D_1 through D_k of Figure 1, we see that an FIR at stage i is computed once every $2 \cdot D_{i-1}$ cycles, and that total computational requirement never exceeds the computational capability of the cell array. Thus the same K cell array can be time-multiplexed (or shared) over different stages of the whole decimation computation.

The overall decimation algorithm is mapped as follows: The first stage is mapped from left to right; the second stage is mapped from right to left; and so on. The two end cells of the array have appropriate micro-control to store the outgoing data at appropriate locations in the memory and use it instead of the input data during subsequent computations. Each cell has a schedule period of $2 \cdot D_k$ cycles of C_0.

References

[1] "The Warp Computer: Architecture, Implementation, and Performance"; M. Annaratone et al, *IEEE Transactions on Computers, December 1987, pp 1523-1538.*

[2] Multirate Digital Signal Processing; R. E. Crochiere and L. R. Rabiner, Prentice-Hall, 1983.

[3] "A Theory for Automated Synthesis of Architectures for Repetitive Multi-rate Algorithms"; S. R. Deshpande, *Ph.D. Dissertation, May 1990, Department of Electrical and Computer Engineering, University of Texas at Austin, Austin, TX 78712.*

[4] "Why Systolic Architectures?"; H. T. Kung, *IEEE Computer Magazine, January 1982, pp 37-46.*

[5] "Systolic Algorithms for the CMU Warp Processor"; H. T. Kung in Collection of Papers on Warp, Department of Computer Science Carnegie Mellon University, Pittsburgh, PA 15213, January, 1987.

[6] "Optimizing Synchronous Systems"; C. E. Leiserson and J. B. Saxe, *22nd IEEE Symposium On Foundations Of Computer Science, 1981, pp 23-36.*

A SYSTOLIC ARRAY EXPLOITING THE INHERENT PARALLELISMS OF ARTIFICIAL NEURAL NETWORKS

Jai-Hoon Chung, Hyunsoo Yoon, and Seung Ryoul Maeng

Department of Computer Science
Korea Advanced Institute of Science and Technology
373-1 Kusong-Dong, Yusung-Gu
Taejon 305-701, Korea

Abstract -- In this paper, a two-dimensional systolic array for back-propagation neural network is presented. The design is based on the classical systolic algorithm of matrix-by-vector multiplication, and exploits the inherent parallelisms of back-propagation neural networks. This design executes the forward and backward passes in parallel, and exploits the pipelined parallelism of multiple patterns in each pass. The estimated performance of this design shows that the pipelining of multiple patterns is an important factor in VLSI neural network implementations.

Introduction

As simulations of artificial neural networks (ANNs) are time consuming on a sequential computer, dedicated architectures that exploit the parallelisms inherent in the ANNs are required to implement these networks. A systolic array [1] is one of the best solutions to these problems. It can overcome the communication problems generated by the highly interconnected neurons, and can exploit the massive parallelism inherent in the problem. Moreover since the computation of ANNs can be represented by a series of matrix-by-vector multiplications, the classical systolic algorithms can be used to implement them.

There have been several research efforts on systolic algorithms and systolic array architectures to implement ANNs. The approaches can be classified into three groups. One is mapping the systolic algorithms for ANNs onto existing parallel computers such as Warp, MasPar MP-1, and Transputer arrays, another is designing programmable systolic arrays for general ANN models, and the other is designing a VLSI systolic array dedicated to a specific model. All of these implementations exploit the spatial and training pattern parallelism inherent in ANNs, and suggest the systolic ring array or two-dimensional array structures.

The back-propagation neural network [2], in addition to the above two types of parallelism, has one more parallel aspect that the forward and backward passes can be executed in parallel with pipelining of multiple patterns. This paper proposes a systolic design dedicated to back-propagation neural network that can exploit this type of parallelism.

The Back-Propagation Model

Let us consider an L-layer network consisting of N_k neurons at the k-th layer. The operations of the back-propagation model are represented by following equations:

The Forward Pass

$$x_{pj}^{(n)} = i_{pj}, \quad (n = 1) \tag{1.1}$$

$$x_{pj}^{(n)} = \sum_{i=1}^{N_{n-1}} y_{pi}^{(n-1)} w_{ji}^{(n)}(t) - \theta_j, \quad (2 \leq n \leq L) \tag{1.2}$$

$$y_{pj}^{(n)} = x_{pj}^{(n)}, \quad (n = 1) \tag{2.1}$$

$$y_{pj}^{(n)} = f(x_{pj}^{(n)}) = \frac{1}{1 + e^{-x_{pj}^{(n)}}}, \quad (2 \leq n \leq L) \tag{2.2}$$

The Backward Pass

$$\beta_{pj}^{(n)} = y_{pj}^{(n)} - d_{pj}, \quad (n = L) \tag{3.1}$$

$$\beta_{pj}^{(n)} = \sum_{k=1}^{N_{n+1}} \delta_{pk}^{(n+1)} w_{kj}^{(n+1)}(t), \quad (2 \leq n \leq L-1) \tag{3.2}$$

$$\delta_{pj}^{(n)} = \beta_{pj}^{(n)} y_{pj}^{(n)} (1 - y_{pj}^{(n)}) = g(\beta_{pj}^{(n)}), \quad (2 \leq n \leq L) \tag{4}$$

The Weight Increment Update

$$\Delta w_{ji}^{(n)}(u+1) = \Delta w_{ji}^{(n)}(u) + \delta_{pj}^{(n)} y_{pi}^{(n-1)}, \quad (2 \leq n \leq L) \tag{5}$$

$$w_{ji}^{(n)}(t+1) = w_{ji}^{(n)}(t) + \varepsilon \Delta w_{ji}^{(n)}, \quad (2 \leq n \leq L) \tag{6}$$

Terminating Conditions

$$E = \sum_p E_p = \sum_p \sum_{i=1}^{N_L} (y_{pi}^{(L)} - d_{pi})^2 < \alpha \tag{7}$$

Inherent Parallelisms

There are three types of parallelism inherent in back-propagation model. The first is the *spatial parallelism* which can be exploited by executing the operations required in each layer on many processors. In spatial parallelism, the operations are partitioned into groups, and the partitioned operations are executed on different processors in parallel.

The second type of parallelism is the *pattern parallelism* which can be exploited by executing the different patterns on many processors. The pattern parallelism can be classified as follows:

1) Partitioning the training pattern set, independent execution on many processors, and epoch update (*training set parallelism*)
2) Pipelining of multiple training patterns or multiple pattern recalls

The parallelism 1) can be exploited by partitioning the training patterns and execution on many processors independently. The parallelism 2) can be exploited by allowing some processors that completed the parts of operations for a pattern to start execution for the next pattern while the other processors are processing for the previous patterns presented.

The third type of parallelism is the *pipelined execution* of forward and backward passes of different training patterns [3]. The depth of pipelining can be maximized to twice the number of neurons in the network, and it is effective if the size of the training set is sufficiently large compared to the number of neurons.

Updating the Weights

The issue of the weight updating time has been controversial between researchers who update network weights continuously after each training pattern is presented (*continuous updating*) and those who update weights only after some subset, or often after the entire set, of training patterns (*batch updating*) [4]. To exploit the traing pattern pipelining, we use the batch updating method because the weights must not be changed for the forward pass and the backward pass of a pattern as shown in Equations (1)-(6).

The Systolic Array Organization

The computation of artificial neural networks can be represented by a series of matrix-by-vector multiplications interleaved with the non-linear activation functions. The matrix represents the weights associated to the connections between two adjacent layers, and the vector represents the input patterns or the outputs of a hidden layer in the forward pass, or the errors propagated in the backward pass. The resultant vector of a matrix-by-vector multiplication is applied to the non-linear activation function and to the next layer.

In Figure 1, the systolic array organization between two adjacent layers is shown. One layer consists of 5 neurons and the other layer consists of 3 neurons. Each basic cell w_{ji} is a computational element which contains the weight value associated to the connection between neuron i of a layer and neuron j of the next layer. The basic cell executes the sum-of-products operations as shown in Equations (1.2) and (3.2) and weight updating operations as shown in Equation (5) and (6). The weight matrix is shared between the forward and backward passes, and the data paths of the activations forward-propagated and the errors back-propagated are disjoint. The θf unit executes the thresholding and the activation function as shown in Equations (1.2) and (2.2), and the g unit executes the derivative of the activation function as shown in Equation (4). The g unit has a FIFO queue to contain the outputs of a neuron, that are fed back to the basic cells to compute the weight updates in the backward pass.

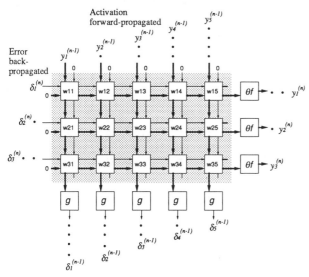

Figure 1. A systolic array organization between two adjacent layers of artificial neural networks.

In Figure 2, an example of a systolic array organization for (5-3-2) 3-layer network is shown. The elements of two weight matrix are set into the two dimensional arrays, the 3-by-5 weight matrix represents the weights associated to the connections between the input layer and the hidden layer, and the 2-by-3 weight matrix represents the weights associated to the connections between the hidden layer and the output layer. The second matrix is transposed in the figure.

Performance Evaluation

In this design, the basic clock cycle time is determined as the maximum time among the time required by one multiplication and one addition, the time for thresholding and activation function (table lookup), and the time required to feed the input patterns continuously. Assuming 100 nsec of cycle time and 16-bit weight precision, the estimated performance measured in *MCUPS (millions of connection updates per second)* is 248 MCUPS for NETtalk text-to-speech neural network [5] for a single training pattern only by the spatial parallelism and the forward/backward pipelining, and for (128-128-128) three-layer network [3] with 64K training patterns pipelined, the performance that can be obtained is 324,000 MCUPS.

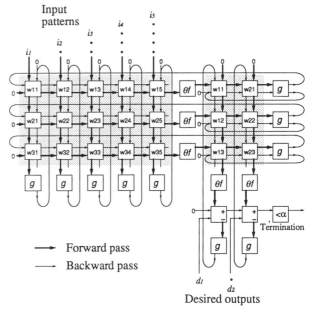

Figure 2. An example of a systolic array organization for (5-3-2) 3-layer artificial neural network.

Conclusions

In this paper, a systolic array for back-propagation neural network that executes the forward and backward passes in parallel, and exploits the pipelining of multiple patterns in each pass has been presented. The estimated performance of this design shows that the pipelining of multiple patterns is an important factor in VLSI implementations of artificial neural networks. To implement a large network with this design, an efficient mapping method that still allows the exploitation of the pipelining is required as a further study.

References

[1] H.T. Kung, "Why Systolic Architectures?," *IEEE Computer*, January 1982, pp.37-46.
[2] D.E. Rumelhart, G.E. Hinton, and R.J. Williams, "Learning Internal Representations by Error Propagation," *Parallel Distributed Processing: Explorations in the Microstructures of Cognition, Vol.1, Foundations*, MIT Press, 1986, pp.318-362.
[3] A. Singer, "Exploiting the Inherent Parallelism of Artificial Neural Networks to Achieve 1300 Million Interconnects per Second," *Proc. of INNC*, Vol.II, July 1990, pp.656-660.
[4] S.E. Fahlman, "Fast-Learning Variations on Back-Propagation: An Empirical Study," *Proc. of the 1988 Connectionist Models Summer School*, Morgan Kaufmann Publishers, June 1988, pp.38-51.
[5] T.J. Sejnowski and C.R. Rosenberg, "Parallel Networks that Learn to Pronounce English Text," *Complex Systems,* Vol.1, 1987, pp.145-168.

A Fault-Tolerant Design of CFM VLSI Parallel Array for Yield Enhancement and Reliability Improvement

Wen-Gang Che Yan-Da Li and Yue Jiao

Department of Automation, Tsinghua University
Beijing 100084,China

ABSTRACT-- We propose a new method in this paper called component-level fault-tolerance. In contrast with using spare PEs in many published results, the redundant computing components are incorporated in every PE to improve the yield and reliability of the PEs and the whole array because every PE's computing units consist of several configurable function components in PE of the CFM array. Taking the advantage of n-element-1-order function structure, which is the basic computing architecture of the computing unit of the CFM, only one redundant component is necessary to design the fault-tolerant PE or array. This design is suitable for both fabrication-time and run-time reconfigurations.

Index terms: Configurable Functional Method, Component Level Fault-Tolerance

1. Introduction

Generally, an architecture of the VLSI parallel array is derived from the structures of its particular algorithm[1][2]. Such a design method is unattractive to the industries and real applications because of their specified architectures[3][4]. The flexible architectures are more feasible because these application-restrictionless structures will make a VLSI parallel array more applicable [3][4]. In addition, these flexibilities may relatively reduce the production overhead. Attempting to set up a near general-purpose parallel array, we introduced a new design strategy called Configuration Function Method[4].

Assume that a PE in a VLSI array has n inputs and m outputs, i.e. $x_0, x_1 ..., x_{n-1}$ and $y_0, y_1 ..., y_{m-1}$, the relationship between inputs and outputs can be expressed as following Fig.(1) and (1):

$$
\begin{bmatrix} y_{m-1} \\ \cdots \\ y_1 \\ y_0 \end{bmatrix} = \begin{bmatrix} f_{m-1}(x_{n-1} \cdots x_1 x_0) \\ \cdots \\ f_1(x_{n-1} \cdots x_1 x_0) \\ f_0(x_{n-1} \cdots x_1 x_0) \end{bmatrix} \quad (1)
$$

Figure(1) PE model with n inputs and m outputs

Suppose element x_i is multiplied by the other elements at one order, we call the product part of (2) n-element-1-order product function. Sum these terms weighted by the coefficients $c_{k,j}$ and assign the summation to y_k, then the (1) is rewritten as:

$$
y_k = \sum_{j=0}^{2^n-1} c_{k,j} \prod_{i=0}^{n-1} (x_i^{b_{j,i}}) \quad Where \ k=0,1,...,m-1. \quad (2)
$$

Therefore y_k is a configurable function because y_k is determined by $c_{k,j}$. Here:

$$
b_{j,i} = 0 \ or \ 1, \ for \ i=0,1,...,n-1 \quad (3.a)
$$

$$
c_{k,j} = 0 \ or \ 1, \ for \ j=0,1,...,2^n-1 \quad (3.b)
$$

We use this computing structure in the PE of the CFM array so that the CFM array is configurable depending on the demands of application. This is because different parallel algorithms can be realized on them using a particular set of $c_{k,j}$ which actually reflects the parallel structure of the algorithms.

2.The Augmented CFM PE and Fault Model

Conventional methods dealing with defects in PEs adopt "PE-level" fault models which are fault-tolerant or reconfigurable among array using spare PEs[5][6][7][8][9][10]. In contrast with the methods proposed previously, a fault-tolerant design at computing component level in every PE of the CFM array is discussed in this paper. This design mainly depends on the special structure of CFM and n-element-1-order product function.

A. The Augmented CFM PE

Besides the primitive CFM PE, only a few of extra hardwares, such as redundant function unit, test and control circuits, are attached to a PE to achieve the reconfiguration. We discuss a three-input three-output (n=2, m=2) CFM PE here[4]. In Fig.(2), the component $x_0 x_1 x_2$ is used as a redundant component for multiplying units. Similarly, an additional add-multiplexer is used for add-multiplexing units.

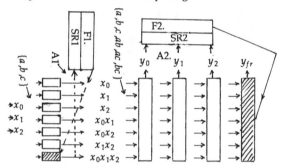

Figure(2) Augmented CFM PE

The process of test and reconfiguration is driven by the following inner PE clock.

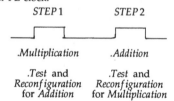

Figure(3) Time-Overlapped Technique for Test and Reconfiguration

B. The Fault Model

Without loss of the generality, we assume:

1) Interconnection between PEs and computing components are fault-free.

2) A failure in a VLSI/WSI affects only a small area of the chip and nature of the failure is precisely unknown.

3) Control circuits, test and reconfiguration circuits are fault-free.

4) A failure occurs only on one of the computing component, one of multiplier or add-multiplexer.

3. Test Scheme

Because of using time-overlapped techniques in test and reconfiguration, the whole processes are time-saved. We discuss the test of multipling components and add-multiplexers respectively.

A. Product Function Test

Instead of the input x_0, x_1 and x_2, a test vector {a, b, c} is fed into the multiplying units. Vector A1 is the tested result, and $A1^*$ is the standard vector. SR1 is a register array triggered by the difference between A1 and $A1^*$. Vector F1 is a vector whose element is the bit of SR1. Similarly, we assume A2, $A2^*$, F2 and SR2 for the add-multiplexers. Consequently, the test process is carried out by the following steps.

Definition: $\sum$FK (K=1,2) = $f K_0 \sum f K_1 \dots \sum f K_l$, where $f K_i$ (i=0,1,...,l) is the element of vector FK.

(1). If A1 = {a, b, c, ab, ac, bc, abc},
      F1 = {0,0,0,0,0,0,0,0} and $\sum$F1=0,
      then no fault.

(2). If A1 ≠ {a, b, c, ab, ac, bc, abc},
      F1 ≠ {0,0,0,0,0,0,0,0} and $\sum$F1=1,
      then one fault.

(3). If A1 ≠ {a, b, c, ab, ac, bc, abc},
      F1 ≠ {0,0,0,0,0,0,0,0} and $\sum$F1>1,
      then two faults or more.

B. Add-Multiplexer Test

The test process of add-multiplexers is similar with that of product component's except that the input vector is {a, b, c, ab, ac, bc}.

4. Reconfiguration

The process of reconfiguration is performed serially as soon as the process of test is completed.

A. Reconfiguration of Product Functions

Because the $x_0 x_1 x_2$ is used as a redundant unit among the multiplying circuits, we design a mask circuit to have this unit generate a subset of $x_0 x_1 x_2$, such as $x_1 x_2$, x_1, etc., to replace the failed.

(1). If $\sum$F1=0,
      then no fault, no reconfiguration.

(2). If $\sum$F1=1,
      then one fault, while SR1 is flagged, the component of $x_0 x_1 x_2$ generates a compensated result the same as the failed to replace the fault depending on F1.

(3). If $\sum$F1>1,
      then completely faied and unrecoverable.

B. Reconfiguration of Add-Multiplexers

The process of add-multiplexer reconfiguration is easier than that of product component reconfiguration for we use a unit as same as add-multiplexer itself.

(1). If $\sum$F2=0,
      then no fault, no reconfiguration.

(2). If $\sum$F2=1,
      then one fault, the fault is replaced by the redundant while SR2 is flagged.

(3). If $\sum$F2>1,
      then completely faied and unrecoverable.

5. Simple Analysis of Reliability

Assume that all the components are independent of each other so that one failure on one of them has no effects on the others, and the reliability of every component is equal to each other. According to our fault model and Fig.(2), the PE reliability is:

$$[\sum_{i=0}^{1} C_N^i \ r_1^{N-i} \ (1-r_1)^i][\sum_{j=0}^{1} C_M^j \ r_2^{M-j} \ (1-r_2)^j] \qquad (4)$$

Therefore, the primitive PE without the attached redundancies has the reliability as (5), where N=6,M=3:

$$R_{PE0} = r_1^6 r_2^3 \qquad (5)$$

Similarly, the augmented PE has the reliability (6), where N=7,M=4:

$$R_{PE} = [r_1^7 + 7 r_1^6 (1-r_1)][r_2^4 + 4 r_2^3 (1-r_2)]$$
$$= [7 r_1^6 - 6 r_1^7][4 r_2^3 - 3 r_2^4] \qquad (6)$$

Comparing (5) and (6) listed above, the improved reliability is (7):

$$R_{PE}/R_{PE0} = [7-6r_1][4-3r_2] \qquad (7)$$

If we consider the time, then the reliability of the product component is $e^{-l_1 t}$, and $e^{-l_2 t}$ for the add-multiplex component. Therefore, the improved reliability of the augmented PE is a function of time.

From the above designs and analyses, our CFM PE or array with the component-level fault-tolerance at least has the following advantages.

1). Only one functional unit is necessary to achieve the fault-tolerance in every PE.

2). While a fault arose, the size of whole array remains unchanged. Therefore, a repartition of the parallel algorithm running on the array can be avoided.

3). Generally speaking, a low level fault-tolerance is more reliable than high level's.

4). CFM array is able to run much more parallel algorithms on it.

References

[1]   H.T. Kung, "Why Systolic Architecture ?", *Computer*, Jan. 1982.

[2]   S.Y. Kung, "VLSI Array Processors", 1987.

[3]   L. Snyder, "Introduction to the Configurable Highly Parallel Computer", *Computer*., 15 (1) Jan. 1982.pp47-56.

[4]   Che Wengang, Li Yanda and Jiao Yue, "A Systolic Realization for 2-D Convolution Using CFM VLSI Parallel Array " accepted by *IEE Proceedings Part E* " *Computers and Digital Techniques* ".

[5]   I. Koren, " A Reconfigurable and Fault-Tolerant VLSI Multiprocessor Array", *Proc. 8th Annu. Symp. Comput. Architec.*, 1981, pp. 425-442.

[6]   I. Koren and O.K. Pradhan, "Yield and Performance Enhancement through Redundancy in VLSI Multiprocessor Systems", *Proc. IEEE*, vol. 74, pp. 699-711, May. 1986.

[7]   R.Negrini, M.G.Sami and R.Stefanelli, "*Fault Tolerance Through Reconfiguration in VLSI and WSI Arrays*" The MIT Press, 1989

[8]   H.T. Kung and M.S. Lam, "Fault-Tolerance and Two Level Pipelining in VLSI Systolic Arrays", *Proc. MIT. Conf. Adv. Res. VLSI*, Jan. 1984, pp. 74-83.

[9]   I. Koren and M.A. Breuer, "On Area and Yield Consideration for Fault-tolerant VLSI Processor Array", *IEEE Trans. on Comput.*, vol. C-33, No. 1,1984, pp. 21-27

[10] M.Wang, M.Cutler and S.Y.H.Su, "On-line Error Detection and Reconfiguration with Two-level Redundancy" *Proc. COMPEURO 87* Hamburg 1987, 703-706, IEEE

The MULTITOP Parallel Computers for ASDEX-Upgrade

G. Raupp, H. Richter

Max-Planck-Institut für Plasmaphysik, EURATOM Association
Boltzmannstr. 2, 8046 Garching, Germany

Abstract

This paper describes a class of multiprocessor architectures which are based upon transputers and fixed or variable interconnection networks. Additionally a transputer control bus and a ´fifth´ link concept is used. The architecture is realized in 7 computers which are used in the fast part of a large experiment control.

I. INTRODUCTION[1]

ASDEX-Upgrade (AUG) is Germany´s largest experiment in the field of plasma physics for controlled nuclear fusion. For this fusion experiment a fast control, superior to standard industry controllers, had to be developped. This holds for I/O speed, number of signals, floating point power and real time behavior related to the costs.

The requirements for the AUG control were: At first place a guaranteed reaction time of 50 ms for overload monitoring of the reactor vessel. A 10 ms response time for plasma generation and its feed forward control (ff). And an under 2 ms reaction time for the feed back control (fb) of the plasma position with several thousand floating point operations included. At second place a large number of signals (about 550) had to be processed. And at third place the costs should be modest.

On the other hand there was 3 years of experience in parallel processing and a first protoype of a "MULTITOP" parallel computer exists [1], [2]. Then it took another 2 years to develop and implement the computer part of the fast AUG control. Seven MULTITOP computers (MTs) were realized in 3 different applications:

In the first application two MT/29 computers with 29 processors each can be used for the ff control. In the second application two MT/14 computers with 14 processors are for the fb control. And in the third application two MT/5 computers with 5 processors are responsible for the overload monitoring. Apart from them there is a MT/33 machine for basic research. For each computer a prototype was built and later replaced by a sucessor in industrial quality.

II. THE MULTITOP ARCHITECTURE

The architecture is based upon 7 principles. Some of them and their combination can be called new:

1. All MT computers are embedded systems for a standard UNIX workstation with VMEbus. The host is used as network- and file server and cares for a graphical user interface.

2. All MT computers consist of a set of processor memory modules (PMs), who communicate by message passing. The processors are transputers [3].

3. There are heterogenous PMs for different tasks. But every system- and user-task is performed by only 2 basic types of boards:

The first is the "CPU0" board with 4 PMs (T800 transputer, 0.5 MB fast SRAM each PM). This type of board is used for computing.

The second is the "MCPU" board with one PM (T800, 4 MB DRAM). This type is used for master functions, as I/O processor and for network switching.

All boards are VMEbus compatible. To interconnect the CPU0 and MCPU boards 3 different backplanes are used: The MT/14 computers have a printed circuit backplane in 9 layer stripline technic to minimize crosstalk and ground bouncing. It incooperates 4 standard INMOS [3] link switches and the slots for 3 CPU0s, 2 MCPUs and their optical converter boards.

The MT/29 computers have two 9 layer printed circuit backplanes. One for 6 CPU0s accompanied by 2 optical boards in a ´triple slot´ unit each. The other backplane is for 5 MCPUs and their converter boards, which are relativly fixed coupled on a jumper basis.

---

[1]We hereby thank our colleagues from the E1 and Informatik division and the UCS company for their valuable remarks and support.

The MT/33 backplane connects a fully reconfigurable network in PAL technic with a MCPU with 32 fifth links and 8 CPU0 boards.

The computers are housed in standard 19″ crates with a total height of 18 units.

4. The PMs are coupled via a fixed or variable interconnection network. The goal was to make the gap between the software (SW) and the underlying hardware (HW) as small as possible. In the case of a variable topology the connections between the processors (links) are switched over several times during one SW cycle to fit the HW to the SW and not vice versa. The interconnection networks are realized either by standard crossbar chips or by a self developped non blocking network with minimum chip count. For this network a patent was applied (H. Richter).

5. Every computing PM (CPU0) has apart its four regular communication links a fifth link. With all fifth links a star can be formed with a MCPU as a master in the center of the star. The MCPUs has many fifth links as counterparts. This gives a fixed topology for system control parallel to the variable user topology. Furthermore it allows better SW synchronization between the PMs among one other and the master by a broadcast function from master to PMs. And vice versa the PMs can access the host screen and file I/O via the master, which is connected to the host. So an efficient debug of parallel programs is possible by this path.

6. All transputers are connected by a "Transputer Control Bus" (TSB). This special bus allows scaleable compute power on a simple ´plug in´ principle. This means, if there are free slots in the crate, more CPU0 boards can be plugged in without any HW changes.

It furthermore allows *topology independent boot* and additional debug of the transputers, because a bus signal can switch the topology of the links from the variable application topology to a fixed topology, which is a chain. So booting and debugging has always the same topology. This mechanism does only apply for the 4 communication links, not for the fifth links star.

With the TSB a HW clock synchronization of the PMs is performed and an additional SW synchronization is possible on a common interrupt basis. This is an important feature for the real time applications to avoid clock and SW drifts.

7. The I/O is strictly done via glass fibre links. The transmission speed is 10 MBit/s at a maximum distance of 1 km. Glass fibre transmission is necessary because of the high electromagnetic radiation and voltage differences between crates in the fusion experiment. The I/O is done massivly parallel, the MT/29 computer for example has 64 input- and 32 output glass fibre links. Each optical converter board has 8 ports, which can be configured as send, receive or bidirectional. They are VMEbus compatible.

Beside this common architecture each MT computer has some special features. The MT/29 has a fixed topology of an input - and an output star. Every CPU0 board in the two stars is combined with 2 optical converter boards for the glass fibre links to form a triple slot unit. All the units are plugable to their master to easily expand the number of I/O channels (together with the TSB). The MT/14 is switched over 3 times every SW cycle from input -, to compute -, to output topology to get maximum benefit of the transputers. The compute topology is a torus. The MT/5 is fixed wired, but with heavily multiplexed I/O signals. The MT/33 realizes in a self developped interconnection network all 32! possible permutations of connections.

III. RESULTS

Up to now there exists half a year of experience in the operation of the computer part of the fast AUG control. The MULTITOP computers have met the requirements with respect to functionality, speed and reliability. The variable interconnection network has been proven as reliable. However for an efficient use of the net the fifth links and the TSB has been shown necessary. The HW costs were moderate. The complex transputer SW could be clearly structured, because of the adaptable underlying HW.

IV. ADDITIONAL INFORMATION

The Annual Report 1990 of the Max-Planck-Institut für Plasmaphysik contains a comprehensive description of the whole fast control of ASDEX-Upgrade.

V. REFERENCES

[1]   H. Richter, Multiprozessor mit Dynamisch Variabler Topologie, University of Munich, Ph.D. dissertation, 1988.

[2]   H. Richter, Multiprocessor with Dynamic Variable Topology, Computer Systems Science and Engineering Vol. 5. No.1 Jan. 1990, Butterworth UK, pp. 29 - 35.

[3]   INMOS company, The Transputer Databook, 1989.

Performance Indices for Parallel Marker-Propagation [†]

R. F. DeMara and D. I. Moldovan

Dept. of Elec. Engr.–Systems, University of Southern California

Los Angeles, California 90089-1115

Abstract

Marker-propagation is an intrinsically parallel reasoning technique which has been used in many AI applications. Recent research has concentrated on how to implement marker-propagation efficiently using existing and special-purpose parallel computer architectures [1] [2] [3]. However, few measures currently exist to quantify the computational work performed in these systems. Existing metrics such as Logical Inferences per Second (LIPS) obscure the computational features of this massively parallel approach. We propose a set of indices for measuring processing throughput and characterizing the associated knowledge bases.

Propagation Mechanisms

Our objective was to develop representative processing indices which could be related to features of the application program. These metrics should be similar in concept to those which already exist for numeric-processing applications, such as the number of 32-bit floating-point operations (FLOPs) performed. In numeric processing the basic floating point operation is a representative metric because it forms a common basis between the algorithm and machine. However, no equivalent measures have been available for marker-propagation.

We have approached this problem by defining metrics capable of quantifying the fundamental operations of "useful work" under the marker-propagation paradigm. We contend that each marker propagation has two primary effects:

1. *Conditioning:* activation or deactivation of the status bit in a node or the execution of an arithmetic operation upon node or link registers,

2. *Spreading:* propagation of the marker to other selected nodes specified by some propagation rule.

During propagation, spreading can be considered transient work while node conditioning produces steady-state effects. This is because spreading consists of search-and-dispersion operations which consume computational resources, but have no lasting impact on the knowledge base. On the other hand, conditioning operations modify the state of the knowledge base during execution. Our set of indices for marker-propagation are listed in Table 1.

[†]This research has been funded by National Science Foundation Grants MIP-89/02426 and MIP-90/09109

Propagation Indices		Significance
$\mathcal{A}$	marker activation	changes to knowledge base
$\mathcal{M}_{inj}$	marker injection	new markers introduced
$\mathcal{S}(t)$	scattering function	time behavior behavior of routing and replication
$\mathcal{I}$	marker inspection	searching work performed
$\mathcal{D}$	marker dispersion	spreading work performed

Table 1: Indices for Marker-Propagation.

Marker Activation

The *activation*, $\mathcal{A}$, of an injected marker is a measure of the effect it has on the knowledge base. It is a computation-oriented metric. This quantity is the sum of the activation status bits flipped at a node (these bits that indicate membership in a hypothesis, i.e. the nodes become marked by them) and the number of FLOPs induced (arithmetic computations to compute marker value). Thus given a marker propagation m_i we measure:

$$\mathcal{A}_{m_i} = \sum_{\substack{prop\ cycle \\ m_i}} (\text{bit inversions}) + \sum_{\substack{prop\ cycle \\ m_i}} \left(\begin{array}{c} \text{FLOPs} \\ \text{performed} \end{array} \right) \tag{1}$$

Thus the number of activations per second is a microscopic indicator of the amount of computational work performed, in the absence of communication expenses.

Marker Scattering Function

The marker *scattering function*, $\mathcal{S}_{m_i}^j$, denotes the time behavior of two figures-of-merit for work done by spreading. It reflects the communication work performed. The first component, called the *inspection*, is defined as:

$$\mathcal{I}_{m_i} = \sum_{\substack{prop\ cycle \\ m_i}} \sum_{\substack{nodes \\ encountered}} \left(\begin{array}{c} relations \\ searched \end{array} \right) \tag{2}$$

It is the amount of work performed by searching the relations in accordance with the propagation rule. The second component of the scattering function is the *dispersion*. The dispersion measures the amount of new propagations that are recursively created:

$$\mathcal{D}_{m_i} = \sum_{\substack{prop\ cycle \\ m_i}} \left(\overset{new}{activations} \right) \quad (3)$$

This is precisely the amount of successful inspections:

$$= \sum_{\substack{prop\ cycle \\ m_i}} \sum_{\substack{nodes \\ encountered}} (\mathcal{I}_{relation} = \mathcal{I}_{prop\ rule}) \quad (4)$$

Thus the scattering function for the jth propagation is:

$$\mathcal{S}_{m_i}^j = \mathcal{I}_{m_i}^j + \mathcal{D}_{m_i}^j \quad (5)$$

The motivation for separating the components of the spreading tasks is to isolate the effects of particular algorithm features. The distinction may be significant in the context of the architecture used. For example, suppose we wanted to assess the suitability of Content Addressable Memory (CAM) in a marker-propagation machine. These metrics could be tabulated for two application algorithms φ_1 and φ_2 as part of the design study. The algorithm $\varphi_1 = \langle large\ \mathcal{I}_{m_i},\ small\ \mathcal{D}_{m_i} \rangle$ would benefit from a CAM for fast searching during the inspection process. On the other hand, $\varphi_2 = \langle small\ \mathcal{I}_{m_i},\ large\ \mathcal{D}_{m_i} \rangle$ would benefit from tighter coupling between nodes, but derive little benefit from a CAM.

Lazy Markers

Under the marker-passing model, it is not possible to know a-priori how long a marker-propagation instruction will take. A major problem this can cause is low resource utilization while waiting for the final markers to complete propagation. We have called this the *lazy marker effect*. We refer to the responsible marker as a *lazy marker*. To quantify this effect, it is necessary to specify precisely what it means to wait a "long" time. This may be readily defined in terms of the scattering function $\mathcal{S}(t)$. Here

$$\mathcal{S}(t) \quad \forall\ t_{injection} \leq t \leq t_{end\ propagation} \quad (6)$$

indicates the markers propagating through the network during the propagation period. Lazy effects occur at the tail-end of this distribution. We know that $\mathcal{S}(t = t_{injection}) = 1$ and $\mathcal{S}(t = t_{end}) = 0$. However, $\mathcal{S}(t)$ is not a strictly bitonic function. Thus we define lazy markers in terms of the length of the propagation cycle spent waiting for the last 1% of the markers to finish.

$$t_{lazy} \triangleq t_{99-100} \quad such\ that\ \mathcal{S}(t) \leq 1\%\ of\ \mathcal{S}_{max} \quad (7)$$

The t_{lazy} parameter must be considered relative to the overall propagation period T_{prop}. Consider the bursty nature of reasoning algorithms which generate ratios of $(T_{prop} - t_{lazy})/T_{prop} \approx 0.75$. This indicates for at least 25% of the time, the machine is operating at less than 1% of capacity. In this case, the price one has to pay to provide massively parallel support for marker-propagation is low processor utilization on average.

Network Indices		Significance
F_{out}	fan-out	outgoing relations per node
F_{in}	fan-in	incoming relations per node
$l_{reltype}$	path length	relations in a chain of links of type $reltype$
M	network size	concepts in hierarchy
$N_{reltype}$	composition	(number of $reltype$ links) $\div M$
T	topology	$T \epsilon \{tree, mesh, random\ replicated\ subgraph\}$

Table 2: Networked Representation Indices.

Networked Knowledge Bases

Marker-propagation algorithms typically represent knowledge using semantic networks. Semantic networks are directed graphs of concepts (graph nodes), their properties (graph links), and the hierarchical relationship (partial ordering) between them. Semantic networks are difficult to characterize because their interconnections are essentially random. To characterize them, we use the average and maximum values of the parameters in Table 2.

We have observed a common feature that we call the *replicated subgraph*. This refers to the existence of large forests of similar trees, with each tree representing an specific event sequence. This regularity of structure provides insight into likely hot spots during computation.

We feel three cases are especially representative for benchmarking marker-propagation performance:

1. vary M (e.g. 256, 1K, 4K, 16K nodes) with T fixed (tests the scale-up of a similar but larger knowledge base),

2. vary $l_{reltype}$ with F_{in}, F_{out} fixed (examines effect of changing the critical path), and

3. vary M and F_{out} with $l_{reltype}$ fixed (examines impact of network bushyness).

We are currently analyzing knowledge bases in terms of these parameters. First, we would like to obtain some idea of what typical ranges are for these indices. Second, this will allow construction of a small set of standardized, synthetic knowledge bases for benchmarking purposes.

References

[1] R. F. DeMara and D. I. Moldovan, "The SNAP-1 Parallel AI Prototype," in *Proceedings of Eighteenth Annual International Symposium on Computer Architecture*, 1991.

[2] M. Evett, J. Hendler and L. Spector, "PARKA: Parallel Knowledge Representation on the Connection Machine," Technical Report UMIACS-TR-90-22, University of Maryland, 1990.

[3] T. Higuchi, *et al*, "The Prototype of a Semantic Network Machine IXM". *Proceedings of 1989 International Conference on Parallel Processing*.

PERFORMANCE PENALTY BY COMMUNICATION IN MULTIPROCESSOR SYSTEMS

Walter H. Burkhardt

Institut für Informatik, Universität Stuttgart

D-7000 Stuttgart-80, West Germany

Tel: (049)-711-781-6391, e-mail: b_hardt@ifi.uni-stuttgart.dbp.de

Abstract: Performance degradation by communication has been investigated for matrix operations and other programs in multiprocessor systems theoretically for several configurations and by experiment on a linear Transputer and on a crossbar connected system. Results show severe limitation and degradation, following a negative exponential, depending on the speed and amount of communication.

1. Introduction.

The problem of input/output and communication and their influence has not been given adequate consideration in most previous investigations of parallel processor systems [BUR87]. Raw estimates diverge from a zero influence by supposed overlapping with processing [BEL89] to several orders of magnitude of performance degradation [HAC86].

2. Theoretical Investigation.

2.1. Linear Model.

A linear array is the simplest form of a multiprocessor system, with n identical processors, each with its own memory. The input and output enters and leaves at one end. The model has been detailed in [BUR89], results are reported here only.

With p the computing power of one processor module, and q be the communications requirements factor, as a fraction of p, for transferring data between input and output to one processor in the row with overlapping with computation for the fraction (1-r). The processor with the index i requires then for communication the processing power:

$$q_i = q * [(r + 1) * (n - i) + 1]. \qquad (1)$$

The total ideal computational power of the array $P = n * p$ decreases to P' by the total communications requirement Q:

$$Q = q * n * [n * (r+1) - (r-1)] / 2. \qquad (2)$$
$$P' = n*[p + q*(r-1)/2 - q*n*(r+1)/2] \qquad (3)$$

Maximally achievable systems power Pmax is obtained at the optimal number of Nmax processors:

$$Nmax = (2*p - q + q*r)/(2*q*r + 2*q) \qquad (4)$$
$$Pmax = (2*p - q + q*r)^2 / (8*q*r + 8*q) \qquad (5)$$

2.1. Overlap of Data Transfer

With total overlap with processsing:

Total computing power $P'_o = n*(p - q/2) - q*n*n/2$ (3a)
Optimal number of processors: $Nmax_o = p/q - 1/2$ (4a)
Maximal systems power: $Pmax_o = p^2/(2*q) - p/2 + q/8$ (5a)

Other cases and configurations appear in [BUR90b]; they give similar results with the optimal number of processors still as a function of p/q.

2.2. Communication Overhead.

As soon as the data has entered the first processor in the row, it cannot be distinguished from other communication. Communication and data transfer behave similarly.

3. Experimental Results.

3.1. Transputer System.

The configuration of the Transputer system consists of eight processor modules, connected as a linear string. One module has the 32-bit Transputer processor, 1 Mbyte local memory, the clock and 5 simple integrated circuits [BOD89].

3.1.1. Programs Performed.

Previous investigations have shown little exploitable parallelism in programs in procedural languages, aside from vector- or matrix-loops [MAK84]. The assumption arose that the serial pattern of thinking in such languages, the language grid, could be the reason. These limitations should then disappear with programs in a language of the next higher level of declarative languages [BUR65]. Functional Programming languages belong to them. Programs on this level are free from side-effects and extremely terse.

The application programs, that were run on the Transputer system, are written in a language derived from KRC [TUR82] with the most important additional features of global and local functions for reasons of modularity. They were: The matrix inversion (Mat20D1), quicksort program (Qs600D1), Hartley transform (Hart50D1), Knapsack problem(Knap13D1), Differential quotient (Dif1500D1), Fibonacci-Series (Fib20D1), Towers of Hanoi (Han10D1), and the Queens problem (Queen7D1). The number after the shorthand description of the problem indicates the size of the data set, D1 refers to the linear dimension of the Transputer array. For listings see [BUR89]. They are translated by a compiler into an intermediate language with the detection of exploitable parallel paths [BUR68, BOD89].

3.1.2. Measurements and Results.

The run-time of the various programs was measured with the distribution of the computational load on the different modules. The Fibonacci program reaches the best performance with a speedup of 4.5; the Towers of Hanoi program the worst with a speedup of only 1.75, running on the eight processor system. The other problems range in between. The low value for the Towers problem indicates a lack for parallelizability.

Analyzing the results by a quadratic polynomial, an exponetial, a power, and a logarithmic regression program [BUR89] gives the communications factors from 14% to 18%. This translates into an optimal number of modules of about six to seven and maximal systems power of between two and five. For other configurations, see [BUR90a].

3.2. Crossbar System.

The system consists of n = 5 processor modules Z8000, m = 4 memory modules, connected by a decentralized crossbar switch [BÜD82, BUR87], running a matrix inversion program. The matrix has 100 non-zero elements. The parallel portion of it was run 1000 times in five runs each on a variable number of processors with adjustable amounts of data for input and output coming from, or going to, memory or to floppy disk drives. Media speed or amount of data can be modified.

3.2.1. Measurements and Results.

The processors used a 4k Byte cache with the Set4-cache-strategy with FIFO mode and the loading of a block of one word at a miss [HAL86]. Table 1 shows the speedup.

Table 1: Measured Results as Speedups.

Sectors:	0	0.144	0.25	.333	0.5	1	2	4
Procs.: 1	1	.906	.861	.795	.717	.544	.363	.272
2	1.99	1.92	1.73	1.59	1.44	1.08	.726	.362
3	2.98	2.66	2.58	2.23	2.15	1.30	.932	.353
4	3.96	3.44	3.40	2.56	2.76	1.41	.917	.361
5	4.94	4.12	4.22	3.06	2.93	1.42	.885	.359

The results obtained were reduced to equation 3a from §2.1.1 above by the least square method. Table 2 shows the parameters for the model for an excellent fit, reaching in nearly all cases a regression coefficient R of close to 1, despite the simplicity of the model.

Table 2: Results of data analysis

SETUP	p	q	Nmax	Pmax	R
0-SECT	1.001	6.09E-03	164.8	82.73	1.0000
.144-SECT	.9963	5.66E-02	13.55	6.906	.9999
.25-SECT	.8842	1.31E-02	67.30	29.54	.9999
.333-SECT	.9466	.1165	7.29	3.356	.9975
.5-SECT	.9079	.1011	8.483	3.636	.9971
1-SECT	.7490	.1554	4.321	1.450	.9981
2-SECT	.5375	.1204	3.976	.9463	.9963
4-SECT	.2716	6.88E-02	3.446	.4087	.9574

An interesting function is $Nmax = f(x)$, where x is the amount of data. It calculates to a steep negative power curve with a regression coefficient of 85.56%. Increasing the communications factor for the 0-SECT case from 0 to .04 improves the regression coefficient to 99.80% and the function changes to $Nmax = 3.85 * x^{(-1.15)}$, see Fig. 1.

Likewise, the function for Pmax follows the function $Pmax = 1.41 * x^{(-1.25)}$, where x is the number of sectors of data transmitted, with a regression coefficient of 99.92%, see Fig. 2.

The decline of the computational power follows the function: $P = 1.00 * \exp(-.428 * x)$ with a regression coefficient of R = 99.77%.

Bibliography.

BEL89: Bell, G.: The future of high performance computers in science and engineering. Comm. ACM, Vol. 32, pp. 1091-1101 (1989).

BOD89: Bodenschatz, W.: Multi-Tansputer-Maschine zur parallelen Reduktion von Funktionalsprachenprogrammen. Doctoral thesis, Universität Stuttgart, 1989.

BÜD82: Büdenbender, H.-W.: Aufbau und Leistungsverhalten eines modularen Multiprozessor-Systems mit dezentral gesteuertem Kreuzschienenschalter. Doctoral Diss. Universität Stuttgart (Febr. 1982).

BUR65: Burkhardt, W. H.: Universal programming languages and processors. Proceedings Fall Joint Computer Conference 1965, Vol. I, pp. 1-21 and Vol. II, pp. 163-164.

BUR68: Burkhardt, W. H.: Automation of program speedup on parallel-processor computers. Computing, Vol. 3, pp. 297-310 (1968).

BUR87: Burkhardt, W. H.: Aspects of multiprocessor systems. Proceedings of the COMPEURO '87-Conference, pp. 99-105 (1987).

BUR89: Burkhardt, W. H.: Performance degradation by input/output in multiprocessors systems. Proceedings 32nd Midwest Symposium on Circuits and Systems, 1989.

BUR90a: Burkhardt, W. H.: Limitations to Parallel Processing. Proc. International Phoenix Conference on Computers and Communications, 1990.

BUR90 Burkhardt, W. H.: Quantification of Performance, Connectivity and Load-Distribution in Multiprocessor Systems. Proceedings 33rd Midwest Symposium on Circuits and Systems, 1990.

HAC86: Hack, J.J.: Peak vs. Sustained Performance in Highly Concurrent Vector Machines. IEEE Computer, Sept. 1986, pp. 11-19.

HAL86: Haller, W.: Leistungsverhalten von privaten Cache-Speichern für Instruktionen in einem Multiprozessorsystem. Doctoral Dissertation Universität Stuttgart, Dec. 1986.

MAK84: Makram, E. N.: Automatic analysis for the detection of inherent concurrency in programs for parallel computing. Diplomarbeit, Universität Stuttgart (May 1984).

TUR82: Turner, D. A.: Recursion equations as a programming language. In Functional programming and its applications, pp. 1-22, Cambridge University Press 1982.

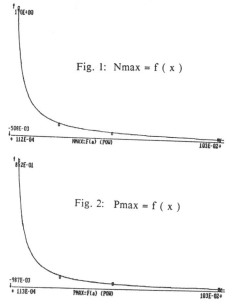

Fig. 1: Nmax = f (x)

Fig. 2: Pmax = f (x)

Simulation Results for a Scalable Coherent Cache System with Incomplete Directory State *

Joseph E. Hoag and Eugene D. Brooks III

Massively Parallel Computing Initiative (MPCI)

Lawrence Livermore National Laboratory

Livermore, CA 94550

Abstract —— A scheme which employs cache grouping and incomplete directory state in order to reduce the cost of maintaining directory state in a shared memory coherent cache system was introduced in earlier work by the authors and others. In this paper we report on detailed simulation studies of a cache grouping scheme employing multistaged network interconnects. We examine the effects of cache group size and support for multicast and combining in the network.

We find that the performance of the cache grouping scheme equals or betters that of a full-directory scheme. The multicast and return reply combination features are most effective when used in concert. We also learn that the system performance is relatively insensitive to cache group size. Perhaps most importantly, we find that there are significant performance gains associated with the use of a coherent cache, and that these performance gains increase as the size of the system is scaled upward.

1 Introduction

The idea of cache grouping and incomplete directory state was introduced in earlier works by the authors [1] and simultaneously by Gupta, Weber and Mowry [2]. Gupta et al. used the Stanford DASH architecture [3] as a model. The DASH architecture features clusters of processors connected by a mesh of buses. They performed event-driven DASH simulations, using Tango [4] to generate multiprocessor references.

We used Cerberus [5], a register-transfer level multiprocessor simulator, for our studies. Cerberus simulates RISC processors connected by a multistage interconnection network, and can be configured in a variety of topologies. Cerberus has its own compiler, assembler and loader, and faithfully models the movement of packets through the interconnection network.

The cache group scheme introduced by the authors involves three techniques that can be used to enforce coherence without using a disproportionate amount of controller memory:

- *Cache Groups*: The primary idea is the concept of *cache groups*; we advocate the use of a directory scheme, with a bit for each cache group instead of one per processor. This allows us to bound the memory overhead requirements to $O(log_2(N))$, as opposed to

$O(N)$ for a full-directory scheme [6]. Gupta, Weber and Mowry call this mechanism *coarse vectors*.

- *Multicasting*: We modified the network switchnodes to be able to route a packet to both outputs simultaneously. Multicasting enables the interconnection network to quickly propagate invalidation requests to cache groups.

- *Return Reply Combination*: Return reply combination, as we describe it, is a low-cost method of combining invalidation acknowledgements from caches in a switchnode. Unlike some other combining schemes [7], it does not require additional memory at the switchnode.

These features are described in greater detail in [1] and [8].

In order to test the viability of the cache grouping scheme, we performed simulations on the Cerberus multiprocessor simulator equipped with a coherent cache model. The codes that we simulated were a Gaussian elimination code (*gauss*), a network simulator (*psim*), an iterative relaxation code using a nine-point stencil (*relax*), and the synchronization portion of *gauss* (*flag*). These four codes span a variety of memory access patterns, and represent the behavior of sundry types of codes that are typically run on a multiprocessor.

This paper is by necessity very concise. For a more detailed description of the cache grouping scheme, the protocols associated with the scheme, the Cerberus multiprocessor simulator, and our simulation results, see [8].

2 Summary of Simulation Results

We learned through simulation that the cache grouping scheme equals or betters a full-directory scheme, though the cache grouping scheme requires much less memory overhead for tracking cache line location(s). The inability of the cache group scheme to track multiple outstanding copies of a cache line on a one-to-one basis did not result in a significant loss of performance, since most of our codes did not exhibit a high degree of one-to-many data sharing.

Detailed simulation showed that system performance was not sensitive to cache group size, so that cache group size could be set to whatever is convenient to the hardware of the system. This is significant, since it allows one to

*Work performed under the auspices of the U. S. Department of Energy by the Lawrence Livermore National Laboratory under contract No. W-7405-ENG-48.

bound the memory overhead due to coherence control to $O(log_2(N))$, where N is the number of processors in the system, without damaging the performance of the system.

Multicasting and return reply combination were most effective when used together. They made the most difference for codes that exhibited a high degree of one-to-many data sharing. Another important finding is that top-level return reply combination appears to be sufficient; there is little to be gained through the implementation of multilevel combination in the switchnodes.

Though the use of a coherent cache means additional network traffic for coherence control, its overall result is improved performance. The implementation of a coherent cache, using the cache grouping scheme, showed a significant performance improvement (up to 30%) over a similar system with no coherent cache. It should be noted that our simulations were limited to what would be fairly small runs on a "real" machine; time would not permit larger runs. Trends indicated improved performance with larger problem size. We also learned that the performance increase due to a coherent cache grew with the number of simulated processors. This leads us to believe that "massively parallel" machines of 1000 or more processors would show a much greater improvement through the addition of a coherent shared memory cache than the 30% increase that we saw using 32 simulated processors.

We also looked into some alternative coherence schemes, namely the one-read scheme and the minimal state scheme. The one-read scheme, which allows only one read-only copy of a cache line to be outstanding at a time, performs very poorly due to the amount of traffic it generates. The minimal state scheme [9], which uses broadcasts to effect coherence operations, fails for the same reason. We find that two capabilities are essential for any scalable coherence scheme: the ability to track cache line location for one-to-one data sharing, and the ability to grant multiple readable copies of a cache line.

Speed of execution, however, is not the only advantage offered by a coherent shared memory cache. The Lawrence Livermore National Laboratory spends hundreds of millions of dollars every year on computation. It is estimated that 80% of that figure is spent on software development and maintenance. Our experience with scalable shared memory multiprocessors without coherent shared memory caches is that a significant amount of effort is required to shape codes to run efficiently in parallel [10]. A lot of explicit localization of memory and decoupling of data structures is typically required. A coherent shared memory cache, supported in the hardware, would greatly cut down on the software costs of efficiently parallelizing existing code. The added hardware to support such a cache need not be overly expensive. The added performance of a coherent shared memory cache, as well as the decreased software development time on such a machine, would be well worth the costs associated with it.

References

[1] Eugene D. Brooks III and Joseph E. Hoag, "A Scalable Coherent Cache System with Incomplete Directory State," Proceedings of the 1990 International Conference on Parallel Processing, Vol. I, Penn State University Press, pp. 553-554, 1990.

[2] Anoop Gupta, Wolf-Dietrich Weber and Todd Mowry, "Reducing Memory and Traffic Requirements for Scalable Directory-Based Cache Coherence Schemes," Proceedings of the 1990 International Conference on Parallel Processing, Vol. I, Penn State University Press, pp. 312-321, 1990.

[3] D. Lenoski, J. Laudon, K. Gharachorloo, A. Gupta, J. Hennessy, M. Horowitz and M. Lam. "Design of Scalable Shared-Memory Multiprocessors: The DASH Approach." In Proceedings of COMPCON '90, pp. 62-67, 1990.

[4] H. Davis, S. Goldschmidt and J. Hennessy. "Tango: A Multiprocessor Simulation and Tracing System." Stanford Technical Report – in preparation, 1989.

[5] E.D. Brooks III, T.S. Axelrod and G.A. Darmohray, "The Cerberus Multiprocessor Simulator." In G. Rodrigue, editor, Parallel Processing for Scientific Computing, pp.384-390, SIAM, 1989.

[6] L.M. Censier and P. Feautrier, "A New Solution to Coherence Problems in Multicache Systems", IEEE Transactions on Computers, C-27(12):1112-1118, 1978.

[7] George S. Almasi and Allan Gottlieb, Highly Parallel Computing, Benjamin/Cummings Publishing Co., 1989, pp. 434-441.

[8] Joseph E. Hoag, "The Cache Group Scheme for Hardware-controlled Cache Coherence and The General Need for Hardware Coherence Control in Large-scale Multiprocessors," MS Thesis, Computer Science Department, University of California at Davis, March 1991.

[9] James Archibald and Jean-Loup Baer, "An Economical Solution to the Cache Coherence Problem", Proceedings of the 11th International Symposium on Computer Architecture, SIGARCH Newsletter, Vol. 12, Num. 3, IEEE, pp. 355-362, June 1984.

[10] Silvio Picano, Eugene D. Brooks III, and Joseph E. Hoag, "MIMD Implementations of a Network Simulator on a Large Scale, Shared Memory Multiprocessor," Lawrence Livermore National Laboratory Tech. Rep. UCRL-JC-105468, Livermore, CA, November 1990.

A Hardware Synchronization Technique
for Structured Parallel Computing

Zhiwei Xu
Department of Computer Science
Polytechnic University
Farmingdale, NY 11735
zxu@tasha.poly.edu

ABSTRACT

We present a hardware technique for efficient support of generalized barrier synchronization which allow multiple barriers to execute at a time, each synchronizing a subset of processes. We formally define the requirements for the needed synchronization mechanism, which must be *orderly* (i.e., it guarantees correct execution order), *fair* (i.e., barriers are served in a first-come-first-serve fashion), and *efficient* (the synchronization overhead should be small.) The hardware synchronization mechanism is shown to satisfy these three criteria.

1. Background and Problem Statement

We have recently proposed a methodology of *structured parallel computing*, which applies the philosophy behind the successful structured techniques for sequential computing to deal with parallelism. This methodology advocates that parallel computing algorithms should be implemented by determinate and terminating parallel programs using simple, structured constructs, and that parallel computers should support these structured constructs with small overheads.

With this methodology, a parallel program consists of several sequential processes. Each process has some computing codes and uses a new construct, called *structured regions*, for interprocess communication and synchronization. An example program segment is illustrated in Fig.1, which contains four sequential processes P1, P2, P3, and P4, and four structured regions C1, C2, C3, and C4. When process P1 finishes the computing code P11 and proceeds to structured region C1, it has to wait till P2 reaches C1. Then both P1 and P2 enter C1 to execute its communication operations specified by a parallel assignment statement. Afterwards they leave C1 to execute P12 and P22, respectively. Similarly, when P3 finishes P32, it has to wait till P2, P3, and P4 finish P22, P32, and P42. Only then can they enter C3. The processes which are to execute structured region C is called the *participants* of C, and they are *partners* of each other with respect to C. For instance, the participants of region C3 are processes P2, P3, and P4.

With the above definition, any structured region can be viewed as a parallel assignment statement enclosed in an entry and an exit barriers. The barriers realize synchronization, while the parallel assignment implements communication. The parallel assignment statement must follow the *single-assignment rule*: no variable can appear more than once in the left hand side of the parallel assignment. This way, no ambiguity can arise when multiple processes enter the region to simultaneously execute the parallel assignment. The major advantage of this structured region construct is that it provides a structured way to specify communication and synchronization operations as enclosed in a single construct.

We are designing the architecture of a structured parallel computer that aims to providing efficient support for synchronization and communication in structured parallel programs. The architecture consists of N Processing Elements (PEs), interconnected by a point-to-point communication network. A SYNCH instruction and a synchronizer hardware circuit realizes barriers needed for structured regions. When a process executes a SYNCH instruction, it will wait until all other partner processes also execute the same instruction. The synchronizer circuit will then generates a signal to allow all these partner processes to proceed.

In a program using structured region, multiple barriers can be executed at a time, each synchronizing a subset of processes. For instance, in the program of Fig.1, two barriers for structured regions C1 and C2 may be active simultaneously, one synchronizing P1 and P2 and the other P3 and P4. Thus our architecture must support multiple, partial barriers. This raises new problems. We need a synchronization logic that is *orderly*, *fair*, and *efficient*, as defined below.

The structured regions in a parallel program form a partially ordered set, where the partial order relation is defined by the sequence orders specified in the program, i.e., region C_i is greater than region C_j if C_i must be executed before C_j. When C_i is the minimum of the regions that are greater than C_j, we call C_i a *parent* of C_j and C_j a *child* of C_i. An *antichain* is a set of regions in which no region is greater than any other (i.e., they are *incomparable*).

We say a synchronization mechanism is *orderly*, if it guarantees that structured regions are executed according to the partial order relation of the program. In other words, the PEs can enter a child region only after they have entered all its parents regions. A mechanism is *fair*, if for any regions C_i and C_j, if C_i becomes executable (i.e., all participant processes of C_i have reached their entry points) before C_j does, then C_i will be executed no later than C_j. In other word, PEs will enter C_i no later than entering C_j. Finally, a mechanism is *efficient*, if a region will be executed at most t_s cycles after it becomes executable, where t_s is a small constant independent of the number of PEs.

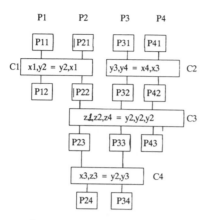

Figure 1. The structure of a parallel program

2. The Synchronization Logic

The synchronization logic is shown in Fig.2. At the beginning of program execution, both the synchronization status bit flip-flop (SB) and the synchronization signal flip-flop (SS) in every PE are reset to zero. A PE executes a SYNCH instruction when it wants to enter a structured region. This instruction sets SB to 1, and then keeps monitoring the SS flip-flop. When SS becomes 1, indicating that the region becomes executable, the PE resets both SB and SS, which terminates the SYNCH instruction. Then the PE can start execute the communication operations in the structured region.

The hardware synchronizer includes, among other components: (1) A synchronization word memory M which contains the synchronization words of all structured regions. The *synchronization word* w of a structured region C is an N-bit binary vector $w = (w_1 \, w_2 \, ... \, w_N)$ where $w_i = 1$ if and only if PE_i is a participant of region C. For instance, the synchronization words of regions C1 and C3 in Fig.1 are (1100) and (0111), respectively. The memory location $M(i)$ contains the synchronization word of structured region C_i. (2) A counter memory D where, at the beginning of program execution, $D(i)$ contains the number of parents of structured region C_i. (3) An N-bit synchronization status word (S) which copies the content of the N SB's and reflects which PEs have executed a SYNCH instruction. When S matches the synchronization word of a structured region, all the participants of that region have executed a SYNCH instruction, thus that region becomes executable. (4) An N-bit synchronization signal word W, the content of which signals the N SS's. (5) An antichain buffer $B = (B(1),B(2),...,B(m))$, where the $B(i)$'s contain the synchronization words of an antichain that is potentially executable.

The matching circuit generates a boolean vector $Q = (q_1, q_2, ... , q_m)$ which indicates whether the synchronization status word S matches the synchronization words in B, i.e., $q_i = 1$ if and only if $B(i) = S$. The multiplier circuit generates the desired synchronization signal word $W = BQ$ by a boolean matrix-vector multiplication.

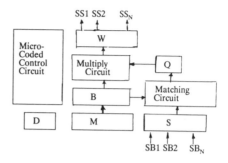

Figure 2. Schematic of the Synchronizer circuit

The micro-coded control circuit repeatedly executes the following sequence of operations:

1. compute Q by matching S with all words in B.
2. Generate W by multiplying B and Q.
3. Update the antichain buffer B:
   for each i such that $q_i = 1$ do begin
       ($B(i)$ contains the synchronization word of C_k)
       clear $B(i)$
       for each child region C_j of region C_k begin
         decrement $D(j)$
         if $D(j) = 0$ then load $M(j)$ to B
       end
   end

We note that PEs can enter a structured region after its synchronization word is loaded into the antichain buffer B. Initially, B contains words of those regions with no parents. Thereafter, a region's word is loaded into B only when $D(i)$ becomes zero, which happens only after PEs have entered all its parent regions. Thus the mechanism is orderly.

As far as fairness is concerned, the worst case is when C_i becomes executable before C_j does, but only the word of C_j is in B. C_i is waiting for the Update step to load its word into B. A detailed analysis shows that the complexity of this step is cm, where m is the maximum size of antichains and c is a small constant. The situation that C_i's word is not yet in B and all participant processes have reached their entry points of C_i can happen only when $t < cm$, assuming t cycles between two synchronization points. In fact, PEs will enter C_i at most $cm - t$ cycles after entering C_j. The value $cm - t$ also constitutes the dominating overhead. This overhead may be masked because updating can be performed by the synchronizer circuit while PEs are executing computing or communication operations. Although m can reach as large as $N/2$ in pathological cases, many parallel algorithms exhibit small m. Thus for medium-grain parallel algorithms (e.g., 50 cycles of communication and computing operations between two synchronization points), the updating overhead can be completely masked. And the synchronization mechanism is both fair and efficient.

References: Omitted due to page limits.

A STUDY OF SINGLE-BIT SPIN-WAITING ALTERNATIVES

David W. Opitz
Computer Science Department
University of Wisconsin - Madison
Madison, WI 53706

Abstract -- Different methods for spin-waiting for a lock were simulated and compared with each other, and the results indicate that a form of hardware queueing offers a highly efficient management for the waiting processors. The performance of other methods, except for queueing of processors and a form of backing off requests, quickly degraded for many processors.

Introduction

As the demands on computing power rise, shared memory multiprocessors are being built with an increasing number of processors and are often used to reduce the amount of time to execute a single large job by partitioning it into pieces which run on separate processors. This means that more time and effort will be spent coordinating the various parts of the parallel program at run-time. It is therefore essential that the costs of synchronization be kept to a minimum.

Multiprocessors with many processors will tend to use interconnects in which the latency of a global request is proportional to the log of the number of processors. The ratio of the latency of global requests to the latency of local requests found in a processor's cache is very large and is increasing. This large latency makes it essential that synchronization be efficient in the number of times that the interconnect is traversed. Also, frequent use of the interconnect can consume a large portion of the bandwidth, which could have otherwise been used to satisfy different independent requests, and can lead to hot-spots and tree saturation.

Synchronization mechanisms can be classified along two dimensions, one being the amount of state that the scheme manipulates in order to provide synchronization and the other being the way in which waiting processors are managed. Single-bit methods can basically only provide mutual exclusion because the only state that is retained is whether or not the lock is currently being held. Methods such as eventcounts/sequencers, counting semaphores, and Fetch-and-Add essentially use an integer to encode their synchronization state, while Fetch-and-Φ and Cedar operations can be applied to any arbitrary variable. These primitives are inherently more powerful than those that manipulate only one bit.

Synchronization methods can also be classified by the way in which waiting processors are managed. Wait management attempts to improve the efficiency of how a particular synchronization scheme operates. Efficient wait management schemes should be quick in terms of transferring control of the lock and should use the interconnect as little as possible. This paper compares wait management schemes for single-bit synchronization methods for multiprocessors with up to 256 processors.

A Comparison of Single-Bit Methods

Six types of single-bit synchronization methods were simulated on a model described later. The first method simulated was repeated test&set instructions until the lock was acquired. A test&set instruction is an atomic read-modify-write instruction which tests a lock value to see if it is unset and then sets it. If the value was unset, this processor obtains the lock then sends an unset instruction when it is done with the lock. If the lock was set, it spins. An improvement to this, known as backoff, is for each processor to delay issuing a test&set until the estimated chance of success is high [1]. Static or dynamic delays can be inserted either after the lock has been released or after every separate access to the lock. The method simulated here had an exponentially increasing backoff delay after each failed attempt up to a maximum of twice the number of processors. A read and test&set are often combined into one primitive called a test&test&set so that architectures with coherent caches can spin on a cache copy. The bandwidth associated with spinning can be avoided because no interconnect cycles are used while spinning. Two methods were simulated, one in which the cached copy was updated by a distributed-write and one in which it was invalidated.

Spinning processors can be efficiently managed by placing them in a queue. When a processor finishes with the lock, it taps the processor with the next highest sequence number, and that processor now owns the lock. This requires $O(N)$ operations over the interconnect for all N processors to gain access to the lock, while all the methods described above require $O(N^2)$ operations. A software version was implemented where each processor does an atomic Fetch-and-Add (read-and-increment) to obtain a unique sequence number, which defines its place in the queue [2]. A hardware queuing version, analogous to the QOLB (Queue On Lock Bit) operation proposed in [3], was also implemented. This is a non-blocking operation on a syncbit address that adds the issuing processor to the syncbit queue if the processor is not already in the queue. A test&set will automatically fail without testing and setting the syncbit whenever it is issued by a processor that is not at the head of the queue. The QOLB operation generates at most one asynchronous operation over the interconnect to put the processor in the queue, and requires at most one additional asynchronous operation to notify the next processor that it is now at the head of the queue. One big advantage to this operation is that with the acquisition of the lock comes the relevant line of data. The QOLB operation can be implemented regardless of the cache coherency used.

Results

The model of synchronization accesses on a shared-memory multiprocessor consisted of up to 256 processors connected to the same number of shared memory modules through an omega network. Each processor was assumed to have immediate access to its own cache memory. Cache coherency was maintained either by a software, a distributed-write, or an invalidation-based scheme, depending on the synchronization method. All synchronization requests were assumed to operate on a single synchronization variable in memory, and memory can handle one request per cycle. Simulations of each method were run for varying means of work section length as well as number of processors. For this experiment, the processors captured the lock, computed a short critical section, released the lock, computed a work section, and then recursively did this over again. The length of the work section was randomly chosen from an exponentially distributed amount of time with varying means. This reflects, for example, conditional execution outside the critical section or nonsymmetric thread functions.

When there was little contention for the lock, the hardware queueing method gave the best performance because of its extremely low overhead latency of acquiring the lock, while the repeated test&sets and backoff methods also performed well. When many processor were contending for the lock, the overhead was due to both the latency of obtaining or passing the lock as well as contention in the network. The hardware queueing method had virtually no overhead due to the fact that the only overhead involved was the contention in the network when initially placing them into the queue. The backoff method didn't do as well as repeated test&sets with few processors, but performed much better than all of the other software methods with many processors. Other backoff methods not presented here, such as backing off a small random number of cycles after each failed attempt, performed better with few processors, but quickly degraded with a larger number of processors. Software queueing also scaled well to many processors, however, there was a large overhead associated with each passing of the lock as well as when initially obtaining the unset lock. The other methods did not perform well with a large number of processors.

Performance was also measured when all the processors arrived at the lock at the same time with each processor achieving the lock once, such as when released from a barrier. The hardware queue had much less overhead than the other methods because its only overhead was initially setting up the queue. Backoff had additional overhead from the fact that the final few processors who hadn't received the lock were backing off a large number of cycles, even though there was no longer contention for the lock. The other methods

performance, besides software queueing, degraded with a large number of processors.

The increased overhead associated with the latency of traversing the interconnect due to contention among the processors was also measured. The queuing methods had the smallest overhead latency due to the fact that processors only contended for the lock one time and were then placed in a queue. Test&set had the highest latency overhead which helps show why repeated test&sets don't scale well to a large number of processors. Using the interconnect can also interfere with processors doing independent work. The less a method uses the interconnect, the better, so the number of messages which were sent by all processors per critical section access was measured. The queueing methods only sent messages to initially get into the queue and to pass the lock to the next processor. Backoff didn't send as many as the remaining three methods because of the exponentially increasing backoff delay after each failure.

Conclusion and Future Work

In the simulations run, hardware queueing had virtually no overhead associated with it and can be implemented with any cache coherency. Backoff showed good performance for scaling up to a large number of processors, in fact it showed better performance than software queueing when using an invalidation-based protocol. It should be noted that backoff needs an atomic read-modify-write instruction while software queueing needs a read-and-increment instruction. Different forms of backoff, such as using smaller dynamic delays, can be used on systems with fewer processors. The other methods had performance degrade rapidly with a large number of processors.

Synchronization can be the bottleneck for massively parallel programs. It is the job of the architect to add hardware that will reduce the effects of bottlenecks. In the Wisconsin Multicube, most of the hardware needed by QOLB is provided by the cache coherency scheme, but in other systems this may not be the case.

For future work, simulations with real workloads need to be done for these methods with a large number of processors. Also, more work needs to be done on how various atomic instructions affect the performance of these methods. For acknowledgements, I would like to thank Gurindar Sohi and Yul Kim for their many insightful comments.

References

[1] A. Agarwal, and M. Cherian, "Adaptive Backoff Synch. Techniques", Int. Symp. of Comp. Arch., (June, 1989), pp. 396-406.

[2] T. Anderson, "The Performance Implication of Spin-Waiting Alternatives for Shared Memory Multiprocessors", ICPP, (1989), pp. 170-174.

[3] J.R. Goodman, M. Vernon, and P.J. Woest, "Eff. Synch. Prim. for Large-Scale Cache-Coherent Mult.", ASPLOS III, (April, 1989), pp. 64-75.

1991 International Conference on Parallel Processing

MAPPING MULTIPLE PROBLEM INSTANCES INTO A SINGLE SYSTOLIC ARRAY
WITH APPLICATION TO CONCURRENT ERROR DETECTION

C. N. Zhang

Department of Computer Science

University of Regina

Regine, Saskatchewan

Canada S4S 0A2

H. D. Cheng

School of Computer Science

Technical University of Nova Scotia

Halifax, Nova Scotia

Canada B3J 2X4

Abstract

The conventional space-time mapping approach has been extended for mapping multiple problem instances into a single systolic array. A theory relating concurrent error detection (CED) and space-time mapping in systolic arrays is also proposed. Procedures which guarantee to have minimum area for mapping multiple problem instances and designing CED systolic array are presented.

1. Mapping Multiple Problem Instances into A Single Systolic Array

Consider a three-index (i-j-k) nested loop algorithm:

Algorithm 1 (matrix multiplication: $C = A \times B$)

```
for i:=1 to n do
    for j:=1 to n do
        for k:=1 to n do
            begin
                a(i,j,k) := a(i,j-1,k);
                b(i,j,k) := b(i-1,j,k);
                c(i,j,k) := a(i,j,k-1) + a(i,j,k) b(i,j,k);
            end
```

where $a(i,0,k) = a_{ik}$, $b(i,j,k) = b_{kj}$, $c(i,j,0) = 0$, $c(i,j,n) = c_{ij}$ for all i,j, and k.

The computational structure of the algorithm can be represented by a pair (D,S) where S is a set of computation sites $S = \{(i, j, k)\}$ where $1 \le i \le n$, $1 \le j \le n$, $1 \le k \le n$ and a dependency matrix D [2,3,5]. The basic ideal of space-time mapping approach is to find a valid 3×3 integer transformation matrix $T = [t_{ij}]$ such that the result matrix $\tilde{D} = TD$ is a systolic matrix (all elements of the first row of $\tilde{D}$ are negative) and the resulting systolic computational space $\{(t,x,y)\} \subset Z^3$ (Z is the linear integer space) where t represents the time, x and y represent x-coordinate and y-coordinate in x-y plan. We call such transformation T is valid with respect to (D, S) denoted by $(D, S) \xrightarrow{T} (\tilde{D}, \tilde{S})$ [5].

One important performance parameter for the designed systolic array is the pipeline period l_t which is the time interval in clock units between two successive input data. Similarly we can define l_x and l_y which are the space intervals (number of idle processing elements (PE)) between two successive active PEs in x-coordinate and y-coordinate of the x-y plan, respectively. We proven that l_t, l_x, and l_y depends on the transformation matrix T only and, they can by calculated by the following formulas:

$$l_t = \left| \frac{|T|}{\gcd(|T_{11}|, |T_{12}|, |T_{13}|)} \right|$$

$$l_x = \left| \frac{|T|}{\gcd(|T_{21}|, |T_{22}|, |T_{23}|)} \right|$$

$$\text{and } l_y = \left| \frac{|T|}{\gcd(|T_{31}|, |T_{32}|, |T_{33}|)} \right| \text{ where } |T| \text{ is the deter-}$$

minant of matrix T and T_{ij} is the (i,j)th co-factor of matrix T. The chip area of the resulting systolic implementation is defined by $area = (x_{max} - x_{min} + 1) \times (y_{max} - y_{min} + 1)$ where $x_{max} = \max \{x \mid (t,x,y) \in \tilde{S} \}$, $x_{min} = \min \{x \mid (t,x,y) \in \tilde{S} \}$, $y_{max} = \max \{y \mid (t,x,y) \in \tilde{S} \}$ and $y_{min} = \min \{y \mid (t,x,y) \in \tilde{S} \}$.

Construct an identical version of Algorithm 1 as follows.

Algorithm 1A (an identical version of Algorithm 1)

```
for i' := 0.5 to n − 0.5 do
    for j' := 1.5 to n + 0.5 do
        for k' := 1 to n do
            begin
                a'(i', j',k') := a'(i', j' − 1, k');
                b'(i', j',k') := b'(i' − 1, j', k');
                c'(i', j',k') := c'(i', j', k' − 1) + a'(i', j',k') b'(i'j'j');
            end;
```

where $a'(i', 0.5, k') = a'_{i',k'}$, $b'(-0.5, j', k') = b'_{k'j'}$, $c'(i', j, 0) = 0$ and $c'(i', j' ,n) = c'_{i',j'}$ for all i', j' and k'.

It is clear that the Algorithm 1 and Algorithm 1A have the same dependency matrix D, but they have the different sets of computational sites represented by S and S' respectively. Note that $S = \{(i, j, k)\} \subset Z^3$ but $S' = \{(i', j', k')\} \notin Z^3$, thus, $S \cap S' = \emptyset$. We call such computation (D, S') is an identical version of (D, S). In general, given an algorithm represented by (D, S) where $S = \{(i, j, k)\} \subset Z^3$, we can define a set of its redundant versions as follows.

Definition 1. $\{(D, S^m)\}$ is a set of identical versions of (D, S) if and only if

(i) at least one of d_i^m, d_j^m, and d_k^m is not integer for all m = 1, 2, ... , h and

(ii) $S^m \cap S^n = \emptyset$ if $m \ne n$.

where $S^m = \{(i^m, j^m, k^m)\}$ and $i^m = i + d_i^m$, $j^m = j + d_j^m$ and $k^m = k + d_k^m$, m = 1, 2, $\cdots$, h.

Definition 2. Suppose $\{(D, S^m)\}$ is a set of redundant versions of (D, S) and T is a valid transformation of (D, S). T is also a valid transformation of $\{(D, S^m)\}$ m = 1, 2, $\cdots$, h, (h > 1) if and only if $\tilde{S}^m = TS^m \subset Z^3$ for all m = 1, 2, $\cdots$, h denoted by $\{(D, S^m)\} \xrightarrow{T} \{(\tilde{D}, \tilde{S}^m)\}$.

We call such transformation T having mapping degree h+1. Suppose T has mapping degree h+1. Since T is a one to one mapping function ($|T| \ne 0$), we have $\tilde{S} \cap \tilde{S}^m = \emptyset$ for all m, and $\tilde{S}^n \cap \tilde{S}^m = \emptyset$ if $n \ne m$. Let $\tilde{S}^* = \tilde{S} \cup \tilde{S}^1 \cup \cdots \tilde{S}^h$. It is obvious that $\tilde{S}^* \subset Z^3$. Thus, the total $1 + h$ problem instances represented by (D, S) and $\{(D, S^m)\}$, m = 1, 2, ... , h, can be computed in a single systolic array $(D, \tilde{S}^*)$ simultaneously. We show that the capability of mapping multiple problem instances depends on the transformation T only and have the following result.

Theorem 1. T has mapping degree $1 + h$ if and only if max $\{l_t, l_x, l_y\} \ge 1+h$.

I-668

Since the problem of choosing a set of redundant versions of (D, S), $\{(D, S^m)\}$, $m = 1, 2, \ldots, h$, is equivalent to the problem of choosing an integer set $\{(d_t^m, d_x^m, d_y^m)\}$

$$\left(\begin{bmatrix} t + d_t^m \\ x + d_x^m \\ y + d_y^m \end{bmatrix} = T \begin{bmatrix} i + d_i^m \\ j + d_j^m \\ k + d_k^m \end{bmatrix} \right) m = 1, 2, \ldots, h. \text{ Note that the}$$

value $\max_{1 \le m \le h} \{d_x^m\}$ indicates the number of additional columns of PEs required; the value $\max_{1 \le m \le h} \{d_y^m\}$ indicates the number of additional rows of PEs required, and $\max_{1 \le m} \{d_t^m\}$ represents the number of additional clock delays required. The following procedure can be used to select a set $\{(d_t^m, d_x^m, d_y^m)\}$, $m = 1, 2, \ldots, h$, such that the resulting systolic array has the minimum area.

procedure 1. (construct a systolic array with minimum area)

case 1. $l_t = h + 1$.
    choose $d_t^m = m$, $d_x^m = 0$ and $d_y = 0$ $m = 1, 2, \ldots, h$ (no additional PEs required).

case 2. $l_t < h + 1$, $l_x = l_y = h + 1$.
    if $n_x \le n_y$ then choose $d_t^m = 0$, $d_x^m = 0$ and $d_y = m$ $m = 1, 2, \ldots, h$ (it requires additional h rows of PEs) else choose $d_t^m = 0$, $d_x^m = m$ and $d_y = 0$, $m = 1, 2, \ldots, h$ (it requires additional h columns of PEs)

case 3. $l_t < h + 1$, $l_x < h + 1$, $l_y = h + 1$
    choose $d_t^m = 0$, $d_x^m = 0$ and $d_y^m = m$ $m = 1, 2, \ldots, h$ (it requires additional h rows of PEs)

case 4. $l_t < h + 1$, $l_x = h + 1$, $l_y < h + 1$
    choose $d_t^m = 0$, $d_x^m = m$ and $d_y^m = 0$ $m = 1, 2, \ldots, h$ (it requires h columns of PEs)

For example, consider the transformation
$$T = \begin{bmatrix} 1 & 1 & 1 \\ -1 & 1 & 0 \\ 0 & 0 & -1 \end{bmatrix} \text{ for Algorithm 1. We have}$$
$l_t = 2$, $l_x = 2$ and $l_y = 2$. According to procedure 1, we choose $d_t^1 = 1$, $d_x^1 = 0$ and $d_y^1 = 0$. The corresponding systolic array implementation is shown in Fig. 1 which computes two problem instances simultaneously.

2. A General Model for CED Using Redundancy

The fact that in the generalized systolic mapping approach, all PEs in a systolic array may not perform the computation every cycle, open up other possibility to use them. These unused processors (in a cycle) can be used to perform a redundant computation to achieve concurrent error detection (CED) [13]. Let computation (D, S') be a redundant version of (D, S) same as the identical version defined before except that the algorithm of (D, S') compute the same problem instance as the (D, S) does. Suppose T is a valid transformation for both (D, S) and (D, S'), that is,

$(D, S) \xrightarrow{T} (\tilde{D}, \tilde{S})$ and $(D, S') \xrightarrow{T} (\tilde{D}, \tilde{S}')$ where $S = \{(i, j, k)\}$, $S' = \{(i', j', k')\}$, $\tilde{S} = \{(t, x, y)\}$ $\tilde{S}' = \{(t', x', y')\}$. We call $(\tilde{D}, \tilde{S}')$ is a concurrent redundant computation (CRC) of $(\tilde{D}, \tilde{S})$. In particular, S' is related to S by $i' = i + d_i$, $j' = j + d_j$, $k' = k + d_k$ and $\tilde{S}'$ is related to $\tilde{S}$ by $t' = t + d_t$, $x' = x + d_x$, $y' = y + d_y$ where d_i, d_j, d_k, d_t, d_x and d_y are small constant. Since mapping an algorithm (S, D) and its redundant version (D, S') into a systolic array is the special case of mapping two problem instances into a single systolic array $(h = 1)$. Thus, we have

Theorem 2. Suppose $(D, S) \xrightarrow{T} (\tilde{D}, \tilde{S})$. There is a CRC of $(\tilde{D}, \tilde{S})$ if and only if $\max \{l_t, l_x, l_y\} \ge 1$.

Since d_t, d_x, and d_y are small constant. The corresponding sites that generate identical results are local to one another, that is, separated by a small constant in space and/or time coordinates. As a result, comparison logic can be inserted between these corresponding sites such that results from (t, x, y) and (t', x', y') are compared directly to detect functional errors. Moreover, based on the single failure assumption, any such failure can be detected. Interestingly, this approach provides a unifying theory relating CED and the space-time mapping. Note that if $\tilde{S}'$ is chosen by $d_x = 0$ and $d_y = 0$ then the original computation $(\tilde{D}, \tilde{S})$ and its redundant computation $(\tilde{D}, \tilde{S}')$ will be performed in the same PE. Some errors, e.g. permanent function error of the PE, may be masked. In the full paper, we present a CED design procedure and prove that the resulting CED systolic array has minimum area and has the capability of detecting any single error.

In general, if $(D, S) \xrightarrow{T} (\tilde{D}, \tilde{S})$ and $\max \{l_t, l_x, l_y\} \ge h+1$, then according to Theorem 1 we can construct a systolic array which computes $h+1$ problem instances simultaneously. Let h of them are different problem instances and two of them are the same problem instance. Consequently, this systolic array pays the cost of decreasing $\eta = \dfrac{throughput}{area}$ by a fact of $\dfrac{1}{h+1}$ but achieves the capability of CED. Formally, it is stated by the following theorem.

Theorem 3.

$$\max_{\substack{\max(l_t, l_x, l_y) \ge 1+h \\ \text{with CED}}} \left\{ \frac{throughput}{area} \right\} \ge \frac{h}{h+1} \max_{l_t = l_x = l_y = 1} \left\{ \frac{throughput}{area} \right\}.$$

References

[1] Jacob, A., Banerjee, P., Chen, C-Y., Fuchs, W., Kuo, S-Y. and Reddy, A.. Fault Tolerance Techniques for Systolic Arrays. IEEE Computer, July 1987, pp

[2] W. L. Miranker. Space-time Representations of Computational Structures. Computing, 32, 1984, pp 93-114.

[3] D. I. Moldovan. On the Design of Algorithms for VLSI Systolic Arrays. Proc. IEEE, 71(1), 1983, pp 113-120.

[4] C-C. Wu, T-S. Wu. Concurrent Error Correction in Unidirectional Linear Arithmetic Arrays. Proc. 17-th Intl. Symp. on Fault-Tolerant Computing, 1987, pp 136-141.

[5] C. N. Zhang, Concurrent Error Detection in Systolic Arrays. IFIP-IEEE Int'l Workshop on Defect and Fault Tolerance in VLSI System, 1990. 65-74.

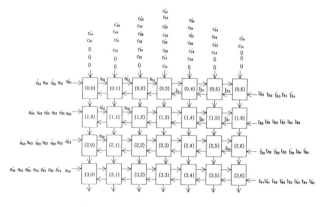

Fig. 1. Systolic Array $(\tilde{D}, \tilde{S} \cup \tilde{S}')$ with $d_t^1 = 1$, $d_x^1 = 0$, $d_y^1 = 0$.

A Parallel Reconfiguration Algorithm for WSI/VLSI Processor Arrays

Jung H. Kim and Phill K. Rhee

The Center for Advanced Computer Studies

University of SW Louisiana

Lafayette, LA 70504-4330

ABSTRACT

This paper addresses a parallel reconfiguration algorithm for VLSI/WSI processor arrays. The aim of the proposed algorithm is to reduce the time complexity of reconfiguration algorithm through parallel implementation.

1. Introduction

The main objective of this paper is to present a parallel reconfiguration algorithm for large scale reconfigurable processor arrays to reduce the reconfiguration time. In the proposed approach, the initial structure of a logical array is decided considering fault distributions in the initialization step. In the rule-based reconfiguration step, a logical cell which could not be assigned to a fault-free cell in the initialization step, is matched to a spare cell by repetitive shifts of logical cells. Reconfiguration rules are devised based on the connection capability of interconnection resources, and they utilize only local information of neighboring cells to guide repetitive shifts. Since the proposed reconfiguration rules do not require global information of the array, several unassigned logical cells can be matched to spare cells simultaneously. The proposed parallel algorithm of the reconfiguration step can limit the time complexity as a constant in terms of the array size, i.e., reconfiguration time does not depend upon the size of the array. The detailed description of definitions and reconfiguration rules can be found in [Rhee90].

2. Parallel Rule-Based Reconfiguration

The proposed algorithm can be executed in parallel to minimize reconfiguration time, by constructing several assignment paths for unassigned logical cells simultaneously. *Global map* of the array is defined as the information of the locations of logical cells in the array and δ^rs. δ^r (*Inverse row range*) of a physical cell denotes the range of a logical row which must not be connected below and above the cell. We assume that tasks in the parallel machine can create other tasks, can communicate with each other, and can access a shared memory. A task involved in the construction of an assignment path stores pointers to its previous and next tasks of the assignment path in its local memory. Since each assignment path modifies a logical array in a different way, the modification information must be stored independently. If a cell P is visited by an assignment path i, the logical cell of P and associated δ^rs must be updated accordingly in the global map. If P is visited by an assignment path j again, then the global map will lose the previously updated information of the assignment path i. Hence, if there is only one global map, the construction of the assignment path j must be suspended until the assignment path i is completed or backtracked from P. However, if a cell can only be in one assignment path, the degree of parallelism is limited too severely and the parallel implementation can not achieve much.

Let α be the maximum number of temporary global maps allowed in the parallel machine, β be the maximum number of assignment paths constructed in parallel for an unassigned logical cell, γ be the number of unassigned logical cells after the initialization step. For the independent construction of assignment paths, $\beta \times \gamma = \alpha$. Since α is restricted by the limited memory, β may not be large enough to allow high parallelism. To obtain high parallelism with a limited memory space, the array is divided into partitions called *regions*. Each region has its *regional map* which is the partition of the global map. The size of region is decided such that an assignment path visits limited number of regions. In each region α temporary regional maps are allowed. Since an assignment path visits only limited number of regions, β can be decided such that $\beta \times \gamma \gg \alpha$. *Temporary regional map* is referenced as Φ_i^A, where A indicates the region name and i indicates that the assignment path i visits the temporary regional map. Smaller number of temporary regional maps than that of the assignment paths may cause an assignment path collision as well as a deadlock among assignment paths, since a region may be requested by more than α assignment paths.

2.1. Tasks and Temporary Regional Maps (Φs)

Tasks are initialized for the parallel execution. Φs of regions are initialized by R-tasks.

1) *C-tasks, I-tasks, and R-tasks:* Suppose that a task $\mathbf{T}_0$ has the reconfiguration information of the array in the initialization step, i.e., the global map, the locations of faulty cells, unassigned logical cells, domains, etc. $\mathbf{T}_0$ creates following tasks: 1) *C-task* which is allocated to each physical cell, 2) *I-task* which is allocated to each unassigned logical cell, and 3) *R-task* which is allocated to each region. R-tasks update the global map simultaneously. Note that the created tasks can be used repetitively in as many reconfigurations as they are requested.

2) *Temporary Regional Map* (Φ): R-task of each region creates α temporary regional map Φs by copying from the global map. If a Φ is allocated to an assignment path i, the Φ is named as Φ_i to prevent another assignment path from using it. Only the assignment path i can modify or release a Φ_i. Φ has the counter which is initially set to 0. The counter indicates whether an assignment path i is in a region or not. That is, if the counter of Φ_i is reduced to 0, the assignment path i is no more in the region, thus releases Φ_i. Φ_i of a region is called *available* if either Φ_i is already in the region or an unnamed Φ exists in the region.

2.1) Suppose that an assignment path i visits a cell Q in region A and L_P is a new unassigned cell in the assignment path i.

Case 1 (a Φ is available in region A):

Case 1.1 (Φ_i^A has already been in the region): The counter of the Φ_i^A is increased by 1. If any of affected cells is in another region B, one of the followings occurs: i) If a Φ^B is available, the counter of the Φ_i^B is increased by 1. ii) If a Φ^B is not available, an assignment path collision occurs, and it is discussed later in the parallel construction of assignment paths.

Case 1.2 (an unnamed Φ exists in region A): An unnamed Φ is named as Φ_i^A, and do similarly as Case 1.

Case 2 (Φ is not available in region A): An assignment path collision occurs, and it is discussed later in the parallel construction of assignment paths.

2.2) If an assignment path i is backtracked from a cell Q whose previous logical cell was L_Q in region A, then

i) Q decreases the counter of the Φ_i^A by 1, ii) the path i backtracks to previous Φ_i^A by assigning L_Q to Q, and iii) if the counter of the Φ_i^A is 0, the Φ_i^A is released from the assignment path i, and it becomes an unnamed Φ^A so that it can be used by another assignment path. If Q has an affected cell located in another region B, the counter of Φ_i^B is decreased by 1.

2.2. Initiation of Assignment Paths

Each I-task creates β assignment paths simultaneously.

1) An I-task I_c allocated to an unassigned logical cell L_C, decides the spare cell list of L_C. If S_1 is the first priority spare cell, I_c allocates S_1 to the assignment path 1 for L_C. Assignment paths of L_C are distinguishable by identification numbers. Assignment paths initiated from other unassigned logical cells are distinguishable by the unassigned logical cell coordinates. I_c decides the assignment trial list of L_C for the assignment path 1, and selects the first priority cell, say Q, for the assignment path 1. I_c requests C-task of Q to assign L_C to Q. At the same time as the I_c allocates S_1 to the assignment path 1, I_c decides the first priority cell for the assignment path 2, and sends a request to assign L_C and so on, until β assignment paths are initiated.

2) In many cases, full list of spare cells for an unassigned logical cell is not necessary, especially when the size of the array is large. A spare cell located at the nearest location of an unassigned logical cell will be matched with the highest probability. The *spare region* of an unassigned logical cell is defined as the area covered by cells whose Manhattan distance from the domain-1 of the unassigned logical cell is equal to or less than r, where r is a system parameter. Since a spare cell is decided within the spare region, an assignment path can be constructed within the spare region. Therefore, the time complexity of the parallel rule-based reconfiguration step is constant independent of the number of cells in the array. That is, reconfiguration time is not dependent upon the number of cells in the array, but it is dependent upon only the system parameter "r" which can be readily controlled.

3) If an assignment path is failed, I_c accesses a new spare cell from the spare list of L_C and initiates an assignment path. I_c knows that an assignment path is failed, if all cells in the assignment trial list return negative responses.

2.3. Parallel Construction of Assignment Paths

Maximum $\beta \times \gamma$ assignment paths are being constructed simultaneously in the parallel machine. Each assignment path is constructed independently utilizing corresponding Φs. Suppose that a cell P_i located in region A is the first priority cell for an unassigned logical cell L_U in an assignment path i.

1) *Assignment path construction:* If Φ_i^A is available to the assignment path i, the assignment safety of L_U to P_i is checked by $\mathbf{T_i}$ (C-task of P_i), utilizing the information in the Φ_is. If an affected cell of P_i is located in another region, say B, Φ_i^B also must be available.

1.1) If the assignment of L_U to P_i is safe, L_U is assigned to P_i in Φ_i^A, and L_V which was previously the logical cell of P_i in Φ_i^A becomes a new unassigned cell in the assignment path i. $\mathbf{T_i}$ requests the assignment of L_V to the C-task of the first priority cell of P_i for the continuous construction of assignment path i.

1.2) If the assignment of L_U to P_i is not safe, $\mathbf{T_i}$ sends a negative response to $\mathbf{T_h}$ (C-task of P_h), where L_U was previously assigned to P_h. P_i is deleted from the assignment trial list of L_U and $\mathbf{T_h}$ tries the next priority cell. If $\mathbf{T_h}$ receives negative responses from all cells in the assignment trial list of the assignment path i, backtracking is required from $\mathbf{T_h}$ to its previous task.

2) *Assignment path collision:* Suppose that region A has already been allocated to α assignment paths, but the assignment path i is not one of them. If an assignment path i visits P_i in region A, Φ^A for the assignment path i is not available. That is, assignment path collision occurs at P_i. $\mathbf{T_i}$ sends back a suspensive response, implying that the request must be suspended. If $\mathbf{T_h}$ receives the message, $\mathbf{T_h}$ tries the next priority cell in the assignment trial list of L_U. But, $\mathbf{T_h}$ does not delete P_i in the assignment trial list so that P_i can be tried again later. If P_i is the last cell in the assignment trial list, a deadlock may occur.

3) *Deadlock among assignment paths:* A deadlock among Φs may occur if the following necessary conditions for a deadlock meet: i) *Mutual Exclusion*: A temporary local map is allocated to exactly one assignment path, ii) *Nonpreemption*: A temporary local map is released only by an assignment path which holds it, iii) *Hold and Wait*: An assignment path holds temporary local maps already allocated to it, while waiting for an additional temporary local map, and iv) *Circular Wait*: There must be a circular chain of two or more assignment paths, each of which is waiting for a temporary local map held by the next assignment path in the chain.

Our strategy to prevent a deadlock is to eliminate circular wait condition: i) an assignment path with smaller identification number has higher priority in accessing a Φ, and ii) if a C-task receives a suspensive response from the C-task of the last cell in the assignment trial list, the C-task sends only the limited number of assignment requests.

3. References

[Rhee90] P. K. Rhee "On the Reconfiguration for VLSI/WSI Processor Arrays", *Ph.D. Thesis*, Center for Advanced Computer Studies, University of SW Louisiana, 1990.

Choosing the Right Grains for Data Flow Machines

Roni Hardon *and* Shlomit S. Pinter

Dept. of Electrical Engineering

Technion — Israel Institute of Technology

Haifa 32000, Israel

Abstract — Data flow computers execute data flow graph programs by dividing the graph into instruction templates which are scheduled as early as possible. However, implementing such a scheme involves communication overheads which affect the running time of the program. In this work we present a model for data flow machines which includes both communication and execution times. With this model we derive bounds on the execution time of programs represented as DAGs or as trees. In addition, for tree like programs we provide a quadratic time (in the size of the tree) complexity algorithm for optimally partitioning a program into sets of instruction templates, providing minimum execution time.

1 Introduction

We provide a general model of communication and computation which is applicable to most data flow machines, and consider programs which can be represented as Directed Acyclic Graphs (DAGs). A DAG can either be a whole program for a "static" data flow machine or a basic block of a "dynamic" machine. Note that when the destination of every operation is unique the program is represented as a tree. Graph nodes (operations) are scheduled as soon as their operands arrive, and executed as soon as there is a free execution unit to handle them.

The problem of optimally scheduling a DAG on our data flow machine model is a generalization of the precedence constrained scheduling problem which is NP-complete [3]. There is a polynomial time algorithm for solving the precedence constrained scheduling problem when the precedence graph is a tree [4]. We provide both a lower and a matching upper bound for the generalized problem of scheduling computation trees in our model, (with an unlimited number of processors) and we show that the obvious solution of one-instruction per packet (*minimal grain size*) is not optimal due to communication (and matching) time overhead. Algorithms for similar problems on different machine models do not provide an optimal solution for our data flow model even for the simple case where the computation graph is a tree. Thus, we provide an $O(n^2)$ time complexity algorithm which generate an optimal partitioning of a program tree into execution grains, for a tree with n nodes.

2 The Model

Our model of a data flow computer consists of one control unit, several execution units, and three data structures (see Figure 1). The instruction templates are waiting for their arguments to arrive in order to be passed to the *ready instructions store*. There the instructions get fetched and executed. The results are passed to the *result store* and afterwards moved into their destination.

A classification of data flow machines with respect to communication networks is given in [1]. Our model fits the two stage machines such as in [2], as well as the two level and one level machines [1].

A program for a data flow machine is represented as a DAG. A node in the graph represents an operation, an arc in the graph represents a data dependency between two operations. An instruction template is composed of an operation and its arguments. A *grain* is a set of (one or more) instruction templates (a subgraph of the graph).

The arguments of the grain are the arguments for the contained instructions whose dependency arcs are not in the grain. The machine schedules grains to the execution units. Scheduling grains saves some of the communication overhead and reduces program execution time.

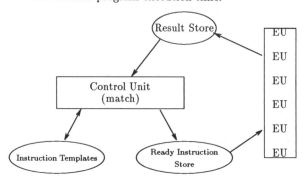

Figure 1: data flow machine model

Model assumptions:

- The execution time of a single instruction v is $ex(v)$.

- The size of each grain is linear in the number of the arguments to the grain.

- Passing one argument through the grain distribution network (from the instruction ready store to the execution units) takes t_f time units.

- Writing the result plus the match operation and moving a grain to the ready instruction store takes t_m time units. (If moving a grain to the ready instruction store takes non constant time it can be added to t_f.)

- The program's output (which is the output of some node(s)) is delivered through the network and takes t_m time units as well.

Let t_0 be the time in which the last argument of a k argument grain, P, arrives at the ready instruction store. Then the execution of P can start at $t_1 = t_0 + k \times t_f$ and ends at $t_2 = t_1 + \sum_{v \in P} ex(v)$. The computed result will arrive (for use in the next grain) at $t_3 = t_2 + t_m$.

Denoting the time t_0 by $\mathrm{WS}(grain)$, $t_2 - t_0$ by $t(grain)$, and t_3 by $T(grain)$ we get: $T(P) = \mathrm{WS}(P) + t(P) + t_m$.

2.1 Formalization

The problem is: Given a DAG $G = (V, E)$, numbers t_f, t_m, a number of execution units p and a mapping $ex : V \to N$, find a set of connected sub-DAGs (i.e. their underlying graphs are each connected) $G_i = (V_i, E_i)$ covering all internal nodes (not leaves) of G (each sub-DAG represents a grain), such that the number of active grains at time t, which is the number of grains G_i satisfying $\mathrm{WS}(G_i) \leq t < \mathrm{WS}(G_i) + t(G_i)$, is less than or equal to p, and such that $T(\mathrm{DAG}) \equiv \max_i \{T(G_i)\}$ is minimized. (Note that the V_i's are not required to be disjoint.) The following notation is used in the above definition:

- $V_i \subseteq V$; $\cup_i \{V_i\} = V \setminus \{v \mid v \text{ is a leaf}\}$

- $E_i \equiv \{(u,v) \in E \mid u,v \in V_i\}$

- $d_{in}(G_i) \equiv \mid \{u \mid (u,v) \in E , u \notin V_i , v \in V_i\} \mid$

- $WS(G_i) \equiv$
  $\max_U \left\{ T(U) \,\middle|\, \begin{array}{l} U \in \{V_j\}, \ U \neq V_i \text{ and there is an} \\ \text{arc in E from } u \in U, u \notin V_i, \text{ to } v \in V_i \end{array} \right\}$

- $t(G_i) \equiv t_f \times d_{in}(G_i) + \sum_{v \in V_i} ex(v)$

- $T(G_i) \equiv WS(G_i) + t(G_i) + t_m$

3 Results

We consider the case with an unlimited number of execution units. A lower bound for the execution time of trees is derived and an algorithm which produces optimal instructions scheduling is presented.

To get the lower bound for trees let h be the height of the tree, and consider the "heaviest" chain from leaves to the root to be the set of nodes G_* composed of all nodes on a directed path from a leaf to the root for which $t_f \times d_{in}(G_*) + \sum_{v \in G_*} ex(v)$ is maximal. Consider the chain G_* as one grain and denote the grain cover using G_* by S_1. In the full paper we prove that the completion time of S_1, is at least $t(G_*) + t_m$, and this is a lower bound for the completion time of the tree.

3.1 Algorithm for binary trees

Assuming an unlimited number of processors, all we have to do is calculate the best possible $T(grain)$ for the grain containing the root of the tree.

Given a tree $G = (V,E)$, numbers t_f, t_m, and a mapping $ex : V \to N$, traverse the tree from leaves to root (in postorder), computing for each node as follows :

ST(v) — a subtree rooted at v which is the grain calculating v and delivering its result at the best possible time.

WS(v) — the start time of ST(v). (The completion time of the worst son of ST(v).)

$t(v)$ — the number of time units needed to execute the grain ST(v).

$T(v)$ — the completion time of ST(v).

1. For all leaves v define $ST(v) \equiv \phi$, $T(v) \equiv 0$, $WS(v) \equiv 0$, $t(v) \equiv \infty$.

2. For a node v having two sons x, y (x and y have been evaluated due to postorder):

   (a) Define
   $$S_*(v) \equiv \{v\} \cup ST(x) \cup ST(y)$$

   (b) Sort the $(k+1)$ nodes of $S_*(v)$ in decreasing order of their completion times to get[1]:
   $$T(v_0 = v) > T(v_1) \geq T(v_2) \geq \cdots \geq T(v_k)$$

   (c) For all $i \in [1 \ldots k]$ calculate :
   - $ST_i(v) = \{z \mid z \in S_*(v), T(z) > T(v_i)\}$
     $ST_{k+1}(v) = S_*(v)$
   - $WS_i(v) = T(v_i)$
     $WS_{k+1}(v) = \max\{WS(x), WS(y)\}$

For all $i \in [1 \ldots (k+1)]$ calculate :
- $t_i(v) = t_f \times d_{in}(ST_i(v)) + \sum_{s \in ST_i(v)} ex(s)$
- $T_i(v) = WS_i(v) + t_i(v) + t_m$

(d) Choose
- $T(v) = T_{i \cdot}(v) = \min_{i=1}^{k+1}\{T_i(v)\}$
- $WS(v) = WS_{i \cdot}(v)$
- $t(v) = t_{i \cdot}(v)$
- $ST(v) = ST_{i \cdot}(v)$

Now we have a grain associated with each node. However, grains might not be disjoint. For trees we do not need nodes to be calculated in several grains. To get the required grains, traverse the tree from the root to the leaves collecting the required grains. The root's grain is required. For every direct son x of a required grain, the grain rooted at x is required.

Theorem 1 *Let $G=(V,E)$ be a program graph which is a binary tree. For every node v, $T(v)$— the completion time of the grain rooted at v calculated by the algorithm — is the smallest over all possible grain covers having a grain rooted at v.*

The optimality of the algorithm is given in the full paper where we also discuss some extensions and show that the time complexity of the algorithm is $O(n^2)$, where $|V| = n$.

The algorithm and the optimality proof are given for binary trees, however it seems that this result is applicable to non binary trees as well. The same problem regarding DAGs is harder and this solution is not optimal even for binary DAGs, however binary DAGs seem easier than non binary DAGs. Considering a finite number of processors is again a harder problem and the algorithm given does not find the optimal solution for this case. We consider these cases as interesting problems for future work.

References

[1] A.H.Veen. Dataflow machine architecture. *ACM Computing Surveys*, 18(4):365–396, December 1986.

[2] J.B.Dennis and D.P.Misunas. A preliminary architecture for a basic data flow processor. In *2nd annual Symposium on Computer Architecture*, pages 126–132, 1974.

[3] M.R.Garey and D.S.Johnson. *Computers and Intractability: a Guide to the Theory of NP-completeness.* w.h.freeman and company, 1979.

[4] T.C.Hu. Parallel sequencing and assembly line problems. *Operations Res.*, 9:841–848, 1961.

[5] C.H.Papadimitriou and M.Yannakakis. Towards an architecture independent analysis of parallel algorithms. *SIAM J.Computing*, 19(2):322–328, April 1990.

[6] J.L.Gaudiot and M.D.Ercegovac. Performance evaluation of a simulated data flow computer with low resolution actors. *Journal of Parallel and Distributed Computing*, 2(4):321–351, 1985.

EFFICIENT PRODUCTION SYSTEM EXECUTION ON THE PESA ARCHITECTURE

Franz Schreiner
Gerhard Zimmermann
University of Kaiserslautern, SFB 124
Postfach 3049 W-6750 Germany

Abstract -- This paper sketches simulation results of the parallel distributed memory architecture PESA-I. The speedup of six different production systems is analysed on several alternative hardware configurations. PESA-I allows a very fine grain data distribution not achievable by shared memory multiprocessors implementations. Thus, parallelism at a lower level can be exploited. This advantage is illustrated by a performance comparison of several parallel production system architectures.

Introduction

The current tendency towards knowledge based systems relies on easy methods to express correlations between different data sets. In OPS5 [1], the correlations are expressed by the condition parts of rules. The declarative knowledge builds the working memory (WM). Each of its elements (WME) belongs to a class and can be identified by a time tag. The execution of the knowledge based system is achieved by the recognize act cycle. Firstly, the rule base is checked against the WM- all executable rules form the conflict set(CS). Next, the conflict resolution selects one of these rules. Finally, the action part of this rule gets executed by the act phase and changes the WM. Thus, executability of rules changes, the cycle repeats. Caused by the large number of tests, match is the most expensive part of this cycle. Parallelism is present in each part of recognize act.

Figure 1 shows an example rule and its intermediate representation by the RETE algorithm [2]. The RETE graph stores the elements affected by positive tests associated with the constants occurring in single conditions elements (constant tests) in memory nodes as well as those combinations of elements fulfilling several condition by consistent variable bindings (intertests). This is reasonable. Just a small part of the working memory changes by a rule's execution. Thus, repeating tests with identical input in different cycles is avoided at the price of permanent storing the memory nodes and the conflict set. The WM-changes by the last rule execution are attached with a sign and update the sets. They and their derivates form tokens by attaching the appropriate test node address. Tokens activate RETE's nodes in data-flow manner and report CS-changes.

Expert systems require large RETE graphs. Nevertheless, just a small part of a graph gets activated by the few WM-changes filtering through the graph. But a single Intertest activation causes a high effort. An Intertest node is inseparable from the memory nodes at its input arcs. At one time, it is activated by just one token in the following way: Firstly the activating token must stored/deleted (depending on its sign)

in/from the memory node to which it belongs. Secondly the comparison(s) described by the testnode must be applied on the activating token and each entry of the other (partner) memory node. The filling of memory nodes changes, they are stored as lists. Those lists contain typically 20-30 elements. Hence, a testnode activation provides two important sources of parallelism

P0: If the token has negative sign, its copy must be found and deleted in the activating memory node.

P1: In each case, <u>all</u> elements of the partner memory must be addressed and tested.

Furtheron, the intertest may produce several positive results and may activate the following Intertest node(s). Thus, the range of parallelism is smaller than that of P0/P1, but the parallelism type should be noticed:

P2: Manyfold activation of the same Intertest.

Self evidently, the following parallelism types are present too:

P3: Parallel Execution of different Intertests,

P4: Parallel Execution of different Constant Tests.

Both kinds of parallelism can be used by distributing the respective nodes to different processing elements. In contrast to P3, P4 does not require a permanent storage of data. P3 tolerates the concurrent use of the intra node parallelism types P0 and P1. The number of processors cannot be infinite. Hence, if a fixed assignment of testnodes to processors is made, concurrently active nodes should be assigned to different PEs. Intertests succeeding one another are often active almost concurrently. Positive results of the predecessor can activate the succeeding node while the predecessor is still performing the same Intertest with other partner elements. This kind of pipelining can be exploitet by the assignment of succeeding nodes to different PEs.

The non-match parallelism types are mentioned in the sequel. The first concerns the act phase, whose actions can be executed in parallel (P5). To access the necessary data, the act phase administers the working memory. The actions are very simple and should be finished quickly. Deletions of WMEs and the retrival of bound variables introduced in the LHS require search for a definite class entry by its time tag. A class can be represented by a list. A fixed class assignment to PEs allows a better distribution of tasks to PEs. A clever class distribution guarantees the utilization of P5 in many cases. If the WM is small, it could be assigned redundantly to each PE, but the performance does not rise significantly [5]. A RHS action is directly followed by the constant tests of respective WM change (elements to be deleted are tested beforehand). It is useful to perform the constant tests by the same PE which performs the RHS-action. The WME is addressed already. The constant tests require no permanent data storage. The complexity of the combination act and constant testing (=ACT) gets closer to that of conflict resolution. An utilization of P5 guarantees the utilization of P4 as well.

Conflict Resolution maintains the CS and delivers the dominant instantiation to ACT. Just the dominant instantiation has to be computed. It is not necessary to provide the order between all elements of the CS. The CS is partitioned into conflict subsets (CSS). Thus, parallelism is present if elements are inserted or deleted (P6).

The use of several CSSs requires an arbiter, because n CSSs produce n local dominant instantiations. This arbiter should be located at a special PE. It stores the first arriving entry. An element arriving later can have a higher priority, in this case it replaces its predecessor, otherwise it is discarded. This method runs fast. Initial experiments showed us that the CSS's operations consume a little bit more time than other operations, although the number of CS changes is not very high. Thus, we decided to deliver the local CSS-dominants at a definite point of time during the match phase to the arbiter. If furthermore, the arbiter consideres all CS-changes, it can determine the dominant rule itself. By this way, the arbiter (called CR processor) often has finished computing the dominant instantiation while all CSSs are still busy. The next cycle can be started, the CSS maintenance overlaps with the next cycle. The arbiter's result is invalid if it is deleted again during the match phase. Then the CR PE signals the CSSs to send their local dominants after they have performed all CS changes.

The architecture

The proposed PESA-I architecture is shown in figure 2. Most of the measurements were performed with the depicted configuration, but others show also a reasonable performance. The architecture is scalable. The communication principle is broadcasting on the busses of the match pipeline and multicasting on bus zero. The

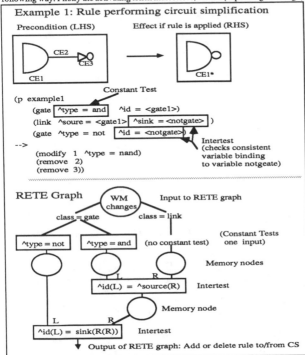

Example 1: Rule performing circuit simplification

Precondition (LHS)   Effect if rule is applied (RHS)

Constant Test

(p example1
   (gate ^type = and   ^id = <gate1>)
   (link ^soure = <gate1> ^sink = <notgate>)
   (gate ^type = not   ^id = <notgate>)
 -->
   (modify 1 ^type = nand)
   (remove 2)
   (remove 3))

Intertest (checks consistent variable binding to variable notgeate)

RETE Graph

WM changes

Input to RETE graph

class = gate   class = link

^type = not ^type = and (no constant test)

(Constant Tests one input)

Memory nodes

^id(L) = ^source(R) Intertest

Memory node

^id(L) = sink(R(R)) Intertest

Output of RETE graph: Add or delete rule to/from CS

multicasting messages on bus zero are sent to one of the groups ACT, Conflict Resolution, I/O or Match. Other groups neglect this message. All PEs are of identical type especially designed for PESA-I.

Programs and data are assigned to the PEs in the following way: A class of the WME is assigned to a fixed ACT PE which contains all associated RHS actions and constant tests. The match pipeline contains the Intertest nodes. The match pipeline consists of stages. PESA does not restrict the depth of branches of the RETE graph, because the match pipeline can be used several times. An Intertest is assigned to one stage. All PEs of this stage contain the appropriate test program and a part of test partners. Activating tokens are queued in a buffer. Their first word points to the Intertest program. This has the following structure: Firstly, the token must be stored/deleted by the PE matching the address field of the token. All PEs address the partner memory of the activating token and perform the Intertest with its entries. If an Intertest succeeds, a new token is formed out of the two partners and passed to a sendbuffer. The sign remains unchanged, the tag is contained in the Intertest program and the address is computed by adding the activating tokens address and the PEs address. While this token activates the succeeding Intertest, the predecessor may still be active. CS changes may have to be passed via other stage(s) to the conflict resolution group.

Handling of negative tokens

In RETE, a WM change of identical class and attribute values requires nearly the same amount of computation, independent whether it was initiated by remove or make actions. No significant difference beside store/delete of tokens is present in the handling tokens of different sign. Especially, the Intertests between negative tokens and all partners are performed. In general, this is not necessary. All memory nodes could be looked up for entries containing the deleted WME's time tag. The instantiations present in the conflict set can be cleared from invalid entries in the same way. Hence, PESA exclusively performs the constant tests if a WME is removed. In case of a succesful test, just the time tag is sent to all succeeding memory nodes.

Simulation Results

The results we present here are gained by a detailed two stage simulation concerning a single PE and the entire architecture. Both levels interpret the code of our compiler.

A unique unit of measurement is important to compare architectures. For OPS5 working memory element changes per second (WMEC/s), production executions per second (P/s) and the speedup against sequential implementations are common. Our

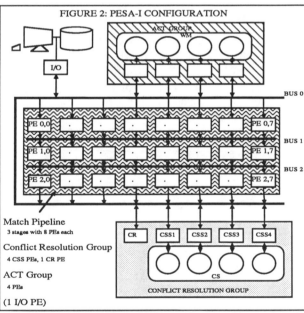

FIGURE 2: PESA-I CONFIGURATION

Match Pipeline
3 stages with 8 PEs each

Conflict Resolution Group
4 CSS PEs, 1 CR PE

ACT Group
4 PEs

(1 I/O PE)

sequential reference system is the OPS83 compiler (to avoid competing with interpreters) running on an Apollo 3010 workstation (MC 68020, 12MHz, 4 MB main memory). The PESA PEs and busses run with 10MHz, therefore the preconditions are similar. By the exchange of programs and measurements we are able to normalize this performance to that of other systems. The following production systems were used:

- A: the n-queens problem, here n = 8 (19 productions)
- B: implementation of the mincut algorithm (42 productions)
- C: circuit simplification program (8 productions)
- D: scheduling of a bridge tournament (16 productions)
- E: Waltz's edge labelling algorithm (48 productions)

F: simulation of a cafeteria (74 productions)

Program C terminates after 40 cycles, E after 211. Program D runs very slow. The simulation time corresponds to this behaviour. Therefore, just six cycles of this program were simulated (one of them produces 882 CS changes). For the others, the first 200 cycles are executed. The OPS83 runs were performed without any trace information, the time to load the system is not included. The bold-face rows of table one show the performance of the PESA configuration depicted if fig 2. WMEC/s-Results of other configurations are listed below. Because the match phase consumes about 90% of the runtime, the speedup is essentially caused by P0/P1 (approx factor 6 with 8 PEs/row) and the changed negative update (~factor 1.5), while the gain by P3 is low or gets sucked by the transportation overhead (max 1.5). The rendering match gain is caused by the PE and its special instruction set (~faction 3-6). ACT and CSA utilize the overlap with match, the parallelism type P4, P5 and P6 (~factor 2-3 with 4 PEs) and the special PE structure (~factor 4-6 for CSA instructions. The reader interested in more details is referred to [5].

Comparison to other architectures

Several architecture for production system have been proposed. I just mention those with the highest performance. PSM [3], a shared memory match coprocessor, indicates that a speedup of 16 of the pure match time is possible with 32 processors (=

	A	B	C	D	E	F
# WM-changes	788	464	411	58	262	848
runtime OPS83 (s)	1,03	1,62	1,51	2,33	1,82	1,19
runtime PESA(ms)	**23,3**	**42,7**	**17,4**	**35,7**	**36**	**28,6**
speedup	**44,2**	**37,9**	**86,8**	**65,3**	**50,6**	**41,6**
P/s	**8584**	**4684**	**2299**	**168**	**5861**	**6986**
WMEC/s	**33800**	**10864**	**23623**	**1623**	**7286**	**29620**
2row/8PE	25745	9913	22279	–	7115	–
3row/16PE	–	–	26975	–	–	–
3row/4PE	32289	9421	17166	–	6625	–
3row/2PE	29870	6280	10966	–	5288	–
4row/8PE	30373	10044	26411	–	6923	–
5row/8PE	29085	10215	–	–	7380	–
1CSA PE	31199	7840	–	–	6629	–
1ACT PE	–	9760	14471	–	–	–
1CSA, 1ACT PE	–	7159	14340	–	–	–

9400 WMEC/s, 3800 P/s) on the average. The use of parallel rule firing and 64 PEs achieves 16000 WMEC/s. The shared memory does not allow the use of the parallelism types P0/P1. These are used more intensive in the CUPID [4] architecture, an array of processors with local memories, coupled by a communication network. The applied D-RETE algorithm creates and deletes nodes dynmically. While we store a new element in a memory node corresponding to an Intertest, CUPID creates a new test node for this element. A match speedup of 18 (with 64 PEs) compared to a VAX 11/785 when using program 'C', is possible. With many PEs (no number available) a speedup of 'D' by a factor of 21.8 is estimated.

Conclusion

The quoted results indicate PESA-I to be the currently fastest architecture for production systems, although the number of PEs is not large. Speedups of 50 were measured by a detailed simulation of several production systems. They are achieved by a use of several kinds of parallelism and a special PE structure and instruction set. PESA's speedup is under some restrictions valid for all times, because it is achieved by the use of parallelism and not of technology parameters. The PE can be realized with any technology. Its advantages consist of the structure and the adapted instruction set. All phases of the recognize act cycle are implemented. The detail of the simulation was useful to detect interdependencies of implementation details. The machine is independent from any host, all subgroups use the same type of PE. In case of single rule firing the architecture schedules itself. During execution all actions are evoked in a data-flow manner. The architecture is scaleable.

References

[1] L. Brownston, R. Farrell, E. Kant, and N. Martin. "Programming Expert Systems in OPS 5: An Introduction to Rule-Based Programming". Addison-Wesley, 1985.

[2] C. L. Forgy. "Rete: A Fast Algorithm for the Many Pattern/Many Object Pattern Match Problem". Artificial Intelligence 19 (Sep 1982), p 17-37.

[3] A. Gupta, C. Forgy, A. Newell, and R. Wedig. "Parallel Algorithms and Architectures for Rule-Based Systems", 13th Intern. Symp. on Computer Architecture, Tokyo, 1986

[4] M. A. Kelly and R. E. Seviora. "Performance of OPS 5 Pattern Matching on CUPID", Proc of the Euromicro 89, Microprocessing and Microprogramming Vol 27, No 1-5 August 1989, pp 397-404.

[5] F. Schreiner. "PESA-1: A Parallel and Distributed Architecture for Production Systems". Ph. D. Thesis, University of Kaiserslautern 1991.

IMPLEMENTATION OF THE HOPFIELD NEURAL NETWORK ON THE MASPAR SYSTEM

Jeffrey S. Clary   Suraj Kothari

Department of Computer Science
Iowa State University
Ames, Iowa 50011

Abstract – Four different implementations of the Hopfield neural network were developed on an 8192-processor MasPar, a single instruction, multiple data (SIMD) machine that became commercially available in 1990. We briefly describe the implementations and report Mflop ratings to compare their relative performance. Factors that affect performance are discussed.

Introduction

Single instruction, multiple data (SIMD) machines like the MasPar [4, 5], are especially well suited for implementation of neural networks. A matrix-vector multiply is the most significant computation in all the common neural network models [2]. Methods are known for performing this operation efficiently on SIMD machines.

MasPar System

The MasPar system is a SIMD parallel computer with from 1K to 16K processing elements (PEs). This study was done on a MasPar with 8K PEs. Each PE has 16 Kbytes of RAM. The Array Control Unit (ACU) performs operations on singular data and broadcasts instructions to the PEs to manipulate plural data.

Communication between the ACU and PEs is provided, as well as among PEs. The PEs are arranged in a 2-dimensional array. Each PE may directly communicate with its eight nearest neighbors using *xnet* communication. Arbitrary communications among PEs are provided by the usually slower *router* communication.

Programming for this project was done using the MasPar Parallel Application Language (MPL), based on Kernighan and Ritchie C. MPL provides the functionality of C, with extensions for parallel variables (called *plural*), parallel expressions, parallel control statements, and interprocessor communications.

Hopfield Networks

A Hopfield network consists of N *neurons*, each with a state X_i and a threshold Θ_i, and an $N \times N$ matrix of connection weights. The connection weight between neurons i and j is denoted W_{ij}. The state X_i of each neuron is one of $\{1, -1\}$ for a bipolar network or $\{1, 0\}$ for a binary network.

"Running" a bipolar Hopfield network means *updating* each state X_i according to the following update rule:

$$X_i \leftarrow \left(\begin{array}{l} 1 \; if \; \sum_{j=1}^{N} W_{ij}X_j - \Theta_i > 0 \\ -1 \; if \; \sum_{j=1}^{N} W_{ij}X_j - \Theta_i \; <= 0 \end{array} \right)$$

If none of the states X_i change during an update, the network is in a *stable state*. Applications of Hopfield networks seek to make the stable states correspond to solutions to a given problem. For example, an associative memory application seeks to make the stable states correspond to the patterns being stored.

Implementations

The four different implementations described below illustrate various ways of distributing data items to processors, and for implementing the update rule. Since most of the update rule is just a matrix-vector multiplication, the lessons learned here apply to many other applications.

For convenience, indexing was implemented as $[0..N - 1]$.

Version 1 (N^2 PEs)

Version 1 uses N^2 processing elements, one for each element in the weight matrix, with some PEs containing states and thresholds. To perform an update, each processor retrieves the X_j for its weight W_{ij} using router communicaton, and calculates $W_{ij}X_j$. The results are summed to the state PEs using a segmented sum operation.

Version 2 (N PEs with router communication)

Version 2 uses only N processors. Each PE i stores X_i and Θ_i, as well as an array of weights $W_{i0}...W_{i,N-1}$. To perform an update, each PE uses router communication to retrieve the X_j for each of its W_{ij} elements in turn, and calculates the sum of the $W_{ij}X_j$ terms.

Version 3 (N PEs with xnet communication)

Version 3 is similar to Version 2, but uses xnet nearest neighbor communication instead of router communication to retrieve each X_j. The distribution of data to the PEs is identical to that in Version 2. However, to give the data distribution a regular pattern for the use of xnet communication, this implementation restricts the number of neurons to a multiple of $NXPROC$, the "width" of the MasPar PE array.

Version 4 ($NPROC$ PEs)

Version 4 uses all the processors available, with each PE storing several weights, and some PEs storing one state and threshold.

X_i and Θ_i are stored on PE i. The distribution of weights is best understood by picturing the MasPar PE array repeatedly tiled over a part of the weight matrix until all of it has been covered. Figure 1 illustrates the layout of a 6-neuron Hopfield network on a hypothetical 3×3 PE machine. This version requires N to be a multiple of both $NXPROC$ and $NYPROC$, the number of columns and number of rows of the MasPar PE array, respectively.

Figure 1: Version 4 – Initial Data Distribution

Tile	1			2		
PE #	0	1	2	0	1	2
W_{ij}	W_{00}	W_{01}	W_{02}	W_{03}	W_{04}	W_{05}
X_i	X_0	X_1	X_2	X_0	X_1	X_2
PE #	3	4	5	3	4	5
W_{ij}	W_{10}	W_{11}	W_{12}	W_{13}	W_{14}	W_{15}
X_i	X_3	X_4	X_5	X_3	X_4	X_5
PE #	6	7	8	6	7	8
W_{ij}	W_{20}	W_{21}	W_{22}	W_{23}	W_{24}	W_{25}
Tile	3			4		
PE #	0	1	2	0	1	2
W_{ij}	W_{30}	W_{31}	W_{32}	W_{33}	W_{34}	W_{35}
X_i	X_0	X_1	X_2	X_0	X_1	X_2
PE #	3	4	5	3	4	5
W_{ij}	W_{40}	W_{41}	W_{42}	W_{43}	W_{44}	W_{45}
X_i	X_3	X_4	X_5	X_3	X_4	X_5
PE #	6	7	8	6	7	8
W_{ij}	W_{50}	W_{51}	W_{52}	W_{53}	W_{54}	W_{55}

Updating involves iterating over the "tiles," calculating part of the product for each, as described in the algorithm below:

1. ACU sets $VTILES = N/NYPROC$
2. ACU sets $HTILES = N/NXPROC$
3. For $v = 0$ to $VTILES - 1$
   a. Each PE sets $PARTSUM = 0$
   b. For $h = 0$ to $HTILES - 1$
      i. Each PE on row h of the MasPar array copies X_i into X_j down its column of the MasPar array
      ii. Each PE sets $PARTSUM = PARTSUM + W_{ij}X_j$
   c. Each row of PEs on the MasPar array sums $PARTSUM$ into SUM on its eastmost PE on the MasPar array
   d. Each PE on the eastmost column of the MasPar array sends its SUM to the PE with the corresponding Θ_i using router communication
4. Each state PE compares Θ_i SUM and updates X_i accordingly

Performance

Mflop Calculation

A count of floating-point operations performed per Hopfield network update was obtained from the update rule, as follows:

- N^2 multiplies $(W_{ij}X_j)$
- N^2 adds to sum the above products
- N compares with Θ_i

The update time and the formula $2N^2 + N$ was used to calculate Mflops for each version. Single-precision arithmetic was used.

Version 1

Version 1, limited to 90 neurons on the 8K MasPar used in this study, could not be directly compared to the others. A disappointing performance of about 9.5 Mflops results from two factors. First, the router operation used to move the X_j values to the PEs with weights W_{ij} is expensive. Router communication is generally slow, and in this case each X_j must be sent to N PEs. These sends are serialized. Second, the system function *scan*, used to sum the $W_{ij}X_j$ terms, is expensive. While these operations are executing, no floating-point operations are performed. These two operations together account for more than 90% of the update time.

Versions 2 and 3

These two implementations use only N PEs, so have a higher maximum number of neurons (3840) than Version 1. In fact, the number of neurons is limited by the memory per PE, not by the number of PEs, because each PE i must store weights $W_{i0}...W_{i,N-1}$.

Since the programs for this paper were run on an 8K-processor MasPar, these two versions always use less than 50% of the PEs. Therefore, the Mflop ratings for these two versions were several times smaller than the one for Version 4. On a 16K-processor MasPar, the inefficiency would be even worse.

The difference between the performance of Version 2 and Version 3 is also dramatic. Recall that in both versions each active PE retrieves an X_j from every other active PE in turn. Version 2 uses general router communication, while Version 3 takes advantage of the regular pattern of communication to pass each X_j using xnet communication, much as in a systolic array processor. As shown in Figure 2, as N increases, the advantage of xnet over router communication increases. Using general communication when generality was not needed decreased performance.

Version 4

The final implementation, Version 4, avoids many of the problems of the previous three. This version always uses all the PEs available, but is not limited by needing N^2 PEs. PE memory limits N to 5504 in this implementation.

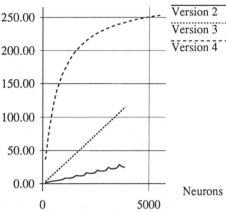

Figure 2: Hopfield Network Update Mflops (Versions 2,3,4)

A sum operation like the scan described for Version 1 is required for this version. However, the data distribution and restrictions on N in this version allow the implementation of a relatively fast sum using xnet communication. The router is used to send the sums to the state PEs as described in the algorithm above, but this only involves $NYPROC$ PEs and is only done $N/NYPROC$ times, so it does not affect performance greatly.

As expected, Version 4 outperforms the others by a large margin, especially at larger values of N. The superior performance is achieved by keeping all of the PEs busy most of the time and by making use of the faster xnet communication.

Conclusions

The implementations described above illustrate the two important requirements for achieving high performance on the Maspar:

- Keep as many PEs busy as possible
- Minimize communication overhead

Version 4 satisfies these requirements far better than any of the other implementations, because it is designed to use all the PEs available, and to use fast xnet communication.

We have concluded that implementing Hopfield networks on a SIMD computer is very natural and efficient. Each neuron computes an identical function, and the fully-connected network makes for a very regular communication pattern.

Acknowledgments – Thanks to the Scalable Computer Laboratory at Iowa State University for the use of the MasPar computer as well as to Michael Carter and Naresh Nayar for ideas which contributed to Version 4.

References

[1] V.C. Barbosa, and P.M.V. Lima, "On the Distributed Parallel Simulation of Hopfield's Neural Networks," *Software–Practice and Experience*, vol. 20(10), 967-983, October, 1990.

[2] B.M. Forrest, D. Roweth, N. Stroud, D.J. Wallace, and G.V. Wilson, "Implementing Neural Network Models on Parallel Computers," *The Computer Journal*, vol. 30, no. 5, pp. 413-419, 1987.

[3] J.J Hopfield, and D.W. Tank, "Neural Computation of Decisions in Optimization Problems," *Biological Cybernetics*, Springer-Verlag, 1985, pp. 141-152.

[4] *MasPar Parallel Application Language (MPL) User Guide*, MasPar Computer Corporation, Sunnyvale, CA, 1990.

[5] *MasPar Parallel Application Language (MPL) Reference Manual*, MasPar Computer Corporation, Sunnyvale, CA, 1990.

LOOP STAGGERING, LOOP STAGGERING AND COMPACTING: RESTRUCTURING TECHNIQUES
FOR THRASHING PROBLEM

Guohua Jin   Xuejun Yang   Fujie Chen

Department of Computer Science
Changsha Institute of Technology
Changsha, Hunan, 410073 , P.R.China

Abstract -- Executing nested loops in parallel with low run-time overhead is very important for parallel processing systems to achieve high performance. However,with caches or local memories in memory hierarchies, "thrashing problem" may arise whenever data moves back and forth between the caches or local memories in different processors. In this paper, we present two restructuring techniques, with which we can not only eliminate the thrashing phenomena significantly,but also exploit the additional parallelism existing in outer serial loop.

Introduction

Parallel processing systems with complex memory hierarchies have become very common today. Cache as a means to bridge the gap between fast processor cycles and slow main memory access times is introduced in most multiprocessor systems. Local memory (or programmable cache) in some supercomputers is another example. In such parallel processing systems, a problem called"cache or local memory thrashing " may arise whenever data moves back and forth between the caches or local memories in different processors[1]. In this paper,we present an approach to solve the thrashing problem. Based on thorough study of the relationship between the array element accesses and its enclosed loop indices in the nested loop construct,we partition iteration space into a set of equivalent classes at first.Then,we discuss two new restructuring techniques called loop staggering,loop staggering and compacting.With which we can not only eliminate the thrashing phenomena significantly,but also exploit additional parallelism existing in outer serial loop.Loop staggering benifits the dynamic loop scheduling strategies,whereas loop staggering and compacting is good for static loop scheduling strategies.They both solve the thrashing problem satisfactorily.

To simplify our discussion,we adopt the same assumption as Fang's. More complicated machine model can be viewed as an extension of the simple model with more levels of local memories in the memory hierarchy.

Preliminaries

Because there is no cache or local memory thrashing problem when the outermost loop is parallel[1],in this paper,we consider the nested loop constructs of the following form for simplicity:

```
DO I=1,N₁
    DOALL J=1,N₂
        DO K=1,N₃
        ......
        A(a₁I+b₁J+c₁K+d₁,a₂I+b₂J+c₂K+d₂)=...
        ......
        ...=A(a₁I+b₁J+c₁K+d₁,a₂I+b₂J+c₂K+d₂)+...
        ......
        ENDDO
    ENDDOALL
ENDDO
```

Fig. 1   Nested Loop Construct

Where $N_1*N_2*N_3$ is the iteration space of the nested loop construct, N_1*N_2 is the reduced iteration space which consists of N_1*N_2 threads T_{ij} ($1 \le i \le N_1$, $1 \le j \le N_2$). As the cache size is large enough,dependences between statements within a thread are ignored.

Restructuring Techniques

Different from Fang's method, we solve the thrashing problem by means of restructuring techniques, with which we can transform the nested construct in Fig.1 into a nested construct with the parallel loop outermost. Here we first introduce some definitions and properties for the construct in Fig. 1.

Definition1: Let T_{ij}, $T_{i'j'}$ be threads,$1 \le i,i' \le N_1$,$1 \le j,j' \le N_2$, if $\exists$ k,k', $1 \le k,k' \le N_3$

$$a_1 i + b_1 j + c_1 k + d_1 = a_1 i' + b_1 j' + c_1 k' + d_1$$
$$a_2 i + b_2 j + c_2 k + d_2 = a_2 i' + b_2 j' + c_2 k' + d_2$$

then there is data dependence between T_{ij} and $T_{i'j'}$,denoted by $T_{ij} \, \psi \, T_{i'j'}$.

From definition1, we have:

Lemma 1: let $b_1 a_2 - a_1 b_2 \ne 0$, if $T_{ij} \, \psi \, T_{i'j'}$ then there exist k,k', $1 \le k,k' \le N_3$

$$i' - i = \frac{(b_1 c_2 - c_1 b_2)}{(b_1 a_2 - a_1 b_2)} (k-k') \,,\, j' - j = \frac{(a_2 c_1 - a_1 c_2)}{(b_1 a_2 - a_1 b_2)} (k-k')$$

Definition2: For a given pair (i,j) in the reduced iteration spaceN_1*N_2,S_{ij} denotes a set of iteration pair (i',j') in this space,which satisfies :

$$S_{ij} = \{(i',j') | T_{ij} \, \psi \, T_{i'j'} \}$$

From lemma 1 and definition 2,we have following theorems.

Theorem 1: S_{ij} defined above is an equivalent class of the reduced iteration space N_1*N_2, therefore relation ψ partitions the reduced iteration space.

Theorem 2: Let $b_1 a_2 - a_1 b_2 \ne 0$, S_{ij} defined above satisfies :

$$S_i = \left\{ (i+r*L_1, j+r*L_2) \,\middle|\, r=0,\pm1,\pm2,\cdots; 1 \le i \le N_1, 1 \le j \le N_2, 1 \le i+r*L_1 \le N_1, 1 \le j+r*L_2 \le N_2, 1-N_3 \le r*L_3 \le N_3 - 1 \right\}$$

where $L_1 = \varphi(b_1 c_2 - c_1 b_2), L_2 = \varphi(a_2 c_1 - a_1 c_2)$ are the numerators in the unreducible fraction numbers with common denominator of fraction numbers:

$$\frac{b_1 c_2 - c_1 b_2}{b_1 a_2 - a_1 b_2} \text{ and } \frac{a_2 c_1 - a_1 c_2}{b_1 a_2 - a_1 b_2} \text{ respectively,} (L_1 > 0), L_3 = \varphi(b_1 a_2 - a_1 b_2)$$

is the denominator in the unreducible fraction numbers.

From theorem 2,we know the relation ψ staggers the reduced iteration space into follow forms if $L_2 < 0$:

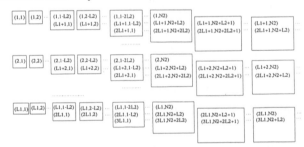

In the above staggered reduced iteration space, each column forms an equivalent class.The thread determined by an index pair in an equivalent class never accesses any element of array A,which is accessed in the thread determined by other index pairs in the different equivalent classes.In other words,if the threads determined by the index pairs in an equivalent class are not assigned to the other processor,there is no data of array A moving between processors.To simplify our discussion,we change the above staggered reduced iteration space into more regular staggered space as following form:

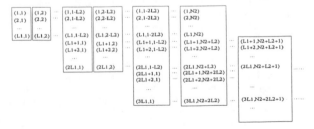

Same with original staggered reduced iteration space, in this more regular staggered iteration space, the thread determined by an index pair in one column also never accesses any element of A,which is accessed in the thread determined by other index pairs in the different columns.So, there is no memory access from one processor to caches or local memories of the other processor, if all these threads are assigned to a same physical processor.

The regular staggered reduced iteration space(RSRIS) can also be regarded as consisting of many blocks which form a flight of steps as Fig.2 shows. In order to explain the restructuring techniques clearly,we define it as follows.

Definition3: RSRIS can be discribed by a 4-tuple (N_1,N_2,L_1,L_2),where N_1,N_2 are the upper bounds corresponding to the outer loops I,J of the nested loop construct in Fig.1,L_1,L_2 are the height and the width of each step in the RSRIS, $L_1=\psi(b_1c_2-c_1b_2)$ and $L_2=\psi(a_2c_1-a_1c_2)$ as indicated before.

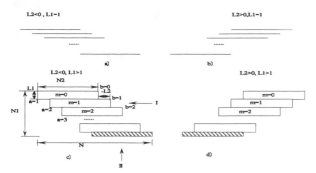

Fig. 2 Regular staggered reduced iteration space

As to RSRIS, there are some parameters which are useful in the rest of this paper.

Theorem3: Let II,I be current column number and current line number of RSRIS, N the number of columns in RSRIS, m the block number of the current line in RSRIS. Let a,b be step number of the flights of steps on the two sides of RSRIS. If $L_2<0$ then

(1) $\quad N=N_2 L_2*(\left\lceil \dfrac{N_1}{L_1} \right\rceil - 1)$

(2) $\quad m=\left\lfloor \dfrac{I-1}{L_1} \right\rfloor$

(3) $\quad a=\begin{cases} \left\lceil \dfrac{N_1}{L_1} \right\rceil & II>-L_2*\left(\left\lceil \dfrac{N_1}{L_1}\right\rceil -1\right) \\ \left\lceil \dfrac{II}{-L_2} \right\rceil & \text{Otherwise} \end{cases}$

(4) $\quad b=\left\lceil \dfrac{(II-N_2)^*}{-L_2} \right\rceil$

where $\quad x^*=\begin{cases} x & x>0 \\ 0 & x\le 0 \end{cases}$

Therefore, we can restructure the nested loop construct in Fig.1 into following form by staggering the reduced iteration space.

```
N=N2-L2 *(⌈N1/L1⌉ -1 )
DOALL II=1,N
    IF II>-L2*(⌈N1/L1⌉ -1)
    THEN  a=⌈N1/L1⌉
    ELSE  a=⌈II/-L2⌉
    b=⌈(II-N2)*/-L2⌉
    DO I=1+b*L1,a*L1
        J=II+b*L2+⌊(I-(1+b*L1))/L1⌋*L2
        DO K=1,N3
            ......
            A(a1*I+b1*J+c1*K+d1,a2*I+b2*J+c2*K+d2)=......
            ......
            ......=A(a1*I+b1*J+c1*K+d1,a2*I+b2*J+c2*K+d2)+......
            ......
        ENDDO
    ENDDO
ENDDOALL
```

With the second restructuring technique, we first partition the RSRIS into segments, each segment have N_2 columns as Fig. 3(a). There are $\left\lceil N/N_2 \right\rceil$ such segments. Then we compact the RSRIS with followin algorithm

```
Algorithm COMPACT(S1,S2)
//S1:inputed RSRIS;  S2: compacted RSRIS;  MERGE merges the i-th column in
segment s and the i-th column in S1' into S2 as the i-th column//
    cut the first segment s from S1
    COMPACT(S1-s,S1')
    MERGE(s,S1',S2)
END
```

After compacting , we get compacted RSRIS in Fig. 3(b)

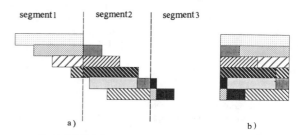

Fig. 3 RSRIS before compacting and after compacting

Therefore, we can restructure the nested loop construct in Fig. 1 into following form by staggering and compacting the reduced iteration space.

```
DOALL II=1,N2
    DO I=1,N1
        M=⌊(I-1)/L1⌋
        J=(II+M*L2) MOD N2
        DO K=1,N3
            ......
            A(a1I+b1J+c1K+d1,a2I+b2J+c2K+d2)= ---
            ......
            ---=A(a1I+b1J+c1K+d1,a2I+b2J+c2K+d2)+ ---
        ENDDO
    ENDDO
ENDDOALL
```

Theorem4: If $(i',j'),(i'',j'') \in S_{ij}$,then $(i',j'),(i'',j'')$ belong to the same column in the compacted RSRIS.

By definition 2 and theorem 4, there is also no memory access from one processor to caches or local memories of the other processor, so long as the threads corresponding to each column of the compacted RSRIS are assigned into a same physical processor.

Conclusions

Exploiting loop parallelism with minimal overhead is extremely important for achieving high speedup and efficiency rates on parallel machines.In this paper,we presented loop staggering,loop staggering and compacting,loop restructuring techniques that transforms the nested loop construct with serial loop outermost into the nested loop construct with parallel loop outermost, and therefore not only solve the cache or local memory thrashing problem but also exploit the additional parallelism existing in outer serial loop efficiently.In this paper,we only discussed a simple machine model and a simple program construct.There are lots of work to be done in this field. Although it is very difficult to deal with the cache or local memory thrashing problem by means of restructuring techniques,we think it is important and necessary to minimize the run-time overhead and believe it is the best way.

Reference

[1] Z.Fang, " Cache or local memory thrashing and compiler strategy in parallel processing systems",ICPP'90 ,pp.271-275.

[2] C.D. Polychronopoulos, " Parallel programming and compilers ", 1988, by Kluwer Academic Publishers.

[3] M.Wolfe and U.Banerjee, "Data dependence and its application to parallel processing",International J. of Parallel Programming,Vol.16,No.2,1987.

Memory models, Compiler Optimizations and P-RISC

K.Gopinath
Computer Science Department
Indian Institute of Science
Bangalore 560012 India

1 Introduction

Nikhil and Arvind[1] have recently proposed the P-RISC architecture that can tolerate variable latencies that result when large number of processors are used to build a multiprocessor. Except for HEP and their successors, it has been only recently that variable latencies are being explicitly incorporated into the architectural models. Another approach for designing large shared-memory multiprocessors has been to start from building blocks that have not been designed with variable latencies explicitly in mind but to impose memory consistency models to solve the variable latencies problem.

A close study of the· two approaches shows areas of similarity. For example, variable latencies can also be handled in the second approach[4] but the issue of contention has been the number of threads required to keep the processors busy. While Weber and Gupta[4] have found that 4 contexts are all that are needed to keep the processors busy, this study has been criticised[5] on the grounds that more contexts are needed when programmed in a non-strict functional language because of the additional available parallelism. One aspect that has not received much attention is a fair comparison of the latency tolerant aspects of these two architectures assuming similar endowments in terms of resources. In this short paper, we will qualitatively study the performance of the P-RISC and DASH[3] architectures under this assumption for a wavefront computation that does relaxation.

2 Overview of P-RISC and DASH

A close study of the P-RISC and DASH architectures can show the rough equivalence of the hardware for DASH and proposed models for P-RISC. Due to lack of space, we do not discuss this any further.

In P-RISC there are no coherence problems as we assume no caches or write-buffers. However, one of the fundamental unsolved problems of the P-RISC architecture is the pipelinability of the instruction set. Since P-RISC does not have registers, the pipelining becomes difficult unless the frame memory and instruction memory are multi-ported which is costly. If we have to avoid this cost, then memory banking has to be implemented so that different threads use different memory banks just as register banking is implemented in [2]. However, for programs like wavefront we are considering below, many of the available threads would use the same banks of frame and instruction memory and hence it will be difficult to pipeline the instructions without introducing bubbles in the pipeline. Having multiple register contexts as in HEP is not an attractive proposition as this increases the cost of a fork. Similarly, caching of the frame memory is problematic. For the purposes of analysis, we assume that the frame and instruction memory has two read ports and one write port though this is expensive.

In P-RISC, all non-local or heap memory references become split-phase transactions and we essentially have sequential consistency. This does not necessarily reduce the performance of P-RISC-like architectures if the number of threads in each processor is high. If the parallelism profile is such that there is only a few threads at certain times, the performance may be impacted. For producer and consumer types of problems on architectures like DASH, the write of a value produced is typically a cache miss and will result in a cache-coherency transaction but later writes to the same cache line will not stall the processor unless the write-buffer is full. This can result in a net performance gain for DASH-like architectures since the later writes are quite cheap but have to be retired at memory speeds in the case of P-RISC.

3 Compiler optimizations

3.1 Wavefront/Skewing Transformations

The power of non-strict languages is illustrated by the wavefront example. However, compiler optimizations can be helpful in rearranging the computation in such a way that cycles are not wasted in checking if operands are available. Using the skewing/wavefront transformations, the computation can be arranged in such a way that any computation that is attempted has the operands ready. The wavefront method has the maximal parallelism but the parallelism may be difficult to exploit unless very light-weight tasks are available (as in P-RISC). However, some of the parallelism is lost if the computation for each element of the wavefront is executed sequentially. The skewing transformation is more suitable for architectures like DASH as the computation per iteration is constant, the cache behaviour is more predictable and compiler optimizations may be able to exploit these features.

3.2 Prefetching and low-latency synchronization operations

Architectures like DASH and PLUS have operations like prefetching that can be managed by the compiler to improve latency tolerance. However, prefetching may result in too much coherence traffic if a cacheline is updated while it is used by other nodes concurrently. Similarly, synchronization operations can be initiated and verified later for completion as in the PLUS architecture.

4 Wavefront example

```
for i from 1 to n do //relaxation
  for j from 1 to n do
    m[i,j] = (m[i-1,j] + m[i,j-1] + m[i-1,j-1])/3;
```

The above can be transformed into a wavefront computation as follows:

```
for t from 1 to n do
  for j from 1 to t do
    m[t-j+1,j] = (m[t-j,j] + m[t-j+1,j-1] + m[t-j,j-1])/3;
for t from n-1 to 1 do
  for j from 1 to t do
    m[n-j+1,n-t+j] = (m[n-j,n-t+j] +
      m[n-j+1,n-t+j] + m[n-j,n-t+j])/3;
```

The skewing transformation output is as follows

```
for i from 1 to n do //relaxation
  for j from 1+i to n+i do
    m[i,j-i] = (m[i-1,j-i] + m[i,j-i-1] + m[i-1,j-i-1])/3;
```

4.1 Wavefront on DASH

Assume we have a processor array of p by p and that the array m of n by n is distributed equally across each processor. This requires that each node has a copy of its grid as the shared array m may be capable of being stored in a linear order in only one or two nodes as the node memory size is substantial (16-64MB). If this is not done, one or two nodes will experience lot of memory activity and may become a bottleneck. In addition, it is necessary to write back the contents of the copies at the end of computation to the main array.

Let $l = n/p$ (assuming $p|n$). Each processor has a grid l by l. For a first level analysis, a grid cannot get started unless the left and top grids are completely done. A more aggressive implementation would have two synchronizing variables that are set as soon as the computation at the ends of the wavefront are finished. This can enable other computations in other nodes to start. However this analysis is omitted here.

```
wait(left); wait(top)
for each wavefront
```

```
for all elements in wavefront
  ld ld ld comp st
  (* fence *)
signal(right); signal(bottom)
```

Assuming that the 64K data cache at the first level is sufficient to store all the elements of the l by l grid, the first three loads are likely to be cache hits because they have been just recently computed, there is no contention in the direct mapped cache between elements of the array unless $l > 16K$ (as we have already assumed that each node will have a copy of its grid stored linearly). There will be a cache miss for the store and this will generate a cache-coherency transaction for the first word of the cache line. The processor is not stalled on a write to a shared variable if the processor has ownership to the line. If the latter condition is not true, a cache-coherency transaction ensues and the write-buffer will be stalled but the processor can continue until an access is attempted that cannot be satisfied past the first level store-through cache. However, for the next words of the cache line, no cache coherency transaction will take place as the cache line is already present in the cache with the dirty state. The cache-coherency transaction involved is the reading of the directory information to see if the cache line is being cached elsewhere. If the cache is big enough to hold the complete subgrid, the DASH architecture has the advantage of having to read from memory only once per cache line whereas P-RISC has to go to memory each time. Again, each store need not be retired from the write buffer before the next three loads are attempted as the loads will be satisfied from the first level cache. Using release consistency model, there is only one full fence just before the signals are executed.

For the computation of elements at the left and top edges, non-local communication is necessary. The producer prefetch primitives in DASH can be used to make this communication effective by exploiting the fact that multiple values can be sent at least in one dimension (depending on the row major or column major format) because the linesize is more than one. However, this is quite difficult to effect in practice due to the wavefront nature of computation and requires sophisticated compiler optimizations. Under the simpler model, the computation time for a grid in DASH is as follows:

* Synchronization time (waits) : t_{sw}

* First wavefront with only one element: There are three non-local accesses (t_{nl}) followed by the computation (t_C) and then the store. This first store will be a miss and a 16 byte line is brought into the second-level cache on an exclusive basis after a cache coherency transaction (time t_c). Time = $3 * t_{nl} + t_C + t_c$

* i th wavefront with i elements ($1 < i \le l$): There are two non-local + 1 local accesses, followed by the computation and then store for each of the end elements whereas there are three local accesses for the interior elements. There is a store miss for each of the elements on the left side of the grid but the interior points miss whenever the column modulo 4 = 0 (assuming double precision data). Each of the misses has an associated cache coherency transaction. The access of the local values can be overlapped with the store as it can be satisfied from the first-level cache. Time = $2 * t_{nl} + t_C + t_c$ for left end element, $2 * t_{nl} + t_C$ for top end element and $(i - 2) * t_C + (\lfloor i/4 \rfloor) * t_c$ for interior elements

* $l + i$ th wavefront with $l - i$ elements ($1 \le i < l$): Here all the loads are local; hence 3 local accesses followed by the computation and then the store. The store misses are like in the previous case. Time = $(l - i) * t_C + (\lfloor (l - i)/4 \rfloor) * t_c$

* Fence delay. This is at worst case the time to empty the 4 deep write buffer (there is no sharing of cachelines at this stage for invalidates to be awaited). Time = t_{wb}

* Synchronization time (signals). Time = t_{ss}

If l is a multiple of 4, the total time (T) for each grid is given by $t_{sw} + (4 * l + 1) * t_{nl} + l * l * t_C + (l * l/4 - l/4) * t_c + t_{wb} + t_{ss}$

The total time for the complete relaxation is $(2 * p - 1) * T - t_{sw} - t_{ss}$.

For DASH, the numbers are: $t_{nl} = 61$ processor clocks, t_C is 2 FP adds, 1 divide, 2 index calculations and looping overhead (approx. 40 clocks), t_{wb} is 4 bus clocks (= 6 clocks), $t_c = 18$ clocks, $t_{ss} = 32$ clocks, $t_{sw} = 58$ clocks.

4.2 Wavefront on P-RISC

There are two approaches to this computation on P-RISC. If the non-strict aspect is to be exploited, then all the loops and all the computations are unfolded but the progress of any computation depends on the availability of the operands. This causes large amount of queuing in the I-structure memory and when a memory element has valid data, it has to dequeue those computations that have been already enqueued. The other approach is the use of the wavefront/skewing transformations so that no queuing in I-structure memory is needed.

Since the detailed design of P-RISC has not yet been reported in the literature, we will confine ourselves to some rough calculations. Under this model, the computation time for a grid is as follows:

* Synchronization time (wait for full) : t_{sw}

* First wavefront with only one element: There are three non-local accesses followed by the computation and then the store. Time = $r * t_{nl} + t_C + t_S$ where $1 < r < 3$ and reflects the amount of overlap possible because of split-phase transactions.

* i th wavefront with i elements ($1 < i \le l$): There are two non-local + 1 local accesses, followed by the computation and then store for each of the end elements whereas there are three local accesses for the interior elements. Assume that the access of the local values (time = t_l) can be overlapped with the access of the non-local values because they take considerably longer. Time = $q * t_{nl} + t_C + t_S$ for each end element (where $1 < q < 2$ and reflects the amount of overlap possible because of split-phase transactions) and $s * t_l + t_C + t_S$ for each interior element (where $1 < s < 3$). We can create independent threads for each computation but the limitations of reading and writing into frame memory simultaneously as needed by pipelining prevent the realisation of the corresponding speedup. Let t be the effective speedup that is obtained by the use of running threads (likely to be in the range of 2-5(?)). Then the time for all wavefront elements is $i/t * (s * t_l + t_C + t_S)$ assuming the end elements are initiated first so that non-local accesses are not in the critical path.

* $l + i$ th wavefront with $l - i$ elements ($1 \le i < l$): Here all the loads are local; hence 3 local accesses followed by the computation and then the store. Time = $s * t_l + t_C + t_S$ for each element. Total time for all the elements = $(l - i)/t * (s * t_l + t_C + t_S)$

* Synchronization time (signals). Time = t_{ss}

Approximate total time for a grid = $t_{sw} + r * t_{nl} + t_C + t_S + \sum_{i=1}^{l} i/t * (s * t_l + t_C + t_S) + \sum_{i=1}^{l-1}((l-i)/t * (s * t_l + t_C + t_S)) + t_{ss}$ $= t_{sw} + r * t_{nl} + l * l(s/t * t_l + t_C + t_S) + t_C + t_S + t_{ss}$ The second sigma term is dropped in case we exploit the non-strict features of P-RISC computation.

Using numbers comparable to DASH ($t_l = 6$, $t_S = 6$, s=2, t=2), we notice that the $l * l$ term has a coefficient of approximately 44 in DASH but it is approximately 26 (full utilization of non-strict features of P-RISC computation) and 52 (wait for the grid to be completely finished before the right and bottom grids start) for P-RISC. We have not analysed the more aggressive DASH implementation of wavefronting here which could have an advantage over P-RISC but the amount of synchronization needed is large. Our analysis has been generous to P-RISC by giving benefit of doubt to P-RISC with respect to realizability of a pipelined P-RISC using multi-ported memory but at the same time not incorporating any penalties for P-RISC for using the more costly memory. In addition, we have assumed that the I-structure memory does not have any speed degradation and that there is some controller that wakes up threads that are waiting for a memory value to be full instead of being done by the processor.

References

[1] Nikhil and Arvind, "Can dataflow subsume von Neumann computing?", '89 International Symposium on Computer Architecture

[2] Burton Smith, "A Pipelined, Shared Resource MIMD Computer", '78 International Conference on Parallel Processing

[3] Gharachorloo, et al. "Memory Consistency and Event Ordering in Scalable Shared-Memory Multiprocessors," '90 International Symposium on Computer Architecture.

Inverted Memory

S. Bhattacharya, C. T. Liang, W. T. Tsai

University of Minnesota, Minneapolis, MN 55455

Abstract

Vectorizing and associative processing involves parallel access to data elements. However, conventional RAM limits accessing one word at a time only. Thus parallel data access using RAM storage require multiple memory modules leading to a number of disadvantages and restrictions over vectorizing. This paper proposes a new memory organization, Inverted Memory (IM), in which words can be formed either row-wise or column-wise at the control of an external input. IM design does not require significant overhead compared to RAM. IM allows unrestricted vectorizing and can be used in associative processing.

1. Introduction

Vectorizing and Associative Processing are two useful strategies in parallel processing. Vectorization involves parallel access to data elements. However, the standard word organized Random Access Memory (RAM) limits accessing one word at a time only. Hence, parallel data access using RAM is not straightforward. *Skewing* [1] is a popular method to implement parallel access to matrix data elements using RAM. Essence of the skewing techniques lies in distributing matrix data elements into *different* memory modules. Associative Processing allows parallel search on data elements using Content Addressable Memories (CAM). Fully word parallel CAM is expensive in terms of hardware cost. A cost-effective alternative is found in bit-serial CAM which can be implemented [2] using Multi Dimensional Access memory (MDA). [3] developed an implementation of MDA using *parallel* RAM modules. Thus MDA realization also requires parallel RAM modules. Hence, both vectorizing and bit-serial associative processing requires careful data distribution to parallel RAM modules.

However, the above approaches have several disadvantages: using multiple RAM chips require higher address/data/control bus bandwidth, higher fan in/out of the connecting busses, high order of interleaving and complicated address generation. Data distribution and redistribution (known as *gather-and-scatter* operations in vector processing) to parallel memory modules requires substantial time overhead. The time taken by the gather-and-scatter steps are often prohibitive and overshoots the time saved by vectorizing. For this reason, vectorizing is not always appropriate [4]. Similar data distribution overhead also bottlenecks MDA implementation using RAM modules.

The idea in inverted memory (IM) is to provide built-in hardware row-coulmn access ability. In RAM words are formed strictly row-wise. One particular cell of storage is dedicated to deliver a particular bit-position data of one particular memory word only. This assumption is relaxed in IM where external control input can dictate the way memory words would be formed. A set of cells form a word being guided by an external control input. In IM words can be formed either row-wise (horizontal) or column-wise (vertical). In horizontal mode it works exactly similar to a RAM. In the vertical mode memory words are formed using vertical strips (hence it is called Inverted Memory).

The built-in capability of IM to perform row/column access gives advantages in parallel data access. This eliminates the need (and the associated disadvantages) of distributing the data to different memory modules. A data matrix can be stored in a single IM module (leading to reduced bus bandwidth, less fan in/out and reduced order of memory interleaving) and still provide parallel row/column access. Also this eliminates the gather-and-scatter steps of vectorizing. Therefore, *IM allows unbounded vectorizing*. In this regard IM is better than any data skewing method. IM is a class of MDA but unlike [3] IM does not require multiple memory modules. With an additional set of *external* comparators IM can implement bit-serial CAM and be used in associative processing. IM provides a new design of MDA. This design is better than [3] since it requires less memory interleaving and avoids time for data distributions. It is shown that in terms of hardware IM needs no significant overhead than RAM. Section 2 introduces IM and its design overhead. Section 3 presents applications of IM and Section 4 concludes the paper with suggestions for future extensions.

2. Introduction to Inverted Memory

The conventional RAM maps an address (given in Memory Address Register, MAR) to a data (available in the Memory Data Register, MDR) while a CAM maps a data to an address. Inverted Memory is a memory organization which belongs to the category of the conventional RAM. However, an extra control *tag* is used in the memory operation. These mapping properties can be specified as:

data = Memory(*address*), in a conventional RAM.
address = Memory(*data*), in a CAM.
data = Memory(*address, tag*), in Inverted Memory.

The *tag* is a 1-bit control flag, as shown in Fig. 1b. It decides *how bits in the memory* are grouped to form a word. Let the memory unit be a two dimensional array of cells. Each cell is a 1-bit of storage. A memory of size $N \times M$ thus has N rows each of which has M cells. An address consists of $\log_2 N$ bits (for pointing to one of the N rows) and every word is of M bits. Let $b_{i,j}$ denote the cell at the i-th row and j-th column. The role of the control *tag* is expressed in the following:

   i) *tag*=0: $b_{i,j}$ forms the j-th *bit* of the i-th *word*,
   ii) *tag*=1: $b_{i,j}$ forms the (i mod M)-th *bit* of the [j+(i div M)$\times M$]-th *word*,

where $0 \le j \le M-1$, $0 \le i \le N-1$ and the memory size is $N \times M$. Fig. 2 shows an 8×4 memory with the control *tag*. With control tag = 0 eight words are formed one in each row. But with control tag = 1 words are formed by vertical strips of four bits as shown in Fig. 2b. In Fig. 2 cell A (B, C) is bit-2 (bit-3, bit-4) of word-4 (word-2, word-6) in horizontal mode while it is bit-1 (bit-3, bit-3) of word-5 (word-2, word-7) in the vertical mode. Intuitively, this memory organization has a row-column interchanging ability. With control tag=0 the organization is similar to the conventional RAM. Data can be written into (or read from) this memory unit in a way identical to the conventional RAM. However, if the control *tag* is toggled to 1 then the memory contents can be read out in an inverted fashion. The control *tag* should be under program control so that the user can utilize the features of the IM.

Design Overhead: No significant additional hardware, e.g., address decoder, is needed for IM. Formation of the words from individual bits is controlled by the *Tag* control input. Typically, two multiplexor are used to feed to the MAR/MDR register. One set of address select wires and data formation wires are run in a way exactly identical to RAM construction. These wires are in effect for *Tag*=0 indicating a normal RAM-mode operation. The second set of address wires and data formation wires connect in column-wise fashion. These wires are in effect for *Tag*=1 indicating a vertical inverted mode word formation. Fig. 4a and Fig. 4b show the address selection and data formation circuitry respectively for a (4×4) IM. Each memory cell, storing one bit of data, is augmented by OR gates at its data input, select input. The cell output is also duplicated into two lines separated by isolation diodes. Fig. 3 shows the logical view of a cell in IM design. It may be noted that the cell in Fig. 3 is only a logical view. In actual design the OR gates can be avoided using wired-OR gates or tri-state logic. Similarly isolation diodes can be implemented using multiple collector transistors etc. Thus, each cell in IM design is realistic and is not excessively complicated or enlarged. We address this layout design of IM as IM_1. In an alternative and more efficient layout of IM (called as IM_2) cell pairs with symmetric row-column positions are placed together. As a result any address, data wire available to one also can be used in the other. In this design the *tag* control wire is the only additional track that needs to pass through every cell. Prototype IM chips have been implemented using XILINX Logic Cell Array (LCA) architecture [5]. It has been observed that the design complexity of IM is not significantly more than that of an equivalent sized RAM. Table 1 compares design complexity of IM over RAM.

3. Applications of Inverted Memory

IM storage is extended for generalized matrix handling. Matrices of any size are considered while data elements can be of multiple bits. Inherent row-column interchanging feature of IM leads to efficient matrix applications, unrestricted vectorizing. Bit-serial CAM can be designed using IM. These results are developed in this section.

3.1. Generalized Matrix Storage in IM

Arbitrary Size: Previous examples assumed that the IM size (i.e., the number of rows and number of columns) and the matrix size are equal. Also each matrix data elements was considered 1-bit only. This range is extended here. If the IM size is smaller than the matrix size then the matrix needs to be partitioned into several memory chips. A $P \times Q$ matrix can be stored with $N \times M$ IMs using (P/N) sets of IM modules (i.e., dividing along rows) where each set consists of (Q/M) IM chips (dividing along columns). Using one $N \times M$ IM ($N > M$) necessitates (N/M) memory-reads to read-out one column, even though reading out any row can be done in a single step. For this reason, IMs of square sizes (i.e., number of rows = number of columns) are most efficient to use. In other cases there would be a constant slowdown factor as mentioned above.

Multi-Bit Data: IM can handle multiple bit data values by using one separate memory module per bit-position. If every element of the matrix is a k-bit value then k distinct memory modules are used, one for each bit-position. Within every IM module the row-column transposing property are retained. The *Tag* control input for all the *k* IM modules are connected together and controlled as a single control line. When *Tag*=0 *all k* IM modules reads/writes row-wise, while making *Tag*=1 would imply all the *k* IM modules read/write column-wise. It may be noted that using one memory chip per bit-plane would involve only a constant number of parallel memory modules, unlike the data skewing schemes which require a number of parallel memory modules proportional to the size of the concerned matrix.

3.2. Matrix Applications Using IM

Inherent row-column interchanging property of IM supports easy matrix transposing. Data can be written into IM row-wise from the matrix with *tag* =0. Then data can be read out changing *tag* as 1. Let $D1_{r,c}$ ($D2_{r,c}$) denote the input (output) matrix data element at the *r*-th row and *c*-th column. Let IM_k denote the *k*-th memory module. Then the *Data Write In* phase is: bit_i of $D1_{r,c}$ is written into ($bit_{c\,modN}$, $word_r$, $IM_{c\,divN}$, bit-plane *i*), with *Tag*=0. While the *Data Read Out* phase is: bit_l of $word_j$ of IM module *k* [$0 \le k \le (P/N)-1$] in bit-plane *i* is read as bit-*i* of $D2_{(k \times N)+l,\,j}$. Matrix transposing in IM requires no extra time. Similarly, matrix symmetry can also be checked.

In a matrix multiplication operation it is required to compute $C=A \times B$, where *A* and *B* matrices are stored in a distinct set of IM modules. Following the properties of IM storage it is known that any row/column of either matrices can be accessed in single memory operation. Let both the *A* and *B* matrices be of size $N \times N$. The equation for computing the elements of the product matrix *C* is: $C_{i,j} = \sum (A_{i,k} \times B_{k,j})$, for $0 \le k \le N-1$. For computing one element of the product matrix one row of *A* matrix and one column of *B* matrix are needed. These can be read out in single memory operation since the matrices are stored in IM modules. Thus matrix multiplication using IM requires no data access overhead. It may be noted that no data skewing is needed while using IM. Also, the number of memory modules accessed in parallel get reduced. This saves in data reorganization time and reduces the order of memory interleaving. Suppose, skewing (de-skewing) requires rewriting the entire matrix. For $N \times N$ matrices skewing and de-skewing would require $2 \times N$ memory operations. Let *T*, *V* be the uni-processor execution time for matrix multiplication and speedup factor by vectorizing respectively. Hence, the tradeoff whether or not to use vectorizing would be

$$[\ 2 \times N + T \times V < T\ ], \text{ i.e., } [\ 2 \times N < T \times (1-V)\ ]$$

For large values of *N* the term $2 \times N$ might be dominant and nullify the above inequality. Hence vectorizing would not be suitable always [4]. Using IM such restrictions does not arise since the term $2 \times N$ is no longer present. Vectorizing is always advantageous. In other words, IM allows unrestricted vectorizing.

3.3. Configuring IM as a Bit-Slice CAM

An $N \times N$ IM is considered. When it is operated with *Tag*=1, it can read vertical strips of bits (i.e., bit-slices). This particular property is used to design a bit-serial CAM based on IM. In a bit-serial CAM comparison is done at every bit-slice position and the results of comparisons are made output. Equivalent effect can be obtained by first reading out a bit-slice and then performing comparison operations. Comparisons are done after reading out a bit-slice, i.e., between the MDR contents and the comparand register. Fig. 5 shows this structure. Bit-slice CAM design using IM can be extended for the case when number of words is not equal to the number of bits, i.e., the IM is not square. For example, with an IM(8×4) there are 4 bit-slices of 8 bits each. Thus MAR has 3 bits of which 2 bits are supplied by the bit-slice number. The other bit can be set/reset to 1/0 to access two different nibbles in each bit-slice.

	RAM	IM_1	IM_2
Number of cells	$N \times M$	$N \times M$	$N \times M$
Complexity of each cell	one storage element	one storage element (using wired-OR logic)	one storage element with two 2-input AND gates
Number of address decoders ($\log_2 N$-to-N)	one	one	one
Number of address select lines	N	$N + 2 \times N$	N
Number of data collect lines	M	$M + 4 \times M$	M
Complexity of MDR	M-bit register with Read/Write multiplexed I/O	M-bit register with Read/Write multiplexed I/O	M-bit register with Read/Write multiplexed I/O

Table 1. Layout complexity comparison of RAM with IM_1 and IM_2.

4. Conclusion

Inverted Memory (IM) is proposed as a memory organization for shared memory multiprocessors. IM provides built-in hardware row-column interchanging ability. This property makes IM suitable for matrix related applications in vectorizing. IM can also offer a low cost bit-slice CAM design. A prototype IM has been designed using [5]. The idea in IM, i.e., externally controlling a set of memory cells to form as a word, can be used to design other versions of IM, e.g., for diagonal access or block access. The

present design of IM focuses only on the row-column access since this represent majority of the matrix applications. Future work includes demonstration of memory fault diagnostic features of IM and compiler support for IM in a vectorizing application environment.

Reference

[1] Kichul, K., Kumar, V. K. Prasanna, "Parallel Memory System for Image Processing", *Proc. IEEE Computer Society Conf. on Computer Vision and Pattern Recognition*, pp 654-659, 1989.

[2] Rudolf, J. A., "A Production Implementation of an Associative Array Processor STARAN", *Proc. AFIPS Fall Joint Comp. Conf.*, no. 41, AFIPS Press, Montvale, N.J., pp 229-241, 1972.

[3] Batcher, K. E., "The Multidimensional Access Memory in STARAN", *IEEE Transactions on Computers*, Vol. C-26, no. 2, pp 174-177, Feb 1977.

[4] Stone, H. S., "High-performance Computer Architecture", *Addison-Wesley Pub. Co.*, 2nd ed., Reading, Mass., 1990.

[5] The Programmable Gate Array Data Book, *XILINX, 2100 Logic Drive, Sab Jose, CA 95124*.

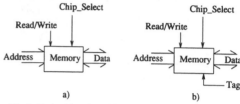

Fig. 1. External interface of RAM and IM: a) RAM, b) IM.

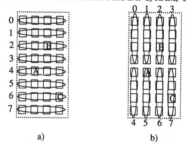

Fig. 2. Two different word formation styles in IM:
a) Horizontal (when Tag=0), b) Vertical (when Tag=1).

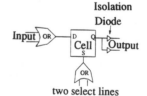

Fig. 3. Cell design of Inverted Memory.

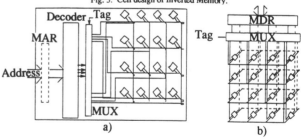

Fig. 4. IM(4x4) design: a) Address selection, b) Word formation.

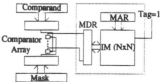

Fig. 5. Bit-slice CAM design using IM.

CARTESIAN PRODUCT NETWORKS

Abdou Youssef

Department of Electrical Engineering and Computer Science

The George Washington University

Washington, DC 20052

Abstract –This paper will study product graphs as interconnection networks. The topological properties of product networks will be presented, and generic, divide-and-conquer algorithms for point-to-point routing, broadcasting and permuting will be designed. Finally, linear arrays, rings, meshes, toruses and trees will be embedded on product networks.

Introduction

In this paper a unified framework is developed in which most existing networks and many new ones can be studied. The class of cartesian product graphs which will be called here product networks, will provide this common framework. Multidimensional meshes, multidimensional toruses, binary and generalized k-ary hypercubes and others belong to this class. Other useful product network that can be extended with fixed node degree will be proposed. The paper will study the topological properties of product networks and develop various generic routing and embedding algorithms for them.

Product Networks

In this section product graphs will be reviewed and their topological properties will be presented.

Let $G_1 = (V_1, E_1)$ and $G_2 = (V_2, E_2)$ be two graphs. The product graph of G_1 and G_2, denoted $G_1 G_2 = (V_1 V_2, E)$, is a graph where the set of nodes is the product set $V_1 V_2 = \{x_1 x_2 \mid x_1 \in V_1 \text{ and } x_2 \in V_2\}$ and $E = \{\langle x_1 x_2, y_1 y_2 \rangle \mid (x_1 = y_1 \text{ and } \langle x_2, y_2 \rangle \in E_2) \text{ or } (x_2 = y_2 \text{ and } \langle x_1, y_1 \rangle \in E_1)\}$. The graphs G_1 and G_2 are called the factors of $G_1 G_2$.

As can be observed, $G_1 G_2$ consists of $|V_2|$ copies of G_1 where every set of the $|V_1|$ corresponding nodes of these copies form a G_2 graph. The copy of G_2 in $G_1 G_2$ that corresponds to a node $x_1 \in V_1$ is denoted $x_1 G_2$. Its set of nodes is $\{x_1 x_2 \mid x_2 \in V_2\}$ and its set of edges is $\{\langle x_1 x_2, x_1 y_2 \rangle \mid \langle x_2, y_2 \rangle \in E_2\}$. Similarly, The copy of G_1 in $G_1 G_2$ that corresponds to a node $x_2 \in V_2$ is denoted $G_1 x_2$.

Note that the product $G_1 G_2 G_3 ... G_n = (V, E)$ of n networks $G_1 = (V_1, E_1)$, $G_2 = (V_2, E_2)$, ..., $G_n = (V_n, E_n)$ can be derived. Clearly, $V = V_1 V_2 V_3 ... V_n$ and $E = \{\langle x_1 x_2 ... x_n, y_1 y_2 ... y_n \rangle \mid$ there exists an i such that $\langle x_i, y_i \rangle \in E_i$ and for every $j \neq i$ we have $x_j = y_j\}$. If all G_i's are equal (to some G), $G_1 ... G_n$ is denoted G^n.

Denote by L_p, R_p and K_p a linear array of p nodes, a ring of p nodes, and a complete graph of p nodes, respectively. It can be seen that a $p_1 \times p_2 \times ... \times p_n$ mesh (resp., torus) is $L_{p_1} L_{p_2} ... L_{p_n}$ (resp., $R_{p_1} R_{p_2} ... R_{p_n}$). Similarly, an n-cube Q_n is the product network k_2^n. The n-cube of radix $r \geq 2$ is K_r^n.

A network is said to be *indefinitely extendable* if it can be extended without increasing the node degree. Linear arrays, rings, and meshes are indefinitely extendable but hypercubes are not. As $degree(G_1 G_2) = degree(G_1) + degree(G_2)$, we conclude that if G_1 is indefinitely extendable then the product $G_1 G_2$ is indefinitely extendable. Note that G_2 does need not be indefinitely extendable. Thus, the **line-hyeprcube** $L_r Q_n$, the **ring-hypercube** $R_r Q_n$ and the **mesh-hypercube** $L_p L_q Q_n$ are indefinitely extendable to $L_s Q_n$, $R_s Q_n$ and $L_{p'} L_{q'} Q_n$ respectively, for any $s > r$, $p' \geq p$, $q' \geq q$. Hence, the product operator provides an excellent cube augmentation scheme. It has also other desirable properties as summarized below.

Denote by $d_G(x, y)$ the distance from x to y in G, by D_G the diameter of G, by $\overline{d}_G$ the average distance of G, and by C_G the connectivity of G, that is, 1+the largest number of nodes that can be deleted without disconnecting G.

Theorem 1. Let G_1 and G_2 be two graphs.
1) $d_{G_1 G_2}(xx', yy') = d_{G_1}(x, y) + d_{G_2}(x', y')$.
2) $D(G_2 G_2) = D(G_1) + D(G_2)$.
3) $\overline{d}_{G_1 G_2} = \overline{d}_{G_1} + \overline{d}_{G_2}$.
4) $C_{G_1 G_2} \geq C_{G_1} + C_{G_2}$.
Proof. See [5].

In part (4) above, the equality holds for some products but not for others. As a corollary of this theorem, the diameter, average distance and connectivity of meshes, toruses and hypercubes can be rediscovered and these same measures can be concluded for the line-hypercube, ring-hypercube and mesh-hypercube.

Routing

Let G_1 and G_2 be two networks such that each G_i is endowed with a point-to-point routing algorithm $\text{ROUTE}_{G_i}(s, d)$ which sends a message from s to d, a broadcasting algorithm $\text{BROADCAST}_{G_i}(s, M)$ which broadcasts a message M from s to all other nodes, and a permuting algorithm $\text{PERMUTE}_{G_i}(f)$ which routes a message from every node x to node $f(x)$, where f is a permutation. These algorithms will be used to devise corresponding algorithm for $G_1 G_2$ in a divide-and-conquer fashion.

procedure $\text{ROUTE}_{G_1 G_2}(s_1 s_2, d_1 d_2)$
begin
    $\text{ROUTE}_{G_1}(s_1 s_2, d_1 s_2)$;
    $\text{ROUTE}_{G_2}(d_1 s_2, d_1 d_2)$;
end

procedure $\text{BROADCAST}_{G_1 G_2}(s_1 s_2, M)$
begin
    1. $\text{BROADCAST}_{G_1 s_2}(s_1 s_2, M)$;
    2. **forall** nodes x in G_1 **do in parallel**
        $\text{BROADCAST}_{x G_2}(x s_2, M)$;
end

Note that if each ROUTE_{G_i} and BROADCAST_{G_i} is optimal in time, then $\text{ROUTE}_{G_1 G_2}$ and $\text{BROADCAST}_{G_1 G_2}$ are optimal.

Next, the more elaborate permutation routing will be addressed. Our approach is based on Clos routing [3]. We review Clos networks first. A Clos network $C(p, q)$ has pq input terminals, pq output terminals and three columns of switches. The 1st and 3rd columns have p $q \times q$ crossbar switches each. The 2nd column has q $p \times p$ crossbar switches. The switches in each column are labeled $0, 1, ...$. The input ports and the output ports of every switch x are labeled xy ($y = 0, 1, ...$) so that xy is the y-th port of switch x. The interconnection between the first two columns links output port xy in the 1st column to input port yx in the 2nd column. Similarly, the interconnection between the last two columns links output port yx in the 2nd column to input port xy in the 3rd column. It has been shown in [1] that every permutation of pq elements is realizable by $C(p, q)$ in $O((pq)^2)$ time, that is, for every permutation f, there are switch settings for all the switches so that the source-destination paths $i \to f(i)$ are established without conflict. Call the algorithm that determines the switch setting $CLOS_{pq}(f)$.

Let xy be an arbitrary source and $xy \to f(xy)$ the corresponding source-destination path established by $CLOS_{pq}(f)$. The detailed parts of this path are:

$$xy \to xy' \to y'x \to y'x' \to x'y' \to x'y'' = f(xy)$$

for some x', y' and y''. Denote by f_x the permutation that maps y to y' in switch x of the 1st column, by $g_{y'}$ the permutation that maps x to x' in switch y' of the 2nd column, and by $h_{x'}$ the permutation that maps y' to y'' in switch x' of the 3rd column.

Suppose that f is to be routed in $G_1 G_2$, where G_1 has p nodes labeled $0, 1, ..., p-1$, and G_2 has q nodes labeled $0, 1, ..., q-1$. By corresponding every node xy in $G_1 G_2$ to input/output terminal xy of $C(p,q)$, every xG_2 to switch x in the 1st/3rd column of $C(p,q)$, and every $G_1 y$ to switch y in the 2nd column of $C(p,q)$, we obtain an algorithm for $G_1 G_2$:

Procedure PERMUTE$_{G_1 G_2}(f)$
1. Let $CLOS_{pq}(f)$ determine the f_x's, $g_{y'}$'s and $h_{x'}$'s.
2. PERMUTE$_{xG_2}(f_x)$ for all x in parallel
3. PERMUTE$_{G_1 y'}(g_{y'})$ for all y' in parallel
4. PERMUTE$_{x'G_2}(h_{x'})$ for all x' in parallel

Proof of correctness: Let M_{xy} be the message to be send from the arbitrary node xy to node $f(xy)$. Let x', y' and y'' be as before. After routing f_x on xG_2, M_{xy} is at node xy'. After routing $g_{y'}$ on $G_1 y'$, M_{xy} is at node $g_{y'}(x)y' = x'y'$. Finally, after routing $h_{x'}$ on $x'G_2$, the message M_{xy} is at node $x'h_{x'}(y') = x'y'' = f(xy)$.

The major drawback of this algorithm is the inefficiency of the CLOS algorithm. However, we have developed a new approach to self-routing on CLOS networks [4]. This approach is based on the observation that if the 1st column in a Clos network is set to some configuration, the resulting network becomes self-routed using destination addresses. Accordingly, the approach seeks, for every given family of permutations, a configuration to which to set the first column so that the resulting delta network realizes all the permutations of the family. Such configuration of the first column is called the compatibility factor. Compatibility factors were found in [4] for several important families of permutations. These include the families of permutations required by FFT, bitonic sorting, tree computations, multidimensional mesh/torus computations, multigrid computations [2] as well as the Omega permutations.

It will be argued next that if the compatibilty factor is known, then PERMUTE$_{G_1 G_2}$ becomes a self-routing algorithm and step 1 is bypassed. The f_x's are clearly the compatibility factor and hence known. $g_{y'}(x) = x'$ =the first part of the destination address $x'y'' = f(xy)$ of the message M_{xy}. Thus, node xy' can alone determine its intermediate destination $x'y'$. $h_{x'}(y') = y''$ =the second part of the destination address of M_{xy}. Thus, node $x'y'$ can alone determine the destination $x'y''$.

A noteworthy special case is the Omega-realizable permutations. Their compatibility factor is the identity permutation. That is, the f_x's for every Omega-realizable permutation f are identity permutations and therefore, do not have to be routed. Thus, step 1 and 2 can be bypassed in this case while steps 3 and 4 are executable in a distributed manner.

Next we evaluate the communication complexity of permuting using PERMUTE and taking the number of conflict-free steps as the complexity measure.

Theorem 2. Let $P(G)$ be the minimum number of steps needed for permuting in G. then,
1) $P(G_1 G_2) \leq$ MIN$(2P(G_1) + P(G_2), 2P(G_2) + P(G_1))$.
2) $P(G_1 G_2 ... G_k) \leq 2\sum_{i=1}^{k} P(G_i) -$ MAX$_{i=1}^{k}(P(G_i))$

3) For Ω-realizable permutations, $P(G_1 G_2 ... G_k) \leq \sum_{i=1}^{k} P(G_i)$
The proof follows directly from the algorithm and the above discussion. We can then conclude: $P(p \times p$ mesh$) \leq 3(p-1)$, $P(K_r^n) \leq 2n - 1$ and $P(L_r Q_n) \leq$ MIN$(2r + 2n - 4, 4n + r - 5)$.

Embedding in Product Networks

An embedding of a guest graph $G = (V_g, E_g)$ on a host graph (i.e., network) $H = (V_h, E_h)$ is a mapping f from V_g to V_h. The standard goodness measure of a mapping f is the dilation cost: $dilation(f) =$ MAX$\{d_{G_h}(f(x), f(y)) \mid (x,y) \in E_g\}$. We will limit ourselves to one-to-one embeddings, that is, to cases where f is one-to-one. In this section, embeddings for lines, rings, meshes, toruses and trees will be constructed on product networks using the embeddings of these structures on the factor networks. We first start with a general theorem.

Theorem 3. If for $i = 1, ..., k$ the graph G_i can be embedded on the graph H_i with a mapping f_i of dilation d_i, then $G_1 G_2 ... G_k$ can be embedded on $H_1 H_2 ... H_k$ with dilation MAX$(d_1, ..., d_k)$ with the mapping $f(x_1 ... x_k) = f_1(x_1) ... f_k(x_k)$.

Therefore, if a linear array L_{p_i} (resp., ring R_{p_i}) can be embedded on H_i with dilation cost d_i, for every i, then the k-dimensional $p_1 \times p_2 \times ... \times p_k$ mesh (resp., torus), can be embedded on $H_1 H_2 ... H_k$ with dilation cost MAX$(d_1, ..., d_k)$. In particular, by using the standard embedding of linear arrays and rings in meshes and toruses, one can achieve an embedding of a linear array or a ring in the product graph.

Next we embed a tree in $G_1 G_2$. Assume that T_i is a tree of hight h_i that can be embedded in G_i with dilation d_i $(i = 1, 2)$. Let x_2 be a the root node of T_2 in G_2. Embed the tree $T_1 x_2$ in $G_1 x_2$, then embed the tree $x_1 T_2$ of root $x_1 x_2$ in $x_1 G_2$ for all nodes x_1 of G_1. The resulting embedded structure $T_1 x_2 \cup \cup_{x_1 \in G_1}\{x_1 T_2\}$ is clearly a tree of hight $h_1 + h_2$ and the dilation cost of the embedding is MAX(d_1, d_2).

Conclusions

A theory of product interconnection networks has been developed in this paper. Product networks were shown to include many of the existing networks as well as other useful, indefinitely extendable networks. It was shown that as the number of nodes grows multiplicatively, the degree, diameter, average distamce and connectivity grow additively. Finally, product networks were constructively shown to yield to a divide-and-conquer approach of routing and embedding.

References

[1] V. E. Benes, *Mathematical theory on connecting networks and telephone traffic,* Academic Press, New York, 1965.

[2] T. F. Chan and Y. Saad, "Multigrid algorithms on the Hypercube multiprocessor," *IEEE Trans. Comput.,* vol. C-35, pp. 969–977, Nov. 1986.

[3] C. Clos, "A Study of Non-Blocking Networks," *Bell System Technical Journal 32,* pp. 494–603.

[4] A. Youssef and B. Arden, "A New Approach to Fast Control of $r^2 \times r^2$ 3-Stage Benes Networks of $r \times r$ Crossbar Switches," *Proc. of the 17th Annual Int'l Symp. on Computer Architecture,* pp. 50–59, May 1990.

[5] A. Youssef, "Product Networks: A Unified Theory of Fixed Interconnection Networks," Technical Report GWU-IIST-90-38, The George Washington University, Dec. 1990.

The VEDIC network for multicomputers[1]

VIPIN CHAUDHARY, BIKASH SABATA, and J. K. AGGARWAL

Computer and Vision Research Center

The University of Texas at Austin

1 Introduction

In this paper we introduce a clan of hierarchical networks for interconnecting multicomputers. The network is described by eight topological parameters, varying which generates different families of networks. By a suitable assignment of the parameters, most of commonly known networks are realizable [1]. Because of the generality of this formulation we refer to the network as the **VEDIC** (*Virtually Every Data InterConnection*) network. The size of the network, its diameter, degree, and the number of links are evaluated in terms of the parameters of the network. A self routing distributed routing algorithm is developed. A procedure to generate and expand the **VEDIC** network is described. All the families of networks generated are fault tolerant too.

2 The VEDIC network

VEDIC interconnection networks are regular or irregular, hierarchical networks formed by interconnecting rings of various sizes. The lowest level, i.e. level 0, of the hierarchy is a ring consisting of n nodes. The next level consists of rings, each with equal or fewer number of nodes than the lower ring. Each of the higher level rings necessarily has at least one node common with a ring in the level immediately below that level. These rings formed at the higher level can also have subsets of other rings (at the same level) common to them. The next level is formed by constructing rings using subsets of rings at the immediate lower level (not using any subset of the rings at levels lower than the immediate lower level). Figure 1 gives an example of a VEDIC network. The VEDIC network is represented as $\aleph[n, l, m, k, q, w]$, where

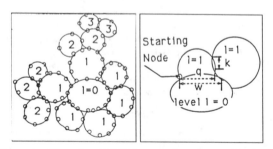

Figure 1: *Example of* VEDIC *networks. Figure (b) shows the definitions of* w, q, k *in the network*

n is the number of nodes in the ring at level 0, l is the number of levels of the network, m is the maximum difference between the number of nodes at adjacent levels, k is the number of nodes common to two rings at the same level, q is the number of nodes common to two rings at adjacent levels, and w is the distance between two rings at the same level that have common nodes. The distance between rings is defined as the distance between the starting nodes of the two rings. Different variations of m, k, q, and w generate families of networks. The network in Figure 2 is $\aleph[4, 3, 0, 1, 1, 1]$. Note that the maximum possible values for l, m, q, and w are functions of n and m. For ease of

[1]This research was supported in part by IBM.

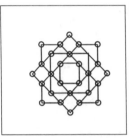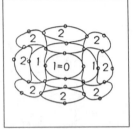

Figure 2: *Example of a network with n=4, q=1, w=1 & k=1*

understanding we restrict our discussion here to networks of the form $\aleph_0[n, l, m, k, 1, 1]$, where m and k have the same values for all the levels. In this case the rings at adjacent levels can have only one node in common, and the rings at the same level that have k common nodes originate from adjacent nodes of the ring at the immediate lower level. An example of this is shown in figure 2. These restrictions are eliminated and the network further generalized in later discussion.

2.1 Properties of the VEDIC Network

Unique indexing. In a multiprocessor each processor has a distinct processor identification, (or address) so that messages can be routed from one processor to another. Thus, each node in the VEDIC network should be assigned a unique address.

Every node of the network is distinguished by the level of the loop it belongs to, the number of the loop at that level, and the number of the node in that loop. The count on each loop proceeds in the clockwise direction. The rings are referred to by the address of the starting node of the ring. A node at level p has its address denoted by a tuple $[p, i_0, i_1, \ldots, i_{p-2}, i_{p-1}, j]$, where p denotes the level of the network at which the node is located, i_b denotes the number of the node at level b where the loop at level $b + 1$ starts, and j denotes the number of the node on the loop i_{p-1} at level p. The address of a node can be recognized to have three components; the level p, then the ring address given by $[p - 1, i_0, i_1, \ldots, i_{p-1}]$ and finally the location j on the ring. The nodes common to two rings are assigned addresses corresponding to the ring with a lower lexicographical address.

By invoking an inductive argument it can easily be established that each node has a unique address. The total number of nodes in the network is computed to be:

$$N = n \left[1 + \sum_{j=1}^{l} \prod_{i=1}^{j} (n - im - k - 1) \right].$$

The total number of edges (links between nodes) in the network can be calculated similarly.

Degree. The nodes of the VEDIC network need not have the same degree. Therefore, it may be an irregular network. But, it can be shown that there is an upper bound of eight on the degree of each node of the network.

Diameter. The evaluation of the upper bound of the diameter of the network follows from the indexing scheme used to address the nodes in the network. The distance between the two nodes n_1 and n_2, located at levels l_1 and l_2 respectively, evaluated to be bounded by

$$\left\lceil \frac{l_1}{2}[(n-1) - \frac{m}{2}(l_1+1)] \right\rceil + \left\lceil \frac{n}{2} \right\rceil + \left\lceil \frac{l_1}{2}[(n-1) - \frac{m}{2}(l_1+1)] \right\rceil.$$

This distance is maximized when $l_1 = l_2 = L$. Thus, the upper bound on the diameter of a network $[n, l, m, k, 1, 1]$ with N nodes is $D = \lceil L[(n-1) - \frac{m}{2}(L+1)] \rceil + \lceil \frac{n}{2} \rceil$.

Expansibility. The VEDIC network is easily expandable. However, not all these networks can be arbitrarily expanded. For a network $\aleph_0[n, l, m, k, 1, 1]$ there is a bound on the number of nodes in the network. Even if we do not bound the number of levels, i.e., l, the network does not grow infinitely (since the number of nodes in the rings at higher levels keep decreasing by m, and will eventually become zero) unless m is always zero. In order to add new nodes to a complete network (a network to which no more nodes can be added) we will have to change certain parameters, i.e., n or m.

Fault tolerance. The VEDIC network is fault tolerant since there exist at least two paths between any two nodes. Except for the nodes at the highest level there may (and usually so) exist more than two paths between them.

3 Routing in the VEDIC network

We describe a distributed control routing algorithm for the VEDIC network. The algorithm needs only the local information, i.e., the destination address of the message. Figure 3 gives the format of the messages sent by one node to another. The field **T** is bit to denote the direction of traversal of the message (**T**= 0 if message going down levels, and **T**= 1 if message is going up levels). The message is first sent to the lowest common

T	Final Destination	Message : MSG

Figure 3: *The format of the messages sent by the processors.*

level between the source and the destination nodes, and then routed to the destination *via.* the appropriate rings. To start with, three cases can be distinguished: the source and destination nodes are in the same level; the source is at a higher level than the destination; the source is at a lower address than the destination. Therefore, if a node after receiving the message, realizes that the destination and the node have the same ring address, then it would send the message towards the destination node on the ring. If the node is at level $p = 0$ which is lower than the level of the destination and the message reached the node on its way down ($T = 0$), then the message is sent towards the proper ring that will take the message to the final destination node. On its way downwards, in any intermediate level, the message is either send down one level, (if it is possible) or on the ring (anti-clockwise) towards a ring on the lower level, or towards a ring on the higher level if the ring is common to source and destination addresses. If the node is not at level zero but is lower than the level of the destination and is traveling upwards ($T = 1$), then we either send it up a level (if it is the correct higher level ring), or send it clockwise towards the ring at the higher level. Note that it is not possible for the message to be traveling upwards and the destination node is at a level lower than the present node because we always start by going to the lowest common level. It can be shown by induction that the message will reach its destination.

Consider the case when the node and the destination are at the same level but on different rings. In such a case it is entirely possible that it might be quicker to go around the rings of destination level rather than to go down levels and up again. This will happen only when the number of nodes along the rings at the level of the node is less than the nodes on the way down and up. But, to evaluate this we will need that each processor know the parameters of the entire network. In both the algorithms discussed we have assumed that the node knows only its own address. Thus, to make the routing more efficient as we have described, the parameters of the network need to be known to each node. This will lead to problems when we have to expand the network. Thus, in case of expansion of the network we will need to broadcast the new network parameters.

4 Generalizing the network

The network can be generalized by allowing q and w to take values in the range $1 \le q \le Q$, and $1 \le w \le W$, respectively. These values can vary with the rings at different levels. Also, the values of m and k may vary with levels (however with the added constraint that $q - w \le k$ because of the definitions). These relaxations on the values of m, k, q, and w allow us to represent very complex networks. Such networks and further generalizations of these can generate most other networks in the literature. The indexing of the nodes in the network is similar to the indexing of the restricted network. We need to make special note of the fact that k and m can vary with levels, and q and w can vary with rings as well as levels.

These generalized networks have a bound on the degree of each node which is independent of the total number of nodes in the network. Thus, we cannot generate networks, e.g., hypercube, whose degree is a function of the number of nodes in the network. In order to incorporate a change in the bound of degrees of the node we introduce two new parameters x and y in our network. The generalized VEDIC network is now represented as $\aleph[n, l, m, k, q, w, x, y]$. The parameter x denotes the upper bound on the number of rings at the same level that a node can belong to. The upper bound on the number of rings starting from a node is given by the parameter y. With these generalizations it is now possible to generate networks with arbitrarily high degrees.

5 Conclusion and discussion

The VEDIC network can generate other well known networks by fixing some of the parameters [2]. The VEDIC network in its most general form is a very powerful framework for studying the properties of other networks. This shows the versatility of the network in modeling a general network. Also, since all the networks are studied in the same framework, the interrelationships between the networks becomes apparent.

The families of networks generated by this network on varying certain parameters can be investigated in detail. Also, the dependence of fault tolerance on the various parameters needs further evaluation.

References

[1] T. Feng, "A survey of interconnection networks," *Computer*, Dec. 1981, pp. 12-27.

[2] V. Chaudhary, B. Sabata, and J. K. Aggarwal, "The VEDIC network for multicomputers," *TR-91-7-71, Computer and Vision Research Center*, The University of Texas at Austin.

Block Shift Network:
A New Interconnection Network for Efficient Parallel Computation

Yi Pan and Henry Y. H. Chuang
Department of Computer Science
University of Pittsburgh
Pittsburgh, PA 15260

ABSTRACT

A new class of networks, called *Block Shift Network* (BSN), is proposed for constructing very large multicomputer systems for efficient parallel processing. The topological properties of the network are analyzed, several basic data movement operations on BSN are described, and their time complexities presented. A distinct advantage of the BSN over the hypercube is that while the degree (the number of links incident on each node) of the hypercube is logarithmically proportional to the network size, the degree of the BSN is constant.

1. Introduction

A significant drawback of the hypercube is that it is not truly expandable because the number of links per node increases linearly with the dimensions. On the other hand, mesh type structures have the problem of either large diameter (e.g., plain mesh) or traffic bottleneck (e.g., mesh of tree). A careful look at algorithms on a hypercube reveals that only a small percentage of the links are used frequently. Most of the time a large number of links are idle. For example, in the matrix multiplication in[2], the 2-D convolution algorithms in[4], and the Hough transform in[6], half of the data exchange in the hypercube uses dimension 0, one quarter uses dimension 1, and so on. This suggests that in most cases we do not need link connections along all dimensions of a hypercube, and it is possible to construct networks which have fewer links than the hypercube and still have as good performance.

In this note, we first describe a new class of networks called *block shift networks (BSN)*, which have a constant number of links per node and surpass the hypercube in several properties while keeping most of its advantages. We then briefly discuss the topological properties of BSN and the time complexities of the major basic data movement operations.

2. Block Shift Networks

A hypercube of size $n = logN$ has connections along all its n dimensions. In order to reduce the connection complexity, links can be provided on only certain number of dimensions. Thus, we have to choose certain dimensions on which connections should be provided. First we introduce three connection schemes and two access modes. We then use these to define our Block Shift Networks (BSNs).

1. *Rightmost connection*: Suppose that b dimensions are to be connected, the rightmost connection scheme selects the rightmost b dimensions. In other words, only nodes with addresses differing in the rightmost b bits have direct link connections.

2. *Leftmost connection*: Assume that we can have links on b dimensions in its address, leftmost connection scheme selects the leftmost b dimensions. That is, only nodes with addresses differing in the leftmost b bits have direct link connections.

3. *Even connection*: Assume that we can have links on at most b dimensions in an n dimensional network. Using a even connections scheme, links are provided along dimensions distributed evenly over the n dimensions. We refer the connection scheme as *k even rightmost* if we divide an address into k sections, and connections are provided on the dimensions of the rightmost b/k bits in each section.

Given a particular connection scheme, data can be routed from a node to another in one step or multiple steps depending on how the nodes are connected. Thus, we have the following access modes for the rightmost connection scheme.

1. If two processors whose addresses differ only in their rightmost b positions are directly connected, we call this *concurrent access* mode. That is, in a concurrent access mode, the rightmost b bits in the address can be changed in one step to reach another address. When b is equal to n, the connection scheme corresponds to the fully connected topology.

2. If two processors whose addresses differ only in their rightmost b positions can reach each other by changing the b bits one by one, we call this *sequential access* mode. Thus, for two processors with addresses differing in all of the b rightmost bits, one needs b unit-routes to send a message to the other. If only one bit is different in the b bits, only one step is needed. Obviously, when b is equal to n the connection scheme corresponds to a hypercube topology.

For the other two connection schemes, similar access modes apply. Therefore, there are six combinations of connection topologies.

Basicly, the Block Shift Network consists of three groups of edges. The first group of edges connects nodes to their counterparts with addresses shifted cyclically b positions left. That is, they connect the processor at address $a_{n-1}a_{n-2}\cdots a_1 a_0$ to processor at $a_{n-b-1}a_{n-b-2}\cdots a_1 a_0 a_{n-1}\cdots a_{n-b}$. These connections will be called LEFT-SHIFT links and the data transfers over these links LEFT-SHIFT operations. Similarly, the second group of edges connects nodes to those with addresses shifted cyclically b positions right, and they will be called RIGHT-SHIFT links, and The data transfers over these links RIGHT-SHIFT operations. The last group of edges contains the links in one of the six combinations of connection topologies defined above, and are called R-change links. The data transfers over these links will be called R-change operations. In the following, we assume that a BSN has 2^n nodes and a given value for the parameter b, and we will focus on the rightmost connection scheme with concurrent access mode.

Conceptually, in a BSN there are 2^{n-b} blocks each with 2^b nodes. Within each block is a complete graph, and blocks are connected through either the RIGHT-SHIFT or the LEFT-SHIFT links. If we use the sequential access mode, then instead of a complete graph, each block is a hypercube with 2^b nodes. Clearly, the BSN is a hierarchical structure with nodes connected tightly within blocks and blocks connected loosely. This property matches the communication requirements of most parallel application algorithms[5]. The BSN has superior topological properties over other networks.

3. Topological Properties

The three most useful metrics for comparing interconnection networks are degree, diameter, and average distance. In this section, we shall analyze these topological properties of the BSN and compare with those of other popular interconnection networks.

3.1. Network Degree

Network degree is an important parameter in determining the cost of a system. Network degree d is defined as the maximum number of output (or input) links in a network node. For a BSN, $d=2^b+1$ in a network of $N=2^n$ nodes, as explained in the following. Each node contains at most one LEFT-SHIFT link since left-shifting an address by b positions results in on new address. Similarly, each node has one RIGHT-SHIFT link. For the R-change links, since the rightmost b bits of a node address have 2^b different patterns and one of them must belong to the processor itself, there are 2^b-1 connections to other processors. Therefore, the network degree of a BSN network is $(2^b - 1) + 2 = 2^b+1$, independent of n, while the degree of a hypercube node increases linearly with the network size.

3.2. Diameter

Given a source address and a destination address, we can shift the source address b positions left using a LEFT-SHIFT link, then using a R-change to change the b least significant bits so that they match the corresponding b bits in the destination address. We need at most $\lceil n/b \rceil$ such shifts and changes. In the worst case when the source and destination addresses are completely different, e.g., one consisting of all 0's and the other all 1's, exactly $\lceil n/b \rceil$ shifts and changes are necessary. Since a shift or a change requires one hop, a total of $2\lceil n/b \rceil$ hops are needed in the worst case. Therefore, the diameter of a BSN with $N=2^n$ nodes is $2\lceil n/b \rceil$.

While the diameter of a BSN increases with $O(logN)$ as in hypercube, and complete binary tree (CBT), the diameter of the 2D mesh and 2D torus increases with $O(N^{1/2})$. Clearly, as the value b of a BSN increases, its diameter decreases and the number of its links increases. Since the system cost increases as the number of links per node increases, links per node should be kept small. For the purpose of comparison, consider a BSN having the same link complexity as a hypercube; i.e., when b is equal to $loglogN-1=logn-1$. In this case, the diameter of the BSN is $2\lceil logN/(loglogN-1) \rceil$, while the diameter of the hypercube is $logN$. Clearly, when $n>8$, the diameter of the BSN is smaller than that of the hypercube. Networks with small diameters are desirable because the diameter is indicative of the maximum delay incurred in interprocessor communications. Obviously, the diameter of a BSN is smaller than those of ring, CBT, and mesh.

3.3. Average distance

The distance between two nodes is the length of the shortest path between them. In a hypercube, the distance between two nodes is equal to the Hamming distance (the number of different bit positions) between them. The average distance is measured in terms of the average number of hops between two nodes. For a hypercube with $N=2^n$ nodes, the average distance is $n/2$. The average distance of a BSN as derived in[7] is equal to

$$3 \left[n/b - \frac{1/2^b}{1 - 1/2^b} \right] -2qp^{n/b-1} .$$

The average distance traveled by a message is a popular criterion in evaluating a multicomputer topology. For a meaningful comparison between networks with different number of output ports per node, some normalization is desired. After considering realistic limitations on the number of pins and the amount of power available to drive communication lines, it has been proposed[3] that, in the context of single-chip computers, a constant bandwidth per node be assumed. If the total bandwidth available per node is fixed, then the bandwidth available per port is inversely

proportional to the number of ports for that node. Therefore, we normalize the average distance $\bar{d}$ by multiplying it with the number of ports per node[1]. For a hypercube with $N=2^n$ nodes, the normalized average distance is $n^2/2$ since each node in a hypercube has n ports. Similarly, the normalized average distance of a complete binary tree is about $6n$ and that of a BSN is roughly $3(2^b)n/b$. For a torus, the normalized average distance is still $O(N^{1/2})$. Therefore, the BSN and the CBT outperform the hypercube and the torus in this aspect. When $b=1$ or $b=2$, the BSN and the CBT have the same performance. For other values of b, CBT is better.

4. Basic Data Movement Operations

The previous section shows that the BSN topology has many advantages over the hypercube, the complete binary tree, and the mesh. For an interconnection network to be attractive, we still have to show that most application algorithms can be mapped onto the network naturally and executed efficiently. This can be accomplished by considering the basic data movement operations in the BSN.

Table 1 shows the communication times of five basic data movement operations. The time complexities in the table indicate that not only do all these operations have the same order of magnitude in time as those for a hypercube of the same size, but also the constant factors in the time complexities are very close to those of their counterparts for the hypercube.

Communication Times of some BSN and Hypercube Data Movement.

Algorithms	Hypercube	BSN
Broadcast	$logN$	$2\lceil logN/b \rceil - 1$
ASCEND/DESCEND	$logN$	$(1 + 1/b)logN$
Semigroup	$logN$	$(1 + 1/b)logN$
Data Circulation	$N-1$	$(1 + 4/b)N$
Min Finding	$logN$	$(1 + 1/b)logN$

References

1. D. P. Agrawal, V. K. Janakiram, and G. C. Pathak, "Evaluating the performance of multicomputer configurations," *IEEE Computer*, vol. 19, pp. 23-37, May 1986.

2. E. Dekel, D. Nassimi, and S. Sahni, "Parallel matrix and graph algorithms," *SIAM J. Comput.*, vol. 10, no. 4, pp. 657-675, 1981.

3. A. M. Despain and D. A.P atterson, "X-Tree: A tree structured multiprocessor computer architecture," *Proc. 5th Annu. Sump. Comput. Architecture*, pp. 144-151, Aug. 1978.

4. Z. Fang, X. Li, and L. Ni, "Parallel algorithms for 2-D convolution," *IEEE Trans. on Computers*, vol. 30, no. 2, pp. 184-194, Feb. 1989.

5. K. Hwang and J. Ghosh, "Hypernet: A communication-efficient architecture for constructing massively parallel computers," *IEEE Trans. on Computers*, vol. C-36, no. 12, pp. 1450-1467, Dec. 1987.

6. Yi Pan and Henry H. Y. Chuang, "Parallel Hough Transform Algorithms on SIMD Hypercube Arrays," *1990 International Conference on Parallel Processing*, pp. 83-86, St. Charles, IL, Aug. 13-17, 1990.

7. Yi Pan and Henry Y. H. Chuang, "Block Shift Network: A new interconnection network for efficient parallel computation," TR-91-11, Department of Computer Science, University of Pittsburgh, 1991.

LAYERED NETWORKS, A NEW CLASS OF MULTISTAGE INTERCONNECTION NETWORKS

Donald B. Bennett and Stephen M. Sohn*
Unisys Defense Systems, Inc.
St. Paul, MN 55164

Abstract -- We developed Layered networks to connect processors and memories in a shared memory massive multiprocessor. We emphasized latency time, fault tolerance, low contention and low cost growth at the expense of characteristics that might be more important, for example, in message-passing systems. We overcame limitations found in existing networks with respect to cost, speed (latency), fault tolerance, routing patterns and fast routing algorithms. We characterized Layered networks' behavior using simulation.

Introduction

We introduce Layered[1] networks by showing how they relate to existing well-known networks. We compare performance, topology and growth properties with baseline-equivalent networks. We present simulation results and hypothesize the non-blocking property for a subclass called "Fully-layered" networks. We conclude with speculation about some open issues.

Previously known networks

We recommend [1] for the new reader to survey known networks.

Investigators have studied multistage interconnection networks topologically equivalent to the Baseline network [2] such as the Omega network [3], [4]. The Modified Data Manipulator network, or Reverse Banyan network, Figure 1, most closely resembles

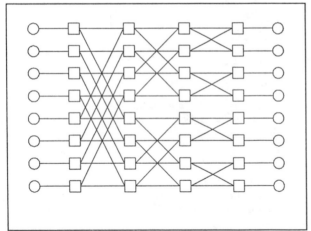

Figure 1 - The Reverse Banyan network is one of the baseline-equivalent networks.

a Layered network from among the classical, Baseline-equivalent networks. Note that we show only one node per switch rather than the customary two. Although such networks have distributed routing algorithms, good cost and access delay growth, and support for fetch-and-op operations [5], the blocking property inherent in such networks imposes uncertain and often lengthy delays, even with one node per switch. Kruskal and Snir nicely capture properties of the equivalent networks [6].

---

[1]Kruskal and Snir [6] used the term *layered* as a generalization of the notion of "stage" in a multistage network. In contrast, layers and stages are quite different concepts in our Layered networks.

The Cantor and Benes networks have low cost and the non-blocking property. However, they lack a fast (O(log N)) routing method.

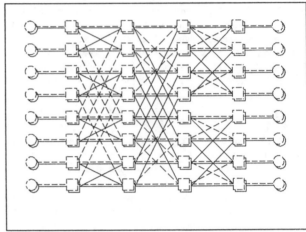

Figure 2 - The second layer's interconnects are rotated one stage.

General concept

A Layered network is composed of one or more copies of banyan networks superimposed on a switch-by-switch basis. Each copy is called a layer. The pattern of edges between stages is rotated by one stage in each layer as shown in Figure 2. The corresponding switches in the layers are merged to form the (larger) switches of the Layered network so as to create many routing choices at each switch.

Several properties of Layered networks are immediately clear. For a fixed number p of layers, growth is O(N log N), the same as a baseline-equivalent network. The number of lines is larger by a factor of p. There is a maximum of log N distinct layers. Routing algorithms can route any one of p bits in each request address.

Layered and baseline equivalent networks may be constructed using a logarithm base b other than 2; we have determined a few characteristics of higher order networks. However, most of our work is concerned with order 2, binary networks.

Node and Network Contention

Two unrelated requests entering a network may conflict so that the network or the destination node cannot properly deal with them simultaneously. We call those situations network and node contention respectively. While no network can resolve node contention, networks vary in their ability to control network contention. Layered networks can deal effectively with network contention.

Cost and Performance Comparisons

Cost growth of the Layered network is the same as the baseline-equivalent networks for a fixed number of layers. The number of stages is the same, but the larger switches might operate more slowly in a given technology. Nevertheless, because all switches are of the same design, network delay is O(log N), the same as the baseline-equivalent networks. These values are the same as, or lower than, values for other kinds of networks, given reasonable assumptions about implementations (e.g. bounded power). Of course, the Layered network is more complex by a factor equal to the number of layers, plus the cost of merging the switch layers. A larger switch may cost on the order of the square of the size of a smaller switch. The 2p wire cost and $2p^2$ switch cost factors in a Layered network serve to reduce network contention by a typical factor of $10(p-1)$.

Switch Setting (Routing)

Each switch setting is determined by comparing a request's destination address with its current location in the network, and with locally competing requests. To avoid a centralized controller, each switch routes the requests using only the information contained in the requests arriving at the switch.

A Layered network switch can choose to route an incoming request on any outgoing layer. It uses this flexibility to arrange competing requests to minimize local blocking while maximizing all requests' routing progress. Reduction of the Hamming distance from the switch to the destination is used as a measure of routing progress.

Simulation of Layered Networks

We have simulated layered networks of many sizes, bases and orders. The results show that multiple layers dramatically improve performance at all size levels, and that network contention can be reduced to any level desired.

Description of the Simulator.

The simulator first constructs a network according to input parameters and then simulates each switch's behavior given a set of specified or pseudorandom input requests. The pseudorandom generator can cause hot spots [7], that is, multiple simultaneous requests to the same node, at specified rates. The generator normally generates a request at each node for every network cycle, therefore simulating a worst case loading scenario.

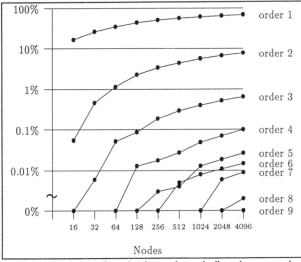

Figure 3 - Simulations show that layers dramatically reduce network contention.

Binary Network Results.

Figure 3 shows simulation results for base 2 Layered networks. It is apparent that the number of routing failures (non-routes) depends strongly on the number of layers as well as the number of nodes in the network. The simulations show that each additional layer reduces the number of non-routes by approximately an order of magnitude for a given network size. The low number of non-routes for a two-layer network taken with the low cost for construction show this network to be very cost-effective.

Speculation on Open Issues

Fully-layered networks are Layered networks with all log N distinct layers. Our simulations have never found blocked requests in Fully-layered and log N - 1 layered binary networks, nor have we been able to construct a blocking combination of requests. While we have observed this, we have not been able to prove it. Fully-layered networks with order p = 2, 3 and 4 completely route all patterns in simulation, but those with of p = 5 or greater do not. Counterexamples exist for $p \geq 5$, but neither proof nor counterexample has been found for $2 \leq p \leq 4$.

Variations on the Layered network topology are potentially interesting. The number of straight-through paths might be reduced to as low as one. The number of requestors on the nodes might be increased from one to as many as 2p. The number of layers actually needed for non-blocking varies by stage, and it may be possible to prune the network without degrading its performance.

References

[1] G.M. Masson, G.C. Gingher, and S. Nakamura, "A Sampler of Circuit Switching Networks," Computer, (June 1979), pages 32-48.

[2] C.-l. Wu and T.-y. Feng, "The Universality of the Shuffle-Exchange Network," IEEE Transactions on Computers, (May 1981), pages 324-332.

[3] G.F. Pfister, et. al., "The IBM Research Parallel Processor Prototype (RP3): Introduction and Architecture," Proceedings of the 1985 International Conference on Parallel Processing, pages 764-771.

[4] G.F. Pfister and V.A. Norton, "'Hot Spot' Contention and Combining in Multistage Interconnection Networks", Proceedings of the 1985 International Conference on Parallel Processing, pages 790-797.

[5] A. Gottlieb and J.T. Schwartz, "Networks and Algorithms for Very-Large-Scale Parallel Computation," Computer, January 1982, pages 27-35.

[6] C. P. Kruskal and M Snir, "A Unified Theory of Interconnection Network Structure," Theoretical Computer Science 48, North-Holland, (1986), pp 75-94.

[7] M. Kumar and G. F. Pfister, "The Onset of Hotspot Contention," Proc. 1986 Intl. Conf. on Parallel Processing, pp. 28-34.

On the Construction of Fault-Tolerant Cube-Connected Cycles Networks

Jehoshua Bruck   Robert Cypher   Ching-Tien Ho

IBM Almaden Research Center

San Jose, CA 95120

bruck@ibm.com, cypher@ibm.com, ho@ibm.com

1 Introduction

This paper addresses the issue of adding fault-tolerance to Cube-Connected Cycles (CCC) [9] networks based on the graph model of fault-tolerance introduced by Hayes [5]. In this model, a parallel computer is viewed as a graph in which the processors correspond to nodes and the communication links correspond to edges. A desired interconnection network, called a *target graph*, is first selected. Then a *fault-tolerant* graph is defined such that after the removal of a bounded number of faulty edges and/or nodes, the remaining graph is guaranteed to still contain the target graph as a subgraph. Previous fault-tolerance work on this model was studied in, for instance, [5] and [3].

We present constructions for fault-tolerant graphs in which the target graph is a CCC. Let $CCC(n, x)$ denote the CCC with 2^n cycles, each of which contains x nodes. We have two main results. First, we give a family of fault-tolerant graphs, G_n, which consist of $(n+1)2^n$ nodes. The graph G_n can tolerate any single edge failure when the target graph is $CCC(n, n+1)$. In fact, the graph G_n can tolerate the simultaneous failure of both an arbitrary cycle-edge and an arbitrary cube-edge when the target graph is $CCC(n, n+1)$. The target graph $CCC(n, n+1)$ also contains $(n+1)2^n$ nodes, so all of the nodes are utilized. All nodes in G_n have degree at most 4 except when n is even, in which case there is one node per cycle with degree 5. Surprisingly, the same graph G_n can also tolerate any single node fault when the target graph is $CCC(n, n)$. Second, we construct a family of fault-tolerant graphs, $H_n(y)$, which contain $(n+1)2^n$ nodes and have degree $y+2$. The graph $H_n(y)$ can tolerate *any* $2y-1$ cube-edge faults when the target graph is $CCC(n, n+1)$.

Several other authors have studied the issue of fault-tolerance in CCC networks ([1],[2],[10]). Banerjee [2] gave a fault-tolerant graph with $n(2^n + 1)$ nodes which can tolerate any single edge or node fault when the target graph is $CCC(n, n)$. His fault-tolerant graph has degree 4. However, when there are either no faults or one edge fault, his construction does not utilize n healthy nodes. In contrast, both of our constructions utilize all of the nodes when there are either no faults or only edge faults. Tzeng et al. [10] presented a degree-4 fault-tolerant network for a CCC target graph. That fault-tolerant network can tolerate either a single cube-edge fault or a single node fault. Note that no cycle-edge faults can be tolerated. Moreover, their model assumes that a (healthy or faulty) node can short-cut two incident cycle-edges. In contrast, we make no such assumption and yet our first construction can yield a degree-4 graph which tolerates the simultaneous failure of any cycle-edge and any cube-edge. Due to space limitation, all proofs are omitted and can be found in [4].

2 Preliminaries

Definition 1 A $CCC(n, x)$ is an n-dimensional cube-connected cycles (CCC) with x nodes per cycle, where $x \geq n$. Each node is labeled (c, l), $0 \leq c < 2^n$ and $0 \leq l < x$, where c identifies a cycle and l identifies a node within the cycle. Each node (c, l) is connected to node $(c, l - 1 \bmod x)$ and $(c, l + 1 \bmod x)$ through *cycle-edges*. Furthermore, if $0 \leq l < n$ then node (c, l) is also connected to node $(c \oplus 2^l, l)$ through a *cube-edge*, where "$\oplus$" is the bitwise exclusive-or operator.

Definition 2 [7] An n-dimensional folded hypercube, denoted by F_n, is an n-dimensional hypercube enhanced with an extra edge between each pair of nodes of the form i and $\bar{i}$, where $\bar{i}$ is the bitwise complement of i.

Definition 3 An $FCCC(n, x)$ is an n-dimensional *folded* CCC with x nodes per cycle, where $x \geq n+1$. It can be derived from $CCC(n, x)$ by adding a *cube-edge* between each pair of nodes of the form (c, n) and $(\bar{c}, n)$ where $0 \leq c < 2^n$.

For convenience, throughout the paper let $f(c, l) = c \oplus 2^l$ if $0 \leq l < n$, and $f(c, l) = \bar{c}$ if $l = n$. Thus each node (c, l), where $0 \leq c < 2^n$ and $0 \leq l \leq n$, is connected to node $(f(c, l), l)$ in an $FCCC(n, x)$ through a cube-edge.

3 Fault-Tolerant Cycles

A graph is said to be k-node Hamiltonian (resp. k-edge Hamiltonian) if after the removal of p vertices (resp. p edges), $0 < p \leq k$, it remains Hamiltonian. Several constructions of k-node and k-edge Hamiltonian graphs have been presented by other researchers [5,11,8]. However, none of these constructions yield graphs which are both 1-node and 1-edge Hamiltonian and which have the minimum number of edges. In this section we give such an optimal result. This optimal result will be used in our first fault-tolerant CCC construction.

We denote the family of fault-tolerant cycles by C_n. The topology of C_n, where $n \geq 4$, is defined according to three cases: (1) $n \bmod 4 = 2$, (2) $n \bmod 4 = 0$, and (3) n is odd. The construction rules for the three cases are easily seen from Figure 1-(a) to 1-(c), respectively. It is easy to verify that each node of C_n has degree 3, except when n is odd (i.e., case 3) in which case one node has degree 4. We now state the main theorem of this section. It follows from a lower bound argument given by Hayes [5] that the fault-tolerant cycle C_n has the minimum number of edges.

Theorem 1 *The graph C_n is 1-edge Hamiltonian and 1-node Hamiltonian.*

4 Fault-Tolerant CCC, Construction 1

In this section we define a new fault-tolerant graph G_n which is obtained by taking a folded hypercube F_n and replacing each node with a copy of the fault-tolerant cycle C_{n+1}. The specific construction rule of G_n, its degree, and fault-tolerant properties are described next.

The graph G_n, where $n \geq 3$, is defined as follows. First, take the fault-tolerant cycle C_{n+1}, select any Hamiltonian cycle (HC) in C_{n+1}, and starting at an arbitrary node label the nodes along the HC in order with the numbers $0, 1, \cdots, n$. Then take the graph $\text{FCCC}(n, n + 1)$ and add cycle-edges between each pair of nodes of the form (c, l) and (c, l') whenever there is an edge in C_{n+1} between the nodes labeled l and l'. Note that all nodes in G_n have degree at most 4 except when n is even, in which case there is one node per cycle with degree 5.

Theorem 2 *The graph G_n can tolerate the simultaneous failure of any single cycle-edge and any single cube-edge and is still guaranteed to contain $\text{CCC}(n, n + 1)$ as a subgraph. Furthermore, the graph G_n can also tolerate the failure of any single node and is still guaranteed to contain $\text{CCC}(n, n)$ as a subgraph.*

The above techniques can be used to obtain edge fault-tolerant graphs for target graphs of the form $\text{CCC}(n, x)$ where $x > n + 1$. The resulting fault-tolerant graphs have degree 4. However, it should be noted that following an edge fault, the resulting graph may not contain the exact graph $\text{CCC}(n, x)$, as a noncontiguous set of nodes in each cycle may be connected to cube-edges.

5 Fault-Tolerant CCC, Construction 2

In this section we construct a family of fault-tolerant CCC's, denoted $H_n(y)$, with $(n + 1)2^n$ nodes and degree $y + 2$. $H_n(y)$ can tolerate up to $2y - 1$ *cube-edge* faults such that the remaining graph contains $\text{CCC}(n, n + 1)$ as a subgraph. The construction of $H_n(y)$ is based on the idea of "dimensional substitution" used in [10] and the idea of the "wildcard dimension" [6] in the folded CCC.

Definition 4 Nodes in $H_n(y)$ are labeled as (c, l) where $0 \leq c < 2^n$ and $0 \leq l \leq n$. Each node (c, l) is connected to node $(c, l - 1 \bmod n + 1)$ and node $(c, l + 1 \bmod n + 1)$ through *cycle-edges*, and to nodes $(f(c, (l - i) \bmod n + 1), l)$ for all $i \in \{0, 1, \cdots, y - 1\}$ through *cube-edges*.

Intuitively, each node (c, l), in addition to having its two original cycle-edges and one original cube-edge in dimension l, has $y - 1$ additional cube-edges in dimensions $l - 1, l - 2, \cdots, l - y + 1$ (modulo $n + 1$), respectively. Note that $H_n(1) \equiv \text{FCCC}(n, n + 1)$. The main theorem is stated below. Note that $H_n(2)$ is a degree-4 fault-tolerant CCC which can tolerate any 3 cube-edge faults.

Theorem 3 *The family of graphs $H_n(y)$, where $y \geq 1$, can tolerate any $2y - 1$ cube-edge faults such that the remaining graph contains $CCC(n, n + 1)$ as a subgraph.*

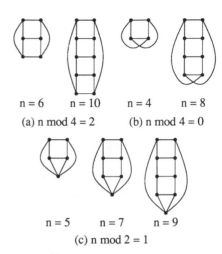

n = 6   n = 10   n = 4   n = 8

(a) n mod 4 = 2   (b) n mod 4 = 0

n = 5   n = 7   n = 9

(c) n mod 2 = 1

Figure 1: The topology of the fault-tolerant cycles.

References

[1] P. Banerjee, S.-Y. Kuo, and W.K. Fuchs. Reconfigurable cube-connected cycles architectures. In *Proc. 16th Int'l Symp. Fault-Tolerant Computing*, pages 286–291, July 1986.

[2] P. Banerjee. The cubical ring connected cycles: a fault-tolerant parallel computation network. *IEEE Trans. Computers*, 37(5):632–636, May 1988.

[3] J. Bruck, R. Cypher, and C.-T. Ho. *Fault-Tolerant Parallel Architectures with Minimal Numbers of Spares*. Technical Report, IBM RJ 8029, March 1991.

[4] J. Bruck, R. Cypher, and C.-T. Ho. *On the Construction of Fault-Tolerant Cube-Connected Cycles Networks*. Technical Report, IBM RJ, 1991.

[5] J.P. Hayes. A graph model for fault tolerant computing systems. *Computer*, C-25:875–883, September 1976.

[6] C.-T. Ho. Full bandwidth communications on folded hypercubes. In *1990 International Conf. on Parallel Processing, Vol. I*, pages 276–280, Penn State, 1990.

[7] S. Latifi and A. El-Amawy. On folded hypercubes. In *1989 International Conf. on Parallel Processing, Vol. I*, pages 180–187, Penn State, 1989.

[8] M. Paoli, W.W. Wong, and C.K. Wong. Minimum k-hamiltonian graphs, II. *J. of Graph Theory*, 10:79–95, 1986.

[9] F.P. Preparata and J.E. Vuillemin. The Cube-Connected Cycles: a versatile network for parallel computation. *Comm. of ACM*, 300–309, May 1981.

[10] N.-F. Tzeng, S. Bhattacharya, and P.-J. Chuang. Fault-tolerant cube-connected cycles structures through dimensional substitution. In *1990 International Conf. on Parallel Processing, Vol. I*, pages 433–440, Penn State, 1990.

[11] W.W. Wong and C.K. Wong. Minimum k-hamiltonian graphs. *J. of Graph Theory*, 8:155–165, 1984.

Binary Hypermesh Networks for Parallel Processing

Douglas M. Blough and *Wei-Kang Tsai**
Department of Electrical and Computer Engineering
University of California
Irvine, CA 92717

1 Introduction

Communication time is a primary factor in determining the speedup obtainable from a parallel computer. Hence, selection of a communication network can have dramatic impact on performance. In this paper, we propose a new class of networks for parallel processing called binary hypermeshes. We show that the binary hypermeshes are capable of providing fast communication between processors by presenting upper bounds on the transmission time required for point-to-point, broadcast, and multicast messages and show that these times are within a small constant factor of any other proposed network.

2 Binary Hypermesh Topology

The foundation of the binary hypermesh networks is a torus of processors which is augmented with binary trees of switching elements that provide additional paths between processors. There are two types of binary hypermeshes, referred to as Class 1 and Class 2 hypermeshes. Both classes of hypermeshes contain a binary tree of switching elements for each column of the torus. The Class 1 hypermeshes have a single binary tree of switching elements that connects the roots of the column trees while the Class 2 hypermeshes have a separate binary tree for each row of the torus. Hence, Class 2 hypermeshes can be constructed by superimposing a torus and a mesh of trees network.

We refer to the networks described above as *basic* binary hypermeshes. In addition, we have proposed in [1] a class of networks referred to as *reduced* binary hypermeshes. These reduced binary hypermeshes start with the basic hypermesh topology and remove the lowest level of switching elements in all the binary trees. By careful modification of the network (see [1] for details), the average distance between processors in the network can actually be reduced while eliminating many switching elements. It is proven in [1] that these reduced networks have smaller average distance than the basic hypermeshes while using fewer than half of the switching elements. Figure 1 shows examples of Class 1 and Class 2 reduced hypermesh networks. It should be noted that the same reduction techniques can be applied to quad hypermesh networks [2,3] with even more dramatic results. For quad hypermesh networks, more than 3/4 of the switching elements can be eliminated while reducing the average distance between processors.

3 Communication Performance

3.1 Worst-Case Performance

We are concerned with three types of communication – one-to-one, broadcast, and multicast. In [1], it was shown that the communication delay between two processors engaged in a one-to-one communication is $O(\log n)$. We also showed in [1] that broadcast can be accomplished in the binary hypermesh networks in $O(\log n)$ time. The final communication type we are interested in is multicast in which a single processor sends distinct information to each other processor. Since the amount of data is $\Theta(kn)$, where k is the length of each data item, any network of sequential processors requires $\Omega(kn)$ time for multicast. Even if a single processor can send data over multiple links simultaneously the data must be collected and distributed to the multiple ports, requiring $\Theta(kn)$ time. Multicast can be performed in $O(kn)$ time in the binary hypermesh by pipelining the messages and routing via shortest paths. The details of this pipelining approach are given in [1]. All of the communication times discussed are within a constant factor of the times required in the mesh of trees, the hypercube, the cube-connected cycle, the quad hypermesh, the hypernet, and the pyramid. In addition, the one-to-one and broadcast times are orders of magnitude shorter than the times required by those operations performed on a mesh or torus.

3.2 Average-Case Performance

Table 1 compares the average distance travelled by messages in the mesh of trees network to that of the Class 2 reduced binary hypermesh networks. While the Class 2 reduced hypermeshes had the smallest average distance, the other hypermesh networks were only marginally worse. The data for the mesh of trees is exact, based on a derivation of the average distance contained in [1] while the data for the hypermeshes was obtained by averaging the distances between 10,000 source-destination pairs chosen at random.

Note that the binary hypermesh has much smaller average distance than the mesh of trees when n is relatively small. Even when n is larger, the reduced binary

*D. Blough was supported in part by NSF under Grant CCR-9010547. W.-K. Tsai was supported in part by NSF under Grant ECS-8910328.

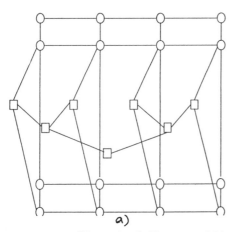 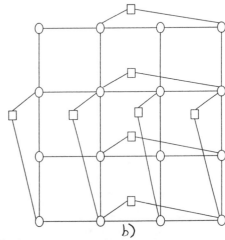

Figure 1: a) Class 1 and b) Class 2 Reduced Binary Hypermesh Networks

n	Mesh of Trees	Class 2 Reduced Binary Hypermesh
16	5.33	1.49
64	8.63	3.77
256	12.30	7.64
1024	16.14	12.33
4096	20.07	17.02
65536	28.02	25.59
1048576	36.00	33.92

Table 1: Average Message Distance

hypermesh network has considerably shorter average distance. For example, in a 1024 processor system, the average path is about 25% shorter in the reduced binary hypermesh network. These results come despite the fact that the reduced binary hypermesh uses fewer than 1/2 of the switching elements required in the mesh of trees network.

4 Algorithm Performance

Binary hypermesh networks are efficient at executing *local communicating* algorithms, *i.e.* algorithms that require only local message exchange between processors and $O(1)$ messages per processor. This class includes most of the classical iterative relaxation algorithms such as the SOR algorithms for discretized partial differential equations and the low-pass filter algorithms for signal processing. The class also includes many image processing algorithms. For the hypercube network, $O(\log n)$ links are unused when executing algorithms in this class and the cost/performance ratio is very poor.

The *generic inner product* algorithms are those that require, in addition to local communication, only

generic inner product. A generic parallel inner product computation is an operation which takes two arbitrary vectors, x and y, and performs a computation task of the form $(x_1 \circ y_1) * (x_2 \circ y_2) * (\dots) * \dots$, where $\circ$ and $*$ are two operations left unspecified and each component of the vectors x and y could also be a vector of $O(1)$ sub-components. Examples of the (parallel) generic inner product algorithms include the usual inner product, Dijsktra's algorithm for shortest path computation, Gaussian elimination with pivot, and the cyclic reduction algorithms. Any member of this class of algorithms can be executed in time $O(\log n)$ on binary hypermesh networks (see [1] for details).

5 Conclusion

We have presented a scalable, communication-efficient network for parallel processing. Several interesting issues remain to be considered. The first is to determine the most efficient VLSI layout for the binary hypermesh. It would also be interesting and useful to study the performance of symbolic algorithms on this network.

References

[1] D.M. Blough and W.-K. Tsai, "Binary Hypermesh Networks for Parallel Processing," UC Irvine Tech. Rep. No. ECE–91–02, Jan. 1991.

[2] W.K. Tsai, Y.C. Kim, and N. Bagherzadeh, "A Hierarchical Mesh Architecture," *Proc. 4th Ann. Par. Proc. Symp.*, pp. 459–477, 1990.

[3] W.K. Tsai, N. Bagherzadeh, and Y.C. Kim, "Hypermesh: A Combined Quad-Tree and Mesh Network for Parallel Processing" *Proc. Int. Conf. Par. Proc.*, 1991.

Hypermesh: A Combined Quad Tree and Mesh Network for Parallel Processing

Wei K. Tsai, Nader Bagherzadeh, and Young C. Kim*
Department of Electrical and Computer Engineering
University of California, Irvine
Irvine, CA 92717

1 Introduction

The hypermesh has almost all the advantages of the pyramid while the wire density of the hypermesh is smaller than the corresponding pyramid. The next closest architectures are the mesh-of-trees. and the enhanced mesh architecture. The mesh-of-tree architecture has $2\sqrt{N}$ binary trees and the network diameter is still $\log N$ while the wire density is much higher than the hypermesh and the average distance is about the same as the hypermesh, where N is the number of processors. The enhanced mesh has an idealistic $O(1)$ diameter while only $O(1)$ number of links per PE. Another competing architecture is the hypernet. It has been shown that the hypernet performs favorably for various algorithms when compared to the mesh, the hypercube, or the binary tree.

2 Synthesis

A m-D hypermesh is characterized by (m, h), where m is the dimension of the mesh and h is the number of hierarchy levels in the switching network. Let b denote the number of branches per father in the tree of switching network, and b is related to d by $b = 2^d$. We will call the links connecting PEs ordinary links, and the links connecting between a PE and a SE or between a SE and another SE switching links. First we state the assumptions:

1. The number of PEs along each dimension axis is equal to n, thus, the total number of PEs is n^m. For example, in the 2-D hypermesh, the number of column PEs is the same as the number of row PEs, and this number is n, the total number of PEs in the network is hence n^2.

2. $n = 2^h$, where h is the number of hierarchy levels, $h = 1, 2, 3, \ldots$.

The synthesis procedure is as follows:

1. Take an m-D mesh with torus connections and connect the PEs at the edges to the corresponding PEs at the opposite edge, to form a torus.

2. Add an SE to every b number of PEs and connect the SE with switching links to the b number of PEs. These switches will be called the 1-st level switches.

3. Complete the tree of SEs by recursively adding a $(i + 1)$-th level SE to every b number of i-th level SEs and connecting the $(i + 1)$-th level SE to the corresponding i-th level switches.

Figure 1 shows a 2-D Quad Hypermesh.

3 Traffic patterns based comparison

We assume that each processor has $O(1)$ number links/ports and each link has $O(1)$ number of wires. We will call such a PE a *realistic PE*.

For comparison purposes, we classify the traffic requirement in an interconnection network, and based on this classification, we identify proper networks for different communication requirements. The following is a modified classification originally due to Bertsekas and Tsitsiklis [1]: Single-node Local Exchange; Single-node Global Exchange; Single-node Broadcast; Single-node Accumulation; Single-node Scatter/Gather; Total exchange. For convenience, we will call each of the above data traffic requirement a *data traffic pattern*.

From a cursory survey on parallel algorithms [1,2], it is clear that a large portion of numerical and symbolic (non-numerical) algorithms require data traffic ranging from $O(1)$ single-node local exchanges to $O(1)$ single-node broadcast/accumulation. There exists less number of algorithms requiring $O(1)$ single-node scatter/gather or $\Omega(N)$ global exchanges. Algorithms that require total exchange are least in number.

The most important global data traffic patterns are single-node broadcast and accumulation. Suppose that single-node scatter/gather traffic patterns are required. A realistic PE has to process $\Omega(N)$ number of distinct data with $\Omega(Nm)$ time, where m is the size of the data item. This complexity compare unfavorably with the $O(m \log N)$ time for the single-node broadcast and accumulation, using a tree network. Indeed, most algorithms requiring global communications can be designed such that the global communication can be combined with computation to form inner product computations–inner product computation can then be implemented using only single-node accumulation.

*W.K. Tsai was supported by NSF grant number ECS-8910328, N. Bagherzadeh was supported by a University of California MICRO project in collaboration with Rockwell Corporation, Y.C. Kim was supported by Korea Science and Engineering Foundation.

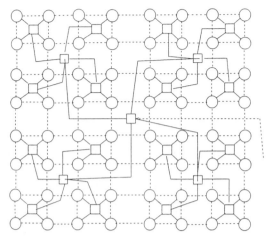

Figure 1: The 2-D Quad Hypermesh Network

○ Processing Element (PE)

□ Switching Element (SE)

Single-node scatter in the hypercube network with realistic PEs has communication time complexity of $\Omega(Nm)$ [3]. For the hypermesh, the best communication strategy is pipelining using the mesh interconnection. The pipelining algorithm will only need $O(mN)$ time [4]. Thus, even for the single-node scatter problem, the hypermesh performs just as well as the hypercube. Since the hypercube has a lot more wires, the hypermesh is obviously better. Similar argument can be made for the single-node gather problem.

4 General comparison

Table 1: Worst case distances and average distances

structure	worst case distance	average distance
binary tree	$2\log N$	$2\log N - 2 + \frac{2}{N}$
mesh	$\sqrt{N}$	$\frac{\sqrt{N}}{2}$
hypercube	$\log N$	$\frac{\log N}{2}$
mesh of tree	$\log N$	$2\log N$
hypertree	$3\log N - \frac{((\log N))_2}{2}$	$\frac{5}{4}\log N + \frac{4}{3N} - \frac{((\log N))_2}{12}$
hypernet	$2\log N - 1$	$\frac{5}{4}\log N - 1$
hypermesh	$\log N$	$\log N(1 - \frac{2\log^2 N}{3N} - \frac{1}{3N})$

Note that $((x))_2 \equiv x \bmod 2$.

5 Algorithms Performance

Any algorithm that can efficiently utilize the tree switch network without causing congestion in the roots of the trees will be appropriate to be realized in a hypermesh network. These algorithms include semigroup computation, local communicating algorithms, and the multi-level algorithms, see [4]. In this paper, we will focus on multi-level algorithms. The *multi-level* algorithms are those in which the only communication required is *local communication*, or *local inter-level com-*

munication. The local inter-level communication is defined only for those computing structures that consist of multiple levels of computation grid (mesh), and message communication will only occur between local processors on the same level and between local processors on two consecutive levels, and all messages between two processors will be $O(1)$ in number and message size. The processors on two consecutive levels that need to communicate are also directly connected and each processor is connected to only $O(1)$ number of neighboring processors. The multi-level algorithms include the famous multigrid algorithms for partial differential equations, the multi-level aggregation algorithms for general problems, and the multi-level image processing algorithms. For these algorithms, the best architecture is the pyramid network.

A more detailed disccusion of the architecture and algorithm performance can be found in [4].

References

[1] Bertsekas, D.P. and J.N. Tsitsiklis, *Parallel and Distributed Computation: Numerical Methods,* Englewood Cliffs, N.J.: Prentice Hall, 1989.

[2] Akl, S.G., *The Design and Analysis of Parallel Algorithms,* Englewood Cliffs, N.J.: Prentice Hall, 1989.

[3] S.L. Johnson and C.-T. Ho, "Optimum broadcasting and personalized communication in hypercubes,", *IEEE Trans. Computers,* pp. 1249-68, Vol. 38, No. 9, Sept. 1989.

[4] Wei K. Tsai, Nader Bagherzadeh and Young C. Kim, "Hypermesh: A Combined Quad Tree and Mesh Network for Parallel Processing," ECE Department Technical Report No. 91-05, UC-Irvine, Irvine, CA 92717, April, 1991.

C2SC: A FOUR-PATH
FAULT-TOLERANT INTERCONNECTION NETWORK

F. N. Sibai
Electrical Engineering Dept.

A. Abonamah
Mathematical Sciences Dept.

University of Akron
Akron, Ohio 44325

Abstract-- The Complementary Two-Stage Cube (C2SC) network is presented as a reliable multiple-path multi-stage interconnection network (MIN) with comparable hardware cost to other reliable MINs. The C2SC network augments the Generalized Cube (GC) network with two extra stages each added to one side of the GC to improve its fault-tolerance and robustness. The C2SC allows four possible paths between any source to any destination module. This is desirable to bypass paths along which faults may occur. We show that the C2SC network is significantly more robust than the Extra Stage Cube (ESC) network, a prominent fault-tolerant MIN, at comparable hardware cost. We also present a simple routing scheme which defines the four possible routes connecting any source to any destination.

I. Introduction

One important component of a highly parallel computer is an efficient network for information transfer between the computer's resources. Multistage Interconnection Networks (MINs) such as the baseline [1], Delta [2], Generalized Cube [2], indirect binary n-cube [1], Omega [1-2], STARAN flip [1], and SW-banyan (S=F=2) [1], have been proposed to interconnect the resources of parallel and distributed systems. While MINs have many interesting properties such as a good balance between performance and cost, topological equivalence, simple routing schemes, and distributed control, they are *unique path networks*. That is, they allow only a *single* path between any pair of input/output modules. If there is a fault on that path, no communication is possible and the network is paralyzed. Therefore, in order for MINs to fulfill their role as a reliable and efficient interconnection mechanism for parallel and distributed systems, they must be fault-tolerant.

Herein, we investigate the capabilities and functional characteristics of the Complementary Two-Stage Cube (C2SC), a fault-tolerant network which is a derivative of the Generalized Cube. The C2SC network is similar to the Extra Stage Cube (ESC) network [2]. Like the ESC, the C2SC augments the Generalized Cube by adding extra hardware to recover from link and interchange box failure. It consists of a Generalized Cube network with two added stages at the input and output sides whose purpose is to *pad* the Generalized Cube by providing alternate paths in the presence of faults. While in the ESC the links of the extra stage are connected like stage 0, in the C2SC the links of the extra stages are interconnected using *new* interconnection functions which are derived from the *complementary Generalized Cube*. The hardware cost of the C2SC is *comparable* to that of the ESC. However, the fault tolerance of the two networks are different. The ESC is *single* fault-tolerant and robust in the presence of multiple faults. On the other hand, the C2SC is *double fault-tolerant* (under a certain condition) *and significantly more robust* in the presence of multiple faults. The faults, however, are not allowed at the extra stages unless a Mux/Demux scheme as the one used in the ESC [2] is adopted. Hence, the C2SC is more fault-tolerant and robust than its counterpart the ESC at comparable cost which makes it more attractive.

II. The Complementary-Generalized Cube Network

In this section, we describe the structure of the Complementary Generalized Cube, or Co-Generalized Cube (CGC), network. The CGC is a MIN type network which has the same hardware cost as the Generalized Cube network. However, its connection capability is far inferior. That is, only certain source to destination connections are possible. Thus, the CGC used by itself is not a very useful network; however, when it is used to augment the Generalized Cube, a more reliable network is produced.

The CGC network is described by n interconnection functions, co-cube$_i$, defined by

$$co\text{-}cube_i(p_{n-1}...p_1p_0) = (p_{n-1}'...p_{i-1}'p_ip_{i+1}'...p_0')$$

where $0 <= i < n$, $P = p_{n-1}...p_1p_0$ is the binary representation of an I/O line label, $0 <= p_{n-1}...p_1p_0 < N$, and $n=\log_2 N$. In other words, two inputs at stage i are connected to an interchange box if and only if they differ in n-1 bit positions, except the ith bit position.

III. The Complementary Two-Stage Cube Network

In our fault model, both links and interchange boxes in stages n-1, ... ,1 may fail. Input and output ports are assumed to be operational. Further, the interchange boxes and links of stage n (in the Complementary One-Stage Cube; stage n+1 in the Complementary Two-Stage Cube) and 0 are assumed to be fault-free. Our model could be easily extended to allow faults in these stages by adding Muxes/Demuxes which are used to bypass the faulty components. We assume that there exists a mechanism for detecting and locating faults. Of course, an interchange box or a link failure renders it unusable.

The Complementary Two-Stage Cube (C2SC) network is constructed from the Generalized Cube network with two extra stages added, one at the input side and the other at the output side of the network, for the purpose of improving the network's reliability. The extra stage added at the input side of the MIN implements the co-cube$_{n-1}$ interconnection function, whereas the extra stage added at the output side implements the co-cube$_0$ interconnection function. Fig. 1 shows the C2SC network for N = 8. The two extra stages are disabled in the absence of link or exchange box faults in the network and are enabled in the presence of at least one fault. We show that the C2SC network has a higher fault-tolerance and robustness than the ESC network at the small price of an extra stage. As the network size N is increased for large systems the added cost becomes more and more negligible. Thus, for the negligible cost of an additional stage over the ESC network, the C2SC network can tolerate more faults in the network.

IV. Fault-Tolerance and Robustness of the C2SC Network

Below, we define the degrees of fault-tolerance and robustness of an interconnection network , and discuss the robustness of the C2SC. Next, the C2SC and ESC networks are compared in terms of robustness and hardware cost.

Definition : The degree of fault tolerance is defined as the number of faults that can be tolerated by the MIN while still providing a fault-free connection from every source to every destination. It indicates how fault tolerant the network is.

Definition: The degree of robustness of a MIN also provides an indication of the network's fault tolerance capability, but is a looser indicator than the degree of fault tolerance. It indicates the number of faults tolerated by the network such that only certain instances of those faults are allowed. In other words, the locations of those faults are restricted to provide for the maximum number of such faults to be simultaneously tolerated by the network.

The robustness of the C2SC network is considerably higher than that of the ESC network. Let SR(N) denote the switch robustness of the network of size N. Then, SR(N) for the C2SC is given by

$$SR(N) = \begin{cases} 0 & \text{if } N=2 \\ N(n+2)/2 - N\left[2^{-(n+3)/2} + \sum_{i=0}^{(n-1)/2} 2^{-i}\right] & \text{if } N > 2 \text{ and } n \text{ is odd} \\ N(n+2)/2 - N\sum_{i=0}^{n/2} 2^{-i} & \text{if } N > 2 \text{ and } n \text{ is even} \end{cases}$$

The percent additional hardware required by the C2SC network over the ESC network is given by

$$H(N) = \frac{\text{No. of boxes req. by C2SC - No. of boxes req. by ESC}}{\text{No. of boxes req. by C2SC}}$$

$$= \frac{100\%}{\log_2 N + 2}$$

Table I summarizes the degree of switch robustness for the C2SC and ESC networks with sizes ranging from 2 to 128. The percent difference in switch robustness as well as the percent additional hardware (H(N)) are also given for comparison purposes between the two MINs. Thus for the network of size 32, the switch robustness of the C2SC is 40.74% higher than that of the ESC network of the same size at the expense of only 14.29% additional hardware.

In the C2SC network, four paths connect each source to every destination. For this reason, the C2SC network is a multiple-path MIN. In addition to improving the network's performance and bandwidth, by providing alternate paths to a busy path, it also results in a substantial increase in the fault tolerance and robustness capabilities of the network. Next, theorems pertaining to the proposed network are stated (the proofs are omitted due to space).

Theorem 1: In the C2SC network with both extra stages disabled, there exists *one and only one* path between any source and any destination.
Theorem 2: In the C2SC network with one of the extra stages disabled, there exist exactly *two* paths from any source to any destination.
Theorem 3: In the C2SC network with stages n+1 and 0 enabled, there are exactly *four* paths connecting any source S to any destination D.
Theorem 4: The degree of fault tolerance of the C2SC network is 2, provided that no faults are allowed in stages 0 and n + 1, and that at most one fault is allowed to occur in each of stages 1 and n. If no faults are allowed to occur in stages 0, 1, n, and n + 1, then the degree of fault tolerance is 3.

V. Routing Tag Generation

Generating one routing tag for the C2SC network is achieved by padding the source and destination tags with zeros and taking the bitwise exclusive-or of both tags. The other three routing tags for the three remaining paths can be generated from the first routing tag. The routing tags RT1...RT4 for all four possible paths in the C2SC network can be generated as follows. Below, we assume that the source and destination tags are represented by $s_1 s_2 ... s_n$ and $d_1 d_2 ... d_n$ for a network of size $N = 2^n$.

- *case 1*: N = 2

		0	s_1	0
		0	d_1	0
RT1	=	0	$(s_1 \oplus d_1)$	0
RT2	=	1	$(s_1 \oplus d_1)$	1
RT3	=	0	$(s_1 \oplus d_1)'$	1
RT4	=	1	$(s_1 \oplus d_1)'$	0

- *case 2*: N > 2

0	s_1	s_2	...	s_{n-1}	s_n	0
0	d_1	d_2	...	d_{n-1}	d_n	0

$RT1 = 0 \; (s_1 \oplus d_1) \; (s_2 \oplus d_2) ... (s_{n-1} \oplus d_{n-1}) \; (s_n \oplus d_n) \; 0$
$RT2 = 1 \; (s_1 \oplus d_1)' \; (s_2 \oplus d_2) ... (s_{n-1} \oplus d_{n-1}) \; (s_n \oplus d_n)' \; 1$
$RT3 = 0 \; (s_1 \oplus d_1)' \; (s_2 \oplus d_2)' ... (s_{n-1} \oplus d_{n-1})' \; (s_n \oplus d_n) \; 1$
$RT4 = 1 \; (s_1 \oplus d_1) \; (s_2 \oplus d_2)' ... (s_{n-1} \oplus d_{n-1})' \; (s_n \oplus d_n)' \; 0$

Fig. 1 shows the four possible paths connecting source 5 ($s_1 s_2 s_3 = 101$) to destination 2 ($d_1 d_2 d_3 = 010$) in the C2SC with N=8 corresponding to the four routing tags: RT1=01110, RT2=10101, RT3=00011, RT4=11000.

VI. Conclusion

We have presented the C2SC network as a reliable multiple-path multi-stage interconnection network with comparable cost to other prominent reliable MINs and introduced the Co-Generalized Cube network on which the outermost stages of the C2SC network are based. A simple routing tag generation scheme for the C2SC was also given.

Future work includes developing reconfiguration algorithms to reconfigure it in the presence of one or multiple faults under one-to-one, one-to-many, and many-to-many connections.

References

[1] Wu, C. L., and Feng, T. Y., "On a Class of Multistage Interconnection Networks," *IEEE Trans. Comp.*, vol. C-29 (8), Aug. 1980, pp. 694-702.

[2] Adams, G., Agrawal, D. P., and Siegel, H.J., "A Survey and Comparison of Fault-Tolerant Multistage Interconnection Networks," *IEEE Computer*, vol. 20(6), June 1987, pp. 14-27.

Table I. Comparison of Switch Robustness and Hardware Costs for the C2SC and ESC Networks

N = size of network	SR(N) C2SC Network	SR(N) ESC Network	% Difference in SR(N)	% Additional Hardware: H(N)
2	0	0	0.00%	33.33%
4	2	1	50.00%	25.00%
8	7	4	42.00%	20.00%
16	20	12	40.00%	16.67%
32	54	32	40.74%	14.29%
64	136	80	41.18%	12.50%
128	332	192	42.17%	11.11%

Fig. 1   The Four Paths Connecting Source 5 (101) to Destination 2 (010) in the C2SC Network with N= 8

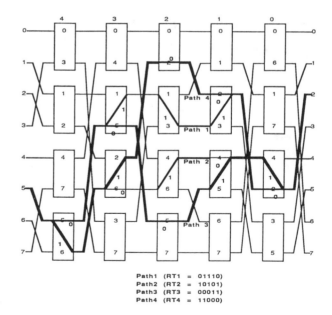

Path1 (RT1 = 01110)
Path2 (RT2 = 10101)
Path3 (RT3 = 00011)
Path4 (RT4 = 11000)

On Multistage Networks with Ternary Switching Elements

Krishnan Padmanabhan

Computing Systems Research Laboratory
AT&T Bell Laboratories
Murray Hill, New Jersey 07974

Abstract

Ternary shuffle-exchange architectures are investigated in this paper as an alternative to the augmented data manipulator (ADM) type networks for multiprocessor interconnections. These architectures use a 3×3 switch as the building block and are aimed at applications requiring higher performance levels than that of the multistage Omega type networks. A generalization of the k-ary shuffle-exchange architecture to non-power-of-k system sizes permits the construction of ternary switching networks that are fundamentally different than the ADM topology. The new networks require fewer stages and provide superior performance in an MIMD environment.

Ternary Shuffle-Exchange Architecture

We generate a $3N \times 3N$ ternary shuffle-exchange network with $N = 2^n$, and use only one input link at each first stage switch and one output link at each last stage switch. The latter technique is similar to that employed in the ADM type networks [1,3]. (Throughout this paper we will use the terms "Gamma" and "ADM" networks interchangeably, assuming full crossbars for the switches.) The architecture of binary shuffle-exchange networks for non-power-of-two system sizes has been presented in [2]. We will first outline an extension of that work to non-binary shuffles and switching elements, and then consider the ternary architecture as a special case.

Consider an integer N', which is a multiple, but not necessarily a power, of $k \geq 2$. N' can be expressed as $N' = P + M$, where $P = k^p$ and $k \leq M \leq (k-1)P$. The multistage k-ary shuffle-exchange network with N' inputs and outputs consists of $\lceil \log_k N' \rceil = p+1$ stages, with each stage composed of the k-ary shuffle of N' elements followed by N'/k crossbar switches of size $k \times k$. Each such switch can be controlled by a single k-ary digit, and the network as a whole can be controlled by a distributed tag based algorithm. The following theorem specifies the control tag to get from input i to output j. (Proof is omitted due to space constraints.)

Theorem: Any input i, $0 \leq i < N'$, in the k-ary shuffle-exchange network can set up a path to output j, $0 \leq j < N'$, by using the control tag
$$T_1 = (j + kMi) \bmod N'.$$
In addition, other control tags (and paths) may be available, specified by
$$T_c = T_1 + (c-1)N' \quad \text{if} \quad T_c < kP, \quad 1 < c \leq k.$$

The maximum number of disjoint paths between any input and output is given by $\lceil k^{p+1}/N' \rceil$. As in the case of $k = 2$, not all input output pairs might be connected by this maximum number of paths.

We now consider a specific instance of this architecture, which is a $3N \times 3N$ ternary shuffle-exchange network. Such a network consists of
$$s = \lceil \log_3 3N \rceil = 1 + \lceil \log_3 N \rceil = 1 + \lceil n \log_3 2 \rceil$$
ternary shuffle-exchange stages. An example of a 24×24 network constructed in this manner is shown in Fig. 1b. If we use only one fixed input at each first stage switch and only one output at each last stage switch then we obtain an $N \times N$ connection structure that is similar to the ADM structure (Fig. 1a). We will explore this structure in the rest of the paper, and compare the two architectures.

Referring to Fig. 1b, the N chosen inputs and outputs are numbered using $\lceil \log_3 N \rceil$ ternary digits. However, in the computation of the control tag (Theorem), the full $\lceil \log_3 3N \rceil = 1 + \lceil \log_3 N \rceil$ digit address is required for each source and destination. Without loss of generality we can assume that the N inputs with the most significant digit = 0 and the N outputs with the least significant digit = 0 are the chosen terminals. These are indicated by the solid lines in Fig. 1. For this architecture, $N' = P + M$, where P is the largest power of 3 that is less than $N' = 3N$. The values of N', P, and M are given by:
$$N' = 3N = P + M, \quad P = 3^p = 3^{\lfloor \log_3 3N \rfloor}, \quad \text{and} \quad M = 3N - 3^p, \quad 3 \leq M \leq 2 \cdot 3^p.$$

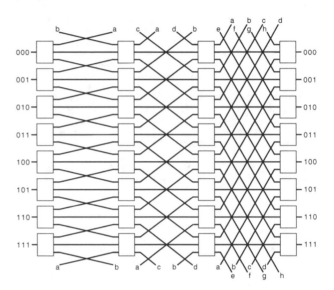

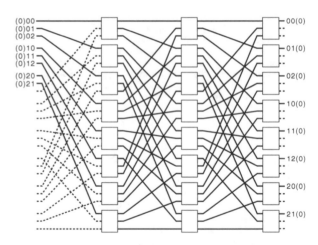

Fig 1. (a) Structure of the 8×8 inverse augmented data manipulator (IADM) and Gamma networks, and (b) The 8×8 ternary shuffle-exchange network.

From Theorem 1, one control tag to get from input i to output j is given by $T_1 = (j + 3Mi) \bmod 3N$. In addition up to two more control tags may exist for this input-output pair, given by
$$T_2 = T_1 + 3N \quad \text{if} \quad T_2 < 3^{p+1}, \quad \text{and} \quad T_3 = T_1 + 6N \quad \text{if} \quad T_3 < 3^{p+1}.$$

Note that the number of stages $s = p+1$, and this is also the number of digits in the control tag. As N increases it takes fewer and fewer stages to implement the shuffle-exchange network than the ADM network — asymptotically the former requires 63% of the number of stages in the latter, but for reasonable network sizes (≈ 1024), this fraction is about 70%.

In the case of the 8×8 network in Fig. 1, $N' = 24 = 9 + 15$, so that $p = 2$ and $M = 15$. Control tags are given by $(j + 45i) \bmod 24$. Each

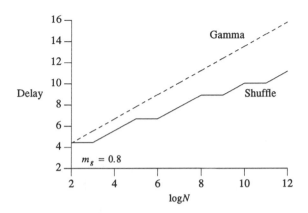

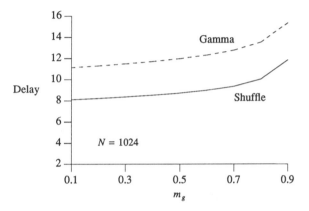

Fig. 2. Message delay as a function of (a) system size and (b) message generation rate under the lossless system model.

input has at most two paths to get to a destination. The control tags can be generated in a simple manner, as in the case of the binary shuffle-exchange networks [2]. Note that at each input i, $3Mi$ is a constant and $0 \leq 3Mi \bmod 3N < 3N$. Let $C = 3Mi \bmod 3N$. Then $T_1 = j+C-3N$ if $j+C-3N \geq 0$ and $T_1 = j+C$ if $j+C-3N < 0$. Thus T_1 reduces to a binary choice between the results of two additions (that can be done in parallel), based on the sign of one of the sums. Additional tags can be generated by adding $3N$ and $6N$ to T_1, and each is invalidated if a carry-out results during the addition.

Performance

We consider the performance of the ternary shuffle-exchange and ADM architectures under the following operating conditions:
1. The mode of communication is packet or message switched, with the entire message contained in a single packet.
2. Equiprobable addressing: Every output is equally likely to be the destination of a generated message.
3. Conflict resolution at any switch output is egalitarian.
4. Sources are identical, but are independent of each other.

Results of an analytical model for the two networks are presented in Figs. 2-3. The analysis itself is omitted due to space constraints.

Message delays under a "lossless" model that assumes buffers of infinite capacity at switch outputs are plotted as functions of network size and message generation rate per cycle (m_g) in Fig. 2. The variation with respect to size in both networks is similar to that of the number of stages, s. We see from Fig. 2a that for a 1024×1024 network, the Gamma network results in about 35% longer message delays than the shuffle-exchange architecture; this difference is less for smaller networks. At low traffic ($m_g < 0.5$) the queueing delays at all stages except the last are very small and the message delays result from just transiting the network stages. Even at high traffic ($m_g \approx 0.9$) the increase in delays is only about 25% (Fig. 2b). The shuffle-exchange architecture maintains the performance advantage throughout the range of message generation rates.

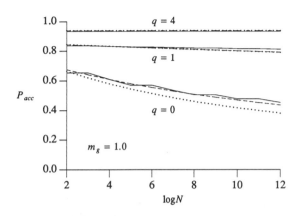

Fig. 3. Message acceptance probability as a function of system and buffer sizes under the lossy system model. Solid: Shuffle-exchange, Dashed: Gamma-unrestricted, Dotted: Gamma-restricted.

The acceptance probability of a message under a "lossy" system model that assumes buffers of finite capacity q is shown in Fig. 3. Two different routing schemes have been investigated for the ADM type networks, — "natural" or "restricted" routing and "unrestricted" routing [4]. The essential difference between the two schemes is that dynamic rerouting of messages is permitted when possible in the unrestricted routing scheme to avoid blocking. Messages that encounter full buffers (and which cannot be rerouted) are assumed dropped from the system and the acceptance probability is the probability of a message being successfully delivered to the destination. The step form of the curve for the number of stages in the shuffle-exchange network is evident here also in the unbuffered network. For instance, networks of size 256 and 512 have the same number of stages and hence the same acceptance probability. We see that in the absence of buffers at switch outputs, the acceptance probability in a shuffle-exchange architecture (without rerouting of blocked messages) is roughly the same as that in a Gamma network *with* dynamic rerouting of messages. Both of these are somewhat better than the latter architecture without message rerouting. As the buffer size increases, the acceptance probability is almost identical in the two architectures, and dynamic rerouting of messages has only a negligible effect in the Gamma network.

Thus, the shuffle-exchange network provides the same level of acceptance probabilities for messages as the Gamma network, but with fewer stages. In addition, because very low loss of messages is ensured by relatively small buffers in all cases, it is reasonable to use the lossless model to compare message delays. Under that model, the shuffle-exchange architecture is seen to result in substantially lower message delays.

Conclusion

A shuffle-exchange architecture based upon ternary switching elements has been presented in this paper as an alternative to the augmented data manipulator type networks, including the Gamma network. The ADM structure is useful in applications in which the stochastic performance and connection capability of the Omega type networks are not sufficient. The new interconnection structure is aimed at similar applications, but in this paper we have concentrated only on performance in an MIMD environment.

References

[1] McMillen, R.J., and Siegel, H.J. "MIMD Communication Using the Augmented Data Manipulator Network," *Proc. 7th Ann. Int. Symp. Computer Architecture*, May 1980, pp 51-60.
[2] Padmanabhan, K. "A Generalization of the Binary Shuffle-Exchange Architecture for Non-Power-of-Two System Sizes," *Proc. 1990 Int. Conf. Parallel Processing*, Aug. 1990, pp I363-I371.
[3] Parker, D.S., and Raghavendra, C.S. "The Gamma Network," *IEEE Trans. Computers*, Vol. C-33, No. 4, Apr. 1984, pp 367-373.
[4] Varma, A., and Raghavendra, C.S. "Performance Analysis of a Redundant-Path Interconnection Network," *Proc. 1985 Int. Conf. Parallel Processing*, Aug. 1985, pp 474-479.

Self Synchronizing Data Transfer Scheme for Multicomputers

Liao Guoning, Joar Østby, Oddvar Søråsen, Yngvar Lundh
Department of Informatics
University of Oslo
POB 1080 Blindern
0316 Oslo 3, Norway

Abstract–A line transmission scheme for data communications in multicomputers is presented. An 8bit/10bit transmission code is proposed to efficiently encode data and control symbols. A line signal conditioning scheme for the 8bit/10bit code which can achieve both bit and frame synchronization is also presented.

1 Introduction

The interconnection network for multicomputers must meet the requirements of high speed and low latency[1]. Thus, the data transmission from one node to another plays an important role in designing such interconnection networks, and influences the performance of the multicomputer system. Optical fibers are used because optical transmission offers advantages for such requirements. The computers in a multicomputer system exchange information in the form of data packets, the clock signal and line control symbols must be recovered from the data stream for proper operations on the the data packets. This is done by bit and frame synchronization.

In order to achieve bit synchronization, scrambling, bit insertion or block coding techniques can be used. Scramblers[2] try to "randomize" the bit stream, thus eliminating long strings of equal bits that might impair the receiver's synchronization. The difficulty of framing, on the other hand, makes scramblers unsuitable for data packets transmission in a multicomputer system. A bit insertion[3] code guarantees transitions by inserting extra bits into each code word. However, bit insertion code is not DC balanced which might pose problems to the optical receiver circuits. Block coding[4] guarantees both minimum transitions and a balanced number of 1's and 0's in the data stream. But block coding has bit inflation like bit insertion and is more complex.

In this paper, firstly a special transmission code, 8bit/10bit, is proposed for data encoding. The 8bit/10bit code is simple because it is a bit insertion code. However, it is a DC balanced code and it provides great flexibility to encode control symbols for the management and monitoring of the data link. Secondly, a line signal conditioning scheme which can achieve both bit and frame synchronization is presented.

2 8bit/10bit code

Walker[5] et.al proposed a line code that is not only DC balanced but also guarantees at least one transition for every 20 bits. The code is constructed by prepending 4 extra bits in front of every 16 data bits. The first two bits are master transition and the rest two bits are used to indicate the 16 bits be data or control symbols.

Multicomputers transmit information in the form of packets with a defined field structure for addresses, information, line control, and error control, for example, CRC. The number of bits in each of these fields and in the entire packet is generally a multiple of eight bits. So a simple and byte oriented encoding scheme is of great interest.

The 8bit/10bit code is constructed similarly by inserting 2 bits in front of every 8-bit data word. These inserted bits, either 01 or 10, guarantee at least one transition for every code word. The basic serial encoding scheme is shown in Figure 1.

Figure 1: The 8bit/10bit encoding scheme.

The encoding scheme not only guarantees one transition for each symbol frame, but also guarantees that the average encoded signal has a balanced DC bias. The basic approach is described as follows:

A summing logic circuit monitors the DC bias of the code words being transmitted by accumulating the number of 1's and 0's. If the number is positive and the next symbol also has a positive bias, then this entire next code word is inverted, including the leading $1 \rightarrow 0$ transition. The new code word now has a negative bias that will compensate the DC bias. If this compensation is not enough, the next code word might be inverted and so forth. Similarly, if the bias is negative, i.e. more 0's than 1's, the opposite action takes place. Thus at any time the summer compares the number of excess 1's transmitted so far and decrements or increments this number according to the next code word. The bias monitor will thus guarantee an average DC-free signal with only a minor momentary transitions in the positive or negative domain.

A special case is formed by those code words which have an equal number of 1's and 0's. They provide a special opportunity, since they can be inverted with no effect on the DC bias. Therefore if they are to be treated as normal data, we will insert a $1 \rightarrow 0$ transition to them as usual, indicating that they are to be treated as normal data at the receiving end. However these code words could also be defined as control symbols, simply by inserting a $0 \rightarrow 1$ transition. A control symbol is thus defined as having an equal number of 1's and 0's and a $0 \rightarrow 1$ transition inserted.

3 The bit and frame sychronization for the 8bit/10bit transmission code

The first thing that has to be done to the bit and frame synchronization is to recover the clock from the data stream. A good method to accomplishing this is by using a phase-locked loop (PLL).

The phase lock system shown in Figure 2 has the advantage that both bit and frame synchronizations are recovered by the same circuit. The bit synchonization is achieved by extracting the clock information from inside the data stream. The PLL is initialized by the training sequence of "0111110000". The rising edge of the *word clock*, which is derived from by dividing the bit clock by ten, is always tracking the first bit of the 8bit/10bit code word. Frame synchronization is obtained by using both clocks and a serial-in-parallel-out register.

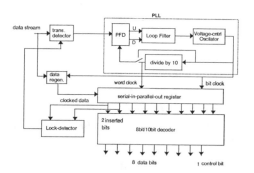

Figure 2: The bit and frame synchronization circuitry.

The system is initialized by a training sequence. This puts certain restrictions on the synchronization symbols that can be used. In particular, it requires the system to go into a complete initialization sequence each time the synchronization is lost. To maitain bit and frame sychronizations, sync symbols, which have the same code format as training sequences, are transmitted durin idle period.

Initially, the transition-detect circuit should function as an and-gate to detect only $0 \rightarrow 1$ transitions. Thus there is only one pulse every ten bit-intervals. The word clock is used to obtain the phase difference during the initialization. After synchronization has been achieved, the circuit should detect both $1 \rightarrow 0$ and $0 \rightarrow 1$ transitions, i.e. function as an exclusive-or gate, because part of the normal data codes have a $1 \rightarrow 0$ transition inserted. A control signal is used to switch the transition detector from one working condition to another.

If the phase of the data stream is ahead of word clock, then a pulse appears at the output U (pull-up the local clock) of the phase-frequency detector (PFD), and the duration of the pulse is dependent on the phase difference. If the phase of the word clock is ahead, then a pulse appears at the output D (pull-down the local clock).

Either bit clock or word clock can be used to sample the phase difference with the data stream. In the latter case, a window signal is needed to let only the inserted transitions pass through the transition detector. Otherwise, the word clock may have too many transitions to follow, thus losing synchronization.

The transition detector switches its working mode according to a control signal, which indicates whether or not the PLL system is in lock. The control signal also applies to the bit clock or word clock respectively to the PFD. The lock detector checks the first two inserted bits of every code word at the output of the serial-in parallel-out register. These two bits should be either $0 \rightarrow 1$ or $1 \rightarrow 0$ for every code word when the PLL is in lock, any violence may suggest that lock be lost. So only a correlation calculation is needed and the 01 or 10 transitions are accumulated for a certain number, then a lock condition is assured.

One of the outstanding features of the 8bit/10bit encoding scheme comes from the simplicity to decode the 8bit/10bit code. The two inserted bits and the number of 1's and 0's of the code word determines the re-inversion of the code word, and decides whether a code word represents a control symbol or normal data.

The inserted two bits are first checked. If they are a "10" combination, then the remaining 8-bit data should be output without inversion, accompanied by another bit set to logical "1" indicating an 8 bits normal data. If the inser ted two bits are a "01" combination, then the 8 data bits are checked to see whether the number of 1's is equal to the number of 0's. If yes, the data should be output without inversion, accompanied by the other bit set to logical "0" indicating a control symbol. Otherwise, the 8 bits data are inverted and output as normal data.

4   Conclusion

A serial line transmission scheme using 8bit/10bit code for self synchronization has been presented. It can readily implemented by circuit integration.

In a multicomputer interconnection network, fast and robust data links are important. The 8bit/10bit transmission code is defined in such a way that by properly inserting a guaranteed transition to every group of 8-bit data, the data and control symbols are easily encoded and decoded, which is key to a high performance data link. Both bit and frame synchronization can be obtained by the presented scheme.

The 8bit/10bit code has 20 per cent of bit inflation. When optical fibers are used as transmission medium, the bit inflation of the 8bit/10bit code will not be a much waste and the simplicity of the transmitter and receiver has been given top priority. Thus, a fast and low cost line signal conditioning circuit can be constructed.

Acknowlegement

This work has been supported by the Royal Norwegian Council for Scientific and Industrial Research (NTNF) under a postdoctoral fellowship, and in part by NTNF under the "complex system" project.

References

[1]William C. Athas and Charles L. Seitz, Mulitcomputers: Message passing concurrent computers, Computer, Aug. 1988, pp9-24.

[2]D.G. Leeper, " A universal digital data scrambler", Bell Syst. Tech. J., vol-52, no. 10, Dec. 1973, pp1851-1865.

[3]Joseph E. Berthold, High Speed Integrated Electronics for Communications Systems, Proc. of IEEE, Vol-78, No.3, March 1990, pp486-511.

[4]A.X. Widmer, P.A. Franaszek, A DC-balanced, partioned-block, 8B/10B transmission code, IBM J. Res. Develop., vol-27, No. 5, September 1983, pp440-451.

[5]R. Walker, T. Hornak, C.-S. Yen, J. Doernberg, K. Springer, A 1.5 Gb/s link interface chipset for computer data transmission, Technical Report, (c) Copyright Hewlett-Packard Company 1990.

A PARALLEL COMPUTER ARCHITECTURE BASED ON OPTICAL SHUFFLE INTERPROCESSOR CONNECTIONS

Barry G. Douglass

Electrical Engineering Department

Texas A&M University

College Station, TX 77843

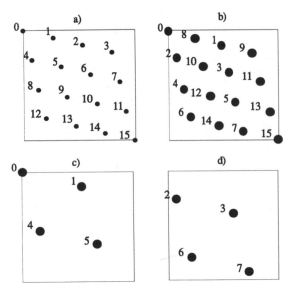

Fig. 1. Square array images with source points and shuffled destination points.

Abstract: For Multiprocessor based parallel architectures, various designs exist for interprocessor connection networks with efficient switching hardware, but all suffer from the $O(N^2)$ wiring area constraint in VLSI. This paper proposes a multiprocessor array design using free space optics as a global communications pathway, thereby overcoming the VLSI wiring area constraint, using a new simple optics scheme to produce a recirculating shuffle-exchange network.

A Shuffle Structure for Optical Imaging

Optical computers, which use optical switching elements instead of electronic (transistor) elements, show promise for off chip massive connection capability using free space optics. Electronically controlled spatial light modulators [1], such as the multiple quantum well electro-optic device, coupled with standard VLSI implemented processors, can be used to build large arrays of processors, shuffle connected to form a high speed massively parallel structure. A perfect shuffle connection scheme, is constructed by partitioning the set of inputs into two halves, then interlacing the two halves in alternating fashion at the outputs. If the number of inputs, N, is a power of two, and $p(i)$ is the output connected to input i, $0 \leq i \leq N-1$, then $p(i) = 2i - \lfloor 2i/N \rfloor N + \lfloor 2i/N \rfloor$. Consider how the shuffle connection can be implemented when the inputs are arranged in a two dimensional array instead of a straight line. Let (r,q) be the output in row r, column q. Let (j,k) be the input in row j, column k. Now let $(r(j,k), q(j,k))$ be the output connected to input (j,k). Further, let $N = 2^n$ where n is an even integer. The array of positions are a square. Let the array be numbered in both directions from 0 to $\sqrt{N}-1$ so there are N positions in the array occupied by source and destination points. Then, $r(j,k) = 2j - \lfloor 2j/\sqrt{N} \rfloor \sqrt{N} + \lfloor 2k/\sqrt{N} \rfloor$, and

$q(j,k) = 2k - \lfloor 2k/\sqrt{N} \rfloor \sqrt{N} + \lfloor 2j/\sqrt{N} \rfloor$. This will produce a shuffle where the row and column indices correspond to a one dimensional index i by the formula $i = k + j\sqrt{N}$, and $p = q + r\sqrt{N}$. Since the formulas for r and q are identical except for the exchange of indices, the structure of the connections is symmetrical about both diagonals.

To arrange this square array for shuffling by imaging, the partitioning is extended into a two dimensional process. For the linear case the image is partitioned into two halves. For the case of a square array, there are $\sqrt{N}$ rows and columns, and each is a separate partition. If each point is separated from its neighbors by a distance α, then the length of each row (or column) must be equal to the total length divided by the number of partitions, or $(N-1)\alpha/\sqrt{N}$. The last point in the first row is positioned at a distance $\alpha(1 - 1/\sqrt{N})$ before the end of the row. The first point in the next row is positioned $\alpha/\sqrt{N}$ after the start of the row. Each successive row is offset in this fashion from the previous row, and by symmetry the columns are also offset in the same fashion. With this layout a simple imaging process is applied to achieve the perfect shuffle. The image is divided into four quadrants. Each quadrant is magnified to twice the original size, and the four resulting images are superimposed on each other. The combined projections will have the image points rearranged in a perfect shuffle configuration.

Consider the position of each point in the vertical direction, that is from row to row. Let x be the vertical position, with $x = 0$ corresponding to the point $(j,k) = (0,0)$. The next point, $(j,k) = (0,1)$, is at $x = \alpha/\sqrt{N}$, and so on until the first point in the next row is at $x = \alpha$. The last point, $(j,k) = (\sqrt{N}-1, \sqrt{N}-1)$, is at $x = \alpha(N-1)/\sqrt{N}$. In terms of the one-dimensional index i defined above, the points are numbered row by row. In terms of this index, the position in the x direction of each point is in order of increasing value of i, such that point i is at position $x = (i \times \alpha)/\sqrt{N}$. Therefore in terms of the vertical axis a perfect shuffle can be constructed from the one-dimensional process by partitioning the array into an upper and lower part, dividing the image along the line $x = \alpha(N-1)/(2\sqrt{N})$, doubling the size of the two halves, and projecting the two image halves onto each other. Since, as shown above, the connection process is the same in the vertical and horizontal axes, a division of the image along the vertical axis as well results in a complete perfect shuffle. This process is illustrated in Figure 1. Figure 1a) shows the layout of the source image, while Figure 1b) shows the layout of the projected image required to achieve a shuffle. Figures 1c), and 1d) show two of the four quadrants of the source image magnified to twice their original size. The other two quadrants are the inverse of the two shown. When these four images are overlaid they produce Figure 1b).

Imaging Connections With Free Space Optics

To perform the above optically, a square illuminated flat surface is divided into four equal parts along the horizontal and vertical centerlines. The image of each quadrant of the surface is magnified and projected back onto the original image. In the *directed beam* model the surface is illuminated by a global source which produces directed, coherent

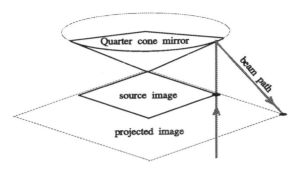

Fig. 2. Quarter cone mirror showing the projected image.

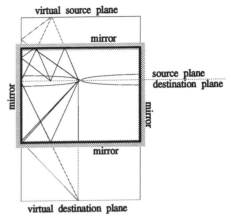

Fig. 3. A single plane projection system.

light beams, such as a laser source placed behind the surface and directed to form a parallel beam passing through every point in the surface perpendicular to it. Those points in the surface which are used as signal sources have spatial light modulators, which selectively block or admit the passage of the light beam depending on the logic state. All inactive points at the surface will have an optical mask to block the passage of light. Seen from the other side, the surface appears lit by individual coherent light sources directed vertically to the plane.

Light beams from the source plane are directed perpendicular to the surface, and therefore to achieve a perfect shuffle a mirror reflects the light back onto the surface at the proper location. The mirror must be a convex surface designed to magnify the projections to twice the original size, split into four overlapping parts. If a cone shaped mirror with a 120° internal angle is placed with its axis perpendicular to the surface, with its apex touching the surface, then each light beam will reflect back to the surface at a point twice the distance from the apex of the cone. This results in a projected image centered at the same point (the cone apex) and twice the size of the source image, as shown in Figure 2. To create four overlapping quarter images, the cone is divided into four quadrants which are swapped in both directions. The resulting mirror consists of four quarter cones, each centered at one of the four corners of the square image. Each quarter cone projects its corresponding source image back onto the entire square surface. It can be shown that any hyperboloid of revolution with the cone as its asymptote will also produce the same projected image.

As the size of the spatial light modulators decreases, approaching the wavelength of the light, diffraction becomes significant and the light beams are no longer exclusively directed vertically to the plane. A more appropriate *radiant source* model assumes that each source point at the surface behaves as a radiant illuminated source, with light emanating in all directions. Any image projection scheme for this model will require a focusing system. Each point in the source image must be focused onto the destination point at the same surface. This means passing the image through a focusing lens or mirror. Now consider a system with a separate source and destination plane. Placing a lens of the proper focal length exactly one third of the distance between the two planes will result in a projected image twice the size of the source image, and inverted. Inverting the image back can be accomplished by offsetting the lens horizontally so that the light can be reflected off a flat mirror, perpendicular to the image planes, thereby reinverting the projection at the destination plane. Since the image is two-dimensional, completely inverting it requires two orthogonal flat mirrors. Now once again, instead of a single magnified image, four overlapping images are necessary. This is obtained by dividing the lens into four quarters, and swapping the quarters. Each quarter lens is centered at one of the four corners of

the source image, and the inverting mirrors form a mirror box with the two image planes at the ends and the mirrors forming the sides.

To project the destination image back onto the source plane, the above structure is modified so that the source, destination, and lens planes coincide. Flat mirrors are placed at the two ends of the box, and the length of the box from end to end is halved. Now since the coincident source and image planes contain the electronic hardware, this hardware is spaced with gaps so that light can pass in between the electronic components and through the lens. The source and destination planes are on opposite sides of the surface. The source side is illuminated by a single, bright point light originating at the corresponding end of the box. Light is selectively reflected from the surface, under the control of electronically modulated reflecting spatial light modulators. As shown in Figure 3, this light strikes the mirror at one end of the box, reflects back through the gaps in the source surface, through the lens, reflecting off the mirror on the far end and refocusing on the opposite face. The light illuminating the source plane is placed at the center of the source face, aiming toward the mirrored end wall. The reflection from the wall illuminates the source face, and any illuminating light which passes through the surface will refocus on the four corners of the destination face.

References

1. G. D. Boyd, D. Miller, D. Chemla, S. McCall, A. Gossard, J. English, "Multiple Quantum Well Reflection Modulator", *Applied Physics Letters*, Vol. 50, 1987.

2. K-H. Brenner, and A. Huang, "Optical Implementations of the Perfect Shuffle Interconnection", *Applied Optics*, Vol. 27, no. 1, 15 January, 1988. pp. 135-137.

3. T. J. Cloonan, "Topological equivalence of optical crossover networks and modified data manipulator networks", *Applied Optics*, Vol. 28, no. 13, 1 July, 1989. pp. 2494-2498.

4. A. W. Lohmann, W. Stork, and G. Stucke, "Optical Perfect Shuffle", *Applied Optics*, Vol. 25, no. 10, 15 May, 1986. pp. 1530-1531.

5. Shing-Hong Lin, Thomas F. Krile and John F. Walkup "2-D Optical Multistage Interconnection Networks", *Digital Optical Computing*, SPIE vol. 752, 1987, pp. 209-216.

PARALLEL MODULAR ARITHMETIC ON A PERMUTATION NETWORK[†]

A. Yavuz Oruç and V. Peris

Electrical Engineering Department and Institute of Advanced Computer Studies

University of Maryland, College Park, MD 20742

and

M. Yaman Oruç 3613 Marlbrough Way

College Park, MD 20740

ABSTRACT

A permutation network is a device which can interconnect a set of n terminals in all $n!$ ways, and has been extensively studied in connection with interprocessor communications in parallel processors. This paper presents an alternative application of such networks, namely performing modular arithmetic in parallel. A theorem of Cayley states that every finite group is isomorphic to a permutation group. Based on this fact, this paper establishes that modulo m addition, subtraction, multiplication, and division can be performed on a permutation network with $O(\log_2 m)$ inputs, $O((\log_2 m) \log_2 \log_2 m)$, cost and $O(\log_2 \log_2 m)$, depth.

1 Introduction

Modular arithmetic is used extensively in many computations. Despite this, conventional algorithms for carrying multiplication, division, powering in modular arithmetic are cumbersome, and do not lend themselves to parallelization very easily. A more serious problem with these algorithms is that they are structurally different, and this leads to different hardware structures for different algorithms; for example, multiplication and addition cannot use the same hardware unless a serial multiplication algorithm is used, but this means reduction in speed which is often not tolerable.

In this paper, we propose a new scheme which unifies modular arithmetic operations using a permutation network. This scheme relies on a fundamental result of Cayley on isomorphisms between finite groups and permutation groups [Fr73]. Using this fact, we show that the operations of modulo addition, subtraction, multiplication, and division can be reduced to realizing permutations on a permutation network.

2 Modulo Multiplication and Division

In this section we describe how to multiply two numbers modulo m by composing permutations. The division modulo m can be carried out by multiplying the dividend with the inverse of the divisor, and therefore will not be considered separately. For the notation and basic facts which follow, the reader is referred to a standard text in elementary number theory.

Let $\phi(m)$ (so-called "Euler's phi" function) denote the number of positive integers which are less than or equal to, and relatively prime to m. It is well-known that $\phi(m)$ residue classes relatively prime to m form an abelian group of order $\phi(m)$, denoted $(G_m, \otimes)$ where, for any $g_1, g_2 \in G_m$, $\otimes$ is defined as $g_1 \otimes g_2 = g_1 g_2 \bmod m$. Let g be a residue class which is relatively prime to m, and define the order of $g \in G_m$ with respect to modulus m as the smallest exponent e such that $g^e \equiv 1 \pmod{m}$. If $e = \phi(m)$ then g is called a primitive root of G_m. In particular, if m is a

prime then G_m has $m-1$ elements which are the residue classes $\{n_1 : n_1 \equiv 1 \pmod{m}\}, \{n_2 : n_2 \equiv 2 \pmod{m}\}, \ldots, \{n_{m-1} : n_{m-1} \equiv m-1 \pmod{m}\}$, and every primitive root of G_m is a residue class of order $m-1$. A theorem of Gauss states that if $d|m-1$ where m is a prime then there are $\phi(d)$ residue classes of order d modulo m [Sh85, p. 72]. Thus, G_m has $\phi(m-1)$ distinct primitive roots. Let $g \in G_m$, where m is a prime, be any one of these primitive roots. It can be shown that $g, g^2, \ldots, g^{m-1}$ are all distinct. Since G_m has $m-1$ elements, it follows that any element of G_m can be generated by g, and therefore $(G_m, \otimes)$ is a cyclic group, simply corresponds to multiplication modulo m.

Now, to carry out this multiplication, we exploit Cayley's theorem on isomorphisms between finite groups and permutation groups [Fr73, pp. 59-62]. This theorem states that every finite group $(G, \otimes)$ is isomorphic to a permutation group, and furthermore, its proof provides an explicit construction of such a permutation group which is called the regular representation of G. This representation is given by $R = \{\rho_\alpha : \alpha \in G\}$ where ρ_α is defined as $\rho_\alpha : g \rightarrow \alpha \otimes g$, for all $g \in G$. In particular, the regular representation of $(G_m, \otimes)$ is given by $(R_m, \bullet)$ where $R_m = \{\rho_i : 1 \le i \le m-1\}$ and $\rho_i : j \rightarrow i \otimes j$ for all $j; 1 \le j \le m-1$.

Let us define the degree of a permutation group G, as the cardinality of the set of elements upon which the permutations in G act. While the regular representation of $(G_m, \otimes)$ facilitates a means for converting multiplications of integers to compositions of permutations, the degree of R_m grows linearly with m and this may render its use impractical for large values of m. This is because we ultimately need to realize R_m by a permutation network whose complexity grows at least at the rate of $m \log_2 m$. This problem can be avoided if we can construct another permutation group which is isomorphic to $(G_m, \otimes)$ and of order $m-1$, but degree less than $m-1$. The following theorem shows that this is possible.

Theorem 1: If m is a prime and $m-1 = m_1^{q_1} m_2^{q_2} \ldots m_k^{q_k}$ where $m_1, m_2, \ldots, m_k$ are relatively prime then $(G_m, \otimes)$ is isomorphic to a cyclic group of permutations of order $m-1$ and degree $\sum_{i=1}^k m_i^{q_i}$.
Proof: See OVO90. ||

Let m, be a prime, where $m-1$ has a prime factorization $m-1 = m_1^{q_1} m_2^{q_2} \ldots m_k^{q_k}$. Let $\sum_{i=1}^k m_i^{q_i}$ be called the degree of m, and denoted $Deg(m)$. It must be clear that $Deg(m)$ increases at a slower rate than m. To see how slowly it increases, a plot of $Deg(m)$ versus $\log_2 m$ is shown in Figure 2, where each point in the graph corresponds to an actual prime in the interval $[2^r, 2^{r+1}]$, for a given positive integer $r; 0 \le r \le 24$ and whose degree is minimal. These primes along with their degrees were computed using *Mathematica*, and can be found in [OVO90]. The plot in Figure 1 indicates that, for $0 \le r \le 24$, the $Deg(m)$ grows roughly at the same rate as $\log_2 m$. Therefore, for $0 \le m \le 2^{24}$, there exists at least one prime m in the interval $[2^r, 2^{r+1}]$ whose degree is about $\log_2 m$. This, when combined with Theorem 1, implies that the degree of H_{m-1} in the same interval also increases at the same rate as a function of m. Thus, we have

---

[*]This work is supported in part by the National Science Foundation under Grant No: CCR-8708864, and in part by the Minta Martin Fund of the School of Engineering at the University of Maryland.

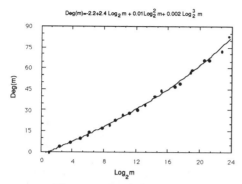

Figure 1: $Deg(m)$ v.s. $\log_2 m$.

Theorem 2: For each integer $r; 0 \leq r \leq 24$, the degree of $(H_m, \bullet)$ is $c \log_2 m$), where m is a prime which falls between 2^r and 2^{r+1}, and $c \leq 3$. ||

We conjecture that this statement is true for all integers $r \geq 0$, but the proof of this remains to be given.

Now, consider the problem of constructing an isomorphism between G_m and $(H_{m-1}, \bullet)$. Since both groups are cyclic, all we have to do is to map a generator of G_m onto a generator of $(H_{m-1}, \bullet)$. Theorem 1 already specifies a generator for $(H_{m-1}, \bullet)$ in terms of a set of pairwise disjoint cycles $c_i; 1 \leq i \leq k$ whose lengths are determined by the prime factors of $m - 1$. Thus, all we have to do is to determine a generator for G_m. By the earlier discussion, m has $\phi(m-1)$ distinct primitive roots, and any one of these can be used as a generator of G_m. In general, there are methods to find the primitive roots of m when m is prime, see for example, [Gu80. pp. 52-55]. In fact, a theorem of Gauss shows that primitive roots exists only for the numbers $2, 4, m^q$ and $2m^q$ where $m \geq 3$ is an odd prime and $q \geq 1$ [Sh85, p. 92]. Figure 2 gives one such isomorphism for $(G_m, \otimes)$ and $(H_{m-1}, \bullet)$ for $m = 11$. The isomorphism η between G_{11} and H_{10} is defined by $\eta : (a^i) \rightarrow \rho^i; 0 \leq i < 10$, where a is any of the numbers, $2,6,7,8$, and $\rho = \begin{pmatrix} 1234567 \\ 2145673 \end{pmatrix}$. With the isomorphism defined, the multiplication over G_{11} can be carried out by composing permutations in H_{10}. For example, $5 \otimes 7 = 2$ in G_{11} corresponds to $\begin{pmatrix} 1234567 \\ 1273456 \end{pmatrix} \bullet \begin{pmatrix} 1234567 \\ 2156734 \end{pmatrix} = \begin{pmatrix} 1234567 \\ 2145673 \end{pmatrix}$.

3 Modulo Addition and Subtraction

Now, consider addition and subtraction modulo m. It is well-known that the set $\{0, 1, \ldots, m-1\}$ with the addition operation modulo m forms a cyclic group of order m. Denote this group by $(\mathcal{Z}_m, \oplus)$. Since, for any two integers a, b, $a - b \bmod m = a + (m-b) \bmod m$, we need only to consider addition over $\mathcal{Z}_m$. In fact, once we construct an isomorphism between $(\mathcal{Z}_m, \oplus)$ and a permutation group, then computing the additive inverse of a number $b \in \mathcal{Z}_m$, i.e., $-b$, corresponds to computing the inverse of a permutation in the permutation group, and this is trivial.

Since $(\mathcal{Z}_m, \oplus)$ is cyclic as in the case of multiplicative groups, we can construct an isomorphism between $(\mathcal{Z}_m, \oplus)$ and a cyclic group of permutations of order m. Let this group be denoted by $(H_m, \bullet)$ as before. This time, the element 1 is a generator of $(\mathcal{Z}_m, \oplus)$ and therefore the isomorphism between $(\mathcal{Z}_m, \oplus)$ and $(H_m, \bullet)$ is defined by $\eta : (1^i) \rightarrow \rho^i; 0 \leq i < m$, where 1^i means "add 1 i times modulo m," and ρ is a generator of $(H_m, \bullet)$. An illustration of this isomorphism can be found in [OVO90].

4 Network Realization

Now, that the isomorphisms between $(\mathcal{Z}_m, \oplus)$ and $(H_m, \bullet)$, and $(G_m, \otimes)$ and $(H_{m-1}, \bullet)$ are defined, we can realize $(\mathcal{Z}_m, \oplus)$ and $(G_m, \otimes)$ on a *permutation network*. Such a network is typically used in telephone switching and in parallel computers. Here we will use it to compose permutations, and by the results of the pre-

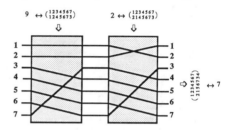

Figure 2: Isomorphism between G_{11} and H_{10}

Figure 3: Computing $9 \otimes 2 \bmod 11$ on a permutation network.

ceding sections this amounts to adding or multiplying numbers. To illustrate this with an example, Figure 3 shows how to multiply $9 \otimes 2 \bmod 11$. The permutation network on the left is programmed to realize the permutation corresponding to 9 in Figure 1, while the network on the right realizes the permutation corresponding to 2. The two networks combined together realize the permutation corresponding to $7 = 9 \otimes 2 \bmod 11$. It is seen that the product of the two numbers which are being multiplied is coded in the form of a permutation which the network represents by a set of paths between its inputs and outputs. A similar example can be given for addition as well.

The following results summarize the complexity of performing modulo arithmetic on a permutation network.

Theorem 3: For $r; 0 \leq r \leq 24$, there exists at least one prime $m; 2^r \leq m \leq 2^{r+1}$, for which $(H_{m-1}, \bullet)$ can be realized by a permutation network with about $(\log_2 m) \log_2 \log_2 m$ cost and $\log_2 \log_2 m$ depth.
Proof: See [OVO90] ||

Corollary 3.1: For any $m; 1 \leq m \leq 2^{24}$, any two integers can be added or multiplied modulo m using a permutation network with about $(\log_2 m) \log_2 \log_2 m$ switches and $\log_2 \log_2 m$ depth.

5 Concluding Remarks

We have described a method to perform parallel modulo arithmetic on a permutation network. The obvious advantage of our approach over conventional arithmetic algorithms stems from the fact that we can perform all our arithmetic on a single permutation network, rather than a different circuit for each operation.

BIBLIOGRAPHY

[Fr73] J. B. Fraleigh, *A First Course In Abstract Algebra*, Addison-Wesley, Fifth printing, Reading MA, 1973.

[OVO90] A. Yavuz Oruç, V. Peris and M. Yaman Oruç, "Fast modular arithmetic on a permutation network," Technical Report, UMIACS-TR-90-44, UMIACS, University of Maryland, College Park, MD.

[Sh85] D. Shanks, *Solved and Unsolved Problems in Number Theory*, Chelsea Publishing Company, New York, 1985.

Multicasting In Optical Bus Connected Processors Using Coincident Pulse Techniques

Chunming Qiao, Rami G. Melhem, Donald M. Chiarulli and Steven P. Levitan

Department of Computer Science and Department of Electrical Engineering
University of Pittsburgh, Pittsburgh, PA 15260

INTRODUCTION : Multicasting are common to many parallel algorithms in image processings and scientific applications. Traditional addressing mechanisms for multicasting, such as separate-addressing, multi-destination addressing and source routing have been mainly developed for point-to-point networks and are inefficient, especially in bus connected systems. Most recent research work exploits *tree forwarding* on broadcast networks which uses group identifiers when multicasting [1,4]. This requires explicit group formations of communicating processors and therefore is not efficient for dynamic communication environments where group formations change frequently.

In this paper, we will use coincident pulse techniques as addressing mechanisms among optical bus connected multiprocessors. Coincident pulse techniques are based on two properties of optical pulse transmission, namely unidirectional propagation and predictable propagation delay per unit length. When used as addressing mechanisms for multicasting, the address information of multiple destinations is multiplexed at the source processor and demultiplexed along the waveguide on which it propagates. The address decoding, which is a usual bottleneck in traditional addressing mechanisms, is done at the destination through the detection of a coincidence of light pulses.

MULTICASTING WITH UNARY ADDRESSING : To explain coincident pulse addressing techniques, we consider the optical bus connected multiprocessor system shown in Figure 1. Each processor transmits on the lower half segment of a bus, while receiving from the upper half segment. The optical bus consists of three waveguides, one for carrying messages, one for carrying *reference pulses* and one for carrying *select pulses*, which we call the *message waveguide*, the *reference waveguide* and the *select waveguide* respectively.

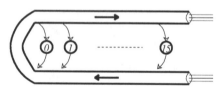

Figure 1. the Folded Bus model

Let w be the pulse duration in seconds, and let c_b be the velocity of light in these waveguides. Define a unit delay to be the spatial length of a single optical pulse, that is $w \times c_b$. Starting with the fact that all three waveguides have equal intrinsic propagation delays, we add one unit delay (shown as one loop) on the reference waveguide and the message waveguide between any two adjacent receivers. Figure 2 shows only the upper receiving segments of the select waveguide and the reference waveguide for a 16 processor system.

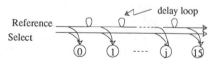

Figure 2. A unary addressing implementation

This research is supported in part by AFOSR-89-0469 and in part by MIP 89 01053

The basic idea of using coincident pulse techniques as an addressing mechanism is as follows. Addressing a destination processor is done by the source processor which sends a reference pulse and a select pulse with appropriate delays, so that after these two pulses propagate through their corresponding waveguides, a coincidence of the two occurs at the desired destination. More specifically, let T_{ref} be the time when processor i transmits its reference pulse and $T_{sel}(j)$ be the time when it transmits a select pulse. With delays added on the reference waveguide as in Figure 2, these two pulses will coincide at processor j if and only if

$$T_{sel}(j) = T_{ref} + j \qquad (1)$$

where $0 \le i, j < N$ and N is the total number of processors in the system.

This means that for a given reference pulse transmitted at time t, the presence of a select pulse at time $t + j$ will address processor j while the absence of a select pulse at that time will not. Therefore, addressing several destinations can be accomplished by having corresponding select pulses in the address frame. In essence, the address of a destination processor is unary encoded by the source processor using the relative transmission time of a reference pulse and a select pulse. The address information of multiple destinations is a multiplexing of address information of each individual destination. We define an *address frame* to be the address information in the form of a sequence of either the presence or the absence of select pulses relative to a given reference pulse. Since we have $0 \le (T_{sel}(j) - T_{ref}) < N$, an address frame has a length of N units long with unary addressing.

After transmitting an address frame, the source processor also sends a fixed length *message frame*. If a processor detects a coincidence, it reads the message frame which follows the address frame synchronously. Multicasting is accomplished by sending one address frame, which addresses several processors, followed by a message frame.

MULTICASTING WITH TWO LEVEL ADDRESSING : There are two reasons why we want to reduce the length of address frames by using two level addressing. One has to do with efficiency. Unary addressing could be very inefficient in a large multiprocessing system where the address frame is much longer than the message frame. The other reason has to do with simultaneous transmissions of messages by all processors (called bus pipelining [2]). To prevent overlapping of messages, it is required that the optical path between any two adjacent processors be at least as long as the transmitted message. Given a fixed optical path length between two adjacent processors on a bus, say D units, the address frame length has to be at most D units. That is, if unary addressing is used, the system size, N, is limited to D processors.

A two level addressing implementation divides a linear system into logical clusters. Addressing of a single destination is accomplished by using one level of unary addressing to select a particular cluster and another level of unary addressing to select an individual processor within the selected cluster. Assuming that $N = n^2$ processors are linearly connected. If every n consecutive processors constitute one logical cluster, two level addressing in this linear system is logically equivalent to addressing a two dimensional array. Hereafter, terms such as "row", "column" will be used logically. Two select waveguides are used, one for carrying row select pulses ($W1$) and the other for column select pulses ($W2$). Additional unit delays are also added to these two waveguides such that each $W1$ pulse corresponds to one row and each $W2$ pulse corresponds to one column. The quantity and the location of these delays added on $W1$ and $W2$ waveguide can be determined by solving a set of underconstrained equations. Figure 3 shows a two level addressing implementation of a 9 processor system that is logically addressed as a 3×3 array. Note that only the upper segments of the reference, $W1$ and $W2$ waveguides are shown and taps are omitted.

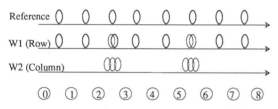

Figure 3. A two level addressing implementation

A coincidence is said to occur at a given processor only if all *three* pulses, namely a reference pulse, a $W1$ pulse and a $W2$ pulse, coincide with each other at that processor. Conditions for coincidence to occur can be similarly established. An address frame with the two level addressing contains a train of $W1$ pulses and a train of $W2$ pulses and has a length of $2 \times n - 1$ units long.

Multicasting to a group of processors can be accomplished by sending a reference pulse, one train of $W1$ pulses with one or more pulses present and one train of $W2$ pulses with one or more pulses present along with a message frame. However, by doing so, coincidences may also occur at unintended processors, which we call *shadows*. For example, let r_i and c_i be, respectively, the row number and the column number for processor i and similarly, r_j and c_j for processor j. When both processor i and j are addressed, the $W1$ train consists of two pulses, one for row r_i and another for row r_j. Similarly, the $W2$ train also consists of two pulses, one for column c_i and another for column c_j. In addition to processor i and j, the processor at row r_i and column c_j also detects a coincidence and picks up a copy of the multicasted message. So does the processor at row r_j and column c_i.

SHADOW REDUCTION AND AVOIDANCE : Shadows are created because of the unintended couplings of a $W1$ pulse with a $W2$ pulse. One way to reduce shadows is to further identify the intended pairs by using additional select waveguides, called *check* waveguides for carrying select pulses called *check* pulses. Check pulses are arranged such that they do not coincide with the reference pulse at places where shadows were created. Only processors at which a coincidence of the reference pulse and each of all select pulses are addressed. Simulations have been done and it is shown that significant numbers of shadows can be reduced with two additional select waveguides for most multicasting applications [6].

One way to avoid shadows when multicasting to a group of processors is to partition the whole group into several subgroups such that each subgroup is multicasted within one cycle without creating any shadows. More formally, assume a group of m processors is a set of processors denoted by $G = \{P_1, P_2, ..., P_m\}$. Define a *Shadow Free (SF)* partition of the set G to be a number of subgroups S_1 through S_g such that for all $1 \le i, j \le g$ the following conditions are satisfied. (a). $S_i \cap S_j = \phi$ if $i \ne j$, (b). $\bigcup_{i=1}^{g} S_i = G$ and (c). each S_i can be multicasted in one cycle without any shadows.

A SF partition is called a *maximal SF* partition if multicasting to more than one of the subgroups within one cycle will create a shadow. Therefore the number of subgroups, namely g, of a maximal partition is the number of cycles needed to complete the multicasting to the whole group G.

An incremental SF partition algorithm has been proposed [6]. With two additional select waveguides, it generates a maximal SF partition of any G with at most $min(\lceil \frac{n}{2} \rceil, \lceil \frac{m}{2} \rceil)$ subgroups.

Although multiple cycles are needed, speed ups over unary addressing are still obtained. The time needed to multicast g subgroups, one in each cycle, is $g \times (2 \times n - 1) \times c_b$ in the two level addressing implementation. However, $N \times c_b$ is needed in a unary addressing implementation. Therefore, the speed up is at least $\frac{n}{2 \times g}$. Noting that the worst case speed up is 1. Figure 4 shows that with three select waveguides, the average number of subgroups, g, is only about 5 for a group of $m = 50$ processor in a 2,500 processor system (where n is equal to $m = 50$). A speed up of 5 over unary addressing is obtained.

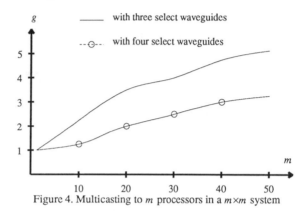

Figure 4. Multicasting to m processors in a $m \times m$ system

Several theorems have been established and it has been found that many useful multicasting applications such as those found in finite element analysis, image processings and matrix multiplications can be done in one cycle [6]. Finally, we note that in this paper, many technological aspects of pulse generation, coincidence detection, power distribution and other related issues have not been discussed. They can be found in [3, 5].

References

1. A. Frank, L. Wittie, and A. Bernstein, "Multicast Communication on Network Computers," *IEEE Software*, vol. 2, no. 3, pp. 49-61, May 1985.

2. Z. Guo, R. Melhem, R. Hall, D. Chiarulli, and S. Levitan, "Array Processors with Pipelined Optical Busses," *Proceeding of the 3rd Symposium on the Frontiers of Massively Parallel Computations*, pp. 333-342, 1990.

3. S. Levitan, D. Chiarulli, and R. Melhem, "Coincident Pulse Techniques for Multiprocessor Interconnection Structures," *Applied Optics*, vol. 29, no. 14, pp. 2024-2039, 1990.

4. P. McKinley and J. Liu, "Multicast Tree Construction in Bus Based Networks," *Communications of ACM*, vol. 33, no. 1, pp. 29-42, Jan. 1990.

5. M. Nassehi, F. Tobagi, and M. Marhic, "Fiber Optic Configurations for Local Area Networks," *IEEE Journal on Selected Areas in Communication*, vol. SAC-3, no. 6, pp. 941-949, Nov. 1985.

6. C. Qiao, R. Melhem, D. Chiarulli, and S. Levitan, "Multicasting in Optical Bus Connected Processors Using Coincident Pulse Techniques," *Tech. Rep. 91-3, CS Dept., University of Pittsburgh*, (Jan. 1991).

Fault–Tolerant Ring Embedding in de Bruijn Networks

(extended abstract)

Robert Rowley   Bella Bose

Department of Computer Science
Oregon State University
Corvallis, Oregon 97331

Abstract — We present an algorithm for embedding a ring in a de Bruijn multiprocessor network in the event of multiple node (processor) failures. The algorithm guarantees that a $(2^n\text{-}n\text{-}1)$-node ring will be found in a binary de Bruijn network with a single faulty node, where 2^n is the total number of nodes in the network.

1. Introduction

The directed de Bruijn graph has $N = r^n$ nodes represented by n-tuples $\{x_0...x_{n-1} \mid x_i \in D\}$, where $D = \{0, ..., r-1\}$. There is an edge from $x_0...x_{n-1}$ to $x_1...x_{n-1}\alpha$ for each $\alpha \in D$. Every node has indegree and outdegree r.

The undirected version of B_n has much to recommend it as a model for a processor interconnection network [Sam 89, Ber 89]. Its features include the ability to efficiently emulate several useful topologies, such as an N-node ring, a complete r-ary tree, and a shuffle-exchange graph.

Samatham and Pradhan [Sam 89] have demonstrated that a complete binary tree with $2^{n-1}-1$ nodes is contained in a binary (r=2) de Bruijn graph with a single faulty node. In this paper we consider the related problem of finding a fault-free cycle in a de Bruijn graph with faulty nodes. In particular, we show that a $(2^n\text{-}n\text{-}1)$-node ring can be found in a binary de Bruijn network with a single node failure. Edge failures are briefly discussed in Section 3.

2. Ring Embedding With Node Failures

In this section we present a general method of embedding a fault-free ring in B_n for any radix r. Our approach exploits the correspondence between circuits (closed paths with no edge appearing more than once) in B_{n-1} and cycles in B_n. This relationship arises from the fact that, if the edge $(x_1...x_{n-1}, x_2...x_n)$ in B_{n-1} is labeled $x_1...x_n$, then adjacent edges in B_{n-1} correspond to adjacent nodes in B_n, and vice versa. In other words, B_n is the line graph of B_{n-1}. A graph is said to be *balanced* if, for every node x, the indegree of x is the same as its outdegree. A balanced graph remains balanced if a circuit is removed. A graph that is both balanced and connected contains an Euler circuit (i.e., a circuit that traverses every edge in the graph).

Let F_c be a set of node-disjoint cycles in B_n that together include all of the faulty nodes, and let B_n–F_c be the graph obtained from B_n by removing the cycles in F_c. Note that B_n–F_c corresponds to a balanced subgraph of B_{n-1} whose components are Eulerian. Therefore, the connected components of B_n–F_c are Hamiltonian. We choose a Hamiltonian cycle in the largest component of B_n–F_c to be our fault-free ring. The following algorithm is based on a well-known technique for constructing an Euler circuit (see, e.g., [Gou 88]). Nodes are 'removed' from B_n by marking them, and initially all nodes in F_c are marked.

Cycle(z): Finds a Hamiltonian cycle in the component of B_n–C containing node z, where C is a set of node disjoint cycles in B_n.

   1. $C_z := (z)$; mark z; x := z;

   2. While x has an unmarked neighbor y

      2a. make y the successor of x in C_z;

      2b. mark y;

      2c. x := y;

   3. For each node x in C_z with an unmarked neighbor y

      3a. $C_y := \text{Cycle}(y)$;

      3b. $C_z := C_z$ JOIN C_y;

   4. return(C_z);

When Step 2 terminates, the path $C_z = (z, ..., x)$ is a cycle. The JOIN operation in Step 3b is accomplished as follows. If x is followed by y' in C_z, and y is preceded by $\hat{x}$ in C_y, then there is an edge from $\hat{x}$ to y'. The cycles are joined by making y the successor of x, and y' the successor of $\hat{x}$.

The cycle set F_c can be comprised of cycles, known as *necklaces*, obtained by cyclically shifting the radix-r representation of the faulty nodes. For example, if nodes 1011 and 1212 fail, we can take F_c to be {(1011, 0111, 1110, 1101), (1212, 2121)}

The maximum length of a necklace in B_n is n, so the number of nodes in F_c does not exceed n·f, where f is the number of faults. The connectivity of B_n is r-1, so B_n–F_c may be disconnected if n·f ≥ r-1. In practice, the size of the ring seems to be about r^n – n·f, even when n·f greatly exceeds the connectivity. The simulated performance of the Cycle algorithm on a 1024-node binary de Bruijn graph is shown in Table 1 in the appendix.

For the special case when r = 2 and there is a single faulty node x, it can be shown that the largest component of B_n-N(x) contains at least 2^n - n - 1 nodes, where N(x) is the necklace containing x. This leads to the following theorem.

THEOREM 1. A binary de Bruijn graph B_n with a single faulty node admits a ring with at least 2^n-n-1 nodes.

3. Edge Failures

One way to cope with edge failures is to assume that every node with a faulty incident edge is itself faulty. In this section we briefly describe a more efficient strategy for the binary de Bruijn graph with a single faulty edge.

An nth order binary *m-sequence* is a sequence of period 2^n-1 defined by

$$s_{n+i} = a_{n-1}s_{n-1+i} + \ldots + a_0s_i \pmod 2 \quad i \geq 0,$$

for $a_i \in \{0,1\}$ [McE 87]. An m-sequence corresponds to a cycle of length 2^n-1 in B_n. For example, the m-sequence $s_{3+i} = s_{2+i} + s_i \pmod 2$, with initial conditions 001, is 0011101001.... The corresponding cycle C in B_3 is (001, 011, 111, 110, 101, 010, 100).

If S = s_0s_1... is an m-sequence of period 2^n-1, then the sequence S' whose i^{th} element is s_i+1 (mod 2) also has period 2^n-1. The cycles corresponding to S and S' have no common edges. In the example above, the cycle edge-disjoint to C is (001, 010, 101, 011, 110, 100, 000).

THEOREM 2. A binary de Bruijn graph B_n with a single faulty edge admits a ring with at least 2^n-1 nodes.

In fact, r-ary m-sequences exist for every r that is a power of a prime. These r-ary sequences can be used to construct r edge-disjoint cycles of length r^n-1, in a manner analogous to that described above for r=2.

References

[Ber 89] J-C. Bermond and C. Peyrat, "de Bruijn and Kautz networks: a competitor for the hypercube?," in *Hypercube and Distributed Computers,* edited by F. André and J.P. Verjus, Holland: Elsevier, 1989.

[Gou 88] R. Gould, *Graph Theory*, Menlo Park, CA: Benjamin/Cummings, 1988, p. 129.

[McE 87] R. J. McEliece, *Finite Fields for Computer Scientists and Engineers*, Norwell, Mass: Kluwer Academic, 1987.

[Sam 89] M. R. Samatham and D. K. Pradhan, "The de Bruijn multiprocessor network: a versatile parallel processing and sorting network for VLSI," *IEEE Trans. Comput.*, vol. C-38 (April 1989), pp. 567-581.

Appendix

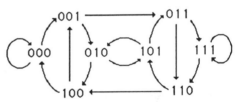

Figure 1. Binary de Bruijn graph with 8 nodes.

Faults	Smallest Ring	Average Ring	Largest Ring
1	1013	1014.2	1023
2	1002	1004.4	1019
3	992	994.7	1014
4	981	985.2	1004
5	972	975.7	998
10	918	929.8	964
20	818	844.2	903

Table 1. Size of fault-free rings in a binary 2^{10}-node de Bruijn graph with random faults (5,000 simulations).

ON EMBEDDING VIRTUAL INCOMPLETE HYPERCUBES INTO WDM-BASED HIGH-SPEED OPTICAL NETWORKS

Swie Tsing Tan and David H.C. Du
Computer Science Department
University of Minnesota
Minneapolis, MN 55455

Abstract -- Embedding a virtual extended incomplete hypercube into a high-speed Wavelength-Division-Multiplexing (WDM) optical network is proposed in this paper. The proposed scheme has the following advantages: (a) it allows a variable number of nodes in the network, (b) it requires only a minor effort to reconfigure the interconnection whenever a node is added or deleted from the network, (c) it possesses a simple and distributed routing strategy, (d) the aggregate throughput of the network increases as more nodes are connected to the network, and (e) each node is connected to either $n-1$ or n links where $n = \lceil \log_2 N \rceil$ and N is the total number of nodes in the network.

1. Introduction

The emergence of lightwave technology offers the network architect a huge bandwidth in the order of terabits per second in a single optical fiber. The current peak transmission rate of an electronic transmitter, however, is just in the order of hundreds of megabits per second. This substantial speed incompatibility between the speed of an electronic component and the speed of an optical component results in electro-optical interface being a performance bottleneck. One common approach to this speed incompatibility problem is to partition the optical bandwidth into wavelengths which are speed-wise compatible to the electronic devices connected to the optical fiber. This approach is called Wavelength-Division-Multiplexing (WDM) [1-3].

The WDM-based optical network which is discussed in this paper is a multichannel and multihop network. To simplify the discussion, we assume that each optical channel is assigned to a distinct wavelength. The multichannel multihop network allows more than one node to transmit the data concurrently by modulating their respective wavelengths in the spectrum of the optical fiber. Thus, theoretically, the throughput of a multichannel multihop network increases as more nodes are connected to the network. We will use the terms optical channel, channel and link interchangeably since they denote the same object.

Another advantage of a multichannel multihop optical network is that two physically non-adjacent nodes can be virtually adjacent by connecting them through a channel. This is because an optical channel may span more than one segment of optical fibers. To eliminate the bottleneck in the transmission speed due to wavelength switching at each intermediate node, the same wavelength is used in each segment of optical fiber for the same channel. Furthermore, as Chlamtac pointed out in [4], in order to transmit the data across the network at "the speed of light", an *optical switch box* is attached to each node. The purpose of the optical switch box is to receive and terminate the wavelengths that are destined to the attached node, to bypass the remaining wavelengths with no delay, and to transmit the data through a selected wavelength. Detail discussion on the optical switch box can be found in [4-5].

The purpose of this paper is to propose embedding extended virtual incomplete hypercubes into WDM-based optical networks. An *Incomplete Hypercube* (IH) [6] is a hypercube topology with some of the nodes removed. For the purpose of the proposed scheme, we require the removed nodes to be selected from the highest address nodes. Removing these nodes varies the number of nodes which are adjacent to each of the remaining nodes. Depending on the number of nodes in an IH, a node may be adjacent to only one other node while the remaining links are unused. We extend the interconnection by connecting some of the unused links to selected nodes such that: (a) the interconnection is transparent leading to a simple and distributed routing scheme, and (b) each node is adjacent to at least $n-1$ other nodes in an N-node IH where $n = \lceil \log_2 N \rceil$ is the dimension of the IH. The newly connected pair of unused links will be called an *extra link*. Due to page limitations, we

will illustrate the idea of embedding an IH* into a physical topology by an example in Section 2. A detailed discussion can be found in [8]. The proposed extended IHs is different from the scheme which was proposed in [7]. The scheme in [7] requires: (a) the sender to calculate and determine the shortest path to the destination node, and (b) a node may be adjacent to only two nodes while the remaining links are still unused.

The remainder of this paper is organized as follows. The algorithm to create an extended IH is described in Section 2, followed by the routing algorithm for the extended IH in Section 3. Simulation results are presented in Section 4. Section 5 concludes the discussion of this paper.

2. Extended Incomplete Hypercube (IH*)

Let $a_{n-1}...a_1 a_0$, $a_i \in \{0,1\}$, be the address of a node in an n-dimensional IH. Let $link^j_i$ be the i-th link ($0 \le i \le n-1$) of node j ($0 \le j \le N-1$). Node A is the i-th adjacent node of node B iff $link^A_i$ is connected to $link^B_i$ and the address of node A differs from that of node B at the i-th bit position. The algorithm to create an IH* from an IH is given in Algorithm 1.

Algorithm 1:

> **repeat**
> > select $node_i$ with the maximum number of unused links.
> > **if** $link^i_j$ is an unused link **then**
> > > connect $link^i_j$ with another unused $link^p_q$ where the address of $node_i$ differs from that of $node_p$ at positions $n-1$ and j.
> >
> **until** no other pair of unused links can be connected.

Example 1: A 3-dimensional IH* with 5 nodes is shown in figure 1a. The links are clearly indicated by $l_0, l_1, l_2, l_3, l_4, l^*_5$ and l^*_6. The extra links l^*_5 and l^*_6 are the result of connecting the unused links $link^{100}_0$ with $link^{001}_2$ and $link^{100}_1$ with $link^{010}_2$ respectively. Figure 1b shows a physical topology. A possible mapping of each node in a 5-node IH* into a node in the physical topology in figure 1b is shown in figure 1c. Note that link l_2 spans to three segments of optical fibers in figure 1c. The virtual connectivity of the network is shown in figure 1d.

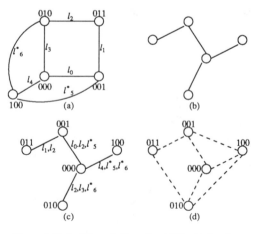

Figure 1. Embedding a 3-dimensional IH* with 5 nodes.

Adding a node in an IH* requires only a minor effort to reconfigure the new interconnection. Reconfiguring the new interconnection is as simple as retuning some of the transceivers to a new set of channels. This retuning process usually takes only a few microseconds. Each new node will be assigned an address. This address is a virtual address, therefore the address of a node can be changed if necessary. The address of the new node is selected from either the address of one of the previously deleted nodes or the next higher address in the network. As an example, if a new node is added to the network in figure 1b then the address of this node will be "101". In this example, the extra link l^*_5 will be disconnected, $link^{100}_0$ will be connected to $link^{101}_0$, $link^{101}_2$ will be connected to $link^{001}_2$, and a new extra link will be created by connecting $link^{101}_1$ to $link^{011}_2$. Observe that only one of the existing links is affected in this case. Another important point to note is that even though the dimension of the IH* may be changed as the result of adding some new nodes, no immediate major changes are required in the network. These changes are performed gradually as more nodes are added to the network.

Deleting a node in an IH* is also straightforward. We assume that although a node is deleted, the optical fibers which were used to connect to the deleted node are coupled together. That is, the channels in that segments of optical fibers are not disconnected. As an example, if node "001" in figure 1c is deleted then the two segments of optical fibers which were used to connected nodes "011" with "001" and nodes "001" with "000" will be coupled to connect nodes "011" with "000". Thus link l_2 is not affected. It should be easy to see that if node "100" were deleted from the network in figure 1c then the remaining links are not affected. If the node and its optical fibers are removed altogether then the channels that go through the removed segments of optical fibers must be rerouted to other segments of optical fibers. The rerouting process is done by modulating the wavelengths of the affected channels in the spectrum of the new segments of optical fibers.

3. The Routing Algorithm for IH*

The routing algorithm for an IH* is simple and distributed. It is similar to the routing algorithm for a hypercube, except for a minor modification. This minor modification is to take the advantage of using the extra link. Each extra link connects a pair of nodes whose addresses differ in two bit positions. One of the differing bits is the leftmost bit position and the other differing bit is in the remaining $n-1$ bit positions. Thus there is a potential that the number of hops between a pair of nodes may be reduced by one if an extra link is used in that path.

The algorithm to route a packet in an IH* is given in Algorithm 2. Let $node_{cur}$ and $node_{dst}$ be the current node and the destination node respectively. The link which connects two nodes whose addresses differ in one bit position is called a default link in Algorithm 2.

Algorithm 2:

    if ($node_{cur} = node_{dst}$) **then** receive the data; **exit**;
    find the positions of the differing bits between $node_{cur}$ and $node_{dst}$;
    if an extra link exists **and** two differing bits can be corrected **then**
        select this extra link; **exit**;
    if a default link can correct a differing bit **then**
        select this default link; **exit**;
    if an extra link can only fix one of the differing bits **then**
        select this extra link; **exit**;
    if no differing bit can be fixed **then**
        select any default link; **exit**;

Note that the links from the first two **if** statements are mostly selected. These links are the favored links since they reduce the number of hops to the destination node. The links from the remaining **if** statements are selected whenever the favored links are not available. As an example, to route a packet from node "011" to node "100" in figure 1c can be accomplished in two hops. The first hop will deliver the packet from node "011" to node "001" by link l_1. The second hop will deliver the packet from node "001" to the destination node "100" by an extra link l^*_5.

4. Simulation Results

We simulated and compared the performance between an IH* and a hypercube. To simplify the simulations, we assumed that the length of each link is uniform and each link is assigned to a different wavelength. In practice, however, these assumptions are not valid since: (a) the physical length between a pair of virtually adjacent nodes is unlikely to be uniform, (b) the average physical distance between a pair of virtually adjacent nodes must be minimized, (c) the channels must be carefully distributed to each segment of optical fibers so that no segment of optical fiber is heavily loaded while others are lightly loaded.

The average number of hops between every pair of nodes in an N-node IH* is calculated and compared with that of hypercubes. The result is presented in figure 2. Simulation results show that the performance of IH* is slightly better than that of hypercube, but IH* has a major advantage over hypercube since IH* allows an arbitrary number of nodes to be connected to the network. Other simulation results can be found in [8].

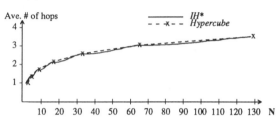

Figure 2. The average number of hops between every pair of nodes.

5. Conclusion

Due to the fact that the number of nodes in the network can be variable, a new scheme is proposed that allows a variable number of nodes to be embedded into a physical topology. The proposed scheme is similar to a hypercube but it allows an arbitrary number of nodes to be connected to the network. This topological similarity allows numerous algorithms which were designed for hypercube topology to be run in the proposed scheme. Another interesting scheme which is based on a unidirectional topology has been proposed and presented in [9].

6. References

[1]   I. Chlamtac, A. Ganz and G. Karmi, "Lightnet: Lightpath Based Solutions for Wide Bandwidth WANs," IEEE Infocom '90, vol. III, pp. 1014-1021.

[2]   Charles A. Brackett, "Dense Wavelength Division Multiplexing Networks: Principle and Applications," IEEE Journal on Selected Areas in Comm., vol. 8, no. 6, August 1990, pp. 948-964.

[3]   Joseph A. Bannister, "The Wavelength-Division Optical Network: Architectures, Topologies and Protocols," PhD. Thesis, Computer Science Dept., UCLA, 1990.

[4]   I. Chlamtac, A. Ganz and G. Karmi, "Purely Optical Networks for Terabit Communication," IEEE Infocom '89, vol. III, pp. 887-896.

[5]   Ronald J. Vetter, Kenneth Williams and David Du, "Topological Design of Optically Switched WDM Networks," Tech. Report, CSci Dept., U of Minnesota, Minneapolis, 1991.

[6]   Howard P. Katseff, "Incomplete Hypercubes," IEEE Trans. on Computers, vol. c-37, May 1988, pp. 604-608.

[7]   Hsing-Lung Chen and Nian-Feng Tzeng, "Enhanced Incomplete Hypercubes," Proc. of Int. Conf. on Parallel Processing 1989, vol. 1, pp. I-270-277.

[8]   Swie T. Tan and David Du, "On Embedding Virtual Incomplete Hypercubes into WDM-Based High-Speed Optical Networks," Tech. Report, CSci Dept., U of Minnesota, Minneapolis, 1991.

[9]   Swie T. Tan and David Du, "On Embedding Virtual Unidirectional Incomplete Hypercubes into WDM-Based High-Speed Optical Networks," Tech. Report, CSci Dept., U of Minnesota, Minneapolis, 1991.

Embedding Large Binary Trees to Hypercube Multiprocessors

Ming-Yi Fang and Wen-Tsuen Chen

Department of Computer Science
National Tsing Hua University
Hsinchu, Taiwan 30043, Republic of China.

Abstract - In this paper, many-to-one embeddings of large binary trees to hypercube multiprocessors are investigated. An efficient embedding procedure, which maps binary trees into a hypercube with constant overhead, is presented. The result can also be extended to the embedding of augmented binary trees which incorporate extra links to connect nodes that lie in the same level of the tree so as to reduce average path distance, and to provide redundant paths for fault tolerance as well.

I. INTRODUCTION

Efficient embeddings into the hypercube have been investigated for several simple network structures. A survey of several known results is given in [3]. Most of the previous results [1]-[5] have dealt with one-to-one embeddings only, which consider the one-to-one association of the processors of a network with the processors of a hypercube network. However, in practice, we are often faced with hypercubes of limited size. It is therefore desirable to study other useful mappings, such as the many-to-one mappings in which multiple nodes are mapped to a single node. Many-to-one embeddings are highly demanded in the situation where the guest graph has more nodes than the host graph.

In this paper, we consider a many-to-one mapping in which only the leaves of a binary tree are mapped to distinct nodes. Such embeddings commonly arise from the execution of tree-like programs in which only one level of nodes are activated in each step of the execution. As these nodes represent active processes, nodes lying at the other levels are waiting processes. Then, only the leaf nodes have to be mapped to distinct nodes. Here, the host graph used in the mapping is always a hypercube while the guest graphs considered for embedding can be the complete binary trees, the X-trees, the ringed trees, or the Hypertrees.

II. DEFINITIONS AND NOTATIONS

An **n-dimension boolean hypercube** has 2^n nodes. Two nodes with binary addresses $a_{n-1}...a_0$ and $b_{n-1}...b_0$ have an edge joining them whenever their binary addresses differ in a single bit position. A hypercube of dimension n is also called an n-cube, and is denoted by Q_n.

We consider only simple undirected graph $G = (V, E)$ consisting of a node set $V(G)$ and an edge set $E(G)$. An embedding $<\varphi, \rho>$ of a **guest** graph G into a **host** graph H is a mapping from each node v in G to a node $\varphi(v)$ in H, and from each edge (u, v) in G to a path $\rho(\varphi(u), \varphi(v))$ in H, which starts at node $\varphi(u)$ and ends at node $\varphi(v)$. In this paper, we consider a mapping function φ that is a many-to-one mapping, i.e., we allow multiple nodes in G to be mapped onto a single node in H.

The cost of an embedding $<\varphi, \rho>$ is measured in terms of dilation, congestion, and expansion. The **dilation** of an edge $(u,v) \in G$ is the length of the path $\rho(\varphi(u), \varphi(v))$ in H and the dilation of the embedding $<\varphi, \rho>$ is the maximum dilation over all edges in G. The **congestion** of an edge $e \in E(H)$ is the number of edges in G whose image in H contains the edge e and the congestion of $<\varphi, \rho>$ is the maximum congestion over all edges of H. The **expansion** of an embedding is defined to be $|V(H)|/|V(G)|$, which is a measure of processor utilization. When embedding is performed in a many-to-one manner, we use load factor instead of expansion to measure the processor utilization. The **load factor** is defined as the maximum number of nodes in G that are mapped to the same node in H.

In the embedding algorithm, we also use an identification operation. Two nodes are said to be **identified** (or **merged**) if we replace two nodes u and v in the graph with a new node w such that every node that is adjacent to u or v is also adjacent to w.

III. A MANY-TO-ONE EMBEDDING

Here, we consider the many-to-one embedding of an n-level complete binary tree T_n into a hypercube Q_k of size 2^k, $1 \le k \le n-1$. First, we assume k = n-1. The result can be easily generalized to all other cases for $1 \le k \le n-2$.

First, we observe that a T_n contains almost twice as many nodes as a Q_{n-1}, while 2^{n-1} nodes of the T_n are at the leaf level, which are as many as the number of nodes in Q_{n-1}. Thus, our basic idea is that we view the leaf nodes of the T_n as the base nodes and attempt to fold the internal nodes onto the leaf nodes such that the resulting graph is a subgraph of Q_{n-1}. To make the folding operation uniform and to ensure the folded graph is a subgraph of Q_{n-1}, we have to properly stretch the T_n prior to the folding operation.

Therefore, the overall embedding procedure proceeds as follows. First, we stretch the T_n to make the tree sparse by introducing dilation-two edges among internal nodes. Then, the internal nodes of the T_n are uniformly folded upon the leaf nodes. Finally, each leaf node is mapped to a distinct node of the Q_{n-1}.

We are now ready to describe the embedding procedure in detail. We begin with the stretching operation. The main purpose of the operation is to make the T_n, $n \ge 3$, sparse by adding an articulation node (A-node) to every edge between level 0 and level n-3 which stretches an edge into a dilation-two edge. As a result, two nodes are added between level 0 and level 1, and four nodes between level 1 and level 2 etc., and a total of 2^{n-2}-2 nodes are added. Note that we view the A-nodes as virtual nodes and ignore the number of levels introduced by them.

After stretching the T_n, a second operation, called **folding** or **identification**, is used to merge multiple nodes into a single node. The merging is conducted so as to identify a limited number of nodes with a single leaf node. In other words, we regard the leaf node as the base nodes and attempt to distribute the internal nodes evenly over the leaf nodes. Informally, we first partition a stretched T_n into four subtrees of the same size, each denoted by B_0, B_1,..., B_3 respectively, as shown in Figure 1. An identification operation deals with the identification of the root of a subtree and its two immediate A-nodes with the leaf nodes. That is, the root of a subtree is identified with its right immediate A-node, which in turn identified with the leftmost leaf node of B_3. While the other A-node is identified with the leftmost leaf node of B_1. Similarly, the above identification operation is applied to all

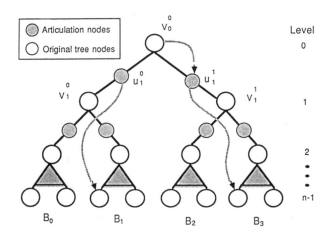

Figure 1.. Identification of internal nodes with leaf nodes

subtrees of height greater than 2 that contain A-nodes. For subtrees not having A-nodes, i.e. subtrees of height at most 2, we simply identify the root node with its right immediate descendant. We shall show that for a given leaf node there are at most two internal nodes identified with it.

To be specific, the folding operation is formally described as follows. To begin with, we associate a label with each node in the tree. Nodes at each level are labelled from left to right in turn. Namely, the root node is labelled by v_0^0, nodes at level 1 labelled by v_1^0, v_1^1, nodes at level 2 labelled by v_2^0, v_2^1,..., v_2^3, and so on. Similarly, A-nodes between level 0 and 1 are labelled by u_1^0, u_1^1, A-nodes between level 1 and 2 are labelled by u_2^0, u_2^1,..., u_2^3, etc, as shown in Figure 1. Then, the folding operation, described by two cases, is defined as follows, in which the identification of a sequence of nodes $u, v,..., w$ is denoted by $u \rightarrow v \rightarrow ... \rightarrow w$.

(1) For a subtree rooted at node v_i^j, $0 \leq i \leq n-4$, $0 \leq j \leq 2^i-1$,

$$v_i^j \rightarrow u_{i+1}^{2j+1} \rightarrow v_{n-1}^k \text{ where } k = j \cdot 2^{n-i-1} + 3 \cdot 2^{n-i-3},$$

and $u_{i+1}^{2j} \rightarrow v_{n-1}^{k'}$ where $k' = j \cdot 2^{n-i-1} + 2^{n-i-3}$.

(2) For a subtree rooted at node v_i^j, $n-3 \leq i \leq n-2$, $0 \leq j \leq 2^i-1$,

$$v_i^j \rightarrow v_{i+1}^{2j+1}.$$

Finally, we have to find a one-to-one mapping of the leaf nodes to the nodes in the Q_{n-1}. The mapping process is straightforward. We simply map a leaf node v_{n-1}^j to the node with address j in the Q_{n-1}, $0 \leq j \leq 2^n-1$, which completes the embedding procedure.

We now turn to analyze the cost of the embedding in terms of dilation, congestion, and load factor. Let the stretched T_n be denoted by S_n. Then we have:
Theorem 1. There exists an embedding of the T_n into the Q_{n-1} with dilation 2, congestion 2, and load factor 3.

Note that the optimal load factor of embedding a T_n in a Q_{n-1} is $\lceil |T_n|/|Q_{n-1}| \rceil = 2$. As a result, the load factor induced by the embedding procedure is optimal within a factor of two.

To deal with the embedding of the T_n in a smaller Q_m, where m < n-1, we first embed the T_n in a Q_{n-1}, and then successively fold the Q_{n-1} until it reaches the size of Q_m. In

such a case, both the load factor and optimal load factor increase by a factor of 2^{m-n}. Thus, we have
Corollary 1. The T_n can be embedded in the Q_m, m ≤ n-1, with dilation 2 and optimal load factor within a factor of 2.

VI. AUGMENTED BINARY TREES

In this section, we shall study the embedding of augmented binary trees which make use of additional links to enhance the basic binary tree.

The key to the embedding is based on the observation that the nodes lying at the k-*th* level of the T_n form a subcube of dimension k, as stated in the following lemma:
Lemma 1. The nodes at the i-*th* level of the T_n, $n \geq 3$, embedded in a Q_{n-1}, are connected as a subcube Q_i, $0 \leq i \leq n-1$.

An n-level **X-tree**, denoted by X_n, is constructed from the T_n by connecting adjacent nodes at every level of the T_n. Since nodes lying in the same level of T_n form a subcube and a linear array is a subgraph of the hypercube, we can easily obtain:
Theorem 2. An X_n can be embedded in a Q_{n-1} with dilation 2 and load factor 3.

Similarly, an n-level **full-ringed tree**, denoted by F_n, is constructed from the T_n by connecting adjacent nodes at every level into the ring topology. Since hypercubes are Hamiltonian, we have:
Corollary 2. An F_n can be embedded on a Q_{n-1} with dilation 2 and load factor 3.

An n-level **Hypertree**, denoted by H_n, is constructed from the T_n by choosing a set of n-cube connections to connect nodes lying in the same level of the tree. Since nodes in every level of T_n form a subcube, we have:
Corollary 3. An H_n can be embedded on a Q_{n-1} with dilation 2 and load factor 3.

V. CONCLUSIONS

We have proposed an embedding procedure for embedding a family of binary trees in the hypercube. The embedding is preceded by a prior stretching of the tree edges, followed by folding internal nodes onto the leaf node, which leads to a many-to-one mapping with optimal dilation, optimal congestion and optimal load factor all within a factor of two. Augmenting the basic binary tree by connecting nodes in every level results in the X-tree, the ringed tree, and the Hypertree, which are also shown to be embeddable in the hypercube with $O(1)$ dilation cost and $O(1)$ load factor cost.

REFERENCES

[1] S.N. Bhatt, F.R.K. Chung, F.T. Leighton, and A.L. Rosenberg, "Optimal Simulations of Tree Machines", Proc. of the 27th annual symp. on the Found. of Comp. Sci., (1986), pp. 274-282.

[2] S.R. Deshpande and R.M. Jenevein, "Scalability of a Binary Tree on a Hypercube", Proc. IEEE 1986 Int. Conf. Parallel Processing, (1986), pp. 661-668.

[3] M. Livingston and Q.F. Stout, "Embeddings in Hypercubes", in Proc. 6th Intern. Conf. on Mathe. Modeling, (Aug., 1987).

[4] A.S. Wagner, "Embedding Arbitrary Trees in a Hypercube", J. of Parallel and Distributed Computing 7, (1989), 503-520.

[5] A.Y. Wu, "Embedding of Tree Networks into Hypercubes", J. of Parallel and Distributed Computing 2, (1985), 238-249.

Simulation of SIMD Algorithms on Faulty Hypercubes[1]

Sing-Ban Tien and C.S. Raghavendra
Department of Electrical Engineering-Systems
University of Southern California, Los Angeles, CA 90089-0781

Abstract. To achieve fault-tolerance in hypercubes, efficient data communication should be supported even in the presence of faulty processors. In this paper a simulation scheme is developed to support communication steps for SIMD hypercube algorithms. We consider an n-dimensional faulty hypercube, with arbitrary $(n-1)$ faulty processors, and simulate SIMD data communications of any algorithm with a small constant slowdown factor. Specifically, the communication time is at most 4 times when the number of faults is no more than $\lceil n/2 \rceil$. Computation time is only doubled as fault-free processors take over tasks of faulty processors. Our approach to communication in faulty hypercube is based on a key concept called **free dimension**

1 Introduction

Multiprocessor systems have been made feasible by the advances in VLSI technology; several machines are commercially available, for example Connection Machine and Hypercube computers. Many algorithms have been designed for hypercubes, namely BPC routing [2], PDE algorithm [4]. Most of these algorithms are SIMD type: in each step every node either performs computation or communicates along the same dimension with its neighbor.

An important issue in hypercube parallel computers is fault tolerance, as the probability of a failure can be high with large number of nodes. One common approach in providing fault tolerance is through introducing spare nodes and/or links. However, adding spare nodes and links increases the cost of hypercube computers.

A cost effective alternative is by allowing performance degradation to make spare nodes and links unnecessary. The approach is to move tasks from faulty processors to fault-free processors. In this direction, Hastad et. al. [3] proposed a probabilistic scheme. The results show that with high probability an n-dimensional hypercube can simulate fault-free hypercube operations only with a constant factor of slowdown. However, since the approach is probabilistic, even with a small number of faults, say $n-1$, there is a chance of system failure. A deterministic scheme proposed in [1], finds the largest fault-free subcube and then runs the algorithms in that subcube. The scheme can run any algorithm on faulty hypercubes. However, the slowdown factor can be significant; only 2 faulty nodes can reduce the dimension of the largest fault-free subcube by 2 in the worst case. Thus the slowdown is by a factor of 4.

In this paper, we present a *deterministic* scheme for simulating SIMD algorithms on faulty hypercubes (which operates in SIMD mode). We introduce a key concept, called **free dimension** – a dimension is free if for all pairs of nodes across this dimension, at most one node is faulty. Otherwise the dimension is *occupied*. Based on this concept we obtain the following results. In an n-dimensional hypercube, with arbitrary $n-1$ faulty nodes, our scheme can simulate any SIMD hypercube algorithm, with a small constant slowdown factor. Specifically, the computation time is doubled and communication time is increased by at most 3 times if partial fault model of nodes, the communi-

---
[1]This research is supported by the NSF Grant MIP8452003

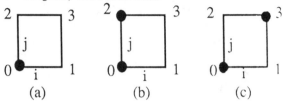

Figure 1: Three cases of fault patterns.

cation function of the faulty node is intact, is assumed. In total fault model, the communication function is lost rendering computational function useless, it is 4 times when the number of faulty nodes $f \le \lceil n/2 \rceil$ and bounded by a factor of 9, when $\lceil n/2 \rceil < f < n$. In the latter case, the factor can be 4 if a specific condition is satisfied. (The condition is the existence of free and uncovered dimensions as explained later.) Since worst case scenarios are rare, our simulation scheme is expected to perform with a slowdown factor of 2 to 4. In this paper, only node failures are considered. Link failures can be absorbed as node failures easily. We assume total fault model in the following, since the results for partial fault model can be easily obtained from this analysis. The proofs of the theorems in this paper can be found in [5].

2 Reassigning Faulty Nodes and Their Links

In order to run an SIMD algorithm on a faulty hypercube, the tasks of all faulty processors need to be reassigned to some fault-free processors. New communication paths between the new location of the faulty node's task and the neighbors of the faulty node need to be established too. In the case that one neighbor of the faulty node is also faulty, the path goes to the new location of the neighbor.

Theorem 2.1 *In Q_n with a set of faulty nodes F, the number of dimensions occupied by F is at most $|F|-1$ provided $|F| \le n$, and hence at least $n - |F| + 1$ free dimensions exist.*

From Theorem 2.1, we know that as long as $|F| \le n$ there is at least one free dimension. The tasks of the faulty nodes can then be assigned to their neighbors across a free dimension, However, n faults can disconnect a node. Hence, in the following $|F| < n$ is assumed.

New communication paths are established as follows. Consider the 3 possible fault cases for simulating the communication along a dimension j as depicted in Fig. 1, where dimension i is free and faulty nodes are marked by a black dot. In Fig. 1(a) node 0 is faulty and its task is relocated in node 1. The communication between the task of faulty node 0 and the task of node 1 along dimension i can now be achieved within the same node 1. Communication with node 2 along dimension j can be achieved by going through node 3 with dilation 2. In Fig. 1(b), both the tasks of node 0 and 2 are assigned to node 1 and 3, respectively. For the task of faulty node 0 (2), communication with the task of node 1 (3) along dimension i is the same as above within a node, whereas communication with the task of node 2 along dimension j can now be done in one step by communicating with node 3. In Fig. 1(c), communication

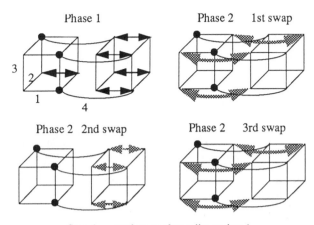

Phase 1   Phase 2   1st swap

3
2
1   4

Phase 2   2nd swap   Phase 2   3rd swap

Step 1 -- exchange along dimension 1

Figure 2: A BPC permutation simulated on a faulty hypercube.

along dimension i can be achieved within node 1 and 3. For communication along j, a path between node 1 and node 2 need to be established. It can not be routed within the 2-subcube. However, a path of length 4 can be found through an adjacent fault-free 2-subcube. There exists at least a fault-free adjacent 2-subcube due to the following. There are 2 faults in the 2-subcube and the 2-subcube has $(n-2)$ adjacent 2-subcubes. If every adjacent subcube contains a faulty node, then the total number of faulty nodes is $2 + n - 2 = n > |F|$.

3   Simulation of SIMD Algorithms

We separate the processing steps (and the communication steps) into two phases. In the first phase, all fault-free processors perform processing (communication) for their own tasks. In the second phase, the ones having the tasks of their faulty neighbors perform processing (communication) for the tasks of the faulty nodes. Therefore the processing time is doubled, since each node assumes at most 2 tasks. For communication, if it is in the situation of Fig. 1(a) or (b), it takes at most 4 times as much – one swap for the first phase and at most 3 swaps for the second phase. In Fig 1(a) we need to perform swaps along dimension (i, j, i) to exchange messages between 1 and 2. (An example is given later.) However, if the situation is as in Fig. 1(c), it can be 8 swaps in the second phase. This is due to the fact that the path is shared by two pairs of tasks – those in nodes 1 and 3 and those in nodes 0 and 2 and each pair takes 4 hops for communication. But if the two messages can be combined, then it takes only 4 times as much. The total time taken in the 2nd phase is 8 times as that in the first phase only if all such paths in the second phase of a communication step along dimension j, $1 \leq j \leq n$ are node-disjoint; i.e., no further congestion on the 4-hop paths.

Theorem 3.1 *All the paths established in the second phase of the communication steps for simulating the fault-free communication along dimension j, are node-disjoint.*

Definition 3.1 *Dimension i is covered if there are two faulty nodes in F with distance 2 and differing in exactly two bits, bit i and another bit. Otherwise it is uncovered.*

An example that a dimension is covered is that shown in Fig. 1(c), where dimensions i and j are both covered. We show in the following, when $|F|$ is less that $n/2$, there exists an uncovered free dimension.

Theorem 3.2 *If $|F| < \lceil n/2 \rceil + 1$, then there must be a free and uncovered dimension.*

Using such an uncovered free dimension, simulating the communication takes at most 4 times as much as that in the fault-free case, since Fig 1(c) does not occur. We summarize our results in the following theorem.

Theorem 3.3 *In an n-dimensional SIMD hypercube with at most $n-1$ faulty nodes, any SIMD hypercube algorithms can be simulated with a constant slowdown factor.*

The actual slowdown factor will depend on how many times an algorithm communicates along covered dimensions. Since worst case situation is rare, it is very likely with $n - 1$ faults a free and uncovered dimension exists. Therefore, slowdown factor is likely to be at most 4 upto $n-1$ faulty nodes. A parallel algorithm runing on faulty hypercube with complexity $O(n)$ for finding free dimensions and detailed implementation issues can be found in [5].

Example. Consider a 4 dimensional hypercube containing 3 faulty nodes as shown in Fig. 2. Notice that dimension 3 is occupied and dimension 1 and 2 are covered. Dimension 4 is a free and uncovered dimension. We like to perform a BPC routing defined as follows. Every source node has its destination obtained by complementing the 1st and the 3rd bit. In fault-free situation, we only need to swap along dimension 1 once and then along dimension 3 once. To simulate this routing, the tasks of faulty nodes are assigned to their 4th dimensional neighbors. As shown in Fig 2, the communication are conducted in 2 phases. In the first phase, all fault-free nodes swap along dimension 1 and in the second phase those fault-free nodes assuming some faulty nodes' tasks take 3 swaps to perform the exchange. Similar operations are performed for exchanging messages along dimension 3.

4   Conclusions

In this paper, we present a technique for simulating SIMD algorithms on faulty hypercubes. When the number of faulty nodes, $|F|$, is less than n, the number of dimensions, our simulation scheme has a small constant slowdown factor. By our scheme, many important algorithms can be simulated in a faulty hypercube such as sorting, BPC routing, and other applications.

References

[1] Bernd Becker and Hans-Ulrich Simon. How robust is the n-cube? *Information and Computation*, 77, 1988.

[2] Rajendra Boppana and C.S. Raghavendra. Optimal self-routing of linear-complement permutations in hypercubes. In *Proc. Distributed Memory Comput. Conference*, 1990.

[3] Johan Hastad, Tom Leighton, and Mark Newman. Fast computation using faulty hypercubes. In *Proc. 21th ACM Symposium Theory of Comp.*, 1989.

[4] O.A. Mcbryan and E.F. De Velde. Hypercube algorithms and implementations. *SIAM Journal Sci. Stat. Comput.*, 8(2), Mar. 1987.

[5] Sing-Ban Tien and C.S. Raghavendra. Simulation of simd alogrithms on faulty hypercubes. Technical report, Univ. of So. Calif., EE dept., 1991.

A Batcher Double Omega network with Combining

Hideharu Amano and Gaye Kalidou

Dept. of Electrical Engineering

Keio University

3-14-1, Hiyoshi Yokohama 223 JAPAN

1   Introduction

Multistage Interconnection Networks (MINs) have been well researched as an interconnection mechanism between processors and memory modules in parallel machines. Most of them are blocking networks like the Omega-network, and packets are transferred by using the store-and-forward manner in a certain size of data width (8bit-64bit). The implementation of conventional MINs have been difficult work. Each switching element needs some storage for buffering, and mechanisms for handshake and conflict resolution. To implement the combine mechanism[1], parallel comparators and a sophisticated controller are required in each element. The pin neck problem also introduces difficulties to implement a lot of switching elements into one LSI chip.

We propose a novel MIN architecture for parallel machines. Packets are transferred in the serial and synchronized manner Although such a strategy limits the bandwidth of MIN, the structure and control can be much simplified. Moreover, the simple structure of the switching element enables the use of higher frequency system clock and a plenty of switching elements in order to achieve the non-blocking feature.

Here, we introduce such a type of MIN called the Batcher Double Omega network with Combining, or BDOC in short.

2   A Batcher Double Omega network

2.1   A Batcher-Banyan network

A Batcher-Banyan network has been well researched as a switching fabric for fast packet switching based on the ATM (Asynchronous Transfer Mode) because it can overcome the problem of internal conflict if the destination of all packets are different from each other [2]. Packets for the identical destination still collide with each other in the Banyan network. In order to eliminate the packets for the same destination, a trap network is required at the end of the Batcher sorter. Since packets are sorted by the Batcher sorter, the packets for the same destination can be checked by comparison with neighboring outputs at the end of the sorter.

However, empty inputs are generated by the trapping network and they still cause conflicts in the Banyan network even if all packets have different destinations. In order to maintain the conflict free feature in the Banyan network, a concentrator is required. Although the role of the concentrator is just shifting the packets and eliminate the empty inputs, the implementation is difficult because the number of the trapped packets cannot be predicted. Usually, the Batcher sorter is used again for the concentrator, and the fabric actually becomes a Batcher-Batcher-Banyan network.

The Batcher Double Omega network (BDO) is the alternative approach to overcome the problem around the trap network and concentrator.

2.2   The Batcher Double Omega network

The Batcher Double Omega network consists of a cascade connection of Batcher sorter, a double sized Omega network, and the double sized inverse Omega network (Fig.1). The Omega network is called the distributor, while the inverse Omega network is called the concentrator. In the distributor, the packet is routed by using the new header which is computed by the adder at the input. The header for the distributor (Z_i) is simply the sum of the destination (D_i) of the packet and the input label (S_i) of the distributor:

$$Z_i = D_i + S_i. \tag{1}$$

The header Z_i is stripped at the input of the concentrator, and the original destination D_i is used for the routing in the concentrator again.

Property 1: The distributor is conflict free.

Proof

Let $N = 2^l$ be the size of the Batcher sorter, then the size of the Omega is $2N = 2^{l+1}$. Let $S_i = si_{l-1}si_{l-2}\cdots si_0$ and $S_j = sj_lsj_{l-1}\cdots sj_0$ be the source tag, i.e. the binary representation of two input numbers. Let $Z_i = zi_lzi_{l-1}\cdots zi_0$ and $Z_j = zj_lzj_{l-1}\cdots zj_0$ be the destination tag, i.e. the binary representation of two destination numbers. The packet from S_i to Z_i goes through the position $si_{l-k}si_{l-k-1}\cdots si_00zi_lzi_{l-1}\cdots zi_{l-k+1}$ at the output of the stage k.

In the distributor, let Z_i be the destination of S_i, and Z_j be the destination of S_j respectively. There are the following relations because the input packets are sorted:

$$S_i < S_j \rightarrow D_i \leq D_j \; D_i \leq S_i, D_j \leq S_j. \tag{2}$$

If $si_{l-k}si_{l-k-1}\cdots si_0 = sj_{l-k}sj_{l-k-1}\cdots sj_0$, then $si_{l-1}si_{l-2}\cdots si_{l-k+1} < sj_{l-1}sj_{l-2}\cdots sj_{l-k+1}$, that is, $S_j - S_i > 2^{l-k+1}$.

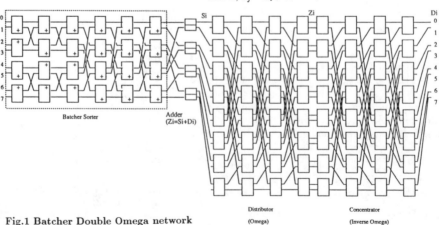

Batcher Sorter   Adder
(Zi=Si+Di)

Fig.1 Batcher Double Omega network

Distributor   Concentrator

(Omega)   (Inverse Omega)

From expression (1), (2) and (3),

$$Z_j - Z_i \geq S_j - S_i > 2^{l-k+1}.$$

$$zj_l zj_{l-1} \cdots zj_{l-k+1} > zi_l zi_{l-1} \cdots zi_{l-k+1}.$$

$$si_{l-k} si_{l-k-1} \cdots si_0 \neq sj_{l-k} sj_{l-k-1} \cdots sj_0$$
$$\cup zi_l zi_{l-1} \cdots zi_{l-k+1} \neq zj_l zj_{l-1} \cdots zj_{l-k+1}.$$

It means that the path from S_i to Z_i and the path from S_j to Z_j go through different numbers at every stage k.

Property 2: The concentrator is conflict free if the destination of all packets are different from each other.

Proof

In the concentrator, packets are routed from the inputs Z_i and Z_j to the original destination D_i and D_j ($D_i < D_j$, that is $Z_i < Z_j$) respectively. Since the concentrator is an Inverse-Omega network, the packet from D_i to Z_i goes through the position $di_{k-1} di_{k-2} \cdots di_0 zi_l zi_{l-1} \cdots zi_{l-k}$ at the input of the stage k.

If $di_{k-1} di_{k-2} \cdots di_0 = dj_{k-1} dj_{k-2} \cdots dj_0$, then $di_{l-1} di_{l-2} \cdots di_{l-k+1} < dj_{l-1} dj_{l-2} \cdots dj_{l-k+1}$, that is, $D_j - D_i \geq 2^{l-k+1}$. From the expression (2),

$$Z_j - Z_i > D_j - D_i > 2^{l-k+1}.$$

$$zj_l zj_{l-1} \cdots zj_{l-k+1} > zi_l zi_{l-1} \cdots zi_{l-k+1}.$$

$$di_{k-1} di_{k-2} \cdots di_0 \neq di_{k-1} di_{k-2} \cdots di_0$$
$$\cup zi_l zi_{l-1} \cdots zi_{l-k+1} \neq zj_l zj_{l-1} \cdots zj_{l-k+1}$$

Lemma 2.1: In the concentrator, only packets for the same destination collide each other if one of the packets disappears after collision.

Proof

The position $di_{k-1} di_{k-2} \cdots di_0 zi_l zi_{l-1} \cdots zi_{l-k}$ and $dj_{k-1} dj_{k-2} \cdots dj_0 zj_l zj_{l-1} \cdots zj_{l-k}$ are identical only if $D_i = D_j$. Therefore, two paths share the same position only the destination is identical.

In the concentrator, the packet provides the conflict bit in the header. Once the packet collides with the other packet (i.e. for the same destination) and is misrouted, the bit is set to one. The packet whose conflict bit is set is treated as a dead packet, and never interferes the other packets. By using this control, the concentrator can select only one packet for a destination. Moreover, the combining can be also implemented easily using this property.

2.3 Control of the element

Since processors issue both read and write requests to memory modules, the MIN is required to be bi-directional. In the conventional MINs, the network for the return path requires the same amount of hardware as that for the request path.

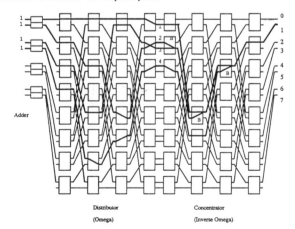

Fig.2 Packets Combining

In the BDOC, the return path can be formed easily by using the same control signals for the forward path. Since each switching element provides only a few bits storage, the packet goes through the BDOC like a stream. In a sense, the BDOC behaves like a set of long shift registers. If the length of the packet is longer than the total amount of the registers from an input to output, the source and destination are connected with a stream and forms a virtual circuit. The return path can be formed by taking advantage of this circuit.

The packet transfer is divided into the address transfer phase and data transfer phase. Each switching element provides three paths, one for the address transfer, and two for the bi-directional data transfer. In the address transfer phase, the packet including routing tag, the type of the operation, and memory address goes through the address path. The state of the switching element is fixed in this phase.

2.4 Combining packets

Lemma 2.1 guarantees that only packets for the same memory module collide with each other in the concentrator. However, packets can be combined only if the whole memory address is exactly matched each other. To distinguish it, the whole address tags in packets are compared in each switching element of the concentrator. If two address tags are equal, the switching element is set to the backward broadcast mode as shown in Fig.2. In this example, four packets for module 1 are combined with each other. Switching elements which is marked B are set to the backward broadcast mode. After the address phase, the tree-like path for data phase remains in switching elements, and combined packets are returned through this path. The copy of the packets are easily performed using the tree. This technique enables the following types of combining:

- Read-read combining

- Write-read combining

These types of combine are useful for the broadcasting or multicasting of data.

If conflict packets are not for the same address, one packet is routed to the requested direction. The control is performed based on the priority bit in the packet header. The collision bit in the header of the packet which is routed to the wrong direction is set to one, and it will never disturb other packets in the later stage.

For such miss-routed packets, the NAK (Not AcKnowledge) packet is returned through the data path, and the packet is transferred again. The priority bit of the packet is used to avoid retransmission too many times. The Test&Set combine is performed by using this mechanism. In this case, only one packet is selected and accesses the memory address. Other packets are combined and reply packets are multicast using the backward multicast tree. The BDO with combining mechanism is called a BDOC (Batcher Double Banyan with Combining).

Two double sized Omega networks in the BDOC takes a role of the trap network and the second Batcher sorter in the Batcher-Batcher-Omega network. The number of the switching element required for two Omega networks is $2N(\log_2 N - 1)$, while the Batcher network needs $\frac{1}{4} N \log_2 N(\log_2 N + 1)$. If the size N grows beyond 2^8, the number of elements for BDOC is less than that of the Batcher-Batcher-Omega network.

REFERENCES

[1] G.F.Phister, and V.A.Norton, "Hot Spot Contention and Combining in Multistage Interconnection Networks," IEEE Trans. on Comput. vol. c-34, No.10, Oct. 1985.

[2] M.J.Narasimha, "The Batcher-banyan self-routing network: Universality and Simplification," IEEE Trans. Commun., vol.36, No 10, pp.1175-1178, Oct.1988.

Stability and Performance of Alternative Two-level Interconnection Networks *

Shyamal Chowdhury and Mark A. Holliday

Department of Computer Science

Duke University, Durham, NC 27706

Abstract

The ability to construct a multicomputer with multiple processors on a single chip introduces a two-level network with a local network on chip and a global network between chips. An important issue in two-level networks is how alternative numbers of processors per chip changes the per link traffic in the global network. In the case of a hypercube global network we consider three communication schemes. We derive system stability conditions for each scheme under very general assumptions. Assuming the packets sent by a processor form a Poisson process, we also derive the worst-case mean end-to-end packet delay for the three schemes. We then use these results to compare alternative numbers of processors per chip for the system size of a given multicomputer.

1   Introduction

With advances in VLSI technology, it has become possible to have multiple processors on a single chip. A multicomputer with thousands of processors can be built by interconnecting several such chips. Processors residing on the same chip communicate with each other by using a network local to the chip. Processors on different chips communicate by using a global network that connects the chips together. Thus, such a multicomputer has a two-level interconnection network. Quantifying the effect of variations in the number of processors per chip on the traffic demand presented to the global network arises as an important question. In this paper we address this question in the case where the global network is a hypercube.

Each chip contains i processors and a router. The i processors on a chip use the on-chip router to form the local network. The routers of the different chips are connected to form a global hypercube network of dimension n. Fixed-length data packets are used for communication. We consider three communication schemes. In time division multiplexing, TDM, time is divided into communication cycles. Each communication cycle is divided into n slots (since the network has dimension n, there are n adjacent chips and associated output links). Each slot is associated with one of the chip's links. During one slot, one packet can be transfered on that link. If multiple packets need to use the same link (and thus, slot), only one of those packet can be transfered per cycle and the packets are scheduled in first-come-first-served order.

The second communication scheme is statistical multiplexing, STM [1], which maintains the n slots, but decouples a slot from a link. That is, any slot can be used to traverse any adjacent link. We call the third scheme the multiple driver scheme (MD). It can be viewed as STM combined with a driver for each output link. TDM and STM have a single communication line driver per chip so only one link can be used per slot. In MD, by having multiple drivers, multiple links adjacent to the given chip can be traversed by packets during the same slot. In fact, a chip can in the same slot send packets to all of its neighbors.

2   System Stability

A hypercube of dimension n has 2^n nodes, and is called an n-cube. Each node is directly connected by bidirectional links to n other nodes. When a packet is sent from one node to another it goes through at most $n - 1$ intermediate nodes. We assume that when a packet is sent from one node to another, the routing algorithm divides the packet stream equally among the alternative shortest paths. We assume that each router has an unbounded buffer to hold packets which are to be sent to other nodes. We assume packet destinations are uniformly distributed over all other nodes. We make

*This work was supported in part by the National Science Foundation (Grant CCR-8821809).

no assumption about the distribution of time between each sending of a packet by a processor. We denote by L the average traffic (number of packets) carried by each link per time unit. Our first result gives the value of L given the average number of packets sent by each processor to every other processor per time unit. The value of L is independent of the communication scheme used.

Lemma 1 *If λ is the average number of packets per time unit sent by each processor to every other processor, then the average link traffic per time unit is $L = 2^{n-1}i^2\lambda$ where n is the dimension of the global hypercube network and i is the number of processors per node.*

Each link would have a stable operating condition if the average number of packets per time unit arriving at the link for transmission is less than the capacity of the link. If the number of packets arriving at the link per time unit is greater than or equal to the capacity of the link the system loses stability. That is, the queue of packets builds up without limit. Unlike the average link traffic, L, system stability and system performance depend on the communication scheme. The following three theorems (see [2] for the proofs) give the maximum allowable value of λ for stable operation in the three schemes.

Let s be the time taken to send a packet between adjacent nodes; that is, the slot length. Let c be the length of the communication cycle. Consider two directly-connected nodes A and B. A sends packets to B after time intervals of length c if there are packets to send. At most one packet is sent during a communication cycle. During the rest of the cycle, A sends packets to the other nodes directly connected to it. For an n-cube $c = ns$.

Theorem 1 *Let λ be the average number of packets per time unit sent by each processor to every other processor. Each node in the global hypercube network sends packets to neighboring nodes using the TDM scheme. Then the system is stable if $\lambda 2^{n-1}i^2ns < 1$, where n is the dimension of the hypercube, i is the number of processors per chip, and s is the time taken to send a packet.*

Suppose we want to build a multicomputer with a total of N processors. We would like to know how placing different numbers of processors per chip affects system stability. We have an n-cube as the global interconnection network, and i processors in each chip. So $i2^n = N$. That is $n = (logN - log\,i)/log2$. By the stability condition in Theorem 1, the upper bound on the average number of packets per time unit sent by each processor to every other processor is as follows.

$$\bar{\lambda}_1 = \frac{1}{2^{n-1}i^2ns} = \frac{2}{Nins} = \frac{2log2}{Ni(logN - log\,i)s}.$$

The region of stable operation for STM is the same as that for TDM since for very heavy traffic the behavior of STM is identical to that of TDM. The main difference between the two schemes appears in system performance for low link traffic. We can write a theorem similar to Theorem 1 for the stability condition of STM. Also, the same upper bound on the average number of packets sent by each processor to every other processor holds: $\bar{\lambda}_2 = 2log2/(Ni(logN - log\,i)s)$.

Theorem 2 *Let λ be the average number of packets per time unit sent by each processor to every other processor. Each node in the global hypercube network sends packets to neighboring nodes using the STM scheme. Then the system is stable if $\lambda 2^{n-1}i^2ns < 1$, where n is the dimension of the hypercube, i is the number of processors per chip, and s is the time taken to send a packet.*

In MD each node is given the capability of sending packets simultaneously to all of its neighbors. So the node's capacity to transmit packets along any link will increase n-fold. The price we pay for this is a more complicated chip: a chip that must have n drivers to drive the communication links to its neighbors simultaneously.

Theorem 3 *Let λ be the average number of packets per time unit sent by each processor to every other processor. Each node in the global hypercube network is capable of simultaneously sending packets to neighboring nodes. Then the system is stable if $\lambda 2^{n-1} i^2 s < 1$, where n is the dimension of the hypercube, i is the number of processors per chip, and s is the time taken to send a packet.*

By Theorem 3, the upper bound on the average number of packets per time unit that can be sent by a procesor to every other processor without losing system stability is $\bar{\lambda}_3 = 1/(2^{n-1} i^2 s) = 2/Nis$. The improvement in stability is given by the ratio $\bar{\lambda}_3/\bar{\lambda}_1 = n = (\log N - \log i)/\log 2$. If $N = 2^K$ and $i = 2^k$ then the improvement in stability can be written as $\bar{\lambda}_3/\bar{\lambda}_1 = K - k$.

3 System Performance

To assess the overall system performance, we need to determine the average communication delay for two processors that are residing on nodes that are farthest apart in the hypercube network. In this section we provide an approximate analytical model to determine the average communication delay. We model the communication delay at each router with an $M/D/1$ queueing system model. Unlike the stability question above, we assume that the packets sent by each processor to every other processor in the system constitute a Poisson stream of rate λ. The packets sent by a processor to other processors in the same node do not contribute to the delay suffered by the packets in the global network.

In TDM the total delay suffered by a packet at a router consists of three components. If, at the time of the arrival of a tagged packet X, the router has no other packet to send in the direction of the destination of X, then X has to wait until the appropriate slot for the transmission of X comes up. If the arrival time of X is random then the time for which X has to wait is uniformly distributed over 0 to c. The time required to transmit X equals s. If at the time of arrival of X, there are other packets to be sent in the direction of the destination of X, then X must wait until these packets are all transmitted. We call this component of the delay suffered by X its waiting time at a router, and denote it by W_1. The value of W_1 can be determined by modeling the router as a queueing system. For the purpose of determining the waiting time, the service time of this queueing system can be considered to be constant of length c, since the time between the service initiations of two packets going in the same direction in two consecutive communication cycles is c.

For the first scheme, let T_1 be a random variable representing the time taken to send a packet from one node to a node farthest from it in the global hypercube. Applying the Pollaczek-Khinchin formula([3], p.191) we have

$$\mathbf{E}[T_1] = \frac{nc}{2} + n\mathbf{E}[W_1] + ns = \frac{nc}{2} + \frac{\lambda 2^{n-1} i^2 n^3 s^2}{2(1 - \lambda 2^{n-1} i^2 ns)} + ns. \quad (1)$$

In STM the delay suffered by a packet X at a router consists of two components. The time required to transmit X is equal to s. Also, X must wait until all the packets that arrived at the router before X have been transmitted. Let us denote the waiting time of X at a router by W_2. The mean value of W_2 can be determined by modeling the router as an $M/D/1$ queueing system. The service time of the queueing system is s. The arrival rate is nL since in the communication driver we are multiplexing the traffic for the n outgoing lines. Let T_2 denote the total delay suffered by a packet sent by a node to a node farthest from it in the global network. Applying the Pollaczek-Khinchin formula we have

$$\mathbf{E}[T_2] = ns + n\mathbf{E}[W_2] = ns + \frac{\lambda 2^{n-1} i^2 n^2 s^2}{2(1 - \lambda 2^{n-1} i^2 ns)}. \quad (2)$$

Note that $\mathbf{E}[T_1] > \mathbf{E}[T_2]$ since $\mathbf{E}[T_1]$ has an extra first term and an extra n in the numerator of the second term.

In MD we have a communication line driver for each outgoing line. The delay suffered by a tagged packet X at a router consists of two components. The time required to transmit X is equal to s. Also, X must wait until all the packets that arrived at the router before X have been transmitted. Let us denote the waiting time of X at a router by W_3. The mean value of W_3 can be determined by modeling

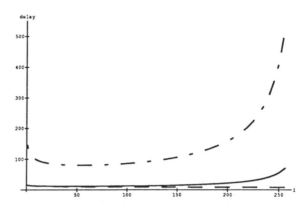

Figure 1: Mean end-to-end packet delay in the worst case of two nodes farthest apart. The dot-dashed line is TDM. The solid line is STM. The dashed line is MD. $s = 1$ time unit, $N = 2^{16}$, and $\lambda = 1.4 \times 10^{-8}$. i is varied: $i = 2^k, 0 \le k \le 8$.

the router as an $M/D/1$ queueing system. The service time of the queueing system is s. The arrival rate is L. Let T_3 denote the total delay suffered by a packet sent by a node to a node farthest from it in the global network. Applying the Pollaczek-Khinchin formula we have

$$\mathbf{E}[T_3] = ns + n\mathbf{E}[W_3] = ns + \frac{\lambda 2^{n-1} i^2 ns^2}{2(1 - \lambda 2^{n-1} i^2 s)}. \quad (3)$$

Note that $\mathbf{E}[T_2] > \mathbf{E}[T_3]$ since $\mathbf{E}[T_2]$ has an extra n in the numerator of the second term and an extra n in the denominator of the second term (which reduces the size of the denominator).

For comparing packet delays we chose a $\lambda = 1.4 \times 10^{-8}$. In Figure 1 we plot $\mathbf{E}[T_1]$, $\mathbf{E}[T_2]$, and $\mathbf{E}[T_3]$ for $s = 1$ time unit, $N = 2^{16}$, and $\lambda = 1.4 \times 10^{-8}$. i is varied: $i = 2^k, 0 \le k \le 8$.

There are two interesting results. First, as the number of processors per chip increases, the packet delays initially decrease. This is because the dimension n of the hypercube decreases so the worst case number of hops in a path between two nodes decreases. However, when queueing starts to dominate, the packet delay starts to increase again. The queueing effect is most noticeable for TDM. Second, is that, as expected, TDM does worse than STM which, in turn, does worse than MD. However, the difference between TDM and STM is much larger than the difference between STM and MD. At lower rates of packets sent, the difference between STM and MD will be even smaller.

References

[1] A. S. Tanenbaum, *Computer Networks*. Englewood Cliffs, NJ: Prentice-Hall, 1981.

[2] S. Chowdhury and M. Holliday, "Stability and performance of alternative two-level interconnection networks," Tech. Rep. CS-1991-12, Dept. of Computer Science, Duke University, Durham, NC, April 1991.

[3] L. Kleinrock, *Queueing Systems, Volume I: Theory*. John Wiley, 1975.

Fibonacci Cubes - Properties of An Asymmetric Interconnection Topology

Wen-Jing Hsu

Department of Computer Science

Michigan State University

Oct. 11, 1990

1 Introduction

We present a new interconnection topology called the Fibonacci cube, which is inspired by the famous Fibonacci numbers[1]. The following main results are reported:

1. We show that the Fibonacci cube has attractive recurrent structures (*self-similarity*, Section 2) such that a Fibonacci cube can be decomposed into subgraphs which are also Fibonacci cubes by themselves. In particular, we present a *Decomposition Lemma* which, in some sense, is the bridge between the forms and functions of the Fibonacci cube.

2. The Fibonacci cube encompasses certain topologies that are commonly used. Specifically, we show that the linear array, certain types of mesh and trees (*Fibonacci mesh and Fibonacci trees*), and Boolean cubes can be directly embedded in the Fibonacci cube; on the other hand, the Fibonacci cube can be embedded in the Boolean cube.

3. For a cube with N nodes, we show that the diameter, the edge connectivity and the node connectivity of the Fibonacci cube are in the order of $O(logN)$, which are similar to the Boolean cube. We also show how to construct the shortest path and independent paths between nodes.

4. We show that common system communication primitives can be implemented efficiently. Assuming that each node can send or receive along one incident link at each step, we show that Single node broadcast and Single node accumulation can be done in $O(logN)$ time; Single node gather, Single node scatter, Multiple node broadcast, and Multinode accumulation can all be done in $O(N)$ time; moreover, the Total Exchange operation can be done in $O(NlogN)$ time.

2 Basic Properties of Fibonacci Cubes

An interconnection topology is represented by a graph $G = (V, E)$, where V (the set of *vertices*, or, alternatively, *nodes*) denotes the memory modules or processors and E (edges) the communication links between them. If $G_2 \cap G_3 = (\phi, \phi)$ (i.e., the two graphs G_1 and G_2 are disjoint), then we write $G_1 = G_2 \uplus G_3$, instead of $G_2 \cup G_3$, to emphasize that G_1 consists of two disjoint subgraphs. Also, for convenience, write $\uplus_{1=i}^{m} G_i = m \cdot G$, if the graphs are all isomorphic, i.e., $G_i \cong G$ for $1 \leq m$.

By *Zeckendorf's theorem* [2], any natural number can be uniquely represented as a sum of Fibonacci numbers. Specifically, let i be an integer, and $0 \leq i < f_n$, where $n \geq 3$. $(b_{n-1}, ..., b_3, b_2)_F$ denotes the *Fibonacci representation* (or, alternatively, the *Fibonacci code*) of i, if $b_j \cdot b_{j-1} = 0$ for $n - 1 \geq j \geq 3$, b_j is either 0 or 1 for $2 \leq j \leq (n - 1)$ and $i = \sum_{j=2}^{n-1} b_j \cdot f_j$.

Example Here the integers between 0 and 20 are written in this notation: $0 = (000000)_F$, $1 = (000001)_F$, $2 = (000010)_F$, $3 = (000100)_F$, $4 = (000101)_F$, $5 = (001000)_F$, $6 = (001001)_F$, $7 = (001010)_F$, $8 = (010000)_F$, $9 = (010001)_F$, $10 = (010010)_F$, $11 = (010100)_F$, $12 = (010101)_F$, $13 = (100000)_F$, $14 = (100001)_F$, $15 = (100010)_F$, $16 = (100100)_F$, $17 = (100101)_F$, $18 = (101000)_F$, $19 = (101001)_F$, $20 = (101010)_F$.

Let $I = (b_{n-1}, ..., b_3, b_2)$ and $J = (c_{n-1}, ..., c_3, c_2)$ denote two binary numbers. The *Hamming distance* between I and J, denoted by $H(I, J)$, is the number of bits where the two binary numbers differ. It is straightforward to define the Fibonacci cube based on the Fibonacci codes.

Definition [Fibonacci Cube of order n]

The *Fibonacci cube of order n*, denoted by Γ_n, is a graph (V_n, E_n), where $V_n = \{0, 1, ..f_n - 1\}$ and $(i, j) \in E(N)$ if and only if $H(I_F, J_F) = 1$. Define $\Gamma_0 = (\phi, \phi)$.

Example Fig. 1 shows the Fibonacci cubes Γ_i for $0 \leq i \leq 7$. In the Fibonacci cube of size 8, for example, Node $0 = (0000)_F$ has a link with each of the following nodes: $1 = (0001)_F$, $2 = (0010)_F$, $3 = (0100)_F$, $5 = (1000)_F$.

The preceding definition can be modified and applied to define a Fibonacci cube with arbitrary $N \geq 1$ nodes [6].

2.1 Self-similarity of the Fibonacci cube

The following lemma shows that the Fibonacci cube Γ_n contains two disjoint subgraphs that are isomorphic to Γ_{n-1} and Γ_{n-2} respectively (with proper re-naming of the vertices in Γ_{n-2}); moreover, there are exactly f_{n-2} edges linking the two subgraphs together.

Lemma 1 *Let* $\Gamma_n = (V_n, E_n)$ *denote the Fibonacci cube of order n, where $n \geq 2$. Let* $LINK(n) = \{(i, j) : |i - j| = f_{n-1}, (i, j) \in E_n\}$. *Let* $LOW(n)$ *and* $HIGH(n)$ *be the subgraphs induced by the set of vertices* $\{0, 1, ..., f_{n-1} - 1\}$ *and* $\{f_{n-1}, ..., f_n - 1\}$ *respectively. Then* Γ_n *can be decomposed into two and* $HIGH(n) \cong \Gamma_{n-2}$; *the two subgraphs are connected exactly by*

[proof] ([6]).

Clearly Lemma 1 is useful for deriving substructures or embeddings of other types of graphs. It is also the formal basis of some (divide-and-conquer) algorithms [4] for the Fibonacci cube. Here we will present a general *decomposition lemma*. Recall that $\uplus_{1=i}^{m} G_i = m \cdot G$ if $G_i \cong G$ for all $1 \leq i \leq m$.

Example In Fig. 1 we see that Γ_6 contains a subgraph Γ_5 and a subgraph Γ_4 (after renaming vertices). Since Γ_5 can be decomposed into a subgraph Γ_4 and a subgraph Γ_3, so Γ_6 contains two disjoint Γ_4 and one Γ_3. Using the notations introduced, we will write $\Gamma_6 \supseteq (2 \cdot \Gamma_4 \uplus \Gamma_3)$ to denote that Γ_6 contains two disjoint subgraphs which are isomorphic to Γ_4 and another disjoint subgraph which is isomorphic to Γ_3.

Theorem 1 *Assume that* $2 \leq k \leq n$. *The Fibonacci cube of order n (Γ_n) admits the following decompositions:*

1. $\Gamma_n \supseteq (f_k \cdot \Gamma_{n-k+1} \uplus f_{k-1} \cdot \Gamma_{n-k})$

2. $\Gamma_n \supseteq (f_{n-k+1} \cdot \Gamma_k \uplus f_{n-k} \cdot \Gamma_{k-1})$

[proof] ([6]).

Because of the rich properties of the Fibonacci numbers, there are many conceivable ways to decompose the Fibonacci cube. See [6].

3 Topological Properties of Fibonacci Cube

We shall analyze the Fibonacci Cube in terms of connectivity: degree of vertices, diameter, shortest path and independent paths between vertices, edge connectivity and node connectivity. We refer the readers to [6] for proofs.

Lemma 2 *The Fibonacci cube of order n has* $\frac{2(n-1)f_n - nf_{n-1}}{5}$ *edges, where f_n denotes the n-th Fibonacci number.*

Corollary 1 *Let $A(N)$ (resp., $B(N)$) denote the number of edges in the Fibonacci cube (resp., Boolean cube) with N vertices. Then $0.79 < \frac{A(N)}{B(N)} < 0.80$ for sufficiently large N.*

Note that, in general, there are more Fibonacci numbers than powers of 2 between two given numbers.

Lemma 3 *Let $d_n(i)$ denote the degree of a node i in the Fibonacci cube Γ_n, for $n \geq 3$. Then* $\lfloor \frac{n-2}{3} \rfloor \leq d_n(i) \leq (n - 2)$.

The *distance* between two nodes i and j (denoted $\Delta(i, j)$) in a graph G is the number of edges on the *shortest* path connecting i and j. The *diameter* of a graph G is the maximum distance between two nodes in G.

---

[1]Throughout the paper, we let f_n denote the n-th Fibonacci number which is given by $f_0 = 0$, $f_1 = 1$, and, for $n \geq 2$, $f_n = f_{n-1} + f_{n-2}$ [2].

Lemma 4 *Let $\Delta(i, j)$ denote the distance between Node i and Node j in the Fibonacci cube Γ_n. Let $H(I_F, J_F)$ denote the Hamming distance between I_F and J_F, where $I_F = (b_{n-1}, b_{n-2}, \ldots b_2)_F$ and $J_F = (c_{n-1}, c_{n-2}, \ldots c_2)_F$ denote the Fibonacci codes of i and j respectively. Then $\Delta(i, j) = H(I_F, J_F)$.*

Theorem 2 *The Fibonacci cube Γ_n is a connected graph for all $n \geq 1$. Let Δ_n denote the diameter of Γ_n, where $n \geq 2$. Then $\Delta_n = n - 2$.*

The *edge connectivity* (resp., *node connectivity*) of a connected graph G, denoted $K_e(G)$ (resp., $K_n(G)$), is the minimum number of edges (resp., vertices) whose removal will disconnect G. The following theorem is from [6], which shows that both the node connectivity and the edge connectivity are logarithmic functions of the size of the Fibonacci cube.

Theorem 3 *Let $K_e(\Gamma_n)$ and $K_n(\Gamma_n)$ denote the edge connectivity and node connectivity of Γ_n respectively. Then $\lfloor \frac{n}{8} \rfloor \leq K_n(\Gamma_n) \leq K_e(\Gamma_n) \leq \lfloor \frac{n-2}{3} \rfloor$.*

4 Embeddings of Other Structures

Here we show how to embed other structures such as the linear array, the mesh, certain types of trees, and even Boolean cubes on the Fibonacci cube.

Definition

A graph G_1 is said to be *directly embedded* [2] in G_2, denoted $G_1 \sqsubseteq G_2$ (or, alternatively, $G_2 \sqsupseteq G_1$), if and only if there is a subgraph G_3 in G_2, such that $G_1 \cong G_3$.

A fundamental relation between the Fibonacci cube and the Boolean cube is given below.

Lemma 5 *Let Γ_n and B_n denote the Fibonacci cube of order n and the Boolean cube of order n respectively. Then*

1. *$\Gamma_n \sqsubseteq B_{n-2}$ for all $n \geq 2$.*

2. *$B_n \sqsubseteq \Gamma_{2n+1}$ for all $n \geq 1$.*

Example The graph B_4 is shown in Fig. 3(a). In Fig. 3(b), an embedding of f_6 in B_4 is shown, where vertices marked with a cross correspond to the "forbidden" codes in the Fibonacci representation.

To discuss the embeddings of other topologies, it is convenient to introduce the following classes of (Fibonacci) interconnections. Notice that it is impossible to directly embed an arbitrary graph in the Fibonacci cube, because the graph Γ_n has exactly f_n nodes and a fixed interconnection pattern. Also, since there are more Fibonacci numbers than powers of 2 between two given numbers, the class of (Fibonacci) graphs proposed are no less general than, say, $2^k \times 2^k$ meshes or full binary trees.

Definition

1. The *Fibonacci linear array of order n*, denoted by L_n, where $n > 0$, is a linear array of size f_n. Define $L_0 = (\phi, \phi)$.

2. Let $n_1, n_2 > 0$ denote integers. The *Fibonacci mesh of order (n_1, n_2)*, denoted by $M_{(n_1, n_2)}$, is a $f_{n_1} \times f_{n_2}$ mesh. Define $M_{0,n} = M_{n,0} = M_{0,0} = (\phi, \phi)$.

3. The *Fibonacci tree of order n*, denoted T_n, is a graph (V_n, E_n). Define $T_0 = (\phi, \phi)$, $T_1 = (\{0\}, \phi)$, and for $n \geq 2$, $T_n = T_{n-1} \cup T'_{n-2} \cup T''_n$, where $T'_{n-2} = (\{i + f_{n-1} : i \in V_{n-2}\}, \{(i + f_{n-1}, j + f_{n-1}) : (i, j) \in E_{n-2}\})$ and $T''_n = (\{0, f_{n-1}\}, \{(0, f_{n-1})\})$.

Example Fig. 4 shows T_i, for $1 \leq i \leq 7$. Notice that T_i can be decomposed into a copy of T_{i-1} and a disjoint T_{i-2}; note that the two subgraphs are linked only by a single edge (through the roots), which distinguishes between the Fibonacci tree and the Fibonacci cube.

Lemma 6 *Let $Height(T_n)$ denote the height of the Fibonacci tree T_n. Then $Height(T_0) = 0$, $Height(T_1) = 1$. For $n \geq 2$, $Height(T_n) = \lceil \frac{n}{2} \rceil$.*

Theorem 4 *Let $B_n, \Gamma_n, L_n, M_{(n_1, n_2)}$, and T_n be defined as before.*

1. *$\Gamma_n \sqsupseteq L_n$ for all $n \geq 1$.*

2. *$\Gamma_n \sqsupseteq T_n$ for all $n \geq 1$.*

3. *$\Gamma_{2n+1} \sqsupseteq (M_{(n,n)} \uplus M_{(n+1,n+1)})$ and $\Gamma_{2n} \sqsupseteq (M_{(n,n+1)} \uplus M_{(n,n-1)})$ for all $n \geq 1$.*

4. *$B_n \sqsupseteq (4 \cdot \Gamma_n)$, for all $n \geq 2$.*

5. *$B_n \sqsubseteq \Gamma_{2n+1}$ for all $n \geq 1$.*

In [7] it is shown that the embedding of a Hamiltonian cycle cannot be achieved in general. However, it is shown that when $n = 3 \cdot k$ for some integer k, the Hamiltonian cycle does exist; and when $n = 3 \cdot k \mp 1$ or $3 \cdot k + 2$, there is a cycle which includes all nodes except one. Similar results were reported for embeddings of Fibonacci tori (Fibonacci mesh with wraparound) [7].

5 Communication Primitives

To further evaluate the Fibonacci cube, we next show that certain useful operations can be efficiently performed on the Fibonacci cube. The following communication patterns are often needed when solving problems [1]:

Lemma 7 *On a Fibonacci cube of order n, the send/receive operation between any two nodes can be done in at most $n - 2$ steps, which is optimal.*

Theorem 5 *Assume that each node can send or receive over a single link at each step (rather than to send or receive simultaneously over all incident links). Then*

1. *Broadcasting from Node 0 to all nodes can be performed in $n - 2$ steps on Γ_n, which is optimal.*

2. *The Single-node broadcast and the Single-node gather can be performed in $\lceil \frac{n}{2} \rceil + n - 3$ steps on Γ_n, which is optimal to within a multiplicative (1.5) factor.*

3. *Assume that a node can send or receive along one incident link at each step. Multinode broadcast (accumulation) can be done in $f_n - 1$ steps, which is optimal.*

4. *Single node scatter and Single node gather can be done in $f_n - 1$ steps, which is optimal.*

5. *The Total Exchange operation on Γ_n can be done in $O(n \cdot f_n)$ steps.*

[proof] ([6].)

Theorem 6 *Assume that multiple incident links of a node can be active simultaneously. Then*

1. *Broadcasting operation from Node 0 can be done in $\lceil \frac{n}{2} \rceil - 1$ steps, which is optimal.*

2. *The Single-node broadcast (gather) requires at most $2 \cdot \lfloor \frac{n}{2} \rfloor$ steps, which is optimal to within a constant additive factor.*

[proof] ([6]).

6 Discussion and Conclusion

We have presented the Fibonacci cube, analyzed its structural properties, and studied its functionalities. The structural issues that were analyzed include the recursive decompositions, diameter, node/edge connectivity, and relations with other structures; the functional issues include common system communication primitives. We also showed that various types of structures can be directly embedded in the Fibonacci cube. These preliminary results show that the Fibonacci cube has very attractive recurrent structures. Clearly, there are many questions left unanswered. See [6] for prominent open problems.

Acknowledgments The author wishes to thank Professors L. M. Kelly and C. V. Page for their kind comments. C. T. Ho of IBM Alamaden Research Center is acknowledged for pointing out $B_n \sqsubseteq \Gamma_{2n+1}$.

References

[1] D. P. Bertsekas and J. N. Tsitsiklis, *PARALLEL AND DISTRIBUTED COMPUTATION: Numerical Methods*, Prentice-Hall, Inc. 1989. (Chapter 1, pp. 27-68).

[2] R. L. Graham, D. E. Knuth, and O. Patashnik, *CONCRETE MATHEMATICS*, Addison-Wesley Publishing Co., 1989 (Chapter 6: Special Numbers).

[3] J. P Hayes, *COMPUTER ARCHITECTURE AND ORGANIZATION*, 2nd ed., McGraw-Hill, 1988. Ch. 6.

[4] E. Horowitz and S. Sahni, *Fundamentals of Computer Algorithms*, Computer Science Press, 1978. Ch. 4.

TABLE OF CONTENTS - FULL PROCEEDINGS

(R): Regular Paper
(C): Concise Paper
(P): Poster Paper

Volume I: Architecture
Volume II: Software
Volume III: Algorithms and Applications

A1

A3

A5

A7

POSTER SESSION

A10

AREA 6: Advances in Parallel Algorithms and Applications

A13